Annals of Mathematics Studies
Number 197

The Mathematics of Shock Reflection-Diffraction and von Neumann's Conjectures

Gui-Qiang G. Chen
Mikhail Feldman

PRINCETON UNIVERSITY PRESS
PRINCETON AND OXFORD
2018

Published by Princeton University Press, 41 William Street,
Princeton, New Jersey 08540

In the United Kingdom: Princeton University Press, 6 Oxford Street,
Woodstock, Oxfordshire OX20 1TR

press.princeton.edu

Library of Congress Cataloging-in-Publication Data

Names: Chen, Gui-Qiang, 1963– | Feldman, Mikhail, 1960–
Title: The mathematics of shock reflection-diffraction and von Neumann's conjectures / Gui-Qiang G. Chen and Mikhail Feldman.
Description: Princeton : Princeton University Press, 2017. | Series: Annals of mathematics studies ; number 197 | Includes bibliographical references and index.
Identifiers: LCCN 2017008667| ISBN 9780691160542 (hardcover : alk. paper) | ISBN 9780691160559 (pbk. : alk. paper)
Subjects: LCSH: Shock waves–Diffraction. | Shock waves–Free boundaries. | Shock waves–Mathematics. | von Neumann conjectures. | Analysis of PDEs.
Classification: LCC QC168.85.S45 C44 2017 | DDC 531/.1133–dc23 LC record available at https://lccn.loc.gov/2017008667

British Library Cataloging-in-Publication Data is available

This book has been composed in LaTeX. [sjp]

The publisher would like to acknowledge the authors of this volume for providing the camera-ready copy from which this book was printed.

Printed on acid-free paper ∞

10 9 8 7 6 5 4 3 2 1

Contents

Preface

The purpose of this research monograph is to survey some recent developments in the analysis of shock reflection-diffraction, to present our original mathematical proofs of von Neumann's conjectures for potential flow, to collect most of the related results and new techniques in the analysis of partial differential equations (PDEs) achieved in the last decades, and to discuss a set of fundamental open problems relevant to the directions of future research in this and related areas.

Shock waves are fundamental in nature, especially in high-speed fluid flows. Shocks are generated by supersonic or near-sonic aircraft, explosions, solar wind, and other natural processes. They are governed by the Euler equations for compressible fluids or their variants, generally in the form of nonlinear conservation laws – nonlinear PDEs of divergence form. The Euler equations describing the motion of a perfect fluid were first formulated by Euler [112, 113, 114] in 1752 (based in part on the earlier work of Bernoulli [15]), and were among the first PDEs for describing physical processes to be written down.

When a shock hits an obstacle (steady or flying), shock reflection-diffraction configurations take shape. One of the most fundamental research directions in mathematical fluid dynamics is the analysis of shock reflection-diffraction by wedges, with focus on the wave patterns of the reflection-diffraction configurations formed around the wedge. The complexity of such configurations was first reported by Ernst Mach [206] in 1878, who observed two patterns of shock reflection-diffraction configurations that are now named the Regular Reflection (RR) and the Mach Reflection (MR). The subject remained dormant until the 1940s when von Neumann [267, 268, 269], as well as other mathematical and experimental scientists, began extensive research on shock reflection-diffraction phenomena, owing to their fundamental importance in applications. It has since been found that the phenomena are much more complicated than what Mach originally observed, and various other patterns of shock reflection-diffraction configurations may occur. On the other hand, the shock reflection-diffraction configurations are core configurations in the structure of global entropy solutions of the two-dimensional Riemann problem, while the Riemann solutions themselves are local building blocks and determine local structures, global attractors, and large-time asymptotic states of general entropy solutions of multidimensional hyperbolic systems of conservation laws. In this sense, we have to understand the shock reflection-diffraction configurations, in order to understand fully the global entropy solutions of multidimensional hyperbolic systems of conservation laws.

Diverse patterns of shock reflection-diffraction configurations have attracted many asymptotic/numerical analysts since the middle of the 20th century. However, most of the fundamental issues involved, such as the structure and transition criteria of the different patterns, have not been understood. This is partially because physical and numerical experiments are hampered by various difficulties and have not yielded clear transition criteria between the different patterns. In light of this, a natural approach for understanding fully the shock reflection-diffraction configurations, especially with regard to the transition criteria, is via rigorous mathematical analysis. To achieve this, it is essential to establish the global existence, regularity, and structural stability of shock reflection-diffraction configurations: That is the main topic of this book.

Mathematical analysis of shock reflection-diffraction configurations involves dealing with several core difficulties in the analysis of nonlinear PDEs. These include nonlinear PDEs of mixed hyperbolic-elliptic type, nonlinear degenerate elliptic PDEs, nonlinear degenerate hyperbolic PDEs, free boundary problems for nonlinear degenerate PDEs, and corner singularities (especially when free boundaries meet the fixed boundaries), among others. These difficulties also arise in many further fundamental problems in continuum mechanics, differential geometry, mathematical physics, materials science, and other areas, including transonic flow problems, isometric embedding problems, and phase transition problems. Therefore, any progress in solving these problems requires new mathematical ideas, approaches, and techniques, all of which will both be very helpful for solving other problems with similar difficulties and open up new research directions.

Our efforts in the analysis of shock reflection-diffraction configurations for potential flow started 18 years ago when both of us were at Northwestern University, USA. We soon realized that the first step to achieving our goal should be to develop new free boundary techniques for multidimensional transonic shocks, along with other analytical techniques for nonlinear degenerate elliptic PDEs. After about two years of struggle, we developed such techniques, and these were published in [49] in 2003 and subsequent papers [42, 50, 51, 53]. With this groundwork, we first succeeded in developing a rigorous mathematical approach to establish the global existence and stability of regular shock reflection-diffraction solutions for large-angle wedges in [52] in 2005, the complete version of which was published electronically in 2006 and in print form in [54] in 2010. Since 2005, we have continued our efforts to solve von Neumann's sonic conjecture (*i.e.*, the existence of global regular reflection-diffraction solutions up to the sonic wedge angle with the supersonic reflection-diffraction configuration, containing a transonic reflected-diffracted shock), as well as von Neumann's detachment conjecture (*i.e.*, the necessary and sufficient condition for the existence of global regular reflection-diffraction solutions, even beyond the sonic angle, up to the detachment angle with the subsonic reflection-diffraction configuration, containing a transonic reflected-diffracted shock) (*cf.* [55, 57]). The results of these efforts were announced in [56, 58], and their detailed proofs constitute the main part of this book.

Some efforts have also been made by several groups of researchers on related models, including the unsteady small disturbance equation (USD), the pressure gradient equations, and the nonlinear wave system, as well as for some partial results for the potential flow equation and the full Euler equations. For the sake of completeness, we have made remarks and notes about these contributions throughout the book, and have tried to collect a detailed list of appropriate references in the bibliography.

Based on these results, along with our recent results on von Neumann's conjectures for potential flow, mathematical understanding of shock reflection-diffraction, especially for the global regular reflection-diffraction configurations, has reached a new height, and several new mathematical approaches and techniques have been developed. Moreover, new research opportunities and many new, challenging, and important problems have arisen during this exploration. Given these developments, we feel that it is the right time to publish this research monograph.

During the process of assembling this work, we have received persistent encouragement and invaluable suggestions from many leading mathematicians and scientists, especially John Ball, Luis Caffarelli, Alexander Chorin, Demetrios Christodoulou, Peter Constantin, Constantine Dafermos, Emmanuele DiBenedetto, Xiaxi Ding, Weinan E, Björn Engquist, Lawrence Craig Evans, Charles Fefferman, Edward Fraenkel, James Glimm, Helge Holden, Jiaxing Hong, Carlos Kenig, Sergiu Klainerman, Peter D. Lax, Tatsien Li, Fanhua Lin, Andrew Majda, Cathleen Morawetz, Luis Nirenberg, Benoît Perthame, Richard Schoen, Henrik Shahgholian, Yakov Sinai, Joel Smoller, John Toland, Neil Trudinger, and Juan Luis Vázquez. The materials presented herein contain direct and indirect contributions from many leading experts – teachers, colleagues, collaborators, and students alike, including Myoungjean Bae, Sunčica Canić, Yi Chao, Jun Chen, Shuxing Chen, Volker Elling, Beixiang Fang, Jingchen Hu, Feimin Huang, John Hunter, Katarina Jegdić, Siran Li, Tianhong Li, Yachun Li, Gary Lieberman, Tai-Ping Liu, Barbara Keyfitz, Eun Heui Kim, Jie Kuang, Stefano Marchesani, Ho Cheung Pang, Matthew Rigby, Matthew Schrecker, Denis Serre, Wancheng Sheng, Marshall Slemrod, Eitan Tadmor, Dehua Wang, Tian-Yi Wang, Yaguang Wang, Wei Xiang, Zhouping Xin, Hairong Yuan, Tong Zhang, Yongqian Zhang, Yuxi Zheng, and Dianwen Zhu, among others. We are grateful to all of them.

A significant portion of this work was done while the authors attended the Spring 2011 Program "*Free Boundary Problems: Theory and Applications*" at the Mathematical Sciences Research Institute in Berkeley, California, USA, and the 2014 Program "*Free Boundary Problems and Related Topics*" at the Isaac Newton Institute for Mathematical Sciences in Cambridge, UK. A part of the work was also supported by Keble College, University of Oxford, and a UK EPSRC Science and Innovation Award to the Oxford Centre for Nonlinear PDE (EP/E035027/1) when Mikhail Feldman visited Oxford in 2010.

The work of Gui-Qiang G. Chen was supported in part by the National Science Foundation under Grants DMS-0935967 and DMS-0807551, a UK EP-

SRC Science and Innovation Award to the Oxford Centre for Nonlinear PDE (EP/E035027/1), a UK EPSRC Award to the EPSRC Centre for Doctoral Training in PDEs (EP/L015811/1), the National Natural Science Foundation of China (under joint project Grant 10728101), and the Royal Society–Wolfson Research Merit Award (UK). The work of Mikhail Feldman was supported in part by the National Science Foundation under Grants DMS-0800245, DMS-1101260, and DMS-1401490, the Vilas Award from the University of Wisconsin-Madison, and the Simons Foundation via the Simons Fellows Program. Kurt Ballstadt deserves our special thanks for his effective assistance during the preparation of the manuscript. We are indebted to Princeton University Press, especially Vickie Kearn (Executive Editor) and Betsy Blumenthal and Lauren Bucca (Editorial Assistants), for their professional assistance.

Finally, we remark in passing that further supplementary materials to this research monograph will be posted at:

http://people.maths.ox.ac.uk/chengq/books/Monograph-CF-17/index.html
https://www.math.wisc.edu/~feldman/Monograph-CF-17/monograph.html

Part I

Shock Reflection-Diffraction, Nonlinear Conservation Laws of Mixed Type, and von Neumann's Conjectures

Chapter One

Shock Reflection-Diffraction, Nonlinear Partial Differential Equations of Mixed Type, and Free Boundary Problems

Shock waves are steep fronts that propagate in compressible fluids when convection dominates diffusion. They are fundamental in nature, especially in high-speed fluid flows. Examples include transonic shocks around supersonic or near-sonic flying bodies (such as aircraft), transonic and/or supersonic shocks formed by supersonic flows impinging onto solid wedges, bow shocks created by solar wind in space, blast waves caused by explosions, and other shocks generated by natural processes. Such shocks are governed by the Euler equations for compressible fluids or their variants, generally in the form of nonlinear conservation laws – nonlinear partial differential equations (PDEs) of divergence form. When a shock hits an obstacle (steady or flying), shock reflection-diffraction phenomena occur. One of the most fundamental research directions in mathematical fluid mechanics is the analysis of shock reflection-diffraction by wedges; see Ben-Dor [12], Courant-Friedrichs [99], von Neumann [267, 268, 269], and the references cited therein. When a plane shock hits a two-dimensional wedge head-on (*cf.* Fig. 1.1), it experiences a reflection-diffraction process; a fundamental question arisen is then what types of wave patterns of shock reflection-diffraction configurations may be formed around the wedge.

An archetypal system of PDEs describing shock waves in fluid mechanics, widely used in aerodynamics, is that of the Euler equations for potential flow (*cf.* [16, 95, 99, 139, 146, 221]). The Euler equations for describing the motion of a perfect fluid were first formulated by Euler [112, 113, 114] in 1752, based in part on the earlier work of D. Bernoulli [15], and were among the first PDEs for describing physical processes to be written down. The n-dimensional Euler equations for potential flow consist of the conservation law of mass and the Bernoulli law for the density and velocity potential (ρ, Φ):

$$\begin{cases} \partial_t \rho + \mathrm{div}_{\mathbf{x}}(\rho \nabla_{\mathbf{x}} \Phi) = 0, \\ \partial_t \Phi + \dfrac{1}{2}|\nabla_{\mathbf{x}} \Phi|^2 + h(\rho) = B_0, \end{cases} \tag{1.1}$$

where $\mathbf{x} \in \mathbb{R}^n$, B_0 is the Bernoulli constant determined by the incoming flow

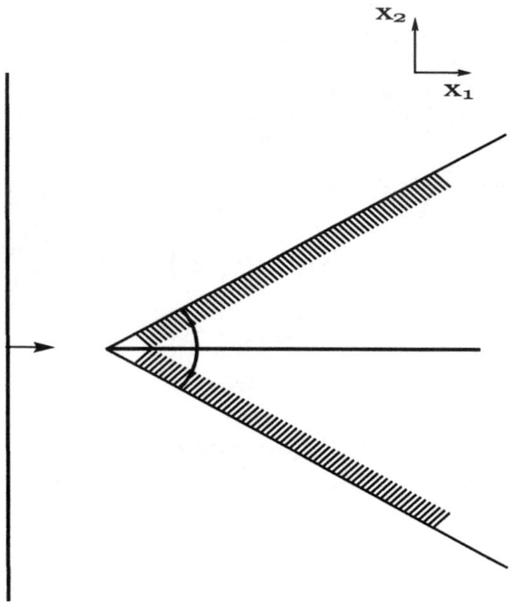

Figure 1.1: A plane shock hits a two-dimensional wedge in $\mathbb{R}^2$ head-on

and/or boundary conditions,

$$h'(\rho) = \frac{p'(\rho)}{\rho} = \frac{c^2(\rho)}{\rho},$$

and $c(\rho) = \sqrt{p'(\rho)}$ is the sonic speed (*i.e.*, the speed of sound).

The first equation in (1.1) is a transport-type equation for density ρ for a given $\nabla_{\mathbf{x}}\Phi$, while the second equation is the Hamilton-Jacobi equation for the velocity potential Φ coupling with density ρ through function $h(\rho)$.

For polytropic gases,

$$p(\rho) = \kappa\rho^\gamma, \qquad c^2(\rho) = \kappa\gamma\rho^{\gamma-1}, \qquad \gamma > 1,\ \kappa > 0.$$

Without loss of generality, we may choose $\kappa = \frac{1}{\gamma}$ so that

$$h(\rho) = \frac{\rho^{\gamma-1}-1}{\gamma-1}, \qquad c^2(\rho) = \rho^{\gamma-1}. \tag{1.2}$$

This can be achieved by noting that (1.1) is invariant under scaling:

$$(t, \mathbf{x}, B_0) \mapsto (\alpha^2 t, \alpha\mathbf{x}, \alpha^{-2}B_0)$$

with $\alpha^2 = \kappa\gamma$. In particular, Case $\gamma = 1$ can be considered as the limit of $\gamma \to 1+$ in (1.2):

$$h(\rho) = \ln\rho, \qquad c(\rho) = 1. \tag{1.3}$$

Henceforth, we will focus only on Case $\gamma > 1$, since Case $\gamma = 1$ can be handled similarly by making appropriate changes in the formulas so that the results of the main theorems for $\gamma > 1$ (below) also hold for $\gamma = 1$.

From the Bernoulli law, the second equation in (1.1), we have

$$\rho(\partial_t\Phi, |\nabla_{\mathbf{x}}\Phi|^2) = h^{-1}(B_0 - (\partial_t\Phi + \frac{1}{2}|\nabla_{\mathbf{x}}\Phi|^2)). \tag{1.4}$$

Then system (1.1) can be rewritten as the following time-dependent potential flow equation of second order:

$$\partial_t\rho(\partial_t\Phi, |\nabla_{\mathbf{x}}\Phi|^2) + \nabla_{\mathbf{x}} \cdot \big(\rho(\partial_t\Phi, |\nabla_{\mathbf{x}}\Phi|^2)\nabla_{\mathbf{x}}\Phi\big) = 0 \tag{1.5}$$

with $\rho(\partial_t\Phi, |\nabla_{\mathbf{x}}\Phi|^2)$ determined by (1.4). Equation (1.5) is a nonlinear wave equation of second order. Notice that equation (1.5) is invariant under a symmetry group formed of space-time dilations.

For a steady solution $\Phi = \varphi(\mathbf{x})$, *i.e.*, $\partial_t\Phi = 0$, we obtain the celebrated steady potential flow equation, especially in aerodynamics (*cf.* [16, 95, 99]):

$$\nabla_{\mathbf{x}} \cdot \big(\rho(|\nabla_{\mathbf{x}}\varphi|^2)\nabla_{\mathbf{x}}\varphi\big) = 0, \tag{1.6}$$

which is a second-order nonlinear PDE of mixed elliptic-hyperbolic type. This is a simpler case of the nonlinear PDE of mixed type for self-similar solutions, as shown in (1.12)–(1.13) later.

When the effects of vortex sheets and the deviation of vorticity become significant, the full Euler equations are required. The full Euler equations for compressible fluids in $\mathbb{R}^{n+1}_+ = \mathbb{R}_+ \times \mathbb{R}^n, t \in \mathbb{R}_+ := (0,\infty)$ and $\mathbf{x} \in \mathbb{R}^n$, are of the following form:

$$\begin{cases} \partial_t\,\rho + \nabla_{\mathbf{x}} \cdot (\rho\mathbf{v}) = 0, \\ \partial_t(\rho\mathbf{v}) + \nabla_{\mathbf{x}} \cdot (\rho\mathbf{v} \otimes \mathbf{v}) + \nabla_{\mathbf{x}}p = 0, \\ \partial_t\big(\rho(\frac{1}{2}|\mathbf{v}|^2 + e)\big) + \nabla_{\mathbf{x}} \cdot \big(\rho\mathbf{v}(\frac{1}{2}|\mathbf{v}|^2 + e + \frac{p}{\rho})\big) = 0, \end{cases} \tag{1.7}$$

where ρ is the density, $\mathbf{v} \in \mathbb{R}^n$ the fluid velocity, p the pressure, and e the internal energy. Two other important thermodynamic variables are temperature θ and entropy S. Here, $\mathbf{a} \otimes \mathbf{b}$ denotes the tensor product of vectors $\mathbf{a}$ and $\mathbf{b}$.

Choose (ρ, S) as the independent thermodynamical variables. Then the constitutive relations can be written as $(e, p, \theta) = (e(\rho, S), p(\rho, S), \theta(\rho, S))$, governed by

$$\theta dS = de + pd\tau = de - \frac{p}{\rho^2}d\rho,$$

as introduced by Gibbs [129].

For a polytropic gas,

$$p = (\gamma - 1)\rho e, \qquad e = c_v\theta, \qquad \gamma = 1 + \frac{R}{c_v}, \tag{1.8}$$

or equivalently,

$$p = p(\rho, S) = \kappa \rho^\gamma e^{S/c_v}, \qquad e = e(\rho, S) = \frac{\kappa}{\gamma - 1}\rho^{\gamma-1} e^{S/c_v}, \tag{1.9}$$

where $R > 0$ may be taken to be the universal gas constant divided by the effective molecular weight of the particular gas, $c_v > 0$ is the specific heat at constant volume, $\gamma > 1$ is the adiabatic exponent, and $\kappa > 0$ may be chosen as any constant through scaling.

The full Euler equations in the general form presented here were originally derived by Euler [112, 113, 114] for mass, Cauchy [29, 30] for linear and angular momentum, and Kirchhoff [165] for energy.

The nonlinear equations (1.5) and (1.7) fit into the general form of hyperbolic conservation laws:

$$\partial_t \mathbf{A}(\partial_t \mathbf{u}, \nabla_{\mathbf{x}} \mathbf{u}, \mathbf{u}) + \nabla_{\mathbf{x}} \cdot \mathbf{B}(\partial_t \mathbf{u}, \nabla_{\mathbf{x}} \mathbf{u}, \mathbf{u}) = 0, \tag{1.10}$$

or

$$\partial_t \mathbf{u} + \nabla_{\mathbf{x}} \cdot \mathbf{f}(\mathbf{u}) = 0, \qquad \mathbf{u} \in \mathbb{R}^m, \ \mathbf{x} \in \mathbb{R}^n, \tag{1.11}$$

where $\mathbf{A} : \mathbb{R}^m \times \mathbb{R}^{n\times m} \times \mathbb{R}^m \mapsto \mathbb{R}^m$, $\mathbf{B} : \mathbb{R}^m \times \mathbb{R}^{n\times m} \times \mathbb{R}^m \mapsto (\mathbb{R}^m)^n$, and $\mathbf{f} : \mathbb{R}^m \mapsto (\mathbb{R}^m)^n$ are nonlinear mappings. Besides (1.5) and (1.7), most of the nonlinear PDEs arising from physical or engineering science can also be formulated in accordance with form (1.10) or (1.11), or their variants. Moreover, the second-order form (1.10) of hyperbolic conservation laws can be reformulated as a first-order system (1.11). The hyperbolicity of system (1.11) requires that, for all $\boldsymbol{\xi} \in \mathbb{S}^{n-1}$, matrix $[\boldsymbol{\xi} \cdot \nabla_{\mathbf{u}} \mathbf{f}(\mathbf{u})]_{m\times m}$ have m real eigenvalues $\lambda_j(\mathbf{u}, \boldsymbol{\xi}), j = 1, 2, \cdots, m$, and be diagonalizable. See Lax [171], Glimm-Majda [139], and Majda [210].

The complexity of shock reflection-diffraction configurations was first reported in 1878 by Ernst Mach [206], who observed two patterns of shock reflection-diffraction configurations that are now named the Regular Reflection (RR: two-shock configuration; see Fig. 1.2) and the Simple Mach Reflection (SMR: three-shock and one-vortex-sheet configuration; see Fig. 1.3); see also [12, 167, 228]. The problem remained dormant until the 1940s when von Neumann [267, 268, 269], as well as other mathematical/experimental scientists, began extensive research on shock reflection-diffraction phenomena, owing to their fundamental importance in various applications (see von Neumann [267, 268] and Ben-Dor [12]; see also [11, 132, 152, 160, 166, 205, 248, 249] and the references cited therein).

It has since been found that there are more complexity and variety of shock reflection-diffraction configurations than what Mach originally observed: The Mach reflection can be further divided into more specific sub-patterns, and many other patterns of shock reflection-diffraction configurations may occur, for example, the Double Mach Reflection (see Fig. 1.4), the von Neumann Reflection, and the Guderley Reflection; see also [12, 99, 139, 143, 159, 243, 257, 258, 259, 263, 267, 268] and the references cited therein.

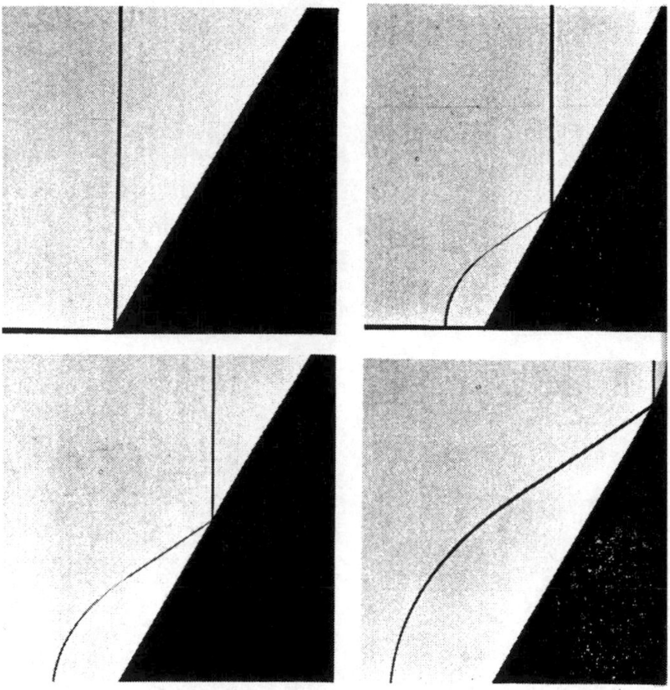

Figure 1.2: Regular Reflection for large-angle wedges. From Van Dyke [263, pp. 142].

The fundamental scientific issues arising from all of this are

(i) *The structure of shock reflection-diffraction configurations*;

(ii) *The transition criteria between the different patterns of shock reflection-diffraction configurations*;

(iii) *The dependence of the patterns upon the physical parameters such as the wedge angle* θ_w*, the incident-shock Mach number* M_I *(a measure of the strength of the shock), and the adiabatic exponent* $\gamma \geq 1$.

Careful asymptotic analysis has been made for various reflection-diffraction configurations in Lighthill [199, 200], Keller-Blank [162], Hunter-Keller [158], and Morawetz [221], as well as in [128, 148, 155, 255, 267, 268] and the references cited therein; see also Glimm-Majda [139]. Large or small scale numerical simulations have also been made; *e.g.*, [12, 139], [104, 105, 149, 170, 232, 240], and [133, 134, 135, 160, 273] (see also the references cited therein).

On the other hand, most of the fundamental issues for shock reflection-diffraction phenomena have not been understood, especially the global structure

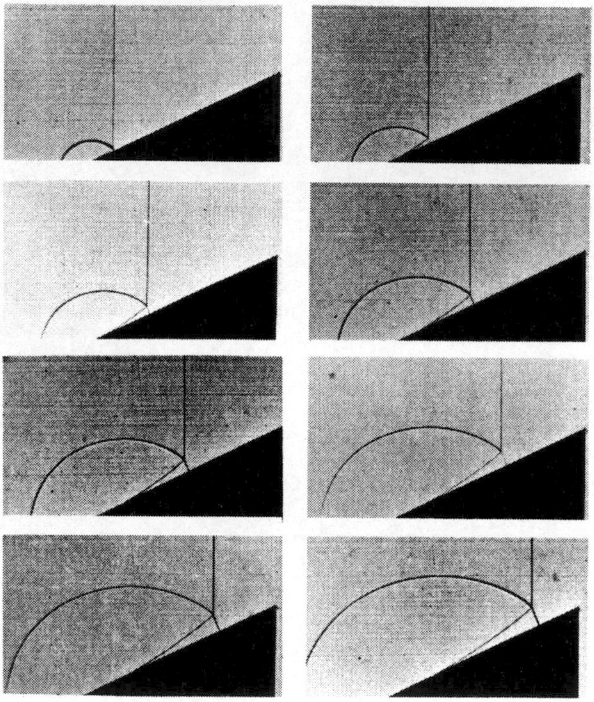

Figure 1.3: Simple Mach Reflection when the wedge angle becomes small. From Van Dyke [263, pp. 143].

and transition between the different patterns of shock reflection-diffraction configurations. This is partially because physical and numerical experiments are hampered by various difficulties and have not thusfar yielded clear transition criteria between the different patterns. In particular, numerical dissipation or physical viscosity smears the shocks and causes the boundary layers that interact with the reflection-diffraction configurations and may cause spurious Mach steams; *cf.* Woodward-Colella [273]. Furthermore, some different patterns occur in which the wedge angles are only fractions of a degree apart; a resolution has challenged even sophisticated modern numerical and laboratory experiments. For this reason, it is almost impossible to distinguish experimentally between the sonic and detachment criteria, as was pointed out by Ben-Dor in [12] (also *cf.* Chapter 7 below). On account of this, a natural approach to understand fully the shock reflection-diffraction configurations, especially the transition criteria, is via rigorous mathematical analysis. To carry out this analysis, it is essential to establish first the global existence, regularity, and structural stability of shock reflection-diffraction configurations: That is the main topic of this book.

Furthermore, the shock reflection-diffraction configurations are core configurations in the structure of global entropy solutions of the two-dimensional Rie-

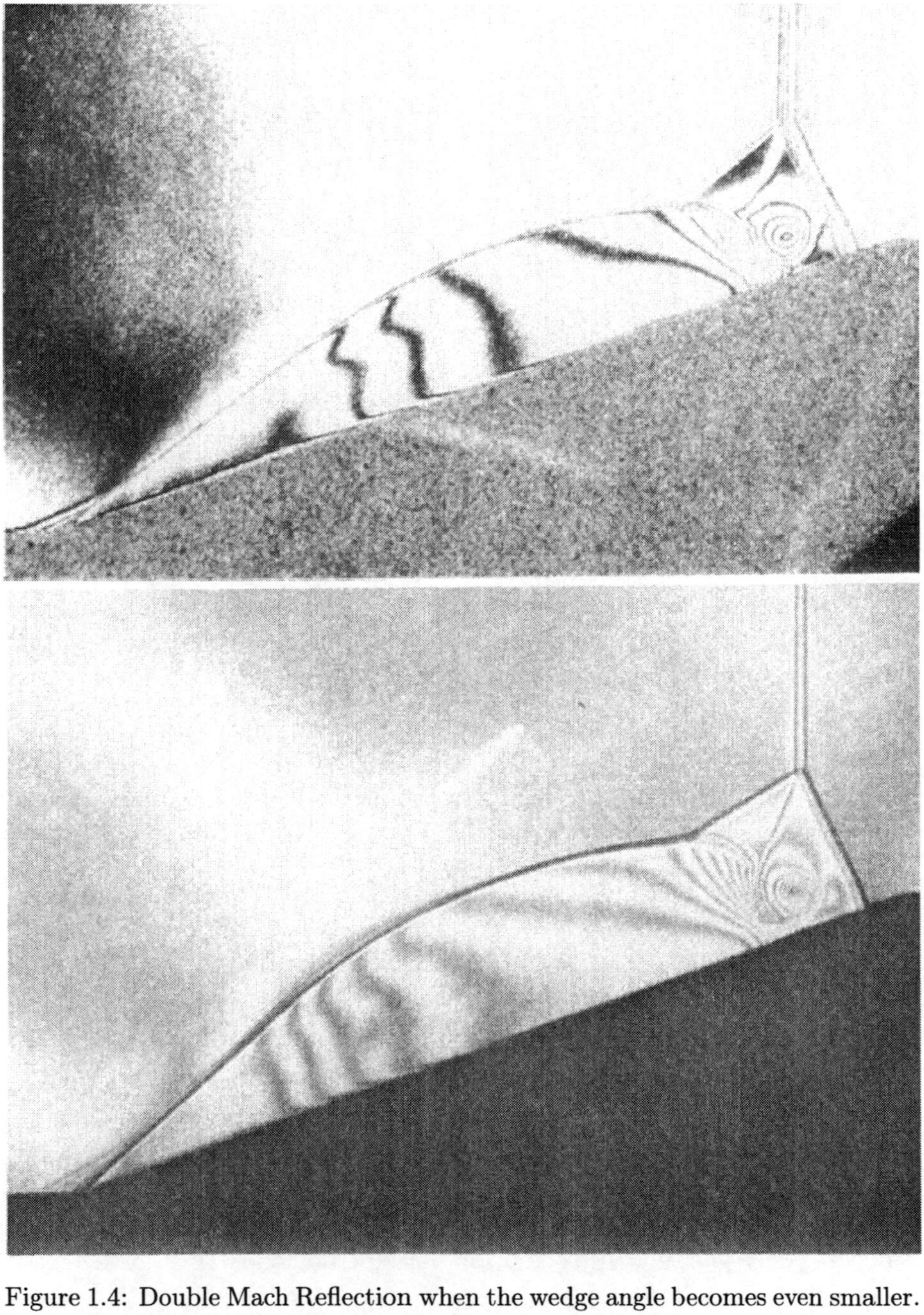

Figure 1.4: Double Mach Reflection when the wedge angle becomes even smaller. From Ben-Dor [12, pp. 67].

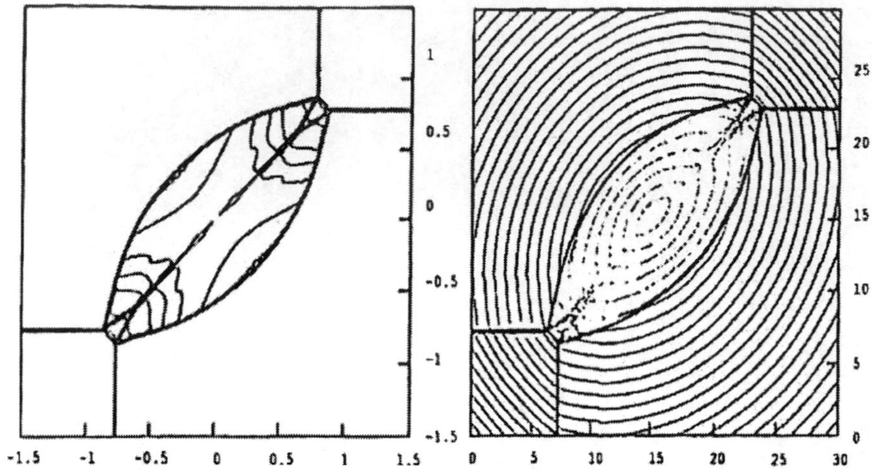

Density contour curves Self-Mach number contour curves

Figure 1.5: Riemann solutions: Simple Mach Reflection; see [33]

mann problem for hyperbolic conservation laws (see Figs. 1.5–1.6), while the Riemann solutions are building blocks and determine local structures, global attractors, and large-time asymptotic states of general entropy solutions of multidimensional hyperbolic systems of conservation laws (see [31]–[35], [138, 139, 169, 175, 181, 233, 235, 236, 286], and the references cited therein). Consequently, we have to understand the shock reflection-diffraction configurations in order to fully understand global entropy solutions of the multidimensional hyperbolic systems of conservation laws.

Mathematically, the analysis of shock reflection-diffraction configurations involves several core difficulties that we have to face for the mathematical theory of nonlinear PDEs:

(i) Nonlinear PDEs of Mixed Elliptic-Hyperbolic Type: The first is that the underlying nonlinear PDEs change type from hyperbolic to elliptic in the shock reflection-diffraction configurations, so that the nonlinear PDEs are of mixed hyperbolic-elliptic type.

This can be seen as follows: Since both the system and the initial-boundary conditions admit a symmetry group formed of space-time dilations, we seek self-similar solutions of the problem:

$$\rho(t,\mathbf{x}) = \rho(\boldsymbol{\xi}), \quad \Phi(t,\mathbf{x}) = t\phi(\boldsymbol{\xi}),$$

depending only upon $\boldsymbol{\xi} = \frac{\mathbf{x}}{t} \in \mathbb{R}^2$. For the Euler equation (1.5) for potential flow, the corresponding pseudo-potential function $\varphi(\boldsymbol{\xi}) = \phi(\boldsymbol{\xi}) - \frac{|\boldsymbol{\xi}|^2}{2}$ satisfies the following potential flow equation of second order:

$$\operatorname{div}\left(\rho(|D\varphi|^2,\varphi)D\varphi\right) + 2\rho(|D\varphi|^2,\varphi) = 0 \tag{1.12}$$

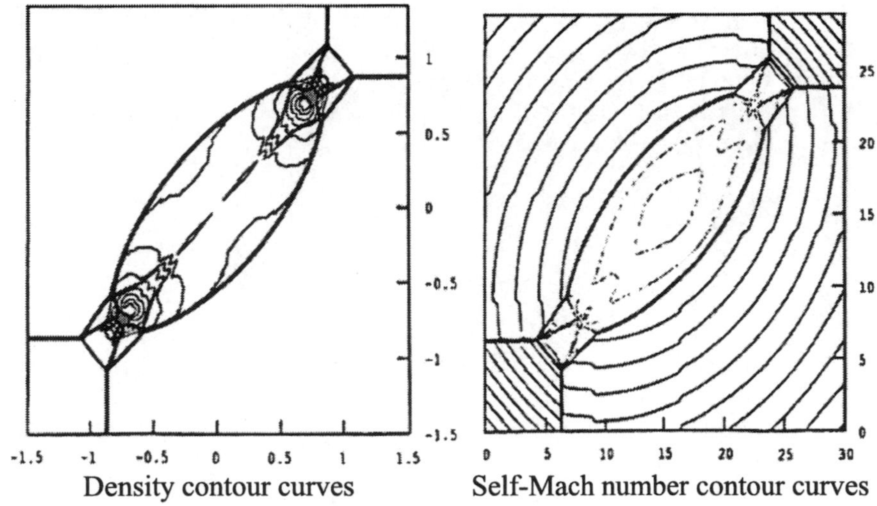

Figure 1.6: Riemann solutions: Double Mach reflection; see [33]

with

$$\rho(|D\varphi|^2, \varphi) = \left(\rho_0^{\gamma-1} - (\gamma-1)(\varphi + \frac{1}{2}|D\varphi|^2)\right)^{\frac{1}{\gamma-1}}, \tag{1.13}$$

where div and D represent the divergence and the gradient, respectively, with respect to the self-similar variables $\boldsymbol{\xi} = (\xi_1, \xi_2)$, that is, $D := (D_1, D_2) = (D_{\xi_1}, D_{\xi_2})$. Then the sonic speed becomes:

$$c = c(|D\varphi|^2, \varphi, \rho_0^{\gamma-1}) = \left(\rho_0^{\gamma-1} - (\gamma-1)(\frac{1}{2}|D\varphi|^2 + \varphi)\right)^{\frac{1}{2}}. \tag{1.14}$$

Equation (1.12) can be written in the following non-divergence form of nonlinear PDE of second order:

$$\sum_{i,j=1}^{2} a_{ij}(\varphi, D\varphi) D_{ij}\varphi = f(\varphi, D\varphi), \tag{1.15}$$

where $[a_{ij}(\varphi, D\varphi)]_{1\le i,j\le 2}$ is a symmetric matrix and $D_{ij} = D_iD_j, i, j = 1, 2$. The type of equation that (1.12) or (1.15) is depends on the values of solution φ and its gradient $D\varphi$. More precisely, equation (1.15) is elliptic on a solution φ when the two eigenvalues $\lambda_j(\varphi, D\varphi), j = 1, 2$, of the symmetric matrix $[a_{ij}(\varphi, D\varphi)]$ have the same sign on φ:

$$\lambda_1(\varphi, D\varphi)\lambda_2(\varphi, D\varphi) > 0. \tag{1.16}$$

Correspondingly, equation (1.15) is (strictly) hyperbolic on a solution φ if the two eigenvalues of the matrix have the opposite signs on φ:

$$\lambda_1(\varphi, D\varphi)\lambda_2(\varphi, D\varphi) < 0. \tag{1.17}$$

The more complicated case is that of the mixed elliptic-hyperbolic type for which $\lambda_1(\varphi, D\varphi)\lambda_2(\varphi, D\varphi)$ changes its sign when the values of φ and $D\varphi$ change in the physical domain under consideration.

In particular, equation (1.12) is a *nonlinear second-order conservation law of mixed elliptic-hyperbolic type*. It is *elliptic* if

$$|D\varphi| < c(|D\varphi|^2, \varphi, \rho_0^{\gamma-1}), \tag{1.18}$$

and *hyperbolic* if

$$|D\varphi| > c(|D\varphi|^2, \varphi, \rho_0^{\gamma-1}). \tag{1.19}$$

The types normally change with $\boldsymbol{\xi}$ from hyperbolic in the far field to elliptic around the wedge vertex, which is the case that the corresponding physical velocity $\nabla_{\mathbf{x}}\Phi$ is bounded.

Similarly, for the full Euler equations, the corresponding self-similar solutions are governed by a nonlinear system of conservation laws of composite-mixed hyperbolic-elliptic type, as shown in (18.3.1) in Chapter 18.

Such nonlinear PDEs of mixed type also arise naturally in many other fundamental problems in continuum physics, differential geometry, elasticity, relativity, calculus of variations, and related areas.

Classical fundamental linear PDEs of mixed elliptic-hyperbolic type include the following:

The Lavrentyev-Bitsadze equation for an unknown function $u(x, y)$:

$$u_{xx} + \text{sign}(x)u_{yy} = 0. \tag{1.20}$$

This becomes the wave equation (hyperbolic) in half-plane $x < 0$ and the Laplace equation (elliptic) in half-plane $x > 0$, and changes the type from elliptic to hyperbolic via a jump discontinuous coefficient $\text{sign}(x)$.

The Keldysh equation for an unknown function $u(x, y)$:

$$xu_{xx} + u_{yy} = 0. \tag{1.21}$$

This is hyperbolic in half-plane $x < 0$, elliptic in half-plane $x > 0$, and degenerates on line $x = 0$. This equation is of parabolic degeneracy in domain $x \leq 0$, for which the two characteristic families are quadratic parabolas lying in half-plane $x < 0$ and tangential at contact points to the degenerate line $x = 0$. Its degeneracy is also determined by the classical elliptic or hyperbolic Euler-Poisson-Darboux equation:

$$u_{\tau\tau} \pm u_{yy} + \frac{\beta}{\tau}u_\tau = 0 \tag{1.22}$$

with $\beta = -\frac{1}{4}$, where $\tau = \frac{1}{2}|x|^{\frac{1}{2}}$, and signs "$\pm$" in (1.22) are determined by the corresponding half-planes $\pm x > 0$.

The Tricomi equation for an unknown function $u(x, y)$:

$$u_{xx} + xu_{yy} = 0. \tag{1.23}$$

This is hyperbolic when $x < 0$, elliptic when $x > 0$, and degenerates on line $x = 0$. This equation is of hyperbolic degeneracy in domain $x \leq 0$, for which the two characteristic families coincide *perpendicularly* to line $x = 0$. Its degeneracy is also determined by the classical elliptic or hyperbolic Euler-Poisson-Darboux equation (1.22) with $\beta = \frac{1}{3}$, where $\tau = \frac{2}{3}|x|^{\frac{3}{2}}$.

For linear PDEs of mixed elliptic-hyperbolic type such as (1.20)–(1.23), the transition boundary between the elliptic and hyperbolic phases is known *a priori*. One of the classical approaches to the study of such mixed-type linear equations is the fundamental solution approach, since the optimal regularity and/or singularities of solutions near the transition boundary are determined by the fundamental solution (see [17, 37, 39, 41, 275, 278]).

For nonlinear PDEs of mixed elliptic-hyperbolic type such as (1.12), the transition boundary between the elliptic and hyperbolic phases is *a priori* unknown, so that most of the classical approaches, especially the fundamental solution approach, no longer work. New ideas, approaches, and techniques are in great demand for both theoretical and numerical analysis.

(ii) Free Boundary Problems: Following the discussion in **(i)**, above, the analysis of shock reflection-diffraction configurations can be reduced to the analysis of a *free boundary problem*, as we will show in §2.4, in which the reflected-diffracted shock, defined as the transition boundary from the hyperbolic to elliptic phase, is a free boundary that cannot be determined prior to the determination of the solution.

The subject of free boundary problems has its origin in the study of the Stefan problem, which models the melting of ice (*cf.* Stefan [250]). In that problem, the moving-in-time boundary between water and ice is not known *a priori*, but is determined by the solution of the problem. More generally, free boundary problems are concerned with sharp transitions in the variables involved in the problems, such as the change in the temperature between water and ice in the Stefan problem, and the changes in the velocity and density across the shock wave in the shock reflection-diffraction configurations. Mathematically, this rapid transition is simplified to be seen as occurring infinitely fast across a curve or surface of discontinuity or constraint in the PDEs governing the physical or other processes under consideration. The location of these curves and surfaces, called free boundaries, is required to be determined in the process of solving the free boundary problem. Free boundaries subdivide the domain into subdomains in which the governing equations (usually PDEs) are satisfied. On the free boundaries, the free boundary conditions, derived from the models, are prescribed. The number of conditions on the free boundary is such that the PDE governing the problem, combined with the free boundary conditions, allows us to determine both the location of the free boundary and the solution in the whole domain. That is, more conditions are required on the free boundary than in the case of the fixed boundary value problem for the same PDEs in a fixed domain. Great progress has been made on free boundary problems for linear PDEs. Further developments, especially in terms of solving such problems for nonlinear PDEs of mixed type, ask for new mathematical approaches and techniques. For a better sense of these, see Chen-Shahgholian-Vázquez [67] and the references cited therein.

(iii) Estimates of Solutions to Nonlinear Degenerate PDEs: The third difficulty concerns the degeneracies that are along the sonic arc, since the sonic arc is another transition boundary from the hyperbolic to elliptic phase in the shock reflection-diffraction configurations, for which the corresponding nonlinear PDE becomes a nonlinear degenerate hyperbolic equation on its one side and a nonlinear degenerate elliptic equation on the other side; both of these degenerate on the sonic arc. Also, unlike the reflected-diffracted shock, the sonic arc is not a free boundary; its location is explicitly known. In order to construct a global regular reflection-diffraction configuration, we need to determine the unknown velocity potential in the subsonic (elliptic) domain such that the reflected-diffracted shock and the sonic arc are parts of its boundary. Thus, we can view our problem as a free boundary problem for an elliptic equation of second order with ellipticity degenerating along a part of the fixed boundary. Moreover, the solution should satisfy two Rankine-Hugoniot conditions on the transition boundary of the elliptic region, which includes both the shock and the sonic arc. While this over-determinacy gives the correct number of free boundary conditions on the shock, the situation is different on the sonic arc that is a fixed boundary. Normally, only one condition may be prescribed for the elliptic problem. Therefore, we have to prove that the other condition is also satisfied on the sonic arc by the solution. To achieve this, we exploit the detailed structure of the elliptic degeneracy of the nonlinear PDE to make careful estimates of the solution near the sonic arc in the properly weighted and scaled $C^{2,\alpha}$–spaces, for which the nonlinearity plays a crucial role.

(iv) Corner Singularities: Further difficulties include the singularities of solutions at the corner formed by the reflected-diffracted shock (free boundary) and the sonic arc (degenerate elliptic curve), at the wedge vertex, as well as at the corner between the reflected shock and the wedge at the reflection point for the transition from the supersonic to subsonic regular reflection-diffraction configurations when the wedge angle decreases. For the latter, it requires uniform *a priori* estimates for the solutions as the sonic arc shrinks to a point; the degenerate ellipticity then changes to the uniform ellipticity when the wedge angle decreases across the sonic angle up to the detachment angle, as described in §2.4–§2.6.

These difficulties also arise in many further fundamental problems in continuum physics (fluid/solid), differential geometry, mathematical physics, materials science, and other areas, such as transonic flow problems, isometric embedding problems, and phase transition problems; see [9, 10, 16, 93, 68, 69, 95, 99, 139, 147, 168, 181, 220, 270, 286] and the references cited therein. Therefore, any progress in shock reflection-diffraction analysis requires new mathematical ideas, approaches, and techniques, all of which will be very useful for solving other problems with similar difficulties and open up new research opportunities.

We focus mainly on the mathematics of shock reflection-diffraction and von Neumann's conjectures for potential flow, as well as offering (in Parts I–IV) new analysis to overcome the associated difficulties. The mathematical approaches and techniques developed here will be useful in tackling other nonlinear problems involving similar difficulties. One of the recent examples of this is the Prandtl-Meyer problem for supersonic flow impinging onto solid wedges, another longstanding open problem in mathematical fluid mechanics, which has been treated in Bae-Chen-Feldman [5, 6].

In Part I, we state our main results and give an overview of the main steps of their proofs.

In Part II, we present some relevant results for nonlinear elliptic equations of second order (for which the structural conditions and some regularity of coefficients are not required), convenient for applications in the rest of the book, and study the existence and regularity of solutions of certain boundary value problems in the domains of appropriate structure for an equation with ellipticity degenerating on a part of the boundary, which include the boundary value problems used in the construction of the iteration map in the later chapters. We also present basic properties of the self-similar potential flow equation, with focus on the two-dimensional case.

In Part III, we first focus on von Neumann's sonic conjecture – that is, the conjecture concerning the existence of regular reflection-diffraction solutions up to the sonic angle, with a supersonic shock reflection-diffraction configuration containing a transonic reflected-diffracted shock, and then provide its whole detailed proof and related analysis. We treat this first on account of the fact that the presentation in this case is both foundational and relatively simpler than that in the case beyond the sonic angle.

Once the analysis for the sonic conjecture is done, we present, in Part IV, our proof of von Neumann's detachment conjecture – that is, the conjecture concerning the existence of regular reflection-diffraction solutions, even beyond the sonic angle up to the detachment angle, with a subsonic shock reflection-diffraction configuration containing a transonic reflected-diffracted shock. This is more technically involved. To achieve it, we make the whole iteration again, starting from the normal reflection when the wedge angle is $\frac{\pi}{2}$, and prove the results for both the supersonic and subsonic regular reflection-diffraction configurations by going over the previous arguments with the necessary additions (instead of writing all the details of the proof up to the detachment angle from the beginning). We present the proof in this way to make it more readable.

In Part V, we present the mathematical formulation of the shock reflection-diffraction problem for the full Euler equations and uncover the role of the potential flow equation for the shock reflection-diffraction even in the realm of the full Euler equations. We also discuss further connections and their roles in developing new mathematical ideas, techniques, and approaches for solving further open problems in related scientific areas.

Chapter Two

Mathematical Formulations and Main Theorems

In this chapter, we first analyze the potential flow equation (1.5) and its planar shock-front solutions, and then formulate the shock reflection-diffraction problem into an initial-boundary value problem. Next we employ the self-similarity of the problem to reformulate the initial-boundary value problem into a boundary value problem in the self-similar coordinates. To solve von Neumann's conjectures, we further reformulate the boundary value problem into a free boundary problem for a nonlinear second-order conservation law of mixed hyperbolic-elliptic type. Finally, we present the main theorems for the existence, regularity, and stability of regular reflection-diffraction solutions of the free boundary problem.

2.1 THE POTENTIAL FLOW EQUATION

The time-dependent potential flow equation of second order for the velocity potential Φ takes the form of (1.5) with $\rho(\partial_t\Phi, \nabla_{\mathbf{x}}\Phi)$ determined by (1.4), which is a nonlinear wave equation.

Definition 2.1.1. *A function $\Phi \in W^{1,1}_{\rm loc}(\mathbb{R}_+ \times \mathbb{R}^2)$ is called a weak solution of equation* (1.5) *in a domain $\mathcal{D} \subset \mathbb{R}_+ \times \mathbb{R}^2$ if Φ satisfies the following properties*:

(i) $B_0 - \big(\partial_t\Phi + \frac{1}{2}|\nabla_{\mathbf{x}}\Phi|^2\big) \ge h(0+)$ *a.e. in* $\mathcal{D}$;

(ii) $(\rho(\partial_t\Phi, |\nabla_{\mathbf{x}}\Phi|^2), \rho(\partial_t\Phi, |\nabla_{\mathbf{x}}\Phi|^2)|\nabla_{\mathbf{x}}\Phi|) \in (L^1_{\rm loc}(\mathcal{D}))^2$;

(iii) *For every* $\zeta \in C^\infty_c(\mathcal{D})$,

$$\int_{\mathcal{D}} \Big(\rho(\partial_t\Phi, |\nabla_{\mathbf{x}}\Phi|^2)\partial_t\zeta + \rho(\partial_t\Phi, |\nabla_{\mathbf{x}}\Phi|^2)\nabla_{\mathbf{x}}\Phi \cdot \nabla_{\mathbf{x}}\zeta\Big)\, d\mathbf{x}dt = 0.$$

In the study of a piecewise smooth weak solution of (1.5) with jump for $(\partial_t\Phi, \nabla_{\mathbf{x}}\Phi)$ across an oriented surface $\mathcal{S}$ with unit normal $\mathbf{n} = (n_t, \mathbf{n}_{\mathbf{x}}), \mathbf{n}_{\mathbf{x}} = (n_1, n_2)$, in the $(t, \mathbf{x})$–coordinates, the requirement of the weak solution of (1.5) in the sense of Definition 2.1.1 yields the *Rankine-Hugoniot jump condition* across $\mathcal{S}$:

$$[\rho(\partial_t\Phi, |\nabla_{\mathbf{x}}\Phi|^2)]n_t + [\rho(\partial_t\Phi, |\nabla_{\mathbf{x}}\Phi|^2)\nabla_{\mathbf{x}}\Phi] \cdot \mathbf{n}_{\mathbf{x}} = 0, \qquad (2.1.1)$$

where the square bracket, $[w]$, denotes the jump of quantity w across the oriented surface $\mathcal{S}$; that is, assuming that $\mathcal{S}$ subdivides $\mathcal{D}$ into subregions $\mathcal{D}^+$ and $\mathcal{D}^-$

so that, for every $(t,\mathbf{x}) \in \mathcal{S}$, there exists $\varepsilon > 0$ such that $(t,\mathbf{x}) \pm s\mathbf{n} \in \mathcal{D}^{\pm}$ if $s \in (0,\varepsilon)$, define

$$[w](t,\mathbf{x}) := \lim_{\substack{(\tau,\mathbf{y})\to(t,\mathbf{x})\\(\tau,\mathbf{y})\in\mathcal{D}^+}} w(\tau,\mathbf{y}) - \lim_{\substack{(\tau,\mathbf{y})\to(t,\mathbf{x})\\(\tau,\mathbf{y})\in\mathcal{D}^-}} w(\tau,\mathbf{y}).$$

Notice that $\Phi \in W^{1,1}$ is required in Definition 2.1.1, which implies the continuity of Φ across a shock-front $\mathcal{S}$ for piecewise smooth solutions:

$$[\Phi]_{\mathcal{S}} = 0. \tag{2.1.2}$$

In fact, the continuity of Φ in (2.1.2) can also be derived for the piecewise smooth solution Φ (without assumption $\Phi \in W^{1,1}$) by requiring that Φ keep the validity of the equations:

$$\nabla_{\mathbf{x}} \times \mathbf{v} = 0, \quad \partial_t \mathbf{v} = \nabla_{\mathbf{x}}(\partial_t\Phi) \tag{2.1.3}$$

in the sense of distributions. This is tantamount to requiring that

$$(\partial_t\Phi, \mathbf{v}) = (\partial_t\Phi, \nabla_{\mathbf{x}}\Phi). \tag{2.1.4}$$

The condition on $\mathbf{v}$ is that

$$(\partial_t, \nabla_{\mathbf{x}}) \times (\partial_t\Phi, \mathbf{v}) = 0,$$

which is equivalent to (2.1.3). By definition, the equations in (2.1.4) for piecewise smooth solutions are understood as

$$\iint \Phi\partial_{x_i}\psi \, dtd\mathbf{x} = -\iint v_i\psi \, dtd\mathbf{x}, \qquad i = 1,2,$$

for any test function $\psi \in C_0^\infty((0,\infty)\times\mathbb{R}^2)$. Using the Gauss-Green formula in the two regions of continuity of $(\partial_t\Phi, \nabla_{\mathbf{x}}\Phi)$ separated by $\mathcal{S}$ in the standard fashion, we obtain

$$\int_{\mathcal{S}} [\Phi]\psi \, d\sigma = 0,$$

where $d\sigma$ is the surface measure on $\mathcal{S}$. This implies the continuity of Φ across a shock-front $\mathcal{S}$ in (2.1.2).

The discontinuity $\mathcal{S}$ of $(\partial_t\Phi, \nabla_{\mathbf{x}}\Phi)$ is called a *shock* if Φ further satisfies the physical *entropy condition*: The corresponding density function $\rho(\partial_t\Phi, \nabla_{\mathbf{x}}\Phi)$ increases across $\mathcal{S}$ in the relative flow direction with respect to $\mathcal{S}$ (*cf.* [94, 99]).

Definition 2.1.2. *Let Φ be a piecewise smooth weak solution of* (1.5) *with jump for $(\partial_t\Phi, \nabla_{\mathbf{x}}\Phi)$ across an oriented surface $\mathcal{S}$. The discontinuity $\mathcal{S}$ of $(\partial_t\Phi, \nabla_{\mathbf{x}}\Phi)$ is called a shock if Φ further satisfies the physical entropy condition: The corresponding density function $\rho(\partial_t\Phi, \nabla_{\mathbf{x}}\Phi)$ increases across $\mathcal{S}$ in the relative flow direction with respect to $\mathcal{S}$ (cf.* [94, 99]).

The jump condition in (2.1.1) from the conservation of mass and the continuity of Φ in (2.1.2) are the conditions that are actually used in practice, especially in aerodynamics, resulting from the *Rankine-Hugoniot conditions* for the time-dependent potential flow equation (1.5). The empirical evidence for this is that entropy solutions of (1.1) or (1.5) are fairly close to the corresponding entropy solutions of the full Euler equations, provided that the strengths of shock-fronts are small, the curvatures of shock-fronts are not too large, and the amount of vorticity is small in the region of interest. In fact, for the solutions containing a weak shock, especially in aerodynamic applications, the potential flow equation (1.5) and the full Euler flow model (1.7) match each other well up to the third order of the shock strength. Furthermore, we will show in Chapter 18 that, for the shock reflection-diffraction problem, the Euler equations (1.1) for potential flow are actually an *exact* match in an important region of the shock reflection-diffraction configurations to the full Euler equations (1.7). See also Bers [16], Glimm-Majda [139], and Morawetz [220, 221, 222].

Planar shock-front solutions are special piecewise smooth solutions given by the explicit formulae:

$$\Phi = \begin{cases} \Phi^+, & x_1 > st, \\ \Phi^-, & x_1 < st \end{cases} \tag{2.1.5}$$

with

$$\Phi^\pm = a_0^\pm t + u^\pm x_1 + v^\pm x_2. \tag{2.1.6}$$

Then the continuity condition (2.1.2) of Φ across the shock-front implies

$$[v] = 0, \qquad [a_0] + s[u] = 0. \tag{2.1.7}$$

The jump condition (2.1.1) yields

$$s[\rho] = [\rho u], \tag{2.1.8}$$

since $\mathbf{n} = \frac{1}{\sqrt{s^2+1}}(-s, 1, 0)$.

The relation between ρ and Φ via the Bernoulli law is

$$[a_0 + \frac{1}{2}u^2] + \frac{1}{\gamma - 1}[\rho^{\gamma-1}] = 0, \tag{2.1.9}$$

where we have used (1.2) for polytropic gases. From now on, we focus on $\gamma > 1$.

Combining (2.1.7)–(2.1.9), we conclude

$$\begin{cases} [u] = -\sqrt{\dfrac{2[\rho][\rho^{\gamma-1}]}{(\gamma-1)(\rho^+ + \rho^-)}}, \\ [a_0] = -\dfrac{[\rho u][u]}{[\rho]}, \\ s = \dfrac{[\rho u]}{[\rho]}. \end{cases} \tag{2.1.10}$$

This implies that the shock speed s is

$$s = u_+ + \rho_- \sqrt{\frac{2[\rho^{\gamma-1}]}{(\gamma-1)[\rho^2]}}. \tag{2.1.11}$$

The entropy condition is

$$\rho_+ < \rho_- \qquad \text{if } u^\pm > 0. \tag{2.1.12}$$

2.2 MATHEMATICAL PROBLEMS FOR SHOCK REFLECTION-DIFFRACTION

When a plane shock in the $(t, \mathbf{x})$–coordinates, $t \in \mathbb{R}_+ := [0, \infty), \mathbf{x} = (x_1, x_2) \in \mathbb{R}^2$, with left state $(\rho, \nabla_{\mathbf{x}}\Phi) = (\rho_1, u_1, 0)$ and right state $(\rho_0, 0, 0), u_1 > 0$, $\rho_0 < \rho_1$, hits a symmetric wedge

$$W := \{\mathbf{x} : |x_2| < x_1 \tan\theta_{\mathrm{w}}, x_1 > 0\}$$

head-on (see Fig. 1.1), it experiences a reflection-diffraction process. Then system (1.1) in $\mathbb{R}_+ \times (\mathbb{R}^2 \setminus W)$ becomes

$$\begin{cases} \partial_t \rho + \mathrm{div}_{\mathbf{x}}(\rho \nabla_{\mathbf{x}}\Phi) = 0, \\ \partial_t \Phi + \dfrac{1}{2}|\nabla_{\mathbf{x}}\Phi|^2 + \dfrac{\rho^{\gamma-1} - \rho_0^{\gamma-1}}{\gamma-1} = 0, \end{cases} \tag{2.2.1}$$

where we have used the Bernoulli constant $B_0 = \frac{\rho_0^{\gamma-1}-1}{\gamma-1}$ determined by the right state $(\rho_0, 0, 0)$. From (2.1.10), we find that $u_1 > 0$ is uniquely determined by (ρ_0, ρ_1) and $\gamma > 1$:

$$u_1 = \sqrt{\frac{2(\rho_1 - \rho_0)(\rho_1^{\gamma-1} - \rho_0^{\gamma-1})}{(\gamma-1)(\rho_1 + \rho_0)}} > 0, \tag{2.2.2}$$

where we have used that $\rho_0 < \rho_1$.

Then the shock reflection-diffraction problem can be formulated as the following problem:

Problem 2.2.1 (Initial-Boundary Value Problem). *Seek a solution of system* (2.2.1) *for* $B_0 = \frac{\rho_0^{\gamma-1}-1}{\gamma-1}$ *with the initial condition at* $t = 0$:

$$(\rho, \Phi)|_{t=0} = \begin{cases} (\rho_0, 0) & \textit{for } |x_2| > x_1 \tan\theta_{\mathrm{w}},\ x_1 > 0, \\ (\rho_1, u_1 x_1) & \textit{for } x_1 < 0, \end{cases} \tag{2.2.3}$$

and the slip boundary condition along the wedge boundary ∂W:

$$\nabla_{\mathbf{x}}\Phi \cdot \boldsymbol{\nu}|_{\mathbb{R}_+ \times \partial W} = 0, \tag{2.2.4}$$

where $\boldsymbol{\nu}$ *is the exterior unit normal to* ∂W *(see Fig.* 2.1*).*

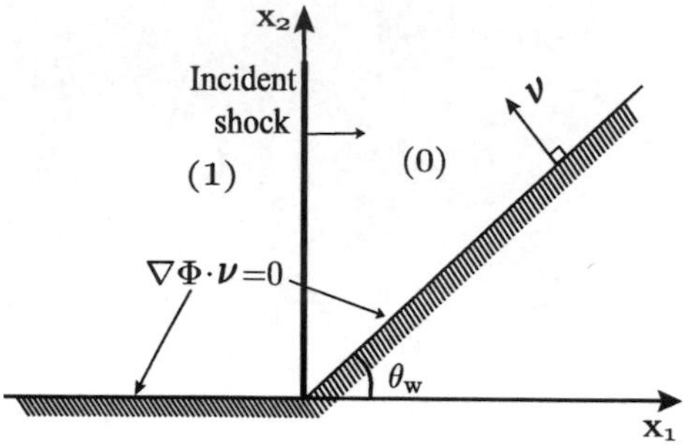

Figure 2.1: Initial-boundary value problem

Notice that the initial-boundary value problem (**Problem 2.2.1**) is invariant under the self-similar scaling:

$$(t, \mathbf{x}) \mapsto (\alpha t, \alpha \mathbf{x}), \quad (\rho, \Phi) \mapsto (\rho, \frac{\Phi}{\alpha}) \qquad \text{for} \quad \alpha \neq 0. \tag{2.2.5}$$

That is, if $(\rho, \Phi)(t, \mathbf{x})$ satisfy (2.2.1)–(2.2.4), so do $(\rho, \frac{\Phi}{\alpha})(\alpha t, \alpha \mathbf{x})$ for any constant $\alpha \neq 0$.

Therefore, we seek self-similar solutions with the following form:

$$\rho(t, \mathbf{x}) = \rho(\boldsymbol{\xi}), \quad \Phi(t, \mathbf{x}) = t\,\phi(\boldsymbol{\xi}) \qquad \text{for} \quad \boldsymbol{\xi} = (\xi_1, \xi_2) = \frac{\mathbf{x}}{t}. \tag{2.2.6}$$

We then see that the pseudo-potential function $\varphi = \phi - \frac{|\boldsymbol{\xi}|^2}{2}$ satisfies the following Euler equations for self-similar solutions:

$$\begin{cases} \operatorname{div}(\rho\, D\varphi) + 2\rho = 0, \\ (\gamma - 1)(\frac{1}{2}|D\varphi|^2 + \varphi) + \rho^{\gamma-1} = \rho_0^{\gamma-1}, \end{cases} \tag{2.2.7}$$

where div and D represent the divergence and the gradient, respectively, with respect to the self-similar variables $\boldsymbol{\xi}$.

This implies that the pseudo-potential function $\varphi(\boldsymbol{\xi})$ is governed by the following potential flow equation of second order:

$$\operatorname{div}\big(\rho(|D\varphi|^2, \varphi)D\varphi\big) + 2\rho(|D\varphi|^2, \varphi) = 0 \tag{2.2.8}$$

with

$$\rho(|D\varphi|^2, \varphi) = \big(\rho_0^{\gamma-1} - (\gamma - 1)(\varphi + \frac{1}{2}|D\varphi|^2)\big)^{\frac{1}{\gamma-1}}. \tag{2.2.9}$$

We consider (2.2.8) with (2.2.9) for functions φ satisfying

$$\rho_0^{\gamma-1} - (\gamma - 1)\big(\varphi + \frac{1}{2}|D\varphi|^2\big) \geq 0. \tag{2.2.10}$$

Definition 2.2.2. *A function $\varphi \in W^{1,1}_{\rm loc}(\Omega)$ is called a weak solution of equation* (2.2.8) *in domain Ω if φ satisfies* (2.2.10) *and the following properties:*

(i) *For $\rho(|D\varphi|^2, \varphi)$ determined by* (2.2.9),

$$(\rho(|D\varphi|^2, \varphi), \rho(|D\varphi|^2, \varphi)|D\varphi|) \in (L^1_{loc}(\overline{\Omega}))^2;$$

(ii) *For every $\zeta \in C^\infty_c(\Omega)$,*

$$\int_\Omega \Big(\rho(|D\varphi|^2, \varphi)D\varphi \cdot D\zeta - 2\rho(|D\varphi|^2, \varphi)\zeta\Big)\, d\boldsymbol{\xi} = 0.$$

We will also use the non-divergence form of equation (2.2.8) for $\phi = \varphi + \frac{|\boldsymbol{\xi}|^2}{2}$:

$$(c^2 - \varphi^2_{\xi_1})\phi_{\xi_1\xi_1} - 2\varphi_{\xi_1}\varphi_{\xi_2}\phi_{\xi_1\xi_2} + (c^2 - \varphi^2_{\xi_2})\phi_{\xi_2\xi_2} = 0, \tag{2.2.11}$$

where the sonic speed $c = c(|D\varphi|^2, \varphi, \rho_0^{\gamma-1})$ is determined by (1.14).

Equation (2.2.8) or (2.2.11) is a nonlinear second-order PDE of mixed elliptic-hyperbolic type. It is elliptic if and only if (1.18) holds, which is equivalent to the following condition:

$$|D\varphi| < c_*(\varphi, \rho_0, \gamma) := \sqrt{\frac{2}{\gamma+1}(\rho_0^{\gamma-1} - (\gamma - 1)\varphi)}. \tag{2.2.12}$$

Throughout the rest of this book, for simplicity, we drop term "pseudo" and simply call φ as a *potential function* and $D\varphi$ as a *velocity*, respectively, when no confusion arises.

Shocks are discontinuities in the velocity functions $D\varphi$. That is, if Ω^+ and $\Omega^- := \Omega \setminus \overline{\Omega^+}$ are two non-empty open subsets of $\Omega \subset \mathbb{R}^2$, and $\mathcal{S} := \partial\Omega^+ \cap \Omega$ is a C^1–curve where $D\varphi$ has a jump, then $\varphi \in W^{1,1}_{\rm loc}(\Omega) \cap C^1(\overline{\Omega^\pm}) \cap C^2(\Omega^\pm)$ is a global weak solution of (2.2.8) in Ω in the sense of Definition 2.2.2 if and only if φ is in $W^{1,\infty}_{\rm loc}(\Omega)$ and satisfies equation (2.2.8) in $\Omega^\pm$ and the Rankine-Hugoniot condition on $\mathcal{S}$:

$$\big[\rho(|D\varphi|^2, \varphi)D\varphi \cdot \boldsymbol{\nu}\big]_{\mathcal{S}} = 0. \tag{2.2.13}$$

Note that the condition that $\varphi \in W^{1,\infty}_{\rm loc}(\Omega)$ implies the continuity of φ across shock $\mathcal{S}$:

$$[\varphi]_{\mathcal{S}} = 0. \tag{2.2.14}$$

The plane incident shock solution in the $(t, \mathbf{x})$–coordinates with the left and right states:

$$(\rho, \nabla_{\mathbf{x}}\Phi) = (\rho_0, 0, 0),\ (\rho_1, u_1, 0)$$

corresponds to a continuous weak solution φ of (2.2.8) in the self-similar coordinates $\boldsymbol{\xi}$ with the following form:

$$\varphi = \begin{cases} \varphi_0 & \text{for } \xi_1 > \xi_1^0, \\ \varphi_1 & \text{for } \xi_1 < \xi_1^0, \end{cases} \tag{2.2.15}$$

where

$$\varphi_0(\boldsymbol{\xi}) = -\frac{|\boldsymbol{\xi}|^2}{2}, \tag{2.2.16}$$

$$\varphi_1(\boldsymbol{\xi}) = -\frac{|\boldsymbol{\xi}|^2}{2} + u_1(\xi_1 - \xi_1^0), \tag{2.2.17}$$

and

$$\xi_1^0 = \rho_1\sqrt{\frac{2(\rho_1^{\gamma-1} - \rho_0^{\gamma-1})}{(\gamma-1)(\rho_1^2 - \rho_0^2)}} = \frac{\rho_1 u_1}{\rho_1 - \rho_0} > 0 \tag{2.2.18}$$

is the location of the incident shock in the $\boldsymbol{\xi}$–coordinates, uniquely determined by (ρ_0, ρ_1, γ) through (2.2.13), which is obtained from (2.1.11) and (2.2.2) owing to the fact that $\xi_1^0 = s$ here. Since the problem is symmetric with respect to the ξ_1–axis, it suffices to consider the problem in half-plane $\xi_2 > 0$ outside the half-wedge:

$$\Lambda := \{\boldsymbol{\xi} \,:\, \xi_1 \in \mathbb{R},\, \xi_2 > \max(\xi_1 \tan\theta_{\mathrm{w}},\, 0)\}. \tag{2.2.19}$$

Then the initial-boundary value problem (2.2.1)–(2.2.4) in the $(t, \mathbf{x})$–coordinates can be formulated as a boundary value problem in the self-similar coordinates $\boldsymbol{\xi}$.

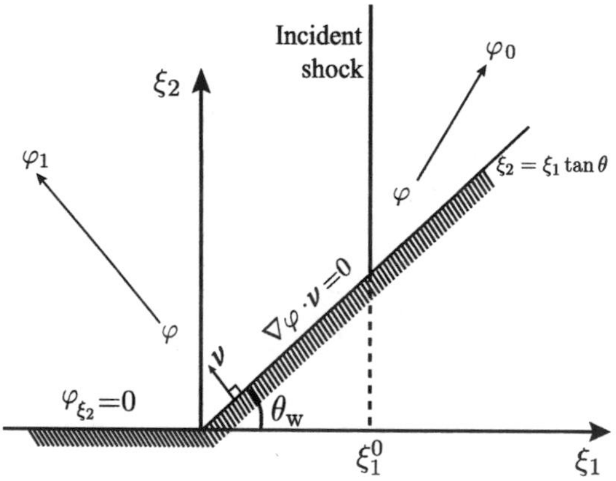

Figure 2.2: Boundary value problem

Problem 2.2.3 (Boundary Value Problem; see Fig. 2.2). *Seek a solution* φ *of equation* (2.2.8) *in the self-similar domain* Λ *with the slip boundary condition:*

$$D\varphi\cdot\boldsymbol{\nu}|_{\partial\Lambda} = 0 \tag{2.2.20}$$

and the asymptotic boundary condition at infinity:

$$\varphi \to \bar{\varphi} = \begin{cases} \varphi_0 & \text{for } \boldsymbol{\xi}\in\Lambda,\ \xi_1 > \xi_1^0, \\ \varphi_1 & \text{for } \boldsymbol{\xi}\in\Lambda,\ \xi_1 < \xi_1^0, \end{cases} \qquad \text{when } |\boldsymbol{\xi}| \to \infty, \tag{2.2.21}$$

where (2.2.21) *holds in the sense that* $\lim_{R\to\infty} \|\varphi - \overline{\varphi}\|_{C^{0,1}(\Lambda\setminus B_R(\mathbf{0}))} = 0$.

This is a boundary value problem for the second-order nonlinear conservation law (2.2.8) of mixed elliptic-hyperbolic type in an unbounded domain. The main feature of this boundary value problem is that $D\varphi$ has a jump at $\xi_1 = \xi_1^0$ at infinity, which is not conventional, coupling with the wedge corner for the domain. The solutions with complicated patterns of wave configurations as observed experimentally should be the global solutions of this boundary value problem: **Problem 2.2.3**.

2.3 WEAK SOLUTIONS OF PROBLEM 2.2.1 AND PROBLEM 2.2.3

Note that the boundary condition (2.2.20) for **Problem 2.2.3** implies

$$\rho D\varphi \cdot \boldsymbol{\nu}|_{\partial\Lambda} = 0. \tag{2.3.1}$$

Conditions (2.2.20) and (2.3.1) are equivalent if $\rho \neq 0$. Since $\rho \neq 0$ for the solutions under consideration, we use condition (2.3.1) instead of (2.2.20) in the definition of weak solutions of **Problem 2.2.3**.

Similarly, we write the boundary condition (2.2.4) for **Problem 2.2.1** as

$$\rho \nabla_{\mathbf{x}}\Phi \cdot \boldsymbol{\nu}|_{\mathbb{R}_+\times\partial W} = 0. \tag{2.3.2}$$

Condition (2.3.1) is the conormal condition for equation (2.2.8). Also, (2.3.2) is the spatial conormal condition for equation (1.5). This yields the following definitions:

Definition 2.3.1 (Weak Solutions of **Problem 2.2.1**). *A function*

$$\Phi \in W^{1,1}_{\mathrm{loc}}(\mathbb{R}_+ \times (\mathbb{R}^2 \setminus W))$$

is called a weak solution of **Problem 2.2.1** *if* Φ *satisfies the following properties*:

(i) $B_0 - \left(\partial_t\Phi + \frac{1}{2}|\nabla_{\mathbf{x}}\Phi|^2\right) \geq h(0+)$ *a.e. in* $\mathbb{R}_+ \times (\mathbb{R}^2 \setminus W)$;

(ii) *For* $\rho(\partial_t\Phi, \nabla_{\mathbf{x}}\Phi)$ *determined by* (1.4),

$$(\rho(\partial_t\Phi, |\nabla_{\mathbf{x}}\Phi|^2), \rho(\partial_t\Phi, |\nabla_{\mathbf{x}}\Phi|^2)|\nabla_{\mathbf{x}}\Phi|) \in (L^1_{\mathrm{loc}}(\overline{\mathbb{R}}_+ \times \overline{\mathbb{R}^2 \setminus W}))^2;$$

(iii) *For every* $\zeta \in C_c^\infty(\overline{\mathbb{R}_+} \times \mathbb{R}^2)$,

$$\int_0^\infty \int_{\mathbb{R}^2\setminus W} \left(\rho(\partial_t\Phi, |\nabla_{\mathbf{x}}\Phi|^2)\partial_t\zeta + \rho(\partial_t\Phi, |\nabla_{\mathbf{x}}\Phi|^2)\nabla\Phi\cdot\nabla\zeta\right) d\mathbf{x}dt$$
$$+ \int_{\mathbb{R}^2\setminus W} \rho|_{t=0}\zeta(0,\mathbf{x})d\mathbf{x} = 0,$$

where

$$\rho|_{t=0} = \begin{cases} \rho_0 & \text{for } |x_2| > x_1 \tan\theta_{\rm w},\ x_1 > 0, \\ \rho_1 & \text{for } x_1 < 0. \end{cases}$$

Remark 2.3.2. *Since ζ does not need to be zero on ∂W, the integral identity in Definition* 2.3.1 *is a weak form of equation* (1.5) *and the boundary condition* (2.3.2).

Definition 2.3.3 (Weak solutions of **Problem 2.2.3**). *A function $\varphi \in W^{1,1}_{\rm loc}(\Lambda)$ is called a weak solution of* **Problem 2.2.3** *if φ satisfies* (2.2.21) *and the following properties:*

(i) $\rho_0^{\gamma-1} - (\gamma-1)\left(\varphi + \frac{1}{2}|D\varphi|^2\right) \geq 0$ *a.e. in* Λ;

(ii) *For $\rho(|D\varphi|^2, \varphi)$ determined by* (2.2.9),

$$\left(\rho(|D\varphi|^2,\varphi), \rho(|D\varphi|^2,\varphi)|D\varphi|\right) \in (L^1_{loc}(\overline{\Lambda}))^2;$$

(iii) *For every $\zeta \in C_c^\infty(\mathbb{R}^2)$,*

$$\int_\Lambda \Big(\rho(|D\varphi|^2,\varphi)D\varphi\cdot D\zeta - 2\rho(|D\varphi|^2,\varphi)\zeta\Big)\, d\boldsymbol{\xi} = 0.$$

Remark 2.3.4. *Since ζ does not need to be zero on $\partial\Lambda$, the integral identity in Definition* 2.3.3 *is a weak form of equation* (2.2.8) *and the boundary condition* (2.3.1).

Remark 2.3.5. *From Definition* 2.3.3, *we observe the following fact: If $B \subset \mathbb{R}^2$ is an open set, and φ is a weak solution of* **Problem 2.2.3** *satisfying $\varphi \in C^2(B\cap\Lambda)\cap C^1(B\cap\overline{\Lambda})$, then φ satisfies equation* (2.2.8) *in the classical sense in $B\cap\Lambda$, the boundary condition* (2.3.1) *on $B\cap\partial\Lambda\setminus\{\mathbf{0}\}$, and $D\varphi(\mathbf{0}) = \mathbf{0}$.*

2.4 STRUCTURE OF SOLUTIONS: REGULAR REFLECTION-DIFFRACTION CONFIGURATIONS

We now discuss the structure of solutions φ of **Problem 2.2.3** corresponding to shock reflection-diffraction.

Since φ_1 does not satisfy the slip boundary condition (2.2.20), the solution must differ from φ_1 in $\{\xi_1 < \xi_1^0\}\cap\Lambda$ so that a shock diffraction by the wedge occurs. We now describe two of the most important configurations: the supersonic and subsonic regular reflection-diffraction configurations, as shown in Fig. 2.3 and Fig. 2.4, respectively. From now on, we will refer to these two configurations as a *supersonic reflection configuration* and a *subsonic reflection configuration* respectively, whose corresponding solutions are called the *supersonic reflection solution* and the *subsonic reflection solution* respectively, when no confusion arises.

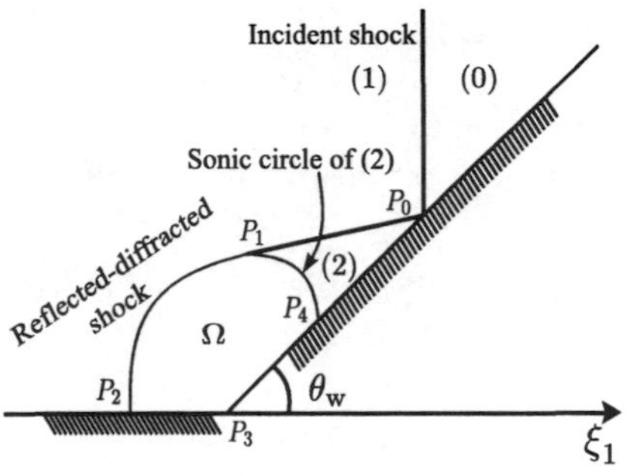

Figure 2.3: Supersonic regular reflection-diffraction configuration

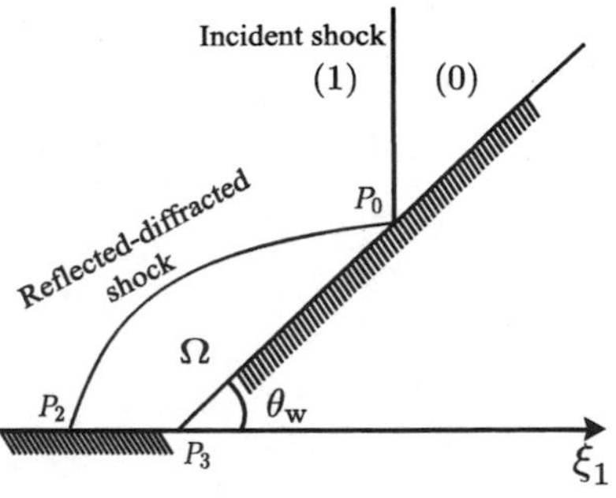

Figure 2.4: Subsonic regular reflection-diffraction configuration

In Figs. 2.3 and 2.4, the vertical line is the incident shock $\mathcal{S}_0 = \{\xi_1 = \xi_1^0\}$ that hits the wedge at point $P_0 = (\xi_1^0, \xi_1^0 \tan\theta_{\rm w})$, and state (0) and state (1), ahead of and behind $\mathcal{S}_0$, are given by φ_0 and φ_1 defined in (2.2.16) and (2.2.17), respectively. Thus, we only need to describe the solution in subregion $P_0P_2P_3$ between the wedge and the reflected-diffracted shock. The solution is expected to be C^1 in $P_0P_2P_3$. Now we describe its structure. Below, φ denotes the potential of the solution in $P_0P_2P_3$, while we use the uniform states φ_0 and φ_1 to describe the solution outside $P_0P_2P_3$, ahead of and behind the incident shock $\mathcal{S}_1$, respectively. In particular, $D\varphi(P_0)$ denotes the limit at P_0 of the gradient of the solution in $P_0P_2P_3$.

2.4.1 Definition of state (2)

Since φ is C^1 in region $P_0P_2P_3$, it should satisfy the boundary condition $D\varphi\cdot\boldsymbol{\nu} = 0$ on the wedge boundary P_0P_3 including the endpoints, as well as the Rankine-Hugoniot conditions (2.2.13)–(2.2.14) at P_0 across the reflected shock separating φ from φ_1. Let

$$(u_2, v_2) := D\varphi(P_0) + \boldsymbol{\xi}_{P_0}.$$

Then $v_2 = u_2\tan\theta_{\rm w}$ by $D\varphi\cdot\boldsymbol{\nu} = 0$. Moreover, using (2.2.17), in addition to the previous properties, we see by a direct calculation that the uniform state with the pseudo-potential:

$$\varphi_2(\boldsymbol{\xi}) = -\frac{|\boldsymbol{\xi}|^2}{2} + u_2(\xi_1 - \xi_1^0) + u_2\tan\theta_{\rm w}(\xi_2 - \xi_1^0\tan\theta_{\rm w}), \tag{2.4.1}$$

called *state* (2), satisfies the boundary condition on the wedge boundary:

$$D\varphi_2\cdot\boldsymbol{\nu} = 0 \qquad \text{on } \partial\Lambda\cap\{\xi_1 = \xi_2\cot\theta_{\rm w}\}, \tag{2.4.2}$$

and the Rankine-Hugoniot conditions (2.2.13) on the flat shock $\mathcal{S}_1$ determined by (2.2.14):

$$\mathcal{S}_1 := \{\varphi_1 = \varphi_2\} \tag{2.4.3}$$

which passes through P_0 between states (1) and (2).

We note that the constant velocity $u_2 > 0$ is determined by $(\rho_0, \rho_1, \gamma, \theta_{\rm w})$ from the algebraic equation expressing (2.2.13) for φ_1 and φ_2 across $\mathcal{S}_1$, where we have used (2.2.2) to eliminate u_1 from the list of parameters and have noted that $\boldsymbol{\nu}_{\mathcal{S}_1} = \frac{(u_1-u_2,-u_2\tan\theta_{\rm w})}{|(u_1-u_2,-u_2\tan\theta_{\rm w})|}$.

Thus, state (2) is defined by the following requirements: It is a uniform state with pseudo-potential $\varphi_2(\boldsymbol{\xi})$ such that $D\varphi_2 \cdot \boldsymbol{\nu} = 0$ on the wedge boundary, and the Rankine-Hugoniot conditions (2.2.13)–(2.2.14) for φ_1 and φ_2 hold at P_0. As we will discuss in §2.5, such a state (2) exists for any wedge angle $\theta_{\rm w} \in [\theta_{\rm w}^{\rm d}, \frac{\pi}{2}]$, for some $\theta_{\rm w}^{\rm d} = \theta_{\rm w}^{\rm d}(\rho_0, \rho_1, \gamma) > 0$. From the discussion above, it is apparent that the existence of state (2) is a necessary condition for the existence of regular reflection-diffraction configurations as shown in Figs. 2.3–2.4.

From now on, we fix the data, *i.e.*, parameters (ρ_0, ρ_1, γ). Thus, the parameters of state (2) depend only on $\theta_{\rm w}$.

State (2) can be either pseudo-subsonic or pseudo-supersonic at P_0. This determines the subsonic or supersonic type of regular reflection-diffraction configurations, as shown in Figs. 2.3–2.4.

We note that the uniform state (2) is pseudo-subsonic within its sonic circle with center $\mathcal{O}_2 = (u_2, u_2\tan\theta_{\rm w})$ and radius $c_2 = \rho_2^{(\gamma-1)/2} > 0$, the sonic speed of state (2), and that φ_2 is pseudo-supersonic outside this circle.

Thus, if state (2) is pseudo-supersonic at P_0, P_0 lies outside the sonic circle $B_{c_2}(\mathcal{O}_2)$ of state (2). It can be shown (see §7.5) that line $\mathcal{S}_1$ intersects $\partial B_{c_2}(\mathcal{O}_2)$ at two points and, denoting by P_1 the point that is closer to P_0, we find that P_1 lies in Λ and segment P_0P_1 lies in Λ and outside of $B_{c_2}(u_2, v_2)$; see Fig. 2.3. Denote by P_4 the point of intersection of $\partial B_{c_2}(\mathcal{O}_2)$ with the wedge boundary $\{\xi_2 = \xi_1\tan\theta_{\rm w},\ \xi_1 > 0\}$ such that arc P_1P_4 lies between $\mathcal{S}_1$ and $\{\xi_2 = \xi_1\tan\theta_{\rm w},\ \xi_1 > 0\}$.

2.4.2 Supersonic regular reflection-diffraction configurations

This is the case when state (2) is supersonic at P_0.

The supersonic reflection configuration as shown in Fig. 2.3 consists of three uniform states: (0), (1), (2), plus a non-uniform state in domain $\Omega = P_1P_2P_3P_4$. As described above, the solution is equal to state (0) and state (1) ahead of and behind the incident shock $\mathcal{S}_0$, away from subregion $P_0P_2P_3$. The solution is equal to state (2) in subregion $P_0P_1P_4$. Note that state (2) is supersonic in $P_0P_1P_4$.

The non-uniform state in Ω is subsonic, *i.e.*, the potential flow equation (2.2.8) for φ is elliptic in Ω.

We denote the boundary parts of Ω by

$$\Gamma_{\rm shock} := P_1P_2,\quad \Gamma_{\rm sym} := P_2P_3,\quad \Gamma_{\rm wedge} := P_3P_4,\quad \Gamma_{\rm sonic} := P_1P_4, \qquad (2.4.4)$$

where $\Gamma_{\rm shock}$ is the curved part of the reflected shock, $\Gamma_{\rm sonic}$ is the sonic arc, and $\Gamma_{\rm wedge}$ (the wedge boundary) and $\Gamma_{\rm sym}$ are the straight segments, respectively.

Note that the curved part of the reflected shock Γ_{shock} separates the supersonic flow outside Ω from the subsonic flow in Ω, *i.e.*, Γ_{shock} is a *transonic* shock.

2.4.3 Subsonic regular reflection-diffraction configurations

This is the case when state (2) is subsonic or sonic at P_0.

The subsonic reflection configuration as shown in Fig. 2.4 consists of two uniform states – (0) and (1) – in the regions described above, and a non-uniform state in domain $\Omega = P_0P_2P_3$. The non-uniform state in Ω is subsonic, *i.e.*, the potential flow equation (2.2.8) for φ is elliptic in Ω. Moreover, solution φ in Ω matches with φ_2 at P_0 as follows:

$$\varphi(P_0) = \varphi_2(P_0), \qquad D\varphi(P_0) = D\varphi_2(P_0).$$

The boundary parts of Ω in this case are

$$\Gamma_{\text{shock}} := P_0P_2, \;\; \Gamma_{\text{sym}} := P_2P_3, \;\; \Gamma_{\text{wedge}} := P_0P_3. \tag{2.4.5}$$

Similar to the previous case, Γ_{shock} is a transonic shock. We unify the notations with supersonic reflection configurations by introducing points P_1 and P_4 for subsonic reflection configurations via setting

$$P_1 := P_0, \quad P_4 := P_0, \quad \overline{\Gamma_{\text{sonic}}} := \{P_0\}. \tag{2.4.6}$$

Note that, with this convention, (2.4.5) coincides with (2.4.4).

In Part III, we develop approaches, techniques, and related analysis to establish the global existence of a supersonic reflection configuration up to the sonic angle, or the critical angle in the attached case (defined in §2.6).

In Part IV, we develop the theory further to establish the global existence of regular reflection-diffraction configurations up to the detachment angle, or the critical angle in the attached case. In particular, this will imply the existence of both supersonic and subsonic reflection configurations.

2.5 EXISTENCE OF STATE (2) AND CONTINUOUS DEPENDENCE ON THE PARAMETERS

We note that state (2), the uniform state (2.4.1), satisfies (2.4.2) and the Rankine-Hugoniot condition with state (1) on $\mathcal{S}_1 = \{\varphi_1 = \varphi_2\}$ as defined in (2.4.3):

$$\rho_2 D\varphi_2 \cdot \boldsymbol{\nu} = \rho_1 D\varphi_1 \cdot \boldsymbol{\nu} \qquad \text{on } \mathcal{S}_1, \tag{2.5.1}$$

where $\boldsymbol{\nu}$ is the unit normal on $\mathcal{S}_1$.

From the regular reflection-diffraction configurations as described in §2.4.2–§2.4.3, the existence of state (2) is a necessary condition for the existence of such a solution. We note that $\mathcal{S}_1$, defined in (2.4.3), is a straight line, which is concluded from the explicit expressions of $\varphi_j, j = 1, 2$, and the fact that $(u_1, 0) \neq (u_2, v_2)$. The last statement holds since φ_1 does not satisfy (2.4.2).

State (2), (u_2, v_2) in (2.4.1), is obtained as a solution of the algebraic system involving the slope of $\mathcal{S}_1$ (*i.e.*, the direction of $\boldsymbol{\nu}$) and the equality in (2.5.1); see §7.4 below.

This algebraic system has solutions for some but not all $\theta_{\rm w} \in (0, \frac{\pi}{2})$. More precisely, there exist the sonic angle $\theta_{\rm w}^{\rm s}$ and the detachment angle $\theta_{\rm w}^{\rm d}$ satisfying

$$0 < \theta_{\rm w}^{\rm d} < \theta_{\rm w}^{\rm s} < \frac{\pi}{2}$$

such that there are two states (2), weak and strong with $\rho_2^{\rm wk} < \rho_2^{\rm sg}$, for all $\theta_{\rm w} \in (\theta_{\rm w}^{\rm d}, \frac{\pi}{2})$, but $\rho_2^{\rm wk} = \rho_2^{\rm sg}$ at $\theta_{\rm w} = \theta_{\rm w}^{\rm d}$. Moreover, the strong state (2) is always subsonic at the reflection point $P_0(\theta_{\rm w})$, while the weak state (2) is:

(i) supersonic at the reflection point $P_0(\theta_{\rm w})$ for $\theta_{\rm w} \in (\theta_{\rm w}^{\rm s}, \frac{\pi}{2})$;

(ii) sonic at $P_0(\theta_{\rm w})$ for $\theta_{\rm w} = \theta_{\rm w}^{\rm s}$;

(iii) subsonic at $P_0(\theta_{\rm w})$ for $\theta_{\rm w} \in (\theta_{\rm w}^{\rm d}, \hat{\theta}_{\rm w}^{\rm s})$, for some $\hat{\theta}_{\rm w}^{\rm s} \in (\theta_{\rm w}^{\rm d}, \theta_{\rm w}^{\rm s}]$.

Moreover, the weak state (2)$= (u_2, v_2)$ depends continuously on $\theta_{\rm w}$ in $[\theta_{\rm w}^{\rm d}, \frac{\pi}{2}]$. For details of this, see Theorem 7.1.1 in Chapter 7.

As for the weak and strong states for each $\theta_{\rm w} \in (\theta_{\rm w}^{\rm d}, \frac{\pi}{2})$, there has been a long debate to determine which one is physical for the local theory; see Courant-Friedrichs [99], Ben-Dor [12], and the references cited therein. It has been conjectured that the strong reflection-diffraction configuration is non-physical. Indeed, when the wedge angle $\theta_{\rm w}$ tends to $\frac{\pi}{2}$, the weak reflection-diffraction configuration tends to the unique normal reflection as proved in Chen-Feldman [54]; however, the strong reflection-diffraction configuration does not (see Chapter 7 below).

In the existence results of regular reflection-diffraction solutions below, we always use the *weak* state (2).

2.6 VON NEUMANN'S CONJECTURES, PROBLEM 2.6.1 (FREE BOUNDARY PROBLEM), AND MAIN THEOREMS

If the weak state (2) is supersonic, on which equation (2.2.8) is hyperbolic, the propagation speeds of the solution are finite, and state (2) is completely determined by the local information: state (1), state (0), and the location of point P_0. That is, any information from the region of shock reflection-diffraction, such as the disturbance at corner P_3, cannot travel towards the reflection point P_0. However, if the weak state (2) is subsonic, on which equation (2.2.8) is elliptic, the information can reach P_0 and interact with it, potentially creating a new type of shock reflection-diffraction configurations. This argument motivated the conjecture by von Neumann in [267, 268], which can be formulated as follows:

von Neumann's Sonic Conjecture: *There exists a supersonic regular reflection-diffraction configuration when* $\theta_{\rm w} \in (\theta_{\rm w}^{\rm s}, \frac{\pi}{2})$, *i.e., the supersonicity of the weak state* (2) *at* $P_0(\theta_{\rm w})$ *implies the existence of a supersonic regular reflection-diffraction configuration to* **Problem 2.2.3** *as shown in Fig.* 2.3.

Another conjecture states that the global regular reflection-diffraction configuration is possible whenever the local regular reflection at the reflection point P_0 is possible, even beyond the sonic angle $\theta_{\rm w}^{\rm s}$ up to the detachment angle $\theta_{\rm w}^{\rm d}$:

von Neumann's Detachment Conjecture: *There exists a global regular reflection-diffraction configuration for any wedge angle* $\theta_{\rm w} \in (\theta_{\rm w}^{\rm d}, \frac{\pi}{2})$, *i.e., the existence of state* (2) *implies the existence of a regular reflection-diffraction configuration to* **Problem 2.2.3**. *Moreover, the type* (*subsonic or supersonic*) *of the reflection-diffraction configuration is determined by the type of the weak state* (2) *at* $P_0(\theta_{\rm w})$, *as shown in Figs.* 2.3–2.4.

It is clear that the supersonic and subsonic reflection configurations are not possible without a local two-shock configuration at the reflection point on the wedge, so that it is the necessary criterion for the existence of supersonic and subsonic reflection configurations.

There has been a long debate in the literature whether there still exists a global regular reflection-diffraction solution beyond the sonic angle $\theta_{\rm w}^{\rm s}$ up to the detachment angle $\theta_{\rm w}^{\rm d}$; see Ben-Dor [12] and the references cited therein. As shown in Fig. 18.7 for the full Euler case, the difference on the physical parameters between the sonic conjecture and the detachment conjecture is only fractions of a degree apart in terms of the wedge angles; a resolution has challenged even sophisticated modern numerical and laboratory experiments. In Part IV (Chapters 15–17), we rigorously prove the global existence of regular reflection-diffraction configurations, beyond the sonic angle up to the detachment angle. This indicates that the necessary criterion is also sufficient for the existence of supersonic and subsonic reflection configurations, at least for potential flow.

To solve von Neumann's conjectures, we reformulate **Problem 2.2.3** into the following free boundary problem:

Problem 2.6.1 (Free Boundary Problem). *For* $\theta_{\rm w} \in (\theta_{\rm w}^{\rm d}, \frac{\pi}{2})$, *find a free boundary* (*curved reflected-diffracted shock*) $\Gamma_{\rm shock}$ *in region* $\Lambda \cap \{\xi_1 < \xi_{1P_1}\}$ (*where we use* (2.4.6) *for subsonic reflections*) *and a function* φ *defined in region* Ω *as shown in Figs.* 2.3–2.4 *such that* φ *satisfies*:

(i) *Equation* (2.2.8) *in* Ω;

(ii) $\varphi = \varphi_1$ *and* $\rho D\varphi \cdot \boldsymbol{\nu} = \rho_1 D\varphi_1 \cdot \boldsymbol{\nu}$ *on the free boundary* $\Gamma_{\rm shock}$ *separating the elliptic phase from the hyperbolic phase*;

(iii) $\varphi = \varphi_2$ *and* $D\varphi = D\varphi_2$ *on* $\Gamma_{\rm sonic}$ *for the supersonic reflection configuration as shown in Fig.* 2.3 *and at* P_0 *for the subsonic reflection configuration as shown in Fig.* 2.4;

(iv) $D\varphi \cdot \boldsymbol{\nu} = 0$ *on* $\Gamma_{\text{wedge}} \cup \Gamma_{\text{sym}}$,

where $\boldsymbol{\nu}$ *is the interior unit normal to* Ω *on* $\Gamma_{\text{shock}} \cup \Gamma_{\text{wedge}} \cup \Gamma_{\text{sym}}$.

We remark that condition (iii) is equivalent to the Rankine-Hugoniot conditions for φ across Γ_{sonic}. The sonic arc Γ_{sonic} is a weak discontinuity of φ (which is different from a strong discontinuity such as Γ_{shock}); that is, if the state from one side is sonic, the Rankine-Hugoniot conditions require the gradient of φ to be continuous across Γ_{sonic}.

Furthermore, since condition (ii) is the Rankine-Hugoniot conditions for φ across Γ_{shock}, we can extend solution φ of **Problem 2.6.1** from Ω to Λ so that the extended function (still denoted) φ is a weak solution of **Problem 2.2.3** (at least when Γ_{shock} and φ are sufficiently regular). More specifically,

Definition 2.6.2. *Let* φ *be a solution of* **Problem** 2.6.1 *in region* Ω. *Define the extension of* φ *from* Ω *to* Λ *by setting:*

$$\varphi = \begin{cases} \varphi_0 & \text{for } \xi_1 > \xi_1^0 \text{ and } \xi_2 > \xi_1 \tan\theta_{\text{w}}, \\ \varphi_1 & \text{for } \xi_1 < \xi_1^0 \text{ and above curve } P_0P_1P_2, \\ \varphi_2 & \text{in region } P_0P_1P_4, \end{cases} \tag{2.6.1}$$

where we have used the notational convention (2.4.6) *for subsonic reflections. In particular, for subsonic reflections, region* $P_0P_1P_4$ *is one point, and curve* $P_0P_1P_2$ *is* P_0P_2. *See Figs.* 2.3 *and* 2.4.

We note that ξ_1^0 used in (2.6.1) is the location of the incident shock; *cf.* (2.2.15) and (2.2.18). Also, the extension by (2.6.1) is well-defined because of the requirement $\Gamma_{\text{shock}} \subset \Lambda \cap \{\xi_1 < \xi_{1P_1}\}$ in **Problem 2.6.1**.

From now on, using Definition 2.6.2, we consider solutions φ of **Problem 2.6.1** to be defined in Λ.

It turns out that another key obstacle for establishing the existence of regular reflection-diffraction configurations is the additional possibility that, for some wedge angle $\theta_{\text{w}}^{\text{c}} \in (\theta_{\text{w}}^{\text{d}}, \frac{\pi}{2})$, the reflected-diffracted shock P_0P_2 may strike the wedge vertex P_3, an additional sub-type of regular reflection-diffraction configurations in which the reflected-diffracted shock is attached to the wedge vertex P_3, *i.e.*, $P_2 = P_3$. Indeed, in such a case, we establish the existence of such a global solution of regular reflection-diffraction configurations as shown in Figs. 2.5–2.6 for any wedge angle $\theta_{\text{w}} \in (\theta_{\text{w}}^{\text{c}}, \frac{\pi}{2})$.

Observe that some experimental results (*cf.* [263, Fig. 238, Page 144]) suggest that solutions with an attached shock to the wedge vertex may exist for the Mach reflection case. We are not aware of experimental or numerical evidence of the existence of regular reflection-diffraction configurations with an attached shock to the wedge vertex. However, it is possible that such solutions may exist, as shown in Figs. 2.5–2.6. Thus, it is not surprising that two different cases on the parameters of the initial data in **Problem 2.2.1** are considered separately in our study.

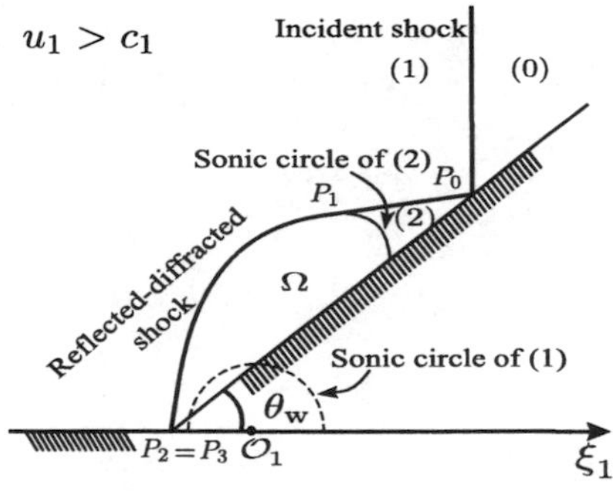

Figure 2.5: Attached supersonic regular reflection-diffraction configuration

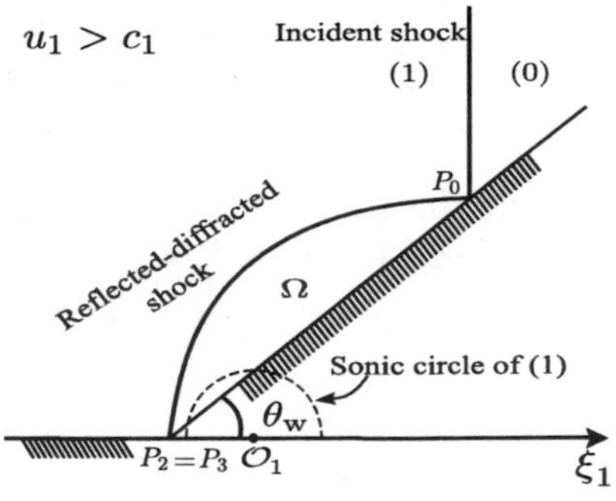

Figure 2.6: Attached subsonic regular reflection-diffraction configuration

We show that the solutions with an attached shock do not exist when the initial data of **Problem 2.2.1**, equivalently parameters (ρ_0, ρ_1, γ) in **Problem 2.6.1** which also define u_1 by (2.2.2), satisfy $u_1 \leq c_1$. Moreover, in this case, the regular reflection-diffraction solution of **Problem 2.6.1** exists for each $\theta_w \in (\theta_w^d, \frac{\pi}{2})$, as von Neumann conjectured [267, 268]. In the other case, $u_1 > c_1$, we assert the existence of a regular reflection-diffraction configuration for **Problem 2.6.1** for any $\theta_w \in (\theta_w^c, \frac{\pi}{2})$, where either $\theta_w^c = \theta_w^d$ or $\theta_w^c \in (\theta_w^d, \frac{\pi}{2})$ is the *hitting* wedge angle in the sense that a solution with $P_2 = P_3$ exists for $\theta_w = \theta_w^c$.

We also note that both cases $u_1 \leq c_1$ and $u_1 > c_1$ exist, for the corresponding parameters (ρ_0, ρ_1, γ), where $\rho_1 > \rho_0$ by the entropy condition on $\mathcal{S}_1$. Indeed, since $c_1 = \rho_1^{\frac{\gamma-1}{2}}$, and u_1 is a function of (ρ_0, ρ_1) for fixed $\gamma > 1$, determined by (2.2.2), then, for each $\rho_0 > 0$, there exists $\rho_1^* > \rho_0$ such that

(i) $u_1 < c_1$ for any $\rho_1 \in (\rho_0, \rho_1^*)$;

(ii) $u_1 = c_1$ for $\rho_1 = \rho_1^*$;

(iii) $u_1 > c_1$ for any $\rho_1 \in (\rho_1^*, \infty)$.

This is verified via a straightforward but lengthy calculation by both noting that (2.2.2) implies that $u_1 = 0 < c_1$ for $\rho_1 = \rho_0$ and showing that $\partial_{\rho_1}(u_1 - c_1) \geq C(\rho_0) > 0$ for $\rho_1 > \rho_0$.

Therefore, Case $u_1 \leq c_1$ (resp. Case $u_1 > c_1$) corresponds to the weaker (resp. stronger) incident shocks.

In Parts II–III, we focus on von Neumann's sonic conjecture, that is, the existence of supersonic reflection configurations up to the sonic angle θ_w^s for $\theta_w \in (\theta_w^s, \frac{\pi}{2})$, or the critical angle $\theta_w^c \in (\theta_w^s, \frac{\pi}{2})$ for $\theta_w \in (\theta_w^c, \frac{\pi}{2})$. We establish two existence theorems: Theorem 2.6.3, which corresponds to the case of a relatively weaker incident shock, and Theorem 2.6.5, which corresponds to the case of a relatively stronger incident shock. We also establish a regularity theorem, Theorem 2.6.6, for supersonic reflection solutions. We stress that, in what follows, φ_2 always denotes the *weak* state (2). Furthermore, in all of the theorems below, we always assume that $\rho_1 > \rho_0 > 0$ and $\gamma > 1$.

Theorem 2.6.3 (Existence of Supersonic Reflection Configurations for $u_1 \le c_1$). *Consider all* (ρ_0, ρ_1, γ) *such that* $u_1 \le c_1$. *Then there is* $\alpha = \alpha(\rho_0, \rho_1, \gamma) \in (0, \frac{1}{2})$ *so that, when* $\theta_{\rm w} \in (\theta_{\rm w}^{\rm s}, \frac{\pi}{2})$, *there exists a solution* φ *of* **Problem 2.6.1** *such that*

$$\Phi(t, \mathbf{x}) = t\,\varphi(\frac{\mathbf{x}}{t}) + \frac{|\mathbf{x}|^2}{2t} \qquad \text{for } \frac{\mathbf{x}}{t} \in \Lambda,\ t > 0$$

with

$$\rho(t, \mathbf{x}) = \Big(\rho_0^{\gamma-1} - (\gamma-1)\big(\Phi_t + \frac{1}{2}|\nabla_{\mathbf{x}}\Phi|^2\big)\Big)^{\frac{1}{\gamma-1}}$$

is a global weak solution of **Problem 2.2.1** *in the sense of Definition* 2.3.3, *which satisfies the entropy condition (cf. Definition* 2.1.2). *Furthermore,*

(a) $\varphi \in C^\infty(\Omega) \cap C^{1,\alpha}(\overline{\Omega})$;

(b) φ *has structure* (2.6.1) *in* $\Lambda \setminus \overline{\Omega}$;

(c) φ *is* $C^{1,1}$ *across part* $\Gamma_{\rm sonic}$ *of the sonic circle including endpoints* P_1 *and* P_4;

(d) *The reflected-diffracted shock* $P_0P_1P_2$ *is* $C^{2,\beta}$ *up to its endpoints for any* $\beta \in [0, \frac{1}{2})$ *and* C^∞ *except* P_1;

(e) *The relative interior of the reflected-diffracted shock* $P_0P_1P_2$ *lies in* $\{\xi_2 > \xi_1 \tan\theta_{\rm w},\ \xi_2 > 0\}$, *i.e., the domain bounded by the wedge and the symmetry line* $\{\xi_2 = 0\}$.

Moreover, φ *satisfies the following properties:*

(i) *Equation* (2.2.8) *is strictly elliptic in* $\overline{\Omega} \setminus \overline{\Gamma_{\rm sonic}}$:

$$|D\varphi| < c(|D\varphi|^2, \varphi) \qquad \text{in } \overline{\Omega} \setminus \overline{\Gamma_{\rm sonic}}; \tag{2.6.2}$$

(ii) *In* Ω,

$$\varphi_2 \le \varphi \le \varphi_1. \tag{2.6.3}$$

See Fig. 2.3.

Remark 2.6.4. *In fact,* φ *in Theorem* 2.6.3 *is an admissible solution in the sense of Definition* 8.1.1 *below so that* φ *satisfies the further conditions listed in Definition* 8.1.1.

Now we address Case $u_1 > c_1$. In this case, the results of Theorem 2.6.3 hold for any wedge angle $\theta_{\rm w}$ from $\frac{\pi}{2}$ until either $\theta_{\rm w}^{\rm s}$ or $\theta_{\rm w}^{\rm c} \in (\theta_{\rm w}^{\rm s}, \frac{\pi}{2})$ when the shock hits the wedge vertex P_3.

Theorem 2.6.5 (Existence of Supersonic Reflection Configurations when $u_1 > c_1$). *Consider all* (ρ_0, ρ_1, γ) *such that* $u_1 > c_1$. *There are* $\theta_{\mathrm{w}}^{\mathrm{c}} \in [\theta_{\mathrm{w}}^{\mathrm{s}}, \frac{\pi}{2})$ *and* $\alpha \in (0, \frac{1}{2})$ *depending only on* (ρ_0, ρ_1, γ) *so that the results of Theorem* 2.6.3 *hold for each wedge angle* $\theta_{\mathrm{w}} \in (\theta_{\mathrm{w}}^{\mathrm{c}}, \frac{\pi}{2})$. *If* $\theta_{\mathrm{w}}^{\mathrm{c}} > \theta_{\mathrm{w}}^{\mathrm{s}}$, *then, for the wedge angle* $\theta_{\mathrm{w}} = \theta_{\mathrm{w}}^{\mathrm{c}}$, *there exists an attached solution as shown in Fig.* 2.5 (*i.e., a solution* φ *of* **Problem 2.6.1** *with the properties as those in Theorem* 2.6.3 *except that* $P_3 = P_2$) *with the regularity:*

$$\varphi \in C^\infty(\Omega) \cap C^{2,\alpha}(\overline{\Omega} \setminus (\Gamma_{\text{sonic}} \cup \{P_2\})) \cap C^{1,1}(\overline{\Omega} \setminus \{P_2\}) \cap C^{0,1}(\overline{\Omega}),$$

and the reflected-diffracted shock $P_0P_1P_2$ *is Lipschitz up to its endpoints,* $C^{2,\beta}$ *for any* $\beta \in [0, \frac{1}{2})$ *except point* P_2, *and* C^∞ *except points* P_1 *and* P_2.

Since solution φ of **Problem 2.6.1** constructed in Theorems 2.6.3 and 2.6.5 satisfies the $C^{1,1}$–continuity across $\overline{P_1P_4}$, (2.6.2)–(2.6.3), and further estimates including (11.2.23)–(11.2.24), (11.2.38)–(11.2.40), (11.4.38)–(11.4.39), and (11.5.2), as well as Propositions 11.2.8 and 11.4.6, then the regularity results of Theorem 14.2.8 and Corollary 14.2.11 apply. More precisely, we have

Theorem 2.6.6 (Regularity of Solutions up to Γ_{sonic}). *Any solution* φ *in Theorems* 2.6.3 *and* 2.6.5 *satisfies the following:*

(i) φ *is* $C^{2,\alpha}$ *up to the sonic arc* $\overline{\Gamma_{\text{sonic}}}$ *away from point* P_1 *for any* $\alpha \in (0,1)$. *That is, for any* $\alpha \in (0,1)$ *and any given* $\boldsymbol{\xi}^0 \in \overline{\Gamma_{\text{sonic}}} \setminus \{P_1\}$, *there exist* $C < \infty$ *depending only on* $(\rho_0, \rho_1, \gamma, \alpha)$ *and* $\operatorname{dist}(\boldsymbol{\xi}^0, \Gamma_{\text{shock}})$, *and* $d > 0$ *depending only on* (ρ_0, ρ_1, γ) *and* $\operatorname{dist}(\boldsymbol{\xi}^0, \Gamma_{\text{shock}})$ *such that*

$$\|\varphi\|_{2,\alpha;\overline{B_d(\boldsymbol{\xi}^0)\cap\Omega}} \leq C.$$

(ii) *For any* $\boldsymbol{\xi}^0 \in \overline{\Gamma_{\text{sonic}}} \setminus \{P_1\}$,

$$\lim_{\substack{\boldsymbol{\xi}\to\boldsymbol{\xi}^0\\ \boldsymbol{\xi}\in\Omega}} (D_{rr}\varphi - D_{rr}\varphi_2) = \frac{1}{\gamma+1},$$

where (r,θ) *are the polar coordinates with the center at* (u_2, v_2).

(iii) $D^2\varphi$ *has a jump across* $\overline{\Gamma_{\text{sonic}}}$: *For any* $\boldsymbol{\xi}^0 \in \overline{\Gamma_{\text{sonic}}} \setminus \{P_1\}$,

$$\lim_{\substack{\boldsymbol{\xi}\to\boldsymbol{\xi}^0\\ \boldsymbol{\xi}\in\Omega}} D_{rr}\varphi \;-\; \lim_{\substack{\boldsymbol{\xi}\to\boldsymbol{\xi}^0\\ \boldsymbol{\xi}\in\Lambda\setminus\Omega}} D_{rr}\varphi \;=\; \frac{1}{\gamma+1}.$$

(iv) $\lim_{\substack{\boldsymbol{\xi}\to P_1\\ \boldsymbol{\xi}\in\Omega}} D^2\varphi$ *does not exist.*

From Chapter 4 to Chapter 14, we develop approaches, techniques, and related analysis to complete the proofs of these theorems in detail, which provide a solution to **von Neumann's Sonic Conjecture**. We give an overview of these techniques in Chapter 3.

In Part IV, we extend Theorems 2.6.3 and 2.6.5 beyond the sonic angle to include the wedge angles $\theta_{\rm w} \in (\theta_{\rm w}^{\rm d}, \theta_{\rm w}^{\rm s}]$. Therefore, we establish the global existence of regular reflection-diffraction configurations for any wedge angle between $\frac{\pi}{2}$ and the detachment angle $\theta_{\rm w}^{\rm d}$, or the critical angle $\theta_{\rm w}^{\rm c}$ in the attached case. As in Theorems 2.6.3 and 2.6.5, we need to analyze two separate cases: Case $u_1 \le c_1$ and Case $u_1 > c_1$, since the reflected-diffracted shock may hit the wedge vertex in the latter case. Below, we use the notations introduced in §2.4.2–§2.4.3, and φ_2 denotes the *weak* state (2).

The theorems that follow assert the global existence of regular reflection-diffraction configurations for any wedge angle between $\frac{\pi}{2}$ and the detachment angle $\theta_{\rm w}^{\rm d}$, or the critical angle $\theta_{\rm w}^{\rm c}$ in the attached case. The type of regular reflection-diffraction configuration for $\theta_{\rm w} \in (\theta_{\rm w}^{\rm d}, \frac{\pi}{2})$ is supersonic if $|D\varphi_2(P_0)| > c_2$ and subsonic if $|D\varphi_2(P_0)| \le c_2$. In particular, the regular reflection-diffraction configuration is supersonic for all $\theta_{\rm w} \in (\theta_{\rm w}^{\rm s}, \frac{\pi}{2})$, as we have presented in Theorems 2.6.3 and 2.6.5. Also, as we have discussed in (iii) of §2.5 and will prove in Theorem 7.1.1(vi) later, there exists $\hat{\theta}_{\rm w}^{\rm s} \in (\theta_{\rm w}^{\rm d}, \theta_{\rm w}^{\rm s}]$ such that $|D\varphi_2(P_0)| < c_2$ for all $\theta_{\rm w} \in (\theta_{\rm w}^{\rm d}, \hat{\theta}_{\rm w}^{\rm s})$, and then the regular reflection-diffraction configuration is subsonic for these wedge angles.

Theorem 2.6.7 (Global Solutions up to the Detachment Angle when $u_1 \le c_1$). *Consider all (ρ_0, ρ_1, γ) such that $u_1 \le c_1$. Then there is $\alpha = \alpha(\rho_0, \rho_1, \gamma) \in (0, \frac{1}{2})$ so that, when $\theta_{\rm w} \in (\theta_{\rm w}^{\rm d}, \frac{\pi}{2})$, there exists a solution φ of* **Problem 2.6.1** *such that*

$$\Phi(t, \mathbf{x}) = t\,\varphi(\frac{\mathbf{x}}{t}) + \frac{|\mathbf{x}|^2}{2t} \qquad \text{for } \frac{\mathbf{x}}{t} \in \Lambda,\ t > 0$$

with

$$\rho(t, \mathbf{x}) = \Big(\rho_0^{\gamma-1} - (\gamma-1)\big(\Phi_t + \frac{1}{2}|\nabla_{\mathbf{x}}\Phi|^2\big)\Big)^{\frac{1}{\gamma-1}}$$

is a global weak solution of **Problem 2.2.1** *in the sense of Definition* 2.3.3, *which satisfies the entropy condition (cf. Definition* 2.1.2), *and the type of reflection configurations (supersonic or subsonic) is determined by $\theta_{\rm w}$:*

- *If $|D\varphi_2(P_0)| > c_2$, then φ has the* supersonic *reflection configuration and satisfies all the properties in Theorem* 2.6.3, *which is the case for any wedge angle $\theta_{\rm w} \in (\theta_{\rm w}^{\rm s}, \frac{\pi}{2})$;*

- *If $|D\varphi_2(P_0)| \le c_2$, then φ has the* subsonic *reflection configuration and satisfies*

$$\varphi \in C^\infty(\Omega) \cap C^{2,\alpha}(\overline{\Omega} \setminus \{P_0, P_3\}) \cap C^{1,\alpha}(\overline{\Omega}),$$

$$\varphi = \begin{cases} \varphi_0 & \text{for } \xi_1 > \xi_1^0 \text{ and } \xi_2 > \xi_1 \tan\theta_{\rm w}, \\ \varphi_1 & \text{for } \xi_1 < \xi_1^0 \text{ and above curve } P_0P_2, \\ \varphi_2(P_0) & \text{at } P_0, \end{cases} \tag{2.6.4}$$

$D\varphi(P_0) = D\varphi_2(P_0)$, and the reflected-diffracted shock $\Gamma_{\rm shock}$ is $C^{1,\alpha}$ up to its endpoints and C^∞ except P_0. Also, the relative interior of shock $\Gamma_{\rm shock}$

lies in $\{\xi_2 > \xi_1 \tan\theta_w, \xi_2 > 0\}$, *i.e., a domain bounded by the wedge and the symmetry line* $\{\xi_2 = 0\}$.

Furthermore, φ *satisfies the following properties:*

(i) *Equation* (2.2.8) *is strictly elliptic:*

$$|D\varphi| < c(|D\varphi|^2, \varphi) \tag{2.6.5}$$

in $\overline{\Omega} \setminus \{P_0\}$ *if* $|D\varphi_2(P_0)| = c_2$ *and in* $\overline{\Omega}$ *if* $|D\varphi_2(P_0)| < c_2$;

(ii) *In* Ω,

$$\varphi_2 \le \varphi \le \varphi_1. \tag{2.6.6}$$

Note that the regular reflection-diffraction solution has a subsonic reflection configuration for any $\theta_w \in (\theta_w^d, \hat{\theta}_w^s)$, *where* $\hat{\theta}_w^s$ *is from* (iii) *in* §2.5.

Moreover, the optimal regularity theorem, Theorem 2.6.6, *applies to any global regular reflection solutions of supersonic reflection configuration.*

Remark 2.6.8. *Solution* φ *in Theorem* 2.6.7 *is also an admissible solution in the sense of Definition* 15.1.1 *in the supersonic case and of Definition* 15.1.2 *in the subsonic case, so that* φ *satisfies the further conditions listed in Definitions* 15.1.1–15.1.2, *respectively, in Chapter* 15.

Now we address Case $u_1 > c_1$. In this case, the results of Theorem 2.6.7 hold from the wedge angle $\frac{\pi}{2}$ until either θ_w^d or the critical angle $\theta_w^c \in (\theta_w^d, \frac{\pi}{2})$ when the shock hits the wedge vertex P_3.

Theorem 2.6.9 (Global Solutions up to the Detachment Angle when $u_1 > c_1$). *Consider all* (ρ_0, ρ_1, γ) *such that* $u_1 > c_1$. *Then there are* $\theta_w^c \in [\theta_w^d, \frac{\pi}{2})$ *and* $\alpha \in (0, \frac{1}{2})$ *depending only on* (ρ_0, ρ_1, γ) *so that the results of Theorem* 2.6.7 *hold for each wedge angle* $\theta_w \in (\theta_w^c, \frac{\pi}{2})$.

If $\theta_w^c > \theta_w^d$, *then, for the wedge angle* $\theta_w = \theta_w^c$, *there exists an attached solution* φ *as shown in Fig.* 2.5–2.6, *i.e., a solution* φ *of* **Problem 2.6.1** *with the properties as in Theorem* 2.6.7 *except that* $P_2 = P_3$. *Moreover, the attached solution* φ *has the following two cases:*

- *If* $|D\varphi_2(P_0)| > c_2$ *for* $\theta_w = \theta_w^c$ *(the supersonic case), the reflected-diffracted shock of the attached solution satisfies all of the properties listed in Theorem* 2.6.5;

- *If* $|D\varphi_2(P_0)| \le c_2$ *for* $\theta_w = \theta_w^c$ *(the subsonic case),*

$$\varphi \in C^\infty(\Omega) \cap C^{2,\alpha}(\overline{\Omega} \setminus \{P_0, P_3\}) \cap C^{1,\alpha}(\overline{\Omega} \setminus \{P_3\}) \cap C^{0,1}(\overline{\Omega}).$$

The reflected-diffracted shock is Lipschitz up to its endpoints, $C^{1,\alpha}$ *except point* P_2, *and* C^∞ *except its endpoints* P_0 *and* P_2.

Remark 2.6.10. *We emphasize that all the results in the main theorems – Theorem* 2.6.3, *Theorems* 2.6.5–2.6.7, *and Theorem* 2.6.9 – *hold when* $\gamma = 1$, *which can be handled similarly with appropriate changes in the formulas in their respective proofs.*

Remark 2.6.11. *In Chen-Feldman-Xiang* [60], *the strict convexity of self-similar transonic shocks has also been proved in the regular shock reflection-diffraction configurations in Theorem* 2.6.3, *Theorems* 2.6.5–2.6.7, *and Theorem* 2.6.9.

In Part IV, we further develop analytical techniques to complete the proofs of these theorems and further results, which provides a solution to **von Neumann's Detachment Conjecture**. The main challenge is the analysis of the transition from the supersonic to subsonic reflection configurations, which requires uniform *a priori* estimates for the solutions at the corner between the reflected-diffracted shock and the wedge when the wedge angle $\theta_{\rm w}$ decreases across the sonic angle $\theta_{\rm w}^{\rm s}$ up to the detachment angle $\theta_{\rm w}^{\rm d}$, or the critical angle $\theta_{\rm w}^{\rm c}$ in the attached case.

Chapter Three

Main Steps and Related Analysis in the Proofs of the Main Theorems

In this chapter, we give an overview of the main steps and related analysis in the proofs of the main theorems for the existence of global regular reflection-diffraction solutions. We first discuss the proof of the existence and properties of supersonic regular reflection-diffraction configurations up to the sonic angle in Theorems 2.6.3 and 2.6.5. Then we discuss the proofs of the existence and properties of regular reflection-diffraction configurations beyond the sonic angle, up to the detachment angle, in Theorems 2.6.7 and 2.6.9. The detailed proofs and analysis developed for the main theorems will be given in Parts III–IV.

3.1 NORMAL REFLECTION

When the wedge angle $\theta_{\rm w} = \frac{\pi}{2}$, the incident shock reflects normally (see Fig. 3.1). The reflected shock is also a plane at $\xi_1 = \bar{\xi}_1 < 0$. Then the velocity of state (2) is zero, $\bar{u}_2 = \bar{v}_2 = 0$, state (1) is of form (2.2.17), and state (2) is of the form:

$$\varphi_2(\boldsymbol{\xi}) = -\frac{|\boldsymbol{\xi}|^2}{2} + u_1(\bar{\xi}_1 - \xi_1^0), \tag{3.1.1}$$

where $\xi_1^0 = \frac{\rho_1 u_1}{\rho_1 - \rho_0} > 0$, which is the position of the incident shock in the self-similar coordinates $\boldsymbol{\xi}$.

The position: $\xi_1 = \bar{\xi}_1 < 0$ of the reflected shock and density $\bar{\rho}_2$ of state (2) can be determined uniquely from the Rankine-Hugoniot condition (2.2.13) at the reflected shock and the Bernoulli law (2.2.7).

3.2 MAIN STEPS AND RELATED ANALYSIS IN THE PROOF OF THE SONIC CONJECTURE

In this section, we always discuss the global solutions of **Problem 2.6.1** in §2.6 for the wedge angles $\theta_{\rm w} \in (\theta_{\rm w}^{\rm s}, \frac{\pi}{2})$, where $\theta_{\rm w}^{\rm s}$ is the sonic angle. Then the expected solutions are of the supersonic reflection configuration as described in §2.4.2.

To solve the free boundary problem (**Problem 2.6.1**) for the wedge angles $\theta_{\rm w} \in (\theta_{\rm w}^{\rm s}, \frac{\pi}{2})$, we first define a class of admissible solutions that are of the

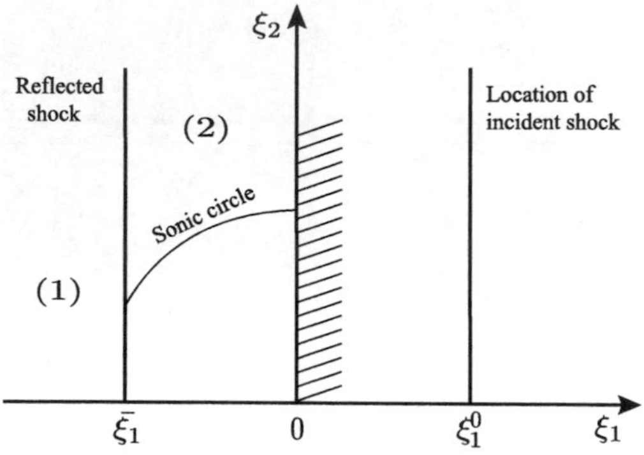

Figure 3.1: Normal reflection

structure of §2.4.2 and satisfy some additional properties as discussed below. Then we make *a priori* estimates of admissible solutions. Finally, based on the *a priori* estimates, we obtain the existence of admissible solutions as fixed points of the iteration procedure by employing the Leray-Schauder degree theory. We now discuss these steps in more detail.

3.2.1 Admissible solutions

To solve the free boundary problem (**Problem 2.6.1**), we first define a class of admissible solutions φ that are the solutions with supersonic reflection configuration as described in §2.4.2, which is the case when the wedge angle θ_w is between θ_w^s and $\frac{\pi}{2}$.

Let $\gamma > 1$ and $\rho_1 > \rho_0 > 0$ be given constants, and let $\xi_1^0 > 0$ and $u_1 > 0$ be defined by (2.2.18). Let the incident shock be defined by $\mathcal{S}_0 := \{\xi_1 = \xi_1^0\}$, and let state (0) and state (1) ahead of and behind $\mathcal{S}_0$ be given by (2.2.16)–(2.2.17), respectively, so that the Rankine-Hugoniot condition (2.2.13) holds on $\mathcal{S}_0$. As we will show in Theorem 7.1.1 (see also the discussion in §2.5), there exists $\theta_w^s \in (0, \frac{\pi}{2})$ such that, when the wedge angle $\theta_w \in (\theta_w^s, \frac{\pi}{2})$, there is a unique weak state (2) of form (2.4.1) so that

(i) $u_2 > 0$. Then $v_2 = u_2 \tan\theta_w > 0$, and $\mathcal{S}_1 := \{\varphi_1 = \varphi_2\}$ is a line. Lines $\mathcal{S}_0$ and $\mathcal{S}_1$ meet the wedge boundary $\{\xi_2 = \xi_1 \tan\theta_w\}$ at point $P_0 \equiv P_0(\theta_w) = (\xi_1^0, \xi_1^0 \tan\theta_w)$.

(ii) The entropy condition, $\rho_2 > \rho_1$, holds.

(iii) The Rankine-Hugoniot condition (2.2.13) holds for φ_1 and φ_2 along line $\mathcal{S}_1$.

(iv) φ_2 is supersonic at the reflection point P_0, *i.e.*, $|D\varphi_2(P_0)| > c_2$.

(v) u_2 and ρ_2 depend continuously on $\theta_w \in (\theta_w^s, \frac{\pi}{2})$.

(vi) $\lim_{\theta_w \to \frac{\pi}{2}-}(u_2(\theta_w), \rho_2(\theta_w)) = (0, \bar{\rho}_2)$, where $\bar{\rho}_2$ is the unique density of state (2) for the normal reflection solution.

Using the properties of the uniform state solutions of (2.2.8), we show that, for each $\theta_w \in (\theta_w^s, \frac{\pi}{2})$, line $\mathcal{S}_1 = \{\varphi_1 = \varphi_2\}$ necessarily intersects with boundary $\partial B_{c_2}(u_2, v_2)$ of the sonic circle of state (2) in two points. Let P_1 be the nearest point of intersection of $\mathcal{S}_1$ with $\partial B_{c_2}(u_2, v_2)$ to $P_0 = (\xi_1^0, \xi_1^0 \tan\theta_w)$; see Fig. 2.3. Then P_1 necessarily lies within Λ, so does the whole segment P_0P_1.

With this, for any $\theta_w \in (\theta_w^s, \frac{\pi}{2})$, we define the points, segments, and curves shown in Fig. 2.3 as follows:

- Line $\mathcal{S}_1 := \{\varphi_1 = \varphi_2\}$.
- Point $P_0 := (\xi_1^0, \xi_1^0 \tan\theta_w)$.
- Point P_1 is the unique point of intersection of $\mathcal{S}_1$ with $\partial B_{c_2}(u_2, v_2)$ such that state (2) is supersonic at any $\boldsymbol{\xi} = (\xi_1, \xi_2) \in \mathcal{S}_1$ satisfying $\xi_2 > \xi_{2P_1}$.
- Point $P_3 := \mathbf{0} := (0, 0)$.
- Point $P_4 := (q_2 + c_2)(\cos\theta_w, \sin\theta_w)$, where $q_2 := \sqrt{u_2^2 + v_2^2}$; that is, P_4 is the upper point of intersection of the sonic circle of state (2) with the wedge boundary $\{\xi_2 = \xi_1 \tan\theta_w\}$. From the definition,
$$\xi_{1P_1} < \xi_{1P_4}.$$
- Line segment $\Gamma_{\text{wedge}} := P_3P_4 \subset \{\xi_2 = \xi_1 \tan\theta_w\}$.
- Γ_{sonic} is the upper arc P_1P_4 of the sonic circle of state (2), that is,
$$\Gamma_{\text{sonic}} := \{(\xi_1, v_2 + \sqrt{c_2^2 - (\xi_1 - u_2)^2}) \ : \ \xi_{1P_1} \le \xi_1 \le \xi_{1P_4}\}.$$

Now we define the admissible solutions of **Problem 2.6.1** for the wedge angles $\theta_w \in (\theta_w^s, \frac{\pi}{2})$. The admissible solutions are of the structure of supersonic reflection configuration described in §2.4.2. These conditions are listed in conditions (i)–(iii) of Definition 3.2.1. We also add conditions (iv)–(v) of Definition 3.2.1. This is motivated by the fact that, for the wedge angles sufficiently close to $\frac{\pi}{2}$, the solutions of **Problem 2.6.1** which satisfy conditions (i)–(iii) of Definition 3.2.1 also satisfy conditions (iv)–(v) of Definition 3.2.1, as we will prove in Appendix 8.3.

Definition 3.2.1. *Fix a wedge angle $\theta_w \in (\theta_w^s, \frac{\pi}{2})$. A function $\varphi \in C^{0,1}(\Lambda)$ is called an admissible solution of the regular shock reflection-diffraction problem if φ is a solution of* **Problem 2.6.1** *and satisfies the following:*

(i) *There exists a relatively open curve segment Γ_{shock} (without self-intersection) whose endpoints are $P_1 = (\xi_{1P_1}, \xi_{2P_1})$ and $P_2 = (\xi_{1P_2}, 0)$ with*
$$\xi_{1P_2} < \min\{0,\ u_1 - c_1\}, \qquad \xi_{1P_2} \le \xi_{1P_1},$$
such that Γ_{shock} satisfies that:

- *For the sonic circle* $\partial B_{c_1}(u_1, 0)$ *of state* (1),

$$\Gamma_{\rm shock} \subset \big(\Lambda \setminus \overline{B_{c_1}(u_1,0)}\big) \cap \{\xi_{1P_2} \le \xi_1 \le \xi_{1P_1}\}; \qquad (3.2.1)$$

- $\Gamma_{\rm shock}$ *is* C^2 *in its relative interior, and curve* $\Gamma^{\rm ext}_{\rm shock} := \Gamma_{\rm shock} \cup \Gamma^-_{\rm shock} \cup \{P_2\}$ *is* C^1 *at its relative interior (including* P_2*), where* $\Gamma^-_{\rm shock}$ *is the reflection of* $\Gamma_{\rm shock}$ *with respect to the* ξ_1*-axis.*

Let $\Gamma_{\rm sym} := P_2P_3$ *be the line segment. Then* $\Gamma_{\rm sonic}$, $\Gamma_{\rm sym}$, *and* $\Gamma_{\rm wedge}$ *do not have common points except their common endpoints* $\{P_3, P_4\}$. *We* require *that there be no common points between* $\Gamma_{\rm shock}$ *and curve* $\overline{\Gamma_{\rm sym}} \cup \overline{\Gamma_{\rm wedge}} \cup \overline{\Gamma_{\rm sonic}}$ *except their common endpoints* $\{P_1, P_2\}$. *Thus,* $\overline{\Gamma_{\rm shock}} \cup \overline{\Gamma_{\rm sym}} \cup \overline{\Gamma_{\rm wedge}} \cup \overline{\Gamma_{\rm sonic}}$ *is a closed curve without self-intersection. Denote by* Ω *the open bounded domain restricted by this curve. Note that* $\Omega \subset \Lambda$ *and* $\partial\Omega \cap \partial\Lambda = \overline{\Gamma_{\rm sym}} \cup \overline{\Gamma_{\rm wedge}}$.

(ii) φ *satisfies*

$$\begin{aligned}
&\varphi \in C^{0,1}(\Lambda) \cap C^1(\overline{\Lambda} \setminus \overline{P_0P_1P_2}),\\
&\varphi \in C^3(\overline{\Omega} \setminus (\overline{\Gamma_{\rm sonic}} \cup \{P_2, P_3\})) \cap C^1(\overline{\Omega}),\\
&\varphi = \begin{cases} \varphi_0 & \text{for } \xi_1 > \xi_1^0 \text{ and } \xi_2 > \xi_1 \tan\theta_{\rm w},\\ \varphi_1 & \text{for } \xi_1 < \xi_1^0 \text{ and above curve } P_0P_1P_2,\\ \varphi_2 & \text{in } P_0P_1P_4. \end{cases}
\end{aligned} \qquad (3.2.2)$$

(iii) *Equation* (2.2.8) *is strictly elliptic in* $\overline{\Omega} \setminus \overline{\Gamma_{\rm sonic}}$:

$$|D\varphi| < c(|D\varphi|^2, \varphi) \qquad \text{in } \overline{\Omega} \setminus \overline{\Gamma_{\rm sonic}}. \qquad (3.2.3)$$

(iv) *In* Ω,

$$\varphi_2 \le \varphi \le \varphi_1. \qquad (3.2.4)$$

(v) *Let* $\mathbf{e}_{S_1}$ *be the unit vector parallel to* S_1 *oriented so that* $\mathbf{e}_{S_1} \cdot D\varphi_2(P_0) > 0$; *that is,*

$$\mathbf{e}_{S_1} = \frac{P_1 - P_0}{|P_1 - P_0|} = -\frac{(v_2,\, u_1 - u_2)}{\sqrt{(u_1-u_2)^2 + v_2^2}}. \qquad (3.2.5)$$

Then

$$\partial_{\mathbf{e}_{S_1}}(\varphi_1 - \varphi) \le 0 \qquad \text{in } \Omega, \qquad (3.2.6)$$

$$\partial_{\xi_2}(\varphi_1 - \varphi) \le 0 \qquad \text{in } \Omega. \qquad (3.2.7)$$

Remark 3.2.2. *Condition* (3.2.4) *of Definition* 3.2.1 *implies that* $\Omega \subset \{\varphi_2 < \varphi_1\}$, *i.e., that* Ω *lies between line* S_1 *and the wedge boundary; see Fig.* 3.2.

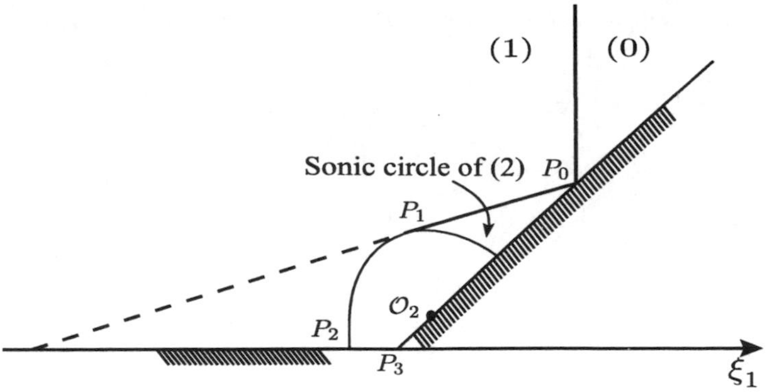

Figure 3.2: Location of domain Ω

Remark 3.2.3 (Cone of monotonicity directions). *Conditions* (3.2.6) *and* (3.2.7) *imply that, if* φ *is an admissible solution of* **Problem 2.6.1** *in the sense of Definition* 3.2.1, *then*

$$\partial_{\mathbf{e}}(\varphi_1 - \varphi) \leq 0 \qquad in\ \overline{\Omega}, \ \ for\ all\ \mathbf{e} \in \mathrm{Cone}(\mathbf{e}_{\mathcal{S}_1}, \mathbf{e}_{\xi_2}),\ \mathbf{e} \neq 0, \tag{3.2.8}$$

where, for $\mathbf{e}, \mathbf{g} \in \mathbb{R}^2 \setminus \{0\}$ *with* $\mathbf{e}, \mathbf{g} \neq 0$ *and* $\mathbf{e} \neq c\mathbf{g}$,

$$\mathrm{Cone}(\mathbf{e}, \mathbf{g}) := \{a\mathbf{e} + b\mathbf{g} \ : \ a, b \geq 0\}. \tag{3.2.9}$$

We denote by $\mathrm{Cone}^0(\mathbf{e}, \mathbf{g})$ *the interior of* $\mathrm{Cone}(\mathbf{e}, \mathbf{g})$.

Remark 3.2.4 (Γ_{shock} does not intersect with Γ_{wedge} and the sonic circle of state (1)). *The property that* $\Gamma_{\text{shock}} \subset \Lambda \setminus \overline{B_{c_1}(u_1, 0)}$ *of* (3.2.1) *implies that* Γ_{shock} *intersects with neither* Γ_{wedge} *nor the sonic circle* $\partial B_{c_1}(u_1, 0)$ *of state* (1) *with*

$$B_{c_1}(u_1, 0) \cap \Lambda \subset \Omega. \tag{3.2.10}$$

Remark 3.2.5 (φ matches with φ_2 on Γ_{sonic}). *From Definition* 3.2.1(ii), *it follows that*

$$\varphi = \varphi_2, \quad D\varphi = D\varphi_2 \qquad on\ \overline{\Gamma_{\text{sonic}}}.$$

Note that the Rankine-Hugoniot conditions (2.2.13)–(2.2.14) imply the following equalities on Γ_{shock}:

$$\rho(|D\varphi|^2, \varphi)\partial_{\boldsymbol{\nu}}\varphi = \rho_1 \partial_{\boldsymbol{\nu}}\varphi_1, \tag{3.2.11}$$

$$\partial_{\boldsymbol{\tau}}\varphi = \partial_{\boldsymbol{\tau}}\varphi_1, \tag{3.2.12}$$

$$\varphi = \varphi_1, \tag{3.2.13}$$

where, on the left-hand sides of (3.2.11)–(3.2.12), $D\varphi$ is evaluated on the Ω–side of Γ_{shock}.

Throughout the rest of this section, we always assume that φ is an admissible solution of **Problem 2.6.1**.

3.2.2 Strict monotonicity cone for $\varphi_1 - \varphi$ and its geometric consequences

First, we prove that, for any $\mathbf{e} \in \mathrm{Cone}^0(\mathbf{e}_{\mathcal{S}_1}, \mathbf{e}_{\xi_2})$,

$$\partial_{\mathbf{e}}(\varphi_1 - \varphi) < 0 \qquad \text{in } \overline{\Omega}, \tag{3.2.14}$$

where $\mathrm{Cone}^0(\mathbf{e}, \mathbf{g})$ is the interior of $\mathrm{Cone}(\mathbf{e}, \mathbf{g})$ defined by (3.2.9) for $\mathbf{e}, \mathbf{g} \in \mathbb{R}^2 \setminus \{0\}$. For the proof, we derive an equation for $w = \partial_{\mathbf{e}}(\varphi_1 - \varphi)$ in Ω, and employ the maximum principle and boundary conditions on $\partial\Omega$, including the conditions on Γ_{sonic} in Remark 3.2.5.

This implies that Γ_{shock} *is a graph in the directions of the cone.* That is, for $\mathbf{e} \in \mathrm{Cone}^0(\mathbf{e}_{\mathcal{S}_1}, \mathbf{e}_{\xi_2})$ with $\mathbf{e}^\perp$ being orthogonal to $\mathbf{e}$ and oriented so that $\mathbf{e}^\perp \cdot \mathbf{e}_{\mathcal{S}_1} < 0$ and $|\mathbf{e}| = |\mathbf{e}^\perp| = 1$, coordinates (S, T) with basis $\{\mathbf{e}, \mathbf{e}^\perp\}$, and $P_k = (S_{P_k}, T_{P_k})$, $k = 1, \dots, 4$, with $T_{P_2} < T_{P_1} < T_{P_4}$, there exists $f_{\mathbf{e},\text{sh}} \in C^1(\mathbb{R})$ such that

(i) $\Gamma_{\text{shock}} = \{S = f_{\mathbf{e},\text{sh}}(T) \,:\, T_{P_2} < T < T_{P_1}\}$ and
$\Omega \subset \{S < f_{\mathbf{e},\text{sh}}(T) \,:\, T \in \mathbb{R}\}$.

(ii) In the (S,T)–coordinates, $P_k = (f_{\mathbf{e},\text{sh}}(T_{P_k}), T_{P_k}), k = 1, 2$.

(iii) The tangent directions to Γ_{shock} are between the directions of line $\mathcal{S}_1$ and $\{t\mathbf{e}_{\xi_2} \,:\, t \in \mathbb{R}\}$, which are the tangent lines to Γ_{shock} at points P_1 and P_2, respectively. That is, for any $T \in (T_{P_2}, T_{P_1})$,

$$\begin{aligned} -\infty < \frac{\mathbf{e}_{\mathcal{S}_1} \cdot \mathbf{e}}{\mathbf{e}_{\mathcal{S}_1} \cdot \mathbf{e}^\perp} &= f'_{\mathbf{e},\text{sh}}(T_{P_1}) \\ &\le f'_{\mathbf{e},\text{sh}}(T) \le f'_{\mathbf{e},\text{sh}}(T_{P_2}) = \frac{\mathbf{e}_{\xi_2} \cdot \mathbf{e}}{\mathbf{e}_{\xi_2} \cdot \mathbf{e}^\perp} < \infty. \end{aligned}$$

Note that the last property gives an estimate of the Lipschitz constant of Γ_{shock} for an admissible solution in terms of the parameters of states (0), (1), and (2).

3.2.3 Monotonicity cone for $\varphi - \varphi_2$ and its consequences

We prove that, for any $\mathbf{e} \in \mathrm{Cone}^0(\mathbf{e}_{\mathcal{S}_1}, -\boldsymbol{\nu}_{\mathrm{w}})$,

$$\partial_{\mathbf{e}}(\varphi - \varphi_2) \ge 0 \qquad \text{in } \overline{\Omega}, \tag{3.2.15}$$

where $\mathrm{Cone}^0(\mathbf{e}_{\mathcal{S}_1}, -\boldsymbol{\nu}_{\mathrm{w}})$ is defined by (3.2.9), and $\boldsymbol{\nu}_{\mathrm{w}}$ is the unit normal on Γ_{wedge}, interior to Ω.

As a consequence of this, we conclude that, in the local coordinates (x, y) with x the normal directional coordinate into Ω with respect to the sonic arc Γ_{sonic},

$$\partial_x(\varphi - \varphi_2) \ge 0$$

in a uniform neighborhood of Γ_{sonic}. This is important for the regularity estimates near Γ_{sonic}; see §3.2.5.2.

3.2.4 Uniform estimates for admissible solutions

We next discuss several uniform estimates for admissible solutions. Some of these estimates hold for any wedge angle $\theta_{\rm w} \in (\theta_{\rm w}^{\rm s}, \frac{\pi}{2})$. The universal constant C in these estimates depends only on the data: (ρ_0, ρ_1, γ).

In the other estimates, we have to restrict the range of wedge angles as follows: Fix any $\theta_{\rm w}^* \in (\theta_{\rm w}^{\rm s}, \frac{\pi}{2})$, and consider admissible solutions with $\theta_{\rm w} \in [\theta_{\rm w}^*, \frac{\pi}{2})$. In the case of Theorem 2.6.5, for some estimates, we need to restrict the wedge angles further by considering only $\theta_{\rm w}^* \in (\theta_{\rm w}^{\rm c}, \frac{\pi}{2})$, where $\theta_{\rm w}^{\rm c} \in [\theta_{\rm w}^{\rm s}, \frac{\pi}{2})$ is defined in §2.6 (also see §3.2.4.3). Then we obtain the uniform estimates for admissible solutions with the wedge angles $\theta_{\rm w} \in [\theta_{\rm w}^*, \frac{\pi}{2}]$. The universal constant C in these estimates depends only on the data and $\theta_{\rm w}^*$.

The proofs are achieved by employing the conditions of admissible solutions in Definition 3.2.1 and the monotonicity properties discussed in §3.2.2–§3.2.3. The arguments are based on the maximum principle via the strict ellipticity of the equation in Ω.

3.2.4.1 Uniform estimate of the size of Ω, the Lipschitz norm of the potential, and the density from above and below

In estimating $\mathrm{diam}(\Omega)$, a difficulty is that we cannot exclude the possibility that the ray:

$$\mathcal{S}_1^+ = \{P_0 + t(P_1 - P_0) \ : \ t > 0\}$$

does not intersect with the ξ_1–axis for $\theta_{\rm w} \in [\theta_{\rm w}^{\rm s}, \frac{\pi}{2})$. Then $\mathrm{diam}(\Omega)$ would not be estimated by the coordinates of the points of intersection of $\mathcal{S}_1^+$ with the ξ_1–axis.

By using the potential flow equation (2.2.8) and the conditions of admissible solutions in Definition 3.2.1, including the strict ellipticity in Ω, we show that there exists $C > 0$ such that, if φ is an admissible solution of **Problem 2.6.1** with $\theta_{\rm w} \in (\theta_{\rm w}^{\rm s}, \frac{\pi}{2})$, then

$$\Omega \subset B_C(\mathbf{0}), \tag{3.2.16}$$

$$\|\varphi\|_{0,1,\overline{\Omega}} \le C, \tag{3.2.17}$$

$$a\rho_1 \le \rho \le C \qquad \text{in } \Omega \text{ with } a = \Big(\frac{2}{\gamma+1}\Big)^{\frac{1}{\gamma-1}} > 0, \tag{3.2.18}$$

$$\rho_1 < \rho \le C \qquad \text{on } \overline{\Gamma_{\rm shock}} \cup \{P_3\}. \tag{3.2.19}$$

These properties allow us to obtain the uniform $C^{2,\alpha}$–estimates of φ in Ω away from $\Gamma_{\rm shock} \cup \Gamma_{\rm sonic} \cup \{P_3\}$. With this, we obtain certain (preliminary) compactness properties of admissible solutions. In particular, we show that the admissible solutions tend to the normal reflection as the wedge angles tend to $\frac{\pi}{2}$, where the convergence is understood in the appropriate sense that implies the convergence of $\Gamma_{\rm shock}$ to the normal reflected shock $\Gamma_{\rm shock}^{\rm norm}$.

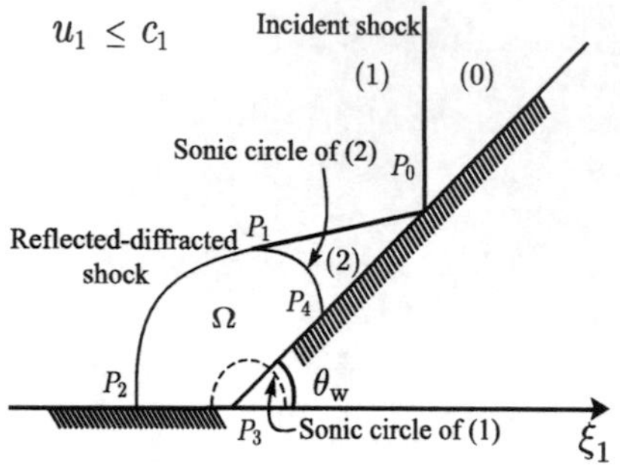

Figure 3.3: Supersonic regular reflection-diffraction configuration when $u_1 \leq c_1$

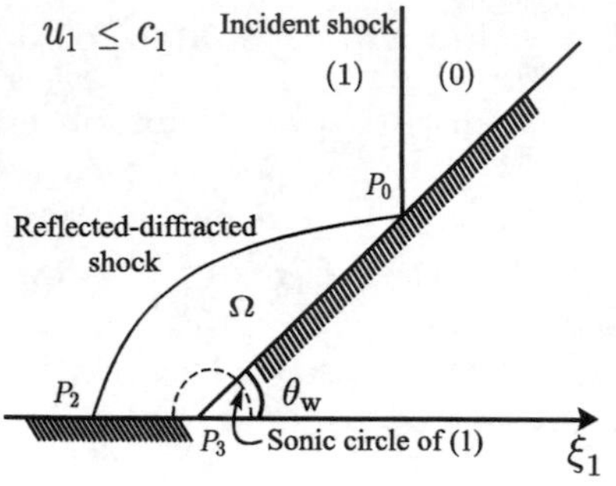

Figure 3.4: Subsonic regular reflection-diffraction configuration when $u_1 \leq c_1$

3.2.4.2 Separation of $\Gamma_{\rm shock}$ and $\Gamma_{\rm sym}$

There exists $\mu > 0$ depending only on the data such that, for any admissible solution φ of **Problem 2.6.1** with $\theta_{\rm w} \in (\theta_{\rm w}^{\rm s}, \frac{\pi}{2})$,

$$f_{\rm sh}(\xi_1) \geq \min(\frac{c_1}{2}, \mu(\xi_1 - \xi_1^{P_2})) \qquad \text{for all } \xi_1 \in [\xi_1^{P_2}, \min\{\xi_1^{P_1}, 0\}],$$

where $\xi_2 = f_{\rm sh}(\xi_1)$ represents $\Gamma_{\rm shock}$ when $\xi_1 \in [\xi_1^{P_2}, \min\{\xi_1^{P_1}, 0\}]$.

3.2.4.3 Uniform positive lower bound for the distance between $\Gamma_{\rm shock}$ and $\Gamma_{\rm wedge}$

We now extend the set of admissible solutions by including the normal reflection as the unique admissible solution for $\theta_{\rm w} = \frac{\pi}{2}$. This is justified by the fact that all the admissible solutions converge to the normal reflection solution as the wedge angles tend to $\frac{\pi}{2}$; see §3.2.4.1. Fix $\theta_{\rm w}^* \in (\theta_{\rm w}^{\rm s}, \frac{\pi}{2})$.

If $u_1 \leq c_1$, which is determined by (ρ_0, ρ_1, γ), then there exists $C > 0$ such that

$$\text{dist}(\Gamma_{\rm shock}, \Gamma_{\rm wedge}) > \frac{1}{C} \tag{3.2.20}$$

for any admissible solution of **Problem 2.6.1** with $\theta_{\rm w} \in [\theta_{\rm w}^*, \frac{\pi}{2}]$. In this case, the reflected-diffracted shock does not hit the wedge vertex P_3, since point P_2 should be away from the sonic circle of state (1), as shown in Figs. 3.3–3.4.

Without assuming the condition that $u_1 \leq c_1$, we show the uniform lower bound of the distance between $\Gamma_{\rm shock}$ and $\Gamma_{\rm wedge}$ away from P_3 for any $\theta_{\rm w} \in [\theta_{\rm w}^*, \frac{\pi}{2}]$: For any small $r > 0$, there exists $C_r > 0$ such that

$$\text{dist}(\Gamma_{\rm shock}, \Gamma_{\rm wedge} \setminus B_r(P_3)) > \frac{1}{C_r} \tag{3.2.21}$$

for every admissible solution with $\theta_{\rm w} \in [\theta_{\rm w}^*, \frac{\pi}{2}]$.

Recall that estimates (3.2.20)–(3.2.21) hold for the wedge angles $\theta_{\rm w} \in [\theta_{\rm w}^*, \frac{\pi}{2}]$, and the constants in these estimates depend on $\theta_{\rm w}^*$. However, for the application in §3.2.4.4, we need an estimate of the distance between $\Gamma_{\rm shock}$ and $\Gamma_{\rm wedge}$ which holds for all the wedge angles $\theta_{\rm w} \in [\theta_{\rm w}^{\rm s}, \frac{\pi}{2}]$. We do not assume that $u_1 \le c_1$, which implies that our estimate has to be made away from P_3, as we discussed earlier. Moreover, for $\theta_{\rm w} = \theta_{\rm w}^{\rm s}$, $\Gamma_{\rm shock}$ and $\Gamma_{\rm wedge}$ meet at P_0, which implies that our estimate has to be made away from P_0. Then we obtain the following estimate: For every small $r > 0$, there exists $C_r > 0$ depending on $(\rho_0, \rho_1, \gamma, r)$ such that

$$\text{dist}\,(\Gamma_{\rm shock}, \Gamma_{\rm wedge} \setminus (B_r(P_0) \cup B_r(P_3))) \ge \frac{1}{C_r} \tag{3.2.22}$$

for any admissible solution of **Problem 2.6.1** with the wedge angle $\theta_{\rm w} \in (\theta_{\rm w}^{\rm s}, \frac{\pi}{2})$.

If $u_1 > c_1$, the *critical angle* $\theta_{\rm w}^{\rm c}$ in Theorem 2.6.5 is defined as follows:

$$\theta_{\rm w}^{\rm c} = \inf \mathcal{A},$$

where

$$\mathcal{A} := \left\{ \theta_{\rm w}^* \in (\theta_{\rm w}^{\rm s}, \frac{\pi}{2}] : \begin{array}{c} \exists\, \varepsilon > 0 \text{ so that } \text{dist}(\Gamma_{\rm shock}, \Gamma_{\rm wedge}) \ge \varepsilon \text{ for} \\ \text{any admissible solution with } \theta_{\rm w} \in [\theta_{\rm w}^*, \frac{\pi}{2}] \end{array} \right\}.$$

Since the normal reflection solution is the unique admissible solution for $\theta_{\rm w} = \frac{\pi}{2}$, the set of admissible solutions with angles $\theta_{\rm w} \in [\theta_{\rm w}^*, \frac{\pi}{2}]$ is non-empty for any $\theta_{\rm w}^* \in (\theta_{\rm w}^{\rm s}, \frac{\pi}{2}]$. Moreover, since $\text{dist}(\Gamma_{\rm shock}, \Gamma_{\rm wedge}) > 0$ for the normal reflection solution, we conclude that $\frac{\pi}{2} \in \mathcal{A}$, *i.e.*, $\mathcal{A} \neq \emptyset$. Furthermore, we show that $\theta_{\rm w}^{\rm c} < \frac{\pi}{2}$ by using that $\Gamma_{\rm shock} \to \Gamma_{\rm shock}^{\rm norm}$ as $\theta_{\rm w} \to \frac{\pi}{2}$; see §3.2.4.1. Then it follows directly from the definition of $\theta_{\rm w}^{\rm c}$ that, for each $\theta_{\rm w}^* \in (\theta_{\rm w}^{\rm c}, \frac{\pi}{2})$, there exists $C > 0$ such that

$$\text{dist}(\Gamma_{\rm shock}, \Gamma_{\rm wedge}) \ge \frac{1}{C} \tag{3.2.23}$$

for any admissible solution with $\theta_{\rm w} \in [\theta_{\rm w}^*, \frac{\pi}{2}]$.

We note that property (3.2.21) is employed in the proof of Theorem 2.6.5 to show the existence of the *attached* solution for $\theta_{\rm w} = \theta_{\rm w}^{\rm c}$ when $\theta_{\rm w}^{\rm c} > \theta_{\rm w}^{\rm s}$.

3.2.4.4 Uniform positive lower bound for the distance between $\Gamma_{\rm shock}$ and the sonic circle of state (1)

Employing the detail structure of the potential flow equation (2.2.8) for a solution φ that is close to a uniform state near its sonic circle, and the property of admissible solutions that $\varphi \le \varphi_1$ holds in Ω by (3.2.4), we use estimate (3.2.22) to prove that there exists $C > 0$ such that

$$\text{dist}(\Gamma_{\rm shock}, B_{c_1}(\mathcal{O}_1)) > \frac{1}{C} \tag{3.2.24}$$

for any admissible solution φ of **Problem 2.6.1** with $\theta_{\rm w} \in (\theta_{\rm w}^{\rm s}, \frac{\pi}{2})$, where $\mathcal{O}_1 = (u_1, 0)$ is the center of the sonic circle of state (1).

Estimate (3.2.24) is crucial, especially since it is employed for the ellipticity estimate in §3.2.4.5 below and the uniform estimate of the lower bound of the gradient jump across $\Gamma_{\rm shock}$ in the radial direction with respect to the sonic circle of state (1); see (3.2.29).

3.2.4.5 Uniform estimate of the ellipticity of equation (2.2.8) in Ω up to $\Gamma_{\rm shock}$

Set the Mach number

$$M^2 = \frac{|D\varphi|^2}{c^2} = \frac{|D\varphi|^2}{\rho_0^{\gamma-1} - (\gamma-1)(\varphi + \frac{1}{2}|D\varphi|^2)}, \tag{3.2.25}$$

where we have used (2.2.9) for the second equality. Note that, for an admissible solution of **Problem 2.6.1**, by (3.2.2),

$$M \in C(\overline{\Omega}) \cap C^2(\overline{\Omega} \setminus (\overline{\Gamma_{\rm sonic}} \cup \{P_2, P_3\})).$$

We conclude that there exists $\mu > 0$ depending only on (ρ_0, ρ_1, γ) such that, if φ is an admissible solution of **Problem 2.6.1** with $\theta_{\rm w} \in (\theta_{\rm w}^{\rm s}, \frac{\pi}{2})$, then

$$M^2(\boldsymbol{\xi}) \le 1 - \mu\,{\rm dist}(\boldsymbol{\xi}, \Gamma_{\rm sonic}) \qquad \text{for all } \boldsymbol{\xi} \in \Omega(\varphi). \tag{3.2.26}$$

To achieve this, we show that a maximum of $M^2 + \mu d$, which is close to one, cannot be attained on $\overline{\Omega} \setminus \overline{\Gamma_{\rm sonic}}$, where $\mu > 0$ is a small constant and $d(\boldsymbol{\xi})$ is an appropriate function comparable with ${\rm dist}(\boldsymbol{\xi}, \Gamma_{\rm sonic})$.

First, the maximum of $M^2 + \mu d$ cannot be attained on $\Omega \cup \Gamma_{\rm wedge} \cup \Gamma_{\rm sym}$ if $1 - M^2 \ge 0$ is sufficiently small; see §5.2–§5.3 below. Also, we explicitly check that $M = 0$ at P_3 so that, by choosing μ small, we conclude that $M^2 + \mu d$ is small at P_3.

Thus, it remains to show that the maximum of $M^2 + \mu d$ cannot be attained on $\Gamma_{\rm shock} \cup \{P_2\}$. Crucially, the result of §3.2.4.4 on the positive lower bound on the distance between $\Gamma_{\rm shock}$ and the sonic circle of state (1) is employed, since it allows us to estimate that state (1) is *sufficiently hyperbolic* on the other side of $\Gamma_{\rm shock}$. Then, assuming that the maximum of $M^2+\mu d$ is attained at $P \in \Gamma_{\rm shock}$, we use the first-order conditions at the maximum point, $\partial_{\boldsymbol{\tau}}(M^2 + \mu d)(P) = 0$ and $\partial_{\boldsymbol{\nu}}(M^2 + \mu d)(P) \le 0$ (where $\boldsymbol{\nu}$ is the interior normal to $\Gamma_{\rm shock}$), the fact that the equation holds at P, and the Rankine-Hugoniot condition:

$$\partial_{\boldsymbol{\tau}}\Big((\rho D\varphi - \rho_1 D\varphi_1) \cdot \frac{D\varphi - D\varphi_1}{|D\varphi - D\varphi_1|}\Big) = 0 \qquad \text{on } \Gamma_{\rm shock},$$

to obtain the four relations at P for the three components of $D^2\varphi$, which leads to a contradiction. Thus, the maximum of $M^2 + \mu d$ cannot be attained on $\Gamma_{\rm shock}$. The maximum at P_2 is handled similarly, since P_2 can be regarded as an interior point of Γ_{shock} after extending the solution by even reflection with respect to the ξ_1–axis. This completes the proof of (3.2.26).

Write equation (2.2.8) in the form:

$$\operatorname{div}\mathcal{A}(D\varphi,\varphi,\boldsymbol{\xi})+\mathcal{B}(D\varphi,\varphi,\boldsymbol{\xi})=0 \tag{3.2.27}$$

with $\mathbf{p}=(p_1,p_2)\in\mathbb{R}^2$ and $z\in\mathbb{R}$, where

$$\mathcal{A}(\mathbf{p},z,\boldsymbol{\xi})\equiv\mathcal{A}(\mathbf{p},z):=\rho(|\mathbf{p}|^2,z)\mathbf{p},\quad \mathcal{B}(\mathbf{p},z,\boldsymbol{\xi})\equiv\mathcal{B}(\mathbf{p},z):=2\rho(|\mathbf{p}|^2,z)$$

with function $\rho(|\mathbf{p}|^2,z)$ defined by (2.2.9). We restrict $(\mathbf{p},z)$ in a set such that (2.2.9) is defined, *i.e.*, satisfying $\rho_0^{\gamma-1}-(\gamma-1)(z+\frac{1}{2}|\mathbf{p}|^2)\geq 0$.

As a corollary of (3.2.26), we employ (3.2.18) to conclude that, for any $\theta_{\text{w}}^*\in(\theta_{\text{w}}^{\text{s}},\frac{\pi}{2})$, there exists $C>0$ depending only on (ρ_0,ρ_1,γ) such that, if φ is an admissible solution of **Problem 2.6.1** with $\theta_{\text{w}}\in[\theta_{\text{w}}^*,\frac{\pi}{2})$, equation (3.2.27) satisfies the strict ellipticity condition:

$$\frac{\operatorname{dist}(\boldsymbol{\xi},\Gamma_{\text{sonic}})}{C}|\boldsymbol{\kappa}|^2\leq\sum_{i,j=1}^{2}\mathcal{A}^i_{p_j}(D\varphi(\boldsymbol{\xi}),\varphi(\boldsymbol{\xi}),\boldsymbol{\xi})\kappa_i\kappa_j\leq C|\boldsymbol{\kappa}|^2 \tag{3.2.28}$$

for any $\boldsymbol{\xi}\in\Omega$ and $\boldsymbol{\kappa}=(\kappa_1,\kappa_2)\in\mathbb{R}^2$. Note that the ellipticity degenerates on Γ_{sonic}.

3.2.5 Regularity and related uniform estimates

We consider $\theta_{\text{w}}^*\in(\theta_{\text{w}}^{\text{s}},\frac{\pi}{2})$ for Case $u_1\leq c_1$ and $\theta_{\text{w}}^*\in(\theta_{\text{w}}^{\text{c}},\frac{\pi}{2})$ for Case $u_1>c_1$. Then, from §3.2.4.3, we obtain the uniform estimate:

$$\operatorname{dist}(\Gamma_{\text{shock}},\Gamma_{\text{wedge}})\geq\frac{1}{C}$$

for any admissible solution with $\theta_{\text{w}}\in[\theta_{\text{w}}^*,\frac{\pi}{2})$. This fixes the geometry of Ω for such a solution.

With the geometry of Ω and the strict ellipticity controlled, we can conclude the estimates in the properly scaled and weighted $C^{k,\alpha}$–spaces. We perform the estimates separately away from Γ_{sonic} where the equation is uniformly elliptic, and near Γ_{sonic} where the ellipticity degenerates.

3.2.5.1 Weighted $C^{k,\alpha}$–estimates away from Γ_{sonic}

Away from the ε-neighborhood of Γ_{sonic}, we use the uniform ellipticity to estimate admissible solutions with the bounds independent of the solution and the wedge angle $\theta_{\text{w}}\in[\theta_{\text{w}}^*,\frac{\pi}{2})$. Also, in order to avoid the difficulty related to the corner at point P_2 of intersection of Γ_{shock} and Γ_{sym}, we extend the elliptic domain Ω by reflection over the symmetry line and use the even extension of the solution. From the structure of the potential flow equation (2.2.8), it follows that (2.2.8) is satisfied in the extended domain, and the Rankine-Hugoniot conditions (2.2.13)–(2.2.14) are satisfied on the extended shock. Now the boundary part $\Gamma_{\rm sym}$ lies in the interior of the extended domain $\Omega^{\rm ext}$, and P_2 is the interior point of the extended shock $\Gamma^{\rm ext}_{\rm shock}$.

In the argument below, we consider the points of $\Omega^{\rm ext}$ which are on the distance, $d > 0$, from the original and reflected sonic arcs, and for which the constants in the estimates depend on d.

We use the estimates obtained in §3.2.4.2–§3.2.4.3 to control the geometry of domain Ω. Then, for any $\alpha \in (0,1)$, the $C^{2,\alpha}$–estimates in the interior of $\Omega^{\rm ext}$ and near $\Gamma_{\rm wedge}$ and the reflected $\Gamma^-_{\rm wedge}$ (away from corner P_3) follow from the standard elliptic theory, where we use the homogeneous Neumann boundary condition on $\Gamma_{\rm wedge}$, the uniform estimate of the distance between $\Gamma_{\rm shock}$ and $\Gamma_{\rm wedge}$, and the Lipschitz estimates of the solution.

For the estimates of the shock curve $\Gamma_{\rm shock}$ and the solution near $\Gamma_{\rm shock}$ (away from the ε–neighborhood of $\Gamma_{\rm sonic}$), we first show that the function:

$$\bar{\phi} := \varphi_1 - \varphi$$

is uniformly monotone in a uniform neighborhood of $\Gamma_{\rm shock}$ in the radial direction with respect to the center of the sonic circle $\mathcal{O}_1$ of state (1), *i.e.*, there exist $\delta, \sigma > 0$ such that

$$\partial_r \bar{\phi} \le -\delta \qquad \text{in } \mathcal{N}_\sigma(\Gamma^{\rm ext}_{\rm shock}) \cap \Omega^{\rm ext}. \tag{3.2.29}$$

Note that $\bar{\phi} = 0$ on $\Gamma_{\rm shock}$, by the Rankine-Hugoniot condition (2.2.14), and that $\bar{\phi} > 0$ in $\Omega^{\rm ext}$. Using that φ is an admissible solution of **Problem 2.6.1**, we show that the extended shock $\Gamma^{\rm ext}_{\rm shock}$ is a graph in the radial direction in the polar coordinates (r, θ) with center $\mathcal{O}_1$. With this, working in the (r, θ)–coordinates, we inductively make the $C^{k,\alpha}$–estimates of $\Gamma^{\rm ext}_{\rm shock}$ and φ near $\Gamma^{\rm ext}_{\rm shock}$, for $k = 1, 2, \dots$, as follows: φ satisfies the uniformly elliptic equation in $\Omega^{\rm ext}$ (away from the original and reflected sonic arcs) and an oblique boundary condition on $\Gamma^{\rm ext}_{\rm shock}$ from the Rankine-Hugoniot conditions. The nonlinear equation and boundary condition are given by smooth functions. Now we use the estimates due to Lieberman [192] (stated in §4.3 below) for two-dimensional elliptic equations with nonlinear boundary conditions, which show that the regularity of the solution is higher than that of the boundary. More precisely, from (3.2.17) and (3.2.29), we obtain the Lipschitz estimates of $\Gamma^{\rm ext}_{\rm shock}$ and of φ in $\Omega^{\rm ext}$. Then, from Theorem 4.3.2, we obtain the $C^{1,\alpha}$–estimates of φ near $\Gamma^{\rm ext}_{\rm shock}$ for some $\alpha \in (0,1)$. Moreover, by (3.2.29) and the fact that $\varphi = \varphi_1$ on $\Gamma^{\rm ext}_{\rm shock}$, we obtain the $C^{1,\alpha}$–estimates of $\Gamma^{\rm ext}_{\rm shock}$. Now, using Corollary 4.3.5, we obtain the $C^{2,\alpha}$–estimates of φ near $\Gamma^{\rm ext}_{\rm shock}$, which in turn implies the $C^{2,\alpha}$–estimates of $\Gamma^{\rm ext}_{\rm shock}$. We repeat this argument inductively for $k = 2, 3, \dots$.

Finally, we obtain the $C^{1,\alpha}$–estimates near corner P_3 for sufficiently small $\alpha > 0$ by using the results of Lieberman [189], stated in Theorem 4.3.13 below. For that, we work on the original domain Ω instead of the extended domain $\Omega^{\rm ext}$, and use the homogeneous Neumann conditions on $\Gamma_{\rm wedge} \cup \Gamma_{\rm sym}$. This is crucial, because the angle at the corner point P_3 of $\partial\Omega$ is less than π in this way, which allows us to obtain the $C^{1,\alpha}$–estimates.

Combining all the above estimates, we obtain

$$\varphi \in C^k(\overline{\Omega} \setminus (\mathcal{N}_d(\Gamma_{\text{sonic}}) \cup \{P_3\})) \cap C^{1,\alpha}(\overline{\Omega} \setminus \mathcal{N}_d(\Gamma_{\text{sonic}}))$$

and

$$\overline{\Gamma_{\text{shock}}} \setminus \mathcal{N}_d(\Gamma_{\text{sonic}}) \in C^k \qquad \text{for } k = 1, 2, \dots,$$

with uniform estimates.

3.2.5.2 Weighted and scaled $C^{2,\alpha}$–estimates near Γ_{sonic}

Near Γ_{sonic}, *i.e.*, in $\mathcal{N}_{\varepsilon_1}(\Gamma_{\text{sonic}}) \cap \Omega$ for sufficiently small $\varepsilon_1 > 0$, it is convenient to work in the coordinates flattening Γ_{sonic}. We consider the polar coordinates (r, θ) with respect to $\mathcal{O}_2 = (u_2, v_2)$, note that Γ_{sonic} is an arc of the circle with radius $r = c_2$ and center $\mathcal{O}_2$, and define

$$(x, y) = (c_2 - r, \theta).$$

Then there exists $\varepsilon_0 \in (0, \varepsilon_1)$ such that, for any $\varepsilon \in (0, \varepsilon_0]$,

$$\begin{aligned}
&\Omega_\varepsilon := \Omega \cap \mathcal{N}_{\varepsilon_1}(\Gamma_{\text{sonic}}) \cap \{x < \varepsilon\} = \{0 < x < \varepsilon,\ \theta_{\text{w}} < y < \hat{f}(x)\},\\
&\Gamma_{\text{sonic}} = \partial\Omega_\varepsilon \cap \{x = 0\},\\
&\Gamma_{\text{wedge}} \cap \partial\Omega_\varepsilon = \{0 < x < \varepsilon,\ y = \theta_{\text{w}}\},\\
&\Gamma_{\text{shock}} \cap \partial\Omega_\varepsilon = \{0 < x < \varepsilon,\ y = \hat{f}(x)\}
\end{aligned} \tag{3.2.30}$$

for some $\hat{f}(x)$ defined on $(0, \varepsilon_0)$. We perform the estimates in terms of the function:

$$\psi = \varphi - \varphi_2.$$

Note that $\psi(0, y) \equiv 0$, since $\varphi = \varphi_2$ on Γ_{sonic} by Definition 3.2.1(ii).

We write the potential flow equation (2.2.8) in terms of ψ in the (x, y)–coordinates. Then (3.2.28) implies that there exists $\delta > 0$ so that, for each admissible solution,

$$\psi_x \leq \frac{2 - \delta}{1 + \gamma} x \qquad \text{in } \Omega_\varepsilon.$$

Combining this with the estimate that $\psi_x \geq 0$ shown in §3.2.3, we obtain

$$|\psi_x| \leq Cx \qquad \text{in } \mathcal{N}_\varepsilon(\Gamma_{\text{sonic}}) \cap \Omega.$$

From this, using the monotonicity cone of ψ discussed in §3.2.3, we obtain

$$|\psi_y| \leq Cx.$$

Now, since $|D\psi| \le Cx$, we can modify equation (2.2.8) to obtain that any admissible solution ψ satisfies an equation in $\mathcal{N}_\varepsilon(\Gamma_{\text{sonic}})$:

$$\sum_{i,j=1}^{2} A_{ij}(D\psi,\psi,x)D_{ij}\psi + \sum_{i=1}^{2} A_i(D\psi,\psi,x)D_i\psi = 0 \tag{3.2.31}$$

with smooth $(A_{ij}, A_i)(\mathbf{p}, z, x)$ (independent of y), which is of the degenerate ellipticity structure:

$$\lambda|\boldsymbol{\xi}|^2 \le A_{11}(\mathbf{p},z,x)\frac{\xi_1^2}{x} + 2A_{12}(\mathbf{p},z,x)\frac{\xi_1\xi_2}{\sqrt{x}} + A_{22}(\mathbf{p},z,x)\xi_2^2 \le \frac{1}{\lambda}|\boldsymbol{\xi}|^2 \tag{3.2.32}$$

for $(\mathbf{p}, z) = (D\psi, \psi)(x, y)$ and for any $(x, y) \in \Omega_\varepsilon$, where we recall (3.2.30).

We use (3.2.31)–(3.2.32) for the estimates in the weighted and scaled $C^{2,\alpha}$–norms with the weights depending on x, which reflect the ellipticity structure. One way to define these norms is as follows: For any $(x_0, y_0) \in \Omega_\varepsilon$ and $\rho \in (0,1)$, let

$$\begin{aligned} \tilde{R}_\rho^{(x_0,y_0)} &:= \Big\{(s,t) \;:\; |s-x_0| < \frac{\rho}{4}x_0, |t-y_0| < \frac{\rho}{4}\sqrt{x_0}\Big\}, \\ R_\rho^{(x_0,y_0)} &:= \tilde{R}_\rho^{(x_0,y_0)} \cap \Omega. \end{aligned} \tag{3.2.33}$$

Rescale ψ from $R_\rho^{(x_0,y_0)}$ to the portion of the square with side-length 2ρ, *i.e.*, define the rescaled function:

$$\psi^{(x_0,y_0)}(S,T) = \frac{1}{x_0^2}\psi(x_0 + \frac{x_0}{4}S, y_0 + \frac{\sqrt{x_0}}{4}T) \qquad \text{in } Q_\rho^{(x_0,y_0)}, \tag{3.2.34}$$

where

$$Q_\rho^{(x_0,y_0)} := \Big\{(S,T) \in (-\rho,\rho)^2 \;:\; (x_0 + \frac{x_0}{4}S, \hat{y}_0 + \frac{\sqrt{x_0}}{4}T) \in \Omega\Big\}.$$

The *parabolic* norm of $\|\psi\|^{(\text{par})}_{2,\alpha,\Omega_\varepsilon}$ is the supremum over $(x_0, y_0) \in \Omega_\varepsilon$ of norms $\|\psi^{(x_0,y_0)}\|_{2,\alpha,\overline{Q_1^{(x_0,y_0)}}}$. Note that the estimate in norm $\|\cdot\|^{(\text{par})}_{2,\alpha,\Omega_\varepsilon}$ implies the $C^{1,1}$–estimate in Ω_ε.

In order to estimate $\|\psi\|^{(\text{par})}_{2,\alpha,\Omega_\varepsilon}$, we need to obtain the $C^{2,\alpha}$–estimates of the rescaled functions $\psi^{(x_0,y_0)}$. By the standard covering argument, it suffices to consider three cases:

(i) The interior rectangle: $R_{1/10}^{(x_0,y_0)} \subset \Omega$ for $(x_0, y_0) \in \Omega_\varepsilon$;

(ii) Rectangle $R_{1/2}^{(x_0,y_0)}$ centered at $(x_0, y_0) \in \Gamma_{\text{wedge}} \cap \partial\Omega_\varepsilon$ (on the wedge);

(iii) Rectangle $R_{1/2}^{(x_0,y_0)}$ centered at $(x_0, y_0) \in \Gamma_{\text{shock}} \cap \partial\Omega_\varepsilon$ (on the shock).

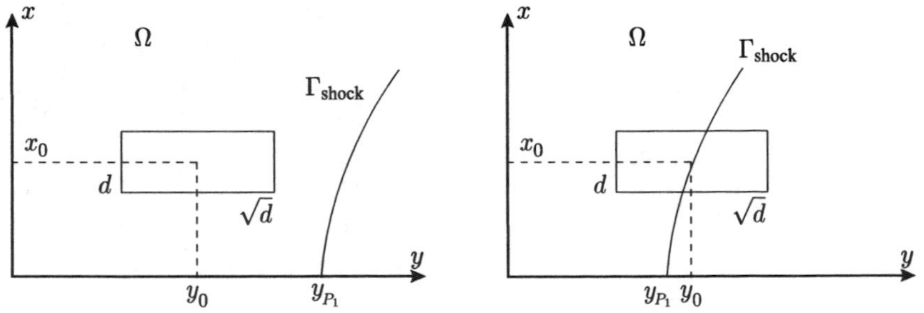

Figure 3.5: Rectangles in Cases (i) and (iii)

See Fig. 3.5 for $d = \frac{x_0}{4}$.

The gradient estimate $|D\psi| \leq Cx$ and the property that $\psi(0, y) \equiv 0$ imply

$$|\psi| \leq Cx^2,$$

so that

$$\|\psi^{(x_0,y_0)}\|_{L^\infty(\overline{Q_1^{(x_0,y_0)}})} \leq C \qquad \text{for any } (x_0, y_0) \in \Omega_\varepsilon.$$

Also, writing equation (3.2.31) in terms of the rescaled function $\psi^{(x_0,y_0)}$ and using the ellipticity structure (3.2.32), we see that $\psi^{(x_0,y_0)}$ satisfies a uniformly elliptic homogeneous equation in $Q_1^{(x_0,y_0)}$, *with the ellipticity constants and certain Hölder norms of the coefficients independent of* (x_0, y_0). Then the $C^{2,\alpha}$–estimates of $\psi^{(x_0,y_0)}$ in the smaller square $Q_{1/20}^{(x_0,y_0)}$ in Case (i) follow from the interior elliptic estimates. In Case (ii), in addition to the equation, we use the boundary condition $\partial_\nu \psi = 0$ on Γ_{wedge}, which holds under rescaling.

In Case (iii), we need to make the estimates up to Γ_{shock}, *i.e.*, the free boundary, for which only the Lipschitz estimates are *a priori* available. Thus, we rescale the region as in Cases (i)–(ii) to obtain the uniformly elliptic equation, and then follow the argument in §3.2.5.1 for the estimates near Γ_{shock}.

Owing to the non-isotropic rescaling (3.2.34), some difference from the estimates in §3.2.5.1 appears because:

(a) The Lipschitz estimate for ψ, combined with (3.2.34), does *not* imply the uniform Lipschitz estimate of $\psi^{(x_0,y_0)}$ with respect to $(x_0, y_0) \in \Gamma_{\text{shock}} \cap \partial\Omega_\varepsilon$. The estimate blows up as $d = \frac{x_0}{4} \to 0$, *i.e.*, for the rectangles close to Γ_{sonic}. Thus we have, *a priori*, only the L^∞ bound of $\psi^{(x_0,y_0)}$ uniform with respect to $(x_0, y_0) \in \Gamma_{\text{shock}} \cap \partial\Omega_\varepsilon$.

(b) The boundary condition for ψ on Γ_{shock} is uniformly oblique up to P_1 (*i.e.*, up to $x = 0$). However, the obliqueness of the rescaled condition for $\psi^{(x_0,y_0)}$ on Γ_{shock} degenerates as $d \to 0$. On the other hand, we can show that the rescaled boundary condition has an *almost tangential structure* with the constants uniform with respect to $(x_0, y_0) \in \Gamma_{\text{shock}} \cap \partial\Omega_\varepsilon$.

For these reasons, we cannot use the estimates of [192] (stated in §4.3 below) for the oblique derivative problem, with the bounds depending on the $C^{0,1}$–norm of the solution. Instead, we employ the estimates for the problem with *almost tangential structure*, when only the L^∞ bound of the solution is *a priori* known; see Theorems 4.2.4 and 4.2.8. These results give the gain-in-regularity similar to the estimates in [192], *i.e.*, we obtain the $C^{1,\alpha}$–estimate of the solution for the C^1–boundary with the Lipschitz estimate and the $C^{2,\alpha}$–estimate of the solution for the $C^{1,\alpha}$–boundary. This allows us to obtain the $C^{2,\alpha}$–estimates of $\Gamma_{\rm shock}$ and $\psi^{(x_0,y_0)}$ in Case (iii).

3.2.6 Existence of the supersonic regular reflection-diffraction configurations up to the sonic angle

Once the *a priori* estimates are established, we obtain a solution to **Problem 2.6.1** as a fixed point of an iteration map. The existence of a fixed point follows from the Leray-Schauder degree theory (*cf.* §3.4).

In order to apply the degree theory, the iteration set should be bounded and open in an appropriate function space (actually, in its product with the parameter space, *i.e.*, interval $[\theta_{\rm w}^*, \frac{\pi}{2}]$ of the wedge angles), the iteration map should be defined and continuous on the closure of the iteration set, and any fixed point of the iteration map should not occur on the boundary of the iteration set. We choose this function space according to the norms and the other quantities in the *a priori* estimates. Then the *a priori* estimates allow us to conclude that the fixed point cannot occur on the boundary of the iteration set, if the bounds defining the iteration set are chosen appropriately large or small, depending on the context and the *a priori* estimates. This can be done for any $\theta_{\rm w}^* \in (\theta_{\rm w}^{\rm s}, \frac{\pi}{2})$ if $u_1 \le c_1$ and for any $\theta_{\rm w}^* \in (\theta_{\rm w}^{\rm c}, \frac{\pi}{2})$ if $u_1 > c_1$.

In our case, there is an extra issue of connecting the admissible solutions with the normal reflection solution in the setup convenient for the application of the degree theory. We use the strict monotonicity properties of the admissible solutions (proved as a part of the *a priori* estimates) in our definition of the iteration set. These strict monotonicity properties can be made uniform for any wedge angle $\theta_{\rm w}$ away from $\frac{\pi}{2}$ and any point away from the appropriate parts of the boundary of the elliptic region by using the compactness of the set of admissible solutions, which is a corollary of the *a priori* estimates. However, the monotonicities become nonstrict when the wedge angle $\theta_{\rm w}$ is $\frac{\pi}{2}$, *i.e.*, at the normal reflection solution. Then, for the wedge angles near $\frac{\pi}{2}$, we use the fact that *the admissible solutions converge to the normal reflection solution as* $\theta_{\rm w}$ *tends to* $\frac{\pi}{2}$.

From this fact, we can derive the estimates similar to Chen-Feldman [54] for the admissible solutions and the *approximate* solutions for $\theta_{\rm w}$ near $\frac{\pi}{2}$. Then, for the wedge angle $\theta_{\rm w}$ near $\frac{\pi}{2}$, the iteration set $\mathcal{K}_{\theta_{\rm w}}$ is a small neighborhood of the normal reflection solution, where the norms used and the size of neighborhood are related to the estimates of Chapters 9–11. For the wedge angle $\theta_{\rm w}$ away from $\frac{\pi}{2}$, the iteration set $\mathcal{K}_{\theta_{\rm w}}$ is defined by the bounds in the appropriate norms related to the *a priori* estimates and by the lower bounds of certain directional derivatives, corresponding to the strict monotonicity properties so that the actual solution cannot be on the boundary of the iteration set according to the *a priori* estimates. These two definitions are combined into one setup, with the bounds depending continuously on the wedge angle $\theta_{\rm w}$.

Also, since the elliptic domain Ω depends on the solution, we define the iteration set in terms of the functions on the unit square $Q^{\rm iter} = (0,1)^2 := (0,1) \times (0,1)$ and, for each such function and wedge angle $\theta_{\rm w}$, define the elliptic domain Ω of the *approximate solution* and a smooth invertible map from $Q^{\rm iter}$ to Ω, where Ω is of the same structure as the elliptic region of supersonic reflection configurations; see §2.4.2 and Fig. 2.3. This defines the iteration set:

$$\mathcal{K} = \cup_{\theta_{\rm w}\in[\theta_{\rm w}^*, \frac{\pi}{2}]}\mathcal{K}_{\theta_{\rm w}} \times \{\theta_{\rm w}\}$$

with $\mathcal{K}_{\theta_{\rm w}} \subset \mathcal{C}$, where $\mathcal{C}$ is a weighted and scaled $C^{2,\alpha}$–type space on $Q^{\rm iter}$ so that $\mathcal{C} \subset C^{2,\alpha}(Q^{\rm iter}) \cap C^{1,\alpha}(\overline{Q^{\rm iter}})$. For each $(u, \theta_{\rm w}) \in \mathcal{K}$, the map from $Q^{\rm iter}$ to $\Omega(u, \theta_{\rm w})$ can be extended to the smooth and smoothly invertible map $\overline{Q^{\rm iter}} \mapsto \overline{\Omega}$, where the sides of square $Q^{\rm iter}$ are mapped to the boundary parts $\Gamma_{\rm sonic}$, $\Gamma_{\rm shock}$, $\Gamma_{\rm sym}$, and $\Gamma_{\rm wedge}$ of Ω.

The iteration map $\mathcal{I}$ is defined as follows: Given $(u, \theta_{\rm w}) \in \overline{\mathcal{K}}$, define the corresponding *elliptic domain* $\Omega = \Omega(u, \theta_{\rm w})$ by both mapping from the unit square $Q^{\rm iter}$ to the *physical plane* and determining the iteration $\Gamma_{\rm shock}$ depending on $(u, \theta_{\rm w})$, and set up a boundary value problem in Ω for an elliptic equation that is degenerate near $\Gamma_{\rm sonic}$. Let $\hat{\varphi}$ be the solution of the boundary value problem in Ω. Then we define $\hat{u}$ on $Q^{\rm iter}$ by mapping $\hat{\varphi}$ back in such a way that the gain-in-regularity of the solution is preserved. This requires some care, since the original mapping between $Q^{\rm iter}$ and the *physical domain* is defined by u and hence has a lower regularity. Then the iteration map is defined by

$$\mathcal{I}(u, \theta_{\rm w}) = \hat{u}.$$

The boundary value problem in the definition of $\mathcal{I}$ is defined so that, at the fixed point $u = \hat{u}$, its solution satisfies the potential flow equation (2.2.8) with the ellipticity cutoff in a small neighborhood of $\Gamma_{\rm sonic}$, and both the Rankine-Hugoniot conditions (2.2.13)–(2.2.14) on $\Gamma_{\rm shock}$ and $D\hat{\varphi}\cdot\boldsymbol{\nu} = 0$ on $\Gamma_{\rm wedge}\cup\Gamma_{\rm sym}$. On the sonic arc $\Gamma_{\rm sonic}$ that is a fixed boundary, we can prescribe only one condition, the Dirichlet condition $\hat{\varphi} = \varphi_2$. However, it is not sufficient to have the potential flow equation (2.2.8) satisfied across $\Gamma_{\rm sonic}$. Indeed, the Rankine-Hugoniot conditions (2.2.13)–(2.2.14) need to be satisfied for $\hat{\varphi}$ and φ_2 on $\Gamma_{\rm sonic}$, and condition (2.2.13) implies that $D\varphi = D\varphi_2$ on $\Gamma_{\rm sonic}$, since φ_2 is sonic on $\Gamma_{\rm sonic}$. Thus, we need to prove that the last property holds for the solution of the iteration problem (at least for the fixed point). In this proof, we use the elliptic degeneracy of the iteration equation in Ω near $\Gamma_{\rm sonic}$ by obtaining the estimates of $\hat{\psi} = \hat{\varphi} - \varphi_2$ in the norms of $\|\cdot\|^{\rm (par)}_{2,\alpha,\mathcal{N}_\varepsilon(\Gamma_{\rm sonic})\cap\Omega}$ introduced in §3.2.5.2. These estimates imply that $D\hat{\psi} = 0$ on $\Gamma_{\rm sonic}$, *i.e.*, $D\varphi = D\varphi_2$ on $\Gamma_{\rm sonic}$.

Furthermore, the other conditions required in the definition of an admissible solution φ (including the inequalities, $\varphi_2 \leq \varphi \leq \varphi_1$, and the monotonicity properties) are satisfied for $\hat{\varphi}$ for any wedge angle $\theta_{\rm w}$ away from $\frac{\pi}{2}$ and for any point away from the appropriate parts of the boundary of the elliptic domain.

Then we prove the following facts:

(i) Any fixed point $u = \mathcal{I}(u, \theta_{\rm w})$, mapped to the *physical plane*, is an admissible solution φ. For that, we remove the ellipticity cutoff and prove the inequalities and monotonicity properties mentioned above for the regions and the wedge angles where they are not readily known from the definition of the iteration set. The fact that these estimates need to be proved only in the localized regions is crucial. This localization is achieved by using the uniform bounds and monotonicity properties which are a part of the *a priori* estimates.

(ii) The iteration set is open. We prove this by showing the existence of a solution for the iteration boundary value problem determined by any (v, θ) in a sufficiently small neighborhood of any $(u, \theta_{\rm w}) \in \mathcal{K}$.

(iii) The iteration map is compact. We prove this by using the gain-in-regularity of the solution of the iteration boundary value problem.

(iv) Any fixed point of the iteration map cannot occur on the boundary of the iteration set. This is shown by using the *a priori* estimates, which can be applied since the fixed point is, by (i) above, an admissible solution.

(v) The normal reflection solution $u^{(\rm norm)}$, expressed on the unit square, is in the iteration set, which shows that the iteration set is non-empty for $\theta_{\rm w} = \frac{\pi}{2}$.

Now the Leray-Schauder degree theory (see §3.4) guarantees that the fixed point index:

$$\mathbf{Ind}(\mathcal{I}^{(\theta_{\rm w})}, \overline{\mathcal{K}}(\theta_{\rm w}))$$

of the iteration map on the iteration set (for given $\theta_{\rm w}$) is independent of the wedge angle $\theta_{\rm w} \in [\theta_{\rm w}^*, \frac{\pi}{2}]$.

It remains to show that $\mathbf{Ind}(\mathcal{I}^{(\theta_{\rm w})}, \overline{\mathcal{K}}(\theta_{\rm w}))$ is nonzero. In fact, at $\theta_{\rm w} = \frac{\pi}{2}$, we show that $\mathcal{I}_{\frac{\pi}{2}}(v) = u^{(\rm norm)}$ for any $v \in \mathcal{K}_{\frac{\pi}{2}}$. This implies that $\mathbf{Ind}(\mathcal{I}^{(\frac{\pi}{2})}, \overline{\mathcal{K}}(\frac{\pi}{2})) = 1$.

Then, for any $\theta_{\rm w} \in [\theta_{\rm w}^*, \frac{\pi}{2}]$, $\mathbf{Ind}(\mathcal{I}^{(\theta_{\rm w})}, \overline{\mathcal{K}}(\theta_{\rm w})) = 1$, which implies that a fixed point exists. Moreover, the fixed point is an admissible solution of **Problem 2.6.1**.

Since $\theta_{\rm w}^*$ is arbitrary in interval $(\theta_{\rm w}^{\rm s}, \frac{\pi}{2})$ if $u_1 \leq c_1$ and in $(\theta_{\rm w}^{\rm c}, \frac{\pi}{2})$ if $u_1 > c_1$, we obtain the existence of admissible solutions in the intervals of the wedge angles $\theta_{\rm w}$ indicated in Theorems 2.6.3 and 2.6.5.

Moreover, for Case $u_1 > c_1$, if $\theta_{\rm w}^{\rm c} > \theta_{\rm w}^{\rm s}$, then, from the definition of $\theta_{\rm w}^{\rm c}$ in §3.2.4.3, there exists a sequence $\theta_{\rm w}^{(i)} \in [\theta_{\rm w}^{\rm c}, \frac{\pi}{2})$ with $\lim_{i\to\infty} \theta_{\rm w}^{(i)} = \theta_{\rm w}^{\rm c}$ and a corresponding admissible solution $\varphi^{(i)}$ with the wedge angle $\theta_{\rm w}^{(i)}$ such that

$$\lim_{i\to\infty} \operatorname{dist}(\Gamma_{\rm shock}^{(i)}, \Gamma_{\rm wedge}^{(i)}) = 0.$$

Taking the uniform limit in a subsequence of $\varphi^{(i)}$ and employing the geometric properties of the free boundary (shock) proved in §3.2.4.3, including (3.2.21), and the regularity of admissible solutions and involved shocks, we obtain an attached solution for the wedge angle $\theta_w = \theta_w^c$ as asserted in Theorem 2.6.5.

In Part III, we give the detailed proofs of the steps described above for the main theorems for von Neumann's sonic conjecture, as well as related further finer estimates and analysis of the solutions.

3.3 MAIN STEPS AND RELATED ANALYSIS IN THE PROOF OF THE DETACHMENT CONJECTURE

In this section we discuss the solutions of **Problem 2.6.1** in §2.6 for the full range of wedge angles for which state (2) exists, *i.e.*, for any $\theta_w \in (\theta_w^d, \frac{\pi}{2})$, where θ_w^d is the detachment angle. We make the whole iteration again, starting from the normal reflection, and prove the results for both the supersonic and subsonic reflection configurations. We follow the procedure discussed in §3.2 with the changes described below.

The difference with §3.2 is from the fact that, depending on $\theta_w \in (\theta_w^d, \frac{\pi}{2})$, the expected solutions have the structure of either supersonic or subsonic reflection configurations described in §2.4.2 and §2.4.3, respectively; it is of supersonic (resp. subsonic) structure if state (2) is supersonic (resp. subsonic or sonic) at P_0, *i.e.*, $|D\varphi_2(P_0)| > c_2$ (resp. $|D\varphi_2(P_0)| \leq c_2$), where we recall that P_0 and (u_2, v_2, c_2) depend only on θ_w.

Then we will use the following terminology: $\theta_w \in (\theta_w^d, \frac{\pi}{2})$ is a *supersonic* (resp. *subsonic*, or *sonic*) wedge angle if $|D\varphi_2(P_0)| > c_2$ (resp. $|D\varphi_2(P_0)| < c_2$, or $|D\varphi_2(P_0)| = c_2$) for θ_w. Note that the sonic angle θ_w^s, introduced above, is a sonic wedge angle according to this terminology; moreover, θ_w^s is the supremum of the set of sonic wedge angles (even though it is not clear if more sonic wedge angles other than θ_w^s exist).

Note that, if $\theta_w^{(i)} \in (\theta_w^d, \frac{\pi}{2})$ is a sequence of supersonic wedge angles, and $\theta_w^{(i)} \to \theta_w^{(\infty)}$ for a sonic wedge angle $\theta_w^{(\infty)}$, then ${P_0}^{(i)} \to {P_0}^{(\infty)}$, ${P_1}^{(i)} \to {P_0}^{(\infty)}$, ${P_4}^{(i)} \to {P_0}^{(\infty)}$, and $\overline{\Gamma_{\text{sonic}}}^{(i)}$ shrinks to point ${P_0}^{(\infty)}$. Thus, we define that, for the subsonic/sonic wedge angles, $P_1 = P_4 := P_0$ and $\overline{\Gamma_{\text{sonic}}} := \{P_0\}$. That is, $P_0 = P_1 = P_4$ for the subsonic/sonic wedge angles.

Now we comment on the steps in §3.2 with the changes necessary in the present case.

3.3.1 Admissible solutions of Problem 2.6.1

The definition of admissible solutions of Problem 2.6.1 in §3.2.1 has included only the supersonic reflection solutions. Now we need to define admissible solutions of both supersonic and subsonic reflection configurations.

For the supersonic wedge angles $\theta_w \in (\theta_w^d, \frac{\pi}{2})$, we define the admissible solutions by Definition 3.2.1.

For the subsonic/sonic wedge angles $\theta_w \in (\theta_w^d, \frac{\pi}{2})$, we define the admissible solutions which correspond to the subsonic configuration described in §2.4.3 and shown on Fig. 2.4, which are elliptic in Ω as in Definition 3.2.1(iii), and satisfy conditions (iv)–(v) of Definition 3.2.1. Moreover, we require the property similar to that in Remark 3.2.5 to be held for the subsonic reflection configurations. Since $\overline{\Gamma_{\text{sonic}}} = \{P_0\}$ in this case, Definition 3.2.1(ii) for the subsonic reflection solutions is changed into the following:

(ii) φ satisfies (2.6.4) and

$$\begin{aligned}&\varphi \in C^{0,1}(\Lambda) \cap C^1(\overline{\Lambda} \setminus \overline{\Gamma_{\text{shock}}}),\\&\varphi \in C^3(\overline{\Omega} \setminus \{P_0, P_2, P_3\}) \cap C^1(\overline{\Omega}),\end{aligned} \tag{3.3.1}$$

together with

$$\varphi(P_0) = \varphi_2(P_0), \quad D\varphi(P_0) = D\varphi_2(P_0). \tag{3.3.2}$$

3.3.2 Strict monotonicity cones for $\varphi_1 - \varphi$ and $\varphi - \varphi_2$

All of the results discussed in §3.2.2–§3.2.3 hold without change. In the proofs, the only difference is that, for subsonic reflection solutions, $\overline{\Gamma_{\text{shock}}}$ is only one point, P_0. However, we use (3.3.2) instead of Remark 3.2.5 in this case, and then the argument works without change.

3.3.3 Uniform estimates for admissible solutions

We discuss the extensions of the estimates stated in §3.2.4 to the present case.

Some of the estimates hold for any $\theta_{\text{w}} \in (\theta_{\text{w}}^{\text{d}}, \frac{\pi}{2})$, in which the universal constant C depends only on (ρ_0, ρ_1, γ).

In the other estimates, we have to restrict the range of angles by fixing any $\theta_{\text{w}}^* \in (\theta_{\text{w}}^{\text{d}}, \frac{\pi}{2})$ and considering the admissible solutions with $\theta_{\text{w}} \in [\theta_{\text{w}}^*, \frac{\pi}{2})$. The universal constant C in these estimates depends only on $(\rho_0, \rho_1, \gamma, \theta_{\text{w}}^*)$. Note that both the supersonic and subsonic reflection configurations occur if $\theta_{\text{w}}^* \in (\theta_{\text{w}}^{\text{d}}, \theta_{\text{w}}^{\text{s}}]$. We need to consider such θ_{w}^*, since we will prove the existence of solutions up to $\theta_{\text{w}}^{\text{d}}$.

3.3.3.1 Basic estimates of (φ, ρ, Ω), the distance between Γ_{shock} and the sonic circle of state (1), and separation of Γ_{shock} and Γ_{sym}

The estimates in §3.2.4.1–§3.2.4.2 and §3.2.4.4 hold without change in the present case.

Specifically, the estimates of (φ, ρ, Ω) in §3.2.4.1 hold for admissible solutions (supersonic and subsonic) with $\theta_{\text{w}}^* \in (\theta_{\text{w}}^{\text{d}}, \theta_{\text{w}}^{\text{s}}]$ for some $C > 0$. The proofs are the same as those in the previous case; indeed, the only difference is that, in the subsonic reflection case, we use (3.3.2) instead of Remark 3.2.5.

Then we obtain (3.2.24) with uniform C for any admissible solution for $\theta_{\text{w}} \in (\theta_{\text{w}}^{\text{d}}, \frac{\pi}{2})$.

The estimate in §3.2.4.2 is extended to all $\theta_{\text{w}} \in (\theta_{\text{w}}^{\text{d}}, \frac{\pi}{2})$ without change in its proof, since the supersonic and subsonic admissible solutions are of similar structures near Γ_{sym}.

3.3.3.2 The distance between Γ_{shock} and Γ_{wedge}

We note that estimates (3.2.20)–(3.2.21) of the distance between Γ_{shock} and Γ_{wedge} discussed in §3.2.4.3 cannot hold for the subsonic reflection configurations. Indeed, in this case, $\overline{\Gamma_{\text{shock}}} \cap \overline{\Gamma_{\text{wedge}}} = \{P_0\}$, *i.e.*, $\text{dist}(\Gamma_{\text{shock}}, \Gamma_{\text{wedge}}) = 0$, even if $u_1 \leq c_1$. Thus, we need to consider the distance between Γ_{shock} and Γ_{wedge} away from P_0, as we have done in estimate (3.2.22). Then the estimates in §3.2.4.3 in the present case have the following two forms:

If $u_1 \leq c_1$, then, for every small $r > 0$, there exists $C_r > 0$ such that

$$\text{dist}(\Gamma_{\text{shock}}, \Gamma_{\text{wedge}} \setminus B_r(P_0)) > \frac{1}{C_r} \tag{3.3.3}$$

for any admissible solution (supersonic and subsonic) of **Problem 2.6.1** with $\theta_{\text{w}} \in (\theta_{\text{w}}^{\text{d}}, \frac{\pi}{2})$. Note that, if θ_{w} is supersonic, $P_0 \notin \overline{\Gamma_{\text{wedge}}}$. Thus, choosing r sufficiently small, we see that $\Gamma_{\text{wedge}} \setminus B_r(P_0) = \Gamma_{\text{wedge}}$ so that, for such θ_{w} and r, estimate (3.3.3) coincides with (3.2.20). Moreover, in this case, the reflected-diffracted shock does not hit the wedge vertex P_3 as shown in Figs. 3.3–3.4.

Without assuming the condition that $u_1 \leq c_1$, we show the uniform lower bound of the distance between Γ_{shock} and Γ_{wedge} away from P_0 and P_3, *i.e.*, extending estimate (3.2.22) to the present case. That is, for any small $r > 0$, there exists $C_r > 0$ such that

$$\text{dist}\,(\Gamma_{\text{shock}},\ \Gamma_{\text{wedge}} \setminus (B_r(P_0) \cup B_r(P_3))) \geq \frac{1}{C_r} \tag{3.3.4}$$

for every admissible solution with $\theta_{\text{w}} \in (\theta_{\text{w}}^{\text{d}}, \frac{\pi}{2})$.

If $u_1 > c_1$, the wedge angle $\theta_{\text{w}}^{\text{c}}$ in Theorem 2.6.9 is defined as follows: As in §3.2.4.3, we extend the set of admissible solutions by including the normal reflection as the unique admissible solution for $\theta_{\text{w}} = \frac{\pi}{2}$. Let $r_1 := \inf_{\theta_{\text{w}} \in (\theta_{\text{w}}^{\text{d}}, \frac{\pi}{2})} |\Gamma_{\text{wedge}}^{(\theta_{\text{w}})}|$, which can be shown that $r_1 > 0$. Then we replace the definition of set $\mathcal{A}$ in §3.2.4.3 by

$$\mathcal{A} := \left\{ \theta_{\text{w}}^* \in (\theta_{\text{w}}^{\text{d}}, \frac{\pi}{2}] \ : \ \begin{array}{l} \text{For each } r \in (0, r_1), \text{ there exists } \varepsilon > 0 \text{ such that} \\ \text{dist}(\Gamma_{\text{shock}}, \Gamma_{\text{wedge}} \setminus B_r(P_0)) \geq \varepsilon \text{ for all admissible} \\ \text{solutions with the wedge angles } \theta_{\text{w}} \in [\theta_{\text{w}}^*, \dfrac{\pi}{2}] \end{array} \right\}.$$

Since the normal reflection solution is the unique admissible solution for $\theta_{\text{w}} = \frac{\pi}{2}$, the set of admissible solutions with angles $\theta_{\text{w}} \in [\theta_{\text{w}}^*, \frac{\pi}{2}]$ is non-empty for any $\theta_{\text{w}}^* \in (\theta_{\text{w}}^{\text{d}}, \frac{\pi}{2}]$. Moreover, since $\text{dist}(\Gamma_{\text{shock}}, \Gamma_{\text{wedge}}) > 0$ for the normal reflection solution, then $\frac{\pi}{2} \in \mathcal{A}$, *i.e.*, $\mathcal{A} \neq \emptyset$. Thus, we have

$$\theta_{\text{w}}^{\text{c}} = \inf \mathcal{A}.$$

Similarly to §3.2.4.3, we find that $\theta_{\text{w}}^{\text{c}} < \frac{\pi}{2}$. Therefore, for any $\theta_{\text{w}}^* \in (\theta_{\text{w}}^{\text{c}}, \frac{\pi}{2})$ and $r \in (0, r_1)$, there exists $C_r > 0$ such that, for any admissible solution φ with $\theta_{\text{w}} \in [\theta_{\text{w}}^*, \frac{\pi}{2})$,

$$\text{dist}(\Gamma_{\text{shock}}, \Gamma_{\text{wedge}} \setminus B_r(P_0)) \geq \frac{1}{C_r}. \tag{3.3.5}$$

We note that, while the estimates of this section are weaker than the estimates in §3.2.4.3, since Γ_{wedge} is replaced by $\Gamma_{\text{wedge}} \setminus B_r(P_0)$, the present estimates are used in the same way as the estimates in §3.2.4.3. Specifically, (3.2.20) and (3.2.23) are used in §3.2.5.1 to obtain the weighted $C^{k,\alpha}$–estimates away from Γ_{sonic}. Clearly, (3.3.3) and (3.3.5) can be used for that purpose as well. Similarly, one can replace (3.2.21) by (3.3.4) in the proof that, for the *attached* solution for $\theta_{\text{w}} = \theta_{\text{w}}^{\text{c}}$, the relative interior of Γ_{wedge} is disjoint from Γ_{shock}.

3.3.3.3 Uniform estimate of the ellipticity of equation (2.2.8) in Ω up to Γ_{shock}

We estimate the Mach number defined by (3.2.25).

First, we prove that (3.2.26) holds for all the supersonic admissible solutions with any supersonic wedge angle $\theta_{\text{w}} \in (\theta_{\text{w}}^{\text{d}}, \frac{\pi}{2})$, with uniform $\mu > 0$.

For the subsonic admissible solutions, we obtain the following estimate of the Mach number:

$$M^2(\boldsymbol{\xi}) \leq \max(1 - \hat{\zeta}, \frac{|D\varphi_2(P_0)|^2}{c_2^2} - \hat{\mu}|\boldsymbol{\xi} - P_0|) \qquad \text{for all } \boldsymbol{\xi} \in \overline{\Omega(\varphi)},$$

where the positive constants $\hat{\zeta}$ and $\hat{\mu}$ depend only on (ρ_0, ρ_1, γ).

From these estimates, we obtain the following ellipticity properties of the potential flow equation (2.2.8), written in the form of (3.2.27): There exist $\hat{\zeta} > 0$ and $C > 0$ depending only on (ρ_0, ρ_1, γ) such that, if φ is an admissible solution of **Problem 2.6.1** with $\theta_{\text{w}} \in (\theta_{\text{w}}^{\text{d}}, \frac{\pi}{2})$, then

(i) For any supersonic wedge angle θ_{w}, (3.2.28) holds;

(ii) For any subsonic/sonic wedge angle θ_{w},

$$\begin{aligned} &\frac{1}{C}\min(c_2 - |D\varphi_2(P_0)| + |\boldsymbol{\xi} - P_0|,\ \hat{\zeta})|\boldsymbol{\kappa}|^2 \\ &\leq \sum_{i,j=1}^{2} \mathcal{A}_{p_j}^i(D\varphi(\boldsymbol{\xi}), \varphi(\boldsymbol{\xi}))\kappa_i\kappa_j \leq C|\boldsymbol{\kappa}|^2 \end{aligned} \tag{3.3.6}$$

for any $\boldsymbol{\xi} \in \overline{\Omega}$ and $\boldsymbol{\kappa} = (\boldsymbol{\kappa}_1, \boldsymbol{\kappa}_2) \in \mathbb{R}^2$.

Note that, if θ_{w} is a subsonic wedge angle, then $|D\varphi_2(P_0)| < c_2$ so that (3.3.6) shows the uniform ellipticity of (2.2.8) for φ in Ω. However, this ellipticity degenerates near P_0 as the subsonic wedge angles tend to a sonic angle. If θ_{w} is a sonic angle, $|D\varphi_2(P_0)| = c_2$ and $\overline{\Gamma_{\text{sonic}}} = \{P_0\}$ so that (3.3.6) coincides with (3.2.28) in this case.

3.3.4 Regularity and related uniform estimates

3.3.4.1 Weighted $C^{k,\alpha}$–estimates away from $\Gamma_{\rm sonic}$ or the reflection point

Now all the preliminary results used for the estimates in §3.2.5.1 are extended to all the admissible solutions (supersonic and subsonic) with $\theta_{\rm w} \in (\theta_{\rm w}^{\rm d}, \frac{\pi}{2})$, where there is some difference in the estimates of the distance between $\Gamma_{\rm shock}$ and $\Gamma_{\rm wedge}$. However, the estimates obtained there are sufficient, as discussed in §3.3.3.2. Then we obtain the weighted $C^{k,\alpha}$–estimates away from $\Gamma_{\rm sonic}$ or the reflection point for any admissible solutions (supersonic and subsonic) with $\theta_{\rm w} \in (\theta_{\rm w}^{\rm d}, \frac{\pi}{2})$ by the same argument as that in §3.2.5.1.

3.3.4.2 Weighted and scaled $C^{k,\alpha}$–estimates near $\Gamma_{\rm sonic}$ or the reflection point

The main difference between the structure of supersonic and subsonic admissible solutions is near $\overline{\Gamma_{\rm sonic}}$, since $\overline{\Gamma_{\rm sonic}}$ is an arc for the supersonic wedge angles, and $\overline{\Gamma_{\rm sonic}} = \{P_0\}$ is one point for the subsonic and sonic wedge angles. Thus, the main difference from the argument in §3.2 is in the estimates near $\Gamma_{\rm sonic}$ or the reflection point, *i.e.*, near $\overline{\Gamma_{\rm sonic}}$.

Similarly to (3.2.30), we define and characterize Ω_ε, which now works for both supersonic and subsonic reflection solutions. We work in the (x, y)–coordinates introduced in §3.2.5.2, and note that $\overline{\Gamma_{\rm sonic}} \subset \{(x, y) \,:\, x = x_{P_1}\}$ for any wedge angle, where $x_{P_1} = 0$ for supersonic and sonic wedge angles, and $x_{P_1} > 0$ for subsonic wedge angles. Then, for appropriately small $\varepsilon_1 > \varepsilon_0 > 0$, we find that, for any $\varepsilon \in (0, \varepsilon_0]$,

$$
\begin{aligned}
&\Omega_\varepsilon := \Omega \cap \mathcal{N}_{\varepsilon_1}(\overline{\Gamma_{\rm sonic}}) \cap \{x < x_{P_1} + \varepsilon\} \\
&\quad\;\; = \{x_{P_1} < x < x_{P_1} + \varepsilon,\ \theta_{\rm w} < y < \hat{f}(x)\}, \\
&\Gamma_{\rm sonic} = \partial\Omega_\varepsilon \cap \{x = x_{P_1}\}, \\
&\Gamma_{\rm wedge} \cap \partial\Omega_\varepsilon = \{x_{P_1} < x < x_{P_1} + \varepsilon,\ y = \theta_{\rm w}\}, \\
&\Gamma_{\rm shock} \cap \partial\Omega_\varepsilon = \{x_{P_1} < x < x_{P_1} + \varepsilon,\ y = \hat{f}(x)\}
\end{aligned}
\tag{3.3.7}
$$

for some $\hat{f}(x)$ defined on $(x_{P_1},\, x_{P_1} + \varepsilon_0)$ and satisfying

$$
\begin{cases}
\hat{f}(x_{P_1}) = y_{P_1} > y_{P_4} = \theta_{\rm w} & \text{for supersonic reflection solutions,} \\
\hat{f}(x_{P_1}) = y_{P_0} = y_{P_1} = y_{P_4} = \theta_{\rm w} & \text{otherwise,}
\end{cases}
\tag{3.3.8}
$$

and

$$
0 < \omega \le \frac{d\hat{f}}{dx} < C \qquad \text{for any } x \in (x_{P_1},\, x_{P_1} + \varepsilon_0). \tag{3.3.9}
$$

To obtain the estimates near $\overline{\Gamma_{\rm sonic}}$, we consider four separate cases depending on the Mach number $\frac{|D\varphi_2|}{c_2}$ at P_0:

(a) Supersonic: $\frac{|D\varphi_2(P_0)|}{c_2} \geq 1 + \delta$;

(b) Supersonic-almost-sonic: $1 + \delta > \frac{|D\varphi_2(P_0)|}{c_2} > 1$;

(c) Subsonic-almost-sonic: $1 \geq \frac{|D\varphi_2(P_0)|}{c_2} \geq 1 - \delta$;

(d) Subsonic: $\frac{|D\varphi_2(P_0)|}{c_2} \leq 1 - \delta$.

We derive the uniform estimates in Ω_ε for any $\theta_{\mathrm{w}} \in [\theta^*_{\mathrm{w}}, \frac{\pi}{2})$, where ε is independent of θ_{w}. Recall that $P_1 = P_0$ in the subsonic case. The choice of constants (ε, δ) will be described below with the following properties: δ is chosen small, depending on (ρ_0, ρ_1, γ), so that the estimates in Cases (b)–(c) work in Ω_ε for some $\varepsilon > 0$; then ε is further reduced so that all the estimates in Cases (a)–(d) work in Ω_ε.

We first present a general overview of the estimates. In Cases (a)–(b), equation (2.2.8) is degenerate elliptic in Ω near $\Gamma_{\mathrm{sonic}} = P_1P_4$; see Fig. 2.3. In Case (c), the equation is uniformly elliptic in $\overline{\Omega}$, but the ellipticity constant is small near P_0 in Fig. 2.4. Thus, in Cases (a)–(c), we use the local elliptic degeneracy, which allows us to find a comparison function in each case, to show the appropriately fast decay of $\varphi - \varphi_2$ near P_1P_4 in Cases (a)–(b) and near P_0 in Case (c). Similarly to the argument of §3.2.5.2, we perform the local non-isotropic rescaling (different in each of Cases (a)–(c)) near each point of Ω_ε so that the rescaled functions satisfy a uniformly elliptic equation and the uniform L^∞–estimates, which follow from the decay of $\varphi - \varphi_2$ obtained above. Then we obtain the *a priori* estimates in the weighted and scaled $C^{2,\alpha}$–norms, which are different in each of Cases (a)–(c), but they imply the standard $C^{1,1}$–estimates. This is an extension of the methods in §3.2.5.2. In the uniformly elliptic case, Case (iv), the solution is of subsonic reflection configuration (*cf.* Fig. 2.4) and the estimates are more technically challenging than those in Cases (a)–(c), owing to the fact that the lower *a priori* regularity (Lipschitz) of the free boundary presents a new difficulty in Case (d) and the uniform ellipticity does not allow a comparison function that shows the sufficiently fast decay of $\varphi - \varphi_2$ near P_0. Thus, we prove the C^α–estimates of $D(\varphi - \varphi_2)$ near P_0 by deriving the corresponding elliptic equations and oblique boundary conditions for appropriately chosen directional derivatives of $\varphi - \varphi_2$.

Now we discuss the estimates in Cases (a)–(c) in more detail.

The techniques described in §3.2.5.2, for $\theta_{\mathrm{w}} \in [\theta^*_{\mathrm{w}}, \frac{\pi}{2})$ with $\theta^*_{\mathrm{w}} \in (\theta^{\mathrm{s}}_{\mathrm{w}}, \frac{\pi}{2})$, cannot be extended to all the supersonic reflection solutions. The reason for this is that, if the length of Γ_{sonic} is very small, rectangles $R^{(x_0,y_0)}_\rho$ specified in Cases (i)–(iii) in §3.2.5.2 do not *fit into* Ω in the following sense: The argument in §3.2.5.2 uses the property that the rectangles in Cases (i)–(ii) do not intersect with Γ_{shock} and the rectangles in Cases (i) and (iii) do not intersect with Γ_{wedge} (*cf.* Fig. 3.5) so that rectangles $R^{(x_0,y_0)}_\rho$ fit into Ω. From (3.3.7)–(3.3.9), rectangles $R^{(x_0,y_0)}_{1/2}$ in Cases (ii)–(iii) fit into Ω if $\sqrt{x_0} \lesssim y_{P_1} - y_{P_4}$, and do not fit into Ω in the opposite case; see Fig. 3.6. Thus, all the rectangles $R^{(x_0,y_0)}_{1/2}$ with

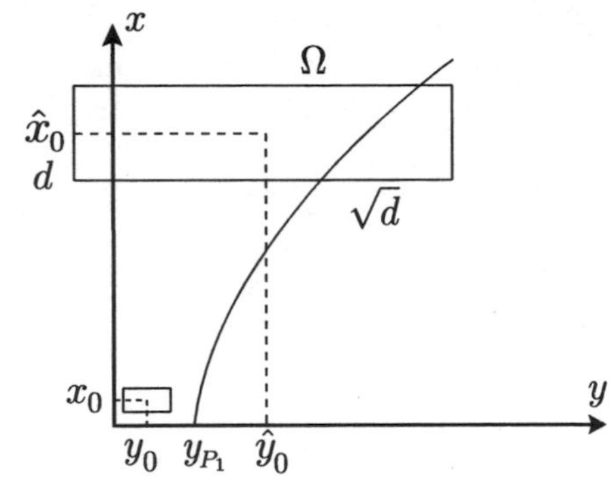

Figure 3.6: Rectangles when the sonic arc is short

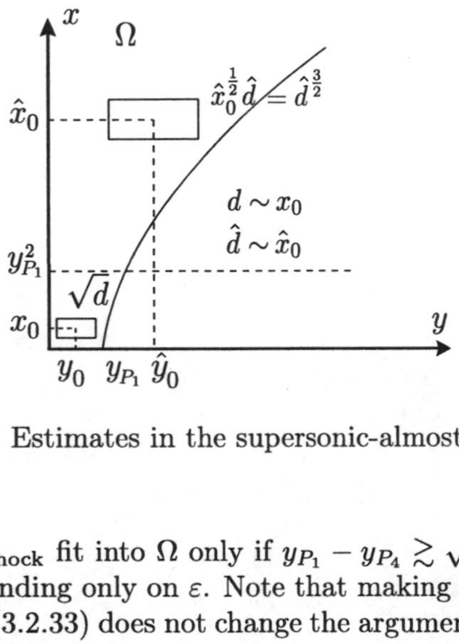

Figure 3.7: Estimates in the supersonic-almost-sonic case

$(x_0, y_0) \in \Gamma_{\rm wedge} \cup \Gamma_{\rm shock}$ fit into Ω only if $y_{P_1} - y_{P_4} \gtrsim \sqrt{\varepsilon}$, *i.e.*, when $\Gamma_{\rm sonic}$ is sufficiently long, depending only on ε. Note that making the rectangles smaller by choosing $\rho < \frac{1}{2}$ in (3.2.33) does not change the argument. The condition that $\frac{|D\varphi_2(P_0)|}{c_2} \geq 1+\delta$ implies a positive lower bound $b > 0$ on the length of $\Gamma_{\rm sonic}$, depending on $\delta > 0$. We fix $\delta > 0$ below. Then the estimates in §3.2.5.2 apply to any $\theta_{\rm w} \in (\theta_{\rm w}^{\rm d}, \frac{\pi}{2})$ satisfying $\frac{|D\varphi_2(P_0)|}{c_2} \geq 1+\delta$, and these estimates are obtained in Ω_ε with $\varepsilon \sim b^2$. This describes the estimates in Case (a) (supersonic).

In Case (b) (supersonic-almost-sonic), when $y_{P_1} - y_{P_4}$ is very small, we use (3.3.7)–(3.3.9) to note that there exists $k > 1$ so that the rectangles:

$$\hat{R}_{(x_0,y_0)} := \Big\{|x - x_0| < \frac{x_0^{3/2}}{10k},\ |y - y_0| < \frac{x_0}{10k}\Big\} \cap \Omega \tag{3.3.10}$$

for $(x_0, y_0) \in (\Gamma_{\rm wedge} \cup \Gamma_{\rm shock}) \cap \partial\Omega_\varepsilon$ fit into Ω in the sense described above. Note that the ratio of the lengths in the $x-$ and y–directions of $\hat{R}_{(x_0,y_0)}$ is $\sqrt{x_0}$, *i.e.*, the same as for the rectangles in (3.2.33). This implies that, rescaling $\hat{R}_{(x_0,y_0)}$ to the portion of square $(-1,1)^2 := (-1,1) \times (-1,1)$:

$$\hat{Q}_{(x_0,y_0)} := \{(S,T) \in (-1,1)^2 \,:\, (x_0 + x_0^{\frac{3}{2}} S, y_0 + \frac{x_0}{10k} T) \in \Omega\},$$

we obtain a uniformly elliptic equation for the function:

$$\psi^{(x_0,y_0)}(S,T) := \frac{1}{x_0^m} \psi(x_0 + x_0^{\frac{3}{2}} S, y_0 + \frac{x_0}{10k} T) \qquad \text{in } \hat{Q}_{(x_0,y_0)} \tag{3.3.11}$$

with any positive integer m. Thus, if the uniform L^∞ bound is obtained for functions $\psi^{(x_0,y_0)}$, we can follow the argument in §3.2.5.2 by using the rectangles in (3.3.10). However, if $m = 2$ is used, the resulting estimates, rescaled back into the (x,y)–variables, are weaker than the estimates obtained in §3.2.5.2, where we have used the rectangles in (3.2.33), and such estimates are not sufficient for the rest of the argument. In fact, we need to use $m = 4$. This requires the estimate: $\psi(x,y) \le Cx^4$, in order to obtain the uniform L^∞ bound of $\psi^{(x_0,y_0)}$. However, Theorem 2.6.6 implies that $\psi \in C^{2,\alpha}(\overline{\Omega_\varepsilon} \setminus \{P_1\})$ with $\psi_{xx} = \frac{1}{\gamma+1} > 0$ on Γ_{sonic} so that, recalling that $\psi = \psi_x = 0$ on Γ_{sonic}, we conclude that the estimate, $\psi(x,y) \le Cx^4$, does not hold near Γ_{sonic}. For this reason, we decompose Ω_ε into two subdomains; see Fig. 3.7. For $\theta_{\text{w}} \in (\theta_{\text{w}}^{\text{d}}, \frac{\pi}{2})$ satisfying the condition of Case (b), define

$$b_{\text{so}} := y_{P_1} - y_{P_4}.$$

As we have discussed above, in $\Omega_{b_{\text{so}}^2}$, we can use the argument in §3.2.5.2 to obtain the estimates described there. Furthermore, for each $m = 2, 3, \dots$, if δ is small in the condition of Case (b) depending only on $(\rho_0, \rho_1, \gamma, m)$, we obtain

$$0 \le \psi(x,y) \le Cx^m \qquad \text{in } \Omega_\varepsilon \cap \{x > \frac{b_{\text{so}}^2}{10}\}, \tag{3.3.12}$$

where $C > 0$ and $\varepsilon \in (0, \varepsilon_0]$ depend only on $(\rho_0, \rho_1, \gamma, m)$. The main point here is that $C > 0$ and ε are independent of b_{so}. Estimate (3.3.12) is proved by showing that

$$0 \le \psi(x,y) \le C(x + Mb_{\text{so}}^2)^m \qquad \text{in } \Omega_\varepsilon$$

with C, M, and ε depending only on $(\rho_0, \rho_1, \gamma, m)$. We use $m = 4$ in (3.3.12). This fixes δ for Cases (a)–(b). Then, as we have discussed above, we obtain the estimates in $\Omega_\varepsilon \cap \{x > \frac{b_{\text{so}}^2}{2}\}$ by using the rectangles in (3.3.10) and the rescaled functions (3.3.11) with $m = 4$. Combining this with the estimates in $\Omega_{b_{\text{so}}^2}$, we complete the uniform estimates in Ω_ε for Case (b).

If $\theta_{\text{w}} \in (\theta_{\text{w}}^{\text{d}}, \frac{\pi}{2})$ satisfies the condition of Case (c), we argue similar to Case (b), by changing the size of the rectangles (*i.e.*, the scaling) according to the geometry of the domain; see Fig. 3.8. Specifically, for each $m = 2, 3, \dots$, if δ is small depending only on $(\rho_0, \rho_1, \gamma, m)$ in the condition of Case (c), we obtain

$$0 \le \psi(x,y) \le C(x - x_{P_0})^m \qquad \text{in } \Omega_\varepsilon \tag{3.3.13}$$

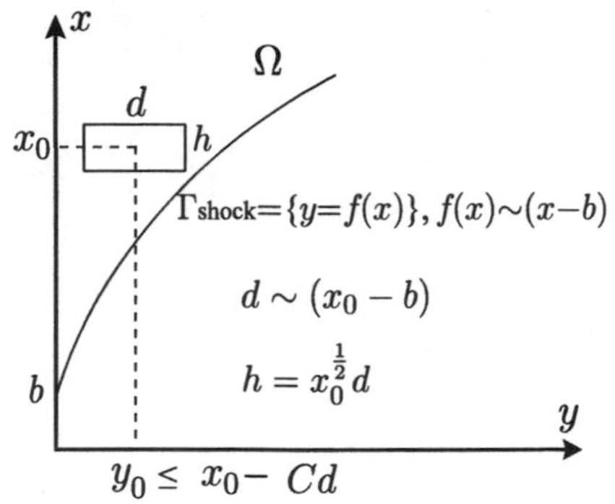

Figure 3.8: Estimates in the subsonic-almost-sonic case

with C, M, and ε depending only on $(\rho_0, \rho_1, \gamma, m)$. Recall that $x_{P_0} = x_{P_1} > 0$ in the subsonic case. Also, for sufficiently large $k > 1$, the rectangles:

$$\hat{R}_{(x_0,y_0)} := \Big\{|x - x_0| < \frac{\sqrt{x_0}}{10k}(x - x_{P_0}),\ |y - y_0| < \frac{1}{10k}(x - x_{P_0})\Big\} \cap \Omega$$

for $(x_0, y_0) \in (\Gamma_{\text{wedge}} \cup \Gamma_{\text{shock}}) \cap \partial\Omega_\varepsilon$ fit into Ω in the sense described above. The ratio of the side lengths in the x– and y–directions of $\hat{R}_{(x_0,y_0)}$ is $\sqrt{x_0}$, as in the previous cases. Thus, rescaling $\hat{R}_{(x_0,y_0)}$ to the portion of square $(-1,1)^2$:

$$\hat{Q}_{(x_0,y_0)} := \{(S,T) \in (-1,1)^2 : (x_0 + \frac{\sqrt{x_0}}{10k}(x - x_{P_0})S, y_0 + \frac{1}{10k}(x - x_{P_0})T) \in \Omega\},$$

we obtain a uniformly elliptic equation in $\hat{Q}_{(x_0,y_0)}$ for the function:

$$\psi^{(z_0)}(S,T) := \frac{1}{(x - x_{P_0})^m}\psi(x_0 + \frac{\sqrt{x_0}}{10k}(x - x_{P_0})S, y_0 + \frac{1}{10k}(x - x_{P_0})T). \quad (3.3.14)$$

We use $m = 5$ in (3.3.13), which fixes δ for Cases (c)–(d). Then, repeating the argument of the previous cases for the rescaled functions (3.3.14) with $m = 5$, we obtain the uniform estimates of ψ in $C^{2,\alpha}(\overline{\Omega_\varepsilon})$ with $\varepsilon(\rho_0, \rho_1, \gamma)$.

Next we consider Case (d). If $\theta_{\text{w}}^* \in (\theta_{\text{w}}^{\text{d}}, \frac{\pi}{2})$ is fixed, and $\theta_{\text{w}} \in (\theta_{\text{w}}^{\text{d}}, \frac{\pi}{2})$ satisfies the condition of Case (d) with δ fixed above, we use the uniform ellipticity (independent of θ_{w}) in the estimates. The main steps of these estimates are described in §16.6.1. We note the following points of this argument:

- We use the fact that φ_2 in (3.3.2) is the *weak* state (2);
- We use the monotonicity cone of $\varphi_1 - \varphi$ (*cf.* §3.3.2), and the convexity of the shock polar;
- We obtain the estimates in $C^{1,\alpha}$ up to P_0, which is a weaker regularity than that in Cases (a)–(c);

- The constants in the estimates depend on $\theta_{\rm w}^*$, in addition to (ρ_0, ρ_1, γ), and blow up as $\theta_{\rm w}^* \to \theta_{\rm w}^{\rm d}+$.

3.3.5 Existence of the regular reflection-diffraction configuration up to the detachment angle

Let $\theta_{\rm w}^* \in (\theta_{\rm w}^{\rm d}, \frac{\pi}{2})$. We show that there exists an admissible solution for any wedge angle $\theta_{\rm w} \in (\theta_{\rm w}^*, \frac{\pi}{2}]$.

We follow the argument described in §3.2.6 with the changes necessary to handle both cases of supersonic and subsonic reflection solutions in the argument. This includes the following three steps:

1. As in §3.2.6, the iteration set $\mathcal{K}$ consists of pairs $(u, \theta_{\rm w})$, for a function u on the unit square $Q^{\rm iter}$ and $\theta_{\rm w} \in [\theta_{\rm w}^*, \frac{\pi}{2}]$:

$$\mathcal{K} = \cup_{\theta_{\rm w} \in [\theta_{\rm w}^*, \frac{\pi}{2}]} \mathcal{K}_{\theta_{\rm w}} \times \{\theta_{\rm w}\}$$

with $\mathcal{K}_{\theta_{\rm w}} \subset \mathcal{C}$, where $\mathcal{C}$ is a weighted and scaled $C^{2,\alpha}$–type space on $Q^{\rm iter}$ for some $\alpha = \alpha(\rho_0, \rho_1, \gamma) \in (0,1)$, which satisfies

$$\mathcal{C} \subset C^{2,\alpha}(Q^{\rm iter}) \cap C^{1,\alpha}(\overline{Q^{\rm iter}}).$$

For each $(u, \theta_{\rm w}) \in \mathcal{K}$, the elliptic domain Ω of the *approximate solution* and a smooth invertible map $\mathcal{G}_{u,\theta_{\rm w}} : Q^{\rm iter} \mapsto \Omega$ are defined. As in §3.2.6, for any supersonic wedge angle $\theta_{\rm w} \in (\theta_{\rm w}^*, \frac{\pi}{2}]$, region Ω is of the same structure as an elliptic region of supersonic reflection solutions; see §2.4.2 and Fig. 2.3. Map $\mathcal{G}_{u,\theta_{\rm w}} : Q^{\rm iter} \mapsto \Omega(u, \theta_{\rm w})$ can be extended to the smooth and smoothly invertible map $\overline{Q^{\rm iter}} \mapsto \overline{\Omega}$, where the sides of square $Q^{\rm iter}$ are mapped to the boundary parts $\Gamma_{\rm sonic}, \Gamma_{\rm shock}, \Gamma_{\rm sym}$, and $\Gamma_{\rm wedge}$ of Ω. However, for any subsonic/sonic wedge angle $\theta_{\rm w}$, $\Omega(u, \theta_{\rm w})$ is of the structure described in §2.4.3 and Fig. 2.4, *i.e.*, has a triangular shape $P_0P_2P_3$. Thus, map $\mathcal{G}_{u,\theta_{\rm w}} : \overline{Q^{\rm iter}} \mapsto \overline{\Omega}$ is smooth but not invertible; one of the sides of $Q^{\rm iter}$ is now mapped into point $\overline{\Gamma_{\rm sonic}} = \{P_0\}$.

2. The singularity of mapping $\mathcal{G}_{u,\theta_{\rm w}} : \overline{Q^{\rm iter}} \mapsto \overline{\Omega}$, described above, affects the choice of the function space $\mathcal{C}$ introduced above. The norm in $\mathcal{C}$ is a weighted and scaled $C^{2,\alpha}$–type norm on $Q^{\rm iter}$ such that

- If $(u, \theta_{\rm w}) \in \mathcal{K}$ and $v \in \mathcal{C}$, then, expressing v as a function w on $\Omega(u, \theta_{\rm w})$ by $w = v \circ \mathcal{G}_{u,\theta_{\rm w}}^{-1}$, we obtain that $w \in C^{1,\alpha}(\overline{\Omega}) \cap C^{2,\alpha}(\Omega)$ and some more detailed properties.

- If φ is an admissible solution for the wedge angle $\theta_{\rm w}$, there exists $u \in \mathcal{K}_{\theta_{\rm w}}$, which is related to φ through map $\mathcal{G}_{u,\theta_{\rm w}}$. The *a priori* estimates of the admissible solutions for all the cases described in §3.3.4.1–§3.3.4.2 imply the estimates for u in a norm which is stronger than the norm of $\mathcal{C}$. This allows us to define an iteration map which is compact in the norm of $\mathcal{C}$ and to show that there is no fixed point of the iteration map on the boundary of the iteration set.

3. The properties of the potential flow equation (2.2.8) for admissible solutions, near $\Gamma_{\rm sonic}$ or the reflection point, are different for $\theta_{\rm w}$ belonging to the different cases (a)–(d) in §3.3.4.2. This affects the definition of the equation in the boundary value problem used in the definition of the iteration map for the corresponding angle $\theta_{\rm w}$. Also, in solving this problem and deriving the estimates of its solutions, we employ techniques similar to the estimates of admissible solutions in §3.3.4.1–§3.3.4.2 for Cases (a)–(d). This allows us to define the iteration map and obtain its compactness.

3.4 APPENDIX: THE METHOD OF CONTINUITY AND FIXED POINT THEOREMS

For completeness, we now present several fundamental theorems regarding the method of continuity and fixed point theorems that are used in this book.

Theorem 3.4.1 (Method of Continuity). *Let $\mathcal{B}$ be a Banach space and $\mathcal{V}$ a normed linear space, and let $\mathbf{L}_0$ and $\mathbf{L}_1$ be bounded linear operators from $\mathcal{B}$ into $\mathcal{V}$. Suppose that there is a constant C such that, for any $\tau \in [0,1]$,*

$$\|\mathbf{x}\|_{\mathcal{B}} \le C\|\big((1-\tau)\mathbf{L}_0 + \tau\mathbf{L}_1\big)\mathbf{x}\|_{\mathcal{V}} \qquad \text{for any } \mathbf{x} \in \mathcal{B}. \tag{3.4.1}$$

Then $\mathbf{L}_1$ maps $\mathcal{B}$ onto $\mathcal{V}$ if and only if $\mathbf{L}_0$ maps $\mathcal{B}$ onto $\mathcal{V}$.

Definition 3.4.2. *Let X and Y be metric spaces. A map $h : X \mapsto Y$ is compact provided that*

(i) *h is continuous;*

(ii) *$f(A)$ is compact whenever $A \subset X$ is bounded.*

Theorem 3.4.3 (Leray-Schauder Fixed Point Theorem). *Let $\mathbf{T}$ be a compact mapping of a Banach space $\mathcal{B}$ into itself. Suppose that there exists a constant M such that, for all $\mathbf{x} \in \mathcal{B}$ and $\tau \in [0,1]$ satisfying $\mathbf{x} = \tau\mathbf{T}\mathbf{x}$,*

$$\|\mathbf{x}\|_{\mathcal{B}} \le M. \tag{3.4.2}$$

Then $\mathbf{T}$ has a fixed point.

Theorems 3.4.1 to 3.4.3 can be found as Theorem 5.2 and 11.3 in [131].

Theorem 3.4.4 (Schauder Fixed Point Theorem). *Let $\mathcal{K}$ be a closed and convex subset of a Banach space, and let $\mathbf{J} : \mathcal{K} \mapsto \mathcal{K}$ be a continuous map such that $\mathbf{J}(\mathcal{K})$ is precompact. Then $\mathbf{J}$ has a fixed point.*

More details can be found in [131], including Corollary 11.2.

Next we present some further basic definitions and facts in the Leray-Schauder degree theory.

Definition 3.4.5. *Let G be an open bounded set in a Banach space X. Denote by $V(G, X)$ the set of all the maps $\mathbf{f} : \bar{G} \mapsto X$ satisfying the following:*

(i) $\mathbf{f}$ *is compact;*

(ii) $\mathbf{f}$ *has no fixed points on boundary ∂G.*

Definition 3.4.6. *Two maps $\mathbf{f}, \mathbf{g} \in V(G, X)$ are called compactly homotopic on ∂G if there exists a map $\mathbf{H}$ with the following properties:*

(i) $\mathbf{H} : \bar{G} \times [0, 1] \mapsto X$ *is compact;*

(ii) $\mathbf{H}(\mathbf{x}, \tau) \neq \mathbf{x}$ *for any* $(\mathbf{x}, \tau) \in \partial G \times [0, 1]$;

(iii) $\mathbf{H}(\mathbf{x}, 0) = \mathbf{f}(\mathbf{x})$ *and* $\mathbf{H}(\mathbf{x}, 1) = \mathbf{g}(\mathbf{x})$ *on* $\bar{G}$.

We write $\partial G : \mathbf{f} \tilde{=} \mathbf{g}$. This map $\mathbf{H}$ is called a compact homotopy.

Then we have the following Leray-Schauder degree theory.

Theorem 3.4.7. *Let G be an open bounded set in a Banach space X. Then, to each map $\mathbf{f} \in V(G, X)$, an integer number $\mathbf{Ind}(\mathbf{f}, G)$ can be uniquely assigned such that*

(i) *If $\mathbf{f}(\mathbf{x}) \equiv \mathbf{x}_0$ for any $\mathbf{x} \in \bar{G}$ and some fixed $\mathbf{x}_0 \in G$, then $\mathbf{Ind}(\mathbf{f}, G) = 1$;*

(ii) *If $\mathbf{Ind}(\mathbf{f}, G) \neq 0$, there exists $\mathbf{x} \in G$ such that $\mathbf{f}(\mathbf{x}) = \mathbf{x}$;*

(iii) $\mathbf{Ind}(\mathbf{f}, G) = \sum_{j=1}^{n} \mathbf{Ind}(\mathbf{f}, G_j)$, *whenever $\mathbf{f} \in V(G, X) \cap \left(\cup_{j=1}^{n} V(G_j, X) \right)$, where $\{G_j\}$ is a regular partition of G, i.e., G_j are pairwise disjoint and $\bar{G} = \cup_{j=1}^{n} \bar{G}_j$;*

(iv) *If $\partial G : \mathbf{f} \tilde{=} \mathbf{g}$, then $\mathbf{Ind}(\mathbf{f}, G) = \mathbf{Ind}(\mathbf{g}, G)$.*

The integer number $\mathbf{Ind}(\mathbf{f}, G)$ is called the fixed point index of f on G.

We also need to consider the case in which set G varies with the homotopy parameter t; see §13.6($A4^*$) in [283].

Theorem 3.4.8 (Generalized Homotopy Invariance of the Fixed Point Index). *Let X be a Banach space, $t_2 > t_1$. Let $U \subset X \times [t_1, t_2]$, and let $U_t := \{\mathbf{x} : (\mathbf{x}, t) \in U\}$. Then*

$$\mathbf{Ind}(\mathbf{h}(\cdot, t), U_t) = const. \qquad \text{for all } t \in [t_1, t_2],$$

provided that U is bounded and open in $X \times [t_1, t_2]$, and operator $\mathbf{h} : \overline{U} \mapsto X$ is compact with $\mathbf{h}(\mathbf{x}, t) \neq \mathbf{x}$ on ∂U.

Note that set U is open with respect to the subspace topology on $X \times [t_1, t_2]$. That is, U is an intersection of an open set in $X \times \mathbb{R}$ with $X \times [t_1, t_2]$.

More details about the degree theory can be found in Chapters 12–13 in [283].

Part II

Elliptic Theory and Related Analysis for Shock Reflection-Diffraction

Chapter Four

Relevant Results for Nonlinear Elliptic Equations of Second Order

In this chapter, we present some relevant results for nonlinear elliptic equations of second order (for which the structural conditions and some regularity of the coefficients are not required) with focus on the two-dimensional case, and study the existence and regularity of solutions of certain boundary value problems in domains of appropriate structure for an equation with ellipticity degenerating on a part of the boundary. These results will be applied in solving von Neumann's conjectures in Parts III–IV.

Through this chapter, we use the following notations: $\mathbf{x} = (x_1, x_2)$ or $\mathbf{X} = (X_1, X_2)$ for the coordinates in $\mathbb{R}^2$, $\mathbb{R}^2_+ := \{x_2 > 0\}$, and $B_r := B_r(\mathbf{0})$ and $B_r^+ := B_r^+(\mathbf{0}) \cap \mathbb{R}^2_+$ for $r > 0$.

4.1 NOTATIONS: HÖLDER NORMS AND ELLIPTICITY

In this section we first introduce some notations, including the Hölder norms and notions of ellipticity, which will be used in subsequent developments.

4.1.1 Hölder norms

Let $\Omega \subset \mathbb{R}^2$ be an open bounded set. We now introduce the Hölder norms in Ω. For $\alpha \in (0,1)$ and $m \in \{0, 1, \dots\}$, define

$$\begin{aligned}
\|u\|_{m,0,\Omega} &:= \sum_{0\le|\boldsymbol{\beta}|\le m} \sup_{\mathbf{x}\in\Omega} |D^{\boldsymbol{\beta}}u(\mathbf{x})|, \\
[u]_{m,\alpha,\Omega} &:= \sum_{|\boldsymbol{\beta}|=m} \sup_{\mathbf{x},\mathbf{y}\in\Omega,\mathbf{x}\neq\mathbf{y}} \frac{|D^{\boldsymbol{\beta}}u(\mathbf{x}) - D^{\boldsymbol{\beta}}u(\mathbf{y})|}{|\mathbf{x}-\mathbf{y}|^{\alpha}}, \\
\|u\|_{m,\alpha,\Omega} &:= \|u\|_{m,0,\Omega} + [u]_{m,\alpha,\Omega},
\end{aligned} \tag{4.1.1}$$

where $D^{\boldsymbol{\beta}} = D_1^{\beta_1} D_2^{\beta_2}$, $D_i = \partial_{x_i}$, and $\boldsymbol{\beta} = (\beta_1, \beta_2)$ is a multi-index with $\beta_j \in \mathbb{N} \cup \{0\}$ and $|\boldsymbol{\beta}| = \beta_1 + \beta_2$ (note that $D^{(0,0)}u = u$), where $\mathbb{N}$ is the set of natural numbers, *i.e.*, all positive integers.

We denote by $C_{m,\alpha,\Omega} := C^{m,\alpha}(\overline{\Omega})$ the space of functions on Ω with finite norm $\|\cdot\|_{m,\alpha,\Omega}$. We also write $\|u\|_{m,\Omega}$ for $\|u\|_{m,0,\Omega}$, $m = 0, 1, \dots$.

Let $\Sigma \subset \partial\Omega$. We define partially interior norms in $\Omega \cup \Sigma$: For $\mathbf{x}, \mathbf{y} \in \Omega$, let

$$d_{\mathbf{x}} = \mathrm{dist}(\mathbf{x}, \partial\Omega \setminus \Sigma), \quad d_{\mathbf{x},\mathbf{y}} = \min(d_{\mathbf{x}}, d_{\mathbf{y}}).$$

For $\alpha \in (0,1)$, $m \in \{0,1,\dots\}$, and $\sigma \in \mathbb{R}$, define

$$\begin{aligned}
\|u\|^{(\sigma)}_{m,0,\Omega\cup\Sigma} &:= \sum_{|\boldsymbol{\beta}|=m} \sup_{\mathbf{x}\in\Omega} \left(d_{\mathbf{x}}^{m+\sigma}|D^{\boldsymbol{\beta}}u(\mathbf{x})|\right),\\
[u]^{(\sigma)}_{m,\alpha,\Omega\cup\Sigma} &:= \sum_{|\boldsymbol{\beta}|=m} \sup_{\mathbf{x},\mathbf{y}\in\Omega,\mathbf{x}\neq\mathbf{y}} \left(d_{\mathbf{x},\mathbf{y}}^{m+\alpha+\sigma}\frac{|D^{\boldsymbol{\beta}}u(\mathbf{x}) - D^{\boldsymbol{\beta}}u(\mathbf{y})|}{|\mathbf{x}-\mathbf{y}|^{\alpha}}\right),\\
\|u\|^{(\sigma)}_{m,\alpha,\Omega\cup\Sigma} &:= \sum_{k=0}^{m} \|u\|^{(\sigma)}_{k,0,\Omega\cup\Sigma},\\
\|u\|^{(\sigma)}_{m,\alpha,\Omega\cup\Sigma} &:= \|u\|^{(\sigma)}_{m,0,\Omega\cup\Sigma} + [u]^{(\sigma)}_{m,\alpha,\Omega\cup\Sigma}.
\end{aligned} \tag{4.1.2}$$

Furthermore, we use the notations:

$$\begin{aligned}
&[u]^{*}_{m,0,\Omega\cup\Sigma} = [u]^{(0)}_{m,0,\Omega\cup\Sigma}, \quad \|u\|^{*}_{m,0,\Omega\cup\Sigma} = \|u\|^{(0)}_{m,0,\Omega\cup\Sigma},\\
&[u]^{*}_{m,\alpha,\Omega\cup\Sigma} = [u]^{(0)}_{m,\alpha,\Omega\cup\Sigma}, \quad \|u\|^{*}_{m,\alpha,\Omega\cup\Sigma} = \|u\|^{(0)}_{m,\alpha,\Omega\cup\Sigma},
\end{aligned} \tag{4.1.3}$$

where the right-hand sides are the norms in (4.1.2). We denote by $C_{m,\alpha,\Omega\cup\Sigma} := C^{m,\alpha}(\Omega \cup \Sigma)$ the space of functions with finite norm $\|\cdot\|^{*}_{m,\alpha,\Omega\cup\Sigma}$.

We also define the interior norms in Ω: For $m \in \mathbb{N} \cup \{0\}$, $\alpha \in (0,1)$, and $\sigma \in \mathbb{R}$,

$$\|u\|^{(\sigma)}_{m,\alpha,\Omega}, \ \|u\|^{*}_{m,\alpha,\Omega} \quad \text{are norms (4.1.2)–(4.1.3) with } \Sigma = \emptyset. \tag{4.1.4}$$

Next, we introduce the Hölder norms in Ω weighted by the distance to $\Sigma \subset \partial\Omega$. Set

$$\delta_{\mathbf{x}} := \mathrm{dist}(\mathbf{x}, \Sigma), \quad \delta_{\mathbf{x},\mathbf{y}} := \min(\delta_{\mathbf{x}}, \delta_{\mathbf{y}}) \qquad \text{for } \mathbf{x}, \mathbf{y} \in \Omega.$$

Then, for $k \in \mathbb{R}$, $\alpha \in (0,1)$, and $l, m \in \mathbb{N} \cup \{0\}$, define

$$\begin{aligned}
[u]^{(k),\Sigma}_{l,0,\Omega} &:= \sum_{|\boldsymbol{\beta}|=l} \sup_{\mathbf{x}\in\Omega} \left(\delta_{\mathbf{x}}^{\max\{l+k,0\}}|D^{\boldsymbol{\beta}}u(\mathbf{x})|\right),\\
[u]^{(k),\Sigma}_{m,\alpha,\Omega} &:= \sum_{|\boldsymbol{\beta}|=m} \sup_{\mathbf{x},\mathbf{y}\in\Omega,\mathbf{x}\neq\mathbf{y}} \left(\delta_{\mathbf{x},\mathbf{y}}^{\max\{m+\alpha+k,0\}}\frac{|D^{\boldsymbol{\beta}}u(\mathbf{x}) - D^{\boldsymbol{\beta}}u(\mathbf{y})|}{|\mathbf{x}-\mathbf{y}|^{\alpha}}\right),\\
\|u\|^{(k),\Sigma}_{m,0,\Omega} &:= \sum_{l=0}^{m} [u]^{(k),\Sigma}_{l,0,\Omega},\\
\|u\|^{(k),\Sigma}_{m,\alpha,\Omega} &:= \|u\|^{(k),\Sigma}_{m,0,\Omega} + [u]^{(k),\Sigma}_{m,\alpha,\Omega}.
\end{aligned} \tag{4.1.5}$$

We denote by $C^{(k),\Sigma}_{m,\alpha,\Omega}$ the closure of space $C^{\infty}(\Omega)$ under norm $\|\cdot\|^{(k),\Sigma}_{m,\alpha,\Omega}$.

Remark 4.1.1. *If $m \geq -k \geq 1$ and k is an integer, any function $u \in C^{(k),\Sigma}_{m,\alpha,\Omega}$ is $C^{|k|-1,1}$ up to Σ, but not necessarily $C^{|k|}$ up to Σ.*

We also note

Lemma 4.1.2. *Let $0 \leq \alpha_1 < \alpha_2 < 1$, and let m be a nonnegative integer. Then*

$$\|u\|^{(-m+1-\alpha_1),\Sigma}_{m,\alpha_1,\Omega} \leq \max\{(\operatorname{diam}\Omega)^{\alpha_2-\alpha_1}, 1\}\|u\|^{(-m+1-\alpha_2),\Sigma}_{m,\alpha_2,\Omega}.$$

This follows directly from the expressions of norms in (4.1.5) by a straightforward estimate.

4.1.2 Notions of ellipticity

We adopt the following notions of strict ellipticity and uniform ellipticity. Let $\Omega \subset \mathbb{R}^2$ be open, $u \in C^2(\Omega)$, and

$$\mathcal{N}(u) = \sum_{i,j=1}^{2} A_{ij}(Du, u, \mathbf{x})D_{ij}u + A(Du, u, \mathbf{x}), \tag{4.1.6}$$

where $A_{ij}(\mathbf{p}, z, \mathbf{x})$ and $B(\mathbf{p}, z, \mathbf{x})$ are continuous on $\mathbb{R}^2 \times \mathbb{R} \times \overline{\Omega}$, $D_{ij} = D_iD_j$, $\mathbf{p} = (p_1, p_2) \in \mathbb{R}^2$, and $z \in \mathbb{R}$. Operator $\mathcal{N}$ is elliptic with respect to u in Ω if the matrix of coefficients of the second-order terms, $[A_{ij}(Du(\mathbf{x}), u(\mathbf{x}), \mathbf{x})]$, is nonnegative for every $\mathbf{x} \in \Omega$, *i.e.*,

$$\sum_{i,j=1}^{2} A_{ij}(Du(\mathbf{x}), u(\mathbf{x}), \mathbf{x})\mu_i\mu_j \geq 0 \quad \text{for any } \mathbf{x} \in \Omega \text{ and } \boldsymbol{\mu} = (\mu_1, \mu_2) \in \mathbb{R}^2. \tag{4.1.7}$$

The ellipticity is strict when inequality (4.1.7) becomes strict for any $\mathbf{x} \in \Omega$ and $\boldsymbol{\mu} = (\mu_1, \mu_2) \in \mathbb{R}^2 \setminus \{\mathbf{0}\}$, which means that matrix $[A_{ij}(Du(\mathbf{x}), u(\mathbf{x}), \mathbf{x})]$ is positive for every $\mathbf{x} \in \Omega$; otherwise, the ellipticity is degenerate. Furthermore, $\mathcal{N}$ is uniformly elliptic with respect to u in Ω if there exists a positive constant $\lambda > 0$ such that, for any $\mathbf{x} \in \Omega$ and $\boldsymbol{\mu} = (\mu_1, \mu_2) \in \mathbb{R}^2$,

$$\lambda|\boldsymbol{\mu}|^2 \leq \sum_{i,j=1}^{2} A_{ij}(Du(\mathbf{x}), u(\mathbf{x}), \mathbf{x})\mu_i\mu_j \leq \lambda^{-1}|\boldsymbol{\mu}|^2.$$

The following standard comparison principle for operator $\mathcal{N}$ follows from [131, Theorem 10.1]:

Lemma 4.1.3. *Let $\Omega \subset \mathbb{R}^2$ be an open bounded set. Let $u, v \in C(\overline{\Omega}) \cap C^2(\Omega)$ be such that operator $\mathcal{N}$ is elliptic in Ω with respect to either u or v, and coefficients $A_{ij}(\mathbf{p}, z, \mathbf{x})$ and $A(\mathbf{p}, z, \mathbf{x})$ are independent of z. Let $\mathcal{N}u \leq \mathcal{N}v$ in Ω and $u \geq v$ on $\partial\Omega$. Then $u \geq v$ in Ω.*

4.2 QUASILINEAR UNIFORMLY ELLIPTIC EQUATIONS

In this section, we present some estimates of solutions of the boundary value problems for quasilinear elliptic equations in dimension two, which are applied in later chapters. The main features of these estimates are that the structural conditions (other than the ellipticity) and some regularity of the coefficients are not required, and the $C^{k,\alpha}$–estimates of the solution, $k = 1, 2$, depend only on its C^0–norm, independent of the $C^{0,1}$–norm of the solution. These features are necessary for our applications to the shock reflection-diffraction problems.

Consider a quasilinear elliptic equation of the form:

$$\mathcal{N}(u) = f(\mathbf{x}) \tag{4.2.1}$$

with

$$\mathcal{N}(u) := \sum_{i,j=1}^{2} A_{ij}(Du, u, \mathbf{x}) D_{ij}u + A(Du, u, \mathbf{x}), \tag{4.2.2}$$

where $A_{ij} = A_{ij}(\mathbf{p}, z, \mathbf{x})$ with $A_{ij} = A_{ji}$, and $A = A(\mathbf{p}, z, \mathbf{x})$ with $A(\mathbf{0}, 0, \mathbf{x}) = 0$, for $(\mathbf{p}, z, \mathbf{x}) \in \mathbb{R}^2 \times \mathbb{R} \times \Omega$, $\Omega \subset \mathbb{R}^2$, and $i, j = 1, 2$.

We study boundary value problems with the following three types of boundary conditions:

(i) the Dirichlet condition;

(ii) the oblique derivative condition;

(iii) the *almost tangential derivative* condition.

We restrict to the two-dimensional case that allows for equations of a general structure. For equation (4.2.1), we require only the strict ellipticity and certain regularity of the coefficients. In particular, we do not require equation (4.2.1) to satisfy any structural conditions. This is important for our applications, because our iteration equation does not satisfy the structural conditions used in higher dimensions; *cf.* [131, Chapter 15] and [197].

The main point at which the structural conditions are needed in higher dimensions is for the gradient estimates; *cf.* [131, Chapter 15] for the interior estimates for the Dirichlet problem and [197] for the oblique derivative problem. The interior gradient estimates and global gradient estimates for the Dirichlet problem, without requiring the structural conditions, were obtained in the earlier work in the two-dimensional case; see Trudinger [262] and the references therein. However, it is not clear how this approach can be extended to the oblique and *almost tangential derivative* problems. We also note some related results by Lieberman [191, 192] for fully nonlinear equations with the boundary conditions in the two-dimensional case, in which the Hölder estimates for the gradient of a solution depend on both the bounds of the solution and its gradient. In this section, we present the $C^{2,\alpha}$–estimates of the solution only in terms of its C^0–norm.

For simplicity, we restrict to the case of the quasilinear equation (4.2.1) with boundary conditions which either are linear or are nonlinear but are of a certain linear structure, which is the case for later applications in this book. We first present the interior estimate in the form that will be used in later parts of this book. Then we give a proof of the $C^{2,\alpha}$–estimates for the *almost tangential derivative* problem. Since the proofs for the Dirichlet and oblique derivative problems are similar to that for the *almost tangential derivative* problem, we will only sketch these proofs.

Let $\Omega \subset \mathbb{R}^2$ be an open bounded set. Let $A_{ij}(\mathbf{p}, z, \mathbf{x})$, $A_i(\mathbf{p}, z, \mathbf{x})$, and $f(\mathbf{x})$ satisfy the following conditions: There exist constants $\lambda > 0$ and $\alpha \in (0,1)$ such that

$$\lambda|\boldsymbol{\mu}|^2 \leq \sum_{i,j=1}^{2} A_{ij}(Du(\mathbf{x}), u(\mathbf{x}), \mathbf{x})\mu_i\mu_j \leq \lambda^{-1}|\boldsymbol{\mu}|^2 \tag{4.2.3}$$

for any $\mathbf{x} \in \Omega$, $\boldsymbol{\mu} = (\mu_1, \mu_2) \in \mathbb{R}^2$, and

$$\|(A_{ij}, D_{(\mathbf{p},z)}A)(\mathbf{p}, z, \cdot)\|_{0,\alpha,\overline{\Omega}} \leq \lambda^{-1} \qquad \text{for all } (\mathbf{p}, z) \in \mathbb{R}^2 \times \mathbb{R}, \tag{4.2.4}$$

$$\|D_{(\mathbf{p},z)}A_{ij}\|_{0,0,\mathbb{R}^2\times\mathbb{R}\times\overline{\Omega}} \leq \lambda^{-1}, \tag{4.2.5}$$

$$A(\mathbf{0}, 0, \mathbf{x}) \equiv 0 \qquad \text{for all } \mathbf{x} \in \Omega. \tag{4.2.6}$$

Theorem 4.2.1. *Let $A_{ij}(\mathbf{p}, z, \mathbf{x})$, $A_i(\mathbf{p}, z, \mathbf{x})$, and $f(\mathbf{x})$ satisfy* (4.2.3)–(4.2.6) *in $\Omega = B_{2r}$ for $r \in (0,1]$. Then, for any solution $u \in C^{2,\alpha}(B_{2r})$ of* (4.2.1) *in B_{2r} satisfying*

$$\|u\|_{0,B_{2r}} + \|f\|_{0,\alpha,B_{2r}} \leq M, \tag{4.2.7}$$

there exists $C > 0$ depending only on (λ, M, α) such that

$$\|u\|_{2,\alpha,B_r} \leq \frac{C}{r^{2+\alpha}} \left(\|u\|_{0,B_{2r}} + r^2\|f\|_{0,\alpha,B_{2r}}\right). \tag{4.2.8}$$

Proof. We start by considering Case $r = 1$.

We first show the following estimate for $A(Du(\mathbf{x}), u(\mathbf{x}), \mathbf{x})$ (to be used later in the proof). For any $\beta \in [0, \alpha]$ and $\mathcal{D} \subset B_2$,

$$\|A(Du(\cdot), u(\cdot), \cdot)\|_{0,\beta,\overline{\mathcal{D}}} \leq C\|u\|_{1,\beta,\overline{\mathcal{D}}}. \tag{4.2.9}$$

Indeed, $|A(Du, u, \mathbf{x})| \leq C(|Du| + |u|)$ by (4.2.4) and (4.2.6). Furthermore, denoting $H(\mathbf{x}) := A(Du(\mathbf{x}), u(\mathbf{x}), \mathbf{x})$, we have

$$\begin{aligned}|H(\mathbf{x}) - H(\hat{\mathbf{x}})| &\leq |A(Du(\mathbf{x}), u(\mathbf{x}), \mathbf{x}) - A(Du(\hat{\mathbf{x}}), u(\hat{\mathbf{x}}), \mathbf{x})| \\ &\quad + |A(Du(\hat{\mathbf{x}}), u(\hat{\mathbf{x}}), \mathbf{x}) - A(Du(\hat{\mathbf{x}}), u(\hat{\mathbf{x}}), \hat{\mathbf{x}})| \\ &=: I_1 + I_2.\end{aligned}$$

Then

$$\begin{aligned}I_1 &\leq \left(\|D_{\mathbf{p}}A\|_{L^\infty(\mathbb{R}^2\times\mathbb{R}\times B_2)}[Du]_{0,\beta,\mathcal{D}} + \|D_zA\|_{0,\mathbb{R}^2\times\mathbb{R}\times B_2}[u]_{0,\beta,\mathcal{D}}\right)|\mathbf{x} - \hat{\mathbf{x}}|^\beta \\ &\leq C\left([Du]_{0,\beta,\mathcal{D}} + [u]_{0,\beta,\mathcal{D}}\right)|\mathbf{x} - \hat{\mathbf{x}}|^\beta,\end{aligned}$$

since $\|D_{(\mathbf{p},z)}A\|_{L^\infty(\mathbb{R}^2\times\mathbb{R}\times B_2)} \le \lambda^{-1}$ by (4.2.4). Moreover, using (4.2.6), we have

$$A(\mathbf{p},z,\mathbf{x}) = a_1p_1 + a_2p_2 + a_0z,$$

where

$$(a_1,a_2,a_0)(\mathbf{x}) = \int_0^1 (D_{p_1},D_{p_2},D_z)A(t\mathbf{p},tz,\mathbf{x})dt.$$

Thus, setting $\mathbf{p} = Du(\hat{\mathbf{x}})$ and $z = u(\hat{\mathbf{x}})$, we have

$$\begin{aligned} I_2 &\le \sup_{t\in[0,1]} \big([D_\mathbf{p}A(t\mathbf{p},tz,\cdot)]_{0,\beta,B_2}|Du(\hat{\mathbf{x}})| + [D_zA(t\mathbf{p},tz,\cdot)]_{0,\beta,B_2}|u(\hat{\mathbf{x}})|\big)|\mathbf{x}-\hat{\mathbf{x}}|^\beta \\ &\le C\big([Du]_{0,0,\mathcal{D}} + [u]_{0,0,\mathcal{D}}\big)|\mathbf{x}-\hat{\mathbf{x}}|^\beta, \end{aligned}$$

by (4.2.4). Now (4.2.9) is proved.

We consider (4.2.1) as a linear elliptic equation:

$$\sum_{i,j=1}^2 a_{ij}(\mathbf{x})D_{ij}u = f(\mathbf{x}) - A(Du(\mathbf{x}),u(\mathbf{x}),\mathbf{x}) \qquad \text{in } B_{3/2}$$

with coefficients $a_{ij}(\mathbf{x}) = A_{ij}(Du(\mathbf{x}),u(\mathbf{x}),\mathbf{x})$. The strict ellipticity and L^∞ bounds of a_{ij} follow from (4.2.3)–(4.2.4). We use the interior norms (4.1.4). By [131, Theorem 12.4], there exists $\beta\in(0,1)$ depending only on λ such that

$$\begin{aligned} [u]^*_{1,\beta,B_2} &\le C(\lambda)\big(\|u\|_{0,B_2} + \|f-A\|^{(2)}_{0,B_2}\big) \\ &\le C(\lambda)\big(\|u\|_{0,B_2} + \|Du\|^{(2)}_{0,B_2} + \|f\|^{(2)}_{0,B_2}\big), \end{aligned}$$

where $A = A(Du(\mathbf{x}),u(\mathbf{x}),\mathbf{x})$, and we have used (4.2.4) and (4.2.6) to obtain the last inequality. Then, applying the interpolation inequality (*cf.* [131, (6.82)]):

$$[u]^*_{1,0,B_2} \le \varepsilon[u]^*_{1,\beta,B_2} + C(\varepsilon,\beta)\|u\|_{0,B_2},$$

using the fact that $\|Du\|^{(2)}_{0,B_2} = [u]^{(1)}_{1,0,B_2} \le 2[u]^*_{1,0,B_2}$ and $\beta = \beta(\lambda)$, and choosing small $\varepsilon = \varepsilon(\lambda)$, we obtain

$$\|u\|^*_{1,\beta,B_2} \le C(\lambda)\big(\|u\|_{0,B_2} + \|f\|^{(2)}_{0,B_2}\big). \tag{4.2.10}$$

We can assume that $\beta\le\alpha$. We obtain $a_{ij}\in C^\beta(\overline{B_{3/2}})$ with

$$\|a_{ij}\|_{0,\alpha\beta,\overline{B_{3/2}}} \le C,$$

by (4.2.4)–(4.2.5), (4.2.7), and (4.2.10).

Then the local estimates for linear elliptic equations yield

$$\|u\|_{2,\beta,\overline{B_{5/4}}} \le C\big(\|u\|_{0,\overline{B_{3/2}}} + \|f\|_{0,\beta,\overline{B_{3/2}}} + \|A(Du,u,\mathbf{x})\|_{0,\beta,\overline{B_{3/2}}}\big).$$

From this, using (4.2.9) with $\mathcal{D} = B_{3/2}$ and (4.2.10) to estimate $\|u\|_{1,\beta,\overline{B_{3/2}}}$, we have

$$\|u\|_{2,\beta,\overline{B_{5/4}}} \le C\big(\|u\|_{0,\overline{B_{3/2}}} + \|f\|_{0,\beta,\overline{B_{3/2}}}\big).$$

With this estimate, we obtain that $\|a_{ij}\|_{0,\alpha,\overline{B_{5/4}}} \le C(\lambda, M)$. Then the local estimates for linear elliptic equations in $B_{5/4}$ yield (4.2.8) for $r=1$.

In the general case $r \in (0,1)$, we rescale by defining

$$v(\mathbf{x}) = \frac{1}{r}u(r\mathbf{x}).$$

Then v is a solution of the equation of form (4.2.1) in B_2 with the modified ingredients $(\hat{A}_{ij}, \hat{A}, \hat{B})$ and $\hat{f}$ defined by

$$\hat{A}_{ij}(\mathbf{p}, z, \mathbf{x}) = A_{ij}(\mathbf{p},\ rz,\ r\mathbf{x}),\ \ \hat{A}(\mathbf{p}, z, \mathbf{x}) = rA(\mathbf{p},\ rz,\ r\mathbf{x}),\ \ \hat{f}(\mathbf{x}) = rf(r\mathbf{x}).$$

It follows that $(\hat{A}_{ij}, \hat{A})$ satisfy (4.2.3)–(4.2.6) in B_2 with the unchanged constants (λ, M, α). Therefore, estimate (4.2.8) with $r=1$ holds for $(v, \hat{f})$. Writing it in terms of (u, f, h) and using that $r \le 1$, we conclude (4.2.8). □

Remark 4.2.2. *Note that ellipticity* (4.2.3) *of equation* (4.2.1) *in Theorem* 4.2.1 *and the following theorems are assumed only on solution u. We will use this for application to the potential flow equation* (2.2.8) *with* (2.2.9), *which is of mixed type.*

Remark 4.2.3. *From the proof, we also obtain the following estimate under the conditions of Theorem* 4.2.1: *There exist $\beta \in (0,1)$ and $\hat{C} > 0$ depending only on λ such that*

$$\|u\|_{1,\beta,B_r} \le \frac{\hat{C}}{r^{1+\beta}}\left(\|u\|_{0,B_{2r}} + r^2\|f\|_{0,\alpha,B_{2r}}\right). \tag{4.2.11}$$

Indeed, estimate (4.2.10) *implies* (4.2.11) *for $r=1$. Then the scaling performed at the end of the proof of Theorem* 4.2.1 *yields* (4.2.11) *for any $r \in (0,1]$.*

Now we consider the boundary value problems. We start with the *almost tangential derivative* problem. The first result is the $C^{1,\alpha}$–estimates up to the boundary. For these estimates, the C^1–boundary suffices.

For $\Phi \in C(\mathbb{R})$ and $R > 0$, we set

$$\Omega_R := B_R \cap \{x_2 > \varepsilon\Phi(x_1)\}, \qquad \Gamma_R := B_R \cap \{x_2 = \varepsilon\Phi(x_1)\}. \tag{4.2.12}$$

Theorem 4.2.4. *Let $\lambda \in (0,1)$ and $r \in (0,1]$. Let $\Phi \in C^1(\mathbb{R})$ satisfy*

$$\|\Phi\|_{1,\mathbb{R}} \le \lambda^{-1}, \qquad \Phi(0) = 0. \tag{4.2.13}$$

Let $\Omega = \Omega_{2r}$ be of structure (4.2.12). *Let $A_{ij}(\mathbf{p}, z, \mathbf{x})$ and $A(\mathbf{p}, z, \mathbf{x})$ satisfy* (4.2.3), (4.2.6), *and*

$$\|(A_{ij}, D_{(\mathbf{p},z)}A)\|_{0,0,\mathbb{R}^2\times\mathbb{R}\times\Omega} \le \lambda^{-1}. \tag{4.2.14}$$

Let $B(p_2, z, \mathbf{x})$, $f(\mathbf{x})$, *and* $h(\mathbf{x})$ *satisfy*

$$B(0,0,\mathbf{x}) \equiv 0 \qquad \textit{for all } \mathbf{x} \in \Gamma_{2r}, \tag{4.2.15}$$

$$|\partial_{p_2} B(p_2, z, \mathbf{x})| \le \varepsilon \qquad \textit{for all } (p_2, z) \in \mathbb{R}^2,\ \mathbf{x} \in \Gamma_{2r}, \tag{4.2.16}$$

$$\|D_{(p_2,z)} B(p_2, z, \cdot)\|_{1,\Gamma_{2r}} \le \lambda^{-1} \qquad \textit{for all } (p_2, z) \in \mathbb{R} \times \mathbb{R}, \tag{4.2.17}$$

$$f \in L^\infty(\Omega_{2r}), \quad h \in C^1(\Gamma_{2r}). \tag{4.2.18}$$

Then there exist $\varepsilon, \beta \in (0,1)$ *and* $C > 0$ *depending only on* λ *such that, for* $u \in C^2(\Omega_{2r}) \cap C^{1,\beta}(\Omega_{2r} \cup \Gamma_{2r})$ *satisfying* (4.2.1) *in* Ω_{2r} *and*

$$u_{x_1} = B(u_{x_2}, u, \mathbf{x}) + h(\mathbf{x}) \qquad \textit{on } \Gamma_{2r}, \tag{4.2.19}$$

we have

$$\|u\|_{1,\beta,\Omega_{9r/5}} \le \frac{C}{r^{1+\beta}} \left(\|u\|_{0,\Omega_{2r}} + r^2 \|f\|_{0,\Omega_{2r}} + r\|h\|_{1,0,\Gamma_{2r}} \right). \tag{4.2.20}$$

Proof. From (4.2.13), if ε is small, depending only on λ, then

$$B_R(\mathbf{x}_0) \cap \{x_2 > \varepsilon\Phi(x_1)\} \quad \text{is connected for any } \mathbf{x}_0 = (x_1^0, \Phi(x_1^0)). \tag{4.2.21}$$

We assume that (4.2.21) holds from now on. In particular, Ω_R is connected for any R. We divide the proof into six steps.

1. We first prove Theorem 4.2.4 for Case $r = 1$.

We extend $B(p_2, z, \mathbf{x})$ from $\mathbb{R} \times \mathbb{R} \times \Gamma_2$ to $\mathbb{R} \times \mathbb{R} \times \Omega_2$, and h from Γ_2 to Ω_2, so that (4.2.15)–(4.2.18) are satisfied for the extended functions in their domains, where the bound in (4.2.17) may change, but still depends on λ. Specifically, if $\mathbf{x} = (x_1, x_2) \in \Omega_2 \cap B_{2-2\varepsilon}$, and $\varepsilon\lambda^{-2}$ is small, then $(x_1, \varepsilon\Phi(x_1)) \in B_2$, which implies that $(x_1, \varepsilon\Phi(x_1)) \in \Gamma_2$. It follows that, by setting

$$B(p_2, z, \mathbf{x}) := B(p_2, z, x_1, 0), \quad h(\mathbf{x}) := h(x_1, 0) \tag{4.2.22}$$

for any $(p_2, z) \in \mathbb{R} \times \mathbb{R}$ and $\mathbf{x} = (x_1, x_2) \in \Omega_2 \cap B_{2-2\varepsilon}$, we obtain B and h extended to $\Omega_2 \cap B_{2-2\varepsilon}$ so that they satisfy all the required properties. Then, for simplicity of notation, we assume without loss of generality that $B(p_2, z, \mathbf{x})$ and $h(\mathbf{x})$ are defined in $\mathbb{R} \times \mathbb{R} \times \Omega_2$ and Ω_2, respectively, and satisfy

$$B(0,0,\mathbf{x}) \equiv 0 \qquad \text{for all } \mathbf{x} \in \Omega_2, \tag{4.2.23}$$

$$|\partial_{p_2} B(p_2, z, \mathbf{x})| \le \varepsilon \qquad \text{for all } (p_2, z) \in \mathbb{R}^2,\ \mathbf{x} \in \Omega_2, \tag{4.2.24}$$

$$\|D_{(p_2,z)} B(p_2, z, \cdot)\|_{1,\Omega_2} \le \lambda^{-1} \qquad \text{for all } (p_2, z) \in \mathbb{R} \times \mathbb{R}, \tag{4.2.25}$$

$$h \in C^1(\overline{\Omega_2}). \tag{4.2.26}$$

The universal constant C in the argument below depends only on λ, unless otherwise specified.

2. As in [131, §13.2], we define $w_i := D_i u$ for $i = 1, 2$. Then we conclude from equation (4.2.1) that w_1 and w_2 are weak solutions of the following equations of divergence form:

$$D_1\Big(\frac{A_{11}}{A_{22}}D_1 w_1 + \frac{A_{12}+A_{21}}{A_{22}}D_2 w_1\Big) + D_{22}w_1 = D_1\Big(\frac{f}{A_{22}} - \frac{A}{A_{22}}\Big), \tag{4.2.27}$$

and

$$D_{11}w_2 + D_2\Big(\frac{A_{12}+A_{21}}{A_{11}}D_1 w_2 + \frac{A_{22}}{A_{11}}D_2 w_2\Big) = D_2\Big(\frac{f}{A_{11}} - \frac{A}{A_{11}}\Big), \tag{4.2.28}$$

where $(A_{ij}, A) = (A_{ij}, A)(Du, u, \mathbf{x})$.

Condition (4.2.3) implies that equations (4.2.27)–(4.2.28), considered as linear equations for w_1 and w_2 respectively, are elliptic with ellipticity constant λ. Also, (4.2.3), (4.2.6), and (4.2.14) imply that, in Ω_2,

$$A_{22}(Du, u, \mathbf{x}) \geq \lambda, \qquad |A_{ij}(Du, u, \mathbf{x})| \leq C, \tag{4.2.29}$$

$$\Big|\Big(\frac{f-A}{A_{ii}}\Big)(Du, u, \mathbf{x})\Big| \leq C\big(|f| + |u| + |Du|\big). \tag{4.2.30}$$

From (4.2.19), we have

$$w_1 = g \qquad \text{on } \Gamma_2, \tag{4.2.31}$$

where

$$g(\mathbf{x}) := B(D_2 u(\mathbf{x}), u(\mathbf{x}), \mathbf{x}) + h(\mathbf{x}) \qquad \text{in } \Omega_2. \tag{4.2.32}$$

We first obtain the following Hölder estimates of $D_1 u$, for which $d_{\mathbf{x}} := \operatorname{dist}(\mathbf{x}, \partial\Omega_2 \setminus \Gamma_2)$:

Lemma 4.2.5. *There exist $\beta \in (0, \alpha]$ and $C > 0$ depending only on λ such that, for any $\mathbf{x}_0 \in \Omega_2 \cup \Gamma_2$,*

$$\begin{aligned} &d^{\beta}_{\mathbf{x}_0}[w_1]_{0,\beta,B_{\frac{d_{\mathbf{x}_0}}{16}}(\mathbf{x}_0)\cap\Omega_2} \\ &\leq C\Big(\|(Du, f)\|_{0,0,B_{\frac{d_{\mathbf{x}_0}}{2}}(\mathbf{x}_0)\cap\Omega_2} + d^{\beta}_{\mathbf{x}_0}[g]_{0,\beta,B_{\frac{d_{\mathbf{x}_0}}{2}}(\mathbf{x}_0)\cap\Omega_2}\Big), \end{aligned} \tag{4.2.33}$$

where g is defined in (4.2.32).

Proof. For $\mathbf{x}_0 \in \{x_2 = \varepsilon\Phi(x_1)\}$ and $R > 0$, denote

$$\Omega_R(\mathbf{x}_0) = B_R(\mathbf{x}_0) \cap \{x_2 > \varepsilon\Phi(x_1)\}. \tag{4.2.34}$$

We first prove that, for $\hat{\mathbf{x}} \in \Gamma_2$ such that $\Omega_{2R}(\hat{\mathbf{x}}) \subset \Omega_2$,

$$R^{\beta}[w_1]_{0,\beta,\Omega_R(\hat{\mathbf{x}})} \leq C\big(\|(Du, Rf)\|_{0,0,\Omega_{2R}(\hat{\mathbf{x}})} + R^{\beta}[g]_{0,\beta,\Omega_{2R}(\hat{\mathbf{x}})}\big). \tag{4.2.35}$$

We rescale u, w_1, and f from $\Omega_{2R}(\hat{\mathbf{x}})$ into $\hat{\Omega}_1$, where

$$\hat{\Omega}_\rho = B_\rho \cap \{X_2 > \varepsilon\hat{\Phi}(X_1)\}, \qquad \hat{\Gamma}_\rho = B_\rho \cap \{X_2 = \varepsilon\hat{\Phi}(X_1)\}$$

with $\hat{\Phi}(X_1) = \frac{1}{2R}\big(\Phi(x_1 + 2RX_1) - \Phi(x_1)\big)$.

Note that

$$\|\hat{\Phi}'\|_{L^\infty(\mathbb{R})} = \|\Phi'\|_{L^\infty(\mathbb{R})} \le \lambda^{-1}, \qquad \hat{\Phi}(0) = 0.$$

For $\mathbf{X} \in \hat{\Omega}_1$, define

$$\begin{aligned} &\hat{u}(\mathbf{X}) = \frac{1}{2R}u(\hat{\mathbf{x}} + 2R\mathbf{X}), \qquad \hat{f}(\mathbf{X}) = 2Rf(\hat{\mathbf{x}} + 2R\mathbf{X}), \\ &(\hat{A}_{ij}, \hat{A})(\mathbf{X}) = (A_{ij}, 2RA)(D\hat{u}(\mathbf{X}), 2R\hat{u}(\mathbf{X}), \hat{\mathbf{x}} + 2R\mathbf{X}), \\ &\hat{w}_i = D_{X_i}\hat{u}. \end{aligned} \tag{4.2.36}$$

Then $\hat{w}_1$ satisfies an equation of form (4.2.27) in $\hat{\Omega}_1$ with (A_{ij}, A, f) replaced by $(\hat{A}_{ij}(\mathbf{X}), \hat{A}(\mathbf{X}), \hat{f})$. Note that $\hat{A}_{ij}(\mathbf{X}), i, j = 1, 2$, satisfy (4.2.3) in $\hat{\Omega}_1$, and (4.2.29)–(4.2.30) imply

$$\left|\frac{\hat{f} - \hat{A}}{\hat{A}_{22}}\right| \le C\big(|\hat{f}| + 2R|\hat{u}| + |D\hat{u}|\big) \qquad \text{in } \hat{\Omega}_1$$

with unchanged constants λ and C. By the elliptic version of [196, Theorem 6.33] stated in the parabolic setting (it can also be obtained by using [196, Lemma 4.6] instead of [131, Lemma 8.23] in the proofs of [131, Theorems 8.27 and 8.29] to achieve $\alpha = \alpha_0$ in [131, Theorem 8.29]), we find $\tilde{\beta} \in (0, 1)$ depending only on λ, and $C = C(\lambda, \mathrm{Lip}[\varepsilon\hat{\Phi}])$ such that, for $\beta = \min\{\tilde{\beta}, \alpha\}$,

$$[\hat{w}_1]_{0,\beta,\hat{\Omega}_{1/2}} \le C\big(\|(R\hat{u}, D\hat{u}, \hat{f})\|_{0,0,\hat{\Omega}_1} + [\hat{w}_1]_{0,\beta,\hat{\Gamma}_1}\big).$$

Since $\mathrm{Lip}[\varepsilon\hat{\Phi}] \le \varepsilon\lambda^{-1}$ and $\varepsilon < 1$, it follows that $C = C(\lambda)$.

Rescaling back and using (4.2.31), we obtain (4.2.35).

If $\hat{\mathbf{x}} \in \Omega_2$ and $B_{2R}(\hat{\mathbf{x}}) \subset \Omega_2$, an argument similar to the proof of (4.2.35) by using the interior estimates [131, Theorem 8.24] yields

$$R^\beta[w_1]_{0,\beta,B_R(\hat{\mathbf{x}})} \le C\|(u, Du, Rf)\|_{0,0,B_{2R}(\hat{\mathbf{x}})}. \tag{4.2.37}$$

Now let $\mathbf{x}_0 \in \Omega_2$. If $B_{\frac{d_{\mathbf{x}_0}}{8}}(\mathbf{x}_0) \subset \Omega_2$, we apply (4.2.37) with $\hat{\mathbf{x}} = \mathbf{x}_0$ and $R = \frac{d_{\mathbf{x}_0}}{16} \le 1$ to obtain (4.2.33). Otherwise, there exists $\mathbf{x}_0' \in \overline{B_{\frac{d_{\mathbf{x}_0}}{8}}(\mathbf{x}_0)} \cap \Gamma_2$ such that

$$B_{\frac{d_{\mathbf{x}_0}}{16}}(\mathbf{x}_0) \cap \Omega_2 \subset \Omega_{\frac{d_{\mathbf{x}_0}}{2}}(\mathbf{x}_0') \subset \Omega_2.$$

Then, applying (4.2.35) with $\hat{\mathbf{x}} = \mathbf{x}_0'$ and $R = \frac{d_{\mathbf{x}_0}}{2} \le 1$ and using the inclusions stated above, we obtain (4.2.33). □

3. Next, we make the Hölder estimates for Du. We first note that g defined by (4.2.32) satisfies

$$|g| \le C\big(\varepsilon|Du| + |u|\big) + |h| \qquad \text{in } \Omega_2, \tag{4.2.38}$$

$$|Dg| \le C\big(\varepsilon|D^2u| + |Du| + |u|\big) + |Dh| \qquad \text{in } \Omega_2, \tag{4.2.39}$$

$$[g]_{0,\beta,\mathcal{D}} \le C\big(\varepsilon[Du]_{0,\beta,\mathcal{D}} + \|u\|_{1,0,\mathcal{D}}\big) + [h]_{0,\beta,\mathcal{D}} \qquad \text{for any } \mathcal{D} \subset \Omega_2. \tag{4.2.40}$$

Indeed, (4.2.38) follows directly from (4.2.23)–(4.2.25).

To show (4.2.39), we differentiate (4.2.32) with respect to x_i to obtain

$$D_{x_i}g = B_{p_2}D_{i2}u + B_zD_iu + B_{x_i} + h_{x_i}, \qquad i = 1, 2, \tag{4.2.41}$$

where $(B, DB) = (B, DB)(Du, u, \mathbf{x})$. From (4.2.24) and the bound:

$$\|B_z\|_{0,\mathbb{R}^2\times\mathbb{R}\times\Omega_2} \le \lambda^{-1},$$

which follows from (4.2.25), we see that the first two terms on the right-hand side of (4.2.41) are estimated by the right-hand side of (4.2.39). Thus, it only remains to estimate $B_{x_i}(Du, u, \mathbf{x})$. We first note that $B_{x_i}(0, 0, \mathbf{x}) \equiv 0$ by (4.2.23). Also, (4.2.25) implies that $\|D_{\mathbf{x}}(B_{p_2}, B_z)\|_{0,\mathbb{R}^2\times\mathbb{R}\times\Omega_2} \le \lambda^{-1}$. Then $|D_{\mathbf{x}}B(Du, u, \mathbf{x})| \le C\big(|Du| + |u|\big)$, which completes the proof of (4.2.39).

To show (4.2.40), we denote

$$g_1(\mathbf{x}) := B(D_2u, u, \mathbf{x}),$$

and estimate $[g_1]_{0,\beta,\mathcal{D}}$. For $\mathbf{x}, \hat{\mathbf{x}} \in \mathcal{D}$, we have

$$\begin{aligned} |g_1(\mathbf{x}) - g_1(\hat{\mathbf{x}})| &\le |B(D_2u(\mathbf{x}), u(\mathbf{x}), \mathbf{x}) - B(D_2u(\hat{\mathbf{x}}), u(\hat{\mathbf{x}}), \mathbf{x})| \\ &\qquad + |B(D_2u(\hat{\mathbf{x}}), u(\hat{\mathbf{x}}), \mathbf{x}) - B(D_2u(\hat{\mathbf{x}}), u(\hat{\mathbf{x}}), \hat{\mathbf{x}})| \\ &=: I_1 + I_2. \end{aligned}$$

Then

$$\begin{aligned} I_1 &\le \big(\|B_{p_2}\|_{0,\mathbb{R}^2\times\mathbb{R}\times\Omega_2}[Du]_{0,\beta,\mathcal{D}} + \|B_z\|_{0,\mathbb{R}^2\times\mathbb{R}\times\Omega_2}[u]_{0,\beta,\mathcal{D}}\big)\,|\mathbf{x} - \hat{\mathbf{x}}|^\beta \\ &\le \big(\varepsilon[Du]_{0,\beta,\mathcal{D}} + C[u]_{0,\beta,\mathcal{D}}\big)|\mathbf{x} - \hat{\mathbf{x}}|^\beta, \end{aligned}$$

by (4.2.24)–(4.2.25). Furthermore, using (4.2.23), we have

$$B(p_2, z, \mathbf{x}) = a_1p_2 + a_2z,$$

where $(a_1, a_2) = \int_0^1 D_{(p_2,z)}B(tp_2, tz, \mathbf{x})dt$. Thus, setting $p_2 = D_2u(\hat{\mathbf{x}})$ and $z = u(\hat{\mathbf{x}})$, we have

$$\begin{aligned} I_2 &\le \sup_{t\in[0,1]} \big([B_{p_2}(tp_2, tz, \cdot)]_{0,\beta,\Omega_2}|D_2u(\hat{\mathbf{x}})| + [B_z(tp_2, tz, \cdot)]_{0,\beta,\Omega_2}|u(\hat{\mathbf{x}})|\big)|\mathbf{x} - \hat{\mathbf{x}}|^\beta \\ &\le C\big([Du]_{0,0,\mathcal{D}} + [u]_{0,0,\mathcal{D}}\big)|\mathbf{x} - \hat{\mathbf{x}}|^\beta, \end{aligned}$$

by (4.2.25). Now (4.2.40) is proved.

Lemma 4.2.6. *Let β be as in Lemma 4.2.5. Then there exist $\varepsilon_0(\lambda) > 0$ and $C(\lambda) > 0$ such that, if $0 \le \varepsilon \le \varepsilon_0$,*

$$d^{\beta}_{\mathbf{x}_0}[Du]_{0,\beta,B_{\frac{d_{\mathbf{x}_0}}{32}}(\mathbf{x}_0)\cap\Omega_2} \le C\Big(\|u\|_{1,0,B_{\frac{d_{\mathbf{x}_0}}{2}}(\mathbf{x}_0)\cap\Omega_2} + \varepsilon d^{\beta}_{\mathbf{x}_0}[Du]_{0,\beta,B_{\frac{d_{\mathbf{x}_0}}{2}}(\mathbf{x}_0)\cap\Omega_2} + \|f\|_{0,0,\Omega_2} + \|h\|_{1,0,\Omega_2}\Big) \tag{4.2.42}$$

for any $\mathbf{x}_0 \in \Omega_2 \cup \Gamma_2$.

Proof. The Hölder norm of $D_1 u$ has been estimated in Lemma 4.2.5. It remains to estimate the Hölder norm of $D_2 u$. We follow the proof of [131, Theorem 13.1].

Fix $\mathbf{x}_0 \in \Omega_2 \cup \Gamma_2$. In order to prove (4.2.42), it suffices to show that, for every $\hat{\mathbf{x}} \in B_{\frac{d_{\mathbf{x}_0}}{32}}(\mathbf{x}_0) \cap \Omega_2$ and every $R > 0$ such that $B_R(\hat{\mathbf{x}}) \subset B_{\frac{d_{\mathbf{x}_0}}{16}}(\mathbf{x}_0)$, we have

$$\int_{B_R(\hat{\mathbf{x}})\cap\Omega_2} |D^2 u|^2 d\mathbf{x} \le \frac{L^2}{d^{2\beta}_{\mathbf{x}_0}} R^{2\beta}, \tag{4.2.43}$$

where L is the right-hand side of (4.2.42) (*cf.* [131, Theorem 7.19] and [196, Lemma 4.11]).

In order to prove (4.2.43), we consider two separate cases: (i) $B_{2R}(\hat{\mathbf{x}})\cap\Gamma_2 \neq \emptyset$ and (ii) $B_{2R}(\hat{\mathbf{x}}) \cap \Gamma_2 = \emptyset$.

We first consider Case (i). Let $B_{2R}(\hat{\mathbf{x}}) \cap \Gamma_2 \neq \emptyset$. Since $B_R(\hat{\mathbf{x}}) \subset B_{\frac{d_{\mathbf{x}_0}}{16}}(\mathbf{x}_0)$, then

$$R \le \frac{d_{\mathbf{x}_0}}{16}. \tag{4.2.44}$$

Let $\eta \in C^1_0(B_{2R}(\hat{\mathbf{x}}))$ and $\zeta = (w_1 - g)\eta^2$. Note that $\zeta \in W^{1,2}_0(B_{2R}(\hat{\mathbf{x}}) \cap \Omega_2)$ by (4.2.31). We use ζ as a test function in the weak form of (4.2.27):

$$\int_{\Omega_2} \big(\sum_{i,j=1}^{2} a_{ij} D_i w_1 D_j \zeta\big)\, d\mathbf{x} = \int_{\Omega_2} \frac{1}{A_{22}}(f - A) D_1\zeta\, d\mathbf{x}, \tag{4.2.45}$$

where

$$a_{ii} = \frac{A_{ii}}{A_{22}}, \quad a_{12} = 0, \quad a_{21} = \frac{A_{12} + A_{21}}{A_{22}},$$

and $(A_{ij}, A) = (A_{ij}, A)(Du(\mathbf{x}), u(\mathbf{x}), \mathbf{x})$. We apply (4.2.3), (4.2.29)–(4.2.30), and (4.2.39) to obtain

$$\begin{aligned} &\int_{\Omega_2} |Dw_1|^2 \eta^2\, d\mathbf{x} \\ &\le C\int_{\Omega_2} \Big(\big((\delta+\varepsilon)|Dw_1|^2 + \varepsilon|D^2u|^2\big)\eta^2 \\ &\qquad + (\frac{1}{\delta} + 1)\big((|D\eta|^2 + |f|\eta^2)(w_1 - g)^2 + |(Du, u, f, Dh)|^2\eta^2\big)\Big)\, d\mathbf{x}, \end{aligned} \tag{4.2.46}$$

where C depends only on λ, and the sufficiently small constant $\delta > 0$ will be chosen below. Since

$$|Dw_1|^2 = (D_{11}u)^2 + (D_{12}u)^2, \tag{4.2.47}$$

it remains to estimate $|D_{22}u|^2$. Using the ellipticity property (4.2.3) and (4.2.29), we can express $D_{22}u$ from equation (4.2.1) and use (4.2.30) to obtain

$$|D_{22}u|^2 \leq C(\lambda)|(D_{11}u, D_{12}u, Du, u, f)|^2.$$

Combining this with (4.2.46)–(4.2.47), we have

$$\begin{aligned}
&\int_{\Omega_2} |D^2u|^2\eta^2\, d\mathbf{x}\\
&\leq C\int_{\Omega_2}\Big((\varepsilon+\delta)|D^2u|^2\eta^2\\
&\qquad + (\frac{1}{\delta}+1)\left(|(D\eta,\eta)|^2(w_1-g)^2 + |(Du,u,f,Dh)|^2\eta^2\right)\Big)\, d\mathbf{x}.
\end{aligned}$$

Choose $\varepsilon_0 = \delta = \frac{1}{4C}$, where C is from the last inequality. Then, when $\varepsilon \in (0, \varepsilon_0)$, we have

$$\int_{\Omega_2} |D^2u|^2\eta^2\, d\mathbf{x} \leq C\int_{\Omega_2}\left(|(D\eta,\eta^2)|^2(w_1-g)^2 + |(Du,u,f,Dh)|^2\eta^2\right)\, d\mathbf{x}. \tag{4.2.48}$$

Now we make a more specific choice of η: In addition to $\eta \in C_0^1(B_{2R}(\hat{\mathbf{x}}))$, we assume that $\eta \equiv 1$ on $B_R(\hat{\mathbf{x}})$, $0 \leq \eta \leq 1$ on $\mathbb{R}^2$, and $|D\eta| \leq \frac{10}{R}$. Also, since $B_{2R}(\hat{\mathbf{x}}) \cap \Gamma_2 \neq \emptyset$, we fix some $\mathbf{x}^* \in B_{2R}(\hat{\mathbf{x}}) \cap \Gamma_2$. Then $|\mathbf{x}-\mathbf{x}^*| \leq 4R$ for any $\mathbf{x} \in B_{2R}(\hat{\mathbf{x}})$. Moreover, $(w_1-g)(\mathbf{x}^*) = 0$ by (4.2.31). Since $B_{2R}(\hat{\mathbf{x}}) \subset B_{\frac{d_{\mathbf{x}_0}}{16}}(\mathbf{x}_0)$, we find from (4.2.33), (4.2.44), and (4.2.40) with $\mathcal{D} = B_{\frac{d_{\mathbf{x}_0}}{2}}(\mathbf{x}_0) \cap \Omega_2$ that, for any $\mathbf{x} \in B_{2R}(\hat{\mathbf{x}}) \cap \Omega_2$,

$$\begin{aligned}
|(w_1-g)(\mathbf{x})| &= |(w_1-g)(\mathbf{x}) - (w_1-g)(\mathbf{x}^*)|\\
&\leq |w_1(\mathbf{x}) - w_1(\mathbf{x}^*)| + |g(\mathbf{x}) - g(\mathbf{x}^*)|\\
&\leq \frac{C}{d_{\mathbf{x}_0}^{\beta}}\left(\|(Du,f)\|_{0,0,B_{\frac{d_{\mathbf{x}_0}}{2}}(\mathbf{x}_0)\cap\Omega_2} + d_{\mathbf{x}_0}^{\beta}[g]_{0,\beta,B_{\frac{d_{\mathbf{x}_0}}{2}}(\mathbf{x}_0)\cap\Omega_2}\right)|\mathbf{x}-\mathbf{x}^*|^{\beta}\\
&\quad + [g]_{0,\beta,B_{\frac{d_{\mathbf{x}_0}}{2}}(\mathbf{x}_0)\cap\Omega_2}|\mathbf{x}-\mathbf{x}^*|^{\beta}\\
&\leq C\Big(\frac{1}{d_{\mathbf{x}_0}^{\beta}}\|(Du,f)\|_{0,0,B_{\frac{d_{\mathbf{x}_0}}{2}}(\mathbf{x}_0)\cap\Omega_2} + \varepsilon[Du]_{0,\beta,B_{\frac{d_{\mathbf{x}_0}}{2}}(\mathbf{x}_0)\cap\Omega_2}\\
&\qquad + \|u\|_{0,0,B_{\frac{d_{\mathbf{x}_0}}{2}}(\mathbf{x}_0)\cap\Omega_2} + [h]_{0,\beta,B_{\frac{d_{\mathbf{x}_0}}{2}}(\mathbf{x}_0)\cap\Omega_2}\Big)R^{\beta}.
\end{aligned}$$

Using this estimate and our choice of η, we obtain from (4.2.48) that

$$\begin{aligned}
&\int_{B_R(\hat{\mathbf{x}})\cap\Omega_2} |D^2u|^2 d\mathbf{x}\\
&\le C\Big(\frac{1}{d_{\mathbf{x}_0}^{2\beta}}\|(Du,f)\|^2_{0,0,B_{\frac{d_{\mathbf{x}_0}}{2}}(\mathbf{x}_0)\cap\Omega_2} + \|h\|^2_{0,\beta,\Omega_2} + \varepsilon^2[Du]^2_{0,\beta,B_{\frac{d_{\mathbf{x}_0}}{2}}(\mathbf{x}_0)\cap\Omega_2}\Big)R^{2\beta}\\
&\quad + C\|u\|^2_{1,0,B_{\frac{d_{\mathbf{x}_0}}{2}}(\mathbf{x}_0)\cap\Omega_2}(R^{2\beta}+R^2) + C\|Dh\|^2_{0,0,B_{\frac{d_{\mathbf{x}_0}}{2}}\cap\Omega_2}R^2,
\end{aligned}$$

which implies (4.2.43) for Case (i).

Now we consider Case (ii): $\hat{\mathbf{x}} \in \Omega_2$ and $R > 0$ satisfy $B_R(\hat{\mathbf{x}}) \subset B_{\frac{d_{\mathbf{x}_0}}{32}}(\mathbf{x}_0)$ and $B_{2R}(\hat{\mathbf{x}}) \cap \Gamma_2 = \emptyset$. Then $B_{2R}(\hat{\mathbf{x}}) \subset B_{\frac{d_{\mathbf{x}_0}}{16}}(\mathbf{x}_0) \cap \Omega_2$. Let $\eta \in C_0^1(B_{2R}(\hat{\mathbf{x}}))$ and $\zeta = \eta^2(w_1 - w_1(\hat{\mathbf{x}}))$. Note that $\zeta \in W_0^{1,2}(\Omega_2)$, since $B_{2R}(\hat{\mathbf{x}}) \subset \Omega_2$. Thus, we can use ζ as a test function in (4.2.45). Performing estimates similar to those that we have made for obtaining (4.2.48), we have

$$\int_{\Omega_2} |D^2u|^2\eta^2 d\mathbf{x} \le C(\lambda)\int_{\Omega_2}\left(|(D\eta,\eta)|^2(w_1 - w_1(\hat{\mathbf{x}}))^2 + |(Du,u,f)|^2\eta^2\right)d\mathbf{x}. \tag{4.2.49}$$

Choose $\eta \in C_0^1(B_{2R}(\hat{\mathbf{x}}))$ so that $\eta \equiv 1$ on $B_R(\hat{\mathbf{x}})$, $0 \le \eta \le 1$ on $\mathbb{R}^2$, and $|D\eta| \le \frac{10}{R}$. Note that, for any $\mathbf{x} \in B_{2R}(\hat{\mathbf{x}})$,

$$|w_1(\mathbf{x}) - w_1(\hat{\mathbf{x}})| \le C\Big(\frac{1}{d_{\mathbf{x}_0}^{\beta}}\|(Du,f)\|_{0,0,B_{\frac{d_{\mathbf{x}_0}}{2}}(\mathbf{x}_0)\cap\Omega_2} + \varepsilon[Du]_{0,\beta,B_{\frac{d_{\mathbf{x}_0}}{2}}(\mathbf{x}_0)\cap\Omega_2}\Big)R^\beta$$

by (4.2.33), since $B_{2R}(\hat{\mathbf{x}}) \subset B_{\frac{d_{\mathbf{x}_0}}{16}}(\mathbf{x}_0)\cap\Omega_2$. Now we obtain (4.2.43) from (4.2.49) in a way similar to that for Case (i). Then Lemma 4.2.6 is proved. □

4. Now we make the $C^{1,\beta}$–estimate of u. In the next lemma, we use the partially interior norms (4.1.3) in $\Omega_2 \cup \Gamma_2$. Note that the distance function related to the norms is $d_{\mathbf{x}} = \operatorname{dist}(\mathbf{x}, \partial\Omega_2 \setminus \Gamma_2)$, *i.e.*, the function used in Lemmas 4.2.5–4.2.6.

Lemma 4.2.7. *Let β and ε_0 be as in Lemma* 4.2.6. *Then, for $\varepsilon \in (0, \varepsilon_0)$, there exists $C(\lambda)$ such that*

$$[u]^*_{1,\beta,\Omega_2\cup\Gamma_2} \le C\left(\|u\|^*_{1,0,\Omega_2\cup\Gamma_2} + \varepsilon[u]^*_{1,\beta,\Omega_2\cup\Gamma_2} + \|f\|_{0,0,\Omega_2} + \|h\|_{1,0,\Omega_2}\right). \tag{4.2.50}$$

This can be seen as follows: Estimate (4.2.50) follows directly from Lemma 4.2.6 and an argument similar to the proof of [131, Theorem 4.8]. Let $\hat{\mathbf{x}}, \tilde{\mathbf{x}} \in \Omega_2$ with $d_{\hat{\mathbf{x}}} \le d_{\tilde{\mathbf{x}}}$ (so that $d_{\hat{\mathbf{x}},\tilde{\mathbf{x}}} = d_{\hat{\mathbf{x}}}$) and let $|\hat{\mathbf{x}} - \tilde{\mathbf{x}}| \le \frac{d_{\hat{\mathbf{x}}}}{64}$. Then $\tilde{\mathbf{x}} \in B_{\frac{d_{\mathbf{x}_0}}{32}}(\mathbf{x}_0)\cap\Omega_2$

and, by Lemma 4.2.6 applied to $\mathbf{x}_0 = \hat{\mathbf{x}}$, we have

$$
\begin{aligned}
&d^{1+\beta}_{\hat{\mathbf{x}},\tilde{\mathbf{x}}} \frac{|Du(\hat{\mathbf{x}}) - Du(\tilde{\mathbf{x}})|}{|\hat{\mathbf{x}} - \tilde{\mathbf{x}}|^\beta} \\
&\le C\Big(d_{\hat{\mathbf{x}}}\|u\|_{1,0,B_{\frac{d_{\hat{\mathbf{x}}}}{2}}(\hat{\mathbf{x}})\cap\Omega_2} + \varepsilon d^{1+\beta}_{\hat{\mathbf{x}}}[Du]_{0,\beta,B_{\frac{d_{\hat{\mathbf{x}}}}{2}}(\hat{\mathbf{x}})\cap\Omega_2} \\
&\qquad + \|f\|_{0,0,\Omega_2} + \|h\|_{1,0,\Omega_2}\Big) \\
&\le C\big(\|u\|^*_{1,0,\Omega_2\cup\Gamma_2} + \varepsilon[u]^*_{1,\beta,\Omega_2\cup\Gamma_2} + \|f\|_{0,0,\Omega_2} + \|h\|_{1,0,\Omega_2}\big),
\end{aligned}
$$

where the last inequality holds since $2d_{\mathbf{x}} \ge d_{\hat{\mathbf{x}}}$ for any $\mathbf{x} \in B_{\frac{d_{\hat{\mathbf{x}}}}{2}}(\hat{\mathbf{x}}) \cap \Omega_2$. If $\hat{\mathbf{x}}, \tilde{\mathbf{x}} \in \Omega_2$ with $d_{\hat{\mathbf{x}}} \le d_{\tilde{\mathbf{x}}}$ and $|\hat{\mathbf{x}} - \tilde{\mathbf{x}}| \ge \frac{d_{\hat{\mathbf{x}}}}{64}$, then

$$
\begin{aligned}
d^{1+\beta}_{\hat{\mathbf{x}},\tilde{\mathbf{x}}} \frac{|Du(\hat{\mathbf{x}}) - Du(\tilde{\mathbf{x}})|}{|\hat{\mathbf{x}} - \tilde{\mathbf{x}}|^\beta} &\le 64\big(d_{\hat{\mathbf{x}}}|Du(\hat{\mathbf{x}})| + d_{\tilde{\mathbf{x}}}|Du(\tilde{\mathbf{x}})|\big) \\
&\le 64\,\|u\|^*_{1,0,\Omega_2\cup\Gamma_2}.
\end{aligned}
$$

5. Now we can complete the proof of Theorem 4.2.4. For sufficiently small $\varepsilon_0 > 0$ depending only on λ, when $\varepsilon \in (0, \varepsilon_0)$, we use Lemma 4.2.7 to obtain

$$
[u]^*_{1,\beta,\Omega_2\cup\Gamma_2} \le C(\lambda)\big(\|u\|^*_{1,0,\Omega_2\cup\Gamma_2} + \|f\|_{0,0,\Omega_2} + \|h\|_{1,0,\Omega_2}\big). \tag{4.2.51}
$$

We use the interpolation inequality [131, Eq. (6.89)] to estimate

$$
\|u\|^*_{1,0,\Omega_2\cup\Gamma_2} \le C(\beta,\delta)\|u\|_{0,\Omega_2} + \delta[u]^*_{1,\beta,\Omega_2\cup\Gamma_2} \qquad \text{for } \delta > 0.
$$

Since $\beta = \beta(\lambda)$, we choose sufficiently small $\delta(\lambda) > 0$ to find

$$
\|u\|^*_{1,\beta,\Omega_2\cup\Gamma_2} \le C(\lambda)\big(\|u\|_{0,0,\Omega_2} + \|f\|_{0,0,\Omega_2} + \|h\|_{1,0,\Omega_2}\big) \tag{4.2.52}
$$

from (4.2.51). Since h has been obtained by extension from the boundary so that $\|h\|_{1,0,\Omega_2} = \|h\|_{1,0,\Gamma_2}$, estimate (4.2.20) is proved for $r = 1$.

6. Now let $r \in (0, 1)$. We rescale into the case when $r = 1$, by defining

$$
v(\mathbf{x}) = \frac{1}{r}u(r\mathbf{x}).
$$

Then v is a solution of the equation of form (4.2.1) in Ω_2, defined by the modified boundary function $\hat{\Phi}(t) = \frac{1}{r}\Phi(rt)$, and the boundary condition of form (4.2.19) on the corresponding boundary Γ_2, with the modified ingredients $(\hat{A}_{ij}, \hat{A}, \hat{B})$ and the right-hand sides $(\hat{f}, \hat{h})$, defined by

$$
\begin{aligned}
&\hat{A}_{ij}(\mathbf{p}, z, \mathbf{x}) = A_{ij}(\mathbf{p}, rz, r\mathbf{x}), \quad \hat{A}(p, z, \mathbf{x}) = rA(\mathbf{p}, rz, r\mathbf{x}), \\
&\hat{B}(p_2, z, \mathbf{x}) = B(p_2, rz, r\mathbf{x}), \quad \hat{f}(\mathbf{x}) = rf(r\mathbf{x}), \quad \hat{h}(\mathbf{x}) = h(r\mathbf{x}).
\end{aligned}
$$

It follows that $(\hat{\Phi}, \hat{A}_{ij}, \hat{A}, \hat{B})$ satisfy (4.2.3), (4.2.6), and (4.2.14)–(4.2.17) with $r = 1$ and the unchanged constant λ. Thus, if ε is small, depending only on λ, estimate (4.2.20) with $r = 1$ holds for $(v, \hat{f}, \hat{h})$. Writing this in terms of (u, f, h), we obtain (4.2.20). □

Now we assume the $C^{1,\alpha}$–regularity of the boundary, and more regularity in the ingredients of the equation and the boundary condition than those in Theorem 4.2.4, to prove the $C^{2,\alpha}$–estimates for the solutions of the *almost tangential derivative problem.*

Theorem 4.2.8. *Let $\lambda > 0, \alpha \in (0,1)$, $M > 0$, and $r \in (0,1]$. Let $\Phi \in C^1(\mathbb{R})$ satisfy*

$$\|\Phi\|_{C^{1,\alpha}(\mathbb{R})} \leq \lambda^{-1}, \qquad \Phi(0) = 0. \tag{4.2.53}$$

Assume that $A_{ij}(\mathbf{p}, z, \mathbf{x})$, $A(\mathbf{p}, z, \mathbf{x})$, and $B(p_2, z, \mathbf{x})$ satisfy (4.2.3)–(4.2.6) *in $\Omega = \Omega_{2r}$, and* (4.2.15)–(4.2.17), *and*

$$\|D_{(p_2,z)}B(p_2, z, \cdot)\|_{1,\alpha,\Gamma_{2r}} \leq \lambda^{-1} \qquad \textit{for all } (p_2, z) \in \mathbb{R} \times \mathbb{R}, \tag{4.2.54}$$

$$\|D^2_{(p_2,z)}B\|_{1,0,\mathbb{R}\times\mathbb{R}\times\Gamma_{2r}} \leq \lambda^{-1}. \tag{4.2.55}$$

Then there exist $\varepsilon \in (0,1)$ and $C > 0$ depending only on (λ, α, M) such that, for $u \in C^{2,\alpha}(\Omega_{2r} \cup \Gamma_{2r})$ satisfying (4.2.1), (4.2.19), *and*

$$\|u\|_{0,\Omega_{2r}} + \|f\|_{0,\alpha,\Omega_{2r}} + \|h\|_{1,\alpha,\Gamma_{2r}} \leq M, \tag{4.2.56}$$

we have

$$\|u\|_{2,\alpha,\Omega_r} \leq \frac{C}{r^{2+\alpha}}\big(\|u\|_{0,\Omega_{2r}} + r^2\|f\|_{0,\alpha,\Omega_{2r}} + r\|h\|_{1,\alpha,\Gamma_{2r}}\big). \tag{4.2.57}$$

Proof. We first prove the theorem for Case $r = 1$ and then extend the results to the general case $r \in (0,1]$ by scaling. The universal constant C in the argument below depends only on (λ, α, M), unless otherwise specified. We divide the proof into five steps.

1. For $r = 1$, we extend $B(p_2, z, \mathbf{x})$ from $\mathbb{R}\times\mathbb{R}\times\Gamma_2$ to $\mathbb{R}\times\mathbb{R}\times\Omega_2$, and h from Γ_2 to Ω_2, so that the conditions of the theorem still hold for the extension. The extension can be done in the same way as that at the beginning of the proof of Theorem 4.2.4: Extend to $\mathbf{x} \in \Omega_2 \cap B_{2-2\varepsilon}$ by (4.2.22) and then, for simplicity of notation, assume without loss of generality that $B(p_2, z, \mathbf{x})$ and $h(\mathbf{x})$ are defined in $\mathbb{R} \times \mathbb{R} \times \Omega_2$ and Ω_2 respectively, and satisfy (4.2.15)–(4.2.16) and (4.2.54)–(4.2.55) with Ω_2 instead of Γ_2. That is, $B(p_2, z, \mathbf{x})$ satisfies (4.2.23)–(4.2.25) and

$$\|D_{(p_2,z)}B(p_2, z, \cdot)\|_{1,\alpha,\Omega_2} \leq C\lambda^{-1} \qquad \text{for all } (p_2, z) \in \mathbb{R} \times \mathbb{R}, \tag{4.2.58}$$

$$\|D^2_{(p_2,z)}B\|_{1,\mathbb{R}\times\mathbb{R}\times\Omega_2} \leq \lambda^{-1}, \tag{4.2.59}$$

$$\|h\|_{1,\alpha,\Omega_2} \leq C\|h\|_{1,\alpha,\Gamma_2}, \tag{4.2.60}$$

where C depends only on α and the $C^{1,\alpha}$–norm of Γ_2, *i.e.*, only on (α, M).

Note that the conditions of Theorem 4.2.8 imply the conditions of Theorem 4.2.4 with the same constant λ. Thus, u satisfies (4.2.20). Then, using (4.2.56) with $r = 1$, *i.e.*,

$$\|u\|_{0,\Omega_2} + \|f\|_{0,\alpha,\Omega_2} + \|h\|_{1,\alpha,\Gamma_2} \leq M,$$

we obtain

$$\|u\|_{1,\beta,\Omega_{9/5}} \leq C. \tag{4.2.61}$$

2. We may assume that $\beta \leq \alpha$. Now we consider (4.2.27) as a linear elliptic equation:

$$\sum_{i,j=1}^{2} D_i(a_{ij}(\mathbf{x})D_j w_1) = D_1 F \qquad \text{in } \Omega_{9/5}, \tag{4.2.62}$$

where $a_{ij}(\mathbf{x}) = \frac{A_{ij}}{A_{22}}$, $a_{12}(\mathbf{x}) = 0$, $a_{21}(\mathbf{x}) = \frac{A_{12}+A_{21}}{A_{22}}$, and $F(\mathbf{x}) = -\frac{A-f}{A_{22}}$ with $(A_{ij}, A) = (A_{ij}, A)(Du(\mathbf{x}), u(\mathbf{x}), \mathbf{x})$. Then $a_{ij}, i, j = 1, 2$, satisfy the ellipticity in $\Omega_{9/5}$ with the same constant λ as in (4.2.3). Also, combining (4.2.61) with (4.2.3)–(4.2.5) implies that

$$\|a_{ij}\|_{0,\beta,\Omega_{9/5}} \leq C(\lambda, M). \tag{4.2.63}$$

From now on, $d_{\mathbf{x}}$ denotes the distance related to the partially interior norms in $\Omega_{9/5} \cup \Gamma_{9/5}$, *i.e.*, for $\mathbf{x} \in \Omega_{9/5}$, $d_{\mathbf{x}} := \text{dist}(\mathbf{x}, \partial\Omega_{9/5} \setminus \Gamma_{9/5}) \equiv \text{dist}(\mathbf{x}, \partial\Omega_{9/5} \cap \partial B_{9/5})$. Now, similarly to the proof of Lemma 4.2.5, we rescale equation (4.2.62) and the Dirichlet condition (4.2.31) from subdomains $\Omega_R(\mathbf{x}_1') \subset \Omega_{9/5}$ and $\Omega_R(\hat{\mathbf{x}}) \subset \Omega_{9/5}$ with $R \leq 1$ to $B = B_1^+$ or $B = B_1$, respectively, by defining

$$(\hat{w}_1, \hat{g}, \hat{a}_{ij})(\mathbf{X}) = (w_1, g, a_{ij})(\hat{\mathbf{x}} + R\mathbf{X}), \;\; \hat{F}(\mathbf{X}) = RF(\hat{\mathbf{x}} + R\mathbf{X}) \qquad \text{for } \mathbf{X} \in B.$$

Then

$$\sum_{i,j=1}^{2} D_i(\hat{a}_{ij}(\mathbf{x})D_j \hat{w}_1) = D_1 \hat{F} \qquad \text{in } B,$$

the strict ellipticity of this rescaled equation is the same as that for (4.2.62), and $\|\hat{a}_{ij}\|_{0,\beta,B} \leq C$ for $C = C(\lambda, M)$ in (4.2.63), where we have used $R \leq 1$. This allows us to apply the local $C^{1,\beta}$ interior and boundary estimates for the Dirichlet problem [131, Theorem 8.32, Corollary 8.36] to the rescaled problems in balls $B^+_{\frac{3d_{\mathbf{x}_0}}{8}}(\mathbf{x}_0')$ and $B_{\frac{d_{\mathbf{x}_0}}{8}}(\mathbf{x}_0)$ as in Lemma 4.2.5. Then scaling back, multiplying by $d_{\mathbf{x}_0}$, and applying the covering argument as in Lemma 4.2.5, we see that, for any $\mathbf{x}_0 \in \Omega_{9/5} \cup \Gamma_{9/5}$,

$$\begin{aligned}
&d_{\mathbf{x}_0}^{2+\beta}[w_1]_{1,\beta,\mathcal{D}_1} + d_{\mathbf{x}_0}^2\|w_1\|_{1,0,\mathcal{D}_1} \\
&\leq C\Big(d_{\mathbf{x}_0}\|w_1\|_{0,0,\mathcal{D}_2} + d_{\mathbf{x}_0}\|F\|_{0,0,\mathcal{D}_2} + d_{\mathbf{x}_0}^{1+\beta}[F]_{0,\beta,\mathcal{D}_2} + d_{\mathbf{x}_0}\|w_1\|_{0,0,\Gamma_2} \\
&\qquad + d_{\mathbf{x}_0}^2\|Dw_1\|_{0,0,\Gamma_2} + d_{\mathbf{x}_0}^{2+\beta}[Dw_1]_{0,\beta,\Gamma_2}\Big) \\
&\leq C\Big(d_{\mathbf{x}_0}\|u\|_{1,0,\mathcal{D}_2} + \|f\|_{0,\beta,\mathcal{D}_2} + d_{\mathbf{x}_0}^{1+\beta}\|u\|_{1,\beta,\mathcal{D}_2} + d_{\mathbf{x}_0}\|g\|_{0,0,\mathcal{D}_2} \\
&\qquad + d_{\mathbf{x}_0}^2\|Dg\|_{0,0,\mathcal{D}_2} + d_{\mathbf{x}_0}^{2+\beta}[Dg]_{0,\beta,\mathcal{D}_2}\Big),
\end{aligned} \tag{4.2.64}$$

where

$$\mathcal{D}_1 = B_{\frac{d_{\mathbf{x}_0}}{16}}(\mathbf{x}_0) \cap \Omega_{\frac{9}{5}}, \quad \mathcal{D}_2 = B_{\frac{d_{\mathbf{x}_0}}{2}}(\mathbf{x}_0) \cap \Omega_{\frac{9}{5}}, \quad \Gamma_2 = \partial\mathcal{D}_2 \cap \{x_2 = 0\},$$

and we have also used $d_{\mathbf{x}_0} < 2$, $w_1 = D_1 u$, the boundary condition (4.2.19) with (4.2.32), and the definition of F to obtain the last inequality. In addition, we have also used that $A(Du, u, \mathbf{x})$ satisfies estimate (4.2.9) for any $\beta \in [0, \alpha]$ and $\mathcal{D} \subset \Omega_2$, which follows from the present assumptions, by repeating the proof of (4.2.9).

Now we prove that, for any $\mathcal{D} \subset \Omega_{9/5}$,

$$[g]_{1,\beta,\mathcal{D}} \le C\big(\varepsilon[D^2u]_{0,\beta,\mathcal{D}} + \|u\|_{2,0,\mathcal{D}} + [h]_{1,\beta,\mathcal{D}}\big). \tag{4.2.65}$$

To obtain this estimate, we use expression (4.2.41). From (4.2.55) and (4.2.61), we obtain that functions $\mathbf{x} \mapsto D_{(p_2,z)}B(D_2u(\mathbf{x}), u(\mathbf{x}), \mathbf{x})$ satisfy

$$\|D_{(p_2,z)}B(D_2u(\cdot), u(\cdot), \cdot)\|_{0,\beta,\Omega_{9/5}} \le C.$$

From this and (4.2.24), we obtain that $[B_{p_2}D_{i2}u + B_z D_i u + h_{x_i}]_{0,\beta,\mathcal{D}}$ is estimated by the right-hand side of (4.2.65).

Thus, it remains to estimate $[B_{x_i}(D_2u(\cdot), u(\cdot), \cdot)]_{0,\beta,\mathcal{D}}$. Using that $B_{x_i}(0,0,\mathbf{x}) \equiv 0$ by (4.2.23), we have

$$B_{x_i}(p_2, z, \mathbf{x}) = a_1 p_2 + a_2 z,$$

where $(a_1, a_2) = \int_0^1 D_{(p_2,z)}B_{x_i}(tp_2, tz, \mathbf{x})dt$. Therefore, we employ (4.2.58)–(4.2.59) and (4.2.61) to obtain

$$[B_{x_i}(D_2u(\cdot), u(\cdot), \cdot)]_{0,\beta,\mathcal{D}} \le C\|Du\|_{2,0,\mathcal{D}}, \qquad i = 1, 2.$$

Now (4.2.65) is proved.

Recall that $Dw_1 = (D_{11}u, D_{12}u)$. Expressing $D_{22}u$ from equation (4.2.1) and then using (4.2.3)–(4.2.6), (4.2.9), and (4.2.52) to estimate the Hölder norms of $D_{22}u$, in terms of the norms of $(D_{11}u, D_{12}u, Du, u, f)$, as follows:

$$\begin{aligned}
&d_{\mathbf{x}_0}^{2+\beta}[D_{22}u]_{0,\beta,\mathcal{D}_1} + d_{\mathbf{x}_0}^2\|D_{22}u\|_{0,0,\mathcal{D}_1} \\
&\quad \le C\Big(d_{\mathbf{x}_0}^{2+\beta}[(D_{11}, D_{12}u, f)]_{0,\beta,\mathcal{D}_1} + d_{\mathbf{x}_0}^2\|(D_{11}, D_{12}u, f)\|_{0,0,\mathcal{D}_1} \\
&\qquad + d_{\mathbf{x}_0}^2\|u\|_{1,\beta,\mathcal{D}_1}\Big),
\end{aligned}$$

and using (4.2.38)–(4.2.39) and (4.2.65) to estimate the terms involving g in (4.2.64), we obtain from (4.2.64) that, for every $\mathbf{x}_0 \in \Omega_{9/5} \cup \Gamma_2$,

$$\begin{aligned}
&d_{\mathbf{x}_0}^{2+\beta}[D^2u]_{0,\beta,\mathcal{D}_1} + d_{\mathbf{x}_0}^2\|D^2u\|_{0,0,\mathcal{D}_1} \\
&\le C\Big(d_{\mathbf{x}_0}\|u\|_{1,0,\mathcal{D}_2} + d_{\mathbf{x}_0}^{1+\beta}[u]_{1,\beta,\mathcal{D}_2} + d_{\mathbf{x}_0}\|u\|_{1,0,\mathcal{D}_2} + \|f\|_{0,\beta,\mathcal{D}_2} \\
&\quad + d_{\mathbf{x}_0}^{2+\beta}\big(\varepsilon[D^2u]_{0,\beta,\mathcal{D}_2} + \|D^2u\|_{0,0,\mathcal{D}_2} + [h]_{1,\beta,\mathcal{D}_2}\big) + \varepsilon d_{\mathbf{x}_0}^2\|D^2u\|_{0,0,\mathcal{D}_2}\Big).
\end{aligned}$$

Then the argument in the proof of Lemma 4.2.7, combined with $d_{\mathbf{x}_0} \le 2$, implies

$$\|u\|^*_{2,\beta,\Omega_{9/5}\cup\Gamma_{9/5}} \\ \le C\big(\|u\|^*_{2,0,\Omega_{9/5}\cup\Gamma_{9/5}} + \varepsilon\|u\|^*_{2,\beta,\Omega_{9/5}\cup\Gamma_{9/5}} + \|f\|_{0,\beta,\Omega_{9/5}} + \|h\|_{1,\beta,\Omega_{9/5}}\big). \tag{4.2.66}$$

Using the interpolation inequality – For any $\varepsilon > 0$, there exists $C_\varepsilon > 0$ such that

$$\|u\|^*_{2,0,\Omega_{9/5}\cup\Gamma_{9/5}} \le \varepsilon\|u\|^*_{2,\beta,\Omega_{9/5}\cup\Gamma_{9/5}} + C_\varepsilon\|u\|_{0,\Omega_{9/5}},$$

reducing ε_0 if necessary, and using (4.2.60), we conclude

$$\|u\|^*_{2,\beta,\Omega_{9/5}\cup\Gamma_{9/5}} \le C(\lambda, M)\big(\|u\|_{0,\Omega_2} + \|f\|_{0,\beta,\Omega_2} + \|h\|_{1,\beta,\Gamma_2}\big). \tag{4.2.67}$$

This implies

$$\|u\|_{1,\alpha,\Omega_{8/5}} \le C(\lambda, M)\big(\|u\|_{0,\Omega_2} + \|f\|_{0,\beta,\Omega_2}\big) \le C(\lambda, M),$$

that is, we obtain (4.2.61) with α in place of β. Now we can repeat the argument, which leads from (4.2.61) to (4.2.67) with β replaced by α in $\Omega_{8/5}$, to obtain

$$\|u\|^*_{2,\alpha,\Omega_{8/5}\cup\Gamma_{8/5}} \le C(\lambda, M, \alpha)\big(\|u\|_{0,\Omega_2} + \|f\|_{0,\alpha,\Omega_2} + \|h\|_{1,\alpha,\Gamma_2}\big), \tag{4.2.68}$$

where we have used norm (4.1.3). This implies (4.2.57). Theorem 4.2.8 is proved for $r = 1$.

3. Now let $r \in (0, 1)$. By Theorem 4.2.4, estimate (4.2.20) holds. Then, using (4.2.56), we have

$$\|u\|_{1,\beta,\overline{\Omega}_{9r/5}} \le C(\lambda, M).$$

We rescale by defining

$$v(\mathbf{x}) = \frac{1}{\rho}u(\rho\mathbf{x}) \qquad \text{with } \rho = \frac{9r}{10}.$$

Then v is a solution of the equation of form (4.2.1) in Ω_2, defined by the modified boundary function $\hat{\Phi}(t) = \frac{1}{r}\Phi(rt)$, and the boundary condition of form (4.2.19) on the corresponding boundary Γ_2 with the modified ingredients $(\hat{A}_{ij}, \hat{A}, \hat{B})$ and the right-hand sides $(\hat{f}, \hat{h})$, defined by

$$\hat{A}_{ij}(\mathbf{p}, z, \mathbf{x}) = A_{ij}(\mathbf{p},\ \rho z,\ \rho\mathbf{x}), \quad \hat{A}(\mathbf{p}, z, \mathbf{x}) = \rho A(\mathbf{p},\ \rho z,\ \rho\mathbf{x}),$$
$$\hat{B}(p_2, z, \mathbf{x}) = B(p_2, \rho z, \rho\mathbf{x}), \quad \hat{f}(\mathbf{x}) = \rho f(\rho\mathbf{x}), \quad \hat{h}(\mathbf{x}) = h(\rho\mathbf{x}).$$

It follows that $(\hat{\Phi}, \hat{A}_{ij}, \hat{A}, \hat{B})$ satisfy the conditions of Theorem 4.2.8 with $r = 1$ and the unchanged constants (λ, M, α). Therefore, if ε is small, depending only on (λ, M, α), estimate (4.2.57) with $r = 1$ holds for $(v, \hat{f}, \hat{h})$. Writing this in terms of (u, f, h) and using that $r \le 1$, we conclude (4.2.57). □

Next we show the *a priori* estimates for the Dirichlet problem. Below, we use the notations:

$$B_R^+ := B_R(\mathbf{0}) \cap \{x_2 > 0\}, \quad \Sigma_R := B_R(\mathbf{0}) \cap \{x_2 = 0\} \qquad \text{for } R > 0.$$

Theorem 4.2.9. *Let $\lambda > 0$ and $\alpha \in (0,1)$. Let $\Phi \in C^{2,\alpha}(\mathbb{R})$ satisfy*

$$\|\Phi\|_{2,\alpha,\mathbb{R}} \leq \lambda^{-1}, \quad \Phi(0) = 0, \tag{4.2.69}$$

and let $\Omega_R := B_R \cap \{x_2 > \Phi(x_1)\}$ for $R \in (0, 2]$. Let $u \in C^2(\Omega_R) \cap C(\overline{\Omega_R})$ satisfy (4.2.1) *in Ω_R and*

$$u = g \qquad \text{on } \Gamma_R := \partial\Omega_R \cap \{x_2 = \Phi(x_1)\}, \tag{4.2.70}$$

where $A_{ij} = A_{ij}(Du, \mathbf{x})$, $A = A(Du, u, \mathbf{x})$, and $f = f(\mathbf{x})$ satisfy the assumptions of Theorem 4.2.8, *and $g = g(\mathbf{x})$ satisfies*

$$\|g\|_{2,\alpha,\overline{\Omega_R}} \leq M. \tag{4.2.71}$$

Assume that $\|u\|_{0,\Omega_R} \leq M$. Then

$$\|u\|_{2,\alpha,\overline{\Omega_{R/2}}} \leq C(\lambda, M, \alpha, R)\big(\|u\|_{0,\overline{\Omega_R}} + \|f\|_{0,\alpha,\overline{\Omega_R}} + \|g\|_{2,\alpha,\overline{\Omega_R}}\big). \tag{4.2.72}$$

Proof. We flatten the boundary by the change of variables:

$$\mathcal{A}(x_1, x_2) = (x_1, x_2 - \Phi(x_1)).$$

Assumption (4.2.69) implies that $\mathcal{A} : \mathbb{R}^2 \mapsto \mathbb{R}^2$ is a $C^{2,\alpha}$–diffeomorphism with $\|(\mathcal{A}, \mathcal{A}^{-1})\|_{2,\alpha,\mathbb{R}^2} \leq C(\lambda, R)$, and that $B^+_{R/K} \subset \mathcal{A}(\Omega_R)$ with $K = K(\lambda, R) > 1$. Then, changing further the variables to rescale $B^+_{R/K}$ to B_2^+, and noticing that the ingredients and the right-hand sides of the equation and the boundary condition in the new variables satisfy the same conditions as those for the original equation with the constants depending only on (λ, M, R, α), we reduce the problem to Case $R = 2$ with $\Omega_R = B_2^+$.

Furthermore, replacing u by $u - g$, we can assume without loss of generality that $g \equiv 0$. Thus, $D_1 u = 0$ on Γ_2. Then estimate (4.2.72) in our case $\Omega_R = B_2^+$ and $g \equiv 0$ follows from Theorem 4.2.8. Changing the variables back, we obtain (4.2.72) with the norm of the left-hand side in domain $\Omega_{\mu R}$ for some $\mu \in (0, \frac{1}{2})$ depending only on (λ, M, R). Then, from a covering argument and using the interior estimates (4.2.8) in Theorem 4.2.1, we obtain the full estimate (4.2.72). □

We now derive the estimates for a class of oblique derivative problems.

Theorem 4.2.10. *Let $\lambda > 0, \alpha \in (0,1)$, and $M > 0$. Let $\Phi \in C^{2,\alpha}(\mathbb{R})$ satisfy* (4.2.69), *and let*

$$\Omega_R := B_R \cap \{x_2 > \Phi(x_1)\}, \quad \Gamma_R := \partial\Omega_R \cap \{x_2 = \Phi(x_1)\} \quad \textit{for } R \in (0, 2).$$

Let $u \in C^{2,\alpha}(\Omega_R \cup \Gamma_R)$ *satisfy*

$$\sum_{i,j=1}^{2} A_{ij}(Du, u, \mathbf{x})D_{ij}u + A(Du, u, \mathbf{x}) = f \qquad \text{in } \Omega_R, \tag{4.2.73}$$

$$\mathbf{b} \cdot Du + b_0 u = h \qquad \text{on } \Gamma_R. \tag{4.2.74}$$

Assume that $A_{ij} = A_{ij}(Du, u, \mathbf{x})$ *and* $A = A(Du, u, \mathbf{x})$ *satisfy* (4.2.3)–(4.2.6), *and that* $\mathbf{b} = (b_1, b_2)(\mathbf{x})$ *and* $b_0 = b_0(\mathbf{x})$ *satisfy the following obliqueness condition and* $C^{1,\alpha}$*–bounds:*

$$\mathbf{b} \cdot \boldsymbol{\nu} \geq \lambda \qquad \text{on } \Gamma_R, \tag{4.2.75}$$

$$\|(\mathbf{b}, b_0)\|_{1,\alpha,\Gamma_R} \leq \lambda^{-1}. \tag{4.2.76}$$

Assume that (4.2.56) *holds in* $\Omega_R \cup \Gamma_R$. *Then there exists* $C > 0$ *depending only on* (λ, α, M) *such that*

$$\|u\|_{2,\alpha,\Omega_{R/2}} \leq \frac{C}{R^{2+\alpha}} \big(\|u\|_{0,\Omega_R} + R^2 \|f\|_{0,\alpha,\Omega_R} + R\|h\|_{1,\alpha,\Gamma_R} \big). \tag{4.2.77}$$

In addition, there exist $\beta \in (0,1)$ *and* $\hat{C} > 0$ *depending only on* λ *such that*

$$\|u\|_{1,\beta,\Omega_{R/2}} \leq \frac{\hat{C}}{R^{1+\beta}} \big(\|u\|_{0,\Omega_R} + R^2 \|f\|_{0,\Omega_R} + R\|h\|_{0,1,\Gamma_R} \big). \tag{4.2.78}$$

Proof. We divide the proof into five steps.

1. We first consider Case $R = 2$. Then, by flattening boundary $\Gamma_R = \Gamma_2$ as at the beginning of the proof of Theorem 4.2.9, we can assume without loss of generality that

$$R = 2, \quad \Omega_R = B_2^+, \quad \Gamma_R = \Sigma_2,$$

that $u \in C^{2,\alpha}(B_2^+ \cup \Sigma_2)$ satisfies (4.2.73)–(4.2.74) with $\Omega_R = B_2^+$ and $\Gamma_R = \Sigma_2$, and (4.2.56) holds, and that the ingredients of the equation, the boundary condition, and the right-hand sides satisfy the same assumptions as those for the original problem. We also note that (4.2.75) now is of the form:

$$b_2(\mathbf{x}) \geq \lambda \qquad \text{for all } \mathbf{x} \in \Sigma_2. \tag{4.2.79}$$

We need to prove that

$$\|u\|_{2,\alpha,B_1^+} \leq C\big(\|u\|_{0,B_2^+} + \|f\|_{0,\alpha,B_2^+} + \|h\|_{1,\alpha,\Sigma_2}\big), \tag{4.2.80}$$

$$\|u\|_{1,\beta,B_1^+} \leq \hat{C}\big(\|u\|_{0,B_2^+} + \|f\|_{1,B_2^+} + \|h\|_{1,0,\Sigma_2}\big), \tag{4.2.81}$$

where C depends only on (λ, M, α) in (4.2.80), and $\hat{C}$ and $\beta \in (0,1)$ depend only on λ in (4.2.81).

Using (4.2.76)–(4.2.79), we can extend $(\mathbf{b}, b_0)$ to B_2^+ so that

$$b_2(\mathbf{x}) \geq \frac{\lambda}{2} \qquad \text{for all } \mathbf{x} \in B_2^+, \tag{4.2.82}$$

$$\|(\mathbf{b}, b_0)\|_{1,\alpha,B_2^+} \leq \lambda^{-1}. \tag{4.2.83}$$

Note that, in order to prove (4.2.80), it suffices to prove that there exist K and C depending only on (λ, M, α) such that

$$\|u\|_{2,\alpha,B^+_{1/K}} \leq C\big(\|u\|_{0,B_2^+} + \|f\|_{0,\alpha,B_2^+} + \|h\|_{1,\alpha,\Sigma_2}\big). \tag{4.2.84}$$

Indeed, if (4.2.84) is proved, (4.2.80) is obtained by the standard covering technique, the scaled estimates (4.2.84), and the interior estimates (4.2.8) in Theorem 4.2.1.

2. To obtain (4.2.84), we first make a linear change of variables to normalize the problem so that

$$b_1(\mathbf{0}) = 0, \quad b_2(\mathbf{0}) = 1 \tag{4.2.85}$$

for the modified problem.

This can be done as follows: Let

$$\mathbf{X} = \tilde{\Psi}(\mathbf{x}) := \frac{1}{b_2(\mathbf{0})}(b_2(\mathbf{0})x_1 - b_1(\mathbf{0})x_2, x_2).$$

Then

$$\mathbf{x} = \tilde{\Psi}^{-1}(\mathbf{X}) = (X_1 + b_1(\mathbf{0})X_2, b_2(\mathbf{0})X_2), \quad |D\tilde{\Psi}| + |D\tilde{\Psi}^{-1}| \leq C(\lambda),$$

where the estimate follows from (4.2.82)–(4.2.83). Then

$$v(\mathbf{X}) := u(\mathbf{x}) \equiv u(X_1 + b_1(\mathbf{0})X_2, b_2(\mathbf{0})X_2)$$

is a solution of the equation of form (4.2.73) in domain $\tilde{\Psi}(B_2^+)$ and the boundary condition of form (4.2.74) on the boundary part $\tilde{\Psi}(\Sigma_1)$ such that (4.2.3)–(4.2.6), (4.2.56), and (4.2.82)–(4.2.83) are satisfied with constant $\hat{\lambda} > 0$ depending only on λ, (4.2.85) holds, and $\|v\|_{0,\tilde{\Psi}(B_2^+)} \leq M$, which can be verified by a straightforward calculation.

Moreover, $\tilde{\Psi}(B_2^+) \subset \mathbb{R}^2_+ := \{X_2 > 0\}$ and $\tilde{\Psi}(\Sigma_1) = \partial\tilde{\Psi}(B_2^+) \cap \{X_2 = 0\}$. Since $|D\tilde{\Psi}| + |D\tilde{\Psi}^{-1}| \leq C(\lambda)$, there exists $K_1 = K_1(\lambda) > 0$ such that, for any $r > 0$, $B_{r/K_1} \subset \tilde{\Psi}(B_r) \subset B_{K_1 r}$. Therefore, it suffices to prove

$$\|v\|_{2,\alpha,B^+_{r/2}} \leq C\big(\|v\|_{0,B_r^+} + \|\hat{f}\|_{0,\alpha,B_r^+} + \|\hat{h}\|_{1,\alpha,\Sigma_r}\big)$$

for some $r \in (0, \frac{1}{K_1})$, where $\hat{f}$ and $\hat{h}$ are the right-hand sides in the equation and the boundary condition for v. This estimate implies (4.2.84) with $K = \frac{2K_1}{r}$.

3. As a result of the reduction performed in Step 2, it suffices to prove that *there exist $\varepsilon \in (0,1)$ and C depending only on (λ, α, M) such that, if u satisfies*

(4.2.73)–(4.2.74) *in* $B^+_{2\varepsilon}$ *and on* $\Sigma_{2\varepsilon}$ *respectively, if* (4.2.3)–(4.2.6), (4.2.56), (4.2.82)–(4.2.83) *in* $B^+_{2\varepsilon}$, $\|u\|_{0,B^+_{2\varepsilon}} \le M$, *and* (4.2.85) *hold, then*

$$\|u\|_{2,\alpha,B^+_\varepsilon} \le C\big(\|u\|_{0,B^+_{2\varepsilon}} + \|f\|_{0,\alpha,B^+_{2\varepsilon}} + \|h\|_{1,\alpha,\Sigma_{2\varepsilon}}\big). \tag{4.2.86}$$

This implies (4.2.80), and the corresponding $C^{1,\beta}$–estimate leads to (4.2.81).

We now prove (4.2.86). For $\varepsilon > 0$ to be chosen later, we rescale from $B^+_{2\varepsilon}$ into B^+_2 by defining

$$v(\mathbf{x}) = \frac{1}{\varepsilon}\big(u(\varepsilon\mathbf{x}) - u(\mathbf{0})\big) \qquad \text{for } \mathbf{x} \in B^+_2. \tag{4.2.87}$$

Then v satisfies

$$\sum_{i,j=1}^{2} \tilde{A}_{ij}(Dv, v, \mathbf{x})D_{ij}v + \tilde{A}(Dv, v, \mathbf{x}) = \tilde{f} \qquad \text{in } B^+_2, \tag{4.2.88}$$

$$D_2 v = \tilde{\mathbf{b}} \cdot Dv + \tilde{b}_0 v + b_0(\varepsilon\mathbf{x})u(\mathbf{0}) + \tilde{h} \qquad \text{on } \Sigma_2, \tag{4.2.89}$$

where

$$(\tilde{A}_{ij}, \tilde{A}_i)(\mathbf{p}, z, \mathbf{x}) = (A_{ij}, \varepsilon A_i)(\mathbf{p}, \varepsilon z + u(\mathbf{0}), \varepsilon\mathbf{x}) \qquad \text{for } i = 1, 2,$$
$$\tilde{\mathbf{b}}(\mathbf{x}) = (-b_1(\varepsilon\mathbf{x}), -b_2(\varepsilon\mathbf{x}) + 1), \qquad \tilde{b}_0(\mathbf{x}) = -\varepsilon b_0(\varepsilon\mathbf{x}),$$
$$\tilde{f}(\mathbf{x}) = \varepsilon f(\varepsilon\mathbf{x}), \qquad \tilde{h}(\mathbf{x}) = h(\varepsilon\mathbf{x}),$$

so that $(\tilde{A}_{ij}, \tilde{A}_i)$ satisfy (4.2.3)–(4.2.6) in B^+_2. Using (4.2.76), (4.2.85), and $\varepsilon \le 1$, and extending $(\tilde{\mathbf{b}}, \tilde{b}_0)$ from Σ_2 by $(\tilde{\mathbf{b}}, \tilde{b}_0)(\mathbf{x}) := (\tilde{\mathbf{b}}, \tilde{b}_0)(x_1, 0)$, we have

$$\|(\tilde{\mathbf{b}}, \tilde{b}_0)\|_{1,\alpha,B^+_2} \le C\varepsilon \qquad \text{for some } C = C(\lambda). \tag{4.2.90}$$

Similarly, $(v, \tilde{f}, \tilde{h})$ satisfy (4.2.56) in B^+_2.

Now we follow the proofs of Theorems 4.2.4 and 4.2.8 with the only difference that we now work with $w_2 = D_2 v$ and equation (4.2.28), instead of w_1 and (4.2.27).

More specifically, the sketch of argument is as follows: We first find from (4.2.88) that w_2 is a weak solution of the equation:

$$D_{11}w_2 + D_2\Big(\frac{2\tilde{A}_{12}}{\tilde{A}_{11}}D_1 w_2 + \frac{\tilde{A}_{22}}{\tilde{A}_{11}}D_2 w_2\Big) = D_2\Big(\frac{\tilde{f} - \tilde{A}}{\tilde{A}_{11}}\Big) \qquad \text{in } B^+_2. \tag{4.2.91}$$

From (4.2.89), we have

$$w_2 = \tilde{g} \qquad \text{on } \Sigma_2, \tag{4.2.92}$$

where

$$\tilde{g} := \tilde{\mathbf{b}} \cdot Dv + \tilde{b}_0 v + b_0(\varepsilon\mathbf{x})u(\mathbf{0}) + \tilde{h} \qquad \text{in } B^+_2.$$

Using equation (4.2.91) and the Dirichlet boundary condition (4.2.92) for w_2 and following the proof of Lemma 4.2.5, we can show the existence of $\beta \in (0, \alpha]$ and C depending only on λ such that, for any $\mathbf{x}_0 \in B_2^+ \cup \Sigma_2$,

$$
\begin{aligned}
& d_{\mathbf{x}_0}^{\beta}[w_2]_{0,\beta,B_{\frac{d_{\mathbf{x}_0}}{16}}(\mathbf{x}_0)\cap B_2^+} \\
& \le C\Big(\|(Dv,\tilde{f})\|_{0,B_{\frac{d_{\mathbf{x}_0}}{2}}(\mathbf{x}_0)\cap B_2^+} + d_{\mathbf{x}_0}^{\beta}[\tilde{g}]_{0,\beta,B_{\frac{d_{\mathbf{x}_0}}{2}}(\mathbf{x}_0)\cap B_2^+}\Big).
\end{aligned} \tag{4.2.93}
$$

Next, we obtain the Hölder estimates of Dv if ε is sufficiently small. We first note that, by (4.2.90), $\tilde{g}$ satisfies

$$
|D\tilde{g}| \le C\varepsilon\big(|D^2v| + |Dv| + |v| + \|u\|_{0,B_{2\varepsilon}^+}\big) + |D\tilde{h}| \qquad \text{in } B_2^+, \tag{4.2.94}
$$

$$
[\tilde{g}]_{0,\beta,B_{\frac{d_{\mathbf{x}_0}}{2}}(\mathbf{x}_0)\cap B_2^+} \le C\varepsilon\Big(\|v\|_{1,\beta,B_{\frac{d_{\mathbf{x}_0}}{2}}(\mathbf{x}_0)\cap B_2^+} + \|u\|_{0,B_{2\varepsilon}^+}\Big) + [\tilde{h}]_{0,\beta,B_2^+} \tag{4.2.95}
$$

for some $C = C(\lambda)$. The term, $\varepsilon\|u\|_{0,B_{2\varepsilon}^+}$, in (4.2.94)–(4.2.95) comes from term $b_0(\varepsilon\mathbf{x})u(\mathbf{0})$ in the definition of $\tilde{g}$. We follow the proof of Lemma 4.2.6, but we now employ the integral form of equation (4.2.91) with a test function $\zeta = \eta^2(w_2 - \tilde{g})$ and $\zeta = \eta^2(w_2 - w_2(\hat{\mathbf{x}}))$ to obtain an integral estimate of $|Dw_2|$, *i.e.*, $|D_{ij}v|$ for $i + j > 2$, and then use the elliptic equation (4.2.88) to estimate the remaining derivative $D_{11}v$. In these estimates, we use (4.2.93)–(4.2.95). We obtain that, for sufficiently small ε depending only on λ,

$$
\begin{aligned}
& d_{\mathbf{x}_0}^{\beta}[Dv]_{0,\beta,B_{\frac{d_{\mathbf{x}_0}}{32}}(\mathbf{x}_0)\cap B_2^+} \\
& \le C\Big(\|v\|_{1,B_{\frac{d_{\mathbf{x}_0}}{2}}(\mathbf{x}_0)\cap B_2^+} + \varepsilon d_{\mathbf{x}_0}^{\beta}[Dv]_{0,\beta,B_{\frac{d_{\mathbf{x}_0}}{2}}(\mathbf{x}_0)\cap B_2^+} \\
& \qquad + \varepsilon d_{\mathbf{x}_0}^{\beta}\|u\|_{0,B_{2\varepsilon}^+} + \|f\|_{0,0,B_{2\varepsilon}^+} + \|h\|_{1,0,B_{2\varepsilon}^+}\Big)
\end{aligned} \tag{4.2.96}
$$

for any $\mathbf{x}_0 \in B_2^+ \cup \Sigma_2$, with $C = C(\lambda)$. Using (4.2.96), we follow the proof of Lemma 4.2.7 to obtain

$$
\begin{aligned}
[v]^*_{1,\beta,B_2^+\cup\Sigma_2} \le C\Big(& \|v\|^*_{1,0,B_2^+\cup\Sigma_2} + \varepsilon[v]^*_{1,\beta,B_2^+\cup\Sigma_2} + \varepsilon\|u\|_{0,B_{2\varepsilon}^+} \\
& + \|f\|_{0,0,B_{2\varepsilon}^+} + \|h\|_{1,0,B_{2\varepsilon}^+}\Big),
\end{aligned} \tag{4.2.97}
$$

where we have used norms (4.1.3). Now we choose sufficiently small $\varepsilon > 0$ depending only on λ such that $C\varepsilon \le \frac{1}{2}$ in (4.2.97). Then we obtain

$$
[v]^*_{1,\beta,B_2^+\cup\Sigma_2} \le C(\lambda)\big(\|v\|^*_{1,0,B_2^+\cup\Sigma_2} + \|u\|_{0,B_{2\varepsilon}^+} + \|f\|_{0,0,B_{2\varepsilon}^+} + \|h\|_{1,0,B_{2\varepsilon}^+}\big).
$$

We employ the interpolation inequality, similar to the proof of (4.2.52), to obtain

$$
\|v\|^*_{1,\beta,B_2^+\cup\Sigma_2} \le C(\lambda)\big(\|v\|_{0,B_2^+} + \|u\|_{0,B_{2\varepsilon}^+} + \|f\|_{0,0,B_{2\varepsilon}^+} + \|h\|_{1,0,B_{2\varepsilon}^+}\big). \tag{4.2.98}
$$

4. We can assume that $\beta \le \alpha$ for the rest of the argument. Problem (4.2.91)–(4.2.92) can be regarded as a Dirichlet problem for the linear elliptic equation:

$$\sum_{i,j=1}^{2} D_i(a_{ij}(\mathbf{x})D_jw_2) = D_2F \qquad \text{in } B^+_{9/5},$$

where $a_{ij}(\mathbf{x}) = \frac{\tilde{A}_{ij}}{\tilde{A}_{22}}$, $a_{12}(\mathbf{x}) = 0$, $a_{21}(\mathbf{x}) = \frac{\tilde{A}_{12}+\tilde{A}_{21}}{\tilde{A}_{22}}$, and $F(\mathbf{x}) = \frac{\tilde{f}-\tilde{A}}{\tilde{A}_{22}}$ with $(\tilde{A}_{ij}, \tilde{A}) = (\tilde{A}_{ij}, \tilde{A})(Dv(\mathbf{x}), v(\mathbf{x}), \mathbf{x})$. Then $\|(a_{ij}, F)\|_{0,\beta,B^+_{9/5}} \le C(\lambda, M)$, from (4.2.4)–(4.2.6), (4.2.56), and (4.2.98). Note that the equation and the estimates of its coefficients are similar to (4.2.62)–(4.2.63). Thus, we employ (4.2.90) and argue as in Step 2 of the proof of Theorem 4.2.8 to obtain estimate (4.2.65) for $\tilde{g}$. Then we obtain the following estimate similar to (4.2.68):

$$\|v\|^*_{2,\alpha,B^+_{8/5}\cup\Sigma_{8/5}} \le C(\lambda, M, \alpha)\big(\|v\|_{0,B^+_2} + \|\tilde{f}\|_{0,\alpha,B^+_2} + \|\tilde{h}\|_{1,\alpha,\Sigma_2}\big),$$

where we have used norm (4.1.3). By (4.2.87) with $\varepsilon = \varepsilon(\lambda)$ as chosen above, the last estimate implies (4.2.86) with $C = C(\lambda, M, \alpha)$.

As we have discussed in Steps 1–2, this implies (4.2.84), so that (4.2.80) holds. Moreover, by the same change of variables, estimate (4.2.81) with $\hat{C} = \hat{C}(\lambda)$ is obtained directly from (4.2.98) by using $\varepsilon = \varepsilon(\lambda)$.

As we have shown in Step 1, this implies (4.2.77)–(4.2.78) for $R = 2$.

5. To obtain (4.2.77)–(4.2.78) for any $R \in (0, 2]$, we perform the scaling as in Step 3 of Theorem 4.2.8; see also Step 6 of Theorem 4.2.4. □

Corollary 4.2.11. *Under the assumptions of Theorem* 4.2.10, *solution u satisfies the following estimates*:

$$\|u\|^*_{2,\alpha,\Omega_R\cup\Gamma_R} \le C(\lambda, M, \alpha)\big(\|u\|_{0,\Omega_R} + \|f\|^{(2)}_{0,\alpha,\Omega_R\cup\Gamma_R} + \|h\|^{(1)}_{1,\alpha,\Gamma_R}\big), \tag{4.2.99}$$

$$\|u\|^*_{1,\beta,\Omega_R\cup\Gamma_R} \le C(\lambda)\big(\|u\|_{0,\Omega_R} + \|f\|^{(2)}_{0,0,\Omega_R\cup\Gamma_R} + \|h\|^{(1)}_{0,1,\Gamma_R}\big) \tag{4.2.100}$$

for $\beta = \beta(\lambda) \in (0, 1)$, where we have used norms (4.1.2)–(4.1.3).

Proof. To obtain (4.2.99), we use estimate (4.2.77) in $B_{d_{\mathbf{x}}/4}(\mathbf{x}) \cap \Omega_R$ for each $\mathbf{x} \in \Gamma_R$, and the interior estimate (4.2.8) in $B_{d_{\mathbf{x}}/4}(\mathbf{x}) \subset \Omega_R$, where $d_{\mathbf{x}} := \text{dist}(\mathbf{x}, \partial\Omega_R \setminus \Gamma_R)$ and we have used radius $\frac{d_{\mathbf{x}}}{4}$ in the estimates; finally we argue as in the proof of [131, Theorem 4.8] (*cf.* the proof of Lemma 4.2.7 above).

By a similar argument, using the boundary estimate (4.2.78) and the interior estimate (4.2.11), we obtain (4.2.100). □

Next we consider a version of Theorem 4.2.10 where some nonlinearity in the boundary conditions is included.

Theorem 4.2.12. *Let $\lambda > 0, \alpha \in (0, 1)$, and $M > 0$. Let $\Phi \in C^{2,\alpha}(\mathbb{R})$ satisfy* (4.2.69), *and let*

$$\Omega_R := B_R \cap \{x_2 > \Phi(x_1)\}, \quad \Gamma_R := \partial\Omega_R \cap \{x_2 = \Phi(x_1)\} \qquad \textit{for } R > 0.$$

Let $u \in C^{2,\alpha}(\Omega_R \cup \Gamma_R)$ satisfy equation (4.2.73) in Ω_R and the boundary condition:

$$B(Du, u, \mathbf{x}) = 0 \qquad on\ \Gamma_R. \tag{4.2.101}$$

Assume that equation (4.2.73) satisfies (4.2.3)–(4.2.6) in Ω_R. Assume that there exist $\delta \in (0,1)$, a nonhomogeneous linear operator:

$$L(\mathbf{p}, z, \mathbf{x}) := \mathbf{b}(\mathbf{x}) \cdot \mathbf{p} + b_0(\mathbf{x})z - h(\mathbf{x}) \qquad for\ (\mathbf{p}, z, \mathbf{x}) \in \mathbb{R}^2 \times \mathbb{R} \times \Gamma_R,$$

and a function $v \in C^{2,\alpha}(\overline{\Gamma_R})$ such that

$$\begin{aligned} &|B(\mathbf{p}, z, \mathbf{x}) - L(\mathbf{p}, z, \mathbf{x})| \le \delta\big(|\mathbf{p} - Dv(\mathbf{x})| + |z - v(\mathbf{x})|\big), \\ &\big|D_{(\mathbf{p},z)}(B - L)(\mathbf{p}, z, \mathbf{x})\big| \le \delta \end{aligned} \tag{4.2.102}$$

for any $(\mathbf{p}, z, \mathbf{x}) \in \mathbb{R}^2 \times \mathbb{R} \times \Gamma_R$. Assume that B satisfies

$$\|D_{(\mathbf{p},z)}B(\mathbf{p}, z, \cdot)\|_{C^{1,\alpha}(\overline{\Gamma_R})} \le \lambda^{-1} \qquad for\ all\ (\mathbf{p}, z) \in \mathbb{R}^2 \times \mathbb{R}, \tag{4.2.103}$$

$$\|D^2_{(\mathbf{p},z)}B\|_{C^1(\mathbb{R}^2 \times \mathbb{R} \times \overline{\Gamma_R})} \le \lambda^{-1}, \tag{4.2.104}$$

that $\mathbf{b}$ satisfies (4.2.75)–(4.2.76) with $c = b_0$, that

$$\|v\|_{2,\alpha,\Gamma_R} \le M,$$

and that (4.2.56) holds in Ω_R. Then there exist $\delta_0 > 0$ and $C < \infty$ depending only on (λ, α, M, R) such that, if $\delta \in (0, \delta_0)$, u satisfies

$$\|u\|_{2,\alpha,\Omega_{R/2}} \le C. \tag{4.2.105}$$

Proof. Repeating the argument in the first part of the proof of Theorem 4.2.9, we flatten the boundary and reduce Problem (4.2.73) with (4.2.101) to the case:

$$R = 2, \quad \Omega_R = B_2^+, \quad \Gamma_R = \Sigma_2.$$

We work in this setting for the rest of proof, which is divided into three steps. The universal constant C below depends only on (λ, α, M), unless otherwise specified.

1. First we may assume that $v \equiv 0$. Indeed, introducing $\tilde{u} := u - v$, and substituting $u = \tilde{u} + v$ and its derivatives into (4.2.73) and (4.2.101), we find that $\tilde{u}$ satisfies the problem of form (4.2.73) and (4.2.101) with (A_{ij}, A, B) satisfying the same properties as the problem for u, except that (4.2.102) holds with $v = 0$ and the new function h in operator L is $h_{\rm old} - \mathbf{b} \cdot Dv - b_0 v$, which has a similar estimate. Also, from the regularity of v, estimate (4.2.105) for $\tilde{u}$ implies a similar estimate for u.

Then it suffices to prove (4.2.105) for $\tilde{u}$. That is, keeping the original notation for u and the ingredients of the problem, we can assume that $v \equiv 0$ so that (4.2.102) is of the form:

$$|(B - L)(\mathbf{p}, z, \mathbf{x})| \le \delta\,(|\mathbf{p}| + |z|), \quad \big|D_{(\mathbf{p},z)}(B - L)(\mathbf{p}, z, \mathbf{x})\big| \le \delta \tag{4.2.106}$$

for any $(\mathbf{p}, z, \mathbf{x}) \in \mathbb{R}^2 \times \mathbb{R} \times \Sigma_2$. Furthermore, we extend (B, L) from Σ_2 to B_2^+ by $(B, L)(\mathbf{p}, z, \mathbf{x}) := (B, L)(\mathbf{p}, z, (x_1, 0))$ for $\mathbf{x} = (x_1, x_2)$. Then it follows that these extensions satisfy (4.2.106) for any $(\mathbf{p}, z, \mathbf{x}) \in \mathbb{R}^2 \times \mathbb{R} \times B_2^+$.

We can rewrite the boundary condition (4.2.101) as

$$\mathbf{b}(\mathbf{x}) \cdot Du + b_0(\mathbf{x})u = h(\mathbf{x}) + H(\mathbf{x}), \tag{4.2.107}$$

where $H(\mathbf{x}) = L(Du(\mathbf{x}), u(\mathbf{x}), \mathbf{x}) - B(Du(\mathbf{x}), u(\mathbf{x}), \mathbf{x})$. Let $\beta = \beta(\lambda) \in (0, 1)$ be the constant from (4.2.78) for the present constants λ and $R = 2$. Combining (4.2.106) with the other assumptions on (B, L), *i.e.*, (4.2.76), (4.2.103)–(4.2.104), and arguing as in the proof of (4.2.38)–(4.2.39), we conclude that H satisfies (4.2.38)–(4.2.39) in B_2^+, without term involving h. Let $d_{\mathbf{x}} := \text{dist}(\mathbf{x}, \partial B_2^+ \setminus \Sigma_2)$. Then, for each $\mathbf{x} \in B_2^+$, multiplying (4.2.38) and (4.2.39) for H at $\mathbf{x}$ by $d_{\mathbf{x}}$ and $d_{\mathbf{x}}^2$, respectively, we obtain

$$\|H\|^{(1)}_{0,1,B_2^+ \cup \Sigma_2} \le \delta[u]^*_{1,\beta,B_2^+ \cup \Sigma_2} + C\|u\|^*_{1,0,B_2^+ \cup \Sigma_2} \tag{4.2.108}$$

for C depending only on (λ, M, β). Here and hereafter, we use norms (4.1.2)–(4.1.3). Since u satisfies equation (4.2.73) and the boundary condition (4.2.107), our assumptions allow us to apply Corollary 4.2.11. From (4.2.100), we have

$$\|u\|^*_{1,\beta,B_2^+ \cup \Sigma_2} \le C\big(\|u\|_{0,B_2^+} + \|f\|^{(2)}_{0,0,B_2^+ \cup \Sigma_2} + \|(h, H)\|^{(1)}_{0,1,\Sigma_2}\big),$$

where $\beta \in (0, 1)$ depends only on λ. Now we apply (4.2.108) to obtain

$$\begin{aligned}\|u\|^*_{1,\beta,B_2^+ \cup \Sigma_2} &\le C\big(\delta\|u\|^*_{1,\beta,B_2^+ \cup \Sigma_2} + \|u\|^*_{1,0,B_2^+ \cup \Sigma_2}\big)\\ &\quad + C\big(\|u\|_{0,B_2^+} + \|f\|^{(2)}_{0,0,B_2^+ \cup \Sigma_2} + \|h\|^{(1)}_{0,1,\Sigma_2}\big).\end{aligned} \tag{4.2.109}$$

Using the interpolation inequality [131, Eq. (6.89)]:

$$\|u\|^*_{1,0,B_2^+ \cup \Sigma_2} \le \delta[u]^*_{1,\beta,B_2^+ \cup \Sigma_2} + C(\beta, \delta)\|u\|_{0,B_2^+}$$

on the right-hand side of estimate (4.2.109), choosing small δ depending only on (λ, M) (which also fixes constant C), and using (4.2.56), we have

$$\|u\|^*_{1,\beta,B_2^+ \cup \Sigma_2} \le C\big(\|u\|_{0,B_2^+} + \|f\|^{(2)}_{0,0,B_2^+ \cup \Sigma_2} + \|h\|^{(1)}_{0,1,\Sigma_2}\big) \le C. \tag{4.2.110}$$

Now we consider two separate cases: $\alpha \le \beta$ and $\alpha > \beta$.

2. If $\alpha \le \beta$, we argue as in the proof of (4.2.65) by using estimate (4.2.110) and the assumptions with (4.2.76), (4.2.103)–(4.2.104), and (4.2.106) to obtain (4.2.65) for H in B_2^+, without term involving h. Then, for each $\mathbf{x} \in B_2^+$, multiplying (4.2.65) for H with $\mathcal{D} = B_{d_{\mathbf{x}}/4}(\mathbf{x})$ by $d_{\mathbf{x}}^{1+\alpha}$ and combining these estimates as in the proof of Lemma 4.2.7, we have

$$[H]^{(1)}_{1,\alpha,B_2^+ \cup \Sigma_2} \le \delta[u]^*_{2,\alpha,B_2^+ \cup \Sigma_2} + C\|u\|^*_{2,0,B_2^+ \cup \Sigma_2}. \tag{4.2.111}$$

Since u satisfies equation (4.2.73) and the boundary condition (4.2.107), we apply estimate (4.2.99) in Corollary 4.2.11 to obtain

$$\|u\|^*_{2,\alpha,B_2^+\cup\Sigma_2} \le C\big(\|u\|_{0,B_2^+} + \|f\|^{(2)}_{0,\alpha,B_2^+\cup\Sigma_2} + \|(h,H)\|^{(1)}_{1,\alpha,\Sigma_2}\big).$$

From this, using (4.2.108) and (4.2.111), we have

$$\begin{aligned}&\|u\|^*_{2,\alpha,B_2^+\cup\Sigma_2}\\ &\le C\big(\delta[u]^*_{2,\alpha,B_2^+\cup\Sigma_2} + \|u\|^*_{2,0,B_2^+\cup\Sigma_2} + \|u\|_{0,B_2^+} + \|f\|^{(2)}_{0,\alpha,B_2^+\cup\Sigma_2} + \|h\|^{(1)}_{1,\alpha,\Sigma_2}\big).\end{aligned}$$

Using the interpolation inequality [131, Eq.(6.89)]:

$$\|u\|^*_{2,0,B_2^+\cup\Sigma_2} \le \delta[u]^*_{2,\alpha,B_2^+\cup\Sigma_2} + C(\alpha,\delta)\|u\|_{0,B_2^+}$$

on the right-hand side of the last estimate, choosing δ small, and using (4.2.56), we obtain (4.2.105).

3. If $\alpha > \beta$, we obtain (4.2.111) and the subsequent estimates with β instead of α, so that $\|u\|^*_{2,\beta,B_2^+\cup\Sigma_2} \le C$ holds. Now we can obtain (4.2.111) with α, and then the subsequent estimates imply (4.2.105). □

Remark 4.2.13. *The conditions for $A(\mathbf{p},z,\mathbf{x})$ in (4.2.4)–(4.2.6) in Theorems 4.2.1, 4.2.8–4.2.10, and 4.2.12 can be replaced by the following conditions:*

$$\begin{aligned}&A(\mathbf{p},z,\mathbf{x}) = A_1(\mathbf{p},z,\mathbf{x})p_1 + A_2(\mathbf{p},z,\mathbf{x})p_2 + A_0(\mathbf{p},z,\mathbf{x})z,\\ &\|A_i(\mathbf{p},z,\cdot)\|_{0,\alpha,\Omega} \le \lambda^{-1} \qquad \text{for all } (\mathbf{p},z)\in\mathbb{R}^2\times\mathbb{R},\\ &\|D_{(\mathbf{p},z)}A_i\|_{0,\mathbb{R}^2\times\mathbb{R}\times\overline{\Omega}} \le \lambda^{-1},\end{aligned} \tag{4.2.112}$$

where Ω is the respective domain considered in the conditions of the respective theorems. Then the conclusions of these theorems remain unchanged. This can be seen from the proofs of these theorems, which may be briefly summarized as follows: Condition (4.2.6) follows from (4.2.112) and, furthermore, conditions (4.2.4)–(4.2.6) for $A(\cdot)$ are used in the proofs in the following way:

(i) *To obtain that $|A(\mathbf{p},z,\mathbf{x})| \le C(\lambda)(|\mathbf{p}|+|z|)$. Clearly, (4.2.112) also implies this estimate.*

(ii) *To obtain (4.2.9) with constant C depending only on (λ, M). Under assumptions (4.2.112), we can obtain (4.2.9) with constant C depending only on $(\lambda, M, \|u\|_{1,\overline{\mathcal{D}}})$. However, in the proof, we obtain the estimate that $\|u\|_{1,\beta,\overline{\mathcal{D}}} \le C(\lambda, M)$, before we apply (4.2.9). With this estimate, conditions (4.2.112) imply (4.2.9) with C depending only on (λ, M). This allows us to complete the proof of the corresponding theorem without change.*

Remark 4.2.14. *Conditions* (4.2.6) *and* (4.2.14) *in Theorem* 4.2.4 *can be replaced by*

$$\begin{aligned} &A(\mathbf{p}, z, \mathbf{x}) = A_1(\mathbf{p}, z, \mathbf{x})p_1 + A_2(\mathbf{p}, z, \mathbf{x})p_2 + A_0(\mathbf{p}, z, \mathbf{x})z, \\ &\|A_i\|_{0,\mathbb{R}^2\times\mathbb{R}\times\Omega} \leq \lambda^{-1}. \end{aligned} \tag{4.2.113}$$

Then the conclusion of Theorem 4.2.4 *remains unchanged.*

This can be seen from the proof of Theorem 4.2.4. *Indeed,* (4.2.6) *and* (4.2.14) *are used only to obtain* (4.2.30). *Clearly,* (4.2.113), *combined with the strict ellipticity, also implies* (4.2.30).

Next we discuss an estimate of a solution of an oblique derivative problem near a corner, in the case that the equation in the domain is linear, but the boundary condition is nonlinear. In this case, we can allow a lower regularity of the coefficients for the lower-order terms of the equation.

Let $\lambda > 0, \alpha \in (0,1)$, $M > 0$, and

$$\Phi(0) = 0, \quad |\Phi'(0)| \leq \lambda^{-1}, \quad \|\Phi\|^{(-1-\alpha),\{0\}}_{2,\alpha,(0,\infty)} \leq M. \tag{4.2.114}$$

Denote

$$\mathcal{R}^{+,\Phi}_r = B_r(\mathbf{0}) \cap \{x_1 > 0, x_2 > \Phi(x_1)\} \qquad \text{for } r > 0. \tag{4.2.115}$$

We denote the boundary parts of $\mathcal{R}^{+,\Phi}_r$ by

$$\Gamma^{(l)}_r = \left(\partial\mathcal{R}^{+,\Phi}_r \cap \{x_1 = 0\}\right)^0, \; \Gamma^{(n)}_r = \left(\partial\mathcal{R}^{+,\Phi}_r \cap \{x_2 = \Phi(x_1)\}\right)^0, \; \mathbf{x}_0 = \mathbf{0}, \tag{4.2.116}$$

where $(\cdot)^0$ denotes the relative interior of a segment of a curve or line. From (4.2.114), it follows that there exists R_0 depending only on (λ, M, α) such that

$$\Gamma^{(n)}_r \quad \text{is connected for all } 0 < r \leq R_0. \tag{4.2.117}$$

For $R \in (0, R_0)$, we consider the following problem:

$$\sum_{i,j=1}^{2} a_{ij}(\mathbf{x})D_{ij}u + \sum_{i=1}^{2} a_i(\mathbf{x})D_i u + a_0(\mathbf{x})u = f \qquad \text{in } \mathcal{R}^{+,\Phi}_R, \tag{4.2.118}$$

$$B(Du, u, \mathbf{x}) = 0 \qquad \text{on } \Gamma^{(n)}_R, \tag{4.2.119}$$

$$\mathbf{b}^{(l)}(\mathbf{x}) \cdot Du + b^{(l)}_0(\mathbf{x})u = h^{(l)}(\mathbf{x}) \qquad \text{on } \Gamma^{(l)}_R. \tag{4.2.120}$$

Theorem 4.2.15. *Let $\lambda \in (0,1)$, $\kappa > 0$, and $M > 0$. Let Φ satisfy* (4.2.114). *Let $\mathcal{R}^{+,\Phi}_R$ be defined by* (4.2.115) *for some $R \in (0, R_0)$, where R_0 is such that* (4.2.117) *holds. Then there exists $\alpha_1 \in (0,1)$ depending only on (λ, κ) so that, for each $\alpha, \beta \in (0, \alpha_1)$ with $\beta \leq \alpha$, there exist $\delta \in (0,1)$ and $C > 0$ depending only on $(\kappa, \lambda, M, R, \alpha, \beta)$ with the following property: Let $u \in C^{2,\alpha}(\mathcal{R}^{+,\Phi}_R \cup$*

$\Gamma_R^{(n)}) \cap C^{1,\alpha}(\mathcal{R}_R^{+,\Phi} \cup \Gamma_R^{(l)} \cup \overline{\Gamma_R^{(n)}})$ *satisfy* (4.2.118)–(4.2.120). *Assume that the coefficients of* (4.2.118) *satisfy the ellipticity:*

$$\sum_{i,j=1}^{2} a_{ij}(\mathbf{x})\mu_i\mu_j \geq \lambda|\boldsymbol{\mu}|^2 \qquad \text{for all } \mathbf{x} \in \mathcal{R}_R^{+,\Phi},\ \boldsymbol{\mu} = (\mu_1, \mu_2) \in \mathbb{R}^2, \tag{4.2.121}$$

and

$$\|a_{ij}\|_{0,\alpha,\mathcal{R}_R^{+,\Phi}}^{(-\beta),\,\overline{\Gamma_R^{(l)}}} + \|(a_i, a_0)\|_{0,\alpha,\mathcal{R}_R^{+,\Phi}}^{(1-\alpha),\,\overline{\Gamma_R^{(l)}}} \leq M. \tag{4.2.122}$$

Assume that the boundary condition (4.2.119) *satisfies*

$$\|D_{(\mathbf{p},z)}B(\mathbf{p}, z, \cdot)\|_{1,\alpha,\Gamma_R^{(n)}}^{(-\alpha),\mathbf{x}_0} \leq M \qquad \text{for all } (\mathbf{p}, z) \in \mathbb{R}^2 \times \mathbb{R}, \tag{4.2.123}$$

$$\|D^2_{(\mathbf{p},z)}B\|_{C^1(\mathbb{R}^2\times\mathbb{R}\times\overline{\Gamma_R^{(n)}})} \leq M. \tag{4.2.124}$$

Let there be a nonhomogeneous linear operator:

$$L^{(n)}(\mathbf{p}, z, \mathbf{x}) = \mathbf{b}^{(n)}(\mathbf{x}) \cdot \mathbf{p} + b_0^{(n)}(\mathbf{x})z - h^{(n)}(\mathbf{x}),$$

defined for $(\mathbf{p}, z, \mathbf{x}) \in \mathbb{R}^2 \times \mathbb{R} \times \Gamma_R^{(n)}$, *and a function* v *satisfying*

$$\|v\|_{2,\alpha,\mathcal{R}_R^{+,\Phi}}^{(-1-\alpha),\,\Gamma_R^{(l)}} \leq M,$$

such that (4.2.102) *holds for any* $(\mathbf{p}, z, \mathbf{x}) \in \mathbb{R}^2 \times \mathbb{R} \times \Gamma_R^{(n)}$. *Assume that*

$$\lambda \leq \mathbf{b}^{(l)} \cdot \boldsymbol{\nu} \leq \lambda^{-1} \ \text{ on } \Gamma_R^{(l)}, \qquad \lambda \leq \mathbf{b}^{(n)} \cdot \boldsymbol{\nu} \leq \lambda^{-1} \ \text{ on } \Gamma_R^{(n)}, \tag{4.2.125}$$

$$\|(\mathbf{b}^{(l)}, b_0^{(l)})\|_{0,\alpha,\Gamma_R^{(l)}} + \|(\mathbf{b}^{(n)}, b_0^{(n)})\|_{1,\alpha,\Gamma_R^{(n)}}^{(-\alpha),\{\mathbf{x}_0\}} \leq M, \tag{4.2.126}$$

$$\left|\frac{\mathbf{b}^{(l)}}{|\mathbf{b}^{(l)}|} - \frac{\mathbf{b}^{(n)}}{|\mathbf{b}^{(n)}|}\right| \geq \kappa \ \text{ at } \mathbf{x}_0 = \mathbf{0}. \tag{4.2.127}$$

Furthermore, assume that

$$\|u\|_{0,\mathcal{R}_R^{+,\Phi}} + \|f\|_{0,\alpha,\mathcal{R}_R^{+,\Phi}}^{(1-\alpha),\,\Gamma_R^{(l)}} + \|h^{(l)}\|_{0,\alpha,\Gamma_R^{(l)}} + \|h^{(n)}\|_{1,\alpha,\Gamma_R^{(n)}}^{(-\alpha),\{\mathbf{x}_0\}} \leq M. \tag{4.2.128}$$

Then u *satisfies*

$$\|u\|_{2,\alpha,\mathcal{R}_{R/2}^{+,\Phi}}^{(-1-\alpha),\,\Gamma_{R/2}^{(l)}} \leq C. \tag{4.2.129}$$

Before establishing Theorem 4.2.15, we prove the local versions in some cases of intermediate Schauder estimates for linear equations.

Lemma 4.2.16. *Let* $\alpha, \beta, \sigma, \lambda \in (0, 1)$ *with* $\beta \leq \alpha$.

(i) *Let* $u \in C^{(-1-\sigma),\,\Sigma_1}_{2,\alpha,B_1^+}$ *satisfy equation* (4.2.118) *in half-ball* $B_2^+ = B_2 \cap \{x_1 > 0\}$ *and the boundary condition:*

$$\mathbf{b}(\mathbf{x}) \cdot Du + b_0(\mathbf{x})u = h(\mathbf{x}) \qquad \text{on } \Sigma_2 := B_2 \cap \{x_1 = 0\}, \tag{4.2.130}$$

where the equation and boundary conditions satisfy ellipticity (4.2.121) *in* B_2^+, *obliqueness* (4.2.75) *on* Σ_2, *and*

$$\|a_{ij}\|^{(-\beta),\,\Sigma_2}_{0,\alpha,B_2^+} + \|(a_i, a)\|^{(1-\sigma),\,\Sigma_2}_{0,\alpha,B_2^+} + \|\mathbf{b}\|_{0,\sigma,\Sigma_2} \le \lambda^{-1}.$$

Then u *satisfies*

$$\|u\|^{(-1-\sigma),\,\Sigma_1}_{2,\alpha,B_1^+} \le C\big(\|u\|_{0,B_2^+} + \|f\|^{(1-\sigma),\Sigma_2}_{0,\alpha,B_2^+} + \|h\|_{0,\sigma,\Sigma_2}\big), \tag{4.2.131}$$

where C *depends only on* $(\lambda, \alpha, \beta, \sigma)$.

(ii) *Let* $k \in \{0, 1\}$. *Let* $u \in C^{(-k-\sigma),\,\Sigma_1}_{2,\alpha,B_1^+}$ *satisfy equation* (4.2.118) *in half-ball* $B_2^+ = B_2 \cap \{x_1 > 0\}$ *and the boundary condition:*

$$u = h(\mathbf{x}) \qquad \text{on } \Sigma_2 := B_2 \cap \{x_1 = 0\}. \tag{4.2.132}$$

We assume that the coefficients of equation (4.2.118) *satisfy the same properties as in part* (i) *above. Then* u *satisfies*

$$\|u\|^{(-k-\sigma),\,\Sigma_1}_{2,\alpha,B_1^+} \le C\big(\|u\|_{0,B_2^+} + \|f\|^{(2-k-\sigma),\Sigma_2}_{0,\alpha,B_2^+} + \|h\|_{k,\sigma,\Sigma_2}\big), \tag{4.2.133}$$

where C *depends only on* $(\lambda, k, \alpha, \beta, \sigma)$.

Proof. We first prove part (i). The universal constant C below depends only on $(\lambda, \alpha, \beta, \sigma)$, unless otherwise specified.

In the proof, we will use the following partially interior Hölder norms, related to norms (4.1.5): For open $\Omega \subset \mathbb{R}^2$ and $E \subset \partial\Omega$, let

$$\begin{aligned} &\delta_{\mathbf{x}} := \operatorname{dist}(\mathbf{x}, \Sigma), \quad \delta_{\mathbf{x},\mathbf{y}} := \min(\delta_{\mathbf{x}}, \delta_{\mathbf{y}}), \\ &d_{\mathbf{x}} := \operatorname{dist}(\mathbf{x}, \partial\Omega \setminus \Sigma), \quad d_{\mathbf{x},\mathbf{y}} := \min(d_{\mathbf{x}}, d_{\mathbf{y}}) \end{aligned}$$

for $\mathbf{x}, \mathbf{y} \in \Omega$. Then, for nonnegative integers l and m, and for $k \in \mathbb{R}$ and $\alpha \in (0, 1)$, define the weights:

$$\begin{aligned} &d^{(k;\,l)}_{\mathbf{x}} := \min\big(d^l_{\mathbf{x}},\, d^{\min\{l,-k\}}_{\mathbf{x}}\delta^{\max\{l+k,0\}}_{\mathbf{x}}\big), \\ &d^{(k;\,m;\,\alpha)}_{\mathbf{x},\mathbf{y}} := \min\big(d^{m+\alpha}_{\mathbf{x},\mathbf{y}},\, d^{\min\{m+\alpha,-k\}}_{\mathbf{x},\mathbf{y}}\delta^{\max\{m+\alpha+k,0\}}_{\mathbf{x},\mathbf{y}}\big) \end{aligned}$$

and the norms:

$$
\begin{aligned}
&[u]^{*,(k),\Sigma}_{l,0,\Omega} := \sum_{|\beta|=l} \sup_{\mathbf{x}\in\Omega} \left(d_{\mathbf{x}}^{(k;\,l)} |D^\beta u(\mathbf{x})| \right), \\
&[u]^{*,(k),\Sigma}_{m,\alpha,\Omega} := \sum_{|\beta|=m} \sup_{\mathbf{x},\mathbf{y}\in\Omega,\mathbf{x}\neq\mathbf{y}} \left(d_{\mathbf{x},\mathbf{y}}^{(k;\,m;\,\alpha)} \frac{|D^\beta u(\mathbf{x}) - D^\beta u(\mathbf{y})|}{|\mathbf{x}-\mathbf{y}|^\alpha} \right), \\
&\|u\|^{*,(k),\Sigma}_{m,0,\Omega} := \sum_{l=0}^{m} [u]^{*,(k),\Sigma}_{l,0,\Omega}, \\
&\|u\|^{*,(k),\Sigma}_{m,\alpha,\Omega} := \|u\|^{*,(k),\Sigma}_{m,0,\Omega} + [u]^{*,(k),\Sigma}_{m,\alpha,\Omega}.
\end{aligned}
\tag{4.2.134}
$$

We first discuss the motivation for the choice of weights in norms (4.2.134). In the following discussion, we use the fact that $\min\{l,-k\}+\max\{l+k,0\}=l$ and $\min\{m+\alpha,-k\}+\max\{m+\alpha+k,0\}=m+\alpha$.

To fix the notations, we consider the case that $\Omega = B_2^+$ and $\Sigma = \Sigma_2$. First, the homogeneity of weights in norms $\|\cdot\|^{*,(k),\Sigma}_{m,\alpha,\Omega}$ and its seminorms is the same as for norms $\|\cdot\|^{*}_{m,\alpha,\Omega\cup\Sigma}$ in (4.1.3). Also, away from Σ for which $d_{\mathbf{x}} < \delta_{\mathbf{x}}$ and $d_{\mathbf{x},\mathbf{y}} < \delta_{\mathbf{x},\mathbf{y}}$, the weights in norms $\|\cdot\|^{*,(k),\Sigma}_{m,\alpha,\Omega}$ in (4.2.134) coincide with the weights in norms $\|\cdot\|^{*}_{m,\alpha,\Omega\cup\Sigma}$ in (4.1.3). On the other hand, for $\mathbf{x},\mathbf{y}\in B_1^+$, *i.e.*, away from $\Gamma := \partial\Omega\setminus\Sigma$, we see that $\delta_{\mathbf{x}}, \delta_{\mathbf{x},\mathbf{y}} < 1$ and $d_{\mathbf{x}}, d_{\mathbf{x},\mathbf{y}} > 1$, and then

$$[u]^{(-1-\sigma),\,\Sigma_1}_{2,\alpha,B_1^+} \le [u]^{*,(-1-\sigma),\,\Sigma_2}_{2,\alpha,B_2^+}.$$

Thus, to obtain (4.2.131), it suffices to show

$$\|u\|^{*,(-1-\sigma),\,\Sigma_2}_{2,\alpha,B_2^+} \le C\big(\|u\|_{0,B_2^+} + \|f\|^{(1-\sigma),\Sigma_2}_{0,\alpha,B_2^+} + \|h\|_{0,\sigma,\Sigma_2}\big). \tag{4.2.135}$$

Now we prove (4.2.135).

Fix an open region $\mathcal{D}$ with $\partial\mathcal{D}\in C^{2,\alpha}$ such that

$$B^+_{8/5} \subset \mathcal{D} \subset B^+_{9/5}.$$

Such $\mathcal{D}$ clearly exists. Let $\eta\in C^\infty(R^2)$ be such that $\eta\equiv 1$ in $B_1(\mathbf{0})$, and $\eta\equiv 0$ in $\mathbb{R}^2\setminus B_{6/5}(\mathbf{0})$.

Denote by L the operator on the left-hand side of (4.2.118). Then $v=\eta u$ satisfies

$$Lv = \hat{f} \qquad \text{in } \mathcal{D}, \tag{4.2.136}$$

where $\hat{f} := \eta f - uL\eta - \sum_{i,j=1}^2 a_{ij}(D_i\eta D_j u + D_j\eta D_i u)$.

In order to determine the boundary condition for v on $\partial\mathcal{D}$, we note that $\partial\mathcal{D}\cap(\operatorname{supp}\eta)\subset\Sigma_{6/5}$, so that functions $\eta\mathbf{b}, \eta b_0$, and ηh are well-defined on $\partial\mathcal{D}$, and the boundary condition for v is

$$\hat{\mathbf{b}}\cdot Dv + b_0 v = \hat{h} \qquad \text{on } \partial\mathcal{D},$$

where $\hat{\mathbf{b}} = \zeta\mathbf{b} + (1-\zeta)\boldsymbol{\nu}$, $\hat{h} = \eta h - u\mathbf{b}\cdot D\eta$, $\boldsymbol{\nu}$ is the interior normal to $\partial\mathcal{D}$, and $\zeta \in C^\infty(\mathbb{R}^2)$ is such that $\zeta \equiv 1$ in $B_{7/5}(\mathbf{0})$ and $\zeta \equiv 0$ in $\mathbb{R}^2 \setminus B_{8/5}(\mathbf{0})$. Note that this boundary condition has the same obliqueness and regularity properties of the coefficients with the modified constants which depend only on λ, since $\mathcal{D}$ is fixed. Now, from [186, Corollary] with parameters $a = 2+\alpha$ and $b = 1+\sigma$, we obtain

$$\|v\|^{(-1-\sigma),\,\partial\mathcal{D}}_{2,\alpha,\mathcal{D}} \le C\big(\|v\|_{0,\mathcal{D}} + \|\hat{f}\|^{(1-\sigma),\partial\mathcal{D}}_{0,\alpha,\mathcal{D}} + \|\hat{h}\|_{0,\sigma,\partial\mathcal{D}}\big), \tag{4.2.137}$$

which implies

$$\begin{aligned}\|u\|^{(-1-\sigma),\,\Sigma_1}_{2,\alpha,B_1^+} &\le C\big(\|u\|_{1,B_2^+} + \|f\|^{(1-\sigma),\Sigma_2}_{0,\alpha,B_2^+} + \|h\|_{0,\sigma,\Sigma_2}\big)\\ &\le C\big(\hat{C}(\delta)\|u\|_{0,B_2^+} + \delta[u]^{(-1-\sigma),\,\Sigma_2}_{2,\alpha,B_2^+} + \|f\|^{(1-\sigma),\Sigma_2}_{0,\alpha,B_2^+} + \|h\|_{0,\sigma,\Sigma_2}\big),\end{aligned}$$

where we have used the interpolation inequality to obtain the last expression. Note that $\delta > 0$ will be chosen below.

Furthermore, by the scaling argument as in Step 6 of the proof of Theorem 4.2.4, we see that, for any $r \in (0,1]$,

$$\begin{aligned}\|u\|^{(-1-\sigma),\,\Sigma_r}_{2,\alpha,B_r^+} \le \frac{C}{r^{1+\sigma}}\Big(&\hat{C}(\delta)\|u\|_{0,B_{2r}^+} + r^{1+\sigma}\delta[u]^{(-1-\sigma),\,\Sigma_{2r}}_{2,\alpha,B_{2r}^+}\\ &+ r\|f\|^{(1-\sigma),\Sigma_{2r}}_{0,\alpha,B_{2r}^+} + r\|h\|_{0,\sigma,\Sigma_{2r}}\Big).\end{aligned} \tag{4.2.138}$$

Similarly, for any $B_{2r}(\mathbf{x}_0) \subset B_2^+$, we apply Theorem 4.2.1 to equation (4.2.136), rewrite the resulting estimate (4.2.8) in terms of (u, f), and use the interpolation inequality on the right-hand side to obtain

$$\begin{aligned}&\|u\|_{2,\alpha,B_r^+(\mathbf{x}_0)}\\ &\le \frac{C}{r^{2+\alpha}}\big(\hat{C}(\delta)\|u\|_{0,B_{2r}^+(\mathbf{x}_0)} + \sigma r^{2+\alpha}[u]_{2,\alpha,B_{2r}^+(\mathbf{x}_0)} + r^2\|f\|_{0,\alpha,B_{2r}^+(\mathbf{x}_0)}\big).\end{aligned} \tag{4.2.139}$$

Next, we note that

(i) For any $\mathbf{x}$ and $\mathbf{y}$ in the boundary half-balls $B^+_{d_{\mathbf{x}_0}/10}(\mathbf{x}_0)$ with $x_0 \in \Sigma_2$, $\delta_{\mathbf{x}} \le d_{\mathbf{x}}$ so that $d^{(-1-\sigma;\,2+\alpha)}_{\mathbf{x},\mathbf{y}} = d^{1+\sigma}_{\mathbf{x},\mathbf{y}}\delta^{1+\alpha-\sigma}_{\mathbf{x},\mathbf{y}}$.

(ii) For any $\mathbf{x}$ and $\mathbf{y}$ in the interior balls $B_{d_{\mathbf{x}_0}/50}(\mathbf{x}_0) \subset B_2^+$ with $\frac{d_{\mathbf{x}_0}}{20} \le \delta_{\mathbf{x}_0}$, $\frac{d^{(-1-\sigma;\,2+\alpha)}_{\mathbf{x},\mathbf{y}}}{d^{2+\alpha}_{\mathbf{x},\mathbf{y}}} \in [(\frac{1}{20})^{1+\alpha-\sigma}, 1]$.

Also, $\frac{d_{\mathbf{x}}}{d_{\mathbf{x}_0}} \in (\frac{1}{2}, 2)$ for any $\mathbf{x}$ in the half-balls and balls in (i) and (ii), while these half-balls and balls cover all B_2^+.

Thus, for the half-balls described in (i), we use (4.2.138) in $B^+_{d_{\mathbf{x}_0}/10}(\mathbf{x}_0)$ with $r = \frac{d_{\mathbf{x}_0}}{10}$, and multiply this by $d^{1+\sigma}_{\mathbf{x}_0}$. For the balls described in (ii), we use

(4.2.139) in $B_{d_{\mathbf{x}_0}/50}(\mathbf{x}_0)$ with $r = \frac{d_{\mathbf{x}_0}}{50}$, and multiply this by $d_{\mathbf{x}_0}^{2+\alpha}$. Combining these estimates with the argument as in the proof of [131, Theorem 4.8] (*cf.* also the proof of Lemma 4.2.7 above), we have

$$\|u\|^{*,(-1-\sigma),\,\Sigma_2}_{2,\alpha,B_2^+} \le C\big(\hat{C}(\delta)\|u\|_{0,B_2^+} + \delta[u]^{*,(-1-\sigma),\,\Sigma_2}_{2,\alpha,B_2^+} + \|f\|^{(1-\sigma),\Sigma_2}_{0,\alpha,B_2^+} + \|h\|_{0,\sigma,\Sigma_2}\big).$$

Choosing δ small, we obtain (4.2.135) so that (4.2.131) holds.

Now we prove part (ii). We follow the proof of part (i) and use domain $\mathcal{D}$ and function η defined above. Then $v = \eta u$ satisfies equation (4.2.136) in $\mathcal{D}$, and the Dirichlet boundary condition:

$$v = \hat{h} \qquad \text{on } \partial\mathcal{D},$$

where $\hat{h} := \eta h$. We use that $\partial\mathcal{D} \cap (\operatorname{supp}\eta) \subset \Sigma_{6/5}$ to verify these boundary conditions. Now we apply the estimate for the Dirichlet problem in [130, Theorem 5.1], with parameters $\gamma = a = 2+\alpha$ and $b = k+\sigma$, to obtain

$$\|v\|^{(-k-\sigma),\,\partial\mathcal{D}}_{2,\alpha,\mathcal{D}} \le C\big(\|v\|_{0,\mathcal{D}} + \|\hat{f}\|^{(2-k-\sigma),\partial\mathcal{D}}_{0,\alpha,\mathcal{D}} + \|\hat{h}\|_{k,\sigma,\partial\mathcal{D}}\big). \tag{4.2.140}$$

Now we follow the argument in the proof of part (i) starting after (4.2.137), and use (4.2.140) instead of (4.2.137) to conclude (4.2.133). □

Proof of Theorem 4.2.15. From (4.2.114), there exists $\hat{R} \in (0, R]$ depending only on (λ, α, M) such that

$$\|\Phi'\|_{L^\infty((0,\hat{R}))} \le 2\lambda^{-1}. \tag{4.2.141}$$

Denote

$$\mathcal{R}_r^+ := \mathcal{R}_r^{+,\Phi} \qquad \text{for } \Phi \equiv 0.$$

By flattening boundary $\Gamma_R^{(n)}$ via the map: $\mathcal{A}(x_1, x_2) = (x_1, x_2 - \Phi(x_1))$, it follows from (4.2.114)–(4.2.115) that $\mathcal{A}$ is a diffeomorphism on $\mathbb{R}^2$, and there exists $R' \in (0, \hat{R}]$ depending only on (λ, α, M) such that $\mathcal{A}(\mathcal{R}_R^{+,\Phi}) \cap B_{R'}(\mathbf{0}) = \mathcal{R}_{R'}^+$. Moreover, the equations and boundary conditions, written in the new coordinates, have the same properties as in the original coordinates, with the modified constants. More precisely, it follows from (4.2.141) that the constants in (4.2.121) and (4.2.125) in the new coordinates depend only on λ, and the constant in (4.2.127) depends only on (λ, κ), while the constants in the other conditions depend only on (λ, α, M) by (4.2.114).

Then it suffices to prove the results in $\mathcal{R}_{R'}^+$ with α_1 depending only on the constants in conditions (4.2.121), (4.2.125), and (4.2.127), *i.e.*, on (λ, κ), and with C depending on the constants in all the conditions, *i.e.*, on $(\lambda, \kappa, \alpha, M)$, where we have used that R' depends also on these constants. Indeed, by (4.2.114) and (4.2.141), we obtain the estimate:

$$\|u\|^{(-1-\alpha),\,\overline{\Gamma_\rho^{(l)}}}_{2,\alpha,\mathcal{R}_\rho^{+,\Phi}} \le C$$

in the original coordinates for some $\rho := \rho(R, \lambda) = \rho(\lambda, \alpha, M)$ with unchanged α, and for C depending only on $(\lambda, \kappa, \alpha, M)$. Then, performing a covering argument by employing the interior estimates in Theorem 4.2.1, and using Theorem 4.2.12 in the balls of radius $r(\lambda, R)$ centered at boundaries $\Gamma_R^{(l)} \cup \Gamma_R^{(n)}$, we recover the full estimate (4.2.129).

Thus, it remains to prove the results in $\mathcal{R}_{R'}^+$. Assuming without loss of generality that $R' \le 2$, we can rescale to $\mathcal{R}_2^+$ so that, in the new coordinates, the constants in the assumptions for the equation and boundary conditions are unchanged. Since R' depends only on (λ, α, M), we rescale the resulting estimate back from $\mathcal{R}_2^+$ into $\mathcal{R}_{R'}^+$ to obtain the unchanged Hölder exponent, and the modified constant C depending only on $(\lambda, \kappa, \alpha, M)$. Then we reduce the case to the following:

$$R = 2, \quad \Phi \equiv 0, \quad \mathcal{R}_R^{+,\Phi} = \mathcal{R}_2^+.$$

For the rest of the proof, we work in this setting. We divide the proof into four steps.

1. Similarly to the proof of Theorem 4.2.12, introducing $\tilde{u} := u - v$, rewriting the problem in terms of $\tilde{u}$, and noting that the problem is of the same structure with similar estimates for its ingredients, we reduce the problem to the case that $v \equiv 0$. Thus, we assume without loss of generality that (4.2.106) holds, instead of (4.2.102).

For $0 < \rho < r \le 2$, denote

$$\begin{aligned} &\mathcal{R}_{[\rho,r]}^+ = \mathcal{R}_2^+ \cap \{\rho < |\mathbf{x}| < r\}, \\ &\Gamma_{[\rho,r]}^{(l)} = \Gamma_2^{(l)} \cap \{\rho < |\mathbf{x}| < r\}, \qquad \Gamma_{[\rho,r]}^{(n)} = \Gamma_2^{(n)} \cap \{\rho < |\mathbf{x}| < r\}, \end{aligned}$$

and

$$\Gamma_r = \Gamma_r^{(l)} \cup \Gamma_r^{(n)} \cup \{\mathbf{x}_0\}, \qquad \Gamma_{[\rho,r]} = \Gamma_2 \cap \{\rho < |\mathbf{x}| < r\}. \tag{4.2.142}$$

2. We first show that u satisfies the following estimate in $\mathcal{R}_{[1/5,9/5]}^+$:

$$\|u\|_{2,\alpha,\mathcal{R}_{[1/2,5/4]}^+}^{(-1-\alpha),\,\Gamma_{[1/2,5/4]}} \le C. \tag{4.2.143}$$

Indeed, in $\mathcal{D} = \mathcal{R}_{[1/3,3/2]}^+ \setminus \mathcal{N}_{1/20}(\Gamma_{[1/3,3/2]}^{(l)})$, we see from (4.2.122) that

$$\|(a_{ij}, a_i, a)\|_{0,\alpha,\overline{\mathcal{D}}} \le C,$$

where $\mathcal{N}_r(\cdot)$ denotes the r–neighborhood of a set. Thus, applying the properly scaled Theorem 4.2.12 in $B_{1/100}(\mathbf{x}) \cap \mathcal{D}$ for each $\mathbf{x} \in \Gamma_{[1/2,5/4]}^{(n)}$ and the standard interior Schauder estimates for linear elliptic equations in each $B_{1/200}(\mathbf{x}) \subset \mathcal{D}$, we have

$$\|u\|_{2,\alpha,\mathcal{R}_{[1/2,5/4]}^+ \setminus \mathcal{N}_{1/10}(\Gamma_{[1/2,5/4]}^{(l)})} \le C.$$

Applying (4.2.131), scaled from B_2^+ into half-balls $B_{1/4}(\mathbf{x}) \cap \mathcal{R}_2^+$ for each $\mathbf{x} \in \Gamma^{(l)}_{[1/2,5/4]}$, we complete the proof of (4.2.143).

3. Now we make the estimates near the corner. Similarly to the proof of Theorem 4.2.12, we rewrite the boundary conditions (4.2.119) as

$$\mathbf{b}^{(n)}(\mathbf{x}) \cdot Du + b_0^{(n)}(\mathbf{x})u = h^{(n)}(\mathbf{x}) + H^{(n)}(\mathbf{x}) \qquad \text{on } \Gamma_2^{(n)}, \tag{4.2.144}$$

where $H^{(n)}(\mathbf{x}) := L^{(n)}(Du, u, \mathbf{x}) - B^{(n)}(Du, u, \mathbf{x})$. Using the assumptions on $(B^{(n)}, L^{(n)})$, *i.e.*, (4.2.106), (4.2.123)–(4.2.124), and (4.2.126), and arguing as in the proof of (4.2.38) and (4.2.40), we obtain that, for each $\rho \in (0, 2]$,

$$\|H^{(n)}\|_{0,\alpha,\Gamma_\rho^{(n)}} \le \|H^{(n)}\|_{0,\alpha,\mathcal{R}_\rho^+} \le \delta \|u\|_{1,\alpha,\mathcal{R}_\rho^+} + C\|u\|_{1,0,\mathcal{R}_\rho^+}.$$

Now, u satisfies the linear elliptic equation (4.2.118) and the linear oblique boundary conditions (4.2.120) and (4.2.144). Also, (4.2.143) implies that u satisfies the condition:

$$\partial_{\boldsymbol{\nu}} u = h^{(\Sigma)} \qquad \text{on } \Sigma_{6/5} := \partial\mathcal{R}_{6/5}^+ \setminus \overline{\Gamma_{6/5}}, \tag{4.2.145}$$

where $\Gamma_{6/5}$ is defined by (4.2.142) with

$$\|h^{(\Sigma)}\|_{0,\alpha,\Sigma_{6/5}} \le C.$$

Furthermore, we can modify the coefficients of the conditions on $\Gamma^{(n)} \cup \Gamma^{(l)}$ within $\Gamma_{6/5} \setminus B_1$, without changing notations, so that, at corners $\{\mathbf{x}^{(n)}, \mathbf{x}^{(l)}\}$ where $\Sigma_{6/5}$ meets $\Gamma^{(n)}_{6/5}$ and $\Gamma^{(l)}_{6/5}$, respectively, the resulting coefficients satisfy $\mathbf{b}^{(l)}(\mathbf{x}^{(l)}) = \boldsymbol{\nu}_{\Gamma^{(l)}}$ and $\mathbf{b}^{(n)}(\mathbf{x}^{(n)}) = \boldsymbol{\nu}_{\Gamma^{(n)}}$, obliqueness (4.2.125) is satisfied on $\Gamma_{6/5}$ with the same constant, and (4.2.126) is satisfied with an updated constant depending only on (λ, α). We note that all the properties are achieved by defining the modified function $\mathbf{b}^{(l)}$ at $\mathbf{x}_r = \Gamma^{(l)} \cap \partial B_r$ by $\eta(r)\mathbf{b}^{(l)} + (1-\eta(r))\boldsymbol{\nu}_{\Gamma^{(l)}}$ for $r \in [1, \frac{6}{5}]$, where $\eta(\cdot)$ is smooth and monotone on $\mathbb{R}$ with $\eta \equiv 1$ on $(-\infty, 1]$ and $\eta \equiv 0$ on $[\frac{6}{5}, \infty)$. On $\Gamma^{(n)}$, the definition is similar.

Then, from (4.2.128), (4.2.143), and the estimates of the modified coefficients discussed above, u satisfies these modified conditions on $\Gamma_{6/5}$ with the modified right-hand sides $(h^{(l)}, h^{(n)})$ satisfying (4.2.128) on $\Gamma_{6/5}$ with the constant depending only on (λ, α). Now, at corners $\{\mathbf{x}^{(n)}, \mathbf{x}^{(l)}\}$, all the boundary conditions are in the normal direction to the respective boundary curves, so that a condition similar to (4.2.127) is satisfied at the corner points $\mathbf{x}^{(n)}$ and $\mathbf{x}^{(l)}$ with uniform constant $\sqrt{2}$, from the definition of domain $\mathcal{R}^+_{6/5}$.

Lemma 1.3 in [193] implies the existence of $\alpha_1 \in (0,1)$ depending only on (κ, λ) such that, if $\alpha \in (0, \alpha_1)$, then

$$\begin{aligned}\|u\|^{(-1-\alpha),\,\partial\mathcal{R}^+_{6/5}}_{2,\alpha,\mathcal{R}^+_{6/5}} \le C\big(&\|u\|_{0,\mathcal{R}^+_{6/5}} + \|f\|^{(1-\alpha),\,\Gamma^{(l)}_{3/2}}_{\alpha,\mathcal{R}^+_{6/5}} + \|h^{(l)}\|_{0,\alpha,\Gamma^{(l)}_{6/5}} \\ &+ \|h^{(n)} + H^{(n)}\|_{0,\alpha,\Gamma^{(n)}_{6/5}} + \|h^{(\Sigma)}\|_{0,\alpha,\Sigma_{6/5}}\big),\end{aligned}$$

where $C = C(\kappa, \lambda, M, \alpha, \beta)$.

Using the estimates of $H^{(n)}$ and $h^{(\Sigma)}$ obtained above, we have

$$\|u\|^{(-1-\alpha),\,\partial\mathcal{R}^+_{6/5}}_{2,\alpha,\mathcal{R}^+_{6/5}} \le C\big(\delta[u]_{1,\alpha,\mathcal{R}^+_{6/5}} + \|u\|_{1,0,\mathcal{R}^+_{6/5}} + \|u\|_{0,\mathcal{R}^+_2} + \|f\|_{0,\alpha,\mathcal{R}^+_2} + \|h^{(l)}\|_{0,\alpha,\Gamma^{(l)}_2} + \|h^{(n)}\|_{0,\alpha,\Gamma^{(n)}_2} + 1\big).$$

Using the interpolation inequality:

$$\|u\|_{1,0,\mathcal{R}^+_{6/5}} \le \delta[u]_{1,\alpha,\mathcal{R}^+_{6/5}} + C(\alpha,\delta)\|u\|_{0,\mathcal{R}^+_{6/5}}$$

on the right-hand side of the last estimate, choosing δ small, and then using (4.2.128), we obtain that $\|u\|^{(-1-\alpha),\,\partial\mathcal{R}^+_{6/5}}_{2,\alpha,\mathcal{R}^+_{6/5}} \le C$, which implies

$$\|u\|^{(-1-\alpha),\,\partial\mathcal{R}^+_{7/6}}_{2,\alpha,\mathcal{R}^+_{7/6}} \le C. \tag{4.2.146}$$

4. In order to complete the proof of (4.2.105), we need to prove the higher regularity near $\Gamma^{(n)}_1$. Let $\hat{\mathbf{x}} \in \Gamma^{(n)}_1$. Denote $d_{\mathbf{x}} := \operatorname{dist}(\mathbf{x}, \Gamma^{(n)}_2)$. Then $d_{\hat{\mathbf{x}}} = \operatorname{dist}(\hat{\mathbf{x}}, \mathbf{x}_0) = |\hat{\mathbf{x}}|$. Rescale u from half-ball $B_{\frac{d_{\hat{\mathbf{x}}}}{2}}(\hat{\mathbf{x}}) \cap \mathcal{R}^+_2$ to half-ball B^+_2 by defining

$$v^{(\hat{\mathbf{x}})}(\mathbf{X}) = \frac{1}{d^{1+\alpha}_{\hat{\mathbf{x}}}}\Big(u(\hat{\mathbf{x}} + \frac{d_{\mathbf{x}}}{4}R(\mathbf{X})) - u(\mathbf{x}_0) - \frac{d_{\mathbf{x}}}{4}Du(\mathbf{x}_0)\cdot R(\mathbf{X})\Big),$$

where $R(\mathbf{X}) = X_1\boldsymbol{\nu} + X_2\boldsymbol{\tau}$. Then $\|v^{(\hat{\mathbf{x}})}\|_{1,\alpha,B^+_2} \le \|u\|_{1,\alpha,B_{d_{\hat{\mathbf{x}}}/2}(\hat{\mathbf{x}})\cap\mathcal{R}^+_2} \le C$ by (4.2.143) and (4.2.146). The equation in B^+_2 and the boundary condition on Σ_2 for $v^{(\hat{\mathbf{x}})}$, obtained from (4.2.118)–(4.2.119), satisfy the conditions of Theorem 4.2.12. Thus, $v^{(\hat{\mathbf{x}})}$ satisfies (4.2.105). Then, writing this estimate in terms of $u(\mathbf{x})$ and following an argument similar to the proof of Lemma 4.2.7, we conclude the proof of (4.2.129). □

4.3 ESTIMATES FOR LIPSCHITZ SOLUTIONS OF ELLIPTIC BOUNDARY VALUE PROBLEMS

In this section we state and prove some results by Lieberman [192], as well as some related results, which are used in the later sections. The results in [192] are obtained for fully nonlinear elliptic equations with nonlinear first-order boundary conditions. We state these results for quasilinear equations with less general assumptions than those in [192] (for convenience in later applications in this book). The difference from the results in §4.2 is that the $C^{k,\alpha}$–estimates of solutions in this section, $k = 1, 2$, are obtained for more general nonlinear boundary conditions, without assumptions on the linear structure as in §4.2. However, now the estimates depend on the $C^{0,1}$–norm of the solution, while, in §4.2, the estimates depend only on the L^∞–norm of solutions.

Remark 4.3.1. *In later chapters, the estimates of this section are used for the a priori estimates of the regular reflection-diffraction solutions, where the boundary conditions, derived from the Rankine-Hugoniot conditions, are not of a linear structure as required in §4.2, but the Lipschitz estimates of the solutions are available. The estimates in §4.2 are used for solving the iteration problem, where the boundary condition is defined so that it is of a linear structure, but the Lipschitz estimate of the solution is not available a priori.*

Some calculations and notations below are from the proof of [192, Theorem 2.1], given in [111, Appendix 5.1].

4.3.1 Estimates near the Lipschitz boundary

Theorem 4.3.2. *Let $R > 0$, $\lambda \in (0,1)$, $\gamma \in [0,1)$, and $K > 0$. Let $\Phi \in C^1(\mathbb{R})$ satisfy*

$$\|\Phi\|_{0,1,\mathbb{R}} \le \lambda^{-1}, \qquad \Phi(0) = 0, \tag{4.3.1}$$

let $\Omega_r := B_r \cap \{x_2 > \Phi(x_1)\}$ and $\Gamma_r := B_r \cap \{x_2 = \Phi(x_1)\}$ for $r > 0$, and let $d(\mathbf{x}) := \mathrm{dist}(\mathbf{x}, \Gamma_R)$ for $\mathbf{x} \in \Omega_R$. Assume that $u \in C^3(\Omega_R) \cap C^1(\overline{\Omega_R})$ is a solution of

$$\sum_{i,j=1}^{2} A_{ij}(Du, u, \mathbf{x}) D_{ij} u + A(Du, u, \mathbf{x}) = 0 \qquad \text{in } \Omega_R, \tag{4.3.2}$$

$$B(Du, u, \mathbf{x}) = 0 \qquad \text{on } \Gamma_R. \tag{4.3.3}$$

Let $A_{ij}(\mathbf{p}, z, \mathbf{x})$, $A(\mathbf{p}, z, \mathbf{x})$, and $z \in \mathbb{R}$ satisfy that, for any $\mathbf{x} \in \Omega_R$ and $|\mathbf{p}| + |z| \le 2K$,

$$\begin{aligned} \lambda |\boldsymbol{\mu}|^2 \le \sum_{i,j=1}^{2} & A_{ij}(Du(\mathbf{x}), u(\mathbf{x}), \mathbf{x}) \mu_i \mu_j \\ & \le \lambda^{-1} |\boldsymbol{\mu}|^2 \qquad \text{for all } \boldsymbol{\mu} = (\mu_1, \mu_2) \in \mathbb{R}^2, \end{aligned} \tag{4.3.4}$$

$$|(A_{ij}, A)(\mathbf{p}, z, \mathbf{x})| \le \lambda^{-1}, \tag{4.3.5}$$

$$|D_{(\mathbf{p},z)}(A_{ij}, A)(\mathbf{p}, z, \mathbf{x})| + [d(\mathbf{x})]^{\gamma} |D_{\mathbf{x}}(A_{ij}, A)(\mathbf{p}, z, \mathbf{x})| \le \lambda^{-1}, \tag{4.3.6}$$

and let $B(\mathbf{p}, z, \mathbf{x})$ satisfy

$$|D_{\mathbf{p}} B(Du(\mathbf{x}), u(\mathbf{x}), \mathbf{x})| \ge \lambda \qquad \text{for all } \mathbf{x} \in \overline{\Omega_R}, \tag{4.3.7}$$

$$\|B\|_{2,\{|\mathbf{p}|+|z| \le 2K,\ \mathbf{x} \in \overline{\Omega_R}\}} \le \lambda^{-1}. \tag{4.3.8}$$

Assume that u satisfies

$$|u| + |Du| \le K \qquad \text{on } \Omega_R \cup \Gamma_R. \tag{4.3.9}$$

Then there exist $\beta \in (0,1]$ depending only on (λ, K, γ), and $C > 0$ depending only on (R, λ, K, γ), such that

$$\|u\|_{1,\beta,\overline{\Omega}_{R/2}} \le C \tag{4.3.10}$$

and

$$\|u\|^{(-1-\beta),\Gamma_{R/2}}_{2,\beta,\Omega_{R/2}} \le C. \tag{4.3.11}$$

Proof. We may assume without loss of generality that

$$A_{12}(\mathbf{p}, z, \mathbf{x}) = A_{21}(\mathbf{p}, z, \mathbf{x}), \tag{4.3.12}$$

which can be achieved by replacing A_{12} and A_{21} by $\frac{1}{2}(A_{12} + A_{21})$. We divide the proof into six steps.

1. First we show that the proof may be reduced to Case $R = 1$.

Indeed, if $R \le 1$, we reduce to $R = 1$ by scaling, *i.e.*, rewriting the problem in terms of $\tilde{u}(\mathbf{x}) := \frac{u(R\mathbf{x})-u(\mathbf{0})}{R}$ on Ω_1. Then the rescaled domain $\tilde{\Omega}_1$ has the boundary part $\tilde{\Gamma}_1 = \partial\tilde{\Omega}_1 \cap B_1$ with the same Lipschitz constant as for Γ_R, $\tilde{u}(\mathbf{x})$ satisfies (4.3.9) in $\tilde{\Omega}_1$ with the same K, and the equation and boundary conditions written in terms of $\tilde{u}(\mathbf{x})$ satisfy the assumptions in the theorem with the same constants (λ, γ, M), since $R \le 1$. When estimates (4.3.10)–(4.3.11), proved for $\tilde{u}$ in $\tilde{\Omega}_{1/2}$, rewritten in terms of u and β remain unchanged, constant C depends on R.

If $R > 1$, we cover $\Omega_{R/2}$ by the balls of radius $\frac{1}{2}$ centered on $\Gamma_{R/2}$ and by the interior balls $B_{1/4}(\mathbf{x}) \subset \Omega_R$. Combining estimates (4.3.11)–(4.3.12) for the boundary balls with the interior estimates in Theorem 4.2.1 for the interior balls, we obtain (4.3.10)–(4.3.11) with β independent of R.

Thus, for the rest of the proof, we fix $R = 1$. Then $d(\mathbf{x}) = \mathrm{dist}(\mathbf{x}, \Gamma_1)$ for $\mathbf{x} \in \Omega_1$. All the constants in this proof depend only on (λ, K, γ). We also use the notations:

$$(a_{ij}, a, g)(\mathbf{x}) := (A_{ij}, A, B)(Du(\mathbf{x}), u(\mathbf{x}), \mathbf{x}).$$

2. By [214], there exist $\delta > 0$ and $L_0 > 1$ (depending only on λ) and a function $w \in C^2(\Omega_1) \cap C^0(\overline{\Omega_1})$ such that

$$\begin{aligned} &\sum_{i,j=1}^{2} a_{ij} D_{ij} w(\mathbf{x}) \le -[d(\mathbf{x})]^{\delta-2}, \\ &\frac{1}{L_0}[d(\mathbf{x})]^{\delta} \le w(\mathbf{x}) \le L_0[d(\mathbf{x})]^{\delta}, \quad |\nabla w(\mathbf{x})| \le L_0[d(\mathbf{x})]^{\delta-1}, \end{aligned} \tag{4.3.13}$$

where the upper and lower bounds of $w(x)$ follow from [214, Eq. (53)] and Lemma 3.6 with $k = 0$. With these bounds, the other estimates follow from [214, Theorem 3.7] and [215, Definition 3.1].

Then it follows that there exists $L_1 = L_1(\lambda) > 1$ such that, for any $\varepsilon \in (0, 1]$, $\mathbf{x} \in \Omega_1$,

$$\begin{aligned} &\sum_{i,j=1}^{2} a_{ij} D_{ij} w^{\varepsilon}(\mathbf{x}) \le -\frac{\varepsilon}{L_1}[d(\mathbf{x})]^{\varepsilon\delta-2}, \\ &\frac{[d(\mathbf{x})]^{\varepsilon\delta}}{L_1} \le w^{\varepsilon}(\mathbf{x}) \le L_1[d(\mathbf{x})]^{\varepsilon\delta}, \quad |\nabla w^{\varepsilon}(\mathbf{x})| \le \varepsilon L_1[d(\mathbf{x})]^{\varepsilon\delta-1}. \end{aligned} \tag{4.3.14}$$

Denote

$$v(\mathbf{x}) = g(\mathbf{x}) + Sw^\varepsilon(\mathbf{x}) + S|\mathbf{x}|^2, \tag{4.3.15}$$

where

$$S = \hat{d}^{-\varepsilon\delta}\varepsilon^{-\frac{1}{2}}, \tag{4.3.16}$$

and $\varepsilon, \hat{d} \in (0,1]$ will be determined below. The assumptions in the theorem imply $v \in C^2(\Omega) \cap C(\overline{\Omega})$.

It suffices to prove the estimates in $\Omega^{(\hat{d})} := \Omega_1 \cap \{d(\mathbf{x}) \le \hat{d}\}$, where $\hat{d} = \hat{d}(\lambda, \gamma, K) \in (0,1)$ will be determined below. Then, combining this with the interior estimates of Theorem 4.2.1 in balls $B_{\hat{d}/4}(\mathbf{x})$ for $\mathbf{x} \in \Omega_1 \cap \{d(\mathbf{x}) \ge \frac{\hat{d}}{2}\}$, we obtain (4.3.10)–(4.3.11) with $R = 1$.

3. We now derive the boundary conditions for v in $\Omega^{(\hat{d})} = \Omega_1 \cap \{d(\mathbf{x}) \le \hat{d}\}$. Since

$$\partial\Omega^{(\hat{d})} = \overline{\Gamma}_1 \cup \big(\partial\Omega^{(\hat{d})} \cap \partial B_1\big) \cup \big(\overline{\Omega}_1 \cap \{d(\mathbf{x}) = \hat{d}\}\big),$$

we employ (4.3.3) and (4.3.14)–(4.3.16), use the fact that $g(\mathbf{x}) \ge -\lambda^{-1}$ in $\overline{\Omega}_1$ by (4.3.8)–(4.3.9), and choose small $\varepsilon > 0$ depending only on (λ, K) to compute:

$$\begin{aligned} &v = S(w^\varepsilon + x^2) \ge 0 \qquad \text{on } \Gamma_1,\\ &v(\mathbf{x}) \ge g(\mathbf{x}) + S \ge -\lambda^{-1} + \varepsilon^{-\frac{1}{2}} \ge 0 \qquad \text{on } \partial\Omega_1 \cap \partial B_1,\\ &v \ge g + Sw^\varepsilon \ge g + \varepsilon^{-\frac{1}{2}}\hat{d}^{-\varepsilon\delta}\frac{\hat{d}^{\varepsilon\delta}}{L_1} \ge -\lambda^{-1} + \frac{\varepsilon^{-\frac{1}{2}}}{L_1} \ge 0 \quad \text{on } \overline{\Omega}_1 \cap \{d(\mathbf{x}) = \hat{d}\}. \end{aligned}$$

Thus, for $\varepsilon > 0$ chosen small as above,

$$v \ge 0 \qquad \text{on } \partial\Omega^{(\hat{d})}. \tag{4.3.17}$$

4. We now derive the differential inequality for v in $\Omega^{(\hat{d})} = \Omega_1 \cap \{d(\mathbf{x}) \le \hat{d}\}$. Notice that

$$D_i g = D_i(B(Du, u, \mathbf{x})) = B_{\mathbf{p}} \cdot D(D_i u) + B_z D_i u + B_{x_i} \qquad \text{for } i = 1, 2. \tag{4.3.18}$$

We write these expressions and equation (4.3.2) as a linear algebraic system for $(D_{11}u, D_{12}u, D_{22}u)$:

$$\begin{bmatrix} a_{11} & 2a_{12} & a_{22} \\ B_{p_1} & B_{p_2} & 0 \\ 0 & B_{p_1} & B_{p_2} \end{bmatrix} \begin{bmatrix} D_{11}u \\ D_{12}u \\ D_{22}u \end{bmatrix} = \begin{bmatrix} 0 \\ D_1 g \\ D_2 g \end{bmatrix} - \begin{bmatrix} a \\ B_z D_1 u + B_{x_1} \\ B_z D_2 u + B_{x_2} \end{bmatrix}. \tag{4.3.19}$$

The determinant of the matrix in this system is

$$a_{11}(B_{p_2})^2 - 2a_{12}B_{p_1}B_{p_2} + a_{22}(B_{p_1})^2 \ge \lambda|D_{\mathbf{p}}B|^2 \ge \lambda^3, \tag{4.3.20}$$

where we have used (4.3.7), and (4.3.4) with $\boldsymbol{\mu} = (-B_{p_2}, B_{p_1})$. Then we have

$$\begin{bmatrix} D_{11}u \\ D_{12}u \\ D_{22}u \end{bmatrix} = Q(\mathbf{x})Dg + R(\mathbf{x}), \tag{4.3.21}$$

where Q is a 3×2 matrix, and

$$|(Q, R)(\mathbf{x})| \leq C, \tag{4.3.22}$$

by (4.3.8)–(4.3.9) and (4.3.20).

Now we compute

$$\begin{aligned}\sum_{i,j=1}^{2} a_{ij} D_{ij} g = &\sum_{i,j,k} B_{p_k} a_{ij} D_{kij} u \\ &+ \sum_{i,j,k} a_{ij} \big(B_{p_k p_l} D_{ik} u D_{jl} u + B_{p_k z} D_{ik} u D_j u + B_{p_k z} D_{jk} u D_i u \\ &\qquad + B_{p_k x_j} D_{ik} u + B_{p_k x_i} D_{jk} u \big) \\ &+ \sum_{i,j} a_{ij} \big(B_z D_{ij} u + 2 B_{z x_i} D_j u + B_{x_i x_j} \big).\end{aligned} \tag{4.3.23}$$

Taking the partial derivative with respect to x_k on both sides of equation (4.3.2), we have

$$\begin{aligned}&\sum_{i,j=1}^{2} \big(a_{ij} D_{kij} u + \partial_{p_l} A_{ij}\, D_{ij} u\, D_{kl} u + \partial_z A_{ij}\, D_{ij} u\, D_k u + \partial_{\mathbf{x}} A_{ij}\, D_{ij} u \big) \\ &\quad + \sum_{l=1}^{2} A_{p_l} D_{kl} u + A_z D_k u + A_{x_k} = 0.\end{aligned} \tag{4.3.24}$$

We replace the term of $B_{p_k} a_{ij} D_{kij} u$ in (4.3.23) by its expression from (4.3.24). In the resulting equation (which does not contain the third derivatives of u), we replace $D_{ij} u$ by their expression from (4.3.21) to obtain the equation for g:

$$\sum_{i,j=1} \big(a_{ij} D_{ij} g + m_{ij} D_i g D_j g \big) + [d(\mathbf{x})]^{-\gamma} \big(\sum_{i=1}^{2} m_i D_i g + m \big) = 0 \quad \text{in } \Omega^{(\hat{d})}, \tag{4.3.25}$$

where

$$|(m_{ij}, m_i, m)(\mathbf{x})| \leq C \qquad \text{in } \Omega^{(\hat{d})}, \tag{4.3.26}$$

by (4.3.5)–(4.3.6), (4.3.8)–(4.3.9), (4.3.12), and (4.3.22) so that C depends only on (λ, γ, K).

Now we derive the differential inequality for v given by (4.3.15). Denote

$$q_i = \sum_{j=1}^{2} \big(m_{ij} v_j - S(m_{ij} + m_{ji})((w^\varepsilon)_{x_j} + 2x_j) \big) + [d(\mathbf{x})]^{-\gamma} m_i,$$

where S is defined by (4.3.16). Then, substituting $D^k g = D^k v - S\big(D^k(w^\varepsilon) + D^k(x^2)\big)$ into (4.3.25) and using (4.3.14), we obtain that, in $\Omega^{(\hat{d})} = \Omega_1 \cap \{d(\mathbf{x}) \le \hat{d}\}$,

$$
\begin{aligned}
&\sum_{i,j=1}^{2} a_{ij} D_{ij} v + \sum_{i=1}^{2} q_i D_i v \\
&= \sum_{i,j=1}^{2} \big(S a_{ij}(w^\varepsilon + |\mathbf{x}|^2)_{x_i x_j} - S^2 m_{ij}((w^\varepsilon)_{x_i} + 2x_i)((w^\varepsilon)_{x_j} + 2x_j)\big) \\
&\quad + [d(\mathbf{x})]^{-\gamma}\Big(\sum_{i=1}^{2} S m_i((w^\varepsilon)_{x_i} + 2x_i) - m\Big) \qquad (4.3.27)\\
&\le -\frac{\varepsilon}{L_1} S[d(\mathbf{x})]^{\varepsilon\delta-2} + CS + C\big(\varepsilon^2 S^2 [d(\mathbf{x})]^{2\varepsilon\delta-2} + S^2[d(\mathbf{x})]^2\big) \\
&\quad + C\big(\varepsilon S[d(\mathbf{x})]^{\varepsilon\delta-\gamma-1} + [d(\mathbf{x})]^{-\gamma}\big) \\
&= I_1 + I_2,
\end{aligned}
$$

where

$$
\begin{aligned}
I_1 &= -\frac{\varepsilon}{L_1} S[d(\mathbf{x})]^{\varepsilon\delta-2} + C\varepsilon^2 S^2 [d(\mathbf{x})]^{2\varepsilon\delta-2} + C\varepsilon S[d(\mathbf{x})]^{\varepsilon\delta-\gamma-1}, \\
I_2 &= CS + CS^2[d(\mathbf{x})]^2 + C[d(\mathbf{x})]^{-\gamma}.
\end{aligned}
$$

We first estimate I_1. Using (4.3.16), $0 < d(\mathbf{x}) < \hat{d} < 1$ in $\Omega^{(\hat{d})}$, and $\gamma < 1$, we have

$$
\begin{aligned}
I_1 &= \frac{S}{L_1}\varepsilon[d(\mathbf{x})]^{\varepsilon\delta-2}\Big(-1 + C\sqrt{\varepsilon}\big[\frac{d(\mathbf{x})}{\hat{d}}\big]^{\varepsilon\delta} + C[d(\mathbf{x})]^{1-\gamma}\Big) \\
&\le \frac{S}{L_1}\varepsilon[d(\mathbf{x})]^{\varepsilon\delta-2}\left(-1 + C\sqrt{\varepsilon} + C\hat{d}^{1-\gamma}\right).
\end{aligned}
$$

To estimate the last expression, we choose $(\varepsilon, \hat{d})$ so that $\varepsilon \in (0,1)$ is small enough with $C\sqrt{\varepsilon} \le \frac{1}{4}$ and all the previous requirements are satisfied. Now $\varepsilon > 0$ is fixed for the rest of proof. Then we choose $\hat{d} = \hat{d}(\lambda, K, \gamma)$ small so that $C\hat{d}^{1-\gamma} \le \frac{1}{4}$. We obtain

$$
I_1 \le -\frac{S}{2L_1}\varepsilon[d(\mathbf{x})]^{\varepsilon\delta-2}.
$$

Now we estimate $I_1 + I_2$. We compare $\frac{I_1}{3}$ with each of the three terms in I_2. For the first term, we use $2 - \varepsilon\delta > 1$ and recall that C is a universal constant independent of ε and L_1 to obtain that, if $\hat{d}$ is small,

$$\begin{aligned}\frac{I_1}{3}+CS &\le -\frac{S}{6L_1}\varepsilon[d(\mathbf{x})]^{\varepsilon\delta-2}+CS\\ &= \frac{S}{6L_1}\varepsilon[d(\mathbf{x})]^{\varepsilon\delta-2}\big(-1+C\varepsilon^{-1}[d(\mathbf{x})]^{2-\varepsilon\delta}\big)\\ &\le \frac{S}{6L_1}\varepsilon[d(\mathbf{x})]^{\varepsilon\delta-2}\big(-1+C\varepsilon^{-1}\hat{d}\big)<0.\end{aligned}$$

For the next term, we employ (4.3.16), $\varepsilon\delta<1$, and $0<d(\mathbf{x})<\hat{d}$ to obtain that, if $\hat{d}$ is small,

$$\begin{aligned}\frac{I_1}{3}+C[d(\mathbf{x})]^2S^2 &\le -\frac{S}{6L_1}\varepsilon[d(\mathbf{x})]^{\varepsilon\delta-2}+CS^2[d(\mathbf{x})]^2\\ &= \frac{S}{6L_1}\varepsilon[d(\mathbf{x})]^{\varepsilon\delta-2}\Big(-1+C\varepsilon^{-\frac{3}{2}}[d(\mathbf{x})]^{4-2\varepsilon\delta}\big(\frac{d(\mathbf{x})}{\hat{d}}\big)^{\varepsilon\delta}\Big)\\ &\le \frac{S}{6L_1}\varepsilon[d(\mathbf{x})]^{\varepsilon\delta-2}\left(-1+C\varepsilon^{-\frac{3}{2}}\hat{d}^2\right)<0.\end{aligned}$$

For the last term, we use $2-\varepsilon\delta-\gamma>0$ and choose $\hat{d}$ small to find

$$\begin{aligned}\frac{I_1}{3}+C[d(\mathbf{x})]^{-\gamma} &\le -\frac{S\varepsilon}{6L_1}[d(\mathbf{x})]^{\varepsilon\delta-2}+C[d(\mathbf{x})]^{-\gamma}\\ &= \frac{S\varepsilon}{6L_1}[d(\mathbf{x})]^{\varepsilon\delta-2}\left(-1+C\varepsilon^{-\frac{1}{2}}[d(\mathbf{x})]^{2-\varepsilon\delta-\gamma}\right)<0.\end{aligned}$$

Therefore, we have

$$\sum_{i,j=1}^{2}a_{ij}D_{ij}v+\sum_{i=1}^{2}q_iD_iv<0 \qquad \text{in } \Omega^{(\hat{d})}.$$

Combining this with (4.3.17) and recalling that $v\in C^2(\Omega)\cap C(\overline{\Omega})$, we see that $v(\mathbf{x})=g(\mathbf{x})+Lw^{\varepsilon}(\mathbf{x})+Lx^2\ge 0$ in $\Omega^{(\hat{d})}$. A similar argument shows that $\tilde{v}(\mathbf{x})=-g(\mathbf{x})+Lw^{\varepsilon}(\mathbf{x})+Lx^2\ge 0$ in $\Omega^{(\hat{d})}$. This implies

$$|g(\mathbf{x})|\le C[d(\mathbf{x})]^{\beta}+C|\mathbf{x}|^2 \qquad \text{in } \Omega^{(\hat{d})}$$

with $\beta=\min\{\varepsilon\delta,1\}$. Shifting the origin to the other points in $\Gamma_{7/8}$ and taking the infimum of the right-hand sides of the resulting estimates, we have

$$|g(\mathbf{x})|\le C[d(\mathbf{x})]^{\beta}+[d(\mathbf{x})]^2\le 2C[d(\mathbf{x})]^{\beta} \qquad \text{in } \Omega^{(\hat{d})}\cap B_{5/6}(\mathbf{0}).$$

Also, the interior estimates in Theorem 4.2.1 yield

$$\|u\|_{2,\alpha,\overline{\Omega_{3/4}}\cap\{d(\mathbf{x})\ge\hat{d}/2\}}\le C.$$

Combining the last two estimates, we have

$$|g(\mathbf{x})| \le C[d(\mathbf{x})]^{\beta} \qquad \text{in } \Omega_{3/4}. \tag{4.3.28}$$

5. We now estimate Dg and D^2u. Let $\mathbf{x}_0 \in \Omega_{3/4}$. Denote $r := d(\mathbf{x}_0)$. Denote $V(\mathbf{x}) = r^{1-\beta}g(\mathbf{x})$. We note that $d(\mathbf{x}) \ge \frac{9r}{10}$ for any $\mathbf{x} \in B_{r/10}(\mathbf{x}_0)$ and that $\beta + \gamma < 2$. Then, from (4.3.25) and (4.3.28),

$$|a_{ij}D_{ij}V| \le \frac{C}{r}(|DV|^2 + 1), \quad |V| \le Cr \qquad \text{in } B_{r/10}(\mathbf{x}_0).$$

By [262, Part (i) on Page 75], we obtain that $|DV(\mathbf{x}_0)| \le C$, which implies

$$|Dg(\mathbf{x})| \le C[d(\mathbf{x})]^{\beta-1} \qquad \text{in } \Omega_{3/4}.$$

By (4.3.21), we have

$$|D^2u(\mathbf{x})| \le C[d(\mathbf{x})]^{\beta-1} \qquad \text{in } \Omega_{3/4}. \tag{4.3.29}$$

Let $\eta \in C^\infty(\mathbb{R}^2)$ satisfy $\eta \equiv 1$ on $B_{1/2}(\mathbf{0})$ and $\eta \equiv 0$ on $\mathbb{R}^2 \setminus B_{5/8}(\mathbf{0})$. Then, using (4.3.29) and (4.3.9) with $R = 1$, we have

$$|D^2(\eta u)(\mathbf{x})| \le C[\text{dist}(\mathbf{x}, \partial\Omega_{3/4})]^{\beta-1} \qquad \text{in } \Omega_{3/4}.$$

Combining this with (4.3.9), we obtain that $\|\eta u\|^{(-1-\beta),\partial\Omega_{3/4}}_{2,0,\Omega_{3/4}} \le C$. Now (4.3.10) with $R = 1$ follows from

$$\|\eta u\|_{1,\beta,\Omega_{3/4}} = \|\eta u\|^{(-1-\beta),\partial\Omega_{3/4}}_{1,\beta,\Omega_{3/4}} \le \|\eta u\|^{(-1-\beta),\partial\Omega_{3/4}}_{2,0,\Omega_{3/4}},$$

where the last inequality holds by [130, Lemma 2.1].

6. Estimate (4.3.11) follows from (4.3.10) by the standard argument, which we now sketch below. By the covering argument, we obtain (4.3.10) in $\Omega_{3R/4}$, instead of $\Omega_{R/2}$. We continue to consider Case $R = 1$, without loss of generality.

For each $\mathbf{x}_0 \in \Omega_{1/2}$, denote $\rho := \frac{d(\mathbf{x}_0)}{10}$ and

$$U(\mathbf{X}) = \frac{u(\mathbf{x}_0 + \rho\mathbf{X}) - u(\mathbf{x}_0) - \rho Du(\mathbf{x}_0)\cdot\mathbf{X}}{\rho^{\beta+1}} \qquad \text{for all } \mathbf{X} \in B_1(\mathbf{0}).$$

By (4.3.10) in $\Omega_{3/4}$, we find that $\|U\|_{1,\beta,B_1(\mathbf{0})} \le C$, independent of $\mathbf{x}_0$. Since u satisfies equation (4.3.2), U satisfies the equation of similar form with ingredients $(\hat{A}_{ij}, \hat{A})(\mathbf{p}, z, \mathbf{X})$ in $B_1(\mathbf{0})$, where

$$\begin{aligned}&(\hat{A}_{ij}, \hat{A})(\mathbf{p}, z, \mathbf{X})\\ &= (A_{ij}, \rho^{1-\beta}A)(Du(\mathbf{x}_0) + \rho^{\beta}\mathbf{p},\ u(\mathbf{x}_0) + \rho^{1+\beta}z + \rho Du(\mathbf{x}_0)\cdot\mathbf{X},\ \mathbf{x}_0 + \rho\mathbf{X}).\end{aligned}$$

In particular, by (4.3.6), $\gamma \le 1$, and $\frac{1}{2}d(\mathbf{x}_0) \le d(\mathbf{x}) \le 2d(\mathbf{x}_0)$ for any $\mathbf{x} \in B_\rho(\mathbf{x}_0)$, we note that, for any $\mathbf{X}, \hat{\mathbf{X}} \in B_1(\mathbf{0})$ and $(\mathbf{p}, z) \in \mathbb{R}^2 \times \mathbb{R}$ with $|Du(\mathbf{x}_0) + \rho^{\beta}\mathbf{p}| + |u(\mathbf{x}_0) + \rho Du(\mathbf{x}_0)\cdot\mathbf{X} + \rho^{\beta+1}z| \le 2K$, we have

$$|\hat{A}_{ij}(\mathbf{p}, z, \mathbf{X}) - \hat{A}_{ij}(\mathbf{p}, z, \hat{\mathbf{X}})| \le C\rho^{-\gamma}|\rho\mathbf{X} - \rho\hat{\mathbf{X}}| \le C|\mathbf{X} - \hat{\mathbf{X}}|,$$

and the same holds for $\hat{A}(\cdot)$. Combining this with the estimate of $D_{(\mathbf{p},z)}(A_{ij}, A)$ in (4.3.6), estimate (4.3.9), and $\|U\|_{1,\beta,B_1(0)} \le C$, we have

$$\|(a_{ij}, a)\|_{0,\beta,B_1(\mathbf{0})} \le C,$$

where $(a_{ij}, a)(\mathbf{X}) := (\hat{A}_{ij}, \hat{A})(DU(\mathbf{X}), U(\mathbf{X}), \mathbf{X})$.

Since U satisfies the linear equation:

$$\sum_{i,j=1}^{2} a_{ij}(\mathbf{X})D_{ij}U + a(\mathbf{X}) = 0 \qquad \text{in } B_1(\mathbf{0}),$$

which is elliptic by (4.3.4), and $\|U\|_{1,\beta,B_1(\mathbf{0})} \le C$, we conclude

$$\|U\|_{2,\beta,B_{1/2}(\mathbf{0})} \le C.$$

Since this holds for each $\mathbf{x}_0 \in \Omega_{1/2}$ with the same constant C, and (4.3.10) also holds, then (4.3.11) follows. □

We obtain the $C^{2,\alpha}$–regularity of the solution under stronger assumptions, including the $C^{1,\alpha}$–boundary. We start by noting that the localized intermediate Schauder estimates for linear boundary value problems shown in Lemma 4.2.16, in the case that the boundary conditions are prescribed on the flat boundary parts, can be extended to the case of $C^{1,\alpha}$–boundaries. We state this only for the linear Dirichlet problem which will be used below, since the estimates for the oblique derivative problem are extended similarly.

Lemma 4.3.3. *Let $\alpha, \beta, \sigma, \lambda \in (0,1)$ with $\beta \le \alpha$. Let*

$$\|\Phi\|_{C^{1,\sigma}(\mathbb{R})} \le \lambda^{-1}, \quad \Phi(0) = 0,$$

and let $\Omega_r := B_r \cap \{x_2 > \Phi(x_1)\}$ and $\Gamma_r := B_r \cap \{x_2 = \Phi(x_1)\}$ for $r > 0$. Let $R > 0$, and let $u \in C^{(-1-\sigma),\,\Gamma_R}_{2,\alpha,\Omega_R}$ satisfy equation (4.2.118) in Ω_R with

$$u = h(\mathbf{x}) \qquad on\ \Gamma_R, \tag{4.3.30}$$

where equation (4.2.118) satisfies ellipticity (4.2.121) in Ω_R, and

$$\|a_{ij}\|^{(-\beta),\,\Gamma_R}_{0,\alpha,\Omega_R} + \|(a_i, a)\|^{(1-\sigma),\,\Gamma_R}_{0,\alpha,\Omega_R} \le \lambda^{-1}. \tag{4.3.31}$$

Then u satisfies

$$\|u\|^{(-1-\sigma),\,\Gamma_{R/2}}_{2,\alpha,\Omega_{R/2}} \le C\big(\|u\|_{0,\Omega_R} + \|f\|^{(1-\sigma),\Gamma_R}_{0,\alpha,\Omega_R} + \|h\|_{1,\sigma,\Gamma_R}\big), \tag{4.3.32}$$

where C depends only on $(\lambda, \alpha, \beta, \sigma, R)$.

Proof. The proof consists of two steps.

1. First consider Case $\alpha \le \sigma$. We flatten the boundary by the change of variables: $\mathcal{A}(x_1, x_2) = (x_1, x_2 - \Phi(x_1))$ and reduce the problem to Case $R = 2$, as in the proof of Theorem 4.2.9. Using the $C^{1,\sigma}$–regularity of Φ with $\sigma \ge \alpha$, we note that, under this change of variables, the ellipticity and regularity of the coefficients remain unchanged, with the modified constants depending only on $(\lambda, \sigma, \alpha, \beta, R)$. Now the assertion follows from Lemma 4.2.16(ii) with $k = 1$.

2. Next we consider Case $\alpha > \sigma$. First, we note that, reducing α by setting $\alpha := \sigma$, then the conditions of the lemma are satisfied for this reduced α. Thus, by the result of Step 1, we obtain (4.3.32) with σ instead of α. Also, by a covering argument, we obtain a similar estimate with region $\Omega_{3R/4}$ on the left-hand side. In particular,

$$\|u\|_{1,\sigma,\,\Omega_{3R/4}} \le CM. \tag{4.3.33}$$

Here and hereafter, M denotes the sum of norms on the right-hand side of (4.3.32), and $C = C(\lambda, \alpha, \beta, \sigma, R)$.

It remains to prove the higher interior regularity of the solution by using the higher interior regularity of the coefficients in (4.3.31), determined by $\alpha > \sigma$. This is done by following the argument in Step 6 of the proof of Theorem 4.3.2. Specifically, for each $\mathbf{x}_0 \in \Omega_{R/2}$, denote $\rho := \frac{d(\mathbf{x}_0)}{10}$ and

$$U(\mathbf{X}) := \frac{u(\mathbf{x}_0 + \rho\mathbf{X}) - u(\mathbf{x}_0) - \rho Du(\mathbf{x}_0)\cdot\mathbf{X}}{\rho^{\sigma+1}} \qquad \text{for all } \mathbf{X} \in B_1(\mathbf{0}).$$

From (4.3.33), we find that $\|U\|_{1,\sigma,B_1(\mathbf{0})} \le CM$. Also, U satisfies an equation of form (4.2.118) in $B_1(\mathbf{0})$, with coefficients $(\hat{a}_{ij}, \hat{a}_i, \hat{a})$ and the right-hand side $\hat{f}$ defined by

$$\begin{aligned}
&(\hat{a}_{ij},\, \hat{a}_i,\, \hat{a})(\mathbf{X}) = (a_{ij},\, \rho a_i,\, \rho^2 a)(\mathbf{x}),\\
&\hat{f}(\mathbf{X}) = \rho^{1-\sigma}\big(f(\mathbf{x}) - \sum_{i=1}^{2} a_i(\mathbf{x})u_i(\mathbf{x}_0) - a(\mathbf{x}_0)(u(\mathbf{x}_0) + \rho Du(\mathbf{x}_0)\cdot\mathbf{X})\big),
\end{aligned}$$

where $\mathbf{x} = \mathbf{x}_0 + \rho\mathbf{X}$ and $\mathbf{X} \in B_1(\mathbf{0})$. Recalling that $\rho = \frac{d(\mathbf{x}_0)}{10}$, we have

$$\begin{aligned}
&\|(\hat{a}_{ij},\, \hat{a}_i,\, \hat{a})\|_{0,\alpha,B_1} \le \|a_{ij}\|^{(-\beta),\,\Gamma_R}_{0,\alpha,\Omega_R} + \|(a_i, a)\|^{(1-\sigma),\,\Gamma_R}_{0,\alpha,\Omega_R} \le \lambda^{-1},\\
&\|\hat{f}\|_{0,\alpha,B_1} \le C\big(\|f\|^{(1-\sigma),\Gamma_R}_{0,\alpha,\Omega_R} + \|(a_i, a)\|^{(1-\sigma),\,\Gamma_R}_{0,\alpha,\Omega_R}\|u\|_{1,0,\Omega_{3R/4}}\big) \le CM.
\end{aligned}$$

Thus, the interior Schauder estimates imply that $\|U\|_{2,\sigma,B_{1/2}(\mathbf{0})} \le CM$. Combining with (4.3.33) and arguing as in the proof of [131, Theorem 4.8] (*cf.* the proof of Lemma 4.2.7 above), we obtain (4.3.32). □

The next theorem is a version of [192, Corollary 1.4].

Theorem 4.3.4. *Let the assumptions of Theorem 4.3.2 be satisfied with $\gamma = 0$. In addition, assume that $\alpha, \sigma \in (0,1)$,*

$$\|\Phi\|_{C^{1,\sigma}(\mathbb{R})} \leq \lambda^{-1}, \quad \Phi(0) = 0, \tag{4.3.34}$$

and

$$\|(A_{ij}, A)\|_{C^{1,\alpha}(\{|\mathbf{p}|+|z|\leq 2K,\ \mathbf{x}\in\overline{\Omega_R}\})} + \|B\|_{C^{2,\alpha}(\{|\mathbf{p}|+|z|\leq 2K,\ \mathbf{x}\in\overline{\Omega_R}\})} \leq \lambda^{-1}. \tag{4.3.35}$$

Then

$$\|u\|_{C^{2,\sigma}(\overline{\Omega}_{R/4})} \leq C, \tag{4.3.36}$$

where C depends only on $(\lambda, K, \alpha, \sigma, R)$.

Proof. As in the proof of Theorem 4.3.2, we may assume $R = 1$, without loss of generality. All the constants in this proof depend only on $(\lambda, K, \alpha, \sigma)$. Furthermore, we may also assume without loss of generality that $\sigma \geq \alpha$. Indeed, if $\alpha > \sigma$, (4.3.35) holds with σ instead of α, *i.e.*, we can set $\alpha := \sigma$.

From (4.3.11) in Theorem 4.3.2 with $R = 1$ and by a covering argument, we have

$$\|u\|^{(-1-\beta),\Gamma_{3/4}}_{2,\beta,\Omega_{3/4}} \leq C. \tag{4.3.37}$$

We write equation (4.3.25) with $\gamma = 0$ as

$$\sum_{i,j=1}^{2} a_{ij} D_{ij} g + \sum_{i=1}^{2} \tilde{m}_i D_i g + m = 0, \tag{4.3.38}$$

with $\tilde{m}_i = \sum_{j=1}^{2} m_{ij} D_j g + m_i, i = 1,2$. Note that this equation is satisfied in the whole region Ω_1 and is also uniformly elliptic. Using (4.3.35) and (4.3.37), we see that coefficients (Q, R) in (4.3.21) and (a_{ij}, m) in (4.3.38), $i,j = 1,2$, have $C^{\alpha\beta}(\overline{\Omega_{3/4}})$–bounds, and $\tilde{m}_i, i = 1,2$, have a $C^{1-\beta,\Gamma_{3/4}}_{0,\alpha\beta,\Omega_{3/4}}$–bound depending only on (λ, α, β). Since β depends only on (λ, K), it follows that these estimates of the coefficients depend only on (λ, K, α).

Moreover, $g(\mathbf{x}) := B(Du(\mathbf{x}), u(\mathbf{x}), \mathbf{x})$ satisfies the linear elliptic equation (4.3.38) in $\Omega_{3/4}$ with the coefficients discussed above, and the boundary condition: $g = 0$ on $\Gamma_{3/4}$. Also, $|g| \leq C$ in Ω_1 by (4.3.35) and (4.3.37). Then we employ Lemma 4.3.3 to obtain

$$\|g\|^{(-1-\alpha\beta),\Gamma_{1/2}}_{2,\alpha\beta,\Omega_{1/2}} \leq C.$$

In particular, from (4.3.21) with $C^{\alpha\beta}(\overline{\Omega_{3/4}})$–coefficients (Q, R), we have

$$\|u\|_{C^{2,\alpha\beta}(\overline{\Omega}_{1/2})} \leq C.$$

From this, we obtain the estimates of coefficients (Q, R) in (4.3.21) in the $C^{1,\alpha\beta}(\overline{\Omega}_{1/2})$–norm, and hence in the $C^{\sigma}(\overline{\Omega}_{1/2})$–norm, and the estimates of $\tilde{m}_i$

in (4.3.38) in the $C^{\alpha\beta}(\overline{\Omega_{1/2}})$–norm. We also recall that the estimates of (a_{ij}, m) in (4.3.38) in the $C^{\alpha\beta}(\overline{\Omega_{3/4}})$–norm have been obtained earlier. Then we use Lemma 4.3.3 to obtain

$$\|g\|^{(-1-\sigma),\Gamma_{1/2}}_{2,\alpha\beta,\Omega_{1/4}} \leq C,$$

that is, $\|g\|_{1,\sigma,\Omega_{1/4}} \leq C$. Now, from (4.3.21) with $C^{\sigma}(\overline{\Omega_{1/2}})$–coefficients (Q, R), we conclude (4.3.36). □

Corollary 4.3.5. *Let the assumptions of Theorem* 4.3.2 *be satisfied with* $\gamma = 0$. *In addition, we assume that* $\alpha \in (0,1)$, $k \geq 1$ *integer,*

$$\|\Phi\|_{k,\alpha,\mathbb{R}} \leq \lambda^{-1}, \qquad \Phi(0) = 0, \tag{4.3.39}$$

and

$$\|(A_{ij}, A)\|_{C^{k,\alpha}(\{|\mathbf{p}|+|z|\leq 2K,\ \mathbf{x}\in\overline{\Omega_R}\})} + \|B\|_{C^{k+1,\alpha}(\{|\mathbf{p}|+|z|\leq 2K,\ \mathbf{x}\in\overline{\Omega_R}\})} \leq \lambda^{-1}. \tag{4.3.40}$$

Then

$$\|u\|_{k+1,\alpha,\overline{\Omega}_{R/2}} \leq C, \tag{4.3.41}$$

where C *depends only on* $(\lambda, K, k, \alpha, R)$.

Proof. As in the proof of Theorem 4.3.2, we may assume that $R = 1$, without loss of generality. We do induction over k.

If $k = 1$, this is Theorem 4.3.4.

Assume that $k \geq 2$, and estimate (4.3.41) with $R = 1$ holds for $k-1$ instead of k. Then, by a covering argument, we obtain that $\|u\|_{k,\alpha,\overline{\Omega}_{3/4}} \leq C$, which yields the estimates of coefficients (Q, R) in (4.3.21) and $(a_{ij}, \tilde{m}_i, m)$ in (4.3.38) in the $C^{k-1+\alpha}(\overline{\Omega_{3/4}})$–norm. Arguing as in the proof of Theorem 4.3.4, we note that g satisfies the elliptic equation (4.3.38) in $\Omega_{3/4}$, $g = 0$ on $\Gamma_{3/4}$, and $|g| \leq C$ in Ω_1. The ellipticity of $[a_{ij}]$, the estimates of coefficients $(a_{ij}, \tilde{m}_i, m)$ in $C^{k-1+\alpha}(\overline{\Omega_{3/4}})$, and the boundary regularity (4.3.39) with $k \geq 2$ allow us to use the standard local Schauder estimates for the Dirichlet problem (see *e.g.* [131, the argument after Theorem 6.19]) to obtain

$$\|g\|_{k,\alpha,\Omega_{1/2}} \leq C.$$

Now, from (4.3.21) with $C^{k-1+\alpha}(\overline{\Omega_{1/2}})$–coefficients (Q, B), we obtain (4.3.41) for k. □

Theorem 4.3.6. *The assertions of Theorems* 4.3.2 *and* 4.3.4 *and Corollary* 4.3.5 *remain true when condition* (4.3.7) *is replaced by*

$$|D_{\mathbf{p}}B(\mathbf{p}, z, \mathbf{x})| \geq \lambda \tag{4.3.42}$$

for any $(\mathbf{p}, z, \mathbf{x}) \in (\mathbb{R}^2 \times \mathbb{R} \times \overline{\Gamma_R}) \cap \{|\mathbf{p}| + |z| \leq 3K\}$.

Proof. From (4.3.8), there exists $\hat{R} \in (0, R)$ depending only on (λ, M) such that

$$|D_{\mathbf{p}}B(\mathbf{p}, z, \mathbf{x})| \geq \frac{\lambda}{2}$$

for any $(\mathbf{p}, z, \mathbf{x}) \in \big(\mathbb{R}^2 \times \mathbb{R} \times \{\mathbf{x} \in \Omega_R \,:\, \mathrm{dist}(\mathbf{x}, \Gamma_1) \leq \hat{R}\}\big) \cap \{|\mathbf{p}| + |z| \leq 2K\}$. Now we can apply Theorems 4.3.2 and 4.3.4 and Corollary 4.3.5 in $\Omega_{\hat{R}}$. □

4.3.2 Estimates of solutions at a corner for the oblique derivative problem

Next we present the results in Lieberman [192, Theorem 2.1] on the $C^{1,\alpha}$–regularity of solutions of the oblique derivative problem at a corner. Some of the argument and notations below follow the presentation of the results in [111, Appendix 5.1]. In addition, we prove some related estimates under weaker assumptions than those in [192, Theorem 2.1] and [111, Appendix 5.1]. We need to do this in order to apply these results in the situation when one side of the corner is a free boundary with *a priori* unknown regularity, and the functional independence of the boundary conditions is known only for the solution on the free boundary, *i.e.*, not for all $(\mathbf{p}, z, \mathbf{x})$.

We first consider the case that the solution is *a priori* known to be C^1 up to the corner.

Proposition 4.3.7. *Let $R > 0$, $\beta \in (0,1)$, $\gamma \in [0,1)$, $\lambda \in (0,1]$, and $K, M \geq 1$. Let $\hat{\Omega} \subset \mathbb{R}^2$ be an open domain contained within $\mathbf{x}_0 + \{\mathbf{x} \,:\, x_2 > \tau|x_1|\}$ for some $\mathbf{x}_0 \in \partial\hat{\Omega}$ and $\tau > 0$, and let*

$$\partial\hat{\Omega} \cap B_R(\mathbf{x}_0) = \Gamma^1 \cup \Gamma^2,$$

where Γ^k, $k = 1, 2$, are two Lipschitz curves intersecting only at $\mathbf{x}_0$. Let

$$\Omega \equiv \Omega_R := \hat{\Omega} \cap B_R(\mathbf{x}_0).$$

Assume that Γ^2 is $C^{1,\sigma}$ up to the endpoints for some $\sigma \in (0,1)$ with the bound:

$$\|\overline{\Gamma^2}\|_{1,\sigma} \leq M. \tag{4.3.43}$$

Let $u \in C^1(\overline{\Omega}) \cap C^2(\Omega \cup \Gamma^2) \cap C^3(\Omega)$ satisfy

$$\|u\|_{0,1,\overline{\Omega}} \leq K. \tag{4.3.44}$$

Assume that u is a solution of the boundary value problem:

$$\sum_{i,j=1}^{2} A_{ij}(Du, u, \mathbf{x})D_{ij}u + A(Du, u, \mathbf{x}) = 0 \qquad \text{in } \Omega, \tag{4.3.45}$$

$$B^{(1)}(Du, u, \mathbf{x}) = h(\mathbf{x}) \qquad \text{on } \Gamma^1, \tag{4.3.46}$$

$$B^{(2)}(Du, u, \mathbf{x}) = 0 \qquad \text{on } \Gamma^2, \tag{4.3.47}$$

where functions $(A_{ij}, A, B^{(k)})$ *are defined in the set:*

$$V = \{(\mathbf{p}, z, \mathbf{x}) \in \mathbb{R}^2 \times \mathbb{R} \times \overline{\Omega} \; : \; |\mathbf{p}| + |z| \le 2K\}. \tag{4.3.48}$$

Assume that $(A_{ij}, A) \in C(\overline{V}) \cap C^1(\overline{V} \setminus \{\mathbf{x}_0\})$, $B^{(1)} \in C^2(\overline{V})$, $B^{(2)} \in C^1(\overline{V})$, *and* $h \in C(\overline{\Gamma^1})$ *with*

$$\|(A_{ij}, A)\|_{0,V} + \|D_{(\mathbf{p},z)}(A_{ij}, A)\|_{0,V} \le M, \tag{4.3.49}$$

$$|D_{\mathbf{x}}(A_{ij}, A)(\mathbf{p}, z, \mathbf{x})| \le M|\mathbf{x} - \mathbf{x}_0|^{-\gamma} \qquad \text{for all } (\mathbf{p}, z, \mathbf{x}) \in V, \tag{4.3.50}$$

$$\|B^{(1)}\|_{2,V} + \|B^{(2)}\|_{1,V} \le M, \tag{4.3.51}$$

$$|h(\mathbf{x}) - h(\mathbf{x}_0)| \le \frac{\lambda^{-1}}{R^\beta}|\mathbf{x} - \mathbf{x}_0|^\beta \qquad \text{for all } \mathbf{x} \in \Gamma^1. \tag{4.3.52}$$

Assume that equation (4.3.45) *is uniformly elliptic on solution* u *in* Ω*:*

$$\lambda|\boldsymbol{\mu}|^2 \le \sum_{i,j=1}^{2} A_{ij}(Du(\mathbf{x}), u(\mathbf{x}), \mathbf{x})\mu_i\mu_j \le \frac{1}{\lambda}|\boldsymbol{\mu}|^2 \tag{4.3.53}$$

for any $\mathbf{x} \in \Omega$ *and* $\boldsymbol{\mu} = (\mu_1, \mu_2) \in \mathbb{R}^2$, *and that the boundary condition* $B^{(2)}$ *is oblique for* u *on* Γ^2*:*

$$D_{\mathbf{p}}B^{(2)}(Du(\mathbf{x}), u(\mathbf{x}), \mathbf{x}) \cdot \boldsymbol{\nu} \ge \lambda \qquad \text{for any } \mathbf{x} \in \Gamma^2, \tag{4.3.54}$$

where $\boldsymbol{\nu}$ *is the interior unit normal to* Γ^2. *Moreover, assume*

$$|D_{\mathbf{p}}B^{(1)}(Du(\mathbf{x}), u(\mathbf{x}), \mathbf{x})| \ge \lambda \qquad \text{for any } \mathbf{x} \in \overline{\Omega}_2. \tag{4.3.55}$$

Furthermore, assume the functional independence of $B^{(1)}$ *and* $B^{(2)}$ *for* u *on* Γ^2*:*

$$|\det G(Du(\mathbf{x}), u(\mathbf{x}), \mathbf{x})| \ge \lambda \qquad \text{for any } \mathbf{x} \in \Gamma^2, \tag{4.3.56}$$

where $G(\mathbf{p}, z, \mathbf{x})$ *is the matrix with columns* $D_pB^{(k)}(\mathbf{p}, z, \mathbf{x})$, $k = 1, 2$.

Then there exist $\alpha \in (0, \beta]$ *and* C *depending only on* (λ, K, M), *and* $R' \in (0, R]$ *depending on* $(\lambda, \gamma, K, M, \sigma)$, *such that, for any* $\mathbf{x} \in \overline{\Omega \cap B_{R'}(\mathbf{x}_0)}$,

$$|B^{(1)}(Du(\mathbf{x}), u(\mathbf{x}), \mathbf{x}) - B^{(1)}(Du(\mathbf{x}_0), u(\mathbf{x}_0), \mathbf{x}_0)| \le C|\mathbf{x} - \mathbf{x}_0|^\alpha. \tag{4.3.57}$$

Proof. We use the notations and some calculations from the proof of Theorem 4.3.2. In particular, we define

$$(a_{ij}, a, g^{(k)})(\mathbf{x}) := (A_{ij}, A, B^{(k)})(Du(\mathbf{x}), u(\mathbf{x}), \mathbf{x}). \tag{4.3.58}$$

Constants α, μ, and C, below, depend only on (λ, K, M), and constant R' depends only on $(\lambda, \gamma, K, M, \sigma)$, unless otherwise specified. We may also assume as before, without loss of generality, that (4.3.12) holds. We divide our proof into four steps.

1. We use the polar coordinates (r, θ) centered at $\mathbf{x}_0$. Under the assumptions on Ω given in Proposition 4.3.7 and (4.3.43), reducing R depending only on M, there exists $\theta_1 \in (0, \pi)$ so that, choosing the direction of θ and ray $\theta = 0$ appropriately, we have

$$\begin{aligned} &\Gamma^k \cap B_R(\mathbf{x}_0) = \{(r, \phi^k(r)) \;:\; 0 < r < R\} \qquad \text{for } k = 1, 2, \\ &\|\phi^2\|_{C^{1,\sigma}([0,R])} \le CM, \quad -\frac{\theta_1}{2} \le \phi^1(\cdot) < \phi^2(\cdot) \le \frac{\theta_1}{2}, \end{aligned} \tag{4.3.59}$$

$$\Omega = \{0 < r < R,\ \phi^1(r) < \theta < \phi^2(r)\}. \tag{4.3.60}$$

Denote

$$w(r, \theta) = r^\alpha h(\theta), \quad h(\theta) = 1 - \mu e^{-\bar{L}\theta}, \tag{4.3.61}$$

where $\alpha \in (0, \frac{1}{2}]$ and $\bar{L}, \mu > 0$ to be chosen. Take μ so small that $h(\theta) \ge \frac{1}{2}$ for any $\theta \in [-\frac{\pi}{2}, \frac{\pi}{2}]$. Then the explicit calculation in [111, Page 97] shows, by using (4.3.43) and (4.3.53), that, for $\mathbf{x} \in \Omega$, rotating the coordinates to be along the angular and radial directions, and denoting $(a_{rr}, a_{r\theta}, a_{\theta\theta})$ as the corresponding components of a_{ij}, we have

$$\sum_{i,j=1}^{2} a_{ij} D_{ij} w \le \Big(e^{-\bar{L}\theta} \mu \bar{L}\big(2|a_{r\theta}| - \bar{L} a_{\theta\theta}\big) + \alpha\, a_{\theta\theta}\Big) r^{\alpha-2} \qquad \text{at } \mathbf{x}.$$

Note that $\frac{2\sup|a_{r\theta}+1|}{\inf a_{rr}} \le \frac{3}{\lambda^2}$. If $\bar{L} = \frac{3}{\lambda^2}$, then, for every $\mu > 0$ small, there exist $\alpha_0 \in (0, \mu]$ depending only on (λ, μ), and $R' \in (0, R)$ depending only on (λ, M, μ), so that, for every $\alpha \in (0, \alpha_0)$,

$$\begin{aligned} &\sum_{i,j=1}^{2} a_{ij}(\mathbf{x}) D_{ij} w(\mathbf{x}) \le -\frac{\mu \bar{L}}{2} e^{-\bar{L}\theta} r^{\alpha-2}, \\ &|Dw(\mathbf{x})| \le C(\alpha + \mu) r^{\alpha-1}, \quad \frac{1}{2} r^\alpha \le w(\mathbf{x}) \le r^\alpha \qquad \text{in } \Omega. \end{aligned} \tag{4.3.62}$$

The further calculation in [111, Page 97] shows that, for any $T : \Gamma^2 \cap B_{R'} \mapsto \mathbb{R}$,

$$w_{\boldsymbol{\nu}} + T w_{\boldsymbol{\tau}} \le -\Big(\mu \bar{L} e^{-\bar{L}\theta} - C(R')^\sigma - T(\alpha + C(R')^\sigma \bar{L}\mu)\Big) r^{\alpha-1} \qquad \text{on } \Gamma^2 \cap B_{R'}, \tag{4.3.63}$$

where $\boldsymbol{\nu}$ is the inner normal to Ω on Γ_2.

Define

$$v(\mathbf{x}) = g^{(1)}(\mathbf{x}) - g^{(1)}(\mathbf{x}_0) + L w(\mathbf{x}), \tag{4.3.64}$$

where

$$L = \frac{1}{\sqrt{\mu}(R')^\alpha}. \tag{4.3.65}$$

Then our assumptions imply that $v \in C(\overline{\Omega}) \cap C^1(\Omega \cup \Gamma^2) \cap C^2(\Omega)$.

2. We first derive the boundary conditions for v on $\partial(\Omega \cap B_{R'}(\mathbf{x}_0)) \setminus \Gamma^2$, *i.e.*, on $\Gamma_1 \cap B_{R'}(\mathbf{x}_0)$ and $\partial B_{R'}(\mathbf{x}_0) \cap \Omega$.

Notice that $|B^{(1)}(Du(\mathbf{x}), u(\mathbf{x}), \mathbf{x})| \leq \|B^{(1)}\|_{L^\infty(V)} \leq C$ for any $\mathbf{x} \in \overline{\Omega}$, by (4.3.44) and (4.3.51). Also, $w(\mathbf{x}) \geq \frac{1}{2}(R')^\alpha$ on $\partial B_{R'}(\mathbf{x}_0) \cap \Omega$, by (4.3.62). Thus, using (4.3.65) and choosing μ sufficiently small,

$$v \geq -2\|B^{(1)}\|_{L^\infty(V)} + \frac{(R')^\alpha}{2\sqrt{\mu}(R')^\alpha} \geq 0 \qquad \text{on } \partial B_{R'}(\mathbf{x}_0) \cap \Omega. \tag{4.3.66}$$

For $\mathbf{x} \in \Gamma_1 \cap B_{R'}(\mathbf{x}_0)$, we choose $\alpha \leq \beta$ and employ (4.3.46) and (4.3.52) to obtain that, on $\Gamma^1 \cap B_{R'}(\mathbf{x}_0)$,

$$\begin{aligned} v(\mathbf{x}) &= h(\mathbf{x}) - h(\mathbf{x}_0) + \frac{w(\mathbf{x})}{\sqrt{\mu}(R')^\alpha} \\ &\geq -\lambda^{-1}\left(\frac{r}{R}\right)^\beta + \frac{1}{\sqrt{\mu}}\left(\frac{r}{R'}\right)^\alpha \geq 0, \end{aligned} \tag{4.3.67}$$

by using that $r < R' < R$ and choosing μ so small that $\mu^{-1/2} \geq \lambda^{-1}$.

3. Next, using (4.3.55) and following the calculations in Step 4 of the proof of Theorem 4.3.2, we find that D^2u and $g^{(1)}$ satisfy (4.3.21) with $(Q(\mathbf{x}), R(\mathbf{x}))$ satisfying (4.3.22) in Ω and that, similar to (4.3.25), $g^{(1)}$ satisfies

$$\sum_{i,j=1}^{2} \left(a_{ij}D_{ij}g^{(1)} + m_{ij}D_ig^{(1)}D_jg^{(1)}\right) + r^{-\gamma}\Big(\sum_{i=1}^{2} m_iD_ig^{(1)} + m\Big) = 0 \quad \text{in } \Omega, \tag{4.3.68}$$

where, similar to (4.3.26),

$$|(m_{ij}, m_i, m)(\mathbf{x})| \leq C \qquad \text{for } \mathbf{x} \in \Omega.$$

Now, setting

$$q_i := \sum_{j=1}^{2} \left(m_{ij}v_j - L(m_{ij} + m_{ji})w_j\right) + r^{-\gamma}m_i, \tag{4.3.69}$$

we perform a calculation similar to (4.3.27): Substituting $D^kg^{(1)} = D^kv - LD^kw$, $k = 1, 2$, into (4.3.68) and using (4.3.62) and (4.3.65), we find that, in $\Omega \cap B_{R'}(\mathbf{x}_0)$,

$$\begin{aligned} &\sum_{i,j=1}^{2} a_{ij}D_{ij}v + \sum_{i=1}^{2} q_iD_iv \\ &= L\sum_{i,j=1}^{2} \left(a_{ij}D_{ij}w - Lm_{ij}D_iwD_jw\right) + r^{-\gamma}\Big(L\sum_{i=1}^{2} m_iD_iw - m\Big) \\ &\leq -\frac{L\mu}{C}r^{\alpha-2} + CL^2(\alpha^2 + \mu^2)r^{2\alpha-2} + CL(\alpha + \mu)r^{\alpha-\gamma-1} + Cr^{-\gamma} \\ &= \Big(-\frac{\sqrt{\mu}}{C} + C\frac{\alpha^2 + \mu^2}{\mu}\big(\frac{r}{R'}\big)^\alpha + C\frac{\alpha + \mu}{\sqrt{\mu}}r^{1-\gamma}\Big)\big(\frac{r}{R'}\big)^\alpha r^{-2} + Cr^{-\gamma}. \end{aligned}$$

Using that $\alpha \leq \mu$ (where α will further be chosen below), $\gamma \in [0,1)$, and $\frac{r}{R'} \in (0,1)$, we have

$$\sum_{i,j=1}^{2} a_{ij} D_{ij} v + \sum_{i=1}^{2} q_i D_i v \leq \Big(-\frac{\sqrt{\mu}}{C} + C\mu + C\sqrt{\mu}(R')^{1-\gamma} \Big) \big(\frac{r}{R'}\big)^{\alpha} r^{-2} + Cr^{-\gamma}$$
$$\leq -\frac{\sqrt{\mu}}{2C} \big(\frac{r}{R'}\big)^{\alpha} r^{-2} + Cr^{-\gamma},$$

where the last inequality has been obtained by choosing μ so small that $-\frac{\sqrt{\mu}}{4C} + C\mu < 0$ and then taking $R' = R'(\lambda, M, K, \gamma)$ small so that $-\frac{\sqrt{\mu}}{4C} + C\sqrt{\mu}(R')^{1-\gamma} < 0$. Now $\mu \in (0,1)$ is fixed for the rest of the proof.

Next, noting that $\alpha, \gamma \in (0,1)$ so that $\alpha + \gamma < 2$, we find that, in $\Omega \cap B_{R'}(\mathbf{x}_0)$,

$$\sum_{i,j=1}^{2} a_{ij} D_{ij} v + \sum_{i=1}^{2} q_i D_i v \leq r^{-\gamma} \left(-\frac{\sqrt{\mu}}{2C} (R')^{-\alpha} r^{-2+\gamma+\alpha} + C \right)$$
$$\leq r^{-\gamma} \left(-\frac{\sqrt{\mu}}{2C} (R')^{-\alpha} (R')^{-2+\gamma+\alpha} + C \right)$$
$$= r^{-\gamma} \left(-\frac{\sqrt{\mu}}{2C} (R')^{-2+\gamma} + C \right).$$

Since $\mu > 0$ is fixed and $\gamma < 1$, we choose R' small to conclude

$$\sum_{i,j=1}^{2} a_{ij} D_{ij} v + \sum_{i=1}^{2} q_i D_i v < 0 \qquad \text{in } \Omega \cap B_{R'}(\mathbf{x}_0). \tag{4.3.70}$$

4. We now derive a differential inequality for v on the $C^{1,\sigma}$–curve Γ^2. First, we seek $T = T(\mathbf{x})$ on Γ_2 so that

$$|g^{(1)}_{\boldsymbol{\nu}} + T g^{(1)}_{\boldsymbol{\tau}}| \leq C \qquad \text{on } \Gamma^2. \tag{4.3.71}$$

Since $g^{(2)}_{\boldsymbol{\tau}} = 0$ on Γ_2 and equation (4.3.45) holds, it suffices to find (T, y_1, y_2) at every point $\mathbf{x} \in \Gamma^2$ so that, after substituting (4.3.18) for $(g^{(1)}, g^{(2)})$ into the expression:

$$g^{(1)}_{\boldsymbol{\nu}} + T g^{(1)}_{\boldsymbol{\tau}} + y_1 \mathcal{N}(u) + y_2 g^{(2)}_{\boldsymbol{\tau}},$$

all the coefficients of $D_{ij}u$ vanish, where $\mathcal{N}(u)$ is the left-hand side of (4.3.45). By explicit calculation, this happens if

$$[y_1 \; y_2 \; 1] \begin{bmatrix} a_{\boldsymbol{\tau}\boldsymbol{\tau}} & 2a_{\boldsymbol{\tau}\boldsymbol{\nu}} & a_{\boldsymbol{\nu}\boldsymbol{\nu}} \\ B^{(2)}_{\mathbf{p}_{\boldsymbol{\tau}}} & B^{(2)}_{\mathbf{p}_{\boldsymbol{\nu}}} & 0 \\ TB^{(1)}_{\mathbf{p}_{\boldsymbol{\tau}}} & TB^{(1)}_{\mathbf{p}_{\boldsymbol{\nu}}} + B^{(1)}_{\mathbf{p}_{\boldsymbol{\tau}}} & B^{(1)}_{\mathbf{p}_{\boldsymbol{\nu}}} \end{bmatrix} = 0,$$

where we have used (4.3.12). Then the determinant must vanish, which gives

$$a_{\boldsymbol{\nu}\boldsymbol{\nu}} (B^{(1)}_{\mathbf{p}_{\boldsymbol{\nu}}} B^{(2)}_{\mathbf{p}_{\boldsymbol{\tau}}} - B^{(1)}_{\mathbf{p}_{\boldsymbol{\tau}}} B^{(2)}_{\mathbf{p}_{\boldsymbol{\nu}}}) T = \tilde{m}, \tag{4.3.72}$$

where $\tilde{m} = -a_{\nu\nu}B^{(1)}_{\mathbf{p}_\tau}B^{(2)}_{\mathbf{p}_\tau} + B^{(1)}_{\mathbf{p}_\nu}(a_{\tau\tau}B^{(2)}_{\mathbf{p}_\nu} - 2a_{\nu\tau}B^{(2)}_{\mathbf{p}_\tau})$. From (4.3.49) and (4.3.51), we see that $|\tilde{m}(\mathbf{x})| \le C$ on Γ^2. The assumptions in (4.3.53) and (4.3.56) allow us to solve for T to obtain $|T(\mathbf{x})| \le C$ on Γ^2. From (4.3.54), $B^{(2)}_{\mathbf{p}_\nu} \ge \lambda$ so that

$$y_1 = -\frac{B^{(1)}_{\mathbf{p}_\nu}}{a_{\nu\nu}}, \qquad y_2 = -\frac{a_{\nu\nu}B^{(1)}_{\mathbf{p}_\tau} - 2a_{\tau\nu}B^{(1)}_{\mathbf{p}_\nu} + Ta_{\nu\nu}B^{(1)}_{\mathbf{p}_\nu}}{a_{\nu\nu}\,B^{(2)}_{\mathbf{p}_\nu}}, \tag{4.3.73}$$

which implies that $|(y_1, y_2)| \le C$ on Γ^2. It follows that (4.3.71) holds. With this, using (4.3.63) with $\bar{L} + |T(\mathbf{x})| \le C$ and $\theta \in [-\frac{\pi}{2}, \frac{\pi}{2}]$ by (4.3.59)–(4.3.60), and using (4.3.65), we obtain that, on $\Gamma^2 \cap B_{R'}(\mathbf{x}_0)$,

$$\begin{aligned} v_\nu + Tv_\tau &= L(w_\nu + Tw_\tau) + g^{(1)}_\nu + Tg^{(1)}_\tau \\ &\le -\frac{1}{\sqrt{\mu}}\Big\{\frac{\mu}{C} - C(R')^\sigma - C(\alpha + C(R')^\sigma\mu)\Big\}\frac{r^{\alpha-1}}{(R')^\alpha} + C. \end{aligned}$$

Now we choose α and R' small so that the first term of the expression in the braces dominates the other terms. For this, we recall that $\mu \in (0,1)$ has been fixed. First choose $\alpha \le \frac{\mu}{10C^2}$. Combining this with the previous requirements on α, we fix $\alpha \in (0, \beta]$ such that it depends only on (λ, M, K) and is independent of (γ, σ). Now, choosing R' small so that

$$C\big(1 + \mu C\big)(R')^\sigma \le \frac{\mu}{10C},$$

we obtain that, for $r \in (0, R')$,

$$v_\nu + Tv_\tau \le -\frac{\sqrt{\mu}r^{\alpha-1}}{2C(R')^\alpha} + C \le -\frac{\sqrt{\mu}}{2C}\frac{(R')^{\alpha-1}}{(R')^\alpha} + C = -\frac{\sqrt{\mu}}{2C}(R')^{-1} + C.$$

Since $\mu > 0$ is fixed, we choose $R' = R'(\lambda, M, \gamma, \sigma, K)$ small to obtain

$$v_\nu + Tv_\tau < 0 \qquad \text{on } \Gamma^2 \cap B_{R'}(\mathbf{x}_0). \tag{4.3.74}$$

Therefore, $v \in C(\overline{\Omega}) \cap C^1(\Omega \cup \Gamma^2) \cap C^2(\Omega)$ satisfies the differential inequality (4.3.70) in $\Omega \cap B_{R'}(\mathbf{x}_0)$ and the boundary conditions (4.3.74) and (4.3.66)–(4.3.67) on $\partial(\Omega \cap B_{R'}(\mathbf{x}_0))$. Note that matrix $[a_{ij}]$ satisfies the uniform ellipticity. Then, by the maximum principle, $v \ge 0$ in $\Omega \cap B_{R'}(\mathbf{x}_0)$, *i.e.*,

$$B^{(1)}(Du(\mathbf{x}), u(\mathbf{x}), \mathbf{x}) - B^{(1)}(Du(\mathbf{x}_0), u(\mathbf{x}_0), \mathbf{x}_0) + Lw(\mathbf{x}) \ge 0 \ \text{ in } \Omega \cap B_{R'}(\mathbf{x}_0).$$

By a similar argument,

$$-\big(B^{(1)}(Du(\mathbf{x}), u(\mathbf{x}), \mathbf{x}) - B^{(1)}(Du(\mathbf{x}_0), u(\mathbf{x}_0), \mathbf{x}_0)\big) + Lw(\mathbf{x}) \ge 0 \ \text{ in } \Omega \cap B_{R'}(\mathbf{x}_0).$$

Combining these estimates and using (4.3.62), we obtain (4.3.57). □

Remark 4.3.8. *Note that, in Proposition* 4.3.7, *we assume the functional independence* (4.3.56) *only on* $\{(\mathbf{p}, z, \mathbf{x}) = (Du(\mathbf{x}), u(\mathbf{x}), \mathbf{x}) \,:\, \mathbf{x} \in \Gamma_2\}$. *Also, the* $C^{1,\sigma}$*–regularity of* Γ^1 *is not required in Proposition* 4.3.7.

Next we note the following fact for the functional independence.

Proposition 4.3.9. *Suppose that* $\Omega = \Omega_R$ *is as in Proposition* 4.3.7 *and that* $u \in C^1(\overline{\Omega})$ *satisfies* (4.3.44). *Assume that functions* $B^{(k)}$, $k = 1, 2$, *are defined in* V *given by* (4.3.48) *and satisfy*

$$|B^{(k)}(\hat{\mathbf{Y}}) - B^{(k)}(\tilde{\mathbf{Y}})| \leq M|\hat{\mathbf{Y}} - \tilde{\mathbf{Y}}| \qquad \text{for all } \hat{\mathbf{Y}}, \tilde{\mathbf{Y}} \in V. \tag{4.3.75}$$

Moreover, denoting

$$h^{(k)}(\mathbf{p}) = B^{(k)}(\mathbf{p}, u(\mathbf{x}_0), \mathbf{x}_0) \qquad \text{for } k = 1, 2,$$

and noting that functions $h^{(k)}$ *are defined on* $B_K(Du(\mathbf{x}_0))$, *we assume that* $h^{(k)} \in C^{1,\alpha}(\overline{B_K(Du(\mathbf{x}_0))})$ *with*

$$\|h^{(k)}\|_{1,\alpha,\overline{B_K(Du(\mathbf{x}_0))}} \leq M \tag{4.3.76}$$

and

$$|\det H(Du(\mathbf{x}_0))| \geq M^{-1}, \tag{4.3.77}$$

where $\alpha \in (0, 1)$, *and* $H(\mathbf{p})$ *is the matrix with columns* $D_{\mathbf{p}} h^{(k)}(\mathbf{p})$, $k = 1, 2$.

Let $W \subset \overline{\Omega}$ *satisfy*

$$W \cap \partial B_r(\mathbf{x}_0) \neq \emptyset, \quad \overline{W} \cap \partial B_r(\mathbf{x}_0) \subset \overline{W \cap B_r(\mathbf{x}_0)} \qquad \text{for all } r \in (0, R), \tag{4.3.78}$$

where $B_r(\mathbf{x}_0)$ *is the open ball of radius* r *centered at* $\mathbf{x}_0$. *For* $k = 1, 2$, *let*

$$\begin{aligned} &|B^{(k)}(Du(\mathbf{x}), u(\mathbf{x}), \mathbf{x}) - B^{(k)}(Du(\mathbf{x}_0), u(\mathbf{x}_0), \mathbf{x}_0)| \\ &\leq M|\mathbf{x} - \mathbf{x}_0|^\alpha \qquad \text{for all } \mathbf{x} \in W. \end{aligned} \tag{4.3.79}$$

Then there exists C *depending only on* (K, M, R, α) *such that*

$$|Du(\mathbf{x}) - Du(\mathbf{x}_0)| \leq C|\mathbf{x} - \mathbf{x}_0|^\alpha \qquad \text{for all } \mathbf{x} \in W. \tag{4.3.80}$$

Proof. In this proof, constants C and C_k depend only on (K, M, R, α).

From (4.3.79), using (4.3.44) and (4.3.75), we obtain that, for $k = 1, 2$,

$$|B^{(k)}(Du(\mathbf{x}), u(\mathbf{x}_0), \mathbf{x}_0) - B^{(k)}(Du(\mathbf{x}_0), u(\mathbf{x}_0), \mathbf{x}_0)| \leq C|\mathbf{x} - \mathbf{x}_0|^\alpha,$$

which is

$$|h^{(k)}(Du(\mathbf{x})) - h^{(k)}(Du(\mathbf{x}_0))| \leq C|\mathbf{x} - \mathbf{x}_0|^\alpha \qquad \text{for any } \mathbf{x} \in W. \tag{4.3.81}$$

From (4.3.76)–(4.3.77), by the inverse function theorem, there exists $\rho \in (0, K)$ depending only on (K, M, α) such that the map:

$$\mathbf{h}(\mathbf{p}) := (h^{(1)}, h^{(2)})(\mathbf{p})$$

is one-to-one on $B_\rho(Du(\mathbf{x}_0))$ onto an open set U, and

$$|\mathbf{h}^{-1}(\mathbf{q}_1) - \mathbf{h}^{-1}(\mathbf{q}_2)| \le C|\mathbf{q}_1 - \mathbf{q}_2| \qquad \text{for all } \mathbf{q}_1, \mathbf{q}_2 \in U. \tag{4.3.82}$$

Since $u \in C^1(\overline{\Omega})$, there exists $r > 0$ such that

$$|Du(\mathbf{x}) - Du(\mathbf{x}_0)| \le \rho \qquad \text{for all } \mathbf{x} \in W \cap B_r(\mathbf{x}_0). \tag{4.3.83}$$

Then

$$Du(\mathbf{x}) = \mathbf{h}^{-1}(\mathbf{h}(Du(\mathbf{x}))) \qquad \text{for all } \mathbf{x} \in W \cap B_r(\mathbf{x}_0),$$

so that, from (4.3.81)–(4.3.82) and $u \in C^1(\Omega)$,

$$|Du(\mathbf{x}) - Du(\mathbf{x}_0)| \le C|\mathbf{h}(Du(\mathbf{x})) - \mathbf{h}(Du(\mathbf{x}_0))| \le C_1|\mathbf{x} - \mathbf{x}_0|^\alpha \tag{4.3.84}$$

for any $\mathbf{x} \in \overline{W \cap B_r(\mathbf{x}_0)}$.

Let R' be the supremum of all $r \in (0, R)$ such that (4.3.83) holds. Since $\rho \in (0, K)$, the regularity that $u \in C^1(\Omega)$ and $h^{(k)} \in C^{1,\alpha}(\overline{B_K(Du(\mathbf{x}_0))})$ and properties (4.3.78) of W imply that (4.3.83)–(4.3.84) hold in $\overline{W \cap B_{R'}(\mathbf{x}_0)}$.

If $R' = R$, (4.3.84) in $\overline{W \cap B_{R'}(\mathbf{x}_0)}$ implies (4.3.80).

If $R' < R$, we obtain from (4.3.76) and (4.3.84), with $r = R'$, that

$$|Du(\mathbf{x}) - Du(\mathbf{x}_0)| \le MC_1(R')^\alpha \qquad \text{for all } \mathbf{x} \in \overline{W \cap B_{R'}(\mathbf{x}_0)}.$$

If $MC_1(R')^\alpha < \rho$, we use (4.3.78) and $u \in C^1(\overline{\Omega})$ to conclude that (4.3.83) holds with some $r > R'$. This contradicts the definition of R'. Therefore, (4.3.84) holds in $\overline{W \cap B_r(\mathbf{x}_0)}$ with $r = R' \ge \big(\frac{\rho}{MC_1}\big)^{1/\alpha}$, where we recall that ρ depends only on (K, M, α). From this, using (4.3.44), we obtain (4.3.80). ☐

Remark 4.3.10. *Examples for such sets $W \subset \Omega$ satisfying (4.3.78) include:*

(i) $W = \overline{\Omega}$;

(ii) $W = \{(r, \theta) \,:\, 0 \le r \le R,\ \theta = f(r)\} \subset \overline{\Omega}$, *where $f(\cdot)$ is continuous. In particular, if $\Omega = \Omega_R$ is as in Proposition 4.3.7 and (4.3.43) holds, then, from (4.3.59), $W = \overline{\Gamma^2} \cap B_R(\mathbf{x}_0)$ satisfies (4.3.78).*

To obtain the $C^{1,\alpha}$–regularity at the corner, we now prove

Proposition 4.3.11. *Let $R, \lambda > 0$, $\alpha \in (0, 1]$, $\gamma \in [0, 1)$, and $M \ge 1$.*

(i) *Let $\Omega = \Omega_R$ be the domain between two Lipschitz curves as in Proposition 4.3.7. Assume that*

$$\Gamma^k \in C^1 \qquad \text{with } \|\Gamma^k\|_{C^{0,1}} \le M, \qquad k = 1, 2, \tag{4.3.85}$$

in the sense that, for $k = 1, 2$, there exist $c^{(k)} > 0$ and $f^{(k)} \in C^1((0, c^{(k)}))$ with $\|f^{(k)}\|_{C^{0,1}([0,c^{(k)}])} \le M$ such that, in the appropriate basis in $\mathbb{R}^2$,

$$\Omega \subset \{x_2 > f^{(k)}(x_1)\}, \qquad \overline{\Gamma^k} = \{x_2 = f^{(k)}(x_1) \,:\, 0 \le x_1 \le c^{(k)}\}.$$

Moreover, curves Γ^1 and Γ^2 are separated in the sense that, for $k = 1, 2$,

$$B_{\frac{d(\mathbf{x})}{M}}(\mathbf{x}) \cap \partial\Omega = B_{\frac{d(\mathbf{x})}{M}}(\mathbf{x}) \cap \Gamma^k \qquad \text{for all } \mathbf{x} \in \Gamma^k \cap B_{\frac{3R}{4}}(\mathbf{x}_0), \quad (4.3.86)$$

where $d(\mathbf{x}) := |\mathbf{x} - \mathbf{x}_0|$. Assume that $u \in C^1(\overline{\Omega}) \cap C^3(\Omega)$ satisfies (4.3.44) and is a solution of (4.3.45)–(4.3.47) with $h \equiv 0$. Let $(A_{ij}, A)(\mathbf{p}, z, \mathbf{x})$ satisfy (4.3.49)–(4.3.50) and ellipticity (4.3.53). For $k = 1, 2$, let

$$\|B^{(k)}\|_{2,\{|\mathbf{p}|+|z|\le 2K,\ \mathbf{x}\in\overline{\Omega}\}} \le M, \quad (4.3.87)$$

$$|D_{\mathbf{p}}B^{(k)}(Du(\mathbf{x}), u(\mathbf{x}), \mathbf{x})| \ge \lambda \qquad \text{for all } \mathbf{x} \in \Omega. \quad (4.3.88)$$

Moreover, assume that

$$|Du(\mathbf{x}) - Du(\mathbf{x}_0)| \le M|\mathbf{x} - \mathbf{x}_0|^\alpha \qquad \text{for all } \mathbf{x} \in \Omega. \quad (4.3.89)$$

Then there exist $\beta \in (0, \alpha]$ depending only on (λ, K, M, α), and $C > 0$ depending only on $(\lambda, K, M, R, \alpha)$, such that $u \in C^{1,\beta}(\overline{\Omega \cap B_{R/2}(\mathbf{x}_0)})$ with

$$\|u\|_{1,\beta,\Omega\cap B_{R/2}(\mathbf{x}_0)} \le C. \quad (4.3.90)$$

(ii) *If, in addition to the previous assumptions,*

$$\|\overline{\Gamma}^k\|_{1,\sigma} \le M, \qquad k = 1, 2, \quad (4.3.91)$$

for some $\sigma \in (0, 1)$ in the sense that, for $k = 1, 2$, functions f_k introduced above satisfy $\|f^{(k)}\|_{1,\sigma,[0,c^{(k)}]} \le M$, and if assumptions (4.3.49)–(4.3.50) and (4.3.87) for (A_{ij}, A) and $B^{(k)}$ are replaced by

$$\|((A_{ij}, A)(\mathbf{0}, 0, \cdot),\ D^m_{(\mathbf{p},z)}(A_{ij}, A)(\mathbf{p}, z, \cdot))\|^{(-\delta),\{\mathbf{x}_0\}}_{1,\delta,\Omega} \le M \quad (4.3.92)$$

for any $|\mathbf{p}| + |z| \le 2K$ and $m = 1, 2$, and

$$\|B^{(k)}\|_{2,\delta,\{|\mathbf{p}|+|z|\le 2K,\ \mathbf{x}\in\overline{\Omega}\}} \le M \qquad \text{for } k = 1, 2, \quad (4.3.93)$$

with some $\delta \in (0, 1)$, then

$$\|u\|^{(-1-\alpha),\{\mathbf{x}_0\}}_{2,\sigma,\Omega\cap B_{R/2}(\mathbf{x}_0)} \le C, \quad (4.3.94)$$

where C depends only on $(\lambda, K, M, R, \alpha, \sigma, \delta)$.

Proof. In the proof, constants R', C, and C_k depend only on $(\lambda, K, M, R, \alpha)$.

We first note that it suffices to prove the estimates in $\Omega \cap B_{R'/2}(\mathbf{x}_0)$ for some $R' \in (0, R)$ depending only on $(\lambda, K, M, R, \alpha)$. Indeed, estimate (4.3.90) in $\Omega \cap (B_{R/2}(\mathbf{x}_0) \setminus B_{R'/2}(\mathbf{x}_0))$ is obtained by a covering argument and by using:

(a) estimates (4.3.10) from Theorem 4.3.2 in $\Omega \cap B_{R'/(10M)}(\mathbf{x})$ for each $\mathbf{x} \in \Gamma^k \cap (B_{R/2}(\mathbf{x}_0) \setminus B_{R'/2}(\mathbf{x}_0))$, $k = 1, 2$, where we have used condition (4.3.86);

(b) the interior estimates (as in Theorem 4.2.1) applied in the interior balls $B_{R'/(50M)}(\mathbf{x}) \subset \Omega$.

These estimates imply

$$\|u\|_{1,\beta,\Omega\cap(B_{R/2}(\mathbf{x}_0)\setminus B_{R'/2}(\mathbf{x}_0))} \leq C. \tag{4.3.95}$$

In the conditions of assertion (ii), for the balls described in (b), we use Theorem 4.3.4 to obtain

$$\|u\|_{2,\sigma,\Omega\cap(B_{R/2}(\mathbf{x}_0)\setminus B_{R'/2}(\mathbf{x}_0))} \leq C. \tag{4.3.96}$$

Thus, it remains to prove the estimates in $\Omega \cap B_{R'}(\mathbf{x}_0)$, for sufficiently small $R' \in (0, \frac{R}{2})$ which will be chosen. We divide the proof into two steps.

1. We first prove (4.3.90) in $\Omega \cap B_{R'}(\mathbf{x}_0)$. There are two cases.

Case 1: $k \in \{1, 2\}$ *and* $\mathbf{x} = (x_1, x_2) \in \Gamma^k \cap B_{R'}(\mathbf{x}_0)$. Then $x_2 = f^{(k)}(x_1)$ in the appropriate basis. Let $r_{\mathbf{x}} := \min(\frac{d(\mathbf{x})}{4M}, 1)$, where $d(\mathbf{x}) = |\mathbf{x} - \mathbf{x}_0|$. Define the function:

$$v(\mathbf{X}) = \frac{u(\mathbf{x} + r_{\mathbf{x}}\mathbf{X}) - u(\mathbf{x}) - r_{\mathbf{x}} Du(\mathbf{x}) \cdot \mathbf{X}}{r_{\mathbf{x}}^{1+\alpha}} \tag{4.3.97}$$

for $\mathbf{X} \in \tilde{\Omega}_{\mathbf{x}} = \{\mathbf{X} \in B_1(\mathbf{0}) \ : \ \mathbf{x} + r_{\mathbf{x}}\mathbf{X} \in \Omega\}$. Then (4.3.86) implies that, in the appropriate basis in $\mathbb{R}^2$, $\tilde{\Omega}_{\mathbf{x}} = B_1(\mathbf{0}) \cap \{X_2 > \Phi(X_1)\}$ with $\Phi(X_1) = \frac{1}{r_{\mathbf{x}}}\big(f^{(k)}(x_1 + r_{\mathbf{x}}X_1) - f^{(k)}(x_1)\big)$. Thus, $\Phi(0) = 0$ and $\|\Phi\|_{C^{0,1}(\mathbb{R})} \leq M$, by (4.3.85). Furthermore, using (4.3.89), we have

$$\|v\|_{1,0,\tilde{\Omega}_{\mathbf{x}}} \leq C_1, \qquad (v, Dv)(\mathbf{0}) = (0, \mathbf{0}). \tag{4.3.98}$$

Also, v satisfies an equation of form (4.3.45) in $\tilde{\Omega}_{\mathbf{x}}$, and a condition that comes from rescaling (4.3.46) (with $h \equiv 0$) or (4.3.47) on $\tilde{\Gamma}_{\mathbf{x}} = B_1(\mathbf{0}) \cap \{X_2 = \Phi(X_1)\}$. The condition is of a form similar to the original conditions; specifically, the corresponding functions $(\hat{A}_{ij}, \hat{A}, \hat{B}^{(k)})$ in the equation and the boundary condition for v are

$$(\hat{A}_{ij},\ \hat{A},\ \hat{B}^k)(\mathbf{p}, z, \mathbf{X}) = (A_{ij},\ r_{\mathbf{x}}^{1-\alpha}A,\ r_{\mathbf{x}}^{-\alpha}B^k)(\mathcal{Y}), \tag{4.3.99}$$

where $\mathcal{Y} = (Du(\mathbf{x}) + r_{\mathbf{x}}^{\alpha}\mathbf{p},\ u(\mathbf{x}) + r_{\mathbf{x}}Du(\mathbf{x}) \cdot \mathbf{X} + r_{\mathbf{x}}^{\alpha+1}z,\ \mathbf{x} + r_{\mathbf{x}}\mathbf{X})$. Denote

$$V = \{(\mathbf{p}, z, \mathbf{X}) \ : \ |\mathbf{p}| + |z| \leq 2C_1,\ \mathbf{X} \in \overline{\tilde{\Omega}_{\mathbf{x}}}\},$$

where C_1 is from (4.3.98). Since $|Du(\mathbf{x})| + |u(\mathbf{x})| \leq K$ and $r_{\mathbf{x}} \leq \frac{d(\mathbf{x})}{4} \leq \frac{R'}{4}$, we choose R' small to conclude

$$\mathcal{Y}(\mathbf{p}, z, \mathbf{x}) \in \{(\mathbf{q}, w, \mathbf{y}) \in \mathbb{R}^2 \times \mathbb{R} \times \overline{\Omega} \ : \ |\mathbf{q}| + |w| \leq 2K\} \qquad \text{if } (\mathbf{p}, z, \mathbf{X}) \in V.$$

Noting that

$$B^{(k)}(Du(\mathbf{x}_0), u(\mathbf{x}_0), \mathbf{x}_0) = 0,\ |\mathbf{x} - \mathbf{x}_0| = 10r_{\mathbf{x}},\ r_{\mathbf{x}} \leq 1, \quad |\mathbf{X}| \leq 1 \text{ in } \tilde{\Omega}_{\mathbf{x}},$$

we see that $(\hat{A}_{ij}, \hat{A})$ satisfy condition (4.3.53) on $\tilde{\Omega}_{\mathbf{x}}$, $\hat{B}^{(k)}$ satisfies (4.3.88) on $\tilde{\Omega}_{\mathbf{x}}$ with unchanged constants, and

$$\begin{aligned}&\|(\hat{A}_{ij}, \hat{A})\|_{0,V} + \|D_{(\mathbf{p},z)}(\hat{A}_{ij}, \hat{A})\|_{0,V} \le M \qquad \text{for } i,j = 1,2,\\ &\|\hat{B}^{(k)}\|_{2,V} \le M \qquad \text{for } k = 1,2,\end{aligned} \tag{4.3.100}$$

from (4.3.49)–(4.3.50) and (4.3.87). Then, from Theorem 4.3.2, there exist $\hat{\beta} \in (0,1)$ and $C > 0$ depending only on (λ, K, M, α) such that

$$\|v\|_{1,\hat{\beta},\tilde{\Omega}_{\mathbf{x}}\cap B_{1/2}(\mathbf{0})} \le C. \tag{4.3.101}$$

Case 2: $\mathbf{x} \in \Omega$ *and* $\mathrm{dist}(\mathbf{x}, \partial\Omega) \ge \frac{d(\mathbf{x})}{10M}$. Let $r_{\mathbf{x}} := \frac{d(\mathbf{x})}{10M}$. Defining $v(\mathbf{X})$ by (4.3.97) as in Case 1, v is now defined on $B_1(\mathbf{0})$ and satisfies the same rescaled elliptic equation in $B_1(\mathbf{0})$ as in Case 1, with $(\hat{A}_{ij}, \hat{A})$ satisfying (4.3.100) for $\tilde{\Omega}_{\mathbf{x}} = B_1(\mathbf{0})$. The interior estimates (*e.g.*, Theorem 4.2.1) imply that, for any $\kappa \in (0,1)$,

$$\|v\|_{2,\kappa,B_{1/2}(\mathbf{0})} \le C(\kappa), \tag{4.3.102}$$

where $C(\kappa)$ depends only on $(\lambda, K, M, R, \alpha, \kappa)$.

Now, defining $\beta = \min\{\hat{\beta}, \alpha\}$, we note that estimates (4.3.89) and (4.3.101)–(4.3.102) hold with exponent β, which also fixes constant C in (4.3.102), depending on $(\lambda, K, M, R, \alpha)$. These estimates imply (4.3.90) on $\Omega \cap B_{R'}(\mathbf{x}_0)$ by a standard argument; *cf.* the proof of Lemma 4.2.7 or the proof of [131, Theorem 4.8]. Combining this with (4.3.95), we obtain the full estimate (4.3.90).

2. Now we prove (4.3.94) on $\Omega \cap B_{R'}(\mathbf{x}_0)$ under the assumptions of part (ii). We consider Cases 1–2 as in Step 1 above. In Case 1, by using the notations introduced there, assumption (4.3.91) implies that $\|\Phi\|_{C^{1,\sigma}(\mathbb{R})} \le M$, and assumptions (4.3.92)–(4.3.93) imply that

$$\begin{aligned}&\|(\hat{A}_{ij}, \hat{A})\|_{1,\delta,V} \le C(M, K, \alpha, \delta) \qquad \text{for } i,j = 1,2,\\ &\|\hat{B}^{(k)}\|_{2,\delta,V} \le M \qquad \text{for } k = 1,2.\end{aligned} \tag{4.3.103}$$

Thus, for the boundary value problem introduced in Case 1, we can use Theorem 4.3.4 to obtain the stronger estimate:

$$\|v\|_{2,\sigma,\tilde{\Omega}_{\mathbf{x}}\cap B_{1/2}(\mathbf{0})} \le C, \tag{4.3.104}$$

instead of (4.3.101).

In Case 2, we employ estimate (4.3.102) with $\kappa = \sigma$. These estimates and (4.3.89) imply (4.3.94).

Finally, we extend estimate (4.3.94) from $\Omega \cap B_{R'}(\mathbf{x}_0)$ to $\Omega \cap B_{R/2}(\mathbf{x}_0)$ by using (4.3.96). □

Remark 4.3.12. *Condition* (4.3.92) *implies* (4.3.49)–(4.3.50) *with* $\gamma = 1 - \delta$ *and an updated constant* M, *depending only on* M *in* (4.3.92) *and* K.

Combining Proposition 4.3.7 with Proposition 4.3.11, we obtain the following $C^{1,\alpha}$–regularity:

Theorem 4.3.13. *Let $R > 0$, $\gamma \in [0,1)$, $\lambda \in (0,1]$, $\sigma \in (0,1)$, and $K, M \geq 1$.*

(i) *Let $\Omega \equiv \Omega_R$ be as in Proposition 4.3.7 with*

$$\|\overline{\Gamma}^k\|_{1,\sigma} \leq M \qquad \text{for } k = 1, 2. \tag{4.3.105}$$

We also assume that curves Γ^1 and Γ^2 satisfy (4.3.86). Let $u \in C^1(\overline{\Omega}) \cap C^2(\Omega \cup \Gamma^1 \cup \Gamma^2) \cap C^3(\Omega)$ be a solution of (4.3.45)–(4.3.47) with $h \equiv 0$ in (4.3.46), satisfying (4.3.44). Let $(A_{ij}, A)(\mathbf{p}, z, \mathbf{x})$ satisfy (4.3.49)–(4.3.50) and ellipticity (4.3.53). Assume that $B^{(k)}(\mathbf{p}, z, \mathbf{x})$ satisfy (4.3.87)–(4.3.88) for $k = 1, 2$, and

$$D_{\mathbf{p}} B^{(k)}(Du(\mathbf{x}), u(\mathbf{x}), \mathbf{x}) \cdot \boldsymbol{\nu} \geq \lambda \qquad \text{for all } \mathbf{x} \in \Gamma^k, \tag{4.3.106}$$

$$|\det G(Du(\mathbf{x}), u(\mathbf{x}), \mathbf{x})| \geq \lambda \qquad \text{for all } \mathbf{x} \in \Gamma^1 \cup \Gamma^2, \tag{4.3.107}$$

where $G(\mathbf{p}, z, \mathbf{x})$ is the matrix with columns $D_p B^{(k)}(\mathbf{p}, z, \mathbf{x})$, $k = 1, 2$.

Then there exist $\beta \in (0,1)$ depending only on (λ, K, M), and $C > 0$ depending only on $(\lambda, K, M, \gamma, R, \sigma)$, such that $u \in C^{1,\beta}(\overline{\Omega \cap B_{R/2}})$, and (4.3.90) holds.

(ii) *Replace conditions (4.3.49)–(4.3.50) and (4.3.87) on $(A_{ij}, A, B^{(k)})(\mathbf{p}, z, \mathbf{x})$ by conditions (4.3.92)–(4.3.93) for some $\delta \in (0,1)$ with all the other assumptions unchanged. Then u satisfies (4.3.94) with $\alpha \in (0,1)$ depending only on (λ, K, M) and $C > 0$ depending only on $(\lambda, K, M, R, \delta, \sigma)$.*

Proof. By Proposition 4.3.7 with $h \equiv 0$, we have

$$|B^{(k)}(Du(\mathbf{x}), u(\mathbf{x}), \mathbf{x}) - B^{(k)}(Du(\mathbf{x}_0), u(\mathbf{x}_0), \mathbf{x}_0)| \leq C|\mathbf{x} - \mathbf{x}_0|^\alpha \tag{4.3.108}$$

for any $\mathbf{x} \in \Omega \cap B_{R'}(\mathbf{x}_0)$, $k = 1, 2$, where α and C depend only on (λ, K, M), and R' depends only on $(\lambda, K, M, \gamma, \sigma)$. Now, we apply Proposition 4.3.9 with $W = \Omega \cap B_{R'}(\mathbf{x}_0)$, and note that (4.3.75)–(4.3.76) hold, by (4.3.87), and that (4.3.77) holds, by (4.3.107). Then we have

$$|Du(\mathbf{x}) - Du(\mathbf{x}_0)| \leq C|\mathbf{x} - \mathbf{x}_0|^\alpha \qquad \text{for all } \mathbf{x} \in \Omega \cap B_{R'}(\mathbf{x}_0), \tag{4.3.109}$$

where $C = C(\lambda, K, M, \gamma, \sigma)$.

Now estimate (4.3.90) for some $\beta \in (0, \alpha]$ depending only on (λ, K, M, γ) and the fact that C also depends on (R, σ) directly follow from Proposition 4.3.11(i) applied in $\Omega \cap B_{R'}(\mathbf{x}_0)$. Then we use Theorems 4.2.1 and 4.3.2 to extend this estimate from $\Omega \cap B_{R'/2}(\mathbf{x}_0)$ to $\Omega \cap B_{R/2}(\mathbf{x}_0)$ by arguing as in the proof of (4.3.95).

If $(A_{ij}, A, B^{(k)})(\mathbf{p}, z, \mathbf{x})$ satisfy (4.3.92)–(4.3.93) for some $\delta \in (0,1)$, we use Remark 4.3.12 and apply Proposition 4.3.7 with $h \equiv 0$ to obtain (4.3.108) with

α and C depending only on (λ, K, M), and R' depending only on $(\lambda, K, M, \delta, \sigma)$. Note that α is independent of δ. From this, arguing as above, we obtain (4.3.109) with $C = C(\lambda, K, M, \delta, \sigma)$. Using (4.3.109), we can apply Proposition 4.3.11(ii) to obtain (4.3.94) in $\Omega \cap B_{R'}(\mathbf{x}_0)$. Then we apply Theorems 4.2.1 and 4.3.4 to extend this estimate from $\Omega \cap B_{R'/2}(\mathbf{x}_0)$ to $\Omega \cap B_{R/2}(\mathbf{x}_0)$ by arguing as in the proof of (4.3.96). □

Next we obtain a refined version of estimate (4.3.94) in Theorem 4.3.13(ii): Under the conditions of Theorem 4.3.13(ii), the Hölder exponent α in (4.3.94) can be chosen independently of $\|u\|_{C^{0,1}(\Omega)}$, constant γ, and some higher regularity norms of the ingredients of the equation and the oblique boundary conditions (even though these norms still affect estimate (4.3.94) through constant C). This form of (4.3.94) will be used in §12.7.2 and §17.4 via Lemma 4.5.12.

Remark 4.3.14. *In what follows, constant λ is used in the conditions that affect the Hölder exponent of the resulting estimates, and the other constants $(M, K, \dots)$ are used for the conditions that do not affect the Hölder exponent, so that the Hölder exponent in the resulting estimates depends only on λ. This sometimes leads to a redundancy in the conditions (e.g., in* (4.3.92) *and* (4.3.110) *of Theorem* 4.3.15; *we include both, since the weaker condition* (4.3.110) *has bound λ, and the stronger condition* (4.3.92) *does not).*

Theorem 4.3.15. *Assume that all the conditions of Theorem* 4.3.13(ii) *hold: $\Omega \equiv \Omega_R$ is as in Proposition* 4.3.7, *and curves Γ^1 and Γ^2 satisfy* (4.3.86) *and* (4.3.105). *Furthermore, $u \in C^1(\overline{\Omega}) \cap C^2(\Omega \cup \Gamma^1 \cup \Gamma^2) \cap C^3(\Omega)$ is a solution of* (4.3.45)–(4.3.47) *with $h \equiv 0$ and satisfies* (4.3.44), *where functions $(A_{ij}, A, B^{(k)})(\mathbf{p}, z, \mathbf{x})$ satisfy* (4.3.53), (4.3.88), (4.3.92)–(4.3.93), *and* (4.3.106)–(4.3.107). *Moreover, assume*

$$\|A_{ij}\|_{0,\{|\mathbf{p}|+|z|\le 2K,\ \mathbf{x}\in\overline{\Omega}\}} \le \lambda^{-1} \qquad \text{for } i,j = 1,2, \tag{4.3.110}$$

$$\|D_{\mathbf{p}}B^{(k)}\|_{0,\{|\mathbf{p}|+|z|\le 2K,\ \mathbf{x}\in\overline{\Omega}\}} \le \lambda^{-1} \qquad \text{for } k = 1,2; \tag{4.3.111}$$

see Remark 4.3.14. *Then there exist $\hat{\alpha} \in (0,1)$ depending only on λ, and $C > 0$ depending only on $(\lambda, K, M, R, \delta, \sigma)$ such that*

$$\|u\|^{(-1-\hat{\alpha}),\{\mathbf{x}_0\}}_{2,\sigma,\Omega\cap B_{R/2}} \le C. \tag{4.3.112}$$

Proof. The assumptions above allow us to apply Theorem 4.3.13(ii) to obtain (4.3.94) with $\alpha \in (0,1)$ depending only on (λ, K, M), and $C > 0$ depending only on $(\lambda, K, M, R, \delta, \sigma)$. We denote this α by α_1.

Then the main part of proof is to use estimate (4.3.94) and the additional assumptions (4.3.110)–(4.3.111) to refine the argument for Proposition 4.3.7 (with $h \equiv 0$) to obtain (4.3.57) with α depending only on λ, which is $\hat{\alpha}$ in (4.3.112). We do that in Step 1 now.

1. Note that $B^{(k)}$ are the same as in Proposition 4.3.7. We set $h \equiv 0$. Then the assumptions of Proposition 4.3.7 are satisfied with $\gamma = 1 - \delta$; we use this γ and follow the proof of Proposition 4.3.7 with the changes specified below.

In the remaining part of Step 1, constants α, μ, $\bar{L}$, and C_λ depend only on λ, the universal constant C depends only on (λ, γ, K, M), and R' depends on $(\lambda, \gamma, K, M, \sigma)$, unless otherwise specified. We use the notations introduced in the proof of Proposition 4.3.7, especially the notations in (4.3.58).

Then, using (4.3.55) and following the calculations in Step 3 of the proof of Theorem 4.3.2, we find that D^2u and $B^{(1)}$ satisfy (4.3.19) and (4.3.21) with $(Q, R)(\mathbf{x})$ satisfying (4.3.22) in Ω. Using (4.3.19), (4.3.94), and assumptions (4.3.92)–(4.3.93), we have

$$|Dg^{(1)}(\mathbf{x})| \le C|Du(\mathbf{x})| + C \le Cr^{\alpha_1 - 1} \qquad \text{for any } \mathbf{x} \in \Omega_{R/2}, \tag{4.3.113}$$

where $r = |\mathbf{x}|$.

Furthermore, $g^{(1)}$ satisfies equation (4.3.68) in Ω with

$$|(m_{ij}, m_i, m)(\mathbf{x})| \le C \qquad \text{in } \Omega.$$

Denote

$$\tilde{q}_i = \sum_{j=1}^{2} m_{ij} D_j g^{(1)} + r^{-\gamma} m_i. \tag{4.3.114}$$

Then equation (4.3.68) takes the form:

$$\sum_{i,j=1}^{2} a_{ij} D_{ij} g^{(1)} + \sum_{i=1}^{2} \tilde{q}_i D_i g^{(1)} + r^{-\gamma} m = 0. \tag{4.3.115}$$

Moreover, from (4.3.113)–(4.3.114), we have

$$|\tilde{q}_i(\mathbf{x})| \le Cr^{-\gamma_1} \qquad \text{in } \Omega_{R/2}, \tag{4.3.116}$$

where $\gamma_1 = \max\{1 - \alpha_1, \gamma\} \in (0, 1)$. Note that this is the main point where we have used (4.3.94) to improve the estimates in the proof of Proposition 4.3.7: We have rewritten equation (4.3.68) as a linear elliptic equation with $|(a_{ij}, m)| \le C$ and the coefficients of the first-order terms satisfying (4.3.116).

Now we define $w(\mathbf{x})$ by (4.3.61) with $\mu = \frac{1}{2}$ and $\bar{L} = \frac{3}{\lambda^2}$. Then there exist $\alpha_0 \in (0, \frac{1}{2}]$ depending only on λ, and $R' \in (0, \frac{R}{2})$ depending only on (λ, M) so that, for every $\alpha \in (0, \alpha_0)$ and $T : \Gamma^2 \cap B_{R'} \mapsto \mathbb{R}$, we obtain (4.3.62)–(4.3.63) with $\mu = \frac{1}{2}$. We will fix α later, depending only on λ, and further reduce R' in the following calculations.

Define

$$v(\mathbf{x}) = g^{(1)}(\mathbf{x}) + \frac{4M}{(R')^\alpha} w(\mathbf{x}) \qquad \text{in } \Omega_{R'}. \tag{4.3.117}$$

Then, by (4.3.58), (4.3.62), and (4.3.93), we have

$$\begin{aligned} v &\ge -\|g^{(1)}\|_{L^\infty(V)} + \frac{1}{2}\frac{4M}{(R')^\alpha}(R')^\alpha \\ &= -\|g^{(1)}\|_{L^\infty(V)} + 2M \ge 0 \qquad \text{on } \Omega \cap \partial B_{R'}(\mathbf{x}_0). \end{aligned} \tag{4.3.118}$$

Also, since $h \equiv 0$ so that $B^{(1)}|_{\Gamma_1} \equiv 0$, we have

$$v(\mathbf{x}) = \frac{4M}{(R')^\alpha} w(\mathbf{x}) \geq 0 \qquad \text{on } \Gamma^1 \cap B_{R'}(\mathbf{x}_0). \tag{4.3.119}$$

Now, using (4.3.115)–(4.3.116) and (4.3.62) with $\mu = \frac{1}{2}$, we calculate in $\Omega \cap B_{R'}(\mathbf{x}_0)$:

$$\begin{aligned}
\sum_{i,j=1}^{2} a_{ij} D_{ij} v + \sum_{i=1}^{2} \tilde{q}_i D_i v &= \frac{4M}{(R')^\alpha} \Big(\sum_{i,j=1}^{2} a_{ij} w_{ij} + \sum_{i=1}^{2} \tilde{q}_i w_i \Big) - r^{-\gamma} m \\
&\leq \frac{4M}{(R')^\alpha} \Big(-\frac{1}{C} r^{\alpha-2} + C r^{-\gamma_1+\alpha-1} + C r^{-\gamma} (R')^\alpha \Big) \\
&= \frac{4M}{(R')^\alpha} r^{\alpha-2} \Big(-\frac{1}{C} + C r^{1-\gamma_1} + C r^{2-\alpha-\gamma} (R')^\alpha \Big).
\end{aligned}$$

Using that $\alpha, \gamma, \gamma_1 \in (0,1)$ (so that $2 - \alpha - \gamma > 1 - \gamma > 0$) and $r \in (0, R')$, and choosing R' small depending only on (C, γ, γ_1), *i.e.*, on $(\lambda, K, M, \gamma, \delta)$, we have

$$\sum_{i,j=1}^{2} a_{ij} D_{ij} v + \sum_{i=1}^{2} \tilde{q}_i D_i v < 0 \qquad \text{in } \Omega \cap B_{R'}(\mathbf{x}_0). \tag{4.3.120}$$

It remains to derive a differential inequality for v on Γ^2. Repeating the argument starting from (4.3.71), we obtain (4.3.72), where

$$\tilde{m} = -a_{\nu\nu} B^{(1)}_{\mathbf{p}_\tau} B^{(2)}_{\mathbf{p}_\tau} + B^{(1)}_{\mathbf{p}_\nu} \big(a_{\tau\tau} B^{(2)}_{\mathbf{p}_\nu} - 2 a_{\nu\tau} B^{(2)}_{\mathbf{p}_\tau} \big).$$

Using (4.3.110)–(4.3.111), we obtain

$$|\tilde{m}| \leq C_\lambda \qquad \text{on } \Gamma^2,$$

where we recall that constant C_λ depends only on λ. Then assumptions (4.3.53) and (4.3.56) allow us to solve for T with $|T(\mathbf{x})| \leq C_\lambda$ on Γ^2. From (4.3.54), $B^{(2)}_{\mathbf{p}_\nu} \geq \lambda$. Using this, we obtain (4.3.73), which yields $|(y_1, y_2)| \leq C_\lambda$ on Γ^2 by using (4.3.110)–(4.3.111) and (4.3.53) again. It follows that (4.3.71) holds. With this, using (4.3.63) with $\mu = \frac{1}{2}$ and $\bar{L} = \frac{3}{\lambda^2}$, $|T(\mathbf{x})| \leq C_\lambda$, and recalling that $\theta \in [-\frac{\pi}{2}, \frac{\pi}{2}]$ by (4.3.59)–(4.3.60), we obtain that, on $\Gamma^2 \cap B_{R'}(\mathbf{x}_0)$, $v(\mathbf{x})$ in (4.3.117) satisfies

$$\begin{aligned}
v_\nu + T v_\tau &= \frac{4M}{(R')^\alpha} (w_\nu + T w_\tau) + g^{(1)}_\nu + T g^{(1)}_\tau \\
&\leq -4M \Big\{ \big(\frac{1}{C_\lambda} - C(R')^\sigma - C_\lambda \big(\alpha + C(R')^\sigma \big) \Big\} \frac{r^{\alpha-1}}{(R')^\alpha} + C.
\end{aligned}$$

Now we choose α and R' small so that the first term of the expression in the braces dominates the other terms. For that, we first choose $\alpha = \frac{1}{10C_\lambda^2}$ (which fixes $\alpha \in (0,1)$ depending only on λ) and then choose R' so small that $C(1+C_\lambda)(R')^\sigma \le \frac{1}{10C_\lambda}$ to deduce that, for $r \in (0, R')$,

$$v_{\boldsymbol{\nu}} + Tv_{\boldsymbol{\tau}} \le -\frac{M}{C_\lambda}\frac{r^{\alpha-1}}{(R')^\alpha} + C \le -\frac{M}{C_\lambda}\frac{(R')^{\alpha-1}}{(R')^\alpha} + C = -\frac{M}{C_\lambda}(R')^{-1} + C.$$

Now, choosing $R' = R'(\lambda, M, \gamma, \sigma, K)$ small, we obtain (4.3.74). Then, following the rest of the argument in the proof of Proposition 4.3.7, we conclude the proof of (4.3.57) with $B^{(1)}(Du(\mathbf{x}_0), u(\mathbf{x}_0), \mathbf{x}_0) = 0$, since $h \equiv 0$, and with α chosen above so that $\alpha = \alpha(\lambda)$, $C = C(\lambda, \delta, K, M)$, and $R' = R'(\lambda, \delta, K, M, \sigma)$, where we recall that $\gamma = 1 - \delta$.

We also obtain a similar estimate for $B^{(2)}(Du(\mathbf{x}), u(\mathbf{x}), \mathbf{x})$. Denote by $\hat{\alpha}$ the exponent α chosen above so that $\hat{\alpha} = \hat{\alpha}(\lambda)$.

2. With both $B^{(k)}$, $k = 1, 2$, satisfying (4.3.57) with $\alpha = \hat{\alpha}$, we apply Proposition 4.3.9 with $W = \Omega \cap B_{R'}(\mathbf{x}_0)$ and note that (4.3.75)–(4.3.76) hold by (4.3.93), and that (4.3.77) holds by (4.3.107). Then we have

$$|Du(\mathbf{x}) - Du(\mathbf{x}_0)| \le C|\mathbf{x} - \mathbf{x}_0|^{\hat{\alpha}} \qquad \text{for any } \mathbf{x} \in \Omega \cap B_{R'}(\mathbf{x}_0),$$

where C and R' depend only on $(\lambda, K, M, \varepsilon_0, \delta, \sigma)$. Using (4.3.44), we have

$$|Du(\mathbf{x}) - Du(\mathbf{x}_0)| \le C|\mathbf{x} - \mathbf{x}_0|^{\hat{\alpha}} \qquad \text{for any } \mathbf{x} \in \Omega \cap B_{R}(\mathbf{x}_0)$$

with an updated constant C, depending only on R, in addition to the previous parameters. Now we conclude the proof by applying Proposition 4.3.11(ii). □

Next, we consider the case that the solution is *a priori* assumed to be only Lipschitz up to the corner. Then, following [192, Theorem 2.1], we show the $C^{1,\alpha}$–regularity up to the corner under stronger assumptions on the functional independence of the boundary conditions.

Theorem 4.3.16. *Let $R > 0$, $\lambda \in (0,1]$, $\sigma, \delta \in (0,1)$, and $K, M \ge 1$. Let $\Omega := \Omega_R$ be as in Proposition 4.3.7, and let curves Γ^1 and Γ_2 satisfy (4.3.86) and (4.3.105). Let $u \in C^{0,1}(\overline{\Omega}) \cap C^2(\overline{\Omega} \setminus \{\mathbf{x}_0\}) \cap C^3(\Omega)$ satisfy (4.3.44) and be a solution of (4.3.45)–(4.3.47) with $h \equiv 0$ in (4.3.46). Let $(A_{ij}, A, B^{(k)})(\mathbf{p}, z, \mathbf{x})$ satisfy (4.3.53), (4.3.88), (4.3.92)–(4.3.93), (4.3.106), (4.3.110)–(4.3.111), and the following functional independence condition: Map $\mathcal{H} : \mathbb{R}^2 \mapsto \mathbb{R}^2$ defined by $\mathcal{H}(\mathbf{p}) := (B^{(1)}, B^{(2)})(\mathbf{p}, u(\mathbf{x}_0), \mathbf{x}_0)$ satisfies*

$$\mathcal{H} : \mathbb{R}^2 \mapsto \mathbb{R}^2 \text{ is one-to-one and onto,} \tag{4.3.121}$$

$$\mathcal{H} \in C^1(\mathbb{R}^2; \mathbb{R}^2), \qquad \|D\mathcal{H}\|_{L^\infty(\mathbb{R}^2)} \le \lambda^{-1}, \tag{4.3.122}$$

$$|\det D\mathcal{H}(\mathbf{p})| \ge \lambda \qquad \text{for all } \mathbf{p} \in \mathbb{R}^2. \tag{4.3.123}$$

Then there exist $\hat{\alpha} \in (0,1)$ depending only on λ, and $C > 0$ depending only on $(\lambda, K, M, \delta, \sigma, R)$, such that (4.3.112) holds.

Proof. In the argument below, constants α and $\tilde{R}$ depend only on λ, whereas constants C, R', and R'' depend only on $(\lambda, K, M, \delta, R, \sigma)$. We also set $\gamma = 1-\delta$ based on Remark 4.3.12. Then the proof consists of five steps.

1. We combine (4.3.123) with (4.3.44) and (4.3.111) to obtain (4.3.107) on $(\Gamma^1 \cup \Gamma^2) \cap B_{\tilde{R}}(\mathbf{x}_0)$ for some small $\tilde{R} \in (0, R)$ depending only on (K, λ, R). In order to prove (4.3.112) with $\tilde{R}$ instead of R, it suffices to show that $u \in C^1(\overline{\Omega \cap B_{\tilde{R}}(\mathbf{x}_0)})$: Indeed, we can then apply Theorem 4.3.15 in $\Omega \cap B_{\tilde{R}}(\mathbf{x}_0)$ and extend (4.3.112) in $\Omega \cap B_{\tilde{R}/2}(\mathbf{x}_0)$ to the similar estimate in $\Omega \cap B_{R/2}(\mathbf{x}_0)$ by applying the interior estimates (Theorem 4.2.1) and the estimates near the $C^{1,\sigma}$–boundaries (Theorem 4.3.4), as in the proof of (4.3.96).

Therefore, it remains to show that $u \in C^1(\overline{\Omega \cap B_{\tilde{R}}(\mathbf{x}_0)})$.

2. As a preliminary step, we show that, for sufficiently small $\tilde{R}$,

$$|D^2 u(\mathbf{x})| \leq \frac{C}{|\mathbf{x} - \mathbf{x}_0|} \qquad \text{for any } \mathbf{x} \in \Omega \cap B_{\tilde{R}}(\mathbf{x}_0). \tag{4.3.124}$$

Now the argument repeats the proof of Proposition 4.3.11: For $\mathbf{x} \in \Omega \cap B_{\tilde{R}}(\mathbf{x}_0)$, we denote $r_{\mathbf{x}} := \frac{|\mathbf{x}-\mathbf{x}_0|}{M}$ and define

$$v(\mathbf{X}) := \frac{u(\mathbf{x} + r_{\mathbf{x}}\mathbf{X}) - u(\mathbf{x})}{r_{\mathbf{x}}} \tag{4.3.125}$$

for $\mathbf{X} \in \tilde{\Omega}_{\mathbf{x}} = \{\mathbf{X} \in B_1(\mathbf{0}) \ : \ \mathbf{x} + r_{\mathbf{x}}\mathbf{X} \in \Omega\}$. Then (4.3.44) implies that $v(\mathbf{X})$ satisfies (4.3.98) with $C_1 = K$. We argue as in the proof of Proposition 4.3.11 by considering Cases 1–2 and deriving the equation and boundary condition for v, which have the same properties in the present case as in Proposition 4.3.11. Indeed, Ω and Γ^k now have the same properties as in Proposition 4.3.11(ii). Thus, as in Step 2 of the proof of Proposition 4.3.11, we find that, in Case 1, *i.e.*, when $\mathbf{x} \in \Gamma^k \cap B_{\tilde{R}}(\mathbf{x}_0)$, in the appropriate basis in $\mathbb{R}^2$, $\tilde{\Omega}_{\mathbf{x}} = B_1(\mathbf{0}) \cap \{X_2 > \Phi(X_1)\}$ with $\Phi(0) = 0$ and $\|\Phi\|_{C^{1,\sigma}(\mathbb{R})} \leq M$. Also, v satisfies an equation of form (4.3.45) in $\tilde{\Omega}_{\mathbf{x}}$, a condition of form (4.3.47) on $\tilde{\Gamma}_{\mathbf{x}} = B_1(\mathbf{0}) \cap \{X_2 = \Phi(X_1)\}$, and the corresponding functions $(\hat{A}_{ij}, \hat{A}, \hat{B}^{(k)})$ in the equation and the boundary condition for v are

$$(\hat{A}_{ij}, \hat{A}, \hat{B}^{(k)})(\mathbf{p}, z, \mathbf{X}) = (A_{ij}, r_{\mathbf{x}}A, B^{(k)})(\mathbf{p},\, u(\mathbf{x}) + r_{\mathbf{x}}z,\, \mathbf{x} + r_{\mathbf{x}}\mathbf{X}). \tag{4.3.126}$$

Then $\hat{A}_{ij}$ and $\hat{B}^{(k)}$ satisfy (4.3.53) and (4.3.88) on $\tilde{\Omega}_{\mathbf{x}}$ with the unchanged constants, respectively. Moreover, since $|u(\mathbf{x})| \leq K$ by (4.3.44), then (4.3.92)–(4.3.93) imply

$$\begin{aligned} &\|(\hat{A}_{ij}, \hat{A})\|_{C^{1,\delta}(\{|\mathbf{p}|+|z|\leq 2K,\ \mathbf{X}\in\overline{\tilde{\Omega}_{\mathbf{x}}}\})} \leq M, \\ &\|\hat{B}^{(k)}\|_{2,\delta,\{|\mathbf{p}|+|z|\leq 2K,\ \mathbf{x}\in\overline{\tilde{\Omega}_{\mathbf{x}}}\}} \leq M \qquad \text{for } k = 1, 2, \end{aligned} \tag{4.3.127}$$

by using $r_{\mathbf{x}} \leq \tilde{R}$ and further reducing $\tilde{R}$ if necessary.

Thus, following the argument in Proposition 4.3.11, we obtain estimate (4.3.104) in Case 1 and (4.3.102) in Case 2. This implies (4.3.124).

3. We now show that, for $k = 1, 2$,

$$|B^{(k)}(Du(\mathbf{x}), u(\mathbf{x}), \mathbf{x})| \leq C|\mathbf{x} - \mathbf{x}_0|^\alpha \qquad \text{for all } \mathbf{x} \in \Omega \cap B_{R'}(\mathbf{x}_0). \tag{4.3.128}$$

We first show (4.3.128) with $k = 1$. We follow the proof of Proposition 4.3.7, with R replaced by $\tilde{R}$, and $h \equiv 0$. Let w be defined by (4.3.61), with the parameters chosen so that (4.3.62)–(4.3.63) are satisfied. We use the notations in (4.3.58) and define

$$v(\mathbf{x}) = g^{(1)}(\mathbf{x}) + Lw(\mathbf{x}), \tag{4.3.129}$$

where L is defined by (4.3.65). Then $v \in L^\infty(\Omega) \cap C^1(\overline{\Omega} \setminus \{\mathbf{x}_0\}) \cap C^2(\Omega)$, where we have used (4.3.44) to conclude that $v \in L^\infty(\Omega)$. By (4.3.46) with $h \equiv 0$, and (4.3.62), we have

$$v(\mathbf{x}) = Lw(\mathbf{x}) > 0 \qquad \text{on } \Gamma^1 \cap B_{R'}(\mathbf{x}_0). \tag{4.3.130}$$

Furthermore, using that (4.3.107) holds on $(\Gamma^1 \cup \Gamma^2) \cap B_{\tilde{R}}(\mathbf{x}_0)$, as we have discussed in Step 1, we can repeat the calculations in the proof of Proposition 4.3.7. Then we obtain that v satisfies (4.3.66), and (4.3.70) with q_i defined by (4.3.69), as well as (4.3.74) if parameters (μ, α, R') are chosen appropriately. However, since v is not known to be continuous up to $\mathbf{x}_0 \in \partial\Omega \cap B_{R'}(\mathbf{x}_0)$, we cannot apply the maximum principle in $\Omega \cap B_{R'}(\mathbf{x}_0)$.

Then we construct a comparison function which becomes infinite at $\mathbf{x}_0$. Specifically, we show that, for some $R'' \in (0, R]$, there exist both a function $W \in C^\infty(\overline{\Omega} \setminus \{\mathbf{x}_0\})$ and a constant $\beta > 0$ such that

$$\sum_{i,j=1}^{2} a_{ij} D_{ij} W + \sum_{i=1}^{2} q_i D_i W < 0 \qquad \text{in } \Omega \cap B_{R''}(\mathbf{x}_0), \tag{4.3.131}$$

$$W_{\boldsymbol{\nu}} + T W_{\boldsymbol{\tau}} < 0 \qquad \text{on } \Gamma^2 \cap B_{R''}(\mathbf{x}_0), \tag{4.3.132}$$

$$W \geq \frac{1}{2}|\mathbf{x} - \mathbf{x}_0|^{-\beta} \qquad \text{in } \Omega \cap B_{R''}(\mathbf{x}_0), \tag{4.3.133}$$

where $(a_{ij}, q_i)(\mathbf{x})$ are the same as in (4.3.70), and $T(\mathbf{x})$ is the same as in (4.3.74). The proof largely repeats the existence proof of w satisfying (4.3.62)–(4.3.63), given in [111, Page 97]. We only sketch the proof here.

We again use the polar coordinates (r, θ), centered at $\mathbf{x}_0$, chosen so that (4.3.59)–(4.3.60) hold. Then we set

$$W(r, \theta) = r^{-\beta} \bar{h}(\theta), \qquad \bar{h}(\theta) = 1 - \bar{\mu} e^{-\hat{L}\theta}, \tag{4.3.134}$$

where the positive constants $\beta, \bar{\mu}, \hat{L}$, and R'' will be fixed below. More specifically, we will choose $(\beta, \bar{\mu}, R'')$ depending on $(\lambda, K, M, R, \gamma, \sigma)$ and $\hat{L}$, and eventually choose a large $\hat{L}$ depending only on $(\lambda, K, M, R, \gamma, \sigma)$.

For each $\hat{L} \geq 1$, we choose $\bar{\mu} = \frac{1}{2}e^{-\frac{\pi}{2}\hat{L}}$ so that $\bar{h}(\theta) \geq \frac{1}{2}$ for all $\theta \in [-\frac{\pi}{2}, \frac{\pi}{2}]$. Then (4.3.133) holds. We fix such a constant $\bar{\mu}$ from now on.

Now we show (4.3.131). Fix a point $\mathbf{x} \in \Omega \cap B_{R''}(\mathbf{x}_0)$, and rotate the Cartesian coordinates in $\mathbb{R}^2$ so that (x_1, x_2) become the radial and tangential coordinates at $\mathbf{x}$. We denote by a_{rr}, $a_{r\theta}$, *etc.* the coefficients of equation (4.3.131) in these rotated coordinates. As usual, C is a universal constant that may be different at each occurrence, depending only on $(\lambda, K, M, R, \gamma, \sigma)$. We use that $\beta \in (0,1)$ and $\frac{1}{2} \leq h(\theta) \leq 1$ to compute at $\mathbf{x}$:

$$
\begin{aligned}
&\sum_{i,j=1}^{2} a_{ij} D_{ij} W \\
&= a_{rr} W_{rr} + 2a_{r\theta}(r^{-1}W_{r\theta} - r^{-2}W_\theta) + a_{\theta\theta}(r^{-2}W_{\theta\theta} + r^{-1}W_r) \\
&= \Big(\bar{\mu}\hat{L}e^{-\hat{L}\theta}\big(-a_{\theta\theta}\hat{L} - 2a_{r\theta}(\beta+1)\big) + \beta\big((\beta+1)a_{rr} - a_{\theta\theta}\big)\bar{h}(\theta)\Big)r^{-\beta-2} \\
&\leq \Big(\bar{\mu}\hat{L}e^{-\hat{L}\theta}\big(-\frac{\hat{L}}{C} + C\big) + C\beta\Big)r^{-\beta-2}.
\end{aligned}
$$

Then, for any $\hat{L} \geq 10C^2$, using that $\bar{\mu} = \frac{1}{2}e^{-\frac{\pi}{2}\hat{L}} > 0$, we can choose $\beta(\hat{L})$ so small that

$$
\sum_{i,j=1}^{2} a_{ij} D_{ij} W \leq -\bar{\mu}\frac{\hat{L}^2}{2C}e^{-\hat{L}\theta}r^{-\beta-2}.
$$

Next, we estimate q_i by using expression (4.3.69) with L given by (4.3.65) and $|(m_{ij}, m_i)| \leq C$. First we note that $|Dv| \leq \frac{C}{r}$ by (4.3.129), since u satisfies (4.3.44), (4.3.124), and $L|Dw| \leq \frac{C}{r}$ by (4.3.62) and (4.3.65). Using the last estimate in the second term of (4.3.69) as well, and noting that $\gamma < 1$, we conclude that

$$
|q_i| \leq \frac{C}{r}.
$$

Then we have

$$
\begin{aligned}
\sum_{i,j=1}^{2} a_{ij} D_{ij} W + \sum_{i=1}^{2} q_i D_i W &\leq -\bar{\mu}\frac{\hat{L}^2}{2C}e^{-\hat{L}\theta}r^{-\beta-2} + \frac{C}{r}|DW| \\
&\leq \big(-\bar{\mu}\frac{\hat{L}^2}{2C}e^{-\hat{L}\theta} + C(\beta + \bar{\mu}\hat{L}e^{-\hat{L}\theta})\big)r^{-\beta-2} \\
&= \big(\bar{\mu}\hat{L}e^{-\hat{L}\theta}(-\frac{\hat{L}}{2C} + C) + C\beta\big)r^{-\beta-2}.
\end{aligned}
$$

We first choose $\hat{L}$ large so that $-\frac{\hat{L}}{2C} + C \leq -\frac{\hat{L}}{4C}$, *i.e.*, $\hat{L} \geq 4C^2$. Then, for such $\hat{L}$, using that $\bar{\mu} = \frac{1}{2}e^{-\frac{\pi}{2}\hat{L}} > 0$ and $\theta \in [-\frac{\pi}{2}, \frac{\pi}{2}]$, we can choose $\beta(\hat{L})$ small so that the last expression is negative, *i.e.*, that (4.3.131) holds.

To show (4.3.132), we use (4.3.59)–(4.3.60) to see that, at $\mathbf{x} \in \Gamma^2$, choosing $\boldsymbol{\nu}$ to be the interior unit normal to Γ_2 and $\boldsymbol{\tau}$ the unit tangent vector pointing away from the corner,

$$W_{\boldsymbol{\nu}} = -\frac{1+O(r^\sigma)}{r}W_\theta + O(r^\sigma)W_r, \quad W_{\boldsymbol{\tau}} = \big(1+O(r^\sigma)\big)W_r + \frac{O(r^\sigma)}{r}W_\theta,$$

where $O(r^\sigma)$ denotes any quantity satisfying that $|O(r^\sigma)| \le Cr^\sigma$. Then, differentiating (4.3.134) and using that $|T(\mathbf{x})| \le C$, we have

$$\begin{aligned} W_{\boldsymbol{\nu}} + TW_{\boldsymbol{\tau}} &\le \big(-\bar{\mu}\hat{L}e^{-\hat{L}\theta} + C\beta + Cr^\sigma\big)r^{-\beta-1} \\ &\le \big(-\bar{\mu}\hat{L}e^{-\hat{L}\theta} + C\beta + C(R'')^\sigma\big)r^{-\beta-1}. \end{aligned}$$

Therefore, for any $\hat{L} \ge 1$, using that $\bar{\mu} = \frac{1}{2}e^{-\frac{\pi}{2}\hat{L}} > 0$, we can choose both β and R'' small such that the last expression is negative for each $\theta \in [-\frac{\pi}{2}, \frac{\pi}{2}]$ so that (4.3.132) holds. Thus, we fix $\hat{L}$ sufficiently large to satisfy all the conditions stated above, which fixes $(\bar{\mu}, \beta, R'')$. This determines W satisfying (4.3.131)–(4.3.133).

Now we note that, from the proof of Proposition 4.3.7 applied above to function v defined by (4.3.129), it follows that R' can be replaced by any $\hat{R} \in (0, R']$ without change of (μ, α), and the resulting function v (which changes because we now use $\hat{R}$ in (4.3.65)) satisfies (4.3.70) in $\Omega \cap B_{\hat{R}}(\mathbf{x}_0)$, as well as (4.3.66), (4.3.74), and (4.3.130) on the corresponding boundary parts of $\Omega \cap B_{\hat{R}}(\mathbf{x}_0)$. Thus, replacing R' by $\min\{R', R''\}$, we can assume without loss of generality that $R'' \ge R'$.

Now, for any $\varepsilon \in (0,1)$, we can define

$$v_\varepsilon = v + \varepsilon W \qquad \text{in } \Omega \cap B_{R'}(\mathbf{x}_0).$$

Then $v_\varepsilon \in C^1(\overline{\Omega} \setminus \{\mathbf{x}_0\}) \cap C^2(\Omega)$ satisfies (4.3.70) and (4.3.74), and $v_\varepsilon \ge 0$ on $\big(\Omega \cap \partial B_{R'}(\mathbf{x}_0)\big) \cup \big(\Gamma^1 \cap B_{R'}(\mathbf{x}_0)\big)$. Also, since $v \in L^\infty(\Omega)$ and W satisfies (4.3.133), it follows that there exists $r_\varepsilon > 0$ such that $v_\varepsilon > 0$ in $B_{r_\varepsilon}(\mathbf{x}_0)$. Therefore, for any $\rho \in (0, r_\varepsilon)$, applying the maximum principle in $\Omega \cap (B_{R'}(\mathbf{x}_0) \setminus \overline{B_\rho(\mathbf{x}_0)})$, we see that $v_\varepsilon \ge 0$ in that domain. Sending $\rho \to 0^+$, we find that $v_\varepsilon \ge 0$ in $\Omega \cap B_{R'}(\mathbf{x}_0)$. Sending $\varepsilon \to 0^+$, we conclude that $v \ge 0$ in $\Omega \cap B_{R'}(\mathbf{x}_0)$, *i.e.*,

$$B^{(1)}(Du(\mathbf{x}), u(\mathbf{x}), \mathbf{x}) + Lw(\mathbf{x}) \ge 0 \qquad \text{in } \Omega \cap B_{R'}(\mathbf{x}_0).$$

By a similar argument,

$$-B^{(1)}(Du(\mathbf{x}), u(\mathbf{x}), \mathbf{x}) + Lw(\mathbf{x}) \ge 0 \qquad \text{in } \Omega \cap B_{R'}(\mathbf{x}_0).$$

Combining these estimates together, we conclude (4.3.128) for $k = 1$. By a similar argument, we also conclude (4.3.128) for $k = 2$.

4. Using (4.3.121), we can define

$$\mathbf{p}_0 := \mathcal{H}^{-1}(\mathbf{0}).$$

In this step, we now show that

$$|Du(\mathbf{x}) - \mathbf{p}_0| \le C|\mathbf{x} - \mathbf{x}_0|^\alpha \qquad \text{for all } \mathbf{x} \in \Omega \cap B_{R'}(\mathbf{x}_0) \tag{4.3.135}$$

with $\alpha > 0$ depending only on (λ, K, M, γ), and $C > 0$ depending also on (σ, R).

Denote $\mathbf{B}(\mathbf{p}, z, \mathbf{x}) := (B^{(1)}, B^{(2)})(\mathbf{p}, z, \mathbf{x})$. Let $\mathbf{x} \in \Omega \cap B_{R'}(\mathbf{x}_0)$. Then

$$\mathcal{H}(Du(\mathbf{x})) = \mathbf{B}(Du(\mathbf{x}), u(\mathbf{x}), \mathbf{x}) + \big(\mathbf{B}(Du(\mathbf{x}), u(\mathbf{x}_0), \mathbf{x}_0) - \mathbf{B}(Du(\mathbf{x}), u(\mathbf{x}), \mathbf{x})\big).$$

From (4.3.128), $|\mathbf{B}(Du(\mathbf{x}), u(\mathbf{x}), \mathbf{x})| \le C|\mathbf{x} - \mathbf{x}_0|^\alpha$. From (4.3.44) and (4.3.87),

$$|\mathbf{B}(Du(\mathbf{x}), u(\mathbf{x}_0), \mathbf{x}_0) - \mathbf{B}(Du(\mathbf{x}), u(\mathbf{x}), \mathbf{x})| \le C|\mathbf{x} - \mathbf{x}_0|,$$

which implies

$$|\mathcal{H}(Du(\mathbf{x}))| \le C|\mathbf{x} - \mathbf{x}_0|^\alpha.$$

Also, from (4.3.121)–(4.3.123) and the inverse function theorem, map $\mathcal{H}^{-1}$ is in $C^1(\mathbb{R}^2; \mathbb{R}^2)$ with $\|D\mathcal{H}^{-1}\|_{L^\infty(\mathbb{R}^2)} \le C$. Then

$$|Du(\mathbf{x}) - \mathbf{p}_0| = |\mathcal{H}^{-1}(\mathcal{H}(Du(\mathbf{x}))) - \mathcal{H}^{-1}(\mathbf{0})| \le C|\mathcal{H}(Du(\mathbf{x})) - \mathbf{0}| \le C|\mathbf{x} - \mathbf{x}_0|^\alpha.$$

Thus, (4.3.135) is proved.

5. Now we show that u is differentiable at $\mathbf{x}_0$ with $Du(\mathbf{x}_0) = \mathbf{p}_0$.

We recall that Ω is as in Proposition 4.3.7. In particular, Ω is contained in $\mathbf{x}_0 + \{\mathbf{x} : x_2 > \tau|x_1|\}$ for $\mathbf{x}_0 \in \partial\Omega$ and $\tau > 0$, and is between curves Γ^1 and Γ^2 passing through $\mathbf{x}_0$, and (4.3.86) holds. Then, using (4.3.105), we can choose a coordinate system such that $\mathbf{x}_0 = \mathbf{0}$ and, for small $R' = R'(M, \sigma)$,

$$\Omega \cap \{x_1 < R'\} = \{\mathbf{x} : 0 < x_1 < R', \ f_1(x_1) < x_2 < f_2(x_1)\}$$

with $\|f_k\|_{C^{1,\sigma}([0,R'])} \le C$ and $f_k(0) = 0$, $k = 1, 2$. Let $\mathbf{x} \in \Omega \cap B_{R'}$. Then $\mathbf{x} = (x_1, x_2)$ in our coordinate system with $x_1 \in (0, R')$ and $x_2 = \kappa f_1(x_1) + (1-\kappa) f_2(x_1)$ for some $\kappa \in (0,1)$, and $\bar{f}(s) := \kappa f_1(s) + (1-\kappa) f_2(s)$ satisfies $\|\bar{f}\|_{C^{1,\sigma}([0,R'])} \le C$, $\bar{f}(0) = 0$, and $(s, \bar{f}(s)) \in \Omega$ for each $s \in [0, \bar{x}_1]$.

With this, using that $\mathbf{x}_0 = \mathbf{0}$ in our coordinates and applying (4.3.135), we have

$$\begin{aligned}|u(\mathbf{x}) - u(\mathbf{0}) - \mathbf{p}_0 \cdot \mathbf{x}| &= \Big|\int_0^{x_1} (Du(s, \bar{f}(s)) - \mathbf{p}_0) \cdot (1, \bar{f}'(s)) ds\Big| \le C|x_1|^{1+\alpha} \\ &\le C|\mathbf{x}|^{1+\alpha}.\end{aligned}$$

Therefore, $Du(\mathbf{0}) = \mathbf{p}_0$, *i.e.*, $Du(\mathbf{x}_0) = \mathbf{p}_0$.

Combining this with (4.3.135), we conclude that Du is continuous at $\mathbf{x}_0$. Thus, $u \in C^1(\overline{\Omega \cap B_{\tilde{R}}(\mathbf{x}_0)})$. Now the argument in Step 1 implies (4.3.112). □

We note one case (important for our applications in later chapters) in which conditions (4.3.121)–(4.3.123) are satisfied:

Lemma 4.3.17. *Let λ, $\kappa \in (0,1)$. Let Ω, $\mathbf{x}_0$, and Γ_k, $k = 1,2$, be as in Theorem 4.3.16. Let $B^{(k)} \in C^1(\mathbb{R}^2 \times \mathbb{R} \times \Gamma_k)$, $k = 1,2$, satisfy (4.2.102) with some $\delta \geq 0$ and functions $L^{(k)}(\mathbf{p}, z, \mathbf{x}) = \mathbf{b}^{(k)}(\mathbf{x}) \cdot \mathbf{p} + b_0^{(k)}(\mathbf{x})z - h^{(k)}(\mathbf{x})$ and $v^{(k)}(\mathbf{x})$, where $\mathbf{b}^{(k)} = (b_1^{(k)}, b_2^{(k)}), b_0^{(k)}, h^{(k)}, v^{(k)} \in C(\overline{\Gamma}_k)$,*

$$|\mathbf{b}^{(k)}(\mathbf{x}_0)| \leq \lambda^{-1} \qquad \text{for } k = 1,2, \tag{4.3.136}$$

and

$$\left| \frac{\mathbf{b}^{(1)}}{|\mathbf{b}^{(1)}|} - \frac{\mathbf{b}^{(2)}}{|\mathbf{b}^{(2)}|} \right| \geq \kappa \qquad \text{at } \mathbf{x}_0. \tag{4.3.137}$$

If δ in (4.2.102) is small, depending only on (λ, κ), then, for any $z_0 \in \mathbb{R}$, map $\mathcal{H} : \mathbb{R}^2 \mapsto \mathbb{R}^2$ defined by $\mathcal{H}(\mathbf{p}) := (B^{(1)}, B^{(2)})(\mathbf{p}, z_0, \mathbf{x}_0)$ satisfies (4.3.121)–(4.3.123) with some $\tilde{\lambda}(\lambda, \kappa) \in (0,1)$.

Proof. Fix $z_0 \in \mathbb{R}$. Denote $\mathbf{L}(\mathbf{p}) = (L^{(1)}, L^{(2)})(\mathbf{p}, z_0, \mathbf{x}_0)$. Let $\mathcal{L} : \mathbb{R}^2 \mapsto \mathbb{R}^2$ be defined by $\mathcal{L}(\mathbf{p}) := \mathbf{L}(\mathbf{p}) - \mathbf{L}(\mathbf{0})$ for $\mathbf{p} \in \mathbb{R}^2$. Then $\mathcal{L}$ is a linear map, namely, $\mathcal{L}(\mathbf{p}) = \mathcal{B}\mathbf{p}$, where $\mathcal{B}$ is the 2×2 matrix with rows $\mathbf{b}^{(k)}(\mathbf{x}_0)$. Thus, (4.3.136)–(4.3.137) imply that $\mathcal{L}$ satisfies (4.3.121)–(4.3.123) with $\hat{\lambda} \in (0,1)$ depending only on (λ, κ). In particular, the constant matrix $\mathcal{B}$ is invertible, with $\|\mathcal{B}^{-1}\| \leq C(\kappa, \lambda)$.

Using the second inequality in (4.2.102) for $B^k(\mathbf{p}, z_0, \mathbf{x}_0)$, $k = 1,2$, and noting that $D_{\mathbf{p}}\mathbf{L}(\mathbf{p}, z, \mathbf{x}_0) = \mathcal{B}$ for any $(\mathbf{p}, z)$, it follows that, if δ is small depending only on (λ, κ), $\mathcal{H}$ satisfies (4.3.122)–(4.3.123) with $\tilde{\lambda} = \frac{\hat{\lambda}}{2}$.

Therefore, it remains to prove (4.3.121) for $\mathcal{H}$. In the following calculations, we use that $\mathcal{L}$ satisfies (4.3.121)–(4.3.123). Then, for any $\hat{\mathbf{p}}, \tilde{\mathbf{p}} \in \mathbb{R}^2$, we have

$$\hat{\mathbf{p}} - \tilde{\mathbf{p}} = \mathcal{L}^{-1}\big(\mathcal{H}(\hat{\mathbf{p}}) - \mathcal{H}(\tilde{\mathbf{p}}) - \mathbf{q}(\hat{\mathbf{p}}, \tilde{\mathbf{p}})\big),$$

where

$$\mathbf{q}(\hat{\mathbf{p}}, \tilde{\mathbf{p}}) = \mathcal{H}(\hat{\mathbf{p}}) - \mathcal{H}(\tilde{\mathbf{p}}) - \mathcal{L}(\hat{\mathbf{p}} - \tilde{\mathbf{p}}). \tag{4.3.138}$$

Since $\mathcal{L}(\mathbf{p}) = \mathcal{B}\mathbf{p}$ for any $\mathbf{p} \in \mathbb{R}^2$, where $\mathcal{B}$ is a constant matrix,

$$\begin{aligned} |\mathbf{q}(\hat{\mathbf{p}}, \tilde{\mathbf{p}})| &= \left| \int_0^1 \big(D\mathcal{H}(t\hat{\mathbf{p}} + (1-t)\tilde{\mathbf{p}}) - \mathcal{B}\big)(\hat{\mathbf{p}} - \tilde{\mathbf{p}})dt \right| \\ &\leq \sum_{k=1}^{2} \left| \int_0^1 \big(D_{\mathbf{p}}B^{(k)}(t\hat{\mathbf{p}} + (1-t)\tilde{\mathbf{p}}, z_0, \mathbf{x}_0) - \mathbf{b}^{(k)}(\mathbf{x}_0)\big) \cdot (\hat{\mathbf{p}} - \tilde{\mathbf{p}})dt \right| \\ &\leq \delta|\hat{\mathbf{p}} - \tilde{\mathbf{p}}|, \end{aligned}$$

where we have used the second inequality in (4.2.102) and the fact that

$$D_{\mathbf{p}}L^{(k)}(\mathbf{p}, z, \mathbf{x}_0) = \mathbf{b}^{(k)}(\mathbf{x}_0) \qquad \text{for any } (\mathbf{p}, z).$$

If δ is small, depending only on (λ, κ), then

$$|\mathcal{L}^{-1}\big(\mathbf{q}(\hat{\mathbf{p}}, \tilde{\mathbf{p}})\big)| \leq C|\mathbf{q}(\hat{\mathbf{p}}, \tilde{\mathbf{p}})| \leq \frac{1}{2}|\hat{\mathbf{p}} - \tilde{\mathbf{p}}| \qquad \text{for all } \hat{\mathbf{p}}, \tilde{\mathbf{p}} \in \mathbb{R}^2, \tag{4.3.139}$$

so

$$\left|\mathcal{L}^{-1}\big(\mathcal{H}(\hat{\mathbf{p}})-\mathcal{H}(\tilde{\mathbf{p}})\big)\right| \ge \frac{1}{2}|\hat{\mathbf{p}}-\tilde{\mathbf{p}}|,$$

which implies that $\mathcal{H}$ is one-to-one on $\mathbb{R}^2$.

Now we show that $\mathcal{H}(\mathbb{R}^2)=\mathbb{R}^2$. Let $\mathbf{v}\in\mathbb{R}^2$. Then $\mathcal{H}(\mathbf{p})=\mathbf{v}$ if and only if $\mathbf{p}=G_{\mathbf{v}}(\mathbf{p})$ is a fixed point of map $G_{\mathbf{v}}(\mathbf{p})=\mathcal{L}^{-1}\big(\mathbf{v}+(\mathcal{L}(\mathbf{p})-\mathcal{H}(\mathbf{p}))\big)$. We show that $G_{\mathbf{v}}:\mathbb{R}^2\mapsto\mathbb{R}^2$ is a contraction map if δ is small, depending only on (λ,κ), so that (4.3.139) holds. Indeed, we then see that

$$|G_{\mathbf{v}}(\hat{\mathbf{p}})-G_{\mathbf{v}}(\tilde{\mathbf{p}})| = \left|\mathcal{L}^{-1}\big(\mathbf{q}(\hat{\mathbf{p}},\tilde{\mathbf{p}})\big)\right| \le \frac{1}{2}|\hat{\mathbf{p}}-\tilde{\mathbf{p}}|.$$

Therefore, a fixed point of $G_{\mathbf{v}}$ exists for any $\mathbf{v}\in\mathbb{R}^2$, which implies that $\mathcal{H}(\mathbb{R}^2)=\mathbb{R}^2$. □

Next, we prove the regularity of the solution at a corner without assuming its Lipschitz bound *a priori*. Instead, we assume the linear growth of the solution from the corner.

Theorem 4.3.18. *Let $R>0$, $\kappa>0$, $\lambda\in(0,1]$, $\sigma,\hat{\delta}\in(0,1)$, and $L,M\ge 1$. Let $\Omega\equiv\Omega_R$ be as in Theorem 4.3.16, satisfying (4.3.86) and (4.3.105). Let $u\in C(\overline{\Omega})\cap C^2(\overline{\Omega}\setminus\{\mathbf{x}_0\})\cap C^3(\Omega)$ be a solution of (4.3.45)–(4.3.47) with $h\equiv 0$ in (4.3.46) and satisfy*

$$|u(\mathbf{x}_0)|\le L, \tag{4.3.140}$$

$$|u(\mathbf{x})-u(\mathbf{x}_0)|\le L|\mathbf{x}-\mathbf{x}_0| \qquad \text{for all } \mathbf{x}\in\Omega. \tag{4.3.141}$$

Let $(A_{ij},A,B^{(k)})(\mathbf{p},z,\mathbf{x})$ satisfy (4.3.53), (4.3.88), (4.3.92)–(4.3.93), (4.3.106), and (4.3.110)–(4.3.111) with $K=\infty$, and $\hat{\delta}$ instead of δ.

Furthermore, assume that $B^{(k)}(\mathbf{p},z,\mathbf{x})$, for $k=1,2$, satisfy (4.2.102) with some $\delta\ge 0$ and functions $L^{(k)}(\mathbf{p},z,\mathbf{x})=\mathbf{b}^{(k)}(\mathbf{x})\cdot\mathbf{p}+b_0^{(k)}(\mathbf{x})z-h^{(k)}(\mathbf{x})$ and $v^{(k)}(\mathbf{x})$ for $\mathbf{b}^{(k)},b_0^{(k)},h^{(k)},v^{(k)}\in C(\overline{\Gamma}_k)$ satisfying (4.3.136) and

$$\|v^{(k)}\|^{(-1-\alpha),\{\mathbf{x}_0\}}_{2,\alpha,\Omega}+\|(\mathbf{b}^{(k)},h^{(k)})\|^{(-\alpha),\{\mathbf{x}_0\}}_{1,\alpha,\Gamma^{(k)}}\le M \tag{4.3.142}$$

for $k=1,2$. Moreover, let (4.3.137) hold.

Then there exists $\delta_0>0$ depending only on $(\kappa,\lambda,L,M,\hat{\delta},\sigma,R)$ such that, if $\delta\in(0,\delta_0)$, u satisfies (4.3.112) with $\hat{\alpha}\in(0,1)$ depending only on (κ,λ), and $C>0$ depending only on $(\kappa,\lambda,L,M,\hat{\delta},\sigma,R)$.

Proof. We divide the proof into three steps. In the first two steps, we prove estimate (4.3.112) in a smaller region $\Omega\cap B_{\tilde{R}/4}(\mathbf{x}_0)$ for $\tilde{R}\in(0,R)$, depending only on $(\lambda,L,M,\hat{\delta},R,\sigma)$. In the last step, we extend this estimate to $\Omega\cap B_{R/2}(\mathbf{x}_0)$.

1. In this step, we show that $u\in C^{0,1}(\overline{\Omega})$ and satisfies

$$\|Du\|_{L^\infty(\Omega\cap B_{\tilde{R}/2}(\mathbf{x}_0))}\le\hat{K}, \tag{4.3.143}$$

where $\tilde{R} \in (0, R)$, and $\hat{K}$ depends only on $(\lambda, L, M, \hat{\delta}, R, \sigma)$.

For the proof of (4.3.143), we follow the argument in Step 2 of the proof of Theorem 4.3.16. We use $\tilde{R}$, $r_{\mathbf{x}}$, and $\tilde{\Omega}_{\mathbf{x}}$ introduced there, and denote $d(\mathbf{x}) := |\mathbf{x} - \mathbf{x}_0|$. Then $v(\mathbf{X})$, defined by (4.3.125), satisfies

$$\|v\|_{L^\infty(\tilde{\Omega}_{\mathbf{x}})} \le 3ML, \tag{4.3.144}$$

under the present assumptions. Indeed, for each $\mathbf{x} \in \Omega \cap B_{\tilde{R}}(\mathbf{x}_0)$ and $\mathbf{X} \in \tilde{\Omega}_{\mathbf{x}}$, $r_{\mathbf{x}} \le d(\mathbf{x})$ and $|\mathbf{X}| \le 1$ so that $|\mathbf{x} + r_{\mathbf{x}}\mathbf{X} - \mathbf{x}_0| \le |\mathbf{x} - \mathbf{x}_0| + r_{\mathbf{x}} \le 2d(\mathbf{x})$, which implies

$$|v(\mathbf{X})| \le \frac{|u(\mathbf{x} + r_{\mathbf{x}}\mathbf{X}) - u(\mathbf{x}_0)|}{d(\mathbf{x})/M} + \frac{|u(\mathbf{x}) - u(\mathbf{x}_0)|}{d(\mathbf{x})/M} \le M(2L + L) = 3\hat{M}L.$$

Thus, (4.3.144) is proved.

Now we show estimates (4.3.104) and (4.3.102) in Cases 1 and 2, respectively, considered in the proof of Proposition 4.3.11.

First, as in Step 2 of the proof of Theorem 4.3.16, we note that, in Case 1, the following properties of the rescaled domain, equation, and boundary conditions hold: Domain $\tilde{\Omega}_{\mathbf{x}} := \{\mathbf{X} \in B_1(\mathbf{0}) \ : \ \mathbf{x} + r_{\mathbf{x}}\mathbf{X} \in \Omega\}$ is of the form:

$$\tilde{\Omega}_{\mathbf{x}} = B_1(\mathbf{0}) \cap \{X_2 > \Phi(X_1)\} \qquad \text{with } \Phi(0) = 0 \text{ and } \|\Phi\|_{C^{1,\sigma}(\mathbb{R})} \le M;$$

v satisfies an equation of form (4.3.45) in $\tilde{\Omega}_{\mathbf{x}}$ and a condition of form (4.3.47) on $\tilde{\Gamma}_{\mathbf{x}} = B_1(\mathbf{0}) \cap \{X_2 = \Phi(X_1)\}$, where the corresponding functions $(\hat{A}_{ij}, \hat{A}, \hat{B}^{(k)})$ in the equation and the boundary condition for v are defined by (4.3.126) and satisfy (4.3.53) and (4.3.88) on $\tilde{\Omega}_{\mathbf{x}}$ with the same λ, and (4.3.127) (with $\hat{\delta}$ instead of δ) with the unchanged constants, M and K as in (4.3.92)–(4.3.93), so that $K = \infty$ in the present case.

However, since the Lipschitz continuity of v has not been proved yet, we cannot use Theorem 4.3.4 to obtain (4.3.104) in Case 1. Instead, we use Theorem 4.2.12. Then we need to check that the conditions of Theorem 4.2.12 are satisfied.

The properties of $(\hat{A}_{ij}, \hat{A}, \hat{B}^{(k)})$ discussed above imply that it suffices to check that (4.2.102) is satisfied for $\hat{B}^{(k)}$, $k = 1, 2$. Since condition (4.2.102) is satisfied for $B^{(k)}(\mathbf{p}, z, \mathbf{x})$ with $L^{(k)}(\mathbf{p}, z, \mathbf{x}) = \mathbf{b}^{(k)}(\mathbf{x}) \cdot \mathbf{p} + b_0^{(k)}(\mathbf{x})z - h^{(k)}(\mathbf{x})$ and $v^{(k)}(\mathbf{x})$, using (4.3.126) and $r_{\mathbf{x}} \le 1$, we obtain that, in the $\mathbf{X}$-coordinates, condition (4.2.102) is satisfied with the same constant δ for $\hat{B}^{(k)}(\mathbf{p}, z, \mathbf{X})$, where the corresponding $\hat{L}^{(k)}(\mathbf{p}, z, \mathbf{X})$ and $\hat{v}^{(k)}$ are defined by

$$\hat{L}^{(k)}(\mathbf{p}, z, \mathbf{X}) = L^{(k)}(\mathbf{p},\ u(\mathbf{x}) + r_{\mathbf{x}}z,\ \mathbf{x} + r_{\mathbf{x}}\mathbf{X}), \ \ \hat{v}^{(k)}(\mathbf{X}) = \frac{1}{r_{\mathbf{x}}}(v^{(k)} - u)(\mathbf{x} + r_{\mathbf{x}}\mathbf{X}).$$

Then

$$\hat{L}^{(k)}(\mathbf{p}, z, \mathbf{X}) = \hat{\mathbf{b}}^{(k)}(\mathbf{X}) \cdot \mathbf{p} + \hat{b}_0^{(k)}(\mathbf{X})z - \hat{h}^{(k)}(\mathbf{X}),$$

where

$$(\hat{\mathbf{b}}^{(k)}, \hat{b}_0^{(k)})(\mathbf{X}) = (\mathbf{b}^{(k)}, r_{\mathbf{x}} b_0^{(k)})(\mathbf{x} + r_{\mathbf{x}}\mathbf{X}),$$
$$\hat{h}^{(k)}(\mathbf{X}) = h^{(k)}(\mathbf{x} + r_{\mathbf{x}}\mathbf{X}) - u(\mathbf{x}) b_0^{(k)}(\mathbf{x} + r_{\mathbf{x}}\mathbf{X}).$$

From (4.3.142), (4.3.144), and $r_{\mathbf{x}} \le 1$,

$$\|\hat{v}^{(k)}\|_{2,\alpha,\tilde{\Omega}_{\mathbf{x}}} + \|(\hat{\mathbf{b}}^{(k)}, \hat{h}^{(k)})\|_{1,\alpha,\tilde{\Gamma}_{\mathbf{x}}} \le C(M, L) \qquad \text{for } k = 1, 2.$$

It follows that the conditions of Theorem 4.2.12 for the rescaled problem are satisfied with the constants depending only on $(\lambda, L, M, \hat{\delta}, \sigma, R)$, independent of $\mathbf{x} \in \Gamma^k$. Then there exist δ_0 and C depending only on these constants such that, when $\delta \in (0, \delta_0)$, (4.3.104) holds.

The argument for the interior estimate (4.3.102) in Case 2 works without change, because the estimate of Theorem 4.2.1 depends only on its L^∞–norm of the solution, which is estimated in (4.3.144). Thus, in the present case, for any $\kappa \in (0, 1)$, we obtain (4.3.102) with $C(\kappa)$ depending only on $(\lambda, L, M, R, \alpha, \kappa)$.

Estimates (4.3.102) and (4.3.104), with C depending only on $(\lambda, L, M, \hat{\delta}, \sigma, R)$, obtained for each $v(\cdot) = v^{(\mathbf{x})}(\cdot)$ for every $\mathbf{x}$ considered in Cases 1–2 for Proposition 4.3.11, imply (4.3.143).

2. Combining (4.3.143) with (4.3.140), we obtain

$$\|u\|_{C^{0,1}(\Omega \cap B_{\tilde{R}/2}(\mathbf{x}_0))} \le C(\lambda, L, M, \hat{\delta}, R, \sigma).$$

With this estimate, we note that all the conditions of Theorem 4.3.16 are satisfied in $\Omega \cap B_{\tilde{R}/2}(\mathbf{x}_0)$, where (4.3.121)–(4.3.123) hold in the present case with some $\tilde{\lambda}(\lambda, \kappa) > 0$ by Lemma 4.3.17. Now we can apply Theorem 4.3.16 in $\Omega \cap B_{\tilde{R}/2}(\mathbf{x}_0)$, with λ replaced by $\min\{\lambda, \tilde{\lambda}\}$, to obtain (4.3.112) in $\Omega \cap B_{\tilde{R}/4}(\mathbf{x}_0)$. Note that $\hat{\alpha}$ depends on (κ, α), since $\tilde{\lambda}$ depends on these constants.

3. Finally, we extend this estimate from $\Omega \cap B_{\tilde{R}/4}(\mathbf{x}_0)$ to $\Omega \cap B_{R/2}(\mathbf{x}_0)$, arguing as in the proof of (4.3.96) with the following difference: From (4.3.140)–(4.3.141), $\|u\|_{L^\infty(\Omega \cap (B_R(\mathbf{x}_0) \setminus B_{\tilde{R}/2}(\mathbf{x}_0)))} \le L(R+1)$. With this, we can apply Theorem 4.2.1 in the interior balls $B_{\tilde{R}/50\tilde{M}}(\mathbf{x}) \subset \Omega$. However, in the boundary half-balls $\Omega \cap B_{\tilde{R}/10M}(\mathbf{x})$ for $\mathbf{x} \in \Gamma^k \cap (B_{R/2}(\mathbf{x}_0) \setminus B_{R'/2}(\mathbf{x}_0))$, we cannot apply Theorem 4.3.4 (since the Lipschitz bound of u is not available). Instead, we apply Theorem 4.2.12 by noting that its conditions are satisfied by the assumptions of this theorem. □

Corollary 4.3.19. *The assertion of Theorem* 4.3.18 *also holds if assumption* (4.3.92) *is replaced by the following*:

(i) $A(\mathbf{p}, z, \mathbf{x}) = (A_1(\mathbf{p}, z, \mathbf{x}), A_2(\mathbf{p}, z, \mathbf{x})) \cdot \mathbf{p} + A_0(\mathbf{p}, z, \mathbf{x})z$;

(ii) *For any* $(\mathbf{p}, z) \in \mathbb{R}^2 \times \mathbb{R}$ *and* $m = 1, 2$,

$$\|((A_{ij}, A_i)(\mathbf{p}, z, \cdot),\ D^m_{(\mathbf{p},z)}(A_{ij}, A_i)(\mathbf{p}, z, \cdot))\|^{(-\delta),\{\mathbf{x}_0\}}_{1,\delta,\Omega} \le M. \qquad (4.3.145)$$

Proof. We follow the proof of Theorem 4.3.18, except that we now apply Theorem 4.2.12 with the conditions modified in Remark 4.2.13. Thus, we obtain (4.3.143) with $\tilde{R} \in (0, R)$ and $\hat{K}$ depending only on $(\kappa, \lambda, L, M, \hat{\delta}, R, \sigma)$. Then we notice that conditions (i)–(ii) imply (4.3.92) for $K = 4\hat{K}$, with the constant on the right-hand side depending only on $(M, \hat{K})$, hence on $(\kappa, \lambda, L, M, \hat{\delta}, R, \sigma)$. With this, we can complete the proof as in Theorem 4.3.18. □

4.4 COMPARISON PRINCIPLE FOR A MIXED BOUNDARY VALUE PROBLEM IN A DOMAIN WITH CORNERS

We present a comparison principle for elliptic equations with degenerate ellipticity near a part of the boundary of the domain with corners and with mixed boundary conditions.

4.4.1 Linear case

We first define the obliqueness of the boundary conditions at a corner.

Definition 4.4.1. *Let $\Omega \subset \mathbb{R}^2$ be open, $P \in \partial\Omega$, and $R > 0$, and let $\partial\Omega \cap B_R(P)$ be a union of two C^1–curves Γ_1 and Γ_2 with common endpoint P. Assume that, for each $i = 1, 2$, a vector field $\mathbf{b}^{(i)} = (b_1^{(i)}, b_2^{(i)})$ is defined on Γ_i, is continuous up to P, and points into Ω, i.e., $\mathbf{b}^{(i)} \cdot \boldsymbol{\nu}^{(i)} \geq 0$ on Γ_i, where $\boldsymbol{\nu}^{(i)}$ is the interior unit normal on Γ_i to Ω. We say that the boundary conditions:*

$$\mathbf{b}^{(i)} \cdot Du + b_0^{(i)} u = g_i \qquad \text{on } \Gamma_i, \quad \text{for } i = 1, 2,$$

are oblique at P if vectors $\mathbf{b}^{(i)}(P), i = 1, 2$, lie on the same side of a supporting $C^{1,\alpha}$–curve Σ at P as Ω, i.e., there exist a $C^{1,\alpha}$–curve Σ and a constant $\lambda > 0$ such that

$$\begin{aligned} &\Sigma \cap \Omega = \emptyset, && P \in \Sigma^0, \\ &\mathbf{b}^{(i)} \cdot \boldsymbol{\nu}_\Sigma \geq \lambda, && \text{at } P \quad \text{for } i = 1, 2, \end{aligned} \tag{4.4.1}$$

where $\boldsymbol{\nu}_\Sigma$ is a normal to Σ at P, oriented so that $\boldsymbol{\nu}_\Sigma \cdot \boldsymbol{\nu}^{(i)} \geq 0$ at P for $i = 1, 2$, and Σ^0 is the relative interior of Σ. We also say that the λ-obliqueness holds at P for a particular choice of λ in (4.4.1).

Lemma 4.4.2 (Comparison Principle). *Let $\Omega \subset \mathbb{R}^2$ be a bounded domain whose boundary $\partial\Omega$ is a union of curves Γ_k, $k = 0, \dots, 3$, such that curve Γ_k has endpoints P_k and P_{k+1} for each $k = 1, 2, 3$, and Γ_0 has endpoints P_1 and P_4, where Γ_k refers to the corresponding curve segment that includes the endpoints. Assume that curves $\Gamma_k, k = 1, 2, 3$, are C^2 at their relative interiors and $C^{1,\alpha}$ up to their endpoints for some $\alpha \in (0, 1)$. Furthermore, assume that curves Γ_k do not intersect each other at the points of their relative interiors Γ_k^0. Moreover, assume that the angles (from the Ω–side) between the curves meeting at points*

P_2 and P_3 are less than π. Let $u \in C(\overline{\Omega}) \cap C^1(\overline{\Omega} \setminus \overline{\Gamma}_0) \cap C^2(\Omega)$ satisfy

$$\begin{array}{ll} \mathcal{L}(u) \leq 0 & \text{in } \Omega, \\ u \geq 0 & \text{on } \Gamma_0, \\ \mathbf{b}^{(k)} \cdot Du + b_0^{(k)} u \leq 0 & \text{on } \Gamma_k \quad \text{for } k = 1, 2, 3, \end{array} \tag{4.4.2}$$

where

$$\mathcal{L}(u) := \sum_{i,j=1}^{2} a_{ij} D_{ij} u + \sum_{i=1}^{2} a_i D_i u + a_0 u. \tag{4.4.3}$$

Assume that coefficients (a_{ij}, a_i) are continuous in $\overline{\Omega}$ and that operator $\mathcal{L}$ in (4.4.3) is strictly elliptic in $\overline{\Omega} \setminus \Gamma_0$ in the sense that $\sum_{i,j=1}^{2} a_{ij}(\mathbf{x}) \xi_i \xi_j > 0$ for every $\boldsymbol{\xi} \in \mathbb{R}^2 \setminus \{\mathbf{0}\}$ and $\mathbf{x} \in \overline{\Omega} \setminus \Gamma_0$. Assume that

$$(\mathbf{b}^{(1)}, b_0^{(1)}) \in C(\overline{\Gamma}_1 \setminus \{P_1\}), \;\; (\mathbf{b}^{(2)}, b_0^{(2)}) \in C(\overline{\Gamma}_2), \;\; (\mathbf{b}^{(3)}, b_0^{(3)}) \in C(\overline{\Gamma}_3 \setminus \{P_4\}),$$

and the strict obliqueness of the boundary conditions hold on Γ_k, $k = 1, 2, 3$, in the following sense:

$$\begin{array}{ll} \mathbf{b}^{(1)} \cdot \boldsymbol{\nu}^{(1)} > 0 & \text{on } \overline{\Gamma}_1 \setminus \{P_1\}, \\ \mathbf{b}^{(2)} \cdot \boldsymbol{\nu}^{(2)} > 0 & \text{on } \overline{\Gamma}_2, \\ \mathbf{b}^{(3)} \cdot \boldsymbol{\nu}^{(3)} > 0 & \text{on } \overline{\Gamma}_3 \setminus \{P_4\}, \end{array}$$

where $\boldsymbol{\nu}^{(k)}$ is the interior unit normal on Γ_k to Ω. Moreover, assume that the boundary conditions are oblique at corners $\{P_2, P_3\}$ in the sense of Definition 4.4.1. Furthermore, assume

$$a_0 \leq 0 \quad \text{in } \Omega, \qquad b_0^{(k)} \leq 0 \quad \text{on } \Gamma_k \quad \text{for } k = 1, 2, 3.$$

Then $u \geq 0$ in Ω.

Proof. If u is not a constant in Ω, then a negative minimum of u over $\overline{\Omega}$ cannot be attained:

(i) In the interior of Ω, by the strong maximum principle for linear elliptic equations;

(ii) In the relative interiors of Γ_k for $k = 1, 2, 3$, by Hopf's lemma, using the C^2–regularity of Γ_k^0, the obliqueness of the boundary conditions on Γ_k, and the non-positivity of the zero-order term on Γ_k for $k = 1, 2, 3$;

(iii) At corners P_2 and P_3, by the result in Lieberman [189, Lemma 2.2], via the standard argument as in [131, Theorem 8.19], where we have used the obliqueness of the boundary conditions at P_2 and P_3. Note that curves Γ_k have to be flattened in order to apply [189, Lemma 2.2] near P_2 and P_3, which can be done by using the $C^{1,\alpha}$–regularity of Γ_k.

Since $u \geq 0$ on Γ_0, we conclude the proof. □

4.4.2 Nonlinear case

Next, we consider the corresponding nonlinear problems and related comparison principle.

Definition 4.4.3. *Let $\Omega \subset \mathbb{R}^2$ be open, $P \in \partial\Omega$, and $R > 0$. Assume that $\partial\Omega \cap B_R(P)$ is a union of two C^1-curves Γ_1 and Γ_2 with common endpoint P. Assume that, for $i = 1, 2$, a function $B^{(i)}(\mathbf{p}, z, \mathbf{x})$ is defined on $\mathbb{R}^2 \times \mathbb{R} \times \Gamma_i$, and $B^{(i)}$ and $D_{(\mathbf{p},z)}B^{(i)}$ are continuous on $\mathbb{R}^2 \times \mathbb{R} \times (\Gamma_i \cup \{P\})$, and the vector field $D_{\mathbf{p}}B^{(i)}$ points into Ω, i.e., $D_{\mathbf{p}}B^{(i)}(\mathbf{p}, z, \mathbf{x}) \cdot \boldsymbol{\nu}^{(i)}(\mathbf{x}) \geq 0$ on Γ_i, for each $(\mathbf{p}, z, \mathbf{x}) \in \mathbb{R}^2 \times \mathbb{R} \times \Gamma_i$, where $\boldsymbol{\nu}^{(i)}$ is the interior unit normal on Γ_i to Ω. We say that the boundary conditions:*

$$B^{(i)}(Du, u, \mathbf{x}) = 0 \qquad \text{on } \Gamma_i, \quad \text{for } i = 1, 2,$$

are oblique at P if, for each $(\mathbf{p}, z) \in \mathbb{R}^2 \times \mathbb{R}$, vectors $D_{\mathbf{p}}B^{(i)}(\mathbf{p}, z, P), i = 1, 2$, lie on the same side of a supporting $C^{1,\alpha}$-curve Σ at P; that is, there exist a C^1-curve Σ and a constant $\lambda > 0$ such that $\Sigma \cap \Omega = \emptyset$, $P \in \Sigma^0$, and, for $(\mathbf{p}, z) \in \mathbb{R}^2 \times \mathbb{R}$,

$$\lambda \leq D_{\mathbf{p}}B^{(i)}(\mathbf{p}, z, P) \cdot \boldsymbol{\nu}_\Sigma \leq \frac{1}{\lambda} \qquad \text{for } i = 1, 2, \tag{4.4.4}$$

where $\boldsymbol{\nu}_\Sigma$ is the unit normal to Σ at P, oriented so that $\boldsymbol{\nu}_\Sigma \cdot \boldsymbol{\nu}^{(i)} \geq 0$ at P for $i = 1, 2$, and Σ^0 is the relative interior of Σ. We also say that the λ-obliqueness holds at P for a particular choice of λ in (4.4.4).

Lemma 4.4.4 (Comparison Principle)**.** *Let $\Omega \subset \mathbb{R}^2$, Γ_k, and P_k be as in Lemma 4.4.2. Let $u, v \in C(\overline{\Omega}) \cap C^1(\overline{\Omega} \setminus \overline{\Gamma}_0) \cap C^2(\Omega)$ satisfy*

$$\begin{aligned} &\mathcal{N}(u) \leq \mathcal{N}(v) && \text{in } \Omega, \\ &u \geq v && \text{on } \Gamma_0, \\ &B^{(k)}(Du, u, \mathbf{x}) \leq B^{(k)}(Dv, v, \mathbf{x}) && \text{on } \Gamma_k \ \text{ for } k = 1, 2, 3, \end{aligned}$$

where $\mathcal{N}(u)$ is defined in (4.2.2) in which (A_{ij}, A) are functions of $(\mathbf{p}, z, \mathbf{x}) \in \mathbb{R}^2 \times \mathbb{R} \times \Omega$. Assume that (A_{ij}, A) and $D_{(\mathbf{p},z)}(A_{ij}, A)$ are continuous on $\mathbb{R}^2 \times \mathbb{R} \times \overline{\Omega}$, and that the equation is strictly elliptic in $\overline{\Omega} \setminus \Gamma_0$ in the sense that matrix $[A_{ij}](\mathbf{p}, z, \mathbf{x})$ is strictly positive definite for each $(\mathbf{p}, z, \mathbf{x}) \in \mathbb{R}^2 \times \mathbb{R} \times (\overline{\Omega} \setminus \Gamma_0)$. Assume that

$$\begin{aligned} &(B^{(1)}, D_{(\mathbf{p},z)}B^{(1)}) \in C(\mathbb{R}^2 \times \mathbb{R} \times (\overline{\Gamma}_1 \setminus \{P_1\})), \\ &(B^{(2)}, D_{(\mathbf{p},z)}B^{(2)}) \in C(\mathbb{R}^2 \times \mathbb{R} \times \overline{\Gamma}_2), \\ &(B^{(3)}, D_{(\mathbf{p},z)}B^{(3)}) \in C(\mathbb{R}^2 \times \mathbb{R} \times (\overline{\Gamma}_3 \setminus \{P_4\})), \end{aligned}$$

and that the boundary conditions on Γ_k, $k = 1, 2, 3$, are oblique in the following sense:

$$\begin{aligned} &D_{\mathbf{p}}B^{(1)}(\mathbf{p}, z, \mathbf{x}) \cdot \boldsymbol{\nu}^{(1)}(\mathbf{x}) > 0 && \text{for } (\mathbf{p}, z, \mathbf{x}) \in \mathbb{R}^2 \times \mathbb{R} \times (\overline{\Gamma}_1 \setminus \{P_1\}), \\ &D_{\mathbf{p}}B^{(2)}(\mathbf{p}, z, \mathbf{x}) \cdot \boldsymbol{\nu}^{(2)}(\mathbf{x}) > 0 && \text{for } (\mathbf{p}, z, \mathbf{x}) \in \mathbb{R}^2 \times \mathbb{R} \times \overline{\Gamma}_2, \\ &D_{\mathbf{p}}B^{(3)}(\mathbf{p}, z, \mathbf{x}) \cdot \boldsymbol{\nu}^{(3)}(\mathbf{x}) > 0 && \text{for } (\mathbf{p}, z, \mathbf{x}) \in \mathbb{R}^2 \times \mathbb{R} \times (\overline{\Gamma}_3 \setminus \{P_4\}), \end{aligned}$$

where $\boldsymbol{\nu}^{(k)}$ is the interior unit normal on Γ_k to Ω. Moreover, assume that the boundary conditions are oblique at corners $\{P_2, P_3\}$ in the sense of Definition 4.4.3. *Furthermore, assume that*

$$\begin{aligned} D_z A \le 0 \qquad & \text{in } \mathbb{R}^2 \times \mathbb{R} \times \Omega, \\ D_z B^{(k)} \le 0 \qquad & \text{on } \mathbb{R}^2 \times \mathbb{R} \times \Gamma_k \ \text{ for } k = 1, 2, 3. \end{aligned}$$

Then $u \ge v$ in Ω.

Proof. From the conditions, $u - v$ solves a linear problem of form (4.4.2) with the coefficients satisfying the conditions of Lemma 4.4.2. Now the assertion follows directly from Lemma 4.4.2. □

Remark 4.4.5. *In most of our applications of Lemma* 4.4.4, $\Gamma_0 = \Gamma_{\text{sonic}}$, $\Gamma_1 = \Gamma_{\text{shock}}$, $\Gamma_2 = \Gamma_{\text{sym}}$, *and* $\Gamma_3 = \Gamma_{\text{wedge}}$.

Remark 4.4.6. *Assume that $P_1 = P_4$, i.e., $\overline{\Gamma_0}$ is one point. In this case, in Lemmas* 4.4.2 *and* 4.4.4, *the condition that $u \ge v$ on Γ_0 is replaced by the one-point condition that $u \ge v$ at $P_1 = P_4$. Then Lemmas* 4.4.2 *and* 4.4.4 *still hold without change of the proofs.*

Note that, in this case, sides Γ_1 and Γ_3 meet at corner $P_1 = P_4$. We do not *assume that the boundary conditions on $\Gamma_1 \cup \Gamma_3$ satisfy the obliqueness at P_1 in the sense of Definitions* 4.4.1 *and* 4.4.3 *for Lemmas* 4.4.2 *and* 4.4.4, *respectively. Indeed, we use the one-point Dirichlet condition: $u \ge v$ at $P_1 = P_4$ for the proofs of the comparison principles.*

4.5 MIXED BOUNDARY VALUE PROBLEMS IN A DOMAIN WITH CORNERS FOR UNIFORMLY ELLIPTIC EQUATIONS

4.5.1 Linear problem

In this section, we consider a domain

$$\Omega = \{\mathbf{x} \in \mathbb{R}^2 \,:\, 0 < x_1 < h,\ 0 < x_2 < f_{\text{bd}}(x_1)\}, \tag{4.5.1}$$

where

$$\begin{aligned} & f_{\text{bd}} \in C^1([0,h]), \quad f_{\text{bd}}(0) = t_0, \quad \|f_{\text{bd}}'\|_{L^\infty([0,h])} \le M_{\text{bd}}, \\ & f_{\text{bd}}(x_1) \ge \min(t_0 + t_1 x_1,\, t_2) \qquad \text{for all } x_1 \in (0,h), \end{aligned} \tag{4.5.2}$$

for some $h, t_0, t_1, t_2, M_{\text{bd}} > 0$. We denote the boundary vertices and segments as follows:

$$\begin{aligned} & P_1 = (0, f_{\text{bd}}(0)), \ \ P_2 = (h, f_{\text{bd}}(h)), \ \ P_3 = (h, 0), \ \ P_4 = \mathbf{0}, \\ & \overline{\Gamma_0} = \partial\Omega \cap \{x_1 = 0\}, \quad \overline{\Gamma_1} = \partial\Omega \cap \{x_2 = f_{\text{bd}}(x_1)\}, \\ & \overline{\Gamma_2} = \partial\Omega \cap \{x_1 = h\}, \quad \overline{\Gamma_3} = \partial\Omega \cap \{x_2 = 0\}, \end{aligned} \tag{4.5.3}$$

and Γ_k, $k = 0, \dots, 3$, are the relative interiors of the segments defined above. Then $\partial\Omega = \cup_{k=0}^{3}(\Gamma_k \cup \{P_{k+1}\})$.

Consider the following mixed boundary value problem for the linear elliptic equation:

$$\sum_{i,j=1}^{2} a_{ij} D_{ij} u + \sum_{i=1}^{2} a_i D_i u + a_0 u = f \qquad \text{in } \Omega, \tag{4.5.4}$$

$$\mathbf{b}^{(k)} \cdot Du + b_0^{(k)} u = g_k \qquad \text{on } \Gamma_k \ \text{ for } k = 1, 2, 3, \tag{4.5.5}$$

$$u = 0 \qquad \text{on } \Gamma_0. \tag{4.5.6}$$

Assume that there exist constants $\lambda > 0$, $\kappa > 0$, and $M < \infty$ such that

$$\lambda |\boldsymbol{\mu}|^2 \le \sum_{i,j=1}^{2} a_{ij}(\mathbf{x}) \mu_i \mu_j \le \lambda^{-1} |\boldsymbol{\mu}|^2 \tag{4.5.7}$$

for any $\mathbf{x} \in \Omega$, $\boldsymbol{\mu} = (\mu_1, \mu_2) \in \mathbb{R}^2$, and

$$a_0 \le 0 \qquad \text{in } \Omega, \tag{4.5.8}$$

$$\|a_{ij}\|_{0,\alpha,\Omega} + \|(a_i, a_0)\|_{0,\alpha,\Omega}^{(1-\alpha),\Gamma_0} \le M, \qquad i, j = 1, 2. \tag{4.5.9}$$

Assume that the boundary conditions satisfy

Obliqueness: $\lambda \le \mathbf{b}^{(k)} \cdot \boldsymbol{\nu}^{(k)} \le \lambda^{-1}$ on Γ_k for $k = 1, 2, 3$, (4.5.10)

λ *– obliqueness at corners* $\{P_2, P_3\}$ *in the sense of Definition* 4.4.1, (4.5.11)

$$b_0^{(k)} \le 0 \qquad \text{on } \Gamma_k \ \text{ for } k = 1, 2, 3, \tag{4.5.12}$$

$$\left| \frac{\mathbf{b}^{(k)}}{|\mathbf{b}^{(k)}|}(P_k) \pm \frac{\mathbf{b}^{(k-1)}}{|\mathbf{b}^{(k-1)}|}(P_k) \right| \ge \kappa \qquad \text{for } \ k = 2, 3, \tag{4.5.13}$$

$$\max_{k \in \{1,3\}} \|(\mathbf{b}^{(k)}, b_0^{(k)})\|_{C^{(-\alpha),\partial\Gamma_k}_{1,\alpha,\Gamma_k}} + \|(\mathbf{b}^{(2)}, b_0^{(2)})\|_{0,\alpha,\overline{\Gamma_2}} \le M, \tag{4.5.14}$$

where $\boldsymbol{\nu}^{(k)}$ is the interior unit normal on Γ_k to Ω, and $\partial\Gamma_k$ denotes the endpoints of Γ_k.

We consider the following norm in Ω: Let $\varepsilon > 0$ be such that $\mathcal{N}_\varepsilon(\Gamma_0) \cap \mathcal{N}_\varepsilon(\Gamma_2) = \emptyset$. Also, denote by $\Omega_{(-\varepsilon/2)} := \Omega \setminus (\mathcal{N}_{\varepsilon/2}(\Gamma_0) \cup \mathcal{N}_{\varepsilon/2}(\Gamma_2))$. Then, for $m = 0, 1, \dots$, and $\alpha \in (0, 1)$,

$$\|u\|_{*,m,\alpha,\Omega} := \|u\|_{m,\alpha,\mathcal{N}_\varepsilon(\Gamma_0)\cap\Omega}^{(-m+2-\alpha),\overline{\Gamma_0}} + \|u\|_{m,\alpha,\mathcal{N}_\varepsilon(\Gamma_2)\cap\Omega}^{(-m+1-\alpha),\overline{\Gamma_2}} + \|u\|_{m,\alpha,\Omega_{(-\varepsilon/2)}}.$$

Denote $C^{*,m,\alpha}(\Omega) := \{u \in C^m(\Omega) \ : \ \|u\|_{*,m,\alpha,\Omega} < \infty\}$. Then $C^{*,m,\alpha}(\Omega)$ with norm $\|\cdot\|_{*,m,\alpha,\Omega}$ is a Banach space. Also, the choice of different $\varepsilon > 0$, satisfying the properties described above, determines an equivalent norm on $C^{*,m,\alpha}(\Omega)$.

Similarly, define

$$\|g\|_{*,1,\alpha,\Gamma_1} := \|g\|_{1,\alpha,\mathcal{N}_\varepsilon(\Gamma_0)\cap\Gamma_1}^{(1-\alpha),\{P_1\}} + \|g\|_{1,\alpha,\mathcal{N}_\varepsilon(\Gamma_2)\cap\Gamma_1}^{(-\alpha),\{P_2\}} + \|g\|_{1,\alpha,\Omega_{(-\varepsilon/2)}\cap\Gamma_1},$$

$$\|g\|_{*,1,\alpha,\Gamma_3} := \|g\|_{1,\alpha,\mathcal{N}_\varepsilon(\Gamma_0)\cap\Gamma_3}^{(1-\alpha),\{P_4\}} + \|g\|_{1,\alpha,\mathcal{N}_\varepsilon(\Gamma_2)\cap\Gamma_3}^{(-\alpha),\{P_3\}} + \|g\|_{1,\alpha,\Omega_{(-\varepsilon/2)}\cap\Gamma_3},$$

and $C^{*,1,\alpha}(\Gamma_k) := \{g \in C^1(\Gamma_k) \;:\; \|g\|_{*,1,\alpha,\Gamma_k} < \infty\}$ for $k = 1, 3$.

Lemma 4.5.1. *Let $h, t_0, t_1, t_2, M_{\rm bd} > 0$, and let Ω be a domain of structure (4.5.1)–(4.5.3). Let $\kappa, \lambda \in (0,1]$ and $M < \infty$ be constants. Then there exists $\alpha_1(\kappa, \lambda, M_{\rm bd}) \in (0,1)$ such that, for every $\alpha \in (0, \alpha_1]$, there is $C(\Omega, \lambda, \kappa, M, \alpha)$ so that the following holds: Let*

$$\|f_{\rm bd}\|^{(-1-\alpha),\,\{0,h\}}_{2,\alpha,(0,h)} \leq M. \tag{4.5.15}$$

Let (4.5.7)–(4.5.14) be satisfied. Then any solution $u \in C^{\alpha}(\overline{\Omega}) \cap C^{1,\alpha}(\overline{\Omega} \setminus \overline{\Gamma}_0) \cap C^{2,\alpha}(\Omega)$ of Problem (4.5.4)–(4.5.6) satisfies

$$\|u\|_{*,2,\alpha,\Omega} \leq C\big(\|f\|_{*,0,\alpha,\Omega} + \sum_{j=1,3} \|g_j\|_{*,1,\alpha,\Gamma_j} + \|g_2\|_{0,\alpha,\Gamma_2}\big). \tag{4.5.16}$$

Proof. In this proof, constants α_1 and C depend only on the parameters listed in the statement, unless otherwise specified.

From (4.5.1), we obtain the existence of $\theta_0 \in (0, \frac{\pi}{2})$ depending only on $M_{\rm bd}$ such that angles θ_k at vertices P_k, $k = 1, \dots, 4$, satisfy

$$\theta_0 \leq \theta_k \leq \frac{\pi}{2} - \theta_0.$$

From (4.5.2), there exists $r > 0$ depending only on $(t_0, t_1, t_2, M_{\rm bd}, M)$ such that, for $i = 1, \dots, 4$,

(i) $\partial\Omega \cap B_{10r}(P_i) \subset \Gamma_{i-1} \cup \Gamma_i$, where Γ_4 denotes Γ_0;

(ii) $\partial\Omega \cap B_{10r}(P) \subset \Gamma_i$ for any $P \in \Gamma_i$ such that $\mathrm{dist}(P, \partial\Gamma_i) > 10r$, where $\partial\Gamma_i$ denotes the endpoints of Γ_i.

Fix r in such a way.

We first estimate the solution near corners $\{P_i\}_{i=1}^4$.

We now estimate the solution in $B_{2r}(P_i) \cap \Omega$ for $i = 1, 4$. Consider $i = 1$ to fix the notations. We show that there exists $\alpha_1 \in (0,1)$ such that, for any $\alpha \in (0, \alpha_1)$,

$$\begin{aligned} \|u\|^{(-\alpha),\Gamma_0}_{1,\alpha,\Omega\cap B_{3r/4}(P_1)} \leq C\big(&\|u\|_{0,\Omega\cap B_{2r}(P_1)} + \|f\|^{(2-\alpha),\Gamma_0}_{0,\Omega\cap B_{2r}(P_1)} \\ &+ \|g_1\|^{(1-\alpha),\{P_1\}}_{0,\alpha,\Gamma_1\cap B_{2r}(P_1)}\big). \end{aligned} \tag{4.5.17}$$

This is a localized version of the estimates of [194, Theorem 1] for the mixed derivative problem.

Estimate (4.5.17) is obtained from the global estimate of [194, Theorem 1] by an argument similar to the proof of Lemma 4.2.16. The dependence of the constants in (4.5.17) is as follows: α_1 depends only on the ellipticity, the obliqueness, and angle θ_1, and hence on $(\lambda, M_{\rm bd})$. Constant C depends on these parameters, as well as (α, r) and the right-hand sides of estimates (4.5.9) and

(4.5.14) for $(\mathbf{b}^{(1)}, b_0^{(1)})$. Thus, C in (4.5.17) depends only on (λ, M, α) and r (where $M_{\rm bd}$ is not included since $M \geq M_{\rm bd}$). Since r depends only on the parameters of domain Ω in (4.5.2), we see that $C = C(\Omega, \lambda, M, \alpha)$. That is, the dependence of constants α_1 and C is as asserted in this lemma.

Furthermore, we improve (4.5.17) by showing a higher regularity away from Γ_0, with the use of the corresponding higher regularity of the coefficients in (4.5.9). Specifically, following Step 4 of the proof of Theorem 4.2.15, we obtain

$$\begin{aligned}\|u\|^{(-\alpha),\Gamma_0}_{2,\alpha,\Omega\cap B_r(P_1)} \leq C\big(&\|u\|_{0,\Omega\cap B_{2r}(P_1)} + \|f\|^{(2-\alpha),\Gamma_0}_{\alpha,\Omega\cap B_{2r}(P_1)} \\ &+ \|g_1\|^{(1-\alpha),\{P_1\}}_{1,\alpha,\Gamma_1\cap B_{2r}(P_1)}\big).\end{aligned} \tag{4.5.18}$$

Near P_4, we obtain a similar estimate, namely (4.5.18), with (P_1, Γ_1, g_1) replaced by (P_4, Γ_3, g_3).

The estimates near P_2 and P_3 are obtained via a similar argument by using the localized version of [193, Lemma 1.3] and then improving the regularity away from Γ_2. The resulting estimate near P_2 is

$$\begin{aligned}\|u\|^{(-1-\alpha),\Gamma_2}_{2,\alpha,\Omega\cap B_r(P_2)} \leq C\big(&\|u\|_{0,\Omega\cap B_{2r}(P_2)} + \|f\|^{(1-\alpha),\Gamma_2}_{\alpha,\Omega\cap B_{2r}(P_2)} \\ &+ \|g_1\|^{(-\alpha),\{P_2\}}_{1,\alpha,\Gamma_1\cap B_{2r}(P_2)} + \|g_2\|_{0,\alpha,\Gamma_2}\big).\end{aligned} \tag{4.5.19}$$

Near P_3, we obtain a similar estimate, with (P_2, Γ_1, g_1) replaced by (P_3, Γ_3, g_3).

Next we make the estimates near Γ_m, $m = 0, \dots, 3$, away from the corners. If $P \in \Gamma_m$ and $B_{r/2}(P) \cap \partial\Gamma_m = \emptyset$, we obtain the following estimates in $B_r(P) \cap \Omega$. For $m = 0$, we have (4.2.133) with $k = 0$ and $\sigma = \alpha$, rescaled from B_2^+ to $B_{r/4}(P) \cap \Omega$. For $m = 2$, we have (4.2.131) with $\sigma = \alpha$, rescaled from B_2^+ to $B_{r/4}(P) \cap \Omega$. For $m = 1, 3$, we use the standard local Schauder estimates for the oblique derivative problem near the $C^{2,\alpha}$–boundaries in $B_{r/4}(P) \cap \Omega$, where we note that

$$\|(a_{ij}, a_i, a_0)\|_{0,\alpha,B_{r/4}(P)\cap\Omega} \leq C, \quad \|(\mathbf{b}^{(m)}, b_0^{(m)})\|_{1,\alpha,B_{r/4}(P)\cap\Omega} \leq C.$$

Similarly, in the interior, *i.e.*, in $B_{r/2}(P) \subset \Omega$, $\|(a_{ij}, a_i, a_0)\|_{0,\alpha,B_{r/4}(P)} \leq C$, so that the standard interior local Schauder estimates for the elliptic equation can be used.

Combining all the estimates discussed above with a scaling technique similar to that of Step 4 of the proof of Theorem 4.2.15, and noticing the regularity of $\partial\Omega$, we obtain that there exists α_1 depending only on $(\kappa, \lambda, \theta_0)$, hence on $(\kappa, \lambda, M_{\rm bd})$, such that, for each $\alpha \in (0, \alpha_1)$, there is $C > 0$ depending only on $(\Omega, \lambda, \kappa, M, \alpha)$ so that any solution $u \in C^{\alpha}(\overline{\Omega}) \cap C^{1,\alpha}(\overline{\Omega} \setminus \overline{\Gamma_0}) \cap C^{2,\alpha}(\Omega)$ of Problem (4.5.4)–(4.5.6) satisfies

$$\|u\|_{*,2,\alpha,\Omega} \leq C\big(\|u\|_{0,\Omega} + \|f\|_{*,0,\alpha,\Omega} + \sum_{j=1,3} \|g_j\|_{*,1,\alpha,\Gamma_j} + \|g_2\|_{0,\alpha,\Gamma_2}\big). \tag{4.5.20}$$

Thus, in order to show (4.5.16), we need to estimate $\|u\|_{0,\Omega}$ by the right-hand side of (4.5.16).

Suppose that such an estimate is false. Then, for $m = 1, 2, \ldots$, there exists a sequence of problems of form (4.5.4)–(4.5.6) with coefficients $(a_{ij}^m, a_i^m, \mathbf{b}^{(k),m})$, the right-hand sides (f^m, g_k^m), and solutions $u^m \in C^{*,2,\alpha}(\Omega)$, where the assumptions on $(a_{ij}^m, \mathbf{b}^{(k),m})$ stated above are satisfied with the uniform constants κ, λ, α, and M so that

$$\|f^m\|_{*,0,\alpha,\Omega} + \sum_{j=1,3} \|g_j^m\|_{*,1,\alpha,\Gamma_j} + \|g_2^m\|_{C^\alpha(\overline{\Gamma_2})} \to 0 \qquad \text{as } m \to \infty,$$

but $\|u^m\|_{0,\Omega} = 1$ for $m = 1, 2, \ldots$. Then, from (4.5.20), we obtain that $\|u^m\|_{*,2,\alpha} \leq C$ with C independent of m. Passing to a subsequence (without change of notation), we find that $u^m \to u^\infty$ in $C^{*,2,\frac{\alpha}{2}}$, $a_{ij}^m \to a_{ij}^\infty$ in $C^{\frac{\alpha}{2}}(\overline{\Omega})$, $(a_i^m, a_0^m) \to (a_i^\infty, a_0^\infty)$ in $C_{0,\alpha/2,\Omega}^{(1-\frac{\alpha}{2}),\Gamma_0}$, and $\mathbf{b}^{(k),m} \to \mathbf{b}^{(k),\infty}$ in $C_{1,\alpha/2,\Gamma_k}^{(-\frac{\alpha}{2}),\partial\Gamma_k}$ for $k = 1, 3$, and in $C^\alpha(\overline{\Gamma_2})$ for $k = 2$. Moreover, a_{ij}^∞ and $\mathbf{b}^{(k),\infty}$ satisfy (4.5.4)–(4.5.14). Then $\|u^\infty\|_{0,\Omega} = 1$, and u^∞ is a solution of the homogeneous problem (4.5.4)–(4.5.6) with coefficients $(a_{ij}^\infty, \mathbf{b}^{(k),\infty})$. Obviously, $v = 0$ is another solution of the same problem. This is in contradiction to the uniqueness of a solution in $C(\overline{\Omega}) \cap C^1(\overline{\Omega} \setminus \overline{\Gamma}_0) \cap C^2(\Omega)$ of Problem (4.5.4)–(4.5.6), where the uniqueness follows from Lemma 4.4.2. Therefore, (4.5.16) is proved. □

We first prove the existence of solutions in the case that Ω is a unit square.

Proposition 4.5.2. *Let Ω be a square $Q = (0,1) \times (0,1)$, which corresponds to $h = 1$ and $f_{\rm bd} \equiv 1$ in* (4.5.1). *We use the notations in* (4.5.3). *Let $\kappa, \lambda \in (0, 1]$, $M < \infty$, and $\alpha \in (0, \alpha_1]$ be constants, where $\alpha_1(\kappa, \lambda, 0) \in (0, 1)$ is determined in Lemma* 4.5.1. *Let* (4.5.7)–(4.5.14) *be satisfied. Then, for every $(f, g_1, g_2, g_3) \in C^{*,0,\alpha}(Q) \times C^{*,1,\alpha}(\Gamma_1) \times C^\alpha(\overline{\Gamma_2}) \times C^{*,1,\alpha}(\Gamma_3)$, there exists a unique solution $u \in C^{*,2,\alpha}(Q)$ of Problem* (4.5.4)–(4.5.6). *Moreover, u satisfies* (4.5.16), *where C depends only on $(\kappa, \lambda, M, \alpha)$.*

Proof. We first consider **Problem $\mathcal{P}_0$** defined as follows:

$$\Delta u = f \ \text{ in } Q; \qquad \partial_\nu u|_{\Gamma_k} = g_k,\ k = 1, 2, 3; \qquad u|_{\Gamma_0} = 0.$$

By [53, Theorem 3.2], for any $f \in C^\alpha(\overline{Q})$ and $g_k \in C^\alpha(\overline{\Gamma_k})$ with $k = 1, 2, 3$, there exists a unique weak solution $u \in H^1(Q)$ of **Problem $\mathcal{P}_0$**. Moreover, $u \in C^\alpha(\overline{Q}) \cap C^{1,\alpha}(\overline{Q} \setminus \overline{\Gamma}_0)$.

Next, it is easy to see that each

$$(f, g_1, g_2, g_3) \in \mathcal{Y}^\alpha := C^{*,0,\alpha}(Q) \times C^{*,1,\alpha}(\Gamma_1) \times C^\alpha(\overline{\Gamma_2}) \times C^{*,1,\alpha}(\Gamma_3)$$

can be approximated by a sequence of C^α–functions $(f^{(m)}, g_1^{(m)}, g_2^{(m)}, g_3^{(m)})$, which are uniformly bounded in the $\mathcal{Y}^\alpha$–norm, such that $f^{(m)} \to f$ uniformly on every $K \Subset Q$, and $g_i^{(m)} \to g_i$ uniformly on every $K \Subset \Gamma_k$ as $m \to \infty$ for

$i = 0, \ldots, 3$. Indeed, for $m = 1, 2, \ldots$, we define that, for each $(x_1, x_2) \in Q$,

$$f^{(m)}(x_1, x_2) = \begin{cases} f(x_1, x_2) & \text{if } x_1 \in [\frac{1}{10m}, 1 - \frac{1}{10m}], \\ f(\frac{1}{10m}, x_2) & \text{if } x_1 \in (0, \frac{1}{10m}), \\ f(1 - \frac{1}{10m}, x_2) & \text{if } x_1 \in (1 - \frac{1}{10m}, 1), \end{cases}$$

and $g_j^{(m)}(x_1), j = 1, 3$, are defined similarly, while $g_2^{(m)} := g_2 \in C^\alpha(\overline{\Gamma_2})$. It is easy to see that $(f^{(m)}, g_1^{(m)}, g_2^{(m)}, g_3^{(m)})$ satisfy the properties asserted above.

From the existence of a unique solution $u^{(m)}$ of **Problem** $\mathcal{P}_0$ with the right-hand side $(f^{(m)}, g_1^{(m)}, g_2^{(m)}, g_3^{(m)})$, and using estimate (4.5.16), we find that there exists C such that $\|u^{(m)}\|_{*,2,\alpha,Q} \leq C$ for all m. Then a subsequence $u^{(m_j)}$ converges to $u \in C^{*,2,\alpha}(Q)$ in $C^{*,2,\frac{\alpha}{2}}(Q)$ which is a solution of **Problem** $\mathcal{P}_0$ for the right-hand sides (f, g_1, g_2, g_3). This leads to the existence of a solution $u \in C^{*,2,\alpha}$ for any $(f, g_1, g_2, g_3) \in \mathcal{Y}^\alpha$. By Lemma 4.5.1, u satisfies (4.5.16), which also implies the uniqueness of the solution in $C^{*,2,\alpha}$ for **Problem** $\mathcal{P}_0$.

Now the existence of a unique solution $u \in C^{*,2,\alpha}(Q)$ of Problem (4.5.4)–(4.5.6), denoted by **Problem** $\mathcal{P}$, for any $(f, g_1, g_2, g_3) \in \mathcal{Y}^\alpha$, follows by the method of continuity (*cf.* Theorem 3.4.1), applied to the family of **Problems** $t\mathcal{P} + (1-t)\mathcal{P}_0$ for $t \in [0, 1]$, where we have used that all of these problems satisfy estimate (4.5.16) with uniform C by Lemma 4.5.1. □

Now we consider the case of general domain Ω.

Proposition 4.5.3. *Let $h, t_0, t_1, t_2, M_{\rm bd} > 0$, and let Ω be a domain of structure* (4.5.1)–(4.5.3). *Let $\kappa, \lambda \in (0, 1)$, $M < \infty$, and $\alpha \in (0, \alpha_1]$ be constants, where $\alpha_1(\kappa, \lambda, M_{\rm bd}) \in (0, 1)$ is determined in Lemma* 4.5.1. *Let $f_{\rm bd}$ satisfy* (4.5.15). *Let* (4.5.7)–(4.5.14) *be satisfied. Then, for every $(f, g_1, g_2, g_3) \in C^{*,0,\alpha}(\Omega) \times C^{*,1,\alpha}(\Gamma_1) \times C^\alpha(\overline{\Gamma_2}) \times C^{*,1,\alpha}(\Gamma_3)$, there exists a unique solution $u \in C^{*,2,\alpha}(\Omega)$ of Problem* (4.5.4)–(4.5.6). *Moreover, u satisfies* (4.5.16), *where C depends only on $(\Omega, \kappa, \lambda, M, \alpha)$.*

Proof. Let Q be the unit square as in Proposition 4.5.2. We denote by $P_k(Q)$ and $\Gamma_k(Q)$ its boundary parts in (4.5.3). Define mapping $F : \Omega \mapsto Q$ by $F(x_1, x_2) = (\frac{x_1}{h}, \frac{x_2}{f_{\rm bd}(x_1)})$. Then (4.5.1) and (4.5.15) imply that F is a diffeomorphism $\overline{\Omega} \mapsto \overline{Q}$, with $F(\Gamma_k) = \Gamma_k(Q)$ and $F(P_{k+1}) = P_{k+1}(Q)$ for $k = 0, \ldots, 3$, and

$$\|F\|^{(-1-\alpha), \Gamma_0 \cup \Gamma_2}_{2,\alpha,\Omega} + \|F^{-1}\|^{(-1-\alpha), \Gamma_0(Q) \cup \Gamma_2(Q)}_{2,\alpha,Q} \leq C,$$

where $C = C(M, t_0, t_2)$. It is easy to check that, if $u \in C^{*,2,\alpha}(\Omega)$ is a solution of (4.5.4)–(4.5.6) (which is called **Problem** $\mathcal{P}_1$), then the function on Q defined by $v(\mathbf{x}) := u(F^{-1}(\mathbf{x}))$ satisfies $v \in C^{*,2,\alpha}(Q)$. Moreover, v is a solution of a problem on Q with structure (4.5.4)–(4.5.6), which is called **Problem** $\mathcal{P}_2$. It is easy to see that, for **Problem** $\mathcal{P}_2$, the equation and boundary conditions satisfy the assumptions of Proposition 4.5.2. If $(f, g_1, g_2, g_3) \in C^{*,0,\alpha}(\Omega) \times C^{*,1,\alpha}(\Gamma_1) \times C^{\alpha}(\overline{\Gamma_2}) \times C^{*,1,\alpha}(\Gamma_3)$ and $(\hat{f}, \hat{g}_k)(\mathbf{x}) := (f, g_k)(F^{-1}(\mathbf{x}))$, then

$$(\hat{f}, \hat{g}_1, \hat{g}_2, \hat{g}_3) \in C^{*,0,\alpha}(Q) \times C^{*,1,\alpha}(\Gamma_1(Q)) \times C^{\alpha}(\overline{\Gamma_2(Q)}) \times C^{*,1,\alpha}(\Gamma_3(Q)).$$

Moreover, if $v \in C^{*,2,\alpha}(Q)$ is a solution of **Problem** $\mathcal{P}_2$ with the right-hand sides $(\hat{f}, \hat{g}_1, \hat{g}_2, \hat{g}_3)$, then $u(\mathbf{x}) = v(F(\mathbf{x}))$ on Ω satisfies $u \in C^{*,2,\alpha}(\Omega)$ and is a solution of **Problem** $\mathcal{P}_1$ with the right-hand sides (f, g_1, g_2, g_3). Then the existence and uniqueness assertion in Proposition 4.5.2 implies the existence and uniqueness assertion we want here. The estimate of the solution follows from Lemma 4.5.1. □

Next we discuss some cases in which the regularity at corners $\{P_1, P_4\}$ can be improved from C^{α} to $C^{1,\alpha}$. We prove the local estimates in the corner-shaped domains in two cases.

Consider the corner domain:

$$\mathcal{R}_s^+ = B_s(\mathbf{0}) \cap \{x_1 > 0, x_2 > f_{\rm ob}(x_1)\}, \quad s > 0, \quad f_{\rm ob}(0) = 0, \tag{4.5.21}$$

where $f_{\rm ob} \in C^1([0, \infty))$. Denote the boundary parts of $\mathcal{R}_s^+$:

$$\Gamma_s^{(\rm d)} := \partial\mathcal{R}_s^+ \cap \{x_1 = 0\}, \; \Gamma_s^{(\rm ob)} := \partial\mathcal{R}_s^+ \cap \{x_2 = f_{\rm ob}(x_1)\}, \; \hat{P} = \mathbf{0}. \tag{4.5.22}$$

We now consider equation (4.5.4) in $\mathcal{R}_s^+$ with the boundary conditions:

$$\mathbf{b} \cdot Du + b_0 u = g \qquad \text{on } \Gamma_s^{(\rm ob)}, \tag{4.5.23}$$

$$u = 0 \qquad \text{on } \Gamma_s^{(\rm d)}. \tag{4.5.24}$$

The first case corresponds with the conditions near point P_4, for which $\Gamma_s^{(\rm ob)}$ is flat.

Lemma 4.5.4. *Let $s > 0$, and let $\mathcal{R}_s^+$ be the domain in (4.5.21) with*

$$f_{\rm ob} \equiv 0 \qquad \text{on } [0, \infty).$$

Let $\alpha \in (0, 1)$, $\lambda \in (0, 1]$, and $M < \infty$ be constants. Let (4.5.4) satisfy (4.5.7) and (4.5.9) with $\Omega = \mathcal{R}_s^+$, and

$$\|(a_{ij}, a_i, a_0)\|_{C^{\alpha}(\overline{\mathcal{R}_s^+})} \le M, \qquad i, j = 1, 2, \tag{4.5.25}$$

$$a_{12}(\mathbf{0}) = a_{21}(\mathbf{0}) = 0. \tag{4.5.26}$$

Let the coefficients of (4.5.23) satisfy $\mathbf{b} \equiv (0, 1)$ and $b_0 \equiv 0$, and let $g \equiv 0$, i.e., condition (4.5.23) is

$$u_{\nu} = 0 \qquad \text{on } \Gamma_s^{(\rm ob)}.$$

Furthermore, let $f \in C^{\alpha}(\mathcal{R}_s^+)$. Then any solution $u \in C(\overline{\mathcal{R}_s^+}) \cap C^1(\overline{\mathcal{R}_s^+ \setminus \Gamma_s^{(\rm d)}}) \cap C^2(\mathcal{R}_s^+)$ of (4.5.4) and (4.5.23)–(4.5.24) is in $C^{2,\alpha}(\overline{\mathcal{R}_{s/2}^+})$ with the estimate:

$$\|u\|_{C^{2,\alpha}(\overline{\mathcal{R}_{s/2}^+})} \le C\big(\|u\|_{L^{\infty}(\mathcal{R}_s^+)} + \|f\|_{C^{\alpha}(\mathcal{R}_s^+)}\big), \tag{4.5.27}$$

where C depends only on (λ, M, α).

Proof. It suffices to prove (4.5.27) in $\mathcal{R}^+_{\varrho/2}$ for some $\varrho > 0$ depending only on (λ, M). Indeed, in order to complete the proof of (4.5.27), it then suffices to derive the $C^{2,\alpha}$–estimate in domain $\mathcal{R}^+_{s/2} \setminus \mathcal{R}^+_{\varrho/4}$. These estimates are obtained by combining the standard interior estimates for elliptic equations and the local estimates near the boundary for the Dirichlet and oblique derivative problems. The small constant $\varrho \in (0, \frac{s}{2})$ will be chosen below. The proof is divided into four steps.

1. Rescale u by

$$v(\mathbf{x}) = u(\varrho\mathbf{x}) \qquad \text{for } \mathbf{x} \in \mathcal{R}^+_2.$$

Then $v \in C(\overline{\mathcal{R}^+_2}) \cap C^1(\overline{\mathcal{R}^+_2} \setminus \overline{\{x_1 = 0\}}) \cap C^2(\mathcal{R}^+_2)$ satisfies

$$\|v\|_{L^\infty(\mathcal{R}^+_2)} = \|u\|_{L^\infty(\mathcal{R}^+_{2\varrho})}, \tag{4.5.28}$$

and is a solution of

$$\sum_{i,j=1}^{2} \hat{a}^{(\varrho)}_{ij} D_{ij}v + \hat{a}^{(\varrho)}_i D_i v + \hat{a}^{(\varrho)}_0 v = \hat{f}^{(\varrho)} \qquad \text{in } \mathcal{R}^+_2, \tag{4.5.29}$$

$$v = 0 \qquad \text{on } \partial\mathcal{R}^+_2 \cap \{x_1 = 0\}, \tag{4.5.30}$$

$$v_{\boldsymbol{\nu}} \equiv D_2 v = 0 \qquad \text{on } \partial\mathcal{R}^+_2 \cap \{x_2 = 0\}, \tag{4.5.31}$$

where

$$\begin{aligned} &\hat{a}^{(\varrho)}_{ij}(\mathbf{x}) = a_{ij}(\varrho\mathbf{x}), \quad \hat{a}^{(\varrho)}_i(\mathbf{x}) = \varrho\, a_i(\varrho\mathbf{x}) \qquad \text{for } i, j = 1, 2,\\ &\hat{a}^{(\varrho)}_0(\mathbf{x}) = \varrho^2 a_0(\varrho\mathbf{x}), \quad \hat{f}^{(\varrho)}(\mathbf{x}) = \varrho^2 f(\varrho\mathbf{x}) \qquad \text{for } \mathbf{x} \in \mathcal{R}^+_2. \end{aligned} \tag{4.5.32}$$

Thus, $\hat{a}^{(\varrho)}_{ij}, i, j = 1, 2$, satisfy (4.5.8), (4.5.26), and (4.5.7) with the unchanged constant $\lambda > 0$. Moreover, since $\varrho \le 1$,

$$\begin{aligned} &\|(\hat{a}^{(\varrho)}_{ij}, \hat{a}^{(\varrho)}_i, \hat{a}^{(\varrho)}_0)\|_{C^\alpha(\overline{\mathcal{R}^+_2})} \le \|(\hat{a}_{ij}, \hat{a}_i, a_0)\|_{C^\alpha(\overline{\mathcal{R}^+_s})} \le M,\\ &\|\hat{f}^{(\varrho)}\|_{C^\alpha(\overline{\mathcal{R}^+_2})} \le \|f\|_{C^\alpha(\overline{\mathcal{R}^+_s})}. \end{aligned} \tag{4.5.33}$$

Denote $Q := \{\mathbf{x} \in \mathcal{R}^+_2 \;:\; \operatorname{dist}(\mathbf{x}, \partial\mathcal{R}^+_2) > \frac{1}{50}\}$. The interior estimates for the elliptic equation (4.5.29) imply

$$\|v\|_{C^{2,\alpha}(\overline{Q})} \le C\big(\|v\|_{L^\infty(\mathcal{R}^+_2)} + \|\hat{f}^{(\varrho)}\|_{C^\alpha(\overline{\mathcal{R}^+_2})}\big).$$

The local estimates for the Dirichlet problem (4.5.29)–(4.5.30) imply

$$\|v\|_{C^{2,\alpha}(\overline{B_{1/10}(\mathbf{x}) \cap \mathcal{R}^+_2})} \le C\big(\|v\|_{L^\infty(\mathcal{R}^+_2)} + \|\hat{f}^{(\varrho)}\|_{C^\alpha(\overline{\mathcal{R}^+_2})}\big) \tag{4.5.34}$$

for every $\mathbf{x} \in \{x_1 = 0, \frac{1}{2} \le x_2 \le \frac{3}{2}\}$. The local estimates for the oblique derivative problem (4.5.29) and (4.5.31) imply (4.5.34) for every $\mathbf{x} \in \{\frac{1}{2} \le x_1 \le \frac{3}{2}, x_2 = 0\}$. Combining these estimates, we have

$$\|v\|_{C^{2,\alpha}(\overline{\mathcal{R}^+_{3/2} \setminus \mathcal{R}^+_{1/2}})} \le C\big(\|v\|_{L^\infty(\mathcal{R}^+_2)} + \|\hat{f}^{(\varrho)}\|_{C^\alpha(\overline{\mathcal{R}^+_2})}\big). \tag{4.5.35}$$

2. We modify domain $\mathcal{R}_1^+$ by mollifying the corner at $(0,1)$ and denote the resulting domain by D. That is, D denotes an open domain such that

$$D \subset \mathcal{R}_1^+, \qquad D \setminus B_{1/10}((0,1)) = \mathcal{R}_1^+ \setminus B_{1/10}((0,1)),$$

and $\partial D \cap B_{1/5}((0,1))$ is a $C^{2,\alpha}$–curve. Then we prove the fact that, for any $\hat{f} \in C^\alpha(\overline{D})$, there exists a unique solution $w \in C^{2,\alpha}(\overline{D})$ of the problem:

$$\begin{aligned} &\textstyle\sum_{i=1}^2 \hat{a}_{ii}^{(\varrho)} D_{ii} w + \hat{a}_1^{(\varrho)} D_1 w = \hat{f} && \text{in } D, \\ &w = 0 && \text{on } \partial D \cap \{x_1 = 0, x_2 > 0\}, \\ &w_\nu \equiv D_2 w = 0 && \text{on } \partial D \cap \{x_1 > 0, x_2 = 0\}, \\ &w = v && \text{on } \partial D \cap \{x_1 > 0, x_2 > 0\} \end{aligned} \tag{4.5.36}$$

with

$$\|w\|_{C^{2,\alpha}(\overline{D})} \le C\big(\|v\|_{L^\infty(\mathcal{R}_2^+)} + \|\hat{f}\|_{C^\alpha(\overline{D})}\big). \tag{4.5.37}$$

This can be seen as follows: Denote $B_s^+ := B_s(\mathbf{0}) \cap \{x_1 > 0\}$ for any $s > 0$. Denote by D^+ the even extension of D from $\{x_1 > 0, x_2 > 0\}$ into $\{x_1 > 0\}$, *i.e.*,

$$D^+ := D \cup \{(x_1, 0) \;:\; 0 < x_1 < 1\} \cup \{\mathbf{x} \;:\; (x_1, -x_2) \in D\}.$$

Then $B_{7/8}^+ \subset D^+ \subset B_1^+$, and ∂D^+ is a $C^{2,\alpha}$–curve.

Extend $F = (v, g, \hat{a}_{11}^{(\varrho)}, \hat{a}_{22}^{(\varrho)}, \hat{a}_1^{(\varrho)}, \hat{a}_0^{(\varrho)})$ from $\overline{\mathcal{R}_2^+}$ to $\overline{B_2^+}$ by setting

$$F(x_1, -x_2) = F(x_1, x_2) \qquad \text{for } (x_1, x_2) \in \overline{\mathcal{R}_2^+}.$$

Then it follows from (4.5.30)–(4.5.31) and (4.5.35) that, denoting by $\hat{v}$ the restriction of the extended function v to ∂D^+, we obtain that $\hat{v} \in C^{2,\alpha}(\partial D^+)$ with

$$\|\hat{v}\|_{C^{2,\alpha}(\partial D^+)} \le C\big(\|v\|_{L^\infty(\mathcal{R}_2^+)} + \|\hat{f}\|_{C^\alpha(\overline{D})}\big). \tag{4.5.38}$$

Also, the extended function $\hat{f}$ satisfies that $\hat{f} \in C^\alpha(\overline{D^+})$ with $\|\hat{f}\|_{C^\alpha(\overline{D^+})} = \|\hat{f}\|_{C^\alpha(\overline{D})}$. For $\varrho \le 1$, the extended functions $(\hat{a}_{11}^{(\varrho)}, \hat{a}_{22}^{(\varrho)}, \hat{a}_1^{(\varrho)})$ satisfy (4.5.7) and

$$\begin{aligned} \|(\hat{a}_{11}^{(\varrho)}, \hat{a}_{22}^{(\varrho)}, \hat{a}_1^{(\varrho)})\|_{C^\alpha(\overline{B_2^+})} &= \|(\hat{a}_{11}^{(\varrho)}, \hat{a}_{22}^{(\varrho)}, \hat{a}_1^{(\varrho)})\|_{C^\alpha(\overline{\mathcal{R}_2^+})} \\ &\le \sum_{i,j=1}^2 \|(a_{ij}, a_i)\|_{C^\alpha(\overline{\mathcal{R}_s^+})}. \end{aligned}$$

Then, by [131, Theorem 6.8], there exists a unique solution $w \in C^{2,\frac{\alpha}{2}}(D^+)$ of the Dirichlet problem:

$$\sum_{i=1}^2 \hat{a}_{ii}^{(\varrho)} D_{ii} w + \hat{a}_1^{(\varrho)} D_1 w = \hat{f} \qquad \text{in } D^+, \tag{4.5.39}$$

$$w = \hat{v} \qquad \text{on } \partial D^+, \tag{4.5.40}$$

and w satisfies

$$\|w\|_{C^{2,\alpha}(\overline{D^+})} \le C\big(\|\hat{v}\|_{C^{2,\alpha}(\partial D^+)} + \|\hat{f}\|_{C^\alpha(\overline{D^+})}\big). \qquad (4.5.41)$$

From the structure of equation (4.5.39) and the symmetry of the domain, it follows that $\hat{w}$, defined by $\hat{w}(x_1, x_2) = w(x_1, -x_2)$ in D^+, is also a solution of (4.5.39)–(4.5.40), since the coefficients and right-hand sides are obtained by even extension. By uniqueness for Problem (4.5.39)–(4.5.40), we have

$$w(x_1, x_2) = w(x_1, -x_2) \qquad \text{in } D^+.$$

It follows that $D_2 w(x_1, 0) = 0$ for all $x_1 \in (0,1)$. Thus, w restricted to D is a solution of (4.5.36), where we have used (4.5.30) to see that $w = 0$ on $\partial D \cap \{x_1 = 0, x_2 > 0\}$. Moreover, (4.5.38) and (4.5.41) imply (4.5.37). The uniqueness of solution $w \in C^{2,\alpha}(\overline{D})$ of (4.5.36) follows from the maximum principle and Hopf's lemma, as in Steps (i) and (ii) of the proof of Lemma 4.4.2.

3. Now we prove the existence of a solution $w \in C^{2,\alpha}(\overline{D})$ of the problem:

$$\begin{aligned}
&\textstyle\sum_{i,j=1}^2 \hat{a}_{ij}^{(\varrho)} D_{ij} w + \sum_{j=1}^2 \hat{a}_j^{(\varrho)} D_j w + \hat{a}_0^{(\varrho)} w = \hat{f}^{(\varrho)} && \text{in } D,\\
&w = 0 && \text{on } \partial D \cap \{x_1 = 0\},\\
&w_\nu \equiv D_2 w = 0 && \text{on } \partial D \cap \{x_2 = 0\},\\
&w = v && \text{on } \partial D \cap \{x_1 > 0, x_2 > 0\}.
\end{aligned} \qquad (4.5.42)$$

Moreover, we prove that w satisfies

$$\|w\|_{C^{2,\alpha}(\overline{D})} \le C\big(\|v\|_{L^\infty(\mathcal{R}_2^+)} + \|\hat{f}^{(\varrho)}\|_{C^\alpha(\overline{D})}\big). \qquad (4.5.43)$$

We obtain such a function w as a fixed point of map $K : C^{2,\alpha}(\overline{D}) \mapsto C^{2,\alpha}(\overline{D})$ defined as follows: Let $W \in C^{2,\alpha}(\overline{D})$. Define

$$\hat{f} = -2\hat{a}_{12}^{(\varrho)} D_{12} W - \hat{a}_2^{(\varrho)} D_2 W - \hat{a}_0^{(\varrho)} W + \hat{f}^{(\varrho)}. \qquad (4.5.44)$$

By (4.5.25)–(4.5.26) and (4.5.32), we have

$$\|(a_{12}^{(\varrho)}, a_2^{(\varrho)}, a_0^{(\varrho)})\|_{C^\alpha(\overline{D})} \le M\varrho^\alpha, \qquad (4.5.45)$$

which implies

$$\hat{f} \in C^\alpha(\overline{D}).$$

Then, by the results of Step 2, there exists a unique solution $w \in C^{2,\alpha}(\overline{D})$ of (4.5.36) with $\hat{f}$ defined by (4.5.44). We set $K[W] = w$.

Now we prove that, if $\varrho > 0$ is sufficiently small, K is a contraction map. Let $W^{(i)} \in C^{2,\alpha}(\overline{D})$ and $w^{(i)} := K[W^{(i)}]$ for $i = 1, 2$. Then $w := w^{(1)} - w^{(2)}$ is a solution of (4.5.36) with

$$\begin{aligned}
&\hat{f} = -2\hat{a}_{12}^{(\varrho)} D_{12}(W^{(1)} - W^{(2)}) - \hat{a}_2^{(\varrho)} D_2(W^{(1)} - W^{(2)}) - \hat{a}_0^{(\varrho)}(W^{(1)} - W^{(2)}),\\
&v \equiv 0.
\end{aligned}$$

By (4.5.45), we have

$$\|\hat{f}\|_{\alpha,\overline{D}} \leq C\varrho^{\alpha}\|W^{(1)} - W^{(2)}\|_{2,\alpha,\overline{D}}.$$

Thus, we apply (4.5.37) to obtain

$$\begin{aligned}\|w^{(1)} - w^{(2)}\|_{2,\alpha,\overline{D}} &\leq C\varrho^{\alpha}\|W^{(1)} - W^{(2)}\|_{2,\alpha,\overline{D}}\\ &\leq \frac{1}{2}\|W^{(1)} - W^{(2)}\|_{2,\alpha,\overline{D}},\end{aligned}$$

where the last inequality holds if $\varrho > 0$ is sufficiently small, depending only on (M, α). We fix such ϱ. Then map K has a fixed point $w \in C^{2,\alpha}(\overline{D})$, which is a solution of (4.5.42).

4. Since v satisfies (4.5.29)–(4.5.31), v satisfies (4.5.42). Now it follows from the uniqueness of solutions in $C(\overline{D}) \cap C^1(\overline{D} \setminus \overline{\{x_1 = 0\}}) \cap C^2(D)$ of Problem (4.5.42) that $w = v$ in D. Therefore, $v \in C^{2,\alpha}(\overline{D})$, which implies that $u \in C^{2,\alpha}(\overline{\mathcal{R}^+_{\varrho/2}})$. □

The next case we consider is modeled by the conditions near P_1: Consider domain $\mathcal{R}^+_s$, defined by (4.5.21), with curved boundary segment $\Gamma^{(\text{ob})}_s$:

$$\|f_{\text{ob}}\|^{(-1-\alpha),\{0\}}_{2,\alpha,(0,s)} \leq M_{\text{ob}} \qquad \text{for } \alpha \in (0,1). \tag{4.5.46}$$

For equation (4.5.4), the uniform ellipticity (4.5.7) is replaced by the ellipticity that is uniform, but with a certain degeneracy near $\Gamma^{(d)}_s$: There exist $\lambda \in (0, 1]$ and $\delta > 0$ such that

$$\lambda|\boldsymbol{\mu}|^2 \leq \sum_{i,j=1}^{2} a_{ij}(\mathbf{x})\frac{\mu_i\mu_j}{\big(\max\{x_1, \delta\}\big)^{2-\frac{i+j}{2}}} \leq \lambda^{-1}|\boldsymbol{\mu}|^2 \tag{4.5.47}$$

for any $\mathbf{x} \in \mathcal{R}^+_s$ and $\boldsymbol{\mu} \in \mathbb{R}^2$, and that the conditions for $\mathbf{x} \in \mathcal{R}^+_s$:

$$|a_{ii}(\mathbf{x}) - a_{ii}(\hat{P})| + |a_{12}(\mathbf{x})| \leq \lambda^{-1}|\mathbf{x}|^{\alpha}, \tag{4.5.48}$$

$$|a_i(\mathbf{x})| \leq \lambda^{-1}|\mathbf{x}|^{\alpha-1}, \tag{4.5.49}$$

$$a_0(\mathbf{x}) \leq 0, \tag{4.5.50}$$

$$\|a_{ij}\|_{C^{\alpha}(\overline{\mathcal{R}^+_s})} + \|(a_i, a_0)\|_{C^{(1-\alpha),\{\hat{P}\}}_{0,\alpha,\mathcal{R}^+_s}} \leq M, \tag{4.5.51}$$

and (4.5.8)–(4.5.9) hold with $\Omega = \mathcal{R}^+_s$. The boundary condition (4.5.23) with $\mathbf{b} = (b_1, b_2)$ satisfies that, on $\Gamma^{(\text{ob})}_s$,

$$\text{Obliqueness: } \lambda \leq \mathbf{b} \cdot \boldsymbol{\nu} \leq \lambda^{-1}, \tag{4.5.52}$$

$$|(\mathbf{b}(\mathbf{x}), b_0(\mathbf{x}))| \leq \lambda^{-1}, \tag{4.5.53}$$

$$b_0 \leq 0, \tag{4.5.54}$$

$$b_1 \leq -\lambda, \tag{4.5.55}$$

$$\|(\mathbf{b}, b_0)\|_{C^{(-\alpha),\{\hat{P}\}}_{1,\alpha,\Gamma^{(\text{ob})}_s}} \leq M. \tag{4.5.56}$$

Lemma 4.5.5. *Let $s > 0$ and $M_{\rm ob} \geq 1$, and let $\mathcal{R}_s^+$ be the domain in* (4.5.21) *such that* (4.5.46) *holds. Let $\alpha \in (0,1)$ and $\lambda \in (0,1]$. Then there exists $\delta_0 \in (0,1)$ depending only on $(M_{\rm ob}, \alpha, \lambda)$ such that, for every $\delta \in (0, \delta_0)$, if* (4.5.47)–(4.5.50) *and* (4.5.52)–(4.5.55) *hold, and $f \in C^{(1-\alpha),\{\hat{P}\}}_{\alpha,\mathcal{R}_s^+}$, $g \in C^{(-\alpha),\{\hat{P}\}}_{1,\alpha,\Gamma_s^{(\rm ob)}}$, and one of the following assumptions hold:*

(i) $g(\hat{P}) = 0$,

(ii) (4.5.51) *and* (4.5.56) *hold for some M,*

then any solution $u \in C(\overline{\mathcal{R}_s^+}) \cap C^1(\overline{\mathcal{R}_s^+} \setminus \overline{\Gamma_s^{(d)}}) \cap C^2(\mathcal{R}_s^+)$ of (4.5.4) *and* (4.5.23)–(4.5.24) *satisfies*

$$
\begin{aligned}
&\Big|u(\mathbf{x}) - \frac{g(\hat{P})}{b_1(\hat{P})}x_1\Big| \\
&\leq C\Big(\|u\|_{L^\infty(\mathcal{R}_s^+)} + \|f\|^{(1-\alpha),\{\hat{P}\}}_{0,\alpha,\mathcal{R}_s^+} + \|g\|^{(-\alpha),\{\hat{P}\}}_{1,\alpha,\Gamma_s^{(\rm ob)}}\Big)|\mathbf{x}|^{1+\alpha} \qquad \text{in } \mathcal{R}_s^+,
\end{aligned}
\tag{4.5.57}
$$

where constant C depends only on $(s, M_{\rm ob}, \alpha, \lambda, \delta)$ for Case (i), *and depends on M in addition to the previous parameters for Case* (ii).

In particular, for Case (ii), *u is in $C^{1,\alpha}(\overline{\mathcal{R}_{s/2}^+}) \cap C^{2,\alpha}(\overline{\mathcal{R}_{s/2}^+} \setminus \{\hat{P}\})$ and satisfies*

$$
\|u\|^{(-1-\alpha),\{\hat{P}\}}_{2,\alpha,\mathcal{R}_{s/2}^+} \leq C\Big(\|u\|_{L^\infty(\mathcal{R}_s^+)} + \|f\|^{(1-\alpha),\{\hat{P}\}}_{0,\alpha,\mathcal{R}_s^+} + \|g\|^{(-\alpha),\{\hat{P}\}}_{1,\alpha,\Gamma_s^{(\rm ob)}}\Big) \tag{4.5.58}
$$

with C depending only on $(s, M_{\rm ob}, \alpha, \lambda, \delta, M)$.

Proof. We divide the proof into five steps.

1. In Steps 1–4, we prove (4.5.57). To achieve this, it suffices to prove (4.5.57) in $\mathcal{R}_\varrho^+$ for some $\varrho > 0$ and C depending only on $(M_{\rm ob}, \alpha, \lambda, \delta)$ for Case (i), and on $(M_{\rm ob}, \alpha, \lambda, \delta, M)$ for Case (ii). Indeed, in order to extend (4.5.57) from $\mathcal{R}_\varrho^+$ to $\mathcal{R}_s^+$, we need only the L^∞–bound of u in $\mathcal{R}_s^+$. Then, since term $\|u\|_{L^\infty(\mathcal{R}_s^+)}$ is present on the right-hand side of (4.5.57), we readily extend (4.5.57) from $\mathcal{R}_\varrho^+$ to $\mathcal{R}_s^+$, where C now depends on s, in addition to the previous parameters.

2. We may assume without loss of generality that $g(\hat{P}) = 0$. Otherwise, (4.5.51) and (4.5.56) hold by assumption (ii), and then u may be replaced by $\tilde{u} = u - \frac{g(\hat{P})}{b_1(\hat{P})}x_1$ so that $\tilde{u}$ satisfies (4.5.4) and (4.5.23)–(4.5.24) with the right-hand sides $\tilde{f} = f - \frac{g(\hat{P})}{b_1(\hat{P})}(a_1(\mathbf{x}) + a_0(\mathbf{x})x_1)$ in (4.5.4) and $\tilde{g} = g - \frac{g(\hat{P})}{b_1(\hat{P})}(b_1(\mathbf{x}) + b_0(\mathbf{x})x_1)$ in (4.5.23). Since $\hat{P} = \mathbf{0}$, we obtain that $\tilde{g}(\hat{P}) = 0$. Moreover, using (4.5.51) and (4.5.55)–(4.5.56), we find that estimate (4.5.57) for $(\tilde{u}, \tilde{f}, \tilde{g})$ implies (4.5.57) for (u, f, g), with C depending on M, in addition to $(M_{\rm ob}, \alpha, \lambda, \delta)$.

To summarize, it suffices to prove (4.5.57) in $\mathcal{R}_\varrho^+$ for some $\varrho > 0$ and C, depending only on $(M_{\rm ob}, \alpha, \lambda, \delta)$, under the assumption that $g(\hat{P}) = 0$.

In Steps 3–4 below, the positive constants C, N, and ϱ depend only on $(M_{\rm ob}, \lambda, \delta, \alpha)$, and constant $\delta_0 > 0$ depends only on $(M_{\rm ob}, \alpha, \lambda)$.

3. Since $g(\hat{P}) = 0$, we use (4.5.46) to obtain

$$|g(\mathbf{x})| \le C\|g\|^{(-\alpha),\{\hat{P}\}}_{1,\alpha,\Gamma_s^{(\rm ob)}}|x_1|^\alpha \qquad \text{on } \Gamma_s^{(\rm ob)}. \tag{4.5.59}$$

Fix $\delta \in (0, \delta_0)$. We work in $\mathcal{R}_\varrho^+$ for ϱ to be defined. We change the variables in such a way that the second-order part of equation (4.5.4) at $\hat{P}$ becomes the Laplacian. Denote

$$\mu = \sqrt{\hat{a}_{11}(\hat{P})/\hat{a}_{22}(\hat{P})}. \tag{4.5.60}$$

Then, using (4.5.47) and $\hat{P} = \mathbf{0}$, we have

$$\lambda\sqrt{\delta} \le \mu \le \frac{\sqrt{\delta}}{\lambda}. \tag{4.5.61}$$

In particular, $\mu < 1$, by choosing δ_0 small.

Now we introduce the variables:

$$\mathbf{X} = (X_1, X_2) := (\frac{x_1}{\mu}, x_2).$$

Expressing domain $\mathcal{R}_s^+$ in (4.5.21) and its boundary $\Gamma_s^{(\rm ob)}$ in (4.5.22) in the $\mathbf{X}$-coordinates, and restricting them to $B_r(\mathbf{0})$ for $r \in (0, s)$, we obtain the following regions: For $r \in (0, \infty)$,

$$\hat{\mathcal{R}}_r^+ = \{X_1 > 0, X_2 > F(X_1)\} \cap B_r(\mathbf{0}),$$
$$\hat{\Gamma}_r^{(\rm ob)} := \{X_1 > 0,\ X_2 = F(X_1)\} \cap B_r(\mathbf{0}),$$

where

$$F(X_1) = f_{\rm ob}(\mu X_1). \tag{4.5.62}$$

From (4.5.21), (4.5.46), and (4.5.61)–(4.5.62),

$$F(0) = 0, \qquad |F'(X_1)| \le \frac{M_{\rm ob}}{\lambda}\sqrt{\delta} \quad \text{for } X_1 \in [0, \frac{s}{\mu}]. \tag{4.5.63}$$

We now consider our problem in domain $\hat{\mathcal{R}}_\varrho^+$ in the $\mathbf{X}$–coordinates.

We first write u in the $\mathbf{X}$–coordinates. Introduce the function:

$$v(\mathbf{X}) := u(\mathbf{x}) = u(\mu X_1, X_2).$$

Since $\mu < 1$, and u is defined in $\mathcal{R}_\varrho^+$, then v is defined in $\hat{\mathcal{R}}_\varrho^+$. Since u satisfies equation (4.5.4) and the boundary conditions (4.5.23)–(4.5.24), v satisfies

$$Av := \frac{1}{\mu^2}\tilde{a}_{11}v_{X_1X_1} + \frac{2}{\mu}\tilde{a}_{12}v_{X_1X_2} + \tilde{a}_{22}v_{X_2X_2} + \frac{1}{\mu}\tilde{a}_1v_{X_1} + \tilde{a}_2v_{X_2} + \tilde{a}_0v = \tilde{f} \tag{4.5.64}$$

in $\hat{\mathcal{R}}_\varrho^+$,

$$Bv := \frac{1}{\mu}\tilde{b}_1 v_{X_1} + \tilde{b}_2 v_{X_2} + \tilde{b}_0 v = \tilde{g} \qquad \text{on } \hat{\Gamma}_\varrho^{(\mathrm{ob})}, \tag{4.5.65}$$

and

$$v = 0 \qquad \text{on } \{X_1 = 0,\ X_2 > F(0)\} \cap B_\varrho, \tag{4.5.66}$$

where $(\tilde{a}_{ij}, \tilde{a}_i, \tilde{b}_i, \tilde{f}, \tilde{g})(\mathbf{X}) = (a_{ij}, a_i, b_i, f, g)(\mu X_1, X_2)$. In particular, from (4.5.48)–(4.5.50) and (4.5.60), we have

$$\tilde{a}_{22}(\mathbf{0}) = \frac{1}{\mu^2}\tilde{a}_{11}(\mathbf{0}), \quad \tilde{a}_{12}(\mathbf{0}) = \tilde{a}_2(\mathbf{0}) = 0, \quad \tilde{a}_0 \le 0, \tag{4.5.67}$$

$$|\tilde{a}_{ii}(\mathbf{X}) - \tilde{a}_{ii}(\mathbf{0})| \le C|\mathbf{X}|^\alpha \qquad \text{for } i = 1, 2, \tag{4.5.68}$$

$$|\tilde{a}_{12}(\mathbf{X})| + \mu^{1-\alpha}|\mathbf{X}||(\tilde{a}_1, \tilde{a}_2)(\mathbf{X})| \le C|\mathbf{X}|^\alpha. \tag{4.5.69}$$

From (4.5.53) and (4.5.55), we have

$$\tilde{b}_1 \le -\lambda, \quad |\tilde{\mathbf{b}}| \le \lambda^{-1} \qquad \text{on } \hat{\Gamma}_\varrho^{(\mathrm{ob})}. \tag{4.5.70}$$

Also, (4.5.52) and the expression of F in (4.5.62) imply

$$(\frac{1}{\mu}\tilde{b}_1, \tilde{b}_2) \cdot \boldsymbol{\nu}_F > 0 \qquad \text{on } \hat{\Gamma}_\varrho^{(\mathrm{ob})},$$

where $\boldsymbol{\nu}_F$ is the normal to $\hat{\Gamma}_\varrho^{(\mathrm{ob})}$. Therefore, condition (4.5.65) is oblique.

4. We use the polar coordinates (r, θ) on the $\mathbf{X}$–plane, *i.e.*,

$$\mathbf{X} = (r\cos\theta, r\sin\theta).$$

From (4.5.63), it follows that, if $\delta_0 > 0$ is small, $|\theta| \le C\sqrt{\delta}$ on $\Gamma_\varrho^{(\mathrm{ob})}$ so that, on $\Gamma_\varrho^{(\mathrm{ob})}$,

$$\partial_\theta\big(X_2 - F(X_1)\big) = \partial_\theta\big(r\sin\theta - F(r\cos\theta)\big) \ge r\big(\cos\theta - C\sqrt{\delta}|\sin\theta|\big) \ge \frac{r}{2},$$

$$\big|\partial_r\big(X_2 - F(X_1)\big)\big| = \big|\partial_r\big(r\sin\theta - F(r\cos\theta)\big)\big| \le |\sin\theta + |F'(r\cos\theta)| \le C\sqrt{\delta}.$$

Thus, there exists a function $\theta_F \in C^1([0, \varrho])$ such that

$$\hat{\mathcal{R}}_\varrho^+ = \{0 < r < \varrho,\ \theta_F(r) < \theta < \frac{\pi}{2}\}, \qquad \hat{\Gamma}_\varrho^{(\mathrm{ob})} = \{0 < r < \varrho,\ \theta = \theta_F(r)\}. \tag{4.5.71}$$

Also, by (4.5.63),

$$|\theta_F(r)| \le C\sqrt{\delta}. \tag{4.5.72}$$

Choosing sufficiently small $\delta_0 > 0$ and $\varrho > 0$, and large $N = N(\alpha, \lambda) \ge 1$, we show that, for any $\delta \in (0, \delta_0)$, the function:

$$w(r, \theta) = Lr^{1+\alpha}\cos(G(\theta)) \tag{4.5.73}$$

is a positive supersolution of (4.5.64)–(4.5.66) in $\mathcal{R}_\varrho^+$, where

$$\begin{aligned} G(\theta) &= \frac{3+\alpha}{2}(\theta - \frac{\pi}{4}), \\ L &= N\Big(\|u\|_{L^\infty(\mathcal{R}_s^+)} + \|f\|^{(1-\alpha),\{\hat{P}\}}_{0,\alpha,\mathcal{R}_s^+} + \|g\|^{(-\alpha),\{\hat{P}\}}_{1,\alpha,\Gamma_s^{(\mathrm{ob})}}\Big). \end{aligned} \tag{4.5.74}$$

From (4.5.71)–(4.5.72), we find that, for sufficiently small δ_0 depending only on (α, λ, M) and any $\delta \in (0, \delta_0)$,

$$-\frac{\pi}{2} + \frac{(1-\alpha)\pi}{16} \le G(\theta) \le \frac{\pi}{2} - \frac{(1-\alpha)\pi}{16} \qquad \text{for any } (r,\theta) \in \mathcal{R}_\varrho^+.$$

In particular, we have

$$\cos(G(\theta)) \ge \sin\big(\frac{1-\alpha}{16}\pi\big) > 0 \qquad \text{for any } (r,\theta) \in \mathcal{R}_\varrho^+ \setminus \{\mathbf{X} = \mathbf{0}\}, \tag{4.5.75}$$

which implies

$$w > 0 \qquad \text{in } \mathcal{R}_\varrho^+.$$

By (4.5.71)–(4.5.72), we find that, if $\delta \in (0, \delta_0)$ with small $\delta_0 > 0$,

$$\cos(\theta_F(r)) \ge 1 - C\delta_0 > 0, \quad |\sin(\theta_F(r))| \le C\sqrt{\delta_0} \qquad \text{for all } r \in (0, \varrho).$$

Now, possibly reducing δ_0 further, we show that w is a supersolution of (4.5.65). Using (4.5.61), (4.5.65), and (4.5.70), the estimates of $(\theta_F, G(\theta_F))$ derived above, and the fact that $\theta = \theta_F$ on $\hat{\Gamma}_\varrho^{(\mathrm{ob})}$, we have

$$\begin{aligned} Bw \le\; & L\frac{\tilde{b}_1}{\mu}r^\alpha\Big((\alpha+1)\cos(\theta_F)\cos(G(\theta_F)) + \frac{3+\alpha}{2}\sin(\theta_F)\sin(G(\theta_F))\Big) \\ & + CLr^\alpha|\tilde{b}_2| + CLr^{\alpha+1}|\tilde{b}_3| \\ \le & - Lr^\alpha\Big(\frac{\lambda^2}{\sqrt{\delta_0}}\Big((1 - C\delta_0)\sin(\frac{1-\alpha}{16}\pi) - C\sqrt{\delta_0}\Big) - C\Big) \\ \le & - Lr^\alpha, \end{aligned}$$

if δ_0 is sufficiently small. We now fix δ_0 such that it satisfies all the smallness assumptions above. Then we have

$$Bw \le -Lr^\alpha \le -N\|g\|^{(-\alpha),\{\hat{P}\}}_{1,\alpha,\Gamma_r^{(\mathrm{ob})}} r^\alpha \le g,$$

where we have used (4.5.59) and chosen N sufficiently large to achieve the last inequality.

Finally, we show that w is a supersolution of equation (4.5.64) in $\mathbf{X} \in \{X_1 > 0,\ X_2 > F(X_1)\} \cap B_\varrho$ if ϱ is small. Denote by A_0 the operator obtained by fixing the coefficients of A in (4.5.64) at $\mathbf{X} = \mathbf{0}$. Then $A_0 = \tilde{a}_{22}(\mathbf{0})\Delta$ by (4.5.67). By (4.5.47), we obtain that $\tilde{a}_{22}(\mathbf{0}) \ge \lambda$. Now, by explicit calculation and using

(4.5.61), (4.5.67)–(4.5.69), and (4.5.75), we find that, for any $\delta \in (0, \delta_0)$ and $\mathbf{X} \in \{X_1 > 0,\ X_2 > F(X_1)\} \cap B_\varrho$,

$$\begin{aligned}
Aw(r,\theta) =& a_2(\mathbf{0})\Delta w(r,\theta) + (A - A_0)w(r,\theta)\\
\le& L\tilde{a}_{22}(\mathbf{0})r^{\alpha-1}\big((\alpha+1)^2 - (\frac{3+\alpha}{2})^2\big)\cos(G(\theta))\\
&+ CLr^{\alpha-1}\Big(\frac{1}{\mu^2}|\tilde{a}_{11}(\mathbf{X}) - \tilde{a}_{11}(\mathbf{0})| + |\tilde{a}_{22}(\mathbf{X}) - \tilde{a}_{22}(\mathbf{0})|\Big)\\
&+ \frac{CL}{\mu}\Big(r^{\alpha-1}|\tilde{a}_{12}(\mathbf{X})| + r^\alpha|\tilde{a}_1(\mathbf{X})|\Big) + CLr^\alpha\Big(|\tilde{a}_2(\mathbf{X})| + r|\tilde{a}_0(\mathbf{X})|\Big)\\
\le& -Lr^{\alpha-1}\Big(\frac{(1-\alpha)(5+3\alpha)}{4}\lambda\sin\big(\frac{1-\alpha}{16}\pi\big) - C\frac{\varrho^\alpha}{\delta}\Big)\\
\le& -\frac{(1-\alpha)(5+3\alpha)}{8}Lr^{\alpha-1}\lambda\sin\big(\frac{1-\alpha}{16}\pi\big),
\end{aligned}$$

where we have chosen sufficiently small $\varrho > 0$ to obtain the last inequality. Note that such a small constant ϱ depends on $(\delta, \alpha, \lambda, C)$, and hence on $(M_{\rm ob}, \alpha, \lambda, \delta)$ as stated in Step 2. We now fix ϱ. Choosing N in (4.5.74) sufficiently large, we have

$$Aw \le -\|f\|^{(1-\alpha),\{\hat{P}\}}_{0,\alpha,\mathcal{R}^+_\varrho} r^{\alpha-1} \le f.$$

Now the maximum principle and Hopf's lemma imply that

$$v \le w \le Cr^{\alpha+1} \qquad \text{in } \mathcal{R}^+_\varrho,$$

by the argument that repeats Steps (i)–(ii) of the proof of Lemma 4.4.2. A similar estimate can be obtained for $-v$. Thus, $|v| \le w$ in $\mathcal{R}^+_\varrho$. Then, from (4.5.73)–(4.5.74) and $v(\mathbf{X}) = u(\mu X_1, X_2)$, we obtain (4.5.57) in $\mathcal{R}^+_\varrho$ in the case that $g(\hat{P}) = 0$. This implies the full estimate (4.5.57) in the general case, according to Steps 1–2.

5. We now prove (4.5.58) under the assumption that (4.5.51) and (4.5.56) hold. Similar to Step 2, it suffices to consider the case that $g(\hat{P}) = 0$. In the following argument, the positive constants C and L depend only on $(s, M_{\rm ob}, \lambda, \delta, \alpha, M)$.

Estimate (4.5.58) can be obtained from (4.5.57), combined with rescaling from balls $B_{\frac{d_{\mathbf{x}}}{10}}(\mathbf{x}) \cap \mathcal{R}^+_s$ for $\mathbf{x} \in \overline{\mathcal{R}^+_{s/2}} \setminus \{\hat{P}\}$ into the unit ball and the standard interior estimates for the linear elliptic equations and the local estimates for the linear Dirichlet and oblique derivative problems in smooth domains. Specifically, from (4.5.21) and (4.5.46), it follows that there exists $L > 1$, depending only on $M_{\rm ob}$, such that

$$\begin{aligned}
&B_{\frac{d_{\mathbf{x}}}{L}}(\mathbf{x}) \cap (\partial\mathcal{R}^+_s \setminus \Gamma^{(\rm ob)}_s) = \emptyset \qquad \text{for any } \mathbf{x} \in \Gamma^{(\rm ob)}_{s/2},\\
&B_{\frac{d_{\mathbf{x}}}{L}}(\mathbf{x}) \cap (\partial\mathcal{R}^+_s \setminus \Gamma^{(d)}_s) = \emptyset \qquad \text{for any } \mathbf{x} \in \Gamma^{(d)}_{s/2},
\end{aligned}$$

where $d_{\mathbf{x}} := \mathrm{dist}(\mathbf{x}, \hat{P})$. Also, for any $\mathbf{x} \in \mathcal{R}^+_{s/2}$, we have at least one of the following three cases:

(a) $B_{\frac{d_{\mathbf{x}}}{10L}}(\mathbf{x}) \subset \mathcal{R}^+_s$;

(b) $\mathbf{x} \in B_{\frac{d_{\hat{\mathbf{x}}}}{2L}}(\hat{\mathbf{x}})$ for some $\hat{\mathbf{x}} \in \Gamma^{(d)}_s$;

(c) $\mathbf{x} \in B_{\frac{d_{\hat{\mathbf{x}}}}{2L}}(\hat{\mathbf{x}})$ for some $\hat{\mathbf{x}} \in \Gamma^{(\mathrm{ob})}_s$.

Thus, it suffices to make the $C^{2,\alpha}$–estimates of u in the following subdomains:

(i) $B_{\frac{d_{\hat{\mathbf{x}}}}{20L}}(\hat{\mathbf{x}})$ when $B_{\frac{d_{\hat{\mathbf{x}}}}{10L}}(\hat{\mathbf{x}}) \subset \mathcal{R}^+_s$;

(ii) $B_{\frac{d_{\hat{\mathbf{x}}}}{2L}}(\hat{\mathbf{x}}) \cap \mathcal{R}^+_s$ for $\hat{\mathbf{x}} \in \Gamma^{(d)}_{s/2}$;

(iii) $B_{\frac{d_{\hat{\mathbf{x}}}}{2L}}(\hat{\mathbf{x}}) \cap \mathcal{R}^+_s$ for $\hat{\mathbf{x}} \in \Gamma^{(\mathrm{ob})}_{s/2}$.

We discuss only Case (iii), since the other cases are simpler and can be handled similarly.

Let $\hat{\mathbf{x}} \in \Gamma^{(\mathrm{ob})}_{s/2}$. Denote $\hat{d} := \frac{d_{\hat{\mathbf{x}}}}{2L} > 0$. Without loss of generality, we can assume that $\hat{d} \leq 1$.

We rescale near $\hat{\mathbf{x}} = (\hat{x}_1, \hat{x}_2)$, *i.e.*, define the coordinates:

$$\mathbf{X} := \frac{\mathbf{x} - \hat{\mathbf{x}}}{\hat{d}}.$$

Since $B_{\hat{d}}(\hat{\mathbf{x}}) \cap (\partial\mathcal{R}^+_s \setminus \Gamma^{(\mathrm{ob})}_s) = \emptyset$, then, for $\rho \in (0,1)$, the domain obtained by rescaling $\mathcal{R}^+_s \cap B_{\rho\hat{d}}(\hat{\mathbf{x}})$ is

$$\Omega^{\hat{\mathbf{x}}}_\rho := B_\rho \cap \{X_2 > F(X_1)\},$$

where $F(X_1) = \frac{f_{\mathrm{ob}}(\hat{x}_1 + \hat{d}X_1) - f_{\mathrm{ob}}(\hat{x}_1)}{\hat{d}}$. Then $\|F\|_{2,\alpha,(-1,1)} \leq \|f\|^{(-1-\alpha),\{0\}}_{2,\alpha,(0,s)} \leq M_{\mathrm{ob}}$.

Define

$$v(\mathbf{X}) = \frac{u(\hat{\mathbf{x}} + \hat{d}\mathbf{X})}{\hat{d}^{1+\alpha}} \qquad \text{for } \mathbf{X} \in \Omega^{\hat{\mathbf{x}}}_1. \tag{4.5.76}$$

Then, by (4.5.57),

$$\|v\|_{0,\Omega^{\hat{\mathbf{x}}}_1} \leq C\Big(\|u\|_{0,\mathcal{R}^+_s} + \|f\|^{(1-\alpha),\{\hat{P}\}}_{0,\alpha,\mathcal{R}^+_s} + \|g\|^{(-\alpha),\{\hat{P}\}}_{1,\alpha,\Gamma^{(\mathrm{ob})}_s}\Big), \tag{4.5.77}$$

since $g(\hat{P}) = 0$ in the case under consideration.

Since u satisfies equation (4.5.4) in $\mathcal{R}^+_s$ and the oblique derivative condition (4.5.23) on $\Gamma^{(\mathrm{ob})}_s$, v satisfies an equation and an oblique derivative condition of a similar form in $\Omega^{\hat{\mathbf{x}}}_1$ and on $\Gamma^{(\mathrm{ob})}_{\hat{\mathbf{x}}} := \partial\Omega^{\hat{\mathbf{x}}}_1 \cap \{X_2 = F(X_1)\}$, respectively, whose coefficients satisfy properties (4.5.47) and (4.5.51) with the same constants as for the original equations, since $\hat{d} \le 1$. Moreover, (4.5.51) and (4.5.56) imply that the coefficients of the rescaled equation and the boundary condition are in $C^{\alpha}(\overline{\Omega_1^{\hat{\mathbf{x}}}})$ and $C^{1,\alpha}(\overline{\Gamma_{\hat{\mathbf{x}}}^{(\mathrm{ob})}})$, respectively, with the norms bounded by M. Furthermore, the right-hand sides $(\hat{f}, \hat{g})$ of these equations and the oblique condition for v are determined by

$$\hat{f}(\mathbf{X}) = \hat{d}^{1-\alpha} f(\hat{\mathbf{x}} + \hat{d}\mathbf{X}), \qquad \hat{g}(\mathbf{X}) = \hat{d}^{-\alpha} g(\hat{\mathbf{x}} + \hat{d}\mathbf{X}).$$

Since $g(\hat{P}) = 0$, we have

$$g(\hat{\mathbf{x}}) \le \|g\|^{(-\alpha),\{\hat{P}\}}_{1,\alpha,\Gamma_s^{(\mathrm{ob})}} |\hat{\mathbf{x}}|^{\alpha} \le C \|g\|^{(-\alpha),\{\hat{P}\}}_{1,\alpha,\Gamma_s^{(\mathrm{ob})}} \hat{d}^{\alpha}.$$

Using this estimate and $\hat{d} \le 1$, we have

$$\|\hat{f}\|_{0,\alpha,\Omega_1^{\hat{\mathbf{x}}}} \le C \|f\|^{(1-\alpha),\{\hat{P}\}}_{0,\alpha,\mathcal{R}_s^+}, \qquad \|\hat{g}\|_{1,\alpha,\Gamma_{\hat{\mathbf{x}}}^{(\mathrm{ob})}} \le C \|g\|^{(-\alpha),\{\hat{P}\}}_{1,\alpha,\Gamma_s^{(\mathrm{ob})}}.$$

We also note the $C^{2,\alpha}$–regularity of the boundary $\Gamma_{\hat{\mathbf{x}}}^{(\mathrm{ob})} := \partial\Omega_1^{\hat{\mathbf{x}}} \cap \{X_2 = F(X_1)\}$: $\|F\|_{2,\alpha,(-1,1)} \le M_{\mathrm{ob}}$. Then, from the standard local estimates for linear oblique derivative problems, we have

$$\begin{aligned}\|v\|_{2,\alpha,\Omega_{1/2}^{\hat{\mathbf{x}}}} &\le C\Big(\|v\|_{0,\Omega_1^{\hat{\mathbf{x}}}} + \|\hat{f}\|_{0,\alpha,\overline{\Omega_1^{\hat{\mathbf{x}}}}} + \|\hat{g}\|_{1,\alpha,\Gamma_{\hat{\mathbf{x}}}^{(\mathrm{ob})}}\Big)\\ &\le C\Big(\|u\|_{0,\mathcal{R}_s^+} + \|f\|^{(1-\alpha),\{\hat{P}\}}_{0,\alpha,\mathcal{R}_s^+} + \|g\|^{(-\alpha),\{\hat{P}\}}_{1,\alpha,\Gamma_s^{(\mathrm{ob})}}\Big)\end{aligned}$$

with C depending only on (λ, M, δ).

We obtain similar estimates for Cases (i)–(ii) by using the interior estimates for elliptic equations for Case (i) and the local estimates for the Dirichlet problem for linear elliptic equations for Case (ii).

Writing the above estimates in terms of u and using the fact that the whole domain $\mathcal{R}^+_{s/2}$ is covered by the subdomains in (i)–(iii), we obtain (4.5.58) by an argument similar to that for the proof of [131, Theorem 4.8]; see also the proof of Lemma 4.2.7. □

Now we obtain the following existence result for solutions of Problem (4.5.4)–(4.5.6) with global $C^{1,\alpha}$–regularity:

Proposition 4.5.6. *Let $h, t_0, t_1, t_2, M_{\mathrm{bd}} > 0$, and let Ω be a domain of structure* (4.5.1)–(4.5.3). *Let $\varepsilon \in (0, \frac{1}{2}\min\{h, t_0\})$, $\kappa, \lambda \in (0, 1]$, $M < \infty$, and $\alpha \in (0, \alpha_1]$ be constants, where $\alpha_1(\kappa, \lambda, M_{\mathrm{bd}}) \in (0, 1)$ is determined in Lemma* 4.5.1. *Let f_{bd} satisfy* (4.5.15). *Let $\delta \in (0, \delta_0)$ for $\delta_0(\lambda, \alpha)$ determined in Lemma* 4.5.5. *Assume that the coefficients of equation* (4.5.4) *satisfy the ellipticity condition* (4.5.47) *for all $\mathbf{x} \in \Omega$ and $\boldsymbol{\mu} \in \mathbb{R}^2$ and the properties:*

$$\|a_{ij}\|_{0,\alpha,\Omega} + \|(a_i, a_0)\|^{(1-\alpha),\{P_1\}}_{0,\alpha,\Omega} \le M, \tag{4.5.78}$$

$$a_{12}(P_j) = a_{21}(P_j) = 0, \qquad j = 1, 4, \tag{4.5.79}$$

and (4.5.8). *Assume that the coefficients of the boundary conditions* (4.5.5) *satisfy* (4.5.10)–(4.5.14) *and the following additional properties:*

$$b_1^{(1)}(P_1) \le -\lambda, \tag{4.5.80}$$

$$b_1^{(3)} \equiv 0, \quad b_2^{(3)} \equiv 1, \quad b_0^{(3)} \equiv 0 \qquad \text{on } \Gamma_3 \cap \{x_1 < \varepsilon\}. \tag{4.5.81}$$

Let $f \in C_{0,\alpha,\Omega}^{(1-\alpha),\{P_1\}\cup\overline{\Gamma_2}}$, $g_2 \in C^\alpha(\overline{\Gamma_2})$, *and* $g_i \in C_{1,\alpha,\Gamma_i}^{(-\alpha),\partial\Gamma_i}$ *for* $i = 1, 3$, *where* $\partial\Gamma_i$ *denotes the endpoints of* Γ_i. *Moreover, let*

$$g_3 \equiv 0 \qquad \text{on } \Gamma_3 \cap \{x_1 < \varepsilon\}. \tag{4.5.82}$$

Then there exists a unique solution $u \in C_{2,\alpha,\Omega}^{(-1-\alpha),\{P_1\}\cup\overline{\Gamma_2}}$ *of Problem* (4.5.4)–(4.5.6) *with*

$$\|u\|_{2,\alpha,\Omega}^{(-1-\alpha),\{P_1\}\cup\overline{\Gamma_2}} \le C\Big(\|f\|_{0,\alpha,\Omega}^{(1-\alpha),\{P_1\}\cup\overline{\Gamma_2}} + \sum_{i=1,3} \|g_i\|_{1,\alpha,\Gamma_i}^{(-\alpha),\partial\Gamma_i} + \|g_2\|_{0,\alpha,\Gamma_2}\Big), \tag{4.5.83}$$

where C *depends only on* $(\Omega, \kappa, \lambda, M, \varepsilon, \alpha)$.

Proof. Fix α and δ satisfying the conditions above. Since λ and δ determine the ellipticity in the present case, then, from Proposition 4.5.2, there exists $\beta_1(\kappa, \lambda, \delta)$ such that, for each $\beta \in (0, \beta_1)$, there is a solution $u \in C^{*,2,\alpha,\beta}(\Omega)$ of Problem (4.5.4)–(4.5.6) satisfying (4.5.16). Fix $\beta = \frac{\beta_1}{2}$. Then C in (4.5.16) depends only on $(\kappa, \lambda, M, \alpha)$. Applying Lemma 4.5.4 in $B_\varepsilon(P_4) \cap \Omega$ and Lemma 4.5.5 in $B_\varepsilon(P_1) \cap \Omega$, we obtain the higher regularity of u near $\{P_1, P_4\}$. Note that our assumptions allow us to apply Lemmas 4.5.4–4.5.5, that assumption (4.5.82) especially determines the size of region where we apply Lemma 4.5.4, and that restriction $\varepsilon \in (0, \frac{1}{2}\min\{h, t_0\})$ implies that regions $B_\varepsilon(P_j) \cap \Omega$, $j = 1, 4$, are of structure (4.5.21) in the appropriate coordinate system. Also, we apply the standard local Schauder estimates for the Dirichlet problem in $\Omega_{\varepsilon/4}(\mathbf{x}) := B_{\varepsilon/4}(\mathbf{x}) \cap \Omega$ for any $\mathbf{x} \in \Gamma_0$ with $\operatorname{dist}(\mathbf{x}, \{P_1, P_4\}) > \frac{\varepsilon}{2}$, where we have used that $\|f\|_{0,\alpha,\Omega_{\varepsilon/4}(\mathbf{x})} \le C(\varepsilon, \alpha)\|f\|_{0,\alpha,\Omega}^{(1-\alpha),\{P_1\}\cup\overline{\Gamma_2}}$ and

$$\|(a_1, a_0)\|_{0,\alpha,\Omega_{\varepsilon/4}(\mathbf{x})} \le C(\varepsilon, \alpha)M$$

by (4.5.78) to obtain

$$\|u\|_{2,\alpha,\Omega_{\varepsilon/8}(\mathbf{x})} \le C(\lambda, M, \varepsilon, \alpha)\big(\|u\|_{0,\Omega} + \|f\|_{0,\alpha,\Omega}^{(1-\alpha),\{P_1\}\cup\overline{\Gamma_2}}\big).$$

Combining these estimates allows us to improve (4.5.16) to (4.5.83). The uniqueness of the solution follows from (4.5.83). □

4.5.2 Nonlinear problem

In this section, we consider a nonlinear problem in the domain of structure (4.5.1)–(4.5.2), which will be used later. We use the notations introduced in (4.5.3) for its sides and vertices.

The nonlinear problem under consideration is

$$\sum_{i,j=1}^{2} A_{ij}(Du,\mathbf{x})D_{ij}u + \sum_{i=1}^{2} A_i(Du,\mathbf{x})D_iu = 0 \qquad \text{in } \Omega, \tag{4.5.84}$$

$$B(Du,u,\mathbf{x}) = 0 \qquad \text{on } \Gamma_1, \tag{4.5.85}$$

$$\mathbf{b}^{(2)}(\mathbf{x})\cdot Du = g_2 \qquad \text{on } \Gamma_2, \tag{4.5.86}$$

$$\mathbf{b}^{(3)}(\mathbf{x})\cdot Du = 0 \qquad \text{on } \Gamma_3, \tag{4.5.87}$$

$$u = 0 \qquad \text{on } \Gamma_0. \tag{4.5.88}$$

Now we list the assumptions for Problem (4.5.84)–(4.5.88). Let $\alpha \in (0,\frac{1}{2})$, $\beta \in (\frac{1}{2},1)$, $\lambda \in (0,1]$, $\delta > 0$, $\kappa > 0$, $\varepsilon \in (0,\frac{h}{10})$, $\sigma \in (0,1)$, and $M < \infty$.

Assumptions on the equation:

(i) Ellipticity degenerating on Γ_0: For any $\mathbf{x} \in \Omega$ and $\mathbf{p},\boldsymbol{\mu} \in \mathbb{R}^2$,

$$\lambda\,\mathrm{dist}(\mathbf{x},\Gamma_0)|\boldsymbol{\mu}|^2 \le \sum_{i,j=1}^{2} A_{ij}(\mathbf{p},\mathbf{x})\mu_i\mu_j \le \lambda^{-1}|\boldsymbol{\mu}|^2; \tag{4.5.89}$$

Anisotropic uniform ellipticity near Γ_0: For any $\mathbf{x} \in \Omega\cap\{x_1 < \frac{\varepsilon}{2}\}$ and $\mathbf{p},\boldsymbol{\mu} = (\mu_1,\mu_2) \in \mathbb{R}^2$,

$$\lambda|\boldsymbol{\mu}|^2 \le \sum_{i,j=1}^{2} A_{ij}(\mathbf{p},\mathbf{x})\frac{\mu_i\mu_j}{\big(\max(x_1,\delta)\big)^{2-\frac{i+j}{2}}} \le \lambda^{-1}|\boldsymbol{\mu}|^2; \tag{4.5.90}$$

(ii) Functions $(A_{ij},A_i)(\mathbf{p},\mathbf{x})$ are independent of $\mathbf{p}$ on $\Omega\cap\{x_1 \ge \varepsilon\}$, *i.e.*, $(A_{ij},A_i)(\mathbf{p},\mathbf{x}) = (A_{ij},A_i)(\mathbf{x})$ for any $\mathbf{x} \in \Omega\cap\{x_1 \ge \varepsilon\}$ and $\mathbf{p} \in \mathbb{R}^2$, with

$$\|A_{ij}\|_{L^\infty(\Omega\cap\{x_1\ge\varepsilon\})} \le \lambda^{-1}, \quad \|(A_{ij},A_i)\|_{C^{(-\alpha),\Gamma_2}_{1,\alpha,\Omega\cap\{x_1\ge\varepsilon\}}} \le M; \tag{4.5.91}$$

(iii) There exists $\beta \in (\frac{1}{2},1)$ such that, for any $\mathbf{p} \in \mathbb{R}^2$,

$$\begin{aligned}&\|(A_{ij},A_i)(\mathbf{p},\cdot)\|_{C^\beta(\overline{\Omega\cap\{x_1<2\varepsilon\}})}\\ &\quad + \|(D_{\mathbf{p}}A_{ij},D_{\mathbf{p}}A_i)(\mathbf{p},\cdot)\|_{L^\infty(\Omega\cap\{x_1<2\varepsilon\})} \le M;\end{aligned} \tag{4.5.92}$$

(iv) $(A_{ij},A_i) \in C^{1,\alpha}(\mathbb{R}^2\times(\overline{\Omega}\setminus(\overline{\Gamma_0}\cup\overline{\Gamma_2})))$ and, for any $s \in (0,\frac{h}{4})$,

$$\|(A_{ij},A_i)\|_{C^{1,\alpha}(\mathbb{R}^2\times(\overline{\Omega}\cap\{s\le x_1\le h-s\}))} \le M\Big(\frac{h}{s}\Big)^M; \tag{4.5.93}$$

(v) For any $(\mathbf{p},(x_1,0)) \in \mathbb{R}^2 \times (\Gamma_3 \cap \{x_1 < \varepsilon\})$,

$$\begin{aligned}&(A_{11},A_{22},A_1)((p_1,-p_2),(x_1,0))\\&=(A_{11},A_{22},A_1)((p_1,p_2),(x_1,0))\end{aligned} \tag{4.5.94}$$

and, for any $(\mathbf{p},(x_1,x_2)) \in \mathbb{R}^2 \times (\Omega \cap \{x_1 < \varepsilon\})$,

$$|A_{ii}(\mathbf{p},(x_1,x_2)) - A_{ii}(\mathbf{0},(0,x_2))| \le M|x_1|^\beta, \qquad i=1,2; \tag{4.5.95}$$

(vi) For any $\mathbf{p} \in \mathbb{R}^2$ and $(0,x_2) \in \Gamma_0$,

$$(A_{12},A_{21})(\mathbf{p},(0,x_2)) = 0; \tag{4.5.96}$$

(vii) For any $\mathbf{p} \in \mathbb{R}^2$ and $\mathbf{x} \in \Omega \cap \{x_1 < \frac{\varepsilon}{2}\}$,

$$A_1(\mathbf{p},\mathbf{x}) \le -\lambda. \tag{4.5.97}$$

Assumptions on the nonlinear boundary condition (4.5.85):

(a) Uniform obliqueness: For any $(\mathbf{p},z,\mathbf{x}) \in \mathbb{R}^2 \times \mathbb{R} \times \Gamma_1$,

$$D_{\mathbf{p}}B(\mathbf{p},z,\mathbf{x}) \cdot \boldsymbol{\nu}^{(1)}(\mathbf{x}) \ge \lambda, \tag{4.5.98}$$

where $\boldsymbol{\nu}^{(1)}$ is the interior unit normal on Γ_1 to Ω;

(b) Regularity of the coefficients: For any $(\mathbf{p},z) \in \mathbb{R}^2 \times \mathbb{R}$,

$$\|(B(\mathbf{0},0,\cdot),\ D^k_{(\mathbf{p},z)}B(\mathbf{p},z,\cdot))\|_{C^3(\overline{\Omega})} \le M, \quad k=1,2,3, \tag{4.5.99}$$

$$\|D_{\mathbf{p}}B(\mathbf{p},z,\cdot)\|_{C^0(\overline{\Omega})} \le \lambda^{-1}, \tag{4.5.100}$$

$$D_zB(\mathbf{p},z,\mathbf{x}) \le -\lambda \qquad \text{for all } \mathbf{x} \in \Gamma_1, \tag{4.5.101}$$

$$D_{p_1}B(\mathbf{p},z,\mathbf{x}) \le -\lambda \qquad \text{for all } \Gamma_1 \cap \{x_1 < \varepsilon\}; \tag{4.5.102}$$

(c) Almost-linear structure: There exist $v \in C^3(\overline{\Gamma}_1)$ and a nonhomogeneous linear operator:

$$L(\mathbf{p},z,\mathbf{x}) = \mathbf{b}^{(1)}(\mathbf{x}) \cdot \mathbf{p} + b_0^{(1)}(\mathbf{x})z + g_1(\mathbf{x}),$$

defined for $\mathbf{x} \in \Gamma_1$ and $(\mathbf{p},z) \in \mathbb{R}^2 \times \mathbb{R}$, satisfying

$$\|v\|_{C^3(\overline{\Omega})} + \|(\mathbf{b}^{(1)},\ b_0^{(1)},\ g_1)\|_{C^3(\overline{\Gamma_1})} \le M, \tag{4.5.103}$$

such that, for any $(\mathbf{p},z,\mathbf{x}) \in \mathbb{R}^2 \times \mathbb{R} \times \Gamma_1$,

$$\begin{aligned}&|B(\mathbf{p},z,\mathbf{x}) - L(\mathbf{p},z,\mathbf{x})| \le \sigma\big(|\mathbf{p} - Dv(\mathbf{x})| + |z - v(\mathbf{x})|\big),\\&\big|D_{\mathbf{p}}B(\mathbf{p},z,\mathbf{x}) - \mathbf{b}^{(1)}(\mathbf{x})\big| + \big|D_zB(\mathbf{p},z,\mathbf{x}) - b_0^{(1)}(\mathbf{x})\big| \le \sigma.\end{aligned} \tag{4.5.104}$$

Assumptions on the linear boundary conditions (4.5.86)–(4.5.87): The coefficients $\mathbf{b}^{(k)}$ of the boundary conditions on Γ_k, $k = 2, 3$, satisfy

$$\text{Obliqueness: } \mathbf{b}^{(k)} \cdot \boldsymbol{\nu}^{(k)} \ge \lambda \ \text{ on } \Gamma_k, \tag{4.5.105}$$

$$b_1^{(3)} \le 0 \ \text{ on } \Gamma_3, \qquad \mathbf{b}^{(3)} = (0,1) \ \text{ on } \Gamma_3 \cap \{0 < x_1 < \varepsilon\}, \tag{4.5.106}$$

$$\|\mathbf{b}^{(k)}\|_{C^3(\overline{\Omega})} \le \lambda^{-1} \qquad \text{for } k = 2, 3, \tag{4.5.107}$$

where $\boldsymbol{\nu}^{(k)}$ is the interior unit normal on Γ_k to Ω, and $\partial\Gamma_k$ denotes the endpoints of Γ_k.

Assumptions at corners P_k, $k = 2, 3$:

$$\begin{aligned}&\text{The boundary conditions on } \Gamma_{k-1} \text{ and } \Gamma_k \text{ satisfy the}\\ &\quad \lambda\text{-obliqueness at corner } P_k \text{ in the sense of Definition 4.4.3,}\end{aligned} \tag{4.5.108}$$

$$\left|\frac{\mathbf{b}^{(k-1)}}{|\mathbf{b}^{(k-1)}|} - \frac{\mathbf{b}^{(k)}}{|\mathbf{b}^{(k)}|}\right| \ge \kappa \qquad \text{at } P_k. \tag{4.5.109}$$

Assumptions on the right-hand sides:

$$\|g_2\|_{C^3(\Gamma_2)} \le M, \tag{4.5.110}$$

$$B(\mathbf{0}, 0, \cdot) \equiv 0 \qquad \text{on } \Gamma_1 \cap \{x_1 < \varepsilon\}. \tag{4.5.111}$$

We first establish the *a priori* estimates for solutions of (4.5.84)–(4.5.88).

Remark 4.5.7. *Some of the estimates below are proved under the assumptions of possibly nonstrictly positive t_0 in* (4.5.1)*. In the case that $f_{\rm bd}(0) = t_0 = 0$, $P_1 = P_4$ so that the boundary part Γ_0 becomes one point. Then condition* (4.5.6) *becomes the one-point Dirichlet condition. We will consider such problems in* §4.8.

We start with the L^∞–estimates. One important point for our applications to the degenerate elliptic equations is that these estimates are independent of parameter δ in (4.5.90). Another important point is that these estimates are independent of $t_0 \ge 0$ in (4.5.1).

Lemma 4.5.8. *Let $\lambda \in (0,1)$ and $M < \infty$. Let Ω be a domain of structure* (4.5.1)–(4.5.3) *with $h, t_1, t_2, M_{\rm bd} > 0$ and $t_0 \ge 0$. Assume that $h \in (\lambda, \frac{1}{\lambda})$. Let $\varepsilon \in (0, \frac{h}{2})$. Let equation* (4.5.84) *be strictly elliptic in $\overline{\Omega} \setminus \overline{\Gamma_0}$, and let*

$$(A_{ij}, A_i) \in C^1(\mathbb{R}^2 \times \Omega) \ \text{ with } \sup_{(\mathbf{p},\mathbf{x}) \in \mathbb{R}^2 \times \Omega} |(A_{11}, A_1)(\mathbf{p}, \mathbf{x})| \le M, \tag{4.5.112}$$

$$\frac{A_{11}(\mathbf{p}, \mathbf{x})}{x_1} \ge \lambda \qquad \text{for all } \mathbf{x} = (x_1, x_2) \in \Omega, \ \mathbf{p} \in \mathbb{R}^2, \tag{4.5.113}$$

and (4.5.97) *hold.*

Let the boundary condition (4.5.85) *satisfy $B \in C^1(\mathbb{R}^2 \times (\overline{\Gamma_1} \setminus \{P_1\}))$ with*

$$\|B(\mathbf{0}, 0, \cdot)\|_{0,\Gamma_1} + \|B_{(p_1,z)}\|_{0,\mathbb{R}^2 \times \mathbb{R} \times \Gamma_1} \le M, \tag{4.5.114}$$

the strict obliqueness:

$$D_{\mathbf{p}}B(\mathbf{p},z,\mathbf{x})\cdot\boldsymbol{\nu}^{(1)}>0 \qquad \text{for all } (\mathbf{p},z,\mathbf{x})\in\mathbb{R}^2\times\mathbb{R}\times(\overline{\Gamma}_1\setminus\{P_1\}), \tag{4.5.115}$$

and (4.5.101)–(4.5.102). *Let coefficients* $\mathbf{b}^{(k)}$ *of the boundary conditions on* $\Gamma_2\cup\Gamma_3$ *satisfy*

$$\mathbf{b}^{(2)}\in C(\overline{\Gamma}_2),\quad \mathbf{b}^{(3)}\in C(\overline{\Gamma}_3\setminus\{P_4\}),\quad \|\mathbf{b}^{(2)}\|_{0,\Gamma_2}\le M, \tag{4.5.116}$$

the obliqueness properties:

$$\mathbf{b}^{(2)}\cdot\boldsymbol{\nu}^{(2)}\ge\lambda \quad \text{on } \overline{\Gamma}_2, \qquad \mathbf{b}^{(3)}\cdot\boldsymbol{\nu}^{(3)}>0 \quad \text{on } \overline{\Gamma}_3\setminus\{P_4\}, \tag{4.5.117}$$

and (4.5.106). *Also, we assume that* λ*-obliqueness* (4.5.108) *holds at* $P_2\cup P_3$.

Let $f\in C(\overline{\Omega})$ *and* $g_k\in C(\overline{\Gamma_k})$, $k=2,3$, *with*

$$\|f\|_{0,\Omega}+\|g_2\|_{0,\Gamma_2}+\|g_3\|_{0,\Gamma_3}\le M. \tag{4.5.118}$$

Let $u\in C^2(\Omega)\cap C^1(\overline{\Omega}\setminus\overline{\Gamma_0})\cap C(\overline{\Omega})$ *be a solution of* (4.5.84)–(4.5.88), *in which equation* (4.5.84) *and the boundary condition* (4.5.87) *are now nonhomogeneous with the right-hand sides* f *and* g_3, *respectively. Then there exists* C *depending only on* (λ,M,ε) *so that*

$$\|u\|_{0,\overline{\Omega}}\le C, \tag{4.5.119}$$

$$|u(\mathbf{x})|\le Cx_1 \qquad \text{in } \Omega. \tag{4.5.120}$$

Proof. It suffices to prove (4.5.120), since it implies (4.5.119). Constants C and C_1 in this proof are positive and depend only on (λ,M,ε).

Consider equation (4.5.84) and the boundary condition (4.5.85) as a linear equation and boundary condition:

$$\mathcal{L}u=\sum_{i,j=1}^{2}a_{ij}(\mathbf{x})D_{ij}u+\sum_{i=1}^{2}a_i(\mathbf{x})D_iu=f \qquad \text{in } \Omega, \tag{4.5.121}$$

$$\hat{\mathcal{B}}^{(1)}u:=\hat{\mathbf{b}}^{(1)}(\mathbf{x})\cdot Du+\hat{b}_0^{(1)}(\mathbf{x})u=\hat{g}_1 \qquad \text{on } \Gamma_1, \tag{4.5.122}$$

where

$$\begin{aligned}
&(a_{ij},\,a_i)(\mathbf{x})=(A_{ij},\,A_i)(Du(\mathbf{x}),\mathbf{x}) \qquad \text{for } i,j=1,2,\\
&(\hat{\mathbf{b}}^{(1)},\hat{b}_0^{(1)})(\mathbf{x})=\int_0^1 D_{(\mathbf{p},z)}B(tDu(\mathbf{x}),tu(\mathbf{x}),\mathbf{x})dt,\\
&\hat{g}_1(\mathbf{x})=B(\mathbf{0},0,\mathbf{x}).
\end{aligned} \tag{4.5.123}$$

Then our assumptions imply that a_{ij} satisfy the strict ellipticity condition and the bounds:

$$\begin{aligned}
&\lambda x_1\le a_{11}(\mathbf{x})\le M && \text{for all } \mathbf{x}\in\Omega,\\
&|a_1(\mathbf{x})|\le M && \text{for all } \mathbf{x}\in\Omega,\\
&a_1(\mathbf{x})\le-\lambda && \text{for all } \mathbf{x}\in\Omega\cap\{0<x_1<\tfrac{\varepsilon}{2}\},
\end{aligned} \tag{4.5.124}$$

where the last property follows from (4.5.97). Also, from (4.5.101)–(4.5.102) and (4.5.114)–(4.5.115), $(\hat{\mathbf{b}}^{(1)}, \hat{b}_0^{(1)}, \hat{g}_1)$ satisfy the strict obliqueness:

$$\hat{\mathbf{b}}^{(1)} \cdot \boldsymbol{\nu}^{(1)} > 0 \qquad \text{on } \Gamma_1$$

and

$$\|(\hat{\mathbf{b}}^{(1)}, \hat{g}_1)\|_{L^\infty(\Gamma_1)} \le M, \qquad \hat{b}_1^{(1)} \le -\lambda, \;\; \hat{b}_0^{(1)} \le -\lambda \;\text{ on } \Gamma_1. \tag{4.5.125}$$

We use the comparison function:

$$v(\mathbf{x}) = w(x_1),$$

where $w(x_1)$ with

$$w' > 0, \quad w'' < 0 \qquad \text{on } (0, h) \tag{4.5.126}$$

will be determined below in such a way that $v(\mathbf{x})$ is a supersolution of equation (4.5.121) with the boundary conditions (4.5.86)–(4.5.88) and (4.5.122).

First we consider equation (4.5.121). We employ (4.5.124) and (4.5.126) to compute that

$$\mathcal{L}v - f = a_{11}w'' + a_1 w' - f \le \lambda x_1 w'' + Mw' + M. \tag{4.5.127}$$

Now we work in $\Omega \cap \{x_1 \ge \frac{\varepsilon}{2}\}$ and $\Omega \cap \{x_1 < \frac{\varepsilon}{2}\}$, separately.

We first consider $\Omega \cap \{x_1 \ge \frac{\varepsilon}{2}\}$. From (4.5.126)–(4.5.127), we obtain that $\mathcal{L}v \le f$ in $\Omega \cap \{x_1 \ge \frac{\varepsilon}{2}\}$ if

$$\lambda\varepsilon w'' + Mw' + M = 0.$$

Solving this equation with condition: $w(0) = 0$, we have

$$w(x_1) = -x_1 + C_1(1 - e^{-\hat{M}x_1}), \quad w'(x_1) = -1 + \hat{M}C_1 e^{-\hat{M}x_1}, \tag{4.5.128}$$

where $\hat{M} = \frac{M}{\lambda\varepsilon} > 0$, and C_1 is an arbitrary constant, which will be chosen below. We define $w(x_1)$ by (4.5.128) on the whole interval $(0, h)$. Then, from the expression of $w'(\cdot)$ and its derivative, it follows that (4.5.126) holds if $C_1 = C_1(\lambda, M, \varepsilon)$ is sufficiently large, where we recall that $h \in (\lambda, \frac{1}{\lambda})$. This justifies the estimates above.

We now show that $\mathcal{L}v - f \le 0$ on $\Omega \cap \{x_1 < \frac{\varepsilon}{2}\}$, when C_1 is large. On $\Omega \cap \{x_1 < \frac{\varepsilon}{2}\}$, $a_1 \le -\lambda$ by (4.5.124). Thus, using that $a_{11} > 0$ by (4.5.124), and employing (4.5.126) and the fact that $x_1 \in (0, \frac{\varepsilon}{2})$ in the present case, we find from (4.5.124) and (4.5.127)–(4.5.128) that

$$\mathcal{L}v - f \le a_1 w' - f \le -\lambda w' + M < 0 \qquad \text{on } \Omega \cap \{0 < x_1 < \frac{\varepsilon}{2}\},$$

if $C_1 > 0$ is sufficiently large. Therefore, $\mathcal{L}v - f < 0$ in Ω.

Next consider (4.5.86). Note that $\boldsymbol{\nu}^{(2)} = -(1,0)$ so that (4.5.117) implies that $b_1^{(2)} \leq -\lambda$. Then, using (4.5.126), we have

$$\mathcal{B}^{(2)}v := \mathbf{b}^{(2)} \cdot Dv = b_1^{(2)} w'(x_1) \leq -\lambda w'(h).$$

Thus, $\mathcal{B}^{(2)}v \leq g_2$ on Γ_2 if

$$w'(h) \geq \lambda^{-1} \|g_2\|_{L^\infty}.$$

By (4.5.128), the last inequality holds if C_1 is sufficiently large.

Furthermore, from (4.5.126) and (4.5.128), $w(0) = 0$ and $w' > 0$ on $(0,h)$ so that

$$v(\mathbf{x}) = w(x_1) \geq 0 \qquad \text{in } \Omega.$$

Next we show that v is a supersolution of (4.5.87). We use (4.5.106) and (4.5.126) to compute:

$$\mathcal{B}^{(3)}v := \mathbf{b}^{(3)} \cdot Dv = b_1^{(3)} w'(x_1) \leq 0 \qquad \text{on } \Gamma_3.$$

Finally, consider the boundary condition (4.5.122) on Γ_1: From (4.5.122),

$$\hat{\mathcal{B}}^{(1)}v(\mathbf{x}) = \hat{b}_1^{(1)}(\mathbf{x}) w'(x_1) + \hat{b}_0^{(1)}(\mathbf{x}) w(x_1) \leq \hat{b}_1^{(1)}(\mathbf{x}) w'(x_1) \qquad \text{for } \mathbf{x} \in \Gamma_1,$$

where we have used (4.5.125) and $w \geq 0$. Then, using (4.5.114), (4.5.123), and (4.5.128), and choosing C_1 large, we have

$$\hat{\mathcal{B}}^{(1)}v(\mathbf{x}) \leq \hat{b}_1^{(1)}(\mathbf{x}) w'(x_1) \leq -M \leq \hat{g}_1(\mathbf{x}) \qquad \text{on } \Gamma_1.$$

Thus, $\mathcal{L}v \leq f$ in Ω, $\hat{\mathcal{B}}^{(1)}v \leq \hat{g}_1$ on Γ_1, $\mathcal{B}^{(i)}v \leq g_i$ on Γ_i for $i = 2,3$, and $v \geq 0$ on Γ_0. Also, u satisfies the same equation and boundary conditions with equalities. Now we can apply the comparison principle in Lemma 4.4.2 to this problem. Indeed, we use the assumptions regarding the strict ellipticity of the equation and the strict obliqueness of the boundary conditions on Γ_i, $i = 1, \dots, 3$. Therefore, we only need to check the obliqueness at the corner points P_2 and P_3. The obliqueness at P_3 follows directly from assumption (4.5.108). Also, by (4.5.108), the λ-obliqueness at corner P_2 is satisfied for the nonlinear boundary condition (4.5.85) on Γ_1 and (4.5.86) on Γ_2. From Definition 4.4.3 and (4.5.123), the λ-obliqueness at P_2 is still satisfied when the nonlinear condition (4.5.85) is replaced by the linear condition (4.5.122). Then, by Lemma 4.4.2 applied to $v - u$, we conclude that $u \leq v$ in Ω. A similar argument shows that $-u \leq v$ in Ω. Therefore, we have

$$|u(\mathbf{x})| \leq v(\mathbf{x}) = w(x_1) \qquad \text{in } \Omega.$$

From (4.5.126) and (4.5.128),

$$0 = w(0) \leq w(x_1) \leq \|w'\|_{L^\infty([0,h])} x_1 = (\hat{M}C_1 - 1)x_1 \qquad \text{for } x_1 \in [0,h).$$

Then

$$|u(\mathbf{x})| \leq (\hat{M}C_1 - 1)x_1 \qquad \text{in } \Omega.$$

Now (4.5.120) is proved. □

We will also need the following version of Lemma 4.5.8.

Lemma 4.5.9. *In the conditions of Lemma* 4.5.8, *assumptions* (4.5.97) *and* (4.5.113) *are replaced by the following assumption:*

$$A_{11}(\mathbf{p}, \mathbf{x}) \geq \lambda \qquad \textit{for all } \mathbf{x} \in \Omega, \ \mathbf{p} \in \mathbb{R}^2. \tag{4.5.129}$$

Then the assertion of Lemma 4.5.8 *still holds.*

Proof. We argue similar to the proof of Lemma 4.5.8, with the changes outlined below. We prove only (4.5.120), since that implies (4.5.119).

We rewrite equation (4.5.84) and the boundary condition (4.5.85) as a linear equation (4.5.121) and boundary condition (4.5.122). Then (4.5.124) is now replaced by

$$\lambda \leq a_{11}(\mathbf{x}) \leq M, \quad |a_1(\mathbf{x})| \leq M \qquad \text{for all } \mathbf{x} \in \Omega. \tag{4.5.130}$$

The properties of coefficients $\hat{\mathbf{b}}^{(1)}$ of (4.5.122) are the same as in the proof of Lemma 4.5.8.

We use the comparison function $v(\mathbf{x}) = w(x_1)$. In order to find $w(\cdot)$ so that $v(\mathbf{x})$ is a supersolution of equation (4.5.121), we require w to satisfy (4.5.126) and follow the calculation in (4.5.127). In the present case, using (4.5.130), we have

$$\mathcal{L}v - f = a_{11}w'' + a_1 w' - f \leq \lambda w'' + Mw' + M \qquad \text{in } \Omega.$$

Then it suffices to determine w by solving the equation:

$$\lambda w'' + Mw' + M = 0 \qquad \text{on } (0, h)$$

with condition $w(0) = 0$. It has a solution $w(x_1)$ as in (4.5.128) with $\varepsilon = 1$ for an arbitrary constant C_1. Thus, (4.5.126) holds if $C_1 > 0$ is sufficiently large. This justifies our estimates above, which implies that $v(\mathbf{x}) = w(x_1)$ is a supersolution of equation (4.5.121) in Ω.

Then, repeating the argument for Lemma 4.5.8, we show that $v(\cdot)$ is a supersolution of the boundary conditions if constant C_1 is chosen appropriately large, which leads to (4.5.120). □

Now we prove the *a priori* estimates of solutions of (4.5.84)–(4.5.88). We first estimate a solution away from Γ_0.

Lemma 4.5.10. *Let* $\kappa > 0$, $\lambda > 0$, $M < \infty$, $\alpha \in (0, 1)$, $\beta \in [\frac{1}{2}, 1)$, *and* $s \in (0, \frac{\lambda}{10})$. *Then there exist* $\alpha_1 \in (0, \frac{1}{2})$ *depending only on* (κ, λ) *and* $\sigma > 0$ *depending only on* $(\kappa, \lambda, M, \alpha, \beta, s)$ *such that the following holds:*

Let Ω *be a domain of structure* (4.5.1)–(4.5.3) *with* $h, t_1, t_2 \in [\lambda, \frac{1}{\lambda}]$, $M_{\text{bd}} \leq \frac{1}{\lambda}$, *and* $t_0 \geq 0$. *Let* f_{bd} *satisfy* (4.5.15), *and let* $\varepsilon \in (0, \frac{h}{10})$. *Let* (4.5.89), (4.5.91)–(4.5.93), (4.5.98)–(4.5.100), (4.5.103)–(4.5.105), *and* (4.5.107)–(4.5.110) *hold with the constants fixed above, including* $\sigma = \sigma(s)$.

Let $u \in C^2(\Omega) \cap C^1(\overline{\Omega} \setminus \overline{\Gamma_0}) \cap C(\overline{\Omega})$ *satisfy* (4.5.84)–(4.5.87) *and*

$$\|u\|_{L^\infty(\Omega)} \le K.$$

Then

$$\|u\|^{(-1-\alpha_1),\,\overline{\Gamma_2}}_{2,\alpha_1,\Omega^{(s)}} \le C_s, \tag{4.5.131}$$

where $\Omega^{(s)} := \Omega \cap \{x_1 > s\}$, *and* C_s *depends only on* $(\kappa, \lambda, M, K, \alpha, \beta, s)$.

Proof. In the proof below, all the following constants C, C_k, and C_s depend only on $(\kappa, \lambda, M, K, \alpha, \beta, s)$. The proof is divided into two steps.

1. Fix $s \in (0, \frac{\lambda}{10})$. Since $\mathrm{dist}(\mathbf{x}, \Gamma_0) \ge \frac{s}{4}$ for any $\mathbf{x} \in \Omega^{(s/4)}$, then, from (4.5.89), equation (4.5.84) is uniformly elliptic in $\Omega^{(s/4)}$. That is, for any $\mathbf{x} \in \Omega^{(s/4)}$ and $\mathbf{p}, \boldsymbol{\mu} \in \mathbb{R}^2$,

$$\frac{s}{4}\lambda|\boldsymbol{\mu}|^2 \le \sum_{i,j=1}^{2} A_{ij}(\mathbf{p}, \mathbf{x})\mu_i\mu_j \le \lambda^{-1}|\boldsymbol{\mu}|^2. \tag{4.5.132}$$

To do further estimates, we note that, from (4.5.1) with $t_0 \ge 0$ and $t_1, t_2 \in [\lambda, \frac{1}{\lambda}]$,

$$f_{\mathrm{bd}} \ge l_s := \min\{t_1 s,\, t_2\} \ge \lambda \min\{s, 1\} > 0 \qquad \text{on } [s, h]. \tag{4.5.133}$$

We note that l_s is independent of t_0.

From (4.5.133), for $\rho := \frac{1}{4}\min\{\lambda, s\} \le \frac{1}{4}\min\{t_1 s,\, t_2, s\}$ and $k = 1, 3$, we have

$$B_\rho(\mathbf{x}) \cap \partial\Omega = B_\rho(\mathbf{x}) \cap (\Gamma_k \cup \Gamma_2) \qquad \text{for all } \mathbf{x} \in \Gamma_k \cap \partial\Omega^{(s)}.$$

From $h \in [\lambda, \frac{1}{\lambda}]$, we find that $\rho \le \frac{h}{4}$. Also, from (4.5.1), we see that, for any $R < \rho$, $\Omega \cap B_R(P_2)$ is of structure (4.2.115)–(4.2.116) with $\Gamma_r^{(n)} = \Gamma_1 \cap B_R(P_2)$ and $\Gamma_r^{(l)} = \Gamma_2 \cap B_R(P_2)$. Then, by (4.5.15), there exists $R_0 \in (0, \rho]$ depending only on (λ, M, α, s) such that (4.2.117) holds. Using (4.5.15), we find that, for $R_1 = \frac{R_0}{8\sqrt{1+M^2}}$ and $k = 1, 2, 3$,

$$B_{R_1}(\mathbf{x}) \cap \partial\Omega = B_{R_1}(\mathbf{x}) \cap \Gamma_k \tag{4.5.134}$$

for $\mathbf{x} \in (\Gamma_k \cap \{x_1 \ge \frac{s}{4}\}) \setminus (B_{R_0/8}(P_2) \cup B_{R_0/8}(P_3))$, where we have used that $R_1 \le \frac{\rho}{8} \le \frac{s}{32}$.

Also, combining (4.5.91) with (4.5.92), we have

$$\|(A_{ij}, A_i)(\mathbf{p}, \cdot)\|^{(-\alpha),\Gamma_2}_{0,\beta,\Omega} + \|D_{\mathbf{p}}(A_{ij}, A_i)(\mathbf{p}, \cdot)\|_{L^\infty(\Omega)} \le C, \tag{4.5.135}$$

and $\beta \in [\frac{1}{2}, 1)$.

We now make the estimates of u in Ω_s, in which equation (4.5.84) is uniformly elliptic by (4.5.132). We use R_0 and R_1 defined above and, in the argument below, we further reduce R_0 depending only on (λ, M, κ), keeping $R_1 = \frac{R_0}{8\sqrt{1+M^2}}$, so that (4.5.134) holds.

(a) We apply the interior estimates in Theorem 4.2.1, rescaled into the balls:

$$B_{R_1/100}(\mathbf{x}) \subset \Omega \qquad \text{with } \mathbf{x} \in \Omega^{(s)}.$$

We then obtain that $\|u\|_{C^{2,\beta}(\overline{B_{R_1/200}(\mathbf{x})})} \leq C$, where we have used (4.5.135). Thus, we estimate u in the interior of $\Omega^{(s)}$ and near $\partial\Omega^{(s)} \cap \{x_1 = s\}$, away from Γ_1 and Γ_3.

(b) We employ estimate (4.2.131) in Lemma 4.2.16, rescaled into half-balls $B_{R_1/10}(\mathbf{x}) \cap \Omega$ with $\mathbf{x} \in \Gamma_2 \setminus (B_{R_0}(P_2) \cup B_{R_0}(P_3))$ for the estimates near Γ_2 away from $P_2 \cup P_3$, where we have used the fact that the equation near Γ_2 and the boundary condition on Γ_2 are linear, with the regularity of the coefficients given in (4.5.91) and (4.5.107). Then we obtain that $\|u\|^{(-1-\gamma),\,\Gamma_2}_{2,\gamma,B_{R_1/20}(\mathbf{x})\cap\Omega} \leq C_\gamma$ for any $\gamma \in (0,1)$.

(c) We apply estimate (4.2.77) in Theorem 4.2.10, rescaled into the half-balls:

$$B_{R_1/10}(\mathbf{x}) \cap \Omega \qquad \text{with } \mathbf{x} \in \Gamma_3 \cap \{s < x_1 < 1 - s\},$$

with the regularity of coefficients in (4.5.135), to obtain

$$\|u\|_{C^{2,\beta}(\overline{B_{R_1/20}(\mathbf{x})\cap\Omega})} \leq C(M, \lambda, \beta, s).$$

Thus, we obtain the estimates on $\Gamma_3 \cap \partial\Omega^{(s)}$ away from P_3.

(d) Now we obtain the estimates near corners $\{P_2, P_3\}$, *i.e.*, in $B_{R_0}(P_2) \cap \Omega$ and $B_{R_0}(P_3) \cap \Omega$, by reducing R_0 depending only on (λ, M, κ), since $h \in (\lambda, \frac{1}{\lambda})$. We recall that regions $B_{R_0}(P_2) \cap \Omega$ and $B_{R_0}(P_3) \cap \Omega$ are of the structure of $\mathcal{R}^{+,\Phi}_{R_0}$ in (4.2.115) with $\Phi(x_1) = f_{\text{bd}}(h - x_1)$ for P_2, and $\Phi \equiv 0$ for P_3. Therefore, in both cases,

$$\|\Phi\|^{(-1-\alpha),\{0\}}_{2,\alpha,(0,R_0)} \leq M,$$

where we have used (4.5.15) near P_2.

We also note that, since $R_0 < \rho \leq \frac{h}{4}$, $B_{R_0}(P_k) \subset \{x_1 \geq \frac{h}{2}\}$ for $k = 2, 3$, so that, by (4.5.89), equation (4.5.84) is uniformly elliptic in $B_{R_0}(P_2) \cap \Omega$ and $B_{R_0}(P_3) \cap \Omega$ with constant $\tilde{\lambda} := \frac{\lambda}{2} \min\{h, 1\}$. The assumption that $h \in (\lambda, \frac{1}{\lambda})$ implies that $\tilde{\lambda}$ depends only on λ.

With this, we will perform the estimates in two steps. First, since we do not have the Lipschitz bound of u near the corners, we apply Theorem 4.2.15. Then we obtain the C^{1,α_2}–estimate near the corners, with α_2 depending only on $(\kappa, \lambda, \alpha)$, where the dependence on α is from the requirement $\alpha_2 \in (0, \alpha)$. However, we show that the Hölder exponent in the corner can be estimated in terms of (κ, λ) only, but independent of α. Thus, we refine the previous estimate as follows: Using the Lipschitz bound provided by that estimate, we can apply Theorem 4.3.15 to obtain the C^{1,α_1}–estimate of u near the corners with α_1 depending only on (κ, λ). Now we prove these estimates:

(i) In this sub-step, we obtain the Lipschitz estimate of u in $B_{R_0}(P_k)\cap\Omega$, $k = 2, 3$. We apply Theorem 4.2.15, with Γ_2 corresponding to $\Gamma^{(l)}$ in Theorem 4.2.15 for both corners $\{P_2, P_3\}$.

Our assumptions now imply that all the conditions of Theorem 4.2.15 are satisfied, where we only need to note that (4.5.98) and the second estimate in (4.5.104) with $\sigma \le \frac{\lambda}{2}$ imply that $\mathbf{b}^{(1)}$ satisfies (4.2.125) with constant $\tilde{\lambda}$. Indeed, $\frac{\lambda}{2} \le \mathbf{b}^{(1)} \cdot \boldsymbol{\nu}_{\Gamma_2} \le (\frac{\lambda}{2})^{-1}$ on Γ_2, and $\tilde{\lambda} \le \frac{\lambda}{2}$. Let $\hat{\alpha}_1 := \alpha_1(\tilde{\lambda}, \kappa)$ determined in Theorem 4.2.15 with κ from (4.5.109), and let $\alpha_2 := \frac{1}{2}\min\{\hat{\alpha}_1, \alpha\}$. Then, further reducing σ depending only on $(\kappa, \lambda, M, \alpha)$, we obtain from Theorem 4.2.15 that $\|u\|^{(-1-\alpha_2),\,\overline{\Gamma_2}}_{2,\alpha_2,B_{R_0/2}(P_k)\cap\Omega} \le \hat{C}$ for $k = 2, 3$, where $\hat{C} = \hat{C}(\kappa, \lambda, M, K, \alpha, R_0) = \hat{C}(\kappa, \lambda, M, K, \alpha, s)$. In particular, $\|u\|_{1,0,B_{R_0/2}(P_k)\cap\Omega} \le \hat{C}$.

(ii) We now apply Theorem 4.3.15 in $B_{R_0/2}(P_k) \cap \Omega$. As we have discussed above, domains $B_{R_0}(P_k) \cap \Omega$, for $k = 2, 3$, are of the structure of $\mathcal{R}^{+,\Phi}_{R_0}$ in (4.2.115) with $\|\Phi\|_{1,\alpha,(0,R_0)} \le M$. The estimate of the solution in $C^1(B_{R_0/2}(P_k) \cap \Omega)$ is obtained in (a). The assumptions of Theorem 4.3.15 on the equations and boundary conditions are satisfied as described below.

For equation (4.5.84) in $B_{R_0/2}(P_k) \cap \Omega$, $k = 2, 3$, the ellipticity with constant $\tilde{\lambda} > 0$ depending only on λ is shown above. The regularity estimate (4.3.92) holds by (4.5.91), where we also note that the equation is linear in $B_{R_0/2}(P_k) \cap \Omega$ so that $D^2_{(\mathbf{p},z)}(A_{ij}, A) = 0$.

For the boundary conditions, the obliqueness condition (4.3.106) holds with constant λ by (4.5.98) and (4.5.105). Regularity (4.3.93) holds by (4.5.99) and (4.5.107). Condition (4.3.88) is satisfied in $B_{R_0/2}(P_k)\cap\Omega$, $k = 2, 3$, with constant $\frac{\lambda}{2}$ after further reducing R_0 depending only on (λ, α, M). Indeed, for $B(Du, u, \mathbf{x})$, this follows from (4.5.98)–(4.5.99), while, for $G_k(Du, \mathbf{x}) := \mathbf{b}^{(k)}(\mathbf{x}) \cdot Du$, $k = 2, 3$, this follows from (4.5.105) and (4.5.107).

Furthermore, the functional independence (4.3.107) holds in $B_{R_0/2}(P_2) \cap \Omega$ with constant $\frac{\kappa}{2}$ by (4.5.109) with $k = 2$, combined with (4.5.104) for sufficiently small σ depending only on (κ, λ), and by reducing R_0 depending only on (λ, M, κ), where we have used (4.5.99) and (4.5.107). Similarly, (4.3.107) holds in $B_{R_0/2}(P_3) \cap \Omega$ with constant $\frac{\kappa}{2}$ by (4.5.109) for $k = 3$, by using (4.5.107) for small R_0.

Finally, (4.3.110) holds with constant λ by the first estimate in (4.5.91), and (4.3.111) holds with constant λ for $B(\cdot)$ by (4.5.100) and the conditions: $\mathbf{b}^{(k)}(\mathbf{x}) \cdot Du$, $k = 2, 3$, by (4.5.107).

We note that, from the discussion above, the conditions of Theorem 4.3.15 with the bounds depending only on λ are now satisfied with the constants depending only on parameters (λ, κ). Thus, applying Theorem 4.3.15 and recalling that Γ_1 is a $C^{1,\alpha}$–curve by (4.5.15), and Γ_2 and Γ_3 are flat, there exists $\hat{\alpha} \in (0,1)$ depending only on (λ, κ) such that

$$\|u\|^{(-1-\hat{\alpha}),\,\overline{\Gamma_2}}_{2,\alpha,B_{R_0/4}(P_k)\cap\Omega} \le \hat{C} \qquad \text{for } k = 2,3,$$

where $\hat{C} = \hat{C}(\lambda, \kappa, M, K, \alpha, R_0) = \hat{C}(\lambda, \kappa, M, K, \alpha, s)$. Then, using the higher regularity of Γ_1 in (4.5.15), recalling that Γ_2 and Γ_3 are flat, and employing the argument in Step 4 of the proof of Theorem 4.2.15 by using Theorem 4.3.4 instead of Theorem 4.2.12, we find that, for any $\gamma \in (0,1)$,

$$\|u\|^{(-1-\hat{\alpha}),\,\overline{\Gamma_2}}_{2,\gamma,B_{R_0/4}(P_k)\cap\Omega} \le \hat{C} \quad \text{for } k = 2,3, \text{ with } \hat{C} = \hat{C}(\lambda, \kappa, M, K, \alpha, s, \gamma).$$

We now fix α_1 in the lemma to be $\hat{\alpha}$ from the last estimate.

Therefore, combining the estimates obtained above and noting that $\alpha_1 \le \frac{1}{2} \le \beta$, we see that there exists α_1 depending only on (λ, κ) such that

$$\|u\|^{(-1-\alpha_1),\,\overline{\Gamma_2}}_{2,\alpha_1,\Omega^{(s)}\setminus\mathcal{N}_{R_1/100}(\Gamma_1\setminus B_{R_0/4}(P_2))} \le C_s, \tag{4.5.136}$$

where $\mathcal{N}_r(\cdot)$ denotes the r–neighborhood.

Remark 4.5.11. *Note that, in the proof of* (4.5.136)*, we have introduced a smallness requirement for σ in the argument of the estimate near P_2, but at this point, σ depends only on (κ, λ), i.e., is independent of s.*

It remains to obtain the estimates near $\Gamma_1 \cap \partial\Omega^{(s)}$ away from P_2, *i.e.*, near $(\Gamma_1 \cap \{x_1 \ge s\}) \setminus B_{R_0/4}(P_2)$, where we recall (4.5.134). To do this, we will introduce a further smallness requirement for σ, which will depend on s, among the other parameters.

2. We now estimate u in $\mathcal{N}_{R_1/10}((\Gamma_1 \cap \{x_1 \ge s\}) \setminus B_{R_0/4}(P_2)) \cap \overline{\Omega}$. Since the boundary condition on Γ_1 is oblique and is of structure (4.5.104), we intend to apply Theorem 4.2.12 with Remark 4.2.13, followed by Theorem 4.3.4, similar to the estimates near corners $\{P_2, P_3\}$ in Step 1(d). For the application of Theorem 4.2.12, we then need to choose σ in (4.5.104) sufficiently small, depending especially on the ellipticity of the equation, which degenerates near Γ_0, *i.e.*, for small x_1. This will introduce the dependence of σ on s.

We use R_0 and R_1 chosen above so that (4.5.134) holds. Fix

$$\mathbf{x} = (x_1, f_{\mathrm{bd}}(x_1)) \in (\Gamma_1 \cap \{x_1 \ge s\}) \setminus B_{R_0/4}(P_2).$$

Note that, by (4.5.15) and $R_1 = \frac{R_0}{8\sqrt{1+M^2}}$, it follows that

$$(\Gamma_1 \cap \{x_1 \ge s\}) \setminus B_{R_0/4}(P_2) \subset \Gamma_1 \cap \{h - 2R_1 \ge x_1 \ge s\}.$$

Also, by (4.5.15), $h \in (\lambda, \frac{1}{\lambda})$, and $M_{\text{bd}} \le \frac{1}{\lambda}$,

$$\|f_{\text{bd}}\|_{2,\alpha,(s/2,h-R_1)} \le M_s, \tag{4.5.137}$$

where M_s depends only on $(M, \lambda, \alpha, s, R_1)$, and hence on $(M, \lambda, \kappa, \alpha, s)$. We also recall that $R_1 \le \frac{\rho}{8} \le \frac{s}{32}$. Then domain $B_{R_1}(\mathbf{x}) \cap \Omega$ satisfies the conditions of Theorem 4.2.12 in the appropriately rotated coordinate system. Using the uniform ellipticity (4.5.132) with the constant depending on s, obliqueness (4.5.98), regularity (4.5.91)–(4.5.92) and (4.5.99), and structure (4.5.103)–(4.5.104), we can choose σ in (4.5.104) depending only on $(\kappa, \lambda, M, K, \alpha, s)$ such that Theorem 4.2.12 with Remark 4.2.13 can be applied in $B_{R_1}(\mathbf{x})$ for any $\mathbf{x} = (x_1, f_{\text{bd}}(x_1)) \in (\Gamma_1 \cap \{x_1 \ge s\}) \setminus B_{R_0/4}(P_2)$. Thus, we have

$$\|u\|_{C^{2,\alpha}(\overline{B_{R_1/2(\mathbf{x})} \cap \Omega})} \le C \qquad \text{for all } \mathbf{x} \in (\Gamma_1 \cap \{x_1 \ge s\}) \setminus B_{R_0/4}(P_2)$$

with $C = C(\kappa, \lambda, M, K, \alpha, s)$. This implies that $\|u\|_{C^{1,0}(\overline{B_{R_1/2(\mathbf{x})} \cap \Omega})} \le C$. Therefore, we can apply Theorem 4.3.4 in $B_{R_1/2}(\mathbf{x}) \cap \Omega$. Its conditions are satisfied in $B_{R_1/2}(\mathbf{x}) \cap \Omega$ by (4.5.89) and (4.5.93) (since $\mathbf{x} \in (\Gamma_1 \cap \{x_1 \ge s\}) \setminus B_{R_0/4}(P_2)$) and (4.5.98)–(4.5.99), where the last two properties also imply that (4.3.7) holds in $B_{R_1/2}(\mathbf{x}) \cap \Omega$ after reducing R_1 depending only on (λ, M). Moreover, (4.5.137) implies that $\|f_{\text{bd}}\|_{1,\gamma,(s/2,h-R_1)} \le M_s$ for any $\gamma \in (0,1)$, which means that (4.3.34) is satisfied for any Hölder exponent in $(0,1)$. Therefore, Theorem 4.3.4 implies that, for any $\gamma \in (0,1)$,

$$\|u\|_{C^{2,\gamma}(\overline{B_{R_1/4(\mathbf{x})} \cap \Omega})} \le C \qquad \text{for all } \mathbf{x} \in (\Gamma_1 \cap \{x_1 \ge s\}) \setminus B_{R_0/4}(P_2)$$

with $C = C(\kappa, \lambda, M, K, \alpha, s)$. Combining this with (4.5.136), we obtain (4.5.131). □

Now we show the *a priori* estimates for Problem (4.5.84)–(4.5.88). We use the structural conditions near Γ_0 in order to remove the dependence of σ on s, which we have in Lemma 4.5.10.

Lemma 4.5.12. *Let $\kappa > 0$, $\lambda > 0$, $M < \infty$, $\alpha \in (0,1)$, $\beta \in [\frac{1}{2}, 1)$, and $\varepsilon \in (0, \frac{\lambda}{10})$. Then there exist $\alpha_1 \in (0, \frac{1}{2})$ depending only on (κ, λ), and $\sigma, \delta_0 > 0$ depending only on $(\kappa, \lambda, M, \alpha, \beta, \varepsilon)$, such that the following holds:*

Let Ω be a domain of structure (4.5.1)–(4.5.3) with $h, t_1, t_2 \in (\lambda, \frac{1}{\lambda})$, $M_{\text{bd}} \le \frac{1}{\lambda}$, and $t_0 \ge 0$. Let f_{bd} satisfy (4.5.15), and $\delta \in (0, \delta_0)$. Let (4.5.89)–(4.5.111) hold. Let $u \in C^2(\Omega) \cap C^1(\overline{\Omega} \setminus \overline{\Gamma_0}) \cap C(\overline{\Omega})$ be a solution of (4.5.84)–(4.5.88). Then u satisfies (4.5.119)–(4.5.120) with constant C depending only on $(\lambda, M, \varepsilon)$ and, for each $s \in (0, \frac{h}{10})$, there exists C_s depending only on $(\kappa, \lambda, M, \alpha, \beta, \varepsilon, s)$, but independent of δ, so that

$$\|u\|^{(-1-\alpha_1),\,\overline{\Gamma_2}}_{2,\alpha_1,\Omega^{(s)}} \le C_s, \tag{4.5.138}$$

where $\Omega^{(s)} := \Omega \cap \{x_1 > s\}$. Moreover, if $t_0 > 0$ in addition to the previous assumptions, then

$$\|u\|^{(-1-\alpha_1),\,\{P_1\}\cup\overline{\Gamma_2}}_{2,\alpha_1,\Omega} \le \hat{C}_\delta, \tag{4.5.139}$$

where $\hat{C}_\delta$ *depends only on* $(\kappa, \lambda, M, \alpha, \beta, \varepsilon, \delta, t_0)$.

Proof. We first note that our assumptions allow us to apply Lemma 4.5.8, so that u satisfies the L^∞–estimates (4.5.119)–(4.5.120) with constant C depending only on $(\lambda, M, \varepsilon)$. Then it remains to show the regularity. We divide the proof into four steps.

In the proof, we assume without loss of generality that $A_{12} = A_{21}$. Otherwise, we replace A_{12} and A_{21} by $\frac{1}{2}(A_{12} + A_{21})$ so that the modified A_{ij} satisfy the same assumptions and u is a solution of the modified equation.

1. We first prove (4.5.138). In this step, constants C, C_k, and C_s depend only on $(\kappa, \lambda, M, \alpha, \beta, s)$, but are independent of δ. Also, from the conditions, $\varepsilon < \frac{\lambda}{10} \le \frac{1}{10}\min\{h, 1\}$.

We note that an estimate similar to (4.5.138) is obtained in Lemma 4.5.10, but now we need σ to be independent of s and constant δ in (4.5.90). As we have discussed in Step 2 of the proof of Lemma 4.5.10, this dependence comes from the application of Theorem 4.2.12 with Remark 4.2.13, since we need then to choose σ in (4.5.104) sufficiently small, depending on the ellipticity of the equation among the other parameters. In the present lemma, the ellipticity is uniform up to Γ_0, but depends on δ, or we can consider the uniform ellipticity in $\Omega^{(s)}$ independent of δ but depending on s. In both cases, we do not obtain the required estimate by employing Theorem 4.2.12. Instead, for small $\varepsilon^* \in (0, \varepsilon)$ to be chosen depending only on the parameters in the assumptions, we use the non-isotropy of the ellipticity condition (4.5.90) and property (4.5.102) to apply Theorem 4.2.8 for the estimates near $\Gamma_1 \cap \{x_1 < \varepsilon^*\}$, after the rescaling that makes the equation locally uniformly elliptic, and then the boundary condition becomes almost tangential. On $\Gamma_1 \cap \{x_1 \ge \frac{\varepsilon^*}{2}\}$, we obtain the estimate from Lemma 4.5.10 with $\sigma = \sigma(\frac{\varepsilon^*}{2})$, hence independent of (δ, s). Combining these estimates, we obtain (4.5.138).

More precisely, let $\varepsilon^* \in (0, \varepsilon)$ be chosen depending only on $(\kappa, \lambda, M, \alpha, \beta, \varepsilon)$ below. Let α_1 and σ be determined in Lemma 4.5.10 for $s = \frac{\varepsilon^*}{2}$, where Lemma 4.5.10 is applied with parameters $(\kappa, \lambda, M, \alpha, \beta)$ in this lemma, and with K equal to the right-hand side of the L^∞–estimate (4.5.119) for u, which depends only on $(\lambda, M, \varepsilon)$. Thus, $\alpha_1 = \alpha_1(\kappa, \lambda)$ and $\sigma = \sigma(\kappa, \lambda, M, \alpha, \beta, \varepsilon)$.

If $s \ge \frac{\varepsilon^*}{2}$, then (4.5.138) with C_s depending only on $(\kappa, \lambda, M, \alpha, \beta, \varepsilon)$ follows from Lemma 4.5.10 applied with $s = \frac{\varepsilon^*}{2}$ and the parameters described in the previous paragraph. Therefore, we assume that $s < \varepsilon^*$ from now on.

We follow the argument in Step 1 of the proof of Lemma 4.5.10 by using L^∞–estimate (4.5.119) for u obtained above and noting Remark 4.5.11. Then, for fixed σ above, we obtain estimate (4.5.136) with $R_0 \le \frac{1}{4}\min\{\lambda, s\}$ depending only on (λ, M, κ, s) and satisfying $R_0 \le \frac{1}{4}\min\{t_1 s, t_2, s\}$ such that $R_1 = \frac{R_0}{8\sqrt{1+M^2}}$ satisfies (4.5.134), and with C_s depending only on $(\kappa, \lambda, M, \alpha, \beta, \varepsilon, s)$.

Now it remains to estimate u near $\Gamma_1 \cap \{s < x_1 < \varepsilon^*\}$, *i.e.*, in $\mathcal{N}_{R_1/10}(\Gamma_1 \cap \{s < x_1 < \varepsilon^*\})$. Specifically, we show that

$$\|u\|_{C^{2,\beta}(\overline{B_{R_1/10}(\hat{\mathbf{x}}) \cap \Omega})} \le C(\lambda, M, \beta, s) \tag{4.5.140}$$

for all $\hat{\mathbf{x}} \in \Gamma_1 \cap \{s < x_1 \le \varepsilon^*\}$.

We first note that, using (4.5.102), we can rewrite the boundary condition (4.5.85) on $\Gamma_1 \cap \{\frac{s}{2} < x_1 < \varepsilon\}$ in the form:

$$u_{x_1} = \tilde{B}(u_{x_2}, u, \mathbf{x}) \qquad \text{on } \Gamma_1 \cap \{0 < x_1 < \varepsilon\}, \tag{4.5.141}$$

where

$$\tilde{B}(0, 0, \mathbf{x}) = 0 \qquad \text{for all } \mathbf{x} \in \Gamma_1 \cap \{0 < x_1 < \varepsilon\}, \tag{4.5.142}$$

by (4.5.111). Using (4.5.99) and (4.5.102), we have

$$\|D^k_{(p_2,z)} \tilde{B}(p_2, z, \cdot)\|_{C^3(\Gamma_1 \cap \{0 \le x_1 \le \varepsilon\})} \le C(\lambda) M \tag{4.5.143}$$

for any $(p_2, z) \in \mathbb{R} \times \mathbb{R}$, $k = 1, 2, 3$.

From (4.5.133), in which we can assume without loss of generality that $l_s \le \frac{1}{2}$, we conclude that, if $\varepsilon^* \in (0, 1)$ is small, depending on M in (4.5.15), then, for each $\hat{\mathbf{x}} = (\hat{x}_1, f_{\rm bd}(\hat{x}_1)) \in \Gamma_1 \cap \{s < x_1 < \varepsilon^*\}$,

$$\begin{aligned} (\hat{\mathbf{x}} + Q_s^{\hat{\mathbf{x}}}) \cap \Omega &= \{\mathbf{x} \in \hat{\mathbf{x}} + Q_s^{\hat{\mathbf{x}}} \ : \ x_2 < f_{\rm bd}(x_1)\}, \\ (\hat{\mathbf{x}} + Q_s^{\hat{\mathbf{x}}}) \cap \Gamma_1 &= \{(x_1, f_{\rm bd}(x_1)) \ : \ x_2 = f_{\rm bd}(x_1)\}, \end{aligned} \tag{4.5.144}$$

where $Q_s^{\hat{\mathbf{x}}} := (-l_s \hat{x}_1, l_s \hat{x}_1) \times (-l_s\sqrt{\hat{x}_1}, l_s\sqrt{\hat{x}_1})$. We rescale u in $\hat{\mathbf{x}} + Q_s^{\hat{\mathbf{x}}}$ as follows: Let

$$d^{(\delta)}_{\mathbf{x}} = \max(\delta, x_1) \qquad \text{for } \mathbf{x} = (x_1, x_2) \in \overline{\Omega}. \tag{4.5.145}$$

Define the change of variables:

$$\mathbf{x} = (\hat{x}_1 + d^{(\delta)}_{\hat{\mathbf{x}}} X_1, \ f_{\rm bd}(\hat{x}_1) - \sqrt{d^{(\delta)}_{\hat{\mathbf{x}}}} X_2) \qquad \text{for } \mathbf{X} = (X_1, X_2) \in \mathbb{R}^2,$$

so that

$$\mathbf{x} \in (\hat{\mathbf{x}} + Q_s^{\hat{\mathbf{x}}}) \cap \Omega \quad \text{if and only if} \quad \mathbf{X} = (X_1, X_2) \in \hat{\Omega}^{(\hat{\mathbf{x}})},$$

where

$$\hat{\Omega}^{(\hat{\mathbf{x}})} := \left\{ \mathbf{X} : \begin{array}{l} X_1 \in \big(-\frac{l_s \hat{x}_1}{d^{(\delta)}_{\hat{\mathbf{x}}}}, \frac{l_s \hat{x}_1}{d^{(\delta)}_{\hat{\mathbf{x}}}} \big), \ X_2 \in \Big(-l_s \sqrt{\frac{\hat{x}_1}{d^{(\delta)}_{\hat{\mathbf{x}}}}}, l_s \sqrt{\frac{\hat{x}_1}{d^{(\delta)}_{\hat{\mathbf{x}}}}} \Big) \\ X_2 \ge \sqrt{d^{(\delta)}_{\hat{\mathbf{x}}}} f^{(\hat{\mathbf{x}})}_{\rm bd}(X_1) \end{array} \right\} \tag{4.5.146}$$

with $f^{(\hat{\mathbf{x}})}_{\rm bd}(X_1) = \frac{1}{d^{(\delta)}_{\hat{\mathbf{x}}}} \Big(f_{\rm bd}(\hat{x}_1) - f_{\rm bd}(\hat{x}_1 + d^{(\delta)}_{\hat{\mathbf{x}}} X_1) \Big)$.

Denote $\Gamma_1^{(\hat{\mathbf{x}})} := \partial \Omega^{(\hat{\mathbf{x}})} \cap \{X_2 = f^{(\hat{\mathbf{x}})}_{\rm bd}(X_1)\}$. From (4.5.144),

$$\Gamma_1^{(\hat{\mathbf{x}})} = \Big\{ (X_1, \sqrt{d^{(\delta)}_{\hat{\mathbf{x}}}} f^{(\hat{\mathbf{x}})}_{\rm bd}(X_1)) \ : \ X_1 \in (-\frac{l_s \hat{x}_1}{d^{(\delta)}_{\hat{\mathbf{x}}}}, \frac{l_s \hat{x}_1}{d^{(\delta)}_{\hat{\mathbf{x}}}}) \Big\}, \tag{4.5.147}$$

and

$$\mathbf{x} \in (\hat{\mathbf{x}} + Q_s^{\hat{\mathbf{x}}}) \cap \Gamma_1 \qquad \text{if and only if} \quad \mathbf{X} = (X_1, X_2) \in \Gamma_1^{(\hat{\mathbf{x}})}.$$

From the definition of $f_{\rm bd}^{(\hat{\mathbf{x}})}$ and (4.5.15), we have

$$\|f_{\rm bd}\|^{(-1-\alpha),\,\{h\}}_{1,\beta,(s/2,h)} \le M_s,$$

where M_s depends only on (M, β, s). Then

$$f_{\rm bd}^{(\hat{\mathbf{x}})}(0) = 0, \quad \|f_{\rm bd}^{(\hat{\mathbf{x}})}\|_{C^{1,\beta}(-\frac{\hat{x}_1}{10d_{\hat{\mathbf{x}}}^{(\delta)}}, \frac{\hat{x}_1}{10d_{\hat{\mathbf{x}}}^{(\delta)}})} \le CM_s, \tag{4.5.148}$$

where C depends only on β. From the first equality in (4.5.148),

$$\mathbf{0} \in \Gamma_1^{(\hat{\mathbf{x}})}. \tag{4.5.149}$$

From (4.5.148), noting that

$$d_{\mathbf{x}}^{(\delta)} = \max(\delta, x_1) \le \max\{\varepsilon^*, \delta_0\} \qquad \text{for } \mathbf{x} \in \Omega \cap \{x_1 < \varepsilon^*\},$$

and choosing δ_0 and ε^* small, depending on (β, M), we have

$$\sqrt{d_{\hat{\mathbf{x}}}^{(\delta)}}|f_{\rm bd}^{(\hat{\mathbf{x}})}(X_1)| \le \frac{l_s\hat{x}_1}{d_{\hat{\mathbf{x}}}^{(\delta)}} \le l_s\sqrt{\frac{\hat{x}_1}{d_{\hat{\mathbf{x}}}^{(\delta)}}} \qquad \text{for } X_1 \in (-\frac{l_s\hat{x}_1}{d_{\hat{\mathbf{x}}}^{(\delta)}}, \frac{l_s\hat{x}_1}{d_{\hat{\mathbf{x}}}^{(\delta)}}),$$

where we have used that $\hat{x}_1 \le d_{\hat{\mathbf{x}}}^{(\delta)}$ in the last inequality. With this, restricting v to the region:

$$\Omega^{(\hat{\mathbf{x}})} := \Big((-\frac{l_s\hat{x}_1}{d_{\hat{\mathbf{x}}}^{(\delta)}}, \frac{l_s\hat{x}_1}{d_{\hat{\mathbf{x}}}^{(\delta)}}) \times (-\frac{l_s\hat{x}_1}{d_{\hat{\mathbf{x}}}^{(\delta)}}, \frac{l_s\hat{x}_1}{d_{\hat{\mathbf{x}}}^{(\delta)}})\Big) \cap \hat{\Omega}^{(\hat{\mathbf{x}})},$$

we have

$$\Omega^{(\hat{\mathbf{x}})} = \left\{\mathbf{X} : \begin{array}{l} X_1 \in \big(-\frac{l_s\hat{x}_1}{d_{\hat{\mathbf{x}}}^{(\delta)}}, \frac{l_s\hat{x}_1}{d_{\hat{\mathbf{x}}}^{(\delta)}}\big), \\ X_2 \in \big(-\frac{l_s\hat{x}_1}{d_{\hat{\mathbf{x}}}^{(\delta)}}, \sqrt{d_{\hat{\mathbf{x}}}^{(\delta)}} f_{\rm bd}^{(\hat{\mathbf{x}})}(X_1)\big) \end{array}\right\} \tag{4.5.150}$$

and $\Gamma_1^{(\hat{\mathbf{x}})} = \Omega^{(\hat{\mathbf{x}})} \cap \{X_2 = f_{\rm bd}^{(\hat{\mathbf{x}})}(X_1)\}$.

Now define

$$v(\mathbf{X}) = \frac{1}{d_{\hat{\mathbf{x}}}^{(\delta)}} u(\hat{x}_1 + d_{\hat{\mathbf{x}}}^{(\delta)} X_1,\ f_{\rm bd}(\hat{x}_1) - \sqrt{d_{\hat{\mathbf{x}}}^{(\delta)}} X_2) \qquad \text{in } \Omega^{(\hat{\mathbf{x}})}. \tag{4.5.151}$$

Estimate (4.5.120) for $u(\cdot)$ implies that

$$\|v\|_{L^\infty(\Omega^{(\hat{\mathbf{x}})})} \le C,$$

where C is independent of $\hat{x}_1$.

Furthermore, v satisfies an equation of form (4.5.84) in $\Omega^{(\hat{\mathbf{x}})}$, and the boundary condition of form (4.5.141) on $\Gamma_1^{(\hat{\mathbf{x}})}$ with modified ingredients $(\hat{A}_{ij}, \hat{A}_i, \hat{B})$ defined as follows:

$$\begin{aligned}
&\hat{A}_{ij}(\mathbf{p}, \mathbf{X}) \\
&= (-1)^{i+j}(d_{\hat{\mathbf{x}}}^{(\delta)})^{\frac{i+j}{2}-2}\, A_{ij}(p_1,\, -\sqrt{d_{\hat{\mathbf{x}}}^{(\delta)}}p_2,\, \hat{x}_1 + d_{\hat{\mathbf{x}}}^{(\delta)}X_1,\, f_{\rm bd}(\hat{x}_1) - \sqrt{d_{\hat{\mathbf{x}}}^{(\delta)}}X_2), \\
&\hat{A}_i(\mathbf{p}, \mathbf{X}) \\
&= (-1)^{i+1}(d_{\hat{\mathbf{x}}}^{(\delta)})^{\frac{i-1}{2}}\, A_i(p_1,\, -\sqrt{d_{\hat{\mathbf{x}}}^{(\delta)}}p_2,\, \hat{x}_1 + d_{\hat{\mathbf{x}}}^{(\delta)}X_1,\, f_{\rm bd}(\hat{x}_1) - \sqrt{d_{\hat{\mathbf{x}}}^{(\delta)}}X_2), \\
&\hat{B}(p_2, z, \mathbf{X}) = \tilde{B}(-\sqrt{d_{\hat{\mathbf{x}}}^{(\delta)}}p_2,\, d_{\hat{\mathbf{x}}}^{(\delta)}z,\, \hat{x}_1 + d_{\hat{\mathbf{x}}}^{(\delta)}X_1,\, f_{\rm bd}(\hat{x}_1) - \sqrt{d_{\hat{\mathbf{x}}}^{(\delta)}}X_2).
\end{aligned}$$

From (4.5.142), choosing $\varepsilon^* \le \frac{\varepsilon}{2}$, we have

$$\hat{B}(0, 0, \mathbf{X}) = 0 \qquad \text{on } \Gamma_1^{(\hat{\mathbf{x}})}.$$

Also, since $l_s \le \frac{1}{2}$, it follows from (4.5.90) that $\hat{A}_{ij}, i, j = 1, 2$, satisfy (4.2.3) with constant $\frac{\lambda}{2}$ in $\Omega^{(\hat{\mathbf{x}})}$, independent of the rescaling base point $\hat{\mathbf{x}}$. Furthermore, from (4.5.135), it follows that (4.2.113) holds for $\hat{A}_i$ in $\Omega^{(\hat{\mathbf{x}})}$. Also, from the properties of $\tilde{B}$, it follows that $\hat{B}$ satisfies (4.5.143) on $\Gamma_1^{(\hat{\mathbf{x}})}$ with constant $C(\lambda, M)$. Furthermore,

$$\begin{aligned}
&\hat{B}_{p_2}(p_2, z, \mathbf{X}) \\
&= -\sqrt{d_{\hat{\mathbf{x}}}^{(\delta)}}\,\tilde{B}_{p_2}(-\sqrt{d_{\hat{\mathbf{x}}}^{(\delta)}}p_2,\, d_{\hat{\mathbf{x}}}^{(\delta)}z,\, \hat{x}_1 + d_{\hat{\mathbf{x}}}^{(\delta)}X_1,\, f_{\rm bd}(\hat{x}_1) - \sqrt{d_{\hat{\mathbf{x}}}^{(\delta)}}X_2).
\end{aligned}$$

Thus, using (4.5.143), we have

$$|\hat{B}_{p_2}(p_2, z, \mathbf{X})| \le C\sqrt{d_{\hat{\mathbf{x}}}^{(\delta)}} \le C\sqrt{\max\{\varepsilon^*, \delta_0\}} \quad \text{for all } (p_2, z, \mathbf{X}) \in \mathbb{R} \times \mathbb{R} \times \Gamma_1^{(\hat{\mathbf{x}})}. \tag{4.5.152}$$

That is, $\hat{B}$ satisfies (4.2.16) with $\sqrt{d_{\hat{\mathbf{x}}}^{(\delta)}}$ on the right-hand side and, from (4.5.145), this quantity can be made small for all $\hat{\mathbf{x}} \in \Gamma_1 \cap \{0 < x_1 \le \varepsilon^*\}$ by making ε^* and δ_0 sufficiently small and recalling that $\delta \in (0, \delta_0)$. Furthermore, if $\hat{x}_1 \ge s$ and $\delta_0 \in (0, 1)$, then

$$s \le \frac{\hat{x}_1}{\delta} \le \frac{\hat{x}_1}{d_{\hat{\mathbf{x}}}^{(\delta)}} \le 1, \tag{4.5.153}$$

from (4.5.145). Also, we use the boundary structure (4.5.147) and (4.5.150) with (4.5.148). Then Theorem 4.2.4 and Remark 4.2.14 imply that, if $\varepsilon^* \in (0, \frac{h}{10})$ and $\delta_0 \in (0, 1)$ are small in (4.5.152), depending only on (λ, M), there exists $\gamma = \gamma(\lambda, M) \in (0, 1)$ such that, for any $\hat{\mathbf{x}} \in \Gamma_1 \cap \{s < x_1 \le \varepsilon^*\}$,

$$\|v\|_{C^{1,\gamma}(\overline{\Omega_{1/2}^{(\hat{\mathbf{x}})}})} \le C\Big(\frac{\hat{x}_1}{10 d_{\hat{\mathbf{x}}}^{(\delta)}}\Big)^{-(\gamma+1)} \|v\|_{C(\overline{\Omega^{(\hat{\mathbf{x}})}})} \le C(\lambda, M, s), \tag{4.5.154}$$

where $\Omega_\rho^{(\hat{\mathbf{x}})} := \big((-\rho\frac{l_s\hat{x}_1}{d_{\hat{\mathbf{x}}}^{(\delta)}}, \rho\frac{l_s\hat{x}_1}{d_{\hat{\mathbf{x}}}^{(\delta)}}) \times (-\rho\frac{l_s\hat{x}_1}{d_{\hat{\mathbf{x}}}^{(\delta)}}, \rho\frac{l_s\hat{x}_1}{d_{\hat{\mathbf{x}}}^{(\delta)}})\big) \cap \Omega^{(\hat{\mathbf{x}})}$ for $\rho \in (0,1]$.

However, we cannot apply Theorem 4.2.8 with Remark 4.2.13 to obtain a $C^{2,\beta}$–estimate, since $(\hat{A}_{11}, \hat{A}_{12})$ do not satisfy (4.2.4)–(4.2.5) for the Hölder exponent β with the constant independent of $\delta, s > 0$. Instead, we use (4.5.154) and perform further rescaling:

For $\hat{\mathbf{x}} \in \Gamma_1 \cap \{s < x_1 \le \varepsilon^*\}$, let $v(\mathbf{X})$ be the corresponding function (4.5.151). Define

$$v^{(\varrho)}(\mathbf{X}) = \frac{1}{\varrho^{1+\gamma}}\big(v(\varrho\mathbf{X}) - v(\mathbf{0}) - \varrho Dv(\mathbf{0})\cdot\mathbf{X}\big) \qquad \text{in } \tfrac{1}{\rho}\Omega_\rho^{(\hat{\mathbf{x}})}. \tag{4.5.155}$$

From (4.5.147) and (4.5.150),

$$\frac{1}{\rho}\Omega_\rho^{(\hat{\mathbf{x}})} = \Big\{(-\frac{l_s\hat{x}_1}{d_{\hat{\mathbf{x}}}^{(\delta)}}, \frac{l_s\hat{x}_1}{d_{\hat{\mathbf{x}}}^{(\delta)}}) \times (-\frac{l_s\hat{x}_1}{d_{\hat{\mathbf{x}}}^{(\delta)}}, \frac{l_s\hat{x}_1}{d_{\hat{\mathbf{x}}}^{(\delta)}}) \;:\; X_2 \ge \sqrt{d_{\hat{\mathbf{x}}}^{(\delta)}}\, f_{\rm bd}^{(\hat{\mathbf{x}},\rho)}(X_1)\Big\},$$

where $f_{\rm bd}^{(\hat{\mathbf{x}},\rho)}(X_1) = \frac{1}{\rho} f_{\rm bd}^{(\hat{\mathbf{x}})}(\rho X_1)$. It follows that $f_{\rm bd}^{(\hat{\mathbf{x}},\rho)}$ satisfies (4.5.148) with unchanged constants for any $\rho \in (0,1)$. Also, $\Gamma_1^{(\hat{\mathbf{x}},\rho)} := \partial(\frac{1}{\rho}\Omega_\rho^{(\hat{\mathbf{x}})}) \cap \{X_2 = f_{\rm bd}^{(\hat{\mathbf{x}},\rho)}(X_1)\}$ is of form (4.5.147) with function $f_{\rm bd}^{(\hat{\mathbf{x}},\rho)}$.

From (4.5.154),

$$\|v^{(\varrho)}\|_{L^\infty(\frac{1}{\rho}\Omega_\rho^{(\hat{\mathbf{x}})})} \le C(\lambda, M, s).$$

Furthermore, $v^{(\varrho)}$ satisfies an equation of form (4.5.84) with the right-hand side $\hat{f}^{(\varrho)}(\mathbf{X})$ in domain $\frac{1}{\rho}\Omega_\rho^{(\hat{\mathbf{x}})}$, and the boundary condition of form (4.5.141) on $\Gamma_1^{(\hat{\mathbf{x}},\rho)}$ with the modified ingredients $(\hat{A}_{ij}^{(\varrho)}, \hat{A}_i^{(\varrho)}, \hat{f}^{(\varrho)}, \hat{B}^{(\varrho)})$ defined as follows:

$$\begin{aligned}
&\hat{A}_{ij}^{(\varrho)}(\mathbf{p}, \mathbf{X}) = \hat{A}_{ij}(\varrho^\gamma\mathbf{p} + Dv(\mathbf{0}),\ \varrho\mathbf{X}),\\
&\hat{A}_i^{(\varrho)}(\mathbf{p}, \mathbf{X}) = \varrho\hat{A}_i(\varrho^\gamma\mathbf{p} + Dv(\mathbf{0}),\ \varrho\mathbf{X}),\\
&\hat{f}^{(\varrho)}(\mathbf{X}) = -\varrho^{1-\gamma}\sum_{i=1}^{2}\hat{A}_i(\varrho^\gamma\mathbf{p} + Dv(\mathbf{0}),\ \varrho\mathbf{X})D_iv(\mathbf{0}),\\
&\hat{B}^{(\varrho)}(p_2, z, \mathbf{X}) = \frac{1}{\varrho^\gamma}\Big(\hat{B}(\varrho^\gamma p_2 + v_{X_2}(\mathbf{0}),\ \hat{Z},\ \varrho\mathbf{X}) - v_{X_1}(\mathbf{0})\Big),
\end{aligned} \tag{4.5.156}$$

where $\hat{Z} = \varrho^{1+\gamma}z + \varrho Dv(\mathbf{0})\cdot\mathbf{X} + v(\mathbf{0})$.

Since $\hat{A}_{ij}, i,j = 1,2$, satisfy (4.2.3) with constant $\frac{\lambda}{2}$ in $\Omega^{(\hat{\mathbf{x}})}$, then $\hat{A}_{ij}^{(\varrho)}$ satisfy the same property in $\frac{1}{\rho}\Omega_\rho^{(\hat{\mathbf{x}})}$. This and $\hat{A}_{12}^{(\varrho)} = \hat{A}_{21}^{(\varrho)}$ imply

$$\|A_{ij}^{(\varrho)}\|_{L^\infty(\mathbb{R}^2\times(\frac{1}{\rho}\Omega_\rho^{(\hat{\mathbf{x}})}))} \le C.$$

Also, from the definition of $\hat{A}_{ij}^{(\varrho)}$ and using (4.5.135), it follows that, for every $\mathbf{p} \in \mathbb{R}^2$,

$$[A_{ij}^{(\varrho)}(\mathbf{p},\cdot)]_{C^\beta(\overline{\frac{1}{\rho}\Omega_\rho^{(\hat{\mathbf{x}})}})} + \|D_\mathbf{p}A_{ij}^{(\varrho)}(\mathbf{p},\cdot)\|_{L^\infty(\frac{1}{\rho}\Omega_\rho^{(\hat{\mathbf{x}})})} \le C\varrho^\beta(d_{\hat{\mathbf{x}}}^{(\delta)})^{\frac{i+j}{2}-2} \le C(\lambda, M, \beta)$$

when $\varrho \le s^{\frac{1}{\beta}}$. Then we fix $\varrho = \frac{1}{2}\min\{s^{\frac{1}{\beta}}, 1\}$. Moreover, from (4.5.135),

$$[A_i^{(\varrho)}(\mathbf{p},\cdot)]_{C^\beta(\overline{\frac{1}{\rho}\Omega_\rho^{(\hat{\mathbf{x}})}})} + \|D_{\mathbf{p}}A_i^{(\varrho)}(\mathbf{p},\cdot)\|_{L^\infty(\frac{1}{\rho}\Omega_\rho^{(\hat{\mathbf{x}})})} \le C(\lambda, M, \beta).$$

Next we discuss the properties of $\hat{B}^{(\varrho)}(\cdot)$. Note that

$$v_{X_1}(\mathbf{0}) = \hat{B}(v_{X_2}(\mathbf{0}),\ v(\mathbf{0}),\ \mathbf{0})$$

by (4.5.149) and the boundary condition for v on $\Gamma_1^{(\hat{\mathbf{x}})}$. Also, recall that $\hat{B}$ satisfies (4.5.143) on $\Gamma_1^{(\hat{\mathbf{x}})}$ with $C(\lambda, M)$. Then there exists $C(\lambda, M)$ such that, for $k = 1, 2, 3$,

$$\|D^k_{(p_2,z)}\hat{B}^{(\varrho)}(p_2, z, \cdot)\|_{C^3(\overline{\hat{\Gamma}_1^{(\hat{\mathbf{x}})}})} \le C \qquad \text{for any } (p_2, z) \in \mathbb{R}\times\mathbb{R}. \qquad (4.5.157)$$

Also, (4.5.152) holds for $\hat{B}^{(\varrho)}$ with the same constant as for $\hat{B}$. Furthermore, $\hat{B}^{(\varrho)}(0, 0, \mathbf{X}) = 0$ for any $\mathbf{X} \in \Gamma_1^{(\hat{\mathbf{x}},\rho)}$, since $\hat{B}$ has the property on $\Gamma_1^{(\hat{\mathbf{x}})}$. Also, (4.5.135), (4.5.154), and the expression of $f^{(\varrho)}$ in (4.5.156) imply

$$\|f^{(\varrho)}\|_{C^\beta(\overline{\frac{1}{\rho}\Omega_\rho^{(\hat{\mathbf{x}})}})} \le C(\lambda, M, \beta).$$

Therefore, we have shown that $(\hat{A}_{ij}^{(\varrho)}, \hat{A}_i^{(\varrho)}, \hat{B}^{(\varrho)})$ satisfy all the conditions in Theorem 4.2.8 and Remark 4.2.13 in $\frac{1}{\rho}\Omega_\rho^{(\hat{\mathbf{x}})}$, for β in place of α and $\|h\|_{C^{1,\beta}} \equiv 0$, with the constants depending only on (λ, M, β). Recall that (4.5.152) holds for $\hat{B}^{(\varrho)}$ with the same constant as for $\hat{B}$. Thus, further reducing ε^* and δ_0 depending only on (λ, M, β), we have

$$\begin{aligned}\|v^{(\varrho)}\|_{C^{2,\beta}(\overline{\frac{1}{\rho}\Omega_{\rho/2}^{(\hat{\mathbf{x}})}})} &\le C\Big(\frac{\hat{x}_1}{10d_{\hat{\mathbf{x}}}^{(\delta)}}\Big)^{-(\beta+2)}\Big(\|v^{(\varrho)}\|_{C(\overline{\frac{1}{\rho}\Omega_\rho^{(\hat{\mathbf{x}})}})} + \|\hat{B}^{(\varrho)}(\mathbf{0},\cdot)\|_{C^{1,\beta}(\overline{\hat{\Gamma}_1^{(\hat{\mathbf{x}})}})}\Big)\\ &\le C(\lambda, M, \beta, s),\end{aligned}$$

where we have used (4.5.153) in the second inequality. Combining this estimate with (4.5.154) and expressing them in terms of u, we find that $\hat{r} > 0$ depends only on (t_1, t_2, M, s), hence on (λ, M, s), such that

$$\|u\|_{C^{2,\beta}(\overline{B_{\hat{r}}(\hat{\mathbf{x}})\cap\Omega})} \le C(\lambda, M, \beta, s) \qquad \text{for all } \hat{\mathbf{x}} \in \Gamma_1 \cap \{s < x_1 \le \varepsilon^*\}.$$

Since R_1 satisfies (4.5.134), we combine the last estimate with the standard interior estimates to obtain (4.5.140).

Now combine (4.5.140) with estimate (4.5.131) for $s = \frac{\varepsilon^*}{2}$, where $K = K(\lambda, M, \varepsilon)$ is from (4.5.119) in the application of Lemma 4.5.10, as we have discussed at the beginning of Step 1. Then we obtain (4.5.138) for any $s \in (0, \min\{\frac{h}{10}, 1\})$.

2. It remains to prove (4.5.139) under the assumption that $t_0 > 0$. The main part is to estimate $u(\cdot)$ near corners $\{P_1, P_4\}$, which we do from now on.

Furthermore, the universal constant $C \geq 1$ depends only on $(\lambda, M, \alpha, \beta, \varepsilon, \delta)$, unless otherwise specified; constants $C_{t_0} \geq 1$ and $\varrho \in (0,1)$ depend only on $(\lambda, M, \alpha, \beta, \varepsilon, \delta, t_0)$; the small constant $\delta_0 > 0$ depends only on $(\lambda, M, \alpha, \beta, \varepsilon)$. We always assume that $\delta \in (0, \delta_0)$ in the proof.

In this step, we first prove that $u \in C^{2,\beta}(\overline{B_\varrho(P_4) \cap \Omega})$ for sufficiently small $\varrho > 0$, so that

$$\|u\|_{C^{2,\beta}(\overline{B_\varrho(P_4)\cap\Omega})} \leq C_{t_0}, \tag{4.5.158}$$

where exponent β is from (4.5.135).

We follow the proof of Lemma 4.5.4. We use $\mathcal{R}^+_\varrho$ defined by (4.5.21). Since $\Gamma_3 \subset \{x_2 = 0\}$, then $\Omega \cap B_{4\varrho}(P_4)$ is $\mathcal{R}^+_{4\varrho}$ with $f_{\rm ob} \equiv 0$ on $[0, \infty)$, if we choose $\varrho \leq \frac{1}{8}\min\{\lambda, t_0\}$ which implies $\varrho \leq \frac{1}{8}\min\{t_0, t_2, h\}$. Also, we define $B^+_\varrho := B_\varrho(\mathbf{0}) \cap \{x_1 > 0\}$ as in Step 2 of the proof of Lemma 4.5.4, and consider the function:

$$v(\mathbf{x}) = \frac{1}{\varrho}u(\varrho\mathbf{x}) \qquad \text{for } \mathbf{x} \in \mathcal{R}^+_2.$$

Then, by (4.5.120), v satisfies

$$\|v\|_{L^\infty(\mathcal{R}^+_2)} \leq C. \tag{4.5.159}$$

Moreover, v is a solution of

$$\sum_{i,j=1}^2 \hat{A}^{(\varrho)}_{ij} D_{ij}v + (\hat{A}^{(\varrho)}_1, \hat{A}^{(\varrho)}_2) \cdot Dv = 0 \qquad \text{in } \mathcal{R}^+_2, \tag{4.5.160}$$

$$v = 0 \qquad \text{on } \partial\mathcal{R}^+_2 \cap \{x_1 = 0\}, \tag{4.5.161}$$

$$v_{\boldsymbol{\nu}} \equiv D_2 v = 0 \qquad \text{on } \partial\mathcal{R}^+_2 \cap \{x_2 = 0\}, \tag{4.5.162}$$

with $(\hat{A}^{(\varrho)}_{ij}, \hat{A}^{(\varrho)}_i) = (A^{(\varrho)}_{ij}, A^{(\varrho)}_i)(Dv, \mathbf{x})$ defined by

$$\hat{A}^{(\varrho)}_{ij}(\mathbf{p}, \mathbf{x}) = A_{ij}(\mathbf{p}, \varrho\mathbf{x}), \quad \hat{A}^{(\varrho)}_i(\mathbf{p}, \mathbf{x}) = \varrho A_i(\mathbf{p}, \varrho\mathbf{x}),$$

where we have used (4.5.111) and assumed that $\varrho < \varepsilon$ to have the homogeneous equation and boundary condition (4.5.160)–(4.5.162).

Then, by (4.5.90), $(\hat{A}^{(\varrho)}_{ij}, \hat{A}^{(\varrho)}_i)$ satisfy that, for any $\mathbf{x} \in \mathcal{R}^+_2$ and $\mathbf{p} \in \mathbb{R}^2$,

$$\delta\lambda|\boldsymbol{\mu}|^2 \leq \sum_{i,j=1}^2 \hat{A}^{(\varrho)}_{ij}(\mathbf{p}, \mathbf{x})\mu_i\mu_j \leq \lambda^{-1}|\boldsymbol{\mu}|^2 \qquad \text{for any } \boldsymbol{\mu} \in \mathbb{R}^2, \tag{4.5.163}$$

and, by (4.5.92),

$$\|(A^{(\varrho)}_{ij},\, D_{\mathbf{p}}A^{(\varrho)}_{ij})\|_{L^\infty(\mathbb{R}^2\times\overline{\mathcal{R}^+_2})} \leq M, \tag{4.5.164}$$

$$\|(A^{(\varrho)}_i,\, D_{\mathbf{p}}A^{(\varrho)}_i)\|_{L^\infty(\mathbb{R}^2\times\overline{\mathcal{R}^+_2})} \leq \varrho M, \tag{4.5.165}$$

$$[(A^{(\varrho)}_{ij},\, A^{(\varrho)}_i)(\mathbf{p}, \cdot)]_{C^\beta(\overline{\mathcal{R}^+_2})} \leq M\varrho^\beta, \tag{4.5.166}$$

since $0 < \varrho \le 1$. Moreover, by (4.5.94), we find that, for any $(\mathbf{p}, (x_1, 0)) \in \mathbb{R}^2 \times (\partial\mathcal{R}_2^+ \cap \{x_2 = 0\})$,

$$(A_{11}^{(\varrho)}, A_{22}^{(\varrho)}, A_1^{(\varrho)})((p_1, -p_2), (x_1, 0)) = (A_{11}^{(\varrho)}, A_{22}^{(\varrho)}, A_1^{(\varrho)})((p_1, p_2), (x_1, 0)). \tag{4.5.167}$$

Furthermore, (4.5.92) and (4.5.96) imply that

$$|(A_{12}, A_{21})(\mathbf{p}, \mathbf{x})| \le M x_1^\beta \tag{4.5.168}$$

for any $(\mathbf{p}, \mathbf{x}) \in \mathbb{R}^2 \times \overline{\Omega \cap \{x_1 < \varepsilon\}}$, so that

$$|(A_{12}^{(\varrho)}, A_{21}^{(\varrho)})(\mathbf{p}, \mathbf{x})| \le M \varrho^\beta x_1^\beta \qquad \text{for any } (\mathbf{p}, \mathbf{x}) \in \mathbb{R}^2 \times \overline{\mathcal{R}_2^+}. \tag{4.5.169}$$

Using (4.5.159) and (4.5.163)–(4.5.166), and combining the estimates in Theorems 4.2.1 and 4.2.9–4.2.10 with Remark 4.2.13 and the argument that has led to (4.5.35), we have

$$\|v\|_{C^{2,\beta}(\overline{\mathcal{R}_{3/2}^+ \setminus \mathcal{R}_{1/2}^+})} \le C, \tag{4.5.170}$$

where C is independent of ϱ.

We now use domain D introduced in Step 2 of the proof of Lemma 4.5.4. Recall that $D \subset \mathcal{R}_1^+$. We prove that, for any $g \in C^\beta(\overline{D})$ with $\|g\|_{C^\beta(\overline{D})} \le 1$, there exists a unique solution $w \in C^{2,\beta}(\overline{D})$ of the following problem:

$$\sum_{i=1}^{2} \hat{A}_{ii}^{(\varrho)} D_{ii} w + \hat{A}_1^{(\varrho)} D_1 w = g \qquad \text{in } D, \tag{4.5.171}$$

$$w = 0 \qquad \text{on } \partial D \cap \{x_1 = 0, x_2 > 0\}, \tag{4.5.172}$$

$$w_{\boldsymbol{\nu}} \equiv D_2 w = 0 \qquad \text{on } \partial D \cap \{x_1 > 0, x_2 = 0\}, \tag{4.5.173}$$

$$w = v \qquad \text{on } \partial D \cap \{x_1 > 0, x_2 > 0\}, \tag{4.5.174}$$

with $(A_{ii}^{(\varrho)}, A_1^{(\varrho)}) = (A_{ii}^{(\varrho)}, A_1^{(\varrho)})(Dw, \mathbf{x})$. Moreover, we show

$$\|w\|_{C^{2,\beta}(\overline{D})} \le C. \tag{4.5.175}$$

For that, in a way similar to Step 2 of the proof of Lemma 4.5.4, we consider the even reflection D^+ of D and the even reflection of $(v, g, \hat{A}_{11}^{(\varrho)}, \hat{A}_{22}^{(\varrho)}, \hat{A}_1^{(\varrho)})$ from $\overline{\mathcal{R}_2^+}$ to $\overline{B_2^+}$, without change of notation, where the even reflection of $(\hat{A}_{11}^{(\varrho)}, \hat{A}_{22}^{(\varrho)}, \hat{A}_1^{(\varrho)})$, which depends on $(\mathbf{p}, \mathbf{x})$, is defined by

$$\begin{aligned} \hat{A}_{ii}^{(\varrho)}((p_1, p_2), x_1, -x_2) &= \hat{A}_{ii}^{(\varrho)}((p_1, -p_2), x_1, x_2), \\ \hat{A}_1^{(\varrho)}((p_1, p_2), x_1, -x_2) &= \hat{A}_1^{(\varrho)}((p_1, -p_2), x_1, x_2) \end{aligned} \tag{4.5.176}$$

for any $\mathbf{x} = (x_1, x_2) \in \overline{\mathcal{R}_2^+}$ and $\mathbf{p} = (p_1, p_2) \in \mathbb{R}^2$.

Also, denote by $\hat{v}$ the restriction of the extended function v to ∂D^+. It follows from (4.5.161)–(4.5.162) and (4.5.170) that $\hat{v} \in C^{2,\beta}(\partial D^+)$ with

$$\|\hat{v}\|_{C^{2,\beta}(\partial D^+)} \le C. \tag{4.5.177}$$

Moreover, the extended function g satisfies that $g \in C^\beta(\overline{D^+})$ with $\|g\|_{C^\beta(\overline{D^+})} = \|g\|_{C^\beta(\overline{D})} \le 1$. The extended functions $(\hat{A}_{11}^{(\varrho)}, \hat{A}_{22}^{(\varrho)}, \hat{A}_1^{(\varrho)})$ satisfy (4.5.163) and (4.5.166) in D^+ with the same constants as in $\mathcal{R}_2^+$. Also, using (4.5.164)–(4.5.165) and (4.5.167) in $\mathcal{R}_2^+$, we see that the extended functions $(\hat{A}_{11}^{(\varrho)}, \hat{A}_{22}^{(\varrho)}, \hat{A}_1^{(\varrho)})$ are Lipschitz with respect to $\mathbf{p}$ in D^+ and satisfy

$$\|(A_{ij}^{(\varrho)}, A_i^{(\varrho)}, D_\mathbf{p} A_{ij}^{(\varrho)}, D_\mathbf{p} A_i^{(\varrho)})(\mathbf{p}, \cdot)\|_{L^\infty(\overline{D^+})} \le M.$$

We consider the Dirichlet problem:

$$\sum_{i=1}^{2} \hat{A}_{ii}^{(\varrho)} D_{ii} w + \hat{A}_1^{(\varrho)} D_1 w = g \qquad \text{in } D^+, \tag{4.5.178}$$

$$w = \hat{v} \qquad \text{on } \partial D^+, \tag{4.5.179}$$

with $(A_{ii}^{(\varrho)}, A_1^{(\varrho)}) := (A_{ii}^{(\varrho)}, A_1^{(\varrho)})(Dw, \mathbf{x})$. By the maximum principle,

$$\|w\|_{L^\infty(D^+)} \le \|\hat{v}\|_{L^\infty(\partial D^+)} + C\|g\|_{L^\infty(D^+)} \le \|\hat{v}\|_{L^\infty(\partial D^+)} + C.$$

Using (4.5.177), we obtain an estimate of $\|w\|_{L^\infty(D^+)}$. Now, using Theorems 4.2.1 and 4.2.9 and the estimates of $\|g\|_{C^\beta(\overline{D^+})}$ and $\|\hat{v}\|_{C^{2,\beta}(\partial D^+)}$ discussed above, we obtain the *a priori* estimate for the $C^{2,\beta}$–solution w of (4.5.178)–(4.5.179):

$$\|w\|_{C^{2,\beta}(\overline{D^+})} \le C. \tag{4.5.180}$$

Moreover, for every $\hat{w} \in C^{1,\beta}(\overline{D^+})$, the existence of a unique solution $w \in C^{2,\beta}(\overline{D^+})$ of the linear Dirichlet problem, obtained by substituting $\hat{w}$ into the coefficients of (4.5.178), follows from [131, Theorem 6.8]. Now, by a standard application of the Leray-Schauder theorem, there exists a solution $w \in C^{2,\beta}(\overline{D^+})$ of the Dirichlet problem (4.5.178)–(4.5.179), which satisfies (4.5.180). The uniqueness of this solution follows from the comparison principle.

From the structure of equation (4.5.178) in D^+, specifically from (4.5.167), (4.5.176), and the symmetry of the right-hand sides obtained by even extension, it follows that $\hat{w}$, defined by $\hat{w}(x_1, x_2) = w(x_1, -x_2)$, is also a solution of (4.5.178)–(4.5.179). By uniqueness for Problem (4.5.178)–(4.5.179), we find that $w(x_1, x_2) = w(x_1, -x_2)$ in D^+. Thus, w restricted to D is a solution of (4.5.171)–(4.5.174), where (4.5.172) follows from (4.5.161) and (4.5.179). Moreover, (4.5.180) implies (4.5.175).

The uniqueness of a solution $w \in C^{2,\beta}(\overline{D})$ of (4.5.171)–(4.5.174) follows from the standard comparison principle.

Now we prove the existence of a solution $w \in C^{2,\beta}(\overline{D})$ of the problem:

$$\begin{aligned}
&\textstyle\sum_{i,j=1}^2 \hat{A}_{ij}^{(\varrho)} D_{ij}w + \sum_{i=1}^2 \hat{A}_i^{(\varrho)} D_i w = 0 && \text{in } D,\\
&w = 0 && \text{on } \partial D \cap \{x_1 = 0, x_2 > 0\},\\
&w_\nu \equiv D_2 w = 0 && \text{on } \partial D \cap \{x_1 > 0, x_2 = 0\},\\
&w = v && \text{on } \partial D \cap \{x_1 > 0, x_2 > 0\},
\end{aligned} \tag{4.5.181}$$

where $(A_{ij}^{(\varrho)}, A_i^{(\varrho)}) := (A_{ij}^{(\varrho)}, A_i^{(\varrho)})(Dw, \mathbf{x})$. Moreover, we prove that w satisfies

$$\|w\|_{C^{2,\beta}(\overline{D})} \le C. \tag{4.5.182}$$

Define

$$\mathcal{R}(N) := \left\{ W \in C^{2,\beta}(\overline{D}) \;:\; \|W\|_{C^{2,\beta}(\overline{D})} \le N \right\}, \tag{4.5.183}$$

where N will be determined later. We obtain such w as a fixed point of map $K : \mathcal{R}(N) \mapsto \mathcal{R}(N)$ (if ϱ is small and N is large) defined as follows:

For $W \in \mathcal{R}(N)$, define

$$g = -2\hat{A}_{12}^{(\varrho)}(DW, \mathbf{x}) W_{x_1x_2} - \hat{A}_2^{(\varrho)}(DW, \mathbf{x}) W_{x_2}. \tag{4.5.184}$$

By (4.5.164)–(4.5.166) and (4.5.169), we can estimate $\|g\|_{C^\beta(\overline{D})}$. First, $G(\mathbf{x}) := \hat{A}_{12}^{(\varrho)}(DW(\mathbf{x}), \mathbf{x})$ satisfies

$$\|G\|_{C(\overline{D})} \le C\varrho^\beta, \qquad \|DG\|_{C(\overline{D})} \le CN,$$

where constant C depends only on the parameters described at the beginning of Step 2, which are fixed for this argument, so we do not specify the dependence on them below. Then, by the interpolation inequality, for any $\varepsilon > 0$,

$$\|G\|_{C^\beta(\overline{D})} \le \varepsilon \|DG\|_{C(\overline{D})} + \hat{C}(\varepsilon, \beta)\|G\|_{C(\overline{D})} \le CN\varepsilon + \hat{C}(\varepsilon, \beta) C\varrho^\beta,$$

where $\hat{C}(\varepsilon, \beta)$ depends only on ε and β. Thus, choosing ε small depending on N so that $CN\varepsilon \le \frac{1}{4N^2}$, and then choosing ϱ small depending on (ε, β, N), we obtain that $\|G\|_{C^\beta(\overline{D})} \le \frac{1}{2N^2}$. With this, we have

$$\|g\|_{C^\beta(\overline{D})} \le \frac{1}{2N^2} N + CN(1+N)\varrho^\beta \le 1,$$

if $\varrho \le \varrho_0$ with $\varrho_0(N, \beta, \delta)$ small. Then, as we have proved above, there exists a unique solution $w \in C^{2,\beta}(\overline{D})$ of (4.5.171)–(4.5.174) with g defined by (4.5.184). Moreover, w satisfies (4.5.175). Choosing N to be constant C in (4.5.175), we conclude that $w \in \mathcal{R}(N)$. Thus, N is chosen depending only on $(\lambda, M, \alpha, \beta, \varepsilon, \delta)$. Now our choice of $\varrho \le \varrho_0$ and the other smallness conditions stated above determines ϱ in terms of $(\delta, \lambda, M, \alpha, \beta, \varepsilon, \delta, t_0)$. We define $K[W] := w$ so that $K : \mathcal{R}(N) \mapsto \mathcal{R}(N)$.

Now the existence of a fixed point of K follows from the Schauder fixed point theorem in the following setting: From its definition, $\mathcal{R}(N)$ is a compact and convex subset in $C^{2,\beta/2}(\overline{D})$. Map $K : \mathcal{R}(N) \mapsto \mathcal{R}(N)$ is continuous in $C^{2,\beta/2}(\overline{D})$. Indeed, if $W_k \in \mathcal{R}(N)$ for $k = 1, \dots$, and $W_k \to W$ in $C^{2,\beta/2}(\overline{D})$, it is easy to see that $W \in \mathcal{R}(N)$. Define g_k and g_∞ by (4.5.184) for W_k and W, respectively. Then $g_k \to g_\infty$ in $C^{\beta/2}(\overline{D})$ by (4.5.164)–(4.5.166). Let $w_k = K[W_k]$. Then $w_k \in \mathcal{R}(N)$, and $\mathcal{R}(N)$ is bounded in $C^{2,\beta}(\overline{D})$. Thus, for any subsequence w_{k_l}, there exists a further subsequence $w_{k_{l_m}}$ converging in $C^{2,\beta/2}(\overline{D})$. The limit function $\tilde{w}$ is a solution of (4.5.171)–(4.5.174) with g_∞ on the right-hand side of (4.5.171). By the uniqueness of solutions in $\mathcal{R}(N)$ to (4.5.171)–(4.5.174), we find that $\tilde{w} = K[W]$. Then it follows that the whole sequence $K[W_k]$ converges to $K[W]$, so that $K : \mathcal{R}(N) \mapsto \mathcal{R}(N)$ is continuous in $C^{2,\beta/2}(\overline{D})$. Therefore, there exists $w \in \mathcal{R}(N)$, which is a fixed point of K. This function w is a solution of (4.5.181).

Since v satisfies (4.5.160)–(4.5.162) and $D \subset \mathcal{R}_1^+$, it follows from the uniqueness of solutions in $C(\overline{D}) \cap C^1(\overline{D} \setminus \overline{\{x_1 = 0\}}) \cap C^2(D)$ of Problem (4.5.181) that $w = v$ in D. Thus, $v \in C^{2,\beta}(\overline{D})$ and satisfies (4.5.175). This implies that u satisfies (4.5.158).

3. In this step, we show the following estimate near corner P_1: There exist (δ_0, C, ϱ) such that, if $\delta \in (0, \delta_0)$,

$$\|u\|^{(-1-\beta),\{P_1\}}_{2,\beta,\Omega\cap B_\varrho(P_1)} \le C_{t_0}. \tag{4.5.185}$$

The dependence of constants $(\delta_0, C, C_{t_0}, \varrho)$ on the parameters in the problem has been discussed at the beginning of Step 2.

We assume that $\varrho < \frac{1}{10}\min\{t_0, \lambda, \varepsilon\}$ so that $\varrho < \frac{1}{10}\min\{t_0, t_2, h, \varepsilon\}$. Then, by shifting the origin into P_1 and inverting the direction of the x_2–axis, $\Omega \cap B_{4\varrho}(P_1)$ becomes $\mathcal{R}^+_{4\varrho}$ defined by (4.5.21) with

$$f_{\rm ob}(x_1) = f_{\rm bd}(0) - f_{\rm bd}(x_1) \qquad \text{on } [0, \infty),$$

and parts Γ_0 and Γ_1 within $\partial\Omega \cap B_{4\varrho}(P_1)$ are mapped into $\Gamma^{(\rm d)}_{4\varrho}$ and $\Gamma^{(\rm ob)}_{4\varrho}$, respectively. We now work in these coordinates.

As in Step 1 of the proof of Lemma 4.5.8, we consider equation (4.5.84) and the boundary condition (4.5.85) as a linear equation and boundary condition (4.5.121)–(4.5.122), in which we now have that $f = 0$ by (4.5.84), and $\hat{g}_1 = 0$ within $\Gamma^{(\rm ob)}_{4\varrho}$ by (4.5.111) for $\varrho \le \varepsilon$. Also, since $u \in C^1(\overline{\Omega})$ satisfies (4.5.88), (4.5.85) with (4.5.102), and (4.5.111), $Du(P_1) = \mathbf{0}$ in the original coordinates, which is $Du(\hat{P}) = \mathbf{0}$ in the new coordinates, where we recall that $\hat{P} = \mathbf{0}$. Then (4.5.90), (4.5.92), and (4.5.95)–(4.5.102) imply that conditions (4.5.47)–(4.5.50) and (4.5.52)–(4.5.55) are satisfied for Problem (4.5.121)–(4.5.122) in the present case with β in place of α, where we have used that $Du(\hat{P}) = \mathbf{0}$ to derive (4.5.48) from (4.5.95). Using that $f = 0$ and $\hat{g}_1 = 0$, we obtain estimate (4.5.57) for u with α replaced by β and with $f = g = 0$, if δ is sufficiently small, depending only on $(\lambda, M, \alpha, \varepsilon)$:

$$|u(\mathbf{x})| \le C\|u\|_{L^\infty(\mathcal{R}_\varrho^+)}|\mathbf{x}|^{1+\beta} \le C|\mathbf{x}|^{1+\beta} \qquad \text{in } \mathcal{R}_\varrho^+, \tag{4.5.186}$$

where we have used the L^∞–bound (4.5.119) of u.

We note that $d_{\mathbf{x}}^{(\delta)}$ defined by (4.5.145) satisfies

$$d_{\mathbf{x}}^{(\delta)} \equiv \delta \qquad \text{for all } \mathbf{x} \in \overline{\Omega} \cap \{x_1 < \delta\}.$$

We assume that $\delta_0 \le \varepsilon$ and rescale u in $\mathcal{R}_\varrho^+$ by combining (4.5.151) with (4.5.155), which now takes the form:

$$w(\mathbf{x}) := \frac{1}{\varrho^{1+\beta}} u(\sqrt{\delta}\,\varrho x_1, \varrho x_2) \qquad \text{for } \mathbf{x} = (x_1, x_2) \in \hat{\mathcal{R}}_1^+, \tag{4.5.187}$$

where

$$\hat{\mathcal{R}}_r^+ := B_r(\mathbf{0}) \cap \{x_1 > 0, x_2 > \sqrt{\delta} F(x_1)\} \qquad \text{for } r > 0 \tag{4.5.188}$$

with

$$F(x_1) = \frac{1}{\sqrt{\delta}\,\varrho} f_{\rm ob}(\sqrt{\delta}\,\varrho x_1) = \frac{1}{\sqrt{\delta}\,\varrho}\big(f_{\rm bd}(0) - f_{\rm bd}(\sqrt{\delta}\,\varrho x_1)\big).$$

From (4.5.15),

$$F(0) = 0, \qquad \|F\|_{2,\alpha,(0,2)}^{(-1-\alpha),\{0\}} \le 2M. \tag{4.5.189}$$

Hence, $\hat{\mathcal{R}}_r^+$ is of structure (4.5.21), where $\hat{\Gamma}_r^{(\rm d)}$ and $\hat{\Gamma}_r^{(\rm ob)}$ are denoted as its boundary parts from (4.5.22). We note that (4.5.187) is well-defined if $\delta_0 < 1$, since, in this case, $(\varrho\sqrt{\delta}x_1, \varrho x_2) \in \mathcal{R}_\varrho^+$ if $(x_1, x_2) \in \hat{\mathcal{R}}_1^+$. Thus, defining $\beta' := \frac{1+\beta}{2}$ so that $\frac{1}{2} < \beta' < \beta$ (since $\beta \in (\frac{1}{2}, 1)$), and choosing $\varrho > 0$ small, depending only on (β, δ), we use (4.5.186) to obtain

$$|w(\mathbf{x})| \le C\delta^{-\frac{1}{2}}\varrho^{\beta-\beta'}|\mathbf{x}|^{1+\beta'} \le |\mathbf{x}|^{1+\beta'} \qquad \text{in } \mathcal{R}_\varrho^+. \tag{4.5.190}$$

Furthermore, w satisfies an equation of form (4.5.84) in $\mathcal{R}_\varrho^+$, and a boundary condition of form (4.5.141) on $\hat{\Gamma}_1^{(\rm ob)} = \partial\hat{\mathcal{R}}_1^+ \cap \{x_2 = 0\}$ with modified ingredients $(\hat{A}_{ij}, \hat{A}_i, \hat{f}, \hat{B})$ defined as follows:

$$\hat{A}_{ij}(\mathbf{p}, \mathbf{x}) = \delta^{\frac{i+j}{2}-2}\, A_{ij}(\frac{\varrho^\beta}{\sqrt{\delta}} p_1,\ \varrho^\beta p_2,\ \sqrt{\delta}\,\varrho x_1,\ \varrho x_2),$$

$$\hat{A}_i(\mathbf{p}, \mathbf{x}) = \delta^{\frac{i-2}{2}}\,\varrho\, A_i(\frac{\varrho^\beta}{\sqrt{\delta}} p_1,\ \varrho^\beta p_2,\ \sqrt{\delta}\,\varrho x_1,\ \varrho x_2),$$

$$\hat{B}(p_2, z, \mathbf{x}) = \sqrt{\delta}\,\varrho^{-\beta}\tilde{B}(\varrho^\beta p_2, \varrho^{1+\beta} z,\ \sqrt{\delta}\,\varrho x_1,\ \varrho x_2).$$

Choosing $\delta_0 < \varepsilon$, we obtain from (4.5.142) that

$$\hat{B}(0, 0, \mathbf{x}) = 0.$$

Also, $w = 0$ on $\hat{\Gamma}_{\varrho}^{(\mathrm{d})} = \partial\hat{\mathcal{R}}_1^+ \cap \{x_1 = 0\}$.

From (4.5.90), it follows that $\hat{A}_{ij}, i, j = 1, 2$, satisfy (4.2.3) with constant λ in $\mathcal{R}_1^+$. This also implies that $\|\hat{A}_{ij}\|_{L^\infty(\mathbb{R}^2\times\hat{\mathcal{R}}_1^+)} \leq C(\lambda)$, where we have used the symmetry: $A_{12} = A_{21}$. Using this estimate, (4.5.92), and the definitions of $(\hat{A}_{ij}, \hat{A}_i)$, and choosing ϱ small depending only on (δ, β), we obtain that, for each $\mathbf{p} \in \mathbb{R}^2$,

$$\|(\hat{A}_{ij}, \hat{A}_i)(\mathbf{p}, \cdot)\|_{C^\beta(\overline{\hat{\mathcal{R}}_1^+})} + \|D_{\mathbf{p}}(\hat{A}_{ij}, \hat{A}_i)(\mathbf{p}, \cdot)\|_{L^\infty(\hat{\mathcal{R}}_1^+)} \leq 1. \tag{4.5.191}$$

Since $\varrho, \delta_0 < 1$, then $|D^k_{(\mathbf{p},z,\mathbf{x})}\hat{B}| \leq |D^k_{(\mathbf{p},z,\mathbf{x})}\tilde{B}|$, $k = 1, 2, \dots$. Combining this with (4.5.143), and again choosing ϱ small depending only on (δ, β), we have

$$\|D^k_{(p_2,z)}\hat{B}(p_2, z, \cdot)\|_{C^3(\overline{\hat{\Gamma}_1^{(\mathrm{ob})}})} \leq C(\lambda, \beta, M) \tag{4.5.192}$$

for any $(p_2, z) \in \mathbb{R} \times \mathbb{R}$ and $k = 0, 1, 2, 3$, where we have used (4.5.142) to obtain the estimate for $k = 0$.

Also, $\hat{B}_{p_2}(p_2, z, x_1) = \sqrt{\delta}\tilde{B}_{p_2}(\sqrt{\delta}p_2,\ \delta z,\ \delta x_1)$. Thus, using (4.5.143), we have

$$|\hat{B}_{p_2}(p_2, z, x_1)| \leq C(\lambda, \beta, M)\sqrt{\delta} \qquad \text{for all } (p_2, \mathbf{x}) \in \mathbb{R} \times \hat{\Gamma}_{\varrho}^{(\mathrm{ob})}. \tag{4.5.193}$$

That is, $\hat{B}$ satisfies (4.2.16) with $\sqrt{\delta}$ on the right-hand side.

Then, for small $\varrho(\delta, \beta)$, $(\hat{A}_{ij}, \hat{A}_i, \hat{B})$ satisfy all the conditions of Theorem 4.2.8 and Remark 4.2.13 in $\mathcal{R}_1^+$, with the constants depending only on (λ, M, β). At this point, recalling also our previous requirements on the smallness of ϱ, we fix ϱ for the rest of Step 3.

Now we follow the argument in Step 5 of the proof of Lemma 4.5.5. Consider Cases (i)–(iii) in $\hat{\mathcal{R}}_{1/2}^+$, which are defined by the same conditions. We discuss only Case (iii), since the other cases are simpler and can be handled similarly. On the other hand, in Case (iii), we use that the boundary condition is *almost tangential*, instead of using its obliqueness as in Step 5 of the proof of Lemma 4.5.5. The details are as follows:

We define $L > 0$ and $d_{\mathbf{x}}$ as in Step 5 of the proof of Lemma 4.5.5 with respect to $\hat{\mathcal{R}}_{1/2}^+$. Then, in this case, $d_{\mathbf{x}} := \mathrm{dist}(\mathbf{x}, P_1) = |\mathbf{x}|$, since we have shifted P_1 to the origin. Also, using (4.5.189), we have $L = L(M)$.

In Case (iii), we fix $\hat{\mathbf{x}} \in \hat{\Gamma}_{1/2}^{(\mathrm{ob})}$ and define $v(\mathbf{x})$, similar to (4.5.76), by

$$v(\mathbf{x}) = \frac{1}{\hat{d}^{1+\beta'}} w(\hat{\mathbf{x}} + \hat{d}\mathbf{x}) \qquad \text{for } \mathbf{x} \in \Omega_1^{\hat{\mathbf{x}}},$$

where $\hat{d} = \frac{d_{\hat{\mathbf{x}}}}{2L}$ and, for $r \in (0, 1]$,

$$\Omega_r^{\hat{\mathbf{x}}} := B_r(\mathbf{0}) \cap \{x_2 > \sqrt{\delta}\,F^{(\hat{\mathbf{x}})}(x_1)\},$$

where $F^{(\hat{\mathbf{x}})}(x_1) := \frac{F(\hat{x}_1+\hat{d}x_1)-F(\hat{x}_1)}{\hat{d}}$ and $F(x_1)$ is from (4.5.188). By (4.5.189),

$$F^{(\hat{\mathbf{x}})}(0) = 0, \qquad \|F^{(\hat{\mathbf{x}})}\|_{2,0,(-2,2)} \le C(L)M, \tag{4.5.194}$$

so that $\|F^{(\hat{\mathbf{x}})}\|_{1,\beta,(-2,2)} \le C(M)$, where we have used that $L = L(M)$. By (4.5.190), we have

$$\|v\|_{L^\infty(\Omega_1^{\hat{\mathbf{x}}})} \le C$$

with C independent of $\hat{\mathbf{x}}$. Also, v satisfies an equation of form (4.5.84) in $\Omega_1^{\hat{\mathbf{x}}}$ and the boundary condition of form (4.5.141) on

$$\Gamma_1^{\hat{\mathbf{x}}} = \partial\Omega_1^{\hat{\mathbf{x}}} \cap \{x_2 = \sqrt{\delta}\, F^{(\hat{\mathbf{x}})}(x_1)\},$$

with ingredients:

$$\begin{aligned}
(\hat{A}_{ij}^{(\hat{\mathbf{x}})}, \hat{A}_i^{(\hat{\mathbf{x}})})(\mathbf{p}, \mathbf{x}) &= (\hat{A}_{ij}, \hat{d}\hat{A}_i)(\hat{d}^{\beta'}\mathbf{p}, \hat{\mathbf{x}} + \hat{d}\mathbf{x}),\\
\hat{B}^{(\hat{\mathbf{x}})}(p_2, z, \mathbf{x}) &= \frac{1}{\hat{d}^{\beta'}}\hat{B}(\hat{d}^{\beta'}p_2, d^{1+\beta'}z, \hat{\mathbf{x}} + \hat{d}\mathbf{x}).
\end{aligned}$$

From the properties of $(\hat{A}_{ij}, \hat{A}_i, \hat{B})$, it follows that $(\hat{A}_{ij}^{(\hat{\mathbf{x}})}, \hat{A}_i^{(\hat{\mathbf{x}})}, \hat{B}^{(\hat{\mathbf{x}})})$ satisfy the uniform ellipticity, the *almost tangentiality* property (4.5.193), and the regularity estimates (4.5.191)–(4.5.192) with the unchanged constants, since $\hat{B}(0, 0, \mathbf{x}) \equiv 0$ and $\hat{d} < 1$. Also, for boundary $\Gamma_1^{\hat{\mathbf{x}}}$, we note (4.5.194). Then we can choose δ_0 small, depending only on these constants, *i.e.*, $(h, t_1, t_2, \lambda, M, \beta)$, such that, for any $\delta \in (0, \delta_0)$, we can employ Theorem 4.2.8 and Remark 4.2.13 in $\Omega_1^{\hat{\mathbf{x}}}$ to obtain

$$\|v\|_{C^{2,\beta}(\Omega_{1/2}^{\hat{\mathbf{x}}})} \le C\big(\|v\|_{L^\infty(\Omega_1^{\hat{\mathbf{x}}})} + \|\hat{B}^{(\varrho)}(0, 0, \cdot)\|_{C^{1,\beta}(\overline{\Gamma_1^{(\hat{\mathbf{x}})}})}\big) \le C(\lambda, M, \beta, \delta).$$

Cases (i)–(ii) in Step 5 of the proof of Lemma 4.5.5 are handled similarly, by using Theorem 4.2.1 for Case (i) and Theorem 4.2.9 for Case (ii).

Combining these estimates as in Step 5 of the proof of Lemma 4.5.5, we find that, for w in (4.5.187),

$$\|w\|_{2,\beta,\mathcal{R}_{1/2}^+}^{(-1-\beta),\{0\}} \le C.$$

This implies (4.5.185) with ϱ fixed above.

4. In this step, we prove (4.5.139). Let σ be fixed as in Step 1. Let δ_0 be small, depending only on $(\lambda, M, \alpha, \beta, \varepsilon)$, to satisfy the conditions in Steps 1–3. Let ϱ be small, depending on $(\lambda, M, \alpha, \varepsilon, \delta, t_0)$, to satisfy the conditions of Steps 2–3. Then we combine estimate (4.5.138) with $s = \frac{\varrho}{100}$ so that constant C_s depends only on $(\kappa, \lambda, M, \alpha, \varepsilon, \delta, t_0)$, estimates (4.5.158) and (4.5.185), and the estimates of Theorem 4.2.9 with Remark 4.2.13 applied in half-balls $B_{\varrho/10}(\hat{\mathbf{x}}) \cap \{x_1 > 0\}$ for any $\hat{\mathbf{x}} = (0, \hat{x}_2)$ with $\hat{x}_2 \in (\frac{\varrho}{4},\ t_0 - \frac{\varrho}{4})$, where the constants in (4.2.72) with $f = g = 0$ depend only on $(\lambda, M, \alpha, \beta, \varepsilon)$, since ϱ depends on these parameters, and the ellipticity constant is δ. Finally, we conclude (4.5.139). □

Next we prove the existence of solutions.

Proposition 4.5.13. *Let $\kappa > 0$, $\lambda > 0$, $M < \infty$, $\alpha \in (0,1)$, $\beta \in [\frac{1}{2},1)$, and $\varepsilon \in (0,\frac{\lambda}{10})$. Then there exist $\alpha_1 \in (0,\frac{1}{2})$ depending only on (κ,λ), and $\sigma, \delta_0 > 0$ depending only on $(\kappa,\lambda,M,\alpha,\beta,\varepsilon)$, such that the following holds: Let Ω be a domain of structure (4.5.1)–(4.5.3) with $h, t_1, t_2, M_{\rm bd} \in (\lambda, \frac{1}{\lambda})$ and $t_0 > 0$. Let $f_{\rm bd}$ satisfy (4.5.15), and let $\varepsilon \in (0,\frac{h}{10})$ and $\delta \in (0,\delta_0)$. Let (4.5.89)–(4.5.111) hold.*

Then there exists a unique solution $u \in C(\overline{\Omega}) \cap C^1(\overline{\Omega}\backslash\overline{\Gamma_0}) \cap C^2(\Omega)$ of (4.5.84)–(4.5.88). Moreover, $u \in C(\overline{\Omega}) \cap C^{1,\alpha_1}(\overline{\Omega}\backslash\overline{\Gamma_0}) \cap C^{2,\alpha_1}(\overline{\Omega}\backslash(\overline{\Gamma_0}\cup\overline{\Gamma_2}))$ and satisfies (4.5.119)–(4.5.120), (4.5.139), as well as (4.5.138) for each $s \in (0,\frac{h}{10})$. The constants in these estimates depend only on the parameters described in Lemma 4.5.12. In particular, both constant C in (4.5.119)–(4.5.120) and C_s in (4.5.138) are independent of δ and t_0.

Proof. We employ a nonlinear method of continuity for this proof.

Let α_1 be sufficiently small in order to satisfy the conditions of Lemma 4.5.1, Proposition 4.5.6, and Lemma 4.5.12 with $\frac{\kappa}{2}$, $\frac{\lambda}{2}$, and $M_{\rm bd} = \frac{2}{\lambda}$, where κ and λ are given in the assumptions. Then $\alpha_1 = \alpha_1(\kappa,\lambda)$. Fix $\alpha \in (0,1)$. Let δ_0 be chosen to satisfy the conditions of Proposition 4.5.6 and Lemma 4.5.12, and let σ be as in Lemma 4.5.12 so that these constants depend only on $(\kappa,\lambda,M,\alpha,\beta,\varepsilon)$.

For $t \in [0,1]$, denote by $\mathcal{P}_t$ the operator defined for $u \in C^1(\overline{\Omega}) \cap C^2(\Omega)$ by

$$\mathcal{P}_t(u) = \begin{pmatrix} \sum_{i,j=1}^2 A_{ij}(Du,\mathbf{x})u_{x_ix_j} + \sum_{i=1}^2 A_i(Du,\mathbf{x})u_{x_i} \\ \big(B(Du,u,\mathbf{x}) - (1-t)B(\mathbf{0},0,\mathbf{x})\big)|_{\Gamma_1} \\ \big(\mathbf{b}^{(2)}(\mathbf{x})\cdot Du - tg_2\big)|_{\Gamma_2} \\ \mathbf{b}^{(3)}(\mathbf{x})\cdot Du|_{\Gamma_3} \end{pmatrix},$$

where $(A_{ij}, A_i, B, b_i^{(k)}, g_2)$ are from (4.5.84)–(4.5.87). Note that we do not include condition (4.5.88) in the operator, since it will be included in the definition of spaces below. With that, the existence of a solution of (4.5.84)–(4.5.88) is equivalent to the existence of u such that $\mathcal{P}_1(u) = 0$.

Now we define the spaces:

$$\mathcal{C}_D = \Big\{u \in C^{(-1-\alpha_1),\{P_1\}\cup\overline{\Gamma_2}}_{2,\alpha_1,\Omega} \;:\; u_\nu|_{\Gamma_3\cap\{0<x_1<\varepsilon\}} = 0,\ u|_{\Gamma_0} = 0\Big\},$$

$$\mathcal{C}_T = \left\{ \begin{array}{l} (f,g_1,g_2,g_3) \in C^{(1-\alpha_1),\{P_1\}\cup\overline{\Gamma_2}}_{0,\alpha_1,\Omega} \times C^{(-\alpha_1),\partial\Gamma_1}_{1,\alpha_1,\Gamma_1} \times C^{\alpha_1}(\overline{\Gamma_2}) \times C^{(-\alpha_1),\partial\Gamma_3}_{1,\alpha_1,\Gamma_3} \\ \text{with } g_3|_{\Gamma_3\cap\{0<x_1<\varepsilon\}} = 0 \end{array} \right\}.$$

Sets $\mathcal{C}_D$ and $\mathcal{C}_T$ are closed subspaces of the Banach spaces $C^{(-1-\alpha_1),\{P_1\}\cup\overline{\Gamma_2}}_{2,\alpha_1,\Omega}$ and $C^{(1-\alpha_1),\{P_1\}\cup\overline{\Gamma_2}}_{0,\alpha_1,\Omega} \times C^{(-\alpha_1),\partial\Gamma_1}_{1,\alpha_1,\Gamma_1} \times C^{\alpha_1}(\overline{\Gamma_2}) \times C^{(-\alpha_1),\partial\Gamma_3}_{1,\alpha_1,\Gamma_3}$, respectively. Therefore, $\mathcal{C}_D$ and $\mathcal{C}_T$ are the Banach spaces with respect to these norms. Also, $u \in \mathcal{C}_D$ implies that u satisfies (4.5.88).

Our assumptions in this proposition imply that $(t, u) \mapsto \mathcal{P}_t(u)$ is a Fréchet-differentiable mapping from $[0,1] \times \mathcal{C}_D$ to $\mathcal{C}_T$. In particular, (4.5.106) implies that, if $u \in \mathcal{C}_D$ and $\mathcal{P}_t(u) = (f, g_1, g_2, g_3)$, then $g_3 = 0$ on $\Gamma_3 \cap \{0 < x_1 < \varepsilon\}$.

Denote

$$\mathcal{T} = \{t \in [0,1] \, : \, \mathcal{P}_t(u_t) = 0 \text{ for some } u_t \in \mathcal{C}_D\}.$$

We need to show that $1 \in \mathcal{T}$. For that, we show that $\mathcal{T} = [0,1]$.

From the explicit definition of $\mathcal{P}_t(\cdot)$, we see that $\mathcal{P}_0(0) = \mathbf{0}$ so that $0 \in \mathcal{T}$; that is, $\mathcal{T}$ is non-empty.

Next, we show that $\mathcal{T}$ is relatively open in $[0,1]$. Let $t_0 \in \mathcal{T}$. Then there exists a corresponding $u_{t_0} \in \mathcal{C}_D$ such that $\mathcal{P}_{t_0}(u_{t_0}) = \mathbf{0}$. From assumptions (4.5.89)–(4.5.111), it follows that the linearization on $u = u_{t_0}$ of the boundary value problem:

$$\mathcal{P}_{t_0}(u) = \mathbf{0}, \qquad u = 0 \text{ on } \Gamma_0$$

is a boundary value problem of form (4.5.4)–(4.5.6), satisfying the conditions of Proposition 4.5.6 with $\alpha = \alpha_1$. In addition to the ellipticity, obliqueness, non-positive zero-order terms, and properties of the boundary conditions at corners $P_1 \cup P_4$, we note the following properties of the linearization:

(i) (4.5.78) with $\alpha = \alpha_1$, which follows from the fact that the equation is linear on $\Omega \cap \{x_1 \geq \varepsilon\}$ with the coefficients satisfying (4.5.91), as well as from (4.5.92) with $u \in C^{(-1-\alpha_1),\{P_1\}\cup\overline{\Gamma_2}}_{2,\alpha_1,\Omega}$;

(ii) (4.5.14) with $\alpha = \alpha_1$, which follows from $u \in C^{(-1-\alpha_1),\{P_1\}\cup\overline{\Gamma_2}}_{2,\alpha_1,\Omega}$ and (4.5.99);

(iii) (4.5.79) which follows from (4.5.96);

(iv) (4.5.80) from (4.5.102);

(v) (4.5.81) from (4.5.106) and $b_0^{(3)} = 0$ in (4.5.87).

Also, (4.5.82) holds for the linearized problem from the definition of space $\mathcal{C}_T$. Then Proposition 4.5.6 with $\alpha = \alpha_1$ implies that the partial Fréchet derivative $D_u\mathcal{P}_{t_0}(u_{t_0}) \, : \, \mathcal{C}_D \mapsto \mathcal{C}_T$ is an isomorphism. Thus, by the implicit function theorem, for each $t \in [0,1]$ sufficiently close to t_0, Problem $\mathcal{P}_t(u) = \mathbf{0}$ has a solution $u_t \in \mathcal{C}_D$; that is, $\mathcal{T}$ is open.

Finally, for each $t \in [0,1]$, Problem $\mathcal{P}_t u = \mathbf{0}$ with $u|_{\Gamma_0} = 0$ has the form of (4.5.84)–(4.5.88) and satisfies (4.5.90)–(4.5.111) with the constants independent of t. In fact, this applies with the same constants as for the original problem for all the conditions with one exception that, for $B_t(\mathbf{p}, z, x_1) := B(\mathbf{p}, z, \mathbf{x}) - (1-t)B(\mathbf{0}, 0, \mathbf{x})$, estimate (4.5.99) holds with constant $2M$ instead of M. Then estimate (4.5.139) of Lemma 4.5.12 implies that $\mathcal{T}$ is closed. Indeed, let $\mathcal{T} \ni t_i \to t_\infty$. Then $t_\infty \in [0,1]$. Also, for each i, there exists $u_i \in \mathcal{C}_D$ such that $\mathcal{P}_{t_i}(u_i) = \mathbf{0}$. Estimate (4.5.139) holds for each i with the uniform constant. Thus, there exists a subsequence u_{i_j} converging in the norm of $C^{(-1-\alpha_1/2),\{P_1\}\cup\overline{\Gamma_2}}_{2,\alpha_1/2,\Omega}$, and its limit u_∞ satisfies $u_\infty \in \mathcal{C}_D$ and $\mathcal{P}_{t_\infty}(u_\infty) = \mathbf{0}$, so that $t_\infty \in \mathcal{T}$.

Thus, $\mathcal{T} = [0,1]$, which implies the existence of a solution $u \in C^{(-1-\alpha_1),\{P_1\}\cup\overline{\Gamma_2}}_{2,\alpha_1,\Omega}$ for the original problem (4.5.84)–(4.5.88). Furthermore, Lemmas 4.5.8 and 4.5.12 imply that this solution satisfies the estimates asserted.

The uniqueness of the solution follows directly from Lemma 4.4.4. □

4.6 HÖLDER SPACES WITH PARABOLIC SCALING

We now define a family of norms convenient for the analysis of degenerate elliptic equations in the open domain $\mathcal{D} \subset \{(x_1,x_2) : x_1 > 0\} \subset \mathbb{R}^2$, with ellipticity degenerating at $\{x_1 = 0\}$.

For $\alpha \in (0,1)$, denote

$$\delta^{(\text{par})}_{\alpha}(\mathbf{x},\tilde{\mathbf{x}}) := \left(|x_1-\tilde{x}_1|^2 + \max(x_1,\tilde{x}_1)|x_2-\tilde{x}_2|^2\right)^{\frac{\alpha}{2}}. \tag{4.6.1}$$

For a nonnegative integer m, and real constants $\sigma > 0$ and $\alpha \in (0,1)$, define the *parabolic* Hölder norms, weighted and scaled by the distance to $\{x_1 = 0\}$:

$$\begin{aligned}
\|u\|^{(\sigma),(\text{par})}_{m,0,\mathcal{D}} &:= \sum_{0\le k+l\le m} \sup_{z\in\mathcal{D}} \left(x_1^{k+\frac{l}{2}-\sigma}|\partial^k_{x_1}\partial^l_{x_2}u(\mathbf{x})|\right),\\
[u]^{(\sigma),(\text{par})}_{m,\alpha,\mathcal{D}} &:= \sum_{k+l=m} \sup_{\mathbf{x},\tilde{\mathbf{x}}\in\mathcal{D},\mathbf{x}\neq\tilde{\mathbf{x}}} \Bigg(\min\left(x_1^{\alpha+k+\frac{l}{2}-\sigma}, \tilde{x}_1^{\alpha+k+\frac{l}{2}-\sigma}\right)\\
&\qquad\qquad \times \frac{|\partial^k_{x_1}\partial^l_{x_2}u(\mathbf{x}) - \partial^k_{x_1}\partial^l_{x_2}u(\tilde{\mathbf{x}})|}{\delta^{(\text{par})}_{\alpha}(\mathbf{x},\tilde{\mathbf{x}})}\Bigg),\\
\|u\|^{(\sigma),(\text{par})}_{m,\alpha,\mathcal{D}} &:= \|u\|^{(\sigma),(\text{par})}_{m,0,\mathcal{D}} + [u]^{(\sigma),(\text{par})}_{m,\alpha,\mathcal{D}},
\end{aligned} \tag{4.6.2}$$

where k and l are nonnegative integers.

For integer $m \ge 0$ and constant $\alpha \in [0,1)$, denote by $C^{m,\alpha}_{\sigma,(\text{par})}(\mathcal{D})$ the completion of space $\{u \in C^\infty(\mathcal{D}) : \|u\|^{(\sigma),(\text{par})}_{m,\alpha,\mathcal{D}} < \infty\}$ under norm $\|\cdot\|^{(\sigma),(\text{par})}_{m,\alpha,\mathcal{D}}$.

Related to the estimates of the degenerate elliptic equations in later sections, we characterize spaces $C^{m,\alpha}_{\sigma,(\text{par})}(\mathcal{D})$ in terms of the $C^{m,\alpha}$–norms of the rescaled functions in the parabolic rectangles.

Fix $\varepsilon > 0$. Denote $\mathcal{D}_\varepsilon := \mathcal{D} \cap \{0 < x_1 < \varepsilon\}$. We define a scaled version of function $u(\mathbf{x})$ in the parabolic rectangles: For $\mathbf{x} \in \mathcal{D}_\varepsilon$ and $\rho \in (0,1]$,

$$R_{\mathbf{x},\rho} := \left\{(s,t) : |s-x_1| < \frac{\rho}{4}x_1,\ |t-x_2| < \frac{\rho}{4}\sqrt{x_1}\right\} \cap \mathcal{D}_\varepsilon. \tag{4.6.3}$$

Note that, for $\rho \in (0,1]$,

$$R_{\mathbf{x},\rho} \subset \mathcal{D}_\varepsilon \cap \{(s,t) : \frac{3x_1}{4} < s < \frac{5x_1}{4}\}. \tag{4.6.4}$$

For $\rho \in (0,1]$, denote $Q_\rho := (-\rho,\rho) \times (-\rho,\rho)$. Then, rescaling $R_{\mathbf{x},\rho}$, we obtain

$$Q^{(\mathbf{x})}_\rho := \left\{(S,T) \in Q_\rho : (x_1 + \frac{x_1}{4}S, x_2 + \frac{\sqrt{x_1}}{4}T) \in \mathcal{D}_\varepsilon\right\}. \tag{4.6.5}$$

Denote by $u^{(\mathbf{x})}(S,T)$ the following function in $Q_\rho^{(\mathbf{x})}$:

$$u^{(\mathbf{x})}(S,T) := \frac{1}{x_1^\sigma} u(x_1 + \frac{x_1}{4}S, x_2 + \frac{\sqrt{x_1}}{4}T) \qquad \text{for } (S,T) \in Q_\rho^{(\mathbf{x})}. \tag{4.6.6}$$

Lemma 4.6.1. *For any* $\varepsilon \in (0,1)$ *and* $\rho \in (0,1]$, *we have*

$$C^{-1} \sup_{\mathbf{x}\in\mathcal{D}_{3\varepsilon/4}} \|u^{(\mathbf{x})}\|_{m,\alpha,Q_\rho^{(\mathbf{x})}} \le \|u\|^{(\sigma),(\text{par})}_{m,\alpha,\mathcal{D}_\varepsilon} \le C \sup_{\mathbf{x}\in\mathcal{D}_\varepsilon} \|u^{(\mathbf{x})}\|_{m,\alpha,Q_\rho^{(\mathbf{x})}}, \tag{4.6.7}$$

where C *depends only on* (m,α,σ,ρ) *and is independent of* $\mathcal{D}$ *and* $\varepsilon > 0$.

Proof. Most of the asserted estimates follow directly from the definitions. We prove only the estimate of $[u]^{(\sigma),(\text{par})}_{m,\alpha,\mathcal{D}}$ in terms of the right-hand side of (4.6.7).

For simplicity of notation, we consider only Case $\rho = 1$, since Case $\rho \in (0,1)$ can be handled similarly. We also use the notation for the rectangles in (4.6.3):

$$R_{\mathbf{x}} := R_{\mathbf{x},1}.$$

The universal constant C below depends only on (m,α,σ) and $\rho = 1$.

For integers $k, l \ge 0$, differentiating (4.6.6), we obtain

$$\frac{1}{4^{k+l}} x_1^{k+\frac{l}{2}-\sigma} \partial_{x_1}^k \partial_{x_2}^l u(\mathbf{x}) = \partial_S^k \partial_T^l u^{(\mathbf{x})}(\mathbf{0}),$$

which implies

$$\|u\|^{(\sigma),(\text{par})}_{m,0,\mathcal{D}_\varepsilon} \le C \sup_{\mathbf{x}\in\mathcal{D}_\varepsilon} \|u^{(\mathbf{x})}\|_{m,\alpha,Q_1^{(\mathbf{x})}} =: M. \tag{4.6.8}$$

Let $k + l = m$ and $\alpha \in (0,1)$. Let $\mathbf{x} = (x_1,x_2), \tilde{\mathbf{x}} = (\tilde{x}_1,\tilde{x}_2) \in \mathcal{D}$, and let

$$x_1 \ge \tilde{x}_1.$$

If $\tilde{\mathbf{x}} \in R_{\mathbf{x}}$, then, from (4.6.5)–(4.6.6), there exist $(S_j,T_j) \in Q_1^{(\mathbf{x})}$, $j = 1,2$, such that $\mathbf{x} = \mathbf{x} + (\frac{x_1}{4}S_1, \frac{\sqrt{x_1}}{4}T_1)$ and $\tilde{\mathbf{x}} = \mathbf{x} + (\frac{x_1}{4}S_2, \frac{\sqrt{x_1}}{4}T_2)$, where $(S_1,T_1) = \mathbf{0}$. With this, we have

$$\frac{x_1^{\alpha+k+\frac{l}{2}-\sigma} |\partial_{x_1}^k \partial_{x_2}^l u(\mathbf{x}) - \partial_{x_1}^k \partial_{x_2}^l u(\tilde{\mathbf{x}})|}{4^{k+l+\alpha}\left(|x_1-\tilde{x}_1|^2 + x_1|x_2-\tilde{x}_2|^2\right)^{\frac{\alpha}{2}}} = \frac{|\partial_S^k \partial_T^l u^{(\mathbf{x})}(\mathbf{0}) - \partial_S^k \partial_T^l u^{(\tilde{\mathbf{x}})}(S_2,T_2)|}{|(S_2,T_2)|^\alpha}.$$

Also, $\tilde{\mathbf{x}} \in R_{\mathbf{x}}$ implies $\tilde{x}_1 \in (\frac{3}{4}x_1, \frac{5}{4}x_1)$. It follows that

$$\min(x_1^{\alpha+k+\frac{l}{2}-\sigma}, \tilde{x}_1^{\alpha+k+\frac{l}{2}-\sigma}) \frac{|\partial_{x_1}^k \partial_{x_2}^l u(\mathbf{x}) - \partial_{x_1}^k \partial_{x_2}^l u(\tilde{\mathbf{x}})|}{\delta_\alpha^{(\text{par})}(\mathbf{x},\tilde{\mathbf{x}})} \le C \|u^{(\mathbf{x})}\|_{m,\alpha,Q_1^{(\mathbf{x})}}.$$

If $\tilde{\mathbf{x}} \notin R_{\mathbf{x}}$, there are two cases:

(i) $x_1 - \tilde{x}_1 \ge \frac{x_1}{4}$;

(ii) $0 < x_1 - \tilde{x}_1 < \frac{x_1}{4}$, but $\sqrt{x_1}|x_2 - \tilde{x}_2| \geq \frac{x_1}{4}$.

Thus, in both Cases (i) and (ii),

$$\delta_\alpha^{(\mathrm{par})}(\mathbf{x}, \tilde{\mathbf{x}}) \geq \left(\frac{x_1}{4}\right)^\alpha \geq \left(\frac{\tilde{x}_1}{4}\right)^\alpha. \tag{4.6.9}$$

Also, $0 < \tilde{x}_1 \leq x_1 \leq \varepsilon < 1$. Then, using (4.6.8),

$$\begin{aligned} \frac{|\partial_{x_1}^k \partial_{x_2}^l u(\mathbf{x}) - \partial_{x_1}^k \partial_{x_2}^l u(\tilde{\mathbf{x}})|}{\delta_\alpha^{(\mathrm{par})}(\mathbf{x}, \tilde{\mathbf{x}})} &\leq C \left(\frac{|\partial_{x_1}^k \partial_{x_2}^l u(\mathbf{x})|}{x_1^\alpha} + \frac{|\partial_{x_1}^k \partial_{x_2}^l u(\tilde{\mathbf{x}})|}{\tilde{x}_1^\alpha} \right) \\ &\leq C \max(x_1^{-\alpha-k-\frac{l}{2}+\sigma}, \tilde{x}_1^{-\alpha-k-\frac{l}{2}+\sigma}) M, \end{aligned} \tag{4.6.10}$$

where M is from (4.6.8). This implies

$$\min(x_1^{\alpha+k+\frac{l}{2}-\sigma}, \tilde{x}_1^{\alpha+k+\frac{l}{2}-\sigma}) \frac{|\partial_{x_1}^k \partial_{x_2}^l u(\mathbf{x}) - \partial_{x_1}^k \partial_{x_2}^l u(\tilde{\mathbf{x}})|}{\delta_\alpha^{(\mathrm{par})}(\mathbf{x}, \tilde{\mathbf{x}})} \leq CM.$$

The proof is completed. □

We also note the following properties:

Lemma 4.6.2. *Let $\mathcal{D}$ be an open bounded subset of $\mathbb{R}^2$. Let m be a nonnegative integer, $\alpha \in (0,1)$, and $\sigma > 0$. Then there exists C such that, for any $\varepsilon > 0$,*

$$\|u\|_{m,\alpha,\mathcal{D}}^{(\sigma),(\mathrm{par})} \leq C \left(\|u\|_{m,\alpha,\mathcal{D}\cap\{x_1>\varepsilon\}}^{(\sigma),(\mathrm{par})} + \|u\|_{m,\alpha,\mathcal{D}\cap\{x_1<2\varepsilon\}}^{(\sigma),(\mathrm{par})} \right) \tag{4.6.11}$$

for any $u \in C_{\sigma,(\mathrm{par})}^{m,\alpha}(\mathcal{D})$.

Proof. The assertion obviously holds for norms $\|\cdot\|_{m,0,\mathcal{D}}^{(\sigma),(\mathrm{par})}$. Then we only need to estimate $[u]_{m,\alpha,\mathcal{D}}^{(\sigma),(\mathrm{par})}$ in terms of the right-hand side of (4.6.11). More precisely, it suffices to consider points $\mathbf{x}$ and $\tilde{\mathbf{x}}$ with $x_1 > 2\varepsilon$ and $\tilde{x}_1 < \varepsilon$, since the other cases are included in the norms on the right-hand side of (4.6.11). Therefore, $x_1 - \tilde{x}_1 > \frac{1}{2}x_1$ so that we can follow the argument in Case (i) in the last part of the proof of Lemma 4.6.1. We first see that (4.6.9) holds, and then we obtain the first inequality in (4.6.10) and, from that,

$$\begin{aligned} &\min(x_1^{\alpha+k+\frac{1}{2}-\sigma}, \tilde{x}_1^{\alpha+k+\frac{1}{2}-\sigma}) \frac{|\partial_{x_1}^k \partial_{x_2}^l u(\mathbf{x}) - \partial_{x_1}^k \partial_{x_2}^l u(\tilde{\mathbf{x}})|}{\delta_\alpha^{(\mathrm{par})}(\mathbf{x}, \tilde{\mathbf{x}})} \\ &\leq \|u\|_{m,0,\mathcal{D}}^{(\sigma),(\mathrm{par})} \leq \|u\|_{m,0,\mathcal{D}\cap\{x_1>\varepsilon\}}^{(\sigma),(\mathrm{par})} + \|u\|_{m,0,\mathcal{D}\cap\{x_1<2\varepsilon\}}^{(\sigma),(\mathrm{par})}. \end{aligned}$$

□

Lemma 4.6.3. *Let $\mathcal{D}$ be an open bounded subset of $\mathbb{R}^2$ with a Lipschitz boundary. Let m_1 and m_2 be nonnegative integers, $\alpha_1, \alpha_2 \in [0,1)$, and $m_1 + \alpha_1 > m_2 + \alpha_2$. Let $\sigma_1 > \sigma_2 > 0$. Then $C_{\sigma_1,(\mathrm{par})}^{m_1,\alpha_1}(\mathcal{D})$ is compactly imbedded into $C_{\sigma_2,(\mathrm{par})}^{m_2,\alpha_2}(\mathcal{D})$.*

Proof. We note first that, for every $\varepsilon > 0$, norm $\|\cdot\|^{(\sigma),(\mathrm{par})}_{m,\alpha,\mathcal{D}\cap\{x_1>\varepsilon\}}$ is equivalent to the standard Hölder norm $\|\cdot\|_{m,\alpha,\mathcal{D}\cap\{x>\varepsilon\}}$. This follows directly from the definitions (though the constants in the equivalence of the norms would blow up as $\varepsilon \to 0$).

Let $\|u_i\|^{(\sigma_1),(\mathrm{par})}_{m_1,\alpha_1,\mathcal{D}} \le C$ for $i = 1, 2, \ldots$. Let $j = 1$. From the equivalence of the norms mentioned above and the standard compactness results on the Hölder spaces, there exists a subsequence of $\{u_i\}$, denoted as $\{u_{1k}\}_{k=1}^\infty$, which is a Cauchy sequence in $C^{m_2,\alpha_2}_{\sigma_2,(\mathrm{par})}(\mathcal{D}\cap\{x_1 > \frac{1}{2j}\})$. Then, for $j = 2$, we can select a further subsequence, denoted as $\{u_{2k}\}_{k=1}^\infty$, with the properties stated above (for $j = 2$), and continue the process for $j = 3, 4, \ldots$.

Now, using the properties of the selected subsequences, for each j, there exists $N(j) > 0$ so that, for any $k, l \ge N(j)$,

$$\|u_{jk} - u_{jl}\|^{(\sigma_2),(\mathrm{par})}_{m_2,\alpha_2,\mathcal{D}\cap\{x_1>\frac{1}{2j}\}} \le j^{\sigma_2-\sigma_1}.$$

Moreover, we can choose these $N(j)$ so that $N(j+1) > N(j)$ for each j. Then it follows that $\{u_{jN(j)}\}_{j=1}^\infty$ is a subsequence of the original sequence $\{u_i\}$.

Also, for any $j, k, l = 1, 2, \ldots$,

$$\begin{aligned}\|u_{jk} - u_{jl}\|^{(\sigma_2),(\mathrm{par})}_{m_2,\alpha_2,\mathcal{D}\cap\{x_1<\frac{1}{j}\}} &\le \|u_{jk}\|^{(\sigma_2),(\mathrm{par})}_{m_2,\alpha_2,\mathcal{D}\cap\{x_1<\frac{1}{j}\}} + \|u_{jl}\|^{(\sigma_2),(\mathrm{par})}_{m_2,\alpha_2,\mathcal{D}\cap\{x_1<\frac{1}{j}\}}\\ &\le j^{\sigma_2-\sigma_1}\left(\|u_{jk}\|^{(\sigma_1),(\mathrm{par})}_{m_1,\alpha_1,\mathcal{D}} + \|u_{jl}\|^{(\sigma_1),(\mathrm{par})}_{m_1,\alpha_1,\mathcal{D}}\right)\\ &\le Cj^{\sigma_2-\sigma_1}.\end{aligned}$$

From the estimates above and Lemma 4.6.2, there exists C so that, for each $j_1 > j_2 \ge 1$,

$$\|u_{j_1N(j_1)} - u_{j_2N(j_2)}\|^{(\sigma_2),(\mathrm{par})}_{m_2,\alpha_2,\mathcal{D}} \le Cj_2^{\sigma_2-\sigma_1}.$$

That is, $\{u_{jN(j)}\}_{j=1}^\infty$ is a Cauchy sequence in $C^{m_2,\alpha_2}_{\sigma_2,(\mathrm{par})}(\mathcal{D})$ for $\sigma_2 < \sigma_1$. ☐

We also note the following fact:

Lemma 4.6.4. *Let $0 \le \alpha_1 < \alpha_2 < 1$, $\sigma > 0$, and let m be a nonnegative integer. Then*

$$\|u\|^{(\sigma),(\mathrm{par})}_{m,\alpha_1,\mathcal{D}} \le 9\|u\|^{(\sigma),(\mathrm{par})}_{m,\alpha_2,\mathcal{D}}.$$

Proof. It suffices to show that

$$[u]^{(\sigma),(\mathrm{par})}_{m,\alpha_1,\mathcal{D}} \le 8\|u\|^{(\sigma),(\mathrm{par})}_{m,0,\mathcal{D}} + [u]^{(\sigma),(\mathrm{par})}_{m,\alpha_2,\mathcal{D}}. \tag{4.6.12}$$

For all nonnegative integers k and l with $k + l = m$, and $\alpha \in (0, 1)$, and for any $\mathbf{x}, \tilde{\mathbf{x}} \in \mathcal{D}$ with $\mathbf{x} \neq \tilde{\mathbf{x}}$, define

$$A^{(\alpha)}_{kl}(\mathbf{x}, \tilde{\mathbf{x}}) := \min(x_1^{\alpha+k+\frac{l}{2}-\sigma}, \tilde{x}_1^{\alpha+k+\frac{l}{2}-\sigma})\frac{|\partial^k_{x_1}\partial^l_{x_2}u(\mathbf{x}) - \partial^k_{x_1}\partial^l_{x_2}u(\tilde{\mathbf{x}})|}{\delta^{(\mathrm{par})}_\alpha(\mathbf{x}, \tilde{\mathbf{x}})}.$$

To prove (4.6.12), we need to estimate $A^{(\alpha_1)}_{kl}(\mathbf{x},\tilde{\mathbf{x}})$ for all $k,l,\mathbf{x}$, and $\tilde{\mathbf{x}}$ as above.

We use rectangles (4.6.3), with the simplified notation $R_{\mathbf{x}} := R_{\mathbf{x},1}$. We can assume without loss of generality that

$$x_1 \geq \tilde{x}_1.$$

If $\tilde{\mathbf{x}} \in R_{\mathbf{x}}$, it follows that $\delta_1^{(\mathrm{par})}(\mathbf{x},\tilde{\mathbf{x}}) \leq \frac{\sqrt{2}}{4}x_1$. Also, $\tilde{\mathbf{x}} \in R_{\mathbf{x}}$ with $x_1 \geq \tilde{x}_1$ implies $x_1 \geq \tilde{x}_1 \geq \frac{3}{4}x_1$. Then we have

$$\delta_1^{(\mathrm{par})}(\mathbf{x},\tilde{\mathbf{x}}) \leq \frac{\sqrt{2}}{3}\tilde{x}_1 = \frac{\sqrt{2}}{3}\min(x_1,\tilde{x}_1).$$

Now we note that $0 \leq \alpha_1 < \alpha_2 < 1$ implies

$$A^{(\alpha_2)}_{kl}(\mathbf{x},\tilde{\mathbf{x}}) \geq \left(\frac{(x_1,\,\tilde{x}_1)}{\delta_1^{(\mathrm{par})}(\mathbf{x},\tilde{\mathbf{x}})}\right)^{\alpha_2-\alpha_1} A^{(\alpha_1)}_{kl}(\mathbf{x},\tilde{\mathbf{x}}).$$

Therefore, when $\tilde{\mathbf{x}} \in R_{\mathbf{x}}$, we have

$$A^{(\alpha_1)}_{kl}(\mathbf{x},\tilde{\mathbf{x}}) \leq \left(\frac{\sqrt{2}}{3}\right)^{\alpha_2-\alpha_1} A^{(\alpha_2)}_{kl}(\mathbf{x},\tilde{\mathbf{x}}) \leq [u]^{(\sigma),(\mathrm{par})}_{m,\alpha_2,\mathcal{D}},$$

where we have used that $\alpha_2-\alpha_1 \in (0,1)$.

If $\tilde{\mathbf{x}} \notin R_{\mathbf{x}}$, we have shown that inequalities (4.6.9) hold. Then we have

$$\begin{aligned} A^{(\alpha_1)}_{kl}(\mathbf{x},\tilde{\mathbf{x}}) &\leq \frac{(\max(x_1,\tilde{x}_1))^{\alpha_1}}{\delta_\alpha^{(\mathrm{par})}(\mathbf{x},\tilde{\mathbf{x}})}\Big(|x_1|^{k+\frac{l}{2}-\sigma}|\partial^k_{x_1}\partial^l_{x_2}u(\mathbf{x})| + |\tilde{x}_1|^{k+\frac{l}{2}-\sigma}|\partial^k_{x_1}\partial^l_{x_2}u(\tilde{\mathbf{x}})|\Big) \\ &\leq 4^{\alpha_1}\big(2\|u\|^{(\sigma),(\mathrm{par})}_{m,0,\mathcal{D}}\big) \leq 8\|u\|^{(\sigma),(\mathrm{par})}_{m,0,\mathcal{D}}, \end{aligned}$$

since $\alpha_1 \in (0,1)$.

Combining the two cases considered above, we conclude (4.6.12). □

Next we consider the functions on a unit square: $\Omega \equiv (0,1)\times(0,1)$. We use the notations for the sides and vertices of Ω, introduced in Proposition 4.5.2. We introduce the following spaces: For a set $\mathcal{S} \in \{\Omega,\Gamma_1,\Gamma_3\}$, and for integer $k \geq 0$ and constants $\sigma > 0$ and $\alpha \in (0,1)$, we choose $\varepsilon \in (0,1]$ and denote

$$\|u\|^{*,\sigma}_{k,\alpha,\mathcal{S}} := \|u\|^{(-k+1-\alpha),\Gamma_2}_{k,\alpha,\mathcal{S}\cap\{x_1>\varepsilon/10\}} + \|u\|^{(\sigma),(\mathrm{par})}_{k,\alpha,\mathcal{S}\cap\{x_1<\varepsilon/5\}}. \tag{4.6.13}$$

Of course, for a different choice of $\varepsilon \in (0,1]$, we obtain an equivalent norm (4.6.13).

Define

$$C^{k,\alpha}_{*,\sigma}(\mathcal{S}) := \Big\{u \in C^{k-1}(\overline{\mathcal{S}}) \cap C^k(\mathcal{S}) \;:\; \|u\|^{*,\sigma}_{k,\alpha,\mathcal{S}} < \infty\Big\}. \tag{4.6.14}$$

Lemma 4.6.5. *Let $\alpha_1,\alpha_2 \in (0,1)$, and $\alpha_1 > \alpha_2$. Let $\sigma \geq 0$. Then $C^{2,\alpha_1}_{*,\sigma}(\Omega)$ is dense in $C^{2,\alpha_2}_{*,\sigma}(\Omega)$.*

Proof. Let $u \in C^{2,\alpha_2}_{*,\sigma_2}(\Omega)$. We approximate u by the following functions $u_\varepsilon \in C^\infty(\overline{\Omega} \setminus \Gamma_0)$ for $\varepsilon > 0$:

We first extend u to a larger region. Using Theorem 13.9.5, we extend u through the boundary parts $\Gamma_1 = \{x_2 = 1\} \cap \partial\Omega$ and $\Gamma_3 = \{x_2 = 0\} \cap \partial\Omega$ so that $u \in C^{2,\alpha_2}_{*,\sigma}((0,1) \times (-1,2))$. In particular, we obtain that $u \in W^{2,\frac{1}{1-\alpha_2}}((0,1) \times (-1,2)) \cap C^{1,\alpha_2}([\frac{1}{2},1] \times [-1,2])$. Then the standard extension of order 2 of u through the boundary part $\{x_1 = 1\}$ yields the extended function u that is in $W^{2,\frac{1}{1-\alpha_2}}((0,2) \times (-1,2)) \cap C^{1,\alpha_2}([\frac{1}{2},2] \times [-1,2])$; moreover, $u \in C^{(-1-\alpha_2),\{x_1=1\}}_{2,\alpha_2}((0,1) \times (-1,2)) \cap C^{(-1-\alpha_2),\{x_1=1\}}_{2,\alpha_2}((1,2) \times (-1,2))$. In particular, $|D^2 u(\mathbf{x})| \le C|x_1 - 1|^{\alpha_2 - 1}$ for $\mathbf{x} \in ((\frac{1}{2},2) \times (-1,2)) \setminus \{x_1 = 1\}$.

Let $\zeta \in C^\infty_c(\mathbb{R}^2)$ with $\zeta \ge 0$ and $\int_{\mathbb{R}^2} \zeta \, d\mathbf{x} = 1$. We introduce the following elliptic and parabolic rescaling on ζ: For $r > 0$, let

$$\zeta^{(\text{ell})}_r(\mathbf{x}) = \frac{1}{r^2}\zeta(\frac{\mathbf{x}}{r}), \qquad \zeta^{(\text{par})}_r(\mathbf{x}) = \frac{1}{r^{3/2}}\zeta(\frac{x_1}{r}, \frac{x_2}{\sqrt{r}}).$$

Let $\eta \in C^\infty(\mathbb{R})$ with $\eta \equiv 1$ on $(\frac{3}{4},\infty)$, $\eta \equiv 0$ on $(-\infty,\frac{1}{4})$, and $\eta \ge 0$ on $\mathbb{R}$. Then, for u extended as described above, we define that, for $\varepsilon \in (0,\frac{1}{10})$,

$$u_\varepsilon(\mathbf{x}) = \eta(x_1)\int_{\mathbb{R}^2} u(\hat{\mathbf{x}})\zeta^{(\text{ell})}_\varepsilon(\mathbf{x} - \hat{\mathbf{x}})d\hat{\mathbf{x}} + \big(1 - \eta(x_1)\big)\int_{\mathbb{R}^2} u(\hat{\mathbf{x}})\zeta^{(\text{par})}_{2\varepsilon x_1}(\mathbf{x} - \hat{\mathbf{x}})d\hat{\mathbf{x}}$$

for $\mathbf{x} \in \Omega$. This can also be written as

$$\begin{aligned} u_\varepsilon(\mathbf{x}) &= \eta(x_1)\int_{\mathbb{R}^2} u(\mathbf{x} - \varepsilon\mathbf{X})\zeta(\mathbf{X})d\mathbf{X} \\ &\quad + \big(1 - \eta(x_1)\big)\int_{\mathbb{R}^2} u(x_1 - \varepsilon x_1 X_1, x_2 - \sqrt{\varepsilon x_1}X_2)\zeta(\mathbf{X})\,d\mathbf{X}. \end{aligned}$$

Now a standard calculation by using the properties of the extended function u yields that $u_\varepsilon \in C^{2,\alpha_1}_{*,\sigma}(\Omega)$ and $\|u_\varepsilon - u\|^{*,\sigma}_{2,\alpha_2,\Omega} \to 0$ as $\varepsilon \to 0^+$. □

4.7 DEGENERATE ELLIPTIC EQUATIONS

4.7.1 Mixed boundary value problem in a domain with corners for degenerate elliptic equations: The nonlinear case

In this section, we establish the existence of solutions for Problem (4.5.84)–(4.5.88) under the conditions that include the case that the equation is degenerate elliptic with ellipticity degenerating near Γ_0. That is, we require δ to be small in (4.5.90), including Case $\delta = 0$. Then we consider either degenerate elliptic or close to degenerate elliptic equations.

Furthermore, we consider the domain of structure (4.5.1)–(4.5.3) with $t_0 \ge 0$. In Case $t_0 = 0$, *i.e.*, $f_{\text{bd}}(0) = 0$, the boundary part Γ_0 becomes a single point, *i.e.*, $P_1 = P_4 = \mathbf{0}$ and

$$\overline{\Gamma_0} = P_1 = P_4 \qquad \text{if } t_0 = 0.$$

In this case, we interpret (4.5.88) as a one-point Dirichlet condition. To simplify the notation, we denote point $P_1 = P_4$ by P_0:

$$P_0 := P_1 = P_4 = \mathbf{0} \qquad \text{when } t_0 = 0 \text{ in (4.5.1)–(4.5.3).} \tag{4.7.1}$$

Remark 4.7.1. *Let $t_0 = 0$. Then the oblique derivative conditions are prescribed on sides Γ_1 and Γ_3 that meet at P_0, where we have used the notation in* (4.7.1). *If the obliqueness holds at P_0 in the sense of Definition* 4.4.3, *and the equation is uniformly elliptic at and near P_0, Lieberman's Harnack estimate at the corner* [189, Lemma 2.2], *combined with* (4.5.101), *would not allow us to prescribe a one-point Dirichlet condition at P_0. Thus, it is important that* (4.5.102), (4.5.106), *and the structure of Ω imply that conditions* (4.5.85) *and* (4.5.88) *do* not *satisfy the obliqueness at P_0. In Proposition* 4.7.2 *below, we show that, under these conditions and for the small ellipticity constant of the equation at and near P_0, we can prescribe the one-point Dirichlet condition at P_0. The case of similar boundary conditions, but with the uniform ellipticity of the equation near P_0 (without smallness assumption on the ellipticity constant), will be considered in §4.8.*

Now we show the existence of solutions.

Proposition 4.7.2. *Let $\kappa > 0$, $\lambda \in (0, \frac{1}{2})$, $M < \infty$, $\alpha \in (0,1)$, $\beta \in [\frac{1}{2}, 1)$, and $\varepsilon \in (0, \frac{\lambda}{10})$. Then there exist $\alpha_1 \in (0, \frac{1}{2})$ depending only on (κ, λ), and $\sigma, \delta_0 > 0$ depending only on $(\kappa, \lambda, M, \alpha, \beta, \varepsilon)$ such that the following hold: Let Ω be a domain of structure* (4.5.1)–(4.5.3) *with $h, t_1, t_2, M_{bd} \in (\lambda, \frac{1}{\lambda})$ and $t_0 \geq 0$. Let f_{bd} satisfy* (4.5.15), *and let $\varepsilon \in (0, \frac{h}{10})$. Let $\delta \in [0, \delta_0)$. Let* (4.5.89)–(4.5.111) *be satisfied. Then there exists a unique solution $u \in C(\overline{\Omega}) \cap C^1(\overline{\Omega} \setminus \overline{\Gamma_0}) \cap C^2(\Omega)$ of* (4.5.84)–(4.5.88). *Moreover, u satisfies* (4.5.119)–(4.5.120) *with constant C depending only on $(\lambda, M, \varepsilon)$. Furthermore, $u \in C(\overline{\Omega}) \cap C^{1,\alpha_1}(\overline{\Omega} \setminus \overline{\Gamma_0}) \cap C^{2,\alpha_1}(\overline{\Omega} \setminus (\overline{\Gamma_0} \cup \overline{\Gamma_2}))$ and satisfies* (4.5.138) *for each $s \in (0, \frac{h}{10})$, with constant C_s depending only on $(\kappa, \lambda, M, \alpha, \beta, \varepsilon, s)$.*

Proof. Let α_1, σ, and δ_0 be the constants defined in Proposition 4.5.13 for $(\kappa, M, \alpha, \beta)$, and $(\lambda^2\sqrt{\lambda^2+1}, \frac{\varepsilon}{2})$ instead of (λ, ε). Note that α_1 depends only on (κ, λ) given in this proposition.

Let $\delta \in (0, \delta_0)$. We assume that (4.5.89)–(4.5.111) are satisfied with σ fixed above, and with $(\kappa, \lambda, M, \alpha, \beta, \varepsilon)$ given in this proposition.

We approximate (4.5.84)–(4.5.88) in Ω by the problems in domains $\Omega^{(\gamma)}$ of structure (4.5.1) and (4.5.3) with $t_0^{(\gamma)} > 0$. More precisely, for each $\gamma \in (0, \min\{\frac{\varepsilon}{2}, \delta_0 - \delta\})$, denote

$$\Omega^{(\gamma)} := \Omega \cap \{x_1 > \gamma\}, \qquad \Gamma_0^{(\gamma)} := \partial\Omega^{(\gamma)} \cap \{x_1 = \gamma\},$$
$$\Gamma_k^{(\gamma)} := \Gamma_k \cap \{x_1 > \gamma\} \qquad \text{for } k = 1, 2, 3.$$

Denoting the left-hand side of equation (4.5.84) by $\mathcal{N}(Du, \mathbf{x})$, we consider the following boundary value problem in $\Omega^{(\gamma)}$ for $u_\gamma \in C^1(\overline{\Omega^{(\gamma)}}) \cap C^2(\Omega^{(\gamma)})$:

$$\begin{aligned} &\mathcal{N}(Du, \mathbf{x}) = 0 && \text{in } \Omega^{(\gamma)}, \\ &B(Du, u, \mathbf{x}) = 0 && \text{on } \Gamma_1^{(\gamma)}, \\ &\mathbf{b}^{(k)} \cdot Du + b_0^{(k)} u = g_k && \text{on } \Gamma_k^{(\gamma)},\ k = 2, 3, \\ &u = 0 && \text{on } \Gamma_0^{(\gamma)}, \end{aligned} \tag{4.7.2}$$

where $g_3 \equiv 0$.

We shift the coordinates replacing x_1 by $x_1 - \gamma$, so that $\{x_1 = \gamma\}$ becomes $\{x_1 = 0\}$. In the new coordinates, domain $\Omega^{(\gamma)}$ is of structure (4.5.1) with constants $(t_k^{(\gamma)}, h^{(\gamma)})$ and function $f_{\rm bd}^{(\gamma)}$, where

$$\begin{aligned} &t_0^{(\gamma)} = f_{\rm bd}(\gamma) \geq \min\{t_1\gamma + t_0,\, t_2\}, \quad (t_1^{(\gamma)}, t_2^{(\gamma)}) = (t_1, t_2), \\ &h^{(\gamma)} = h - \gamma, \quad f_{\rm bd}(x_1)^{(\gamma)} = f_{\rm bd}(x_1 + \gamma). \end{aligned}$$

Thus, $t_0^{(\gamma)} \geq \min\{t_1\gamma,\, t_2\} \geq \lambda\gamma > 0$. Also, $h^{(\gamma)} \geq \frac{9}{10}h$, $M_{\rm bd}^{(\gamma)} = M_{\rm bd}$, and $f_{\rm bd}^{(\gamma)}$ satisfies (4.5.15) on $(0, h^{(\gamma)})$.

Moreover, in the shifted coordinates, the boundary value problem (4.7.2) for u_γ satisfies conditions (4.5.89)–(4.5.111) in $\Omega^{(\gamma)}$ for each $\gamma \in (0, \min\{\frac{\varepsilon}{2},\, \delta_0 - \delta\})$, with constants (κ, M, α) given in the proposition, and with $(\lambda, \delta, \varepsilon)$ replaced by $(\lambda^2\sqrt{\lambda^2+1},\, \delta + \gamma,\, \frac{\varepsilon}{2})$. Indeed, condition (4.5.90) in the shifted coordinates is satisfied with $\delta + \gamma$ instead of δ, where we note that $\delta + \gamma \in (0, \delta_0)$. The lower bound in condition (4.5.89) with a modified constant follows from (4.5.90), since $\mathrm{dist}(\mathbf{x}, \Gamma_0) \leq \frac{1}{\lambda\sqrt{\lambda^2+1}} x_1$ by (4.5.2) with $M_{\rm bd} \in [\lambda, \frac{1}{\lambda}]$, and the upper bound in (4.5.89) is unchanged. Furthermore, since $\gamma < \frac{\varepsilon}{2}$, conditions (4.5.91)–(4.5.111) are satisfied with $\frac{\varepsilon}{2}$ instead of ε, and the other constants are unchanged, for Problem (4.7.2) in the shifted coordinates. We note that the modified constant λ depends only on the original constant λ.

Finally, since $\lambda \in (0, \frac{1}{2})$, $\hat{\lambda} := \lambda^2\sqrt{\lambda^2+1} > \frac{\lambda}{2}$. Also, since $\varepsilon \in (0, \frac{\lambda}{10})$, $h \geq \lambda$, and $\gamma \leq \frac{\varepsilon}{2}$, then $h^{(\gamma)} = h - \gamma > \frac{h}{2}$. Thus, $(h^{(\gamma)}, t_1^{(\gamma)}, t_2^{(\gamma)}, M_{bd}^{(\gamma)}) \in [\hat{\lambda}, \frac{1}{\hat{\lambda}}]$.

Therefore, by Proposition 4.5.13, there exists a solution $u_\gamma \in C(\overline{\Omega^{(\gamma)}}) \cap C^1(\overline{\Omega^{(\gamma)}} \setminus \overline{\Gamma_0^{(\gamma)}}) \cap C^2(\Omega^{(\gamma)})$ of (4.7.2) and, for each $s \in (0, \frac{h}{10})$, solution u_γ satisfies (4.5.138) in $\Omega^{(\gamma)} \cap \{x_1 > s\}$ in the shifted coordinates.

Then, for each $\gamma \in (0, \min\{\frac{\varepsilon}{2},\, \delta_0 - \delta\})$, we can apply Lemma 4.5.8 (with $\frac{\varepsilon}{2}$ instead of ε) to Problem (4.8.8) to obtain that u_γ satisfies (4.5.120) in $\Omega^{(\gamma)}$. Writing this estimate in the original coordinates, we obtain that, for each $\gamma \in (0, \min\{\frac{\varepsilon}{2},\, \delta_0 - \delta\})$,

$$|u_\gamma(\mathbf{x})| \leq C(x_1 - \gamma) \leq Cx_1 \qquad \text{in } \Omega \cap \{x_1 > \gamma\}, \tag{4.7.3}$$

where constant C depends only on $(\lambda, M, \varepsilon)$.

Since each u_γ satisfies (4.5.138) in $(\Omega^{(\gamma)})^{(s)} \equiv \Omega^{(\gamma+s)}$ for each $s \in (0, \frac{h}{10})$, we have

$$\|u_\gamma\|^{(-1-\alpha_1),\,\overline{\Gamma_2}}_{2,\alpha_1,\Omega^{(\gamma+s)}} \leq C_s, \tag{4.7.4}$$

where C_s depends only on $(\kappa, \lambda, \alpha, M, s)$.

Define

$$h_1 = \min\{\frac{\varepsilon}{2}, \frac{h}{10}\}.$$

From (4.7.4) with $s = \frac{h_1}{2}$, there exists a sequence $\gamma_j \to 0^+$ such that u_{γ_j} converges in $C^{(-1-\frac{\alpha_1}{2}),\,\overline{\Gamma_2}}_{2,\alpha_1/2,\Omega^{(h_1)}}$. Similarly, from (4.7.4) with $s = \frac{h_1}{4}, \frac{h_1}{8}, \dots$, there exists a subsequence of $\{u_{\gamma_j}\}$, converging in $C^{(-1-\frac{\alpha_1}{2}),\,\overline{\Gamma_2}}_{2,\alpha_1/2,\Omega^{(h_1/2)}}$, then a further subsequence converging in $C^{(-1-\frac{\alpha_1}{2}),\,\overline{\Gamma_2}}_{2,\alpha_1/2,\Omega^{(h_1/4)}}$, *etc.* By the diagonal procedure, there exists a sequence $\hat{\gamma}_k \to 0$ such that $u_{\hat{\gamma}_k}$ converges to a function u in $C^{(-1-\frac{\alpha_1}{2}),\,\overline{\Gamma_2}}_{2,\alpha_1/2,K}$ for each compact $K \subset \overline{\Omega} \setminus \Gamma_0$. Since each $u_{\hat{\gamma}_k}$ is a solution of (4.7.2) satisfying the uniform estimates (4.7.3)–(4.7.4), it follows that u satisfies $u \in C^{1,\alpha_1}(\overline{\Omega}\setminus\overline{\Gamma_0}) \cap C^{2,\alpha_1}(\overline{\Omega}\setminus(\overline{\Gamma_0}\cup\overline{\Gamma_2}))$ and is a solution of (4.5.84)–(4.5.87) satisfying

$$|u(\mathbf{x})| \leq Cx_1 \qquad \text{in } \Omega, \tag{4.7.5}$$

where $C = C(\lambda, M, \varepsilon)$ is from (4.7.3). Then (4.7.5), combined with the property that $u \in C^{1,\alpha_1}(\overline{\Omega} \setminus \overline{\Gamma_0})$, implies that $u \in C(\overline{\Omega})$, and (4.5.88) holds. Therefore, $u \in C(\overline{\Omega}) \cap C^{1,\alpha_1}(\overline{\Omega}\setminus\overline{\Gamma_0}) \cap C^3(\Omega)$ and is a solution of (4.5.84)–(4.5.88) satisfying (4.5.138) for each $s \in (0, \frac{h}{10})$, where we have used the interior regularity of solutions of (4.5.84). □

Remark 4.7.3. *Note that the existence result in Proposition* 4.7.2 *applies to the case that the ellipticity of the equation either degenerates on* Γ_0 *or is close to degenerate near* Γ_0. *In §4.8, we consider Case* $t_0 = 0$, *when the equation is uniformly elliptic at* $\overline{\Omega}$.

4.7.2 Regularity in the Hölder spaces with parabolic scaling for a class of degenerate elliptic equations

In this section, we make the estimates in the norms defined in (4.6.2) with $\sigma = 2$ for a nonlinear mixed problem for an elliptic equation with ellipticity degenerating near the boundary part Γ_0, under the condition that the solution has a quadratic growth from Γ_0. In Chapter 12, we will have this case for the iteration problem.

We consider domain $\Omega \subset \mathbb{R}^2$ of the form:

$$\Omega = \{\mathbf{x} \,:\, x_1 > 0,\ 0 < x_2 < f(x_1)\}, \tag{4.7.6}$$

where $f \in C^{1,\alpha}(\overline{\mathbb{R}_+})$ and $f > 0$ on $\mathbb{R}_+$. We denote the boundary parts of Ω as follows:

$$\Gamma_0 = \partial\Omega \cap \{x_1 = 0\}, \;\; \Gamma_n = \partial\Omega \cap \{x_2 = 0\}, \;\; \Gamma_f = \partial\Omega \cap \{x_1 > 0, \; x_2 = f(x_1)\}.$$

For $r > 0$, denote

$$\Omega_r = \Omega \cap \{x_1 < r\}, \;\; \Gamma_{n,r} = \Gamma_n \cap \{x_1 < r\}, \;\; \Gamma_{f,r} = \Gamma_f \cap \{x_1 < r\}.$$

For $r > 0$, we consider a boundary value problem of the form:

$$\sum_{i,j=1}^{2} A_{ij}(Du, u, \mathbf{x}) D_{ij} u + \sum_{i=1}^{2} A_i(Du, u, \mathbf{x}) D_i u = 0 \qquad \text{in } \Omega_r, \tag{4.7.7}$$

$$B(Du, u, \mathbf{x}) = 0 \qquad \text{on } \Gamma_{f,r}, \tag{4.7.8}$$

$$D_2 u = 0 \qquad \text{on } \Gamma_{n,r}, \tag{4.7.9}$$

$$u = 0 \qquad \text{on } \Gamma_0. \tag{4.7.10}$$

We assume that (4.7.7) satisfies the following degenerate ellipticity condition: For any $\mathbf{p} \in \mathbb{R}^2$, $z \in \mathbb{R}$, $\mathbf{x} \in \Omega_r$, and $\boldsymbol{\kappa} = (\kappa_1, \kappa_2) \in \mathbb{R}^2$,

$$\lambda|\boldsymbol{\kappa}|^2 \le \sum_{i,j=1}^{2} A_{ij}(\mathbf{p}, z, \mathbf{x}) \frac{\kappa_i \kappa_j}{x_1^{2-\frac{i+j}{2}}} \le \frac{1}{\lambda}|\boldsymbol{\kappa}|^2. \tag{4.7.11}$$

Note that the ellipticity degenerates near $x_1 = 0$. We show the following regularity result for solutions with quadratic growth from Γ_{sonic}:

Theorem 4.7.4. *Let Ω be of form* (4.7.6). *Let $r > 0$, $M \ge 1$, and $l, \lambda \in (0,1)$. Let $\beta \in (0,1)$, and let $f \in C^{1,\beta}([0,r])$ satisfy*

$$\|f\|_{C^{(-1-\beta),\{0\}}_{2,\beta,(0,r)}} \le M, \qquad f \ge l \;\; \text{on } \mathbb{R}_+, \tag{4.7.12}$$

and let Problem (4.7.7)–(4.7.10) *be given in Ω_r. Let functions (A_{ij}, A_i), $j = 1, 2$, satisfy ellipticity* (4.7.11) *and the following regularity properties:*

$$\|(A_{11}, A_{12})\|_{0,1,\mathbb{R}^2\times\mathbb{R}\times\overline{\Omega_r}} \le M, \tag{4.7.13}$$

$$|D_{x_2} A_{11}(\mathbf{p}, z, \mathbf{x})| \le M\sqrt{x_1} \qquad \text{on } \mathbb{R}^2 \times \mathbb{R} \times \Omega_r, \tag{4.7.14}$$

$$\|(A_{22}, A_1, A_2)\|_{0,\mathbb{R}^2\times\mathbb{R}\times\overline{\Omega_r}} + \|D_{(\mathbf{p},z)}(A_{22}, A_1, A_2)\|_{0,\mathbb{R}^2\times\mathbb{R}\times\overline{\Omega_r}} \le M, \tag{4.7.15}$$

$$\sup_{(\mathbf{p},z)\in\mathbb{R}^2\times\mathbb{R}, \mathbf{x}\in\Omega_r} \big|(x_1 D_{x_1}, \sqrt{x_1} D_{x_2})(A_{22}, A_1, A_2)(\mathbf{p}, z, \mathbf{x})\big| \le M. \tag{4.7.16}$$

Let B satisfy

$$\|B\|_{3,\mathbb{R}^2\times\mathbb{R}\times\Gamma_{f,r}} \le M, \tag{4.7.17}$$

$$\partial_{p_1} B(\mathbf{p}, z, \mathbf{x}) \le -M^{-1} \qquad \text{for all } (\mathbf{p}, z) \in \mathbb{R}^2 \times \mathbb{R}, \; \mathbf{x} \in \Gamma_{f,r}, \tag{4.7.18}$$

$$B(\mathbf{0}, 0, \mathbf{x}) = 0 \qquad \text{for all } \mathbf{x} \in \Gamma_{f,r}. \tag{4.7.19}$$

Let $u \in C(\overline{\Omega_r}) \cap C^2(\overline{\Omega_r} \setminus \overline{\Gamma_0})$ *be a solution of* (4.7.7)–(4.7.10) *satisfying*

$$|u(\mathbf{x})| \le Mx_1^2 \qquad \text{in } \Omega_r. \tag{4.7.20}$$

Then, for any $\alpha \in (0,1)$, *there exist* $r_0 \in (0,1]$ *and* $C > 0$ *depending only on* (M, λ, α) *such that*

$$\|u\|^{(2),(\mathrm{par})}_{2,\alpha,\Omega_\varepsilon} \le C, \tag{4.7.21}$$

where

$$\varepsilon = \min\{\frac{r}{2}, r_0, l^2\}. \tag{4.7.22}$$

Proof. In this proof, the universal constant C depends only on (M, λ, β), and constants $C(\alpha)$ and r_0 depend only on $(M, \lambda, \beta, \alpha)$. We divide the proof into five steps.

1. Let $r_0 > 0$ to be chosen below. Denote $r' = \min\{\frac{r}{2}, r_0\}$ so that $\varepsilon = \min\{r', l^2\}$ for ε defined by (4.7.22).

For $\mathbf{x} \in \Omega_{r'}$ and $\rho \in (0,1)$, define

$$\tilde{R}_{\mathbf{x},\rho} := \Big\{(s,t) \,:\, |s - x_1| < \frac{\rho}{4}x_1, |t - x_2| < \frac{\rho}{4}\sqrt{x_1}\Big\}, \quad R_{\mathbf{x},\rho} := \tilde{R}_{\mathbf{x},\rho} \cap \Omega_r. \tag{4.7.23}$$

Then, for any $\mathbf{x} \in \Omega_{r'}$ and $\rho \in (0,1)$,

$$R_{\mathbf{x},\rho} \subset \Omega_r \cap \{(s,t) \,:\, \frac{3}{4}x_1 < s < \frac{5}{4}x_1\} \subset \Omega_r. \tag{4.7.24}$$

Let ε be defined by (4.7.22). For any $\mathbf{x} \in \Omega_\varepsilon$, we have at least one of the following three cases:

(i) $R_{\mathbf{x},1/10} = \tilde{R}_{\mathbf{x},1/10}$;

(ii) $\mathbf{x} \in R_{\mathbf{x}_n,1/2}$ for $\mathbf{x}_n = (x_1, 0) \in \Gamma_{n,\varepsilon}$;

(iii) $\mathbf{x} \in R_{\mathbf{x}_f,1/2}$ for $\mathbf{x}_f = (x_1, f(x_1)) \in \Gamma_{f,\varepsilon}$.

Then it suffices to make the local estimates of Du and D^2u in the following rectangles:

(i) $R_{\hat{\mathbf{x}},1/20}$ for $\hat{\mathbf{x}} \in \Omega_\varepsilon$ such that $R_{\hat{\mathbf{x}},1/10} = \tilde{R}_{\hat{\mathbf{x}},1/10}$;

(ii) $R_{\hat{\mathbf{x}},1/4}$ for $\hat{\mathbf{x}} \in \Gamma_{n,\varepsilon}$;

(iii) $R_{\hat{\mathbf{x}},1/4}$ for $\hat{\mathbf{x}} \in \Gamma_{f,\varepsilon}$.

Since $\varepsilon \le l^2$ by (4.7.22), we obtain from (4.7.23) with $\rho = 1$ and $f \ge l$ in (4.7.12) that

$$\begin{aligned} &R_{\mathbf{x},1} \cap \Gamma_{f,r} = \emptyset && \text{for all } \mathbf{x} = (x_1, 0) \in \Gamma_{n,\varepsilon}, \\ &R_{\mathbf{x},1} \cap \Gamma_{n,r} = \emptyset && \text{for all } \mathbf{x} = (x_1, f(x_1)) \in \Gamma_{f,\varepsilon}. \end{aligned} \tag{4.7.25}$$

Furthermore, denoting

$$Q_\rho := (-\rho,\rho)^2,$$

and introducing variables (S,T) by the invertible change of variables:

$$\mathbf{x} = \hat{\mathbf{x}} + \frac{1}{4}(\hat{x}_1 S, \sqrt{\hat{x}_1}T), \tag{4.7.26}$$

we find that there exists $Q^{\hat{\mathbf{x}}}_\rho \subset Q_\rho$ such that rectangle $R_{\hat{\mathbf{x}},\rho}$ in (4.7.12) is expressed as:

$$R_{\hat{\mathbf{x}},\rho} = \Big\{(\hat{x}_1 + \frac{\hat{x}_1}{4}S, \hat{x}_2 + \frac{\sqrt{\hat{x}_1}}{4}T) \;:\; (S,T) \in Q^{\hat{\mathbf{x}}}_\rho\Big\}. \tag{4.7.27}$$

Rescale u in $R_{\hat{\mathbf{x}},\rho}$ by defining

$$u^{(\hat{\mathbf{x}})}(S,T) := \frac{1}{\hat{x}_1^2}u(\hat{x}_1 + \frac{\hat{x}_1}{4}S,\; \hat{x}_2 + \frac{\sqrt{\hat{x}_1}}{4}T) \qquad \text{for } (S,T) \in Q^{\hat{\mathbf{x}}}_\rho. \tag{4.7.28}$$

Then, by (4.7.20) and (4.7.24),

$$\|u^{(\hat{\mathbf{x}})}\|_{0,Q^{(\hat{\mathbf{x}})}_\rho} \le 4M. \tag{4.7.29}$$

Moreover, since u satisfies equation (4.7.7) in $R_{\hat{\mathbf{x}},\rho}$, then $u^{(\hat{\mathbf{x}})}$ satisfies

$$\sum_{i,j=1}^{2} A^{(\hat{\mathbf{x}})}_{ij} D_{ij}u^{(\hat{\mathbf{x}})} + \sum_{i}^{2} A^{(\hat{\mathbf{x}})}_{i} D_i u^{(\hat{\mathbf{x}})} = 0 \qquad \text{in } Q^{\hat{\mathbf{x}}}_\rho, \tag{4.7.30}$$

where $(A^{\hat{\mathbf{x}}}_{ij}, A^{\hat{\mathbf{x}}}_{i})$ are defined by

$$\begin{aligned} A^{(\hat{\mathbf{x}})}_{ij}(\mathbf{p},z,S,T) &= \hat{x}_1^{\frac{i+j}{2}-2} A_{ij}(4\hat{x}_1 p_1, 4\hat{x}_1^{\frac{3}{2}} p_2, \hat{x}_1^2 z, \mathbf{x}),\\ A^{(\hat{\mathbf{x}})}_{i}(\mathbf{p},z,S,T) &= \frac{1}{4}\hat{x}_1^{\frac{i-1}{2}} A_{i}(4\hat{x}_1 p_1, 4\hat{x}_1^{\frac{3}{2}} p_2, \hat{x}_1^2 z, \mathbf{x}), \end{aligned} \tag{4.7.31}$$

with

$$\mathbf{x} = \hat{\mathbf{x}} + \frac{1}{4}(\hat{x}_1 S,\; \sqrt{\hat{x}_1}T). \tag{4.7.32}$$

From (4.7.11) and (4.7.24), it follows that (4.7.30) is uniformly elliptic, *i.e.*, it satisfies that, for $\mathbf{p} \in \mathbb{R}^2$, $z \in \mathbb{R}$, $(S,T) \in Q^{(\hat{\mathbf{x}})}_\rho$, and $\boldsymbol{\kappa} = (\kappa_1,\kappa_2) \in \mathbb{R}^2$,

$$\frac{\lambda}{2}|\boldsymbol{\kappa}|^2 \le \sum_{i,j=1}^{2} A^{(\hat{\mathbf{x}})}_{ij}(\mathbf{p},z,S,T)\kappa_i\kappa_j \le \frac{2}{\lambda}|\boldsymbol{\kappa}|^2. \tag{4.7.33}$$

Also, from (4.7.33) with $A^{(\hat{\mathbf{x}})}_{12} = A^{(\hat{\mathbf{x}})}_{21}$, we obtain

$$|A^{(\hat{\mathbf{x}})}_{ij}(\mathbf{p},z,S,T)| \le C.$$

Combining this with (4.7.15) and using the fact that, in (4.7.31) for $A_i^{(\hat{\mathbf{x}})}$, the corresponding function A_i is multiplied by $\hat{\mathbf{x}}$ in a nonnegative power, we obtain

$$\|(A_{ij}^{(\hat{\mathbf{x}})}, A_i^{(\hat{\mathbf{x}})})\|_{0,\mathbb{R}^2\times\mathbb{R}\times Q_\rho^{(\hat{\mathbf{x}})}} \le C \qquad \text{for } i,j=1,2. \tag{4.7.34}$$

Next we show that

$$\|D_{(\mathbf{p},z,S,T)}(A_{ij}^{(\hat{\mathbf{x}})}, A_i^{(\hat{\mathbf{x}})})\|_{0,\mathbb{R}^2\times\mathbb{R}\times Q_\rho^{(\hat{\mathbf{x}})}} \le C \qquad \text{for } i,j=1,2. \tag{4.7.35}$$

Indeed, let $(\mathbf{p},z)\in\mathbb{R}^2\times\mathbb{R}$ and $(S,T)\in Q_\rho^{(\hat{\mathbf{x}})}$. Then, using that $\mathbf{x}$ in (4.7.31) is given by (4.7.32), we obtain from (4.7.13)–(4.7.14) that

$$\begin{aligned}
|D_{p_1}A_{11}^{(\hat{\mathbf{x}})}(\mathbf{p},z,S,T)| &= 4\hat{x}_1^{-1}\hat{x}_1|(D_{p_1}A_{11})(4\hat{x}_1p_1, 4\hat{x}_1^{\frac{3}{2}}p_2, \hat{x}_1^2z, \mathbf{x})| \le C,\\
|D_TA_{11}^{(\hat{\mathbf{x}})}(\mathbf{p},z,S,T)| &= \frac{1}{4}\hat{x}_1^{-1}\sqrt{\hat{x}_1}|(D_{x_2}A_{11})(4\hat{x}_1p_1, 4\hat{x}_1^{\frac{3}{2}}p_2, \hat{x}_1^2z, \mathbf{x})|\\
&\le C\hat{x}_1^{-1}\hat{x}_1 = C.
\end{aligned}$$

The estimates of $D_{(p_2,z,S)}A_{11}^{(\hat{\mathbf{x}})}$ are obtained similarly. This confirms (4.7.35) for $A_{11}^{(\hat{\mathbf{x}})}$. Also, (4.7.35) for $A_{12}^{(\hat{\mathbf{x}})}$ is obtained similarly.

Next we show (4.7.35) for $A_{22}^{(\hat{\mathbf{x}})}$. Using (4.7.15)–(4.7.16) and (4.7.31)–(4.7.32), we estimate

$$\begin{aligned}
|D_{p_1}A_{22}^{(\hat{\mathbf{x}})}(\mathbf{p},z,S,T)| &= 4\hat{x}_1|(D_{p_1}A_{22})(4\hat{x}_1p_1, 4\hat{x}_1^{\frac{3}{2}}p_2, \hat{x}_1^2z, \mathbf{x})| \le C,\\
|D_SA_{22}^{(\hat{\mathbf{x}})}(\mathbf{p},z,S,T)| &= \frac{1}{4}\hat{x}_1|(D_{x_1}A_{22})(4\hat{x}_1p_1, 4\hat{x}_1^{\frac{3}{2}}p_2, \hat{x}_1^2z, \mathbf{x})| \le C\hat{x}_1\hat{x}_1^{-1} = C,\\
|D_TA_{22}^{(\hat{\mathbf{x}})}(\mathbf{p},z,S,T)| &= \frac{1}{4}\sqrt{\hat{x}_1}|(D_{x_2}A_{22})(4\hat{x}_1p_1, 4\hat{x}_1^{\frac{3}{2}}p_2, \hat{x}_1^2z, \mathbf{x})| \le C.
\end{aligned}$$

The estimates of $D_{(p_2,z)}A_{22}^{(\hat{\mathbf{x}})}$ are obtained similarly. This confirms (4.7.35) for $A_{22}^{(\hat{\mathbf{x}})}$. Also, (4.7.35) for $A_i^{(\hat{\mathbf{x}})}$, $i=1,2$, is obtained similarly.

2. We first consider Case (i) in Step 1. Let $\hat{\mathbf{x}}\in\Omega_\varepsilon$ be such that $R_{\mathbf{x},1/10} = \tilde{R}_{\mathbf{x},1/10}$. Then $Q_{1/10}^{(\hat{\mathbf{x}})} = Q_{1/10}$ in (4.7.27), *i.e.*, for any $\rho\in(0,\frac{1}{10}]$,

$$R_{\hat{\mathbf{x}},\rho} = \Big\{\hat{\mathbf{x}} + \frac{1}{4}(\hat{x}_1S, \sqrt{\hat{x}_1}T) \;:\; (S,T)\in Q_\rho\Big\}.$$

Using the bounds in (4.7.29) and (4.7.33)–(4.7.35) in $Q_{1/10}$, and employing Theorem 4.2.1 with Remark 4.2.13, where conditions (4.2.4)–(4.2.5) for $A_{ij}^{(\hat{\mathbf{x}})}$ and (4.2.112) for $A_i^{(\hat{\mathbf{x}})}$ in $Q_{1/10}$ are satisfied in our case with any $\alpha\in(0,1)$ by (4.7.34)–(4.7.35), we obtain that, for each $\alpha\in(0,1)$, $\|u^{(\hat{\mathbf{x}})}\|_{2,\alpha,Q_{1/20}} \le C(\alpha)$, that is,

$$\|u^{(\hat{\mathbf{x}})}\|_{2,\alpha,Q_{1/20}^{(\hat{\mathbf{x}})}} \le C(\alpha). \tag{4.7.36}$$

3. We then consider Case (ii) in Step 1. Let $\hat{\mathbf{x}} \in \Gamma_{n,\varepsilon}$. By (4.7.25), for any $\rho \in (0,1]$,

$$R_{\hat{\mathbf{x}},\rho} = \Big\{\hat{\mathbf{x}} + \frac{1}{4}(\hat{x}_1 S, \sqrt{\hat{x}_1}T) \;:\; (S,T) \in Q_\rho \cap \{T > 0\}\Big\},$$

so that

$$Q_\rho^{(\hat{\mathbf{x}})} = Q_\rho \cap \{T > 0\}.$$

Define $u^{(\hat{\mathbf{x}})}(S,T)$ by (4.7.28) for $(S,T) \in Q_1 \cap \{T > 0\}$. Then, by (4.7.20) and (4.7.24), we have

$$\|u^{(\hat{\mathbf{x}})}\|_{C(Q_1\cap\{T\geq 0\})} \leq 4M. \tag{4.7.37}$$

Moreover, $u^{(\hat{\mathbf{x}})}$ satisfies equation (4.7.30) in $Q_1 \cap \{T > 0\}$. As in Step 2, we conclude that (4.7.30) satisfies ellipticity (4.7.33) in $Q_1 \cap \{T > 0\}$ and properties (4.2.4)–(4.2.5) and (4.2.112) for any $\alpha \in (0,1)$ in $Q_1 \cap \{T > 0\}$. Moreover, since u satisfies (4.7.9), it follows that

$$\partial_T u^{(\hat{\mathbf{x}})} = 0 \qquad \text{on } \{T = 0\} \cap Q_1.$$

Then, from Theorem 4.2.10 and Remark 4.2.13, we obtain that, for each $\alpha \in (0,1)$, $\|u^{(\hat{\mathbf{x}})}\|_{2,\alpha,Q_{1/2}\cap\{T\geq 0\}} \leq C$; that is,

$$\|u^{(\hat{\mathbf{x}})}\|_{2,\alpha,Q_{1/2}^{(\hat{\mathbf{x}})}} \leq C. \tag{4.7.38}$$

4. We now consider Case (iii) in Step 1. Let $\hat{\mathbf{x}} \in \Gamma_{f,\varepsilon}$. By (4.7.23) and (4.7.25), for any $\rho \in (0,1]$,

$$R_{\hat{\mathbf{x}},\rho} = \Big\{\hat{\mathbf{x}} + \frac{1}{4}(\hat{x}_1 S, \sqrt{\hat{x}_1}T) \;:\; (S,T) \in Q_\rho \cap \{T < F_{\hat{\mathbf{x}}}(S)\}\Big\} \tag{4.7.39}$$

with $F_{\hat{\mathbf{x}}}(S) = 4\frac{f(\hat{x}_1 + \frac{\hat{x}_1}{4}S) - \hat{f}(\hat{x}_1)}{\sqrt{\hat{x}_1}}$. That is, for any $\rho \in (0,1]$,

$$Q_\rho^{(\hat{\mathbf{x}})} = Q_\rho \cap \{T < F_{\hat{\mathbf{x}}}(S)\}.$$

Since $\varepsilon \leq r'$ by (4.7.22), $\hat{x}_1 \in (0, r')$. Then we obtain

$$\begin{aligned}
&F_{\hat{\mathbf{x}}}(0) = 0,\\
&\|F_{\hat{\mathbf{x}}}\|_{1,[-1,1]} \leq \frac{2\|f'\|_{0,[0,r]}\hat{x}_1}{\sqrt{\hat{x}_1}} \leq 2M\sqrt{r'},\\
&\|F''_{\hat{\mathbf{x}}}\|_{0,[-1,1]} = \sup_{x_1\in(\frac{3}{4}\hat{x}_1,\frac{5}{4}\hat{x}_1)} \frac{\hat{x}_1^2|f''(x_1)|}{16\sqrt{\hat{x}_1}} \leq \|f\|^{(-1-\beta),\{0\}}_{2,\beta,(0,r)}\sqrt{\hat{x}_1} \leq M\sqrt{r'}.
\end{aligned}$$

Since $r_0 \leq 1$ so that $r' \leq 1$, we have

$$\|F_{\hat{\mathbf{x}}}\|_{1,\alpha,[-1,1]} \leq C(M,\alpha). \tag{4.7.40}$$

By (4.7.20) and (4.7.24), $u^{(\hat{\mathbf{x}})}(S,T)$ defined by (4.7.28) in $Q_1^{(\hat{\mathbf{x}})} = Q_1 \cap \{T < F_{\hat{\mathbf{x}}}(S)\}$ satisfies

$$\|u^{(\hat{\mathbf{x}})}\|_{0,Q_1\cap\{T\le F_{\hat{\mathbf{x}}}(S)\}} \le 4M. \tag{4.7.41}$$

Similar to Steps 2–3, $u^{(\hat{\mathbf{x}})}$ satisfies equation (4.7.30), which satisfies ellipticity (4.7.33) and properties (4.2.4)–(4.2.5) and (4.2.112) in $Q_1 \cap \{T < F_{\hat{\mathbf{x}}}(S)\}$.

Moreover, u satisfies (4.7.8) on $\Gamma_{f,r}$, which implies that $u^{(\hat{\mathbf{x}})}$ satisfies

$$B^{\hat{\mathbf{x}}}(Du^{(\hat{\mathbf{x}})}, u^{(\hat{\mathbf{x}})}, S, T) = 0 \qquad \text{on } \{T = F_{\hat{\mathbf{x}}}(S)\} \cap Q_1, \tag{4.7.42}$$

where

$$B^{(\hat{\mathbf{x}})}(\mathbf{p}, z, S, T) = B(4\hat{x}_1 p_1,\ 4\hat{x}_1^{\frac{3}{2}} p_2,\ x_0^2 z,\ \mathbf{x})$$

with $\mathbf{x} = \hat{\mathbf{x}} + \frac{1}{4}(\hat{x}_1 S, \sqrt{\hat{x}_1}T)$. Using (4.7.18)–(4.7.19), this condition can be written in the form:

$$\partial_S u^{(\hat{\mathbf{x}})} = \hat{B}^{\hat{\mathbf{x}}}(\partial_T u^{(\hat{\mathbf{x}})}, u^{(\hat{\mathbf{x}})}, S, T) \qquad \text{on } \{T = F_{\hat{\mathbf{x}}}(S)\} \cap Q_1,$$

where $\hat{B}^{\hat{\mathbf{x}}}(p_2, z, S, T)$ satisfies

$$\partial_{p_2}\hat{B}^{\hat{\mathbf{x}}}(\mathbf{p}, z, S, T) = -\sqrt{\hat{x}_1}\frac{B_{p_2}}{B_{p_1}}(4\hat{x}_1 p_1,\ 4\hat{x}_1^{\frac{3}{2}} p_2,\ \hat{x}_1^2 z,\ \hat{x}_1(1+\frac{S}{4}),\ \hat{x}_2 + \frac{\sqrt{\hat{x}_1}}{4}T).$$

Thus, from (4.7.17)–(4.7.18), using that $\hat{x}_1 \in (0,\varepsilon)$ and $0 < \varepsilon \le r_0$, we have

$$|\partial_{p_2}\hat{B}^{\hat{\mathbf{x}}}| \le \sqrt{r_0}M \qquad \text{for all } (p_2, z) \in \mathbb{R}^2,\ (S,T) \in \{T = F_{\hat{\mathbf{x}}}(S)\} \cap Q_1. \tag{4.7.43}$$

Also, computing $D^k_{(p_2,z,S,T)}\hat{B}^{\hat{\mathbf{x}}}$, $k = 1,2,3$, in a similar way and using (4.7.17)–(4.7.18), we have

$$\|\hat{B}^{\hat{\mathbf{x}}}\|_{3,\Gamma_1^{\hat{\mathbf{x}}}} \le CM.$$

Now, for any $\alpha \in (0,1)$, if r_0 in (4.7.43) is sufficiently small, depending only on (M, λ, α), we can employ Theorem 4.2.8 to obtain

$$\|u^{(\hat{\mathbf{x}})}\|_{2,\alpha,Q_{1/2}\cap\{T\le F_{z_0}(S)\}} \le C,$$

which is

$$\|u^{(\hat{\mathbf{x}})}\|_{2,\alpha,Q_{1/2}^{\hat{\mathbf{x}}}} \le C. \tag{4.7.44}$$

5. Combining (4.7.36), (4.7.38), and (4.7.44), we have

$$\|u^{(\hat{\mathbf{x}})}\|_{2,\alpha,Q_{1/100}^{\hat{\mathbf{x}}}} \le C \qquad \text{for all } \hat{\mathbf{x}} \in \overline{\Omega_\varepsilon} \setminus \Gamma_{\text{sonic}},$$

where we have used (4.6.5). Then Lemma 4.6.1 implies (4.7.21). ☐

4.8 UNIFORMLY ELLIPTIC EQUATIONS IN A CURVED TRIANGLE-SHAPED DOMAIN WITH ONE-POINT DIRICHLET CONDITION

In order to analyze the subsonic reflection configurations, we need to consider Problem (4.5.84)–(4.5.88) in the domain of structure (4.5.1)–(4.5.3) with $t_0 = 0$, *i.e.*, $f_{\rm bd}(0) = 0$. In this case, $P_1 = P_4 = \mathbf{0}$, so that Ω has the triangular shape with one curved side P_1P_2. Also, $\overline{\Gamma_0}$ is now one point $P_1 = P_4$ which is origin $\mathbf{0}$, so (4.5.1) becomes the one-point Dirichlet condition. To simplify the notations, we denote point $P_1 = P_4$ by P_0:

$$P_0 := P_1 = P_4 = \mathbf{0} \qquad \text{when } t_0 = 0 \text{ in (4.5.1)–(4.5.3)}. \tag{4.8.1}$$

Moreover, we need to consider the case that equation (4.5.84) is uniformly elliptic in Ω. Note that the oblique derivative conditions are prescribed on sides Γ_1 and Γ_3 that meet at P_0. Then, if the obliqueness holds at P_0 in the sense of Definition 4.4.3, Lieberman's Harnack estimate [189, Lemma 2.2] at the corner, combined with (4.5.101), would not allow us to prescribe a one-point Dirichlet condition at P_0. Therefore, it is important that (4.5.102), (4.5.106), and the structure of Ω imply that conditions (4.5.85) and (4.5.88) do *not* satisfy the obliqueness at P_0. We will show that these conditions in fact allow us to prescribe the one-point Dirichlet condition at P_0.

Another important point is that, in the view of applications to the case for subsonic reflection configurations, we will not require condition (4.5.97). For this reason, we cannot use Lemma 4.5.8 for the L^∞–estimates, but can use Lemma 4.5.9, since the equation is uniformly elliptic in Ω.

Now we state and prove the results. We follow the procedure of §4.5, and consider first the corresponding linear problem.

4.8.1 Linear problems with one-point Dirichlet condition

We consider Problem (4.5.4)–(4.5.6) in the domain of structure (4.5.1)–(4.5.3) with $t_0 = 0$. Then, using (4.8.1), we can write (4.5.6) in the form:

$$u(P_0) = 0. \tag{4.8.2}$$

Throughout this section, we assume that equation (4.5.4) and the boundary conditions (4.5.5) satisfy properties (4.5.7)–(4.5.15), (4.5.80) ($P_0 = P_1 = P_4$ in this case), (4.5.106), and

$$\|\mathbf{b}^{(k)}\|_{L^\infty(\Gamma_k)} \le \lambda^{-1} \qquad \text{for } k = 1, 2, 3, \tag{4.8.3}$$

where we introduce condition (4.8.3), in addition to (4.5.14), for the reasons described in Remark 4.3.14.

We note that the conditions listed above imply that the following condition similar to (4.5.13) holds at P_0, where Γ_1 and Γ_3 meet:

$$\left| \frac{\mathbf{b}^{(1)}}{|\mathbf{b}^{(1)}|}(P_0) \pm \frac{\mathbf{b}^{(3)}}{|\mathbf{b}^{(3)}|}(P_0) \right| \ge \lambda^2. \tag{4.8.4}$$

Indeed, using (4.5.80), (4.5.106), and (4.8.3), and denoting by L the left-hand side of (4.8.4), we have

$$L = \left| \frac{\mathbf{b}^{(1)}}{|\mathbf{b}^{(1)}|}(P_0) \pm (0,1) \right| \geq \frac{|b_1^{(1)}|}{|\mathbf{b}^{(1)}|}(P_0) \geq \lambda^2.$$

Then (4.8.4) is proved.

Lemma 4.8.1. *Let $t_0 = 0$, let $h, t_1, t_2, M_{\rm bd} > 0$, and let Ω be a domain of structure* (4.5.1)–(4.5.3) *and* (4.8.1). *Let $\kappa > 0$, $\lambda > 0$, $M < \infty$, and $\varepsilon \in (0, \frac{h}{10})$ be constants. Then there exists $\alpha_1(\kappa, \lambda, M_{\rm bd}) \in (0,1)$ such that, for every $\alpha \in (0, \alpha_1]$, there is $C(\Omega, \lambda, \kappa, M, \alpha)$ so that the following holds: Let* (4.5.7)–(4.5.15) *and* (4.5.80) *with $P_0 = P_1 = P_4$,* (4.5.106), *and* (4.8.3) *be satisfied. Then any solution $u \in C^{1,\alpha}(\overline{\Omega}) \cap C^{2,\alpha}(\Omega)$ of Problem* (4.5.4)–(4.5.5) *and* (4.8.2) *satisfies*

$$\|u\|^{(-1-\alpha),\{P_0\}\cup\Gamma_2}_{2,\alpha,\Omega} \leq C\big(\|f\|^{(1-\alpha),\{P_0\}\cup\Gamma_2}_{0,\alpha,\Omega} + \sum_{j=1,3} \|g_j\|^{(-\alpha),\partial\Gamma_j}_{1,\alpha,\Gamma_1} + \|g_2\|_{C^\alpha(\overline{\Gamma_2})}\big), \tag{4.8.5}$$

where $\partial\Gamma_k$ denotes the endpoints of Γ_k.

Proof. From (4.5.1) with $t_0 = 0$ and (4.5.15), there exists $\theta_0 \in [\frac{\pi}{2}, \pi)$ depending on $M_{\rm bd}$ such that angles θ_1 at P_0, and θ_k at P_k for $k = 2, 3$, satisfy

$$0 < \theta_j \leq \theta_0, \qquad j = 1, 2, 3.$$

Now we follow the proof of Lemma 4.5.1, with the only difference that the localized version of the estimates of [193, Lemma 1.3] is used near all the three corner points, *i.e.*, in $B_{2r}(P_i) \cap \Omega$ for $i = 0, 2, 3$, where we use (4.8.4) for the estimate near P_0. We also use a scaling technique similar to Step 2 of the proof of Proposition 4.3.11 to obtain the estimates in the weighted norms near P_0. Note that, for the estimate near P_0, we use the oblique conditions on $\Gamma_1 \cup \Gamma_3$ (but do not use the Dirichlet condition at P_0). Then we have

$$\begin{aligned} &\|u\|^{(-1-\alpha),\{P_0\}\cup\Gamma_2}_{2,\alpha,\Omega} \\ &\leq C\big(\|u\|_{0,\Omega} + \|f\|^{(1-\alpha),\{P_0\}\cup\Gamma_2}_{0,\alpha,\Omega} + \sum_{j=1,3} \|g_j\|^{(-\alpha),\partial\Gamma_j}_{1,\alpha,\Gamma_1} + \|g_2\|_{C^\alpha(\overline{\Gamma_2})}\big), \end{aligned} \tag{4.8.6}$$

where $C = C(\Omega, \lambda, \kappa, M, \alpha)$. Thus, it remains to estimate $\|u\|_{0,\Omega}$ by the right-hand side of (4.8.5) multiplied by $C(\Omega, \lambda, \kappa, M, \alpha)$. For that, we employ an argument similar to the one after (4.5.20) in the proof of Lemma 4.5.1, by using (4.8.6) and the uniqueness for Problem (4.5.4)–(4.5.5) and (4.8.2), which follows from Lemma 4.4.2 and Remark 4.4.6. ☐

Remark 4.8.2. *In the proof of Lemma* 4.8.1, *the Dirichlet condition* (4.8.2) *at P_0 has been used only through the application of the comparison principle in Lemma* 4.4.2 *with Remark* 4.4.6.

Next we note that any solution of Problem (4.5.4)–(4.5.6) in the domain of structure (4.5.1)–(4.5.3) with $t_0 \geq 0$ satisfies the estimates similar to (4.5.131) in Lemma 4.5.10.

Lemma 4.8.3. *Let $\kappa > 0$, $\lambda > 0$, and $M < \infty$ be constants. Let Ω be a domain of structure (4.5.1)–(4.5.3) with $h, t_1, t_2 \in (\lambda, \frac{1}{\lambda})$, $M_{\rm bd} \leq \frac{1}{\lambda}$, and $t_0 \geq 0$. Let $f_{\rm bd}$ satisfy (4.5.15). Then there exists $\alpha_1(\kappa, \lambda) \in (0,1)$ such that the following holds: Let $\alpha \in (0, \alpha_1]$. Let (4.5.7)–(4.5.15) and (4.8.3) be satisfied. Let $u \in C^2(\Omega) \cap C^1(\overline{\Omega} \setminus \{P_0\})$ satisfy (4.5.4)–(4.5.5). Then*

$$\|u\|^{(-1-\alpha),\,\overline{\Gamma_2}}_{2,\alpha,\Omega^{(s)}} \leq C_s\big(\|u\|_{0,\Omega^{(s/2)}} + \|f\|^{(1-\alpha),\Gamma_2}_{0,\alpha,\Omega^{(s/2)}} + \sum_{j=1,3} \|g_j\|^{(-\alpha),\{P_2\}}_{1,\alpha,\Gamma_j^{(s/2)}} + \|g_2\|_{C^\alpha(\overline{\Gamma_2})}\big), \tag{4.8.7}$$

where $\Omega^{(s)} = \Omega \cap \{x_1 > s\}$, $\Gamma_j^{(s)} = \Gamma_j \cap \{x_1 > s\}$ for $j = 1, 3$, and C_s depends only on $(\kappa, \lambda, M, K, \alpha, s)$.

Proof. The proof follows that for Lemma 4.5.10, in which we replace the estimates for the nonlinear equations and boundary conditions by the corresponding linear estimates used in the proof of Lemma 4.8.1. □

Next we prove the existence of solutions of Problem (4.5.4)–(4.5.5) and (4.8.2) in the domain of structure (4.5.1)–(4.5.3) with $t_0 = 0$.

Proposition 4.8.4. *Let $\kappa > 0$, $\lambda > 0$, and $M < \infty$ be constants. Let $t_0 = 0$, $h, t_1, t_2 \in (\lambda, \frac{1}{\lambda})$, and $M_{\rm bd} \leq \frac{1}{\lambda}$, and let Ω be a domain of structure (4.5.1)–(4.5.3) and (4.8.1). Then there exists $\alpha_1 = \alpha_1(\kappa, \lambda) \in (0,1)$ such that the following holds: Let $\varepsilon \in (0, \frac{h}{10})$, and let $\alpha \in (0, \alpha_1]$. Let $f_{\rm bd}$ satisfy (4.5.15). Let $(a_{ij}, a_i, \mathbf{b}^{(k)})$ satisfy (4.5.7)–(4.5.15) and (4.5.80) with $P_0 = P_1 = P_4$, (4.5.106), and (4.8.3). Then, for every $(f, g_1, g_2, g_3) \in C^{(1-\alpha),\{P_0\}\cup\overline{\Gamma_2}}_{0,\alpha,\Omega} \times C^{(-\alpha),\partial\Gamma_1}_{1,\alpha,\Gamma_1} \times C^\alpha(\overline{\Gamma_2}) \times C^{(-\alpha),\partial\Gamma_3}_{1,\alpha,\Gamma_3}$, there exists a unique solution $u \in C^{(-1-\alpha),\{P_0\}\cup\overline{\Gamma_2}}_{2,\alpha,\Omega}$ of Problem (4.5.4)–(4.5.5) and (4.8.2). Moreover, u satisfies (4.8.5), where C depends only on $(\Omega, \kappa, \lambda, M, \alpha)$.*

Proof. We assume that $\alpha \in (0, \alpha_1)$, where $\alpha_1 := \alpha_1(\kappa, \lambda)$ will be fixed below. We divide the proof into two steps.

1. We first replace equation (4.5.4) by the corresponding Poisson equation, and consider **Problem $\mathcal{P}_0$** defined as follows:

$$\Delta u = f \;\text{ in } \Omega; \quad \mathbf{b}^{(k)} \cdot Du + b_0^{(k)} u = g_k \;\text{ on } \Gamma_k,\; k = 1, 2, 3; \quad u(P_0) = 0.$$

Clearly, **Problem $\mathcal{P}_0$** satisfies all the conditions in the proposition. We first prove that, for every $(f, g_1, g_2, g_3) \in C^{(1-\alpha),\{P_0\}\cup\overline{\Gamma_2}}_{0,\alpha,\Omega} \times C^{(-\alpha),\partial\Gamma_1}_{1,\alpha,\Gamma_1} \times C^\alpha(\overline{\Gamma_2}) \times C^{(-\alpha),\partial\Gamma_3}_{1,\alpha,\Gamma_3}$, there exists a unique solution $u \in C^{(-1-\alpha),\{P_0\}\cup\overline{\Gamma_2}}_{2,\alpha,\Omega}$ of **Problem $\mathcal{P}_0$**.

Note that, while $f \in C^{(1-\alpha),\{P_0\}\cup\overline{\Gamma_2}}_{0,\alpha,\Omega}$ may be unbounded, the assumptions on g_k imply that $g_k \in L^\infty(\Gamma_k)$. We first prove the existence of a solution of **Problem** $\mathcal{P}_0$ for the right-hand sides (f, g_1, g_2, g_3) satisfying the additional condition that $f \in L^\infty(\Omega)$.

We approximate **Problem** $\mathcal{P}_0$ by the problems of structure (4.5.4)–(4.5.6) in domains $\Omega^{(\delta)}$ of structure (4.5.1)–(4.5.3) with $t_0^{(\delta)} > 0$. Specifically, for each $\delta \in (0, \frac{h}{10})$, denote

$$\begin{aligned} &\Omega^{(\delta)} := \Omega \cap \{x_1 > \delta\}, \quad \Gamma_0^{(\delta)} := \partial\Omega^{(\delta)} \cap \{x_1 = \delta\}, \\ &\Gamma_k^{(\delta)} := \Gamma_k \cap \{x_1 > \delta\} \qquad \text{for } k = 1, 2, 3. \end{aligned}$$

We consider the following boundary value problem $\mathcal{P}_0^{(\delta)}$ for $u_\delta \in C^1(\overline{\Omega^{(\delta)}}) \cap C^2(\Omega^{(\delta)})$:

$$\Delta u = f \ \text{ in } \Omega^{(\delta)}; \quad u = 0 \ \text{ on } \Gamma_0^{(\delta)}; \quad \mathbf{b}^{(k)} \cdot Du + b_0^{(k)} u = g_k \ \text{ on } \Gamma_k^{(\delta)}, \ k = 1, 2, 3. \tag{4.8.8}$$

We shift the coordinates, replacing x_1 by $x_1 + \delta$, so that $\{x_1 = \delta\}$ becomes $\{x_1 = 0\}$. In the new coordinates, domain $\Omega^{(\delta)}$ is of structure (4.5.1), with constants $t_k^{(\delta)}$ and $h^{(\delta)}$, and function $f_{\text{bd}}^{(\delta)}$, where

$$\begin{aligned} &t_0^{(\delta)} = f_{\text{bd}}(\delta) \geq \min\{t_1\delta + t_0, \, t_2\}, \quad (t_1^{(\delta)}, t_2^{(\delta)}) = (t_1, t_2), \\ &h^{(\delta)} = h - \delta, \quad f_{\text{bd}}(x_1)^{(\delta)} = f_{\text{bd}}(x_1 + \delta). \end{aligned}$$

Then $t_0^{(\delta)} \geq \min\{t_1\delta, \, t_2\} > 0$. Also, $h^{(\delta)} \geq \frac{9}{10}h$, $M_{\text{bd}}^{(\delta)} = M_{\text{bd}}$, and $f_{\text{bd}}^{(\delta)}$ satisfies (4.5.15) on $(0, h^{(\delta)})$. Thus, $\Omega^{(\delta)}$ for $\delta \in (0, \frac{h}{10})$ satisfies the conditions of Proposition 4.5.3 with $\frac{\lambda}{2}$ instead of λ.

Moreover, in the shifted coordinates, the boundary value problem (4.8.8) for u_δ is of the structure of Problem (4.5.4)–(4.5.6) in $\Omega^{(\delta)}$. It is also easy to see that conditions (4.5.7)–(4.5.14) are satisfied for Problem (4.8.8) in $\Omega^{(\delta)}$ for each δ, with constants $(\kappa, \lambda, M, \alpha)$ given in the proposition. Furthermore, the right-hand sides (f, g_1, g_2, g_3) restricted to $\{x_1 > \delta\}$ satisfy the conditions of Proposition 4.5.3 for each $\delta \in (0, \frac{h}{10}]$.

Thus, by Proposition 4.5.3 applied with $\frac{\lambda}{2}$ instead of λ, and the other constants from this proposition unchanged, there exists $\alpha_1 = \alpha_1(\kappa, \lambda) \in (0, 1)$ such that, if $\alpha \in (0, \alpha_1]$, for each $\delta \in (0, \frac{h}{10}]$, there is a solution $u_\delta \in C(\overline{\Omega^{(\delta)}}) \cap C^1(\overline{\Omega^{(\delta)}} \setminus \overline{\Gamma}_0^{(\delta)}) \cap C^2(\Omega^{(\delta)})$ of (4.8.8), which satisfies (4.5.16) in domain $\Omega^{(\delta)}$ in the shifted coordinates.

Furthermore, from (4.5.80), using (4.5.14)–(4.5.15), we obtain the existence of $\hat{\varepsilon} \in (0, \varepsilon]$ depending only on $(\lambda, \alpha, M, \varepsilon)$ such that

$$b_1^{(1)} \leq -\frac{\lambda}{2} \qquad \text{on } \Gamma_1 \cap \{x_1 < \hat{\varepsilon}\}.$$

Then, using the uniform ellipticity (4.5.7), for each $\delta \in (0, \frac{\hat{\varepsilon}}{2})$, we can apply Lemma 4.5.9 (with $\hat{\varepsilon}$ instead of ε) to Problem (4.8.8), so that u_δ satisfies

(4.5.120) in $\Omega^{(\delta)}$. Writing this estimate in the original coordinates, we obtain that, for each $\delta \in (0, \frac{\hat{\varepsilon}}{2})$,

$$|u_\delta(\mathbf{x})| \le C(x_1 - \delta) \le Cx_1 \qquad \text{in } \Omega \cap \{x_1 > \delta\}, \tag{4.8.9}$$

where C depends only on $(\lambda, M, \hat{\varepsilon})$, while we keep $(f, g_1, g_2, g_3) \in L^\infty$ fixed. Thus, $C = C(\lambda, \alpha, M, \varepsilon)$, since $\hat{\varepsilon} = \varepsilon(\lambda, \alpha, M, \varepsilon)$.

Applying Lemma 4.8.3 with $\frac{\lambda}{2}$ instead of λ, and (κ, M, α) given in this proposition, which possibly requires the reduction of α_1 depending only on (κ, λ), and using that $\|u_\delta\|_{0,\Omega^\delta} \le Ch$ for C from (4.8.9), we obtain that u_δ satisfies (4.8.7) in $(\Omega^{(\delta)})^{(s)} \equiv \Omega^{(\delta+s)}$ for each $s \in (0, \frac{h}{10})$, which implies

$$\begin{aligned}&\|u_\delta\|^{(-1-\alpha),\,\overline{\Gamma_2}}_{2,\alpha,\Omega^{(\delta+s)}}\\ &\le C_s\big(Ch + \|f\|^{(1-\alpha),\Gamma_2}_{0,\alpha,\Omega^{(s/2)}} + \sum_{j=1,3} \|g_j\|^{(-\alpha),\{P_2\}}_{1,\alpha,\Gamma_j^{(s/2)}} + \|g_2\|_{C^\alpha(\overline{\Gamma_2})}\big) =: \tilde{C}_s,\end{aligned} \tag{4.8.10}$$

where C_s depends only on $(\kappa, \lambda, \alpha, M, s)$, and $\tilde{C}_s$ depends on the same parameters if we keep (f, g_1, g_2, g_3) fixed.

Define $h_1 = \min\{\frac{\varepsilon}{2}, \frac{h}{10}\}$. From (4.8.10), we have

$$\|u_\delta\|^{(-1-\alpha),\,\overline{\Gamma_2}}_{2,\alpha,\Omega^{(2s)}} \le \tilde{C}_s \qquad \text{for all } s \in (0, \frac{h_1}{2}],\ \delta \in (0, s]. \tag{4.8.11}$$

Thus, from (4.8.11) with $s = \frac{h_1}{2}$, there exists a sequence $\delta_j \to 0^+$ such that u_{δ_j} converges in $C^{(-1-\frac{\alpha}{2}),\,\overline{\Gamma_2}}_{2,\alpha/2,\Omega^{(h_1)}}$. Similarly, from (4.8.11) with $s = \frac{h_1}{4}, \frac{h_1}{8}, \dots$, there exists a subsequence of $\{u_{\delta_j}\}$, converging in $C^{(-1-\frac{\alpha}{2}),\,\overline{\Gamma_2}}_{2,\alpha/2,\Omega^{(h_1/2)}}$, then a further subsequence converging in $C^{(-1-\frac{\alpha}{2}),\,\overline{\Gamma_2}}_{2,\alpha/2,\Omega^{(h_1/4)}}$, *etc.* By the diagonal procedure, there exists a sequence $\hat{\delta}_k \to 0$ such that $u_{\hat{\delta}_k}$ converges in $C^{(-1-\frac{\alpha}{2}),\,\overline{\Gamma_2}}_{2,\,\alpha/2,\,K}$ for each compact $K \subset \overline{\Omega}\backslash\Gamma_0$. Since each $u_{\hat{\delta}_k}$ is a solution of (4.8.8) satisfying the uniform estimates (4.8.9)–(4.8.11), it follows that the limit function u is a solution of

$$\Delta u = f \ \text{ in } \Omega, \qquad \mathbf{b}^{(k)} \cdot Du + b_0^{(k)} u = g_k \ \text{ on } \Gamma_k,\ k = 1, 2, 3,$$

and satisfies $u \in C^{1,\alpha}(\overline{\Omega} \setminus \overline{\Gamma_0}) \cap C^{2,\alpha}(\overline{\Omega} \setminus (\overline{\Gamma_0} \cup \overline{\Gamma_2}))$ and

$$|u(\mathbf{x})| \le Cx_1 \qquad \text{in } \Omega, \tag{4.8.12}$$

where $C = C(\lambda, \alpha, M)$ is from (4.8.9). Then (4.8.12), combined with the fact that $u \in C^{1,\alpha}(\overline{\Omega} \setminus \{P_0\})$, implies that $u \in C(\overline{\Omega})$ with $u(P_0) = 0$. Therefore, $u \in C(\overline{\Omega}) \cap C^{1,\alpha}(\overline{\Omega} \setminus \{P_0\}) \cap C^3(\Omega)$ and is a solution of **Problem** $\mathcal{P}_0$, where we have used (4.8.12) (that is satisfied by u) and the interior regularity of harmonic functions.

Now we note that Theorem 4.3.18 (with the conditions as in Corollary 4.3.19) applies to **Problem** $\mathcal{P}_0$ near corner P_0. Indeed, (4.5.1)–(4.5.3) with

$t_0 = 0$, (4.8.1), and (4.5.15) imply that region $\Omega \cap B_r(P_0)$ for sufficiently small $r = r(M, \alpha, \lambda) > 0$ is of the structure described in Proposition 4.3.7, (4.3.86) holds for $M = \frac{\sqrt{t_1^2+1}}{t_1} \leq \frac{\sqrt{\lambda^2+1}}{\lambda^2}$, and (4.3.105) holds with $\sigma = \alpha$ and M as in this proposition. The ellipticity of the equation, obliqueness of the boundary conditions, and regularity of the ingredients and the right-hand sides required in Theorem 4.3.18 (with the conditions as in Corollary 4.3.19) follow from the assumptions. Condition (4.2.102) for the boundary conditions on both Γ_1 and Γ_3 is satisfied with $\delta = 0$, $v \equiv 0$, and the nonhomogeneous linear operator $L(\cdot, \cdot, \cdot)$ equal to the original linear boundary condition minus the right-hand side, so that (4.3.142) is satisfied. Also, (4.3.137) holds by (4.8.4) with $\kappa = \frac{\lambda^2}{\sqrt{2}}$. Furthermore, u satisfies (4.8.12) so that constant L in Theorem 4.3.18 is now constant C in (4.8.12). Thus, reducing α_1 if necessary depending only on λ so that the resulting constant α_1 depends only on (κ, λ), we obtain from Theorem 4.3.18 (with the conditions as in Corollary 4.3.19) that $u \in C^{1,\alpha}(\overline{\Omega}) \cap C^{2,\alpha}(\Omega)$. Then estimate (4.8.5) for u follows from Lemma 4.8.1. This estimate also implies the uniqueness of the solution.

Therefore, we have proved the existence of $\alpha_1 = \alpha_1(\kappa, \lambda) \in (0, 1)$ such that, for each $\alpha \in (0, \alpha_1]$, there exists a solution $u \in C^{1,\alpha}(\overline{\Omega}) \cap C^{2,\alpha}(\Omega)$ of **Problem** $\mathcal{P}_0$ for the general functions:

$$(f, g_1, g_2, g_3) \in \mathcal{Y} := C^{(1-\alpha),\{P_0\}\cup\overline{\Gamma_2}}_{0,\alpha,\Omega} \times C^{(-\alpha),\partial\Gamma_1}_{1,\alpha,\Gamma_1} \times C^{\alpha}(\overline{\Gamma_2}) \times C^{(-\alpha),\partial\Gamma_3}_{1,\alpha,\Gamma_3}$$

with the additional assumption of the boundedness of f.

The existence of solution $u \in C^{1,\alpha}(\overline{\Omega}) \cap C^{2,\alpha}(\Omega)$ of **Problem** $\mathcal{P}_0$, for general $(f, g_1, g_2, g_3) \in \mathcal{Y}$, follows by the approximation procedure via repeating the one in Proposition 4.5.2 and employing estimate (4.8.5). Here, in the present case, the bounded approximations of f are defined as follows: For $(x_1, x_2) \in \Omega$ and $m = 1, 2, \ldots,$

$$f^{(m)}(x_1, x_2) = \begin{cases} f(x_1, x_2) & \text{if } x_1 \in [\frac{h}{10m}, h - \frac{h}{10m}], \\ f(\frac{h}{10m}, x_2) & \text{if } x_1 \in (0, \frac{h}{10m}), \\ f(h - \frac{h}{10m}, x_2) & \text{if } x_1 \in (h - \frac{h}{10m}, h), \end{cases}$$

and $g_k^{(m)} = g_k$ for $k = 1, 2, 3$, sequence $(f^{(m)}, g_1^{(m)}, g_2^{(m)}, g_3^{(m)})$ is uniformly bounded in $\mathcal{Y}$. Estimate (4.8.5) for u follows from Lemma 4.8.1, which also implies the uniqueness of the solution.

Therefore, the proposition is proved for **Problem** $\mathcal{P}_0$.

2. Now the existence and uniqueness of solutions for Problem (4.5.4)–(4.5.5) and (4.8.2) satisfying the assumptions of the proposition (denoted this problem as **Problem** $\mathcal{P}$) follow by the method of continuity, applied to the family of **Problems** $t\mathcal{P} + (1-t)\mathcal{P}_0$ **for** $t \in [0, 1]$, where we have used that all of these problems satisfy estimate (4.8.5) with uniform C by Lemma 4.8.1. □

Remark 4.8.5. *The proof of Proposition* 4.8.4 *is based crucially on assumptions* (4.5.15), (4.5.80) (*with* $P_0 = P_1 = P_4$), *and* (4.5.106); *they allow us to employ Lemma* 4.5.9 *to obtain* (4.8.9) *and* (4.8.12), *which leads to the* $C^{1,\alpha}$*–regularity at* P_0 *by Theorem* 4.3.18 *with Corollary* 4.3.19. *It is also easy to see from* (4.5.15) *and* (4.5.80) *that the obliqueness at the corner point* P_0 *in the sense of Definition* 4.4.1 *is* not *satisfied.*

4.8.2 Nonlinear problems with one-point Dirichlet condition

We consider Problem (4.5.84)–(4.5.88) in the domain of structure (4.5.1)–(4.5.3) with $t_0 = 0$. We continue to use notation (4.8.1). Then the Dirichlet condition (4.5.88) becomes the one-point Dirichlet condition (4.8.2).

Our assumptions on the ingredients of the problem are the following: Fix constants $\alpha \in (0, \frac{1}{2})$, $\beta \in (0,1)$, $\lambda \in (0,1]$, $\delta > 0$, $\kappa > 0$, $\varepsilon \in (0, \frac{h}{10})$, $\sigma \in (0,1)$, and $M < \infty$. Then we assume that

$$\big(\lambda \operatorname{dist}(\mathbf{x}, \Gamma_0) + \delta\big)|\boldsymbol{\mu}|^2 \le \sum_{i,j=1}^{2} A_{ij}(\mathbf{p}, \mathbf{x})\mu_i\mu_j \le \lambda^{-1}|\boldsymbol{\mu}|^2 \tag{4.8.13}$$

for all $\mathbf{x} \in \Omega$ and $\mathbf{p}, \boldsymbol{\mu} \in \mathbb{R}^2$,

$$\|((A_{ij}, A)(\mathbf{0}, \cdot),\ D^m_{\mathbf{p}}(A_{ij}, A)(\mathbf{p}, \cdot))\|^{(-\alpha),\{P_0\}}_{1,\alpha,\Omega\cap\{x_1<2\varepsilon\}} \le M \tag{4.8.14}$$

for all $\mathbf{p} \in \mathbb{R}^2$ and $m = 1, 2$, as well as (4.5.91), (4.5.93), and (4.5.98)–(4.5.110).

We note that the conditions listed above imply that, if σ is small, depending on λ, the following condition similar to (4.5.109) holds at P_0 where Γ_1 and Γ_3 meet:

$$\left|\frac{\mathbf{b}^{(1)}}{|\mathbf{b}^{(1)}|}(P_0) \pm \frac{\mathbf{b}^{(3)}}{|\mathbf{b}^{(3)}|}(P_0)\right| \ge \frac{3\lambda^2}{8} \tag{4.8.15}$$

with $\mathbf{b}^{(1)}$ from (4.5.103)–(4.5.104). Indeed, using (4.5.100), (4.5.102), and (4.5.104) with small $\sigma = \sigma(\lambda)$, we have

$$\|\mathbf{b}^{(1)}\|_{L^\infty(\Gamma_k)} \le \frac{2}{\lambda}, \qquad b_1^{(1)}(P_0) \le -\frac{3\lambda}{4}.$$

Also, $\mathbf{b}^{(3)}(P_0) = (0,1)$ by (4.5.106). With this, we prove (4.8.15) by repeating the corresponding calculation in the proof of (4.8.4). Now (4.8.15) is proved.

We first show an *a priori estimate.* Note that, unlike Lemma 4.5.12, here σ depends on δ, *i.e.*, we only consider the uniformly elliptic case, but do not consider the limit for these estimates as $\delta \to 0+$.

Lemma 4.8.6. *Let* $\kappa > 0$, $\lambda > 0$, $\delta > 0$, $M < \infty$, $\alpha \in (0,1)$, *and* $\varepsilon \in (0, \frac{\lambda}{10})$. *Then there exists* $\sigma > 0$ *depending only on* $(\kappa, \lambda, \delta, M, \alpha, \varepsilon)$ *such that the following hold: Let* Ω *be a domain of structure* (4.5.1)–(4.5.3) *with* $h, t_1, t_2 \in (\lambda, \frac{1}{\lambda})$, $M_{\rm bd} \le \frac{1}{\lambda}$, *and* $t_0 = 0$. *Let* $f_{\rm bd}$ *satisfy* (4.5.15). *Let* (4.8.13)–(4.8.14),

(4.5.91), (4.5.93), *and* (4.5.98)–(4.5.110) *hold. Let* $u \in C^2(\Omega) \cap C^1(\overline{\Omega} \setminus \{P_0\}) \cap C(\overline{\Omega})$ *be a solution of Problem* (4.5.84)–(4.5.87) *and* (4.8.2). *Then* u *satisfies* (4.5.119)–(4.5.120) *with* C *depending only on* $(\lambda, M, \delta, \varepsilon)$. *Moreover,*

$$\|u\|^{(-1-\alpha_1),\ \{P_0\}\cup\overline{\Gamma_2}}_{2,\alpha_1,\Omega} \leq \hat{C} \tag{4.8.16}$$

with $\alpha_1 = \alpha_1(\kappa, \lambda, \delta) \in (0, \frac{1}{2})$, *where* $\hat{C}$ *depends on* $(\kappa, \lambda, \delta, M, \alpha, \varepsilon)$.

Proof. We divide the proof into two steps.

1. Condition (4.8.13) (which implies the uniform ellipticity with constant δ) and condition (4.5.102), together with the obliqueness of the boundary conditions and the regularity of the ingredients of the equation and boundary conditions, imply that Lemma 4.5.9 can be employed to obtain (4.5.119)–(4.5.120) with constant C depending only on $(\lambda, M, \delta, \varepsilon)$.

2. We now prove (4.8.16). First, using (4.8.15) and the uniform ellipticity (4.8.13) (with the ellipticity constant $\delta > 0$), together with the obliqueness of the boundary conditions on $\Gamma_1 \cup \Gamma_3$, then the almost-linear structure (4.5.104) of the boundary condition on Γ_1, the regularity of the ingredients of the equation and boundary conditions, and the structure of Ω allow us to apply Theorem 4.3.18 (with the conditions as in Corollary 4.3.19) near P_0. More precisely, we have shown in Step 1 of the proof of Proposition 4.8.4 that there exists $R(M, \alpha, \lambda) > 0$ such that $\Omega \cap B_R(P_0)$ satisfies the requirements of Theorem 4.3.18. Reducing R if necessary, then $R \in (0, \varepsilon]$ so that $R = R(M, \alpha, \lambda, \varepsilon)$. We use estimate (4.5.120) of u obtained in Step 1 to satisfy assumption (4.3.141) with L depending only on $(\lambda, M, \delta, \varepsilon)$. Also, (4.3.140) is satisfied by (4.8.2). Furthermore, assumption (4.3.137) of Theorem 4.3.18 with $\kappa = \frac{3\lambda^2}{8}$ follows from (4.8.15). The other assumptions of Theorem 4.3.18 (with the conditions as in Corollary 4.3.19) follow directly from the assumptions of this lemma, where we recall that the ellipticity constant is δ, by (4.8.13). Then we find that there exist $\alpha_2 = \alpha_2(\lambda, \delta) \in (0, \frac{1}{2})$ and positive constants $(\sigma_1, \hat{C}_1)$ depending only on $(\lambda, \delta, M, \alpha, \varepsilon)$ such that, if (4.5.104) is satisfied with $\sigma \in (0, \sigma_1]$,

$$\|u\|^{(-1-\alpha_2),\ \{P_0\}}_{2,\alpha_2,B_R(P_0)\cap\Omega} \leq \hat{C}_1.$$

Next, using the structure of Ω, we note that $\Omega \cap \{x_1 < s^*\} \subset B_R(P_0) \cap \Omega$ for $s^* = \frac{R}{\sqrt{1+\lambda^{-2}}}$. Then we have

$$\|u\|^{(-1-\alpha_2),\ \{P_0\}}_{2,\alpha_2,\Omega\cap\{x_1<s^*\}} \leq \hat{C}_1. \tag{4.8.17}$$

Also, from its definition, $s^* = s^*(\lambda, M, \alpha, \varepsilon)$.

Now we note that (4.8.13) with $\delta > 0$ implies (4.5.89). With this, it follows that our conditions and the L^∞–bound of u obtained in Step 1 allow us to apply Lemma 4.5.10 with $K = K(\lambda, M, \delta, \varepsilon)$. We use $s = \frac{1}{2}s^*(\lambda, \delta, M, \alpha, \varepsilon)$ in Lemma 4.5.10 to obtain $\alpha_3 = \alpha_3(\kappa, \lambda) \in (0, \frac{1}{2})$ and $\sigma_2 = \sigma_2(\kappa, \lambda, \delta, M, \alpha, \varepsilon) > 0$ such that, if (4.5.104) is satisfied with $\sigma \in (0, \sigma_2]$, then

$$\|u\|^{(-1-\alpha_3),\,\overline{\Gamma_2}}_{2,\alpha_3,\Omega\cap\{x_1>s^*/2\}} \leq \hat{C}_2,$$

where $\hat{C}_2$ depends only on $(\kappa, \lambda, \delta, M, \alpha, \varepsilon)$. Combining this estimate with (4.8.17), choosing $\alpha_1 = \min\{\alpha_2, \alpha_3\}$, and assuming that (4.5.104) is satisfied with $\sigma = \min\{\sigma_1, \sigma_2\}$, we obtain (4.8.16). □

Now we prove the existence of solutions.

Proposition 4.8.7. *Let $\kappa > 0$, $\lambda > 0$, $\delta > 0$, $M < \infty$, $\alpha \in (0,1)$, and $\varepsilon \in (0, \frac{\lambda}{10})$. Then there exist $\alpha_1 \in (0, \frac{1}{2})$ depending only on $(\kappa, \lambda, \delta)$, and $\sigma > 0$ depending only on $(\kappa, \lambda, \delta, M, \alpha, \varepsilon)$, such that the following holds: Let Ω be a domain of structure* (4.5.1)–(4.5.3) *with $h, t_1, t_2 \in (\lambda, \frac{1}{\lambda})$, $M_{\rm bd} \leq \frac{1}{\lambda}$, and $t_0 = 0$. Let $f_{\rm bd}$ satisfy* (4.5.15). *Let* (4.8.13)–(4.8.14), (4.5.91), (4.5.93), *and* (4.5.98)–(4.5.110) *hold. Then there exists a unique solution $u \in C(\overline{\Omega}) \cap C^1(\overline{\Omega} \setminus \{P_0\}) \cap C^2(\Omega)$ of* (4.5.84)–(4.5.87) *and* (4.8.2). *Moreover, $u \in C(\overline{\Omega}) \cap C^{1,\alpha_1}(\overline{\Omega} \setminus \{P_0\}) \cap C^{2,\alpha_1}(\overline{\Omega} \setminus (\{P_0\} \cup \overline{\Gamma_2}))$ satisfies* (4.5.119)–(4.5.120) *and* (4.8.16), *where the constants in these estimates depend only on the parameters described in Lemma* 4.8.6.

Proof. We use a nonlinear method of continuity by following the proof of Proposition 4.5.13. We only comment on the proof and highlight the differences.

Let α_1 be sufficiently small to satisfy the conditions of Proposition 4.8.4 with λ replaced by $\frac{1}{2}\min\{\lambda, \delta\}$, and with $(\frac{\kappa}{2}, \frac{\delta}{2})$ and $M_{\rm bd} = \frac{2}{\lambda}$; and the conditions of Lemma 4.8.6 with $(\frac{\kappa}{2}, \frac{\lambda}{2}, \frac{\delta}{2})$ and $M_{\rm bd} = \frac{2}{\lambda}$, where $(\kappa, \lambda, \delta)$ are given in this proposition. Hence, $\alpha_1 = \alpha_1(\kappa, \lambda, \delta)$. Fix $\alpha \in (0,1)$. Let σ be as in Lemma 4.5.12, which depends only on $(\kappa, \lambda, \delta, M, \alpha, \varepsilon)$.

In the proof, we define **Problems** $\mathcal{P}_t(u)$, spaces $\mathcal{C}_D$ and $\mathcal{C}_T$, and set $\mathcal{T}$ in the same way as in Proposition 4.5.13, with only notational change: Both Γ_0 and P_1 are replaced by P_0, in both the definitions and the argument. We employ Proposition 4.8.4 to show that $\mathcal{T}$ is open, and Lemma 4.8.6 to show that $\mathcal{T}$ is close.

Further details of the proof of Proposition 4.5.13 are easily adjusted to the present case. □

Chapter Five

Basic Properties of the Self-Similar Potential Flow Equation

5.1 SOME BASIC FACTS AND FORMULAS FOR THE POTENTIAL FLOW EQUATION

We first show some facts that hold for sufficiently regular solutions of **Problem 2.6.1**, as well as for *approximate solutions* which will be considered in Chapter 12. Therefore, we do not assume that φ is an admissible solution; that is, φ is not required to satisfy any equation in Ω or any boundary conditions on $\partial\Omega$, unless otherwise specified. Instead, we use the following notations through this section: Ω denotes a domain in $\mathbb{R}^2$ which is of the structure as described in §2.4.2, where we assume that curve Γ_{shock} is C^2 in its relative interior; furthermore, φ denotes a function in Ω satisfying $\varphi \in C^3(\overline{\Omega} \setminus (\overline{\Gamma_{\text{sonic}}} \cup \{P_2, P_3\})) \cap C^1(\overline{\Omega})$.

For a smooth function G on Ω and vectors $\mathbf{a}, \mathbf{b} \in \mathbb{R}^2$, we denote

$$D^2G[\mathbf{a}, \mathbf{b}] \equiv D^2G[\mathbf{a}, \mathbf{b}](\boldsymbol{\xi}) := \sum_{i,j=1}^{2} D_{ij}G(\boldsymbol{\xi})a_i b_j \qquad \text{for } \boldsymbol{\xi} \in \Omega.$$

Note that the right-hand side in the above equality is invariant under the orthogonal transform so that we can use the partial derivatives of G and components $(\mathbf{a}, \mathbf{b})$ with respect to any orthonormal basis in $\mathbb{R}^2$. In particular, we often use the basis, $\{\boldsymbol{\nu}, \boldsymbol{\tau}\}$, at $\boldsymbol{\xi} = (\xi_1, \xi_2) \in \Gamma_{\text{shock}}$.

Denote

$$\phi(\boldsymbol{\xi}) := (\varphi - \varphi_0)(\boldsymbol{\xi}) = \varphi(\boldsymbol{\xi}) + \frac{|\boldsymbol{\xi}|^2}{2}, \quad \bar{\phi}(\boldsymbol{\xi}) := (\varphi_1 - \varphi)(\boldsymbol{\xi}), \quad \psi(\boldsymbol{\xi}) := (\varphi - \varphi_2)(\boldsymbol{\xi}), \tag{5.1.1}$$

where φ_0, φ_1, and φ_2 are the potentials of states (0), (1), and (2) defined by (2.2.16), (2.2.17), and (2.4.1), respectively. Since φ_0, φ_1, and φ_2 are uniform states, we have

$$D^2\phi = -D^2\bar{\phi} = D^2\psi.$$

Denote

$$\hat{\rho}(s) = \left(\rho_0^{\gamma-1} + (\gamma-1)s\right)^{\frac{1}{\gamma-1}}.$$

Then, by (2.2.9),

$$\rho(|D\varphi|^2, \varphi) = \hat{\rho}(-(\varphi + \frac{1}{2}|D\varphi|^2)). \tag{5.1.2}$$

In the calculations below, $\hat{\rho}$ and $\hat{\rho}'$ are always evaluated at $-(\varphi + \frac{1}{2}|D\varphi|^2)$, and we often drop the argument. From (1.14),

$$c^2(|D\varphi|^2, \varphi) = \frac{\hat{\rho}}{\hat{\rho}'}. \tag{5.1.3}$$

5.1.1 Partial derivatives of the density and sonic speed

By explicit calculation from (5.1.2), we have

$$\begin{aligned}\partial_{\xi_i}\rho(|D\varphi|^2, \varphi) &= -(\varphi_{\xi_i} + D\varphi \cdot D\varphi_{\xi_i})\hat{\rho}' = -(\varphi_{\xi_i} + D\varphi \cdot D^2\varphi\, \mathbf{e}_i)\hat{\rho}' \\ &= -\big(\varphi_{\xi_i} + D^2\varphi[\mathbf{e}_i, D\varphi]\big)\hat{\rho}',\end{aligned} \tag{5.1.4}$$

where $\mathbf{e}_1 = (1,0)^\top$, $\mathbf{e}_2 = (0,1)^\top$, and $D^2\varphi[\mathbf{e},\mathbf{v}] := \mathbf{e}^\top D^2\varphi \mathbf{v}$ for any vectors $\mathbf{e}$ and $\mathbf{v}$.

Using that $D^2\varphi = D^2\phi - I$, we can write (5.1.4) as

$$\partial_{\xi_i}\rho(|D\varphi|^2, \varphi) = -\hat{\rho}' D^2\phi[\mathbf{e}_i, D\varphi]. \tag{5.1.5}$$

Similarly, from (1.14), we have

$$\begin{aligned}\partial_{\xi_i}c^2(|D\varphi|^2, \varphi) &= -\,(\gamma-1)\big(\varphi_{\xi_i} + D^2\varphi[\mathbf{e}_i, D\varphi]\big) \\ &= -\,(\gamma-1)\hat{\rho}' D^2\phi[\mathbf{e}_i, D\varphi].\end{aligned} \tag{5.1.6}$$

In particular, on Γ_{shock},

$$\rho_{\boldsymbol{\tau}} = -\big(\varphi_{\boldsymbol{\tau}} + D^2\varphi[\boldsymbol{\tau}, D\varphi]\big)\hat{\rho}' = -D^2\phi[\boldsymbol{\tau}, D\varphi]\hat{\rho}', \tag{5.1.7}$$

$$\rho_{\boldsymbol{\nu}} = -\big(\varphi_{\boldsymbol{\nu}} + D^2\varphi[\boldsymbol{\nu}, D\varphi]\big)\hat{\rho}' = -D^2\phi[\boldsymbol{\nu}, D\varphi]\hat{\rho}'. \tag{5.1.8}$$

Also, since $D^2\phi = -D^2\bar{\phi} = D^2\psi$, we can replace $D^2\phi$ by either $-D^2\bar{\phi}$ or $D^2\psi$ on the right-hand sides of (5.1.5)–(5.1.6).

5.1.2 An elliptic equation for $\partial_{\mathbf{e}}\phi$ in Ω, provided that φ is a subsonic potential

In this subsection, we assume that φ satisfies equation (2.2.8) (hence its non-divergence form (2.2.11)) in Ω and that equation (2.2.8) is strictly elliptic on φ in $\overline{\Omega} \setminus \overline{\Gamma_{\text{sonic}}}$.

Let $\mathbf{e} \in \mathbb{R}^2$ be a unit vector and $\mathbf{e}^\perp$ the unit vector orthogonal to $\mathbf{e}$. Denote by (S,T) the coordinates with basis $\{\mathbf{e}, \mathbf{e}^\perp\}$. Then $\partial_{\mathbf{e}}\phi = \partial_S\phi$. Since equation (2.2.11) is invariant with respect to the orthogonal coordinate transforms, it is of the same form in the (S,T)–coordinates as in the $\boldsymbol{\xi}$–coordinates, that is,

$$(c^2 - \varphi_S^2)\phi_{SS} - 2\varphi_S\varphi_T\phi_{ST} + (c^2 - \varphi_T^2)\phi_{TT} = 0. \tag{5.1.9}$$

We differentiate equation (5.1.9) with respect to S and use (5.1.6) to obtain

$$\partial_S c^2 = -(\gamma-1)D^2\phi[\mathbf{e}, D\varphi] = -(\gamma-1)(\varphi_S\phi_{SS} + \varphi_T\phi_{ST}).$$

Also, we use that $\varphi_{ST} = \phi_{ST}$. Then we have the following equation for $w := \partial_{\mathbf{e}}\phi = \partial_S\phi$:

$$\begin{aligned}
&(c^2 - \varphi_S^2)w_{SS} - 2\varphi_S\varphi_T w_{ST} + (c^2 - \varphi_T^2)w_{TT} \\
&\quad - \big((\gamma-1)\varphi_S(\phi_{SS} + \phi_{TT}) + (\gamma-1)\varphi_T\phi_{ST} + 2\varphi_S(\phi_{SS} - 1)\big)w_S \\
&\quad - \big(2\varphi_T(\phi_{SS} - 1) + 2\phi_{ST}\varphi_S + (\gamma+1)\varphi_T\phi_{TT}\big)w_T = 0.
\end{aligned} \tag{5.1.10}$$

Now we assume that (2.2.11) is strictly elliptic in a region $\mathcal{D} \subset \overline{\Omega}$. Equation (5.1.10) has the same coefficients of the second derivative terms as in (5.1.9), *i.e.*, in (2.2.11). Thus, (5.1.10) is strictly elliptic in $\mathcal{D}$.

Furthermore, in some cases, we want to avoid the dependence of the coefficients on the second derivatives of φ. Since (2.2.11) is strictly elliptic in $\mathcal{D}$, then, using its form (5.1.9), we obtain that $c^2 - \varphi_T^2 > 0$ in $\mathcal{D}$. Using (5.1.9) to express ϕ_{TT} through (ϕ_{SS}, ϕ_{ST}) that are substituted by (w_S, w_T) below, we obtain the following nonlinear equation for w:

$$\begin{aligned}
&(c^2 - \varphi_S^2)w_{SS} - 2\varphi_S\varphi_T w_{ST} + (c^2 - \varphi_T^2)w_{TT} \\
&\quad - \Big(\big((\gamma-1)\frac{\varphi_S^2 - \varphi_T^2}{c^2 - \varphi_T^2} + 2\big)\varphi_S w_S \\
&\qquad + (\gamma-1)\big(\frac{2\varphi_S^2}{c^2 - \varphi_T^2} + 1\big)\varphi_T w_T - 2\varphi_S\Big)w_S \\
&\quad - \Big(\big((\gamma+1)\frac{c^2 - \varphi_S^2}{c^2 - \varphi_T^2} + 2\big)\varphi_T w_S \\
&\qquad + 2\big(\frac{2(\gamma+1)\varphi_S^2}{c^2 - \varphi_T^2} + 1\big)\varphi_S w_T - (\gamma+1)\varphi_T\Big)w_T = 0.
\end{aligned} \tag{5.1.11}$$

Equation (5.1.11) has the same coefficients of the second derivative terms as (2.2.11). Thus, (5.1.11) is strictly elliptic in $\mathcal{D}$.

5.1.3 Tangential derivative of φ on Γ_{shock} through the Rankine-Hugoniot conditions

In this subsection, we assume that $\Omega \subset \mathbb{R}^2$ is an open region, $\Gamma_{\text{shock}} \subset \partial\Omega$, and Γ_{shock} is a relatively open segment of curve, which is locally C^2. Let $\varphi \in C^2(\Omega \cup \Gamma_{\text{shock}})$ satisfy the Rankine-Hugoniot conditions (2.2.13)–(2.2.14) with the uniform state φ_1 across Γ_{shock}. Then, on Γ_{shock},

$$\rho(|D\varphi|^2, \varphi)\partial_{\boldsymbol{\nu}}\varphi = \rho_1\partial_{\boldsymbol{\nu}}\varphi_1, \tag{5.1.12}$$

$$\varphi = \varphi_1, \tag{5.1.13}$$

where $\rho_1 > 0$ is the density of state φ_1. Moreover, we assume that

$$\partial_{\boldsymbol{\nu}}\varphi_1 > \partial_{\boldsymbol{\nu}}\varphi > 0 \qquad \text{on } \Gamma_{\text{shock}}. \tag{5.1.14}$$

Thus, by (5.1.12),

$$\rho(|D\varphi|^2, \varphi) > \rho_1 \qquad \text{on } \overline{\Gamma_{\text{shock}}}. \tag{5.1.15}$$

Then, assuming that φ satisfies (5.1.12)–(5.1.15) on Γ_{shock} and using the notations in (5.1.1), we have the following:

- From (5.1.13)–(5.1.14) and (2.2.17), the normal to Γ_{shock} is in the direction:

$$\tilde{\boldsymbol{\nu}} = D\varphi_1 - D\varphi = (u_1 - \phi_{\xi_1}, -\phi_{\xi_2})^\top, \tag{5.1.16}$$

and the tangent vector to Γ_{shock} is in the direction:

$$\tilde{\boldsymbol{\tau}} = (\phi_{\xi_2}, u_1 - \phi_{\xi_1})^\top. \tag{5.1.17}$$

Then the unit normal and tangent vectors to Γ_{shock} are

$$\boldsymbol{\nu} = \frac{\tilde{\boldsymbol{\nu}}}{|\tilde{\boldsymbol{\nu}}|}, \qquad \boldsymbol{\tau} = \frac{\tilde{\boldsymbol{\tau}}}{|\tilde{\boldsymbol{\tau}}|},$$

respectively. We can also express vectors $\boldsymbol{\nu}$ and $\boldsymbol{\tau}$ at $P \in \Gamma_{\text{shock}}$ as

$$\boldsymbol{\nu} = \frac{D(\varphi_1 - \varphi)}{|D(\varphi_1 - \varphi)|} = \frac{D\bar{\phi}}{|D\bar{\phi}|}, \quad \boldsymbol{\tau} = (\tau_1, \tau_2) = \frac{(-\partial_{\xi_2}\bar{\phi}, \partial_{\xi_1}\bar{\phi})}{|D\bar{\phi}|}, \tag{5.1.18}$$

where $\bar{\phi}$ is defined by (5.1.1). Using the Rankine-Hugoniot conditions (5.1.12)–(5.1.13) along Γ_{shock}, we have

$$|\tilde{\boldsymbol{\nu}}| = |D\varphi_1 - D\varphi| = |D\varphi_1 \cdot \boldsymbol{\nu} - D\varphi \cdot \boldsymbol{\nu}| = \frac{\rho - \rho_1}{\rho_1}\varphi_{\boldsymbol{\nu}} > 0. \tag{5.1.19}$$

- Since $\tilde{\boldsymbol{\nu}}_{\xi_i} = (D\varphi_1)_{\xi_i} - D\varphi_{\xi_i} = D^2\varphi_1\,\mathbf{e}_i - D^2\varphi\,\mathbf{e}_i = -\mathbf{e}_i - D^2\varphi\,\mathbf{e}_i$, we have

$$\tilde{\boldsymbol{\nu}}_{\boldsymbol{\tau}} = -\boldsymbol{\tau} - D^2\varphi\,\boldsymbol{\tau}. \tag{5.1.20}$$

Taking the tangential derivative to both sides of (5.1.12) with $\boldsymbol{\nu}$ replaced by $\tilde{\boldsymbol{\nu}}$ for convenience:

$$(\rho D\varphi - \rho_1 D\varphi_1) \cdot \tilde{\boldsymbol{\nu}} = 0,$$

we have

$$(\rho D^2\varphi\,\boldsymbol{\tau} + \rho_{\boldsymbol{\tau}} D\varphi - \rho_1\boldsymbol{\tau}) \cdot \tilde{\boldsymbol{\nu}} + (\rho D\varphi - \rho_1 D\varphi_1) \cdot (-D^2\varphi\,\boldsymbol{\tau} - \boldsymbol{\tau}) = 0.$$

Noting that $\boldsymbol{\tau} \cdot \tilde{\boldsymbol{\nu}} = 0$ and using (5.1.3), (5.1.7), and (5.1.19), we have

$$\begin{aligned} &D^2\varphi\big[\boldsymbol{\tau},\, \rho\tilde{\boldsymbol{\nu}} + \rho_1 D\varphi_1 - \rho(1 + \frac{\rho - \rho_1}{\rho_1}\frac{\varphi_{\boldsymbol{\nu}}^2}{c^2})D\varphi\big] \\ &= \frac{\rho(\rho - \rho_1)}{\rho_1 c^2}\varphi_{\boldsymbol{\tau}}\varphi_{\boldsymbol{\nu}}^2 + (\rho - \rho_1)\varphi_{\boldsymbol{\tau}} \\ &= \frac{\rho - \rho_1}{\rho_1}\varphi_{\boldsymbol{\tau}}\Big(\rho_1 + \rho\frac{\varphi_{\boldsymbol{\nu}}^2}{c^2}\Big). \end{aligned}$$

That is, for $\mathbf{g} = \rho\tilde{\boldsymbol{\nu}} + \rho_1 D\varphi_1$,

$$D^2\varphi[\boldsymbol{\tau}, \mathbf{g}] = \rho\Big(1 + \frac{\rho-\rho_1}{\rho_1}\frac{\varphi_{\boldsymbol{\nu}}^2}{c^2}\Big)D^2\varphi[\boldsymbol{\tau}, D\varphi] + \frac{\rho-\rho_1}{\rho_1}\varphi_{\boldsymbol{\tau}}\Big(\rho_1 + \rho\frac{\varphi_{\boldsymbol{\nu}}^2}{c^2}\Big). \tag{5.1.21}$$

Using that $D^2\phi = D^2\varphi + I$, we can rewrite (5.1.21) in terms of ϕ to obtain the following lemma:

Lemma 5.1.1. *Let $\Omega \subset \mathbb{R}^2$ be an open set, and let $\Gamma_{\rm shock}$ be a relatively open segment of curve, which is C^2 locally. Let φ satisfy* (5.1.12)–(5.1.14), *where φ_1 is a uniform state with density ρ_1. Then*

$$D^2\phi[\boldsymbol{\tau}, \mathbf{h}] = 0 \qquad \text{on } \Gamma_{\rm shock}, \tag{5.1.22}$$

where

$$\mathbf{h} = \frac{\rho-\rho_1}{\rho_1 c^2}\left(-\rho(c^2 - \varphi_{\boldsymbol{\nu}}^2)\varphi_{\boldsymbol{\nu}}\boldsymbol{\nu} + (\rho\varphi_{\boldsymbol{\nu}}^2 + \rho_1 c^2)\varphi_{\boldsymbol{\tau}}\boldsymbol{\tau}\right). \tag{5.1.23}$$

Remark 5.1.2. *Note that* (5.1.22)–(5.1.23) *still hold if $\boldsymbol{\nu}$ is replaced by $-\boldsymbol{\nu}$, or $\boldsymbol{\tau}$ by $-\boldsymbol{\tau}$.*

We also use the expression of function $\mathbf{h}$ in basis $\{\mathbf{e}_1, \mathbf{e}_2\}$ corresponding to the $\boldsymbol{\xi}$–coordinates:

$$\begin{aligned}\mathbf{h} = \frac{\rho-\rho_1}{\rho_1 c^2}\Big(&\left(-\rho(c^2 - \varphi_{\boldsymbol{\nu}}^2)\varphi_{\boldsymbol{\nu}}\nu_1 + (\rho\varphi_{\boldsymbol{\nu}}^2 + \rho_1 c^2)\varphi_{\boldsymbol{\tau}}\tau_1\right)\mathbf{e}_1\\ &+ \left(-\rho(c^2 - \varphi_{\boldsymbol{\nu}}^2)\varphi_{\boldsymbol{\nu}}\nu_2 + (\rho\varphi_{\boldsymbol{\nu}}^2 + \rho_1 c^2)\varphi_{\boldsymbol{\tau}}\tau_2\right)\mathbf{e}_2\Big),\end{aligned} \tag{5.1.24}$$

where $\boldsymbol{\tau} = (\tau_1, \tau_2)$ and $\boldsymbol{\nu} = (\nu_1, \nu_2)$ are the unit tangent and normal vectors to $\Gamma_{\rm shock}$, respectively.

5.1.4 Oblique derivative condition for the directional derivatives of ϕ on $\Gamma_{\rm sym} \cup \Gamma_{\rm wedge}$.

Lemma 5.1.3. *Let φ be a solution of* **Problem 2.6.1** *with the structure of supersonic reflection configuration as in §2.4.2, or subsonic reflection configuration as in §2.4.3. Moreover, assume that equation* (2.2.11) *is strictly elliptic for φ in $\Omega \cup \Gamma_{\rm wedge} \cup \Gamma_{\rm sym}$. Let $\phi = \varphi + \frac{|\boldsymbol{\xi}|^2}{2}$, and $w := \partial_{\mathbf{e}}\phi$ for $\mathbf{e} \in \mathbb{R}^2 \setminus \{0\}$. Then*

(i) *Let $\mathbf{e}$ be non-orthogonal to $\Gamma_{\rm wedge}$, i.e., $\mathbf{e} = a_1\boldsymbol{\nu} + a_2\boldsymbol{\tau}$, where $\boldsymbol{\nu}$ and $\boldsymbol{\tau}$ are the unit normal and tangent vectors to $\Gamma_{\rm wedge}$, and a_1 and a_2 are constants with $a_2 \neq 0$. Then $w := \partial_{\mathbf{e}}\phi$ satisfies the following oblique derivative condition on $\Gamma_{\rm wedge}$:*

$$\partial_{\boldsymbol{\nu}} w + \frac{a_1(c^2 - \varphi_{\boldsymbol{\nu}}^2)}{a_2(c^2 - \varphi_{\boldsymbol{\tau}}^2)}\partial_{\boldsymbol{\tau}} w = 0. \tag{5.1.25}$$

(ii) *Let* $\mathbf{e}$ *be non-orthogonal to* Γ_{sym}*, i.e.,* $\mathbf{e} = a_1\boldsymbol{\nu} + a_2\boldsymbol{\tau}$*, where* $\boldsymbol{\nu}$ *and* $\boldsymbol{\tau}$ *are the unit normal and tangent vectors to* Γ_{sym}*, and* a_1 *and* a_2 *are constants with* $a_2 \neq 0$*. Then* (5.1.25) *holds on* Γ_{sym}*.*

Proof. Using that $\varphi_{\boldsymbol{\nu}} = 0$ on Γ_{wedge}, we have

$$\partial_{\boldsymbol{\nu}}\phi = 0 \qquad \text{on } \Gamma_{\text{wedge}}. \tag{5.1.26}$$

Now we work in the (S, T)–coordinates with basis $\{\boldsymbol{\nu}, \boldsymbol{\tau}\}$ so that $\Gamma_{\text{wedge}} \subset \{S = 0\}$. Then (5.1.26) becomes

$$\partial_S\phi = 0 \qquad \text{on } \Gamma_{\text{wedge}} \subset \{S = 0\}. \tag{5.1.27}$$

Differentiating condition (5.1.27) in the T-direction (tangential to Γ_{wedge}), we have

$$\partial_{ST}\phi = 0 \qquad \text{on } \Gamma_{\text{wedge}}. \tag{5.1.28}$$

Recall that equation (2.2.11) in the (S, T)–variables is of form (5.1.9). Moreover, since the equation is strictly elliptic in $\Omega \cup \Gamma_{\text{wedge}}\Gamma_{\text{sym}}$, then

$$c^2 - \varphi_S^2 > 0$$

in that region. Also recall that $\varphi_S = \varphi_{\boldsymbol{\nu}}$ and $\varphi_T = \varphi_{\boldsymbol{\tau}}$ on Γ_{wedge}. Thus equation (5.1.9), combined with (5.1.28), implies

$$\partial_{SS}\phi = -\frac{c^2 - \varphi_{\boldsymbol{\tau}}^2}{c^2 - \varphi_{\boldsymbol{\nu}}^2}\partial_{TT}\phi \qquad \text{on } \Gamma_{\text{wedge}}.$$

Now we use (5.1.28) to compute on Γ_{wedge} that

$$\begin{aligned}\partial_S w &= a_1\partial_{SS}\phi + a_2\partial_{ST}\phi = a_1\partial_{SS}\phi = -a_1\frac{c^2 - \varphi_{\boldsymbol{\tau}}^2}{c^2 - \varphi_{\boldsymbol{\nu}}^2}\partial_{TT}\phi \\ &= -\frac{a_1(c^2 - \varphi_{\boldsymbol{\tau}}^2)}{a_2(c^2 - \varphi_{\boldsymbol{\nu}}^2)}(a_1\partial_{ST}\phi + a_2\partial_{TT}\phi) = -\frac{a_1(c^2 - \varphi_{\boldsymbol{\tau}}^2)}{a_2(c^2 - \varphi_{\boldsymbol{\nu}}^2)}\partial_T w.\end{aligned}$$

This implies (5.1.25). Part (i) is proved.

Part (ii) is proved similarly, using that

$$\partial_{\boldsymbol{\nu}}\phi = \partial_{\boldsymbol{\nu}}\varphi = 0 \qquad \text{on } \Gamma_{\text{sym}}.$$

□

More generally, we have

Lemma 5.1.4. *Let* Ω *be an open region, and let* L *be a line in* $\mathbb{R}^2$ *passing through the origin so that* $\Gamma = L \cap \partial\Omega$ *is a non-empty segment. Let* $\varphi \in C^2(\Omega \cup \Gamma^0)$*, where* Γ^0 *denotes the relative interior of* Γ*, with*

$$\partial_{\boldsymbol{\nu}}\varphi = 0 \qquad \text{on } \Gamma^0,$$

and let $\phi = \varphi + \frac{|\boldsymbol{\xi}|^2}{2}$ *satisfy the equation:*

$$\sum_{i,j=1}^{2} a_{ij} D_{ij}\phi = 0 \qquad \text{in } \Omega, \tag{5.1.29}$$

where $a_{ij} \in C(\overline{\Omega})$ *satisfy*

$$\sum_{i,j=1}^{2} a_{ij}\nu_i\nu_j \geq \kappa \qquad \text{on } \Gamma^0 \tag{5.1.30}$$

for some constant $\kappa > 0$, *and* $\boldsymbol{\nu} = (\nu_1, \nu_2)$ *is the unit normal to* Γ^0. *Let* $\mathbf{e} \in \mathbb{R}^2 \setminus \{0\}$ *be non-orthogonal to* L. *Then* $w := \partial_{\mathbf{e}}\phi$ *satisfies the following oblique derivative condition on* Γ^0:

$$\partial_{\boldsymbol{\nu}} w + b\partial_{\boldsymbol{\tau}} w = 0 \qquad \text{on } \Gamma^0, \tag{5.1.31}$$

where $b \in C(\Gamma^0)$.

Proof. Note that equation (5.1.29) remains the same form with some new coefficients $\hat{a}_{ij}$ under a rotation of the coordinates, and these coefficients $\hat{a}_{ij}$ satisfy (5.1.30). Then the rest of the proof follows the proof of Lemma 5.1.3. Indeed, in the (S,T)–coordinates defined in that proof, $\boldsymbol{\nu} = (1,0)$ so that condition (5.1.30) becomes $\hat{a}_{11} \geq \kappa > 0$. Also, (5.1.28) holds on Γ^0. Therefore, equation (5.1.29) implies

$$\partial_{SS}\phi = -\frac{\hat{a}_{22}}{\hat{a}_{11}}\partial_{TT}\phi \qquad \text{on } \Gamma^0.$$

Then we obtain (5.1.31) with

$$b = \frac{a_1\hat{a}_{22}}{a_2\hat{a}_{11}} \qquad \text{on } \Gamma^0,$$

where $a_1 = \mathbf{e}\cdot\boldsymbol{\nu}$ and $a_2 = \mathbf{e}\cdot\boldsymbol{\tau}$ so that $\mathbf{e} = a_1\boldsymbol{\nu} + a_2\boldsymbol{\tau}$, and $a_2 \neq 0$ since $\mathbf{e}$ is not orthogonal to Γ. □

5.2 INTERIOR ELLIPTICITY PRINCIPLE FOR SELF-SIMILAR POTENTIAL FLOW

For $\gamma \geq 1$, the coefficients of the self-similar potential flow equation for φ, written in either the divergence form (2.2.8) or the non-divergence form (2.2.11), depend on the potential function φ itself, besides $D\varphi$, which is a significant difference from the steady case.

We now discuss an extension of the interior ellipticity principle of Elling-Liu [110] in $\mathbb{R}^2$, as well as the corresponding ellipticity principle for flat boundaries in §5.3. We let $\gamma \geq 1$ and use the notation:

$$M := \frac{|D\varphi|}{c(D\varphi,\varphi)}$$

as the pseudo-Mach number.

Theorem 5.2.1. *Let $\Omega \subset \mathbb{R}^2$ be an open bounded domain.*

(i) *Let $\varphi \in C^3(\Omega)$ satisfy* (2.2.11) *with $M \leq 1$ and $\rho > 0$ in Ω. Then either $M \equiv 0$ in Ω or M does not attain its maximum in Ω.*

(ii) *More generally, for any $d > 0$, there exists $C_0 > 0$ depending only on (γ, d) such that, if $\operatorname{diam}(\Omega) \leq d$, for any $\delta \geq 0$, $\hat{c} \geq 1$, and $b \in C^2(\Omega)$ with $|Db| + \hat{c}|D^2 b| \leq \frac{\delta}{\hat{c}}$, and for any solution $\varphi \in C^3(\Omega)$ of* (2.2.11) *satisfying $M \leq 1$, $\rho(|D\varphi|^2, \varphi) > 0$, and $c(|D\varphi|^2, \varphi) \leq \hat{c}$ in Ω, then either*

$$M^2 \leq C_0 \delta \qquad \text{in } \Omega$$

or $M^2 + b$ does not attain its maximum in Ω.

Proof. In the proof below, for $\sigma \in \mathbb{R}$, $O(\sigma)$ denotes any expression that can be estimated as $|O(\sigma)| \leq C|\sigma|$, where C depends only on γ. We follow the calculation in the proof of [110, Theorem 2.1], keeping more terms in its exact form.

Note that assertion (i) is essentially assertion (ii) with $\delta = 0$, except that (ii) has the assumption of boundedness of $c(|D\varphi|^2, \varphi)$ in Ω, which is not present in (i). Thus, we first prove (ii) for $\delta > 0$, and then give the proof of (i).

1. Let $\delta > 0$. Let b satisfy the conditions in (ii). Let the maximum of $M^2 + b$ be attained at $\hat{P} = \hat{\boldsymbol{\xi}} \in \Omega$, and let

$$M^2 > C_1 \delta \qquad \text{at } \hat{P}$$

for a constant $C_1 \geq 1$ to be chosen. We will arrive at a contradiction if C_1 is large, depending only on γ.

Our assumptions imply that $|D\varphi| = Mc > 0$ at $\hat{P}$ since $M(\hat{P}) > 0$. Also, since $C_1 \geq 1$, our assumptions imply that $\frac{\delta}{M^2} \leq 1$ at $\hat{P}$.

In the calculation below, all the equations and inequalities hold only at $\hat{P}$, unless otherwise specified. Also, we use expression (1.14) and the notations: $\varphi_i = \varphi_{\xi_i}, \varphi_{ij} = \varphi_{\xi_i \xi_j}$, and $\varphi_{ijk} = \varphi_{\xi_i \xi_j \xi_k}$ in the calculation for simplicity.

2. Since equation (2.2.11) is rotationally invariant, we can assume without loss of generality that $\varphi_{\xi_1} = |D\varphi|$ and $\varphi_{\xi_2} = 0$ at $\hat{P}$. Then the first-order condition at the maximum point implies that, at $\hat{P}$,

$$\begin{aligned} 0 = \partial_{\xi_1}\Big(\frac{|D\varphi|^2}{c^2(|D\varphi|^2, \varphi)} + b\Big) &= \frac{2c^2 \varphi_1 \varphi_{11} + (\gamma - 1)(\varphi_1 + \varphi_1 \varphi_{11})\varphi_1^2}{c^4} + \partial_{\xi_1} b \\ &= \frac{M\big(2\varphi_{11} + (\gamma - 1)(1 + \varphi_{11})M^2\big)}{c} + \partial_{\xi_1} b. \end{aligned}$$

Using that $M \leq 1$, we have

$$\varphi_{11} = \frac{(1-\gamma)M^2}{2 + (\gamma - 1)M^2} + O(\frac{\delta}{M^2}). \tag{5.2.1}$$

Remark 5.2.2. *Equality* (5.2.1) *holds with* $O(\frac{\delta}{M})$ *instead of* $O(\frac{\delta}{M^2})$, *since* $O(\frac{\delta}{M})$ *is a stronger estimate of the error term than* $O(\frac{\delta}{M^2})$ *owing to the fact that* $M \leq 1$. *We have adapted the weaker error term* $O(\frac{\delta}{M^2})$ *here and below, since the weak error form is good enough to carry through our analysis later on and some estimates below hold only with* $O(\frac{\delta}{M^2})$.

Similarly,

$$
\begin{aligned}
0 = \partial_{\xi_2}\Big(\frac{|D\varphi|^2}{c^2(|D\varphi|^2,\varphi)} + b\Big) &= \frac{2c^2\varphi_1\varphi_{12} + (\gamma-1)\varphi_1\varphi_{12}\varphi_1^2}{c^4} + \partial_{\xi_2} b \\
&= \frac{M\big(2+(\gamma-1)M^2\big)\varphi_{12}}{c} + \partial_{\xi_2} b.
\end{aligned}
$$

Thus, we have

$$
\varphi_{12} = O(\frac{\delta}{M^2}). \tag{5.2.2}
$$

Now we note that equation (2.2.11) in Ω can be written in the following form:

$$
c^2\Delta\varphi - \sum_{i,j=1}^{2} \varphi_i\varphi_j\varphi_{ij} = |D\varphi|^2 - 2c^2. \tag{5.2.3}
$$

Then, at $\hat{P}$, we have

$$
(1-M^2)\varphi_{11} + \varphi_{22} = M^2 - 2,
$$

and, using (5.2.1),

$$
\varphi_{22} = \frac{(3-\gamma)M^2 - 4}{2+(\gamma-1)M^2} + O(\frac{\delta}{M^2}). \tag{5.2.4}
$$

In particular, we have

$$
|\varphi_{11}| + |\varphi_{22}| \leq C(\gamma), \tag{5.2.5}
$$

where we have used that $\frac{\delta}{M^2} \leq 1$.

3. We now use the second-order conditions at the maximum point. First we compute at a generic point for $k = 1, 2$:

$$
\begin{aligned}
&\partial_{\xi_k\xi_k}\Big(\frac{|D\varphi|^2}{c^2}\Big) \\
&= \partial_{\xi_k}\Big(\frac{2}{c^2}\sum_i \varphi_i\varphi_{ki} + \frac{\gamma-1}{c^4}\big(\varphi_k + \sum_i \varphi_i\varphi_{ki}\big)|D\varphi|^2\Big) \\
&= \frac{2}{c^4}\Big(\big(\sum_i \varphi_i\varphi_{kki} + \sum_i \varphi_{ki}^2\big)c^2 + (\gamma-1)\big(\varphi_k + \sum_j \varphi_j\varphi_{kj}\big)\sum_i \varphi_i\varphi_{ki}\Big) \\
&\quad + \frac{\gamma-1}{c^4}\Big(\big(\varphi_{kk} + \sum_i \varphi_{ki}^2 + \sum_i \varphi_i\varphi_{kki}\big)|D\varphi|^2 + 2\big(\varphi_k + \sum_i \varphi_i\varphi_{ki}\big)\sum_j \varphi_j\varphi_{kj}\Big) \\
&\quad + \frac{2(\gamma-1)^2}{c^6}\big(\varphi_k + \sum_i \varphi_i\varphi_{ki}\big)^2|D\varphi|^2.
\end{aligned}
$$

Then, at $\hat{P}$, using that $\varphi_1 = |D\varphi| = Mc$ and $\varphi_2 = 0$, and employing (5.2.2) and (5.2.5), we obtain

$$\begin{aligned}
0 &\geq \partial_{\xi_1\xi_1}\Big(\frac{|D\varphi|^2}{c^2} + b\Big) \\
&= \frac{1}{c^2}\Big(Mc(2+(\gamma-1)M^2)\varphi_{111} + O(\frac{\delta}{M^2}) \\
&\qquad + \big(\varphi_{11} + 2(\gamma-1)M^2(1+\varphi_{11})\big)\big(2\varphi_{11} + (\gamma-1)M^2(1+\varphi_{11})\big)\Big),
\end{aligned}$$

where we have used that $\frac{\delta}{M^2} \leq 1$ so that $O(\frac{\delta}{M^2})$ can be used instead of $O\big(\frac{\delta^2}{M^4}\big)$. Now, from (5.2.1), we have

$$2\varphi_{11} + (\gamma-1)M^2(1+\varphi_{11}) = O(\frac{\delta}{M^2}).$$

Thus, using (5.2.5), we have

$$\frac{1}{c^2}\Big(Mc(2+(\gamma-1)M^2)\varphi_{111} + O(\frac{\delta}{M^2})\Big) \leq 0,$$

which implies

$$Mc\varphi_{111} \leq O(\frac{\delta}{M^2}). \tag{5.2.6}$$

Similarly,

$$\begin{aligned}
0 &\geq \partial_{\xi_2\xi_2}\Big(\frac{|D\varphi|^2}{c^2} + b\Big) \\
&= \frac{1}{c^2}\Big(Mc(2+(\gamma-1)M^2)\varphi_{122} \\
&\qquad + \big(2+(\gamma-1)M^2\big)\varphi_{22}^2 + (\gamma-1)M^2\varphi_{22} + O(\frac{\delta}{M^2})\Big),
\end{aligned}$$

which implies

$$cM\varphi_{122} \leq -\varphi_{22}^2 - \frac{(\gamma-1)M^2}{2+(\gamma-1)M^2}\varphi_{22} + O(\frac{\delta}{M^2}).$$

Now, from (5.2.4), we obtain

$$cM\varphi_{122} \leq -\frac{\big((\gamma-3)M^2+4\big)\big(4-2M^2\big)}{\big(2+(\gamma-1)M^2\big)^2} + O(\frac{\delta}{M^2}) \leq -\frac{4}{(\gamma+1)^2} + O(\frac{\delta}{M^2}),$$

where we have used that $\gamma \geq 1$ and $0 \leq M \leq 1$ to derive the last inequality. Therefore, we have

$$cM\varphi_{122} \leq -\frac{4}{(\gamma+1)^2} + O(\frac{\delta}{M^2}). \tag{5.2.7}$$

4. We differentiate equation (5.2.3) with respect to ξ_1 and use $\varphi_1 = |D\varphi| = Mc$ and $\varphi_2 = 0$ at $\hat{P}$ to obtain

$$c^2(\varphi_{111} + \varphi_{122}) - (\gamma - 1)\varphi_1(1 + \varphi_{11})\Delta\varphi - \varphi_1^2\varphi_{111} - 2\varphi_1(\varphi_{11}^2 + \varphi_{12}^2)$$
$$= 2\varphi_1\varphi_{11} + 2(\gamma - 1)\varphi_1(1 + \varphi_{11}),$$

which, using also (5.2.2), implies

$$(c^2 - \varphi_1^2)\varphi_{111} + c^2\varphi_{122}$$
$$= (\gamma - 1)Mc(1 + \varphi_{11})\Delta\varphi + 2Mc\varphi_{11}^2 + 2\gamma Mc\varphi_{11}$$
$$+ 2(\gamma - 1)Mc + cMO(\frac{\delta}{M^2}).$$

Since $M \leq 1$, we obtain from (5.2.6) that $(c^2 - \varphi_1^2)\varphi_{111} \leq \frac{c}{M}O(\frac{\delta}{M^2})$. Then we have

$$cM\varphi_{122} \geq M^2\Big((\gamma - 1)(1 + \varphi_{11})\Delta\varphi + 2\varphi_{11}^2 + 2\gamma\varphi_{11} + 2(\gamma - 1)\Big) + O(\frac{\delta}{M^2}).$$

Substituting expressions (5.2.1) and (5.2.4) into the right-hand side of the inequality above, we conclude after a tedious but direct calculation that

$$cM\varphi_{122} \geq O(\frac{\delta}{M^2}). \tag{5.2.8}$$

This contradicts (5.2.7) if $\frac{\delta}{M^2} \leq \frac{1}{C_1}$ for sufficiently large $C_1 \geq 1$, depending only on γ.

Therefore, we have shown that the maximum of $M^2 + b$ cannot be attained in Ω unless $\dfrac{\delta}{M^2} \geq \dfrac{1}{C_1}$ at $\hat{P}$, that is, $M^2(\hat{P}) \leq C_1\delta$. Since $\mathrm{osc}_\Omega b \leq \dfrac{\delta}{\hat{c}}\mathrm{diam}(\Omega) \leq \delta\,\mathrm{diam}(\Omega)$, and $\hat{P}$ is a maximum point of $M^2 + b$, it follows that, at any point in Ω,

$$M^2 \leq M^2(\hat{P}) + \mathrm{osc}_\Omega b \leq C_0\delta,$$

where $C_0 = C_1 + \mathrm{diam}(\Omega)$. Assertion (ii) is proved for $\delta > 0$.

5. Now we prove assertion (i). Let the maximum of M^2 be attained at $\hat{P} = \hat{\boldsymbol{\xi}} \in \Omega$, and let M^2 be not identically zero in Ω. Then

$$M^2 > 0 \qquad \text{at } \hat{P}.$$

It also follows that $|D\varphi| = Mc > 0$ at $\hat{P}$, since $c = \rho^{\frac{\gamma-1}{2}} > 0$ in Ω.

Now we follow the proof of (ii) for $\delta > 0$. The only difference is that, in all of the expressions, the $O(\cdot)$–terms are now replaced by zero. In particular, instead of (5.2.7), we have

$$cM\varphi_{122} \leq -\frac{4}{(\gamma + 1)^2},$$

and, instead of (5.2.8), we obtain

$$cM\varphi_{122} \geq 0.$$

These two inequalities clearly contradict each other, thus assertion (i) is proved. □

5.3 ELLIPTICITY PRINCIPLE FOR SELF-SIMILAR POTENTIAL FLOW WITH SLIP CONDITION ON THE FLAT BOUNDARY

We consider a domain $\Omega \subset \mathbb{R}^2$ with a flat boundary part $\Gamma \subset \partial\Omega$, and a solution $\varphi \in C^3(\Omega \cup \Gamma)$ of (2.2.11) satisfying

$$\partial_\nu \varphi = 0 \qquad \text{on } \Gamma. \tag{5.3.1}$$

Theorem 5.3.1. *Let $\Omega \subset \mathbb{R}^2$ be an open bounded domain and $\Gamma \subset \partial\Omega$ a relatively open flat segment.*

(i) *Let $\varphi \in C^3(\Omega \cup \Gamma)$ satisfy (2.2.11) in Ω, and (5.3.1) on Γ, with $M \le 1$ and $\rho > 0$ in $\Omega \cup \Gamma$. Then either $M \equiv 0$ in $\Omega \cup \Gamma$ or M does not attain its maximum in $\Omega \cup \Gamma$.*

(ii) *More generally, for any $d > 0$, there exists $C_0 > 0$ depending only on (γ, d) such that, if $\operatorname{diam}(\Omega) \le d$, for any $\delta \ge 0$, $\hat{c} \ge 1$, and $b \in C^2(\Omega)$ with $\partial_\nu b = 0$ on Γ and $|Db| + \hat{c}|D^2 b| \le \frac{\delta}{\hat{c}}$ in Ω, and for any $\varphi \in C^3(\Omega \cup \Gamma)$ satisfying (2.2.11) in Ω and (5.3.1) on Γ with $M \le 1$, $\rho(|D\varphi|^2, \varphi) > 0$, and $c(|D\varphi|^2, \varphi) \le \hat{c}$ in $\Omega \cup \Gamma$, then either*

$$M^2 \le C_0 \delta \qquad \text{in } \Omega \cap \Gamma$$

or $M^2 + b$ does not attain its maximum in $\Omega \cap \Gamma$.

Proof. The proof consists of two steps. We continue to use $\varphi_k := \varphi_{\xi_k}$ and form (5.2.3) of equation (2.2.11) in the proof.

1. First consider the case that $\Omega = B_{d/2}(\mathbf{0}) \cap \{\xi_1 > 0\}$ with $\Gamma = B_{d/2}(\mathbf{0}) \cap \{\xi_1 = 0\}$, for some $d > 0$. Then (5.3.1) is of the form:

$$\varphi_1 = 0 \qquad \text{on } \Gamma.$$

Taking the tangential derivative along Γ, we have

$$\varphi_{12} = 0, \quad \varphi_{122} = 0 \qquad \text{on } \Gamma.$$

From this, using (5.1.6),

$$\partial_{\xi_1}(c^2(|D\varphi|^2, \varphi)) = -(\gamma - 1)\big(\varphi_1 + \varphi_{11}\varphi_1 + \varphi_{12}\varphi_2\big) = 0 \qquad \text{on } \Gamma.$$

Using all of the vanishing properties obtained above and taking ∂_{ξ_1} to both sides of equation (5.2.3), we obtain that $c^2\varphi_{111} = 0$ on Γ, since all the other terms vanish on Γ. Then, using that $c = \rho^{\frac{\gamma-1}{2}} > 0$ in $\Omega \cup \Gamma$, we obtain

$$\varphi_{111} = 0 \qquad \text{on } \Gamma.$$

From all of the properties on Γ obtained above, we see that the even extension $\varphi(\xi_1, \xi_2) = \varphi(-\xi_1, \xi_2)$ of φ into $B_{d/2}(\mathbf{0})$ satisfies $\varphi \in C^3(B_{d/2}(\mathbf{0}))$. Moreover, using the explicit form (5.2.3), it follows that the extended function φ is a solution of (5.2.3) in $B_{d/2}(\mathbf{0})$. Now, the conditions of Theorem 5.2.1 are satisfied. Thus, we obtain Theorem 5.3.1 in the present special case of Ω and Γ, with the same constant $C_0(\gamma, d)$ as in Theorem 5.2.1.

2. Now we consider the general case of Ω and Γ. We only give the argument for the proof of (ii), since the proof of (i) is similar.

By an orthogonal coordinate transformation and a shift of the origin, we reduce to the case that $\Gamma \subset \{\xi_1 = 0\}$ and $\boldsymbol{\nu} = \mathbf{e}_1$ on Γ, where the invariance of equation (2.2.11) with respect to the rotation and translation of coordinates has been used. Let $d > 0$ be such that $\mathrm{diam}(\Omega) \leq d$. Let $C_0 = C_0(\gamma, d)$ be from Theorem 5.2.1. Assume that there exists $P \in \overline{\Omega \cap \Gamma}$ such that

$$M^2(P) = \max_{\boldsymbol{\xi} \in \overline{\Omega \cap \Gamma}} M^2(\boldsymbol{\xi}), \qquad M^2(P) > C_0\delta.$$

If $P \in \Omega$, we arrive at a contradiction with Theorem 5.2.1. Thus, $P \in \Gamma$. Since Γ is a relatively open segment, there exists $r \in (0, \frac{d}{2})$ such that $B_r(P) \cap \Omega = B_r(P) \cap \{\xi_1 > 0\}$. Also, since $\Gamma \subset \{\xi_1 = 0\}$, then, shifting the origin into P, we obtain that $P = \mathbf{0}$ and $B_r(P) \cap \Omega = B_r(\mathbf{0}) \cap \{\xi_1 > 0\}$, and the maximum of φ over $\overline{B_r(\mathbf{0}) \cap \{\xi_1 > 0\}}$ is attained at $\mathbf{0}$. Using that $r < \frac{d}{2}$, we arrive at a contradiction to the result of Step 1. This proves (ii). $\square$

Part III

Proofs of the Main Theorems for the Sonic Conjecture and Related Analysis

Chapter Six

Uniform States and Normal Reflection

In this chapter, we analyze the uniform states and normal reflection in the self-similar coordinates for potential flow.

6.1 UNIFORM STATES FOR SELF-SIMILAR POTENTIAL FLOW

Let $\rho_0 > 0$ and $\mathcal{O}_\pm = (u_\pm, v_\pm)$ be fixed constants. Let φ^+ and φ^- represent the uniform (physical) states, *i.e.*,

$$\varphi^\pm(\boldsymbol{\xi}) = -\frac{|\boldsymbol{\xi}|^2}{2} + (u_\pm, v_\pm) \cdot (\boldsymbol{\xi} - \boldsymbol{\xi}^\pm), \tag{6.1.1}$$

where $\boldsymbol{\xi}^\pm$ are constant vectors. Set

$$\rho_\pm^{\gamma-1} := \rho_0^{\gamma-1} - (\gamma - 1)\Big(- (u_\pm, v_\pm) \cdot \boldsymbol{\xi}^\pm + \frac{1}{2}|(u_\pm, v_\pm)|^2\Big) > 0, \tag{6.1.2}$$

where we assume that ρ_0 is chosen so that the right-hand side is positive. Then $\varphi^\pm$ are the solutions of equation (2.2.8) with constant densities $\rho_\pm$ defined by (2.2.9) and corresponding constant sonic speeds $c_\pm^2 = \rho_\pm^{\gamma-1}$, respectively.

Lemma 6.1.1. *Let $P, \boldsymbol{\tau} \in \mathbb{R}^2$ with $|\boldsymbol{\tau}| = 1$, and let $L := \{P + s\boldsymbol{\tau} \,:\, s \in \mathbb{R}\}$ be a line. Let $\boldsymbol{\nu}$ be a unit vector orthogonal to $\boldsymbol{\tau}$. Then $D\varphi^\pm \cdot \boldsymbol{\nu}$ is constant along line L. Moreover, $|D\varphi^\pm \cdot \boldsymbol{\nu}|(\boldsymbol{\xi}) = \mathrm{dist}(\mathcal{O}_\pm, L)$ for any $\boldsymbol{\xi} \in L$; see Fig.* 6.1.

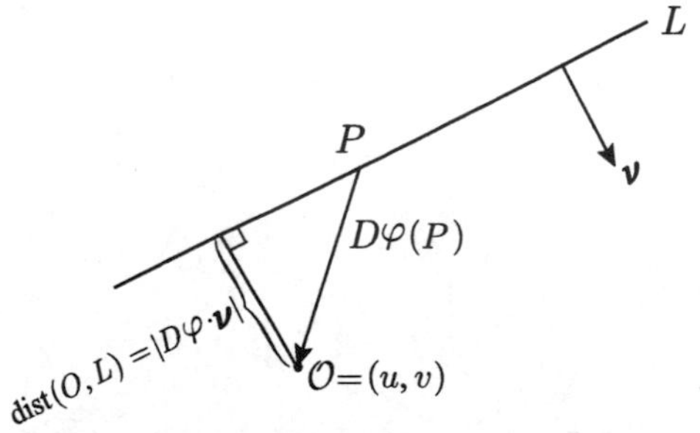

Figure 6.1: $|D\varphi^\pm \cdot \boldsymbol{\nu}| = \mathrm{dist}(\mathcal{O}_\pm, L)$

Proof. Let $Q \in L$. Then $L = \{Q + s\boldsymbol{\tau} \ : \ s \in \mathbb{R}\}$. Moreover,

$$D\varphi^{\pm}(\boldsymbol{\xi}) = (u_{\pm} - \xi_1, v_{\pm} - \xi_2) = \mathcal{O}_{\pm} - \boldsymbol{\xi}.$$

If $\boldsymbol{\xi} \in L$, then $\boldsymbol{\xi} = Q + s\boldsymbol{\tau}$ for some $s \in \mathbb{R}$. Since $\boldsymbol{\nu} \cdot \boldsymbol{\tau} = 0$, then

$$D\varphi^{\pm}(\boldsymbol{\xi}) \cdot \boldsymbol{\nu} = (\mathcal{O}_{\pm} - Q) \cdot \boldsymbol{\nu},$$

independent of s, *i.e.*, $\boldsymbol{\xi} \in L$. Furthermore, since $Q \in L$, and $\boldsymbol{\nu}$ is the unit vector orthogonal to L, then $|(\mathcal{O}_{\pm} - Q) \cdot \boldsymbol{\nu}| = \text{dist}(\mathcal{O}_{\pm}, L)$. □

Note that, if $(u_+, v_+) \neq (u_-, v_-)$, set $\{\varphi^+ = \varphi^-\}$ is a line. Also, if $\rho_+ \neq \rho_-$, $(u_+, v_+) \neq (u_-, v_-)$ by (1.13).

Lemma 6.1.2. *Assume that $\rho_+ > \rho_-$ so that $(u_+, v_+) \neq (u_-, v_-)$. Denote*

$$\mathcal{S} := \{\boldsymbol{\xi} \ : \ \varphi^+(\boldsymbol{\xi}) = \varphi^-(\boldsymbol{\xi})\}.$$

Assume that $\varphi^{\pm}$ satisfy the Rankine-Hugoniot condition on $\mathcal{S}$:

$$\rho_+ D\varphi^+ \cdot \boldsymbol{\nu} = \rho_- D\varphi^- \cdot \boldsymbol{\nu} \qquad \text{on } \mathcal{S}.$$

Then the following holds:

(i) $D\varphi^{\pm} \cdot \boldsymbol{\nu} \neq 0 \quad$ *on* $\mathcal{S}$.

(ii) $\varphi := \min(\varphi^+, \varphi^-)$ *is a weak solution of* (2.2.8)–(2.2.9) *satisfying the entropy condition on shock $\mathcal{S}$. That is, choosing the unit normal $\boldsymbol{\nu}$ on $\mathcal{S}$ so that $D\varphi^+ \cdot \boldsymbol{\nu} > 0$, we have*

$$D\varphi^- \cdot \boldsymbol{\nu} > D\varphi^+ \cdot \boldsymbol{\nu} > 0 \qquad \text{on } \mathcal{S}.$$

Thus, $\varphi = \varphi^-$ in the upstream half-plane $\{\boldsymbol{\xi} \in \mathbb{R}^2 \ : \ (\boldsymbol{\xi} - \boldsymbol{\xi}_{\mathcal{S}}) \cdot \boldsymbol{\nu} < 0\}$, where $\boldsymbol{\xi}_{\mathcal{S}} \in \mathcal{S}$ is arbitrary and fixed, and $\varphi = \varphi^+$ in the downstream half-plane $\{\boldsymbol{\xi} \in \mathbb{R}^2 \ : \ (\boldsymbol{\xi} - \boldsymbol{\xi}_{\mathcal{S}}) \cdot \boldsymbol{\nu} > 0\}$.

(iii) *In addition, we have*

$$\overline{B_{c_-}(\mathcal{O}_-)} \subset \{\varphi^- > \varphi^+\}, \tag{6.1.3}$$

$$B_{c_+}(\mathcal{O}_+) \cap \mathcal{S} \neq \emptyset, \tag{6.1.4}$$

$$\mathcal{O}_+ \in \{\varphi^- > \varphi^+\}, \tag{6.1.5}$$

$$\mathcal{O}_+\mathcal{O}_- \perp \mathcal{S}, \tag{6.1.6}$$

where $B_a(P)$ is the ball with center at P and radius $a > 0$; see also Fig. 6.2.

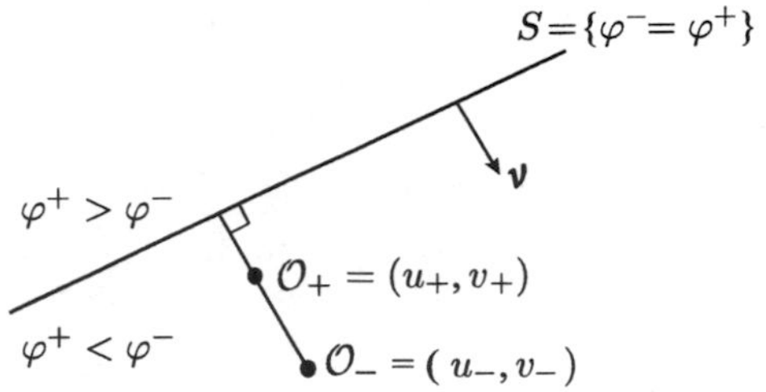

Figure 6.2: Line $\mathcal{S} = \{\varphi^+ = \varphi^-\}$ and $\mathcal{O}_+\mathcal{O}_- \perp \mathcal{S}$

Proof. We divide the proof into four steps.

1. The Rankine-Hugoniot conditions (2.2.13)–(2.2.14) imply that, at any $P \in \mathcal{S}$,

$$\rho_+\varphi^+_{\boldsymbol{\nu}} = \rho_-\varphi^-_{\boldsymbol{\nu}}, \tag{6.1.7}$$
$$\varphi^+ = \varphi^-, \tag{6.1.8}$$
$$\varphi^+_{\boldsymbol{\tau}} = \varphi^-_{\boldsymbol{\tau}}. \tag{6.1.9}$$

Also, $\rho_\pm \neq 0$ and, by Lemma 6.1.1,

$$\varphi^\pm_{\boldsymbol{\nu}} = const. \qquad \text{on } \mathcal{S}.$$

Thus, if $\varphi^+_{\boldsymbol{\nu}} = 0$ on $\mathcal{S}$, $\varphi^-_{\boldsymbol{\nu}} = 0$ by (6.1.7). This, combined with (6.1.9), yields $D\varphi^+ = D\varphi^-$ on $\mathcal{S}$, which implies

$$D\phi^+ = D\phi^- \qquad \text{on } \mathcal{S}$$

for $\phi = \varphi + \frac{|\boldsymbol{\xi}|^2}{2}$. Since $\phi^\pm$ are linear functions, then

$$D\phi^+ = D\phi^- \qquad \text{in } \mathbb{R}^2.$$

Using $\phi^+ = \phi^-$ on $\mathcal{S}$ by (6.1.8), it follows that $\phi^+ = \phi^-$ in $\mathbb{R}^2$, which is in contradiction to the fact that $\rho_+ > \rho_-$. Thus, $\varphi^\pm_{\boldsymbol{\nu}} \neq 0$ on $\mathcal{S}$. Assertion (i) is proved.

2. By (6.1.7), $\varphi^+_{\boldsymbol{\nu}}$ and $\varphi^-_{\boldsymbol{\nu}}$ on $\mathcal{S}$ have the same sign. Choose the direction of $\boldsymbol{\nu}$ so that $\varphi^\pm_{\boldsymbol{\nu}} > 0$ on $\mathcal{S}$. Then (6.1.7) and $\rho_+ > \rho_-$ imply that $D\varphi^- \cdot \boldsymbol{\nu} > D\varphi^+ \cdot \boldsymbol{\nu} > 0$ on $\mathcal{S}$. Assertion (ii) is proved.

3. It remains to prove (6.1.3)–(6.1.6). Note that assertion (i) and Lemma 6.1.1 imply that $\mathcal{O}_\pm \notin \mathcal{S}$. Let $P \in \mathcal{S}$ be such that $P\mathcal{O}_- \perp \mathcal{S}$. Since $D\varphi^\pm(P) = \mathcal{O}_\pm - P$ by (6.1.1), $\varphi^-_{\boldsymbol{\tau}}(P) = D\varphi^-(P) \cdot \boldsymbol{\tau} = 0$. By (6.1.9), $\varphi^+_{\boldsymbol{\tau}}(P) = 0$ so that $P\mathcal{O}_+ \perp \mathcal{S}$. That is,

$$P\mathcal{O}_\pm \perp \mathcal{S}. \tag{6.1.10}$$

Moreover, since $(\mathcal{O}_\pm - P)\cdot\boldsymbol{\nu} = \varphi^\pm_{\boldsymbol{\nu}}(P) > 0$, (6.1.10) implies that $\varphi^\pm_{\boldsymbol{\nu}}(P) = |P\mathcal{O}_\pm|$. Thus, from the entropy condition shown above, $|P\mathcal{O}_-| > |P\mathcal{O}_+|$. Note that we have shown that

$$D\varphi^\pm(P) = \mathcal{O}_\pm - P = |P\mathcal{O}_\pm|\boldsymbol{\nu}.$$

Also, recall that $\phi^\pm$ are linear functions, and $\varphi_-(P) = \varphi_+(P)$ since $P \in \mathcal{S}$. Then we compute:

$$\begin{aligned}\varphi_-(\mathcal{O}_-) - \varphi_+(\mathcal{O}_-) &= \phi_-(\mathcal{O}_-) - \phi_+(\mathcal{O}_-)\\ &= (\phi_- - \phi_+)(\mathcal{O}_-) - (\phi_- - \phi_+)(P)\\ &= D(\phi_- - \phi_+)(P)\cdot(\mathcal{O}_- - P)\\ &= D(\varphi_- - \varphi_+)(P)\cdot(\mathcal{O}_- - P)\\ &= \big(|P\mathcal{O}_-|\boldsymbol{\nu} - |P\mathcal{O}_+|\boldsymbol{\nu}\big)\cdot\big(|P\mathcal{O}_-|\boldsymbol{\nu}\big)\\ &= \big(|P\mathcal{O}_-| - |P\mathcal{O}_+|\big)|P\mathcal{O}_-| > 0.\end{aligned}$$

Hence, $\mathcal{O}_- \in \{\varphi^- > \varphi^+\}$. Then $(\mathcal{O}_\pm - P)\cdot\boldsymbol{\nu} = \varphi^\pm_{\boldsymbol{\nu}}(P) > 0$ with $P \in \mathcal{S} = \{\varphi^+ = \varphi^-\}$ implies that $\mathcal{O}_+ \in \{\varphi^- > \varphi^+\}$, which yields (6.1.5). Also, (6.1.10) leads to (6.1.6).

Furthermore, since $\mathcal{O}_- \in \{\varphi^- > \varphi^+\}$, and P is the nearest point to $\mathcal{O}_-$ on $\mathcal{S} = \partial\{\varphi^- > \varphi^+\}$, then, in order to prove (6.1.3), it suffices to show that φ_- is supersonic at P. Also, (6.1.4) follows if φ^+ is shown to be subsonic at P.

4. It remains to show that $|D\varphi_-(P)| > c_-$ and $|D\varphi_+(P)| < c_+$. The Bernoulli law in (2.2.7), or equivalently, (2.2.9), implies that, at P,

$$\rho_+^{\gamma-1} + (\gamma-1)\big(\frac{1}{2}|D\varphi^+|^2 + \varphi^+\big) = \rho_-^{\gamma-1} + (\gamma-1)\big(\frac{1}{2}|D\varphi^-|^2 + \varphi^-\big).$$

Using (6.1.8)–(6.1.9), this can be reduced to

$$\rho_+^{\gamma-1} + \frac{\gamma-1}{2}|\varphi^+_{\boldsymbol{\nu}}|^2 = \rho_-^{\gamma-1} + \frac{\gamma-1}{2}|\varphi^-_{\boldsymbol{\nu}}|^2 =: B_0 \qquad \text{at } P. \tag{6.1.11}$$

Consider the functions:

$$\tilde\rho(s) = \big(B_0 - \frac{\gamma-1}{2}s^2\big)^{\frac{1}{\gamma-1}} \quad\text{and}\quad \Phi(s) = s\tilde\rho(s) \qquad \text{on } [0, \sqrt{\frac{2B_0}{\gamma-1}}]. \tag{6.1.12}$$

Then $\rho_\pm = \tilde\rho(\varphi^\pm_{\boldsymbol{\nu}}(P))$, and condition (6.1.7) at P is equivalent to

$$\Phi(\varphi^-_{\boldsymbol{\nu}}(P)) = \Phi(\varphi^+_{\boldsymbol{\nu}}(P)). \tag{6.1.13}$$

By explicit differentiation, we have

$$\begin{aligned}&\tilde\rho'(s) < 0 \qquad \text{on } (0, q_{\max}),\\ &\Phi'(s)\begin{cases} > 0 & \text{if } s \in (0, q_*),\\ < 0 & \text{if } s \in (q_*, q_{\max}),\end{cases}\end{aligned} \tag{6.1.14}$$

where $q_{\max} = \sqrt{\frac{2B_0}{\gamma-1}}$ and $q_* = \sqrt{\frac{2B_0}{\gamma+1}}$.

Therefore, if (6.1.13) holds and $\varphi^-_{\boldsymbol{\nu}}(P) > \varphi^+_{\boldsymbol{\nu}}(P) > 0$, then

$$\varphi^-_{\boldsymbol{\nu}}(P) > q_* > \varphi^+_{\boldsymbol{\nu}}(P) > 0. \tag{6.1.15}$$

Also, by an explicit calculation,

$$\tilde{\rho}(q_*)^{\gamma-1} = q_*^2. \tag{6.1.16}$$

Then, using $\tilde{\rho}'(s) < 0$, we have

$$|\varphi^-_{\boldsymbol{\nu}}(P)|^2 > q_*^2 = \tilde{\rho}(q_*)^{\gamma-1} > \tilde{\rho}(\varphi^-_{\boldsymbol{\nu}}(P))^{\gamma-1} = \rho_-^{\gamma-1} = c_-^2,$$
$$|\varphi^+_{\boldsymbol{\nu}}(P)|^2 < q_*^2 = \tilde{\rho}(q_*)^{\gamma-1} < \tilde{\rho}(\varphi^+_{\boldsymbol{\nu}}(P))^{\gamma-1} = \rho_+^{\gamma-1} = c_+^2.$$

Since $\varphi^\pm_{\boldsymbol{\tau}}(P) = 0$, this implies that $|D\varphi_-(P)| > c_-$ and $|D\varphi_+(P)| < c_+$. □

In the next lemma, we show that, for any point P of the shock curve separating the upstream uniform state φ^- with density ρ_- from a downstream state φ^+ (possibly non-uniform) such that the entropy condition holds at P, the gradient jump across the shock at P depends only on ρ_- and $\partial_{\boldsymbol{\nu}}\varphi^-(P)$, and strictly increases with respect to $\partial_{\boldsymbol{\nu}}\varphi^-(P)$ when ρ_- is fixed. Also, by Lemma 6.1.1 and the entropy condition, $\partial_{\boldsymbol{\nu}}\varphi^-(P) = \text{dist}(L_P, \mathcal{O}_-)$, where L_P is the tangent line to the shock at P, so we can express the gradient jump at P through ρ_- and $\text{dist}(L_P, \mathcal{O}_-)$.

Lemma 6.1.3. *Let $\Omega \subset \mathbb{R}^2$ be open. Let a smooth curve $\mathcal{S}$ subdivide Ω into two open subdomains Ω^+ and Ω^-. Let $\varphi \in C^{0,1}(\Omega)$ be a weak solution of equation (2.2.8) in Ω such that $\varphi \in C^2(\Omega^\pm) \cap C^1(\Omega^\pm \cup \mathcal{S})$. Denote by $\varphi^\pm := \varphi|_{\Omega^\pm}$. Suppose that φ is a constant state in Ω^- with density ρ_- and sonic speed $c_- = \rho_-^{\frac{\gamma-1}{2}}$, i.e.,*

$$\varphi^-(\boldsymbol{\xi}) = -\frac{\boldsymbol{\xi}^2}{2} + (u_-, v_-) \cdot \boldsymbol{\xi} + A,$$

where (u_-, v_-) is a constant vector and A is the constant such that $\rho_- = \rho(|(u_-, v_-)|^2, A)$ for $\rho(\cdot,\cdot)$ from (2.2.9). Let $P \in \mathcal{S}$ be such that

(i) *φ^- is supersonic at P, i.e., $|D\varphi^-| > c_-$ at P;*

(ii) *$D\varphi^- \cdot \boldsymbol{\nu} > D\varphi^+ \cdot \boldsymbol{\nu} > 0$ at P, where $\boldsymbol{\nu}$ is the unit normal vector to $\mathcal{S}$ oriented from Ω^- to Ω^+.*

Let L_P be the tangent line to $\mathcal{S}$ at P. Let $d(P)$ be the distance between line L_P and center $\mathcal{O}_- = (u_-, v_-)$ of the sonic circle of state φ^-, where $d(P) > c_-$ by (6.1.3); see Fig. 6.3. Then $(\varphi^-_{\boldsymbol{\nu}} - \varphi^+_{\boldsymbol{\nu}})(P)$ depends only on ρ_- and $d(P)$.

More precisely, for each $\rho_- > 0$, there exists a unique function

$$\hat{H} \in C([c_-, \infty)) \cap C^\infty((c_-, \infty))$$

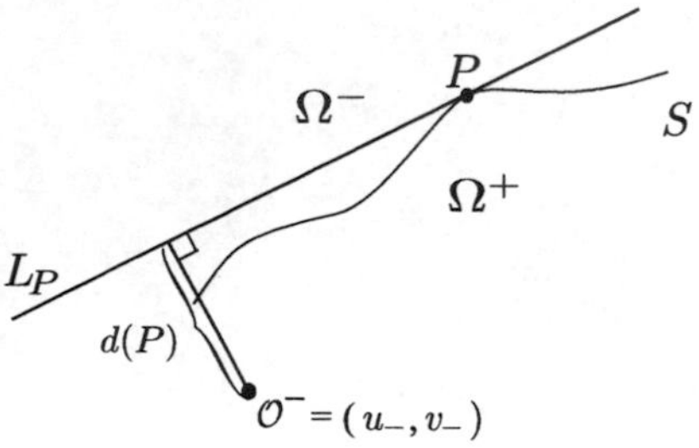

Figure 6.3: Distance $d(P)$ between L_P and $\mathcal{O}_- = (u_-, v_-)$

such that, for any Ω, $\mathcal{S}$, $\varphi^{\pm}$, *and* P, *as above,*

$$(\varphi_{\boldsymbol{\nu}}^- - \varphi_{\boldsymbol{\nu}}^+)(P) = \hat{H}(d(P)).$$

Moreover, $\hat{H}$ *satisfies*

$$\hat{H}(c_-) = 0, \qquad \hat{H}'(d) > 0 \quad \textit{for all } d \in (c_-, \infty). \tag{6.1.17}$$

Proof. We divide the proof into four steps.

1. The Rankine-Hugoniot conditions (2.2.13)–(2.2.14) imply that equations (6.1.7)–(6.1.9) hold at P, where ρ_- is the constant density of φ^- and $\rho_+ = \rho(|D\varphi^+(P)|^2, \varphi^+(P))$. Moreover, the Bernoulli law in (2.2.7), or equivalently, (2.2.9), combined with (6.1.8)–(6.1.9), implies that (6.1.11) holds at P.

Since $\boldsymbol{\nu}$ is orthogonal to L_P, and $D\varphi^- \cdot \boldsymbol{\nu} > 0$ at P for $D\varphi^- = \mathcal{O}_1 - P$, we have

$$d(P) = (\mathcal{O}_1 - P) \cdot \boldsymbol{\nu}(P) = \partial_{\boldsymbol{\nu}} \varphi^-(P).$$

Denote $d := d(P)$. Then, at P, we have

$$\varphi_{\boldsymbol{\nu}}^+ = (\varphi_{\boldsymbol{\nu}}^+ - \varphi_{\boldsymbol{\nu}}^-) + \varphi_{\boldsymbol{\nu}}^- = (\phi_{\boldsymbol{\nu}}^+ - \phi_{\boldsymbol{\nu}}^-) + d = d - w,$$

where

$$w := \phi_{\boldsymbol{\nu}}^- - \phi_{\boldsymbol{\nu}}^+ = (u_-, v_-) \cdot \boldsymbol{\nu} - \phi_{\boldsymbol{\nu}}^+ \qquad \text{at } P. \tag{6.1.18}$$

By conditions (i)–(ii),

$$d \in (c_-, \infty), \quad w \in (0, d). \tag{6.1.19}$$

With these notations, (6.1.7) and (6.1.11) imply

$$G(w, d) := \rho_-^{\gamma-1}\Big(\frac{d^{\gamma-1}}{(d-w)^{\gamma-1}} - 1\Big) + \frac{\gamma-1}{2}\big((d-w)^2 - d^2\big) = 0. \tag{6.1.20}$$

We show that, for each $d \in (c_-, \infty)$, this equation has a unique solution $\hat{w} \in (0, d)$ and, defining $\hat{H}(\cdot)$ by $\hat{H}(d) = \hat{w}$, $\hat{H} \in C^\infty((c_-, \infty))$.

2. We now prove that, for any $d \in (c_-, \infty)$, equation (6.1.20) has a unique solution $\hat{w} \in (0, d)$. Moreover, $\frac{\partial G}{\partial w}(\hat{w}, d) > 0$.

In order to show this, we note from the explicit definition of $G(w, d)$ and relation $\rho_-^{\gamma-1} = c_-^2$ that

$$
\begin{aligned}
& G \in C^\infty(\{c_- \le d < \infty,\ 0 \le w < d\}),\\
& G(0, d) = 0 \qquad \text{for all } d \in (c_-, \infty),\\
& \frac{\partial G}{\partial w}(0, d) = \frac{\gamma - 1}{d}(c_-^2 - d^2) < 0 \qquad \text{for all } d \in (c_-, \infty),\\
& \lim_{w \to d-} G(w, d) = \infty \qquad \text{for all } d \in (c_-, \infty),\\
& \frac{\partial^2 G}{\partial w^2} = (\gamma - 1)\Big(\frac{\gamma \rho_-^{\gamma-1} d^{\gamma-1}}{(d - w)^{\gamma+1}} + 1\Big) > 0 \quad \text{for all } d \in (c_-, \infty) \text{ and } w \in [0, d).
\end{aligned}
$$

Combining these facts implies the assertion.

3. From Step 2 and the implicit function theorem, there exists $\hat{H} \in C^\infty((c_-, \infty))$ such that, for any $d \in (c_-, \infty)$, the unique solution $w \in (0, d)$ of (6.1.20) is $w = \hat{H}(d)$.

We also note that

$$
\frac{\partial G}{\partial w} = \frac{\gamma - 1}{(d - w)^\gamma}\big(d^{\gamma-1}\rho_-^{\gamma-1} - (d - w)^{\gamma+1}\big).
$$

Since, by Step 2, $\frac{\partial G}{\partial w}(d, H(d)) > 0$, we have

$$
d^{\gamma-1}\rho_-^{\gamma-1} - (d - w)^{\gamma+1} > 0 \qquad \text{for all } d \in (c_-, \infty),\ w = \hat{H}(d). \tag{6.1.21}
$$

4. Now we show that $\hat{H}'(d) > 0$ for any $d \in (c_-, \infty)$.

Differentiating the equality, $G(\hat{H}(d), d) = 0$, with respect to d, we have

$$
\begin{aligned}
& \frac{\gamma - 1}{(d - w)^\gamma}\Big(-w\big(d^{\gamma-2}\rho_-^{\gamma-2} + (d - w)^\gamma\big) + \hat{H}'(d)\big(d^{\gamma-1}\rho_-^{\gamma-1} - (d - w)^{\gamma+1}\big)\Big)\\
& = 0 \qquad \text{for all } d \in (c_-, \infty) \text{ and } w = \hat{H}(d).
\end{aligned}
$$

That is,

$$
\hat{H}'(d) = \frac{w\big(d^{\gamma-2}\rho_-^{\gamma-2} + (d - w)^\gamma\big)}{d^{\gamma-1}\rho_-^{\gamma-1} - (d - w)^{\gamma+1}} \qquad \text{for all } d \in (c_-, \infty) \text{ and } w = \hat{H}(d).
$$

Then (6.1.21) and $0 < \hat{H}(d) < d$ imply the assertion.

Finally, the assertions of Steps 2–4 yield the result stated in Lemma 6.1.3. □

Lemma 6.1.4. *Let $\Omega \subset \mathbb{R}^2$, S, φ, and $\varphi^\pm$ be as in Lemma 6.1.3. Let $P_k \in S$, $k = 1, 2$, be such that*

(i) *φ^- is supersonic at P_k, i.e., $|D\varphi^-| > c_- := c(|D\varphi^-|^2, \varphi^-)$ at P_k;*

(ii) *$D\varphi^- \cdot \boldsymbol{\nu} > D\varphi^+ \cdot \boldsymbol{\nu} > 0$ at P_k for the unit normal vector $\boldsymbol{\nu}$ to $\mathcal{S}$ oriented from Ω^- to Ω^+;*

(iii) *$\boldsymbol{\nu}(P_1) = \boldsymbol{\nu}(P_2)$;*

(iv) *$d(P_1) > d(P_2)$ for $d(P_k) := \mathrm{dist}(L_{P_k}, \mathcal{O}_-)$, where $\mathcal{O}_- = (u_-, v_-)$, and L_{P_k} is the tangent line to $\mathcal{S}$ at P_k.*

Let $\phi^\pm(\boldsymbol{\xi}) = \frac{1}{2}|\boldsymbol{\xi}|^2 + \varphi^\pm(\boldsymbol{\xi})$. Then

$$\phi^+_{\boldsymbol{\nu}}(P_1) < \phi^+_{\boldsymbol{\nu}}(P_2).$$

Proof. By assumptions (i)–(ii), we can apply Lemma 6.1.3 to obtain

$$\phi^+_{\boldsymbol{\nu}}(P_k) = (u_-, v_-) \cdot \boldsymbol{\nu}(P_k) - \hat{H}(d(P_k)) \qquad \text{for } k = 1, 2,$$

where we note by assumption (i) that $d(P_k) > c_-$. Now, using (6.1.17) and assumptions (iii)–(iv), we have

$$\phi^+_{\boldsymbol{\nu}}(P_2) - \phi^+_{\boldsymbol{\nu}}(P_1) = \hat{H}(d(P_1)) - \hat{H}(d(P_2)) > 0.$$

□

6.2 NORMAL REFLECTION AND ITS UNIQUENESS

We consider the shock reflection when the wedge angle $\theta_{\rm w}$ is $\frac{\pi}{2}$. Then the incident shock reflects normally (see Fig. 3.1), and the reflected shock is also a plane at $\xi_1 = \bar{\xi}_1 < 0$, which will be defined below. Thus, we seek $\bar{\xi}_1 < 0$ and a uniform state (2) with potential:

$$\varphi_2(\boldsymbol{\xi}) = -\frac{|\boldsymbol{\xi}|^2}{2} + (\bar{u}_2, \bar{v}_2) \cdot \boldsymbol{\xi} + C,$$

which satisfies $\partial_{\boldsymbol{\nu}}\varphi_2 = 0$ on $\Gamma_{\rm wedge} = \{\xi_1 = 0\}$ and the Rankine-Hugoniot conditions (2.2.13)–(2.2.14) for (φ_1, φ_2) on $\mathcal{S}_1 = \{\xi_1 = \bar{\xi}_1\}$, where φ_1 for state (1) is of form (2.2.17). Then, from (2.2.14) on $\{\xi_1 = \bar{\xi}_1\}$, we obtain that $\partial_{\xi_2}\varphi_2 = \partial_{\xi_2}\varphi_1$ on that line, so that $\bar{v}_2 = v_1 = 0$ by (2.2.17). Also, from the condition on $\Gamma_{\rm wedge}$, $\bar{u}_2 = 0$. Again using (2.2.14) on $\{\xi_1 = \bar{\xi}_1\}$ and (2.2.17), we obtain that state (2) is of the form:

$$\varphi_2(\boldsymbol{\xi}) = -\frac{|\boldsymbol{\xi}|^2}{2} + u_1(\bar{\xi}_1 - \xi_1^0), \tag{6.2.1}$$

where $\xi_1^0 = \frac{\rho_1 u_1}{\rho_1 - \rho_0} > 0$ by (2.2.18).

Now the Rankine-Hugoniot condition (2.2.13) on $\{\xi_1 = \bar{\xi}_1\}$ implies

$$\bar{\xi}_1 = -\frac{\rho_1 u_1}{\bar{\rho}_2 - \rho_1}, \tag{6.2.2}$$

where $\bar{\rho}_2$ is the density of state (2). In particular, we see that $\bar{\xi}_1 < 0$ if and only if $\bar{\rho}_2 > \rho_1$.

We use the Bernoulli law in (2.2.7) and expressions (2.2.17) and (6.2.1) to see that

$$\rho_0^{\gamma-1} = \rho_1^{\gamma-1} + (\gamma-1)\big(\frac{1}{2}u_1^2 - u_1\xi_1^0\big) = \bar{\rho}_2^{\gamma-1} + (\gamma-1)u_1(\bar{\xi}_1 - \xi_1^0),$$

so that we obtain the equation for $\bar{\rho}_2$:

$$\bar{\rho}_2^{\gamma-1} = \rho_1^{\gamma-1} + (\gamma-1)\big(\frac{1}{2}u_1^2 + \frac{\rho_1 u_1^2}{\bar{\rho}_2 - \rho_1}\big). \tag{6.2.3}$$

Now we show that there is a unique solution $\bar{\rho}_2$ of (6.2.3) such that

$$\bar{\rho}_2 > \rho_1.$$

Indeed, for fixed $\gamma > 1$ and $\rho_1, u_1 > 0$, and for $F(\bar{\rho}_2)$ that is the right-hand side of (6.2.3), we have

$$\lim_{s\to\infty} F(s) = \rho_1^{\gamma-1} + \frac{\gamma-1}{2}u_1^2 > \rho_1^{\gamma-1}, \qquad \lim_{s\to\rho_1+} F(s) = \infty,$$

$$F'(s) = -\frac{(\gamma-1)\rho_1 u_1^2}{(s-\rho_1)^2} < 0 \qquad \text{for } s > \rho_1.$$

Thus, there exists a unique $\bar{\rho}_2 \in (\rho_1, \infty)$ satisfying $\bar{\rho}_2^{\gamma-1} = F(\bar{\rho}_2)$, *i.e.*, (6.2.3). Then the position of the reflected shock $\bar{\xi}_1 < 0$ is uniquely determined by (6.2.2).

Note that, on $\mathcal{S}_1 = \{\xi_1 = \bar{\xi}_1\}$, choosing the orientation of the normal as $\boldsymbol{\nu} = (1,0)$, we find that $\partial_{\boldsymbol{\nu}}\varphi_1 = u_1 - \bar{\xi}_1$ and $\partial_{\boldsymbol{\nu}}\varphi_2 = -\bar{\xi}_1 > 0$, so that $\partial_{\boldsymbol{\nu}}\varphi_1 > \partial_{\boldsymbol{\nu}}\varphi_2 > 0$. This allows us to apply Lemma 6.1.2 with $\varphi^- = \varphi_1$ and $\varphi^+ = \varphi_2$. Then $\mathcal{O}^+ = \mathbf{0}$. From (6.1.4), we conclude that, for the sonic speed $\bar{c}_2 = \sqrt{\bar{\rho}_2^{\gamma-1}}$ of state (2),

$$|\bar{\xi}_1| < \bar{c}_2. \tag{6.2.4}$$

This leads to the following theorem:

Theorem 6.2.1. *There exists a unique solution to the normal reflection with the unique state* (2) *whose velocity* $(\bar{u}_2, \bar{v}_2) = \mathbf{0}$ *and density* $\bar{\rho}_2 \in (\rho_1, \infty)$, *and the unique location of the reflected shock:*

$$\bar{\xi}_1 = -\frac{\rho_1 u_1}{\bar{\rho}_2 - \rho_1} \in (-\bar{c}_2, 0).$$

6.3 THE SELF-SIMILAR POTENTIAL FLOW EQUATION IN THE COORDINATES FLATTENING THE SONIC CIRCLE OF A UNIFORM STATE

Let $\rho_0 > 0$ be a fixed constant. Let φ be a C^2–solution of equation (2.2.8) with (2.2.9) in a domain $\mathcal{D} \subset \mathbb{R}^2$.

Let $(\hat{u}, \hat{v})$ be a constant vector, $\hat{P} = \hat{\boldsymbol{\xi}} \in \mathbb{R}^2$, and

$$\varphi_{\rm un}(\boldsymbol{\xi}) = -\frac{|\boldsymbol{\xi}|^2}{2} + (\hat{u}, \hat{v}) \cdot (\boldsymbol{\xi} - \hat{\boldsymbol{\xi}}).$$

Assume that the above constants satisfy condition (6.1.2). Then $\varphi_{\rm un}$ is the pseudo-potential of the uniform state, which is a solution of (2.2.8)–(2.2.9) with the constant velocity $(\hat{u}, \hat{v})$ and density $\rho_{\rm un}$ defined by

$$\rho_{\rm un}^{\gamma-1} + (\gamma - 1)\Big(- (\hat{u}, \hat{v}) \cdot \hat{\boldsymbol{\xi}} + \frac{\hat{u}^2 + \hat{v}^2}{2}\Big) = \rho_0^{\gamma-1}.$$

Introduce the function:

$$\psi = \varphi - \varphi_{\rm un} \qquad \text{in } \mathcal{D}. \tag{6.3.1}$$

Since both φ and $\varphi_{\rm un}$ satisfy (2.2.9),

$$\rho^{\gamma-1}(|D\varphi|^2, \varphi) + (\gamma - 1)(\varphi + \frac{1}{2}|D\varphi|^2) = \rho_{\rm un}^{\gamma-1} + (\gamma - 1)(\varphi_{\rm un} + \frac{1}{2}|D\varphi_{\rm un}|^2).$$

Then we obtain the following expression for the sonic speed in terms of ψ:

$$c^2(D\psi, \psi, \boldsymbol{\xi}) = c_{\rm un}^2 - (\gamma - 1)\Big(D\varphi_{\rm un} \cdot D\psi + \frac{1}{2}|D\psi|^2 + \psi\Big). \tag{6.3.2}$$

Denote $O_{\rm un} := (\hat{u}, \hat{v})$, so that $O_{\rm un}$ is the center of the sonic circle of state $\varphi_{\rm un}$. Introduce the polar coordinates (r, θ) with respect to $O_{\rm un}$:

$$(\xi_1 - \hat{u}, \xi_2 - \hat{v}) = r(\cos\theta, \sin\theta). \tag{6.3.3}$$

Then

$$\varphi_{\rm un}(r, \theta) = -\frac{r^2}{2} + C_0 \tag{6.3.4}$$

with $C_0 = \frac{\hat{u}^2+\hat{v}^2}{2} - (\hat{u}, \hat{v}) \cdot \hat{\boldsymbol{\xi}}$.

In $\mathcal{D} \setminus \{O_{\rm un}\}$, we introduce the coordinates:

$$(x, y) = (c_{\rm un} - r, \theta). \tag{6.3.5}$$

Note that, in the (x, y)–coordinates, the sonic circle of state $\varphi_{\rm un}$ is $\{x = 0\}$, and $\varphi_{\rm un}$ is subsonic on $\{0 < x < c_{\rm un}\}$ and supersonic on $\{x < 0\}$.

Substituting $\varphi = \psi + \varphi_{\rm un}$ into the non-divergence form (2.2.11) of equation (2.2.8), writing the resulting equation in the (x, y)–coordinates (6.3.5), and using (6.3.2) and (6.3.4), we obtain that ψ satisfies the following equation in $\mathcal{D} \setminus \{O_{\rm un}\}$:

$$\big(2x - (\gamma+1)\psi_x + O_1\big)\psi_{xx} + O_2\psi_{xy} + \big(\frac{1}{c_{\rm un}} + O_3\big)\psi_{yy} - (1 + O_4)\psi_x + O_5\psi_y = 0, \tag{6.3.6}$$

where

$$
\begin{aligned}
O_1(D\psi,\psi,x) &= -\frac{x^2}{c_{\text{un}}} + \frac{\gamma+1}{2c_{\text{un}}}(2x-\psi_x)\psi_x - \frac{\gamma-1}{c_{\text{un}}}\Big(\psi + \frac{1}{2(c_{\text{un}}-x)^2}\psi_y^2\Big),\\
O_2(D\psi,\psi,x) &= -\frac{2}{c_{\text{un}}(c_{\text{un}}-x)^2}(\psi_x + c_{\text{un}} - x)\psi_y,\\
O_3(D\psi,\psi,x) &= \frac{1}{c_{\text{un}}(c_{\text{un}}-x)^2}\Big(x(2c_{\text{un}}-x) - (\gamma-1)\big(\psi + (c_{\text{un}}-x)\psi_x + \frac{1}{2}\psi_x^2\big)\\
&\qquad - \frac{\gamma+1}{2(c_{\text{un}}-x)^2}\psi_y^2\Big),\\
O_4(D\psi,\psi,x) &= \frac{1}{c_{\text{un}}-x}\Big(x - \frac{\gamma-1}{c_{\text{un}}}\big(\psi + (c_{\text{un}}-x)\psi_x + \frac{1}{2}\psi_x^2\\
&\qquad + \frac{\gamma+1}{2(\gamma-1)(c_{\text{un}}-x)^2}\psi_y^2\big)\Big),\\
O_5(D\psi,\psi,x) &= -\frac{2}{c_{\text{un}}(c_{\text{un}}-x)^3}(\psi_x + c_{\text{un}} - 2x)\psi_y.
\end{aligned}
\tag{6.3.7}
$$

We also write equation (6.3.6)–(6.3.7) as

$$
\sum_{i,j=1}^{2} A_{ij}(D\psi,\psi,x,y)D_{ij}\psi + \sum_{i=1}^{2} A_i(D\psi,\psi,x,y)D_i\psi = 0, \tag{6.3.8}
$$

where $D_1 = D_x$, $D_2 = D_y$, $D_{ij} = D_iD_j$, and $A_{12} = A_{21}$.

Chapter Seven

Local Theory and von Neumann's Conjectures

In this chapter, we describe the local theory of shock reflection and von Neumann's conjectures for shock reflection-diffraction configurations.

7.1 LOCAL REGULAR REFLECTION AND STATE (2)

In this section, we follow von Neumann's detachment criterion to derive the necessary condition for the existence of regular reflection-diffraction configurations, as described in §2.6.

The incident shock $\mathcal{S}_0 = \{\xi_1 = \xi_1^0\}$ separates the upstream uniform state (0) determined by pseudo-potential φ_0 from the downstream uniform state (1) determined by pseudo-potential φ_1. Here the potential functions (φ_0, φ_1) and positive constants (ξ_1^0, u_1) are determined by (2.2.16)–(2.2.18), where $\rho_1 > \rho_0 > 0$ are the densities of states (0) and (1) determined by (2.2.9). For each wedge angle $\theta_{\rm w} \in (0, \frac{\pi}{2})$, the wedge in the upper half-plane is defined by

$$W := \{\boldsymbol{\xi} \,:\, 0 < \xi_2 < \xi_1 \tan\theta_{\rm w},\ \xi_1 > 0\}. \tag{7.1.1}$$

The incident shock $\mathcal{S}_0$ intersects with the wedge boundary $\{\xi_2 = \xi_1 \tan\theta_{\rm w}\}$ at the point:

$$P_0 := (\xi_1^0, \xi_1^0 \tan\theta_{\rm w}). \tag{7.1.2}$$

We study the local reflection at point P_0; that is, we seek a straight shock $\mathcal{S}_1$ passing through P_0 and a uniform state φ_2, so that $\mathcal{S}_1$ separates the upstream state φ_1 from the downstream state φ_2 with the velocity parallel to the wedge boundary, thus satisfying the boundary condition:

$$\partial_{\boldsymbol{\nu}_{\rm w}} \varphi_2 = 0 \qquad \text{on } \Gamma_{\rm wedge} := \{\xi_2 = \xi_1 \tan\theta_{\rm w}\} \tag{7.1.3}$$

for the unit normal $\boldsymbol{\nu}_{\rm w}$ to $\Gamma_{\rm wedge}$:

$$\boldsymbol{\nu}_{\rm w} = (-\sin\theta_{\rm w}, \cos\theta_{\rm w}). \tag{7.1.4}$$

Then the Rankine-Hugoniot conditions (2.2.13)–(2.2.14) and the entropy condition are satisfied on $\mathcal{S}_1$. From (2.2.14), it follows that $\mathcal{S}_1 = \{\varphi_1 = \varphi_2\}$, so that the unit normal to $\mathcal{S}_1$ in the direction of the downstream region is $\boldsymbol{\nu} = \frac{D(\varphi_1 - \varphi_2)}{|D(\varphi_1 - \varphi_2)|}$. It also follows that $D\varphi_2 \cdot \boldsymbol{\tau} = D\varphi_1 \cdot \boldsymbol{\tau}$ on $\mathcal{S}_1$.

In order to find such a two-shock configuration, it suffices to find the uniform state (2) with pseudo-potential φ_2, which satisfies all the conditions described above only at point P_0 – then it satisfies these conditions at any $P \in \mathcal{S}_1$, a fact that is easily checked by a direct calculation.

Thus, we seek a uniform state φ_2 that satisfies the Rankine-Hugoniot conditions with state (1):

$$\begin{aligned}&\varphi_2(P_0) = \varphi_1(P_0),\\ &\rho_2 D\varphi_2 \cdot \boldsymbol{\nu} = \rho_1 D\varphi_1 \cdot \boldsymbol{\nu}, \quad D\varphi_2 \cdot \boldsymbol{\tau} = D\varphi_1 \cdot \boldsymbol{\tau} \qquad \text{at } P_0,\end{aligned} \tag{7.1.5}$$

the entropy condition:

$$D\varphi_1 \cdot \boldsymbol{\nu} > D\varphi_2 \cdot \boldsymbol{\nu} > 0 \qquad \text{at } P_0 \tag{7.1.6}$$

for $\boldsymbol{\nu} = \frac{D(\varphi_1 - \varphi_2)}{|D(\varphi_1-\varphi_2)|}$, $\boldsymbol{\tau} \perp \boldsymbol{\nu}$, $|\boldsymbol{\tau}| = 1$, and $\rho_k = \rho(|D\varphi_k|^2, \varphi_k)$, with $\rho(\cdot)$ given by (2.2.9) and $k = 1, 2$, and the boundary condition:

$$D\varphi_2 \cdot \boldsymbol{\nu}_{\rm w} = 0 \qquad \text{at } P_0, \tag{7.1.7}$$

where $\boldsymbol{\nu}_{\rm w}$ is given by (7.1.4).

In particular, substituting the expression of $\boldsymbol{\nu}_{\rm w}$ into (7.1.5), we have

$$g^{\rm sh}(D\varphi_2(\boldsymbol{\xi}), \varphi_2(\boldsymbol{\xi}), \boldsymbol{\xi}) = 0 \qquad \text{at } \boldsymbol{\xi} = P_0, \tag{7.1.8}$$

where

$$g^{\rm sh}(\mathbf{p}, z, \boldsymbol{\xi}) := \big(\rho(|\mathbf{p}|^2, z)\mathbf{p} - \rho_1 D\varphi_1(\boldsymbol{\xi})\big) \cdot \frac{D\varphi_1(\boldsymbol{\xi}) - \mathbf{p}}{|D\varphi_1(\boldsymbol{\xi}) - \mathbf{p}|} \tag{7.1.9}$$

for $\rho(|\mathbf{p}|^2, z)$ defined by (2.2.9).

Theorem 7.1.1 (Local Theory). *Let $\rho_1 > \rho_0 > 0$. Then there exists $\theta_{\rm w}^{\rm d} = \theta_{\rm w}^{\rm d}(\rho_0, \rho_1, \gamma) \in (0, \frac{\pi}{2})$, called the* detachment angle, *such that*

(a) *For each $\theta_{\rm w} \in (\theta_{\rm w}^{\rm d}, \frac{\pi}{2})$, there are exactly two states (2) satisfying (7.1.5)–(7.1.7): the* weak *reflection state $\varphi_2^{\rm wk} := \varphi_2^{{\rm wk},\theta_{\rm w}}$ and the* strong *reflection state $\varphi_2^{\rm sg} := \varphi_2^{{\rm sg},\theta_{\rm w}}$, such that*

$$\rho_1 < \rho_2^{{\rm wk},\theta_{\rm w}} < \rho_2^{{\rm sg},\theta_{\rm w}}.$$

Moreover,

$$\lim_{\theta_{\rm w}\to\theta_{\rm w}^{\rm d}+} (\rho_2^{{\rm wk},\theta_{\rm w}}, D\phi_2^{{\rm wk},\theta_{\rm w}}) = \lim_{\theta_{\rm w}\to\theta_{\rm w}^{\rm d}+} (\rho_2^{{\rm sg},\theta_{\rm w}}, D\phi_2^{{\rm sg},\theta_{\rm w}}); \tag{7.1.10}$$

(b) *For the detachment angle $\theta_{\rm w} = \theta_{\rm w}^{\rm d}$, there exists the unique state (2), $\phi_2^{\theta_{\rm w}^{\rm d}}(\boldsymbol{\xi})$, satisfying (7.1.5)–(7.1.7) so that $(\rho_2^{\theta_{\rm w}^{\rm d}}, D\phi_2^{\theta_{\rm w}^{\rm d}})$ is equal to the limit in (7.1.10), which can be interpreted as the coincidence of the weak and strong states (2) when $\theta_{\rm w} = \theta_{\rm w}^{\rm d}$.*

Here $\rho_2^{\mathrm{wk},\theta_{\mathrm{w}}}$ *and* $\rho_2^{\mathrm{sg},\theta_{\mathrm{w}}}$ *are the constant densities for the weak and strong states* (2), $\phi_2^{\mathrm{wk},\theta_{\mathrm{w}}}(\boldsymbol{\xi}) = \varphi_2^{\mathrm{wk},\theta_{\mathrm{w}}}(\boldsymbol{\xi})+\frac{|\boldsymbol{\xi}|^2}{2}$ *and* $\phi_2^{\mathrm{sg},\theta_{\mathrm{w}}}(\boldsymbol{\xi}) = \varphi_2^{\mathrm{sg},\theta_{\mathrm{w}}}(\boldsymbol{\xi})+\frac{|\boldsymbol{\xi}|^2}{2}$ *with the pseudo-potentials* $\varphi_2^{\mathrm{wk},\theta_{\mathrm{w}}}(\boldsymbol{\xi})$ *and* $\varphi_2^{\mathrm{sg},\theta_{\mathrm{w}}}(\boldsymbol{\xi})$ *of the weak and strong states* (2), *respectively. Note that* $D\phi_2^{\mathrm{wk},\theta_{\mathrm{w}}}$ *and* $D\phi_2^{\mathrm{sg},\theta_{\mathrm{w}}}$ *are constant vectors.*

Furthermore, denoting by $c_2^{\mathrm{wk},\theta_{\mathrm{w}}}$ *and* $c_2^{\mathrm{sg},\theta_{\mathrm{w}}}$ *the sonic speeds of the weak and strong states* (2) *respectively, we have*

(i) $\rho_2^{\mathrm{wk},\theta_{\mathrm{w}}}$, $\rho_2^{\mathrm{sg},\theta_{\mathrm{w}}}$, $D\phi_2^{\mathrm{wk},\theta_{\mathrm{w}}}$, *and* $D\phi_2^{\mathrm{sg},\theta_{\mathrm{w}}}$ *depend continuously on* $\theta_{\mathrm{w}} \in [\theta_{\mathrm{w}}^{\mathrm{d}}, \frac{\pi}{2})$ *and* C^∞*-smoothly on* $\theta_{\mathrm{w}} \in (\theta_{\mathrm{w}}^{\mathrm{d}}, \frac{\pi}{2})$. *Moreover,* $\rho_2^{\mathrm{wk},\theta_{\mathrm{w}}}$ *and* $D\phi_2^{\mathrm{wk},\theta_{\mathrm{w}}}$ *depend continuously on* $\theta_{\mathrm{w}} \in [\theta_{\mathrm{w}}^{\mathrm{d}}, \frac{\pi}{2}]$ *and* C^∞*-smoothly on* $\theta_{\mathrm{w}} \in (\theta_{\mathrm{w}}^{\mathrm{d}}, \frac{\pi}{2}]$;

(ii) $\lim_{\theta_{\mathrm{w}}\to\frac{\pi}{2}-}(\rho_2^{\mathrm{wk},\theta_{\mathrm{w}}}, D\phi_2^{\mathrm{wk},\theta_{\mathrm{w}}}) = (\bar{\rho}_2, \mathbf{0})$, *where* $\bar{\rho}_2$ *is the density of state* (2) *for the normal reflection in Theorem* 6.2.1, *so that the normal state* (2) *may be considered as the weak state* (2) *for* $\theta_{\mathrm{w}} = \frac{\pi}{2}$;

(iii) *For any* $\theta_{\mathrm{w}} \in [\theta_{\mathrm{w}}^{\mathrm{d}}, \frac{\pi}{2})$,

$$|D\varphi_1(P_0(\theta_{\mathrm{w}}))| > c_1; \tag{7.1.11}$$

(iv) *For any* $\theta_{\mathrm{w}} \in (\theta_{\mathrm{w}}^{\mathrm{d}}, \frac{\pi}{2})$,

$$|D\varphi_2^{\mathrm{sg},\theta_{\mathrm{w}}}(P_0)| < c_2^{\mathrm{sg},\theta_{\mathrm{w}}}; \tag{7.1.12}$$

(v) *There exists* $\theta_{\mathrm{w}}^{\mathrm{s}} \in (\theta_{\mathrm{w}}^{\mathrm{d}}, \frac{\pi}{2})$, *called the* sonic wedge angle, *such that*

$$\begin{aligned} &|D\varphi_2^{\mathrm{wk},\theta_{\mathrm{w}}}(P_0)| > c_2^{\mathrm{wk},\theta_{\mathrm{w}}} && \textit{for all } \theta_{\mathrm{w}} \in (\theta_{\mathrm{w}}^{\mathrm{s}}, \frac{\pi}{2}),\\ &|D\varphi_2^{\mathrm{wk},\theta_{\mathrm{w}}}(P_0)| = c_2^{\mathrm{wk}} && \textit{for } \theta_{\mathrm{w}} = \theta_{\mathrm{w}}^{\mathrm{s}}; \end{aligned}$$

(vi) *There exists* $\tilde{\theta}_{\mathrm{w}}^{\mathrm{s}} \in (\theta_{\mathrm{w}}^{\mathrm{d}}, \theta_{\mathrm{w}}^{\mathrm{s}}]$ *such that* $|D\varphi_2^{\mathrm{wk},\theta_{\mathrm{w}}}(P_0)| < c_2^{\mathrm{wk},\theta_{\mathrm{w}}}$ *for any* $\theta_{\mathrm{w}} \in [\theta_{\mathrm{w}}^{\mathrm{d}}, \tilde{\theta}_{\mathrm{w}}^{\mathrm{s}})$;

(vii) *For any* $\theta_{\mathrm{w}} \in (\theta_{\mathrm{w}}^{\mathrm{d}}, \frac{\pi}{2})$, *the weak reflection satisfies*

$$D_{\mathbf{p}}g^{\mathrm{sh}}(D\varphi_2^{\mathrm{wk},\theta_{\mathrm{w}}}(P_0), \varphi_2^{\mathrm{wk},\theta_{\mathrm{w}}}(P_0), P_0) \cdot D\varphi_2^{\mathrm{wk},\theta_{\mathrm{w}}}(P_0) < 0,$$

and the strong reflection satisfies

$$D_{\mathbf{p}}g^{\mathrm{sh}}(D\varphi_2^{\mathrm{sg},\theta_{\mathrm{w}}}(P_0), \varphi_2^{\mathrm{sg},\theta_{\mathrm{w}}}(P_0), P_0) \cdot D\varphi_2^{\mathrm{sg},\theta_{\mathrm{w}}}(P_0) > 0;$$

(viii) *Let* φ_2 *be a pseudo-potential of either the weak or strong state* (2). *Then, for any* $\boldsymbol{\xi} \in \mathcal{S}_1 = \{\varphi_1 = \varphi_2\}$,

$$D_{\mathbf{p}}g^{\mathrm{sh}}(D\varphi_2(\boldsymbol{\xi}), \varphi_2(\boldsymbol{\xi}), \boldsymbol{\xi}) \cdot D^\perp\varphi_2(\boldsymbol{\xi}) < 0,$$

where $D^\perp\varphi_2(\boldsymbol{\xi})$ *is the* $\frac{\pi}{2}$*–rotation of* $D\varphi_2(\boldsymbol{\xi})$ *in the direction chosen so that* $D\varphi_1(\boldsymbol{\xi})\cdot D^\perp\varphi_2(\boldsymbol{\xi}) < 0$, *that is,* $D^\perp\varphi_2(\boldsymbol{\xi})$ *is either* $(-\partial_{\xi_2}\varphi_2(\boldsymbol{\xi}), \partial_{\xi_1}\varphi_2(\boldsymbol{\xi}))$ *or* $(\partial_{\xi_2}\varphi_2(\boldsymbol{\xi}), -\partial_{\xi_1}\varphi_2(\boldsymbol{\xi}))$, *for the sign chosen so that* $D\varphi_1(\boldsymbol{\xi}) \cdot D^\perp\varphi_2(\boldsymbol{\xi}) < 0$.

For a possible two-shock configuration satisfying the corresponding boundary condition on the wedge boundary $\xi_2 = \xi_1 \tan\theta_{\rm w}$, the three state pseudo-potentials $\varphi_j, j = 0, 1, 2$, must be of form (2.2.16)–(2.2.17) and (2.4.1).

We prove this theorem in the next subsections. We first consider the case of large-angle wedges in §7.2. After that, we discuss the shock polar for steady potential flow in §7.3. Then, in §7.4, we use the shock polar to prove the existence of the weak and strong state (2) for any wedge angle $\theta_{\rm w} \in (\theta_{\rm w}^{\rm d}, \frac{\pi}{2})$, and to identify the sonic angle $\theta_{\rm w}^{\rm s}$.

7.2 LOCAL THEORY OF SHOCK REFLECTION FOR LARGE-ANGLE WEDGES

In this section, we assume that $\theta_{\rm w} \in (\frac{\pi}{2} - \sigma_1, \frac{\pi}{2})$, where $\sigma_1 > 0$ is small, as determined below, depending only on (ρ_0, ρ_1, γ). We construct the weak state (2) satisfying (7.1.5)–(7.1.7) close to state (2) of the normal reflection in §6.2 and show that (7.1.10) holds.

Let $P_0 = (\xi_1^0, \xi_1^0 \tan\theta_{\rm w})$ be the reflection point (*i.e.*, the intersection point of the incident shock with $\Gamma_{\rm wedge}$). We show below that, if σ_1 is small, then $\mathcal{S}_1$ intersects with the ξ_1–axis, and the point of intersection $(\tilde{\xi}_1, 0)$ is close to point $(\bar{\xi}_1, 0)$ from Theorem 6.2.1. Denote by $\theta_{\rm sh}$ the angle between line $\mathcal{S}_1$ and the ξ_1-axis.

We look for $\varphi_2(\cdot)$ in the form:

$$\varphi_2(\boldsymbol{\xi}) = -\frac{|\boldsymbol{\xi}|^2}{2} + (u_2, v_2) \cdot \boldsymbol{\xi} + C,$$

where (u_2, v_2, C) are constants to be determined. Note that $\varphi_1(\boldsymbol{\xi})$ is defined by (2.2.17). Using that $\varphi_1 = \varphi_2$ on $\mathcal{S}_1$, we find that $(\varphi_1 - \varphi_2)(\tilde{\xi}_1, 0) = 0$. This implies

$$\varphi_2(\boldsymbol{\xi}) = -\frac{|\boldsymbol{\xi}|^2}{2} + (u_2, v_2) \cdot \boldsymbol{\xi} + u_1(\tilde{\xi}_1 - \xi_1^0) - u_2\tilde{\xi}_1. \tag{7.2.1}$$

Furthermore, (7.1.4) and (7.1.7) yield

$$v_2 = u_2 \tan\theta_{\rm w}. \tag{7.2.2}$$

Also, since $\mathcal{S}_1$ passes through points $P_0 = (\xi_1^0, \xi_1^0 \tan\theta_{\rm w})$ and $(\tilde{\xi}_1, 0)$, and has angle $\theta_{\rm sh}$ with the ξ_1–axis, we have

$$\tilde{\xi}_1 = \xi_1^0 - \xi_1^0 \frac{\tan\theta_{\rm w}}{\tan\theta_{\rm sh}}. \tag{7.2.3}$$

The Bernoulli law (2.2.7) becomes

$$\rho_0^{\gamma-1} = \rho_2^{\gamma-1} + (\gamma-1)\big(\frac{1}{2}(u_2^2 + v_2^2) + (u_1 - u_2)\tilde{\xi}_1 - u_1\xi_1^0\big). \tag{7.2.4}$$

We can express $\boldsymbol{\tau} = (\cos\theta_{\rm sh}, \sin\theta_{\rm sh})$. Since $\boldsymbol{\nu} = \frac{D(\varphi_1-\varphi_2)}{|D(\varphi_1-\varphi_2)|} = \frac{(u_2-u_1,v_2)}{|(u_2-u_1,v_2)|}$, we have

$$(u_2 - u_1, v_2)\cdot(\cos\theta_{\rm sh}, \sin\theta_{\rm sh}) = 0, \tag{7.2.5}$$

so that, using (7.2.2),

$$u_2 = u_1 \frac{\cos\theta_{\rm w}\cos\theta_{\rm sh}}{\cos(\theta_{\rm w}-\theta_{\rm sh})}. \tag{7.2.6}$$

From (7.2.5), the Rankine-Hugoniot condition (2.2.13) along the reflected shock is

$$[\rho\, D\varphi]\cdot(\sin\theta_{\rm sh}, -\cos\theta_{\rm sh}) = 0,$$

that is,

$$\rho_1(u_1-\tilde{\xi}_1)\sin\theta_{\rm sh} = \rho_2\big(u_2\frac{\sin(\theta_{\rm sh}-\theta_{\rm w})}{\cos\theta_{\rm w}} - \tilde{\xi}_1\sin\theta_{\rm sh}\big). \tag{7.2.7}$$

Combining (7.2.3)–(7.2.7), we obtain the following system for $(\rho_2, \theta_{\rm sh}, \tilde{\xi}_1)$:

$$(\tilde{\xi}_1-\xi_1^0)\cos\theta_{\rm w} + \xi_1^0\sin\theta_{\rm w}\cot\theta_{\rm sh} = 0, \tag{7.2.8}$$

$$\frac{\rho_2^{\gamma-1}-\rho_0^{\gamma-1}}{\gamma-1} + \frac{u_1^2\cos^2\theta_{\rm sh}}{2\cos^2(\theta_{\rm w}-\theta_{\rm sh})} + \frac{u_1\sin\theta_{\rm w}\sin\theta_{\rm sh}}{\cos(\theta_{\rm w}-\theta_{\rm sh})}\tilde{\xi}_1 - u_1\xi_1^0 = 0, \tag{7.2.9}$$

$$\big(u_1\cos\theta_{\rm sh}\tan(\theta_{\rm sh}-\theta_{\rm w}) - \tilde{\xi}_1\sin\theta_{\rm sh}\big)\rho_2 - \rho_1(u_1-\tilde{\xi}_1)\sin\theta_{\rm sh} = 0. \tag{7.2.10}$$

Lemma 7.2.1. *There exist positive constants σ_1, ε, and C depending only on (ρ_0,ρ_1,γ) such that, for any $\theta_{\rm w}\in[\frac{\pi}{2}-\sigma_1,\frac{\pi}{2})$, system (7.2.8)–(7.2.10) has a unique solution satisfying $|(\rho_2,\theta_{\rm sh},\tilde{\xi}_1)-(\bar{\rho}_2,\frac{\pi}{2},\bar{\xi}_1)|<\varepsilon$, where $\bar{\rho}_2$ and $\bar{\xi}_1$ are the parameters of the normal reflection state (2) from Theorem 6.2.1. Moreover, $(\rho_2,\theta_{\rm sh},\tilde{\xi}_1)$, as functions of $\theta_{\rm w}$, are in $C^\infty([\frac{\pi}{2}-\sigma_1,\frac{\pi}{2}])$ and*

$$|\rho_2-\bar{\rho}_2| + |\frac{\pi}{2}-\theta_{\rm sh}| + |\tilde{\xi}_1-\bar{\xi}_1| + |c_2-\bar{c}_2| \le C(\frac{\pi}{2}-\theta_{\rm w}), \tag{7.2.11}$$

where c_2 is the sonic speed of state (2).

Proof. We compute the Jacobian J of the left-hand side of (7.2.8)–(7.2.10) in terms of $(\rho_2,\theta_{\rm sh},\tilde{\xi}_1)$ at the normal reflection state $(\bar{\rho}_2,\frac{\pi}{2},\bar{\xi}_1)$ in §3.1 for state (2) when $\theta_{\rm w}=\frac{\pi}{2}$ to obtain

$$J = -\xi_1^0\big(\bar{\rho}_2^{\gamma-2}(\bar{\rho}_2-\rho_1) - u_1\bar{\xi}_1\big) < 0,$$

since $\bar{\rho}_2>\rho_1$ and $\bar{\xi}_1<0$. Then, by the implicit function theorem, we conclude the proof. ∎

Reducing $\sigma_1>0$ if necessary, we find that, for any $\sigma\in(0,\sigma_1)$,

$$\tilde{\xi}_1 < 0, \tag{7.2.12}$$

from (6.2.2) and (7.2.11). Since $\theta_{\rm w} \in (\frac{\pi}{2} - \sigma_1, \frac{\pi}{2})$, $\theta_{\rm sh} \in (\frac{\pi}{4}, \frac{3\pi}{4})$ if σ_1 is small, which implies $\sin\theta_{\rm sh} > 0$. We conclude from (7.2.8), (7.2.12), and $\xi_1^0 > 0$ that $\tan\theta_{\rm w} > \tan\theta_{\rm sh} > 0$. Therefore, we have

$$\frac{\pi}{4} < \theta_{\rm sh} < \theta_{\rm w} < \frac{\pi}{2}. \tag{7.2.13}$$

Now, given $\theta_{\rm w}$, we define φ_2 as follows: We have shown that there exists a unique solution $(\rho_2, \theta_{\rm sh}, \tilde{\xi}_1)$ close to $(\bar{\rho}_2, \frac{\pi}{2}, \bar{\xi}_1)$ of system (7.2.8)–(7.2.10), and define u_2 by (7.2.6), v_2 by (7.2.2), and φ_2 by (7.2.1). Now (7.2.2) and (7.2.6) imply (7.2.5). This, combined with (2.2.17) and (7.2.1), implies that line $\mathcal{S}_1 = \{\boldsymbol{\xi} : \varphi_1(\boldsymbol{\xi}) = \varphi_2(\boldsymbol{\xi})\}$ is given by the equation:

$$\xi_1 = \xi_2 \cot\theta_{\rm sh} + \tilde{\xi}_1.$$

Now (7.2.10) implies that the Rankine-Hugoniot condition (2.2.13) holds on $\mathcal{S}_1$. Furthermore, (7.2.8) implies (7.2.3). From (7.2.3), using that $\mathcal{S}_1 = \{\xi_1 = \xi_2 \cot\theta_{\rm sh} + \tilde{\xi}_1\}$, we find that point $P_0 = (\xi_1^0, \xi_1^0 \tan\theta_{\rm w})$ lies on $\mathcal{S}_1$. Also, (7.2.9) and (7.2.6) imply (7.2.4), *i.e.*, (1.13), or equivalently (2.2.9), holds. This can be stated as the Bernoulli law:

$$\rho_2^{\gamma-1} + (\gamma-1)\big(\frac{1}{2}|D\varphi_2|^2 + \varphi_2\big) = \rho_0^{\gamma-1}. \tag{7.2.14}$$

Thus, we have shown that φ_2 defined above satisfies (7.1.5)–(7.1.7).

We also notice from (7.2.2) and (7.2.6) that

$$|u_2| + |v_2| \le C(\frac{\pi}{2} - \theta_{\rm w}), \tag{7.2.15}$$

by using (7.2.11) and (7.2.13). Then, reducing σ_1 if necessary, we obtain from (7.2.1) and (7.2.11) that, for $\theta_{\rm w} \in [\frac{\pi}{2} - \sigma_1, \frac{\pi}{2})$ and $P_0 = (\xi_1^0, \xi_1^0 \tan\theta_{\rm w})$,

$$|D\varphi_2(P_0)| > c_2.$$

Furthermore, from (6.2.4) and the continuity of $(\rho_2, \tilde{\xi}_1)$ with respect to $\theta_{\rm w}$ on $(\frac{\pi}{2} - \sigma_1, \frac{\pi}{2}]$, it follows that, if $\sigma > 0$ is small,

$$|\tilde{\xi}_1| < c_2. \tag{7.2.16}$$

Therefore, we have

Proposition 7.2.2. *There exist positive constants σ_1 and C depending only on (ρ_0, ρ_1, γ) such that, for any $\theta_{\rm w} \in [\frac{\pi}{2} - \sigma_1, \frac{\pi}{2})$, there exists a state (2) of form (7.2.1). Its parameters $(u_2, v_2, \tilde{\xi}_1)$, and the corresponding density and the sonic speed (ρ_2, c_2), determined by (1.13) and (1.14), satisfy (7.2.11) and (7.2.15)–(7.2.16). Moreover, φ_2 is supersonic at P_0, i.e., $|D\varphi_2(P_0)| > c_2$.*

7.3 THE SHOCK POLAR FOR STEADY POTENTIAL FLOW AND ITS PROPERTIES

The steady potential flow is governed by the following equations for the velocity potential φ:

$$\begin{cases} \operatorname{div}(\rho D\varphi) = 0, \\ (u, v) = D\varphi, \\ \frac{1}{2}(u^2 + v^2) + h(\rho) = K_0 \qquad \text{(Bernoulli's law)}, \end{cases} \tag{7.3.1}$$

with the Bernoulli constant K_0, where $h(\rho)$ is given by (1.2).

Suppose that a point P_0 lies on a smooth shock curve $\mathcal{S}$ separating two smooth states, and $\{\boldsymbol{\nu}, \boldsymbol{\tau}\}$ are the unit normal and tangent vectors to $\mathcal{S}$ at P_0. Furthermore, assume that the upstream velocity and density are $\mathbf{v}_\infty = (u_\infty, 0)$ and ρ_∞, and the downstream velocity and density are $\mathbf{v} = (u_\mathcal{O}, v_\mathcal{O})$ and $\rho_\mathcal{O}$. Then the condition that the equations in (7.3.1) hold in the weak sense across $\mathcal{S}$ implies the following Rankine-Hugoniot conditions at P_0:

$$\rho_\mathcal{O}\mathbf{v} \cdot \boldsymbol{\nu} = \rho_\infty \mathbf{v}_\infty \cdot \boldsymbol{\nu}, \tag{7.3.2}$$

$$(\mathbf{v}_\infty - \mathbf{v}) \cdot \boldsymbol{\tau} = 0, \tag{7.3.3}$$

$$\frac{\rho_\mathcal{O}^{\gamma-1}}{\gamma - 1} + \frac{1}{2}|\mathbf{v}|^2 = \frac{\rho_\infty^{\gamma-1}}{\gamma - 1} + \frac{1}{2}u_\infty^2. \tag{7.3.4}$$

If the upstream and downstream states are uniform, $\mathcal{S}$ is a straight line $\mathcal{S}_\mathcal{O}$, and $\{\boldsymbol{\nu}, \boldsymbol{\tau}\}$ are the unit normal and tangent vectors to $\mathcal{S}_\mathcal{O}$, respectively. Thus, if $(\rho_\mathcal{O}, u_\mathcal{O}, v_\mathcal{O})$ is a solution of (7.3.2)–(7.3.4) with some unit vectors $\{\boldsymbol{\nu}, \boldsymbol{\tau}\}$ such that $\boldsymbol{\nu} \cdot \boldsymbol{\tau} = 0$, then $\rho_\mathcal{O}$ and $(u_\mathcal{O}, v_\mathcal{O})$ are the density and velocity of the downstream uniform state, separated by the straight shock $\mathcal{S}_\mathcal{O} \perp \boldsymbol{\nu}$ from the upstream uniform state with density ρ_∞ and velocity $(u_\infty, 0)$, which will be denoted as the incoming uniform state (ρ_∞, u_∞) below.

We also note that, if $\mathcal{S}_\mathcal{O}$ forms an angle β with the positive v–axis,

$$\boldsymbol{\nu} = (\cos\beta, -\sin\beta) \quad \text{and} \quad \boldsymbol{\tau} = (\sin\beta, \cos\beta) \tag{7.3.5}$$

are a unit normal and a unit tangent vector to $\mathcal{S}_\mathcal{O}$, respectively. Of course, replacing $\boldsymbol{\nu}$ by $-\boldsymbol{\nu}$ and/or $\boldsymbol{\tau}$ by $-\boldsymbol{\tau}$ does not change system (7.3.2)–(7.3.4). However, if $u_\infty > 0$ is assumed, then the uniform flow (ρ_∞, u_∞) is the upstream state, and $\boldsymbol{\nu}$ in (7.3.5) is the unit normal to $\mathcal{S}_\mathcal{O}$ toward downstream. In fact, the choice of directions of the normal $\boldsymbol{\nu}$ and the tangent vector $\boldsymbol{\tau}$ to $\mathcal{S}_\mathcal{O}$ in (7.3.5) for $\mathbf{v} \neq \mathbf{v}_\infty$ is uniquely determined by the following requirements:

$$\boldsymbol{\nu} = \frac{\mathbf{v}_\infty - \mathbf{v}}{|\mathbf{v}_\infty - \mathbf{v}|}, \qquad \boldsymbol{\tau} \cdot \mathbf{v}_\infty = \boldsymbol{\tau} \cdot \mathbf{v} \geq 0. \tag{7.3.6}$$

Therefore, we have

Definition 7.3.1. *Let* $\rho_\infty > 0$ *and* $u_\infty > 0$, *and let* (ρ_∞, u_∞) *be the incoming uniform state. Then* $\rho_\mathcal{O}$ *and* $(u_\mathcal{O}, v_\mathcal{O})$ *are the density and velocity of the downstream uniform state behind a straight oblique shock* $\mathcal{S}_\mathcal{O}$ *of angle* $\beta \in [0, \frac{\pi}{2})$ *with the positive* v*–axis if* $(\rho_\mathcal{O}, u_\mathcal{O}, v_\mathcal{O})$ *is a solution of* (7.3.2)–(7.3.4) *with vectors* $\{\boldsymbol{\nu}, \boldsymbol{\tau}\}$ *defined by* (7.3.5).

The shock polar is the collection of all the downstream velocities $(u_\mathcal{O}, v_\mathcal{O})$ for a fixed incoming uniform state (ρ_∞, u_∞).

7.3.1 Convexity and further properties of the steady shock polar

We now present some further properties of the steady shock polar.

Lemma 7.3.2. *Fix* $\gamma \geq 1$ *and the incoming constant state* (ρ_∞, u_∞) *with* $u_\infty > c_\infty := \rho_\infty^{\frac{\gamma-1}{2}}$. *Then, for any* $\beta \in [0, \cos^{-1}(\frac{c_\infty}{u_\infty})]$, *there exists a unique state* $(u_\mathcal{O}, v_\mathcal{O}) \in \mathbb{R}^+ \times \mathbb{R}^+$ *so that* $(u_\mathcal{O}, v_\mathcal{O})$ *becomes the downstream velocity behind a straight oblique shock* $\mathcal{S}_\mathcal{O}$ *of angle* β *with the positive* v*–axis in the sense of Definition* 7.3.1, *and satisfies the entropy condition* $|(u_\mathcal{O}, v_\mathcal{O})| \leq u_\infty$. *In particular,*

- *For* $\beta \in [0, \cos^{-1}(\frac{c_\infty}{u_\infty}))$, *the entropy inequality is strict:* $|(u_\mathcal{O}, v_\mathcal{O})| < u_\infty$;
- *For* $\beta = \cos^{-1}(\frac{c_\infty}{u_\infty})$, *the unique downstream state has velocity* $(u_\infty, 0)$;
- *For* $\beta \in (\cos^{-1}(\frac{c_\infty}{u_\infty}), \frac{\pi}{2})$, *there is no solution (downstream state) satisfying* $|(u_\mathcal{O}, v_\mathcal{O})| \leq u_\infty$.

Furthermore, the following properties hold:

(a) *The collection of all* $(u_\mathcal{O}, v_\mathcal{O})$ *for* $\beta \in [0, \cos^{-1}(\frac{c_\infty}{u_\infty})]$ *forms a convex curve in the upper half-plane* $\{v \geq 0\}$ *of the* (u, v)*–plane. The curve:* $[0, \cos^{-1}(\frac{c_\infty}{u_\infty})] \mapsto \mathbb{R}^2$, *defined by* $\beta \mapsto (u_\mathcal{O}, v_\mathcal{O})$, *is called a shock polar for potential flow. The shock polar is in* $C([0, \cos^{-1}(\frac{c_\infty}{u_\infty})]; \mathbb{R}^2) \cap C^\infty((0, \cos^{-1}(\frac{c_\infty}{u_\infty})); \mathbb{R}^2)$.

(b) *Function* $\beta \mapsto u_\mathcal{O}$ *satisfies* $u'_\mathcal{O}(\beta) > 0$ *on* $(0, \cos^{-1}(\frac{c_\infty}{u_\infty})]$.

(c) *The shock polar in the upper half-plane* $\{v \geq 0\}$ *is the graph:*

$$\{(u, f_{\text{polar}}(u)) \ : \ \hat{u}_0 \leq u \leq u_\infty\},$$

where $\hat{u}_0 \in (0, u_\infty)$ *is the unique solution of the equation on* $(0, u_\infty)$:

$$\big(\rho_\infty^{\gamma-1} + \frac{\gamma-1}{2}(u_\infty^2 - \hat{u}_0^2)\big)^{\frac{1}{\gamma-1}} \hat{u}_0 = \rho_\infty u_\infty. \tag{7.3.7}$$

Also f_{polar} *satisfies* $f_{\text{polar}} \in C([\hat{u}_0, u_\infty]) \cap C^\infty((\hat{u}_0, u_\infty))$ *and*

$$f_{\text{polar}}(\hat{u}_0) = f_{\text{polar}}(u_\infty) = 0. \tag{7.3.8}$$

Moreover, f_{polar} *is strictly concave:*

$$f''_{\text{polar}}(u) < 0 \qquad \text{for all } u \in (\hat{u}_0, u_\infty). \tag{7.3.9}$$

(d) *For a downstream state* $(u_{\mathcal{O}}, v_{\mathcal{O}})$*, define the wedge angle:*

$$\theta_{\rm w} = \tan^{-1}\big(\frac{v_{\mathcal{O}}}{u_{\mathcal{O}}}\big).$$

Since $(u_{\mathcal{O}}, v_{\mathcal{O}})$ *is uniquely defined by angle* β*, the above equation defines a function* $\theta_{\rm w} = \Theta_{\rm w}(\beta)$*, where* $\Theta_{\rm w} \in C([0, \cos^{-1}(\frac{c_\infty}{u_\infty})]) \cap C^\infty((0, \cos^{-1}(\frac{c_\infty}{u_\infty})))$*. Then there exist the* detachment wedge angle $\theta_{\rm d} \in (0, \frac{\pi}{2})$ *and the* detachment shock angle $\beta_{\rm d} \in (0, \cos^{-1}(\frac{c_\infty}{u_\infty}))$ *such that*

(i) $\Theta_{\rm w}(\beta_{\rm d}) = \theta_{\rm d}$*, and* $\beta_{\rm d}$ *is uniquely determined in* $(0, \cos^{-1}(\frac{c_\infty}{u_\infty}))$ *by this equation;*

(ii) *For any* $\theta_{\rm w} \in [0, \theta_{\rm d})$*, there exist two values of* β *such that* $\Theta_{\rm w}(\beta) = \theta_{\rm w}$*: The strong reflection* $\beta^{\rm sg} \in [0, \beta_{\rm d})$ *and the weak reflection* $\beta^{\rm wk} \in (\beta_{\rm d}, \cos^{-1}(\frac{c_\infty}{u_\infty})]$*;*

(iii) $\Theta_{\rm w}(\cdot)$ *is strictly increasing on* $(0, \beta_{\rm d})$ *and strictly decreasing on* $(\beta_{\rm d}, \cos^{-1}(\frac{c_\infty}{u_\infty}))$ *with*

$$\Theta_{\rm w}(0) = \Theta_{\rm w}(\cos^{-1}(\frac{c_\infty}{u_\infty})) = 0.$$

(e) *There exists the sonic value* $\beta_{\rm s} \in (\beta_{\rm d}, \cos^{-1}(\frac{c_\infty}{u_\infty}))$ *such that*

$$\begin{aligned} |(u_{\mathcal{O}}, v_{\mathcal{O}})| &< c(|(u_{\mathcal{O}}, v_{\mathcal{O}})|^2) \qquad \text{if } \beta \in (0, \beta_{\rm s}), \\ |(u_{\mathcal{O}}, v_{\mathcal{O}})| &> c(|(u_{\mathcal{O}}, v_{\mathcal{O}})|^2) \qquad \text{if } \beta \in (\beta_{\rm s}, \cos^{-1}(\frac{c_\infty}{u_\infty})), \end{aligned} \tag{7.3.10}$$

where $c_\infty = c(u_\infty^2)$*, and the sonic speed* $c(q^2) \geq 0$ *and density* $\rho(q^2) \geq 0$ *are defined by*

$$c^2(q^2) = \rho^{\gamma-1}(q^2) := \rho_\infty^{\gamma-1} + \frac{\gamma-1}{2}(u_\infty^2 - q^2). \tag{7.3.11}$$

We also denote by $\theta_{\rm s}$ *the wedge angle, introduced in* (d) *above, corresponding to* $\beta_{\rm s}$*, i.e.,* $\theta_{\rm s} := \Theta(\beta_{\rm s})$*.*

(f) *Denote* $u_{\rm d} := u_{\mathcal{O}}(\beta_{\rm d})$ *as the detachment value of* u*, and* $u_{\rm s} := u_{\mathcal{O}}(\beta_{\rm s})$ *the sonic value of* u*. Then*

$$\hat{u}_0 < u_{\rm d} < u_{\rm s} < u_\infty. \tag{7.3.12}$$

Furthermore,

$$\mathbf{v} = (u, f_{\rm polar}(u)) \;\; \text{is} \;\; \begin{cases} \text{strong reflection if } u \in [\hat{u}_0, u_{\rm d}), \\ \text{weak reflection if } u \in (u_{\rm d}, u_\infty]. \end{cases} \tag{7.3.13}$$

Also, for $\mathbf{v} = (u, f_{\rm polar}(u))$*,*

$$|\mathbf{v}| < c(|\mathbf{v}|^2) \;\; \text{if } u \in [\hat{u}_0, u_{\rm s}), \qquad |\mathbf{v}| > c(|\mathbf{v}|^2) \;\; \text{if } u \in (u_{\rm s}, u_\infty]. \tag{7.3.14}$$

In particular, all the strong reflections are subsonic.

(g) *The continuous dependence on the parameters of the incoming flow* (ρ_∞, u_∞):

(i) $\hat{u}_0$ *is a* C^∞*–function of* (ρ_∞, u_∞) *on the domain:*

$$\{(\rho_\infty, u_\infty) \,:\, \rho_\infty > 0,\; u_\infty > c_\infty\};$$

(ii) *Denote by* $f_{\rm polar}(\cdot\,; \rho_\infty, u_\infty)$ *the shock polar function* $f_{\rm polar}(\cdot)$ *for the incoming flow* (ρ_∞, u_∞). *Then the function:* $(u, \rho_\infty, u_\infty) \mapsto f_{\rm polar}(u; \rho_\infty, u_\infty)$ *is* C^∞ *on the domain:*

$$\{(u, \rho_\infty, u_\infty) \,:\, \rho_\infty > 0,\; u_\infty > c_\infty,\; \hat{u}_0(\rho_\infty, u_\infty) < u < u_\infty\};$$

(iii) $(u_{\rm d}, u_{\rm s})$, *defined in assertion* (f), *have the property that*

$$(u_{\rm d}, u_{\rm s}) \text{ depend continuously on } (\rho_\infty, u_\infty). \tag{7.3.15}$$

(h) *Any* $\mathbf{v} = (u_{\mathcal{O}}, v_{\mathcal{O}}) \neq \mathbf{v}_\infty$ *on the shock polar satisfies*

$$g(\mathbf{v}) := \big(\rho(|\mathbf{v}|^2)\mathbf{v} - \rho_\infty \mathbf{v}_\infty\big) \cdot \frac{\mathbf{v}_\infty - \mathbf{v}}{|\mathbf{v}_\infty - \mathbf{v}|} = 0, \tag{7.3.16}$$

where $\rho(|\mathbf{v}|^2)$ *is defined by* (7.3.11). *Moreover,* $Dg \neq 0$ *on the shock polar and, for all* $u \in (\hat{u}_0, u_\infty)$,

$$Dg(u, f_{\rm polar}(u)) = \frac{|Dg(u, f_{\rm polar}(u))|}{\sqrt{1 + (f'_{\rm polar}(u))^2}}(f'_{\rm polar}(u),\; -1). \tag{7.3.17}$$

Furthermore, for any $\mathbf{v} = (u_{\mathcal{O}}, v_{\mathcal{O}})$ *on the shock polar,*

$$\begin{aligned} &Dg(\mathbf{v}) \cdot \mathbf{v} < 0 \;\; \text{if and only if } \mathbf{v} \text{ is a weak reflection,}\\ &Dg(\mathbf{v}) \cdot \mathbf{v} > 0 \;\; \text{if and only if } \mathbf{v} \text{ is a strong reflection,} \end{aligned} \tag{7.3.18}$$

and

$$Dg(\mathbf{v}) \cdot \mathbf{v}^\perp < 0 \qquad \text{for all } \beta \in (0, \cos^{-1}(\frac{c_\infty}{u_\infty})], \tag{7.3.19}$$

where $\mathbf{v}^\perp$ *is uniquely determined by*

$$\mathbf{v}^\perp \cdot \mathbf{v} = 0, \qquad \mathbf{v}^\perp \cdot \mathbf{v}_\infty \leq 0, \qquad |\mathbf{v}^\perp| = |\mathbf{v}|.$$

Here we have used the definition of weak and strong reflections given in assertion (d)(ii) *above.*

Proof. The convexity of the shock polar curve will be shown by adjusting the proof of [108, Theorem 1] to the case of steady flow.

Fix $\gamma > 1$ and set $\mathbf{v}_\infty := (u_\infty, 0)$. Let $\mathcal{S}_{\mathcal{O}}$ be a straight oblique shock with angle $\frac{\pi}{2} - \beta$ from the horizontal ground that is the u-axis, and let $\rho_{\mathcal{O}}$ and

$\mathbf{v} = (u_{\mathcal{O}}, v_{\mathcal{O}})$ be the density and velocity behind shock $\mathcal{S}_{\mathcal{O}}$. In this case, the unit normal $\boldsymbol{\nu}$ towards the downstream and the unit tangent $\boldsymbol{\tau}$ obtained from rotating $\boldsymbol{\nu}$ by $\frac{\pi}{2}$ counterclockwise are given in (7.3.5). On $\mathcal{S}_{\mathcal{O}}$, state $(\rho_{\mathcal{O}}, \mathbf{v})$ satisfies the Rankine-Hugoniot conditions (7.3.2)–(7.3.4). We divide the proof into nine steps.

1. We first reduce the general case to Case $\rho_\infty = 1$. From the explicit expressions, if $(\rho_{\mathcal{O}}, \mathbf{v})$ satisfies (7.3.2)–(7.3.4) for the incoming constant state (ρ_∞, u_∞), then, defining $\hat{u}_\infty$ and $(\hat{\rho}_{\mathcal{O}}, \hat{\mathbf{v}})$ by

$$u_\infty = c_\infty \hat{u}_\infty, \quad \mathbf{v} = c_\infty \hat{\mathbf{v}}, \quad \rho_{\mathcal{O}} = \rho_\infty \hat{\rho}_{\mathcal{O}}$$

with $c_\infty = \rho_\infty^{\frac{\gamma-1}{2}}$, we see that $(\hat{\rho}_{\mathcal{O}}, \hat{\mathbf{v}})$ satisfies (7.3.2)–(7.3.4) for the incoming constant state $(\hat{\rho}_\infty, \hat{u}_\infty) = (1, \hat{u}_\infty)$. Moreover, the direction of shock $\mathcal{S}_{\mathcal{O}}$ does not change. Indeed, from (7.3.6), we have

$$\boldsymbol{\nu} = \frac{\mathbf{v} - \mathbf{v}_\infty}{|\mathbf{v} - \mathbf{v}_\infty|} = \frac{\hat{\mathbf{v}} - \hat{\mathbf{v}}_\infty}{|\hat{\mathbf{v}} - \hat{\mathbf{v}}_\infty|} = \hat{\boldsymbol{\nu}}.$$

Therefore, from now on, we assume without loss of generality that $\rho_\infty = 1$. Then $c_\infty = 1$, and condition $u_\infty > c_\infty$ becomes

$$u_\infty > 1,$$

and (7.3.2)–(7.3.4) can be written as

$$\rho_{\mathcal{O}} \mathbf{v} \cdot \boldsymbol{\nu} = \mathbf{v}_\infty \cdot \boldsymbol{\nu}, \tag{7.3.20}$$

$$(\mathbf{v}_\infty - \mathbf{v}) \cdot \boldsymbol{\tau} = 0, \tag{7.3.21}$$

$$\frac{\rho_{\mathcal{O}}^{\gamma-1}}{\gamma-1} + \frac{1}{2}(\mathbf{v} \cdot \boldsymbol{\nu})^2 = \frac{1}{\gamma-1} + \frac{1}{2}(\mathbf{v}_\infty \cdot \boldsymbol{\nu})^2 = \frac{1}{\gamma-1} + \frac{1}{2}(u_\infty \cos\beta)^2, \tag{7.3.22}$$

where $h(\rho)$ is given by (1.2). In particular, (7.3.22) is obtained from the Bernoulli law in (7.3.1) and (7.3.21). By (7.3.21), the angle between vector $\mathbf{v} - \mathbf{v}_\infty$ and the horizontal axis in Fig. 7.1 is β.

By the expression of $\{\boldsymbol{\nu}, \boldsymbol{\tau}\}$ above, we have

$$\begin{aligned} &\mathbf{v}_\infty \cdot \boldsymbol{\nu} = u_\infty \cos\beta, && \mathbf{v}_\infty \cdot \boldsymbol{\tau} = u_\infty \sin\beta, \\ &\mathbf{v} \cdot \boldsymbol{\nu} = u_{\mathcal{O}} \cos\beta - v_{\mathcal{O}} \sin\beta, && \mathbf{v} \cdot \boldsymbol{\tau} = u_{\mathcal{O}} \sin\beta + v_{\mathcal{O}} \cos\beta. \end{aligned} \tag{7.3.23}$$

2. Fix $\beta \in [0, \frac{\pi}{2})$. Seek a solution $(\mathbf{v}, \rho_{\mathcal{O}})$ of (7.3.20)–(7.3.22) satisfying the entropy condition:

$$|(u_{\mathcal{O}}, v_{\mathcal{O}})| \leq u_\infty$$

and the conditions on β for the existence of such a solution.

Since $(u_{\mathcal{O}}, v_{\mathcal{O}}) = (\mathbf{v} \cdot \boldsymbol{\nu})\boldsymbol{\nu} + (\mathbf{v} \cdot \boldsymbol{\tau})\boldsymbol{\tau}$, and $\mathbf{v} \cdot \boldsymbol{\tau} = u_\infty \sin\beta$ by (7.3.21) and (7.3.23), it remains to express $\mathbf{v} \cdot \boldsymbol{\nu}$ in terms of u_∞ and β. From (7.3.20) and (7.3.22), $q_{\boldsymbol{\nu}} := \mathbf{v} \cdot \boldsymbol{\nu}$ satisfies

$$\rho(q_{\boldsymbol{\nu}}^2) q_{\boldsymbol{\nu}} = u_\infty \cos\beta, \tag{7.3.24}$$

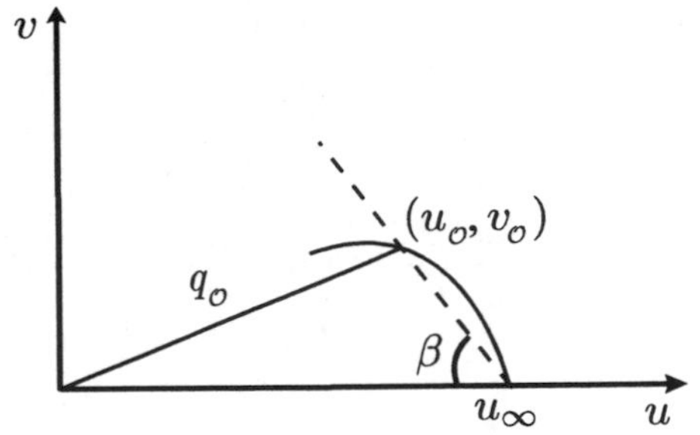

Figure 7.1: The shock polar for potential flow

where $\rho(q_\nu^2) = \left(1+\frac{\gamma-1}{2}(u_\infty^2\cos^2\beta - q_\nu^2)\right)^{\frac{1}{\gamma-1}}$. In order to satisfy the entropy condition, we seek a solution q_ν satisfying that $0 \le q_\nu < u_\infty \cos\beta$.

Introducing the function:

$$\Phi_K(q) := \rho_K(q^2)q = \left(K - \frac{\gamma-1}{2}q^2\right)^{\frac{1}{\gamma-1}} q \qquad \text{on } [0, \sqrt{\frac{2K}{\gamma-1}}], \tag{7.3.25}$$

we rewrite equation (7.3.24) as

$$\Phi_K(q) = \Phi_K(u_\infty \cos\beta) \qquad \text{for } K := 1 + \tfrac{\gamma-1}{2}u_\infty^2\cos^2\beta. \tag{7.3.26}$$

We compute

$$\begin{aligned}
&\Phi_K(0) = \Phi_K(\sqrt{\frac{2K}{\gamma-1}}) = 0, \\
&\Phi_K'(q) = \Big(K - \frac{\gamma-1}{2}q^2\Big)^{\frac{2-\gamma}{\gamma-1}}\Big(K - \frac{\gamma+1}{2}q^2\Big), \\
&\Phi_K' > 0 \ \text{ on } [0, q_c], \qquad \Phi_K' < 0 \ \text{ on } [q_c, \sqrt{\frac{2K}{\gamma-1}}],
\end{aligned} \tag{7.3.27}$$

where $q_c = \sqrt{\frac{2K}{\gamma+1}}$. Moreover, $q = u_\infty \cos\beta$ obviously satisfies equation (7.3.26). Then there exists a unique solution q_ν such that $0 \le q_\nu \ne u_\infty\cos\beta$. Moreover, $q_\nu < u_\infty \cos\beta$ if $u_\infty\cos\beta > q_c$. The condition that $u_\infty \cos\beta > q_c$ is

$$u_\infty^2\cos^2\beta > \frac{2(1+\frac{\gamma-1}{2}u_\infty^2\cos^2\beta)}{\gamma+1}, \quad \textit{i.e.,} \quad u_\infty\cos\beta > 1.$$

Thus, from (7.3.27), we obtain the following properties: For any $\beta \in [0, \cos^{-1}(\frac{1}{u_\infty}))$, there exists a unique $q_\nu \in (0, u_\infty\cos\beta)$ satisfying (7.3.25). For $\beta = \cos^{-1}(\frac{1}{u_\infty})$, we obtain that $u_\infty \cos\beta = q_c = 1$, and hence the only solution of equation (7.3.25) is $q_{\boldsymbol{\nu}} = u_\infty \cos\beta$. Also, $q_{\boldsymbol{\nu}}$ is a continuous function of β on $[0, \cos^{-1}(\frac{1}{u_\infty})]$.

Once $q_{\boldsymbol{\nu}}$ is determined, we find (u_O, v_O) from the last two equations in (7.3.23) by setting that $\mathbf{v}\cdot\boldsymbol{\nu} = q_{\boldsymbol{\nu}}$ and $\mathbf{v}\cdot\boldsymbol{\tau} = u_\infty \sin\beta$, where the last equality follows from (7.3.21). Also, from the continuity of $q_{\boldsymbol{\nu}}(\beta)$ on $[0, \cos^{-1}(\frac{1}{u_\infty})]$, we see that (u_O, v_O) depends continuously on $\beta \in [0, \cos^{-1}(\frac{1}{u_\infty})]$.

Therefore, we have shown that, for every $\beta \in [0, \cos^{-1}(\frac{1}{u_\infty})]$, there exists (u_O, v_O) satisfying (7.3.20)–(7.3.22) and the entropy condition, and that curve $\beta \mapsto (u_O, v_O)$ has the regularity stated in assertion (a). Also, since $q_{\boldsymbol{\nu}} < u_\infty \cos\beta$ for all $\beta \in [0, \cos^{-1}(\frac{1}{u_\infty}))$, $(u_O, v_O)\cdot\boldsymbol{\nu} = q_{\boldsymbol{\nu}} < u_\infty\cos\beta = (u_\infty, 0)\cdot\boldsymbol{\nu}$ for these angles. Using (7.3.21), we show that, for $\beta \mapsto (u_O, v_O)$, the entropy inequality is strict: $|(u_O, v_O)| < u_\infty$. Moreover, since

$$(q_{\boldsymbol{\nu}})_{|\beta=\cos^{-1}(\frac{1}{u_\infty})} = u_\infty \cos\beta = 1,$$

it follows that

$$(u_O, v_O)_{|\beta=\cos^{-1}(\frac{1}{u_\infty})} = (u_\infty, 0).$$

Furthermore, by the implicit function theorem, we obtain from equation (7.3.25) that

$$\begin{aligned}&\text{function } (\rho_\infty, u_\infty, \beta) \mapsto (u_O, v_O) \text{ is } C^\infty \text{ on the set:}\\ &\{(\rho_\infty, u_\infty, \beta) \;:\; \rho_\infty > 0,\ u_\infty > c_\infty,\ 0 < \beta < \cos^{-1}(\frac{c_\infty}{u_\infty})\}.\end{aligned} \tag{7.3.28}$$

Now, plugging (7.3.23) into (7.3.20)–(7.3.21), we have

$$(u_O, v_O) = (u_\infty(1 - (1 - \frac{1}{\rho_O})\cos^2\beta), u_\infty(1 - \frac{1}{\rho_O})\cos\beta\sin\beta). \tag{7.3.29}$$

Combining the entropy condition with (7.3.21)–(7.3.22) implies

$$\rho_O > \rho_\infty = 1 \tag{7.3.30}$$

for each β. Then, by (7.3.29), curve $\beta \mapsto (u_O, v_O)$ for $\beta \in (0, \cos^{-1}(\frac{c_\infty}{u_\infty}))$ lies in the first quadrant of the (u, v)–plane and is bounded.

Thus, we have proved the existence of (u_O, v_O, ρ_O), satisfying the conditions in Definition 7.3.1, the entropy condition for all $\beta \in [0, \cos^{-1}(\frac{c_\infty}{u_\infty})]$, and the properties of the shock polar stated in assertion (a).

3. Next, we prove assertion (b). Set $M_{\boldsymbol{\nu}} := \frac{\mathbf{v}\cdot\boldsymbol{\nu}}{c_O}$ and $M_{\infty,\boldsymbol{\nu}} = \mathbf{v}_\infty\cdot\boldsymbol{\nu}$ for $c_O = \rho_O^{\frac{\gamma-1}{2}}$. Also, since $\beta \in [0, \cos^{-1}(\frac{1}{u_\infty})]$, we employ (7.3.23) to conclude $M_{\infty,\boldsymbol{\nu}} > 1$.

Furthermore, we note that (7.3.27) implies that the unique solution $q_{\boldsymbol{\nu}} \in (0, u_\infty\cos\beta)$ of (7.3.26) satisfies $q_{\boldsymbol{\nu}} < q_c$ for every $\beta \in [0, \cos^{-1}(\frac{1}{u_\infty}))$. Then, from (7.3.25),

$$\rho_K^{\gamma-1}(q_{\boldsymbol{\nu}}) > \rho_K^{\gamma-1}(q_c) = q_c^2 > q_{\boldsymbol{\nu}}^2,$$

where the equality holds, since $q_c = \sqrt{\frac{2K}{\gamma+1}}$. Note that, for a fixed β, $\mathbf{v} \cdot \boldsymbol{\nu} = q_{\boldsymbol{\nu}}$ for K defined in (7.3.26). Since $c_{\mathcal{O}}^2 = \rho_{\mathcal{O}}^{\gamma-1} = \rho_K^{\gamma-1}(q_{\boldsymbol{\nu}})$, we can rewrite the above inequality as

$$(\mathbf{v} \cdot \boldsymbol{\nu})^2 < c_{\mathcal{O}}^2, \tag{7.3.31}$$

which implies that $M_{\boldsymbol{\nu}} < 1$. Thus, we have shown that

$$0 < M_{\boldsymbol{\nu}} < 1 < M_{\infty,\boldsymbol{\nu}}.$$

Plugging these expressions into (7.3.20) and (7.3.22), and then combining the resulting expressions, we obtain the equation:

$$\mathfrak{g}(M_{\boldsymbol{\nu}}) = \mathfrak{g}(M_{\infty,\boldsymbol{\nu}}), \tag{7.3.32}$$

where

$$\mathfrak{g}(M) = (M^2 + \frac{2}{\gamma-1})M^{-\frac{2(\gamma-1)}{\gamma+1}}.$$

We note that $\mathfrak{g}'(M) = \frac{4}{\gamma+1}(M - \frac{1}{M})M^{-\frac{2(\gamma-1)}{\gamma+1}} < 0$ for $0 < M < 1$, and $\mathfrak{g}'(M) > 0$ for $M > 1$. Thus, differentiating equation (7.3.32) with respect to β and using $0 < M_{\boldsymbol{\nu}} < 1 < M_{\infty,\boldsymbol{\nu}} = u_\infty \cos\beta$, we obtain that, on $(0, \cos^{-1}(\frac{1}{u_\infty})]$,

$$\frac{dM_{\boldsymbol{\nu}}}{d\beta} = \frac{\mathfrak{g}'(M_{\infty,\boldsymbol{\nu}})}{\mathfrak{g}'(M_{\boldsymbol{\nu}})}\frac{dM_{\infty,\boldsymbol{\nu}}}{d\beta} = -\frac{\mathfrak{g}'(M_{\infty,\boldsymbol{\nu}})}{\mathfrak{g}'(M_{\boldsymbol{\nu}})}u_\infty \sin\beta > 0. \tag{7.3.33}$$

Using the definition of $M_{\boldsymbol{\nu}}$, we can write equation (7.3.24) as follows:

$$\rho_{\mathcal{O}}^{\frac{\gamma+1}{2}} M_{\boldsymbol{\nu}} = u_\infty \cos\beta.$$

Differentiating this equation with respect to β and using (7.3.33), we have

$$\frac{d\rho_{\mathcal{O}}}{d\beta} < 0 \qquad \text{for all } \beta \in (0, \cos^{-1}(\frac{1}{u_\infty})]. \tag{7.3.34}$$

From (7.3.29), we have

$$\frac{du_{\mathcal{O}}}{d\beta} = u_\infty \left((1 - \frac{1}{\rho_{\mathcal{O}}})\sin(2\beta) - \frac{\cos^2\beta}{\rho_{\mathcal{O}}^2}\frac{d\rho_{\mathcal{O}}}{d\beta}\right).$$

Thus, using (7.3.30) and (7.3.34), we conclude assertion (b).

This also implies assertion (c), except for the convexity part. In particular, the definition of $\hat{u}_0$ in (c) is from the fact that $\hat{u}_0 := (u_{\mathcal{O}})_{|\beta=0}$ satisfies equation (7.3.24) for $\beta = 0$. Noting that, for $\beta = 0$, $\boldsymbol{\nu} = (1, 0)$ such that $u_{\mathcal{O}} = \mathbf{v} \cdot \boldsymbol{\nu}$, we conclude that equation (7.3.24) for $\beta = 0$ is (7.3.7).

This also implies the smooth dependence of $\hat{u}_0$ on (ρ_∞, u_∞) as stated in part (i) of assertion (g). Since the problem has been reduced to Case $\rho_\infty = 1$ by a smooth transformation, it suffices to consider the continuity of $\hat{u}_0$ with respect to u_∞, while keeping $\rho_\infty = 1$.

This continuity can be seen as follows: Note that $q_{\boldsymbol{\nu}} = \hat{u}_0$ is the unique solution of equation (7.3.24) for $\beta = 0$, satisfying $q_{\boldsymbol{\nu}} \in (0, q_c(u_\infty)_{|\beta=0})$. Then the C^∞–smoothness of $u_\infty \mapsto \hat{u}_0$ follows from the implicit function theorem by using form (7.3.26) of equation (7.3.24), considering $\Phi_K(q) - \Phi_K(u_\infty)$ with K defined in (7.3.26) and $\beta = 0$ as a function of (u_∞, q), and using that $\frac{\partial \Phi_K}{\partial q} > 0$ for any $q \in (0, q_c(u_\infty)_{|\beta=0})$ by (7.3.27).

4. Now we prove the strict convexity of the shock polar, *i.e.*, the convexity of curve $(u_{\mathcal{O}}(\beta), v_{\mathcal{O}}(\beta))$ for $\beta \in [0, \cos^{-1}(\frac{c_\infty}{u_\infty})]$.

Let $\mathbf{v} = (u, v)$ denote a point on the shock polar curve, corresponding to $\beta \in (0, \cos^{-1}(\frac{c_\infty}{u_\infty}))$, *i.e.*, $\mathbf{v} \neq \mathbf{v}_\infty$. By (7.3.21), the unit normal $\boldsymbol{\nu}$ to $\mathcal{S}_{\mathcal{O}}$ is defined by (7.3.6) up to the orientation. Plugging this expression into (7.3.20) and expressing $\rho_{\mathcal{O}}$ from (7.3.4), we obtain equation (7.3.16) with $\rho_\infty = 1$. On the other hand, if $\boldsymbol{\nu}$ and $\boldsymbol{\tau}$ for $\mathbf{v} \neq \mathbf{v}_\infty$ are defined by (7.3.6), then (7.3.21) holds. Under this condition, (7.3.20) with ρ defined by (7.3.23) is equivalent to (7.3.16). This means that the shock polar curve is the zero level set of function g on the first quadrant of the (u, v)–plane. Also, (7.3.21) holds on the shock polar. Then, differentiating g and using (7.3.21), we obtain that, on the shock polar,

$$
\begin{aligned}
g_{\mathbf{v}} \cdot \boldsymbol{\nu} &= \rho\Big(1 - \big(\frac{\mathbf{v} \cdot \boldsymbol{\nu}}{c}\big)^2\Big), \\
g_{\mathbf{v}} \cdot \boldsymbol{\tau} &= -(\mathbf{v}_\infty \cdot \boldsymbol{\tau})\Big(\frac{\rho \mathbf{v} \cdot \boldsymbol{\nu}}{c^2} + \frac{\rho - 1}{|\mathbf{v}_\infty - \mathbf{v}|}\Big).
\end{aligned}
\tag{7.3.35}
$$

As we have discussed earlier, $M_{\boldsymbol{\nu}} < 1$, which implies

$$
g_{\mathbf{v}} \cdot \boldsymbol{\nu} > 0.
$$

Then we can define a vector $\mathbf{q}$ by

$$
\mathbf{q} = \frac{g_{\mathbf{v}}}{g_{\mathbf{v}} \cdot \boldsymbol{\nu}} = \boldsymbol{\nu} + \frac{g_{\mathbf{v}} \cdot \boldsymbol{\tau}}{g_{\mathbf{v}} \cdot \boldsymbol{\nu}} \boldsymbol{\tau}.
$$

We claim that

$$
\mathbf{q} \times \mathbf{q}_\beta < 0 \qquad \text{for } \beta \in (0, \cos^{-1}\big(\tfrac{1}{u_\infty}\big)).
$$

Set $A := -\frac{g_{\mathbf{v}} \cdot \boldsymbol{\tau}}{g_{\mathbf{v}} \cdot \boldsymbol{\nu}}$. Noting that $\boldsymbol{\nu} = (\cos\beta, -\sin\beta)$ and $\boldsymbol{\tau} = (\sin\beta, \cos\beta)$, we find that $\mathbf{q}_\beta = -(1 + A_\beta)\boldsymbol{\tau} - A\boldsymbol{\nu}$, which yields

$$
\mathbf{q} \times \mathbf{q}_\beta = -(1 + A^2 + A_\beta).
$$

By (7.3.35), on the shock polar, A can be written as

$$
A = \frac{u_\infty \sin\beta}{1 - M_{\boldsymbol{\nu}}^2}\Big(\frac{M_{\boldsymbol{\nu}}}{c} + \frac{1}{u_\infty \cos\beta}\Big) \qquad \text{for } M_{\boldsymbol{\nu}} := \frac{\mathbf{v} \cdot \boldsymbol{\nu}}{c},
$$

where we have used that (7.3.20)–(7.3.21) hold on the shock polar so that $|\mathbf{v}_\infty - \mathbf{v}| = (1 - \frac{\rho_\infty}{\rho})(\mathbf{v}_\infty \cdot \boldsymbol{\nu}) = (1 - \frac{\rho_\infty}{\rho})u_\infty \cos\beta$. From (7.3.34), we obtain

$$\frac{dc_O}{d\beta} < 0 \qquad \text{on } [0, \cos^{-1}\tfrac{1}{u_\infty}).$$

Then, using (7.3.33), we obtain that $A_\beta \geq 0$, which yields

$$\mathbf{q} \times \mathbf{q}_\beta \leq -1 \qquad \text{for } \beta \in (0, \cos^{-1}(\tfrac{1}{u_\infty})].$$

From the definition of $\mathbf{q}$, we see that $\mathbf{q} \neq \mathbf{0}$. Then we finally obtain

$$\frac{\mathbf{q}}{|\mathbf{q}|} \times \Big(\frac{\mathbf{q}}{|\mathbf{q}|}\Big)_\beta = \frac{\mathbf{q} \times \mathbf{q}_\beta}{|\mathbf{q}|^2} \leq -\frac{1}{|\mathbf{q}|^2} < 0 \tag{7.3.36}$$

on the shock polar curve.

Fix a point $P_0 = (u_O, v_O)$ on the curve that $g(\mathbf{v}) = \mathbf{0}$, and let P_0 correspond to β_0 for $\beta_0 \in (0, \cos^{-1}(\frac{1}{u_\infty}))$.

Connect $(u_\infty, 0)$ to P_0 by a line L. Let L^* be the line perpendicular to L and passing through P_0. Then vectors $\{\boldsymbol{\nu}, \boldsymbol{\tau}\}$, introduced above, are parallel to lines L and L^*, respectively. We introduce a coordinate system (s,t) with basis $\{\boldsymbol{\nu}, \boldsymbol{\tau}\}$. Then $g_s = -g_{\mathbf{v}} \cdot \boldsymbol{\nu} < 0$ at P_0. Thus, we can apply the implicit function theorem to see that there exists a function f_{P_0} so that

$$\{g(\mathbf{v}) = 0, |\mathbf{v} - P_0| < \varepsilon_0\} = \{(f_{P_0}(t), t) : |t - t_{P_0}| < \varepsilon_1\}$$

for some small positive constants ε_0 and ε_1. Note that $\frac{\mathbf{q}}{|\mathbf{q}|}$ is the unit normal to the shock polar at P_0 with $\frac{\mathbf{q}}{|\mathbf{q}|} \cdot \boldsymbol{\nu} > 0$. Then $\frac{\mathbf{q}}{|\mathbf{q}|}$ can be expressed as

$$\frac{\mathbf{q}}{|\mathbf{q}|} = \frac{(-f'_{P_0}(t), 1)}{\sqrt{1 + (f'_{P_0})^2(t)}} \qquad \text{for } |t - t_{P_0}| < \varepsilon_1,$$

from which

$$\frac{\mathbf{q}}{|\mathbf{q}|} \times \Big(\frac{\mathbf{q}}{|\mathbf{q}|}\Big)_t = \frac{f''_{P_0}(t)}{1 + (f'_{P_0})^2(t)}.$$

From the definition of the (s,t)–coordinates,

$$\Big(\frac{\mathbf{q}}{|\mathbf{q}|}\Big)_t (P_0) = \frac{1}{B}\Big(\frac{\mathbf{q}}{|\mathbf{q}|}\Big)_\beta (P_0),$$

where $B = |P_0 - (u_\infty, 0)| > 0$. Then (7.3.36) implies that

$$f''_{P_0}(t_{P_0}) < 0 \qquad \text{for } \beta_0 \in (0, \cos^{-1}\tfrac{1}{u_\infty}).$$

This implies the strict convexity (7.3.9). One can repeat this argument to lead to the same conclusion about the shock polar for Case $\gamma = 1$.

Now we prove assertion (d). We have shown in Step 2 that curve $\beta \mapsto (u_{\mathcal{O}}, v_{\mathcal{O}})$ for $\beta \in (0, \cos^{-1}(\frac{c_\infty}{u_\infty}))$ lies in the first quadrant of the (u,v)–plane and is bounded. This implies that we can find $\theta_{\rm d} \in (0, \frac{\pi}{2})$ such that line $\frac{v}{u} = \tan\theta_{\rm d}$ is tangential to the shock polar curve, and the whole curve lies between line $\frac{v}{u} = \tan\theta_{\rm d}$ and the u–axis in the first quadrant of the (u,v)–plane. Combining the strict convexity of the shock polar with its structure of the graph as described in assertion (c), property (7.3.8), we obtain that line $\frac{v}{u} = \tan\theta_{\rm d}$ has exactly one contact point with the shock polar and, for any $\theta_{\rm w} \in [0, \theta_{\rm d})$, line $\frac{v}{u} = \tan\theta_{\rm w}$ intersects with the curve of $(u_{\mathcal{O}}, v_{\mathcal{O}})$ at exactly two different points. Applying the monotonicity shown in assertion (b) and using (7.3.8) again, we obtain the existence of $\Theta(\cdot)$ and its properties (i)–(iii) in assertion (d). Now the regularity that $\Theta_{\rm w} \in C([0, \cos^{-1}(\frac{c_\infty}{u_\infty})]) \cap C^\infty((0, \cos^{-1}(\frac{c_\infty}{u_\infty})))$ follows from $\Theta_{\rm w}(\beta) = \tan^{-1}\big(\frac{v_{\mathcal{O}}}{u_{\mathcal{O}}}\big)(\beta)$, the regularity of $(u_{\mathcal{O}}, v_{\mathcal{O}})(\cdot)$ in assertion (a), and the fact that $u_{\mathcal{O}} \in [\hat{u}_0, u_\infty]$ with $\hat{u}_0 > 0$.

Now assertions (a)–(d) are proved.

5. Next we prove assertion (h). We first show (7.3.18). As shown in Step 4, equation (7.3.16) holds and $Dg \neq 0$ on the shock polar. Then the strict convexity of the shock polar and its structure described in assertion (c) imply that $Dg(\mathbf{v}) \cdot \mathbf{v} = 0$ if and only if $\mathbf{v}$ is the detachment point that corresponds to angle $\beta_{\rm d}$ introduced in (d) and the sign of $Dg(\mathbf{v}) \cdot \mathbf{v}$ as a continuous function of $\beta \in [0, \cos^{-1}(\frac{c_\infty}{u_\infty})]$ changes across $\beta = \beta_{\rm d}$. Therefore, it remains to check the sign of $Dg(\mathbf{v}) \cdot \mathbf{v}$ for some $\beta \in [0, \beta_{\rm d})$.

At $\beta = 0$, as shown in Step 2, we have

$$\boldsymbol{\nu} = (1,0), \quad \mathbf{v} = (\hat{u}_0, 0), \quad 0 < \hat{u}_0 < c.$$

Then, from (7.3.35),

$$Dg(\mathbf{v}) \cdot \mathbf{v} = \rho\Big(1 - \frac{\hat{u}_0^2}{c^2}\Big) > 0 \qquad \text{at } \beta = 0.$$

Thus,

$$Dg(\mathbf{v}) \cdot \mathbf{v} \begin{cases} > 0 & \text{for all } \beta \in [0, \beta_{\rm d}), \\ < 0 & \text{for all } \beta \in (\beta_{\rm d}, \cos^{-1}(\frac{c_\infty}{u_\infty})]. \end{cases}$$

This implies (7.3.18).

Furthermore, since $g(\mathbf{v}) = 0$ and $Dg(\mathbf{v}) \neq 0$ on the shock polar:

$$\{(u, f_{\rm polar}(u)) \ : \ \hat{u}_0 \leq u \leq u_\infty\},$$

then either (7.3.17) holds on the whole shock polar or the right-hand side of (7.3.17) is equal to the minus left-hand side of (7.3.17) on the whole shock polar. Thus, it suffices to check the sign at one point. Choose any $\theta_{\rm w} \in (0, \theta_{\rm d})$. As we have discussed at the end of Step 4, line $\frac{v}{u} = \tan\theta_{\rm w}$ intersects with the curve of $(u_{\mathcal{O}}, v_{\mathcal{O}})$ at exactly two different points which lie in the first quadrant.

Let $\mathbf{v} = (u_{\mathcal{O}}, f_{\text{polar}}(u_{\mathcal{O}}))$ be the nearest to the origin of intersection. It follows that the segment connecting the origin to $\mathbf{v}$ lies outside of the open convex set:

$$\mathcal{A} := \{(u,v) \; : \; \hat{u}_0 < u < u_\infty,\, 0 < v < f_{\text{polar}}(u)\};$$

the line containing that segment (*i.e.*, line $v = u\tan\theta_{\text{w}}$) has a non-empty intersection with $\mathcal{A}$, and $\mathbf{v} \in \partial\mathcal{A}$. This implies that $f'_{\text{polar}}(u_{\mathcal{O}}) > \frac{f_{\text{polar}}(u_{\mathcal{O}})}{u_{\mathcal{O}}}$. Then

$$\mathbf{v} \cdot (f'_{\text{polar}}(u_{\mathcal{O}}), -1) = (u_{\mathcal{O}}, f_{\text{polar}}(u_{\mathcal{O}})) \cdot (f'_{\text{polar}}(u_{\mathcal{O}}), -1) > 0.$$

Since $Dg(\mathbf{v}) \cdot \mathbf{v} > 0$, we conclude (7.3.17) at $\mathbf{v}$, and hence also on the whole shock polar.

6. We continue the proof of assertion (h). It remains to show (7.3.19). We can write (7.3.35) as

$$g_{\mathbf{v}} = \rho_{\mathcal{O}}\boldsymbol{\nu} - \frac{\rho_{\mathcal{O}}}{c_{\mathcal{O}}^2}(\mathbf{v}\cdot\boldsymbol{\nu})\mathbf{v} - \frac{\rho_{\mathcal{O}} - 1}{|\mathbf{v} - \mathbf{v}_\infty|}(\mathbf{v}\cdot\boldsymbol{\tau})\boldsymbol{\tau}.$$

Let $\beta \neq 0$, *i.e.*, $\beta \in (0, \cos^{-1}(\frac{c_\infty}{u_\infty})]$. Then $\mathbf{v} = (u_{\mathcal{O}}, v_{\mathcal{O}})$ and $\mathbf{v}_\infty = (u_\infty, 0)$ with $u_{\mathcal{O}}, v_{\mathcal{O}}, u_\infty > 0$. Thus, from the definition of $\mathbf{v}^\perp$, it follows that

$$\mathbf{v}^\perp = (-v_{\mathcal{O}}, u_{\mathcal{O}}),$$

that is, $\mathbf{v}^\perp$ is a counterclockwise rotation of $\mathbf{v}$. Then, from (7.3.5),

$$\boldsymbol{\tau} = \boldsymbol{\nu}^\perp,$$

so that $\mathbf{v}^\perp\cdot\boldsymbol{\nu} = -\mathbf{v}\cdot\boldsymbol{\tau}$ and $\mathbf{v}^\perp\cdot\boldsymbol{\tau} = \mathbf{v}\cdot\boldsymbol{\nu}$. We calculate

$$g_{\mathbf{v}}\cdot\mathbf{v}^\perp = -(\mathbf{v}\cdot\boldsymbol{\tau})\Big(\rho_{\mathcal{O}} + \frac{\rho_{\mathcal{O}} - 1}{|\mathbf{v} - \mathbf{v}_\infty|}(\mathbf{v}\cdot\boldsymbol{\nu})\Big) < 0,$$

where we have used that $\rho_{\mathcal{O}} \geq 1$ by (7.3.22) and the entropy condition, and

$$\mathbf{v}\cdot\boldsymbol{\nu} > 0, \quad \mathbf{v}\cdot\boldsymbol{\tau} > 0. \tag{7.3.37}$$

The first inequality in (7.3.37) follows by (7.3.20), since $\rho_{\mathcal{O}}\mathbf{v}_{\boldsymbol{\nu}} > 0$ and $\mathbf{v}_\infty\cdot\boldsymbol{\nu} = u_\infty\cos\beta > 0$, where we have used (7.3.5). The second inequality in (7.3.37) for $\beta \in (0, \cos^{-1}(\frac{c_\infty}{u_\infty})]$ follows by (7.3.5) and $u_{\mathcal{O}}, v_{\mathcal{O}} > 0$. Now assertion (h) is proved.

7. It remains to prove assertions (c)–(g). In this step, we first prove assertion (e).

Assertion (a) implies that the function:

$$\beta \mapsto (q_{\mathcal{O}}, c_{\mathcal{O}}) \quad \text{is in } C([0,\cos^{-1}(\tfrac{1}{u_\infty})];\mathbb{R}^2) \cap C^\infty((0,\cos^{-1}(\tfrac{1}{u_\infty}));\mathbb{R}^2).$$

Then we need to show that there exists a unique solution $\beta_{\text{s}} \in [0, \cos^{-1}(\frac{1}{u_\infty})]$ of the equation:

$$c_{\mathcal{O}} = q_{\mathcal{O}} \tag{7.3.38}$$

for $q_O = |(u_O, v_O)|$, where q_O and c_O are functions of β (for fixed u_∞ and $\rho_\infty = 1$).

Differentiating the Bernoulli law in (7.3.1), we obtain that $\frac{d\rho_O}{dq_O} = -\frac{\rho_O q_O}{c_O^2}$. Thus, (7.3.34) implies that

$$\frac{dq_O}{d\beta} > 0 \qquad \text{for all } \beta \in (0, \cos^{-1}(\frac{1}{u_\infty})].$$

Moreover, (7.3.34) implies that $\frac{dc_O}{d\beta} < 0$. Also, $q_O, c_O > 0$. Then

$$\frac{d(\frac{q_O}{c_O})}{d\beta} > 0 \qquad \text{for all } \beta \in (0, \cos^{-1}(\frac{1}{u_\infty})]. \tag{7.3.39}$$

Since $\mathbf{v}_{|\beta=0} = (\hat{u}_0, 0)$ by (7.3.29), and $\boldsymbol{\nu}_{|\beta=0} = (1, 0)$, we conclude that $\hat{u}_0 = (\mathbf{v} \cdot \boldsymbol{\nu})_{|\beta=0}$. Then (7.3.31) implies that

$$\frac{q_O}{c_O}\Big|_{\beta=0} = \frac{\hat{u}_0}{c_{O|\beta=0}} < 1.$$

Since $(u_O, v_O)_{|\beta=\cos^{-1}(\frac{1}{u_\infty})} = (u_\infty, 0)$, as discussed in Step 2, and $u_\infty > c_\infty$ by the assumption, we obtain

$$\frac{q_O}{c_O}\Big|_{\beta=\cos^{-1}(\frac{1}{u_\infty})} = \frac{u_\infty}{c_\infty} > 1.$$

This implies that there exists a unique $\beta_{\mathrm{s}} \in (0, \cos^{-1}(\frac{c_\infty}{u_\infty}))$ so that (7.3.10) holds.

It remains to show that $\beta_{\mathrm{s}} > \beta_{\mathrm{d}}$, *i.e.*, that the strong reflections are subsonic. From (7.3.35), using the notations that $\mathbf{v}_{\boldsymbol{\nu}} = \mathbf{v} \cdot \boldsymbol{\nu}$ and $\mathbf{v}_{\boldsymbol{\tau}} = \mathbf{v} \cdot \boldsymbol{\tau}$, and recalling that $\rho_\infty = 1$, we have

$$g_{\mathbf{v}} \cdot \mathbf{v} = \rho_O \mathbf{v}_{\boldsymbol{\nu}}\Big(1 - \frac{|\mathbf{v}|^2}{c_O^2}\Big) - \mathbf{v}_{\boldsymbol{\tau}}^2 \frac{\rho_O - 1}{|\mathbf{v}_\infty - \mathbf{v}|} \qquad \text{at } \beta_{\mathrm{s}}.$$

We note that $\rho_O > 1$ by (7.3.22) and the entropy condition, where the inequality is strict since $\beta_{\mathrm{s}} \in (0, \cos^{-1}(\frac{c_\infty}{u_\infty}))$. Then $\rho_O \mathbf{v}_{\boldsymbol{\nu}} > 0$ by (7.3.37). Thus, $g_{\mathbf{v}} \cdot \mathbf{v} < 0$ if $|\mathbf{v}|^2 \geq c_O^2$. Then, using assertion (h), proved above, we see that $|\mathbf{v}|^2 \geq c_O^2$ can hold only for (strictly) weak reflections. This implies that $\beta_{\mathrm{s}} > \beta_{\mathrm{d}}$.

8. We now prove assertion (f) with the following steps:

- (7.3.12) follows from the inequality: $0 < \beta_{\mathrm{d}} < \beta_{\mathrm{s}} < \cos^{-1}(\frac{c_\infty}{u_\infty})$, proved in assertions (d) and (e), by using assertion (b) and recalling that $u_O(0) = \hat{u}_0$ and $u_O\big(\cos^{-1}(\frac{c_\infty}{u_\infty})\big) = u_\infty$;
- (7.3.13) follows from assertions (b) and (d);
- (7.3.14) follows from assertions (b) and (e).

9. It remains to prove assertion (g). Since we have reduced the problem to Case $\rho_\infty = 1$ by a smooth explicit transformation, it suffices to prove the continuity and regularity stated in (g) with respect to the other parameters stated there, while keeping $\rho_\infty = 1$ fixed.

Part (i) of assertion (g) has been shown at the end of Step 3 above.

Part (ii) of assertion (g) follows from part (i), property (7.3.28), assertion (b), the regularity of $\beta \mapsto u_O$ in assertion (a), and the regularity of f_{polar} in assertion (c).

Part (iii) of assertion (g) can be seen as follows: We first show the continuity of $u_{\mathrm{d}}(u_\infty)$. Recall that $(u_{\mathrm{d}}, v_{\mathrm{d}})$ are the coordinates of the unique contact point of the tangent line $L_{\mathrm{d}} = \{v = u \tan\theta_{\mathrm{d}}\}$ from the origin to graph $\{v = f_{\mathrm{polar}}(u) \,:\, \hat{u}_0 \le u \le u_\infty\}$ of the strictly concave and positive function $f_{\mathrm{polar}} \in C^\infty((\hat{u}_0, u_\infty)) \cap C([\hat{u}_0, u_\infty])$ satisfying $f_{\mathrm{polar}}(\hat{u}_0) = f_{\mathrm{polar}}(u_\infty) = 0$. These properties, combined with the continuity property of parts (i)–(ii) of assertion (g), imply the continuous dependence of u_{d} on u_∞. Indeed, if $u_\infty^{(j)} \to u_\infty^*$, then, from (i)–(ii) of assertion (g) and the concavity of $f_{\mathrm{polar}}^{(j)}$ and f_{polar}^*, it is easy to see that lines $L_d^{(j)}$ converge to the tangent line L_{d}^* from the origin to graph $\{v = f_{\mathrm{polar}}^*(u) \,:\, \hat{u}_0^* \le u \le u_\infty^*\}$ and, for any subsequence of the contact points of $L_{\mathrm{d}}^{(j)}$ with graphs $\{v = f_{\mathrm{polar}}^{(j)}(u) \,:\, \hat{u}_0^{(j)} \le u \le u_\infty^{(j)}\}$, we can extract a further subsequence converging to a contact point of the limiting line and graph. Such a contact point is unique, since $f_{\mathrm{polar}}^*(\cdot)$ is strictly concave and positive in $(\hat{u}_0^*, u_\infty^*)$ with $f_{\mathrm{polar}}^*(\hat{u}_0^*) = f_{\mathrm{polar}}^*(u_\infty^*) = 0$; we denote the unique contact point by $(u_{\mathrm{d}}^*, v_{\mathrm{d}}^*)$. Thus, $u_{\mathrm{d}}^{(j)} \to u_{\mathrm{d}}^*$.

Now we show the continuity of $u_{\mathrm{s}}(u_\infty)$. From (7.3.28) and assertion (b), it suffices to show that β_{s} depends continuously on u_∞, where $\beta_{\mathrm{s}} \in (\beta_{\mathrm{d}}, \cos^{-1}(\frac{c_\infty}{u_\infty}))$ is defined in assertion (e). Recall that $\beta_{\mathrm{s}} \in (0, \cos^{-1}(\frac{1}{u_\infty}))$ is a solution of equation (7.3.38). Considering q_O and c_O as functions of (u_∞, β) (with $\rho_\infty = 1$ fixed) and using (7.3.28) and (7.3.39) (where the derivative is now understood as the partial derivative), we apply the implicit function theorem to obtain the continuous dependence of $u_\infty \mapsto \beta_{\mathrm{s}}$, which yields the continuity of $u_{\mathrm{s}}(u_\infty)$. □

7.3.2 The limit of the detachment angle as the upstream velocity tends to the sonic speed for steady flow

Denote by $\hat{\theta}_{\mathrm{w}}^{\mathrm{d}}(u_\infty)$ the detachment angle for the steady incoming uniform flow (ρ_∞, u_∞).

Lemma 7.3.3. *Fix $\rho_\infty > 0$. Then, for $u_\infty > c_\infty$,*

$$\lim_{u_\infty \to c_\infty+} \hat{\theta}_{\mathrm{w}}^{\mathrm{d}}(u_\infty) = 0.$$

Proof. In this proof, we denote $\hat{u}_0$ defined in Lemma 7.3.2(c) for the steady incoming uniform flow (ρ_∞, u_∞) by $\hat{u}_0^{(u_\infty)}$, while ρ_∞ is fixed through this proof.

1. We first show that

$$\lim_{u_\infty\to c_\infty+} \hat{u}_0^{(u_\infty)} = c_\infty. \tag{7.3.40}$$

Let $u_\infty > c_\infty$. Then $\hat{u}_0^{(u_\infty)} \in (0, u_\infty)$ satisfies (7.3.7), which can be written as

$$\hat{\rho}(u_\infty^2, (\hat{u}_0^{(u_\infty)})^2)\hat{u}_0^{(u_\infty)} = \rho_\infty u_\infty, \tag{7.3.41}$$

where

$$\hat{\rho}(s^2, t^2) = \big(\rho_\infty^{\gamma-1} + \frac{\gamma-1}{2}(s^2 - t^2)\big)^{\frac{1}{\gamma-1}}.$$

Denote

$$\Phi(s,t) := \hat{\rho}(s^2, t^2)t.$$

We calculate:

$$\frac{\partial\Phi}{\partial t}(s,t) = \hat{\rho}^{2-\gamma}(s^2,t^2)\Big(\rho_\infty^{\gamma-1} + \frac{\gamma-1}{2}s^2 - \frac{\gamma+1}{2}t^2\Big).$$

Then, for each $s \geq c_\infty$, $\Phi(s,\cdot)$ is:

(i) defined on $(0, \sqrt{\frac{2}{\gamma-1}\rho_\infty^{\gamma-1} + s^2})$;

(ii) C^∞ smooth on its domain;

(iii) $\frac{\partial\Phi}{\partial t}(s,\cdot) > 0$ on $(0, \sqrt{\frac{2}{\gamma+1}(\rho_\infty^{\gamma-1} + \frac{\gamma-1}{2}s^2)}\,)$, and
$\frac{\partial\Phi}{\partial t}(s,\cdot) < 0$ on $(\sqrt{\frac{2}{\gamma+1}(\rho_\infty^{\gamma-1} + \frac{\gamma-1}{2}s^2)},\ \sqrt{\frac{2}{\gamma-1}\rho_\infty^{\gamma-1} + s^2}\,)$.

We also note that $\rho_\infty = \hat{\rho}(u_\infty^2, u_\infty^2)$. Thus, equation (7.3.41) can be written as

$$\Phi(u_\infty, \hat{u}_0^{(u_\infty)}) = \Phi(u_\infty, u_\infty).$$

If $u_\infty > c_\infty = \rho_\infty^{\frac{\gamma-1}{2}}$, $u_\infty > \sqrt{\frac{2}{\gamma+1}(\rho_\infty^{\gamma-1} + \frac{\gamma-1}{2}u_\infty^2)}$, *i.e.*, $\frac{\partial\Phi}{\partial t}(u_\infty, u_\infty) < 0$. From properties (i)–(iii) above, and since $\Phi(u_\infty, 0) = 0$ and $\Phi(u_\infty, u_\infty) = \rho_\infty u_\infty > 0$, we conclude that, for each $u_\infty > c_\infty$, there exists a unique solution $\hat{u}_0^{(u_\infty)} \in (0, u_\infty)$ of (7.3.41).

Let $u_\infty^{(i)} \to c_\infty$ as $i \to \infty$. Then there exists a subsequence (still denoted) $\hat{u}_0^{(u_\infty^{(i)})}$ converging to some constant $\hat{u}_0^* \in [0, c_\infty]$. Taking the limit in (7.3.41), using the smoothness of $\Phi(\cdot,\cdot)$ in a neighborhood of $\{c_\infty\} \times [0, c_\infty]$, and recalling that $\rho_\infty = \hat{\rho}(u_\infty^2, u_\infty^2)$, we have

$$\Phi(c_\infty, \hat{u}_0^*) = \Phi(c_\infty, c_\infty).$$

Note that $c_\infty = \rho_\infty^{\frac{\gamma-1}{2}} = \sqrt{\frac{2}{\gamma+1}(\rho_\infty^{\gamma-1} + \frac{\gamma-1}{2}c_\infty^2)}$. Then it follows from property (iii) of $\Phi(s,\cdot)$ that $\hat{u}_0^* = c_\infty$. Therefore, (7.3.40) is proved.

2. Let $u_\infty^{(i)} \to c_\infty$ as $i \to \infty$. Assume that there exists $\delta > 0$ such that

$$\hat{\theta}_{\mathrm{w}}^{\mathrm{d}}(u_\infty^{(i)}) \in (\delta, \frac{\pi}{2}) \qquad \text{for all } i. \tag{7.3.42}$$

Then we denote by $(u_{\mathrm{d}}^{(i)}, v_{\mathrm{d}}^{(i)})$ the coordinates of the corresponding detachment point on the shock polar $f_{\mathrm{polar}}^{(u_\infty)}(\cdot)$, *i.e.*, the unique $u_{\mathrm{d}}^{(i)} \in [\hat{u}_0^{(u_\infty^{(i)})}, u_\infty^{(i)}]$ with $u_{\mathrm{d}}^{(i)} \tan(\hat{\theta}_{\mathrm{w}}^{\mathrm{d}}(u_\infty^{(i)})) = f_{\mathrm{polar}}^{(u_\infty)}(u_{\mathrm{d}}^{(i)})$, and by β_i the corresponding β–angle defined in §7.3.1. From this, using $u_\infty^{(i)} \to c_\infty > 0$, (7.3.40), and (7.3.42), we have

$$\tan \beta_i = \frac{u_{\mathrm{d}}^{(i)} \tan(\hat{\theta}_{\mathrm{w}}^{\mathrm{d}}(u_\infty^{(i)}))}{u_\infty^{(i)} - u_{\mathrm{d}}^{(i)}} \to \infty,$$

which implies

$$\beta_i \to \frac{\pi}{2}-, \tag{7.3.43}$$

since $\beta_i \in (0, \frac{\pi}{2})$.

On the other hand, since $(u_\infty^{(i)}, 0)$ and $(u_{\mathrm{d}}^{(i)}, v_{\mathrm{d}}^{(i)})$ satisfy the Rankine-Hugoniot conditions (7.3.20)–(7.3.21) on the straight-line shock:

$$\{(u, v) \,:\, v = u \cot(\beta_{\mathrm{w}}^{\mathrm{d}}(u_\infty^{(i)}))\},$$

where $\beta_{\mathrm{w}}^{\mathrm{d}}$ is angle β from Definition 7.3.1 for the detachment point of the shock polar for the incoming flow (u_∞, ρ_∞) (where ρ_∞ has been fixed at the beginning of the proof). Then the equality of the tangential to the shock components of $(u_\infty^{(i)}, 0)$ and $(u_{\mathrm{d}}^{(i)}, v_{\mathrm{d}}^{(i)})$ implies

$$u_\infty^{(i)} \sin \beta_i = q_O^{(i)} \sin(\beta_i + \hat{\theta}_{\mathrm{w}}^{\mathrm{d}}(u_\infty^{(i)})),$$

where $q_O^{(i)} = \big|(u_{\mathrm{d}}^{(i)}, v_{\mathrm{d}}^{(i)})\big|$. Moreover, from the Rankine-Hugoniot conditions (7.3.20)–(7.3.21), we know that $u_\infty^{(i)} \ge q_O^{(i)}$ for any i. From this and (7.3.43), we have

$$\liminf_{i \to \infty} \sin(\beta_i + \hat{\theta}_{\mathrm{w}}^{\mathrm{d}}(u_\infty^{(i)})) = \liminf_{i \to \infty} \Big(\frac{u_\infty^{(i)}}{q_O^{(i)}} \sin \beta_i\Big) \ge 1,$$

which implies

$$\lim_{i \to \infty} \sin(\beta_i + \hat{\theta}_{\mathrm{w}}^{\mathrm{d}}(u_\infty^{(i)})) = 1.$$

This is impossible, since both $\hat{\theta}_{\mathrm{w}}^{\mathrm{d}}(u_\infty^{(i)}) \in (\delta, \frac{\pi}{2})$ and (7.3.43) hold. Thus, (7.3.42) is impossible. This completes the proof. □

7.4 LOCAL THEORY FOR SHOCK REFLECTION: EXISTENCE OF THE WEAK AND STRONG STATE (2) UP TO THE DETACHMENT ANGLE

Proof of Theorem 7.1.1. We divide the proof into five steps.

1. For any angle $\theta_w \in (0, \frac{\pi}{2})$, the existence of the weak and strong state (2) can be proved by employing the steady shock polar.

We start by fixing $\theta_w \in (0, \frac{\pi}{2})$. This determines the reflection point $P_0 = (\xi_1^0, \xi_1^0 \tan\theta_w)$ that is the intersection point of the incident shock with the wedge boundary $\Gamma_{\text{wedge}} = \{\xi_2 = \xi_1 \tan\theta_w\}$, and the incident shock is line $\{\xi_1 = \xi_1^0\}$ with $\xi_1^0 = \frac{\rho_1 u_1}{\rho_1 - \rho_0} > u_1$; see §6.2.

Note that we can rewrite (7.1.5) with $\rho(\cdot)$ given by (2.2.9) as

$$\begin{aligned}
&\rho_2 D\varphi_2(P_0)\cdot\boldsymbol{\nu} = \rho_1 D\varphi_1(P_0)\cdot\boldsymbol{\nu},\\
&D\varphi_2(P_0)\cdot\boldsymbol{\tau} = D\varphi_1(P_0)\cdot\boldsymbol{\tau},\\
&\frac{\rho_2^{\gamma-1}}{\gamma-1} + \frac{1}{2}|D\varphi_2(P_0)|^2 = \frac{\rho_1^{\gamma-1}}{\gamma-1} + \frac{1}{2}|D\varphi_1(P_0)|^2,
\end{aligned} \tag{7.4.1}$$

where $\boldsymbol{\nu}$, $\boldsymbol{\tau}$, and $\rho_k, k = 1, 2$, are defined in (7.1.5). Note that the condition that $\varphi_2(P_0) = \varphi_1(P_0)$ has been used to write the Bernoulli law (2.2.9), as the last line in (7.4.1).

Also, with possibly changing $\boldsymbol{\tau}$ to $-\boldsymbol{\tau}$, we may assume that

$$\boldsymbol{\tau}\cdot D\varphi_1(P_0) \geq 0.$$

We consider (7.4.1) as an algebraic system for $(\rho_2, D\varphi_2(P_0))$. Note that, if $(\rho_2, D\varphi_2(P_0))$ satisfy (7.4.1), the pseudo-potential φ_2 of the self-similar uniform state uniquely determined by $\varphi_2(P_0) := \varphi_1(P_0)$ and $D\varphi_2(P_0)$, *i.e.*,

$$\varphi_2(\boldsymbol{\xi}) = \varphi_2(P_0) + D\varphi_2(P_0)\cdot(\boldsymbol{\xi} - P_0) - \frac{1}{2}(\boldsymbol{\xi} - P_0)^2$$

satisfies (7.1.5), and its density, calculated by (2.2.9), is $\rho_2 = \rho(|D\varphi_2|^2, \varphi_2)$. The last assertion follows from (7.4.1), since $\rho_1 = \rho(|D\varphi_1|^2, \varphi_1)$ and $\varphi_2(P_0) = \varphi_1(P_0)$. Therefore, we have proved the following claim:

Claim 7.4.1. *Every state* (2) *for a self-similar wedge angle* $\theta_w \in (0, \frac{\pi}{2})$ *is determined by a solution* $(\rho_2, D\varphi_2(P_0))$ *of algebraic system* (7.4.1) *with* $P_0 = (\xi_1^0, \xi_1^0 \tan\theta_w)$.

Therefore, from now on, we study the existence and multiplicity of solutions of the algebraic system (7.4.1) with $P_0 = (\xi_1^0, \xi_1^0 \tan\theta_w)$.

Fix a self-similar wedge angle $\theta_w \in (0, \frac{\pi}{2})$. Denote

$$\boldsymbol{\tau}_w^{(0)} := -(\cos\theta_w, \sin\theta_w). \tag{7.4.2}$$

Then $\boldsymbol{\tau}_w^{(0)}$ is a unit tangent vector to the wedge boundary Γ_{wedge} and

$$\boldsymbol{\tau}_w^{(0)}\cdot D\varphi_1(P_0) > 0 \qquad \text{for all } \theta_w \in (0, \frac{\pi}{2}). \tag{7.4.3}$$

Fact (7.4.3) can be seen as follows: Since $D\varphi_1(P_0) = (u_1 - \xi_1^0, -\xi_1^0 \tan\theta_{\rm w})$ by (2.2.17), and $\xi_1^0 > u_1$ by (2.2.18), we have

$$\begin{aligned}\boldsymbol{\tau}_{\rm w}^{(0)} \cdot D\varphi_1(P_0) &= (\cos\theta_{\rm w}, \sin\theta_{\rm w}) \cdot (\xi_1^0 - u_1, \xi_1^0 \tan\theta_{\rm w}) \\ &= \frac{\xi_1^0}{\cos\theta_{\rm w}} - u_1 \sin\theta_{\rm w} > 0.\end{aligned} \tag{7.4.4}$$

Thus, we can define the coordinate system with the origin at P_0 and basis $\{\mathbf{e}(P_0), \mathbf{e}^{\perp}(P_0)\}$ such that

$$\begin{aligned}&\mathbf{e}(P_0) := \frac{D\varphi_1(P_0)}{|D\varphi_1(P_0)|}, \\ &\mathbf{e}^{\perp}(P_0) \perp \mathbf{e}(P_0), \qquad \mathbf{e}^{\perp}(P_0) \cdot \boldsymbol{\nu}_{\rm w} > 0,\end{aligned} \tag{7.4.5}$$

where $\mathbf{e}^{\perp}(P_0) \cdot \boldsymbol{\nu}_{\rm w} \neq 0$ by (7.4.3). Then we see that conditions (7.4.1) coincide with the Rankine-Hugoniot conditions (7.3.2)–(7.3.3) and the Bernoulli law (7.3.4) for the steady shock with incoming steady flow:

$$(\rho_\infty, u_\infty) = (\rho_1, |D\varphi_1(P_0)|). \tag{7.4.6}$$

Definition 7.4.2. *Fix the self-similar uniform state $\varphi_1(\boldsymbol{\xi})$ with density ρ_1 and sonic speed c_1. For $\boldsymbol{\xi} \in \mathbb{R}^2 \setminus \overline{B_{c_1}(\mathcal{O}_1)}$, we denote by $f_{\rm polar}^{(\boldsymbol{\xi})}(\cdot)$ the steady shock polar considered in Lemma* 7.3.2 *for the steady incoming flow with density and velocity $(\rho_\infty, u_\infty) = (\rho_1, |D\varphi_1(\boldsymbol{\xi})|)$; see Fig.* 7.2.

We use the notations introduced in Lemma 7.3.2 below. One difference is that $\hat{\theta}_{\rm w}$ is denoted as the wedge angle for steady flow, introduced in Lemma 7.3.2(d), which will *not* be the same as the wedge angle $\theta_{\rm w}$ for self-similar flow in our application.

Note that, if state (2) exists for some $\theta_{\rm w} \in (0, \frac{\pi}{2})$, the boundary condition (7.1.7) and property (7.4.3) imply that

$$D\varphi_2(P_0) = |D\varphi_2(P_0)| \boldsymbol{\tau}_{\rm w}^{(0)}.$$

Thus, angle $\hat{\theta}_{\rm w}$ between $D\varphi_1(P_0)$ and $D\varphi_2(P_0)$ is

$$\hat{\theta}_{\rm w} = \hat{\theta}_{\rm w}(\theta_{\rm w}) := \cos^{-1}(\frac{D\varphi_1(P_0) \cdot \boldsymbol{\tau}_{\rm w}^{(0)}}{|D\varphi_1(P_0)|}) \in (0, \frac{\pi}{2}), \tag{7.4.7}$$

where the last inclusion follows from both (7.4.3) and the fact that $D\varphi_1(P_0) = (u_1 - \xi_1^0, -\xi_1^0 \tan\theta_{\rm w})$ is not parallel to $\boldsymbol{\tau}_{\rm w}^{(0)}$, since $u_1 \neq 0$. From Lemma 7.3.2(d), it follows that $(\rho_2, D\varphi_2(P_0))$ satisfying (7.4.1)–(7.4.5) (or its non-existence) is determined by the shock polar $f_{\rm polar}^{(P_0)}(\cdot)$ and the *steady wedge angle* $\hat{\theta}_{\rm w}$ defined above. More precisely, by Lemma 7.3.2(c)–(d) and Definition 7.4.2 of $f_{\rm polar}^{(\boldsymbol{\xi})}(\cdot)$,

$$D\varphi_2(P_0) = u\,\mathbf{e}(P_0) + f_{\rm polar}^{(P_0)}(u)\mathbf{e}^{\perp}(P_0), \tag{7.4.8}$$

where u is determined by the intersection point (u,v) of the shock polar curve:

$$\{(u,v)\ :\ \hat{u}_0^{(|D\varphi_1(P_0)|)} \le u \le |D\varphi_1(P_0)|,\ v = f_{\rm polar}^{(P_0)}(u)\}, \tag{7.4.9}$$

with line $\{(u,v)\ :\ v = u\tan\hat{\theta}_{\rm w}\}$. Note that, for the steady shock polar described above, $\hat{\theta}_{\rm w}$ in (7.4.7) is the angle introduced in Lemma 7.3.2(d) (denoted by $\theta_{\rm w}$ there). Thus, we have

Definition 7.4.3. *Angle $\hat{\theta}_{\rm w} = \hat{\theta}_{\rm w}(\theta_{\rm w})$, defined by (7.4.7) for $(P_0, \boldsymbol{\tau}_{\rm w}^{(0)})(\theta_w)$, is called a steady wedge angle corresponding to the self-similar wedge angle $\theta_{\rm w} \in (0, \frac{\pi}{2})$.*

Furthermore, using Lemma 7.3.2(d), we see that the existence or non-existence of solutions $(\rho_2, D\varphi_2(P_0))$ of (7.4.1), when (7.1.11) holds, is determined as follows:

(a) If $\hat{\theta}_{\rm w} < \hat{\theta}_{\rm w}^{\rm d}(\rho_1, |D\varphi_1(P_0)|)$, there are two solutions;

(b) If $\hat{\theta}_{\rm w} = \hat{\theta}_{\rm w}^{\rm d}(\rho_1, |D\varphi_1(P_0)|)$, there is one solution;

(c) If $\hat{\theta}_{\rm w} > \hat{\theta}_{\rm w}^{\rm d}(\rho_1, |D\varphi_1(P_0)|)$, there is no solution.

Here $\hat{\theta}_{\rm w}^{\rm d}(\rho_\infty, u_\infty)$ denotes the detachment angle for the steady incoming flow (ρ_∞, u_∞) defined in Lemma 7.3.2(d) (denoted by $\theta_{\rm d}$ there).

2. By Proposition 7.2.2, there exists $\sigma_1 > 0$ such that, for any self-similar wedge angle $\theta_{\rm w} \in (\frac{\pi}{2}-\sigma_1, \frac{\pi}{2})$, there exists a solution $(\rho_2, D\varphi_2(P_0))$ of (7.4.1) such that $|D\varphi_2(P_0)| > c_2$, *i.e.*, the steady uniform flow $(\rho_2, D\varphi_2(P_0))$ is supersonic: $|D\varphi_2(P_0)| > c_2$.

We note that, from Lemma 7.3.2(d)–(f), it follows that the unique solution for the steady detachment wedge angle $\hat{\theta}_{\rm w}^{\rm d}$ is $\mathbf{v} = (u_{\rm d}, f_{\rm polar}(u_{\rm d}))$ which is subsonic by (7.3.12). Since, for any self-similar wedge angle $\theta_{\rm w} \in (\frac{\pi}{2} - \sigma_1, \frac{\pi}{2})$, the steady uniform flow $(\rho_2, D\varphi_2({P_0}^{(\theta_{\rm w})}))$ is supersonic as we have shown above, it follows from Lemma 7.3.2(d)–(f) that, in this case, there exist two solutions $(\rho_2^{\rm wk}, D\varphi_2^{\rm wk}(P_0))$ and $(\rho_2^{\rm sg}, D\varphi_2^{\rm sg}(P_0))$ of (7.4.1) with $\rho_2^{\rm wk} < \rho_2^{\rm sg}$, and solution $\varphi_2^{\rm sg}$ is subsonic at P_0, *i.e.*, $|D\varphi_2^{\rm sg}(P_0)| < c_2^{\rm sg} = (\rho_2^{\rm sg})^{\frac{\gamma-1}{2}}$. Note that $(\rho_2, D\varphi_2(P_0))$ obtained in §7.2 is the weak reflection solution $(\rho_2^{\rm wk}, D\varphi_2^{\rm wk}(P_0))$, since it is supersonic.

This implies that, for any self-similar wedge angle $\theta_{\rm w} \in (\frac{\pi}{2} - \sigma_1, \frac{\pi}{2})$, there exist two solutions $\varphi_2^{\rm wk}$ and $\varphi_2^{\rm sg}$ for state (2).

Moreover, since $D\varphi_2^{\rm wk}(P_0) > c_2^{\rm wk}$ for $\theta_{\rm w} \in (\frac{\pi}{2} - \sigma_1, \frac{\pi}{2})$ as discussed above, then, by Lemma 7.3.2(d)–(e), we obtain that, for any $\theta_{\rm w} \in (\frac{\pi}{2} - \sigma_1, \frac{\pi}{2})$,

$$\hat{\theta}_{\rm w}(\theta_{\rm w}) < \hat{\theta}_{\rm w}^{\rm s}(\rho_1, |D\varphi_1({P_0}^{(\theta_{\rm w})})|) < \hat{\theta}_{\rm w}^{\rm d}(\rho_1, |D\varphi_1({P_0}^{(\theta_{\rm w})})|), \tag{7.4.10}$$

where $\hat{\theta}_{\rm w}^{\rm s}(\rho_\infty, u_\infty)$ denotes the sonic angle for the steady incoming flow (ρ_∞, u_∞) defined in Lemma 7.3.2(e).

3. We next prove the existence of the self-similar detachment angle $\theta_{\rm w}^{\rm d} \in (0, \frac{\pi}{2})$ and assertions (i)–(iv) and (vi) of Theorem 7.1.1.

We first note that the incident shock $\mathcal{S}_0 = \{\xi_1 = \xi_1^0\}$ is the shock between states (0) and (1) with $\rho_0 < \rho_1$. Denoting by $\boldsymbol{\nu}_{\mathcal{S}_0} := (-1, 0)$ the normal to $\mathcal{S}_0$, we have

$$D\varphi_0(\boldsymbol{\xi}) \cdot \boldsymbol{\nu}_{\mathcal{S}_0} = \xi_1^0 > \xi_1^0 - u_1 = D\varphi_1(\boldsymbol{\xi}) \cdot \boldsymbol{\nu}_{\mathcal{S}_0} \qquad \text{for all } \boldsymbol{\xi} \in \mathcal{S}_0.$$

Applying Lemma 6.1.2 with $\varphi^- = \varphi_0$ and $\varphi^+ = \varphi_1$, we obtain from (6.1.4) that $B_{c_1}(\mathcal{O}_1) \cap \mathcal{S}_0 \neq \emptyset$, that is,

$$c_1 > \xi_1^0 - u_1.$$

Thus, denoting

$$\theta_{\rm w}^+ := \tan^{-1}\Big(\frac{\sqrt{c_1^2 - (\xi_1^0 - u_1)^2}}{\xi_1^0}\Big) > 0,$$

we obtain that the steady uniform flow $(\rho_1, |D\varphi_1({P_0}^{(\theta_{\rm w}^+)})|)$ is sonic, and

$$|D\varphi_1({P_0}^{(\theta_{\rm w})})| > c_1^{(\theta_{\rm w})} \qquad \text{for all } \theta_{\rm w} \in (\theta_{\rm w}^+, \frac{\pi}{2}).$$

On the other hand,

$$\hat{\theta}_{\rm w}^+ := \hat{\theta}_{\rm w}(\theta_{\rm w}^+) = \angle \mathcal{O}_1 {P_0}^{(\theta_{\rm w}^+)} P_3 > 0.$$

Using Lemma 7.3.3 and the continuous dependence of $\hat{\theta}_{\rm w}$ on $\theta_{\rm w}$, we conclude that there exists $\delta > 0$ such that

$$\hat{\theta}_{\rm w}(\theta_{\rm w}) > \hat{\theta}_{\rm w}^{\rm d}(\rho_1, |D\varphi_1({P_0}^{(\theta_{\rm w})})|) \qquad \text{for all } \theta_{\rm w} \in (\theta_{\rm w}^+, \theta_{\rm w}^+ + \delta).$$

Combining with (7.4.10), we obtain the existence of $\theta_{\rm w}^{(\rm d_1)} \in (\theta_{\rm w}^+ + \delta, \frac{\pi}{2} - \sigma_1)$ such that

$$\hat{\theta}_{\rm w}(\theta_{\rm w}^{(\rm d_1)}) = \hat{\theta}_{\rm w}^{\rm d}(\rho_1, |D\varphi_1({P_0}^{(\theta_{\rm w}^{(\rm d_1)})})|). \tag{7.4.11}$$

Defining $\theta_{\rm w}^{\rm d}$ to be the supremum of all such angles $\theta_{\rm w}^{(\rm d_1)}$, we find that $\theta_{\rm w}^{\rm d} \in [\theta_{\rm w}^+ + \delta, \frac{\pi}{2} - \sigma_1]$. Then it follows from Lemma 7.3.2(d) that there exist two states for state (2) for each $\theta_{\rm w} \in (\theta_{\rm w}^{\rm d}, \frac{\pi}{2})$, and there exists a unique state (2) when $\theta_{\rm w} = \theta_{\rm w}^{\rm d}$.

We now show the continuous dependence of the weak and strong states (2) on $\theta_{\rm w}$, stated in (7.1.10) and Theorem 7.1.1(i).

We first note the C^∞–dependence of $\hat{\theta}_{\rm w}$ on $\theta_{\rm w} \in [\theta_{\rm w}^{\rm d}, \frac{\pi}{2})$. Indeed, it follows from the explicit formula of $\hat{\theta}_{\rm w}(\theta_{\rm w})$ in (7.4.7), using that $D\varphi_1(P_0) = -(\xi_1^0 - u_1, \xi_1^0 \tan\theta_{\rm w})$ by (2.2.17), and $|D\varphi_1(P_0)| \geq \xi_1^0 - u_1 > 0$ by (7.4.4).

Now the continuous dependence of the weak and strong states for (2) on $\theta_{\rm w}$ can be seen by letting $\theta_{\rm w}^{(i)} \in (\theta_{\rm w}^{\rm d}, \frac{\pi}{2})$ with $\theta_{\rm w}^{(i)} \to \theta_{\rm w}^{(\infty)} \in [\theta_{\rm w}^{\rm d}, \frac{\pi}{2})$. Recall that, for each $\theta_{\rm w} \in (\theta_{\rm w}^{\rm d}, \frac{\pi}{2})$, the strictly convex shock polar curve, defined by (7.4.9) and denoted by $\Gamma(\theta_{\rm w})$, intersects with line $L(\theta_{\rm w}) = \{v = u \tan(\hat{\theta}_{\rm w}(\theta_{\rm w}))\}$ at exactly two points, with coordinates $(u_{\mathcal{O}}^{\rm wk}, v_{\mathcal{O}}^{\rm wk})$ and $(u_{\mathcal{O}}^{\rm sg}, v_{\mathcal{O}}^{\rm sg})$. Moreover, $D\varphi_2^{\rm wk}(P_0)$ and $D\varphi_2^{\rm sg}(P_0)$, written in coordinates (7.4.8), are $D\varphi_2^{\rm wk}(P_0) = (u_{\mathcal{O}}^{\rm wk}, v_{\mathcal{O}}^{\rm wk})$ and $D\varphi_2^{\rm sg}(P_0) = (u_{\mathcal{O}}^{\rm sg}, v_{\mathcal{O}}^{\rm sg})$, *i.e.*,

$$D\varphi_2^{\rm wk}(P_0) = u_{\mathcal{O}}^{\rm wk}\mathbf{e}(P_0) + v_{\mathcal{O}}^{\rm wk}\mathbf{e}^{\perp}(P_0), \;\; D\varphi_2^{\rm sg}(P_0) = u_{\mathcal{O}}^{\rm sg}\mathbf{e}(P_0) + v_{\mathcal{O}}^{\rm sg}\mathbf{e}^{\perp}(P_0). \tag{7.4.12}$$

If $\theta_{\rm w} = \theta_{\rm w}^{\rm d}$, $L(\theta_{\rm w})$ touches $\Gamma(\theta_{\rm w})$ at exactly one point, with the coordinates given by the components of $D\varphi^{\rm wk}(P_0) = D\varphi^{\rm sg}(P_0)$ in coordinates (7.4.8). From the continuity of $\hat{\theta}_{\rm w}(\cdot)$, it follows that lines $L(\theta_{\rm w}^{(i)})$ converge to $L(\theta_{\rm w}^{(\infty)})$. Then the continuity stated in assertions (ii)–(iii) of Lemma 7.3.2(g) implies that

$$(u_{\mathcal{O}}^{\rm wk}, v_{\mathcal{O}}^{\rm wk})(\hat{\theta}_{\rm w}^{(i)}) \to (u_{\mathcal{O}}^{\rm wk}, v_{\mathcal{O}}^{\rm wk})(\hat{\theta}_{\rm w}^{(\infty)}), \quad (u_{\mathcal{O}}^{\rm sg}, v_{\mathcal{O}}^{\rm sg})(\hat{\theta}_{\rm w}^{(i)}) \to (u_{\mathcal{O}}^{\rm sg}, v_{\mathcal{O}}^{\rm sg})(\hat{\theta}_{\rm w}^{(\infty)}),$$

where

$$(u_{\mathcal{O}}^{\rm wk}, v_{\mathcal{O}}^{\rm wk})(\hat{\theta}_{\rm w}^{(\infty)}) \neq (u_{\mathcal{O}}^{\rm sg}, v_{\mathcal{O}}^{\rm sg})(\hat{\theta}_{\rm w}^{(\infty)}) \qquad \text{if } \theta_{\rm w}^{\infty} \in (\theta_{\rm w}^{\rm d}, \tfrac{\pi}{2}),$$

and

$$(u_{\mathcal{O}}^{\rm wk}, v_{\mathcal{O}}^{\rm wk})(\hat{\theta}_{\rm w}^{(\infty)}) = (u_{\mathcal{O}}^{\rm sg}, v_{\mathcal{O}}^{\rm sg})(\hat{\theta}_{\rm w}^{(\infty)}) \qquad \text{if } \theta_{\rm w}^{\infty} = \theta_{\rm w}^{\rm d}.$$

Then, using (7.4.12) where vectors $(\mathbf{e}({P_0}^{(\theta_{\rm w})}), \mathbf{e}^{\perp}({P_0}^{(\theta_{\rm w})}))$ depend C^∞-smoothly on $\theta_{\rm w}$, and using that $D\phi_2 = D\varphi_2(P_0) + P_0 = (u_{\mathcal{O}}, v_{\mathcal{O}}) + (\xi_1^0, \xi_1^0\tan\theta_{\rm w})$ for any $\theta_{\rm w}$ and for both the weak and strong states (2), we obtain the continuous dependence of $(\rho_2^{\rm wk}, \rho_2^{\rm sg}, D\phi_2^{\rm wk}, D\phi_2^{\rm sg})$ on $\theta_{\rm w} \in [\theta_{\rm w}^{\rm d}, \frac{\pi}{2})$. We have also shown (7.1.10).

To show the C^∞–dependence on $\theta_{\rm w} \in (\theta_{\rm w}^{\rm d}, \frac{\pi}{2})$, we consider the function:

$$G(u, \theta_{\rm w}) := u\tan(\hat{\theta}_{\rm w}(\theta_{\rm w})) - f_{\rm polar}^{({P_0}^{(\theta_{\rm w})})}(u),$$

where $f_{\rm polar}^{(\boldsymbol{\xi})}(\cdot)$ is from Definition 7.4.2. Then, for fixed $\theta_{\rm w} \in (\theta_{\rm w}^{\rm d}, \frac{\pi}{2})$,

$$G(u, \theta_{\rm w}) = 0 \qquad \text{only for } u = u_{\mathcal{O}}^{\rm sg}(\theta_{\rm w}) \;\text{ and }\; u = u_{\mathcal{O}}^{\rm wk}(\theta_{\rm w}).$$

From the smoothness of $\hat{\theta}_{\rm w}(\cdot)$ shown above and assertions (i)–(ii) of Lemma 7.3.2(g), we have

$$G \in C^\infty(\{(u, \theta_{\rm w}) \,:\, \theta_{\rm w}^{\rm d} \le \theta_{\rm w} < \frac{\pi}{2}, \hat{u}_0(\rho_1, |D\varphi_1({P_0}^{(\theta_{\rm w})})|) < u < |D\varphi_1({P_0}^{(\theta_{\rm w})})|\}).$$

Since $f_{\rm polar}^{({P_0}^{(\theta_{\rm w})})}(\cdot)$ is strictly concave in the interior of its domain, we have

$$\partial_u G(u_{\mathcal{O}}^{\rm wk}(\theta_{\rm w}), \theta_{\rm w}) \neq 0, \;\; \partial_u G(u_{\mathcal{O}}^{\rm sg}(\theta_{\rm w}), \theta_{\rm w}) \neq 0 \qquad \text{for all } \theta_{\rm w} \in (\theta_{\rm w}^{\rm d}, \frac{\pi}{2}),$$

since $L(\theta_{\rm w}) = \{v = u\tan(\hat{\theta}_{\rm w}(\theta_{\rm w}))\}$ intersects with the graph of $f_{\rm polar}^{({P_0}^{(\theta_{\rm w})})}(\cdot)$ at two points, and hence intersects transversally at both points. Now, from the implicit function theorem, we obtain the C^∞–dependence of $(u_{\mathcal{O}}^{\rm wk}, u_{\mathcal{O}}^{\rm sg})$ on $\theta_{\rm w} \in (\theta_{\rm w}^{\rm d}, \frac{\pi}{2})$. Since $v_{\mathcal{O}} = f_{\rm polar}^{(P_0)}(u_{\mathcal{O}})$ for both the weak and strong reflections, with

$P_0 = (\xi_1^0, \xi_1^0 \tan\theta_{\rm w})$, we use Definition 7.4.2 and assertion (ii) of Lemma 7.3.2(g) to obtain the C^∞–dependence of $(u_{\mathcal{O}}^{\rm wk}, v_{\mathcal{O}}^{\rm wk}, u_{\mathcal{O}}^{\rm sg}, v_{\mathcal{O}}^{\rm sg})$ on $\theta_{\rm w} \in (\theta_{\rm w}^{\rm d}, \frac{\pi}{2})$. Finally, using (7.4.12) with the argument as in the proof of continuity above, we conclude the proof of the C^∞–dependence of $(\rho_2^{\rm wk}, \rho_2^{\rm sg}, D\phi_2^{\rm wk}, D\phi_2^{\rm sg})$ on $\theta_{\rm w} \in (\theta_{\rm w}^{\rm d}, \frac{\pi}{2})$. This completes the proof of assertion (i) of Theorem 7.1.1.

Now we prove assertion (ii) of Theorem 7.1.1. First, we note that assertion (iv) holds, since the existence of two states for (2) for a self-similar wedge angle $\theta_{\rm w}$, and equivalently for the steady wedge angle $\hat{\theta}_{\rm w}(\theta_{\rm w})$, implies (7.1.12) as follows: By (b) and (d) of Lemma 7.3.2, we conclude that $u_{\mathcal{O}}^{\rm sg} \in [\hat{u}_0, u_{\rm d})$, and then (7.3.12)–(7.3.14) in Lemma 7.3.2(f) imply (7.1.12). Noting that state (2) constructed in Proposition 7.2.2 is supersonic at P_0, we conclude that it is a weak state (2). Now Proposition 7.2.2 implies the C^∞–dependence of $(\rho_2^{\rm wk}, D\phi_2^{\rm wk})$ on $\theta_{\rm w} \in (\frac{\pi}{2} - \sigma_1, \frac{\pi}{2}]$. Combining this with the C^∞–dependence on $\theta_{\rm w} \in (\theta_{\rm w}^{\rm d}, \frac{\pi}{2})$ proved above, we conclude the C^∞–dependence of $(\rho_2^{\rm wk}, D\phi_2^{\rm wk})$ on $\theta_{\rm w} \in (\theta_{\rm w}^{\rm d}, \frac{\pi}{2}]$.

Therefore, assertions (i)–(ii) and (iv) of Theorem 7.1.1 hold. Also, assertion (iii) of Theorem 7.1.1 follows from the fact that $\theta_{\rm w}^{\rm d} > \theta_{\rm w}^{+}$.

Furthermore, from the continuity of $\hat{\theta}_{\rm w}(\cdot)$ and $P_0(\cdot)$ on $\theta_{\rm w} \in [\theta_{\rm w}^{\rm d}, \frac{\pi}{2})$ and assertion (iii) of Lemma 7.3.2(g), there exists $\delta > 0$ such that, for any $\theta_{\rm w} \in [\theta_{\rm w}^{\rm d}, \frac{\pi}{2} - \sigma_1]$,

$$\hat{\theta}_{\rm w}^{\rm d}(\rho_1, |D\varphi_1(P_0{}^{(\theta_{\rm w})})|) - \hat{\theta}_{\rm w}^{\rm s}(\rho_1, |D\varphi_1(P_0{}^{(\theta_{\rm w})})|) \geq \delta > 0. \tag{7.4.13}$$

This, combined with (7.4.11), Lemma 7.3.2(f), and the continuity on $\theta_{\rm w} \in [\theta_{\rm w}^{\rm d}, \frac{\pi}{2})$ in assertion (i) of Theorem 7.1.1, implies assertion (vi) of Theorem 7.1.1.

Therefore, we have shown assertions (i)–(iv) and (vi) of Theorem 7.1.1.

4. Next, we show the existence of the self-similar sonic angle, *i.e.*, assertion (v) of Theorem 7.1.1. Denote by $\mathcal{S}$ the set of all angles $\theta_{\rm w}^{\rm s_1} \in [\theta_{\rm w}^{\rm d}, \frac{\pi}{2})$ such that $|D\varphi_2^{\rm wk}(P_0)| > c_2$ for any $\theta_{\rm w} \in (\theta_{\rm w}^{\rm s_1}, \frac{\pi}{2})$. From Proposition 7.2.2 and Theorem 7.1.1(iv), we conclude that $(\frac{\pi}{2} - \sigma_1, \frac{\pi}{2}) \subset \mathcal{S}$. Thus, $\mathcal{S} \neq \emptyset$. Let $\theta_{\rm w}^{\rm s} := \inf \mathcal{S}$. We show that $\theta_{\rm w}^{\rm s}$ satisfies all the properties given in assertion (v) of Theorem 7.1.1.

First, we note that $\theta_{\rm w}^{\rm s} > \theta_{\rm w}^{\rm d}$, since

- (7.4.11) holds for $\theta_{\rm w}^{\rm d}$;
- (7.4.10) holds for any $\theta_{\rm w} \in (\theta_{\rm w}^{\rm s}, \frac{\pi}{2})$;
- (7.4.13) holds for any $\theta_{\rm w} \in [\theta_{\rm w}^{\rm d}, \frac{\pi}{2} - \sigma_1]$.

Now, using assertion (i) of Theorem 7.1.1, we find that $|D\varphi_2^{\rm wk}(P_0)| = c_2^{\rm wk}$ for $\theta_{\rm w} = \theta_{\rm w}^{\rm s}$.

Now $\theta_{\rm w}^{\rm s}$ satisfies all the properties asserted in Theorem 7.1.1(v).

5. We show assertions (vii)–(viii) of Theorem 7.1.1. Let φ_2 be the pseudo-potential of either weak or strong state (2). Since $\rho_1 = \rho(|D\varphi_1|^2, \varphi_1)$ for $\rho(\cdot)$ given by (2.2.9) and $\varphi_1(P_0) = \varphi_2(P_0)$ by (7.1.5), then, for function (7.1.9), we have

$$g^{\rm sh}(\mathbf{p}, \varphi_2(P_0), P_0) = \big(\rho(|\mathbf{p}|^2, P_0)\mathbf{p} - \rho_1 D\varphi_1(P_0)\big) \cdot \frac{D\varphi_1(P_0) - \mathbf{p}}{|D\varphi_1(P_0) - \mathbf{p}|}, \tag{7.4.14}$$

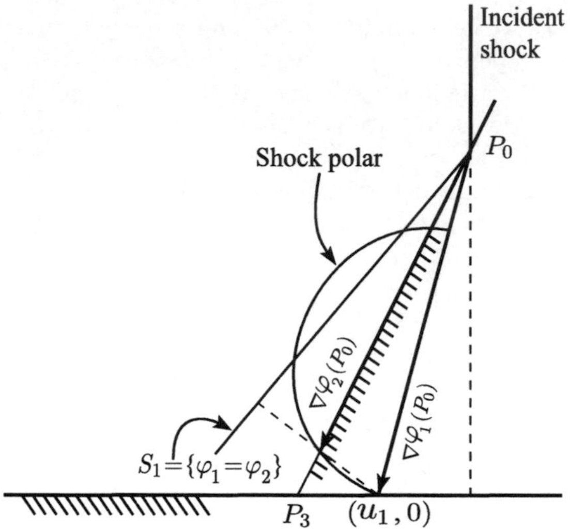

Figure 7.2: $D\varphi_2(P_0)$ lies on the shock polar of the steady incoming flow $(\rho_\infty, u_\infty) = (\rho_1, |D\varphi(P_0)|)$

where $\rho(|\mathbf{p}|^2, P_0) = \big(\rho_1^{\gamma-1} + \frac{\gamma-1}{2}(|D\varphi_1(P_0)|^2 - |\mathbf{p}|^2)\big)^{\frac{1}{\gamma-1}}$. Thus, function $\mathbf{p} \mapsto g^{\rm sh}(\mathbf{p}, \varphi_2(P_0), P_0)$ coincides with function (7.3.16) for the steady incoming flow $(\rho_\infty, u_\infty) = (\rho_1, |D\varphi_1(P_0)|)$. Next, using that $D\varphi_1(P_0) \cdot D\varphi_2(P_0) > 0$ by the Rankine-Hugoniot conditions (7.4.1) at P_0, by rotating the coordinates $\boldsymbol{\xi}$ so that the new ξ_1–axis is in the direction of $D\varphi_1(P_0)$, and reflecting ξ_2 if necessary, we conclude that $D\varphi_1(P_0) = (|D\varphi_1(P_0)|, 0)$, and $\mathbf{v} := D\varphi_2(P_0)$ has positive components in the new coordinates. Then, using that $D\varphi_1(P_0)\cdot\boldsymbol{\tau} = D\varphi_2(P_0)\cdot\boldsymbol{\tau}$ by (7.1.5), we find that $\mathbf{v} := D\varphi_2(P_0)$ (expressed in the new coordinates) lies on the shock polar of the steady incoming flow $(\rho_\infty, u_\infty) = (\rho_1, |D\varphi_1(P_0)|)$; see Fig. 7.2.

Moreover, for any fixed self-similar wedge angle $\theta_{\rm w} \in (\theta_{\rm w}^{\rm d}, \frac{\pi}{2})$ and the corresponding weak and strong reflection solutions $\varphi_2^{\rm wk}$ and $\varphi_2^{\rm sg}$, velocities $D\varphi_2^{\rm w}(P_0)$ and $D\varphi_2^{\rm s}(P_0)$, which lie on the shock polar described in the previous paragraph, correspond to the weak and strong steady reflection solutions for the steady wedge angle $\hat\theta_{\rm w}(\theta_{\rm w})$ introduced in Definition 7.4.3. Indeed, the corresponding steady flows have both the same steady wedge angle $\hat\theta_{\rm w}(\theta_{\rm w})$ with densities $\rho_2^{\rm wk}$ and $\rho_2^{\rm sg}$, respectively. Since $\rho_2^{\rm wk} < \rho_2^{\rm sg}$, the assertion follows. Therefore, we have proved the following lemma:

Lemma 7.4.4. *For any $\theta_{\rm w} \in (\theta_{\rm w}^{\rm d}, \frac{\pi}{2})$ and the corresponding $P_0 = P_0(\theta_{\rm w})$, denoting by φ_2 the pseudo-potential of either weak or strong state* (2), *we have*

(i) *Function $\mathbf{p} \mapsto g^{\rm sh}(\mathbf{p}, \varphi_2(P_0), P_0)$ coincides with function* (7.3.16) *for the steady incoming flow $(\rho_\infty, u_\infty) = (\rho_1, |D\varphi_1(P_0)|)$;*

(ii) *Change the coordinates in* $\mathbb{R}^2$ *to the ones determined on the basis of* (7.4.5), *so that* $D\varphi_1(P_0) = (|D\varphi_1(P_0)|, 0)$, *and* $D\varphi_2(P_0)$ *has positive components. Then* $\mathbf{v} := D\varphi_2(P_0)$ *(expressed in the new coordinates) lies on the shock polar of the steady incoming flow* $(\rho_\infty, u_\infty) = (\rho_1, |D\varphi_1(P_0)|)$; *see Fig.* 7.2.

(iii) *For weak and strong self-similar reflection solutions* $\varphi_2^{\rm wk}$ *and* $\varphi_2^{\rm sg}$, *velocities* $D\varphi_2^{\rm wk}(P_0)$ *and* $D\varphi_2^{\rm sg}(P_0)$ *correspond to the weak and strong steady reflection solutions respectively on the steady shock polar described in* (ii) *above with the steady wedge angle* $\hat{\theta}_{\rm w}(\theta_{\rm w})$ *introduced in Definition* 7.4.3.

Now assertion (vii) of Theorem 7.1.1 directly follows from Lemma 7.4.4(ii)–(iii) and (7.3.18) in Lemma 7.3.2(h).

Moreover, it is easy to check that, if two self-similar uniform states with pseudo-potentials φ_1 and φ_2 satisfy (7.1.5)–(7.1.6) at some point $P_0 \in \mathcal{S}_1 := \{\varphi_1 = \varphi_2\}$, then (7.1.5)–(7.1.6) with P_0 replaced by $\boldsymbol{\xi}$ hold at any $\boldsymbol{\xi} \in \mathcal{S}_1$. Repeating the argument that leads to (7.4.14), we see that, for any $\boldsymbol{\xi} \in \mathcal{S}_1$,

$$g^{\rm sh}(\mathbf{p}, \varphi_2(\boldsymbol{\xi}), \boldsymbol{\xi}) = \big(\rho(|\mathbf{p}|^2, \boldsymbol{\xi})\mathbf{p} - \rho_1 D\varphi_1(\boldsymbol{\xi})\big) \cdot \frac{D\varphi_1(\boldsymbol{\xi}) - \mathbf{p}}{|D\varphi_1(\boldsymbol{\xi}) - \mathbf{p}|}, \tag{7.4.15}$$

where

$$\rho(|\mathbf{p}|^2, \boldsymbol{\xi}) = \Big(\rho_1^{\gamma-1} + \frac{\gamma-1}{2}(|D\varphi_1(\boldsymbol{\xi})|^2 - |\mathbf{p}|^2)\Big)^{\frac{1}{\gamma-1}}. \tag{7.4.16}$$

Then, arguing as in the proof of Lemma 7.4.4, we have

Lemma 7.4.5. *Let* φ_1 *and* φ_2 *be the pseudo-potentials of two self-similar uniform states, and let* (7.1.5)–(7.1.6) *hold at some point* $P_0 \in \mathcal{S}_1 := \{\varphi_1 = \varphi_2\}$. *Then, for every* $\boldsymbol{\xi} \in \mathcal{S}_1$, *we have*

(i) (7.1.5)–(7.1.6) *hold with* P_0 *replaced by* $\boldsymbol{\xi} \in \mathcal{S}_1$;

(ii) *Function* $\mathbf{p} \mapsto g^{\rm sh}(\mathbf{p}, \varphi_2(\boldsymbol{\xi}), \boldsymbol{\xi})$ *coincides with function* (7.3.16) *for the steady incoming flow* $(\rho_\infty, u_\infty) = (\rho_1, |D\varphi_1(\boldsymbol{\xi})|)$;

(iii) *There exists an orthogonal transformation of* $\mathbb{R}^2$ *such that, in the new coordinates (still denoted* $\boldsymbol{\xi}$), $D\varphi_1(\boldsymbol{\xi}) = (|D\varphi_1(\boldsymbol{\xi})|, 0)$, *and* $D\varphi_2(\boldsymbol{\xi})$ *has positive components. Then* $\mathbf{v} := D\varphi_2(\boldsymbol{\xi})$ *(expressed in the new coordinates) lies on the shock polar of the steady incoming flow* $(\rho_\infty, u_\infty) = (\rho_1, |D\varphi_1(\boldsymbol{\xi})|)$.

Now assertion (viii) of Theorem 7.1.1 directly follows from (7.3.19) in Lemma 7.3.2(h).

Theorem 7.1.1 is proved. □

Next we show some corollaries from Theorem 7.1.1 and their proofs.

Corollary 7.4.6. *Let* φ_1 *be a uniform state. Let* $\Omega \subset \mathbb{R}^2$ *be open, and let* Ω *be subdivided by a* C^1*-smooth curve* $\mathcal{S}$ *into two open subdomains* Ω_1 *and* Ω_2. *Let* $\varphi \in C^{0,1}(\Omega)$ *be a weak solution of* (2.2.8) *in* Ω *such that* $\varphi \in C^2(\Omega_k) \cap C^1(\Omega_k \cup \mathcal{S})$

for $k = 1, 2$. *Assume that* $\mathcal{S} \subset \mathbb{R}^2 \setminus \overline{B_{c_1}(\mathcal{O}_1)}$ *and that* $\varphi = \varphi_1$ *on* Ω_1. *Assume that the following entropy condition holds on* $\mathcal{S}$:

$$\partial_{\boldsymbol{\nu}}\varphi_1 > \partial_{\boldsymbol{\nu}}\varphi > 0 \qquad \text{on } \mathcal{S},$$

where $\boldsymbol{\nu}$ *is the unit normal to* $\mathcal{S}$ *pointing into* Ω_2. *Let* $\boldsymbol{\xi} \in \mathcal{S}$. *Denote*

$$\begin{aligned} &\mathbf{e}(\boldsymbol{\xi}) := \frac{D\varphi_1(\boldsymbol{\xi})}{|D\varphi_1(\boldsymbol{\xi})|}, \\ &\mathbf{e}^{\perp}(\boldsymbol{\xi}) \text{ as the rotation of } \mathbf{e}(\boldsymbol{\xi}) \text{ by } \frac{\pi}{2} \text{ such that } D\varphi_2(\boldsymbol{\xi}) \cdot \mathbf{e}^{\perp}(\boldsymbol{\xi}) \geq 0. \end{aligned} \tag{7.4.17}$$

Then, for $u = D\varphi_2(\boldsymbol{\xi}) \cdot \mathbf{e}(\boldsymbol{\xi})$ *and* $v = D\varphi_2(\boldsymbol{\xi}) \cdot \mathbf{e}^{\perp}(\boldsymbol{\xi})$,

$$v = f^{(\boldsymbol{\xi})}_{\rm polar}(u), \tag{7.4.18}$$

where $f^{(\boldsymbol{\xi})}_{\rm polar}(\cdot)$ *is introduced in Definition* 7.4.2.

Proof. The argument is the same as that in the proof of (7.4.8) in Step 1 of the proof of Theorem 7.1.1 for point $\boldsymbol{\xi}$ (instead of P_0). □

Corollary 7.4.7. *Let* φ_1 *and* φ_2 *be pseudo-potentials of two self-similar uniform states with densities* ρ_1 *and* ρ_2, *respectively, and let* (7.1.5)–(7.1.6) *hold at some point* $P_0 \in \mathcal{S}_1 := \{\varphi_1 = \varphi_2\}$. *Let* $g^{\rm sh}$ *be defined by* (7.1.9). *Let* $\boldsymbol{\xi} \in \mathcal{S}_1$ *satisfy* $|D\varphi_2(\boldsymbol{\xi})| \geq c_2$, *where* c_2 *is the sonic speed of the uniform state* φ_2. *Then*

$$D_{\mathbf{p}} g^{\rm sh}(D\varphi_2(\boldsymbol{\xi}), \varphi_2(\boldsymbol{\xi}), \boldsymbol{\xi}) \cdot D\varphi_2(\boldsymbol{\xi}) < 0. \tag{7.4.19}$$

Proof. By Lemma 7.4.5, $\mathbf{v} = D\varphi_2(\boldsymbol{\xi})$, expressed in the appropriately chosen coordinates, lies on the shock polar of the steady incoming flow $(\rho_\infty, u_\infty) = (\rho_1, |D\varphi_1(\boldsymbol{\xi})|)$. Comparing (7.4.16) with (7.3.11), we see that the condition that $|D\varphi_2(\boldsymbol{\xi})| \geq c_2$ implies that the steady flow with velocity $\mathbf{v} = D\varphi_2(\boldsymbol{\xi})$ (expressed in the coordinates described in Lemma 7.4.5(iii)) is supersonic, *i.e.*, $|\mathbf{v}| \geq c(|\mathbf{v}|^2)$, where $c(\cdot)$ is given by (7.3.11) with $(\rho_\infty, u_\infty) = (\rho_1, |D\varphi_1(\boldsymbol{\xi})|)$. Now Lemma 7.3.2(f) implies that $\mathbf{v}$ is a weak reflection solution for the steady shock polar of the incoming flow $(\rho_\infty, u_\infty) = (\rho_1, |D\varphi_1(\boldsymbol{\xi})|)$.

Also, $D\varphi_1(\boldsymbol{\xi}) \neq D\varphi_2(\boldsymbol{\xi})$ by (7.1.6). Now (7.4.19) follows from Lemma 7.4.5(ii) and (7.3.18) in Lemma 7.3.2(h). □

We also note the following property of state (2):

Lemma 7.4.8. *Let* $\theta_{\rm w} \in (\theta^{\rm d}_{\rm w}, \frac{\pi}{2})$, *and let* φ_2 *be a state* (2) (*weak or strong*) *written in the form of* (2.4.1). *Then center* $\mathcal{O}_2 = (u_2, u_2 \tan\theta_{\rm w})$ *of the sonic circle of* φ_2 *lies within the relative interior of segment* $P_0P_3 \subset \{\xi_2 = \xi_1 \tan\theta_{\rm w}\}$, *where* $P_3 = \mathbf{0}$.

Proof. Recall that $P_0 = (\xi_1^0, \xi_1^0 \tan\theta_w)$.

Points P_0, P_3, and $\mathcal{O}_2$ lie on line $\{\xi_2 = \xi_1 \tan\theta_w\}$, which implies

$$\mathcal{O}_2 - P_0 = t(P_3 - P_0) \qquad \text{for } t \in \mathbb{R}.$$

Thus, it remains to show that $t \in (0,1)$.

Denoting by $\mathcal{O}_1 = (u_1, 0)$ the center of state (1) and using (7.1.5) and the fact that $D\varphi_k(P_0) = \mathcal{O}_k - P_0$ for $k = 1, 2$, we can write (7.1.6) as

$$|P_0\mathcal{O}_1| > |P_0\mathcal{O}_2|.$$

Also, by (2.2.18),

$$\xi_1^0 > u_1 > 0,$$

and hence $|P_0P_3| = \xi_1^0 \sec\theta_w > \sqrt{(\xi_1^0)^2 \tan^2\theta_w + (\xi_1^0 - u_1)^2} = |P_0\mathcal{O}_1|$, which implies

$$|P_0P_3| > |P_0\mathcal{O}_2|.$$

Furthermore, $P_0 \in \mathcal{S}_1$, and hence $\mathcal{O}_2 \neq P_0 = (\xi_1^0, \xi_1^0 \tan\theta_w)$, on account of (6.1.5) applied with $\varphi^- = \varphi_1$, $\varphi^+ = \varphi_2$, and $\mathcal{S} = \mathcal{S}_1$. Then we have

$$|P_0P_3| > |P_0\mathcal{O}_2| \neq 0,$$

which implies that $|t| \in (0,1)$.

It remains to show that $t > 0$. We first note that $D\varphi_1(P_0) \cdot D\varphi_2(P_0) > 0$ by (7.1.5) so that

$$(\mathcal{O}_1 - P_0) \cdot (\mathcal{O}_2 - P_0) > 0.$$

Again using that $\xi_1^0 > u_1 > 0$ by (2.2.18), we calculate

$$(\mathcal{O}_1 - P_0) \cdot (P_3 - P_0) = \xi_1^0(\xi_1^0 - u_1) + (\xi_1^0)^2 \tan^2\theta_w > 0.$$

The last two inequalities imply $t > 0$, which concludes the proof. □

From Lemma 7.4.8, we conclude

Corollary 7.4.9. *For any* $\theta_w \in (\theta_w^d, \frac{\pi}{2})$, *writing state* (2) *in the form of* (2.4.1), *we have*

$$u_2 > 0, \quad v_2 > 0.$$

7.5 BASIC PROPERTIES OF THE WEAK STATE (2) AND THE DEFINITION OF SUPERSONIC AND SUBSONIC WEDGE ANGLES

In this section, for any wedge angle $\theta_w \in (\theta_w^d, \frac{\pi}{2})$, we consider the *weak* state (2) associated with the wedge angle (that is, state (2) means the weak state (2)), and denote φ_2 and (u_2, v_2, ρ_2, c_2) as the pseudo-potential and the quantities related to the weak state (2) in the form of (2.4.1).

Definition 7.5.1. *We call the wedge angle $\theta_{\rm w} \in (\theta_{\rm w}^{\rm d}, \frac{\pi}{2})$ supersonic if state (2) is supersonic at P_0, i.e., if $|D\varphi_2(P_0)| > c_2$ for $\theta_{\rm w}$; sonic if state (2) is sonic at P_0, i.e., if $|D\varphi_2(P_0)| = c_2$ for $\theta_{\rm w}$; and subsonic if state (2) is subsonic at P_0, i.e., if $|D\varphi_2(P_0)| < c_2$ for $\theta_{\rm w}$.*

We also define the wedge angle $\theta_{\rm w} = \frac{\pi}{2}$ to be supersonic (since, as $\theta_{\rm w} \to \frac{\pi}{2}-$, $P_0(\theta_{\rm w}) \to \infty$, while $B_{c_2(\theta_{\rm w})}(\mathcal{O}_2(\theta_{\rm w})) \to B_{\bar{c}_2}(\mathbf{0})$ in the Hausdorff metric, which implies that $|D\varphi_2 P_0(\theta_{\rm w})| = |P_0(\theta_{\rm w}) - \mathcal{O}_2(\theta_{\rm w})| \to \infty > \bar{c}_2$).

Remark 7.5.2. *By Theorem* 7.1.1(v)*, any $\theta_{\rm w} \in (\theta_{\rm w}^{\rm s}, \frac{\pi}{2})$ is a supersonic wedge angle.*

Remark 7.5.3. *The sonic angle $\theta_{\rm w}^{\rm s}$ introduced in Theorem* 7.1.1(v) *is also a sonic angle in the terms of Definition* 7.5.1*. The existence (or non-existence) of other sonic angles is unclear; however, if they existed, $\theta_{\rm w}^{\rm s}$ from Theorem* 7.1.1(v) *would be the largest sonic angle, on account of Remark* 7.5.2.

Remark 7.5.4. *Let $\tilde{\theta}_{\rm w}^{\rm s}$ be an angle introduced in assertion* (vi) *of Theorem* 7.1.1*. Then, by Theorem* 7.1.1(vi)*, any $\theta_{\rm w} \in (\theta_{\rm w}^{\rm d}, \tilde{\theta}_{\rm w}^{\rm s})$ is a subsonic wedge angle.*

Next we introduce some points and lines shown on Figs. 2.3 and 2.4. We use region Λ defined by (2.2.19) and note the following remarks.

Remark 7.5.5. *Let $\theta_{\rm w} \in (\theta_{\rm w}^{\rm d}, \frac{\pi}{2})$ be a supersonic wedge angle. Then*

(i) *Line $\mathcal{S}_1 = \{\varphi_1 = \varphi_2\}$ necessarily intersects with the sonic circle $\partial B_{c_2}(u_2, v_2)$ of state* (2) *in two points.*

This follows from (6.1.4) *of Lemma* 6.1.2*, which can be applied with $\varphi^- = \varphi_1$ and $\varphi^+ = \varphi_2$ by properties* (7.1.5)–(7.1.6) *of state* (2)*, where we note that* (7.1.5) *at point P_0 for the uniform states* (1) *and* (2) *implies that the equalities in* (7.1.5) *hold at every point of $\mathcal{S}_1 = \{\varphi_1 = \varphi_2\}$.*

(ii) *Let P_1 be the nearest point of intersection of $\mathcal{S}_1$ with $\partial B_{c_2}(u_2, v_2)$ to $P_0 = (\xi_1^0, \xi_1^0 \tan\theta_{\rm w})$. Then P_1 necessarily lies within Λ.*

Indeed, by (6.1.6) *of Lemma* 6.1.2 *(applied again with $\varphi^- = \varphi_1$ and $\varphi^+ = \varphi_2$), it follows that $\mathcal{O}_1\mathcal{O}_2 \perp \mathcal{S}_1$. Denote by Q the point of intersection of line $L_{\mathcal{O}}$ through $\mathcal{O}_1$ and $\mathcal{O}_2$ with $\mathcal{S}_1$. Noting that $D\varphi_i(P_0) = \mathcal{O}_i - P_0$ so that* (6.1.6) *implies that $D\varphi_i(P_0) \cdot \boldsymbol{\nu} = |Q\mathcal{O}_i|$ for $i = 1, 2$, we use* (7.1.6) *to obtain $|Q\mathcal{O}_1| > |Q\mathcal{O}_2|$. Then, since $\mathcal{O}_1, \mathcal{O}_2 \in \{\varphi_1 > \varphi_2\}$ by* (6.1.3) *and* (6.1.5)*, and $Q \in \mathcal{S}_1 = \{\varphi_1 = \varphi_2\}$, we conclude that $\mathcal{O}_2$ lies on segment $Q\mathcal{O}_1$. From this and* (2.2.19)*, using that $\mathcal{O}_1 = (u_1, 0) \notin \overline{\Lambda}$ and $\mathcal{O}_2 = (u_2, u_2 \tan\theta_{\rm w})$ with $u_2 > 0$ and $\theta_{\rm w} \in (0, \frac{\pi}{2})$ so that $\mathcal{O}_2 \in \partial\Lambda$, we see that $Q \in \Lambda$. Since $\theta_{\rm w}$ is a supersonic wedge angle, i.e., $P_0 \notin \overline{B_{c_2}(u_2, v_2)}$, it follows that P_1 is on segment P_0Q. Using* (2.2.19)*, we conclude that $P_1 \in \Lambda$.*

(iii) *From the last argument, it also follows that the whole segment P_0P_1 (including endpoint P_1, but excluding endpoint P_0) lies within Λ.*

Remark 7.5.6. *Let $\theta_w \in (\theta_w^s, \frac{\pi}{2})$. Then θ_w is supersonic by Remark 7.5.2. Let $P_1(\theta_w)$ be the point introduced in Remark 7.5.5 for θ_w. Let $P_1(\frac{\pi}{2})$ be the corresponding point for the normal reflection, i.e., $P_1(\frac{\pi}{2})$ is the unique point of intersection within $\Lambda(\frac{\pi}{2})$ of the vertical shock (between states (1) and (2)) and the sonic circle of state (2). That is, using the notations in §6.2, $P_1(\frac{\pi}{2}) = (\bar{\xi}_1, \sqrt{\bar{c}_2^2 - \bar{\xi}_1^2})$. Then, by (7.2.11),*

$$P_1(\theta_w) \to P_1(\frac{\pi}{2}) \qquad \text{as } \theta_w \to \frac{\pi}{2}- .$$

Definition 7.5.7. *If $\theta_w \in (\theta_w^d, \frac{\pi}{2}]$, the following lines, points, and segments are defined for the wedge angle θ_w in terms of the data and the parameters of state (2):*

- *Line $\mathcal{S}_1 = \{\varphi_1 = \varphi_2\}$. We note that, for $\theta_w \in (\theta_w^d, \frac{\pi}{2})$, line $\mathcal{S}_1$ is not parallel to the wedge boundary $\{\xi_1 = \xi_2 \cot\theta_w\}$; otherwise, $\boldsymbol{\nu} = \boldsymbol{\nu}_w$ in (7.1.6), which contradicts (7.1.7). For $\theta_w = \frac{\pi}{2}$, $\mathcal{S}_1(\frac{\pi}{2}) = \{\xi_1 = \bar{\xi}_1\}$, by (2.2.16) and (6.2.1).*

- *Point P_1 is defined as follows: If θ_w is a supersonic wedge angle, $P_1 \in \Lambda$ is the point described in Remarks 7.5.5–7.5.6. If θ_w is a subsonic/sonic wedge angle, we set $P_1 := P_0$.*

- *Point $P_3 = \mathbf{0}$.*

- *Point P_4 for supersonic wedge angles is defined by*

$$P_4 = (q_2 + c_2)(\cos\theta_w, \sin\theta_w) \qquad \text{for } q_2 = \sqrt{u_2^2 + v_2^2}.$$

  That is, P_4 is the upper point of intersection of the sonic circle of state (2) with the wedge boundary $\{\xi_1 = \xi_2 \cot\theta_w\}$. Also, from the definitions,

$$\xi_{1P_1} < \xi_{1P_4}.$$

  For supersonic wedge angles $\theta_w \in (\theta_w^d, \frac{\pi}{2})$, we note that, since $\mathcal{O}_2 = (u_2, v_2) \in P_0P_3$ (by Lemma 7.4.8) and $|D\varphi_2(P_0)| > c_2$, we have

$$P_4 \in P_0\mathcal{O}_2 \subset P_0P_3 \subset \{\xi_1 = \xi_2 \cot\theta_w\}. \tag{7.5.1}$$

  Furthermore, since $\mathcal{O}_2 = (u_2, v_2) \in P_0P_3$, we obtain from the definition of P_4 that

$$\mathcal{O}_2 \subset (P_3P_4)^0 \subset \{\xi_1 = \xi_2 \cot\theta_w\}. \tag{7.5.2}$$

- *For subsonic/sonic wedge angles θ_w, we set $P_4 := P_0$.*

- *The relatively open line segment $\Gamma_{\text{wedge}} := P_3P_4 \subset \{\xi_1 = \xi_2 \cot\theta_w\}$. Note that, for subsonic/sonic wedge angles, the previous definition implies that $\Gamma_{\text{wedge}} = P_0P_3$.*

- *For supersonic wedge angles, denote by* Γ_{sonic} *the upper arc* P_1P_4 *of the sonic circle of state* (2), *that is,*

$$\Gamma_{\text{sonic}} = \{(\xi_1, f_{\text{so}}(\xi_1)) \, : \, \xi_{1P_1} \le \xi_1 \le \xi_{1P_4}\}$$

with $f_{\text{so}}(\xi_1) = v_2 + \sqrt{c_2^2 - (\xi_1 - u_2)^2}$. *For subsonic/sonic wedge angles,* Γ_{sonic} *denotes point* $P_0 = P_1 = P_4$.

The property that $\xi_{1P_1} < \xi_{1P_4}$ *for supersonic wedge angles, as claimed above, is from* (2.2.19), *since* $\mathcal{O}_2, P_4 \in \{\xi_1 = \xi_2 \cot\theta_{\text{w}}\} \subset \partial\Lambda$, $P_1 \in \Lambda$, *and* $|\mathcal{O}_2P_1| = |\mathcal{O}_2P_4| = c_2$, *where* $\mathcal{O}_2 := (u_2, v_2)$ *is the center of sonic circle of state* (2).

Remark 7.5.8. *If* $\theta_{\text{w}} \in (\theta_{\text{w}}^{\text{d}}, \frac{\pi}{2})$, *then*

$$\xi_{2P_0} > v_2, \quad \xi_{2P_1} > v_2,$$

where we recall that $P_0 = P_1$ *for subsonic/sonic wedge angles.*

The first inequality follows from Lemma 7.4.8. *This also implies the second inequality for subsonic/sonic wedge angles. To show the second inequality for supersonic wedge angles, we note that Lemma* 7.4.8 *and* (7.5.1) *imply that* $\mathcal{O}_2$ *lies in the relative interior of* P_3P_4, *which concludes the proof.*

We now prove

Lemma 7.5.9. *The points introduced in Definition* 7.5.7 *depend continuously on* θ_{w}:

(i) P_0 *depends continuously on* $\theta_{\text{w}} \in [\theta_{\text{w}}^{\text{d}}, \frac{\pi}{2})$;

(ii) P_1 *and* P_4 *continuously depend on* $\theta_{\text{w}} \in [\theta_{\text{w}}^{\text{d}}, \frac{\pi}{2}]$;

(iii) $P_1 \neq P_3$ *and* $P_4 \neq P_3$ *for any* $\theta_{\text{w}} \in [\theta_{\text{w}}^{\text{d}}, \frac{\pi}{2}]$.

Proof. The proof consists of four steps.

1. Since $P_0 = (\xi_1^0, \xi_1^0 \tan\theta_{\text{w}})$, (i) follows.

2. Next, we show the continuity of $P_1(\theta_{\text{w}})$ on interval $[\theta_{\text{w}}^{\text{d}}, \frac{\pi}{2}]$. To do this, we first show the continuity of P_1 on set $\mathcal{A}$ of all supersonic angles $\theta_{\text{w}} \in [\theta_{\text{w}}^{\text{d}}, \frac{\pi}{2})$. Since $\mathcal{S}_1 = \{\varphi_1 = \varphi_2\}$, the unit vector $\boldsymbol{\nu}_{\mathcal{S}_1} = \frac{(u_1-u_2,-v_2)}{|(u_1-u_2,-v_2)|}$ is orthogonal to $\mathcal{S}_1$. Moreover, note that the denominator in the last expression of $\boldsymbol{\nu}_{\mathcal{S}_1}$ is nonzero for any $\theta_{\text{w}} \in [\theta_{\text{w}}^{\text{d}}, \frac{\pi}{2})$, by Corollary 7.4.9, and also for $\theta_{\text{w}} = \frac{\pi}{2}$, since $(u_2, v_2)_{\theta_{\text{w}}=\frac{\pi}{2}} = \mathbf{0}$ by Theorem 7.1.1(ii) and $u_1 > 0$. Now the continuity of the parameters of the weak state (2) in Theorem 7.1.1(i) and the fact that $(u_2, v_2) = D\phi_2^{\text{wk}}$ imply that $\boldsymbol{\nu}_{\mathcal{S}_1}$ depends continuously on $\theta_{\text{w}} \in [\theta_{\text{w}}^{\text{d}}, \frac{\pi}{2}]$. Notice that, for any supersonic angle $\theta_{\text{w}} \in [\theta_{\text{w}}^{\text{d}}, \frac{\pi}{2})$, point P_1 is the nearest intersection point of $\mathcal{S}_1$ with $\partial B_{c_2}(\mathcal{O}_2)$ to P_0, and $\mathcal{S}_1 = \{\varphi_1 = \varphi_2\}$ is the line containing P_0 and perpendicular to $\boldsymbol{\nu}_{\mathcal{S}_1}$. Then, noting that (7.1.2) and Theorem 7.1.1(i) imply that all of P_0, $\boldsymbol{\nu}_{\mathcal{S}_1}$, c_2, and $\mathcal{O}_2 = (u_2, v_2)$ depend continuously on $\theta_{\text{w}} \in [\theta_{\text{w}}^{\text{d}}, \frac{\pi}{2})$, and that $\mathcal{A}$ is open, we obtain the continuous dependence of P_1 on $\theta_{\text{w}} \in \mathcal{A}$.

Also, combining this with Remark 7.5.6, we obtain the continuity of $P_1(\theta_{\rm w})$ on the set of all supersonic angles $\theta_{\rm w} \in [\theta_{\rm w}^{\rm d}, \frac{\pi}{2}]$.

Since set $\mathcal{B}$ of all the subsonic and sonic angles $\theta_{\rm w}$ satisfies $\mathcal{B} \subset [\theta_{\rm w}^{\rm d}, \theta_{\rm w}^{\rm s}] \subset (0, \frac{\pi}{2})$, and $P_1 = P_0$ on $\mathcal{B}$, the continuity of $P_1(\theta_{\rm w})$ on $\mathcal{B}$ follows from assertion (i).

It remains to show that, if $\theta_{\rm w}^{(i)}$ are supersonic angles for $i = 1, 2, \dots$, such that $\theta_{\rm w}^{(i)} \to \theta_{\rm w}^{(\infty)}$ for which $\theta_{\rm w}^{(\infty)}$ is a sonic wedge angle, then

$$P_1(\theta_{\rm w}^{(i)}) \to P_0(\theta_{\rm w}^{(\infty)}). \tag{7.5.3}$$

Indeed, from the definition of P_1 for supersonic angles in Remark 7.5.5, we estimate

$$|P_0P_1| \le \sqrt{\big(c_2 + \mathrm{dist}(P_0, \partial B_{c_2}(\mathcal{O}_2))\big)^2 - c_2^2} \qquad \text{for each } \theta_{\rm w}^{(i)}. \tag{7.5.4}$$

Also, $\theta_{\rm w}^{(\infty)}$ is a sonic wedge angle so that

$$\mathrm{dist}(P_0, \partial B_{c_2}(\mathcal{O}_2))(\theta_{\rm w}^{(i)}) \to \mathrm{dist}(P_0, \partial B_{c_2}(\mathcal{O}_2))(\theta_{\rm w}^{(\infty)}) = 0,$$

and $c_2(\theta_{\rm w})$ is uniformly bounded on $[\theta_{\rm w}^{\rm d}, \frac{\pi}{2}]$ for the weak states (2) by Theorem 7.1.1(i). Thus, from (7.5.4), we have

$$|P_0P_1|(\theta_{\rm w}^{(i)}) \to 0 \qquad \text{as } i \to \infty.$$

This, combined with the continuity of $P_0(\theta_{\rm w})$ and $\boldsymbol{\nu}_{\mathcal{S}_1}(\theta_{\rm w})$ at $\theta_{\rm w} = \theta_{\rm w}^{(\infty)}$, implies (7.5.3).

Therefore, we have proved the continuous dependence of P_1 on $[\theta_{\rm w}^{\rm d}, \frac{\pi}{2}]$.

3. Next we show the continuity of $P_4(\theta_{\rm w})$ on the angle interval $[\theta_{\rm w}^{\rm d}, \frac{\pi}{2}]$.

On the set of supersonic angles $\theta_{\rm w} \in [\theta_{\rm w}^{\rm d}, \frac{\pi}{2}]$, $P_4(\theta_{\rm w}) = \hat{q}(\cos\theta_{\rm w}, \sin\theta_{\rm w})$, where $\hat{q} = \sqrt{u_2^2 + v_2^2} + c_2$. This implies the continuity of $P_4(\theta_{\rm w})$ on the set of all the supersonic angles $\theta_{\rm w} \in [\theta_{\rm w}^{\rm d}, \frac{\pi}{2}]$. The remaining part of the proof follows an argument similar to that in Step 2, with the obvious modifications, by using that $|P_0P_4| = \mathrm{dist}(P_0, \partial B_{c_2}(\mathcal{O}_2))$ for any supersonic angle $\theta_{\rm w}$. Then assertion (ii) is proved.

4. Now we prove assertion (iii).

We first show that $P_1 \ne P_3$.

Consider first Case $\theta_{\rm w} \in [\theta_{\rm w}^{\rm d}, \frac{\pi}{2})$. Recall that $P_3 = \mathbf{0}$ lies on the wedge boundary $\{\xi_2 = \xi_1 \tan\theta_{\rm w}\}$. If $P_1 = P_3$, then, using that $P_0 \in \mathcal{S}_1 \cap \{\xi_2 = \xi_1 \tan\theta_{\rm w}\}$ and $P_1 \in \mathcal{S}_1$, we obtain that $\mathcal{S}_1$ coincides with $\{\xi_2 = \xi_1 \tan\theta_{\rm w}\}$. Then $\boldsymbol{\nu} = \boldsymbol{\nu}_{\rm w}$ in (7.1.6), which contradicts (7.1.7). Therefore, $P_1 \ne P_3$.

If $\theta_{\rm w} = \frac{\pi}{2}$, $\mathcal{O}_2 = \mathbf{0}$, by Theorem 7.1.1(ii), so that $P_3 = \mathcal{O}_2$. Then $P_1 \in \partial B_{\bar{c}_2}(P_3)$, so $P_1 \ne P_3$, since $\bar{c}_2 > c_1 > 0$.

Moreover, $|P_4P_3| \ge c_2 > 0$ for any $\theta_{\rm w} \in [\theta_{\rm w}^{\rm d}, \frac{\pi}{2}]$. □

Furthermore, we note the following properties of state (2).

Lemma 7.5.10. *For any* $\theta_{\rm w} \in (\theta_{\rm w}^{\rm d}, \frac{\pi}{2})$,

$$B_{c_1}(\mathcal{O}_1) \cap \Lambda \subset \{\varphi_1 > \varphi_2\} \cap \Lambda, \tag{7.5.5}$$

$$\angle P_3\mathcal{O}_2\mathcal{O}_1 < \frac{\pi}{2}. \tag{7.5.6}$$

Moreover, if $\theta_{\rm w} \in (\theta_{\rm w}^{\rm d}, \frac{\pi}{2})$ *is any supersonic wedge angle,*

$$\overline{B_{c_1}(\mathcal{O}_1)} \cap \overline{\Gamma_{\rm sonic}} = \emptyset. \tag{7.5.7}$$

Proof. (7.5.5) follows from (6.1.3). Now we prove (7.5.6) and (7.5.7).

From (6.1.6), it follows that $\mathcal{O}_1\mathcal{O}_2 \perp \mathcal{S}_1$. Denote by Q the point of intersection of line $\mathcal{O}_1\mathcal{O}_2$ with $\mathcal{S}_1$. Then, in triangle $P_0\mathcal{O}_2Q$, angle $\angle\mathcal{O}_2QP_0$ is $\frac{\pi}{2}$, so that $\angle Q\mathcal{O}_2P_0 < \frac{\pi}{2}$. Thus, $\angle P_3\mathcal{O}_2\mathcal{O}_1 = \angle Q\mathcal{O}_2P_0 < \frac{\pi}{2}$, which shows (7.5.6).

Furthermore, let $\theta_{\rm w} \in (\theta_{\rm w}^{\rm d}, \frac{\pi}{2})$ be any supersonic wedge angle. Then, from (7.5.1) and (7.5.6), $\angle P_4\mathcal{O}_2\mathcal{O}_1 = \pi - \angle P_3\mathcal{O}_2\mathcal{O}_1 > \frac{\pi}{2}$. Since $\Gamma_{\rm sonic}$ and $\mathcal{O}_1$ are on the opposite sides of line $\{\xi_2 = \xi_1 \tan\theta_{\rm w}\}$, which contains points $\mathcal{O}_2$ and P_4, it follows that, for every $P \in \overline{\Gamma_{\rm sonic}}$, $\angle P\mathcal{O}_2\mathcal{O}_1 \geq \angle P_4\mathcal{O}_2\mathcal{O}_1 > \frac{\pi}{2}$. From triangle $P\mathcal{O}_2\mathcal{O}_1$, $|\mathcal{O}_1P| > |\mathcal{O}_2P|$. Since $P \in \Gamma_{\rm sonic}$, then $|\mathcal{O}_2P| = c_2$. Also, $\rho_1 < \rho_2$, so that $c_1 < c_2$. Then $|\mathcal{O}_1P| > |\mathcal{O}_2P| = c_2$ implies that $P \notin \overline{B_{c_1}(\mathcal{O}_1)}$. Therefore, (7.5.7) holds. □

Lemma 7.5.11. *There exists* $\delta > 0$ *depending only on the data such that, for any* $\theta_{\rm w} \in [\theta_{\rm w}^{\rm d}, \frac{\pi}{2}]$,

$$|D(\varphi_1 - \varphi_2^{(\theta_{\rm w})})| \geq \delta.$$

Note that $D(\varphi_1 - \varphi_2^{(\theta_{\rm w})})$ *is a constant vector (independent of* $\boldsymbol{\xi}$*).*

Proof. The parameters of the weak state (2) depend continuously on $\theta_{\rm w} \in [\theta_{\rm w}^{\rm d}, \frac{\pi}{2}]$ by Theorem 7.1.1(i)–(ii). Thus, it suffices to show that

$$|D(\varphi_1 - \varphi_2^{(\theta_{\rm w})})| > 0 \qquad \text{for each } \theta_{\rm w} \in [\theta_{\rm w}^{\rm d}, \tfrac{\pi}{2}].$$

If $\theta_{\rm w}^{\rm d} \leq \theta_{\rm w} < \frac{\pi}{2}$, then, by Corollary 7.4.9, $v_2 > 0$. Thus, we have

$$|D(\varphi_1 - \varphi_2^{(\theta_{\rm w})})| = \sqrt{(u_1 - u_2)^2 + v_2^2} > 0.$$

If $\theta_{\rm w} = \frac{\pi}{2}$ (which is the normal reflection case), then, by Theorem 6.2.1, $(u_2, v_2) = \mathbf{0}$. Thus, using $u_1 > 0$, we obtain

$$|D(\varphi_1 - \varphi_2^{(\theta_{\rm w})})| = \sqrt{(u_1 - u_2)^2 + v_2^2} = u_1 > 0.$$

□

Let $\mathbf{e}_{\mathcal{S}_1}$ be the unit vector parallel to $\mathcal{S}_1$, oriented so that $\mathbf{e}_{\mathcal{S}_1} \cdot D\varphi_2(P_0) > 0$, that is,

$$\mathbf{e}_{\mathcal{S}_1} = -\frac{(v_2,\, u_1 - u_2)}{\sqrt{(u_1 - u_2)^2 + v_2^2}}. \tag{7.5.8}$$

Lemma 7.5.12. *If $\theta_{\mathrm{w}} \in (\theta_{\mathrm{w}}^{\mathrm{d}}, \frac{\pi}{2})$, then $\mathbf{e}_{\mathcal{S}_1}$ is not orthogonal to $\Gamma_{\mathrm{wedge}} \cup \Gamma_{\mathrm{sym}}$.*

Proof. From its definition, $\mathbf{e}_{\mathcal{S}_1} \parallel \mathcal{S}_1$. If $\mathcal{S}_1$ were orthogonal to Γ_{wedge}, then, since $D\varphi_2(P_0)$ is parallel to Γ_{wedge} and $\varphi_1 = \varphi_2$ on $\mathcal{S}_1$, it would follow that $D\varphi_1(P_0)$ is parallel to Γ_{wedge}. However, $D\varphi_1(P_0) = -(\xi_{1P_0}, \xi_{2P_0}) + (u_1, 0)$, where (ξ_{1P_0}, ξ_{2P_0}) is parallel to Γ_{wedge} and $(u_1, 0)$ is not parallel to Γ_{wedge}. Thus, $\mathbf{e}_{\mathcal{S}_1}$ is not orthogonal to Γ_{wedge}.

Furthermore, by Corollary 7.4.9, $v_2 > 0$. Then, by (7.5.8) with $v_2 > 0$, we see that $\mathbf{e}_{\mathcal{S}_1}$ is not orthogonal to $\Gamma_{\mathrm{sym}} \subset \{\xi_2 = 0\}$. □

Remark 7.5.13. *For any supersonic wedge angle $\theta_{\mathrm{w}} \in (\theta_{\mathrm{w}}^{\mathrm{d}}, \frac{\pi}{2})$, we can express $\mathbf{e}_{\mathcal{S}_1}$ as*

$$\mathbf{e}_{\mathcal{S}_1} = \frac{P_1 - P_0}{|P_1 - P_0|}. \tag{7.5.9}$$

Remark 7.5.14. *For $\theta_{\mathrm{w}} = \frac{\pi}{2}$, we also define $\mathbf{e}_{\mathcal{S}_1}$ by (7.5.8) for $(u_2, v_2) = \mathbf{0}$. Then*

$$\mathbf{e}_{\mathcal{S}_1} = -(0,1) = -\mathbf{e}_{\xi_2} \qquad \text{for } \theta_{\mathrm{w}} = \frac{\pi}{2}.$$

7.6 VON NEUMANN'S SONIC AND DETACHMENT CONJECTURES

The local theory indicates that there are two possible states for state (2), and there has been a long debate on the issue of which one is physical for the local theory; see Courant-Friedrichs [99], Ben-Dor [12], and the references cited therein.

Notice that the shock reflection-diffraction is not a local problem. Therefore, we take a different point of view, namely that the selection of state (2) should be determined by the global feature of the problem and, more precisely, by the stability of the configuration with respect to the wedge angle θ_{w}.

Stability/Continuity Criterion to Select the Correct State (2) (Chen-Feldman [52]): *Since the normal reflection solution is unique when the wedge angle $\theta_{\mathrm{w}} = \frac{\pi}{2}$, the global regular reflection-diffraction configurations should be required to converge to the unique normal reflection solution when $\theta_{\mathrm{w}} \to \frac{\pi}{2}$, provided that such global configurations can be constructed.*

Employing this stability/continuity criterion, we conclude that the choice of state (2) should be φ_2^{wk}, provided that such a global configuration can be constructed, as is confirmed in the main theorems of this book.

As indicated in the previous sections, in general, φ_2^{wk} may be supersonic or subsonic.

As described in §2.6, **von Neumann's Sonic Conjecture** asserts that *there exists a global supersonic reflection configuration when $\theta_{\mathrm{w}} \in (\theta_{\mathrm{w}}^{\mathrm{s}}, \frac{\pi}{2})$ for $\theta_{\mathrm{w}}^{\mathrm{s}} > \theta_{\mathrm{w}}^{\mathrm{d}}$ such that $|D\varphi_2^{\mathrm{wk}}(P_0)| > c_2^{\mathrm{wk}}$ at the reflection point P_0.* In Chapters 8–13, we present the detailed proof of this conjecture. Also, in Chapter 14, we study the further regularity properties of solutions.

Another conjecture, **von Neumann's Detachment Conjecture**, states that global regular reflection is possible whenever the local regular reflection at the reflection point is possible. That is, *there exists a regular reflection-diffraction configuration for any wedge angle* $\theta_{\rm w} \in (\theta_{\rm w}^{\rm d}, \frac{\pi}{2})$, *i.e., the existence of state* (2) *implies the existence of a regular reflection-diffraction solution to* **Problem 2.2.3**, *of the structure shown in Figs.* 2.3–2.6. *In particular, there exists a global subsonic reflection configuration when* $\theta_{\rm w}$ *is beyond the sonic angle up to the detachment angle as shown in Figs.* 2.4 *and* 2.6.

It is clear that the regular reflection-diffraction configurations are not possible without a local two-shock configuration at the reflection point on the wedge (*i.e.*, the existence of state (2)), so this is the necessary criterion. In Chapters 15–17, we prove this conjecture, which indicates that the detachment criterion is also sufficient to ensure a global regular reflection-diffraction configuration.

The local theory developed in this chapter indicates that, for any $\theta_{\rm w} \in (\theta_{\rm w}^{\rm d}, \frac{\pi}{2})$, there are two possible solutions for state (2). In the case of subsonic wedge angles, both states (2) are subsonic at the reflection point P_0. In Chapters 15–17, we show that the weak state (2) is physically admissible indeed, while the strong state (2) will not ensure the existence of a global regular reflection-diffraction in general. On the other hand, the regime between the wedge angles $\theta_{\rm w}^{\rm s}$ and $\theta_{\rm w}^{\rm d}$ is normally very narrow, and the two are only fractions of a degree apart; see also Fig. 18.7, below.

Chapter Eight

Admissible Solutions and Features of Problem 2.6.1

8.1 DEFINITION OF ADMISSIBLE SOLUTIONS

Let $\gamma > 1$ and $\rho_1 > \rho_0 > 0$ be given constants. We use the terminology and notations introduced in §7.5.

Starting from this chapter, right through to Chapter 14, we focus on the wedge angles in the angle interval $(\theta_{\rm w}^{\rm s}, \frac{\pi}{2})$, where $\theta_{\rm w}^{\rm s}$ is the sonic angle introduced in Theorem 7.1.1(v). From Definition 7.5.1 and Remark 7.5.2, any wedge angle $\theta_{\rm w} \in (\theta_{\rm w}^{\rm s}, \frac{\pi}{2})$ is supersonic.

Now we define the admissible solutions of **Problem 2.6.1** for the wedge angles $\theta_{\rm w} \in (\theta_{\rm w}^{\rm s}, \frac{\pi}{2})$, for which we establish the existence of such solutions starting in this chapter and carrying on to Chapter 13. The admissible solutions are of the structure of supersonic reflection configuration described in §2.4.2. These properties are listed in (i)–(iii) of Definition 8.1.1. We also add conditions (iv)–(v) of Definition 8.1.1. This is motivated by the fact that, for any wedge angle sufficiently close to $\frac{\pi}{2}$, the solution of **Problem 2.6.1** which satisfies (i)–(iii) of Definition 8.1.1 also satisfies (iv)–(v) of Definition 8.1.1, as we will show in Appendix 8.3, below.

Definition 8.1.1. *Fix a wedge angle $\theta_{\rm w} \in (\theta_{\rm w}^{\rm s}, \frac{\pi}{2})$. A function $\varphi \in C^{0,1}(\Lambda)$ is called an admissible solution of the regular shock reflection-diffraction problem if φ is a solution of* **Problem 2.6.1** *and satisfies the following:*

(i) *There exists a relatively open curve segment $\Gamma_{\rm shock}$ (without self-intersection) whose endpoints are P_1 and $P_2 = (\xi_{1P_2}, 0)$ with*

$$\xi_{1P_2} < \min\{0,\ u_1 - c_1\}, \qquad \xi_{1P_2} \le \xi_{1P_1}, \tag{8.1.1}$$

so that

- *For the sonic circle $C_1 = \partial B_{c_1}(u_1, 0)$ of state (1),*

$$\Gamma_{\rm shock} \subset \left(\Lambda \setminus \overline{B_{c_1}(u_1,0)}\right) \cap \{\xi_{1P_2} \le \xi_1 \le \xi_{1P_1}\}; \tag{8.1.2}$$

- *$\Gamma_{\rm shock}$ is C^2 in its relative interior. Moreover, denote $\Gamma_{\rm shock}^{\rm ext} := \Gamma_{\rm shock} \cup \Gamma_{\rm shock}^- \cup \{P_2\}$, where $\Gamma_{\rm shock}^-$ is the reflection of $\Gamma_{\rm shock}$ with respect to the ξ_1-axis. Then curve $\Gamma_{\rm shock}^{\rm ext}$ is C^1 at its relative interior* *including P_2 in the sense that, for any P in the relative interior of $\Gamma^{\rm ext}_{\rm shock}$, there exist $r > 0$, $f \in C^1(\mathbb{R})$, and an orthonormal coordinate system $(S,T) \in \mathbb{R}^2$ such that $\Gamma_{\rm shock} \cap B_r(P) = \{S = f(T)\} \cap B_r(P)$. Furthermore, if $P \neq P_2$, $f \in C^2(\mathbb{R})$ can be chosen for sufficiently small r.*

Let $\Gamma_{\rm sonic}$ and $\Gamma_{\rm wedge}$ be the arc and line segment introduced in Definition 7.5.7, respectively. Let $\Gamma_{\rm sym} := P_2P_3$ be the line segment. Note that $\Gamma_{\rm sonic}$, $\Gamma_{\rm sym}$, and $\Gamma_{\rm wedge}$ do not have common points except their common endpoints $\{P_3, P_4\}$. We require that there be no common points between $\Gamma_{\rm shock}$ and curve $\overline{\Gamma_{\rm sym}} \cup \overline{\Gamma_{\rm wedge}} \cup \overline{\Gamma_{\rm sonic}}$ except their common endpoints $\{P_1, P_2\}$. Thus, $\overline{\Gamma_{\rm shock}} \cup \overline{\Gamma_{\rm sym}} \cup \overline{\Gamma_{\rm wedge}} \cup \overline{\Gamma_{\rm sonic}}$ is a closed curve without self-intersection. Denote by Ω the open bounded domain restricted by this curve. Note that $\Omega \subset \Lambda$, $\partial\Omega = \overline{\Gamma_{\rm shock}} \cup \overline{\Gamma_{\rm sym}} \cup \overline{\Gamma_{\rm wedge}} \cup \overline{\Gamma_{\rm sonic}}$, and $\partial\Omega \cap \partial\Lambda = \overline{\Gamma_{\rm sym}} \cup \overline{\Gamma_{\rm wedge}}$.

(ii) *φ satisfies*

$$\begin{aligned} &\varphi \in C^{0,1}(\Lambda) \cap C^1(\overline{\Lambda} \setminus \overline{P_0P_1P_2}), \\ &\varphi \in C^3(\overline{\Omega} \setminus (\overline{\Gamma_{\rm sonic}} \cup \{P_2, P_3\})) \cap C^1(\overline{\Omega}), \\ &\varphi = \begin{cases} \varphi_0 & \text{for } \xi_1 > \xi_1^0 \text{ and } \xi_2 > \xi_1 \tan\theta_{\rm w}, \\ \varphi_1 & \text{for } \xi_1 < \xi_1^0 \text{ and above curve } P_0P_1P_2, \\ \varphi_2 & \text{in } P_0P_1P_4. \end{cases} \end{aligned} \tag{8.1.3}$$

(iii) *Equation* (2.2.8) *is strictly elliptic in $\overline{\Omega} \setminus \overline{\Gamma_{\rm sonic}}$:*

$$|D\varphi| < c(|D\varphi|^2, \varphi) \qquad \text{in } \overline{\Omega} \setminus \overline{\Gamma_{\rm sonic}}. \tag{8.1.4}$$

(iv) *In Ω,*

$$\varphi_2 \le \varphi \le \varphi_1. \tag{8.1.5}$$

(v) *Let $\mathbf{e}_{S_1}$ be defined by* (7.5.8)*. Then, in Ω,*

$$\partial_{\mathbf{e}_{S_1}}(\varphi_1 - \varphi) \le 0, \tag{8.1.6}$$

$$\partial_{\xi_2}(\varphi_1 - \varphi) \le 0. \tag{8.1.7}$$

Remark 8.1.2 (C^1–smoothness across $\Gamma_{\rm sonic}$). *By Definition* 8.1.1(ii), *solution φ is C^1 across $\Gamma_{\rm sonic}$ so that*

$$\varphi = \varphi_2, \quad D\varphi = D\varphi_2 \qquad \text{on } \Gamma_{\rm sonic}.$$

Remark 8.1.3 ($\Gamma_{\rm sym} \cup \{P_2\}$ are interior points). *Let Ω^- (resp. $\Gamma^-_{\rm sonic}$) be the reflection of Ω (resp. $\Gamma_{\rm sonic}$) with respect to the ξ_1–axis, and*

$$\Omega^{\rm ext} = \Omega \cup \Omega^- \cup \Gamma_{\rm sym}, \qquad \Gamma^{\rm ext}_{\rm sonic} = \Gamma_{\rm sonic} \cup \Gamma^-_{\rm sonic}.$$

Then $\Gamma^{\rm ext}_{\rm shock} \subset \partial\Omega^{\rm ext}$, *where* $\Gamma^{\rm ext}_{\rm shock}$ *is defined in Definition* 8.1.1(i). *Let* $\varphi^{\rm ext}$ *be the even extension of* φ *into* $\Omega^{\rm ext}$, *i.e.,* $\varphi^{\rm ext}(\xi_1, \pm\xi_2) = \varphi(\xi_1,\xi_2)$ *for* $\xi_2 > 0$. *Using* (8.1.3) *and* $\partial_{\boldsymbol\nu}\phi = 0$ *on* $\Gamma_{\rm sym}$, *we have*

$$\varphi^{\rm ext} \in C^2(\overline{\Omega^{\rm ext}} \setminus (\overline{\Gamma^{\rm ext}_{\rm sonic}} \cup \{P_2, P_3\})) \cap C^1(\overline{\Omega^{\rm ext}}). \tag{8.1.8}$$

Then $\phi^{\rm ext} := \varphi^{\rm ext} + \frac{|\boldsymbol\xi|^2}{2} \equiv \varphi^{\rm ext} - \varphi_0^{\rm ext}$ *is the even extension of* ϕ *into* $\Omega^{\rm ext}$ *with* $\phi^{\rm ext} \in C^2(\overline{\Omega^{\rm ext}} \setminus (\overline{\Gamma^{\rm ext}_{\rm sonic}} \cup \{P_2, P_3\})) \cap C^1(\overline{\Omega^{\rm ext}})$. *It can be checked by an explicit calculation that* $\varphi^{\rm ext}$ *and* $\phi^{\rm ext}$ *satisfy equations* (2.2.8) *and the non-divergence form* (2.2.11) *in* $\Omega^{\rm ext}$ *and that the equations are strictly elliptic in* $\overline{\Omega^{\rm ext}} \setminus \overline{\Gamma^{\rm ext}_{\rm sonic}}$. *Moreover,* $\varphi_1(\xi_1, -\xi_2) = \varphi_1(\xi_1,\xi_2)$ *in* $\mathbb{R}^2$ *so that* $\varphi^{\rm ext}$ *satisfies*

$$\varphi^{\rm ext} = \varphi_1, \quad \rho(|D\varphi^{\rm ext}|^2, \varphi^{\rm ext})D\varphi^{\rm ext} \cdot \boldsymbol\nu = \rho_1 D\varphi_1 \cdot \boldsymbol\nu \qquad \text{on } \Gamma^{\rm ext}_{\rm shock}.$$

Remark 8.1.4 (Velocity jump across $\Gamma_{\rm shock}$). *Condition* (8.1.2) *implies that* φ_1 *is pseudo-supersonic on* $\Gamma_{\rm shock}$. *Then Definition* 8.1.1(iii) *implies that* $D\varphi \neq D\varphi_1$ *on* $\Gamma_{\rm shock}$. *A similar argument, using* $\xi_{1P_2} < u_1 - c_1$ *from* (8.1.1), *implies the gradient jump at* P_2. *Also, using* (8.1.3), $D\varphi(P_1) = D\varphi_2(P_1) \neq D\varphi_1(P_1)$. *Thus, we have*

$$D\varphi \neq D\varphi_1 \qquad \text{on } \overline{\Gamma_{\rm shock}}. \tag{8.1.9}$$

In the next remark, we use the following notation: For $\mathbf{e}, \mathbf{g} \in \mathbb{R}^2 \setminus \{0\}$ with $\mathbf{e} \neq c\mathbf{g}$,

$$\begin{aligned} &\text{Cone}(\mathbf{e},\mathbf{g}) := \{a\mathbf{e} + b\mathbf{g} \ : \ a, b \geq 0\}, \\ &\text{Cone}^0(\mathbf{e},\mathbf{g}) \ \text{ is the interior of } \ \text{Cone}(\mathbf{e},\mathbf{g}). \end{aligned} \tag{8.1.10}$$

Remark 8.1.5 (Cone of monotonicity directions). *Conditions* (8.1.6)–(8.1.7) *imply that, for any admissible solution* φ *of* **Problem 2.6.1** *in the sense of Definition* 8.1.1,

$$\partial_{\mathbf{e}}(\varphi_1 - \varphi) \leq 0 \qquad \text{in } \overline{\Omega}, \ \text{for all } \mathbf{e} \in \text{Cone}(\mathbf{e}_{S_1}, \mathbf{e}_{\xi_2}) \text{ with } \mathbf{e} \neq \mathbf{0}. \tag{8.1.11}$$

Remark 8.1.6 ($\Gamma_{\rm shock}$ does not intersect with $\Gamma_{\rm wedge}$ and the sonic circle of state (1)). $\Gamma_{\rm shock} \subset \Lambda \setminus \overline{B_{c_1}(u_1, 0)}$ *in condition* (8.1.2) *implies that* $\Gamma_{\rm shock}$ *does not intersect with* $\Gamma_{\rm wedge}$ *and the sonic circle of state* (1). *Furthermore, from this property and Lemma* 7.5.10, *we have*

$$B_{c_1}(u_1, 0) \cap \Lambda \subset \Omega. \tag{8.1.12}$$

Note that the Rankine-Hugoniot conditions (2.2.13)–(2.2.14) imply the following equalities on $\Gamma_{\rm shock}$:

$$\rho(|D\varphi|^2, \varphi)\partial_{\boldsymbol\nu}\varphi = \rho_1 \partial_{\boldsymbol\nu}\varphi_1, \tag{8.1.13}$$

$$\partial_{\boldsymbol\tau}\varphi = \partial_{\boldsymbol\tau}\varphi_1, \tag{8.1.14}$$

$$\varphi = \varphi_1, \tag{8.1.15}$$

where, on the left-hand sides of (8.1.13)–(8.1.14), $D\varphi$ is evaluated on the Ω–side of $\Gamma_{\rm shock}$.

We also note the following property:

Lemma 8.1.7 (Directions of pseudo-velocities on Γ_{shock}: The entropy condition). *If φ is a solution of* **Problem 2.6.1** *satisfying conditions* (i)–(iii) *of Definition* 8.1.1, *then*

$$\partial_{\boldsymbol{\nu}}\varphi_1 > \partial_{\boldsymbol{\nu}}\varphi > 0 \qquad \text{on } \Gamma_{\text{shock}}, \tag{8.1.16}$$

where $\boldsymbol{\nu}$ is the unit normal to Γ_{shock}, interior to Ω.

Proof. We first notice that $D\varphi_1(P_2) = (u_1 - \xi_{1P_2}, 0) = (u_1 + |\xi_{1P_2}|, 0)$, and $\boldsymbol{\nu}_{\text{sh}}(P_2) = (1,0)$, since $\Gamma_{\text{shock}}^{\text{ext}}$ is C^1 in part (i) of Definition 8.1.1. Then

$$\partial_{\boldsymbol{\nu}}\varphi_1(P_2) > 0.$$

Suppose that $\partial_{\boldsymbol{\nu}}\varphi_1(\hat{P}) = 0$ at some point $\hat{P} \in \Gamma_{\text{shock}}$. Then the Rankine-Hugoniot condition (8.1.13) at $\hat{P}$ implies that either $\partial_{\boldsymbol{\nu}}\varphi = 0$ or $\rho(|D\varphi|^2, \varphi) = 0$ at $\hat{P}$. In the latter case, $c(|D\varphi|^2, \varphi) = \rho^{\frac{\gamma-1}{2}}(|D\varphi|^2, \varphi) = 0$ at $\hat{P}$, and then the ellipticity condition (8.1.4) in Definition 8.1.1 implies that $D\varphi(\hat{P}) = 0$. Thus, in either case, $\partial_{\boldsymbol{\nu}}\varphi(\hat{P}) = 0 = \partial_{\boldsymbol{\nu}}\varphi_1$ at $\hat{P}$. This, combined with (8.1.14), implies

$$D\varphi(\hat{P}) = D\varphi_1(\hat{P}).$$

Therefore, by (8.1.13), we have

$$\rho(|D\varphi|^2, \varphi) = \rho_1 \qquad \text{at } \hat{P},$$

which implies that $c(|D\varphi|^2, \varphi) = c_1$ at $\hat{P}$. Now the ellipticity condition (8.1.4) implies that $|D\varphi_1(\hat{P})| \le c_1$, that is, $\hat{P} \in \overline{B_{c_1}(u_1, 0)}$. This contradicts (8.1.2). Thus, $\partial_{\boldsymbol{\nu}}\varphi_1 \neq 0$ on Γ_{shock}. Also, $\partial_{\boldsymbol{\nu}}\varphi_1$ is continuous on $\Gamma_{\text{shock}} \cup \{P_2\}$ since $\Gamma_{\text{shock}}^{\text{ext}} \in C^1$. Since $\partial_{\boldsymbol{\nu}}\varphi_1(P_2) > 0$, we conclude that $\partial_{\boldsymbol{\nu}}\varphi_1 > 0$ on Γ_{shock}. Also, $\rho_1 > 0$ and $\rho(|D\varphi|^2, \varphi) \ge 0$ since $\varphi \in C^1(\overline{\Omega})$ is a solution of (2.2.8) in Ω, *i.e.*, satisfies (2.2.10) in $\overline{\Omega}$. Now (8.1.13) implies

$$\partial_{\boldsymbol{\nu}}\varphi > 0, \quad \rho(|D\varphi|^2, \varphi) > 0 \qquad \text{on } \Gamma_{\text{shock}}.$$

Then expression (2.2.9) of the density, the Rankine-Hugoniot conditions (8.1.13)–(8.1.15), and the strict ellipticity of equation (8.1.4) in Ω imply the first inequality in (8.1.16). □

Corollary 8.1.8. *Let φ be a solution of* **Problem 2.6.1** *satisfying conditions* (i)–(iii) *of Definition* 8.1.1. *Let* **h** *be defined by* (5.1.23) (*or equivalently, by* (5.1.24)) *for φ on Γ_{shock}. Then*

$$\mathbf{h}\cdot\boldsymbol{\nu} = -\frac{\rho - \rho_1}{\rho_1 c^2}\rho(c^2 - \varphi_{\boldsymbol{\nu}}^2)\varphi_{\boldsymbol{\nu}} < 0 \qquad \text{on } \Gamma_{\text{shock}}, \tag{8.1.17}$$

where $\rho = \rho(|D\varphi|^2, \varphi)$. Moreover, assertion (5.1.22) *of Lemma* 5.1.1 *holds.*

Proof. From Lemma 8.1.7, (5.1.14)–(5.1.15) hold.

The first equality in (8.1.17) follows from (5.1.23). The inequality in (8.1.17) follows from (5.1.15), (8.1.4), and (8.1.16).

Also, since (5.1.14) holds, φ satisfies all the conditions of Lemma 5.1.1. □

Next, we show that condition (iv) of Definition 8.1.1 in fact holds with strict inequalities. This follows from the more general fact below (which we will use later). Let $\mathcal{D} \subset \Lambda$ be open. Denote

$$\begin{aligned}\Gamma_{\rm wedge}(\mathcal{D}) &:= \big(\partial\mathcal{D} \setminus (\overline{\partial\mathcal{D} \setminus \{\xi_2 = \xi_1 \tan\theta_{\rm w}\}})\big)^0,\\ \Gamma_{\rm sym}(\mathcal{D}) &:= \big(\partial\mathcal{D} \setminus (\overline{\partial\mathcal{D} \setminus \{\xi_2 = 0\}})\big)^0,\end{aligned} \tag{8.1.18}$$

where these sets may be empty, and $(\cdot)^0$ denotes the relative interior of the subset of a line. Note also that, by definition,

$$\Gamma_{\rm wedge}(\mathcal{D}) \subset \partial\mathcal{D} \cap \{\xi_2 = \xi_1 \tan\theta_{\rm w}\}, \quad \Gamma_{\rm sym}(\mathcal{D}) \subset \partial\mathcal{D} \cap \{\xi_2 = 0\}.$$

Furthermore, it is easy to show that every point of $\Gamma_{\rm wedge}(\mathcal{D}) \cup \Gamma_{\rm sym}(\mathcal{D})$ has an interior ball with respect to $\mathcal{D}$.

Lemma 8.1.9. *Let $\mathcal{D} \subset \Lambda$ be open. Let $\varphi \in C^1(\overline{\mathcal{D}}) \cap C^2(\mathcal{D})$ satisfy equation* (2.2.8) *in $\mathcal{D}$ and $\partial_\nu\varphi = 0$ on $\Gamma_{\rm wedge}(\mathcal{D}) \cup \Gamma_{\rm sym}(\mathcal{D})$ if any of the boundary parts are non-empty. Let equation* (2.2.8) *be strictly elliptic on φ in $\mathcal{D} \cup \Gamma_{\rm wedge}(\mathcal{D}) \cup \Gamma_{\rm sym}(\mathcal{D})$. Assume that* (8.1.5) *holds in $\mathcal{D}$. Then*

$$\varphi < \varphi_1 \quad \text{in } \mathcal{D} \cup \Gamma_{\rm wedge}(\mathcal{D}) \cup \Gamma_{\rm sym}(\mathcal{D}) \quad \text{if } \Gamma_{\rm wedge}(\mathcal{D}) \neq \emptyset; \tag{8.1.19}$$

$$\varphi > \varphi_2 \quad \text{in } \mathcal{D} \cup \Gamma_{\rm wedge}(\mathcal{D}) \cup \Gamma_{\rm sym}(\mathcal{D}) \quad \text{if } \Gamma_{\rm sym}(\mathcal{D}) \neq \emptyset. \tag{8.1.20}$$

If, in addition to the previous assumptions, $\Lambda \cap B_r(\mathbf{0}) \subset \mathcal{D}$ for some $r > 0$ and equation (2.2.8) *is strictly elliptic on φ at $\mathbf{0}$, then*

$$\varphi_2 < \varphi < \varphi_1 \qquad \text{in } \mathcal{D} \cup \Gamma_{\rm wedge}(\mathcal{D}) \cup \Gamma_{\rm sym}(\mathcal{D}) \cup \{\mathbf{0}\}. \tag{8.1.21}$$

Proof. We divide the proof into three steps.

1. We first show (8.1.19). Note that equality $\varphi \equiv \varphi_1$ in $\mathcal{D}$ is not possible if $\Gamma_{\rm wedge}(\mathcal{D}) \neq \emptyset$, since φ_1 does not satisfy that $\partial_\nu\varphi = 0$ on $\Gamma_{\rm wedge}(\mathcal{D})$.

Note that φ satisfies the non-divergence form (2.2.11) of (2.2.8). $\bar\phi := \varphi_1 - \varphi = \phi_1 - \phi$ satisfies $D^2\bar\phi = -D^2\phi$, since ϕ_1 is a linear function. Thus equation (2.2.11), considered as a linear equation for $D^2\phi$, is satisfied with $D^2\phi$ replaced by $D^2\bar\phi$, and the equation is strictly elliptic in $\mathcal{D} \cup \Gamma_{\rm wedge}(\mathcal{D}) \cup \Gamma_{\rm sym}(\mathcal{D})$. Also, since $\phi_1 = u_1\xi_1 + const.$ and $\partial_\nu\varphi = 0$ on $\Gamma_{\rm wedge}(\mathcal{D}) \cup \Gamma_{\rm sym}(\mathcal{D})$, we have

$$\partial_\nu\bar\phi = 0 \ \text{ on } \Gamma_{\rm sym}(\mathcal{D}), \qquad \partial_\nu\bar\phi = -u_1 \sin\theta_{\rm w} < 0 \ \text{ on } \Gamma_{\rm wedge}(\mathcal{D}).$$

Moreover, $\bar\phi \geq 0$ in $\mathcal{D}$ since $\varphi_1 \geq \varphi$ in $\mathcal{D}$. Then, by the strong maximum principle and Hopf's lemma, the minimum value $\bar\phi = 0$ cannot be achieved in $\mathcal{D} \cup$

$\Gamma_{\text{wedge}}(\mathcal{D})\cup\Gamma_{\text{sym}}(\mathcal{D})$ unless $\bar{\phi}\equiv 0$ in $\mathcal{D}$, since every point of $\Gamma_{\text{wedge}}(\mathcal{D})\cup\Gamma_{\text{sym}}(\mathcal{D})$ has an interior ball with respect to $\mathcal{D}$. However, $\bar{\phi}\equiv 0$ in $\mathcal{D}$ is impossible, since $\varphi\equiv\varphi_1$ in $\mathcal{D}$ is not possible as we have shown above in that case. Therefore, $\bar{\phi}>0$ on $\mathcal{D}\cup\Gamma_{\text{wedge}}(\mathcal{D})\cup\Gamma_{\text{sym}}(\mathcal{D})$, which is (8.1.19).

2. Assertion (8.1.20) is proved similarly by considering $\psi=\varphi-\varphi_2=\phi-\phi_2$, and by noting that equality $\varphi\equiv\varphi_2$ in $\mathcal{D}$ is impossible if $\Gamma_{\text{sym}}(\mathcal{D})\neq\emptyset$ (since φ_2 does not satisfy that $\partial_{\boldsymbol{\nu}}\varphi=0$ on $\Gamma_{\text{sym}}(\mathcal{D})$) and that $\partial_{\boldsymbol{\nu}}\psi=-\partial_{\boldsymbol{\nu}}\psi_2=-v_2<0$ on $\Gamma_{\text{sym}}(\mathcal{D})$, where the last inequality holds by Corollary 7.4.9.

3. If, in addition to the previous assumptions, $\Lambda\cap B_r(\mathbf{0})\subset\mathcal{D}$ for some $r>0$, then both $\Gamma_{\text{wedge}}(\mathcal{D})$ and $\Gamma_{\text{sym}}(\mathcal{D})$ are non-empty. Since $\varphi\in C^1(\overline{\mathcal{D}})$ and $\partial_{\boldsymbol{\nu}}\varphi=0$ on $\Gamma_{\text{wedge}}(\mathcal{D})\cup\Gamma_{\text{sym}}(\mathcal{D})$, we conclude

$$D\phi(\mathbf{0})=\mathbf{0}.$$

We perform the reflection about the ξ_1–axis, as in Remark 8.1.3, to the extended domain $\mathcal{D}^{\text{ext}}$. Then equation (2.2.11) in $\mathcal{D}^{\text{ext}}$ is strictly elliptic on φ in $\mathcal{D}^{\text{ext}}\cup\{\mathbf{0}\}$, and hence in a neighborhood of $\mathbf{0}$. Since $\phi_1(\boldsymbol{\xi})=u_1\xi_1+const.$ is independent of ξ_2, the extended function ϕ_1 is of the same form $\phi_1(\boldsymbol{\xi})=u_1\xi_1+const.$ in domain $\mathcal{D}^{\text{ext}}$.

We now consider the extended function $\bar{\phi}=\phi_1-\phi$. Note that domain $\mathcal{D}^{\text{ext}}$ satisfies the interior sphere condition at $\mathbf{0}\in\partial\mathcal{D}^{\text{ext}}$ with $\boldsymbol{\nu}(\mathbf{0})=(-1,0)$, since $\mathbf{0}\in\partial B_r(r,0)$. Then, using (2.2.19), we have

$$B_r(-\frac{r}{2},0)\cap\{\xi_2>0\}\subset B_r(\mathbf{0})\cap\Lambda\subset\mathcal{D},$$

which implies that $B_r(-\frac{r}{2},0)\subset\mathcal{D}^{\text{ext}}$. If the minimum value $\bar{\phi}=0$ is attained at $\mathbf{0}$, then, by Hopf's lemma, $\partial_{\boldsymbol{\nu}}\bar{\phi}(\mathbf{0})>0$, *i.e.*, $\partial_{\xi_1}\bar{\phi}(\mathbf{0})<0$. On the other hand, from $D\phi(\mathbf{0})=\mathbf{0}$ and $\phi_1(\boldsymbol{\xi})=u_1\xi_1+const.$, we find that $\partial_{\xi_1}\bar{\phi}(\mathbf{0})=u_1>0$. This contradiction shows that the minimum value $\bar{\phi}=0$ cannot be attained at $\mathbf{0}$; hence the first inequality in (8.1.21) is proved. The second inequality in (8.1.21) is proved similarly by considering $\psi=\varphi-\varphi_2$. □

Corollary 8.1.10. *Let φ be an admissible solution of* **Problem 2.6.1**. *Then*

$$\varphi_2<\varphi<\varphi_1 \qquad \text{in } \Omega\cup\Gamma_{\text{wedge}}\cup\Gamma_{\text{sym}}\cup\{\mathbf{0}\}. \tag{8.1.22}$$

8.2 STRICT DIRECTIONAL MONOTONICITY FOR ADMISSIBLE SOLUTIONS

We consider a solution φ of **Problem 2.6.1** for the wedge angle $\theta_{\text{w}}\in[\theta_{\text{w}}^{\text{s}},\frac{\pi}{2})$. For further applications, some results of this section are proved for the solutions of **Problem 2.6.1** only satisfying *a priori* some conditions stated in Definition 8.1.1.

In this section, the universal positive constant C depends only on the data, which may be different at different occurrence. Throughout this section, we use the notations introduced in Definition 7.5.7. We note that, from (7.1.6) and assertion (i) of Theorem 7.1.1 for the weak state (2), there exists C such that, for any $\theta_{\rm w} \in [\theta_{\rm w}^{\rm s}, \frac{\pi}{2}]$,

$$\begin{aligned} &|(u_2, v_2)| \le C, \quad 0 < \rho_1 < \rho_2 \le C, \quad 0 < c_1 < c_2 \le C, \\ &|\xi_1^0| + |P_1| + |P_3| + |P_4| \le C. \end{aligned} \tag{8.2.1}$$

8.2.1 Directions of strict monotonicity of $\varphi_1 - \varphi$

We first prove several properties of the directional derivatives of ϕ and $\bar{\phi} = \varphi_1 - \varphi$.

Lemma 8.2.1. *Let $\varphi = \phi - \frac{|\boldsymbol{\xi}|^2}{2}$ be a solution of* **Problem 2.6.1** *satisfying conditions* (i)–(iii) *of Definition* 8.1.1. *Let $\mathbf{e} \in \mathbb{R}^2 \setminus \{0\}$. Then $\partial_{\mathbf{e}}\phi$ is not a constant in Ω.*

Proof. Denote by (S, T) the coordinates with basis $\{\mathbf{e}, \mathbf{e}^{\perp}\}$. Let $a \in \mathbb{R}$, and let $\partial_{\mathbf{e}}\phi \equiv a$ in Ω. Then $\phi_S \equiv a$ in Ω, which implies that $\phi_{SS} \equiv 0$ and $\phi_{ST} \equiv 0$ in Ω. Since both equation (2.2.8) and equation (5.1.9) are strictly elliptic in $\overline{\Omega} \setminus \overline{\Gamma_{\rm sonic}}$, it follows that $\phi_{TT} \equiv 0$ in Ω. Thus, there exist constants (u, v) and C such that

$$\phi(\boldsymbol{\xi}) = (u, v) \cdot \boldsymbol{\xi} + C \qquad \text{in } \Omega.$$

Now, using that φ is C^1 across $\Gamma_{\rm sonic}$ by Definition 8.1.1(ii), we see that $(\varphi, D\varphi) = (\varphi_2, D\varphi_2)$ on $\Gamma_{\rm sonic}$. This implies that $(u, v) = (u_2, v_2)$ so that

$$\varphi = \varphi_2 \qquad \text{in } \Omega.$$

This is a contradiction, since φ_2 does not satisfy (2.2.20) on $\Gamma_{\rm sym}$. Therefore, $\partial_{\mathbf{e}}\phi$ is not a constant in Ω. □

Corollary 8.2.2. *Let φ, ϕ, and $\mathbf{e}$ be as those in Lemma* 8.2.1. *Then, for $\bar{\phi} = \varphi_1 - \varphi$, $\partial_{\mathbf{e}}\bar{\phi}$ is not a constant in Ω. Furthermore, for $\psi = \varphi - \varphi_2$, $\partial_{\mathbf{e}}\psi$ is not a constant in Ω.*

Proof. Since ϕ_1 is a linear function, $\partial_{\mathbf{e}}\bar{\phi} = \partial_{\mathbf{e}}(\phi_1 - \phi)$ is not constant in Ω by Lemma 8.2.1. The argument for $\partial_{\mathbf{e}}\psi$ is similar. □

Lemma 8.2.3. *Let $\varphi = \phi - \frac{|\boldsymbol{\xi}|^2}{2}$ be a solution of* **Problem 2.6.1** *satisfying conditions* (i)–(iii) *of Definition* 8.1.1. *Let $\mathbf{e} \in \mathbb{R}^2 \setminus \{0\}$ and $w = \partial_{\mathbf{e}}\phi$ in Ω. Let $m = \min_{\overline{\Omega}} w$ and $M = \max_{\overline{\Omega}} w$, where $M > m$ by Lemma* 8.2.1. *Then*

(i) *$m < w < M$ in Ω;*

(ii) *If $\mathbf{e}$ is not orthogonal to $\Gamma_{\rm wedge}$, $m < w < M$ on $\Gamma_{\rm wedge}$;*

(iii) *If $\mathbf{e}$ is not orthogonal to $\Gamma_{\rm sym}$, $m < w < M$ on $\Gamma_{\rm sym}$,*

where segments $\Gamma_{\rm wedge}$ and $\Gamma_{\rm sym}$ do not include their endpoints.

Proof. Let $\mathbf{e} = (a_1, a_2)$ in the $\boldsymbol{\xi}$–coordinates. Then

$$w = \mathbf{e} \cdot D\phi.$$

Now we prove each item.

For (i), we know that $w = \partial_{\mathbf{e}}\phi$ satisfies equation (5.1.10) in Ω, and the equation is strictly elliptic in $\overline{\Omega} \setminus \overline{\Gamma_{\text{sonic}}}$ and has continuous coefficients in $\Omega \cup \Gamma_{\text{wedge}} \cup \Gamma_{\text{sym}}$ by Definition 8.1.1(ii)–(iii), where we recall that segments Γ_{wedge} and Γ_{sym} do not include their endpoints. Also, w is not constant in Ω by Lemma 8.2.1. Then the strong maximum principle implies that w cannot attain its minimum or maximum in Ω.

For (ii), Lemma 5.1.3(i) implies that $w = \partial_{\mathbf{e}}\phi$ satisfies the oblique derivative condition (5.1.25) on Γ_{wedge} whose coefficients are continuous on Γ_{wedge} by Definition 8.1.1(ii)–(iii). Since w satisfies the elliptic equation (5.1.10) in Ω with continuous coefficients in $\Omega \cup \Gamma_{\text{wedge}} \cup \Gamma_{\text{sym}}$, and w is not a constant in Ω, we conclude from (5.1.25) and Hopf's lemma that w cannot attain its minimum or maximum over $\overline{\Omega}$ on Γ_{wedge}.

The proof of (iii) is similar to the proof of (ii), and is achieved by using Lemma 5.1.3(ii). □

Lemma 8.2.4. *Let* φ *be a solution of* **Problem 2.6.1** *satisfying conditions* (i)–(iv) *of Definition* 8.1.1. *Let* $w = \partial_{\mathbf{e}}\phi$ *in* Ω *for* $\mathbf{e} \in \mathbb{R}^2 \setminus \{\mathbf{0}\}$. *Let* $\mathbf{h} = h_{\boldsymbol{\nu}}\boldsymbol{\nu} + h_{\boldsymbol{\tau}}\boldsymbol{\tau}$ *is defined by* (5.1.23). *If the extremum of* w *over* $\overline{\Omega}$ *is attained at* $\hat{P} \in \Gamma_{\text{shock}}$, *then* $\mathbf{h}(\hat{P}) = k\mathbf{e}$ *for some* $k \in \mathbb{R}$, *where curve* Γ_{shock} *does not include its endpoints.*

Proof. Notice that, for $\mathbf{e} \in \mathbb{R}^2 \setminus \{\mathbf{0}\}$,

$$w = \partial_{\mathbf{e}}\phi = \mathbf{e} \cdot D\phi.$$

Since w is not constant in Ω by Lemma 8.2.1, we may assume that the minimum of w over $\overline{\Omega}$ is attained at $\hat{P} = \hat{\boldsymbol{\xi}} \in \Gamma_{\text{shock}}$, since the maximum case can be argued similarly. Then $w(\hat{\boldsymbol{\xi}}) = \min_{\boldsymbol{\xi} \in \overline{\Omega}} w(\boldsymbol{\xi})$. Thus, we have the following equalities at $\boldsymbol{\xi} = \hat{\boldsymbol{\xi}}$:

- Since $\partial_{\boldsymbol{\tau}} w = 0$ at $\hat{P}$, and $\partial_{\boldsymbol{\tau}} w = \boldsymbol{\tau} \cdot D(\mathbf{e} \cdot D\phi) = D^2\phi[\mathbf{e}, \boldsymbol{\tau}]$, we have

$$D^2\phi[\mathbf{e}, \boldsymbol{\tau}] = 0 \qquad \text{at } \hat{P}. \tag{8.2.2}$$

  Writing (8.2.2) in basis $\{\boldsymbol{\nu}(\hat{P}), \boldsymbol{\tau}(\hat{P})\}$, we have

$$\phi_{\boldsymbol{\nu}\boldsymbol{\tau}} b_{\boldsymbol{\nu}} + \phi_{\boldsymbol{\tau}\boldsymbol{\tau}} b_{\boldsymbol{\tau}} = 0 \qquad \text{at } \hat{P}, \tag{8.2.3}$$

  where $(b_{\boldsymbol{\nu}}, b_{\boldsymbol{\tau}})$ are the coordinates of $\mathbf{e}$ with basis $\{\boldsymbol{\nu}(\hat{P}), \boldsymbol{\tau}(\hat{P})\}$, *i.e.*, $\mathbf{e} = b_{\boldsymbol{\nu}}\boldsymbol{\nu}(\hat{P}) + b_{\boldsymbol{\tau}}\boldsymbol{\tau}(\hat{P})$.

- Writing (5.1.22) at $\hat{P}$ in basis $\{\boldsymbol{\nu}(\hat{P}), \boldsymbol{\tau}(\hat{P})\}$, we obtain

$$\phi_{\boldsymbol{\nu}\boldsymbol{\tau}} h_{\boldsymbol{\nu}} + \phi_{\boldsymbol{\tau}\boldsymbol{\tau}} h_{\boldsymbol{\tau}} = 0 \qquad \text{at } \hat{P}, \tag{8.2.4}$$

  where $\mathbf{h} = h_{\boldsymbol{\nu}}\boldsymbol{\nu} + h_{\boldsymbol{\tau}}\boldsymbol{\tau}$ is defined by (5.1.23).

Assume that

$$\mathbf{h} \neq k\mathbf{e} \qquad \text{for any } k \in \mathbb{R}. \tag{8.2.5}$$

Then (8.2.3)–(8.2.4) imply that

$$\phi_{\boldsymbol{\nu}\boldsymbol{\tau}} = \phi_{\boldsymbol{\tau}\boldsymbol{\tau}} = 0 \qquad \text{at } \hat{P}. \tag{8.2.6}$$

Equation (2.2.11), written in the (S, T)–coordinates with basis $\{\boldsymbol{\nu}(\hat{P}), \boldsymbol{\tau}(\hat{P})\}$, is of form (5.1.9). Equality (8.2.6) becomes

$$\phi_{ST}(\hat{P}) = \phi_{TT}(\hat{P}) = 0.$$

Since equation (5.1.9) is strictly elliptic in $\Omega \cup \Gamma_{\text{shock}}$, we obtain that $\phi_{SS}(\hat{P}) = 0$. Thus, $D^2\phi = 0$ at $\hat{P}$.

Then, at $\hat{P}$,

$$\partial_{\boldsymbol{\nu}} w = D^2\phi[\mathbf{e}, \boldsymbol{\nu}] = 0. \tag{8.2.7}$$

On the other hand, w satisfies equation (5.1.10), which is uniformly elliptic and has continuous coefficients in $\overline{\Omega} \cap \overline{B_r(\hat{P})}$ for some small $r > 0$, $\hat{P}$ is a minimum point of w over $\overline{\Omega}$, and w is not constant in Ω by Lemma 8.2.1. Then Hopf's lemma implies that $\partial_{\boldsymbol{\nu}} w > 0$ at $\hat{P}$. Therefore, we have arrived at a contradiction with (8.2.7). This completes the proof. □

Lemma 8.2.5. *Let φ be a solution of* **Problem 2.6.1** *satisfying conditions* (i)–(iv) *of Definition* 8.1.1. *Let $\partial_{\mathbf{e}}\bar{\phi} \geq 0$ in Ω for some $\mathbf{e} \in \mathbb{R}^2 \setminus \{0\}$, where $\bar{\phi} = \varphi_1 - \varphi$. Then $\partial_{\mathbf{e}}\bar{\phi} > 0$ on Γ_{shock}, where curve Γ_{shock} does not include its endpoints.*

Proof. By Corollary 8.2.2, $\partial_{\mathbf{e}}\bar{\phi}$ is not constant in Ω. We need to show that the minimum value $\partial_{\mathbf{e}}\bar{\phi} = 0$ over $\overline{\Omega}$ cannot be attained on Γ_{shock}.

Since $\bar{\phi} = \varphi_1 - \varphi = \phi_1 - \phi$ and $D\phi_1(\boldsymbol{\xi}) = (u_1, 0)$ for any $\boldsymbol{\xi} \in \mathbb{R}^2$, we need to show that the maximum of $w := \partial_{\mathbf{e}}\phi$ over $\overline{\Omega}$ cannot be attained at $\hat{\boldsymbol{\xi}} \in \Gamma_{\text{shock}}$ if $\partial_{\mathbf{e}}\bar{\phi}(\hat{\boldsymbol{\xi}}) = 0$.

Suppose that there exists $\hat{\boldsymbol{\xi}} \in \Gamma_{\text{shock}}$ such that

$$w(\hat{\boldsymbol{\xi}}) = \max_{\boldsymbol{\xi} \in \overline{\Omega}} w(\boldsymbol{\xi}), \qquad \partial_{\mathbf{e}}\bar{\phi}(\hat{\boldsymbol{\xi}}) = 0.$$

Then, by Lemma 8.2.4,

$$\mathbf{h}(\hat{\boldsymbol{\xi}}) = k\mathbf{e} \qquad \text{for some } k \in \mathbb{R},$$

where $\mathbf{h} = h_{\boldsymbol{\nu}}\boldsymbol{\nu} + h_{\boldsymbol{\tau}}\boldsymbol{\tau}$ is defined by (5.1.23).

Furthermore, $\partial_{\mathbf{e}}\bar{\phi}(\hat{\boldsymbol{\xi}}) = 0$, *i.e.*, $\mathbf{e} \cdot D\bar{\phi}(\hat{\boldsymbol{\xi}}) = 0$. From this, we use (5.1.14) and (5.1.18) to see that

$$\mathbf{e} = l\boldsymbol{\tau}(\hat{\boldsymbol{\xi}}) \qquad \text{for some } l \in \mathbb{R}.$$

Thus, $\mathbf{h}(\hat{\boldsymbol{\xi}}) = kl\boldsymbol{\tau}(\hat{\boldsymbol{\xi}})$, that is,

$$\mathbf{h} \cdot \boldsymbol{\nu} = 0 \qquad \text{at } \boldsymbol{\xi} = \hat{\boldsymbol{\xi}}.$$

This contradicts (8.1.17). □

Now we can establish the strict monotonicity of $\bar{\phi}$ in the appropriate directions.

Proposition 8.2.6. *Let φ be an admissible solution of* **Problem 2.6.1** *in the sense of Definition* 8.1.1. *Then*

$$\partial_{\mathbf{e}_{S_1}}(\varphi_1 - \varphi) < 0 \qquad \text{in } \overline{\Omega} \setminus \overline{\Gamma_{\text{sonic}}}, \tag{8.2.8}$$

where $\mathbf{e}_{S_1}$ is defined by (7.5.8).

Proof. Denote

$$w := \partial_{-\mathbf{e}_{S_1}}(\varphi_1 - \varphi) = \partial_{-\mathbf{e}_{S_1}}\bar{\phi}.$$

We need to show that $w > 0$ in $\overline{\Omega} \setminus \overline{\Gamma_{\text{sonic}}}$.

Note that w is not constant in Ω by Corollary 8.2.2. From Definition 8.1.1(v), $w \geq 0$ in $\overline{\Omega}$. We need to show that the minimum value $w = 0$ cannot be attained within $\overline{\Omega} \setminus \overline{\Gamma_{\text{sonic}}}$.

From Lemma 7.5.12, $\mathbf{e}_{S_1}$ is not orthogonal to $\Gamma_{\text{wedge}} \cup \Gamma_{\text{sym}}$. Then combining the fact that $w \geq 0$ in Ω with Lemmas 8.2.3 and 8.2.5 implies that $w > 0$ in $\Omega \cup \Gamma_{\text{shock}} \cup \Gamma_{\text{wedge}} \cup \Gamma_{\text{sym}}$. Thus, it remains to show that $w > 0$ at points $\{P_2, P_3\}$. We recall that φ is C^1 up to P_2 and P_3.

The condition that $\partial_{\boldsymbol{\nu}}\varphi = 0$ on Γ_{sonic} and (2.2.17) imply that $\partial_{\xi_2}\varphi_1 = \partial_{\xi_2}\varphi = 0$ at P_2. Then

$$\partial_{\xi_2}\bar{\phi}(P_2) = 0.$$

Also, $\partial_{\xi_1}\varphi_1(P_2) = u_1 - \xi_{1P_2} > 0$ since $\xi_{1P_2} < 0$ by (8.1.1). Furthermore, considering the reflection with respect to the ξ_1-axis as in Remark 8.1.3, we find that the shock normal $\boldsymbol{\nu}_{\text{sh}}(P_2) = (1, 0)$, and (8.1.13) is satisfied at P_2 so that

$$0 < \partial_{\xi_1}\varphi = \frac{\rho_1}{\rho(|D\varphi|^2, \varphi)}\partial_{\xi_1}\varphi_1 < \partial_{\xi_1}\varphi_1 \qquad \text{at } P_2,$$

where we have used (5.1.15) in the last inequality. Then $\partial_{\xi_1}\bar{\phi}(P_2) > 0$. Now, using (7.5.8), we compute at P_2 that

$$w = \frac{v_2\partial_{\xi_1}\bar{\phi} + (u_1 - u_2)\partial_{\xi_2}\bar{\phi}}{\sqrt{(u_1 - u_2)^2 + v_2^2}} = \frac{v_2\partial_{\xi_1}\bar{\phi}}{\sqrt{(u_1 - u_2)^2 + v_2^2}} > 0.$$

We estimate $w(P_3)$. The condition that $\varphi_{\boldsymbol{\nu}} = 0$ on $\Gamma_{\text{wedge}} \cup \Gamma_{\text{sym}}$ implies $D\varphi(P_3) = \mathbf{0}$. Thus, using (7.5.8), we compute at $P_3 = \mathbf{0}$ to obtain

$$w = \partial_{-\mathbf{e}_{S_1}}(\varphi_1 - \varphi) = \partial_{-\mathbf{e}_{S_1}}\varphi_1 = \frac{u_1 v_2}{\sqrt{(u_1 - u_2)^2 + v_2^2}} > 0.$$

□

Corollary 8.2.7. *Let φ be an admissible solution of* **Problem 2.6.1** *in the sense of Definition* 8.1.1 *for $\theta_{\text{w}} \in (\theta_{\text{w}}^{\text{s}}, \frac{\pi}{2})$. Then*

(i) $\overline{\Gamma_{\text{shock}}} \setminus \{P_1\} \subset \{\varphi_2 < \varphi_1\}$;

(ii) $\varphi > \varphi_2$ *on* $\overline{\Omega} \setminus \overline{\Gamma_{\text{sonic}}}$.

Proof. Since $\overline{\Gamma_{\text{shock}}} \subset \{\varphi_2 \le \varphi_1\}$ from Definition 8.1.1(iv), then, in order to prove assertion (i), we need to show that no point of $\overline{\Gamma_{\text{shock}}} \setminus \{P_1\}$ lies on line $\mathcal{S}_1 := \{\varphi_1 = \varphi_2\}$. On the contrary, we suppose that $Q \in (\overline{\Gamma_{\text{shock}}} \setminus \{P_1\}) \cap \mathcal{S}_1$.

We first consider the case that $Q \ne P_2$, *i.e.*, when $Q \in \Gamma_{\text{shock}} \cap \mathcal{S}_1$. Then $\Gamma_{\text{shock}} \subset \{\varphi_2 \le \varphi_1\}$ touches $\mathcal{S}_1 = \{\varphi_2 = \varphi_1\}$ at Q. That is, denoting by $\boldsymbol{\nu}$ the unit normal to Γ_{shock} at Q interior for Ω, and by $\boldsymbol{\tau}$ the unit tangent vector to Γ_{shock} at Q, we see that $\boldsymbol{\tau} = \pm \mathbf{e}_{\mathcal{S}_1}$. Hence, we can choose $\boldsymbol{\tau} = \mathbf{e}_{\mathcal{S}_1}$. Moreover, since $\varphi_1 - \varphi \equiv 0$ on Γ_{shock}, then, by Proposition 8.2.6, $\Gamma_{\text{shock}} \cap \mathcal{S}_1$ cannot include the segments of nonzero length. Thus, there exists $\delta > 0$ such that $\Gamma_{\text{shock}} \cap \{Q + s\mathbf{e}_{\mathcal{S}_1} \; : \; |s| < \delta\} = \{Q\}$. From this, and since Γ_{shock} is a C^1–curve, there exist small constants $\varepsilon, \sigma > 0$ so that $Q_\varepsilon := Q + \varepsilon\boldsymbol{\nu} \in \Omega$ and $Q_\varepsilon - s\mathbf{e}_{\mathcal{S}_1} \in \Omega$ for $s \in [0, \sigma)$, but $Q_\varepsilon - \sigma\mathbf{e}_{\mathcal{S}_1} \in \Gamma_{\text{shock}}$. Then Proposition 8.2.6 implies that $(\varphi_1 - \varphi)(Q_\varepsilon) < (\varphi_1 - \varphi)(Q_\varepsilon - \sigma\mathbf{e}_{\mathcal{S}_1})$. On the other hand, $Q_\varepsilon \in \Omega \subset \{\varphi \le \varphi_1\}$ by Definition 8.1.1(iv), and $Q_\varepsilon - \sigma\mathbf{e}_{\mathcal{S}_1} \in \Gamma_{\text{shock}} \subset \{\varphi = \varphi_1\}$ so that $(\varphi_1 - \varphi)(Q_\varepsilon) \ge 0 = (\varphi_1 - \varphi)(Q_\varepsilon - \sigma\mathbf{e}_{\mathcal{S}_1})$, which contradicts the previous estimate. Therefore, $\Gamma_{\text{shock}} \cap \mathcal{S}_1 = \emptyset$.

The remaining case is that $P_2 \in \mathcal{S}_1$. Recall that $P_2 \subset \{\xi_1 < 0, \; \xi_2 = 0\}$. Since $P_0 \in \mathcal{S}_1 \cap \{\xi_1 > 0 \; \xi_2 > 0\}$, and $\mathbf{e}_{\mathcal{S}_1}$ is parallel to $\mathcal{S}_1$, (7.5.8) with $u_1, u_2, v_2 > 0$ implies that $\mathcal{S}_1$ intersects half-line $\{\xi_1 < 0, \; \xi_2 = 0\}$ only if $u_1 > u_2$. Also, we note that the normal to $\mathcal{S}_1 = \{\varphi_1 = \varphi_2\}$ in the direction to region $\{\varphi_1 > \varphi_2\}$ is $\boldsymbol{\nu}_{\mathcal{S}_1} = \frac{(u_1 - u_2, -v_2)}{\sqrt{(u_1-u_2)^2 + v_2^2}}$. Finally, since $\Gamma_{\text{shock}} \subset \{\varphi_1 \ge \varphi_2\}$ is C^1 up to P_2 (because $\Gamma_{\text{shock}}^{\text{ext}}$ is C^1), it follows that the tangent line L_{P_2} to Γ_{shock} at P_2 is parallel to vector $\mathbf{e}_{\mathcal{S}_1} + s\boldsymbol{\nu}_{\mathcal{S}_1}$ for some $s \le 0$. We calculate

$$(\mathbf{e}_{\mathcal{S}_1} + s\boldsymbol{\nu}_{\mathcal{S}_1}) \cdot (1, 0) = -\frac{v_2 - s(u_1 - u_2)}{\sqrt{(u_1 - u_2)^2 + v_2^2}} < 0,$$

since $u_1 > u_2$, $v_2 > 0$, and $s \le 0$. Thus, line L_{P_2} is not vertical. However, L_{P_2} is the tangent line to $\Gamma_{\text{shock}}^{\text{ext}}$ at P_2, which must be vertical. This contradiction yields the proof of assertion (i).

Now we prove assertion (ii). Using $\psi = \varphi - \varphi_2$ and $\phi = \varphi - \varphi_0$, we have $D^2\psi = D^2(\phi + \varphi_0 - \varphi_2) = D^2\phi$ because $\varphi_0 - \varphi_2 = \phi_0 - \phi_2$ is a linear function. Thus, equation (2.2.11) is satisfied with $D^2\phi$ replaced by $D^2\psi$. This equation (considered as a linear equation for ψ) is uniformly elliptic on each compact subset of $\overline{\Omega} \setminus \overline{\Gamma_{\text{sonic}}}$, by Definition 8.1.1(iii) and since $\varphi \in C^1(\overline{\Omega})$. Also, $\psi \ge 0$ in Ω by (8.1.5).

If the minimum value $\psi(P) = 0$ is attained at some point $P \in \Omega$, $\psi = 0$ everywhere in Ω by the strong maximum principle. This contradicts Corollary 8.2.2.

On Γ_{wedge}, $\partial_{\boldsymbol{\nu}}\psi = \partial_{\boldsymbol{\nu}}\varphi - \partial_{\boldsymbol{\nu}}\varphi_2 = 0$. Thus, if the minimum value $\psi(P) = 0$ is attained at some point $P \in \Gamma_{\text{wedge}}$, then $\psi = 0$ everywhere in Ω by Hopf's lemma and the strong maximum principle. This leads to a contradiction, as in the previous case.

On $\overline{\Gamma_{\rm sym}}$, the interior unit normal to Ω is $\mathbf{e}_{\xi_2}$. Thus, using the regularity: $\varphi \in C^1(\overline{\Omega})$, and the boundary condition: $\partial_{\boldsymbol{\nu}}\varphi = 0$ on $\Gamma_{\rm sym} \subset \{\xi_2 = 0\}$, we find that $\partial_{\boldsymbol{\nu}}\phi = \partial_{\boldsymbol{\nu}}\varphi + \partial_{\xi_2}(\frac{|\boldsymbol{\xi}|^2}{2}) = 0$ on $\overline{\Gamma_{\rm sym}}$ so that $\partial_{\boldsymbol{\nu}}\psi = \partial_{\boldsymbol{\nu}}\phi - \partial_{\xi_2}(u_2\xi_1 + v_2\xi_2) = -v_2 < 0$, where we have used that $\theta_{\rm w} < \frac{\pi}{2}$. This implies that the minimum of ψ cannot be attained on $\overline{\Gamma_{\rm sym}}$.

Also, from part (i) as we have proved above, and by the fact that $\varphi = \varphi_1$ on $\Gamma_{\rm shock}$, we have

$$\varphi = \varphi_1 > \varphi_2 \qquad \text{on } \Gamma_{\rm shock} \setminus \{P_1\}.$$

Now assertion (ii) is proved. □

Proposition 8.2.8. *Let φ be an admissible solution of* **Problem 2.6.1** *in the sense of Definition* 8.1.1. *Then*

$$\partial_{\xi_2}(\varphi_1 - \varphi) < 0 \qquad \textit{in } \overline{\Omega} \setminus \overline{\Gamma_{\rm sym}}. \tag{8.2.9}$$

Proof. Denote

$$w := -\partial_{\xi_2}\bar{\phi},$$

where $\bar{\phi} = \varphi_1 - \varphi$. Then w is not constant in Ω by Corollary 8.2.2.

From Definition 8.1.1(v), $w \geq 0$ in $\overline{\Omega}$. We need to show that the minimum value $w = 0$ cannot be attained within $\overline{\Omega} \setminus \overline{\Gamma_{\rm sym}}$.

Note that $\mathbf{e}_{-\xi_2} = (0, -1)$ is not orthogonal to $\Gamma_{\rm wedge}$. Then, repeating the argument in the proof of Proposition 8.2.6, we conclude that $w > 0$ in $\Omega \cup \Gamma_{\rm shock} \cup \Gamma_{\rm wedge}$.

Thus, it remains to show that $w > 0$ on $\overline{\Gamma_{\rm sonic}}$. We recall that, by (8.1.3), $(\varphi, D\varphi) = (\varphi_2, D\varphi_2)$ on $\overline{\Gamma_{\rm sonic}}$. Then

$$-\partial_{\xi_2}(\varphi_1 - \varphi) = -\partial_{\xi_2}(\varphi_1 - \varphi_2) = v_2 > 0 \qquad \text{on } \overline{\Gamma_{\rm sonic}}.$$

□

Since $\phi_1 - \phi = \varphi_1 - \varphi = 0$ on $\Gamma_{\rm shock}$ and $\partial_{\xi_2}(\phi_1 - \phi) = \partial_{\xi_2}(\varphi_1 - \varphi) < 0$ in $\overline{\Omega} \setminus \overline{\Gamma_{\rm sym}}$ by Proposition 8.2.8 with $\phi_1 = u_1\xi_2 + const.$, we employ (8.1.2)–(8.1.3) to obtain

Corollary 8.2.9 ($\Gamma_{\rm shock}$ is a graph in the vertical direction)**.** *The strict inequality $\xi_{1P_1} > \xi_{1P_2}$ holds in* (8.1.1). *Moreover, there exists*

$$f_{\rm sh} \in C^2((\xi_{1P_2}, \xi_{1P_1})) \cap C^1((\xi_{1P_2}, \xi_{1P_1}])$$

such that

$$\Gamma_{\rm shock} = \{(\xi_1, f_{\rm sh}(\xi_1)) \,:\, \xi_{1P_2} \leq \xi_1 \leq \xi_{1P_1}\} \tag{8.2.10}$$

and

$$\Omega = \left\{ \boldsymbol{\xi} \in \Lambda \,: \quad \begin{array}{l} \xi_{1P_2} < \xi_1 < \xi_{1P_4},\ \xi_2 < f_{\rm sh}(\xi_1) \text{ for } \xi_1 \in (\xi_{1P_2}, \xi_{1P_1}] \\ \xi_2 < f_{\rm so}(\xi_1) \text{ for } \xi_1 \in [\xi_{1P_1}, \xi_{1P_4}) \end{array} \right\} \tag{8.2.11}$$

with $f_{\rm so}(\xi_1) = v_2 + \sqrt{c_2^2 - (\xi_1 - u_2)^2}$. *Furthermore, for* $\xi_1 \in (\xi_{1P_2}, \xi_{1P_1})$,

$$f_{\rm sh}(\xi_1) > \max(\xi_1 \tan\theta_{\rm w},\ \sqrt{\max(0,\ c_1^2 - (\xi_1 - u_1)^2)}\,). \tag{8.2.12}$$

Next, Propositions 8.2.6 and 8.2.8, combined with $\Gamma_{\rm sonic} \cap \Gamma_{\rm sym} = \emptyset$, imply

Corollary 8.2.10 (Cone of strict monotonicity directions for $\varphi_1 - \varphi$)**.** *Let* φ *be an admissible solution of* **Problem 2.6.1** *in the sense of Definition* 8.1.1. *Then, for all* $\mathbf{e} \in {\rm Cone}^0(\mathbf{e}_{\mathcal{S}_1}, \mathbf{e}_{\xi_2})$,

$$\partial_{\mathbf{e}}(\varphi_1 - \varphi) < 0 \qquad \text{in } \overline{\Omega}, \tag{8.2.13}$$

where ${\rm Cone}^0(\mathbf{e}_{\mathcal{S}_1}, \mathbf{e}_{\xi_2})$ *is defined by* (8.1.10).

In the next lemma, we show that, for any $\theta_{\rm w} \in (\theta_{\rm w}^{\rm s}, \frac{\pi}{2})$, the interior unit normal to $\Gamma_{\rm wedge}$ with respect to Λ:

$$\boldsymbol{\nu}_{\rm w} = (-\sin\theta_{\rm w}, \cos\theta_{\rm w}), \tag{8.2.14}$$

is in ${\rm Cone}^0(\mathbf{e}_{\mathcal{S}_1}, \mathbf{e}_{\xi_2})$. We use the terminology in §7.5. In particular, we recall vector $\mathbf{e}_{\mathcal{S}_1}$ defined for $\theta_{\rm w} \in (\theta_{\rm w}^{\rm d}, \frac{\pi}{2}]$ by (7.5.8) and Remark 7.5.14. From this, ${\rm Cone}^0(\mathbf{e}_{\mathcal{S}_1}, \mathbf{e}_{\xi_2})$ is a half-plane when $\theta_{\rm w} = \frac{\pi}{2}$, since ${\rm Cone}^0(\mathbf{e}_{\mathcal{S}_1}, \mathbf{e}_{\xi_2}) = {\rm Cone}^0(-\mathbf{e}_{\xi_2}, \mathbf{e}_{\xi_2})$; see also (9.2.20). To be precise, we define this to be the left half-plane:

$${\rm Cone}^0(\mathbf{e}_{\mathcal{S}_1}, \mathbf{e}_{\xi_2})|_{\theta_{\rm w}=\frac{\pi}{2}} := \{\xi_1 < 0\}. \tag{8.2.15}$$

Then, by (7.5.8) and Remark 7.5.14, it follows that, for any $R > 0$,

$$B_R(\mathbf{0}) \cap {\rm Cone}^0(\mathbf{e}_{\mathcal{S}_1}, \mathbf{e}_{\xi_2})_{|\theta_{\rm w}=\theta_{\rm w}^{(i)}} \to B_R(\mathbf{0}) \cap {\rm Cone}^0(\mathbf{e}_{\mathcal{S}_1}, \mathbf{e}_{\xi_2})_{|\theta_{\rm w}=\frac{\pi}{2}} \tag{8.2.16}$$

in the Hausdorff metric, as $\theta_{\rm w}^{(i)} \to \frac{\pi}{2}-$.

We also define $\boldsymbol{\tau}_{\rm w}$ to be the unit tangent vector to $\Gamma_{\rm wedge}$ in the direction of $D\varphi_2(P_3)$, *i.e.*, from P_3 to P_4:

$$\boldsymbol{\tau}_{\rm w} = \frac{P_3P_4}{|P_3P_4|} = (\cos\theta_{\rm w}, \sin\theta_{\rm w}). \tag{8.2.17}$$

Note that $\boldsymbol{\nu}_{\rm w}$ and $\boldsymbol{\tau}_{\rm w}$ can be defined when $\theta_{\rm w} = \frac{\pi}{2}$ by (8.2.14) and (8.2.17). Then $\boldsymbol{\nu}_{\rm w} = -\mathbf{e}_{\xi_1}$ and $\boldsymbol{\tau}_{\rm w} = \mathbf{e}_{\xi_2}$ when $\theta_{\rm w} = \frac{\pi}{2}$.

Lemma 8.2.11. *For any wedge angle* $\theta_{\rm w} \in (\theta_{\rm w}^{\rm d}, \frac{\pi}{2}]$,

$$\boldsymbol{\tau}_{\rm w} \cdot \mathbf{e}_{\mathcal{S}_1} < 0, \tag{8.2.18}$$

$$\boldsymbol{\nu}_{\rm w} \in {\rm Cone}^0(\mathbf{e}_{\mathcal{S}_1}, \mathbf{e}_{\xi_2}). \tag{8.2.19}$$

Moreover, if (S, T) *are the coordinates in* $\mathbb{R}^2$ *with basis* $\{\boldsymbol{\nu}_{\rm w}, \boldsymbol{\tau}_{\rm w}\}$ *and the origin at* P_3, *then*

$$\begin{array}{ll} T(P_2) < T(P_3) < T(P_1) < T(P_4) < T(P_0) & \text{for supersonic } \theta_{\rm w} \in (\theta_{\rm w}^{\rm d}, \frac{\pi}{2}), \\ T(P_2) < T(P_3) < T(P_0) & \text{for subsonic } \theta_{\rm w} \in (\theta_{\rm w}^{\rm d}, \frac{\pi}{2}), \\ T(P_2) = T(P_3) < T(P_1) < T(P_4) & \text{for } \theta_{\rm w} = \frac{\pi}{2}. \end{array} \tag{8.2.20}$$

Proof. We divide the proof into three steps.

1. Prove (8.2.18). If $\theta_{\rm w} \in (\theta^{\rm d}_{\rm w}, \frac{\pi}{2})$, we use $v_2 = u_2 \tan\theta_{\rm w}$ with $u_2 > 0$ to find from (8.2.17) that

$$\boldsymbol{\tau}_{\rm w} = (\cos\theta_{\rm w}, \sin\theta_{\rm w}) = \frac{(u_2, v_2)}{|(u_2, v_2)|}.$$

Now, using (7.5.8),

$$\boldsymbol{\tau}_{\rm w} \cdot \mathbf{e}_{\mathcal{S}_1} = \frac{-u_1 v_2}{|(u_1 - u_2, -v_2)||(u_2, v_2)|} < 0,$$

which is (8.2.18).

If $\theta_{\rm w} = \frac{\pi}{2}$, we have

$$\boldsymbol{\tau}_{\rm w} \cdot \mathbf{e}_{\mathcal{S}_1} = \mathbf{e}_{\xi_2} \cdot (-\mathbf{e}_{\xi_2}) = -1.$$

Now (8.2.18) is proved.

2. Prove (8.2.19). For $\theta_{\rm w} = \frac{\pi}{2}$, $\boldsymbol{\nu}_{\rm w} = (0, -1)$ so that (8.2.19) follows directly from definition (8.2.15) of $Cone^0(\mathbf{e}_{\mathcal{S}_1}, \mathbf{e}_{\xi_2})_{|\theta_{\rm w}=\frac{\pi}{2}}$.

Now let $\theta_{\rm w} \in (\theta^{\rm d}_{\rm w}, \frac{\pi}{2})$. Then $v_2 \neq 0$. Thus, (7.5.8) shows that $\mathbf{e}_{\mathcal{S}_1} \neq c\mathbf{e}_{\xi_2}$. Also, $|(u_1 - u_2, v_2)| \neq 0$. Therefore, there exist a and b such that

$$\boldsymbol{\nu}_{\rm w} = a|(u_1 - u_2, v_2)|\mathbf{e}_{\mathcal{S}_1} + b\mathbf{e}_{\xi_2}.$$

We need to show that $a, b > 0$.

From (7.5.8), (8.2.14), (8.2.17), and $\mathbf{e}_{\xi_2} = (0, 1)$, we have

$$a = \frac{\sin\theta_{\rm w}}{v_2} > 0,$$
$$b = \frac{v_2\cos\theta_{\rm w} + (u_1 - u_2)\sin\theta_{\rm w}}{v_2} = -|(u_1 - u_2, v_2)|\frac{\boldsymbol{\tau}_{\rm w} \cdot \mathbf{e}_{\mathcal{S}_1}}{v_2}.$$

From (8.2.18), we see that $b > 0$. This implies (8.2.19).

3. Prove (8.2.20). First let $\theta_{\rm w} \in (\theta^{\rm d}_{\rm w}, \frac{\pi}{2})$ be supersonic. Since $P_2 = (0, a)$ with $a < 0$ in the $\boldsymbol{\xi}$–coordinates, $T_{P_2} = a\cos\theta_{\rm w} < 0 = T_{P_3}$. Also, we use that P_3, $\mathcal{O}_2$, P_4, and P_0 lie on line $\{\xi_1 = \xi_2 \cot\theta_{\rm w}\}$ (*i.e.*, on the T–axis) to obtain

$$\mathcal{O}_2 = P_3 + |(u_2, v_2)|\boldsymbol{\tau}_{\rm w}, \quad P_4 = \mathcal{O}_2 + c_2\boldsymbol{\tau}_{\rm w}, \quad P_0 = \mathcal{O}_2 + |\mathcal{O}_2 P_0|\boldsymbol{\tau}_{\rm w} \tag{8.2.21}$$

with $|\mathcal{O}_2 P_0| > c_2$. This implies

$$T_{\mathcal{O}_2} = |(u_2, v_2)| > 0 = T_{P_3}, \quad T_{P_4} = T_{\mathcal{O}_2} + c_2 > T_{\mathcal{O}_2},$$
$$T_{P_0} = T_{\mathcal{O}_2} + |\mathcal{O}_2 P_0| > T_{\mathcal{O}_2} + c_2 = T_{P_4}.$$

Furthermore, $|\mathcal{O}_2 P_1| = c_2$ and $\angle P_1 \mathcal{O}_2 P_0 \in (0, \frac{\pi}{2})$, since segment $P_1 P_0$ is outside circle $B_{c_2}(\mathcal{O}_2)$ and is not tangential to the circle by assertion (6.1.4) of Lemma 6.1.2 applied with $\varphi^- = \varphi_1$ and $\varphi^+ = \varphi_2$. Thus, we have

$$T_{P_1} = T_{\mathcal{O}_2} + |\mathcal{O}_2 P_1|\cos(\angle P_1 \mathcal{O}_2 P_0) = T_{\mathcal{O}_2} + c_2\cos(\angle P_1 \mathcal{O}_2 P_0) \in (T_{\mathcal{O}_2}, T_{P_4}),$$

since $\angle P_1 O_2 P_0 \in (0, \frac{\pi}{2})$. Now (8.2.20) is proved for supersonic wedge angles.

For subsonic angles $\theta_w \in (\theta_w^d, \frac{\pi}{2})$, the argument is similar, except now we take into account that $P_0 = P_1 = P_4$.

If $\theta_w = \frac{\pi}{2}$, then $T_{P_2} = \xi_{2P_2} = 0 = T_{P_3}$. The rest of the argument is as above. □

Remark 8.2.12. Let (S,T) be the coordinates in $\mathbb{R}^2$ with basis $\{\boldsymbol{\nu}_w, \boldsymbol{\tau}_w\}$ and the origin at P_3. Using (8.2.20), we define the function:

$$f_{\boldsymbol{\nu}_w,so}(T) = \sqrt{c_2^2 - (T - |(u_2, v_2)|)^2} \qquad \text{on } (T_{P_1}, T_{P_4}),$$

and the linear functions:

$$L_{\boldsymbol{\nu}_w,w}(T) = 0, \quad L_{\boldsymbol{\nu}_w,sym}(T) = -(T - T_{P_3}) \tan\theta_w,$$

and recall that $T_{P_3} = 0$, where we have used that $|T-(u_2, v_2)| < c_2$ on (T_{P_1}, T_{P_4}). Then

$$\begin{aligned}
\Gamma_{sonic} &= \{S = f_{\boldsymbol{\nu}_w,so}(T) \ : \ T_{P_1} < T < T_{P_4}\},\\
\Gamma_{wedge} &= \{S = L_{\boldsymbol{\nu}_w,w}(T) \ : \ T_{P_3} < T < T_{P_4}\},\\
\Gamma_{sym} &= \{S = L_{\boldsymbol{\nu}_w,sym}(T) \ : \ T_{P_2} < T < T_{P_3}\}.
\end{aligned}$$

On Γ_{shock}, the assertion of Corollary 8.2.10 can be expressed as follows:

Corollary 8.2.13. *Let φ be an admissible solution of* **Problem 2.6.1** *in the sense of Definition* 8.1.1. *Then, for all* $\mathbf{e} \in \mathrm{Cone}^0(\mathbf{e}_{\mathcal{S}_1}, \mathbf{e}_{\xi_2})$,

$$\boldsymbol{\nu} \cdot \mathbf{e} < 0 \qquad \text{on } \Gamma_{shock}, \tag{8.2.22}$$

where $\mathrm{Cone}^0(\mathbf{e}_{\mathcal{S}_1}, \mathbf{e}_{\xi_2})$ *is defined by* (8.1.10), *and* $\boldsymbol{\nu}$ *is the interior unit normal to* Γ_{shock} *with respect to* Ω.

Proof. We first see that $\mathbf{e} \cdot \boldsymbol{\nu} = \frac{\mathbf{e} \cdot D(\varphi_1 - \varphi)}{|D(\varphi_1 - \varphi)|}$ on Γ_{shock}, by (5.1.18). Then (8.2.22) follows from (8.2.13). □

In the next property, we use the following fact: For $\mathbf{g}^{(1)}, \mathbf{g}^{(2)} \in \mathbb{R}^2 \setminus \{0\}$ with $\mathbf{g}^{(1)} \neq c\mathbf{g}^{(2)}$, it follows from definition (8.1.10) of $\mathrm{Cone}(\mathbf{g}^{(1)}, \mathbf{g}^{(2)})$ that

$$\begin{aligned}
&\mathbf{e}^{\perp} \cdot \mathbf{g}^{(k)} \neq 0 \qquad \text{for all } \mathbf{e} \in \mathrm{Cone}^0(\mathbf{g}^{(1)}, \mathbf{g}^{(2)}),\ k = 1, 2,\\
&(\mathbf{e}^{\perp} \cdot \mathbf{g}^{(1)})(\mathbf{e}^{\perp} \cdot \mathbf{g}^{(2)}) < 0.
\end{aligned} \tag{8.2.23}$$

Corollary 8.2.14 (Γ_{shock} is a graph for a cone of directions). *Let φ be an admissible solution of* **Problem 2.6.1**. *Let* $\mathbf{e} \in \mathrm{Cone}^0(\mathbf{e}_{\mathcal{S}_1}, \mathbf{e}_{\xi_2})$ *with* $|\mathbf{e}| = 1$, *and let* $\mathbf{e}^{\perp}$ *be orthogonal to* $\mathbf{e}$ *and oriented, so that* $|\mathbf{e}^{\perp}| = 1$ *and* $\mathbf{e}^{\perp} \cdot \mathbf{e}_{\mathcal{S}_1} < 0$, *which is possible by* (8.2.23). *Let (S,T) be the coordinates with basis* $\{\mathbf{e}, \mathbf{e}^{\perp}\}$. *Denote (S_{P_k}, T_{P_k}) the (S,T)–coordinates of points P_k, $k = 1, \dots, 4$, and note that $T_{P_2} < T_{P_1}$. Then there exists* $f_{\mathbf{e}} := f_{\mathbf{e},sh} \in C^1(\mathbb{R})$ *such that*

(i) $\Gamma_{\text{shock}} = \{S = f_{\mathbf{e}}(T) \,:\, T_{P_2} < T < T_{P_1}\}$ *and* $\Omega \subset \{S < f_{\mathbf{e}}(T) \,:\, T \in \mathbb{R}\}$;

(ii) *In the* (S,T)*–coordinates,* $P_1 = (f_{\mathbf{e}}(T_{P_1}), T_{P_1})$ *and* $P_2 = (f_{\mathbf{e}}(T_{P_2}), T_{P_2})$;

(iii) *For any* $P \in \Gamma_{\text{shock}}$*, there exists* $r > 0$ *such that*

$$\begin{aligned}
&\big(P - \text{Cone}^0(\mathbf{e}_{\mathcal{S}_1}, \mathbf{e}_{\xi_2})\big) \subset \{S < f_{\mathbf{e}}(T) \,:\, T \in \mathbb{R}\},\\
&\big(P - \text{Cone}^0(\mathbf{e}_{\mathcal{S}_1}, \mathbf{e}_{\xi_2})\big) \cap B_r(P) \subset \Omega,\\
&\big(P + \text{Cone}^0(\mathbf{e}_{\mathcal{S}_1}, \mathbf{e}_{\xi_2})\big) \subset \{S > f_{\mathbf{e}}(T) \,:\, T \in \mathbb{R}\},\\
&\big(P + \text{Cone}^0(\mathbf{e}_{\mathcal{S}_1}, \mathbf{e}_{\xi_2})\big) \cap \Omega = \emptyset.
\end{aligned}$$

That is, at P*,* $\text{Cone}^0(\mathbf{e}_{\mathcal{S}_1}, \mathbf{e}_{\xi_2})$ *is below the graph of* $f_{\mathbf{e}}$*, and* $-\text{Cone}^0(\mathbf{e}_{\mathcal{S}_1}, \mathbf{e}_{\xi_2})$ *is above the graph of* $f_{\mathbf{e}}$.

Moreover, $f_{\mathbf{e}}$ *satisfies the following:*

(a) *The directions of tangent lines to* Γ_{shock} *are between the directions of line* $\mathcal{S}_1$ *and* $\{t\mathbf{e}_{\xi_2} \,:\, t \in \mathbb{R}\}$*, which are tangent lines to* Γ_{shock} *at points* P_1 *and* P_2*, respectively. That is, for any* $T \in (T_{P_2}, T_{P_1})$*,*

$$-\infty < \frac{\mathbf{e}_{\mathcal{S}_1} \cdot \mathbf{e}}{\mathbf{e}_{\mathcal{S}_1} \cdot \mathbf{e}^{\perp}} = f'_{\mathbf{e}}(T_{P_1}) \le f'_{\mathbf{e}}(T) \le f'_{\mathbf{e}}(T_{P_2}) = \frac{\mathbf{e}_{\xi_2} \cdot \mathbf{e}}{\mathbf{e}_{\xi_2} \cdot \mathbf{e}^{\perp}} < \infty. \tag{8.2.24}$$

(b) *In particular, when* $\mathbf{e} = \boldsymbol{\nu}_{\text{w}}$ *(cf.* (8.2.14)*), region* Ω *in the* (S,T)*–coordinates is of the following form:*

$$\begin{aligned}
&f_{\mathbf{e},\text{sh}}(T) > \max(L_{\mathbf{e},\text{w}}(T), L_{\mathbf{e},\text{sym}}(T)) \ge 0 \qquad \textit{for } T \in [T_{P_2}, T_{P_1}],\\
&f_{\mathbf{e},\text{so}}(T) > \max(L_{\mathbf{e},\text{w}}(T), L_{\mathbf{e},\text{sym}}(T)) \qquad\quad\;\; \textit{for } T \in [T_{P_1}, T_{P_4}),\\
&\Lambda = \big\{(S,T) \in \mathbb{R}^2 \,:\, T \in \mathbb{R},\; S > \max(L_{\mathbf{e},\text{w}}(T), L_{\mathbf{e},\text{sym}}(T))\big\},\\
&\Omega = \left\{ (S,T) \in \mathbb{R}^2 \,:\, \begin{array}{l} T_{P_2} < T < T_{P_4},\\ -(T - T_{P_3})\tan\theta_{\text{w}} < S < f_{\mathbf{e},\text{sh}}(T)\\ \qquad\qquad\qquad\qquad \textit{for } T \in (T_{P_2}, T_{P_3}],\\ 0 < S < f_{\mathbf{e},\text{sh}}(T) \;\; \textit{for } T \in (T_{P_3}, T_{P_1}],\\ 0 < S < f_{\mathbf{e},\text{so}}(T) \;\; \textit{for } T \in (T_{P_1}, T_{P_4}) \end{array} \right\},
\end{aligned} \tag{8.2.25}$$

where $f_{\mathbf{e},\text{sh}} = f_{\mathbf{e},\text{so}}$ *at* T_{P_1}*, and the notations introduced in Remark* 8.2.12 *have been used.*

Proof. Since $\varphi = \varphi_1$ on Γ_{shock} and $\varphi < \varphi_1$ in Ω by Corollary 8.1.10, then Corollary 8.2.10, regularity $\varphi \in C^1(\overline{\Omega})$ given in (8.1.3), and the implicit function theorem imply the existence of $f_{\mathbf{e}} \in C^1([T_{P_2}, T_{P_1}])$ such that (i) holds. Thus, (ii) follows from the definition of P_1 and P_2.

We extend $f_{\mathbf{e}}$ from $[T_{P_2}, T_{P_1}]$ to $\mathbb{R}$ by the tangent lines to $S = f_{\mathbf{e}}(T)$ at the endpoints, *i.e.*, we define

$$f_{\mathbf{e}}(T) = \begin{cases} f'_{\mathbf{e}}(T_{P_2})(T - T_{P_2}) + f_{\mathbf{e}}(T_{P_2}) & \text{for } T < T_{P_2}, \\ f'_{\mathbf{e}}(T_{P_1})(T - T_{P_1}) + f_{\mathbf{e}}(T_{P_1}) & \text{for } T > T_{P_1}. \end{cases}$$

Note that the tangent lines to Γ_{shock} at P_1 and P_2 are in directions $\mathbf{e}_{\mathcal{S}_1}$ and $\mathbf{e}_{\xi_2}$, respectively. Also, the extended function $f_{\mathbf{e}}$ is in $C^1(\mathbb{R})$.

From Corollary 8.2.10 (combined with the extension of $f_{\mathbf{e}}$ beyond $[T_{P_2}, T_{P_1}]$ defined above), it follows that Γ_{shock} intersects only once with each line $t \mapsto P + t\mathbf{e}$, where $P \in \Gamma_{\text{shock}}$ and $\mathbf{e} \in \text{Cone}^0(\mathbf{e}_{\mathcal{S}_1}, \mathbf{e}_{\xi_2})$. Moreover, $\partial_{\mathbf{e}}(\varphi_1 - \varphi) < 0$ in $\overline{\Omega}$ by Corollary 8.2.10 and $\Omega \subset \{\varphi_1 > \varphi\} \cap \Lambda$ with $\Gamma_{\text{shock}} \subset \{\varphi_1 = \varphi\}$. Using this and the fact that $\Gamma_{\text{shock}} \subset \partial\Omega$ is a C^1–curve without self-intersection, which does not intersect with the other boundary parts in the points of its relative interior as assumed in Definition 8.1.1(i), we conclude (iii).

Now we prove (8.2.24). We work in the (S, T)–coordinates. For any $P = (S_P, T_P) \in \mathbb{R}^2$ and $\mathbf{g} \neq c\mathbf{e}$, line $\{P + t\mathbf{g} : t \in \mathbb{R}\}$ is the graph: $\{(S, T) : S = l_{\mathbf{g}}(T),\ T \in \mathbb{R}\}$, where

$$l_{\mathbf{g}}(T) = S_P + \frac{\mathbf{g} \cdot \mathbf{e}}{\mathbf{g} \cdot \mathbf{e}^{\perp}}(T - T_P).$$

Let $T \in (T_{P_2}, T_{P_1})$. Then $P = (f_{\mathbf{e}}(T), T) \in \Gamma_{\text{shock}}$. Recall that $\mathbf{e}^{\perp}$ is chosen so that $\mathbf{e}^{\perp} \cdot \mathbf{e}_{\mathcal{S}_1} < 0$. Also, $\mathbf{e} \in \text{Cone}^0(\mathbf{e}_{\mathcal{S}_1}, \mathbf{e}_{\xi_2})$ implies $\mathbf{e} = a\mathbf{e}_{\mathcal{S}_1} + b\mathbf{e}_{\xi_2}$ for some $a, b > 0$. Then

$$\begin{aligned} b\mathbf{e}_{\xi_2} \cdot \mathbf{e} &= |\mathbf{e}|^2 - a\mathbf{e}_{\mathcal{S}_1} \cdot \mathbf{e} > -a\mathbf{e}_{\mathcal{S}_1} \cdot \mathbf{e}, \\ b\mathbf{e}_{\xi_2} \cdot \mathbf{e}^{\perp} &= -a\mathbf{e}_{\mathcal{S}_1} \cdot \mathbf{e}^{\perp} > 0, \end{aligned}$$

so that

$$\frac{\mathbf{e}_{\xi_2} \cdot \mathbf{e}}{\mathbf{e}_{\xi_2} \cdot \mathbf{e}^{\perp}} > \frac{\mathbf{e}_{\mathcal{S}_1} \cdot \mathbf{e}}{\mathbf{e}_{\mathcal{S}_1} \cdot \mathbf{e}^{\perp}}.$$

It follows from (iii) that, for any small $\tau > 0$,

$$l_{\mathbf{e}_{\xi_2}}(T + \tau) \geq f(T + \tau) \geq l_{\mathbf{e}_{\mathcal{S}_1}}(T + \tau),$$

that is,

$$f(T) + \frac{\mathbf{e}_{\xi_2} \cdot \mathbf{e}}{\mathbf{e}_{\xi_2} \cdot \mathbf{e}^{\perp}}\tau \geq f(T + \tau) \geq f(T) + \frac{\mathbf{e}_{\mathcal{S}_1} \cdot \mathbf{e}}{\mathbf{e}_{\mathcal{S}_1} \cdot \mathbf{e}^{\perp}}\tau.$$

This implies (8.2.24).

Finally, combining assertions (i)–(iii) with the facts about state (2) stated in Definition 7.5.7 and Lemma 8.2.11, we conclude (8.2.25). □

8.2.2 Directional derivatives of ϕ in a cone of directions cannot attain the maximum on Γ_{shock}

In §8.2.1, we have shown that $\bar{\phi} = \varphi_1 - \varphi$ is strictly monotone in the cone of directions, $\text{Cone}^0(\mathbf{e}_{\mathcal{S}_1}, \mathbf{e}_{\xi_2})$. In order to prove this, we have demonstrated in Lemma 8.2.3 that the maximum or minimum of $\partial_{\mathbf{e}}\phi$ (and hence of $\bar{\phi}$) for any $\mathbf{e} \in \mathbb{R}^2 \setminus \{0\}$ cannot be attained within $\Omega \cup \Gamma_{\text{wedge}} \cup \Gamma_{\text{sym}}$. On the other hand, in order to prove the same property on Γ_{shock}, we used Lemma 8.2.5, which is specific to $\bar{\phi}$ because of the assumption of non-positivity of $\partial_{\mathbf{e}}\bar{\phi}$ in the cone of directions $\mathbf{e} \in \text{Cone}^0(\mathbf{e}_{\mathcal{S}_1}, \mathbf{e}_{\xi_2})$ and the fact that $\bar{\phi} = 0$ on Γ_{shock}.

In this section, we partially extend the properties proved in Lemma 8.2.3 to Γ_{shock}. We first refine the result of Lemma 8.2.4.

Lemma 8.2.15. *Let φ be a solution of* **Problem 2.6.1** *satisfying conditions* (i)–(iv) *of Definition* 8.1.1. *For a fixed $\mathbf{e} \in \mathbb{R}^2 \setminus \{0\}$, define $w := \partial_{\mathbf{e}}\phi$ in Ω. If the maximum of w over $\overline{\Omega}$ is attained at $\hat{P} \in \Gamma_{\text{shock}}$ and $\boldsymbol{\nu}(\hat{P}) \cdot \mathbf{e} < 0$, then $\phi_{\boldsymbol{\tau}\boldsymbol{\tau}}(\hat{P}) < 0$, where $\boldsymbol{\nu}$ denotes the interior unit normal to Γ_{shock} with respect to Ω, and curve Γ_{shock} does not include its endpoints.*

Proof. Let $\mathbf{e} = (a_1, a_2)$ in the $\boldsymbol{\xi}$–coordinates. Then $w = \mathbf{e} \cdot D\phi$. By Lemma 8.2.4,

$$\mathbf{h}(\hat{P}) = k\mathbf{e} \qquad \text{for some } k \in \mathbb{R},$$

where $\mathbf{h}$ is defined by (5.1.23). From (8.1.17), $\mathbf{h} \cdot \boldsymbol{\nu} < 0$ at $\hat{P}$. Thus, $k\mathbf{e} \cdot \boldsymbol{\nu}(\hat{P}) < 0$. Now the assumption that $\boldsymbol{\nu}(\hat{P}) \cdot \mathbf{e} < 0$ implies

$$k > 0.$$

Writing (5.1.22) at $\hat{P}$ in basis $\{\boldsymbol{\nu}(\hat{P}), \boldsymbol{\tau}(\hat{P})\}$, we obtain (8.2.4):

$$\phi_{\boldsymbol{\nu}\boldsymbol{\tau}} = -\frac{h_{\boldsymbol{\tau}}}{h_{\boldsymbol{\nu}}} \phi_{\boldsymbol{\tau}\boldsymbol{\tau}} \qquad \text{at } \hat{P}. \tag{8.2.26}$$

Equation (2.2.11), written in the (S, T)–coordinates with basis $\{\boldsymbol{\nu}(\hat{P}), \boldsymbol{\tau}(\hat{P})\}$, combined with (8.2.26), implies

$$\phi_{\boldsymbol{\nu}\boldsymbol{\nu}} = \frac{1}{c^2 - \varphi_{\boldsymbol{\nu}}^2} \Big(-2\varphi_{\boldsymbol{\nu}}\varphi_{\boldsymbol{\tau}} \frac{h_{\boldsymbol{\tau}}}{h_{\boldsymbol{\nu}}} - (c^2 - \varphi_{\boldsymbol{\tau}}^2) \Big) \phi_{\boldsymbol{\tau}\boldsymbol{\tau}} \qquad \text{at } \hat{P}, \tag{8.2.27}$$

where we have used that $|D\varphi| < c$ in $\overline{\Omega} \setminus \overline{\Gamma_{\text{sonic}}}$.

We have

$$w_{\boldsymbol{\nu}} = (\mathbf{e} \cdot \boldsymbol{\nu})\phi_{\boldsymbol{\nu}\boldsymbol{\nu}} + (\mathbf{e} \cdot \boldsymbol{\tau})\phi_{\boldsymbol{\nu}\boldsymbol{\tau}} = D^2\phi[\boldsymbol{\nu}, \mathbf{e}] = \frac{1}{k} D^2\phi[\boldsymbol{\nu}, \mathbf{h}] \qquad \text{at } \hat{P},$$

so that

$$w_{\boldsymbol{\nu}} = \frac{1}{k}(h_{\boldsymbol{\nu}}\phi_{\boldsymbol{\nu}\boldsymbol{\nu}} + h_{\boldsymbol{\tau}}\phi_{\boldsymbol{\nu}\boldsymbol{\tau}}) \qquad \text{at } \hat{P}.$$

Then, substituting the expressions of $h_{\boldsymbol{\nu}}$ and $h_{\boldsymbol{\tau}}$ given by (5.1.23), and the expressions of $\psi_{\boldsymbol{\nu}\boldsymbol{\nu}}$ and $\psi_{\boldsymbol{\nu}\boldsymbol{\tau}}$ given by (8.2.26)–(8.2.27), we obtain (after a somewhat tedious but straightforward calculation):

$$w_{\boldsymbol{\nu}} = \frac{c^2}{k\rho\varphi_{\boldsymbol{\nu}}(c^2-\varphi_{\boldsymbol{\nu}}^2)}\big(\rho^2\varphi_{\boldsymbol{\nu}}^2(c^2-|D\varphi|^2)+\rho_1^2c^2\varphi_{\boldsymbol{\tau}}^2\big)\phi_{\boldsymbol{\tau}\boldsymbol{\tau}} \qquad \text{at } \hat{P}. \tag{8.2.28}$$

Since w satisfies equation (5.1.10), which is uniformly elliptic and has continuous coefficients in $\overline{\Omega}\cap B_r(\hat{P})$ for some small $r>0$, and since $\hat{P}$ is a maximum point of w over $\overline{\Omega}$ and w is not constant in Ω by Lemma 8.2.1, then Hopf's lemma implies

$$w_{\boldsymbol{\nu}}(\hat{P})<0.$$

Thus, the right-hand side of (8.2.28) is negative. We employ that $\varphi_{\boldsymbol{\nu}}>0$ on Γ_{shock} (by Lemma 8.1.7), $|D\varphi|<c$ at $\hat{P}$, and $k>0$ to conclude

$$\phi_{\boldsymbol{\tau}\boldsymbol{\tau}}<0 \;\text{ at } \hat{P}.$$

□

Now we prove the main result of this subsection.

Proposition 8.2.16. *Let* $\varphi=\phi-\frac{|\boldsymbol{\xi}|^2}{2}$ *be an admissible solution of* **Problem 2.6.1** *in the sense of Definition* 8.1.1. *Let* $\mathbf{e}\in \text{Cone}^0(\mathbf{e}_{S_1},\mathbf{e}_{\xi_2})$. *Then the maximum of* $\partial_{\mathbf{e}}\phi$ *over* $\overline{\Omega}$ *cannot be attained on* Γ_{shock}.

Proof. We can assume without loss of generality that $|\mathbf{e}|=1$. Denote

$$w=\partial_{\mathbf{e}}\phi.$$

We need to show that w cannot attain its maximum over $\overline{\Omega}$ on Γ_{shock}. We divide the proof into three steps.

1. Suppose that there exists $\hat{P}=\hat{\boldsymbol{\xi}}\in\Gamma_{\text{shock}}$ such that

$$w(\hat{P})=\max_{\boldsymbol{\xi}\in\overline{\Omega}} w(\boldsymbol{\xi}).$$

Let (S,T) be the coordinates with basis $\{\mathbf{e},\mathbf{e}^{\perp}\}$, where $\mathbf{e}^{\perp}$ is as in Corollary 8.2.14. Then, by Corollary 8.2.14, Γ_{shock} is a graph in the S–direction, *i.e.*, there exists $f_{\mathbf{e}}\in C^1(\mathbb{R})$ such that $\Gamma_{\text{shock}}=\{S=f_{\mathbf{e}}(T)\;:\;T_{P_2}<T<T_{P_1}\}$, $\Omega\subset\{S<f_{\mathbf{e}}(T)\;:\;T\in\mathbb{R}\}$, and $f_{\mathbf{e}}$ satisfies all the other properties in Corollary 8.2.14.

Furthermore, by Lemma 8.2.15,

$$\phi_{\boldsymbol{\tau}\boldsymbol{\tau}}(\hat{P})<0. \tag{8.2.29}$$

Let $\hat{P}=(\hat{S},\hat{T})$ in the (S,T)–coordinates. Then $\hat{T}\in(T_{P_2},T_{P_1})$. From the definition of $f_{\mathbf{e}}$, $\bar{\phi}(f_{\mathbf{e}}(T),T)=0$ holds on $T\in(T_{P_2},T_{P_1})$ for $\bar{\phi}=\phi_1-\phi$.

Differentiating this equality twice and using $\boldsymbol{\nu} = \frac{D\bar{\phi}}{|D\bar{\phi}|}$ and $\boldsymbol{\tau} = \boldsymbol{\nu}^{\perp}$, which imply that $\partial_{\boldsymbol{\tau}}\bar{\phi}(\hat{P}) = 0$, and $\partial_{\mathbf{e}}\bar{\phi}(\hat{P}) > 0$ by Corollary 8.2.10, we have

$$f_{\mathbf{e}}''(\hat{T}) = -\frac{D^2\bar{\phi}[D^{\perp}\bar{\phi}, D^{\perp}\bar{\phi}]}{(\partial_{\mathbf{e}}\bar{\phi})^3}(\hat{P}) = -\frac{(\partial_{\boldsymbol{\nu}}\bar{\phi})^2\bar{\phi}_{\boldsymbol{\tau}\boldsymbol{\tau}}}{(\partial_{\mathbf{e}}\bar{\phi})^3}(\hat{P}),$$

where $D^{\perp}\bar{\phi} = (-D_T\bar{\phi}, D_S\bar{\phi})$, and we have also used the fact that the expression of $D^2\bar{\phi}[D^{\perp}\bar{\phi}, D^{\perp}\bar{\phi}]$ is invariant with respect to the orthogonal change of variables that can be expressed in basis $\{\boldsymbol{\nu}, \boldsymbol{\tau}\}$. Recall that ϕ_1 is a linear function so that $D^2\bar{\phi} = D^2(\phi_1 - \phi) = -D^2\phi$. Then $\bar{\phi}_{\boldsymbol{\tau}\boldsymbol{\tau}}(\hat{P}) > 0$ by (8.2.29). Also, $\partial_{\mathbf{e}}\bar{\phi}(\hat{P}) > 0$ by Corollary 8.2.10. Thus, we have

$$f_{\mathbf{e}}''(\hat{T}) > 0.$$

The tangent line to $\Gamma_{\rm shock}$ at $\hat{P}$ is $\{S = l(T)\}$, where

$$l(T) = f_{\mathbf{e}}'(\hat{T})(T - \hat{T}) + \hat{S}.$$

Denote

$$F(T) = f_{\mathbf{e}}(T) - l(T).$$

Let $T^* \in [T_{P_2}, T_{P_1}]$ be a point at which the maximum of $F(T)$ over $[T_{P_2}, T_{P_1}]$ is attained. Let $S^* = f_{\mathbf{e}}(T^*)$ and $P^* = (S^*, T^*)$. Then $P^* \in \overline{\Gamma_{\rm shock}}$.

Since $l(T)$ is a linear function, $F''(\hat{T}) = f_{\mathbf{e}}''(\hat{T}) > 0$. Using that $\hat{T} \in (T_{P_2}, T_{P_1})$, we find that the maximum of F over $[T_{P_2}, T_{P_1}]$ cannot be attained at $\hat{T}$. Thus, $\hat{T} \neq T^*$ and

$$F(\hat{T}) < F(T^*). \tag{8.2.30}$$

Note that

$$F'(\hat{T}) = f_{\mathbf{e}}'(\hat{T}) - f_{\mathbf{e}}'(\hat{T}) = 0.$$

If $T^* \in (T_{P_2}, T_{P_1})$, then $F'(T^*) = 0$, since T^* is a maximum point. Suppose that $T^* = T_{P_2}$ or $T^* = T_{P_1}$. Since $l(T)$ is a linear function, we obtain from (8.2.24) that

$$F'(T_{P_1}) \le F'(T) \le F'(T_{P_2}) \qquad \text{for all } T \in [T_{P_2}, T_{P_1}]. \tag{8.2.31}$$

From this, since $F'(\hat{T}) = 0$, $T_{P_2} < T_{P_1}$, and T^* is a maximum point, we find that $F'(T^*) = 0$ if $T^* = T_{P_2}$ or $T^* = T_{P_1}$. Indeed, if $T^* = T_{P_2}$, then $F'(T_{P_2}) \le 0$, since the maximum over $[T_{P_2}, T_{P_1}]$ is attained at T_{P_2}. If $F'(T_{P_2}) < 0$, we use (8.2.31) to see that $F'(T) \le F'(T_{P_2}) < 0$ for any $T \in [T_{P_2}, T_{P_1}]$, which is in contradiction to the fact that $F'(\hat{T}) = 0$. Case $T^* = T_{P_1}$ is similar. Thus, in all cases, $F'(T^*) = 0 = F'(\hat{T})$, which implies

$$f_{\mathbf{e}}'(T^*) = f_{\mathbf{e}}'(\hat{T}).$$

Therefore, the tangent lines to $\Gamma_{\rm shock}$ at $\hat{P}$ and P^*, denoted by $L_{\hat{P}}$ and L_{P^*}, are parallel to each other.

Moreover, using that $\Omega \subset \{S < f_{\mathbf{e}}(T) \ : \ T \in \mathbb{R}\}$ and denoting by $B := f'_{\mathbf{e}}(T^*) = f'_{\mathbf{e}}(\hat{T})$, we have

$$\boldsymbol{\nu}_{\rm sh}(\hat{P}) = \boldsymbol{\nu}_{\rm sh}(P^*) = \frac{1}{\sqrt{1+B^2}}\left(-\mathbf{e} + B\mathbf{e}^{\perp}\right). \tag{8.2.32}$$

2. Denote by $\mathcal{O}_1$ the center of sonic circle of state (1). Then $\mathcal{O}_1 = (0, u_1)$ in the $\boldsymbol{\xi}$–coordinates. In this step, we show that $\text{dist}(\mathcal{O}_1, L_{P^*}) > \text{dist}(\mathcal{O}_1, L_{\hat{P}})$.

Since $\boldsymbol{\nu}(\hat{P}) = \boldsymbol{\nu}(P^*)$ and $\boldsymbol{\tau}(\hat{P}) = \boldsymbol{\tau}(P^*)$ from the results of Step 1, in the calculations below, $\boldsymbol{\nu}_{\rm sh}$ denotes both $\boldsymbol{\nu}(\hat{P})$ and $\boldsymbol{\nu}(P^*)$, and $\boldsymbol{\tau}_{\rm sh}$ denotes both $\boldsymbol{\tau}(\hat{P})$ and $\boldsymbol{\tau}(P^*)$.

By Lemma 8.1.7, we see that $(\mathcal{O}_1 - P)\cdot\boldsymbol{\nu}(P) = D\varphi_1(P)\cdot\boldsymbol{\nu}(P) > 0$ for any $P \in \Gamma_{\rm shock}$. Then, since $\hat{P} \in L_{\hat{P}}$ and $P^* \in L_{P^*}$, and $\boldsymbol{\nu}_{\rm sh}$ is the normal to both $L_{\hat{P}}$ and L_{P^*}, we obtain by Lemma 6.1.1 that

$$\text{dist}(\mathcal{O}_1, L_{\hat{P}}) = \boldsymbol{\nu}_{\rm sh}\cdot(\mathcal{O}_1 - \hat{P}), \qquad \text{dist}(\mathcal{O}_1, L_{P^*}) = \boldsymbol{\nu}_{\rm sh}\cdot(\mathcal{O}_1 - P^*).$$

We now find the expression of $\mathcal{O}_1 - P^*$ in terms of $\mathcal{O}_1 - \hat{P}$.

From the definition of (S,T)–coordinates and the shock function $f_{\mathbf{e}}$ in Step 1, we have

$$P^* = \hat{P} + \big(f_{\mathbf{e}}(T^*) - f_{\mathbf{e}}(\hat{T})\big)\mathbf{e} + \big(T^* - \hat{T}\big)\mathbf{e}^{\perp}.$$

Using the definitions of functions $F(T)$ and $l(T)$ in Step 1, we rewrite this expression as

$$P^* = \hat{P} + \big(F(T^*) - F(\hat{T})\big)\mathbf{e} + \big(T^* - \hat{T}\big)\big(f'_{\mathbf{e}}(\hat{T})\mathbf{e} + \mathbf{e}^{\perp}\big).$$

Since $\Gamma_{\rm shock} = \{S = f_{\mathbf{e}}(T) \ : \ T_{P_2} < T < T_{P_1}\}$, we have

$$\boldsymbol{\tau}_{\rm sh} = \boldsymbol{\tau}(\hat{P}) = \frac{1}{\sqrt{1+(f'_{\mathbf{e}}(\hat{T}))^2}}\big(f'_{\mathbf{e}}(\hat{T})\mathbf{e} + \mathbf{e}^{\perp}\big).$$

Combining the last two equations, we obtain

$$P^* = \hat{P} + (F(T^*) - F(\hat{T}))\mathbf{e} + M\boldsymbol{\tau}_{\rm sh},$$

where $M = (T^* - \hat{T})\sqrt{1+(f'_{\mathbf{e}}(\hat{T}))^2}$. This implies

$$\mathcal{O}_1 - P^* = (\mathcal{O}_1 - \hat{P}) - (F(T^*) - F(\hat{T}))\mathbf{e} - M\boldsymbol{\tau}_{\rm sh}.$$

Then

$$\boldsymbol{\nu}_{\rm sh}\cdot(\mathcal{O}_1 - P^*) = \boldsymbol{\nu}_{\rm sh}\cdot(\mathcal{O}_1 - \hat{P}) - (F(T^*) - F(\hat{T}))\mathbf{e}\cdot\boldsymbol{\nu}_{\rm sh} > \boldsymbol{\nu}_{\rm sh}\cdot(\mathcal{O}_1 - \hat{P}),$$

where the inequality holds since $F(T^*) - F(\hat{T}) > 0$ by (8.2.30) and $\mathbf{e}\cdot\boldsymbol{\nu}_{\rm sh} < 0$ by Corollary 8.2.13. Thus, we have shown that

$$\text{dist}(\mathcal{O}_1, L_{P^*}) > \text{dist}(\mathcal{O}_1, L_{\hat{P}}). \tag{8.2.33}$$

3. By (8.1.2), Lemma 8.1.7, and (8.2.32)–(8.2.33), we can apply Lemma 6.1.4 to points $\hat{P}$ and P^* to obtain

$$\partial_{\boldsymbol{\nu}}\phi(P^*) < \partial_{\boldsymbol{\nu}}\phi(\hat{P}) \qquad \text{for } \boldsymbol{\nu} := \boldsymbol{\nu}_{\text{sh}}(\hat{P}) = \boldsymbol{\nu}_{\text{sh}}(\hat{P}^*).$$

Also, since $\boldsymbol{\tau}_{\text{sh}}(P^*) = \boldsymbol{\tau}_{\text{sh}}(\hat{P}) =: \boldsymbol{\tau}_{\text{sh}}$, and ϕ_1 is a linear function, we use (8.1.14) to obtain

$$\partial_{\boldsymbol{\tau}}\phi(P^*) = \boldsymbol{\tau}_{\text{sh}} \cdot D\phi_1(P^*) = \boldsymbol{\tau}_{\text{sh}} \cdot D\phi_1(\hat{P}) = \partial_{\boldsymbol{\tau}}\phi(\hat{P}).$$

Furthermore, $\mathbf{e} \cdot \boldsymbol{\nu}_{\text{sh}} < 0$, by Corollary 8.2.13. Then we have

$$\begin{aligned}
\mathbf{e} \cdot D\phi(P^*) &= \partial_{\boldsymbol{\nu}}\phi(P^*)\, \boldsymbol{\nu}_{\text{sh}} \cdot \mathbf{e} + \partial_{\boldsymbol{\tau}}\phi(P^*)\boldsymbol{\tau}_{\text{sh}} \cdot \mathbf{e} \\
&= \partial_{\boldsymbol{\nu}}\phi(P^*)\, \boldsymbol{\nu}_{\text{sh}} \cdot \mathbf{e} + \partial_{\boldsymbol{\tau}}\phi(\hat{P})\boldsymbol{\tau}_{\text{sh}} \cdot \mathbf{e} \\
&> \partial_{\boldsymbol{\nu}}\phi(\hat{P})\, \boldsymbol{\nu}_{\text{sh}} \cdot \mathbf{e} + \partial_{\boldsymbol{\tau}}\phi(\hat{P})\boldsymbol{\tau}_{\text{sh}} \cdot \mathbf{e} \\
&= \mathbf{e} \cdot D\phi(\hat{P}),
\end{aligned}$$

which is to say that $w(P^*) > w(\hat{P})$. This contradicts the assumption that the maximum of w over $\overline{\Omega}$ is attained at $\hat{P}$. ☐

8.2.3 Directions of monotonicity of $\varphi - \varphi_2$

Proposition 8.2.17. *Let φ be an admissible solution of* **Problem 2.6.1** *in the sense of Definition* 8.1.1. *Then*

$$\partial_{\mathbf{e}_{\mathcal{S}_1}}(\varphi - \varphi_2) > 0 \qquad \text{in } \overline{\Omega} \setminus \overline{\Gamma_{\text{sonic}}}, \tag{8.2.34}$$

where $\mathbf{e}_{\mathcal{S}_1}$ is defined by (7.5.8).

Proof. Since $\mathbf{e}_{\mathcal{S}_1} \parallel \mathcal{S}_1$ and $\mathcal{S}_1 = \{\varphi_1 = \varphi_2\} = \{\phi_1 = \phi_2\}$, and ϕ_1 and ϕ_2 are linear functions, we find that $\partial_{\mathbf{e}_{\mathcal{S}_1}}\phi_1 = \partial_{\mathbf{e}_{\mathcal{S}_1}}\phi_2$ in $\mathbb{R}^2$. Thus, by Proposition 8.2.6, we obtain

$$\partial_{\mathbf{e}_{\mathcal{S}_1}}(\varphi - \varphi_2) = \partial_{\mathbf{e}_{\mathcal{S}_1}}(\phi - \phi_2) = \partial_{\mathbf{e}_{\mathcal{S}_1}}(\phi - \phi_1) = \partial_{\mathbf{e}_{\mathcal{S}_1}}(\varphi - \varphi_1) > 0$$

in $\overline{\Omega} \setminus \overline{\Gamma_{\text{sonic}}}$. ☐

Next, we show that $\varphi - \varphi_2$ is monotone in the direction orthogonal to Γ_{wedge}. More precisely, $\varphi - \varphi_2$ is non-decreasing in the $\boldsymbol{n}_{\text{w}}$–direction in Ω, where

$$\boldsymbol{n}_{\text{w}} = -\boldsymbol{\nu}_{\text{w}} = (\sin\theta_{\text{w}}, -\cos\theta_{\text{w}}). \tag{8.2.35}$$

We first prove some preliminary facts. We will use vector $\boldsymbol{\tau}_{\text{w}}$ defined by (8.2.17).

Lemma 8.2.18. *Let* φ *be a solution of* **Problem 2.6.1** *satisfying* (8.1.7) *and conditions* (i)–(iv) *of Definition* 8.1.1. *Then*

$$\phi(P_2) \le \phi(P) \qquad \textit{for all } P \in \overline{\Omega}, \tag{8.2.36}$$

$$\partial_{\xi_1}\phi(P) \ge 0 \qquad \textit{for all } P \in \overline{\Gamma_{\rm sym}}. \tag{8.2.37}$$

Proof. We divide the proof into two steps.

1. We first note that $\partial_{\xi_2}(\phi_1 - \phi) = \partial_{\xi_2}(\varphi_1 - \varphi) \le 0$ in Ω by (8.1.7), where $\partial_{\xi_2}\phi_1 = 0$ by (2.2.17). Then

$$\partial_{\xi_2}\phi \ge 0 \qquad \text{in } \Omega. \tag{8.2.38}$$

Let $\mathbb{R}_+ = \{t > 0\}$. By (8.2.11), we have

$$\Omega \subset \cup_{P \in \overline{\Gamma_{\rm sym}} \cup \Gamma_{\rm wedge}} (P + \mathbb{R}_+ \mathbf{e}_{\xi_2}).$$

Thus, by (8.2.38),

$$\min_{\overline{\Omega}}\phi = \min_{\overline{\Gamma_{\rm sym} \cup \Gamma_{\rm wedge}}}\phi. \tag{8.2.39}$$

On the other hand, we can express the unit tangent vector to $\Gamma_{\rm wedge}$ as

$$\boldsymbol{\tau}_{\rm w} = A\Big(\cot\theta_{\rm w}\boldsymbol{n}_{\rm w} + \frac{1}{\sin\theta_{\rm w}}\mathbf{e}_{\xi_2}\Big) \qquad \text{for } A = \frac{\sin\theta_{\rm w}}{\sqrt{\cos^2\theta_{\rm w} + 1}} > 0.$$

From that, using $\phi_{\boldsymbol{\nu}} = 0$ on $\Gamma_{\rm wedge}$ and (8.2.38), we have

$$\partial_{\boldsymbol{\tau}_{\rm w}}\phi = A\Big(\cot\theta_{\rm w}\partial_{\boldsymbol{n}_{\rm w}}\phi + \frac{1}{\sin\theta_{\rm w}}\partial_{\mathbf{e}_{\xi_2}}\phi\Big) = \frac{A}{\sin\theta_{\rm w}}\partial_{\mathbf{e}_{\xi_2}}\phi \ge 0 \qquad \text{on } \Gamma_{\rm wedge}.$$

Thus, $\phi(P) \ge \phi(P_3)$ for any $P \in \Gamma_{\rm wedge}$. Then (8.2.39) implies

$$\min_{\overline{\Omega}}\phi = \min_{\overline{\Gamma_{\rm sym}}}\phi. \tag{8.2.40}$$

Clearly, (8.2.40), combined with (8.2.37), implies (8.2.36). Therefore, it remains to prove (8.2.37).

2. To achieve this, it is convenient to consider the extension of ϕ by even reflection into domain $\Omega^{\rm ext}$, as in Remark 8.1.3. Then the extended function $\phi \in C^1(\overline{\Omega^{\rm ext}})$. Also, $D\phi(P_3) = 0$, since $\phi_{\boldsymbol{\nu}_{\rm sym}}(P_3) = 0$ and $\phi_{\boldsymbol{n}_{\rm w}}(P_3) = 0$, where $\boldsymbol{\nu}_{\rm sym} = (0, 1)$. Furthermore, (8.2.38) and the extension of ϕ into $\xi_2 < 0$ by even reflection imply

$$\xi_2\,\partial_{\xi_2}\phi(\boldsymbol{\xi}) \ge 0 \qquad \text{for all } \boldsymbol{\xi} \in \Omega^{\rm ext}. \tag{8.2.41}$$

Moreover, equation (2.2.8) (considered as a linear equation with respect to ϕ) is strictly elliptic near P_3. Since $\phi \in C^1(\overline{\Omega^{\rm ext}})$, the coefficients of (2.2.8) are in $C(\overline{\Omega^{\rm ext}})$, which implies that (2.2.8) is uniformly elliptic in a neighborhood of P_3 in $\overline{\Omega^{\rm ext}}$.

Now we prove (8.2.37). Recall that $|D\phi(P_3)| = 0$. Suppose that there exists $P^* \in \overline{\Gamma_{\rm sym}} \setminus \{P_3\}$ satisfying that $\partial_{\xi_1}\phi(P^*) < 0$.

If $\partial_{\xi_1}\phi(P) \le 0$ for any $P \in [P^*, P_3]$ (in this argument, $[P^*, P_3] \subset \Gamma_{\rm sym}$ denotes the straight segment connecting P^* to P_3), we then use (8.2.41) to conclude that there exists $r > 0$ such that $\hat{B}_r(P_3) := B_r(P_3) \cap \{\xi_1 < 0\} \subset \Omega$ and $\phi(P_3) \le \phi(\boldsymbol{\xi})$ for any $\boldsymbol{\xi} \in \hat{B}_r(P_3)$. Since $D\phi(P_3) = 0$ and $P_3 = \mathbf{0} \in \partial\hat{B}_r(P_3)$, we obtain a contradiction to Hopf's lemma at P_3 applied in domain $\hat{B}_r(P_3)$, where we note that this domain satisfies the interior sphere condition at P_3.

Therefore, there exists $P^{**} \in (P^*, P_3)$ satisfying that $\partial_{\xi_1}\phi(P^{**}) > 0$. Then the minimum of ϕ over $[P^*, P^{**}]$ is attained at some point $Q \in (P^*, P^{**})$ with $\phi(Q) < \phi(P^*)$ and $\phi(Q) < \phi(P^{**})$. Using (8.2.41), we conclude that Q is a point of local minimum of ϕ in $\Omega^{\rm ext}$, and ϕ is not constant on $\Gamma_{\rm sym}$. This contradicts the strong maximum principle, since Q is an interior point of $\Omega^{\rm ext}$ and equation (2.2.8) is uniformly elliptic near $\Gamma_{\rm sym}$ by Definition 8.1.1(iii). Therefore, (8.2.37) is proved, which implies (8.2.36). □

Now we prove the monotonicity property:

Proposition 8.2.19. *Let φ be a regular reflection-diffraction solution in the sense of Definition* 8.1.1. *Then*

$$\partial_{\boldsymbol{n}_{\rm w}}(\varphi - \varphi_2) \ge 0 \qquad \text{in } \overline{\Omega}. \tag{8.2.42}$$

Proof. We divide the proof into three steps.

1. By Definition 8.1.1(ii), $D\varphi = D\varphi_2$ on $\Gamma_{\rm sonic}$. Also, $\partial_{\boldsymbol{n}_{\rm w}}\varphi = \partial_{\boldsymbol{n}_{\rm w}}\varphi_2 = 0$ on $\Gamma_{\rm wedge}$ by (2.2.20) and (7.1.3).

At P_2, $\partial_{\xi_1}\phi(P_2) \ge 0$ by (8.2.37), and $\partial_{\xi_2}\phi(P_2) = 0$ by the boundary condition (2.2.20). Thus, using (8.2.35), we have

$$\boldsymbol{n}_{\rm w} \cdot D\phi(P_2) = \partial_{\xi_1}\phi(P_2)\sin\theta_{\rm w} - \partial_{\xi_2}\phi(P_2)\cos\theta_{\rm w} = \partial_{\xi_1}\phi(P_2)\sin\theta_{\rm w} \ge 0.$$

Since $D\phi_2(P) = (u_2, v_2) = \sqrt{u_2^2 + v_2^2}\,\boldsymbol{\tau}_{\rm w}$ for any $P \in \mathbb{R}^2$, $\boldsymbol{n}_{\rm w} \cdot D\phi_2(P_2) = 0$. Hence, we have

$$\partial_{\boldsymbol{n}_{\rm w}}(\varphi - \varphi_2)(P_2) = \partial_{\boldsymbol{n}_{\rm w}}(\phi - \phi_2)(P_2) \ge 0.$$

Then $\partial_{\boldsymbol{n}_{\rm w}}(\varphi - \varphi_2) \ge 0$ on $\overline{\Gamma_{\rm sonic}} \cup \overline{\Gamma_{\rm wedge}} \cup \{P_2\}$. It remains to show that the minimum of $\partial_{\boldsymbol{n}_{\rm w}}(\varphi - \varphi_2)$ over $\overline{\Omega}$ cannot be attained within $\overline{\Omega} \setminus (\overline{\Gamma_{\rm sonic}} \cup \overline{\Gamma_{\rm wedge}} \cup \{P_2\})$.

2. Denote

$$w := \partial_{\boldsymbol{n}_{\rm w}}\phi.$$

Since $\partial_{\boldsymbol{n}_{\rm w}}(\varphi - \varphi_2) = \partial_{\boldsymbol{n}_{\rm w}}(\phi - \phi_2)$, and $\partial_{\boldsymbol{n}_{\rm w}}\phi_2$ is constant in $\mathbb{R}^2$, it suffices to show that the minimum of w over $\overline{\Omega}$ cannot be attained within $\overline{\Omega} \setminus (\overline{\Gamma_{\rm sonic}} \cup \overline{\Gamma_{\rm wedge}} \cup \{P_2\})$.

3. Since w satisfies equation (5.1.10), which is uniformly elliptic and has continuous coefficients on any compact subset of Ω, and w is non-constant in Ω by Lemma 8.2.1, it follows that the minimum of w over $\overline{\Omega}$ cannot be attained in the interior of Ω.

Since $\theta_{\rm w} \in (0, \frac{\pi}{2})$, $\boldsymbol{n}_{\rm w}$ is not orthogonal to $\Gamma_{\rm sym} \subset \{\xi_2 = 0\}$. Then Lemma 8.2.3(iii) implies that w cannot attain its minimum over $\overline{\Omega}$ on $\Gamma_{\rm sym}$.

Also, by assertion (8.2.19) in Lemma 8.2.11 and Proposition 8.2.16, the maximum of $\partial_{-\boldsymbol{n}_{\rm w}}\phi$ over $\overline{\Omega}$ cannot be attained on $\Gamma_{\rm shock}$. In other words, the minimum of w over $\overline{\Omega}$ cannot be attained on $\Gamma_{\rm shock}$.

Therefore, we have shown in Step 3 that w cannot attain its minimum over $\overline{\Omega}$ on $\Omega \cup \Gamma_{\rm sym} \cup \Gamma_{\rm shock}$. Note that

$$\Omega \cup \Gamma_{\rm sym} \cup \Gamma_{\rm shock} = \overline{\Omega} \setminus (\overline{\Gamma_{\rm sonic}} \cup \overline{\Gamma_{\rm wedge}} \cup \{P_2\}).$$

This, combining with the conclusion of Step 2, completes the proof. □

Combining Proposition 8.2.17 with Proposition 8.2.19, we have

Corollary 8.2.20 (Cone of monotonicity directions for $\varphi - \varphi_2$). *Let φ be an admissible solution of* **Problem 2.6.1** *in the sense of Definition* 8.1.1. *Then, for all* $\mathbf{e} \in \mathrm{Cone}(\mathbf{e}_{\mathcal{S}_1}, \boldsymbol{n}_{\rm w})$,

$$\partial_{\mathbf{e}}(\varphi - \varphi_2) \geq 0 \qquad in\ \overline{\Omega}, \tag{8.2.43}$$

where $\mathrm{Cone}(\mathbf{e}_{\mathcal{S}_1}, \boldsymbol{n}_{\rm w})$ *is defined by* (8.1.10).

8.3 APPENDIX: PROPERTIES OF SOLUTIONS OF PROBLEM 2.6.1 FOR LARGE-ANGLE WEDGES

In this section we show that, if $\theta_{\rm w}$ is sufficiently close to $\frac{\pi}{2}$ depending only on (ρ_0, ρ_1, γ), and a solution φ of **Problem 2.6.1** is of supersonic reflection configuration as described in §2.4.2 and is sufficiently close to φ_2 in Ω, then solution φ satisfies conditions (iv)–(v) of Definition 8.1.1. This motivates Definition 8.1.1 for the class of admissible solutions.

The main result of this section is the following:

Proposition 8.3.1. *There exists $\sigma > 0$ depending only on (ρ_0, ρ_1, γ) such that, if φ is a solution of* **Problem 2.6.1** *for $\theta_{\rm w} \in (\frac{\pi}{2} - \sigma, \frac{\pi}{2})$, which satisfies conditions* (i)–(iii) *of Definition* 8.1.1 *and*

$$\|\varphi - \varphi_2^{(\theta_{\rm w})}\|_{C^1(\overline{\Omega})} \leq C_0\sigma \tag{8.3.1}$$

for some constant $C_0 > 0$ independent of σ, then φ satisfies conditions (iv)–(v) *of Definition* 8.1.1.

To prove Proposition 8.3.1, we first recall some facts proved earlier. Denote by φ_2^0 the pseudo-potential for state (2) for the normal reflection (*i.e.*, $\theta_{\rm w} = \frac{\pi}{2}$) corresponding to (ρ_0, ρ_1, γ). Then φ_2^0 is defined by (6.2.1), and the Rankine-Hugoniot conditions (2.2.13) hold between states (1) and (2) along the flat reflected shock $\mathcal{S}_1^0 = \{\varphi_1 = \varphi_2^0\} \equiv \{\xi_1 = \bar{\xi}_1\}$.

Denote by $\bar{\rho}_2$ and $\bar{c}_2$ the density and sonic speed of the normal reflection solution, and denote by $\bar{P}_1$ the point in half-plane $\{\xi_2 > 0\}$ of intersection of the sonic circle $\partial B_{\bar{c}_2}$ of the normal reflection solution with the flat reflected shock $\{\xi_1 = \bar{\xi}_1\}$. Let $\bar{P}_2 = (\bar{\xi}_1, 0)$, and denote by $\Gamma^{\rm norm}_{\rm shock}$ the line segment $\bar{P}_1\bar{P}_2$ that is the transonic part of the flat reflected shock $\mathcal{S}_1^0$ in half-plane $\{\xi_2 > 0\}$. For $r > 0$, denote by $N_r(\Gamma^{\rm norm}_{\rm shock})$ the r–neighborhood of $\Gamma^{\rm norm}_{\rm shock}$. Also, for $\theta_{\rm w} \in (\frac{\pi}{2}-\sigma, \frac{\pi}{2})$, let φ_2 be the pseudo-potential of state (2). Let $\Gamma^{\rm flat}_{\rm shock}$ be the transonic part of shock $\mathcal{S}_1 = \{\varphi_1 = \varphi_2\}$, *i.e.*, the line segment between P_1 and point $\hat{P}_2 := \mathcal{S}_1 \cap \{\xi_2 = 0\}$. Moreover, the normal to $\Gamma^{\rm flat}_{\rm shock}$ in the direction to the region for state (2) is $\boldsymbol{\nu}_{\mathcal{S}_1} = \frac{(u_1-u_2, -v_2)}{\sqrt{(u_1-u_2)^2+v_2^2}}$. Then

$$-\bar{c}_2 < \bar{\xi}_1 < 0, \qquad \bar{\rho}_2 > \rho_1, \tag{8.3.2}$$

and there exists $\sigma > 0$ such that, for any $\theta_{\rm w} \in (\frac{\pi}{2} - \sigma, \frac{\pi}{2})$,

$$\text{state (2) exists and } (u_2, v_2, \rho_2) \text{ depends continuously on } \theta_{\rm w}, \tag{8.3.3}$$

$$|(u_2, v_2)| + |(\bar{\rho}_2, \bar{c}_2) - (\rho_2, c_2)| \le C\sigma, \tag{8.3.4}$$

$$|P_1 - \bar{P}_1| + |\hat{P}_2 - \bar{P}_2| \le C\sigma, \tag{8.3.5}$$

where the universal constant C may be different at different occurrence, but depends only on (ρ_0, ρ_1, γ).

Property (8.3.2) follows from Theorem 6.2.1. Properties (8.3.3)–(8.3.4) follow from Theorem 7.1.1 (see also §3.1–§3.2 of [54]). Property (8.3.5) easily follows from (8.3.2) and (8.3.4). Furthermore, it holds that

$$\partial_{\boldsymbol{\nu}}\varphi_1 > \partial_{\boldsymbol{\nu}}\varphi_2 > 0 \qquad \text{on } \Gamma^{\rm flat}_{\rm shock}, \tag{8.3.6}$$

which follows from (7.1.6).

In the rest of this section, we prove Proposition 8.3.1.

8.3.1 $\varphi \le \varphi_1$ if φ is a smooth solution in $\overline{\Omega}$

Lemma 8.3.2. *Let φ be a solution of* **Problem 2.6.1** *for the wedge angle $\theta_{\rm w} \in (\theta^{\rm s}_{\rm w}, \frac{\pi}{2})$, which satisfies conditions* (i)–(iii) *of Definition* 8.1.1. *Then*

$$\varphi \le \varphi_1 \qquad \textit{in } \overline{\Omega}.$$

Proof. We use $\bar{\phi} := \varphi_1 - \varphi$. Since $\boldsymbol{\nu}_{\rm w} = (-\sin\theta_{\rm w}, \cos\theta_{\rm w})$,

$$\partial_{\boldsymbol{\nu}_{\rm w}}(\varphi_1 - \varphi) = \partial_{\boldsymbol{\nu}_{\rm w}}\phi_1 = -u_1 \sin\theta_{\rm w} < 0 \qquad \text{on } \Gamma_{\rm wedge}.$$

Since $\varphi \in C^1(\overline{\Omega})$, the same holds at $P_3 = \mathbf{0}$, which is an endpoint of $\Gamma_{\rm wedge}$. Also, at any point $\boldsymbol{\xi} \in \Gamma_{\rm wedge} \cup \{P_3\}$, $\boldsymbol{\nu}_{\rm w} = (-\sin\theta_{\rm w}, \cos\theta_{\rm w})$ points into Ω, *i.e.*, there exists $\varepsilon_0 > 0$ such that $\boldsymbol{\xi} + \varepsilon\boldsymbol{\nu}_{\rm w} \in \Omega$ for $\varepsilon \in (0, \varepsilon_0)$. Then the minimum of $\bar{\phi}$ over $\overline{\Omega}$ cannot be achieved on $\Gamma_{\rm wedge} \cup \{P_3\}$.

Furthermore, $\bar{\phi}$ satisfies:

- The elliptic equation (2.2.11) in Ω (considered as a linear equation with respect to $D^2\phi$) with $D^2\phi$ replaced by $D^2\bar{\phi}$, which can be done since $D^2\phi = -D^2\bar{\phi}$;
- $\partial_\nu \bar{\phi} = 0$ on $\Gamma_{\rm sym}$;
- $\partial_\nu \bar{\phi} = -u_1 \sin\theta_{\rm w} < 0$ on $\Gamma_{\rm wedge}$;
- $\bar{\phi} = 0$ on $\overline{\Gamma_{\rm shock}}$;
- $\bar{\phi} = \varphi_1 - \varphi = \varphi_1 - \varphi_2 > 0$ on $\overline{\Gamma_{\rm sonic}}$.

Then, by the strong maximum principle and Hopf's lemma, if the minimum of $\bar{\phi}$ over $\overline{\Omega}$ is achieved in $\Omega \cup \Gamma_{\rm sym}$, $\bar{\phi}$ is a constant in Ω; in this case, the condition on $\Gamma_{\rm wedge}$ is not satisfied. Thus, the minimum of $\bar{\phi}$ over $\overline{\Omega}$ cannot be achieved in $\Omega \cup \Gamma_{\rm sym}$. Also, the minimum of $\bar{\phi}$ over $\overline{\Omega}$ cannot be achieved on $\Gamma_{\rm wedge} \cup \{P_3\}$ as we have shown above. Therefore, we have

$$\min_{\overline{\Omega}} \bar{\phi} = \min_{\overline{\Gamma_{\rm shock} \cup \Gamma_{\rm sonic}}} \bar{\phi} = 0.$$

□

8.3.2 $\varphi \geq \varphi_2$ if φ is a solution close to the normal reflection

Now we show that $\varphi \geq \varphi_2$ in Ω, if the conditions of Proposition 8.3.1 are satisfied with small σ. For this, it is convenient to rewrite the potential flow equation in Ω and the boundary conditions on $\partial\Omega$ in terms of $\psi = \varphi - \varphi_2$.

8.3.2.1 Shifting coordinates

It is more convenient to change the coordinates in the self-similar plane by shifting the origin to the center of sonic circle of state (2). We define

$$\boldsymbol{\xi}_{\rm new} := \boldsymbol{\xi} - (u_2, v_2).$$

For simplicity of notation, throughout the rest of this section, we always work in the new coordinates without changing notation $\boldsymbol{\xi}$.

Rewriting the background solutions (2.2.16)–(2.2.17) and (2.4.1) in the shifted coordinates, we have

$$\varphi_0(\boldsymbol{\xi}) = -\frac{1}{2}|\boldsymbol{\xi}|^2 - (u_2, v_2)\cdot\boldsymbol{\xi} - \frac{1}{2}q_2^2, \tag{8.3.7}$$

$$\varphi_1(\boldsymbol{\xi}) = -\frac{1}{2}|\boldsymbol{\xi}|^2 + (u_1 - u_2, -v_2)\cdot\boldsymbol{\xi} - \frac{1}{2}q_2^2 + u_1(u_2 - \xi_1^0), \tag{8.3.8}$$

$$\varphi_2(\boldsymbol{\xi}) = -\frac{1}{2}|\boldsymbol{\xi}|^2 - \frac{1}{2}q_2^2 + (u_1 - u_2)\hat{\xi}_1 + u_1(u_2 - \xi_1^0), \tag{8.3.9}$$

where $q_2^2 = u_2^2 + v_2^2$.

We first note that, from (8.3.4),

$$u_1 - u_2 > 0 \qquad \text{for } \theta_w \in (\frac{\pi}{2} - \sigma, \frac{\pi}{2})$$

if σ is small, depending only on the data. Since $\mathbf{e}_{\xi_1} = (1, 0)$ and $D(\varphi_1 - \varphi_2) = (u_1 - u_2, -v_2)$, it follows that $\partial_{\mathbf{e}_{\xi_1}}(\varphi_1 - \varphi_2) = u_1 - u_2 > 0$ in $\mathbb{R}^2$ for such θ_w. Then, after possibly further reducing σ, it follows from (8.3.1) that $\partial_{\mathbf{e}_{\xi_1}}(\varphi_1 - \varphi) > 0$ in $\overline{\Omega}$. Since $\varphi = \varphi_1$ on Γ_{shock} by the Rankine-Hugoniot condition (5.1.12), and $\varphi \geq \varphi_1$ in Ω by Lemma 8.3.2, it follows that there exists a function $f \in C^1(\mathbb{R})$ such that

$$\Gamma_{\text{shock}} = \{\xi_1 = f(\xi_2) \;:\; 0 < \xi_2 < \xi_{2P_1}\}, \qquad \Omega \subset \{\xi_1 > f(\xi_2) \;:\; \xi_2 \in \mathbb{R}\}.$$

$$\|f - l\|_{C^1([\xi_{2P_2}, \xi_{2P_1}])} \leq C\sigma, \tag{8.3.10}$$

where $\xi_1 = l(\xi_2)$ is line $\mathcal{S}_1 = \{\varphi_1 = \varphi_2\}$. That is,

$$l(\xi_2) = \xi_2 \cot\theta_s + \hat{\xi}_1, \tag{8.3.11}$$

and

$$\hat{\xi}_1 = \tilde{\xi}_1 - u_2 + v_2 \cot\theta_{sh} < 0 \tag{8.3.12}$$

if $\sigma = \frac{\pi}{2} - \theta_w > 0$ is sufficiently small, since $|(u_2, v_2)|$ is small and $\tilde{\xi}_1 < 0$ by (6.2.2) in this case. Since $u_2 = v_2 \cot\theta_w > 0$, it follows from (7.2.13) that

$$\hat{\xi}_1 > \tilde{\xi}_1. \tag{8.3.13}$$

Another condition on f comes from the fact that, since φ satisfies Definition 8.1.1(i)–(ii), the curved part and straight part of the reflected-diffracted shock should match at least up to the first-order at P_1. From its definition, $P_1 = \boldsymbol{\xi}_{P_1}$ is the intersection point of line $\xi_1 = l(\xi_2)$ and the sonic circle $|\boldsymbol{\xi}|^2 = c_2^2$ with $\xi_{2P_1} > 0$, *i.e.*, $\boldsymbol{\xi}_{P_1}$ is the unique point for small $\sigma > 0$ satisfying

$$l(\xi_{2P_1})^2 + \xi_{2P_1}^2 = c_2^2, \qquad \xi_{1P_1} = l(\xi_{2P_1}), \qquad \xi_{2P_1} > 0. \tag{8.3.14}$$

The existence and uniqueness of such a point $\boldsymbol{\xi}_{P_1}$ follows from $-c_2 < \tilde{\xi}_1 < 0$, which holds from (7.2.13), (7.2.16), (8.3.12), and the smallness of $|(u_2, v_2)|$. Then f satisfies

$$f(\xi_{2P_1}) = l(\xi_{2P_1}), \qquad f'(\xi_{2P_1}) = l'(\xi_{2P_1}) = \cot\theta_{sh}. \tag{8.3.15}$$

Note also that, for small $\sigma > 0$, we obtain from (7.2.16), (8.3.12)–(8.3.13), and $l'(\xi_2) = \cot\theta_{sh} > 0$ that

$$-c_2 < \tilde{\xi}_1 < \hat{\xi}_1 < \xi_{1P_1} < 0, \qquad c_2 - |\tilde{\xi}_1| \geq \frac{\bar{c}_2 - |\bar{\xi}_1|}{2} > 0. \tag{8.3.16}$$

Thus, in the new shifted coordinates, domain Ω is expressed as

$$\Omega = B_{c_2}(\mathbf{0}) \cap \{\xi_2 > -v_2\} \cap \{f(\xi_2) < \xi_1 < \xi_2 \cot\theta_w\}. \tag{8.3.17}$$

Furthermore, equations (2.2.8)–(2.2.9) and the Rankine-Hugoniot conditions (2.2.13)–(2.2.14) on $\Gamma_{\rm shock}$ do not change under the shift of coordinates. That is, φ satisfies both (2.2.8)–(2.2.9) in Ω (so that the equation is elliptic on φ) and the following boundary conditions on $\Gamma_{\rm shock}$: The continuity of the pseudo-potential function across the shock

$$\varphi = \varphi_1 \qquad \text{on } \Gamma_{\rm shock}, \tag{8.3.18}$$

and the gradient jump condition

$$\rho(|D\varphi|^2, \varphi) D\varphi \cdot \boldsymbol{\nu}_{\rm sh} = \rho_1 D\varphi_1 \cdot \boldsymbol{\nu}_{\rm sh} \qquad \text{on } \Gamma_{\rm shock}, \tag{8.3.19}$$

where $\boldsymbol{\nu}_{\rm sh}$ is the interior unit normal to $\Gamma_{\rm shock} \subset \partial\Omega$.

The boundary conditions on the other parts of $\partial\Omega$ are:

$$\varphi = \varphi_2 \qquad \text{on } \Gamma_{\rm sonic} = \partial\Omega \cap \partial B_{c_2}(\mathbf{0}), \tag{8.3.20}$$

$$\varphi_{\boldsymbol{\nu}} = 0 \qquad \text{on } \Gamma_{\rm wedge} = \partial\Omega \cap \{\xi_2 = \xi_1 \tan\theta_{\rm w}\}, \tag{8.3.21}$$

$$\varphi_{\boldsymbol{\nu}} = 0 \qquad \text{on } \Gamma_{\rm sym} = \partial\Omega \cap \{\xi_2 = -v_2\}. \tag{8.3.22}$$

Moreover, substituting $\tilde{\xi}_1$ in (8.3.12) into equation (7.2.8) and using (7.2.2) and (7.2.5), we have

$$\rho_2 \hat{\xi}_1 = \rho_1 \Big(\hat{\xi}_1 - \frac{(u_1 - u_2)^2 + v_2^2}{u_1 - u_2} \Big), \tag{8.3.23}$$

which expresses the Rankine-Hugoniot conditions on $\mathcal{S}_1$ in terms of $\hat{\xi}_1$. We use this equality below.

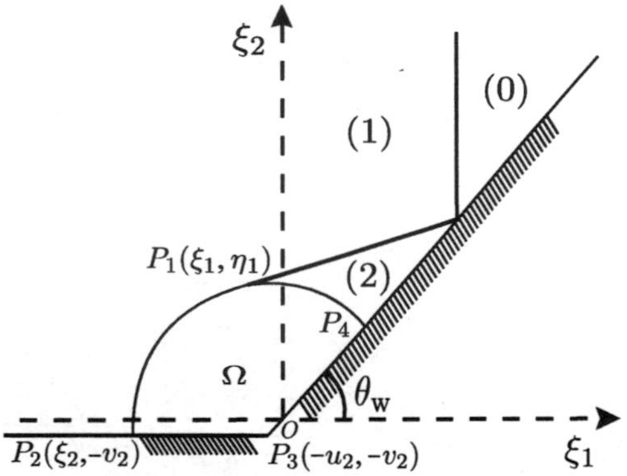

Figure 8.1: Supersonic reflection configurations in the new coordinates

8.3.2.2 The equations and boundary conditions in terms of $\psi = \varphi - \varphi_2$

We consider the function: $\psi = \varphi - \varphi_2$. It follows from (2.2.8)–(2.2.9), (7.2.14), and (8.3.9), by explicit calculation, that ψ satisfies the following equation in Ω:

$$\begin{aligned}&\big(c^2(D\psi,\psi,\boldsymbol{\xi}) - (\psi_{\xi_1}-\xi_1)^2\big)\psi_{\xi_1\xi_1} + \big(c^2(D\psi,\psi,\boldsymbol{\xi}) - (\psi_{\xi_2}-\xi_2)^2\big)\psi_{\xi_2\xi_2}\\ &-2(\psi_{\xi_1}-\xi_1)(\psi_{\xi_2}-\xi_2)\psi_{\xi_1\xi_2} = 0, \qquad (8.3.24)\end{aligned}$$

and the expressions of the density and sonic speed in Ω in terms of ψ are

$$\rho(D\psi,\psi,\boldsymbol{\xi}) = \Big(\rho_2^{\gamma-1} + (\gamma-1)\big(\boldsymbol{\xi}\cdot D\psi - \frac{1}{2}|D\psi|^2 - \psi\big)\Big)^{\frac{1}{\gamma-1}}, \qquad (8.3.25)$$

$$c^2(D\psi,\psi,\boldsymbol{\xi}) = c_2^2 + (\gamma-1)\Big(\boldsymbol{\xi}\cdot D\psi - \frac{1}{2}|D\psi|^2 - \psi\Big), \qquad (8.3.26)$$

where ρ_2 is the density of state (2).

From (8.3.9) and (8.3.20)–(8.3.21), we obtain

$$\psi = 0 \qquad \text{on } \Gamma_{\text{sonic}} = \partial\Omega \cap \partial B_{c_2}(\mathbf{0}), \qquad (8.3.27)$$

$$\psi_{\boldsymbol{\nu}} = 0 \qquad \text{on } \Gamma_{\text{wedge}} = \partial\Omega \cap \{\xi_2 = \xi_1 \tan\theta_{\text{w}}\}, \qquad (8.3.28)$$

$$\psi_{\xi_2} = -v_2 \qquad \text{on } \Gamma_{\text{sym}} = \partial\Omega \cap \{\xi_2 = -v_2\}. \qquad (8.3.29)$$

From (8.3.8)–(8.3.9), the Rankine-Hugoniot conditions in terms of ψ take the following form: The continuity of the pseudo-potential function across (8.3.18) is written as

$$\psi - \frac{1}{2}q_2^2 + \hat{\xi}_1(u_1-u_2) + u_1(u_2-\xi_1^0) = \xi_1(u_1-u_2) - \xi_2 v_2 - \frac{1}{2}q_2^2 + u_1(u_2-\xi_1^0) \quad (8.3.30)$$

on Γ_{shock}, that is,

$$\xi_1 = \frac{\psi(\boldsymbol{\xi}) + v_2\xi_2}{u_1-u_2} + \hat{\xi}_1 \qquad \text{on } \Gamma_{\text{shock}}, \qquad (8.3.31)$$

where $\hat{\xi}_1$ is defined by (8.3.12). The gradient jump condition (8.3.19) is

$$\rho(D\psi,\psi)\,(D\psi - \boldsymbol{\xi})\cdot\boldsymbol{\nu}_{\text{sh}} = \rho_1\,(u_1-u_2-\xi_1, -v_2-\xi_2)\cdot\boldsymbol{\nu}_{\text{sh}} \qquad (8.3.32)$$

on Γ_{shock}, where $\rho(D\psi,\psi)$ is defined by (8.3.25) and $\boldsymbol{\nu}_{\text{sh}}$ is the interior unit normal to Ω on Γ_{shock}. If $|(u_2, v_2, D\psi)| < \frac{u_1}{50}$, the unit normal $\boldsymbol{\nu}_{\text{sh}}$ can be expressed as

$$\boldsymbol{\nu}_{\text{sh}} = \frac{D(\varphi_1-\varphi)}{|D(\varphi_1-\varphi)|} = \frac{(u_1-u_2-\psi_{\xi_1}, -v_2-\psi_{\xi_2})}{\sqrt{(u_1-u_2-\psi_{\xi_1})^2 + (v_2+\psi_{\xi_2})^2}}, \qquad (8.3.33)$$

where we have used (8.3.8)–(8.3.9) and (6.3.1) to obtain the last expression.

Now we rewrite the jump condition (8.3.32) in a more convenient form for ψ satisfying (8.3.18) when $\sigma > 0$ and $\|\psi\|_{C^1(\bar{\Omega})}$ are sufficiently small.

Lemma 8.3.3. *There exists $\sigma > 0$ small, depending only on (ρ_0, ρ_1, γ), such that, if θ_w and φ satisfy the conditions of Proposition 8.3.1, φ satisfies the following condition on Γ_{shock}:*

$$\rho_2'(c_2^2 - |\hat{\xi}_1|^2)\psi_{\xi_1} + \big(\frac{\rho_2 - \rho_1}{u_1} - \rho_2'\hat{\xi}_1\big)(\xi_2\psi_{\xi_2} - \psi)$$
$$+E_1(D\psi, \psi, \xi_2) \cdot D\psi + E_2(D\psi, \psi, \xi_2)\psi = 0, \tag{8.3.34}$$

where functions $E_1 = (E_{11}, E_{12})(\mathbf{p}, z, \xi_2)$ and $E_2(\mathbf{p}, z, \xi_2)$ are smooth on $\mathbb{R}^2 \times \mathbb{R} \times \mathbb{R}$ and satisfy

$$|E_i(\mathbf{p}, z, \xi_2)| \leq C\left(|\mathbf{p}| + |z| + \sigma\right), \tag{8.3.35}$$

$$|(D_{(\mathbf{p},z,\xi_2)}E_i,\ D^2_{(\mathbf{p},z,\xi_2)}E_i)| \leq C, \tag{8.3.36}$$

for $i = 1, 2$.

Proof. We use the notations introduced in §6.2. In particular, $\bar{\rho}_2$ and $\bar{c}_2$ denote the density and sonic speed of the normal reflection.

We first discuss the smallness assumptions for $\sigma > 0$ and $\|\psi\|_{C^1(\bar{\Omega})}$. By (7.2.11) and (7.2.15), it follows that, if σ is small, depending only on the data, then

$$\frac{5\bar{c}_2}{6} \leq c_2 \leq \frac{6\bar{c}_2}{5}, \quad \frac{5\bar{\rho}_2}{6} \leq \rho_2 \leq \frac{6\bar{\rho}_2}{5}, \quad \sqrt{u_2^2 + v_2^2} \leq \frac{u_1}{50}. \tag{8.3.37}$$

We also require that $\|\psi\|_{C^1(\bar{\Omega})}$ be sufficiently small so that, if (8.3.37) holds, expressions (8.3.25) and (8.3.33) are well-defined in Ω, and ξ_1 defined by the right-hand side of (8.3.31) satisfies $|\xi_1| \leq \frac{7\bar{c}_2}{5}$ for $\xi_2 \in (-v_2, c_2)$, which is the range of ξ_2 on Γ_{shock}. Since (8.3.37) holds and $\Omega \subset B_{c_2}(0)$ by (8.3.17), it suffices to assume

$$\|\psi\|_{C^1(\bar{\Omega})} \leq \min\Big\{\frac{\bar{\rho}_2^{\gamma-1}}{50(1+4\bar{c}_2)}, \min\{1, \bar{c}_2\}\frac{u_1}{50}\Big\} =: \delta^*. \tag{8.3.38}$$

For the rest of the proof, we assume that (8.3.37)–(8.3.38) hold. Under these conditions, we can substitute the right-hand side of (8.3.33) for $\boldsymbol{\nu}_{\text{sh}}$ into (8.3.32). Thus, we rewrite (8.3.32) as

$$F(D\psi, \psi, u_2, v_2, \boldsymbol{\xi}) = 0 \qquad \text{on } \Gamma_{\text{shock}}, \tag{8.3.39}$$

where, denoting $\mathbf{p} = (p_1, p_2) \in \mathbb{R}^2$ and $z \in \mathbb{R}$,

$$F(\mathbf{p}, z, u_2, v_2, \boldsymbol{\xi}) = \big(\tilde{\rho}\,(\mathbf{p} - \boldsymbol{\xi}) - \rho_1\,(u_1 - u_2 - \xi_1, -v_2 - \xi_2)\big) \cdot \hat{\boldsymbol{\nu}}, \tag{8.3.40}$$

with $\tilde{\rho} := \tilde{\rho}(\mathbf{p}, z, \boldsymbol{\xi})$ and $\hat{\boldsymbol{\nu}} := \hat{\boldsymbol{\nu}}(\mathbf{p}, u_2, v_2)$ defined by

$$\tilde{\rho}(\mathbf{p}, z, \boldsymbol{\xi}) = \Big(\rho_2^{\gamma-1} + (\gamma-1)\big(\mathbf{p}\cdot\boldsymbol{\xi} - \frac{|\mathbf{p}|^2}{2} - z\big)\Big)^{\frac{1}{\gamma-1}}, \tag{8.3.41}$$

$$\hat{\boldsymbol{\nu}}(\mathbf{p}, u_2, v_2) = \frac{(u_1 - u_2 - p_1, -v_2 - p_2)}{\sqrt{(u_1 - u_2 - p_1)^2 + (v_2 + p_2)^2}}. \tag{8.3.42}$$

From the explicit definitions of $\tilde{\rho}$ and $\hat{\boldsymbol{\nu}}$, it follows from (8.3.37) that

$$\tilde{\rho} \in C^\infty(\overline{B_{\delta^*}(0) \times (-\delta^*, \delta^*) \times B_{2\bar{c}_2}(\mathbf{0})}), \quad \hat{\boldsymbol{\nu}} \in C^\infty(\overline{B_{\delta^*}(\mathbf{0}) \times B_{u_1/50}(\mathbf{0})}),$$

where $B_R(\mathbf{0})$ denotes the ball in $\mathbb{R}^2$ with center $\mathbf{0}$ and radius R and, for $k \in \mathbf{N}$ (the set of nonnegative integers), the C^k–norms of $\tilde{\rho}$ and $\hat{\boldsymbol{\nu}}$ over the regions specified above are bounded by the constants depending only on $(u_1, \bar{\rho}_2, \bar{c}_2, \gamma, k)$, that is, the C^k–norms depend only on the data and k. Then

$$F \in C^\infty(\overline{B_{\delta^*}(\mathbf{0}) \times (-\delta^*, \delta^*) \times B_{u_1/50}(\mathbf{0}) \times B_{2\bar{c}_2}(\mathbf{0})}) \tag{8.3.43}$$

with its C^k–norm depending only on the data and k.

Furthermore, since ψ satisfies (8.3.18) so that (8.3.31) holds, we can substitute the right-hand side of (8.3.31) for ξ_1 into (8.3.39). Thus, we rewrite (8.3.32) as

$$\Psi(D\psi, \psi, u_2, v_2, \xi_2) = 0 \qquad \text{on } \Gamma_{\text{shock}}, \tag{8.3.44}$$

where

$$\Psi(\mathbf{p}, z, u_2, v_2, \xi_2) = F(\mathbf{p}, z, u_2, v_2, \frac{z + v_2\xi_2}{u_1 - u_2} + \hat{\xi}_1, \xi_2). \tag{8.3.45}$$

If $\xi_2 \in (-\frac{6\bar{c}_2}{5}, \frac{6\bar{c}_2}{5})$ and $|z| \le \delta^*$, it follows from (8.3.16) and (8.3.37)–(8.3.38) that $\left|\frac{z+v_2\xi_2}{u_1-u_2} + \hat{\xi}_1\right| \le \frac{7\bar{c}_2}{5}$. That is, $(\frac{z+v_2\xi_2}{u_1-u_2} + \hat{\xi}_1,\, \xi_2) \in B_{2\bar{c}_2}(\mathbf{0})$ if $\xi_2 \in (-\frac{6\bar{c}_2}{5}, \frac{6\bar{c}_2}{5})$ and $|z| \le \delta^*$. Thus, from (8.3.43) and (8.3.45), $\Psi \in C^\infty(\overline{\mathcal{A}})$ with $\|\Psi\|_{C^k(\overline{\mathcal{A}})}$ depending only on the data and $k \in \mathbf{N}$, where $\mathcal{A} = B_{\delta^*}(\mathbf{0}) \times (-\delta^*, \delta^*) \times B_{u_1/50}(\mathbf{0}) \times (-\frac{6\bar{c}_2}{5}, \frac{6\bar{c}_2}{5})$.

Using the explicit expression of Ψ given by (8.3.40)–(8.3.42) and (8.3.45), we calculate

$$\begin{aligned}
&\Psi(\mathbf{0}, 0, u_2, v_2, \xi_2)\\
&= -\frac{(u_1 - u_2)\rho_2\hat{\xi}_1}{\sqrt{(u_1 - u_2)^2 + v_2^2}} - \rho_1\Big(\sqrt{(u_1 - u_2)^2 + v_2^2} - \frac{(u_1 - u_2)\hat{\xi}_1}{\sqrt{(u_1 - u_2)^2 + v_2^2}}\Big).
\end{aligned}$$

Now, using (8.3.23), we have

$$\Psi(\mathbf{0}, 0, u_2, v_2, \xi_2) = 0 \qquad \text{for any } (u_2, v_2, \xi_2) \in B_{\frac{u_1}{50}}(\mathbf{0}) \times (-\frac{6\bar{c}_2}{5}, \frac{6\bar{c}_2}{5}).$$

Then, denoting $p_0 = z$ and $\mathcal{X} = (\mathbf{p}, p_0, u_2, v_2, \xi_2) \in \mathcal{A}$, we have

$$\Psi(\mathcal{X}) = \sum_{i=0}^{2} p_i D_{p_i}\Psi(\mathbf{0}, 0, u_2, v_2, \xi_2) + \sum_{i,j=0}^{2} p_i p_j g_{ij}(\mathcal{X}), \tag{8.3.46}$$

where $g_{ij}(\mathcal{X}) = \int_0^1 (1-t) D_{p_i p_j}\Psi(t\mathbf{p}, tp_0, u_2, v_2, \xi_2)dt$ for $i, j = 0, 1, 2$. Thus, $g_{ij} \in C^\infty(\overline{\mathcal{A}})$ and $\|g_{ij}\|_{C^k(\overline{\mathcal{A}})} \le \|\Psi\|_{C^{k+2}(\overline{\mathcal{A}})}$, depending only on the data and $k \in \mathbf{N}$.

Next, denoting $\rho_2' := \hat{\rho}'(\rho_2^{\gamma-1}) = \frac{\rho_2}{c_2^2} > 0$, we compute from the explicit expression of Ψ given by (8.3.40)–(8.3.42) and (8.3.45) that

$$D_{(\mathbf{p},z)}\Psi(\mathbf{0},0,0,0,\xi_2) = \big(\rho_2'(c_2^2-\hat{\xi}_1^2),\ \big(\frac{\rho_2-\rho_1}{u_1}-\rho_2'\hat{\xi}_1\big)\xi_2,\ \rho_2'\hat{\xi}_1-\frac{\rho_2-\rho_1}{u_1}\big).$$

Note that, for $i=0,1,2$,

$$\partial_{p_i}\Psi(\mathbf{0},0,u_2,v_2,\xi_2) = \partial_{p_i}\Psi(\mathbf{0},0,0,0,\xi_2) + h_i(u_2,v_2,\xi_2)$$

with

$$\|h_i\|_{C^k(\overline{B_{u_1/50}(\mathbf{0})\times(-6\bar{c}_2/5,6\bar{c}_2/5)})} \le \|\Psi\|_{C^{k+2}(\overline{\mathcal{A}})} \qquad \text{for } k\in\mathbf{N},$$
$$|h_i(u_2,v_2,\xi_2)| \le \|D^2\Psi\|_{C(\overline{\mathcal{A}})}\big(|u_2|+|v_2|\big).$$

Then we obtain from (8.3.46) that, for any $\mathcal{X}=(\mathbf{p},z,u_2,v_2,\xi_2)\in\mathcal{A}$,

$$\Psi(\mathcal{X}) = \rho_2'(c_2^2-\hat{\xi}_1^2)p_1+\big(\frac{\rho_2-\rho_1}{u_1}-\rho_2'\hat{\xi}_1\big)(\xi_2p_2-z)+\hat{E}_1(\mathcal{X})\cdot\mathbf{p}+\hat{E}_2(\mathcal{X})z, \quad (8.3.47)$$

where $\hat{E}_1\in C^\infty(\overline{\mathcal{A}};\mathbb{R}^2)$ and $\hat{E}_2\in C^\infty(\overline{\mathcal{A}})$ with

$$\|\hat{E}_i\|_{C^k(\overline{\mathcal{A}})} \le \|\Psi\|_{C^{k+2}(\overline{\mathcal{A}})}, \qquad i=1,2, \quad k\in\mathbf{N},$$
$$|\hat{E}_i(\mathbf{p},z,u_2,v_2,\xi_2)| \le C(|\mathbf{p}|+|z|+|u_2|+|v_2|) \text{ for any } (\mathbf{p},z,u_2,v_2,\xi_2)\in\mathcal{A},$$

for C, depending only on $\|D^2\Psi\|_{C(\overline{\mathcal{A}})}$.

From now on, we fix (u_2,v_2) to equal the velocity of state (2) and write $E_i(\mathbf{p},z,\xi_2)$ for $\hat{E}_i(\mathbf{p},z,u_2,v_2,\xi_2)$. We conclude that, if (8.3.37) holds and $\psi\in C^1(\overline{\Omega})$ satisfies (8.3.38), then $\psi=\varphi-\varphi_2$ satisfies (8.3.18)–(8.3.19) on Γ_{shock} if and only if ψ satisfies conditions (8.3.31) on Γ_{shock}:

$$\begin{aligned}&\rho_2'(c_2^2-\hat{\xi}_1^2)\psi_{\xi_1}+\big(\frac{\rho_2-\rho_1}{u_1}-\rho_2'\hat{\xi}_1\big)(\xi_2\psi_{\xi_2}-\psi)\\&\qquad+E_1(D\psi,\psi,\xi_2)\cdot D\psi+E_2(D\psi,\psi,\xi_2)\psi=0,\end{aligned} \qquad (8.3.48)$$

and $E_i(\mathbf{p},z,\xi_2), i=1,2$, are smooth on $\overline{B_{\delta^*}(\mathbf{0})\times(-\delta^*,\delta^*)\times(-\frac{6\bar{c}_2}{5},\frac{6\bar{c}_2}{5})}$ and satisfy that, for any $(\mathbf{p},z,\xi_2)\in B_{\delta^*}(\mathbf{0})\times(-\delta^*,\delta^*)\times(-\frac{6\bar{c}_2}{5},\frac{6\bar{c}_2}{5})$,

$$|E_i(\mathbf{p},z,\xi_2)| \le C\left(|\mathbf{p}|+|z|+\sigma\right), \qquad (8.3.49)$$

$$|(D_{(\mathbf{p},z,\xi_2)},D^2_{(\mathbf{p},z,\xi_2)})E_i)| \le C, \qquad (8.3.50)$$

where we have used (7.2.15) in the derivation of (8.3.49), and C depends only on the data. □

Lemma 8.3.4. *Let θ_{w},Ω, and φ be as those in Lemma* 8.3.3. *Write* (8.3.34) *in the form:*

$$a_1\psi_{\xi_1}+a_2\psi_{\xi_2}+a_3\psi=0 \qquad \textit{on } \Gamma_{\text{shock}}, \qquad (8.3.51)$$

where

$$a_1 = \rho_2'(c_2^2 - \hat{\xi}_1^2) + E_{11}, \qquad a_2 = (\frac{\rho_2 - \rho_1}{u_1} - \rho_2'\hat{\xi}_1)\xi_2 + E_{12},$$
$$a_3 = -(\frac{\rho_2 - \rho_1}{u_1} - \rho_2'\hat{\xi}_1) + E_2, \tag{8.3.52}$$

for $(E_{1k}, E_2) = (E_{1k}, E_2)(D\psi(\boldsymbol{\xi}), \psi(\boldsymbol{\xi}), \xi_2)$, *where* $E_1 = (E_{11}, E_{12})$ *and* E_2 *are from Lemma* 8.3.3. *Then, if* σ *is sufficiently small, depending only on* (ρ_0, ρ_1, γ), *the linear homogeneous boundary condition* (8.3.51) *is oblique:*

$$(a_1, a_2) \cdot \boldsymbol{\nu}_{\rm sh} \geq \frac{1}{2}\rho_2'(c_2^2 - \hat{\xi}_1^2) > 0, \tag{8.3.53}$$

where $\boldsymbol{\nu}_{\rm sh}$ *is the normal to* $\Gamma_{\rm shock}$, *interior with respect to* Ω. *Moreover,*

$$a_1(P_2) > 0, \quad a_2(P_2) < 0, \tag{8.3.54}$$
$$a_3 \leq -\frac{1}{2}(\frac{\rho_2 - \rho_1}{u_1} - \rho_2'\hat{\xi}_1) < 0 \qquad \text{on } \Gamma_{\rm shock}, \tag{8.3.55}$$

where we have used that $\hat{\xi}_1 < 0$.

Proof. Denote by $\boldsymbol{\nu}_{\mathcal{S}_1}$ the unit normal to line $\mathcal{S}_1 = \{\varphi_1 = \varphi_2\}$ towards the region of state (2). Then $\boldsymbol{\nu}_{\mathcal{S}_1} = \frac{D(\varphi_1 - \varphi_2)}{|D(\varphi_1 - \varphi_2)|} = \frac{(u_1 - u_2, -v_2)}{\sqrt{(u_1-u_2)^2 + v_2^2}}$. We employ (7.2.15) to compute

$$\begin{aligned}
&\big(\rho_2'(c_2^2 - \hat{\xi}_1^2), (\frac{\rho_2 - \rho_1}{u_1} - \rho_2'\hat{\xi}_1)\xi_2\big) \cdot \boldsymbol{\nu}_{\mathcal{S}_1} \\
&= \frac{1}{\sqrt{(u_1 - u_2)^2 + v_2^2}}\Big(\rho_2'(c_2^2 - \hat{\xi}_1^2)(u_1 - u_2) + (\frac{\rho_2 - \rho_1}{u_1} - \rho_2'\hat{\xi}_1)\xi_2 v_2\Big) \\
&\geq \frac{3}{4}\rho_2'(c_2^2 - \hat{\xi}_1^2) > 0,
\end{aligned}$$

if $\frac{\pi}{2} - \theta_{\rm w}$ is small and $\xi_2 \in [\xi_{2P_2}, \xi_{2P_1}]$. From (8.3.33), we have

$$\|\boldsymbol{\nu}_{\rm sh} - \boldsymbol{\nu}_{\mathcal{S}_1}\|_{L^\infty(\Gamma_{\rm shock})} \leq C\|D\psi\|_{C(\overline{\Omega})}.$$

Using this and (8.3.52) with (8.3.35), we conclude that, if $\sigma > 0$, (8.3.53) holds.

Similarly, using (8.3.52) with (8.3.35), noting that $\hat{\xi}_1 < 0$ and $\xi_2(P_2) = -v_2 < 0$, and choosing σ small, we obtain (8.3.54)–(8.3.55). □

Lemma 8.3.5. *Let* φ *be as that in Proposition* 8.3.1. *Then, if* σ *is sufficiently small,*

$$\varphi \geq \varphi_2 \qquad \text{in } \Omega.$$

Proof. We have shown that $\psi = \varphi - \varphi_2$ satisfies equation (8.3.24) in Ω and the boundary conditions (8.3.27)–(8.3.29) and (8.3.51).

Equation (8.3.24) is strictly elliptic in Ω, since its ellipticity constants for ψ are equal to the ellipticity constants for φ, and φ satisfies Definition 8.1.1(iii).

Also, by Lemma 8.3.4, ψ satisfies the linear homogeneous condition (8.3.51) on $\Gamma_{\rm shock}$, and (8.3.53)–(8.3.55) hold.

Now we apply the comparison principle of Lemma 4.4.2 with $\Gamma_0 = \Gamma_{\rm sonic}$, $\Gamma_1 = \Gamma_{\rm shock}$, $\Gamma_2 = \Gamma_{\rm sym}$, and $\Gamma_3 = \Gamma_{\rm wedge}$. We note that the obliqueness condition at corners P_2 and P_3 is satisfied. Indeed, for corner P_3, the Neumann condition is prescribed at both straight sides $\Gamma_{\rm wedge}$ and $\Gamma_{\rm sym}$, which meet at angle $\pi - \theta_{\rm w} \in (0, \pi)$, so that the obliqueness at P_3 holds. At corner P_2, the Neumann condition is prescribed on $\Gamma_{\rm sym} \subset \{\xi_2 = -v_2\}$ with the interior normal $\boldsymbol{\nu}_{\rm sym} = (0,1)$, and condition (8.3.4) holds on $\Gamma_{\rm shock}$, which is close to line $\{\xi_1 = \xi_1(P_2)\}$ by (7.2.11), (7.2.15), and (8.3.10)–(8.3.12). Coefficients a_1 and a_2 are continuous near P_2 by (8.3.52), Lemma 8.3.3, and Definition 8.1.1(ii). Also, (8.3.53) implies that vector (a_1, a_2) on $\Gamma_{\rm shock}$ points into Ω. Then (8.3.54) and $\boldsymbol{\nu}_{\rm sym} = (0,1)$ imply that, if σ is small, (4.4.1) is satisfied with Σ as a straight line through P_2 parallel to vector $(a_1, 2a_2)(P_2)$. Now, by Lemma 4.4.2, we conclude that $\psi \geq 0$ in Ω. □

Corollary 8.3.6. *φ satisfies condition* (iv) *of Definition* 8.1.1.

8.3.3 Cone of monotonicity for solutions close to the normal reflection

In the next argument, we use vector $\mathbf{h}$ introduced in Lemma 5.1.1, specifically form (5.1.24) of $\mathbf{h}$. Denote by $\bar{\mathbf{h}} = (\bar{h}_1, \bar{h}_2)$ and $\bar{P}_1 = (\bar{\xi}_1, \sqrt{\bar{c}_2^2 - |\bar{\xi}_1|^2})$ the corresponding vector and point in the normal reflection case. Thus, $\Gamma^{\rm norm}_{\rm shock} = \{(\bar{\xi}_1, \xi_2) \; : \; 0 < \xi_2 < \sqrt{\bar{c}_2^2 - |\bar{\xi}_1|^2}\}$, and $\boldsymbol{\nu} = (1,0)$ and $\boldsymbol{\tau} = (0,1)$ at any $P \in \Gamma^{\rm norm}_{\rm shock}$. Furthermore, $\varphi = -\frac{|\boldsymbol{\xi}|^2}{2} + const.$ in Ω. Then, from (5.1.24), we have

$$\bar{\mathbf{h}} = -\frac{\bar{\rho}_2 - \rho_1}{\rho_1}(\hat{h}_1, \hat{h}_2) \qquad \text{at } P = (\bar{\xi}_1, \xi_2) \in \Gamma^{\rm norm}_{\rm shock}, \tag{8.3.56}$$

where $\hat{h}_1 = \bar{\rho}_2(\bar{c}_2^2 - |\bar{\xi}_1|^2)\bar{\xi}_1$ and $\hat{h}_2 = -(\bar{\rho}_2|\bar{\xi}_1|^2 + \rho_1\bar{c}_2^2)\xi_2$.

Lemma 8.3.7. *Let φ be as in Proposition* 8.3.1. *Let $\mathbf{e} = (a_1, a_2) \in \mathbb{R}^2 \setminus \{\mathbf{0}\}$ with $|\mathbf{e}| = 1$ be uniformly separated from $\hat{h} = (\hat{h}_1, \hat{h}_2)$ on $\overline{\Gamma^{\rm norm}_{\rm shock}}$ in the sense that there exists $\mu > 0$ such that*

$$d := \det\begin{bmatrix} \hat{h}_1(P) & \hat{h}_2(P) \\ a_1 & a_2 \end{bmatrix} \; \textit{satisfies } |d| \geq \mu \textit{ for any } P \in \Gamma^{\rm norm}_{\rm shock}. \tag{8.3.57}$$

Then there exists $\sigma > 0$ depending only on $(\rho_0, \rho_1, \gamma, \mu)$ such that $w = \partial_{\mathbf{e}}\phi$ for $\phi = \varphi - \varphi_0$ in Ω satisfies that $m < w < M$ on $\Gamma_{\rm shock}$, where $m = \min_{\overline{\Omega}} w$ and $M = \max_{\overline{\Omega}} w$.

Proof. By Lemma 8.2.1, $\partial_{\mathbf{e}}\phi$ is not a constant in Ω. Then we need to show that the extremum of w over $\overline{\Omega}$ cannot be attained on $\Gamma_{\rm shock}$. We divide the proof into three steps.

1. Suppose that $P^* = \boldsymbol{\xi}^* \in \Gamma_{\rm shock}$ is a point of extremum of w over $\overline{\Omega}$. Then, at P^*, we have three equalities involving $D^2\phi$:

(i) Equation (2.2.11),

(ii) The tangential derivative of the Rankine-Hugoniot condition (5.1.22),

(iii) $\partial_{\boldsymbol{\tau}} w{=}0$,

where the last equality is true since P^* is a point of extremum. Also, (ii) has been obtained by Lemma 5.1.1, which can be applied since condition (5.1.14) is satisfied for small σ by (7.1.6) and (8.3.1).

Denoting $\mathbf{e} = (a_1, a_2)$, we then see that $w = a_1\partial_1\phi + a_2\partial_2\phi$. Also, denote by $\boldsymbol{\tau}(P^*) = (\tau_1, \tau_2)$ the unit tangent vector to $\Gamma_{\rm shock}$ at P^*. Thus, the three equalities at P^* mentioned above are written as:

$$\begin{aligned} &(c^2-\varphi_{\xi_1}^2)\phi_{\xi_1\xi_1} - 2\varphi_{\xi_1}\varphi_{\xi_2}\phi_{\xi_1\xi_2} + (c^2-\varphi_{\xi_2}^2)\phi_{\xi_2\xi_2} = 0,\\ &\tau_1 h_1\phi_{11} + (\tau_1 h_2 + \tau_2 h_1)\phi_{12} + \tau_2 h_2\phi_{22} = 0, \\ &\tau_1 a_1\phi_{11} + (\tau_1 a_2 + \tau_2 a_1)\phi_{12} + \tau_2 a_2\phi_{22} = 0. \end{aligned} \tag{8.3.58}$$

Regarding (8.3.58) as a linear algebraic system for ϕ_{ij}, $1 \le i \le j \le 2$, we show that the only solution is $D^2\phi = 0$ if $\sigma > 0$ is sufficiently small, *i.e.*, the determinant of the linear system (8.3.58) is nonzero for small $\sigma > 0$.

2. For that, we compute first the determinant of the linear system (8.3.58) in the normal reflection case. Then $\boldsymbol{\tau} = (0,1)$, $\varphi = -\frac{|\boldsymbol{\xi}|^2}{2} + const.$, $\bar{P}_1 = (\bar{\xi}_1, \sqrt{\bar{c}_2^2 - |\bar{\xi}_1|^2})$, and $\Gamma_{\rm shock}^{\rm norm} = \{(\bar{\xi}_1, \xi_2) \,:\, 0 < \xi_2 < \sqrt{\bar{c}_2^2 - |\bar{\xi}_1|^2}\}$. Now, using (8.3.56), we compute the determinant of (8.3.58) at $P = (\bar{\xi}_1, \xi_2) \in \Gamma_{\rm shock}^{\rm norm}$ in the normal reflection case and obtain

$$\begin{aligned} D_0 &:= \det \begin{bmatrix} \bar{c}_2^2 - |\bar{\xi}_1|^2 & -2\bar{\xi}_1\xi_2 & \bar{c}_2^2 - \xi_2^2 \\ 0 & \hat{h}_1 & \hat{h}_2 \\ 0 & a_1 & a_2 \end{bmatrix} \\ &= (\bar{c}_2^2 - |\bar{\xi}_1|^2)\Big(\frac{\bar{\rho}_2 - \rho_1}{\rho_1}\Big)^2 \det \begin{bmatrix} \hat{h}_1 & \hat{h}_2 \\ a_1 & a_2 \end{bmatrix}. \end{aligned}$$

From this, we employ (8.3.57) to obtain

$$|D_0| \ge (\bar{c}_2^2 - |\bar{\xi}_1|^2)\Big(\frac{\bar{\rho}_2 - \rho_1}{\rho_1}\Big)^2 \mu \qquad \text{for any } P \in \Gamma_{\rm shock}^{\rm norm},$$

where, by (8.3.2), the right-hand side is a positive constant.

3. Now, if $\theta_{\rm w} \in (\frac{\pi}{2} - \sigma)$, the coefficients of the linear algebraic system (8.3.58) differ from the coefficients of the system for the normal reflection by no more than $C\sigma$, which follows from (1.14), (8.3.1), and (8.3.4). Then, if σ is sufficiently small, the determinant of (8.3.58) is positive. Thus, at point $\hat{P}$ of extremum of w, $D^2\phi(\hat{P}) = 0$. Then, at $\hat{P}$,

$$\partial_{\boldsymbol{\nu}} w = D^2\phi[\mathbf{e}, \boldsymbol{\nu}] = 0.$$

On the other hand, since w satisfies equation (5.1.10), which is uniformly elliptic in $\overline{\Omega} \cap \overline{B_r(\hat{P})}$ for sufficiently small $r > 0$, and $\hat{P}$ is a point of extremum of w over $\overline{\Omega}$, Hopf's lemma implies that $\partial_{\boldsymbol{\nu}} w > 0$ at $\hat{P}$, where we have used that, by Lemma 8.2.1, w is not a constant in $\overline{\Omega}$. Therefore, we arrive at a contradiction, which implies that a point of extremum of w cannot be on Γ_{shock}. □

Lemma 8.3.8. *Let φ be as that in Proposition* 8.3.1. *Then, if $\sigma > 0$ is sufficiently small, depending on (ρ_0, ρ_1, γ),*

$$\partial_{\mathbf{e}_{S_1}}(\varphi_1 - \varphi) \le 0 \qquad \text{in } \overline{\Omega}, \tag{8.3.59}$$

where $\mathbf{e}_{S_1}$ is defined by (7.5.8).

Proof. Let $w := \partial_{-\mathbf{e}_{S_1}}\phi$. We need to show that $w \ge 0$ in Ω. We divide the proof into three steps.

1. We first show that condition (8.3.57) holds for $(a_1, a_2) = \mathbf{e}_{S_1}$. From (7.5.8) and (8.3.4), $|\mathbf{e}_{S_1} - (0, -1)| \le C\sigma$. Then

$$\det\begin{bmatrix} \hat{h}_1(P) & \hat{h}_2(P) \\ a_1 & a_2 \end{bmatrix} \ge -\hat{h}_1 - C\sigma.$$

Note that, by (8.3.2) and (8.3.56), $-\hat{h}_1$ is a positive constant. Thus, if σ is small, depending only on (ρ_0, ρ_1, γ), condition (8.3.57) is satisfied with $\mu = -\frac{\hat{h}_1}{2} > 0$.

Then, further reducing σ if necessary, depending only on (ρ_0, ρ_1, γ), since now μ depends on these constants, we see that the conclusion of Lemma 8.3.7 holds.

2. By Lemma 7.5.12, $\mathbf{e}_{S_1}$ is not orthogonal to $\Gamma_{\text{wedge}} \cup \Gamma_{\text{sym}}$. Then, by Lemmas 8.2.3 and 8.3.7, and using that w is not constant in Ω, we conclude that the extremum of w over $\overline{\Omega}$ cannot be attained on $\Omega \cup \Gamma_{\text{shock}} \cup \Gamma_{\text{wedge}} \cup \Gamma_{\text{sym}}$. Since $\partial_{\mathbf{e}_{S_1}}\phi_1 = \partial_{\mathbf{e}_{S_1}}(u_1\xi_1) = const.$ and $\varphi_1 - \varphi = \phi_1 - \phi$, we conclude that the extremum of $\partial_{\mathbf{e}_{S_1}}(\varphi_1 - \varphi)$ over $\overline{\Omega}$ cannot be attained on $\Omega \cup \Gamma_{\text{shock}} \cup \Gamma_{\text{wedge}} \cup \Gamma_{\text{sym}}$. That is, the extremum of $\partial_{\mathbf{e}_{S_1}}(\varphi_1 - \varphi)$ over $\overline{\Omega}$ can be attained only on $\{P_2, P_3\} \cup \overline{\Gamma_{\text{sonic}}}$.

3. In the proof of Proposition 8.2.6, we have shown that $\partial_{\mathbf{e}_{S_1}}(\varphi_1 - \varphi) \le 0$ at P_2 and P_3. This argument works in the present case. Thus, it remains to show that $\partial_{\mathbf{e}_{S_1}}(\varphi_1 - \varphi) \le 0$ on $\overline{\Gamma_{\text{sonic}}}$.

Then $D\varphi = D\varphi_2$ on $\overline{\Gamma_{\text{sonic}}}$; see also Remark 8.1.2. Thus, on $\overline{\Gamma_{\text{sonic}}}$,

$$\partial_{\mathbf{e}_{S_1}}(\varphi_1 - \varphi) = \partial_{\mathbf{e}_{S_1}}(\phi_1 - \phi_2) = \partial_{\mathbf{e}_{S_1}}\big((u_1 - u_2)\xi_1 - v_2\xi_2\big) = 0,$$

where we have used (7.5.8). This completes the proof. □

Lemma 8.3.9. *Let φ be as that in Proposition* 8.3.1. *Then, if $\sigma > 0$ is sufficiently small, depending on (ρ_0, ρ_1, γ),*

$$\partial_{\xi_2}(\varphi_1 - \varphi) \leq 0 \qquad \textit{in } \overline{\Omega}. \tag{8.3.60}$$

Proof. Note that $\partial_{\xi_2}(\varphi_1 - \varphi) = \partial_{\mathbf{e}_{\xi_2}}(\varphi_1 - \varphi)$ for $\mathbf{e}_{\xi_2} = (0, 1)$. Then $\mathbf{e}_{\xi_2}$ is not orthogonal to Γ_{wedge}. Arguing as in the proof of Lemma 8.3.8, and using that $\partial_{\xi_2}(\varphi_1 - \varphi) = -v_2 < 0$ on Γ_{sonic} as shown in the proof of Proposition 8.2.8, we see that, for sufficiently small $\sigma > 0$, it suffices to check that $\partial_{\xi_2}(\varphi_1 - \varphi) \leq 0$ on Γ_{sym}. To show this, we notice that, since $\partial_{\xi_2}\varphi = 0$ on Γ_{sym} by the boundary conditions, and $\partial_{\xi_2}\varphi_1 = \partial_{\xi_2}(-\frac{|\boldsymbol{\xi}|^2}{2} + u_1\xi_1) = 0$ on $\Gamma_{\text{sym}} \subset \{\xi_2 = 0\}$, we find that $\partial_{\xi_2}(\varphi_1 - \varphi) = 0$ on Γ_{sym}. ☐

Lemmas 8.3.8–8.3.9 imply that φ satisfies condition (v) of Definition 8.1.1. This and Corollary 8.3.6 imply Proposition 8.3.1.

Chapter Nine

Uniform Estimates for Admissible Solutions

In this chapter, we always assume that φ is an admissible solution of **Problem 2.6.1** in the sense of Definition 8.1.1. We make several key uniform estimates for the admissible solution φ with respect to the wedge angle θ_w.

9.1 BOUNDS OF THE ELLIPTIC DOMAIN Ω AND ADMISSIBLE SOLUTION φ IN Ω

In this section, we first make some basic uniform estimates for any admissible solution φ for the wedge angle $\theta_w \in (\theta_w^s, \frac{\pi}{2})$. The universal constant C in this section depends only on the data, *i.e.*, (ρ_0, ρ_1, γ), unless otherwise specified.

Lemma 9.1.1. *For* $\phi = \varphi + \frac{1}{2}|\boldsymbol{\xi}|^2$,

$$\sup_{\Omega} |\phi| = \sup_{\overline{\Gamma_{\text{sonic}}} \cup \overline{\Gamma_{\text{shock}}} \cup \{P_3\}} |\phi|.$$

Proof. Since φ is smooth in Ω, then φ satisfies equation (2.2.11) in Ω. Equation (2.2.11) is strictly elliptic in $\overline{\Omega} \setminus \overline{\Gamma_{\text{sonic}}}$ by Definition 8.1.1(iii). Then we conclude that $\sup_{\Omega} |\phi| = \sup_{\partial\Omega} |\phi|$, by writing (2.2.11) as a linear equation:

$$a_{11}(\boldsymbol{\xi})\phi_{\xi_1\xi_1} + a_{12}(\boldsymbol{\xi})\phi_{\xi_1\xi_2} + a_{22}(\boldsymbol{\xi})\phi_{\xi_2\xi_2} = 0 \tag{9.1.1}$$

with strict ellipticity inside Ω, by considering the domains:

$$\Omega_\epsilon = \{P \in \Omega \,:\, \text{dist}(P, \partial\Omega) > \epsilon\} \qquad \text{for } \epsilon > 0,$$

in which (9.1.1) is uniformly elliptic, and then by sending $\epsilon \to 0+$ and using that $\phi \in C(\overline{\Omega})$.

Moreover, the boundary condition (2.2.20) implies that ϕ satisfies $\partial_{\boldsymbol{\nu}}\phi = 0$ on $\Gamma_{\text{wedge}} \cup \Gamma_{\text{sym}}$. Also, for any $P \in \Gamma_{\text{wedge}} \cup \Gamma_{\text{sym}}$ (where the segments do not include the endpoints), equation (9.1.1) is uniformly elliptic in some neighborhood of P by Definition 8.1.1(iii). Thus, the extremum of ϕ over $\overline{\Omega}$ cannot be attained on $\Gamma_{\text{wedge}} \cup \Gamma_{\text{sym}}$, unless ϕ is constant in Ω. However, for an admissible solution φ for $\theta_w < \frac{\pi}{2}$, ϕ cannot be constant in Ω by Lemma 8.2.1. Therefore, the extremum of ϕ over $\overline{\Omega}$ cannot be attained on $\Gamma_{\text{wedge}} \cup \Gamma_{\text{sym}}$. $\square$

Now we prove the estimates on $\mathrm{diam}(\Omega)$ and $\sup_\Omega |\varphi|$. The main difficulty is that, if θ_{w} is sufficiently small, we cannot exclude the possibility that ray $\mathcal{S}_1^+ := \{P_0 + t(P_1 - P_0) \,:\, t > 0\}$ does not intersect with the ξ_1–axis. Thus, we cannot obtain a uniform bound of $\mathrm{diam}(\Omega)$ by the coordinates of the intersection points of $\mathcal{S}_1^+$ with the ξ_1–axis.

Proposition 9.1.2. *There exists $C > 0$ such that*

$$\Omega \subset B_C(\mathbf{0}), \tag{9.1.2}$$
$$|\varphi| + |\phi| \le C \qquad \text{in } \Omega. \tag{9.1.3}$$

Proof. We divide the proof into three steps.

1. In the first step, we show a lower bound for ϕ in Ω. By (8.2.36) in Lemma 8.2.18, it suffices to bound ϕ at P_2. Moreover, we also obtain the bound of ξ_{1P_2}.

We note that $P_2 = (\xi_{1P_2}, 0)$ and $\boldsymbol{\nu}(P_2) = (1, 0)$. Then the Rankine-Hugoniot condition (8.1.13) at P_2 implies

$$\rho(\partial_{\xi_1}\phi - \xi_1) = \rho_1(u_1 - \xi_1) \qquad \text{at } \boldsymbol{\xi} = (\xi_{1P_2}, 0).$$

Since $\phi(P_2) = \min_{P\in\overline{\Omega}} \phi(P)$ by (8.2.36),

$$\partial_{\xi_1}\phi(P_2) = \partial_{\boldsymbol{\nu}}\phi(P_2) \ge 0.$$

We also note that $\rho > \rho_1 > 0$ on $\Gamma_{\text{shock}} \cup \{P_2\}$ by (8.1.13) and (8.1.16). Combining this with the last two facts implies

$$\rho_1 u_1 - (\rho - \rho_1)(-\xi_{1P_2}) = \rho\partial_{\xi_1}\phi \ge 0 \qquad \text{at } P_2.$$

Since $\xi_{1P_2} < 0$, we have

$$(\rho - \rho_1)|\xi_{1P_2}| \le \rho_1 u_1. \tag{9.1.4}$$

Now we recall that the sonic circle of state (1) is $\mathcal{C}_1 = \partial B_{c_1}(u_1, 0)$. Since $u_1 > 0$, we find that either $0 > \xi_{1P_2} > -c_1$ or $\mathrm{dist}(P_2, \mathcal{C}_1) > u_1$.

In the first case, we have the bound that $|\xi_{1P_2}| \le c_1$.

In the second case, $\mathrm{dist}(P_2, \mathcal{C}_1) > u_1$, and then Lemma 6.1.3 implies that $\partial_{\boldsymbol{\nu}}\varphi_1(P_2) - \partial_{\boldsymbol{\nu}}\varphi(P_2) \ge \frac{1}{C}$. Combining this with the facts that $\varphi(P_2) = \varphi_1(P_2)$ and $\partial_{\boldsymbol{\tau}}\varphi(P_2) = \partial_{\boldsymbol{\tau}}\varphi_1(P_2) = 0$, and using the Bernoulli law (2.2.9), we conclude that $\rho(P_2) - \rho_1 > \frac{1}{C}$. Then (9.1.4) implies that $|\xi_{1P_2}| \le C$.

Therefore, we obtain that $|\xi_{1P_2}| \le C$ in both cases. Now $\phi(P_2) = \phi_1(P_2) = u_1(\xi_{1P_2} - \xi_1^0)$ implies that $|\phi(P_2)| \le C$.

2. Prove (9.1.2). From Step 1, $|\phi| \le C$ at its minimum point. Thus, $\phi \ge -C$ in Ω. Then, on Γ_{shock}, $u_1(\xi_1 - \xi_1^0) = \phi_1(\boldsymbol{\xi}) = \phi(\boldsymbol{\xi}) \ge -C$, which implies that $\xi_1 \ge -C$ by using (8.2.1). From this, using that $\xi_1 \le \xi_{1P_1}$ on Γ_{shock}, we obtain that $|\xi_1| \le C$ on Γ_{shock}. Thus, by Definition 8.1.1(i), we conclude that $|\xi_1| \le C$ in Ω.

It remains to bound ξ_2 in Ω. By Definition 8.1.1(i), $\xi_2 > 0$ in Ω. Thus, we only need to estimate ξ_2 from above.

Since u_2 is a continuous function of $\theta_{\rm w} \in [\theta_{\rm w}^{\rm s}, \frac{\pi}{2}]$ with $u_{2|\theta_{\rm w}=\frac{\pi}{2}} = 0$, we conclude from (7.5.8) that there exists $\hat{\theta} \in [\theta_{\rm w}^{\rm s}, \frac{\pi}{2})$ such that $\mathbf{e}_{\mathcal{S}_1}\cdot(0,1) \le 0$ for any $\theta_{\rm w} \in [\hat{\theta}, \frac{\pi}{2})$. For such $\theta_{\rm w}$, we obtain that $\sup_\Omega \xi_2 \le \sup_{\Gamma_{\rm sonic}} \xi_2 \le u_2 + c_2 \le C$.

Next, we estimate $\sup_\Omega \xi_2$ for $\theta_{\rm w} \in (\theta_{\rm w}^{\rm s}, \hat{\theta})$. By (8.1.22), $\Omega \subset \{\varphi_2 < \varphi_1\}$. Thus, by (2.2.17) and (2.4.1),

$$\xi_2 < \frac{(u_1 - u_2)(\xi_1 - \xi_1^0)}{u_2 \tan\theta_{\rm w}} + \xi_1^0 \tan\theta_{\rm w} \qquad \text{in } \Omega.$$

From this, using that $\theta_{\rm w}^{\rm s} < \theta_{\rm w} < \hat{\theta}$, with $\theta_{\rm w}^{\rm s} > 0$ and $\hat{\theta} < \frac{\pi}{2}$, which depend only on (ρ_0, ρ_1, γ), we conclude that $|\xi_2| \le C|\xi_1| + C$ in Ω so that (9.1.2) follows.

3. Now, $\sup_\Omega |\phi| = \sup_{\Gamma_{\rm sonic}\cup\Gamma_{\rm shock}\cup\{P_3\}} |\phi|$ by Lemma 9.1.1. Since $\phi = \phi_1$ on $\Gamma_{\rm shock}$, $\phi = \phi_2$ on $\Gamma_{\rm sonic}$, and $\phi_1 \ge \phi \ge \phi_2$ at P_3, then (8.2.1) and (9.1.2) imply that $|\phi| \le C$ in Ω. Thus, using (9.1.2) again, we see that $|\varphi| \le C$ in Ω. Now (9.1.3) is proved. □

Corollary 9.1.3. *There exists $C > 0$ such that*

$$\|(\varphi, \phi, \psi)\|_{C^{0,1}(\overline{\Omega})} \le C. \tag{9.1.5}$$

Proof. By Definition 8.1.1(iii), equation (2.2.8) is elliptic in Ω. Then (1.18) holds, which implies (2.2.12). From (2.2.12), we use (9.1.3) to obtain that $|D\varphi| \le C$ in Ω. From this and (9.1.2), we find that $|D\phi| = |D\varphi + \boldsymbol{\xi}| \le C$ in Ω. Combining these derivative estimates with (9.1.3), we obtain the estimate of $\|\varphi\|_{C^{0,1}(\overline{\Omega})}$ in (9.1.5). Also, using this estimate, we have

$$\|\psi\|_{C^{0,1}(\overline{\Omega})} \le \|\varphi\|_{C^{0,1}(\overline{\Omega})} + \|\varphi_2\|_{C^{0,1}(\overline{\Omega})} \le C,$$

where we have used (9.1.2) and Theorem 7.1.9(i). □

Lemma 9.1.4. *There exists $C > 0$ such that, for any admissible solution φ of* **Problem 2.6.1**,

$$\begin{aligned} &\Big(\frac{2}{\gamma+1}\Big)^{\frac{1}{\gamma-1}} \rho_1 \le \rho \le C \qquad && \text{in } \Omega, \\ &\rho_1 < \rho \le C && \text{on } \overline{\Gamma_{\rm shock}} \cup \{P_3\}. \end{aligned} \tag{9.1.6}$$

Proof. The upper bound in (9.1.6) follows from (2.2.9) and (9.1.5).

Now we prove the lower bounds. From the Bernoulli law in (2.2.7) applied to φ in Ω and to φ_1 in $\mathbb{R}^2$, and using $\varphi \in C^1(\overline{\Omega})$ by (8.1.3), we have

$$\rho^{\gamma-1} + \frac{\gamma-1}{2}\big(|D\varphi|^2 + \varphi\big) = \rho_1^{\gamma-1} + \frac{\gamma-1}{2}\big(|D\varphi_1|^2 + \varphi_1\big) \qquad \text{in } \overline{\Omega}. \tag{9.1.7}$$

Since $\varphi \le \varphi_1$ in Ω by (8.1.5), we have

$$\rho^{\gamma-1} + \frac{\gamma-1}{2}|D\varphi|^2 \ge \rho_1^{\gamma-1} + \frac{\gamma-1}{2}|D\varphi_1|^2 \qquad \text{in } \overline{\Omega}. \tag{9.1.8}$$

Using (9.1.8) and the fact that $|D\varphi|^2 \le c^2 = \rho^{\gamma-1}$ in Ω by (8.1.4), we have

$$\rho^{\gamma-1} + \frac{\gamma-1}{2}\rho^{\gamma-1} \ge \rho_1^{\gamma-1},$$

that is, $\rho \ge \left(\frac{2}{\gamma+1}\right)^{\frac{1}{\gamma-1}}\rho_1$ in Ω.

On $\overline{\Gamma_{\text{shock}}}$, $|D\varphi|^2 \le c^2 = \rho^{\gamma-1}$ by (8.1.4) and $|D\varphi_1|^2 > c_1^2 = \rho_1^{\gamma-1}$ by (8.1.2). Thus, (9.1.8) implies

$$\rho^{\gamma-1} + \frac{\gamma-1}{2}\rho^{\gamma-1} > \rho_1^{\gamma-1} + \frac{\gamma-1}{2}\rho_1^{\gamma-1},$$

that is, $\rho > \rho_1$ on $\overline{\Gamma_{\text{shock}}}$.

At P_3, $|D\varphi| = 0$, by combining (2.2.20) on $\Gamma_{\text{wedge}} \cup \Gamma_{\text{sym}}$ with $\varphi \in C^1(\overline{\Omega})$. Also, $|D\varphi_1(P_3)| = u_1 > 0$. Therefore, (9.1.8) implies that $\rho > \rho_1$ at P_3. □

9.2 REGULARITY OF ADMISSIBLE SOLUTIONS AWAY FROM $\Gamma_{\text{shock}} \cup \Gamma_{\text{sonic}} \cup \{P_3\}$

9.2.1 An elliptic equation satisfied by the admissible solutions in Ω

Any admissible solution φ of **Problem 2.6.1** satisfies the potential flow equation (2.2.8) in Λ. The equation is of mixed type and, by Definition 8.1.1(ii)–(iii), it is uniformly elliptic for φ on any compact subset of $\overline{\Omega} \setminus (\overline{\Gamma_{\text{shock}}} \cup \overline{\Gamma_{\text{sonic}}} \cup \{P_3\})$. It is more convenient to show that φ satisfies a uniformly elliptic equation with smooth ingredients on these subsets by making an appropriate modification of the potential flow equation (2.2.8).

Write equation (2.2.8) in the form:

$$\operatorname{div}\mathcal{A}(D\varphi, \varphi) + \mathcal{B}(D\varphi, \varphi) = 0, \tag{9.2.1}$$

where

$$\mathcal{A}(\mathbf{p}, z) := \rho(|\mathbf{p}|^2, z)\mathbf{p}, \quad \mathcal{B}(\mathbf{p}, z) := 2\rho(|\mathbf{p}|^2, z) \tag{9.2.2}$$

with function $\rho(|\mathbf{p}|^2, z)$ defined by (2.2.9), for which we restrict to such $(\mathbf{p}, z)$ that (2.2.9) is defined, *i.e.*, satisfying $\rho_0^{\gamma-1} - (\gamma-1)(z + \frac{1}{2}|\mathbf{p}|^2) \ge 0$, $\mathbf{p} \in \mathbb{R}^2$, and $z \in \mathbb{R}$.

Lemma 9.2.1. *Let $M \ge 2$. Denote*

$$\mathcal{K}_M := \left\{ (\mathbf{p}, z) \in \mathbb{R}^2 \times \mathbb{R} \, : \, \begin{array}{l} |\mathbf{p}| + |z| \le M, \ \rho(|\mathbf{p}|^2, z) \ge \frac{1}{M} \\ |\mathbf{p}|^2 \le (1 - \frac{1}{M})c^2(|\mathbf{p}|^2, z) \end{array} \right\},$$

where $\rho(|\mathbf{p}|^2, z)$ is defined by (2.2.9):

$$\rho(s, z) = \left(\rho_0^{\gamma-1} - (\gamma-1)(\frac{1}{2}s + z)\right)^{\frac{1}{\gamma-1}},$$

and $c^2 = \rho^{\gamma-1}$. Then there exist $\tilde{\mathcal{A}}(\mathbf{p}, z)$ and $\tilde{\mathcal{B}}(\mathbf{p}, z)$ on $\mathbb{R}^2 \times \mathbb{R}$ satisfying the following properties:

(i) *If* $|(\mathbf{p}, z) - (\tilde{\mathbf{p}}, \tilde{z})| < \varepsilon$ *for some* $(\tilde{\mathbf{p}}, \tilde{z}) \in \mathcal{K}_M$,

$$\tilde{\mathcal{A}}(\mathbf{p}, z) = \mathcal{A}(\mathbf{p}, z), \qquad \tilde{\mathcal{B}}(\mathbf{p}, z) = \mathcal{B}(\mathbf{p}, z); \tag{9.2.3}$$

(ii) *For any* $(\mathbf{p}, z) \in \mathbb{R}^2 \times \mathbb{R}$ *and* $\boldsymbol{\kappa} = (\boldsymbol{\kappa}_1, \boldsymbol{\kappa}_2) \in \mathbb{R}^2$,

$$\sum_{i,j=1}^{2} \tilde{\mathcal{A}}^i_{p_j}(\mathbf{p}, z)\kappa_i\kappa_j \geq \lambda|\boldsymbol{\kappa}|^2; \tag{9.2.4}$$

(iii) *For* $k = 1, \ldots,$

$$|\tilde{\mathcal{B}}| \leq C_0, \quad |D^k_{(\mathbf{p},z)}(\tilde{\mathcal{A}}, \tilde{\mathcal{B}})| \leq C_k \qquad \text{on } \mathbb{R}^2 \times \mathbb{R}, \tag{9.2.5}$$

where the positive constants ε, λ, *and* C_k *with* $k = 0, 1, \ldots$ *depend only on* (ρ_0, γ, M). *In particular, if* $\mathcal{D} \subset \mathbb{R}^2$ *is open, and* $\varphi \in C^2(\mathcal{D})$ *is a solution of* (2.2.8) *in* $\mathcal{D}$ *such that*

$$\|\varphi\|_{C^{0,1}(\mathcal{D})} \leq M, \tag{9.2.6}$$

$$\rho(|D\varphi|^2, \varphi) \geq \frac{1}{M}, \tag{9.2.7}$$

$$\frac{|D\varphi|^2}{c^2(|D\varphi|^2, \varphi)} \leq 1 - \frac{1}{M} \tag{9.2.8}$$

for $M \geq 2$, *then* φ *satisfies the equation:*

$$\mathrm{div}\tilde{\mathcal{A}}(D\varphi, \varphi) + \tilde{\mathcal{B}}(D\varphi, \varphi) = 0 \qquad \text{in } \mathcal{D}, \tag{9.2.9}$$

with functions $\tilde{\mathcal{A}}$ *and* $\tilde{\mathcal{B}}$ *introduced above.*

Proof. In this proof, constants C, C_k, and δ are positive and depend only on (ρ_0, γ, M).

Let $\bar{\rho}(s, z)$ be a nonnegative smooth function on $(s, z) \in \mathbb{R} \times \mathbb{R}$ with $\partial_s\bar{\rho} \leq 0$. Then, by a simple explicit calculation, we find that the equation:

$$\mathrm{div}\big(\bar{\rho}(|D\varphi|^2, \varphi)D\varphi\big) + 2\bar{\rho}(|D\varphi|^2, \varphi) = 0, \tag{9.2.10}$$

considered as (9.2.9), has

$$\tilde{\mathcal{A}}^i_{p_j}(\mathbf{p}, z) = \bar{\rho}\delta_{ij} + 2p_ip_j\partial_s\bar{\rho} = \bar{\rho}\delta_{ij} - 2p_ip_j|\partial_s\bar{\rho}|$$

with $(\bar{\rho}, \partial_s\bar{\rho}) = (\bar{\rho}, \partial_s\bar{\rho})(|\mathbf{p}|^2, z)$. Thus, denoting $\bar{\Phi}(s, z) = \bar{\rho}(s^2, z)s$ and noting that

$$\partial_s\bar{\Phi}(s, z) = \bar{\rho}(s^2, z) + 2s^2\partial_s\bar{\rho}(s^2, z) = \bar{\rho}(s^2, z) - 2s^2|\partial_s\bar{\rho}(s^2, z)|,$$

we have

$$\sum_{i,j=1}^{2} \tilde{\mathcal{A}}^i_{p_j}(\mathbf{p}, z)\kappa_i\kappa_j = \Big(\bar{\rho} - 2\frac{(\mathbf{p}\cdot\boldsymbol{\kappa})^2}{|\boldsymbol{\kappa}|^2}|\partial_s\bar{\rho}|\Big)|\boldsymbol{\kappa}|^2$$
$$\geq \big(\bar{\rho} - 2|\mathbf{p}|^2|\partial_s\bar{\rho}|\big)|\boldsymbol{\kappa}|^2 = \partial_s\bar{\Phi}(|\mathbf{p}|, z)|\boldsymbol{\kappa}|^2,$$

where the inequality becomes an equality if $\boldsymbol{\kappa} = a\mathbf{p}$ with $a \in \mathbb{R}$. Therefore, ellipticity (9.2.4) holds if there exists $\tilde{\lambda} > 0$ such that

$$\partial_s\bar{\Phi}(s, z) \geq \tilde{\lambda} \qquad \text{for all } s \geq 0, z \in \mathbb{R}.$$

We note that (2.2.8) is of form (9.2.10) with

$$\bar{\rho}_*(s, z) = \big(\rho_0^{\gamma-1} - (\gamma - 1)(z + \frac{s}{2})\big)^{\frac{1}{\gamma-1}}.$$

We compute

$$\partial_s\bar{\rho}_* = -\frac{1}{2}\bar{\rho}_*^{2-\gamma},$$
$$\partial_s\bar{\Phi}_*(s, z) = \bar{\rho}_*^{2-\gamma}(s^2, z)\Big(\rho_0^{\gamma-1} - (\gamma - 1)z - \frac{\gamma+1}{2}s^2\Big).$$

Denoting $s_{\rm so}(z) := \sqrt{\frac{2}{\gamma+1}\big(\rho_0^{\gamma-1} - (\gamma-1)z\big)}$ as the sonic speed, we can rewrite the last expression as

$$\partial_s\bar{\Phi}_*(s, z) = \frac{\gamma+1}{2}\bar{\rho}_*^{2-\gamma}(s^2, z)\big(s_{\rm so}^2(z) - s^2\big). \tag{9.2.11}$$

Then $\partial_s\bar{\Phi}_*(s, z) > 0$ for $s \in [0, s_{\rm so}(z))$. Denoting

$$\tilde{\mathcal{K}}_M := \{(|\mathbf{p}|, z) \ :\ (\mathbf{p}, z) \in \mathcal{K}_M\} \subset \mathbb{R}^2,$$

we find from the definition of $\mathcal{K}_M$ and the fact that $c^2(s^2, z) \leq s_{\rm so}^2(z)$ for $\gamma > 1$ that, for all $(s, z) \in \tilde{\mathcal{K}}_M$,

$$s^2 \leq \big(1 - \frac{1}{M}\big)c^2(s^2, z) \leq \big(1 - \frac{1}{M}\big)s_{\rm so}^2(z),$$
$$\frac{1}{M} \leq \bar{\rho}_*(s^2, z) \leq C,$$
$$\frac{1}{C} \leq c^2(s^2, z) \leq s_{\rm so}^2(z) \leq C.$$

The estimate of $\bar{\rho}_*$ above and the explicit expression of $\bar{\rho}_*(s, z)$ imply that

$$-C_1 \leq z \leq \frac{1}{\gamma-1}\rho_0^{\gamma-1} - \delta \qquad \text{if } (s, z) \in \tilde{\mathcal{K}}_M.$$

Therefore, we have proved that

$$\tilde{\mathcal{K}}_M \subset \Big\{(s,z) \ : \ -C_1 < z < \frac{1}{\gamma-1}\rho_0^{\gamma-1} - \delta, \ s^2 \le \big(1-\frac{1}{M}\big)s_{\rm so}^2(z)\Big\} =: \mathcal{S}_M.$$

Then we need to modify $\bar{\rho}_*(s,z)$ outside $\mathcal{S}_M$ so that the corresponding function $\bar{\rho}_{\rm mod}(s,z)$ is smooth and that $\Phi_{\rm mod}(s,z) := \bar{\rho}_{\rm mod}(s^2,z)s$ satisfies

$$\partial_s\Phi_{\rm mod}(s,z) \ge \frac{1}{C} \qquad \text{for all } s\ge 0, z\in\mathbb{R}.$$

We first define $\bar{\rho}_{\rm mod}(s,z)$ for all $s \ge 0$ and $z \in [-2C_1, \ \frac{1}{\gamma-1}\rho_0^{\gamma-1} - \frac{\delta}{2}]$. In fact, we first define $\partial_s\bar{\rho}_{\rm mod}(s,z)$ on this set.

Choose $\eta \in C^\infty(\mathbb{R})$ satisfying $0\le\eta\le 1$ on $\mathbb{R}$, $\eta\equiv 1$ on $(-\infty,0)$, and $\eta\equiv 0$ on $(1,\infty)$.

Fix $z \in [-2C_1, \ \frac{1}{\gamma-1}\rho_0^{\gamma-1} - \frac{\delta}{2}]$. We define

$$\begin{aligned}
&\partial_s\bar{\rho}_{\rm mod}(s,z) := \partial_s\bar{\rho}_*(s,z)\eta\Big(\frac{2M(s^2-(1-\frac{1}{M})s_{\rm so}^2(z))}{s_{\rm so}^2(z)}\Big),\\
&\bar{\rho}_{\rm mod}(s,z) := \bar{\rho}_*(0,z) + \int_0^s \partial_s\bar{\rho}_{\rm mod}(t,z)\,dt,\\
&\Phi_{\rm mod}(s,z) := \bar{\rho}_{\rm mod}(s^2,z)s
\end{aligned}$$

for all $s \ge 0$.

Using that $\partial_s\bar{\rho}_*(s,z) < 0$ for $(s,z)\in\mathcal{S}_M$, we obtain that, for each $z \in [-2C_1, \ \frac{1}{\gamma-1}\rho_0^{\gamma-1} - \frac{\delta}{2}]$,

$$\begin{aligned}
&\partial_s\bar{\rho}_*(s,z) \le \partial_s\bar{\rho}_{\rm mod}(s,z) \le 0 && \text{for all } s\in[0,\infty),\\
&\partial_s\bar{\rho}_{\rm mod}(s,z) = \partial_s\bar{\rho}_*(s,z) && \text{for all } s^2 \le (1-\tfrac{1}{M})s_{\rm so}^2(z),\\
&\partial_s\bar{\rho}_{\rm mod}(s,z) = 0 && \text{for all } s^2 \ge (1-\tfrac{1}{2M})s_{\rm so}^2(z).
\end{aligned} \tag{9.2.12}$$

From the definition of $\bar{\rho}_{\rm mod}(\cdot)$ above, we find that, for each z as above,

$$\begin{aligned}
&\bar{\rho}_*(s,z) \le \bar{\rho}_{\rm mod}(s,z) \le \bar{\rho}_*(0,z) && \text{for all } s\in[0,\infty),\\
&\bar{\rho}_{\rm mod}(s,z) = \bar{\rho}_*(s,z) && \text{for all } s^2 \le (1-\tfrac{1}{M})s_{\rm so}^2(z),\\
&\bar{\rho}_{\rm mod}(s,z) = C_2 \ge \bar{\rho}_*((1-\tfrac{1}{2M})s_{\rm so}^2(z),z) && \text{for all } s^2 \ge (1-\tfrac{1}{2M})s_{\rm so}^2(z).
\end{aligned}$$

We also note from the explicit expression that

$$s_{\rm so}^2(z) \ge \frac{\delta}{\gamma+1} \qquad \text{for all } z\in(-\infty, \ \frac{1}{\gamma-1}\rho_0^{\gamma-1} - \frac{\delta}{2}],$$

which implies that, for such z,

$$\bar{\rho}_{\rm mod}(s,z) \ge \bar{\rho}_*(s,z) \ge \big(\frac{\gamma+1}{2}s_{\rm so}^2(z)\big)^{\frac{1}{\gamma-1}} \ge \frac{1}{C} \qquad \text{for all } s\ge 0.$$

Now we show the ellipticity. Using the first estimate in (9.2.12), we see that, for all $z \in (-2C_1, \frac{1}{\gamma-1}\rho_0^{\gamma-1} - \frac{\delta}{2}]$ and $s \in [0, (1 - \frac{1}{2M})s_{\rm so}^2(z)]$,

$$\begin{aligned}\partial_s \Phi_{\rm mod}(s, z) &= \bar{\rho}_{\rm mod}(s^2, z) + 2s^2 \partial_s \bar{\rho}_{\rm mod}(s^2, z)\\ &\geq \bar{\rho}_*(s^2, z) + 2s^2 \partial_s \bar{\rho}_*(s^2, z)\\ &= \partial_s \Phi_*(s, z)\\ &\geq \frac{\gamma+1}{2} \bar{\rho}_*^{2-\gamma}(s^2, z) \frac{1}{M}\\ &\geq \frac{1}{C},\end{aligned}$$

where we have used (9.2.11). For $z \in (-2C_1, \frac{1}{\gamma-1}\rho_0^{\gamma-1} - \frac{\delta}{2}]$ and $s \geq (1 - \frac{1}{2M})s_{\rm so}^2(z)$, $\partial_s \bar{\rho}_{\rm mod} = 0$ so that

$$\partial_s \Phi_{\rm mod}(s, z) = \bar{\rho}_{\rm mod}((1 - \frac{1}{2M})s_{\rm so}^2(z),\ z) \geq \frac{1}{C},$$

where we have used the positive lower bound of $\bar{\rho}_{\rm mod}$ obtained above.

Combining the last two estimates, we have

$$\partial_s \Phi_{\rm mod} \geq \frac{1}{C} \qquad \text{on } \Big\{ -2C_1 \leq z \leq \frac{1}{\gamma-1}\rho_0^{\gamma-1} - \frac{\delta}{2},\ s \geq 0 \Big\}.$$

Note also that $\bar{\rho}_{\rm mod} \in C^\infty([0, \infty) \times [-2C_1, \frac{1}{\gamma-1}\rho_0^{\gamma-1} - \frac{\delta}{2}])$. Furthermore,

$$\partial_s \Phi_{\rm mod} = \bar{\rho}_{\rm mod}((1 - \frac{1}{2M})s_{\rm so}^2(z),\ z)$$

for any $(s^2, z) \in ((1 - \frac{1}{2M})s_{\rm so}^2(z), \infty) \times [-2C_1, \frac{1}{\gamma-1}\rho_0^{\gamma-1} - \frac{\delta}{2}]$. From this and the estimates of $\bar{\rho}_{\rm mod}$ obtained above, it is easy to see that $|\bar{\rho}_{\rm mod}| \leq C$ and $|D^k(\Phi_{\rm mod}, \bar{\rho}_{\rm mod})| \leq C_k$ for any $(s^2, z) \in [0, \infty) \times [-2C_1, \frac{1}{\gamma-1}\rho_0^{\gamma-1} - \frac{\delta}{2}]$ for $k = 1, 2, \dots$.

Thus, it remains to extend $\bar{\rho}_{\rm mod}$ to all z, without modifying $\bar{\rho}_{\rm mod}$ for $z \in [-2C_1, \frac{1}{\gamma-1}\rho_0^{\gamma-1} - \delta]$.

Let $\zeta \in C^\infty(\mathbb{R})$ satisfying $-2C_1 \leq \zeta \leq \frac{1}{\gamma-1}\rho_0^{\gamma-1} - \frac{\delta}{2}$ on $\mathbb{R}$, and let $\eta(z) = z$ on $(-C_1, \frac{1}{\gamma-1}\rho_0^{\gamma-1} - \delta)$. Then we define

$$\bar{\rho}_{\rm mod}^{\rm new}(s, z) := \bar{\rho}_{\rm mod}(s, \zeta(z)).$$

The resulting function $\bar{\rho}_{\rm mod}^{\rm new}$ is defined on $s \in [0, \infty)$ and $z \in \mathbb{R}$, and equation (9.2.10) with $\bar{\rho} = \bar{\rho}_{\rm mod}^{\rm new}$ satisfies all of the asserted properties. ☐

9.2.2 Regularity away from $\Gamma_{\rm shock} \cup \Gamma_{\rm sonic} \cup \{P_3\}$ and compactness

Lemma 9.2.2. *For any $\alpha \in (0, 1)$ and $r > 0$, there exists $C > 0$ depending only on $(\rho_0, \rho_1, \gamma, \alpha, r)$ such that, if φ is an admissible solution of* **Problem 2.6.1** *for the wedge angle $\theta_{\rm w} \in (\theta_{\rm w}^{\rm s}, \frac{\pi}{2})$, then*

(i) *If* $P \in \Omega$ *such that* $B_{4r}(P) \subset \Omega$,

$$\|\varphi\|_{C^{2,\alpha}(\overline{B_r(P)})} \leq C; \tag{9.2.13}$$

(ii) *If* $P \in \Gamma_{\rm sym}$ *such that* $B_{4r}(P) \cap \Omega$ *is a half-ball* $B_{4r}(P) \cap \{\xi_2 > 0\}$,

$$\|\varphi\|_{C^{2,\alpha}(\overline{B_r^+(P)})} \leq C, \tag{9.2.14}$$

where $B_r^+(P) := B_r(P) \cap \{\xi_2 > 0\}$;

(iii) *If* $P \in \Gamma_{\rm wedge}$ *such that* $B_{4r}(P) \cap \Omega$ *is a half-ball,*

$$\|\varphi\|_{C^{2,\alpha}(\overline{B_r(P)\cap\Omega})} \leq C. \tag{9.2.15}$$

Proof. Let φ be an admissible solution. We divide the proof into two steps.

1. Let $B_{4r}(P) \subset \Omega$. Equation (2.2.8) on φ is elliptic in Ω. We use the ellipticity principle in Theorem 5.2.1 with the function:

$$b(\boldsymbol{\xi}) = \delta\, \tilde{b}(\boldsymbol{\xi}),$$

where $\delta > 0$ is a small constant to be fixed later, and $\tilde{b} \in C^2(B_{4r}(P))$ such that

$$\tilde{b}(\boldsymbol{\xi}) = 0 \ \text{ on } \partial B_{4r}(P), \qquad \tilde{b}(\boldsymbol{\xi}) \geq r \ \text{ in } B_{2r}(P), \qquad \|\tilde{b}\|_{C^2(B_{4r}(P))} \leq \frac{C}{r^2}$$

for some constant C. The existence of such $\tilde{b}$ is easily obtained by first choosing $\hat{b}(\boldsymbol{\xi}) = \max(3r - |\boldsymbol{\xi} - P|,\ 0)$ and then smoothing it as $\tilde{b} = \hat{b} * \eta_\varepsilon$, where η_ε is the standard mollifier, and $\varepsilon = \frac{r}{100}$. Using the uniform bounds in (9.1.6), we apply Theorem 5.2.1(ii) in ball $B_{4r}(P)$ (*i.e.*, $d = 4r$) with function b defined above. Then, choosing δ small depending on (r, γ), the constants in (9.1.6), and constant C_0 in Theorem 5.2.1(ii) (hence $\delta = \delta(\rho_0, \rho_1, \gamma)$), we obtain the following ellipticity estimate for equation (2.2.8) on φ in the smaller ball $B_{2r}(P)$: There exists $M > 0$ depending only on r and (ρ_0, ρ_1, γ) such that

$$\frac{|D\varphi|^2}{c^2(D\varphi, \varphi)} \leq 1 - \frac{1}{M} \qquad \text{in } B_{2r}(P). \tag{9.2.16}$$

With this and (9.1.5)–(9.1.6), we use Lemma 9.2.1 to modify the coefficients in (9.2.1), as $(\tilde{\mathcal{A}}, \tilde{\mathcal{B}})(\mathbf{p}, z)$, so that they are defined for all $(\mathbf{p}, z) \in \mathbb{R}^2 \times \mathbb{R}$ and satisfy (9.2.3)–(9.2.5) for $\mathcal{D} = B_{2r}(P)$, with constants λ and $C_k, k = 0, 1, 2, \dots$, depending only on (ρ_0, ρ_1, γ), and φ still satisfies equation (9.2.9) in $B_{2r}(P)$.

Now write equation (9.2.9) in the non-divergence form with separation of the nonhomogeneous part:

$$\sum_{i,j=1}^{2} \tilde{\mathcal{A}}^i_{p_j}(D\varphi, \varphi) D_{ij}\varphi + \big(\hat{\mathcal{B}}(D\varphi, \varphi) - \tilde{\mathcal{B}}(\mathbf{0}, 0)\big) = -\tilde{\mathcal{B}}(\mathbf{0}, 0), \tag{9.2.17}$$

where $\hat{\mathcal{B}}(D\varphi,\varphi) = \tilde{\mathcal{B}}(D\varphi,\varphi)+\tilde{A}_\varphi(D\varphi,\varphi)\cdot D\varphi$. Then, using the properties of its coefficients discussed above and estimate (9.1.5), we can apply Theorem 4.2.1 (rescaled from B_1 to B_r) to obtain (i).

2. Let $P \in \Gamma_{\text{sym}}$ so that $B_{4r}(P)\cap\Omega$ is half-ball $B^+_{4r}(P) = B_{4r}(P)\cap\{\xi_2 > 0\}$. Then equation (2.2.8) on φ is elliptic in $B^+_{4r}(P)$, and the boundary condition $\varphi_{\boldsymbol{\nu}} = 0$ holds on $B_{4r}(P)\cap\partial\Lambda = B_{4r}(P)\cap\{\xi_2 = 0\}$. We use the ellipticity principle with the slip boundary condition, Theorem 5.3.1. We note that function b, constructed in Step 1, is radial, and has the structure: $b(\boldsymbol{\xi}) = f(|\boldsymbol{\xi} - P|)$ with $f \in C^2(\overline{\mathbb{R}_+})$. Thus, for $P \in \Gamma_{\text{sym}} \subset \{\xi_2 = 0\}$, we find that $\partial_{\boldsymbol{\nu}} b = 0$ on segment $B_{4r}(P)\cap\{\xi_2 = 0\}$. Applying Theorem 5.3.1(ii) in $B^+_{4r}(P)$ with this b, and the uniform bounds in (9.1.6), we obtain, as in Step 1, the uniform ellipticity estimates (9.2.16) in $B^+_{2r}(P)$ with $M > 0$ depending only on r and the data. Then, arguing as in Step 1, we can modify coefficients $(\mathcal{A},\mathcal{B})$ in equation (9.2.1) so that φ satisfies equation (9.2.9) in $B^+_{2r}(P)$ with the coefficients satisfying (9.2.3)–(9.2.5) for $\mathcal{D} = B^+_{2r}(P)$ and the constants depending only on r and the data for all $(\mathbf{p}, z, \boldsymbol{\xi}) \in \mathbb{R}^2 \times \mathbb{R} \times B^+_{2r}(P)$.

Then, writing equation (9.2.9) in the form of (9.2.17), recalling the boundary condition that $\partial_{\boldsymbol{\nu}}\varphi = 0$ on $B_{2r}(P) \cap \{\xi_2 = 0\}$, and using the uniform bounds in (9.1.5), we can apply Theorem 4.2.10 to obtain (9.2.14) from (4.2.77) for any $\alpha \in (0, 1)$. Now Case (ii) is proved.

The proof of Case (iii) is similar. □

Now we need to define weak solutions of **Problem 2.6.1**.

Definition 9.2.3 (Weak solutions of Problem 2.6.1). *Let $\theta_{\rm w} \in [\theta^{\rm s}_{\rm w}, \frac{\pi}{2}]$, and let P_1, $\overline{\Gamma_{\text{sonic}}}$, and Γ_{wedge} be from Definition* 7.5.7 *for $\theta_{\rm w}$. Let $\Gamma_{\text{shock}} \subset \overline{\Lambda} \cap \{\xi_1 \le \xi_{1P_1}\}$ be a Lipschitz curve with endpoints P_1 and $P_2 \in \partial\Lambda \cap \{\xi_2 = 0\}$. Let Ω be the interior of the region bounded by Γ_{shock}, $\overline{\Gamma_{\text{sym}}}$, Γ_{wedge}, and $\overline{\Gamma_{\text{sonic}}}$.*

A function $\varphi \in W^{1,1}_{\text{loc}}(\Omega)$ is called a weak solution of **Problem 2.6.1** *if its extension to Λ introduced in Definition* 2.6.2 *is a weak solution of Problem* 2.2.3 *in the sense of Definition* 2.3.3.

This definition is motivated by the following fact (which follows directly from the definitions): Let φ be an admissible solution of **Problem 2.6.1**, then φ is a weak solution of **Problem 2.6.1** in the sense of Definition 9.2.3.

We note the following features of Definition 9.2.3:

(a) The relative interior of Γ_{shock} may have common points with Γ_{wedge} and Γ_{sym};

(b) Set Ω is not necessarily connected;

(c) If $\theta_{\rm w} = \theta^{\rm s}_{\rm w}$, then region $P_0P_1P_4$ in (2.6.1) is one point;

(d) If $\theta_{\rm w} = \frac{\pi}{2}$, then region $P_0P_1P_4$ in (2.6.1) is the region between the vertical lines $\{\xi_1 \in (\xi_{1P_1(\frac{\pi}{2})}, 0)\}$ and above Γ_{sonic}, where we have used Lemma 7.5.9 to define $P_1(\frac{\pi}{2})$. Note that $\xi_{1P_1(\frac{\pi}{2})} = \bar{\xi}_1$ for $\bar{\xi}_1$ defined in Theorem 6.2.1.

Remark 9.2.4. *In some arguments below, we consider the limit of a sequence of admissible solutions $\varphi^{(i)}$ of* **Problem 2.6.1** *with the wedge angles $\theta_{\rm w}^{(i)} \in (\theta_{\rm w}^{\rm s}, \frac{\pi}{2})$. Then each solution $\varphi^{(i)}$ is defined in the corresponding domain $\Lambda^{(i)}$ as described in Definition* 2.6.2, *which is different for each different i. We define the uniform limit of such functions as follows:*

First, we extend $\varphi^{(i)}$ to $\mathbb{R}^2$ in the following way: It suffices to extend $\phi^{(i)} = \varphi^{(i)} + \frac{|\boldsymbol{\xi}|^2}{2}$. Note that $\phi^{(i)}$ on $\Lambda^{(i)} \setminus \Omega^{(i)}$ is equal to one of the linear functions $\phi_k^{(i)}$, for $k = 0, 1, 2$, in the respective regions of the constant states (0), (1), *and* (2) *as in* (8.1.3) *and, by* (9.1.3), *$\phi^{(i)}$ is Lipschitz in $\overline{\Omega^{(i)}}$ and continuous across $\partial\Omega^{(i)} \setminus \partial\Lambda^{(i)}$ so that $\phi^{(i)}$ is Lipschitz in $\Lambda^{(i)}$. Then, by the standard results, we can extend $\phi^{(i)}$ to $\mathbb{R}^2$ as a Lipschitz function with the same Lipschitz constant as in $\Lambda^{(i)}$. For each admissible solution, fix such an extension, which defines an extension of $\varphi^{(i)}$. Note that, using* (8.1.3), *Corollary* 9.1.3, *and estimate* (9.1.2), *and by the continuity (hence the boundedness) of the parameters of state* (2) *on $[\theta_{\rm w}^{\rm s}, \frac{\pi}{2}]$, we conclude that there exist $C < \infty$ and $C(K) < \infty$ for every compact $K \subset \mathbb{R}^2$ such that, for any admissible solution φ of* **Problem 2.6.1** *with $\theta_{\rm w} \in (\theta_{\rm w}^{\rm s}, \frac{\pi}{2})$, the extension of φ satisfies*

$$\|D\phi\|_{L^\infty(\mathbb{R}^2)} \leq C, \qquad \|\varphi\|_{C^{0,1}(K)} \leq C(K). \tag{9.2.18}$$

We will consider these extensions in the convergence statements below.

The equi-Lipschitz property (9.2.18) *implies that the limit assertions below are independent of the particular choice of the Lipschitz extension of $\varphi^{(i)}$ with the properties described above, in the light of the following fact: Let compact sets $K_i \subset \overline{\Lambda^{(i)}}$ for $i = 1, 2, \dots$, and $K_\infty \subset \overline{\Lambda^{(\infty)}}$ satisfy $K_i \to K_\infty$ in the Hausdorff metric. Let $\|\varphi^{(i)}\|_{C^{0,1}(K_i)} \leq C$ for all i, and let extensions $\varphi_{\rm ext}^{(i)}$ of $\varphi^{(i)}$ satisfy $\|D\varphi_{\rm ext}^{(i)}\|_{L^\infty(\overline{B_R})} \leq C$, where R is such that $\cup_{i=1}^\infty K_i \subset B_R$ (here we have used that $\cup_{i=1}^\infty K_i$ is bounded since $K_i \to K_\infty$). Then $\varphi_{\rm ext}^{(i)} \to \varphi_\infty$ uniformly on K_∞ if and only if this convergence holds for any other such extensions $\tilde{\varphi}_{\rm ext}^{(i)}$ of $\varphi^{(i)}$.*

In a similar way, we define the uniform convergence of the equi-Lipschitz functions f_i defined on intervals $[a_i, b_i] \subset \mathbb{R}$ with $a_i \to a$ and $b_i \to b$.

When taking the limits below, we allow the case that $\theta_{\rm w}^{(i)} \to \frac{\pi}{2}-$. In order to include this case, we introduce the notion of weak solutions to **Problem 2.6.1** for $\theta_{\rm w} = \frac{\pi}{2}$. The only exception is that, in Definition 2.3.3, condition (2.2.21) at infinity should be replaced by the condition of the convergence to the normal reflection at infinity, that is,

$$\varphi \to \varphi^{\rm norm} = \begin{cases} \varphi_1 & \text{for } \xi_1 < \bar{\xi}_1,\ \xi_2 > 0, \\ \varphi_2^{\rm norm} & \text{for } 0 > \xi_1 > \bar{\xi}_1,\ \xi_2 > 0, \end{cases} \qquad \text{when } |\boldsymbol{\xi}| \to \infty, \tag{9.2.19}$$

in the same sense as in **Problem 2.6.1**. We also note that line $\mathcal{S}_1$, points $\{P_1, P_3, P_4\}$, the boundary curves $\Gamma_{\rm sonic} \cup \Gamma_{\rm wedge}$, and vector $\mathbf{e}_{\mathcal{S}_1}$ for $\theta_{\rm w} = \frac{\pi}{2}$

have been introduced in Definition 7.5.7 and Remark 7.5.14 with

$$\mathcal{S}_1(\frac{\pi}{2}) = \{\xi_1 = \bar{\xi}_1\}, \qquad \mathbf{e}_{\mathcal{S}_1}(\frac{\pi}{2}) = -\mathbf{e}_{\xi_2} = (0, -1). \tag{9.2.20}$$

Then $\mathbf{e}_{\mathcal{S}_1}(\theta_{\rm w})$ is continuous on $[\theta_{\rm w}^{\rm s}, \frac{\pi}{2}]$. Also, by (7.2.11) and (7.2.15), it follows that, as $\theta_{\rm w} \to \frac{\pi}{2}-$,

$$\mathcal{S}_1(\theta_{\rm w}) \to \mathcal{S}_1(\frac{\pi}{2}) \tag{9.2.21}$$

in the Hausdorff metric on compact subsets of $\mathbb{R}^2$. Finally, we recall that $\text{Cone}^0(\mathbf{e}_{\mathcal{S}_1}, \mathbf{e}_{\xi_2})$ for $\theta_{\rm w} = \frac{\pi}{2}$ is defined by (8.2.15), so that (8.2.16) holds.

Furthermore, in taking the limits below, we allow the case that $\theta_{\rm w}^{(i)} \to \theta_{\rm w}^{\rm s}$. Then, as in Definition 7.5.7, $P_1(\theta_{\rm w}^{\rm s}) = P_4(\theta_{\rm w}^{\rm s}) = P_0(\theta_{\rm w}^{\rm s})$.

Corollary 9.2.5. *Let $\{\varphi^{(i)}\}$ be a sequence of admissible solutions of* **Problem 2.6.1** *with the corresponding wedge angles $\theta_{\rm w}^{(i)} \in (\theta_{\rm w}^{\rm s}, \frac{\pi}{2})$ such that*

$$\theta_{\rm w}^{(i)} \to \theta_{\rm w}^{(\infty)} \in [\theta_{\rm w}^{\rm s}, \frac{\pi}{2}].$$

Then there exists a subsequence $\{\varphi^{(i_j)}\}$ converging uniformly to a function $\varphi^{(\infty)} \in C^{0,1}_{loc}(\overline{\Lambda^{(\infty)}})$ in any compact subset of $\overline{\Lambda^{(\infty)}}$, which is a weak solution of **Problem 2.6.1** *for the wedge angle $\theta_{\rm w}^{(\infty)}$, where $\Lambda^{(\infty)} := \Lambda(\theta_{\rm w}^{(\infty)})$. Moreover, under the notations in Definition 7.5.7 and Corollary 8.2.14, with superscripts i and ∞ indicating which of solutions $\varphi^{(i)}$ and $\varphi^{(\infty)}$ these objects are related, $\varphi^{(\infty)}$ is of the following structure:*

Let $\mathbf{g} \in \text{Cone}^0(\mathbf{e}_{\mathcal{S}_1^{(\infty)}}, \mathbf{e}_{\xi_2})$, let $\mathbf{g}^\perp$ be orthogonal to $\mathbf{g}$ and oriented so that $\mathbf{g}^\perp \cdot \mathbf{e}_{\mathcal{S}_1^{(\infty)}} < 0$, let $|\mathbf{g}| = |\mathbf{g}^\perp| = 1$, and let (S, T) be the coordinates with basis $\{\mathbf{g}, \mathbf{g}^\perp\}$. Then

(i) *Sequence $\{{P_2}^{(i_j)}\}$ converges to a limit ${P_2}^{(\infty)}$, so that ${P_2}^{(\infty)} = (\xi_{1P_2^{(\infty)}}, 0)$ with $\xi_{1P_2^{(\infty)}} < \xi_{1P_1^{(\infty)}}$ in the $\boldsymbol{\xi}$-coordinates.*

(ii) *Let $f^{(i_j)}_{\mathbf{g},\rm sh}$ be the functions in Corollary 8.2.14 for $\varphi^{(i_j)}$. Then functions $f^{(i_j)}_{\mathbf{g},\rm sh}$ are uniformly bounded in $C^{0,1}([T_{P_2^{(\infty)}}, T_{P_1^{(\infty)}}])$ and converge uniformly on $[T_{P_2^{(\infty)}}, T_{P_1^{(\infty)}}]$ to a limit $f^{(\infty)}_{\mathbf{g},\rm sh} \in C^{0,1}([T_{P_2^{(\infty)}}, T_{P_1^{(\infty)}}])$.*

(iii) *Let $\mathbf{e} = \boldsymbol{\nu}_{\rm w}^{(\infty)}$, and let (S, T) be the coordinates with basis $\{\boldsymbol{\nu}_{\rm w}^{(\infty)}, \boldsymbol{\tau}_{\rm w}^{(\infty)}\}$, where $\boldsymbol{\nu}_{\rm w}^{(\infty)}$ and $\boldsymbol{\tau}_{\rm w}^{(\infty)}$ are defined by (8.2.14) and (8.2.17) with $\theta_{\rm w}^{(\infty)}$, and Lemma 8.2.11 has also been used. Then*

$$f^{(\infty)}_{\mathbf{e},\rm sh}(T) \geq \max(0, -T \tan\theta_{\rm w}^{(\infty)}) \qquad \textit{for } T \in (T_{P_2^{(\infty)}}, T_{P_1^{(\infty)}}).$$

Denote

$$\widehat{\Omega^{(\infty)}} = \left\{ (S,T) : \begin{array}{r} T_{P_2^{(\infty)}} \le T \le T_{P_4^{(\infty)}}, \\ -(T - T_{P_3})\tan\theta_{\rm w}^{(\infty)} \le S \le f_{\mathbf{e},\rm sh}^{(\infty)}(T) \\ \text{for } T \in [T_{P_2^{(\infty)}}, T_{P_3^{(\infty)}}], \\ 0 \le S \le f_{\mathbf{e},\rm sh}^{(\infty)}(T) \;\; \text{for } T \in (T_{P_3^{(\infty)}}, T_{P_1^{(\infty)}}], \\ 0 \le S \le f_{\mathbf{e},\rm so}^{(\infty)}(T) \;\; \text{for } T \in (T_{P_1^{(\infty)}}, T_{P_4^{(\infty)}}] \end{array} \right\}, \tag{9.2.22}$$

where $T_{P_1^{(\infty)}} = T_{P_4^{(\infty)}}$ *when* $\theta_{\rm w}^{(\infty)} = \theta_{\rm w}^{\rm s}$*, i.e., the corresponding interval is not present in* (9.2.22).

Denote by $\Omega^{(\infty)}$ *the interior of* $\widehat{\Omega^{(\infty)}}$*. Denote*

$$\Gamma_{\rm shock}^{(\infty)} := \{S = f_{\mathbf{e}}^{(\infty)}(T) \;:\; T_{P_2^{(\infty)}} < T < T_{P_1^{(\infty)}}\}. \tag{9.2.23}$$

Let $\Gamma_{\rm wedge}^{(\infty)}$ *be as defined in Definition* 7.5.7 *for* $\theta_{\rm w} = \theta_{\rm w}^{(\infty)}$*, and let* $\Gamma_{\rm sym}^{(\infty)} = \{(\xi_1, 0) \;:\; \xi_{1P_2^{(\infty)}} < \xi_1 < \xi_{1P_3^{(\infty)}}\}$ *in the* $\boldsymbol{\xi}$*-coordinates. Denote by* $\Gamma_{\rm wedge}^{(\infty),0}$ *(resp.* $\Gamma_{\rm sym}^{(\infty),0}$*) the relative interior of* $\Gamma_{\rm wedge}^{(\infty)} \setminus \Gamma_{\rm shock}^{(\infty)}$ *(resp.* $\Gamma_{\rm sym}^{(\infty)} \setminus \Gamma_{\rm shock}^{(\infty)}$*). Then*

$$\varphi^{(\infty)} \in C^\infty\big(\Omega^{(\infty)} \cup \Gamma_{\rm sym}^{(\infty),0} \cup \Gamma_{\rm wedge}^{(\infty),0}\big), \tag{9.2.24}$$

$$\varphi^{(\infty)} = \varphi_1 \quad \text{on } \Gamma_{\rm shock}^{(\infty)}, \tag{9.2.25}$$

$$\text{(2.2.8) is strictly elliptic for } \varphi^{(\infty)} \text{ in } \Omega^{(\infty)} \cup \Gamma_{\rm sym}^{(\infty),0} \cup \Gamma_{\rm wedge}^{(\infty),0}, \tag{9.2.26}$$

$$\varphi^{(i_j)} \to \varphi^{(\infty)} \quad \text{in } C^2 \text{ on compact subsets of } \Omega^{(\infty)}, \tag{9.2.27}$$

$$\partial_{\mathbf{g}}(\varphi_1 - \varphi^{(\infty)}) \le 0 \;\; \text{in } \Omega^{(\infty)} \;\; \text{for all } \mathbf{g} \in {\rm Cone}^0(\mathbf{e}_{S_1^{(\infty)}}, \mathbf{e}_{\xi_2}), \tag{9.2.28}$$

$$B_{c_1}(\mathcal{O}_1) \cap \Lambda^{(\infty)} \subset \Omega^{(\infty)}, \tag{9.2.29}$$

where $\mathcal{O}_1$ *is the center of sonic circle of state* (1)*, i.e.,* $\mathcal{O}_1 = (u_1, 0)$ *in the* $\boldsymbol{\xi}$*-coordinates.*

(iv) $\varphi^{(\infty)}$ *and* $\Gamma_{\rm wedge}^{(\infty)}$ *satisfy that*

$$\{\varphi_1 > \varphi^{(\infty)}\} \cap \Gamma_{\rm wedge}^{(\infty)} \qquad \text{is dense in } \Gamma_{\rm wedge}^{(\infty)}. \tag{9.2.30}$$

(v) *If* $\theta_{\rm w}^{(\infty)} > \theta_{\rm w}^{(s)}$*, then* $\varphi^{(\infty)}$ *is equal to the constant states* φ_0*,* φ_1*, and* $\varphi_2^{(\infty)}$ *in their respective subdomains of* $\Lambda^{(\infty)} \setminus \overline{\Omega^{(\infty)}}$*, as in* (8.1.3)*. If* $\theta_{\rm w}^{(\infty)} = \theta_{\rm w}^{(s)}$*, then* $\varphi^{(\infty)}$ *is equal to the constant states in their respective domains as in* (2.6.4).

Proof. Let $\{\varphi^{(i)}\}_{i=1}^\infty$ be admissible solutions of **Problem 2.6.1** with

$$\theta_{\rm w}^{(i)} \to \theta_{\rm w}^{(\infty)} \in [\theta_{\rm w}^{\rm s}, \frac{\pi}{2}].$$

From (9.2.18), there exist a subsequence (still denoted) $\{\varphi^{(i)}\}_{i=1}^{\infty}$ and a function $\varphi^{(\infty)} \in C^{0,1}_{\rm loc}(\overline{\Lambda^{(\infty)}})$ such that

$$\varphi^{(i)} \to \varphi^{(\infty)} \quad \text{uniformly in any compact subset of } \overline{\Lambda^{(\infty)}}. \tag{9.2.31}$$

We divide the remaining proof into four steps.

1. *Prove assertions* (i)–(ii). Fix a unit vector $\mathbf{g} \in \mathrm{Cone}^0(\mathbf{e}_{\mathcal{S}_1^{(\infty)}}, \mathbf{e}_{\xi_2})$. Let $\mathbf{g}^\perp$ be the orthogonal unit vector to $\mathbf{g}$ which is oriented so that

$$\mathbf{g}^\perp \cdot \mathbf{e}_{\mathcal{S}_1^{(\infty)}} < 0, \qquad \mathbf{e}_{\xi_2} \cdot \mathbf{g}^\perp > 0. \tag{9.2.32}$$

Such a vector $\mathbf{g}^\perp$ exists by (8.2.23). Using the smooth dependence of the parameters of state (2) from $\theta_{\rm w}$, we have

$$\mathbf{e}_{\mathcal{S}_1^{(i)}} \to \mathbf{e}_{\mathcal{S}_1^{(\infty)}} \qquad \text{as } i \to \infty.$$

Thus, from (9.2.32), there exists $N > 0$ such that, for all $i > N$, $\mathbf{g} \in \mathrm{Cone}^0(\mathbf{e}_{\mathcal{S}_1^{(i)}}, \mathbf{e}_{\xi_2})$ and

$$\mathbf{e}_{\mathcal{S}_1^{(i)}} \cdot \mathbf{g}^\perp \le \frac{1}{2}\mathbf{e}_{\mathcal{S}_1^{(\infty)}} \cdot \mathbf{g}^\perp < 0. \tag{9.2.33}$$

Let (S,T) be the coordinates with basis $\{\mathbf{g}, \mathbf{g}^\perp\}$. Then, by Corollary 8.2.14, there exist $T_{P_2{}^{(i)}} < T_{P_1{}^{(i)}}$ and functions $f_{\mathbf{g}}^{(i)}$ such that

$$\Gamma_{\rm shock}^{(i)} = \{S = f_{\mathbf{g}}^{(i)}(T) \ : \ T_{P_2{}^{(i)}} < T < T_{P_1{}^{(i)}}\}, \ \Omega^{(i)} \subset \{S < f_{\mathbf{g}}^{(i)}(T) \ : \ T \in \mathbb{R}\},$$

and all the other properties in Corollary 8.2.14 are satisfied. Extend $f_{\mathbf{g}}^{(i)}$ from $(T_{P_2{}^{(i)}}, T_{P_1{}^{(i)}})$ to $\mathbb{R}$ by setting $f_{\mathbf{g}}^{(i)}(T) = f_{\mathbf{g}}^{(i)}(T_{P_2{}^{(i)}})$ for $T < T_{P_2{}^{(i)}}$, and $f_{\mathbf{g}}^{(i)}(T) = f_{\mathbf{g}}^{(i)}(T_{P_1{}^{(i)}})$ for $T > T_{P_1{}^{(i)}}$. Now (8.2.24), (9.2.32)–(9.2.33), and $|\mathbf{e}_{\mathcal{S}_1^{(i)}}| = |\mathbf{e}_{\xi_2}| = |\mathbf{g}| = 1$ imply that

$$\{f_{\mathbf{g}}^{(i)}\} \qquad \text{are equi-Lipschitz on } \mathbb{R}. \tag{9.2.34}$$

Furthermore, using (9.1.2) applied to each $\varphi^{(i)}$ and the fact that $f_{\mathbf{g}}^{(i)}$ is constant outside $(T_{P_2{}^{(i)}}, T_{P_1{}^{(i)}})$, we obtain that

$$\{f_{\mathbf{g}}^{(i)}\} \qquad \text{are uniformly bounded on } \mathbb{R}. \tag{9.2.35}$$

Using (9.1.2), there exist subsequences (still denoted) $T_{P_1{}^{(i)}}$ and $T_{P_2{}^{(i)}}$, and corresponding limits $T_{P_1{}^{(\infty)}}$ and $T_{P_2{}^{(\infty)}}$, such that

$$(T_{P_1{}^{(i)}}, T_{P_2{}^{(i)}}) \to (T_{P_1{}^{(\infty)}}, T_{P_2{}^{(\infty)}}) \qquad \text{with } T_{P_2{}^{(\infty)}} \le T_{P_1{}^{(\infty)}}. \tag{9.2.36}$$

Then, using (9.2.34)–(9.2.35) and passing to a further subsequence, we conclude that there exists $f^{(\infty)} \in C^{0,1}([T_{P_2{}^{(\infty)}} - 1, T_{P_1{}^{(\infty)}} + 1])$ such that

$$f_{\mathbf{g}}^{(i)} \to f_{\mathbf{g}}^{(\infty)} \qquad \text{uniformly on } [T_{P_2{}^{(\infty)}} - 1, T_{P_1{}^{(\infty)}} + 1]. \tag{9.2.37}$$

Denote

$$P_2{}^{(\infty)} := f_{\mathbf{g}}^{(\infty)}(T_{P_2{}^{(\infty)}})\mathbf{g} + T_{P_2{}^{(\infty)}}\mathbf{g}^{\perp}. \tag{9.2.38}$$

Now we consider the case that $\theta_{\rm w}^{\infty} > \theta_{\rm w}^{\rm s}$ and show that, as $i \to \infty$,

$$\begin{aligned} &P_k^{(i)} \to P_k^{(\infty)} && \text{for } k = 0,1,2,3,4, \\ &\Gamma_{\rm sonic}^{(i)} \to \Gamma_{\rm sonic}^{(\infty)} && \text{in the Hausdorff metric,} \end{aligned} \tag{9.2.39}$$

where all the limiting objects, except $P_2{}^{(\infty)}$, are from Definition 7.5.7 for $\theta_{\rm w}^{(\infty)}$, and point $P_2{}^{(\infty)}$ is defined by (9.2.38). We work in the $\boldsymbol{\xi}$–coordinates. Since $P_0{}^{(i)} = (\xi_1^0, \xi_1^0 \tan\theta_{\rm w}^{(i)})$ and $P_3{}^{(i)} = \mathbf{0}$ for all i, the convergence of these points follows. Furthermore, $P_1{}^{(i)}$ is a point of intersection of line

$$\mathcal{S}_1 = \{\boldsymbol{\xi} \,:\, (u_1 - u_2^{(i)})(\xi_1 - \xi_1^0) - v_2^{(i)}(\xi_2 - \xi_1^0\tan\theta_{\rm w}^{(i)}) = 0\}$$

and the sonic circle $C_2^{(i)} = \partial B_{c_2^{(i)}}(u_2^{(i)}, v_2^{(i)})$ of state (2). Similarly, $P_4{}^{(i)}$ is a point of intersection of the wedge boundary $\{\boldsymbol{\xi} \,:\, \xi_2 = \xi_1\tan\theta_{\rm w}^{(i)}\}$ and $C_2^{(i)}$. Moreover, $\Gamma_{\rm sonic}^{(i)}$ is the smaller arc of $C_2^{(i)}$ between points $P_4{}^{(i)}$ and $P_1{}^{(i)}$. Thus, the convergence of $P_4{}^{(i)}$, $P_1{}^{(i)}$, and $\Gamma_{\rm sonic}^{(i)}$ follows from the smooth dependence of the parameters of state (2) from $\theta_{\rm w}$.

If $\theta_{\rm w}^{\infty} = \theta_{\rm w}^{\rm s}$, a similar argument shows that $P_k^{(i)} \to P_k^{(\infty)}$ for $k = 0,2,3$. Also, we now know that $P_k^{(i)} \to P_0{}^{(\infty)}$ for $k = 1,4$, by arguing as in the previous case and using the definition of the sonic angle.

Finally, $P_2{}^{(i)} = f_{\mathbf{g}}^{(i)}(T_{P_2{}^{(i)}})\mathbf{g} + T_{P_2{}^{(i)}}\mathbf{g}^{\perp}$, by Corollary 8.2.14(ii). Then (9.2.36)–(9.2.38) imply that

$$P_2{}^{(i)} \to f_{\mathbf{g}}^{(\infty)}(T_{P_2{}^{(\infty)}})\mathbf{g} + T_{P_2{}^{(\infty)}}\mathbf{g}^{\perp} = P_2{}^{(\infty)}.$$

Now (9.2.39) is proved.

Next, we show that $T_{P_2{}^{(\infty)}} < T_{P_1{}^{(\infty)}}$ strictly, *i.e.*, that the limit of $\Gamma_{\rm shock}^{(i)}$ does not degenerate to a point. We work in the $\boldsymbol{\xi}$–coordinates. By Remark 7.5.8, $\xi_{2P_1{}^{(i)}} > v_2^{(i)}$. Passing to the limit, $\xi_{2P_1{}^{(\infty)}} \ge v_2^{(\infty)} > 0$. Since $\xi_{2P_2{}^{(\infty)}} = 0$, then $P_2{}^{(\infty)} \neq P_1{}^{(\infty)}$. On the other hand, from (9.2.39) and Corollary 8.2.14(ii) (applied to $f_{\mathbf{g}}^{(i)}$), we find that $P_1{}^{(\infty)} = f^{(\infty)}(T_{P_1{}^{(\infty)}})\mathbf{g} + T_{P_1{}^{(\infty)}}\mathbf{g}^{\perp}$. Also, from its definition, $P_2{}^{(\infty)} = f^{(\infty)}(T_{P_2{}^{(\infty)}})\mathbf{g} + T_{P_2{}^{(\infty)}}\mathbf{g}^{\perp}$. Thus, $T_{P_2{}^{(\infty)}} \neq T_{P_1{}^{(\infty)}}$, that is, $T_{P_2{}^{(\infty)}} < T_{P_1{}^{(\infty)}}$.

Now assertions (i) and (ii) are proved.

2. *Prove assertion* (iii). In the following argument, when $\theta_{\rm w}^{\infty} = \theta_{\rm w}^{\rm s}$, we set $\Gamma_{\rm sonic}^{(\infty)} = \emptyset$ and $\overline{\Gamma_{\rm sonic}^{(\infty)}} = \{P_0{}^{(\infty)}\}$.

Let $\mathbf{e}$ and coordinates (S,T) be as those defined in the statement of assertion (iii). By Lemma 8.2.11, $\mathbf{e} = \boldsymbol{\nu}_{\rm w}^{(\infty)} \in {\rm Cone}^0(\mathbf{e}_{\mathcal{S}_1^{(\infty)}}, \mathbf{e}_{\xi_2})$ with $\boldsymbol{\tau}_{\rm w}^{(\infty)} \cdot \mathbf{e}_{\mathcal{S}_1}^{(\infty)} < 0$.

Then, using assertions (i)–(ii) as proved in Step 1, we obtain a function $f^{(\infty)}_{\mathbf{e},\mathrm{sh}}$ on $(T_{P_2^{(\infty)}}, T_{P_1^{(\infty)}})$.

From (8.2.25) in Corollary 8.2.14 (under the notations introduced there and in Remark 8.2.12), applied to $\varphi^{(i)}$ with $\mathbf{e}^{(i)} = \boldsymbol{\nu}^{(i)}_{\mathrm{w}}$, using (9.2.37) and (9.2.39), and noting that $\boldsymbol{\nu}^{(i)}_{\mathrm{w}} \to \boldsymbol{\nu}^{(\infty)}_{\mathrm{w}}$ as $i \to \infty$ by (8.2.14), we have

$$f^{(\infty)}_{\mathbf{e},\mathrm{sh}}(T) \geq \max(0, -T \tan\theta^{(\infty)}_{\mathrm{w}}) \qquad \text{for } T \in (T_{P_2}, T_{P_1}).$$

It follows that

$$\begin{aligned} &\partial\widehat{\Omega^{(\infty)}} = \overline{\Gamma^{(\infty)}_{\mathrm{shock}}} \cup \overline{\Gamma^{(\infty)}_{\mathrm{sonic}}} \cup \overline{\Gamma^{(\infty)}_{\mathrm{wedge}}} \cup \overline{\Gamma^{(\infty)}_{\mathrm{sym}}}, \\ &\widehat{\Omega^{(\infty)}} \setminus \overline{\Omega^{(\infty)}} = (\Gamma^{(\infty)}_{\mathrm{wedge}} \setminus \overline{\Gamma^{(\infty),0}_{\mathrm{wedge}}}) \cup (\Gamma^{(\infty)}_{\mathrm{sym}} \setminus \overline{\Gamma^{(\infty),0}_{\mathrm{sym}}}), \\ &\partial\Omega^{(\infty)} \subset \partial\widehat{\Omega^{(\infty)}}, \qquad \partial\widehat{\Omega^{(\infty)}} \setminus \partial\Omega^{(\infty)} = \widehat{\Omega^{(\infty)}} \setminus \overline{\Omega^{(\infty)}}, \end{aligned} \tag{9.2.40}$$

where sets $\partial\widehat{\Omega^{(\infty)}}$ and $\Omega^{(\infty)}$ are defined in (9.2.22). Since $\Gamma^{(\infty)}_{\mathrm{shock}}$ is the graph of a Lipschitz function, $\overline{\Gamma_{\mathrm{sonic}}}^{(\infty)}$ is an arc of circle (which becomes a point if $\theta^{\infty}_{\mathrm{w}} = \theta^{\mathrm{s}}_{\mathrm{w}}$), and $\Gamma^{(\infty)}_{\mathrm{wedge}}$ and $\Gamma^{(\infty)}_{\mathrm{sym}}$ are line segments, we have

$$|\partial\Omega^{(\infty)}| = |\partial\widehat{\Omega^{(\infty)}}| = 0. \tag{9.2.41}$$

Let P be an interior point of $\Omega^{(\infty)}$. Then, from (9.2.37) and (9.2.39), there exist $r > 0$ and N such that $B_{4r}(P) \subset \Omega^{(i)}$ for all $i > N$. Fix some $\alpha \in (0,1)$. Then, from Lemma 9.2.2(i), estimate (9.2.13) holds for each $\varphi^{(i)}$ with $i > N$. Combining these uniform $C^{2,\alpha}(\overline{B_r(P)})$–estimates with (9.2.31), we have

$$\varphi^{(i)} \to \varphi^{(\infty)} \qquad \text{in } C^{2,\frac{\alpha}{2}}(\overline{B_r(P)}), \tag{9.2.42}$$

and $\varphi^{(\infty)} \in C^{2,\alpha}(\overline{B_r(P)})$ is a solution of (2.2.8) in $B_r(P)$. Also, since Theorem 5.2.1 and the bounds in (9.1.5)–(9.1.6) provide the uniform ellipticity estimates with respect to i for equation (2.2.8) on $\varphi^{(i)}$ in $B_{2r}(P)$, we conclude that equation (2.2.8) is uniformly elliptic on $\varphi^{(\infty)}$ in $B_r(P)$. Then the further regularity, $\varphi^{(\infty)} \in C^{\infty}(\overline{B_r(P)})$, follows from the linear elliptic theory.

If $P \in \Gamma^{(\infty),0}_{\mathrm{wedge}}$, then, using the (S,T)–coordinates with basis $\{\boldsymbol{\nu}^{(\infty)}_{\mathrm{w}}, \boldsymbol{\tau}^{(\infty)}_{\mathrm{w}}\}$, we find that $P = (S_P, T_P)$ with $S_P = L^{(\infty)}_{\mathbf{e},\mathrm{w}}(T_P) < f^{(\infty)}_{\mathbf{e},\mathrm{sh}}(T_P)$. Using the continuity of $f^{(\infty)}_{\mathbf{e},\mathrm{sh}}$, there exists $r > 0$ such that $B_{3r}(P) \cap \Lambda^{(\infty)}$ is a half-ball and $B_{4r}(P) \cap \Lambda^{(\infty)} \subset \Omega^{(\infty)}$. Then there exists N such that, for each $i > N$, there is $P^{(i)} \in \Gamma^{(i),0}_{\mathrm{wedge}}$ so that $B_{r/2}(P) \subset B_r(P^{(i)})$ and $B_{4r}(P^{(i)}) \cap \Lambda^{(i)} \subset \Omega^{(i)}$ for all $i > N$, *i.e.*, $B_{4r}(P^{(i)}) \cap \Omega^{(i)}$ is a half-ball. Fix some $\alpha \in (0,1)$. From Lemma 9.2.2(iii), estimate (9.2.15) holds for each $\varphi^{(i)}$ with $i > N$. Then, for any compact $\mathcal{K} \subset B_{r/2}(P) \cap \Lambda^{(\infty)}$, there exists $N_{\mathcal{K}}$ such that $\mathcal{K} \subset B_r(P^{(i)})$ for each $i \geq N_{\mathcal{K}}$. Thus, the uniform convergence $\varphi^{(i)} \to \varphi^{(\infty)}$ on $\mathcal{K}$, combined with estimates (9.2.15) for $\varphi^{(i)}$ on $B_r(P^{(i)}) \cap \Omega^{(i)}$, implies

$$\varphi^{(i)} \to \varphi^{(\infty)} \qquad \text{in } C^{2,\frac{\alpha}{2}}(\mathcal{K}),$$

and

$$\|\varphi^{(\infty)}\|_{C^{2,\alpha}(\mathcal{K})} \le \|\varphi^{(i)}\|_{C^{2,\alpha}(\mathcal{K})} \le \|\varphi^{(i)}\|_{C^{2,\alpha}(\overline{B_r(P^{(i)})\cap\Omega^{(i)}})} \le C$$

with C independent of $\mathcal{K}$, where we have used that $B_{4r}(P^{(i)})\cap\Omega^{(i)}$ is a half-ball. This implies that

$$\|\varphi^{(\infty)}\|_{C^{2,\alpha}(\overline{B_{r/2}(P)\cap\Omega^{(\infty)}})} \le C,$$

and that $\varphi^{(\infty)}$ is a smooth solution of (2.2.8) in $B_r(P)\cap\Lambda^{(\infty)}$ and satisfies $\partial_\nu\varphi^{(\infty)} = 0$ on $\Gamma^{(\infty)}_{\text{wedge}}\cap B_{r/2}(P)$.

Since Theorem 5.2.1 (also see [110, Theorem 3.1]) and the bounds in (9.1.5)–(9.1.6) provide the uniform ellipticity estimates with respect to i for equation (2.2.8) on $\varphi^{(i)}$ in $B_{2r}(P^{(i)})\cap\Omega^{(i)}$, we conclude that equation (2.2.8) is uniformly elliptic on $\varphi^{(\infty)}$ in $B_{r/2}(P)\cap\Omega^{(\infty)}$. Then the further regularity, $\varphi^{(\infty)}\in C^\infty(\overline{B_r(P)}\cap\Lambda)$, follows from the linear elliptic theory for the oblique derivative problem.

Case $P\in\Gamma^{(\infty),0}_{\text{sym}}$ is similar. Now (9.2.24) and (9.2.26)–(9.2.27) are proved.

Furthermore, (9.2.25) follows from the uniform convergence of both $\varphi^{(i)}\to\varphi^{(\infty)}$ and $f^{(i)}_{\boldsymbol{\nu}_{\text{w}}^{(\infty)},\text{sh}}\to f^{(\infty)}_{\boldsymbol{\nu}_{\text{w}}^{(\infty)},\text{sh}}$, and the fact that $\varphi^{(i)}=\varphi_1$ on $\Gamma^{(i)}_{\text{shock}}$.

Now we prove property (9.2.28). Fix $\mathbf{g}\in\text{Cone}^0(\mathbf{e}_{\mathcal{S}_1^{(\infty)}},\mathbf{e}_{\xi_2})$. First we note that $\mathbf{g}\in\text{Cone}^0(\mathbf{e}_{\mathcal{S}_1^{(i_k)}},\mathbf{e}_{\xi_2})$ for large k, since $\mathbf{e}_{\mathcal{S}_1^{(i_k)}}\to\mathbf{e}_{\mathcal{S}_1^{(\infty)}}$. Thus, $\varphi_1-\varphi^{(i_k)}$ is monotonically non-increasing in direction $\mathbf{g}$ in $\Omega^{(i_k)}$ for such k. Then we see that $\varphi_1-\varphi^{(\infty)}$ is monotonically non-increasing in direction $\mathbf{g}$ in $\Omega^{(\infty)}$, by employing the uniform convergence of $\varphi^{(i_k)}$ to $\varphi^{(\infty)}$ on compact subsets of $\Lambda^{(\infty)}$, the equi-Lipschitz estimate and uniform convergence of $f^{(i_k)}_{\mathbf{g},\text{sh}}$ to $f^{(\infty)}_{\mathbf{g},\text{sh}}$ on $[T_{P_2^{(\infty)}},T_{P_1^{(\infty)}}]$, and the definition of $\Omega^{(\infty)}$ in part (iii). Combining this with (9.2.24), we conclude (9.2.28).

Property (9.2.29) follows from (8.1.12) of $\varphi^{(i)}$, (9.2.37), (9.2.39), and the definition of $\Omega^{(\infty)}$ in (9.2.22).

3. *Prove assertion* (v) *and show that* $\varphi^{(\infty)}$ *is a weak solution of* **Problem 2.6.1**. In this proof, in order to fix the notations, we focus on the case that $\theta_{\text{w}}^\infty>\theta_{\text{w}}^{\text{s}}$. The other case, $\theta_{\text{w}}^\infty=\theta_{\text{w}}^{\text{s}}$, is similar and simpler.

We first prove assertion (v) for $\varphi^{(\infty)}$ in $\Lambda\setminus\overline{\Omega^{(\infty)}}$. For that, we consider two separate cases: $\theta_{\text{w}}^{(\infty)}<\frac{\pi}{2}$ and $\theta_{\text{w}}^{(\infty)}=\frac{\pi}{2}$.

First consider Case $\theta_{\text{w}}^{(\infty)}<\frac{\pi}{2}$. Denote by $\mathcal{S}^{(\infty)}_{P_0P_1}$ the segment between points ${P_0}^{(\infty)}$ and ${P_1}^{(\infty)}$ on line $\mathcal{S}_1^{(\infty)}=\{\varphi_1=\varphi_2^{(\infty)}\}$. Let $P\in\Lambda^{(\infty)}\setminus(\overline{\Omega^{(\infty)}}\cup\mathcal{S}^{(\infty)}_{P_0P_1}\cup\{\xi_1=\xi_1^0\})$. By (9.2.40), it follows that $P\in\Lambda^{(\infty)}\setminus(\widehat{\Omega^{(\infty)}}\cup\mathcal{S}^{(\infty)}_{P_0P_1}\cup\{\xi_1=\xi_1^0\})$. Then, using (9.2.36)–(9.2.38), we find that there exist $k\in\{0,1,2\}$, large integer N, and small $r>0$ such that, for all $i>N$, $B_{2r}(P)\subset\Lambda^{(i)}$ and $\varphi^{(i)}\equiv\varphi_k^{(i)}$ in $B_{2r}(P)$, where we have used the convention that $\varphi_k=\varphi_k^{(i)}=\varphi_k^{(\infty)}$ for $k=0,1$. Sending $i\to\infty$ and using (9.2.31), we arrive at assertion (v).

Now we consider Case $\theta_{\rm w}^{(\infty)} = \frac{\pi}{2}$. In this case, we denote $\mathcal{S}_{P_0P_1}^{(\infty)}$ as half-line $\{\boldsymbol{\xi} \in \mathcal{S}_1^{(\infty)} \,:\, \xi_2 > \xi_{2P_1(\infty)}\}$. Let $P \in \Lambda^{(\infty)} \setminus (\overline{\Omega^{(\infty)}} \cup \mathcal{S}_{P_0P_1}^{(\infty)})$. Using (9.2.20)–(9.2.21), we can follow the same argument as in the previous case (the only difference is now that $k \in \{1,2\}$).

Furthermore, in both cases with the choice of P and r as above, (9.2.42) holds and $B_{2r}(P) \subset \overline{\Lambda^{(\infty)}}$. Combining this with the fact that the convergence in (9.2.42) holds for a subsequence for any $B_{4r}(P) \subset \Omega^{(\infty)}$, considering a countable collection of balls $B_{4r}(P) \subset \Omega^{(\infty)}$ covering $\Omega^{(\infty)}$, and using a diagonal procedure, we conclude that, for a subsequence,

$$(\varphi^{(i)}, D\varphi^{(i)}) \to (\varphi^{(\infty)}, D\varphi^{(\infty)})$$

everywhere in $\Lambda^{(\infty)} \setminus (\partial\Omega^{(\infty)} \cup \mathcal{S}_{P_0P_1}^{(\infty)} \cup \{\xi_1 = \xi_1^0\})$. Then, using (9.2.41),

$$(\varphi^{(i)}, D\varphi^{(i)}) \to (\varphi^{(\infty)}, D\varphi^{(\infty)}) \qquad a.e. \text{ in } \Lambda^{(\infty)}. \tag{9.2.43}$$

Each $\varphi^{(i)}$ is a weak solution of **Problem 2.6.1** in the sense of Definition 9.2.3. Using (9.1.5)–(9.1.6), (9.2.43), and the dominated convergence theorem, we can take the limit in the conditions of Definition 2.3.3 to conclude that $\varphi^{(\infty)}$ is a weak solution of **Problem 2.6.1**.

4. *It remains to prove assertion* (iv). Note that, on $\Gamma_{\rm sonic}^{(\infty)}$, $\varphi^{(\infty)} = \varphi_2^{(\infty)} < \varphi_1$. Then, using (9.2.22) and (9.2.28) with $\mathbf{g} = \boldsymbol{\nu}_{\rm w}^{(\infty)}$, we have

$$\varphi^{(\infty)} < \varphi_1 \qquad \text{on } \Gamma_{\rm wedge}^{(\infty)} \cap \{T_{P_1(\infty)} < T \le T_{P_4(\infty)}\}.$$

Thus, it remains to show (9.2.30) on $\Gamma_{\rm wedge}^{(\infty)} \cap \{T_{P_3(\infty)} < T \le T_{P_1(\infty)}\}$.

On the contrary, if (9.2.30) is false on $\Gamma_{\rm wedge}^{(\infty)} \cap \{T_{P_3(\infty)} < T \le T_{P_1(\infty)}\}$, then there exist T_1 and T_2 satisfying

$$T_{P_3(\infty)} < T_1 < T_2 \le T_{P_1(\infty)},$$

such that

$$\varphi^{(\infty)} = \varphi_1 \quad \text{on } \Gamma_{\rm wedge}^{(\infty)} \cap \{(S,T) : T_1 < T < T_2\} \equiv \{(0,T) : T_1 < T < T_2\}.$$

Using (9.2.22), (9.2.28) with $\mathbf{g} = \boldsymbol{\nu}_{\rm w}^{(\infty)}$, and $\varphi^{(\infty)} = \varphi_1$ on $\Gamma_{\rm shock}^{(\infty)}$, we have

$$\varphi^{(\infty)} = \varphi_1 \qquad \text{in } \{(S,T) \,:\, T_1 < T < T_2,\, 0 < S < f_{\boldsymbol{\nu}_{\rm w}^{(\infty)},{\rm sh}}^{(\infty)}(T)\}.$$

Also, from part (v) as proved above,

$$\varphi^{(\infty)} = \varphi_1 \qquad \text{in } \{(S,T) \,:\, T_{P_3(\infty)} < T < T_{P_1(\infty)},\, S > f_{\boldsymbol{\nu}_{\rm w}^{(\infty)},{\rm sh}}^{(\infty)}(T)\}.$$

Since $\varphi^{(\infty)} \in C^{0,1}(\Lambda^{(\infty)})$, it follows that

$$\varphi^{(\infty)} = \varphi_1 \qquad \text{in } \{(S,T) \,:\, T_1 < T < T_2,\, S > 0\}.$$

This implies that $\varphi^{(\infty)}$ is not a weak solution of **Problem 2.6.1** in the sense of Definition 9.2.3, since it does not satisfy the boundary condition that $\varphi_{\boldsymbol{\nu}} = 0$ on $\Gamma^{(\infty)}_{\text{wedge}} \cap \{(S,T) \;:\; T_1 < T < T_2\}$, even in the weak sense. Indeed, if $\zeta \in C_c^\infty(\mathbb{R}^2)$ with $\text{supp}(\zeta) \subset \{(S,T) \;:\; T_1 < T < T_2\}$ in the (S,T)–coordinates, then $\varphi^{(\infty)} = \varphi_1$ on $\text{supp}(\zeta)$ so that, calculating the left-hand side of the equality in Definition 2.3.3(iii) in the $\boldsymbol{\xi}$–coordinates, we have

$$\begin{aligned}
&\int_{\Lambda^{(\infty)}} \Big(\rho(|D\varphi^{(\infty)}|^2, \varphi^{(\infty)})D\varphi^{(\infty)} \cdot D\zeta - 2\rho(|D\varphi^{(\infty)}|^2, \varphi^{(\infty)})\zeta\Big)\, d\boldsymbol{\xi} \\
&\quad = \int_{\Lambda^{(\infty)}} \Big(\rho(|D\varphi_1|^2, \varphi_1)D\varphi_1 \cdot D\zeta - 2\rho(|D\varphi_1|^2, \varphi_1)\zeta\Big)\, d\boldsymbol{\xi} \\
&\quad = \int_{\Lambda^{(\infty)}} \big(\rho_1 D\varphi_1 \cdot D\zeta - 2\rho_1\zeta\big)\, d\boldsymbol{\xi} \\
&\quad = \rho_1 u_1 \sin\theta_{\text{w}}^{(\infty)} \int_{\Gamma^{(\infty)}_{\text{wedge}}} \zeta\, dl > 0,
\end{aligned}$$

if ζ is chosen to be nonnegative in $\mathbb{R}^2$ and positive on some segment on $\Gamma^{(\infty)}_{\text{wedge}} \cap (\{0\} \times (T_1, T_2))$. This is a contradiction, since $\varphi^{(\infty)}$ is a weak solution of **Problem 2.6.1** as we have proved in Step 3. □

Lemma 9.2.6. *Let $\{\varphi^{(i)}\}$ be a sequence of admissible solutions of* **Problem 2.6.1** *with the wedge angles $\theta_{\text{w}}^{(i)} \in (\theta_{\text{w}}^{\text{s}}, \frac{\pi}{2})$ such that $\theta_{\text{w}}^{(i)} \to \frac{\pi}{2}$. Then*

(i) *$\{\varphi^{(i)}\}$ converges to the normal reflection solution, uniformly in compact subsets of $\overline{\Lambda^{(\infty)}} \equiv \overline{\Lambda(\frac{\pi}{2})} \equiv \{\boldsymbol{\xi} \;:\; \xi_2 \le 0\}$, as $\theta_{\text{w}}^{(i)} \to \frac{\pi}{2}$;*

(ii) *For sufficiently large N,*

$$\Gamma^{(i)}_{\text{shock}} = \{(f^{(i)}_{-\mathbf{e}_{\xi_1},\text{sh}}(\xi_2), \xi_2) \;:\; 0 < \xi_2 < \xi_2(P_1{}^{(i)})\} \qquad \textit{for all } i \ge N,$$

where $f^{(i)}_{-\mathbf{e}_{\xi_1},\text{sh}}$ are equi-Lipschitz on $[0, \xi_2(P_1(\frac{\pi}{2}))]$,

$$f^{(i)}_{-\mathbf{e}_{\xi_1},\text{sh}} \to f^{(\infty)}_{-\mathbf{e}_{\xi_1},\text{sh}} \qquad \textit{uniformly on } \; [0, \xi_2(P_1(\frac{\pi}{2}))],$$

and $f^{(\infty)}_{-\mathbf{e}_{\xi_1},\text{sh}}$ is the constant function $f^{(\infty)}_{-\mathbf{e}_{\xi_1},\text{sh}}(\cdot) \equiv \bar{\xi}_1$ with $\bar{\xi}_1 < 0$ defined by (6.2.2).

Proof. By Corollary 9.2.5, there exists a subsequence $\{\varphi^{(i_j)}\}$ converging uniformly in compact subsets of $\overline{\Lambda^{(\infty)}}$ to a function $\varphi^{(\infty)} \in C^{0,1}_{\text{loc}}(\overline{\Lambda^{(\infty)}})$, which is a weak solution of **Problem 2.6.1** for the wedge angle $\theta_{\text{w}} = \frac{\pi}{2}$. Then, by (8.2.15)–(8.2.16), it follows that $-\mathbf{e}_{\xi_1} \in \text{Cone}^0(\mathbf{e}_{\mathcal{S}_1}^{(\theta_{\text{w}}^{(i_j)})}, \mathbf{e}_{\xi_2})$ for all $j > N$ for sufficiently large N. Expressing $\Gamma^{(i_j)}_{\text{shock}} = (f^{(i_j)}_{-\mathbf{e}_{\xi_1},\text{sh}}(\xi_2), \xi_2)$ by Corollary 8.2.14, we find from Corollary 9.2.5 that $f^{(i_j)}_{-\mathbf{e}_{\xi_1},\mathrm{sh}}$ are equi-Lipschitz and converge uniformly on $[0, \xi_2(P_1(\frac{\pi}{2}))]$ to a function $f^{(\infty)}_{-\mathbf{e}_{\xi_1},\mathrm{sh}}$. Using (8.2.15)–(8.2.16) and the properties in Corollary 8.2.14(iii) for each $f^{(i_j)}_{-\mathbf{e}_{\xi_1},\mathrm{sh}}$, we find that $f^{(\infty)}_{-\mathbf{e}_{\xi_1},\mathrm{sh}}$ is a constant function. Furthermore, since $f^{(i_j)}_{-\mathbf{e}_{\xi_1},\mathrm{sh}}(\xi_2(P_1(\theta_{\mathrm{w}}^{(i_j)}))) = \xi_1(P_1(\theta_{\mathrm{w}}^{(i_j)}))$, we employ Remark 7.5.6 and the equi-Lipschitz property of $f^{(i_j)}_{-\mathbf{e}_{\xi_1},\mathrm{sh}}$ to conclude

$$f^{(\infty)}_{-\mathbf{e}_{\xi_1},\mathrm{sh}}(\xi_2(P_1(\frac{\pi}{2}))) = \xi_1(P_1(\frac{\pi}{2})) = \bar{\xi}_1,$$

where $\bar{\xi}_1$ is defined by (6.2.2). Thus, $f^{(\infty)}_{-\mathbf{e}_{\xi_1},\mathrm{sh}}(\xi_2) = \bar{\xi}_1$ for all ξ_2, and $\bar{\xi}_1 < 0$ by (6.2.2). Then assertion (ii) is proved for a subsequence $\{\varphi^{(i_j)}\}$.

Now we prove assertion (i) for a subsequence $\{\varphi^{(i_j)}\}$. From Corollary 9.2.5(v), it suffices to prove that $\varphi^{(\infty)} = \varphi_2^{(\frac{\pi}{2})}$ in $\overline{\Omega^{(\infty)}}$, below.

Since assertion (ii) has been proved, Corollary 9.2.5(iii), combined with (8.2.15), implies that

$$\widehat{\Omega^{(\infty)}} = \left\{ \boldsymbol{\xi} \in \mathbb{R}^2 \;:\; \begin{array}{l} \bar{\xi}_1 \le \xi_1 \le 0 \;\text{ for } \xi_2 \in [0, \xi_2(P_1(\frac{\pi}{2}))], \\ f^{(\infty)}_{-\mathbf{e}_{\xi_1},\mathrm{so}}(\xi_2) < \xi_1 < 0 \\ \qquad \text{for } \xi_2 \in (\xi_2(P_1(\frac{\pi}{2})), \xi_2(P_4(\frac{\pi}{2}))] \end{array} \right\}.$$

In particular, $\Omega^{(\infty)}$ coincides with the elliptic domain of the normal reflection, *i.e.*, $\Omega^{(\infty)}$ is the domain between the vertical lines $\xi_1 = \bar{\xi}_1 < 0$ and $\xi_1 = 0$, and is bounded by the horizontal line $\xi_2 = 0$ from below and the sonic arc $\partial B_{\bar{c}_2}(\mathbf{0}) \cap \{\xi_2 > 0, \ \bar{\xi}_1 < \xi_1 < 0\}$ from above. Since each $\varphi_1 - \varphi^{(i)}$ is monotonically non-increasing in $\Omega^{(i)}$ in the directions in $\mathrm{Cone}^0(\mathbf{e}_{S_1}, \mathbf{e}_{\xi_2})|_{\theta_{\mathrm{w}}=\theta_{\mathrm{w}}^{(i)}}$, then, using (8.2.15)–(8.2.16) and the uniform convergence of $\varphi^{(i)}$ to $\varphi^{(\infty)}$, we obtain that $\varphi_1 - \varphi^{(\infty)}$ is monotonically non-increasing in $\Omega^{(\infty)}$ in every direction in $\mathrm{Cone}^0(-\mathbf{e}_{\xi_2}, \mathbf{e}_{\xi_2})|_{\theta_{\mathrm{w}}=\frac{\pi}{2}}$. Using (9.2.24), we conclude that $\partial_{\xi_2}(\varphi_1 - \varphi^{(\infty)}) \equiv 0$ in $\Omega^{(\infty)}$, *i.e.*, $(\varphi_1 - \varphi^{(\infty)})(\boldsymbol{\xi}) = g(\xi_1)$ in $\Omega^{(\infty)}$ for some $g \in C([\bar{\xi}_1, 0]) \cap C^\infty((\bar{\xi}_1, 0])$, where we have used (9.2.24) and the structure of $\Omega^{(\infty)}$ as shown above. Also, by Corollary 9.2.5, $\varphi^{(\infty)}$ is a solution of the potential flow equation (2.2.8) in $\Omega^{(\infty)}$.

Since $D^2(\varphi - \varphi_0) = D^2(\varphi - \varphi_1)$, equation (2.2.11) with $g(\xi_1)$ instead of $\phi(\boldsymbol{\xi})$ holds in $\Omega^{(\infty)}$, and this equation, considered as a linear equation for g, is strictly elliptic in $\Omega^{(\infty)}$ by (9.2.26). It follows that $g'' = 0$ on $(\bar{\xi}_1, 0)$. Thus, $\varphi_1 - \varphi^{(\infty)}$ in Ω is a linear function of form $(\varphi_1 - \varphi^{(\infty)})(\boldsymbol{\xi}) = a\xi_1 + b$. Using that $\varphi_1 = -\frac{|\boldsymbol{\xi}|^2}{2} + u_1\xi_1 + const.$, it follows that $\varphi^{(\infty)} = -\frac{|\boldsymbol{\xi}|^2}{2} + \hat{a}\xi_1 + \hat{b}$. Moreover, from Corollary 9.2.5(v), it follows that $\varphi^{(\infty)} = \varphi_2^{(\frac{\pi}{2})}$ on Γ_{sonic}. Since $\varphi_2^{(\frac{\pi}{2})} = -\frac{|\boldsymbol{\xi}|^2}{2} + const.$, $\varphi^{(\infty)} - \varphi_2^{(\frac{\pi}{2})}$ is a linear function in $\overline{\Omega^{(\infty)}}$, which vanishes on Γ_{sonic}. Thus, $\varphi^{(\infty)} - \varphi_2^{(\frac{\pi}{2})}$ is identically zero in $\overline{\Omega^{(\infty)}}$. Therefore, $\varphi^{(\infty)}$ is the normal reflection solution.

Repeating the argument above, from any subsequence of $\varphi^{(i)}$, one can extract a further subsequence converging to the normal reflection solution in the sense of assertions (i)–(ii), as shown above. Then, from the uniqueness of the normal reflection solution, it follows that the whole sequence $\varphi^{(i)}$ converges to the normal reflection solution in the sense of assertions (i)–(ii). □

More generally, from (8.2.15)–(8.2.16), it follows that there exist $\sigma^* > 0$ and C, depending on (ρ_0, ρ_1, γ), such that $-\mathbf{e}_{\xi_1} \in \mathrm{Cone}^0(\mathbf{e}_{\mathcal{S}_1}^{(\theta_{\rm w})}, \mathbf{e}_{\xi_2})$ for any $\theta_{\rm w} \in (\frac{\pi}{2}-\sigma^*, \frac{\pi}{2})$. By Corollary 8.2.14, for any admissible solution φ of **Problem 2.6.1** with the wedge angle $\theta_{\rm w} \in (\frac{\pi}{2} - \sigma, \frac{\pi}{2})$,

$$\Gamma_{\rm shock} = \{(f_{-\mathbf{e}_{\xi_1},{\rm sh}}(\xi_2), \xi_2) \;:\; 0 < \xi_2 < \xi_2(P_1(\theta_{\rm w}))\},$$

where $\|f_{-\mathbf{e}_{\xi_1},{\rm sh}}\|_{C^{0,1}((0,\xi_2(P_1)))} \leq C$. Then we have

Corollary 9.2.7. *For any $\varepsilon > 0$, there exists $\sigma > 0$ such that, for any $\theta_{\rm w} \in (\frac{\pi}{2}-\sigma, \frac{\pi}{2})$ and any admissible solution φ of* **Problem 2.6.1** *with the wedge angle $\theta_{\rm w}$,*

$$|f_{-\mathbf{e}_{\xi_1},{\rm sh}}(\xi_2) - \bar{\xi}_1| \leq \varepsilon \qquad \text{on } (0, \xi_2(P_1(\theta_{\rm w}))),$$

where $\bar{\xi}_1 < 0$ is defined by (6.2.2).

Proof. Noting that $P_1(\theta_{\rm w}) \to P_1(\frac{\pi}{2})$ as $\theta_{\rm w} \to \frac{\pi}{2}$ by Remark 7.5.5 and using the equi-Lipschitz property of $f_{-\mathbf{e}_{\xi_1},{\rm sh}}$ for admissible solutions with $\theta_{\rm w} \in (\frac{\pi}{2}-\sigma^*, \frac{\pi}{2})$, it follows that, if the assertion is false, there exists a sequence $\varphi^{(i)}$ of admissible solutions of **Problem 2.6.1** with the wedge angles $\theta_{\rm w}^{(i)} \to \frac{\pi}{2}$, which does not satisfy the convergence property as stated in Lemma 9.2.6(ii). This contradiction yields Corollary 9.2.7. □

9.3 SEPARATION OF $\Gamma_{\rm shock}$ FROM $\Gamma_{\rm sym}$

In this section, we use the expression of $\Gamma_{\rm shock}$ as a graph in the vertical direction given in Corollary 8.2.9 with the corresponding function $f_{\rm sh}$ from (8.2.10).

Proposition 9.3.1. *There exists $\mu > 0$ depending only on the data such that*

$$f'_{\rm sh}(\xi_1) \geq \mu \qquad \text{for any } \xi_1 \in (\xi_{1P_2}, \min\{\xi_{1P_1}, 0\}) \cap \{0 < f_{\rm sh}(\xi_1) < \frac{c_1}{2}\}$$

for any admissible solution φ of **Problem 2.6.1** *with $\theta_{\rm w} \in (\theta_{\rm w}^{\rm s}, \frac{\pi}{2})$, where $f_{\rm sh}$ is defined in* (8.2.10).

Proof. We use the C^1–regularity of $f_{\rm sh}$ shown in Corollary 8.2.9. Now we divide the remaining proof into two steps.

1. Prove

$$f'_{\rm sh}(\xi_1) > 0 \qquad \text{for any } \xi_1 \in (\xi_{1P_2}, \min\{\xi_{1P_1}, 0\}) \cap \{0 < f_{\rm sh}(\xi_1) < \frac{c_1}{2}\}. \tag{9.3.1}$$

For $P = (\xi_{1P}, \xi_{2P}) \in \Gamma_{\text{shock}}$, denote by L_P the tangent line to Γ_{shock} at P. By (8.2.10), $L_P = \{(\xi_{1P}+s,\ f_{\text{sh}}(\xi_{1P})+sf'_{\text{sh}}(\xi_{1P}))\ :\ s \in \mathbb{R}\}$. Denote by $G(\xi_{1P})$ the ξ_2–coordinate of the point of intersection of L_P with line $\mathcal{O}_1 + \mathbb{R}\mathbf{e}_{\xi_2}$. Then

$$G(\xi_{1P}) = f_{\text{sh}}(\xi_{1P}) + (u_1 - \xi_{1P})f'_{\text{sh}}(\xi_{1P}). \tag{9.3.2}$$

By Lemma 8.1.7 and (6.1.3) in Lemma 6.1.2 (applied with $\varphi^- = \varphi_1$), it follows that L_P does not intersect with $B_{c_1}(\mathcal{O}_1)$. This implies

$$|G(\xi_{1P})| > c_1 \qquad \text{for all } \xi_{1P} \in (\xi_{1P_2}, \xi_{1P_1}). \tag{9.3.3}$$

By Definition 8.1.1(i), $f_{\text{sh}}(\xi_{1P_2}) = \xi_{2P_2} = 0$ and $f_{\text{sh}} > 0$ on (ξ_{1P_2}, ξ_{1P_1}), where the last property follows from (8.1.2). Then there exists $\hat{P} = \hat{\boldsymbol{\xi}} \in \Gamma_{\text{shock}}$ such that $\hat{\xi}_1 \in (\xi_{1P_2}, \min\{\xi_{1P_1}, 0\})$ and $f'_{\text{sh}}(\hat{\xi}_1) > 0$. By (9.3.2), $G(\hat{\xi}_1) > 0$. Thus, by (9.3.3),

$$G(\hat{\xi}_1) > c_1.$$

Let $\bar{P} = \bar{\boldsymbol{\xi}} \in \Gamma_{\text{shock}}$ be such that $\bar{\xi}_1 \in (\xi_{1P_2}, \min\{\xi_{1P_1}, 0\})$, $\bar{\xi}_2 = f_{\text{sh}}(\bar{\xi}_1) \le \frac{c_1}{2}$, and $f'_{\text{sh}}(\bar{\xi}_1) < 0$. Then $G(\bar{\xi}_1) < \frac{c_1}{2}$ by (9.3.2). Thus, by (9.3.3),

$$G(\bar{\xi}_1) < -c_1.$$

Since $f_{\text{sh}} \in C^1((\xi_{1P_2}, \xi_{1P_1}])$, $G(\xi_1)$ depends continuously on $\xi_1 \in (\xi_{1P_2}, \xi_{1P_1}]$. Therefore, there exists $\xi_1^* \in (\xi_{1P_2}, \xi_{1P_1}]$ between $\bar{\xi}_1$ and $\hat{\xi}_1$ such that $G(\xi_1^*) = 0$. This contradicts (9.3.3). Now (9.3.1) is proved.

2. From (9.3.1)–(9.3.2), it follows that

$$G(\xi_1) > 0 \qquad \text{for any } \xi_1 \in (\xi_{1P_2}, \min\{\xi_{1P_1}, 0\})$$

such that $f_{\text{sh}}(\xi_1) \in (0, \frac{c_1}{2})$. By (9.3.3), $G(\xi_1) > c_1$ for all such ξ_1. Also we recall that, by (9.1.2), $\xi_{1P_2} \ge -C$, depending only on the data. Then, from (9.3.2),

$$f'_{\text{sh}}(\xi_1) = \frac{G(\xi_1) - f_{\text{sh}}(\xi_1)}{u_1 - \xi_1} \ge \frac{c_1 - \frac{c_1}{2}}{u_1 + C} = \mu.$$

This completes the proof. ∎

Corollary 9.3.2. *For any admissible solution φ of* **Problem 2.6.1** *with $\theta_{\text{w}} \in (\theta_{\text{w}}^{\text{s}}, \frac{\pi}{2})$,*

$$f_{\text{sh}}(\xi_1) \ge \min(\frac{c_1}{2}, \mu(\xi_1 - \xi_{1P_2})) \qquad \textit{for all } \xi_1 \in [\xi_{1P_2}, \min\{\xi_{1P_1}, 0\}],$$

where $\mu > 0$ is the constant in Proposition 9.3.1.

9.4 LOWER BOUND FOR THE DISTANCE BETWEEN Γ_{shock} AND Γ_{wedge}

In this section, we obtain the estimates of the lower bound for the distance between Γ_{shock} and Γ_{wedge} in several cases.

We start by proving some preliminary facts.

Lemma 9.4.1. *For every $a > 0$, there exists $C > 0$ depending only on $(a, \rho_0, \rho_1, \gamma)$ such that, if φ is an admissible solution of* **Problem 2.6.1** *with the wedge angle $\theta_{\text{w}} \in (\theta_{\text{w}}^{\text{s}}, \frac{\pi}{2})$, and there are Q_1 and $Q_2 \in \overline{\Gamma_{\text{shock}}}$ so that*

$$\operatorname{dist}(Q_1, Q_2) \geq a,$$
$$\operatorname{dist}(Q_j, L_{\Gamma_{\text{wedge}}}) \geq a \qquad \textit{for } j = 1, 2,$$

where $L_{\Gamma_{\text{wedge}}} = \{\boldsymbol{\xi} \,:\, \xi_1 = \xi_2 \cot\theta_{\text{w}}\}$, then

$$\operatorname{dist}(\Gamma_{\text{shock}}[Q_1, Q_2],\ \Gamma_{\text{wedge}}) \geq \frac{1}{C},$$

where $\Gamma_{\text{shock}}[Q_1, Q_2]$ denotes the segment of curve Γ_{shock} between points Q_1 and Q_2.

Proof. The proof is achieved by employing the maximum principle in the case that $\Gamma_{\text{shock}}[Q_1, Q_2]$ is very close to Γ_{wedge} in order to lead to a contradiction.

In this proof, the universal constants C and R depend only on the data and a, *i.e.*, on $(\rho_0, \rho_1, \gamma, a)$. We divide the proof into three steps.

1. *Choose a sequence of admissible solutions.* If the claim in the proposition is false, then there exist a sequence $\{\theta_{\text{w}}^{(i)}\} \subset (\theta_{\text{w}}^{\text{s}},\ \frac{\pi}{2})$ and corresponding sequences of admissible solutions $\{\varphi^{(i)}\}$ of **Problem 2.6.1** and points $\{Q_1^{(i)}, Q_2^{(i)}\} \subset \Gamma_{\text{shock}}^{(i)}$ and $Q^{(i)} \in \Gamma_{\text{shock}}^{(i)}[Q_1^{(i)}, Q_2^{(i)}]$ such that

$$\begin{aligned}
&\operatorname{dist}(Q_1^{(i)}, Q_2^{(i)}) \geq a, \qquad \operatorname{dist}(Q_j^{(i)}, L_{\Gamma_{\text{wedge}}^{(i)}}) \geq a \quad \text{for } j = 1, 2,\\
&\operatorname{dist}(Q^{(i)},\ \Gamma_{\text{wedge}}^{(i)}) \to 0.
\end{aligned} \tag{9.4.1}$$

Passing to a subsequence (without change of the index), we have

$$\theta_{\text{w}}^{(i)} \to \theta_{\text{w}}^{(\infty)} \in [\theta_{\text{w}}^{\text{s}}, \frac{\pi}{2}].$$

Denote $\Gamma_{\text{wedge}}^{(\infty)} := \Gamma_{\text{wedge}}(\theta_{\text{w}}^{(\infty)})$ and $\Lambda^{(\infty)} := \Lambda(\theta_{\text{w}}^{(\infty)})$. Let

$$\boldsymbol{\nu}^{(\infty)} = \boldsymbol{\nu}_{\text{w}}^{(\infty)}$$

be the unit normal on $\Gamma_{\text{wedge}}^{(\infty)}$, interior to $\Lambda^{(\infty)}$. Thus, $\boldsymbol{\nu}_{\text{w}}^{(\infty)} \in \operatorname{Cone}^0(\mathbf{e}_{\mathcal{S}_1^{(\infty)}}, \mathbf{e}_{\xi_2})$ by Lemma 8.2.11. Then $\boldsymbol{\nu}_{\text{w}}^{(\infty)} \in \operatorname{Cone}^0(\mathbf{e}_{\mathcal{S}_1^{(i)}}, \mathbf{e}_{\xi_2})$ for all $i > N$, where N is large. From Corollary 8.2.14(i), it follows that, in the (S,T)–coordinates with basis $\{\boldsymbol{\nu}_{\rm w}^{(\infty)}, \boldsymbol{\tau}_{\rm w}^{(\infty)}\}$, we see that, for $j=1,2$,

$$\begin{aligned} Q_j^{(i)} &= (f^{(i)}_{\boldsymbol{\nu}^{(\infty)},{\rm sh}}(T_{Q_j^{(i)}}), T_{Q_j^{(i)}}) \qquad \text{with } T_{P_2^{(i)}} \le T_{Q_j^{(i)}} \le T_{P_1^{(i)}},\\ Q^{(i)} &= (f^{(i)}_{\boldsymbol{\nu}^{(\infty)},{\rm sh}}(T_{Q^{(i)}}), T_{Q^{(i)}}) \qquad \text{with } T_{Q_1^{(i)}} < T_{Q^{(i)}} < T_{Q_2^{(i)}}, \end{aligned} \tag{9.4.2}$$

where $T_{Q_1^{(i)}} < T_{Q_2^{(i)}}$ by switching $Q_1^{(i)}$ and $Q_2^{(i)}$ if necessary.

By passing to a subsequence (without change of the index) and using (9.1.2), we have

$$(Q_1^{(i)}, Q_2^{(i)}, Q^{(i)}) \to (Q_1^{(\infty)}, Q_2^{(\infty)}, Q^{(\infty)}).$$

It follows from (9.4.1) that

$$\begin{aligned} &{\rm dist}(Q_1^{(\infty)}, Q_2^{(\infty)}) \ge a, \qquad {\rm dist}(Q_j^{(\infty)}, L_{\Gamma^{(\infty)}_{\rm wedge}}) \ge a \quad \text{for } j=1,2,\\ &{\rm dist}(Q^{(\infty)},\ \Gamma^{(\infty)}_{\rm wedge}) = 0. \end{aligned} \tag{9.4.3}$$

2. *Choose a solution, a domain, and coordinates for employing the maximum principle.* Using Corollary 9.2.5 and passing to a subsequence (without change of the index), and employing the Lipschitz extensions of $\varphi^{(i)}$ as in Remark 9.2.4, we have

$$\begin{aligned} &\varphi^{(i)} \to \varphi^{(\infty)} \qquad \text{uniformly on each compact subset of } \overline{\Lambda}(\theta_{\rm w}^{\infty}),\\ &f^{(i)}_{\boldsymbol{\nu}^{(\infty)},{\rm sh}} \to f^{(\infty)}_{\boldsymbol{\nu}^{(\infty)},{\rm sh}} \qquad \text{uniformly on } [T_{P_2^{(\infty)}}, T_{P_1^{(\infty)}}], \end{aligned}$$

where $\varphi^{(\infty)}$ is a weak solution of **Problem 2.6.1** in $\Lambda^{(\infty)}$, and $\varphi^{(\infty)}$ is of the structure described in Corollary 9.2.5. In particular,

$$\|(f^{(i)}_{\boldsymbol{\nu}^{(\infty)},{\rm sh}},\ f^{(\infty)}_{\boldsymbol{\nu}^{(\infty)},{\rm sh}})\|_{C^{0,1}([T_{P_2^{(\infty)}}, T_{P_1^{(\infty)}}])} \le C. \tag{9.4.4}$$

Passing to the limit in (9.4.2), we obtain that, in the (S,T)–coordinates with basis $\{\boldsymbol{\nu}_{\rm w}^{(\infty)}, \boldsymbol{\tau}_{\rm w}^{(\infty)}\}$,

$$\begin{aligned} Q^{(\infty)} &= (f^{(\infty)}_{\boldsymbol{\nu}^{(\infty)},{\rm sh}}(T_{Q^{(\infty)}}), T_{Q^{(\infty)}}),\\ Q_j^{(\infty)} &= (f^{(\infty)}_{\boldsymbol{\nu}^{(\infty)},{\rm sh}}(T_{Q_j^{(\infty)}}), T_{Q_j^{(\infty)}}) \qquad \text{for } j=1,2, \end{aligned}$$

and, noting that $Q^{(\infty)} \ne Q_j^{(\infty)}$ for $j=1,2$, by (9.4.3),

$$T_{P_2^{(\infty)}} \le T_{Q_1^{(\infty)}} < T_{Q^{(\infty)}} < T_{Q_2^{(\infty)}} \le T_{P_1^{(\infty)}}. \tag{9.4.5}$$

Now, by Corollary 9.2.5(iii),

$$Q^{(\infty)}, Q_j^{(\infty)} \in \Gamma^{(\infty)}_{\rm shock}.$$

Also, by (9.2.22) and (9.4.3),

$$
\begin{aligned}
&f^{(\infty)}_{\boldsymbol{\nu}^{(\infty)},\mathrm{sh}}(T_{Q_j^{(\infty)}}) = \mathrm{dist}(Q_j^{(\infty)}, L_{\Gamma^{(\theta_{\mathrm{w}}^{(\infty)})}_{\mathrm{wedge}}}) \geq a \qquad \text{for } j=1,2,\\
&f^{(\infty)}_{\boldsymbol{\nu}^{(\infty)},\mathrm{sh}}(T_{Q^{(\infty)}}) = \mathrm{dist}(Q^{(\infty)}, L_{\Gamma^{(\theta_{\mathrm{w}}^{(\infty)})}_{\mathrm{wedge}}}) = 0.
\end{aligned}
\tag{9.4.6}
$$

From this, using (9.2.22) and (9.4.5), we have

$$T_{Q^{(\infty)}} \in [T_{P_3{}^{(\infty)}}, T_{P_1{}^{(\infty)}}). \tag{9.4.7}$$

Next we show that

$$T_{Q^{(\infty)}} \in (T_{P_3{}^{(\infty)}}, T_{P_1{}^{(\infty)}}). \tag{9.4.8}$$

Indeed, assume that (9.4.8) is false. Then (9.4.7) implies that $T_{Q^{(\infty)}} = T_{P_3{}^{(\infty)}}$. Thus, (9.4.4)–(9.4.6) imply

$$T_{P_2{}^{(\infty)}} + \frac{1}{C} \leq T_{Q_1^{(\infty)}} + \frac{1}{C} \leq T_{P_3{}^{(\infty)}} = T_{Q^{(\infty)}} \leq T_{Q_2^{(\infty)}} - \frac{1}{C} \leq T_{P_1{}^{(\infty)}} - \frac{1}{C}.$$

Then, for all $i \geq N$ for sufficiently large N, we have

$$T_{P_2{}^{(i)}} + \frac{1}{2C} \leq T_{P_3{}^{(i)}} \leq T_{P_1{}^{(i)}} - \frac{1}{2C}. \tag{9.4.9}$$

Thus, $|P_3{}^{(i)} - P_2{}^{(i)}| \geq \frac{1}{2C}$. Since $\xi_{2P_3{}^{(i)}} = \xi_{2P_2{}^{(i)}} = 0$ in the $\boldsymbol{\xi}$–coordinates, we have

$$\xi_{1P_3{}^{(i)}} - \xi_{1P_2{}^{(i)}} = |P_3{}^{(i)} - P_2{}^{(i)}| \geq \frac{1}{2C}.$$

Now (8.2.10) and Corollary 9.3.2 imply that there exists $R > 0$ such that

$$\Gamma^i_{\mathrm{shock}} \cap \big(B_R(P_3{}^{(i)}) \cap \{\xi_1 < 0\}\big) = \emptyset \qquad \text{for all } i \geq N.$$

This and (8.2.25) imply that $f^{(i)}_{\boldsymbol{\nu}^{(\infty)},\mathrm{sh}}(T_{P_3{}^{(i)}}) \geq R$. Sending this to the limit as $i \to \infty$ and using the uniform convergence of $f^{(i)}_{\boldsymbol{\nu}^{(\infty)},\mathrm{sh}} \to f^{(\infty)}_{\boldsymbol{\nu}^{(\infty)},\mathrm{sh}}$, we obtain that $f^{(\infty)}_{\boldsymbol{\nu}^{(\infty)},\mathrm{sh}}(T_{P_3{}^{(\infty)}}) \geq R$. That is, $f^{(\infty)}_{\boldsymbol{\nu}^{(\infty)},\mathrm{sh}}(T_{Q^{(\infty)}}) \geq R$. This contradicts (9.4.6). Now (9.4.8) is proved.

From (9.4.6) and (9.4.8),

$$Q^{(\infty)} \in \Gamma^{(\infty)}_{\mathrm{wedge}}.$$

Then, by (9.2.30), (9.4.6), and the continuity of $f^{(\infty)}_{\boldsymbol{\nu}^{(\infty)},\mathrm{sh}}(\cdot)$, there exist $\hat{T}_1 \in (T_{P_3{}^{(\infty)}}, T_{Q^{(\infty)}})$ and $\hat{T}_2 \in (T_{Q^{(\infty)}}, T_{P_1{}^{(\infty)}})$ such that the points:

$$B_j = (0, \hat{T}_j) \in \Gamma^{\infty}_{\mathrm{wedge}}, \quad \hat{Q}_j^{(\infty)} := (f^{(\infty)}_{\boldsymbol{\nu}^{(\infty)},\mathrm{sh}}(\hat{T}_j), \hat{T}_j), \qquad j = 1,2, \tag{9.4.10}$$

satisfy

$$(\varphi_1 - \varphi^{(\infty)})(B_j) > 0, \;\; f^{(\infty)}_{\boldsymbol{\nu}^{(\infty)},\mathrm{sh}}(\hat{T}_j) > 0 \qquad \text{for } j = 1,2.$$

Denote

$$\hat{a} := \min_{j=1,2} f^{(\infty)}_{\boldsymbol{\nu}^{(\infty)},\mathrm{sh}}(\hat{T}_j), \tag{9.4.11}$$

$$b := \min_{j=1,2} (\varphi_1 - \varphi^{(\infty)})(B_j). \tag{9.4.12}$$

Then $\hat{a} > 0$ and $b > 0$.

Also, note that

$$\begin{aligned} &T_{P_2^{(\infty)}} \le T_{Q_1^{(\infty)}} < \hat{T}_1 < T_{Q^{(\infty)}} < \hat{T}_2 < T_{Q_2^{(\infty)}} \le T_{P_1^{(\infty)}}, \\ &\mathrm{dist}(\hat{Q}_j^{(\infty)}, L_{\Gamma^{(\infty)}_{\mathrm{wedge}}}) = f^{(\infty)}_{\boldsymbol{\nu}^{(\infty)},\mathrm{sh}}(\hat{T}_j) \ge \hat{a} \qquad \text{for } j = 1, 2. \end{aligned} \tag{9.4.13}$$

Denote

$$\hat{Q}_j^{(i)} := (f^{(i)}_{\boldsymbol{\nu}^{(\infty)},\mathrm{sh}}(\hat{T}_j), \hat{T}_j) \qquad \text{for } j = 1, 2.$$

Then $T_{Q^{(\infty)}}, T_j \in (T_{P_3^{(i)}}, T_{P_1^{(i)}})$ for all sufficiently large i, which implies

$$\hat{Q}_j^{(i)} \in \Gamma^{(i)}_{\mathrm{shock}}.$$

Moreover, from the uniform convergence $f^{(i)}_{\boldsymbol{\nu}^{(\infty)},\mathrm{sh}} \to f^{(\infty)}_{\boldsymbol{\nu}^{(\infty)},\mathrm{sh}}$, it follows that

$$(\hat{Q}_1^{(i)}, \hat{Q}_2^{(i)}) \to (\hat{Q}_1^{(\infty)}, \hat{Q}_2^{(\infty)}) \qquad \text{as } i \to \infty.$$

This and (9.4.13) imply that

$$\lim_{i\to\infty} \mathrm{dist}(\hat{Q}_j^{(i)}, L_{\Gamma^{(i)}_{\mathrm{wedge}}}) = \mathrm{dist}(\hat{Q}_j^{(\infty)}, L_{\Gamma^{(\infty)}_{\mathrm{wedge}}}) \ge \hat{a} \qquad \text{for } j = 1, 2, \tag{9.4.14}$$

where $L_{\Gamma^{(i)}_{\mathrm{wedge}}} = \{\boldsymbol{\xi} \,:\, \xi_1 = \xi_2 \cot\theta_{\mathrm{w}}^{(i)}\}$.

Now we write points $Q^{(i)}$ and $\hat{Q}_j^{(i)}$ in the (s,t)–coordinates with basis $\{\boldsymbol{\nu}^{(i)}, (\boldsymbol{\nu}^{(i)})^\perp\}$ and the origin at P_3, where

$$\boldsymbol{\nu}^{(i)} = \boldsymbol{\nu}_{\mathrm{w}}^{(i)} = (-\sin\theta_{\mathrm{w}}^{(i)},\, \cos\theta_{\mathrm{w}}^{(i)}), \quad (\boldsymbol{\nu}^{(i)})^\perp = \boldsymbol{\tau}_{\mathrm{w}}^{(i)} = (\cos\theta_{\mathrm{w}}^{(i)}, \sin\theta_{\mathrm{w}}^{(i)}).$$

In the coordinates, $\Lambda^{(i)} \subset \{s > 0\}$ and $\Gamma^{(i)}_{\mathrm{wedge}} \subset \{s = 0\}$. Using (8.2.25), we have

$$\Omega^{(i)} = \left\{ (s,t) \in \mathbb{R}^2 \,:\, \begin{array}{l} t_{P_2^{(i)}} < t < t_{P_4^{(i)}}, \\ -(t - t_{P_3^{(i)}})\tan\theta_{\mathrm{w}}^{(i)} < s < f^{(i)}_{\boldsymbol{\nu}^{(i)},\mathrm{sh}}(t) \\ \hfill \text{for } t \in (t_{P_2^{(i)}}, t_{P_3^{(i)}}], \\ 0 < s < f^{(i)}_{\boldsymbol{\nu}^{(i)},\mathrm{sh}}(t) \;\; \text{for } t \in (t_{P_3^{(i)}}, t_{P_1^{(i)}}], \\ 0 < s < f_{\boldsymbol{\nu}^{(i)},\mathrm{so}}(t) \;\; \text{for } t \in (t_{P_1^{(i)}}, t_{P_4^{(i)}}) \end{array} \right\} \tag{9.4.15}$$

and

$$\Gamma^{(i)}_{\mathrm{shock}} = \left\{ (f^{(i)}_{\boldsymbol{\nu}^{(i)},\mathrm{sh}}(t), t) \,:\, t_{P_2^{(i)}} < t < t_{P_1^{(i)}} \right\} \tag{9.4.16}$$

with $f^{(i)}_{\boldsymbol{\nu}^{(i)},\mathrm{sh}} > 0$ on $(t_{P_2^{(i)}}, t_{P_1^{(i)}})$, since φ_i is an admissible solution of **Problem 2.6.1**. From the definition and properties of points $Q^{(i)}$ and $\hat{Q}^{(i)}_j$ expressed in the (S,T)–coordinates with basis $\{\boldsymbol{\nu}^{(\infty)}_{\mathrm{w}}, \boldsymbol{\tau}^{(\infty)}_{\mathrm{w}}\}$ (*cf.* (9.4.2) and (9.4.13)), and the convergence: $\boldsymbol{\nu}^{(i)}_{\mathrm{w}} \to \boldsymbol{\nu}^{(\infty)}_{\mathrm{w}}$, it follows that

$$Q^{(i)} = (f^{(i)}_{\boldsymbol{\nu}^{(i)},\mathrm{sh}}(t_{Q^{(i)}}), t_{Q^{(i)}}), \qquad \hat{Q}^{(i)}_j = (f^{(i)}_{\boldsymbol{\nu}^{(i)},\mathrm{sh}}(t^{(i)}_j), t^{(i)}_j) \ \text{ for } j = 1, 2,$$

where, restricting to $i \geq N$ for sufficiently large N, we have

$$\begin{aligned}
&t_{P_2^{(i)}} < t^{(i)}_1 < t_{Q^{(i)}} < t^{(i)}_2 < t_{P_1^{(i)}},\\
&\lim_{i\to\infty} t^{(i)}_j = \hat{T}_j \qquad \text{for } j = 1, 2,\\
&f^{(i)}_{\boldsymbol{\nu}^{(i)},\mathrm{sh}}(t_{Q^{(i)}}) = \mathrm{dist}(Q^{(i)}, L_{\Gamma^{(i)}_{\mathrm{wedge}}}) \to \mathrm{dist}(Q^{(\infty)}, L_{\Gamma^{(\infty)}_{\mathrm{wedge}}}) = 0,\\
&f^{(i)}_{\boldsymbol{\nu}^{(i)},\mathrm{sh}}(t^{(i)}_j) = \mathrm{dist}(\hat{Q}^{(i)}_j, L_{\Gamma^{(i)}_{\mathrm{wedge}}}) \geq \frac{\hat{a}}{2} \qquad \text{for } j = 1, 2.
\end{aligned} \tag{9.4.17}$$

Thus, for any sufficiently large i, there exists $t^{(i)} \in (t^{(i)}_1, t^{(i)}_2)$ such that

$$f^{(i)}_{\boldsymbol{\nu}^{(i)},\mathrm{sh}}(t^{(i)}) = \min_{t\in[t^{(i)}_1, t^{(i)}_2]} f^{(i)}_{\boldsymbol{\nu}^{(i)},\mathrm{sh}}(t) \leq \frac{\hat{a}}{4} < \min_{j=1,2} f^{(i)}_{\boldsymbol{\nu}^{(i)},\mathrm{sh}}(t^{(i)}_j). \tag{9.4.18}$$

Denote

$$s^{(i)} := f^{(i)}_{\boldsymbol{\nu}^{(i)},\mathrm{sh}}(t^{(i)}), \qquad R^{(i)} = (s^{(i)}, t^{(i)}).$$

Then $R^{(i)} \in \Gamma^{(i)}_{\mathrm{shock}}$. By (8.2.25), $s^{(i)} > 0$. Also, by (9.4.17) and $0 < s^{(i)} \leq f^{(i)}_{\boldsymbol{\nu}^{(i)},\mathrm{sh}}(t_{Q^{(i)}})$, we have

$$s^{(i)} \to 0. \tag{9.4.19}$$

In the (s,t)–coordinates with basis $\{\boldsymbol{\nu}^{(i)}_{\mathrm{w}}, \boldsymbol{\tau}^{(i)}_{\mathrm{w}}\}$, denote $\mathcal{D}^{(i)} := (0, s^{(i)}) \times (t^{(i)}_1, t^{(i)}_2)$. Then it follows from (8.2.25) for $\varphi^{(i)}$ by using (9.4.18) that, for sufficiently large i,

$$\mathcal{D}^{(i)} \subset \Omega^{(i)}, \qquad R^{(i)} \in \partial\mathcal{D}^{(i)} \cap \partial\Omega^{(i)},$$

and $\partial\mathcal{D}^{(i)}$ is smooth (flat) at and near $R^{(i)}$. Then, denoting by $\boldsymbol{\nu}_{\partial\mathcal{D}^{(i)}} = \boldsymbol{\nu}_{\partial\mathcal{D}^{(i)}}(R^{(i)})$ the interior unit normal with respect to $\mathcal{D}^{(i)}$ at $R^{(i)}$, we have

$$\boldsymbol{\nu}_{\partial\mathcal{D}^{(i)}} = \boldsymbol{\nu}_{\Gamma^{(i)}_{\mathrm{shock}}}(R^{(i)}) = -\boldsymbol{\nu}^{(i)}_{\mathrm{w}}. \tag{9.4.20}$$

3. *Application of the maximum principle in $\mathcal{D}^{(i)}$ for sufficiently large i.* Consider the function:

$$\bar{\phi}^{(i)} = \varphi_1 - \varphi^{(i)} \qquad \text{in } \mathcal{D}^{(i)}.$$

Since $D^2\bar{\phi}^{(i)} = -D^2\phi^{(i)}$ for $\phi^{(i)} = \varphi^{(i)} + \frac{|\boldsymbol{\xi}|^2}{2}$, equation (2.2.11) (considered as a linear equation for $D^2\phi^{(i)}$) is satisfied with $D^2\phi^{(i)}$ replaced by $D^2\bar{\phi}^{(i)}$, and is strictly elliptic in $\overline{\Omega^{(i)}} \setminus \Gamma^{(i)}_{\text{sonic}}$. Also, from (9.4.17)–(9.4.19), we see that $\overline{\mathcal{D}^{(i)}} \subset \Omega^{(i)}$ for all large i, so that $\text{dist}(\overline{\mathcal{D}^{(i)}}, \Gamma^{(i)}_{\text{sonic}}) > 0$. Thus, $\bar{\phi}$ satisfies the linear equation:

$$a_{11}D_{ss}\bar{\phi}^{(i)} + 2a_{12}D_{st}\bar{\phi}^{(i)} + a_{22}D_{tt}\bar{\phi}^{(i)} = 0 \qquad \text{in } \mathcal{D}^{(i)}, \tag{9.4.21}$$

with continuous coefficients in $\overline{\mathcal{D}^{(i)}}$, which is uniformly elliptic in $\overline{\mathcal{D}^{(i)}}$.

Next, we show the following properties of $\bar{\phi}^{(i)}$ on $\partial\mathcal{D}^{(i)}$ for sufficiently large i:

$$\bar{\phi}^{(i)} \geq 0 \qquad \text{on } \{s^{(i)}\} \times (t_1^{(i)}, t_2^{(i)}), \tag{9.4.22}$$

$$\partial_{\boldsymbol{\nu}}\bar{\phi}^{(i)} = -u_1 \sin\theta_{\text{w}}^{(i)} \qquad \text{on } \{0\} \times (t_1^{(i)}, t_2^{(i)}), \tag{9.4.23}$$

$$\bar{\phi}^{(i)} \geq \frac{b}{2} \qquad \text{on } (0, s^{(i)}) \times \{t_j^{(i)}\} \ \text{ for } j = 1, 2, \tag{9.4.24}$$

where b is from (9.4.12).

Indeed, (9.4.22) holds because $\varphi^{(i)}$ is an admissible solution of **Problem 2.6.1**.

To show (9.4.23), we note that $\{0\}\times(t_1^{(i)}, t_2^{(i)}) = \Gamma^{(i)}_{\text{wedge}} \cap \partial\mathcal{D}^{(i)}$, which implies that $\partial_{\boldsymbol{\nu}}\varphi^{(i)} = 0$ on $\{0\} \times (t_1^{(i)}, t_2^{(i)})$. Then we calculate in the $\boldsymbol{\xi}$–coordinates to obtain that $\partial_{\boldsymbol{\nu}}\bar{\phi}^{(i)} = \partial_{\boldsymbol{\nu}}\varphi_1 = -u_1 \sin\theta_{\text{w}}^{(i)}$ on $\Gamma^{(i)}_{\text{wedge}}$, which implies (9.4.23).

Next we show (9.4.24). Denoting $B_j^{(i)} := (0, t_j^{(i)})$ for $j = 1, 2$ in the (s, t)–coordinates with basis $\{\boldsymbol{\nu}_{\text{w}}^{(i)}, \boldsymbol{\tau}_{\text{w}}^{(i)}\}$, we have

$$B_j^{(i)} \in \Gamma^{(i)}_{\text{wedge}}.$$

Using (9.4.17) and the convergence: $\boldsymbol{\nu}_{\text{w}}^{(i)} \to \boldsymbol{\nu}_{\text{w}}^{(\infty)}$, we see that $B_j^{(i)} \to B_j$ for B_j defined in (9.4.10). From the uniform convergence that $\varphi^{(i)} \to \varphi^{(\infty)}$, (9.4.4), and (9.4.12), it follows that $\bar{\phi}(B_j^{(i)}) \geq \frac{2}{3}b$ for large i. Then, from the equi-Lipschitz property of $\varphi^{(i)}$ and (9.4.19), we conclude (9.4.24) for large i.

Now, define the function:

$$v(s,t) = -u_1 \sin\theta_{\text{w}}^{(i)}(s - s^{(i)}).$$

Note that $v(R^{(i)}) = v(s^{(i)}, t^{(i)}) = 0$. Also, for i sufficiently large, v satisfies

$$a_{11}D_{ss}v + 2a_{12}D_{st}v + a_{22}D_{tt}v = 0 \qquad \text{in } \mathcal{D}^{(i)}, \tag{9.4.25}$$

$$v = 0 \qquad \text{on } \{s^{(i)}\} \times (t_1^{(i)}, t_2^{(i)}), \tag{9.4.26}$$

$$\partial_{\boldsymbol{\nu}}v = -u_1 \sin\theta_{\text{w}}^{(i)} \qquad \text{on } \{0\} \times (t_1^{(i)}, t_2^{(i)}), \tag{9.4.27}$$

$$0 < v < \frac{b}{2} \qquad \text{on } (0, s^{(i)}) \times \{t_j^{(i)}\} \ \text{ for } j = 1, 2, \tag{9.4.28}$$

where we obtain (9.4.28) by noting that $0 < v(s,t) < u_1 s^{(i)} \sin\theta_{\text{w}}^{(i)}$ on $(0, s^{(i)}) \times \{t_j^{(i)}\}$ and $s^{(i)} \to 0$.

From (9.4.21)–(9.4.28), by the comparison principle,

$$\bar{\phi}^{(i)} \geq v \qquad \text{in } \mathcal{D}^{(i)}. \tag{9.4.29}$$

On the other hand, since $R^{(i)} \in \Gamma^{(i)}_{\text{shock}}$, the Rankine-Hugoniot condition (8.1.13), combined with (9.4.20), implies

$$\rho D\varphi^{(i)} \cdot \boldsymbol{\nu}_{\partial\mathcal{D}^{(i)}} = -\rho_1 D\varphi_1 \cdot \boldsymbol{\nu}^{(i)}_{\text{w}} \qquad \text{at } R^{(i)}.$$

Since $\varphi_1 = -\frac{|\boldsymbol{\xi}|^2}{2} + u_1\xi_1 + const.$, we have

$$-(D\varphi_1 \cdot \boldsymbol{\nu}^{(i)}_{\text{w}})(R^{(i)}) = s^{(i)} + u_1 \sin\theta^{(i)}_{\text{w}} > 0,$$

by using $s^{(i)} > 0$ and $\theta^{(i)}_{\text{w}} \in (\theta^{\text{s}}_{\text{w}}, \frac{\pi}{2})$. Combining this with the bound of the density in Lemma 9.1.4, we have

$$D\varphi^{(i)} \cdot \boldsymbol{\nu}_{\partial\mathcal{D}^{(i)}} = \frac{\rho_1}{\rho}(-D\varphi_1 \cdot \boldsymbol{\nu}^{(i)}_{\text{w}}) \geq \frac{1}{C}u_1 \sin\theta^{(i)}_{\text{w}} \qquad \text{at } R^{(i)}.$$

Thus, using (9.4.20) again, we compute at $R^{(i)}$:

$$\begin{aligned} D\bar{\phi}^{(i)}(R^{(i)}) \cdot \boldsymbol{\nu}_{\partial\mathcal{D}^{(i)}} &= -D\varphi_1(R^{(i)}) \cdot \boldsymbol{\nu}^{(i)}_{\text{w}} - D\varphi^{(i)}(R^{(i)}) \cdot \boldsymbol{\nu}_{\partial\mathcal{D}^{(i)}} \\ &= s^{(i)} + u_1 \sin\theta^{(i)}_{\text{w}} - D\varphi^{(i)}(R^{(i)}) \cdot \boldsymbol{\nu}_{\partial\mathcal{D}^{(i)}} \\ &\leq s^{(i)} + u_1 \sin\theta^{(i)}_{\text{w}} - \frac{1}{C}u_1 \sin\theta^{(i)}_{\text{w}}. \end{aligned}$$

Then, using (9.4.19), we obtain that $D\bar{\phi}^{(i)}(R^{(i)}) \cdot \boldsymbol{\nu}_{\partial\mathcal{D}^{(i)}} < u_1 \sin\theta^{(i)}_{\text{w}}$ if i is large. That is,

$$D\bar{\phi}^{(i)}(R^{(i)}) \cdot \boldsymbol{\nu}_{\partial\mathcal{D}^{(i)}} < Dv(R^{(i)}) \cdot \boldsymbol{\nu}_{\partial\mathcal{D}^{(i)}}.$$

Since $R^{(i)} \in \Gamma^{(i)}_{\text{shock}} \cap \{s = s^{(i)}\}$, we see that $\bar{\phi}^{(i)}(R^{(i)}) = v(R^{(i)}) = 0$. Then the last inequality contradicts (9.4.29). □

In order to state the next property, we note that, from the continuous dependence of $P_3 = \mathbf{0}$ and P_4 on $\theta_w \in [\theta^{\text{d}}_{\text{w}}, \frac{\pi}{2}]$ by Lemma 7.5.9(ii), it follows that

$$r_0 := \min_{\theta_{\text{w}} \in [\theta^{\text{s}}_{\text{w}}, \frac{\pi}{2}]} |P_3 P_4{}^{(\theta_{\text{w}})}| > 0. \tag{9.4.30}$$

Moreover, if φ is an admissible solution for $\theta_{\text{w}} \in [\theta^{\text{s}}_{\text{w}}, \frac{\pi}{2})$, then $\Gamma^{(\theta_{\text{w}})}_{\text{wedge}} = P_3 P_4{}^{(\theta_{\text{w}})}$.

Lemma 9.4.2. *Let $r \in (0, r_0)$. There exists $C_r > 0$ depending only on $(r, \rho_0, \rho_1, \gamma)$ such that, for any admissible solution of* **Problem 2.6.1** *with the wedge angle $\theta_{\text{w}} \in (\theta^{\text{s}}_{\text{w}}, \frac{\pi}{2})$,*

$$\sup_{P \in \Gamma_{\text{shock}} \cap B_r(P_3)} \text{dist}(P, L_{\Gamma_{\text{wedge}}}) > \frac{1}{C_r} \qquad \text{if } |P_2 - P_3| \leq \frac{r}{10}, \tag{9.4.31}$$

and

$$\sup_{P \in \Gamma_{\text{shock}} \cap B_r(P_4)} \text{dist}(P, L_{\Gamma_{\text{wedge}}}) > \frac{1}{C_r} \qquad \text{if } |P_1 - P_4| \leq \frac{r}{10}. \tag{9.4.32}$$

Proof. We first prove (9.4.32).

If the statement is false, then there exist a sequence $\{\theta_{\rm w}^{(i)}\} \subset (\theta_{\rm w}^{\rm s}, \frac{\pi}{2})$ and a corresponding admissible solution sequence $\{\varphi^{(i)}\}$ of **Problem 2.6.1** such that

$$|{P_1}^{(i)} - {P_4}^{(i)}| \le \frac{r}{10}, \qquad \sup_{P \in \Gamma_{\rm shock}^{(i)} \cap B_r({P_4}^{(i)})} \operatorname{dist}(P, L_{\Gamma_{\rm wedge}^{(i)}}) \le \frac{1}{i}.$$

Using Corollary 9.2.5 and passing to a subsequence (without change of the index), and employing the Lipschitz extensions of $\varphi^{(i)}$ as in Remark 9.2.4, we conclude that

$$\varphi^{(i)} \to \varphi^{(\infty)} \qquad \text{uniformly on each compact subset of } \overline{\Lambda}(\theta_{\rm w}^{\infty}),$$

and

$$|{P_1}^{(\infty)} - {P_4}^{(\infty)}| \le \frac{r}{10}, \qquad \sup_{P \in \Gamma_{\rm shock}^{(\infty)} \cap B_r({P_4}^{(\infty)})} \operatorname{dist}(P, L_{\Gamma_{\rm wedge}^{(\infty)}}) = 0, \tag{9.4.33}$$

where $\varphi^{(\infty)}$ is a weak solution of **Problem 2.6.1** in $\Lambda^{(\infty)}$, with the structure described in Corollary 9.2.5. In particular, we obtain from (9.2.23) and (9.4.33) that ${P_1}^{(\infty)} \in L_{\Gamma_{\rm wedge}^{(\infty)}}$, which implies

$${P_1}^{(\infty)} = {P_4}^{(\infty)}. \tag{9.4.34}$$

By Remark 8.2.12,

$$\Gamma_{\rm wedge}^{(\infty)} = \{(0,T) \,:\, T_{{P_3}^{(\infty)}} < T < T_{{P_4}^{(\infty)}}\}.$$

Then, using (9.2.23) with $\mathbf{e} = \boldsymbol{\nu}_{\rm w}^{(\infty)}$, we obtain from (9.4.33)–(9.4.34) that

$$\Gamma_{\rm wedge}^{(\infty)} \cap \{(0,T) \,:\, T_{{P_4}^{(\infty)}} - r < T < T_{{P_4}^{(\infty)}}\} \subset \Gamma_{\rm shock}^{(\infty)}.$$

From this, since $\varphi^{(\infty)} = \varphi_1$ on $\Gamma_{\rm shock}^{(\infty)}$, we find that $\varphi^{(\infty)} = \varphi_1$ on $\Gamma_{\rm wedge}^{(\infty)} \cap \{(0,T) \,:\, T_{{P_4}^{(\infty)}} - r < T < T_{{P_4}^{(\infty)}}\}$. This contradicts (9.2.30).

The proof of (9.4.31) is similar. □

Corollary 9.4.3. *Let $r \in (0, r_0)$. There exists $C_r > 0$ depending only on $(r, \rho_0, \rho_1, \gamma)$ such that, for any admissible solution of* **Problem 2.6.1** *with the wedge angle $\theta_{\rm w} \in (\theta_{\rm w}^{\rm s}, \frac{\pi}{2})$, there is $Q \in \Gamma_{\rm shock}$ so that, in the (S,T)-coordinates with basis $\{\boldsymbol{\nu}_{\rm w}, \boldsymbol{\tau}_{\rm w}\}$,*

$$\operatorname{dist}(Q, L_{\Gamma_{\rm wedge}}) \ge \frac{1}{C_r}, \qquad T_Q \le T_{P_3} + \frac{r}{2}, \tag{9.4.35}$$

where C depends only on $(\rho_0, \rho_1, \gamma, r)$.

Proof. To prove (9.4.35), we first consider the case that $|P_3 - P_2| \geq \frac{r}{20}$. Then, using (8.2.25), we estimate

$$f_{\boldsymbol{\nu}_{\rm w}}(T_{P_2}) \geq \frac{r}{20}\sin\theta_{\rm w} \geq \frac{r}{20}\sin\theta_{\rm w}^{\rm s} > 0,$$

so that

$$\operatorname{dist}(P_2, L_{\Gamma_{\rm wedge}}) = f_{\boldsymbol{\nu}_{\rm w}}(T_{P_2}) \geq \frac{r}{20}\sin\theta_{\rm w}^{\rm s}.$$

Moreover, $T_{P_2} \leq T_{P_3}$ by Lemma 8.2.11. Then (9.4.35) is satisfied with $Q = P_2$ in this case.

Otherwise, $|P_3 - P_2| < \frac{r}{20}$. From Lemma 9.4.2 applied with $\frac{r}{2}$ instead of r, there exists $Q \in \overline{\Gamma_{\rm shock} \cap B_{r/2}(P_3)}$ with $\operatorname{dist}(Q, L_{\Gamma_{\rm wedge}}) \geq \frac{1}{C}$. Clearly, $\overline{\Gamma_{\rm shock} \cap B_{r/2}(P_3)} \subset \Gamma_{\rm shock} \cap \{T \leq T_{P_3} + \frac{r}{2}\}$. Now (9.4.35) is proved. □

9.4.1 Uniform positive lower bound for the distance between $\Gamma_{\rm shock}$ and $\Gamma_{\rm wedge}$ when $u_1 < c_1$

We first prove the lower bound for the distance between $\Gamma_{\rm shock}$ and $\Gamma_{\rm wedge}$ when (ρ_0, ρ_1, γ) satisfy $u_1 < c_1$. The bound is uniform for the wedge angles $\theta_{\rm w} \in [\theta_{\rm w}^*, \frac{\pi}{2})$ where $\theta_{\rm w}^* \in (\theta_{\rm w}^{\rm s}, \frac{\pi}{2})$. Note that, for the sonic angle, $P_1 = P_4 = P_0 \in \overline{\Gamma_{\rm wedge}}$. For this reason, in order to have a uniform bound with respect to the wedge angles up to the sonic angle, we would need to remove a neighborhood of P_0. We will do that in §9.4.4 and §15.4, in order to obtain the estimates for all the wedge angles up to the sonic angle and the detachment angle, respectively.

Remark 9.4.4. *When $u_1 = c_1$, the uniform positive lower bound will also be obtained in §9.5 (Corollary 9.5.7) below. On the other hand, when $u_1 > c_1$, experimental results suggest that $\Gamma_{\rm shock}$ may hit the wedge vertex P_3 in certain cases, and hence the general positive lower bound for the distance between $\Gamma_{\rm shock}$ and $\Gamma_{\rm wedge}$ does not seem to exist in this case. This is the reason for the difference in the conclusions of Theorems 2.6.3 and 2.6.5. However, in the next subsections, we will obtain the positive lower bound of the distance between $\Gamma_{\rm shock}$ and $\Gamma_{\rm wedge}$ away from P_3 for any (ρ_0, ρ_1, γ), including for Case $u_1 \geq c_1$.*

Proposition 9.4.5. *Assume that $u_1 < c_1$. Let $\theta_{\rm w}^* \in (\theta_{\rm w}^{\rm s}, \frac{\pi}{2})$. Then there exists $C > 0$ depending only on $(\rho_0, \rho_1, \gamma, \theta_{\rm w}^*)$ such that, for any admissible solution of* **Problem 2.6.1** *with $\theta_{\rm w} \in [\theta_{\rm w}^*, \frac{\pi}{2})$,*

$$\operatorname{dist}(\Gamma_{\rm shock}, \Gamma_{\rm wedge}) > \frac{1}{C}.$$

Proof. Let φ be an admissible solution for a wedge angle $\theta_{\rm w} \in [\theta_{\rm w}^*, \frac{\pi}{2})$. Let $\{\boldsymbol{\nu}_{\rm w}, \boldsymbol{\tau}_{\rm w}\}$ be the unit normal and tangent vectors to $\Gamma_{\rm wedge}$, defined by (8.2.14) and (8.2.17), respectively. Let (S, T) be the coordinates with basis $\{\boldsymbol{\nu}_{\rm w}, \boldsymbol{\tau}_{\rm w}\}$ and the origin at P_3.

Let $L_{\Gamma_{\rm wedge}} = \{\boldsymbol{\xi} : \xi_1 = \xi_2 \cot\theta_{\rm w}\}$ as that defined in Lemma 9.4.1. Let $f_{\boldsymbol{\nu}_{\rm w}} = f_{\boldsymbol{\nu}_{\rm w},{\rm sh}}$ be the function from (8.2.25).

Using (8.1.2) and the condition that $u_1 < c_1$, we obtain that $\xi_{1P_2} \le -(c_1 - u_1)$. Then, using that $|P_2P_3| = |\xi_{1P_2}|$, changing $\boldsymbol{\xi}$ to the (S,T)–coordinates, and using (8.2.25), we have

$$f_{\boldsymbol{\nu}_{\rm w}}(T_{P_2}) = |P_2P_3|\sin\theta_{\rm w} \ge (c_1 - u_1)\sin\theta_{\rm w} > 0.$$

Using that $\theta_{\rm w} \in [\theta_{\rm w}^*, \frac{\pi}{2}) \subset (\theta_{\rm w}^{\rm s}, \frac{\pi}{2})$, we obtain

$$\text{dist}(P_2, L_{\Gamma_{\rm wedge}}) = f_{\boldsymbol{\nu}_{\rm w}}(T_{P_2}) \ge (c_1 - u_1)\sin\theta_{\rm w}^{\rm s} > 0, \tag{9.4.36}$$

i.e., the positive lower bound depends only on the data, *i.e.*, (ρ_0, ρ_1, γ).

Next, by Lemma 7.5.9(ii), $\text{dist}(P_1, L_{\Gamma_{\rm wedge}})$ depends continuously on $\theta_{\rm w} \in [\theta_{\rm w}^{\rm d}, \frac{\pi}{2}]$. Also $\text{dist}(P_1, L_{\Gamma_{\rm wedge}}) > 0$ for each $\theta_{\rm w} \in [\theta_{\rm w}^*, \frac{\pi}{2})$. Then

$$\min_{\theta_{\rm w}\in[\theta_{\rm w}^*, \frac{\pi}{2}]} \text{dist}(P_1, L_{\Gamma_{\rm wedge}})(\theta_{\rm w}) \ge \frac{1}{C}, \tag{9.4.37}$$

where C depends only on $(\rho_0, \rho_1, \gamma, \theta_{\rm w}^*)$.

Finally, using Corollary 8.2.14(i) and (8.2.25) with $\mathbf{e} = \boldsymbol{\nu}_{\rm w}$, we find that, in the (S,T)–coordinates, $P_k = (f_{\boldsymbol{\nu}_{\rm w}}(T_{P_k}), T_{P_k})$ for $k = 1, 2$, and

$$|P_1 - P_2| \ge T_{P_1} - T_{P_2} \ge T_{P_1} - T_{P_3}. \tag{9.4.38}$$

Also, $T_{P_1} - T_{P_3}$ is strictly positive and depends continuously on $\theta_{\rm w} \in [\theta_{\rm w}^{\rm d}, \frac{\pi}{2}]$ by Lemma 7.5.9(ii) and Lemma 8.2.11. Then

$$|P_1 - P_2| \ge \min_{\theta_{\rm w}\in[\theta_{\rm w}^{\rm d}, \frac{\pi}{2}]} (T_{P_1} - T_{P_3}) \ge \frac{1}{C}, \tag{9.4.39}$$

where $C > 0$ depends only on (ρ_0, ρ_1, γ).

Now we can apply Lemma 9.4.1 with $Q_1 = P_1$ and $Q_2 = P_3$, and with $a > 0$ being the smallest of the lower bounds in (9.4.36)–(9.4.37) and (9.4.39), so that a depends only on $(\rho_0, \rho_1, \gamma, \theta_{\rm w}^*)$. Also, note that $\Gamma_{\rm shock}[P_1, P_2] = \Gamma_{\rm shock}$. Then, applying Lemma 9.4.1, we conclude the proof. □

Remark 9.4.6. *In the proof of Proposition* 9.4.5 (*cf. equation* (9.4.39)), *we have shown that*

$$\hat{r}^0 := \min_{\theta_{\rm w}\in[\theta_{\rm w}^{\rm d}, \frac{\pi}{2}]} (T_{P_1^{(\theta_{\rm w})}} - T_{P_3}) > 0,$$

where we use the coordinates (S,T) with basis $\{\boldsymbol{\nu}_{\rm w}{}^{(\theta_{\rm w})}, \boldsymbol{\tau}_{\rm w}{}^{(\theta_{\rm w})}\}$ for each $\theta_{\rm w}$.

9.4.2 Lower bound for the distance between $\Gamma_{\rm shock}$ and $\Gamma_{\rm wedge}$ for large-angle wedges

When the requirement that $u_1 < c_1$ is dropped, the lower bound for the distance between $\Gamma_{\rm shock}$ and $\Gamma_{\rm wedge}$ cannot be established in general. In this and the next subsections, we prove some partial results for this. We fix (ρ_0, ρ_1, γ) without the condition on $u_1 < c_1$.

Lemma 9.4.7. *There exists $\sigma_1 > 0$ such that, if φ is an admissible solution of* **Problem 2.6.1** *with the wedge angle θ_w,*

$$\Gamma_{shock} \subset \{\xi_2 \le \frac{\bar{\xi}_1}{2}\} \qquad \text{if } \theta_w \in [\frac{\pi}{2} - \sigma_1, \frac{\pi}{2}),$$

where $\bar{\xi}_1 < 0$ is defined by (6.2.2). In particular, we have

$$\mathrm{dist}(\Gamma_{shock}, \Gamma_{wedge}) \ge \frac{|\bar{\xi}_1|}{2} \qquad \text{if } \theta_w \in [\frac{\pi}{2} - \sigma_1, \frac{\pi}{2}).$$

Proof. Indeed, this follows from Corollary 9.2.7 with $\varepsilon = \frac{|\bar{\xi}_1|}{2}$, and the fact that $\Gamma_{wedge} \subset \{\xi_2 \ge 0\}$. □

9.4.3 Lower bound for the distance between Γ_{shock} and Γ_{wedge} away from P_3.

In the next proposition, we use r_0 defined by (9.4.30).

Proposition 9.4.8. *Let $\theta_w^* \in (\theta_w^s, \frac{\pi}{2})$. For every $r \in (0, \frac{r_0}{10})$, there exists $C_r > 0$ depending only on $(\rho_0, \rho_1, \gamma, \theta_w^*, r)$ such that*

$$\mathrm{dist}(\Gamma_{shock}, \Gamma_{wedge} \setminus B_r(P_3)) \ge \frac{1}{C_r}$$

for any admissible solution of **Problem 2.6.1** *with the wedge angle $\theta_w \in [\theta_w^*, \frac{\pi}{2})$.*

Proof. In this proof, the universal constant C depends only on $(\rho_0, \rho_1, \gamma, \theta_w^*, r)$, unless otherwise specified. It suffices to consider $r \in (0, \frac{r^*}{10}]$ for $r^* = \min\{r_0, \hat{r}_0\}$, where $\hat{r}_0$ is defined in Remark 9.4.6. Fix such a constant r.

Let φ be an admissible solution for a wedge angle $\theta_w \in [\theta_w^*, \frac{\pi}{2})$. We use coordinates (S, T) with basis $\{\boldsymbol{\nu}_w, \boldsymbol{\tau}_w\}$ and the origin at P_3, as in the proof of Proposition 9.4.5.

First, we apply Corollary 9.4.3 to obtain $Q \in \Gamma_{shock}$ such that (9.4.35) holds.

Next, we note that (9.4.37) holds in the present case, which can be seen from its proof. Also, since $r < \frac{r^*}{10} \le \frac{\hat{r}_0}{10}$, we use Remark 9.4.6 and (9.4.35) to estimate

$$|P_1 - Q| \ge T_{P_1} - T_Q \ge T_{P_1} - T_{P_3} - \frac{r}{2} \ge \frac{\hat{r}_0}{2}. \tag{9.4.40}$$

Thus, we can apply Lemma 9.4.1 with $Q_1 = P_1$ and $Q_2 = Q$, and with $a > 0$ being the smallest of the lower bounds in (9.4.35), (9.4.37), and (9.4.40), so that a depends only on $(\rho_0, \rho_1, \gamma, \theta_w^*, r)$. Then we obtain

$$\mathrm{dist}(\Gamma_{shock}[Q, P_1], \Gamma_{wedge}) \ge \frac{1}{C}. \tag{9.4.41}$$

Using Corollary 8.2.14(i), (8.2.25), and $T_Q < T_{P_1}$, we have

$$\Gamma_{shock}[Q, P_1] = \Gamma_{shock} \cap \{T > T_Q\}.$$

From (9.4.35), we see that $\Gamma_{\rm shock} \cap \{T > T_{P_3} + \frac{r}{2}\} \subset \Gamma_{\rm shock}[Q, P_1]$. Then, by (9.4.41), we have

$$\mathrm{dist}(\Gamma_{\rm shock} \cap \{T > T_{P_3} + \frac{r}{2}\},\, \Gamma_{\rm wedge}) \geq \frac{1}{C}. \tag{9.4.42}$$

On the other hand,

$$\Gamma_{\rm wedge} \setminus B_r(P_3) = \{(0, T) \,:\, T_{P_3} + r \leq T < T_{P_4}\}.$$

Thus

$$\mathrm{dist}(\Gamma_{\rm shock} \cap \{T \leq T_{P_3} + \frac{r}{2}\},\, \Gamma_{\rm wedge} \setminus B_r(P_3)) \geq \frac{r}{2}.$$

Combining this with (9.4.42), we conclude the proof. □

9.4.4 Lower bound for the distance between $\Gamma_{\rm shock}$ and $\Gamma_{\rm wedge}$ away from P_0 and P_3.

The lower bounds in Propositions 9.4.5 and 9.4.8 are obtained for admissible solutions with the wedge angles $\theta_{\rm w} \in [\theta_{\rm w}^*, \frac{\pi}{2})$, and the bounds depend on $\theta_{\rm w}^*$. However, for some applications below, we need to obtain the lower bounds independent of $\theta_{\rm w}^*$, which hold for admissible solutions with $\theta_{\rm w} \in (\theta_{\rm w}^{\rm s}, \frac{\pi}{2})$.

We do not assume that $u_1 < c_1$, which implies that our estimate has to be made away from P_3, as we have discussed earlier. Moreover, for $\theta_{\rm w} = \theta_{\rm w}^{\rm s}$, $\Gamma_{\rm shock}$ and $\Gamma_{\rm wedge}$ meet at P_0, which implies that our estimate has to be made away from P_0. Then we obtain the following estimate:

Proposition 9.4.9. *Fix $\rho_1 > \rho_0 > 0$ and $\gamma > 1$. For every $r \in (0, \frac{r_0}{10})$, there exists $C_r > 0$ depending only on $(\rho_0, \rho_1, \gamma, r)$ such that*

$$\mathrm{dist}\,(\Gamma_{\rm shock},\, \Gamma_{\rm wedge} \setminus (B_r(P_0) \cup B_r(P_3))) \geq \frac{1}{C_r} \tag{9.4.43}$$

for any admissible solution of **Problem 2.6.1** *with the wedge angle $\theta_{\rm w} \in (\theta_{\rm w}^{\rm s}, \frac{\pi}{2})$.*

Proof. In this proof, the universal constants C and C_k are positive and depend only on $(\rho_0, \rho_1, \gamma, r)$. It suffices to consider $r \in (0, \frac{r^*}{10}]$ for $r^* = \min\{r_0, \hat{r}_0\}$, where $\hat{r}_0$ is defined in Remark 9.4.6. Fix such a constant r.

Let φ be an admissible solution for a wedge angle $\theta_{\rm w} \in (\theta_{\rm w}^{\rm s}, \frac{\pi}{2})$. We use coordinates (S, T) with basis $\{\boldsymbol{\nu}_{\rm w}, \boldsymbol{\tau}_{\rm w}\}$ and the origin at P_3, as in the proof of Proposition 9.4.5. The proof consists of four steps.

1. We show that there exist $C, C_1 > 0$ such that, if $\mathrm{dist}(P_1, L_{\Gamma_{\rm wedge}}) \leq \frac{1}{C_1}$, then

$$\begin{aligned} &|P_0 - P_4| \leq \frac{r}{20}, \\ &\exists\ \hat{Q} \in \Gamma_{\rm shock}\ \textit{with}\ \mathrm{dist}(\hat{Q},\, L_{\Gamma_{\rm wedge}}) \geq \frac{1}{C}\ \textit{and}\ T_{\hat{Q}} \geq T_{P_0} - \frac{3r}{4}. \end{aligned} \tag{9.4.44}$$

To prove (9.4.44), we first define the set:

$$\mathcal{A} = \{\theta_{\rm w} \in [\theta_{\rm w}^{\rm s}, \frac{\pi}{2}) \ : \ |P_0 - P_4| \geq \frac{r}{20}\} \cup \{\frac{\pi}{2}\}.$$

Then, using the continuous dependence of points P_0 and P_4 on $\theta_{\rm w} \in [\theta_{\rm w}^{\rm d}, \frac{\pi}{2})$, which holds by Lemma 7.5.9(i)–(ii), we conclude that $\mathcal{A}$ is closed. Also, $\mathcal{A}$ includes only the supersonic wedge angles (since $P_0 = P_4$ for the sonic angle), so that $\text{dist}(P_1, L_{\Gamma_{\rm wedge}}) > 0$ for all $\theta_{\rm w} \in \mathcal{A}$. Using the continuous dependence of $\text{dist}(P_1, L_{\Gamma_{\rm wedge}})$ on $\theta_{\rm w} \in [\theta_{\rm w}^{\rm d}, \frac{\pi}{2}]$, we have

$$\text{dist}(P_1, L_{\Gamma_{\rm wedge}}) \geq \frac{2}{C_1} \qquad \text{for all } \theta_{\rm w} \in \mathcal{A}.$$

Also, define

$$\mathcal{B} := \{\theta_{\rm w} \in [\theta_{\rm w}^{\rm s}, \frac{\pi}{2}] \ : \ |P_1 - P_4| \geq \frac{r}{20}\}.$$

Using the continuous dependence of points P_1 and P_4 on $\theta_{\rm w} \in [\theta_{\rm w}^{\rm d}, \frac{\pi}{2}]$, which holds by Lemma 7.5.9(ii), we conclude that $\mathcal{B}$ is closed. Then, noting that $\text{dist}(P_1, L_{\Gamma_{\rm wedge}}) > 0$ for any $\theta_{\rm w} \in \mathcal{B}$ and using the continuous dependence of $\text{dist}(P_1, L_{\Gamma_{\rm wedge}})$ on $\theta_{\rm w} \in [\theta_{\rm w}^{\rm d}, \frac{\pi}{2}]$, we obtain (possibly increasing C_1):

$$\text{dist}(P_1, L_{\Gamma_{\rm wedge}}) \geq \frac{2}{C_1} \qquad \text{for all } \theta_{\rm w} \in \mathcal{B}.$$

If $\theta_{\rm w} \notin \mathcal{A} \cup \mathcal{B}$, then $|P_1 - P_4| \leq \frac{r}{20}$. We can apply Lemma 9.4.2 with $\frac{r}{2}$ instead of r to obtain the existence of $\hat{Q} \in \overline{\Gamma_{\rm shock} \cap B_{r/2}(P_4)}$ with $\text{dist}(\hat{Q}, L_{\Gamma_{\rm wedge}}) \geq \frac{1}{C}$ from (9.4.32). Moreover, using that $|P_0 - P_4| \leq \frac{r}{20}$, we see that $\hat{Q} \in \overline{\Gamma_{\rm shock} \cap B_{3r/4}(P_0)}$, so that $T_{\hat{Q}} \geq T_{P_0} - \frac{3r}{4}$. Thus, (9.4.44) holds in this case.

That is, we have shown that, if $\text{dist}(P_1, L_{\Gamma_{\rm wedge}}) \leq \frac{1}{C_1}$ for some $\theta_{\rm w}$, then $\theta_{\rm w} \notin \mathcal{A} \cup \mathcal{B}$ so that (9.4.44) holds for such $\theta_{\rm w}$.

2. Next we apply Corollary 9.4.3 to obtain $Q \in \Gamma_{\rm shock}$ such that (9.4.35) holds with C_r depending only on $(\rho_0, \rho_1, \gamma, r)$.

3. If $\theta_{\rm w}$ is such that

$$\text{dist}(P_1, L_{\Gamma_{\rm wedge}}) \geq \frac{1}{C_1} \tag{9.4.45}$$

for C_1 from (9.4.44), we argue as in the proof of Proposition 9.4.8 by using (9.4.45) instead of (9.4.37). Note that, in this argument, we use the lower bounds in (9.4.35), (9.4.40), and (9.4.45), which depend only on $(\rho_0, \rho_1, \gamma, r)$, where, for (9.4.39), $\frac{1}{C} = \hat{r}_0$, from Remark 9.4.6 by (9.4.38). Thus, we obtain the estimate of Proposition 9.4.8 with C_r depending only on $(\rho_0, \rho_1, \gamma, r)$. This implies (9.4.43).

4. If $\theta_{\rm w}$ is such that

$$\text{dist}(P_1, L_{\Gamma_{\rm wedge}}) \leq \frac{1}{C_1}, \tag{9.4.46}$$

then (9.4.44) holds. Now we can apply Lemma 9.4.1 with $Q_1 = \hat{Q}$ from (9.4.44) and $Q_2 = Q$ from (9.4.35), which holds, as we have shown in Step 2. Moreover, since $r < \frac{r^*}{10} \leq \frac{\hat{r}_0}{10}$, we use (9.4.35), (9.4.44), and Remark 9.4.6 to estimate

$$|\hat{Q} - Q| \geq T_{\hat{Q}} - T_Q \geq T_{P_0} - \frac{3r}{4} - T_{P_3} - \frac{r}{2} \geq \hat{r}_0 - \frac{5r}{4} \geq \frac{\hat{r}_0}{2}. \tag{9.4.47}$$

Thus, in the application of Lemma 9.4.1 with $Q_1 = \hat{Q}$ and $Q_2 = Q$, we use $a > 0$ that is the smallest of the lower bounds in (9.4.35), (9.4.44), and (9.4.47), so that a depends only on $(\rho_0, \rho_1, \gamma, r)$. We then obtain

$$\mathrm{dist}(\Gamma_{\mathrm{shock}}[Q, \hat{Q}], \Gamma_{\mathrm{wedge}}) \geq \frac{1}{C}. \tag{9.4.48}$$

Using Corollary 8.2.14(i) and noting that $T_Q < T_{\hat{Q}}$ by (9.4.47), we have

$$\Gamma_{\mathrm{shock}}[Q, \hat{Q}] = \Gamma_{\mathrm{shock}} \cap \{T_Q < T < T_{\hat{Q}}\}.$$

Then, from (9.4.35) and (9.4.44),

$$\Gamma_{\mathrm{shock}} \cap \{T_{P_3} + \frac{r}{2} \leq T \leq T_{P_0} - \frac{3r}{4}\} \subset \Gamma_{\mathrm{shock}}[Q, \hat{Q}],$$

which, by (9.4.48), implies

$$\mathrm{dist}(\Gamma_{\mathrm{shock}} \cap \{T_{P_3} + \frac{r}{2} \leq T \leq T_{P_0} - \frac{3r}{4}\},\ \Gamma_{\mathrm{wedge}}) \geq \frac{1}{C}. \tag{9.4.49}$$

On the other hand,

$$\Gamma_{\mathrm{wedge}} \setminus (B_r(P_3) \cup B_r(P_4)) = \{(0, T)\ :\ T_{P_3} + r \leq T \leq T_{P_4} - r\}.$$

Thus, using that $T_{P_0} \geq T_{P_4}$,

$$\mathrm{dist}\big(\Gamma_{\mathrm{shock}} \setminus \{T_{P_3} + \frac{r}{2} \leq T \leq T_{P_0} - \frac{3r}{4}\},\ \Gamma_{\mathrm{wedge}} \setminus (B_r(P_3) \cup B_r(P_4))\big) \geq \frac{r}{4}.$$

Combining this with (9.4.49), we conclude the proof. □

9.5 UNIFORM POSITIVE LOWER BOUND FOR THE DISTANCE BETWEEN Γ_{shock} AND THE SONIC CIRCLE OF STATE (1)

In this section, we keep the notation that $\mathcal{O}_1 = (u_1, 0)$ as the center of the sonic circle of state (1) and recall that $P_3 = \mathbf{0}$. The universal constant C depends only on the data, *i.e.*, on (ρ_0, ρ_1, γ), unless otherwise specified. We first prove a preliminary fact.

Lemma 9.5.1. *For every non-empty compact set* $K \subset B_{c_1}(\mathcal{O}_1) \setminus \{P_3\}$, *there exists* $C = C(K) > 0$ *such that* $K \cap \overline{\Lambda^{(\theta_w)}} \subset \Omega(\varphi)$ *and*

$$\inf_{K \cap \overline{\Lambda^{(\theta_w)}}} (\varphi_1 - \varphi) \geq \frac{1}{C(K)}, \tag{9.5.1}$$

for any wedge angle $\theta_w \in (\theta_w^s, \frac{\pi}{2})$ *satisfying* $K \cap \overline{\Lambda^{(\theta_w)}} \neq \emptyset$ *and any admissible solution* φ *of* **Problem 2.6.1** *with the wedge angle* θ_w.

Proof. Since $K \subset B_{c_1}(\mathcal{O}_1)$, the fact that $K \cap \overline{\Lambda^{(\theta_w)}} \subset \Omega(\varphi)$ follows directly from (8.1.2). Thus, it suffices to prove (9.5.1).

On the contrary, if (9.5.1) does not hold, then, since $\varphi \leq \varphi_1$ in Ω by (8.1.5) for any admissible solution, there exist a compact $K \subset B_{c_1}(u_1, 0)$, a sequence of admissible solutions $\{\varphi^{(i)}\}$ of **Problem 2.6.1** for the wedge angles $\{\theta_w^{(i)}\} \subset (\theta_w^s, \frac{\pi}{2})$ extended to $\Lambda^{(\theta_w^{(i)})}$ as in Definition 2.6.2, and a corresponding sequence of points $\{Q^{(i)}\} \subset K \cap \overline{\Lambda^{(\theta_w^{(i)})}}$ such that

$$(\varphi_1 - \varphi^{(i)})(Q^{(i)}) \to 0. \tag{9.5.2}$$

Passing to a subsequence (without change of the index), we have

$$\theta_w^{(i)} \to \theta_w^{(\infty)} \in [\theta_w^s, \frac{\pi}{2}], \qquad Q^{(i)} \to Q^{(\infty)} \in K \cap \overline{\Lambda^{(\infty)}}.$$

For each i, choose and fix a Lipschitz extension of $\varphi^{(i)}$ from $\overline{\Lambda^{(\theta_w^{(i)})}}$ to $\mathbb{R}^2$ as in Remark 9.2.4, *i.e.*, satisfying (9.2.18) for any compact set K in $\mathbb{R}^2$ with the constants independent of i. Passing to a further subsequence and using Corollary 9.2.5, we obtain that, for the extended functions $\varphi^{(i)}$,

$$\varphi^{(i)} \to \varphi^{(\infty)} \qquad \text{uniformly on compact subsets of } \mathbb{R}^2,$$

where $\varphi^{(\infty)}$ is a weak solution of **Problem 2.6.1** in $\Lambda^{(\infty)}$. Moreover, $\varphi^{(\infty)}$ in $\Lambda^{(\infty)}$ is of the structure described in Corollary 9.2.5. Using the uniform convergence and equicontinuity of the extended function $\varphi^{(i)}$, we find from (9.5.2) that

$$\varphi^{(\infty)}(Q^{(\infty)}) = \varphi_1(Q^{(\infty)}). \tag{9.5.3}$$

Since the compact set $K \subset B_{c_1}(u_1, 0) \setminus \{\mathbf{0}\}$, then $K \subset B_{c_1 - 2\epsilon}(u_1, 0) \setminus \overline{B_{2\epsilon}(\mathbf{0})}$ for some $\epsilon > 0$ satisfying $c_1 > 5\varepsilon$. Denote

$$\mathcal{D}_\epsilon := \left(B_{c_1-\epsilon}(u_1, 0) \setminus \overline{B_\epsilon(\mathbf{0})}\right) \cap \Lambda^{(\infty)}.$$

Since $Q^{(\infty)} \in K \cap \overline{\Lambda^{(\infty)}} \subset \left(B_{c_1-2\epsilon}(u_1, 0) \setminus \overline{B_{2\epsilon}(\mathbf{0})}\right) \cap \overline{\Lambda^{(\infty)}}$, it follows from the form of $\Lambda^{(\infty)}$ that

$$\left(B_{c_1-2\epsilon}(u_1, 0) \setminus \overline{B_{2\epsilon}(\mathbf{0})}\right) \cap \Lambda^{(\infty)} \neq \emptyset,$$

which implies, after possibly reducing ε, that $\mathcal{D}_\epsilon \neq \emptyset$. From this, we infer that $\Gamma_{\text{wedge}}(\mathcal{D}_\epsilon) \neq \emptyset$, where we have used and will hereafter use the notations in (8.1.18). Indeed, since $\theta_{\rm w} \in (0, \frac{\pi}{2})$, it follows from (7.5.6) that the nearest point R to $\mathcal{O}_1 = (u_1, 0)$ on line $\{\xi_2 = \xi_1 \tan\theta_{\rm w}\}$ lies in the relative interior of $P_3\mathcal{O}_2 \subset \Gamma_{\text{wedge}}$, where the last inclusion follows from Lemma 7.4.8 and the definition of P_4. Then, reducing ε, we conclude that $R \notin \overline{B_\epsilon(\mathbf{0})}$ so that $R \in \Gamma_{\text{wedge}}(\mathcal{D}_\epsilon)$.

Note that (9.2.29) implies that $\overline{\mathcal{D}}_{\epsilon/4} \subset \Omega^{(\infty)} \cup \Gamma^{(\infty),0}_{\text{sym}} \cup \Gamma^{(\infty),0}_{\text{wedge}}$. Then, from (9.2.24) and (9.2.26), we infer that $\varphi^{(\infty)} \in C^3\left(\overline{\mathcal{D}_{\epsilon/2}}\right)$ and that equation (2.2.8) is strictly elliptic for $\varphi^{(\infty)}$ in $\overline{\mathcal{D}_{\epsilon/2}}$. Since $\Gamma_{\text{wedge}}(\mathcal{D}_\epsilon) \neq \emptyset$ (as shown above) and $\Gamma_{\text{wedge}}(\mathcal{D}_\epsilon) \subset \Gamma_{\text{wedge}}(\mathcal{D}_{\epsilon/2})$, then $\Gamma_{\text{wedge}}(\mathcal{D}_{\epsilon/2}) \neq \emptyset$. Now (8.1.19) in Lemma 8.1.9 implies that $\varphi_1 > \varphi^{(\infty)}$ on $\mathcal{D}_{\epsilon/2} \cup \Gamma_{\text{wedge}}(\mathcal{D}_{\epsilon/2}) \cup \Gamma_{\text{sym}}(\mathcal{D}_{\epsilon/2})$. Since $Q^{(\infty)} \in \overline{\mathcal{D}_\epsilon} \subset \mathcal{D}_{\epsilon/2} \cup \Gamma_{\text{wedge}}(\mathcal{D}_{\epsilon/2}) \cup \Gamma_{\text{sym}}(\mathcal{D}_{\epsilon/2})$, we arrive at a contradiction to (9.5.3). $\square$

Now we prove the main technical result of this section. In order to clarify the conditions for this, we note that, if $u_1 < c_1$, then $B_{c_1}(\mathcal{O}_1) \cap \Lambda(\theta_{\rm w}) \neq \emptyset$ for any $\theta_{\rm w} \in (0, \frac{\pi}{2}]$. However, if $u_1 \geq c_1$, $B_{c_1}(\mathcal{O}_1) \cap \Lambda(\theta_{\rm w}) = \emptyset$ for all $\theta_{\rm w} \in [\arcsin(\frac{c_1}{u_1}), \frac{\pi}{2}]$, and $B_{c_1}(\mathcal{O}_1) \cap \Lambda(\theta_{\rm w}) \neq \emptyset$ for any $\theta_{\rm w} \in (0, \arcsin(\frac{c_1}{u_1}))$. Now we consider the case that $B_{c_1-\varepsilon_0}(\mathcal{O}_1) \cap \Lambda(\theta_{\rm w}) \neq \emptyset$ for some small $\varepsilon_0 > 0$. Then, when $c_1 - \varepsilon_0 \leq u_1$, we have

$$B_{c_1-\varepsilon_0}(\mathcal{O}_1) \cap \Lambda(\theta_{\rm w}) \neq \emptyset \qquad \text{for all } \theta_{\rm w} \in (0, \arcsin(\frac{c_1-\varepsilon_0}{u_1})),$$

$$B_{c_1-\varepsilon_0}(\mathcal{O}_1) \cap \Lambda(\theta_{\rm w}) = \emptyset \qquad \text{for all } \theta_{\rm w} \in [\arcsin(\frac{c_1-\varepsilon_0}{u_1}), \frac{\pi}{2}].$$

We also recall that u_1 and c_1 are determined by (ρ_0, ρ_1, γ).

Lemma 9.5.2. *For any $\varepsilon_0 \in (0, \frac{c_1}{2})$, there exists $C > 0$ such that, if $\theta_{\rm w} \in (\theta_{\rm w}^{\rm s}, \frac{\pi}{2})$ satisfies $B_{c_1-\varepsilon_0} \cap \Lambda(\theta_{\rm w}) \neq \emptyset$, and φ is a corresponding admissible solution of* **Problem 2.6.1** *with the wedge angle $\theta_{\rm w}$, then*

$$\text{dist}(\Gamma_{\text{shock}}, B_{c_1}(\mathcal{O}_1)) \geq \frac{1}{C}.$$

Proof. In this proof, C denotes a universal constant that depends only on $(\rho_0, \rho_1, \gamma, \varepsilon_0)$. Denote

$$d := \text{dist}(\Gamma_{\text{shock}}, B_{c_1}(\mathcal{O}_1)). \tag{9.5.4}$$

Then $d > 0$ by (8.1.2); see also Remark 8.1.6. We now show that d has a positive lower bound $d \geq \frac{1}{C}$ for any admissible solution φ of **Problem 2.6.1**. We divide the proof into five steps.

1. We first note that $B_{c_1-\varepsilon_0} \cap \Lambda(\theta_{\rm w}) \neq \emptyset$ implies

$$B_{c_1}(\mathcal{O}_1) \cap \Gamma_{\text{wedge}} \neq \emptyset. \tag{9.5.5}$$

Indeed, from (7.5.6), we obtain that the nearest point to $\mathcal{O}_1$ on $\{\xi_2 = \xi_1 \tan\theta_{\rm w}\}$ lies within segment $P_3\mathcal{O}_2$. This and $B_{c_1-\varepsilon_0} \cap \Lambda(\theta_{\rm w}) \neq \emptyset$ imply (9.5.5).

For $\varepsilon \in (0, \varepsilon_0)$ and $\delta > -c_1$, denote

$$\mathcal{D}_{\varepsilon,\delta} := \big(B_{c_1+\delta}(\mathcal{O}_1) \setminus \overline{B_{c_1-\varepsilon}(\mathcal{O}_1)}\big) \cap \Omega.$$

Using (8.1.12) and (9.5.5), we find that, if $\varepsilon \in (0, \varepsilon_0)$, and d is from (9.5.4),

$$\mathcal{D}_{\varepsilon,d} \neq \emptyset, \quad \partial\mathcal{D}_{\varepsilon,d} \cap \Gamma_{\text{wedge}} \neq \emptyset, \quad \partial\mathcal{D}_{\varepsilon,d} \cap \Gamma_{\text{shock}} \neq \emptyset. \tag{9.5.6}$$

Furthermore, if $\varepsilon \in (0, \varepsilon_0)$ and $\delta \in (-\varepsilon, d)$, then

$$\mathcal{D}_{\varepsilon,\delta} \neq \emptyset, \quad \partial\mathcal{D}_{\varepsilon,\delta} \cap \Gamma_{\text{wedge}} \neq \emptyset, \quad \partial\mathcal{D}_{\varepsilon,\delta} \cap \Gamma_{\text{shock}} = \emptyset, \tag{9.5.7}$$

$$\mathcal{D}_{\varepsilon,\delta} = \big(B_{c_1+\delta}(\mathcal{O}_1) \setminus \overline{B_{c_1-\varepsilon}(\mathcal{O}_1)}\big) \cap \Lambda, \tag{9.5.8}$$

where we have used (8.1.2) and (9.5.4) to obtain (9.5.8). Thus, $\varphi \in C^3(\mathcal{D}_{\varepsilon,d}) \cap C^1(\overline{\mathcal{D}_{\varepsilon,d}})$ by (8.1.3). Note also that $\mathcal{O}_1 \notin \overline{\Lambda}$ implies that $\mathcal{O}_1 \notin \overline{\mathcal{D}_{\varepsilon,d}}$. Then we can write equation (2.2.8) for φ in $\mathcal{D}_{\varepsilon,d}$ as in §6.3, with respect to state (1). That is, for $(\hat{u}, \hat{v}) = (u_1, 0)$, $\varphi_{\text{un}} = \varphi_1$, and $c_{\text{un}} = c_1$, we define function (6.3.1), coordinates (6.3.3) and (6.3.5), and rewrite equation (2.2.8) in $\mathcal{D}_{\varepsilon,d}$ as (6.3.6)–(6.3.7). Also, in the (x, y)–coordinates with $(\hat{u}, \hat{v}) = (u_1, 0)$ and $c_{\text{un}} = c_1$, we use (9.5.8) to obtain

$$\mathcal{D}_{\varepsilon,\delta} = \Lambda \cap \{-\delta < x < \varepsilon\} \qquad \text{for } \varepsilon \in (0, \varepsilon_0) \text{ and } \delta \in (-\varepsilon, d]. \tag{9.5.9}$$

Note that, by Corollary 8.1.10, function ψ defined by (6.3.1) with $\varphi_{\text{un}} = \varphi_1$ satisfies that $\psi < 0$ in $\mathcal{D}_{\varepsilon,d}$. Since it is more convenient to work with a positive function, define

$$w = -\psi = \varphi_1 - \varphi.$$

Then equation (6.3.6) in terms of w takes the form:

$$\mathcal{N}(w) = 0, \tag{9.5.10}$$

where

$$\begin{aligned}\mathcal{N}(w) := &\big(2x + (\gamma+1)w_x + O_1^-\big)w_{xx} + O_2^- w_{xy} \\ &+ \big(\frac{1}{c_1} + O_3^-\big)w_{yy} - (1 + O_4^-)w_x + O_5^- w_y\end{aligned} \tag{9.5.11}$$

with $O_k^-(Dw, w, x) := \mathcal{O}_k(-Dw, -w, x)$ for $k = 1, \dots, 5$, where $\mathcal{O}_k$ are defined by (6.3.7). Note that the plus sign $(+)$ of term $(\gamma+1)w_x$ in the coefficient of w_{xx} makes equation (9.5.10) very different from equation (6.3.6), as we will see below. Furthermore, separating the zero-order terms, we have

$$\begin{aligned}O_1^-(Dw, w, x) &= \frac{\gamma-1}{c_1}w + \widehat{O_1^-}(Dw, x), \\ O_4^-(Dw, w, x) &= \frac{\gamma-1}{c_1(c_1 - x)}w + \widehat{O_4^-}(Dw, x).\end{aligned} \tag{9.5.12}$$

Denote

$$\begin{aligned}\mathcal{N}_1(V) := &(2x + (\gamma+1)V_x + \widehat{O_1^-} + \frac{\gamma-1}{c_1}w)V_{xx} + O_2^- V_{xy} \\ &+ (\frac{1}{c_1} + O_3^-)V_{yy} - (1 + O_4^-)V_x + O_5^- V_y,\end{aligned} \tag{9.5.13}$$

where $\widehat{O_1^-} = \widehat{O_1^-}(DV, x)$ and $O_k^- = O_k^-(DV, V, x)$ for $k = 2, \dots, 5$. The only difference between $\mathcal{N}_1(V)$ and $\mathcal{N}(V)$ is term $\frac{\gamma-1}{c_1}w$ in the coefficient of V_{xx} in $\mathcal{N}_1(V)$. By (9.5.12), the corresponding term in $\mathcal{N}(V)$ is $\frac{\gamma-1}{c_1}V$.

2. Now we show the following fact:

Claim 9.5.3. *Let $\varepsilon \in (0, \varepsilon_0)$ and $\delta \in [0, d]$, and let $U \in C^2(\mathcal{D}_{\varepsilon,\delta})$ be a subsolution of $\mathcal{N}_1(U) \geq 0$ in $\mathcal{D}_{\varepsilon,\delta}$. Assume that U is independent of y, i.e., $U = U(x)$, and that $U_x > 0$ and $U_{xx} > 0$ in $\mathcal{D}_{\varepsilon,\delta}$. Then $U - w$ cannot attain a positive maximum in the interior of $\mathcal{D}_{\varepsilon,\delta}$.*

We first note that $\mathcal{N}(w) = \mathcal{N}_1(w)$ so that

$$\mathcal{N}_1(w) = 0 \qquad \text{in } \mathcal{D}_{\varepsilon,\delta}.$$

For the sake of brevity, we write (9.5.13) as

$$\mathcal{N}_1(V) = \sum_{i,j=1}^{2} a_{ij} D_{ij} V + \sum_{i=1}^{2} a_i D_i V,$$

where $a_{12} = a_{21}$ and $(D_1, D_2) = (\partial_x, \partial_y)$. Note explicitly that

$$\begin{aligned}a_{11} &= a_{11}(DV, x, y) = 2x + (\gamma+1)V_x + \widehat{O_1^-}(DV, x) + \frac{\gamma-1}{c_1}w(x, y), \\ a_1 &= a_1(DV, V, x) = -1 - O_1^-(DV, V, x) = -1 - \frac{\gamma-1}{c_1(c_1 - x)}V - \widehat{O_4^-}(DV, x).\end{aligned} \tag{9.5.14}$$

We now write $a_{ij}(V)$, representing $a_{ij}(DV, V, x, y)$, and do the same for a_i. Since $\mathcal{D}_{\varepsilon,\delta} \subset \Omega$, equation (9.5.10) is elliptic on w in $\mathcal{D}_{\varepsilon,\delta}$, which implies

$$[a_{ij}(w)]_{i,j=1}^2 \quad \text{is positive definite in } \mathcal{D}_{\varepsilon,\delta}.$$

Assume that $U - w$ attains a positive maximum at $Q \in \mathcal{D}_{\varepsilon,\delta}$. Then

$$Dw = DU, \quad D^2 w \geq D^2 U \qquad \text{at } Q.$$

Using the explicit form of a_{11} and a_1 in (9.5.13), we have

$$a_{11}(w) = a_{11}(U), \quad a_1(w) = a_1(U) + \frac{\gamma-1}{c_1(c_1 - x)}(U - w) \qquad \text{at } Q,$$

where we have used (9.5.14) to obtain the second equality. Also, since $U = U(x)$,

$$U_y = U_{xy} = U_{yy} = 0.$$

Then we obtain that, at Q,

$$\begin{aligned}
0 &\le \mathcal{N}_1(U) - \mathcal{N}_1(w) \\
&= \sum_{i,j=1}^{2} a_{ij}(w)D_{ij}(U-w) + \sum_{i=1}^{2} a_i(w)D_i(U-w) - \frac{\gamma-1}{c_1(c_1-x)}(U-w)U_x \\
&= \sum_{i,j=1}^{2} a_{ij}(w)D_{ij}(U-w) - \frac{\gamma-1}{c_1(c_1-x)}(U-w)U_x \\
&=: I_1 - I_2 < 0,
\end{aligned}$$

where the last inequality is proved as follows: $I_1 \le 0$ because $[a_{ij}(w)] > 0$ (positive definite) and $D^2(w-U)(Q) \ge 0$, and $I_2 > 0$ because $(U-w)(Q) > 0$, $U_x > 0$, and $x < \varepsilon < \varepsilon_0 < c_1$ in $\mathcal{D}_{\varepsilon,\delta}$. Therefore, we arrive at a contradiction, which implies Claim 9.5.3.

3. Now we show a lower bound of $w := \varphi_1 - \varphi$.

Claim 9.5.4. *Fix $\alpha \in (\frac{1}{2}, 1)$. There exist $\varepsilon \in (0, \varepsilon_0)$ and $A > 0$ depending only on $(\rho_0, \rho_1, \gamma, \alpha)$ such that*

$$(\varphi_1 - \varphi)(x,y) \ge Ax^{1+\alpha} \qquad \textit{in } \mathcal{D}_{\varepsilon,0}.$$

First, we note that $0 < x < \varepsilon$ in $\mathcal{D}_{\varepsilon,0}$ by (9.5.9). Then the function:

$$U(x,y) = U(x) = Ax^{1+\alpha}$$

satisfies $U \in C^\infty(\mathcal{D}_{\varepsilon,0}) \cap C^{1,\alpha}(\overline{\mathcal{D}_{\varepsilon,0}})$. We divide the proof into four sub-steps.

3.1. We first show that there exists $\varepsilon \in (0, \varepsilon_0)$ depending only on $(\rho_0, \rho_1, \gamma, \alpha)$ such that, for any $A \in (0,1)$, U is a subsolution of (9.5.13) in $\mathcal{D}_{\varepsilon,0}$.

Note that

$$U_x = (1+\alpha)Ax^\alpha > 0, \quad U_{xx} = \alpha(1+\alpha)Ax^{\alpha-1} > 0 \qquad \text{in } \mathcal{D}_{\varepsilon,0}.$$

Also, $w := \varphi_1 - \varphi > 0$ in $\mathcal{D}_{\varepsilon,0}$. Furthermore, $U_y = U_{xy} = U_{yy} = 0$. Then

$$\begin{aligned}
\mathcal{N}_1(U) &= \Big(2x + (\gamma+1)U_x + \widehat{O_1^-}(DU,x) + \frac{\gamma-1}{c_1}w\Big)U_{xx} \\
&\quad - \big(1 + O_4^-(DU,U,x)\big)U_x \\
&\ge \big(2x + (\gamma+1)U_x + \widehat{O_1^-}(DU,x)\big)U_{xx} - \big(1 + O_4^-(DU,U,x)\big)U_x \\
&= (1+\alpha)Ax^{\alpha-1}\Big(x(2\alpha - 1 + \frac{\widehat{O_1^-}}{x} - O_4^-) + (\gamma+1)(1+\alpha)Ax^\alpha\Big) \\
&\ge (1+\alpha)Ax^\alpha\Big(2\alpha - 1 + \frac{\widehat{O_1^-}}{x} - O_4^-\Big).
\end{aligned}$$

Using (6.3.7) and (9.5.12), and choosing $\varepsilon < 1$, we obtain that, in $\mathcal{D}_{\varepsilon,0}$,

$$|\widehat{O_1^-}(DU, x)| \le C(|DU|^2 + x^2), \quad |O_4^-(DU, U, x)| \le C(|DU| + |U| + x).$$

Here and hereafter, C denotes a universal constant that depends on $(\rho_0, \rho_1, \gamma, \alpha)$ only. Thus, from the explicit expressions of U and DU, and using $\alpha > \frac{1}{2}$, we conclude that, for any $A \in (0,1)$ and $\varepsilon > 0$ sufficiently small, depending only on $(\rho_0, \rho_1, \gamma, \alpha)$, when $x \in \mathcal{D}_{\varepsilon,0}$,

$$\frac{|\widehat{O_1^-}(DU, x)|}{x} \le C\big(x + A^2x^{2\alpha-1}\big) \le C\big(\varepsilon + \varepsilon^{2\alpha-1}\big) \le \frac{2\alpha-1}{4},$$
$$|O_4^-(DU, U, x)| \le CAx^\alpha \le \frac{2\alpha-1}{4}.$$

Thus, $\mathcal{N}_1(U) > 0$, *i.e.*, $U = Ax^{1+\alpha}$ is a subsolution of (9.5.13) in $\mathcal{D}_{\varepsilon,0}$. Therefore, by Claim 9.5.3, $Ax^{1+\alpha} - w$ cannot attain a positive maximum in $\mathcal{D}_{\varepsilon,0}$.

3.2. Since $\mathcal{D}_{\varepsilon,0} \subset B_{c_1}(u_1, 0)$, we use (9.5.9) to obtain

$$\partial\mathcal{D}_{\varepsilon,0} = \overline{\partial\mathcal{D}_{\varepsilon,0} \cap \Omega} \cup \overline{\partial\mathcal{D}_{\varepsilon,0} \cap \Gamma_{\text{wedge}}} \cup (\partial\mathcal{D}_{\varepsilon,0} \cap \Gamma_{\text{sym}})^0, \tag{9.5.15}$$

where $(\Gamma)^0$ denotes the relative interior of Γ which is a subset of a line or smooth curve. In order to obtain (9.5.15), we have used (7.5.7) in Lemma 7.5.10 to obtain that $\partial\mathcal{D}_{\varepsilon,0} \cap \overline{\Gamma_{\text{sonic}}} = \emptyset$.

Part $(\partial\mathcal{D}_{\varepsilon,0} \cap \Gamma_{\text{sym}})^0$ of decomposition (9.5.15) may be empty: Specifically, it is empty if $u_1 \ge c_1$. On the other hand, the first two parts in decomposition (9.5.15) are non-empty by (9.5.7).

We first consider $\overline{\partial\mathcal{D}_{\varepsilon,0} \cap \Gamma_{\text{wedge}}}$ in the $\boldsymbol{\xi}$–coordinates. Since $\partial_{\boldsymbol{\nu}}\varphi_1 = -u_1 \sin\theta_{\text{w}}$ and $\partial_{\boldsymbol{\nu}}\varphi = 0$ on Γ_{wedge} for the interior unit normal $\boldsymbol{\nu}$ on Γ_{wedge} to Ω, then

$$\partial_{\boldsymbol{\nu}} w = -u_1 \sin\theta_{\text{w}} \qquad \text{on } (\partial\mathcal{D}_{\varepsilon,0} \cap \Gamma_{\text{wedge}})^0.$$

Writing $U(x) = Ax^{1+\alpha}$ in the $\boldsymbol{\xi}$–coordinates, we have

$$U(\boldsymbol{\xi}) = A\Big(c_1 - \sqrt{(\xi_1 - u_1)^2 + {\xi_2}^2}\Big)^{1+\alpha}.$$

Since $\theta_{\text{w}}^{\text{d}} > 0$, there exists C such that $\text{dist}((u_1, 0), \Lambda(\theta_{\text{w}})) > \frac{1}{C}$ for any $\theta_{\text{w}} \in [\theta_{\text{w}}^{\text{d}}, \frac{\pi}{2}]$. Then there exists C_1 depending only on the data and α such that

$$\|U\|_{C^1(\overline{B_{c_1}(\mathcal{O}_1) \cap \Lambda(\theta_{\text{w}})})} \le C_1 A \qquad \text{for all } \theta_{\text{w}} \in [\theta_{\text{w}}^*, \frac{\pi}{2}].$$

Thus, choosing A such that it satisfies $0 < A < \frac{u_1 \sin\theta_{\text{w}}^{\text{d}}}{2C_1}$, and using $\theta_{\text{w}} \in [\theta_{\text{w}}^{\text{d}}, \frac{\pi}{2}]$, we find that, on $\overline{\partial\mathcal{D}_{\varepsilon,0} \cap \Gamma_{\text{wedge}}}$,

$$\partial_{\boldsymbol{\nu}}(U - w) \ge -\|U\|_{C^1(\overline{B_{c_1}(\mathcal{O}_1)})} + u_1 \sin\theta_{\text{w}} \ge u_1\big(-\frac{1}{2}\sin\theta_{\text{w}}^{\text{d}} + \sin\theta_{\text{w}}\big) > 0. \tag{9.5.16}$$

The strict inequality implies that the maximum of $U - w$ cannot be attained on $\partial\mathcal{D}_{\varepsilon,0} \cap \Gamma_{\rm wedge}$.

Next consider $(\partial\mathcal{D}_{\varepsilon,0} \cap \Gamma_{\rm sym})^0$ with the assumption that it is non-empty, since this step can be skipped if it is empty. Since $\partial_\nu\varphi = \partial_\nu\varphi_1 = 0$ on $\Gamma_{\rm sym} \subset \{\xi_2 = 0\}$, we can perform the reflection about the ξ_1–axis as in Remark 8.1.3 to obtain C^2–functions φ and φ_1 in the extended domain $\Omega^{\rm ext}$. Since $\mathcal{O}_1 = (u_1, 0) \in \{\xi_2 = 0\}$, coordinates (6.3.3) and (6.3.5) with center $(\hat{u}, \hat{v}) = (u_1, 0)$ satisfy that, if $P_+ = (\xi_1, \xi_2)$ is (x, y) in coordinates (6.3.5), $P_- := (\xi_1, -\xi_2)$ is $(x, 2\pi - y)$ in coordinates (6.3.5). Moreover, $\Omega^{\rm ext} \cap \{\xi_2 = 0\} \subset \{\xi_1 < 0\} \subset \{y = \pi\}$. Thus, the extended function $w = \varphi_1 - \varphi$ is in $C^2(\Omega^{\rm ext})$ in the (x, y)–coordinates. Moreover, $w(x, y) = w(x, 2\pi - y)$ in $\Omega^{\rm ext}$.

Now, from the explicit structure of equation (9.5.10) given by (6.3.6)–(6.3.7), it follows that the extended function w satisfies (9.5.10) in $\Omega^{\rm ext}$ expressed in the (x, y)–coordinates. Furthermore, similar to (9.5.9), the extended domain $\mathcal{D}_{\varepsilon,\delta}$ is $\Lambda^{\rm ext} \cap \{-\delta < x < \varepsilon\}$. Then Claim 9.5.3 and the assertion in Step 3.1 hold in the extended domain $\mathcal{D}_{\varepsilon,\delta}$. Now, the points of $(\partial\mathcal{D}_{\varepsilon,0} \cap \Gamma_{\rm sym})^0$ are interior points of the extended domain $\mathcal{D}_{\varepsilon,\delta}$. Therefore, the positive maximum of $Ax^{1+\alpha} - w$ cannot be attained on $(\partial\mathcal{D}_{\varepsilon,0} \cap \Gamma_{\rm sym})^0$.

Finally, we consider $\overline{\partial\mathcal{D}_{\varepsilon,0} \cap \Omega}$. From (9.5.9), we have

$$\overline{\partial\mathcal{D}_{\varepsilon,0} \cap \Omega} = \overline{\Omega \cap \{x = 0\}} \cup \overline{\Omega \cap \{x = \varepsilon\}}.$$

Since $w = \varphi_1 - \varphi \geq 0$ on Ω by (8.1.5), and $U(x) = Ax^{1+\alpha} = 0$ on $\{x = 0\}$, we have

$$w \geq U(x) \qquad \text{on } \overline{\Omega \cap \{x = 0\}}.$$

Furthermore, $\overline{\Omega \cap \{x = \varepsilon\}} = \overline{\Omega \cap \partial B_{c_1 - \varepsilon}(u_1, 0)} \subset B_{c_1}(u_1, 0)$. Also, reducing ε if necessary, depending only on (ρ_0, ρ_1, γ), we have $\mathbf{0} \notin \partial B_{c_1-\varepsilon}(u_1, 0)$, where we note that point $\mathbf{0}$ may be either inside $B_{c_1-\varepsilon}(u_1, 0)$ or outside $\overline{B_{c_1-\varepsilon}(u_1, 0)}$, depending on (ρ_0, ρ_1, γ). Thus, $\mathbf{0} \notin \{x = \varepsilon\}$. Then it follows from Lemma 9.5.1 that

$$w \geq \frac{1}{C} \qquad \text{on } \overline{\Omega \cap \{x = \varepsilon\}},$$

where C depends only on $(\rho_0, \rho_1, \gamma, \varepsilon)$. Moreover, since the choice of ε above depends only on $(\rho_0, \rho_1, \gamma, \alpha)$, then constant C depends only on $(\rho_0, \rho_1, \gamma, \alpha)$. Since $U = A\varepsilon^{1+\alpha}$ on $\{x = \varepsilon\}$, then reducing A, depending only on (C, ε, α), *i.e.*, on $(\rho_0, \rho_1, \gamma, \alpha)$, we find that $U \leq \frac{1}{C}$ on $\{x = \varepsilon\}$. That is,

$$w \geq U(x) \qquad \text{on } \overline{\Omega \cap \{x = \varepsilon\}}.$$

3.3. Therefore, we have discussed all the parts of decomposition (9.5.15). It follows that $w \geq U(x)$ on $\overline{\partial\mathcal{D}_{\varepsilon,0} \cap \Omega}$, and the maximum of $U - w$ cannot be attained on $(\partial\mathcal{D}_{\varepsilon,0} \cap \Gamma_{\rm wedge})^0 \cup (\partial\mathcal{D}_{\varepsilon,0} \cap \Gamma_{\rm sym})^0$. Also, by Step 3.1, the positive maximum of $U - w$ cannot be attained in $\mathcal{D}_{\varepsilon,0}$. Furthermore, $\overline{\partial\mathcal{D}_{\varepsilon,0} \cap \Omega} \neq \emptyset$. Thus, $w \geq U = Ax^{1+\alpha}$ on $\overline{\mathcal{D}_{\varepsilon,0}}$. Claim 9.5.4 is proved.

4. Now we further improve the lower bound of $w := \varphi_1 - \varphi$.

Claim 9.5.5. *Let $\varepsilon > 0$ be the constant from Claim 9.5.4 for $\alpha = \frac{3}{4}$ (we choose $\alpha = \frac{3}{4}$ from now on). Then there exist $d_0 \in (0,1)$, $\sigma \in (0,\varepsilon)$, and $k \in (0,1)$ depending only on $(\rho_0, \rho_1, \gamma, \varepsilon_0)$ such that, if d defined by (9.5.4) satisfies $d < d_0$, the corresponding function $w = \varphi_1 - \varphi$ satisfies*

$$w(x,y) \geq (x+d)^2 + k(x+d) \qquad \text{in } \mathcal{D}_{\sigma,d}.$$

We prove Claim 9.5.5 by following a procedure similar to that of the proof of Claim 9.5.4 and using the result of Claim 9.5.4. Since $\mathcal{O}_1 \notin \overline{\Lambda}$, $\mathcal{D}_{\sigma,d} \subset \Lambda$, and the (x,y)–coordinates are defined by (6.3.3) and (6.3.5) with $(\hat{u},\hat{v}) = \mathcal{O}_1$, then the function:

$$\tilde{U}(x,y) = \tilde{U}(x) = (x+d)^2 + k(x+d)$$

satisfies $\tilde{U} \in C^\infty(\overline{\mathcal{D}_{\sigma,d}})$. We divide the proof into two sub-steps.

4.1. We show first that, for sufficiently small $d_0, k > 0$ and $\sigma \in (0,\varepsilon)$ depending only on $(\rho_0, \rho_1, \gamma, \varepsilon_0)$, $\tilde{U}$ is a subsolution of (9.5.13) in $\mathcal{D}_{\sigma,d}$ if $d \leq d_0$.

Notice that, in $\mathcal{D}_{\sigma,d}$,

$$\tilde{U}_x = 2(x+d) + k > 0, \qquad \tilde{U}_{xx} = 2 > 0.$$

Also, $w > 0$ in $\mathcal{D}_{\sigma,d}$. Furthermore, $\tilde{U}_y = \tilde{U}_{xy} = \tilde{U}_{yy} = 0$. Then

$$\begin{aligned}
\mathcal{N}_1(\tilde{U}) &= \Big(2x + (\gamma+1)\tilde{U}_x + \widehat{O_1^-}(D\tilde{U},x) + \frac{\gamma-1}{c_1}w\Big)\tilde{U}_{xx} \\
&\quad - \big(1 + O_4^-(D\tilde{U},\tilde{U},x)\big)\tilde{U}_x \\
&\geq \Big(2x + (\gamma+1)\tilde{U}_x + \widehat{O_1^-}(D\tilde{U},x)\Big)\tilde{U}_{xx} - \big(1 + O_4^-(D\tilde{U},\tilde{U},x)\big)\tilde{U}_x \\
&= 2\big(1 + 2(\gamma+1)\big)(x+d) + k\big(2(\gamma+1) - 1\big) - 4d + 2\widehat{O_1^-} \\
&\quad - \big(2(x+d) + k\big)O_4^-.
\end{aligned}$$

Using (6.3.7) and (9.5.12) and noting that $k \in (0,1)$ and $|x| < 1$ in $\mathcal{D}_{\sigma,d}$, we obtain that, in $\mathcal{D}_{\sigma,d}$,

$$\begin{aligned}
|\widehat{O_1^-}(D\tilde{U},x)| &\leq C(|D\tilde{U}|^2 + x^2) \leq C\big((x+d)^2 + k^2\big), \\
|(2(x+d)+k)O_4^-(D\tilde{U},\tilde{U},x)| &\leq C(2(x+d)+k)(|D\tilde{U}| + |\tilde{U}| + x) \\
&\leq C\big((x+d)^2 + k^2\big),
\end{aligned}$$

where C denotes a universal constant that depends only on the data and ε_0. Then, using also $\gamma \geq 1$, we have

$$\mathcal{N}_1(\tilde{U}) \geq 10(x+d) + 3k - 4d - C\big((x+d)^2 + k^2\big) \qquad \text{in } \mathcal{D}_{\sigma,d}.$$

We choose

$$k := 4d_0, \quad \sigma := d_0. \tag{9.5.17}$$

Since $-d < x < \sigma$ in $\mathcal{D}_{\sigma,d}$ and $0 \le \mathbf{x} + d < 2d_0$, we have

$$\mathcal{N}_1(\tilde{U}) \ge (10 - Cd_0)(x+d) + 4(2 - Cd_0)d_0 \qquad \text{in } \mathcal{D}_{\sigma,d}.$$

Choosing d_0 small, depending on C above, *i.e.*, on (ρ_0, ρ_1, γ), we conclude that $\mathcal{N}_1(\tilde{U}) > 0$, which implies that $\tilde{U}$ is a subsolution of (9.5.13) in $\mathcal{D}_{\sigma,d}$. Therefore, by Claim 9.5.3, $\tilde{U} - w$ cannot attain a positive maximum in $\mathcal{D}_{\sigma,d}$.

4.2. It remains to estimate $\tilde{U} - w$ on $\partial\mathcal{D}_{\sigma,d}$. Similar to (9.5.15), we have

$$\partial\mathcal{D}_{\sigma,d} = \overline{\partial\mathcal{D}_{\sigma,d} \cap \Omega} \cup \overline{\partial\mathcal{D}_{\sigma,d} \cap \Gamma_{\text{wedge}}} \cup (\partial\mathcal{D}_{\sigma,d} \cap \Gamma_{\text{sym}})^0, \tag{9.5.18}$$

where we have used (7.5.7) in Lemma 7.5.10 and reduced σ if necessary depending only on the data to obtain $\overline{B_{c_1+\sigma}(\mathcal{O}_1)} \cap \overline{\Gamma_{\text{sonic}}} = \emptyset$, so that $\partial\mathcal{D}_{\sigma,d} \cap \overline{\Gamma_{\text{sonic}}} = \emptyset$.

Now all the parts of this decomposition, except $\overline{\partial\mathcal{D}_{\sigma,d} \cap \Omega}$, are considered similarly to the corresponding parts of (9.5.15) in Step 3.2 of the proof of Claim 9.5.4. We briefly comment on each part.

For $(\partial\mathcal{D}_{\sigma,d} \cap \Gamma_{\text{wedge}})^0$, write $\tilde{U}$ in the $\boldsymbol{\xi}$–coordinates as follows:

$$\tilde{U}(\boldsymbol{\xi}) = \Big(c_1 - \sqrt{(\xi_1 - u_1)^2 + \xi_2^2}\Big)^2 + k\Big(c_1 - \sqrt{(\xi_1 - u_1)^2 + \xi_2^2}\Big),$$

and use that $\operatorname{dist}((u_1, 0), \Lambda(\theta_{\text{w}})) > \frac{1}{C}$ for any $\theta_{\text{w}} \in [\theta_{\text{w}}^{\text{d}}, \frac{\pi}{2}]$. Then we have

$$\|\tilde{U}\|_{C^1(\overline{\mathcal{D}_{\sigma,d} \cap \Lambda(\theta_{\text{w}})})} \le C_1\big((\sigma + d)^2 + \sigma + d + k\big) \qquad \text{for all } \theta_{\text{w}} \in [\theta_{\text{w}}^{\text{d}}, \frac{\pi}{2}].$$

Thus, recalling (9.5.17) with $d_0 \in (0,1)$, we have

$$\|\tilde{U}\|_{C^1(\overline{\mathcal{D}_{\sigma,d} \cap \Lambda(\theta_{\text{w}})})} \le Cd_0.$$

Choosing d_0 small, we obtain that $\partial_{\boldsymbol{\nu}}(\tilde{U} - w) > 0$ on $(\partial\mathcal{D}_{\sigma,d} \cap \Gamma_{\text{wedge}})^0$, by an argument similar to (9.5.16). This implies that the maximum of $\tilde{U} - w$ cannot be attained on $(\partial\mathcal{D}_{\sigma,d} \cap \Gamma_{\text{wedge}})^0$. Points $(\partial\mathcal{D}_{\sigma,d} \cap \Gamma_{\text{sym}})^0$ can be considered as the interior points by even reflection with respect to the ξ_1–axis expressed in the (x, y)–coordinates. Thus, the positive maximum of $\tilde{U} - w$ cannot be attained on $(\partial\mathcal{D}_{\sigma,d} \cap \Gamma_{\text{sym}})^0$ by Step 4.1.

Then it remains to consider $\overline{\partial\mathcal{D}_{\sigma,d} \cap \Omega}$. From (9.5.9), we have

$$\overline{\partial\mathcal{D}_{\sigma,d} \cap \Omega} = \overline{\Lambda \cap \{x = -d\}} \cup \overline{\Lambda \cap \{x = \sigma\}}.$$

Since $w = \varphi_1 - \varphi \ge 0$ on Ω by (8.1.5), and $\tilde{U}(x) = 0$ on $\{x = -d\}$, we have

$$w \ge \tilde{U}(x) \qquad \text{on } \overline{\Omega \cap \{x = -d\}} = \Lambda \cap \{x = -d\}.$$

Next, using Claim 9.5.4 (with $\alpha = \frac{3}{4}$) and (9.5.17), in which we choose d_0 small, depending only on the data and ε_0, we conclude that, on $\overline{\Omega \cap \{x = \sigma\}} = \overline{\Lambda \cap \{x = \sigma\}}$,

$$w \ge A\sigma^{1+\frac{3}{4}} \ge 5\sigma^2 = \sigma^2 + k\sigma = \tilde{U}.$$

Thus, we have considered all the parts of decomposition (9.5.18). Now repeating the argument in Step 3.3 of the proof of Claim 9.5.4, we complete the proof of Claim 9.5.5.

5. Now we conclude the proof of Lemma 9.5.2. By (9.5.4), $\Gamma_{\text{shock}} \subset \mathbb{R}^2 \setminus B_{c_1+d}(\mathcal{O}_1)$, and there exists $Q \in \overline{\Gamma_{\text{shock}}}$ such that $Q \in \partial B_{c_1+d}(\mathcal{O}_1)$. Also, from (6.1.3) and the continuous dependence of the parameters of state (2) on $\theta_{\text{w}} \in [\theta_{\text{w}}^{\text{d}}, \frac{\pi}{2}]$, it follows that there exists $a > 0$ depending only on the data such that $\text{dist}(\mathcal{S}_1, B_{c_1}(\mathcal{O}_1)) \geq a$ for any $\theta_{\text{w}} \in [\theta_{\text{w}}^{\text{d}}, \frac{\pi}{2}]$. Thus, $\text{dist}(P_1, B_{c_1}(\mathcal{O}_1)) \geq a$, *i.e.*, either $d \geq a$ (in which case the lemma is proved) or $Q \neq P_1$. Point P_2 can be considered as a relative interior point of Γ_{shock} by reflection across the ξ_1–axis, *i.e.*, $Q \in \Gamma_{\text{shock}}^{\text{ext}}$, by using the notations in Remark 8.1.3. Moreover, since $\mathcal{O}_1 \in \{\xi_2 = 0\}$, $\Gamma_{\text{shock}}^{\text{ext}} \subset \mathbb{R}^2 \setminus B_{c_1+d}(\mathcal{O}_1)$. Then

$$\boldsymbol{\nu}_{\Gamma_{\text{shock}}}(Q) = \boldsymbol{\nu}_{\partial B_{c_1+d}(\mathcal{O}_1)}(Q) = \frac{Q\mathcal{O}_1}{|Q\mathcal{O}_1|}.$$

Using the polar coordinates (6.3.3) with $(\hat{u}, \hat{v}) = (u_1, 0) = \mathcal{O}_1$, we have

$$\partial_{\boldsymbol{\nu}} w(Q) = -\frac{1}{c_1 + d}\partial_r w(Q) = \frac{1}{c_1 + d}\partial_x w(Q). \tag{9.5.19}$$

Let k and d_0 be the constants in Claim 9.5.5, and let $d < d_0$ (otherwise, the lemma is proved). Since $Q \in B_{c_1+d}(\mathcal{O}_1) \cap \Gamma_{\text{shock}}^{\text{ext}}$, then $Q \in \{x = -d\}$ and $w(Q) = \varphi_1(Q) - \varphi(Q) = 0$. Thus, $w(Q) = \tilde{U}(Q)$, where $\tilde{U}(x, y) = (x + d)^2 + k(x + d)$ in the (x, y)–coordinates. Using Claim 9.5.5 and the fact that Q is in the relative interior of $\Gamma_{\text{shock}}^{\text{ext}}$, *i.e.*, $B_r(Q) \cap \{x > -d\} \subset \mathcal{D}_{\sigma,d}$ for some small $r > 0$, we see that $w \geq \tilde{U}$ in $B_r(Q) \cap \{x > -d\}$, so that

$$\partial_x w(Q) \geq \partial_x \tilde{U}(Q) = 2d + k \geq k,$$

which implies

$$\partial_{\boldsymbol{\nu}}(\varphi_1 - \varphi)(Q) \geq \frac{k}{c_1 + d_0},$$

on account of (9.5.19) and by recalling that $w = \varphi_1 - \varphi$. Lemma 6.1.3 applied with $\varphi_- = \varphi_1$ and $\varphi_+ = \varphi$ implies that

$$\text{dist}(Q, B_{c_1}(\mathcal{O}_1)) \geq \hat{d} := \hat{H}^{-1}(\frac{k}{c_1 + d_0}) - c_1, \tag{9.5.20}$$

where $\hat{H}^{-1}(\cdot) : [0, \infty) \mapsto [c_-, \infty)$ is the inverse function of $\hat{H}(\cdot)$. Then, by (6.1.17), $\hat{d} > 0$, since $k > 0$. Note that $\hat{H}(\cdot)$ in Lemma 6.1.3, applied with $\varphi_- = \varphi_1$ and $\varphi_+ = \varphi$, depends only on (ρ_0, ρ_1, γ). By the dependence of (k, d_0) in Claim 9.5.5, it follows that $\hat{d}$ depends only on $(\rho_0, \rho_1, \gamma, \varepsilon_0)$. Then (9.5.20) leads to the expected result. □

Now we show a uniform positive lower bound for the distance between Γ_{shock} and the sonic circle of state (1).

Proposition 9.5.6. *There exists $C > 0$ such that*

$$\operatorname{dist}(\Gamma_{\text{shock}}, B_{c_1}(\mathcal{O}_1)) \geq \frac{1}{C}$$

for any admissible solution of **Problem 2.6.1** *with* $\theta_{\text{w}} \in (\theta_{\text{w}}^{\text{s}}, \frac{\pi}{2})$.

Proof. Recall that u_1 and c_1 are determined by (ρ_0, ρ_1, γ). We consider three separate cases: $u_1 < c_1$, $u_1 = c_1$, and $u_1 > c_1$.

Case 1: $u_1 < c_1$. Let $\varepsilon_0 = \frac{1}{2}(c_1 - u_1)$. Since $\mathcal{O}_1 = (u_1, 0)$, it follows that $B_{c_1-\varepsilon_0}(\mathcal{O}_1) \cap \Lambda(\theta_{\text{w}}) \neq \emptyset$ for any $\theta_{\text{w}} \in (0, \frac{\pi}{2})$. Thus, the result in this case follows directly from Lemma 9.5.2 applied with $\varepsilon_0 = \frac{1}{2}(c_1 - u_1)$.

Case 2: $u_1 = c_1$. Then $B_{c_1}(\mathcal{O}_1) \subset \{\xi_1 > 0\}$ and $\mathbf{0} \in \partial B_{c_1}(\mathcal{O}_1)$. From Corollary 9.2.7 applied with $\varepsilon = \frac{|\bar{\xi}_1|}{2}$, there exists $\sigma > 0$ depending only on (ρ_0, ρ_1, γ) such that $\Gamma_{\text{shock}} \subset \{\xi_1 < -\frac{|\bar{\xi}_1|}{2}\}$ for any admissible solution for the wedge angle $\theta_{\text{w}} \in (\frac{\pi}{2} - \sigma, \frac{\pi}{2})$. Thus, $\operatorname{dist}(\Gamma_{\text{shock}}, B_{c_1}(\mathcal{O}_1)) \geq \frac{|\bar{\xi}_1|}{2}$ for any admissible solution with $\theta_{\text{w}} \in (\frac{\pi}{2} - \sigma, \frac{\pi}{2})$.

It remains to consider an admissible solution with the wedge angle $\theta_{\text{w}} \in (\theta_{\text{w}}^{\text{s}}, \frac{\pi}{2} - \sigma]$. Since $u_1 = c_1$, there exists ε_0, depending only on the data, such that $B_{c_1-\varepsilon_0}(\mathcal{O}_1) \cap \Lambda(\theta_{\text{w}}) \neq \emptyset$ for any $\theta_{\text{w}} \in (0, \frac{\pi}{2} - \sigma]$. Thus, applying Lemma 9.5.2 with ε_0 just determined, we conclude the proof in this case.

Case 3: $u_1 > c_1$. To motivate the ongoing argument, we note that

$$B_{c_1-\sigma}(\mathcal{O}_1) \cap \Lambda(\theta_{\text{w}}) \neq \emptyset \qquad \text{for all } \theta_{\text{w}} \in (0, \arcsin(\frac{c_1 - \sigma}{u_1})),$$

$$B_{c_1-\sigma}(\mathcal{O}_1) \cap \Lambda(\theta_{\text{w}}) = \emptyset \qquad \text{for all } \theta_{\text{w}} \in [\arcsin(\frac{c_1 - \sigma}{u_1}), \frac{\pi}{2}]$$

for any small $\sigma > 0$. Thus, if $\theta_{\text{w}} \geq \arcsin(\frac{c_1}{u_1})$, then $B_{c_1-\sigma}(\mathcal{O}_1) \cap \Lambda(\theta_{\text{w}}) = \emptyset$ for any $\sigma \geq 0$. That is, we cannot apply Lemma 9.5.2 in this case. Then, for any wedge angle $\theta_{\text{w}} \geq \arcsin(\frac{c_1-\sigma}{u_1})$ for $\sigma > 0$ determined below, we obtain the lower bound of $\operatorname{dist}(\Gamma_{\text{shock}}, B_{c_1}(\mathcal{O}_1))$ by the argument which follows:

Let $r_2^{(\theta_{\text{w}})} := |\mathcal{O}_2^{(\theta_{\text{w}})} P_4^{(\theta_{\text{w}})}|$. Then $r_2^{(\theta_{\text{w}})} > 0$ for each $\theta_{\text{w}} \in [\theta_{\text{w}}^{\text{s}}, \frac{\pi}{2}]$ by Lemma 7.4.8 and the definition of $P_4^{(\theta_{\text{w}})}$ in Definition 7.5.7. Then $\min_{\theta_{\text{w}} \in [\theta_{\text{w}}^{\text{s}}, \frac{\pi}{2}]} r_2^{(\theta_{\text{w}})} > 0$ from the continuous dependence of points $\mathcal{O}_2 = (u_2, v_2)$ and P_4 on $\theta_{\text{w}} \in [\theta_{\text{w}}^{\text{s}}, \frac{\pi}{2}]$, where we have used Theorem 7.1.1(i) and Lemma 7.5.9. Thus,

$$\hat{r} := \frac{1}{10} \min\{\sqrt{u_1^2 - c_1^2},\, r_0,\, \min_{\theta_{\text{w}} \in [\theta_{\text{w}}^{\text{s}}, \frac{\pi}{2}]} r_2^{(\theta_{\text{w}})}\} > 0$$

for r_0 in (9.4.30). From Proposition 9.4.9 applied with $r = \hat{r}$, there exists $\delta > 0$ depending only on the data such that, for any admissible solution with $\theta_{\text{w}} \in (\theta_{\text{w}}^{\text{s}}, \frac{\pi}{2})$,

$$\operatorname{dist}(\Gamma_{\text{shock}}, \Gamma_{\text{wedge}} \setminus (B_{\hat{r}}(P_3) \cup B_{\hat{r}}(P_4))) \geq \delta.$$

From this, working in the (S, T)–coordinates with basis $\{\boldsymbol{\nu}_{\rm w}, \boldsymbol{\tau}_{\rm w}\}$ introduced in Lemma 8.2.11, and using (8.2.25), we find that, for any admissible solution with the wedge angle $\theta_{\rm w} \in (\theta_{\rm w}^{\rm s}, \frac{\pi}{2})$,

$$\Gamma_{\rm shock} \subset \Big\{S \geq F^{(\theta_{\rm w})}(T) \;:\; T_{P_2} < T < T_{P_1}^{(\theta_{\rm w})}\Big\}, \tag{9.5.21}$$

where $F^{(\theta_{\rm w})}(\cdot)$ is defined by noting that $T_{P_3} = 0$ and $\hat{r} \leq \frac{1}{10}|T_{P_4}^{(\theta_{\rm w})} - T_{P_3}|$ by (9.4.30) as follows:

$$F^{(\theta_{\rm w})}(T) = \begin{cases} -T\tan\theta_{\rm w}, & -\infty < T \leq 0, \\ 0, & 0 < T \leq \hat{r}, \\ \frac{\delta}{\hat{r}}(T - \hat{r}), & \hat{r} < T \leq 2\hat{r}, \\ \delta, & 2\hat{r} < T \leq T_{P_4}^{(\theta_{\rm w})} - 2\hat{r}, \\ \frac{\delta}{\hat{r}}(T_{P_4}^{(\theta_{\rm w})} - \hat{r} - T), & T_{P_4}^{(\theta_{\rm w})} - 2\hat{r} < T \leq T_{P_4}^{(\theta_{\rm w})} - \hat{r}, \\ 0, & T \leq T_{P_4}^{(\theta_{\rm w})} - \hat{r} < T < \infty. \end{cases}$$

The expression above defines $F^{(\theta_{\rm w})}(T)$ for all $(T, \theta_{\rm w}) \in \mathbb{R} \times [\theta_{\rm w}^{\rm s}, \frac{\pi}{2}]$. Note that, using Lemma 7.5.9(ii) and the definition of the (S, T)–coordinates, we find that $T_{P_4}^{(\theta_{\rm w})}$ depends continuously on $\theta_{\rm w} \in [\theta_{\rm w}^{\rm s}, \frac{\pi}{2}]$. If $\theta_{\rm w}^{\rm s} > \frac{1}{2}\arcsin(\frac{c_1}{u_1})$, we extend the function: $\theta_{\rm w} \mapsto T_{P_4}^{(\theta_{\rm w})}$ to domain $[\frac{1}{2}\arcsin(\frac{c_1}{u_1}), \frac{\pi}{2}]$ by defining $T_{P_4}^{(\theta_{\rm w})} := T_{P_4}^{(\theta_{\rm w}^{\rm s})}$ on $\theta_{\rm w} \in [\frac{1}{2}\arcsin(\frac{c_1}{u_1}), \theta_{\rm w}^{\rm s}]$. Then $\theta_{\rm w} \mapsto T_{P_4}^{(\theta_{\rm w})}$ is continuous on $[\frac{1}{2}\arcsin(\frac{c_1}{u_1}), \frac{\pi}{2}]$. Now $F^{(\theta_{\rm w})}(T)$ is defined on $(T, \theta_{\rm w}) \in \mathbb{R} \times [\frac{1}{2}\arcsin(\frac{c_1}{u_1}), \frac{\pi}{2}]$.

Also, for each $\theta_{\rm w}$, $\mathcal{O}_1 = (-u_1\sin\theta_{\rm w}, u_1\cos\theta_{\rm w})$ in the (S, T)–coordinates, which implies

$$B_{c_1}(\mathcal{O}_1) \subset \Big\{S < G^{(\theta_{\rm w})}(T),\; |T - u_1\cos\theta_{\rm w}| \leq c_1\Big\}, \tag{9.5.22}$$

where

$$G^{(\theta_{\rm w})}(T) = \begin{cases} -u_1\sin\theta_{\rm w} + \sqrt{c_1^2 - (T - u_1\cos\theta_{\rm w})^2} & \text{if } |T - u_1\cos\theta_{\rm w}| \leq c_1, \\ -u_1\sin\theta_{\rm w} & \text{otherwise.} \end{cases}$$

Note that $F(T, \theta_{\rm w}) := F^{(\theta_{\rm w})}(T)$ and $G(T, \theta_{\rm w}) := G^{(\theta_{\rm w})}(T)$ are continuous functions of $(T, \theta_{\rm w})$ on $\mathbb{R} \times [\frac{1}{2}\arcsin(\frac{c_1}{u_1}), \frac{\pi}{2}]$, where we have used the continuous dependence of $T_{P_4}^{(\theta_w)}$ on $\theta_w \in [\frac{1}{2}\arcsin(\frac{c_1}{u_1}), \frac{\pi}{2}]$, as discussed above.

From the explicit expressions,

$$F(T, \theta_{\rm w}) \geq 0 \geq G(T, \theta_{\rm w}) \qquad \text{on } \mathbb{R} \times [\arcsin(\frac{c_1}{u_1}), \frac{\pi}{2}], \tag{9.5.23}$$

and also

$$\begin{aligned} &G < 0 \qquad \text{on } \big(\mathbb{R} \times [\arcsin(\frac{c_1}{u_1}), \frac{\pi}{2}]\big) \setminus \{(\sqrt{u_1^2 - c_1^2},\; \arcsin(\frac{c_1}{u_1}))\}, \\ &G(\sqrt{u_1^2 - c_1^2},\; \arcsin(\frac{c_1}{u_1})) = 0. \end{aligned} \tag{9.5.24}$$

Then we have

$$F > G \qquad \text{on } \big(\mathbb{R} \times [\arcsin(\frac{c_1}{u_1}), \frac{\pi}{2}]\big) \setminus \{(\sqrt{u_1^2 - c_1^2},\ \arcsin(\frac{c_1}{u_1}))\}.$$

Now we show that $F > G$ at $(T, \theta_{\rm w}) = (\sqrt{u_1^2 - c_1^2}, \arcsin(\frac{c_1}{u_1}))$. For $\theta_{\rm w} \in [\theta_{\rm w}^{\rm s}, \frac{\pi}{2})$, denote by $\hat{Q} = \hat{Q}^{(\theta_{\rm w})}$ the nearest point on line $\{\xi_2 = \xi_1 \tan\theta_{\rm w}\}$ to center $\mathcal{O}_1$ of the sonic circle of state (1). From (7.5.1) and (7.5.6), we obtain that $\hat{Q}$ lies within segment $P_3\mathcal{O}_2$. Using (7.5.1)–(7.5.2), we have

$$|\hat{Q}P_4| > |\mathcal{O}_2 P_4| = r_2 \geq 10\hat{r}. \tag{9.5.25}$$

Since $\hat{Q} = (0, u_1 \cos\theta_{\rm w})$ in the (S, T)–coordinates, we obtain from the previous inequality that

$$u_1 \cos\theta_{\rm w} = T_{\hat{Q}^{(\theta_{\rm w})}} < T_{P_4}^{(\theta_{\rm w})} - 10\hat{r}. \tag{9.5.26}$$

If $\theta_{\rm w}^{\rm s} \leq \arcsin(\frac{c_1}{u_1})$, we use (9.5.26) with $\theta_{\rm w} = \arcsin(\frac{c_1}{u_1})$ to obtain

$$\sqrt{u_1^2 - c_1^2} = u_1 \cos(\arcsin(\frac{c_1}{u_1})) < T_{P_4}^{(\arcsin(\frac{c_1}{u_1}))} - 10\hat{r}.$$

If $\theta_{\rm w}^{\rm s} > \arcsin(\frac{c_1}{u_1})$, then, recalling that we have extended $\theta_{\rm w} \mapsto T_{P_4}^{(\theta_{\rm w})}$ to $\theta_{\rm w} \in [\arcsin(\frac{c_1}{u_1}), \frac{\pi}{2}]$ so that $T_{P_4}^{(\arcsin(\frac{c_1}{u_1}))} = T_{P_4}^{(\theta_{\rm w}^{\rm s})}$ in the definition of F, we obtain similar estimates by using (9.5.26) with $\theta_{\rm w} = \theta_{\rm w}^{\rm s}$:

$$\sqrt{u_1^2 - c_1^2} = u_1 \cos(\arcsin(\frac{c_1}{u_1})) < u_1 \cos\theta_{\rm w}^{\rm s} < T_{P_4}^{(\theta_{\rm w}^{\rm s})} - 10\hat{r} = T_{P_4}^{(\arcsin(\frac{c_1}{u_1}))} - 10\hat{r}.$$

Thus, in both cases,

$$\sqrt{u_1^2 - c_1^2} < T_{P_4}^{(\arcsin(\frac{c_1}{u_1}))} - 10\hat{r}.$$

Using this and $\sqrt{u_1^2 - c_1^2} > 2\hat{r}$ from the choice of $\hat{r}$, we employ the explicit definition of F and (9.5.24) to obtain

$$F(\sqrt{u_1^2 - c_1^2},\ \arcsin(\frac{c_1}{u_1})) = \delta > 0 = G(\sqrt{u_1^2 - c_1^2},\ \arcsin(\frac{c_1}{u_1})).$$

Combining this with (9.5.23)–(9.5.24), we have

$$F > G \qquad \text{on } \mathbb{R} \times [\arcsin(\frac{c_1}{u_1}), \frac{\pi}{2}]. \tag{9.5.27}$$

Also, using (9.1.2), there exists $\hat{T} > 1$ such that, for every admissible solution φ with $\theta_{\rm w} \in (\theta_{\rm w}^{\rm s}, \frac{\pi}{2})$,

$$T_{P_1}^{(\theta_{\rm w})}, T_{P_2}^{(\theta_{\rm w})}, u_1 \cos\theta_{\rm w} \pm c_1 \in (-\hat{T}, \hat{T})$$

in the corresponding (S,T)–coordinates. Then, recalling that F and G are continuous on $\mathbb{R}\times[\frac{1}{2}\arcsin(\frac{c_1}{u_1}),\frac{\pi}{2}]$, we obtain from (9.5.27) that there exists a constant $\sigma>0$ such that

$$F(T,\theta_{\rm w})\geq G(T,\theta_{\rm w})+\sigma \quad \text{for all } (T,\theta_{\rm w})\in[-2\hat{T},2\hat{T}]\times[\arcsin(\frac{c_1-\sigma}{u_1}),\frac{\pi}{2}],$$

and σ depends only on F, G, and (u_1,c_1), *i.e.*, on (ρ_0,ρ_1,γ). This, combined with (9.5.21)–(9.5.22), implies that, for any admissible solution with the wedge angle $\theta_{\rm w}$,

$$\operatorname{dist}(\Gamma_{\rm shock},B_{c_1}(\mathcal{O}_1))\geq\sigma \qquad \text{if } \theta_{\rm w}\in[\arcsin(\frac{c_1-\sigma}{u_1}),\frac{\pi}{2}].$$

If $\theta_{\rm w}^{\rm s}\geq\arcsin(\frac{c_1-\sigma}{u_1})$, the proof for Case 3 is completed.

If $\theta_{\rm w}^{\rm s}<\arcsin(\frac{c_1-\sigma}{u_1})$, we notice that

$$B_{c_1-\sigma}(\mathcal{O}_1)\cap\Lambda(\theta_{\rm w})\neq\emptyset \qquad \text{for all } \theta_{\rm w}\in(0,\arcsin(\frac{c_1-\sigma}{u_1})).$$

Therefore, for $\theta_{\rm w}\in[\theta_{\rm w}^{\rm s},\arcsin(\frac{c_1-\sigma}{u_1}))$, we apply Lemma 9.5.2 with $\varepsilon_0=\sigma$ to complete the proof for Case 3. □

Now we extend the result of Proposition 9.4.5 to Case $u_1=c_1$.

Corollary 9.5.7. *Let $\gamma>1$ and $\rho_1>\rho_0>0$ be such that $u_1=c_1$. Let $\theta_{\rm w}^*\in(\theta_{\rm w}^{\rm s},\frac{\pi}{2})$. Then there exists $C>0$ such that*

$$\operatorname{dist}(\Gamma_{\rm shock},\Gamma_{\rm wedge})\geq\frac{1}{C}$$

for any admissible solution of **Problem 2.6.1** *with $\theta_{\rm w}\in[\theta_{\rm w}^*,\frac{\pi}{2})$.*

Proof. In this proof, the universal constant $C>0$ depends only on (ρ_0,ρ_1,γ).

By Lemma 9.4.7, it suffices to consider only admissible solutions with the wedge angles $\theta_{\rm w}\in[\theta_{\rm w}^*,\frac{\pi}{2}-\sigma_1)$, where $\sigma_1>0$ depends only on (ρ_0,ρ_1,γ).

Since $u_1=c_1$, choosing $r=2c_1\sin\sigma_1$, we obtain

$$\Gamma_{\rm wedge}(\theta_{\rm w})\cap B_r(\mathbf{0})\subset B_{c_1}(\mathcal{O}_1)\equiv B_{c_1}(c_1,0) \qquad \text{for all } \theta_{\rm w}\in(0,\frac{\pi}{2}-\sigma_1).$$

Then, using (8.1.2) and Proposition 9.5.6, we have

$$\operatorname{dist}(\Gamma_{\rm shock},\Gamma_{\rm wedge}\cap B_r(\mathbf{0}))\geq\operatorname{dist}(\Gamma_{\rm shock},B_{c_1}(\mathcal{O}_1))\geq\frac{1}{C}.$$

On the other hand, from Proposition 9.4.8 applied with r, as defined above, we conclude

$$\operatorname{dist}(\Gamma_{\rm shock},\Gamma_{\rm wedge}\setminus B_r(\mathbf{0}))\geq\frac{1}{C}.$$

This completes the proof. □

9.6 UNIFORM ESTIMATES OF THE ELLIPTICITY CONSTANT IN $\overline{\Omega} \setminus \overline{\Gamma_{\text{sonic}}}$

Set the Mach number

$$M^2 = \frac{|D\varphi|^2}{c^2} = M_1^2 + M_2^2 \qquad \text{with } M_1^2 = \frac{\varphi_{\boldsymbol{\nu}}^2}{c^2} \text{ and } M_2^2 = \frac{\varphi_{\tau}^2}{c^2}.$$

Notice that, for an admissible solution of **Problem 2.6.1**, $M, M_1, M_2 \in C(\overline{\Omega}) \cap C^2(\overline{\Omega} \setminus (\overline{\Gamma_{\text{sonic}}} \cup \{P_2, P_3\}))$, by (8.1.3). Also, by (8.1.4),

$$M^2 < 1 \qquad \text{in } \overline{\Omega} \setminus \overline{\Gamma_{\text{sonic}}}. \tag{9.6.1}$$

In this section, we improve estimate (9.6.1).

We now prove two preliminary lemmas. We first show that, for the hyperbolic-elliptic shock, if the elliptic part is strictly elliptic, the gradient jump across Γ_{shock} can be estimated in terms of the ellipticity.

Lemma 9.6.1. *Fix $\gamma > 1$ and $\rho_- > 0$. Then, for any $\delta \in (0,1)$ and $R > 0$, there exists $\kappa > 0$ such that the following holds: Let $\Omega \subset \mathbb{R}^2$, $\mathcal{S}$, φ, and $\varphi^{\pm}$ be as in Lemma 6.1.3 (note, in particular, that φ^- is a uniform state, and φ^+ is not assumed to be uniform). Let $P \in \mathcal{S}$ be such that*

(i) *$|PO_-| \le R$, where $O_- = \mathbf{v}^-$ is the center of sonic circle of state φ^-;*

(ii) *$D\varphi^- \cdot \boldsymbol{\nu} > D\varphi^+ \cdot \boldsymbol{\nu} > 0$ at P, where $\boldsymbol{\nu}$ is a unit normal to $\mathcal{S}$ oriented from Ω^- to Ω^+;*

(iii) *$M_+^2 \le 1 - \delta$ at P, where M_+^2 is the Mach number of φ^+.*

Then

$$D\varphi^- \cdot \boldsymbol{\nu} - D\varphi^+ \cdot \boldsymbol{\nu} \ge \kappa \qquad \text{at } P. \tag{9.6.2}$$

Proof. In the proof, constants C and C_1 below are positive and depend only on (ρ_-, γ, R). We use functions $\tilde{\rho}$ and Φ, as defined in (6.1.12), and constants q_* and $q_{\max}$, as defined in the line after (6.1.14).

First, assumption (ii), combined with (6.1.14), implies that (6.1.15).

Note that constant B_0 in definition (6.1.12) of $\tilde{\rho}$ is determined by the second equality in (6.1.11). Thus, using that $|D\varphi^-(P)|^2 = |PO_-|^2 \le R^2$, we have

$$\rho_-^{\gamma-1} \le B_0 \le \rho_-^{\gamma-1} + R^2.$$

Now, by an explicit calculation, we find that $\tilde{\rho}'(s) \le -\frac{1}{C}$ for any $s \in (0, q_*)$. Also, $B_0^{\frac{1}{\gamma-1}} = \tilde{\rho}(0) > \tilde{\rho}(s) > \tilde{\rho}(q_*) = q_*^{\frac{2}{\gamma-1}} > 0$ for any $s \in (0, q_*)$. Then $M_+^2 \le 1 - \delta$ and (6.1.16) yield

$$\varphi_{\boldsymbol{\nu}}^+(P) \le q_* - \frac{\delta}{C_1}.$$

Combining this with (6.1.15) implies (9.6.2) with $\kappa = \frac{\delta}{C_1}$. □

Next we show that, for the hyperbolic-elliptic shock, if the upstream state is strictly hyperbolic, and the shock direction is almost orthogonal to the flow, the downstream state is strictly elliptic with a quantitative estimate of the ellipticity.

Lemma 9.6.2. *For any $\gamma > 1$, $\rho_- > 0$, and $\delta \in (0, \frac{1}{2})$, there exist $\alpha, \zeta > 0$ such that the following holds: Let $\Omega \subset \mathbb{R}^2$, $\mathcal{S}$, φ, and $\varphi^\pm$ be as in Lemma 6.1.3 (again, φ^- is a uniform state, and φ^+ is not assumed to be uniform). Let $P \in \mathcal{S}$ be such that*

(i) *$D\varphi^- \cdot \boldsymbol{\nu} > D\varphi^+ \cdot \boldsymbol{\nu} > 0$ at P, where $\boldsymbol{\nu}$ is a unit normal to $\mathcal{S}$ oriented from Ω^- to Ω^+;*

(ii) *$|D\varphi^-(P)|^2 \geq (1+\delta)c_-^2$ at P;*

(iii) *$|\varphi_{\boldsymbol{\tau}}^+|^2 \leq \alpha|\varphi_{\boldsymbol{\nu}}^+|^2$ at P.*

Let M_+^2 be the Mach number of φ^+. Then

$$M_+^2 \leq 1 - \zeta \qquad \text{at } P. \tag{9.6.3}$$

Proof. In the proof, for simplicity, we write φ and ρ for φ^+ and ρ_+. Also, we write M^2, M_1^2, and M_2^2 for M_+^2, $(M_+)_1^2$, and $(M_+)_2^2$, respectively.

By assumption (i),

$$\rho \geq \rho_- \qquad \text{at } P.$$

Since $\rho\varphi_{\boldsymbol{\nu}} = \rho_-\varphi_{\boldsymbol{\nu}}^-$ along Γ_{shock}, we have

$$|D\varphi^-|^2 - |\varphi_{\boldsymbol{\nu}}^-|^2 \leq \alpha\varphi_{\boldsymbol{\nu}}^2 = \alpha\big(\frac{\rho_-}{\rho}\big)^2|\varphi_{\boldsymbol{\nu}}^-|^2,$$

that is,

$$|D\varphi^-|^2 \leq \Big(1 + \alpha\big(\frac{\rho_-}{\rho}\big)^2\Big)|\varphi_{\boldsymbol{\nu}}^-|^2.$$

Then we have

$$|\varphi_{\boldsymbol{\nu}}^-|^2 \geq \frac{1}{1+\alpha\big(\frac{\rho_-}{\rho}\big)^2}|D\varphi^-|^2 \geq \frac{1+\delta}{1+\alpha\big(\frac{\rho_-}{\rho}\big)^2}\rho_-^{\gamma-1}. \tag{9.6.4}$$

The Bernoulli law in (2.2.7) and the Rankine-Hugoniot conditions (6.1.8)–(6.1.9) imply

$$\frac{1}{2}\Big(1 - \big(\frac{\rho_-}{\rho}\big)^2\Big)|\varphi_{\boldsymbol{\nu}}^-|^2 = \frac{1}{\gamma-1}(\rho^{\gamma-1} - \rho_-^{\gamma-1}) = \rho_*^{\gamma-2}(\rho - \rho_-)$$

for some $\rho_* \in (\rho_-, \rho)$. That is,

$$|\varphi_{\boldsymbol{\nu}}^-|^2 = \frac{2\rho^2\rho_*^{\gamma-2}}{\rho+\rho_-}. \tag{9.6.5}$$

Set $t = \frac{\rho_-}{\rho} \in (0,1)$. Combining (9.6.4) with (9.6.5), we have

$$(\frac{\rho_-}{\rho_*})^{\gamma-2}(t^2+t) \le \frac{2(1+\alpha t^2)}{1+\delta}. \tag{9.6.6}$$

We now consider two separate cases: $1 < \gamma \le 2$ and $\gamma > 2$.

When $1 < \gamma \le 2$, $(\frac{\rho_-}{\rho_*})^{\gamma-2} \ge 1$, since $\rho_- < \rho_*$. Thus, (9.6.6) implies

$$(1 - \frac{2\alpha}{1+\delta})t^2 + t - \frac{2}{1+\delta} \le 0.$$

Set $s = t - 1 < 0$. Then

$$s^2 + bs + c \le 0,$$

where $b = \frac{3(1+\delta)-4\alpha}{1+\delta-2\alpha}$ and $c = \frac{2(\delta-\alpha)}{1+\delta-2\alpha}$. Thus, we have

$$s \le \frac{-2c}{\sqrt{b^2-4c}+b} \le -\frac{c}{b} \le -\frac{1}{3}(\delta-\alpha)$$

when $0 < \alpha < \delta \le \frac{1}{2}$, which implies

$$\frac{\rho_-}{\rho} \le 1 - \frac{1}{3}(\delta-\alpha). \tag{9.6.7}$$

By (6.1.7) and (9.6.5),

$$\varphi_\nu^2 = \Big(\frac{\rho_-}{\rho}\Big)^2 |\varphi_\nu^-|^2 = \frac{2\rho_-^2\rho_*^{\gamma-2}}{\rho+\rho_-} = 2\Big(\frac{\rho_*}{\rho}\Big)^{\gamma-2}\rho^{\gamma-1}\frac{t^2}{t+1}.$$

Also, since $\gamma - 2 \le 0$ and $\rho_- \le \rho_*$, it follows that $(\frac{\rho_*}{\rho})^{\gamma-2} \le (\frac{\rho_-}{\rho})^{\gamma-2} = t^{\gamma-2}$. Then

$$M_1^2 \le \frac{2t^{\gamma+1}}{t+1} =: g(t),$$

which implies

$$M^2 = M_1^2 + M_2^2 \le (1+\alpha)M_1^2 \le (1+\alpha)g(t).$$

Since $t \in (0, 1 - \frac{1}{3}(\delta-\alpha))$ with δ and α small, and $g'(t) > 0$ for $t > 0$ with $\min_{t\in[\frac{1}{2},1]} g'(t) = B(\gamma) > 0$, we have

$$g(t) \le g(1) - \frac{1}{3}(\delta-\alpha)\min_{t\in[\frac{1}{2},1]} g'(t) = 1 - \frac{B}{3}(\delta-\alpha).$$

Therefore, we obtain

$$M^2 \le (1+\alpha)\big(1 - \frac{B}{3}(\delta-\alpha)\big) \le 1 - \zeta, \tag{9.6.8}$$

if ζ is chosen as

$$0 < \zeta \le \frac{B}{3}(\delta - \alpha)(1+\alpha) - \alpha,$$

where the right-hand side is positive if α is chosen as

$$0 < \alpha \le \frac{\delta}{4}\min\{\frac{B}{3}, 1\}.$$

Thus, with the choice of parameters (α, ζ) specified above, (9.6.3) is proved when $1 < \gamma \le 2$.

When $\gamma > 2$,

$$\left(\frac{\rho_-}{\rho_*}\right)^{\gamma-2} \ge \left(\frac{\rho_-}{\rho}\right)^{\gamma-2} \qquad \text{for } \rho_- < \rho_* \le \rho,$$

which, by (9.6.6) and $t \in (0,1)$, implies the inequality:

$$t^{\gamma-2}(t^2+t) \le \frac{2(1+\alpha t^2)}{1+\delta} \le \frac{2(1+\alpha)}{1+\delta}. \tag{9.6.9}$$

Denote

$$h(t) = t^{\gamma-2}(t^2+t).$$

Then

$$h(1) = 2, \qquad h'(t) = \gamma t^{\gamma-1} + (\gamma-1)t^{\gamma-2} > 0 \quad \text{on } (0,1],$$
$$\max_{t\in[0,1]} h'(t) = h'(1) = 2\gamma - 1 > 1.$$

Now (9.6.9) is that $h(t) \le \frac{2(1+\alpha)}{1+\delta} < 2$ if $\alpha < \delta$. Then $1 - K(\delta-\alpha) > 0$ if $0 < \alpha < \delta \le \frac{1}{2}$ and $K \in (0,1)$. Thus, if $t > 1 - K(\delta - \alpha)$,

$$h(t) \ge h(1) - K(\delta-\alpha)\max_{t\in[0,1]} h'(t) = 2 - K(2\gamma-1)(\delta-\alpha),$$

which contradicts (9.6.9) if $K < \frac{2}{1+\delta}$. Therefore, choosing $K = \frac{1}{1+\delta}$, we have

$$t \le 1 - K(\delta-\alpha) \le 1 - \frac{1}{2}(\delta - \alpha). \tag{9.6.10}$$

By (6.1.7) and (9.6.5),

$$M_1^2 = \left(\frac{\rho_*}{\rho}\right)^{\gamma-2}\frac{2t^2}{t+1} \le \frac{2t^2}{t+1},$$

where we have used $\gamma > 2$ and $\rho_* < \rho$ in the last inequality. Thus, assuming that $0 < \alpha \le \frac{\delta}{100}$, we employ (9.6.10) and $\left(\frac{2t^2}{t+1}\right)' > 0$ for $t > 0$ to obtain

$$\begin{aligned} M^2 \le (1+\alpha)M_1^2 &\le (1+\alpha)\frac{2t^2}{t+1} \\ &\le \frac{2(1+\alpha)(1-\frac{1}{2}(\delta-\alpha))^2}{2-\frac{1}{2}(\delta-\alpha)} \\ &\le \frac{(1+\alpha)(1-(\delta-\alpha)+\frac{1}{2}\delta^2)}{1-\frac{1}{4}(\delta-\alpha)} \\ &\le \frac{1-(\delta-\alpha)+\frac{1}{2}\delta^2(1+\alpha)+\alpha}{1-\frac{1}{4}(\delta-\alpha)} \\ &\le 1-\frac{\delta}{2}. \end{aligned}$$

Choose $0 < \zeta \le \frac{\delta}{2}$. Then

$$M^2 \le 1-\zeta.$$

Thus, with the choice of parameters (α, ζ) specified above, (9.6.3) is proved when $\gamma > 2$. □

Now we can prove the main technical result in this section.

Proposition 9.6.3. *There exists $\mu > 0$ depending only on (ρ_0, ρ_1, γ) such that, if φ is an admissible solution of* **Problem 2.6.1** *with $\theta_w \in (\theta_w^s, \frac{\pi}{2})$, and $g \in C^1(\overline{\Omega})$ with $g(P_1) = 0$ and $|Dg| \le 1$ in Ω, then the maximum of $M^2 + \mu g$ over $\overline{\Omega}$ cannot be attained on $\Gamma_{\rm shock} \cup \{P_2\}$.*

Proof. In this proof, constants C, μ, α, ζ, and δ depend only on (ρ_0, ρ_1, γ). The proof consists of three steps.

1. By Proposition 9.5.6,

$$\mathrm{dist}(\Gamma_{\rm shock}, B_{c_1}(\mathcal{O}_1)) \ge \frac{1}{C}$$

for any admissible solution of **Problem 2.6.1** with $\theta_w \in (\theta_w^s, \frac{\pi}{2})$. Since $D\varphi_1(\boldsymbol{\xi}) = (u_1, 0) - \boldsymbol{\xi}$, then there exists $\delta > 0$ such that, for all such solutions,

$$|D\varphi_1|^2 \ge (1+\delta)c_1^2 \qquad \text{on } \Gamma_{\rm shock}. \tag{9.6.11}$$

Let φ be an admissible solution of **Problem 2.6.1** with $\theta_w \in (\theta_w^s, \frac{\pi}{2})$. Note that, from Lemma 9.1.4,

$$\rho \ge a\rho_1 > 0 \qquad \text{in } \overline{\Omega}.$$

Let $g \in C^1(\overline{\Omega})$ with $g(P_1) = 0$ and $|Dg| \le 1$ in Ω. Denote

$$d(\boldsymbol{\xi}) = \mu g(\boldsymbol{\xi})$$

for $\mu > 0$ to be chosen. Since $\mathrm{diam}(\Omega) \le C$ by Proposition 9.1.2,

$$d(\boldsymbol{\xi}) \le \mu\big(g(P_1) + \|Dg\|_{L^\infty(\Omega)}\mathrm{diam}(\Omega)\big) \le C\mu \qquad \text{for all } \boldsymbol{\xi} \in \overline{\Omega}. \tag{9.6.12}$$

Assume that the maximum of $M^2 + d$ over $\overline{\Omega}$ is attained at $P_{\max} \in \Gamma_{\rm shock} \cup \{P_2\}$. Note that $M^2 = 1$ on $\Gamma_{\rm sonic}$, since φ is C^1 across $\Gamma_{\rm sonic}$ by (8.1.3). Thus, we have

$$(M^2 + d)(P_{\max}) \ge (M^2 + d)(P_1) = 1.$$

Then, using (9.6.12),

$$M^2(P_{\max}) \ge 1 - d(P_{\max}) \ge 1 - C\mu. \tag{9.6.13}$$

Let $\alpha, \zeta > 0$ be the constants in Lemma 9.6.2 for δ defined by (9.6.11). Without loss of generality, we may assume that $\alpha, \zeta < \frac{1}{2}$. From (9.6.13), choosing μ sufficiently small, we have

$$M^2(P_{\max}) \ge 1 - \frac{\zeta}{2}. \tag{9.6.14}$$

This implies

$$M_2^2 \ge \alpha M_1^2 \qquad \text{at } P_{\max} \tag{9.6.15}$$

by (8.1.16), (9.6.11), and Lemma 9.6.2. Equivalently,

$$\varphi_{\boldsymbol{\tau}}^2 \ge \alpha \varphi_{\boldsymbol{\nu}}^2 \qquad \text{at } P_{\max}.$$

In particular, $\varphi_{\boldsymbol{\tau}} \ne 0$ by (8.1.16). This implies

$$P_{\max} \ne P_2,$$

since $\boldsymbol{\tau}_{\rm sh}(P_2) = \boldsymbol{\nu}_{\rm sym}(P_2)$. Therefore, $(D\varphi \cdot \boldsymbol{\tau}_{\rm sh})(P_2) = (D\varphi \cdot \boldsymbol{\nu}_{\rm sym})(P_2) = 0$ by (2.2.20), where we have used that $\varphi \in C^1(\overline{\Omega})$.

It remains to consider the case that $P_{\max} \in \Gamma_{\rm shock}$. Since the maximum of $M^2 + d$ over Ω is achieved at $P_{\max}$, it follows that, at $P_{\max}$,

$$(M^2 + d)_{\boldsymbol{\tau}} = 0, \qquad i.e., \quad (M^2)_{\boldsymbol{\tau}} = -d_{\boldsymbol{\tau}}, \tag{9.6.16}$$

$$\partial_{\boldsymbol{\nu}}(M^2 + d) \le 0. \tag{9.6.17}$$

Now we compute the derivatives of the Mach number along $\Gamma_{\rm shock}$ and then apply these calculations at (9.6.16)–(9.6.17) to arrive at a contradiction.

2. *Derivatives of the Mach number along* $\Gamma_{\rm shock}$. Since

$$(|D\varphi|^2)_{\xi_i} = 2D\varphi \cdot D^2\varphi\, \mathbf{e}_i = 2D^2\varphi[\mathbf{e}_i, D\varphi],$$

we have

$$(|D\varphi|^2)_{\boldsymbol{\tau}} = 2D^2\varphi[\boldsymbol{\tau}, D\varphi], \qquad (|D\varphi|^2)_{\boldsymbol{\nu}} = 2D^2\varphi[\boldsymbol{\nu}, D\varphi].$$

From (5.1.6),

$$(c^2)_{\boldsymbol{\tau}} = -(\gamma-1)\big(D^2\varphi[\boldsymbol{\tau}, D\varphi] + \varphi_{\boldsymbol{\tau}}\big), \quad (c^2)_{\boldsymbol{\nu}} = -(\gamma-1)\big(D^2\varphi[\boldsymbol{\nu}, D\varphi] + \varphi_{\boldsymbol{\nu}}\big).$$

Thus, we have

$$(M^2)_{\boldsymbol{\tau}} = \frac{2c^2 D^2\varphi[\boldsymbol{\tau}, D\varphi] + (\gamma-1)|D\varphi|^2\big(D^2\varphi[\boldsymbol{\tau}, D\varphi] + \varphi_{\boldsymbol{\tau}}\big)}{c^4},$$

that is,

$$(M^2)_{\boldsymbol{\tau}} = \frac{\big(2+(\gamma-1)M^2\big)D^2\varphi[\boldsymbol{\tau}, D\varphi] + (\gamma-1)M^2\varphi_{\boldsymbol{\tau}}}{c^2}. \tag{9.6.18}$$

Similarly, we have

$$(M^2)_{\boldsymbol{\nu}} = \frac{\big(2+(\gamma-1)M^2\big)D^2\varphi[\boldsymbol{\nu}, D\varphi] + (\gamma-1)M^2\varphi_{\boldsymbol{\nu}}}{c^2}. \tag{9.6.19}$$

3. *The maximum point $P_{\max}$.* From now on, all the functions are estimated at $P_{\max}$. Then, from (9.6.16) and (9.6.18), we have

$$D^2\varphi[\boldsymbol{\tau}, D\varphi] = -\frac{(\gamma-1)M^2\varphi_{\boldsymbol{\tau}} + c^2 d_{\boldsymbol{\tau}}}{2+(\gamma-1)M^2} =: B_1. \tag{9.6.20}$$

Consider equation (5.1.21). Then the right-hand side of (5.1.21) is

$$\begin{aligned}
&\frac{\rho-\rho_1}{\rho_1}\varphi_{\boldsymbol{\tau}}(\rho_1+\rho M_1^2) - \rho\Big(1+\frac{\rho-\rho_1}{\rho_1}M_1^2\Big)\frac{(\gamma-1)M^2\varphi_{\boldsymbol{\tau}} + c^2 d_{\boldsymbol{\tau}}}{2+(\gamma-1)M^2}\\
&= \varphi_{\boldsymbol{\tau}}\Big(\frac{\rho-\rho_1}{\rho_1}(\rho_1+\rho M_1^2) - \frac{(\gamma-1)\rho M^2}{2+(\gamma-1)M^2}\big(1+\frac{\rho-\rho_1}{\rho_1}M_1^2\big)\Big)\\
&\quad - \frac{c^2\rho(1+\frac{\rho-\rho_1}{\rho_1}M_1^2)d_{\boldsymbol{\tau}}}{2+(\gamma-1)M^2}\\
&= \varphi_{\boldsymbol{\tau}}\Big(-\frac{(\gamma-1)\rho M^2}{2+(\gamma-1)M^2} + \frac{\rho-\rho_1}{\rho_1}\frac{2\rho M_1^2}{2+(\gamma-1)M^2} + \rho - \rho_1\Big)\\
&\quad - \frac{c^2\rho(1+\frac{\rho-\rho_1}{\rho_1}M_1^2)d_{\boldsymbol{\tau}}}{2+(\gamma-1)M^2}.
\end{aligned}$$

Thus, (5.1.21) yields

$$\begin{aligned}
D^2\varphi[\boldsymbol{\tau}, \mathbf{g}] &= \frac{\big(2(\rho-\rho_1)(1+\frac{\rho}{\rho_1}M_1^2) - (\gamma-1)\rho_1 M^2\big)\varphi_{\boldsymbol{\tau}} - c^2\rho(1+\frac{\rho-\rho_1}{\rho_1}M_1^2)d_{\boldsymbol{\tau}}}{2+(\gamma-1)M^2}\\
&=: B_2.
\end{aligned} \tag{9.6.21}$$

Note that, under the orthogonal transformation:

$$O: (\mathbf{e}_1, \mathbf{e}_2) \mapsto (\boldsymbol{\nu}, \boldsymbol{\tau}),$$

the left-hand sides of equations (9.6.20)–(9.6.21) are invariant. Then (9.6.20) implies

$$\varphi_{\boldsymbol{\nu}\boldsymbol{\tau}}\varphi_{\boldsymbol{\nu}} + \varphi_{\boldsymbol{\tau}\boldsymbol{\tau}}\varphi_{\boldsymbol{\tau}} = B_1, \tag{9.6.22}$$

and (9.6.21) implies

$$\varphi_{\boldsymbol{\nu}\boldsymbol{\tau}}g_1 + \varphi_{\boldsymbol{\tau}\boldsymbol{\tau}}g_2 = B_2, \tag{9.6.23}$$

where

$$\begin{aligned} g_1 &:= \mathbf{g}\cdot\boldsymbol{\nu} = \rho|\tilde{\boldsymbol{\nu}}| + \rho_1 D\varphi_1\cdot\boldsymbol{\nu} = \frac{\rho}{\rho_1}(\rho-\rho_1)\varphi_{\boldsymbol{\nu}} + \rho\varphi_{\boldsymbol{\nu}} = \frac{\rho^2}{\rho_1}\varphi_{\boldsymbol{\nu}},\\ g_2 &= \rho_1 D\varphi_1\cdot\boldsymbol{\tau} = \rho_1\varphi_{\boldsymbol{\tau}}. \end{aligned}$$

Then the Jacobian of system (9.6.22)–(9.6.23) is

$$J := \varphi_{\boldsymbol{\nu}}g_2 - \varphi_{\boldsymbol{\tau}}g_1 = \frac{\rho_1^2-\rho^2}{\rho_1}\varphi_{\boldsymbol{\tau}}\varphi_{\boldsymbol{\nu}} = (\rho_1-\rho)\big(1+\frac{\rho}{\rho_1}\big)\varphi_{\boldsymbol{\nu}}\varphi_{\boldsymbol{\tau}}.$$

Recall that $\varphi_{\boldsymbol{\nu}}, \varphi_{\boldsymbol{\tau}} \neq 0$, by (8.1.16) and (9.6.15). Also, $\rho > \rho_1$ on Γ_{shock}, by Lemma 9.1.4. Thus, $J \neq 0$. Then

$$\begin{aligned} \varphi_{\boldsymbol{\nu}\boldsymbol{\tau}} &= \frac{g_2B_1 - \varphi_{\boldsymbol{\tau}}B_2}{J} = \frac{\rho_1(\rho_1B_1 - B_2)}{(\rho_1^2-\rho^2)\varphi_{\boldsymbol{\nu}}},\\ \varphi_{\boldsymbol{\tau}\boldsymbol{\tau}} &= \frac{-g_1B_1 + \varphi_{\boldsymbol{\nu}}B_2}{J} = \frac{\rho_1B_2 - \rho^2B_1}{(\rho_1-\rho)(\rho_1+\rho)\varphi_{\boldsymbol{\tau}}}. \end{aligned}$$

Since

$$\begin{aligned} \rho_1B_1 - B_2 &= -\frac{2(\rho-\rho_1)(\rho_1+\rho M_1^2)\varphi_{\boldsymbol{\tau}} + \rho_1c^2d_{\boldsymbol{\tau}}\big(\rho_1 - \rho(1+\frac{\rho-\rho_1}{\rho_1}M_1^2)\big)}{\rho_1(2+(\gamma-1)M^2)}\\ &= -\frac{(\rho-\rho_1)(\rho_1+\rho M_1^2)}{\rho_1(2+(\gamma-1)M^2)}(2\varphi_{\boldsymbol{\tau}} - c^2d_{\boldsymbol{\tau}}), \end{aligned}$$

we have

$$\varphi_{\boldsymbol{\nu}\boldsymbol{\tau}} = \frac{(\rho_1+\rho M_1^2)(2\varphi_{\boldsymbol{\tau}} - c^2d_{\boldsymbol{\tau}})}{(\rho+\rho_1)(2+(\gamma-1)M^2)}. \tag{9.6.24}$$

Similarly, we have

$$\begin{aligned} &\rho_1B_2 - \rho^2B_1\\ &= \frac{(\rho-\rho_1)\Big(\big(2(\rho_1+\rho M_1^2) + (\gamma-1)M^2(\rho+\rho_1)\big)\varphi_{\boldsymbol{\tau}} + c^2\rho(1-M_1^2)d_{\boldsymbol{\tau}}\Big)}{2+(\gamma-1)M^2} \end{aligned}$$

and

$$\varphi_{\boldsymbol{\tau}\boldsymbol{\tau}} = -\frac{\big((\gamma-1)(\rho+\rho_1)M^2 + 2(\rho_1+\rho M_1^2)\big)\varphi_{\boldsymbol{\tau}} + c^2\rho(1-M_1^2)d_{\boldsymbol{\tau}}}{\varphi_{\boldsymbol{\tau}}(\rho+\rho_1)(2+(\gamma-1)M^2)}. \tag{9.6.25}$$

Equation (2.2.8) is also invariant under the orthogonal transformation O so that

$$(c^2-\varphi_{\boldsymbol{\nu}}^2)\varphi_{\boldsymbol{\nu\nu}}-2\varphi_{\boldsymbol{\nu}}\varphi_{\boldsymbol{\tau}}\varphi_{\boldsymbol{\tau\nu}}+(c^2-\varphi_{\boldsymbol{\tau}}^2)\varphi_{\boldsymbol{\tau\tau}}=|D\varphi|^2-2c^2. \tag{9.6.26}$$

Then we have

$$\begin{aligned}\varphi_{\boldsymbol{\nu\nu}}&=\frac{1}{c^2-\varphi_{\boldsymbol{\nu}}^2}\Big(2\varphi_{\boldsymbol{\nu}}\varphi_{\boldsymbol{\tau}}\varphi_{\boldsymbol{\tau\nu}}-(c^2-\varphi_{\boldsymbol{\tau}}^2)\varphi_{\boldsymbol{\tau\tau}}+|D\varphi|^2-2c^2\Big)\\&=\frac{1}{1-M_1^2}\bigg(\frac{4M_2^2(\rho_1+\rho M_1^2)+(1-M_2^2)\big((\gamma-1)(\rho+\rho_1)M^2+2(\rho_1+\rho M_1^2)\big)}{(\rho+\rho_1)\big(2+(\gamma-1)M^2\big)}\\&\qquad+M^2-2-c^2d_{\boldsymbol{\tau}}\frac{2M_2^2(\rho_1+\rho M_1^2)-\rho(1-M_2^2)(1-M_1^2)}{\varphi_{\boldsymbol{\tau}}(\rho+\rho_1)\big(2+(\gamma-1)M^2\big)}\bigg).\end{aligned}$$

Therefore, we obtain

$$\begin{aligned}\varphi_{\boldsymbol{\nu\nu}}=&\frac{(\rho+\rho_1)\big(2(M^2+M_1^2-1)+(\gamma-1)M^2(M_1^2-1)\big)+2\rho(M_1^2M_2^2-1)}{(\rho+\rho_1)(1-M_1^2)\big(2+(\gamma-1)M^2\big)}\\&-\frac{c^2d_{\boldsymbol{\tau}}\Big(2M_2^2(\rho_1+\rho M_1^2)-\rho(1-M_2^2)(1-M_1^2)\Big)}{\varphi_{\boldsymbol{\tau}}(\rho+\rho_1)(1-M_1^2)\big(2+(\gamma-1)M^2\big)}.\end{aligned} \tag{9.6.27}$$

Furthermore, using (9.6.19), inequality (9.6.17) can be written as

$$\big(2+(\gamma-1)M^2\big)D^2\varphi[\boldsymbol{\nu},D\varphi]+(\gamma-1)M^2\varphi_{\boldsymbol{\nu}}\le-c^2d_{\boldsymbol{\nu}},$$

or, equivalently,

$$\big(2+(\gamma-1)M^2\big)(\varphi_{\boldsymbol{\nu\nu}}M_1^2+\varphi_{\boldsymbol{\nu\tau}}M_1M-2)+(\gamma-1)M^2M_1^2+c^2\varphi_{\boldsymbol{\nu}}d_{\boldsymbol{\nu}}\le 0.$$

The substitution of $\varphi_{\boldsymbol{\nu\nu}}$ and $\varphi_{\boldsymbol{\nu\tau}}$ into the above inequality yields

$$\begin{aligned}&\frac{(\rho+\rho_1)M_1^2\big(2(M^2+M_1^2-1)+(\gamma-1)M^2(M_1^2-1)\big)+2\rho M_1^2(M_1^2M_2^2-1)}{(\rho+\rho_1)(1-M_1^2)}\\&\quad+\frac{2M_2^2(\rho_1+\rho M_1^2)}{\rho+\rho_1}+(\gamma-1)M^2M_1^2+c^2\varphi_{\boldsymbol{\nu}}d_{\boldsymbol{\nu}}\\&\quad-\frac{c^2d_{\boldsymbol{\tau}}}{\varphi_{\boldsymbol{\tau}}}\Big(\frac{M_1^2\big((2\rho_1^2-\rho^2)M_2^2+\rho^2\big)}{(\rho_1^2-\rho^2)(1-M_1^2)}+\frac{\rho_1^2M_2^2}{\rho_1^2-\rho^2}\Big)\le 0,\end{aligned}$$

that is,

$$\begin{aligned}&2\rho M_1^2(M^2+M_1^2+M_1^2M_2^2-2)+2\rho_1M_1^2(M^2+M_2^2-1)\\&+2M_2^2(1-M_1^2)(\rho_1+\rho M_1^2)+c^2\varphi_{\boldsymbol{\nu}}(\rho+\rho_1)(1-M_1^2)d_{\boldsymbol{\nu}}\\&-\frac{c^2d_{\boldsymbol{\tau}}}{\varphi_{\boldsymbol{\tau}}(\rho_1+\rho)}\Big(M_1^2\big((\rho_1^2-\rho^2)M_2^2+\rho^2\big)+\rho_1^2M_2^2\Big)\le 0.\end{aligned}$$

Therefore, we have

$$\Delta := 2(2\rho+\rho_1)M_1^2(M^2-1) + 2\rho_1 M_2^2 + c^2\varphi_{\boldsymbol{\nu}}(\rho+\rho_1)(1-M_1^2)d_{\boldsymbol{\nu}} \\ -\frac{c^2\big((\rho_1^2-\rho^2)M_1^2M_2^2+\rho^2M_1^2+\rho_1^2M_2^2\big)}{\varphi_{\boldsymbol{\tau}}(\rho_1+\rho)}d_{\boldsymbol{\tau}} \le 0. \tag{9.6.28}$$

Property (9.6.15) implies that

$$M_2^2 \ge \alpha(M^2 - M_2^2).$$

Thus, recalling that $\alpha \in (0, \frac{1}{2})$ and choosing $\mu > 0$ small,

$$M_2^2 \ge \frac{\alpha}{2}M^2 \ge \frac{\alpha}{2}(1-C\mu) \ge \frac{\alpha}{4},$$

where we have used (9.6.13). Equivalently,

$$\varphi_{\boldsymbol{\tau}}^2 \ge \frac{\alpha}{4}c^2.$$

Therefore, (9.6.13) and (9.6.15) imply

$$M^2 - 1 \ge -C\mu, \quad M_1^2 \le M^2 \le 1 \qquad \text{at } P_{\max},$$

where we have used that $M^2 \le 1$ in Ω by (8.1.4). Then we find from (9.6.28) that

$$\begin{aligned}\Delta \ge & -2(2\rho+\rho_1)C\mu + 2\rho_1\frac{\alpha}{4} - C\mu - C\frac{\mu}{\sqrt{\alpha}}\\ \ge & \frac{1}{2}\rho_1\big(\alpha - 4(2\frac{\rho}{\rho_1}+1)C\mu\big) - C\frac{\mu}{\sqrt{\alpha}}\\ \ge & \frac{1}{2}\rho_1\alpha\big(1 - C\frac{\mu}{\alpha}\big) - C\frac{\mu}{\sqrt{\alpha}}\\ \ge & \frac{1}{2}\rho_1\alpha\big(1 - C\frac{\mu}{\alpha\sqrt{\alpha}}\big) > 0,\end{aligned}$$

if μ is chosen sufficiently small (recall that $\alpha > 0$ has been chosen).

This contradicts (9.6.28), which implies that the maximum of $M^2 + d$ for such a choice of $\mu > 0$ cannot be achieved on Γ_{shock}. □

Noting that $\boldsymbol{\xi} \mapsto \text{dist}(\boldsymbol{\xi}, \Gamma_{\text{sonic}})$ is Lipschitz, but not smooth in Ω, we now construct a function $g(\boldsymbol{\xi})$ that is smooth and comparable with $\text{dist}(\boldsymbol{\xi}, \Gamma_{\text{sonic}})$ in Ω.

Lemma 9.6.4. *There exist $g \in C(\mathbb{R}^2 \times [\theta_{\text{w}}^{\text{s}}, \frac{\pi}{2}])$ with $g(\cdot, \theta_{\text{w}}) \in C^\infty(\mathbb{R}^2)$ for each $\theta_{\text{w}} \in (\theta_{\text{w}}^{\text{s}}, \frac{\pi}{2}]$, and constants $C_0, C_1 < \infty$ depending only on (ρ_0, ρ_1, γ) such that, for any $\boldsymbol{\xi} \in \Omega(\varphi)$,*

$$\begin{aligned}&\frac{1}{C_0}\text{dist}(\boldsymbol{\xi}, \Gamma_{\text{sonic}}(\theta_{\text{w}})) \le g(\boldsymbol{\xi}, \theta_{\text{w}}) \le C_0\,\text{dist}(\boldsymbol{\xi}, \Gamma_{\text{sonic}}(\theta_{\text{w}})),\\ &|D_{\boldsymbol{\xi}}g(\boldsymbol{\xi}, \theta_{\text{w}})| \le 1, \qquad |D_{\boldsymbol{\xi}}^2 g(\boldsymbol{\xi}, \theta_{\text{w}})| \le C_1\end{aligned} \tag{9.6.29}$$

for any admissible solution φ *of* **Problem 2.6.1** *with* $\theta_{\rm w} \in (\theta_{\rm w}^{\rm s}, \frac{\pi}{2})$. *Furthermore, for any* $\theta_{\rm w} \in (\theta_{\rm w}^{\rm s}, \frac{\pi}{2})$,

$$\partial_\nu g(\cdot, \theta_{\rm w}) = 0 \qquad \text{on } \{\xi_2 = 0\} \cup \{\xi_2 = \xi_1 \tan\theta_{\rm w}\}. \tag{9.6.30}$$

Proof. Fix an admissible solution φ of **Problem 2.6.1** with corresponding wedge angle $\theta_{\rm w}$. In the following argument, we use the notations from Definition 7.5.7.

First, we note that $\Omega \subset \{\varphi_1 > \varphi_2\} \cap \Lambda$ by Definition 8.1.1(iv). Let $\mathcal{C}_2 = \partial B_{c_2}(\mathcal{O}_2)$ be the sonic circle of state (2) with center $\mathcal{O}_2 = (u_2, v_2)$. Line $\mathcal{S}_1 = \{\varphi_1 = \varphi_2\}$ intersects with $\mathcal{C}_2$ at point P_1, and segment P_0P_1 lies outside $\mathcal{C}_2$, as indicated in Remark 7.5.5. Moreover, $\mathcal{S}_1$ is not tangential to $\mathcal{C}_2$ as shown in Remark 7.5.5. Denote by Q another point of intersection of $\mathcal{S}_1$ and $\mathcal{C}_2$. It follows that P_1 lies between P_0 and Q on $\mathcal{S}_1$. Denote by $Q' := \frac{1}{2}(P_1 + Q)$ and let $\hat{c} = |\mathcal{O}_2 Q'|$. Then $\hat{c} = \sqrt{c_2^2 - \frac{1}{4}|P_1Q|^2} < c_2$. Denote by $Q'' := \mathcal{O}_2 + \hat{c}\boldsymbol{\tau}_{\rm w}$, where $\boldsymbol{\tau}_{\rm w}$ is the unit tangent vector to $\Gamma_{\rm wedge}$ defined by (8.2.17). Then $Q'' \in \Gamma_{\rm wedge}$ since $\Gamma_{\rm wedge} = P_3P_4 \ni \mathcal{O}_2$ and $P_4 = \mathcal{O}_2 + c_2\boldsymbol{\tau}_{\rm w}$ with $\hat{c} < c_2$. Let $\mathcal{D} = P_1P_4Q''Q'$ be the domain bounded by arcs $\Gamma_{\rm sonic} = P_1P_4$ and $Q''Q'$ (the smaller arc of $\partial B_{\hat{c}}(\mathcal{O}_2)$ with these endpoints), and segments P_1Q' and P_4Q'', *i.e.*, $\mathcal{D} \subset B_{c_2}(\mathcal{O}_2) \setminus B_{\hat{c}}(\mathcal{O}_2)$. Using that $\Omega \subset \{\varphi_1 > \varphi_2\} \cap \Lambda$, it follows that

$$\Omega \cap \{\boldsymbol{\xi} \,:\, \mathrm{dist}(\boldsymbol{\xi}, \mathcal{C}_2) < c_2 - \hat{c}\} \subset \mathcal{D}. \tag{9.6.31}$$

Let $P \in \mathcal{D}$, and let $A \in \mathcal{C}_2$ be the endpoint of the radius of $B_{c_2}(\mathcal{O}_2)$ passing through P. Then either $A \in \Gamma_{\rm sonic}$ or A on arc $P_1\hat{Q}$, where $\hat{Q} \in \mathcal{C}_2$ is the endpoint of the radius of $B_{c_2}(\mathcal{O}_2)$ passing through Q', *i.e.*, $\hat{Q}$ is the midpoint of arc P_1Q.

When $A \in \Gamma_{\rm sonic}$, $\mathrm{dist}(P, \Gamma_{\rm sonic}) = \mathrm{dist}(P, \mathcal{C}_2) = |AP|$. In the case that A is in arc $P_1\hat{Q}$, $\mathrm{dist}(P, \Gamma_{\rm sonic}) = |P_1P|$ and $\mathrm{dist}(P, \mathcal{C}_2) = |AP| < |P_1P|$. From the elementary geometry,

$$\frac{|P_1P|}{\sin(\angle P_1AP)} = \frac{|AP|}{\sin(\angle AP_1P)}.$$

Also, $0 < \beta := \angle\hat{Q}P_1Q < \angle AP_1Q < \angle AP_1P < \frac{\pi}{2}$. Note that $\beta = \angle\hat{Q}P_1Q$ is independent of P. Thus, for any $\boldsymbol{\xi} \in \mathcal{D}$,

$$\sin\beta \,\mathrm{dist}(\boldsymbol{\xi}, \Gamma_{\rm sonic}) \le \mathrm{dist}(\boldsymbol{\xi}, \mathcal{C}_2) \le \mathrm{dist}(\boldsymbol{\xi}, \Gamma_{\rm sonic}). \tag{9.6.32}$$

Let $h \in C^\infty([0,\infty))$ satisfy

$$h(s) = \begin{cases} s & \text{if } s \in [0, \frac{1}{2}], \\ 1 & \text{if } s \ge 1, \end{cases} \qquad 0 \le h' \le 2 \quad \text{on } [0,\infty).$$

Then the function:

$$\hat{g}_{\theta_{\rm w}}(\boldsymbol{\xi}) = \frac{1}{2}(c_2 - \hat{c})\, h(\frac{\mathrm{dist}(\boldsymbol{\xi}, \mathcal{C}_2)}{c_2 - \hat{c}}) \tag{9.6.33}$$

depends only on the parameters of state (2) for the wedge angle $\theta_{\rm w}$, and satisfies $\hat{g}_{\theta_{\rm w}}(\cdot) \in C^\infty(\mathbb{R}^2)$.

Since the parameters of state (2) depend smoothly on $\theta_{\rm w} \in [\theta_{\rm w}^{\rm s}, \frac{\pi}{2}]$, $g(\boldsymbol{\xi}, \theta_{\rm w}) = \hat{g}_{\theta_{\rm w}}(\boldsymbol{\xi})$ satisfies $g \in C(\mathbb{R}^2 \times [\theta_{\rm w}^{\rm s}, \frac{\pi}{2}])$ and $g(\cdot, \theta_{\rm w}) \in C^\infty(\mathbb{R}^2)$ for each $\theta_{\rm w} \in [\theta_{\rm w}^{\rm s}, \frac{\pi}{2}]$. Also, from (9.6.32) and the properties of $h(\cdot)$, it follows that, for any admissible solution φ for the wedge angle $\theta_{\rm w}$, g satisfies the properties in (9.6.29) with the constants:

$$C_0 = \frac{4\max\{1, \operatorname{diam}(\Omega)\}}{\sin\beta}, \qquad C_1 = C_1(\frac{1}{\hat{c}}).$$

Furthermore, from the construction of $\hat{g}$, $\beta > 0$ depends smoothly on the parameters of state (2) for $\theta_{\rm w}$, and hence on $\theta_{\rm w} \in [\theta_{\rm w}^{\rm s}, \frac{\pi}{2}]$. Thus, $\beta(\theta_{\rm w})$ and $\hat{c}(\theta_{\rm w})$ have positive lower bounds on $[\theta_{\rm w}^{\rm s}, \frac{\pi}{2}]$. Also, $\operatorname{diam}(\Omega)$ has a uniform upper bound by Proposition 9.1.2. Now (9.6.29) holds for every admissible solution φ for $\theta_{\rm w} \in (\theta_{\rm w}^{\rm s}, \frac{\pi}{2})$ with the uniform constant C_0.

Also, (9.6.30) is satisfied, since, for each $\theta_{\rm w} \in [\theta_{\rm w}^{\rm s}, \frac{\pi}{2})$, we see that

- Following from $\xi_{2P_0} > \xi_{2\mathcal{O}_2} = v_2$, $\xi_{2Q'} > \xi_{2\mathcal{O}_2}$ in such a way that, from the construction of $\mathcal{D}$, $\operatorname{dist}(\overline{\mathcal{D}}, \{\xi_2 = 0\}) \geq \delta > 0$. Now, using (9.6.31) and the definition of g, we find that $g(\cdot, \theta_{\rm w})$ is constant in the δ–neighborhood of $\{\xi_2 = 0\}$, *i.e.*, $\partial_{\boldsymbol{\nu}} g(\cdot, \theta_{\rm w}) = 0$ on $\{\xi_2 = 0\}$.

- Since $\mathcal{O}_2 \in \{\xi_2 = \xi_1 \tan\theta_{\rm w}\}$, $\partial_{\boldsymbol{\nu}}\big(\operatorname{dist}(\cdot, \mathcal{C}_2)\big) = 0$ on $\{\xi_2 = \xi_1 \tan\theta_{\rm w}\}$, which implies that $\partial_{\boldsymbol{\nu}} g(\cdot, \theta_{\rm w}) = 0$ on $\{\xi_2 = \xi_1 \tan\theta_{\rm w}\}$.

□

Proposition 9.6.5. *There exists $\tilde{\mu} > 0$ depending only on (ρ_0, ρ_1, γ) such that, for any admissible solution φ of* **Problem 2.6.1** *with $\theta_{\rm w} \in (\theta_{\rm w}^{\rm s}, \frac{\pi}{2})$,*

$$M^2(\boldsymbol{\xi}) \leq 1 - \tilde{\mu}\operatorname{dist}(\boldsymbol{\xi}, \Gamma_{\rm sonic}) \qquad \textit{for all} \ \ \boldsymbol{\xi} \in \overline{\Omega(\varphi)}.$$

Proof. Let $g = g(\boldsymbol{\xi}, \theta_{\rm w})$ be the function constructed in Lemma 9.6.4. Let φ be an admissible solution of **Problem 2.6.1** with $\theta_{\rm w} \in (\theta_{\rm w}^{\rm s}, \frac{\pi}{2})$. We now show that the maximum of $M^2 + \mu g$ over $\overline{\Omega}$ cannot be attained on $\overline{\Omega} \setminus \Gamma_{\rm sonic}$ if $\mu > 0$ is small, depending only on (ρ_0, ρ_1, γ).

Note also that, performing the even reflection as in Remark 8.1.3, φ satisfies equation (2.2.8) in the extended domain $\Omega^{\rm ext}$ and the Rankine-Hugoniot conditions (8.1.13)–(8.1.14) on $\Gamma_{\rm shock}^{\rm ext}$. Also, since $\partial_{\boldsymbol{\nu}} g = 0$ on $\Gamma_{\rm sym} \subset \{\xi_2 = 0\}$ by (9.6.30), then, for each $\theta_{\rm w}$, $g(\cdot, \theta_{\rm w})$ extended by even reflection to $\Omega^{\rm ext}$ is in $C^2(\Omega^{\rm ext})$.

Now, by Theorems 5.2.1 and 5.3.1 applied to the extended functions in $\Omega^{\rm ext}$, the maximum of $M^2 + \mu g$ cannot be attained in $\Omega^{\rm ext}$ and on $\Gamma_{\rm wedge}^{\rm ext}$ if μ is sufficiently small depending only on the data, where we have used the uniform bounds in (9.1.2), (9.1.6), and (9.6.29), and property (9.6.30) on $\Gamma_{\rm wedge}$. Thus, the maximum of $M^2 + \mu g$ cannot be attained in $\Omega \cup \Gamma_{\rm sym} \cup \Gamma_{\rm wedge}$.

By Proposition 9.6.3, reducing μ if necessary (depending only on the data), we find that the maximum of $M^2 + \mu g_{\theta_w}$ cannot be attained on $\Gamma_{\text{shock}} \cup P_2$.

It remains to estimate $M^2 + \mu g$ at $P_3 = \mathbf{0}$. Since $\varphi \in C^1(\overline{\Omega})$ and $\partial_{\boldsymbol{\nu}}\varphi = 0$ on $\Gamma_{\text{sym}} \cup \Gamma_{\text{wedge}}$, then $D\varphi(P_3) = 0$. Also, by (9.1.6),

$$c(P_3) = \rho^{\frac{\gamma-1}{2}}(P_3) \geq \big(\frac{2}{\gamma+1}\rho_1\big)^{\frac{\gamma-1}{2}} > 0.$$

Thus, $M^2(P_3) = 0$. Moreover, we obtain that $\mu g(P_3) \leq 1$ by further reducing μ, if necessary, with μ depending only on (ρ_0, ρ_1, γ), by (9.1.2) and (9.6.29). Therefore, $M^2 + \mu g \leq 1$ at P_3.

On the other hand, $M^2 + \mu g = 1$ on Γ_{sonic}, since $M^2 = 1$ on Γ_{sonic} by (8.1.3), and $g(\cdot, \theta_w) = 0$ on Γ_{sonic} by (9.6.29).

Therefore, $M^2 + \mu g \leq 1$ in $\overline{\Omega}$. Combining this with (9.6.29), we complete the proof with $\tilde{\mu} = \frac{\mu}{C_0}$. □

In Corollary 9.6.6 below, we use form (9.2.1)–(9.2.2) of equation (2.2.8).

Corollary 9.6.6. *There exists $C > 0$ depending only on (ρ_0, ρ_1, γ) such that, if φ is an admissible solution of* **Problem 2.6.1** *with $\theta_w \in (\theta_w^s, \frac{\pi}{2})$, equation* (2.2.8) *for solution φ is elliptic in $\overline{\Omega} \setminus \Gamma_{\text{sonic}}$ with degeneracy near Γ_{sonic}. More precisely, for any $\boldsymbol{\xi} \in \overline{\Omega}$ and $\boldsymbol{\kappa} = (\kappa_1, \kappa_2) \in \mathbb{R}^2$,*

$$\frac{\operatorname{dist}(\boldsymbol{\xi}, \Gamma_{\text{sonic}})}{C}|\boldsymbol{\kappa}|^2 \leq \sum_{i,j=1}^{2} \mathcal{A}^i_{p_j}(D\varphi(\boldsymbol{\xi}), \varphi(\boldsymbol{\xi}))\kappa_i\kappa_j \leq C|\boldsymbol{\kappa}|^2, \tag{9.6.34}$$

where we have used the notations in (9.2.1)–(9.2.2).

Proof. From (9.2.2) and (2.2.9), we obtain that, for any $\boldsymbol{\xi} \in \overline{\Omega}$ and $\boldsymbol{\kappa} \in \mathbb{R}^2$,

$$c^2(|D\varphi|^2, \varphi)\Big(1 - \frac{|D\varphi|^2}{c^2(|D\varphi|^2, \varphi)}\Big)|\boldsymbol{\kappa}|^2 \leq \sum_{i,j=1}^{2} \mathcal{A}^i_{p_j}(D\varphi, \varphi)\kappa_i\kappa_j \leq c^2(|D\varphi|^2, \varphi)|\boldsymbol{\kappa}|^2,$$

where φ is evaluated at $\boldsymbol{\xi}$. Furthermore, by Proposition 9.6.5,

$$\frac{|D\varphi|^2}{c^2(|D\varphi|^2, \varphi)} \leq 1 - \frac{1}{2}\tilde{\mu}\operatorname{dist}(\boldsymbol{\xi}, \Gamma_{\text{sonic}}) \qquad \text{in } \Omega,$$

where $\tilde{\mu} > 0$ depends only on (ρ_0, ρ_1, γ). Combining these estimates with (9.1.6) and $c^2 = \rho^{\gamma-1}$, we conclude (9.6.34). □

Chapter Ten

Regularity of Admissible Solutions away from the Sonic Arc

In this chapter, we focus on the regularity of admissible solutions away from the sonic arc Γ_{sonic}.

Throughout this chapter, we always fix $\gamma > 1$ and $\rho_1 > \rho_0 > 0$, and use the notations introduced in Definition 7.5.7. In order to obtain the regularity near the symmetry line Γ_{sym}, it is convenient to extend solution φ to φ^{ext} by even reflection into $\{\xi_2 < 0\}$ as in Remark 8.1.3 and to use the extended sets Ω^{ext}, $\Gamma^{\text{ext}}_{\text{shock}}$, and $\Gamma^{\text{ext}}_{\text{sonic}}$ defined in Remark 8.1.3. We also define

$$\Gamma^{\text{ext}}_{\text{wedge}} := \Gamma_{\text{wedge}} \cup \Gamma^{-}_{\text{wedge}} \cup \{P_3\},$$

where $\Gamma^{-}_{\text{wedge}}$ is the reflection of Γ_{wedge} with respect to the ξ_1–axis.

10.1 Γ_{shock} AS A GRAPH IN THE RADIAL DIRECTIONS WITH RESPECT TO STATE (1)

Let (r, θ) be the polar coordinates with respect to center $\mathcal{O}_1 = (u_1, 0)$ of the sonic circle of state (1), *i.e.*, (r, θ) are defined by (6.3.3) with $(\hat{u}, \hat{v}) = (u_1, 0)$.

Lemma 10.1.1. *There exist $\varepsilon, \delta > 0$ depending only on (ρ_0, ρ_1, γ) such that, for any admissible solution φ with the wedge angle $\theta_{\text{w}} \in (\theta^{\text{s}}_{\text{w}}, \frac{\pi}{2})$,*

$$\partial_r(\varphi_1 - \varphi^{\text{ext}}) \leq -\delta \qquad \textit{in } \; \mathcal{N}_\varepsilon(\Gamma^{\text{ext}}_{\text{shock}}) \cap \Omega^{\text{ext}},$$

where $\mathcal{N}_\varepsilon(A)$ is the ε–neighborhood of set A in the $\boldsymbol{\xi}$–coordinates, and φ^{ext}, Ω^{ext}, and $\Gamma^{\text{ext}}_{\text{shock}}$ are defined in Remark 8.1.3.

Proof. Since $\varphi^{\text{ext}}(\xi_1, -\xi_2) = \varphi(\xi_1, \xi_2)$ and $\varphi^{\text{ext}}_1(\xi_1, -\xi_2) = \varphi_1(\xi_1, \xi_2)$ for $(\xi_1, \xi_2) \in \Omega$, and $\mathcal{O}_1$ lies on the ξ_1–axis, it suffices to show that

$$\partial_r(\varphi_1 - \varphi) \leq -\delta \qquad \text{in } \mathcal{N}_\varepsilon(\Gamma_{\text{shock}}) \cap \Omega.$$

From the Bernoulli law (2.2.7) applied to φ in Ω and to φ_1 in $\mathbb{R}^2$, using $\varphi \in C^1(\overline{\Omega})$ by (8.1.3), we have

$$\rho^{\gamma-1} + (\gamma - 1)\big(\frac{1}{2}|D\varphi|^2 + \varphi\big) = \rho_1^{\gamma-1} + (\gamma - 1)\big(\frac{1}{2}|D\varphi_1|^2 + \varphi_1\big) \quad \text{in } \overline{\Omega}. \; (10.1.1)$$

Now, since $\varphi \leq \varphi_1$ in Ω by (8.1.5),

$$\rho^{\gamma-1} + \frac{\gamma-1}{2}|D\varphi|^2 \geq \rho_1^{\gamma-1} + \frac{\gamma-1}{2}|D\varphi_1|^2 \qquad \text{in } \overline{\Omega}. \tag{10.1.2}$$

Next, choosing $\hat{d} = \frac{1}{2C}$ that depends only on (ρ_0, ρ_1, γ) for C from Proposition 9.5.6, we have

$$\text{dist}(\mathcal{N}_{\hat{d}}(\Gamma_{\text{shock}}), B_{c_1}(\mathcal{O}_1)) \geq \frac{1}{C} - \hat{d} = \hat{d}.$$

Thus, for any $P \in \mathcal{N}_{\hat{d}}(\Gamma_{\text{shock}}) \cap \Omega$,

$$|D\varphi_1(P)| = |P\mathcal{O}_1| \geq c_1 + \hat{d} \qquad \text{in } \mathcal{N}_{\hat{d}}(\Gamma_{\text{shock}}) \cap \Omega.$$

That is, choosing

$$\delta = \frac{\hat{d}^2}{\gamma+1},$$

we have

$$|D\varphi_1(P)|^2 \geq c_1^2 + (\gamma+1)\delta = \rho_1^{\gamma-1} + (\gamma+1)\delta \qquad \text{in } \mathcal{N}_{\hat{d}}(\Gamma_{\text{shock}}) \cap \Omega.$$

Also, $|D\varphi|^2 \leq c^2 = \rho^{\gamma-1}$ in Ω by (8.1.4). Then, from (10.1.2), we obtain

$$\rho^{\gamma-1} \geq \rho_1^{\gamma-1} + (\gamma-1)\delta \qquad \text{in } \mathcal{N}_{\hat{d}}(\Gamma_{\text{shock}}) \cap \Omega.$$

Combining this with (10.1.1), we have

$$|D\varphi_1|^2 - |D\varphi|^2 \geq 2\delta - (\varphi_1 - \varphi) \qquad \text{in } \mathcal{N}_{\hat{d}}(\Gamma_{\text{shock}}) \cap \Omega. \tag{10.1.3}$$

Since $\varphi = \varphi_1$ on Γ_{shock}, we employ Corollary 9.1.3 to find that, for any $\varepsilon > 0$,

$$|\varphi_1 - \varphi| \leq C\varepsilon \qquad \text{in } \mathcal{N}_{\varepsilon}(\Gamma_{\text{shock}}) \cap \Omega.$$

Combining this with (10.1.3), we obtain that, for any $\varepsilon \in (0, \hat{d})$,

$$|D\varphi_1|^2 - |D\varphi|^2 \geq 2\delta - C\varepsilon \qquad \text{in } \mathcal{N}_{\varepsilon}(\Gamma_{\text{shock}}) \cap \Omega.$$

Choosing $\varepsilon = \min\{\frac{\delta}{C}, \hat{d}\}$, we have

$$|D\varphi_1|^2 - |D\varphi|^2 \geq \delta \qquad \text{in } \mathcal{N}_{\varepsilon}(\Gamma_{\text{shock}}) \cap \Omega.$$

From this, we use Corollary 9.1.3 to conclude

$$|D\varphi_1| - |D\varphi| \geq \frac{\delta}{C} \qquad \text{in } \mathcal{N}_{\varepsilon}(\Gamma_{\text{shock}}) \cap \Omega.$$

Since $|D\varphi_1| = -\partial_r\varphi_1$, the last inequality implies the assertion (with $\frac{\delta}{C}$ instead of δ). $\square$

Corollary 10.1.2. *There exist $\hat{\delta}, \hat{\delta}_1 > 0$ depending only on (ρ_0, ρ_1, γ) such that, if φ is an admissible solution with $\theta_{\rm w} \in (\theta_{\rm w}^{\rm s}, \frac{\pi}{2}]$, then, on $\overline{\Gamma_{\rm shock}}$,*

$$\partial_{\boldsymbol{\nu}}(\varphi_1 - \varphi) \geq \hat{\delta}, \tag{10.1.4}$$

$$\partial_{\boldsymbol{\nu}}\varphi_1 > \partial_{\boldsymbol{\nu}}\varphi \geq \hat{\delta}_1. \tag{10.1.5}$$

Proof. Since $\varphi < \varphi_1$ in $\Omega^{\rm ext}$, $\varphi = \varphi_1$ on $\Gamma^{\rm ext}_{\rm shock} \subset \partial\Omega^{\rm ext}$, and $\boldsymbol{\nu}$ is the interior normal on $\Gamma^{\rm ext}_{\rm shock}$ to $\Omega^{\rm ext}$, then

$$\partial_{\boldsymbol{\nu}}(\varphi_1 - \varphi) = |D(\varphi_1 - \varphi)| \qquad \text{on } \overline{\Gamma^{\rm ext}_{\rm shock}}.$$

Now (10.1.4) follows from Lemma 10.1.1.

Next, we show (10.1.5). From the Rankine-Hugoniot condition (8.1.13) on $\Gamma_{\rm shock}$, we obtain that $\partial_{\boldsymbol{\nu}_{\rm sh}}\varphi_1 = \frac{\rho}{\rho_1}\partial_{\boldsymbol{\nu}_{\rm sh}}\varphi$. Then, from (10.1.4),

$$(\frac{\rho}{\rho_1} - 1)\partial_{\boldsymbol{\nu}_{\rm sh}}\varphi \geq \hat{\delta} \qquad \text{on } \overline{\Gamma_{\rm shock}}.$$

Lemma 9.1.4 implies that $\frac{\rho}{\rho_1} > 1$ on $\Gamma_{\rm shock}$. Therefore, we employ the bound: $\rho < C$ in (9.1.6) to obtain (10.1.5). □

Corollary 10.1.3 ($\Gamma_{\rm shock}$ is a graph in the radial directions with respect to $\mathcal{O}_1$). *For any admissible solution φ with $\theta_{\rm w} \in (\theta_{\rm w}^{\rm s}, \frac{\pi}{2})$, there exists a function $f_{\mathcal{O}_1,{\rm sh}} \in C^1(\mathbb{R})$ such that*

$$\begin{aligned} &\Omega^{\rm ext} \cap \{(r,\theta) \,:\, r > 0,\ \theta_{P_1} < \theta < \theta_{P_1{}^-}\} \\ &\quad = \Lambda \cap \{(r,\theta) \,:\, \theta_{P_1} < \theta < \theta_{P_1{}^-},\ r < f_{\mathcal{O}_1,{\rm sh}}(\theta)\}, \\ &\Gamma^{\rm ext}_{\rm shock} = \{(r,\theta) \,:\, \theta_{P_1} < \theta < \theta_{P_1{}^-},\ r = f_{\mathcal{O}_1,{\rm sh}}(\theta)\}, \end{aligned} \tag{10.1.6}$$

where (r_P, θ_P) are the (r,θ)-coordinates of point P, and ${P_1}^-$ denotes the reflection of P_1 with respect to the ξ_1-axis in the $\boldsymbol{\xi}$-coordinates with $\theta_{P_1{}^-} = 2\pi - \theta_{P_1} > \pi$.

Proof. The existence of $f_{\mathcal{O}_1,{\rm sh}}$ follows from Lemma 10.1.1. The C^1-regularity follows from (10.1.1) and Lemma 10.1.1, since $\Gamma^{\rm ext}_{\rm shock} = \{\varphi^{\rm ext} = \varphi_1\}$, as shown in Remark 8.1.3. □

Lemma 10.1.4. *There exists $C > 0$ such that the following holds: Let φ be an admissible solution of* **Problem 2.6.1** *with $\theta_{\rm w} \in (\theta_{\rm w}^{\rm s}, \frac{\pi}{2})$. Let $f_{\mathcal{O}_1,{\rm sh}}$ be the extended shock function of φ in the (r,θ)-coordinates; cf.* (10.1.6). *Then*

$$|f'_{\mathcal{O}_1,{\rm sh}}| \leq C \qquad \text{on } (\theta_{P_1}, \theta_{P_1{}^-}). \tag{10.1.7}$$

Proof. In this proof, constants L and C depend only on (ρ_0, ρ_1, γ).

Let $\bar{\phi} = \varphi_1 - \varphi$ in Ω. Since φ_1 defined by (2.2.17) is an even function with respect to ξ_2, we extend $\bar{\phi}$ into $\Omega^{\rm ext}$ by even reflection and employ Remark 8.1.3 to obtain

$$\bar{\phi}^{\rm ext} = \varphi_1 - \varphi^{\rm ext} \in C^2(\overline{\Omega^{\rm ext}} \setminus (\overline{\Gamma^{\rm ext}_{\rm sonic}} \cup \{P_2, P_3\})) \cap C^1(\overline{\Omega^{\rm ext}}). \tag{10.1.8}$$

From (8.1.5) and (8.1.15), we have

$$\bar{\phi}^{\text{ext}} > 0 \quad \text{in } \Omega^{\text{ext}}, \qquad \bar{\phi}^{\text{ext}} = 0 \quad \text{on } \Gamma^{\text{ext}}_{\text{shock}}. \tag{10.1.9}$$

Also, from (2.2.17), (9.1.2), and (9.1.5), we obtain

$$\|\bar{\phi}^{\text{ext}}\|_{C^{0,1}(\overline{\Omega^{\text{ext}}})} \le L. \tag{10.1.10}$$

From (8.1.2), we have

$$\text{dist}(\Gamma^{\text{ext}}_{\text{shock}}, \mathcal{O}_1) > c_1.$$

Then, in the $\frac{c_1}{2}$–neighborhood of $\Gamma^{\text{ext}}_{\text{shock}}$, denoted as $\mathcal{N}_{\frac{c_1}{2}}(\Gamma^{\text{ext}}_{\text{shock}})$, the C^1–norms of the coordinate transform $\boldsymbol{\xi} \mapsto (r, \theta)$ and its inverse are bounded by a constant C depending only on $c_1 = \rho_1^{\gamma-1}$. Thus, from (10.1.10),

$$|\partial_{(r,\theta)}\bar{\phi}^{\text{ext}}| \le CL \qquad \text{in } \mathcal{N}_{\frac{c_1}{2}}(\Gamma^{\text{ext}}_{\text{shock}}) \cap \Omega^{\text{ext}}. \tag{10.1.11}$$

Moreover, (10.1.9) and Lemma 10.1.1 imply

$$f'_{O_1,\text{sh}}(\theta) = -\frac{\partial_\theta \bar{\phi}^{\text{ext}}(f_{O_1,\text{sh}}(\theta),\ \theta)}{\partial_r \bar{\phi}^{\text{ext}}(f_{O_1,\text{sh}}(\theta),\ \theta)} \qquad \text{on } (\theta_{P_1}, \theta_{P_1 -}).$$

Therefore, using (10.1.6), (10.1.11), and Lemma 10.1.1, we conclude (10.1.7). □

10.2 BOUNDARY CONDITIONS ON Γ_{shock} FOR ADMISSIBLE SOLUTIONS

Let φ be an admissible solution with $\theta_{\text{w}} \in [\theta^{\text{s}}_{\text{w}}, \frac{\pi}{2})$. Then φ satisfies (8.1.13)–(8.1.15) on Γ_{shock}. From (8.1.14),

$$\boldsymbol{\nu}_{\text{sh}} = \frac{D(\varphi_1 - \varphi)}{|D(\varphi_1 - \varphi)|}.$$

Thus, (8.1.13) can be written as

$$g^{\text{sh}}(D\varphi, \varphi, \boldsymbol{\xi}) = 0 \qquad \text{on } \Gamma_{\text{shock}}, \tag{10.2.1}$$

where $g^{\text{sh}}(\mathbf{p}, z, \boldsymbol{\xi})$ is defined by (7.1.9). We now regularize $g^{\text{sh}}(\mathbf{p}, z, \boldsymbol{\xi})$ in such a way that admissible solutions satisfy (10.2.1) with the modified/regularized function $g^{(\text{sh})}_{\text{mod}}(\mathbf{p}, z, \boldsymbol{\xi})$.

From Proposition 9.6.5 and Lemma 10.1.1,

$$\frac{|D\varphi|}{c(|D\varphi|^2, \varphi)} \le 1 - \delta \qquad \text{on } \Omega \setminus \mathcal{N}_\varepsilon(\Gamma_{\text{sonic}}),$$

$$|D(\varphi_1 - \varphi^{\text{ext}})| \ge \delta \qquad \text{on } \mathcal{N}_\varepsilon(\Gamma^{\text{ext}}_{\text{shock}}) \cap \Omega^{\text{ext}}$$

for $\varepsilon, \delta > 0$ depending only on $(\rho_0, \rho_1, \gamma, \theta^*_{\rm w})$. In particular, choosing

$$\eta \in C^\infty(\mathbb{R}) \text{ with } \eta(t) = \begin{cases} \frac{\delta}{2} & \text{on } (-\infty, \frac{\delta}{2}], \\ t & \text{on } [\frac{3\delta}{4}, \infty), \\ \eta'(t) \geq 0 & \text{on } \mathbb{R}, \end{cases} \tag{10.2.2}$$

so that $\eta(t) \geq \frac{\delta}{2}$ on $\mathbb{R}$, we find that, for any admissible solution φ,

$$\boldsymbol{\nu}_{\rm sh} = \frac{D(\varphi_1 - \varphi)}{\eta\left(|D(\varphi_1 - \varphi)|\right)}.$$

Motivated by these properties, we now modify $g^{\rm sh}(\mathbf{p}, z, \boldsymbol{\xi})$.

Lemma 10.2.1. *Let $M \geq 2$, and let $\mathcal{K}_M$ and $\tilde{\mathcal{A}}(\mathbf{p}, z)$ be the set and function defined in Lemma 9.2.1, respectively. Let $\delta > 0$, and let $\eta(\cdot)$ be the function in (10.2.2). Let*

$$g^{(\rm sh)}_{\rm mod}(\mathbf{p}, z, \boldsymbol{\xi}) = \left(\tilde{\mathcal{A}}(\mathbf{p}, z) - \rho_1 D\varphi_1(\boldsymbol{\xi})\right) \cdot \frac{D\varphi_1(\boldsymbol{\xi}) - \mathbf{p}}{\eta\left(|D\varphi_1(\boldsymbol{\xi}) - \mathbf{p}|\right)}. \tag{10.2.3}$$

Then there exist positive constants ε, $\delta_{\rm bc}$, and C_k with $k = 1, 2, \dots$, depending only on $(M, \delta, \rho_0, \rho_1, \gamma)$, such that

(i) *For any $\boldsymbol{\xi} \in \mathbb{R}^2$ and any $(\tilde{\mathbf{p}}, \tilde{z}) \in \mathcal{K}_M$ with $|\tilde{\mathbf{p}} - D\varphi_1(\boldsymbol{\xi})| \geq \delta$,*

$$g^{(\rm sh)}_{\rm mod}(\mathbf{p}, z, \boldsymbol{\xi}) = g^{\rm sh}(\mathbf{p}, z, \boldsymbol{\xi}) \qquad \text{if } |(\mathbf{p}, z) - (\tilde{\mathbf{p}}, \tilde{z})| < \varepsilon; \tag{10.2.4}$$

(ii) *For $\mathcal{R}_M = \{(\mathbf{p}, z, \boldsymbol{\xi}) : |\mathbf{p}| \leq 2M, |z| \leq 2M, |\boldsymbol{\xi}| \leq 2M\}$,*

$$\|g^{(\rm sh)}_{\rm mod}\|_{C^k(\overline{\mathcal{R}_M})} \leq C_k \qquad \text{for } k = 1, 2, \dots; \tag{10.2.5}$$

(iii) *For any $(\mathbf{p}, z) \in \mathcal{K}_M$ with $|\mathbf{p} - D\varphi_1(\boldsymbol{\xi})| \geq \delta$,*

$$D_{\mathbf{p}} g^{(\rm sh)}_{\rm mod}(\mathbf{p}, z, \boldsymbol{\xi}) \cdot \frac{D\varphi_1(\boldsymbol{\xi}) - \mathbf{p}}{|D\varphi_1(\boldsymbol{\xi}) - \mathbf{p}|} \geq \delta_{\rm bc}. \tag{10.2.6}$$

Proof. Property (10.2.4) follows from (9.2.3) and (9.2.5). Estimate (10.2.5) follows from (9.2.5). Thus, it remains to show (10.2.6).

Fix $(\tilde{\mathbf{p}}, \tilde{z}, \tilde{\boldsymbol{\xi}})$ satisfying the conditions in (iii). Then, from (10.2.2),

$$\frac{D\varphi_1(\tilde{\boldsymbol{\xi}}) - \mathbf{p}}{\eta(|D\varphi_1(\tilde{\boldsymbol{\xi}}) - \mathbf{p}|)} = \frac{D\varphi_1(\tilde{\boldsymbol{\xi}}) - \mathbf{p}}{|D\varphi_1(\tilde{\boldsymbol{\xi}}) - \mathbf{p}|} \qquad \text{for all } \mathbf{p} \in B_{\frac{\delta}{8}}(\tilde{\mathbf{p}}),$$

which implies that $\frac{|D\varphi_1(\tilde{\boldsymbol{\xi}}) - \mathbf{p}|}{\eta(|D\varphi_1(\tilde{\boldsymbol{\xi}}) - \mathbf{p}|)} = 1$ for any $\mathbf{p} \in B_{\delta/8}(\tilde{\mathbf{p}})$. Denoting $\mathbf{e} := \frac{D\varphi_1(\tilde{\boldsymbol{\xi}}) - \tilde{\mathbf{p}}}{\eta(|D\varphi_1(\tilde{\boldsymbol{\xi}}) - \tilde{\mathbf{p}}|)}$, then $|\mathbf{e}| = 1$ and

$$(\mathbf{e} \cdot D_{\mathbf{p}})\Big(\frac{D\varphi_1(\tilde{\boldsymbol{\xi}}) - \mathbf{p}}{\eta(|D\varphi_1(\tilde{\boldsymbol{\xi}}) - \mathbf{p}|)}\Big)\Big|_{\mathbf{p}=\tilde{\mathbf{p}}} = \frac{1}{2} D_{\mathbf{p}}\Big(\Big|\frac{D\varphi_1(\tilde{\boldsymbol{\xi}}) - \mathbf{p}}{\eta(|D\varphi_1(\tilde{\boldsymbol{\xi}}) - \mathbf{p}|)}\Big|^2\Big)\Big|_{\mathbf{p}=\tilde{\mathbf{p}}} = 0.$$

From this, we use (10.2.3) and calculate at $(\mathbf{p}, z, \boldsymbol{\xi}) = (\tilde{\mathbf{p}}, \tilde{z}, \tilde{\boldsymbol{\xi}})$ to obtain

$$D_{\mathbf{p}} g^{(\mathrm{sh})}_{\mathrm{mod}}(\tilde{\mathbf{p}}, \tilde{z}, \tilde{\boldsymbol{\xi}}) \cdot \mathbf{e} = \sum_{i,j=1}^{2} \tilde{\mathcal{A}}^i_{p_j}(\tilde{\mathbf{p}}, \tilde{z}) \Big(\frac{D\varphi_1(\tilde{\boldsymbol{\xi}}) - \tilde{\mathbf{p}}}{\eta(|D\varphi_1(\tilde{\boldsymbol{\xi}}) - \tilde{\mathbf{p}}|)} \Big)_i e_j$$
$$= \sum_{i,j=1}^{2} \tilde{\mathcal{A}}^i_{p_j}(\tilde{\mathbf{p}}, \tilde{z}) e_i e_j \geq \lambda > 0,$$

where we have used ellipticity (9.2.4). This implies (10.2.6) with $\delta_{\mathrm{bc}} = \lambda$ from (9.2.4). □

10.3 LOCAL ESTIMATES NEAR Γ_{shock}

We continue to consider the solutions extended by reflection into $\{\xi_2 < 0\}$. Again, in this section, (r, θ) still denotes the polar coordinates with respect to center $\mathcal{O}_1 = (u_1, 0)$ of the sonic circle of state (1).

Proposition 10.3.1. *For each $d > 0$, there exist $s > 0$ and $C_k, \hat{C}_k < \infty$ for $k = 2, 3, \ldots$, such that the following holds: Let φ be an admissible solution of* **Problem 2.6.1** *with the wedge angle $\theta_{\mathrm{w}} \in (\theta^{\mathrm{s}}_{\mathrm{w}}, \frac{\pi}{2})$. Let $f_{\mathcal{O}_1,\mathrm{sh}}$ be the extended shock function of φ in the (r, θ)–coordinates (cf. (10.1.6)). If $P = (r_P, \theta_P) \in \Gamma^{\mathrm{ext}}_{\mathrm{shock}}$ satisfies*

$$\mathrm{dist}(P, \Gamma^{\mathrm{ext}}_{\mathrm{sonic}}) \geq d, \qquad \mathrm{dist}(P, \Gamma^{\mathrm{ext}}_{\mathrm{wedge}}) \geq d,$$

then

$$|D^k f_{\mathcal{O}_1,\mathrm{sh}}(\theta_P)| \leq C_k(d) \qquad \text{for } k = 2, 3, \ldots, \tag{10.3.1}$$
$$|D^k_{\boldsymbol{\xi}} \varphi| \leq \hat{C}_k(d) \qquad \text{on } B_s(P) \cap \Omega^{\mathrm{ext}} \ \text{ for } k = 2, 3, \ldots, \tag{10.3.2}$$

where $B_s(P)$ is the ball with radius s and center at P in the $\boldsymbol{\xi}$–coordinates.

Proof. In this proof, constants C, M, ε, δ, β, λ, and Λ are positive and depend only on $(\rho_0, \rho_1, \gamma, d)$. Furthermore, from Remark 8.1.3, equation (2.2.8) holds in Ω^{ext} for $\varphi = \varphi^{\mathrm{ext}}$, and the Rankine-Hugoniot condition (8.1.13) holds on $\Gamma^{\mathrm{ext}}_{\mathrm{shock}}$. We divide the proof into three steps.

1. Let ε and δ be from Lemma 10.1.1. If we choose

$$R := \min\{\frac{d}{2}, \varepsilon, \frac{c_1}{2}\} > 0,$$

then R depends only on $(\rho_0, \rho_1, \gamma, d)$, and ball $B_R(P)$ in the $\boldsymbol{\xi}$–coordinates satisfies

$$B_R(P) \cap \Gamma^{\mathrm{ext}}_{\mathrm{wedge}} = \emptyset, \tag{10.3.3}$$
$$\mathrm{dist}(B_R(P), \Gamma^{\mathrm{ext}}_{\mathrm{sonic}}) \geq \frac{d}{2}, \tag{10.3.4}$$
$$\mathrm{dist}(B_R(P), \mathcal{O}_1) \geq \frac{c_1}{2}, \tag{10.3.5}$$
$$\partial_r(\varphi_1 - \varphi) \leq -\delta \qquad \text{in } B_R(P) \cap \Omega^{\mathrm{ext}}, \tag{10.3.6}$$

where, by the choice of R, property (10.3.3) follows from $\mathrm{dist}(P, \Gamma^{\mathrm{ext}}_{\mathrm{wedge}}) \geq d$, property (10.3.4) follows from $\mathrm{dist}(P, \Gamma^{\mathrm{ext}}_{\mathrm{sonic}}) \geq d$, property (10.3.5) follows from (8.1.2), and property (10.3.6) follows from Lemma 10.1.1.

Our first step is to employ Theorem 4.3.2 to establish the $C^{1,\alpha}$–regularity, so that we need to verify its assumptions. We work in the (r, θ)–coordinates.

For $P = (r_p, \theta_P)$, by Corollary 10.1.3, after shifting the (r, θ)–coordinates and inverting the r–direction, *i.e.*, changing (r, θ) to $(\tilde{r}, \tilde{\theta}) = (r_p - r, \theta - \theta_p)$, domain $B_R(P) \cap \Omega^{\mathrm{ext}}$ is of the form as in Theorem 4.3.2, and (4.3.1) holds with λ depending only on (ρ_0, ρ_1, γ) by Lemma 10.1.4.

To verify the other assumptions in Theorem 4.3.2, we first work in the $\boldsymbol{\xi}$–coordinates.

By (10.3.4) and Proposition 9.6.5, there exists $\delta > 0$ such that

$$\frac{|D\varphi|^2}{c^2(|D\varphi|^2, \varphi)} \leq 1 - \delta \qquad \text{in } B_R(P) \cap \Omega^{\mathrm{ext}}.$$

Using this and (9.1.5)–(9.1.6), we see that there exists a sufficiently large M such that φ satisfies (9.2.6)–(9.2.8) in $\mathcal{D} = B_R(P) \cap \Omega^{\mathrm{ext}}$. Thus, by Lemma 9.2.1, we can modify coefficients $\mathcal{A}$ and B in (9.2.1) so that they are defined for all $(\mathbf{p}, z) \in \mathbb{R}^2 \times \mathbb{R}$ and satisfy (9.2.3)–(9.2.5) with λ and C_k depending only on $(\rho_0, \rho_1, \gamma, d)$, and φ satisfies the modified equation (9.2.9) in $\Omega^{\mathrm{ext}} \cap B_R(P)$.

From (10.3.6),

$$|D(\varphi_1 - \varphi)| \geq \delta \qquad \text{in } \Omega^{\mathrm{ext}} \cap B_R(P). \tag{10.3.7}$$

Then, applying Lemma 10.2.1 with M as above and δ from (10.3.7) to obtain the corresponding function $g^{(\mathrm{sh})}_{\mathrm{mod}}(\mathbf{p}, z, \boldsymbol{\xi})$, we find from (10.2.4) that φ satisfies

$$g^{(\mathrm{sh})}_{\mathrm{mod}}(D\varphi, \varphi, \boldsymbol{\xi}) = 0 \qquad \text{on } \Gamma^{\mathrm{ext}}_{\mathrm{shock}} \cap B_R(P). \tag{10.3.8}$$

Furthermore, since $\dfrac{D(\varphi_1 - \varphi)}{|D(\varphi_1 - \varphi)|}$ is a unit vector, then, from (10.2.6), we have

$$|D_{\mathbf{p}} g^{(\mathrm{sh})}_{\mathrm{mod}}(D\varphi, \varphi, \boldsymbol{\xi})| \geq \delta_{\mathrm{bc}} > 0 \qquad \text{in } \Omega^{\mathrm{ext}} \cap B_R(P). \tag{10.3.9}$$

Using the regularity in (9.2.5) and (10.2.5) of the modified equation and the boundary condition, and writing the equation in the non-divergence form (4.2.1) with $f \equiv 0$, we obtain that conditions (4.3.5)–(4.3.6) and (4.3.8) are satisfied. Furthermore, ellipticity (4.3.4) holds by (9.2.4). The nondegeneracy of the boundary condition (4.3.7) holds by (10.3.9). Finally, (4.3.9) holds by (9.1.5).

Now we change to the polar coordinates (r, θ) with respect to $\mathcal{O}_1 = (u_1, 0)$. By (10.3.5), this change is smooth and nondegenerate, with the norms controlled in terms of the data. Thus, conditions (4.3.4)–(4.3.9) hold in the (r, θ)–coordinates, with the constants depending only on $(\rho_0, \rho_1, \gamma, d)$.

2. We can now apply Theorem 4.3.2 to obtain $\beta \in (0,1)$ and $C > 0$ depending only on $(\rho_0, \rho_1, \gamma, d)$ such that

$$\|\varphi\|_{C^{1,\beta}(\overline{\Omega^{\rm ext}\cap B_{R/2}(P)})} \le C$$

in the (r,θ)–coordinates, and hence also in the $\boldsymbol{\xi}$–coordinates, by (10.3.5). From this, we use (10.3.6) and the smoothness of φ_1 to obtain

$$\|f_{O_1,{\rm sh}}\|_{C^{1,\beta}(\overline{I_1})} \le C,$$

where $I_1 \subset \mathbb{R}$ is an open interval such that

$$\Gamma^{\rm ext}_{\rm shock} \cap B_{R/2}(P) = \{r = f_{O_1,{\rm sh}}(\theta) \;:\; \theta \in I_1\}.$$

3. Now we use Corollary 4.3.5 inductively, for the open intervals $I_k \subset \mathbb{R}$, $k = 1, 2, \dots$, such that $\Gamma^{\rm ext}_{\rm shock} \cap B_{R/2^k}(P) = \{r = f_{O_1,{\rm sh}}(\theta) \;:\; \theta \in I_k\}$. Then $\theta_P \in I_{k+1} \subset I_k$ for $k = 1, 2, \dots$. Note that condition (4.3.40) is satisfied in $\Omega^{\rm ext} \cap B_R$ for each k by (9.2.5) and (10.2.5), where we recall that the equation is written in the non-divergence form. Therefore, the induction goes as follows: Suppose that, for some k,

$$\|f_{O_1,{\rm sh}}\|_{C^{k,\beta}(\overline{I_k})} \le C.$$

This estimate for $k = 1$ has been obtained in Step 2, which verifies condition (4.3.39) for domain $\Omega^{\rm ext} \cap B_{R/2^k}$. Thus, by Corollary 4.3.5,

$$\|\varphi\|_{C^{k+1,\beta}(\overline{\Omega^{\rm ext}\cap B_{R/2^{k+1}}(P)})} \le C.$$

From this, we apply (10.3.6) and the smoothness of φ_1 to conclude

$$\|f_{O_1,{\rm sh}}\|_{C^{k+1,\beta}(\overline{I_{k+1}})} \le C.$$

This completes the proof. □

10.4 THE CRITICAL ANGLE AND THE DISTANCE BETWEEN $\Gamma_{\rm shock}$ AND $\Gamma_{\rm wedge}$

From Proposition 9.4.5 and Corollary 9.5.7, we see that, when $u_1 \le c_1$, there is a uniform positive lower bound for the distance between $\Gamma_{\rm shock}$ and $\Gamma_{\rm wedge}$ for admissible solutions with $\theta_{\rm w} \in [\theta^*_{\rm w}, \frac{\pi}{2})$ for any $\theta^*_{\rm w} \in (\theta^{\rm s}_{\rm w}, \frac{\pi}{2})$. However, when $u_1 > c_1$, we only know from Lemma 9.4.7 that $\Gamma_{\rm shock}$ cannot hit $\Gamma_{\rm wedge}$ for the wedge angles sufficiently close to $\frac{\pi}{2}$, so that we cannot rule out the possibility that, for a limit of a sequence of admissible solutions, the limiting shock may hit $\Gamma_{\rm wedge}$. Moreover, from Proposition 9.4.8, it follows that, if $\Gamma_{\rm shock}$ does hit $\Gamma_{\rm wedge}$ for the limiting solution described in Corollary 9.2.5, then the contact point is only the wedge vertex P_3. These issues lead to the difference in the results in our main theorems, Theorems 2.6.3 and 2.6.5.

We also note that, while we have not known yet whether there exists a case that $\Gamma_{\rm shock}$ of its solution is attached to $\Gamma_{\rm wedge}$, the experimental results [263, Fig. 238, Page 144] suggest that such *attached* solutions may indeed exist.

Thus, we expect the existence of admissible solutions only for the wedge angles $\theta_{\rm w} \in (\theta_{\rm w}^{\rm c}, \frac{\pi}{2})$, where the *critical angle* $\theta_{\rm w}^{\rm c}$ is defined as follows:

Definition 10.4.1. *Fix $\gamma > 1$ and $\rho_1 > \rho_0 > 0$. Define the set:*

$$\mathcal{A} := \left\{ \theta_{\rm w}^* \in (\theta_{\rm w}^{\rm s}, \frac{\pi}{2}] \;:\; \begin{array}{c} \exists\, \varepsilon > 0 \;\; \textit{such that} \;\; \mathrm{dist}(\Gamma_{\rm shock}, \Gamma_{\rm wedge}) \geq \varepsilon \\ \textit{for any admissible solution with } \theta_{\rm w} \in [\theta_{\rm w}^*, \frac{\pi}{2}] \end{array} \right\},$$

for which the normal reflection solution as the unique admissible solution for $\theta_{\rm w} = \frac{\pi}{2}$ is included. Since $\mathrm{dist}(\Gamma_{\rm shock}, \Gamma_{\rm wedge}) > 0$ for the normal reflection, $\frac{\pi}{2} \in \mathcal{A}$, i.e. $\mathcal{A} \neq \emptyset$. Now we define the critical angle

$$\theta_{\rm w}^{\rm c} = \inf \mathcal{A}.$$

Directly from Definition 10.4.1, $\theta_{\rm w}^{\rm c} \in [\theta_{\rm w}^{\rm s}, \frac{\pi}{2}]$. Furthermore, we have

Lemma 10.4.2. *Fix $\gamma > 1$ and $\rho_1 > \rho_0 > 0$. Then the critical angle $\theta_{\rm w}^{\rm c}$ satisfies the following properties:*

(i) $\theta_{\rm w}^{\rm c} < \frac{\pi}{2}$, *i.e.*, $\theta_{\rm w}^{\rm c} \in [\theta_{\rm w}^{\rm s}, \frac{\pi}{2})$;

(ii) *If* $u_1 \leq c_1$, $\theta_{\rm w}^{\rm c} = \theta_{\rm w}^{\rm s}$;

(iii) *If $\theta_{\rm w}^{\rm c} > \theta_{\rm w}^{\rm s}$, there exists a sequence of admissible solutions $\varphi^{(i)}$ with the wedge angles $\theta_{\rm w}^{(i)} \in [\theta_{\rm w}^{\rm c}, \frac{\pi}{2}]$ such that*

$$\lim_{i \to \infty} \theta_{\rm w}^{(i)} = \theta_{\rm w}^{\rm c}, \qquad \lim_{i \to \infty} \mathrm{dist}(\Gamma_{\rm shock}^{(i)}, \Gamma_{\rm wedge}^{(i)}) = 0;$$

(iv) *For any $\theta_{\rm w}^* \in (\theta_{\rm w}^{\rm c}, \frac{\pi}{2})$, there exists $C > 0$ such that, for any admissible solution φ with $\theta_{\rm w} \in [\theta_{\rm w}^*, \frac{\pi}{2})$,*

$$\mathrm{dist}(\Gamma_{\rm shock}, \Gamma_{\rm wedge}) \geq \frac{1}{C}.$$

Proof. Assertion (i) follows from Lemma 9.4.7. Assertion (ii) follows from Proposition 9.4.5 and Corollary 9.5.7. Assertions (iii)–(iv) follow directly from Definition 10.4.1. □

10.5 REGULARITY OF ADMISSIBLE SOLUTIONS AWAY FROM $\Gamma_{\rm sonic}$

In this section, we consider admissible solutions for the wedge angles up to the critical angle $\theta_{\rm w}^{\rm c}$ introduced in Definition 10.4.1. In particular, we establish the estimates for admissible solutions with the wedge angles $\theta_{\rm w} \in (\theta_{\rm w}^{\rm c}, \frac{\pi}{2})$.

First we show the regularity of $\Gamma_{\rm shock}$ including endpoint P_2, and the regularity of solutions near $\Gamma_{\rm shock}$.

Proposition 10.5.1. *Fix $\theta_{\rm w}^* \in (\theta_{\rm w}^{\rm c}, \frac{\pi}{2})$. Then, for each $d > 0$, there exist $s > 0$ and $C_k, \hat{C}_k < \infty$ for $k = 2, 3, \dots$, such that the following holds: Let φ be an admissible solution of* **Problem 2.6.1** *with the wedge angle $\theta_{\rm w} \in [\theta_{\rm w}^*, \frac{\pi}{2})$. Let $f_{O_1,{\rm sh}}$ be the extended shock function of φ in the (r,θ)–coordinates (cf.* (10.1.6)). *If $d > 0$ and $P = (r_P, \theta_P) \in \Gamma_{\rm shock}^{\rm ext}$ satisfy $\operatorname{dist}(P, \Gamma_{\rm sonic}^{\rm ext}) \ge d$, then*

$$|D^k f_{O_1,{\rm sh}}(\theta_P)| \le C_k(d) \qquad \textit{for } k = 2,3,\dots, \tag{10.5.1}$$

$$|D^k_{\boldsymbol{\xi}} \varphi| \le \hat{C}_k(d) \qquad \textit{in } B_s(P) \cap \Omega^{\rm ext} \ \textit{ for } k = 2,3,\dots, \tag{10.5.2}$$

where $B_s(P)$ is the ball in the $\boldsymbol{\xi}$–coordinates.

Proof. By Lemma 10.4.2(iv), $\operatorname{dist}(\Gamma_{\rm shock}^{\rm ext}, \Gamma_{\rm wedge}^{\rm ext}) \ge \frac{1}{C}$, where C depends only on the data and $\theta_{\rm w}^*$. In particular, $\operatorname{dist}(P, \Gamma_{\rm wedge}^{\rm ext}) \ge \frac{1}{C}$. Now, applying Proposition 10.3.1 with $\hat{d} = \min\{d, \frac{1}{C}\}$ instead of d, we complete the proof. □

Next, we show the regularity near the wedge vertex.

Lemma 10.5.2. *There exists $\alpha \in (0,1)$ such that, for any $\theta_{\rm w}^* \in (\theta_{\rm w}^{\rm c}, \frac{\pi}{2})$, there are $s > 0$ and $C < \infty$ so that, for any admissible solution φ of* **Problem 2.6.1** *with the wedge angle $\theta_{\rm w} \in [\theta_{\rm w}^*, \frac{\pi}{2})$,*

$$\|\varphi\|_{C^{1,\alpha}(\overline{B_s(P_3)\cap\Omega})} \le C. \tag{10.5.3}$$

Proof. In this proof, α depends only on (ρ_0, ρ_1, γ). The universal constant C depends only on $(\rho_0, \rho_1, \gamma, \theta_{\rm w}^*)$.

Since P_2 and $\mathcal{O}_2$ lie on $\Gamma_{\rm wedge}$, $\operatorname{dist}(P_2, \Gamma_{\rm sonic}) = |P_2 - P_1| = |\Gamma_{\rm wedge}|$. Thus, for any $\theta_{\rm w} \in (\theta_{\rm w}^{\rm s}, \frac{\pi}{2})$, we have

$$\operatorname{dist}(P_2, \Gamma_{\rm sonic}) \ge r_0, \tag{10.5.4}$$

where $r_0 > 0$ is from (9.4.30) and depends only on (ρ_0, ρ_1, γ).

Fix $\theta_{\rm w}^* \in (\theta_{\rm w}^{\rm c}, \frac{\pi}{2})$. Let φ be an admissible solution with $\theta_{\rm w} \in [\theta_{\rm w}^*, \frac{\pi}{2})$. Then, by Lemma 10.4.2(iv), there exists $s > 0$ depending only on $(\rho_0, \rho_1, \gamma, \theta_{\rm w}^*)$ such that $\operatorname{dist}(\Gamma_{\rm shock}, \Gamma_{\rm wedge}) > 4s$. Further reducing s if necessary, depending on the same parameters, we see that $s \le \frac{r_0}{4}$, so that $4s \le \operatorname{dist}(P_2, \Gamma_{\rm sonic})$. Now, from the properties of Ω in Definition 8.1.1(i), we have

$$\Omega \cap B_{3s}(P_3) = \Lambda \cap B_{3s}(P_3),$$

which implies that $\Omega \cap B_{3s}(P_3)$ is the sector of $B_{3s}(P_3)$ with angle $\pi - \theta_{\rm w} \in (\frac{\pi}{2}, \pi - \theta_{\rm w}^*)$.

Since $\operatorname{dist}(\Omega \cap B_{2s}(P_3), \Gamma_{\rm sonic}) \ge \frac{r_0}{4}$ by (10.5.4), Corollary 9.6.6 implies that equation (2.2.8) is uniformly elliptic on solution φ in domain $\Omega \cap B_{2s}(P_3) = \Lambda \cap B_{3s}(P_3)$, with the ellipticity constants depending only on $(\rho_0, \rho_1, \gamma, r_0)$, *i.e.*, on (ρ_0, ρ_1, γ). Moreover, the $C^{0,1}(\Omega)$–bound of φ is given in Corollary 9.1.5. In particular, writing equation (2.2.8) in the form of (9.2.1) and arguing as in Step 1 of the proof of Lemma 9.2.2, we conclude that φ satisfies equation (9.2.9) in $\Lambda \cap B_{2s}(P_3)$, for which properties (9.2.3)–(9.2.5) hold with constants $\lambda, C_0, C_1, \dots$, depending only on (ρ_0, ρ_1, γ). Thus, the requirements of Theorem 4.3.13 hold for the equation with the constants depending on (ρ_0, ρ_1, γ).

On the sides of sector $\Lambda \cap B_{3s}(P_3)$, *i.e.*, on $\Gamma_{\text{wedge}} \cap B_{3s}(P_3)$ and $\Gamma_{\text{sym}} \cap B_{3s}(P_3)$, the boundary condition: $\frac{\partial\varphi}{\partial\nu} = 0$ is satisfied. The requirements of Theorem 4.3.13 for the boundary conditions clearly hold for these boundary conditions; specifically, (4.3.88) and (4.3.106) hold with $\lambda = 1$, and (4.3.107) holds with $\lambda = \sin\theta_{\text{w}} \geq \sin\theta_{\text{w}}^{\text{d}}$. Thus, the constants for the boundary conditions depend also on (ρ_0, ρ_1, γ).

Now (10.5.3) follows from Theorem 4.3.13 applied in domain $\Lambda \cap B_{3s}(P_3)$. Then the Hölder exponent α in (10.5.3) depends on (ρ_0, ρ_1, γ), and C depends on θ_{w}^* in addition to (ρ_0, ρ_1, γ), since s depends on these parameters. □

Now we have the following estimate away from Γ_{sonic}:

Corollary 10.5.3. *There exists $\alpha \in (0,1)$ such that, for any $\theta_{\text{w}}^* \in (\theta_{\text{w}}^{\text{c}}, \frac{\pi}{2})$ and $d > 0$, there is $C < \infty$ so that, for any admissible solution φ of* **Problem 2.6.1** *with $\theta_{\text{w}} \in [\theta_{\text{w}}^*, \frac{\pi}{2})$,*

$$\|\varphi\|^{(-1-\alpha),\{P_3\}}_{2,\alpha,\Omega\setminus\mathcal{N}_d(\Gamma_{\text{sonic}})} \leq C. \tag{10.5.5}$$

Proof. By Lemma 10.4.2(iv), estimate (10.5.5) follows directly from Lemma 9.2.2, Proposition 10.5.1, and Lemma 10.5.2, as well as a covering argument, combined with the standard rescaling technique for obtaining the weighted estimates near P_3. □

10.6 REGULARITY OF THE LIMIT OF ADMISSIBLE SOLUTIONS AWAY FROM Γ_{sonic}

In this section, we use the notations introduced in Corollary 9.2.5. Also, recall that Γ_{shock}, Γ_{sonic}, Γ_{wedge}, and Γ_{sym} denote the relative interiors of these segments, *i.e.*, without endpoints.

Proposition 10.6.1. *Fix $\theta_{\text{w}}^* \in (\theta_{\text{w}}^{\text{c}}, \frac{\pi}{2})$. Let $\{\varphi^{(i)}\}$ be a sequence of admissible solutions of* **Problem 2.6.1** *with the wedge angles $\theta_{\text{w}}^{(i)} \in [\theta_{\text{w}}^*, \frac{\pi}{2})$ such that*

$$\theta_{\text{w}}^{(i)} \to \theta_{\text{w}}^{(\infty)} \in [\theta_{\text{w}}^*, \frac{\pi}{2}).$$

Let $\varphi^{(i)} \to \varphi^{(\infty)} \in C^{0,1}_{\text{loc}}(\overline{\Lambda^{(\infty)}})$ uniformly in compact subsets of $\overline{\Lambda^{(\infty)}}$, where $\varphi^{(\infty)}$ is a weak solution of **Problem 2.6.1**, *and let the convergence properties in Corollary* 9.2.5(i)–(iii) *hold for the whole sequence. Fix a unit vector $\mathbf{g} \in \text{Cone}^0(\mathbf{e}_{\mathcal{S}_1^{(\infty)}}, \mathbf{e}_{\xi_2})$, use the (S,T)-coordinates as in Corollary* 9.2.5, *let $f^{(\infty)}_{\mathbf{g},\text{sh}}$ be the corresponding shock function, and let $\Gamma^{(\infty)}_{\text{shock}}$ be defined by* (9.2.23). *Then*

(i) $\Gamma^{(\infty)}_{\text{shock}}$ *is disjoint with* $\Gamma^{(\infty)}_{\text{sym}} \cup \overline{\Gamma^{(\infty)}_{\text{wedge}}} \cup \Gamma^{(\infty)}_{\text{sonic}}$. *In particular,* $\Gamma^{(\infty),0}_{\text{wedge}}$, $\Gamma^{(\infty),0}_{\text{sym}}$, $\widehat{\Omega^{(\infty)}}$, *and* $\Omega^{(\infty)}$ *introduced in Corollary* 9.2.5(iii) *satisfy that* $\Gamma^{(\infty),0}_{\text{wedge}} = \Gamma^{(\infty)}_{\text{wedge}}$, $\Gamma^{(\infty),0}_{\text{sym}} = \Gamma^{(\infty)}_{\text{sym}}$, *and* $\widehat{\Omega^{(\infty)}} = \Omega^{(\infty)}$, *and that* $\Omega^{(\infty)}$ *is connected;*

(ii) *In the* (S,T) *coordinates with basis* $\{\boldsymbol{\nu}_{\text{w}}^{(\infty)}, \boldsymbol{\tau}_{\text{w}}^{(\infty)}\}$,

$$f^{(\infty)}_{\boldsymbol{\nu}_{\text{w}}^{(\infty)},\text{sh}}(T) > \max(0, -T\tan\theta_{\text{w}}^{(\infty)}) \qquad \textit{for } T \in (T_{P_2{}^{(\infty)}}, T_{P_1{}^{(\infty)}}),$$

and $\Omega^{(\infty)}$ *is of the form:*

$$\Omega^{(\infty)} = \left\{ (S,T) \in \mathbb{R}^2 \;:\; \begin{array}{c} T_{P_2{}^{(\infty)}} \le T \le T_{P_4{}^{(\infty)}}, \\ -(T - T_{P_3})\tan\theta_{\text{w}}^{(\infty)} < S < f^{(\infty)}_{\boldsymbol{\nu}_{\text{w}}^{(\infty)},\text{sh}}(T) \\ \qquad\qquad \textit{for } T \in [T_{P_2{}^{(\infty)}}, T_{P_3{}^{(\infty)}}], \\ 0 < S < f^{(\infty)}_{\boldsymbol{\nu}_{\text{w}}^{(\infty)},\text{sh}}(T) \;\; \textit{for } T \in (T_{P_3{}^{(\infty)}}, T_{P_1{}^{(\infty)}}], \\ 0 < S < f^{(\infty)}_{\boldsymbol{\nu}_{\text{w}}^{(\infty)},\text{so}}(T) \;\; \textit{for } T \in (T_{P_1{}^{(\infty)}}, T_{P_4{}^{(\infty)}}] \end{array} \right\}, \tag{10.6.1}$$

where $f^{(\infty)}_{\boldsymbol{\nu}_{\text{w}}^{(\infty)},\text{so}}$ *is defined in Remark* 8.2.12;

(iii) *There exists a function* $f^{(\infty)}_{O_1,\text{sh}} \in C^\infty((\theta_{P_1{}^{(\infty)}}, \theta_{(P_1{}^{-})^{(\infty)}}))$ *such that, in polar coordinates,*

$$\Gamma^{(\infty)}_{\text{shock}} = \{r = f^{(\infty)}_{O_1,\text{sh}}(\theta) \;:\; \theta_{P_1{}^{(\infty)}} < \theta < \theta_{(P_1{}^{-})^{(\infty)}}\}, \tag{10.6.2}$$

where we have used the notation from Remark 8.1.3 *and Corollary* 10.1.3;

(iv) $\varphi^{(\infty)} \in C^\infty(\overline{\Omega^{(\infty)}} \setminus (\overline{\Gamma^{(\infty)}_{\text{sonic}}} \cup \{P_3{}^{(\infty)}\}))$;

(v) $\varphi^{(\infty)} \in C^{1,\alpha}(\overline{\Omega^{(\infty)}} \setminus \overline{\Gamma^{(\infty)}_{\text{sonic}}})$ *with* $\alpha \in (0,1)$ *depending only on* (ρ_0, ρ_1, γ), *but independent of* θ_{w}^*.

Proof. We divide the proof into five steps.

1. Proof of (i). By Lemma 10.4.2(iv), $\text{dist}(\Gamma^{(i)}_{\text{shock}}, \Gamma^{(i)}_{\text{wedge}}) \ge \frac{1}{C}$ for all i. Also, $P_3{}^{(i)} = \mathbf{0}$, $P_4{}^{(i)} \to P_4{}^{(\infty)}$ since $\theta_{\text{w}}^{(i)} \to \theta_{\text{w}}^{(\infty)}$, and $\Gamma^{(i)}_{\text{wedge}}$ is segment $P_3{}^{(i)}P_4{}^{(i)}$. Thus, $\Gamma^{(i)}_{\text{wedge}} \to \Gamma^{(\infty)}_{\text{wedge}}$ in the Hausdorff metric. Moreover, $f^{(i)}_{\mathbf{g},\text{sh}} \to f^{(\infty)}_{\mathbf{g},\text{sh}}$ uniformly by Corollary 9.2.5(ii). Therefore, $\Gamma^{(i)}_{\text{shock}} \to \Gamma^{(\infty)}_{\text{shock}}$ in the Hausdorff metric. Then it follows that $\text{dist}(\Gamma^{(\infty)}_{\text{shock}}, \Gamma^{(\infty)}_{\text{wedge}}) \ge \frac{1}{C}$.

Similarly, $\Gamma^{(\infty)}_{\text{shock}}$ does not intersect with $\Gamma^{(\infty)}_{\text{sym}}$, which is known by using Corollary 9.3.2 applied to each $\varphi^{(i)}$ and the convergence: $P_2{}^{(i)} \to P_2{}^{(\infty)}$ that holds by Corollary 9.2.5 (i).

Finally, $\Gamma_{\rm shock}^{(\infty)}$ does not intersect with $\Gamma_{\rm sonic}^{(\infty)}$, since $\Gamma_{\rm shock}^{(i)} \subset \{\xi_{1P_2} \le \xi_1 \le \xi_{1P_1}\}$ by (8.1.2), $\Gamma_{\rm sonic}^{(i)} \subset \{\xi_1 \ge \xi_{1P_1}\}$ from the definition of $\Gamma_{\rm sonic}$ in Definition 7.5.7, and the same properties hold for the limiting solution.

Also, the properties proved above imply that $\widehat{\Omega^{(\infty)}} = \Omega^{(\infty)}$, and $\Omega^{(\infty)}$ is connected. Now (i) is proved.

2. Now (ii) follows directly from (i) and Corollary 9.2.5(iii).

3. Proof of (iii). We note that

$$(\Gamma_{\rm shock}^{\rm ext})^{(i)} = \{r = f_{O_1,{\rm sh}}^{(i)}(\theta) \;:\; \theta_{{P_1}^{(i)}} < \theta < \theta_{({P_1}^-)^{(i)}}\} \tag{10.6.3}$$

in the (r,θ)–coordinates by (10.1.6). Also, ${P_1}^{(i)} \to P_1(\theta_{\rm w}^{(\infty)}) =: {P_1}^{(\infty)}$, $({P_1}^-)^{(i)} \to ({P_1}^-)(\theta_{\rm w}^{(\infty)}) =: ({P_1}^-)^{(\infty)}$, and $\mathcal{O}_1^{(i)} \to \mathcal{O}_1(\theta_{\rm w}^{(\infty)})$ as $i \to \infty$. Moreover, (10.1.7) and (10.5.1) hold for every $f_{O_1,{\rm sh}}^{(i)}$, $i = 1, 2, \dots$. Then there exists a subsequence of $\{f_{O_1,{\rm sh}}^{(i)}\}$ converging uniformly on compact subsets of $(\theta_{{P_1}^{(\infty)}}, \theta_{({P_1}^-)^{(\infty)}})$, whose limit is denoted by $f_{O_1,{\rm sh}}^{(\infty)}$. Therefore, $f_{O_1,{\rm sh}}^{(\infty)}$ satisfies (10.1.7) and (10.5.1) on $(\theta_{{P_1}^{(\infty)}}, \theta_{({P_1}^-)^{(\infty)}})$ so that

$$f_{O_1,{\rm sh}}^{(\infty)} \in C^\infty((\theta_{{P_1}^{(\infty)}}, \theta_{({P_1}^-)^{(\infty)}})).$$

On the other hand, denoting $\boldsymbol{\nu}_{\rm w}^{(\infty)} := \boldsymbol{\nu}_{\rm w}(\theta_{\rm w}^{(\infty)})$, we obtain that $\boldsymbol{\nu}_{\rm w}^{(\infty)} \in {\rm Cone}^0(\mathbf{e}_{\mathcal{S}_1^{(\infty)}}, \mathbf{e}_{\xi_2})$ by Lemma 8.2.11. Then $\boldsymbol{\nu}_{\rm w}^{(\infty)} \in {\rm Cone}^0(\mathbf{e}_{\mathcal{S}_1^{(i)}}, \mathbf{e}_{\xi_2})$ for sufficiently large i so that, for such i, we have

$$(\Gamma_{\rm shock}^{\rm ext})^{(i)} = \{S = f_{\boldsymbol{\nu}_{\rm w}^{(\infty)}}^{(i)}(T) \;:\; T_{P_1}^{(i)} < T < T_{P_1}^{(i)}\},$$

by Corollary 8.2.14(i). Also, $(\Gamma_{\rm shock}^{\rm ext})^{(\infty)}$ is defined by (9.2.23) with $f_{\rm e}^{(\infty)} = f_{\boldsymbol{\nu}_{\rm w}^{(\infty)}}^{(\infty)}$. Then Corollary 9.2.5(ii) implies that

$$\Gamma_{\rm shock}^{(i)} \to \Gamma_{\rm shock}^{(\infty)} \qquad \text{in the Hausdorff metric as } i \to \infty.$$

Then (10.6.3) and the uniform convergence, $\{f_{O_1,{\rm sh}}^{(i)}\} \to \{f_{O_1,{\rm sh}}^{(\infty)}\}$, on compact subsets of $(\theta_{{P_1}^{(\infty)}}, \theta_{({P_1}^-)^{(\infty)}})$ imply that (10.6.2) holds. Also, we have shown that $f_{O_1,{\rm sh}}^{(\infty)} \in C^\infty((\theta_{{P_1}^{(\infty)}}, \theta_{({P_1}^-)^{(\infty)}}))$. Now (iii) is proved.

4. Proof of (iv). Let $P \in \Gamma_{\rm shock}^{(\infty)} \cup \{P_2\}$ (extended as in Remark 8.1.3) and ${\rm dist}(P, \Gamma_{\rm sonic}) = 2d > 0$. Let $s > 0$ be the constant from Proposition 10.5.1 for distance d from $\Gamma_{\rm sonic}$. We can also assume without loss of generality that $s < \frac{d}{2}$. Then the convergence, $\Gamma_{\rm sonic}^{(i)} \to \Gamma_{\rm sonic}^{(\infty)}$ and $\Gamma_{\rm shock}^{(i)} \to \Gamma_{\rm shock}^{(\infty)}$, in the Hausdorff metric implies that there exists N such that $B_{s/2}(P) \subset B_s(P^{(i)})$ for some $P^{(i)} \in \Gamma_{\rm shock}^{(i)}$, and ${\rm dist}(P^{(i)}, \Gamma_{\rm sonic}^{(i)}) > d$. For every compact $\mathcal{K} \subset B_{s/2}(P) \cap \Omega^{(\infty)}$, there exists $N_{\mathcal{K}}$ such that $\mathcal{K} \subset B_s(P^{(i)}) \cap \Omega^{(i)}$ for each $i \ge N_{\mathcal{K}}$. Then the uniform convergence $\varphi^{(i)} \to \varphi^{(\infty)}$ on $\mathcal{K}$, combined with estimates (10.5.2), for $\varphi^{(i)}$ on $B_s(P^{(i)})\cap\Omega^{(i)}$, implies the same estimates for $\varphi^{(\infty)}$ on $\mathcal{K}$: For each $j = 1, 2, \dots$,

$$\|\varphi^{(\infty)}\|_{C^j(\mathcal{K})} \leq \limsup_{i\to\infty} \|\varphi^{(i)}\|_{C^j(\mathcal{K})} \leq \limsup_{i\to\infty} \|\varphi^{(i)}\|_{C^j(\overline{B_s(P^{(i)})\cap\Omega^{(i)}})} \leq C_j,$$

independent of $\mathcal{K}$, where we have used that $\operatorname{dist}(P^{(i)}, \Gamma^{(i)}_{\text{sonic}}) > d$ and $s = s(d)$. This implies

$$\|\varphi^{(\infty)}\|_{C^j(\overline{B_{s/2}(P)\cap\Omega^{(\infty)}})} \leq C_j.$$

That is, (10.5.2) holds for the extended function $\varphi^{(\infty)}$. Combining this with (9.2.24) and the structure of $\Omega^{(\infty)}$ described in assertion (i) shown above, we obtain (iv).

5. Proof of (v). By Lemma 10.4.2(iv), $\operatorname{dist}(\Gamma^{(i)}_{\text{shock}}, \Gamma^{(i)}_{\text{wedge}}) \geq \frac{1}{C}$ for all i. Let $s = \frac{1}{4C}$. Then, from the argument in the proof of Lemma 10.5.2, it follows that $\Lambda^{(i)} \cap B_{3s}(P_3) \subset \Omega^{(i)}$ for each i. Combining this with the uniform convergence $\Gamma^{(i)}_{\text{wedge}} \to \Gamma^{(\infty)}_{\text{wedge}}$ implies that, for every compact $K \subset B_s(P_3)\cap\Omega^{(\infty)}$, there exists N such that $\mathcal{K} \subset B_s(P_3) \cap \Omega^{(i)}$ for each $i \geq N$. By (9.2.27) in Corollary 9.2.5, $\varphi^{(i)} \to \varphi^{(\infty)}$ holds in $C^{1,\alpha}(\mathcal{K})$, where $\alpha \in (0, 1)$ is from Lemma 10.5.2. Then (10.5.3) in Lemma 10.5.2 for $\varphi^{(i)}$ implies that there exists $C > 0$ independent of $\mathcal{K}$ such that

$$\|\varphi^{(\infty)}\|_{C^{1,\alpha}(\mathcal{K})} \leq \lim_{i\to\infty} \|\varphi^{(i)}\|_{C^{1,\alpha}(\mathcal{K})} \leq \lim_{i\to\infty} \|\varphi^{(i)}\|_{C^{1,\alpha}(\overline{B_s(P_3)\cap\Omega^{(i)}})} \leq C.$$

This implies

$$\|\varphi^{(\infty)}\|_{C^{1,\alpha}(\overline{B_s(P_3)\cap\Omega^{(\infty)}})} \leq C.$$

Combining this with (iv) yields (v). □

Chapter Eleven

Regularity of Admissible Solutions near the Sonic Arc

In this chapter, we establish the regularity of admissible solutions of **Problem 2.6.1** near the sonic arc Γ_{sonic}. To achieve this, it is more convenient to consider the function:

$$\psi = \varphi - \varphi_2 \qquad \text{in } \Omega.$$

From (8.1.3) and (8.1.5),

$$\psi \in C^1(\overline{\Omega}), \tag{11.0.1}$$
$$\psi \geq 0 \qquad \text{in } \Omega, \tag{11.0.2}$$
$$\psi = 0 \qquad \text{on } \Gamma_{\text{sonic}}. \tag{11.0.3}$$

We always fix $\rho_1 > \rho_0 > 0$, $\gamma > 1$, and $\theta_{\text{w}}^* \in (\theta_{\text{w}}^{\text{s}}, \frac{\pi}{2})$ for the sonic angle $\theta_{\text{w}}^{\text{s}} = \theta_{\text{w}}^{\text{s}}(\rho_0, \rho_1, \gamma)$ throughout this chapter.

11.1 THE EQUATION NEAR THE SONIC ARC AND STRUCTURE OF ELLIPTIC DEGENERACY

Fix $\theta_{\text{w}} \in (\theta_{\text{w}}^{\text{s}}, \frac{\pi}{2}]$. From §6.1 applied with $\varphi_{\text{un}} = \varphi_2$, $(\hat{u}, \hat{v}) = (u_2, v_2)$, and $c_{\text{un}} = c_2$, we have the following: Let (r, θ) be the polar coordinates with respect to $\mathcal{O}_2 = (u_2, v_2)$:

$$(\xi_1 - u_2, \xi_2 - v_2) = (r\cos\theta, r\sin\theta). \tag{11.1.1}$$

Let (x, y) be the coordinates in Ω defined by

$$(x, y) = (c_2 - r, \theta). \tag{11.1.2}$$

Then, in the (x, y)–coordinates, for sufficiently small $\varepsilon > 0$,

$$\Omega \cap \mathcal{N}_\varepsilon(\Gamma_{\text{sonic}}) \subset \{x > 0\}, \qquad \Gamma_{\text{sonic}} = \{(0, y) \,:\, y_{P_4} < y < y_{P_1}\}, \tag{11.1.3}$$

as shown in Proposition 11.2.3 below. Furthermore, ψ in the (x, y)–coordinates satisfies (6.3.6)–(6.3.7), which now takes the form in $\Omega \cap \{x < \frac{c_2}{2}\}$:

$$(2x - (\gamma+1)\psi_x + O_1)\psi_{xx} + O_2\psi_{xy} + (\frac{1}{c_2} + O_3)\psi_{yy} - (1 + O_4)\psi_x + O_5\psi_y = 0 \tag{11.1.4}$$

with $O_k = O_k(D\psi, \psi, x)$, where $O_k(\mathbf{p}, z, x)$ with $\mathbf{p} = (p_1, p_2) \in \mathbb{R}^2$ and $z, x \in \mathbb{R}$ are defined by

$$\begin{aligned}
O_1(\mathbf{p}, z, x) &= -\frac{x^2}{c_2} + \frac{\gamma+1}{2c_2}(2x - p_1)p_1 - \frac{\gamma-1}{c_2}\Big(z + \frac{p_2^2}{2(c_2-x)^2}\Big), \\
O_2(\mathbf{p}, z, x) &= -\frac{2(p_1 + c_2 - x)p_2}{c_2(c_2-x)^2}, \\
O_3(\mathbf{p}, z, x) &= \frac{1}{c_2(c_2-x)^2}\Big(x(2c_2 - x) - (\gamma-1)(z + (c_2-x)p_1 + \frac{p_1^2}{2}) \\
&\qquad - \frac{(\gamma+1)p_2^2}{2(c_2-x)^2}\Big), \\
O_4(\mathbf{p}, z, x) &= \frac{1}{c_2 - x}\Big(x - \frac{\gamma-1}{c_2}(z + (c_2-x)p_1 + \frac{p_1^2}{2} + \frac{(\gamma+1)p_2^2}{2(\gamma-1)(c_2-x)^2})\Big), \\
O_5(\mathbf{p}, z, x) &= -\frac{(p_1 + 2c_2 - 2x)p_2}{c_2(c_2-x)^3}.
\end{aligned} \tag{11.1.5}$$

Equation (11.1.4) with (11.1.5) can be rewritten as

$$\sum_{i,j=1}^{2} \hat{A}_{ij}(D\psi, \psi, x)D_{ij}\psi + \sum_{i=1}^{2} \hat{A}_i(D\psi, \psi, x)D_i\psi = 0, \tag{11.1.6}$$

where $(D_1, D_2) = (D_x, D_y)$, $D_{ij} = D_iD_j$, and $\hat{A}_{12} = \hat{A}_{21}$. Note that $(\hat{A}_{ij}, \hat{A}_i) = (\hat{A}_{ij}, \hat{A}_i)(\mathbf{p}, z, x)$, independent of y, with $\mathbf{p} = (p_1, p_2) \in \mathbb{R}^2$ and $x, y, z \in \mathbb{R}$.

Lemma 11.1.1. *Let $L > 1$ be fixed. Then, for any $\delta \in (0, 1)$ and $M > 0$, there exists $\varepsilon > 0$ such that, when*

$$\frac{1}{L} \le c_2 \le L,$$

equation (11.1.4)–(11.1.5), *written as* (11.1.6), *satisfies*

$$\frac{\delta}{2}|\boldsymbol{\kappa}|^2 \le \sum_{i,j=1}^{2} \hat{A}_{ij}(\mathbf{p}, z, x)\frac{\kappa_i\kappa_j}{x^{2-\frac{i+j}{2}}} \le \frac{2}{\delta}|\boldsymbol{\kappa}|^2 \tag{11.1.7}$$

for any $\boldsymbol{\kappa} = (\kappa_1, \kappa_2) \in \mathbb{R}^2$ and $(\mathbf{p}, z, x) \in \mathbb{R}^2 \times \mathbb{R} \times \mathbb{R}$ satisfying $0 < x < \varepsilon$, $|p_1| \le \frac{2-\delta}{1+\gamma}x$, $|p_2| \le Mx$, and $|z| \le Mx^2$.

Proof. From the explicit expressions (11.1.5) and the restrictions on $(\mathbf{p}, z, x)$, we estimate

$$|O_1| \le C_1x^2, \qquad |O_k| \le C_1x \quad \text{for } k = 2, \dots, 5. \tag{11.1.8}$$

Then, combining this with (11.1.4), using once again the restriction on p_1, and choosing $\varepsilon > 0$ small depending only on (c_2, γ), we obtain (11.1.7). □

11.2 STRUCTURE OF THE NEIGHBORHOOD OF Γ_{sonic} IN Ω AND ESTIMATES OF $(\psi, D\psi)$

In order to obtain the estimates near Γ_{sonic}, we need to analyze the structure of the neighborhood of Ω near Γ_{sonic}. We continue using the polar coordinates (11.1.1). Also, for a point P, we denote by (r_P, θ_P) its polar coordinates with $\theta_P \in [0, 2\pi)$.

Let $\theta_{\text{w}} \in (\theta_{\text{w}}^{\text{s}}, \frac{\pi}{2}]$. By Remark 7.5.5 and Definition 7.5.7, P_1 is a point of intersection of line $\mathcal{S}_1$ with circle $\partial B_{c_2}(\mathcal{O}_2)$ such that segment P_0P_1 is away from $B_{c_2}(\mathcal{O}_2)$. Note that $\mathcal{S}_1$ is not orthogonal to Γ_{wedge}, since $\mathcal{S}_1 \perp \mathcal{O}_1\mathcal{O}_2$, $\mathcal{O}_1 \notin \{\xi_2 = \xi_1 \tan\theta_{\text{w}}\}$, and $\mathcal{O}_2 \in \{\xi_2 = \xi_1 \tan\theta_{\text{w}}\}$. Then it follows that $\mathcal{S}_1$ intersects with $\partial B_{|\mathcal{O}_2P_0|}(\mathcal{O}_2)$ at two points. One of the points of intersection is P_1 as noted above.

Definition 11.2.1. *Define the following points:*

- *Denote by $\bar{P}_1$ the second point of intersection of $\mathcal{S}_1$ with $\partial B_{c_2}(\mathcal{O}_2)$. Then $|P_0\bar{P}_1| > |P_0P_1|$. Also, by* (6.1.4)–(6.1.5) *of Lemma* 6.1.2 *applied with $\varphi^- = \varphi_1$ and $\varphi^+ = \varphi_2$, it follows that the smaller arc $P_1\bar{P}_1$ of $\partial B_{c_2}(\mathcal{O}_2)$ lies within $\{\varphi_2 > \varphi_1\}$.*

- *Denote by Q the midpoint of the smaller arc $P_1\bar{P}_1$ of $\partial B_{c_2}(\mathcal{O}_2)$. Then $Q \in \{\varphi_2 > \varphi_1\}$.*

- *Denote by Q' the midpoint of the line segment* (*chord*) *$P_1\bar{P}_1$.*

From the definition of $\bar{P}_1$ and the fact that $|P_0\bar{P}_1| > |P_0P_1|$,

$$\theta_{\bar{P}_1} > \theta_{P_1}, \qquad \theta_{\bar{P}_1} - \theta_{P_1} < \pi.$$

From the definition of Q, we have

$$r_Q = c_2, \qquad \theta_Q = \frac{\theta_{\bar{P}_1} + \theta_{P_1}}{2}.$$

Furthermore, the tangent line to $\partial B_{c_2}(\mathcal{O}_2)$ at Q is parallel to $\mathcal{S}_1$, which implies that $\mathbf{e}_{\mathcal{S}_1} = \pm\mathbf{e}_y(Q)$. To determine the sign, we notice that $|P_0\bar{P}_1| > |P_0P_1|$ and $r_{P_1} = r_{\bar{P}_1} = c_2$ imply that $\theta_{\bar{P}_1} - \theta_{P_1} \in (0, \pi)$ so that, by using (7.5.9),

$$\mathbf{e}_{\mathcal{S}_1} = \frac{\bar{P}_1 - P_1}{|\bar{P}_1 - P_1|} = \mathbf{e}_y(Q),$$

which is

$$\mathbf{e}_{\mathcal{S}_1} = (-\sin\theta_Q, \cos\theta_Q). \tag{11.2.1}$$

Note that, from the definition of Q,

$$\theta_{\text{w}} = \theta_{P_4} < \theta_{P_1} < \theta_Q < \pi.$$

Introduce the function:

$$\bar{\phi}_0 := \varphi_1 - \varphi_2,$$

and note that, in both the polar coordinates (11.1.1) and the (x,y)–coordinates (11.1.2), this function has the expression:

$$\begin{aligned}\bar{\phi}_0 := \varphi_1 - \varphi_2 &= (u_1 - u_2) r\cos\theta - v_2 r \sin\theta + \hat{C}\\ &= (u_1 - u_2)(c_2 - x)\cos y - v_2(c_2 - x)\sin y + \hat{C},\end{aligned} \tag{11.2.2}$$

where $\hat{C} = (\varphi_1 - \varphi_2)(\mathcal{O}_2) = -((u_1 - u_2)\cos\theta_{\rm w} - v_2 \sin\theta_{\rm w})|\mathcal{O}_2 P_0|$, and distance $|\mathcal{O}_2 P_0|$ is in the Euclidean coordinates.

Furthermore, we have

Lemma 11.2.2. *There exist $\varepsilon_1 > \varepsilon_0 > 0$, $\delta > 0$, $\omega > 0$, $C > 0$, $M \geq 2$, and a continuous function $m(\cdot)$ on $[0,\infty)$ satisfying $m(0) = 0$ and $m(t) > 0$ for $t > 0$, depending only on (ρ_0, ρ_1, γ), such that, for any $\theta_{\rm w} \in [\theta_{\rm w}^{\rm s}, \frac{\pi}{2}]$,*

(i) *The following inclusions hold:*

$$\Lambda \cap \mathcal{N}_{\varepsilon_1}(\Gamma_{\rm sonic}) \subset \{\theta_{\rm w} < \theta < \theta_Q - \delta\}, \tag{11.2.3}$$

$$\{\varphi_2 < \varphi_1\} \cap \mathcal{N}_{\varepsilon_1}(\Gamma_{\rm sonic}) \cap \{\theta > \theta_{P_1}\} \subset \{x > 0\}; \tag{11.2.4}$$

(ii) *In $\{\theta_{\rm w} < \theta < \theta_Q - \delta\} \cap \{-\varepsilon_1 < x < \varepsilon_1\}$,*

$$\begin{aligned}&\frac{2}{M}(\theta - \theta_{\rm w}) + m(\frac{\pi}{2} - \theta_{\rm w}) \leq \partial_x(\varphi_1 - \varphi_2) \leq \frac{M}{2},\\ &\frac{2}{M} \leq -\partial_y(\varphi_1 - \varphi_2) \leq \frac{M}{2};\end{aligned} \tag{11.2.5}$$

(iii) *In $\{-\varepsilon_1 < x < \varepsilon_1\}$,*

$$|(D^2_{(x,y)}, D^3_{(x,y)})(\varphi_1 - \varphi_2)| \leq C; \tag{11.2.6}$$

(iv) *$\phi_1(\boldsymbol{\xi}) = \varphi_1(\boldsymbol{\xi}) + \frac{1}{2}|\boldsymbol{\xi}|^2 = u_1\xi_1 + const.$ satisfies*

$$\frac{2}{M} \leq -\partial_y \phi_1 \leq \frac{M}{2} \qquad in\ \{\theta_{\rm w} < \theta < \theta_Q - \frac{\delta}{2}\} \cap \{-\varepsilon_1 < x < \varepsilon_1\}; \tag{11.2.7}$$

(v) *There exists $\hat{f}_0 \in C^\infty([-\varepsilon_0, \varepsilon_0])$ such that, for each $\varepsilon', \varepsilon'' \in [-\varepsilon_0, \varepsilon_0]$ with $\varepsilon' \leq \varepsilon''$,*

$$\begin{aligned}&\{\varphi_2 < \varphi_1\} \cap \Lambda \cap \mathcal{N}_{\varepsilon_1}(\Gamma_{\rm sonic}) \cap \{\varepsilon' \leq x \leq \varepsilon''\}\\ &\quad = \{(x,y)\,:\, \varepsilon' \leq x \leq \varepsilon'',\ \ \theta_{\rm w} < y < \hat{f}_0(x)\},\end{aligned} \tag{11.2.8}$$

$$\begin{aligned}&\mathcal{S}_1 \cap \mathcal{N}_{\varepsilon_1}(\Gamma_{\rm sonic}) \cap \{\varepsilon' \leq x \leq \varepsilon''\}\\ &\quad = \{(x,y)\,:\, \varepsilon' \leq x \leq \varepsilon'',\ \ y = \hat{f}_0(x)\},\end{aligned} \tag{11.2.9}$$

$$2\omega \leq \hat{f}_0' \leq C \qquad on\ (-\varepsilon_0, \varepsilon_0); \tag{11.2.10}$$

(vi) *For any* $\theta_{\rm w} \in [\theta_{\rm w}^{\rm s}, \frac{\pi}{2}]$,

$$\delta \le \theta_Q - \theta_{P_4} = \frac{\pi}{2} - \tilde{m}(\frac{\pi}{2} - \theta_{\rm w}), \tag{11.2.11}$$

where $\tilde{m}(\cdot)$ *has the same properties as* $m(\cdot)$ *above.*

Proof. In this proof, we use the polar coordinates (r, θ) centered at $\mathcal{O}_2 = (u_2, v_2)$. Also, when $\theta_{\rm w} = \theta_{\rm w}^{\rm s}$, $P_0 = P_1 = P_4$, which do not cause any changes in the argument. We divide the proof into five steps.

1. We note that $\theta_{P_1} < \theta_Q$. Also, $\theta \in [\theta_{\rm w}, \theta_{P_1}]$ on $\Gamma_{\rm sonic}$, and $\theta \ge \theta_{\rm w}$ on Λ. Then, since P_1 and Q depend continuously on $\theta_{\rm w} \in [\theta_{\rm w}^{\rm s}, \frac{\pi}{2}]$, there exist $\varepsilon_1 > 0$ and $\delta > 0$ such that (11.2.3) holds.

Furthermore, since $\theta_Q = \theta_{Q'}$, and points P_1 and Q' lie on line $\mathcal{S}_1 = \{\varphi_1 = \varphi_2\}$, then (11.2.3) implies that

$$\{\varphi_2 < \varphi_1\} \cap \mathcal{N}_{\varepsilon_1}(\Gamma_{\rm sonic}) \cap \{\theta > \theta_{P_1}\} \subset \mathcal{O}_2 P_1 Q' \subset B_{c_2}(\mathcal{O}_2) = \{x > 0\}.$$

This shows (11.2.4).

2. Note that, for any $\theta_{\rm w} \in [\theta_{\rm w}^{\rm s}, \frac{\pi}{2})$, point P_0 is well-defined. Then, from triangle $P_0\mathcal{O}_2Q'$ in which $\angle P_0Q'\mathcal{O}_2 = \frac{\pi}{2}$ and $\angle P_0\mathcal{O}_2Q' = \theta_Q - \theta_{P_4}$, we have

$$0 < \theta_Q - \theta_{P_4} < \frac{\pi}{2} \qquad \text{for any } \theta_{\rm w} \in [\theta_{\rm w}^{\rm s}, \frac{\pi}{2}). \tag{11.2.12}$$

Furthermore, since $\theta_Q > \theta_{P_1}$ for each $\theta_{\rm w}$, then, using the continuous dependence of Q and P_4 on $\theta_{\rm w}$ and reducing $\delta > 0$ if necessary, we see that, for any $\theta_{\rm w} \in [\theta_{\rm w}^{\rm s}, \frac{\pi}{2}]$,

$$\theta_Q - \theta_{P_4} \ge \delta.$$

Also, $\theta_Q = \pi$ and $\theta_{P_4} = \frac{\pi}{2}$ when $\theta_{\rm w} = \frac{\pi}{2}$. Thus, we define

$$\tilde{m}(\frac{\pi}{2} - \theta_{\rm w}) := \frac{\pi}{2} - (\theta_Q - \theta_{P_4}) \qquad \text{for } \theta_{\rm w} \in [\theta_{\rm w}^{\rm s}, \frac{\pi}{2}],$$

so that $\tilde{m}(\cdot)$ is defined and continuous on $[0, \frac{\pi}{2} - \theta_{\rm w}^{\rm s}]$ with

$$\tilde{m}(0) = 0, \qquad 0 < \tilde{m}(t) \le \frac{\pi}{2} - \delta \quad \text{for } t > 0. \tag{11.2.13}$$

We can arbitrarily extend $\tilde{m}$ to a continuous function on $[0, \infty)$, satisfying the inequalities given above for all $t > 0$. Then (11.2.11) holds.

3. Note that $\sqrt{(u_1 - u_2)^2 + v_2^2}$ is a continuous function of $\theta_{\rm w} \in [\theta_{\rm w}^{\rm s}, \frac{\pi}{2}]$ and is positive for every such $\theta_{\rm w}$, since $u_2 = v_2 = 0$ in the normal reflection case $\theta_{\rm w} = \frac{\pi}{2}$, and $v_2 > 0$ for $\theta_{\rm w} \in [\theta_{\rm w}^{\rm s}, \frac{\pi}{2})$. Then there exists $\tilde{C} > 0$ such that

$$\frac{1}{\tilde{C}} \le \sqrt{(u_1 - u_2)^2 + v_2^2} \le \tilde{C} \qquad \text{for any } \theta_{\rm w} \in [\theta_{\rm w}^{\rm s}, \frac{\pi}{2}]. \tag{11.2.14}$$

Using expression (11.2.2) and noting that (7.5.8) and (11.2.1) imply

$$(\sin\theta_Q, \cos\theta_Q) = \frac{(v_2, u_2 - u_1)}{\sqrt{(u_1-u_2)^2+v_2^2}},$$

we compute to see

$$\begin{aligned}
&\partial_r(\varphi_1 - \varphi_2)\\
&= (u_1 - u_2)\cos\theta - v_2\sin\theta = -\sqrt{(u_1-u_2)^2+v_2^2}\cos(\theta_Q - \theta)\\
&= -\sqrt{(u_1-u_2)^2+v_2^2}\big(\cos(\theta_Q-\theta_{\rm w})\cos(\theta-\theta_{\rm w}) + \sin(\theta_Q-\theta_{\rm w})\sin(\theta-\theta_{\rm w})\big).
\end{aligned}$$

Since $\partial_x = -\partial_r$ by (11.1.2), we note that $\theta_{\rm w} = \theta_{P_4}$, and employ (11.2.11) and (11.2.14) to obtain (11.2.5) for $\partial_x(\varphi_1 - \varphi_2)$ with large $C > 0$, small $\varepsilon_1 > 0$, and $m(\cdot) = \frac{1}{C}\sin\delta\sin(\tilde{m}(\cdot))$. The properties of $\tilde{m}(\cdot)$ in (11.2.13) imply that $m(\cdot)$ satisfies the properties asserted.

4. Similarly, from (11.2.2),

$$\frac{1}{r}\partial_\theta(\varphi_1 - \varphi_2) = -(u_1 - u_2)\sin\theta - v_2\cos\theta = -\sqrt{(u_1-u_2)^2+v_2^2}\sin(\theta_Q - \theta).$$

Since $c_2(\theta_{\rm w}) \geq \frac{2}{C}$ for some large C and any $\theta_{\rm w} \in [\theta_{\rm w}^{\rm s}, \frac{\pi}{2}]$, then $r \in (c_2 - \varepsilon, c_2)$ implies that $r \geq \frac{1}{C}$ for any $\theta_{\rm w} \in [\theta_{\rm w}^{\rm s}, \frac{\pi}{2}]$ if $\varepsilon > 0$ is sufficiently small. Then, from (11.1.2), (11.2.11), and (11.2.14), we obtain (11.2.5) for $\partial_y(\varphi_1 - \varphi_2)$.

Taking the derivatives of $\varphi_1 - \varphi_2$ with respect to (r, θ) up to the third order, we obtain (11.2.6).

Now, from (11.2.3) and (11.2.5), there exist $\varepsilon_0 > 0$, $\hat{f}_0 \in C^\infty([-\varepsilon_0, \varepsilon_0])$, and $C, \omega > 0$ such that (11.2.8)–(11.2.10) hold for any $\theta_{\rm w} \in [\theta_{\rm w}^{\rm s}, \frac{\pi}{2}]$.

5. To show (11.2.7), we note that, since $\phi_1(\boldsymbol{\xi}) = u_1\xi_1 + const.$,

$$\frac{1}{r}\partial_\theta\phi_1 = -u_1\sin\theta.$$

Also, $\theta_Q \in (\theta_{\rm w}^{\rm s}, \pi)$ for all $\theta_{\rm w} \in [\theta_{\rm w}^{\rm s}, \frac{\pi}{2}]$. Thus, for every such $\theta_{\rm w}$, we have

$$0 < u_1\min(\sin\theta_{\rm w}^{\rm s}, \sin\delta) \leq -\frac{1}{r}\partial_\theta\phi_1 \leq u_1 \qquad \text{in } \{\theta_{\rm w} < \theta < \theta_Q - \frac{\delta}{2}\}. \quad (11.2.15)$$

Moreover, the continuous dependence of $c_2 > 0$ on $\theta_{\rm w} \in [\theta_{\rm w}^{\rm s}, \frac{\pi}{2}]$ implies that there exists $\hat{\delta} > 0$ depending only on the data such that $c_2 \in [\hat{\delta}, \frac{1}{\hat{\delta}}]$ for all such $\theta_{\rm w}$, so that $\hat{\delta} \leq \frac{c_2}{2} \leq r(P) \leq 2c_2 \leq \frac{1}{\hat{\delta}}$ in $\{-\varepsilon_1 < x < \varepsilon_1\}$. Combining with (11.2.15), we conclude (11.2.7). ☐

Proposition 11.2.3. *Let ε_0 and ε_1 be the constants defined in Lemma* 11.2.2. *Let φ be an admissible solution with $\theta_{\rm w} \in (\theta_{\rm w}^{\rm s}, \frac{\pi}{2}]$ and $\Omega = \Omega(\varphi)$. For $\varepsilon \in (0, \varepsilon_0)$, denote*

$$\Omega_\varepsilon := \Omega \cap \mathcal{N}_{\varepsilon_1}(\overline{\Gamma_{\rm sonic}}) \cap \{x < \varepsilon\}. \quad (11.2.16)$$

Then, for each $\varepsilon \in (0, \varepsilon_0)$,

$$\Omega \cap \mathcal{N}_\varepsilon(\Gamma_{\text{sonic}}) \subset \Omega_\varepsilon, \tag{11.2.17}$$

$$\Omega_\varepsilon = \Omega_\varepsilon \cap \{x > 0\}, \tag{11.2.18}$$

$$\Omega_\varepsilon \subset \{\varphi_2 < \varphi_1\} \cap \Lambda \cap \mathcal{N}_{\varepsilon_1}(\Gamma_{\text{sonic}}) \cap \{0 < x < \varepsilon\}. \tag{11.2.19}$$

Proof. For any $\varepsilon < \varepsilon_0 < \varepsilon_1$, $\Gamma_{\text{sonic}} \subset \partial B_{c_2}(\mathcal{O}_2)$ implies that

$$\mathcal{N}_\varepsilon(\Gamma_{\text{sonic}}) \subset \mathcal{N}_\varepsilon(\partial B_{c_2}(\mathcal{O}_2)) = \{|x| < \varepsilon\} \subset \{x < \varepsilon\}.$$

Then (11.2.17) follows from (11.2.16).

Note that $\Omega \subset \{\varphi_2 < \varphi_1\} \cap \Lambda$ by conditions (i) and (iv) of Definition 8.1.1, and $\Lambda \subset \{\theta > \theta_{\text{w}}\}$. Then it follows from Definition 8.1.1(i) that

$$\Omega \cap \{\theta_{\text{w}} < \theta \le \theta_{P_1}\} \subseteq B_{c_2}(\mathcal{O}_2) \cap \{\theta_{\text{w}} < \theta \le \theta_{P_1}\} \subset \{x > 0\}.$$

Combining this with (11.2.4) and noting that $\Omega \cap \mathcal{N}_{\varepsilon_1}(\Gamma_{\text{sonic}}) \cap \{\theta > \theta_{P_1}\}$ is a subset of the left-hand side of (11.2.4), we obtain (11.2.18).

Furthermore, we combine the fact that $\Omega \subset \{\varphi_2 < \varphi_1\} \cap \Lambda$ with (11.2.16) and (11.2.18) to conclude (11.2.19). □

Lemma 11.2.4. *Let* ε_0 *and* ε_1 *be the constants defined in Lemma* 11.2.2. *Then there exists* C *such that, for any admissible solution with* $\theta_{\text{w}} \in (\theta_{\text{w}}^{\text{s}}, \frac{\pi}{2}]$ *and any* $\varepsilon \in (0, \varepsilon_0)$,

$$\Gamma_{\text{shock}} \cap \partial\Omega_\varepsilon \subset B_{C\varepsilon}(P_1).$$

Proof. Let $\{\boldsymbol{\nu}_{\text{w}}, \boldsymbol{\tau}_{\text{w}}\}$ be the unit normal and tangent vectors to Γ_{wedge} defined in (8.2.14) and (8.2.17). Let (S, T) be the coordinates with basis $\{\boldsymbol{\nu}_{\text{w}}, \boldsymbol{\tau}_{\text{w}}\}$. Then, by Corollary 8.2.14 and Lemma 8.2.11, Γ_{shock} is a graph in the S–direction: There exists $f_{\boldsymbol{\nu}_{\text{w}}} \in C^{1,\alpha}(\mathbb{R})$ such that

$$\Gamma_{\text{shock}} = \{S = f_{\boldsymbol{\nu}_{\text{w}}}(T) \, : \, T_{P_2} < T < T_{P_1}\}, \quad \Omega \subset \{S < f_{\boldsymbol{\nu}_{\text{w}}}(T) \, : \, T \in \mathbb{R}\},$$

and $f_{\boldsymbol{\nu}_{\text{w}}}$ satisfies all the other properties in Corollary 8.2.14. In particular, it follows from (8.2.24) that

$$\text{Lip}[f_{\boldsymbol{\nu}_{\text{w}}}] \le L,$$

where L is a continuous function of θ_{w} on $[\theta_{\text{w}}^{\text{s}}, \frac{\pi}{2}]$ and the parameters of states (1) and (2), all of which are continuous functions of (ρ_0, ρ_1, γ). Thus, we can choose L depending only on (ρ_0, ρ_1, γ), but independent of $\theta_{\text{w}} \in (\theta_{\text{w}}^{\text{s}}, \frac{\pi}{2}]$.

Set $Q_\varepsilon := P_1 + \varepsilon \mathbf{e}_{\mathcal{S}_1}$, where we have used (7.5.8). Then (11.2.19) implies that

$$T_P \ge T_{Q_\varepsilon} \qquad \text{for all } P \in \Omega_\varepsilon.$$

It follows that

$$\Gamma_{\text{shock}} \cap \partial\Omega_\varepsilon \subset \{S = f_{\boldsymbol{\nu}_{\text{w}}}(T) \, : \, T_{Q_\varepsilon} < T < T_{P_1}\}. \tag{11.2.20}$$

From the definition of Q_ε, we have

$$|T_{P_1} - T_{Q_\varepsilon}| \le \varepsilon.$$

From (11.2.20) and the uniform Lipschitz bound $\mathrm{Lip}[f_{\nu_{\mathrm{w}}}(\cdot)] \le L$ by (8.2.24), we conclude that, for any $P \in \Gamma_{\mathrm{shock}} \cap \partial\Omega_\varepsilon$,

$$|PP_1| \le \sqrt{1+L^2}\,\varepsilon.$$

□

In the next two lemmas, we estimate the gradient of $\psi = \varphi - \varphi_2$ near Γ_{sonic}, specifically in Ω_ε for sufficiently small $\varepsilon > 0$.

Lemma 11.2.5. *There exist $\varepsilon \in (0, \varepsilon_0)$ and $\delta > 0$ depending only on (ρ_0, ρ_1, γ) such that, if φ is an admissible solution with $\theta_{\mathrm{w}} \in (\theta_{\mathrm{w}}^{\mathrm{s}}, \frac{\pi}{2})$, then*

$$\psi_x \le \frac{2-\delta}{\gamma+1}x \qquad \text{in } \Omega_\varepsilon. \tag{11.2.21}$$

Proof. In this proof, all constants δ, ε, and C depend only on (ρ_0, ρ_1, γ). Let φ be an admissible solution of **Problem 2.6.1** with $\theta_{\mathrm{w}} \in [\theta_{\mathrm{w}}^{\mathrm{s}}, \frac{\pi}{2})$.

Using the smooth dependence of the parameters of state (2) on $\theta_{\mathrm{w}} \in [\theta_{\mathrm{w}}^{\mathrm{s}}, \frac{\pi}{2}]$, the coordinate transform: $\boldsymbol{\xi} \mapsto (x, y)$ is bi-Lipschitz in $B_{c_2}(\mathcal{O}_2) \setminus B_{\frac{c_2}{2}}(\mathcal{O}_2)$ with the constants depending only on (ρ_0, ρ_1, γ). Thus, using Corollary 9.6.6, there exists $\delta > 0$ such that, for any $(x, y) \in \Omega(\varphi) \cap \{0 < x < \frac{c_2}{2}\}$ and $\boldsymbol{\kappa} = (\kappa_1, \kappa_2) \in \mathbb{R}^2$,

$$2\delta x|\boldsymbol{\kappa}|^2 \le \sum_{i,j=1}^{2} \hat{A}_{ij}(D\psi(x,y),\, \psi(x,y),\, x)\kappa_i\kappa_j \le \frac{1}{\delta}|\boldsymbol{\kappa}|^2,$$

by employing form (11.1.6) of equation (11.1.4). Then

$$\hat{A}_{11}(D\psi(x,y),\, \psi(x,y),\, x) \ge 2\delta x,$$

which is

$$2x - (\gamma+1)\psi_x + O_1(D\psi, \psi, x) \ge 2\delta x.$$

Using the expression of O_1 in (11.1.5) and employing (11.0.2), (11.1.3), and $x < c_2$, we have

$$O_1 \le \frac{\gamma+1}{c_2} x\psi_x,$$

so that

$$2x - (\gamma+1)\psi_x + \frac{\gamma+1}{c_2} x\psi_x \ge 2\delta x. \tag{11.2.22}$$

Since c_2 is positive and depends continuously on $\theta_{\mathrm{w}} \in [\theta_{\mathrm{w}}^{\mathrm{s}}, \frac{\pi}{2}]$, there exists $C > 0$ such that $c_2(\theta_{\mathrm{w}}) \ge \frac{1}{C}$ for all $\theta_{\mathrm{w}} \in [\theta_{\mathrm{w}}^{\mathrm{s}}, \frac{\pi}{2}]$. Then there exists $\varepsilon \in (0, \varepsilon_0)$ such that (11.2.22) implies (11.2.21) with δ given in (11.2.22). □

Lemma 11.2.6. *There exist $\varepsilon \in (0, \varepsilon_0)$, $\delta > 0$, and $C > 0$ depending only on (ρ_0, ρ_1, γ) such that, if φ is an admissible solution with $\theta_{\rm w} \in (\theta_{\rm w}^{\rm s}, \frac{\pi}{2})$, then, in Ω_ε,*

$$0 \le \psi_x \le \frac{2-\delta}{\gamma+1}x, \tag{11.2.23}$$

$$|\psi_y| \le Cx. \tag{11.2.24}$$

Proof. In the proof, we use that $\partial_{\boldsymbol{n}_{\rm w}}\psi \ge 0$ and $\partial_{\mathbf{e}_{\mathcal{S}_1}}\psi \ge 0$ in $\overline{\Omega}$ by Corollary 8.2.20, and then express these differentiations in terms of the (x, y)–differentiations. We divide the proof into four steps.

1. At point $P \neq \mathcal{O}_2$ in the $\boldsymbol{\xi}$–plane, let $\{\mathbf{e}_x, \mathbf{e}_y\} = \{\mathbf{e}_x(P), \mathbf{e}_y(P)\}$ be unit vectors in the (x, y)–directions, respectively. Using the polar coordinates (r, θ) with respect to $\mathcal{O}_2 = (u_2, v_2)$ defined by (11.1.1), we have

$$\mathbf{e}_x = -\mathbf{e}_r = -(\cos\theta, \sin\theta), \qquad \mathbf{e}_y = \frac{1}{r}\mathbf{e}_\theta = (-\sin\theta, \cos\theta).$$

Using (11.2.1), we have

$$\mathbf{e}_{\mathcal{S}_1} = (-\sin\theta_Q, \cos\theta_Q) = \mathbf{e}_y(Q).$$

From (8.2.35) and $\theta_{P_4} = \theta_{\rm w}$, we have

$$\boldsymbol{n}_{\rm w} = (\sin\theta_{P_4}, -\cos\theta_{P_4}) = -\mathbf{e}_y(P_4).$$

Now, for point $P = r(\cos\theta, \sin\theta)$, we employ the expressions obtained above to compute:

$$\begin{aligned} &\mathbf{e}_{\mathcal{S}_1} \cdot \mathbf{e}_x = \sin(\theta_Q - \theta), \qquad &&\mathbf{e}_{\mathcal{S}_1} \cdot \mathbf{e}_y = \cos(\theta_Q - \theta),\\ &\boldsymbol{n}_{\rm w} \cdot \mathbf{e}_x = \sin(\theta - \theta_{P_4}), &&\boldsymbol{n}_{\rm w} \cdot \mathbf{e}_y = -\cos(\theta - \theta_{P_4}). \end{aligned} \tag{11.2.25}$$

From (11.2.3) and (11.2.19), we have

$$\Omega_\varepsilon \subset \{\theta_{P_4} < \theta < \theta_Q - \delta\}. \tag{11.2.26}$$

Now we use that $\partial_{\boldsymbol{n}_{\rm w}}\psi \ge 0$ and $\partial_{\mathbf{e}_{\mathcal{S}_1}}\psi \ge 0$ in $\overline{\Omega}$ by Corollary 8.2.20. Thus, using that $\partial_x = \partial_{\mathbf{e}_x}$ and $\partial_y = \frac{1}{c_2 - x}\partial_{\mathbf{e}_y}$, we find that, in Ω,

$$\begin{aligned} &(\boldsymbol{n}_{\rm w} \cdot \mathbf{e}_x)\psi_x + (\boldsymbol{n}_{\rm w} \cdot \mathbf{e}_y)\frac{\psi_y}{c_2 - x} \ge 0,\\ &(\mathbf{e}_{\mathcal{S}_1} \cdot \mathbf{e}_x)\psi_x + (\mathbf{e}_{\mathcal{S}_1} \cdot \mathbf{e}_y)\frac{\psi_y}{c_2 - x} \ge 0. \end{aligned} \tag{11.2.27}$$

Then, using (11.2.11) and (11.2.25)–(11.2.26), we have

$$\big(\sin(\theta - \theta_{P_4})\cos(\theta_Q - \theta) + \cos(\theta - \theta_{P_4})\sin(\theta_Q - \theta)\big)\psi_x \ge 0 \qquad \text{in } \Omega_\varepsilon,$$

that is, $\sin(\theta_Q - \theta_{P_4})\psi_x \geq 0$. Now, using (11.2.11), we conclude that $\psi_x \geq 0$, which implies

$$|\psi_x| \leq Cx,$$

by using (11.2.21).

2. In Ω_ε, from (11.2.11) and (11.2.25)–(11.2.27), we have

$$-\tan(\theta_Q - \theta)\psi_x \leq \frac{\psi_y}{c_2 - x} \leq \tan(\theta - \theta_{P_4})\psi_x. \tag{11.2.28}$$

However, this does not immediately imply (11.2.24) for $\theta_{\rm w}$ near $\frac{\pi}{2}$, since $\theta_Q = \pi$ when $\theta_{\rm w} = \frac{\pi}{2}$ (*i.e.*, in the normal reflection case), which implies that $\theta_Q - \theta \to \frac{\pi}{2}$ when $\theta_{\rm w} = \frac{\pi}{2}$ and $\theta \to \theta_{\rm w}$. Then, to prove (11.2.24), we consider the following two cases:

Case 1: $\theta_{\rm w} \in (\theta_{\rm w}^{\rm s}, \frac{\pi+2\theta_{\rm w}^{\rm s}}{4})$. In this case, for each $\theta_{\rm w} \in [\theta_{\rm w}^{\rm s}, \frac{\pi+2\theta_{\rm w}^{\rm s}}{4}]$,

$$0 < \theta_Q - \theta_{P_4} = \frac{\pi}{2} - \angle QP_0P_3 < \frac{\pi}{2}.$$

Thus, we use the continuous dependence of Q and P_1 on $\theta_{\rm w} \in [\theta_{\rm w}^{\rm s}, \frac{\pi+2\theta_{\rm w}^{\rm s}}{4}]$ to conclude that there exists $\hat{\delta} > 0$ depending only on the data so that

$$\hat{\delta} \leq \theta_Q - \theta_{P_4} \leq \frac{\pi}{2} - \hat{\delta} \qquad \text{for all } \theta_{\rm w} \in [\theta_{\rm w}^{\rm s}, \frac{\pi + 2\theta_{\rm w}^{\rm s}}{4}].$$

Then (11.2.24) follows from (11.2.28).

Case 2: $\theta_{\rm w} \in [\hat{\theta}_{\rm w}^*, \frac{\pi}{2})$ for $\hat{\theta}_{\rm w}^* = \frac{\pi+2\theta_{\rm w}^{\rm s}}{4}$. By (8.1.3),

$$\psi \in C^1(\overline{\Lambda} \setminus \overline{P_0P_1P_2}) \cap C^1(\overline{\Omega}) \cap C^3(\Omega),$$

and $\psi = 0$ in $P_0P_1P_4$. Thus, we have

$$\psi_y \in C(\overline{\Omega_\varepsilon}) \cap C^2(\Omega_\varepsilon)$$

and $\psi_y = 0$ on $\Gamma_{\rm sonic}$. Note that $\Gamma_{\rm sonic} \subset \partial\Omega_\varepsilon$, by (11.2.16).

We first show (11.2.24) on $\partial\Omega_\varepsilon \setminus \Gamma_{\rm sonic}$. After reducing ε if necessary, $\Omega_\varepsilon \cap \Gamma_{\rm sym} = \emptyset$. By using Definition 8.1.1(i), it follows from (11.2.16) that

$$\partial\Omega_\varepsilon = \overline{\Gamma_{\rm sonic}} \cup (\Gamma_{\rm wedge} \cap \partial\Omega_\varepsilon) \cup (\Gamma_{\rm shock} \cap \partial\Omega_\varepsilon) \cup (\overline{\Omega} \cap \{x = \varepsilon\}).$$

Lemma 11.2.4 and the fact that $\inf_{\theta_{\rm w} \in [\hat{\theta}_{\rm w}^*, \frac{\pi}{2}]}(\theta_{P_1} - \theta_{\rm w}) > 0$ imply that there exist $\varepsilon, \delta > 0$, depending only on the data, such that

$$\Gamma_{\rm shock} \cap \partial\Omega_\varepsilon \subset \{\theta \geq \theta_{\rm w} + \delta\}.$$

Also, using (11.2.3) and recalling that $\theta_{\rm w} = \theta_{P_4}$, we obtain that $\theta_{P_4} + \delta \leq \theta \leq \theta_Q - \delta$ on $\Gamma_{\rm shock} \cap \partial\Omega_\varepsilon$. Combining these facts with (11.2.11), we have

$$\delta \leq \theta_Q - \theta \leq \frac{\pi}{2} - \delta, \quad \delta \leq \theta - \theta_{P_4} \leq \frac{\pi}{2} - \delta \qquad \text{on } \Gamma_{\rm shock} \cap \partial\Omega_\varepsilon,$$

where $\delta > 0$ depends only on the data. Now, using (11.2.23), (11.2.28), and the boundedness of $c_2(\theta)$ on $[\hat{\theta}^*_{\rm w}, \frac{\pi}{2}]$, and reducing ε if necessary so that (11.2.21) can be used, we obtain (11.2.24) on $\Gamma_{\rm shock} \cap \partial\Omega_\varepsilon$.

Notice that the Lipschitz estimates in Corollary 9.1.3 and the uniform bound $c_2(\theta_{\rm w}) \geq \delta > 0$ for all $\theta_{\rm w} \in [\hat{\theta}^*_{\rm w}, \frac{\pi}{2}]$ lead to the uniform bounds of the coordinate transform: $\boldsymbol{\xi} \mapsto (x, y)$ on $\overline{\Omega_\varepsilon}$ and its inverse transform with respect to $\theta_{\rm w}$ in the C^k–norms, $k = 0, 1, 2 \dots$. Then we conclude that $|\psi_y| \leq C$ in $\Omega_{2\varepsilon}$, which implies that $|\psi_y| \leq \tilde{C}x$ on $\overline{\Omega} \cap \{x = \varepsilon\}$ with $\tilde{C} = \frac{C}{\varepsilon}$.

Also, $\partial_{\boldsymbol{\nu}} = \frac{1}{c_2 - x}\partial_y$ on $\Gamma_{\rm wedge} \cap \partial\Omega_\varepsilon$. Thus, by the boundary condition, $\psi_y = 0$ on $\Gamma_{\rm wedge} \cap \partial\Omega_\varepsilon$.

Therefore, we have shown that, for some $M > 0$ depending only on the data and ε,

$$|\psi_y| \leq Mx \qquad \text{on } \partial\Omega_\varepsilon. \tag{11.2.29}$$

We do not fix ε at this point, since we will further reduce ε in the forthcoming argument. Then, after ε is fixed, depending only on the data, $M > 0$ will be determined, depending only on the data.

3. It remains to show that $\pm\psi_y - Mx$ cannot attain the maximum in the interior of Ω_ε. We denote

$$w = \psi_y$$

and derive a PDE for w in Ω_ε by differentiating equation (11.1.4)–(11.1.5) with respect to y. Note that there is no explicit dependence on y in the coefficients of this equation. We use notation $D = (D_1, D_2) = (D_x, D_y)$, and denote by $a_{ij}(x, y)$ the coefficients of $D_{ij}\psi$ with $i \leq j$ in (11.1.4)–(11.1.5), *i.e.*, with $(\psi, D\psi)(x, y)$ substituted into the coefficients. Then functions $a_{ij}(x, y)$ are smooth in the interior of Ω_ε. Noting that $\psi_{xy} = w_x$ and $\psi_{yy} = w_y$, we obtain the following equation:

$$\begin{aligned} a_{11}w_{xx} + a_{12}w_{xy} + a_{22}w_{yy} - \big(1 + O_4 - (O_2)_y\big)w_x + \big(O_5 + (O_3)_y\big)w_y \\ - \big((\gamma+1)\psi_x - O_1\big)_y\psi_{xx} - (O_4)_y\psi_x + (O_5)_y w = 0, \end{aligned} \tag{11.2.30}$$

where $O_k := O_k(D\psi(x, y), \psi(x, y), x)$ with functions $O_k(\mathbf{p}, z, x)$ defined in (11.1.5).

We will rewrite this equation as a linear second-order equation for w by computing the $(O_k)_y$–terms and redistributing some sub-terms among the terms. In this calculation, we will compute the exact expressions of the resulting coefficients of w_x and w, but will not emphasize the precise form of the coefficient of w_y. All the coefficients will be smooth in the interior of Ω_ε.

First, we do the following preliminary estimates. Note that

$$\begin{aligned} 0 < a_{11} &= 2x - (\gamma+1)\psi_x + O_1 \\ &= 2x - (\gamma+1)\psi_x - \frac{x^2}{c_2} + \frac{\gamma+1}{2c_2}(2x - \psi_x)\psi_x \\ &\quad - \frac{\gamma-1}{c_2}\big(\psi + \frac{\psi_y^2}{2(c_2 - x)^2}\big). \end{aligned} \tag{11.2.31}$$

Then, in Ω_ε, we employ $\psi \geq 0$ and $0 \leq \psi_x \leq \frac{2-\delta}{\gamma+1}x$ to obtain

$$0 < a_{11} \leq 2x - (\gamma+1)\psi_x - \frac{x^2}{c_2} + \frac{\gamma+1}{2c_2}(2x - \psi_x)\psi_x \leq Cx. \tag{11.2.32}$$

Next, using (11.2.31) and recalling that $\psi \geq 0$, $0 \leq \psi_x \leq Cx$, and $x \in (0, \varepsilon)$ in Ω_ε, we have

$$|\psi_y|^2 \leq C(x + x\psi_x + \psi_x^2) \leq Cx,$$

which implies

$$|\psi_y| \leq C\sqrt{x}. \tag{11.2.33}$$

Now we rewrite the terms of equation (11.2.30) in a convenient form. We use the expressions in (11.1.5) for the O_k–terms, and note that $\partial_y(\psi_x, \psi_y) = (w_x, w_y)$. Then we rewrite the terms of the equation as in (a)–(e) below:

(a) The first term is

$$-\big(1 + O_4 - (O_2)_y\big)w_x = -(1+O_4)w_x - \frac{2}{c_2(c_2-x)^2}w_x^2\psi_y + b_1^{(y)}w_y,$$

where $b_1^{(y)} = -\frac{2(\psi_x + c_2 - x)}{c_2(c_2-x)^2}$. Since $\psi_y = w$, we can rewrite the last expression as

$$\begin{aligned}-\big(1 + O_4 - (O_2)_y\big)w_x &= -(1+O_4)w_x + b_1^{(y)}w_y - \frac{2w_x^2}{c_2(c_2-x)^2}w \\ &= b_1^{(x)}w_x + b_1^{(y)}w_y + d_1 w.\end{aligned}$$

Note that $|O_4| \leq Cx$ by $|\psi_x| \leq Cx$ and (11.2.33). Then, reducing ε if necessary, we have

$$b_1^{(x)} \leq 0, \quad d_1 \leq 0 \qquad \text{in } \Omega_\varepsilon.$$

(b) The next term of (11.2.30) is

$$\big(O_5 + (O_3)_y\big)w_y = b_2^{(y)}w_y.$$

(c) For the further term of (11.2.30), using the expression of O_1, we have

$$\begin{aligned}&-\big((\gamma+1)\psi_x - O_1\big)_y\psi_{xx} \\ &= -\frac{\gamma+1}{c_2}(c_2 - x + \psi_x)w_x\psi_{xx} - \frac{\gamma-1}{c_2}\big(w + \frac{1}{(c_2-x)^2}\psi_y w_y\big)\psi_{xx}.\end{aligned} \tag{11.2.34}$$

In order to rewrite further this term, we note that $a_{11} > 0$ in Ω_ε. Thus, we express ψ_{xx} from equation (11.1.4). Using $\psi_{xy} = w_x$ and $\psi_{yy} = w_y$, we obtain

$$\psi_{xx} = \frac{1}{a_{11}}\Big(- O_2 w_x - (\frac{1}{c_2} + O_3)w_y + (1 + O_4)\psi_x - O_5 w\Big).$$

Note that $\frac{1}{a_{11}}$ is smooth in the interior of Ω_ε (even though it becomes unbounded as $x \to 0+$). Substituting this expression of ψ_{xx} into (11.2.34), we compute its terms:

$$-w_x\psi_{xx} = \frac{1}{a_{11}}\Big(O_2 w_x^2 + (\frac{1}{c_2} + O_3)w_x w_y - (1+O_4)\psi_x w_x + O_5 w w_x\Big).$$

Using the expressions of the O_k–terms, we have

$$\begin{aligned} O_2 w_x^2 &= -\frac{2(c_2+\psi_x - x)w_x^2}{c_2(c_2-x)^2} w = d_{3,1} w, \\ O_5 w w_x &= -\frac{(2c_2 + \psi_x - 2x)w^2}{c_2(c_2-x)^3} w_x = b_{3,1}^{(x)} w_x, \end{aligned}$$

where $d_{3,1} \le 0$ and $b_{3,1}^{(x)} \le 0$ in Ω_ε if ε is chosen small.

Furthermore, another term of (11.2.34) is

$$-w\psi_{xx} = \frac{1}{a_{11}}\Big(O_2 w_x w + (\frac{1}{c_2} + O_3)w w_y - (1+O_4)\psi_x w + O_5 w^2\Big)$$

with

$$\begin{aligned} O_2 w_x w &= -\frac{2(c_2+\psi_x - x)w^2}{c_2(c_2-x)^2} w_x = b_{3,2}^{(x)} w_x, \\ O_5 w^2 &= -\frac{(2c_2 + \psi_x - 2x)w^2}{c_2(c_2-x)^3} w = d_{3,2} w, \end{aligned}$$

where $d_{3,2} \le 0$ and $b_{3,2}^{(x)} \le 0$ in Ω_ε if ε is chosen small.

We have shown above that $|O_4| \le Cx$ in Ω_ε. Then, using that $\psi_x \ge 0$ as shown above, we conclude that $-(1+O_4)\psi_x \le -\frac{1}{2}\psi_x$ in Ω_ε if ε is chosen small.

Therefore, the substitution of the expressions of the terms into (11.2.34) leads to

$$-\big((\gamma+1)\psi_x - O_1\big)_y \psi_{xx} = b_3^{(x)} w_x + b_3^{(y)} w_y + d_3 w,$$

where

$$\begin{aligned} b_3^{(x)} &= \frac{1}{a_{11}}\left(\frac{\gamma+1}{c_2}(c_2 - x + \psi_x)\Big(b_{3,1}^{(x)} - (1+O_4)\psi_x\Big) + \frac{\gamma-1}{c_2} b_{3,2}^{(x)}\right) \le -\frac{1}{Cx}\psi_x, \\ d_3 &= \frac{1}{a_{11}}\left(\frac{\gamma+1}{c_2}(c_2 - x + \psi_x)d_{3,1} + \frac{\gamma-1}{c_2}\Big(d_{3,2} - (1+O_4)\psi_x\Big)\right) \le -\frac{1}{Cx}\psi_x, \end{aligned}$$

by using (11.2.32) and the fact that $d_{3,k}, b_{3,k}^{(x)} \le 0$ for $k = 1, 2$.

(d) Another term of (11.2.30) is

$$\begin{aligned} -(O_4)_y\psi_x &= \frac{(\gamma-1)\psi_x}{c_2(c_2-x)} w + \frac{\gamma-1}{c_2}(1 + \frac{1}{c_2 - x}\psi_x)\psi_x w_x + \frac{(\gamma+1)\psi_x\psi_y}{c_2(c_2-x)^3} w_y \\ &= d_4 w + b_4^{(x)} w_x + b_4^{(y)} w_y. \end{aligned}$$

Then the estimate that $0 \le \psi_x \le Cx$ implies

$$|d_4| \le C\psi_x, \quad |b_4^{(x)}| \le C\psi_x \qquad \text{in } \Omega_\varepsilon. \tag{11.2.35}$$

(e) The last term of (11.2.30) is

$$(O_5)_y w = -\frac{w^2}{c_2(c_2-x)^3} w_x - \frac{(2c_2+\psi_x-2x)w}{c_2(c_2-x)^3} w_y = b_5^{(x)} w_x + b_5^{(y)} w_y,$$

where $b_5^{(x)} \le 0$.

Now, we combine all the terms in (a)–(e) to rewrite equation (11.2.30) as

$$a_{11}w_{xx} + a_{12}w_{xy} + a_{22}w_{yy} + b^{(x)}w_x + b^{(y)}w_y + dw = 0, \tag{11.2.36}$$

where, in Ω_ε with sufficiently small $\varepsilon > 0$,

$$b^{(x)} = b_1^{(x)} + b_3^{(x)} + b_4^{(x)} + b_5^{(x)} \le (-\frac{1}{Cx} + C)\psi_x \le 0,$$
$$d = d_1 + d_3 + d_4 \le (-\frac{1}{Cx} + C)\psi_x \le 0.$$

From now on, we fix such a constant ε, which also fixes M in (11.2.29), as we have discussed above.

4. We now consider equation (11.2.36) as a linear equation, *i.e.*, the coefficients are fixed by substituting $\psi(x,y)$, $w(x,y)$, and their first derivatives. Denote by L the operator of this equation. Then L is linear, and

$$L(\pm w) = 0.$$

The coefficients of this operator are smooth in the interior of Ω_ε. Also, the coefficients of the second-order terms of this operator are the same as those for the potential flow equation (11.1.4) for solution ψ. Thus, L is strictly elliptic in the interior of Ω_ε. Since the coefficients are smooth there, it follows that L is uniformly elliptic on each compact subset of the open set Ω_ε. Since $d \le 0$, it follows that L satisfies the strong maximum principle on every compact subset of Ω_ε.

Now, we show that Mx is a supersolution of L. Indeed,

$$L(Mx) = M(b^{(x)} + dx) \le 0 \qquad \text{in } \Omega_\varepsilon,$$

since $\Omega_\varepsilon \subset \{0 < x < \varepsilon\}$ and $b^{(x)}, d \le 0$.

Thus, $L(\pm w - Mx) \ge 0$ in Ω_ε. Since L satisfies the strong maximum principle on every compact subset of Ω_ε, it follows that, if a positive maximum of $\pm w - Mx$ is attained in the interior of Ω_ε, then $\pm w - Mx = const.$ in Ω_ε. From this, since $\pm w - Mx \in C(\overline{\Omega_\varepsilon})$, and $\pm w - Mx \le 0$ on $\partial\Omega_\varepsilon$, we find that $\pm w - Mx \le 0$ in Ω_ε. This contradicts the assumption that the maximum of $\pm w - Mx$ is positive. Thus, the positive maximum of $\pm w - Mx$ is attained in the interior of Ω_ε. Combining this with (11.2.29) yields (11.2.24). □

Lemma 11.2.7. *There exists $\varepsilon \in (0, \varepsilon_0)$ depending only on (ρ_0, ρ_1, γ) such that, for any admissible solution φ with $\theta_{\rm w} \in (\theta_{\rm w}^{\rm s}, \frac{\pi}{2})$,*

$$\begin{aligned} &\frac{2}{M}(\theta - \theta_{\rm w}) + m(\frac{\pi}{2} - \theta_{\rm w}) - Cx \le \partial_x(\varphi_1 - \varphi) \le M, \\ &\frac{1}{M} \le -\partial_y(\varphi_1 - \varphi) \le M \end{aligned} \tag{11.2.37}$$

in Ω_ε, where M and $m(\cdot)$ are from (11.2.5).

Proof. Since $\varphi_1 - \varphi = \varphi_1 - \varphi_2 - \psi$, estimates (11.2.37) follow from (11.2.3), (11.2.5), (11.2.19), and (11.2.23)–(11.2.24) if $\varepsilon > 0$ is chosen sufficiently small. □

Now we can describe more precisely the structure of set Ω_ε defined in Proposition 11.2.3 for sufficiently small ε.

Proposition 11.2.8. *Choose ε_0 to be the smallest of ε determined in Lemmas 11.2.5–11.2.7. Then there exist $\omega, C > 0$ depending only on (ρ_0, ρ_1, γ) such that, for any admissible solution φ with $\theta_{\rm w} \in (\theta_{\rm w}^{\rm s}, \frac{\pi}{2})$, there is a function $\hat{f} \in C^1([0, \varepsilon_0])$ so that, for every $\varepsilon \in (0, \varepsilon_0]$, region Ω_ε defined by* (11.2.16) *satisfies*

$$\begin{aligned} &\Omega_\varepsilon = \{(x,y) : 0 < x < \varepsilon, \ \theta_{\rm w} < y < \hat{f}(x)\}, \\ &\Gamma_{\rm shock} \cap \partial\Omega_\varepsilon = \{(x,y) : 0 < x < \varepsilon, \ y = \hat{f}(x)\}, \\ &\Gamma_{\rm wedge} \cap \partial\Omega_\varepsilon = \{(x,y) : 0 < x < \varepsilon, \ y = \theta_{\rm w}\}, \\ &\Gamma_{\rm sonic} = \Gamma_{\rm sonic} \cap \partial\Omega_\varepsilon = \{(0,y) : \theta_{\rm w} < y < \hat{f}(0)\}, \end{aligned} \tag{11.2.38}$$

and

$$\hat{f}(0) = y_{P_1} > y_{P_4} = \theta_{\rm w}, \tag{11.2.39}$$

$$0 < \omega \le \frac{d\hat{f}}{dx} \le C \qquad \text{for all } x \in (0, \ \varepsilon_0). \tag{11.2.40}$$

Proof. In this proof, all the constants depend on (ρ_0, ρ_1, γ).

We continue to use the polar coordinates with the center at $\mathcal{O}_2$. Then, from Lemma 9.4.7, we see that there exists $\mu > 0$ such that $\Gamma_{\rm shock} \subset \{\theta > \theta_{\rm w} + \mu\}$ for any admissible solution with $\theta_{\rm w} \in [\frac{\pi}{2} - \sigma_1, \frac{\pi}{2})$. From Lemma 11.2.7, there exists $\tilde{M} > 0$ such that, for any admissible solution φ with $\theta_{\rm w} \in (\theta_{\rm w}^{\rm s}, \frac{\pi}{2})$, we find that, on $\Gamma_{\rm shock}$,

$$\begin{aligned} &\frac{2}{\tilde{M}} \le \partial_x(\varphi_1 - \varphi) \le \frac{M}{2}, \\ &\frac{2}{M} \le -\partial_y(\varphi_1 - \varphi) \le \frac{M}{2}. \end{aligned} \tag{11.2.41}$$

Also, $\varphi_1 > \varphi_2 = \varphi$ on $\Gamma_{\rm sonic}$, where the inequality follows from the definition of $\Gamma_{\rm sonic}$ (see Definition 7.5.7), and the equality follows from Definition 8.1.1(ii).

With these properties and the fact that $\varphi_1 > \varphi$ in Ω and $\varphi_1 = \varphi$ on Γ_{shock}, we employ (11.2.8), (11.2.19), Lemma 11.2.6, and the structure of Ω in Definition 8.1.1(i) to conclude that there exists $\hat{f} \in C^1([0, \varepsilon_0])$ such that (11.2.38) holds for each $\varepsilon \in (0, \varepsilon_0]$. Then $\hat{f}$ satisfies (11.2.39) and $\frac{d\hat{f}}{dx}(x) = -\frac{\partial_x(\varphi_1-\varphi)}{\partial_y(\varphi_1-\varphi)}(x, \hat{f}(x))$. Using the expression of $\frac{d\hat{f}}{dx}(x)$ given above and Lemma 11.2.7, we conclude (11.2.40). □

Corollary 11.2.9. *Let ε_0 be as in Proposition* 11.2.8. *There exists $C > 0$ depending only on (ρ_0, ρ_1, γ) such that, for any admissible solution φ with $\theta_{\rm w} \in (\theta_{\rm w}^{\rm s}, \frac{\pi}{2})$,*

$$0 < \psi \leq Cx^2 \qquad in\ \Omega_{\varepsilon_0}. \tag{11.2.42}$$

Proof. Since $\psi = 0$ on $\partial\Omega_{\varepsilon_0} \cap \{x = 0\} \equiv \Gamma_{\text{sonic}}$, and (11.2.23) holds in Ω_{ε_0} for ε_0 defined in Proposition 11.2.8, we employ structure (11.2.38) and (11.2.40) of Ω_{ε_0} to obtain (11.2.42). □

Corollary 11.2.10. *There exist $\varepsilon, \lambda > 0$ depending only on (ρ_0, ρ_1, γ) such that, if φ is an admissible solution with $\theta_{\rm w} \in (\theta_{\rm w}^{\rm s}, \frac{\pi}{2})$, then equation* (11.1.4) *with* (11.1.5), *written as* (11.1.6), *satisfies the following scaled ellipticity with respect to solution $\psi = \varphi - \varphi_2$ in Ω_ε:*

$$\lambda|\boldsymbol{\kappa}|^2 \leq \sum_{i,j=1}^{2} \hat{A}_{ij}(D\psi(x,y), \psi(x,y), x)\frac{\kappa_i\kappa_j}{x^{2-\frac{i+j}{2}}} \leq \frac{1}{\lambda}|\boldsymbol{\kappa}|^2, \tag{11.2.43}$$

for any $(x, y) \in \Omega_\varepsilon$ and $\boldsymbol{\kappa} = (\kappa_1, \kappa_2) \in \mathbb{R}^2$.

Proof. Note that $\frac{1}{C} \leq c_2 \leq C$ for c_2 in (11.1.4) for $\theta_{\rm w} \in [\theta_{\rm w}^{\rm s}, \frac{\pi}{2}]$, where C depends only on (ρ_0, ρ_1, γ).

Let $\varepsilon > 0$ be smaller than the small constant ε in Lemmas 11.2.5–11.2.6 and Proposition 11.2.8. Then, employing (11.2.23)–(11.2.24), (11.2.42), and Lemma 11.1.1, and reducing ε further if necessary, we conclude the proof. □

From (11.2.23)–(11.2.24) and (11.2.42), it follows that there exist $\varepsilon \in (0, \varepsilon_0)$, $\delta > 0$, and $L < \infty$ depending only on (ρ_0, ρ_1, γ) such that, if φ is an admissible solution with $\theta_{\rm w} \in (\theta_{\rm w}^{\rm s}, \frac{\pi}{2})$, and $\psi = \varphi - \varphi_2$, then

$$0 \leq \partial_x\psi(x,y) \leq \frac{2-\delta}{1+\gamma}x, \quad |D\psi(x,y)| \leq Lx, \quad |\psi(x,y)| \leq Lx^2 \tag{11.2.44}$$

for any $(x, y) \in \Omega_\varepsilon$.

Lemma 11.2.11. *For any constant $M > 0$, there exist $\varepsilon \in (0, \frac{\varepsilon_0}{2})$ and C depending only on $(\rho_0, \rho_1, \gamma, M)$ such that, for any $\theta_{\rm w} \in (\theta_{\rm w}^{\rm s}, \frac{\pi}{2})$, the corresponding equation* (11.1.4) *with* (11.1.5), *written as* (11.1.6), *satisfies*

$$\begin{aligned} &|(\hat{A}_{11}, \hat{A}_{12}, \hat{A}_2)(\mathbf{p}, z, x)| \leq Cx, \\ &|(\hat{A}_{22}, \hat{A}_1)| + |D_{(\mathbf{p},z,x)}(\hat{A}_{ij}, \hat{A}_i)| \leq C \end{aligned} \tag{11.2.45}$$

on $\{(\mathbf{p}, z, x) : |\mathbf{p}| \le Mx,\ |z| \le Mx^2,\ x \in (0, \varepsilon)\}$.

Proof. Since $c_2^* := \inf_{\theta_{\rm w}\in[\theta_{\rm w}^{\rm s},\frac{\pi}{2}]} c_2(\theta_{\rm w}) > 0$, then, choosing $\varepsilon < \frac{c_2^*}{10}$, the results follow directly from the explicit expressions (11.1.4)–(11.1.5) of $(\hat{A}_{ij}, \hat{A}_i)$. □

From (11.2.44) and Lemma 11.2.11, we can modify $(\hat{A}_{ij}, \hat{A}_i)(\mathbf{p}, z, x)$ outside set $\{|\mathbf{p}| \le Lx,\ |z| \le Lx^2,\ x \in (0, \varepsilon)\}$ in such a way that (11.2.45) holds for all $(\mathbf{p}, z, x) \in \mathbb{R}^2 \times \mathbb{R} \times (0, \varepsilon)$ and the admissible solutions satisfy the resulting equation.

Corollary 11.2.12. *There exist $\varepsilon \in (0, \frac{\varepsilon_0}{2})$ and C depending only on (ρ_0, ρ_1, γ) such that any admissible solution $\psi = \varphi - \varphi_2$ with $\theta_{\rm w} \in (\theta_{\rm w}^{\rm s}, \frac{\pi}{2})$ satisfies the equation of form* (11.1.6) *whose coefficients $(A_{ij}^{(\rm mod)}, A_i^{(\rm mod)})$ satisfy*

$$(A_{ij}^{(\rm mod)}, A_i^{(\rm mod)}) = (\hat{A}_{ij}, \hat{A}_i) \tag{11.2.46}$$

on $\{(\mathbf{p}, z, x) : |\mathbf{p}| \le Lx,\ |z| \le Lx^2,\ 0 < x < \varepsilon\}$, *where* $(A_{ij}, A_i)(\mathbf{p}, z, x)$ *are from Lemma* 11.2.11, *L is the constant in* (11.2.44), *and*

$$\begin{aligned} &|(A_{11}^{(\rm mod)}, A_{12}^{(\rm mod)}, A_2^{(\rm mod)})(\mathbf{p}, z, x)| \le Cx \qquad \text{on } \mathbb{R}^2 \times \mathbb{R} \times (0, \varepsilon),\\ &\|(A_{22}^{(\rm mod)}, A_1^{(\rm mod)})\|_{L^\infty(\mathbb{R}^2\times\mathbb{R}\times(0,\varepsilon))} \le C,\\ &\|D_{(\mathbf{p},z,x)}(A_{ij}^{(\rm mod)}, A_i^{(\rm mod)})\|_{L^\infty(\mathbb{R}^2\times\mathbb{R}\times(0,\varepsilon))} \le C. \end{aligned} \tag{11.2.47}$$

Proof. In this proof, the universal constant $C > 0$ depends only on (ρ_0, ρ_1, γ). Without loss of generality, we may assume that $L \ge 1$.

Let $\eta \in C^\infty(\mathbb{R})$ so that $0 \le \eta \le 1$ on $\mathbb{R}$ with $\eta(t) = t$ for all $|t| \le L$, $|\eta(t)| \le 2L$ on $\mathbb{R}$, and $\eta(t) = 0$ for all $|t| \ge 2L$. We note that $|\eta'(t)| \le C$ and $|t\eta'(t)| \le C$ on $\mathbb{R}$, where C depends only on L, and hence on (ρ_0, ρ_1, γ). From this, $f(t, x) := x\eta(\frac{t}{x})$ satisfies that $f \in C^\infty(\mathbb{R} \times (0, \varepsilon))$ with

$$\begin{aligned} &f(t, x) = t \ \text{ if } |t| \le Lx, \qquad f(t, x) = 0 \ \text{ if } |t| \ge 2Lx,\\ &|f(t, x)| \le 2Lx \qquad \text{for all } (t, x) \in \mathbb{R} \times (0, \varepsilon),\\ &\|f\|_{L^\infty(\mathbb{R}\times(0,\varepsilon))} \le C,\\ &\|D_{(t,x)}f\|_{L^\infty(\mathbb{R}\times(0,\varepsilon))} \le C, \end{aligned}$$

where the last estimate follows from $|D_t f(t, x)| = |\eta'(\frac{t}{x})|$ and $|D_x f(t, x)| = |\frac{t}{x}\eta'(\frac{t}{x})|$. Similar facts hold for $f_1(t, x) := x^2\eta(\frac{t}{x^2})$.

Then, using Lemma 11.2.11 with $M = 10L$, we find that the functions:

$$(A_{ij}^{(\rm mod)}, A_i^{(\rm mod)})(\mathbf{p}, z, x) := (\hat{A}_{ij}, \hat{A}_i)(x(\eta(\frac{p_1}{x}), \eta(\frac{p_2}{x})),\ x^2\eta(\frac{z}{x^2}),\ x)$$

satisfy all the assertions. □

11.3 PROPERTIES OF THE RANKINE-HUGONIOT CONDITION ON $\Gamma_{\rm shock}$ NEAR $\Gamma_{\rm sonic}$

Next we write the Rankine-Hugoniot condition (8.1.13) with (8.1.14) as an oblique condition on $\Gamma_{\rm shock}$, analyze its properties near $\Gamma_{\rm sonic}$, and then write it in the (x,y)–coordinates (11.1.1)–(11.1.2) in terms of $\psi = \varphi - \varphi_2$.

We first work in the $\boldsymbol{\xi}$–coordinates. Let φ be an admissible solution with $\theta_{\rm w} \in (\theta_{\rm w}^{\rm s}, \frac{\pi}{2})$. Then φ satisfies (8.1.13)–(8.1.14) on $\Gamma_{\rm shock}$. From (8.1.14), it follows that $\boldsymbol{\nu}_{\rm sh} = \frac{D(\varphi_1-\varphi)}{|D(\varphi_1-\varphi)|}$, so that (8.1.13) can be written as

$$\mathcal{M}_0(D\psi, \psi, \boldsymbol{\xi}) = 0 \qquad \text{on } \Gamma_{\rm shock}, \tag{11.3.1}$$

where

$$\mathcal{M}_0(\mathbf{p}, z, \boldsymbol{\xi}) = g^{\rm sh}(D\varphi_2(\boldsymbol{\xi}) + \mathbf{p},\ \varphi_2(\boldsymbol{\xi}) + z,\ \boldsymbol{\xi}) \tag{11.3.2}$$

and $g^{\rm sh}(\mathbf{p}, z, \boldsymbol{\xi})$ is defined by (7.1.9).

We note that $g^{\rm sh}(\mathbf{p}, z, \boldsymbol{\xi})$ is not defined for all $(\mathbf{p}, z, \boldsymbol{\xi}) \in \mathbb{R}^2 \times \mathbb{R} \times \mathbb{R}^2$, and its domain is clear from its explicit expression in (7.1.9). This determines the domain of $\mathcal{M}_0(\mathbf{p}, z, \boldsymbol{\xi})$ by (11.3.2). In the $\boldsymbol{\xi}$–coordinates with the origin shifted to center $\mathcal{O}_2$ of the sonic center of state (2), the domain of $\mathcal{M}_0(\mathbf{p}, z, \boldsymbol{\xi})$ is

$$\mathcal{D}(\mathcal{M}_0) := \left\{ (\mathbf{p}, z, \boldsymbol{\xi}) \ : \ \begin{array}{l} \rho_2^{\gamma-1} + (\gamma-1)(\boldsymbol{\xi}\cdot\mathbf{p} - \frac{|\mathbf{p}|^2}{2} - z) > 0 \\ |\mathbf{p} - (u_1 - u_2, -v_2)| > 0 \end{array} \right\}. \tag{11.3.3}$$

From (7.1.5)–(7.1.6) at P_0, it follows that these properties hold at every point of $\mathcal{S}_1 = \{\varphi_1 = \varphi_2\}$, especially at P_1. Also, $\rho(|D\varphi_2(\mathbf{0},0,P_1)|^2, \varphi_2(\mathbf{0},0,P_1)) = \rho_2 > 0$, where $\rho(|\mathbf{p}|^2, z)$ is defined by (2.2.9). Then, from Theorem 7.1.1(i), it follows that there exist $\delta > 0$ and C depending only on (ρ_0, ρ_1, γ) such that, for any $\theta_{\rm w} \in [\theta_{\rm w}^{\rm s}, \frac{\pi}{2}]$,

$$\begin{aligned} &B_\delta(\mathbf{0}, 0, P_1) \subset \mathcal{D}(\mathcal{M}_0), \\ &|D^k \mathcal{M}_0(\mathbf{p}, z, \boldsymbol{\xi})| \le C \quad \text{for all } (\mathbf{p}, z, \boldsymbol{\xi}) \in B_\delta(\mathbf{0}, 0, P_1), k = 1, 2, 3. \end{aligned} \tag{11.3.4}$$

We first show the following properties of $g^{\rm sh}(\mathbf{p}, z, \boldsymbol{\xi})$ and $\mathcal{M}_0(\mathbf{p}, z, \boldsymbol{\xi})$:

Lemma 11.3.1. *There exists $\delta > 0$ depending only on (ρ_0, ρ_1, γ) such that, for any $\theta_{\rm w} \in [\theta_{\rm w}^{\rm s}, \frac{\pi}{2}]$,*

$$D_{\mathbf{p}} g^{\rm sh}(D\varphi_2(P_1), \varphi_2(P_1), P_1) \cdot D\varphi_2(P_1) \le -\delta, \tag{11.3.5}$$

$$D_{\mathbf{p}} g^{\rm sh}(D\varphi_2(P_1), \varphi_2(P_1), P_1) \cdot D^\perp\varphi_2(P_1) \le -\delta, \tag{11.3.6}$$

where $D^\perp\varphi_2(P_1) = (\partial_{\xi_2}, -\partial_{\xi_1})\varphi_2(P_1)$.

Proof. By definition, $P_1 \in \partial B_{c_2}(\mathcal{O}_2)$ so that

$$|D\varphi_2(P_1)| = c_2.$$

Now, (11.3.5) follows from Corollary 7.4.7, applied at $\boldsymbol{\xi} = P_1$, by using the continuous dependence of the parameters of state (2) and the coordinates of point P_1 on the wedge angle $\theta_{\rm w} \in [\theta_{\rm w}^{\rm s}, \frac{\pi}{2}]$.

To prove (11.3.6), we apply Theorem 7.1.1(viii) with $P_1 \in \mathcal{S}_1$. The only thing we need to check is that the orientation of $D^{\perp}\varphi_2(P_1)$ coincides with the one given in Theorem 7.1.1(viii), that is,

$$D\varphi_1(P_1) \cdot (\partial_{\xi_2}\varphi_2(P_1), -\partial_{\xi_1}\varphi_2(P_1)) < 0. \tag{11.3.7}$$

Recall that

$$D\varphi_k(P_1) = \mathcal{O}_k - P_1, \qquad k = 1, 2.$$

Thus, $(\partial_{\xi_2}, -\partial_{\xi_1})\varphi_2(P_1)$ is the clockwise rotation of $P_1\mathcal{O}_2$ by $\frac{\pi}{2}$. Using the notations in Definition 11.2.1, it follows from $\mathcal{O}_2 \in \Gamma_{\rm wedge}$ and property (6.1.6) in Lemma 6.1.2 that, in triangle $P_1\mathcal{O}_1Q'$, angle $\angle P_1Q'\mathcal{O}_1$ is $\frac{\pi}{2}$, and $\mathcal{O}_2$ lies on segment $\mathcal{O}_1Q'$ so that $\angle\mathcal{O}_1P_1\mathcal{O}_2 \in (0, \frac{\pi}{2})$. Let $\beta \in (0, \frac{\pi}{2})$ be the angle between vector $\mathcal{O}_1 - \mathcal{O}_2$ and the positive direction of the ξ_1–axis. Rotate the coordinate axes by angle β clockwise, and denote by (S, T) the rotated coordinates that have basis $\{\hat{\mathbf{e}}_1, \hat{\mathbf{e}}_2\}$ with $\hat{\mathbf{e}}_1 = \frac{\mathcal{O}_2\mathcal{O}_1}{|\mathcal{O}_2\mathcal{O}_1|} = (\cos\beta, -\sin\beta)$ and $\hat{\mathbf{e}}_2 = (\sin\beta, \cos\beta)$ in the original coordinates. From the above argument, taking into account the locations of the points as described in Definition 7.5.7 and denoting $a_k := |Q'\mathcal{O}_k|$ for $k = 1, 2$, and $b := |P_1Q'|$, we find that, in the (S, T)–coordinates,

$$\mathcal{O}_k - P_1 = (a_k, -b) = a_k\hat{\mathbf{e}}_1 - b\hat{\mathbf{e}}_2 \qquad \text{for } k = 1, 2,\ a_1 > a_2 > 0,\ b > 0.$$

Thus, denoting $\{\hat{\mathbf{e}}_1, \hat{\mathbf{e}}_2\}$ the basis in the original (non-rotated) $\boldsymbol{\xi}$–coordinates, we have

$$D\varphi_k(P_1) = \mathcal{O}_k - P_1 = a_k\hat{\mathbf{e}}_1 - b\hat{\mathbf{e}}_2 = (a_k\cos\beta - b\sin\beta)\mathbf{e}_1 - (a_k\sin\beta + b\cos\beta)\mathbf{e}_2.$$

Now we calculate

$$\begin{aligned} &D\varphi_1(P_1) \cdot (\partial_{\xi_2}\varphi_2, -\partial_{\xi_1}\varphi_2)(P_1) \\ &= -(a_1\cos\beta - b\sin\beta)(a_2\sin\beta + b\cos\beta) \\ &\quad + (a_1\sin\beta + b\cos\beta)(a_2\cos\beta - b\sin\beta) \\ &= b(a_2 - a_1) < 0. \end{aligned}$$

Then (11.3.7) is proved, which implies (11.3.6). □

To state the next properties, we define the unit normal and tangent vectors to $\mathcal{S}_1$:

$$\begin{aligned} \boldsymbol{\nu}_{\mathcal{S}_1} &= \frac{D(\varphi_1 - \varphi_2)}{|D(\varphi_1 - \varphi_2)|} = \frac{(u_1 - u_2,\ -v_2)}{\sqrt{(u_1 - u_2)^2 + v_2^2}}, \\ \boldsymbol{\tau}_{\mathcal{S}_1} &= \boldsymbol{\nu}_{\mathcal{S}_1}^{\perp} = -\frac{(v_2,\ u_1 - u_2)}{\sqrt{(u_1 - u_2)^2 + v_2^2}}. \end{aligned} \tag{11.3.8}$$

Note that $\boldsymbol{\tau}_{\mathcal{S}_1} = \mathbf{e}_{\mathcal{S}_1}$, as defined by (7.5.8). Also we have

$$\boldsymbol{\tau}_{\mathcal{S}_1} \cdot D\varphi_1(P_0) > 0. \tag{11.3.9}$$

Indeed, $P_0 = (\xi_1^0, \xi_1^0 \tan\theta_{\rm w})$ and $\mathcal{O}_1 = (u_1, 0)$ so that $D\varphi_1(P_0) = P_0\mathcal{O}_1 = (u_1 - \xi_1^0, -\xi_1^0 \tan\theta_{\rm w})$. Thus, using (11.3.8) and $v_2 = u_2 \tan\theta_{\rm w}$, we have

$$\boldsymbol{\tau}_{\mathcal{S}_1} \cdot D\varphi_1(P_0) = \frac{u_1 \tan\theta_{\rm w}}{\sqrt{(u_1 - u_2)^2 + v_2^2}}(\xi_1^0 - u_2) > 0,$$

where the last inequality follows, since $\xi_1^0 > u_2$ by Lemma 7.4.8, and $u_1 > 0$.

Lemma 11.3.2. *There exists $\delta > 0$ depending only on (ρ_0, ρ_1, γ) such that, for any $\theta_{\rm w} \in (\theta_{\rm w}^{\rm s}, \frac{\pi}{2}]$,*

$$D_z g^{\rm sh}(D\varphi_2(P_1), \varphi_2(P_1), P_1) = -\rho_2^{2-\gamma} D\varphi_2(P_1) \cdot \boldsymbol{\nu}_{\mathcal{S}_1} \le -\delta. \tag{11.3.10}$$

Proof. The expression of $D_z g^{\rm sh}(D\varphi_2(P_1), \varphi_2(P_1), P_1)$ given in (11.3.10) follows from (7.1.9) through an explicit calculation by using (2.2.9) and $c^2 = \rho^{\gamma-1}$.

To obtain the estimate, we note that $P_0, P_1 \in \mathcal{S}_1$ so that $D\varphi_2(P_1) \cdot \boldsymbol{\nu}_{\mathcal{S}_1} = D\varphi_2(P_0) \cdot \boldsymbol{\nu}_{\mathcal{S}_1}$ by Lemma 6.1.1. It remains to show that $D\varphi_2(P_0) \cdot \boldsymbol{\nu}_{\mathcal{S}_1} \le -\delta$ for some $\delta(\rho_0, \rho_1, \gamma) > 0$. We use (7.1.6), in which $\boldsymbol{\nu} = \boldsymbol{\nu}_{\mathcal{S}_1}$ by (7.1.5). Since $\boldsymbol{\nu}_{\mathcal{S}_1}$ depends continuously on the parameters of state (2), we conclude that $\rho_2^{2-\gamma} D\varphi_2(P_1) \cdot \boldsymbol{\nu}_{\mathcal{S}_1}$ depends continuously on the wedge angle $\theta_{\rm w} \in [\theta_{\rm w}^{\rm s}, \frac{\pi}{2}]$. This completes the proof. □

Lemma 11.3.3. *There exists $\delta > 0$ depending only on (ρ_0, ρ_1, γ) such that, for any $\theta_{\rm w} \in [\theta_{\rm w}^{\rm s}, \frac{\pi}{2}]$,*

$$D_{\mathbf{p}} g^{\rm sh}(D\varphi_2(P_1), \varphi_2(P_1), P_1) \cdot \boldsymbol{\nu}_{\mathcal{S}_1} \ge \delta. \tag{11.3.11}$$

Proof. From (2.2.9) and $c^2 = \rho^{\gamma-1}$,

$$D_{\mathbf{p}}\rho(|\mathbf{p}|^2, z) = -\frac{\rho}{c^2}(|\mathbf{p}|^2, z)\mathbf{p}. \tag{11.3.12}$$

Since φ_1 and φ_2 satisfy the Rankine-Hugoniot conditions (7.1.5) for all $\boldsymbol{\xi} \in \mathcal{S}_1$, then, at P_1, $\frac{D(\varphi_1 - \varphi_2)}{|D(\varphi_1 - \varphi_2)|} = \boldsymbol{\nu}_{\mathcal{S}_1}$ and

$$\rho_2 D\varphi_2(P_1) - \rho_1 D\varphi_1(P_1) = (\rho_2 - \rho_1)(D\varphi_2(P_1) \cdot \boldsymbol{\tau}_{\mathcal{S}_1})\boldsymbol{\tau}_{\mathcal{S}_1}.$$

Combining this with the calculation at $(\mathbf{p}, z, \boldsymbol{\xi}) = (D\varphi_2(P_1), \varphi_2(P_1), P_1)$:

$$(\boldsymbol{\nu}_{\mathcal{S}_1} \cdot D_{\mathbf{p}}) \frac{D\varphi_1 - \mathbf{p}}{|D\varphi_1 - \mathbf{p}|} = -\frac{1}{|D\varphi_1 - \mathbf{p}|}\boldsymbol{\nu}_{\mathcal{S}_1} - \frac{(\boldsymbol{\nu}_{\mathcal{S}_1} \cdot D_{\mathbf{p}})|D\varphi_1 - \mathbf{p}|}{|D\varphi_1 - \mathbf{p}|}\boldsymbol{\nu}_{\mathcal{S}_1},$$

we conclude, again at $(\mathbf{p}, z, \boldsymbol{\xi}) = (D\varphi_2(P_1), \varphi_2(P_1), P_1)$, that

$$(\rho_2 D\varphi_2 - \rho_1 D\varphi_1)(\boldsymbol{\nu}_{\mathcal{S}_1} \cdot D_{\mathbf{p}}) \frac{D\varphi_1 - \mathbf{p}}{|D\varphi_1 - \mathbf{p}|} = 0.$$

With this, we calculate that, at $(\mathbf{p}, z, \boldsymbol{\xi}) = (D\varphi_2(P_1), \varphi_2(P_1), P_1)$,

$$\begin{aligned}(\boldsymbol{\nu}_{\mathcal{S}_1} \cdot D_{\mathbf{p}})g^{\rm sh} &= \big(\boldsymbol{\nu}_{\mathcal{S}_1} \cdot D_{\mathbf{p}}\rho(|\mathbf{p}|^2, z)\big)D\varphi_2 \cdot \boldsymbol{\nu}_{\mathcal{S}_1} + \rho_2\boldsymbol{\nu}_{\mathcal{S}_1} \cdot \boldsymbol{\nu}_{\mathcal{S}_1} \\ &= -\frac{\rho_2}{c_2^2}(D\varphi_2 \cdot \boldsymbol{\nu}_{\mathcal{S}_1})^2 + \rho_2,\end{aligned} \tag{11.3.13}$$

by using (7.1.9) and (11.3.12). Since P_1 lies on the sonic circle of state (2), we find that $D\varphi_2(P_1) = c_2\mathbf{e}_{P_1}$ for $\mathbf{e}_{P_1} := \frac{D\varphi_2(P_1)}{|D\varphi_2(P_1)|} = \frac{O_2-P_1}{|O_2-P_1|}$. We note that vectors $\boldsymbol{\nu}_{\mathcal{S}_1}$ and $\mathbf{e}_{P_1}$ are not parallel to each other, since $\mathbf{e}_{P_1}$ is in the radial direction for $\partial B_{c_2}(0)$ at P_1, and line $\mathcal{S}_1$ passes through point P_1 and is also non-tangential to $\partial B_{c_2}(0)$ at P_1 by (6.1.4), where Lemma 6.1.2 is applied with $(\varphi^-, \varphi^+) = (\varphi_1, \varphi_2)$. Then $\big(\boldsymbol{\nu}_{\mathcal{S}_1} \cdot \mathbf{e}_{P_1}\big)^2 < 1$, since $|\boldsymbol{\nu}_{\mathcal{S}_1}| = |\mathbf{e}_{P_1}| = 1$. Using this and $D\varphi_2(P_1) = c_2\mathbf{e}_{P_1}$, we obtain

$$(\boldsymbol{\nu}_{\mathcal{S}_1} \cdot D_{\mathbf{p}})g^{\rm sh}(D\varphi_2(P_1), \varphi_2(P_1), P_1) = \rho_2\big(1 - (\boldsymbol{\nu}_{\mathcal{S}_1} \cdot \mathbf{e}_{P_1})^2\big) > 0. \tag{11.3.14}$$

Moreover, expression $\rho_2\big(1 - (\boldsymbol{\nu}_{\mathcal{S}_1} \cdot \mathbf{e}_{P_1})^2\big)$ depends only on the parameters of state (2), and the dependence is continuous on $\theta_{\rm w} \in [\theta_{\rm w}^{\rm s}, \frac{\pi}{2}]$. Now (11.3.11) is proved. □

Corollary 11.3.4. *There exists $\delta > 0$ depending only on (ρ_0, ρ_1, γ) such that, for any $\theta_{\rm w} \in [\theta_{\rm w}^{\rm s}, \frac{\pi}{2}]$,*

$$D_{\mathbf{p}}\mathcal{M}_0(\mathbf{0}, 0, P_1) \cdot D\varphi_2(P_1) \le -\delta, \tag{11.3.15}$$

$$D_{\mathbf{p}}\mathcal{M}_0(\mathbf{0}, 0, P_1) \cdot D^{\perp}\varphi_2(P_1) \le -\delta, \tag{11.3.16}$$

$$D_z\mathcal{M}_0(\mathbf{0}, 0, P_1) = -\rho_2^{2-\gamma}D\varphi_2(P_1) \cdot \boldsymbol{\nu}_{\mathcal{S}_1} \le -\delta, \tag{11.3.17}$$

$$D_{\mathbf{p}}\mathcal{M}_0(\mathbf{0}, 0, P_1) \cdot \boldsymbol{\nu}_{\mathcal{S}_1} \ge \delta, \tag{11.3.18}$$

where $D^{\perp}\varphi_2(P_1) = (\partial_{\xi_2}, -\partial_{\xi_1})\varphi_2(P_1)$.

These results follow directly from Lemmas 11.3.1–11.3.3.

However, the boundary condition (11.3.1)–(11.3.2) is not convenient for our purposes because it is nonhomogeneous; $\mathcal{M}_0(\mathbf{0}, 0, \boldsymbol{\xi}) \neq 0$ in general. Now we define $\mathcal{M}_1(\mathbf{p}, z, \boldsymbol{\xi})$ so that both $\mathcal{M}_1(\mathbf{0}, 0, \boldsymbol{\xi}) \equiv 0$ and (11.3.1) hold with $\mathcal{M}_1$ for admissible solutions.

Since $\mathcal{S}_1 = \{\varphi_1 = \varphi_2\}$ and $P_1 \in \mathcal{S}_1$, it follows that

$$(\varphi_1 - \varphi_2)(\boldsymbol{\xi}) = |(u_1 - u_2, -v_2)|\,\boldsymbol{\nu}_{\mathcal{S}_1} \cdot (\boldsymbol{\xi} - \boldsymbol{\xi}_{P_1}) \qquad \text{for all } \boldsymbol{\xi} \in \mathbb{R}^2,$$

where we have used that $D(\varphi_1 - \varphi_2) = (u_1 - u_2, -v_2)$ and (11.3.8). We write this as

$$\boldsymbol{\xi} \cdot \boldsymbol{\nu}_{\mathcal{S}_1} = \boldsymbol{\xi}_{P_1} \cdot \boldsymbol{\nu}_{\mathcal{S}_1} + \frac{(\varphi_1 - \varphi_2)(\boldsymbol{\xi})}{\sqrt{(u_1 - u_2)^2 + v_2^2}} \qquad \text{on } \Gamma_{\rm shock}. \tag{11.3.19}$$

If φ is a solution of **Problem 2.6.1** with regular reflection-diffraction configuration, then, in addition to (11.3.1), it satisfies that $\varphi = \varphi_1$ on Γ_{shock}, that is,

$$\varphi_1 - \varphi_2 = \psi \qquad \text{on } \Gamma_{\text{shock}}.$$

Using (11.3.19), we have

$$\boldsymbol{\xi} \cdot \boldsymbol{\nu}_{S_1} = \boldsymbol{\xi}_{P_1} \cdot \boldsymbol{\nu}_{S_1} + \frac{\psi(\boldsymbol{\xi})}{\sqrt{(u_1 - u_2)^2 + v_2^2}} \qquad \text{on } \Gamma_{\text{shock}}.$$

Then we modify $\mathcal{M}_0(\mathbf{p}, z, \boldsymbol{\xi})$ given by (11.3.2) via replacing $\boldsymbol{\xi} = (\boldsymbol{\xi} \cdot \boldsymbol{\nu}_{S_1})\boldsymbol{\nu}_{S_1} + (\boldsymbol{\xi} \cdot \boldsymbol{\tau}_{S_1})\boldsymbol{\tau}_{S_1}$ by the following expression:

$$F(z, \boldsymbol{\xi}) := \Big(\boldsymbol{\xi}_{P_1} \cdot \boldsymbol{\nu}_{S_1} + \frac{z}{\sqrt{(u_1 - u_2)^2 + v_2^2}}\Big)\boldsymbol{\nu}_{S_1} + (\boldsymbol{\xi} \cdot \boldsymbol{\tau}_{S_1})\boldsymbol{\tau}_{S_1}. \tag{11.3.20}$$

Note that $|D(\varphi_1 - \varphi_2)| = |(u_1 - u_2, -v_2)| \geq \frac{1}{C}$ and (11.3.19) imply that, for each $\theta_{\text{w}} \in [\theta_{\text{w}}^{\text{s}}, \frac{\pi}{2}]$,

$$F(z, \boldsymbol{\xi}) = \boldsymbol{\xi} \qquad \text{if and only if} \quad z = (\varphi_1 - \varphi_2)(\boldsymbol{\xi}), \tag{11.3.21}$$

$$|D^k F(z, \boldsymbol{\xi})| \leq C \qquad \text{for } k = 1, 2, 3, \tag{11.3.22}$$

for all $(\boldsymbol{\xi}, z) \in \mathbb{R}^2 \times \mathbb{R}$, where C depends only on (ρ_0, ρ_1, γ).

Now we can define

$$\mathcal{M}_1(\mathbf{p}, z, \boldsymbol{\xi}) := \mathcal{M}_0(\mathbf{p}, z, F(z, \boldsymbol{\xi})) \tag{11.3.23}$$

for all $(\mathbf{p}, z, \boldsymbol{\xi}) \in \mathbb{R}^2 \times \mathbb{R} \times \mathbb{R}^2$ satisfying $(\mathbf{p}, z, F(z, \boldsymbol{\xi})) \in \mathcal{D}(\mathcal{M}_0)$. From (11.3.4), (11.3.22), and $F(0, P_1) = P_1$, there exist $\delta > 0$ and C depending only on (ρ_0, ρ_1, γ) such that, for any $\theta_{\text{w}} \in [\theta_{\text{w}}^{\text{s}}, \frac{\pi}{2}]$,

$$(\mathbf{p}, z, F(z, \boldsymbol{\xi})) \in \mathcal{D}(\mathcal{M}_0), \quad |(D, D^2, D^3)\mathcal{M}_1(\mathbf{p}, z, \boldsymbol{\xi})| \leq C \tag{11.3.24}$$

for all $(\mathbf{p}, z, \boldsymbol{\xi}) \in B_\delta(\mathbf{0}, 0, P_1)$.

From (11.3.20)–(11.3.21),

$$\mathcal{M}_1(\mathbf{p}, z, \boldsymbol{\xi}) = \mathcal{M}_0(\mathbf{p}, z, \boldsymbol{\xi}) \tag{11.3.25}$$

for all $(\mathbf{p}, z, \boldsymbol{\xi}) \in \mathcal{D}(\mathcal{M}_0)$ with $z = (\varphi_1 - \varphi_2)(\boldsymbol{\xi})$.

From this, since $\mathcal{D}(\mathcal{M}_0)$ is an open set, we have

$$D_{\mathbf{p}}\mathcal{M}_1(\mathbf{p}, z, \boldsymbol{\xi}) = D_{\mathbf{p}}\mathcal{M}_0(\mathbf{p}, z, \boldsymbol{\xi}) \tag{11.3.26}$$

for all $(\mathbf{p}, z, \boldsymbol{\xi}) \in \mathcal{D}(\mathcal{M}_0)$ with $z = (\varphi_1 - \varphi_2)(\boldsymbol{\xi})$.

Lemma 11.3.5. *For any $\boldsymbol{\xi} \in \mathbb{R}^2$ such that $(\mathbf{0}, 0, F(0, \boldsymbol{\xi})) \in \mathcal{D}(\mathcal{M}_0)$,*

$$\mathcal{M}_1(\mathbf{0}, 0, \boldsymbol{\xi}) = 0. \tag{11.3.27}$$

Proof. Since $P_1 \in \mathcal{S}_1$, $(\boldsymbol{\xi}_{P_1} \cdot \boldsymbol{\nu}_{\mathcal{S}_1})\boldsymbol{\nu}_{\mathcal{S}_1} \in \mathcal{S}_1$ so that $(\boldsymbol{\xi}_{P_1} \cdot \boldsymbol{\nu}_{\mathcal{S}_1})\boldsymbol{\nu}_{\mathcal{S}_1} + a\boldsymbol{\tau}_{\mathcal{S}_1} \in \mathcal{S}_1$ for any $a \in \mathbb{R}$. With this, for any $\boldsymbol{\xi} \in \mathbb{R}^2$,

$$F(0,\boldsymbol{\xi}) = (\boldsymbol{\xi}_{P_1} \cdot \boldsymbol{\nu}_{\mathcal{S}_1})\boldsymbol{\nu}_{\mathcal{S}_1} + (\boldsymbol{\xi} \cdot \boldsymbol{\tau}_{\mathcal{S}_1})\boldsymbol{\tau}_{\mathcal{S}_1} \in \mathcal{S}_1.$$

Thus, using the explicit expressions of $\mathcal{M}_0$ and $\mathcal{M}_1$,

$$\begin{aligned}\mathcal{M}_1(\mathbf{0},0,\boldsymbol{\xi}) &= \mathcal{M}_0(\mathbf{0},0,F(0,\boldsymbol{\xi})) \\ &= \big(\rho_2 D\varphi_2(F(0,\boldsymbol{\xi})) - \rho_1 D\varphi_1(F(0,\boldsymbol{\xi}))\big) \cdot \boldsymbol{\nu}_{\mathcal{S}_1} \\ &= 0,\end{aligned}$$

where the last equality holds since states (1) and (2) satisfy the Rankine-Hugoniot conditions (8.1.13)–(8.1.14) across $\mathcal{S}_1$ at point $F(0,\boldsymbol{\xi}) \in \mathcal{S}_1$. □

Lemma 11.3.6. *There exists $\delta > 0$ depending only on (ρ_0, ρ_1, γ) such that, for any $\theta_{\rm w} \in [\theta_{\rm w}^{\rm s}, \frac{\pi}{2}]$,*

$$D_{\mathbf{p}}\mathcal{M}_1(\mathbf{0},0,P_1) \cdot D\varphi_2(P_1) \le -\delta, \tag{11.3.28}$$

$$D_{\mathbf{p}}\mathcal{M}_1(\mathbf{0},0,P_1) \cdot D^{\perp}\varphi_2(P_1) \le -\delta, \tag{11.3.29}$$

$$D_{\mathbf{p}}\mathcal{M}_1(\mathbf{0},0,P_1) \cdot \boldsymbol{\nu}_{\mathcal{S}_1} \ge \delta, \tag{11.3.30}$$

$$D_z\mathcal{M}_1(\mathbf{0},0,\boldsymbol{\xi}) \le -\delta \qquad \text{for any } \boldsymbol{\xi} \in \mathbb{R}^2, \tag{11.3.31}$$

so that $(\mathbf{0},0,F(0,\boldsymbol{\xi})) \in \mathcal{D}(\mathcal{M}_0)$, where $D^{\perp}\varphi_2(P_1) = (\partial_{\xi_2}, -\partial_{\xi_1})\varphi_2(P_1)$. In particular, (11.3.31) *holds for $\boldsymbol{\xi} = P_1$.*

Proof. Properties (11.3.28)–(11.3.30) follow from (11.3.26) and Corollary 11.3.4, where we have used that $(\varphi_1 - \varphi_2)(P_1) = 0$. Now we prove (11.3.31).

From (11.3.20),

$$\partial_z F(z,\boldsymbol{\xi}) = \frac{1}{\sqrt{(u_1-u_2)^2 + v_2^2}}\boldsymbol{\nu}_{\mathcal{S}_1} \qquad \text{for all } (z,\boldsymbol{\xi}) \in \mathbb{R} \times \mathbb{R}^2.$$

Also, from (7.1.9), via an explicit calculation by using (2.2.9), $c^2 = \rho^{\gamma-1}$, and $\frac{D(\varphi_1-\varphi_2)}{|D(\varphi_1-\varphi_2)|} = \boldsymbol{\nu}_{\mathcal{S}_1}$, we have

$$\partial_z\mathcal{M}_0(\mathbf{0},0,\boldsymbol{\xi}) = -\rho_2^{2-\gamma} D\varphi_2(\boldsymbol{\xi}) \cdot \boldsymbol{\nu}_{\mathcal{S}_1}.$$

Furthermore, since $F(0,\boldsymbol{\xi}) \in \mathcal{S}_1$ for any $\boldsymbol{\xi} \in \mathbb{R}^2$ and $P_1 \in \mathcal{S}_1$, then

$$\begin{aligned}&D\varphi_2(F(0,\boldsymbol{\xi})) - D\varphi_2(P_1) = P_1 - F(0,\boldsymbol{\xi}) \\ &\quad = ((P_1 - F(0,\boldsymbol{\xi})) \cdot \boldsymbol{\tau}_{s_1})\boldsymbol{\tau}_{\mathcal{S}_1} = ((P_1 - \boldsymbol{\xi}) \cdot \boldsymbol{\tau}_{s_1})\boldsymbol{\tau}_{\mathcal{S}_1}.\end{aligned}$$

It follows that

$$\partial_z\mathcal{M}_0(\mathbf{0},0,F(0,\boldsymbol{\xi})) = -\rho_2^{2-\gamma} D\varphi_2(P_1) \cdot \boldsymbol{\nu}_{\mathcal{S}_1} \qquad \text{for all } \boldsymbol{\xi} \in \mathbb{R}^2.$$

Also, using (7.1.9), (11.3.2), and $\boldsymbol{\nu}_{\mathcal{S}_1} = \frac{D(\varphi_1-\varphi_2)}{|D(\varphi_1-\varphi_2)|}$, we have

$$\mathcal{M}_0(\mathbf{0},0,\boldsymbol{\xi}) = \big(\rho_2 D\varphi_2(\boldsymbol{\xi}) - \rho_1 D\varphi_1(\boldsymbol{\xi})\big)\cdot \boldsymbol{\nu}_{\mathcal{S}_1},$$

which implies

$$D_{\boldsymbol{\xi}}\mathcal{M}_0(\mathbf{0},0,\boldsymbol{\xi}) = -(\rho_2-\rho_1)\boldsymbol{\nu}_{\mathcal{S}_1},$$

since $D\varphi_2(\boldsymbol{\xi}) = (u_2,v_2)-\boldsymbol{\xi}$. From these expressions and (11.3.23), we find that, for any $\boldsymbol{\xi}\in\mathbb{R}^2$,

$$\begin{aligned} D_z\mathcal{M}_1(\mathbf{0},0,\boldsymbol{\xi}) &= D_z\mathcal{M}_0(\mathbf{0},0,F(0,\boldsymbol{\xi})) + D_{\boldsymbol{\xi}}\mathcal{M}_0(\mathbf{0},0,F(0,\boldsymbol{\xi}))\cdot\partial_z F(0,\boldsymbol{\xi}) \\ &= -\rho_2^{2-\gamma}D\varphi_2(P_1)\cdot\boldsymbol{\nu}_{\mathcal{S}_1} - (\rho_2-\rho_1)\boldsymbol{\nu}_{\mathcal{S}_1}\cdot\partial_z F(0,\boldsymbol{\xi}) \\ &= -\rho_2^{2-\gamma}D\varphi_2(P_1)\cdot\boldsymbol{\nu}_{\mathcal{S}_1} - \frac{\rho_2-\rho_1}{\sqrt{(u_1-u_2)^2+v_2^2}} < -\delta \end{aligned}$$

for some $\delta>0$ depending only on (ρ_0,ρ_1,γ). In the last inequality above, we have used that $D\varphi_2(P_1)\cdot\boldsymbol{\nu}_{\mathcal{S}_1} > 0$ by (7.1.6), and $\rho_2(\theta_{\rm w}) > \rho_1$ for each $\theta_{\rm w}\in[\theta_{\rm w}^{\rm s},\frac{\pi}{2}]$ by Theorem 7.1.1, and that ρ_2 depends continuously on $\theta_{\rm w}$ on $[\theta_{\rm w}^{\rm s},\frac{\pi}{2}]$. Now (11.3.31) is proved. □

From (11.3.1), we employ (8.1.14) and (11.3.25) to obtain

$$\mathcal{M}_1(D\psi,\psi,\boldsymbol{\xi}) = 0 \qquad \text{on } \Gamma_{\rm shock}. \tag{11.3.32}$$

Finally, we write the boundary condition (11.3.32) in the (x,y)–coordinates. Changing the variables in (11.3.32), we find that $\psi(x,y)$ satisfies

$$B_1(D\psi,\psi,x,y) = 0 \qquad \text{on } \Gamma_{\rm shock}\cap\partial\Omega_{\varepsilon_0}, \tag{11.3.33}$$

where ε_0 is from Proposition 11.2.8, and

$$\begin{aligned} &B_1(\mathbf{p},z,x,y) = \mathcal{M}_1((\mathcal{P}_1,\mathcal{P}_2),\ z,\ (c_2-x)\cos y,(c_2-x)\sin y), \\ &(\mathcal{P}_1,\mathcal{P}_2) = -p_1(\cos y,\sin y) + \frac{p_2}{c_2-x}(-\sin y,\cos y). \end{aligned} \tag{11.3.34}$$

Then we obtain the homogeneity of B_1. Indeed, by (11.3.27) and (11.3.34),

$$B_1(\mathbf{0},0,x,y) = 0 \tag{11.3.35}$$

for any $(x,y)\in\mathbb{R}^2$ such that $(\mathbf{0},0,x,y)$ is in the domain of B_1.

Furthermore, from (11.3.24) and since $c_2 \ge (\rho_{\min})^{\frac{\gamma-1}{2}} > 0$ for any $\theta_{\rm w}\in[\theta_{\rm w}^{\rm s},\frac{\pi}{2}]$, there exist $\delta_{\rm bc}>0$ and C depending only on (ρ_0,ρ_1,γ) such that, for any $\theta_{\rm w}\in[\theta_{\rm w}^{\rm s},\frac{\pi}{2}]$, $B(\cdot)$ is well-defined for $(\mathbf{p},z,x,y)\in B_{\delta_{\rm bc}}(\mathbf{0},0,0,y_{P_1})$, and

$$|(D,D^2,D^3)B_1(\mathbf{p},z,x,y)| \le C \tag{11.3.36}$$

if $|(\mathbf{p},z,x)|\le\delta_{\rm bc}$ and $|y-y_{P_1}|\le\delta_{\rm bc}$.

Lemma 11.3.7. *There exist $\hat{\delta}_{\rm bc} > 0$ and C depending only on (ρ_0, ρ_1, γ) such that, for each $\theta_{\rm w} \in [\theta_{\rm w}^{\rm s}, \frac{\pi}{2}]$,*

$$D_{p_1} B_1(\mathbf{p}, z, x, y) \le -\frac{1}{C}, \tag{11.3.37}$$

$$D_{p_2} B_1(\mathbf{p}, z, x, y) \le -\frac{1}{C}, \tag{11.3.38}$$

$$D_z B_1(\mathbf{p}, z, x, y) \le -\frac{1}{C} \tag{11.3.39}$$

for all $(\mathbf{p}, z, x, y) \in \mathbb{R}^2 \times \mathbb{R} \times \mathbb{R} \times \mathbb{R}$ with $|(\mathbf{p}, z, x)| \le \hat{\delta}_{\rm bc}$ and $|y - y_{P_1}| \le \hat{\delta}_{\rm bc}$.

Proof. Using the polar coordinates (r, θ) with center $\mathcal{O}_2$ as in (11.1.1) and recalling that $|\mathcal{O}_2 P_1| = c_2$, we have

$$D\varphi_2(P_1) = \mathcal{O}_2 - P_1 = -c_2(\cos\theta_{P_1}, \sin\theta_{P_1}),$$
$$D^\perp\varphi_2(P_1) = (\partial_{\xi_2}, -\partial_{\xi_1})\varphi_2(P_1) = c_2(-\sin\theta_{P_1}, \cos\theta_{P_1}).$$

Then, using (11.3.34), definition (11.1.2) of the (x, y)–coordinates, and $x_{P_1} = 0$, we have

$$D_{p_1} B_1(\mathbf{0}, 0, 0, y_{P_1}) = \frac{1}{c_2} D_{\mathbf{p}} \mathcal{M}_1(\mathbf{0}, 0, P_1) \cdot D\varphi_2(P_1),$$
$$D_{p_2} B_1(\mathbf{0}, 0, 0, y_{P_1}) = \frac{1}{c_2^2} D_{\mathbf{p}} \mathcal{M}_1(\mathbf{0}, 0, P_1) \cdot D^\perp\varphi_2(P_1).$$

Then (11.3.37)–(11.3.38) follow from (11.3.28)–(11.3.29) and (11.3.36), and (11.3.39) directly follows from (11.3.31), (11.3.34), and (11.3.36). □

Furthermore, we note the following uniform obliqueness of the boundary condition (11.3.33) for any admissible solution of **Problem 2.6.1**.

Lemma 11.3.8. *There exist $\varepsilon > 0$ and C depending only on (ρ_0, ρ_1, γ) such that the following holds: Let φ be an admissible solution for the wedge angle $\theta_{\rm w} \in [\theta_{\rm w}^{\rm s}, \frac{\pi}{2})$. Express Ω and $\Gamma_{\rm shock}$ in the (x, y)–coordinates. Denote by $\boldsymbol{\nu}_{\rm sh}^{(xy)}$ the unit normal on $\Gamma_{\rm shock}$ pointing into Ω in the (x, y)–coordinates. Then*

$$D_{\mathbf{p}} B_1(D\psi, \psi, x, y) \cdot \boldsymbol{\nu}_{\rm sh}^{(xy)} \ge \frac{1}{C} \qquad \text{for } (x, y) \in \Gamma_{\rm shock} \cap \partial\Omega_\varepsilon.$$

Proof. The assertion follows from (11.3.30) in Lemma 11.3.6, and smooth and smoothly invertible change of variables: $\boldsymbol{\xi} \mapsto (x, y)$ with bounds in the C^k–norms independent of $\theta_{\rm w} \in [\theta_{\rm w}^{\rm s}, \frac{\pi}{2}]$, where we have used $c_2(\theta_{\rm w}) \ge \frac{1}{C}$ for any $\theta_{\rm w} \in [\theta_{\rm w}^{\rm s}, \frac{\pi}{2}]$ and Lemma 11.2.6. □

Lemma 11.3.9. *There exist ε and C depending only on (ρ_0, ρ_1, γ) such that any admissible solution ψ satisfies*

$$\partial_x \psi = b(\partial_y \psi, \psi, x, y) \qquad \text{on } \Gamma_{\rm shock} \cap \partial\Omega_\varepsilon, \tag{11.3.40}$$

where $b(p_2, z, x, y)$ *is defined on* $\mathbb{R} \times \mathbb{R} \times \mathbb{R}^2$ *and satisfies*

$$|(D, D^2, D^3)b(p_2, z, x, y)| \le C \qquad \text{on } \mathbb{R} \times \mathbb{R} \times \mathbb{R}^2, \tag{11.3.41}$$

$$b(0, 0, x, y) = 0 \qquad \text{for all } (x, y) \in \mathbb{R}^2. \tag{11.3.42}$$

Proof. Recall that any admissible solution ψ satisfies (11.3.33). We obtain $b(\cdot)$ first by solving (11.3.33) for p_1 for each (p_2, z, x, y) in an s–neighborhood of $(0, 0, x_{P_1}, y_{P_1})$ with $s > 0$ uniform with respect to $\theta_{\rm w}$ via using (11.3.35)–(11.3.37). Then $b(\cdot)$ is defined on ball $B_s((0, 0, x_{P_1}, y_{P_1}))$ for each $\theta_{\rm w} \in [\theta_{\rm w}^{\rm s}, \frac{\pi}{2}]$, and satisfies (11.3.41)–(11.3.42) on its domain. Also, from Lemma 11.2.6, Proposition 11.2.8, and Corollary 11.2.9, we obtain the existence of $\varepsilon \in (0, \varepsilon_0)$ such that every admissible solution satisfies

$$(\psi_y(x, y), \psi(x, y), x, y) \in B_s((0, 0, x_{P_1}, y_{P_1})) \quad \text{for all } (x, y) \in \Gamma_{\rm shock} \cap \partial\Omega_\varepsilon.$$

This shows that (11.3.40) holds. Finally, for each $\theta_{\rm w} \in [\theta_{\rm w}^{\rm s}, \frac{\pi}{2}]$, we choose the extension of $b(\cdot)$ to $\mathbb{R} \times \mathbb{R} \times \mathbb{R}^2$ satisfying (11.3.41)–(11.3.42) with C uniform with respect to $\theta_{\rm w}$. □

11.4 $C^{2,\alpha}$–ESTIMATES IN THE SCALED HÖLDER NORMS NEAR $\Gamma_{\rm sonic}$

In this section, we obtain the main estimates of this chapter – the regularity estimates of both $\psi = \varphi - \varphi_2$ and the shock curve $\{y = \hat{f}(x)\}$ near $\Gamma_{\rm sonic}$.

To achieve these estimates, we will use the norms introduced in §4.6 by (4.6.2). Moreover, we will use the following simplified notations:

$$\|u\|_{m,\alpha,\mathcal{D}}^{(\rm par)} := \|u\|_{m,\alpha,\mathcal{D}}^{(2),(\rm par)}, \qquad C_{(\rm par)}^{m,\alpha}(\mathcal{D}) := C_{2,(\rm par)}^{m,\alpha}(\mathcal{D}).$$

Remark 11.4.1. *From* (4.6.2) *with* $\sigma = 2$, $\|u\|_{2,0,\Omega_\varepsilon}^{(\rm par)} < \infty$ *implies* $\|u\|_{C^{1,1}(\Omega_\varepsilon)} < \infty$. *Thus, the estimates in the* $\|\cdot\|_{2,0,\Omega_\varepsilon}^{(\rm par)}$*–norm imply the* $C^{1,1}$*–estimates.*

Furthermore, for a function $f \in C^2((0, \varepsilon))$, considered as $f(x)$, norm $\|f\|_{2,\alpha,(0,\varepsilon)}^{(\rm par)}$ is defined as in (4.6.2) for $\sigma = 2$, with only the x–variable. Then $\delta_\alpha^{(\rm par)}(x, \tilde{x})$ becomes $|x - \tilde{x}|^\alpha$ and, for $I := (a, b) \subset (0, \varepsilon)$, norm $\|f\|_{2,\alpha,I}^{(\rm par)}$ is defined as

$$\begin{aligned} &\|f\|_{2,0,I}^{(\rm par)} := \sum_{k=0}^{2} \sup_{x \in I} \left(x^{k-2} |\partial_x^k f(x)|\right), \\ &[f]_{2,\alpha,I}^{(\rm par)} := \sup_{x,\tilde{x} \in I, x \neq \tilde{x}} \left(\left(\min(x, \tilde{x})\right)^\alpha \frac{|\partial_x^2 f(x) - \partial_x^2 f(\tilde{x})|}{|x - \tilde{x}|^\alpha}\right), \\ &\|f\|_{2,\alpha,I}^{(\rm par)} := \|f\|_{2,0,I}^{(\rm par)} + [f]_{2,\alpha,I}^{(\rm par)}. \end{aligned} \tag{11.4.1}$$

Remark 11.4.2. *Note that, for $I = (0, \varepsilon)$, estimate $\|f\|^{(\mathrm{par})}_{2,\alpha,(0,\varepsilon)} \le C$ implies the estimate in the standard weighted Hölder norm $\|f\|^{(-2),\{0\}}_{2,\alpha,(0,\varepsilon)} \le C$ and the property that $f(0) = f'(0) = 0$. This follows directly from* (11.4.1).

Now we make the main estimates. First we recall that $\hat{f}_0$ from (11.2.8) is such that $\{y = \hat{f}_0(x)\}$ is line $\mathcal{S}_1$ written in the (x, y)–coordinates near Γ_{sonic}.

From Proposition 11.2.8, for any $\theta_{\mathrm{w}} \in (\theta^{\mathrm{s}}_{\mathrm{w}}, \frac{\pi}{2}]$, we have

$$\Gamma^{(\theta_{\mathrm{w}})}_{\mathrm{sonic}} = \{(0, y) \,:\, \theta_{\mathrm{w}} < y < y^{(\theta_{\mathrm{w}})}_{P_1}\}, \tag{11.4.2}$$

where $y^{(\theta_{\mathrm{w}})}_{P_1} - \theta_{\mathrm{w}} > 0$ for any $\theta_{\mathrm{w}} \in (\theta^{\mathrm{s}}_{\mathrm{w}}, \frac{\pi}{2}]$, and $y^{(\theta_{\mathrm{w}})}_{P_1} - \theta_{\mathrm{w}} \to 0$ as $\theta_{\mathrm{w}} \to \theta^{\mathrm{s}}_{\mathrm{w}}+$.

Proposition 11.4.3. *There exists $\sigma \in (0, \frac{\varepsilon_0}{2})$ depending only on (ρ_0, ρ_1, γ) and, for any $\alpha \in (0, 1)$, there exists $C > 0$ depending only on $(\rho_0, \rho_1, \gamma, \alpha)$ such that the following estimates hold: Let $l_{\mathrm{so}} > 0$. Define*

$$\varepsilon = \min\{\sigma, l^2_{\mathrm{so}}\}. \tag{11.4.3}$$

If $\theta_{\mathrm{w}} \in [\theta^{\mathrm{s}}_{\mathrm{w}}, \frac{\pi}{2})$ satisfies

$$y^{(\theta_{\mathrm{w}})}_{P_1} - \theta_{\mathrm{w}} \ge l_{\mathrm{so}},$$

and φ is an admissible solution with the wedge angle θ_{w}, then $\psi = \varphi - \varphi^{(\theta_{\mathrm{w}})}_2$ satisfies

$$\|\psi\|^{(\mathrm{par})}_{2,\alpha,\Omega_\varepsilon} \le C, \tag{11.4.4}$$

and the shock function $\hat{f}(x)$ from (11.2.38) *satisfies*

$$\|\hat{f} - \hat{f}^{(\theta_{\mathrm{w}})}_0\|^{(\mathrm{par})}_{2,\alpha,(0,\varepsilon)} \le C, \tag{11.4.5}$$

where $\hat{f}^{(\theta_{\mathrm{w}})}_0$ is the function from (11.2.8), *and we have used norm* (11.4.1).

Proof. Since we have obtained estimates (11.2.21), (11.2.23)–(11.2.24), and (11.2.42) of $(D_{(x,y)}\psi, \psi)$, and estimate (11.2.43) for the ellipticity of equation (11.1.4) with (11.1.5), the proof is similar to the one of Theorem 4.7.4, except for the argument near Γ_{shock}, where the regularity is initially known only Lipschitz, but on which the two Rankine-Hugoniot conditions (8.1.13)–(8.1.14) hold.

In the argument below, all constants σ, C, C_i, L, and λ are positive and depend only on (ρ_0, ρ_1, γ). In some estimates, C depends on $\alpha \in (0, 1)$ in addition to (ρ_0, ρ_1, γ), which is written as $C(\alpha)$. Also, from Lemma 9.2.2 and Proposition 10.3.1, we have

$$\varphi,\ \psi \in C^\infty(\overline{\Omega_\varepsilon} \setminus \Gamma_{\mathrm{sonic}}), \qquad \hat{f} \in C^\infty((0, \varepsilon]), \tag{11.4.6}$$

where $\varepsilon > 0$ is from Proposition 11.2.8. We divide the proof into five steps.

1. We choose $\sigma \in (0, \frac{\varepsilon_0}{2})$ so small that the results in §11.2, quoted below, hold in $\Omega_{2\sigma}$ defined by (11.2.16). Then σ depends only on (ρ_0, ρ_1, γ). We reduce

σ below, if necessary, depending only on the same data. Let ε be defined by (11.4.3).

For $z := (x, y) \in \Omega_\varepsilon$ and $\rho \in (0, 1)$, define

$$\begin{aligned} \tilde{R}_{z,\rho} &:= \Big\{(s,t) \;:\; |s-x| < \frac{\rho}{4}x, |t-y| < \frac{\rho}{4}\sqrt{x}\Big\}, \\ R_{z,\rho} &:= \tilde{R}_{z,\rho} \cap \Omega_{2\varepsilon}. \end{aligned} \tag{11.4.7}$$

Then

$$R_{z,\rho} \subset \Omega \cap \{(s,t) \;:\; \frac{3}{4}x < s < \frac{5}{4}x\} \subset \Omega_{2\varepsilon}. \tag{11.4.8}$$

In particular,

$$R_{z,\rho} \cap \Gamma_{\text{sonic}} = \emptyset \qquad \text{for all } z \in \Omega_{2\varepsilon}.$$

If σ is small, depending only on C in (11.2.40) (*i.e.*, on (ρ_0, ρ_1, γ)), we conclude from Proposition 11.2.8 that, if ε is defined by (11.4.3), then, for any $z \in \Omega_\varepsilon$, at least one of the following three cases holds:

(i) $z \in R_{\hat{z},1/20}$ for $R_{\hat{z},1/10} = \tilde{R}_{\hat{z},1/10}$;

(ii) $z \in R_{z_w,1/4}$ for $z_w = (x, 0) \in \Gamma_{\text{wedge}} \cap \{0 < x < \varepsilon\}$;

(iii) $z \in R_{z_s,1/4}$ for $z_s = (x, \hat{f}(x)) \in \Gamma_{\text{shock}} \cap \{0 < x < \varepsilon\}$.

Thus, it suffices to make the local and semi-local estimates of $D\psi$ and $D^2\psi$ (*i.e.*, up to the part of boundary) in the following rectangles with $z_0 := (x_0, y_0)$:

(i) $R_{z_0,1/10}$ for $z_0 \in \Omega_\varepsilon$ satisfying $R_{z_0,1/10} = \tilde{R}_{z_0,1/10}$;

(ii) $R_{z_0,1}$ for $z_0 \in \Gamma_{\text{wedge}} \cap \{x < \varepsilon\}$;

(iii) $R_{z_0,1}$ for $z_0 \in \Gamma_{\text{shock}} \cap \{x < \varepsilon\}$.

Using Proposition 11.2.8, we see from (11.2.40) that

$$\hat{f}(x) \geq \hat{f}(0) = y_{P_1} \geq l_{\text{so}} + \theta_{\text{w}} \qquad \text{for all } x \in (0, \varepsilon).$$

Since $\varepsilon \leq l_{\text{so}}^2$ by (11.4.3), we obtain from (11.2.40) and (11.4.7) with $\rho = 1$ that

$$\begin{aligned} R_{z_w,1} \cap \Gamma_{\text{shock}} = \emptyset \qquad & \text{for all } z_w = (x, 0) \in \Gamma_{\text{wedge}} \cap \{x < \varepsilon\}, \\ R_{z_s,1} \cap \Gamma_{\text{wedge}} = \emptyset \qquad & \text{for all } z_s = (x, \hat{f}(x)) \in \Gamma_{\text{shock}} \cap \{x < \varepsilon\}. \end{aligned} \tag{11.4.9}$$

Remark 11.4.4. *The only place in the proof where the restriction that* $\varepsilon \leq l_{\text{so}}^2$ *is used is for obtaining* (11.4.9). *However, this is a crucial restriction that does not allow one to obtain the uniform estimates in* Ω_ε *with uniform* $\varepsilon > 0$ *up to* $\theta_{\text{w}}^{\text{s}}$ *by using the method for proving Proposition* 11.4.3.

Furthermore, denoting

$$Q_\rho := (-\rho, \rho)^2$$

and introducing variables (S, T) by the invertible change of variables:

$$(x, y) = (x_0 + \frac{x_0}{4} S, y_0 + \frac{\sqrt{x_0}}{4} T), \tag{11.4.10}$$

we find that there exists $Q_\rho^{(z_0)} \subset Q_\rho$ such that rectangle $R_{z_0,\rho}$ in (11.4.7) is expressed as

$$R_{z_0,\rho} = \left\{ (x_0 + \frac{x_0}{4} S, \hat{y}_0 + \frac{\sqrt{x_0}}{4} T) \ : \ (S, T) \in Q_\rho^{(z_0)} \right\}. \tag{11.4.11}$$

2. We first consider Case (i) in Step 1. Then

$$Q_\rho^{(z_0)} = Q_\rho \qquad \text{for all } \rho \in (0, \frac{1}{10}]$$

in (11.4.11). That is, for any $\rho \in (0, \frac{1}{10}]$,

$$R_{z_0,\rho} = \{ (x_0 + \frac{x_0}{4} S, y_0 + \frac{\sqrt{x_0}}{4} T) \ : \ (S, T) \in Q_\rho \}. \tag{11.4.12}$$

Rescale ψ in $R_{z_0,1/10}$ by defining

$$\psi^{(z_0)}(S, T) := \frac{1}{x_0^2} \psi(x_0 + \frac{x_0}{4} S, y_0 + \frac{\sqrt{x_0}}{4} T) \qquad \text{for } \ (S, T) \in Q_{1/10}. \tag{11.4.13}$$

Then, by (11.2.44), we have

$$|\psi^{(z_0)}| \le L. \tag{11.4.14}$$

Remark 11.4.5. *By* (11.2.44), *we have*

$$|\psi_S^{(z_0)}| \le \frac{L}{2}, \quad |\psi_T^{(z_0)}| \le \frac{L}{2\sqrt{x_0}} \qquad \textit{in } Q_{1/10}.$$

Note that we do not have the uniform Lipschitz bound for the rescaled function; see the estimate of $\psi_T^{(z_0)}$ above. Thus, unlike in the argument for Proposition 10.5.1, we will use the regularity estimates under the only assumption of the L^∞-bound of the solution below.

By Corollary 11.2.12, ψ satisfies the equation of form (11.1.6) in $R_{z_0,1/10}$, with coefficients $(A_{ij}^{(\mathrm{mod})}, A_i^{(\mathrm{mod})})(\mathbf{p}, z, x)$ satisfying (11.2.46)–(11.2.47). Changing the variables to (S, T) and dividing by $16x_0$, we see that $\psi^{(z_0)}$ satisfies

$$\begin{aligned} &\sum_{i,j=1}^{2} A_{ij}^{(z_0)}(D\psi^{(z_0)}, \psi^{(z_0)}, S, T) D_{ij}\psi^{(z_0)} \\ &+ \sum_{i=1}^{2} A_i^{(z_0)}(D\psi^{(z_0)}, \psi^{(z_0)}, S, T) D_i\psi^{(z_0)} = 0 \qquad \text{in } Q_{1/10}, \end{aligned} \tag{11.4.15}$$

where $(D_1, D_2) = (D_S, D_T)$, and

$$\begin{aligned} A_{ij}^{(z_0)}(p_1, p_2, z, S, T) &:= x_0^{\frac{i+j}{2}-2} A_{ij}^{(\mathrm{mod})}(4x_0p_1, 4x_0^{\frac{3}{2}}p_2, x_0^2 z, x), \\ A_i^{(z_0)}(p_1, p_2, z, S, T) &:= \frac{1}{4} x_0^{\frac{i-1}{2}} A_i^{(\mathrm{mod})}(4x_0p_1, 4x_0^{\frac{3}{2}}p_2, x_0^2 z, x), \end{aligned} \tag{11.4.16}$$

where $(x, y) = (x_0(1+\frac{S}{4}), y_0 + \frac{\sqrt{x_0}}{4}T)$, and $(A_{ij}^{(z_0)}, A_i^{(z_0)})$ are independent of T.

From definitions (11.4.13) and (11.4.16), using Corollary 11.2.10 and property (11.2.46), we have

$$\frac{\lambda}{4}|\boldsymbol{\kappa}|^2 \le \sum_{i,j=1}^{2} A_{ij}^{(z_0)}(D\psi^{(z_0)}(S,T),\ \psi^{(z_0)}(S,T),\ S,\ T)\kappa_i\kappa_j \le \frac{4}{\lambda}|\boldsymbol{\kappa}|^2 \tag{11.4.17}$$

for all $\boldsymbol{\kappa} = (\kappa_1, \kappa_2) \in \mathbb{R}^2$ and $(S,T) \in Q_{1/10}^{(z_0)}$, where λ is from (11.2.43).

Furthermore, using (11.4.16) and (11.2.47), we have

$$\begin{aligned} &\|(A_{ij}^{(z_0)}, A_i^{(z_0)})\|_{L^\infty(\mathbb{R}^2\times\mathbb{R}\times Q_{1/10}^{(z_0)})} \le C, \\ &\|D_{(\mathbf{p},z,S,T)}(A_{ij}^{(z_0)}, A_i^{(z_0)})\|_{L^\infty(\mathbb{R}^2\times\mathbb{R}\times Q_{1/10}^{(z_0)})} \le C, \end{aligned} \tag{11.4.18}$$

where C depends only on (ρ_0, ρ_1, γ), and we have used that ε depends also on these parameters. Indeed, let $(\mathbf{p}, z, S, T) \in \mathbb{R}^2 \times \mathbb{R} \times Q_{1/10}^{(z_0)}$. Then, using that $(x, y) = (x_0(1+\frac{S}{4}), y_0 + \frac{\sqrt{x_0}}{4}T)$ with $|S| \le \frac{1}{10}$ in (11.4.16) so that x satisfies

$$0 \le x = x_0(1+\frac{S}{4}) \le x_0(1+x_0) \le 2x_0,$$

which especially implies that $(x, y) \in \Omega_{2\varepsilon}$ if $(S,T) \in Q_{1/10}^{(z_0)}$ with $z_0 = (x_0, y_0) \in \Omega_\varepsilon$, we find from (11.2.47) that

$$\begin{aligned} &|A_{11}^{(z_0)}(\mathbf{p}, z, S, T)| = x_0^{-1}|A_{11}^{(\mathrm{mod})}(4x_0p_1, 4x_0^{\frac{3}{2}}p_2, x_0^2 z, x_0(1+\frac{S}{4}))| \le C, \\ &|D_{p_1}A_{11}^{(z_0)}(\mathbf{p}, z, S, T)| = 4|(D_{p_1}A_{11}^{(\mathrm{mod})})(4x_0p_1, 4x_0^{\frac{3}{2}}p_2, x_0^2 z, x_0(1+\frac{S}{4}))| \le C, \\ &D_T A_{ij}^{(z_0)}(\mathbf{p}, z, S, T) = 0. \end{aligned}$$

The other estimates in (11.4.18) are also obtained similarly.

Moreover, $\psi^{(z_0)} \in C^\infty(\overline{Q_{1/10}^{(z_0)}})$ by (11.4.6).

Recalling that $Q_{1/10}^{(z_0)} = Q_{1/10}$ in the present case, we obtain, by (11.4.17)–(11.4.18), that equation (11.4.15) and solution $\psi^{(z_0)}$ satisfy the conditions of Theorem 4.2.1 in $Q_{1/10}$, where (4.2.4) is satisfied with any $\alpha \in (0,1)$ in our case, since $\|D_S(A_{ij}^{(z_0)}, A_i^{(z_0)})\|_{L^\infty(\mathbb{R}^2\times\mathbb{R}\times\overline{Q_{1/10}})} \le C$. Then, using Theorem 4.2.1 and (11.4.14), we see that, for each $\alpha \in (0,1)$,

$$\|\psi^{(z_0)}\|_{C^{2,\alpha}(\overline{Q^{(z_0)}_{1/20}})} = \|\psi^{(z_0)}\|_{C^{2,\alpha}(\overline{Q_{1/20}})} \le C(\alpha). \tag{11.4.19}$$

3. We then consider Case (ii) in Step 1. Let $z_0 = (x_0, 0) \in \Gamma_{\text{wedge}} \cap \{x < \varepsilon\}$. Using Proposition 11.2.8 and (11.4.9), and assuming that ε is sufficiently small, depending only on the data, we obtain that $\overline{R_{z_0,1}} \cap \partial\Omega \subset \Gamma_{\text{wedge}}$ so that, in (11.4.11),

$$Q_\rho^{(z_0)} = Q_\rho \cap \{T > 0\} \qquad \text{for all } \rho \in (0,1],$$

i.e., for any $\rho \in (0,1]$,

$$R_{z_0,\rho} = \Big\{(x_0 + \frac{x_0}{4}S, y_0 + \frac{\sqrt{x_0}}{4}T) \;:\; (S,T) \in Q_\rho \cap \{T>0\}\Big\}. \tag{11.4.20}$$

Define $\psi^{(z_0)}(S,T)$ by (11.4.13) for $(S,T) \in Q_1^{(z_0)} = Q_1 \cap \{T > 0\}$. Then, as in Step 2, (11.4.14) holds in $\overline{Q_1} \cap \{T \ge 0\}$. Moreover, by an argument similar to Step 2, $\psi^{(z_0)}$ satisfies equation (11.4.15) in $Q_1 \cap \{T > 0\}$ with the coefficients in (11.4.16). Since $\partial_\nu\psi = 0$ on Γ_{wedge}, it follows that

$$\partial_T\psi^{(z_0)} = 0 \qquad \text{on } \{T = 0\} \cap Q_1.$$

Then, similarly to the arguments in Step 2, we obtain from Theorem 4.2.10 that, for each $\alpha \in (0,1)$,

$$\|\psi^{(z_0)}\|_{C^{2,\alpha}(\overline{Q_{1/2}}\cap\{T\ge 0\})} \le C(\alpha),$$

that is,

$$\|\psi^{(z_0)}\|_{C^{2,\alpha}(\overline{Q^{(z_0)}_{1/2}})} \le C(\alpha). \tag{11.4.21}$$

4. We now consider Case (iii). Using Proposition 11.2.8, we obtain that, for each $\rho \in (0,1]$, domain $Q_\rho^{(z_0)}$ from (11.4.11) in the present case is of the form:

$$Q_\rho^{(z_0)} = \Big\{(S,T) \in Q_\rho \;:\; T < F^{(z_0)}(S)\Big\}, \tag{11.4.22}$$

where $F^{(z_0)}(S) = \frac{4}{\sqrt{x_0}}\Big(\hat f(x_0 + \frac{x_0}{4}S) - \hat f(x_0)\Big)$. Then, using (11.2.40), we see that, for any $\rho \in (0,1]$,

$$\|F^{(z_0)}\|_{C^{0,1}([-\rho,\rho])} \le C\sqrt{x_0} \le \frac{1}{10}, \tag{11.4.23}$$

where the last inequality is obtained by choosing ε_1 sufficiently small and recalling that $\varepsilon \in (0, \varepsilon_1]$.

From (11.4.23) and $F^{(z_0)}(0) = 0$, we obtain that, for any $\rho \in (0,1]$,

$$F^{(z_0)}(S) > -\frac{\rho}{2} \qquad \text{for all } S \in (-\rho, \rho).$$

This implies from (11.4.7) and (11.4.22) with a further reduction of ε_1 if necessary that

$$Q_1^{(z_0)} = \Big\{(S,T)\ :\ -1 < S < 1,\ -1 < T < F^{(z_0)}(S)\Big\} \tag{11.4.24}$$

and

$$\Gamma_{\rm shock}^{(z_0)} := \Big\{(S,T)\ :\ -1 < S < 1,\ T = F^{(z_0)}(S)\Big\} \subset \partial Q_1^{(z_0)} \tag{11.4.25}$$

satisfy

$$\mathrm{dist}(\Gamma_{\rm shock}^{(z_0)},\ \partial Q_1^{(z_0)} \cap \{T=-1\}) \ge \frac{1}{2}. \tag{11.4.26}$$

As in Steps 2–3, in the present case, $\psi^{(z_0)}(S,T)$ defined by (11.4.13) satisfies (11.4.14) in $Q_1^{(z_0)}$. Moreover, similar to Step 2, $\psi^{(z_0)}$ satisfies equation (11.4.15) in $Q_1^{(z_0)}$ with the coefficients in (11.4.16).

By Lemma 11.3.9, ψ satisfies the boundary condition (11.3.40) with properties (11.3.41)–(11.3.42). Now it follows that $\psi^{(z_0)}$ satisfies the following condition on the boundary part $\Gamma_{\rm shock}^{(z_0)}$ defined in (11.4.25):

$$\partial_S\psi^{(z_0)} = \sqrt{x_0}\,b^{(z_0)}(\partial_T\psi^{(z_0)}, \psi^{(z_0)}, S, T) \qquad \text{on } \Gamma_{\rm shock}^{(z_0)}, \tag{11.4.27}$$

where

$$b^{(z_0)}(p_2, z, S, T) = \frac{1}{4x_0^{\frac{3}{2}}}\, b(4x_0^{\frac{3}{2}}p_2, x_0^2 z, x, y) \qquad \text{on } \Gamma_{\rm shock}^{(z_0)}$$

for $(x,y) = (x_0(1+\frac{S}{4}), y_0 + \frac{\sqrt{x_0}}{4}T)$.

From (11.3.41)–(11.3.42) on $\Omega_{2\varepsilon}$, it follows that $b^{(z_0)}$ on $\Gamma_{\rm shock}^{(z_0)}$ satisfies

$$\|b^{(z_0)}\|_{C^3(\mathbb{R}\times\mathbb{R}\times\overline{\Gamma_{\rm shock}^{(z_0)}})} \le C. \tag{11.4.28}$$

Thus, the right-hand side of (11.4.27):

$$B^{(z_0)}(\partial_T\psi^{(z_0)}, \psi^{(z_0)}, S, T) := \sqrt{x_0}\, b^{(z_0)}(\partial_T\psi^{(z_0)}, \psi^{(z_0)}, S, T)$$

satisfies (4.2.15)–(4.2.17), where the small parameter in (4.2.16) in our case is $\sqrt{x_0}$, which is from the fact that $B^{(z_0)}(\cdot) = \sqrt{x_0}\, b(\cdot)$ and (11.4.28). With this, using structure (11.4.24) with (11.4.23) and (11.4.26) of domain $Q_1^{(z_0)}$, employing that $(\hat{A}_{ij}, \hat{A}_i)$ satisfy (11.4.18) in $Q_1^{(z_0)}$, choosing σ in (11.4.3) sufficiently small, and using $0 < x_0 \le \sigma$ in (11.4.27), we can apply Theorem 4.2.4. Thus, from (4.2.20), using also that $\psi^{(z_0)}$ satisfies (11.4.14) on $Q_1^{(z_0)}$ which leads to a uniform $L^\infty(Q_1^{(z_0)})$–bound, we find from (4.2.20) that

$$\|\psi^{(z_0)}\|_{C^{1,\beta}(\overline{Q_{1/2}^{(z_0)}})} \le C, \tag{11.4.29}$$

where $\beta \in (0,1)$ and C depend only on the bounds in the conditions of Theorem 4.2.4, *i.e.*, on (ρ_0, ρ_1, γ).

This estimate allows us to improve the regularity of the free boundary $\Gamma^{(z_0)}_{\text{shock}}$. We note that, in Case (iii), definition (11.4.7) of $R_{z,\rho}$, combined with the properties that $\varphi_1 > \varphi$ in Ω (by Corollary 8.1.10) and $\varphi = \varphi_1$ on Γ_{shock}, implies that

$$\bar{\phi} > 0 \ \text{ in } R_{z_0,\rho}, \qquad \bar{\phi} = 0 \ \text{ on } \partial R_{z_0,\rho} \cap \Gamma_{\text{shock}}, \tag{11.4.30}$$

where $\bar{\phi} = \varphi_1 - \varphi$.

Define $\bar{\phi}^{(z_0)}(S,T)$ on $Q_1^{(z_0)}$ by (11.4.13) with $\bar{\phi}$ on the right-hand side (instead of ψ). Then we obtain from (11.4.11), (11.4.24)–(11.4.25), and (11.4.30) that

$$\begin{aligned} &\bar{\phi}^{(z_0)} > 0 && \text{in } Q_1^{(z_0)}, \\ &\bar{\phi}^{(z_0)} = 0 && \text{on } \Gamma^{(z_0)}_{\text{shock}}. \end{aligned} \tag{11.4.31}$$

From Lemma 11.2.7, we have

$$|D_T\bar{\phi}^{(z_0)}| \geq C^{-1}x_0^{-\frac{3}{2}}, \quad |D\bar{\phi}^{(z_0)}| \leq Cx_0^{-\frac{3}{2}} \qquad \text{on } Q_1^{(z_0)}. \tag{11.4.32}$$

Now, from (11.4.31)–(11.4.32) and (11.4.24)–(11.4.25), we obtain

$$D_S F^{(z_0)}(S) = -\frac{D_S\bar{\phi}^{(z_0)}}{D_T\bar{\phi}^{(z_0)}}(S, F^{(z_0)}(S)) \qquad \text{for } S \in [-1,1]. \tag{11.4.33}$$

Note that $\bar{\phi} = \bar{\phi}_0 - \psi$, where function $\bar{\phi}_0(x,y)$ is defined by (11.2.2). Define $\bar{\phi}_0^{(z_0)}(S,T)$ by (11.4.13) with $\bar{\phi}_0$ on the right-hand side (instead of ψ). Then $\bar{\phi}^{(z_0)} = \bar{\phi}_0^{(z_0)} - \psi^{(z_0)}$. Since

$$|(D, D_2, D_3)\bar{\phi}_0| \leq C \qquad \text{on } R_{z_0,q},$$

where we have used the uniform bounds for the parameters of state (2) for $\theta_{\mathrm{w}} \in [\theta_{\mathrm{w}}^{\mathrm{s}}, \frac{\pi}{2}]$, and $\varepsilon < \frac{c_2}{2}$ for each $\theta_{\mathrm{w}} \in [\theta_{\mathrm{w}}^{\mathrm{s}}, \frac{\pi}{2}]$, then we have

$$[D\bar{\phi}_0^{(z_0)}]_{0,\beta,Q_1^{(z_0)}} \leq x_0^{\frac{\beta-3}{2}}[D\bar{\phi}_0]_{0,\beta,R_{z_0,q}} \leq Cx_0^{\frac{\beta-3}{2}}.$$

Here we have assumed without loss of generality that $\varepsilon < 1$. Combining this with (11.4.29), and using (11.4.32)–(11.4.33), we have

$$\begin{aligned} [D_S F^{(z_0)}]_{0,\beta,[-\frac{1}{2},\frac{1}{2}]} &\leq \frac{[D_S\bar{\phi}^{(z_0)}]_{0,\beta,Q_1^{(z_0)}}}{\inf_{Q_1^{(z_0)}}|D_T\bar{\phi}^{(z_0)}|} + \frac{\|D_S\bar{\phi}^{(z_0)}\|_{L^\infty(Q_1^{(z_0)})}}{(\inf_{Q_1^{(z_0)}}|D_T\bar{\phi}^{(z_0)}|)^2}[D_T\bar{\phi}^{(z_0)}]_{0,\beta,Q_1^{(z_0)}} \\ &\leq Cx_0^{\frac{3}{2}}[D\bar{\phi}_0^{(z_0)} - D\psi^{(z_0)}]_{0,\beta,Q_1^{(z_0)}} \\ &\leq Cx_0^{\frac{3}{2}}(x_0^{\frac{\beta-3}{2}} + 1) \leq C, \end{aligned}$$

and, using (11.4.23),

$$\|D_S F^{(z_0)}\|_{C^{1,\beta}([-\frac{1}{2},\frac{1}{2}])} \leq C. \tag{11.4.34}$$

Now we can apply Theorem 4.2.8, where the required regularity of the ingredients of the equation and boundary conditions holds by (11.4.18) and (11.3.41), and σ in (11.4.3) is reduced if necessary. Then, using (4.2.57) and the $L^\infty(Q_1^{(z_0)})$–bound of $\psi^{(z_0)}$ which holds by (11.4.14) on $Q_1^{(z_0)}$, we have

$$\|\psi^{(z_0)}\|_{C^{2,\beta}(\overline{Q_{1/3}^{(z_0)}})} \le C. \tag{11.4.35}$$

In particular, $\|\psi^{(z_0)}\|_{C^{1,\alpha}(\overline{Q_{1/3}^{(z_0)}})} \le C$. Now, fixing $\alpha \in (0,1)$, we can repeat the argument for obtaining (11.4.35) from (11.4.29) with α instead of β to obtain

$$\|\psi^{(z_0)}\|_{C^{2,\alpha}(\overline{Q_{1/4}^{(z_0)}})} \le C(\alpha). \tag{11.4.36}$$

5. Combining estimates (11.4.19), (11.4.21), and (11.4.36) together, we obtain

$$\|\psi^{(z_0)}\|_{C^{2,\alpha}(\overline{Q_{1/100}^{(z_0)}})} \le C(\alpha) \qquad \text{for all } z_0 \in \overline{\Omega_\varepsilon} \setminus \Gamma_{\text{sonic}}. \tag{11.4.37}$$

Then Lemma 4.6.1 implies (11.4.4).

Since $\bar\phi_0(x, \hat f_0(x)) = 0$ on $(0,\varepsilon_0)$ by (11.2.8), and $(\bar\phi_0 - \psi)(x, \hat f(x)) = (\varphi_1 - \varphi)(x, \hat f(x)) = 0$ for all $x \in (0,\varepsilon)$ by (11.2.38), then (11.4.5) holds by (11.4.4) and the fact that $|\partial_y \bar\phi_0| \ge \frac{1}{C}$ on $(0,\varepsilon_0)$, which holds by (11.2.5). This completes the proof. □

Now Proposition 11.4.3 implies:

Proposition 11.4.6. *Let $\theta_{\rm w}^* \in (\theta_{\rm w}^{\rm s}, \frac{\pi}{2})$. There exists $\varepsilon \in (0, \frac{\varepsilon_0}{2})$ depending only on $(\rho_0, \rho_1, \gamma, \theta_{\rm w}^*)$ such that, for any $\alpha \in (0,1)$, there exists $C > 0$ depending only on $(\rho_0, \rho_1, \gamma, \alpha)$ so that, if φ is an admissible solution with the wedge angle $\theta_{\rm w} \in [\theta_{\rm w}^*, \frac{\pi}{2})$, then $\psi = \varphi - \varphi_2$ satisfies (11.4.4) and the shock function $\hat f(x)$ from (11.2.38) satisfies (11.4.5).*

Proof. Since $y_{P_1}^{(\theta_{\rm w})} > \theta_{\rm w}$ for each $\theta_{\rm w} \in [\theta_{\rm w}^*, \frac{\pi}{2})$ and $y_{P_1}^{(\theta_{\rm w})}$ depends continuously on $\theta_{\rm w} \in [\theta_{\rm w}^*, \frac{\pi}{2})$, then $l^* := \min_{\theta_{\rm w} \in [\theta_{\rm w}^*, \frac{\pi}{2})} (y_{P_1}^{(\theta_{\rm w})} - \theta_{\rm w}) > 0$. Therefore, we apply Proposition 11.4.3 with $l_{\rm so} = l^*$ to conclude the proof. □

For the wedge angles $\theta_{\rm w} \in (\theta_{\rm w}^{\rm c}, \frac{\pi}{2})$, where $\theta_{\rm w}^{\rm c}$ is the critical angle introduced in Definition 10.4.1, we obtain the following global estimate:

Corollary 11.4.7. *Let $\theta_{\rm w}^* \in (\theta_{\rm w}^{\rm c}, \frac{\pi}{2})$. Let α be the constant determined in Lemma 10.5.2, and ε_0 in Proposition 11.2.8. Let $\varepsilon \in (0, \varepsilon_0]$. Then there exists C depending only on $(\rho_0, \rho_1, \gamma, \theta_{\rm w}^*, \varepsilon)$ such that, for any admissible solution φ with $\theta_{\rm w} \in [\theta_{\rm w}^*, \frac{\pi}{2}]$,*

$$\begin{aligned} &\varphi^{\rm ext} \in C^{1,\alpha}(\overline{\Omega^{\rm ext}}) \cap C^{1,1}(\overline{\Omega^{\rm ext}} \setminus \{P_3\}) \cap C^\infty(\overline{\Omega^{\rm ext}} \setminus (\overline{\Gamma^{\rm ext}_{\rm sonic}} \cup \{P_3\})), \\ &\|\varphi\|^{(-1-\alpha),\{P_3\}}_{2,\alpha,\Omega\setminus\Omega_{\varepsilon/10}} + \|\varphi - \varphi_2\|^{(\rm par)}_{2,\alpha,\Omega_\varepsilon} \le C, \end{aligned} \tag{11.4.38}$$

where we have used the notation introduced in Remark 8.1.3. *Furthermore, the shock function* $f_{O_1,\mathrm{sh}}$ *for* $\Gamma^{\mathrm{ext}}_{\mathrm{shock}}$, *introduced in Corollary* 10.1.3, *satisfies*

$$\begin{aligned} &f_{O_1,\mathrm{sh}} \in C^{1,1}([\theta_{P_1}, \theta_{P_1}-]) \cap C^\infty((\theta_{P_1}, \theta_{P_1}-)), \\ &\|f_{O_1,\mathrm{sh}}\|_{C^{1,1}([\theta_{P_1},\theta_{P_1}-])} \le \hat{C}, \end{aligned} \tag{11.4.39}$$

where $\hat{C}$ *depends only on* $(\rho_0, \rho_1, \gamma, \theta^*_{\mathrm{w}})$.

Proof. We denote by ε^* the small constant ε determined in Proposition 11.4.6. Then ε^* depends only on the data and θ^*_{w}.

If $\varepsilon \in (0, \varepsilon^*]$, then the estimate in (11.4.38) follows directly from Corollary 10.5.3 (applied with $d = \frac{\varepsilon}{2}$) and Proposition 11.4.6, by using (11.2.17).

Now let $\varepsilon \in (\varepsilon^*, \varepsilon_0]$. Note that, for any function v,

$$\|v\|^{(\mathrm{par})}_{2,\alpha,\Omega_\varepsilon} \le C\big(\|v\|^{(\mathrm{par})}_{2,\alpha,\Omega_{\varepsilon^*}} + \|v\|_{2,\alpha,\Omega_\varepsilon\setminus\Omega_{\varepsilon^*/2}}\big),$$

where constant C depends only on ε and ε^*. Combining this with (11.4.38) for ε^*, we obtain (11.4.38) for ε, with C depending on ε, in addition to the data and θ^*_{w}. Then we conclude the estimate in (11.4.38) for all $\varepsilon \in (0, \varepsilon_0]$. To conclude the proof of (11.4.38), it remains to show that $\varphi^{\mathrm{ext}} \in C^\infty(\overline{\Omega^{\mathrm{ext}}} \setminus (\overline{\Gamma^{\mathrm{ext}}_{\mathrm{sonic}}} \cup \{P_3\}))$. Near $\Gamma^{\mathrm{ext}}_{\mathrm{shock}}$, this is shown in (10.5.2) of Proposition 10.5.1. Away from the shock, this is obtained via a similar argument, by using the standard interior estimates (recalling that φ^{ext} satisfies the potential flow equation in Ω^{ext}, *i.e.*, across Γ_{sym}; see Remark 8.1.3), and the estimates for the oblique derivative problem near the flat boundary $\Gamma^{\mathrm{ext}}_{\mathrm{wedge}}$ away from P_3.

Now we prove (11.4.39). We work in domain Ω^{ext}. We first show that there exists $r > 0$ depending on the data such that

$$|D^k_{\boldsymbol{\xi}}\varphi| \le \hat{C} \qquad \text{on } \mathcal{N}_r(\Gamma^{\mathrm{ext}}_{\mathrm{shock}}) \cap \Omega^{\mathrm{ext}}, k = 0, 1, 2, \tag{11.4.40}$$

where C depends only on the data and θ^*_{w}. For $k = 0, 1$, the estimates follow from (9.1.5). Thus, it suffices to prove (11.4.40) for $k = 2$.

We apply (10.5.2) with $d = \frac{\varepsilon^*}{10}$, and set r to be equal to constant s in Proposition 10.5.1, which corresponds to $d = \frac{\varepsilon^*}{10}$. This proves (11.4.40) for the points that are $\frac{\varepsilon^*}{10}$–away from Γ_{sonic}. Then we employ (11.4.4) with $\varepsilon = \varepsilon^*$ to complete the proof of (11.4.40), where we have used the fact that (11.4.4) implies that $\|\psi\|_{C^{1,1}(\Omega^*_\varepsilon)} \le C$ in the (x, y)–coordinates, and hence in the $\boldsymbol{\xi}$–coordinates, since the norms of the coordinate transformation depend only on the parameters of state (2), *i.e.*, on the data.

Now the estimate in (11.4.39) follows from (11.4.40), combined with Lemma 10.1.1, and by using that $\Gamma^{\mathrm{ext}}_{\mathrm{shock}} = \{\varphi_1 - \varphi = 0\}$ as shown in Remark 8.1.3 and that φ_1 is a smooth fixed function. ☐

11.5 THE REFLECTED-DIFFRACTED SHOCK IS $C^{2,\alpha}$ NEAR P_1

In this section, we fix $\theta_{\rm w}^* \in (\theta_{\rm w}^{\rm s}, \frac{\pi}{2})$ and use ε from Proposition 11.4.6 corresponding to $\theta_{\rm w}^*$.

We note that the reflected-diffracted shock $P_0P_1P_2$ has a flat segment P_0P_1, which lies on line $\mathcal{S}_1$. In the (x, y)–coordinates (11.1.1)–(11.1.2) near $\Gamma_{\rm sonic}$ (from both sides of it), $\mathcal{S}_1$ is a graph of a smooth function $y = \hat{f}_0(x)$ on $-\varepsilon < x < \varepsilon$; see (11.2.8) for a more precise statement. Also, by (11.2.8), $\Gamma_{\rm shock} = \{(0, y) : \theta_{\rm w} < y < \hat{f}_0(x)\}$ and $P_1 = (0, \hat{f}_0(x))$ in the (x, y)–coordinates.

On the other hand, $\Gamma_{\rm shock} \cap \partial\Omega_\varepsilon$ is a graph of a function $\hat{f}$ near $\Gamma_{\rm sonic}$, which is contained within $\{0 < x < \varepsilon\}$, as shown in Proposition 11.2.8. Moreover,

$$\hat{f}_0(0) = \hat{f}(0), \quad \hat{f}_0'(0) = \hat{f}'(0), \tag{11.5.1}$$

by Proposition 11.4.6.

Thus, the $C^{2,\alpha}$–regularity of $P_0P_1P_2$ at and near P_1 with $\alpha \in (0, \frac{1}{2})$ follows from the following:

Proposition 11.5.1. *Extend $\hat{f}$ to $(-\varepsilon, \varepsilon)$ by defining $\hat{f}(x) = \hat{f}_0(x)$ for $x \in (-\varepsilon, 0]$. For any $\theta_{\rm w}^* \in (\theta_{\rm w}^{\rm s}, \frac{\pi}{2})$, there exists $\varepsilon > 0$ and, for each $\alpha \in (0, \frac{1}{2})$, there exists $C = C(\theta_{\rm w}^*, \alpha)$ such that, for any admissible solution φ with $\theta_{\rm w} \in [\theta_{\rm w}^*, \frac{\pi}{2})$, the extended function $\hat{f}$ satisfies that $\hat{f} \in C^{2,\alpha}([-\varepsilon, \varepsilon])$ and*

$$\|\hat{f}\|_{C^{2,\alpha}([-\varepsilon,\varepsilon])} \le C \qquad \text{for any } \alpha \in (0, \tfrac{1}{2}). \tag{11.5.2}$$

Proof. Fix $\theta_{\rm w}^* \in (\theta_{\rm w}^{\rm s}, \frac{\pi}{2})$. In this proof, constants C and ε depend on the data and $\theta_{\rm w}^*$. We use the same notations (φ_1, φ_2) for these functions expressed in the (x, y)–coordinates. Choose some $\varepsilon \in (0, \varepsilon_0)$ to be adjusted later. We divide the proof into three steps.

1. First, we show that $\hat{f}$ is twice differentiable at $x = 0$. From Proposition 11.4.6, it follows that $\hat{f} \in C^{1,1}([0, \varepsilon]) \cap C^2((0, \varepsilon])$, and (11.5.1) holds. Moreover,

$$(\varphi_1 - \varphi_2)(x, \hat{f}_0(x)) = 0, \ \ (\varphi_1 - \varphi)(x, \hat{f}(x)) = 0 \qquad \text{for all } x \in (0, \varepsilon), \tag{11.5.3}$$

and

$$\begin{aligned}
&|\partial_y(\varphi_1 - \varphi_2)| \ge \frac{1}{C} \quad \text{ in } \{\theta_{\rm w} < \theta < \theta_Q - \delta\} \cap \{-\varepsilon < x < \varepsilon\}, \\
&|(\varphi - \varphi_2)(x, y)| \le Cx^2 \quad \text{ in } \Omega_\varepsilon, \\
&|D_x(\varphi - \varphi_2)(x, y)| + x^{-\frac{1}{2}}|D_y(\varphi - \varphi_2)(x, y)| \le Cx \quad \text{ in } \Omega_\varepsilon, \\
&|D_{xy}(\varphi - \varphi_2)(x, y)| + x^{-\frac{1}{2}}|D_{yy}(\varphi - \varphi_2)(x, y)| \le C\sqrt{x} \quad \text{ in } \Omega_\varepsilon, \\
&\|\hat{f}\|_{C^{1,1}([0,\varepsilon])} \le C,
\end{aligned} \tag{11.5.4}$$

by (11.2.5) and (11.4.4). Then we subtract the two equalities in (11.5.3) to obtain that, for all $x \in (0, \varepsilon)$,

$$(\varphi_1 - \varphi_2)(x, \hat{f}_0(x)) - (\varphi_1 - \varphi_2)(x, \hat{f}(x)) = (\varphi_2 - \varphi)(x, \hat{f}(x)), \tag{11.5.5}$$

which implies

$$|\hat{f}_0(x) - \hat{f}(x)| \le Cx^2 \qquad \text{on } x \in [0, \varepsilon) \tag{11.5.6}$$

if ε is small, by employing the first and second estimates in (11.5.4). Now, differentiating (11.5.5) and performing estimates similar to the previous argument by using (11.5.4) and (11.5.6), we have

$$|\hat{f}_0'(x) - \hat{f}'(x)| \le Cx^{\frac{3}{2}} + C|\partial_x(\varphi - \varphi_2)(x, \hat{f}(x))| \qquad \text{on } x \in [0, \varepsilon). \tag{11.5.7}$$

Since φ is an admissible solution, it follows that $\psi = \varphi - \varphi_2$ satisfies the boundary condition (11.3.33). Also, from (11.3.35),

$$B_1(\mathbf{0}, 0, x, y) = 0 \qquad \text{for all } (x, y) \in B_\varepsilon(P_1). \tag{11.5.8}$$

Then, applying Lemma 11.3.7, noting that estimate (11.4.4) holds for small ε depending on the data and $\theta_{\rm w}^*$ by Proposition 11.4.6, and using (11.3.36) and reducing ε if necessary, we have

$$a_1(x, y)\psi_x + a_2(x, y)\psi_y + a_0(x, y)\psi = 0 \qquad \text{on } \Gamma_{\rm shock} \cap \{0 < x < \varepsilon\},$$

where $a_i \in C^{0,1}(\overline{\Omega_\varepsilon})$ with $\|a_i\|_{C^{0,1}(\overline{\Omega_\varepsilon})} \le C$ for $i = 0, 1, 2$, and $a_1 \le -\frac{1}{C}$ on $\{0 < x < \varepsilon\}$. Thus, on $\Gamma_{\rm shock} \cap \{0 < x < \varepsilon\}$,

$$|\psi_x| \le C\big(|\psi_y| + |\psi|\big) \le Cx^{\frac{3}{2}}.$$

Therefore, from (11.5.7),

$$|\hat{f}_0'(x) - \hat{f}'(x)| \le Cx^{\frac{3}{2}} \qquad \text{on } x \in [0, \varepsilon). \tag{11.5.9}$$

This implies that $\hat{f}_0'(x) - \hat{f}'(x)$ has the (right) derivative at $x = 0$, which is zero. Since $\hat{f}_0$ is smooth on $(-\varepsilon_0, \varepsilon_0)$, it follows that $\hat{f}$, extended as described above, is twice differentiable at $x = 0$, and

$$\hat{f}''(0) = \hat{f}_0''(0). \tag{11.5.10}$$

2. We now show $\hat{f} \in C^2([-\varepsilon, \varepsilon])$. For this, using (11.5.10) and the smoothness $\hat{f}_0 \in C^\infty(-\varepsilon, \varepsilon)$ and $\hat{f} \in C^\infty(0, \varepsilon)$, it suffices to show that

$$\lim_{x \to 0+} \big(\hat{f}''(x) - \hat{f}_0''(x)\big) = 0.$$

Differentiating (11.5.5) twice and performing the estimates as in the proof of (11.5.7), by using (11.5.4) and (11.5.6)–(11.5.7), we have

$$|\hat{f}_0''(x) - \hat{f}''(x)| \le C\sqrt{x} + C|D^2(\varphi - \varphi_2)(x, \hat{f}(x))| \quad \text{on } x \in [0, \varepsilon). \tag{11.5.11}$$

We now show that

$$|D^2\psi(x, \hat{f}(x))| \le C\sqrt{x}.$$

From (11.4.4), this estimate holds for ψ_{xy} and ψ_{yy}. Thus, it remains to consider ψ_{xx}. Differentiating (11.5.8), we have

$$D_{(x,y)}B_1(\mathbf{0},0,x,y)=0 \qquad \text{for all } (x,y)\in B_\varepsilon(P_1).$$

Then, writing (11.3.33) as

$$B_1(D_{(x,y)}\psi,\psi,x,\hat{f}(x))-B_1(\mathbf{0},0,x,\hat{f}(x))=0 \qquad \text{on } \{0<x<\varepsilon\}$$

with $(D_{(x,y)}\psi,\psi)=(D_{(x,y)}\psi,\psi)(x,\hat{f}(x))$, differentiating with respect to x, and employing (11.3.36) and (11.4.4)–(11.4.5) for small ε, we have

$$\sum_{i,j=1}^{2} b_{ij}(x)D_{ij}\psi+\sum_{i=1}^{2} b_i(x)D_i\psi+b_0(x)\psi=0 \qquad \text{on } \Gamma_{\text{shock}}\cap\{0<x<\varepsilon_0\}, \tag{11.5.12}$$

where $(D_1,D_2)=(D_x,D_y)$, $(b_{ij},b_i)\in C^{0,1}([0,\varepsilon])$ with

$$\|(b_{ij},b_i)\|_{C^{0,1}([0,\varepsilon])}\le C,$$

and $b_{11}(x)=D_{p_1}\hat{\mathcal{M}}(D_{(x,y)}\psi(x,\hat{f}(x)),\psi(x,\hat{f}(x)),x,\hat{f}(x))$. Using (11.4.4), we see that $|(D_{(x,y)}\psi,\psi)(x,\hat{f}(x))|\le Cx$. Thus, reducing ε if necessary and using Lemma 11.3.7, we obtain that $b_{11}(x)\le -\frac{1}{C}$. Then we have

$$|\psi_{xx}|\le C(|D(\psi_y,\psi)|+|\psi|)\le C\sqrt{x} \qquad \text{at } (x,\hat{f}(x)).$$

Thus, by (11.5.11), $|\hat{f}_0''(x)-\hat{f}''(x)|\le C\sqrt{x}$. This implies that $\hat{f}\in C^2([-\varepsilon,\varepsilon])$, as shown above.

3. Now we show that $\hat{f}\in C^{2,\alpha}([-\varepsilon,\varepsilon])$ for each $\alpha\in(0,\frac{1}{2})$. For that, it suffices to prove that $\hat{f}\in C^{2,\alpha}([0,\varepsilon])$, since $\hat{f}\in C^2([-\varepsilon,\varepsilon])$ and $\hat{f}=\hat{f}_0\in C^\infty([-\varepsilon,0])$.

Differentiating (11.5.5) twice, we see that $\hat{f}\in C^{2,\alpha}([0,\varepsilon])$ follows, provided that the function:

$$x\mapsto (D\psi,D^2\psi)(x,\hat{f}(x)) \qquad \text{is in } C^{0,\alpha}([0,\varepsilon]). \tag{11.5.13}$$

Now we prove (11.5.13). From Proposition 11.4.6, it follows that $(D\psi,D\psi_y)\in C^{0,\alpha}(\overline{\Omega_\varepsilon})$ for each $\alpha\in(0,\frac{1}{2})$, with the estimates in these spaces by the constants depending only on $(\rho_0,\rho_1,\gamma,\alpha,\theta_{\rm w}^*)$. Since $\hat{f}\in C^2([0,\varepsilon])$, we conclude that the functions:

$$x\mapsto (D\psi(x,\hat{f}(x)),D\psi_y(x,\hat{f}(x))) \qquad \text{are in } C^{0,\alpha}([0,\varepsilon]).$$

Then we express $\psi_{xx}(x,\hat{f}(x))$ from equation (11.5.12) with

$$\|(b_{ij},b_i)\|_{C^{0,1}([0,\varepsilon])}\le C(\rho_0,\rho_1,\gamma,\alpha,\theta_{\rm w}^*), \qquad b_{11}(x)\le-\lambda,$$

where $\lambda=\lambda(\rho_0,\rho_1,\gamma)>0$. Then the $C^{0,\alpha}([0,\varepsilon])$–regularity of $\psi_{xx}(x,\hat{f}(x))$ follows, with the uniform estimate in this space. Now (11.5.13) is proved. This completes the proof. □

Remark 11.5.2. *From the proof above, it follows that the dependence of constants* (ε, C) *on* $\theta_{\rm w}^*$ *in* (11.5.2) *is only from the dependence of estimate* (11.4.4) *on* $\theta_{\rm w}^*$, *obtained by an application of Proposition* 11.4.6.

11.6 COMPACTNESS OF THE SET OF ADMISSIBLE SOLUTIONS

In Corollary 9.2.5, it has been shown that, from every sequence of admissible solutions of **Problem 2.6.1**, we can extract a subsequence converging uniformly on compact subsets of $\mathbb{R}^2$ to a weak solution of **Problem 2.6.1**. In this section, we show that the limit is still an admissible solution of **Problem 2.6.1**. For this purpose, we always fix $\theta_{\rm w}^* \in (\theta_{\rm w}^{\rm c}, \frac{\pi}{2})$ for $\theta_{\rm w}^{\rm c}$ defined in Definition 10.4.1.

We use the *a priori* estimates for admissible solutions of **Problem 2.6.1**, collected in Corollary 11.4.7.

Proposition 11.6.1. *Let* $\{\varphi^{(i)}\}$ *be a sequence of admissible solutions with the wedge angles* $\theta_{\rm w}^{(i)} \in [\theta_{\rm w}^*, \frac{\pi}{2})$ *satisfying* $\theta_{\rm w}^{(i)} \to \theta_{\rm w}^{(\infty)} \in [\theta_{\rm w}^*, \frac{\pi}{2}]$. *Then*

(i) *There exists a subsequence* $\{\varphi^{(i_j)}\}$ *converging uniformly on compact subsets of* $\overline{\Lambda^{(\infty)}}$ *to a function* $\varphi^{(\infty)} \in C^{0,1}_{\rm loc}(\overline{\Lambda^{(\infty)}})$, *where the convergence is understood in the sense of Remark* 9.2.4. *Moreover,* $\varphi^{(\infty)}$ *is an admissible solution of* **Problem 2.6.1** *for the wedge angle* $\theta_{\rm w}^{(\infty)}$ *if* $\theta_{\rm w}^{(\infty)} \in [\theta_{\rm w}^*, \frac{\pi}{2})$, *and the normal reflection solution if* $\theta_{\rm w}^{(\infty)} = \frac{\pi}{2}$.

(ii) $\Omega(\varphi^{(i_j)}) \to \Omega(\varphi^{(\infty)})$ *in the Hausdorff metric.*

(iii) *If* $\boldsymbol{\xi}_{i_j} \in \overline{\Omega(\varphi^{(i_j)})}$ *and* $\boldsymbol{\xi}_{i_j} \to \boldsymbol{\xi}_\infty$, *then* $\boldsymbol{\xi}_\infty \in \overline{\Omega(\varphi^{(\infty)})}$ *and*

$$\varphi^{(i_j)}(\boldsymbol{\xi}_{i_j}) \to \varphi^{(\infty)}(\boldsymbol{\xi}_\infty), \quad D\varphi^{(i_j)}(\boldsymbol{\xi}_{i_j}) \to D\varphi^{(\infty)}(\boldsymbol{\xi}_\infty),$$

where $D\varphi^{(i_j)}(\boldsymbol{\xi}_{i_j}) := \lim_{\boldsymbol{\xi}\in\Omega^{(i_j)}, \boldsymbol{\xi}\to\boldsymbol{\xi}_{i_j}} D\varphi^{(i_j)}(\boldsymbol{\xi})$ *for* $\boldsymbol{\xi}_{i_j} \in \Gamma_{\rm shock}^{(i_j)}$, *and* $D\varphi^{(\infty)}(\boldsymbol{\xi}_\infty)$ *for* $\boldsymbol{\xi}_\infty \in \Gamma_{\rm shock}^{\infty}$ *is defined similarly.*

Proof. We use the notations that $\Omega^{(i)} := \Omega(\varphi^{(i)})$ and $\Omega^{(\infty)} := \Omega(\varphi^{(\infty)})$, and divide the proof into three steps.

1. The convergence: $\Omega^{(i_j)} \to \Omega^{(\infty)}$ in the Hausdorff metric follows from Corollary 9.2.5(i)–(ii) and the continuity of the parameters of state (2) in $\theta_{\rm w}$. This implies assertion (ii). Next, we prove assertion (i).

By Corollary 9.2.5, there exists a subsequence $\{\varphi^{(i_j)}\}$ converging uniformly on any compact subset of $\overline{\Lambda^{(\infty)}}$ to a function $\varphi^{(\infty)} \in C^{0,1}_{\rm loc}(\overline{\Lambda^{(\infty)}})$ that is a weak solution of **Problem 2.6.1** for the wedge angle $\theta_{\rm w}^{(\infty)}$.

Estimate (11.4.4) in Proposition 11.4.6 implies that $\psi^{(\infty)} \in C^{1,1}(\overline{\Omega_\varepsilon^{(\infty)}})$ with $D\psi^{(\infty)} = 0$ on $\Gamma_{\rm sonic}$. Then $\varphi^{(\infty)} \in C^{1,1}(\overline{\Omega_\varepsilon^{(\infty)}})$ and $D\varphi^{(\infty)} = D\varphi_2^{(\infty)}$ on $\Gamma_{\rm sonic}^{(\infty)}$, which is arc P_1P_4 (here and below, points $P_0, P_1, \dots$ are for angle $\theta_{\rm w}^{(\infty)}$). We also note that the parameters of state (2), as well as points $\{P_1, P_4\}$, depend continuously on $\theta_{\rm w} \in [\theta_{\rm w}^{\rm s}, \frac{\pi}{2}]$.

When $\theta_{\rm w}^{(\infty)} \in [\theta_{\rm w}^{*}, \frac{\pi}{2})$, we use the continuous dependence of P_0 on $\theta_{\rm w} \in [\theta_{\rm w}^{\rm s}, \frac{\pi}{2})$ to conclude that $\varphi^{(\infty)} = \varphi_2^{(\infty)}$ on $P_0P_1P_4$.

When $\theta_{\rm w}^{(\infty)} = \frac{\pi}{2}$, from the continuous dependence of the parameters of state (2) and point P_1 on $\theta_{\rm w} \in (\theta_{\rm w}^{\rm s}, \frac{\pi}{2}]$, it follows that lines $S_1^{i_j}$ converge to the vertical line through $P_{1|\theta_{\rm w}=\frac{\pi}{2}}$, so that $S_1^{i_j}$ converge to the reflected shock of the normal reflection. It follows from the argument above that, when $\theta_{\rm w}^{(\infty)} = \frac{\pi}{2}$, $\varphi^{(\infty)} = \varphi_2^{(\infty)}$ in the unbounded domain $\Omega_{(2)}^{(\infty)}$ between the two vertical lines: The vertical wall $\xi_1 = 0$ and the reflected shock in the normal reflection, bounded from below by the sonic arc $\Gamma_{\rm sonic}^{(\infty)}$ of the normal reflection.

Also, for any $\theta_{\rm w}^{(\infty)} \in [\theta_{\rm w}^{*}, \frac{\pi}{2}]$, it follows from the above argument that $\varphi^{(\infty)}$ is $C^{1,1}$ across $\Gamma_{\rm sonic}^{(\infty)}$.

Combining this with the other properties stated in Corollary 9.2.5 and Proposition 10.6.1, we conclude that $\varphi^{(\infty)}$ satisfies conditions (i)–(ii) of Definition 8.1.1.

Condition (iii) of Definition 8.1.1 for $\varphi^{(\infty)}$ follows from the uniform ellipticity estimates of Corollary 9.6.6 for $\{\varphi^{(i_j)}\}$, combined with (9.2.27) and the fact that $\Omega^{(i_j)} \to \Omega^{(\infty)}$ in the Hausdorff metric.

When $\theta_{\rm w}^{(\infty)} \in [\theta_{\rm w}^{*}, \frac{\pi}{2})$, conditions (iv)–(v) of Definition 8.1.1 for $\varphi^{(\infty)}$ follow from the fact that these conditions are satisfied by each $\varphi^{(i_j)}$, combined with the uniform convergence $\varphi^{(i_j)} \to \varphi^{(\infty)}$ and $\mathbf{e}_{\mathcal{S}_1}(\theta_{\rm w}^{(i_j)}) \to \mathbf{e}_{\mathcal{S}_1}(\theta_{\rm w}^{(\infty)})$. This proves that $\varphi^{(\infty)}$ is an admissible solution when $\theta_{\rm w}^{(\infty)} \in [\theta_{\rm w}^{*}, \frac{\pi}{2})$.

When $\theta_{\rm w}^{(\infty)} = \frac{\pi}{2}$, the equality:

$$\mathbf{e}_{\mathcal{S}_1} = -\frac{(v_2, u_1 - u_2)}{\sqrt{(u_1 - u_2)^2 + v_2^2}}$$

in (7.5.8), the continuous dependence of the parameters of state (2) on $\theta_{\rm w} \in (\theta_{\rm w}^{\rm s}, \frac{\pi}{2}]$, and the fact that $u_2 = v_2 = 0$ for $\theta_{\rm w} = \frac{\pi}{2}$ together imply that

$$\mathbf{e}_{\mathcal{S}_1}(\theta_{\rm w}^{(i_j)}) \to -\mathbf{e}_{\xi_2}.$$

Thus, from the monotonicity properties of $\varphi^{(i_j)}$ stated in Definition 8.1.1(v) and the uniform convergence $\varphi^{(i_j)} \to \varphi^{(\infty)}$ on any compact subset of $\overline{\Omega^{(\infty)}} \setminus \Gamma_{\rm wedge}^{(\infty)}$ (which follows from the convergence in Remark 9.2.4), it follows that the continuous function $\varphi_1 - \varphi^{(\infty)}$ in $\Omega^{(\infty)}$ is monotonically non-increasing in any direction $\mathbf{e} = (a, b)$ with $a < 0$. This implies that $\varphi_1 - \varphi^{(\infty)}$ is a function of ξ_1 in any ball $B_r(\boldsymbol{\xi}^0) \subset \Omega^{(\infty)}$:

$$(\varphi_1 - \varphi^{(\infty)})(\xi_1, \xi_2) = \eta(\xi_1) \qquad \text{in } B_r(\boldsymbol{\xi}^0).$$

Let $\phi^{(\infty)} = \varphi^{(\infty)} + \frac{|\boldsymbol{\xi}|^2}{2}$. Using (2.2.17), we have

$$(\varphi^{(\infty)} - \varphi_1) - \phi^{(\infty)} = \frac{|\boldsymbol{\xi}|^2}{2} - \varphi_1 = -u_1\xi_1 + const.$$

It follows that equation (2.2.11) for $\varphi^{(\infty)}$ still holds when $\phi^{(\infty)}$ is replaced by $\varphi^{(\infty)} - \varphi_1$. Then

$$(c^2 - (\varphi^{(\infty)}_{\xi_1})^2)\eta''(\xi_1) = 0 \qquad \text{in } B_r(\boldsymbol{\xi}^0),$$

where $c^2 = c^2(|D\varphi^{(\infty)}|^2, \varphi^{(\infty)}, \rho_0^{\gamma-1})$, and we have used function (1.14). Since we have proved above that $\varphi^{(\infty)}$ satisfies Definition 8.1.1(iii), then

$$\eta''(\xi_1) = 0 \qquad \text{in } B_r(\boldsymbol{\xi}^0).$$

Also, Ω^∞ is open and connected, since $\varphi^{(\infty)}$ satisfies Definition 8.1.1(i), as shown above. We conclude that

$$\varphi_1 - \varphi^{(\infty)} = A + B\xi_1 \qquad \text{in } \Omega^{(\infty)}.$$

Thus, $\varphi^{(\infty)}$ is a uniform state in $\Omega^{(\infty)}$. In particular, $\varphi^{(\infty)} - \varphi_2^{(\text{norm})}$ is a linear function of $\boldsymbol{\xi}$ in $\Omega^{(\infty)}$. Since $\varphi^{(\infty)}$ is $C^{1,1}$ across arc $\Gamma^{(\infty)}_{\text{sonic}}$ as shown above, and $\varphi^{(\infty)} = \varphi_2^{(\text{norm})}$ in $\Omega^{(\infty)}_{(2)}$, it follows that the linear function $\zeta := \varphi^{(\infty)} - \varphi_2^{(\text{norm})}$ satisfies

$$(\zeta, D\zeta) = (0, \mathbf{0}) \qquad \text{on } \Gamma^{(\infty)}_{\text{sonic}}.$$

From this and the fact that Γ_{sonic} is an arc, we obtain that $\zeta \equiv 0$, so that $\varphi^{(\infty)} = \varphi_2^{(\text{norm})}$ in $\Omega^{(\infty)}$. Combining this with $\varphi^{(\infty)} = \varphi_2^{(\text{norm})}$ in $\Omega^{(\infty)}_{(2)}$ implies that $\varphi^{(\infty)}$ is the normal reflection solution. This completes the proof of assertion (i).

2. It remains to prove assertion (iii). We note that, from Corollary 11.4.7,

$$\|\varphi^{(i_j)}\|_{C^{1,\alpha}(\overline{\Omega^{(i_j)}})} \le C.$$

Let points $\boldsymbol{\xi}_{i_j}$ and $\boldsymbol{\xi}_\infty$ be as in (iii). Then $\boldsymbol{\xi}_\infty \in \overline{\Omega^{(\infty)}}$ by assertion (ii).

Consider first Case $\boldsymbol{\xi}_\infty \in \Omega^{(\infty)}$. Then, using (ii) above, we conclude that there exists $R > 0$ such that $\overline{B_R(\boldsymbol{\xi}_\infty)} \subset \Omega^{(\infty)}$ and $B_R(\boldsymbol{\xi}_{i_j}) \subset \Omega^{(i_j)}$ for all sufficiently large j. Then, defining $\Psi^{(i_j)}(\boldsymbol{\xi}) = \varphi^{(i_j)}(\boldsymbol{\xi} - \boldsymbol{\xi}_{i_j})$, we have

$$\|\Psi^{(i_j)}\|_{C^{1,\alpha}(\overline{B_R(\mathbf{0})})} \le C.$$

Using that $\boldsymbol{\xi}_{i_j} \to \boldsymbol{\xi}_\infty$, and $\varphi^{(i_j)} \to \varphi^\infty$ uniformly on compact subsets of Λ^∞, we see that $\Psi^{(i_j)} \to \Psi^\infty$ in $C^{1,\frac{\alpha}{2}}(\overline{B_{R/2}(\mathbf{0})})$. Then $\Psi^{(i_j)}(\mathbf{0}) \to \Psi^{(\infty)}(\mathbf{0})$ and $D\Psi^{(i_j)}(\mathbf{0}) \to D\Psi^{(\infty)}(\mathbf{0})$ which imply

$$(\varphi^{(i_j)}, D\varphi^{(i_j)})(\boldsymbol{\xi}_{i_j}) \to (\varphi^{(\infty)}, D\varphi^{(\infty)})(\boldsymbol{\xi}_\infty).$$

Next, consider Case $\boldsymbol{\xi}_\infty \in \Gamma^{(\infty)}_{\text{wedge}}$. Then there exists $R > 0$ such that $B_{2R}(\boldsymbol{\xi}_{i_j}) \cap \partial\Omega^{(\infty)} \subset \Gamma^{(i_j)}_{\text{wedge}}$ and $\text{dist}(\boldsymbol{\xi}_{i_j}, \Gamma^{(i_j)}_{\text{wedge}}) < \frac{R}{100}$ for all $j > N$, where N is sufficiently large. Using that $\Gamma^{(i_j)}_{\text{wedge}}$ is a straight line, there exists $C > 0$ such that $\varphi^{(i_j)}$ can be extended from $\Omega^{(i_j)} \cap B_R(\boldsymbol{\xi}_{i_j})$ to $B_R(\boldsymbol{\xi}_{i_j})$ so that the extended function $\varphi^{(i_j)}_{\text{ext}}$ satisfies

$$\|\varphi^{(i_j)}_{\text{ext}}\|_{C^{1,\alpha}(\overline{B_R(\boldsymbol{\xi}_{i_j})})} \le C\|\varphi^{(i_j)}\|_{C^{1,\alpha}(\overline{\Omega^{(i_j)}\cap B_R(\boldsymbol{\xi}_{i_j})})} \le \hat{C} \tag{11.6.1}$$

for all $j > N$. By possibly selecting a further subsequence (without change of notations), we conclude that $\varphi^{(i_j)}_{\text{ext}}$ converges in $C^{1,\frac{\alpha}{2}}$ on the compact subsets of $B_R(\boldsymbol{\xi}_\infty)$. Denote the limit as $\varphi^{(\infty)}_{\text{ext}}$. Then $\|\varphi^{(i_j)}_{\text{ext}}\|_{C^{1,\alpha}(\overline{B_R(\boldsymbol{\xi}_{i_j})})} \le \hat{C}$, by (11.6.1). Note that, from the uniform convergence $\varphi^{(i_j)} \to \varphi^{(\infty)}$ on compact subsets of $\Lambda^{(\infty)}$, $\varphi^{(\infty)}_{\text{ext}} = \varphi^{(\infty)}$ on $\Omega^{(\infty)} \cap B_R(\boldsymbol{\xi}_\infty)$. Then we can argue as in the previous case.

Cases $\boldsymbol{\xi}_\infty \in \Gamma^{(\infty)}_{\text{sym}}$ and $\boldsymbol{\xi}_\infty \in \Gamma^{(\infty)}_{\text{sonic}}$ are treated similarly. In the latter case, we use that $\Gamma^{(i_j)}_{\text{sonic}}$ is an arc with radius $c_2^{(i_j)} \ge \frac{1}{\hat{C}}$ for all i_j by the continuous dependence of the parameters of state (2) on θ_{w}, and that we may assume without loss of generality that $R \le \frac{1}{100\hat{C}}$. Case $\boldsymbol{\xi}_\infty \in \Gamma^{(\infty)}_{\text{shock}}$ is considered similarly by employing estimate (11.4.39) for each $\Gamma^{(i_j)}_{\text{shock}}$.

3. It remains to consider the case that $\boldsymbol{\xi}_\infty$ is a corner point of $\Omega^{(\infty)}$.

When $\boldsymbol{\xi}_\infty = P_2{}^{(\infty)}$, $\Gamma^{(\infty)}_{\text{shock}}$ and $\Gamma^{(\infty)}_{\text{sym}}$ meet at $\boldsymbol{\xi}_\infty$. Thus, there exists $R > 0$ such that $\boldsymbol{\xi}_{i_j} \subset B_{R/10}(P_2{}^{(i_j)})$ and $B_{2R}(P_2{}^{(i_j)}) \cap \partial\Omega^{(i_j)} \subset \Gamma^{(i_j)}_{\text{shock}} \cup \{P_2{}^{(i_j)}\} \cup \Gamma^{(i_j)}_{\text{sym}}$ for all $j > N$, where N is sufficiently large. Extending $\varphi^{(i_j)}$ in $(\Omega^{\text{ext}})^{(i_j)}$ by even reflection as in Remark 8.1.3 and using Proposition 10.5.1, we can extend $\varphi^{(i_j)}$ from $(\Omega^{\text{ext}})^{(i_j)} \cap B_R(P_2{}^{(i_j)})$ to $B_R(P_2{}^{(i_j)})$ so that

$$\begin{aligned}\|\varphi^{(i_j)}_{\text{ext}}\|_{C^{1,\alpha}(\overline{B_R(P_2{}^{(i_j)})})} &\le C\|\varphi^{(i_j)}_{\text{ext}}\|_{C^{1,\alpha}(\overline{(\Omega^{\text{ext}})^{(i_j)}\cap B_R(P_2{}^{(i_j)})})} \\ &= C\|\varphi^{(i_j)}\|_{C^{1,\alpha}(\overline{\Omega^{(i_j)}\cap B_R(P_2{}^{(i_j)})})} \le \hat{C}\end{aligned}$$

for all $j > N$. Then we can argue as in the previous cases.

When $\boldsymbol{\xi}_\infty = P_3{}^{(\infty)}$, $\Gamma^{(\infty)}_{\text{wedge}}$ and $\Gamma^{(\infty)}_{\text{sym}}$ meet at $\boldsymbol{\xi}_\infty$. Thus, there exists $R > 0$ such that $\boldsymbol{\xi}_{i_j} \subset B_{R/10}(P_3{}^{(i_j)})$ and $B_{2R}(P_3{}^{(i_j)}) \cap \partial\Omega^{(i_j)} \subset \Gamma^{(i_j)}_{\text{wedge}} \cup \{P_3{}^{(i_j)}\} \cup \Gamma^{(i_j)}_{\text{sym}}$ for all $i_j > N$, where N is sufficiently large. Since $\Gamma^{(i_j)}_{\text{sym}}$ is a segment on $\{\xi_2 = 0\}$ and $\Gamma^{(i_j)}_{\text{wedge}}$ is a segment of the straight line that meets $\Gamma^{(i_j)}_{\text{sym}}$ at angle $\pi - \theta^{(i_j)}_{\text{w}} \in [\frac{\pi}{2}, \pi - \theta^*_{\text{w}}]$ and $\theta^*_{\text{w}} > 0$, we can extend $\varphi^{(i_j)}$ from $\Omega^{(i_j)} \cap B_R(P_3{}^{(i_j)})$ to half-disc $B_R(P_3{}^{(i_j)}) \cap \{\xi_2 > 0\}$, and then to $B_R(P_3{}^{(i_j)})$, so that

$$\begin{aligned}\|\varphi^{(i_j)}_{\text{ext}}\|_{C^{1,\alpha}(\overline{B_R(P_3{}^{(i_j)})})} &\le C\|\varphi^{(i_j)}_{\text{ext}}\|_{C^{1,\alpha}(\overline{B_R(P_3{}^{(i_j)})\cap\{\xi_2>0\}})} \\ &\le CC_1\|\varphi^{(i_j)}\|_{C^{1,\alpha}(\overline{\Omega^{(i_j)}\cap B_R(P_3{}^{(i_j)})})} \le \hat{C}\end{aligned}$$

for all $i_j > N$. Then we can argue as in the previous cases.

When $\boldsymbol{\xi}_\infty = P_1{}^{(\infty)}$ and $\boldsymbol{\xi}_\infty = P_4{}^{(\infty)}$, we use the (x, y)–coordinates near Γ_{sonic}, introduced in §11.1, where we note that the C^3–norms of the coordinate transform and its inverse depend only on the parameters of state (2) and hence are uniformly bounded for all $\theta_{\text{w}}^{(i)}$. Also, in the (x, y)–coordinates, $\Gamma_{\text{sonic}}^{(i_j)}$ is a segment of line $\{x = 0\}$ and $\Omega^{(i_j)} \subset \{x > 0\}$. Then there exists $R > 0$ such that $B_{2R}(P_1{}^{(i_j)}) \cap \partial\Omega^{(i_j)} \subset \Gamma_{\text{shock}}^{(i_j)} \cup \{P_1{}^{(i_j)}\} \cup \Gamma_{\text{sonic}}^{(i_j)}$ and $B_{2R}(P_4{}^{(i_j)}) \cap \partial\Omega^{(i_j)} \subset \Gamma_{\text{wedge}}^{(i_j)} \cup \{P_4{}^{(i_j)}\} \cup \Gamma_{\text{sonic}}^{(i_j)}$ for all $j > N$, where N is sufficiently large. Let ε be as in Proposition 11.2.8. For each $j > N$, we extend the shock function $\hat{f} = \hat{f}^{(i_j)}$ defined in Proposition 11.2.8 to $(-\varepsilon, \varepsilon)$ by setting $\hat{f}_{\text{ext}}^{(i_j)} = \hat{f}_0^{(i_j)}$ on $[-\varepsilon, 0)$, where $\hat{f}_0 = \hat{f}_0^{(i_j)}$ is the function from (11.2.8). Define

$$(\Omega_\varepsilon^{\text{ext}})^{(i_j)} = \{(x, y) \,:\, -\varepsilon < x < \varepsilon,\;\; \theta_{\text{w}} < y < \hat{f}_{\text{ext}}^{(i_j)}(x)\}.$$

By Proposition 11.2.8,

$$\Omega_\varepsilon^{(i_j)} = (\Omega_\varepsilon^{\text{ext}})^{(i_j)} \cap \{x > 0\}.$$

Also, extend $\varphi^{(i_j)}$ into $(\Omega_\varepsilon^{\text{ext}})^{(i_j)}$ by even reflection in the (x, y)–coordinates, *i.e.*, by setting $\varphi_{\text{ext}}^{(i_j)}(x, y) = \varphi_2^{(i_j)}(-x, y)$ on $(\Omega_\varepsilon^{\text{ext}})^{(i_j)} \cap \{x < 0\}$. From Proposition 11.4.6, we find that $\varphi_{\text{ext}}^{(i_j)} \in C^{1,1}((\Omega_\varepsilon^{\text{ext}})^{(i_j)})$, $f_{\text{ext}}^{(i_j)} \in C^{1,1}((-\varepsilon,\, \varepsilon))$, and there exists $C > 0$ such that, for each $j > N$,

$$\|\varphi_{\text{ext}}^{(i_j)}\|_{C^{1,1}((\Omega_\varepsilon^{\text{ext}})^{(i_j)})} + \|f_{\text{ext}}^{(i_j)}\|_{C^{1,1}((-\varepsilon,\, \varepsilon))} \le C.$$

Also, $P_1{}^{(i_j)} = (0, \hat{f}_0^{(i_j)}(0)) \in \partial(\Omega_\varepsilon^{\text{ext}})^{(i_j)}$ and $P_4{}^{(i_j)} = (0, \hat{\theta}_{\text{w}}^{(i_j)}) \in \partial(\Omega_\varepsilon^{\text{ext}})^{(i_j)}$. Thus, choosing $R < \frac{\varepsilon}{2}$, we have

$$B_{2R}(P_1{}^{(i_j)}) \cap \partial(\Omega_\varepsilon^{\text{ext}})^{(i_j)} = \{-2R < x < 2R,\; y = f_{\text{ext}}^{(i_j)}(x)\},$$
$$B_{2R}(P_4{}^{(i_j)}) \cap \partial(\Omega_\varepsilon^{\text{ext}})^{(i_j)} = \{-2R < x < 2R,\; y = \theta_{\text{w}}\}.$$

Then we can further extend $\varphi_{\text{ext}}^{(i_j)}$ into $B_{2R}(P_1{}^{(i_j)})$ and $B_{2R}(P_4{}^{(i_j)})$ with the uniform estimate:

$$\|\varphi_{\text{ext}}^{(i_j)}\|_{C^{1,1}(B_{2R}(P_1{}^{(i_j)}))} + \|\varphi_{\text{ext}}^{(i_j)}\|_{C^{1,1}(B_{2R}(P_4{}^{(i_j)}))} \le \hat{C} \qquad \text{for each } j > N.$$

Now Cases $\boldsymbol{\xi}_\infty = P_1{}^{(\infty)}$ and $\boldsymbol{\xi}_\infty = P_4{}^{(\infty)}$ can be considered similar to the previous cases. □

Corollary 11.6.2. *Fix (ρ_0, ρ_1, γ) and $\theta_{\text{w}}^* \in (\theta_{\text{w}}^{\text{c}}, \frac{\pi}{2})$. For any $\varepsilon, \sigma > 0$, there exists $\hat{\delta} > 0$ such that, for any admissible solution φ with $\theta_{\text{w}} \in [\theta_{\text{w}}^*, \frac{\pi}{2} - \sigma]$,*

$$(\varphi - \varphi_2) \ge \hat{\delta} \qquad \textit{in } \overline{\Omega} \setminus \mathcal{N}_\varepsilon(\Gamma_{\text{sonic}}), \tag{11.6.2}$$

$$\partial_{\mathbf{e}_{S_1}}(\varphi_1 - \varphi) \le -\hat{\delta} \qquad \textit{in } \overline{\Omega} \setminus \mathcal{N}_\varepsilon(\Gamma_{\text{sonic}}), \tag{11.6.3}$$

$$\partial_{\xi_2}(\varphi_1 - \varphi) \le -\hat{\delta} \qquad \textit{in } \overline{\Omega} \setminus \mathcal{N}_\varepsilon(\Gamma_{\text{sym}}). \tag{11.6.4}$$

Proof. In fact, (11.6.2) follows from Proposition 11.6.1, property (9.1.2), and Corollary 8.2.7(ii); (11.6.3) follows from Proposition 11.6.1, property (9.1.2), and Proposition 8.2.6; and (11.6.4) follows from Proposition 11.6.1, property (9.1.2), and Proposition 8.2.8.

We focus on the proof of (11.6.3), since the arguments for the other assertions are similar. Let (11.6.3) be false for some $\sigma, \varepsilon > 0$. Then, for each $i = 1, 2, \dots$, there exists an admissible solution $\varphi^{(i)}$ with the wedge angle $\theta_{\rm w}^{(i)} \in [\theta_{\rm w}^*, \frac{\pi}{2} - \sigma]$ such that

$$\partial_{\mathbf{e}_{S_1}}(\varphi_1 - \varphi^{(i)})(\boldsymbol{\xi}_i) \geq -\frac{1}{i} \qquad \text{for some } \boldsymbol{\xi}_i \in \overline{\Omega(\varphi_i)} \setminus \mathcal{N}_\varepsilon(\Gamma_{\rm sonic}^{(i)}).$$

From (9.1.2) and Proposition 11.6.1, there exists a subsequence i_j such that $\theta_{\rm w}^{(i_j)} \to \theta_{\rm w}^{(\infty)} \in [\theta_{\rm w}^*, \frac{\pi}{2} - \sigma]$, functions $\varphi^{(i_j)}$ converge as in Proposition 11.6.1(i) to an admissible solution $\varphi^{(\infty)}$ for the wedge angle $\theta_{\rm w}^{(\infty)}$, properties (ii)–(iii) in Proposition 11.6.1 hold, and $\boldsymbol{\xi}_{i_j} \to \boldsymbol{\xi}_\infty \in \overline{\Omega(\varphi^{(\infty)})} \setminus \mathcal{N}_\varepsilon(\Gamma_{\rm sonic}^{(\infty)})$. Then, by Proposition 11.6.1(iii), $\partial_{\mathbf{e}_{S_1}}(\varphi_1 - \varphi^{(\infty)})(\boldsymbol{\xi}_\infty) \geq 0$, which contradicts Proposition 8.2.6. □

Chapter Twelve

Iteration Set and Solvability of the Iteration Problem

In this chapter, we develop an iteration procedure and solve the iteration problem which, combined with the arguments in the next chapter, leads to the existence of an admissible solution for **Problem 2.6.1**.

12.1 STATEMENT OF THE EXISTENCE RESULTS

In this and the next chapter, we give a proof of the following existence assertion:

Proposition 12.1.1. *Let $\gamma > 1$ and $\rho_1 > \rho_0 > 0$. Let $\theta_{\rm w}^{\rm c}$ be the corresponding critical wedge angle as defined in Definition* 10.4.1. *Then, for any $\theta_{\rm w} \in (\theta_{\rm w}^{\rm c}, \frac{\pi}{2})$, there exists an admissible solution of* **Problem 2.6.1**.

In order to prove Proposition 12.1.1, it suffices to prove the existence of admissible solutions with the wedge angles $\theta_{\rm w} \in [\theta_{\rm w}^*, \frac{\pi}{2})$ for each $\theta_{\rm w}^* \in (\theta_{\rm w}^{\rm c}, \frac{\pi}{2})$. Thus, for the rest of this and the next chapter, we always fix $\theta_{\rm w}^* \in (\theta_{\rm w}^{\rm c}, \frac{\pi}{2})$.

12.2 MAPPING TO THE ITERATION REGION

We first discuss the procedure of the mapping of an admissible solution φ for the wedge angle $\theta_{\rm w} \in (\theta_{\rm w}^*, \frac{\pi}{2})$ in the subsonic domain Ω to a function u on a unit square $Q^{\rm iter}$. This procedure is invertible in the sense that, given $\theta_{\rm w} \in (\theta_{\rm w}^*, \frac{\pi}{2})$ and a function u on the unit square $Q^{\rm iter}$, which satisfy the conditions that $(u, \theta_{\rm w}) \in \mathfrak{S}$ where $\mathfrak{S} \subset C^{1,\alpha}(\overline{Q^{\rm iter}}) \times [\theta_{\rm w}^*, \frac{\pi}{2}]$ is introduced in Definition 12.2.6, we are able to recover both domain Ω and the corresponding function φ in Ω. We later define the iteration set as a subset of $\mathfrak{S}$.

For a given $(\varphi, \theta_{\rm w})$, define $u := (\varphi - \tilde{\varphi}_2) \circ \mathfrak{F}$, where $\tilde{\varphi}_2 = \tilde{\varphi}_2^{(\theta_{\rm w})}$ is the appropriately modified version of $\varphi_2^{(\theta_{\rm w})}$, and $\mathfrak{F} : Q^{\rm iter} \to \Omega$ is a $C^{1,\alpha}$–diffeomorphism. We will define $\mathfrak{F}$ in two steps through $\mathfrak{F}^{-1} = F_{(2,\mathfrak{g}_{\rm sh})} \circ F_1$. Here F_1 is a C^∞–mapping, independent of the particular admissible solution but depending on $\theta_{\rm w}$, so that F_1 is defined on a set $\mathcal{Q}_{\rm bd}(\theta_{\rm w}) \subset \Lambda(\theta_{\rm w})$ such that $\Omega \subset \mathcal{Q}_{\rm bd}(\theta_{\rm w})$ for any admissible solution φ for the wedge angle $\theta_{\rm w}$. Moreover, for any admissible solution φ with angle $\theta_{\rm w}$,

$$F_1(\Omega) = \{(s,t) \,:\, 0 < s < \hat{s}(\theta_{\rm w}),\ 0 < t < \mathfrak{g}_{\rm sh}(s)\}$$

with $F_1(\Gamma_{\rm sonic}) = \partial F_1(\Omega) \cap \{s = 0\}$ and $F_1(\Gamma_{\rm shock}) = \partial F_1(\Omega) \cap \{t = \mathfrak{g}_{\rm sh}(s)\}$, among other properties. This mapping $F_{(2,\mathfrak{g}_{\rm sh})}$ depends on φ and is of the $C^{1,\alpha}$–regularity.

Now we define the mappings and functions, discussed above, and show their properties.

12.2.1 Definition and properties of F_1

From Definition 8.1.1, the elliptic region Ω of any admissible solution φ lies within the subregion in Λ, whose boundary consists of lines $\Gamma_{\rm wedge}$, S_1, and $\Gamma_{\rm sym}$, and arc $\Gamma_{\rm sonic}$. In principle, we need to define an appropriate coordinate mapping in this region. However, in fact, it is more convenient to define it in a slightly larger region. In the definition above, we replace line $S_1 = \{\varphi_1 - \varphi_2 = 0\}$ by the line:

$$S_{1,\delta_*} = \{\varphi_1 - \varphi_2 = -\delta_*\} \tag{12.2.1}$$

for small $\delta_* > 0$ chosen below. The precise construction is as follows:

Fix a wedge angle $\theta_{\rm w} \in (\theta_{\rm w}^{\rm s}, \frac{\pi}{2}]$. Let Q be the point defined in Definition 11.2.1. Since the parameters of state (2) depend smoothly on $\theta_{\rm w} \in [\theta_{\rm w}^{\rm s}, \frac{\pi}{2}]$, and (11.2.3) holds for any $\theta_{\rm w} \in [\theta_{\rm w}^{\rm s}, \frac{\pi}{2}]$, there exists $\delta_* > 0$ such that

$$\{\varphi_2 < \varphi_1 + 2\delta_*\} \cap \Lambda \cap \mathcal{N}_{\varepsilon_1}(\Gamma_{\rm sonic}) \cap \{0 < x < \varepsilon\} \subset \{\theta_{\rm w} < \theta < \theta_Q - \delta\} \tag{12.2.2}$$

for any $\theta_{\rm w} \in (\theta_{\rm w}^{\rm s}, \frac{\pi}{2}]$, where ε and ε_1 are the same as those in Lemma 11.2.2, and we have used the same polar and (x, y)–coordinates as in §11.2.

Note that, from the definition of Q, it follows that $(\varphi_1 - \varphi_2)(Q) < 0$. Then, with the choice of δ_* as above, for each $\theta_{\rm w} \in (\theta_{\rm w}^{\rm s}, \frac{\pi}{2}]$, line $S_{1,\delta_*} = S_{1,\delta_*}(\theta_{\rm w})$ is parallel to $S_1(\theta_{\rm w})$ and intersects with the sonic circle of state (2) at two points. Let $\hat{P}_1$ be the point of intersection that is the nearest to P_1, which lies between P_1 and Q:

$$\begin{aligned}&\hat{P}_1 \text{ is the unique intersection point of } S_{1,\delta_*}(\theta_{\rm w}) \text{ with } \partial B_{c_2}(\mathcal{O}_2)\\ &\quad\text{satisfying } \ 0 < \theta_{\rm w} = \theta_{P_4} < \theta_{P_1} < \theta_{\hat{P}_1} < \theta_Q - \delta.\end{aligned} \tag{12.2.3}$$

Denote by $\Gamma_{\rm sonic}^{(\delta_*)}$ the smaller arc $\hat{P}_1 P_4$, *i.e.*, $\Gamma_{\rm sonic}^{(\delta_*)} = \partial B_{c_2}(\mathcal{O}_2) \cap \{\theta_{P_4} < \theta < \theta_{\hat{P}_1}\}$. Note that

$$\Gamma_{\rm sonic} \subset \Gamma_{\rm sonic}^{(\delta_*)}.$$

Definition 12.2.1. *$\mathcal{Q} = \mathcal{Q}^{(\delta_*)}$ denotes the subset of Λ, whose boundary consists of lines $\Gamma_{\rm wedge}$, S_{1,δ_*}, and $\Gamma_{\rm sym}$, and arc $\Gamma_{\rm sonic}^{(\delta_*)}$.*

Note that

$$\mathcal{Q} \subset \{\varphi_2 < \varphi_1 + \delta_*\} \cap \Lambda. \tag{12.2.4}$$

Then the corner points of $\partial\mathcal{Q}$ include $\{\hat{P}_1, P_3, P_4\}$.

Note also that $\mathcal{Q}$ may be unbounded, which happens when $\mathbf{e}_{S_1} \cdot \mathbf{e}_{\xi_2} \geq 0$. We then restrict $\mathcal{Q}$ to

$$\mathcal{Q}_{\rm bd} := \mathcal{Q} \cap \{\xi_1 > -10C\}, \tag{12.2.5}$$

where C is from (9.1.2). In fact, it is possible that $\mathcal{Q} \subset \{\xi_1 > -10C\}$, in which case, $\mathcal{Q}_{\text{bd}} = \mathcal{Q}$.

The main reason to restrict $\mathcal{Q}$ to $\mathcal{Q}_{\text{bd}}$ is the fact that sets $\mathcal{Q}_{\text{bd}} \equiv \mathcal{Q}_{\text{bd}}^{(\theta_{\text{w}})}$ are uniformly bounded with respect to $\theta_{\text{w}} \in [\theta_{\text{w}}^{\text{s}}, \frac{\pi}{2}]$. Specifically, there exists $C_1 > 0$ depending only on the data so that

$$\mathcal{Q}_{\text{bd}}^{(\theta_{\text{w}})} \subset [-10C, \xi_1^0] \times [0, C_1] \qquad \text{for all } \theta_{\text{w}} \in [\theta_{\text{w}}^{\text{s}}, \frac{\pi}{2}], \tag{12.2.6}$$

where ξ_1^0 is the location of the incident shock, defined in (2.2.18), and C is from (12.2.5).

To show (12.2.6), we first note that $\Lambda \subset \{\xi_1 < \xi_1^0,\ \xi_2 > 0\}$ for any θ_{w}, so that

$$\mathcal{Q}_{\text{bd}}^{(\theta_{\text{w}})} \subset [-10C, \xi_1^0] \cap \{\xi_2 > 0\}.$$

Next, we consider $\theta_{\text{w}} \in [\theta_{\text{w}}^{\text{d}}, \frac{\pi}{2})$. Then, using (7.1.4), (7.5.8), and $v_2 = u_2 \tan\theta_{\text{w}}$, we have

$$\mathbf{e}_{\mathcal{S}_1} \cdot \boldsymbol{\nu}_{\text{w}} = \frac{1}{\sqrt{(u_1 - u_2)^2 + v_2^2}} \big(\frac{u_2}{\cos\theta_{\text{w}}} - u_1 \cos\theta_{\text{w}}\big).$$

We show that the last expression is positive. From (7.5.6), noting that $|P_3\mathcal{O}_1| = u_1$ and $\angle\mathcal{O}_1 P_3 \mathcal{O}_2 = \theta_{\text{w}} \in (\theta_{\text{w}}^*, \frac{\pi}{2})$, we obtain that $|P_3\mathcal{O}_2| > u_1 \cos\theta_{\text{w}}$. Also, $\mathcal{O}_2 = (u_2, u_2 \tan\theta_{\text{w}})$ and $P_3 = (0,0)$, so that $|P_3\mathcal{O}_2| = \frac{u_2}{\cos\theta_{\text{w}}}$. Thus, $\frac{u_2}{\cos\theta_{\text{w}}} - u_1\cos\theta_{\text{w}} > 0$, which implies that $\mathbf{e}_{\mathcal{S}_1} \cdot \boldsymbol{\nu}_{\text{w}} > 0$. From this and the fact that $\boldsymbol{\nu}_{\text{w}}$ is the interior normal on Γ_{wedge} to Λ, we have

$$S_{1,\delta_*} \cap \Lambda = \{P_{0,\delta^*} + \tau \mathbf{e}_{\mathcal{S}_1} \ : \ \tau > 0\}, \tag{12.2.7}$$

where P_{0,δ^*} is the point of intersection of S_{1,δ_*} with line $\{\xi_2 = \xi_1 \tan\theta_{\text{w}}\}$. It follows from (7.5.8) that, if $u_1 \geq u_2$, then

$$\mathcal{Q}_{\text{bd}}^{(\theta_{\text{w}})} \subset [-10C, \xi_1^0] \times [0, \max_{P \in \Gamma_{\text{sonic}}^{(\delta_*)}} \xi_2(P)].$$

Since the parameters of state (2) depend continuously on $\theta_{\text{w}} \in [\theta_{\text{w}}^{\text{s}}, \frac{\pi}{2}]$, then there exists C_1 such that

$$\max_{P \in \Gamma_{\text{sonic}}^{(\delta_*)}} \xi_2(P) \leq v_2 + c_2 \leq C_1 \qquad \text{for any } \theta_{\text{w}} \in [\theta_{\text{w}}^{\text{s}}, \tfrac{\pi}{2}].$$

From (7.2.15), there exists $\sigma > 0$ depending only on the data, so that $u_1 > u_2$ for $\theta_{\text{w}} \in [\frac{\pi}{2} - \sigma, \frac{\pi}{2}]$. Since $u_2, v_2 > 0$ for $\theta_{\text{w}} \in [\theta_{\text{w}}^{\text{s}}, \frac{\pi}{2})$ and depend continuously on θ_{w}, there exists $\delta > 0$ such that $v_2 \geq \delta$ for $\theta_{\text{w}} \in [\theta_{\text{w}}^{\text{s}}, \frac{\pi}{2} - \sigma]$. Now, using the explicit form of $\mathbf{e}_{\mathcal{S}_1}$ in (7.5.8) and the fact that point $\hat{P}_0^{(\theta_{\text{w}})} = (\xi_1^0, \xi_1^0 \cot\theta_{\text{w}} + \frac{\delta^*}{v_2})$ lies on S_{1,δ_*}, we find that, for $\theta_{\text{w}} \in [\theta_{\text{w}}^{\text{s}}, \frac{\pi}{2} - \sigma]$ such that $u_2(\theta_{\text{w}}) > u_1$,

$$\mathcal{Q}_{\text{bd}}^{(\theta_{\text{w}})} \subset [-10C, \xi_1^0] \times [0, \xi_1^0 \cot\theta_{\text{w}} + \frac{\delta^*}{v_2} + (10C + \xi_1^0)\frac{u_2 - u_1}{v_2}].$$

From the discussion above, we conclude that

$$\xi_1^0 \cot\theta_{\rm w} + \frac{\delta^*}{v_2^{(\theta_{\rm w})}} + (10C + \xi_1^0)\frac{u_2^{(\theta_{\rm w})} - u_1}{v_2^{(\theta_{\rm w})}}$$

is bounded on $\theta_{\rm w} \in [\theta_{\rm w}^{\rm s}, \frac{\pi}{2} - \sigma]$. This completes the proof of (12.2.6) for $\theta_{\rm w} \in [\theta_{\rm w}^{\rm s}, \frac{\pi}{2})$. Next, the continuity of the parameters of state (2) and point $(\tilde{\xi}_1, 0)$ up to $\theta_{\rm w} = \frac{\pi}{2}$ in Lemma 7.2.1 allows us to extend (12.2.6) to all $\theta_{\rm w} \in (\theta_{\rm w}^{\rm s}, \frac{\pi}{2}]$.

From Definition 8.1.1(iv) and Proposition 9.1.2, we find that, for any admissible solution,

$$\Omega \subset \mathcal{Q} \cap B_C(0) \subset \mathcal{Q}_{\rm bd}.$$

We use the (x, y)–coordinates near $\Gamma_{\rm sonic}^{(\delta_*)}$ as in §11.1, constant ε_1 from Lemma 11.2.2, and constant ε_0 from Proposition 11.2.8. For $\varepsilon \in (0, \varepsilon_0]$, denote

$$\begin{aligned}\mathcal{D}_{\delta_*,\varepsilon}^{(\theta_{\rm w})} &:= \{\varphi_2 < \varphi_1 + \delta_*\} \cap \Lambda \cap \mathcal{N}_{\varepsilon_1}(\Gamma_{\rm sonic}^{(\delta_*)}) \cap \{0 < x < \varepsilon\} \\ &\equiv \mathcal{Q} \cap \mathcal{N}_{\varepsilon_1}(\Gamma_{\rm sonic}^{(\delta_*)}) \cap \{0 < x < \varepsilon\},\end{aligned} \tag{12.2.8}$$

where φ_2 and $\Gamma_{\rm sonic}^{(\delta_*)}$ correspond to the wedge angle $\theta_{\rm w}$. Using (12.2.2) and following the proof of (11.2.8) in Lemma 11.2.2, we obtain that there exists $\hat{f}_{0,\delta_*} \in C^\infty([0, \varepsilon_0])$ such that $\mathcal{D}_{\delta_*,\varepsilon_0}$ is of the form:

$$\begin{aligned}&\mathcal{D}_{\delta_*,\varepsilon_0} = \{(x, y) \,:\, 0 < x < \varepsilon_0, \;\; \theta_{\rm w} < y < \hat{f}_{0,\delta_*}(x)\}, \\ &S_{1,\delta_*} \cap \mathcal{N}_{\varepsilon_1}(\Gamma_{\rm sonic}^{(\delta_*)}) \cap \{0 < x < \varepsilon_0\} \\ &\qquad = \{(x, y) \,:\, 0 < x < \varepsilon_0, \;\; y = \hat{f}_{0,\delta_*}(x)\}.\end{aligned} \tag{12.2.9}$$

Now we introduce the following change of coordinates:

Lemma 12.2.2. *There exist $\delta > 0$, $C > 0$, a smooth and strictly positive function $\hat{s} : (\theta_{\rm w}^{\rm s}, \frac{\pi}{2}] \mapsto \mathbb{R}$ (defined explicitly by $\hat{s}(\theta_{\rm w}) := |P_3P_4| \equiv \sqrt{u_2^2 + v_2^2} + c_2$), a one-to-one map*

$$F_1 \equiv F_1^{(\theta_{\rm w})} : \overline{\mathcal{Q}_{\rm bd}}^{(\theta_{\rm w})} \mapsto [0, \hat{s}(\theta_{\rm w})] \times [0, \infty) \qquad \textit{for each } \theta_{\rm w} \in (\theta_{\rm w}^{\rm s}, \tfrac{\pi}{2}],$$

and a smooth function $g : \mathbb{R} \times \mathbb{R} \mapsto \mathbb{R}$ such that, reducing ε_0 depending only on the data, we obtain that, for each $\theta_{\rm w} \in (\theta_{\rm w}^{\rm s}, \frac{\pi}{2}]$, the following holds with $(F_1, \mathcal{Q}_{\rm bd}) = (F_1^{(\theta_{\rm w})}, \mathcal{Q}_{\rm bd}^{(\theta_{\rm w})})$:

(i) *$F_1(\mathcal{Q}_{\rm bd})$ is open, $F_1(\partial\mathcal{Q}_{\rm bd}) = \partial F_1(\mathcal{Q}_{\rm bd})$, and there exists a continuous function $\eta \equiv \eta^{(\theta_{\rm w})} : [0, \hat{s}(\theta_{\rm w})] \mapsto \mathbb{R}^+$ such that*

$$F_1(\mathcal{Q}_{\rm bd}) = \big\{(s, t) \;:\; 0 < s < \hat{s}(\theta_{\rm w}),\, 0 < t < \eta(s)\big\}, \tag{12.2.10}$$

and the rescaled functions: $t \mapsto \eta^{(\theta_{\rm w})}(\frac{s}{\hat{s}(\theta_{\rm w})})$ depend on $\theta_{\rm w} \in [\theta_{\rm w}^{\rm s}, \frac{\pi}{2}]$ continuously in the $C([0, 1])$–norm.

(ii) $\|F_1\|_{C^3(\overline{\mathcal{Q}_{\rm bd}})} + \|F_1^{-1}\|_{C^3(\overline{F_1(\mathcal{Q}_{\rm bd})})} \le C$, *and* $|\det D(F_1^{-1})| \ge \frac{1}{C}$ *in* $F_1(\mathcal{Q}_{\rm bd})$ *for each* $\theta_{\rm w} \in [\theta_{\rm w}^{\rm s}, \frac{\pi}{2}]$.

(iii) *Smooth dependence on* $\theta_{\rm w}$: *Set* $\mathcal{Q}_{\rm bd}^{\cup} := \cup_{\theta_{\rm w}\in(\theta_{\rm w}^{\rm s},\frac{\pi}{2}]}\mathcal{Q}_{\rm bd}^{(\theta_{\rm w})} \times \{\theta_{\rm w}\}$ *is open and bounded in* $\mathbb{R}^2 \times (\theta_{\rm w}^{\rm s}, \frac{\pi}{2}]$, *and the map:* $(\boldsymbol{\xi}, \theta_{\rm w}) \mapsto F_1^{(\theta_{\rm w})}(\boldsymbol{\xi})$ *is in* $C^\infty(\overline{\mathcal{Q}_{\rm bd}^{\cup}})$. *Set* $F_1(\mathcal{Q}_{\rm bd}^{\cup}) := \cup_{\theta_{\rm w}\in(\theta_{\rm w}^{\rm s},\frac{\pi}{2}]}F_1^{(\theta_{\rm w})}(\mathcal{Q}_{\rm bd}^{(\theta_{\rm w})})\times\{\theta_{\rm w}\}$ *is open and bounded in* $\mathbb{R}^2\times (\theta_{\rm w}^{\rm s}, \frac{\pi}{2}]$, *and the map:* $((s,t),\theta_{\rm w}) \mapsto (F_1^{(\theta_{\rm w})})^{-1}(s,t)$ *is in* $C^\infty(\overline{F_1(\mathcal{Q}_{\rm bd}^{\cup})})$.

(iv) $F_1(P) = (x_P, y_P - y_{P_4})$ *for all* $P \in \mathcal{D}_{\delta_*, 3\varepsilon_0/4}^{(\theta_{\rm w})}$, *where we have used notation* (12.2.8). *In particular,*

$$
\begin{aligned}
&F_1(P_1) = (0, y_{P_1} - y_{P_4}), \quad F_1(P_4) = (0,0),\\
&F_1(\Gamma_{\rm sonic}) = \{s = 0,\ 0 < t < y_{P_1} - y_{P_4}\}.
\end{aligned}
$$

Also, $F_1(\mathcal{Q}_{\rm bd} \setminus \mathcal{D}_{\delta_*, 3\varepsilon_0/4}^{(\theta_{\rm w})}) = F_1(\mathcal{Q}_{\rm bd}) \cap \{s \ge \frac{3\varepsilon_0}{4}\}$.

(v) $F_1(P_3) = (\hat{s}(\theta_{\rm w}), 0)$. *Moreover, for any* $P = (\xi_{1P}, 0) \in \partial\mathcal{Q}_{\rm bd} \cap \{\xi_2 = 0\}$ *with* $\xi_{1P} \le 0$, $F_1(P) = (\hat{s}(\theta_{\rm w}), g^{(\theta_{\rm w})}(\xi_{1P}))$ *and* g *satisfies that* $g \in C^3(\mathbb{R}\times\mathbb{R})$, $g^{(\theta_{\rm w})}(0) = 0$, *and* $(g^{(\theta_{\rm w})})' \le -\delta$ *on* $\mathbb{R}$ *for each* $\theta_{\rm w}$.

(vi) $F_1(P) = |P - P_1|$ *for all* $P \in \Gamma_{\rm wedge}$, *and* $F_1(\Gamma_{\rm wedge}) = \{(s,0)\ :\ s \in (0, \hat{s}(\theta_{\rm w}))\}$. *Moreover,* $\mathbf{e}_t\cdot\boldsymbol{\nu}_{\rm w} \ge \delta$ *on* $\Gamma_{\rm wedge}$, *where we have used notation* (12.2.11).

(vii) $\frac{\partial\phi_1(F_1^{-1}(s,t))}{\partial t} \le -\delta$ *for any* $(s,t) \in F_1(\mathcal{Q}_{\rm bd})$.

(viii) $\mathbf{e}_t\cdot\boldsymbol{\nu}_{\rm sh} \le -\delta$ *on* $\Gamma_{\rm shock}(\varphi)$ *for any admissible solution* φ.

(ix) $\boldsymbol{\nu}_{\rm w} = a_s(P)\mathbf{e}_s(P) + a_t(P)\mathbf{e}_t(P)$ *with* $a_s(P) \le 0$ *and* $a_t(P) > 0$ *for any* $P \in \Gamma_{\rm wedge}$.

Here $\phi_1 = \varphi_1 + \frac{|\boldsymbol{\xi}|^2}{2} \equiv \varphi_1 - \varphi_0$, *and* (s,t) *denote the coordinates of points in* $F_1(\mathcal{Q}_{\rm bd}) \subset (0, \hat{s}(\theta_{\rm w})) \times (0,\infty)$. *Also, we have defined*

$$
\mathbf{e}_s(F_1^{-1}(s,t)) = \frac{\partial_s F_1^{-1}}{|\partial_s F_1^{-1}|}(s,t), \quad \mathbf{e}_t(F_1^{-1}(s,t)) = \frac{\partial_t F_1^{-1}}{|\partial_t F_1^{-1}|}(s,t) \tag{12.2.11}
$$

for all $(s,t) \in F_1(\mathcal{Q}_{\rm bd})$. *Then, by* (i)–(ii), *the* C^2 *vector fields* $\{\mathbf{e}_s, \mathbf{e}_t\}$ *in* $\mathcal{Q}_{\rm bd}$ *are defined.*

Proof. Note that, since $\phi_1 = u_1\xi_1 + const.$, property (vii) is equivalent to:

$$
\frac{\partial F_1^{-1}(s,t)}{\partial t}\cdot\mathbf{e}_{\xi_1} \le -\hat{\delta} \qquad \text{for any } (s,t) \in F_1(\mathcal{Q}_{\rm bd}). \tag{12.2.12}
$$

Now we divide the proof into two steps.

1. In this step, we show that, to construct such a transformation in $\mathcal{Q}_{\rm bd}$, it suffices to construct two unit vector fields $\{\mathbf{e}_s(\cdot), \mathbf{e}_t(\cdot)\}$ in the larger domain $\mathcal{Q}$ defined above, for each $\theta_{\rm w}$, such that

$$\|(\mathbf{e}_s, \mathbf{e}_t)\|_{C^3(\overline{\mathcal{Q}})} \le N, \tag{12.2.13}$$

$$\mathbf{e}_s \text{ and } \mathbf{e}_t \text{ are } C^\infty \text{ functions of } (\boldsymbol{\xi}, \theta_{\rm w}) \in \cup_{\theta_{\rm w}\in(\theta_{\rm w}^{\rm s}, \frac{\pi}{2}]} \mathcal{Q}^{(\theta_{\rm w})} \times \{\theta_{\rm w}\}, \tag{12.2.14}$$

$$|\mathbf{e}_s \cdot \mathbf{e}_t^\perp| \ge \delta \qquad \text{in } \mathcal{Q}, \tag{12.2.15}$$

$$\mathbf{e}_s \equiv \mathbf{e}_x, \quad \mathbf{e}_t \equiv \mathbf{e}_y \qquad \text{in } \mathcal{D}^{(\theta_{\rm w})}_{\delta_*, 3\varepsilon_0/4}, \tag{12.2.16}$$

$$\mathbf{e}_s = -\boldsymbol{\tau}_{\rm w}, \quad \mathbf{e}_t \cdot \boldsymbol{\nu}_{\rm w} \ge \delta \qquad \text{on } \Gamma_{\rm wedge} = P_3P_4, \tag{12.2.17}$$

$$(\mathbf{e}_s, \mathbf{e}_t) = -(\boldsymbol{\tau}_{\rm w}, \mathbf{e}_{\xi_1}) \qquad \text{on } \partial\mathcal{Q} \cap \{\xi_2 = 0\}, \tag{12.2.18}$$

$$\mathbf{e}_s(\boldsymbol{\xi}) \cdot \boldsymbol{\nu}_{S_1} \ge \delta\big(\text{dist}(\boldsymbol{\xi}, \Gamma_{\rm wedge}) + m(\frac{\pi}{2} - \theta_{\rm w})\big), \;\; \mathbf{e}_s \cdot \mathbf{e}_{\xi_2} \le -\delta \qquad \text{in } \mathcal{Q}, \tag{12.2.19}$$

$$\mathbf{e}_t \cdot \boldsymbol{\nu}_{S_1} < 0 \qquad \text{on } S_{1,\delta_*} \cap \overline{\mathcal{Q}}, \tag{12.2.20}$$

$$\mathbf{e}_t(P) \cdot \mathbf{e}_{\xi_1}(P) \le -\hat{\delta} \qquad \text{for all } P \in \mathcal{Q}, \tag{12.2.21}$$

$$\{\mathbf{e}_s, \mathbf{e}_t\} \text{ satisfy assertions (viii)–(ix).} \tag{12.2.22}$$

Here N, δ, and $\hat{\delta}$ are positive constants, $\{\mathbf{e}_x(\cdot), \mathbf{e}_y(\cdot)\}$ is the basis for the (x,y)–coordinates in $\mathcal{N}_{\varepsilon_1}(\Gamma^{(\delta_*)}_{\rm sonic})$ (*i.e.*, the unit vectors in the negative-radial and angular directions with respect to $\mathcal{O}_2$), $m(\cdot)$ is from Lemma 11.2.2, $\boldsymbol{\nu}_{\rm w}$ is the interior unit normal on $\Gamma_{\rm wedge}$ to Λ defined by (7.1.4), and $\boldsymbol{\tau}_{\rm w}$ is the unit vector along $\Gamma_{\rm wedge}$ in the direction from P_3 to P_4, *i.e.*, (8.2.17). Also, $\boldsymbol{\nu}_{S_1} = \frac{D(\varphi_1-\varphi_2)}{|D(\varphi_1-\varphi_2)|}$.

Indeed, assume that the vector fields $\{\mathbf{e}_s, \mathbf{e}_t\}$ satisfying (12.2.13)–(12.2.22) are constructed. Then, from the properties of $\mathbf{e}_s$ in (12.2.16)–(12.2.17) and (12.2.19), it follows that the integral curves of the vector field $\mathbf{e}_s$ can exit $\mathcal{Q}$ only through $\partial\mathcal{Q} \cap \{\xi_2 = 0\}$. Using $\mathcal{Q} \subset \{\xi_2 > 0\}$ and, once again, the property that $\mathbf{e}_s \cdot \mathbf{e}_{\xi_2} \le -\delta$ in $\mathcal{Q}$ in (12.2.19), we conclude that every integral curve of $\mathbf{e}_s$ passing through any point $P \in \mathcal{Q}$ necessarily exits $\mathcal{Q}$ at a finite point. This implies that the integral curves of $\mathbf{e}_s$ passing through the points in $\partial\mathcal{Q}\cap\{\xi_2 = 0\}$ cover $\mathcal{Q}$. In particular, since $\Gamma_{\rm wedge}$ is an integral curve of $\mathbf{e}_s$ by (12.2.18), there exists $\xi_1^0 < 0$ such that the integral curves of $\mathbf{e}_s$ originating in the points in $\{(\xi_1, 0) \;:\; \xi_1^0 < \xi_1 < 0\}$ cover $\mathcal{Q}_{\rm bd}$. Let $\hat{\mathcal{Q}}$ be the union of all these integral curves within $\mathcal{Q}$. Then $\hat{\mathcal{Q}}$ is a bounded domain by (12.2.19) and the boundedness of $\mathcal{Q}_{\rm bd}$. Also, $\mathcal{Q}_{\rm bd} \subset \hat{\mathcal{Q}}$, and $\partial\hat{\mathcal{Q}}$ consists of $\Gamma_{\rm wedge}$, $\Gamma^{(\delta_*)}_{\rm sonic}$, $\{(\xi_1, 0) \;:\; \xi_1^0 < \xi_1 < 0\}$, and a segment $\hat{P}_1\hat{P}$ of S_{1,δ_*}, where $\hat{P} \in S_{1,\delta_*}$ is such that either $\hat{P} = (\xi_1^0, 0)$ or $\xi_{2\hat{P}} > 0$ in which case $\partial\hat{\mathcal{Q}}$ also contains a segment of integral curve of $\mathbf{e}_s$ with one endpoint at $(\xi_1^0, 0)$ and another endpoint $\hat{P}$. We denote this segment by Γ_g. Then, denoting $\boldsymbol{\nu}_{\Gamma_g}$ the unit normal to Γ_g, interior for $\hat{\mathcal{Q}}$, we obtain that, from (12.2.15) and (12.2.18),

$$\mathbf{e}_t \cdot \boldsymbol{\nu}_{\Gamma_g} \le -\delta < 0 \qquad \text{on } \Gamma_g. \tag{12.2.23}$$

Now we consider the integral curves of $\mathbf{e}_t$. By (12.2.16)–(12.2.18), (12.2.20), (12.2.23), and the structure of $\partial\hat{\mathcal{Q}}$, it follows that these integral curves can enter $\hat{\mathcal{Q}}$ only through $\Gamma_{\text{wedge}} = P_3P_4$. Also, since region $\hat{\mathcal{Q}}$ is bounded, it follows from (12.2.21) that, for every $P \in \hat{\mathcal{Q}}$, the integral curve of $\mathbf{e}_t$ passing through P must enter $\hat{\mathcal{Q}}$ at some point of $\partial\hat{\mathcal{Q}}$, and hence at some point of P_3P_4. It follows that the connected segments of integral curves of $\mathbf{e}_t$ originating at all points $Z \in P_3P_4$ cover $\hat{\mathcal{Q}}$. Then, for each $Z \in P_3P_4$, we set $s := |ZP_4|$ on the integral curve originating at Z. Since $\mathbf{e}_t$ is non-characteristic on Γ_{wedge} by (12.2.17), then, using the standard properties of ODEs and the smooth dependence of solutions on the initial data and parameters which hold by (12.2.13)–(12.2.14), we obtain that $\|s^{(\theta_{\rm w})}(\cdot)\|_{C^3(\overline{\hat{\mathcal{Q}}^{(\theta_{\rm w})}})} \le C$, and the function: $(P, \theta_{\rm w}) \mapsto s^{(\theta_{\rm w})}(P)$ is C^∞ on $\mathcal{Q}^{\cup}_{\text{bd}}$, where C depends on the data, N, and δ.

Next, we assign the value of t to the integral curves of $\mathbf{e}_s$ originating at the points in $Z \in \{(\xi_1, 0) \,:\, \xi_1^0 < \xi_1 < 0\}$ and lying within $\hat{\mathcal{Q}}$. From the discussion above and the definition of domain $\hat{\mathcal{Q}}$, these curves cover $\hat{\mathcal{Q}}$. Combining this with (12.2.16), using that vectors $\mathbf{e}_x$ are orthogonal to arc $\Gamma^{(\delta_*)}_{\text{sonic}}$ so that this arc is non-characteristic for $\mathbf{e}_s$, and employing the fact that line $\{\xi_2 = 0\}$ is non-characteristic for $\mathbf{e}_s$ uniformly with respect to $\theta_{\rm w}$ which is seen by combining (12.2.18) with

$$\boldsymbol{\tau}_{\rm w} \cdot \mathbf{e}_{\xi_2} = \sin\theta_{\rm w} \ge \sin\theta^{\rm d}_{\rm w} =: \delta_1 > 0,$$

we conclude that there exists $\hat{\xi}_1 \in (\xi_1^0, 0)$ such that, for each $\xi_1^* \in (\hat{\xi}_1, 0)$, the integral curve of $\mathbf{e}_s$ originating at $P = (\xi_1^*, 0)$ intersects arc $\Gamma^{(\delta_*)}_{\text{sonic}}$ at a point P^* and that, defining $g(\xi_1^*) = \theta_{P^*} - \theta_{\rm w}$ so that $g(\xi_1^*) + \theta_{\rm w} = g(\xi_1^*) + y_{P_4}$ is the y–coordinate of P^*, it follows that $g \in C^\infty([\hat{\xi}_1, 0])$ and satisfies $g(0) = 0$ and $g' \le -\delta_2 < 0$ for some $\delta_2 > 0$ that depends smoothly on $\theta_{\rm w} \in [\theta^{\rm s}_{\rm w}, \frac{\pi}{2}]$ and hence can be chosen independently of $\theta_{\rm w} \in [\theta^{\rm s}_{\rm w}, \frac{\pi}{2}]$ by taking the minimum. Moreover, using that the vector field $\mathbf{e}_s$ depends smoothly on $(\boldsymbol{\xi}, \theta_{\rm w})$ by (12.2.14) and employing the non-characteristic properties of $\Gamma^{(\delta_*)}_{\text{sonic}}$ and line $\{\xi_2 = 0\}$ with respect to $\mathbf{e}_s$ discussed above, we conclude that the function: $(\xi_1, \theta_{\rm w}) \mapsto g^{(\theta_{\rm w})}(\xi_1)$ is in $C^\infty(\cup_{\theta_{\rm w} \in [\theta^{\rm s}_{\rm w}, \frac{\pi}{2}]} [\hat{\xi}_1^{(\theta_{\rm w})}, 0] \times \{\theta_{\rm w}\})$. Then, extending $g^{(\cdot)}(\cdot)$ to a function in $C^\infty(\mathbb{R} \times \mathbb{R})$ satisfying $(g^{(\theta_{\rm w})})'(\cdot) \le -\frac{\delta}{2}$ on $\mathbb{R}$ for each $\theta_{\rm w} \in [\theta^{\rm s}_{\rm w}, \frac{\pi}{2}]$, we set $t = g^{(\theta_{\rm w})}(\xi_1)$ on the integral curve originating at $(\xi_1, 0)$ with $\xi_1^0 < \xi_1 < 0$, which determines a smooth function $t^{(\theta_{\rm w})}(P)$ on $\overline{\hat{\mathcal{Q}}^{(\theta_{\rm w})}}$ satisfying that $\|t^{(\theta_{\rm w})}(\cdot)\|_{C^3(\overline{\hat{\mathcal{Q}}^{(\theta_{\rm w})}})} \le C$, and function $(P, \theta_{\rm w}) \mapsto t^{(\theta_{\rm w})}(P)$ is C^∞ on $\mathcal{Q}^{\cup}_{\text{bd}}$, where we have again used (12.2.13)–(12.2.14) and the non-characteristic properties of line $\{\xi_2 = 0\}$ with respect to $\mathbf{e}_s$.

Now we define

$$F_1(P) = (s(P), t(P)) \qquad \text{for } P \in \mathcal{Q}_{\text{bd}}.$$

Since $\hat{\mathcal{Q}}$ is covered by both the connected segments of $\mathbf{e}_t$–integral curves originating on $\Gamma_{\text{wedge}} = P_3P_4$ (which are non-characteristic for $\mathbf{e}_t$) and the connected segments of $\mathbf{e}_s$–integral curves exiting $\hat{\mathcal{Q}}$ through $\{(\xi_1, 0) \,:\, \xi_1^0 <$

$\xi_1 < 0\}$ (which are non-characteristic for $\mathbf{e}_s$), it follows from (12.2.15) that each $\mathbf{e}_s$–integral curve intersects with any $\mathbf{e}_t$–integral curve at most once in $\hat{\mathcal{Q}}$. Indeed, if an s–curve and a t–curve have two intersection points Z_1 and Z_2 within $\hat{\mathcal{Q}}$, then, from the property of the s–curves and t–curves mentioned above, it follows that the whole segment between Z_1 and Z_2 of the s–curve lies within $\hat{\mathcal{Q}}$ and that there exists a point $\hat{Z}$ on this segment at which some t–curve is tangential to this s–curve. Then (12.2.15) is violated at $\hat{Z}$.

Thus, F_1 is one-to-one on $\hat{\mathcal{Q}}$. Then, from (12.2.13)–(12.2.15) and the properties of the s– and t–integral curves discussed above, it follows that assertions (ii)–(iii) hold and that (12.2.11) holds for the vector fields $\{\mathbf{e}_s, \mathbf{e}_t\}$ given in this step and map F defined above. Assertions (iv)–(vii) follow directly from the corresponding properties of $\{\mathbf{e}_s, \mathbf{e}_t\}$ and function $g^{(\theta_w)}(\cdot)$, as well as the rules of assigning values of s and t to the corresponding integral curves of $\mathbf{e}_s$ and $\mathbf{e}_t$ as described above. Specifically, (iv) follows from (12.2.16); (v) follows from the properties of $g^{(\theta_w)}(\cdot)$ shown above; (vi) follows from (12.2.17); (vii) is equivalent to (12.2.12), which follows from (12.2.21).

It remains to show that (i) holds. Using that $\mathcal{Q}_{\rm bd} \subset \hat{\mathcal{Q}}$ and $\Gamma_{\rm wedge} \subset \partial\mathcal{Q}_{\rm bd}$, it follows that $\mathcal{Q}_{\rm bd}$ is covered by the integral curves of $\mathbf{e}_t$ entering through $\Gamma_{\rm wedge}$. Moreover, from (12.2.5), it follows that $\partial\mathcal{Q}_{\rm bd}$ consists of $\Gamma_{\rm sonic}^{(\delta_*)}$, $\Gamma_{\rm wedge}$, $\{(\xi_1, 0) \;:\; a < \xi_1 < 0\}$ for some $a \in [-10C, 0)$, and curve $\Gamma_{\rm exit}$ in Λ connecting $\hat{P}_1$ and point $(a, 0)$. To describe curve $\Gamma_{\rm exit}$, we note that the point of intersection of line S_{1,δ_*} with the vertical line $\xi_1 = -10C$ has coordinates $(-10C, \xi_2^*)$ for some $\xi_2^* \in \mathbb{R}$. If $\xi_2^* \le 0$, then line S_{1,δ_*} intersects line $\{\xi_2 = 0\}$ within $\{(\xi_1, 0) \;:\; \xi_1 \in [-10C, 0)\}$, *i.e.*, at a point $(a, 0)$ with $a \in [-10C, 0)$ and, in this case, curve $\Gamma_{\rm exit}$ is the segment of line S_{1,δ_*} between points $\hat{P}_1$ and $(a, 0)$. If $\xi_2^* > 0$, then $a = -10C$, and $\Gamma_{\rm exit}$ consists of the segment of line S_{1,δ_*} between points $\hat{P}_1$ and $(-10C, \xi_2^*)$, and the segment of line $\{\xi_1 = -10C\}$ between points $(-10C, 0)$ and $(-10C, \xi_2^*)$. Then, from assertions (iv)–(vi), combined with (12.2.17) and (12.2.20)–(12.2.21), it follows that each integral curve of $\mathbf{e}_t$ in $\mathcal{Q}_{\rm bd}$ enters through $\Gamma_{\rm wedge}$, exists through $\Gamma_{\rm exit}$, and intersects each of these boundary curves only once, and that these intersections are transversal. This and assertion (iii) imply (i).

2. It remains to construct $\{\mathbf{e}_s, \mathbf{e}_t\}$ satisfying (12.2.13)–(12.2.22). We construct these vector fields separately near $\Gamma_{\rm sonic}^{(\delta_*)}$ (denoted by $\{\mathbf{e}_s^{(1)}, \mathbf{e}_t^{(1)}\}$) and away from $\Gamma_{\rm sonic}^{(\delta_*)}$ (denoted by $\{\mathbf{e}_s^{(2)}, \mathbf{e}_t^{(2)}\}$), and then glue $\{\mathbf{e}_s^{(1)}, \mathbf{e}_t^{(1)}\}$ with $\{\mathbf{e}_s^{(2)}, \mathbf{e}_t^{(2)}\}$ by a smooth interpolation.

Let Q be the point introduced in Definition 11.2.1. Then we note that the vector fields $\{\mathbf{e}_s^{(1)}, \mathbf{e}_t^{(1)}\} = \{\mathbf{e}_x, \mathbf{e}_y\}$ in domain $\{\theta_w < \theta < \theta_Q\} \cap \{0 < x < \varepsilon_0\}$ satisfy (12.2.13)–(12.2.17) from their definitions, the first inequality in (12.2.19), (12.2.20) by (11.2.5), and (12.2.21) by (11.2.7). The second inequality in (12.2.19) follows from (12.2.21) by rotating both vectors in (12.2.21) by $\frac{\pi}{2}$ clockwise, where we use the explicit form of $\{\mathbf{e}_s^{(1)}, \mathbf{e}_t^{(1)}\}$. Also, from the second estimate of (11.2.37), property (viii) holds for $\mathbf{e}_t^{(1)} = \mathbf{e}_y$ in

$\{\theta_{\rm w} < \theta < \theta_Q\} \cap \{0 < x < \varepsilon_0\}$ after reducing ε_0 if necessary. Also, property (ix) holds since $\boldsymbol{\nu}_{\rm w} = \mathbf{e}_y = \mathbf{e}_t^{(1)}$. Thus, $\{\mathbf{e}_s^{(1)}, \mathbf{e}_t^{(1)}\} = \{\mathbf{e}_x, \mathbf{e}_y\}$ in domain $\{\theta_{\rm w} < \theta < \theta_Q\} \cap \{0 < x < \varepsilon_0\}$ satisfy (12.2.13)–(12.2.17) and (12.2.19)–(12.2.22).

Now we construct $\{\mathbf{e}_s^{(2)}, \mathbf{e}_t^{(2)}\}$. We note that $\boldsymbol{\nu}_{\rm w} \in \text{Cone}^0(\mathbf{e}_{\mathcal{S}_1}, \mathbf{e}_{\xi_2})$ by Lemma 8.2.11, so that $\{\mathbf{e}_s^{(2,1)}, \mathbf{e}_t^{(2,1)}\} := \{-\boldsymbol{\tau}_{\rm w}, \boldsymbol{\nu}_{\rm w}\}$ satisfy properties (viii) and (ix) of map F_1, where we have used Corollary 8.2.10, (8.2.19), and the compactness argument by using Proposition 11.6.1 to obtain (viii). It is easy to see that $\{\mathbf{e}_s^{(2,1)}, \mathbf{e}_t^{(2,1)}\}$ also satisfy (12.2.13)–(12.2.15), (12.2.17), and (12.2.19)–(12.2.21), but do not satisfy (12.2.18).

On the other hand, $\{\mathbf{e}_s^{(2,2)}, \mathbf{e}_t^{(2,2)}\} := \{-\boldsymbol{\tau}_{\rm w}, -\mathbf{e}_{\xi_1}\}$ satisfy (12.2.13)–(12.2.15), (12.2.17)–(12.2.19), (12.2.21), and (ix), but do not necessarily satisfy (12.2.20) and property (viii) of map F_1. Indeed, if (12.2.20) holds for $\mathbf{e}_t^{(2,2)} = -\mathbf{e}_{\xi_1}$, then rotating vectors $\boldsymbol{\nu}_{\mathcal{S}_1}$ and $-\mathbf{e}_{\xi_1}$ by $\frac{\pi}{2}$ clockwise, we obtain that $\mathbf{e}_{\mathcal{S}_1} \cdot \mathbf{e}_{\xi_2} < 0$. It is possible that $\mathbf{e}_{\mathcal{S}_1} \cdot \mathbf{e}_{\xi_2} \geq 0$ for sufficiently small $\theta_{\rm w}$. In such a case, rotating vectors $\mathbf{e}_{\mathcal{S}_1}$ and $\mathbf{e}_{\xi_2}$ by $\frac{\pi}{2}$ counterclockwise, we obtain that $-\mathbf{e}_{\xi_1} \cdot \boldsymbol{\nu}_{\mathcal{S}_1} \geq 0$, *i.e.*, (12.2.20) does not hold. However, in this case, we see from (12.2.7) that $S_{1,\delta_*} \cap \overline{\mathcal{Q}} \subset \{\xi_2 > \xi_{2P_0}\}$. Thus, (12.2.20) holds on $S_{1,\delta_*} \cap \overline{\mathcal{Q}} \subset \{\xi_2 \leq \xi_{2P_0}\}$ for all $\theta_{\rm w}$. Also, if $\mathbf{e}_{\mathcal{S}_1} \cdot \mathbf{e}_{\xi_2} \geq 0$, then $-\mathbf{e}_{\xi_1} \notin \text{Cone}^0(\mathbf{e}_{\mathcal{S}_1}, \mathbf{e}_{\xi_2})$, so that it is not clear whether property (viii) of map F_1 holds. On the other hand, for any admissible solution, since $\Gamma_{\rm shock}^{\rm ext} \cap \{\xi_2 = 0\} = \{P_2\}$ and $\boldsymbol{\nu}_{\rm sh}(P_2) = (1,0) = \mathbf{e}_{\xi_1}$, we employ the uniform estimates (10.5.1) to conclude that there exists $\delta > 0$ such that, for every admissible solution, $\mathbf{e}_t^{(2,2)} \cdot \boldsymbol{\nu}_{\rm sh} = (-\mathbf{e}_{\xi_1}) \cdot \boldsymbol{\nu}_{\rm sh} \leq -\frac{1}{2}$ in $\{0 \leq \xi_2 < \delta\}$. We also note that $\mathbf{e}_t^{(2,2)} \cdot \boldsymbol{\tau}_{\rm w} = (-\mathbf{e}_{\xi_1}) \cdot \boldsymbol{\tau}_{\rm w} = -\cos\theta_{\rm w} < 0$.

Thus, we have shown that $\{\mathbf{e}_s^{(2,1)}, \mathbf{e}_t^{(2,1)}\}$ satisfy all the required properties in $\{\xi_2 > \frac{\delta}{2}\}$, and $\{\mathbf{e}_s^{(2,2)}, \mathbf{e}_t^{(2,2)}\}$ satisfy all the required properties in $\{0 \leq \xi_2 < \delta\}$.

Now we employ the cutoff function $\eta \in C^\infty(\mathbb{R})$ satisfying $0 \leq \eta \leq 1$ on $\mathbb{R}$, $\eta \equiv 1$ on $(\frac{3\delta}{4}, \infty)$, and $\eta \equiv 0$ on $(-\infty, \frac{\delta}{4})$, and define that, for all $\xi_1 \in \mathbb{R}$, $\xi_2 > 0$,

$$(\mathbf{e}_s^{(2)}, \mathbf{e}_t^{(2)})(\boldsymbol{\xi}) = \eta(\xi_2)(\mathbf{e}_s^{(2,1)}, \mathbf{e}_t^{(2,1)}) + (1 - \eta(\xi_2))(\mathbf{e}_s^{(2,2)}, \mathbf{e}_t^{(2,2)}).$$

Then $\{\mathbf{e}_s^{(2)}, \mathbf{e}_t^{(2)}\}$ satisfy (12.2.13)–(12.2.15) and (12.2.17)–(12.2.21), as well as properties (viii) and (ix) of map F_1, where we have used that $\mathbf{e}_s^{(2,1)} = \mathbf{e}_s^{(2,2)}$ in the proof of (12.2.15).

Using (12.2.9), we can combine $\{\mathbf{e}_s^{(1)}, \mathbf{e}_t^{(1)}\}$ with $\{\mathbf{e}_s^{(2)}, \mathbf{e}_t^{(2)}\}$ by interpolation within domain $\mathcal{Q} \cap \mathcal{N}_{\varepsilon_1}(\Gamma_{\rm sonic}^{(\delta_*)}) \cap \{\frac{\varepsilon_0}{2} < x < \varepsilon_0\}$, working in the (x,y)–coordinates, in the following way:

Let $\zeta \in C^\infty(\mathbb{R})$ satisfy $0 \leq \zeta \leq 1$ on $\mathbb{R}$, $\zeta \equiv 0$ on $(\frac{8\varepsilon_0}{9}, \infty)$, and $\zeta \equiv 1$ on $(-\infty, \frac{7\varepsilon_0}{9})$. We define that, for any $(x,y) \in \mathcal{Q} \cap \mathcal{N}_{\varepsilon_1}(\Gamma_{\rm sonic}^{(\delta_*)}) \cap \{0 < x < \varepsilon_0\}$,

$$(\mathbf{e}_s, \mathbf{e}_t)(x,y) := \zeta(x)(\mathbf{e}_s^{(1)}, \mathbf{e}_t^{(1)})(x,y) + (1 - \zeta(x))(\mathbf{e}_s^{(2)}, \mathbf{e}_t^{(2)})(x,y);$$

otherwise,

$$(\mathbf{e}_s, \mathbf{e}_t)(\boldsymbol{\xi}) := (\mathbf{e}_s^{(2)}, \mathbf{e}_t^{(2)})(\boldsymbol{\xi}).$$

These vector fields satisfy (12.2.13)–(12.2.21) and properties (viii) and (ix) of map F_1. This follows directly from the properties of $\{\mathbf{e}_s^{(1)}, \mathbf{e}_t^{(1)}\}$ and $\{\mathbf{e}_s^{(2)}, \mathbf{e}_t^{(2)}\}$, so that we need only to comment on the proof of (12.2.15).

On $\Gamma_{\text{sonic}}^{(\delta_*)}$, the unit vectors $\mathbf{e}_s^{(1)} = \mathbf{e}_x$ are in $\text{Cone}(\mathbf{e}_x(\hat{P}_1), \mathbf{e}_x(P_4))$, which is convex. We note that $\mathbf{e}_x(P_4) = -\boldsymbol{\tau}_{\text{w}}$. Since $\mathbf{e}_s^{(2)} = -\boldsymbol{\tau}_{\text{w}}$, then $\mathbf{e}_s$ on $\Gamma_{\text{sonic}}^{(\delta_*)}$ also varies within $\text{Cone}(\mathbf{e}_x(\hat{P}_1), \mathbf{e}_x(P_4))$. On the other hand, since $\mathbf{e}_t^{(1)} = \mathbf{e}_y$ are the unit vectors in the counterclockwise tangential to the $\Gamma_{\text{sonic}}^{(\delta_*)}$–directions (with $\mathbf{e}_y(P_4) = \boldsymbol{\nu}_{\text{w}}$) and $\mathbf{e}_t^{(2)} = \boldsymbol{\nu}_{\text{w}}$ on $\Gamma_{\text{sonic}}^{(\delta_*)}$, it follows that $\mathbf{e}_t$ on $\Gamma_{\text{sonic}}^{(\delta_*)}$ varies within the convex cone $\text{Cone}(\mathbf{e}_y(\hat{P}_1), \mathbf{e}_y(P_4))$. It is easy to see that, for any $\mathbf{e} \in \text{Cone}(\mathbf{e}_x(\hat{P}_1), \mathbf{e}_x(P_4))$ and $\mathbf{g} \in \text{Cone}(\mathbf{e}_y(\hat{P}_1), \mathbf{e}_y(P_4))$, $|\mathbf{e} \cdot \mathbf{g}^{\perp}| \geq \delta |\mathbf{e}|\,|\mathbf{g}|$ holds for $\delta = \cos(\angle P_4 \mathcal{O}_2 \hat{P}_1) > 0$. Then (12.2.15) follows on $\Gamma_{\text{sonic}}^{(\delta_*)}$ with constant $\delta = \cos(\angle P_4 \mathcal{O}_2 \hat{P}_1)$. Thus, for sufficiently small ε_0, property (12.2.15) holds with $\delta = \frac{1}{2}\cos(\angle P_4 \mathcal{O}_2 \hat{P}_1)$ in $\mathcal{Q} \cap \mathcal{N}_{\varepsilon_1}(\Gamma_{\text{sonic}}^{(\delta_*)}) \cap \{0 < x < \varepsilon_0\}$. Since $\cos(\angle P_4 \mathcal{O}_2 \hat{P}_1)$ and ε_0 chosen above depend smoothly on $\theta_{\text{w}} \in [\theta_{\text{w}}^{\text{s}}, \frac{\pi}{2}]$, we can choose $\delta > 0$ such that property (12.2.15) holds in $\mathcal{Q} \cap \mathcal{N}_{\varepsilon_1}(\Gamma_{\text{sonic}}^{(\delta_*)}) \cap \{0 < x < \varepsilon_0\}$ for each $\theta_{\text{w}} \in (\theta_{\text{w}}^{\text{s}}, \frac{\pi}{2}]$. Outside this region, $\{\mathbf{e}_s, \mathbf{e}_t\} = \{\mathbf{e}_s^{(2)}, \mathbf{e}_t^{(2)}\}$, so (12.2.15) holds. □

Lemma 12.2.3. *Let ε_1 be as in Lemma* 11.2.2 *and ε_0 as in Proposition* 11.2.8. *Let φ be an admissible solution for the wedge angle $\theta_{\text{w}} \in (\theta_{\text{w}}^*, \frac{\pi}{2})$, or the normal reflection solution for $\theta_{\text{w}} = \frac{\pi}{2}$. Then*

$$\begin{aligned} \tilde{\Omega} &:= F_1(\Omega) = \{(s,t) \,:\, 0 < s < \hat{s}(\theta_{\text{w}}),\ 0 < t < \mathfrak{g}_{\text{sh}}(s)\}, \\ F_1(\Gamma_{\text{shock}}) &= \{(s,t) \,:\, 0 < s < \hat{s}(\theta_{\text{w}}),\ t = \mathfrak{g}_{\text{sh}}(s)\}, \end{aligned} \tag{12.2.24}$$

with

$$\|\mathfrak{g}_{\text{sh}}\|_{C^{0,1}([0,\hat{s}(\theta_{\text{w}})])} \leq C, \tag{12.2.25}$$

$$\|\mathfrak{g}_{\text{sh}}\|_{C^3([\hat{\varepsilon},\hat{s}(\theta_{\text{w}})])} \leq C(\hat{\varepsilon}) \qquad \text{for all } \hat{\varepsilon} \in (0, \varepsilon_0]. \tag{12.2.26}$$

Moreover, denote $\mathcal{D}_{\varepsilon_0} := \{\varphi_2 < \varphi_1\} \cap \Lambda \cap \mathcal{N}_{\varepsilon_1}(\Gamma_{\text{sonic}}) \cap \{0 < x < \varepsilon_0\}$. Then

$$F_1(\mathcal{D}_{\varepsilon_0}) = \{(s,t) \,:\, 0 < s < \varepsilon_0,\ 0 < t < \mathfrak{g}_{\mathcal{S}_1}(s)\}, \tag{12.2.27}$$

$$\|\mathfrak{g}_{\mathcal{S}_1}\|_{C^3([0,\varepsilon_0])} \leq C, \tag{12.2.28}$$

$$C^{-1} \leq \mathfrak{g}_{\mathcal{S}_1}' \leq C \qquad \text{on } [0, \varepsilon_0), \tag{12.2.29}$$

and

$$\|\mathfrak{g}_{\text{sh}} - \mathfrak{g}_{\mathcal{S}_1}\|_{2,\alpha,(0,\varepsilon_0)}^{(\text{par})} \leq C(\alpha) \qquad \text{for any } \alpha \in [0, 1). \tag{12.2.30}$$

Furthermore,

$$C^{-1} \leq \mathfrak{g}_{\text{sh}}(s) \leq \eta^{(\theta_{\text{w}})}(s) - C^{-1} \qquad \text{on } (0, \hat{s}(\theta_{\text{w}})), \tag{12.2.31}$$

where $\eta^{(\theta_w)}(\cdot)$ *is from* (12.2.10). *Here the universal constant* C *depends only on the data and* θ_w^*, *and constant* $C(\hat{\varepsilon})$ (*resp.* $C(\alpha)$) *also depends on* $\hat{\varepsilon}$ (*resp.* α).

Proof. From properties (iv)–(viii) of F_1 in Lemma 12.2.2, it follows that each curve $\{s = const.\}$ in Ω intersects with $\Gamma_{\text{wedge}} \cup \Gamma_{\text{shock}}$ only once. Then, for each $P \in \Gamma_{\text{wedge}}$, we denote $\hat{P}$ as the point of intersection of curve $\{s = s(P)\}$ with Γ_{shock} and define $\mathfrak{g}_{\text{sh}}(s(P)) := t(\hat{P})$. This defines $\mathfrak{g}_{\text{sh}}(\cdot)$ on $[0, \hat{s}(\theta_w)]$ so that $F_1(\Omega)$ is of form (12.2.24). Also, (ii), (vi), and (viii) of Lemma 12.2.2 imply that $\mathfrak{g}_{\text{sh}} \in C^1((0, \hat{s}(\theta_w)))$.

Since $\Gamma_{\text{shock}} = \{\varphi = \varphi_1\}$, we differentiate the following equality in s:

$$(\varphi_1 - \varphi)(F_1^{-1}(s, \mathfrak{g}_{\text{sh}}(s))) = 0 \qquad \text{on } s \in (0, \hat{s}(\theta_w)) \tag{12.2.32}$$

to obtain

$$\mathfrak{g}_{\text{sh}}'(s) = -\frac{|\partial_s F_1^{-1}|\,(\mathbf{e}_s \cdot \boldsymbol{\nu}_{\text{sh}}) \circ F_1^{-1}}{|\partial_t F_1^{-1}|\,(\mathbf{e}_t \cdot \boldsymbol{\nu}_{\text{sh}}) \circ F_1^{-1}}(s, \mathfrak{g}_{\text{sh}}(s)).$$

From Lemma 12.2.2(ii), we have

$$|\partial_t F_1^{-1}| \geq \frac{1}{C^2}, \quad |\partial_s F_1^{-1}| \leq C \qquad \text{in } \overline{F_1(\mathcal{Q}_{\text{bd}})}.$$

Then, using Lemma 12.2.2(viii), we obtain (12.2.25). Taking further derivatives of (12.2.32), and using (10.5.2) and the higher derivative estimates in Lemma 12.2.2(ii), in addition to the estimates discussed above, we conclude (12.2.26).

Properties (12.2.27)–(12.2.29) follow directly from (11.2.8) and (11.2.10) of Lemma 11.2.2, and Lemma 12.2.2(iv). In fact, we obtain

$$\mathfrak{g}_{S_1}(\cdot) = \hat{f}_0(\cdot) - \theta_w,$$

where $\hat{f}_0$ is defined in (11.2.8).

Moreover, again using Lemma 12.2.2(iv), we see that

$$\mathfrak{g}_{\text{sh}}(\cdot) = \hat{f}(\cdot) - \theta_w \qquad \text{on } (0, \varepsilon_0),$$

where $\hat{f}$ is defined in (11.2.38) in Proposition 11.2.8. Then (12.2.30) follows directly from Proposition 11.4.6.

Finally, the lower bound in (12.2.31) follows from Lemma 10.4.2(iv) and Lemma 12.2.2(ii).

The upper bound in (12.2.31) can be seen as follows: Since $\Omega \subset \mathcal{Q}_{\text{bd}}$, then it follows from (12.2.10) and (12.2.24) that

$$\mathfrak{g}_{\text{sh}}(s) \leq \eta(s) \qquad \text{on } (0, \hat{s}(\theta_w)).$$

It remains to show that these two functions are separated. We first note that, in terms of Definition 12.2.1,

$$\Omega \subset \mathcal{Q}^{(0)} \cap \{\xi_1 > -C\}, \tag{12.2.33}$$

where C is from (9.1.2), *i.e.*, the same constant C as in (12.2.5). We also note that, using notation (12.2.1), we have

$$\mathrm{dist}(S_{1,\delta_*}, S_{1,0}) = \frac{\delta^*}{\sqrt{(u_1-u_2)^2+v_2^2}} \geq \frac{1}{\tilde{C}} \qquad \text{for any } \theta_{\mathrm{w}} \in [\theta_{\mathrm{w}}^{\mathrm{s}}, \tfrac{\pi}{2}],$$

where $\tilde{C}$ depends only on the data. Now the upper bound in (12.2.31) follows from (12.2.5), (12.2.33), and Lemma 12.2.2(ii). □

12.2.2 Definition and properties of $F_{(2,\mathfrak{g}_{\mathrm{sh}})}$ and u

In order to define an iteration set, we need to take into account estimate (11.4.4) for $\psi = \varphi - \varphi_2$ near Γ_{sonic}, and perform an iteration in terms of ψ. This is convenient because φ is close to φ_2 near Γ_{sonic}. Also, for any wedge angle close to $\frac{\pi}{2}$, solution φ is close to φ_2 in the whole region $\Omega := \Omega(\varphi)$, as shown in [54]. Thus, for such wedge angles, it is convenient to consider ψ in the whole region Ω. However, φ is not close to φ_2 away from Γ_{sonic} when the wedge angles are away from $\frac{\pi}{2}$, so it is not convenient to consider ψ away from Γ_{sonic}, since it does not have certain monotonicity properties. Specifically, it does not satisfy property (12.2.37) below. Then it is more convenient to consider $\phi = \varphi + \frac{|\boldsymbol{\xi}|^2}{2} \equiv \varphi - \varphi_0$ away from Γ_{sonic} for the wedge angles away from $\frac{\pi}{2}$. This motivates us to define a function $\tilde{\varphi}_2$ below and employ the new function $\varphi - \tilde{\varphi}_2$ to define the iteration set.

Now we define $\tilde{\varphi}_2$. In the following lemma, for each $\theta_{\mathrm{w}} \in (\theta_{\mathrm{w}}^{\mathrm{s}}, \frac{\pi}{2}]$, we use the (x,y)–coordinates near Γ_{sonic} and $\mathcal{D}_{\delta_*,\varepsilon} := \mathcal{D}_{\delta_*,\varepsilon}^{(\theta_{\mathrm{w}})}$ for each $\varepsilon \in (0, \varepsilon_0]$, introduced in (12.2.8).

Lemma 12.2.4. *For each $\theta_{\mathrm{w}} \in (\theta_{\mathrm{w}}^{\mathrm{s}}, \frac{\pi}{2}]$, there exists a function $\tilde{\varphi}_2(\boldsymbol{\xi}) \equiv \tilde{\varphi}_2^{(\theta_{\mathrm{w}})}(\boldsymbol{\xi})$ on $\mathcal{Q}_{\mathrm{bd}}$ such that*

$$\tilde{\varphi}_2^{(\theta_{\mathrm{w}})} = \varphi_2^{(\theta_{\mathrm{w}})} \qquad \text{in } \overline{\mathcal{Q}_{\mathrm{bd}}} \text{ for all } \theta_{\mathrm{w}} \in [\frac{\pi}{2} - \mu, \frac{\pi}{2}], \tag{12.2.34}$$

$$\tilde{\varphi}_2^{(\theta_{\mathrm{w}})} = \varphi_2^{(\theta_{\mathrm{w}})} \qquad \text{in } \mathcal{D}_{\delta_*,\varepsilon_0/2} \text{ for all } \theta_{\mathrm{w}} \in [\theta_{\mathrm{w}}^{\mathrm{s}}, \frac{\pi}{2}], \tag{12.2.35}$$

$$\partial_{\boldsymbol{\nu}_{\mathrm{w}}} \tilde{\varphi}_2 = 0 \qquad \text{on } \Gamma_{\mathrm{wedge}}, \tag{12.2.36}$$

$$\frac{\partial((\varphi_1 - \tilde{\varphi}_2) \circ F_1^{-1})(s,t)}{\partial t} \leq -\delta \qquad \text{for all } (s,t) \in F_1(\mathcal{Q}_{\mathrm{bd}}), \tag{12.2.37}$$

and $\tilde{\varphi}_2^{(\theta_{\mathrm{w}})}(\cdot)$ depends smoothly on θ_{w} in the sense that function $\tilde{\varphi}_2(\theta_{\mathrm{w}}, \boldsymbol{\xi}) := \tilde{\varphi}_2^{(\theta_{\mathrm{w}})}(\boldsymbol{\xi})$ satisfies

$$\|\tilde{\varphi}_2\|_{C^3(\overline{\mathcal{B}})} \leq C, \tag{12.2.38}$$

where $\mathcal{B} = \cup_{\theta_{\mathrm{w}} \in [\theta_{\mathrm{w}}^{\mathrm{s}}, \frac{\pi}{2}]} \{\theta_{\mathrm{w}}\} \times \mathcal{Q}_{\mathrm{bd}}^{(\theta_{\mathrm{w}})}$, and constants $\mu, \delta > 0$ and C depend only on the data.

Proof. We divide the proof into two steps.

1. For each $\theta_{\rm w} \in [\theta_{\rm w}^{\rm s}, \frac{\pi}{2}]$, we express φ_0 and φ_2 in the (x,y)–coordinates in the neighborhood $\mathcal{D}_{\delta_*,\varepsilon_0}$ of $\Gamma_{\rm sonic}$ to introduce the following function on $\mathcal{Q}_{\rm bd}$:

$$\begin{cases} \tilde\varphi_{20}(x,y) := \zeta_1(x)\varphi_0(x,y) + (1-\zeta_1(x))\varphi_2(x,y) & \text{for } (x,y)\in\mathcal{D}_{\delta_*,\varepsilon_0}, \\ \tilde\varphi_{20} = \varphi_0 & \text{otherwise}, \end{cases} \tag{12.2.39}$$

where $\zeta_1 \in C^\infty(\mathbb{R})$ with $0 \le \zeta_1 \le 1$ on $\mathbb{R}$, $\zeta_1 \equiv 0$ on $(-\infty, \frac{\varepsilon_0}{2})$, and $\zeta_1 \equiv 1$ on $(\frac{3\varepsilon_0}{4}, \infty)$.

From (12.2.8), it follows that coordinates (x,y) are smooth in $\mathcal{D}_{\delta_*,\varepsilon_0}$. Since φ_0 and φ_2 are polynomial functions in the $\boldsymbol{\xi}$–coordinates, then we employ the C^∞–smooth dependence of parameters (u_2, v_2) of the weak state (2) on $\theta_{\rm w} \in (\theta_{\rm w}^{\rm d}, \frac{\pi}{2}]$ in Theorem 7.1.1(i) to conclude that $\tilde\varphi_{20}$ satisfies (12.2.38). Also, from its definition, $\tilde\varphi_{20}$ satisfies (12.2.35).

Furthermore, since $\partial_{\boldsymbol{\nu}_{\rm w}}\varphi_0 = 0$ on $\Gamma_{\rm wedge}$ by (2.2.16), $\partial_{\boldsymbol{\nu}_{\rm w}}\varphi_2 = 0$ on $\Gamma_{\rm wedge}$ from its construction in Chapter 7, and $\partial_{\boldsymbol{\nu}_{\rm w}}\varphi_k = \frac{1}{c_2 - x}\partial_y\varphi_k$ on $\Gamma_{\rm wedge}\cap\partial\mathcal{D}_{\delta_*,\varepsilon_0}$ for $k = 0, 2$, it follows from (12.2.39) that (12.2.36) holds for $\tilde\varphi_{20}$.

Now we show that $\tilde\varphi_{20}$ satisfies (12.2.37). From (12.2.39), it follows that $\varphi_1 - \tilde\varphi_{20} = \varphi_1 - \varphi_0 = \phi_1$ in $\mathcal{Q}_{\rm bd}\setminus\mathcal{D}_{\delta_*, 3\varepsilon_0/4}$ so that, from Lemma 12.2.2(iv),

$$(\varphi_1 - \tilde\varphi_{20})(F_1^{-1}(s,t)) = \phi_1(F_1^{-1}(s,t)) \qquad \text{in } F_1(\mathcal{Q}_{\rm bd})\cap\{s > \frac{3\varepsilon_0}{4}\}.$$

Then, from Lemma 12.2.2(vii), it follows that (12.2.37) holds in $F_1(\mathcal{Q}_{\rm bd})\cap\{s > \frac{3\varepsilon_0}{4}\}$.

It remains to show (12.2.37) in $F_1(\mathcal{Q}_{\rm bd})\cap\{s < \frac{3\varepsilon_0}{4}\}$. From (12.2.39) and Lemma 12.2.2(i)–(ii) and (iv), *i.e.*, using that the (s,t)–coordinates coincide with the (x,y)–coordinates in that domain, we see that, in $F_1(\mathcal{Q}_{\rm bd})\cap\{s < \frac{3\varepsilon_0}{4}\}$,

$$\begin{aligned} &(\varphi_1 - \tilde\varphi_{20})(F_1^{-1}(s,t)) \\ &= \zeta_1(s)\phi_1(F_1^{-1}(s,t)) + (1-\zeta_1(s))(\varphi_1-\varphi_2)_{|(x,y)=(s,t+\theta_{\rm w})} \\ &=: I_1 + I_2. \end{aligned} \tag{12.2.40}$$

Since $0 \le \zeta_1 \le 1$, then $\frac{\partial I_1}{\partial t}(s,t) \le -\zeta_1(s)\delta$ by Lemma 12.2.2(vii), and

$$\frac{\partial I_2}{\partial t}(s,t) = (1-\zeta_1(s))\,\partial_y(\varphi_1-\varphi_2)|_{(x,y)=(s,t+\theta_{\rm w})} \le -(1-\zeta_1(s))\delta$$

by (11.2.5). Thus, $\tilde\varphi_{20}$ satisfies (12.2.37).

Therefore, $\tilde\varphi_{20}$ satisfies (12.2.35)–(12.2.38), but has not satisfied (12.2.34).

2. Since $(u_2, v_2) = (0,0)$ at $\theta_{\rm w} = \frac{\pi}{2}$, *i.e.*,

$$\varphi_2^{(\frac{\pi}{2})} = -\frac{|\boldsymbol{\xi}|^2}{2} + const. = \varphi_0 + const.,$$

we have

$$\varphi_1 - \varphi_2^{(\frac{\pi}{2})} = \phi_1 + const.,$$

so that, for any $(s,t) \in F_1(\mathcal{Q}_{\rm bd})$,

$$\frac{\partial((\varphi_1 - \varphi_2^{(\frac{\pi}{2})}) \circ F_1^{-1})(s,t)}{\partial t} = \frac{\partial(\phi_1 \circ F_1^{-1})(s,t)}{\partial t} \le -\delta$$

by Lemma 12.2.2(vii). The continuous dependence of the parameters of state (2) on $\theta_{\rm w}$ and Lemma 12.2.2(iii) imply the existence of $\mu > 0$ such that

$$\frac{\partial((\varphi_1 - \varphi_2^{(\theta_{\rm w})}) \circ F_1^{-1})(s,t)}{\partial t} \le -\frac{\delta}{2}$$

for any $\theta_{\rm w} \in [\frac{\pi}{2} - 2\mu, \frac{\pi}{2}]$ and $(s,t) \in F_1^{(\theta_{\rm w})}(\mathcal{Q}_{\rm bd})$.

Let $\zeta_2 \in C^\infty(\mathbb{R})$ satisfy $0 \le \zeta_2 \le 1$ on $\mathbb{R}$ with $\zeta_2 \equiv 0$ on $(-\infty, \frac{\pi}{2} - 2\mu)$ and $\zeta_2 \equiv 1$ on $(\frac{\pi}{2} - \mu, \infty)$. Then the function:

$$\tilde{\varphi}_2^{(\theta_{\rm w})}(\boldsymbol{\xi}) = \zeta_2(\theta_{\rm w})\varphi_2^{(\theta_{\rm w})}(\boldsymbol{\xi}) + (1 - \zeta_2(\theta_{\rm w}))\tilde{\varphi}_{20}^{(\theta_{\rm w})}(\boldsymbol{\xi})$$

satisfies all the properties required. This completes the proof. □

Now, for $\theta_{\rm w} \in [\theta_{\rm w}^*, \frac{\pi}{2}]$ and region $\tilde{\Omega}_\infty = (0, \hat{s}(\theta_{\rm w})) \times (0, \infty)$, if (12.2.31) holds, we define a map $F_{(2,\mathfrak{g}_{\rm sh},\theta_{\rm w})} \equiv F_{(2,\mathfrak{g}_{\rm sh})} : \tilde{\Omega}_\infty \mapsto \mathbb{R}^2$ by

$$F_{(2,\mathfrak{g}_{\rm sh})}(s,t) = (\frac{s}{\hat{s}(\theta_{\rm w})}, \frac{t}{\mathfrak{g}_{\rm sh}(s)}) \qquad \text{for } (s,t) \in \tilde{\Omega}_\infty. \tag{12.2.41}$$

Then, for $\tilde{\Omega}$ defined by (12.2.24),

$$F_{(2,\mathfrak{g}_{\rm sh})}(\tilde{\Omega}) = (0,1)^2 =: Q^{\rm iter}, \tag{12.2.42}$$

and $F_{(2,\mathfrak{g}_{\rm sh})}$ on $\tilde{\Omega}$ is invertible with $F^{-1}_{(2,\mathfrak{g}_{\rm sh})} : Q^{\rm iter} \mapsto \tilde{\Omega}$ given by

$$F^{-1}_{(2,\mathfrak{g}_{\rm sh})}(s,t) = (\hat{s}(\theta_{\rm w})\, s, \mathfrak{g}_{\rm sh}(\hat{s}(\theta_{\rm w})\, s)\, t). \tag{12.2.43}$$

Clearly, the regularity of $F_{(2,\mathfrak{g}_{\rm sh})}$ and $F^{-1}_{(2,\mathfrak{g}_{\rm sh})}$ is determined by the regularity of $\mathfrak{g}_{\rm sh}(\cdot)$.

Therefore, given an admissible solution φ related to a wedge angle $\theta_{\rm w}$, we obtain a function

$$u = (\varphi - \tilde{\varphi}_2) \circ F_1^{-1} \circ F^{-1}_{(2,\mathfrak{g}_{\rm sh})} \qquad \text{on } Q^{\rm iter}, \tag{12.2.44}$$

where function $\mathfrak{g}_{\rm sh}$ is defined in Lemma 12.2.3, and function $\tilde{\varphi}_2$ and maps F_1^{-1} and $F^{-1}_{(2,\mathfrak{g}_{\rm sh})}$ correspond to $\theta_{\rm w}$.

Define

$$\varepsilon_0' := \frac{\varepsilon_0}{\max_{\theta_{\rm w} \in [\theta_{\rm w}^{\rm s}, \frac{\pi}{2}]} \hat{s}(\theta_{\rm w})}, \tag{12.2.45}$$

where ε_0 is from Lemma 12.2.2, and the continuity of $\hat{s}(\cdot)$ on $[\theta_{\rm w}^{\rm s}, \frac{\pi}{2}]$ has been applied. Then $\varepsilon_0' \in (0, \frac{1}{2})$ depends only on the data.

Proposition 12.2.5. *There exists $M > 0$ depending only on the data and $\theta_{\rm w}^*$ such that the following holds: Let $\bar{\alpha}$ be constant α in Lemma 10.5.2, and let φ be an admissible solution for $\theta_{\rm w} \in [\theta_{\rm w}^*, \frac{\pi}{2}]$. Then function u defined by (12.2.44) satisfies that, for any $\alpha \in [0, \bar{\alpha}]$,*

$$\|u\|^{(-1-\alpha),\{(1,0)\}}_{2,\alpha,Q^{\rm iter}\cap\{s>\varepsilon_0'/10\}} + \|u\|^{(\rm par)}_{2,\alpha,Q^{\rm iter}\cap\{s<\varepsilon_0'\}} \le M, \tag{12.2.46}$$

where ε_0' is from (12.2.45), norms $\|\cdot\|^{(\rm par)}_{2,\alpha,A}$ are with respect to $\{s=0\}$, i.e., (s,t) replace $\mathbf{x}$ in (4.6.2) in our case here, and $(0,1) = F_{(2,\mathfrak{g}_{\rm sh})} \circ F_1(P_3)$ by Lemma 12.2.2(v) and (12.2.41).

Proof. For $\alpha = \bar{\alpha}$, (12.2.46) directly follows from Corollary 11.4.7, combined with the C^3–regularity of map F_1^{-1} and the explicit form (12.2.43) of map $F^{-1}_{(2,\mathfrak{g}_{\rm sh})}$, in which $\mathfrak{g}_{\rm sh}(\cdot)$ has the properties proved in Lemma 12.2.3.

With this, using Lemmas 4.1.2 and 4.6.4, we obtain (12.2.46) for all $\alpha \in [0, \bar{\alpha}]$, with a modified constant depending on the same parameters. □

12.2.3 Inverting the mapping: $(\varphi, \theta_{\rm w}) \mapsto u$

Now we invert the mapping: $(\varphi, \theta_{\rm w}) \mapsto u$ in the sense that, given $\theta_{\rm w}$ and a function u on $Q^{\rm iter}$ satisfying certain properties, we recover $\mathfrak{g}_{\rm sh}$, Ω, and φ on Ω so that (12.2.44) holds.

Fix $\theta_{\rm w} \in [\theta_{\rm w}^*, \frac{\pi}{2})$. From Lemma 12.2.2(iv), it follows that

$$\Gamma^{(\delta_*)}_{\rm sonic} = \partial Q^{(\theta_{\rm w})}_{\rm bd} \cap F_1^{-1}(\{s=0\}).$$

Since, for point $\hat{P}_1$ in (12.2.3),

$$\hat{P}_1 = F_1^{-1}(0, \theta_{\hat{P}_1} - \theta_{\rm w}), \qquad P_4 = F_1^{-1}(\mathbf{0}),$$

we find from (12.2.1), (12.2.3), and Lemma 12.2.4 that

$$\tilde{\varphi}_2^{(\theta_{\rm w})} = \varphi_2^{(\theta_{\rm w})} \qquad \text{on } \Gamma^{(\delta_*)}_{\rm sonic},$$

so that

$$\begin{aligned}
-\delta^* &= (\varphi_1 - \tilde{\varphi}_2^{(\theta_{\rm w})})(\hat{P}_1) \\
&= \inf_{\partial Q^{(\theta_{\rm w})}_{\rm bd} \cap F_1^{-1}(\{s=0\})} (\varphi_1 - \tilde{\varphi}_2^{(\theta_{\rm w})}) \\
&< 0 \\
&= (\varphi_1 - \tilde{\varphi}_2^{(\theta_{\rm w})})(P_1) \\
&< (\varphi_1 - \tilde{\varphi}_2^{(\theta_{\rm w})})(P_4) \\
&= \sup_{\partial Q^{(\theta_{\rm w})}_{\rm bd} \cap F_1^{-1}(\{s=0\})} (\varphi_1 - \tilde{\varphi}_2^{(\theta_{\rm w})}),
\end{aligned}$$

that is,

$$\inf_{\partial\mathcal{Q}_{\rm bd}^{(\theta_{\rm w})}\cap F_1^{-1}(\{s=0\})}(\varphi_1-\tilde{\varphi}_2^{(\theta_{\rm w})})<0<\sup_{\partial\mathcal{Q}_{\rm bd}^{(\theta_{\rm w})}\cap F_1^{-1}(\{s=0\})}(\varphi_1-\tilde{\varphi}_2^{(\theta_{\rm w})}). \quad (12.2.47)$$

Definition 12.2.6. *For $\theta_{\rm w}\in[\theta_{\rm w}^*,\frac{\pi}{2}]$ and $s^*\in(0,\hat{s}(\theta_{\rm w}))$, denote*

$$\mathcal{Q}_{\rm bd}^{(\theta_{\rm w})}(s^*):=\mathcal{Q}_{\rm bd}^{(\theta_{\rm w})}\cap F_1^{-1}(\{s=s^*\}). \quad (12.2.48)$$

Let $u\in C^1(\overline{Q^{\rm iter}})$ and $\theta_{\rm w}\in[\theta_{\rm w}^*,\frac{\pi}{2}]$ satisfy that, for any $s^*\in(0,\hat{s}(\theta_{\rm w})]$,

$$\inf_{\mathcal{Q}_{\rm bd}^{(\theta_{\rm w})}(s^*)}(\varphi_1-\tilde{\varphi}_2^{(\theta_{\rm w})})<u(\frac{s^*}{\hat{s}(\theta_{\rm w})},1)<\sup_{\mathcal{Q}_{\rm bd}^{(\theta_{\rm w})}(s^*)}(\varphi_1-\tilde{\varphi}_2^{(\theta_{\rm w})}). \quad (12.2.49)$$

Let $\alpha\in[0,1)$. Consider the set:

$$\mathfrak{S}\equiv\mathfrak{S}^{(\alpha)}:=\left\{(u,\theta_{\rm w})\in C^{1,\alpha}(\overline{Q^{\rm iter}})\times[\theta_{\rm w}^*,\frac{\pi}{2}]\ :\ \begin{array}{l}(u,\theta_{\rm w})\ \textit{satisfy}\ (12.2.49)\\ (u,Du)(0,\cdot)=(0,\mathbf{0})\end{array}\right\}. \quad (12.2.50)$$

Furthermore, for each $\theta_{\rm w}\in[\theta_{\rm w}^*,\frac{\pi}{2}]$, we define

$$\mathfrak{S}(\theta_{\rm w}):=\{u\in C^{1,\alpha}(\overline{Q^{\rm iter}})\ :\ (u,\theta_{\rm w})\in\mathfrak{S}\}. \quad (12.2.51)$$

Note that $\mathfrak{S}(\theta_{\rm w})$ is non-empty for any $\theta_{\rm w}\in[\theta_{\rm w}^*,\frac{\pi}{2}]$ on account of (12.2.37), (12.2.47), and Lemma 12.2.2(i).

For every $(u,\theta_{\rm w})\in\mathfrak{S}$, we define:

(i) *A function $\mathfrak{h}_{\rm sh}:(0,1)\mapsto\mathbb{R}_+$ by setting $\mathfrak{h}_{\rm sh}(s^*)=t^*$, where $(\hat{s}(\theta_{\rm w})s^*,t^*)$ is the unique point on $F_1(\mathcal{Q}_{\rm bd}^{(\theta_{\rm w})})\cap\{s=\hat{s}(\theta_{\rm w})s^*\}$ such that*

$$(\varphi_1-\tilde{\varphi}_2^{(\theta_{\rm w})})(F_1^{-1}(\hat{s}(\theta_{\rm w})s^*,t^*))=u(s^*,1).$$

Note that such a point exists by (12.2.49) and is unique by (12.2.10) and Lemma 12.2.4.

(ii) *A function $\mathfrak{g}_{\rm sh}:(0,\hat{s}(\theta_{\rm w}))\mapsto\mathbb{R}_+$ by setting $\mathfrak{g}_{\rm sh}(s)=\mathfrak{h}_{\rm sh}(\frac{s}{\hat{s}(\theta_{\rm w})})$.*

(iii) *A map $\mathfrak{F}:Q^{\rm iter}\mapsto\Lambda(\theta_{\rm w})$ defined by*

$$\mathfrak{F}=F_1^{-1}\circ F_{(2,\mathfrak{g}_{\rm sh})}^{-1}. \quad (12.2.52)$$

(iv) *Two sets:*

$$\Gamma_{\rm shock}:=F_1^{-1}\left(\{(s,\mathfrak{g}_{\rm sh}(s))\ :\ 0<s<\hat{s}(\theta_{\rm w})\}\right)\equiv\mathfrak{F}((0,1)\times\{1\}),$$
$$\Omega:=F_1^{-1}\left(\{(s,t)\ :\ 0<s<\hat{s}(\theta_{\rm w}),\ 0<t<\mathfrak{g}_{\rm sh}(s)\}\right)\equiv\mathfrak{F}(Q^{\rm iter}).$$

(v) *A function* φ *in* Ω *by setting*

$$\begin{aligned}\varphi(\boldsymbol{\xi}) &= u(F_{(2,\mathfrak{g}_{\rm sh})} \circ F_1(\boldsymbol{\xi})) + \tilde{\varphi}_2^{(\theta_{\rm w})}(\boldsymbol{\xi}) \\ &\equiv u(\mathfrak{F}^{-1}(\boldsymbol{\xi})) + \tilde{\varphi}_2^{(\theta_{\rm w})}(\boldsymbol{\xi}) \qquad \textit{for all } \boldsymbol{\xi} \in \Omega(u).\end{aligned}$$

(vi) *Let* $b, M > 0$. *Consider* $(u, \theta_{\rm w}) \in \mathfrak{S}^{(\alpha)}$ *satisfying*

$$b < \mathfrak{g}_{\rm sh}^{(u,\theta_{\rm w})}(\cdot) < \frac{1}{b} \qquad \textit{on } (0, \hat{s}(\theta_{\rm w})), \tag{12.2.53}$$

$$\|u\|_{C^{1,\alpha}(\overline{Q^{\rm iter}})} < M, \tag{12.2.54}$$

and define

$$\mathfrak{S}_{b,M}^{(\alpha)} = \big\{(u, \theta_{\rm w}) \in \mathfrak{S} \;:\; (12.2.53)\text{–}(12.2.54) \textit{ hold}\big\}. \tag{12.2.55}$$

We will write $\tilde{\varphi}_2$ for $\tilde{\varphi}_2^{(\theta_{\rm w})}$ when $\theta_{\rm w}$ is fixed.

Lemma 12.2.7. *Let* $\alpha \in [0, 1)$ *and* $\mathfrak{S} = \mathfrak{S}^{(\alpha)}$. *Then, for all* $(u, \theta_{\rm w}) \in \mathfrak{S}$, *the following properties hold:*

(i) *Let* $(u, \theta_{\rm w}) \in \mathfrak{S}$. *Then* $\mathfrak{h}_{\rm sh} \in C^{1,\alpha}([0, 1])$ *and* $\mathfrak{g}_{\rm sh} \in C^{1,\alpha}([0, \hat{s}(\theta_{\rm w})])$.

(ii) *Let* $(u, \theta_{\rm w}) \in \mathfrak{S}$. *Then* $\Gamma_{\rm shock}(u, \theta_{\rm w})$ *and* $\Omega(u, \theta_{\rm w})$ *introduced in Definition* 12.2.6(iv) *satisfy:*

(a). $\Omega \cup \Gamma_{\rm shock} \subset \mathcal{Q}_{\rm bd}^{(\theta_{\rm w})} \subset \Lambda(\theta_{\rm w})$, *and* $\Gamma_{\rm shock}$ *is a* $C^{1,\alpha}$*–curve up to its endpoints* ${P_1}^{(\theta_{\rm w})}$ *and* P_2, *for* ${P_1}^{(\theta_{\rm w})}$ *introduced in Definition* 7.5.7 *and* $P_2 \in \{\xi_2 = 0\}$;

(b). $P_1 = F_1^{-1}(0, \mathfrak{g}_{\rm sh}(\hat{s}(\theta_{\rm w})))$ *and* $P_2 = F_1^{-1}(\hat{s}(\theta_{\rm w}), \mathfrak{g}_{\rm sh}(\hat{s}(\theta_{\rm w})))$;

(c). *Curve* $\Gamma_{\rm shock}$ *is tangential to line* S_1 *at* P_1 *so that*

$$\mathfrak{g}_{\rm sh}^{(u,\theta_{\rm w})}(0) = y_{P_1} - y_{P_4} \equiv y_{P_1} - \theta_{\rm w}, \quad (\mathfrak{g}_{\rm sh}^{(u,\theta_{\rm w})})'(0) = \hat{f}_0'(0), \tag{12.2.56}$$

where $\hat{f}_0 \equiv \hat{f}_0^{(\theta_{\rm w})}$ *is defined in* (11.2.8).

In particular, $\mathfrak{g}_{\rm sh}^{(u,\theta_{\rm w})}(0)$ *and* $(\mathfrak{g}_{\rm sh}^{(u,\theta_{\rm w})})'(0)$ *are uniquely defined by* $\theta_{\rm w}$ *and independent of* u.

(iii) $\partial\Omega(u, \theta_{\rm w})$ *consists of the line segments and curves* $\Gamma_{\rm wedge}$, $\Gamma_{\rm sonic}$, $\Gamma_{\rm shock}(u, \theta_{\rm w})$, *and* $\Gamma_{\rm sym}(u, \theta_{\rm w}) := \{(\xi_1, 0) \;:\; \xi_{1P_2} < \xi_1 < 0\}$ *so that*

$$\begin{aligned}\Gamma_{\rm shock} &= \mathfrak{F}((0,1) \times \{1\}), \quad &\Gamma_{\rm sonic} &= \mathfrak{F}(\{0\} \times (0,1)), \\ \Gamma_{\rm wedge} &= \mathfrak{F}((0,1) \times \{0\}), \quad &\Gamma_{\rm sym} &= \mathfrak{F}(\{1\} \times (0,1)).\end{aligned} \tag{12.2.57}$$

These segments do not intersect at the points of their relative interiors.

(iv) *For* ε_0 *defined in Lemma* 12.2.2, *region* $\Omega\cap\mathcal{D}^{(\theta_{\rm w})}_{\delta_*,\varepsilon}$ *in the* (x,y)*–coordinates is of structure* (11.2.38)–(11.2.39) *with* $\hat f=\mathfrak{g}_{\rm sh}^{(u,\theta_{\rm w})}+\theta_{\rm w}$, *where region* $\mathcal{D}^{(\theta_{\rm w})}_{\delta_*,\varepsilon}$ *is defined by* (12.2.8) *for* δ_* *chosen in* (12.2.2).

(v) *For any* $(u,\theta_{\rm w})\in\mathfrak{S}^{(\alpha)}$, *the corresponding function* $\psi=\varphi-\varphi_2$ *satisfies*

$$\psi=0 \qquad on\ \Gamma_{\rm sonic}.$$

(vi) *Let* $M>0$. *There exists* C *depending only on the data and* (M,α) *such that, for any* $(u,\theta_{\rm w})$ *and* $(\tilde u,\tilde\theta_{\rm w})\in\mathfrak{S}^{(\alpha)}$ *with* $\|(u,\tilde u)\|_{C^{1,\alpha}(\overline{Q^{\rm iter}})}<M$, *the following estimates hold:*

$$\|\mathfrak{h}_{\rm sh}\|_{C^{1,\alpha}([0,1])}+\|\mathfrak{g}_{\rm sh}\|_{C^{1,\alpha}([0,\hat s(\theta_{\rm w})])}\le C, \tag{12.2.58}$$

$$\begin{aligned}&\|\mathfrak{h}_{\rm sh}^{(u,\theta_{\rm w})}-\mathfrak{h}_{\rm sh}^{(\tilde u,\tilde\theta_{\rm w})}\|_{C^{1,\alpha}([0,1])}\\&\quad\le C\big(\|(u-\tilde u)(\cdot,1)\|_{C^{1,\alpha}([0,1])}+|\theta_{\rm w}-\tilde\theta_{\rm w}|\big),\end{aligned} \tag{12.2.59}$$

$$\|\mathfrak{F}_{(u,\theta_{\rm w})}\|_{C^{1,\alpha}(\overline{Q^{\rm iter}})}\le C, \tag{12.2.60}$$

$$\begin{aligned}&\|\mathfrak{F}_{(u,\theta_{\rm w})}-\mathfrak{F}_{(\tilde u,\tilde\theta_{\rm w})}\|_{C^{1,\alpha}(\overline{Q^{\rm iter}})}\\&\quad\le C\big(\|(u-\tilde u)(\cdot,1)\|_{C^{1,\alpha}([0,1])}+|\theta_{\rm w}-\tilde\theta_{\rm w}|\big),\end{aligned} \tag{12.2.61}$$

$$\begin{aligned}&\|(\varphi\circ\mathfrak{F})_{(u,\theta_{\rm w})}-(\varphi\circ\mathfrak{F})_{(\tilde u,\tilde\theta_{\rm w})}\|_{C^{1,\alpha}(\overline{Q^{\rm iter}})}\\&\quad\le C\big(\|u-\tilde u\|_{C^{1,\alpha}(\overline{Q^{\rm iter}})}+|\theta_{\rm w}-\tilde\theta_{\rm w}|\big),\end{aligned} \tag{12.2.62}$$

$$\begin{aligned}&\|(\psi\circ\mathfrak{F})_{(u,\theta_{\rm w})}-(\psi\circ\mathfrak{F})_{(\tilde u,\tilde\theta_{\rm w})}\|_{C^{1,\alpha}(\overline{Q^{\rm iter}})}\\&\quad\le C\big(\|u-\tilde u\|_{C^{1,\alpha}(\overline{Q^{\rm iter}})}+|\theta_{\rm w}-\tilde\theta_{\rm w}|\big).\end{aligned} \tag{12.2.63}$$

(vii) *Let* $\alpha\in[0,1)$, $\sigma\in(1,2]$, *and* $M>0$. *There exist* C *depending only on the data and* (α,σ), *and* C_0 *depending only on the data, such that the following hold: Let* $(u,\theta_{\rm w})\in\mathfrak{S}$ *with* $\theta_{\rm w}\in[\theta^*_{\rm w},\frac{\pi}{2}]$. *Assume that* u *satisfies*

$$\|u\|^{(-1-\alpha),\{1\}\times(0,1)}_{2,\alpha,Q^{\rm iter}\cap\{s>\varepsilon_0'/10\}}+\|u\|^{(\sigma),({\rm par})}_{2,\alpha,Q^{\rm iter}\cap\{s<\varepsilon_0'\}}\le M, \tag{12.2.64}$$

where we have used norm (4.6.2) *with respect to* $\{s=0\}$, *and* ε_0' *is from* (12.2.45). *Then, denoting*

$$\hat\varepsilon\equiv\hat\varepsilon(\theta_{\rm w}):=\varepsilon_0'\hat s(\theta_{\rm w}) \qquad for\ \theta_{\rm w}\in[\theta^*_{\rm w},\frac{\pi}{2}), \tag{12.2.65}$$

we have

$$\|\mathfrak{g}_{\rm sh}\|^{(-1-\alpha),\{\hat s(\theta_{\rm w})\}}_{2,\alpha,[\hat\varepsilon/10,\,\hat s(\theta_{\rm w})]}+\|\mathfrak{g}_{\rm sh}-\mathfrak{g}_{S_1}\|^{(\sigma),({\rm par})}_{2,\alpha,(0,\,\hat\varepsilon)}\le CM. \tag{12.2.66}$$

Furthermore, define $\mathfrak{F}_{(0,\theta_{\rm w})}$ *on* $(0,\varepsilon_0')\times(0,\infty)$ *by* (12.2.52) *with function* $\mathfrak{g}_{\rm sh}=\mathfrak{g}_{S_1}$ *in* (12.2.43), *where* $\mathfrak{g}_{S_1}$ *is defined in* (12.2.27). *Note that this corresponds to* $u \equiv 0$ *on* $Q^{\rm iter} \cap \{s < \varepsilon_0'\}$. *Then*

$$\|\mathfrak{F}_{(0,\theta_{\rm w})}\|_{C^3(\overline{Q^{\rm iter}} \cap \{s \le \varepsilon_0'\})} \le C_0, \tag{12.2.67}$$

$$\begin{aligned}&\|\mathfrak{F}_{(u,\theta_{\rm w})}\|^{(-1-\alpha),\{1\}\times(0,1)}_{2,\alpha,Q^{\rm iter}\cap\{s>\varepsilon_0'/10\}}\\ &\quad+\|\mathfrak{F}_{(u,\theta_{\rm w})} - \mathfrak{F}_{(0,\theta_{\rm w})}\|^{(\sigma),({\rm par})}_{2,\alpha,Q^{\rm iter}\cap\{s<\varepsilon_0'\}} \le C.\end{aligned} \tag{12.2.68}$$

(viii) *Let* M, α, *and* σ *be as in* (vii), *and* $b > 0$. *There exist* $\hat{C}$ *depending on the data and* $(\theta_{\rm w}^*, \alpha, \sigma, b)$, *and* $\hat{C}_0$ *depending only on the data and* b, *such that the following hold: Let* $(u, \theta_{\rm w}) \in \mathfrak{S}$ *with* $\theta_{\rm w} \in [\theta_{\rm w}^*, \frac{\pi}{2}]$ *satisfy* (12.2.53) *and* (12.2.64). *Then*

$$\|\mathfrak{F}^{-1}_{(0,\theta_{\rm w})}\|_{C^3(\overline{\mathcal{Q}_{\rm bd}\cap\mathcal{D}_{\hat\varepsilon}})} \le \hat{C}_0, \tag{12.2.69}$$

$$\|\mathfrak{F}^{-1}_{(u,\theta_{\rm w})}\|^{(-1-\alpha),\Gamma_{\rm sym}}_{2,\alpha,\Omega(u)\setminus\mathcal{D}_{\hat\varepsilon/10}} + \|\mathfrak{F}^{-1}_{(u,\theta_{\rm w})} - \mathfrak{F}^{-1}_{(0,\theta_{\rm w})}\|^{(\sigma),({\rm par})}_{2,\alpha,\Omega(u)\cap\mathcal{D}_{\hat\varepsilon}} \le \hat{C}. \tag{12.2.70}$$

Furthermore, let φ *be the function from Definition* 12.2.6(v) *corresponding to* $(u, \theta_{\rm w})$. *Then*

$$\|\varphi - \tilde\varphi_2^{(\theta_{\rm w})}\|^{(-1-\alpha),\Gamma_{\rm sym}}_{2,\alpha,\Omega\setminus\mathcal{D}_{\hat\varepsilon/10}} + \|\varphi - \tilde\varphi_2^{(\theta_{\rm w})}\|^{(\sigma),({\rm par})}_{2,\alpha,\Omega\cap\mathcal{D}_{\hat\varepsilon}} \le \hat{C}M, \tag{12.2.71}$$

where $\mathcal{D}_\delta := \{\varphi_2 < \varphi_1\} \cap \Lambda \cap \mathcal{N}_{\varepsilon_1}(\Gamma_{\rm sonic}) \cap \{0 < x < \delta\}$, *and* $\tilde\varphi_2^{(\theta_{\rm w})}$ *is defined in Lemma* 12.2.4.

(ix) *The reversion of assertions* (vii) *and* (viii): *Let* $\theta_{\rm w} \in [\theta_{\rm w}^*, \frac{\pi}{2}]$, *and let function* $\mathfrak{g}_{\rm sh} : (0, \hat{s}(\theta_{\rm w})) \mapsto \mathbb{R}^+$ *satisfy* (12.2.66) *with* M *on the right-hand side. Let* Ω *be defined by* $\theta_{\rm w}$ *and* $\mathfrak{g}_{\rm sh}$ *as in Definition* 12.2.6(iv). *Let* $u : Q^{\rm iter} \mapsto \mathbb{R}$ *and* $\varphi : \Omega \mapsto \mathbb{R}$ *be related as in Definition* 12.2.6(v). *Assume that* φ *satisfies* (12.2.71) *with* M *on the right-hand side. Then* u *satisfies* (12.2.64) *with* CM *on the right-hand side, where* C *depends only on the data and* (α, σ).

(x) *Let* $(u, \theta_{\rm w})$ *and* $(\tilde{u}, \tilde\theta_{\rm w})$ *be as in* (vii). *Then, for any open* $K \Subset Q^{\rm iter}$ *satisfying* $K \subset (\delta, 1-\delta) \times (0, 1)$ *for some* $\delta > 0$,

$$\begin{aligned}&\|\mathfrak{F}_{(u,\theta_{\rm w})} - \mathfrak{F}_{(\tilde u,\tilde\theta_{\rm w})}\|_{C^{2,\alpha}(\overline{K})}\\ &\quad\le C(\delta)\big(\|(u-\tilde u)(\cdot,1)\|_{C^{2,\alpha}([\delta,1-\delta])} + |\theta_{\rm w} - \tilde\theta_{\rm w}|\big),\end{aligned} \tag{12.2.72}$$

$$\begin{aligned}&\|(\varphi\circ\mathfrak{F})_{(u,\theta_{\rm w})} - (\varphi\circ\mathfrak{F})_{(\tilde u,\tilde\theta_{\rm w})}\|_{C^{2,\alpha}(\overline{K})}\\ &\quad\le C(K)\big(\|u-\tilde u\|_{C^{2,\alpha}(\overline{K})} + |\theta_{\rm w} - \tilde\theta_{\rm w}|\big),\end{aligned} \tag{12.2.73}$$

$$\begin{aligned}&\|(\psi\circ\mathfrak{F})_{(u,\theta_{\rm w})} - (\psi\circ\mathfrak{F})_{(\tilde u,\tilde\theta_{\rm w})}\|_{C^{2,\alpha}(\overline{K})}\\ &\quad\le C(K)\big(\|u-\tilde u\|_{C^{2,\alpha}(\overline{K})} + |\theta_{\rm w} - \tilde\theta_{\rm w}|\big),\end{aligned} \tag{12.2.74}$$

where $C(\delta)$ (*resp.* $C(K)$) *depends on the data,* (α, σ), *and* δ (*resp.* K).

Proof. Let $(u, \theta_{\rm w}) \in \mathfrak{S}$. Using (12.2.37) and the regularity that $u \in C^{1,\alpha}(\overline{Q^{\rm iter}})$ and $F_1^{-1} \in C^3(\overline{F_1(\mathcal{Q}_{\rm bd})})$, we see that $\mathfrak{h}_{\rm sh} \in C^{1,\alpha}([0,1])$. This also implies that $\mathfrak{g}_{\rm sh} \in C^{1,\alpha}([0, \hat{s}(\theta_{\rm w})])$. We conclude (i).

Then (ii) also follows: From condition (12.2.49) and Definition 12.2.6(i), it follows that

$$(\Omega \cup \Gamma_{\rm shock})(u, \theta_{\rm w}) \subset \mathcal{Q}_{\rm bd}^{(\theta_{\rm w})} \subset \Lambda(\theta_{\rm w}).$$

Also, the $C^{1,\alpha}$–regularity of $\Gamma_{\rm shock}$ follows from (i), the asserted properties at P_1 follow from Definition 12.2.6(i)–(iv) and $(u, Du)(0, \cdot) = (0, \mathbf{0})$, and (12.2.56) follows also from Lemma 12.2.2(iv).

Now the structure of $\partial\Omega(u, \theta_{\rm w})$ in assertion (iii) follows from Lemma 12.2.2(iv)–(vi).

Assertion (iv) follows from Definition 12.2.6(iv) and Lemma 12.2.2(iv), since $y_{P_4} = \theta_{\rm w}$.

Assertion (v) follows from $u(0, \cdot) = 0$ which holds by (12.2.50), by using Definition 12.2.6(v), (12.2.35), and the fact that $\Gamma_{\rm sonic} = \mathfrak{F}(\{0\} \times (0, 1))$ from (12.2.57).

Now we prove (vi). Denote

$$G^{(\theta_{\rm w})}(s,t) = (\varphi_1 - \tilde{\varphi}_2^{(\theta_{\rm w})}) \circ F_1^{-1}(\hat{s}(\theta_{\rm w})s, t)$$

on $\mathcal{S} := \{(\theta_{\rm w}, s, t) \ : \ \theta_{\rm w}^{\rm s} < \theta_{\rm w} < \frac{\pi}{2}, \ (\hat{s}(\theta_{\rm w})s, t) \in F_1^{(\theta_{\rm w})}(\mathcal{Q}_{\rm bd}^{(\theta_{\rm w})})\}$. Then, from Lemma 12.2.2(ii)–(iii) and (12.2.37)–(12.2.38) of Lemma 12.2.4, we have

$$\begin{aligned} &\|G^{(\cdot)}(\cdot,\cdot)\|_{C^3(\overline{\mathcal{S}})} \le C, \\ &\partial_t G^{(\theta_{\rm w})}(s,t) \le -\delta \qquad \text{for all } (\theta_{\rm w}, s, t) \in \mathcal{S}. \end{aligned} \tag{12.2.75}$$

From Definition 12.2.6(i),

$$G^{(\theta_{\rm w})}(s, \mathfrak{h}_{\rm sh}(s)) = u(s,1) \qquad \text{for all } s \in [0,1]. \tag{12.2.76}$$

Differentiating (12.2.76) with respect to s, we have

$$\mathfrak{h}_{\rm sh}'(s)\partial_t G^{(\theta_{\rm w})}(s, \mathfrak{h}_{\rm sh}(s)) = \partial_s u(s,1) - \partial_s G^{(\theta_{\rm w})}(s, \mathfrak{h}_{\rm sh}(s)). \tag{12.2.77}$$

Then, from (12.2.54) and (12.2.75), we obtain

$$[\mathfrak{h}_{\rm sh}']_{C^{1,\alpha}([0,1])} \le C.$$

Using (12.2.53), we conclude the proof of (12.2.58) for $\mathfrak{h}_{\rm sh}$. Then Definition 12.2.6(ii) implies (12.2.58) for $\mathfrak{g}_{\rm sh}$. Furthermore, subtracting (12.2.76)–(12.2.77) for $(u, \theta_{\rm w})$ and using the same identities for $(\tilde{u}, \tilde{\theta}_{\rm w}) \in \mathfrak{S}_{b,M}^{(\alpha)}$, we obtain (12.2.59) via a standard argument by using (12.2.53)–(12.2.54) and (12.2.75) for $(u, \theta_{\rm w})$ and $(\tilde{u}, \tilde{\theta}_{\rm w})$.

Now (12.2.60) (resp. (12.2.61)) follows directly from the definition of $\mathfrak{F}$ in (12.2.43) and (12.2.52), and from (12.2.58) (resp. (12.2.59)) by using Lemma 12.2.2(ii)–(iii).

To show (12.2.62), we note that, from Definition 12.2.6(v),

$$(\varphi \circ \mathfrak{F}_{(u,\theta_{\rm w})})(s,t) = u(s,t) + (\tilde{\varphi}_2^{(\theta_{\rm w})} \circ \mathfrak{F}_{(u,\theta_{\rm w})})(s,t) \qquad \text{for all } (s,t) \in Q^{\rm iter}. \tag{12.2.78}$$

Now (12.2.62) follows directly from (12.2.38) and (12.2.61).

Estimate (12.2.63) follows from (12.2.61)–(12.2.62) and the smooth dependence of the parameters of the weak state (2) with respect to $\theta_{\rm w} \in (\theta_{\rm w}^{\rm d}, \frac{\pi}{2}]$ in Theorem 7.1.1(i)–(ii).

Next we prove (vii). The estimate of the first term on the left in (12.2.66) is obtained by an argument similar to the proof of (12.2.58) by using (12.2.64). To estimate the second term on the left in (12.2.66), we note the fact that function $\mathfrak{g}_{S_1}$ on $(0, \varepsilon_0)$, defined by (12.2.27), coincides with the function in Definition 12.2.6(i) for $u \equiv 0$ with the same angle $\theta_{\rm w}$, considered on $Q^{\rm iter} \cap \{s < \varepsilon_0\}$, where it is well-defined by (12.2.47) if ε_0 is small, depending on the data. Then the estimate of the second term on the left in (12.2.66) can be obtained by subtracting (12.2.76) for $(u, \theta_{\rm w})$ from the same identity for $(\tilde{u} = 0, \theta_{\rm w})$, as well as by using (12.2.53), (12.2.64), and (12.2.75).

Estimate (12.2.67) is obtained from (12.2.28) by an argument similar to the proof of (12.2.60). Now (12.2.68) follows from (12.2.64) and (12.2.66)–(12.2.67) by an argument similar to the proof of (12.2.60)–(12.2.61).

Next we show (viii). By using (12.2.41) and condition (vi), the proof of (12.2.69) (resp. (12.2.70)) is similar to the proof of (12.2.67) (resp. (12.2.68)). Finally, (12.2.71) follows directly from Definition 12.2.6(v) by using (12.2.64) and (12.2.70).

To show (ix), we note that map $\mathfrak{F}_{(u,\theta_{\rm w})}$ depends only on $\theta_{\rm w}$ and $\mathfrak{g}_{\rm sh}$ used in the definition of $F_{(2,\mathfrak{g}_{\rm sh})}$ by (12.2.41), and constant C in (12.2.68) depends only on $\theta_{\rm w}$ and the constant in (12.2.66). Now (ix) follows by expressing $u(s,t)$ from (12.2.78), and by using (12.2.68) and (12.2.71).

It remains to show (x). Let $\delta \in (0, \frac{1}{2})$. From (12.2.66),

$$\|\mathfrak{g}_{\rm sh}\|_{C^{2,\alpha}([\delta,1-\delta])} \le CM, \tag{12.2.79}$$

where C depends only on the data and (α, σ, δ). Then, using (12.2.75) and (12.2.77) and arguing as in the proof of (12.2.59), we have

$$\begin{aligned} &\|\mathfrak{h}_{\rm sh}^{(u,\theta_{\rm w})} - \mathfrak{h}_{\rm sh}^{(\tilde{u},\tilde{\theta}_{\rm w})}\|_{C^{2,\alpha}([\delta,1-\delta])} \\ &\quad \le C\big(\|(u-\tilde{u})(\cdot,1)\|_{C^{2,\alpha}([\delta,1-\delta])} + |\theta_{\rm w} - \tilde{\theta}_{\rm w}|\big), \end{aligned} \tag{12.2.80}$$

where C depends only on the data and (α, σ, δ).

Now let $K \Subset Q^{\rm iter}$ so that $K \subset (\delta, 1-\delta)^2$ for some $\delta \in (0, \frac{1}{2})$. Then, using (12.2.80) and repeating the proof of (12.2.61), we obtain (12.2.72) with C depending only on the data and (α, σ, K) since $\delta = \delta(K)$. Now (12.2.73) follows directly from (12.2.78) by using (12.2.38). Similarly, (12.2.74) follows directly from

$$(\psi \circ \mathfrak{F}_{(u,\theta_{\rm w})})(s,t) = u(s,t) + (\tilde{\varphi}_2^{(\theta_{\rm w})} - \varphi_2^{(\theta_{\rm w})}) \circ \mathfrak{F}_{(u,\theta_{\rm w})}(s,t) \qquad \text{in } Q^{\rm iter} \tag{12.2.81}$$

by using (12.2.38) and Theorem 7.1.1(i) for the weak states (2). □

12.3 DEFINITION OF THE ITERATION SET

12.3.1 Iteration set

Let $Q^{\rm iter} = (0,1)^2$. An iteration set $\mathcal{K} \subset C^{1,\alpha}(\overline{Q}) \times [\theta_{\rm w}^*, \frac{\pi}{2}]$ is defined below for some appropriate $\alpha \in (0,1)$. For $\theta_{\rm w} \in [\theta_{\rm w}^*, \frac{\pi}{2}]$, denote

$$\mathcal{K}_{\theta_{\rm w}} := \{u \in C^{1,\alpha}(\overline{Q}) : (u, \theta_{\rm w}) \in \mathcal{K}\}.$$

We intend to obtain admissible solutions for all the wedge angles $\theta \in [\theta_{\rm w}^*, \frac{\pi}{2}]$, which include the normal reflection solution for $\theta_{\rm w} = \frac{\pi}{2}$ as a starting point.

For $\theta_{\rm w}$ away from $\frac{\pi}{2}$, the admissible solutions are not close to the normal reflection solution, and then we use the strict monotonicity of $\varphi_1 - \varphi$ in the directions $\{\mathbf{e}_{\mathcal{S}_1}, \mathbf{e}_{\xi_2}\}$; see §8.2.

At $\theta_{\rm w} = \frac{\pi}{2}$, the unique admissible solution is the normal reflection solution, so that $\partial_{\mathbf{e}_{\mathcal{S}_1}}(\varphi_1 - \varphi) \equiv 0$, $\partial_{\mathbf{e}_{\xi_2}}(\varphi_1 - \varphi) \equiv 0$, and the shock is flat. Thus, we do not have the strict monotonicity, as in §8.2, with the uniform estimates up to $\theta_{\rm w} = \frac{\pi}{2}$. Then, for $\theta_{\rm w}$ near $\frac{\pi}{2}$, we use the closeness of admissible solutions to the normal reflection solution in $C^{1,\alpha}$.

Therefore, the definition of $\mathcal{K}_{\theta_{\rm w}}$ will be different for $\theta_{\rm w}$ near $\frac{\pi}{2}$, or away from $\frac{\pi}{2}$. This requires us to connect these definitions continuously in $\theta_{\rm w}$ in order to have set $\mathcal{K}$ open in an appropriate norm. This motivates our definition below, especially parts (i) and (iv) of Definition 12.3.2.

We first introduce some notations. Below we use constant ε_0 from Proposition 11.2.8, which depends only on the data.

Definition 12.3.1. (i) *Denote by $u^{(\rm norm)} \in C^\infty(Q^{\rm iter})$ the function in the iteration region, corresponding to the normal reflection solution, so that $u^{(\rm norm)}$ is defined by (12.2.44) for $\theta_{\rm w} = \frac{\pi}{2}$ and $\varphi = \varphi_2^{(\frac{\pi}{2})}$. In fact, $u^{(\rm norm)} \equiv 0$ by (12.2.34).*

(ii) *For integer $k \geq 0$, real $\sigma > 0$ and $\alpha \in (0,1)$, we use norm (4.6.13) for $\mathcal{S} = Q^{\rm iter}$ with $\varepsilon = \varepsilon_0'$:*

$$\|u\|^{*,(\sigma)}_{k,\alpha,Q^{\rm iter}} = \|u\|^{(-k+1-\alpha),\{1\}\times(0,1)}_{k,\alpha,Q^{\rm iter}\cap\{s>\varepsilon_0'/10\}} + \|u\|^{(\sigma),(\rm par)}_{k,\alpha,Q^{\rm iter}\cap\{s<\varepsilon_0'\}},$$

where norm $\|\cdot\|^{(\sigma),(\rm par)}_{k,\alpha,Q^{\rm iter}\cap\{s<\varepsilon_0'\}}$ is defined by (4.6.2), and ε_0' is from (12.2.45). Denote by $C^{k,\alpha}_{*,\sigma}(Q^{\rm iter})$ the space defined by (4.6.14) for $\mathcal{S} = Q^{\rm iter}$.

(iii) *For $\theta_{\rm w} \in (\theta_{\rm w}^{\rm s}, \frac{\pi}{2}]$ and $\sigma \in (0, \varepsilon_0]$, we write $\mathcal{D}_\sigma \equiv \mathcal{D}_\sigma^{(\theta_{\rm w})}$ for $\mathcal{D}^{(\theta_{\rm w})}_{\delta_*,\sigma}$ introduced in (12.2.8) with δ_* chosen and fixed above so that (12.2.2) holds for any $\theta_{\rm w} \in [\theta_{\rm w}^{\rm s}, \frac{\pi}{2}]$.*

(iv) *Let* $(u, \theta_{\rm w}) \in \mathfrak{S}$ *as defined in* (12.2.50), *and let* $\Omega = \Omega(u)$ *and* $\Gamma_{\rm shock} = \Gamma_{\rm shock}(u)$ *be from Definition* 12.2.6. *Let* $\hat{\varepsilon}$ *be defined by* (12.2.65). *For a set* $\mathcal{S} \subset \{\Omega, \Gamma_{\rm shock}, \Gamma_{\rm wedge}\}$, *integer* $k \geq 0$, *and real* $\sigma > 0$ *and* $\alpha \in (0,1)$, *we define*

$$\|v\|^{*,(\sigma)}_{k,\alpha,\mathcal{S}} = \|v\|^{(-k+1-\alpha),\Gamma_{\rm sym}}_{k,\alpha,\mathcal{S}\setminus\mathcal{D}_{\hat{\varepsilon}/10}} + \|v\|^{(\sigma),({\rm par})}_{k,\alpha,\mathcal{S}\cap\mathcal{D}_{\hat{\varepsilon}}},$$

$C^{k,\alpha}_{*,\sigma}(\mathcal{S})$ *is the closure of* $\{v \in C^\infty(\mathcal{S}) \,:\, \|v\|^{*,(\sigma)}_{k,\alpha,\mathcal{S}} < \infty\}$ *under the norm of* $\|\cdot\|^{*,(\sigma)}_{k,\alpha,\mathcal{S}}$.

We note that, from Proposition 11.2.3, specifically from (11.2.16) and (11.2.18)–(11.2.19), it follows that, for any admissible solution with $\theta_{\rm w} \in (\theta^{\rm s}_{\rm w}, \frac{\pi}{2})$,

$$\Omega_\sigma = \Omega \cap \mathcal{D}_\sigma \qquad \text{for all } \sigma \in (0, \varepsilon_0). \tag{12.3.1}$$

Now we define the iteration set.

Let $\alpha \in (0,1)$, the small constants $\delta_1, \delta_2, \delta_3, \varepsilon, \lambda > 0$, and large constants $N_k \geq 1$, $k = 0, 1, \dots, 5$, be chosen and fixed below. We always assume that

$$\alpha \leq \min\{\delta^*, \frac{\bar{\alpha}}{2}\}, \quad \varepsilon < \frac{\varepsilon_0}{2}, \tag{12.3.2}$$

where $\bar{\alpha}$ is from Lemma 10.5.2 and $\delta^* = \frac{1}{8}$.

Definition 12.3.2. *The iteration set* $\mathcal{K} \subset C^{2,\alpha}_{*,1+\delta^*}(Q^{\rm iter}) \times [\theta^*_{\rm w}, \frac{\pi}{2}]$ *is the set of all* $(u, \theta_{\rm w})$ *satisfying the following properties:*

(i) $(u, \theta_{\rm w})$ *satisfies the estimates:*

$$\|u\|^{*,(1+\delta^*)}_{2,\alpha,Q^{\rm iter}} < \eta_1(\theta_{\rm w}), \tag{12.3.3}$$

where $\eta_1 \in C(\mathbb{R})$ *is defined by*

$$\eta_1(\theta_{\rm w}) = \begin{cases} \delta_1 & \text{if } \frac{\pi}{2} - \theta_{\rm w} \leq \frac{\delta_1}{N_1}, \\ N_0 & \text{if } \frac{\pi}{2} - \theta_{\rm w} \geq \frac{2\delta_1}{N_1}, \\ \text{linear} & \text{if } \frac{\pi}{2} - \theta_{\rm w} \in (\frac{\delta_1}{N_1}, \frac{2\delta_1}{N_1}), \end{cases}$$

with $N_0 = \max\{10M, 1\}$ *for constant* M *from* (12.2.46).

(ii) $(u, \theta_{\rm w}) \in \mathfrak{S}$, *as defined in* (12.2.50). *Then* $\mathfrak{g}_{\rm sh}$, Ω, $\Gamma_{\rm shock}$, *and* φ *are defined for* $(u, \theta_{\rm w})$ *in Definition* 12.2.6, *and* $\Gamma_{\rm sym} = \Gamma_{\rm sym}(u, \theta_{\rm w})$ *is defined in Lemma* 12.2.7(iii).

(iii) $\Gamma_{\rm shock}$ *satisfies the estimate:*

$$\operatorname{dist}(\Gamma_{\rm shock}, B_{c_1}(\mathcal{O}_1)) > \frac{1}{N_5}, \tag{12.3.4}$$

with $N_5 = 2C$ *for* C *from Proposition* 9.5.6;

$$\frac{1}{N_2} < \mathfrak{g}_{\rm sh} < \eta^{(\theta_{\rm w})}(s) - \frac{1}{N_2} \qquad on\ (0, \hat{s}(\theta_{\rm w})), \tag{12.3.5}$$

where $N_2 = 2C$ *for* C *from* (12.2.31), *and we have used* $\eta^{(\theta_{\rm w})}(\cdot)$ *from* (12.2.10).

(iv) φ *and* $\psi = \varphi - \varphi_2^{(\theta_{\rm w})}$ *satisfy*

$$\psi > \eta_2(\theta_{\rm w}) \qquad in\ \overline{\Omega} \setminus \mathcal{D}_{\varepsilon/10}, \tag{12.3.6}$$

$$|\partial_x \psi(x,y)| < \frac{2-\mu_0}{1+\gamma} x \qquad in\ \overline{\Omega} \cap (\mathcal{D}_{\varepsilon_0} \setminus \mathcal{D}_{\varepsilon/10}), \tag{12.3.7}$$

$$|\partial_y \psi(x,y)| < N_3 x \qquad in\ \overline{\Omega} \cap (\mathcal{D}_{\varepsilon_0} \setminus \mathcal{D}_{\varepsilon/10}), \tag{12.3.8}$$

$$|(\partial_x, \partial_y)\psi| < N_3 \varepsilon \qquad in\ \overline{\Omega} \cap \mathcal{D}_{\varepsilon}, \tag{12.3.9}$$

$$\|\psi\|_{C^{0,1}(\Omega)} < N_4, \tag{12.3.10}$$

$$\partial_{\mathbf{e}_{S_1}}(\varphi_1 - \varphi) < -\eta_2(\theta_{\rm w}) \qquad in\ \overline{\Omega} \setminus \mathcal{D}_{\varepsilon/10}, \tag{12.3.11}$$

$$\partial_{\xi_2}(\varphi_1 - \varphi) < -\eta_2(\theta_{\rm w}) \qquad in\ \overline{\Omega} \setminus \mathcal{N}_{\varepsilon/10}(\Gamma_{\rm sym}), \tag{12.3.12}$$

$$\partial_{\boldsymbol{\nu}}(\varphi_1 - \varphi) > \mu_1 \qquad on\ \overline{\Gamma_{\rm shock}}, \tag{12.3.13}$$

$$\partial_{\boldsymbol{\nu}} \varphi > \mu_1 \qquad on\ \overline{\Gamma_{\rm shock}}, \tag{12.3.14}$$

where $\mathcal{D}_\varepsilon$ *is from Definition* 12.3.1(iii),

$$\eta_2(\theta_{\rm w}) = \delta_2 \min\big\{\frac{\pi}{2} - \theta_{\rm w} - \frac{\delta_1}{N_1^2}, \frac{\delta_1}{N_1^2}\big\},$$

and constants ε_0, μ_k, N_3, *and* N_4 *are chosen as follows:*

(a) ε_0 *is from Proposition* 11.2.8,

(b) $\mu_0 = \frac{\delta}{2}$, *where* δ *is from Lemma* 11.2.5,

(c) $\mu_1 = \frac{\min\{\hat{\delta}, \hat{\delta}_1\}}{2}$, *where* $\hat{\delta}$ *and* $\hat{\delta}_1$ *are the constants from Corollary* 10.1.2,

(d) $N_3 = 10\max\{C, 1\}$, *where* C *is the constant from estimate* (11.2.24) *in Lemma* 11.2.6,

(e) $N_4 = 10C$, *where* C *is from Corollary* 9.1.3.

In (12.3.7)–(12.3.8), *we have used* φ *and* φ_2 *expressed in the* (x, y)*–coordinates* (11.1.1)–(11.1.2).

(v) *Uniform ellipticity in* $\overline{\Omega} \setminus \mathcal{D}_{\varepsilon/10}$:

$$\frac{|D\varphi|^2}{c^2(|D\varphi|^2, \varphi)}(\boldsymbol{\xi}) < 1 - \lambda\, {\rm dist}(\boldsymbol{\xi}, \Gamma_{\rm sonic}) \qquad for\ all\ \boldsymbol{\xi} \in \overline{\Omega} \setminus \mathcal{D}_{\varepsilon/10},$$

with $\lambda = \frac{\tilde{\mu}}{2}$, *where* $\tilde{\mu}$ *is from Proposition* 9.6.5.

(vi) *Bounds on the density:*

$$\rho_{\min} < \rho(|D\varphi|^2, \varphi) < \rho_{\max} \qquad in\ \overline{\Omega},$$

with $\rho_{\max} > \rho_{\min} > 0$ *defined by* $\rho_{\min} = \frac{a}{2}\rho_1$ *and* $\rho_{\max} = 2C$, *where* $a = \left(\frac{2}{\gamma+1}\right)^{\frac{1}{\gamma-1}}$ *and* C *are the constants from Lemma* 9.1.4.

(vii) *The boundary value problem* (12.3.25)–(12.3.29) *determined by* $(u, \theta_{\rm w})$, *as defined below in* §12.3.3, *has a solution* $\hat{\psi} \in C^2(\Omega(u, \theta_{\rm w})) \cap C^1(\overline{\Omega}(u, \theta_{\rm w}))$ *such that, if*

$$\hat{\varphi} := \hat{\psi} + \varphi_2 \qquad in\ \Omega(u, \theta_{\rm w}),$$

and function $\hat{u}$ *is defined by* (12.2.44) *from function* $\hat{\varphi}$, *where map* $F_{(2,\mathfrak{g}_{\rm sh})}$ *is determined by function* $\mathfrak{g}_{\rm sh} = \mathfrak{g}_{\rm sh}^{(u,\theta_{\rm w})}$ *which corresponds to* $(u, \theta_{\rm w})$ *as in Definition* 12.2.6(i)–(ii), *i.e.,*

$$\hat{u} := (\hat{\varphi} - \tilde{\varphi}_2^{(\theta_{\rm w})}) \circ (F_1^{(\theta_{\rm w})})^{-1} \circ F_{(2,\mathfrak{g}_{\rm sh})}^{-1} \qquad on\ Q^{\rm iter}, \tag{12.3.15}$$

then $\hat{u}$ *satisfies*

$$\|\hat{u} - u\|_{2,\alpha/2,Q^{\rm iter}}^{*,(1+\delta^*)} < \delta_3. \tag{12.3.16}$$

Remark 12.3.3. *The choice of constants* $(\delta^*, \mu_0, \mu_1, \alpha, \varepsilon, \lambda)$, δ_k *for* $k = 1, 2, 3$, *and* N_k *for* $k = 0, \dots, 5$, *below, will keep only the following dependence:* $\delta^* = \frac{1}{8}$; *constants* $(\mu_0, \mu_1, \lambda, N_3, N_4, N_5)$ *are fixed above depending on the data; constants* (N_0, N_2) *are fixed above depending on the data and* $\theta_{\rm w}^*$; *constant* α *will be fixed later depending only on the data and* $\theta_{\rm w}^*$; *and, for the other constants, in addition to the dependence on the data and* $\theta_{\rm w}^*$, *small* ε *depends on* α, *small* δ_1 *depends on* (α, ε), *large* N_1 *depends on* δ_1, *small* δ_2 *depends on* $(\delta_1, N_1, \varepsilon)$, *and then* $\delta_3 > 0$ *is chosen as small as needed, depending on all the other constants.*

Remark 12.3.4. *Let* $(u, \theta_{\rm w}) \in C^{2,\alpha}_{*,1+\delta^*}(Q^{\rm iter}) \times [\theta_{\rm w}^*, \frac{\pi}{2}]$ *satisfy the properties in Definition* 12.3.2(i)–(iii), *let* $\Omega = \Omega(u, \theta_{\rm w})$, $\Gamma_{\rm shock} = \Gamma_{\rm shock}(u, \theta_{\rm w})$, *and* $\varphi = \varphi^{(u,\theta_{\rm w})}$ *be as in Definition* 12.2.6, *and let* $\mathfrak{g}_{\rm sh}$ *be the function determined by* $(u, \theta_{\rm w})$ *in Definition* 12.2.6(i)–(ii). *Then, from* (12.3.3) *and Lemma* 12.2.7(vii), $\mathfrak{g}_{\rm sh}$ *satisfies* (12.2.66) *with* $\sigma = 1 + \delta^*$ *and* $M = C\eta_1(\theta_{\rm w})$, *where* C *depends only on the data and* (α, δ^*). *In particular, since* F_1 *and* F_1^{-1} *are* C^3 *maps, we conclude that* $\Gamma_{\rm shock}$ *is a* $C^{1,\alpha}$*-curve up to its endpoints. Furthermore, using Lemma* 12.2.2(iv), *we have*

$$\Gamma_{\rm shock} \cap \mathcal{D}_{\varepsilon_0/10} = \{(x, f_{\rm sh}(x)) \,:\, 0 < x < \frac{\varepsilon_0}{10}\}$$

in the (x, y)*–coordinate system, and*

$$\|f_{\rm sh} - \hat{f}_0\|_{2,\alpha,(0,\varepsilon_0)}^{(1+\delta^*),({\rm par})} < C\eta_1(\theta_{\rm w}), \tag{12.3.17}$$

where $\hat{f}_0$ is from (11.2.8). Also, since $(u,\theta_{\rm w}) \in \mathcal{K}$ implies that $(u,\theta_{\rm w}) \in \mathfrak{S}$, it follows from the construction of $\Omega = \Omega(u,\theta_{\rm w})$ in Definition 12.2.6 that $\Omega \subset \mathcal{Q}_{\rm bd}^{(\theta_{\rm w})}$. Then, from (12.2.6), there exists $M_{\rm dom}$ such that

$$\Omega(u,\theta_{\rm w}) \subset B_{M_{\rm dom}}(0) \qquad \text{for all } (u,\theta_{\rm w}) \in \mathcal{K}, \tag{12.3.18}$$

and $M_{\rm dom}$ depends only on the data, but is independent of the parameters in the definition of $\mathcal{K}$.

Remark 12.3.5. *Let $(u,\theta_{\rm w})$, Ω, $\Gamma_{\rm shock}$, and φ be as in Remark 12.3.4. Below, we use the norm introduced in Definition 12.3.1(iv). Using (12.3.3), (12.3.5), Lemma 12.2.7(viii), and Lemma 12.2.4, choosing $\delta_1 \le \mu$ with μ from (12.2.34), recalling that $N_0, N_1 \ge 1$, and using (12.3.18), we obtain from (12.2.71) that $\psi = \varphi - \varphi_2 \in C^{2,\alpha}_{*,1+\delta^*}(\Omega)$ with*

$$\|\psi\|^{*,(1+\delta^*)}_{2,\alpha,\Omega} < C\eta_1(\theta_{\rm w}) \le N_0^*, \tag{12.3.19}$$

where $N_0^* = CN_0$, and C and N_0^* depend only on the data and $(\theta_{\rm w}^*, \alpha)$.

Also, from the previous argument, we apply (12.2.71) with $\alpha = 0$ to obtain

$$\|\psi\|^{*,(1+\delta^*)}_{2,0,\Omega} < \tilde{C}\eta_1(\theta_{\rm w}) \le \tilde{N}_0^*, \tag{12.3.20}$$

where $\tilde{N}_0^* = CN_0$, and $\tilde{C}$ and $\tilde{N}_0^*$ depend only on the data and $\theta_{\rm w}^*$.

Furthermore, from (12.3.19),

$$\|\psi\|_{1,\alpha,\Omega} \le \hat{N}_0, \qquad (\psi, D\psi)|_{\Gamma_{\rm sonic}} = (0,\mathbf{0}), \tag{12.3.21}$$

where $\hat{N}_0$ depends on the data and $(\theta_{\rm w}^*, \alpha)$. Also, from now on, we always assume that $\delta_1 \le \mu$ as above, where we note that μ in Lemma 12.2.4 depends only on the data.

Remark 12.3.6. *If $(u,\theta_{\rm w}) \in C^1(\overline{Q^{\rm iter}}) \times [\theta_{\rm w}^*, \frac{\pi}{2}]$ satisfies the properties in Definition 12.3.2(ii), and the corresponding ψ satisfies (12.3.21), then, as in Remark 12.3.4, we have*

$$\Gamma_{\rm shock} \cap \mathcal{D}_{\varepsilon_0/10} = \{(x, f_{\rm sh}(x)) \;:\; 0 < x < \frac{\varepsilon_0}{10}\}.$$

Then, using that $|D(\varphi_1 - \varphi_2)| = \sqrt{(u_1-u_2)^2 + v_2^2} \ge \frac{1}{C}$ for any $\theta_{\rm w} \in [\theta_{\rm w}^{\rm s}, \frac{\pi}{2}]$, we obtain that, in the (x,y)–coordinates,

$$\begin{aligned} &\|f_{\rm sh} - \hat{f}_0\|_{1,\alpha,(0,\varepsilon_0)} \le C\hat{N}_0, \\ &(f_{\rm sh} - \hat{f}_0)(0) = (f_{\rm sh} - \hat{f}_0)'(0) = 0, \end{aligned} \tag{12.3.22}$$

where C depends on the data, and $f_{\rm sh}$ and $\hat{f}_0$ are as in Remark 12.3.4. Also, using the property that $|D(\varphi_1 - \varphi_2)| \ge \frac{1}{C}$ with (12.3.9) and the second property in (12.3.22), we have

$$|(f_{\rm sh} - \hat{f}_0)(x)| + |x(f_{\rm sh} - \hat{f}_0)'(x)| \le \varepsilon C x \qquad \text{for all } x \in (0,\varepsilon), \tag{12.3.23}$$

where C depends only on the data.

Remark 12.3.7. *Since* $\varphi = \varphi_1$ *on* $\Gamma_{\rm shock}$, *i.e.,* $|D_{\boldsymbol\nu}(\varphi_1 - \varphi)| = |D(\varphi_1 - \varphi)|$, *then* (12.3.13) *is equivalent to*

$$|D(\varphi_1 - \varphi)| > \mu_1 \qquad \text{on } \overline{\Gamma_{\rm shock}}.$$

Moreover, since $(\varphi - \varphi_2, D(\varphi - \varphi_2))(P_1) = (0, \mathbf{0})$ *for any admissible solution* φ, *we obtain*

$$|(u_1 - u_2, -v_2)| = |D(\varphi_1 - \varphi_2)| > \mu_1.$$

Remark 12.3.8. *From the definition of constants* $\rho_{\min}$ *and* $\rho_{\max}$ *in Definition* 12.3.2(vi), *using Lemma* 9.1.4 *and* $\rho(|D\varphi|^2, \varphi) = \rho_2$ *on* $\Gamma_{\rm sonic}$ *for any admissible solution* φ, *we have*

$$2\rho_{\min} \le \rho_2 \le \frac{\rho_{\max}}{2}.$$

12.3.2 Closure of the iteration set

We first introduce the following notation:

Definition 12.3.9. (i) *$\mathcal{K}^{\rm ext}$ is the set of all* $(u, \theta_{\rm w}) \in C^{2,\alpha}_{*,1+\delta^*}(Q^{\rm iter}) \times [\theta_{\rm w}^*, \frac{\pi}{2}]$ *that satisfy conditions* (i)–(vi) *of Definition* 12.3.2.

(ii) $\overline{\mathcal{K}}$ *and* $\overline{\mathcal{K}^{\rm ext}}$ *are the closures of* $\mathcal{K}$ *and* $\mathcal{K}^{\rm ext}$ *in* $C^{2,\alpha}_{*,1+\delta^*}(Q^{\rm iter}) \times [\theta_{\rm w}^*, \frac{\pi}{2}]$, *respectively.*

(iii) *For each* $\mathcal{C} \in \{\mathcal{K}, \mathcal{K}^{\rm ext}, \overline{\mathcal{K}}, \overline{\mathcal{K}^{\rm ext}}\}$ *and each* $\theta_{\rm w} \in [\theta_{\rm w}^*, \frac{\pi}{2}]$,

$$\mathcal{C}(\theta_{\rm w}) := \{u \,:\, (u, \theta_{\rm w}) \in \mathcal{C}\}.$$

Note that $\mathcal{C}(\theta_{\rm w}) \subset C^{2,\alpha}_{*,1+\delta^*}(Q^{\rm iter})$.

From its definition,

$$\mathcal{K} \subset \mathcal{K}^{\rm ext}.$$

Lemma 12.3.10. *The sets introduced in Definition* 12.3.9 *have the following properties:*

(i) $\overline{\mathcal{K}^{\rm ext}} \subseteq \mathfrak{S}$: *If* $(u, \theta_{\rm w}) \in \overline{\mathcal{K}^{\rm ext}}$, *then* $(u, \theta_{\rm w})$ *satisfies condition* (ii) *of Definition* 12.3.2.

(ii) *If* $(u, \theta_{\rm w}) \in \overline{\mathcal{K}^{\rm ext}}$, *then* $(u, \theta_{\rm w})$ *satisfies conditions* (i) *and* (iii)–(vi) *of Definition* 12.3.2 *with the nonstrict inequalities in the estimates.*

(iii) *The properties in Remarks* 12.3.4–12.3.8 *hold with the nonstrict inequalities for all* $(u, \theta_{\rm w}) \in \overline{\mathcal{K}^{\rm ext}}$.

Proof. We divide the proof into three steps.

1. Fix $\theta_{\rm w} \in [\theta_{\rm w}^*, \frac{\pi}{2}]$. It follows from (12.2.10) and (12.2.37) that, for any $s \in (0, \hat{s}(\theta_{\rm w}^*))$,

$$\sup_{\mathcal{Q}_{\rm bd}^{(\theta_{\rm w})}(s)} (\varphi_1 - \tilde{\varphi}_2^{(\theta_{\rm w})}) = (\varphi_1 - \tilde{\varphi}_2) \circ F_1^{-1}(s, 0),$$

$$\inf_{\mathcal{Q}_{\rm bd}^{(\theta_{\rm w})}(s)} (\varphi_1 - \tilde{\varphi}_2^{(\theta_{\rm w})}) = (\varphi_1 - \tilde{\varphi}_2) \circ F_1^{-1}(s, \eta(s)).$$

Then, from (12.3.5), employing (12.2.37) with constant δ from there, and using Definition 12.2.6(i)–(ii), we obtain that, for each $(u, \theta_{\rm w}) \in \mathcal{K}^{\rm ext}$,

$$\inf_{\mathcal{Q}_{\rm bd}^{(\theta_{\rm w})}(s)} (\varphi_1 - \tilde{\varphi}_2^{(\theta_{\rm w})}) + \frac{\delta}{N_2} \le u(\frac{s}{\hat{s}(\theta_{\rm w})}, 1) \le \sup_{\mathcal{Q}_{\rm bd}^{(\theta_{\rm w})}(s)} (\varphi_1 - \tilde{\varphi}_2^{(\theta_{\rm w})}) - \frac{\delta}{N_2} \quad (12.3.24)$$

for any $s \in (0, \hat{s}(\theta_{\rm w}))$.

Now let $(u, \theta_{\rm w}) \in \overline{\mathcal{K}^{\rm ext}}$. Then there exists a sequence $(u^{(i)}, \theta_{\rm w}^{(i)}) \in \mathcal{K}^{\rm ext}$ such that $(u^{(i)}, \theta_{\rm w}^{(i)}) \to (u, \theta_{\rm w})$ in $C^{2,\alpha}_{*,1+\delta^*}(Q^{\rm iter}) \times [\theta_{\rm w}^*, \frac{\pi}{2}]$. From (12.2.59) in Lemma 12.2.7(vi), it follows that $\mathfrak{h}_{\rm sh}^{(u^{(i)}, \theta_{\rm w}^{(i)})} \to \mathfrak{h}_{\rm sh}^{(u, \theta_{\rm w})}$ in $C^{1,\alpha}([0,1])$. Since each $(u^{(i)}, \theta_{\rm w}^{(i)})$ with $\mathfrak{g}_{\rm sh}^{(u^{(i)}, \theta_{\rm w}^{(i)})}(s) = \mathfrak{h}_{\rm sh}^{(u^{(i)}, \theta_{\rm w}^{(i)})}(\hat{s}(\theta_{\rm w}^{(i)})s)$ satisfies (12.3.24), then the same inequality holds for $(u, \theta_{\rm w}) \in \overline{\mathcal{K}^{\rm ext}}$ with

$$\mathfrak{g}_{\rm sh}^{(u(s), \theta_{\rm w})} = \mathfrak{h}_{\rm sh}^{(u(s), \theta_{\rm w})}(\hat{s}(\theta_{\rm w})s).$$

Then (12.2.49) (with strict inequalities) holds for any $(u, \theta_{\rm w}) \in \overline{\mathcal{K}^{\rm ext}}$, *i.e.*, $(u, \theta_{\rm w}) \in \mathfrak{S}$. This shows that assertion (i) holds.

2. Now we show (ii). Let $(u, \theta_{\rm w}) \in \overline{\mathcal{K}^{\rm ext}}$. Then there exists a sequence $(u^{(i)}, \theta_{\rm w}^{(i)}) \in \mathcal{K}^{\rm ext}$ such that $(u^{(i)}, \theta_{\rm w}^{(i)}) \to (u, \theta_{\rm w})$ in $C^{2,\alpha}_{*,1+\delta^*}(Q^{\rm iter}) \times [\theta_{\rm w}^*, \frac{\pi}{2}]$. This implies that $(u, \theta_{\rm w})$ satisfies (12.3.3) with the nonstrict inequality, where we have used the continuity of $\eta_1(\cdot)$.

Next, from Lemma 12.2.7(vi), it follows that

$$\mathfrak{h}_{\rm sh}^{(u^{(i)}, \theta_{\rm w}^{(i)})} \to \mathfrak{h}_{\rm sh}^{(u, \theta_{\rm w})} \text{ in } C^{1,\alpha}([0,1]), \qquad \mathfrak{F}_{(u^{(i)}, \theta_{\rm w}^{(i)})} \to \mathfrak{F}_{(u, \theta_{\rm w})} \text{ in } C^{1,\alpha}(\overline{Q^{\rm iter}}).$$

Then, using the expression of $\Gamma_{\rm shock}$ in Definition 12.2.6(iv), we conclude that Definition 12.3.2(iii) holds for $(u, \theta_{\rm w})$ with the nonstrict inequalities in (12.3.4)–(12.3.5).

Also, from the convergence shown above and Definition 12.2.6(v), we have

$$(\varphi \circ \mathfrak{F})_{(u^{(i)}, \theta_{\rm w}^{(i)})} \to (\varphi \circ \mathfrak{F})_{(u, \theta_{\rm w})}, \quad (\psi \circ \mathfrak{F})_{(u^{(i)}, \theta_{\rm w}^{(i)})} \to (\psi \circ \mathfrak{F})_{(u, \theta_{\rm w})}$$

in $C^{1,\alpha}(\overline{Q^{\rm iter}})$. Furthermore, using (12.2.70) for each $(u^{(i)}, \theta_{\rm w}^{(i)})$ and for $(u, \theta_{\rm w})$, which is obtained by applying Lemma 12.2.7(viii) with $b = \frac{1}{N_2}$ for N_2 from Definition 12.3.2(iii), we have

$$(D\varphi \circ \mathfrak{F})_{(u^{(i)}, \theta_{\rm w}^{(i)})} \to (D\varphi \circ \mathfrak{F})_{(u, \theta_{\rm w})}, \quad (D\psi \circ \mathfrak{F})_{(u^{(i)}, \theta_{\rm w}^{(i)})} \to (D\psi \circ \mathfrak{F})_{(u, \theta_{\rm w})}$$

in $C^\alpha(\overline{Q^{\rm iter}})$.

From this, using the expression of $\Gamma_{\rm shock}$ in Definition 12.2.6(iv) and (12.2.59), we obtain

$$(\boldsymbol{\nu}_{\rm sh}\circ\mathfrak{F})_{(u^{(i)},\theta_{\rm w}^{(i)})}\to(\boldsymbol{\nu}_{\rm sh}\circ\mathfrak{F})_{(u,\theta_{\rm w})}\qquad\text{in } C^\alpha([0,1]).$$

Finally, we have

$$\overline{\Omega}\setminus\mathcal{D}_r=\mathfrak{F}(\overline{Q^{\rm iter}}\cap\{(s,t)\in[0,1]^2\;:\;s\geq r\}).$$

From all the convergence and mapping properties discussed above, we use the continuity of $\eta_2(\cdot)$ to obtain that $(u,\theta_{\rm w})$ satisfies the nonstrict inequalities in the remaining estimates in assertion (ii).

3. Assertion (iii) follows directly from (i)–(ii). □

In Lemma 12.3.10, we have characterized $\overline{\mathcal{K}^{\rm ext}}$. The similar characterization of $\overline{\mathcal{K}}$ includes condition (vii) of Definition 12.3.2 with the nonstrict inequality in (12.3.16). We will prove this, after we define the equation and the condition on $\Gamma_{\rm shock}$ in the iteration problem (12.3.25)–(12.3.29) below and study their properties in the next section.

Remark 12.3.11. *For the rest of Chapter* 12 (*except* §12.8), *we will mostly consider* $(u,\theta_{\rm w})\in\overline{\mathcal{K}^{\rm ext}}$. *Then, based on Lemma* 12.3.10(ii)–(iii), *we use the following notational convention: We consider the nonstrict inequalities in the estimates given in Definition* 12.3.2 *and Remarks* 12.3.4–12.3.8, *unless otherwise specified* (*as in* §12.8).

12.3.3 The boundary value problem for the iteration

In order to complete Definition 12.3.2, we need to define the boundary value problem for the iteration, as indicated in Definition 12.3.2(vii). We define it below.

Fix $(u,\theta_{\rm w})\in\overline{\mathcal{K}^{\rm ext}}$. For such $(u,\theta_{\rm w})$, we obtain Ω, $\Gamma_{\rm shock}$, $\Gamma_{\rm sonic}$, and φ as in Definition 12.3.2(ii).

We then set up a boundary value problem for a *new potential* $\hat{\varphi}$ in Ω. We express the problem in terms of functions $\psi=\varphi-\varphi_2$ and $\hat{\psi}=\hat{\varphi}-\varphi_2$ in the $\boldsymbol{\xi}$–coordinates shifted to center $\mathcal{O}_2$ of state (2), without change of the notation. Then the problem is:

$$\mathcal{N}(\hat{\psi}):=\sum_{i,j=1}^{2}A_{ij}D_{ij}\hat{\psi}+\sum_{i,j=1}^{2}A_iD_i\hat{\psi}=0\qquad\text{in }\Omega,\tag{12.3.25}$$

$$\mathcal{M}_{(u,\theta_{\rm w})}(D\hat{\psi},\hat{\psi},\boldsymbol{\xi})=0\qquad\text{on }\Gamma_{\rm shock},\tag{12.3.26}$$

$$\hat{\psi}=0\qquad\text{on }\Gamma_{\rm sonic},\tag{12.3.27}$$

$$\hat{\psi}_{\boldsymbol{\nu}}=0\qquad\text{on }\Gamma_{\rm wedge},\tag{12.3.28}$$

$$\hat{\psi}_{\boldsymbol{\nu}}=-v_2\qquad\text{on }\Gamma_{\rm sym},\tag{12.3.29}$$

where, in the shifted coordinates, $\Gamma_{\rm sym} = \partial\Omega \cap \{\xi_2 = -v_2\}$ and $(A_{ij}, A_i) = (A_{ij}, A_i)(D\hat{\psi}, \boldsymbol{\xi})$ with $A_{12} = A_{21}$. The nonlinear differential operators $\mathcal{N}$ and $\mathcal{M}$ in (12.3.25) and (12.3.26) are defined below.

We also note that condition (12.3.29) is that $\hat{\varphi}_{\boldsymbol{\nu}} = 0$ on $\Gamma_{\rm sym}$, expressed in terms of $\hat{\psi} = \hat{\varphi} - \varphi_2$.

It remains to define the iteration equation (12.3.25) and the boundary condition (12.3.26) on $\Gamma_{\rm shock}$. We will do this over the course of the next two sections.

12.4 THE EQUATION FOR THE ITERATION

In this subsection, we fix $(u, \theta_{\rm w}) \in C^{2,\alpha}_{*,1+\delta^*}(Q^{\rm iter}) \times [\theta^*_{\rm w}, \frac{\pi}{2}]$ satisfying conditions (i)–(ii) and (v)–(vi) of Definition 12.3.2 with the nonstrict inequalities in the estimates, and define equation (12.3.25) such that

(i) It is strictly elliptic in $\overline{\Omega} \setminus \overline{\Gamma_{\rm sonic}}$ with elliptic degeneracy at the sonic arc $\Gamma_{\rm sonic}$;

(ii) For a fixed point $\hat{\psi} = \psi$, equation (12.3.25) coincides with the original equation (2.2.11) written in terms of ψ: In the $\boldsymbol{\xi}$–coordinates with the origin shifted to the center of the sonic circle of state (2), equation (12.3.25) for a fixed point $\hat{\psi} = \psi$ coincides with equation (8.3.24) for the density and sonic speed given by (8.3.25) and (8.3.26), respectively;

(iii) It depends continuously on $\theta_{\rm w}$ in a sense specified below,

where $\Omega, \Gamma_{\rm shock}, \Gamma_{\rm sonic}, \varphi$, and ψ correspond to $(u, \theta_{\rm w})$.

In this subsection, we work in the $\boldsymbol{\xi}$–coordinates, with the origin shifted to the center of the sonic circle of state (2). Thus, the potential flow equation (2.2.11) takes form (8.3.24) with (8.3.25)–(8.3.26).

We define equation (12.3.25) in $\Omega \cap \mathcal{D}_{\varepsilon_{\rm eq}}$ and $\Omega \setminus \mathcal{D}_{\varepsilon_{\rm eq}/10}$ separately, and then combine them over region $\mathcal{D}_{\varepsilon_{\rm eq}} \setminus \mathcal{D}_{\varepsilon_{\rm eq}/10}$ by using the cutoff function, where $\varepsilon_{\rm eq} \in (0, \frac{\varepsilon_0}{2})$ will be determined in Lemma 12.4.2, depending only on the data, for constant ε_0 from Proposition 11.2.8.

Also, from now on, we always assume that ε in Definition 12.3.2 satisfies

$$\varepsilon < \frac{\varepsilon_{\rm eq}}{2}. \tag{12.4.1}$$

It will be convenient to write the potential flow equation (2.2.11) in terms of ψ:

$$\sum_{i,j=1}^{2} A^{\rm potn}_{ij}(D\psi, \psi, \boldsymbol{\xi}) D_{ij}\psi = 0, \tag{12.4.2}$$

where

$$\begin{aligned} A^{\rm potn}_{ij}(\mathbf{p}, z, \boldsymbol{\xi}) = {} & c^2(|D\varphi_2(\boldsymbol{\xi}) + \mathbf{p}|^2, \varphi_2(\boldsymbol{\xi}) + z, \rho_0^{\gamma-1})\delta_{ij} \\ & - (D_i\varphi_2(\boldsymbol{\xi}) + p_i)(D_j\varphi_2(\boldsymbol{\xi}) + p_j) \end{aligned} \tag{12.4.3}$$

with $c = c(|D\varphi|^2, \varphi, \rho_0^{\gamma-1})$ determined by (1.14). From these explicit expressions, we find that, for any $R > 0$ and $k = 1, \ldots,$

$$|A_{ij}^{\rm potn}(\mathbf{p}, z, \boldsymbol{\xi})| \le C_R, \qquad |D^k A_{ij}^{\rm potn}(\mathbf{p}, z, \boldsymbol{\xi})| \le C_{R,k} \tag{12.4.4}$$

for any $|(\mathbf{p}, z, \boldsymbol{\xi})| \le R$, where C_R and $C_{R,k}$ depend on the data and (R, k).

12.4.1 The iteration equation in Ω away from $\Gamma_{\rm sonic}$

In $\Omega \setminus \mathcal{D}_{\varepsilon_{\rm eq}/10}$, we employ equation (2.2.11) for $\hat{\varphi}$, via replacing $D^2\hat{\phi}$ by $D^2\hat{\psi}$ (since $D^2\hat{\phi} = D^2\hat{\psi}$) and substituting the given function φ into the coefficients, to obtain

$$(c^2 - \varphi_{\xi_1}^2)\hat{\psi}_{\xi_1\xi_1} - 2\varphi_{\xi_1}\varphi_{\xi_2}\hat{\psi}_{\xi_1\xi_2} + (c^2 - \varphi_{\xi_2}^2)\hat{\psi}_{\xi_2\xi_2} = 0, \tag{12.4.5}$$

where

$$c^2 = c^2(|D\varphi|^2, \varphi, \boldsymbol{\xi}) = \rho_0^{\gamma-1} - (\gamma - 1)(\frac{1}{2}|D\varphi|^2 + \varphi).$$

This defines coefficients $(A_{ij}^1, A_i^1) = (A_{ij}^1, A_i^1)(\boldsymbol{\xi})$ in $\Omega \setminus \mathcal{D}_{\varepsilon_{\rm eq}/10}$ with $A_{12}^1 = A_{21}^1$ and $A_i^1 \equiv 0$. In terms of equation (12.4.2), we have

$$A_{ij}^1(\boldsymbol{\xi}) = A_{ij}^{\rm potn}(D\psi(\boldsymbol{\xi}), \psi(\boldsymbol{\xi}), \boldsymbol{\xi}).$$

Condition (v) of Definition 12.3.2 and (12.4.1) together imply that the equation is uniformly elliptic in $\Omega \setminus \mathcal{D}_{\varepsilon_{\rm eq}/10}$. Also, from Remark 12.3.5, coefficients (A_{ij}^1, A_i^1) are in $C^{(-\alpha), \Gamma_{\rm sym}}_{1,\alpha,\overline{\Omega}\setminus\mathcal{D}_{\varepsilon_{\rm eq}/10}}$ with

$$\|(A_{ij}^1, A_i^1)\|_{C^{(-\alpha), \Gamma_{\rm sym}}_{1,\alpha,\overline{\Omega}\setminus\mathcal{D}_{\varepsilon_{\rm eq}/10}}} \le C, \tag{12.4.6}$$

where C depends only on the data, $\hat{N}_0$ from (12.3.21), and α from the definition of the iteration set, that is, C depends only on the data and $(\alpha, \theta_{\rm w}^*)$.

12.4.2 The iteration equation in Ω near $\Gamma_{\rm sonic}$

Now we define the equation in region $\Omega \cap \mathcal{D}_{\varepsilon_{\rm eq}}$, *i.e.*, near $\Gamma_{\rm sonic}$. We work in the (x, y)–coordinates (11.1.1)–(11.1.2).

The construction is given below. We use equation (11.1.4) with (11.1.5) for $\hat{\psi}$, written in the form of (11.1.6). We will make cutoffs and substitutions in coefficients $(\hat{A}_{ij}, \hat{A}_i)(\mathbf{p}, z, x, y)$ in (11.1.6) so that:

(a) Any fixed point $\hat{\psi} = \psi$ satisfies the original equation;

(b) In the modified equation, coefficients $(\hat{A}_{ij}, \hat{A}_i)$ are independent of z, *i.e.*, $(\hat{A}_{ij}, \hat{A}_i) = (\hat{A}_{ij}, \hat{A}_i)(\mathbf{p}, x, y)$;

(c) The modified equation has degenerate ellipticity structure (11.1.7) for all $(\mathbf{p}, x, y) \in \mathbb{R}^2 \times (\Omega \cap \mathcal{D}_{\varepsilon_{\rm eq}})$;

(d) $\hat{A}_1(\mathbf{p}, x, y) \approx -1$ for all $(\mathbf{p}, x, y)$;

(e) $\|(\hat{A}_{ij}, \hat{A}_i)\|_{C^{0,1}(\mathbb{R}^2 \times (\Omega \cap \mathcal{D}_{\varepsilon_{\rm eq}}))} \leq C$.

In particular, we cannot substitute the known function ψ instead of the unknown function $\hat{\psi}$ into all the terms of coefficients $\hat{A}_{ij}$ in (11.1.4), because the properties of ψ from Definition 12.3.2 of the iteration set and Remark 12.3.5 do not imply that the resulting equation is elliptic. Indeed, estimate (12.3.19) implies only that $|\psi_x| \leq Cx^{\delta^*}$, so that $\hat{A}_{11}$ (the coefficient of ψ_{xx}) in (11.1.4) can be negative if we substitute such ψ.

The leading nonlinear term in $\hat{A}_{11}$ is $(\gamma+1)\hat{\psi}_x$. Then we do not substitute ψ into term $(\gamma+1)\hat{\psi}_x$ in the coefficient of $\hat{\psi}_{xx}$ in (11.1.4). Instead we make a cutoff of the form:

$$x\zeta_1(\frac{\hat{\psi}_x}{x}),$$

where $\zeta_1(\cdot)$ is a cutoff function such that $\zeta_1(s) = s$ within the expected range of values of $\hat{\psi}_x$ at the fixed point $\hat{\psi} = \psi$ (*i.e.*, for $(\gamma+1)|s| \leq 2-\sigma$ for some small $\sigma > 0$) and $(\gamma+1)\sup\zeta_1 < 2$ to have the positivity of $\hat{A}_{11}$. In fact, this nonlinearity will allow us to make the key estimates of $\hat{\psi}$, so that (11.4.38) will be obtained for $\hat{\varphi}$, which implies

$$|\hat{\psi}_x| \leq Cx, \quad |\hat{\psi}_y| \leq Cx^{\frac{3}{2}} \qquad \text{in } \Omega \cap \mathcal{D}_{\varepsilon_{\rm eq}}. \tag{12.4.7}$$

However, these estimates are not sufficient to remove the cutoff described above, for which we need to have a more precise estimate: $|\hat{\psi}_x| \leq \frac{2-\sigma}{\gamma+1}x$. Then we remove the cutoff for a fixed point $\psi = \hat{\psi}$ by an additional argument, which imposes an extra requirement:

(f) The modified equation for a fixed point $\psi = \hat{\psi}$, written in the $\boldsymbol{\xi}$–coordinates in the form of (12.3.25), has $A_i \equiv 0$ in Ω for $i = 1, 2$.

Note that the cutoff of term $(\gamma+1)\hat{\psi}_x$, described above, satisfies this requirement, since the modification is only for the coefficient of $\hat{\psi}_{xx}$. Using the polar coordinates (11.1.1), we have

$$\hat{\psi}_{xx} = \hat{\psi}_{rr} = \frac{1}{r^2}\big(\xi_1^2\hat{\psi}_{\xi_1\xi_1} + 2\xi_1\xi_2\hat{\psi}_{\xi_1\xi_2} + \xi_2^2\hat{\psi}_{\xi_2\xi_2}\big), \tag{12.4.8}$$

so that the change of the coefficient of $\hat{\psi}_{xx}$, written in the $\boldsymbol{\xi}$–coordinates, does not produce lower-order terms in the potential flow equation (12.4.2).

Now we need to make some substitutions and cutoffs in the terms of O_k, $k = 1, \ldots, 5$, in (11.1.4)–(11.1.5).

First, we substitute ψ for $\hat{\psi}$ in the zero-order terms in O_k, *i.e.*, remove the z–dependence from $(\hat{A}_{ij}, \hat{A}_i)$.

We keep the nonlinearity of the first-order terms in O_k and make a cutoff; otherwise, by substituting $\hat{\psi}$, we would obtain a lower regularity of the ingredients of the equation. To define the cutoff, we note that estimates (12.4.7) and expressions (11.1.5) imply that O_k, $k = 1, \dots, 5$, are small, at least in comparison with the main terms in the coefficients of (11.1.4). This will allow us to use a *milder* cutoff for terms O_k, which will readily be removed after establishing estimates (12.4.7) by using conditions (12.3.7)–(12.3.8) of Definition 12.3.2(iv). More specifically, in order to satisfy requirements (c)–(d) listed above, we need the following estimates for all $(\mathbf{p}, x, y) \in \mathbb{R}^2 \times \Omega_{\varepsilon_{\rm eq}}$:

$$|O_1| \le \sigma x, \qquad |O_2| \le \sigma\sqrt{x}, \qquad |O_k| \le \sigma \ \text{ for } k = 3, 4, 5, \tag{12.4.9}$$

for sufficiently small $\sigma > 0$ depending only on the data.

Furthermore, for deriving the estimates near $\Gamma_{\rm sonic}$ in the iteration, we need to rescale the equation in a similar way, as (4.7.7) has been rescaled to obtain (4.7.30). This puts some extra requirements on (A_{11}, A_{12}), since they are multiplied by x_0^β with $\beta < 0$ in (4.7.31) to define the corresponding coefficients in (4.7.30). To satisfy these and the other requirements (see Lemma 12.4.2 for the full description of the required properties of the iteration equation near $\Gamma_{\rm sonic}$), we modify O_k, $k = 1, \dots, 5$, as follows: Choosing an appropriate cutoff function ζ_1, we replace p_1 by $x^{\frac{3}{4}}\zeta_1(\frac{p_1}{x^{3/4}})$, p_2 by $(\gamma+1)N_3 x\zeta_1(\frac{p_2}{(\gamma+1)N_3 x})$ for N_3 from Definition 12.3.2(iv), and z by ψ.

Now we write these cutoffs explicitly. In the definition below, the cutoff level for $\hat{\psi}_x$ from above is based on estimates (11.2.21) and (11.2.23), as well as condition (12.3.7) for $\mu_0 = \frac{\delta}{2} \in (0, \frac{1}{2})$, where δ is from (11.2.21). Let $\zeta_1 \in C^\infty(\mathbb{R})$ satisfy

$$\zeta_1(s) = \begin{cases} s & \text{if } |s| < \dfrac{2 - \frac{\mu_0}{5}}{1+\gamma}, \\ \dfrac{2 - \frac{\mu_0}{10}}{1+\gamma}\operatorname{sign}(s) & \text{if } |s| > \dfrac{2}{\gamma+1}, \end{cases} \tag{12.4.10}$$

and the following properties:

$$\zeta_1'(s) \ge 0 \qquad \text{on } \mathbb{R}, \tag{12.4.11}$$

$$\zeta_1(-s) = -\zeta_1(s) \qquad \text{on } \mathbb{R}, \tag{12.4.12}$$

$$\zeta_1''(s) \le 0 \qquad \text{on } \{s \ge 0\}. \tag{12.4.13}$$

Clearly, such a smooth function $\zeta_1 \in C^\infty(\mathbb{R})$ exists. Property (12.4.13) will be used only in Proposition 13.4.10.

Then we define the modified equation described above in the (x, y)–coordinates:

$$\begin{aligned} &\Big(2x - (\gamma+1)x\zeta_1(\frac{\hat{\psi}_x}{x}) + \tilde{O}_1^m\Big)\hat{\psi}_{xx} + \tilde{O}_2^m\hat{\psi}_{xy} \\ &\quad + (\frac{1}{c_2} + \tilde{O}_3^m)\hat{\psi}_{yy} - (1 + \tilde{O}_4^m)\hat{\psi}_x + \tilde{O}_5^m\hat{\psi}_y = 0 \end{aligned} \tag{12.4.14}$$

with $\tilde{O}_k^m = \tilde{O}_k^m(D\hat{\psi}(x,y), x, y)$, where $\tilde{O}_k^m(\mathbf{p}, x, y)$ is defined by

$$\begin{aligned}&\tilde{O}_k^m(\mathbf{p}, x, y)\\&= O_k(x^{\frac{3}{4}}\zeta_1(\frac{p_1}{x^{3/4}}),\ (\gamma+1)N_3 x\zeta_1(\frac{p_2}{(\gamma+1)N_3 x}),\ \psi(x,y),\ x)\end{aligned} \tag{12.4.15}$$

for $k = 1, \dots, 5$, $O_k(\mathbf{p}, z, x)$ are given by (11.1.5), and N_3 is from Definition 12.3.2(iv).

Lemma 12.4.1. *Let $\varepsilon_{\rm eq} \in (0, \varepsilon_0)$, and let (12.4.1) hold. If $(u, \theta_{\rm w}) \in \mathfrak{S}$, and the corresponding $\psi = \varphi - \varphi_2$ satisfies (12.3.7)–(12.3.9) with constants μ_0 and N_3 depending only on the data, then*

$$|\psi(x,y)| \le C\varepsilon_{\rm eq} x \qquad \textit{for all } (x,y) \in \Omega \cap \mathcal{D}_{\varepsilon_{\rm eq}}, \tag{12.4.16}$$

and

$$\begin{aligned}&|\tilde{O}_1^m(\mathbf{p}, x, y)| \le C(\varepsilon_{\rm eq} + \sqrt{x})x, \quad |(\tilde{O}_2^m, \tilde{O}_5^m)(\mathbf{p}, x, y)| \le Cx,\\&|(\tilde{O}_3^m, \tilde{O}_4^m)(\mathbf{p}, x, y)| \le Cx^{\frac{3}{4}}\end{aligned} \tag{12.4.17}$$

for all $\mathbf{p} \in \mathbb{R}^2$ and $(x,y) \in \Omega \cap \mathcal{D}_{\varepsilon_{\rm eq}}$, with C depending only on the data.

Proof. In this proof, constant C depends only on the data.

By (12.3.7)–(12.3.9) and (12.4.1), we obtain

$$|D\psi(x,y)| \le C\varepsilon_{\rm eq} \qquad \text{in } \Omega \cap \mathcal{D}_{\varepsilon_{\rm eq}}.$$

Also, combining the property that $|D(\varphi_1 - \varphi_2)| \ge \frac{1}{C}$ with (12.3.7)–(12.3.9), and arguing similarly to Remark 12.3.4, we have

$$|f'_{\rm sh}| \le C \qquad \text{for all } x \in (0, \varepsilon_0).$$

Using these estimates and the fact that $|\psi(0,y)| = 0$ on $\Gamma_{\rm sonic}$, by Lemma 12.2.7(v), we obtain (12.4.16). With this, then (11.1.5), (12.4.10), and (12.4.15) imply (12.4.17). □

We write equation (12.4.14)–(12.4.15) in the form:

$$\hat{\mathcal{N}}^2(\hat{\psi}) = 0, \tag{12.4.18}$$

with

$$\hat{\mathcal{N}}^2(\hat{\psi}) := \hat{A}_{11}\hat{\psi}_{xx} + 2\hat{A}_{12}\hat{\psi}_{xy} + \hat{A}_{22}\hat{\psi}_{yy} + \hat{A}_1\hat{\psi}_x + \hat{A}_2\hat{\psi}_y, \tag{12.4.19}$$

where $\hat{A}_{ij} = \hat{A}_{ij}(D\hat{\psi}, x, y)$, $\hat{A}_i = \hat{A}_i(D\hat{\psi}, x, y)$, and $\hat{A}_{21} = \hat{A}_{12}$. Then coefficients $(\hat{A}_{ij}, \hat{A}_i)(\mathbf{p}, x, y)$ are defined for $\mathbf{p} = (p_1, p_2) \in \mathbb{R}^2$ and $(x,y) \in \Omega \cap \mathcal{D}_{\varepsilon_{\rm eq}}$. We also write this equation in the $\boldsymbol{\xi}$–coordinates in the form:

$$\mathcal{N}^2(\hat{\psi}) = 0 \qquad \text{in } \Omega \cap \mathcal{D}_{\varepsilon_{\rm eq}}, \tag{12.4.20}$$

with

$$\mathcal{N}^2(\hat{\psi}) := A_{11}^2\hat{\psi}_{\xi_1\xi_1} + 2A_{12}^2\hat{\psi}_{\xi_1\xi_2} + A_{22}^2\hat{\psi}_{\xi_2\xi_2} + A_1^2\hat{\psi}_{\xi_1} + A_2^2\hat{\psi}_{\xi_2}, \tag{12.4.21}$$

where $(A_{ij}^2, A_i^2) = (A_{ij}^2, A_i^2)(D\hat{\psi}, \boldsymbol{\xi})$ and $A_{21}^2 = A_{12}^2$, in which $(A_{ij}^2, A_i^2)(\mathbf{p}, \boldsymbol{\xi})$ are defined on $\mathbf{p} \in \mathbb{R}^2$ and $\boldsymbol{\xi} \in \Omega \cap \mathcal{D}_{\varepsilon_{\rm eq}}$.

Lemma 12.4.2. *There exist $\lambda_1 > 0$, $\varepsilon_{\rm eq} \in (0, \frac{\varepsilon_0}{2})$, and $N_{\rm eq} \geq 1$ depending only on the data such that the following holds: Let $(u, \theta_{\rm w}) \in C^1(\overline{Q^{\rm iter}}) \times [\theta_{\rm w}^*, \frac{\pi}{2}]$ satisfy the properties in Definition 12.3.2(ii), and let the corresponding $\psi := \varphi - \varphi_2^{(\theta_{\rm w})}$ satisfy (12.3.7)–(12.3.9) with μ_0 and N_3 depending only on the data. Write equation (12.4.14)–(12.4.15) in the form of (12.4.18). Then $(\hat{A}_{ij}, \hat{A}_i)(\mathbf{p}, x, y)$ have the following properties:*

(i) *For any $(x, y) \in \Omega \cap \mathcal{D}_{\varepsilon_{\rm eq}}$ and $\mathbf{p}, \boldsymbol{\kappa} \in \mathbb{R}^2$,*

$$\lambda_1|\boldsymbol{\kappa}|^2 \leq \sum_{i,j=1}^{2} \hat{A}_{ij}(\mathbf{p}, x, y)\frac{\kappa_i\kappa_j}{x^{2-\frac{i+j}{2}}} \leq \frac{1}{\lambda_1}|\boldsymbol{\kappa}|^2. \tag{12.4.22}$$

(ii) *$(\hat{A}_{ij}, \hat{A}_i, D_{\mathbf{p}}\hat{A}_{ij}, D_{\mathbf{p}}\hat{A}_i)(\mathbf{p}, \cdot, \cdot) \in C^{1,\alpha}(\overline{\Omega \cap \mathcal{D}_{\varepsilon_{\rm eq}}} \setminus \overline{\Gamma_{\rm sonic}})$, and*

$$\|(\hat{A}_{11}, \hat{A}_{12}, \hat{A}_2)\|_{C^{0,1}(\mathbb{R}^2 \times \overline{\Omega \cap \mathcal{D}_{\varepsilon_{\rm eq}}})} \leq N_{\rm eq},$$
$$\|(\hat{A}_{22}, \hat{A}_1)\|_{L^\infty(\mathbb{R}^2 \times \overline{\Omega \cap \mathcal{D}_{\varepsilon_{\rm eq}}})} + \|D_{(\mathbf{p},y)}(\hat{A}_{22}, \hat{A}_1)\|_{L^\infty(\mathbb{R}^2 \times \overline{\Omega \cap \mathcal{D}_{\varepsilon_{\rm eq}}})} \leq N_{\rm eq},$$
$$\sup_{\mathbf{p}\in\mathbb{R}^2, (x,y)\in\Omega\cap\mathcal{D}_{\varepsilon_{\rm eq}}} \Big(x^{\frac{1}{4}}\big|D_x(\hat{A}_{22}, \hat{A}_2)(\mathbf{p}, x, y)\big|\Big) \leq N_{\rm eq},$$

with $\hat{A}_{21} = \hat{A}_{12}$.

(iii) *$\sup_{\mathbf{p}\in\mathbb{R}^2} \|(\hat{A}_{ij}, \hat{A}_i)(\mathbf{p}, \cdot, \cdot)\|_{C^{3/4}(\overline{\Omega \cap \mathcal{D}_{\varepsilon_{\rm eq}}})} \leq N_{\rm eq}$ for $i, j = 1, 2$.*

(iv) *$(\hat{A}_{ij}, \hat{A}_i) \in C^{1,\alpha}(\mathbb{R}^2 \times (\overline{\Omega \cap \mathcal{D}_{\varepsilon_{\rm eq}}} \setminus \overline{\Gamma_{\rm sonic}}))$ and, for each $s \in (0, \frac{\varepsilon_{\rm eq}}{2})$,*

$$\sup_{\mathbf{p}\in\mathbb{R}^2} \|D_{\mathbf{p}}^k(\hat{A}_{ij}, \hat{A}_i)(\mathbf{p}, \cdot, \cdot)\|_{C^{1,\alpha}(\mathbb{R}^2 \times (\overline{\Omega \cap \mathcal{D}_{\varepsilon_{\rm eq}}} \setminus \mathcal{N}_s(\Gamma_{\rm sonic})))} \leq N_{\rm eq} s^{-5}$$

for $k = 1, 2$.

(v) *Let $|D_y\psi(x, y)| \leq M\sqrt{x}$ in $\Omega \cap \mathcal{D}_{\varepsilon_{\rm eq}}$, in addition to the previous assumptions. Then, for all $\mathbf{p} \in \mathbb{R}^2$ and $(x, y) \in \Omega \cap \mathcal{D}_{\varepsilon_{\rm eq}}$,*

$$|D_y(\hat{A}_{11}, \hat{A}_{12})(\mathbf{p}, x, y)| \leq \hat{C}\sqrt{x},$$

where $\hat{C}$ depends only on the data and M.

(vi) *For any* $\mathbf{p}$, *functions* $(\hat{A}_{ij}, \hat{A}_i)(\mathbf{p}, \cdot, \cdot)$ *can be extended from* $\Omega \cap \mathcal{D}_{\varepsilon_{\rm eq}}$ *to* $\Gamma_{\rm sonic} = \{x = 0\} \cap \partial(\Omega \cap \mathcal{D}_{\varepsilon_{\rm eq}})$, *by fixing* $\mathbf{p}$ *and taking a limit in* (x, y) *from* $\Omega \cap \mathcal{D}_{\varepsilon_{\rm eq}} \subset \{x > 0\}$. *Moreover,* $(\hat{A}_{ij}, \hat{A}_i)_{|x=0}$ *are independent of* $\mathbf{p}$. *Explicitly, for any* $\mathbf{p} \in \mathbb{R}^2$ *and* $(0, y) \in \Gamma_{\rm sonic}$,

$$\hat{A}_{ij}(\mathbf{p}, 0, y) = 0 \qquad \textit{for } (i,j) = (1,1),\ (1,2),\ (2,1),$$
$$\hat{A}_{22}(\mathbf{p}, 0, y) = \frac{1}{c_2}, \quad \hat{A}_1(\mathbf{p}, 0, y) = -1, \quad \hat{A}_2(\mathbf{p}, 0, y) = 0.$$

(vii) *For every* $(\mathbf{p}, x, y) \in \mathbb{R}^2 \times \overline{\Omega \cap \mathcal{D}_{\varepsilon_{\rm eq}}}$,

$$(\hat{A}_{11}, \hat{A}_{22}, \hat{A}_1)((p_1, -p_2), x, y) = (\hat{A}_{11}, \hat{A}_{22}, \hat{A}_1)((p_1, p_2), x, y), \tag{12.4.23}$$
$$|\hat{A}_{ii}(\mathbf{p}, x, y) - \hat{A}_{ii}(\mathbf{0}, 0, y)| \le N_{\rm eq}|x|^{\frac{3}{4}} \qquad \textit{for } i = 1, 2, \tag{12.4.24}$$
$$|\hat{A}_{12}(\mathbf{p}, x, y)| \le N_{\rm eq} x, \tag{12.4.25}$$
$$\hat{A}_1(\mathbf{p}, x, y) \le -\frac{1}{2}. \tag{12.4.26}$$

(viii) *Assume also that* (12.4.7) *holds for* ψ. *Then, if* $\varepsilon \in (0, \frac{\varepsilon_{\rm eq}}{2})$ *small, depending on the data and the constants in* (12.4.7), $\hat{\mathcal{N}}^2(\hat{\psi})$ *from* (12.4.20) *for* $\hat{\psi} = \psi$ *has the expression:*

$$\begin{aligned}\hat{\mathcal{N}}^2(\psi) = &\Big(2x - (\gamma+1)x\zeta_1(\frac{\psi_x}{x}) + O_1\Big)\psi_{xx} + O_2\psi_{xy} \\ &+ (\frac{1}{c_2} + O_3)\psi_{yy} - (1 + O_4)\psi_x + O_5\psi_y,\end{aligned} \tag{12.4.27}$$

where $O_k = O_k(D\psi, \psi, x)$, *with* $O_k(\mathbf{p}, z, x)$ *given by* (11.1.5). *That is, the cutoff is removed in these terms* $O_j, j = 1, \ldots, 5$.

Moreover, if ψ *satisfies*

$$|\partial_x\psi(x, y)| \le \frac{2 - \mu_0}{1 + \gamma}x, \quad |\partial_y\psi(x, y)| \le N_3 x \qquad \textit{in } \Omega \cap \mathcal{D}_{\varepsilon/10}, \tag{12.4.28}$$

then, for $\hat{\psi} = \psi$, *we obtain that, in* $\Omega \cap \mathcal{D}_{\varepsilon_{\rm eq}}$,

$$\hat{\mathcal{N}}^2(\psi) = \sum_{i,j=1}^{2} \hat{A}_{ij}^{\rm potn}(D\psi, \psi, x)D_{ij}\psi + \sum_{i=1}^{2} \hat{A}_i^{\rm potn}(D\psi, \psi, x)D_i\psi, \tag{12.4.29}$$

where $(\hat{A}_{ij}^{\rm potn}, \hat{A}_i^{\rm potn})$ *are from* (11.1.6). *That is, if* (12.4.28) *holds,* $\hat{\mathcal{N}}(\psi)$ *coincides in* $\Omega \cap \mathcal{D}_{\varepsilon_{\rm eq}}$ *with the left-hand side of equation* (11.1.4) *with* (11.1.5).

Furthermore, if $\hat{\mathcal{N}}^2(\psi)$ *in the* (x, y)*–coordinates is of the form of either* (12.4.27) *or* (12.4.29), *then its expression of* $\mathcal{N}^2(\psi)$ *in* (12.4.20) *in the* $\boldsymbol{\xi}$*–coordinates has only the second-order terms, i.e.,*

$$A_i^2(D\psi, \boldsymbol{\xi}) = 0 \qquad \textit{in } \Omega \cap \mathcal{D}_{\varepsilon_{\rm eq}} \qquad \textit{for } i = 1, 2. \tag{12.4.30}$$

(ix) *Let ψ be a fixed point, i.e., $\hat{\psi} = \psi$ satisfies equation (12.4.18) in $\Omega \cap \mathcal{D}_{\varepsilon_{\rm eq}}$. Let (12.4.7) hold for $\hat{\psi} = \psi$. Then, if $\varepsilon \in (0, \frac{\varepsilon_{\rm eq}}{2})$ is small in Definition 12.3.2, depending on the data and the constants in (12.4.7), then equation (12.4.14) for ψ in $\Omega \cap \mathcal{D}_{\varepsilon_{\rm eq}}$ is of the form:*

$$\begin{aligned}&\Big(2x - (\gamma+1)x\zeta_1(\frac{\psi_x}{x}) + O_1\Big)\psi_{xx} + O_2\psi_{xy}\\&\quad + (\frac{1}{c_2} + O_3)\psi_{yy} - (1+O_4)\psi_x + O_5\psi_y = 0,\end{aligned} \tag{12.4.31}$$

where $O_k = O_k(D\psi, \psi, x)$, with $O_k(\mathbf{p}, z, x)$ given by (11.1.5). That is, ψ satisfies equation (12.4.14) in $\Omega \cap \mathcal{D}_{\varepsilon_{\rm eq}}$ without the cutoff in these terms $O_j, j = 1, \dots, 5$. Furthermore, equation (12.4.20) for ψ in the $\boldsymbol{\xi}$–coordinates has only the second-order terms, i.e., ψ satisfies

$$A^2_{11}\psi_{\xi_1\xi_1} + 2A^2_{12}\psi_{\xi_1\xi_2} + A^2_{22}\psi_{\xi_2\xi_2} = 0 \qquad \text{in } \Omega \cap \mathcal{D}_{\varepsilon_{\rm eq}}, \tag{12.4.32}$$

where $A^2_{ij} = A^2_{ij}(D\psi, \boldsymbol{\xi})$.

Proof. The universal constant C in this proof depends only on the data. We divide the proof into three steps.

1. The estimates in (12.4.17) and the explicit form (12.4.14) of the equation imply that, if $\varepsilon_{\rm eq}$ is chosen sufficiently small, depending only on the data, then (12.4.22) holds with

$$\lambda_1 = \frac{1}{2}\min\big\{\frac{\mu_0}{10}, \min_{\theta_{\rm w}\in[\theta^{\rm s}_{\rm w}, \frac{\pi}{2}]}\frac{1}{c_2(\theta_{\rm w})}\big\},$$

where $\mu_0 \in (0,1)$ is from Definition 12.3.2(iv).

2. In this step, we prove (ii)–(iv).

We first prove (ii). The regularity:

$$(\hat{A}_{ij}, \hat{A}_i, D_{\mathbf{p}}\hat{A}_{ij}, D_{\mathbf{p}}\hat{A}_i)(\mathbf{p}, \cdot, \cdot) \in C^{1,\alpha}(\overline{\Omega \cap \mathcal{D}_{\varepsilon_{\rm eq}}} \setminus \overline{\Gamma_{\rm sonic}})$$

follows directly from (12.4.14)–(12.4.15) and (12.3.21). Then we show the estimates in (ii).

In the estimates below, we use the fact that, for $\varepsilon_{\rm eq}$ small depending on the data, $O_k(\mathbf{p}, z, x)$ given by (11.1.5) for each $\theta_{\rm w} \in [\theta^{\rm s}_{\rm w}, \frac{\pi}{2}]$ are smooth functions of their arguments, with C^3–norms on any compact subset of $\mathbb{R}^2 \times \mathbb{R} \times [0, 2\varepsilon_{\rm eq}]$ independent of $\theta_{\rm w}$, since $c_2(\theta_{\rm w}) \in [\frac{1}{C}, C]$.

We first estimate $\|\hat{A}_{11}\|_{C^{0,1}(\mathbb{R}^2\times\overline{\Omega\cap\mathcal{D}_{\varepsilon_{\rm eq}}})}$. For this, we estimate the two terms, $x\zeta_1(\frac{p_1}{x})$ and $\tilde{O}^m_1(\mathbf{p}, x, y)$, separately. In the estimates, we use that, for each $k \geq 0$, there exists C_k such that

$$|s|^k|\zeta_1'(s)| \leq C_k \qquad \text{for all } s \in \mathbb{R}, \tag{12.4.33}$$

which holds since $\zeta_1' \in C_c(\mathbb{R})$. From this (with $k=1$), for any $p_1 \in \mathbb{R}$ and $x>0$, we have

$$\begin{aligned}
&\big|x\zeta_1(\frac{p_1}{x})\big| \le Cx,\\
&\big|D_x\big(x\zeta_1(\frac{p_1}{x})\big)\big| = \big|\zeta_1(\frac{p_1}{x}) - \frac{p_1}{x}\zeta_1'(\frac{p_1}{x})\big| \le C,\\
&\big|D_{p_1}\big(x\zeta_1(\frac{p_1}{x})\big)\big| = \big|\zeta_1'(\frac{p_1}{x})\big| \le C.
\end{aligned} \tag{12.4.34}$$

To estimate $\tilde{O}_1^m(\mathbf{p},x,y)$, we first note that, for $O_1(\mathbf{p},z,x)$ defined by (11.1.5),

$$\begin{aligned}
&|D_xO_1(\mathbf{p},z,x)| \le C(x+|p_1|+|p_2|^2), \qquad |D_zO_1(\mathbf{p},z,x)| \le C,\\
&|D_{p_1}O_1(\mathbf{p},z,x)| \le C(x+|p_1|), \qquad |D_{p_2}O_1(\mathbf{p},z,x)| \le C|p_2|.
\end{aligned}$$

Then, denoting

$$\mathcal{X} := \big(x^{\frac{3}{4}}\zeta_1(\frac{p_1}{x^{3/4}}),\ \hat{N}x\zeta_1(\frac{p_2}{\hat{N}x}),\ \psi(x,y),\ x\big) \tag{12.4.35}$$

with $\hat{N} = (\gamma+1)N_3$, we obtain that, for $(\mathbf{p},(x,y)) \in \mathbb{R}^2 \times (\Omega \cap \mathcal{D}_{\varepsilon_{\rm eq}})$,

$$\begin{aligned}
&|D_xO_1(\mathcal{X})| \le Cx^{\frac{3}{4}}, \quad |D_zO_1(\mathcal{X})| \le C,\\
&|D_{p_1}O_1(\mathcal{X})| \le Cx^{\frac{3}{4}}, \quad |D_{p_2}O_1(\mathcal{X})| \le Cx.
\end{aligned}$$

From this, using (12.3.7)–(12.3.9), (12.4.15), and (12.4.33), we have

$$\begin{aligned}
&|D_x\tilde{O}_1^m(\mathbf{p},x,y)|\\
&\le \Big|D_xO_1(\mathcal{X}) + D_x\big(x^{\frac{3}{4}}\zeta_1(\frac{p_1}{x^{3/4}})\big)D_{p_1}O_1(\mathcal{X})\Big|\\
&\quad + \Big|D_x\big(\hat{N}x\zeta_1(\frac{p_2}{\hat{N}x})\big)D_{p_2}O_1(\mathcal{X}) + D_x\psi(x,y)D_zO_1(\mathcal{X})\Big|\\
&\le Cx^{\frac{3}{4}} + C\Big(\Big|x^{-\frac{1}{4}}\zeta_1(x^{-\frac{1}{4}}\frac{p_1}{x^{3/4}})\Big| + \Big|x^{-\frac{1}{4}}\frac{p_1}{x^{3/4}}\zeta_1'(\frac{p_1}{x^{3/4}})\Big|\Big)x^{\frac{3}{4}}\\
&\quad + C\Big|\zeta_1(\frac{p_2}{\hat{N}x}) + \frac{p_2}{\hat{N}x}\zeta_1'(\frac{p_2}{\hat{N}x})\Big|x + C\varepsilon_{\rm eq}\\
&\le C_1(\sqrt{x}+\varepsilon_{\rm eq}) \le 2C_1\sqrt{\varepsilon_{\rm eq}} \le 2C_1,
\end{aligned}$$

where we have used $\varepsilon_{\rm eq} \in (0,1)$. Similarly,

$$\begin{aligned}
&|D_y\tilde{O}_1^m(\mathbf{p},x,y)| = |D_zO_1(\mathcal{X})D_y\psi(x,y)| \le C\varepsilon_{\rm eq} \le C,\\
&|D_{p_1}\tilde{O}_1^m(\mathbf{p},x,y)| = |D_{p_1}O_1(\mathcal{X})\zeta_1'(\frac{p_1}{x^{3/4}})| \le Cx^{\frac{3}{4}} \le C,\\
&|D_{p_2}\tilde{O}_1^m(\mathbf{p},x,y)| = |D_{p_2}O_1(\mathcal{X})\zeta_1'(\frac{p_2}{\hat{N}x})| \le Cx \le C.
\end{aligned}$$

Combining all of this with (12.4.17) and (12.4.34) and using the explicit expression of $\hat{A}_{11}$ in (12.4.14), we obtain that $\|\hat{A}_{11}\|_{C^{0,1}(\mathbb{R}^2\times\overline{(\Omega\cap\mathcal{D}_{\varepsilon_{\rm eq}})})} \le C$.

The estimates for $(\hat{A}_{12}, \hat{A}_2)$ are obtained similarly, especially by using that $|D_{p_1}O_k(\mathbf{p}, z, x)| \leq C|p_2|$ to see that $|D_x\tilde{O}_k^m(\mathbf{p}, x, y)| \leq C$ for $k = 2, 5$.

In the estimates for $(\hat{A}_{22}, \hat{A}_1)$, we find that $|D_{p_1}O_k(\mathbf{p}, z, x)| \leq C$ for $k = 3, 4$, so that, estimating as above, we conclude that $|D_x\tilde{O}_k^m(\mathbf{p}, x, y)| \leq Cx^{-\frac{1}{4}}$ for $k = 3, 4$. Then, following the calculations as above, we obtain the estimates for $(\hat{A}_{22}, \hat{A}_1)$ in (ii). Now (ii) is proved.

Moreover, (iii) directly follows from (ii).

Finally, (iv) follows from the explicit expressions of $(\hat{A}_{ij}, \hat{A}_i)$ in (12.4.14)–(12.4.15). Specifically, we only need to estimate the $C^{1,\alpha}$–norms of the cutoff terms. We obtain them by extending the calculation in the proof of (ii). Continuing the calculation in (12.4.34), and using (12.4.33) with $k = 1, 2$, we have

$$\left|D^2_{(p_1,x)}\left(x\zeta_1(\frac{p_1}{x})\right)\right| \leq Cx^{-1}.$$

Similarly,

$$\left|D^2_{(p_1,x)}\left(x^{\frac{3}{4}}\zeta_1(\frac{p_1}{x^{3/4}})\right)\right| \leq Cx^{-\frac{5}{4}}.$$

By these and similar calculations, there exists C depending on the data such that, for each $s \in (0, \frac{\varepsilon_{\rm eq}}{2})$,

$$\|D^k_{(\mathbf{p},x)}(\hat{A}_{ij}, \hat{A}_i)\|_{L^\infty(\mathbb{R}^2\times(\overline{\Omega\cap\mathcal{D}_{\varepsilon_{\rm eq}}}\setminus\mathcal{N}_s(\Gamma_{\rm sonic})))} \leq C_s s^{-5} \qquad \text{for } k = 1, 2, 3, 4.$$

This implies (iv).

3. In this step, we prove properties (v)–(ix).

To show property (v) for A_{11}, we use the assumption that $|D_y\psi(x, y)| \leq M\sqrt{x}$ on $\Omega \cap \mathcal{D}_{\varepsilon_{\rm eq}}$ with (12.4.14)–(12.4.15) and (11.1.5) to calculate:

$$\begin{aligned}
&|D_yA_{11}(\mathbf{p}, x, y)| = |D_y\tilde{O}_1^m(\mathbf{p}, x, y)| \\
&= \left|D_zO_1(x^{\frac{3}{4}}\zeta_1(\frac{p_1}{x^{3/4}}),\ (\gamma+1)N_3x\zeta_1(\frac{p_2}{(\gamma+1)N_3x}),\ \psi(x, y),\ x)\right||D_y\psi(x, y)| \\
&\leq \hat{C}\sqrt{x}.
\end{aligned}$$

Also, by a similar calculation, $D_yA_{12} \equiv 0$, since $D_zO_2 \equiv 0$ by (11.1.5).

Property (vi) follows from (12.4.14)–(12.4.16), since $\zeta_1(\cdot)$ is bounded so that $\lim_{x\to0+} x\zeta_1(\frac{p_1}{x}) = 0$ to obtain that, for any $(0, y_0) \in \Gamma_{\rm sonic}$, $\mathbf{p} \in \mathbb{R}^2$, and $k = 1, \dots, 5$,

$$\lim_{\Omega\ni(x,y)\to(0,y_0)} \tilde{O}_k^m(\mathbf{p}, x, y) = O_k(\mathbf{0}, 0, 0) = 0,$$

where we have used (11.1.5) in the last equality.

Now we prove assertion (vii). Property (12.4.23) follows from the explicit expressions (12.4.14)–(12.4.15) by using (11.1.5) and (12.4.12).

Next we show (12.4.24). From (11.1.5) and (12.4.14)–(12.4.15), we employ (12.4.16) to conclude that $\hat{A}_{11}(\mathbf{0}, 0, y) = 0$. Also, our assumptions imply that

(12.4.17) holds, so that $|\tilde{O}_1^m(\mathbf{p},x,y)| \leq C\varepsilon_{\rm eq}x$, and then (12.4.24) follows for $i=1$. Similarly, for $i=2$, we have

$$\hat{A}_{22}(\mathbf{p},x,y) - \hat{A}_{22}(\mathbf{0},0,y) = \tilde{O}_k^3(\mathbf{p},x,y),$$

and using (12.4.17) again,

$$|\tilde{O}_3^m(\mathbf{p},x,y)| \leq Cx^{\frac{3}{4}},$$

which shows (12.4.24) for $i=2$.

To show (12.4.25), we first note that, from (12.4.14) and (12.4.18), it follows that $A_{12}(\mathbf{p},x,y) = \frac{1}{2}O_2^m(\mathbf{p},x,y)$. Now (12.4.25) follows from (11.1.5) and (12.4.15).

By choosing sufficiently small $\varepsilon_{\rm eq}$, property (12.4.26) follows from $\hat{A}_1(\mathbf{p},x,y) = -\big(1+\hat{O}_4^m(\mathbf{p},x,y)\big)$ and (12.4.17). Now assertion (vii) is proved.

To show (viii) and (ix), we choose ε so small that (12.4.7) for $\hat{\psi}=\psi$ implies that, for each $(x,y)\in\Omega\cap\mathcal{D}_\varepsilon$,

$$|\psi_x(x,y)| \leq \frac{1}{2(\gamma+1)}x^{\frac{3}{4}}, \qquad |\psi_y(x,y)| \leq \frac{N_3}{2}x.$$

Then, from (12.4.10), it follows that, in $\Omega\cap\mathcal{D}_\varepsilon$,

$$\begin{aligned} &x^{\frac{3}{4}}\zeta_1\Big(\frac{\psi_x(x,y)}{x^{3/4}}\Big) = \psi_x(x,y),\\ &(\gamma+1)N_3x\zeta_1\Big(\frac{\psi_y(x,y)}{(\gamma+1)N_3x}\Big) = \psi_y(x,y). \end{aligned} \tag{12.4.36}$$

From (12.3.7)–(12.3.8), it follows that (12.4.36) holds in $\Omega\cap(\mathcal{D}_{\varepsilon_{\rm eq}}\setminus\mathcal{D}_{\varepsilon/10})$ by using that $\varepsilon_{\rm eq}\leq 1$. Thus, (12.4.36) holds in $\Omega\cap\mathcal{D}_{\varepsilon_{\rm eq}}$. Then, by (12.4.15),

$$O_k^m(D\psi,x,y) = O_k(D\psi,\psi,x) \qquad \text{in } \Omega\cap\mathcal{D}_{\varepsilon_{\rm eq}} \text{ for } k=1,\dots,5.$$

Therefore, (12.4.27) holds, and equation (12.4.18) for ψ coincides with (12.4.31), so that ψ satisfies (12.4.31) in $\Omega\cap\mathcal{D}_{\varepsilon_{\rm eq}}$.

Furthermore, expression (12.4.27) (equivalently, the left-hand side of equation (12.4.31)) differs from the left-hand side of equation (11.1.4) only in the coefficient of ψ_{xx}. Equation (11.1.4) is the potential flow equation (divided by c_2) and is of form (12.4.2) in the $\boldsymbol{\xi}$–coordinates. Equation (12.4.2) has only the second-order terms. Then, from (12.4.8), it follows that (12.4.30) holds, and equation (12.4.31), written in the $\boldsymbol{\xi}$–coordinates, also has only the second-order terms, *i.e.*, is of form (12.4.32).

Finally, if ψ satisfies (12.4.28), then, using (12.3.7)–(12.3.8), we see that (12.4.36) also holds, and $x\zeta_1(\frac{\psi_x(x,y)}{x}) = \psi_x(x,y)$ in $\Omega\cap\mathcal{D}_{\varepsilon_{\rm eq}}$. This implies that (12.4.29) holds. □

Note that (12.3.20) implies that the condition of Lemma 12.4.2(v) holds with the constant depending only on $\tilde{N}_0^*$ and $\varepsilon_{\rm eq}$. Then, from Lemma 12.4.2, recalling Remark 12.3.11, we have

Corollary 12.4.3. *There exists* $\varepsilon_{\rm eq} \in (0, \frac{\varepsilon_0}{2})$ *depending only on the data such that, if* $\varepsilon \in (0, \frac{\varepsilon_{\rm eq}}{2})$ *in Definition* 12.3.2, *then, for any* $(u, \theta_{\rm w}) \in \overline{\mathcal{K}^{\rm ext}}$, *assertions* (i)–(vii) *of Lemma* 12.4.2 *hold with* $\lambda_1 > 0$ *and* $N_{\rm eq}$ *depending only on the data, and with* $\hat{C}$ *depending only on the data and* $\theta_{\rm w}^*$. *Also, assertions* (viii)–(ix) *of Lemma* 12.4.2 *hold if* $\varepsilon \in (0, \frac{\varepsilon_{\rm eq}}{2})$ *is small, depending only on the data and the constants in* (12.4.7).

Next, we combine the equations introduced above by defining the coefficients of (12.3.25) in Ω as follows:

First, we define the cutoff function: Let, for each $\tau > 0$, $\zeta_2 \equiv \zeta_2^{(\tau)} \in C^\infty(\mathbb{R})$ satisfy

$$\zeta_2(s) = \begin{cases} 0 & \text{if } s \le \frac{\tau}{2}, \\ 1 & \text{if } s \ge \tau, \end{cases} \qquad \text{and} \quad 0 \le \zeta_2'(s) \le \frac{10}{\tau} \qquad \text{on } \mathbb{R}. \tag{12.4.37}$$

Then, using the mapping in Lemma 12.2.2, we define that, for each $\theta_{\rm w} \in [\theta_{\rm w}^{\rm s}, \frac{\pi}{2}]$ and $\tau \in (0, \frac{\varepsilon_0}{2})$,

$$\hat{\zeta}_2^{(\tau,\theta_{\rm w})}(\boldsymbol{\xi}) := \zeta_2^{(\tau)}(F_1^{(\theta_{\rm w})}(\boldsymbol{\xi})) \qquad \text{for all } \boldsymbol{\xi} \in \mathcal{Q}_{\rm bd}^{(\theta_{\rm w})}. \tag{12.4.38}$$

Using Lemma 12.2.2(i)–(iv), Definition 12.2.6(iv), and set $\mathcal{Q}_{\rm bd}^{\cup}$ defined in Lemma 12.2.2(iii), we note the following properties of $\hat{\zeta}_2^{(\tau,\theta_{\rm w})}$ for each $\tau \in (0, \frac{\varepsilon_0}{2})$:

$$\text{Function } (\boldsymbol{\xi}, \theta_{\rm w}) \mapsto \hat{\zeta}_2^{(\tau,\theta_{\rm w})}(\boldsymbol{\xi}) \ \text{ is in } \ C^\infty(\overline{\mathcal{Q}_{\rm bd}^{\cup}}), \tag{12.4.39}$$

$$\|\zeta_2^{(\tau,\theta_{\rm w})}(\cdot)\|_{C^3(\overline{\mathcal{Q}_{\rm bd}})} \le C \ \text{ for all } \theta_{\rm w} \in [\theta_{\rm w}^{\rm s}, \frac{\pi}{2}], \tag{12.4.40}$$

$$\zeta_2^{(\tau,\theta_{\rm w})}(\cdot) \equiv \begin{cases} 0 & \text{in } \Omega^{(u,\theta_{\rm w})} \cap \mathcal{D}_{\tau/2}, \\ 1 & \text{in } \Omega^{(u,\theta_{\rm w})} \setminus \mathcal{D}_\tau \end{cases} \quad \text{for all } (u, \theta_{\rm w}) \in \mathfrak{S}. \tag{12.4.41}$$

Now we define the iteration equation in Ω. Fix $(u, \theta_{\rm w}) \in \mathfrak{S}$. Then, possibly reducing $\varepsilon_{\rm eq}$ depending only on the data (which will be fixed in Lemma 12.4.5), and writing $\hat{\zeta}_2^{(\varepsilon_{\rm eq})}$ for $\hat{\zeta}_2^{(\varepsilon_{\rm eq},\theta_{\rm w})}$, we define that, for each $\mathbf{p} \in \mathbb{R}^2$ and $\boldsymbol{\xi} \in \Omega$,

$$\begin{aligned} A_{ij}(\mathbf{p}, \boldsymbol{\xi}) &= \hat{\zeta}_2^{(\varepsilon_{\rm eq})}(\boldsymbol{\xi}) A_{ij}^1(\boldsymbol{\xi}) + \big(1 - \hat{\zeta}_2^{(\varepsilon_{\rm eq})}(\boldsymbol{\xi})\big) c_2 A_{ij}^2(\mathbf{p}, \boldsymbol{\xi}), \\ A_i(\mathbf{p}, \boldsymbol{\xi}) &= \big(1 - \hat{\zeta}_2^{(\varepsilon_{\rm eq})}(\boldsymbol{\xi})\big) c_2 A_i^2(\mathbf{p}, \boldsymbol{\xi}), \end{aligned} \tag{12.4.42}$$

where $A_{ij}^1(\boldsymbol{\xi})$ are defined by (12.4.5).

Remark 12.4.4. *Note that* (A_{ij}^2, A_i^2) *have been multiplied by* c_2, *since equation* (11.1.4)–(11.1.5) *is the potential flow equation divided by* $c_2 = c_2^{(\theta_{\rm w})}$. *Thus, with our definition, equation* (12.3.25) *with coefficients* (12.4.42), *taken at a fixed point* $\psi = \hat{\psi}$ *and without the cutoff, is the potential flow equation* (12.4.2).

We state the properties of (A_{ij}, A_i) in the following lemma.

Lemma 12.4.5. *Let $\alpha \in (0, \frac{1}{8})$ and $M > 0$. Then there exist $\varepsilon_{\rm eq} \in (0, \frac{\varepsilon_0}{2})$, $\lambda_0 > 0$, $N_{\rm eq} \geq 1$ (depending only on the data), and C (depending only on the data, M, and α) such that the following hold:*

Let $\varepsilon \in (0, \frac{\varepsilon_{\rm eq}}{2})$ be used in Definition 12.3.2. Let $(u, \theta_{\rm w}) \in C^1(\overline{Q^{\rm iter}}) \times [\theta_{\rm w}^*, \frac{\pi}{2}]$ satisfy conditions (ii) and (v)–(vi) of Definition 12.3.2 with $(\lambda, \rho_{\min}, \rho_{\max})$ fixed there depending only on the data, and let the corresponding function $\psi = \varphi^{(u,\theta_{\rm w})} - \varphi_2^{(\theta_{\rm w})}$ satisfy (12.3.7)–(12.3.9) for μ_0 and N_3 depending only on the data, and

$$\|\psi\|^{(-1-\alpha),\,\Gamma_{\rm sym}}_{2,\alpha,\Omega\setminus\mathcal{D}_{\varepsilon_{\rm eq}/2}} \leq M, \tag{12.4.43}$$

with M fixed above. Then the coefficient functions $(A_{ij}, A_i)(\mathbf{p}, \boldsymbol{\xi})$ defined by (12.4.42) for $\mathbf{p} \in \mathbb{R}^2$ and $\boldsymbol{\xi} \in \Omega$, $i, j = 1, 2$, satisfy the following:

(i) *For any $\boldsymbol{\xi} \in \Omega$ and $\mathbf{p}, \mu \in \mathbb{R}^2$,*

$$\lambda_0 \operatorname{dist}(\boldsymbol{\xi}, \Gamma_{\rm sonic})|\mu|^2 \leq \sum_{i,j=1}^{2} A_{ij}(\mathbf{p}, \boldsymbol{\xi})\mu_i\mu_j \leq \lambda_0^{-1}|\mu|^2.$$

(ii) *$A_{12} = A_{21}$ and*

$$\|(A_{ij}, A_i)\|_{L^\infty(\mathbb{R}^2\times\Omega)} \leq N_{\rm eq}.$$

(iii) *$A_{ij}(\mathbf{p}, \boldsymbol{\xi}) = A^1_{ij}(\boldsymbol{\xi})$ and $A_i(\mathbf{p}, \boldsymbol{\xi}) = 0$ for any $\boldsymbol{\xi} \in \Omega \setminus \mathcal{D}_{\varepsilon_{\rm eq}}$ and $\mathbf{p} \in \mathbb{R}^2$, where $A^1_{ij}(\boldsymbol{\xi})$ are defined by (12.4.5). In particular,*

$$\|A_{ij}\|_{C^{(-\alpha),\Gamma_{\rm sym}}_{1,\alpha,\overline{\Omega}\setminus\mathcal{D}_{\varepsilon_{\rm eq}}}} \leq C.$$

(iv) *Functions $(A_{ij}, A_i)(\mathbf{p}, \boldsymbol{\xi})$ satisfy that, for each $\mathbf{p} \in \mathbb{R}^2$,*

$$(A_{ij}, A_i)(\mathbf{p}, \cdot),\ D_{\mathbf{p}}(A_{ij}, A_i)(\mathbf{p}, \cdot) \in C^{1,\alpha}(\overline{\Omega} \setminus (\overline{\Gamma_{\rm sonic}} \cup \overline{\Gamma_{\rm sym}})).$$

Moreover, for each $\mathbf{p} \in \mathbb{R}^2$,

$$\|(A_{ij}, A_i)(\mathbf{p}, \cdot)\|^{(-\alpha),\Gamma_{\rm sym}}_{0,3/4,\Omega} + \|D_{\mathbf{p}}(A_{ij}, A_i)(\mathbf{p}, \cdot)\|_{L^\infty(\Omega)} \leq C.$$

(v) *$A_{ij}, A_i, D^k_{\mathbf{p}}(A_{ij}, A_i) \in C^{1,\alpha}(\mathbb{R}^2 \times (\overline{\Omega} \setminus (\overline{\Gamma_{\rm sonic}} \cup \overline{\Gamma_{\rm sym}})))$ for $k = 1, 2$, and, for any $s \in (0, \frac{\varepsilon_0}{2})$,*

$$\|(A_{ij}, A_i)\|_{C^{1,\alpha}(\mathbb{R}^2\times(\overline{\Omega}\setminus\mathcal{N}_s(\Gamma_{\rm sonic}\cup\Gamma_{\rm sym})))} \leq Cs^{-5}.$$

(vi) *$(A_{ij}, A_i)(\mathbf{p}, \boldsymbol{\xi}) = (A^2_{ij}, A^2_i)(\mathbf{p}, \boldsymbol{\xi})$ for any $(\mathbf{p}, \boldsymbol{\xi}) \in \mathbb{R}^2 \times (\Omega \cap \mathcal{D}_{\varepsilon_{\rm eq}/2})$. In particular, equation (12.3.25), written in the (x, y)–coordinates in domain $\Omega \cap \mathcal{D}_{\varepsilon_{\rm eq}/2}$, is (12.4.14).*

(vii) *Coefficients* $A_{ij}(\mathbf{p},\boldsymbol{\xi})$ *can be extended by continuity to all* $\boldsymbol{\xi}\in\overline{\Omega}$ *for each* $\mathbf{p}\in\mathbb{R}^2$*. In particular, for* $\boldsymbol{\xi}\in\overline{\Gamma_{\rm sonic}}$*, the coefficients are of the following explicit form:*

$$\begin{aligned} &A_{11}(\mathbf{p},\boldsymbol{\xi})=c_2^2-\xi_1^2,\quad A_{22}(\mathbf{p},\boldsymbol{\xi})=c_2^2-\xi_2^2,\\ &A_{12}(\mathbf{p},\boldsymbol{\xi})=A_{21}(\mathbf{p},\boldsymbol{\xi})=-\xi_1\xi_2, \end{aligned}\tag{12.4.44}$$

where we have worked in the shifted coordinates $\boldsymbol{\xi}$ *with the origin at* O_2.

(viii) *Let* $A_{ij}^{\rm potn}(D\psi,\psi,\boldsymbol{\xi})$ *be the coefficients of equation* (12.4.2).

(a) *If equation* (12.3.25) *is determined by* $(u,\theta_{\rm w})\in\overline{\mathcal{K}^{\rm ext}}$*, then, for the corresponding* Ω *and* ψ,

$$A_{ij}(D\psi,\boldsymbol{\xi})=A_{ij}^{\rm potn}(D\psi,\psi,\boldsymbol{\xi}),\ \ A_i(D\psi,\boldsymbol{\xi})=0\qquad \text{in } \Omega\setminus\mathcal{D}_{\varepsilon/10}.$$

In particular, if ψ *is a fixed point, i.e.,* $\hat{\psi}=\psi$ *satisfies equation* (12.3.25) *in* Ω*, then* ψ *is a solution of* (12.4.2) *in* $\Omega\setminus\mathcal{D}_{\varepsilon/10}$.

Suppose also that (12.4.7) *holds for the fixed point* $\hat{\psi}=\psi$*. Then, if* ε *is small, depending on the data and the constants in* (12.4.7)*, then*

(b) $A_i(D\psi,\boldsymbol{\xi})\equiv 0$ *in* Ω *for equation* (12.3.25)*. Thus,* ψ *in the* $\boldsymbol{\xi}$*–coordinates satisfies the equation that has only the second-order terms:*

$$A_{11}\psi_{\xi_1\xi_1}+2A_{12}\psi_{\xi_1\xi_2}+A_{22}\psi_{\xi_2\xi_2}=0\qquad \text{in } \Omega,\tag{12.4.45}$$

where $A_{ij}(D\psi,\boldsymbol{\xi})$ *are from* (12.3.25).

(c) *Equation* (12.4.45) *for* ψ *is* (12.4.31) *written in the* (x,y)*–coordinates in* $\Omega\cap\mathcal{D}_{\varepsilon_{\rm eq}/2}$.

Proof. Let $\varepsilon_{\rm eq}$ be as in Lemma 12.4.2. We will reduce it as necessary, depending on the data.

Lemma 12.4.2(i) implies the following ellipticity in the $\boldsymbol{\xi}$–coordinates: By (8.3.37), the change of coordinates $\boldsymbol{\xi}$ to (x,y) in $\Omega\cap\mathcal{D}_{\varepsilon_{\rm eq}}$ and its inverse have C^1–norms bounded by a constant depending only on the data if $\varepsilon_{\rm eq}<\frac{1}{10}\min_{\theta_{\rm w}\in[\theta_{\rm w}^{\rm s},\frac{\pi}{2}]}c_2(\theta_{\rm w})$. Then, from (12.4.22), using (12.3.23) to obtain that $\frac{x}{\operatorname{dist}((x,y),\Gamma_{\rm sonic})}\in(\frac{1}{C},C)$ in the (x,y)–coordinates in $\Omega\cap\mathcal{D}_{\varepsilon_{\rm eq}}$ for C depending on the data, we find that there exists $\tilde{\lambda}>0$ depending only on the data such that, for any $(\mathbf{p},\boldsymbol{\xi})\in\mathbb{R}^2\times(\Omega\cap\mathcal{D}_{\varepsilon_{\rm eq}})$ and $\boldsymbol{\mu}=(\mu_1,\mu_2)\in\mathbb{R}^2$,

$$\tilde{\lambda}\operatorname{dist}(\boldsymbol{\xi},\Gamma_{\rm sonic})|\boldsymbol{\mu}|^2\le\sum_{i,j=1}^2 A_{ij}^2(\mathbf{p},\boldsymbol{\xi})\mu_i\mu_j\le\tilde{\lambda}^{-1}|\boldsymbol{\mu}|^2,\tag{12.4.46}$$

where $A_{ij}^2(\mathbf{p},\boldsymbol{\xi}), i,j=1,2$, are from (12.4.32).

Also, conditions (v)–(vi) of Definition 12.3.2 imply that coefficients A^1_{ij} satisfy the ellipticity condition in $\Omega\setminus\mathcal{D}_{\varepsilon_{\rm eq}/10}$ as asserted in (i), with $\tilde\lambda_0$ depending only on $(\lambda,\rho_{\max},\rho_{\min})$, hence on the data. Combining this with (12.4.46) and using (12.4.42), we obtain (i).

Property (ii) can be seen as follows: Since $A^k_{12}=A^k_{21}$ for $k=1,2$, then $A_{12}=A_{21}$ by (12.4.42). By the symmetry of the coefficients, the ellipticity in assertion (i) implies that $\|A_{ij}\|_{L^\infty(\mathbb{R}^2\times\Omega)}\le\lambda_0^{-1}$. Also, since $A^1_i\equiv 0$, Lemma 12.4.2(ii) implies that $\|A_i\|_{L^\infty(\mathbb{R}^2\times\Omega)}\le N_{\rm eq}$. Now (ii) is proved.

To prove the other properties, we note first that (12.4.43) implies that (12.4.6) holds.

Now property (iii) follows from (12.4.42) and estimate (12.4.6).

Property (iv) follows from (12.4.6), Lemma 12.4.2(ii)–(iii), and the fact that A^1_{ij} are independent of $\mathbf{p}$.

Property (v) follows by using (12.4.43) in the explicit definition (12.4.5) of $(A^1_{ij},A^1_i)(\boldsymbol{\xi})$, combined with (12.4.42) and Lemma 12.4.2(iv).

Property (vi) follows directly from (12.4.42).

Property (vii) follows from property (vi) and Lemma 12.4.2(vi) by changing the variables and noting that $|\boldsymbol{\xi}|=c_2$ on $\Gamma_{\rm sonic}$ (in the shifted coordinates).

Now we show property (viii). Assertion (iii) implies that equation (12.3.25) is the potential flow equation in $\Omega\setminus\mathcal{D}_{\varepsilon_{\rm eq}}$ so that it is of form (12.4.45) in $\Omega\setminus\mathcal{D}_{\varepsilon_{\rm eq}}$. Furthermore, using properties (12.3.7)–(12.3.8) and $\varepsilon_{\rm eq}\le\varepsilon_0\le 1$, we obtain that, in $\Omega\cap(\mathcal{D}_{\varepsilon_{\rm eq}}\setminus\mathcal{D}_{\varepsilon/10})$,

$$
\begin{aligned}
&x\zeta_1\Big(\frac{\psi_x(x,y)}{x}\Big)=\psi_x(x,y),\qquad x^{\frac34}\zeta_1\Big(\frac{\psi_x(x,y)}{x^{3/4}}\Big)=\psi_x(x,y),\\
&(\gamma+1)N_3x\zeta_1\Big(\frac{\psi_y(x,y)}{(\gamma+1)N_3x}\Big)=\psi_y(x,y).
\end{aligned}
\tag{12.4.47}
$$

From this, we conclude that coefficients $(\hat A_{ij},\hat A_i)(D\psi,x,y)$ of (12.4.14) in $\Omega\cap(\mathcal{D}_{\varepsilon_{\rm eq}}\setminus\mathcal{D}_{\varepsilon/10})$ coincide with the coefficients of (11.1.4)–(11.1.5), that is, they are the coefficients of the potential flow equation (divided by c_2) in the (x,y)–coordinates. Thus, the coefficients of equation (12.4.20) in the $\boldsymbol{\xi}$–coordinates coincide with the coefficients of (12.4.2) in $\Omega\cap(\mathcal{D}_{\varepsilon_{\rm eq}}\setminus\mathcal{D}_{\varepsilon/10})$, especially $A^2_k=0$ for $k=1,2$, in that region. Then, using (12.4.42) and the fact that equation (12.4.5) is the potential flow equation (12.4.2) for the fixed point, we conclude that equation (12.3.25) is the potential flow equation (12.4.2) for the fixed point $\psi=\hat\psi$ in $\Omega\cap(\mathcal{D}_{\varepsilon_{\rm eq}}\setminus\mathcal{D}_{\varepsilon/10})$. Finally, from assertion (vi) and Lemma 12.4.2(ix), choosing ε small depending only on the data and the constants in (12.4.7), we conclude that equation (12.3.25) is of form (12.4.31) in the (x,y)–coordinates and form (12.4.45) in the $\boldsymbol{\xi}$–coordinates in $\Omega\cap\mathcal{D}_{\varepsilon_{\rm eq}/2}$. Now property (viii) is proved. □

Combining Lemma 12.4.5 with (12.3.19), (12.3.21), and Definition 12.3.2 where we note that (12.3.19) implies (12.4.43), and recalling Remark 12.3.11, we obtain

Corollary 12.4.6. *There exists $\varepsilon_{\rm eq} \in (0, \frac{\varepsilon_0}{2})$ depending only on the data such that, if $\varepsilon \in (0, \frac{\varepsilon_{\rm eq}}{2})$ in Definition 12.3.2, then, for any $(u, \theta_{\rm w}) \in \overline{\mathcal{K}^{\rm ext}}$, assertions (i)–(vii) of Lemma 12.4.5 hold with $\lambda_0 > 0$ depending only on the data, and C depending only on the data and $(\theta_{\rm w}^*, \alpha)$. Moreover, assertion (viii) of Lemma 12.4.5 holds if $\varepsilon \in (0, \frac{\varepsilon_{\rm eq}}{2})$ is small, depending only on the data and the constants in (12.4.7).*

Next, we show

Lemma 12.4.7. *Let $\varepsilon_{\rm eq} \in (0, \frac{\varepsilon_0}{2})$ be as in Corollary 12.4.6 so that it depends only on the data. If $\varepsilon \in (0, \frac{\varepsilon_{\rm eq}}{2})$ in Definition 12.3.2, then the following hold: Let $(u, \theta_{\rm w}) \in \overline{\mathcal{K}}$ satisfy $\hat{\psi} = \psi$ in Ω for function $\hat{\psi}$ introduced in Definition 12.3.2(vii). Assume also that, in the (x, y)–coordinates,*

$$|\hat{\psi}_x| \leq \frac{2 - \frac{\mu_0}{5}}{1+\gamma} x, \qquad |\hat{\psi}_y| \leq N_3 x$$

hold in $\Omega \cap \mathcal{D}_{\varepsilon/10}$. Then ψ satisfies the potential flow equation (12.4.2) in Ω. Moreover, equation (12.4.2) is strictly elliptic for ψ in $\overline{\Omega} \setminus \overline{\Gamma_{\rm sonic}}$.

Proof. From Lemma 12.3.10, for $(u, \theta_{\rm w}) \in \overline{\mathcal{K}}$, the corresponding ψ satisfies the nonstrict inequalities in (12.3.7)–(12.3.8). Then equation (12.4.14)–(12.4.15) for $\hat{\psi} = \psi$ coincides in $\Omega \cap \mathcal{D}_{\varepsilon_0}$ with equation (11.1.4)–(11.1.5). Thus, writing equation (12.4.14) in the $\boldsymbol{\xi}$–coordinates as (12.4.20) and using coefficients $A_{ij}^{\rm potn}(D\psi, \psi, \boldsymbol{\xi})$ of the potential flow equation (12.4.2), we have

$$c_2 A_{ij}^2(D\psi, \boldsymbol{\xi}) = A_{ij}^{\rm potn}(D\psi, \psi, \boldsymbol{\xi}), \;\; A_i^2(D\psi, \boldsymbol{\xi}) = 0 \qquad \text{in } \Omega \cap \mathcal{D}_{\varepsilon_0};$$

see Remark 12.4.4 regarding coefficient c_2. Furthermore, from (12.4.5), we have

$$A_{ij}^1(\boldsymbol{\xi}) = A_{ij}^{\rm potn}(D\psi, \psi, \boldsymbol{\xi}) \qquad \text{in } \Omega \setminus \mathcal{D}_{\varepsilon_{\rm eq}/10}.$$

Therefore, by (12.4.42), the coefficients of equation (12.3.25) for $\hat{\psi} = \psi$ (satisfying the conditions of this lemma) satisfy that, in Ω,

$$\begin{aligned} A_{ij}(D\psi, \boldsymbol{\xi}) &= \hat{\zeta}_2^{(\varepsilon_{\rm eq})}(\boldsymbol{\xi}) A_{ij}^1(\boldsymbol{\xi}) + \big(1 - \hat{\zeta}_2^{(\varepsilon_{\rm eq})}(\boldsymbol{\xi})\big) c_2 A_{ij}^2(D\psi, \boldsymbol{\xi}) \\ &= A_{ij}^{\rm potn}(D\psi, \psi, \boldsymbol{\xi}), \\ A_i(D\psi, \boldsymbol{\xi}) &= \big(1 - \hat{\zeta}_2^{(\varepsilon_{\rm eq})}(\boldsymbol{\xi})\big) c_2 A_i^2(D\psi, \boldsymbol{\xi}) = 0, \end{aligned} \tag{12.4.48}$$

where we have used that $\varepsilon_{\rm eq} \leq \varepsilon_0$. Then, from equation (12.3.25), we find that ψ is a solution of the potential flow equation (8.3.24) in Ω.

Furthermore, equation (12.3.25) is strictly elliptic for ψ in $\overline{\Omega} \setminus \overline{\Gamma_{\rm sonic}}$, by Lemma 12.4.5(i). Then the same property holds for the potential flow equation for ψ by (12.4.48). □

12.5 ASSIGNING A BOUNDARY CONDITION ON THE SHOCK FOR THE ITERATION

In this section, we work in the $\boldsymbol{\xi}$–coordinates with the origin shifted to center $\mathcal{O}_2$ of the sonic center of state (2). These coordinates have been defined in §8.3.2.1, and the background states in these coordinates are of form (8.3.7)–(8.3.9). Also, we continue to follow the notational convention in Remark 12.3.11.

We introduce the following notations: For a subset $\mathcal{B} \subset \mathbb{R}^2$ and a function $\varphi \in C^1(\mathcal{B})$, denote

$$\mathcal{E}(\varphi, \mathcal{B}) := \{(\mathbf{p}, z, \boldsymbol{\xi}) \,:\, \mathbf{p} = D\varphi(\boldsymbol{\xi}),\ z = \varphi(\boldsymbol{\xi}),\ \boldsymbol{\xi} \in \mathcal{B}\}. \tag{12.5.1}$$

Fix $(u, \theta_{\rm w}) \in C^{2,\alpha}_{*,1+\delta^*}(\mathcal{Q}^{\rm iter}) \times [\theta^*_{\rm w}, \frac{\pi}{2}]$ satisfying conditions (i)–(ii) and (vi) of Definition 12.3.2. Then the corresponding Ω, $\Gamma_{\rm shock}$, φ, and $\psi = \varphi_2 - \varphi$ are defined. We assume that (12.3.10) and (12.3.13) hold.

For a given function ψ, we use the notation defined in (12.5.1) and define $(\mathcal{E}(\psi, \Gamma_{\rm shock}))_\sigma$ to be a σ–neighborhood of $\mathcal{E}(\psi, \Gamma_{\rm shock})$ in the $(\mathbf{p}, z, \boldsymbol{\xi})$–space within set $\{\boldsymbol{\xi} \,:\, \varphi_2(\boldsymbol{\xi}) < \varphi_1(\boldsymbol{\xi})\}$, that is,

$$(\mathcal{E}(\psi, \Gamma_{\rm shock}))_\sigma = \left\{ (\mathbf{p}, z, \boldsymbol{\xi}) \in \mathbb{R}^2 \times \mathbb{R} \times \mathbb{R}^2 \,:\, \begin{array}{r} {\rm dist}(\boldsymbol{\xi}, \Gamma_{\rm shock}) < \sigma \\ |(\mathbf{p}, z) - (D\psi(\boldsymbol{\xi}), \psi(\boldsymbol{\xi}))| < \sigma \end{array} \right\}. \tag{12.5.2}$$

We define a boundary condition on $\Gamma_{\rm shock}$ in (12.3.26) for this $(u, \theta_{\rm w})$ such that

(i) $\mathcal{M}(\mathbf{p}, z, \boldsymbol{\xi})$ is C^3–smooth as a function of $(\mathbf{p}, z, \boldsymbol{\xi})$ in $(\mathcal{E}(\psi, \Gamma_{\rm shock}))_\sigma$ for sufficiently small $\sigma > 0$. The key point here is that the smoothness of $\mathcal{M}$ is independent of the smoothness of $\Gamma_{\rm shock}$.

(ii) $\mathcal{M}(\mathbf{p}, z, \boldsymbol{\xi})$ is homogeneous in the sense that $\mathcal{M}(\mathbf{0}, 0, \boldsymbol{\xi}) = 0$.

(iii) For a fixed point $\hat{\psi} = \psi$, the boundary condition (12.3.26) coincides with the Rankine-Hugoniot condition: $(\rho D\varphi - \rho_1 D\varphi_1) \cdot \boldsymbol{\nu}_{\rm sh} = 0$ on $\Gamma_{\rm shock}$.

In §11.3, we have shown that, if φ is an admissible solution, then $\psi = \varphi - \varphi_2$ satisfies the boundary condition (11.3.1) on $\Gamma_{\rm shock}$ with function $\mathcal{M}_0(\mathbf{p}, z, \boldsymbol{\xi})$ defined by (11.3.2), and condition (11.3.32) on $\Gamma_{\rm shock}$ with function $\mathcal{M}_1$ defined by (11.3.23), which satisfies the homogeneity property (11.3.27).

Function $\mathcal{M}_1$ has some convenient properties when $(\mathbf{p}, z)$ are small. Otherwise, it is more convenient to use function $\mathcal{M}_0$. Thus, below we combine these two conditions and define the corresponding function $\mathcal{M}(\mathbf{p}, z, \boldsymbol{\xi})$.

Furthermore, in view of (12.3.18)–(12.3.19), it suffices to define function $\mathcal{M}(\mathbf{p}, z, \boldsymbol{\xi})$ only for $(\mathbf{p}, z, \boldsymbol{\xi}) \in B_{\hat{M}} \times [-2\hat{M}, 2\hat{M}] \times B_{2M_{\rm dom}}$.

Now we define the boundary condition for the iteration.

Recall that, if φ is a solution of regular reflection-diffraction configuration, then it satisfies the Rankine-Hugoniot conditions:

$$\big(\rho(|D\varphi|^2, \varphi)D\varphi - \rho_1 D\varphi_1\big) \cdot \boldsymbol{\nu}_{\rm sh} = 0 \qquad \text{on } \Gamma_{\rm shock}, \tag{12.5.3}$$

where

$$\boldsymbol{\nu}_{\rm sh} = \frac{D(\varphi_1 - \varphi)}{|D(\varphi_1 - \varphi)|} = \frac{D(\varphi_1 - \varphi_2) - D\psi}{|D(\varphi_1 - \varphi_2) - D\psi|}. \tag{12.5.4}$$

Thus, any solution φ of regular reflection-diffraction configuration satisfies the Rankine-Hugoniot condition on $\Gamma_{\rm shock}$ with $\boldsymbol{\nu}_{\rm sh}$ replaced by the right-hand side of (12.5.4). Expressing this condition in terms of ψ, we obtain the boundary condition (11.3.1)–(11.3.2).

To define the iteration condition (12.3.13), for $(u, \theta_{\rm w}) \in C^{2,\alpha}_{*,1+\delta^*}(\mathcal{Q}^{\rm iter}) \times [\theta^*_{\rm w}, \frac{\pi}{2}]$ satisfying (12.3.13) and conditions (i)–(ii) and (vi) of Definition 12.3.2, we note that the corresponding function φ satisfies that $\varphi = \varphi_1$ on $\Gamma_{\rm shock} = \Gamma_{\rm shock}(u, \theta_{\rm w})$. Thus, if we prescribe

$$\mathcal{M}_0(D\hat{\psi}, \hat{\psi}, \boldsymbol{\xi}) = 0 \qquad \text{on } \Gamma_{\rm shock}, \tag{12.5.5}$$

where $\mathcal{M}_0$ is defined by (11.3.2), then the fixed point $\psi = \hat{\psi}$ satisfies the Rankine-Hugoniot conditions. We later modify the boundary condition (12.5.5) to make it homogeneous. Now let us first discuss some of its properties.

Explicitly, the form of $\mathcal{M}_0(\mathbf{p}, z, \boldsymbol{\xi})$ in the $\boldsymbol{\xi}$–coordinates with the origin shifted to center $\mathcal{O}_2$ of state (2) is

$$\begin{aligned}&\mathcal{M}_0(\mathbf{p}, z, \boldsymbol{\xi})\\ &= \Big(\rho(\mathbf{p},\ z,\ \boldsymbol{\xi})\left(\mathbf{p} + D\varphi_2(\boldsymbol{\xi})\right) - \rho_1 D\varphi_1(\boldsymbol{\xi})\Big) \cdot \frac{D(\varphi_1 - \varphi_2) - \mathbf{p}}{|D(\varphi_1 - \varphi_2) - \mathbf{p}|},\end{aligned} \tag{12.5.6}$$

where

$$\rho(\mathbf{p}, z, \boldsymbol{\xi}) = \Big(\rho_2^{\gamma-1} + (\gamma - 1)(\boldsymbol{\xi} \cdot \mathbf{p} - \frac{|p|^2}{2} - z)\Big)^{\frac{1}{\gamma-1}}, \tag{12.5.7}$$

and $D(\varphi_1 - \varphi_2) = (u_1 - u_2, -v_2)$ is independent of $\boldsymbol{\xi}$. We note that this expression defines $\mathcal{M}_0(\mathbf{p}, z, \boldsymbol{\xi})$ on $\mathcal{D}(\mathcal{M}_0)$ in (11.3.3). For the iteration argument, it suffices to restrict the domain of $\mathcal{M}_0(\mathbf{p}, z, \boldsymbol{\xi})$ to the set:

$$\mathcal{A}_{\mathcal{M}_0} := \left\{ (\mathbf{p}, z, \boldsymbol{\xi}) \ :\ \begin{array}{l} \mathbf{p} \in B_{4\hat{M}},\ z \in (-4\hat{M}, 4\hat{M}),\ \boldsymbol{\xi} \in B_{4M_{\rm dom}} \\ \frac{\rho_{\min}^{\gamma-1}}{2} < \rho_2^{\gamma-1} + (\gamma-1)(\boldsymbol{\xi} \cdot \mathbf{p} - \frac{|\mathbf{p}|^2}{2} - z) < 2\rho_{\max}^{\gamma-1} \\ |\mathbf{p} - (u_1 - u_2, -v_2)| > \frac{\mu_1}{2} \end{array} \right\} \tag{12.5.8}$$

with $\hat{M} = N_4$, where $(N_4, \mu_1, M_{\rm dom})$ are from (12.3.10), (12.3.13), and (12.3.18) respectively, $M_{\rm dom}$ is adjusted for the shifted coordinates but still depends only on the data, and $(\rho_{\min}, \rho_{\max})$ are from Definition 12.3.2(vi). Indeed, if $(u, \theta_{\rm w}) \in C^{2,\alpha}_{*,1+\delta^*}(\mathcal{Q}^{\rm iter}) \times [\theta^*_{\rm w}, \frac{\pi}{2}]$ satisfies (12.3.13) and conditions (i)–(ii) and (vi) of Definition 12.3.2, and if φ, ψ, and $\Gamma_{\rm shock}$ are defined by $(u, \theta_{\rm w})$, then, from the bounds in (12.3.10), (12.3.13), and (12.3.18), we have

$$(\mathcal{E}(\psi, \Gamma_{\rm shock}))_\sigma \subset \mathcal{A}_{\mathcal{M}_0} \cap \{|\boldsymbol{\xi}| \le \frac{3}{2} M_{\rm dom}\}, \tag{12.5.9}$$

where we have used the notation in (12.5.2), and $\sigma > 0$ depends only on the data, since the constants in the properties cited above are determined by the data. Also, from the definitions, it directly follows that, if $\Gamma_{\text{shock}} = \Gamma_{\text{shock}}(\psi)$, and (12.5.5) holds for $\hat{\psi} = \psi$, then $\varphi = \psi + \varphi_2$ satisfies (12.5.3).

From the explicit expression of $\mathcal{M}_0$ in (12.5.6), we have

$$\|\mathcal{M}_0\|_{C^k(\overline{\mathcal{A}_{\mathcal{M}_0}})} \le C_k \qquad \text{for } k = 1, 2, \dots, \tag{12.5.10}$$

where each C_k depends only on the data and μ_1, since constants $\hat{M} = N_4$, M_{dom}, $\rho_{\min}$, and $\rho_{\max}$ depend only on the data.

Also, from Remarks 12.3.7–12.3.8, for small $\sigma > 0$ depending only on the data,

$$A_\sigma := \{(\mathbf{p}, z, \boldsymbol{\xi}) \ : \ |(\mathbf{p}, z)| < \sigma, |\boldsymbol{\xi}| < 4M_{\text{dom}}\} \subset \mathcal{A}_{\mathcal{M}_0}. \tag{12.5.11}$$

The boundary condition (12.5.6) is not homogeneous, *i.e.*, $\mathcal{M}_0(\mathbf{0}, 0, \boldsymbol{\xi}) \neq 0$ in general. Thus, we modify it by defining $\mathcal{M}_1$ from (11.3.23), with $F(z, \boldsymbol{\xi})$ defined by (11.3.20). The boundary condition $\mathcal{M}_1$ is homogeneous, since it satisfies (11.3.27).

From (11.3.20), using that $\boldsymbol{\xi}_{P_1} = (\boldsymbol{\xi}_{P_1} \cdot \boldsymbol{\nu}_{S_1})\boldsymbol{\nu}_{S_1} + (\boldsymbol{\xi}_{P_1} \cdot \boldsymbol{\tau}_{S_1})\boldsymbol{\tau}_{S_1}$, we have

$$F(0, \boldsymbol{\xi}) = \boldsymbol{\xi}_{P_1} + ((\boldsymbol{\xi} - \boldsymbol{\xi}_{P_1}) \cdot \boldsymbol{\tau}_{S_1})\boldsymbol{\tau}_{S_1}.$$

Also, from (11.3.20),

$$|F(z, \boldsymbol{\xi}) - F(0, \boldsymbol{\xi})| \le \frac{|z|}{\sqrt{(u_1 - u_2)^2 + v_2^2}} \le \frac{|z|}{\mu_1},$$

where the last estimate follows from the choice of μ_1 in (12.3.13), since $D\varphi(P_1) = D\varphi_2(P_1)$ for any admissible solution. Combining the last two estimates, we have

$$|F(z, \boldsymbol{\xi})| \le |\boldsymbol{\xi}_{P_1}| + |\boldsymbol{\xi} - \boldsymbol{\xi}_{P_1}| + \frac{|z|}{\mu_1}.$$

Using this and (12.5.11), noting that $|\boldsymbol{\xi}_{P_1}| < M_{\text{dom}}$ for any $\theta_{\text{w}} \in [\theta_{\text{w}}^*, \frac{\pi}{2}]$, by (12.3.18), and using the continuous dependence of $\boldsymbol{\xi}_{P_1}$ on θ_{w} and Remark 12.3.8, we obtain (after possibly reducing σ)

$$(\mathbf{p}, z, F(z, \boldsymbol{\xi})) \in \mathcal{A}_{\mathcal{M}_0} \qquad \text{for any } (\mathbf{p}, z, \boldsymbol{\xi}) \in \mathcal{A}_{\mathcal{M}_1}, \tag{12.5.12}$$

where

$$\mathcal{A}_{\mathcal{M}_1} := \{(\mathbf{p}, z, \boldsymbol{\xi}) \ : \ |(\mathbf{p}, z)| < \sigma, |\boldsymbol{\xi}| < 2M_{\text{dom}}\}. \tag{12.5.13}$$

Thus, $\mathcal{M}_1(\cdot)$ is defined on $\mathcal{A}_{\mathcal{M}_1}$.

Also, using the explicit form of $F(z, \boldsymbol{\xi})$ and (12.5.10), we have

$$\|\mathcal{M}_1\|_{C^k(\overline{\mathcal{A}_{\mathcal{M}_1}})} \le C_k \qquad \text{for } k = 1, 2, \dots, \tag{12.5.14}$$

where each C_k depends only on the data and μ_1.

We also note that the dependence of $(\mathcal{M}_0, \mathcal{M}_1)$ on $\theta_{\rm w}$ is through $(u_2, v_2, \boldsymbol{\xi}_{P_1})$, and these quantities depend on $\theta_{\rm w}$ continuously.

Now we define the operator on the left-hand side of (12.3.26) by combining $\mathcal{M}_0$ and $\mathcal{M}_1$ as follows: Let $\sigma_1 \in (0, \sigma)$ be chosen later. Let $\eta \in C^\infty(\mathbb{R})$ satisfy $\eta \equiv 1$ on $(-\infty, \frac{\sigma_1}{2})$, $\eta \equiv 0$ on (σ_1, ∞), and $\eta' \leq 0$ on $\mathbb{R}$. Define

$$\mathcal{M}(\mathbf{p}, z, \boldsymbol{\xi}) = (1 - \eta(|(\mathbf{p}, z)|))\, \mathcal{M}_0(\mathbf{p}, z, \boldsymbol{\xi}) + \eta(|(\mathbf{p}, z)|)\mathcal{M}_1(\mathbf{p}, z, \boldsymbol{\xi}) \quad (12.5.15)$$

for any $(\mathbf{p}, z, \boldsymbol{\xi}) \in \mathcal{A}_{\mathcal{M}}$, where

$$\mathcal{A}_{\mathcal{M}} = \mathcal{A}_{\mathcal{M}_0} \cap \{(\mathbf{p}, z, \boldsymbol{\xi}) \ : \ |\boldsymbol{\xi}| < 2M_{\rm dom}\}. \quad (12.5.16)$$

Note that the coefficient of $\mathcal{M}_1$ on the right-hand side of (12.5.15) is nonzero only for $(\mathbf{p}, z, \boldsymbol{\xi}) \in \mathcal{A}_{\mathcal{M}_1}$. Then (12.5.15) is well-defined on $\mathcal{A}_{\mathcal{M}}$.

Now, from the properties of $(\mathcal{M}_0, \mathcal{M}_1)$ as proved above and in §11.3, we obtain the following properties of $\mathcal{M}$:

Lemma 12.5.1. *There exists $\sigma > 0$ depending only on the data such that the following hold: Let $\sigma_1 \in (0, \sigma)$. Define functions $\mathcal{M}_0$, $\mathcal{M}_1$, and $\mathcal{M}$ by* (12.5.6), (11.3.23), *and* (12.5.15) *on $(\mathcal{A}_{\mathcal{M}_0}, \mathcal{A}_{\mathcal{M}_1}, \mathcal{A}_{\mathcal{M}})$, respectively. Then*

$$\|\mathcal{M}\|_{C^4(\overline{\mathcal{A}_{\mathcal{M}}})} \leq C, \quad (12.5.17)$$

where C depends only on the data and (μ_1, σ_1).

Furthermore, the following properties hold: Let $(u, \theta_{\rm w}) \in C^{2,\alpha}_{*,1+\delta^*}(Q^{\rm iter}) \times [\theta^*_{\rm w}, \frac{\pi}{2}]$ satisfy (12.3.13) *and conditions* (i)–(ii) *and* (vi) *of Definition* 12.3.2, *and let φ, ψ, and $\Gamma_{\rm shock}$ be defined by $(u, \theta_{\rm w})$. Then*

(i) *For such ψ,*
$$(\mathcal{E}(\psi, \Gamma_{\rm shock}))_\sigma \subset \mathcal{A}_{\mathcal{M}} \subset \mathcal{A}_{\mathcal{M}_0}; \quad (12.5.18)$$

(ii) *Mapping $\theta_{\rm w} \mapsto \mathcal{M}$ is in $C([\theta^*_{\rm w}, \frac{\pi}{2}]; C^4(\overline{\mathcal{A}_{\mathcal{M}}}))$;*

(iii) $\mathcal{M}(D\psi, \psi, \boldsymbol{\xi}) = \mathcal{M}_0(D\psi, \psi, \boldsymbol{\xi})$ *and* $\partial_{\mathbf{p}}\mathcal{M}(D\psi, \psi, \boldsymbol{\xi}) = \partial_{\mathbf{p}}\mathcal{M}_0(D\psi, \psi, \boldsymbol{\xi})$ *on* $\Gamma_{\rm shock}$;

(iv) *Condition $\mathcal{M}(D\hat{\psi}, \hat{\psi}, \boldsymbol{\xi}) = 0$ on $\Gamma_{\rm shock}$ is satisfied by $\hat{\psi} = \psi$ if and only if $\varphi = \varphi_2 + \psi$ satisfies the Rankine-Hugoniot condition* (12.5.3) *on $\Gamma_{\rm shock}(\varphi) = \{\varphi = \varphi_1\}$;*

(v) $\mathcal{M}(\mathbf{0}, 0, \boldsymbol{\xi}) = 0$ *for all $\boldsymbol{\xi} \in B_{2M_{\rm dom}}(\mathbf{0})$.*

Proof. Using (12.5.15), the assertions are proved as follows:

(12.5.17) follows from (12.5.9) which is proved in the lines preceding it;

(i) follows from (12.5.11)–(12.5.12);

(ii) follows from the explicit expressions of $(\mathcal{M}_0, \mathcal{M}_1, \mathcal{M})$ and the continuous dependence of the parameters of state (2) on $\theta_{\rm w} \in [\theta^{\rm s}_{\rm w}, \frac{\pi}{2}]$;

(iii) follows from (11.3.25)–(11.3.26) since $\varphi = \varphi_1$ on Γ_{shock}, *i.e.*, $\psi = \varphi_1 - \varphi_2$ on Γ_{shock};

(iv) follows for $\mathcal{M}_0$ (instead of $\mathcal{M}$) from its definition, because $\varphi = \varphi_1$ on Γ_{shock}; (iv) for $\mathcal{M}$ follows from assertion (iii) as proved above;

(v) follows from Lemma 11.3.5 and (12.5.12). □

We choose constant $\sigma_1 \in (0, \sigma)$ in Lemma 12.5.2 below.

In order to define the operator in (12.3.26), for each $(u, \theta_{\text{w}}) \in \overline{\mathcal{K}^{\text{ext}}}$, we extend $\mathcal{M}$, defined in Lemma 12.5.1, to all $(\mathbf{p}, z, \boldsymbol{\xi}) \in \mathbb{R}^2 \times \mathbb{R} \times \Gamma_{\text{shock}}(u, \theta_{\text{w}})$, in such a way that it satisfies some properties (obliqueness, *etc.*). We do this in the rest of this section.

We first prove some properties of the boundary condition: $\mathcal{M}(D\psi, \psi, \boldsymbol{\xi}) = 0$ on $\Gamma_{\text{shock}} = \Gamma_{\text{shock}}(u, \theta_{\text{w}})$ for (u, θ_{w}) satisfying some subsets of the assumptions in $\overline{\mathcal{K}^{\text{ext}}}$. In view of further applications in Chapter 17, we do not assume that $(u, \theta_{\text{w}}) \in \overline{\mathcal{K}^{\text{ext}}}$; instead, we give the specific conditions needed for each property to hold. In particular, we use those constants in Definition 12.3.2 depending only on the data, *i.e.*, $(\mu_0, \mu_1, \lambda, N_3, N_4, N_5)$, according to Remark 12.3.3, and specify the explicit dependence on the other constants. Also, we use the nonstrict inequalities in the conditions of Definition 12.3.2 as described in Remark 12.3.11.

Lemma 12.5.2. *Let σ be the constant defined in Lemma* 12.5.1. *There exist $\sigma_1 \in (0, \sigma)$ and $\tilde{\varepsilon}_0 > 0$ depending only on the data such that, if σ_1 is used in the definition of η in* (12.5.15), *and $\varepsilon \in (0, \tilde{\varepsilon}_0)$ in Definition* 12.3.2, *then the following holds*:

Let $(u, \theta_{\text{w}}) \in C^1(\overline{Q^{\text{iter}}}) \times [\theta_{\text{w}}^*, \frac{\pi}{2}]$ satisfy conditions (ii) *and* (v)–(vi) *of Definition* 12.3.2 *with $(\lambda, \rho_{\min}, \rho_{\max})$ fixed there depending only on the data, and let the corresponding function $\varphi^{(u,\theta_{\text{w}})}$ satisfy* (12.3.9) *and* (12.3.13)–(12.3.14) *with (μ_1, N_3) depending only on the data. Then, for any $\boldsymbol{\xi} \in \Gamma_{\text{shock}}$,*

$$\delta_{\text{bc}} \le D_{\mathbf{p}}\mathcal{M}(D\psi(\boldsymbol{\xi}), \psi(\boldsymbol{\xi}), \boldsymbol{\xi}) \cdot \boldsymbol{\nu}_{\text{sh}}(\boldsymbol{\xi}) \le \delta_{\text{bc}}^{-1}, \tag{12.5.19}$$

$$D_z\mathcal{M}(D\psi(\boldsymbol{\xi}), \psi(\boldsymbol{\xi}), \boldsymbol{\xi}) \le -\delta_{\text{bc}}, \tag{12.5.20}$$

where $\delta_{\text{bc}} > 0$ depends only on the data, and $\boldsymbol{\nu}_{\text{sh}}$ is the interior normal to Γ_{shock} with respect to Ω.

Proof. From Lemma 12.5.1(iii), it suffices to show (12.5.19) with $\mathcal{M}_0$, instead of $\mathcal{M}$. We show (12.5.19) with $\mathcal{M}_0$ near Γ_{sonic} and away from Γ_{sonic}, separately. We divide the proof into three steps.

1. We first show (12.5.19) at point $P_1 \in \Gamma_{\text{shock}}$, *i.e.*, show that

$$D_{\mathbf{p}}\mathcal{M}_0(D\psi(P_1), \psi(P_1), \boldsymbol{\xi}_{P_1}) \cdot \boldsymbol{\nu}_{\text{sh}}(P_1)$$

is uniformly bounded away from zero and bounded from above.

The uniform upper bound: $|D_{\mathbf{p}}\mathcal{M}_0(D\psi(P_1), \psi(P_1), \boldsymbol{\xi}_{P_1}) \cdot \boldsymbol{\nu}_{\text{sh}}(P_1)| \le C$ for any $\theta_{\text{w}} \in [\theta_{\text{w}}^{\text{s}}, \frac{\pi}{2}]$, with C depending on the data, follows from (11.3.4) and Theorem 7.1.1(i)–(ii).

Now we show the positive lower bound. For this, we find that, from (12.3.21), $(\psi, D\psi)(P_1) = (0, \mathbf{0})$ so that $\boldsymbol{\nu}_{\rm sh}(P_1) = \boldsymbol{\nu}_{S_1}$. Then, from (11.3.30) in Lemma 11.3.6, we obtain that $\delta_{\rm bc}^{(1)} > 0$ depending only on the data such that

$$D_{\mathbf{p}}\mathcal{M}_0(D\psi(P_1), \psi(P_1), \boldsymbol{\xi}_{P_1}) \cdot \boldsymbol{\nu}_{\rm sh}(P_1) \geq \delta_{\rm bc}^{(1)} > 0.$$

Since $(\psi, D\psi)(P_1) = (0, \mathbf{0})$, we use (12.3.9) and (12.5.10) and choose $\tilde{\varepsilon}_0 \in (0, \varepsilon_0)$ small, depending on the data, to obtain

$$D_{\mathbf{p}}\mathcal{M}_0(D\psi(\boldsymbol{\xi}), \psi(\boldsymbol{\xi}), \boldsymbol{\xi}) \cdot \boldsymbol{\nu}_{\rm sh}(\boldsymbol{\xi}) \geq \frac{\delta_{\rm bc}^{(1)}}{2} \qquad \text{for all } \boldsymbol{\xi} \in \Gamma_{\rm shock} \cap \mathcal{D}_{\tilde{\varepsilon}_0}.$$

2. Now we establish the obliqueness of $\mathcal{M}_0$ on $\mathcal{E}(\psi, \Gamma_{\rm shock} \setminus \mathcal{D}_{\tilde{\varepsilon}_0})$.

Since φ, ψ, and $\Gamma_{\rm shock}$ are determined by $(u, \theta_{\rm w})$ (*cf.* Definition 12.2.6), it follows that, for any $\boldsymbol{\xi} \in \Gamma_{\rm shock}$,

$$\boldsymbol{\nu}_{\rm sh} = \frac{D\varphi_1 - D\varphi}{|D\varphi_1 - D\varphi|} = \frac{D(\varphi_1 - \varphi_2) - D\psi}{|D(\varphi_1 - \varphi_2) - D\psi|}.$$

We recall that $\mathcal{M}_0$ in (12.5.5) is defined by (11.3.2) with $g^{\rm sh}(\mathbf{p}, z, \boldsymbol{\xi})$ defined by (7.1.9). Now, using the properties in Definition 12.3.2(v)–(vi) and (12.3.13) with the constants depending only on the data, we can apply Lemma 10.2.1 to obtain from (10.2.5)–(10.2.6) that

$$\delta_{\rm bc}^{(2)} \leq D_{\mathbf{p}}\mathcal{M}_0(D\psi(\boldsymbol{\xi}), \psi(\boldsymbol{\xi}), \boldsymbol{\xi}) \cdot \boldsymbol{\nu}_{\rm sh}(\boldsymbol{\xi}) \leq (\delta_{\rm bc}^{(2)})^{-1} \qquad \text{for all } \boldsymbol{\xi} \in \Gamma_{\rm shock} \setminus \mathcal{D}_{\tilde{\varepsilon}_0},$$

where $\delta_{\rm bc}^{(2)} > 0$ depends only on the data, and we have used that $\tilde{\varepsilon}_0$ depends only on the data. Combining this with Step 1, we obtain (12.5.19) on $\Gamma_{\rm shock}$ for $\mathcal{M}_0$, and hence for $\mathcal{M}$.

3. Now we show (12.5.20). We use (12.5.15) with σ_1 to be chosen below.

We first show that (12.5.20) holds for $\mathcal{M}_0$ on the whole shock $\Gamma_{\rm shock}$. We employ (12.5.6) and recall that $D\psi + D\varphi_2 = D\varphi$ to compute

$$D_z\mathcal{M}_0(D\psi(\boldsymbol{\xi}), \psi(\boldsymbol{\xi}), \boldsymbol{\xi}) = -\rho^{2-\gamma} D\varphi \cdot \boldsymbol{\nu}_{\rm sh} \leq -\delta_{\rm bc}^{(3)} \qquad \text{for all } \boldsymbol{\xi} \in \Gamma_{\rm shock} \tag{12.5.21}$$

with $\delta_{\rm bc}^{(3)} > 0$ depending only on the data, where the last inequality follows from (12.3.14) and Definition 12.3.2(vi), and we have used the fact that $\tilde{\rho}' = \tilde{\rho}^{2-\gamma}$.

Also, from (12.5.14) and Lemma 11.3.6, we obtain that, for sufficiently small $\hat{\sigma}, \delta_{\rm bc}^{(3)} > 0$ depending only on the data,

$$D_z\mathcal{M}_1(\mathbf{p}, z, \boldsymbol{\xi}) \leq -\delta_{\rm bc}^{(3)} \qquad \text{for all } |\boldsymbol{\xi}| \leq M_{\rm dom} \text{ and } |(\mathbf{p}, z)| \leq \hat{\sigma}. \tag{12.5.22}$$

From its definition (12.5.15),

$$\begin{aligned} D_z\mathcal{M}(\mathbf{p}, z, \boldsymbol{\xi}) = & (1 - \eta(|(\mathbf{p}, z)|))\partial_z\mathcal{M}_0(\mathbf{p}, z, \boldsymbol{\xi}) + \eta(|(\mathbf{p}, z)|)\partial_z\mathcal{M}_1(\mathbf{p}, z, \boldsymbol{\xi}) \\ & + \frac{z}{|(\mathbf{p}, z)|}\eta'(|(\mathbf{p}, z)|)\big(\mathcal{M}_1(\mathbf{p}, z, \boldsymbol{\xi}) - \mathcal{M}_0(\mathbf{p}, z, \boldsymbol{\xi})\big). \end{aligned} \tag{12.5.23}$$

Consider this last expression at $(\mathbf{p}, z, \boldsymbol{\xi}) = (D\psi(\boldsymbol{\xi}), \psi(\boldsymbol{\xi}), \boldsymbol{\xi})$ with $\boldsymbol{\xi} \in \Gamma_{\text{shock}}$. Then, from Lemma 12.5.1(iii), the last term vanishes. Furthermore, let $\hat{\sigma}$ be from (12.5.22). Since $\Gamma_{\text{shock}} = \{\varphi = \varphi_1\}$ so that $\boldsymbol{\nu}_{\text{sh}} = \frac{D(\varphi_1-\varphi_2)-D\psi}{|D(\varphi_1-\varphi_2)-D\psi|}$, then, using that $|D(\varphi_1 - \varphi_2)| > \mu_1$ by Remark 12.3.8, there exist $\tilde{\sigma}_1 \in (0,1)$ and $\tilde{C} > 0$, depending only on μ_1, and hence on the data, such that the following property holds: If $\sigma_1 \in (0, \tilde{\sigma}_1)$ and $|D\psi(\boldsymbol{\xi}_P)| \leq \sigma_1$ for some $P \in \Gamma_{\text{shock}}$, then

$$|\boldsymbol{\nu}_{\text{sh}}(P) - \boldsymbol{\nu}_{\mathcal{S}_1}| \leq \tilde{C}\sigma_1.$$

We choose $\sigma_1 \in (0, \tilde{\sigma}_1)$ small, depending only on the data, such that $\tilde{C}\sigma_1 \leq \hat{\sigma}$. We use this σ_1 in the definition of η for (12.5.15). Then, using (12.5.21)–(12.5.22) and the fact that $\eta \equiv 0$ on (σ, ∞), we obtain from (12.5.23) (for which the last term vanishes as shown above) that

$$\partial_z \mathcal{M}(\mathbf{p}, z, \boldsymbol{\xi}) \leq -\min\{\delta_{\text{bc}}^{(2)}, \delta_{\text{bc}}^{(3)}\}.$$

□

From now on, we fix σ_1 in (12.5.15) as in Lemma 12.5.2, depending only on the data.

Next we show that, when $(u, \theta_{\text{w}}) \in C^1(\overline{Q^{\text{iter}}}) \times [\theta_{\text{w}}^*, \frac{\pi}{2}]$ is sufficiently close to $\overline{\mathcal{K}}$, the direction of the oblique derivative condition of Γ_{shock} at P_1 is sufficiently different from the direction of the oblique condition on Γ_{sym} at the same point, *i.e.*, from $\boldsymbol{\nu}_{\text{sym}} = (1, 0)$.

Before we state the next lemma, we note the following fact: Let $(u, \theta_{\text{w}}) \in C^1(\overline{Q^{\text{iter}}}) \times [\theta_{\text{w}}^*, \frac{\pi}{2}]$ satisfy the conditions given in Lemma 12.5.2 and

$$\|u\|_{C^1(\overline{Q^{\text{iter}}}) \cap \{s \geq 1/2\}} \leq N_0. \tag{12.5.24}$$

Let Ω, Γ_{shock}, Γ_{sym}, P_2, P_3, and ψ correspond to (u, θ_{w}), and let

$$\hat{\psi} \in C^2(\Omega(u, \theta_{\text{w}})) \cap C^1(\overline{\Omega}(u, \theta_{\text{w}}))$$

be such that $\hat{u}$ defined by (12.3.15) satisfies (12.3.16) with $\delta_3 \in (0,1)$ and, if $|P_2 - P_3| \geq b$ for some $b > 0$, then

$$|(\hat{\psi} - \psi, D(\hat{\psi} - \psi))|(P_2) \leq \tilde{C}\delta_3, \tag{12.5.25}$$

where $\tilde{C}$ depends only on the data and (N_0, b). The proof is directly from Lemma 12.2.2(ii) and the explicit expressions in (12.2.41) and Lemma 12.2.6(v).

Lemma 12.5.3. *Let $N_0 \geq 1$ and $b \in (0,1)$. Let σ_1 and $\tilde{\varepsilon}_0$ be as in Lemma* 12.5.2, *depending only on the data. Then there exists $\delta_3 > 0$ small, depending only on the data and (N_0, b), such that the following holds*:

Let $\varepsilon \in (0, \tilde{\varepsilon}_0)$. Let $(u, \theta_{\text{w}}) \in C^1(\overline{Q^{\text{iter}}}) \times [\theta_{\text{w}}^*, \frac{\pi}{2}]$ satisfy (12.5.24) *and the conditions in Lemma* 12.5.2. *Let Ω, Γ_{shock}, Γ_{sym}, P_2, P_3, and ψ correspond to*

$(u, \theta_{\rm w})$, and let $|P_2 - P_3| \geq b$. Assume that there exists $\hat{\psi} \in C^2(\Omega) \cap C^1(\overline{\Omega})$ satisfying (12.3.29) *and*

$$\mathcal{M}(D\hat{\psi}, \hat{\psi}, \boldsymbol{\xi}) = 0 \qquad \text{on } \Gamma_{\rm shock}$$

such that function $\hat{u}$ defined by (12.3.15) *satisfies the nonstrict inequality in* (12.3.16) *with $\alpha = 0$, where $\mathcal{M}(D\hat{\psi}, \hat{\psi}, \boldsymbol{\xi})$ is well-defined by Lemma* 12.5.1(i) *with σ_1 chosen in Lemma* 12.5.2, *if δ_3 in* (12.5.25) *is sufficiently small, depending only on the data and b. Then*

$$\left| \frac{D_{\mathbf{p}}\mathcal{M}}{|D_{\mathbf{p}}\mathcal{M}|}(D\psi(P_2),\, \psi(P_2),\, P_2) \pm \boldsymbol{\nu}_{\rm sym} \right| \geq 1. \tag{12.5.26}$$

Proof. Since $\hat{\psi} = \hat{\varphi} - \varphi_2^{(\theta_{\rm w})}$ satisfies (12.3.29), and $\boldsymbol{\nu}_{\rm sym} = (0,1)$, we have

$$D\hat{\varphi} \cdot (0,1) = 0 \qquad \text{on } \Gamma_{\rm sym}.$$

Owing to $\hat{\varphi} \in C^1(\overline{\Omega})$, we have

$$|D\hat{\varphi}(P_2) \cdot (0,1)| = 0. \tag{12.5.27}$$

Recall that $\mathcal{M}_0$ in (12.5.5) is defined by (11.3.2) with $g^{\rm sh}(\mathbf{p}, z, \boldsymbol{\xi})$ defined by (7.1.9) and that we work in the $\boldsymbol{\xi}$–coordinates with the origin shifted from P_3 to $\mathcal{O}_2$. Since $P_2 = (\xi_{1P_2}, 0)$ with $\xi_{1P_2} < 0$, and $D\varphi_1(\boldsymbol{\xi}) = (u_1 - \xi_1, -\xi_2)$ in the non-shifted $\boldsymbol{\xi}$–coordinates, it follows that

$$\frac{D\varphi_1(P_2)}{|D\varphi_1(P_2)|} = (1,0) = \frac{D\hat{\varphi}(P_2)}{|D\hat{\varphi}(P_2)|},$$

where we have used (12.5.27) in the second equality. Then, by explicit calculation via (7.1.9), *i.e.*, computing $D_{\mathbf{p}}g^{\rm sh}(D\hat{\varphi}(P_2),\, \hat{\varphi}(P_2),\, P_2)$, we have

$$D_{\mathbf{p}}\mathcal{M}_0 = (|D_{\mathbf{p}}\mathcal{M}_0|, 0) \qquad \text{at } (\mathbf{p}, z, \boldsymbol{\xi}) = (D\hat{\psi}(P_2),\, \hat{\psi}(P_2),\, P_2) =: \mathcal{X}_1. \tag{12.5.28}$$

Using the fact that $\varphi = \varphi_1$ at P_2, we obtain that, from (11.3.26),

$$D_{\mathbf{p}}\mathcal{M}_0 = D_{\mathbf{p}}\mathcal{M}_1 \qquad \text{at } (\mathbf{p}, z, \boldsymbol{\xi}) = (D\hat{\psi}(P_2),\, \psi(P_2),\, P_2) =: \mathcal{X}_2.$$

Thus, using (12.5.10) and (12.5.14),

$$\begin{aligned}
&|D_{\mathbf{p}}\mathcal{M}_1(\mathcal{X}_1) - D_{\mathbf{p}}\mathcal{M}_0(\mathcal{X}_1)| \\
&\quad = \big|\big(D_{\mathbf{p}}\mathcal{M}_1(\mathcal{X}_1) - D_{\mathbf{p}}\mathcal{M}_1(\mathcal{X}_2)\big) + \big(D_{\mathbf{p}}\mathcal{M}_0(\mathcal{X}_2) - D_{\mathbf{p}}\mathcal{M}_0(\mathcal{X}_1)\big)\big| \\
&\quad \leq C|\mathcal{X}_2 - \mathcal{X}_1| = C|\psi(P_2) - \hat{\psi}(P_2)| \leq C\delta_3,
\end{aligned}$$

where C depends only on the data and (μ_1, N_0, b), and we have used (12.5.25) in the last inequality. From this, combined with (12.5.15) and (12.5.28),

$$|D_{\mathbf{p}}\mathcal{M} - (|D_{\mathbf{p}}\mathcal{M}|, 0)| \leq C\delta_3 \qquad \text{at } (\mathbf{p}, z, \boldsymbol{\xi}) = (D\hat{\psi}(P_2),\, \hat{\psi}(P_2),\, P_2). \tag{12.5.29}$$

Moreover, using (12.3.13) and conditions (v)–(vi) of Definition 12.3.2 for $(u, \theta_{\rm w})$, we can apply Lemma 10.2.1 to obtain that, from (10.2.6),

$$|D_{\mathbf{p}}\mathcal{M}(D\hat{\psi}(P_2),\, \hat{\psi}(P_2),\, P_2)| \ge \frac{1}{C}, \tag{12.5.30}$$

where C depends only on the data. Thus, using $\boldsymbol{\nu}_{\rm sym} = (0,1)$, we have

$$\left|\frac{D_{\mathbf{p}}\mathcal{M}}{|D_{\mathbf{p}}\mathcal{M}|}(D\hat{\psi}(P_2),\, \hat{\psi}(P_2),\, P_2) \pm \boldsymbol{\nu}_{\rm sym}\right| \ge |(1,0) \pm (0,1)| - C\delta_3,$$

where C depends only on (N_0, b) and the data. From this, using (12.5.17) and (12.5.25), we obtain

$$\left|\frac{D_{\mathbf{p}}\mathcal{M}}{|D_{\mathbf{p}}\mathcal{M}|}(D\psi(P_2),\, \psi(P_2),\, P_2) \pm \boldsymbol{\nu}_{\rm sym}\right| \ge \sqrt{2} - C(N_0, b)\delta_3.$$

Choosing δ_3 small, we obtain (12.5.26). □

Corollary 12.5.4. *Let $N_0 \ge 1$ and $b \in (0,1)$. Let σ_1 and $\varepsilon > 0$ be as in Lemma 12.5.2. Then there exist small $\delta_{\mathcal{K}}, \delta_3 > 0$ depending only on the data and (N_0, b) such that the following holds:*

Let $(u, \theta_{\rm w}), (u^{\#}, \theta_{\rm w}^{\#}) \in C^1(\overline{Q^{\rm iter}}) \times [\theta_{\rm w}^*, \frac{\pi}{2}]$ satisfy

$$\|u^{\#} - u\|_{C^1(\overline{Q^{\rm iter}} \cap \{s \ge 1/2\})} + |\theta_{\rm w}^{\#} - \theta_{\rm w}| \le \delta_{\mathcal{K}}, \tag{12.5.31}$$

let $(u, \theta_{\rm w})$ satisfy condition (ii) *of Definition* 12.3.2, *and let $(u^{\#}, \theta_{\rm w}^{\#})$ satisfy the conditions in Lemma* 12.5.3 *(with the constants fixed above). Then ψ and P_2 corresponding to $(u, \theta_{\rm w})$ satisfy*

$$\left|\frac{D_{\mathbf{p}}\mathcal{M}}{|D_{\mathbf{p}}\mathcal{M}|}(D\psi(P_2),\, \psi(P_2),\, P_2) \pm \boldsymbol{\nu}_{\rm sym}\right| \ge \frac{3}{4}. \tag{12.5.32}$$

Proof. Let $\Omega^{\#}$, ${P_2}^{\#}$, $\psi^{\#}$, and $\varphi^{\#}$ be defined by $(u^{\#}, \theta_{\rm w}^{\#})$. Since $P_2 = \mathfrak{F}_{(u,\theta_{\rm w})}(1,1)$ and ${P_2}^{\#} = \mathfrak{F}_{(u^{\#},\theta_{\rm w}^{\#})}(1,1)$, then, using (12.2.61)–(12.2.62) in Lemma 12.2.7 (with $\alpha = 0$), we find that there exists $C_{\mathcal{K}} > 0$ depending only on the data and N_0 such that

$$|P_2 - {P_2}^{\#}| + |D\psi(P_2) - D\psi^{\#}({P_2}^{\#})| \le C_{\mathcal{K}}\delta_{\mathcal{K}}. \tag{12.5.33}$$

Since $(u^{\#}, \theta_{\rm w}^{\#})$ satisfies the conditions in Lemma 12.5.3, then $\psi^{\#}$ and ${P_2}^{\#}$ satisfy (12.5.26). With this, using (12.5.17) and choosing $\delta_{\mathcal{K}}$ small, we conclude (12.5.32). □

Next, we study the boundary condition $\mathcal{M}$ near P_1 in the (x,y)–coordinates. From the change of coordinates, it is of the form:

$$\hat{\mathcal{M}}(\psi_x, \psi_y, \psi, x, y) = 0 \qquad \text{on } \Gamma_{\rm shock} \cap \{0 < x < \varepsilon_0\}, \tag{12.5.34}$$

where

$$\hat{\mathcal{M}}(\mathbf{p}, z, x, y)$$
$$= \mathcal{M}\big(-p_1(\cos y, \sin y) - \frac{p_2}{c_2 - x}(\sin y, \cos y), z,\ (c_2 - x)(\cos y, \sin y)\big). \tag{12.5.35}$$

Lemma 12.5.5. *There exist $\varepsilon_{\rm bc}, \delta_{\rm bc}, C > 0$ depending only on the data such that, for any $\theta_{\rm w} \in [\theta_{\rm w}^*, \frac{\pi}{2}]$ and all $(\mathbf{p}, z, x, y)$ satisfying $|(\mathbf{p}, z)| \le \delta_{\rm bc}$, $0 < x - x_{P_1} \le \varepsilon_{\rm bc}$, and $|y - y_{P_1}| \le \delta_{\rm bc}$,*

$$D_{p_i}\hat{\mathcal{M}}(\mathbf{p}, z, x, y) \le -\frac{1}{C} \qquad \text{for } i = 1, 2, \tag{12.5.36}$$

$$D_z\hat{\mathcal{M}}(\mathbf{p}, z, x, y) \le -\frac{1}{C}. \tag{12.5.37}$$

Proof. We first note that, from (12.5.15) with σ_1 chosen in Lemma 12.5.2 depending on the data and μ_1, as well as (12.5.17) and (11.3.34), it follows that $\hat{\mathcal{M}} \equiv B_1$ in a uniform neighborhood of $(\mathbf{0}, 0, x_{P_1}, y_{P_1})$ with $x_{P_1} = 0$. Then the assertion follows from Lemma 11.3.7. □

Finally, we extend function $\mathcal{M}$ defined in Lemma 12.5.1 to all $(\mathbf{p}, z) \in \mathbb{R}^2 \times \mathbb{R}$ so that the resulting boundary condition is oblique and has the C^3–regularity and the other structural properties as in Lemma 12.5.1.

We now define the extension of $\mathcal{M}$: Let $(u, \theta_{\rm w}) \in \overline{\mathcal{K}^{\rm ext}}$. Let Ω, Γ, φ, and ψ correspond to $(u, \theta_{\rm w})$. We first note that, from (12.5.17) with σ_1 from Lemma 12.5.2, combined with (12.5.9) and (12.5.16), there exist C and $\sigma > 0$ depending only on the data such that, for $k = 1, \dots, 4$,

$$|(\mathcal{M}, D^k\mathcal{M})(\mathbf{p}, z, \boldsymbol{\xi})| \le C \tag{12.5.38}$$

for all $(\mathbf{p}, z, \boldsymbol{\xi}) \in \mathbb{R}^2 \times \mathbb{R} \times \overline{\Omega}$ satisfying $|(\mathbf{p}, z) - (D\psi(\boldsymbol{\xi}), \psi(\boldsymbol{\xi}))| \le 2\sigma$.

We first consider ψ satisfying (12.3.21) and regularize it as follows:

Lemma 12.5.6. *Let $\hat{N}_0 > 0$, $N_2 > 0$, and $\alpha \in (0, \frac{1}{8})$. Let $(u, \theta_{\rm w}) \in C^1(\overline{Q^{\rm iter}}) \times [\theta_{\rm w}^*, \frac{\pi}{2}]$ satisfy condition* (ii) *of Definition* 12.3.2, *let the corresponding $\mathfrak{g}_{\rm sh}^{(u,\theta_{\rm w})}$ and $\varphi^{(u,\theta_{\rm w})}$ satisfy* (12.3.5) *and* (12.3.21) *with $\hat{N}_0$ and N_2 fixed above, and let $\|\mathfrak{g}_{\rm sh}^{(u,\theta_{\rm w})}\|_{C^1([0,\hat{s}(\theta_{\rm w})])} \le \hat{N}_0$. Then, for any $\tilde{\sigma}_1 \in (0, 1)$, there exists $v \equiv v^{(u,\theta_{\rm w})} \in C^4(\overline{\Omega}^{(u,\theta_{\rm w})})$ such that*

(i) *$v^{(u,\theta_{\rm w})}$ satisfies*

$$\|v - \psi\|_{C^1(\overline{\Omega})} \le \tilde{\sigma}_1^2, \tag{12.5.39}$$

$$\|v\|_{C^4(\overline{\Omega})} \le C(\tilde{\sigma}_1), \tag{12.5.40}$$

where $C(\tilde{\sigma}_1)$ depends only on the data and $(\hat{N}_0, N_2, \alpha)$;

(ii) $v^{(u,\theta_{\rm w})}$ *depends continuously on* $(u,\theta_{\rm w})$ *in the sense that, if* $(u^{(j)},\theta_{\rm w}^{(j)})\in C^1(\overline{Q^{\rm iter}})\times[\theta_{\rm w}^*,\frac{\pi}{2}]$ *satisfy the conditions above, and if* $(u^{(j)},\theta_{\rm w}^{(j)})\to(u,\theta_{\rm w})$ *in* $C^{1,\alpha}(\overline{Q^{\rm iter}})\times[\theta_{\rm w}^*,\frac{\pi}{2}]$ *as* $j\to\infty$, *then*

$$v^{(u^{(j)},\theta_{\rm w}^{(j)})}\circ\mathfrak{F}_{(u^{(j)},\theta_{\rm w}^{(j)})}\to v^{(u,\theta_{\rm w})}\circ\mathfrak{F}_{(u,\theta_{\rm w})}\qquad in\ C^{1,\alpha}(\overline{Q^{\rm iter}})\quad as\ j\to\infty.$$

In particular, for a fixed $\tilde\sigma_1\in(0,1)$, *if the map:* $(u,\theta_{\rm w})\mapsto v\equiv v^{(u,\theta_{\rm w})}$ *is restricted to* $(u,\theta_{\rm w})\in\overline{\mathcal{K}^{\rm ext}}$, *then assertions* (i)–(ii) *are true, and* $C(\tilde\sigma_1)$ *in* (12.5.40) *depends only on the data and* $(\theta_{\rm w}^*,\alpha)$ *in Definition* 12.3.2.

Proof. In this proof, the universal constant C depends only on the data and $(\hat N_0,N_2,\alpha)$, and constant $C(\tilde\sigma_1)$ depends only on the data and $(\hat N_0,N_2,\alpha,\tilde\sigma_1)$.

Let $(u,\theta_{\rm w})$ satisfy the properties given in Definition 12.3.2(ii), and let (Ω,φ) be determined by $(u,\theta_{\rm w})$ as in Definition 12.2.6(iii)–(v), and $\psi=\varphi-\varphi_2^{(\theta_{\rm w})}$. Then, by Lemma 12.2.2(i), $\tilde w:=\psi\circ F_1^{-1}$ is defined on

$$F_1(\Omega)=\{(s,t)\,:\,0<s<\hat s(\theta_{\rm w}),\ 0<t<\mathfrak{g}_{\rm sh}(s)\}.$$

In order to obtain (12.5.39), we will define v by mollifying a function that is close to $\tilde w$ in $C^{1,\alpha}(\overline{F_1(\Omega)})$ and $C^{1,\alpha}$–smooth in a neighborhood of $F_1(\Omega)$ to obtain the approximation in $C^{1,\alpha}$ up to $\partial F_1(\Omega)$, and then by mapping back to Ω via F_1. Thus, we construct the approximate function in a larger region, and do this in such a way that allows us to obtain the continuity with respect to $(u,\theta_{\rm w})$, which yields assertion (ii).

Let the large number K be fixed below. Let $\delta\in(0,1)$. Define

$$\tilde w_\delta^{(1)}(s,t):=\tilde w\Big(\frac{s+\frac{\delta}{2K}\hat s}{1+\frac{\delta}{K}},\ \frac{t+\frac{\delta}{2N_2}}{1+\delta}\Big)\tag{12.5.41}$$

on

$$\mathcal{A}_\delta:=\left\{(s,t)\,:\,\begin{array}{c}-\frac{\delta}{2K}\hat s<s<(1+\frac{\delta}{2K})\hat s\\[4pt] -\frac{\delta}{2N_2}<t<(1+\delta)\mathfrak{g}_{\rm sh}\big(\frac{s+\frac{\delta}{2K}\hat s}{1+\frac{\delta}{K}}\big)-\frac{\delta}{2N_2}\end{array}\right\},$$

where we have used that $(s,t)\mapsto\big(\frac{s+\frac{\delta}{2K}\hat s}{1+\frac{\delta}{K}},\ \frac{t+\frac{\delta}{2N_2}}{1+\delta}\big)$ is a one-to-one map of $\mathcal{A}_\delta$ onto $F_1(\Omega)=\{(s,t)\,:\,0<s<\hat s(\theta_{\rm w}),\ 0<t<\mathfrak{g}_{\rm sh}(s)\}$ and have written $\hat s$ for $\hat s(\theta_{\rm w})$.

Using (12.3.21), we obtain

$$\|\tilde w_\delta^{(1)}\|_{C^{1,\alpha}(\overline{\mathcal{A}_\delta})}\le C,\quad \|\tilde w_\delta^{(1)}-\tilde w\|_{C^1(\overline{F_1(\Omega)})}\le C\delta.\tag{12.5.42}$$

Now we prove that, if K is sufficiently large depending on $(\hat N_0,N_2)$ and the data, then there exists $\hat C>0$, depending only on $\min_{[\theta_{\rm w}^*,\frac{\pi}{2}]}\hat s(\cdot)$ (hence on the data) and $(\hat N_0,N_2)$, such that

$$\mathcal{N}_{\frac{\delta}{\hat C}}(F_1(\Omega))\subset\mathcal{A}_\delta\qquad\text{for all }\delta\in(0,1).\tag{12.5.43}$$

To show (12.5.43), we fix $\delta \in (0,1)$ and compare the upper ends of t–intervals in the expressions of both $\mathcal{A}_\delta$ and $F_1(\Omega)$ above. We first note that, for each $s \in (0, \hat{s})$,

$$\Big|\mathfrak{g}_{\rm sh}(\frac{s+\frac{\delta}{2K}\hat{s}}{1+\frac{\delta}{K}}) - \mathfrak{g}_{\rm sh}(s)\Big| \le \frac{C\delta}{K},$$

where we have used the bound that $\|\mathfrak{g}_{\rm sh}^{(u,\theta_{\rm w})}\|_{C^1([0,\hat{s}(\theta_{\rm w})])} \le \hat{N}_0$. Then condition (12.3.5) implies that, for each $s \in (0, \hat{s})$,

$$\Big((1+\delta)\mathfrak{g}_{\rm sh}(\frac{s+\frac{\delta}{2K}\hat{s}}{1+\frac{\delta}{K}}) - \frac{\delta}{2N_2}\Big) - \mathfrak{g}_{\rm sh}(s) \ge \frac{\delta}{2N_2} - \frac{C\delta}{K} \ge \frac{\delta}{4N_2}, \qquad (12.5.44)$$

where the last inequality is obtained by choosing K large.

Now it follows that $F_1(\Omega) \subset \mathcal{A}_\delta$. Then we need to estimate the distance between $\partial F_1(\Omega)$ and $\partial \mathcal{A}_\delta$. Both boundaries consist of three flat segments and one curved part. From the explicit expressions of the sets, it is easy to see that the flat part of $\partial\mathcal{A}_\delta$ is on the distance at least $\delta \min\{\frac{\min_{[\theta_{\rm w}^*,\frac{\pi}{2}]}\hat{s}(\cdot)}{2K}, \frac{1}{2N_2}\}$ (*i.e.*, $\frac{\delta}{C}$) from $\partial F_1(\Omega)$.

Using once again the explicit form of $\partial F_1(\Omega)$ and $\partial\mathcal{A}_\delta$, we see that, to complete the proof of (12.5.43), it remains to estimate the distance between the curved parts of boundaries $\partial F_1(\Omega)$ and $\partial\mathcal{A}_\delta$. By (12.5.44), for each $s \in (0, \hat{s})$, the distance in the t–direction between the curved parts of $\partial\mathcal{A}_\delta$ and $\partial F_1(\Omega)$ is at least $\frac{\delta}{4N_2}$. Then, employing again the bound of the C^1–norm of $\mathfrak{g}_{\rm sh}$, we obtain that the distance between the curved parts of the boundaries is at least $\frac{\delta}{C}$. Therefore, (12.5.43) is proved.

Now, we can define

$$w = \tilde{w}_\delta^{(1)} * \zeta_{\delta/\hat{C}} \qquad \text{in } F_1(\Omega) \qquad (12.5.45)$$

with $\zeta_{\hat{\delta}}(\boldsymbol{\xi}) := \frac{1}{\hat{\delta}^2}\zeta(\frac{\boldsymbol{\xi}}{\hat{\delta}})$, where $\zeta(\cdot)$ is a standard mollifier, that is, $\zeta \in C_0^\infty(\mathbb{R}^2)$ is a nonnegative function with $\mathrm{supp}(\zeta) \subset B_1(\mathbf{0})$ and $\int_{\mathbb{R}^2}\zeta(\boldsymbol{\xi})\,d\boldsymbol{\xi} = 1$. Then, using the first estimate in (12.5.42), we find that $\|\tilde{w}_\delta^{(1)} - w\|_{C^1(\overline{F_1(\Omega)})} \le C_\alpha\delta^\alpha$. From this, using the second estimate in (12.5.42), we obtain

$$\|\tilde{w} - w\|_{C^1(\overline{F_1(\Omega)})} \le C_\alpha\delta^\alpha. \qquad (12.5.46)$$

Defining $v := w \circ F_1$ and recalling that $\tilde{w} = \psi \circ F_1^{-1}$, we apply Lemma 12.2.2(ii) to obtain

$$\|\psi - v\|_{C^1(\overline{\Omega})} \le C_\alpha\delta^\alpha.$$

Then, for any $\tilde{\sigma}_1$, we can choose δ so that (12.5.39) holds. This fixes the choice of δ in (12.5.41), depending only on the data and $(\hat{N}_0, N_2, \alpha, \tilde{\sigma}_1)$. Then (12.5.45) implies that $\|w\|_{C^4(\overline{F_1(\Omega)})} \le C(\tilde{\sigma}_1)$. Since $v = w \circ F_1$, Lemma 12.2.2(ii) implies (12.5.40). This concludes assertion (i).

For each $\tilde{\sigma}_1$, constant δ is fixed, depending only on the data and $(\hat{N}_0, N_2, \alpha, \tilde{\sigma}_1)$, and hence is independent of (u, θ_{w}), satisfying the conditions of this lemma. Then assertion (ii) follows from the explicit construction of v, Lemma 12.2.2(iii), and Lemma 12.2.7(vi). □

Let $\eta \in C^\infty(\mathbb{R})$ be a cutoff function such that $\eta \equiv 1$ on $(-\infty, 1)$, $\eta \equiv 0$ on $(2, \infty)$, and $0 \le \eta \le 1$ on $\mathbb{R}$. Fix $\tilde{\sigma}_1 \in (0,1)$ and denote

$$\eta_{\tilde{\sigma}_1}(t) = \eta(\frac{t}{\tilde{\sigma}_1}).$$

Let $\tilde{\sigma}_1 \in (0,1)$ be defined later, depending only on the data and θ^*_{w}. Then we define $\mathcal{M}_{(u,\theta_{\mathrm{w}})}(\mathbf{p}, z, \boldsymbol{\xi})$ for $(\mathbf{p}, z, \boldsymbol{\xi}) \in \mathbb{R}^2 \times \mathbb{R} \times \overline{\Omega}$ by

$$\mathcal{M}_{(u,\theta_{\mathrm{w}})}(\mathbf{p}, z, \boldsymbol{\xi}) = \eta_{\tilde{\sigma}_1}\mathcal{M}(\mathbf{p}, z, \boldsymbol{\xi}) + (1-\eta_{\tilde{\sigma}_1})\mathcal{L}(\mathbf{p} - Dv(\boldsymbol{\xi}), z - v(\boldsymbol{\xi}), \boldsymbol{\xi}), \tag{12.5.47}$$

where $\eta_{\tilde{\sigma}_1} = \eta_{\tilde{\sigma}_1}(|(\mathbf{p}, z) - (Dv(\boldsymbol{\xi}), v(\boldsymbol{\xi}))|)$ and

$$\begin{aligned}\mathcal{L}(\mathbf{p}, z, \boldsymbol{\xi}) = {} & \mathcal{M}(Dv(\boldsymbol{\xi}), v(\boldsymbol{\xi}), \boldsymbol{\xi}) + D_{\mathbf{p}}\mathcal{M}(Dv(\boldsymbol{\xi}), v(\boldsymbol{\xi}), \boldsymbol{\xi}) \cdot \mathbf{p} \\ & + D_z\mathcal{M}(Dv(\boldsymbol{\xi}), v(\boldsymbol{\xi}), \boldsymbol{\xi})z\end{aligned}$$

with $v(\boldsymbol{\xi})$ defined in Lemma 12.5.6 for (u, θ_{w}) and $\tilde{\sigma}_1$.

Lemma 12.5.7. *Let $N_0, \hat{N}_0, N_2 > 0$, $b \in (0,1)$, and $\alpha \in (0, \frac{1}{8})$. There exist positive constants:*

$(N_1, \varepsilon_{\mathrm{bc}}, \delta_{\mathrm{bc}}, \delta_1, C)$ depending only on the data,

$(\delta_{\mathcal{K}}, \delta_3)$ depending on the data and (N_0, b),

C_α depending only on the data and $(\hat{N}_0, N_2, \alpha)$,

such that the following hold: Let $\varepsilon \in (0, \frac{\varepsilon_{\mathrm{bc}}}{2})$ and $\delta_3 \in (0, \frac{\delta_{\mathcal{K}}}{2})$. Let $(u, \theta_{\mathrm{w}}) \in C^1(\overline{Q^{\mathrm{iter}}}) \times [\theta^*_{\mathrm{w}}, \frac{\pi}{2}]$ satisfy the conditions required in Lemmas 12.5.2 and 12.5.6 with the constants fixed above. Moreover, assume that there exists $(u^\#, \theta^\#_{\mathrm{w}}) \in C^1(\overline{Q^{\mathrm{iter}}}) \times [\theta^*_{\mathrm{w}}, \frac{\pi}{2}]$ satisfying the conditions of Lemma 12.5.3 such that (u, θ_{w}) and $(u^\#, \theta^\#_{\mathrm{w}})$ satisfy (12.5.31). Then $\mathcal{M}_{(u,\theta_{\mathrm{w}})}(\mathbf{p}, z, \boldsymbol{\xi})$ in (12.5.47) with $\tilde{\sigma}_1 = \sqrt{\delta_1}$ satisfies the following properties:

(i) *$\mathcal{M}_{(u,\theta_{\mathrm{w}})} \in C^3(\mathbb{R}^2 \times \mathbb{R} \times \overline{\Omega})$ and, for all $(\mathbf{p}, z) \in \mathbb{R}^2 \times \mathbb{R}$,*

$$\|(\mathcal{M}_{(u,\theta_{\mathrm{w}})}(\mathbf{0}, 0, \cdot),\ D^k_{(\mathbf{p},z)}\mathcal{M}_{(u,\theta_{\mathrm{w}})}(\mathbf{p}, z, \cdot))\|_{C^3(\overline{\Omega})} \le C_\alpha, \quad k = 1, 2, 3.$$

(ii) *$\mathcal{M}_{(u,\theta_{\mathrm{w}})}(\mathbf{p}, z, \boldsymbol{\xi}) = \mathcal{M}(\mathbf{p}, z, \boldsymbol{\xi})$ for all $(\mathbf{p}, z, \boldsymbol{\xi}) \in \mathbb{R}^2 \times \mathbb{R} \times \overline{\Omega}$ satisfying*
$$|\mathbf{p} - D\psi(\boldsymbol{\xi})| + |z - \psi(\boldsymbol{\xi})| < \frac{\sqrt{\delta_1}}{2}.$$

(iii) *For all $(\mathbf{p}, z, \boldsymbol{\xi}) \in \mathbb{R}^2 \times \mathbb{R} \times \overline{\Omega}$,*

$$\left|D_{(\mathbf{p},z)}\mathcal{M}_{(u,\theta_{\mathrm{w}})}(\mathbf{p}, z, \boldsymbol{\xi}) - D_{(\mathbf{p},z)}\mathcal{M}(D\psi(\boldsymbol{\xi}), \psi(\boldsymbol{\xi}), \boldsymbol{\xi})\right| \le C\sqrt{\delta_1}. \tag{12.5.48}$$

(iv) *The uniform obliqueness:*

$$\delta_{\rm bc} \le D_{\mathbf p}\mathcal{M}_{(u,\theta_{\rm w})}(\mathbf{p}, z, \boldsymbol{\xi}) \cdot \boldsymbol{\nu}_{\rm sh} \le \delta_{\rm bc}^{-1} \quad \text{for all } (\mathbf{p}, z, \boldsymbol{\xi}) \in \mathbb{R}^2 \times \mathbb{R} \times \overline{\Gamma_{\rm shock}}.$$

(v) $D_z\mathcal{M}_{(u,\theta_{\rm w})}(\mathbf{p}, z, \boldsymbol{\xi}) \le -\delta_{\rm bc}$ *for all* $(\mathbf{p}, z, \boldsymbol{\xi}) \in \mathbb{R}^2 \times \mathbb{R} \times \overline{\Gamma_{\rm shock}}$.

(vi) *Let* $v \in C^4(\overline{\Omega})$*, and let* $\mathcal{L}(\cdot)$ *be defined in* (12.5.47). *Denote*

$$\mathcal{B}^{(\rm sh)}_{(u,\theta_{\rm w})}(\mathbf{p}, z, \boldsymbol{\xi}) := \mathcal{L}(\mathbf{p} - Dv(\boldsymbol{\xi}), z - v(\boldsymbol{\xi}), \boldsymbol{\xi}). \tag{12.5.49}$$

Then

$$\mathcal{B}^{(\rm sh)}_{(u,\theta_{\rm w})}(\mathbf{p}, z, \boldsymbol{\xi}) = \mathbf{b}^{(\rm sh)}(\boldsymbol{\xi}) \cdot \mathbf{p} + b_0^{(\rm sh)}(\boldsymbol{\xi})z + h^{(\rm sh)}(\boldsymbol{\xi}),$$

and

$$\|(v, \mathbf{b}^{(\rm sh)}, h^{(\rm sh)})\|_{C^3(\overline{\Omega})} \le C_\alpha. \tag{12.5.50}$$

Moreover, for all $(\mathbf{p}, z, \boldsymbol{\xi}) \in \mathbb{R}^2 \times \mathbb{R} \times \overline{\Omega}$,

$$\begin{aligned}
&\left|\mathcal{M}_{(u,\theta_{\rm w})}(\mathbf{p}, z, \boldsymbol{\xi}) - \mathcal{B}^{(\rm sh)}_{(u,\theta_{\rm w})}(\mathbf{p},\ z,\ \boldsymbol{\xi})\right| \\
&\qquad \le C\sqrt{\delta_1}\big(|\mathbf{p} - Dv(\boldsymbol{\xi})| + |z - v(\boldsymbol{\xi})|\big), \\
&\left|D_{(\mathbf{p},z)}\mathcal{M}_{(u,\theta_{\rm w})}(\mathbf{p}, z, \boldsymbol{\xi}) - D_{(\mathbf{p},z)}\mathcal{B}^{(\rm sh)}_{(u,\theta_{\rm w})}(\mathbf{p}, z, \boldsymbol{\xi})\right| \le C\sqrt{\delta_1}.
\end{aligned} \tag{12.5.51}$$

(vii) *The homogeneity properties: If, in addition to the previous assumptions,* $(u, \theta_{\rm w})$ *satisfy Definition* 12.3.2(i) *with constants* (N_0, δ_1) *fixed above, then*

$$\begin{aligned}
&\mathcal{M}_{(u,\theta_{\rm w})}(\mathbf{0}, 0, \boldsymbol{\xi}) = 0 \qquad \text{for all } \boldsymbol{\xi} \in \Gamma_{\rm shock} \ \text{ if } \theta_{\rm w} \in [\frac{\pi}{2} - \frac{\delta_1}{N_1}, \frac{\pi}{2}], \\
&\mathcal{M}_{(u,\theta_{\rm w})}(\mathbf{0}, 0, \boldsymbol{\xi}) = 0 \qquad \text{if } \boldsymbol{\xi} \in \Gamma_{\rm shock} \cap \mathcal{D}_{\varepsilon_{\rm bc}} \ \text{ for all } \theta_{\rm w} \in [\theta_{\rm w}^*, \frac{\pi}{2}],
\end{aligned}$$

where $\mathcal{D}_{\varepsilon_{\rm bc}}$ *is from Definition* 12.3.1(iii), *and* (δ_1, N_1) *are from Definition* 12.3.2.

(viii) *At corner* $P_2 = \overline{\Gamma_{\rm shock}} \cap \overline{\Gamma_{\rm sym}}$*, for all* $(\mathbf{p}, z) \in \mathbb{R}^2 \times \mathbb{R}$,

$$\left|\frac{D_{\mathbf p}\mathcal{M}_{(u,\theta_{\rm w})}}{|D_{\mathbf p}\mathcal{M}_{(u,\theta_{\rm w})}|}(\mathbf{p},\ z,\ P_2) \pm \boldsymbol{\nu}_{\rm sym}\right| \ge \frac{1}{2}, \tag{12.5.52}$$

and

$$\left|\frac{\mathbf{b}^{(\rm sh)}}{|\mathbf{b}^{(\rm sh)}|}(P_2) \pm \boldsymbol{\nu}_{\rm sym}\right| \ge \frac{1}{2}. \tag{12.5.53}$$

(ix) *The* $\frac{1}{2}$*–obliqueness holds at corner* P_2 *in the sense of Definition* 4.4.3 *for the boundary conditions* $\mathcal{M}_{(u,\theta_{\rm w})}(D\hat\psi, \hat\psi, \boldsymbol{\xi}) = 0$ *on* $\Gamma_{\rm shock}$ *and* (12.3.29) *on* $\Gamma_{\rm sym}$.

(x) *Write operator* $\mathcal{M}_{(u,\theta_{\rm w})}(D\hat{\psi},\hat{\psi},\boldsymbol{\xi})$ *in the* (x,y)*-coordinates on* $\Gamma_{\rm shock}\cap\mathcal{D}_{\varepsilon_{\rm bc}}$ *as* $\hat{\mathcal{M}}_{(u,\theta_{\rm w})}(D_{(x,y)}\hat{\psi},\hat{\psi},x,y)$. *Then function* $\hat{\mathcal{M}}_{(u,\theta_{\rm w})}(\mathbf{p},z,x,y)$ *satisfies*

$$\|\hat{\mathcal{M}}_{(u,\theta_{\rm w})}\|_{C^3(\mathbb{R}^2\times\mathbb{R}\times\overline{\Gamma_{\rm shock}\cap\mathcal{D}_{\varepsilon_{\rm bc}}})}\le C_\alpha,\qquad \hat{\mathcal{M}}_{(u,\theta_{\rm w})}(\mathbf{p},z,x,y)=\hat{\mathcal{M}}(\mathbf{p},z,x,y) \tag{12.5.54}$$

for $|(\mathbf{p},z)|\le\frac{\delta_{\rm bc}}{C}$ *and* $(x,y)\in\Gamma_{\rm shock}\cap\mathcal{D}_{\varepsilon_0}$, *where* $\hat{\mathcal{M}}$ *is defined by* (12.5.35). *Moreover, for all* $(\mathbf{p},z,x,y)\in\mathbb{R}^2\times\mathbb{R}\times(\Gamma_{\rm shock}\cap\mathcal{D}_{\varepsilon_{\rm bc}})$ *and* $k=1,2,3$,

$$D_{p_1}\hat{\mathcal{M}}_{(u,\theta_{\rm w})}(\mathbf{p},z,x,y)\le-\delta_{\rm bc},\qquad D_z\hat{\mathcal{M}}_{(u,\theta_{\rm w})}(\mathbf{p},z,x,y)\le-\delta_{\rm bc}. \tag{12.5.55}$$

Furthermore, $\hat{\mathcal{M}}_{(u,\theta_{\rm w})}$ *satisfies the homogeneity property: For any* $(x,y)\in\Gamma_{\rm shock}\cap\mathcal{D}_{\varepsilon_{\rm bc}}$ *and* $\theta_{\rm w}\in[\theta_{\rm w}^*,\frac{\pi}{2}]$,

$$\hat{\mathcal{M}}_{(u,\theta_{\rm w})}(\mathbf{0},0,x,y)=0. \tag{12.5.56}$$

Proof. In this proof, all the constants below are positive and depend only on the parameters specified in the formulation of this lemma above, unless otherwise specified.

Fix $(u,\theta_{\rm w})$ satisfying the required conditions. We use (12.5.47) to extend $\mathcal{M}$ to a function $\mathcal{M}_{(u,\theta_{\rm w})}$ defined on $\mathbb{R}^2\times\mathbb{R}\times\overline{\Gamma_{\rm shock}}$. In the rest of the proof, we show that this extension satisfies all the properties asserted in this lemma. We divide the proof into five steps.

1. Let σ be the constant in (12.5.38). Then, with the choice of $\tilde{\sigma}_1<\frac{\sigma}{2}<\frac{1}{2}$, assertion (i) follows directly from (12.5.38)–(12.5.40) and (12.5.47). Also we note that, from (12.5.39) and (12.5.47) with $\tilde{\sigma}_1=\sqrt{\delta_1}$, it follows that assertion (ii) holds if $\delta_1\le\frac{1}{4}$.

2. Now we prove assertion (vi), and then employ it to prove assertions (iii)–(v) and (viii).

Estimate (12.5.50) follows directly from (12.5.38), (12.5.40), and (12.5.49).

In the remaining part of the proof of (vi), constants $(\tilde{\sigma}_1,C)$ depend only on the data.

We now show estimates (12.5.51). Writing $\mathcal{B}=\mathcal{B}^{(\rm sh)}_{(u,\theta_{\rm w})}$, we obtain from (12.5.47) that

$$\big|(\mathcal{M}_{(u,\theta_{\rm w})}-\mathcal{B})(\mathbf{p},z,\boldsymbol{\xi})\big|=\big|\eta_{\tilde{\sigma}_1}\big(\mathcal{M}(\mathbf{p},z,\boldsymbol{\xi})-\mathcal{L}(\mathbf{p}-Dv(\boldsymbol{\xi}),z-v(\boldsymbol{\xi}),\boldsymbol{\xi})\big)\big|.$$

The right-hand side above is zero outside the set:

$$\mathcal{A}_{\sigma_1}:=\big\{(\mathbf{p},z,\boldsymbol{\xi})\,:\,|(\mathbf{p},z)-(Dv(\boldsymbol{\xi}),v(\boldsymbol{\xi}))|<2\tilde{\sigma}_1,\ \boldsymbol{\xi}\in\Omega\big\},$$

since $\eta_{\tilde{\sigma}_1}\equiv\eta_{\tilde{\sigma}_1}(|(\mathbf{p},z)-(Dv(\boldsymbol{\xi}),v(\boldsymbol{\xi}))|)$. Thus, it suffices to consider $(\mathbf{p},z,\boldsymbol{\xi})\in\mathcal{A}_{\tilde{\sigma}_1}$.

If $\tilde{\sigma}_1$ is sufficiently small, then we employ (12.5.39) to see that $(\mathbf{p}, z, \boldsymbol{\xi}) \in \mathcal{A}_{\tilde{\sigma}_1}$ satisfy the conditions in (12.5.38). Then, for $(\mathbf{p}, z, \boldsymbol{\xi}) \in \mathcal{A}_{\tilde{\sigma}_1}$, we use (12.5.38) to obtain

$$\begin{aligned}\left|(\mathcal{M}_{(u,\theta_{\rm w})} - \mathcal{B})(\mathbf{p}, z, \boldsymbol{\xi})\right| &= |\eta_{\tilde{\sigma}_1}(\mathcal{M}(\mathbf{p}, z, \boldsymbol{\xi}) - \mathcal{L}(\mathbf{p} - Dv(\boldsymbol{\xi}), z - v(\boldsymbol{\xi}), \boldsymbol{\xi}))| \\ &\le C\eta_{\tilde{\sigma}_1}|(\mathbf{p}, z) - (Dv(\boldsymbol{\xi}), v(\boldsymbol{\xi}))|^2 \\ &\le C\tilde{\sigma}_1|(\mathbf{p}, z) - (Dv(\boldsymbol{\xi}), v(\boldsymbol{\xi}))|.\end{aligned}$$

Thus, we have shown that

$$\left|(\mathcal{M}_{(u,\theta_{\rm w})} - \mathcal{B})(\mathbf{p}, z, \boldsymbol{\xi})\right| \le C\tilde{\sigma}_1|(\mathbf{p}, z) - (Dv(\boldsymbol{\xi}), v(\boldsymbol{\xi}))|$$

for any $(\mathbf{p}, z, \boldsymbol{\xi}) \in \mathbb{R}^2 \times \mathbb{R} \times \overline{\Omega}$. Since $\tilde{\sigma}_1 = \sqrt{\delta_1}$, we obtain the first estimate in (12.5.51).

Furthermore,

$$D_{p_i}\mathcal{M}_{(u,\theta_{\rm w})}(\mathbf{p}, z, \boldsymbol{\xi}) - D_{p_i}\mathcal{B}(\mathbf{p}, z, \boldsymbol{\xi}) = J_1 + J_2,$$

where

$$\begin{aligned}J_1 &= \eta_{\tilde{\sigma}_1}(D_{p_i}\mathcal{M}(\mathbf{p}, z, \boldsymbol{\xi}) - D_{p_i}\mathcal{M}(Dv(\boldsymbol{\xi}), v(\boldsymbol{\xi}), \boldsymbol{\xi})), \\ J_2 &= \frac{p_i - D_i v}{|(\mathbf{p}, z) - (Dv(\boldsymbol{\xi}), v(\boldsymbol{\xi}))|}\eta'_{\tilde{\sigma}_1}\left(\mathcal{M}(\mathbf{p}, z, \boldsymbol{\xi}) - \mathcal{L}(\mathbf{p} - Dv(\boldsymbol{\xi}), z - v(\boldsymbol{\xi}), \boldsymbol{\xi})\right).\end{aligned}$$

Then, again using that $(\eta_{\tilde{\sigma}_1}, \eta'_{\tilde{\sigma}_1})(|(\mathbf{p}, z) - (Dv(\boldsymbol{\xi}), v(\boldsymbol{\xi}))|) \equiv 0$ outside set $\{(\mathbf{p}, z) : |(\mathbf{p}, z) - (Dv(\boldsymbol{\xi}), v(\boldsymbol{\xi}))| < 2\tilde{\sigma}_1\}$, employing (12.5.38)–(12.5.39), and choosing $\tilde{\sigma}_1 \le \frac{1}{8}\min\{\sigma, 1\}$, we have

$$\begin{aligned}|J_1| &\le C\eta_{\tilde{\sigma}_1}\,|(\mathbf{p}, z) - (Dv(\boldsymbol{\xi}), v(\boldsymbol{\xi}))| \le C\tilde{\sigma}_1, \\ |J_2| &\le C\eta'_{\tilde{\sigma}_1}\,|(\mathbf{p}, z) - (Dv(\boldsymbol{\xi}), v(\boldsymbol{\xi}))|^2 \le \frac{C}{\tilde{\sigma}_1}\tilde{\sigma}_1^2 \le C\tilde{\sigma}_1.\end{aligned}$$

Thus, $\left|D_{p_i}(\mathcal{M}_{(u,\theta_{\rm w})} - \mathcal{B})\right| \le C\tilde{\sigma}_1$. The estimate for $\left|D_z(\mathcal{M}_{(u,\theta_{\rm w})} - \mathcal{B})\right|$ is similar, which leads to the second estimate in (12.5.51). Now (vi) is proved.

To prove assertion (iii), we note that, from (12.5.49),

$$\begin{aligned}&|D_{(\mathbf{p},z)}(\mathcal{B} - \mathcal{M})(D\psi(\boldsymbol{\xi}), \psi(\boldsymbol{\xi}), \boldsymbol{\xi})| \\ &= |D_{(\mathbf{p},z)}\mathcal{M}(Dv(\boldsymbol{\xi}), v(\boldsymbol{\xi}), \boldsymbol{\xi}) - D_{(\mathbf{p},z)}\mathcal{M}(D\psi(\boldsymbol{\xi}), \psi(\boldsymbol{\xi}), \boldsymbol{\xi})| \le C\tilde{\sigma}_1^2,\end{aligned}$$

where we have used (12.5.38)–(12.5.39) for $\tilde{\sigma}_1 < \frac{\sigma}{2}$. Now (12.5.48) follows from the second estimate in (12.5.51), by using the conditions that $\tilde{\sigma}_1 = \sqrt{\delta_1}$, and $\mathcal{B}(\mathbf{p},\ z,\ \boldsymbol{\xi})$ is independent of $(\mathbf{p},\ z)$.

Assertions (iv)–(v) follow from assertion (iii) and Lemma 12.5.2 by choosing δ_1 and $\tilde{\varepsilon}_0$ small, depending on the data.

Next we show assertion (viii). We choose constants $\delta_{\mathcal{K}}, \delta_3 > 0$ that satisfy the smallness conditions of Corollary 12.5.4, and hence depend only on the data and (N_0, b). Now we can apply Corollary 12.5.4. Then (12.5.52) follows from (12.5.32), combined with assertions (iii)–(iv) for estimating the denominator from below, and choosing δ_1 (and hence $\tilde{\sigma}_1 = \sqrt{\delta_1}$) sufficiently small. Furthermore, using (12.5.47) and (12.5.49), we obtain (12.5.53) from (12.5.52) by choosing $(\mathbf{p}, z) = (Dv(\boldsymbol{\xi}), v(\boldsymbol{\xi}))$.

Therefore, $\mathcal{M}_{(u,\theta_{\rm w})}$ satisfies properties (i)–(vi) and (viii) as asserted.

3. Now we prove assertion (vii). In this step, constants $(C, N_1, \delta_1, \varepsilon_{\rm bc})$ depend only on the data.

Let $(u, \theta_{\rm w})$ satisfy conditions (i)–(ii) of Definition 12.3.2. Consider first Case $\theta_{\rm w} \in [\frac{\pi}{2} - \frac{\delta_1}{N_1}, \frac{\pi}{2}]$. Then, from Definition 12.3.2(i),

$$\|u\|^{*,(1+\delta^*)}_{2,0,Q^{\rm iter}} \le \delta_1 \qquad \text{in } Q^{\rm iter}.$$

Let u_0 be the zero function on $Q^{\rm iter}$: $u_0 \equiv 0$. From (12.2.34) in Lemma 12.2.4, Case $(u, \theta_{\rm w}) = (u_0, \frac{\pi}{2})$ is the normal reflection. From (12.2.47)–(12.2.50), it follows that $(u_0, \frac{\pi}{2}) \in \mathfrak{S}$. Denote by $\mathfrak{g}_{\rm sh}^{(\rm norm)}$ the function from Definition 12.2.6(ii) corresponding to the normal reflection $(u_0, \frac{\pi}{2})$. Then, from the definition of $\mathfrak{g}_{\rm sh}$, it follows that

$$\{(s, \mathfrak{g}_{\rm sh}^{(\rm norm)}(s)) \;:\; 0 < s < \hat{s}(\frac{\pi}{2})\} = F_1^{(\frac{\pi}{2})}(\Gamma_{\rm shock}^{(\rm norm)}),$$

where $\Gamma_{\rm shock}^{(\rm norm)}$ is the straight shock P_1P_2 of the normal reflection. Since

$$\operatorname{dist}(\Gamma_{\rm shock}^{(\rm norm)}, \Gamma_{\rm wedge}^{(\rm norm)}) > 0,$$

it follows from Lemma 12.2.2(ii) and (12.2.57) that

$$\mathfrak{g}_{\rm sh}^{(\rm norm)} \ge b_0 > 0 \qquad \text{on } (0, \hat{s}(\tfrac{\pi}{2})),$$

where b_0 depends only on the data. Now, applying (12.2.59) for $(u, \theta_{\rm w})$ and $(u_0, \frac{\pi}{2})$, and choosing N_1 large, we have

$$\mathfrak{g}_{\rm sh} \ge \frac{b_0}{2} \qquad \text{for } (u, \theta_{\rm w}).$$

Then, choosing N_1 large so that $\frac{1}{N_1} \le \mu$ for μ from Lemma 12.2.4, and using (12.2.34) and Lemma 12.2.7(viii) with $(M, b, \alpha) = (\delta_1, \frac{b_0}{2}, 0)$, we have

$$|\psi| + |D\psi| \le C\delta_1.$$

Thus, choosing $\delta_1 \in (0, \frac{1}{8})$ small, we have

$$|\psi| + |D\psi| \le \frac{\sqrt{\delta_1}}{2} = \frac{\tilde{\sigma}_1}{8} \qquad \text{in } \Omega \ \text{ if } \theta_{\rm w} \in [\frac{\pi}{2} - \frac{\delta_1}{N_1}, \frac{\pi}{2}].$$

At this point, we fix $\delta_1 > 0$ satisfying all the smallness requirements above and in Steps 1–2, and hence depending only on the data. This also fixes $\tilde{\sigma}_1 = \sqrt{\delta_1}$.

On the other hand, for any $\theta_w \in [\theta_w^*, \frac{\pi}{2})$, we employ (12.3.7)–(12.3.9) and note that, under our assumptions, (12.4.16) holds with ε_{bc} instead of ε_{eq}. From this,

$$|\psi| + |D\psi| \le C(x+\varepsilon) \le 2C\varepsilon_{bc} \le \frac{\tilde{\sigma}_1}{8} \quad \text{in } \Omega \cap \mathcal{D}_{\varepsilon_{bc}} \text{ for all } \theta_w \in [\theta_w^*, \frac{\pi}{2}],$$

for sufficiently small ε_{bc}. Then, from (12.5.47), using (12.2.71), (12.5.39), and Definition 12.2.6(v), we have

$$\mathcal{M}_{(u,\theta_w)}(\mathbf{p}, z, \boldsymbol{\xi}) = \mathcal{M}(\mathbf{p}, z, \boldsymbol{\xi}) \tag{12.5.57}$$

on $\{|(\mathbf{p}, z)| \le \frac{\tilde{\sigma}_1}{2}, \ \boldsymbol{\xi} \in \Gamma_{\text{shock}}\}$ if $\theta_w \in [\frac{\pi}{2} - \frac{\delta_1}{N_1}, \frac{\pi}{2}]$, and on $\{|(\mathbf{p}, z)| \le \frac{\tilde{\sigma}_1}{2}, \ \boldsymbol{\xi} \in \Gamma_{\text{shock}} \cap \mathcal{D}_{\varepsilon_{bc}}\}$ for any $\theta_w \in [\theta_w^*, \frac{\pi}{2}]$.

Now assertion (vii) follows from Lemma 12.5.1(v).

4. Now we prove assertion (ix). In this step, the universal constant C depends only on the data and (N_0, b).

In the proof of Lemma 12.5.3, we have shown that, under the conditions of Lemma 12.5.3 (which are satisfied in the present case for $(u^\#, \theta_w^\#)$), properties (12.5.29)–(12.5.30) hold for $(\psi^\#, P_2^\#)$. It follows that

$$\left| \frac{D_{\mathbf{p}}\mathcal{M}}{|D_{\mathbf{p}}\mathcal{M}|}(D\hat{\psi}^\#(P_2^\#), \hat{\psi}^\#(P_2^\#), P_2^\#) - (1,0) \right| \le C\delta_3.$$

Then, using (12.5.17) and (12.5.25), we have

$$\left| \frac{D_{\mathbf{p}}\mathcal{M}}{|D_{\mathbf{p}}\mathcal{M}|}(D\psi^\#(P_2^\#), \psi^\#(P_2^\#), P_2^\#) - (1,0) \right| \le C\delta_3.$$

Also, as in the proof of Corollary 12.5.4, we obtain that (12.5.33) holds for (u, θ_w) and $(u^\#, \theta_w^\#)$. Using this and (12.5.17), we find that, from the last estimate,

$$\left| \frac{D_{\mathbf{p}}\mathcal{M}}{|D_{\mathbf{p}}\mathcal{M}|}(D\psi(P_2), \psi(P_2), P_2) - (1,0) \right| \le C(\delta_{\mathcal{K}} + \delta_3).$$

Finally, using (12.5.48) (shown above), we obtain that, for each $(\mathbf{p}, z) \in \mathbb{R}^2 \times \mathbb{R}$,

$$\left| \frac{D_{\mathbf{p}}\mathcal{M}_{(u,\theta_w)}}{|D_{\mathbf{p}}\mathcal{M}_{(u,\theta_w)}|}(\mathbf{p}, z, P_2) - (1,0) \right| \le C(\delta_{\mathcal{K}} + \delta_3 + \sqrt{\delta_1}).$$

Then, using that $\boldsymbol{\nu}_{\text{sym}} = \mathbf{e}_{\xi_2} = (0,1)$, choosing small $\delta_{\mathcal{K}}$ depending only on the data and (N_0, b), and reducing δ_1 and δ_3 so that $\delta_1^2 + \delta_3 < \delta_{\mathcal{K}}$, we can employ line $t \mapsto P_2 + (-1,1)t$ as curve Σ in Definition 4.4.3 to obtain the $\frac{1}{2}$-obliqueness at P_2.

5. It remains to prove assertion (x). First, (12.5.54) follows from assertions (i)–(ii), since the change of variables $\boldsymbol{\xi} \mapsto (x, y)$ and its inverse have C^2–bounds in $\mathcal{D}_{\varepsilon_0}$ depending only on the lower bound of $c_2(\theta_{\rm w})$, which can be chosen uniformly for any $\theta_{\rm w} \in [\theta_{\rm w}^{\rm d}, \frac{\pi}{2}]$.

Also, using these bounds, (12.5.55) follows from (12.5.38) and Lemma 12.5.5, if we choose sufficiently small $\varepsilon_{\rm bc}$ and $\tilde{\sigma}_1$ in (12.5.39) and (12.5.47), where the smallness of $\tilde{\sigma}_1$ is achieved by the choice of small δ_1.

Finally, (12.5.56) follows from the second equality in assertion (vii) by the change of variables. This completes the proof. □

Lemma 12.5.7(ii), combined with Lemma 12.5.1(iv), implies

Corollary 12.5.8. *Let the parameters and $(u, \theta_{\rm w}) \in C^1(\overline{Q^{\rm iter}}) \times [\theta_{\rm w}^*, \frac{\pi}{2}]$ satisfy all the conditions required in Lemma 12.5.7. Let Ω, $\Gamma_{\rm shock}$, and ψ be determined by $(u, \theta_{\rm w})$, and let $\mathcal{M}_{(u,\theta_{\rm w})}$ be the corresponding function (12.5.47). Then $\hat{\psi} = \psi$ satisfies the boundary condition:*

$$\mathcal{M}_{(u,\theta_{\rm w})}(D\psi, \psi, \boldsymbol{\xi}) = 0 \qquad \text{on } \Gamma_{\rm shock}, \tag{12.5.58}$$

if and only if $\varphi = \varphi_2 + \psi$ satisfies the Rankine-Hugoniot condition (12.5.3) on $\Gamma_{\rm shock}(\varphi) = \{\varphi = \varphi_1\}$.

Corollary 12.5.9. *If parameters $(\varepsilon, \delta_1, \delta_3, \frac{1}{N_1})$ of the iteration set in Definition 12.3.2 and $\delta_{\mathcal{K}} > 0$ are small, depending only on the data and $\theta_{\rm w}^*$, then the following holds: Let $(u, \theta_{\rm w}) \in \overline{\mathcal{K}^{\rm ext}}$ be such that there exists $(u^\#, \theta_{\rm w}^\#) \in \overline{\mathcal{K}}$ satisfying (12.5.31) with $(u, \theta_{\rm w})$. Define function $\mathcal{M}_{(u,\theta_{\rm w})}(\cdot)$ by (12.5.47) with $\tilde{\sigma}_1 = \sqrt{\delta_1}$. Then all the assertions of Lemma 12.5.7 hold for $\mathcal{M}_{(u,\theta_{\rm w})}(\cdot)$, with C and $\delta_{\rm bc}$ depending only on the data, and C_α depending only on the data and $(\theta_{\rm w}^*, \alpha)$.*

Proof. Let (N_0, N_2) be the constants fixed in Definition 12.3.2, depending only on the data and $\theta_{\rm w}^*$, and let $b = \frac{1}{N_2}$. Let $\alpha \in (0, \frac{1}{8})$, and let $\hat{N}_0$ be the constant from (12.3.21), depending only on the data and $(\theta_{\rm w}^*, \alpha)$.

Let parameters $(\varepsilon, \delta_1, \delta_3, \frac{1}{N_1})$ of the iteration set in Definition 12.3.2 and the positive constant $\delta_{\mathcal{K}}$ satisfy the smallness conditions of Lemma 12.5.7 with constants $(N_0, \hat{N}_0, N_2, b)$ fixed above. Then $(\varepsilon, \delta_1, \delta_3, \frac{1}{N_1})$ are small, depending only on the data and $\theta_{\rm w}^*$ (where we note that there is no dependence on α, since $\hat{N}_0$ affects only C_α in Lemma 12.5.7).

Fix $(u, \theta_{\rm w}) \in \overline{\mathcal{K}^{\rm ext}}$ satisfying our assumptions. Then there exists $(u^\#, \theta_{\rm w}^\#) \in \overline{\mathcal{K}}$ satisfying estimate (12.5.31) with $(u, \theta_{\rm w})$ and constant $\frac{\delta_{\mathcal{K}}}{2}$. It follows that there exists $(\tilde{u}^\#, \tilde{\theta}_{\rm w}^\#) \in \mathcal{K}$ that satisfies estimate (12.5.31) with $(u, \theta_{\rm w})$ and constant $\delta_{\mathcal{K}}$.

Then $(u, \theta_{\rm w})$ satisfies all the conditions of Lemma 12.5.7. Now the result as claimed follows from Lemma 12.5.7 by taking into account the dependence of constants $(N_0, \hat{N}_0, N_2)$ described above. □

Corollaries 12.5.8–12.5.9 motivate that, for each $(u, \theta_{\rm w}) \in \overline{\mathcal{K}^{\rm ext}}$, we use operator $\mathcal{M}_{(u,\theta_{\rm w})}$ defined by (12.5.47) in (12.3.26). This completes Definition 12.3.2 of the iteration set.

12.6 NORMAL REFLECTION, ITERATION SET, AND ADMISSIBLE SOLUTIONS

In this section, we show that the normal reflection solution is included in the iteration set and that admissible solutions converge to the normal reflection solution as $\theta \to \frac{\pi}{2}$.

Lemma 12.6.1. *The normal reflection solution $u = u^{(\rm norm)} \equiv 0$ is included in the iteration set for the wedge angle $\theta_{\rm w} = \frac{\pi}{2}$, i.e., $(u^{(\rm norm)}, \frac{\pi}{2}) \in \mathcal{K}$. Moreover, for $(u^{(\rm norm)}, \frac{\pi}{2})$,*

$$\Gamma^{(\rm norm)}_{\rm shock} = \{\boldsymbol{\xi} \, : \, \xi_1 = \bar{\xi}_1, \, 0 < \xi_2 < \sqrt{\bar{c}_2^2 - \bar{\xi}_1^2}\},$$

where $\bar{c}_2^2 = \bar{\rho}_2^{\gamma-1}$ and $\bar{\xi}_1$ are from Theorem 6.2.1; see also Fig. 3.1. This determines $\Omega = \Omega^{(\rm norm)}$. Also, $\varphi^{(\rm norm)} = \varphi_2^{(\frac{\pi}{2})}$ and $\psi^{(\rm norm)} = 0$.

Proof. We check that $(u^{(\rm norm)}, \frac{\pi}{2})$ satisfies the conditions of Definition 12.3.2.

Conditions (i)–(ii) and (v)–(vi) of Definition 12.3.2 and the structure of $\Omega^{(\rm norm)}$, $\Gamma^{(\rm norm)}_{\rm shock}$, $\varphi^{(\rm norm)}$, and $\psi^{(\rm norm)}$ follow for $(u^{(\rm norm)}, \frac{\pi}{2})$ from Definition 12.2.6 and Lemma 12.2.4, where we have used (12.2.34) for $\theta_{\rm w} = \frac{\pi}{2}$.

For condition (iii), $\Gamma_{\rm shock} \subset \Lambda(\theta_{\rm w}) \setminus \overline{B_{c_1}(\mathcal{O}_1)}$ for the normal reflection solution by (6.1.3) in Lemma 6.1.2. Condition (12.3.5) for the normal reflection solution is shown in Lemma 12.2.3.

To check condition (iv) of Definition 12.3.2 for $(u^{(\rm norm)}, \frac{\pi}{2})$, we note that $\psi^{(\rm norm)} = 0$ so that the left-hand sides of (12.3.6)–(12.3.10) vanish. Also, the left-hand side of (12.3.11) vanishes, since $S_1 = \{\varphi_1 = \varphi_2\}$. Since $\mathbf{e}_{S_1} = -\mathbf{e}_{\xi_2}$ for $\theta_{\rm w} = \frac{\pi}{2}$, the left-hand side of (12.3.12) also vanishes. Furthermore, $\eta_2(\frac{\pi}{2}) = -\frac{\delta_1}{N_1^2} < 0$ so that the right-hand sides of (12.3.6)–(12.3.12) are positive. Then the strict inequalities hold in (12.3.6)–(12.3.12). The strict inequalities in (12.3.13)–(12.3.14) for the normal reflection solution follow from the definition of μ_1.

Condition (vii) holds, since $\varphi = \hat{\varphi}$, so that $u = \hat{u}$ in this case. □

Now we show the convergence of admissible solutions to the normal reflection solution as $\theta \to \frac{\pi}{2}$.

Lemma 12.6.2. *If α in the definition of $\mathcal{K}$ satisfies that $\alpha < \bar{\alpha}$ for $\bar{\alpha}$ determined in Proposition 12.2.5, then, for any $\mu > 0$, there exists $\delta > 0$ such that, if $\theta_{\rm w} \in (\frac{\pi}{2} - \delta, \frac{\pi}{2})$, φ is any admissible solution for $\theta_{\rm w}$, and u is defined by (12.2.44), then*

$$\|u - u^{(\rm norm)}\|^{*,(1+\delta^*)}_{2,\alpha,Q^{\rm iter}} < \mu. \tag{12.6.1}$$

Proof. Let $(\varphi^{(i)}, \theta_{\rm w}^{(i)})$ be a sequence of admissible solutions and wedge angles, and let $(u^{(i)}, \theta_{\rm w}^{(i)})$ be the corresponding sequence pulled back to $Q^{\rm iter}$ by (12.2.44). Assume that $\theta_{\rm w}^{(i)} \to \frac{\pi}{2}$. Estimate (12.2.46) in Proposition 12.2.5 for $u^{(i)}$ implies that $\|u^{(i)}\|^{*,2}_{2,\bar\alpha,Q^{\rm iter}} \le M$. With this, we combine Lemma 4.6.3 with the standard results on the compact embedding of the Hölder spaces to conclude that there exists a subsequence (still denoted) $u^{(i)}$ converging in the weaker norm $C^{2,\alpha}_{*,1+\delta^*}(Q^{\rm iter})$ to a function $u^{(\infty)} \in C^{2,\bar\alpha}_{*,2}(Q^{\rm iter})$.

Combining with (12.2.59), we conclude that

$$\mathfrak{h}_{\rm sh}^{(u^{(i)},\theta_{\rm w}^{(i)})} \to \mathfrak{h}_{\rm sh}^{(u^{(\infty)},\frac{\pi}{2})} \qquad \text{in } C^{1,\frac{\alpha}{2}}([0,1]).$$

Then, from Lemma 9.2.6(ii), Lemma 12.2.2(iii), and (12.2.24), it follows that $\mathfrak{h}_{\rm sh}^{(u^{(\infty)},\frac{\pi}{2})} = \mathfrak{h}_{\rm sh}^{(\rm norm)}$ with $\mathfrak{h}_{\rm sh}^{(\rm norm)}(s) = \mathfrak{g}_{\rm sh}^{(\rm norm)}(\hat{s}(\frac{\pi}{2})s)$ for function $\mathfrak{g}_{\rm sh}^{(\rm norm)}$ from Lemma 12.2.3 for the normal reflection with $\theta_{\rm w} = \frac{\pi}{2}$, where we have used Lemma 12.6.1 to apply Lemma 12.2.3 to the normal reflection. Then, using Lemma 9.2.6(i) and (12.2.44) with (12.2.43), and recalling Lemma 12.2.2(iii), we obtain that $u^{(i)} \to u^{(\rm norm)}$ pointwise on $Q^{\rm iter}$. Thus, $u^{(\infty)} = u^{(\rm norm)}$ in $Q^{\rm iter}$, which implies that $u^{(i)} \to u^{(\rm norm)}$ in $C^{2,\alpha}_{*,1+\delta^*}(Q^{\rm iter})$.

Therefore, from every subsequence of $u^{(i)}$, we can extract further subsequence converging to $u^{(\rm norm)}$ in the $\|\cdot\|^{*,(1+\delta^*)}_{2,\alpha,Q^{\rm iter}}$–norm. Since the subsequential limit is unique, it follows that the whole sequence converges to $u^{(\rm norm)}$. This completes the proof. □

12.7 SOLVABILITY OF THE ITERATION PROBLEM AND ESTIMATES OF SOLUTIONS

In this section, we solve the iteration problem (12.3.25)–(12.3.29). In fact, we solve the iteration problem for each $(u, \theta_{\rm w}) \in \overline{\mathcal{K}^{\rm ext}}$ that is sufficiently close to some $(u^\#, \theta_{\rm w}^\#) \in \overline{\mathcal{K}}$. This is needed in order to show that the iteration set is open (*cf.* §12.8) and to define the iteration map on $\overline{\mathcal{K}}$ (*cf.* Chapter 4).

12.7.1 A boundary value problem in the iteration region $Q^{\rm iter}$

We rewrite Problem (12.3.25)–(12.3.29) as a boundary value problem in $Q^{\rm iter}$.

Based on (12.2.57), denote the parts of $\partial Q^{\rm iter}$ by

$$\begin{aligned} &\partial_{\rm shock}Q^{\rm iter} := (0,1)\times\{1\}, \quad \partial_{\rm sonic}Q^{\rm iter} := \{0\}\times(0,1),\\ &\partial_{\rm wedge}Q^{\rm iter} := (0,1)\times\{0\}, \quad \partial_{\rm sym}Q^{\rm iter} := \{1\}\times(0,1). \end{aligned} \tag{12.7.1}$$

Fix $(u, \theta_{\rm w}) \in \overline{\mathcal{K}^{\rm ext}}$ and write the boundary value problem (12.3.25)–(12.3.29) with the condition on $\Gamma_{\rm shock}$ given by Corollary 12.5.8 in the (s,t)–coordinates on $Q^{\rm iter}$. Then the problem is of the form:

$$\tilde{\mathcal{N}}_{(u,\theta_{\rm w})}(D^2\hat{u}, D\hat{u}, \hat{u}, s, t) = 0 \qquad \text{in} \quad Q^{\rm iter}, \tag{12.7.2}$$

$$\tilde{\mathcal{M}}_{(u,\theta_{\rm w})}(D\hat{u}, \hat{u}, s) = 0 \qquad \text{on} \ \partial_{\rm shock}Q^{\rm iter}, \tag{12.7.3}$$

$$\hat{u} = 0 \qquad \text{on} \ \partial_{\rm sonic}Q^{\rm iter}, \tag{12.7.4}$$

$$\mathcal{B}^{(\rm w)}_{(u,\theta_{\rm w})}(D\hat{u}, s) := b_1^{(\rm w)}(s)\partial_s\hat{u} + b_2^{(\rm w)}(s)\partial_t\hat{u} = 0 \qquad \text{on} \ \partial_{\rm wedge}Q^{\rm iter}, \tag{12.7.5}$$

$$\mathcal{B}^{(\rm sym)}_{(u,\theta_{\rm w})}(D\hat{u}, t) := b_1^{(\rm sym)}(t)\partial_s\hat{u} + b_2^{(\rm sym)}(t)\partial_t\hat{u} = g^{(\rm sym)}(t) \qquad \text{on} \ \partial_{\rm sym}Q^{\rm iter}, \tag{12.7.6}$$

where the structure and properties of the equation and boundary conditions are described in the following lemmas. We note that coefficients $(\mathbf{b}^{(\rm w)}, \mathbf{b}^{(\rm sym)})$, function $g^{(\rm sym)}$, and operators $(\tilde{\mathcal{N}}_{(u,\theta_{\rm w})}, \tilde{\mathcal{M}}_{(u,\theta_{\rm w})})$ depend on $(u, \theta_{\rm w})$. Furthermore, the right-hand side of (12.7.5) is zero by using (12.2.44) with (12.2.36) and $\partial_{\boldsymbol{\nu}}\varphi_2 = 0$ on $\Gamma_{\rm wedge}$.

This can be seen as follows: Let $(u, \theta_{\rm w}) \in \overline{\mathcal{K}^{\rm ext}}$. We now change the variables in (12.3.25)–(12.3.29) to write them in the (s,t)–variables in terms of function (12.3.15). Using (12.3.15), the notation in (12.2.52), and map $\mathfrak{F} = \mathfrak{F}_{(u,\theta_{\rm w})}$, we have

$$\begin{aligned}
&\hat{\psi}(\boldsymbol{\xi}) = \hat{u}(\mathfrak{F}^{-1}(\boldsymbol{\xi})) + (\tilde{\varphi}_2 - \varphi_2)(\boldsymbol{\xi}),\\
&D_i\hat{\psi}(\boldsymbol{\xi}) = \sum_{k=1}^{2} D_k\hat{u}(\mathfrak{F}^{-1})D_i\big((\mathfrak{F}^{-1})_k\big) + D_i(\tilde{\varphi}_2 - \varphi_2)(\boldsymbol{\xi}),\\
&D_{ij}\hat{\psi}(\boldsymbol{\xi}) = \sum_{k,l=1}^{2} D_{kl}\hat{u}(\mathfrak{F}^{-1})D_i\big((\mathfrak{F}^{-1})_k\big)D_j\big((\mathfrak{F}^{-1})_l\big)\\
&\qquad\qquad + \sum_{k=1}^{2} D_k\hat{u}(\mathfrak{F}^{-1})D_{ij}\big((\mathfrak{F}^{-1})_k\big) + D_{ij}(\tilde{\varphi}_2 - \varphi_2)(\boldsymbol{\xi}),
\end{aligned} \tag{12.7.7}$$

where $i,j = 1,2$, and functions $(\mathfrak{F}^{-1}, D\mathfrak{F}^{-1}, D^2\mathfrak{F}^{-1})$ are evaluated at $\boldsymbol{\xi} = (\xi_1, \xi_2)$. If $\boldsymbol{\xi} = \mathfrak{F}(s,t)$, then we substitute the right-hand sides for $\hat{\psi}$ and its derivatives into (12.3.25)–(12.3.29) with (12.3.26) given by (12.5.58), make the change of variables $\boldsymbol{\xi} = \mathfrak{F}(s,t)$, and express $(D\mathfrak{F}^{-1}, D^2\mathfrak{F}^{-1})$ at $\mathfrak{F}(s,t)$ in terms of $(D\mathfrak{F},\ D^2\mathfrak{F})$ at $(s,t) \in Q^{\rm iter}$ by using the formulas:

$$\begin{aligned}
&D\mathfrak{F}^{-1}(\mathfrak{F}(s,t)) = (D\mathfrak{F}(s,t))^{-1},\\
&(\partial_i D\mathfrak{F}^{-1})(\mathfrak{F}(s,t)) = -(D\mathfrak{F}(s,t))^{-1}(\partial_i D\mathfrak{F})(s,t)(D\mathfrak{F}(s,t))^{-2}.
\end{aligned} \tag{12.7.8}$$

Therefore, we obtain (12.7.2)–(12.7.6), where we have used that $\tilde{\varphi}_2 = \varphi_2$ on $\Gamma_{\rm sonic}$ by (12.2.35) in order to obtain (12.7.4).

Lemma 12.7.1. *Let* $\alpha \in (0,1)$. *Let* $(u, \theta_{\rm w}) \in \mathfrak{S} \cap (C^{2,\alpha}_{*,1+\delta^*}(Q^{\rm iter}) \times [\theta_{\rm w}^*, \frac{\pi}{2}])$, *and let* Ω *and* ψ *correspond to* $(u, \theta_{\rm w})$ *as in Definition* 12.2.6. *Then equation* (12.7.2), *obtained by substituting expressions* (12.7.7)–(12.7.8) *into equation* (12.3.25), *is of the form:*

$$\sum_{i,j=1}^{2} \mathcal{A}_{ij}(D\hat{u}, s, t) D_{ij}\hat{u} + \sum_{i=1}^{2} \mathcal{A}_i(D\hat{u}, s, t) D_i\hat{u} = f \qquad \text{in } Q^{\text{iter}}, \qquad (12.7.9)$$

with the following properties:

(i) *Functions* $(\mathcal{A}_{ij}, \mathcal{A}_i)(\mathbf{p}, s, t)$ *and* $f(s,t)$ *satisfy* $(\mathcal{A}_{ij}, \mathcal{A}_i) \in C(\mathbb{R}^2 \times Q^{\text{iter}})$ *and* $f \in C(Q^{\text{iter}})$.

(ii) *Let* $\hat{\psi} \in C^2(\Omega)$, *and let* $\hat{u}$ *be determined by* (12.3.15). *Then* $\hat{u} \in C^2(Q^{\text{iter}})$, *and* $\hat{\psi}$ *is a solution of* (12.3.25) *in* Ω *if and only if* $\hat{u}$ *is a solution of* (12.7.2) *in* Q^{iter}.

(iii) *If* $(u^{(k)}, \theta_{\text{w}}^{(k)}), (u, \theta_{\text{w}}) \in \mathfrak{S} \cap (C^{2,\alpha}_{*,1+\delta^*}(Q^{\text{iter}}) \times [\theta_{\text{w}}^*, \frac{\pi}{2}])$ *satisfy that* $(u^{(k)}, \theta_{\text{w}}^{(k)}) \to (u, \theta_{\text{w}})$ *in* $C^{2,\alpha}_{*,1+\delta^*}(Q^{\text{iter}}) \times [\theta_{\text{w}}^*, \frac{\pi}{2}]$, *then*

$$(\mathcal{A}_{ij}^{(u^{(k)}, \theta_{\text{w}}^{(k)})}, \mathcal{A}_i^{(u^{(k)}, \theta_{\text{w}}^{(k)})}) \to (\mathcal{A}_{ij}^{(u, \theta_{\text{w}})}, \mathcal{A}_i^{(u, \theta_{\text{w}})})$$

uniformly on any compact subset of $\mathbb{R}^2 \times Q^{\text{iter}}$, *and*

$$f^{(u^{(k)}, \theta_{\text{w}}^{(k)})} \to f^{(u, \theta_{\text{w}})}$$

uniformly on any compact subset of Q^{iter}.

Proof. We divide the proof into two steps.

1. In this step, we describe more precisely the structure of the coefficients and the right-hand side $(\mathcal{A}_{ij}^{(u,\theta_{\text{w}})}, \mathcal{A}_i^{(u,\theta_{\text{w}})}, f^{(u,\theta_{\text{w}})})$ of (12.7.9) to show the smoothness of their dependence on (u, θ_{w}) and, to avoid unnecessary complicated calculations, we do this without writing the explicit formulas.

From the explicit definition of equation (12.3.25) by (12.4.5), (12.4.14), (12.4.20), and (12.4.42), we conclude that functions (A_{ij}, A_i) in equation (12.3.25) are obtained as follows:

$$\begin{aligned}&(A_{ij}^{(u,\theta_{\text{w}})}, A_i^{(u,\theta_{\text{w}})})(\mathbf{p}, \boldsymbol{\xi}) \\ &\quad = (G_{ij}, G_i)(\mathbf{p}, D\varphi^{(u,\theta_{\text{w}})}(\boldsymbol{\xi}), \varphi^{(u,\theta_{\text{w}})}(\boldsymbol{\xi}), D\psi^{(u,\theta_{\text{w}})}(\boldsymbol{\xi}), \psi^{(u,\theta_{\text{w}})}(\boldsymbol{\xi}), \boldsymbol{\xi}, \theta_{\text{w}})\end{aligned}$$

for any $(u, \theta_{\text{w}}) \in \overline{\mathcal{K}^{\text{ext}}}$, where the dependence on $(\varphi, D\varphi)$ is from (12.4.14), the dependence on $(\psi, D\psi)$ is from (12.4.5), and the dependence on θ_{w} is from the term of $c_2^{(\theta_{\text{w}})}$ in (11.1.5) and (12.4.14), through (12.4.15), by using (11.1.5) and (12.4.14) in the (x,y)–coordinates defined by (11.1.1)–(11.1.2) via $(u_2, v_2, c_2)^{(\theta_{\text{w}})}$. Functions $(G_{ij}, G_i)(\mathbf{p}, \mathbf{q}, z, \hat{\mathbf{q}}, \hat{z}, \boldsymbol{\xi}, \theta_{\text{w}})$ are defined on the set:

$$\mathcal{S} = \mathbb{R}^2 \times \mathbb{R}^2 \times \mathbb{R} \times \mathbb{R}^2 \times \mathbb{R} \times \mathcal{Q}^{\cup}_{\text{bd}},$$

where set $\mathcal{Q}^{\cup}_{\mathrm{bd}} \subset \mathbb{R}^2 \times (\theta_{\mathrm{w}}^{\mathrm{s}}, \frac{\pi}{2})$ is defined in Lemma 12.2.2(iii), in which it has shown that $\mathcal{Q}^{\cup}_{\mathrm{bd}}$ is open. Thus, $\mathcal{S}$ is open.

Now we show that (G_{ij}, G_i) are C^1 on $\mathcal{S}$. From the explicit expressions of the ingredients in (12.4.5), the dependence of (A_{ij}, A_i) on $(u_2, v_2, c_2)^{(\theta_{\mathrm{w}})}$ is C^1, and the dependence of $(u_2, v_2, c_2)^{(\theta_{\mathrm{w}})}$ on $\theta_{\mathrm{w}} \in [\theta_{\mathrm{w}}^{\mathrm{d}}, \frac{\pi}{2}]$ is C^∞, by Theorem 7.1.1(i) for the weak reflection. Thus, (G_{ij}, G_i) depend smoothly on $\theta_{\mathrm{w}} \in [\theta_{\mathrm{w}}^{\mathrm{d}}, \frac{\pi}{2}]$. The smooth dependence of (G_{ij}, G_i) on the other arguments follows from their explicit expressions obtained from (12.4.5), (12.4.14), (12.4.20), and (12.4.42).

To obtain $(\mathcal{A}_{ij}^{(u,\theta_{\mathrm{w}})}, \mathcal{A}_i^{(u,\theta_{\mathrm{w}})})(\mathbf{p}, s, t)$, we substitute expressions (12.7.7) into the equation:

$$G_{11}\hat{\psi}_{\xi_1\xi_1} + 2G_{12}\hat{\psi}_{\xi_1\xi_2} + G_{22}\hat{\psi}_{\xi_2\xi_2} + G_1\hat{\psi}_{\xi_1} + G_2\hat{\psi}_{\xi_2} = 0, \tag{12.7.10}$$

where (G_{ij}, G_i) are

$$(G_{ij}, G_i)(D\hat{\psi}(\boldsymbol{\xi}), D\varphi^{(u,\theta_{\mathrm{w}})}(\boldsymbol{\xi}), \varphi^{(u,\theta_{\mathrm{w}})}(\boldsymbol{\xi}), D\psi^{(u,\theta_{\mathrm{w}})}(\boldsymbol{\xi}), \psi^{(u,\theta_{\mathrm{w}})}(\boldsymbol{\xi}), \boldsymbol{\xi}, \theta_{\mathrm{w}}).$$

We do this by changing the variables $\boldsymbol{\xi} = \mathfrak{F}(s,t)$ and expressing $(D\mathfrak{F}^{-1}, D^2\mathfrak{F}^{-1})$ at $\mathfrak{F}(s,t)$ in terms of $(D\mathfrak{F}, D^2\mathfrak{F})$ at $(s,t) \in Q^{\mathrm{iter}}$ via formulas (12.7.8). We also use the expressions similar to (12.7.7)–(12.7.8) in order to express

$$(D\psi^{(u,\theta_{\mathrm{w}})}, \psi^{(u,\theta_{\mathrm{w}})})(\mathfrak{F}(s,t))$$

in the arguments of $(G_{ij}, G_i)(\cdot)$ in terms of

$$(Du, u)(s,t) \quad \text{and} \quad (D\varphi_2, D\tilde{\varphi}_2, \varphi_2, \tilde{\varphi}_2)(\mathfrak{F}(s,t)),$$

and similarly express $(D\varphi^{(u,\theta_{\mathrm{w}})}, \varphi^{(u,\theta_{\mathrm{w}})})(\mathfrak{F}(s,t))$ in the arguments of $(G_{ij}, G_i)(\cdot)$ by using that $\varphi = \psi + \varphi_2$. Moreover, we obtain the nonhomogeneity coming from the term of $\tilde{\varphi}_2 - \varphi_2$ and its derivatives, when we substitute expressions (12.7.7) into equation (12.7.10). Then, since the coefficients of the equation depend only on $\mathbf{p}$, the separation of the nonhomogeneous part yields

$$f(s,t) = -\sum_{i,j=1}^{2} \left(G_{ij}D_{ij}(\tilde{\varphi}_2 - \varphi_2)\right)(\mathfrak{F}(s,t)) - \sum_{i=1}^{2} \left(G_iD_i(\tilde{\varphi}_2 - \varphi_2)\right)(\mathfrak{F}(s,t)),$$

where (G_{ij}, G_i) are

$$(G_{ij}, G_i)(0, D\varphi^{(u,\theta_{\mathrm{w}})}(\boldsymbol{\xi}), \varphi^{(u,\theta_{\mathrm{w}})}(\boldsymbol{\xi}), D\psi^{(u,\theta_{\mathrm{w}})}(\boldsymbol{\xi}), \psi^{(u,\theta_{\mathrm{w}})}(\boldsymbol{\xi}), \boldsymbol{\xi}, \theta_{\mathrm{w}}).$$

Combining all these calculations, we have

$$\begin{aligned}
&(\mathcal{A}_{ij}^{(u,\theta_{\mathrm{w}})}, \mathcal{A}_i^{(u,\theta_{\mathrm{w}})})(\mathbf{p}, (s,t)) = (\tilde{G}_{ij}, \tilde{G}_i)(\mathbf{p}, u, Du, \mathfrak{F}, D\mathfrak{F}, D^2\mathfrak{F}, (s,t), \theta_{\mathrm{w}}),\\
&f^{(u,\theta_{\mathrm{w}})}(s,t) = F(u, Du, \mathfrak{F}, D\mathfrak{F}, D^2\mathfrak{F}, (s,t), \theta_{\mathrm{w}}),
\end{aligned} \tag{12.7.11}$$

where $\mathfrak{F} = \mathfrak{F}_{(u,\theta_{\rm w})}$, and functions $(u, \mathfrak{F})$ and their derivatives are taken at point (s,t). The functions:

$$(\tilde{G}_{ij}, \tilde{G}_i)(\mathbf{p}, \tilde{z}, \tilde{\mathbf{p}}, R, \mathcal{P}, \mathcal{B}, (s,t), \theta_{\rm w}) \quad \text{and} \quad F = F(\tilde{z}, \tilde{\mathbf{p}}, R, \mathcal{P}, \mathcal{B}, (s,t), \theta_{\rm w})$$

are C^1 on the set:

$$\begin{aligned}\tilde{\mathcal{S}} = \{(\mathbf{p}, \tilde{z}, \tilde{\mathbf{p}}, R, \mathcal{P}, \mathcal{B}, (s,t), \theta_{\rm w}) \in \mathbb{R}^2 \times \mathbb{R} \times \mathbb{R}^2 \times \mathbb{R}^2 \times \mathbb{R}^{2\times 2} \\ \times \mathbb{R}^{2\times 2\times 2} \times Q^{\rm iter} \times [\theta_{\rm w}^{\rm s}, \frac{\pi}{2}] \ : \ \det R \neq 0\}.\end{aligned} \tag{12.7.12}$$

Here and hereafter, we formally add variable $\mathbf{p}$ to the arguments of $F(\cdot)$ to simplify the argument. The smoothness of $(\tilde{G}_{ij}, \tilde{G}_i, F)$ on $\tilde{\mathcal{S}}$ follows from the smoothness of (G_{ij}, G_i) of $\mathcal{S}$ and the property that $\mathfrak{F}(Q^{\rm iter}) \subset \mathcal{Q}_{\rm bd}^{(\theta_{\rm w})}$ holds by Definition 12.2.6(iv) and Lemma 12.2.7(ii).

2. Now we prove the assertions.

Assertion (i) follows from (12.7.11)–(12.7.12) with $(\tilde{G}_{ij}, \tilde{G}_i, F) \in C^\infty(\tilde{\mathcal{S}})$ by using (12.2.67)–(12.2.68).

In assertion (ii), property $\hat{u} \in C^2(Q^{\rm iter})$ follows from (12.3.15) by using (12.2.68) and the fact that $\mathfrak{F}_{(u,\theta_{\rm w})}(Q^{\rm iter}) = \Omega$ (*i.e.*, open sets). Now the property that $\hat{\psi}$ is a solution of (12.3.25) in Ω if and only if u is a solution of (12.7.2) in $Q^{\rm iter}$ follows from the explicit definition of equation (12.7.2) in the form of (12.7.9).

Assertion (iii) follows from (12.7.11) with $(\tilde{G}_{ij}, \tilde{G}_i, F) \in C^\infty(\tilde{\mathcal{S}})$ by using (12.2.73)–(12.2.74), where $\tilde{\mathcal{S}}$ is defined by (12.7.12). ☐

Next, we discuss the properties of the boundary conditions (12.7.3) and (12.7.5)–(12.7.6).

Lemma 12.7.2. *Let $\alpha \in (0,1)$. Let $(u, \theta_{\rm w}) \in \mathfrak{S} \cap (C^{2,\alpha}_{*,1+\delta^*}(Q^{\rm iter}) \times [\theta^*_{\rm w}, \frac{\pi}{2}])$, and let Ω and ψ correspond to $(u, \theta_{\rm w})$ as in Definition 12.2.6. Then the boundary conditions* (12.7.3) *and* (12.7.5)–(12.7.6), *obtained by substituting expressions* (12.7.7)–(12.7.8) *into the boundary conditions* (12.3.26) *and* (12.3.28)–(12.3.29), *satisfy the following properties:*

(i) $\mathcal{M}_{(u,\theta_{\rm w})} \in C(\mathbb{R}^2 \times \mathbb{R} \times \partial_{\rm shock} Q^{\rm iter})$, $\mathcal{B}^{(w)}_{(u,\theta_{\rm w})} \in C(\mathbb{R}^2 \times \partial_{\rm wedge} Q^{\rm iter})$, $\mathcal{B}^{(\rm sym)}_{(u,\theta_{\rm w})} \in C(\mathbb{R}^2 \times \partial_{\rm sym} Q^{\rm iter})$, *and* $g^{(\rm sym)} \in C(\partial_{\rm sym} Q^{\rm iter})$.

(ii) *Let $\hat{\psi} \in C^1(\overline{\Omega})$, and let $\hat{u}$ be determined by* (12.3.15). *Then $\hat{u} \in C^1(\overline{Q^{\rm iter}})$, and $\hat{\psi}$ satisfies* (12.3.26) (*resp.* (12.3.28) *and* (12.3.29)) *if and only if $\hat{u}$ satisfies* (12.7.3) (*resp.* (12.7.5) *and* (12.7.6)).

(iii) *If $(u^{(j)}, \theta^{(j)}_{\rm w}), (u, \theta_{\rm w}) \in \mathfrak{S} \cap (C^{2,\alpha}_{*,1+\delta^*}(Q^{\rm iter}) \times [\theta^*_{\rm w}, \frac{\pi}{2}])$ and $(u^{(j)}, \theta^{(j)}_{\rm w}) \to (u, \theta_{\rm w})$ in $C^{2,\alpha}_{*,1+\delta^*}(Q^{\rm iter}) \times [\theta^*_{\rm w}, \frac{\pi}{2}]$, then*

$\tilde{\mathcal{M}}_{(u^{(j)}, \theta^{(j)}_{\rm w})} \to \tilde{\mathcal{M}}_{(u,\theta_{\rm w})}$ *uniformly on compact subsets of* $\mathbb{R}^2 \times \mathbb{R} \times \partial_{\rm shock} Q^{\rm iter}$,

$(b_k^{(\mathrm{w})})_{(u^{(j)},\theta_{\mathrm{w}}^{(j)})} \to (b_k^{(\mathrm{w})})_{(u,\theta_{\mathrm{w}})}$ *uniformly on compact subsets of* $\mathbb{R}^2 \times \partial_{\mathrm{wedge}} Q^{\mathrm{iter}}$,

$(b_k^{(\mathrm{sym})})_{(u^{(j)},\theta_{\mathrm{w}}^{(j)})} \to (b_k^{(\mathrm{sym})})_{(u,\theta_{\mathrm{w}})}$ *and* $(g^{(\mathrm{sym})})_{(u^{(j)},\theta_{\mathrm{w}}^{(j)})} \to (g^{(\mathrm{sym})})_{(u,\theta_{\mathrm{w}})}$

uniformly on compact subsets of $\mathbb{R}^2 \times \partial_{\mathrm{sym}} Q^{\mathrm{iter}}$,

where $k = 1, 2$, *and segments* $\partial_{\mathrm{shock}} Q^{\mathrm{iter}}$, $\partial_{\mathrm{wedge}} Q^{\mathrm{iter}}$, *and* $\partial_{\mathrm{sym}} Q^{\mathrm{iter}}$ *do not include their endpoints, respectively.*

The proof of Lemma 12.7.2 is obtained in the same way as the proof of Lemma 12.7.1.

12.7.2 Solutions of the iteration problem

In this section, we show that Problem (12.7.2)–(12.7.6) has a unique solution, and study some properties of this solution.

We consider $(u, \theta_{\mathrm{w}}) \in \overline{\mathcal{K}^{\mathrm{ext}}}$. Thus, when the conditions of Definition 12.3.2 are used, we always assume the nonstrict inequalities as described in Remark 12.3.11.

In the next proposition, we use $\bar{\alpha}$ from Proposition 12.2.5.

Proposition 12.7.3. *Let parameters* $(\varepsilon, \delta_3, \frac{1}{N_1})$ *of the iteration set in Definition* 12.3.2 *and* $\delta_{\mathcal{K}} > 0$ *be small, depending only on the data and* θ_{w}^*. *Let* δ_1 *be small, depending on the data and* $(\theta_{\mathrm{w}}^*, \alpha)$, *where* (δ_1, α) *are from Definition* 12.3.2. *Then there exist:*

(i) $\hat{\alpha} \in (0, \min\{\bar{\alpha}, \delta^*\})$ *depending only on the data,*

(ii) $C \geq 1$ *depending only on the data and* $(\theta_{\mathrm{w}}^*, \alpha)$,

(iii) $C_s \geq 1$ *depending only on the data,* $(\theta_{\mathrm{w}}^*, \alpha)$, *and* $s \in (0, \varepsilon_0)$

such that, for each $(u, \theta_{\mathrm{w}}) \in \overline{\mathcal{K}^{\mathrm{ext}}}$ *satisfying*

$$\|u^{\#} - u\|_{C^1(\overline{Q^{\mathrm{iter}}})} + |\theta_{\mathrm{w}}^{\#} - \theta_{\mathrm{w}}| \leq \delta_{\mathcal{K}} \tag{12.7.13}$$

with some $(u^{\#}, \theta_{\mathrm{w}}^{\#}) \in \overline{\mathcal{K}}$, *there is a unique solution* $\hat{\psi} \in C^2(\Omega) \cap C^1(\overline{\Omega} \setminus \overline{\Gamma_{\mathrm{sonic}}}) \cap C(\overline{\Omega})$ *of Problem* (12.3.25)–(12.3.29) *determined by* (u, θ_{w}).

Moreover, $\hat{\psi}$ *satisfies*

$$\|\hat{\psi}\|_{L^\infty(\Omega)} \leq C, \tag{12.7.14}$$

$$|\hat{\psi}(x, y)| \leq Cx \qquad \text{in } \Omega \cap \mathcal{D}_{\varepsilon_0}, \tag{12.7.15}$$

and, for each $s \in (0, \varepsilon_0)$,

$$\|\hat{\psi}\|^{(-1-\hat{\alpha}),\ \overline{\partial_{\mathrm{sym}} Q^{\mathrm{iter}}}}_{2,\hat{\alpha},\Omega \setminus \mathcal{D}_s} \leq C_s. \tag{12.7.16}$$

Proof. We assume that constants $\varepsilon, \delta_1, \delta_3, \frac{1}{N_1}, \delta_{\mathcal{K}} > 0$ satisfy the smallness assumptions of Corollaries 12.4.3, 12.4.6, and 12.5.9, hence depending only on the data and $\theta_{\rm w}^*$. We divide the proof into four steps.

1. We make the change of coordinates $(s,t) = F_1(\boldsymbol{\xi})$, where $F_1(\cdot)$ is defined in Lemma 12.2.2 for each $\theta_{\rm w} \in [\theta_{\rm w}^{\rm s}, \frac{\pi}{2}]$. Then we consider Problem (12.3.25)–(12.3.29) rewritten in the (s,t)–coordinates in domain $F_1(\Omega)$ and its boundary. By Lemma 12.2.2(ii), transform $F_1(\cdot)$ is C^3 with C^3–inverse, with uniform estimates in these norms for $\theta_{\rm w} \in [\theta_{\rm w}^{\rm s}, \frac{\pi}{2}]$. We employ Proposition 4.7.2 to establish the existence and estimates of solutions of this problem.

We check that the conditions of Proposition 4.7.2 are satisfied in the present case with coordinates $(x,y) = (s,t)$.

2. Since $(u, \theta_{\rm w}) \in \mathfrak{S}$ from Definition 12.3.2(ii), then, using Definition 12.2.6(iv), we have

$$F_1(\Omega) = \big\{(s,t) \,:\, 0 < s < \hat{s}(\theta_{\rm w}),\ 0 < t < \mathfrak{g}_{\rm sh}(s)\big\}. \tag{12.7.17}$$

Then, using (12.3.5), we see that $F_1(\Omega)$ is of structure (4.5.1)–(4.5.3) with $h = \hat{s}(\theta_{\rm w})$, $f_{\rm bd} = \mathfrak{g}_{\rm sh}$, $t_0 = \mathfrak{g}_{\rm sh}(0) \ge \frac{1}{N_2}$, $t_2 = \frac{1}{N_2}$, and any positive number t_1. We may fix $t_1 = 1$. Indeed, for any $s \in (0, \hat{s}(\theta_{\rm w}))$, we obtain that $\min(t_1 s + t_0,\, t_2) = \frac{1}{N_2} \le \mathfrak{g}_{\rm sh}(s)$.

In order to estimate $M_{\rm bd}$ in (4.5.2) in the present case, we now estimate $\|\mathfrak{g}_{\rm sh}\|_{C^1([0,\hat{s}(\theta_{\rm w})])}$. From (12.3.3), we obtain that $\|u\|_{C^1(\overline{Q^{\rm iter}})} \le N_0$. Then, by (12.2.58) with $\alpha = 0$, we see that $M_{\rm bd} \le \|\mathfrak{g}_{\rm sh}\|_{C^1([0,\hat{s}(\theta_{\rm w})])} \le C(N_0)$, where $C(N_0)$ depends only on the data and N_0.

We also note from (4.5.3) in the present case that $\Gamma_0 = \Gamma_{\rm sonic}$, $\Gamma_1 = \Gamma_{\rm shock}$, $\Gamma_3 = \Gamma_{\rm sym}$, and $\Gamma_4 = \Gamma_{\rm wedge}$.

Then it follows that Problem (12.3.25)–(12.3.29), rewritten in the (s,t)–coordinates in domain $F_1(\Omega)$ and its boundary, is a problem of structure (4.5.84)–(4.5.88) for $u(s,t) := \hat{\psi}(\boldsymbol{\xi})$, where $\boldsymbol{\xi} = F_1^{-1}(s,t)$.

3. In this step we show that Problem (12.3.25)–(12.3.29) satisfies properties (4.5.89)–(4.5.111) and discuss the dependence of the constants in these properties on the parameters of the iteration set in the present case.

Since the C^3–norms of (F_1, F_1^{-1}) depend only on the data, both equation (4.5.84) in $F_1(\Omega)$ and the boundary condition (4.5.85) on Γ_1, obtained from (12.3.25)–(12.3.26), satisfy the properties in Lemma 12.4.5 and Lemma 12.5.7(i)–(ix) with the dependence of the constants specified in Corollaries 12.4.6 and 12.5.9. This implies that conditions (4.5.89), (4.5.91), (4.5.93), (4.5.98)–(4.5.101), and (4.5.103)–(4.5.104) hold, and that (4.5.108)–(4.5.109) hold at P_2.

Furthermore, from Lemma 12.2.2(iv), $(s,t) = (x, y - y_{P_4})$ for all $(s,t) \in F_1(\Omega) \cap \{s < \varepsilon_0\}$. Then, from Lemma 12.4.2 and Lemma 12.5.7(x) with the dependence of the constants as in Corollaries 12.4.3 and 12.5.9, it follows that the equation in $F_1(\Omega) \cap \{s < \frac{\varepsilon_{\rm eq}}{2}\}$ and the boundary condition on $\Gamma_1 \cap \{s < \frac{\varepsilon_{\rm bc}}{2}\}$ satisfy properties (4.5.90), (4.5.92), (4.5.94)–(4.5.97), (4.5.102), and (4.5.111).

Using again that the C^3–norms of (F_1, F_1^{-1}) depend only on the data, and employing the explicit form of (12.3.28)–(12.3.29) and the fact that $\theta_{\rm w} \in [\theta_{\rm w}^{\rm s}, \frac{\pi}{2})$

with $\theta_{\rm w}^{\rm s} > 0$, we find that the boundary conditions (4.5.86)–(4.5.87) on Γ_2 and Γ_3 satisfy (4.5.105), (4.5.107), and (4.5.110), and that (4.5.108)–(4.5.109) hold at P_3. The constants in all these estimates depend only on $\theta_{\rm w}^{\rm s}$ and the C^3–norms of (F_1, F_1^{-1}), and hence on the data.

Moreover, the property in Lemma 12.2.2(ix) and the property that $(s,t) = (x, y - y_{P_4})$ for all $(s,t) \in F_1(\Omega) \cap \{s < \varepsilon_0\}$ in Lemma 12.2.2(iv) imply (4.5.106).

Therefore, we have shown that properties (4.5.89)–(4.5.111) hold in the present case with $(\kappa, \varepsilon, \lambda, M, \beta, \sigma)$ depending only on the data and the following constants in Lemmas 12.4.2, 12.4.5, and 12.5.7:

- Constant ε in (4.5.89)–(4.5.111) is $\frac{1}{2}\min\{\varepsilon_{\rm eq}, \varepsilon_{\rm bc}\}$, and hence depends only on the data;

- Constant κ depends only on the data. Indeed, we first show (4.5.109) at P_2 and P_3 in the $\boldsymbol{\xi}$–coordinates. Then (4.5.109) at P_2 follows from (12.5.53). Also, (4.5.109) at P_3 follows from a similar inequality for the Neumann conditions (12.3.28)–(12.3.29) at corner P_3 with angle $\pi - \theta_{\rm w} \in (\frac{\pi}{2}, \pi - \theta_{\rm w}^{\rm s}]$, hence with constant $\sqrt{2(1-\cos\theta_{\rm w}^{\rm s})}$ on the right-hand side. Now, to obtain (4.5.109) at P_2 and P_3 in the (s,t)–coordinates, we rewrite these estimates in the (s,t)–variables. Since the change of coordinates is nondegenerate, we obtain similar estimates with the right-hand sides multiplied by a nonzero constant depending on the data.

- Similarly, matching the other conditions in (4.5.89)–(4.5.111) with the properties in Lemmas 12.4.2, 12.4.5, and 12.5.7, we have

$$\beta = \frac{3}{4}, \qquad \lambda = \lambda(\lambda_0, \lambda_1, \delta_{\rm bc}, \rho_0, \rho_1, \gamma), \qquad \sigma = C_0\sqrt{\delta_1}, \tag{12.7.18}$$

  where $C_0 > 0$ depends only on the data. Also, we obtain that M in (4.5.89)–(4.5.111) in the present case depends on the data and constants $(\lambda_1, \varepsilon_{\rm eq}, N_{\rm eq}, M, \alpha, N_0, \hat{N}_0, N_2, b)$ in Lemmas 12.4.2, 12.4.5, and 12.5.7.

Thus, from Corollaries 12.4.3, 12.4.6, and 12.5.9, we obtain the following dependence of the constants in (4.5.89)–(4.5.111) on the parameters of the iteration set:

$\delta = 0$;
α is the same as α in the iteration set;
κ and λ depend only on the data;
ε depends only on the data and $\theta_{\rm w}^*$;
$\beta = \frac{3}{4}$;
$\sigma = C_0\sqrt{\delta_1}$ with $C_0 > 0$ depending on the data;
M depends only on the data and $(\theta_{\rm w}^*, \alpha)$.

We also note that the parameters in structure (4.5.1) of domain $F_1(\Omega)$ discussed above satisfy $h, t_1, t_2, \in (\frac{1}{C}, C)$ and $M_{\rm bd} \leq C$, where C depends only on the data and $\theta_{\rm w}^*$, since the positive constants N_0 and N_2 depend only on these parameters.

4. Now we can apply Proposition 4.7.2 with $\delta = 0$. It follows that, if $\delta_1 = \frac{\sigma^2}{C_0^2} > 0$ is small, depending on the data and $(\theta_{\rm w}^*, \alpha)$ (where δ_1 is a parameter in the definition of iteration set $\mathcal{K}$), we obtain the existence and uniqueness of the solution and the asserted estimates in the (s,t)–coordinates in $F_1(\Omega)$, for which $\hat{\alpha}$ is equal to $\alpha_1 = \alpha_1(\kappa, \lambda)$ from Proposition 4.7.2, and hence depends only on the data; the dependence of κ and λ is discussed above. We reduce $\hat{\alpha}$ if necessary to obtain $\hat{\alpha} \in (0, \min\{\bar{\alpha}, \delta^*\}]$, so that the resulting $\hat{\alpha}$ still depends only on the data and $\theta_{\rm w}^*$, since $\bar{\alpha}$ in Proposition 12.2.5 depends on these parameters. Finally, by the change of variables $\boldsymbol{\xi} = F_1^{-1}(s,t)$, we conclude the proof. ☐

We also note the following:

Lemma 12.7.4. (i) *Let* $(u, \theta_{\rm w}) \in \overline{\mathcal{K}^{\rm ext}}$ *and* $\hat{\psi} \in C^2(\Omega) \cap C^1(\overline{\Omega} \setminus \overline{\Gamma_{\rm sonic}}) \cap C(\overline{\Omega})$. *Then the corresponding* $\hat{u}$ *obtained from* $\hat{\psi}$ *by* (12.3.15) *satisfies* $\hat{u} \in C^2(Q^{\rm iter}) \cap C^1(\overline{Q^{\rm iter}} \setminus \partial_{\rm sonic} Q^{\rm iter}) \cap C(\overline{Q^{\rm iter}})$. *Moreover,* $\hat{\psi}$ *is a solution of Problem* (12.3.25)–(12.3.29) *determined by* $(u, \theta_{\rm w})$ *if and only if* $\hat{u}$ *is a solution of Problem* (12.7.2)–(12.7.6).

(ii) *In particular, if* $(u, \theta_{\rm w})$ *and the parameters in Definition* 12.3.2 *are as in Proposition* 12.7.3, *there exists a unique solution* $\hat{u} \in C^2(Q^{\rm iter}) \cap C^1(\overline{Q^{\rm iter}} \setminus \partial_{\rm sonic} Q^{\rm iter}) \cap C(\overline{Q^{\rm iter}})$ *of Problem* (12.7.2)–(12.7.6) *determined by* $(u, \theta_{\rm w})$. *Furthermore,*

$$|\hat{u}(s,t)| \le Cs \qquad \text{in } Q^{\rm iter}, \tag{12.7.19}$$

$$\|\hat{u}\|^{(-1-\alpha),\, \overline{\Gamma_{\rm sym}}}_{2,\alpha,Q^{\rm iter} \cap \{s > s^*\}} \le C_{s^*} \qquad \text{for all } s^* \in (0, \frac{1}{2}), \tag{12.7.20}$$

where α *is from Definition* 12.3.2, *and* (C, C_{s^*}) *for each* $s^* \in (0, \frac{1}{2})$ *depend only on the data and* $(\theta_{\rm w}^*, \alpha)$.

Proof. Rewrite (12.3.15) as

$$\hat{u} := (\hat{\psi} + \varphi_2^{(\theta_{\rm w})} - \tilde{\varphi}_2^{(\theta_{\rm w})}) \circ \mathfrak{F}_{(u,\theta_{\rm w})} \qquad \text{in } Q^{\rm iter}. \tag{12.7.21}$$

Now the regularity of $\hat{u}$ asserted in (i) follows from the regularity of $\hat{\psi}$, Lemma 12.2.4, and (12.2.68). Problem (12.7.2)–(12.7.6) has been defined by expressing Problem (12.3.25)–(12.3.29) in terms of $\hat{u}$ defined by (12.3.15). This proves (i).

Under the conditions in (ii), the existence and uniqueness of a solution $\hat{\psi} \in C^2(\Omega) \cap C^1(\overline{\Omega} \setminus \overline{\Gamma_{\rm sonic}}) \cap C(\overline{\Omega})$ of Problem (12.3.25)–(12.3.29) follow from Proposition 12.7.3. Now the existence and uniqueness of $\hat{u}$ asserted in (ii) follow from (i). Estimate (12.7.19) follows from (12.7.14)–(12.7.15), and estimate (12.7.20) follows from (12.7.16) by the change of variables (12.7.21) and by using (12.2.68) and Lemma 12.2.4. ☐

Lemma 12.7.5. *Let the parameters in Definition* 12.3.2 *and* $(u^{(i)}, \theta_{\rm w}^{(i)}) \in \overline{\mathcal{K}^{\rm ext}}$ *for* $i = 1, 2, \dots,$ *satisfy the conditions of Proposition* 12.7.3, *and let* $\hat{u}^{(i)}$ *be the* *solution of Problem* (12.7.2)–(12.7.6) *determined by* $(u^{(i)}, \theta_{\rm w}^{(i)})$. *Let* $(u^{(i)}, \theta_{\rm w}^{(i)})$ *converge in* $C^1(\overline{Q^{\rm iter}})$ *to* $(u, \theta_{\rm w}) \in \overline{\mathcal{K}^{\rm ext}}$. *Then there exists a unique solution* $\hat{u} \in C^2(Q^{\rm iter}) \cap C^1(\overline{Q^{\rm iter}} \setminus \partial_{\rm sonic} Q^{\rm iter}) \cap C(\overline{Q^{\rm iter}})$ *of Problem* (12.7.2)–(12.7.6) *determined by* $(u, \theta_{\rm w})$. *Moreover,* $\hat{u}^{(i)} \to \hat{u}$ *in the following norms:*

(i) *uniformly in* $\overline{Q^{\rm iter}}$;

(ii) *in* $C^{1,\beta}(K)$ *for each compact* $K \subset \overline{Q^{\rm iter}} \setminus \partial_{\rm sonic} Q^{\rm iter}$ *and each* $\beta \in [0, \alpha)$;

(iii) *in* $C^{2,\beta}(K)$ *for each compact* $K \subset Q^{\rm iter}$ *and each* $\beta \in [0, \alpha)$,

where α *is from Definition* 12.3.2.

Proof. Fix $\beta \in [0, \alpha)$ and $\bar{s} \in (0, \frac{1}{10})$. From (12.7.20) applied for each $\hat{u}^{(i_j)}$, there exists a subsequence converging in $C^{(-1-\beta),\ \overline{\partial_{\rm sym} Q^{\rm iter}}}_{2,\beta,Q^{\rm iter}\cap\{s\geq\bar{s}\}}$. Then there exists a further subsequence converging in $C^{(-1-\beta),\ \overline{\partial_{\rm sym} Q^{\rm iter}}}_{2,\beta,Q^{\rm iter}\cap\{s\geq\bar{s}/10\}}$ and a yet further subsequence converging in $C^{(-1-\beta),\ \overline{\partial_{\rm sym} Q^{\rm iter}}}_{2,\beta,Q^{\rm iter}\cap\{s\geq\bar{s}/100\}}$, *etc.* By the diagonal procedure, we can select a subsequence $\hat{u}^{(i_k)}$ converging to a function $\hat{u} \in C^2(Q^{\rm iter}) \cap C^1(\overline{Q^{\rm iter}} \setminus \partial_{\rm sonic} Q^{\rm iter})$ in the sense of (ii)–(iii) for the fixed β. Applying (12.7.19) to each $\hat{u}^{(i)}$, we conclude that $\hat{u}$ satisfies (12.7.19) and $\hat{u}^{(i_k)}$ converges to $\hat{u}$ uniformly in $\overline{Q^{\rm iter}}$ so that $\hat{u} \in C^2(Q^{\rm iter}) \cap C^1(\overline{Q^{\rm iter}} \setminus \partial_{\rm sonic} Q^{\rm iter}) \cap C(\overline{Q^{\rm iter}})$. Now, since each $\hat{u}^{(i)}$ is a solution of Problem (12.7.2)–(12.7.6) determined by $(u^{(i)}, \theta_{\rm w}^{(i)})$, we use the properties of Lemma 12.7.1(iii) and Lemma 12.7.2(ii) to conclude that $\hat{u}$ is a solution of Problem (12.7.2)–(12.7.6) determined by $(u, \theta_{\rm w})$. From the uniqueness of solution $\hat{u} \in C^2(Q^{\rm iter}) \cap C^1(\overline{Q^{\rm iter}} \setminus \partial_{\rm sonic} Q^{\rm iter}) \cap C(\overline{Q^{\rm iter}})$ for the problem, shown in Lemma 12.7.4(ii), we conclude that the whole sequence $\hat{u}^{(i)}$ converges to $\hat{u}$ in the sense of (i)–(iii) for any fixed $\beta \in [0, \alpha)$. This completes the proof. □

Corollary 12.7.6. *Let the parameters in Definition* 12.3.2 *be as in Proposition* 12.7.3. *If* $(u, \theta_{\rm w}) \in \overline{\mathcal{K}}$, *then* $(u, \theta_{\rm w})$ *satisfies condition* (vii) *of Definition* 12.3.2 *with the nonstrict inequality in* (12.3.16).

Proof. If $(u, \theta_{\rm w}) \in \overline{\mathcal{K}}$, there exists a sequence $(u^{(i)}, \theta_{\rm w}^{(i)}) \in \mathcal{K}$ such that

$$(u^{(i)}, \theta_{\rm w}^{(i)}) \to (u, \theta_{\rm w}) \qquad \text{in } C^{2,\alpha}_{*,1+\delta^*}(Q^{\rm iter}) \times [\theta_{\rm w}^*, \tfrac{\pi}{2}].$$

Then each $(u^{(i)}, \theta_{\rm w}^{(i)})$ satisfies (12.7.13) with $(u^{\#}, \theta_{\rm w}^{\#}) = (u^{(i)}, \theta_{\rm w}^{(i)})$. Now, from Lemma 12.7.5, there exists a unique solution $\hat{u} \in C^2(Q^{\rm iter}) \cap C^1(\overline{Q^{\rm iter}} \setminus \partial_{\rm sonic} Q^{\rm iter}) \cap C(\overline{Q^{\rm iter}})$ of Problem (12.7.2)–(12.7.6) determined by $(u, \theta_{\rm w})$ such that $\hat{u}^{(i)} \to \hat{u}$ uniformly in $Q^{\rm iter}$. Thus, $\hat{u}^{(i)} - u^{(i)} \to \hat{u} - u$ uniformly in $Q^{\rm iter}$. Since each $\hat{u}^{(i)} - u^{(i)}$ satisfies (12.3.16), the limit satisfies the nonstrict inequality in (12.3.16). □

We now study some properties of $\hat{\psi}$.

Lemma 12.7.7. *If the parameters in Definition* 12.3.2 *and* $\delta_{\mathcal{K}}$ *are as in Proposition* 12.7.3, *and if* δ_3 *is further reduced, depending on the data and* $(\theta_{\rm w}^*, \delta_2)$, *then, for any* $(u^\#, \theta_{\rm w}^\#) \in \overline{\mathcal{K}}$, *there exists* $\delta_{u^\#,\theta_{\rm w}^\#} \in (0, \delta_{\mathcal{K}})$ *such that, for any* $(u, \theta_{\rm w}) \in \overline{\mathcal{K}^{\rm ext}}$ *satisfying*

$$\|u^\# - u\|_{C^1(\overline{Q^{\rm iter}})} + |\theta_{\rm w}^\# - \theta_{\rm w}| \le \delta_{u^\#,\theta_{\rm w}^\#}, \tag{12.7.22}$$

solution $\hat{\psi} \in C^2(\Omega) \cap C^1(\overline{\Omega} \setminus \overline{\Gamma_{\rm sonic}}) \cap C(\overline{\Omega})$ *of Problem* (12.3.25)–(12.3.29) *determined by* $(u, \theta_{\rm w})$ *satisfies*

$$\hat{\psi} \ge 0 \qquad in\ \Omega. \tag{12.7.23}$$

Proof. Fix $(u^\#, \theta_{\rm w}^\#) \in \overline{\mathcal{K}}$. We will use (12.3.6). Then, from the form of the right-hand side of that inequality, we consider two cases: $\theta_{\rm w}^\# \in [\theta_{\rm w}^*, \frac{\pi}{2} - \frac{2\delta_1}{N_1^2})$ and $\theta_{\rm w}^\# \in [\frac{\pi}{2} - \frac{2\delta_1}{N_1^2}, \frac{\pi}{2})$.

1. Let $\theta_{\rm w}^\# \in [\theta_{\rm w}^*, \frac{\pi}{2} - \frac{2\delta_1}{N_1^2})$. Denote $\Omega^\#, \Gamma_{\rm shock}^\#, \varphi^\#$, *etc.* as the objects from Definition 12.2.6 corresponding to $(u^\#, \theta_{\rm w}^\#)$. Then, from (12.3.6), we have

$$(\varphi^\# - \varphi_2) \ge \frac{\delta_1\delta_2}{N_1^2} \qquad \text{in } \overline{\Omega} \setminus \mathcal{D}_{\varepsilon/10}.$$

Using (12.2.35) and (12.2.44) for $\varphi^\#$ and $u^\#$, and the properties of map $\mathfrak{F}_{(u^\#,\theta_{\rm w}^\#)}$ in Lemma 12.2.2(iv) and (12.2.41), we have

$$u^\# \ge \frac{\delta_1\delta_2}{N_1^2} \qquad \text{in } \overline{Q^{\rm iter}} \cap \{s \ge \frac{\tilde{\varepsilon}^\#}{10}\} \quad \text{for } \tilde{\varepsilon}^\# := \frac{\varepsilon}{\hat{s}(\theta_{\rm w}^\#)}.$$

Since $(u^\#, \theta_{\rm w}^\#) \in \overline{\mathcal{K}}$, then $u^\#$ and the corresponding function $\hat{u}^\#$ from (12.3.15) satisfy (12.3.16) with possibly nonstrict inequality. Thus, if $\delta_3 \le \frac{\delta_1\delta_2}{2N_1^2}$,

$$\hat{u}^\# \ge \frac{\delta_1\delta_2}{2N_1^2} \qquad \text{in } \overline{Q^{\rm iter}} \cap \{s \ge \frac{\tilde{\varepsilon}^\#}{10}\}. \tag{12.7.24}$$

Next we show that, if $\delta_{u^\#,\theta_{\rm w}^\#} \in (0, \delta_{\mathcal{K}})$ is small, and $(u, \theta_{\rm w}) \in \overline{\mathcal{K}^{\rm ext}}$ satisfies (12.7.22), then

$$\|\hat{u} - \hat{u}^\#\|_{C(\overline{Q^{\rm iter}} \cap \{s \ge \tilde{\varepsilon}^\#/10\})} \le \frac{\delta_1\delta_2}{4N_1^2}. \tag{12.7.25}$$

Otherwise, there exists a sequence $\overline{\mathcal{K}^{\rm ext}} \ni (u^{(i)}, \theta_{\rm w}^{(i)}) \to (u^\#, \theta_{\rm w}^\#)$ in $C^1(\overline{Q^{\rm iter}}) \times [\theta_{\rm w}^*, \frac{\pi}{2}]$ such that, for the corresponding functions $\hat{u}^{(i)}$ (which exist for all sufficiently large i by Proposition 12.7.3 and Lemma 12.7.4),

$$\|\hat{u}^{(i)} - \hat{u}^\#\|_{C(\overline{Q^{\rm iter}} \cap \{s \ge \tilde{\varepsilon}^\#/10\})} > \frac{\delta_1\delta_2}{4N_1^2}.$$

By Lemma 12.7.5, we can pass to the limit as $i \to \infty$ to obtain

$$0 = \|\hat{u}^{\#} - \hat{u}^{\#}\|_{C(\overline{Q^{\rm iter}} \cap \{s \geq \tilde{\varepsilon}^{\#}/10\})} \geq \frac{\delta_1 \delta_2}{4N_1^2},$$

which is a contradiction. Thus, (12.7.25) holds if $\delta_{u^{\#},\theta_{\rm w}^{\#}} > 0$ is small.

From (12.7.24)–(12.7.25), using the continuity of $\hat{s}(\cdot)$, and reducing $\delta_{u^{\#},\theta_{\rm w}^{\#}}$ if necessary, we obtain that, for each $(u, \theta_{\rm w}) \in \overline{\mathcal{K}^{\rm ext}}$ satisfying (12.7.22),

$$\hat{u} \geq \frac{\delta_1 \delta_2}{4N_1^2} \qquad \text{in } \overline{Q^{\rm iter}} \cap \{s \geq \frac{\tilde{\varepsilon}}{5}\} \quad \text{for } \tilde{\varepsilon} := \frac{\varepsilon}{\hat{s}(\theta_{\rm w})}.$$

Inverting (12.3.15) yields

$$\hat{\psi}(\boldsymbol{\xi}) = \hat{u}(u(\mathfrak{F}^{-1}(\boldsymbol{\xi}))) + (\tilde{\varphi}_2^{(\theta_{\rm w})} - \varphi_2^{(\theta_{\rm w})})(\boldsymbol{\xi}) \qquad \text{for all } \boldsymbol{\xi} \in \Omega(u).$$

Then, using (12.2.35), we have

$$\hat{\psi} \geq \frac{\delta_1 \delta_2}{4N_1^2} \qquad \text{in } \Omega \setminus \mathcal{D}_{\varepsilon/5}. \tag{12.7.26}$$

Recall that $\varepsilon \leq \varepsilon_{\rm p} := \frac{1}{2}\min\{\varepsilon_{\rm eq}, \varepsilon_{\rm bc}\}$. From Lemma 12.4.5(vi), $\hat{\psi}$ on $\Omega \cap \mathcal{D}_{\varepsilon_{\rm p}}$ satisfies equation (12.4.14). Also, $\hat{\psi}$ satisfies the boundary conditions (12.3.26) on $\Gamma_{\rm shock} \cap \mathcal{D}_{\varepsilon_{\rm p}}$ with $\mathcal{M}(\cdot)$ given by (12.5.47), and (12.3.28) on $\Gamma_{\rm wedge} \cap \partial\mathcal{D}_{\varepsilon_{\rm p}}$. Condition (12.3.26) is homogeneous on $\Gamma_{\rm shock} \cap \mathcal{D}_{\varepsilon_{\rm p}}$ by the second equality in Lemma 12.5.7(vii). Also, equation (12.4.14) is strictly elliptic in $\Omega \cap \mathcal{D}_{\varepsilon_{\rm p}}$, and the boundary conditions (12.3.26) and (12.3.28) are oblique. Using that $\hat{\psi} = 0$ on $\Gamma_{\rm sonic}$ and $\hat{\psi} > 0$ on $\Omega \cap \partial\mathcal{D}_{\varepsilon_{\rm p}}$, by (12.7.26), we obtain that $\hat{\psi} > 0$ in $\Omega \cap \mathcal{D}_{\varepsilon_{\rm p}}$ by the comparison principle. Then, using (12.7.26) again, we obtain (12.7.23).

2. Let $\theta_{\rm w}^{\#} \in [\frac{\pi}{2} - \frac{2\delta_1}{N_1^2}, \frac{\pi}{2}]$. Since $N_1 \geq 8$, it follows that $\theta_{\rm w}^{\#} \geq \frac{\pi}{2} - \frac{\delta_1}{4N_1}$ in that interval.

Consider $(u, \theta_{\rm w}) \in \overline{\mathcal{K}^{\rm ext}}$ satisfying (12.7.22). Reducing $\delta_{u^{\#},\theta_{\rm w}^{\#}}$ if necessary, we find that $\theta_{\rm w} \in [\frac{\pi}{2} - \frac{\delta_1}{2N_1}, \frac{\pi}{2}]$. Then the first equality in Lemma 12.5.7(vii) shows that the boundary condition (12.3.26) is homogeneous on $\Gamma_{\rm shock}$. Thus, $\hat{\psi}$ satisfies the strictly elliptic homogeneous equation (12.4.14) in Ω, and the oblique homogeneous boundary conditions (12.3.26) and (12.3.28) on $\Gamma_{\rm shock}$ and $\Gamma_{\rm wedge}$ respectively, and the oblique condition (12.3.29) on $\Gamma_{\rm sym}$ with right-hand side $-v_2 \leq 0$. Moreover, the obliqueness at points P_2 and P_3 is shown as in Step 3 of the proof of Proposition 12.7.3. Then, from the comparison principle in Lemma 4.4.2, we obtain that $\hat{\psi} \geq 0$ in Ω. ☐

Lemma 12.7.8. *If the parameters in Definition* 12.3.2 *are chosen as in Lemma* 12.7.7, *and* $\varepsilon > 0$ *is further reduced if necessary depending on the data, then, for any* $(u, \theta_{\rm w}) \in \overline{\mathcal{K}^{\rm ext}}$,

$$\hat{\psi} \leq Cx^2 \qquad \textit{in } \Omega \cap \mathcal{D}_{\varepsilon_0}, \tag{12.7.27}$$

where C depends only on the data and the constant in (12.7.14), *and hence on the data and* $(\theta_{\rm w}^*, \alpha)$.

Proof. We first show (12.7.27) in $\Omega \cap \mathcal{D}_{\varepsilon/2}$ for $\varepsilon \le \varepsilon_{\rm p} := \frac{1}{2}\min\{\varepsilon_{\rm eq}, \varepsilon_{\rm bc}\}$.

From Lemma 12.4.5(vi), $\hat{\psi}$ on $\Omega \cap \mathcal{D}_{\varepsilon_{\rm p}}$ satisfies equation (12.4.14), which is strictly (but non-uniformly) elliptic. Also, $\hat{\psi}$ satisfies the boundary conditions (12.3.26) on $\Gamma_{\rm shock} \cap \mathcal{D}_{\varepsilon_{\rm p}}$ and (12.3.28) on $\Gamma_{\rm wedge} \cap \partial\mathcal{D}_{\varepsilon_{\rm p}}$, where the boundary conditions are oblique and homogeneous by using the first equality in Lemma 12.5.7(vii). Therefore, all these properties hold with $\varepsilon_{\rm p}$ replaced by ε.

Let

$$v(x,y) = \frac{1}{2}Ax^2$$

for $A \ge \frac{2-\frac{\mu_0}{10}}{1+\gamma} > 0$, whose specific value will be chosen below. We show that v is a supersolution of Problem (12.3.25)–(12.3.28) in $\Omega\cap\mathcal{D}_\varepsilon$, if ε is small, depending only on the data, and the Dirichlet condition is added on the remaining boundary part $\partial(\Omega \cap \mathcal{D}_\varepsilon) \cap \{x = \varepsilon\}$.

From Lemma 12.4.5(vi), equation (12.3.25), written in the (x, y)–coordinates on $\Omega \cap \mathcal{D}_\varepsilon$, is of form (12.4.14). Thus, we substitute v into equation (12.4.14). Using (12.4.10) and $A \ge \frac{2-\frac{\mu_0}{10}}{1+\gamma}$, we obtain

$$\zeta_1(\frac{v_x}{x}) = \zeta(A) = \frac{2-\frac{\mu_0}{10}}{1+\gamma} \qquad \text{on } \{x > 0\}$$

for $\mu_0 \in (0,1)$. Then, using estimate (12.4.17) for $\tilde{O}_1^m$ and $\tilde{O}_4^m$ and notation $O(x)$ for a quantity estimated by $|O(x)| \le C|x|$, and denoting by $\mathcal{N}_{(u,\theta_{\rm w})}$ the left-hand side of (12.4.14), we obtain that, in $\Omega \cap \mathcal{D}_\varepsilon$ in which $x \in (0,\varepsilon)$,

$$\begin{aligned}
&\mathcal{N}_{(u,\theta_{\rm w})}(D^2v, Dv, v, x, y)\\
&= \Big(2x - (2-\frac{\mu_0}{10})x + O((\varepsilon+\sqrt{x})x)\Big)A - \Big(1 + O(x^{\frac{3}{4}})\Big)Ax\\
&\le A\Big(-(1-\frac{\mu_0}{10})x + C\sqrt{\varepsilon}x\Big)\\
&< 0 \qquad \text{in } \Omega\cap\mathcal{D}_\varepsilon
\end{aligned}$$

if ε is small, depending only on the data.

Now we substitute v into (12.3.26) on $\Gamma_{\rm shock}\cap\mathcal{D}_\varepsilon$. We employ the properties in (12.3.26) in the (x, y)–coordinates given in Lemma 12.5.7(x). Using (12.5.54)–

(12.5.56), we find that, for each $x \in (0, \varepsilon)$,

$$\begin{aligned}
&\mathcal{M}_{(u,\theta_{\rm w})}(Dv, v, x, y) \\
&= \mathcal{M}_{(u,\theta_{\rm w})}((Ax, 0), \frac{1}{2}Ax^2, x, 0) - \mathcal{M}_{(u,\theta_{\rm w})}((0,0), 0, x, 0) \\
&= \Big(\int_0^1 (\partial_{p_1}, \partial_z)\mathcal{M}_{(u,\theta_{\rm w})}((Axr, 0), \frac{1}{2}Ax^2 r, x, 0)dr\Big) \cdot (Ax, \frac{1}{2}Ax^2) \\
&\le -\delta_{\rm bc}Ax(1 + \frac{1}{2}x) \\
&< 0.
\end{aligned}$$

Next we substitute v into (12.3.28) on $\Gamma_{\rm wedge} \cap \mathcal{D}_\varepsilon$ to obtain

$$\partial_\nu v = \frac{1}{c_2 - x}\partial_y(\frac{1}{2}Ax^2) = 0 \qquad \text{on } \Gamma_{\rm wedge} \cap \mathcal{D}_\varepsilon \subset \{y = 0\}.$$

Finally, using (12.7.14) (with constant $\hat{C}$) and choosing $A = \max\{\frac{2\hat{C}}{\varepsilon^2}, \frac{2-\frac{\mu_0}{10}}{1+\gamma}\}$, we have

$$v \ge \hat{\psi} \qquad \text{on } \partial(\Omega \cap \mathcal{D}_\varepsilon) \cap \{x = \varepsilon\}.$$

Now, applying the comparison principle in $\Omega\cap\mathcal{D}_\varepsilon$, we obtain that $\hat{\psi} \le v$ in that region, which implies (12.7.27) in $\Omega \cap \mathcal{D}_\varepsilon$. From this, we extend (12.7.27) to $\Omega \cap \mathcal{D}_{\varepsilon_0}$ by using (12.3.7)–(12.3.8). □

Proposition 12.7.9. *If the parameters in Definition* 12.3.2 *are chosen as in Lemma* 12.7.8*, then, for any* $\sigma \in (0, 1)$*, there exist* $\hat{\varepsilon}_{\rm p} \in (0, \frac{\varepsilon_{\rm p}}{2}]$ *(for* $\varepsilon_{\rm p} := \frac{1}{2}\min\{\varepsilon_{\rm eq}, \varepsilon_{\rm bc}\}$*) and* C*, depending only on the data and* $(\theta^*_{\rm w}, \alpha, \sigma)$ *such that the following holds: For any* $(u^\#, \theta^\#_{\rm w}) \in \overline{\mathcal{K}}$*, there is* $\delta_{u^\#, \theta^\#_{\rm w}} > 0$ *small so that, for any* $(u, \theta_{\rm w}) \in \overline{\mathcal{K}^{\rm ext}}$ *satisfying* (12.7.22)*, solution* $\hat{\psi} \in C^2(\Omega) \cap C^1(\overline{\Omega} \setminus \overline{\Gamma_{\rm sonic}}) \cap C(\overline{\Omega})$ *of Problem* (12.3.25)–(12.3.29) *determined by* $(u, \theta_{\rm w})$ *satisfies*

$$\|\hat{\psi}\|^{2,(\rm par)}_{2,\sigma,\Omega\cap\mathcal{D}_{\hat{\varepsilon}_{\rm p}}} \le C. \tag{12.7.28}$$

Proof. We assume that the parameters are chosen such that the previous results hold in this section.

As we have shown in Step 1 of the proof of Lemma 12.7.7, $\hat{\psi}$ on $\Omega \cap \mathcal{D}_{\varepsilon_{\rm p}}$ satisfies equation (12.4.14), the boundary conditions (12.3.26) on $\Gamma_{\rm shock} \cap \mathcal{D}_{\varepsilon_{\rm p}}$ with $\mathcal{M}(\cdot)$ given by (12.5.47), and (12.3.28) on $\Gamma_{\rm wedge} \cap \partial\mathcal{D}_{\varepsilon_{\rm p}}$.

We check that the conditions of Theorem 4.7.4 are satisfied in domain $\Omega\cap\mathcal{D}_{\varepsilon_{\rm p}}$ in the present case.

Since $\varepsilon_{\rm p} < \frac{\varepsilon_0}{2}$, then domain $\Omega \cap \mathcal{D}_{\varepsilon_{\rm p}}$ in the (x, y)–coordinates with shift $y \mapsto y - \theta_{\rm w}$ is of form (4.7.6) with $f = \mathfrak{g}^{(u,\theta_{\rm w})}_{\rm sh}$, by Lemma 12.2.7(iv). Then the regularity condition (4.7.12) of f with $\beta = \alpha$ follows from (12.3.17), where $\delta^* \ge \alpha$, and $\hat{f}_0$ is C^∞–smooth with C^k–norms depending only on the data. The lower bound of $f = \mathfrak{g}^{(u,\theta_{\rm w})}_{\rm sh}$ on $(0, \varepsilon_{\rm p})$ in condition (4.7.12) is $\frac{1}{N_2}$ by (12.3.5).

Thus, $l = \frac{1}{N_2}$ in Theorem 4.7.4 in the present case. We note that N_2 in (12.3.5) depends only on the data and $\theta_{\rm w}^*$.

The conditions of Theorem 4.7.4 for the equation, *i.e.*, (4.7.11) and (4.7.13)–(4.7.16), are satisfied by equation (12.4.14) in $\Omega \cap \mathcal{D}_{\varepsilon_{\rm p}}$, which follows from Lemma 12.4.2(i)–(ii), where we recall that, using form (12.4.18) of equation (12.4.14), $(\hat{A}_{ij}, \hat{A}_i) = (\hat{A}_{ij}, \hat{A}_i)(\mathbf{p}, x, y)$, *i.e.*, independent of z. The constants in these estimates depend on the data.

The boundary condition (12.3.26) on $\Gamma_{\rm shock} \cap \mathcal{D}_{\varepsilon_{\rm p}}$, written in the (x,y)–coordinates in Lemma 12.7.2(x), satisfies conditions (4.7.17)–(4.7.19), which follows from (12.5.54)–(12.5.56). The constants in these estimates depend on the data and $(\theta_{\rm w}^*, \alpha)$. Note that this is the only place in this proof, which makes C in (12.7.28) depend on α.

The boundary condition (12.3.28) on $\Gamma_{\rm wedge} \cap \partial\mathcal{D}_{\varepsilon_{\rm p}}$ is of form (4.7.9) in the (x,y)–coordinates, since $\Gamma_{\rm wedge} \cap \partial\mathcal{D}_{\varepsilon_{\rm p}} \subset \{y = \theta_{\rm w}\}$.

Also, $|\hat{\psi}(x,y)| \le Cx^2$ in $\Omega \cap \mathcal{D}_{\varepsilon_{\rm p}}$, by Lemmas 12.7.7–12.7.8.

We note that the ellipticity, obliqueness, and other constants in the estimates of the ingredients of the equation, the boundary condition, and boundary $\Gamma_{f,\varepsilon} = \Gamma_{\rm shock} \cap \mathcal{D}_\varepsilon$, discussed above, as well as the constants in Lemmas 12.7.7–12.7.8, depend only on the data and $\theta_{\rm w}^*$.

Let $\sigma \in (0,1)$. Then r_0, determined in Theorem 4.7.4 with $\alpha = \sigma$ by the constants discussed above, depends only on the data and $(\theta_{\rm w}^*, \sigma)$. Now, we choose $\hat{\varepsilon}_{\rm p} = \min\{r_0, \frac{\varepsilon_{\rm p}}{2}, \frac{1}{N_2^2}\}$. Then, from Theorem 4.7.4 in $\Omega \cap \mathcal{D}_{\varepsilon_{\rm p}}$ applied with $\alpha = \sigma$, we obtain (12.7.28). □

Proposition 12.7.10. *If parameters ε and $\frac{1}{N_1}$ of the iteration set in Definition* 12.3.2 *are small – depending only on the data and $\theta_{\rm w}^*$, if δ_1 is small – depending only on the data and $(\theta_{\rm w}^*, \alpha)$, and if δ_3 is small – depending only on the data and $(\theta_{\rm w}^*, \delta_2)$, then, for any $(u^\#, \theta_{\rm w}^\#) \in \overline{\mathcal{K}}$, there exists $\delta_{u^\#, \theta_{\rm w}^\#} > 0$ so that, for any $(u, \theta_{\rm w}) \in \overline{\mathcal{K}^{\rm ext}}$ satisfying* (12.7.22), *solution $\hat{\psi}$ of* (12.3.25)–(12.3.29) *determined by $(u, \theta_{\rm w})$ satisfies $\hat{\psi} \in C^{2,\hat{\alpha}}_{*,2}(\Omega)$ with*

$$\|\hat{\psi}\|^{*,(2)}_{2,\hat{\alpha},\Omega} \le C, \tag{12.7.29}$$

where $\hat{\alpha} \in (0, \frac{1}{8})$ is the constant determined in Proposition 12.7.3 *(depending only on the data and $\theta_{\rm w}^*$), and C depends only on the data and $(\theta_{\rm w}^*, \alpha)$.*

Proof. Let the parameters satisfy the conditions in Propositions 12.7.3 and 12.7.9. Let $\hat{\varepsilon}_{\rm p}$ be the constant determined in Proposition 12.7.9 for $\sigma = \hat{\alpha}$. Then estimate (12.7.29) follows from estimate (12.7.16) for $s = \frac{\hat{\varepsilon}_{\rm p}}{2}$, combined with (12.7.28) for $\sigma = \hat{\alpha}$. We have used that $\hat{\varepsilon}_{\rm p}$ depends only on the data and $(\theta_{\rm w}^*, \alpha)$. □

Corollary 12.7.11. *Under the conditions of Proposition* 12.7.10, *choosing α in the definition of the iteration set $\mathcal{K}$ to be $\alpha = \frac{\hat{\alpha}}{2}$ for $\hat{\alpha}$ determined in Proposition* 12.7.10, *then, for any $(u^\#, \theta_{\rm w}^\#) \in \overline{\mathcal{K}}$, there exists $\delta_{u^\#, \theta_{\rm w}^\#} > 0$ so that, for* *any* $(u, \theta_{\rm w}) \in \overline{\mathcal{K}^{\rm ext}}$ *satisfying* (12.7.22), *solution* $\hat{u}$ *of Problem* (12.7.2)–(12.7.6) *determined by* $(u, \theta_{\rm w})$ *satisfies*

$$\|\hat{u}\|^{*,(2)}_{2,\alpha,Q^{\rm iter}} \leq C, \tag{12.7.30}$$

where C *depends only on the data and* $\theta_{\rm w}^*$.

Proof. From Lemma 12.7.4, the unique solution $\hat{u} \in C^2(Q^{\rm iter}) \cap C^1(\overline{Q^{\rm iter}} \setminus \partial_{\rm sonic}Q^{\rm iter}) \cap C(\overline{Q^{\rm iter}})$ of Problem (12.7.2)–(12.7.6) and the unique solution $\hat{\psi} \in C^2(\Omega) \cap C^1(\overline{\Omega} \setminus \overline{\Gamma_{\rm sonic}}) \cap C(\overline{\Omega})$ of (12.3.25)–(12.3.29) are related by (12.3.15) with $\hat{\varphi} = \hat{\psi} + \varphi_2$.

Since $(u, \theta_{\rm w}) \in \overline{\mathcal{K}^{\rm ext}}$ satisfies (12.3.3), we apply Lemma 12.2.7(vii) to $(u, \theta_{\rm w})$ to obtain (12.2.66) for $\mathfrak{g}_{\rm sh}^{(u,\theta_{\rm w})}$ with α from the iteration set, *i.e.*, $\alpha = \frac{\hat{\alpha}}{2}$. Also, (12.7.29) and Lemma 12.2.4 imply that $\hat{\varphi} = \hat{\psi} + \varphi_2$ satisfies (12.2.71) with $\sigma = 1 + \delta_*$ and $\hat{\alpha}$ instead of α, and with a constant on the right-hand side depending only on the data and $\theta_{\rm w}^*$ (since $\hat{\alpha}$ depends only on these constants). Then estimate (12.7.30) follows from (12.3.15) and Lemma 12.2.7(ix). □

We fix $\alpha = \frac{\hat{\alpha}}{2}$ (as in Corollary 12.7.11) as the parameter of the iteration set from now on. This also makes the conditions on the smallness of δ_1, given above, depend only on the data and $\theta_{\rm w}^*$.

12.8 OPENNESS OF THE ITERATION SET

In this section, we assume that the parameters of the iteration set satisfy the requirements of Proposition 12.7.10 and $\alpha = \frac{\hat{\alpha}}{2}$. We show that $\mathcal{K}$ is relatively open in $C^{2,\alpha}_{*,1+\delta^*}(Q^{\rm iter}) \times [\theta_{\rm w}^*, \frac{\pi}{2}]$.

We note that this section is focused on the iteration set $\mathcal{K}$, rather than its closure, so that we do not follow the notational convention in Remark 12.3.11. Instead, we consider the strict inequalities in the estimates as specified in Definition 12.3.2 and Remarks 12.3.4–12.3.8.

We first show that set $\mathcal{K}^{\rm ext}$, introduced in Definition 12.3.9, is open.

Lemma 12.8.1. *Set* $\mathcal{K}^{\rm ext}$ *is relatively open in* $C^{2,\alpha}_{*,1+\delta^*}(Q^{\rm iter}) \times [\theta_{\rm w}^*, \frac{\pi}{2}]$.

Proof. It suffices to show that each of conditions (i)–(vi) in Definition 12.3.2 determines an open subset of $C^{2,\alpha}_{*,1+\delta^*}(Q^{\rm iter}) \times [\theta_{\rm w}^*, \frac{\pi}{2}]$. This can be seen as follows:

- Condition (i) defines a relatively open set, since function $\eta_1(\theta_{\rm w})$ is continuous.

- Condition (ii) uses set $\mathfrak{S}$ defined by (12.2.50), that is, by the inequalities in (12.2.49). We first note that all the three terms in the inequalities in (12.2.49) are continuous functions of $(s^*, \theta_{\rm w})$. For the middle term $u(\frac{s^*}{\hat{s}(\theta_{\rm w})}, 1)$, this continuity follows from the inclusion: $u \in C^{2,\alpha}_{*,1+\delta^*}(Q^{\rm iter})$

and from the continuity and positive lower bound of $\hat{s}(\cdot)$. For the right and left terms, such a continuity follows from the fact that the parameters of state (2) depend continuously on $\theta_{\rm w}$. Also, since $u(0,\cdot)=0$ by (12.2.50), then (12.2.47) implies that the strict inequalities in (12.2.49) hold for all $s^*\in[0,\hat{s}(\theta_{\rm w})]$. Since the $C^{2,\alpha}_{*,1+\delta^*}(Q^{\rm iter})$–norm is stronger than the $C(\overline{Q^{\rm iter}})$–norm, the assertion follows.

- Condition (iii) defines a relatively open set. Indeed, if $(u,\theta_{\rm w})\in\mathcal{K}^{\rm ext}$, then, as we have shown above, if $(\tilde{u},\tilde{\theta}_{\rm w})$ is sufficiently close to $(u,\theta_{\rm w})$ in $C^{2,\alpha}_{*,1+\delta^*}(Q^{\rm iter})$, then $(\tilde{u},\tilde{\theta}_{\rm w})\in\mathfrak{S}$, so that $\tilde{\Gamma}_{\rm shock}$ and $\tilde{\mathfrak{g}}_{\rm sh}$ are defined. Then it follows from (12.2.59) and the continuity of $\hat{s}(\cdot)$ that (12.3.5) holds for $\tilde{\mathfrak{g}}_{\rm sh}$, if $(\tilde{u},\tilde{\theta}_{\rm w})$ is even closer to $(u,\theta_{\rm w})$ in $C^{2,\alpha}_{*,1+\delta^*}(Q^{\rm iter})$. Also, combining this with Definition 12.2.6(iv) and Lemma 12.2.2(ii), we obtain (12.3.4) for $(\tilde{u},\tilde{\theta}_{\rm w})$ sufficiently close to $(u,\theta_{\rm w})$.

- Conditions (iv)–(vi) define a relatively open set. Indeed, if $(u,\theta_{\rm w})\in\mathcal{K}^{\rm ext}$, then, as we have shown above, if $(\tilde{u},\tilde{\theta}_{\rm w})$ is sufficiently close to $(u,\theta_{\rm w})$ in the norm of $C^{2,\alpha}_{*,1+\delta^*}(Q^{\rm iter})$, then $(\tilde{u},\tilde{\theta}_{\rm w})\in\mathfrak{S}$ so that $\Omega^{(\tilde{u},\tilde{\theta}_{\rm w})}$, $\Gamma^{(\tilde{u},\tilde{\theta}_{\rm w})}_{\rm shock}$, $\tilde{\varphi}=\varphi^{(\tilde{u},\tilde{\theta}_{\rm w})}$, and $\tilde{\psi}=\psi^{(\tilde{u},\tilde{\theta}_{\rm w})}$ are well-defined. Also, we note that condition (iii) of Definition 12.3.2 for $(u,\theta_{\rm w})$ implies that $b=\frac{1}{N_2}$ can be used in the conditions of Lemma 12.2.7(viii). Then (12.2.60)–(12.2.62) and (12.2.69)–(12.2.70) imply that, in addition to (12.2.62), the following holds:

$$\begin{aligned}&\|(D\varphi\circ\mathfrak{F})_{(u,\theta_{\rm w})}-(D\varphi\circ\mathfrak{F})_{(\tilde{u},\tilde{\theta}_{\rm w})}\|_{L^\infty(\overline{Q^{\rm iter}})}\\&\quad+\|(D\psi\circ\mathfrak{F})_{(u,\theta_{\rm w})}-(D\psi\circ\mathfrak{F})_{(\tilde{u},\tilde{\theta}_{\rm w})}\|_{L^\infty(\overline{Q^{\rm iter}})}\\&\le C\big(\|u-\tilde{u}\|_{C^1(\overline{Q^{\rm iter}})}+|\theta_{\rm w}-\tilde{\theta}_{\rm w}|\big),\end{aligned}\tag{12.8.1}$$

where C depends only on the data and $\theta^*_{\rm w}$. Similar properties for ψ and $\tilde{\psi}$ are also obtained by using (12.2.63).

Combining these properties with (12.2.61), and assuming that $|\theta_{\rm w}-\tilde{\theta}_{\rm w}|+\|u-\tilde{u}\|_{C^1(\overline{Q^{\rm iter}})}$ is sufficiently small, we obtain the strict inequalities in (iv)–(vi) for $\tilde{\psi}$, $\tilde{\varphi}$, and $\tilde{\Omega}$ corresponding to $(\tilde{u},\tilde{\theta}_{\rm w})$, by using the following features:

(a) $\eta_2(\cdot)$ is continuous;

(b) In the proofs for conditions (12.3.7)–(12.3.8), we use that $\frac{\varepsilon}{10}\le x<\varepsilon_0$ in $\Omega_{\varepsilon_0}\setminus\Omega_{\varepsilon/10}$;

(c) For conditions (12.3.13)–(12.3.14), since $\Gamma_{\rm shock}(u,\theta_{\rm w})$ is defined by Definition 12.2.6(iv), using (12.2.59) and Lemma 12.2.2(ii) yields

$$\begin{aligned}&\|(\boldsymbol{\nu}_{\rm sh}\circ\mathfrak{F})_{(u,\theta_{\rm w})}-(\boldsymbol{\nu}_{\rm sh}\circ\mathfrak{F})_{(\tilde{u},\tilde{\theta}_{\rm w})}\|_{C^\alpha([0,1])}\\&\le C\big(\|(u-\tilde{u})(\cdot,1)\|_{C^{1,\alpha}([0,1])}+|\theta_{\rm w}-\tilde{\theta}_{\rm w}|\big).\end{aligned}\tag{12.8.2}$$

□

Now we are ready to prove the main result of this section.

Proposition 12.8.2. *If parameters $(\varepsilon, \delta_1, \frac{1}{N_1})$ of the iteration set in Definition 12.3.2 are small (depending only on the data and $\theta_{\rm w}^*$), if δ_3 is small (depending only on the data and $(\theta_{\rm w}^*, \delta_2)$), and if $\alpha = \frac{\hat\alpha}{2}$ for $\hat\alpha$ determined in Proposition 12.7.10, then the iteration set $\mathcal{K}$ is relatively open in $C^{2,\alpha}_{*,1+\delta^*}(Q^{\rm iter}) \times [\theta_{\rm w}^*, \frac{\pi}{2}]$.*

Proof. We choose parameters $(\varepsilon, \delta_1, \frac{1}{N_1})$ of the iteration set in Definition 12.3.2, and choose $\delta_{\mathcal{K}}$ that satisfies the conditions of Propositions 12.7.3 and 12.7.9–12.7.10.

Let $(u^\#, \theta_{\rm w}^\#) \in \mathcal{K}$. By Lemma 12.8.1, there exists $\delta_\# > 0$ so that, if $(u, \theta_{\rm w}) \in C^{2,\alpha}_{*,1+\delta^*}(Q^{\rm iter}) \times [\theta_{\rm w}^*, \frac{\pi}{2}]$ satisfies

$$\|u^\# - u\|^{*,(1+\delta^*)}_{2,\alpha,Q^{\rm iter}} + |\theta_{\rm w}^\# - \theta_{\rm w}| \le \delta_\#, \tag{12.8.3}$$

then $(u, \theta_{\rm w}) \in \mathcal{K}^{\rm ext}$. It remains to show that condition (vii) of Definition 12.3.2 holds for $(u, \theta_{\rm w})$ if the parameters are chosen properly.

By Proposition 12.7.3 and Corollary 12.7.11, if the parameters of the iteration set are appropriately chosen, possibly reducing $\delta_\#$, then there exists a unique solution $\hat u$ of Problem (12.7.2)–(12.7.6) determined by $(u, \theta_{\rm w})$ satisfying

$$\|\hat u\|^{*,(2)}_{2,\alpha,Q^{\rm iter}} \le C.$$

Since $(u^\#, \theta_{\rm w}^\#) \in \mathcal{K}$, then there exists a solution $\hat u^\#$ of Problem (12.7.2)–(12.7.6) determined by $(u^\#, \theta_{\rm w}^\#)$ so that (12.3.16) holds for $(u^\#, \hat u^\#)$. Denote

$$\hat\delta = \frac{\delta_3 - \|\hat u^\# - u^\#\|^{*,(1+\delta^*)}_{2,\alpha/2,Q^{\rm iter}}}{10}.$$

Then $\hat\delta > 0$ by (12.3.16).

We show that, if $\delta^\#$ in (12.8.3) is sufficiently small, then

$$\|\hat u^\# - \hat u\|^{*,(1+\delta^*)}_{2,\alpha/2,Q^{\rm iter}} \le \hat\delta. \tag{12.8.4}$$

Indeed, if this is not true, then there exists a sequence $(u^{(i)}, \theta_{\rm w}^{(i)})$, converging in $C^{2,\alpha}_{*,1+\delta^*}(Q^{\rm iter}) \times [\theta_{\rm w}^*, \frac{\pi}{2}]$ to $(u^\#, \theta_{\rm w}^\#)$, such that the corresponding solutions $\hat u^{(i)}$ of (12.7.2)–(12.7.6) (which exist and are unique by Lemma 12.7.4, and $\hat u^{(i)} \in C^{2,\alpha}_{*,2}(Q^{\rm iter})$ by Corollary 12.7.11) satisfy

$$\|\hat u^\# - \hat u^{(i)}\|^{*,(1+\delta^*)}_{2,\alpha/2,Q^{\rm iter}} \ge \hat\delta \qquad \text{for all } i. \tag{12.8.5}$$

Since sequence $\hat u^{(i)}$ is bounded in $C^{2,\alpha}_{*,2}(Q^{\rm iter})$ by Corollary 12.7.11, then there exists a subsequence $\hat u_{i_j}$ converging to a limit $\hat u_\infty$ in $C^{2,\frac{\alpha}{2}}_{*,1+\delta^*}(Q^{\rm iter})$. By Lemma

12.7.5, $\hat{u}_\infty$ is a solution of (12.7.2)–(12.7.6) determined by $(u^\#, \theta_{\rm w}^\#)$. By uniqueness in Lemma 12.7.4, this implies that $\hat{u}_\infty = \hat{u}^\#$. This contradicts (12.8.5).

Thus, (12.8.4) holds if $\delta^\#$ is small. We can assume that $\delta^\# \le \hat{\delta}$. Then

$$
\begin{aligned}
&\|\hat{u} - u\|^{*,(1+\delta^*)}_{2,\alpha/2,Q^{\rm iter}} \\
&\le \|\hat{u} - \hat{u}^\#\|^{*,(1+\delta^*)}_{2,\alpha/2,Q^{\rm iter}} + \|\hat{u}^\# - u^\#\|^{*,(1+\delta^*)}_{2,\alpha/2,Q^{\rm iter}} + \|u^\# - u\|^{*,(1+\delta^*)}_{2,\alpha/2,Q^{\rm iter}} \\
&\le \hat{\delta} + (\delta_3 - 10\hat{\delta}) + \hat{\delta} \\
&= \delta_3 - 8\hat{\delta} < \delta_3.
\end{aligned}
$$

Now, let $\hat{\psi} \in C^2(\Omega) \cap C^1(\overline{\Omega} \setminus \overline{\Gamma_{\rm sonic}}) \cap C(\overline{\Omega})$ be the unique solution of Problem (12.3.25)–(12.3.29) determined by $(u, \theta_{\rm w})$, as ensured in Proposition 12.7.3. Then, by Lemma 12.7.4(i), and by the uniqueness for Problem (12.7.2)–(12.7.6) held by Lemma 12.7.4(ii), it follows that $\hat{\psi}$ and $\hat{u}$ are related by (12.3.15). This, combined with the last estimate, shows that condition (vii) of Definition 12.3.2 holds for $(u, \theta_{\rm w})$. Therefore, $(u, \theta_{\rm w}) \in \mathcal{K}$. □

Chapter Thirteen

Iteration Map, Fixed Points, and Existence of Admissible Solutions up to the Sonic Angle

In this chapter, we define an iteration map on the closure of the iteration set, show that its fixed points are admissible solutions, and prove the existence of fixed points (hence admissible solutions) in the iteration set for each $\theta_{\rm w} \in [\theta_{\rm w}^*, \frac{\pi}{2})$. Since this can be done for any $\theta_{\rm w}^* \in (\theta_{\rm w}^{\rm c}, \frac{\pi}{2})$, we obtain the existence of admissible solutions up to the sonic angle, or the critical angle.

13.1 ITERATION MAP

In this section, we define an iteration map on the closure, $\overline{\mathcal{K}}$, of the iteration set $\mathcal{K}$ in the norm of $C^{2,\alpha}_{*,1+\delta^*}(Q^{\rm iter}) \times [\theta_{\rm w}^*, \frac{\pi}{2}]$; *cf.* §12.3.2.

We now describe our heuristics for the construction. One possible definition of the iteration map would be $\overline{\mathcal{K}} \ni (u, \theta_{\rm w}) \mapsto \hat{u}$, where $\hat{u}$ is the solution of the iteration problem in Definition 12.3.2(vii) from Corollary 12.7.6. However, it is not clear how the compactness of such an iteration map can be obtained. In fact, the estimate in Proposition 12.7.10 has shown that, if $\alpha = \frac{\hat{\alpha}}{2}$ in the iteration set, function $\hat{\varphi}$ corresponding to $\hat{u}$ via (12.3.15) has a gain-in-regularity in comparison with φ. However, as the estimate in Corollary 12.7.11 (where $\alpha = \frac{\hat{\alpha}}{2}$ so that the same estimate with $\hat{\alpha}$ cannot be obtained) has shown, the gain-in-regularity for $\hat{\varphi}$ does not hold for $\hat{u}$, which makes the compactness of map $(u, \theta_{\rm w}) \mapsto \hat{u}$ unclear. The reason for this is that, in (12.3.15), we use map $F^{-1}_{(2,\mathfrak{g}_{\rm sh})}$ for $\mathfrak{g}_{\rm sh}$ determined by $(u, \theta_{\rm w})$ in (12.2.24). Thus, $F^{-1}_{(2,\mathfrak{g}_{\rm sh})}$ has the same regularity as u, *i.e.*, as φ.

In order to fix this, we need to define an iteration map by using map $F^{-1}_{(2,\hat{\mathfrak{g}}_{\rm sh})}$ in (12.3.15), which corresponds to $\Gamma_{\rm shock}(\hat{\varphi})$ instead of $\Gamma_{\rm shock}(\varphi)$. Thus, we need to define the new location of the shock by $\{\hat{\varphi} = \varphi_1\}$. On the other hand, $\hat{\varphi}$ is defined only in $\Omega = \Omega(\varphi)$, while some part of the modified shock may lie outside Ω. Therefore, we need to extend $\hat{\varphi}$ through $\Gamma_{\rm shock}$ before we define the modified shock location. Moreover, this extension must have the same regularity as $\hat{\varphi}$, *i.e.*, a higher regularity than $\Gamma_{\rm shock}$. Furthermore, we need the continuity properties of this extension with respect to curve $\Gamma_{\rm shock}$ in order to have the continuity of the iteration map. We show the existence of such an extension in Theorem 13.9.5 in Appendix 13.9. This motivates the construction that follows:

In the argument below, the universal constant C depends only on the data and $\theta_{\rm w}^*$.

Let $(u, \theta_{\rm w}) \in \overline{\mathcal{K}}$, let (Ω, φ) be defined as in Definition 12.3.2(ii), and let $(\hat{\varphi}, \hat{u})$ be determined by $(u, \theta_{\rm w})$ as in Definition 12.3.2(vii), where we have used Corollary 12.7.6.

Define the functions on $F_1(\Omega)$ as

$$v := (\varphi - \tilde{\varphi}_2) \circ F_1^{-1}, \;\; \hat{v} := (\hat{\varphi} - \tilde{\varphi}_2) \circ F_1^{-1}, \;\; v_1 := (\varphi_1 - \tilde{\varphi}_2) \circ F_1^{-1}. \qquad (13.1.1)$$

Using Definition 12.2.6(v), we have

$$v = u \circ F_{(2,\mathfrak{g}_{\rm sh})} \qquad \text{on } F_1(\Omega),$$

and, by (12.3.15),

$$\hat{v} = \hat{u} \circ F_{(2,\mathfrak{g}_{\rm sh})} \qquad \text{on } F_1(\Omega).$$

Note also that (12.3.3) and the choice of α made above imply that the corresponding $\mathfrak{g}_{\rm sh}$ satisfies (12.2.66) with C and M depending only on the data and $\theta_{\rm w}^*$. Then, using (12.2.41), (12.3.5), and $F_1(\Omega) = F^{-1}_{(2,\mathfrak{g}_{\rm sh})}(Q^{\rm iter})$, we have

$$\begin{aligned} \|v - \hat{v}\|^{*,(1+\delta^*)}_{2,\alpha/2,F_1(\Omega)} &= \|(u - \hat{u}) \circ F_{(2,\mathfrak{g}_{\rm sh})}\|^{*,(1+\delta^*)}_{2,\alpha/2,F^{-1}_{(2,\mathfrak{g}_{\rm sh})}(Q^{\rm iter})} \\ &\le C\|u - \hat{u}\|^{*,(1+\delta^*)}_{2,\alpha/2,Q^{\rm iter}} \\ &< C\delta_3, \end{aligned} \qquad (13.1.2)$$

where the last inequality follows from (12.3.16).

From Definition 12.2.6(iv) and Lemma 12.2.7(iii),

$$\begin{aligned} &F_1(\Omega) = \{(s,t) \,:\, 0 < s < \hat{s}(\theta_{\rm w}),\; 0 < t < \mathfrak{g}_{\rm sh}(s)\}, \\ &F_1(\Gamma_{\rm shock}) = \{(s, \mathfrak{g}_{\rm sh}(s)) \,:\, 0 < t < \hat{s}(\theta_{\rm w})\}, \\ &F_1(\overline{\Gamma_{\rm sonic}}) = \{(0,t) \,:\, 0 \le t \le \mathfrak{g}_{\rm sh}(0)\}, \\ &F_1(\Gamma_{\rm sym}) = \{(\hat{s}(\theta_{\rm w}), t) \,:\, 0 < t < \mathfrak{g}_{\rm sh}(\hat{s}(\theta_{\rm w}))\}. \end{aligned} \qquad (13.1.3)$$

Also, since $\varphi = \varphi_1$ on $\Gamma_{\rm shock}$ by Definition 12.2.6, (12.3.13) implies that $\varphi < \varphi_1$ in Ω near $\Gamma_{\rm shock}$. By (13.1.1),

$$v_1 - v = (\varphi_1 - \varphi) \circ F_1^{-1} \qquad \text{on } F_1(\Omega), \qquad (13.1.4)$$

so that $v = v_1$ on $F_1(\Gamma_{\rm shock})$ and $v < v_1$ in $F_1(\Omega)$ near $F_1(\Gamma_{\rm shock})$. Thus, (13.1.3) implies that, on $F_1(\Gamma_{\rm shock})$,

$$\frac{D(v_1 - v)}{|D(v_1 - v)|}(s, \mathfrak{g}_{\rm sh}(s)) = \frac{(\mathfrak{g}'_{\rm sh}(s), -1)}{\sqrt{(\mathfrak{g}'_{\rm sh}(s))^2 + 1}}.$$

Then

$$\frac{\partial(v_1 - v)}{\partial t} = D(v_1 - v) \cdot (0,1) = -\frac{|D(v_1 - v)|}{\sqrt{(\mathfrak{g}'_{\rm sh})^2 + 1}} \qquad \text{on } F_1(\Gamma_{\rm shock}).$$

From this, we estimate $\frac{\partial(v_1-v)}{\partial t}$. First, using (12.3.13), (13.1.4), and Lemma 12.2.2(ii), we have

$$|D(v_1 - v)| \geq \frac{\mu_1}{C} \qquad \text{on } F_1(\Gamma_{\text{shock}}).$$

Using this and (12.2.66) for $\mathfrak{g}_{\text{sh}}$, we obtain

$$\frac{\partial(v_1 - v)}{\partial t} \leq -\frac{\mu_1}{C} \qquad \text{on } F_1(\Gamma_{\text{shock}}).$$

Then, using (13.1.2) and choosing δ_3 sufficiently small, depending only on the data and (μ_1, M) (hence only on the data and θ_{w}^*), we have

$$\frac{\partial(v_1 - \hat{v})}{\partial t} \leq -\frac{1}{C} \qquad \text{on } F_1(\Gamma_{\text{shock}}). \tag{13.1.5}$$

The definition of $\hat{v}$ in (13.1.1) can be written as $\hat{v} = (\hat{\psi} + \varphi_2 - \tilde{\varphi}_2) \circ F_1^{-1}$. Then, using Lemma 12.2.2(ii), (12.2.35), and (12.2.38), it follows from (12.7.29) that $\hat{v}$ satisfies

$$\|\hat{v}\|^{*,(2)}_{2,\hat{\alpha},F_1(\Omega)} \leq C. \tag{13.1.6}$$

Also, $\mathfrak{g}_{\text{sh}}$ satisfies (12.2.66) with $\sigma = 1 + \delta^*$ and the right-hand side depending only on the data and θ_{w}^* by (12.3.3), since α is now fixed in the iteration set. In particular, $\|\mathfrak{g}_{\text{sh}}\|_{C^{0,1}([0,\hat{s}(\theta_{\text{w}})])} \leq C$.

Then the structure of domain $F_1(\Omega)$ as indicated in (13.1.3) allows us to apply Theorem 13.9.5 to extend $\hat{v}$ defined on $F_1(\Omega)$ to a function $\mathcal{E}^{(\hat{s}(\theta_{\text{w}}))}_{\mathfrak{g}_{\text{sh}}}(\hat{v}) \in C^{2,\alpha}_{*,2}(\mathcal{D}^{\text{ext}})$, where

$$\mathcal{D}^{\text{ext}} = \{(s,t) \,:\, 0 < s < \hat{s}(\theta_{\text{w}}),\ 0 < t < (1+\sigma)\mathfrak{g}_{\text{sh}}(s)\} \tag{13.1.7}$$

with $\sigma > 0$ depending only on $\text{Lip}[\mathfrak{g}_{\text{sh}}]$, and hence only on the data. Now, from (13.1.6) and Theorem 13.9.5, we have

$$\|\mathcal{E}^{(\hat{s}(\theta_{\text{w}}))}_{\mathfrak{g}_{\text{sh}}}(\hat{v})\|^{*,(2)}_{2,\hat{\alpha},\mathcal{D}^{\text{ext}}} \leq C, \tag{13.1.8}$$

where $\mathcal{D}^{\text{ext}}$ is from (13.1.7).

From (12.3.5) and (13.1.7), it follows that

$$\{(s,t) \,:\, 0 < s < \hat{s}(\theta_{\text{w}}),\ 0 < t < \mathfrak{g}_{\text{sh}}(s) + \hat{\sigma}\} \subset \mathcal{D}^{\text{ext}}, \tag{13.1.9}$$

where $\hat{\sigma} > 0$ depends only on the data and θ_{w}^*. From (13.1.5) and (13.1.8), there exists $\zeta \in (0, \frac{\hat{\sigma}}{2})$, depending only on the data and θ_{w}^*, so that

$$\frac{\partial(v_1 - \mathcal{E}^{(\hat{s}(\theta_{\text{w}}))}_{\mathfrak{g}_{\text{sh}}}(\hat{v}))}{\partial t} \leq -\frac{1}{2C} \qquad \text{on } \mathcal{R}_\zeta(F_1^{(\theta_{\text{w}})}(\Gamma_{\text{shock}})), \tag{13.1.10}$$

where

$$\mathcal{R}_\zeta(F_1^{(\theta_{\text{w}})}(\Gamma_{\text{shock}})) := \{(s,t) \,:\, 0 < s < \hat{s}(\theta_{\text{w}}),\ |t - \mathfrak{g}_{\text{sh}}(s)| \leq \zeta\},$$

and $\mathcal{E}^{(\hat{s}(\theta_{\rm w}))}_{\mathfrak{g}_{\rm sh}}(\hat{v})$ is defined in $\mathcal{R}_\zeta(F_1^{(\theta_{\rm w})}(\Gamma_{\rm shock}))$ by (13.1.9). Also, $\mathcal{R}_\zeta(F_1^{(\theta_{\rm w})}(\Gamma_{\rm shock}))$ is an open set, since $\mathfrak{g}_{\rm sh} \in C([0,\hat{s}(\theta_{\rm w})])$. Then, further reducing δ_3 if necessary and using (13.1.10), combined with (13.1.2), (12.3.5), and the fact that $v = v_1$ on $\{(s,t) \, : \, 0 < s < \hat{s}(\theta_{\rm w}),\, t = \mathfrak{g}_{\rm sh}(s)\}$, we conclude that there exists a unique function $\hat{\mathfrak{g}}_{\rm sh}(s)$ on $(0, \hat{s}(\theta_{\rm w}))$ such that

$$\begin{aligned} &\mathcal{R}_\zeta(F_1^{(\theta_{\rm w})}(\Gamma_{\rm shock})) \cap \{\mathcal{E}^{(\hat{s}(\theta_{\rm w}))}_{\mathfrak{g}_{\rm sh}}(\hat{v}) = v_1\} \\ &= \{(s, \hat{\mathfrak{g}}_{\rm sh}(s)) \, : \, 0 < s < \hat{s}(\theta_{\rm w})\} \subset F_1^{(\theta_{\rm w})}(\mathcal{Q}^{(\theta_{\rm w})}_{\rm bd}), \end{aligned} \tag{13.1.11}$$

where

$$\|\hat{\mathfrak{g}}_{\rm sh} - \mathfrak{g}_{\rm sh}\|_{C([0,\hat{s}(\theta_{\rm w})])} < C\delta_3 < \zeta. \tag{13.1.12}$$

Now we define

$$\tilde{u} = \mathcal{E}^{(\hat{s}(\theta_{\rm w}))}_{\mathfrak{g}_{\rm sh}}(\hat{v}) \circ F^{-1}_{(2,\hat{\mathfrak{g}}_{\rm sh})}. \tag{13.1.13}$$

Lemma 13.1.1. *For all* $(u, \theta_{\rm w}) \in \overline{\mathcal{K}}$, $\tilde{u}$ *defined by* (13.1.13) *satisfies*

$$\|\tilde{u}\|^{*,(2)}_{2,\hat{\alpha},Q^{\rm iter}} \le C. \tag{13.1.14}$$

Proof. Using (13.1.8), (13.1.10), and the definition of the modified shock function $\hat{\mathfrak{g}}_{\rm sh}(s)$ in (13.1.11) with $\|v_1\|_{C^3(\overline{F_1^{(\theta_{\rm w})}(\mathcal{Q}^{(\theta_{\rm w})}_{\rm bd})})} \le C$, we have

$$\|\hat{\mathfrak{g}}_{\rm sh}\|^{(-1-\hat{\alpha}),\{\hat{s}(\theta_{\rm w})\}}_{2,\hat{\alpha},[\hat{\varepsilon}/10,\, \hat{s}(\theta_{\rm w})]} + \|\hat{\mathfrak{g}}_{\rm sh} - \mathfrak{g}_{S_1}\|^{(1+\delta^*),({\rm par})}_{2,\hat{\alpha},(0,\hat{\varepsilon})} \le C, \tag{13.1.15}$$

where $\hat{\varepsilon}$ is from Definition 12.3.1(iv). Now, from (13.1.8), (13.1.15), and expression (13.1.13), we use the explicit form (12.2.43) to obtain (13.1.14) by a straightforward calculation. □

From Lemma 13.1.1, using $\alpha = \frac{\hat{\alpha}}{2}$ and the notation from Definition 12.3.9(iii), we define the following iteration map:

Definition 13.1.2. *The iteration map* $\mathcal{I} : \overline{\mathcal{K}} \mapsto C^{2,\alpha}_{*,1+\delta^*}(Q^{\rm iter})$ *is defined by*

$$\mathcal{I}(u, \theta_{\rm w}) = \tilde{u} \tag{13.1.16}$$

for $\tilde{u}$ *defined in* (13.1.13). *For each* $\theta_{\rm w} \in [\theta^*_{\rm w}, \frac{\pi}{2}]$, *we define the map:*

$$\mathcal{I}^{(\theta_{\rm w})} \, : \, \overline{\mathcal{K}}(\theta_{\rm w}) \mapsto C^{2,\alpha}_{*,1+\delta^*}(Q^{\rm iter}),$$

by $\mathcal{I}^{(\theta_{\rm w})}(u) = \mathcal{I}(u, \theta_{\rm w})$.

Lemma 13.1.3. *Let* $\theta_{\rm w} \in [\theta^*_{\rm w}, \frac{\pi}{2}]$. *Let* $\mathcal{I}^{(\theta_{\rm w})}$ *be the iteration map from Definition* 13.1.2. *Let* $\mathcal{I}_1^{(\theta_{\rm w})} : \overline{\mathcal{K}}(\theta_{\rm w}) \mapsto C^{2,\alpha}_{*,1+\delta^*}(Q^{\rm iter})$ *be defined by* $\mathcal{I}_1^{(\theta_{\rm w})}(u) = \hat{u}$ *with* $\hat{u}$ *given in Definition* 12.3.2(vii) *for* $(u, \theta_{\rm w}) \in \overline{\mathcal{K}}$, *where we have used Corollary* 12.7.6. *Then* $u \in \overline{\mathcal{K}}(\theta_{\rm w})$ *is a fixed point of* $\mathcal{I}^{(\theta_{\rm w})}$ *if and only if* u *is a fixed point of* $\mathcal{I}_1^{(\theta_{\rm w})}$.

Proof. Fix $\theta_w \in [\theta_w^*, \frac{\pi}{2}]$. Set $\tilde{u} := \mathcal{I}^{(\theta_w)}(u)$ and $\hat{u} := \mathcal{I}_1^{(\theta_w)}(u)$ for each $u \in \overline{\mathcal{K}}(\theta_w)$. We divide the proof into two steps.

1. We first assume that $u \in \overline{\mathcal{K}}(\theta_w)$ is a fixed point of $\mathcal{I}^{(\theta_w)}$. Denote

$$v := (\varphi - \tilde{\varphi}_2) \circ F_1^{-1}.$$

Then $v = u \circ F_{(2,\mathfrak{g}_{sh})}$. From the definitions of $F_{(2,\mathfrak{g}_{sh})}$ and $\mathfrak{g}_{sh}$, we have

$$u(\frac{s}{\hat{s}(\theta_w)}, 1) = v(s, \mathfrak{g}_{sh}(s)) = v_1(s, \mathfrak{g}_{sh}(s)) \qquad \text{for all } s \in (0, \hat{s}(\theta_w)).$$

Similarly, from the definition of the iteration map through (13.1.13),

$$\tilde{u}(\frac{s}{\hat{s}(\theta_w)}, 1) = \mathcal{E}_{\hat{\mathfrak{g}}_{sh}}^{(\hat{s}(\theta_w))}(\hat{v})(s, \hat{\mathfrak{g}}_{sh}(s)) = v_1(s, \hat{\mathfrak{g}}_{sh}(s)) \qquad \text{for all } s \in (0, \hat{s}(\theta_w)).$$

Since $u = \tilde{u}$ on Q^{iter}, we obtain that $v_1(s, \mathfrak{g}_{sh}(s)) = v_1(s, \hat{\mathfrak{g}}_{sh}(s))$ for all s. Using (12.2.37) and (13.1.1), we have

$$\mathfrak{g}_{sh} \equiv \hat{\mathfrak{g}}_{sh} \qquad \text{on } (0, \hat{s}(\theta_w)).$$

Now, since $\hat{v}$ defined by (13.1.1) satisfies $\hat{v} = \hat{u} \circ F_{(2,\mathfrak{g}_{sh})}$, we conclude from (13.1.13) that $\tilde{u} = \hat{u}$ on Q^{iter}. Thus, $u = \hat{u}$, so that u is a fixed point of $\mathcal{I}_1^{(\theta_w)}$.

2. Now we assume that $u \in \overline{\mathcal{K}}(\theta_w)$ is a fixed point of $\mathcal{I}_1^{(\theta_w)}$. Then $u = \hat{u}$ so that

$$\mathfrak{g}_{sh} \equiv \hat{\mathfrak{g}}_{sh} \qquad \text{on } (0, \hat{s}(\theta_w)).$$

Since $\hat{v}$ defined by (13.1.1) satisfies $\hat{v} = \hat{u} \circ F_{(2,\mathfrak{g}_{sh})}$, we conclude from (13.1.13) that $\tilde{u} = \hat{u}$ on Q^{iter}. Thus, $u = \tilde{u}$, so that u is a fixed point of $\mathcal{I}^{(\theta_w)}$. □

13.2 CONTINUITY AND COMPACTNESS OF THE ITERATION MAP

Below we use $\hat{\alpha}$ defined in Proposition 12.7.10 and $\alpha = \frac{\hat{\alpha}}{2}$ defined in Corollary 12.7.11, both depending only on the data and θ_w^*. Then

$$\alpha' := \hat{\alpha} - \alpha = \frac{\hat{\alpha}}{2} > 0.$$

Lemma 13.2.1. *$C_{*,2}^{2,\alpha+\alpha'}(Q^{\text{iter}})$ is compactly embedded into $C_{*,1+\delta^*}^{2,\alpha}(Q^{\text{iter}})$.*

Proof. The assertion follows from Lemma 4.6.3, combined with the standard results on the Hölder spaces, where we recall that $1 + \delta^* < 2$. □

Lemma 13.2.2. *The iteration map*

$$\mathcal{I} : \overline{\mathcal{K}} \subset C_{*,1+\delta^*}^{2,\alpha}(Q^{\text{iter}}) \times [\theta_w^*, \frac{\pi}{2}] \mapsto C_{*,1+\delta^*}^{2,\alpha}(Q^{\text{iter}})$$

is continuous.

Proof. Let $(u^{(j)}, \theta_{\rm w}^{(j)}) \in \overline{\mathcal{K}}$ converge in the norm of $C^{2,\alpha}_{*,1+\delta^*}(Q^{\rm iter}) \times [\theta_{\rm w}^*, \frac{\pi}{2}]$ to $(u, \theta_{\rm w}) \in \overline{\mathcal{K}}$. In particular, $\theta_{\rm w}^{(j)} \to \theta_{\rm w}$, which implies

$$\lim_{j\to\infty} \hat{s}(\theta_{\rm w}^{(j)}) = \hat{s}(\theta_{\rm w}). \tag{13.2.1}$$

Denote by $\mathfrak{h}_{\rm sh}^{(j)}$ and $\mathfrak{g}_{\rm sh}^{(j)}$ the functions in Definition 12.2.6(i)–(ii) for $(u^{(j)}, \theta_{\rm w}^{(j)})$, and denote by $\mathfrak{h}_{\rm sh}$ and $\mathfrak{g}_{\rm sh}$ the corresponding functions for $(u, \theta_{\rm w})$. From (12.2.59) in Lemma 12.2.7(vi),

$$\mathfrak{h}_{\rm sh}^{(j)} \to \mathfrak{h}_{\rm sh} \qquad \text{in } C^{1,\frac{\alpha}{2}}([0,1]). \tag{13.2.2}$$

Fix a compact set $K \Subset F_1^{(\theta_{\rm w})}(\Omega)$. From the explicit form of $F_1^{(\theta_{\rm w})}(\Omega)$ given by Definition 12.2.6(iv), we have

$$K \subset \{(s,t)\ :\ a < s < b,\ 0 < t < \zeta \mathfrak{g}_{\rm sh}(s)\}$$

for some $0 < a < b < \hat{s}(\theta_{\rm w})$ and $\zeta \in (0,1)$. Then, using (12.2.41), (12.3.5), (13.2.2), and the continuity of $\hat{s}(\cdot)$, we obtain the existence of $\hat{K} \Subset Q^{\rm iter}$ such that $F_{(2,\mathfrak{g}_{\rm sh}^{(j)})}(K) \subset \hat{K}$ for all j. Then, using that $\hat{v} = \hat{u} \circ F_{(2,\mathfrak{g}_{\rm sh})}$, $\hat{v}^{(j)} = \hat{u}^{(j)} \circ F_{(2,\mathfrak{g}_{\rm sh}^{(j)})}$, and $\hat{u}^{(j)} \to \hat{u}$ in C^2 on compact subsets of $Q^{\rm iter}$ by Lemma 12.7.5, we have

$$\hat{v}^{(j)} \to \hat{v} \qquad \text{in } C^{1,\frac{\alpha}{2}}(K) \ \text{ for all compact } \ K \Subset F_1(\Omega). \tag{13.2.3}$$

Next we consider functions $(\hat{\mathfrak{h}}_{\rm sh}^{(j)}, \hat{\mathfrak{h}}_{\rm sh})$, determined by functions $(\hat{\mathfrak{g}}_{\rm sh}^{(j)}, \hat{\mathfrak{g}}_{\rm sh})$ in (13.1.11) by the formula in Definition 12.2.6(ii), and show that

$$\hat{\mathfrak{h}}_{\rm sh}^{(j)} \to \hat{\mathfrak{h}}_{\rm sh} \qquad \text{in } C^{2,\alpha}_{*,1+\delta^*}([0,1]). \tag{13.2.4}$$

For this, using (13.1.6) for each j, as well as (13.2.1)–(13.2.3), we can apply Theorem 13.9.5(iii) to conclude

$$w_j \to \mathcal{E}_{\mathfrak{g}_{\rm sh}}^{(\hat{s}(\theta_{\rm w}))}(\hat{v}) \qquad \text{in } C^{2,\alpha}_{*,1+\delta^*}(\mathcal{D}^{\rm ext}_{\sigma/2}) \tag{13.2.5}$$

for

$$w_j(s,t) := \mathcal{E}_{\mathfrak{g}_{\rm sh}^{(j)}}^{(\hat{s}(\theta_{\rm w}^{(j)}))}(\hat{v}^{(j)})(\frac{\hat{s}(\theta_{\rm w}^{(j)})}{\hat{s}(\theta_{\rm w})}s, t),$$

where

$$\mathcal{D}^{\rm ext}_{\sigma/2} := \{(s,t)\ :\ 0 < s < \hat{s}(\theta_{\rm w}),\ 0 < t < (1+\frac{\sigma}{2})\mathfrak{g}_{\rm sh}(s)\}$$

with $\sigma > 0$ from (13.1.7). From (13.1.9), (13.1.12), and (13.2.2), applied for each j and the limiting function, it follows that, reducing δ_3, we obtain that

$$\{(s, \hat{\mathfrak{g}}_{\rm sh}^{(j)}(\frac{\hat{s}(\theta_{\rm w}^{(j)})}{\hat{s}(\theta_{\rm w})}s))\ :\ 0 < s < \hat{s}(\theta_{\rm w})\} \subset \mathcal{D}^{\rm ext}_{\sigma/2} \qquad \text{for large } j,$$

and the same holds for $\{(s, \hat{\mathfrak{g}}_{\rm sh}(s)) : 0 < s < \hat{s}(\theta_{\rm w})\}$. Then, from (13.1.10) applied for each j and the limiting function, the convergence that $w_j \to \mathcal{E}^{(\hat{s}(\theta_{\rm w}))}_{\mathfrak{g}_{\rm sh}}(v)$ in $C^{2,\alpha}_{*,1+\delta^*}(\mathcal{D}^{\rm ext}_{\sigma/2})$, combined with (13.2.1), implies (13.2.4).

Defining $\tilde{u}^{(j)}$ and $\tilde{u}$ by (13.1.13) for $(\hat{v}^{(j)}, \mathfrak{g}^{(j)}_{\rm sh}, \hat{\mathfrak{g}}^{(j)}_{\rm sh})$ and $(\hat{v}, \mathfrak{g}_{\rm sh}, \hat{\mathfrak{g}}_{\rm sh})$ respectively, and using (13.2.1) and (13.2.4)–(13.2.5), we obtain that $\tilde{u}^{(j)} \to \tilde{u}$ pointwise in the open region $Q^{\rm iter}$. Then, from (13.1.14) for each $\tilde{u}^{(j)}$, using the compactness of $C^{2,\hat{\alpha}}_{*,2}(Q^{\rm iter})$ in $C^{2,\alpha}_{*,1+\delta^*}(Q^{\rm iter})$ shown in Lemma 13.2.1, we conclude that $\tilde{u}^{(j)} \to \tilde{u}$ in $C^{2,\alpha}_{*,1+\delta^*}(Q^{\rm iter})$. This shows the continuity of the iteration map. □

Corollary 13.2.3. *The iteration map*

$$\mathcal{I} : \overline{\mathcal{K}} \subset C^{2,\alpha}_{*,1+\delta^*}(Q^{\rm iter}) \times [\theta^*_{\rm w}, \frac{\pi}{2}] \mapsto C^{2,\alpha}_{*,1+\delta^*}(Q^{\rm iter})$$

is compact. Moreover, there exists $C > 0$ depending only on the data and $\theta^*_{\rm w}$ such that, for each $(u, \theta_{\rm w}) \in \overline{\mathcal{K}}$,

$$\|\mathcal{I}^{(\theta_{\rm w})}(u)\|^{*,(2)}_{2,\alpha+\alpha',Q^{\rm iter}} \le C. \tag{13.2.6}$$

Proof. Lemma 13.1.1 and Definition 13.1.2 imply (13.2.6). Combining this with the property shown in Lemma 13.2.1 and the continuity of $\mathcal{I}$ shown in Lemma 13.2.2, we conclude the compactness of $\mathcal{I}$. □

13.3 NORMAL REFLECTION AND THE ITERATION MAP FOR $\theta_{\rm w} = \frac{\pi}{2}$

In this section, we show that function $u^{(\rm norm)}$ on $Q^{\rm iter}$, which corresponds to the normal reflection $\varphi = \varphi_2^{(\frac{\pi}{2})}$ in Ω for $\theta_{\rm w} = \frac{\pi}{2}$, is the unique value of $\mathcal{I}^{(\frac{\pi}{2})}(\cdot)$ on $\overline{\mathcal{K}}(\frac{\pi}{2})$.

Recall that, by (12.2.34) and (12.2.44),

$$u^{(\rm norm)} \equiv 0 \qquad \text{on } Q^{\rm iter}.$$

Proposition 13.3.1. *For any $u \in \overline{\mathcal{K}}(\frac{\pi}{2})$,*

$$\mathcal{I}^{(\frac{\pi}{2})}(u) = u^{(\rm norm)} \equiv 0.$$

Proof. Let $u \in \overline{\mathcal{K}}(\frac{\pi}{2})$. Then the corresponding function $\hat{\psi}$ is a solution of Problem (12.3.25)–(12.3.29). This problem, in Case $\theta_{\rm w} = \frac{\pi}{2}$, satisfies the following properties:

Equation (12.3.25) is homogeneous and strictly elliptic in Ω by (i) and (viii) of Lemma 12.4.5.

By Lemma 12.5.7(ii), $\hat{\psi}$ satisfies the boundary condition (12.3.26) with function $\mathcal{M}_{(u,\theta_{\rm w})}(\mathbf{p}, z, \boldsymbol{\xi})$ constructed in Lemma 12.5.7. By Lemma 12.5.7(iv), this boundary condition is oblique. Furthermore, by Lemma 12.5.7(vii) with $\theta_{\rm w} = \frac{\pi}{2}$, it follows that $\mathcal{M}_{(u,\theta_{\rm w})}(\mathbf{0}, 0, \boldsymbol{\xi}) = 0$ for any $\boldsymbol{\xi} \in \Gamma_{\rm shock}$. That is, (12.3.26) is homogeneous.

Since $v_2 = 0$ for $\theta_{\rm w} = \frac{\pi}{2}$, the boundary condition (12.3.29) is homogeneous. Also the Neumann boundary conditions (12.3.28)–(12.3.29) are oblique. Furthermore, the obliqueness at points $\{P_2, P_3\}$ is verified as in Step 3 of the proof of Proposition 12.7.3.

Therefore, $\hat{\psi} \equiv 0$ is a solution of (12.3.25)–(12.3.29). By the comparison principle in Lemma 4.4.2, this solution is unique.

Next, we note that Theorem 13.9.5(i) implies that $\mathcal{E}^{(\hat{s}(\theta_{\rm w}))}_{\mathfrak{g}_{\rm sh}}(0) = 0$ for any shock location $\mathfrak{g}_{\rm sh}$. Now, from the definition of the iteration map by using (12.2.34), we conclude that $\mathcal{I}^{(\frac{\pi}{2})}(u) = 0 = u^{(\rm norm)}$. □

13.4 FIXED POINTS OF THE ITERATION MAP FOR $\theta_{\rm w} < \frac{\pi}{2}$ ARE ADMISSIBLE SOLUTIONS

In this section, we always assume that the parameters of the iteration set satisfy the requirements of Propositions 12.7.10 and 12.8.2. The main result in this section is the following:

Proposition 13.4.1. *If the parameters in Definition* 12.3.2 *are chosen so that* $(\delta_1, \varepsilon, \frac{1}{N_1})$ *are small, depending on the data and* $\theta^*_{\rm w}$*, then the following holds: Let* $(u, \theta_{\rm w}) \in \overline{\mathcal{K}}$ *with* $\theta_{\rm w} \in [\theta^*_{\rm w}, \frac{\pi}{2})$*, and let* $u(\cdot)$ *be a fixed point of map* $\mathcal{I}^{(\theta_{\rm w})}$*. Let* φ *be determined by* $(u, \theta_{\rm w})$ *as in Definition* 12.2.6(v)*. Then* φ *is an admissible solution of* **Problem 2.6.1** *in the sense of Definition* 8.1.1.

We recall that $\alpha = \frac{\hat{\alpha}}{2}$ has been fixed as defined in Corollary 12.7.11, where $\hat{\alpha}$ is defined in Proposition 12.7.10. Thus, α depends on the data and $\theta^*_{\rm w}$.

Since $(u, \theta_{\rm w}) \in \overline{\mathcal{K}}$ in this section, we follow the notational convention from Remark 12.3.11.

We note that, in order to prove Proposition 13.4.1, it suffices to show the following properties of φ:

$$\varphi_2 \le \varphi \le \varphi_1 \qquad \text{in } \Omega, \tag{13.4.1}$$

$$\partial_{\mathbf{e}_{S_1}}(\varphi_1 - \varphi) \le 0, \quad \partial_{\xi_2}(\varphi_1 - \varphi) \le 0 \qquad \text{in } \Omega, \tag{13.4.2}$$

$$|\psi_x| \le \frac{2 - \frac{\mu_0}{5}}{1+\gamma} x \qquad \text{in } \Omega \cap \mathcal{D}_{\varepsilon/4}. \tag{13.4.3}$$

Indeed, if (13.4.1)–(13.4.3) are proved, the following argument shows that φ is an admissible solution in the sense of Definition 8.1.1, provided that $\varepsilon > 0$ is chosen small in the definition of $\mathcal{K}$:

(a) By Lemma 13.1.3, $u = \hat{u}$. Thus, using Definition 12.2.6(v) and (12.3.15),

$$\varphi = \hat{\varphi} \quad \text{and} \quad \psi = \hat{\psi}.$$

(b) Domain Ω satisfies the requirements of Definition 8.1.1(i). This follows from Definition 12.3.2(ii)–(iii) and Lemma 12.2.7(iii). Furthermore, the last requirement in (8.1.2) is satisfied by (8.1.21) in Ω and (13.4.2), where we can apply Lemma 8.1.9 since φ satisfies equation (2.2.8) as shown in (c) below.

(c) φ satisfies the potential flow equation (2.2.8) in Ω if ε is small. Indeed, using Proposition 12.7.9 with $\sigma = \hat{\alpha}$, we find from (12.7.28) for $\hat{\psi} = \psi$ that $|\psi_y| \le Cx^{\frac{3}{2}}$ in $\Omega \cap \mathcal{D}_{\hat{\varepsilon}_{\mathrm{p}}}$, with C depending only on the data and θ_{w}^*. Thus, choosing ε small, we obtain that $|\psi_y| \le N_3 x$ in $\Omega \cap \mathcal{D}_{\varepsilon/4}$. Combining this with (13.4.3) and applying Lemma 12.4.7, we conclude that φ satisfies equation (2.2.8) in Ω and that (2.2.8) is elliptic for φ in $\overline{\Omega} \setminus \overline{\Gamma_{\mathrm{sonic}}}$.

(d) φ satisfies the Rankine-Hugoniot conditions on Γ_{shock}. Indeed, by Lemma 13.1.3, the fixed point $u = \mathcal{I}^{(\theta_{\mathrm{w}})}(u)$ of map $\mathcal{I}^{(\theta_{\mathrm{w}})}$ satisfies $u = \mathcal{I}_1^{(\theta_{\mathrm{w}})}(u)$ in Definition 12.3.2(vii). This implies that $\varphi = \hat{\varphi}$ and $\psi = \varphi - \varphi_2 = \hat{\psi}$, that is, $\hat{\psi} = \psi$ is a solution of Problem (12.3.25)–(12.3.29) for (u, θ_{w}). Then $\varphi = \varphi_1$ on Γ_{shock} from the definition of Ω in Definition 12.2.6(iv), and the gradient jump condition (12.5.3) is satisfied by Lemma 12.5.1(iv) and Lemma 12.5.7(ii).

(e) Extending φ by $\varphi_2 = \varphi_2^{(\theta_{\mathrm{w}})}$ into $P_0P_1P_4$, we see that $\varphi \in C^{1,\alpha}(P_0P_2P_3)$, where the $C^{1,\alpha}$–matching across Γ_{sonic} follows from the fact that $\varphi - \varphi_2 \in C^{2,\alpha}_{*,1+\delta^*}(\Omega)$ by Remark 12.3.5. Furthermore, $\varphi_{\boldsymbol{\nu}} = 0$ on $\Gamma_{\mathrm{wedge}} \cup \Gamma_{\mathrm{sym}}$, since (12.3.28)–(12.3.29) hold for $\hat{\psi} = \psi$, where we have used the coordinates with the origin shifted to center O_2 of the sonic center of state (2). Extending φ by φ_1 into $(\Lambda \cap \{\xi_1 < \xi_{1P_0}\}) \setminus P_0P_2P_3$ and recalling that φ satisfies the two Rankine-Hugoniot conditions with φ_1 on Γ_{shock}, and φ_2 satisfies the Rankine-Hugoniot conditions with φ_1 on $P_0P_1 \subset S_1$, it follows that the extended function φ satisfies (8.1.3) and is a weak solution of the potential flow equation (2.2.8) in Λ with $\partial_{\boldsymbol{\nu}}\varphi = 0$ on $\partial\Lambda$. That is, φ is a weak solution of **Problem 2.6.1** and satisfies the requirements of Definition 8.1.1(ii).

(f) The requirement of Definition 8.1.1(iii) holds, that is, the potential flow equation (2.2.8) for φ is strictly elliptic in $\overline{\Omega} \setminus \overline{\Gamma_{\mathrm{sonic}}}$, as shown in (c) above.

(g) The requirement of Definition 8.1.1(iv) follows from (13.4.1).

(h) The requirement of Definition 8.1.1(v) follows from (13.4.2).

Therefore, in the remaining part of §13.4, we focus on proving (13.4.1)–(13.4.3).

Lemma 13.4.2. *If $\varepsilon > 0$ in the definition of $\mathcal{K}$ is sufficiently small, depending only on the data and θ_{w}^*, then, for any $(u, \theta_{\mathrm{w}}) \in \overline{\mathcal{K}}$ such that u is a fixed point of map $\mathcal{I}^{(\theta_{\mathrm{w}})}$, the corresponding equation* (12.3.25) *satisfies assertions* (a)–(c) *of Lemma* 12.4.5(viii).

Proof. Fix $(u, \theta_{\rm w}) \in \overline{\mathcal{K}}$ which is a fixed point. Let $\hat{\psi}$ be the unique solution of (12.3.25)–(12.3.29) defined by $(u, \theta_{\rm w})$.

By Lemma 13.1.3, $u = \hat{u}$. Using (12.3.15) and Definition 12.2.6(v), we obtain that $\varphi = \hat{\varphi}$, so that $\psi = \hat{\psi}$. Therefore, by Proposition 12.7.10, ψ satisfies (12.4.7). Now the assertion follows from Lemma 12.4.5(viii). □

Lemma 13.4.3. *Let $(u, \theta_{\rm w}) \in \overline{\mathcal{K}}$ be such that u is a fixed point of map $\mathcal{I}^{(\theta_{\rm w})}$. Then the corresponding function ψ satisfies*

$$\psi \in C^{1,\alpha}(\overline{\Omega}) \cap C^{2,\alpha}(\overline{\Omega} \setminus (\overline{\Gamma_{\rm sonic}} \cup \{P_2, P_3\})) \cap C^{3,\alpha}(\Omega). \tag{13.4.4}$$

Proof. Since $\psi \in C^{2,\alpha}_{*,1+\delta^*}(\Omega)$, $\psi \in C^{1,\alpha}(\overline{\Omega}) \cap C^{2,\alpha}(\overline{\Omega} \setminus (\overline{\Gamma_{\rm sonic}} \cup \overline{\Gamma_{\rm sym}}))$. Furthermore, $\hat{\psi} = \psi$ satisfies equation (12.4.45) by Lemma 13.4.2, which is strictly elliptic in $\overline{\Omega} \setminus \overline{\Gamma_{\rm sonic}}$ and can be considered as a linear equation for ψ with coefficients $\mathcal{A}_{ij}(\boldsymbol{\xi}) = A_{ij}(D\psi(\boldsymbol{\xi}), \boldsymbol{\xi})$ so that $\mathcal{A}_{ij} \in C^{1,\alpha}(\Omega) \cap C^{\alpha}(\overline{\Omega})$, where we have used Lemma 12.4.5. The equation is strictly elliptic in $\overline{\Omega} \setminus \Gamma_{\rm sonic}$ from Lemma 12.4.5. Then, from the standard interior estimates for linear elliptic equations and the local estimates for linear oblique derivative problems which are applied near the points of the relative interior of $\Gamma_{\rm sym}$, it follows that ψ satisfies (13.4.4). □

13.4.1 $\varphi \geq \varphi_2$ for fixed points

Lemma 13.4.4. *If the parameters in Definition* 12.3.2 *are chosen such that $(\delta_1, \varepsilon, \frac{1}{N_1})$ are small, depending on the data and $\theta^*_{\rm w}$, then, for every $(u, \theta_{\rm w}) \in \overline{\mathcal{K}}$ such that u is a fixed point of map $\mathcal{I}^{(\theta_{\rm w})}$, the corresponding function φ satisfies*

$$\varphi \geq \varphi_2^{(\theta_{\rm w})} \qquad \textit{in } \Omega(u, \theta_{\rm w}).$$

Proof. Recall that, for the fixed points, $u = \mathcal{I}_1^{(\theta_{\rm w})}(u)$, by Lemma 13.1.3. That is, $u = \hat{u}$ in Definition 12.3.2(vii). This implies that

$$(\varphi, \psi) = (\hat{\varphi}, \hat{\psi}) \qquad \text{in } \Omega.$$

Then, from Lemma 12.7.7, we have

$$\varphi - \varphi_2 = \psi = \hat{\psi} \geq 0 \qquad \text{in } \Omega.$$

□

13.4.2 Directional monotonicity of $\varphi_1 - \varphi$ for fixed points

In this section, we prove (13.4.2). We first prove a lemma which will also be used in this section to show that the directional derivatives of $\varphi_1 - \varphi$ satisfy an oblique derivative condition on a part of $\Gamma_{\rm shock}$.

Let φ^+ and φ^- be uniform states defined by (6.1.1), and let $\rho_0 > 0$ be a constant such that the inequality in (6.1.2) holds, and densities $\rho_\pm$ of $\varphi^\pm$ are given by (6.1.2). In particular, $\varphi^\pm$ satisfy equation (2.2.8) with (2.2.9).

Lemma 13.4.5. *Let $\rho_0 > 0$ be fixed. For any $\delta \in (0,1)$, there exists $\mu > 0$ with the following property: Let $\varphi^{\pm}$ be the uniform states introduced above such that*

$$\delta \le \rho_- < \rho_+ \le \frac{1}{\delta}, \qquad |(u_+ - u_-,\, v_+ - v_-)| \ge \delta. \tag{13.4.5}$$

Then $S := \{\boldsymbol{\xi} \,:\, \varphi^+(\boldsymbol{\xi}) = \varphi^-(\boldsymbol{\xi})\}$ is a line. Assume that $\varphi^{\pm}$ satisfy the Rankine-Hugoniot conditions on S:

$$\rho_+ D\varphi^+ \cdot \boldsymbol{\nu} = \rho_- D\varphi^- \cdot \boldsymbol{\nu} \qquad \text{on } S.$$

Noting that $\operatorname{dist}(S, (u_+, v_+)) < c_+$ *by* (6.1.4), *we assume that*

$$c_+ - \operatorname{dist}(S, (u_+, v_+)) \ge \delta. \tag{13.4.6}$$

Let $\mathcal{D} \subset \mathbb{R}^2$ be an open set, and let $\Gamma \subset \partial\mathcal{D}$ be a relatively open segment of curve. Assume that Γ is locally C^2 and

$$\Gamma \subset B_{1/\delta}(\mathcal{O}_+), \tag{13.4.7}$$

where $\mathcal{O}_+ = (u_+, v_+)$ is the center of the sonic circle of state φ^+. Let $\varphi \in C^2(\mathcal{D} \cup \Gamma)$ satisfy

$$\|\varphi - \varphi^+\|_{C^1(\overline{\Gamma})} \le \mu. \tag{13.4.8}$$

Moreover, assume that function $\phi = \varphi + \frac{|\boldsymbol{\xi}|^2}{2}$ satisfies the equation:

$$\sum_{i,j=1}^{2} a_{ij} D_{ij}\phi = 0 \qquad \text{in } \mathcal{D}, \tag{13.4.9}$$

where $a_{ij} \in C(\overline{\mathcal{D}})$ and

$$|a_{ij}| \le \frac{1}{\delta}, \qquad \sum_{i,j=1}^{2} a_{ij}\nu_i^S \nu_j^S \ge \delta \qquad \text{on } \Gamma \tag{13.4.10}$$

with a unit normal $\boldsymbol{\nu}_S = (\nu_1^S, \nu_2^S)$ to line S. Assume that φ and φ^- satisfy the Rankine-Hugoniot conditions on Γ, i.e.,

$$\varphi = \varphi^-, \quad \rho(|D\varphi|^2, \varphi) D\varphi \cdot \boldsymbol{\nu} = \rho_- D\varphi^- \cdot \boldsymbol{\nu} \qquad \text{on } \Gamma, \tag{13.4.11}$$

where $\rho(|D\varphi|^2, \varphi)$ is given by (2.2.9). *Let $\mathbf{e}_S$ be a unit vector along line S, and let $w = \partial_{\mathbf{e}_S}(\varphi^- - \varphi)$. Then w satisfies the following oblique derivative condition:*

$$a_1 w_{\xi_1} + a_2 w_{\xi_2} = 0 \qquad \text{on } \Gamma, \tag{13.4.12}$$

where a_k are continuous functions on $\overline{\Gamma}$ such that

$$(a_1, a_2) \cdot \boldsymbol{\nu} \ne 0 \qquad \text{on } \Gamma. \tag{13.4.13}$$

Proof. In this proof, the universal constant C is positive and depends only on δ, and $O(\mu)$ denotes a universal continuous function $g(\boldsymbol{\xi})$ on $\overline{\Gamma}$ satisfying $|g(\boldsymbol{\xi})| \le C\mu$ for any $\boldsymbol{\xi} \in \Gamma$. We divide the proof into four steps.

1. We note that, since $\rho_\pm$ are determined by (6.1.2), then (13.4.5) implies

$$\rho_+ - \rho_- \ge \frac{1}{C}. \tag{13.4.14}$$

We shift the coordinates so that center $\mathcal{O}_+$ of state φ^+ becomes the origin as in §8.3.2.1. Then we rotate the coordinates so that the new ξ_1–axis is perpendicular to line S and the new ξ_2–axis is parallel to line S. Now $\boldsymbol{\xi} = (\xi_1, \xi_2)$ denotes these new (shifted and rotated) coordinates. Since the potential flow equation (2.2.8) is invariant with respect to the shift and rotation, functions φ and $\varphi^\pm$ written in the new coordinates satisfy equation (2.2.8) with the same ρ_0, and the Rankine-Hugoniot conditions (13.4.11) and property (13.4.14) hold on Γ.

In the new coordinates, functions $\varphi^\pm$ have the expressions:

$$\varphi^+(\boldsymbol{\xi}) = -\frac{|\boldsymbol{\xi}|^2}{2} + const., \qquad \varphi^-(\boldsymbol{\xi}) = -\frac{|\boldsymbol{\xi}|^2}{2} + (\tilde{u}_-, 0)\cdot\boldsymbol{\xi} + const.,$$

where $\tilde{u}_- \in \mathbb{R}$ is a constant satisfying

$$|\tilde{u}_-| = |(u_- - u_+,\, v_- - v_+)|.$$

In particular, $\tilde{u}_- \neq 0$ by (13.4.5). Thus, in the new coordinates, we have

$$\varphi^- - \varphi = -(\varphi + \frac{|\boldsymbol{\xi}|^2}{2}) + \tilde{u}_-\xi_1 + const. = -\phi + \tilde{u}_-\xi_1 + const., \tag{13.4.15}$$

where we have used function ϕ defined in (5.1.1). Note that this form of $\phi = \varphi + \frac{|\boldsymbol{\xi}|^2}{2}$ is not invariant with respect to the coordinate shift in the sense that function ϕ in the new coordinates is *not* the original function ϕ calculated in the original coordinates. In the rest of this proof, we use function $\phi = \varphi + \frac{|\boldsymbol{\xi}|^2}{2}$ defined in the new coordinates.

From the definition of the new $\boldsymbol{\xi}$–coordinates, we find that $S = \{\xi_1 = \xi_{1S}\}$, where ξ_{1S} is a constant. Denoting $\{\mathbf{n}_S, \mathbf{e}_S\}$ the unit normal and tangent vectors to line S, we have

$$\mathbf{e}_S = (0,1), \qquad \mathbf{n}_S = (1,0), \tag{13.4.16}$$

where the orientation of $\{\mathbf{e}_S, \mathbf{n}_S\}$ has been chosen arbitrarily and will be fixed from now on.

2. By (13.4.11),

$$\xi_{1S} = -\frac{\rho_-\tilde{u}_-}{\rho_+ - \rho_-} \neq 0,$$

since $\rho_+ > \rho_- > 0$ and $\tilde{u}_- \neq 0$. Then, by (13.4.5)–(13.4.6) and (13.4.14),

$$|\xi_{1S}| \ge \frac{1}{C}, \qquad c_+ - |\xi_{1S}| \ge \delta. \tag{13.4.17}$$

Also,

$$|\partial_{\xi_1}(\varphi^- - \varphi^+)| = |\tilde{u}_-| \geq \delta, \quad \partial_{\xi_2}(\varphi^- - \varphi^+) = 0 \qquad \text{in } \mathbb{R}^2.$$

Then our assumptions and the implicit function theorem imply that, choosing $\mu \leq \frac{\delta}{2}$, we have

$$\Gamma \subset \{\xi_1 = f(\xi_2) \, : \, \xi_2 \in \mathbb{R}\}, \tag{13.4.18}$$

where $\|f - \xi_{1S}\|_{C^1(\mathbb{R})} \leq \frac{2}{\delta}\mu$. Using that

$$(\boldsymbol{\tau}, \boldsymbol{\nu}) = (\frac{(f', 1)}{\sqrt{1+(f')^2}}, \frac{(1, -f')}{\sqrt{1+(f')^2}}),$$

for which the choice of the orientation of $\{\boldsymbol{\tau}, \boldsymbol{\nu}\}$ is consistent with the orientation of $\{\mathbf{e}_S, \mathbf{n}_S\}$ chosen above, we find that, if $\mu \leq \frac{\delta}{2}$,

$$|\boldsymbol{\nu} - \mathbf{n}_S| \leq \frac{4}{\delta}\mu, \quad |\boldsymbol{\tau} - \mathbf{e}_S| \leq \frac{4}{\delta}\mu \qquad \text{on } \Gamma. \tag{13.4.19}$$

From (13.4.15) and $\partial_{\mathbf{e}_S} = \partial_{\xi_2}$,

$$w = \partial_{\mathbf{e}_S}(\varphi^- - \varphi) = -\partial_{\xi_2}\phi.$$

Since, in the new coordinates, φ satisfies the Rankine-Hugoniot conditions given in (13.4.11), then, by Lemma 5.1.1, the corresponding function $\phi = \varphi + \frac{|\boldsymbol{\xi}|^2}{2}$ satisfies (5.1.22) and (5.1.24) with ρ_- instead of ρ_1. Combining this with (13.4.14) and (13.4.16), we have

$$D^2\phi[\boldsymbol{\tau}, \tilde{\mathbf{h}}] = 0 \qquad \text{on } \Gamma, \tag{13.4.20}$$

where

$$\begin{aligned} \tilde{\mathbf{h}} \equiv & \, \tilde{\mathbf{h}}(|D\varphi|^2, \varphi, \boldsymbol{\nu}, \boldsymbol{\tau}) \\ = & \big(- \rho(c^2 - \varphi_{\boldsymbol{\nu}}^2)\varphi_{\boldsymbol{\nu}}\nu_1 + (\rho\varphi_{\boldsymbol{\nu}}^2 + \rho_- c^2)\varphi_{\boldsymbol{\tau}}\tau_1\big)\mathbf{n}_S \\ & + \big(- \rho(c^2 - \varphi_{\boldsymbol{\nu}}^2)\varphi_{\boldsymbol{\nu}}\nu_2 + (\rho\varphi_{\boldsymbol{\nu}}^2 + \rho_- c^2)\varphi_{\boldsymbol{\tau}}\tau_2\big)\mathbf{e}_S, \end{aligned} \tag{13.4.21}$$

where $\rho = \rho(|D\varphi|^2, \varphi)$, $c = c(|D\varphi|^2, \varphi)$, and $\{\boldsymbol{\tau}, \boldsymbol{\nu}\}$ are the unit tangent and normal vectors to Γ.

Note that, in the new coordinates, (13.4.7) becomes $\Gamma \subset B_{1/\delta}(\mathbf{0})$. This, combined with (13.4.8) (with the choice of $\mu \leq \delta$), implies that $|D\varphi| \leq C$ on Γ. Then, using (13.4.5), we have

$$|\tilde{\mathbf{h}}| \leq C \qquad \text{on } \Gamma.$$

3. Now, since $\boldsymbol{\tau}_\Gamma = O(\mu)\mathbf{n}_S + \big(1 + O(\mu)\big)\mathbf{e}_S$ by (13.4.19), equality (13.4.20) becomes

$$\big(\tilde{h}_1 + O(\mu)\big)\phi_{\xi_1\xi_2} + \big(\tilde{h}_2 + O(\mu)\big)\phi_{\xi_2\xi_2} + O(\mu)\phi_{\xi_1\xi_1} = 0 \qquad \text{on } \Gamma, \tag{13.4.22}$$

where $\tilde{\mathbf{h}} = \tilde{h}_1\mathbf{n}_S + \tilde{h}_2\mathbf{e}_S$ with $(\tilde{h}_1, \tilde{h}_2)$ determined by (13.4.21).

Since we are deriving an equation for $w = \phi_{\xi_2}$, we express $\phi_{\xi_1\xi_1}$ from equation (13.4.9). Since the coordinate changes have been in the shift and rotation only, then, in the new coordinates, equation (13.4.9) is of the same form, for which new coefficients $\hat{a}_{ij}$ satisfy condition (13.4.10) with unchanged constants on the right-hand sides and with $\boldsymbol{\nu}_S$ written in the new coordinates. Then (13.4.10) becomes

$$|\hat{a}_{ij}| \le \frac{1}{\delta}, \quad \hat{a}_{11} \ge \delta \qquad \text{on } \Gamma.$$

Thus, we have

$$\phi_{\xi_1\xi_1} = -\frac{\hat{a}_{12}}{\hat{a}_{11}}\phi_{\xi_1\xi_2} - \frac{\hat{a}_{22}}{\hat{a}_{11}}\phi_{\xi_2\xi_2},$$

where $\left|\frac{\hat{a}_{i2}}{\hat{a}_{11}}\right| \le \frac{1}{\delta^2}$ for $i = 1, 2$. Then (13.4.22) becomes

$$\big(\tilde{h}_1 + O(\mu)\big)\phi_{\xi_1\xi_2} + \big(\tilde{h}_2 + O(\mu)\big)\phi_{\xi_2\xi_2} = 0 \qquad \text{on } \Gamma,$$

that is,

$$\big(\tilde{h}_1 + O(\mu)\big)w_{\xi_1} + \big(\tilde{h}_2 + O(\mu)\big)w_{\xi_2} = 0 \qquad \text{on } \Gamma, \tag{13.4.23}$$

where $\tilde{\mathbf{h}} = \tilde{\mathbf{h}}(|D\varphi|^2, \varphi, \boldsymbol{\nu}, \boldsymbol{\tau})$. Equation (13.4.23) is in the form of (13.4.12), whose coefficients are continuous on $\overline{\Gamma}$.

4. It remains to prove the obliqueness property (13.4.13) of form (13.4.12). Definition (13.4.21), combined with (13.4.8) and (13.4.19), implies

$$\big|\tilde{\mathbf{h}}(|D\varphi|^2, \varphi, \boldsymbol{\nu}, \boldsymbol{\tau}) - \tilde{\mathbf{h}}(|D\varphi^+|^2, \varphi^+, \mathbf{n}_S, \mathbf{e}_S)\big| \le C\mu \qquad \text{on } \Gamma.$$

We compute explicitly from (13.4.21) with $D\varphi^+ = -\boldsymbol{\xi} = -(f(\xi_2), \xi_2)$ on Γ:

$$\begin{aligned}
&\tilde{\mathbf{h}}(|D\varphi^+|^2, \varphi^+, \mathbf{n}_S, \mathbf{e}_S)\\
&= \big(\rho_+(c_+^2 - f^2(\xi_2))f(\xi_2)\big)\mathbf{n}_S - \big((\rho_+ f^2(\xi_2) + \rho_- c_+^2)\xi_2\big)\mathbf{e}_S\\
&= \big(\rho_+(c_+^2 - \xi_{1S}^2)\xi_{1S} + O(\mu)\big)\mathbf{n}_S - \big((\rho_+\xi_{1S}^2 + \rho_- c_+^2)\xi_2 + O(\mu)\big)\mathbf{e}_S,
\end{aligned}$$

where we have used (13.4.18) to obtain the last expression.

Also, recall that, in the new coordinates, (13.4.7) becomes $\Gamma \subset B_{1/\delta}(\mathbf{0})$. This, combined with (13.4.5) and (13.4.17), implies

$$|\rho_+(c_+^2 - \xi_{1S}^2)\xi_{1S}| \ge \frac{1}{C}, \quad |(\rho_+\xi_{1S}^2 + \rho_- c_+^2)\xi_2| \le C \qquad \text{on } \Gamma.$$

Then, recalling that $\boldsymbol{\nu}_\Gamma = \big(1 + O(\mu)\big)\mathbf{n}_S + O(\mu)\mathbf{e}_S$ for $\mathbf{n}_S = (1, 0)$ and denoting by (a_1, a_2) the coefficients of (13.4.23), we have

$$\begin{aligned}
|(a_1, a_2)\cdot\boldsymbol{\nu}_\Gamma| &= |(\tilde{h}_1, \tilde{h}_2)\cdot\mathbf{n}_S| + O(\mu) = |\tilde{h}_1| + O(\mu)\\
&= |\rho_+(c_+^2 - \xi_{1S}^2)\xi_{1S}| + O(\mu) \ge \frac{1}{C} + O(\mu) > 0
\end{aligned}$$

if μ is small. □

Now we prove the monotonicity of $\varphi_1 - \varphi$. From (12.3.19), we have a more precise version of estimate (12.3.21) that, for any $(u, \theta_{\rm w}) \in \overline{\mathcal{K}}$,

$$\|\psi\|_{1,\alpha,\Omega} < C\eta_1(\theta_{\rm w}), \quad (\psi, D\psi) = (0, \mathbf{0}) \qquad \text{on } \Gamma_{\rm sonic}, \tag{13.4.24}$$

where C depends only on the data and $(\theta_{\rm w}^*, \alpha)$.

Lemma 13.4.6. *If the parameters in Definition* 12.3.2 *are chosen so that* δ_1 *and* ε *are small, depending only on the data,* α*, and constants* $(C, \hat{N}_0)$ *in* (12.3.22) *and* (13.4.24) *(hence* δ_1 *and* ε *are small, depending only on the data and* $\theta_{\rm w}^*$*), and if* N_1 *is large, depending on the data and* $\theta_{\rm w}^*$*, then, for every* $(u, \theta_{\rm w}) \in \overline{\mathcal{K}}$ *such that* $\theta_{\rm w} \in [\theta_{\rm w}^*, \frac{\pi}{2})$ *and* u *is a fixed point of map* $\mathcal{I}^{(\theta_{\rm w})}$*, the corresponding function* φ *satisfies*

$$\partial_{\mathbf{e}_{S_1}}(\varphi_1 - \varphi) \le 0 \qquad \text{in } \Omega.$$

Proof. In this proof, the universal constant C depends only on the data, α, and constants $(C, \hat{N}_0)$ in (12.3.22) and (13.4.24) (hence only on the data and $\theta_{\rm w}^*$). Denote

$$\bar{\phi} := \varphi_1 - \varphi, \quad w = \partial_{\mathbf{e}_{S_1}}\bar{\phi}.$$

We now show that $w \le 0$ in Ω through four steps.

1. We work in the $\boldsymbol{\xi}$–coordinates with the origin shifted to center $\mathcal{O}_2$ of state (2).

Since $w = -\partial_{\mathbf{e}_{S_1}}\psi + \partial_{\mathbf{e}_{S_1}}(\varphi_1 - \varphi_2)$, and $\partial_{\mathbf{e}_{S_1}}(\varphi_1 - \varphi_2)$ is constant, it follows from (13.4.4) that

$$w \in C^{\alpha}(\overline{\Omega}) \cap C^{1,\alpha}(\overline{\Omega} \setminus (\overline{\Gamma_{\rm sonic}} \cup \{P_2, P_3\})) \cap C^{2,\alpha}(\Omega). \tag{13.4.25}$$

Now we show that w is a solution of a mixed boundary value problem for an elliptic equation in Ω.

We first derive an elliptic equation for w in Ω. Since $\psi = \varphi - \varphi_2$ satisfies equation (12.4.45) by Lemma 13.4.2, and $D^2(\varphi_1 - \varphi_2) = 0$, it follows that $\bar{\phi} := \varphi - \varphi_0$ is governed by

$$\mathcal{A}_{11}\bar{\phi}_{\xi_1\xi_1} + 2\mathcal{A}_{12}\bar{\phi}_{\xi_1\xi_2} + \mathcal{A}_{22}\bar{\phi}_{\xi_2\xi_2} = 0 \qquad \text{in } \Omega \tag{13.4.26}$$

with

$$\mathcal{A}_{ij}(\boldsymbol{\xi}) = A_{ij}(D\psi(\boldsymbol{\xi}), \boldsymbol{\xi}), \tag{13.4.27}$$

where $A_{ij}(\mathbf{p}, \boldsymbol{\xi})$ are from (12.4.45).

The regularity of $\mathcal{A}_{ij}$ is as follows: We first note that, using (12.4.5), (12.4.42), and the improved regularity of ψ near $\Gamma_{\rm sym}$ in (13.4.4), we improve the estimate in Lemma 12.4.5(iii) to

$$A_{ij}^1 \in C^{1,\alpha}(\overline{\Omega} \setminus (\mathcal{D}_\varepsilon \cup \{P_2, P_3\})).$$

Combining this with Lemma 12.4.5(iii)–(v) and (13.4.4), and using (13.4.24), we have

$$\mathcal{A}_{ij} \in C^{1,\alpha}(\overline{\Omega} \setminus (\Gamma_{\rm sonic} \cup \{P_2, P_3\})), \quad \|\mathcal{A}_{ij}\|_{C^{\alpha^2}(\overline{\Omega})} \le C. \tag{13.4.28}$$

Also, equation (13.4.26) is strictly elliptic in $\overline{\Omega}\setminus\overline{\Gamma_{\rm sonic}}$.

Now we derive the equation for w by differentiating equation (13.4.26) in the direction of $\mathbf{e}_{\mathcal{S}_1}$. It is convenient to rotate the coordinates. Denote by (X,Y) the coordinates with basis $\{\mathbf{e}_{\mathcal{S}_1},\boldsymbol{\nu}_{\mathcal{S}_1}\}$, where $\boldsymbol{\nu}_{\mathcal{S}_1}=\frac{D\bar\phi_0}{|D\bar\phi_0|}$ is a unit vector orthogonal to S_1. Then $\partial_{\mathbf{e}_{\mathcal{S}_1}}\phi=\partial_X\phi$. Equation (13.4.26) is of the same form in the (X,Y)–coordinates:

$$\hat{\mathcal{A}}_{11}\bar\phi_{XX}+2\hat{\mathcal{A}}_{12}\bar\phi_{XY}+\hat{\mathcal{A}}_{22}\bar\phi_{YY}=0\qquad\text{in }\Omega,$$

where $\hat{\mathcal{A}}_{ij}\in C(\overline{\Omega})$ and the equation is strictly elliptic in $\overline{\Omega}\setminus\overline{\Gamma_{\rm sonic}}$ by (13.4.27) and Lemma 12.4.5(i). In particular, $\hat{\mathcal{A}}_{22}>0$ in $\overline{\Omega}\setminus\overline{\Gamma_{\rm sonic}}$. Also, $w=\bar\phi_X$. Now, differentiating the equation with respect to X, substituting the right-hand side of expression $\bar\phi_{YY}=-\frac{1}{\hat{\mathcal{A}}_{22}}\big(\hat{\mathcal{A}}_{11}\bar\phi_{XX}+2\hat{\mathcal{A}}_{12}\bar\phi_{XY}\big)$ for $\bar\phi_{YY}$ into the resulting equation, and then writing the resulting equation in terms of w, we have

$$\begin{aligned}&\hat{\mathcal{A}}_{11}w_{XX}+2\hat{\mathcal{A}}_{12}w_{XY}+\hat{\mathcal{A}}_{22}w_{YY}+(\partial_X\hat{\mathcal{A}}_{11})w_X+2(\partial_X\hat{\mathcal{A}}_{12})w_Y\\&\quad-\frac{\partial_X\hat{\mathcal{A}}_{22}}{\hat{\mathcal{A}}_{22}}(\hat{\mathcal{A}}_{11}w_X+2\hat{\mathcal{A}}_{12}w_Y)=0\qquad\text{in }\Omega.\end{aligned}$$

Writing this equation in the $\boldsymbol{\xi}$–coordinates, we obtain

$$\mathcal{A}_{11}w_{\xi_1\xi_1}+2\mathcal{A}_{12}w_{\xi_1\xi_2}+\mathcal{A}_{22}w_{\xi_2\xi_2}+\mathcal{A}_1w_{\xi_1}+\mathcal{A}_2w_{\xi_2}=0\qquad\text{in }\Omega,\tag{13.4.29}$$

where $\mathcal{A}_{ij}$ are from (13.4.26) and, by (13.4.28),

$$\mathcal{A}_i\in C^\alpha(\overline{\Omega}\setminus(\overline{\Gamma_{\rm sonic}}\cup\{P_2,P_3\})).\tag{13.4.30}$$

This equation is strictly elliptic in $\overline{\Omega}\setminus\overline{\Gamma_{\rm sonic}}$ from a similar property of equation (13.4.26).

Next we derive the boundary condition for w on $\Gamma_{\rm wedge}$. Let $\{\boldsymbol{n}_{\rm w},\boldsymbol{\tau}_{\rm w}\}$ be the unit normal and tangent vectors to $\Gamma_{\rm wedge}$ defined by (8.2.35) and (8.2.17), respectively, and $\boldsymbol{\nu}_{\rm w}=-\boldsymbol{n}_{\rm w}$. Using (12.3.28) and

$$D\varphi_1\cdot\boldsymbol{\nu}_{\rm w}=u_1\mathbf{e}_1\cdot\boldsymbol{\nu}_{\rm w},\quad D\varphi_2\cdot\boldsymbol{\nu}_{\rm w}=0\qquad\text{on }\Gamma_{\rm wedge},$$

we find that, on $\Gamma_{\rm wedge}$,

$$\partial_{\boldsymbol{\nu}_{\rm w}}\bar\phi=-\partial_{\boldsymbol{\nu}_{\rm w}}\psi+\partial_{\boldsymbol{\nu}_{\rm w}}(\varphi_1-\varphi_2)=u_1\mathbf{e}_1\cdot\boldsymbol{\nu}_{\rm w}=-u_1\sin\theta_{\rm w}.$$

Differentiating this equality in direction $\boldsymbol{\tau}_{\rm w}$, we have

$$\partial_{\boldsymbol{\nu}_{\rm w}}\partial_{\boldsymbol{\tau}_{\rm w}}\bar\phi=0\qquad\text{on }\Gamma_{\rm wedge},$$

that is, in the (S,T)–coordinates with basis $\{\boldsymbol{\nu}_{\rm w},\boldsymbol{\tau}_{\rm w}\}$,

$$\bar\phi_{ST}=0\qquad\text{on }\Gamma_{\rm wedge}.\tag{13.4.31}$$

Equation (13.4.26) in the (S, T)–coordinates is of the same form:

$$\tilde{\mathcal{A}}_{11}\bar{\phi}_{SS} + 2\tilde{\mathcal{A}}_{12}\bar{\phi}_{ST} + \tilde{\mathcal{A}}_{22}\bar{\phi}_{TT} = 0 \qquad \text{in } \Omega,$$

where $\tilde{\mathcal{A}}_{ij}, i, j = 1, 2$, satisfy (13.4.28), and the equation is strictly elliptic in $\overline{\Omega} \setminus \overline{\Gamma_{\text{sonic}}}$. In particular, $\tilde{\mathcal{A}}_{11} > 0$ in $\overline{\Omega} \setminus \overline{\Gamma_{\text{sonic}}}$. Since φ (hence $\bar{\phi}$) is C^2 up to Γ_{wedge}, the last equation holds on Γ_{wedge}. Combining this with (13.4.31), we have

$$\bar{\phi}_{SS} = -\frac{\tilde{\mathcal{A}}_{22}}{\tilde{\mathcal{A}}_{11}}\bar{\phi}_{TT} \qquad \text{on } \Gamma_{\text{wedge}}.$$

Also, since $\mathbf{e}_{\mathcal{S}_1}$ is not orthogonal to Γ_{wedge} by Lemma 7.5.12,

$$\mathbf{e}_{\mathcal{S}_1} = a_1\boldsymbol{\nu}_{\text{w}} + a_2\boldsymbol{\tau}_{\text{w}} \qquad \text{with } a_2 \neq 0.$$

Then

$$w = \partial_{\mathbf{e}_{\mathcal{S}_1}}\bar{\phi} = a_1\bar{\phi}_S + a_2\bar{\phi}_T.$$

Now we employ (13.4.31) to compute that, on Γ_{wedge},

$$\begin{aligned}\partial_S w &= a_1\partial_{SS}\bar{\phi} + a_2\partial_{ST}\bar{\phi} = a_1\partial_{SS}\bar{\phi} = -a_1\frac{\tilde{\mathcal{A}}_{22}}{\tilde{\mathcal{A}}_{11}}\partial_{TT}\bar{\phi} \\ &= -\frac{a_1\tilde{\mathcal{A}}_{22}}{a_2\tilde{\mathcal{A}}_{11}}(a_1\partial_{ST}\bar{\phi} + a_2\partial_{TT}\bar{\phi}) = -\frac{a_1\tilde{\mathcal{A}}_{22}}{a_2\tilde{\mathcal{A}}_{11}}\partial_T w =: b_1\partial_T w,\end{aligned}$$

where $b_1 \in C(\overline{\Gamma_{\text{wedge}}} \setminus \{P_4\})$. Therefore, w satisfies the oblique condition on Γ_{wedge}:

$$\partial_{\boldsymbol{\nu}_{\text{w}}} w - b_1\partial_{\boldsymbol{\tau}_{\text{w}}} w = 0 \qquad \text{on } \Gamma_{\text{wedge}}. \tag{13.4.32}$$

Next we derive the condition on Γ_{sym} for w. Since $\mathbf{e}_{\mathcal{S}_1}$ is not orthogonal to Γ_{sym} by Lemma 7.5.12, then we repeat the argument for Γ_{wedge} to obtain that w satisfies the oblique condition on Γ_{sym}:

$$\partial_{\boldsymbol{\nu}_{\text{sym}}} w - b_2\partial_{\boldsymbol{\tau}_{\text{sym}}} w = 0 \qquad \text{on } \Gamma_{\text{sym}}, \tag{13.4.33}$$

with $b_2 \in C(\overline{\Gamma_{\text{sym}}})$.

Furthermore, since $\mathcal{S}_1 = \{\varphi_1 = \varphi_2\}$, $\partial_{\mathbf{e}_{\mathcal{S}_1}}\varphi_1 = \partial_{\mathbf{e}_{\mathcal{S}_1}}\varphi_2$ so that $\partial_{\mathbf{e}_{\mathcal{S}_1}}\bar{\phi} = -\partial_{\mathbf{e}_{\mathcal{S}_1}}\psi$. Since $\psi \in C^{2,\alpha}_{*,1+\delta^*}(\Omega)$ by Remark 12.3.5, it follows that $D\psi = 0$ on Γ_{sonic} so that $\partial_{\mathbf{e}_{\mathcal{S}_1}}\bar{\phi} = -\partial_{\mathbf{e}_{\mathcal{S}_1}}\psi = 0$ on Γ_{sonic}. Then the condition on Γ_{sonic} for w is

$$w = 0 \qquad \text{on } \Gamma_{\text{sonic}}. \tag{13.4.34}$$

In Steps 2–4 below, we derive the boundary condition on Γ_{shock} for w. Specifically, we show that Lemma 13.4.5 can be applied in the following setting: $(\varphi^-, \varphi^+) = (\varphi_1, \varphi_2)$, $\mathcal{D} = \Omega$, and $\Gamma = \Gamma_{\text{shock}}$. Then S in Lemma 13.4.5 is line $\mathcal{S}_1$. Now we prove that the assumptions of Lemma 13.4.5 are satisfied.

2. First, the Rankine-Hugoniot conditions (13.4.11) are satisfied, since u is a fixed point of $\mathcal{I}^{\theta_{\text{w}}}$, as we discussed at the beginning of this chapter.

From the properties of state (2), (13.4.5) is satisfied with $\delta > 0$. Similarly, (13.4.6) is satisfied with

$$\delta = \min_{\theta_{\rm w}\in[\theta_{\rm w}^{\rm d},\frac{\pi}{2}]} \big(c_2^{(\theta_{\rm w})} - {\rm dist}(S_1^{(\theta_{\rm w})}, (u_2^{(\theta_{\rm w})}, v_2^{(\theta_{\rm w})}))\big) > 0,$$

where the last inequality holds by using (6.1.4) and the continuous dependence of the parameters of the weak state (2) on $\theta_{\rm w} \in [\theta_{\rm w}^{\rm d}, \frac{\pi}{2}]$.

Condition (13.4.7) is satisfied by (12.3.18) after possibly further reducing δ. Here we note that (12.3.18) is in the non-shifted coordinates, *i.e.*, when the origin is at center $\mathcal{O}_0$ of state (0). Now we work in the coordinates with the origin shifted to $\mathcal{O}_2$. However, since

$$|\mathcal{O}_2 - \mathcal{O}_0| = |D(\varphi_2 - \varphi_0)| = |(u_2, v_2)| \le C$$

with C depending only on the data and $\theta_{\rm w}^*$, we obtain (12.3.18) and hence (13.4.7) in the shifted coordinates, with a modified constant.

Next, we show that conditions (13.4.9)–(13.4.10) are satisfied. Since $\psi = \varphi - \varphi_2$ is a solution of (12.4.45) by Lemma 13.4.2, and $D^2(\varphi_2 + \frac{|\boldsymbol{\xi}|^2}{2}) = D^2(u_2\xi_1 + v_2\xi_2) = 0$, it follows that $\phi = \varphi + \frac{|\boldsymbol{\xi}|^2}{2}$ satisfies the equation:

$$\mathcal{A}_{11}\phi_{\xi_1\xi_1} + 2\mathcal{A}_{12}\phi_{\xi_1\xi_2} + \mathcal{A}_{22}\phi_{\xi_2\xi_2} = 0 \qquad \text{in } \Omega, \tag{13.4.35}$$

where $\mathcal{A}_{ij}$ are from (13.4.26). This verifies (13.4.9).

We now prove (13.4.10), where $\boldsymbol{\nu}^S = \boldsymbol{\nu}_{S_1}$ in the present case. We use notation $\boldsymbol{\nu}_{S_1} = (\nu_1^{S_1}, \nu_2^{S_1})$ below. Let $\tilde{\varepsilon} > 0$ be fixed below. By (13.4.27) and Lemma 12.4.5(i),

$$\sum_{i,j=1}^{2} \mathcal{A}_{ij}\nu_i^{S_1}\nu_j^{S_1} \ge \lambda_0\frac{\tilde{\varepsilon}}{2} \qquad \text{on } \Gamma_{\rm shock} \setminus \mathcal{D}_{\tilde{\varepsilon}/2}, \tag{13.4.36}$$

with $\lambda_0 > 0$ depending only on the data.

In order to check (13.4.10) on $\Gamma_{\rm shock} \cap \mathcal{D}_{\tilde{\varepsilon}/2}$, we first check it at point P_1. Coefficients $\mathcal{A}_{ij}(\boldsymbol{\xi}) = A_{ij}(D\psi(\boldsymbol{\xi}), \boldsymbol{\xi})$ at $P_1 = \overline{\Gamma_{\rm shock}} \cap \overline{\Gamma_{\rm sonic}}$ are of form (12.4.44), where the $\boldsymbol{\xi}$–coordinates are shifted so that the origin is at center $\mathcal{O}_2$ of the sonic center of state (2), *i.e.*, $\boldsymbol{\xi}_{P_1} = -P_1\mathcal{O}_2$. Then we have

$$\sum_{i,j=1}^{2} \mathcal{A}_{ij}\nu_i^{S_1}\nu_j^{S_1} = c_2^2 - (P_1\mathcal{O}_2 \cdot \boldsymbol{\nu}_{S_1})^2 \qquad \text{at } P_1.$$

Since $P_1 \in \overline{\Gamma_{\rm sonic}}$, it follows that $|P_1\mathcal{O}_2| = c_2$. Also, vectors $P_1\mathcal{O}_2$ and $\boldsymbol{\nu}_{S_1}$ are not parallel to each other, since $P_1\mathcal{O}_2$ is the radial vector for the sonic circle $\partial B_{c_2}(\mathcal{O}_2)$ at P_1, *i.e.*, it is orthogonal to the tangent line to $\partial B_{c_2}(\mathcal{O}_2)$ at P_1, and S_1 intersects with $\partial B_{c_2}(\mathcal{O}_2)$ at P_1, by property (6.1.4) in Lemma 6.1.2, applied with $(\varphi^-, \varphi^+) = (\varphi_1, \varphi_2)$. Then

$$\left|\frac{P_1\mathcal{O}_2}{|P_1\mathcal{O}_2|} \cdot \boldsymbol{\nu}_{S_1}\right| < 1,$$

where the left-hand side depends continuously on the parameters of state (2). Thus, there exists $\sigma > 0$ depending only on the data such that, for any $\theta_{\rm w} \in [\theta_{\rm w}^{\rm s}, \frac{\pi}{2}]$,

$$\Big(\frac{P_1O_2}{|P_1O_2|}\cdot\boldsymbol{\nu}_{S_1}\Big)^2 \le 1-\sigma.$$

Then

$$\sum_{i,j=1}^{2} \mathcal{A}_{ij}\nu_i^{S_1}\nu_j^{S_1} = c_2^2 - c_2^2\Big(\frac{P_1O_2}{|P_1O_2|}\cdot\boldsymbol{\nu}_{S_1}\Big)^2 \ge c_2^2\sigma \qquad \text{at } P_1.$$

To obtain such an estimate for all $\boldsymbol{\xi} \in \Gamma_{\rm shock}\cap\mathcal{D}_{\tilde\varepsilon}$, we obtain from (12.3.22) that, for each $\boldsymbol{\xi} \in \Gamma_{\rm shock}\cap\mathcal{D}_{\tilde\varepsilon}$,

$$|\boldsymbol{\xi} - P_1| \le C\tilde\varepsilon,$$

and, combining this with (13.4.28),

$$|\mathcal{A}_{ij}(\boldsymbol{\xi}) - \mathcal{A}_{ij}(P_1)| \le C|\boldsymbol{\xi}-P_1|^{\alpha^2} \le C\tilde\varepsilon^{\alpha^2} \qquad \text{for all } \boldsymbol{\xi}\in\Gamma_{\rm shock}\cap\mathcal{D}_{\tilde\varepsilon}.$$

Choosing $\tilde\varepsilon$ small, depending only on the data, α, and constants $(C, \hat N_0)$ in (12.3.22) (hence only on the data and $\theta_{\rm w}^*$), we have

$$\sum_{i,j=1}^{2} \mathcal{A}_{ij}(\boldsymbol{\xi})\nu_i^{S_1}\nu_j^{S_1} \ge \frac{\sigma}{2}\min_{\theta_{\rm w}\in[\theta_{\rm w}^{\rm s},\frac{\pi}{2}]} c_2^2(\theta_{\rm w}) \qquad \text{for all } \boldsymbol{\xi}\in\Gamma_{\rm shock}\cap\mathcal{D}_{\tilde\varepsilon}.$$

Combining this with (13.4.36) for fixed $\tilde\varepsilon$ now, we obtain (13.4.10) on $\Gamma_{\rm shock}$ with δ depending on the data, α, and constants $(C, \hat N_0)$ in (12.3.22).

Therefore, we have proved the following claim:

Claim 13.4.7. *There exists $\delta > 0$ depending only on the data and $\theta_{\rm w}^*$ such that properties* (13.4.5)–(13.4.7) *and* (13.4.9)–(13.4.11) *are satisfied, where $(\varphi^-, \varphi^+) = (\varphi_1, \varphi_2)$, $(\mathcal{D}, \Gamma, S) = (\Omega, \Gamma_{\rm shock}, S_1)$, and* (13.4.9) *is* (13.4.35).

The remaining assumption of Lemma 13.4.5 is (13.4.8). To verify this assumption and complete the proof of Lemma 13.4.6, we consider two separate cases: $\theta_{\rm w} \in [\frac{\pi}{2} - \frac{2\delta_1}{N_1^2}, \frac{\pi}{2})$ and $\theta_{\rm w} \in [\theta_{\rm w}^*, \frac{\pi}{2} - \frac{2\delta_1}{N_1^2})$.

3. *The case that the wedge angles are close to $\frac{\pi}{2}$, specifically $\theta_{\rm w} \in [\frac{\pi}{2} - \frac{2\delta_1}{N_1^2}, \frac{\pi}{2})$.* Assuming $N_1 > 2$, then $\theta_{\rm w} \in [\frac{\pi}{2} - \frac{\delta_1}{N_1}, \frac{\pi}{2})$. It follows, by (13.4.24) where $\eta_1(\cdot)$ is introduced in Definition 12.3.2(i), that

$$\|\varphi - \varphi_2\|_{C^{1,\alpha}(\overline{\Omega})} < C\eta_1(\theta_{\rm w}) \le C\delta_1. \tag{13.4.37}$$

Let $\mu = \mu(\delta)$ be the constant in Lemma 13.4.5 for δ from Claim 13.4.7. Thus, from (13.4.37), by choosing δ_1 small, we see that (13.4.8) holds on $\Gamma_{\rm shock}$, where $(\varphi^+, \mathcal{D}) = (\varphi_2, \Omega)$. Then, applying Lemma 13.4.5, we conclude that $w = \partial_{\mathbf{e}_{S_1}}(\varphi_1 - \varphi)$ satisfies an oblique derivative condition:

$$\hat a_1 w_{\xi_1} + \hat a_2 w_{\xi_2} = 0 \qquad \text{on } \Gamma_{\rm shock}, \tag{13.4.38}$$

where $\hat{a}_i \in C(\overline{\Gamma_{\text{shock}}})$.

Thus, w is a solution of the following mixed problem in Ω: w satisfies equation (13.4.29) which is strictly elliptic in $\overline{\Omega} \setminus \overline{\Gamma_{\text{sonic}}}$ with the coefficients satisfying (13.4.28) and (13.4.30), the oblique boundary conditions (13.4.32)–(13.4.33) and (13.4.38) with the coefficients that are continuous in the relative interiors of the corresponding boundary segments, and the Dirichlet condition (13.4.34) on Γ_{sonic}.

Remark 13.4.8. *The comparison principle* (*Lemma* 4.4.2) *cannot be applied directly here, since $\mathcal{A}_i, i = 1, 2$, satisfy only* (13.4.30) *so that they are possibly unbounded near P_2 and P_3, and the rate of their blowup is determined by the elliptic estimates near the corner, i.e., it cannot be controlled. Therefore, we cannot conclude that $w \equiv 0$ directly from the homogeneous problem for w.*

From the equation and the boundary conditions for w discussed above, using (13.4.25), the strong maximum principle in the interior of Ω, and Hopf's lemma at the points of relative interior of Γ_{shock}, Γ_{wedge}, and Γ_{sym}, we obtain that either $w = const.$ in $\overline{\Omega}$ or the maximum of w over $\overline{\Omega}$ cannot occur in $\overline{\Omega} \setminus (\overline{\Gamma_{\text{sonic}}} \cup \{P_2, P_3\})$.

If $w = const.$ in $\overline{\Omega}$, then $w \equiv 0$ by (13.4.34). Otherwise, the maximum of w may occur in $\overline{\Gamma_{\text{sonic}}} \cup \{P_2, P_3\}$. We know that $w = 0$ on $\overline{\Gamma_{\text{sonic}}}$ by (13.4.34).

It remains to estimate w at $\{P_2, P_3\}$.

First consider P_3. From (13.4.4) with the fixed point property $\hat{\psi} = \psi$, it follows that ψ satisfies (12.3.28)–(12.3.29) at $P_3 = \overline{\Gamma_{\text{wedge}}} \cap \overline{\Gamma_{\text{sym}}}$. We also note that, in the coordinates with the origin shifted to $\mathcal{O}_2$, $P_3 = -(u_2, v_2)$ and $\varphi_2(\boldsymbol{\xi}) = -\frac{1}{2}|\boldsymbol{\xi}|^2 + const.$ Thus, for any $P \in \mathbb{R}^2$, we have

$$D\varphi_2(P) = -\boldsymbol{\xi}_P = P\mathcal{O}_2.$$

Since $\mathcal{O}_2, P_3 \in \overline{\Gamma_{\text{wedge}}}$, we find that $P_3\mathcal{O}_2 \perp \boldsymbol{\nu}_{\text{w}}$. Then, using (12.3.28),

$$\partial_{\boldsymbol{\nu}_{\text{w}}}\varphi(P_3) = \partial_{\boldsymbol{\nu}_{\text{w}}}\psi(P_3) + \partial_{\boldsymbol{\nu}_{\text{w}}}\varphi_2(P_3) = P_3\mathcal{O}_2 \cdot \boldsymbol{\nu}_{\text{w}} = 0.$$

Also, since $P_3 = -(u_2, v_2)$ and $\boldsymbol{\nu}_{\text{sym}} = (0, 1)$ on $\Gamma_{\text{sym}} = \partial\Omega \cap \{\xi_2 = -v_2\}$, then $\partial_{\xi_2}\varphi_2(P_3) = v_2$. Using (12.3.29),

$$\partial_{\boldsymbol{\nu}_{\text{sym}}}\varphi(P_3) = \partial_{\boldsymbol{\nu}_{\text{sym}}}\psi(P_3) + \partial_{\boldsymbol{\nu}_{\text{sym}}}\varphi_2(P_3) = -v_2 + v_2 = 0.$$

Moreover, $\boldsymbol{\nu}_{\text{w}}$ is not parallel to $\mathbf{e}_{\xi_2}$, since $\theta_{\text{w}} < \frac{\pi}{2}$. Then

$$D\varphi(P_3) = \mathbf{0}.$$

Using expression (7.5.8) of $\mathbf{e}_{S_1}$ and that $(\varphi_1-\varphi_2)(\boldsymbol{\xi}) = (u_1-u_2, -v_2)\cdot\boldsymbol{\xi}+const.$, we find that, at P_3,

$$\begin{aligned} w &= \partial_{\mathbf{e}_{S_1}}(\varphi_1 - \varphi) \\ &= \partial_{\mathbf{e}_{S_1}}(\varphi_1 - \varphi_2) + \partial_{\mathbf{e}_{S_1}}\varphi_2 - \partial_{\mathbf{e}_{S_1}}\varphi \\ &= \big((u_1 - u_2, -v_2) + (u_2, v_2)\big)\cdot \mathbf{e}_{S_1} \\ &= -\frac{u_1 v_2}{\sqrt{(u_1-u_2)^2+v_2^2}}. \end{aligned}$$

Since $u_1, v_2 > 0$, we have

$$w(P_3) < 0.$$

Now we consider point P_2. Since $P_2 = \overline{\Gamma_{\text{sym}}}\cap\overline{\Gamma_{\text{shock}}}$, $\psi = \hat{\psi}$ satisfies (12.3.29) at P_2. Then, using that $\boldsymbol{\nu}_{\text{sym}} = (0,1)$, we find that, on $\overline{\Gamma_{\text{sym}}}$ (including point P_2),

$$\begin{aligned} \partial_{\xi_2}(\varphi_1 - \varphi) &= \partial_{\xi_2}(\varphi_1 - \varphi_2) - \partial_{\xi_2}\psi \\ &= (u_1 - u_2, -v_2)\cdot(0,1) + v_2 \\ &= 0. \end{aligned} \tag{13.4.39}$$

From Definition 12.2.6, it follows that $\varphi = \varphi_1$ on $\Gamma_{\text{shock}} = \Gamma_{\text{shock}}(u, \theta_{\text{w}})$. Then, using (12.3.13), we have

$$\boldsymbol{\nu}_{\text{sh}} = \frac{D(\varphi_1 - \varphi)}{|D(\varphi_1-\varphi)|} \qquad \text{on } \overline{\Gamma_{\text{shock}}}.$$

Thus, $\partial_{\xi_2}(\varphi_1 - \varphi)(P_2) = 0$ implies

$$\boldsymbol{\nu}_{\text{sh}}(P_2) = (1,0).$$

Then (12.3.13) at P_2 becomes

$$\partial_{\xi_1}(\varphi_1 - \varphi)(P_2) > \mu_1 > 0.$$

Now, using (7.5.8), we find that, at P_2,

$$\begin{aligned} w &= \partial_{\mathbf{e}_{S_1}}(\varphi_1 - \varphi) \\ &= -\frac{v_2\partial_{\xi_1}(\varphi_1 - \varphi) + (u_1-u_2)\partial_{\xi_2}(\varphi_1-\varphi)}{\sqrt{(u_1-u_2)^2+v_2^2}} \\ &= \frac{-v_2\partial_{\xi_1}(\varphi_1-\varphi)}{\sqrt{(u_1-u_2)^2+v_2^2}} \\ &< \frac{-v_2\mu_1}{\sqrt{(u_1-u_2)^2+v_2^2}} \\ &< 0. \end{aligned}$$

Thus, $w \le 0$ in $\overline{\Omega}$. This completes the proof of the expected result for Case $\theta_{\rm w} \in [\frac{\pi}{2} - \frac{2\delta_1}{N_1^2}, \frac{\pi}{2})$.

4. *The case that the wedge angles are away from $\frac{\pi}{2}$, specifically $\theta_{\rm w} \in [\theta_{\rm w}^*, \frac{\pi}{2} - \frac{2\delta_1}{N_1^2})$.* In this case, from Definition 12.3.2(iv),

$$\eta_2(\theta_{\rm w}) \ge \frac{\delta_1}{N_1^2} > 0.$$

Thus, by (12.3.11),

$$w = \partial_{\mathbf{e}_{S_1}}(\varphi_1 - \varphi) < -\frac{\delta_1}{N_1^2} < 0 \qquad \text{in } \overline{\Omega} \setminus \mathcal{D}_{\varepsilon/10}. \tag{13.4.40}$$

It remains to consider $w = \partial_{\mathbf{e}_{S_1}}(\varphi_1 - \varphi)$ on $\overline{\Omega} \cap \mathcal{D}_\varepsilon$. From (13.4.24), it follows that

$$\|\varphi - \varphi_2\|_{C^1(\overline{\Omega \cap \mathcal{D}_\varepsilon})} < C\varepsilon^\alpha. \tag{13.4.41}$$

Let $\mu = \mu(\delta)$ be the constant in Lemma 13.4.5, where δ is from Claim 13.4.7. Then μ depends only on the data, α, and constants $(C, \hat{N}_0)$ in (12.3.22). From (13.4.41), by choosing ε small, depending only on the data, α, and constants $(C, \hat{N}_0)$ in (12.3.22) and (13.4.24), we obtain that (13.4.8) is satisfied on $\Gamma_{\rm shock} \cap \mathcal{D}_\varepsilon$, where $(\varphi^+, \mathcal{D}) = (\varphi_2, \Omega \cap \mathcal{D}_\varepsilon)$. Then, applying Lemma 13.4.5, we conclude that $w = \partial_{\mathbf{e}_{S_1}}(\varphi_1 - \varphi)$ satisfies an oblique derivative condition:

$$\hat{a}_1 w_{\xi_1} + \hat{a}_2 w_{\xi_2} = 0 \qquad \text{on } \Gamma_{\rm shock} \cap \mathcal{D}_\varepsilon, \tag{13.4.42}$$

where $\hat{a}_i \in C(\overline{\Gamma_{\rm shock} \cap \mathcal{D}_\varepsilon})$.

Therefore, w is a solution of the following problem in $\Omega \cap \mathcal{D}_\varepsilon$: w satisfies equation (13.4.29) which is strictly elliptic in $\overline{\Omega \cap \mathcal{D}_\varepsilon} \setminus \overline{\Gamma_{\rm sonic}}$, the oblique boundary conditions (13.4.32) and (13.4.42) with continuous coefficients on $\Gamma_{\rm wedge} \cap \mathcal{D}_\varepsilon$ and $\Gamma_{\rm shock} \cap \mathcal{D}_\varepsilon$ respectively, and the Dirichlet condition (13.4.34) on $\Gamma_{\rm sonic}$. Furthermore, (13.4.40) implies that $w < 0$ on $\partial(\Omega \cap \mathcal{D}_\varepsilon) \cap \{x = \varepsilon\}$. Then, by the maximum principle, $w \le 0$ in $\Omega \cap \mathcal{D}_\varepsilon$.

Combining this with (13.4.40), we conclude that $w \le 0$ in Ω. □

Lemma 13.4.9. *If the parameters in Definition* 12.3.2 *are chosen so that* $(\delta_1, \varepsilon, \frac{1}{N_1})$ *are small, depending on the data and* $\theta_{\rm w}^*$*, then, for every* $(u, \theta_{\rm w}) \in \overline{\mathcal{K}}$ *such that* $\theta_{\rm w} \in [\theta_{\rm w}^*, \frac{\pi}{2})$ *and* u *is a fixed point of map* $\mathcal{I}^{(\theta_{\rm w})}$*, the corresponding function* φ *satisfies*

$$\partial_{\xi_2}(\varphi_1 - \varphi) \le 0 \qquad \textit{in } \Omega.$$

Proof. We follow the proof of Lemma 13.4.6, so that we only sketch the argument and describe the differences in more detail. The universal constant C in this proof depends only on the data, α, constants $(C, \hat{N}_0)$ in (12.3.22) and (13.4.24), and λ in Definition 12.3.2(v) (and hence only on the data and $\theta_{\rm w}^*$). Denote

$$\bar{\phi} := \varphi_1 - \varphi, \quad \tilde{w} = \partial_{\xi_2}\bar{\phi}.$$

We show that $\tilde{w} \leq 0$ in Ω through four steps.

1. Recall that $\bar{\phi}$ satisfies (13.4.4) and is a solution of equation (13.4.26) that is strictly elliptic in $\overline{\Omega} \setminus \overline{\Gamma_{\rm sonic}}$ with the coefficients satisfying (13.4.28). This implies that $\tilde{w}$ satisfies (13.4.25). Also, repeating the argument in Lemma 13.4.6, we conclude the following:

- $\tilde{w}$ satisfies an equation of form (13.4.29) in Ω, where $\mathcal{A}_{ij}, i,j = 1,2$, are from (13.4.26) and $\mathcal{A}_i, i = 1,2$, satisfy (13.4.30) (however, $\mathcal{A}_i, i = 1,2$, are different in the present case from the corresponding coefficients in Lemma 13.4.6);
- $\tilde{w}$ satisfies an oblique boundary condition of form (13.4.32) on $\Gamma_{\rm wedge}$, where b_1 is continuous on $\Gamma_{\rm wedge}$ (hence again, b_1 is different in the present case from the corresponding coefficient in Lemma 13.4.6).

Moreover, since (13.4.39) holds on $\Gamma_{\rm sym}$, we have

$$\tilde{w} = 0 \qquad \text{on } \Gamma_{\rm sym}. \tag{13.4.43}$$

Furthermore, (12.3.19) implies that $D\varphi = D\varphi_2$ on $\Gamma_{\rm sonic}$. Then, on $\Gamma_{\rm sonic}$,

$$\tilde{w} = \partial_{\xi_2}(\varphi_1 - \varphi) = \partial_{\xi_2}(\varphi_1 - \varphi_2) = (u_1 - u_2, v_2) \cdot (0,1) = -v_2.$$

Therefore, the boundary condition for w on $\Gamma_{\rm sonic}$ is

$$\tilde{w} = -v_2 \qquad \text{on } \Gamma_{\rm sonic}. \tag{13.4.44}$$

2. Next we derive the boundary condition on $\Gamma_{\rm shock}$ for $\tilde{w}$. By an argument similar to that in Step 2 of the proof of Lemma 13.4.6, we apply Lemma 13.4.5. However, we need to change the setting in the present case. To motivate the forthcoming argument, we note the following: Since we derive a boundary condition on $\Gamma_{\rm shock} = \{\varphi = \varphi_1\}$, we need to take $\varphi^- = \varphi_1$ in Lemma 13.4.5. Since we derive the condition for function $\tilde{w} = \partial_{\xi_2}(\varphi_1 - \varphi)$, we need to choose a uniform state φ^+ for which $\{\varphi_1 = \varphi^+\}$ is a vertical line $\{\xi_1 = const.\}$. Furthermore, the boundary condition needs to hold on the whole $\Gamma_{\rm shock}$ for large wedge angles and on $\Gamma_{\rm shock}$ near P_2 for all the angles, which can be seen from (12.3.12) (where the original non-shifted coordinates have been used, in which $\{P_2, P_3\}$ lie on $\xi_2 = 0$) such that $\|\varphi^+ - \varphi\|_{\Gamma_{\rm shock} \cap B_\varepsilon(P_2)}$ is small. This leads to the following definition for φ^+:

$$\begin{aligned} \phi^+(\boldsymbol{\xi}) &= \phi(P_2) + D\phi(P_2) \cdot (\boldsymbol{\xi} - \boldsymbol{\xi}_{P_2}), \\ \varphi^+(\boldsymbol{\xi}) &= \phi^+(\boldsymbol{\xi}) - \frac{|\boldsymbol{\xi}|^2}{2}, \end{aligned} \tag{13.4.45}$$

where function $\phi = \varphi + \frac{|\boldsymbol{\xi}|^2}{2}$.

From this definition, we have

$$(\varphi^+, D\varphi^+)(P_2) = (\varphi, D\varphi)(P_2). \tag{13.4.46}$$

Then, using (12.3.13) and (13.4.4), we see that, for any $\boldsymbol{\xi} \in \mathbb{R}^2$,

$$\partial_{\boldsymbol{\nu}}(\varphi_1 - \varphi^+)(\boldsymbol{\xi}) = \partial_{\boldsymbol{\nu}}(\varphi_1 - \varphi)(P_2) > \mu_1 > 0. \tag{13.4.47}$$

In particular, $S := \{\varphi_1 = \varphi^+\}$ is a line. Using (13.4.46) and $\varphi(P_2) = \varphi_1(P_2)$ (which holds since u is a fixed point of $\mathcal{I}^{(\theta_{\rm w})}$), we find that $P_2 \in S$. Furthermore, it follows from (13.4.39) and (13.4.47) that S is a vertical line $\{\xi_1 = const.\}$. Thus, we have

$$S = \{\xi_1 = \xi_{1P_2}\}, \qquad \mathbf{e}_S = (0,1).$$

Also, from (12.3.13) and (13.4.39), it follows that $\boldsymbol{\nu}_{\rm sh}(P_2) = (1,0)$. That is,

$$\boldsymbol{\nu}_S = \boldsymbol{\nu}_{\rm sh}(P_2).$$

Then, since φ and φ_1 satisfy the Rankine-Hugoniot conditions at P_2 (again from the fixed point property), it follows that φ and φ_1 satisfy the Rankine-Hugoniot condition along line S:

$$\rho^+\partial_{\xi_1}\varphi^+ = \rho_1\partial_{\xi_1}\varphi_1 \qquad \text{on } S,$$

where density ρ^+ of the uniform state φ^+ is defined by $\rho^+ = \rho(|D\varphi^+|^2, \varphi^+) > 0$ from (2.2.9). Also, from (12.3.14) applied at P_2 and (13.4.46), we have

$$\partial_{\xi_1}\varphi_1 > \partial_{\xi_1}\varphi^+ > 0 \qquad \text{on } S.$$

Then, using (13.4.47) and the Lipschitz estimate $|D\varphi| \le C$ in $\overline{\Omega}$ which follows from (12.3.18) and (13.4.24), we obtain from the Rankine-Hugoniot condition on S that

$$\rho^+ - \rho_1 = \rho_1\frac{\partial_{\xi_1}\varphi_1 - \partial_{\xi_1}\varphi^+}{\partial_{\xi_1}\varphi^+} \ge \rho_1\frac{\mu_1}{C} =: \delta.$$

Combining this with (13.4.47), we conclude that (13.4.5) holds for φ^+ and $\varphi^- = \varphi_1$.

From (13.4.46), it follows that $\rho^+ = \rho(|D\varphi(P_2)|^2, \varphi(P_2))$. Then, from Definition 12.3.2(vi),

$$\rho_{\min} < \rho^+ < \rho_{\max}. \tag{13.4.48}$$

Furthermore, by Definition 12.3.2(v) and (13.4.46), it follows that

$$\begin{aligned}|D\varphi^+(P_2)| &< c(|D\varphi^+(P_2)|^2, \varphi^+(P_2))\big(1 - \lambda\,{\rm dist}(P_2, \Gamma_{\rm sonic})\big)\\ &= c_+\big(1 - \lambda\,{\rm dist}(P_2, \Gamma_{\rm sonic})\big),\end{aligned}$$

where $c_+ = (\rho^+)^{\gamma-1}$ is the sonic speed of the uniform state φ^+. From this, denoting by $\mathcal{O}_+ := D\phi(P_2)$ which is the center of the sonic circle of the uniform state φ^+, noting that $|D\varphi^+(P_2)| = |P_2\mathcal{O}_+|$, and using (13.4.48), we have

$$c_+ - |P_2\mathcal{O}_+| \ge -\lambda(\rho_{\min})^{(\gamma-1)/2}{\rm dist}(\Gamma_{\rm sonic}, \{\xi_2 = 0\}) \ge \delta,$$

where $\delta > 0$ depends only on the data and λ. Then, using that $P_2 \in S$, we have

$$c_+ - \mathrm{dist}(S, \mathcal{O}_+) \geq c_+ - |P_2\mathcal{O}_+| \geq \delta.$$

Therefore, (13.4.6) holds.

We note that, in the coordinates centered at $\mathcal{O}_2$,

$$\varphi_2(\boldsymbol{\xi}) = -\frac{|\boldsymbol{\xi}|^2}{2} + const.,$$

so that $\phi = \varphi - \varphi_2 + const.$ Thus, $|\mathcal{O}_+| = |D\phi(P_2)| = |D(\varphi - \varphi_2)(P_2)| \leq C$ by (13.4.24). Now (13.4.7) follows from (12.3.18) with an argument as in Step 2 of the proof of Lemma 13.4.6.

Next we show that conditions (13.4.9)–(13.4.10) are satisfied. We have shown in Step 2 of the proof of Lemma 13.4.6 that $\phi = \frac{|\boldsymbol{\xi}|^2}{2} + \varphi$ satisfies equation (13.4.35) with $\mathcal{A}_{ij}$ from (12.4.45). This verifies (13.4.9).

In (13.4.10), vector $\boldsymbol{\nu}_S$ in the present case is $\boldsymbol{\nu}_{\xi_1} = (1,0)$. To prove (13.4.10), we follow the corresponding argument in Step 2 of the proof of Lemma 13.4.6, with the only difference that we now need to show that (13.4.10) holds at point P_1 with $\boldsymbol{\nu}_S = (1,0)$.

This can be seen as follows: Coefficients $\mathcal{A}_{ij}(\boldsymbol{\xi}) = \mathcal{A}_{ij}(D\psi(\boldsymbol{\xi}), \boldsymbol{\xi})$ at $P_1 = \overline{\Gamma_{\rm shock}} \cap \overline{\Gamma_{\rm sonic}}$ are of form (12.4.44), where the $\boldsymbol{\xi}$–coordinates are shifted so that the origin is at $\mathcal{O}_2$. Thus, we see that, at P_1 for $\boldsymbol{\nu}_S = (1,0)$,

$$\begin{aligned}\sum_{i,j=1}^{2} \mathcal{A}_{ij}\nu_i^S\nu_j^S &= \mathcal{A}_{11} = c_2^2 - \xi_{1P_1}^2 = c_2^2 - (P_1\mathcal{O}_2 \cdot (1,0))^2 \\ &= c_2^2\Big(1 - \frac{P_1\mathcal{O}_2}{|P_1\mathcal{O}_2|} \cdot (1,0)\Big),\end{aligned}$$

where $|P_1\mathcal{O}_2| = c_2$ has been used, since $P_1 \in \overline{\Gamma_{\rm sonic}}$. Then, as in Step 2 of the proof of Lemma 13.4.6, in order to complete the proof of (13.4.10) at P_1, we need to show that

$$1 - \frac{P_1\mathcal{O}_2}{|P_1\mathcal{O}_2|} \cdot (1,0) \geq \sigma > 0 \qquad \text{for all } \theta_{\rm w} \in [\theta_{\rm w}^{\rm s}, \frac{\pi}{2}], \tag{13.4.49}$$

where σ depends only on the data.

In order to prove (13.4.49), it suffices to show that $(1,0)$ is not parallel to $P_1\mathcal{O}_2$ for all the wedge angles $\theta_{\rm w} \in [\theta_{\rm w}^*, \frac{\pi}{2}]$. Indeed, from the continuity of the parameters of state (2), including the positions of P_1 and $\mathcal{O}_2$, with respect to $\theta_{\rm w} \in [\theta_{\rm w}^{\rm s}, \frac{\pi}{2}]$, (13.4.49) holds then with some $\sigma > 0$.

We now show that $(1,0)$ is not parallel to $P_1\mathcal{O}_2$. First let $\theta_{\rm w} = \frac{\pi}{2}$. Then $\xi_{2P_1} > 0$ by (6.2.4), so that $\frac{P_1\mathcal{O}_2}{|P_1\mathcal{O}_2|} \cdot (1,0) < 1$.

Let $\theta_{\rm w} \in (0, \frac{\pi}{2})$. Then, in the shifted coordinates, $\mathcal{O}_2 = \mathbf{0}$ and $P_0 = |P_0\mathcal{O}_2|(\cos\theta_{\rm w}, \sin\theta_{\rm w})$. Suppose, contrary to our claim, that $\frac{P_1\mathcal{O}_2}{|P_1\mathcal{O}_2|} \cdot (1,0) = 1$.

Then P_1 is a point of intersection of S_1 with line $\{\xi_2 = \xi_{2\mathcal{O}_2}\} \equiv \{\xi_2 = 0\}$. Using that $\mathbf{e}_{S_1} = -\frac{(v_2, u_1-u_2)}{\sqrt{(u_1-u_2)^2+v_2^2}}$ is a unit vector along S_1, we obtain that such a point of intersection does not exist if $u_1 = u_2$. Otherwise, we have

$$P_1 = |P_0\mathcal{O}_2|(\cos\theta_{\rm w} - \frac{v_2}{u_1-u_2}\sin\theta_{\rm w},\ 0).$$

Recall that $v_2 > 0$. Then, if $u_1 < u_2$, we find that $\xi_{1P_1} > \xi_{1P_0}$ and $\xi_{2P_1} > \xi_{2P_0}$, from which $P_1 \notin \Lambda$, which is a contradiction. The remaining case is $u_1 > u_2$. In that case, in triangle $P_0P_1\mathcal{O}_2$, $\angle P_0\mathcal{O}_2P_1 = \pi - \theta_{\rm w} > \frac{\pi}{2}$, so that $|P_0P_1| > |P_0\mathcal{O}_2|$. Then we obtain a contradiction from the elementary geometry, since segment P_0P_1 is outside the sonic circle $B_{c_2}(\mathcal{O}_2)$, by Remark 7.5.5 and $P_1 \in \partial B_{c_2}(\mathcal{O}_2)$.

Thus, we have shown that $(1, 0)$ is not parallel to $P_1\mathcal{O}_2$. This implies (13.4.49) so that (13.4.10) holds on $\Gamma_{\rm shock}$.

Combining the properties proved above, we conclude that Claim 13.4.7 holds in the present case, with constant δ depending only on the data, α, constants $(C, \hat{N}_0)$ in (12.3.22) and (13.4.24), and λ in Definition 12.3.2(v), and hence depending only on the data and $\theta_{\rm w}^*$.

Fix $\mu = \mu(\delta)$ to be the constant in Lemma 13.4.5, where δ is from the present version of Claim 13.4.7.

In Steps 3–4, we verify the remaining assumption (13.4.8) of Lemma 13.4.5 to complete the proof of Lemma 13.4.9.

3. *The case that the wedge angles are close to* $\frac{\pi}{2}$, *specifically* $\theta_{\rm w} \in [\frac{\pi}{2} - \frac{2\delta_1}{N_1^2}, \frac{\pi}{2})$. In this case, (13.4.37) holds with constant C from (13.4.24). Then we estimate

$$\|\varphi - \varphi^+\|_{C^1(\overline{\Omega})} \le \|\varphi - \varphi_2\|_{C^1(\overline{\Omega})} + \|\varphi^+ - \varphi_2\|_{C^1(\overline{\Omega})} \le C\delta_1 + \|\varphi^+ - \varphi_2\|_{C^1(\overline{\Omega})}.$$

Since (φ^+, φ_2) are uniform states, $D(\varphi^+ - \varphi_2)$ is a constant vector so that, using (13.4.46),

$$\|D(\varphi^+ - \varphi_2)\|_{L^\infty(\overline{\Omega})} = |D(\varphi^+ - \varphi_2)(P_2)| = |D(\varphi - \varphi_2)(P_2)| \le C\delta_1,$$

where the last inequality follows from (13.4.37). Thus, we have

$$\|D(\varphi - \varphi^+)\|_{C(\overline{\Omega})} \le C\delta_1.$$

Then, using that $\varphi(P_2) = \varphi^+(P_2)$ and (12.3.18), we see that $\|\varphi - \varphi^+\|_{C(\overline{\Omega})} \le C\delta_1$, which implies

$$\|\varphi - \varphi^+\|_{C^1(\overline{\Omega})} \le C\delta_1.$$

With this, we repeat the argument in Step 3 of the proof of Lemma 13.4.6 to see that (13.4.8) is satisfied on $\Gamma_{\rm shock}$, if δ_1 is small. Then, applying Lemma 13.4.5, we conclude that $\tilde{w} = \partial_{\xi_2}(\varphi_1 - \varphi)$ satisfies an oblique derivative condition of form (13.4.38) on $\Gamma_{\rm shock}$, where $\hat{a}_i \in C(\overline{\Gamma_{\rm shock}})$.

Therefore, $\tilde{w}$ is a solution of the following mixed problem in Ω: $\tilde{w}$ satisfies an equation of form (13.4.29) which is strictly elliptic in region $\overline{\Omega} \setminus \overline{\Gamma_{\rm sonic}}$ with coefficients satisfying (13.4.28) and (13.4.30), the oblique boundary conditions of form (13.4.32) and (13.4.38) with the coefficients that are continuous in the relative interiors of the corresponding boundary segments, and the Dirichlet conditions (13.4.43)–(13.4.44). Thus, the maximum of $\tilde{w}$ cannot occur in the interior of Ω and relative interiors of Γ_{shock} and Γ_{wedge}. Then the Dirichlet conditions (13.4.43)–(13.4.44) and regularity $\tilde{w} \in C(\overline{\Omega})$ imply that

$$\tilde{w} \le 0 \qquad \text{in } \Omega.$$

4. *The case that the wedge angles are away from $\frac{\pi}{2}$, specifically $\theta_{\rm w} \in [\theta_{\rm w}^*, \frac{\pi}{2} - \frac{2\delta_1}{N_1^2})$.* In this case, from Definition 12.3.2(iv),

$$\eta_2(\theta_{\rm w}) \ge \frac{\delta_1}{N_1^2} > 0.$$

Thus, by (12.3.12),

$$\tilde{w} = \partial_{\xi_2}(\varphi_1 - \varphi) < -\frac{\delta_1}{N_1^2} \qquad \text{in } \overline{\Omega} \setminus \mathcal{N}_{\varepsilon/10}(\{\xi_2 = \xi_{2P_2}\}). \tag{13.4.50}$$

It remains to consider $w = \partial_{\mathbf{e}_{\xi_2}}(\varphi_1 - \varphi)$ on $\overline{\Omega} \cap \mathcal{N}_\varepsilon(\{\xi_2 = \xi_{2P_2}\})$. From (13.4.39) and (13.4.47),

$$\boldsymbol{\nu}_{\rm sh}(P_2) = (0, 1).$$

Using (12.2.66) which holds by Remark 12.3.4, and choosing ε small, depending only on the data and $\theta_{\rm w}^*$, we have

$$\Gamma_{\text{shock}}^\varepsilon := \Gamma_{\text{shock}} \cap \mathcal{N}_\varepsilon(\{\xi_2 = \xi_{2P_2}\}) = \{\xi_1 = f(\xi_2) \,:\, \xi_{2P_2} < \xi_2 < \xi_{2P_2} + \varepsilon\},$$

where $\|f\|_{C^{1,\alpha}([\xi_{2P_2}, \xi_{2P_2}+\varepsilon])} \le C\hat{N}_0$. Then

$$|\boldsymbol{\xi} - P_2| \le C\varepsilon \qquad \text{for all } \boldsymbol{\xi} \in \Gamma_{\text{shock}}^\varepsilon.$$

Using (13.4.24) and (13.4.46), we have

$$\|D\varphi - D\varphi^+\|_{C(\overline{\Gamma_{\text{shock}}^\varepsilon})} \le \|D\varphi - D\varphi(P_2)\|_{C(\overline{\Gamma_{\text{shock}}^\varepsilon})} \le C\varepsilon^\alpha.$$

Using that $\varphi(P_2) = \varphi^+(P_2)$, we obtain that $\|\varphi - \varphi^+\|_{C(\overline{\Gamma_{\text{shock}}^\varepsilon})} \le C\varepsilon^{1+\alpha}$ so that

$$\|\varphi - \varphi^+\|_{C^1(\overline{\Gamma_{\text{shock}}^\varepsilon})} \le C\varepsilon^\alpha.$$

Then, choosing ε small and following the argument for deriving (13.4.42), we obtain that $\tilde{w} = \partial_{\xi_2}(\varphi_1 - \varphi)$ satisfies the oblique derivative condition:

$$\hat{a}_1 \tilde{w}_{\xi_1} + \hat{a}_2 \tilde{w}_{\xi_2} = 0 \qquad \text{on } \Gamma_{\text{shock}}^\varepsilon, \tag{13.4.51}$$

where $\hat{a}_i \in C(\overline{\Gamma_{\text{shock}} \cap \mathcal{D}_\varepsilon})$.

Therefore, $\tilde{w}$ is a solution of the following problem in $\Omega\cap\mathcal{N}_\varepsilon(\{\xi_2=\xi_{2P_2}\})$: $\tilde{w}$ satisfies an equation of form (13.4.29) which is strictly elliptic in $\Omega\cap\mathcal{N}_\varepsilon(\{\xi_2=\xi_{2P_2}\})$, the oblique boundary conditions (13.4.32) and (13.4.51) with continuous coefficients on $\Gamma_{\rm wedge}\cap\mathcal{N}_\varepsilon(\{\xi_2=\xi_{2P_2}\})$ and $\Gamma^\varepsilon_{\rm shock}$, and the Dirichlet condition (13.4.43) on $\Gamma_{\rm sym}$. Furthermore, (12.3.12) implies that $\tilde{w}<0$ on $\Omega\cap\{\xi_2=\xi_{2P_2}+\varepsilon\}$. Then, by the maximum principle, $\tilde{w}\le 0$ in $\Omega\cap\mathcal{N}_\varepsilon(\{\xi_2=\xi_{2P_2}\})$.

Combining this with (12.3.12), we conclude that $\tilde{w}\le 0$ in Ω. □

13.4.3 Removing the cutoff in the equation for fixed points

In this section, we show that the fixed point satisfies the non-modified potential flow equation (1.5). As we discussed above, it suffices to prove (13.4.3), which will be done in the next two lemmas.

First, we bound ψ_x from above. We work in the (x,y)–coordinates in $\Omega\cap\mathcal{D}_{2\varepsilon}$. By Lemma 12.2.7(iv),

$$\Omega\cap\mathcal{D}_\varepsilon=\{0<x<\varepsilon,\ 0<y<f_{\rm sh}(x)\}\qquad\text{for all } \varepsilon\in(0,\varepsilon_0), \tag{13.4.52}$$

where $f_{\rm sh}$ satisfies (12.3.17), and ε_0 is determined in Lemma 12.2.2.

Lemma 13.4.10. *If the parameters in Definition* 12.3.2 *are chosen so that the conditions of Proposition* 12.7.10 *are satisfied, and if ε is further reduced depending only on the data and constant C in* (12.7.29) (*hence depending only on the data and $\theta^*_{\rm w}$*), then, for every $(u,\theta_{\rm w})\in\overline{\mathcal{K}}$ such that $\theta_{\rm w}\in[\theta^*_{\rm w},\frac{\pi}{2})$ and u is a fixed point of map $\mathcal{I}^{(\theta_{\rm w})}$, the corresponding function $\psi=\varphi-\varphi_2$ satisfies*

$$\psi_x\le\frac{2-\frac{\mu_0}{5}}{1+\gamma}x\qquad in\ \ \Omega\cap\{x\le\frac{\varepsilon}{4}\}. \tag{13.4.53}$$

Proof. In this proof, the universal constants C and ε are positive and depend only on the data and constant C in (12.7.29) (hence on the data and $\theta^*_{\rm w}$). We divide the proof into three steps.

1. To simplify notation, we denote

$$A=\frac{2-\frac{\mu_0}{5}}{1+\gamma},\qquad \Omega_s:=\Omega\cap\{x\le s\}\ \text{ for } s>0.$$

Define a function

$$v(x,y):=Ax-\psi_x(x,y)\qquad\text{on } \Omega_{\varepsilon/4}. \tag{13.4.54}$$

From (13.4.4) and since $\psi=\hat{\psi}$ satisfies (12.7.29), it follows that

$$v\in C^{0,1}(\overline{\Omega_{\varepsilon/4}})\cap C^1(\overline{\Omega_{\varepsilon/4}}\setminus\{x=0\})\cap C^2(\Omega_{\varepsilon/4}). \tag{13.4.55}$$

Moreover, it follows from (12.7.29) for $\psi = \hat{\psi}$ that, in $\Omega_{\varepsilon/4}$,

$$
\begin{aligned}
&|\psi(x,y)| \le Cx^2, \quad |\psi_x(x,y)| \le Cx, \quad |\psi_y(x,y)| \le Cx^{\frac{3}{2}},\\
&|\psi_{xx}(x,y)| \le C, \quad |\psi_{xy}(x,y)| \le C\sqrt{x}, \quad |\psi_{yy}(x,y)| \le Cx.
\end{aligned}
\tag{13.4.56}
$$

This implies that

$$
v = 0 \qquad \text{on } \partial\Omega_{\varepsilon/4} \cap \{x = 0\}. \tag{13.4.57}
$$

We now use the fact that ψ satisfies (12.3.26), which can be written as (12.5.34) in the (x,y)–coordinates, where $\hat{\mathcal{M}}$ is of the form of $\hat{\mathcal{M}}_{(u,\theta_{\mathrm{w}})}$ from Lemma 12.5.7(x). Then, from (12.5.55)–(12.5.56) and (13.4.56),

$$
|\psi_x| \le C(|\psi_y| + |\psi|) \le Cx^{\frac{3}{2}} \qquad \text{on } \Gamma_{\text{shock}} \cap \{x < \frac{\varepsilon}{4}\},
$$

so that, choosing ε small,

$$
|\psi_x| < Ax \qquad \text{on } \Gamma_{\text{shock}} \cap \{0 < x < \frac{\varepsilon}{4}\}.
$$

Thus, we have

$$
v \ge 0 \qquad \text{on } \Gamma_{\text{shock}} \cap \{0 < x < \frac{\varepsilon}{4}\}. \tag{13.4.58}
$$

Furthermore, condition (12.3.28) on Γ_{wedge} in the (x,y)–coordinates is

$$
\psi_y = 0 \qquad \text{on } \{0 < x < \frac{\varepsilon}{4},\, y = 0\}.
$$

Since (13.4.4) implies that ψ is C^2 up to Γ_{wedge}, we differentiate the condition on Γ_{wedge} with respect to x, *i.e.*, in the tangential direction to Γ_{wedge}, find that $\psi_{xy} = 0$ on $\{0 < x < \frac{\varepsilon}{4},\, y = 0\}$, which implies that

$$
v_y = 0 \qquad \text{on } \Gamma_{\text{wedge}} \cap \{0 < x < \frac{\varepsilon}{4}\}. \tag{13.4.59}
$$

Furthermore, from (12.3.7) and recalling that $A = \frac{2-\frac{\mu_0}{5}}{1+\gamma}$, we obtain that, on $\partial\Omega_{\varepsilon/4} \cap \{x = \frac{\varepsilon}{4}\}$,

$$
v = \frac{A\varepsilon}{4} - \psi_x \ge \frac{A\varepsilon}{4} - \frac{2-\mu_0}{1+\gamma}\frac{\varepsilon}{4} = \frac{\mu_0}{5(1+\gamma)}\varepsilon > 0,
$$

which implies

$$
v > 0 \qquad \text{on } \partial\Omega_{\varepsilon/4} \cap \{x = \frac{\varepsilon}{4}\}. \tag{13.4.60}
$$

2. Now we show that, if ε is small, v is a supersolution of a linear homogeneous elliptic equation on $\Omega_{\varepsilon/2}$. Since, by Lemma 13.4.2, ψ satisfies equation (12.4.31) in $\Omega_{\varepsilon/4}$, where $O_k = O_k(D\psi, \psi, x)$ with $O_k(\mathbf{p}, z, x)$ given by (11.1.5), we differentiate the equation with respect to x and use the regularity of ψ in (13.4.4) and the definition of v in (13.4.54) to obtain

$$\begin{aligned} &a_{11}v_{xx} + a_{12}v_{xy} + a_{22}v_{yy} \\ &+ (A - v_x)\Big(- 1 + (\gamma+1)\big(\zeta_1(A - \frac{v}{x}) + \zeta_1'(A - \frac{v}{x})(\frac{v}{x} - v_x)\big)\Big) \\ &= E(x,y), \end{aligned} \tag{13.4.61}$$

where

$$a_{11} = 2x - (\gamma+1)x\zeta_1\big(\frac{\psi_x}{x}\big) + \hat{O}_1, \quad a_{12} = \hat{O}_2, \quad a_{22} = \frac{1}{c_2} + \hat{O}_3, \tag{13.4.62}$$

$$\begin{aligned} E(x,y) = {} & \psi_{xx}\partial_x\hat{O}_1 + \psi_{xy}\partial_x\hat{O}_2 + \psi_{yy}\partial_x\hat{O}_3 - \psi_{xx}\hat{O}_4 - \psi_x\partial_x\hat{O}_4 \\ & + \psi_{xy}\hat{O}_5 + \psi_y\partial_x\hat{O}_5, \end{aligned} \tag{13.4.63}$$

with

$$\hat{O}_k(x,y) = O_k(D\psi(x,y), \psi(x,y), x) \qquad \text{for } \ k = 1, \dots, 5, \tag{13.4.64}$$

for $O_k(\mathbf{p}, z, x)$ given by (11.1.5). From these explicit expressions, we employ (13.4.56) to obtain

$$|\hat{O}_1(x,y)| \le Cx^2, \qquad |\hat{O}_k(x,y)| \le Cx \quad \text{for } k = 2, \dots, 5. \tag{13.4.65}$$

From (12.4.10), we have

$$\zeta_1(A) = A.$$

Thus, we can rewrite (13.4.61) in the form:

$$a_{11}v_{xx} + a_{12}v_{xy} + a_{22}v_{yy} + bv_x + cv = -A\big((\gamma+1)A - 1\big) + E(x,y), \tag{13.4.66}$$

with

$$b(x,y) = 1 - (\gamma+1)\big(\zeta_1(A - \frac{v}{x}) + \zeta_1'(A - \frac{v}{x})(\frac{v}{x} - v_x + A)\big), \tag{13.4.67}$$

$$c(x,y) = (\gamma+1)\frac{A}{x}\Big(\zeta_1'(A - \frac{v}{x}) - \int_0^1 \zeta_1'(A - s\frac{v}{x})ds\Big), \tag{13.4.68}$$

where v and v_x are evaluated at point (x,y).

Since ψ satisfies (13.4.4) and v is defined by (13.4.54), we have

$$a_{ij}, b, c \in C\big(\overline{\Omega_{\varepsilon/4}} \setminus \{x = 0\}\big).$$

Combining (13.4.62) with (12.4.10) and (13.4.65), we obtain that, for sufficiently small ε depending only on the data and $\theta_{\rm w}^*$,

$$a_{11} \ge \frac{1}{6}x, \quad a_{22} \ge \frac{1}{2c_2}, \quad |a_{12}| \le \frac{1}{3\sqrt{c_2}}\sqrt{x} \qquad \text{on } \Omega_{\varepsilon/4}.$$

Therefore, $4a_{11}a_{22} - (a_{12})^2 \geq \frac{2}{9c_2}x$ on Ω_ε, which implies that equation (13.4.66) is elliptic on Ω_ε and uniformly elliptic on every compact subset of $\overline{\Omega_{\varepsilon/4}}\setminus\{x=0\}$.

Furthermore, using (12.4.13) and (13.4.68), and noting that $A > 0$ and $x > 0$, we have

$$c(x,y) \leq 0 \qquad \text{for every } (x,y) \in \Omega_\varepsilon \text{ such that } v(x,y) \leq 0. \tag{13.4.69}$$

Now we estimate $E(x,y)$. Using (13.4.64) with (11.1.5) and (12.7.29) for $\psi = \hat{\psi}$, we find that, on Ω_ε,

$$\begin{aligned}
|\partial_x \hat{O}_1| &\leq C\big(x + |D\psi| + x|\psi_{xx}| + |\psi_x\psi_{xx}| + |\psi_y\psi_{xy}| + |D\psi|^2\big) \leq Cx,\\
|\partial_x \hat{O}_{2,5}| &\leq C\big(|D\psi| + |D\psi|^2 + |\psi_y\psi_{xx}| + (1+|\psi_x|)|\psi_{xy}|\big) \leq C\sqrt{x},\\
|\partial_x \hat{O}_{3,4}| &\leq C\big(1 + |\psi| + |D\psi| + (1+|D\psi|)|D^2\psi| + |D\psi|^2\big) \leq C.
\end{aligned}$$

Combining these estimates with (13.4.56) and (13.4.65), we obtain from (13.4.63) that

$$|E(x,y)| \leq Cx \qquad \text{on } \Omega_{\varepsilon/4}.$$

From this and $(\gamma+1)A > 1$, we conclude that the right-hand side of (13.4.66) is strictly negative in Ω_ε if ε is sufficiently small, depending only on the data and θ^*_{w}.

We fix ε satisfying all the requirements above. Then

$$a_{11}v_{xx} + a_{12}v_{xx} + a_{22}v_{yy} + bv_x + cv < 0 \qquad \text{on } \Omega_{\varepsilon/4}, \tag{13.4.70}$$

the equation is elliptic in $\Omega_{\varepsilon/4}$ and uniformly elliptic on compact subsets of $\overline{\Omega_{\varepsilon/4}} \setminus \{x=0\}$, and (13.4.69) holds.

3. Moreover, v satisfies (13.4.55) and the boundary conditions (13.4.57)–(13.4.60). Then it follows that

$$v \geq 0 \qquad \text{on } \Omega_{\varepsilon/4}.$$

Indeed, let $z_0 := (x_0, y_0) \in \overline{\Omega_{\varepsilon/4}}$ be a minimum point of v over $\overline{\Omega_{\varepsilon/4}}$ and $v(z_0) < 0$. Then, by (13.4.57)–(13.4.58) and (13.4.60), either z_0 is an interior point of $\Omega_{\varepsilon/4}$ or $z_0 \in \Gamma_{\text{wedge}} \cap \{0 < x < \frac{\varepsilon}{4}\}$. If z_0 is an interior point of $\Omega_{\varepsilon/4}$, (13.4.70) is violated since (13.4.70) is elliptic, $v(z_0) < 0$, and $c(z_0) \leq 0$ by (13.4.69). Thus, the only possibility is $z_0 \in \Gamma_{\text{wedge}} \cap \{0 < x < \frac{\varepsilon}{4}\}$, *i.e.*, $z_0 = (x_0, 0)$ with $x_0 > 0$. Then, by (13.4.52), there exists $\rho > 0$ such that $B_\rho(z_0) \cap \Omega_{\varepsilon/4} = B_\rho(z_0) \cap \{y > 0\}$. Equation (13.4.70) is uniformly elliptic in $\overline{B_{\rho/2}(z_0) \cap \{y \geq 0\}}$, with coefficients $a_{ij}, b, c \in C(\overline{B_{\rho/2}(z_0) \cap \{y \geq 0\}})$. Since $v(z_0) < 0$ and v satisfies (13.4.55), then, reducing $\rho > 0$ if necessary, we see that $v < 0$ in $B_\rho(z_0) \cap \{y > 0\}$. Then, by (13.4.69),

$$c \leq 0 \qquad \text{on } B_\rho(z_0) \cap \{y > 0\}.$$

Moreover, $v(x,y)$ is not a constant in $\overline{B_{x_0/2}(x_0) \cap \{y \geq 0\}}$, since its negative minimum is achieved at $(x_0, 0)$ and cannot be achieved in any interior point, as shown above. Thus, $\partial_y v(z_0) > 0$ by Hopf's lemma, which contradicts (13.4.59). Therefore, $v \geq 0$ on $\Omega_{\varepsilon/4}$ so that (13.4.53) holds on $\Omega_{\varepsilon/4}$.

Then, using (12.3.7), we obtain (13.4.53) on $\Omega_{\varepsilon/4}$. □

Now we prove the bound of ψ_x from below.

Lemma 13.4.11. *If the parameters in Definition* 12.3.2 *are chosen so that the conditions of Proposition* 12.7.10 *and Lemma* 13.4.6 *are satisfied, and if* ε *is further reduced depending only on the data and constant* C *in* (12.7.29) *(hence depending only on the data and* $\theta_{\rm w}^*$*), then, for every* $(u, \theta_{\rm w}) \in \overline{\mathcal{K}}$ *such that* $\theta_{\rm w} \in [\theta_{\rm w}^*, \frac{\pi}{2})$ *and* u *is a fixed point of map* $\mathcal{I}^{(\theta_{\rm w})}$*, the corresponding function* $\psi = \varphi - \varphi_2$ *satisfies*

$$\psi_x \geq -\frac{2 - \frac{\mu_0}{5}}{1+\gamma} x \qquad \text{in } \Omega \cap \mathcal{D}_{\varepsilon/4}. \tag{13.4.71}$$

Proof. Since $S_1 = \{\varphi_1 = \varphi_2\}$, then $\partial_{\mathbf{e}_{S_1}} \varphi_1 = \partial_{\mathbf{e}_{S_1}} \varphi_2$ so that $\partial_{\mathbf{e}_{S_1}}(\varphi - \varphi_2) = -\partial_{\mathbf{e}_{S_1}}(\varphi_1 - \varphi)$. Thus, using Lemma 13.4.6,

$$\partial_{\mathbf{e}_{S_1}} \psi \geq 0 \qquad \text{in } \Omega. \tag{13.4.72}$$

At point $P \neq \mathcal{O}_2$ in the $\boldsymbol{\xi}$–plane, let $\{\mathbf{e}_x, \mathbf{e}_y\} = \{\mathbf{e}_x(P), \mathbf{e}_y(P)\}$ be unit vectors in the x– and y–directions, respectively. Then, using the polar coordinates (r, θ) with respect to $\mathcal{O}_2$, defined by (11.1.1), we have

$$\mathbf{e}_x = -\mathbf{e}_r = -(\cos\theta, \sin\theta), \qquad \mathbf{e}_y = \mathbf{e}_\theta = (-\sin\theta, \cos\theta).$$

In particular, $\mathbf{e}_x(P)$ and $\mathbf{e}_y(P)$ are orthogonal to each other so that

$$\partial_{\mathbf{e}_{S_1}} \psi = (\mathbf{e}_{S_1} \cdot \mathbf{e}_x)\psi_x + (\mathbf{e}_{S_1} \cdot \mathbf{e}_y)\psi_y \qquad \text{in } \Omega \cap \mathcal{D}_{\varepsilon/4}.$$

From (11.2.5),

$$\frac{2}{M} \leq -\partial_y(\varphi_1 - \varphi_2) \leq \frac{M}{2},$$

where M depends only on the data. Then, denoting

$$a := |D(\varphi_1 - \varphi_2)| = \sqrt{(u_1 - u_2)^2 + v_2^2} > 0,$$

and noting that $D(\varphi_1 - \varphi_2) = (u_1 - u_2, -v_2) = a\boldsymbol{\nu}_{S_1}$, we have

$$0 < \frac{2}{M} \leq -a\boldsymbol{\nu}_{S_1} \cdot \mathbf{e}_y \leq \frac{M}{2}.$$

Rotating vectors $\{\boldsymbol{\nu}_{S_1}, \mathbf{e}_y\}$ by $\frac{\pi}{2}$ clockwise, we obtain $\{\mathbf{e}_{S_1}, -\mathbf{e}_x\}$, where we have used the definition of $\mathbf{e}_{S_1}$ in (7.5.8). Then we have

$$0 < \frac{2}{M} \leq a\mathbf{e}_{S_1} \cdot \mathbf{e}_x \leq \frac{M}{2} \qquad \text{in } \Omega \cap \mathcal{D}_{\varepsilon/4}.$$

Thus, using (13.4.72) and (12.7.29) for $\psi = \hat{\psi}$ (where constant C is independent of ε), we calculate in $\Omega \cap \mathcal{D}_{\varepsilon/4}$:

$$\psi_x = \frac{\partial_{\mathbf{e}_{\mathcal{S}_1}} \psi}{\mathbf{e}_{\mathcal{S}_1} \cdot \mathbf{e}_x} - \frac{\mathbf{e}_{\mathcal{S}_1} \cdot \mathbf{e}_y}{\mathbf{e}_{\mathcal{S}_1} \cdot \mathbf{e}_x} \psi_y \geq -\frac{\mathbf{e}_{\mathcal{S}_1} \cdot \mathbf{e}_y}{\mathbf{e}_{\mathcal{S}_1} \cdot \mathbf{e}_x} \psi_y \geq -\frac{aM}{2} C x^{\frac{3}{2}} \geq -\frac{2 - \frac{\mu_0}{5}}{1+\gamma} x,$$

if $\varepsilon > 0$ is small, where we have used that $x \in (0, \frac{\varepsilon}{4})$. □

Then the following corollary follows directly from Lemmas 13.4.10–13.4.11.

Corollary 13.4.12. *Under the conditions of Lemmas* 13.4.10–13.4.11, *estimate* (13.4.3) *holds.*

Now we prove

Corollary 13.4.13. *Under the conditions of Lemmas* 13.4.10–13.4.11, *for every* $(u, \theta_{\rm w}) \in \overline{\mathcal{K}}$ *such that* u *is a fixed point of map* $\mathcal{I}^{(\theta_{\rm w})}$, *the corresponding function* φ *satisfies*

$$\varphi \leq \varphi_1 \qquad in\ \Omega.$$

Proof. Recall that, as the fixed point solution, φ satisfies the two Rankine-Hugoniot conditions with φ_1 on $\Gamma_{\rm shock}$: (12.5.3) and $\varphi = \varphi_1$ on $\Gamma_{\rm shock}$.

By Corollary 13.4.12 and Lemma 12.4.7, φ is a solution of the potential flow equation (2.2.8) with (2.2.9) in Ω.

Moreover, $\varphi_{\boldsymbol{\nu}} = 0$ on $\Gamma_{\rm wedge} \cup \Gamma_{\rm sym}$, since (12.3.28) holds for $\hat{\psi} = \psi$. Also, $\varphi \in C^{1,\alpha}(\overline{\Omega}) \cap C^2(\Omega)$ by (12.3.19).

Now the corollary follows by repeating the proof of Lemma 8.3.2. □

13.4.4 Completing the proof of Proposition 13.4.1

Note that

- (13.4.1) follows from Lemma 13.4.4 and Corollary 13.4.13;
- (13.4.2) follows from Lemmas 13.4.6 and 13.4.9;
- (13.4.3) follows from Corollary 13.4.12.

Then, from the argument after (13.4.3), the proof of Proposition 13.4.1 is completed.

Remark 13.4.14. *At this point, all the parameters in the iteration set, except* $(N_1, \delta_2, \delta_3)$, *are fixed. Also, as we stated at the beginning of this section, we assume that the parameters of the iteration set satisfy the requirements of Proposition* 12.7.10. *In particular,* δ_3 *is chosen small, depending on the data and* $(\theta^*_{\rm w}, \delta_2)$. *In Lemma* 13.5.1 *below, we will choose* (N_1, δ_2) *depending only on the data and* $\theta^*_{\rm w}$. *From the argument above, this will also fix the choice of* δ_3, *so that the choice of all the parameters of the iteration set will be completed, depending only on the data and* $\theta^*_{\rm w}$.

13.5 FIXED POINTS CANNOT LIE ON THE BOUNDARY OF THE ITERATION SET

We assume that the parameters of the iteration set satisfy the conditions of Propositions 12.7.10, 12.8.2, and 13.4.1 with $\alpha = \frac{\hat{\alpha}}{2}$. Then we show that the fixed points of the iteration map are admissible solutions of **Problem 2.6.1** in the sense of Definition 8.1.1.

In the following lemma, $\partial\mathcal{K}$ denotes the boundary of $\mathcal{K}$ relative to space $C^{2,\alpha}_{*,1+\delta^*}(Q^{\rm iter}) \times [\theta_{\rm w}^*, \frac{\pi}{2}]$. In particular, $(u^{(\rm norm)}, \frac{\pi}{2})$ lies in the interior of $\mathcal{K}$, and the same for $(u, \theta_{\rm w}^*) \in \mathcal{K}$.

Lemma 13.5.1. *If the parameters in Definition* 12.3.2 *are chosen to satisfy the conditions of Proposition* 13.4.1, *and if* N_1 *is further increased – depending only on the data and* δ_1, *and if* δ_2 *is chosen sufficiently small – depending only on* $(\varepsilon, \delta_1, N_1)$ *(hence* (N_1, δ_2) *depend only on the data and* $\theta_{\rm w}^*$*), then, for any* $\theta_{\rm w} \in [\theta_{\rm w}^*, \frac{\pi}{2}]$,

$$\mathcal{I}(u, \theta_{\rm w}) \neq u \qquad \text{for all } (u, \theta_{\rm w}) \in \partial\mathcal{K}.$$

Proof. We divide the proof into three steps.

1. Since $\mathcal{K}$ is relatively open in $C^{2,\alpha}_{*,1+\delta^*}(Q^{\rm iter}) \times [\theta_{\rm w}^*, \frac{\pi}{2}]$, it suffices to prove that, if $u \in \overline{\mathcal{K}}(\theta_{\rm w})$ is a fixed point of $\mathcal{I}^{(\theta_{\rm w})} : \overline{\mathcal{K}}(\theta_{\rm w}) \mapsto C^{2,\alpha}_{*,1+\delta^*}(Q^{\rm iter})$, then $(u, \theta_{\rm w}) \in \mathcal{K}$.

Fix $(u, \theta_{\rm w}) \in \overline{\mathcal{K}}$ such that u is a fixed point of $\mathcal{I}^{(\theta_{\rm w})}$. By Lemma 12.3.10(i), $(u, \theta_{\rm w}) \in \mathfrak{S}$. Let (Ω, φ, ψ) be determined by $(u, \theta_{\rm w})$. By Proposition 13.4.1, φ is an admissible solution of **Problem 2.6.1** with the wedge angle $\theta_{\rm w}$.

Thus, we need to show that all the conditions of Definition 12.3.2 are satisfied (with the strict inequalities in the estimates) for $(u, \theta_{\rm w})$.

By Lemma 12.3.10(i), $(u, \theta_{\rm w})$ satisfies Definition 12.3.2(ii). Also, combining Corollary 12.7.6 with Lemma 13.1.3, we find that $u = \hat{u}$ for $\hat{u}$ from (12.3.15), so that the left-hand side of (12.3.16) vanishes, which implies that Definition 12.3.2(vii) is satisfied with the strict inequality in (12.3.16).

With this, it remains to show that conditions (i) and (iii)–(vi) of Definition 12.3.2 are satisfied (with the strict inequalities in the estimates) for $(u, \theta_{\rm w})$ and the corresponding (Ω, φ, ψ), if the parameters of the iteration set are chosen as stated. We do this in what remains of the proof.

2. By Lemma 12.6.2, there exists a large N_1 such that, for any $(u, \theta_{\rm w})$ corresponding to an admissible solution φ with $0 < \frac{\pi}{2} - \theta_{\rm w} < \frac{3\delta_1}{N_1}$,

$$\|u - u^{(\rm norm)}\|^{*,(1+\delta^*)}_{2,\alpha,Q^{\rm iter}} < \frac{\delta_1}{2}.$$

Then, from the definition of $\eta_1(\cdot)$ in Definition 12.3.2(i), it follows that (12.3.3) holds strictly for any $(u, \theta_{\rm w})$ corresponding to an admissible solution φ, for $\theta_{\rm w} \in (\frac{\pi}{2} - \frac{3\delta_1}{N_1}, \frac{\pi}{2}]$.

On the other hand, for $\theta_{\rm w} \le \frac{\pi}{2} - \frac{2\delta_1}{N_1}$, $\eta_1(\theta_{\rm w}) = 10M$ so that (12.3.3) holds strictly for each $(u, \theta_{\rm w})$ corresponding to an admissible solution. This follows from the choice of M in Definition 12.3.2(i).

Therefore, condition (i) of Definition 12.3.2 holds strictly for $(u, \theta_{\rm w})$.

Now we also fix N_1 determined in the argument above. Then N_1 depends only on the data and δ_1.

Condition (iii) of Definition 12.3.2 holds strictly for admissible solutions from Proposition 9.5.6 and estimate (12.2.31) in Lemma 12.2.3, where we have used the choice of the constants in (12.3.4)–(12.3.5).

Now we consider condition (iv) of Definition 12.3.2. We first show that (12.3.6) holds strictly for admissible solutions if δ_2 in the definition of $\eta_2(\cdot)$ is chosen small, depending on the data and (δ_1, N_1) (which are already fixed). Note that

$$\eta_2(\theta_{\rm w}) < 0 \qquad \text{if } \theta_{\rm w} > \frac{\pi}{2} - \frac{\delta_1}{N_1^2}.$$

From this, since $\varphi \ge \varphi_2$ in Ω for admissible solutions, we obtain that (12.3.6) holds strictly for admissible solutions with $\theta_{\rm w} > \frac{\pi}{2} - \frac{\delta_1}{N_1^2}$. Now, from (11.6.2) of Corollary 11.6.2 (applied with ε equal to $\frac{\varepsilon}{10}$ for parameter ε in the iteration set, and with $\sigma = \frac{\delta_1}{N_1^2}$), it follows that there exists $\delta_2 \in (0,1)$ such that, if $\theta_{\rm w} \in [\theta_{\rm w}^*, \frac{\pi}{2} - \frac{\delta_1}{N_1^2}]$,

$$\varphi - \varphi_2 \ge \frac{2\delta_1\delta_2}{N_1^2} \qquad \text{in } \overline{\Omega} \setminus \mathcal{D}_{\varepsilon/10}.$$

Then, noting that $\eta_2(\theta_{\rm w}) \le \frac{\delta_1\delta_2}{N_1^2}$ for all $\theta_{\rm w}$, we find that the strict inequality in (12.3.6) holds for any admissible solution.

3. In the proofs of the next inequalities, we employ (12.3.1).

The strict inequality (12.3.7) follows from estimates (11.2.21) in Lemma 11.2.5 and the choice of $\mu_0 = \frac{\delta}{2}$.

The strict inequalities (12.3.8)–(12.3.9) follow from estimates (11.2.23)–(11.2.24) in Lemma 11.2.6 with constant C and the choice of $N_3 = 10\max\{C, 1\}$.

The strict inequality in (12.3.10) follows from Corollary 9.1.3.

The strict inequalities (12.3.11)–(12.3.12) are proved similarly to (12.3.6), by using (11.6.3)–(11.6.4) in Corollary 11.6.2, and possibly further reducing δ_2 depending only on (ε, δ_1).

The strict inequalities in (12.3.13)–(12.3.14) hold for admissible solutions from the choice of constant μ_1 in Definition 12.3.2(iv). Now the strict inequalities in all the conditions of Definition 12.3.2(iv) are proved.

The strict inequalities in conditions (v)–(vi) of Definition 12.3.2 for admissible solutions follow from the choice of constants (λ, a, C) described there.

Therefore, we have proved that all the conditions of Definition 12.3.2 are satisfied (with the strict inequalities in the estimates). This completes the proof. □

13.6 PROOF OF THE EXISTENCE OF SOLUTIONS UP TO THE SONIC ANGLE OR THE CRITICAL ANGLE

We note, by Remark 13.4.14, that all the parameters of the iteration set are now fixed.

From Proposition 13.3.1 and Theorem 3.4.7(i), it follows that the fixed point index of map $\mathcal{I}^{(\frac{\pi}{2})}$ on set $\overline{\mathcal{K}}(\frac{\pi}{2})$ is nonzero, more precisely,

$$\mathbf{Ind}(\mathcal{I}^{(\frac{\pi}{2})}, \overline{\mathcal{K}}(\frac{\pi}{2})) = 1.$$

From Lemma 12.8.2, Corollary 13.2.3, and Lemmas 13.2.2 and 13.5.1, we see that the conditions of Theorem 3.4.8 are satisfied so that

$$\mathbf{Ind}(\mathcal{I}^{(\theta_{\rm w})}, \overline{\mathcal{K}}(\theta_{\rm w})) = 1 \qquad \text{for any } \theta_{\rm w} \in [\theta_{\rm w}^*, \frac{\pi}{2}].$$

Then, from Theorem 3.4.7(ii), a fixed point of map $\mathcal{I}^{(\theta_{\rm w})}$ on domain $\overline{\mathcal{K}}(\theta_{\rm w})$ exists for any $\theta_{\rm w} \in [\theta_{\rm w}^*, \frac{\pi}{2}]$. By Proposition 13.4.1, the fixed points are admissible solutions. Thus, the admissible solutions exist for all $\theta_{\rm w} \in [\theta_{\rm w}^*, \frac{\pi}{2}]$. This holds for any $\theta_{\rm w}^* \in (\max\{\theta_{\rm w}^{\rm s}, \theta_{\rm w}^{\rm c}\}, \frac{\pi}{2})$. Then Proposition 12.1.1 is proved.

13.7 PROOF OF THEOREM 2.6.2: EXISTENCE OF GLOBAL SOLUTIONS UP TO THE SONIC ANGLE WHEN $u_1 \leq c_1$

The existence of solution φ of **Problem 2.6.1** in Theorem 2.6.3 follows from Proposition 12.1.1 and Lemma 10.4.2(ii). The regularity of these solutions asserted follows from Corollary 11.4.7 and Proposition 11.5.1.

It remains to show that $(\rho, \Phi)(t, \mathbf{x})$ defined in Theorem 2.6.3 is a weak solution of **Problem 2.2.1** in the sense of Definition 2.3.1.

We first note that, since φ is an admissible solution, it follows from conditions (i)–(ii) of Definition 8.1.1 that φ is a weak solution of **Problem 2.2.3** in the sense of Definition 2.3.3 and that, defining $\phi(\boldsymbol{\xi}) = \varphi(\boldsymbol{\xi}) + \frac{|\boldsymbol{\xi}|^2}{2}$ and extending it by even reflection across the ξ_1–axis to $\mathbb{R}^2 \setminus W$, we have

$$\begin{aligned} &\|D\phi\|_{L^\infty(\mathbb{R}^2 \setminus W)} \leq C, \\ &\phi = \begin{cases} \phi_0 & \text{for } \xi_1 > \xi_1^0, \\ \phi_1 & \text{for } \xi_1 < \xi_1^0 \end{cases} \qquad \text{in } \mathbb{R}^2 \setminus (W \cup B_R(\mathbf{0})) \end{aligned} \tag{13.7.1}$$

for some $R, C > 0$, where $\phi_k(\boldsymbol{\xi}) = \varphi_k(\boldsymbol{\xi}) + \frac{|\boldsymbol{\xi}|^2}{2}$ for $k = 0, 1$, and φ_k defined by (2.2.16)–(2.2.17). Expressing (2.2.9) in terms of ϕ, we have

$$\rho(D\phi, \phi, \boldsymbol{\xi}) = (\rho_0^{\gamma-1} - (\gamma - 1)(\phi - \boldsymbol{\xi} \cdot D\phi + \frac{1}{2}|D\phi|^2))^{\frac{1}{\gamma-1}}. \tag{13.7.2}$$

Furthermore, expressing the weak form of equation in Definition 2.3.3(iii) in terms of ϕ, and recalling that ϕ is extended by even reflection across the ξ_1–axis to $\mathbb{R}^2 \setminus W$, we see that, for every $\zeta \in C_c^\infty(\mathbb{R}^2)$,

$$\int_{\mathbb{R}^2\setminus W} \rho(D\phi, \phi, \boldsymbol{\xi})\Big((D\phi - \boldsymbol{\xi}) \cdot D\zeta - 2\zeta\Big)\, d\boldsymbol{\xi} = 0. \tag{13.7.3}$$

Now expressing $\Phi(t, \mathbf{x})$ defined in Theorem 2.6.3 in terms of ϕ, we have

$$\Phi(t, \mathbf{x}) = t\phi(\frac{\mathbf{x}}{t}). \tag{13.7.4}$$

Then we calculate

$$\nabla_{\mathbf{x}}\Phi(t, \mathbf{x}) = D\phi(\boldsymbol{\xi}), \quad \partial_t\Phi(t, \mathbf{x}) = \phi(\boldsymbol{\xi}) - \boldsymbol{\xi} \cdot D\phi(\boldsymbol{\xi}) \qquad \text{for } \boldsymbol{\xi} = \frac{\mathbf{x}}{t}. \tag{13.7.5}$$

Then, from (13.7.1),

$$\|\nabla_{\mathbf{x}}\Phi\|_{L^\infty([0,\infty)\times(\mathbb{R}^2\setminus W))} \le C. \tag{13.7.6}$$

Also, using (13.7.5) and comparing $\rho(t, \mathbf{x})$ defined in Theorem 2.6.3 with (13.7.2), we see that they coincide. Then

$$\rho(t, \mathbf{x}) = \rho(\frac{\mathbf{x}}{t}), \quad \frac{1}{C} \le \rho(t, \mathbf{x}) \le C \qquad \text{on } [0, \infty) \times (\mathbb{R}^2 \setminus W), \tag{13.7.7}$$

where the last property follows, since φ has been obtained from a fixed point of the iteration map so that condition (vi) of Definition 12.3.2 holds.

Since $\varphi(\boldsymbol{\xi}) = \phi(\boldsymbol{\xi}) - \frac{|\boldsymbol{\xi}|^2}{2}$ satisfies condition (i) of Definition 2.3.3, then we use (13.7.5) and recall (1.2) and $B_0 = \frac{\rho_0^{\gamma-1}-1}{\gamma-1}$ to see that condition (i) of Definition 2.3.1 holds for Φ. From (13.7.6)–(13.7.7), we find that condition (ii) of Definition 2.3.1 holds.

It remains to show that condition (iii) of Definition 2.3.1 holds. Let $\zeta \in C_c^\infty(\overline{\mathbb{R}_+} \times \mathbb{R}^2)$. Then $\operatorname{supp}\zeta \in [0, S] \times B_S$ for some $S > 0$, where $B_S = B_S(\mathbf{0}) \subset \mathbb{R}^2$. We calculate

$$\begin{aligned}
&\int_0^\infty \int_{\mathbb{R}^2\setminus W} \rho(\partial_t\Phi, |\nabla_{\mathbf{x}}\Phi|^2)\Big(\partial_t\zeta + \nabla_{\mathbf{x}}\Phi \cdot \nabla_{\mathbf{x}}\zeta\Big)\, d\mathbf{x}dt \\
&= (\int_0^\varepsilon + \int_\varepsilon^S) \int_{B_S\setminus W} \rho(\partial_t\Phi, |\nabla_{\mathbf{x}}\Phi|^2)\Big(\partial_t\zeta + \nabla_{\mathbf{x}}\Phi \cdot \nabla_{\mathbf{x}}\zeta\Big)\, d\mathbf{x}dt \\
&=: I_1^\varepsilon + I_2^\varepsilon
\end{aligned} \tag{13.7.8}$$

for small $\varepsilon > 0$. Using (13.7.6)–(13.7.7),

$$I_1^\varepsilon \to 0 \qquad \text{as } \varepsilon \to 0+.$$

Furthermore, for $t > 0$, define

$$g(t, \boldsymbol{\xi}) = \zeta(t, t\boldsymbol{\xi}) \equiv \zeta(t, \mathbf{x}) \qquad \text{for } \boldsymbol{\xi} = \frac{\mathbf{x}}{t}.$$

Then, for each $t > 0$, $g(t, \cdot) \in C_c^\infty(\mathbb{R}^2)$ with $\mathrm{supp}(g(t, \cdot)) \subset B_{S/t}$. Also,

$$\nabla_{\mathbf{x}}\zeta(t, \mathbf{x}) = \frac{1}{t}\nabla_{\boldsymbol{\xi}} g(t, \boldsymbol{\xi}), \quad \partial_t\zeta(t, \mathbf{x}) = \partial_t g(t, \boldsymbol{\xi}) - \frac{1}{t}\boldsymbol{\xi} \cdot \nabla_{\boldsymbol{\xi}} g(t, \boldsymbol{\xi}) \qquad \text{for } \boldsymbol{\xi} = \tfrac{\mathbf{x}}{t}.$$

Then making the change of variables: $(t, \mathbf{x}) \mapsto (t, \boldsymbol{\xi}) = (t, \frac{\mathbf{x}}{t})$ in the term I_2^ε in (13.7.8) and using (13.7.5) and (13.7.7), we have

$$I_2^\varepsilon = \int_\varepsilon^\infty \int_{\mathbb{R}^2\setminus W} \rho(\boldsymbol{\xi})\Big(\partial_t g(t, \boldsymbol{\xi}) - \frac{1}{t}\boldsymbol{\xi} \cdot \nabla_{\boldsymbol{\xi}} g(t, \boldsymbol{\xi}) + \frac{1}{t}D\phi(\boldsymbol{\xi})\nabla_{\boldsymbol{\xi}} g(t, \boldsymbol{\xi})\Big)t^2 d\boldsymbol{\xi}\, dt.$$

Integrating by parts with respect to t in the term, $\int_\varepsilon^\infty \int_{\mathbb{R}^2\setminus W} \rho(\boldsymbol{\xi})\partial_t g(t, \boldsymbol{\xi})t^2 d\boldsymbol{\xi}\, dt$, we obtain

$$I_2^\varepsilon = J_1^\varepsilon + J_2^\varepsilon,$$

where

$$J_1^\varepsilon = \int_\varepsilon^\infty t \int_{\mathbb{R}^2\setminus W} \rho(\boldsymbol{\xi})\big((D\phi(\boldsymbol{\xi}) - \boldsymbol{\xi}) \cdot \nabla_{\boldsymbol{\xi}} g(t, \boldsymbol{\xi}) - 2g(t, \boldsymbol{\xi})\big)d\boldsymbol{\xi}\, dt,$$

$$J_2^\varepsilon = -\int_{\mathbb{R}^2\setminus W} \rho(\boldsymbol{\xi})g(\varepsilon, \boldsymbol{\xi})\varepsilon^2 d\boldsymbol{\xi}.$$

Then $J_1^\varepsilon = 0$ by (13.7.3). For J_2^ε, changing variables $\mathbf{x} = \varepsilon\boldsymbol{\xi}$, we have

$$J_2^\varepsilon = -\int_{\mathbb{R}^2\setminus W} \rho(\frac{\mathbf{x}}{\varepsilon})\zeta(\varepsilon, \mathbf{x})d\mathbf{x} = -\int_{B_S\setminus W} \rho(\frac{\mathbf{x}}{\varepsilon})\zeta(\varepsilon, \mathbf{x})d\mathbf{x}.$$

From (13.7.1),

$$\rho(\boldsymbol{\xi}) = \begin{cases} \rho_0 & \text{for } \xi_1 > \xi_1^0, \\ \rho_1 & \text{for } \xi_1 < \xi_1^0 \end{cases} \qquad \text{in } \mathbb{R}^2 \setminus (W \cup B_R(\mathbf{0})). \tag{13.7.9}$$

Then we find that, as $\varepsilon \to 0+$,

$$\rho(\frac{\mathbf{x}}{\varepsilon}) \to \bar{\rho}(\mathbf{x}) = \begin{cases} \rho_0 & \text{for } x_1 > 0, \\ \rho_1 & \text{for } x_1 < 0 \end{cases} \qquad \text{for all } \mathbf{x} \in \mathbb{R}^2 \setminus (W \cup \{x_1 = 0\}).$$

Then, using the density bounds in (13.7.7), we obtain from the dominated convergence theorem that

$$\lim_{\varepsilon\to 0+} J_2^\varepsilon = -\int_{\mathbb{R}^2\setminus W} \bar{\rho}(\mathbf{x})\zeta(0, \mathbf{x})d\mathbf{x}.$$

Thus, we have proved that the left-hand side of (13.7.8) is equal to the right-hand side of the last equality. This proves condition (iii) of Definition 2.3.1. Thus, $(\rho, \Phi)(t, \mathbf{x})$ is a weak solution of **Problem 2.2.1** in the sense of Definition 2.3.3.

Furthermore, this solution satisfies the entropy condition. Indeed, we can see the following:

(i) The entropy condition on the incident shock follows from $u_1 > 0$ and $\rho_1 > \rho_0$;

(ii) The entropy condition on the straight part $\mathcal{S}_1$ of the reflected shock is satisfied, because $\rho_2 > \rho_1$ and (7.1.6) holds at every point of $\mathcal{S}_1$;

(iii) The entropy condition on the curved part Γ_{shock} of the reflected-diffracted shock holds by Lemma 8.1.7.

Then the proof of Theorem 2.6.3 is complete.

13.8 PROOF OF THEOREM 2.6.4: EXISTENCE OF GLOBAL SOLUTIONS WHEN $u_1 > c_1$

We prove Theorem 2.6.5.

It follows from Lemma 10.4.2(i) that $\theta_{\text{w}}^{\text{c}} \in [\theta_{\text{w}}^{\text{s}}, \frac{\pi}{2})$. The existence of admissible solutions with $\theta_{\text{w}} \in (\theta_{\text{w}}^{\text{c}}, \frac{\pi}{2})$ follows from Proposition 12.1.1. The regularity of these solutions as asserted follows from Corollary 11.4.7 and Proposition 11.5.1.

Furthermore, if $\theta_{\text{w}}^{\text{c}} > \theta_{\text{w}}^{\text{s}}$, then, by Lemma 10.4.2(iii), there exists a sequence $\theta_{\text{w}}^{(i)} \in [\theta_{\text{w}}^{\text{c}}, \frac{\pi}{2})$ with $\lim_{i\to\infty} \theta_{\text{w}}^{(i)} = \theta_{\text{w}}^{\text{c}}$ and a corresponding admissible solution $\varphi^{(i)}$ with the wedge angle $\theta_{\text{w}}^{(i)}$ such that

$$\lim_{i\to\infty} \text{dist}(\Gamma_{\text{shock}}^{(i)}, \Gamma_{\text{wedge}}^{(i)}) = 0. \tag{13.8.1}$$

By Corollary 9.2.5, a subsequence (still denoted) $\varphi^{(i)}$ converges to a weak solution of $\varphi^{(\infty)}$ of **Problem 2.6.1** for the wedge angle $\theta_{\text{w}}^{\text{c}}$, with the structure described in Corollary 9.2.5(iii), and $\Gamma_{\text{shock}}^{(i)} \to \Gamma_{\text{shock}}^{(\infty)}$ in the sense of the uniform convergence of the shock functions $f_{\mathbf{e},\text{sh}}^{(i_j)} \to f_{\mathbf{e},\text{sh}}^{(\infty)}$ as described in Corollary 9.2.5(ii) for $\mathbf{e} = \boldsymbol{\nu}_{\text{w}}^{(\infty)}$. From this,

$$\text{dist}(\Gamma_{\text{shock}}^{(\infty)}, \Gamma_{\text{wedge}}^{(\infty)}) = 0.$$

By Corollary 9.2.5(i),

$$\lim_{i\to\infty} P_2{}^{(i)} = P_2{}^{(\infty)} = (\xi_{1P_2{}^{(\infty)}}, 0),$$

where $\xi_{1P_2{}^{(\infty)}} \le 0$, since $P_2{}^{(i)} = (\xi_{1P_2{}^{(i)}}, 0)$ with $\xi_{1P_2{}^{(i)}} < 0$ for each i. Now we show that

$$P_2{}^{(\infty)} = P_3 \equiv \mathbf{0}. \tag{13.8.2}$$

Indeed, on the contrary, suppose that $P_2{}^{(\infty)} \ne \mathbf{0}$. Then $\xi_{1P_2{}^{(\infty)}} < 0$. It follows that $\xi_{1P_2{}^{(i)}} \le \frac{1}{2}\xi_{1P_2{}^{(\infty)}}$ for all $i \ge N$ for some large N. Then, using that $\Gamma_{\text{shock}}^{(i)} \subset \{\xi_{1P_2}^{(i)} \le \xi_1 \le \xi_{1P_1}^{(i)}\}$ by (8.1.2) in Definition 8.1.1, we conclude from

Corollary 9.3.2 that there exists $R > 0$ such that $\mathrm{dist}(\Gamma^{(i)}_{\mathrm{shock}}, \mathbf{0}) \geq 2R$ for all $i \geq N$. Then

$$\mathrm{dist}(\Gamma^{(i)}_{\mathrm{shock}}, \Gamma^{(i)}_{\mathrm{wedge}} \cap B_R(\mathbf{0})) \geq R.$$

Also, using that $\theta^{(i)}_{\mathrm{w}} \in [\theta^{\mathrm{c}}_{\mathrm{w}}, \frac{\pi}{2})$ with $\theta^{\mathrm{c}}_{\mathrm{w}} > \theta^{\mathrm{s}}_{\mathrm{w}}$, we apply Proposition 9.4.8 with $\theta^{*}_{\mathrm{w}} = \theta^{\mathrm{c}}_{\mathrm{w}}$ to obtain

$$\mathrm{dist}(\Gamma^{(i)}_{\mathrm{shock}}, \Gamma^{(i)}_{\mathrm{wedge}} \setminus B_R(P_3)) \geq \frac{1}{C_R} \qquad \text{for all } i \geq N.$$

Combining the last two estimates, we see that $\mathrm{dist}(\Gamma^{(i)}_{\mathrm{shock}}, \Gamma^{(i)}_{\mathrm{wedge}}) \geq \frac{1}{C} > 0$ for all $i > N$, which contradicts (13.8.1). This proves (13.8.2).

Also, again using Proposition 9.4.8 with $\theta^{*}_{\mathrm{w}} = \theta^{\mathrm{c}}_{\mathrm{w}}$, we find that, for each i,

$$\mathrm{dist}(\Gamma^{(i)}_{\mathrm{shock}}, \Gamma^{(i)}_{\mathrm{wedge}} \setminus B_r(\mathbf{0})) \geq \frac{1}{C_r} \qquad \text{for any } r \in (0, |\Gamma^{(i)}_{\mathrm{wedge}}|),$$

where C_r depends only on the data and $(\theta^{*}_{\mathrm{w}}, r)$. Then

$$\mathrm{dist}(\Gamma^{(\infty)}_{\mathrm{shock}}, \Gamma^{(\infty)}_{\mathrm{wedge}} \setminus B_r(\mathbf{0})) \geq \frac{1}{C_r} \qquad \text{for any } r \in (0, |\Gamma^{(\infty)}_{\mathrm{wedge}}|).$$

Combining this with (9.2.22)–(9.2.23) and (13.8.2), we have

$$\begin{aligned} &\overline{\Gamma^{(\infty)}_{\mathrm{shock}}} \cap \overline{\Gamma^{(\infty)}_{\mathrm{wedge}}} = \{P_3\}, \\ &f^{(\infty)}_{\mathbf{e},\mathrm{sh}}(T) > 0 \qquad \text{for } T \in (T_{P_2{}^{(\infty)}}, T_{P_1{}^{(\infty)}}], \end{aligned}$$

and

$$\Omega^{(\infty)} = \left\{ (S,T) : \begin{array}{l} T_{P_2{}^{(\infty)}} < T < T_{P_4{}^{(\infty)}} \\ 0 < S < f^{(\infty)}_{\mathbf{e},\mathrm{sh}}(T) \;\; \text{for } T \in (T_{P_2{}^{(\infty)}}, T_{P_1{}^{(\infty)}}] \\ 0 < S < f^{(\infty)}_{\mathbf{e},\mathrm{so}}(T) \;\; \text{for } T \in (T_{P_1{}^{(\infty)}}, T_{P_4{}^{(\infty)}}) \end{array} \right\},$$

where we recall that Ω^{∞} is the interior of $\widehat{\Omega^{(\infty)}}$ defined by (9.2.22). Thus, Ω^{∞} is of form (8.2.25) of domain Ω for an admissible solution, with ${P_2}^{(\infty)} = {P_3}^{(\infty)} = \mathbf{0}$.

Moreover, it follows that, for each $P \in \Gamma^{(\infty)}_{\mathrm{shock}}$, there exists $r > 0$ such that $\mathrm{dist}(B_r(P), \Gamma^{(\infty)}_{\mathrm{wedge}} \cup \Gamma^{(\infty)}_{\mathrm{sonic}}) > r$. Then, from the uniform convergence: $f^{(i)}_{\mathbf{e},\mathrm{sh}} \to f^{(\infty)}_{\mathbf{e},\mathrm{sh}}$, there exists a large N so that, for each $i > N$, there is $P^{(i)} \in \Gamma^{(i)}_{\mathrm{shock}}$ such that $\mathrm{dist}(B_r(P^{(i)}), \Gamma^{(i)}_{\mathrm{wedge}} \cup \Gamma^{(\infty)}_{\mathrm{sonic}}) > \frac{r}{2}$. Applying Proposition 10.3.1 in $B_r(P^{(i)})$, we obtain that, for each $i \geq N$,

$$\begin{aligned} &|D^k f^{(i)}_{O_1,\mathrm{sh}}(\theta_{P^{(i)}})| \leq C_k(\hat{d}) \qquad \text{for } k = 2, 3, \dots, \\ &|D^k_{\boldsymbol{\xi}} \varphi^{(i)}| \leq \hat{C}_k(\hat{d}) \qquad \text{on } B_s(P^{(i)}) \cap \Omega^{\mathrm{ext}} \;\; \text{for } k = 2, 3, \dots, \end{aligned}$$

where constants s and C_k are independent of i. Passing to the limit, we obtain similar estimates for $(f^{(\infty)}_{O_1,\mathrm{sh}}, \varphi^{(\infty)})$. This shows that $\Gamma^{(\infty)}_{\mathrm{shock}}$ (without endpoints) is a C^∞–curve and that $\varphi^{(\infty)}$ is C^∞ up to $\Gamma^{(\infty)}_{\text{shock}}$. Similarly, the C^∞–regularity of φ in $\overline{\Omega} \setminus (\overline{\Gamma_{\text{shock}}} \cup \overline{\Gamma_{\text{sonic}}} \cup \{P_3\})$, and the $C^{1,1}$–regularity of φ near and up to Γ_{sonic} follows from Lemma 9.2.2 and Proposition 11.4.6 applied to each $\varphi^{(i)}$, where the constants in the estimates are independent of $i \geq N$ so that, passing to the limit, the same estimate holds for $\varphi^{(\infty)}$.

It can be proved in the same way as in §13.7 that, for the detached and attached solutions, the corresponding $(\rho, \Phi)(t, \mathbf{x})$ is a weak solution of **Problem 2.2.1** in the sense of Definition 2.3.3 satisfying the entropy condition.

Now Theorem 2.6.5 is proved.

13.9 APPENDIX: EXTENSION OF THE FUNCTIONS IN WEIGHTED SPACES

In this section, we extend the functions which belong to the weighted $C^{2,\alpha}$–spaces through a Lipschitz boundary, so that the extension is of the same regularity as the original function. We follow the methods of Stein [251, Chapter 6], with modification owing to the fact that we extend the functions that belong to the weighted and scaled $C^{2,\alpha}$–spaces and require the continuity of the extension operator with respect to the changes of the boundary in the sense described below.

Let $a > 0$ and $g \in C^{0,1}([0, a])$ with $g(s) > 0$ on $(0, a)$. Let

$$\begin{aligned}
&\Omega^a_\infty = (0, a) \times (0, \infty),\\
&\Omega^a_g = \{\mathbf{x} = (x_1, x_2) \in \Omega^a_\infty \ : \ 0 < x_2 < g(x_1)\},\\
&\Gamma^a_g = \{(x_1, g(x_1)) \ : \ 0 < x_1 < a\},\\
&\Sigma^0_g = \partial\Omega^a_g \cap \{x_1 = 0\} \equiv \{(0, x_2) \ : \ 0 \leq x_2 \leq g(a)\},\\
&\Sigma^a_g = \partial\Omega^a_g \cap \{x_1 = a\} \equiv \{(a, x_2) \ : \ 0 \leq x_2 \leq g(a)\}.
\end{aligned} \tag{13.9.1}$$

For real $\sigma > 0$, $\alpha \in (0, 1)$, and $\varepsilon \in (0, 1)$ with $\varepsilon < a$, denote

$$\|u\|^{*,(\sigma)}_{2,\alpha,\Omega^a_g} := \|u\|^{(-1-\alpha),\Sigma^a_g}_{2,\alpha,\Omega^a_g \cap \{s > \varepsilon/10\}} + \|u\|^{(\sigma),(\text{par})}_{2,\alpha,\Omega^a_g \cap \{s < \varepsilon\}}, \tag{13.9.2}$$

where the parabolic norms are with respect to $\{s = 0\}$.

Denote by $C^{2,\alpha}_{*,\sigma}(\Omega^a_g)$ the space:

$$C^{2,\alpha}_{*,\sigma}(\Omega^a_g) := \left\{ u \in C^1(\overline{\Omega^a_g}) \cap C^2(\Omega^a_g) \quad : \quad \|u\|^{*,(\sigma)}_{2,\alpha,\Omega^a_g} < \infty \right\}.$$

Given $u \in C^{2,\alpha}_{*,\sigma}(\Omega^a_g)$, we wish to extend it to the larger domain $\Omega^a_{(1+\kappa)g}$ for $\kappa > 0$, depending only on $\text{Lip}[g]$, in such a way that the extended function, denoted as $\mathcal{E}^{(a)}(u)$, satisfies $\mathcal{E}^{(a)}(u) \in C^{2,\alpha}_{*,\sigma}(\Omega^a_{(1+\kappa)g})$ with the estimate of its norm, and that the extension operator $\mathcal{E}^{(a)} \equiv \mathcal{E}^{(a)}_g : \Omega^a_g \mapsto \Omega^a_{(1+\kappa)g}$ has some continuity properties with respect to a and Γ^a_g.

If $g(\cdot) \geq c > 0$ on $[0, a]$, the extension into $\Omega^a_{(1+\kappa)g}$ can be continued to Ω^a_∞ with the same estimate in the weighted spaces by using a cutoff function in the x_2–variable. However, we cannot assume this lower bound below, in order to allow in the case that $g(0) = 0$ in Chapter 17.

In order to define the extension, we first construct a regularized distance function related to boundary Γ^a_g. Below, when no confusion arises, we write $(\Omega_\infty, \Omega_g, \Gamma_g, \mathcal{E}_*)$ for $(\Omega^a_\infty, \Omega^a_g, \Gamma^a_g, \mathcal{E}^{(a)})$.

13.9.1 Regularized distance

Given Ω^a_g, we follow [185] to construct a regularized distance to Γ^a_g in $\Omega^a_\infty \setminus \Omega^a_g$. This construction yields the continuity of the regularized distance with respect to Γ^a_g.

Lemma 13.9.1. *For each $a > 0$ and $g \in C^{0,1}([0, a])$ satisfying*

$$g > 0 \qquad \text{on } (0, a), \tag{13.9.3}$$

there exists a function $\delta_g \in C^\infty(\mathbb{R}^2 \setminus \overline{\Gamma^a_g})$, the regularized distance, such that

(i) $\frac{1}{2}\text{dist}(\mathbf{x}, \Gamma^a_g) \leq \delta_g(\mathbf{x}) \leq \frac{3}{2}\text{dist}(\mathbf{x}, \Gamma^a_g)$ *for all* $\mathbf{x} \in \mathbb{R}^2 \setminus \overline{\Gamma^a_g}$;

(ii) $|D^m \delta_g(\mathbf{x})| \leq C(m)\big(\text{dist}(\mathbf{x}, \Gamma^a_g)\big)^{1-m}$ *for all* $\mathbf{x} \in \mathbb{R}^2 \setminus \overline{\Gamma^a_g}$ *and* $m = 1, 2, \dots,$ *where $C(m)$ depends only on m;*

(iii) *There exists $c > 0$ depending only on* Lip$[g]$ *such that*

$$\delta_g(\mathbf{x}) \geq c\big(x_2 - g(x_1)\big) \qquad \text{for all } \mathbf{x} = (x_1, x_2) \in \overline{\Omega^a_\infty} \setminus \overline{\Omega^a_g};$$

(iv) *If $g_i \in C^{0,1}([0, a_i])$ and $g \in C^{0,1}([0, a])$ satisfy* (13.9.3) *and*

$$\|g_i\|_{C^{0,1}([0,a_i])} \leq L \qquad \text{for all } i,$$

and if $a_i \to a$ and functions $g_i(\frac{a_i}{a}x_1)$ (defined on $[0, a]$) converge to $g(x_1)$ uniformly on $[0, a]$, then functions $\delta_{g_i}(\frac{a_i}{a}x_1, x_2)$ converge to $\delta_g(x_1, x_2)$ in $C^m(K)$ for any $m = 0, 1, 2, \dots,$ and any compact $K \subset \overline{\Omega_\infty} \setminus \overline{\Omega_g}$.

Proof. We divide the proof into four steps.

1. We define δ_g as follows: Let

$$d_g(\mathbf{x}) = \text{dist}(\mathbf{x}, \Gamma^a_g) \qquad \text{for } \mathbf{x} \in \mathbb{R}^2.$$

Let $\phi \in C^\infty(\mathbb{R}^2)$ be nonnegative with $\int \phi(\mathbf{x})\, d\mathbf{x} = 1$ and $\text{supp}(\phi) \subset B_1(\mathbf{0})$. Define

$$G(\mathbf{x}, \tau) = \int_{\mathbb{R}^2} d_g(\mathbf{x} - \frac{\tau}{2}\mathbf{y})\phi(\mathbf{y})d\mathbf{y}. \tag{13.9.4}$$

Then, noting that $|d_g(\mathbf{x}) - d_g(\hat{\mathbf{x}})| \le |\mathbf{x} - \hat{\mathbf{x}}|$, we have

$$\begin{aligned} |G(\mathbf{x},\tau_1) - G(\mathbf{x},\tau_2)| &= \left|\int_{|y|\le 1}\Big(d_g(\mathbf{x} - \frac{\tau_1}{2}\mathbf{y}) - d_g(\mathbf{x} - \frac{\tau_2}{2}\mathbf{y})\Big)\phi(\mathbf{y})d\mathbf{y}\right| \\ &\le \frac{1}{2}|\tau_1 - \tau_2| \int_{|y|\le 1} |\mathbf{y}|\phi(\mathbf{y})d\mathbf{y} \\ &\le \frac{1}{2}|\tau_1 - \tau_2|, \end{aligned} \tag{13.9.5}$$

and similarly,

$$\begin{aligned} |G(\mathbf{x}_1,\tau) - G(\mathbf{x}_2,\tau)| &= \left|\int_{|y|\le 1}\Big(d_g(\mathbf{x}_1 - \frac{\tau}{2}\mathbf{y}) - d_g(\mathbf{x}_2 - \frac{\tau}{2}\mathbf{y})\Big)\phi(\mathbf{y})d\mathbf{y}\right| \\ &\le |\mathbf{x}_1 - \mathbf{x}_2| \int_{|y|\le 1} \phi(\mathbf{y})d\mathbf{y} \\ &\le |\mathbf{x}_1 - \mathbf{x}_2|. \end{aligned} \tag{13.9.6}$$

Now we define the regularized distance $\delta_g(\mathbf{x})$ at $\mathbf{x} \in \mathbb{R}^2$ by the equation:

$$\delta_g(\mathbf{x}) = G(\mathbf{x}, \delta_g(\mathbf{x})). \tag{13.9.7}$$

By (13.9.5) and the contraction mapping theorem, for each $\mathbf{x} \in \mathbb{R}^2$, equation (13.9.7) has a unique solution so that $\delta_g(\mathbf{x})$ is well-defined. Also, $\delta_g(\cdot) \ge 0$, since $G(\cdot) \ge 0$. Furthermore, since (13.9.4) yields that $G(\mathbf{x}, 0) = d_g(\mathbf{x})$, (13.9.5) (with $\tau_1 = \delta_g(\mathbf{x})$ and $\tau_2 = 0$) and (13.9.7) imply

$$|\delta_g(\mathbf{x}) - d_g(\mathbf{x})| \le \frac{1}{2}\delta_g(\mathbf{x}),$$

which leads to assertion (i).

2. Writing $G(\mathbf{x}, \tau)$ as

$$G(\mathbf{x},\tau) = \frac{4}{\tau^2}\int_{\mathbb{R}^2} d_g(\mathbf{y})\phi(\frac{2(\mathbf{x}-\mathbf{y})}{\tau})d\mathbf{y},$$

we see that $G \in C^\infty(\mathbb{R}^2 \times (\mathbb{R} \setminus \{0\}))$. In particular, from (13.9.5)–(13.9.6),

$$|\partial_\tau G| \le \frac{1}{2}, \quad |\partial_{\mathbf{x}} G| \le 1 \qquad \text{on } \mathbb{R}^2 \times (\mathbb{R} \setminus \{0\}). \tag{13.9.8}$$

Now (13.9.7) and the implicit function theorem imply

$$\delta_g \in C^\infty(\mathbb{R}^2 \setminus \overline{\Gamma_g^a}),$$

where we have used that, by assertion (i) proved above, $\delta_g(\mathbf{x}) = 0$ if and only if $\mathbf{x} \in \overline{\Gamma_g^a}$. Furthermore, using that $d_g(\cdot) \in C^{0,1}(\mathbb{R}^2)$, we obtain that, for each

$k, l = 0, 1, \ldots$, with $k + l \geq 1$ (we assume $k > 0$ below; Case $l > 0$ is similar),

$$\begin{aligned}\partial^k_{x_1}\partial^l_{x_2}G(\mathbf{x},\tau) &= \frac{4}{\tau^2}\int_{\mathbb{R}^2}\partial_{x_1}d_g(\mathbf{y})\,\partial^{k-1}_{x_1}\partial^l_{x_2}\phi(\frac{2(\mathbf{x}-\mathbf{y})}{\tau})\,d\mathbf{y}\\ &= \frac{4C(k+l)}{\tau^{k+l+1}}\int_{\mathbb{R}^2}\partial_{x_1}d_g(\mathbf{y})\,\partial^{k-1}_{x_1}\partial^l_{x_2}\phi(\frac{2(\mathbf{x}-\mathbf{y})}{\tau})\,d\mathbf{y}\\ &= \frac{C(k+l)}{\tau^{k+l-1}}\int_{\mathbb{R}^2}\partial_{x_1}d_g(\mathbf{x}-\frac{\tau}{2}\mathbf{y})\,\partial^{k-1}_{x_1}\partial^l_{x_2}\phi(\mathbf{y})d\mathbf{y}.\end{aligned}$$

From this and a similar (but longer) calculation that also involves the τ-derivatives of G, and using the fact that

$$|Dd_g(\cdot)| \leq 1 \qquad a.e.,$$

we have

$$|D^m_{\mathbf{x},\tau}G(\mathbf{x},\tau)| \leq \frac{C(m)}{\tau^{m-1}} \qquad \text{for } m = 1, 2, \ldots. \tag{13.9.9}$$

Combining this with the first estimate in (13.9.8), differentiating (13.9.7), and using the induction in $m = 1, 2, \ldots$, we obtain assertion (ii).

3. We now prove assertion (iii). Denote $L := \mathrm{Lip}[g]$. Using assertion (i), it suffices to show that

$$\mathrm{dist}(\mathbf{x}, \Gamma^a_g) \geq c\big(x_2 - g(x_1)\big) \qquad \text{for all } \mathbf{x} = (x_1, x_2) \in \overline{\Omega^a_\infty} \setminus \overline{\Omega^a_g}, \tag{13.9.10}$$

where $c = c(L)$. We show this as follows: Let $\mathbf{x}^* = (x_1^*, x_2^*) \in \overline{\Omega^a_\infty} \setminus \overline{\Omega^a_g}$. Denote by $\mathcal{C}$ the following cone:

$$\mathcal{C} := \{\mathbf{x} \in \mathbb{R}^2 \;:\; x_2 \geq g(x_1^*) + L|x_1 - x_1^*|\}.$$

The Lipschitz property of g implies that $g(x_1^*) + L|x_1 - x_1^*| \geq g(x_1)$ for all $x_1 \in [0, a]$ so that, for $\mathbf{x} \in \mathcal{C}$, $x_2 \geq g(x_1)$, which implies

$$\mathcal{C} \cap \overline{\Omega^a_\infty} \subset \overline{\Omega^a_\infty} \setminus \overline{\Omega^a_g}.$$

Also, since $\mathbf{x}^* \in \overline{\Omega^a_\infty} \setminus \overline{\Omega^a_g}$, then $x_2^* \geq g(x_1^*)$, which implies that $\mathbf{x}^* \in \mathcal{C}$. It follows that

$$\mathrm{dist}(\mathbf{x}^*, \Gamma^a_g) \geq \mathrm{dist}(\mathbf{x}^*, \partial\mathcal{C}).$$

By a simple geometric argument, $\mathrm{dist}(\mathbf{x}^*, \partial\mathcal{C}) = \frac{x_2^* - g(x_1^*)}{\sqrt{1+L^2}}$. This proves (13.9.10) so that assertion (iii) follows.

4. Now we prove assertion (iv). If the curve segments $\Gamma^a_{g_i}$ correspond to $g_i \in C^{0,1}([0, a_i])$, the conditions on sequence $\{g_i\}$ imply that

$$d_{g_i} \to d_g \qquad \text{uniformly in } \mathbb{R}^2. \tag{13.9.11}$$

Indeed, let $\varepsilon \in (0, \frac{a}{2})$, and let $N = N(\varepsilon)$ be such that

$$|a_i - a| < \varepsilon, \ |g_i(\frac{a}{a_i}x_1) - g(x_1)| < \varepsilon \qquad \text{for all } x_1 \in [0,a], \ i \geq N.$$

Let $\mathbf{x} \in \mathbb{R}^2$ and $i \geq N$, and let $\hat{\mathbf{x}} \in \Gamma^a_{g_i}$ be such that $d_{g_i}(\mathbf{x}) = |\mathbf{x} - \hat{\mathbf{x}}|$. Then $\hat{\mathbf{x}} = (\hat{x}_1, g_i(\hat{x}_1))$ for $\hat{x}_1 \in [0, a_i]$. Let $\tilde{\mathbf{x}} := (\frac{a}{a_i}\hat{x}_1, g_i(\frac{a}{a_i}\hat{x}_1))$ so that $\tilde{\mathbf{x}} \in \Gamma^a_g$. Noting that $a_i \geq a - \varepsilon \geq \frac{a}{2}$, we have

$$|\hat{\mathbf{x}} - \tilde{\mathbf{x}}| \leq \varepsilon\sqrt{(\frac{2}{a})^2 + 1}.$$

Then

$$d_g(\mathbf{x}) \leq |\mathbf{x} - \tilde{\mathbf{x}}| \leq |\mathbf{x} - \hat{\mathbf{x}}| + |\hat{\mathbf{x}} - \tilde{\mathbf{x}}| \leq d_{g_i}(\mathbf{x}) + \varepsilon\sqrt{(\frac{2}{a})^2 + 1},$$

that is, $d_g(\mathbf{x}) - d_{g_i}(\mathbf{x}) \leq C(a)\varepsilon$. The opposite inequality, $d_{g_i}(\mathbf{x}) - d_g(\mathbf{x}) \leq C(a)\varepsilon$, is obtained by a similar argument, with the difference that $\hat{\mathbf{x}} \in \Gamma^a_g$ is first chosen such that $d_g(\mathbf{x}) = |\mathbf{x} - \hat{\mathbf{x}}|$, and then the corresponding point on $\Gamma^a_{g_i}$ is chosen. Combining these estimates, we conclude (13.9.11).

Using (13.9.11), we obtain from (13.9.4) that $G_{g_i} \to G_g$ pointwise on $\mathbb{R}^2 \times \mathbb{R}$. Combining this with the equi-Lipschitz property (13.9.8) for all G_{g_i}, we see that $G_{g_i} \to G_g$ uniformly on compact subsets of $\mathbb{R}^2 \times \mathbb{R}$ and

$$\hat{G}_{g_i} \to G_g \qquad \text{uniformly on compact subsets of } \mathbb{R}^2 \times \mathbb{R},$$

where $\hat{G}_{g_i}(\mathbf{x}, \tau) = G_{g_i}(\frac{a_i}{a}x_1, x_2, \tau)$. Also, from (13.9.8), we find that, for all sufficiently large i,

$$|\partial_\tau \hat{G}_{g_i}| \leq \frac{1}{2}, \ |D^m_{(\mathbf{x},\tau)}\hat{G}_{g_i}(\mathbf{x}, \tau)| \leq \frac{C(m)}{\tau^{m-1}} \ \text{ for } m = 1, 2, \dots \quad \text{on } \mathbb{R}^2 \times (\mathbb{R} \setminus \{0\}).$$

Noting that, by (13.9.7), functions $\hat{\delta}_{g_i}(x_1, x_2) := \delta_{g_i}(\frac{a_i}{a}x_1, x_2)$ satisfy

$$\hat{\delta}_{g_i}(\mathbf{x}) = \hat{G}_{g_i}(\mathbf{x}, \hat{\delta}_{g_i}(\mathbf{x})) \qquad \text{for } i \geq N,$$

we conclude the proof of assertion (iv). □

13.9.2 The extension operator

In the construction of the extension operator, we will employ the regularized distance function constructed above and the following cutoff function:

Lemma 13.9.2. *There exists a function $\psi \in C^\infty_c(\mathbb{R})$ with* $\text{supp}(\psi) \subset [1, 2]$ *such that*

$$\int_{-\infty}^{\infty} \psi(\lambda)\, d\lambda = 1, \qquad \int_{-\infty}^{\infty} \lambda^m \psi(\lambda)\, d\lambda = 0 \quad \textit{for } m = 1, 2. \tag{13.9.12}$$

Proof. We divide the proof into two steps.

1. We first construct a piecewise-constant function $\hat{\psi}$ with $\mathrm{supp}(\hat{\psi}) \subset (1,2)$ and satisfying (13.9.12).

Let $k \in (1, 2^{\frac{1}{6}})$ be determined below. Let

$$\hat{\psi}_1 = \chi_{[k,\,k^2]}, \quad \hat{\psi}_2 = \chi_{[k^3,\,k^4]}, \quad \hat{\psi}_3 = \chi_{[k^5,\,k^6]},$$

where $\chi_{[a,b]}(\cdot)$ is the characteristic function of $[a,b]$. Define

$$\hat{\psi} = \sum_{i=1}^{3} s_i \hat{\psi}_i,$$

where constants $s_i \in \mathbb{R}, i = 1,2,3$, will be defined so that $\hat{\psi}$ satisfies (13.9.12). Note that $\mathrm{supp}(\hat{\psi}) \subset [k, k^6] \subset (1,2)$.

Function $\hat{\psi}$ satisfies (13.9.12) if $s_i, i = 1,2,3$, satisfy the linear algebraic system:

$$A \begin{pmatrix} s_1 \\ s_2 \\ s_3 \end{pmatrix} = \begin{pmatrix} 1 \\ 0 \\ 0 \end{pmatrix}, \tag{13.9.13}$$

where matrix $A = [a_{ij}]_{i,j=1}^{3}$ is defined by

$$a_{ij} = \int_{-\infty}^{\infty} \lambda^{i-1} \hat{\psi}_j(\lambda)\, d\lambda = \int_{k^{2j-1}}^{k^{2j}} \lambda^{i-1} d\lambda = \frac{k^{i(2j-1)}}{i}(k^i - 1).$$

By calculation,

$$\det A = \frac{1}{6} k^{14}(k-1)(k^2-1)(k^3-1)P(k),$$

where $P(k) = k^8 - 2k^6 + 2k^2 - 1$. Since $P(k)$ is a non-zero polynomial, then there exists $k_0 \in (1, 2^{\frac{1}{6}})$ such that $P(k_0) \neq 0$, which implies that $\det A \neq 0$ at k_0.

Fix such k_0. Then we can determine s_i by solving system (13.9.13). Thus, $\hat{\psi}$ is defined and satisfies all the required conditions.

2. Now we modify $\hat{\psi}$ to obtain a smooth function ψ with the properties asserted. Let $\eta \in C_c^\infty(\mathbb{R})$ satisfies $\eta \geq 0$, $\mathrm{supp}(\eta) \subset [-1,1]$, and $\int_{-\infty}^{\infty} \eta(\lambda)\, d\lambda = 1$. Let $\eta_\varepsilon(t) = \frac{1}{\varepsilon}\eta(\frac{t}{\varepsilon})$ for $\varepsilon > 0$. Define

$$\psi = \sum_{i=1}^{3} s_i^{(\varepsilon)} \psi_i^{(\varepsilon)} \qquad \text{with } \psi_i^{(\varepsilon)} = \hat{\psi}_i * \eta_\varepsilon,$$

where constants $\varepsilon > 0$ and $s_i^{(\varepsilon)}, i = 1,2,3$, will be determined below.

Function ψ satisfies (13.9.12) if $s_i^{(\varepsilon)}, i = 1,2,3$, satisfy the linear algebraic system of form (13.9.13) with the same right-hand side, as in (13.9.13), and matrix $A^{(\varepsilon)} = [a_{ij}^{(\varepsilon)}]_{i,j=1}^3$ defined by

$$a_{ij}^{(\varepsilon)} = \int_{-\infty}^{\infty} \lambda^{i-1} \psi_j^{(\varepsilon)}(\lambda)\, d\lambda.$$

Since $0 \leq \hat{\psi}_i \leq 1$, $\psi_i^{(\varepsilon)} \to \hat{\psi}_i$ *a.e.* in $\mathbb{R}$ as $\varepsilon \to 0$, and $0 \leq \psi_i^{(\varepsilon)} \leq 1$ with $\operatorname{supp}(\psi_i^{(\varepsilon)}) \subset [1,2]$ if ε is small. Thus, by the dominated convergence theorem, $a_{ij}^{(\varepsilon)} \to a_{ij}$ as $\varepsilon \to 0$ so that, if $\varepsilon > 0$ is sufficiently small, $\det A^{(\varepsilon)} \neq 0$. Therefore, $s_i^{(\varepsilon)}, i = 1,2,3$, can be determined from the linear algebraic system. Then function ψ is well-defined and satisfies all the required properties. □

Now, for domain Ω_g^a defined by (13.9.1), we define the extension operator $\mathcal{E}^{(a)} \equiv \mathcal{E}_g^{(a)} : C^2(\Omega_g^a \cup \Gamma_g^a) \mapsto C^2(\Omega_{(1+\kappa)g}^a)$. We use the regularized distance δ_g from Lemma 13.9.3 and define $\delta_g^* = \frac{2}{c}\delta_g$ for c from Lemma 13.9.1(iii). Then

$$\delta_g^*(\mathbf{x}) \geq 2\big(x_2 - g(x_1)\big) \qquad \text{in } \overline{\Omega_\infty^a} \setminus \overline{\Omega_g^a}.$$

Definition 13.9.3. *For $u \in C^2(\Omega_g^a \cup \Gamma_g^a)$, we define its extension $\mathcal{E}_g^{(a)}(u)$ on $\Omega_{(1+\kappa)g}^a$, for κ defined below, by setting*

$$\mathcal{E}_g^{(a)}(u)(\mathbf{x}) = \begin{cases} u(\mathbf{x}) & \text{for } \mathbf{x} \in \overline{\Omega_g^a}, \\ \int_{-\infty}^{\infty} u(x_1, x_2 - \lambda \delta_g^*(\mathbf{x}))\psi(\lambda)\, d\lambda & \text{for } \mathbf{x} \in \Omega_{(1+\kappa)g}^a \setminus \overline{\Omega_g^a}. \end{cases} \tag{13.9.14}$$

We note that $x_2 - g(x_1) \in (0, \kappa g(x_1))$ for each $\mathbf{x} \in \Omega_{(1+\kappa)g}^a \setminus \overline{\Omega_g^a}$.

Lemma 13.9.4. *There exists $\kappa \in (0, \frac{1}{3})$ depending only on* Lip$[g]$ *such that the integral on the right-hand side of (13.9.14) is well-defined: For each $\mathbf{x} = (x_1, x_2) \in \Omega_{(1+\kappa)g}^a \setminus \overline{\Omega_g^a}$,*

$$(x_1,\ x_2 - \lambda\delta_g^*(\mathbf{x})) \in \{x_1\} \times [\frac{g(x_1)}{3},\ g(x_1) - (x_2 - g(x_1))] \Subset \Omega_g^a$$

for all $\lambda \in \operatorname{supp}(\psi)$.

Proof. Let $\mathbf{x} = (x_1, x_2) \in \Omega_{(1+\kappa)g}^a \setminus \overline{\Omega_g^a}$. Let $\lambda \in \operatorname{supp}(\psi)$. Since $\operatorname{supp}(\psi) \subset [1,2]$,

$$x_2 - \lambda\delta_g^*(\mathbf{x}) \leq x_2 - 2(x_2 - g(x_1)) = g(x_1) - (x_2 - g(x_1)) < g(x_1).$$

Also, using Lemma 13.9.1(i) and the inequality that $\operatorname{dist}(\mathbf{x}, \Gamma_g^a) \leq x_2 - g(x_1)$, we have

$$\delta_g^*(\mathbf{x}) = \frac{2}{c}\delta_g(\mathbf{x}) \leq \frac{3}{c}\operatorname{dist}(\mathbf{x}, \Gamma_g^a) \leq \frac{3}{c}\big(x_2 - g(x_1)\big).$$

Then, since $\lambda \in [1,2]$, we estimate

$$x_2 - \lambda\delta_g^*(\mathbf{x}) \geq x_2 - \frac{6}{c}(x_2 - g(x_1)) \geq g(x_1) - \frac{6}{c}(x_2 - g(x_1)) \geq \frac{g(x_1)}{3},$$

if $x_2 \leq (1+\frac{c}{9})g(x_1)$. This completes the proof with $\kappa = \frac{c}{9}$ for c from Lemma 13.9.1(iii) depending only on Lip$[g]$. □

Now we prove the regularity of the extension.

Theorem 13.9.5. *Let $a > 0$. Let $g \in C^{0,1}([0,a])$ with $g(s) > 0$ on $(0,a)$. Let κ be from Lemma 13.9.4. Then the extension operator $\mathcal{E}^{(a)} \equiv \mathcal{E}_g^{(a)}$, introduced in Definition 13.9.3, maps $C^2(\Omega_g^a \cup \Gamma_g^a)$ into $C^2(\Omega^a_{(1+\kappa)g})$ with the following properties: For any $\sigma > 0$ and $\alpha \in (0,1)$,*

(i) *$\mathcal{E}^{(a)}$ is a linear continuous operator from $C^{2,\alpha}_{*,\sigma}(\Omega_g^a)$ to $C^{2,\alpha}_{*,\sigma}(\Omega^a_{(1+\kappa)g})$;*

(ii) *There exists C depending only on $(\text{Lip}[g], \alpha, \sigma)$ such that*

$$\|\mathcal{E}^{(a)}(u)\|^{*,(\sigma)}_{2,\alpha,\Omega^a_{(1+\kappa)g}} \leq C\|u\|^{*,(\sigma)}_{2,\alpha,\Omega^a_g}$$

for all $u \in C^{2,\alpha}_{*,\sigma}(\Omega_g^a)$. Note that C is independent of a and ε in (13.9.2);

(iii) *Let $g_i \in C^{0,1}([0,a_i])$ and $g \in C^{0,1}([0,a])$ satisfy (13.9.3) and*

$$\|g_i\|_{C^{0,1}([0,a_i])} \leq L \qquad \textit{for all } i,$$

and let $a_i \to a$ and functions $g_i(\frac{a_i}{a}x_1)$ (defined on $[0,a]$) converge to $g(x_1)$ uniformly on $[0,a]$. Then, if $u_i \in C^{2,\alpha}_{*,\sigma}(\Omega^{a_i}_{g_i})$, $u \in C^{2,\alpha}_{*,\sigma}(\Omega^a_g)$, and $u_i \to u$ uniformly on compact subsets of the open set Ω^a_g, and if there exists $M > 0$ such that $\|u_i\|^{*,(\sigma)}_{2,\alpha,\Omega^{a_i}_{g_i}} \leq M$ for all i, then

$$w_i(\mathbf{x}) := \mathcal{E}^{(a_i)}_{g_i}(u_i)(\frac{a_i}{a}x_1, x_2) \to \mathcal{E}^{(a)}_g(u)(\mathbf{x}) \qquad \textit{in } C^{2,\beta}_{*,\sigma'}(\Omega^a_{(1+\frac{1}{2}\kappa)g})$$

for all $\beta \in (0,\alpha)$ and $\sigma' \in (0,\sigma)$, where we note that w_i is defined on $\Omega^a_{(1+\frac{1}{2}\kappa)g}$ for large i.

We start the proof of Theorem 13.9.5 by showing that operator $\mathcal{E}^{(a)}$ determines the extension in the non-weighted $C^{2,\alpha}$–spaces. We also localize in x_1: For any $b_1, b_2 \in [0,a]$ with $b_1 < b_2$, denote

$$\Omega^a_g(b_1,b_2) := \Omega^a_g \cap \{(x_1,x_2) \in \mathbb{R}^2 \;:\; b_1 < x_1 < b_2\},$$

and similarly define $\Gamma^a_g(b_1,b_2)$ and $(\Omega^a_g \cup \Gamma^a_g)(b_1,b_2)$. We note that the extension operator $\mathcal{E}^{(a)}_g$ in Definition 13.9.3 can be applied to $u \in C^2((\Omega^a_g \cup \Gamma^a_g)(b_1,b_2))$ and defines a function on $\Omega^a_{(1+\kappa)g}(b_1,b_2)$.

Lemma 13.9.6. *Let g and κ be as in Theorem 13.9.5. Then the extension operator $\mathcal{E}^{(a)} \equiv \mathcal{E}_g^{(a)}$ maps $C^2(\Omega_g^a \cup \Gamma_g^a)$ into $C^2(\Omega_{(1+\kappa)g}^a)$ with the following properties: For any $\alpha \in (0,1)$, there exists C depending only on $(\mathrm{Lip}[g], \alpha)$ such that, for any $b_1, b_2 \in [0,a]$ with $b_1 < b_2$,*

(i) *$\mathcal{E}^{(a)}$ is a linear continuous operator:*

$$\mathcal{E}^{(a)} : C^{2,\alpha}(\overline{\Omega_g^a(b_1,b_2)}) \mapsto C^{2,\alpha}(\overline{\Omega_{(1+\kappa)g}^a(b_1,b_2)}).$$

(ii) *For all $u \in C^{2,\alpha}(\overline{\Omega_g^a(b_1,b_2)})$,*

$$\|\mathcal{E}^{(a)}(u)\|_{2,\alpha,\Omega_{(1+\kappa)g}^a(b_1,b_2)} \le C\|u\|_{2,\alpha,\Omega_g^a(b_1,b_2)}.$$

More precisely,

$$\|\mathcal{E}^{(a)}(u)\|_{m,0,\Omega_{(1+\kappa)g}^a(b_1,b_2)} \le C\|u\|_{m,0,\Omega_g^a(b_1,b_2)}, \quad m = 0,1,2, \tag{13.9.15}$$

$$[\mathcal{E}^{(a)}(u)]_{2,\alpha,\Omega_{(1+\kappa)g}^a(b_1,b_2)} \le C[u]_{2,\alpha,\Omega_g^a(b_1,b_2)}. \tag{13.9.16}$$

Note that C is independent of (a, b_1, b_2).

(iii) *Let $g_i \in C^{0,1}([0,a_i])$ and $g \in C^{0,1}([0,a])$ satisfy (13.9.3) and*

$$\|g_i\|_{C^{0,1}([0,a_i])} \le L \qquad \text{for all } i,$$

and let $a_i \to a$ and functions $g_i(\frac{a_i}{a}x_1)$ (defined on $[0,a]$) converge to $g(x_1)$ uniformly on $[0,a]$. Then, if $u_i \in C^{2,\alpha}(\overline{\Omega_{g_i}^{a_i}(\frac{a_i}{a}b_1, \frac{a_i}{a}b_2)})$, $u \in C^{2,\alpha}(\overline{\Omega_g^a(b_1,b_2)})$, and $u_i \to u$ uniformly on compact subsets of the open set $\Omega_g^a(b_1,b_2)$, and if there exist $L, M > 0$ such that $\mathrm{Lip}[g_i] \le L$ and $\|u_i\|_{2,\alpha,\Omega_{g_i}^{a_i}(\frac{a_i}{a}b_1,\frac{a_i}{a}b_2)} \le M$ for all i, then

$$w_i(\mathbf{x}) := \mathcal{E}_{g_i}^{(a_i)}(u_i)(\frac{a_i}{a}x_1, x_2) \to \mathcal{E}_g^{(a)}(u)(\mathbf{x}) \quad \text{in } C^{2,\beta}(\Omega_{(1+\frac{1}{2}\kappa)g}^a)(b_1,b_2)$$

for all $\beta \in (0,\alpha)$, where we note that w_i is defined on $\Omega_{(1+\frac{1}{2}\kappa)g}^a(b_1,b_2)$ for large i.

Proof. We divide the proof into three steps.

1. Let $u \in C^2(\Omega_g^a \cup \Gamma_g^a)$. We show that $\mathcal{E}_g^{(a)}(u) \in C^2(\Omega_{(1+\kappa)g}^a)$. We first note that it follows from (13.9.14) that $\mathcal{E}_g^{(a)}(u)$ is in C^2 in the open set $\Omega_{(1+\kappa)g}^a \setminus \overline{\Omega_g^a}$. Indeed, the continuity of $\mathcal{E}_g^{(a)}(u)$ follows directly from (13.9.14). Furthermore, we obtain the explicit expressions of the first and second partial derivatives of $\mathcal{E}_g^{(a)}(u)$ by differentiating (13.9.14), and show that these expressions define the continuous functions in $\Omega_{(1+\kappa)g}^a \setminus \overline{\Omega_g^a}$. We show that only for $D_{ij}u$ (the argument for the first derivatives is similar). For $\mathbf{x} \in \Omega_{(1+\kappa)g}^a \setminus \overline{\Omega_g^a}$,

$$D_{ij}\mathcal{E}_g^{(a)}(u)(\mathbf{x}) = I_1 + I_2 + I_3 + I_4, \tag{13.9.17}$$

where

$$
\begin{aligned}
I_1 &= \int_{-\infty}^{\infty} D_{ij}u(x_1,\ x_2 - \lambda\delta_g^*(\mathbf{x}))\psi(\lambda)\,d\lambda,\\
I_2 &= \int_{-\infty}^{\infty} D_{i2}u(x_1,\ x_2 - \lambda\delta_g^*(\mathbf{x}))\lambda\,D_i\delta_g^*(\mathbf{x})\psi(\lambda)\,d\lambda\\
&\quad + \int_{-\infty}^{\infty} D_{j2}u(x_1,\ x_2 - \lambda\delta_g^*(\mathbf{x}))\lambda\,D_j\delta_g^*(\mathbf{x})\psi(\lambda)\,d\lambda\\
&=: I_{21} + I_{22},\\
I_3 &= \int_{-\infty}^{\infty} D_{22}u(x_1,\ x_2 - \lambda\delta_g^*(\mathbf{x}))\lambda^2 D_i\delta_g^*(\mathbf{x})D_j\delta_g^*(\mathbf{x})\psi(\lambda)\,d\lambda,\\
I_4 &= \int_{-\infty}^{\infty} D_2u(x_1,\ x_2 - \lambda\delta_g^*(\mathbf{x}))\lambda\,D_{ij}\delta_g^*(\mathbf{x})\psi(\lambda)\,d\lambda.
\end{aligned}
\tag{13.9.18}
$$

Now the assumed regularity of u, and Lemmas 13.9.1(ii), 13.9.2, and 13.9.4 imply the continuity of $D_{ij}u$ in $\Omega^a_{(1+\kappa)g} \setminus \overline{\Omega^a_g}$.

We now show that $\mathcal{E}_g^{(a)}(u)$ and its first and second partial derivatives in $\Omega^a_{(1+\kappa)g} \setminus \overline{\Omega^a_g}$ are continuous up to Γ^a_g, *i.e.*, $\mathcal{E}_g^{(a)}(u) \in C^2(\Omega^a_{(1+\kappa)g} \setminus \Omega^a_g)$, and their limits at Γ^a_g match with the corresponding limits of u and its partial derivatives from Ω^a_g. Again, we discuss the proof in detail only for the second derivatives.

Let $\mathbf{x}^0 \in \Gamma^a_g$. If $\mathbf{x} \in \Omega^a_{(1+\kappa)g} \setminus \overline{\Omega^a_g}$ tends to $\mathbf{x}^0$, then $\delta_g^*(\mathbf{x}) \to 0$ and $|D\delta_g^*(\mathbf{x})| \le C$, by Lemma 13.9.1(ii). Thus, $I_1(\mathbf{x}) \to D_{ij}u(\mathbf{x}^0)$, whereas (13.9.12) implies that $I_2(\mathbf{x}), I_3(\mathbf{x}) \to 0$.

It remains to consider term I_4. Note that I_4 includes $D_{ij}\delta_g^*$, which may blow up near Γ^a_g by Lemma 13.9.1(ii). We rewrite I_4 as follows: Since $\text{supp}(\psi) \subset [1, 2]$, we expand

$$
\begin{aligned}
u_{x_2}(x_1, x_2 - \lambda\delta_g^*) &= u_{x_2}(x_1, x_2 - \delta_g^*)\\
&\quad - (\lambda - 1)\delta_g^* u_{x_2x_2}(x_1, x_2 - \delta_g^*) - (\lambda - 1)\delta_g^* R(\mathbf{x}, \lambda)
\end{aligned}
\tag{13.9.19}
$$

with $\delta_g^* = \delta_g^*(\mathbf{x})$ and

$$
\begin{aligned}
R(\mathbf{x}, \lambda) = \int_0^1 \Big(& u_{x_2x_2}(x_1, x_2 - (1 + t(\lambda - 1))\delta_g^*(\mathbf{x}))\\
& - u_{x_2x_2}(x_1, x_2 - \delta_g^*(\mathbf{x}))\Big)\,dt,
\end{aligned}
\tag{13.9.20}
$$

and substitute (13.9.19) into I_4. Then, using (13.9.12) again, we find that the first and second terms of (13.9.19) give the integrals that vanish. Thus, we have shown that

$$
I_4 = (\lambda - 1)\delta_g^*(\mathbf{x})D_{ij}\delta_g^*(\mathbf{x}) \int_{-\infty}^{\infty} R(\mathbf{x}, \lambda)\lambda\,\psi(\lambda)\,d\lambda. \tag{13.9.21}
$$

Now, to estimate I_4, we note that Lemma 13.9.4, regularity $u \in C^2(\Omega_g^a \cup \Gamma_g^a)$, and the fact that $\mathbf{x}^0 \in \Gamma_g^a$ with $g(x_1^0) > 0$ imply that, for sufficiently small $\varepsilon(\mathbf{x}^0) > 0$, there exists C such that

$$|R(\mathbf{x}, \lambda)| \le C \qquad \text{if } \mathbf{x} \in \Omega_g^a \cup \Gamma_g^a,\ |\mathbf{x} - \mathbf{x}^0| \le \tfrac{1}{C(\varepsilon)},\ \text{and } \lambda \in \mathrm{supp}(\psi),$$

and

$$R(\mathbf{x}, \lambda) \to 0 \qquad \text{as } \mathbf{x} \to \mathbf{x}^0 \quad \text{for each } \lambda \in \mathrm{supp}(\psi).$$

Using that $\delta_g^*(\mathbf{x}) D_{ij} \delta_g^*(\mathbf{x})$ is bounded by Lemma 13.9.1(ii), we see that limit (13.9.21) is zero as $\mathbf{x} \to \mathbf{x}^0$.

Thus, we have shown that

$$\lim_{\mathbf{x}\to\mathbf{x}^0, \mathbf{x}\in\Omega^a_{(1+\kappa)g}\setminus\overline{\Omega^a_g}} D_{ij}\mathcal{E}_g^{(a)}(u)(\mathbf{x}) = \lim_{\mathbf{x}\to\mathbf{x}^0, \mathbf{x}\in\Omega_g^a} D_{ij}u(\mathbf{x}). \tag{13.9.22}$$

Then $\mathcal{E}_g^{(a)}(u)$ is in $C^2(\Omega_g^a \cup \Gamma_g^a) \cup C^2(\Omega^a_{(1+\kappa)g} \setminus \Omega_g^a)$, and the values of $\mathcal{E}_g^{(a)}(u)$ match with its first and second derivatives on the common boundary Γ_g^a of these domains, where Γ_g^a is a Lipschitz graph. Then the standard argument (see, *e.g.*, [251, page 186]) shows that $\mathcal{E}_g^{(a)}(u) \in C^2(\Omega^a_{(1+\kappa)g})$.

2. Now we prove assertions (i)–(ii). The linearity of operator $\mathcal{E}_g^{(a)}$ follows directly from its definition. Then (i) follows from the estimate in (ii), which we now prove. We discuss only the estimates of second derivatives. In the argument below, all the constants depend only on $(\mathrm{Lip}[g], \alpha)$.

From (13.9.20),

$$\sup_{\lambda\in[1,2]} \|R(\cdot, \lambda)\|_{0,0,(\Omega^a_{(1+\kappa)g}\setminus\overline{\Omega^a_g})(b_1,b_2)} \le C\|u\|_{2,0,\Omega_g^a(b_1,b_2)} \qquad \text{for } \lambda \in [1,2].$$

Then (13.9.15) with $m = 2$ follows directly from (13.9.17)–(13.9.18) and (13.9.21) by using Lemma 13.9.1(ii), especially noting that $\delta_g^*(\mathbf{x}) D_{ij} \delta_g^*(\mathbf{x})$ is bounded.

3. Now we show (13.9.16). Let $\mathbf{x}, \bar{\mathbf{x}} \in \Omega^a_{(1+\kappa)g}(b_1, b_2)$. We need to prove that

$$|D_{ij}\mathcal{E}_g^{(a)}(u)(\mathbf{x}) - D_{ij}\mathcal{E}_g^{(a)}(u)(\bar{\mathbf{x}})| \le C[u]_{2,\alpha,\mathcal{D}}|\mathbf{x} - \bar{\mathbf{x}}|^\alpha. \tag{13.9.23}$$

If $\mathbf{x}, \bar{\mathbf{x}} \in \overline{\Omega_g^a(b_1, b_2)}$, then $D_{ij}\mathcal{E}_g^{(a)}(u) = D_{ij}u$ at these points, and the estimates follow.

Thus, we need to consider only the case that

$$\mathbf{x} \in (\Omega^a_{(1+\kappa)g} \setminus \overline{\Omega^a_g})(b_1, b_2).$$

Denote by L the open segment connecting $\mathbf{x}$ to $\bar{\mathbf{x}}$. We consider three separate cases:

$$\begin{aligned}
&(a)\ \ \bar{\mathbf{x}} \in \overline{\Omega^a_g}(b_1, b_2);\\
&(b)\ \ \bar{\mathbf{x}} \in (\Omega^a_{(1+\kappa)g} \setminus \overline{\Omega^a_g})(b_1, b_2),\ \ \min_{\mathbf{x}'\in\overline{L}}(x_2' - g(x_1')) \le |\mathbf{x} - \bar{\mathbf{x}}|;\\
&(c)\ \ \bar{\mathbf{x}} \in (\Omega^a_{(1+\kappa)g} \setminus \overline{\Omega^a_g})(b_1, b_2),\ \ \min_{\mathbf{x}'\in\overline{L}}(x_2' - g(x_1')) > |\mathbf{x} - \bar{\mathbf{x}}|.
\end{aligned} \tag{13.9.24}$$

Cases (a)–(b) will follow from the estimate of $|D_{ij}\mathcal{E}_g^{(a)}(u)(\mathbf{x})-D_{ij}\mathcal{E}_g^{(a)}(u)(\bar{\mathbf{x}})|$ when

$$\bar{\mathbf{x}} \in \Gamma_g^a(b_1,b_2), \quad L \subset (\Omega^a_{(1+\kappa)g} \setminus \overline{\Omega_g^a})(b_1,b_2). \tag{13.9.25}$$

Thus, we consider this case. Then $\delta_g^*(\mathbf{x}) \le C|\mathbf{x}-\bar{\mathbf{x}}|$ and $\mathcal{E}_g^{(a)}(u)(\bar{\mathbf{x}}) = u(\bar{\mathbf{x}})$.

We use (13.9.17)–(13.9.18), (13.9.21), and the estimate of $|I_1(\mathbf{x}) - D_{ij}u(\bar{\mathbf{x}})|$ and $I_m(\mathbf{x})$ for $i,j=1,2$, and $m=2,3,4$.

From (13.9.25),

$$\operatorname{dist}(\mathbf{x}, \Gamma_g^a(b_1,b_2)) \le |\mathbf{x}-\bar{\mathbf{x}}|. \tag{13.9.26}$$

For brevity, we denote $\mathcal{D} := \Omega_g^a(b_1,b_2)$. Using (13.9.26), $\operatorname{supp}(\psi) \subset [1,2]$, and the first property in (13.9.12), we have

$$\begin{aligned} &|I_1(\mathbf{x}) - D_{ij}u(\bar{\mathbf{x}})| \\ &= \left| \int_{-\infty}^{\infty} \big(D_{ij}u(x_1,\ x_2 - \lambda\delta_g^*(\mathbf{x})) - D_{ij}u(\bar{\mathbf{x}})\big)\psi(\lambda)\,d\lambda \right| \\ &\le C[u]_{2,\alpha,\mathcal{D}}\big(|\mathbf{x}-\bar{\mathbf{x}}|^\alpha + (\delta_g(\mathbf{x}))^\alpha\big) \\ &\le C[u]_{2,\alpha,\mathcal{D}}|\mathbf{x}-\bar{\mathbf{x}}|^\alpha. \end{aligned} \tag{13.9.27}$$

Next we estimate the first integral in the expression of I_2. We use (13.9.26) and the orthogonality properties in (13.9.12) to obtain

$$\begin{aligned} &|I_{21}(\mathbf{x})| \\ &= |D_i\delta_g^*(\mathbf{x})| \left| \int_{-\infty}^{\infty} \big(D_{i2}u(x_1,\ x_2 - \lambda\delta_g^*(\mathbf{x})) - D_{i2}u(\mathbf{x})\big)\lambda\psi(\lambda)\,d\lambda \right| \\ &\le C[u]_{2,\alpha,\mathcal{D}}(\delta_g(\mathbf{x}))^\alpha \le C[u]_{2,\alpha,\mathcal{D}}|\mathbf{x}-\bar{\mathbf{x}}|^\alpha. \end{aligned} \tag{13.9.28}$$

The estimates of $I_{22}(\mathbf{x})$ and $I_3(\mathbf{x})$ are similar.

To estimate $I_4(\mathbf{x})$, we use its expression (13.9.21) with the following estimate: From (13.9.20), we obtain that, for $\lambda \in [1,2]$,

$$\sup_{\lambda\in[1,2]} \|R(\cdot,\lambda)\|_{0,0,(\Omega^a_{(1+\kappa)g}\setminus\overline{\Omega_g^a})(b_1,b_2)} \le C[u]_{2,\alpha,\mathcal{D}}(\delta_g(\mathbf{x}))^\alpha. \tag{13.9.29}$$

From this, Lemma 13.9.1(ii), and (13.9.26),

$$|I_4(\mathbf{x})| \le C[u]_{2,\alpha,\mathcal{D}}|\mathbf{x}-\bar{\mathbf{x}}|^\alpha.$$

Thus, (13.9.23) is proved when (13.9.25) holds.

Now we show

Claim 13.9.7. *Given that* (13.9.23) *is true when* (13.9.25) *holds, then* (13.9.23) *holds for Cases* (a)–(b) *of* (13.9.24).

We first consider Case (a) of (13.9.24). Then there exists $\hat{\mathbf{x}} \in L \cap \Gamma_g^a(b_1, b_2)$ such that the open segment connecting $\mathbf{x}$ to $\hat{\mathbf{x}}$ lies within $(\Omega^a_{(1+\kappa)g} \setminus \overline{\Omega^a_g})(b_1, b_2)$. We can apply (13.9.23) for $(\mathbf{x}, \hat{\mathbf{x}})$ and use $D_{ij}\mathcal{E}_g^{(a)}(u)(\hat{\mathbf{x}}) = D_{ij}u(\hat{\mathbf{x}})$ by (13.9.22) so that

$$\begin{aligned}
&|D_{ij}\mathcal{E}_g^{(a)}(u)(\mathbf{x}) - D_{ij}\mathcal{E}_g^{(a)}(u)(\bar{\mathbf{x}})| \\
&\le |D_{ij}\mathcal{E}_g^{(a)}(u)(\mathbf{x}) - D_{ij}\mathcal{E}_g^{(a)}(u)(\hat{\mathbf{x}})| + |D_{ij}u(\hat{\mathbf{x}}) - D_{ij}u(\bar{\mathbf{x}})| \\
&\le C[u]_{2,\alpha,\mathcal{D}}|\mathbf{x} - \bar{\mathbf{x}}|^\alpha.
\end{aligned}$$

Thus, (13.9.23) is proved for Case (a).

Next, we consider Case (b) of (13.9.24). Then there are two cases: Either L intersects Γ_g^a or it does not.

If L intersects Γ_g^a, there exists $\mathbf{x}^{(1)}, \mathbf{x}^{(2)} \in L \cap \Gamma_g^a(b_1, b_2)$ such that the open segments connecting $\mathbf{x}$ to $\mathbf{x}^{(1)}$ and $\bar{\mathbf{x}}$ to $\mathbf{x}^{(2)}$ lie within $(\Omega^a_{(1+\kappa)g} \setminus \overline{\Omega^a_g})(b_1, b_2)$. Then $|\mathbf{x} - \mathbf{x}^{(1)}| + |\mathbf{x}^{(1)} - \mathbf{x}^{(2)}| + |\mathbf{x}^{(2)} - \bar{\mathbf{x}}| = |\mathbf{x} - \bar{\mathbf{x}}|$, and we can apply (13.9.23) for both $(\mathbf{x}, \mathbf{x}^{(1)})$ and $(\bar{\mathbf{x}}, \mathbf{x}^{(2)})$, and use $D_{ij}\mathcal{E}_g^{(a)}(u)(\mathbf{x}^{(k)}) = D_{ij}u(\mathbf{x}^{(k)})$ for $k = 1, 2$, to complete estimate (13.9.23) similarly to Case (a).

If L does not intersect Γ_g^a, let $\hat{\mathbf{x}} = (\hat{x}_1, \hat{x}_2) \in \overline{L}$ be such that

$$\hat{x}_2 - g(\hat{x}_1) = \min_{\mathbf{x}' \in \overline{L}} \{x_2' - g(x_1')\}. \tag{13.9.30}$$

The conditions of Case (b) in (13.9.24) imply that $\overline{L} \subset (\Omega^a_{(1+\kappa)g} \setminus \overline{\Omega^a_g})(b_1, b_2)$, so that $\hat{x}_2 - g(\hat{x}_1) > 0$. Denote $\hat{\mathbf{y}} := (\hat{x}_1, g(\hat{x}_1))$. Then $\hat{\mathbf{y}} \in \Gamma_g^a(b_1, b_2)$, and the open segments connecting $\mathbf{x}$ to $\hat{\mathbf{y}}$ and $\bar{\mathbf{x}}$ to $\hat{\mathbf{y}}$ lie in $(\Omega^a_{(1+\kappa)g} \setminus \overline{\Omega^a_g})(b_1, b_2)$; otherwise, (13.9.30) would be violated. Thus, we can apply (13.9.23) for both $(\mathbf{x}, \hat{\mathbf{y}})$ and $(\bar{\mathbf{x}}, \hat{\mathbf{y}})$. Also, using the conditions of Case (b), we have

$$|\mathbf{x} - \hat{\mathbf{y}}| \le |\mathbf{x} - \hat{\mathbf{x}}| + |\hat{\mathbf{x}} - \hat{\mathbf{y}}| = |\mathbf{x} - \hat{\mathbf{x}}| + |\hat{x}_2 - g(\hat{x}_1)| \le 2|\mathbf{x} - \bar{\mathbf{x}}|,$$

and similarly, $|\bar{\mathbf{x}} - \hat{\mathbf{y}}| \le 2|\mathbf{x} - \bar{\mathbf{x}}|$. Then

$$\begin{aligned}
&|D_{ij}\mathcal{E}_g^{(a)}(u)(\mathbf{x}) - D_{ij}\mathcal{E}_g^{(a)}(u)(\bar{\mathbf{x}})| \\
&\le |D_{ij}\mathcal{E}_g^{(a)}(u)(\mathbf{x}) - D_{ij}\mathcal{E}_g^{(a)}(u)(\hat{\mathbf{y}})| + |D_{ij}\mathcal{E}_g^{(a)}(u)(\hat{\mathbf{y}}) - D_{ij}\mathcal{E}_g^{(a)}(u)(\bar{\mathbf{x}})| \\
&\le C[u]_{2,\alpha,\mathcal{D}}|\mathbf{x} - \bar{\mathbf{x}}|^\alpha.
\end{aligned}$$

Now (13.9.23) is proved for Case (b). Claim 13.9.7 is proved.

4. We consider Case (c) of (13.9.24). Using Lemma 13.9.1(iii), we obtain that, for Case (c),

$$|\mathbf{x} - \bar{\mathbf{x}}| \le C(\mathrm{Lip}[g]) \min_{\mathbf{x}' \in \overline{L}} \delta_g(\mathbf{x}'). \tag{13.9.31}$$

Notice that, for any $\mathbf{x}^{(1)}, \mathbf{x}^{(2)} \in \overline{L}$, letting $\mathbf{y}^{(2)} \in \Gamma_g^a(b_1, b_2)$ be such that $|\mathbf{x}^{(2)} - \mathbf{y}^{(2)}| = \mathrm{dist}(\mathbf{x}^{(2)}, \Gamma_g^a(b_1, b_2))$, we have

$$\begin{aligned}
\mathrm{dist}(\mathbf{x}^{(1)}, \Gamma_g^a(b_1, b_2)) &\le |\mathbf{x}^{(1)} - \mathbf{y}^{(2)}| \le |\mathbf{x}^{(1)} - \mathbf{x}^{(2)}| + |\mathbf{x}^{(2)} - \bar{\mathbf{y}}^{(2)}| \\
&\le |\bar{\mathbf{x}} - \mathbf{x}| + \mathrm{dist}(\mathbf{x}^{(2)}, \Gamma_g^a(b_1, b_2)) \le C\delta_g(\mathbf{x}^{(2)}),
\end{aligned}$$

where we have used (13.9.31) and Lemma 13.9.1(i) in the last estimate. Therefore, we obtain the existence of $C(\mathrm{Lip}[g])$ such that, for Case (c),

$$\delta_g(\mathbf{x}^{(1)}) \le C\delta_g(\mathbf{x}^{(2)}) \qquad \text{for all } \mathbf{x}^{(1)}, \mathbf{x}^{(2)} \in \overline{L}. \tag{13.9.32}$$

We use (13.9.17)–(13.9.18), (13.9.21), and the estimate of $|I_m(\mathbf{x}) - I_m(\bar{\mathbf{x}})|$ for $m = 1, 2, 3, 4$, to obtain

$$\begin{aligned}
&|I_1(\mathbf{x}) - I_1(\bar{\mathbf{x}})| \\
&= \Big|\int_{-\infty}^{\infty} \big(D_{ij}u(x_1,\, x_2 - \lambda\delta_g^*(\mathbf{x})) - D_{ij}u(\bar{x}_1,\, \bar{x}_2 - \lambda\delta_g^*(\bar{\mathbf{x}}))\big)\psi(\lambda)\, d\lambda\Big| \\
&\le C[u]_{2,\alpha,\mathcal{D}}\big(|\mathbf{x} - \bar{\mathbf{x}}|^\alpha + |\delta_g(\mathbf{x}) - \delta_g(\bar{\mathbf{x}})|^\alpha\big) \le C[u]_{2,\alpha,\mathcal{D}}|\mathbf{x} - \bar{\mathbf{x}}|^\alpha,
\end{aligned} \tag{13.9.33}$$

where we have used Lemma 13.9.1(ii) in the last estimate.

Next we estimate the difference quotient for I_2. We discuss only the first term in I_2:

$$I_{21}(\mathbf{x}) - I_{21}(\bar{\mathbf{x}}) = J_{21} + J_{22}, \tag{13.9.34}$$

where

$$J_{21} = D_i\delta_g^*(\mathbf{x}) \int_{-\infty}^{\infty} \big(D_{i2}u(x_1,\, x_2 - \lambda\delta_g^*(\mathbf{x})) - D_{i2}u(\bar{x}_1,\, \bar{x}_2 - \lambda\delta_g^*(\bar{\mathbf{x}}))\big)\lambda\psi(\lambda)\, d\lambda,$$

$$J_{22} = \big(D_i\delta_g^*(\mathbf{x}) - D_i\delta_g^*(\bar{\mathbf{x}})\big) \int_{-\infty}^{\infty} D_{i2}u(\bar{x}_1,\, \bar{x}_2 - \lambda\delta_g^*(\bar{\mathbf{x}}))\lambda\psi(\lambda)\, d\lambda.$$

Then J_{21} is estimated similarly to $|I_1(\mathbf{x}) - I_1(\bar{\mathbf{x}})|$. Now we estimate J_{22} by using the orthogonality properties in (13.9.12):

$$J_{22} = \big(D_i\delta_g^*(\mathbf{x}) - D_i\delta_g^*(\bar{\mathbf{x}})\big) \int_{-\infty}^{\infty} \big(D_{i2}u(\bar{x}_1,\, \bar{x}_2 - \lambda\delta_g^*(\bar{\mathbf{x}})) - D_{i2}u(\bar{\mathbf{x}})\big)\lambda\psi(\lambda)\, d\lambda, \tag{13.9.35}$$

so that, from (13.9.31)–(13.9.32),

$$|J_{22}| \le C\frac{|\mathbf{x} - \bar{\mathbf{x}}|}{\inf_{\mathbf{x}' \in L} \delta(\mathbf{x}')}[u]_{2,\alpha,\mathcal{D}}(\delta_g(\bar{\mathbf{x}}))^\alpha \le C[u]_{2,\alpha,\mathcal{D}}|\mathbf{x} - \bar{\mathbf{x}}|^\alpha. \tag{13.9.36}$$

This completes the estimate of $|I_{21}(\mathbf{x}) - I_{21}(\bar{\mathbf{x}})|$. Then $|I_3(\mathbf{x}) - I_3(\bar{\mathbf{x}})|$ is estimated similarly.

Finally, we estimate

$$I_4(\mathbf{x}) - I_4(\bar{\mathbf{x}}) = J_{41} + J_{42},$$

where

$$J_{41} = (\lambda - 1)D_{ij}\delta_g^*(\mathbf{x}) \int_{-\infty}^{\infty} \big(R(\mathbf{x}, \lambda) - R(\bar{\mathbf{x}}, \lambda)\big)\lambda\psi(\lambda)\, d\lambda,$$

$$J_{42} = (\lambda - 1)\big(D_{ij}\delta_g^*(\mathbf{x}) - D_{ij}\delta_g^*(\bar{\mathbf{x}})\big) \int_{-\infty}^{\infty} R(\bar{\mathbf{x}}, \lambda)\lambda\psi(\lambda)\, d\lambda.$$

From (13.9.20), for $\lambda \in [1,2]$,

$$
\begin{aligned}
&|R(\mathbf{x},\lambda) - R(\bar{\mathbf{x}},\lambda)| \\
&\le C\delta_g(\mathbf{x})\big(\big|u_{x_2x_2}(x_1, x_2 - (1+t(\lambda-1))\delta_g^*(\mathbf{x})) - u_{x_2x_2}(\bar{x}_1,\ \bar{x}_2 - \lambda\delta_g^*(\bar{\mathbf{x}}))\big| \\
&\qquad + \big|u_{x_2x_2}(x_1, x_2 - \delta_g^*(\mathbf{x})) - u_{x_2x_2}(\bar{x}_1, \bar{x}_2 - \delta_g^*(\bar{\mathbf{x}}))\big|\big) \\
&\le C\big(\delta_g(\mathbf{x})|\mathbf{x}-\bar{\mathbf{x}}|^\alpha + (\delta_g(\mathbf{x}))^\alpha|\mathbf{x}-\bar{\mathbf{x}}|\big).
\end{aligned}
$$

With this, term J_{41} is estimated similarly to J_{22} in (13.9.36). To estimate J_{42}, we employ (13.9.29) and the fact that

$$
|\delta_g^*(\mathbf{x}) - D_{ij}\delta_g^*(\bar{\mathbf{x}})| \le \frac{C}{\inf_{\mathbf{x}'\in L}(\delta(\mathbf{x}'))^2}|\mathbf{x}-\bar{\mathbf{x}}|,
$$

and then follow the argument for (13.9.36). Now (13.9.16) is proved.

5. It remains to show assertion (iii). Since $\|u_i\|_{2,\alpha,\Omega_{g_i}^{a_i}(\frac{a_i}{a}b_1,\frac{a_i}{a}b_2)} \le M$ for all i, then, using the estimate of assertion (ii) for $\mathcal{E}_{g_i}^{(a_i)}(u_i)$, $i = 1, 2, \dots$, and the convergence of (a_i, g_i) to (a, g) in the sense described in (iii), it follows that, for all sufficiently large i, functions $w_i(\mathbf{x}) := \mathcal{E}_{g_i}^{(a_i)}(u_i)(\frac{a_i}{a}x_1, x_2)$ are defined on $\Omega^a_{(1+\frac{1}{2}\kappa)g}(b_1,b_2)$ and satisfy

$$
\|w_i\|_{2,\alpha,\Omega^a_{(1+\frac{1}{2}\kappa)g}(b_1,b_2)} \le C\|u_i\|_{2,\alpha,\Omega_{g_i}^{a_i}(\frac{a_i}{a}b_1,\frac{a_i}{a}b_2)} \le CM.
$$

Then, if $\beta < \alpha$, it follows that, from every subsequence of w_i, we can extract a further subsequence that converges in $C^{2,\beta}(\Omega^a_{(1+\frac{1}{2}\kappa)g})(b_1,b_2)$. Thus, it remains to prove that the limit in $C^{2,\beta}(\Omega^a_{(1+\frac{1}{2}\kappa)g})(b_1,b_2)$ of each of such sequences is u, for which it suffices to show that

$$
w_i(\mathbf{x}) \to u(\mathbf{x}) \qquad \text{for all } \mathbf{x} \in \Omega^a_{(1+\frac{1}{2}\kappa)g}(b_1,b_2)\setminus\Gamma^a_g. \tag{13.9.37}
$$

If $\mathbf{x} = (x_1,x_2) \in \Omega^a_g(b_1,b_2)$, $(\frac{a_i}{a}x_1, x_2) \in \Omega^{a_i}_{g_i}(\frac{a_i}{a}b_1, \frac{a_i}{a}b_2)$ for all sufficiently large i. Thus, the uniform convergence $u_i \to u$ on compact subsets of the open set $\Omega^a_g(b_1,b_2)$, combined with $a_i \to a$, implies $w_i(\mathbf{x}) = u_i(\frac{a_i}{a}x_1, x_2) \to u(\mathbf{x})$.

Now let $\mathbf{x} = (x_1,x_2) \in (\Omega^a_{(1+\frac{1}{2}\kappa)g}\setminus\overline{\Omega^a_g})(b_1,b_2)$. Then $(\frac{a_i}{a}x_1, x_2) \in (\Omega^a_{(1+\kappa)g_i}\setminus\overline{\Omega^{a_i}_{g_i}})(\frac{a_i}{a}b_1, \frac{a_i}{a}b_2)$ for all sufficiently large i so that

$$
\begin{aligned}
w_i(\mathbf{x}) &= \mathcal{E}_{g_i}^{(a_i)}(u_i)(\frac{a_i}{a}x_1,\ x_2) \\
&= \int_{-\infty}^{\infty} u_i(\frac{a_i}{a}x_1,\ x_2 - \lambda\delta^*_{g_i}(\frac{a_i}{a}x_1,\ x_2))\psi(\lambda)\,d\lambda.
\end{aligned} \tag{13.9.38}
$$

Then, using Lemma 13.9.4 and the convergence: $(a_i, g_i) \to (a, g)$ given in (iii), it follows that there exists a compact set $K(\mathbf{x}) \subset \Omega^a_g$ such that

$$
(\frac{a_i}{a}x_1,\ x_2 - \lambda\delta^*_{g_i}(\frac{a_i}{a}x_1,\ x_2)) \subset K(\mathbf{x}) \qquad \text{for all } \lambda \in \operatorname{supp}(\psi)
$$

for all sufficiently large i. It also follows that $K(\mathbf{x}) \subset \Omega_{g_i}^{a_i}$ for all large i. Then, using the uniform convergence: $u_i \to u$ on $K(\mathbf{x})$ and Lemma 13.9.1(iv), we find that the integrands of (13.9.38) converge to the integrand of (13.9.14) for each $\lambda \in \mathbb{R}$ as $i \to \infty$. Also, the uniform estimates on u_i, Lemma 13.9.1(i), and $\mathrm{supp}(\psi) \in [1,2]$ imply that the integrands of (13.9.38) are uniformly bounded on $[1,2]$, and the integration is actually over $[1,2]$. Therefore, by the dominated convergence theorem, the integrals in (13.9.38) converge to the integral in (13.9.14). Now (13.9.37) is proved.

This completes the proof. □

Proof of Theorem 13.9.5. The fact that $\mathcal{E}_g^{(a)}$ is linear and maps $C^2(\Omega_g^a \cup \Gamma_g^a)$ into $C^2(\Omega^a_{(1+\kappa)g})$ follows from Lemma 13.9.6. It only remains to prove assertions (ii)–(iii). Then assertion (i) follows from (ii).

We focus first on assertion (ii). For this, we need to show that the estimates for the extension operator, proved in Lemma 13.9.6 in the $C^{2,\alpha}$–norms, can be proved for the weighted norms in (13.9.2). We divide the proof into four steps.

1. We first prove the estimate:

$$\|\mathcal{E}^{(a)}(u)\|^{(-1-\alpha),\Sigma^a_{(1+\kappa)g}}_{2,\alpha,\Omega^a_{(1+\kappa)g}\cap\{s>\frac{\varepsilon}{10}\}} \le C\|u\|^{(-1-\alpha),\Sigma^a_g}_{2,\alpha,\Omega^a_g\cap\{s>\frac{\varepsilon}{10}\}}. \tag{13.9.39}$$

We note that, from the structure of Ω_g^a in (13.9.1), it follows that, for any $\mathbf{x} \in \Omega_g^a$,

$$\mathrm{dist}(\mathbf{x}, \Sigma_g^a) \ge a - x_1 \ge \frac{\mathrm{dist}(\mathbf{x}, \Sigma_g^a)}{\sqrt{1+(\mathrm{Lip}[g])^2}}.$$

Thus, in definition (4.1.5) of $\|\cdot\|^{(-1-\alpha),\Sigma^a_g}_{2,\alpha,\Omega^a_g\cap\{s>\frac{\varepsilon}{10}\}}$ (resp. $\|\cdot\|^{(-1-\alpha),\Sigma^a_{(1+\kappa)g}}_{2,\alpha,\Omega^a_{(1+\kappa)g}\cap\{s>\frac{\varepsilon}{10}\}}$), we can replace $\delta_{\mathbf{x}} := \mathrm{dist}(\mathbf{x}, \Sigma_g^a)$ (resp. $\delta_{\mathbf{x}} := \mathrm{dist}(\mathbf{x}, \Sigma^a_{(1+\kappa)g})$) by $\delta_{\mathbf{x}} := a - x_1$ in both cases – this will change C in (13.9.39) by a multiplicative constant depending only on $(\mathrm{Lip}[g], \alpha)$. With $\delta_{\mathbf{x}} = a - x_1$, we see from (4.1.5) that $u \in C^{(-1-\alpha),\Sigma^a_g}_{2,\alpha,\Omega^a_g\cap\{s>\varepsilon/10\}}$ if and only if there exists a constant M such that, using the notations in Lemma 13.9.6, for any $b_1, b_2 \in (\frac{\varepsilon}{10}, a)$ with $b_1 < b_2$, the following estimates hold:

$$\|u\|_{1,0,\Omega^a_g(b_1,b_2)} \le M, \quad \|u\|_{2,0,\Omega^a_g(b_1,b_2)} \le M(a-b_2)^{\alpha-1},$$
$$\|u\|_{2,\alpha,\Omega^a_g(b_1,b_2)} \le M(a-b_2)^{-1}.$$

Moreover, if such a constant M exists, $\|u\|^{(-1-\alpha),\Sigma^a_g}_{2,\alpha,\Omega^a_g\cap\{s>\frac{\varepsilon}{10}\}} \le CM$, where C is a uniform constant (actually, $C = 9$). Now (13.9.39) follows directly from (13.9.15)–(13.9.16).

2. Now we show the estimate:

$$\|\mathcal{E}^{(a)}(u)\|^{(\sigma),(\mathrm{par})}_{2,\alpha,\Omega^a_{(1+\kappa)g}\cap\{s<\varepsilon\}} \le C\|u\|^{(\sigma),(\mathrm{par})}_{2,\alpha,\Omega^a_g\cap\{s<\varepsilon\}}. \tag{13.9.40}$$

From (4.6.2), it follows that $u \in C^{2,\alpha}_{\sigma,(\mathrm{par})}(\Omega^a_g \cap \{s < \varepsilon\})$ if and only if there exists M such that, for any $\mathbf{x}, \bar{\mathbf{x}} \in \Omega^a_g \cap \{s < \varepsilon\}$ and $i, j = 1, 2$,

$$|u(\mathbf{x})| \le M x_1^{\sigma}, \quad |D_i u(\mathbf{x})| \le M x_1^{d_1(i)}, \tag{13.9.41}$$

$$|D_{ij} u(\mathbf{x})| \le M x_1^{d_2(i+j)}, \tag{13.9.42}$$

$$|D_{ij} u(\mathbf{x}) - D_{ij} u(\bar{\mathbf{x}})| \le M x_1^{d_3(i+j)} \delta^{(\mathrm{par})}_{\alpha}(\mathbf{x}, \bar{\mathbf{x}}), \tag{13.9.43}$$

where

$$d_1(k) = \sigma + \frac{k-3}{2}, \; d_2(k) = \sigma + \frac{k}{2} - 3, \; d_3(k) = \sigma + \frac{k}{2} - 3 - \alpha, \tag{13.9.44}$$

and $\delta^{(\mathrm{par})}_{\alpha}(\cdot)$ is defined by (4.6.1). Moreover, if such M exists, then

$$\|u\|^{(\sigma),(\mathrm{par})}_{2,\alpha,\Omega^a_g \cap \{s<\varepsilon\}} \le CM,$$

where C is uniform (actually, $C = 9$).

To prove estimate (13.9.40), we show that $\mathcal{E}^{(a)}(u)$ satisfies estimates (13.9.41)–(13.9.43) with $M = C\|u\|^{(\sigma),(\mathrm{par})}_{2,\alpha,\Omega^a_g \cap \{s<\varepsilon\}}$ in domain $\Omega^a_{(1+\kappa)g} \cap \{s < \varepsilon\}$, where C depends only on $(\mathrm{Lip}[g], \alpha, \sigma)$. In order to obtain these estimates, we repeat the estimates of Lemma 13.9.6, with the changes due to the fact that u satisfies (13.9.41)–(13.9.43) in $\Omega^a_g \cap \{s < \varepsilon\}$ and that the weighted distances $\delta^{(\mathrm{par})}_1(\mathbf{x}, \bar{\mathbf{x}})$ and $\delta^{(\mathrm{par})}_{\alpha}(\mathbf{x}, \bar{\mathbf{x}}) = (\delta^{(\mathrm{par})}_1(\mathbf{x}, \bar{\mathbf{x}}))^{\alpha}$ are used instead of the standard distance $|\mathbf{x} - \bar{\mathbf{x}}|$ in (13.9.43). Below we sketch the argument and give some details for the estimates of several typical terms.

Similarly to Lemma 13.9.6, we discuss only the estimates of the second derivatives, since the other terms are similar and simpler. We will write $\Omega^a_g(\varepsilon)$ for $\Omega^a_g \cap \{s < \varepsilon\}$ and define $\Gamma^a_g(\varepsilon)$, $(\Omega^a_g \cup \Gamma^a_g)(\varepsilon)$, and $(\Omega^a_{(1+\kappa)g} \setminus \Omega^a_g)(\varepsilon)$ similarly.

We first show (13.9.42) for $\mathcal{E}^{(a)}(u)$ in $\Omega^a_{(1+\kappa)g}(\varepsilon)$, which is

$$|D_{ij}\mathcal{E}^{(a)}(u)(\mathbf{x})| \le M x_1^{d_2(i+j)} \qquad \text{for all } \mathbf{x} \in \Omega^a_{(1+\kappa)g}(\varepsilon). \tag{13.9.45}$$

Since this estimate holds in $\Omega^a_g(\varepsilon)$ by the assumption for this theorem, it suffices to consider $\mathbf{x} \in (\Omega^a_{(1+\kappa)g} \setminus \Omega^a_g)(\varepsilon)$. Then, in order to show (13.9.45), we use (13.9.17)–(13.9.18), and expression (13.9.21) of I_4. We also note that, for $D_{ij}\mathcal{E}^{(a)}_g(u)$, these expressions involve $(D_{ij}u, D_{i2}u, D_{j2}u, D_{22}u, D_2 u)$, and that $d_2(\cdot)$ in (13.9.44) is monotone increasing so that, using $x_1 \in (0, \varepsilon)$ and $\varepsilon < 1$,

$$x_1^{d_2(2+2)} \le \max(x_1^{d_2(i+2)}, x_1^{d_2(j+2)}) \le x_1^{d_2(i+j)}. \tag{13.9.46}$$

In particular, by (13.9.20) and estimate (13.9.43) for u in $\Omega^a_g(\varepsilon)$, the estimate of $R(\mathbf{x}, \lambda)$ in (13.9.21) is

$$\sup_{\lambda \in [1,2]} \|R(\cdot, \lambda)\|_{0,0,\Omega^a_{(1+\kappa)g} \setminus \overline{\Omega^a_g(\varepsilon)}} \le C x_1^{d_2(2+2)} [u]^{\sigma,(\mathrm{par})}_{2,0,\Omega^a_g(\varepsilon)} \le C x_1^{d_2(i+j)} \|u\|^{\sigma,(\mathrm{par})}_{2,\alpha,\Omega^a_g(\varepsilon)},$$

which implies the estimate of I_4 by the right-hand side of (13.9.45). The estimates of I_1, I_2, and I_3 are obtained similarly. Now (13.9.45) follows.

3. Now we show (13.9.43) for $\mathcal{E}^{(a)}(u)$ in $\Omega^a_{(1+\kappa)g}(\varepsilon)$, which is

$$|D_{ij}\mathcal{E}^{(a)}(u)(\mathbf{x}) - D_{ij}\mathcal{E}^{(a)}(u)(\bar{\mathbf{x}})| \le M x_1^{d_3(i+j)}\delta^{(\mathrm{par})}_\alpha(\mathbf{x},\bar{\mathbf{x}}) \tag{13.9.47}$$

for all $\mathbf{x}, \bar{\mathbf{x}} \in \Omega^a_{(1+\kappa)g}(\varepsilon)$.

We first note some basic properties of the weighted *distance* $\delta^{(\mathrm{par})}_1$ in (4.6.1), which follow directly from its definition by a simple argument: Let $\mathbf{x} = (x_1, x_2)$, $\hat{\mathbf{x}} = (\hat{x}_1, \hat{x}_2)$, and $\bar{\mathbf{x}} = (\bar{x}_1, \bar{x}_2)$ with $x_1, \hat{x}_1, \bar{x}_1 \in (0,1)$, and let $L(\mathbf{x}, \hat{\mathbf{x}})$ be the open segment connecting $\mathbf{x}$ to $\hat{\mathbf{x}}$. Then

$$\begin{aligned}
&\delta^{(\mathrm{par})}_1(\mathbf{x}^{(1)}, \mathbf{x}^{(2)}) \le \delta^{(\mathrm{par})}_1(\mathbf{x}, \hat{\mathbf{x}}) \qquad \text{for all } \mathbf{x}^{(1)}, \mathbf{x}^{(2)} \in \overline{L(\mathbf{x}, \hat{\mathbf{x}})};\\
&\delta^{(\mathrm{par})}_1((x_1,\, x_2 + b), (\hat{x}_1,\, \hat{x}_2 + \hat{b}))\\
&\qquad \le \sqrt{2}\Big(\delta^{(\mathrm{par})}_1(\mathbf{x}, \hat{\mathbf{x}}) + \sqrt{\max(x_1, \hat{x}_1)}|b - \hat{b}|\Big);\\
&\delta^{(\mathrm{par})}_1(\mathbf{x}, \hat{\mathbf{x}}) \ge \sqrt{|x_1|}|\mathbf{x} - \hat{\mathbf{x}}|;\\
&\delta^{(\mathrm{par})}_1(\mathbf{x}, \hat{\mathbf{x}}) = \sqrt{|x_1|}|x_2 - \hat{x}_2| \qquad \text{if } x_1 = \hat{x}_1;\\
&\delta^{(\mathrm{par})}_1(\mathbf{x}, \hat{\mathbf{x}}) \le \delta^{(\mathrm{par})}_1(\mathbf{x}, \bar{\mathbf{x}}) + \delta^{(\mathrm{par})}_1(\bar{\mathbf{x}}, \hat{\mathbf{x}}) \qquad \text{if } \hat{x}_1 = \bar{x}_1.
\end{aligned} \tag{13.9.48}$$

We note that, in the proof of the first of the above properties, we have used that

$$\max(x_1^{(1)}, x_1^{(2)}) \le \max(x_1, \hat{x}_1) \qquad \text{if } \mathbf{x}^{(1)}, \mathbf{x}^{(2)} \in \overline{L(\mathbf{x}, \hat{\mathbf{x}})}.$$

Properties (13.9.48), combined with the *parabolic* estimates (13.9.43) of u in $\Omega^a_g(\varepsilon)$, allow us to obtain a proof of (13.9.47) by following the proof of (13.9.23) in Lemma 13.9.6 with just notational changes. Some details are the following:

Estimate (13.9.47) follows from the assumption on u if $\mathbf{x}, \bar{\mathbf{x}} \in \Omega^a_g(\varepsilon)$. Then it suffices to consider

$$\mathbf{x} \in (\Omega^a_{(1+\kappa)g} \setminus \Omega^a_g)(\varepsilon).$$

Similarly to Cases (13.9.24) in the proof of Lemma 13.9.6, we consider separately the cases:

$$\begin{aligned}
&\text{(a)}\;\; \bar{\mathbf{x}} \in (\Omega^a_g \cup \Gamma^a_g)(\varepsilon),\\
&\text{(b)}\;\; \bar{\mathbf{x}} \in \Omega^a_{(1+\kappa)g}(\varepsilon) \setminus \overline{\Omega^a_g(\varepsilon)}, \;\; \min_{\mathbf{x}' \in \overline{L}}(x_2' - g(x_1')) \le |\mathbf{x} - \bar{\mathbf{x}}|,\\
&\text{(c)}\;\; \bar{\mathbf{x}} \in \Omega^a_{(1+\kappa)g}(\varepsilon) \setminus \overline{\Omega^a_g(\varepsilon)}, \;\; \min_{\mathbf{x}' \in \overline{L}}(x_2' - g(x_1')) > |\mathbf{x} - \bar{\mathbf{x}}|,
\end{aligned} \tag{13.9.49}$$

where L denotes the open segment connecting $\mathbf{x}$ to $\bar{\mathbf{x}}$.

Cases (a)–(b) follow from estimate (13.9.47) when

$$\bar{\mathbf{x}} \in \Gamma^a_g(\varepsilon), \qquad L \subset \Omega^a_{(1+\kappa)g}(\varepsilon) \setminus \overline{\Omega^a_g(\varepsilon)}. \tag{13.9.50}$$

The proof of this fact repeats the proof of Claim 13.9.7 by using properties (13.9.48) of the weighted distance.

Now we show that (13.9.47) holds under conditions (13.9.50).

In the estimates below, similarly to the estimate of (13.9.45), we use the fact that, in expression (13.9.17)–(13.9.18) with (13.9.21) for I_4, $D_{ij}\mathcal{E}_g^{(a)}(u)$ is expressed through $(D_{ij}u, D_{i2}u, D_{j2}u, D_{22})$, and that (13.9.46) holds for $d_3(\cdot)$. Thus, the weights in all of the terms are estimated by $x_1^{d_3(i+j)}$, as needed for (13.9.47).

From (13.9.50),

$$\sqrt{x_1}\,\mathrm{dist}(\mathbf{x}, \Gamma_g^a(\varepsilon)) \le \sqrt{x_1}|\mathbf{x} - \bar{\mathbf{x}}| \le \delta_1^{(\mathrm{par})}(\mathbf{x}, \bar{\mathbf{x}}). \tag{13.9.51}$$

We estimate $|I_1(\mathbf{x}) - D_{ij}u(\bar{\mathbf{x}})|$ and $I_m(\mathbf{x})$ for $m = 2, 3, 4$.

To estimate $|I_1(\mathbf{x}) - D_{ij}u(\bar{\mathbf{x}})|$, we use its expression in (13.9.27). From (13.9.43), (13.9.48), and (13.9.51), we obtain that, for $\lambda \in [1, 2]$,

$$\begin{aligned}
&\big|u(x_1,\, x_2 - \lambda\delta_g^*(\mathbf{x})) - D_{ij}u(\bar{\mathbf{x}})\big| \\
&\quad \le C[u]^{(\sigma),(\mathrm{par})}_{2,\alpha,\Omega_g^a(\varepsilon)} x_1^{d_3(i+j)} \big(\delta_1^{(\mathrm{par})}(\mathbf{x}, \bar{\mathbf{x}}) + \sqrt{x_1}\delta_g(\mathbf{x})\big)^\alpha \\
&\quad \le C[u]^{(\sigma),(\mathrm{par})}_{2,\alpha,\Omega_g^a(\varepsilon)} x_1^{d_3(i+j)} \big(\delta_1^{(\mathrm{par})}(\mathbf{x}, \bar{\mathbf{x}})\big)^\alpha.
\end{aligned}$$

Using estimate (13.9.27), we obtain the estimate of $|I_1(\mathbf{x}) - D_{ij}u(\bar{\mathbf{x}})|$ by the right-hand side of (13.9.47).

Next we estimate $|I_{21}(\mathbf{x})|$ by using its expression in (13.9.28). For that, we first estimate for $\lambda \in [1, 2]$:

$$\begin{aligned}
&\big|D_{i2}u(x_1,\, x_2 - \lambda\delta_g^*(\mathbf{x})) - D_{i2}u(\mathbf{x})\big| \\
&\le C[u]^{(\sigma),(\mathrm{par})}_{2,\alpha,\Omega_g^a(\varepsilon)} x_1^{d_3(i+2)} \big(\sqrt{x_1}\delta_g(\mathbf{x})\big)^\alpha \le C[u]^{(\sigma),(\mathrm{par})}_{2,\alpha,\Omega_g^a(\varepsilon)} x_1^{d_3(i+j)} \big(\delta_1^{(\mathrm{par})}(\mathbf{x}, \bar{\mathbf{x}})\big)^\alpha.
\end{aligned}$$

Inserting this estimate into (13.9.28), we obtain the estimate of $|I_{21}(\mathbf{x})|$ by the right-hand side of (13.9.47).

Modifying similarly the estimates of the remaining terms in Step 3 of the proof of Lemma 13.9.6, we conclude the proof of (13.9.47) under conditions (13.9.50), hence for Cases (a)–(b) of (13.9.49).

It remains to show (13.9.47) under condition (c) in (13.9.49). We follow the argument in Step 4 of the proof of Lemma 13.9.6 by modifying the case of the weighted distances and parabolic norms. We continue to use (13.9.48).

From (13.9.31),

$$\delta_1^{(\mathrm{par})}(\mathbf{x}, \bar{\mathbf{x}}) \le C(\mathrm{Lip}[g])\sqrt{\max(x_1, \bar{x}_1)}\,\delta_g(\mathbf{x}') \qquad \text{for all } \mathbf{x}' \in \overline{L}. \tag{13.9.52}$$

We need to estimate $|I_m(\mathbf{x}) - I_m(\bar{\mathbf{x}})|$ for $m = 1, 2, 3, 4$.

First consider $m = 1$. We use the expression of $|I_1(\mathbf{x}) - I_1(\bar{\mathbf{x}})|$ in (13.9.33) and first estimate:

$$\begin{aligned}&\big|D_{ij}u(x_1,\ x_2 - \lambda\delta_g^*(\mathbf{x})) - D_{ij}u(\bar{x}_1,\ \bar{x}_2 - \lambda\delta_g^*(\bar{\mathbf{x}}))\big|\\ &\le C[u]^{(\sigma),(\mathrm{par})}_{2,\alpha,\Omega_g^a(\varepsilon)} x_1^{d_3(i+j)}\Big(\delta_1^{(\mathrm{par})}(\mathbf{x},\bar{\mathbf{x}}) + \sqrt{\max(x_1,\bar{x}_1)}|\delta_g(\mathbf{x}) - \delta_g(\bar{\mathbf{x}})|\Big)^\alpha\\ &\le C[u]^{(\sigma),(\mathrm{par})}_{2,\alpha,\Omega_g^a(\varepsilon)} x_1^{d_3(i+j)}\Big(\delta_1^{(\mathrm{par})}(\mathbf{x},\bar{\mathbf{x}}) + \sqrt{\max(x_1,\bar{x}_1)}|\mathbf{x} - \bar{\mathbf{x}}|\Big)^\alpha\\ &\le C[u]^{(\sigma),(\mathrm{par})}_{2,\alpha,\Omega_g^a(\varepsilon)} x_1^{d_3(i+j)}\big(\delta_1^{(\mathrm{par})}(\mathbf{x},\bar{\mathbf{x}})\big)^\alpha.\end{aligned}$$

Inserting this estimate into (13.9.33), we obtain the estimate of $|I_1(\mathbf{x}) - I_1(\bar{\mathbf{x}})|$ by the right-hand side of (13.9.47).

Next we estimate $|I_{21}(\mathbf{x}) - I_{21}(\bar{\mathbf{x}})|$ expressed as (13.9.34). Term J_{21} is estimated as $|I_1(\mathbf{x}) - I_1(\bar{\mathbf{x}})|$ above. Thus, we estimate J_{22} by using its expression in (13.9.35). Since

$$\begin{aligned}&\big|D_{i2}u(\bar{x}_1,\ \bar{x}_2 - \lambda\delta_g^*(\bar{\mathbf{x}})) - D_{i2}u(\bar{x}_1,\ \bar{x}_2))\big|\\ &\le C[u]^{(\sigma),(\mathrm{par})}_{2,\alpha,\Omega_g^a(\varepsilon)} x_1^{d_3(i+2)}\big(\sqrt{x_1}\delta_g(\mathbf{x})\big)^\alpha,\end{aligned}$$

then, using (13.9.31)–(13.9.32), we have

$$\begin{aligned}|J_{22}| &\le C\frac{|\mathbf{x} - \bar{\mathbf{x}}|}{\inf_{\mathbf{x}'\in L}\delta(\mathbf{x}')}[u]^{(\sigma),(\mathrm{par})}_{2,\alpha,\Omega_g^a(\varepsilon)} x_1^{d_3(i+2)}\big(\sqrt{x_1}\delta_g(\mathbf{x})\big)^\alpha\\ &\le C[u]^{(\sigma),(\mathrm{par})}_{2,\alpha,\Omega_g^a(\varepsilon)} x_1^{d_3(i+j)}\big(\sqrt{x_1}|\mathbf{x} - \bar{\mathbf{x}}|\big)^\alpha\\ &\le C[u]^{(\sigma),(\mathrm{par})}_{2,\alpha,\Omega_g^a(\varepsilon)} x_1^{d_3(i+j)}\big(\delta_1^{(\mathrm{par})}(\mathbf{x},\bar{\mathbf{x}})\big)^\alpha,\end{aligned}$$

that is, J_{22} is bounded by the right-hand side of (13.9.47). This completes the estimate of $|I_2(\mathbf{x}) - I_2(\bar{\mathbf{x}})|$.

The remaining terms (*i.e.*, $|I_m(\mathbf{x}) - I_m(\bar{\mathbf{x}})|$ for $m = 3, 4$) are estimated by following the argument in Step 4 of the proof of Lemma 13.9.6 with the modifications similar to those done above.

This completes the proof of (13.9.47), as well as the proof of (13.9.40). Combined with (13.9.39), this completes the proof of assertion (ii) of Theorem 13.9.5. Then assertion (i) also follows.

4. It remains to prove assertion (iii) of Theorem 13.9.5. From its condition, it follows that, for each $b_1, b_2 \in (0, a)$ with $b_1 < b_2$, we have

$$\|u_i\|_{2,\alpha,\Omega_{g_i}^{a_i}(\frac{a_i}{a}b_1,\frac{a_i}{a}b_2)} \le M(b_1, b_2) \qquad \text{for all } i.$$

Then, from Lemma 13.9.6(iii), we conclude that, for each (b_1, b_2) as above,

$$w_i \to \mathcal{E}_g^{(a)}(u) \qquad \text{in } C^{2,\frac{\alpha}{2}}(\Omega^a_{(1+\frac{1}{2}\kappa)g})(b_1, b_2).$$

In particular, $w_i \to \mathcal{E}_g^{(a)}(u)$ pointwise in $\Omega^a_{(1+\frac{1}{2}\kappa)g}$.

Since $\|u_i\|^{*,(\sigma)}_{2,\alpha,\Omega^{a_i}_{g_i}} \le M$ for all i, then, using the estimate of assertion (ii) of Theorem 13.9.5 for $\mathcal{E}^{(a_i)}_{g_i}(u_i)$, $i = 1, 2, \dots$, and the convergence of (a_i, g_i) to (a, g) in the sense described in (iii), we find that, for all sufficiently large i, functions $w_i(x_1, x_2) := \mathcal{E}^{(a_i)}_{g_i}(u_i)(\frac{a_i}{a}x_1, x_2)$ are defined on $\Omega^a_{(1+\frac{1}{2}\kappa)g}$ and satisfy

$$\|w_i\|^{*,(\sigma)}_{2,\alpha,\Omega^a_{(1+\frac{1}{2}\kappa)g}} \le C\|u_i\|^{*,(\sigma)}_{2,\alpha,\Omega^{a_i}_{g_i}} \le CM.$$

Then, if $\beta < \alpha$, it follows that, from every subsequence of w_i, we can extract a further subsequence that converges in $C^{2,\beta}_{*,\sigma'}(\Omega^a_{(1+\frac{1}{2}\kappa)g})$. Combining with the pointwise convergence in $\Omega^a_{(1+\frac{1}{2}\kappa)g}$ of the whole sequence w_i to $\mathcal{E}^{(a)}_g(u)$, we conclude the proof of assertion (iii). Theorem 13.9.5 is proved. □

We also note that, from Step 1 of the proof of Theorem 13.9.5, we obtain from Definition 13.9.3 that $\mathcal{E}^{(a)}$ also extends the functions in spaces $C^{(-1-\alpha),\Sigma^0_g\cup\Sigma^a_g}_{2,\alpha,\Omega^a_g}$ defined by norms (4.1.5).

Theorem 13.9.8. *Let $a > 0$. Let $g \in C^{0,1}([0, a])$ with $g(s) > 0$ on $(0, a)$. Let κ be the constant in Lemma 13.9.4. Then the extension operator $\mathcal{E}^{(a)} \equiv \mathcal{E}^{(a)}_g$, introduced in Definition 13.9.3, maps $C^2(\Omega^a_g \cup \Gamma^a_g)$ into $C^2(\Omega^a_{(1+\kappa)g})$ with the following properties for any $\alpha \in (0, 1)$:*

(i) *$\mathcal{E}^{(a)}$ is a linear continuous operator:*

$$\mathcal{E}^{(a)} : C^{(-1-\alpha),\Sigma^0_g\cup\Sigma^a_g}_{2,\alpha,\Omega^a_g} \mapsto C^{(-1-\alpha),\Sigma^0_{(1+\kappa)g}\cup\Sigma^a_{(1+\kappa)g}}_{2,\alpha,\Omega^a_{(1+\kappa)g}};$$

(ii) *There exists C depending only on $(\mathrm{Lip}[g], \alpha, \sigma)$ such that*

$$\|\mathcal{E}^{(a)}(u)\|^{(-1-\alpha),\Sigma^0_{(1+\kappa)g}\cup\Sigma^a_{(1+\kappa)g}}_{2,\alpha,\Omega^a_{(1+\kappa)g}} \le C\|u\|^{(-1-\alpha),\Sigma^0_g\cup\Sigma^a_g}_{2,\alpha,\Omega^a_g}$$

for any $u \in C^{(-1-\alpha),\Sigma^0_g\cup\Sigma^a_g}_{2,\alpha,\Omega^a_g}$, where C is independent of a;

(iii) *If $(u, Du) = (0, \mathbf{0})$ on Σ^0_g, then $(\mathcal{E}^{(a)}(u), D\mathcal{E}^{(a)}(u)) = (0, \mathbf{0})$ on $\Sigma^0_{(1+\kappa)g}$;*

(iv) *Let $g_i \in C^{0,1}([0, a_i])$ and $g \in C^{0,1}([0, a])$ satisfy (13.9.3) and*

$$\|g_i\|_{C^{0,1}([0,a_i])} \le L \qquad \textit{for all } i,$$

and let $a_i \to a$ and functions $g_i(\frac{a_i}{a}x_1)$ (defined on $[0, a]$) converge to $g(x_1)$ uniformly on $[0, a]$. If $u_i \in C^{(-1-\alpha),\Sigma^0_{g_i}\cup\Sigma^{a_i}_{g_i}}_{2,\alpha,\Omega^{a_i}_{g_i}}$, $u \in C^{(-1-\alpha),\Sigma^0_g\cup\Sigma^a_g}_{2,\alpha,\Omega^a_g}$, and $u_i \to u$ uniformly on compact subsets of the open set Ω^a_g, and if there exist $L, M > 0$ such that $\mathrm{Lip}[g_i] \le L$ and $\|u_i\|^{(-1-\alpha),\Sigma^0_{g_i}\cup\Sigma^{a_i}_{g_i}}_{2,\alpha,\Omega^{a_i}_{g_i}} \le M$ for all i, then

$$w_i(\mathbf{x}) := \mathcal{E}^{(a_i)}_{g_i}(u_i)(\frac{a_i}{a}x_1, x_2) \to \mathcal{E}^{(a)}_g(u)(\mathbf{x}) \qquad \textit{in } C^{(-1-\beta),\Sigma^0_g\cup\Sigma^a_g}_{2,\beta,\Omega^a_{(1+\frac{1}{2}\kappa)g}}$$

for any $\beta \in (0, \alpha)$ *and* $\sigma' \in (0, \sigma)$*, where we note that* w_i *is defined on* $\Omega^a_{(1+\frac{1}{2}\kappa)g}$ *for large* i*.*

Proof. We combine estimate (13.9.39) for space $C^{(-1-\alpha),\Sigma^a_g}_{2,\alpha,\Omega^a_g\cap\{x_1>\frac{a}{4}\}}$ with similar estimates for space $C^{(-1-\alpha),\Sigma^0_g}_{2,\alpha,\Omega^a_g\cap\{x_1<3a/4\}}$. This concludes (i) and (ii).

Fact (iii) follows directly from definition (13.9.14) of $\mathcal{E}^{(a)}_g(u)$ since $\Sigma^0_g = \overline{\Omega^a_g} \cap \{x_1 = 0\}$.

To prove (iv), we repeat the argument in Step 4 of the proof of Theorem 13.9.5, replacing spaces $C^{2,\alpha}_{*,\sigma}(\Omega^{a_i}_{g_i})$ by spaces $C^{(-1-\alpha),\Sigma^0_{g_i}\cup\Sigma^{a_i}_{g_i}}_{2,\alpha,\Omega^{a_i}_{g_i}}$. $\square$

Chapter Fourteen

Optimal Regularity of Solutions near the Sonic Circle

As indicated in Theorems 2.6.3 and 2.6.5, the global solution φ constructed is at least $C^{1,1}$ near the pseudo-sonic circle Γ_{sonic}. More specifically, the solutions constructed in the proofs of Theorems 2.6.3 and 2.6.5 are admissible solutions of **Problem 2.6.1**, so that estimate (11.4.4) near Γ_{sonic} is satisfied. This gives a weighted and scaled $C^{2,\alpha}$–regularity up to Γ_{sonic}, which implies the standard $C^{1,1}$–regularity of φ up to and across Γ_{sonic}. However, (11.4.4) does *not* imply the standard $C^{2,\alpha}$–regularity of ψ up to Γ_{sonic} or across Γ_{sonic}. We study these problems and present a complete proof of Theorem 2.6.6 in this chapter; also see [4]. In particular, we show that, for any admissible solution φ of **Problem 2.6.1**, φ is not C^2 across Γ_{sonic}, and the jump of the second derivative of φ across Γ_{sonic} in the radial direction of its center is determined solely by the adiabatic exponent, independent of the wedge angles $\theta_{\text{w}} \in (\theta_{\text{w}}^{\text{s}}, \frac{\pi}{2})$. Furthermore, we show the one-sided regularity that φ is $C^{2,\alpha}$ in Ω up to $\overline{\Gamma_{\text{sonic}}} \setminus \{P_1\}$.

14.1 REGULARITY OF SOLUTIONS NEAR THE DEGENERATE BOUNDARY FOR NONLINEAR DEGENERATE ELLIPTIC EQUATIONS OF SECOND ORDER

In order to study the regularity of solutions of **Problem 2.6.1**, we first study the regularity of solutions near a degenerate boundary for a class of nonlinear degenerate elliptic equations of second order.

Let φ be an admissible solution of **Problem 2.6.1**. We use the (x, y)–coordinates (11.1.1)–(11.1.2) in a neighborhood $\Omega \cap \mathcal{N}_\varepsilon(\Gamma_{\text{sonic}})$ of Γ_{sonic}, where ε is chosen small so that (11.1.3) holds. Then $\psi = \varphi - \varphi_2$, written in the (x, y)–coordinates, is a positive solution of equation (11.1.4) with (11.1.5) in $\Omega \cap \mathcal{N}_\varepsilon(\Gamma_{\text{sonic}})$. Also, ψ satisfies (11.0.2)–(11.0.3) and, more precisely, (11.2.42). Furthermore, ψ satisfies the estimates in (11.4.4). This, combined with (11.1.5), implies that terms $O_k(x, y) := O_k(D\psi(x, y), \psi(x, y), x)$ in equation (11.1.4) satisfy estimates (14.1.5)–(14.1.6) below. This motivates the structure of the equation considered in the rest of this section.

14.1.1 Nonlinear degenerate elliptic equations and the regularity theorem

We now study the regularity of positive solutions near the degenerate boundary for the Dirichlet problem for the following class of nonlinear degenerate elliptic equations with the form:

$$\begin{aligned} \mathcal{N}_1\psi &:= (2x - a\psi_x + O_1)\psi_{xx} + O_2\psi_{xy} + (b + O_3)\psi_{yy} - (1 + O_4)\psi_x + O_5\psi_y \\ &= 0 \qquad \text{in } Q^+_{r,R}, \qquad (14.1.1) \\ \psi &> 0 \qquad \text{in } Q^+_{r,R}, \qquad (14.1.2) \\ \psi &= 0 \qquad \text{on } \partial Q^+_{r,R} \cap \{x = 0\}, \qquad (14.1.3) \end{aligned}$$

where $a, b > 0$ are constants and, for $r, R > 0$,

$$Q^+_{r,R} := \{(x, y) \,:\, 0 < x < r,\ |y| < R\} \subset \mathbb{R}^2, \tag{14.1.4}$$

and terms $O_i(x, y), i = 1, \ldots, 5$, are continuously differentiable and

$$\frac{|O_1(x,y)|}{x^2},\ \frac{|O_k(x,y)|}{x} \le N \qquad \text{for } k = 2, \ldots, 5, \tag{14.1.5}$$

$$\frac{|DO_1(x,y)|}{x},\ |DO_k(x,y)| \le N \qquad \text{for } k = 2, \ldots, 5, \tag{14.1.6}$$

in $\{x > 0\}$ for some constant N.

Conditions (14.1.5)–(14.1.6) imply that terms $O_i, i = 1, \cdots, 5$, are *small*; the precise meaning of which can be seen in §14.2 for **Problem 2.6.1** (also see the estimates in §11.4). Thus, the main terms of equation (14.1.1) form the following equation:

$$(2x - a\psi_x)\psi_{xx} + b\psi_{yy} - \psi_x = 0 \qquad \text{in } Q^+_{r,R}. \tag{14.1.7}$$

Equation (14.1.7) is elliptic with respect to ψ in $\{x > 0\}$ if $\psi_x < \frac{2x}{a}$. In this chapter, we consider solutions ψ that satisfy

$$-Mx \le \psi_x \le \frac{2-\beta}{a}x \qquad \text{in } Q^+_{r,R} \tag{14.1.8}$$

for some constants $M \ge 0$ and $\beta \in (0, 1)$. Then (14.1.7) is uniformly elliptic in every subdomain $\{x > \delta\}$ with $\delta > 0$. The same is true for equation (14.1.1) in $Q^+_{r,R}$ if r is sufficiently small.

Remark 14.1.1. *If $\hat{r}$ is sufficiently small, depending only on (a, b, N),* (14.1.5)–(14.1.6) *and* (14.1.8) *imply that equation* (14.1.1) *is uniformly elliptic with respect to ψ in $Q^+_{\hat{r},R} \cap \{x > \delta\}$ for any $\delta \in (0, \frac{\hat{r}}{2})$. We always assume such a choice of $\hat{r}$ hereafter.*

Let $\psi \in C^2(Q^+_{\hat{r},R})$ be a solution of (14.1.1) satisfying (14.1.8). Remark 14.1.1 implies that the interior regularity:

$$\psi \in C^{2,\alpha}(Q^+_{\hat{r},R}) \qquad \text{for any } \alpha \in (0,1), \tag{14.1.9}$$

follows first from the linear elliptic theory in two dimensions (*cf.* [131, Chapter 12]) to conclude the solution in $C^{1,\alpha}$ which leads that the coefficient becomes C^α, and then from the Schauder theory to obtain the $C^{2,\alpha}$–estimate (*cf.* [131, Chapter 6]), where we have used the fact that $O_i \in C^1(\{x > 0\})$, $i = 1, \cdots, 5$. Therefore, we focus on the regularity of ψ near boundary $\{x = 0\} \cap \partial Q^+_{\hat{r},R}$, where the ellipticity of (14.1.1) degenerates.

Theorem 14.1.2 (Regularity Theorem). *Let $a, b, M, N, R > 0$ and $\beta \in (0, \frac{1}{4})$ be constants. Let $\psi \in C(\overline{Q^+_{\hat{r},R}}) \cap C^2(Q^+_{\hat{r},R})$ satisfy* (14.1.2)–(14.1.3), (14.1.8), *and equation* (14.1.1) *in $Q^+_{\hat{r},R}$ with $O_i = O_i(x,y)$ satisfying $O_i \in C^1(\overline{Q^+_{\hat{r},R}})$ and* (14.1.5)–(14.1.6). *Then*

$$\psi \in C^{2,\alpha}(\overline{Q^+_{\hat{r}/2,R/2}}) \qquad \textit{for any } \alpha \in (0,1),$$

with

$$\psi_{xx}(0,y) = \frac{1}{a}, \quad \psi_{xy}(0,y) = \psi_{yy}(0,y) = 0 \qquad \textit{for any } |y| < \tfrac{R}{2}.$$

To prove Theorem 14.1.2, it suffices to show that, for any given $\alpha \in (0,1)$,

$$\psi \in C^{2,\alpha}(\overline{Q^+_{r,R/2}}) \qquad \text{for some } r \in (0, \frac{\hat{r}}{2}), \tag{14.1.10}$$

since ψ belongs to $C^{2,\alpha}(\overline{Q^+_{\hat{r}/2,R/2} \cap \{x > \frac{r}{2}\}})$ by (14.1.9).

Note that, by (14.1.2)–(14.1.3) and (14.1.8), it follows that

$$0 < \psi(x,y) \le \frac{2-\beta}{2a} x^2 \qquad \text{for any } (x,y) \in Q^+_{\hat{r},R}. \tag{14.1.11}$$

The essential part of the proof of Theorem 14.1.2 is to show that, if a solution ψ satisfies (14.1.11), then, for any given $\alpha \in (0, 1)$, there exists $r \in (0, \frac{\hat{r}}{2}]$ such that

$$|\psi(x,y) - \frac{1}{2a} x^2| \le C x^{2+\alpha} \qquad \text{for any } (x,y) \in Q^+_{r,7R/8}. \tag{14.1.12}$$

Notice that, although $\psi^{(0)} \equiv 0$ is a solution of (14.1.1), it satisfies neither (14.1.12) nor the conclusion that $\psi^{(0)}_{xx}(0,y) = \frac{1}{a}$ of Theorem 14.1.2. Therefore, it is necessary to improve first the lower bound of ψ in (14.1.11) to separate our solution from the trivial solution $\psi^{(0)} \equiv 0$.

14.1.2 Quadratic lower bound of ψ

By Remark 14.1.1, equation (14.1.1) is strictly elliptic with respect to ψ inside $Q^+_{\hat{r},R}$. Thus, our idea is to construct a positive subsolution of (14.1.1), which provides our desired lower bound of ψ.

Proposition 14.1.3. *Let ψ satisfy the assumptions of Theorem* 14.1.2. *Then there exist $r \in (0, \frac{\hat{r}}{2}]$ and $\mu > 0$, depending only on $\inf_{Q^+_{\hat{r},R}\cap\{x>\hat{r}/2\}} \psi$ and $(a, b, N, R, \hat{r}, \beta)$, such that*

$$\psi(x, y) \geq \mu x^2 \qquad \textit{on } Q^+_{r,15R/16}.$$

Proof. In this proof, all the constants below depend only on the data, *i.e.*, $(a, b, M, N, R, \hat{r}, \beta)$ and $\inf_{Q^+_{\hat{r},R}\cap\{x>\hat{r}/2\}} \psi$, unless otherwise specified.

Fix y_0 with $|y_0| \leq \frac{15R}{16}$. We now prove that

$$\psi(x, y_0) \geq \mu x^2 \qquad \text{for } x \in (0, r). \tag{14.1.13}$$

We first note that, without loss of generality, we may assume that $R = 2$ and $y_0 = 0$. Otherwise, we set $\tilde{\psi}(x, y) := \psi(x, y_0 + \frac{R}{32}y)$ for any $(x, y) \in Q^+_{\hat{r},2}$. Then $\tilde{\psi} \in C(\overline{Q^+_{\hat{r},2}}) \cap C^2(Q^+_{\hat{r},2})$ satisfies equation (14.1.1) with (14.1.5) and conditions (14.1.2)–(14.1.3) and (14.1.8) in $Q^+_{\hat{r},2}$, with some modified constants (a, b, N, β) and functions O_i, depending only on R and the corresponding quantities in the original equation. Moreover,

$$\inf_{Q^+_{\hat{r},2}\cap\{x>\hat{r}/2\}} \tilde{\psi} = \inf_{Q^+_{\hat{r},R}\cap\{x>\hat{r}/2\}} \psi.$$

Then (14.1.13) for ψ follows from (14.1.13) for $\tilde{\psi}$ with $y_0 = 0$ and $R = 2$. Therefore, we will keep the original notation with $y_0 = 0$ and $R = 2$. Then it suffices to prove

$$\psi(x, 0) \geq \mu x^2 \qquad \text{for } x \in (0, r). \tag{14.1.14}$$

By Remark 14.1.1 and the Harnack inequality, we conclude that, for any $r \in (0, \frac{\hat{r}}{2})$, there exists $\sigma = \sigma(r) > 0$, depending only on r, $(a, b, N, R, \hat{r}, \beta)$, and $\inf_{Q^+_{\hat{r},R}\cap\{x>\hat{r}/2\}} \psi$, such that

$$\psi \geq \sigma \qquad \text{on } Q^+_{\hat{r},3/2} \cap \{x > r\}. \tag{14.1.15}$$

Let $r \in (0, \frac{\hat{r}}{2})$, $k > 0$, and

$$0 < \mu \leq \frac{\sigma(r)}{r^2} \tag{14.1.16}$$

which will be chosen later. Set

$$w(x, y) := \mu x^2(1 - y^2) - kxy^2. \tag{14.1.17}$$

Then, using (14.1.15)–(14.1.16), we obtain that, for any $x \in (0,r)$ and $|y| < 1$,

$$\begin{cases} w(0,y) = 0 \le \psi(0,y), \\ w(r,y) \le \mu r^2 \le \psi(r,y), \\ w(x,\pm 1) = -kx \le 0 \le \psi(x,\pm 1). \end{cases}$$

Therefore, we have

$$w \le \psi \qquad \text{on } \partial Q^+_{r,1}. \tag{14.1.18}$$

Next, we show that w is a strict subsolution $\mathcal{N}_1 w > 0$ in $Q^+_{r,1}$, if the parameters are chosen appropriately. In order to estimate $\mathcal{N}_1 w$, we denote

$$A_0 := \frac{k}{\mu} \tag{14.1.19}$$

and notice that

$$\begin{aligned} w_{yy} &= -2x(\mu x + k) = -2x(\mu x + k)\big((1-y^2) + y^2\big) \\ &= -2\mu x(1-y^2)(x + A_0) - 2ky^2 x\big(\frac{x}{A_0} + 1\big). \end{aligned}$$

Then, by a direct calculation and simplification, we obtain

$$\mathcal{N}_1 w = 2\mu x(1-y^2)I_1 + ky^2 I_2, \tag{14.1.20}$$

where

$$\begin{aligned} I_1 &= 1 - 2\mu a(1-y^2) - O_4 + \frac{O_1}{x} - (x + A_0)\big(b + O_3 + yO_5\big) - \frac{y(2x + A_0)}{x} O_2, \\ I_2 &= (1 + O_4) + 2\mu a(1-y^2) - 2x\big(\frac{x}{A_0} + 1\big)\big(b + O_3 + yO_5\big) - 2y\big(\frac{2x}{A_0} + 1\big)O_2. \end{aligned}$$

Now we choose r and μ so that $\mathcal{N}_1 w \ge 0$ holds. Clearly, $\mathcal{N}_1 w \ge 0$ if $I_1, I_2 \ge 0$. By (14.1.5), we find that, in $Q^+_{r,1}$,

$$\begin{aligned} I_1 &\ge 1 - 2\mu a - C_0 r - (b + N + C_0 r)A_0, \\ I_2 &\ge 1 - C_0 r - \frac{r}{A_0} C_0 r. \end{aligned} \tag{14.1.21}$$

Choose r_0 to satisfy the smallness assumptions stated above and

$$0 < r_0 \le \min\{\frac{1}{4C_0}, \frac{b+N}{C_0}, \frac{1}{8\sqrt{C_0(b+N)}}, \frac{\hat{r}}{2}\}, \tag{14.1.22}$$

where C_0 is the constant in (14.1.21). For such a fixed r_0, we choose μ_0 to satisfy (14.1.16) and

$$\mu_0 \le \frac{1}{8a}, \tag{14.1.23}$$

and A_0 to satisfy

$$4C_0r_0^2 < A_0 < \frac{1}{8(b+N)}, \tag{14.1.24}$$

where we have used (14.1.22) to see that $4C_0r_0^2 < \frac{1}{8(b+N)}$ in (14.1.24). Then k is defined from (14.1.19). From (14.1.21)–(14.1.24),

$$I_1,\ I_2 > 0,$$

which implies

$$\mathcal{N}_1 w > 0 \qquad \text{in } Q^+_{r,1} \tag{14.1.25}$$

whenever $r \in (0,\ r_0]$ and $\mu \in (0,\ \mu_0]$.

By (14.1.18), (14.1.25), Remark 14.1.1, and the comparison principle (Lemma 4.1.3), we have

$$\psi(x,y) \geq w(x,y) = \mu x^2(1-y^2) - kxy^2 \qquad \text{in } Q^+_{r,1}.$$

In particular,

$$\psi(x,0) \geq \mu x^2 \qquad \text{for } x \in [0,\ r]. \tag{14.1.26}$$

This implies (14.1.14), hence (14.1.13). The proof is completed. □

With Proposition 14.1.3, we now make the $C^{2,\alpha}$–estimate of ψ.

14.1.3 $C^{2,\alpha}$–estimate of ψ

If ψ satisfies (14.1.1)–(14.1.3) and (14.1.8), it is expected that ψ is *very close* to $\frac{x^2}{2a}$, which is a solution to (14.1.7). More precisely, we now prove (14.1.12). To achieve this, we study the function:

$$W(x,y) := \frac{x^2}{2a} - \psi(x,y). \tag{14.1.27}$$

By (14.1.1), W satisfies

$$\begin{aligned} \mathcal{N}_2 W := &(x + aW_x + O_1)W_{xx} + O_2 W_{xy} + (b + O_3)W_{yy} \\ &-(2+O_4)W_x + O_5 W_y = \frac{O_1 - xO_4}{a} \qquad \text{in } Q^+_{\hat{r},R}, \end{aligned} \tag{14.1.28}$$

$$W(0,y) = 0 \qquad \text{on } \partial Q^+_{\hat{r},R} \cap \{x=0\}, \tag{14.1.29}$$

$$-\frac{1-\beta}{a}x \leq W_x(x,y) \leq (M + \frac{1}{a})x \qquad \text{in } Q^+_{\hat{r},R}. \tag{14.1.30}$$

Lemma 14.1.4. *Let $(a,b,N,R,\hat{r},\beta)$ and O_i be as in Theorem 14.1.2. Let μ be the constant determined in Proposition 14.1.3. Then there exist $\alpha_1 \in (0,\ 1)$ and $r_1 > 0$ such that, if $W \in C(\overline{Q^+_{\hat{r},R}}) \cap C^2(Q^+_{\hat{r},R})$ satisfies (14.1.28)–(14.1.30),*

$$W(x,y) \leq \frac{1-\mu_1}{2ar^\alpha}x^{2+\alpha} \qquad \textit{in } Q^+_{r,7R/8}, \tag{14.1.31}$$

whenever $\alpha \in (0,\ \alpha_1]$ and $r \in (0, r_1]$ with $\mu_1 := \min\{2a\mu, \frac{1}{2}\}$.

Proof. In the proof below, all the constants depend only on the data, *i.e.*, $(a, b, N, \beta, R, \hat{r})$ and $\inf_{Q^+_{\hat{r},R}\cap\{x>\hat{r}/2\}}\psi$, unless otherwise specified.

By Proposition 14.1.3,

$$W(x,y) \le \frac{1-\mu_1}{2a}x^2 \qquad \text{in } Q^+_{r_0,15R/16}, \tag{14.1.32}$$

where r_0 depends only on $(a, b, N, R, \hat{r}, \beta)$.

Fix y_0 with $|y_0| \le \frac{7R}{8}$. We now prove that

$$W(x, y_0) \le \frac{1-\mu_1}{2ar^\alpha}x^{2+\alpha} \qquad \text{for } x \in (0, r).$$

By a scaling argument similar to the one at the beginning of proof of Lemma 14.1.3, *i.e.*, considering $\tilde{\psi}(x,y) = \psi(x, y_0 + \frac{R}{32}y)$ in $Q^+_{\hat{r},2}$, we can assume without loss of generality that $y_0 = 0$ and $R = 2$. That is, it suffices to prove that

$$W(x,0) \le \frac{1-\mu_1}{2ar^\alpha}x^{2+\alpha} \qquad \text{for } x \in (0,r) \tag{14.1.33}$$

for some $r \in (0, r_0)$ and $\alpha \in (0, \alpha_1)$, under the assumptions that (14.1.28)–(14.1.30) hold in $Q^+_{\hat{r},2}$ and (14.1.32) holds in $Q^+_{r_0,2}$.

For any given $r \in (0, r_0)$, let

$$A_1 r^\alpha = \frac{1-\mu_1}{2a}, \quad B_1 = \frac{1-\mu_1}{2a}, \tag{14.1.34}$$

$$v(x,y) = A_1 x^{2+\alpha}(1-y^2) + B_1 x^2 y^2. \tag{14.1.35}$$

Since (14.1.29) holds on $\partial Q^+_{\hat{r},2} \cap \{x = 0\}$ and (14.1.32) holds in $Q^+_{r_0,2}$, we obtain that, for any $x \in (0, r)$ and $|y| \le 1$,

$$\begin{cases} v(0,y) = 0 = W(0,y), \\ v(r,y) = \big(A_1 r^\alpha(1-y^2) + B_1 y^2\big)r^2 = \frac{1-\mu_1}{2a}r^2 \ge W(r,y), \\ v(x,\pm 1) = B_1 x^2 = \frac{1-\mu_1}{2a}x^2 \ge W(x, \pm 1). \end{cases}$$

Then we have

$$W \le v \qquad \text{on } \partial Q^+_{r,1}. \tag{14.1.36}$$

We now show that $\mathcal{N}_2 v < \mathcal{N}_2 W$ in $Q^+_{r,1}$. From (14.1.28),

$$\mathcal{N}_2 v - \mathcal{N}_2 W = \mathcal{N}_2 v - \frac{O_1 - xO_4}{a}.$$

In order to rewrite the right-hand side in a convenient form, we write the term of v_{yy} in the expression of $\mathcal{N}_2 v$ as $(1-y^2)v_{yy} + y^2 v_{yy}$ and use similar expressions for the terms of v_{xy} and v_y. Then a direct calculation yields

$$\mathcal{N}_2 v - \frac{O_1 - xO_4}{a} = (2+\alpha)A_1 x^{1+\alpha}(1-y^2)J_1 + 2B_1 x y^2 J_2,$$

where

$$J_1 = (1+\alpha)\Big(1 + a\big((2+\alpha)A_1x^\alpha(1-y^2) + 2B_1y^2\big) + \frac{O_1}{x}\Big) - (2+O_4) + T_1,$$
$$J_2 = 1 + a\big((2+\alpha)A_1x^\alpha(1-y^2) + 2B_1y^2\big) + \frac{O_1}{x} - (2+O_4) + T_2,$$
$$T_1 = \frac{1}{(2+\alpha)A_1x^{1+\alpha}}\Big(2x^2(B_1 - A_1x^\alpha)\big((b+O_3) + O_5y\big)$$
$$+ 2O_2xy(2B_1 - (2+\alpha)A_1x^\alpha) - \frac{O_1 - xO_4}{a}\Big),$$
$$T_2 = \frac{(2+\alpha)A_1x^{1+\alpha}}{2B_1x}T_1.$$

Thus, in $Q^+_{r,1}$,

$$\mathcal{N}_2 v - \mathcal{N}_2 W < 0 \qquad \text{if } J_1, J_2 < 0. \tag{14.1.37}$$

By (14.1.5) and (14.1.34), we obtain

$$|T_1| + |T_2| \le Cr^{1-\alpha} \qquad \text{in } Q^+_{r,1},$$

so that, in $Q^+_{r,1}$,

$$J_1 \le (1+\alpha)\big(1 + \frac{2+\alpha}{2}(1-\mu_1)\big) - 2 + Cr^{1-\alpha}, \tag{14.1.38}$$
$$J_2 \le 1 + \frac{2+\alpha}{2}(1-\mu_1) - 2 + Cr^{1-\alpha}. \tag{14.1.39}$$

Choose $\alpha_1 > 0$ depending only on μ_1 so that, when $0 < \alpha \le \alpha_1$,

$$(1+\alpha)\Big(1 + \frac{2+\alpha}{2}(1-\mu_1)\Big) - 2 \le -\frac{\mu_1}{4}. \tag{14.1.40}$$

Such a choice of $\alpha_1 > 0$ is possible because we have the strict inequality in (14.1.40) when $\alpha = 0$, and the left-hand side is an increasing function of $\alpha > 0$ (where we have used $0 < \mu_1 \le \frac{1}{2}$ by reducing μ if necessary). Now, choosing $r_1 > 0$ so that

$$r_1 < \min\big\{\big(\frac{\mu_1}{4C}\big)^{\frac{1}{1-\alpha}},\ r_0\big\} \tag{14.1.41}$$

is satisfied, we use (14.1.38)–(14.1.40) to obtain

$$J_1, J_2 < 0 \qquad \text{in } Q^+_{r,1}.$$

Then, by (14.1.37), we obtain

$$\mathcal{N}_2 v < \mathcal{N}_2 W \qquad \text{in } Q^+_{r,1} \tag{14.1.42}$$

whenever $r \in (0, r_1]$ and $\alpha \in (0,\ \alpha_1]$. By (14.1.36), (14.1.42), Remark 14.1.1, and the standard comparison principle (Lemma 4.1.3), we obtain

$$W \le v \qquad \text{in } Q^+_{r,1}. \tag{14.1.43}$$

In particular, using (14.1.34)–(14.1.35) with $y = 0$, we obtain (14.1.33). □

Using Lemma 14.1.4, we now generalize the result in (14.1.31) for any $\alpha \in (0,1)$.

Proposition 14.1.5. *Let $(a, b, N, R, \hat{r}, \beta)$ and O_i be as in Theorem 14.1.2. Then, for any $\alpha \in (0,1)$, there exist positive constants r and A, depending only on $(a, b, N, R, \hat{r}, \beta)$ and α, so that, if $W \in C(\overline{Q^+_{\hat{r},R}}) \cap C^2(Q^+_{\hat{r},R})$ satisfies (14.1.28)–(14.1.30),*

$$W(x,y) \leq Ax^{2+\alpha} \qquad \text{in } Q^+_{r,3R/4}. \tag{14.1.44}$$

Proof. As argued before, without loss of generality, we may assume that $R = 2$, and it suffices to show that

$$W(x,0) \leq Ax^{2+\alpha} \qquad \text{for } x \in [0,\ r]. \tag{14.1.45}$$

By Lemma 14.1.4, it suffices to prove (14.1.45) for $\alpha > \alpha_1$. Fix any $\alpha \in (\alpha_1, 1)$, and set the following comparison function:

$$u(x,y) = \frac{1-\mu_1}{2ar_1^{\alpha_1} r^{\alpha-\alpha_1}} x^{2+\alpha}(1-y^2) + \frac{1-\mu_1}{2ar_1^{\alpha_1}} x^{2+\alpha_1} y^2. \tag{14.1.46}$$

By Lemma 14.1.4,

$$W \leq u \qquad \text{on } \partial Q^+_{r,1} \quad \text{for } r \in (0, r_1]. \tag{14.1.47}$$

As in the proof of Lemma 14.1.4, we write

$$\begin{aligned} &\mathcal{L}_2 u - \frac{O_1 - xO_4}{a} u \\ &= (2+\alpha)\frac{(1-\mu_1)x^{1+\alpha}}{2ar_1^{\alpha_1} r^{\alpha-\alpha_1}}(1-y^2)\hat{J}_1 + (2+\alpha_1)\frac{(1-\mu_1)x^{1+\alpha_1}}{2ar_1^{\alpha_1}} y^2 \hat{J}_2, \end{aligned}$$

where

$$\begin{aligned} D_0 &= \frac{1-\mu_1}{2}\Big((1-y^2)(2+\alpha)\big(\frac{x}{r}\big)^{\alpha} + y^2(2+\alpha_1)\big(\frac{x}{r}\big)^{\alpha_1}\Big), \\ \hat{J}_1 &= (1+\alpha)\Big(1 + \big(\frac{r}{r_1}\big)^{\alpha_1} D_0\Big) - 2 + \hat{T}_1, \\ \hat{J}_2 &= (1+\alpha_1)\Big(1 + \big(\frac{r}{r_1}\big)^{\alpha_1} D_0\Big) - 2 + \hat{T}_2, \\ \hat{T}_1 &= \frac{2ar_1^{\alpha_1} r^{\alpha-\alpha_1}}{(2+\alpha(1-\mu_1))x^{1+\alpha}}\Big(\mathcal{L}_2 u - ((x+au_x)u_{xx} - 2u_x) - \frac{O_1 - xO_4}{a}\Big), \\ \hat{T}_2 &= \frac{2+\alpha(1-\mu_1)}{(a+\alpha_1)(1-\mu)}\Big(\frac{x}{r}\Big)^{\alpha-\alpha_1} \bar{T}_1. \end{aligned}$$

By (14.1.5), we have

$$|\hat{T}_1| + |\hat{T}_2| \leq Cr^{1-\alpha_1}$$

for some positive constant C depending only on $(a, b, N, \beta, r_1, \alpha_1)$. Thus, we find that, for any $(x, y) \in Q^+_{r,1}$,

$$\max(\hat{J}_1, \hat{J}_2) \leq (1+\alpha)\Big(1 + (2+\alpha)\frac{1-\mu_1}{2}\big(\frac{r}{r_1}\big)^{\alpha_1}\Big) - 2 + Cr^{1-\alpha_1}. \tag{14.1.48}$$

Choosing $r > 0$ sufficiently small, depending only on (r_1, α, C), we obtain

$$\mathcal{L}_2 u - \mathcal{L}_2 W = \mathcal{L}_2 u - \frac{O_1 - xO_4}{a} < 0 \qquad \text{in } Q^+_{r,1}.$$

Then Lemma 4.1.3 implies that

$$W \leq u \qquad \text{in } Q^+_{r,1}.$$

Therefore, (14.1.45) holds with

$$A = \frac{1-\mu_1}{2ar_1^{\alpha_1} r^{\alpha-\alpha_1}}.$$

□

Lemma 14.1.6. *Let $(a, b, N, R, \hat{r}, \beta)$ and O_i be as in Theorem 14.1.2. Then there exist $r_2 > 0$ and $\alpha_2 \in (0, 1)$ such that, for any $W \in C(\overline{Q^+_{\hat{r},R}}) \cap C^2(Q^+_{\hat{r},R})$ satisfying* (14.1.28)–(14.1.30),

$$W(x, y) \geq -\frac{1-\beta}{2ar^\alpha} x^{2+\alpha} \qquad \textit{in } Q^+_{r,7R/8}, \tag{14.1.49}$$

whenever $\alpha \in (0,\ \alpha_2]$ and $r \in (0,\ r_2]$.

Proof. By (14.1.29)–(14.1.30), it can be easily verified that $W(x, y) \geq -\frac{1-\beta}{2a} x^2$ in $Q^+_{\hat{r},R}$. Now, similar to the proof of Lemma 14.1.4, it suffices to prove that, with assumption $R = 2$,

$$W(x, 0) \geq -\frac{1-\beta}{2ar^\alpha} x^{2+\alpha} \qquad \text{for } x \in (0, r)$$

for some $r > 0$ and $\alpha \in (0, \alpha_2)$.

For this, we use the comparison function:

$$v(x, y) := -Lx^{2+\alpha}(1-y^2) - Kx^2y^2 \qquad \text{with } Lr^\alpha = K = \frac{1-\beta}{2a}.$$

Then we follow the same procedure as the proof of Lemma 14.1.4, except we use that $\mathcal{N}_2 v > \mathcal{N}_2 W$, to find that the conditions for the choice of $\alpha, r > 0$ are inequalities (14.1.40)–(14.1.41) with an appropriate constant C and (μ_1, r_1) replaced by (β, r_2). □

Using Lemma 14.1.6, we now generalize the result in (14.1.49) for any $\alpha \in (0, 1)$.

Proposition 14.1.7. *Let $(a, b, N, R, \hat{r}, \beta)$ and O_i be as in Theorem 14.1.2. Then, for any $\alpha \in (0,\ 1)$, there exist positive constants r and B depending only on $(a, b, N, R, \hat{r}, \beta, \alpha)$ so that, for any $W \in C(\overline{Q^+_{\hat{r},R}}) \cap C^2(Q^+_{\hat{r},R})$ satisfying (14.1.28)–(14.1.30),*

$$W(x,y) \geq -Bx^{2+\alpha} \qquad in\ Q^+_{r,3R/4}. \tag{14.1.50}$$

Proof. For fixed $\alpha \in (\alpha_2, 1)$, we set the comparison function:

$$u_-(x,y) = -\frac{1-\beta}{2ar_2^{\alpha_2}r^{\alpha-\alpha_2}}x^{2+\alpha}(1-y^2) - \frac{1-\beta}{2ar_2^{\alpha_2}}x^{2+\alpha_2}y^2.$$

Then, using the argument as in the proof of Proposition 14.1.5, we can choose $r > 0$ appropriately small so that

$$\mathcal{L}_2 u_- - \mathcal{L}_2 W = \mathcal{L}_2 u_- - \frac{O_1 - xO_4}{a} > 0$$

holds for any $(x,y) \in Q^+_{r,1}$. □

With Propositions 14.1.5–14.1.7, we now prove Theorem 14.1.2.

14.1.4 Proof of Theorem 14.1.2

We divide the proof into four steps.

1. Let ψ be a solution of (14.1.1) in $Q^+_{\hat{r},R}$ for $\hat{r}$ as in Remark 14.1.1, and let the assumptions of Theorem 14.1.2 hold. Then ψ satisfies (14.1.9). Thus, it suffices to show that, for any given $\alpha \in (0,1)$, there exists $r > 0$ so that $\psi \in C^{2,\alpha}(\overline{Q^+_{r,R/2}})$, and $\psi_{xx}(0,y) = \frac{1}{a}$ and $\psi_{xy}(0,y) = \psi_{yy}(0,y) = 0$ for any $|y| < \frac{R}{2}$.

Let $W(x,y)$ be defined by (14.1.27). Then, in order to prove Theorem 14.1.2, it suffices to show that, for any given $\alpha \in (0,\ 1)$, there exists $r > 0$ so that

(i) $W \in C^{2,\alpha}(\overline{Q^+_{r,R/2}})$;

(ii) $D^2W(0,y) = 0$ for any $|y| < \frac{R}{2}$.

2. By definition, W satisfies (14.1.28)–(14.1.30). For any given $\alpha \in (0,1)$, there exists $r > 0$ so that both (14.1.44) and (14.1.50) hold in $Q^+_{r,3R/4}$, by Propositions 14.1.5–14.1.7. Fix such $r > 0$.

Furthermore, since W satisfies estimate (14.1.30), we can introduce a cutoff function into the nonlinear term of equation (14.1.28), *i.e.*, modify the nonlinear term away from the values determined by (14.1.30) to make the term bounded in $\frac{W_x}{x}$. Namely, fix $\zeta \in C^\infty(\mathbb{R})$ satisfying

$$\begin{aligned} &-\tfrac{2-\beta}{2a} \leq \zeta \leq M + \tfrac{2}{a} && \text{on } \mathbb{R}, \\ &\zeta(s) = s && \text{on } \big(-\tfrac{1-\beta}{a}, M + \tfrac{1}{a}\big), \\ &\zeta \equiv 0 && \text{on } \mathbb{R} \setminus \big(-\tfrac{2-\beta}{a},\ M + \tfrac{4}{a}\big). \end{aligned} \tag{14.1.51}$$

Then, from (14.1.28) and (14.1.30), it follows that W satisfies

$$\begin{aligned} &x\big(1+a\zeta(\frac{W_x}{x})+\frac{O_1}{x}\big)W_{xx}+O_2W_{xy}+(b+O_3)W_{yy} \\ &\quad -(2+O_4)W_x+O_5W_y=\frac{O_1-xO_4}{a} \qquad \text{in } Q^+_{\hat{r},R}. \end{aligned} \tag{14.1.52}$$

3. For $z:=(x,y)\in Q^+_{r/2,R/2}$, define

$$R_z:=\big\{(s,t)\ :\ |s-x|<\frac{x}{8},\ |t-y|<\frac{\sqrt{x}}{8}\big\}. \tag{14.1.53}$$

Then

$$R_z\subset Q^+_{r,3R/4} \qquad \text{for any } z=(x,y)\in Q^+_{r/2,R/2}. \tag{14.1.54}$$

Fix $z_0=(x_0,y_0)\in Q^+_{r/2,R/2}$. Rescale W in R_{z_0} by defining

$$W^{(z_0)}(S,T):=\frac{1}{x_0^{2+\alpha}}W(x_0+\frac{x_0}{8}S,y_0+\frac{\sqrt{x_0}}{8}T) \qquad \text{for } (S,T)\in Q_1, \tag{14.1.55}$$

where $Q_h=(-h,h)^2$ for $h>0$. Then, by (11.4.8), (14.1.44), and (14.1.50), we have

$$\|W^{(z_0)}\|_{C^0(\overline{Q_1})}\le\frac{1}{ar^\alpha}. \tag{14.1.56}$$

Moreover, since W satisfies equation (14.1.28), $W^{(z_0)}$ satisfies the following equation for $(S,T)\in Q_1$:

$$\begin{aligned} &(1+\frac{S}{8})\Big(1+a\zeta(\frac{8x_0^\alpha W_S^{(z_0)}}{1+\frac{S}{8}})+\tilde{O}_1^{(z_0)}\Big)W_{SS}^{(z_0)}+\tilde{O}_2^{(z_0)}W_{ST}^{(z_0)}+(b+\tilde{O}_3^{(z_0)})W_{TT}^{(z_0)} \\ &\quad -\frac{1}{8}(2+\tilde{O}_4^{(z_0)})W_S^{(z_0)}+\frac{1}{8}\tilde{O}_5^{(z_0)}W_T^{(z_0)}=\frac{(1+\frac{S}{8})}{64ax_0^\alpha}(\tilde{O}_1^{(z_0)}-\tilde{O}_4^{(z_0)}), \end{aligned} \tag{14.1.57}$$

where

$$\begin{aligned} &\tilde{O}_1^{(z_0)}(S,T)=\tfrac{1}{x_0(1+\frac{S}{8})}O_1(x,y), && \tilde{O}_2^{(z_0)}(S,T)=\tfrac{1}{\sqrt{x_0}}O_2(x,y), \\ &(\tilde{O}_3^{(z_0)},\tilde{O}_4^{(z_0)})(S,T)=(O_3,O_4)(x,y), && \tilde{O}_5^{(z_0)}(S,T)=\sqrt{x_0}O_5(x,y) \end{aligned}$$

with $x=x_0(1+\frac{S}{8})$ and $y=y_0+\frac{\sqrt{x_0}}{8}T$. Then, from (14.1.5)–(14.1.6), we find that, for any $(S,T)\in\overline{Q_1}$ and $z_0\in Q^+_{r/2,R/2}$ with $r\le 1$,

$$\begin{aligned} &|\tilde{O}_k^{(z_0)}(S,T)|\le 2N\sqrt{r} && \text{for } k=1,\dots,5, \\ &|D\tilde{O}_k^{(z_0)}(S,T)|\le 2N\sqrt{r} && \text{for } k\ne 2, \\ &|D\tilde{O}_2^{(z_0)}(S,T)|\le 2N. \end{aligned} \tag{14.1.58}$$

Also, denoting the right-hand side of (14.1.57) by $F^{(z_0)}(S,T)$, we obtain from (14.1.58) that, for any $(S,T) \in \overline{Q_1}$ and $z_0 \in Q^+_{r/2,R/2}$,

$$|F^{(z_0)}(S,T)| \le Cr^{1-\alpha}, \qquad |DF^{(z_0)}(S,T)| \le Cr^{\frac{1}{2}-\alpha}, \tag{14.1.59}$$

where C depends only on (N,a).

Now, writing equation (14.1.57) as

$$\sum_{i,j=1}^{2} A_{ij}(DW^{(z_0)}, S, T)\, D_{ij}W^{(z_0)} + \sum_{i=1}^{2} B_i(S,T)\, D_i W^{(z_0)} = F^{(z_0)} \qquad \text{in } Q_1, \tag{14.1.60}$$

we obtain from (14.1.51) and (14.1.57)–(14.1.59) that, if $r > 0$ is sufficiently small, depending only on the data, then (14.1.60) is uniformly elliptic with ellipticity constants depending only on b, but independent of z_0, and $(A_{ij}(\mathbf{p}, S, T), B_i(S,T))$ and $F^{(z_0)}(S,T)$ for $\mathbf{p} \in \mathbb{R}^2$ and $(S,T) \in Q_1$ satisfy

$$\|A_{ij}\|_{C^1(\mathbb{R}^2\times\overline{Q_1})} \le C, \quad \|(B_i, \frac{F^{(z_0)}}{r^{\frac{1}{2}-\alpha}})\|_{C^1(\overline{Q_1})} \le C,$$

where C depends only on the data and is independent of z_0. By Theorem 4.2.1 and (14.1.56), we have

$$\begin{aligned}\|W^{(z_0)}\|_{C^{2,\alpha}(\overline{Q_{1/2}})} &\le C(\|W^{(z_0)}\|_{C^0(\overline{Q_1})} + \|F^{(z_0)}\|_{C^\alpha(\overline{Q_1})})\\ &\le C(\frac{1}{ar^\alpha} + r^{\frac{1}{2}-\alpha}) =: \hat{C},\end{aligned} \tag{14.1.61}$$

where C depends only on the data and α in this case. From (14.1.61),

$$|D_x^i D_y^j W(x_0,y_0)| \le Cx_0^{2+\alpha-i-\frac{j}{2}} \tag{14.1.62}$$

for any $(x_0,y_0) \in Q^+_{r/2,R/2}$ and $0 \le i+j \le 2$.

4. It remains to prove the C^α–continuity of D^2W in $\overline{Q^+_{r/2,R/2}}$.

For two distinct points $z_1 = (x_1,y_1)$ and $z_2 = (x_2,y_2) \in Q^+_{r/2,R/2}$, consider

$$A := \frac{|W_{xx}(z_1) - W_{xx}(z_2)|}{|z_1 - z_2|^\alpha}.$$

Without loss of generality, assume that $x_1 \le x_2$. Then there are two cases:

Case 1: $z_1 \in R_{z_2}$. In this case,

$$x_1 = x_2 + \frac{x_2}{8}S, \quad y_1 = y_2 + \frac{\sqrt{x_2}}{8}T \qquad \text{for some } (S,T) \in Q_1.$$

By (14.1.61),

$$\frac{|W^{(z_2)}_{SS}(S,T) - W^{(z_2)}_{SS}(\mathbf{0})|}{(S^2+T^2)^{\frac{\alpha}{2}}} \le \hat{C},$$

which is

$$\frac{|W_{xx}(x_1,y_1)-W_{xx}(x_2,y_2)|}{\left((x_1-x_2)^2+x_2(y_1-y_2)^2\right)^{\frac{\alpha}{2}}}\le \hat{C}.$$

Since $x_2\in(0,r)$ and $r\le 1$, the last estimate implies

$$\frac{|W_{xx}(x_1,y_1)-W_{xx}(x_2,y_2)|}{\left((x_1-x_2)^2+(y_1-y_2)^2\right)^{\frac{\alpha}{2}}}\le \hat{C}.$$

Case 2: $z_1\notin R_{z_2}$. Then either $|x_1-x_2|>\frac{x_2}{8}$ or $|y_1-y_2|>\frac{\sqrt{x_2}}{8}$. Since $0\le x_2\le r\le 1$, we have

$$|z_1-z_2|^\alpha\ge\left(\frac{x_2}{8}\right)^\alpha.$$

Using (14.1.62) and $x_1\le x_2$, we obtain

$$\frac{|W_{xx}(z_1)-W_{xx}(z_2)|}{|z_1-z_2|^\alpha}\le\frac{|W_{xx}(z_1)|+|W_{xx}(z_2)|}{|z_1-z_2|^\alpha}\le\hat{C}\frac{x_1^\alpha+x_2^\alpha}{x_2^\alpha}\le 2\hat{C}.$$

Thus, $A\le 2\hat{C}$ in both cases, where $\hat{C}$ depends on (α,r) and the data. Since $z_1\ne z_2$ are arbitrary points of $Q^+_{r/2,R/2}$, we obtain

$$[W_{xx}]_{C^\alpha(\overline{Q^+_{r/2,R/2}})}\le 2\hat{C}. \tag{14.1.63}$$

The estimates for (W_{xy},W_{yy}) can be obtained similarly. In fact, for these derivatives, we obtain the stronger estimates: For any $\delta\in(0,\frac{r}{2}]$,

$$[W_{xy}]_{C^\alpha(\overline{Q^+_{\delta,R/2}})}\le\hat{C}\sqrt{\delta},\qquad [W_{yy}]_{C^\alpha(\overline{Q^+_{\delta,R/2}})}\le\hat{C}\delta,$$

where $\hat{C}$ depends on (α,r) and the data, but is independent of $\delta>0$ and z_0.

Therefore, $W\in C^{2,\alpha}(\overline{Q^+_{r,R/2}})$ with $\|W\|_{C^{2,\alpha}(\overline{Q^+_{r,R/2}})}$ depending only on the data, because $r>0$ depends on the data. Moreover, (14.1.62) implies that $D^2W(0,y)=0$ for any $|y|\le\frac{R}{2}$. This concludes the proof of Theorem 14.1.2. □

14.2 OPTIMAL REGULARITY OF SOLUTIONS ACROSS Γ_{sonic}

In this section, we apply the results in §14.1 to study the regularity of admissible solutions (even a larger class of solutions) up to and across Γ_{sonic}. Specifically, we prove that $C^{1,1}$ is actually the optimal regularity of any solution φ across the pseudo-sonic circle P_1P_4 in the class of admissible solutions of **Problem 2.6.1**.

In fact, we define a class of regular reflection-diffraction solutions, which includes admissible solutions of **Problem 2.6.1** but requires less conditions, and prove the following three main results:

(i) There is no regular reflection-diffraction solution that is C^2 across the pseudo-sonic circle.

Furthermore, for admissible solutions of **Problem 2.6.1** or, more generally, for any regular reflection-diffraction solutions satisfying the conclusions of Lemma 11.2.5, Proposition 11.2.8, and estimate (11.4.4), we show that

(ii) φ is $C^{2,\alpha}$ in the pseudo-subsonic region Ω up to the pseudo-sonic circle P_1P_4, excluding endpoint P_1, but $D^2\varphi$ has a jump across $\Gamma_{\text{sonic}} := P_1P_4$ depending only on the adiabatic exponent, independent of the wedge angles $\theta_{\text{w}} \in (\theta_{\text{w}}^{\text{s}}, \frac{\pi}{2})$.

(iii) In addition, $D^2\varphi$ does not have a limit at P_1 from Ω.

In order to state these results, we first define the class of regular reflection-diffraction solutions.

Definition 14.2.1. *Let $\gamma > 1$ and $\rho_1 > \rho_0 > 0$ be constants. Let $\theta_{\text{w}} \in (\theta_{\text{w}}^{\text{s}}, \frac{\pi}{2})$. A function $\varphi \in C^{0,1}(\Lambda)$ is a regular reflection-diffraction solution if φ is a solution of* **Problem 2.2.3** *for the wedge angle θ_{w}, satisfying conditions* (i)–(iii) *of Definition* 8.1.1 *and*

$$\varphi \geq \varphi_2 \qquad on\ \Gamma_{\text{shock}} := P_1P_2. \tag{14.2.1}$$

Remark 14.2.2. *The admissible solutions of* **Problem 2.6.1** *are regular reflection-diffraction solutions. This can be seen directly from a comparison between Definitions* 8.1.1 *and* 14.2.1, *since* (14.2.1) *follows by Definition* 8.1.1(iv).

Remark 14.2.3. *We note that, when $\theta_{\text{w}} = \frac{\pi}{2}$, the regular reflection-diffraction becomes the normal reflection, in which $u_2 = 0$ and the solution is smooth across the sonic circle of state* (2); *see* §6.2. *The condition, $\theta_{\text{w}} \in (0, \frac{\pi}{2})$, in Definition* 14.2.1 *rules out this case.*

Remark 14.2.4. *Since $\varphi = \varphi_1$ on Γ_{shock} by* (8.1.3), *condition* (14.2.1) *of Definition* 14.2.1 *is equivalent to*

$$\Gamma_{\text{shock}} \subset \{\varphi_2 \leq \varphi_1\},$$

that is, Γ_{shock} is below $\mathcal{S}_1$.

Furthermore, we have

Lemma 14.2.5. *For any regular reflection-diffraction solution φ in the sense of Definition* 14.2.1,

$$\varphi > \varphi_2 \qquad in\ \Omega. \tag{14.2.2}$$

Proof. By (2.2.8)–(2.2.9) and (2.4.1), $\psi := \varphi - \varphi_2$ satisfies

$$\begin{aligned}&\big(\tilde{c}^2 - (\psi_{\xi_1} - \xi_1 + u_2)^2\big)\psi_{\xi_1\xi_1} + \big(\tilde{c}^2 - (\psi_{\xi_2} - \xi_2 + u_2\tan\theta_{\text{w}})^2\big)\psi_{\xi_2\xi_2}\\ &\quad - 2(\psi_{\xi_1} - \xi_1 + u_2)(\psi_{\xi_2} - \xi_2 + u_2\tan\theta_{\text{w}})\psi_{\xi_1\xi_2} = 0 \qquad \text{in } \Omega,\end{aligned} \tag{14.2.3}$$

where

$$\tilde{c}^2(D\psi, \psi, \boldsymbol{\xi}) = c_2^2 + (\gamma - 1)\Big((\xi_1 - u_2)\psi_{\xi_1} + (\xi_2 - u_2\tan\theta_{\text{w}})\psi_{\xi_2} - \frac{1}{2}|D\psi|^2 - \psi\Big).$$

We regard that the coefficients of (14.2.3) are computed on ψ as fixed, so that (14.2.3) can be considered as a linear equation with respect to the second derivative of ψ.

Since equation (14.2.3) is elliptic and φ is smooth inside Ω, it follows that (14.2.3) is uniformly elliptic in any compact subset of Ω. Furthermore, we have

$$\begin{aligned} \psi &= 0 && \text{on } \Gamma_{\text{sonic}}, \\ D\psi \cdot (-\sin\theta_{\text{w}}, \cos\theta_{\text{w}}) &= 0 && \text{on } \Gamma_{\text{wedge}}, \\ \psi_{\xi_2} &= -u_2 \tan\theta_{\text{w}} < 0 && \text{on } \partial\Omega \cap \{\xi_2 = 0\}, \\ \psi &\geq 0 && \text{on } \Gamma_{\text{shock}} \quad \text{(by Definition 4.1 (c))}. \end{aligned}$$

Then the strong maximum principle implies

$$\psi > 0 \qquad \text{in } \Omega,$$

which is (14.2.2). This completes the proof. □

Now we first show that any regular reflection-diffraction solution in our case cannot be C^2 across the pseudo-sonic circle $\Gamma_{\text{sonic}} := P_1P_4$.

Theorem 14.2.6. *Let φ be a regular reflection-diffraction solution in the sense of Definition* 14.2.1. *Then φ cannot be C^2 across the pseudo-sonic circle* Γ_{sonic}.

Proof. On the contrary, assume that φ is C^2 across Γ_{sonic}. Then $\psi = \varphi - \varphi_2$ is also C^2 across Γ_{sonic}, where φ_2 is given by (2.4.1). Moreover, since $\psi \equiv 0$ in $P_0P_1P_4$ by (8.1.3), we see that $D^2\psi(\boldsymbol{\xi}) = 0$ at any $\boldsymbol{\xi} \in \Gamma_{\text{sonic}}$.

Furthermore, we use the (x, y)–coordinates (11.1.1)–(11.1.2) in a neighborhood of Γ_{sonic}. Since φ is a regular reflection-diffraction solution, it follows that (11.1.3) holds if ε is sufficiently small. Indeed, this follows from (8.1.3), since $\varphi = \varphi_1$ on $P_0P_1P_2$ by the Rankine-Hugoniot conditions, and from $D(\varphi_1 - \varphi_2)(P_1) \neq 0$ and the fact that segment P_0P_1 intersects with (but is not tangential to) circle $\partial B_{c_2}(O_2)$ at P_1, which holds by (6.1.4)–(6.1.5).

Moreover, $\psi = \varphi - \varphi_2$ satisfies equation (11.1.4) with (11.1.5) in $\Omega \cap \mathcal{N}_\varepsilon(\Gamma_{\text{sonic}})$, which is obtained by substituting $\varphi = \psi + \varphi_2$ into equation (2.2.8) and writing the resulting equation in the (x, y)–coordinates (11.1.2).

Let $(0, y_0)$ be a point in the relative interior of Γ_{sonic}. Then

$$(0, y_0) + Q^+_{r,R} \subset \Omega \cap \mathcal{N}_\varepsilon(\Gamma_{\text{sonic}}) \qquad \text{if } r, R > 0 \text{ are sufficiently small.}$$

By shifting the coordinates $(x, y) \mapsto (x, y - y_0)$, we may assume that $(0, y_0) = (0, 0)$ and $Q^+_{r,R} \subset \Omega \cap \mathcal{N}_\varepsilon(\Gamma_{\text{sonic}})$. Note that the shifting coordinates in the y–direction do not change the expressions in (11.1.5).

Since $\psi \in C^2((\Omega \cap \mathcal{N}_\varepsilon(\Gamma_{\text{sonic}})) \cup \Gamma_{\text{sonic}})$ with $(\psi, D\psi, D^2\psi) \equiv \mathbf{0}$ on Γ_{sonic}, reducing r if necessary, we have

$$|D\psi| \leq \delta x \qquad \text{in } Q^+_{r,R},$$

where $\delta > 0$ is so small that (14.1.8) holds in $\Omega \cap \mathcal{N}_\varepsilon(\Gamma_{\text{sonic}})$ with $\beta = M = 1$, and that terms O_i defined by (11.1.5) satisfy (14.1.5)–(14.1.6) with $M = 1$. Also, from Lemma 14.2.5, we obtain that $\psi = \varphi - \varphi_2 > 0$ in $Q^+_{r,R}$. Now we can apply Proposition 14.1.3 to conclude

$$\psi(x,y) \geq \mu x^2 \qquad \text{on } Q^+_{r,15R/16}$$

for some $\mu, r > 0$. This is in contradiction to the fact that $D^2\psi(0,y) = 0$ for all $y \in (-R, R)$, that is, $D^2\psi(\boldsymbol{\xi}) = 0$ at any $\boldsymbol{\xi} \in \Gamma_{\text{sonic}}$. □

In the following theorem, we study more detailed regularity of ψ near the sonic circle Γ_{sonic} in the case of regular reflection-diffraction solutions with $C^{1,1}$ near Γ_{sonic}, satisfying the additional assumptions (*b*)–(*c*) in Theorem 14.2.7 below. Note that this class includes the admissible solutions, and hence includes the solutions constructed in Theorems 2.6.3–2.6.5. In particular, assumptions (*a*)–(*c*) in Theorem 14.2.7 are satisfied for the admissible solutions by Proposition 11.4.6, Lemma 11.2.5, and Proposition 11.2.8 (combined with estimate (11.4.5)).

Theorem 14.2.7. *Let φ be a regular reflection-diffraction solution in the sense of Definition* 14.2.1 *and satisfy the following properties:*

(a) *φ is $C^{1,1}$ across the sonic circle Γ_{sonic}: There exists $\varepsilon_1 > 0$ such that $\varphi \in C^{1,1}(\overline{P_0P_1P_2P_3} \cap \mathcal{N}_{\varepsilon_1}(\Gamma_{\text{sonic}}))$, where $\mathcal{N}_{\varepsilon_1}(\Gamma_{\text{sonic}})$ is an ε_1-neighborhood of Γ_{sonic};*

(b) *There exists $\delta_0 > 0$ such that, in coordinates* (11.1.2),

$$|\partial_x(\varphi - \varphi_2)(x,y)| \leq \frac{2-\delta_0}{\gamma+1}x \qquad \text{in } \Omega_{\varepsilon_0}; \tag{14.2.4}$$

(c) *There exist $\varepsilon_1 \geq \varepsilon_0 > 0$, $\omega > 0$, and a function $y = \hat{f}(x)$ such that, for $\Omega_{\varepsilon_0} := \Omega \cap \mathcal{N}_{\varepsilon_1}(\Gamma_{\text{sonic}}) \cap \{r > c_2 - \varepsilon_0\}$ in coordinates* (11.1.2),

$$\begin{aligned} &\Omega_{\varepsilon_0} = \{(x,y) : 0 < x < \varepsilon_0,\ 0 < y < \hat{f}(x)\}, \\ &\Gamma_{\text{shock}} \cap \partial\Omega_{\varepsilon_0} = \{(x,y) : 0 < x < \varepsilon_0,\ y = \hat{f}(x)\}, \end{aligned} \tag{14.2.5}$$

and

$$\|\hat{f}\|_{C^{1,1}([0,\,\varepsilon_0])} < \infty\ , \quad \frac{d\hat{f}}{dx} \geq \omega > 0 \ \ \text{for } 0 < x < \varepsilon_0. \tag{14.2.6}$$

Then we have

(i) *φ is $C^{2,\alpha}$ in Ω up to Γ_{sonic} away from point P_1 for any $\alpha \in (0,1)$. That is, for any $\alpha \in (0,1)$ and any given $\boldsymbol{\xi}^0 \in \overline{\Gamma_{\text{sonic}}} \setminus \{P_1\}$, there exists $K < \infty$ depending only on $(\rho_0, \rho_1, \gamma, \varepsilon_0, \alpha)$, $\|\varphi\|_{C^{1,1}(\Omega_{\varepsilon_0})}$, and $d = \operatorname{dist}(\boldsymbol{\xi}^0, \Gamma_{\text{shock}})$ such that*

$$\|\varphi\|_{2,\alpha;\overline{B_{d/2}(\boldsymbol{\xi}^0)\cap\Omega_{\varepsilon_0/2}}} \leq K;$$

(ii) *For any* $\boldsymbol{\xi}^0 \in \Gamma_{\rm sonic} \setminus \{P_1\}$,

$$\lim_{\substack{\boldsymbol{\xi}\to\boldsymbol{\xi}^0\\ \boldsymbol{\xi}\in\Omega}} (D_{rr}\varphi - D_{rr}\varphi_2) = \frac{1}{\gamma+1};$$

(iii) $D^2\varphi$ *has a jump across* $\Gamma_{\rm sonic}$: *For any* $\boldsymbol{\xi}^0 \in \Gamma_{\rm sonic} \setminus \{P_1\}$,

$$\lim_{\substack{\boldsymbol{\xi}\to\boldsymbol{\xi}^0\\ \boldsymbol{\xi}\in\Omega}} D_{rr}\varphi \;-\; \lim_{\substack{\boldsymbol{\xi}\to\boldsymbol{\xi}^0\\ \boldsymbol{\xi}\in\Lambda\setminus\Omega}} D_{rr}\varphi \;=\; \frac{1}{\gamma+1};$$

(iv) *The limit* $\lim_{\substack{\boldsymbol{\xi}\to P_1\\ \boldsymbol{\xi}\in\Omega}} D^2\varphi$ *does not exist.*

Proof. The proof consists of seven steps.

1. Let

$$\psi := \varphi - \varphi_2.$$

By (8.1.3) and (14.2.5), we have

$$\psi(0,y) = \psi_x(0,y) = \psi_y(0,y) = 0 \qquad \text{for any } (0,y) \in \Gamma_{\rm sonic}. \tag{14.2.7}$$

Using (14.2.5)–(14.2.6), we have

$$|\psi(x,y)| \le Cx^2, \quad |D_{(x,y)}\psi(x,y)| \le Cx \qquad \text{for any } (x,y) \in \Omega_{\varepsilon_0}, \tag{14.2.8}$$

where C depends only on $\|\psi\|_{C^{1,1}(\overline{\Omega_{\varepsilon_0}})}$ and $\|\hat{f}\|_{C^1([0,\,\varepsilon_0])}$.

Recall that, in the (x,y)–coordinates (11.1.2), domain Ω_{ε_0} defined in property (c) satisfies (14.2.5), and $\psi(x,y)$ satisfies equation (14.1.1) with

$$O_i(x,y) = O_i(x,y,\psi(x,y),D_{(x,y)}\psi(x,y))$$

given by (11.1.5). Then it follows from (11.1.5) and (14.2.8) that (14.1.5)–(14.1.6) hold with N depending only on ε_0, $\|\psi\|_{C^{1,1}(\overline{\Omega_{\varepsilon_0}})}$, and $\|\hat{f}\|_{C^1([0,\,\varepsilon_0])}$.

2. Now, using (14.2.4) and reducing ε_0 if necessary, we conclude that (14.1.1) is uniformly elliptic on $\Omega_{\varepsilon_0} \cap \{x > \delta\}$ for any $\delta \in (0,\varepsilon_0)$. Moreover, by (c), equation (14.1.1) with (11.1.5), considered as a linear elliptic equation, has C^1–coefficients. Furthermore, since the boundary conditions (2.2.20) hold for φ and φ_2, especially on $\Gamma_{\rm wedge} = \{y = 0\}$, it follows that, in the (x,y)–coordinates, we have

$$\psi_y(x,0) = 0 \qquad \text{for any } x \in (0,\varepsilon_0). \tag{14.2.9}$$

Then, by the standard regularity theory for the oblique derivative problem for linear, uniformly elliptic equations, ψ is C^2 in Ω_{ε_0} up to $\partial\Omega_{\varepsilon_0} \cap \{0 < x < \varepsilon_0,\ y = 0\}$. From this and (c), we have

$$\psi \in C^{1,1}(\overline{\Omega_{\varepsilon_0}}) \cap C^2(\Omega_{\varepsilon_0} \cup \Gamma^{(\varepsilon_0)}_{\rm wedge}), \tag{14.2.10}$$

where $\Gamma^{(\varepsilon_0)}_{\rm wedge} := \Gamma_{\rm wedge} \cap \partial\Omega_{\varepsilon_0} \equiv \{(x,0) : 0 < x < \varepsilon_0\}$.

Reflecting Ω_{ε_0} with respect to the y–axis, *i.e.*, using (14.2.5), we define

$$\hat{\Omega}_{\varepsilon_0} := \{(x,y) : 0 < x < \varepsilon_0, \ -\hat{f}(x) < y < \hat{f}(x)\}. \tag{14.2.11}$$

Extend $\psi(x,y)$ from Ω_{ε_0} to $\hat{\Omega}_{\varepsilon_0}$ by even reflection: $\psi(x,-y) = \psi(x,y)$ for $(x,y) \in \Omega_{\varepsilon_0}$. Using (14.2.9)–(14.2.10), we conclude that the extended function $\psi(x,y)$ satisfies

$$\psi \in C^{1,1}(\overline{\hat{\Omega}_{\varepsilon_0}}) \cap C^2(\hat{\Omega}_{\varepsilon_0}). \tag{14.2.12}$$

Now we use the explicit expressions (11.1.5) and (14.1.1) to find that, if $\psi(x,y)$ satisfies equation (14.1.1) with (11.1.5) in Ω_{ε_0}, $\tilde{\psi}(x,y) := \psi(x,-y)$ also satisfies (14.1.1) with $O_k(D\tilde{\psi}, \tilde{\psi}, x)$ defined by (11.1.5) in Ω_{ε_0}. Thus, in the extended domain $\hat{\Omega}_{\varepsilon_0}$, the extended function $\psi(x,y)$ satisfies (14.1.1) with $O_1, \dots, O_5$ defined by the expressions in (11.1.5) in $\hat{\Omega}_{\varepsilon_0}$.

Moreover, by (8.1.3), it follows that $\psi = 0$ on $\Gamma_{\rm sonic}$. Thus, in the (x,y)–coordinates, for the extended function ψ, we have

$$\psi(0,y) = 0 \qquad \text{for any } \ y \in (-\hat{f}(0), \hat{f}(0)). \tag{14.2.13}$$

Also, using $\varphi \geq \varphi_2$ in Ω,

$$\psi(0,y) \geq 0 \qquad \text{in } \ \hat{\Omega}_{\varepsilon_0}. \tag{14.2.14}$$

3. Let $P = \boldsymbol{\xi}^* \in \Gamma_{\rm sonic} \setminus \{P_1\}$. Then, in the (x,y)–coordinates, $P = (0, y_*)$ with $y_* \in [0, \hat{f}(0))$. Moreover, by (14.2.5)–(14.2.6) and (14.2.11), there exist $r > 0$ and $R > 0$, depending only on $\varepsilon_0, c_2 = c_2(\rho_0, \rho_1, u_1, \theta_{\rm w})$, and $d = \text{dist}(\boldsymbol{\xi}^*, \Gamma_{\rm shock})$, such that

$$(0, y_*) + Q^+_{r,R} \subset \hat{\Omega}_{\varepsilon_0}.$$

Then, in $Q^+_{r,R}$, $\hat{\psi}(x,y) := \psi(x, y - y_*)$ satisfies all the conditions in Theorem 14.1.2. Applying Theorem 14.1.2 and expressing the results in terms of ψ, we obtain that, for any $y_* \in [0, \hat{f}(0))$,

$$\lim_{\substack{(x,y)\to(0,y_*)\\(x,y)\in\Omega}} \psi_{xx}(x,y) = \frac{1}{\gamma+1}, \qquad \lim_{\substack{(x,y)\to(0,y_*)\\(x,y)\in\Omega}} (\psi_{xy}, \psi_{yy})(x,y) = (0,0). \tag{14.2.15}$$

Since $\psi_{rr} = \psi_{xx}$ by (11.1.2), this implies assertions (i)–(ii).

Now assertion (iii) follows from (ii), since $\varphi = \varphi_2$ in $B_\varepsilon(\boldsymbol{\xi}^*) \setminus \Omega$ for small $\varepsilon > 0$ by (8.1.3), and φ_2 is a C^∞–smooth function in $\mathbb{R}^2$.

4. It remains to show assertion (iv). We prove this by contradiction. Assume that assertion (iv) is false, *i.e.*, there exists a limit of $D^2\psi$ at P_1 from Ω. Then our strategy is to choose two different sequences of points converging to P_1 and show that the limits of ψ_{xx} along the two sequences are different, which leads to a contradiction. We note that, in the (x,y)–coordinates, point $P_1 = (0, \ \hat{f}(0))$.

5. A sequence close to Γ_{sonic}. Let $\{y_m^{(1)}\}_{m=1}^{\infty}$ be a sequence such that $y_m^{(1)} \in (0,\ \hat{f}(0))$ and $\lim_{m\to\infty} y_m^{(1)} = \hat{f}(0)$. By (14.2.15), there exists $x_m^{(1)} \in (0,\ \frac{1}{m})$ such that

$$\Big|\psi_{xx}(x_m^{(1)}, y_m^{(1)}) - \frac{1}{\gamma+1}\Big| + |\psi_{xy}(x_m^{(1)}, y_m^{(1)})| + |\psi_{yy}(x_m^{(1)}, y_m^{(1)})| < \frac{1}{m}$$

for each $m = 1, 2, 3, \ldots$. Using (14.2.6), we have

$$y_m^{(1)} < \hat{f}(0) \le \hat{f}(x_m^{(1)}).$$

Then we employ (14.2.5) to obtain that $(x_m^{(1)}, y_m^{(1)}) \in \Omega$ such that

$$\begin{aligned}
&\lim_{m\to\infty} (x_m^{(1)}, y_m^{(1)}) = (0, \hat{f}(0)),\\
&\lim_{m\to\infty} \psi_{xx}(x_m^{(1)}, y_m^{(1)}) = \frac{1}{\gamma+1},\\
&\lim_{m\to\infty} (\psi_{xy}, \psi_{yy})(x_m^{(1)}, y_m^{(1)}) = (0,0).
\end{aligned} \tag{14.2.16}$$

6. The Rankine-Hugoniot conditions on Γ_{shock}. In order to construct another sequence, we first combine the Rankine-Hugoniot conditions on Γ_{shock} into a condition with the following form:

Lemma 14.2.8. *There exists $\varepsilon \in (0, \varepsilon_0)$ such that ψ satisfies*

$$\hat{b}_1(x,y)\psi_x + \hat{b}_2(x,y)\psi_y + \hat{b}_3(x,y)\psi = 0 \qquad \text{on } \Gamma_{\text{shock}} \cap \{0 < x < \varepsilon\}, \tag{14.2.17}$$

where $\hat{b}_k \in C(\overline{\Gamma_{\text{shock}} \cap \{0 \le x \le \varepsilon\}})$ with

$$\hat{b}_1(x,y) \ge \lambda, \ \ |\hat{b}_2(x,y)| \le \frac{1}{\lambda}, \ \ |\hat{b}_3(x,y)| \le \frac{1}{\lambda} \qquad \text{on } \Gamma_{\text{shock}} \cap \{0 < x < \varepsilon\} \tag{14.2.18}$$

for some constant $\lambda > 0$.

Proof. To prove this, we first work in the $\boldsymbol{\xi}$–coordinates. Since

$$\varphi = \varphi_1, \quad \rho D\varphi \cdot \boldsymbol{\nu} = \rho_1 D\varphi_1 \cdot \boldsymbol{\nu} \qquad \text{on } \Gamma_{\text{shock}},$$

$\boldsymbol{\nu}$ is parallel to $D\varphi_1 - D\varphi$ so that

$$(\rho_1 D\varphi_1 - \rho D\varphi)\cdot(D\varphi_1 - D\varphi) = 0 \qquad \text{on } \Gamma_{\text{shock}}. \tag{14.2.19}$$

Since both φ and φ_2 satisfy (2.2.8)–(2.2.9) and $\psi := \varphi - \varphi_2$, we have

$$\begin{aligned}
\rho &= \rho(D\psi, \psi, \boldsymbol{\xi})\\
&= \Big(\rho_2^{\gamma-1} + (\gamma-1)\big((\boldsymbol{\xi} - (u_2, v_2))\cdot D\psi - \frac{1}{2}|D\psi|^2 - \psi\big)\Big)^{\frac{1}{\gamma-1}},\\
c^2 &= c^2(D\psi, \psi, \boldsymbol{\xi}) = c_2^2 + (\gamma-1)\Big((\boldsymbol{\xi} - (u_2, v_2))\cdot D\psi - \frac{1}{2}|D\psi|^2 - \psi\Big).
\end{aligned} \tag{14.2.20}$$

Then, using (2.2.16)–(2.2.17) and $\varphi = \varphi_2 + \psi$, we rewrite (14.2.19) as

$$E(D\psi, \psi, \boldsymbol{\xi}) = 0 \qquad \text{on } \Gamma_{\text{shock}}, \tag{14.2.21}$$

where, for $(\mathbf{p}, p_0, \boldsymbol{\xi}) \in \mathbb{R}^5$,

$$\begin{aligned} & E(\mathbf{p}, p_0, \boldsymbol{\xi}) \\ & = \rho_1\Big((u_1 - \xi_1)(u_1 - u_2 - p_1) + \xi_2(v_2 + p_2)\Big) \\ & \quad + \rho(\mathbf{p}, p_0, \boldsymbol{\xi})\big(\boldsymbol{\xi} - \mathbf{p} - (u_2, v_2)\big) \cdot \big((u_1 - u_2, -v_2) - \mathbf{p}\big), \end{aligned} \tag{14.2.22}$$

$$\rho(\mathbf{p}, p_0, \boldsymbol{\xi}) = \Big(\rho_2^{\gamma-1} + (\gamma - 1)\big((\boldsymbol{\xi} - (u_2, v_2)) \cdot \mathbf{p} - \frac{1}{2}|\mathbf{p}|^2 - p_0\big)\Big)^{\frac{1}{\gamma-1}}, \tag{14.2.23}$$

with $v_2 := u_2 \tan\theta_{\rm w}$.

Since points P_0 and P_1 both lie on $\mathcal{S}_1 = \{\varphi_1 = \varphi_2\}$, we have

$$(u_1 - u_2)(\xi_1^1 - \xi_1^0) - v_2(\xi_2^1 - \xi_2^0) = 0,$$

where $\boldsymbol{\xi}^1$ are the coordinates of P_1. Now, using the condition that $\varphi = \varphi_1$ on Γ_{shock}, *i.e.*, $\psi + \varphi_2 = \varphi_1$ on Γ_{shock}, we have

$$\xi_2 = \frac{(u_1 - u_2)(\xi_1 - \xi_1^1) - \psi(\boldsymbol{\xi})}{v_2} + \xi_2^1 \qquad \text{on } \Gamma_{\text{shock}}. \tag{14.2.24}$$

From (14.2.21) and (14.2.24), we conclude

$$F(D\psi, \psi, \xi_1) = 0 \qquad \text{on } \Gamma_{\text{shock}}, \tag{14.2.25}$$

where

$$F(\mathbf{p}, p_0, \xi_1) = E(\mathbf{p}, p_0, \xi_1, \frac{(u_1 - u_2)(\xi_1 - \xi_1^1) - p_0}{v_2} + \xi_2^1). \tag{14.2.26}$$

Now, from (14.2.22)–(14.2.23), we obtain that, for any $\xi_1 \in \mathbb{R}$,

$$\begin{aligned} F(\mathbf{0}, 0, \xi_1) &= E(\mathbf{0}, 0, \xi_1, \frac{(u_1 - u_2)(\xi_1 - \xi_1^1)}{v_2} + \xi_2^1) \\ &= \rho_1\Big((u_1 - \xi_1^1)(u_1 - u_2) + v_2\xi_2^1\Big) \\ &\quad - \rho_2\Big((u_2 - \xi_1^1)(u_1 - u_2) - v_2(v_2 - \xi_2^1)\Big) \\ &= \Big(\rho_1 D\varphi_1(\boldsymbol{\xi}^1) - \rho_2 D\varphi_2(\boldsymbol{\xi}^1)\Big) \cdot (u_1 - u_2,\ -v_2) \\ &= 0, \end{aligned} \tag{14.2.27}$$

where the last expression is zero, since it represents the right-hand side of the Rankine-Hugoniot condition (2.2.13) at point P_1 of the reflected shock $\mathcal{S}_1 = \{\varphi_1 = \varphi_2\}$ separating state (2) from state (1).

Now we write condition (14.2.25) in the (x, y)–coordinates on $\Gamma_{\text{shock}} \cap \{0 < x < \varepsilon_0\}$. By (11.1.1)–(11.1.2) and (14.2.25), we have

$$\Psi(\psi_x, \psi_y, \psi, x, y) = 0 \qquad \text{on } \Gamma_{\text{shock}} \cap \{0 < x < \varepsilon_0\}, \tag{14.2.28}$$

where

$$\begin{aligned}\Psi(\mathbf{p}, p_0, x, y) = F\big(&- p_1 \cos(y + \theta_{\text{w}}) - \frac{p_2}{c_2 - x} \sin(y + \theta_{\text{w}}),\\ &- p_1 \sin(y + \theta_{\text{w}}) + \frac{p_2}{c_2 - x} \cos(y + \theta_{\text{w}}),\\ &p_0,\ u_2 + (c_2 - x)\cos(y + \theta_{\text{w}})\big).\end{aligned} \tag{14.2.29}$$

From (14.2.27) and (14.2.29), we find that, on $\Gamma_{\text{shock}} \cap \{0 < x < \varepsilon_0\}$,

$$\Psi(\mathbf{0}, 0, x, y) = F(\mathbf{0}, 0, u_2 + (c_2 - x)\cos(y + \theta_{\text{w}})) = 0. \tag{14.2.30}$$

By the explicit definitions in (14.2.22)–(14.2.23), (14.2.26), and (14.2.29), we see that $\Psi(\mathbf{p}, p_0, x, y)$ is C^∞ on $\{|(\mathbf{p}, p_0, x)| < \delta\}$, where $\delta > 0$ depends only on $(u_2, v_2, \rho_2, \boldsymbol{\xi}^0)$, *i.e.*, on the data. Using (14.2.8) and choosing $\varepsilon > 0$ small, we have

$$|x| + |\psi(x, y)| + |(\psi_x, \psi_y)(x, y)| \le \delta \qquad \text{for any } (x, y) \in \overline{\Omega_\varepsilon}.$$

Thus, from (14.2.28)–(14.2.30), it follows that ψ satisfies (14.2.17) on $\Gamma_{\text{shock}} \cap \{0 < x < \varepsilon\}$, where

$$\hat{b}_k(x, y) = \int_0^1 \Psi_{p_k}(t\psi_x(x, y), t\psi_y(x, y), t\psi(x, y), x, y)\, dt \qquad \text{for } k = 0, 1, 2. \tag{14.2.31}$$

Then we obtain that, for $k = 0, 1, 2$,

$$\hat{b}_k \in C(\overline{\Gamma_{\text{shock}} \cap \{0 \le x \le \varepsilon\}}), \qquad |\hat{b}_k| \le \frac{1}{\lambda} \ \text{ on } \Gamma_{\text{shock}} \cap \{0 < x < \varepsilon\},$$

for some $\lambda > 0$.

It remains to show that $\hat{b}_1 \ge \lambda$ for some $\lambda > 0$. For this, since $\hat{b}_1$ is defined by (14.2.31), we first show that $\Psi_{p_1}(\mathbf{0}, 0, 0, y_1) > 0$, where $(x_1, y_1) = (0, \hat{f}(0))$ are the coordinates of P_1.

Since $(0, y_1)$ are the (x, y)–coordinates of $P_1 = \boldsymbol{\xi}^1$, then, by (11.1.1)–(11.1.2),

$$\boldsymbol{\xi}^1 = (\xi_1^1, \xi_2^1) = (u_2 + c_2\cos(y_1 + \theta_{\text{w}}), v_2 + c_2 \sin(y_1 + \theta_{\text{w}})),$$

which implies

$$(\xi_1^1 - u_2)^2 + (\xi_2^1 - v_2)^2 = c_2^2.$$

Also, $c_2^2 = \rho_2^{\gamma-1}$. Then, by an explicit calculation, we obtain

$$\begin{aligned}\Psi_{p_1}(\mathbf{0}, 0, 0, y_1) =& \frac{\rho_1}{c_2}\big((u_1 - \xi_1^1)(\xi_1^1 - u_2) - \xi_2^1(\xi_2^1 - v_2)\big)\\ &- \frac{\rho_2}{c_2}\big((u_2 - \xi_1^1)(\xi_1^1 - u_2) + (v_2 - \xi_2^1)(\xi_2^1 - v_2)\big).\end{aligned} \tag{14.2.32}$$

Now, working on the right-hand side in the $\boldsymbol{\xi}$–coordinates and noting that $D\varphi_1(\boldsymbol{\xi}^1) = (u_1 - \xi_1^1, -\xi_2^1)$ and $D\varphi_2(\boldsymbol{\xi}^1) = (u_2 - \xi_1^1, v_2 - \xi_2^1)$, we rewrite (14.2.32) as

$$\Psi_{p_1}(\mathbf{0}, 0, y_1) = -\frac{1}{c_2}\big(\rho_1 D\varphi_1(\boldsymbol{\xi}^1) - \rho_2 D\varphi_2(\boldsymbol{\xi}^1)\big) \cdot D\varphi_2(\boldsymbol{\xi}^1),$$

where $D = (\partial_{\xi_1}, \partial_{\xi_2})$. Since point P_1 lies on shock $\mathcal{S}_1 = \{\varphi_1 = \varphi_2\}$ separating state (2) from state (1), then, denoting by $\boldsymbol{\tau}_0$ the unit vector along line $\mathcal{S}_1$, we have

$$D\varphi_1(\boldsymbol{\xi}^1) \cdot \boldsymbol{\tau}_0 = D\varphi_2(\boldsymbol{\xi}^1) \cdot \boldsymbol{\tau}_0.$$

Now, using the Rankine-Hugoniot condition (2.2.13) for φ_1 and φ_2 at point P_1, we obtain

$$\rho_1 D\varphi_1(\boldsymbol{\xi}^1) - \rho_2 D\varphi_2(\boldsymbol{\xi}^1) = (\rho_1 - \rho_2)\big(D\varphi_2(\boldsymbol{\xi}^1) \cdot \boldsymbol{\tau}_0\big)\boldsymbol{\tau}_0$$

so that

$$\Psi_{p_1}(\mathbf{0}, 0, 0, y_1) = \frac{1}{c_2}(\rho_2 - \rho_1)\big(D\varphi_2(\boldsymbol{\xi}^1) \cdot \boldsymbol{\tau}_0\big)^2,$$

where $\rho_2 > \rho_1$ by the assumption of our theorem.

Thus, it remains to prove that $D\varphi_2(\boldsymbol{\xi}^1) \cdot \boldsymbol{\tau}_0 \neq 0$. Note that

$$|D\varphi_2(\boldsymbol{\xi}^1)| = c_2 = \rho_2^{(\gamma-1)/2},$$

since $\boldsymbol{\xi}^1$ is on the pseudo-sonic circle. Then, on the contrary, if $D\varphi_2(\boldsymbol{\xi}^1) \cdot \boldsymbol{\tau}_0 = 0$, we use $D\varphi_1(\boldsymbol{\xi}^1) \cdot \boldsymbol{\tau}_0 = D\varphi_2(\boldsymbol{\xi}^1) \cdot \boldsymbol{\tau}_0$ to write the Rankine-Hugoniot condition (2.2.13) at $\boldsymbol{\xi}^1$ in the form:

$$\rho_1 |D\varphi_1(\boldsymbol{\xi}^1)| = \rho_2\, \rho_2^{(\gamma-1)/2} = \rho_2^{(\gamma+1)/2}. \tag{14.2.33}$$

Since both φ_1 and φ_2 satisfy (2.2.8), $\varphi_1(\boldsymbol{\xi}^1) = \varphi_2(\boldsymbol{\xi}^1)$, and $|D\varphi_2(\boldsymbol{\xi}^1)| = c_2$, we have

$$\rho_1^{\gamma-1} + \frac{\gamma-1}{2}|D\varphi_1(\boldsymbol{\xi}^1)|^2 = \rho_2^{\gamma-1} + \frac{\gamma-1}{2}\rho_2^{\gamma-1}.$$

Combining this with (14.2.33), we obtain

$$\frac{2}{\gamma+1}\left(\frac{\rho_1}{\rho_2}\right)^{\gamma-1} + \frac{\gamma-1}{\gamma+1}\left(\frac{\rho_2}{\rho_1}\right)^2 = 1. \tag{14.2.34}$$

Consider the function:

$$g(s) = \frac{2}{\gamma+1}s^{\gamma-1} + \frac{\gamma-1}{\gamma+1}s^{-2} \qquad \text{on } (0, \infty).$$

Since $\gamma > 1$, we have

$$g'(s) < 0 \quad \text{on } (0,1), \qquad g'(s) > 0 \quad \text{on } (1, \infty), \qquad g(1) = 1.$$

Thus, $g(s) = 1$ only for $s = 1$. Therefore, (14.2.34) implies that $\rho_1 = \rho_2$, which contradicts the assumption that $\rho_1 < \rho_2$ in our theorem. This implies that $D\varphi_2(\boldsymbol{\xi}^1) \cdot \boldsymbol{\tau}_0 \neq 0$, so that $\Psi_{p_1}(\mathbf{0}, 0, y_1) > 0$.

Choose $\lambda := \frac{1}{2}\Psi_{p_1}(\mathbf{0}, 0, 0, y_1)$. Then $\lambda > 0$. Since $\Psi(\mathbf{p}, p_0, x, y)$ is C^∞ on $\{|(\mathbf{p}, p_0, x)| < \delta\}$ and, by (14.2.7), $\psi \in C^{1,1}(\overline{\Omega_{\varepsilon_0}})$ with $\psi(\mathbf{0}) = \psi_x(\mathbf{0}) = \psi_y(\mathbf{0}) = 0$, we find that, for small $\varepsilon > 0$,

$$\Psi_{p_1}(t\psi_x(x,y), t\psi_y(x,y), t\psi(x,y), x, y) \geq \lambda$$

for any $(x, y) \in \Gamma_{\text{shock}} \cap \{0 < x < \varepsilon\}$ and $t \in [0, 1]$. Therefore, from (14.2.31), we see that $\hat{b}_1 \geq \lambda$. Lemma 14.2.8 is proved. □

7. A sequence close to Γ_{shock}. Now we construct the sequence close to Γ_{shock}. Since we have assumed that assertion (iv) is false, *i.e.*, $D^2\psi$ has a limit at P_1 from Ω, then (14.2.16) implies

$$\lim_{\substack{(x,y)\to(0,\ \hat{f}(0))\\ (x,y)\in\Omega}} (\psi_{xy}, \psi_{yy})(x, y) = (0, 0), \tag{14.2.35}$$

where $(0,\ \hat{f}(0))$ are the coordinates of P_1 in the (x, y)–plane. Note that, from (14.2.7),

$$\psi_y(x, \hat{f}(x)) = \int_0^x \psi_{xy}(s, \hat{f}(0))ds + \int_{\hat{f}(0)}^{\hat{f}(x)} \psi_{yy}(x, t)dt,$$

and, from (14.2.5), all the points in the paths of integration are within Ω. Furthermore, by (14.2.6), $0 < \hat{f}(x) - \hat{f}(0) < Cx$ with C independent of $x \in (0, \varepsilon_0)$. Now, (14.2.35) implies

$$\lim_{x\to 0+} \frac{\psi_y(x, \hat{f}(x))}{x} = 0. \tag{14.2.36}$$

Also, by Lemma 14.2.8,

$$|\psi_x(x, \hat{f}(x))| = |\frac{\hat{b}_2\psi_y + \hat{b}_3\psi}{\hat{b}_1}| \leq C(|\psi_y| + |\psi|) \qquad \text{for } x \in (0, \varepsilon),$$

where $\varepsilon > 0$ is from Lemma 14.2.8. Then, using (14.2.36) and $|\psi(x, y)| \leq Cx^2$ by (14.2.8), we have

$$\lim_{x\to 0+} \frac{\psi_x(x, \hat{f}(x))}{x} = 0. \tag{14.2.37}$$

Let

$$\mathcal{F}(x) := \psi_x(x,\ \hat{f}(x) - \frac{\omega}{10}x) \qquad \text{for some constant } \omega > 0.$$

Then $\mathcal{F}(x)$ is well-defined and differentiable for $0 < x < \varepsilon_0$ so that

$$\begin{aligned}\mathcal{F}(x) &= \psi_x(x,\ \hat{f}(x) - \frac{\omega}{10}x) \\ &= \psi_x(x, \hat{f}(x)) + \int_0^1 \frac{d}{dt}\psi_x(x,\ \hat{f}(x) - \frac{t\omega}{10}x)\,dt \\ &= \psi_x(x, \hat{f}(x)) - \frac{\omega}{10}x\int_0^1 \psi_{xy}(x, \hat{f}(x) - \frac{t\omega}{10}x)\,dt.\end{aligned}$$

Now (14.2.35) and (14.2.37) imply

$$\lim_{x\to 0+}\frac{\mathcal{F}(x)}{x} = 0. \tag{14.2.38}$$

Using (14.2.10) and $\hat{f} \in C^{1,1}([0, \varepsilon_0])$, we have

$$\mathcal{F} \in C([0, \varepsilon]) \cap C^1((0, \varepsilon)). \tag{14.2.39}$$

Then (14.2.38) and the mean-value theorem imply that there exists a sequence $\{x_k^{(2)}\}$ with $x_k^{(2)} \in (0, \varepsilon)$ and

$$\lim_{k\to\infty} x_k^{(2)} = 0, \qquad \lim_{k\to\infty}\mathcal{F}'(x_k^{(2)}) = 0. \tag{14.2.40}$$

By the definition of $\mathcal{F}(x)$,

$$\psi_{xx}(x,\ g(x)) = \mathcal{F}'(x) - g'(x)\psi_{xy}(x,\ g(x)), \tag{14.2.41}$$

where $g(x) := \hat{f}(x) - \frac{\omega}{10}x$.

On the other hand, $|\hat{f}'(x)|$ is bounded. Then, using (14.2.35) and (14.2.40)–(14.2.41), we have

$$\lim_{k\to\infty}\psi_{xx}(x_k^{(2)}, g(x_k^{(2)})) = \lim_{k\to\infty}\mathcal{F}'(x_k^{(2)}) = 0.$$

Note that $\lim_{x\to 0+} g(x) = \hat{f}(0)$. Denoting $y_k^{(2)} = g(x_k^{(2)})$, we conclude

$$(x_k^{(2)}, y_k^{(2)}) \in \Omega, \quad \lim_{k\to\infty}(x_k^{(2)}, y_k^{(2)}) = (0, \hat{f}(0)), \quad \lim_{k\to\infty}\psi_{xx}(x_k^{(2)}, y_k^{(2)}) = 0.$$

Combining this with (14.2.16), we conclude that ψ_{xx} does not have a limit at P_1 from Ω, which implies assertion (iv). This completes the proof of Theorem 14.2.7. □

Remark 14.2.9. *For the isothermal case, $\gamma = 1$, there exists a global regular reflection-diffraction solution in the sense of Definition* 14.2.1 *when $\theta_{\rm w} \in (\theta_{\rm w}^{\rm s}, \frac{\pi}{2})$. Moreover, the solution has the same properties stated in Theorem* 14.2.7

with $\gamma = 1$. *This can be verified by the limiting properties of the solutions for the homentropic case when* $\gamma \to 1+$. *This is because, when* $\gamma \to 1+$,

$$
\begin{aligned}
&i(\rho) \to \ln\rho, \qquad p(\rho) \to \rho, \qquad c^2(\rho) \to 1 && \textit{in } (1.2),\\
&\rho(|D\varphi|^2, \varphi) \to \rho_0 e^{-(\varphi+\frac{1}{2}|D\varphi|^2)} && \textit{in } (2.2.9),\\
&c_*(\varphi, \rho_0, \gamma) \to 1 && \textit{in } (2.2.12).
\end{aligned}
$$

In this case, the arguments for establishing Theorem 14.2.7 *are even simpler.*

Corollary 14.2.10. *Let* φ *be an admissible solution of* **Problem 2.6.1**. *Then the conclusions of Theorems* 14.2.6–14.2.7 *hold for* φ.

Proof. It suffices to check that any admissible solution of **Problem 2.6.1** satisfies the conditions of Theorems 14.2.6–14.2.7.

By Remark 14.2.2, any admissible solution is a regular reflection-diffraction solution. Then it remains to check whether any admissible solution satisfies conditions (a)–(c) of Theorem 14.2.7.

Condition (a) of Theorem 14.2.7 holds for admissible solutions by (11.4.4), condition (b) holds by Lemma 11.2.5, and condition (c) holds by Proposition 11.2.8 and (11.4.5). □

Part IV

Subsonic Regular Reflection-Diffraction and Global Existence of Solutions up to the Detachment Angle

Chapter Fifteen

Admissible Solutions and Uniform Estimates up to the Detachment Angle

In the next chapters, Chapters 15–17, we prove Theorems 2.6.7–2.6.9, stated in §2.6. To achieve this, we extend the argument in Chapters 8–13 to the present case, where the solutions of both supersonic and subsonic reflection configurations (*cf.* §2.4.2–§2.4.3) are presented.

In this chapter, we define admissible solutions and derive their estimates by extending the estimates of Chapters 8–10 to the present case. An overview of some steps of the argument has been given in §3.3.1–§3.3.4.

15.1 DEFINITION OF ADMISSIBLE SOLUTIONS FOR THE SUPERSONIC AND SUBSONIC REFLECTIONS

In this section, we use the terminology and notations introduced in §7.5. In particular, we use the notions of supersonic, sonic, and subsonic wedge angles; see Definition 7.5.1.

We define the notion of admissible solutions for all the wedge angles $\theta_{\rm w} \in (\theta_{\rm w}^{\rm d}, \frac{\pi}{2})$. The definitions are separate for supersonic wedge angles when the solutions are of the structure of supersonic reflection configuration as in §2.4.2 (see also Fig. 2.3), and for subsonic-sonic wedge angles when the solutions are of the structure of subsonic reflection configuration as in §2.4.3 (see also Fig. 2.4). There is a *continuity* between these definitions through the sonic reflection configuration.

Definition 15.1.1 (Admissible Solutions for the Supersonic Reflection). *Let $\gamma > 1$, $\rho_1 > \rho_0 > 0$, and $u_1 > 0$ be constants. Fix a* supersonic *wedge angle $\theta_{\rm w} \in (\theta_{\rm w}^{\rm d}, \frac{\pi}{2})$. A function $\varphi \in C^{0,1}(\Lambda)$ is called an admissible solution of the regular shock reflection-diffraction problem if φ is a solution to* **Problem 2.6.1** *and satisfies all the conditions of Definition* 8.1.1.

Definition 15.1.2 (Admissible Solutions for the Subsonic Reflection). *Let $\gamma > 1$, $\rho_1 > \rho_0 > 0$, and $u_1 > 0$ be constants. Fix a* subsonic or sonic *wedge angle $\theta_{\rm w} \in (\theta_{\rm w}^{\rm d}, \frac{\pi}{2})$. A function $\varphi \in C^{0,1}(\Lambda)$ is called an admissible solution of the regular shock reflection-diffraction problem if φ is a solution to* **Problem 2.6.1** *satisfying the following:*

(i) *There exists a continuous shock curve* Γ_{shock} *with endpoints* $P_0 = (\xi_1^0, \xi_1^0 \tan\theta_{\rm w})$ *and* $P_2 = (\xi_{1P_2}, 0)$, *and*

$$\xi_{1P_2} < \xi_{1P_3} = 0$$

such that

- Γ_{shock} *is outside the sonic circle* $\partial B_{c_1}(u_1, 0)$:

$$\Gamma_{\text{shock}} \subset \big(\Lambda \setminus \overline{B_{c_1}(u_1,0)}\big) \cap \{\xi_{1P_2} \le \xi_1 \le \xi_1^0\}; \tag{15.1.1}$$

- Γ_{shock} *is* C^2 *in its relative interior. Moreover, for* $\Gamma_{\text{shock}}^{\text{ext}} := \Gamma_{\text{shock}} \cup \Gamma_{\text{shock}}^- \cup \{P_2\}$ *with* Γ_{shock}^- *being the reflection of* Γ_{shock} *with respect to* $\{\xi_2 = 0\}$, $\Gamma_{\text{shock}}^{\text{ext}}$ *is* C^1 *in its relative interior (including* P_2*).*

Denote the line segments $\Gamma_{\text{sym}} := P_2P_3$ *and* $\Gamma_{\text{wedge}} := P_0P_3$. *Note that* $\overline{\Gamma_{\text{sym}}} \cap \overline{\Gamma_{\text{wedge}}} = \{P_3\}$. *We require that there be no common points between* $\overline{\Gamma_{\text{shock}}}$ *and curve* $\overline{\Gamma_{\text{sym}} \cup \Gamma_{\text{wedge}}}$ *except their common endpoints* $\{P_0, P_2\}$. *Thus,* $\overline{\Gamma_{\text{shock}}} \cup \overline{\Gamma_{\text{sym}}} \cup \overline{\Gamma_{\text{wedge}}}$ *is a closed curve without self-intersection. Denote by* Ω *the open bounded domain restricted by this curve. Note that* $\Omega \subset \Lambda$ *and* $\partial\Omega \cap \partial\Lambda := \overline{\Gamma_{\text{sym}}} \cup \overline{\Gamma_{\text{wedge}}}$.

(ii) φ *satisfies* (2.6.4) *and*

$$\begin{aligned} &\varphi \in C^{0,1}(\Lambda) \cap C^1(\overline{\Lambda} \setminus \overline{\Gamma_{\text{shock}}}),\\ &\varphi \in C^3(\overline{\Omega} \setminus \{P_0, P_2, P_3\}) \cap C^1(\overline{\Omega}). \end{aligned} \tag{15.1.2}$$

Furthermore,

$$\varphi(P_0) = \varphi_2(P_0), \quad D\varphi(P_0) = D\varphi_2(P_0). \tag{15.1.3}$$

(iii) *Equation* (2.2.8) *is strictly elliptic in* $\overline{\Omega} \setminus \{P_0\} = \overline{\Omega} \setminus \overline{\Gamma_{\text{sonic}}}$:

$$|D\varphi| < c(|D\varphi|^2, \varphi) \qquad in\ \overline{\Omega} \setminus \overline{\Gamma_{\text{sonic}}}. \tag{15.1.4}$$

(iv) *In* Ω,

$$\varphi_2 \le \varphi \le \varphi_1. \tag{15.1.5}$$

(v) *Let* $\mathbf{e}_{S_1}$ *be defined by* (7.5.8). *Then*

$$\partial_{\mathbf{e}_{S_1}}(\varphi_1 - \varphi) \le 0 \qquad in\ \Omega, \tag{15.1.6}$$

$$\partial_{\xi_2}(\varphi_1 - \varphi) \le 0 \qquad in\ \Omega. \tag{15.1.7}$$

Remark 15.1.3. *Condition* (15.1.3) *for the subsonic reflection solutions corresponds to the property of supersonic reflection solutions discussed in Remark* 8.1.2.

15.2 BASIC ESTIMATES FOR ADMISSIBLE SOLUTIONS UP TO THE DETACHMENT ANGLE

In this section, we make some basic estimates for admissible solutions up to the detachment angle.

Proposition 15.2.1. *Lemmas* 8.1.7 *and* 8.1.9, *Corollary* 8.1.10, *and all the assertions proved in* §8.2.1–§8.2.3 *hold for both supersonic and subsonic admissible solutions for the wedge angles* $\theta_w \in (\theta_w^d, \frac{\pi}{2})$, *with only the following notational changes in the subsonic case:* $\overline{\Gamma_{sonic}}$, P_1, *and* P_4 *need to be replaced by* P_0, *and the conditions of Definition* 8.1.1 *need to be replaced by the corresponding conditions of Definition* 15.1.2 *in the statements and proofs.*

Proof. In the proofs of the assertions in §8.1–§8.2, the only property related to Γ_{sonic} used is that $(\varphi, D\varphi) = (\varphi_2, D\varphi_2)$ on Γ_{sonic}. For the subsonic reflection, the admissible solutions satisfy this property at P_0 by (15.1.3). The other properties employed in the proofs hold for both subsonic and supersonic reflection solutions. □

The formulas derived in §5.1 obviously apply to the subsonic admissible solutions.

Proposition 15.2.2. *All the assertions proved in* §9.1–§9.2 *hold for both supersonic and subsonic admissible solutions, with only the following notational changes in the subsonic case:* $\overline{\Gamma_{sonic}}$, P_1, *and* P_4 *need to be replaced by* P_0, *and the conditions of Definition* 8.1.1 *need to be replaced by the corresponding conditions of Definition* 15.1.2 *in the statements and proofs.*

Moreover, for any subsonic/supersonic admissible solution with the wedge angle $\theta_w \in (\theta_w^d, \frac{\pi}{2})$,

- *The estimates in Proposition* 9.1.2, *Corollary* 9.1.3, *and Lemma* 9.1.4 *hold with the constants depending only on the data, i.e.,* (ρ_0, ρ_1, γ);
- *The estimates in Lemma* 9.2.2 *hold with the constants depending only on* $(\rho_0, \rho_1, \gamma, \alpha, r)$;
- *In Definition* 9.2.3, *for the subsonic/sonic wedge angles, region* $P_0P_1P_4$ *in* (2.6.1) *is one point;*
- *Corollary* 9.2.5 *holds for any sequence of subsonic/supersonic admissible solutions of* **Problem 2.6.1** *with the wedge angles* $\theta_w^{(i)} \in (\theta_w^d, \frac{\pi}{2})$ *such that*

$$\theta_w^{(i)} \to \theta_w^{(\infty)} \in [\theta_w^d, \frac{\pi}{2}].$$

Furthermore, in the assertions and the proof of Corollary 9.2.5, *if the type of admissible solutions is subsonic for the wedge angle* $\theta_w^{(i)}$ *(resp.* $\theta_w^{(\infty)}$*), then* ${P_1}^{(i)} = {P_4}^{(i)} = {P_0}^{(i)}$ *(resp.* ${P_1}^{(\infty)} = {P_4}^{(\infty)} = {P_0}^{(\infty)}$*), in which case*

$\Gamma_{\rm sonic}^{(i)}$ *denotes* $\{{P_4}^{(i)}\}$ *(resp.* $\Gamma_{\rm sonic}^{(\infty)}$ *denotes* $\{{P_4}^{(\infty)}\}$*). In particular, if* $\theta_{\rm w}^{(\infty)}$ *is a subsonic/sonic wedge angle, there is no region of state* (2) *in the solution in assertion* (v) *of Corollary* 9.2.5*, i.e.,* $\varphi^{(\infty)}$ *is equal to the constant states in their respective domains as in* (2.6.4)*.*

Proof. The proofs in §9.1 and the proof of Lemma 9.2.2 work without change in the present case.

In the proof of Corollary 9.2.5, the changes are only notational, as described above, by using the fact that the wedge angles corresponding to the supersonic admissible solutions are an open subset of $[\theta_{\rm w}^{\rm d}, \frac{\pi}{2})$. When $\theta^{(i)}$ are supersonic wedge angles and $\theta^{(\infty)}$ is a sonic wedge angle, we see that ${P_1}^{(i)} \to {P_0}^{(\infty)}$ and ${P_4}^{(i)} \to {P_0}^{(\infty)}$, as in the proof of assertion (v) of Corollary 9.2.5 in this case. □

15.3 SEPARATION OF $\Gamma_{\rm shock}$ FROM $\Gamma_{\rm sym}$

Proposition 9.3.1 and Corollary 9.3.2 hold, without change in the formulation and the proof, for any subsonic/supersonic admissible solution of **Problem 2.6.1** for the wedge angles $\theta_{\rm w} \in [\theta_{\rm w}^{\rm d}, \frac{\pi}{2})$.

15.4 LOWER BOUND FOR THE DISTANCE BETWEEN $\Gamma_{\rm shock}$ AND $\Gamma_{\rm wedge}$ AWAY FROM P_0

For the subsonic reflection, the reflected shock intersects the wedge at point P_0. Thus, we make estimates similar to Propositions 9.4.5 and 9.4.8 for all subsonic/supersonic admissible solutions for the portions of $\Gamma_{\rm shock}$ which are away from P_0 by a distance $r > 0$, with the constants in the estimates depending on r. Of course, for any given r, for the supersonic wedge angles such that $\mathrm{dist}(P_0, P_4) \geq r$, these results are equivalent to Propositions 9.4.5 and 9.4.8.

Note that all the wedge angles $\theta_{\rm w} \in (\theta_{\rm w}^{\rm s}, \frac{\pi}{2})$ are supersonic, so that Lemma 9.4.7 remains without change in the present case.

We first note the following:

Lemma 15.4.1. *Lemmas* 9.4.1–9.4.2 *and Corollary* 9.4.3 *hold for both supersonic and subsonic admissible solutions for the wedge angles* $\theta_{\rm w} \in (\theta_{\rm w}^{\rm d}, \frac{\pi}{2})$*, with only notational changes in the subsonic case:* $\overline{\Gamma_{\rm sonic}}$*,* P_1*, and* P_4 *need to be replaced by* P_0*; that is, property* (9.4.32) *for subsonic/sonic admissible solutions takes the form:*

$$\sup_{P \in \Gamma_{\rm shock} \cap B_r(P_0)} \mathrm{dist}(P, L_{\Gamma_{\rm wedge}}) > \frac{1}{C_r}. \tag{15.4.1}$$

Proof. The proofs of Lemmas 9.4.1–9.4.2 and Corollary 9.4.3 work without changes in the present case by using that Corollaries 8.2.14, 9.2.5, and 9.3.2 are extended to the present case in Propositions 15.2.1–15.2.2 and §15.3. □

15.4.1 Uniform positive lower bound for the distance between Γ_{shock} and Γ_{wedge} away from P_0 when $u_1 < c_1$

We note that, from Lemma 7.5.9, the length of $\Gamma^{(\theta_w)}_{\text{wedge}} = P_3P_4$ is positive and depends continuously on $\theta_w \in [\theta_w^d, \frac{\pi}{2}]$ so that

$$r_1 := \min_{\theta_w \in [\theta_w^d, \frac{\pi}{2}]} |\Gamma^{(\theta_w)}_{\text{wedge}}| > 0. \tag{15.4.2}$$

For simplicity, we will often use $\Gamma_{\text{wedge}} = \Gamma^{(\theta_w)}_{\text{wedge}}$, when no confusion arises.

Proposition 15.4.2. *Assume that $u_1 < c_1$. For every $r \in (0, \frac{r_1}{10})$, there exists $C_r > 0$ depending only on the data and r such that*

$$\text{dist}(\Gamma_{\text{shock}}, \Gamma_{\text{wedge}} \setminus B_r(P_0)) > \frac{1}{C_r}$$

for any subsonic/supersonic admissible solution of **Problem 2.6.1** *with the wedge angle $\theta_w \in (\theta_w^d, \frac{\pi}{2})$.*

Proof. We note that Propositions 15.2.1–15.2.2 and Lemma 15.4.1 extend all the results used in the argument below to the present setting.

In this proof, the universal constants C and C_k are positive and depend only on $(\rho_0, \rho_1, \gamma, r)$.

Arguing as in the proof of (9.4.36), we obtain

$$\text{dist}(P_2, L_{\Gamma_{\text{wedge}}}) = f_{\nu_w}(T_{P_2}) \geq (c_1 - u_1)\sin\theta_w^d > 0, \tag{15.4.3}$$

that is, the positive lower bound depends only on the data, *i.e.*, (ρ_0, ρ_1, γ).

Now we follow Step 1 and Steps 3–4 of the proof of Proposition 9.4.9 by using (15.4.3) instead of estimate (9.4.35) obtained in Step 2 of that proof. More specifically, it suffices to consider $r \in (0, \frac{r^*}{10}]$ for $r^* = \min\{r_1, \hat{r}_0\}$, where $\hat{r}_0$ is defined in Remark 9.4.6. Fix such a constant r.

Let φ be an admissible solution for a wedge angle $\theta_w \in (\theta_w^d, \frac{\pi}{2})$.

Then, arguing as in Step 1 of the proof of Proposition 9.4.9 and changing θ_w^s to θ_w^d in the argument, we conclude that there exist C and C_1 such that (9.4.44) holds if $\text{dist}(P_1, L_{\Gamma_{\text{wedge}}}) \leq \frac{1}{C_1}$.

As in Step 3 of the proof of Proposition 9.4.9, we consider the case that $\text{dist}(P_1, L_{\Gamma_{\text{wedge}}}) \geq \frac{1}{C_1}$, *i.e.*, (9.4.45) holds. We argue as in the proof of Proposition 9.4.5 by using (15.4.3) and (9.4.45) instead of (9.4.36) and (9.4.37), respectively. Also, (9.4.38) holds, and from that, using Remark 9.4.6, we obtain (9.4.39) with the lower bound depending only on the data. Then, following the argument in the proof of Proposition 9.4.5 after (9.4.39), we obtain that $\text{dist}(\Gamma_{\text{shock}}, \Gamma_{\text{wedge}}) > \frac{1}{C}$, where C depends on $(\rho_0, \rho_1, \gamma, r)$ now, since the constants in (9.4.39), (9.4.45), and (15.4.3) depend on these parameters.

Finally, as in Step 4 of the proof of Proposition 9.4.9, we consider the case that $\text{dist}(P_1, L_{\Gamma_{\text{wedge}}}) \leq \frac{1}{C_1}$, *i.e.*, (9.4.46) holds. We follow the argument in that step by using P_2 instead of Q and (15.4.3) instead of the first inequality in (9.4.35). We also note, by (8.2.20), that $T_{P_2} < T_{P_3}$, since $\theta_{\rm w} < \frac{\pi}{2}$. Similarly to (9.4.47), we obtain

$$|\hat{Q} - P_2| \geq T_{\hat{Q}} - T_{P_2} \geq T_{P_0} - \frac{3r}{4} - T_{P_3} \geq r_1 - \frac{3r}{4} \geq \frac{r_1}{2},$$

where we have used that $T_{P_0} - T_{P_3} \geq |\Gamma_{\rm wedge}| \geq r_1$ by (15.4.2). With this, repeating the proof of (9.4.48), we have

$$\operatorname{dist}(\Gamma_{\rm shock}[P_2, \hat{Q}], \Gamma_{\rm wedge}) \geq \frac{1}{C}.$$

Then, arguing as in the proof of (9.4.49) and using again that $T_{P_2} < T_{P_3}$, we obtain

$$\operatorname{dist}(\Gamma_{\rm shock} \cap \{T_{P_3} \leq T \leq T_{P_0} - \frac{3r}{4}\},\ \Gamma_{\rm wedge}) \geq \frac{1}{C}. \tag{15.4.4}$$

On the other hand,

$$\Gamma_{\rm wedge} \setminus B_r(P_4) = \{(0, T)\ :\ T_{P_3} < T \leq T_{P_4} - r\},$$

and then we use $T_{P_0} \geq T_{P_4}$ and (15.4.4) to obtain

$$\operatorname{dist}\big(\Gamma_{\rm shock} \setminus \{T_{P_3} \leq T \leq T_{P_0} - \frac{3r}{4}\},\ \Gamma_{\rm wedge} \setminus B_r(P_4)\big) \geq \frac{r}{4}.$$

Combining this with (15.4.4), we conclude

$$\operatorname{dist}(\Gamma_{\rm shock}, \Gamma_{\rm wedge} \setminus B_r(P_0)) > \frac{1}{C_r}.$$

This completes the proof. □

15.4.2 Lower bound for the distance between $\Gamma_{\rm shock}$ and $\Gamma_{\rm wedge}$ away from P_0 and P_3

In the next proposition, we use constant r_1 defined by (15.4.2).

Proposition 15.4.3. *Fix $\rho_1 > \rho_0 > 0$ and $\gamma > 1$. For every $r \in (0, \frac{r_1}{10})$, there exists $C_r > 0$ such that*

$$\operatorname{dist}\left(\Gamma_{\rm shock},\ \Gamma_{\rm wedge} \setminus (B_r(P_0) \cup B_r(P_3))\right) \geq \frac{1}{C_r}$$

for any subsonic/supersonic admissible solution of **Problem 2.6.1** *with the wedge angle $\theta_{\rm w} \in [\theta_{\rm w}^{\rm d}, \frac{\pi}{2})$.*

Proof. The proof of Proposition 9.4.9 works with only notational changes, *i.e.*, changing $\theta_{\rm w}^{\rm s}$ to $\theta_{\rm w}^{\rm d}$ in the argument, and noting that Propositions 15.2.1–15.2.2 and Lemma 15.4.1 extend to the present setting for all the results used in the argument. □

15.5 UNIFORM POSITIVE LOWER BOUND FOR THE DISTANCE BETWEEN Γ_{shock} AND THE SONIC CIRCLE OF STATE (1)

The results in §9.5 hold for all the subsonic/supersonic admissible solutions with uniform constants. More specifically, we have

Lemma 15.5.1. *For every compact set $K \subset B_{c_1}(\mathcal{O}_1) \setminus \{P_3\}$, there exists $C = C(K) > 0$ such that, for any wedge angle $\theta_w \in [\theta_w^d, \frac{\pi}{2})$ satisfying $K \cap \overline{\Lambda^{(\theta_w)}} \neq \emptyset$ and any admissible solution φ of* **Problem 2.6.1** *with the wedge angle θ_w,*

$$\inf_{K\cap\overline{\Lambda^{(\theta_w)}}} (\varphi_1 - \varphi) \geq \frac{1}{C(K)}.$$

Proof. The proof of Lemma 9.5.1 works without amendment, by using Propositions 15.2.1–15.2.2 to adjust the results in §9.1–§9.2 (especially Corollary 9.2.5) to the present case. □

Lemma 15.5.2. *Fix $\rho_1 > \rho_0 > 0$ and $\gamma > 1$. For any $\varepsilon_0 \in (0, \frac{c_1}{2})$, there exists $C > 0$ such that, if $\theta_w \in [\theta_w^d, \frac{\pi}{2})$ satisfies that $B_{c_1-\varepsilon_0} \cap \Lambda(\theta_w) \neq \emptyset$ and φ is an admissible solution of* **Problem 2.6.1** *with the wedge angle θ_w, then*

$$\text{dist}(\Gamma_{\text{shock}}, B_{c_1}(\mathcal{O}_1)) \geq \frac{1}{C}.$$

Proof. The proof of Lemma 9.5.2 works without change, by using Lemma 15.5.1 instead of Lemma 9.5.1, and using Propositions 15.2.1–15.2.2 to adjust the results in §9.1–§9.2 to the present case. □

Proposition 15.5.3. *Fix $\gamma > 1$ and $\rho_1 > \rho_0 > 0$. There exists $C > 0$ such that*

$$\text{dist}(\Gamma_{\text{shock}}, B_{c_1}(\mathcal{O}_1)) \geq \frac{1}{C}$$

for any supersonic/subsonic admissible solution of **Problem 2.6.1** *with $\theta_w \in [\theta_w^d, \frac{\pi}{2})$.*

Proof. The proof of Proposition 9.5.6 works without change by recalling that P_1 and P_4 are identified with P_0 in the case of subsonic/sonic admissible solutions, and using Proposition 15.4.3 and Lemma 15.5.2 instead of Proposition 9.4.9 and Lemma 9.5.2, respectively. Also, we use r_1 defined by (15.4.2) instead of r_0.

Finally, we note that $\min_{\theta_w\in[\theta_w^d,\frac{\pi}{2}]} r_2^{(\theta_w)} > 0$ by the same argument as in Proposition 9.5.6, where the similar quantity with θ_w^s instead of θ_w^d has been considered. □

Corollary 15.5.4. *Let r_1 be defined by* (15.4.2). *Let $\gamma > 1$ and $\rho_1 > \rho_0 > 0$ be such that $u_1 = c_1$. For every $r \in (0, \frac{r_1}{10})$, there exists $C_r > 0$ such that*

$$\text{dist}\,(\Gamma_{\text{shock}}, \Gamma_{\text{wedge}} \setminus B_r(P_0)) > \frac{1}{C_r}$$

for any subsonic/supersonic admissible solution of **Problem 2.6.1**, *where we write P_4 for P_0 in the case of subsonic admissible solutions.*

Proof. The proof repeats that for Corollary 9.5.7 by using Propositions 15.4.3 and 15.5.3 instead of Propositions 9.4.8 and 9.5.6. □

15.6 UNIFORM ESTIMATES OF THE ELLIPTICITY CONSTANT

We now extend the results of §9.6. We use the notation introduced in §9.6. We first extend Proposition 9.6.3 as follows:

Proposition 15.6.1. *There exist $\mu > 0$ and $\hat{\zeta} > 0$ depending only on (ρ_0, ρ_1, γ) such that, if φ is an admissible solution of* **Problem 2.6.1** *with $\theta_{\rm w} \in (\theta_{\rm w}^{\rm d}, \frac{\pi}{2})$ and $g \in C^1(\overline{\Omega})$ so that $g(P_1) = 0$ and $|Dg| \le 1$ in Ω, then either*

$$M^2 \le 1 - \hat{\zeta} \qquad \text{in } \overline{\Omega}$$

or the maximum of $M^2 + \mu g$ over $\overline{\Omega}$ cannot be attained on $\Gamma_{\rm shock} \cup \{P_2\}$, where M is the Mach number, i.e., $M = \frac{|D\varphi|}{c(|D\varphi|^2, \varphi)}$.

Proof. In this proof, all constants $(C, \mu, \alpha, \zeta, \delta)$ depend only on (ρ_0, ρ_1, γ).

By Proposition 15.5.3,

$$\text{dist}(\Gamma_{\rm shock}, B_{c_1}(\mathcal{O}_1)) > \frac{1}{C}$$

for any admissible solution of **Problem 2.6.1** with $\theta_{\rm w} \in [\theta_{\rm w}^{\rm d}, \frac{\pi}{2})$. Then, using $D\varphi_1(\boldsymbol{\xi}) = (u_1, 0) - \boldsymbol{\xi}$, we obtain that there exists $\delta > 0$ such that

$$|D\varphi_1|^2 \ge (1+\delta)c_1^2 \qquad \text{on } \Gamma_{\rm shock} \tag{15.6.1}$$

for all such solutions.

Let φ be an admissible solution of **Problem 2.6.1** with $\theta_{\rm w} \in [\theta_{\rm w}^{\rm d}, \frac{\pi}{2})$.

Note that, from Lemma 9.1.4 and Proposition 15.2.2,

$$\rho \ge a\rho_1 > 0 \qquad \text{in } \overline{\Omega}.$$

Let $g \in C^1(\overline{\Omega})$ with $g(P_1) = 0$ and $|Dg| \le 1$ in Ω. Denote

$$d(\boldsymbol{\xi}) = \mu g(\boldsymbol{\xi})$$

for $\mu > 0$ to be chosen. Since $\text{diam}(\Omega) \le C$ by Propositions 9.1.2 and 15.2.2,

$$|d(\boldsymbol{\xi})| \le \mu\big(|g(P_1)| + \|Dg\|_{L^\infty(\Omega)}\text{diam}(\Omega)\big) \le C\mu \qquad \text{for all } \boldsymbol{\xi} \in \overline{\Omega}. \tag{15.6.2}$$

Let $\alpha, \zeta > 0$ be the constants as defined in Lemma 9.6.2 for $\rho_- = \rho_1$ and δ from (15.6.1). Then, choosing μ sufficiently small, we obtain from (15.6.2) that

$$|d(\boldsymbol{\xi})| \le C\mu \le \frac{\zeta}{10} \qquad \text{for all } \boldsymbol{\xi} \in \overline{\Omega}. \tag{15.6.3}$$

Assume that the maximum of $M^2 + d$ over $\overline{\Omega}$ is attained at $P_{\max} \in \Gamma_{\text{shock}} \cup \{P_2\}$. Consider first the case:

$$M^2(P_{\max}) \le 1 - \frac{\zeta}{2}.$$

Then, using (15.6.3),

$$(M^2 + d)(\boldsymbol{\xi}) \le (M^2 + d)(P_{\max}) \le 1 - \frac{\zeta}{2} + \frac{\zeta}{10} = 1 - \frac{2\zeta}{5} \qquad \text{for all } \boldsymbol{\xi} \in \overline{\Omega},$$

which implies

$$M^2(\boldsymbol{\xi}) \le 1 - \frac{2\zeta}{5} - d(\boldsymbol{\xi}) \le 1 - \frac{2\zeta}{5} + \frac{\zeta}{10}.$$

Therefore, we obtain the first assertion with $\hat{\zeta} = \frac{3\zeta}{10}$.

It remains to consider the case:

$$M^2(P_{\max}) \ge 1 - \frac{\zeta}{2}.$$

This is inequality (9.6.14) in the proof of Proposition 9.6.3. Therefore, we can proceed from there as in the rest of the proof of Proposition 9.6.3 to arrive at a contradiction, which implies that, in this case, the maximum of $M^2 + d$ over $\overline{\Omega}$ cannot be attained at $P_{\max} \in \Gamma_{\text{shock}} \cup \{P_2\}$. $\square$

Proposition 15.6.2. *There exist $\tilde{\mu} > 0$ and $\hat{\zeta} > 0$ depending only on (ρ_0, ρ_1, γ) such that, if φ is an admissible solution of* **Problem 2.6.1** *with $\theta_{\text{w}} \in (\theta_{\text{w}}^{\text{d}}, \frac{\pi}{2})$, then*

(i) *For any supersonic wedge angle θ_{w}, the Mach number satisfies*

$$M^2(\boldsymbol{\xi}) \le 1 - \tilde{\mu}\, \text{dist}(\boldsymbol{\xi}, \overline{\Gamma_{\text{sonic}}}) \qquad \textit{for all } \boldsymbol{\xi} \in \overline{\Omega(\varphi)}. \tag{15.6.4}$$

(ii) *For any subsonic/sonic wedge angle θ_{w}, the Mach number satisfies*

$$M^2(\boldsymbol{\xi}) \le \max\big(1 - \hat{\zeta},\ \frac{|D\varphi_2(P_0)|^2}{c_2^2} - \tilde{\mu}|\boldsymbol{\xi} - P_0|\big) \qquad \textit{for all } \boldsymbol{\xi} \in \overline{\Omega(\varphi)}. \tag{15.6.5}$$

Proof. We use the convention that $P_1 = P_2 = P_0$ and $\overline{\Gamma_{\text{sonic}}} = P_0$ for the subsonic reflection. Also, the constants in this proof depend only on (ρ_0, ρ_1, γ).

We first note that Lemma 9.6.4 can be extended to all the wedge angles $\theta_{\text{w}} \in (\theta_{\text{w}}^{\text{d}}, \frac{\pi}{2})$, that is, $\theta_{\text{w}}^{\text{s}}$ can be replaced by $\theta_{\text{w}}^{\text{d}}$ in the formulation of the lemma.

The only difference in the proof is that, instead of circle $\partial B_{c_2}(\mathcal{O}_2)$, we now use circle $\partial B_r(\mathcal{O}_2)$ with

$$r = \min\{c_2, |\mathcal{O}_2 P_0|\}.$$

Note that $r = c_2$ for the supersonic/sonic wedge angles, and $r < c_2$ for the subsonic wedge angles. Line $\mathcal{S}_1$ intersects with circle $\partial B_r(\mathcal{O}_2)$ at points $\{P_1, Q\}$. Expression (9.6.33) now takes the form:

$$\hat{g}_{\theta_w}(\boldsymbol{\xi}) = \frac{1}{2}(r - \hat{c})\, h(\frac{\operatorname{dist}(\boldsymbol{\xi}, \partial B_r(\mathcal{O}_2))}{r - \hat{c}}).$$

The rest of the proof of Lemma 9.6.4 is unchanged.

Now we note that (9.1.2) and Proposition 15.2.2, combined with (9.6.29) in the extended version of Lemma 9.6.4, imply the existence of $\mu_0 > 0$ such that, for any admissible solution with $\theta_w \in (\theta_w^d, \frac{\pi}{2})$,

$$\mu_0 |g_{\theta_w}(\boldsymbol{\xi})| \le \frac{1}{4} \qquad \text{in } \overline{\Omega}. \tag{15.6.6}$$

Consider an admissible solution φ with $\theta_w \in (\theta_w^d, \frac{\pi}{2})$. We use function g_{θ_w} from the extended version of Lemma 9.6.4.

The argument in the proof of Proposition 9.6.5 shows that, for small $\mu \in (0, \mu_0)$, the maximum of $M^2 + \mu g_{\theta_w}$ over $\overline{\Omega}$ cannot be attained in $\Omega \cup \Gamma_{\text{sym}} \cup \Gamma_{\text{wedge}}$. Furthermore, following the proof of Proposition 9.6.5, we see that $M(P_3) = 0$.

If φ is a supersonic reflection solution, $M^2 + \mu g_{\theta_w} = 1$ on Γ_{sonic}, since $M^2 = 1$ on Γ_{sonic} by (8.1.3), and $g_{\theta_w} = 0$ on Γ_{sonic} by (9.6.29). Using (15.6.6), we conclude that the maximum of $M^2 + \mu g_{\theta_w}$ is not attained at P_3.

Thus, if the maximum of $M^2 + \mu g_{\theta_w}$ is attained at P_3, then φ is a subsonic reflection solution, and $M^2 + \mu g_{\theta_w} \le \frac{1}{4}$ in $\overline{\Omega}$ by (15.6.6). Another application of (15.6.6) implies that $M^2 \le \frac{1}{2}$ in $\overline{\Omega}$.

Therefore, we have proved that, if the maximum of $M^2 + \mu g_{\theta_w}$ is attained at P_3, then φ is a subsonic reflection solution, and (15.6.5) holds with $\hat{\zeta} = \frac{1}{2}$.

It remains to consider the case that the maximum of $M^2 + \mu g_{\theta_w}$ over $\overline{\Omega}$ is not attained in $\Omega \cup \Gamma_{\text{sym}} \cup \Gamma_{\text{wedge}} \cup \{P_3\}$. Recall that this case necessarily holds if φ is a supersonic reflection solution.

Thus, the maximum of $M^2 + \mu g_{\theta_w}$ over $\overline{\Omega}$ is attained in $\overline{\Gamma_{\text{sonic}}} \cup \Gamma_{\text{shock}} \cup \{P_2\}$ in this case. Now Proposition 15.6.1 and the properties of function g_{θ_w} in Lemma 9.6.4 imply that

$$\text{either} \qquad M^2 \le 1 - \hat{\zeta} \quad \text{in } \overline{\Omega} \tag{15.6.7}$$

$$\text{or} \qquad M^2(\boldsymbol{\xi}) \le M^2(P_1) - \tilde{\mu}\operatorname{dist}(\boldsymbol{\xi}, \overline{\Gamma_{\text{sonic}}}) \quad \text{in } \overline{\Omega}, \tag{15.6.8}$$

where $\hat{\zeta}, \tilde{\mu} > 0$ depend only on the data.

For the supersonic wedge angles, the definition of $P_1 \in \partial B_{c_2}(\mathcal{O}_2)$ in Definition 7.5.7, combined with Definition 8.1.1(ii), implies that $M^2(P_1) = 1$. Therefore, (15.6.7) cannot happen, and (15.6.8) implies assertion (i) in this proposition.

For the subsonic wedge angles, $P_1 = P_0$, and (15.1.3) implies that $M^2(P_0) = \frac{|D\varphi_2(P_0)|^2}{c_2^2}$. Therefore, (15.6.7)–(15.6.8) imply assertion (ii). □

Corollary 15.6.3. *There exist $\hat{\zeta} > 0$ and $C > 0$ depending only on (ρ_0, ρ_1, γ) such that, if φ is an admissible solution of* **Problem 2.6.1** *with $\theta_w \in (\theta_w^d, \frac{\pi}{2})$, then, using the notations in (9.2.1), equation (2.2.8) satisfies the following:*

(i) *For any supersonic wedge angle θ_w,*

$$\frac{\operatorname{dist}(\boldsymbol{\xi}, \Gamma_{\text{sonic}})}{C}|\boldsymbol{\kappa}|^2 \le \sum_{i,j=1}^{2} \mathcal{A}_{p_j}^i (D\varphi(\boldsymbol{\xi}), \varphi(\boldsymbol{\xi}))\kappa_i\kappa_j \le C|\boldsymbol{\kappa}|^2 \tag{15.6.9}$$

for any $\boldsymbol{\xi} \in \overline{\Omega}$ and $\boldsymbol{\kappa} = (\kappa_1, \kappa_2) \in \mathbb{R}^2$.

(ii) *For any subsonic/sonic wedge angle θ_w,*

$$\begin{aligned} &\frac{1}{C}\min(c_2 - |D\varphi_2(P_0)| + |\boldsymbol{\xi} - P_0|,\ \hat{\zeta})|\boldsymbol{\kappa}|^2 \\ &\le \sum_{i,j=1}^{2} \mathcal{A}_{p_j}^i (D\varphi(\boldsymbol{\xi}), \varphi(\boldsymbol{\xi}))\kappa_i\kappa_j \le C|\boldsymbol{\kappa}|^2 \end{aligned} \tag{15.6.10}$$

for any $\boldsymbol{\xi} \in \overline{\Omega}$ and $\boldsymbol{\kappa} = (\kappa_1, \kappa_2) \in \mathbb{R}^2$.

Proof. In this proof, the universal constant C depends only on (ρ_0, ρ_1, γ).

We follow the proof of Corollary 9.6.6 by using Proposition 15.6.2 instead of Proposition 9.6.5. Then, for any supersonic wedge angle, we obtain (15.6.9).

For any subsonic/sonic wedge angle, applying Proposition 15.6.2(ii), we obtain the following lower bound:

$$\sum_{i,j=1}^{2} \mathcal{A}_{p_j}^i (D\varphi, \varphi)\kappa_i\kappa_j \ge c^2(|D\varphi|^2, \varphi)\min(1 - \frac{|D\varphi_2(P_0)|^2}{c_2^2} + \tilde{\mu}|\boldsymbol{\xi} - P_0|,\ \hat{\zeta})|\boldsymbol{\kappa}|^2.$$

Since $|D\varphi_2(P_0)| = |P_0 O_2|$, $\frac{1}{C} \le c(|D\varphi|^2, \varphi) \le C$ by Lemma 9.1.4 and Proposition 15.2.2, and $\frac{1}{C} \le c_2^{(\theta_w)} \le C$ for $\theta_w \in [\theta_w^d, \frac{\pi}{2}]$, we obtain (15.6.10). □

15.7 REGULARITY OF ADMISSIBLE SOLUTIONS AWAY FROM Γ_{sonic}

Proposition 15.7.1. *All the assertions proved in §10.1 and §10.3 hold for both supersonic and subsonic admissible solutions with $\theta_w \in (\theta_w^d, \frac{\pi}{2})$, for which the constants in the estimates depend only on (ρ_0, ρ_1, γ).*

Proof. The proof of Lemma 10.1.1 is based on the properties of admissible solutions that $\varphi < \varphi_1$ in Ω and on the uniform estimates: $\text{dist}(\Gamma_{\text{shock}}, B_{c_1}(\mathcal{O}_1)) \geq \frac{1}{C}$ and $\|\varphi\|_{C^{0,1}} \leq C$. All these properties hold for any admissible solution with $\theta_{\text{w}} \in (\theta_{\text{w}}^{\text{d}}, \frac{\pi}{2})$, with constant C depending only on (ρ_0, ρ_1, γ), as we have shown in Propositions 15.2.2 and 15.5.3. Thus, Lemma 10.1.1 holds for any admissible solution with $\theta_{\text{w}} \in [\theta_{\text{w}}^{\text{d}}, \frac{\pi}{2})$, with constant $\delta > 0$ depending only on (ρ_0, ρ_1, γ). Then the rest of the proof in §10.1 readily follows.

Proposition 10.3.1 is independent of the type of the reflection. Its proof employs Lemma 10.1.1, the ellipticity principle (where we use Proposition 15.6.2 now), and the Lipschitz bounds of the solution in Proposition 15.2.2. Therefore, Proposition 10.3.1 holds in the present case, with the constants depending only on (ρ_0, ρ_1, γ). □

Remark 15.7.2. *The results in §10.2 do not use the definition of admissible solutions, so that they apply to the present case without change.*

In the estimates away from the sonic arc, similarly to Chapter 10, we need to take into account the possibility that Γ_{shock} may hit the wedge vertex P_3 if $u_1 > c_1$. Then, similarly to §10.4, we define the critical angle. However, for the subsonic reflection, Γ_{shock} intersects with Γ_{wedge} at P_0 so that we need to modify Definition 10.4.1 as follows:

Definition 15.7.3. Let $\gamma > 1$ and $\rho_1 > \rho_0 > 0$. Define the set:

$$\mathcal{A} := \left\{ \theta_{\text{w}}^* \in (\theta_{\text{w}}^{\text{d}}, \frac{\pi}{2}] \; : \; \begin{array}{l} \text{For each } r \in (0, r_1), \text{ there exists } \varepsilon > 0 \text{ such} \\ \text{that } \text{dist}(\Gamma_{\text{shock}}, \Gamma_{\text{wedge}} \setminus B_r(P_0)) \geq \varepsilon \text{ for} \\ \text{any admissible solution with } \theta_{\text{w}} \in [\theta_{\text{w}}^*, \frac{\pi}{2}] \end{array} \right\},$$

where the normal reflection is included as the unique admissible solution for $\theta_{\text{w}} = \frac{\pi}{2}$, so that the set of admissible solutions with the wedge angles $\theta_{\text{w}} \in [\theta_{\text{w}}^*, \frac{\pi}{2}]$ is non-empty for any $\theta_{\text{w}}^* \in (\theta_{\text{w}}^{\text{d}}, \frac{\pi}{2}]$. Since $\text{dist}(\Gamma_{\text{shock}}, \Gamma_{\text{wedge}}) > 0$ for the normal reflection, $\frac{\pi}{2} \in \mathcal{A}$, *i.e.*, $\mathcal{A} \neq \emptyset$. Also, r_1 above is defined by (15.4.2). Now we define the critical angle

$$\theta_{\text{w}}^{\text{c}} = \inf \mathcal{A},$$

so that $\theta_{\text{w}}^{\text{c}}$ depends only on (ρ_0, ρ_1, γ).

Directly from this definition, $\theta_{\text{w}}^{\text{c}} \in [\theta_{\text{w}}^{\text{d}}, \frac{\pi}{2}]$.

Lemma 15.7.4. *Let $\gamma > 1$ and $\rho_1 > \rho_0 > 0$. Then the critical angle $\theta_{\text{w}}^{\text{c}}$ has the following properties:*

(i) $\theta_{\text{w}}^{\text{c}} < \frac{\pi}{2}$, *i.e.*, $\theta_{\text{w}}^{\text{c}} \in [\theta_{\text{w}}^{\text{d}}, \frac{\pi}{2})$;

(ii) *If* $u_1 \leq c_1$, $\theta_{\text{w}}^{\text{c}} = \theta_{\text{w}}^{\text{d}}$;

(iii) *If $\theta_{\rm w}^{\rm c} > \theta_{\rm w}^{\rm d}$, there exist $r \in (0, r_1)$ and a sequence of admissible solutions $\varphi^{(i)}$ with the wedge angles $\theta_{\rm w}^{(i)} \in [\theta_{\rm w}^{\rm c}, \frac{\pi}{2})$ such that*

$$\lim_{i\to\infty} \theta_{\rm w}^{(i)} = \theta_{\rm w}^{\rm c}, \qquad \lim_{i\to\infty} \operatorname{dist}(\Gamma_{\rm shock}^{(i)}, \Gamma_{\rm wedge}^{(i)} \setminus B_r(P_0{}^{(i)})) = 0;$$

(iv) *Let r_1 be defined by (15.4.2). For any $\theta_{\rm w}^* \in (\theta_{\rm w}^{\rm c}, \frac{\pi}{2})$ and $r \in (0, r_1)$, there exists $C_r > 0$ such that, for any admissible solution φ with $\theta_{\rm w} \in [\theta_{\rm w}^*, \frac{\pi}{2})$,*

$$\operatorname{dist}(\Gamma_{\rm shock}, \Gamma_{\rm wedge} \setminus B_r(P_0)) \geq \frac{1}{C_r}.$$

Proof. Assertion (i) follows from Lemma 9.4.7. Assertion (ii) follows from Proposition 15.4.2 and Corollary 15.5.4. Assertions (iii)–(iv) follow directly from Definition 15.7.3. □

Proposition 15.7.5. *Let $\theta_{\rm w}^* \in (\theta_{\rm w}^{\rm c}, \frac{\pi}{2})$. All the assertions proved in §10.5–§10.6 hold for both supersonic and subsonic admissible solutions for $\theta_{\rm w} \in [\theta_{\rm w}^*, \frac{\pi}{2})$, with the constants in the estimates depending on $(\rho_0, \rho_1, \gamma, \theta_{\rm w}^*)$ and with the following notational adjustments in the statements in the case that φ is a subsonic reflection solution: Arc $\overline{\Gamma_{\rm sonic}}$ and points $\{P_1, P_4\}$ need to be replaced by P_0. In particular, (10.6.1) when $|D\varphi_2^{(\infty)}(P_0{}^{(\infty)})| \leq c_2^{(\infty)}$ is of the form:*

$$\Omega^{(\infty)} = \left\{ (S,T) \in \mathbb{R}^2 \; : \; \begin{array}{l} T_{P_2{}^{(\infty)}} \leq T \leq T_{P_0{}^{(\infty)}}, \\ -(T - T_{P_3}) \tan\theta_{\rm w}^{(\infty)} < S < f_{\mathbf{e},\rm sh}^{(\infty)}(T) \\ \qquad\qquad \text{for } T \in [T_{P_2{}^{(\infty)}}, T_{P_3{}^{(\infty)}}], \\ 0 < S < f_{\mathbf{e},\rm sh}^{(\infty)}(T) \;\; \text{for } T \in (T_{P_3{}^{(\infty)}}, T_{P_0{}^{(\infty)}}] \end{array} \right\}. \tag{15.7.1}$$

Proof. We fix $\theta_{\rm w}^* \in (\theta_{\rm w}^{\rm c}, \frac{\pi}{2})$.

To prove Proposition 10.5.1 in the present case, we choose $P = (r_P, \theta_P) \in \Gamma_{\rm shock}^{\rm ext}$ with $\operatorname{dist}(P, \Gamma_{\rm sonic}^{\rm ext}) \geq d$, where $\Gamma_{\rm sonic}^{\rm ext}$ is replaced by P_0 in the subsonic reflection case. Then, from Lemma 15.7.4(iv) applied with $r = \frac{d}{2}$, we obtain that $\operatorname{dist}(P, \Gamma_{\rm wedge}^{\rm ext}) \geq \frac{1}{C}$. Now we can argue as in the proof of Proposition 10.5.1 and employ Proposition 15.7.1 to conclude the expected result.

The proof of Lemma 10.5.2 is unchanged in the present case by using Lemma 15.7.4 instead of Lemma 10.4.2.

In the proof of Proposition 10.6.1(i), we update the proof of the fact that $\Gamma_{\rm shock}^{(\infty)}$ is disjoint with $\Gamma_{\rm wedge}^{(\infty)}$ as follows: By Lemma 15.7.4(iv), for any $r \in (0, r_1)$, there exists $C_r > 0$ such that the estimate:

$$\operatorname{dist}(\Gamma_{\rm shock}^{(i)}, \Gamma_{\rm wedge}^{(i)} \setminus B_r(P_0)) \geq \frac{1}{C_r}$$

holds for all i. Also, $P_3{}^{(i)} = \mathbf{0}$, $P_4{}^{(i)} \to P_4{}^{(\infty)}$ since $\theta_{\rm w}^{(i)} \to \theta_{\rm w}^{(\infty)}$, and $\Gamma_{\rm wedge}^{(i)} = P_3{}^{(i)}P_4{}^{(i)}$. Thus, $\Gamma_{\rm wedge}^{(i)} \to \Gamma_{\rm wedge}^{(\infty)}$ in the Hausdorff metric. Also, $f_{\mathbf{g},\rm sh}^{(i)} \to f_{\mathbf{g},\rm sh}^{(\infty)}$

uniformly, by (ii) in Corollary 9.2.5, so that $\Gamma^{(i)}_{\rm shock} \to \Gamma^{(\infty)}_{\rm shock}$ in the Hausdorff metric. Then it follows that

$$\mathrm{dist}(\Gamma^{(\infty)}_{\rm shock}, \Gamma^{(\infty)}_{\rm wedge} \setminus B_r(P_0)) \geq \frac{1}{C_r} \qquad \text{for each } r \in (0, r_1).$$

This implies that $\Gamma^{(\infty)}_{\rm shock}$ is disjoint with $\Gamma^{(\infty)}_{\rm wedge}$.

We also update the proof of Proposition 10.6.1(v) in a similar way. Namely, since $\mathcal{O}_2 \in P_3P_4$ by Lemma 7.4.8 and the definition of P_4 in Definition 7.5.7, then $\mathrm{dist}(P_3, \overline{\Gamma_{\rm sonic}}) = |P_3P_1|$, whereas $\overline{\Gamma_{\rm sonic}} = P_4 = P_0$ for subsonic wedge angles. Then $\mathrm{dist}(P_3, \overline{\Gamma_{\rm sonic}}) \geq c_2$ for supersonic wedge angles, while $\mathrm{dist}(P_3, \overline{\Gamma_{\rm sonic}}) \geq r_1$ for subsonic wedge angles, where $r_1 > 0$ is from (15.4.2). We note that $\hat{c}_2 := \min_{\theta_{\rm w} \in [\theta^{\rm d}_{\rm w}, \frac{\pi}{2}]} c_2(\theta_{\rm w}) > 0$. Thus, for any $\theta_{\rm w} \in [\theta^{\rm d}_{\rm w}, \frac{\pi}{2}]$, we obtain that $\mathrm{dist}(P_3, \overline{\Gamma_{\rm sonic}}) \geq \min(r_1, \hat{c}_2) =: \hat{r} > 0$. From Lemma 15.7.4(iv), there exists C depending only on the data such that

$$\mathrm{dist}(\Gamma^{(i)}_{\rm shock}, \Gamma^{(i)}_{\rm wedge} \setminus B_{\hat{r}/10}({P_0}^{(i)})) \geq \frac{1}{C} \qquad \text{for all } i.$$

Without loss of generality, we can assume that $\frac{1}{C} \leq \frac{\hat{r}}{10}$. Let $s = \frac{1}{4C}$. Then, from the argument in the proof of Lemma 10.5.2, it follows that $\Lambda^{(i)} \cap B_{3s}(P_3) \subset \Omega^{(i)}$ for each i. From that point, we follow the proof of Proposition 10.6.1(v).

The rest of the proof of Proposition 10.6.1 is unchanged in the present case, on account of using Propositions 15.2.1–15.2.2 and 15.7.1, as well as §15.3, in place of the corresponding estimates from Chapters 8–10. □

Chapter Sixteen

Regularity of Admissible Solutions near the Sonic Arc and the Reflection Point

In this chapter, we establish the regularity of admissible solutions near the sonic arc $\Gamma_{\rm sonic}$ (an elliptic degenerate curve) in the supersonic reflection configurations, as shown in Figs. 2.3 and 2.5, and near the shock reflection point P_0 in the subsonic reflection configurations, as shown in Figs. 2.4 and 2.6.

16.1 POINTWISE AND GRADIENT ESTIMATES NEAR $\Gamma_{\rm sonic}$ AND THE REFLECTION POINT

In this section, we extend the estimates in §11.2. In §11.2, the pointwise and gradient estimates have been obtained for $\psi = \varphi - \varphi_2$ near $\Gamma_{\rm sonic}$, where φ is an admissible solution with the supersonic reflection configuration for $\theta_{\rm w} \in (\theta_{\rm w}^{\rm s}, \frac{\pi}{2})$. These estimates are now extended further for the following two classes of admissible solutions:

First, Theorem 7.1.1(v)–(vi) do not exclude the possibility that $\frac{|D\varphi_2(P_0)|}{c_2} > 1$ for some wedge angles $\theta_{\rm w} \in (\tilde{\theta}_{\rm w}^{\rm s}, \theta_{\rm w}^{\rm s})$. We will show that, if such a wedge angle $\theta_{\rm w}$ exists, the estimates of $(\psi, D\psi)$ near $\Gamma_{\rm sonic}$, similar to those for $\theta_{\rm w} \in (\theta_{\rm w}^{\rm s}, \frac{\pi}{2})$, also hold, where the only difference is that the constants will be modified, but still depend only on (ρ_0, ρ_1, γ).

Furthermore, we show that similar estimates hold near the reflection point P_0 for the subsonic reflection solutions that are close to the sonic angle, *i.e.*, for subsonic/sonic admissible solutions φ such that $1 - \delta \le \frac{|D\varphi_2(P_0)|}{c_2} \le 1$ for some small $\delta > 0$.

We first recall that, for the subsonic reflection solutions, the straight line

$$S_1 = \{\varphi_1 = \varphi_2\}$$

is well-defined and contains the reflection point P_0. Also, in the (x, y)–coordinates considered in §11.1,

$$\begin{aligned}
&x_{P_0}^{(\theta_{\rm w})} > 0 \text{ for subsonic wedge angles, } i.e., \text{ when } \tfrac{|D\varphi_2(P_0)|}{c_2}(\theta_{\rm w}) < 1,\\
&x_{P_0}^{(\theta_{\rm w})} = 0 \text{ for sonic wedge angles, } i.e., \text{ when } \tfrac{|D\varphi_2(P_0)|}{c_2}(\theta_{\rm w}) = 1,\\
&x_{P_0}^{(\theta_{\rm w})} < 0 \text{ for supersonic wedge angles, } i.e., \text{ when } \tfrac{|D\varphi_2(P_0)|}{c_2}(\theta_{\rm w}) > 1.
\end{aligned} \tag{16.1.1}$$

Next, we note that Definition 11.2.1 applies to the subsonic reflection case with the following change: Instead of the sonic circle $\partial B_{c_2}(\mathcal{O}_2)$ of state (2), we consider circle $\partial B_{|\mathcal{O}_2 P_0|}(\mathcal{O}_2)$. Notice that S_1 is not orthogonal to Γ_{wedge}, since $S_1 \perp \mathcal{O}_1\mathcal{O}_2$, $\mathcal{O}_1 \notin \{\xi_2 = \xi_1 \tan\theta_{\text{w}}\}$, and $\mathcal{O}_2 \in \{\xi_2 = \xi_1 \tan\theta_{\text{w}}\}$. It follows that S_1 intersects with $\partial B_{|\mathcal{O}_2 P_0|}(\mathcal{O}_2)$ at two points. One of the intersection points is P_0, which follows from the choice of the radius.

Definition 16.1.1. *For any subsonic/sonic wedge angle $\theta_{\text{w}} \in (\theta_{\text{w}}^{\text{d}}, \frac{\pi}{2})$ (in this case, $P_0 = P_1 = P_4$ and $\overline{\Gamma_{\text{sonic}}} = \{P_0\}$), we define the following points:*

(i) *By (6.1.4)–(6.1.5) of Lemma 6.1.2, applied with $\varphi^- = \varphi_1$ and $\varphi^+ = \varphi_2$, it follows that $\mathcal{O}_2 \in \{\varphi_1 > \varphi_2\}$ so that the smaller arc $P_1\overline{P_1}$ lies within $\{\varphi_2 > \varphi_1\}$, where $\overline{P_1}$ is the second point of intersection of S_1 with $\partial B_{|\mathcal{O}_2 P_0|}(\mathcal{O}_2)$;*

(ii) *Denote by Q the midpoint of the smaller arc $P_0\overline{P_1}$ of $\partial B_{|\mathcal{O}_2 P_0|}(\mathcal{O}_2)$ so that $Q \in \{\varphi_2 > \varphi_1\}$;*

(iii) *Denote by Q' the midpoint of the line segment (chord) $P_0\overline{P_1}$.*

Since $\mathcal{O}_2 \in \Gamma_{\text{wedge}}$ by Lemma 7.4.8, $\mathcal{O}_2 \in \{\varphi_1 > \varphi_2\}$, and $\mathcal{O}_1\mathcal{O}_2 \perp S_1$, we conclude that $Q, Q' \in \Lambda$.

With these definitions, we have

Lemma 16.1.2. *There exist constants $\varepsilon_1 > \varepsilon_0 > 0$, $\delta > 0$, $\omega > 0$, $C > 0$, and $M \geq 2$, and a function $m(\cdot) \in C([0,\infty))$ depending only on (ρ_0, ρ_1, γ) such that Lemma 11.2.2 holds for any $\theta_{\text{w}} \in [\theta_{\text{w}}^{\text{d}}, \frac{\pi}{2})$ with the following changes which affect only the subsonic/sonic wedge angles:*

- *$P_0 = P_1 = P_4$ and $\overline{\Gamma_{\text{sonic}}} = \{P_0\}$ for any subsonic/sonic wedge angle θ_{w}, so that $x_{P_1} \geq 0$ for any $\theta_{\text{w}} \in [\theta_{\text{w}}^{\text{d}}, \frac{\pi}{2}]$, with $x_{P_1} = 0$ for any supersonic/sonic wedge angle, and $x_{P_1} > 0$ for any subsonic wedge angle;*

- *Γ_{sonic} should be replaced by $\overline{\Gamma_{\text{sonic}}}$ in every occurrence;*

- *(11.2.4) is replaced by*

$$\{\varphi_2 < \varphi_1\} \cap \mathcal{N}_{\varepsilon_1}(\overline{\Gamma_{\text{sonic}}}) \cap \{\theta > \theta_{P_1}\} \subset \{x > x_{P_1}\}; \tag{16.1.2}$$

- *In (11.2.5)–(11.2.10), $-\varepsilon_k < x < \varepsilon_k$ is replaced by $-\varepsilon_k < x - x_{P_1} < \varepsilon_k$ for $k = 0, 1$. Similarly, $\varepsilon' \leq x \leq \varepsilon''$ is replaced by $\varepsilon' \leq x - x_{P_1} \leq \varepsilon''$.*

Proof. Several points in the proof of Lemma 11.2.2 are based on the continuous dependence of the parameters of state (2) on $\theta_{\text{w}} \in [\theta_{\text{w}}^{\text{s}}, \frac{\pi}{2}]$, and the positivity of certain quantities related to state (2), *e.g.*, c_2, $y_Q - y_{P_1}$, *etc.*, and hence the positivity of their infimum over $\theta_{\text{w}} \in [\theta_{\text{w}}^{\text{s}}, \frac{\pi}{2}]$. In the present case, we use the fact that similar continuous dependence and positivity hold up to the detachment angle, *i.e.*, for $\theta_{\text{w}} \in [\theta_{\text{w}}^{\text{d}}, \frac{\pi}{2}]$. With this and the modifications in Definition 11.2.1 for the subsonic/sonic wedge angles, as described in Definition 16.1.1, the proof of Lemma 11.2.2 applies directly to the present case. □

Next, we extend Proposition 11.2.3 to the case of both supersonic and subsonic reflections.

Proposition 16.1.3. *Let ε_0 and ε_1 be the constants defined in Lemma* 16.1.2. *Let φ be an admissible solution with $\theta_{\rm w} \in (\theta_{\rm w}^{\rm d}, \frac{\pi}{2}]$ and $\Omega = \Omega(\varphi)$. Then, for $\varepsilon \in (0, \varepsilon_0)$, region Ω_ε, defined by*

$$\Omega_\varepsilon := \Omega \cap \mathcal{N}_{\varepsilon_1}(\overline{\Gamma_{\rm sonic}}) \cap \{x < x_{P_1} + \varepsilon\}, \tag{16.1.3}$$

satisfies

$$\Omega \cap \mathcal{N}_\varepsilon(\overline{\Gamma_{\rm sonic}}) \subset \Omega_\varepsilon, \tag{16.1.4}$$

$$\Omega_\varepsilon = \Omega_\varepsilon \cap \{x > x_{P_1}\}, \tag{16.1.5}$$

$$\Omega_\varepsilon \subset \{\varphi_2 < \varphi_1\} \cap \Lambda \cap \mathcal{N}_{\varepsilon_1}(\overline{\Gamma_{\rm sonic}}) \cap \{x_{P_1} < x < x_{P_1} + \varepsilon\}. \tag{16.1.6}$$

We note that

- $x_{P_1} = 0$ *for supersonic/sonic wedge angles, and hence* (16.1.3)–(16.1.6) *are* (11.2.16)–(11.2.19);
- $x_{P_1} = x_{P_0} > 0$ *for subsonic wedge angles.*

Proof. For the supersonic wedge angles $\theta_{\rm w}$, the proof of Proposition 11.2.3 applies without change. Thus, it suffices to focus on the subsonic/sonic wedge angles $\theta_{\rm w}$. Then $x_{P_0} \geq 0$ and $y_{P_0} = \theta_{\rm w}$.

Since $\varepsilon < \varepsilon_0 < \varepsilon_1$, and $P_0 \in \partial B_{c_2 - x_{P_1}}(O_2)$ implies

$$\mathcal{N}_\varepsilon(P_0) \subset \mathcal{N}_\varepsilon(\partial B_{c_2 - x_{P_1}}(O_2)) = \{|x - x_{P_1}| < \varepsilon\} \subset \{x < x_{P_1} + \varepsilon\},$$

then (16.1.4) follows from (16.1.3).

By (i) and (iv) of Definition 15.1.2, $\Omega \subset \{\varphi_2 < \varphi_1\} \cap \Lambda$ and $\Lambda \subset \{\theta > \theta_{\rm w}\}$. Then, since $y_{P_0} = \theta_{\rm w}$ and $P_0 = P_1$ in the subsonic case, it follows that $\Omega \subset \{\varphi_2 < \varphi_1\} \cap \Lambda \cap \{\theta > \theta_{P_1}\}$. Combining this with (16.1.2)–(16.1.3), we obtain (16.1.5).

Furthermore, $\Omega \subset \{\varphi_2 < \varphi_1\} \cap \Lambda$, combined with (16.1.3) and (16.1.5), implies (16.1.6). □

In extending the remaining results in §11.2, we restrict to the following set $\mathcal{I}_{\rm w}(\delta_{P_0})$ of wedge angles:

$$\mathcal{I}_{\rm w}(\delta) := \Big\{\theta_{\rm w} \in [\theta_{\rm w}^{\rm d}, \frac{\pi}{2}) \ : \ \frac{|D\varphi_2(P_0)|}{c_2}(\theta_{\rm w}) \geq 1 - \delta\Big\} \qquad \text{for } \delta \in (0,1), \tag{16.1.7}$$

and $\delta_{P_0} \in (0,1)$ is chosen below depending only on (ρ_0, ρ_1, γ). We note that the regular reflection solutions with $\theta_{\rm w} \in \mathcal{I}_{\rm w}(\delta)$ include all the supersonic reflection solutions, as well as the subsonic reflection solutions that are sufficiently close to sonic at P_0.

Denote

$$x^*_{P_0}(\delta) = \sup\{x^{(\theta_w)}_{P_0} \;:\; \theta_w \in \mathcal{I}_w(\delta)\} \qquad \text{for } \delta \in (0,1). \tag{16.1.8}$$

Recall that $x^{(\theta_w)}_{P_0} < 0$ if $\frac{|D\varphi_2(P_0)|}{c_2}(\theta_w) > 1$, and $x^{(\theta_w)}_{P_0} > 0$ if $\frac{|D\varphi_2(P_0)|}{c_2}(\theta_w) < 1$. Then the continuous dependence of the parameters of state (2) on $\theta_w \in [\theta_w^d, \frac{\pi}{2}]$ and (16.1.1) imply that

$$x^*_{P_0}(\cdot) \in C([0,1]), \quad x^*_{P_0}(\cdot) \text{ is increasing,} \quad \lim_{\delta\to 0+} x^*_{P_0}(\delta) = 0. \tag{16.1.9}$$

Now we can extend the rest of the results in §11.2 to the present case.

Proposition 16.1.4. *There exist $\varepsilon_0, \delta_{P_0} \in (0,1)$ depending only on (ρ_0, ρ_1, γ) such that the results in Lemmas 11.2.4–11.2.7, Lemma 11.2.11, Proposition 11.2.8, and Corollaries 11.2.9–11.2.10 and 11.2.12 hold in Ω_{ε_0} for all admissible solutions with any wedge angle $\theta_w \in \mathcal{I}_w(\delta_{P_0})$, with the following changes (which affect only the admissible solutions with subsonic wedge angles): $x \in (0, \varepsilon_0)$ should be replaced by $x \in (x_{P_1}, x_{P_1} + \varepsilon_0)$ in (11.2.38) and (11.2.40), where we recall our convention that $P_0 := P_1 = P_4$ and $\overline{\Gamma_{\rm sonic}} := \{P_0\}$ for subsonic/sonic wedge angles, and that $x_{P_1} \geq 0$ for any $\theta_w \in [\theta_w^d, \frac{\pi}{2}]$ with $x_{P_1} = 0$ for supersonic/sonic wedge angles and $x_{P_1} > 0$ for subsonic wedge angles.*

More explicitly, for any $\theta_w \in \mathcal{I}_w(\delta_{P_0})$, we define Ω_ε by (16.1.3), which is consistent with definition (11.2.16) of Ω_ε used in the supersonic case. Then, for each $\varepsilon \in (0, \varepsilon_0]$, region Ω_ε is of the form:

$$\begin{aligned}
&\Omega_\varepsilon = \{(x,y) \;:\; x_{P_1} < x < x_{P_1} + \varepsilon,\ \ \theta_w < y < \hat{f}(x)\},\\
&\Gamma_{\rm shock} \cap \partial\Omega_\varepsilon = \{(x,y) \;:\; x_{P_1} < x < x_{P_1} + \varepsilon,\ \ y = \hat{f}(x)\},\\
&\Gamma_{\rm wedge} \cap \partial\Omega_\varepsilon = \{(x,y) \;:\; x_{P_1} < x < x_{P_1} + \varepsilon,\ \ y = \theta_w\},\\
&\overline{\Gamma_{\rm sonic}} \equiv \overline{\Gamma_{\rm sonic}} \cap \partial\Omega_\varepsilon = \{(x_{P_1}, y) \;:\; \theta_w \le y \le \hat{f}(x_{P_1})\},
\end{aligned} \tag{16.1.10}$$

and

$$x_{P_1} = 0 \ \textit{ for the supersonic/sonic wedge angles,} \ \ x_{P_1} > 0 \ \textit{ otherwise,} \tag{16.1.11}$$

$$\begin{cases} \hat{f}(x_{P_1}) = y_{P_1} > y_{P_4} = \theta_w & \textit{for supersonic wedge angles,}\\ \hat{f}(x_{P_1}) = y_{P_0} = y_{P_1} = y_{P_4} = \theta_w & \textit{otherwise,} \end{cases} \tag{16.1.12}$$

$$0 < \omega \le \frac{d\hat{f}}{dx} < C \qquad \textit{for any } x \in (x_{P_1},\, x_{P_1} + \varepsilon_0). \tag{16.1.13}$$

Proof. For $\varepsilon_0 > 0$ specified below, we choose $\delta_{P_0} \in (0,1)$ so that $x^*_{P_0}(\delta_{P_0}) = \frac{\varepsilon_0}{2}$. We note that the existence of such a constant δ_{P_0} follows from (16.1.9) if ε_0 is small; otherwise, we use $\delta = 1$. Moreover, $\delta_{P_0} > 0$ defined in this way depends only on $(\rho_0, \rho_1, \gamma, \varepsilon_0)$. Below we choose a uniform $\varepsilon_0 > 0$, *i.e.*, depending only on (ρ_0, ρ_1, γ), which then fixes a uniform $\delta_{P_0} > 0$.

We now show that, for sufficiently small $\varepsilon_0 > 0$ depending only on (ρ_0, ρ_1, γ), the estimates asserted hold in Ω_{ε_0} for every admissible solution with $\theta_{\rm w} \in \mathcal{I}_{\rm w}(\delta_{P_0})$.

The proofs of the above estimates in Ω_ε for small $\varepsilon > 0$ in the present conditions repeat the proofs of the corresponding results in §11.2, by both using that the uniform bounds proved in the previous chapters have to be extended to the present case (see Propositions 15.2.1–15.2.2, Corollary 15.6.3, Lemma 16.1.2, and Proposition 16.1.3) and employing the continuous dependence of the parameters of state (2) on $\theta_{\rm w} \in [\theta_{\rm w}^{\rm d}, \frac{\pi}{2}]$. Also, in the subsonic reflection case, we employ the modifications in Definition 11.2.1 described before Lemma 16.1.2.

We comment only on the proof of Lemma 11.2.5 under the present assumptions. In the supersonic/sonic reflection case, the proof of Lemma 11.2.5 works without change. In the subsonic reflection case, we rewrite the right-hand side of (15.6.10) as follows: Using that $x_{P_1} = c_2 - |D\varphi_2(P_1)| > 0$ and (16.1.6), we obtain that, for any $\boldsymbol{\xi} \in \Omega_\varepsilon$ and ε small,

$$c_2 - |D\varphi_2(P_1)| + |\boldsymbol{\xi} - P_1| = x_{P_1} + |\boldsymbol{\xi} - P_1| \geq x_{\boldsymbol{\xi}},$$

where the last inequality follows from the definition of coordinates (x, y) and a simple geometric argument by using (16.1.6) and reducing ε if necessary depending only on c_2 (hence it can be chosen uniformly). From (15.6.10), we have

$$\sum_{i,j=1}^{2} \mathcal{A}^i_{p_j}(D\varphi(\boldsymbol{\xi}), \varphi(\boldsymbol{\xi}))\kappa_i\kappa_j \geq \frac{1}{C}\min(x_{\boldsymbol{\xi}}, \hat{\zeta})|\boldsymbol{\kappa}|^2$$

for any $\boldsymbol{\xi} \in \Omega_\varepsilon$ and $\boldsymbol{\kappa} = (\kappa_1, \kappa_2) \in \mathbb{R}^2$.

Reducing ε if necessary to have $\varepsilon \leq \hat{\zeta}$, we have

$$\sum_{i,j=1}^{2} \mathcal{A}^i_{p_j}(D\varphi(\boldsymbol{\xi}), \varphi(\boldsymbol{\xi}))\kappa_i\kappa_j \geq \frac{1}{C}x_{\boldsymbol{\xi}}|\boldsymbol{\kappa}|^2$$

for any $\boldsymbol{\xi} \in \Omega_\varepsilon$ and $\boldsymbol{\kappa} = (\kappa_1, \kappa_2) \in \mathbb{R}^2$. Then we can follow the proof of Lemma 11.2.5 without change. □

16.2 THE RANKINE-HUGONIOT CONDITION ON $\Gamma_{\rm shock}$ NEAR $\Gamma_{\rm sonic}$ AND THE REFLECTION POINT

In this section, we extend the constructions and estimates in §11.3 to include both the supersonic and subsonic reflection solutions that are close to sonic. We thus continue to consider the wedge angles $\theta_{\rm w} \in \mathcal{I}_{\rm w}(\delta_{P_0})$ defined by (16.1.7) for $\delta_{P_0} \in (0, 1)$ chosen depending on the data.

Proposition 16.2.1. *For any $\theta_{\rm w}^* \in (\theta_{\rm w}^{\rm d}, \frac{\pi}{2})$, the results in §11.3 hold for all admissible solutions with any wedge angle $\theta_{\rm w} \in [\theta_{\rm w}^*, \frac{\pi}{2})$, where $P_0 = P_1 = P_4$ and $\overline{\Gamma_{\rm sonic}} = \{P_0\}$ in the subsonic reflection case, definition (16.1.3) of Ω_ε is used in the results and constructions in §11.3, and the positive constants $(\varepsilon, \delta, \delta_{\rm bc}, \hat{\delta}_{\rm bc}, C)$ in the estimates depend only on $(\rho_0, \rho_1, \gamma, \theta_{\rm w}^*)$.*

Proof. Fix $\theta_{\rm w}^* \in (\theta_{\rm w}^{\rm d}, \frac{\pi}{2})$. In the argument below, all the constants depend only on $(\rho_0, \rho_1, \gamma, \theta_{\rm w}^*)$ unless otherwise specified. We divide the proof into four steps.

1. We first show that estimate (11.3.5) holds for $\theta_{\rm w} \in [\theta_{\rm w}^*, \frac{\pi}{2}]$.

If $\theta_{\rm w}$ is a subsonic/sonic wedge angle, *i.e.*, $|D\varphi_2(P_0)| \le c_2$ for $\theta_{\rm w}$, then $P_1 = P_0$ so that

$$(D_{\mathbf{p}} g^{\rm sh}(D\varphi_2(P_1), \varphi_2(P_1), P_1) \cdot D\varphi_2(P_1))(\theta_{\rm w}) < 0 \tag{16.2.1}$$

by Theorem 7.1.1(vii) for the weak state (2). If $\theta_{\rm w}$ is a supersonic wedge angle (*i.e.*, $|D\varphi_2(P_0)| > c_2$ for $\theta_{\rm w}$), then (16.2.1) holds by Corollary 7.4.7. Also, using the continuous dependence on $\theta_{\rm w}$ in Theorem 7.1.1(i) and Lemma 7.5.9(i), we obtain that $(\varphi_2^{(\theta_{\rm w})}, D\varphi_2^{(\theta_{\rm w})})({P_1}^{(\theta_{\rm w})})$ depends continuously on $\theta_{\rm w} \in [\theta_{\rm w}^*, \frac{\pi}{2}]$. Then (16.2.1) that holds for any $\theta_{\rm w} \in [\theta_{\rm w}^*, \frac{\pi}{2}]$ implies (11.3.5) for any $\theta_{\rm w} \in [\theta_{\rm w}^*, \frac{\pi}{2}]$ with constant $\delta > 0$ depending only on the data and $\theta_{\rm w}^*$.

2. We note that (11.3.6) and (11.3.10) hold for any $\theta_{\rm w} \in [\theta_{\rm w}^{\rm d}, \frac{\pi}{2}]$, and their proofs given in Lemmas 11.3.1–11.3.2 work without change for the present range of wedge angles, where we recall that $P_0 = P_1$ for the subsonic/sonic wedge angles.

3. Now we show that (11.3.11) holds for any $\theta_{\rm w} \in [\theta_{\rm w}^{\rm d}, \frac{\pi}{2}]$ with constant $\delta > 0$ depending only on (ρ_0, ρ_1, γ). Fix such an angle $\theta_{\rm w}$. We follow the proof of Lemma 11.3.3 without change up to equation (11.3.13) by using $P_1 := P_0$, for the subsonic/sonic wedge angles in the argument.

Next, from the definition of P_1, we have

$$|D\varphi_2(P_1)| \le c_2 \qquad \text{for all } \theta_{\rm w} \in [\theta_{\rm w}^{\rm d}, \frac{\pi}{2}], \tag{16.2.2}$$

where the equality holds for the supersonic/sonic reflection solutions, and the strict inequality holds for the subsonic reflection solutions. Using that $D\varphi_2(P_1) = |D\varphi_2(P_1)|\mathbf{e}_{P_1}$ for $\mathbf{e}_{P_1} := \frac{D\varphi_2(P_1)}{|D\varphi_2(P_1)|} = \frac{O_2 - P_1}{|O_2 - P_1|}$, we calculate from (11.3.13) and then use (16.2.2) to obtain

$$\begin{aligned} &(\boldsymbol{\nu}_{S_1} \cdot D_{\mathbf{p}}) g^{\rm sh}(D\varphi_2(P_1), \varphi_2(P_1), P_1) \\ &= \rho_2\Big(1 - \frac{|D\varphi_2(P_1)|^2}{c_2^2}\big(\boldsymbol{\nu}_{S_1} \cdot \mathbf{e}_{P_1}\big)^2\Big) \\ &\ge \rho_2\big(1 - \big(\boldsymbol{\nu}_{S_1} \cdot \mathbf{e}_{P_1}\big)^2\big) > 0, \end{aligned} \tag{16.2.3}$$

where the last inequality follows from the facts that $|\boldsymbol{\nu}_{S_1}| = |\mathbf{e}_{P_1}| = 1$, and $\{\boldsymbol{\nu}_{S_1}, \mathbf{e}_{P_1}\}$ are not parallel to each other.

Indeed, for supersonic wedge angles, this has been shown in the paragraph before equation (11.3.14).

For subsonic/sonic wedge angles, P_1 and $\mathcal{O}_2$ lie on line $\{\xi_2 = \xi_1 \tan\theta_{\rm w}\}$ so that $\mathbf{e}_{P_1} = \frac{\mathcal{O}_2 - P_1}{|\mathcal{O}_2 - P_1|}$ is parallel to that line. If $\boldsymbol{\nu}_{S_1}$ and $\mathbf{e}_{P_1}$ are parallel, line $S_1 = \{\varphi_1 = \varphi_2\}$ is perpendicular to line $\{\xi_2 = \xi_1 \tan\theta_{\rm w}\}$ at $P_0 = P_1$. Then, from the Rankine-Hugoniot conditions (7.1.5) at which $(\boldsymbol{\nu}, \boldsymbol{\tau}) = (\boldsymbol{\nu}_{S_1}, \boldsymbol{\tau}_{S_1})$, it follows that $D\varphi_1 \cdot \boldsymbol{\tau}_{S_1} = D\varphi_2 \cdot \boldsymbol{\tau}_{S_1} = 0$ at $P_1 = P_0$, since $D\varphi_2(P_1) = |D\varphi_2(P_1)|\mathbf{e}_{P_1}$ is parallel to line $L_{\rm w} := \{\xi_2 = \xi_1 \tan\theta_{\rm w}\}$, which means that $D\varphi_1(P_1)$ is parallel to line $L_{\rm w}$. This is a contradiction, since $D\varphi_1(P_1) = \mathcal{O}_1 - P_1$ for $P_1 = P_0 \in L_{\rm w}$, but $\mathcal{O}_1 = (u_1, 0) \notin L_{\rm w}$. Thus, vectors $\boldsymbol{\nu}_{S_1}$ and $\mathbf{e}_{P_1}$ are not parallel to each other.

Therefore, (16.2.3) holds for each $\theta_{\rm w} \in [\theta_{\rm w}^{\rm d}, \frac{\pi}{2})$. Now (11.3.11) is proved for any $\theta_{\rm w} \in [\theta_{\rm w}^{\rm d}, \frac{\pi}{2})$.

4. In Steps 1–3, we have established Lemmas 11.3.1–11.3.3 for any $\theta_{\rm w} \in [\theta_{\rm w}^*, \frac{\pi}{2}]$ with uniform constants. The rest of §11.3 is obtained based on Lemmas 11.3.1–11.3.3 by using the uniform estimates of admissible solutions, which hold for any $\theta_{\rm w} \in [\theta_{\rm w}^{\rm d}, \frac{\pi}{2})$ by Proposition 15.2.2. Therefore, all the results in §11.3 hold for $\theta_{\rm w} \in [\theta_{\rm w}^*, \frac{\pi}{2})$. ☐

Corollary 16.2.2. *There exists $\sigma > 0$, depending only on (ρ_0, ρ_1, γ), such that constants $(\varepsilon, \delta, \delta_{\rm bc}, \hat{\delta}_{\rm bc}, C)$ in Proposition* 16.2.1 *can be chosen depending only on (ρ_0, ρ_1, γ) for all $\theta_{\rm w} \in [\theta_{\rm w}^{\rm d}, \frac{\pi}{2})$ satisfying*

$$\frac{|D\varphi_2(P_0)|}{c_2} > 1 - \sigma.$$

Proof. By Theorem 7.1.1(vi) and the continuous dependence of the parameters of (weak) state (2) on $\theta_{\rm w} \in [\theta_{\rm w}^{\rm d}, \frac{\pi}{2}]$ by Theorem 7.1.1(i), there exists $\sigma > 0$ such that $\frac{|D\varphi_2(P_0)|}{c_2} \leq 1 - \sigma$ for all $\theta_{\rm w} \in [\theta_{\rm w}^{\rm d}, \tilde{\theta}_{\rm w}^{\rm s}]$. Then, for this σ, we obtain the assertion as expected by using Proposition 16.2.1 with $\theta_{\rm w}^* = \tilde{\theta}_{\rm w}^{\rm s}$. ☐

16.3 A PRIORI ESTIMATES NEAR $\Gamma_{\rm sonic}$ IN THE SUPERSONIC-AWAY-FROM-SONIC CASE

All the estimates in §11.4 hold in this case without change. Specifically, we have

Proposition 16.3.1. *There exists $\sigma \in (0, \frac{\varepsilon_0}{2})$ depending only on (ρ_0, ρ_1, γ) and, for any $\alpha \in (0, 1)$, there exists $C > 0$ depending only on $(\rho_0, \rho_1, \gamma, \alpha)$ such that the following estimates hold: Let $l_{\rm so} > 0$. Define*

$$\varepsilon = \min\{\sigma, l_{\rm so}^2\}. \tag{16.3.1}$$

If $\theta_{\rm w} \in [\theta_{\rm w}^{\rm d}, \frac{\pi}{2})$ is a supersonic wedge angle that satisfies

$$y_{P_1}^{(\theta_{\rm w})} - \theta_{\rm w} \geq l_{\rm so},$$

and φ *is an admissible solution of* **Problem 2.6.1** *with the wedge angle* $\theta_{\rm w}$*, then* $\psi = \varphi - \varphi_2^{(\theta_{\rm w})}$ *satisfies*

$$\|\psi\|^{(\rm par)}_{2,\alpha,\Omega_\varepsilon} \le C, \tag{16.3.2}$$

where Ω_ε *is defined by* (16.1.3)*. Moreover, the shock function* $\hat{f}(x)$ *from* (11.2.38) *satisfies*

$$\|\hat{f} - \hat{f}_0^{(\theta_{\rm w})}\|^{(\rm par)}_{2,\alpha,(0,\varepsilon)} \le C, \tag{16.3.3}$$

where $\hat{f}_0^{(\theta_{\rm w})}$ *is the function defined in* (11.2.8)*.*

Proof. The same proof of Proposition 11.4.3 works for Proposition 16.3.1 without change by using Proposition 16.2.1. We also note that the constants in Proposition 16.2.1 can be chosen uniformly for all supersonic angles $\theta_{\rm w} \in [\theta_{\rm w}^{\rm d}, \frac{\pi}{2})$ by Corollary 16.2.2. □

Next we note that (16.1.12) and the continuous dependence of the parameters of state (2) on $\theta_{\rm w} \in [\theta_{\rm w}^{\rm d}, \frac{\pi}{2}]$ imply that, for every $\delta \in (0,1)$,

$$l(\delta) := \inf_{\theta_{\rm w}\in \mathcal{J}_{\rm w}(\delta)} (y_{P_1}^{(\theta_{\rm w})} - \theta_{\rm w}) > 0, \tag{16.3.4}$$

where

$$\mathcal{J}_{\rm w}(\delta) := \Big\{\theta_{\rm w} \in [\theta_{\rm w}^{\rm d}, \frac{\pi}{2}] \;:\; \frac{|D\varphi_2(P_0)|}{c_2}(\theta_{\rm w}) \ge 1+\delta\Big\}.$$

Then Proposition 16.3.1 implies

Proposition 16.3.2. *Let* $\delta \in (0,1)$*. There exists* $\varepsilon \in (0, \frac{\varepsilon_0}{2})$ *depending only on* $(\rho_0, \rho_1, \gamma, \delta)$ *such that the following estimates hold: For any* $\alpha \in (0,1)$*, there is* $C > 0$ *depending only on* $(\rho_0, \rho_1, \gamma, \alpha)$ *so that, if* $\theta_{\rm w} \in [\theta_{\rm w}^{\rm d}, \frac{\pi}{2})$ *satisfies*

$$\frac{|D\varphi_2(P_0)|}{c_2}(\theta_{\rm w}) \ge 1+\delta \tag{16.3.5}$$

and φ *is an admissible solution of* **Problem 2.6.1** *with the wedge angle* $\theta_{\rm w}$*, then* $\psi = \varphi - \varphi_2$ *satisfies* (16.3.2)*, and the shock function* $\hat{f}(x)$ *defined in* (11.2.38) *satisfies* (16.3.3)*.*

Proof. We apply Proposition 11.4.3 with $l_{\rm so} = l(\delta)$ defined by (16.3.4) to conclude the proof. □

16.4 A PRIORI ESTIMATES NEAR $\Gamma_{\rm sonic}$ IN THE SUPERSONIC-NEAR-SONIC CASE

The estimates of Proposition 16.3.2 depend on $\delta \in (0,1)$ in (16.3.5). Thus, in order to obtain the uniform estimates in Ω_ε with uniform $\varepsilon > 0$ for all the admissible solutions for $\theta_{\rm w} \in (\theta_{\rm w}^{\rm d}, \frac{\pi}{2})$ satisfying $\frac{|D\varphi_2(P_0)|}{c_2}(\theta_{\rm w}) > 1$, we employ a different technique to obtain the estimates near Γ_{sonic} for all the admissible solutions with $\theta_{\text{w}} \in (\theta_{\text{w}}^{\text{d}}, \frac{\pi}{2})$ satisfying

$$\frac{|D\varphi_2(P_0)|}{c_2}(\theta_{\text{w}}) \in (1, 1+\delta_{P_0}), \tag{16.4.1}$$

where $\delta_{P_0} \in (0,1)$ will be determined below. Combining these estimates with Proposition 16.3.2, we obtain the uniform estimate in Ω_ε for all the supersonic reflection solutions.

16.4.1 Main steps of the estimates near Γ_{shock} in the supersonic-near-sonic case

The main issue here is that, if $\delta := \frac{|D\varphi_2(P_0)|}{c_2}(\theta_{\text{w}}) - 1$ becomes small, then the length of Γ_{sonic} (or, equivalently, $y_{P_1}^{(\theta_{\text{w}})} - \theta_{\text{w}}$ in the (x,y)–coordinates) becomes small, so that Proposition 16.3.1 gives the estimates in Ω_ε with $\varepsilon \to 0$ as $\delta \to 0$. Thus, we do not obtain the uniform estimates near Γ_{sonic} for all the supersonic reflection solutions from Proposition 16.3.1.

In order to obtain these estimates, we argue in this section as follows: If $\delta = \frac{|D\varphi_2(P_0)|}{c_2}(\theta_{\text{w}}) - 1$ is small, then

(i) Using Proposition 16.3.1, we obtain the estimates in $\Omega_{b_{\text{so}}^2}$, where

$$b_{\text{so}}(\theta_{\text{w}}) := y_{P_1}^{(\theta_{\text{w}})} - \theta_{\text{w}}. \tag{16.4.2}$$

(ii) We note that, in the (x,y)–coordinates,

$$\Omega_{\varepsilon_0} \subset \{0 < x < \varepsilon_0,\ 0 < y - \theta_{\text{w}} < b_{\text{so}} + kx\}, \tag{16.4.3}$$

where $k > 1$ can be chosen uniformly for all the supersonic reflection solutions. Then we show that, for each integer $m \geq 3$ and all $b_{\text{so}} \in (0, b_{\text{so}}^*)$ for sufficiently small b_{so}^* depending on $(\rho_0, \rho_1, \gamma, m)$, solution $\psi(x,y)$ has a growth x^m in $\Omega_\varepsilon \cap \{x > \frac{b_{\text{so}}^2}{10}\}$:

$$0 \leq \psi(x,y) \leq Cx^m \qquad \text{in } \Omega_\varepsilon \cap \{x > \frac{b_{\text{so}}^2}{10}\}$$

for some $\varepsilon > 0$ and C depending only on $(\rho_0, \rho_1, \gamma, m)$. In the proof, we use the smallness of b_{so} and the elliptic degeneracy of the equation near Γ_{sonic}. The main point of this estimate is that b_{so} can be arbitrarily close to zero, but C is independent of b_{so}. On the other hand, we note that this estimate cannot be extended to the whole region Ω_ε (*i.e.*, the condition that $x > \frac{b_{\text{so}}^2}{10}$ cannot be dropped), since $D_{xx}\psi(0,y) = \frac{1}{\gamma+1}$ for any $y \in [\theta_{\text{w}}, y_{P_1}^{(\theta_{\text{w}})})$ by Theorem 14.1.2.

(iii) Using the algebraic growth, with $m = 4$, we can obtain the estimates of ψ in $\Omega_\varepsilon \cap \{x > \frac{b_{\rm so}^2}{10}\}$ by using a different scaling from the one in Proposition 16.3.1: From (16.4.3), for $z_0 = (x_0, y_0) \in \Omega_\varepsilon \cap \{\frac{b_{\rm so}^2}{5} < x < \varepsilon\}$ and $\rho \in (0, 1]$, rectangle $R_{z_0,\rho}$ defined by (11.4.7) does not satisfy (11.4.9) in general (*i.e.*, the rectangle does not fit into Ω_ε). On the other hand, we note that

$$\{0 < x < \varepsilon_0,\ 0 < y - \theta_{\rm w} < b_{\rm so} + \frac{x}{k}\} \subset \Omega_{\varepsilon_0},$$

where $k > 1$ can be chosen uniformly for all the supersonic reflection solutions. Since $b_{\rm so}$ can be arbitrarily close to zero, then, for $z_0 = (x_0, y_0)$, we have to use a rectangle with the side length $\sim x_0$ (*i.e.*, proportional to x_0) in the y–direction in order to have a property similar to (11.4.9). Also, from the degenerate ellipticity structure (11.2.43) of the equation for the regular reflection solutions, we need to have a rectangle in which the ratio of the side lengths in the x– and y–directions is $\sim \sqrt{x_0}$, in order to obtain a uniformly elliptic equation after rescaling the rectangle into a square. Thus, for $(x_0, y_0) \in \Omega_\varepsilon \cap \{x > \frac{b_{\rm so}^2}{10}\}$, we consider the rectangle:

$$\hat{R}_{(x_0,y_0)} := \{|x - x_0| < \frac{x_0^{3/2}}{10k},\ |y - y_0| < \frac{x_0}{10k}\} \cap \Omega.$$

Such rectangles *fit into* Ω, *i.e.*, for $(x_0, y_0) \in \Gamma_{\rm wedge} \cap \Omega_\varepsilon$, the corresponding rectangle $\hat{R}_{(x_0,y_0)}$ does not intersect with $\Gamma_{\rm shock}$ and, for $(x_0, y_0) \in \Gamma_{\rm shock} \cap \Omega_\varepsilon$, the corresponding rectangle $\hat{R}_{(x_0,y_0)}$ does not intersect with $\Gamma_{\rm wedge}$. Moreover, note that the ratio of the side lengths in the x– and y–directions of $\hat{R}_{(x_0,y_0)}$ is $10k\sqrt{x_0}$, *i.e.*, the same as for the rectangles in Proposition 16.3.1 up to a multiplicative constant. From this, rescaling $\hat{R}_{(x_0,y_0)}$ to the portion of square $(-1, 1)^2 := (-1, 1) \times (-1, 1)$:

$$\hat{Q}_{(x_0,y_0)} := \{(S, T) \in (-1, 1)^2 \ :\ (x_0 + x_0^{\frac{3}{2}} S, y_0 + \frac{x_0}{10k} T) \in \Omega\},$$

we obtain a uniformly elliptic equation for the function:

$$\psi^{(x_0,y_0)}(S, T) := \frac{1}{x_0^4} \psi(x_0 + x_0^{\frac{3}{2}} S, y_0 + \frac{x_0}{10k} T) \qquad \text{for } (S, T) \in \hat{Q}_{(x_0,y_0)}.$$

The algebraic growth of ψ with $m = 4$ implies the uniform estimates in $L^\infty(\hat{Q}_{(x_0,y_0)})$ of $\psi^{(x_0,y_0)}(\cdot)$ for any $(x_0, y_0) \in \Omega_\varepsilon \cap \{x > \frac{b_{\rm so}^2}{10}\}$. Then, arguing as in Proposition 16.3.1, we obtain the estimates of ψ in $\Omega_\varepsilon \cap \{x > \frac{b_{\rm so}^2}{10}\}$ in the appropriate weighted and scaled $C^{2,\alpha}$–norms, which are in fact stronger than the estimates in Proposition 16.3.1.

In the rest of this section, we assume that φ is an admissible solution with $\theta_{\rm w}$ satisfying (16.4.1) with δ_{P_0} to be specified.

16.4.2 Algebraic growth of ψ in $\Omega_\varepsilon \cap \{x > \frac{b_{\rm so}^2}{10}\}$ in the supersonic-near-sonic case

We work in the (x, y)–coordinates (11.1.1)–(11.1.2) in Ω_ε defined by (11.2.16), where $\varepsilon \in (0, \varepsilon_0)$ will be chosen below. Then $\psi = \varphi - \varphi_2$ satisfies equation (11.1.4)–(11.1.5) in Ω_ε.

For the growth estimates of the solution, we consider equation (11.1.4)–(11.1.5) as a linear equation:

$$\mathcal{L}\psi := \sum_{i,j=1}^{2} a_{ij}(x,y)D_{ij}\psi + \sum_{i}^{2} a_i(x,y)D_i\psi = 0 \qquad \text{in } \Omega_\varepsilon, \tag{16.4.4}$$

which is obtained by plugging ψ into the coefficients of (11.1.4), where $D = (D_1, D_2) := (\partial_x, \partial_y)$ and $D_{ij} = D_iD_j$.

From the explicit expressions of the coefficients (a_{ij}, a_i) given by (11.1.4)–(11.1.5), using Lemma 11.2.6 and Corollary 11.2.9 with Proposition 16.1.4, we obtain the following bounds in Ω_ε for sufficiently small $\varepsilon > 0$:

$$\begin{aligned} &x \le a_{11}(x,y) \le 3x, &&\frac{1}{C} \le a_{22}(x,y) \le C,\\ &|(a_{12}, a_2)(x,y)| \le Cx, &&a_1(x,y) \le 0, \end{aligned} \tag{16.4.5}$$

where ε and C depend only on (ρ_0, ρ_1, γ). Also, equation (16.4.5) is strictly elliptic in Ω_ε by the properties of admissible solutions in Definition 15.1.1 combined with Definition 8.1.1(iii).

By Proposition 16.2.1, the boundary condition for ψ on $\Gamma_{\rm shock}$ is (11.3.33), and $B_1(\mathbf{p}, z, x, y)$ satisfies (11.3.35). Then we can write (11.3.33) as a linear condition:

$$\mathcal{N}\psi := \mathbf{b}(x,y) \cdot D\psi + b_0(x,y)\psi = 0 \qquad \text{on } \Gamma_{\rm shock} \cap \partial\Omega_\varepsilon, \tag{16.4.6}$$

where $\mathbf{b} = (b_1, b_2)$ and b_0 are given by

$$(\mathbf{b}, b_0)(x,y) = \int_0^1 (D_{\mathbf{p}}, D_z)B_1(tD\psi(x,y), t\psi(x,y), x, y)dt.$$

By Proposition 16.2.1 with Corollary 16.2.2, Lemma 11.3.7, and (11.3.36), using Lemma 11.2.6 and Corollary 11.2.9 with Proposition 16.1.4, and reducing ε if necessary, we have

$$-C \le b_i \le -\frac{1}{C} \qquad \text{on } \Gamma_{\rm shock} \cap \partial\Omega_\varepsilon \ \text{ for } i = 0, 1, 2. \tag{16.4.7}$$

Also, the boundary condition (16.4.6) is strictly oblique in the sense that

$$\mathbf{b} \cdot \boldsymbol{\nu}_{\rm sh}^{(xy)} > 0 \qquad \text{on } \Gamma_{\rm shock} \cap \partial\Omega_\varepsilon, \tag{16.4.8}$$

where $\boldsymbol{\nu}_{\rm sh}^{(xy)}$ is the unit normal to curve $\Gamma_{\rm shock}$ expressed in the (x,y)–coordinates and directed into the interior of Ω, as defined in Lemma 11.3.8. The obliqueness condition (16.4.8) is obtained by Proposition 16.2.1 and Lemma 11.3.8.

The boundary condition: $\partial_{\boldsymbol{\nu}}\psi = 0$ on $\Gamma_{\rm wedge}$ in the $\boldsymbol{\xi}$–coordinates, expressed in the (x,y)–coordinates, is

$$D_2\psi = 0 \qquad \text{on } \Gamma_{\rm wedge}\cap\partial\Omega_\varepsilon. \tag{16.4.9}$$

Also, ψ satisfies

$$\psi = 0 \qquad \text{on } \overline{\Gamma_{\rm sonic}}, \tag{16.4.10}$$

$$\psi \le C \qquad \text{on } \partial\Omega_\varepsilon\cap\{x=\varepsilon\}, \tag{16.4.11}$$

where the last bound follows from Corollary 11.2.9 with Proposition 16.1.4.

We also note that, from structure (11.2.38)–(11.2.40) of Ω_ε for any supersonic admissible solution, there exists $k>0$ depending only on (ρ_0,ρ_1,γ) such that

$$\Omega_\varepsilon \subset \{0<x<\varepsilon_0,\ 0<y-\theta_{\rm w}<b_{\rm so}+kx\}. \tag{16.4.12}$$

First we prove the algebraic growth estimate for the solutions of the linear problem (16.4.4), (16.4.6), and (16.4.9)–(16.4.10) in a domain of the appropriate structure.

Lemma 16.4.1. *Let $\varepsilon>0$, $\theta_{\rm w}\in\mathbb{R}$, and domain Ω_ε be of the structure as in* (11.2.38)–(11.2.39), *and let* (11.2.40) *hold. Let $\psi\in C(\overline{\Omega_\varepsilon})\cap C^1(\overline{\Omega_\varepsilon}\cap\{0<x<\varepsilon\})\cap C^2(\Omega_\varepsilon)$ satisfy* (16.4.4), (16.4.6), *and* (16.4.9)–(16.4.10). *Let equation* (16.4.4) *be strictly elliptic in $\overline{\Omega_\varepsilon}\setminus\Gamma_{\rm sonic}$, and let the boundary condition* (16.4.6) *be strictly oblique on $\Gamma_{\rm shock}\cap\partial\Omega_\varepsilon$ in the sense of* (16.4.8). *Let* (16.4.5) *and* (16.4.7) *hold. Denote*

$$b_{\rm so} := y_{P_1}-y_{P_4}$$

with the notations from (11.2.39). *Then, for every integer $m\ge 3$, there exist $\hat\varepsilon$, $b_{\rm so}^*>0$, and $\hat C>0$ depending only on m and the constants in* (11.2.40), (16.4.5), *and* (16.4.7) *such that, if $\varepsilon\le\hat\varepsilon$ and $b_{\rm so}\le b_{\rm so}^*$,*

$$\psi(x,y)\le \hat C\frac{\max_{\partial\Omega_\varepsilon\cap\{x=\varepsilon\}}\psi}{\varepsilon^m}x^m \qquad \text{in } \Omega_\varepsilon\cap\{x>\frac{b_{\rm so}^2}{10}\}. \tag{16.4.13}$$

Proof. We divide the proof into four steps.

1. We note that structure (11.2.38)–(11.2.40) of Ω_ε implies that (16.4.12) holds with k depending only on the constants in (11.2.40).

Fix integer $m\ge 3$. In the proof, constants $(C,L,M,N,\hat\varepsilon)$ depend only on m and the constants in (11.2.40), (16.4.5), and (16.4.7). Also, $L,M,N\ge 1$.

Furthermore, it is convenient to shift the y–coordinate by replacing $y-\theta_{\rm w}$ by y. This does not affect our conditions, but in fact means that we can assume without loss of generality that

$$\theta_{\rm w}=0.$$

Then (16.4.12) is replaced by

$$\Omega_\varepsilon \subset \{0 < x < \varepsilon_0,\ 0 < y < b_{so} + kx\}. \tag{16.4.14}$$

2. We now show that, for an appropriate choice of constants $M, L > 0$ and for sufficiently small b_{so}, the function:

$$v(x, y) = (x + Mb_{so}^2)^m - L(x + Mb_{so}^2)^{m-1}y^2 \tag{16.4.15}$$

is a positive supersolution of equation (16.4.4) in Ω_ε with the boundary condition (16.4.6) on $\Gamma_{\text{shock}} \cap \partial\Omega_\varepsilon$. From this, we show that $\psi \le Bv$ in Ω_ε for sufficiently large B depending on the data and (ε, m). This implies (16.4.13).

In the following estimates, the dependence of the constants is as follows: We denote $M = NL$ and choose large $L, N > 1$ depending only on the data, *i.e.*, the constants in (11.2.40), (16.4.5), and (16.4.7). Then we choose $\varepsilon, b_{so} > 0$ small depending on (L, M, N) (hence on the data), where the smallness condition is explicitly

$$b_{so} \le \frac{1}{LMN}, \qquad \varepsilon \le \frac{1}{LN}, \tag{16.4.16}$$

which implies especially that $b_{so} \le \frac{1}{N}$.

Thus, we employ (16.4.14) to obtain that, for $(x, y) \in \Omega_\varepsilon$,

$$\begin{aligned} \frac{1}{(x + Mb_{so}^2)^{m-1}} v(x, y) &= x + Mb_{so}^2 - Ly^2 \\ &\ge x + Mb_{so}^2 - L(b_{so} + kx)^2 \\ &\ge x + Mb_{so}^2 - 2Lb_{so}^2 - 2Lk^2x^2. \end{aligned}$$

For $L, N \ge 1$, choosing $M = NL \ge 1$, assuming that ε is small so that (16.4.16) holds, choosing $N = 4\max\{1, k^2\}$, and using that $x \in (0, \varepsilon)$ in Ω_ε, we have

$$v(x, y) \ge \frac{1}{2}(x + Mb_{so}^2)^m > 0. \tag{16.4.17}$$

Similar estimates imply that, for $(x, y) \in \Omega_\varepsilon$,

$$\begin{aligned} v_x(x, y) &= (x + Mb_{so}^2)^{m-2}\Big(m(x + Mb_{so}^2) - (m-1)Ly^2\Big) \\ &\ge m(x + Mb_{so}^2)^{m-2}\big(x + Mb_{so}^2 - CL(b_{so} + kx)^2\big) \\ &\ge m(x + Mb_{so}^2)^{m-2}\big(x + Mb_{so}^2 - 2CLb_{so}^2 - 2CLk^2x^2\big), \end{aligned}$$

by using (16.4.14). Choosing N large and ε satisfying (16.4.16), we have

$$v_x(x, y) \ge \frac{m}{2}(x + Mb_{so}^2)^{m-1} > 0. \tag{16.4.18}$$

Similarly,

$$\begin{aligned} v_{xx}(x,y) &= (x+Mb_{\mathrm{so}}^2)^{m-3}\Big(m(m-1)(x+Mb_{\mathrm{so}}^2)-(m-1)(m-2)Ly^2\Big) \\ &\geq m(m-1)(x+Mb_{\mathrm{so}}^2)^{m-3}\Big((x+Mb_{\mathrm{so}}^2)-CL(b_{\mathrm{so}}+kx)^2\Big) \\ &\geq \frac{m(m-1)}{2}(x+Mb_{\mathrm{so}}^2)^{m-2}\geq 0, \end{aligned}$$

by further increasing N if necessary and choosing ε satisfying (16.4.16). Then we conclude

$$|v_{xx}(x,y)|\leq m(m-1)(x+Mb_{\mathrm{so}}^2)^{m-2}. \tag{16.4.19}$$

Next we estimate (v_{xy}, v_{yy}, v_y). For $(x,y)\in\Omega_\varepsilon$, using (16.4.14) and then (16.4.16), which implies particularly that $0\leq b_{\mathrm{so}}+kx\leq\frac{C}{LN}$ in Ω_ε, we have

$$\begin{aligned} |v_{xy}(x,y)| &= |-2(m-1)L(x+Mb_{\mathrm{so}}^2)^{m-2}y| \\ &\leq 2(m-1)L(x+Mb_{\mathrm{so}}^2)^{m-2}(b_{\mathrm{so}}+kx) \\ &\leq \frac{C}{N}(x+Mb_{\mathrm{so}}^2)^{m-2}, \end{aligned} \tag{16.4.20}$$

$$v_{yy}(x,y) = -2L(x+Mb_{\mathrm{so}}^2)^{m-1}, \tag{16.4.21}$$

$$\begin{aligned} |v_y(x,y)| &= |-2L(x+Mb_{\mathrm{so}}^2)^{m-1}y| \\ &\leq 2L(x+Mb_{\mathrm{so}}^2)^{m-1}(b_{\mathrm{so}}+kx) \\ &\leq \frac{C}{N}(x+Mb_{\mathrm{so}}^2)^{m-1}. \end{aligned} \tag{16.4.22}$$

3. Now combining the estimates of (v, Dv, D^2v) above with (16.4.5) and (16.4.7), we show that v is a supersolution of equation (16.4.4) and the boundary condition (16.4.6) as follows:

First, we consider equation (16.4.4). For $(x,y)\in\Omega_\varepsilon$, using (16.4.5), (16.4.16), and the estimates of (v, Dv, D^2v), we see that

(i) $a_{22}(x,y)v_{yy}(x,y)\leq -\frac{L}{C}\big(x+Mb_{\mathrm{so}}^2\big)^{m-1}$;

(ii) $a_1(x,y)v_x(x,y)\leq 0$;

(iii) The absolute value of the other terms of $\mathcal{L}v$ is estimated by $C(x+Mb_{\mathrm{so}}^2)^{m-1}$:

$$|a_{11}(x,y)v_{xx}(x,y)|\leq Cx(x+Mb_{\mathrm{so}}^2)^{m-2}\leq C(x+Mb_{\mathrm{so}}^2)^{m-1},$$

$$|a_{12}(x,y)v_{xy}(x,y)|\leq \frac{C}{N}x(x+Mb_{\mathrm{so}}^2)^{m-2}\leq C(x+Mb_{\mathrm{so}}^2)^{m-1},$$

$$|a_2(x,y)v_y(x,y)|\leq \frac{C}{N}x(x+Mb_{\mathrm{so}}^2)^{m-1}\leq C(x+Mb_{\mathrm{so}}^2)^{m-1}.$$

Therefore, we have

$$\mathcal{L}v \le C(x + Mb_{\rm so}^2)^{m-1} - \frac{L}{C}(x + Mb_{\rm so}^2)^{m-1} < 0$$

by choosing L large. That is, v is a supersolution of equation (16.4.4).

Next we consider the boundary condition (16.4.6). For $(x, y) \in \Gamma_{\rm shock} \cap \partial\Omega_\varepsilon$, using (16.4.7) and estimates (16.4.18) and (16.4.22) of Dv, and noting that $b_3 v \le 0$, we have

$$\mathcal{N}v \le -\frac{1}{C}\left(x + Mb_{\rm so}^2\right)^{m-1} + \frac{C}{N}\left(x + Mb_{\rm so}^2\right)^{m-1} < 0,$$

by increasing N.

4. Now, if $\varepsilon \le \hat{\varepsilon} := \frac{1}{LN}$, then, choosing

$$B = \max\left\{0,\ \frac{2}{\varepsilon^m} \max_{\partial\Omega_\varepsilon \cap \{x=\varepsilon\}} \psi\right\} \ge 0$$

and using (16.4.17), we have

$$Bv \ge \psi \qquad \text{on } \partial\Omega_\varepsilon \cap \{x = \varepsilon\}.$$

Also, from the explicit expression of v and the boundary conditions (16.4.9)–(16.4.10) satisfied by ψ, and using structure (11.2.38)–(11.2.39) of the domain, we have

$$Bv = 0 = \psi \qquad \text{on } \Gamma_{\rm sonic} \subset \{x = 0\},$$
$$v_y = 0 = \psi_y \qquad \text{on } \Gamma_{\rm wedge} \subset \{y = 0\},$$

where we have used the shifted coordinates, *i.e.*, set $\theta_{\rm w} = 0$. From the maximum principle (where we have used the strict ellipticity of equation (16.4.4), the strict obliqueness of the boundary condition (16.4.6), and the property that $b_3 < 0$), it follows that $Bv \ge \psi$ in Ω_ε. Now (16.4.13) is proved. □

In order to apply Lemma 16.4.1 to the estimates for regular reflection solutions, we first note the following facts: From Propositions 11.2.8 and 16.1.4 and the continuous dependence of the parameters of state (2) on $\theta_{\rm w} \in [\theta_{\rm w}^{\rm d}, \frac{\pi}{2}]$, it follows that, for any $\theta_{\rm w} \in (\theta_{\rm w}^{\rm d}, \frac{\pi}{2}]$ satisfying $\frac{|D\varphi_2(P_0)|}{c_2}(\theta_{\rm w}) > 1$, we have

$$\Gamma_{\rm sonic}^{(\theta_{\rm w})} = \{(0, y) \,:\, \theta_{\rm w} < y < y_{P_1}^{(\theta_{\rm w})}\}, \tag{16.4.23}$$

where $y_{P_1}^{(\theta_{\rm w})} - \theta_{\rm w} > 0$ and $y_{P_1}^{(\theta_{\rm w}^{(i)})} - \theta_{\rm w}^{(i)} \to 0$ when $\frac{|D\varphi_2(P_0)|}{c_2}(\theta_{\rm w}^{(i)}) \to 1+$. From this, again using the continuous dependence of the parameters of state (2) on $\theta_{\rm w} \in [\theta_{\rm w}^{\rm d}, \frac{\pi}{2}]$, we find that, for every $\delta \in (0, 1)$, denoting

$$s(\delta) := \sup_{\theta_{\rm w} \in \mathcal{G}_w(\delta)} (y_{P_1}^{(\theta_{\rm w})} - \theta_{\rm w}) \tag{16.4.24}$$

with

$$\mathcal{G}_w(\delta) := \Big\{\theta_{\rm w} \in [\theta_{\rm w}^{\rm d}, \frac{\pi}{2}] \,:\, 1 < \frac{|D\varphi_2(P_0)|}{c_2}(\theta_{\rm w}) < 1+\delta\Big\},$$

we obtain that $s(\delta) > 0$ for any $\delta \in (0,1)$ and

$$\lim_{\delta\to 0+} s(\delta) = 0. \tag{16.4.25}$$

Proposition 16.4.2. *For every integer $m \geq 3$, there exist small $\delta_{P_0}, \varepsilon > 0$ and large C, depending only on $(\rho_0, \rho_1, \gamma, m)$, such that, if $\theta_{\rm w} \in (\theta_{\rm w}^{\rm d}, \frac{\pi}{2})$ satisfies* (16.4.1) *and φ is an admissible solution of* **Problem 2.6.1** *with the wedge angle $\theta_{\rm w}$, then*

$$0 \leq \psi(x,y) \leq Cx^m \qquad in\ \Omega_\varepsilon \cap \{x > \frac{b_{\rm so}^2(\theta_{\rm w})}{10}\}, \tag{16.4.26}$$

where $b_{\rm so}$ is defined by (16.4.2).

Proof. As we have shown at the beginning of §16.4.2, there exists $\varepsilon > 0$ depending only on (ρ_0, ρ_1, γ) such that any supersonic reflection solution ψ satisfies (16.4.4), (16.4.6), and (16.4.9)–(16.4.11) in Ω_ε. Moreover, by Proposition 16.1.4, Ω_ε is of structure (11.2.38)–(11.2.39), and equation (16.4.4) is strictly elliptic in $\overline{\Omega_\varepsilon} \setminus \Gamma_{\rm sonic}$. Also, by Proposition 16.2.1 and Corollary 16.2.2, the boundary condition (16.4.6) is strictly oblique on $\Gamma_{\rm shock} \cap \partial\Omega_\varepsilon$, and estimates (11.2.40), (16.4.5), and (16.4.7) hold with the constants depending only on (ρ_0, ρ_1, γ).

Fix $m \geq 3$. Then this and the constants in (11.2.40), (16.4.5), and (16.4.7) for supersonic reflection solutions fix the corresponding constants $\hat{\varepsilon}$, $b_{\rm so}^* > 0$, and $\hat{C} > 0$ in Lemma 16.4.1. We reduce ε if necessary by replacing it with $\min\{\varepsilon, \hat{\varepsilon}\}$.

Furthermore, from (16.4.24)–(16.4.25), there exists $\delta_{P_0} > 0$ depending only on (ρ_0, ρ_1, γ) such that $s(\delta_{P_0}) \leq b_{\rm so}^*$. Then, using Lemma 16.4.1 and (16.4.11), we conclude the proof. □

16.4.3 Regularity and uniform a priori estimates of solutions

Using the growth estimate in Proposition 16.4.2 (with $m = 4$), we now prove the *a priori* estimates of solutions in the weighted and scaled $C^{2,\alpha}$ spaces. The argument follows the proof of Proposition 11.4.3, but with a different scaling. Since, in §16.5, we will also need similar estimates with another scaling, we first prove a more general version of the estimates in Steps 2–4 of the proof of Proposition 11.4.3, which allows for various scalings, including the one in Proposition 11.4.3 and the two cases that we will need below. Moreover, note that we prove these results for ψ that satisfies only the equation, the boundary conditions, and the growth conditions in a domain of an appropriate shape, *without assumption* on ψ to be a regular reflection solution. This will allow us to use these results later for solving the iteration problem.

Lemma 16.4.3. *Let $\theta_{\rm w} \in \mathbb{R}$ and $\varepsilon^* > 10x_{P_1} \geq 0$, and let Ω_{ε^*} be of structure* (16.1.10) *and* (16.1.13) *with $\hat{f}(0) \geq \theta_{\rm w}$, which includes both the supersonic and* *subsonic reflection cases by Proposition* 16.1.4. *Note that* $x \equiv x_{P_1}$ *on* $\overline{\Gamma_{\text{sonic}}}$. *Denote*

$$d_{\text{so}}(x) := x - x_{P_1}. \tag{16.4.27}$$

Then $d_{\text{so}} > 0$ *in* Ω_{ε^*}, *and* $d_{\text{so}} = 0$ *on* $\overline{\Gamma_{\text{sonic}}}$ *by* (16.1.10).

Let $g, h \in C([x_{P_1}, x_{P_1} + \varepsilon^*])$ *and satisfy that, on* $(x_{P_1}, x_{P_1} + \varepsilon^*]$,

$$g(\cdot) \text{ and } h(\cdot) \quad \text{are monotone increasing,} \tag{16.4.28}$$

$$0 < h(x) \le \frac{d_{\text{so}}(x)}{4\sqrt{x}}, \tag{16.4.29}$$

$$0 < g(x) \le x^{\frac{3}{2}} h(x), \tag{16.4.30}$$

$$\frac{g(x)}{g(x - \frac{d_{\text{so}}(x)}{2})} \le M, \qquad \frac{h(x)}{h(x - \frac{d_{\text{so}}(x)}{2})} \le M. \tag{16.4.31}$$

Let $\varepsilon \in (0, \frac{\varepsilon^*}{2})$, *and let* $\psi \in C(\overline{\Omega}_{2\varepsilon}) \cap C^1(\overline{\Omega}_{2\varepsilon} \cap \{x_{P_1} < x < x_{P_1} + 2\varepsilon\}) \cap C^2(\Omega_{2\varepsilon})$ *satisfy the equation:*

$$\sum_{i,j=1}^{2} A_{ij}(D\psi, \psi, x, y) D_{ij}^2 \psi + \sum_{i=1}^{2} A_i(D\psi, \psi, x, y) D_i \psi = 0, \tag{16.4.32}$$

where $A_{12} = A_{21}$, *with* $(A_{ij}, A_i)(\mathbf{p}, z, x, y)$ *satisfying*

$$\lambda |\boldsymbol{\kappa}|^2 \le \sum_{i,j=1}^{2} A_{ij}(D\psi(x,y), \psi(x,y), x, y) \frac{\kappa_i \kappa_j}{x^{2 - \frac{i+j}{2}}} \le \frac{1}{\lambda} |\boldsymbol{\kappa}|^2 \tag{16.4.33}$$

for all $(x, y) \in \Omega_{2\varepsilon}$ *and* $\boldsymbol{\kappa} = (\kappa_1, \kappa_2) \in \mathbb{R}^2$. *Assume also that, for each* $(\mathbf{p}, z) \in \mathbb{R}^2 \times \mathbb{R}$, $(x, y) \in \Omega_{2\varepsilon}$, *and* $i, j = 1, 2$,

$$\begin{aligned}
&|A_{ij}(\mathbf{p}, z, x, y)| \le M x^{2 - \frac{i+j}{2}}, \\
&\|A_i\|_{L^\infty(\mathbb{R}^2 \times \mathbb{R} \times \Omega_{2\varepsilon})} + \|D_{(\mathbf{p},z)}(A_{ij}, A_i)\|_{L^\infty(\mathbb{R}^2 \times \mathbb{R} \times \Omega_{2\varepsilon})} \le M, \\
&|x^{\frac{i+j-3}{2}} h(x) D_x A_{ij}(\mathbf{p}, z, x, y)| + |x^{\frac{i+j}{2} - 2} h(x) D_y A_{ij}(\mathbf{p}, z, x, y)| \le M, \\
&|x^{\frac{i-1}{2}} h^2(x) D_x A_i(\mathbf{p}, z, x, y)| + |x^{\frac{i}{2} - 1} h^2(x) D_y A_i(\mathbf{p}, z, x, y)| \le M.
\end{aligned} \tag{16.4.34}$$

Let ψ *satisfy the boundary conditions:*

$$\psi_x = b(\psi_y, \psi, x, y) \qquad \text{on } \Gamma_{\text{shock}} \cap \partial\Omega_{2\varepsilon}, \tag{16.4.35}$$

$$\psi = G(x, y) \qquad \text{on } \Gamma_{\text{shock}} \cap \partial\Omega_{2\varepsilon}, \tag{16.4.36}$$

$$\psi_y = 0 \qquad \text{on } \Gamma_{\text{wedge}} \cap \partial\Omega_{2\varepsilon}. \tag{16.4.37}$$

Assume that $b(p_2, z, x, y)$ *satisfies* (11.3.41)–(11.3.42) *on* $\Omega_{2\varepsilon}$.

Denote $\Omega_\delta^N := \{(x,y) : x \in (x_{P_1}, x_{P_1}+\delta),\ y \in (\theta_{\rm w}, \hat{f}(x)+\frac{1}{N})\}$ *for* $N>1$ *and* $\delta \in (0,\varepsilon^*]$. *Assume that* $G \in C^{1,1}(\overline{\Omega_{2\varepsilon}^M})$ *and satisfies*

$$\|G\|_{1,1,\Omega_{2\varepsilon}^M} \le M, \qquad D_yG \le -\frac{1}{M} \ \ in\ \Omega_{2\varepsilon}^M. \tag{16.4.38}$$

Assume that $x_{\rm bot} \in [x_{P_1}, x_{P_1}+\varepsilon)$ *and*

$$|\psi(x,y)| \le Mg(x) \qquad for\ all\ (x,y) \in \Omega_{2\varepsilon} \cap \{d_{\rm so}(x) > \frac{d_{\rm so}(x_{\rm bot})}{2}\}. \tag{16.4.39}$$

For $z_0 = (x_0,y_0) \in \Omega_\varepsilon \cap \{x > x_{\rm bot}\}$ *and* $\rho \in (0,1]$, *consider the rectangle:*

$$\begin{aligned} &\tilde{R}_{z_0,\rho} := \{(x,y) : |x-x_0| < \rho\sqrt{x_0}h(x_0),\ |y-y_0| < \rho h(x_0)\},\\ &\hat{R}_{z_0,\rho} := \tilde{R}_{z_0,\rho} \cap \Omega_{2\varepsilon}. \end{aligned} \tag{16.4.40}$$

We assume that $\hat{f}(\cdot)$, $h(\cdot)$, *and* ε *are such that*

$$\begin{aligned} &\hat{R}_{z_w,1} \cap \Gamma_{\rm shock} = \emptyset \qquad for\ all\ z_w = (x,0) \in \Gamma_{\rm wedge} \cap \{x_{\rm bot} < x < x_{P_1}+\varepsilon\},\\ &\hat{R}_{z_s,1} \cap \Gamma_{\rm wedge} = \emptyset \qquad for\ all\ z_s = (x,\hat{f}(x)) \in \Gamma_{\rm shock} \cap \{x_{\rm bot} < x < x_{P_1}+\varepsilon\}. \end{aligned} \tag{16.4.41}$$

Furthermore, introducing (S,T) *by the invertible change of variables:*

$$(x,y) = (x_0 + \sqrt{x_0}\,h(x_0)S,\, y_0 + h(x_0)T), \tag{16.4.42}$$

we find that there exists $Q_\rho^{(z_0)} \subset Q_\rho := (-\rho,\rho)^2$ *such that*

$$\hat{R}_{z_0,\rho} = \Big\{(x_0 + \sqrt{x_0}\,h(x_0)S,\ y_0 + h(x_0)T) : (S,T) \in Q_\rho^{(z_0)}\Big\}. \tag{16.4.43}$$

Define

$$\psi^{(x_0,y_0)}(S,T) := \frac{1}{g(x_0)}\psi(x_0 + \sqrt{x_0}\,h(x_0)S,\ y_0 + h(x_0)T) \qquad in\ Q_1^{(z_0)}. \tag{16.4.44}$$

Then there exists $\varepsilon_1^* \in (0,\varepsilon^*)$ *such that, if* $\varepsilon \le \frac{\varepsilon_1^*}{2}$ *and* $x_{P_1} \le \frac{\varepsilon_1^*}{2}$,

$$\|\psi^{(z_0)}\|_{C^{2,\alpha}(\overline{Q_{1/100}^{(z_0)}})} \le C \qquad for\ all\ z_0 \in \overline{\Omega}_\varepsilon \cap \{x > x_{\rm bot}\} \tag{16.4.45}$$

for each $\alpha \in (0,1)$, *where* ε_1^* *depends only on* M *and the constants in conditions* (11.3.41), (16.1.13), *and* (16.4.33)–(16.4.34), *and* C *depends on* α *in addition to the previous parameters.*

Proof. We divide the proof into five steps.

1. Note that, in the proof of Proposition 11.4.3, we have considered a specific case: $x_{P_1} = x_{\rm bot} = 0$, $h(x) = \frac{\sqrt{x}}{4}$, $g(x) = \frac{x^2}{4}$, and $M = \max\{4, L\}$, where L is from (11.2.44). This choice of M is done by the argument that $M = 4$ works for (16.4.31) so that $M = \max\{4, L\}$ works for both (16.4.31) and (16.4.39).

In order to prove this lemma, we repeat the proof of Proposition 11.4.3 with only notational changes in the general case from the specific case considered in Proposition 11.4.3, which can be achieved since our conditions imply that the argument works. Below, we sketch this argument and do some explicit calculations to illustrate this.

In this proof, all constants C, ε, and x_{P_1} depend only on M and the constants in the conditions in (16.1.13), (16.4.33)–(16.4.34), and (11.3.41). In some estimates, in addition to these parameters, C depends on $\alpha \in (0,1)$, which is written as $C(\alpha)$ in that case.

2. From (16.4.29) and (16.4.40), it follows that, for all $z_0 := (x_0, y_0) \in \Omega_\varepsilon$ and $\rho \in (0,1]$,

$$
\begin{aligned}
\hat{R}_{z_0,\rho} &\subset \Omega \cap \{(x,y) \,:\, x_0 - \frac{d_{\rm so}(x_0)}{4} < x < x_0 + \frac{d_{\rm so}(x_0)}{4}\}\\
&\subset \Omega \cap \{(x,y) \,:\, \frac{3x_0}{4} < x < \frac{5x_0}{4}\} \subset \Omega_{2\varepsilon}.
\end{aligned}
\tag{16.4.46}
$$

In particular,

$$
\overline{\hat{R}_{z,\rho}} \cap \Gamma_{\rm sonic} = \emptyset \qquad \text{for all } z \in \Omega_{2\varepsilon}.
$$

Next, as in the proof of Proposition 11.4.3, we consider three cases of rectangles $\hat{R}_{z_0,\rho}$ as in Step 1 of the proof of Proposition 11.4.3:

(i) $\hat{R}_{z_0,1/10}$ for $z_0 \in \Omega_\varepsilon \cap \{x > x_{\rm bot}\}$ satisfying $\hat{R}_{z_0,1/10} = \tilde{R}_{z_0,1/10}$;

(ii) $\hat{R}_{z_0,1/2}$ for $z_0 \in \Gamma_{\rm wedge} \cap \{x_{\rm bot} < x < x_{P_1} + \varepsilon\}$;

(iii) $\hat{R}_{z_0,1}$ for $z_0 \in \Gamma_{\rm shock} \cap \{x_{\rm bot} < x < x_{P_1} + \varepsilon\}$.

Then, using structure (16.1.10) and (16.1.13) of $\Omega_{2\varepsilon}$, and the properties in (16.4.41) and (16.4.46) of rectangles $\hat{R}_{z_0,\rho}$, we see that domains $Q_\rho^{(z_0)}$ for the (S,T)–variables, as defined in (16.4.43), take the following form in Cases (i)–(iii):

(i) $Q_\rho^{(z_0)} = Q_\rho$ for $\rho \in (0, \frac{1}{10}]$;

(ii) $Q_\rho^{(z_0)} = Q_\rho \cap \{T > 0\}$ for $\rho \in (0,1]$;

(iii) Using (16.1.10),

$$
Q_\rho^{(z_0)} = \Big\{(S,T) \in Q_\rho \,:\, T < F^{(z_0)}(S)\Big\} \qquad \text{for } \rho \in (0,1],
$$

where

$$
F^{(z_0)}(S) = \frac{\hat{f}(x_0 + \sqrt{x_0}\, h(x_0) S) - \hat{f}(x_0)}{h(x_0)}.
$$

Then, using (16.1.13), for any $\rho \in (0,1]$,

$$\|F^{(z_0)}\|_{C^{0,1}([-\rho,\rho])} \le C\sqrt{x_0} \le \frac{1}{10}, \tag{16.4.47}$$

where the last inequality is obtained by choosing ε_1^* sufficiently small and recalling that $x_{P_1},\varepsilon \in [0,\varepsilon_1^*]$. From (16.4.47) and since $F^{(z_0)}(0)=0$, we obtain that, for any $\rho \in (0,1]$,

$$F^{(z_0)}(S) > -\rho \qquad \text{for all } S \in (-\rho,\rho).$$

Combining this with (16.4.40) and (16.4.43), and further reducing ε_1 if necessary, we have

$$Q_\rho^{(z_0)} = \Big\{(S,T)\ :\ -\rho < S < \rho,\ -\rho < T < F^{(z_0)}(S)\Big\}.$$

3. Then, similarly to Steps 2–4 of the proof of Proposition 11.4.3, we obtain the uniformly elliptic equation for the function in (16.4.44) in $Q_\rho^{(z_0)}$ for Cases (i)–(iii), and the boundary conditions for Cases (ii)–(iii).

Specifically, using that

$$\psi_x(x,y) = \frac{g(x_0)}{\sqrt{x_0}\,h(x_0)}\psi_S^{(z_0)}(S,T), \qquad \psi_y(x,y) = \frac{g(x_0)}{h(x_0)}\psi_T^{(z_0)}(S,T)$$

for (x,y) given by (16.4.42), expressing similarly the second derivatives of ψ, substituting them into equation (16.4.32), and dividing by $\frac{g(x_0)}{[h(x_0)]^2}$, we obtain the equation for $\psi^{(x_0,y_0)}(S,T)$, which is of the form:

$$\begin{aligned}\sum_{i,j=1}^{2} & A_{ij}^{(z_0)}(D\psi^{(z_0)},\psi^{(z_0)},S,T)D_{ij}\psi^{(z_0)}\\ &+\sum_{i=1}^{2} A_i^{(z_0)}(D\psi^{(z_0)},\psi^{(z_0)},S,T)D_i\psi^{(z_0)} = 0\end{aligned} \tag{16.4.48}$$

with

$$\begin{aligned}A_{ij}^{(z_0)}(\mathbf{p},z,S,T) &= x_0^{\frac{i+j}{2}-2}A_{ij}\big(\frac{g(x_0)}{\sqrt{x_0}\,h(x_0)}p_1,\ \frac{g(x_0)}{h(x_0)}p_2,\ g(x_0)z,\ x,\ y\big),\\ A_i^{(z_0)}(\mathbf{p},z,S,T) &= h(x_0)x_0^{\frac{i}{2}-1}A_i\big(\frac{g(x_0)}{\sqrt{x_0}\,h(x_0)}p_1,\ \frac{g(x_0)}{h(x_0)}p_2,\ g(x_0)z,\ x,\ y\big),\end{aligned} \tag{16.4.49}$$

where (x,y) are given by (16.4.42).

Then, similarly to the argument in Step 2 of the proof of Proposition 11.4.3, we find that properties (11.4.17)–(11.4.18) hold in the present case, in the respective (S,T)–regions $Q_\rho^{(z_0)}$ for Cases (i)–(iii). Indeed:

(a) (16.4.33) implies (11.4.17);

(b) (16.4.34), combined with the assumptions of this lemma regarding the scaling, implies (11.4.18) in the present case: For (x, y) given by (16.4.42) with $|S|, |T| \le 1$, we obtain that, by (16.4.29),

$$|x - x_0| < \frac{d_{\rm so}(x_0)}{4} \le \frac{x_0}{4}. \tag{16.4.50}$$

Then, using that (x, y) in (16.4.49) given by (16.4.42), we find from (16.4.34) that

$$\begin{aligned}
|A_{11}^{(z_0)}(\mathbf{p}, z, S, T)| &= \frac{1}{x_0}\big|A_{11}\big(\frac{g(x_0)}{\sqrt{x_0}\,h(x_0)}p_1,\ \frac{g(x_0)}{h(x_0)}p_2,\ g(x_0)z,\ x,\ y\big)\big| \\
&\le \frac{1}{x_0}Cx \le C,
\end{aligned}$$

$$\begin{aligned}
&|D_{p_1}A_{11}^{(z_0)}(\mathbf{p}, z, S, T)| \\
&\quad = \frac{1}{x_0}\frac{g(x_0)}{\sqrt{x_0}\,h(x_0)}|D_{p_1}A_{11}\big(\frac{g(x_0)}{\sqrt{x_0}\,h(x_0)}p_1,\ \frac{g(x_0)}{h(x_0)}p_2,\ g(x_0)z,\ x,\ y\big)| \\
&\quad \le C\frac{g(x_0)}{x_0^{3/2}h(x_0)} \le C,
\end{aligned}$$

where constant C above may change at each occurrence, and we have also used (16.4.30) in the last estimate. Moreover, using (16.4.28)–(16.4.31), we obtain from (16.4.34) and (16.4.50) that

$$\begin{aligned}
&|D_T A_{11}^{(z_0)}(\mathbf{p}, z, S, T)| \\
&\quad = \frac{h(x_0)}{x_0}\Big|D_y A_{11}\big(\frac{g(x_0)}{\sqrt{x_0}\,h(x_0)}p_1,\ \frac{g(x_0)}{h(x_0)}p_2,\ g(x_0)z,\ x,\ y\big)\Big| \\
&\quad \le C\frac{h(x_0)}{x_0}\frac{x}{h(x)} \le C,
\end{aligned}$$

$$\begin{aligned}
&|D_S A_{12}^{(z_0)}(\mathbf{p}, z, S, T)| \\
&\quad = h(x_0)\Big|D_x A_{12}\big(\frac{g(x_0)}{\sqrt{x_0}h(x_0)}p_1,\ \frac{g(x_0)}{h(x_0)}p_2,\ g(x_0)z,\ x,\ y\big)\Big| \\
&\quad \le C\frac{h(x_0)}{h(x)} \le C,
\end{aligned}$$

$$|A_1^{(z_0)}(\mathbf{p}, z, S, T)| \le \frac{Ch(x_0)}{\sqrt{x}} \le C\frac{d_{\rm so}(x_0)}{4\sqrt{x_0 x}} \le C\frac{\sqrt{x_0}}{\sqrt{x}} \le C.$$

The other estimates in (11.4.18) are obtained similarly.

4. Furthermore, from (16.4.39), using (16.4.31) and (16.4.46), and recalling that $g(x)$ is an increasing function, we have

$$|\psi^{(z_0)}| \le C. \tag{16.4.51}$$

Now, repeating directly the argument in Steps 2–3 of the proof of Proposition 11.4.3, we obtain estimate (11.4.19) for Case (i), and estimate (11.4.21) for Case (ii) for each $\alpha \in (0,1)$, with constants $C(\alpha)$.

5. It remains to discuss Case (iii). The rescaled boundary conditions on Γ_{shock} for Case (iii) are the following: Since ψ satisfies (16.4.35), the function in (16.4.44) satisfies

$$\partial_S \psi^{(z_0)} = \sqrt{x_0}\, b^{(z_0)}(\partial_T \psi^{(z_0)}, \psi^{(z_0)}, S, T) \qquad \text{on } \Gamma^{(z_0)}_{\text{shock}}, \tag{16.4.52}$$

where $\Gamma^{(z_0)}_{\text{shock}}$ is given by (11.4.25),

$$b^{(z_0)}(p_2, z, S, T) = \frac{h(x_0)}{g(x_0)}\, b\big(\frac{g(x_0)}{h(x_0)} p_2,\ g(x_0)z,\ x, y\big) \qquad \text{on } \Gamma^{(z_0)}_{\text{shock}}, \tag{16.4.53}$$

and (x,y) are given by (16.4.42). From (11.3.41)–(11.3.42) on $\Omega_{2\varepsilon}$, combined with (16.4.29)–(16.4.30), it follows that $b^{(z_0)}$ on $\Gamma^{(z_0)}_{\text{shock}}$ satisfies (11.4.28).

Now, arguing as in Step 4 of the proof of Proposition 11.4.3, we obtain (11.4.29) with $\beta \in (0,1)$ and C depending only on M and the constants in conditions (16.1.13), (16.4.33)–(16.4.34), and (11.3.41).

This estimate allows us to improve the regularity of the free boundary $\Gamma^{(z_0)}_{\text{shock}}$. To see this, we denote

$$G^{(z_0)}(S,T) = \frac{1}{g(x_0)} G(x_0 + \sqrt{x_0}\, h(x_0) S,\ f(x_0) + h(x_0) T).$$

Assume that $\varepsilon \le \frac{1}{4}$ and $x_{P_1} \le \frac{1}{4}$. Then, from (16.4.38), using (16.4.30) and $x_0 \in (x_{P_1}, x_{P_1} + \varepsilon)$, we have

$$|DG^{(z_0)}| \le M \frac{h(x_0)}{g(x_0)}, \quad D_T G^{(z_0)} \le -\frac{h(x_0)}{M g(x_0)} \le -\frac{1}{M} x_0^{-\frac{3}{2}} \ \text{ in } \Omega^M_{2\varepsilon}. \tag{16.4.54}$$

From this and (11.4.29), and since $x_0 \in (x_{P_1}, x_{P_1} + \varepsilon)$, we obtain that, for sufficiently small ε and x_{P_1} (depending only on M and constants C and β in (11.4.29), hence only on M and the constants in conditions (11.3.41), (16.1.13), and (16.4.33)–(16.4.34)),

$$D_T \bar{\phi}^{(z_0)} \le -\frac{h(x_0)}{M g(x_0)} \le -\frac{1}{2M} \qquad \text{in } \Omega_{2\varepsilon}, \tag{16.4.55}$$

where

$$\bar{\phi}^{(z_0)} = G^{(z_0)} - \psi^{(z_0)}.$$

By (16.4.36),

$$\bar{\phi}^{(z_0)} = 0 \qquad \text{on } \Gamma^{(z_0)}_{\text{shock}}. \tag{16.4.56}$$

Then, using (11.4.24)–(11.4.25), we have

$$\bar{\phi}^{(z_0)} < 0 \qquad \text{in } Q^{(z_0)}_1. \tag{16.4.57}$$

Now we obtain (11.4.33). Indeed, it follows from (16.4.55)–(16.4.57) and (11.4.24)–(11.4.25).

From (11.4.33), we will obtain an improved estimate of $F^{(z_0)}$, *i.e.*, $\Gamma^{(z_0)}_{\text{shock}}$.

First, from (11.4.29), (16.4.30), and (16.4.54), we find that, for sufficiently small ε and x_{P_1},

$$\|D\bar{\phi}^{(z_0)}\|_{L^\infty(\Omega_{2\varepsilon})} \le \frac{Mh(x_0)}{2g(x_0)}. \tag{16.4.58}$$

Next, we note that

$$[DG^{(z_0)}]_{0,\beta,Q_1^{(z_0)}} \le \frac{h(x_0)}{g(x_0)}|h(x_0)|^\beta [DG]_{0,\beta,\Omega^M_{2\varepsilon}} \le C\frac{|h(x_0)|^{\beta+1}}{g(x_0)},$$

where we have assumed without loss of generality that $x_{P_1} + \varepsilon < 1$.

Combining this with (11.4.29) (obtained above for the present case), and using (11.4.33) (also obtained above for the present case), (16.4.38), (16.4.55), and (16.4.58), we have

$$\begin{aligned}
&[D_S F^{(z_0)}]_{0,\beta,[-1/2,1/2]}\\
&\le \frac{[D_S\bar{\phi}^{(z_0)}]_{0,\beta,Q_1^{(z_0)}}}{\inf_{Q_1^{(z_0)}}|D_T\bar{\phi}^{(z_0)}|} + \frac{\|D_S\bar{\phi}^{(z_0)}\|_{L^\infty(Q_1^{(z_0)})}}{(\inf_{Q_1^{(z_0)}}|D_T\bar{\phi}^{(z_0)}|)^2}[D_T\bar{\phi}^{(z_0)}]_{0,\beta,Q_1^{(z_0)}}\\
&\le C\frac{g(x_0)}{h(x_0)}[DG^{(z_0)} - D\psi^{(z_0)}]_{0,\beta,Q_1^{(z_0)}}\\
&\le C\frac{g(x_0)}{h(x_0)}\Big(\frac{|h(x_0)|^{\beta+1}}{g(x_0)} + 1\Big) \le C,
\end{aligned}$$

where we have used (16.4.29)–(16.4.30) in the last inequality. Now, using (16.4.47),

$$\|D_S F^{(z_0)}\|_{C^{1,\beta}([-1/2,1/2])} \le C,$$

which is estimate (11.4.34). Then, repeating the argument in the proof of Proposition 11.4.3 after equation (11.4.34), we obtain (11.4.36) in the present case. This completes the proof. □

Remark 16.4.4. *Let* $\varphi = \varphi_2 + \psi$ *be an admissible solution for the wedge angle* $\theta_{\rm w} \in \mathcal{I}_{\rm w}(\delta_{P_0})$, *where we have used the set in* (16.1.7) *and* δ_{P_0} *from Proposition* 16.1.4. *It follows from Corollaries* 11.2.10 *and* 11.2.12 *with Proposition* 16.1.4 *that* ψ *satisfies an equation of form* (16.4.32) *in the corresponding domain* $\Omega_{2\varepsilon}$ *in* (16.1.10) *with coefficients* $(A_{ij}^{(\rm mod)}, A_i^{(\rm mod)})(\mathbf{p}, z, x)$ *independent of* y, *satisfying the properties in* (16.4.33) *and the first two lines in* (16.4.34). *The remaining conditions in* (16.4.34) *depend on* $h(x)$, *and we will comment on them in each application of Lemma* 16.4.3. *Moreover, by Lemma* 11.3.9 *(via Proposition* 16.2.1 *and Corollary* 16.2.2*),* ψ *satisfies the boundary condition* (16.4.35) *with properties* (11.3.41)*–*(11.3.42). *We note that the constants in the properties discussed above depend only on* (ρ_0, ρ_1, γ).

Remark 16.4.5. *The admissible solution* $\varphi = \varphi_2^{(\theta_w)} + \psi$ *for the wedge angle* θ_w *satisfies* (16.4.36) *with* $G = \varphi_1 - \varphi_2^{(\theta_w)}$, *since* $\varphi = \varphi_1$ *on* Γ_{shock}. *Here* φ_1 *and* $\varphi_2^{(\theta_w)}$ *are expressed in the* (x, y)*–coordinates* (11.1.1)–(11.1.2) *for* θ_w. *We note that* $G = \varphi_1 - \varphi_2^{(\theta_w)}$ *satisfies condition* (16.4.38) *with uniform constant* M *for all* $\theta_w \in [\theta_w^d, \frac{\pi}{2}]$. *This follows from the facts that* $\varphi_1 - \varphi_2^{(\theta_w)}$ *is a linear function in the* (ξ_1, ξ_2)*–coordinates, with the parameters depending continuously on* $\theta_w \in [\theta_w^d, \frac{\pi}{2}]$, *that* $c_2^{(\theta_w)} \ge \frac{1}{C}$ *for all* θ_w, *and from property* (11.2.5) *which holds for all* $\theta_w \in [\theta_w^d, \frac{\pi}{2}]$ *by Lemma* 16.1.2.

Using Lemma 16.4.3, we prove the *a priori* estimates for admissible solutions in the *supersonic-close-to-sonic* case.

Proposition 16.4.6. *There exist* $\delta_{P_0} > 0$ *and* $\varepsilon \in (0, \frac{\varepsilon_0}{2})$ *depending only on* (ρ_0, ρ_1, γ) *and, for any* $\alpha \in (0,1)$, *there exists* $C > 0$ *depending only on* $(\rho_0, \rho_1, \gamma, \alpha)$ *such that the following estimates hold:*

If $\theta_w \in (\theta_w^d, \frac{\pi}{2})$ *is a supersonic wedge angle satisfying* (16.4.1) *and* φ *is an admissible solution of* **Problem 2.6.1** *with the wedge angle* θ_w, *then* $\psi = \varphi - \varphi_2^{(\theta_w)}$ *satisfies*

$$\|\psi\|^{(\text{par})}_{2,\alpha,\Omega_\varepsilon} \le C. \tag{16.4.59}$$

Moreover, the shock function $\hat{f}(x)$ *defined in* (11.2.38) *satisfies*

$$\|\hat{f} - \hat{f}_0^{(\theta_w)}\|^{(\text{par})}_{2,\alpha,(0,\varepsilon)} \le C, \tag{16.4.60}$$

where $\hat{f}_0^{(\theta_w)}$ *is the function defined in* (11.2.8).

Proof. In this proof, all the constants depend only on (ρ_0, ρ_1, γ), unless otherwise specified. We divide the proof into three steps.

1. Let $(\delta_{P_0}, \varepsilon)$ be as in Proposition 16.4.2 with $m = 4$. Moreover, reducing ε if necessary, we can assume that $\varepsilon \le \sigma$, where σ is from Proposition 16.3.1.

Let $\theta_w \in [\theta_w^d, \frac{\pi}{2})$ be a supersonic wedge angle satisfying (16.4.1), and let φ be an admissible solution of **Problem 2.6.1** with the wedge angle θ_w. Let $b_{so} = b_{so}(\theta_w)$ be defined by (16.4.2).

If $b_{so}^2 \ge \varepsilon$, the assertions of Proposition 16.4.6 follow from Proposition 16.3.1.

Thus, we assume that $b_{so}^2 < \varepsilon$. In this case, Proposition 16.3.1, applied with $l_{so} = b_{so}$, implies that, for any $\alpha \in (0,1)$,

$$\|\psi\|^{(\text{par})}_{2,\alpha,\Omega_{b_{so}^2}} \le C, \tag{16.4.61}$$

where C depends only on $(\rho_0, \rho_1, \gamma, \alpha)$.

2. In this step, we estimate ψ in $\Omega_\varepsilon \cap \{\frac{b_{so}^2}{5} < x < \varepsilon\}$.

From the choice of $(\delta_{P_0}, \varepsilon)$, we can apply Proposition 16.4.2 with $m = 4$ and further reduce ε if necessary to obtain the growth estimate in $\Omega_{2\varepsilon}$:

$$0 \le \psi(x,y) \le Cx^4 \qquad \text{in } \Omega_{2\varepsilon} \cap \{x > \frac{b_{so}^2(\theta_w)}{10}\}. \tag{16.4.62}$$

Moreover, since $\Omega_{2\varepsilon}$ is of the structure as in (11.2.38)–(11.2.40), there exists $k > 1$ such that

$$\{0 < x < \varepsilon,\ 0 < y - \theta_{\rm w} < b_{\rm so} + \frac{x}{k}\} \subset \Omega_{2\varepsilon}. \tag{16.4.63}$$

Now we define the rectangles for the estimates in the present case. Definition (16.4.64) is motivated by the following: From (16.4.12), for $z_0 = (x_0, y_0) \in \Omega_\varepsilon \cap \{\frac{b_{\rm so}^2}{5} < x < \varepsilon\}$ and $\rho \in (0,1]$, rectangle $R_{z_0,\rho}$ defined by (11.4.7) does not satisfy (11.4.9) in general (*i.e.*, the rectangle does not fit into Ω_ε). On the other hand, (16.4.63) holds uniformly for $k > 1$. Since $b_{\rm so}$ can be arbitrarily close to zero, then, for $z_0 = (x_0, y_0)$, we have to use a rectangle with the side length $\sim x_0$ in the y–direction in order to have a property similar to (11.4.9). Also, from the degenerate ellipticity structure (11.2.43) of the equation for regular reflection solutions, we need to have a rectangle in which the ratio of the side lengths in the x– and y–directions is $\sim \sqrt{x_0}$, in order to obtain a uniformly elliptic equation after rescaling the rectangle into a square.

Thus, for $z_0 = (x_0, y_0) \in \Omega_\varepsilon \cap \{x > \frac{b_{\rm so}^2}{5}\}$ and $\rho \in (0,1]$, we consider the rectangle:

$$\tilde{R}_{z_0,\rho} := \{(x,y)\ :\ |x - x_0| < \frac{\rho}{10k} x_0^{\frac{3}{2}},\ |y - y_0| < \frac{\rho}{10k} x_0\}, \tag{16.4.64}$$

and set

$$\hat{R}_{z_0,\rho} := \tilde{R}_{z_0,\rho} \cap \Omega_{2\varepsilon}. \tag{16.4.65}$$

Then (11.4.8) holds if ε is small. Also, using (11.2.40) and (16.4.63), and reducing ε further if necessary, we have

$$\begin{aligned} &\hat{R}_{z_w,1} \cap \Gamma_{\rm shock} = \emptyset \qquad \text{for all } z_w = (x,0) \in \Gamma_{\rm wedge} \cap \{x < \varepsilon\}, \\ &\hat{R}_{z_s,1} \cap \Gamma_{\rm wedge} = \emptyset \qquad \text{for all } z_s = (x, \hat{f}(x)) \in \Gamma_{\rm shock} \cap \{x < \varepsilon\}, \end{aligned} \tag{16.4.66}$$

i.e., the rectangles in (16.4.64) fit into domain $\Omega_{2\varepsilon}$.

Now, using Remark 16.4.4, we can apply Lemma 16.4.3 with the following corresponding ingredients:

(i) $\hat{f}(\cdot)$ and ε^* are $\hat{f}(\cdot)$ and ε in Proposition 11.2.8, respectively;

(ii) $x_{P_1} = 0$;

(iii) $h(x) = \frac{x}{10k}$ and $g(x) = \frac{x^4}{10k}$;

(iv) $x_{\rm bot} = \frac{b_{\rm so}^2}{5}$;

(v) $G = \varphi_1 - \varphi_2$.

We note that

(a) (16.4.27) and (ii) imply $d_{\rm so}(x) = x$;

(b) the functions in (iii) satisfy (16.4.28)–(16.4.31) with $M = 16$ for $d_{\rm so}(x) = x$;

(c) the potential flow equation, written as (16.4.32), satisfies (16.4.33)–(16.4.34) by Corollaries 11.2.10 and 11.2.12 with Proposition 16.1.4, and by using that $h(x) = \frac{x}{10k}$;

(d) the boundary condition (16.4.35) holds and satisfies (11.3.41)–(11.3.42) by Lemma 11.3.9 with Proposition 16.2.1 and Corollary 16.2.2;

(e) from (v), property (16.4.38) holds, as shown in Remark 16.4.5;

(f) from (iv), property (16.4.39) holds by Proposition 16.4.2 (with $m = 4$), where we have adjusted constant M depending only on (ρ_0, ρ_1, γ);

(g) (16.4.41) holds by (16.4.66).

This allows us to apply Lemma 16.4.3. Fix $\alpha \in (0,1)$. Then (16.4.45) implies that, for rectangles $\hat{R}_{z_0,\rho}$ defined by (16.4.64) and the corresponding $Q_\rho^{(z_0)}$ and $\psi^{(z_0)}$ for $z_0 = (x_0, y_0)$ defined by

$$\begin{aligned} \hat{R}_{z_0,\rho} &= \Big\{(x_0 + \frac{x_0^{3/2}}{10k}S,\, y_0 + \frac{x_0}{10k}T) \;:\; (S,T) \in Q_\rho^{(z_0)}\Big\}, \\ \psi^{(z_0)}(S,T) &:= \frac{10k}{x_0^4}\psi(x_0 + \frac{x_0^{3/2}}{10k}S,\, y_0 + \frac{x_0}{10k}T) \qquad \text{on } Q_\rho^{(z_0)}, \end{aligned} \tag{16.4.67}$$

the following estimate holds:

$$\|\psi^{(z_0)}\|_{C^{2,\alpha}(\overline{Q_{1/100}^{(z_0)}})} \le C \qquad \text{for all } z_0 \in \overline{\Omega}_\varepsilon \cap \{x > \frac{b_{\rm so}^2}{5}\}, \tag{16.4.68}$$

where C depends only on $(\rho_0, \rho_1, \gamma, \alpha)$, and ε is reduced depending only on (ρ_0, ρ_1, γ).

Denote $\mathcal{D} := \Omega_\varepsilon \cap \{x > \frac{b_{\rm so}^2}{5}\}$. From (16.4.68), we repeat the proof of Lemma 4.6.1 to obtain

$$\begin{aligned} &\sum_{0 \le k+l \le 2} \sup_{z \in \mathcal{D}} \Big(x^{\frac{3k}{2}+l-4} |\partial_x^k \partial_y^l \psi(z)| \Big) \\ &\quad + \sum_{k+l=2} \sup_{z,\tilde{z} \in \mathcal{D}, z \ne \tilde{z}} \bigg(\min(x^{\frac{3}{2}(\alpha+k)+l-4}, \tilde{x}^{\frac{3}{2}(\alpha+k)+l-4}) \\ &\qquad\qquad \times \frac{|\partial_x^k \partial_y^l \psi(z) - \partial_x^k \partial_y^l \psi(\tilde{z})|}{\delta_\alpha^{(\rm par)}(z, \tilde{z})} \bigg) \le C, \end{aligned} \tag{16.4.69}$$

where k and l are nonnegative integers, C depends only on constant C in (16.4.68) and α (hence only on $(\rho_0, \rho_1, \gamma, \alpha)$), and $\delta_\alpha^{(\rm par)}(z, \tilde{z})$ is defined by (4.6.1).

Comparing the left-hand side of (16.4.69) with the expression of norm $\|\psi\|_{2,\alpha,\mathcal{D}}^{2,(\rm par)}$ in (4.6.2) with $m = \sigma = 2$, we see that, in both expressions, each term (the derivative or the difference quotient of the derivatives of ψ) is multiplied by x^β and, for each term, exponent $\beta = \beta(k, l, \alpha)$ is larger in the expression of $\|\psi\|^{2,(\mathrm{par})}_{2,\alpha,\mathcal{D}}$ than the one in the corresponding term on the left-hand side of (16.4.69). For $\varepsilon \leq 1$, it follows that the left-hand side of (16.4.69) is larger than $\|\psi\|^{2,(\mathrm{par})}_{2,\alpha,\mathcal{D}}$. Therefore, we obtain

$$\|\psi\|^{2,(\mathrm{par})}_{2,\alpha,\Omega_\varepsilon\cap\{x>b^2_{\mathrm{so}}/5\}} \leq C, \tag{16.4.70}$$

where C is from (16.4.69).

3. We now show that (16.4.70), combined with (16.4.61), implies (16.4.59). Indeed, using expression (4.6.2) with $m = \sigma = 2$, we see that (16.4.61) and (16.4.70) imply

$$\|\psi\|^{2,(\mathrm{par})}_{2,0,\Omega_\varepsilon} \leq C. \tag{16.4.71}$$

Therefore, it remains to show that there exists C depending only on the data and α such that, for any nonnegative integers k and l with $k + l = 2$, and for any $z, \tilde{z} \in \Omega_\varepsilon$ with $z = (x, y)$ and $\tilde{z} = (\tilde{x}, \tilde{y})$,

$$\min(x^{\alpha+k+\frac{l}{2}-2}, \tilde{x}^{\alpha+k+\frac{l}{2}-2}) \frac{|\partial_x^k\partial_y^l u(z) - \partial_x^k\partial_y^l u(\tilde{z})|}{\delta_\alpha^{(\mathrm{par})}(z,\tilde{z})} \leq C. \tag{16.4.72}$$

If $z, \tilde{z} \in \Omega_{b^2_{\mathrm{so}}}$, or $z, \tilde{z} \in \Omega_\varepsilon \cap \{x > \frac{b^2_{\mathrm{so}}}{5}\}$, (16.4.72) is obtained in (16.4.61) and (16.4.70), respectively.

Therefore, it remains to consider the case:

$$x > b^2_{\mathrm{so}} > \frac{b^2_{\mathrm{so}}}{5} > \tilde{x}.$$

Then $x - \tilde{x} \geq \frac{x}{2}$ so that

$$\delta_\alpha^{(\mathrm{par})}(z,\tilde{z}) \geq \frac{x^\alpha}{2^\alpha} \geq \frac{\tilde{x}^\alpha}{2^\alpha}.$$

This estimate is similar to (4.6.9). Then we conclude estimate (16.4.72) by following the argument in the proof of Lemma 4.6.1 starting from (4.6.9) and by using (16.4.71). Now (16.4.59) is proved.

Then (16.4.60) follows as in Step 5 of the proof of Proposition 11.4.3. □

16.4.4 Uniform estimates for supersonic reflection solutions

First, we state the estimates near Γ_{sonic}:

Corollary 16.4.7. *There exists $\varepsilon > 0$ depending only on (ρ_0, ρ_1, γ) and, for any $\alpha \in (0,1)$, there exists $C > 0$ depending only on $(\rho_0, \rho_1, \gamma, \alpha)$ such that, if $\theta_{\mathrm{w}} \in (\theta^{\mathrm{d}}_{\mathrm{w}}, \frac{\pi}{2})$ is a supersonic wedge angle and φ is an admissible solution of* **Problem 2.6.1** *with the wedge angle θ_{w}, then $\psi = \varphi - \varphi_2^{(\theta_{\mathrm{w}})}$ and the shock function $\hat{f}(x)$ from* (11.2.38) *satisfy estimates* (16.4.59)–(16.4.60).

Proof. We combine Proposition 16.4.6 with Proposition 16.3.2 in which $\delta = \delta_{P_0}$ is chosen, where δ_{P_0} is determined by Proposition 16.4.6. Then $\delta = \delta_{P_0}$ depends only on (ρ_0, ρ_1, γ) so that constants ε and C in the estimates of Proposition 16.3.2 depend only on (ρ_0, ρ_1, γ). □

We also obtain the following global estimate for all the supersonic reflections with the wedge angles $\theta_{\rm w} \in (\theta_{\rm w}^{\rm c}, \frac{\pi}{2})$ for $\theta_{\rm w}^{\rm c}$ from Definition 15.7.3:

Corollary 16.4.8. *Let $\theta_{\rm w}^* \in (\theta_{\rm w}^{\rm c}, \frac{\pi}{2})$. Let α from Lemma* 10.5.2 *be extended to all $\theta_{\rm w} \in [\theta_{\rm w}^*, \frac{\pi}{2})$ by Proposition* 15.7.5. *Let $\varepsilon \in (0, \varepsilon_0]$. Then there exists C depending only on the data and ε such that, if φ is an admissible solution of* **Problem 2.6.1** *with the supersonic wedge angle $\theta_{\rm w} \in [\theta_{\rm w}^*, \frac{\pi}{2})$, then*

$$
\begin{aligned}
&\varphi^{\rm ext} \in C^{1,\alpha}(\overline{\Omega^{\rm ext}}) \cap C^{1,1}(\overline{\Omega^{\rm ext}} \setminus \{P_3\}) \cap C^{\infty}(\overline{\Omega^{\rm ext}} \setminus (\overline{\Gamma^{\rm ext}_{\rm sonic}} \cup \{P_3\})),\\
&\|\varphi\|^{(-1-\alpha),\{P_3\}}_{2,\alpha,\Omega\setminus\overline{\Omega_{\varepsilon/10}}} + \|\varphi - \varphi_2\|^{(\rm par)}_{2,\alpha,\Omega_\varepsilon} \le C.
\end{aligned}
\tag{16.4.73}
$$

Furthermore, the shock function $f_{O_1,{\rm sh}}$ for $\Gamma^{\rm ext}_{\rm shock}$, introduced in Corollary 10.1.3 *and Proposition* 15.7.1, *satisfies*

$$
\begin{aligned}
&f_{O_1,{\rm sh}} \in C^{1,1}([\theta_{P_1}, \theta_{P_1}-]) \cap C^{\infty}((\theta_{P_1}, \theta_{P_1}-)),\\
&\|f_{O_1,{\rm sh}}\|_{C^{1,1}([\theta_{P_1},\theta_{P_1}-])} \le \hat{C},
\end{aligned}
\tag{16.4.74}
$$

where $\hat{C}$ depends only on the data and $\theta_{\rm w}^*$.

Proof. We use Corollary 16.4.7 for the estimates near $\Gamma_{\rm shock}$. Then the proof of Corollary 11.4.7 applies directly by employing Proposition 16.3.2 and the other results which extend the results of the previous chapters to $\theta_{\rm w} \in [\theta_{\rm w}^{\rm d}, \frac{\pi}{2}]$. □

16.5 A PRIORI ESTIMATES NEAR THE REFLECTION POINT IN THE SUBSONIC-NEAR-SONIC CASE

Now we consider the subsonic reflection configuration. First we consider the *subsonic-near-sonic* case to obtain the estimates near

$$\overline{\Gamma_{\rm sonic}} = P_0 = P_1 = P_2$$

for all the admissible solutions with $\theta_{\rm w} \in (\theta_{\rm w}^{\rm d}, \frac{\pi}{2})$ satisfying

$$\frac{|D\varphi_2(P_0)|}{c_2}(\theta_{\rm w}) \in (1 - \delta_{P_0}, 1], \tag{16.5.1}$$

where $\delta_{P_0} \in (0, \sigma)$, and $\sigma > 0$ is from Corollary 16.2.2 which will be determined below by reducing constant δ_{P_0} in Proposition 16.4.6, if necessary, so that the resulting constant works both in the cases of §16.4 and the present section.

16.5.1 Main steps of the estimates near the reflection point in the subsonic-near-sonic case

We argue similarly to the *supersonic-near-sonic* case in §16.4, with different scaling based on the different shape of domain Ω_ε in the present case. We also note that, for each fixed θ_w such that $\frac{|D\varphi_2(P_0)|}{c_2}(\theta_w) \in (1-\delta_{P_0}, 1)$, the equation is uniformly elliptic on Ω_ε. However, we do not have a uniform positive ellipticity constant for all the wedge angles θ_w satisfying (16.5.1), so that we cannot use the uniform ellipticity, since we want to obtain the uniform estimates for all the admissible solutions corresponding to these wedge angles. In fact, we use that the ellipticity constant is small near P_0 for the wedge angles θ_w satisfying (16.5.1).

The outline of the estimates in this section is the following:

(i) From (16.1.10)–(16.1.13), in the subsonic-sonic case, domain Ω_ε in the (x, y)–coordinates is of the form:

$$\begin{aligned} &\{0 < x - x_{P_0} < \varepsilon,\ 0 < y - \theta_w < \frac{1}{k} d_{so}(x)\} \subset \Omega_\varepsilon, \\ &\Omega_\varepsilon \subset \{0 < x - x_{P_0} < \varepsilon,\ 0 < y - \theta_w < k d_{so}(x)\}, \end{aligned} \tag{16.5.2}$$

where $k > 1$ can be chosen uniformly for any subsonic-sonic reflection solution, $P_1 = P_0$ in this case, and

$$d_{so}(x) := x - x_{P_1} = x - x_{P_0} \tag{16.5.3}$$

introduced in (16.4.27). Then we show that, for each integer $m \geq 3$ and sufficiently small b_{so} depending on $(\rho_0, \rho_1, \gamma, m)$, solution $\psi(x, y)$ has a growth $|d_{so}(x)|^m$ in Ω_ε:

$$0 \leq \psi(x, y) \leq C|d_{so}(x)|^m \qquad \text{in } \Omega_\varepsilon$$

for some $\varepsilon > 0$ and C depending only on the data and m. In the proof, we use shape (16.5.2) of Ω_ε and the fact that the ellipticity constant is small near P_0.

(ii) Using the algebraic growth, with $m = 5$ (for a reason to be explained below), we can obtain the estimates of ψ in Ω_ε by using a different scaling from that in Propositions 16.3.1 and (16.4.6). Indeed, from (16.5.2), for $z_0 = (x_0, y_0) \in \overline{\Omega_\varepsilon} \setminus \{P_0\}$, we have to use a rectangle with the side length $\sim \frac{1}{k} d_{so}(x_0)$ in the y–direction in order to have a property similar to (11.4.9). Also, from the degenerate ellipticity structure (11.2.43) of the equation for subsonic reflection solutions, we need to have a rectangle in which the ratio of the side lengths in the x– and y–directions is $\sim \sqrt{x_0}$, in order to obtain a uniformly elliptic equation after rescaling the rectangle into a square.

Thus, for $z_0 = (x_0, y_0) \in \overline{\Omega_\varepsilon} \setminus \{P_0\}$, we consider the rectangle:

$$\hat{R}_{z_0} := \{|x - x_0| < \frac{\sqrt{x_0}}{10k} d_{so}(x_0),\ |y - y_0| < \frac{1}{10k} d_{so}(x_0)\} \cap \Omega.$$

Such rectangles *fit into* Ω; that is, for $z_0 = (x_0, y_0) \in \Gamma_{\text{wedge}} \cap \Omega_\varepsilon$, the corresponding $\hat{R}_{z_0}$ does not intersect with Γ_{shock} and, for $(x_0, y_0) \in \Gamma_{\text{shock}} \cap \Omega_\varepsilon$, the corresponding $\hat{R}_{z_0}$ does not intersect with Γ_{wedge}. Also, note that the ratio of the $x-$ and y–directions of $\hat{R}_{z_0}$ is $\sqrt{x_0}$, as required in Lemma 16.4.3. Rescaling $\hat{R}_{z_0}$ to the portion of square $(-1,1)^2$:

$$\hat{Q}_{z_0} := \{(S,T) \in (-1,1)^2 : \, (x_0 + \frac{\sqrt{x_0}}{10k} d_{\text{so}}(x_0) S, y_0 + \frac{1}{10k} d_{\text{so}}(x_0) T) \in \Omega\},$$

we consider the following function in $(S,T) \in \hat{Q}_{z_0}$:

$$\psi^{(z_0)}(S,T) := \frac{1}{|d_{\text{so}}(x_0)|^5} \psi(x_0 + \frac{\sqrt{x_0}}{10k} d_{\text{so}}(x_0) S, y_0 + \frac{1}{10k} d_{\text{so}}(x_0) T).$$

The algebraic growth of ψ with $m = 5$ implies the uniform estimates in $L^\infty(\hat{Q}_{z_0})$ of $\psi^{(z_0)}(\cdot)$ for all $z_0 \in \Omega_\varepsilon$. Then, from Lemma 16.4.3 with $G = \varphi_1 - \varphi_2$, we obtain the estimates of $\psi^{(z_0)}$ in $C^{2,\alpha}$, independent of $z_0 \in \Omega_\varepsilon$. This implies the estimates of ψ in Ω_ε in the weighted and scaled $C^{2,\alpha}$–norms for any $\alpha \in (0,1)$, which are stronger than the standard $C^{2,\alpha}(\overline{\Omega_\varepsilon})$–norm with $D\psi(P_0) = 0$ and $D^2\psi(P_0) = 0$.

16.5.2 Algebraic growth of ψ in Ω_ε in the subsonic-near-sonic case

Similarly to §16.4.2 for the growth estimates of the solution, we consider equation (11.1.4)–(11.1.5) as a linear equation (16.4.4) which is obtained by plugging ψ into the coefficients of (11.1.4). Equation (16.4.4) is strictly elliptic in $\overline{\Omega_\varepsilon} \setminus \{P_0\}$ by the properties of subsonic admissible solutions in Definition 15.1.2(iii).

By Proposition 16.2.1, the boundary condition for ψ on Γ_{shock} is (11.3.33), and $B_1(\mathbf{p}, z, x, y)$ satisfies (11.3.35). Similarly to that in §16.4.2, we can write (11.3.33) as a linear condition (16.4.6).

Choosing $(\varepsilon, \delta_{P_0})$ and employing Propositions 16.1.4 and 16.2.1 and Corollary 16.2.2 similarly as in §16.4.2, we obtain the bounds in (16.4.5) in Ω_ε, and estimates (16.4.7)–(16.4.8) on $\Gamma_{\text{shock}} \cap \partial\Omega_\varepsilon$ in the present case, where C depends only on (ρ_0, ρ_1, γ).

Also, similarly to that in §16.4.2, ψ satisfies (16.4.9) and (16.4.11), as well as (16.4.10) with $\overline{\Gamma_{\text{sonic}}} = \{P_0\}$.

First, we prove the algebraic growth estimate for the solutions of the linear problem (16.4.4), (16.4.6), and (16.4.9)–(16.4.10) in a domain of the appropriate structure, especially satisfying (16.5.2).

Lemma 16.5.1. *Let $\varepsilon > x_{P_1} \geq 0$ and $\theta_{\text{w}} \in \mathbb{R}$. Let domain Ω_ε be of the structure as in* (16.1.10) *with*

$$\hat{f}(x_{P_1}) = \theta_{\text{w}}, \tag{16.5.4}$$

and let (16.1.13) *hold. Let $\psi \in C(\overline{\Omega_\varepsilon}) \cap C^1(\overline{\Omega_\varepsilon} \cap \{x_{P_1} < x < \varepsilon\}) \cap C^2(\Omega_\varepsilon)$ satisfy* (16.4.4), (16.4.6), *and* (16.4.9)–(16.4.10). *Let equation* (16.4.4) *be strictly elliptic in $\overline{\Omega_\varepsilon} \setminus \Gamma_{\text{sonic}}$, and let the boundary condition* (16.4.6) *be strictly oblique* *on* $\Gamma_{\text{shock}} \cap \partial\Omega_\varepsilon$ *in the sense of* (16.4.8). *Let* (16.4.5) *and* (16.4.7) *hold. Then, for every integer* $m \geq 3$, *there exist* $\hat{\varepsilon} > 0$ *and* $\hat{C} > 0$ *depending only on* m *and the constants in* (16.1.13), (16.4.5), *and* (16.4.7) *such that, if* $\varepsilon \leq \hat{\varepsilon}$,

$$\psi(x,y) \leq \hat{C}\frac{\max_{\partial\Omega_\varepsilon \cap \{x=x_{P_1}+\varepsilon\}}\psi}{\varepsilon^m}(x-x_{P_1})^m \qquad \text{in } \Omega_\varepsilon. \tag{16.5.5}$$

Proof. We note that the structure of Ω_ε, as described in (16.1.10), (16.1.13), and (16.5.4), implies that (16.5.2) holds with k depending only on the constants in (16.1.13). Fix integer $m \geq 3$. In the proof, constants C, L, and $\hat{\varepsilon}$ depend only on m and the constants in (16.1.13), (16.4.5), and (16.4.7). Also, $L, M, N \geq 1$. We divide the proof into three steps.

1. It will be convenient to shift the y–coordinate by replacing $y - \theta_{\text{w}}$ by y. This does not affect the conditions of this lemma. In fact, this means that we can assume without loss of generality that

$$\theta_{\text{w}} = 0.$$

Then (16.5.2) is replaced by

$$\begin{aligned} &\{0 < x - x_{P_0} < \varepsilon_0,\ 0 < y < \frac{1}{k}(x - x_{P_1})\} \subset \Omega_{\varepsilon_0}, \\ &\Omega_{\varepsilon_0} \subset \{0 < x - x_{P_0} < \varepsilon_0,\ 0 < y < k(x - x_{P_1})\}. \end{aligned} \tag{16.5.6}$$

To prove this lemma, we show that, if $L > 0$ is sufficiently large and $\varepsilon > 0$ is sufficiently small, then, for any $\mathbf{x}_{P_1} \in [0, \varepsilon)$, the function:

$$v(x,y) = (x - x_{P_1})^m - \frac{1}{L}(x - x_{P_1})^{m-2}y^2 \tag{16.5.7}$$

is a positive supersolution of equation (16.4.4) in Ω_ε with the boundary condition (16.4.6) on $\Gamma_{\text{shock}} \cap \partial\Omega_\varepsilon$. From this, we show that $\psi \leq Bv$ in Ω_ε for sufficiently large B depending on the data and (ε, m). This implies (16.5.5).

2. We now employ (16.5.6) to estimate that, if $L \geq 2k^2$,

$$v(x,y) \geq (x - x_{P_1})^m - \frac{k^2}{L}(x - x_{P_1})^m \geq \frac{1}{2}(x - x_{P_1})^m \quad \text{for } (x,y) \in \Omega_\varepsilon. \tag{16.5.8}$$

Using (16.5.6), similar estimates imply that, for $(x,y) \in \Omega_\varepsilon$,

$$\begin{aligned} v_x(x,y) &= (x - x_{P_1})^{m-3}\Big(m(x - x_{P_1})^2 - \frac{m-2}{L}y^2\Big) \\ &\geq (x - x_{P_1})^{m-3}\Big(m(x - x_{P_1})^2 - \frac{m-2}{L}k^2(x - x_{P_1})^2\Big). \end{aligned}$$

It follows by choosing $L \geq \frac{2(m-2)}{m}k^2$ that

$$v_x(x,y) \geq \frac{m}{2}(x - x_{P_1})^{m-1} > 0. \tag{16.5.9}$$

Similarly, we have

$$\begin{aligned}
&v_{xx}(x,y)\\
&= (x-x_{P_1})^{m-4}\Big(m(m-1)(x-x_{P_1})^2 - \frac{(m-2)(m-3)}{L}y^2\Big)\\
&\geq (x-x_{P_1})^{m-4}\Big(m(m-1)(x-x_{P_1})^2 - k^2\frac{(m-2)(m-3)}{L}(x-x_{P_1})^2\Big)\\
&\geq \frac{m(m-1)}{2}(x-x_{P_1})^{m-2} \geq 0,
\end{aligned}$$

by further increasing L. Then

$$|v_{xx}(x,y)| \leq m(m-1)(x-x_{P_1})^{m-2}. \qquad (16.5.10)$$

Next, we estimate (v_{xy}, v_{yy}, v_y): For $(x,y) \in \Omega_\varepsilon$, using (16.5.6) which implies especially that $0 \leq y \leq k(x-x_{P_1})$, we have

$$|v_{xy}(x,y)| = \Big|-2\frac{m-2}{L}(x-x_{P_1})^{m-3}y\Big| \leq C(x-x_{P_1})^{m-2}, \qquad (16.5.11)$$

$$v_{yy}(x,y) = -\frac{2}{L}(x-x_{P_1})^{m-2}, \qquad (16.5.12)$$

$$|v_y(x,y)| = \Big|-\frac{2}{L}(x-x_{P_1})^{m-2}y\Big| \leq \frac{C}{L}(x-x_{P_1})^{m-1}. \qquad (16.5.13)$$

3. Now, combining the estimates of (v, Dv, D^2v) above with (16.4.5) and (16.4.7), we show that v is a supersolution of equation (16.4.4) with the boundary condition (16.4.6).

First, we consider equation (16.4.4). Using (16.4.5), (16.4.16), and the estimates of (v, Dv, D^2v), we notice that, for $(x,y) \in \Omega_\varepsilon$,

(i) $a_{22}(x,y)v_{yy}(x,y) \leq -\frac{1}{CL}(x-x_{P_1})^{m-2}$;

(ii) $a_1(x,y)v_x(x,y) \leq 0$;

(iii) The absolute value of the other terms of $\mathcal{L}v$ is estimated by $C\varepsilon(x-x_{P_1})^{m-2}$:

$$|a_{11}(x,y)v_{xx}(x,y)| \leq Cx(x-x_{P_1})^{m-2} \leq C\varepsilon(x-x_{P_1})^{m-2},$$

$$|a_{12}(x,y)v_{xy}(x,y)| \leq Cx(x-x_{P_1})^{m-2} \leq C\varepsilon(x-x_{P_1})^{m-2},$$

$$|a_2(x,y)v_y(x,y)| \leq \frac{C}{N}x(x-x_{P_1})^{m-1} \leq C\varepsilon(x-x_{P_1})^{m-2},$$

so that

$$\mathcal{L}v \leq C\varepsilon(x-x_{P_1})^{m-2} - \frac{1}{CL}(x-x_{P_1})^{m-2} < 0, \qquad (16.5.14)$$

if $\varepsilon > 0$ is small, *i.e.*, $\varepsilon \in (0, \hat{\varepsilon})$ where $\hat{\varepsilon} > 0$ depends only on (C, L). Therefore, we have shown that, under the conditions on (ε, L) given above, v is a supersolution of equation (16.4.4).

Next we consider the boundary condition (16.4.6). For $(x,y) \in \Gamma_{\text{shock}} \cap \partial\Omega_\varepsilon$, we employ (16.4.7) and estimates (16.5.9)–(16.5.13) of Dv, and note that $b_3 v \le 0$ to obtain

$$\mathcal{N}v \le -\frac{1}{C}(x - x_{P_1})^{m-1} + \frac{C}{L}(x - x_{P_1})^{m-1} < 0,$$

by increasing L. At this point, we fix L based on all the previous conditions, and note that L is chosen depending only on (m,k) and the constants in (16.4.5) and (16.4.7), and k depends only on the constants in (16.1.13). Thus, L is chosen depending on the data in the problem and m. This also fixes the choice of $\hat{\varepsilon}$, which has been made after equation (16.5.14), so that $\hat{\varepsilon}$ depends only on the data and m.

If $\varepsilon \in (0, \hat{\varepsilon})$, we choose

$$B = \max\big\{0, \frac{2}{\varepsilon^m} \max_{\partial\Omega_\varepsilon \cap \{x = x_{P_1} + \varepsilon\}} \psi\big\} \ge 0,$$

and use (16.5.8) to obtain

$$Bv \ge \psi \qquad \text{on } \partial\Omega_\varepsilon \cap \{x = x_{P_1} + \varepsilon\}.$$

Also, employing the explicit expression of v, structure (16.1.10) and (16.5.4) of the domain with $x_{P_0} = x_{P_1} \in [0, \varepsilon)$, and the fact that ψ satisfies the boundary conditions (16.4.9)–(16.4.10), we have

$$\begin{aligned} &Bv \ge 0 = \psi && \text{at } \overline{\Gamma_{\text{sonic}}} = \{P_0\} \subset \{x = x_{P_1}\}, \\ &v_y = 0 = \psi_y && \text{on } \Gamma_{\text{wedge}} \subset \{y = 0\}, \end{aligned}$$

where we have used the shifted coordinates, *i.e.*, $\theta_{\text{w}} = 0$. Therefore, from the maximum principle by using the strict ellipticity of equation (16.4.4), and the strict obliqueness of the boundary condition (16.4.6) with the property that $b_3 < 0$, it follows that $Bv \ge \psi$ in Ω_ε. Then (16.5.5) is proved. □

Now we apply Lemma 16.5.1 to the estimates for the subsonic reflection solutions.

Proposition 16.5.2. *For every integer $m \ge 3$, there exist small $\delta_{P_0}, \varepsilon > 0$ and large C, depending only on $(m, \rho_0, \rho_1, \gamma)$, such that, if $\theta_{\text{w}} \in [\theta_{\text{w}}^{\text{d}}, \frac{\pi}{2})$ satisfies (16.5.1) and φ is an admissible solution of* **Problem 2.6.1** *with the wedge angle θ_{w}, then*

$$x_{P_1} \le \frac{\varepsilon}{10}, \tag{16.5.15}$$

$$0 \le \psi(x,y) \le C(x - x_{P_1})^m \qquad \text{in } \Omega_\varepsilon. \tag{16.5.16}$$

Proof. It suffices to show (16.5.15) and the upper bound of ψ in (16.5.16), since $\psi = \varphi - \varphi_2 \ge 0$ in Ω by (15.1.5).

Fix integer $m \geq 3$. Fix $\delta_{P_0}, \varepsilon > 0$ depending only on (ρ_0, ρ_1, γ) such that Propositions 16.1.4 and 16.2.1 and Corollary 16.2.2 apply.

Then, for every admissible solution φ with $\theta_{\rm w} \in \mathcal{I}_{\rm w}(\delta_{P_0})$, region Ω_ε is of structure (16.1.10)–(16.1.13), and $\psi = \varphi - \varphi_2$ satisfies a linear strictly elliptic equation (16.4.4) in Ω_ε with a linear oblique boundary condition (11.3.33) on $\Gamma_{\rm shock} \cap \partial\Omega_\varepsilon$, and the coefficients satisfy (16.4.5) and (16.4.7) with the constants depending only on (ρ_0, ρ_1, γ). Then let $\hat{\varepsilon}$ be the constant from Lemma 16.5.1 determined by m and the constants in estimates (16.1.13), (16.4.5), and (16.4.7) for admissible solutions with wedge angles $\theta_{\rm w} \in \mathcal{I}_{\rm w}(\delta_{P_0})$. We replace ε chosen above by $\min\{\varepsilon, \hat{\varepsilon}\}$. Using property (16.1.9) of $x^*_{P_0}(\delta)$ defined by (16.1.8) and reducing $\delta_{P_0} > 0$ if necessary, depending only on $(\rho_0, \rho_1, \gamma, m)$, we have

$$x_{P_1}(\theta_{\rm w}) \leq \frac{\varepsilon}{10} \qquad \text{for each } \theta_{\rm w} \in \mathcal{I}_{\rm w}(\delta_{P_0}).$$

Then (16.5.15) holds for each $\theta_{\rm w} \in \mathcal{I}_{\rm w}(\delta_{P_0})$, especially for each $\theta_{\rm w} \in [\theta_{\rm w}^{\rm d}, \frac{\pi}{2})$ satisfying (16.5.1). Also, if φ is an admissible solution for the wedge angle $\theta_{\rm w} \in [\theta_{\rm w}^{\rm d}, \frac{\pi}{2})$ satisfying (16.5.1), then, from (16.1.11) in the subsonic case, we see that (16.5.4) holds. Thus, we can apply Lemma 16.5.1 to obtain (16.5.16) for $\psi = \varphi - \varphi_2$, with constant C depending only on $(\rho_0, \rho_1, \gamma, m)$, since the constants in (16.1.13), (16.4.5), and (16.4.7) depend only on (ρ_0, ρ_1, γ). □

16.5.3 Regularity and uniform a priori estimates of solutions in the subsonic-close-to-sonic case

Proposition 16.5.3. *There exist $\delta_{P_0} > 0$ and $\varepsilon \in (0, \frac{\varepsilon_0}{2})$ depending only on (ρ_0, ρ_1, γ) and, for any $\alpha \in (0,1)$, there exists $C > 0$ depending only on $(\rho_0, \rho_1, \gamma, \alpha)$ such that the following estimates hold: If $\theta_{\rm w} \in [\theta_{\rm w}^{\rm d}, \frac{\pi}{2})$ is a wedge angle satisfying* (16.5.1) *and φ is an admissible solution of* **Problem 2.6.1** *with wedge angle $\theta_{\rm w}$, then $\psi = \varphi - \varphi_2^{(\theta_{\rm w})}$ satisfies*

$$\begin{aligned} &\|\psi\|_{C^{2,\alpha}(\overline{\Omega_\varepsilon})} \leq C, \\ &D^m\psi(P_0) = 0 \qquad \textit{for } m = 0, 1, 2, \end{aligned} \tag{16.5.17}$$

where the last assertion, for $m = 0, 1$, is actually given by (15.1.3). *Moreover, the shock function $\hat{f}(x)$ from* (16.1.10) *satisfies*

$$\begin{aligned} &\|\hat{f} - \hat{f}_0^{(\theta_{\rm w})}\|_{C^{2,\alpha}([x_{P_0}, x_{P_0}+\varepsilon])} \leq C, \\ &\hat{f}(x_{P_0}) = \hat{f}_0^{(\theta_{\rm w})}(x_{P_0}), \;\; \hat{f}'(x_{P_0}) = (\hat{f}_0^{(\theta_{\rm w})})'(x_{P_0}), \\ &\hat{f}''(x_{P_0}) = (\hat{f}_0^{(\theta_{\rm w})})''(x_{P_0}), \end{aligned} \tag{16.5.18}$$

where $\hat{f}_0^{(\theta_{\rm w})}$ is the function defined by (11.2.8) *by using Lemma* 16.1.2.

Proof. In this proof, all the constants depend only on (ρ_0, ρ_1, γ), unless otherwise specified. We divide the proof into two steps.

1. Let δ_{P_0} and ε be as in Proposition 16.5.2 with $m = 5$. Then we apply Proposition 16.5.2 with $m = 5$ to obtain the growth estimate in $\Omega_{2\varepsilon}$ by further reducing ε if necessary:

$$0 \leq \psi(x, y) \leq Cx^5 \qquad \text{in } \Omega_{2\varepsilon}. \tag{16.5.19}$$

Moreover, since $\Omega_{2\varepsilon}$ is of the structure as in (16.1.10)–(16.1.13) in the subsonic-sonic case in (16.1.11)–(16.1.12) by (16.5.1), there exists $k > 1$ such that (16.5.2) holds, where $d_{\rm so}(\cdot)$ is defined by (16.5.3).

2. As we have discussed in §16.5.1, for each $z_0 = (x_0, y_0) \in \overline{\Omega_\varepsilon} \setminus \overline{\Gamma_{\rm sonic}}$, we consider the rectangle:

$$\begin{aligned}
&\tilde{R}_{z_0,\rho} := \{|x - x_0| < \frac{\rho\sqrt{x_0}}{10k} d_{\rm so}(x_0),\ |y - y_0| < \frac{\rho}{10k} d_{\rm so}(x_0)\}, \\
&\hat{R}_{z_0,\rho} := \tilde{R}_{z_0,\rho} \cap \Omega_{2\varepsilon}.
\end{aligned} \tag{16.5.20}$$

Then (16.4.46) holds if ε is small. Also, using (16.5.2) and reducing ε further if necessary, we have

$$\begin{aligned}
&\hat{R}_{z_w,1} \cap \Gamma_{\rm shock} = \emptyset \qquad \text{for all } z_w = (x, 0) \in \Gamma_{\rm wedge} \cap \partial\Omega_\varepsilon, \\
&\hat{R}_{z_s,1} \cap \Gamma_{\rm wedge} = \emptyset \qquad \text{for all } z_s = (x, \hat{f}(x)) \in \Gamma_{\rm shock} \cap \partial\Omega_\varepsilon,
\end{aligned} \tag{16.5.21}$$

i.e., the rectangles in (16.4.64) fit into domain $\Omega_{2\varepsilon}$.

Now, using Remark 16.4.4, we can apply Lemma 16.4.3 with the following corresponding ingredients:

(i) $\hat{f}(\cdot)$ and ε^* are function $\hat{f}(\cdot)$ and constant ε in Proposition 16.1.4, respectively;

(ii) $x_{P_1} \geq 0$ is determined by $\theta_{\rm w}$;

(iii) Functions $h(x)$ and $g(x)$ are

$$\begin{aligned}
h(x) &= \frac{1}{10k} d_{\rm so}(x) = \frac{1}{10k}(x - x_{P_1}), \\
g(x) &= \frac{1}{10k}\big(d_{\rm so}(x)\big)^5 = \frac{1}{10k}(x - x_{P_1})^5;
\end{aligned}$$

(iv) $x_{\rm bot} = x_{P_1}$;

(v) $G = \varphi_1 - \varphi_2$.

We note that

(a) the functions in (iii) satisfy (16.4.28)–(16.4.31) with $M = 32$, since $d_{\rm so}(x) = x - x_{P_1}$;

(b) the potential flow equation, written as (16.4.32), satisfies (16.4.33)–(16.4.34) by using Corollaries 11.2.10 and 11.2.12, Proposition 16.1.4, and $h(x) = \frac{1}{10k}(x - x_{P_1})$;

(c) the boundary condition (16.4.35) holds and satisfies (11.3.41)–(11.3.42) by Lemma 11.3.9 with Proposition 16.2.1 and Corollary 16.2.2;

(d) from (v), property (16.4.38) holds, as shown in Remark 16.4.5;

(e) from (iv), property (16.4.39) holds by Proposition 16.5.2 (with $m = 5$), where we have adjusted constant M depending only on (ρ_0, ρ_1, γ);

(f) (16.4.41) holds by (16.5.21).

This allows us to apply Lemma 16.4.3. Fix $\alpha \in (0,1)$. Then (16.4.45) implies that, for rectangles $\hat{R}_{z_0,\rho}$ defined by (16.5.20) and the corresponding $Q_\rho^{(z_0)}$ and $\psi^{(z_0)}$ for $z_0 = (x_0, y_0)$ defined by

$$\hat{R}_{z_0,\rho} = \Big\{\big(x_0 + \frac{\sqrt{x_0}}{10k} d_{\rm so}(x_0)S, y_0 + \frac{1}{10k} d_{\rm so}(x_0)T\big) \ : \ (S,T) \in Q_\rho^{(z_0)}\Big\}, \tag{16.5.22}$$

$$\psi^{(z_0)}(S,T) := \frac{10k}{|d_{\rm so}(x_0)|^5}\psi\big(x_0 + \frac{\sqrt{x_0}}{10k} d_{\rm so}(x_0)S, y_0 + \frac{1}{10k} d_{\rm so}(x_0)T\big) \quad \text{on } Q_\rho^{(z_0)}, \tag{16.5.23}$$

the estimate holds:

$$\|\psi^{(z_0)}\|_{C^{2,\alpha}(\overline{Q_{1/100}^{(z_0)}})} \le C \qquad \text{for all } z_0 \in \overline{\Omega_\varepsilon}, \tag{16.5.24}$$

where C depends on $(\rho_0, \rho_1, \gamma, \alpha)$, if ε and x_{P_1} are small depending only on (ρ_0, ρ_1, γ). In order to have x_{P_1} small as required, we reduce δ_{P_0} in (16.5.1), depending only on (ρ_0, ρ_1, γ).

From (16.5.24), repeating the proof of Lemma 4.6.1 and using the fact that, for all nonnegative integers k and l satisfying $k + l = 2$, real $\alpha \in (0,1)$, and $x_{P_1} \ge 0$, the functions:

$$g_{k,l,\alpha}(x) = (x - x_{P_1})^{k+l+\alpha-5} x^{\frac{k+\alpha}{2}} \qquad \text{decrease on } (x_{P_1}, \infty),$$

which can be checked explicitly by taking derivatives, then we have

$$\begin{aligned} &\sum_{0\le k+l\le 2} \sup_{z\in\Omega_\varepsilon} \Big((x - x_{P_1})^{k+l-5} x^{\frac{k}{2}} |\partial_x^k \partial_y^l \psi(z)| \Big) \\ &+ \sum_{k+l=2} \sup_{z,\tilde{z}\in\Omega_\varepsilon, z\ne\tilde{z}} \Bigg(\big(\max(x,\tilde{x}) - x_{P_1}\big)^{k+l+\alpha-5} \big(\max(x,\tilde{x})\big)^{\frac{k+\alpha}{2}} \\ &\qquad \times \frac{|\partial_x^k\partial_y^l\psi(z) - \partial_x^k\partial_y^l\psi(\tilde{z})|}{\delta_\alpha^{(\rm par)}(z,\tilde{z})} \Bigg) \le C, \end{aligned} \tag{16.5.25}$$

where C depends only on α and constant C in (16.5.24), so that it depends only on (ρ_0, ρ_1, γ), k and l are nonnegative integers, and $\delta_\alpha^{(\text{par})}(z, \tilde{z})$ is defined by (4.6.1).

We note that the norm on the left-hand side of (16.5.25) is stronger than that in $C^{2,\alpha}(\overline{\Omega_\varepsilon})$. Indeed, using that $x_{P_1} \in [0, \varepsilon)$ and $k+l+\alpha-5 < 0$ for any (k, l, α) involved in the expression in (16.5.25), we obtain that, for $x, \tilde{x} \in (x_{P_1}, x_{P_1}+\varepsilon)$,

$$(x - x_{P_1})^{k+l-5} x^{\frac{k}{2}} \geq x^{\frac{3}{2}k+l-5},$$

$$\big(\max(x, \tilde{x}) - x_{P_1}\big)^{k+l+\alpha-5} \big(\max(x, \tilde{x})\big)^{\frac{k+\alpha}{2}} \geq \big(\max(x, \tilde{x})\big)^{\frac{3}{2}(k+\alpha)+l-5}.$$

Now, since $\frac{3}{2}(k+\alpha)+l-5 < 0$ for any (k, l, α) involved in (16.5.25), and since $x, \hat{x} \in (x_{P_1}, x_{P_1}+\varepsilon)$ in (16.5.25), we have

$$(x - x_{P_1})^{k+l-5} x^{\frac{k}{2}} \geq \varepsilon^{\frac{3}{2}k+l-5},$$

$$\big(\max(x, \tilde{x}) - x_{P_1}\big)^{k+l+\alpha-5} \big(\max(x, \tilde{x})\big)^{\frac{k+\alpha}{2}} \geq \varepsilon^{\frac{3}{2}(k+\alpha)+l-5}.$$

Also, assuming without loss of generality that $\varepsilon \leq 1$, we obtain from (4.6.1) that

$$\delta_\alpha^{(\text{par})}(z, \tilde{z}) \leq |z - \tilde{z}|^\alpha.$$

Thus, the expression on the left-hand side of (16.5.25) decreases if the weights are replaced by constants $\varepsilon^{\frac{3}{2}k+l-5}$ and $\varepsilon^{\frac{3}{2}(k+\alpha)+l-5}$, respectively, and the denominators in the difference quotients are replaced by $|z - \tilde{z}|^\alpha$. The resulting expression is equivalent to the $C^{2,\alpha}(\overline{\Omega_\varepsilon})$–norm with the constant depending only on (ε, α), *i.e.*, on $(\rho_0, \rho_1, \gamma, \alpha)$.

Also, from (16.5.25),

$$|D^2\psi(x)| \leq C(x - x_{P_1})^2.$$

Therefore, for $P_0 = P_1$, we conclude that $D^2\psi(P_0) = 0$. This completes the proof of (16.5.17).

Then (16.5.18) follows as in Step 5 of the proof of Proposition 11.4.3. □

Using Proposition 16.5.3, we can obtain the global estimate for the wedge angles $\theta_{\rm w} \in (\theta_{\rm w}^{\rm c}, \frac{\pi}{2})$ satisfying (16.5.1), where $\theta_{\rm w}^{\rm c}$ is from Definition 15.7.3.

Corollary 16.5.4. *Let δ_{P_0} be from Proposition 16.5.3. Let $\theta_{\rm w}^* \in (\theta_{\rm w}^{\rm d}, \frac{\pi}{2})$. Let α from Lemma 10.5.2 be extended to all $\theta_{\rm w} \in [\theta_{\rm w}^*, \frac{\pi}{2})$ by Proposition 15.7.5. Then there exists C depending only on the data and $\theta_{\rm w}^*$ such that, if $\theta_{\rm w} \in [\theta_{\rm w}^{\rm d}, \frac{\pi}{2})$ is a wedge angle satisfying (16.5.1) and φ is an admissible solution of* **Problem 2.6.1** *with the wedge angle $\theta_{\rm w}$, then*

$$\begin{aligned} &\varphi^{\text{ext}} \in C^{1,\alpha}(\overline{\Omega^{\text{ext}}}) \cap C^{2,\alpha}(\overline{\Omega}^{\text{ext}} \setminus \{P_3\}) \cap C^\infty(\overline{\Omega^{\text{ext}}} \setminus \{P_0, P_0^-, P_3\}), \\ &\|\varphi\|_{2,\alpha,\Omega}^{(-1-\alpha),\{P_3\}} \leq C, \qquad\qquad (16.5.26) \\ &D^m(\varphi - \varphi_2)(P_0) = 0 \qquad \textit{for } m = 0, 1, 2. \end{aligned}$$

Furthermore, the shock function $f_{O_1,\mathrm{sh}}$ *for* $\Gamma^{\mathrm{ext}}_{\mathrm{shock}}$, *introduced in Corollary* 10.1.3 *and Proposition* 15.7.1, *satisfies*

$$\begin{aligned} &f_{O_1,\mathrm{sh}} \in C^{2,\alpha}([\theta_{P_1}, \theta_{P_1^-}]) \cap C^{\infty}((\theta_{P_1}, \theta_{P_1^-})), \\ &\|f_{O_1,\mathrm{sh}}\|_{C^{2,\alpha}([\theta_{P_1}, \theta_{P_1^-}])} \le C. \end{aligned} \tag{16.5.27}$$

Proof. We use Corollary 16.4.8 for the estimates near Γ_{sonic}. Then the proof of Corollary 11.4.7 applies directly by means of Proposition 16.3.2 and the other results in this chapter which extend the results of the previous chapters to $\theta_{\mathrm{w}} \in [\theta^{\mathrm{d}}_{\mathrm{w}}, \frac{\pi}{2}]$. □

16.6 A PRIORI ESTIMATES NEAR THE REFLECTION POINT IN THE SUBSONIC-AWAY-FROM-SONIC CASE

We fix $\theta^*_{\mathrm{w}} \in (\theta^{\mathrm{d}}_{\mathrm{w}}, \theta^{\mathrm{s}}_{\mathrm{w}})$ and $\delta_{P_0} \in (0,1)$. In the rest of this section, we assume that φ is an admissible solution with $\theta_{\mathrm{w}} \in (\theta^*_{\mathrm{w}}, \theta^{\mathrm{s}}_{\mathrm{w}})$ satisfying

$$\frac{|D\varphi_2(P_0)|}{c_2} \le 1 - \delta_{P_0}. \tag{16.6.1}$$

By (16.6.1), it follows from Proposition 15.6.2 that there exists $\delta_{\mathrm{ell}} > 0$ depending only on (ρ_0, ρ_1, γ) such that

$$\frac{|D\varphi|}{c(|D\varphi|^2, \varphi)} \le 1 - \delta_{\mathrm{ell}} \qquad \text{in } \overline{\Omega}. \tag{16.6.2}$$

Using this and (9.1.5)–(9.1.6), we see that there exists a sufficiently large M such that φ satisfies (9.2.6)–(9.2.8). Thus, by Lemma 9.2.1, we can modify coefficients $(\mathcal{A}, B)$ in (9.2.1) so that the modified coefficients $(\tilde{\mathcal{A}}, \tilde{B})$ are defined for any $(\mathbf{p}, z) \in \mathbb{R}^2 \times \mathbb{R}$ and satisfy (9.2.3)–(9.2.5) with $\mathcal{D} = \Omega$, where constants (λ, C_k) depend only on $(\rho_0, \rho_1, \gamma, \delta_{P_0})$. Then, by (9.2.3), φ satisfies the modified equation (9.2.9) in Ω. Therefore, we have

Corollary 16.6.1. *There exist* $\varepsilon, \lambda > 0$, $C_k > 0$ *for* $k = 0, 1, \dots$, *and functions* $\tilde{\mathcal{A}}, \tilde{B} \in C^{\infty}(\mathbb{R}^2 \times \mathbb{R})$, *depending only on* $(\rho_0, \rho_1, \gamma, \delta_{P_0})$, *such that* (9.2.3)–(9.2.5) *hold and, if* φ *is an admissible solution of* **Problem 2.6.1** *with* $\theta_{\mathrm{w}} \in (\theta^{\mathrm{d}}_{\mathrm{w}}, \frac{\pi}{2})$ *satisfying* (16.6.1), *then* φ *is a solution of equation* (9.2.9) *in* Ω, *and equation* (9.2.9) *is elliptic by* (9.2.4). *Moreover, equation* (9.2.9) *for* φ *coincides with equation* (2.2.8) *by* (9.2.3) *with* (9.2.2).

Next we discuss the boundary condition on Γ_{shock}. From Lemma 10.1.1 and Proposition 15.7.1, there exist $\hat{\varepsilon}, \hat{\delta} \in (0,1)$ depending only on (ρ_0, ρ_1, γ) such that

$$|D(\varphi_1 - \varphi)| \ge \hat{\delta} \qquad \text{in } \mathcal{N}_{\hat{\varepsilon}}(\Gamma_{\mathrm{shock}}) \cap \Omega \tag{16.6.3}$$

for any admissible solution φ. Then, applying Lemma 10.2.1 with M determined by (9.1.5)–(9.1.6), (16.6.2), and $\delta = \hat{\delta}$ to obtain the corresponding $g^{(\mathrm{sh})}_{\mathrm{mod}}(\mathbf{p}, z, \boldsymbol{\xi})$, we find from (10.2.4) that φ satisfies

$$g^{(\mathrm{sh})}_{\mathrm{mod}}(D\varphi, \varphi, \boldsymbol{\xi}) = 0 \qquad \text{on } \Gamma_{\mathrm{shock}}. \tag{16.6.4}$$

Lemma 16.6.2. *There exists $\hat{\varepsilon} > 0$ depending only on $(\rho_0, \rho_1, \gamma, \delta_{P_0})$ such that, for any admissible solution φ with $\theta_{\mathrm{w}} \in (\theta^{\mathrm{d}}_{\mathrm{w}}, \frac{\pi}{2})$ and any $\boldsymbol{\xi} \in \mathbb{R}^2$,*

$$g^{(\mathrm{sh})}_{\mathrm{mod}}(\mathbf{p}, z, \boldsymbol{\xi}) = g^{\mathrm{sh}}(\mathbf{p}, z, \boldsymbol{\xi}) \quad \text{if } |(\mathbf{p}, z) - (D\varphi(\tilde{\boldsymbol{\xi}}), \varphi(\tilde{\boldsymbol{\xi}}))| \le \hat{\varepsilon} \ \text{ for some } \ \tilde{\boldsymbol{\xi}} \in \overline{\Omega},$$

where $g^{\mathrm{sh}}(\cdot)$ is given by (7.1.9).

Proof. The assertion follows from (10.2.4) in Lemma 10.2.1 and the choice of constants M and $\delta = \hat{\delta}$ in the application of Lemma 10.2.1 for defining $g^{(\mathrm{sh})}_{\mathrm{mod}}$, where we have used that $\varphi \in C^1(\overline{\Omega})$ in order to allow $\tilde{\boldsymbol{\xi}} \in \partial\Omega$. □

Lemma 16.6.3. *Let $L > 0$. Denote*

$$\mathcal{R}_L = \{(\mathbf{p}, z, \boldsymbol{\xi}) \ : \ |\mathbf{p}| \le 2L, \ |z| \le 2L, \ |\boldsymbol{\xi}| \le 2L\}.$$

Then there exists $C_L > 0$ depending only on $(\rho_0, \rho_1, \gamma, \delta_{P_0}, L)$ such that

$$\|g^{(\mathrm{sh})}_{\mathrm{mod}}\|_{C^3(\overline{\mathcal{R}_L})} \le C_L. \tag{16.6.5}$$

In particular, let C be the constant depending only on (ρ_0, ρ_1, γ) such that (9.1.2) *and* (9.1.5) *hold for any admissible solution with $\theta_{\mathrm{w}} \in (\theta^{\mathrm{d}}_{\mathrm{w}}, \frac{\pi}{2})$, where such a constant C exists by Proposition* 15.2.2*. Then* (16.6.5) *holds for $L = C$ with M depending only on $(\rho_0, \rho_1, \gamma, \delta_{P_0})$.*

Proof. The assertion follows from estimate (9.2.5) of $\tilde{A}$ in Corollary 16.6.1 and the explicit expression (10.2.3). □

Lemma 16.6.4. *There exist $\delta_{\mathrm{bc}} > 0$ and $R \ge 0$ depending only on $(\rho_0, \rho_1, \gamma, \delta_{P_0})$ such that, for any admissible solution φ of* **Problem 2.6.1** *with $\theta_{\mathrm{w}} \in [\theta^{\mathrm{d}}_{\mathrm{w}}, \frac{\pi}{2})$ satisfying* (16.6.1)*, the boundary condition* (16.6.4) *satisfies*

$$D_{\mathbf{p}} g^{(\mathrm{sh})}_{\mathrm{mod}}(D\varphi(\boldsymbol{\xi}), \varphi(\boldsymbol{\xi}), \boldsymbol{\xi}) \cdot \boldsymbol{\nu}_{\mathrm{sh}}(\boldsymbol{\xi}) \ge \delta_{\mathrm{bc}} \qquad \text{for all } \boldsymbol{\xi} \in \overline{\Gamma_{\mathrm{shock}}}, \tag{16.6.6}$$

$$D_{\mathbf{p}} g^{(\mathrm{sh})}_{\mathrm{mod}}(D\varphi, \varphi, \boldsymbol{\xi}) \cdot \frac{D\varphi_1 - D\varphi}{|D\varphi_1 - D\varphi|} \ge \delta_{\mathrm{bc}} \qquad \text{for all } \boldsymbol{\xi} \in \overline{\Omega} \cap B_R(P_0), \tag{16.6.7}$$

where $\boldsymbol{\nu}_{\mathrm{sh}}$ is the interior normal on Γ_{shock} to Ω.

Proof. We choose $R := \frac{\hat{\varepsilon}}{2}$, where $\hat{\varepsilon}$ is from (16.6.3). Then, noting that $P_0 \in \overline{\Gamma_{\mathrm{shock}}}$, we obtain

$$|D(\varphi_1 - \varphi)| \ge \hat{\delta} \qquad \text{in } \overline{\Omega} \cap B_R(P_0)$$

by (16.6.3). Also, we recall that $\boldsymbol{\nu}_{\rm sh} = \frac{D\varphi_1 - D\varphi}{|D\varphi_1 - D\varphi|}$ on $\Gamma_{\rm shock}$.

Now properties (16.6.6)–(16.6.7) follow from (10.2.6), (16.6.3), and the choice of constants M and $\delta = \hat{\delta}$ used in the application of Lemma 10.2.1 for defining $g^{(\rm sh)}_{\rm mod}$ in the present case. □

Also, by (2.2.20), the boundary condition on $\Gamma_{\rm wedge}$ is

$$g^{\rm w}(D\varphi) = 0 \qquad \text{on } \Gamma_{\rm wedge}, \tag{16.6.8}$$

where $g^{\rm w}(\mathbf{p}) = \mathbf{p} \cdot \boldsymbol{\nu}_{\rm w}$, and $\boldsymbol{\nu}_{\rm w}$ is the inner unit normal $\boldsymbol{\nu}_{\rm w}$ on $\Gamma_{\rm wedge}$ to Ω, given by (8.2.14).

16.6.1 Main steps of the proof of the a priori estimates near the reflection point in the subsonic-away-from-sonic case

In order to obtain the $C^{1,\alpha}$–estimates near the reflection point P_0, we apply the estimates near the corner for the oblique derivative problem by Lieberman [192, Theorem 2.1], in the version presented in §4.3.2. For that, we perform the following steps, which we first outline below and then perform in detail:

1. Prove the functional independence of functions $g^{(\rm sh)}_{\rm mod}(\mathbf{p}, z, \boldsymbol{\xi})$ and $g^{\rm w}(\mathbf{p})$ on $\Gamma_{\rm shock}$ for $(\mathbf{p}, z) = (D\varphi, \varphi)$ in the sense of (4.3.56). In the proof, we use the cone of monotonicity for $\varphi - \varphi_2$ and the structure of the shock polar (including its convexity). We also use the fact that φ_2 is the *weak* state (2); see (16.6.16).

2. Since only the Lipschitz estimate of $\Gamma_{\rm shock}$ is available so that we do not have the sufficient regularity to apply Proposition 4.3.7 directly with $\Gamma_1 = \Gamma_{\rm wedge}$ and $\Gamma_2 = \Gamma_{\rm shock}$, we perform the hodograph transform to flatten $\Gamma_{\rm shock}$. Then we apply Proposition 4.3.7 to obtain (4.3.57) with $g_1 = g^{\rm w}$ and $x_0 = P_0$ in the hodograph variables.

3. Combining with (16.6.4) and applying Proposition 4.3.9 with $W = \overline{\Gamma_{\rm shock}}$ in the hodograph variables, we obtain the full gradient estimate (4.3.80) on the hodograph plane. Then, changing the variables back, we obtain (4.3.80) for φ on $\Gamma_{\rm shock}$ in the $\boldsymbol{\xi}$–variables. We also obtain (4.3.57) with $g_1 = g^{\rm w}$ in Ω.

4. Choosing a vector $\mathbf{e} \neq \boldsymbol{\nu}_w$ and a function $\hat{g}^{\rm sh}(\mathbf{p}) := (\mathbf{p} - D\varphi_2(P_0)) \cdot \mathbf{e}$ and noting that $\hat{g}^{\rm sh}(D\varphi(P_0)) = 0$ by (15.1.3) so that $|\hat{g}^{\rm sh}(D\varphi(\boldsymbol{\xi}))| \leq |\boldsymbol{\xi} - P_0|^\alpha$ by (4.3.80) for φ on $\Gamma_{\rm shock}$, we apply Proposition 4.3.7 with $\Gamma_1 = \Gamma_{\rm shock}$, $\Gamma_2 = \Gamma_{\rm wedge}$, $g_1 = \hat{g}^{\rm sh}$, and $g_2 = g^{\rm w}$ to obtain (4.3.57) with $g_1 = \hat{g}^{\rm sh}$ in Ω.

5. Combining with (4.3.57) for $g_1 = g^{\rm w}$ and applying Proposition 4.3.9 with $W = \overline{\Omega}$, we obtain the full gradient estimate (4.3.80) in Ω. From that, we prove the $C^{1,\alpha}$–regularity of φ and $\Gamma_{\rm shock}$ up to P_0.

Remark 16.6.5. *In this proof, we crucially use both properties* (15.1.3) *and the fact that* φ_2 *corresponds to the weak reflection. We also note that the estimates hold for* $\theta_{\rm w} \in [\theta_{\rm w}^*, \theta_{\rm w}^{\rm s})$ *satisfying* (16.6.1), *and the constants in these estimates depend on* $\theta_{\rm w}^*$ *in addition to* $(\rho_0, \rho_1, \gamma, \delta_{P_0})$ *and blow up as* $\theta_{\rm w}^* \to \theta_{\rm w}^{\rm d}+$.

Now, in §16.6.2–§16.6.5, we give the details of Steps 1–5.

16.6.2 Functional independence of functions $g^{(\rm sh)}_{\rm mod}(\mathbf{p}, z, \boldsymbol{\xi})$ and $g^{\rm w}(\mathbf{p})$ on $\Gamma_{\rm shock}$ for $(\mathbf{p}, z) = (D\varphi, \varphi)$

Denote by $\boldsymbol{\tau}_{\rm w}^{(0)}$ the unit tangent vector to $\Gamma_{\rm wedge}$ in the following direction:

$$\boldsymbol{\tau}_{\rm w}^{(0)} := \frac{D\varphi_2(P_0)}{|D\varphi_2(P_0)|} = -(\cos\theta_{\rm w}, \sin\theta_{\rm w}). \tag{16.6.9}$$

Note that $\boldsymbol{\tau}_{\rm w}^{(0)} = -\boldsymbol{\tau}_{\rm w}$, where $\boldsymbol{\tau}_{\rm w}$ is defined by (8.2.17).

Lemma 16.6.6. *There exist* $r, M > 0$ *depending only on* $(\rho_0, \rho_1, \gamma, \theta_{\rm w}^*, \delta_{P_0})$ *such that, for any admissible solution* φ *of* **Problem 2.6.1** *with* $\theta_{\rm w} \in [\theta_{\rm w}^*,\ \theta_{\rm w}^{\rm s}]$ *satisfying* (16.6.1),

$$|\det G(D\varphi(\boldsymbol{\xi}), \varphi(\boldsymbol{\xi}), \boldsymbol{\xi})| \geq \frac{1}{M} \qquad \text{for all } \boldsymbol{\xi} \in \Gamma_{\rm shock} \cap B_r(P_0), \tag{16.6.10}$$

where $G(\mathbf{p}, z, \boldsymbol{\xi})$ *is the matrix with columns* $D_{\mathbf{p}} g^{(\rm sh)}_{\rm mod}(\mathbf{p}, z, \boldsymbol{\xi})$ *and* $D_{\mathbf{p}} g^{\rm w}(\mathbf{p}, z, \boldsymbol{\xi})$. *More specifically, we have*

$$\boldsymbol{\tau}_{\rm w}^{(0)} \cdot D_{\mathbf{p}} g^{(\rm sh)}_{\rm mod}(D\varphi(\boldsymbol{\xi}), \varphi(\boldsymbol{\xi}), \boldsymbol{\xi}) \leq -\frac{1}{M} \qquad \text{for all } \boldsymbol{\xi} \in \Gamma_{\rm shock} \cap B_r(P_0), \tag{16.6.11}$$

where $\boldsymbol{\tau}_{\rm w}^{(0)}$ *is defined by* (16.6.9). *Note that* (16.6.11) *implies* (16.6.10), *since* $D_{\mathbf{p}} g^{\rm w} = \boldsymbol{\nu}_{\rm w}$.

Proof. It suffices to prove (16.6.11), since (16.6.10) follows from (16.6.11). In this proof, constants C, r, and M depend only on the data and $\theta_{\rm w}^*$, unless otherwise specified. We divide the proof into three steps.

1. *Properties of the shock polar.* Let φ be an admissible solution. From Corollary 7.4.6, for any $\boldsymbol{\xi} \in \Gamma_{\rm shock}$, we have

$$D\varphi(\boldsymbol{\xi}) = u\,\mathbf{e}(\boldsymbol{\xi}) + f^{(\boldsymbol{\xi})}_{\rm polar}(u)\mathbf{e}^{\perp}(\boldsymbol{\xi}), \tag{16.6.12}$$

where $u = \partial_{\mathbf{e}(\boldsymbol{\xi})}\varphi(\boldsymbol{\xi})$, $f^{(\boldsymbol{\xi})}_{\rm polar}(\cdot)$ is introduced in Definition 7.4.2, and

$$\mathbf{e}(\boldsymbol{\xi}) := \frac{D\varphi_1(\boldsymbol{\xi})}{|D\varphi_1(\boldsymbol{\xi})|}, \tag{16.6.13}$$

$\mathbf{e}^{\perp}(\boldsymbol{\xi})$ is the vector from rotating $\mathbf{e}(\boldsymbol{\xi})$ with $\frac{\pi}{2}$ clockwise.

Note that $f^{(\boldsymbol{\xi})}_{\rm polar}(\cdot)$ is the shock polar for the steady incoming flow with velocity $D\varphi_1(\boldsymbol{\xi})$ and density ρ_1, and $D\varphi_1(\boldsymbol{\xi})$ depends smoothly on $\boldsymbol{\xi}$. Then, from Lemma 7.3.2(c) and (g), we have

Claim 16.6.7. *Fix ρ_1 and u_1. For any compact set $K \subset \mathbb{R}^2 \setminus \overline{B_{c_1}(\mathcal{O}_1)}$, there exist $r > 0$ and $C > 0$ depending only on K such that, denoting by $u^{(\boldsymbol{\xi})}_{\rm s}$ and $u^{(\boldsymbol{\xi})}_{\rm d}$ the u-coordinates for the sonic and detachment points on the shock polar $v = f^{(\boldsymbol{\xi})}_{\rm polar}(u)$ respectively, we have*

(i) $|f^{(\boldsymbol{\xi})}_{\rm polar}(u)| \le C$ *for any* $\boldsymbol{\xi} \in \overline{\mathcal{N}_r(K)}$ *and* $u \in [\hat{u}^{(\boldsymbol{\xi})}_0, |D\varphi_1(\boldsymbol{\xi})|]$;

(ii) $|(f^{(\boldsymbol{\xi})}_{\rm polar})'(u)| \le C$ *for any* $\boldsymbol{\xi} \in \overline{\mathcal{N}_r(K)}$ *and* $u \in [u^{(\boldsymbol{\xi})}_{\rm d}, u^{(\boldsymbol{\xi})}_{\rm s}]$;

(iii) $(u^{(\boldsymbol{\xi})}_{\rm d}, u^{(\boldsymbol{\xi})}_{\rm s})$ *depend continuously on* $\boldsymbol{\xi} \in \mathbb{R}^2 \setminus B_{c_1}(\mathcal{O}_1)$;

(iv) $(\hat{u}^{(P_0(\theta_{\rm w}))}_0, u^{(P_0(\theta_{\rm w}))}_{\rm d}, u^{(P_0(\theta_{\rm w}))}_{\rm s})$ *depend continuously on* $\theta \in (\theta^{\rm d}_{\rm w}, \frac{\pi}{2})$;

(v) *If* $\hat{\boldsymbol{\xi}} \in K$, $\boldsymbol{\xi} \in \overline{B_r(\hat{\boldsymbol{\xi}})}$, *and* $u \in (u^{(\boldsymbol{\xi})}_{\rm d}, u^{(\boldsymbol{\xi})}_{\rm s})$, *then u is in the domain of* $f^{(\hat{\boldsymbol{\xi}})}_{\rm polar}(\cdot)$ *and*

$$|f^{(\boldsymbol{\xi})}_{\rm polar}(u) - f^{(\hat{\boldsymbol{\xi}})}_{\rm polar}(u)| \le Cr \qquad \text{for all } u \in [u^{(\boldsymbol{\xi})}_{\rm d}, u^{(\boldsymbol{\xi})}_{\rm s}]. \tag{16.6.14}$$

Now we choose set K in Claim 16.6.7 to fix constants r and C for the rest of the proof of Lemma 16.6.6. We will apply these estimates near the reflection point $P_0 = (\xi^0_1, \xi^0_1 \tan\theta_{\rm w})$ for $\theta_{\rm w} \in [\theta^*_{\rm w}, \theta^{\rm s}_{\rm w}]$, and hence fix

$$K = \{(\xi^0_1, \xi^0_1 \tan\theta_{\rm w}) \;:\; \theta^*_{\rm w} \le \theta_{\rm w} \le \theta^{\rm s}_{\rm w}\}. \tag{16.6.15}$$

Note that $K \subset \mathbb{R}^2 \setminus \overline{B_{c_1}(\mathcal{O}_1)}$ by Theorem 7.1.10(iii). Now, applying Claim 16.6.7 with K defined by (16.6.15), we have

Corollary 16.6.8. *There exist positive constants r and C, depending only on the data and $\theta^*_{\rm w}$, such that all the assertions of Claim* 16.6.7 *are satisfied for any $\boldsymbol{\xi} \in B_r(P_0(\theta_{\rm w}))$ and $\hat{\boldsymbol{\xi}} = P_0(\theta_{\rm w})$ with any $\theta_{\rm w} \in [\theta^*_{\rm w}, \theta^{\rm s}_{\rm w}]$.*

2. *Reduction to the shock polar at the reflection point.* Since the weak state (2) is chosen for the wedge angles $\theta_{\rm w} \in [\theta^*_{\rm w}, \theta^{\rm s}_{\rm w})$ with $\theta^*_{\rm w} \in (\theta^{\rm d}_{\rm w}, \theta^{\rm s}_{\rm w})$, then, employing Theorem 7.1.1(vii) and using that point $P_0 = (\xi^0_1, \xi^0_1 \tan\theta_{\rm w})$ and the quantities in Theorem 7.1.1(i) depend continuously on $\theta_{\rm w} \in (\theta^{\rm d}_{\rm w}, \frac{\pi}{2})$, we conclude that there exists $M > 0$ depending only on the data and $\theta^*_{\rm w}$ such that

$$\boldsymbol{\tau}^{(0)}_{\rm w} \cdot D_{\mathbf{p}} g^{\rm sh}(D\varphi_2(P_0), \varphi_2(P_0), P_0) \le -\frac{1}{M} \qquad \text{for all } [\theta^*_{\rm w}, \theta^{\rm s}_{\rm w}], \tag{16.6.16}$$

where $g^{\rm sh}$ is the function in (7.1.9). Thus, (16.6.11) holds at the reflection point P_0. We need to show that a similar inequality holds (with a possibly modified constant) on $\Gamma_{\rm shock} \cap B_r(P_0)$ for small $r > 0$.

For that, we use the cone of monotonicity directions for $\varphi - \varphi_2$ from Corollary 8.2.20, which can be applied by Proposition 15.2.1. We first prove the following:

Claim 16.6.9. *Let* $\mathbf{a} := P_0\mathcal{O}_1 \equiv D\varphi_1(P_0)$. *Then* $\mathbf{a} \in \mathrm{Cone}^0(\mathbf{e}_{\mathcal{S}_1}, \boldsymbol{n}_{\rm w})$, *where* $\{\mathbf{e}_{\mathcal{S}_1}, \boldsymbol{n}_{\rm w}\}$ *are defined by* (7.5.8) *and* (8.2.35).

This can be seen by showing that

$$\mathbf{a} = s\,\boldsymbol{n}_{\rm w} + t\,\mathbf{e}_{\mathcal{S}_1} \qquad \text{with } s > 0,\ t > 0.$$

We compute

$$s = \frac{\mathbf{a}\cdot\boldsymbol{\nu}_{\mathcal{S}_1}}{\boldsymbol{n}_{\rm w}\cdot\boldsymbol{\nu}_{\mathcal{S}_1}}, \qquad t = \frac{\mathbf{a}\cdot\boldsymbol{\tau}_{\rm w}^{(0)}}{\mathbf{e}_{\mathcal{S}_1}\cdot\boldsymbol{\tau}_{\rm w}^{(0)}},$$

where $\boldsymbol{\nu}_{\mathcal{S}_1} = \frac{D(\varphi_1-\varphi_2)}{|D(\varphi_1-\varphi_2)|} = \frac{(u_1-u_2,-v_2)}{\sqrt{(u_1-u_2)^2+v_2^2}}$ is the unit normal to $\mathbf{e}_{\mathcal{S}_1}$, and $\boldsymbol{\tau}_{\rm w}^{(0)}$ given by (16.6.9) is the unit vector along $\Gamma_{\rm wedge}$ and hence orthogonal to $\boldsymbol{n}_{\rm w}$.

Recall that, by (7.1.2), the coordinates of P_0 are $(\xi_1^0, \xi_1^0\tan\theta_{\rm w})$, with $\xi_1^0 = \frac{\rho_1 u_1}{\rho_1-\rho_0} > u_1$; see §6.2. This implies that $\xi_{1P_0} > u_1$ and $\xi_{2P_0} > 0$. Then, using $\mathcal{O}_1 = (u_1, 0)$, we obtain

$$\mathbf{a} = P_0\mathcal{O}_1 = -|\mathbf{a}|(\cos\hat\theta, \sin\hat\theta) \qquad \text{for } \hat\theta \in (\theta_{\rm w}, \frac{\pi}{2}). \tag{16.6.17}$$

Thus, using (7.5.8) and (8.2.17), we have

$$\mathbf{a}\cdot\boldsymbol{\tau}_{\rm w}^{(0)} = |\mathbf{a}|\cos(\hat\theta-\theta_{\rm w}) > 0, \qquad \mathbf{e}_{\mathcal{S}_1}\cdot\boldsymbol{\tau}_{\rm w}^{(0)} = \frac{u_1\sin\theta_{\rm w}}{\sqrt{(u_1-u_2)^2+v_2^2}} > 0,$$

so that

$$t = \frac{\mathbf{a}\cdot\boldsymbol{\tau}_{\rm w}^{(0)}}{\mathbf{e}_{\mathcal{S}_1}\cdot\boldsymbol{\tau}_{\rm w}^{(0)}} > 0.$$

Next, using (8.2.35) and $\boldsymbol{\nu}_{\mathcal{S}_1} = \frac{(u_1-u_2,-v_2)}{\sqrt{(u_1-u_2)^2+v_2^2}}$, we have

$$\boldsymbol{n}_{\rm w}\cdot\boldsymbol{\nu}_{\mathcal{S}_1} = \frac{u_1\sin\theta_{\rm w}}{\sqrt{(u_1-u_2)^2+v_2^2}} > 0.$$

In order to estimate $\mathbf{a}\cdot\boldsymbol{\nu}_{\mathcal{S}_1}$, we express $\boldsymbol{\nu}_{\mathcal{S}_1}$ as follows: By Lemma 7.4.8, center $\mathcal{O}_2 = (u_2, u_2\tan\theta_{\rm w})$ of the sonic circle of state (2) lies within the relative interior of segment $P_0P_3 \subset \{\xi_2 = \xi_1\tan\theta_{\rm w}\}$ for $P_3 = \mathbf{0}$. Let $\hat Q = \cos\theta_{\rm w}(u_1\cos\theta_{\rm w}, u_1\sin\theta_{\rm w})$ be the nearest point to $\mathcal{O}_1$ on the wedge boundary $\{\xi_2 = \xi_1\tan\theta_{\rm w}\}$. Then $\hat Q \in P_0P_3$ for $P_3 = \mathbf{0}$ and $P_0 = (\xi_1^0\cos\theta_{\rm w}, \xi_1^0\sin\theta_{\rm w})$, since $u_1 < \xi_1^0$. Note that $\mathcal{O}_2$ cannot lie on segment $P_3\hat Q$ since, in that case, $\mathcal{O}_1$ and $\mathcal{O}_2$ are on the different sides of S_1 (since $S_1 \perp \mathcal{O}_1\mathcal{O}_2$), which contradicts (6.1.3) and (6.1.5). Thus, $\mathcal{O}_2$ lies on segment $\hat Q P_0$. Using again that $\mathcal{O}_1\mathcal{O}_2 \perp S_1$, we have

$$\boldsymbol{\nu}_{\mathcal{S}_1} = -(\cos\bar\theta, \sin\bar\theta),$$

where $\hat\theta \le \bar\theta \le \frac{\pi}{2} + \theta_{\rm w}$ for $\hat\theta$ in (16.6.17) so that $\theta_{\rm w} < \hat\theta < \frac{\pi}{2}$. Then

$$\mathbf{a}\cdot\boldsymbol{\nu}_{\mathcal{S}_1} = \cos(\bar\theta - \hat\theta) > 0.$$

Now we have

$$s = \frac{\mathbf{a} \cdot \boldsymbol{\nu}_{S_1}}{\boldsymbol{n}_{\rm w} \cdot \boldsymbol{\nu}_{S_1}} > 0,$$

and Claim 16.6.9 is proved. □

3. We now continue our proof of Lemma 16.6.6. We use $\{\mathbf{e}(\cdot), \mathbf{e}^{\perp}(\cdot)\}$ defined by (16.6.13). From Claim 16.6.9, since φ_1 is a fixed smooth function, there exists r depending only on the data such that, for any $\theta_{\rm w} \in [\theta_{\rm w}^{\rm d}, \theta_{\rm w}^{\rm s}]$,

$$\mathbf{e}(\boldsymbol{\xi}) \in \mathrm{Cone}^0(\mathbf{e}_{S_1}, \boldsymbol{n}_{\rm w}) \qquad \text{for all } \boldsymbol{\xi} \in B_r(P_0).$$

Then, by Corollary 8.2.20 which holds in the present case by Proposition 15.2.1, we have

$$\partial_{\mathbf{e}(\boldsymbol{\xi})}(\varphi - \varphi_2)(\boldsymbol{\xi}) \geq 0 \qquad \text{for all } \boldsymbol{\xi} \in \overline{\Omega} \cap B_r(P_0). \tag{16.6.18}$$

From this and the smooth dependence of $\mathbf{e}(\cdot)$ and $\varphi_2(\cdot)$ on $\boldsymbol{\xi} \in \overline{\mathcal{N}_r(K)}$ and $\theta_{\rm w} \in [\theta_{\rm w}^{*}, \theta_{\rm w}^{\rm s}]$ for K in (16.6.15), we obtain that, for any $\boldsymbol{\xi} \in \overline{\Gamma_{\rm shock}} \cap B_r(P_0)$,

$$u := \partial_{\mathbf{e}(\boldsymbol{\xi})}\varphi(\boldsymbol{\xi}) \geq \partial_{\mathbf{e}(\boldsymbol{\xi})}\varphi_2(\boldsymbol{\xi}) = \mathbf{e}(\boldsymbol{\xi}) \cdot D\varphi_2(\boldsymbol{\xi}) \geq \partial_{\mathbf{e}(P_0)}\varphi_2(P_0) - Cr. \tag{16.6.19}$$

Remark 16.6.10. *Estimate* (16.6.19) *is crucial for the proof of* (16.6.11). *Indeed, by* (15.1.3) *and* (16.6.16), *we see that inequality* (16.6.11) *holds for* $(\mathbf{p}, z, \boldsymbol{\xi}) = (D\varphi(P_0), \varphi(P_0), P_0)$. *Thus,* (16.6.11) *also holds for* $(\mathbf{p}, z, \boldsymbol{\xi})$ *close to* $(D\varphi(P_0), \varphi(P_0), P_0)$. *However, since only the Lipschitz bound of* φ *is available, we cannot argue that* $D\varphi(\boldsymbol{\xi})$ *is close to* $D\varphi(P_0)$ *if* $\boldsymbol{\xi}$ *is close to* P_0. *Instead, using the cone of monotonicity for* $\varphi - \varphi_2$, *we obtain* (16.6.19), *which amounts to the one-sided closedness of a specific directional derivative of* φ *at* $\boldsymbol{\xi}$ *and* P_0. *Below, using the convexity of the shock polar, we will show that the one-sided estimate* (16.6.19) *is sufficient to complete the proof of* (16.6.11).

We also note that $|D\varphi(\boldsymbol{\xi})| \leq c(|D\varphi(\boldsymbol{\xi})|^2, \varphi(\boldsymbol{\xi}))$ for any $\boldsymbol{\xi} \in \Gamma_{\rm shock}$, since φ is an admissible solution in the sense of Definition 15.1.2 so that (15.1.4) holds. From this, using parts (b), (e), and (g) of Lemma 7.3.2, we find that $u \leq u_{\rm s}^{(\boldsymbol{\xi})}$. Combining this with (16.6.19), we have

$$\partial_{\mathbf{e}(P_0)}\varphi_2(P_0) - Cr \leq u \leq u_{\rm s}^{(\boldsymbol{\xi})}. \tag{16.6.20}$$

We also note that one of the consequences of (16.6.20) is

$$u_{\rm d}^{(\boldsymbol{\xi})} \leq u \leq u_{\rm s}^{(\boldsymbol{\xi})}. \tag{16.6.21}$$

Indeed, recalling that $P_0 = (\xi_1^0, \xi_1^0 \tan\theta_{\rm w})$, we see that $\partial_{\mathbf{e}(P_0)}\varphi_2(P_0)$ is a smooth function of $\theta_{\rm w} \in (\theta_{\rm w}^{\rm d}, \frac{\pi}{2})$. Also, since φ_2 is a weak state (2), it follows from (7.3.13) and Lemma 7.4.4 that

$$\partial_{\mathbf{e}(P_0)}\varphi_2(P_0) > u_{\rm d}^{(P_0)} \qquad \text{for each } \theta_{\rm w} \in (\theta_{\rm w}^{\rm d}, \tfrac{\pi}{2}). \tag{16.6.22}$$

Then, again using the continuous dependence of P_0 on $\theta_{\rm w} \in (\theta_{\rm w}^{\rm d}, \frac{\pi}{2})$, and employing Claim 16.6.7(iv) and Corollary 16.6.8, we have

$$\partial_{\mathbf{e}(P_0)}\varphi_2(P_0) - u_{\rm d}^{(P_0)} \geq \frac{1}{C} \qquad \text{for all } \theta_{\rm w} \in [\theta_{\rm w}^*, \theta_{\rm w}^{\rm s}].$$

Now, from the continuous dependence of $u_{\rm d}^{(\boldsymbol{\xi})}$ on $\boldsymbol{\xi}$ in Claim 16.6.7(iii), recalling that $\boldsymbol{\xi} \in \Gamma_{\rm shock} \cap B_r(P_0) \subset \overline{\mathcal{N}_r(K)}$ with K from (16.6.15), we find that, reducing r if necessary (depending only on the data and $\theta_{\rm w}^*$), the left-hand side of (16.6.20) satisfies

$$\partial_{\mathbf{e}(P_0)}\varphi_2(P_0) - Cr \geq u_{\rm d}^{(\boldsymbol{\xi})},$$

which implies (16.6.21) from (16.6.20).

Denote $\mathbf{p}^{(\boldsymbol{\xi})}(t) := t\,\mathbf{e}(\boldsymbol{\xi}) + f_{\rm polar}^{(\boldsymbol{\xi})}(t)\mathbf{e}^{\perp}(\boldsymbol{\xi})$. Then (16.6.12) implies

$$D\varphi(\boldsymbol{\xi}) = \mathbf{p}^{(\boldsymbol{\xi})}(u).$$

From (16.6.21) and parts (i) and (v) of Claim 16.6.7 with Corollary 16.6.8,

$$\begin{aligned} &|(\mathbf{p}^{(P_0)}(u), \varphi_1(P_0)) - (D\varphi(\boldsymbol{\xi}), \varphi(\boldsymbol{\xi}))| \\ &= |(\mathbf{p}^{(P_0)}(u), \varphi_1(P_0)) - (\mathbf{p}^{(\boldsymbol{\xi})}(u), \varphi_1(\boldsymbol{\xi}))| \leq Cr. \end{aligned} \tag{16.6.23}$$

Now, using Lemma 16.6.3, we obtain that, for any $\boldsymbol{\xi} \in \overline{\Gamma_{\rm shock}} \cap B_r(P_0(\theta_{\rm w}))$,

$$\begin{aligned} &\boldsymbol{\tau}_{\rm w}^{(0)} \cdot D_{\mathbf{p}} g_{\rm mod}^{(\rm sh)}(D\varphi(\boldsymbol{\xi}), \varphi(\boldsymbol{\xi}), \boldsymbol{\xi}) \\ &\leq \boldsymbol{\tau}_{\rm w}^{(0)} \cdot D_{\mathbf{p}} g_{\rm mod}^{(\rm sh)}(\mathbf{p}^{(P_0)}(u), \varphi_1(P_0), P_0) + Cr \\ &= \frac{1}{|D\varphi_2(P_0)|} D\varphi_2(P_0) \cdot D_{\mathbf{p}} g_{\rm mod}^{(\rm sh)}(\mathbf{p}^{(P_0)}(u), \varphi_1(P_0), P_0) + Cr, \end{aligned} \tag{16.6.24}$$

where we have used (16.6.9) in the last equality. Thus, in order to prove (16.6.11), it remains to show a negative upper bound of

$$\frac{1}{|D\varphi_2(P_0)|} D\varphi_2(P_0) \cdot D_{\mathbf{p}} g_{\rm mod}^{(\rm sh)}(\mathbf{p}^{(P_0)}(u), \varphi_1(P_0), P_0)$$

for all u satisfying (16.6.20) for $\theta_{\rm w} \in [\theta_{\rm w}^*, \theta_{\rm w}^{\rm s}]$.

We note that $|D\varphi_2(P_0)| \in (0, \infty)$ for each $\theta_{\rm w} \in [\theta_{\rm w}^{\rm d}, \frac{\pi}{2})$, where the fact that $D\varphi_2(P_0) \neq 0$ follows from Lemma 7.4.4(ii). Then the continuous dependence of both point P_0 and the parameters of state (2) on $\theta_{\rm w} \in [\theta_{\rm d}, \theta_{\rm s}]$ implies that

$$\frac{1}{C} \leq |D\varphi_2(P_0)| \leq C \qquad \text{for all } \theta_{\rm w} \in [\theta_{\rm w}^{\rm d}, \theta_{\rm w}^{\rm s}].$$

Then, to complete the proof of (16.6.11), it remains to show

$$D\varphi_2(P_0) \cdot D_{\mathbf{p}} g_{\rm mod}^{(\rm sh)}(\mathbf{p}^{(P_0)}(u), \varphi_1(P_0), P_0) \leq -\frac{1}{C} \tag{16.6.25}$$

for all u satisfying (16.6.20) for $\theta_{\rm w} \in [\theta_{\rm w}^*, \theta_{\rm w}^{\rm s}]$.

Furthermore, from Lemma 16.6.3, Claim 16.6.7(i)–(ii), and Corollary 16.6.8, we obtain that, for any $u, \tilde{u} \in [u_{\rm d}^{(P_0)}, u_{\rm s}^{(P_0)}]$,

$$|D_{\mathbf{p}} g_{\rm mod}^{(\rm sh)}(\mathbf{p}^{(P_0)}(u), \varphi_1(P_0), P_0) - D_{\mathbf{p}} g_{\rm mod}^{(\rm sh)}(\mathbf{p}^{(P_0)}(\tilde{u}), \varphi_1(P_0), P_0)| \leq C|\tilde{u} - u|.$$

Also, noting that $u_{\rm s}^{(\boldsymbol{\xi})} < u_{\infty}^{(\boldsymbol{\xi})} = |D\varphi_1(\boldsymbol{\xi})|$ and reducing r if necessary, we obtain

$$u_{\rm s}^{(\boldsymbol{\xi})} \leq \frac{1}{2}\big(|D\varphi_1(P_0)| + u_{\rm s}^{(P_0)}\big) \qquad \text{for all } \boldsymbol{\xi} \in B_r(P_0) \text{ and } \theta_{\rm w} \in [\theta_{\rm w}^*, \theta_{\rm w}^{\rm s}],$$

by Claim 16.6.7(iii). Therefore, it suffices to show (16.6.25) for all u satisfying

$$\partial_{\mathbf{e}(P_0)}\varphi_2(P_0) \leq u \leq \frac{1}{2}\big(|D\varphi_1(P_0)| + u_{\rm s}^{(P_0)}\big), \tag{16.6.26}$$

since, with this, using (16.6.22) and possibly reducing r, we obtain a similar estimate (with a modified constant, still depending only on the data and $\theta_{\rm w}^*$) for all u satisfying (16.6.20).

4. *Use of the convexity of the shock polar.* We prove (16.6.25) for all u satisfying (16.6.26). We write $f_{\rm polar}$ below for $f_{\rm polar}^{(P_0)}$.

From (16.6.23), reducing r if necessary, we obtain by Lemma 16.6.2 that

$$D_{\mathbf{p}} g_{\rm mod}^{(\rm sh)}(\mathbf{p}^{(P_0)}(u), \varphi_1(P_0), P_0) = D_{\mathbf{p}} g^{\rm sh}(\mathbf{p}^{(P_0)}(u), \varphi_1(P_0), P_0).$$

Thus, from now on, we use $g^{\rm sh}$ instead of $g_{\rm mod}^{(\rm sh)}$ in the last expression in (16.6.24).

Denote

$$h(u, v) = g^{\rm sh}(u\,\mathbf{e}(P_0) + v\,\mathbf{e}^{\perp}(P_0),\ \varphi_1(P_0),\ P_0),$$

where we have used the notation in (16.6.13). Then, working in the coordinates with basis $\{\mathbf{e}(P_0), \mathbf{e}^{\perp}(P_0)\}$, we see that, in order to prove (16.6.25) for all u satisfying (16.6.26), it suffices to show that

$$D\varphi_2(P_0) \cdot Dh(u, f_{\rm polar}(u)) \leq -\frac{1}{C} \qquad \text{for all } u \text{ in (16.6.26)}. \tag{16.6.27}$$

From Lemma 7.4.4, it follows that $h(u, v)$ is for the boundary condition (7.3.16) for the steady incoming flow with $(\rho_\infty, u_\infty) = (\rho_1, |D\varphi_1(P_0)|)$. Thus, the assertions of Lemma 7.3.2(h) hold for $h(u, v)$. In particular,

$$Dh(u, f_{\rm polar}(u)) \neq 0 \qquad \text{for any } u \in (\hat{u}_0, u_\infty)$$

so that, for each $\theta_{\rm w} \in [\theta_{\rm w}^*, \theta_{\rm w}^{\rm s}]$,

$$\frac{1}{C} \leq |Dh(u, f_{\rm polar}^{(P_0)}(u))| \leq C \qquad \text{for all } u \text{ in (16.6.26)},$$

and, by (7.3.17),

$$Dh(u, f_{\rm polar}(u)) = \frac{|Dh(u, f_{\rm polar}(u))|}{\sqrt{1 + |f'_{\rm polar}(u)|^2}}(f'_{\rm polar}(u),\ -1). \tag{16.6.28}$$

With this, using Lemma 16.6.3, Claim 16.6.7(ii), and Corollary 16.6.8, we conclude that (16.6.27) follows from

$$D\varphi_2(P_0)\cdot(f'_{\rm polar}(u),\ -1)\le -\frac{1}{C}\qquad\text{for all } u \text{ in (16.6.26)}.\tag{16.6.29}$$

Thus, it suffices to show (16.6.29).

By (16.6.12), for each $\theta_{\rm w}\in[\theta_{\rm w}^*,\theta_{\rm w}^{\rm s}]$,

$$D\varphi_2(P_0)=u^*\mathbf{e}(P_0)+f_{\rm polar}(u^*)\mathbf{e}^\perp(P_0),\tag{16.6.30}$$

where $u^*=\partial_{\mathbf{e}(P_0)}\varphi_2(P_0)\in[u_d^{(P_0)}+\frac{1}{C},|D\varphi_1(P_0)|)$. Then, denoting

$$G(u):=D\varphi_2(P_0)\cdot(f'_{\rm polar}(u),\ -1)=u^*f'_{\rm polar}(u)-f_{\rm polar}(u^*),$$

we see that, for each $u\in(\hat{u}_0^{(P_0)},u_\infty^{(P_0)})$,

$$G'(u)=u^*f''_{\rm polar}(u)\le 0,\tag{16.6.31}$$

where we have used that $u^*\ge 0$ and the concavity of $f_{\rm polar}(\cdot)$; see (7.3.9) in Lemma 7.3.2.

Next, using (16.6.9), (16.6.28), and (16.6.30), the left-hand side of (16.6.16) can be written as

$$\frac{|Dh(u^*,f_{\rm polar}(u^*))|}{\sqrt{(1+|f'_{\rm polar}(u^*)|^2)\left(|u^*|^2+|f_{\rm polar}(u^*)|^2\right)}}G(u^*).$$

Thus, (16.6.16), combined with Lemma 16.6.3, Claim 16.6.7, and Corollary 16.6.8, implies

$$G(u^*)\le-\frac{1}{C}.$$

Combining this with (16.6.31), we obtain that

$$G(u)\le-\frac{1}{C}\qquad\text{for any } u\in[u^*,|D\varphi_1(P_0)|],$$

which implies (16.6.29). Lemma 16.6.6 is proved. □

16.6.3 $C^{1,\alpha}$–estimates near P_0

Now we use the functional independence in Lemma 16.6.6 to prove the $C^{1,\alpha}$–estimates near P_0.

We note first that $\Gamma_{\rm shock}$ near P_0 can be written as the graph in the coordinates with basis $\{\mathbf{e},\mathbf{e}^\perp\}$ defined as follows: Denote

$$\mathbf{e}:=\frac{P_0\mathcal{O}_1}{|P_0\mathcal{O}_1|}=\frac{D\varphi_1(P_0)}{|D\varphi_1(P_0)|},\tag{16.6.32}$$

$\mathbf{e}^\perp$ is uniquely defined by $|\mathbf{e}^\perp|=1$, $\mathbf{e}\cdot\mathbf{e}^\perp=0$, and $\mathbf{e}^\perp\cdot\boldsymbol{\nu}_{\rm w}>0$, where $\boldsymbol{\nu}_{\mathrm{w}}$ defined by (7.1.4) is the unit normal vector to Γ_{wedge} pointing into Λ.

Let (r,θ) be the polar coordinates with respect to center $\mathcal{O}_1$ of state (1) as in Lemma 10.1.1. Then $\partial_r(\varphi_1-\varphi)(P_0) = -\partial_{\mathbf{e}}(\varphi_1-\varphi)(P_0)$. Also, for $s \in (0, \frac{|P_0\mathcal{O}_1|}{10})$ and $P \in B_s(P_0)\cap\Omega$, vector $\mathbf{e}_r(P) := \frac{P-\mathcal{O}_1}{|P-\mathcal{O}_1|}$ satisfies

$$\mathbf{e}_r(P) = \mathbf{e}_r(P_0) + \mathbf{g}(P) = -\mathbf{e} + \mathbf{g}(P),$$

where $|\mathbf{g}(P)| \le Cs$, with C depending only on $|P_0\mathcal{O}_1|$, and hence on the data. From Corollary 9.1.3 with Proposition 15.2.2, and Lemma 10.1.1 with Proposition 15.7.1, there exist δ and C depending only on the data such that

$$\partial_{\mathbf{e}}(\varphi_1-\varphi)(P) \ge -\partial_r(\varphi_1-\varphi)(P) - |\mathbf{g}(P)|\|\varphi_1-\varphi\|_{C^{0,1}(\Omega)} \ge \delta - Cs.$$

Then there exists $s^*>0$ depending only on the data such that

$$\partial_{\mathbf{e}}(\varphi_1-\varphi) \ge \frac{\delta}{2} \qquad \text{in } B_{s^*}(P_0)\cap\Omega. \tag{16.6.33}$$

Also, $\varphi=\varphi_1$ on Γ_{shock}, and $\varphi_1>\varphi$ in Ω.

Next, it follows from (15.1.5) in Definition 15.1.2 that Γ_{shock} lies in Λ between lines $\Gamma_{\text{wedge}} \subset \{\xi_2 = \xi_1\tan\theta_{\mathrm{w}}\}$ and $S_1\cap\Lambda = \{\varphi_1=\varphi_2\}\cap\Lambda$. Moreover, if $\mathbf{e}$ is defined by (16.6.32) and $\boldsymbol{\tau}_{S_1}$ defined by (11.3.8), then

$$\text{angle } \hat{\alpha} \text{ between vectors } \mathbf{e} \text{ and } \boldsymbol{\tau}_{S_1} \text{ satisfies } \hat{\alpha}\in(0,\frac{\pi}{2}). \tag{16.6.34}$$

Indeed, using (11.3.9) and Lemma 7.4.4(ii), we obtain that $\hat{\alpha} = \frac{\pi}{2}-\beta$, where $\beta\in(\beta_{\mathrm{d}}, \cos^{-1}(\frac{c_\infty}{u_\infty})]$ is defined in Lemma 7.3.2 on the shock polar of the steady incoming flow $(\rho_\infty, u_\infty) = (\rho_1, |D\varphi_1(P_0)|)$, for the steady downstream velocity $D\varphi_2(P_0)$ in the coordinates described in Lemma 7.4.4(ii). Here, as defined in Lemma 7.3.2(d), $\beta_{\mathrm{d}}\in(0,\cos^{-1}(\frac{c_\infty}{u_\infty}))$ is the detachment angle for the steady shock polar, and we have used the fact that $D\varphi_2$ is the self-similar weak reflection, so that $D\varphi_2(P_0)$ is the steady weak reflection solution for the steady shock polar described above by Lemma 7.4.4(ii)–(iii). Thus, $\hat{\alpha}=\frac{\pi}{2}-\beta\in(0,\frac{\pi}{2})$.

Using (16.6.32) and (16.6.34), we conclude that, in the (S,T)–coordinates with basis $\{\mathbf{e},\mathbf{e}^{\perp}\}$ and the origin at P_0,

$$S_1 = \{S = a_{S_1}T\}, \qquad \{\varphi_1<\varphi_2\}\subset\{S>a_{S_1}T\},$$

where $a_{S_1}>0$. Combining this with (16.6.33) and (15.1.5) in Definition 15.1.2, we conclude that there exists $f_{\mathbf{e}}\in C^1(\mathbb{R})$ such that

$$\begin{aligned}
&\Gamma_{\text{shock}}\cap B_{s^*}(P_0) = \{S=f_{\mathbf{e}}(T),\ T>0\}\cap B_{s^*}(P_0),\\
&\Gamma_{\text{wedge}}\cap B_{s^*}(P_0) = \{S=a_{\mathrm{w}}T,\ T>0\}\cap B_{s^*}(P_0),\\
&\Omega\cap B_{s^*}(P_0) = \{a_{S_1}T<f_{\mathbf{e}}(T)<S<a_{\mathrm{w}}T,\ T>0\}\cap B_{s^*}(P_0),
\end{aligned} \tag{16.6.35}$$

where $a_{\mathrm{w}} := \cot(\angle P_3P_0\mathcal{O}_1)>0$. The last inequality holds since, in the $\boldsymbol{\xi}$–coordinates, $P_3=\mathbf{0}$, $P_0=(\xi_1^0, \xi_1^0\tan\theta_{\mathrm{w}})$, $\mathcal{O}_1=(u_1,0)$, and $\xi_1^0>u_1>0$ by

(2.2.18), so that $\angle P_3P_0\mathcal{O}_1 \in (0, \frac{\pi}{2} - \theta_{\rm w})$. We also note that $\angle P_3P_0\mathcal{O}_1$ depends continuously on $\theta_{\rm w} \in [\theta_{\rm w}^{\rm d}, \theta_{\rm w}^{\rm s}]$, which implies

$$\frac{1}{C} \le a_{\rm w}(\theta_{\rm w}) \le C \qquad \text{for all } \theta_{\rm w} \in [\theta_{\rm w}^{\rm d}, \theta_{\rm w}^{\rm s}]. \tag{16.6.36}$$

Proposition 16.6.11. *There exist $s > 0$, $\alpha \in (0,1)$, and $C > 0$ depending only on $(\rho_0, \rho_1, \gamma, \theta_{\rm w}^*, \delta_{P_0})$ such that, for any admissible solution φ of* **Problem 2.6.1** *with $\theta_{\rm w} \in [\theta_{\rm w}^*,\ \theta_{\rm w}^{\rm s}]$ satisfying* (16.6.1),

$$\|\varphi\|^{(-1-\alpha),\{P_0\}}_{2,\alpha,\Omega\cap B_s(P_0)} \le C, \tag{16.6.37}$$

$$\|f_{\mathbf{e}}\|^{(-1-\alpha),\{0\}}_{2,\alpha,(0,s)} \le C, \tag{16.6.38}$$

where $f_{\mathbf{e}}$ is the function defined in (16.6.35).

Proof. All the constants in this proof depend only on $(\rho_0, \rho_1, \gamma, \theta_{\rm w}^*, \delta_{P_0})$. Let φ be the function as in the statement. We divide the proof into three steps.

1. In this step, we apply the hodograph transform in the $\mathbf{e}$–direction to perform Step 2 in the procedure of §16.6.1.

a) *The equation and boundary conditions for $\bar\phi = \varphi_1 - \varphi$.* We use equation (2.2.11) for φ and $\phi = \varphi + \frac{|\boldsymbol{\xi}|^2}{2}$, with $c = c(|D\varphi|^2, \varphi, \rho_0^{\gamma-1})$ defined by (1.14). Then, using that $D^2\bar\phi = -D^2\phi$ and substituting it into the coefficients of the right-hand side of (2.2.11) for $\varphi = \varphi_1 - \bar\phi$, we see that $\bar\phi$ satisfies the equation of form:

$$\sum_{i,j=1}^{2} A_{ij}(D\bar\phi, \bar\phi, \boldsymbol{\xi}) D_{ij}\bar\phi = 0 \qquad \text{in } \Omega, \tag{16.6.39}$$

where, for $\mathbf{p} \in \mathbb{R}^2$ and $z \in \mathbb{R}$,

$$A_{ij}(\mathbf{p}, z, \boldsymbol{\xi}) = \delta_{ij}\hat c^2(\mathbf{p}, z, \boldsymbol{\xi}) - (\partial_i\varphi_1(\boldsymbol{\xi}) - p_i)(\partial_j\varphi_1(\boldsymbol{\xi}) - p_j), \tag{16.6.40}$$

with $\hat c^2(\mathbf{p}, z, \boldsymbol{\xi}) := \rho_0^{\gamma-1} - (\gamma-1)\big(\frac{1}{2}|D\varphi_1(\boldsymbol{\xi}) - \mathbf{p}|^2 + \varphi_1(\boldsymbol{\xi}) - z\big)$.

Then, from (16.6.2) and the definition of A_{ij},

$$\lambda|\boldsymbol{\kappa}|^2 \le \sum_{i,j=1}^{2} A_{ij}(D\bar\phi, \bar\phi, \boldsymbol{\xi})\kappa_i\kappa_j \le \Lambda|\boldsymbol{\kappa}|^2 \qquad \text{for all } \boldsymbol{\xi} \in \overline{\Omega} \text{ and } \boldsymbol{\kappa} \in \mathbb{R}^2, \tag{16.6.41}$$

and, for any $M > 0$, $k = 1, 2, \dots$, there exist L_k depending only on $(M, k, \rho_0, \rho_1, \gamma)$ such that, for $k = 1, 2, \dots$,

$$|D^k_{(\mathbf{p},z,\boldsymbol{\xi})}A_{ij}| \le L_k \qquad \text{on } \{(\mathbf{p}, z, \boldsymbol{\xi}) \in \mathbb{R}^2 \times \mathbb{R} \times \overline{\Omega}\ :\ |\mathbf{p}| + |z| \le M\}. \tag{16.6.42}$$

Next, we write the boundary conditions (16.6.4) on $\Gamma_{\rm shock}$ and (16.6.8) on $\Gamma_{\rm wedge}$ in terms of $\bar\phi$, by substituting $\varphi = \varphi_1 - \bar\phi$. Defining

$$\hat g^{(k)}(\mathbf{p}, z, \boldsymbol{\xi}) = -g^{(k)}(D\varphi_1(\boldsymbol{\xi}) - \mathbf{p},\ \varphi_1(\boldsymbol{\xi}) - z, \boldsymbol{\xi}) \qquad \text{for } k = 1, 2, \tag{16.6.43}$$

with $(g^{(1)}, g^{(2)}) = (g^{\rm w}, g^{(\rm sh)}_{\rm mod})$, we use notation $(\hat g^{\rm w}, \hat g^{\rm sh}) = (\hat g^{(1)}, \hat g^{(2)})$ to obtain

$$\hat g^{\rm sh}(D\bar\phi, \bar\phi, \boldsymbol\xi) = 0 \quad \text{on } \Gamma_{\rm shock}, \qquad \hat g^{\rm w}(D\bar\phi, \boldsymbol\xi) = 0 \quad \text{on } \Gamma_{\rm wedge}.$$

Therefore, $D_{\mathbf p}\hat g^{\rm sh}(D\bar\phi(\boldsymbol\xi), \bar\phi(\boldsymbol\xi), \boldsymbol\xi) = D_{\mathbf p} g^{\rm sh}(D\varphi(\boldsymbol\xi), \varphi(\boldsymbol\xi), \boldsymbol\xi)$. Then, from Lemma 16.6.4 and possibly reducing s^* by replacing it with $\min\{s^*, R\} > 0$, we have

$$D_{\mathbf p}\hat g^{\rm sh}(D\bar\phi(\boldsymbol\xi), \bar\phi(\boldsymbol\xi), \boldsymbol\xi)\cdot \boldsymbol\nu(\boldsymbol\xi) \ge \delta_{\rm bc} \qquad \text{for all } \boldsymbol\xi \in \Gamma_{\rm shock}, \tag{16.6.44}$$

$$D_{\mathbf p}\hat g^{\rm sh}(D\bar\phi(\boldsymbol\xi), \bar\phi(\boldsymbol\xi), \boldsymbol\xi)\cdot \frac{D\bar\phi(\boldsymbol\xi)}{|D\bar\phi(\boldsymbol\xi)|} \ge \delta_{\rm bc} \qquad \text{for all } \boldsymbol\xi \in \overline{\Omega}\cap B_{s^*}(P_0). \tag{16.6.45}$$

We write equation (16.6.39) and the boundary conditions (16.6.43) in the coordinate variables (S, T) with basis $\{\mathbf e, \mathbf e^\perp\}$ and the origin at P_0 without change of notation.

b) *The partial hodograph transform and the domain in the hodograph variables.* Now, using (16.6.33), we perform the hodograph transform in domain $B_{s^*}(P_0)\cap\Omega$ in the $\mathbf e$–direction. We work in the (S,T)–coordinates with basis $\{\mathbf e, \mathbf e^\perp\}$ and the origin at P_0. Then (16.6.35) holds. We write $\bar\phi$ as a function of (S, T).

From (9.1.5), (16.6.33), and (16.6.35), and the fact that $\varphi = \varphi_1$ on $\Gamma_{\rm shock}$, it follows that

$$\begin{aligned} &\bar\phi(f_{\mathbf e}(T), T) = 0 && \text{for all } T\in(0, s^*),\\ &\frac{\delta}{2} \le \partial_S\bar\phi \le C, && \\ &\bar\phi(S,T) \ge \frac{\delta}{2}(S - f_{\mathbf e}(T)) > 0 && \text{in } \Omega\cap B_{s^*}(P_0). \end{aligned} \tag{16.6.46}$$

Define a map $F : B_{s^*}(P_0)\cap\Omega \mapsto \mathbb R^2$, by

$$F(S,T) = (\bar\phi(S,T),\ T) =: (y_1, y_2) = \mathbf y. \tag{16.6.47}$$

Then $F \in C^1(\overline{B_{s^*}(P_0)\cap\Omega})\cap C^3(\overline{B_{s^*}(P_0)\cap\Omega}\setminus\{P_0\})$ by (15.1.2), and $F(P_0) = \mathbf 0$.

Denote

$$\mathcal D := F(B_{s^*}(P_0)\cap\Omega).$$

Then, from (16.6.33), there exists $v \in C^1(\overline{\mathcal D})\cap C^2(\overline{\mathcal D}\setminus\{\mathbf 0\})$ such that, for $\mathbf y\in\overline{\mathcal D}$,

$$\begin{aligned} &(v(\mathbf y),\ y_2) \in \overline{B_{s^*}(P_0)\cap\Omega},\\ &\bar\phi(S, y_2) = y_1 \qquad \text{if and only if} \qquad v(\mathbf y) = S. \end{aligned} \tag{16.6.48}$$

In particular, $F : B_{s^*}(P_0)\cap\Omega \mapsto \mathcal D$ is one-to-one with

$$F^{-1}(\mathbf y) = (v(\mathbf y),\ y_2) \in B_{s^*}(P_0)\cap\Omega \qquad \text{for } \mathbf y\in\mathcal D \tag{16.6.49}$$

in the (S,T)–coordinates. Using that $P_0 = \mathbf 0$ in the (S,T)–coordinates, (16.6.49) implies that

$$|v| \le s^* \qquad \text{on } \mathcal D. \tag{16.6.50}$$

Differentiating the equation: $\bar{\phi}(v(\mathbf{y}), y_2) = y_1$, which holds for any $\mathbf{y} \in \mathcal{D}$, we have

$$\partial_S \bar{\phi} = \frac{1}{\partial_{y_1} v}, \qquad \partial_T \bar{\phi} = -\frac{1}{\partial_{y_1} v} \partial_{y_2} v, \tag{16.6.51}$$

and

$$\begin{aligned} \partial_{SS}\bar{\phi} &= -\frac{1}{v_{y_1}^3} v_{y_1y_1}, \qquad \partial_{ST}\bar{\phi} = \frac{v_{y_2}}{v_{y_1}^3} v_{y_1y_1} - \frac{1}{v_{y_1}^2} v_{y_1y_2}, \\ \partial_{TT}\bar{\phi} &= -\frac{v_{y_2}^2}{v_{y_1}^3} v_{y_1y_1} + 2\frac{v_{y_2}}{v_{y_1}^2} v_{y_1y_2} - \frac{1}{v_{y_1}} v_{y_2y_2}, \end{aligned} \tag{16.6.52}$$

where the left-hand and right-hand sides are taken at points (S, T) and $F(S, T)$, respectively. Now, if C and δ are the constants from (9.1.5) and (16.6.33), respectively, then, from (16.6.51),

$$\frac{1}{C} \le \partial_{y_1} v \le \frac{C}{\delta}, \qquad |Dv| \le \frac{\sqrt{C^2+1}}{\delta}. \tag{16.6.53}$$

Next, we describe more precisely domain $\mathcal{D}$ in the $\mathbf{y}$–coordinates. We first note that line $S_1 = \{\varphi_1 = \varphi_2\}$ passes through P_0 so that, from Lemma 7.4.8 and (16.6.35), we obtain that, in the (S,T)–coordinates,

$$\begin{aligned} &S_1 \cap \Lambda = \{S = a_{S_1} T,\ S > 0\}, \\ &\{\varphi_1 > \varphi_2\} \cap \Lambda = \{a_{S_1} T < S < a_{\mathrm{w}} T,\ T > 0\}, \end{aligned}$$

where $0 < a_{S_1} < a_{\mathrm{w}}$, and a_{S_1} depends continuously on $\theta_{\mathrm{w}} \in [\theta_{\mathrm{w}}^{\mathrm{d}}, \theta_{\mathrm{w}}^{\mathrm{s}}]$. Since $\Omega \cap B_{s^*} \subset \Lambda \cap \{\varphi_1 > \varphi_2\}$, we have

$$\Omega \cap B_{s^*}(P_0) \subset \{a_{S_1} T < S < a_{\mathrm{w}} T,\ T > 0\} \cap B_{s^*}(P_0).$$

Now, using (16.6.35)–(16.6.36), we have

$$0 < a_{S_1} T \le f_{\mathbf{e}}(T) \le a_{\mathrm{w}} T \le CT \qquad \text{for } T \in (0, \frac{s^*}{\sqrt{1+C^2}}),$$

so that

$$\Omega \cap \{0 < T < \hat{s}\} = \{f_{\mathbf{e}}(T) < S < a_{\mathrm{w}} T,\ 0 < T < \hat{s}\}, \tag{16.6.54}$$

where $\hat{s} := \frac{s^*}{\sqrt{1+C^2}}$. Applying transform (16.6.47) to both sides and using (16.6.35) and (16.6.46), we have

$$\begin{aligned} &\mathcal{D} \cap \{0 < y_2 < \hat{s}\} = \{\mathbf{y} \,:\, \hat{f}_{\mathrm{w}}(y_2) > y_1 > 0,\ 0 < y_2 < \hat{s}\}, \\ &F(\Gamma_{\text{shock}} \cap B_{s^*}(P_0)) \cap \{0 < y_2 < \hat{s}\} = \{(0, y_2) \,:\, 0 < y_2 < \hat{s}\}, \\ &F(\Gamma_{\text{wedge}} \cap B_{s^*}(P_0)) \cap \{0 < y_2 < \hat{s}\} = \{(\hat{f}_{\mathrm{w}}(y_2), y_2) \,:\, 0 < y_2 < \hat{s}\}, \end{aligned} \tag{16.6.55}$$

where $\hat{f}_{\rm w}(y_2) := \bar{\phi}(a_{\rm w}y_2, y_2)$, $\hat{f}_{\rm w} \in C^1([0,\hat{s}])$, and the regularity of $\hat{f}_{\rm w}$ follows from (15.1.2). In particular, $\hat{f}_{\rm w}(0) = 0$, and $\hat{f}_{\rm w} > 0$ on $(0,\hat{s})$ by (16.6.46). Thus, we have

$$\begin{aligned} &\mathcal{D} \cap \{0 < y_2 < \hat{s}\} \subset \{\mathbf{y} \,:\, y_1 > 0,\; y_2 > 0\}, \\ &\mathbf{0} = F(P_0) \in \partial\mathcal{D}. \end{aligned} \tag{16.6.56}$$

Below we use the notations:

$$\begin{aligned} &\mathcal{D}_{\hat{s}} := \mathcal{D} \cap \{0 < y_2 < \hat{s}\}, \\ &\Gamma^{(h)}_{\rm shock} := F(\Gamma_{\rm shock} \cap B_{s^*}(P_0)) \cap \{0 < y_2 < \hat{s}\}, \\ &\Gamma^{(h)}_{\rm wedge} := F(\Gamma_{\rm wedge} \cap B_{s^*}(P_0)) \cap \{0 < y_2 < \hat{s}\}. \end{aligned} \tag{16.6.57}$$

c) *The equation and boundary conditions in the hodograph variables.* In order to obtain the equation for $v(\mathbf{y})$ in $\mathcal{D}$, we substitute the right-hand sides of expressions (16.6.51)–(16.6.52) into equation (16.6.39) to obtain the equation:

$$\sum_{i,j=1}^{2} a_{ij}(Dv, v, \mathbf{y}) D_{ij} v = 0 \qquad \text{in } \mathcal{D}, \tag{16.6.58}$$

with

$$\begin{aligned} &a_{11}(\mathbf{p}, z, \mathbf{y}) = \frac{1}{p_1^3}(A_{11} - 2p_2 A_{12} + p_2^2 A_{22}), \\ &a_{12}(\mathbf{p}, z, \mathbf{y}) = a_{21}(\mathbf{p}, z, \mathbf{y}) = \frac{1}{p_1^2}(A_{12} - p_2 A_{22}), \\ &a_{22}(\mathbf{p}, z, \mathbf{y}) = \frac{p_2}{p_1} A_{22}, \end{aligned} \tag{16.6.59}$$

where $A_{ij} = A_{ij}(\frac{1}{p_1}, -\frac{p_2}{p_1}, y_1, (z, y_2))$.

From the definition of a_{ij}, we find that, for $(\mathbf{p}, z, \mathbf{y})$ satisfying $p_1 \neq 0$,

$$\sum_{i,j=1}^{2} a_{ij}(\mathbf{p}, z, \mathbf{y}) \kappa_i \kappa_j = \frac{1}{p_1^3} \sum_{i,j=1}^{2} A_{ij} \eta_i \eta_j$$

for $(\eta_1, \eta_2) = (\kappa_1, p_1\kappa_2 - p_2\kappa_1)$, so that

$$\sum_{i,j=1}^{2} a_{ij}(Dv, v, \mathbf{y}) \kappa_i \kappa_j = \frac{1}{v_{y_1}^3} \sum_{i,j=1}^{2} A_{ij}(D\bar{\phi}, \bar{\phi}, S, T) \eta_i \eta_j,$$

where $\mathbf{y} = F(S,T)$ and $(\eta_1, \eta_2) = (\kappa_1, v_{y_1}\kappa_2 - v_{y_2}\kappa_1)$.

Then, from (16.6.41) combined with (16.6.46), we have

$$\frac{1}{C}|\boldsymbol{\kappa}|^2 \le \sum_{i,j=1}^{2} a_{ij}(Dv, v, \mathbf{y}) \kappa_i \kappa_j \le C|\boldsymbol{\kappa}|^2 \qquad \text{for all } \mathbf{y} \in \mathcal{D},\ \boldsymbol{\kappa} \in \mathbb{R}^2. \tag{16.6.60}$$

Now we discuss the boundary conditions for v. We write conditions (16.6.43) at point $(S,T) = (v(\mathbf{y}), y_2)$ for $\mathbf{y} \in \mathcal{D}$ by using (16.6.48) and substituting the expressions of $(\bar{\phi}, D\bar{\phi})$ from (16.6.48) and (16.6.51) at such a point into conditions (16.6.43), we obtain the following boundary conditions for v:

$$g_h^{\rm sh}(Dv, v, \mathbf{y}) = 0 \ \text{ on } \Gamma_{\rm shock}^{(h)}, \qquad g_h^{\rm w}(Dv, v, \mathbf{y}) = 0 \ \text{ on } \Gamma_{\rm wedge}^{(h)}, \tag{16.6.61}$$

where

$$g_h^{(k)}(\mathbf{p}, z, \mathbf{y}) = -\hat{g}^{(k)}\big((\frac{1}{p_1}, -\frac{p_2}{p_1}),\ y_1, (z, y_2)\big) \qquad \text{for } k = 1, 2, \tag{16.6.62}$$

with $(g_h^{(1)}, g_h^{(2)}) = (g_h^{\rm w}, g_h^{\rm sh})$, and we have used functions $\hat{g}^{(k)}$ on the right-hand side of (16.6.62) with the arguments written in the (S,T)–coordinates.

Now we show some properties of $g_h^{(k)}(\mathbf{p}, z, \mathbf{y})$. Below we always consider $(\mathbf{p}, z, \mathbf{y})$ with $p_1 \neq 0$.

We first check the obliqueness of the condition on $\Gamma_{\rm shock}^{(h)}$. From (16.6.55) and (16.6.57), we obtain that $\boldsymbol{\nu} = (1, 0)$ on $\Gamma_{\rm shock}^{(h)}$. From (16.6.62), we have

$$\partial_{p_1} g_h^{\rm sh}(\mathbf{p}, z, \mathbf{y}) = \frac{1}{p_1} D_{\mathbf{p}} \hat{g}^{\rm sh}\big((\frac{1}{p_1}, -\frac{p_2}{p_1}),\ y_1, (z, y_2)\big) \cdot (\frac{1}{p_1}, -\frac{p_2}{p_1}).$$

Now, for $\mathbf{y} \in \Gamma_{\rm shock}^{(h)}$, we use that $(S,T) := (v(\mathbf{y}), y_2) \in \Gamma_{\rm shock} \cap B_{s^*}(P_0)$ and (16.6.44) expressed in the (S,T)–coordinates to obtain

$$\begin{aligned} D_{\mathbf{p}} g_h^{\rm sh}(Dv, v, \mathbf{y}) \cdot \boldsymbol{\nu} &= \partial_{p_1} g_h^{\rm sh}(Dv, v, \mathbf{y}) \\ &= \partial_S \bar{\phi}(S,T) D_{\mathbf{p}} \hat{g}^{\rm sh}(D\bar{\phi}, \bar{\phi}, (S,T)) \cdot D\bar{\phi}(S,T) \\ &\geq \delta_{\rm bc} \partial_S \bar{\phi}(S,T) |D\bar{\phi}(S,T)| \geq \frac{\delta_{\rm bc}}{C}, \end{aligned} \tag{16.6.63}$$

where we have used (16.6.46) in the last inequality.

We also check the nondegeneracy of the condition on $\Gamma_{\rm wedge}^{(h)}$. Writing (16.6.8) in the (S,T)–coordinates with basis $\{\mathbf{e}, \mathbf{e}^\perp\}$, we have

$$g^{\rm w}(\mathbf{p}) = p_1 \mathbf{e} \cdot \boldsymbol{\nu}_{\rm w} + p_2 \mathbf{e}^\perp \cdot \boldsymbol{\nu}_{\rm w},$$

where $\mathbf{p} = (p_S, p_T)$. Now, using (16.6.43) in the (S,T)–coordinates and then (16.6.62), and expressing φ_1 in the (S,T)–coordinates, we have

$$g_h^{\rm w}(\mathbf{p}, z, \mathbf{y}) = \frac{1}{p_1} \mathbf{e} \cdot \boldsymbol{\nu}_{\rm w} - \frac{p_2}{p_1} \mathbf{e}^\perp \cdot \boldsymbol{\nu}_{\rm w} - D\varphi_1(z, y_2) \cdot \boldsymbol{\nu}_{\rm w}.$$

Then we calculate at $(\mathbf{p}, z, \mathbf{y})$ with $p_1 \neq 0$:

$$|D_{\mathbf{p}} g_h^{\rm w}| \geq |D_{p_2} g_h^{\rm w}| = \frac{|\mathbf{e}^\perp \cdot \boldsymbol{\nu}_{\rm w}|}{p_1}.$$

Thus, at point $(\mathbf{p}, z, \mathbf{y}) = (Dv(\mathbf{y}), v(\mathbf{y}), \mathbf{y})$ for $\mathbf{y} \in \mathcal{D}$, we employ (16.6.53), which implies especially that $p_1 = v_{y_1}(\mathbf{y}) \neq 0$, to obtain

$$|D_{\mathbf{p}} g_h^{\mathrm{w}}(Dv(\mathbf{y}), v(\mathbf{y}), \mathbf{y})| \geq \frac{1}{C}|\mathbf{e}^{\perp} \cdot \boldsymbol{\nu}_{\mathrm{w}}| = \frac{1}{C}|\mathbf{e} \cdot \boldsymbol{\tau}_{\mathrm{w}}|.$$

It remains to estimate $|\mathbf{e} \cdot \boldsymbol{\tau}_{\mathrm{w}}|$ from below. Using (16.6.32), we have

$$\mathbf{e} = \frac{P_0\mathcal{O}_1}{|P_0\mathcal{O}_1|}, \quad \boldsymbol{\tau}_{\mathrm{w}} = \frac{P_0P_3}{|P_0P_3|}, \qquad \mathcal{O}_1,\ P_3 \in \{\xi_1 = 0\},\ \ P_0 \notin \{\xi_1 = 0\}.$$

Then it follows that $|\mathbf{e} \cdot \boldsymbol{\tau}_{\mathrm{w}}|(\theta_{\mathrm{w}}) > 0$ for each $\theta_{\mathrm{w}} \in [\theta_{\mathrm{w}}^{\mathrm{d}}, \theta_{\mathrm{w}}^{\mathrm{s}}]$. Since $\mathcal{O}_1$ and P_3 are independent of θ_{w}, and P_0 depends continuously on $\theta_{\mathrm{w}} \in [\theta_{\mathrm{w}}^{\mathrm{d}}, \theta_{\mathrm{w}}^{\mathrm{s}}]$, it follows that there exists $\delta_{\mathrm{w}} > 0$ depending only on (ρ_0, ρ_1, γ) such that $|\mathbf{e} \cdot \boldsymbol{\tau}_{\mathrm{w}}| \geq \delta_{\mathrm{w}}$ for any $\theta_{\mathrm{w}} \in [\theta_{\mathrm{w}}^{\mathrm{d}}, \theta_{\mathrm{w}}^{\mathrm{s}}]$. Then

$$|D_{\mathbf{p}} g_h^{\mathrm{w}}(Dv(\mathbf{y}), v(\mathbf{y}), \mathbf{y})| \geq \delta_{\mathrm{w}} > 0 \qquad \text{for all } \mathbf{y} \in \overline{\mathcal{D}}. \tag{16.6.64}$$

We now check the functional independence of functions $(g_h^{\mathrm{sh}}, g_h^{\mathrm{w}})$ on $\Gamma_{\mathrm{shock}}^{(h)}$. We first note that, from (16.6.43) and Lemma 16.6.6, the functional independence holds for $(\hat{g}^{\mathrm{sh}}, \hat{g}^{\mathrm{w}})$ on $\Gamma_{\mathrm{shock}} \cap B_r(P_0)$:

$$|\det \hat{G}(D\bar{\phi}(\boldsymbol{\xi}), \bar{\phi}(\boldsymbol{\xi}), \boldsymbol{\xi})| \geq \frac{1}{M} \qquad \text{for all } \boldsymbol{\xi} \in \Gamma_{\mathrm{shock}} \cap B_r(P_0), \tag{16.6.65}$$

where $\hat{G}(\mathbf{p}, z, \boldsymbol{\xi})$ is the matrix with columns $D_{\mathbf{p}}\hat{g}^{\mathrm{sh}}(\mathbf{p}, z, \boldsymbol{\xi})$ and $D_{\mathbf{p}}\hat{g}^{\mathrm{w}}(\mathbf{p}, z, \boldsymbol{\xi})$, and $r > 0$ is determined in Lemma 16.6.6. We can assume that $s^* \leq r$. Let $G_h(\mathbf{p}, z, \mathbf{y})$ be the matrix with columns $D_{\mathbf{p}} g_h^{\mathrm{sh}}(\mathbf{p}, z, \mathbf{y})$ and $D_{\mathbf{p}} g_h^{\mathrm{w}}(\mathbf{p}, z, \mathbf{y})$. Then, writing $\hat{g}^{(k)}$ in the (S, T)–coordinates and using (16.6.62), we obtain, by a direct calculation, that

$$G_h(\mathbf{p}, z, \mathbf{y}) = \frac{1}{p_1^3}\hat{G}((\frac{1}{p_1}, -\frac{p_2}{p_1}),\ y_1, (z, y_2)).$$

Now, for $\mathbf{y} \in \Gamma_{\mathrm{shock}}^{(h)}$, we use that $(S, T) := (v(\mathbf{y}), y_2) \in \Gamma_{\mathrm{shock}} \cap B_{s^*}(P_0)$, as well as (16.6.65) expressed in the (S, T)–coordinates, to obtain

$$|\det G_h(Dv, v, \mathbf{y})| = |\partial_S \bar{\phi}(S, T)|^3\, |\det \hat{G}(D\bar{\phi}, \bar{\phi}, (S, T))| \geq \frac{1}{CM}. \tag{16.6.66}$$

Finally, we modify $(a_{ij}(\mathbf{p}, z, \mathbf{y}), g_h^{(k)}(\mathbf{p}, z, \mathbf{y}))$ near $p_1 = 0$. Using constant C from (16.6.53), we choose $M_1 = C$ and modify $(a_{ij}, g_h^{(k)})$ via multiplying it by a cutoff function $\zeta \in C^\infty(\mathbb{R})$ such that $\zeta(t) \equiv 0$ on $(-\infty, \frac{1}{4M_1})$ and $\zeta(t) \equiv 1$ on $(\frac{1}{2M_1}, \infty)$ to obtain

$$(a_{ij}^{(\mathrm{mod})}, g_h^{(k),(\mathrm{mod})})(\mathbf{p}, z, \mathbf{y}) = \zeta(p_1)\,(a_{ij}, g_h^{(k)})(\mathbf{p}, z, \mathbf{y}).$$

Then $v(\mathbf{y})$ satisfies equation (16.6.58) and the boundary conditions (16.6.61) with $(a_{ij}^{(\mathrm{mod})}, g_h^{(k),(\mathrm{mod})})$. Also, from the choice of M_1 in the cutoff, we find that functions $(a_{ij}^{(\mathrm{mod})}, g_h^{(k),(\mathrm{mod})})$ satisfy (16.6.60), (16.6.63)–(16.6.64), and (16.6.66). Furthermore, from (16.6.5), (16.6.42), and the explicit expression (16.6.8), using (16.6.43) in the (S,T)–variables, (16.6.59), and (16.6.62), we obtain the following estimates similar to (16.6.42) for any $M > 0$ and $l = 1, 2, \dots$: For any $l = 1, 2, \dots$, there exists C_l depending only on $(M, l, \rho_0, \rho_1, \gamma)$ such that

$$|D^l_{(\mathbf{p},z,\mathbf{y})}(a_{ij}^{(\mathrm{mod})},\ g_h^{(k),(\mathrm{mod})})| \le C_l \tag{16.6.67}$$

on $\{(\mathbf{p}, z, \mathbf{y}) \in \mathbb{R}^2 \times \mathbb{R} \times \overline{\mathcal{D}} \,:\, |\mathbf{p}| + |z| \le M\}$.

d) *Gradient estimates.* Now we are in position to apply Proposition 4.3.7 for $v(\mathbf{y})$ in domain $\mathcal{D}_{\hat{s}}$, with $(\Gamma_1, \Gamma_2) = (\Gamma^{(h)}_{\mathrm{wedge}}, \Gamma^{(h)}_{\mathrm{shock}})$. Indeed, by (16.6.55)–(16.6.57), domain $\mathcal{D}_{\hat{s}} \cap B_{\hat{s}}(0)$ is of the form required in Proposition 4.3.7. Also, since $\Gamma^{(h)}_{\mathrm{shock}}$ is a straight segment, (4.3.43) holds with $L = 0$. The Lipschitz bound of the solution in (4.3.44) holds by (16.6.53). The regularity in (4.3.48) for the ingredients of the equation and boundary condition holds by (16.6.67). Also, $f \equiv 0$ in (4.3.52). The ellipticity in (4.3.53) holds by (16.6.60). The obliqueness of $g^{(2)} = g_h^{\mathrm{sh}}$ in (4.3.54) holds by (16.6.63), the nondegeneracy of $g^{(1)} = g_h^{\mathrm{w}}$ in (4.3.55) follows from (16.6.64), and the functional independence (4.3.56) of $(g^{(1)}, g^{(2)}) = (g_h^{\mathrm{w}}, g_h^{\mathrm{sh}})$ on $\Gamma_2 = \Gamma^{(h)}_{\mathrm{shock}}$ holds by (16.6.66). Thus, from Proposition 4.3.7, there exist $\alpha \in (0,1)$, C, and $s' \in (0, \tilde{s})$ depending only on the data such that, for all $\mathbf{y} \in \overline{\mathcal{D}_{\hat{s}} \cap B_{s'}(\mathbf{0})}$,

$$|g_h^{\mathrm{w}}(Dv(\mathbf{y}), v(\mathbf{y}), \mathbf{y}) - g_h^{\mathrm{w}}(Dv(\mathbf{0}), v(\mathbf{0}), \mathbf{0})| \le C|\mathbf{y}|^\alpha. \tag{16.6.68}$$

Furthermore, we can apply Proposition 4.3.9 with $W = \Gamma^{(h)}_{\mathrm{shock}} \cap B_{s'}(\mathbf{0})$. Indeed, functions $(g_h^{\mathrm{sh}}, g_h^{\mathrm{w}})$ satisfy (4.3.75)–(4.3.76) by (16.6.67). The functional independence (4.3.77) follows from (16.6.66). Conditions (4.3.78) clearly hold for the straight segment $W = \Gamma^{(h)}_{\mathrm{shock}} \cap B_{s'}(\mathbf{0})$ given by (16.6.55) and (16.6.57). Finally, (4.3.79) follows from (16.6.68) and the fact that $g_h^{\mathrm{sh}}(Dv, v, \mathbf{y}) = 0$ on $\Gamma^{(h)}_{\mathrm{shock}}$. Thus, from (4.3.80), we obtain

$$|Dv(\mathbf{y}) - Dv(\mathbf{0})| \le C|\mathbf{y}|^\alpha \qquad \text{for all } y \in \Gamma^{(h)}_{\mathrm{shock}} \cap B_{s'}(\mathbf{0}). \tag{16.6.69}$$

Now we change back to the (S,T)–coordinates. We first show that

$$\Omega \cap B_{s'/C}(P_0) \subset F^{-1}(\mathcal{D}_{\hat{s}} \cap B_{s'}(\mathbf{0})). \tag{16.6.70}$$

To see this, we note that, from (16.6.55), $\|\hat{f}'_{\mathrm{w}}\|_{[0,\hat{s}]} \le \|\bar{\phi}\|_{C^{0,1}(\Omega)} \le C$. Using (16.6.55) and (16.6.57), we obtain

$$\{\mathbf{y} \,:\, \hat{f}_{\mathrm{w}}(y_2) > y_1 > 0,\ 0 < y_2 < \frac{s'}{C}\} \subset \mathcal{D}_{\hat{s}} \cap B_{s'}(\mathbf{0})$$

for C depending on the data and δ_{P_0}. From this, using (16.6.54)–(16.6.55), we conclude that $\Omega \cap \{0 < T < \frac{s'}{C}\} \subset F^{-1}(\mathcal{D}_{\hat{s}} \cap B_{s'}(\mathbf{0}))$, which implies (16.6.70). Then, using (16.6.48), (16.6.51), and (16.6.53), rewriting in terms of $\varphi = \varphi - \bar{\phi}$ expressed in the (S,T)–coordinates, and using the regularity of φ_1, we obtain from (16.6.68)–(16.6.69) that

$$\begin{aligned}
&|g^{\mathrm{w}}(D\varphi(S,T),\varphi(S,T),(S,T)) - g^{\mathrm{w}}(D\varphi(P_0),\varphi(P_0),P_0)| \\
&\quad \le C\big(\bar{\phi}^2(S,T)+T^2\big)^{\frac{\alpha}{2}} \le C_1|(S,T)|^{\alpha} \qquad \text{for all } (S,T)\in\overline{\Omega\cap B_{s'/C}(P_0)}, \\
&|D\varphi(S,T) - D\varphi(P_0)| \le C|(S,T)|^{\alpha} \qquad \text{for all } (S,T)\in\Gamma_{\text{shock}}\cap B_{s'/C}(P_0),
\end{aligned} \tag{16.6.71}$$

where we have used that, in the (S,T)–coordinates, $P_0 = \mathbf{0}$ such that $\bar{\phi}(\mathbf{0}) = 0$, since $P_0 \in \overline{\Gamma_{\text{shock}}}$, which implies that $0 \le \bar{\phi}(S,T) \le |(S,T)|$ by (9.1.5) and the regularity of φ_1.

Rewriting in the $\boldsymbol{\xi}$–coordinates:

$$|\partial_{\boldsymbol{\nu}_{\mathrm{w}}}\varphi(\boldsymbol{\xi}) - \partial_{\boldsymbol{\nu}_{\mathrm{w}}}\varphi(P_0)| \le C_1|\boldsymbol{\xi}-P_0|^{\alpha} \quad \text{for all } \boldsymbol{\xi}\in\overline{\Omega\cap B_{s'/C}(P_0)}, \tag{16.6.72}$$

$$|D\varphi(\boldsymbol{\xi}) - D\varphi(P_0)| \le C|\boldsymbol{\xi}-P_0|^{\alpha} \quad \text{for all } \boldsymbol{\xi}\in\Gamma_{\text{shock}}\cap B_{s'/C}(P_0). \tag{16.6.73}$$

2. Note that

$$\hat{g}^{\mathrm{sh}}(D\varphi) = f(\boldsymbol{\xi}) \qquad \text{on } \Gamma_{\text{shock}}\cap B_{s'/C}(P_0), \tag{16.6.74}$$

where $\hat{g}^{\mathrm{sh}}(\mathbf{p}) = (\mathbf{p} - D\varphi(P_0))\cdot\boldsymbol{\tau}_{\mathrm{w}}$, $f(\boldsymbol{\xi}) := (D\varphi(\boldsymbol{\xi}) - D\varphi(P_0))\cdot\boldsymbol{\tau}_{\mathrm{w}}$, and the equality holds simply because the left-hand side for $D\varphi$ is the same expression as the right-hand side. By (16.6.73),

$$|f(\boldsymbol{\xi})| \le C|\boldsymbol{\xi}-P_0|^{\alpha} \qquad \text{for all } \boldsymbol{\xi}\in\Gamma_{\text{shock}}\cap B_{s'/C}(P_0). \tag{16.6.75}$$

Now we apply Proposition 4.3.7 for φ in domain $\Omega\cap B_{s'/C}(P_0)$, with $\Gamma_1 = \Gamma_{\text{shock}}\cap B_{s'/C}(P_0)$ and $\Gamma_2 = \Gamma_{\text{wedge}}\cap B_{s'/C}(P_0)$, on which we prescribe the boundary conditions (16.6.74) and (16.6.8), respectively. The straight segment Γ_2 obviously satisfies (4.3.43). Also, (4.3.44) follows from (9.1.5) and Proposition 15.2.2. Then the ellipticity and regularity of the ingredients of the equation for φ follow from Corollary 16.6.1. The boundary conditions (16.6.8) and (16.6.74) are linear with constant coefficients, which have the bounds independent of $\theta_{\mathrm{w}} \in [\theta_{\mathrm{w}}^{\mathrm{d}}, \theta_{\mathrm{w}}^{\mathrm{s}}]$. Thus, the conditions in (4.3.51) are satisfied with the constants depending only on the data and δ_{P_0}. Furthermore, $f(\cdot)$ in (16.6.74) satisfies (4.3.52) by (16.6.75). Moreover, from the explicit expressions of the boundary conditions, obliqueness (4.3.54) on $\Gamma_2 = \Gamma_{\text{wedge}}\cap B_{s'/C}(P_0)$, nondegeneracy (4.3.55), and the functional independence (4.3.56) hold with $\lambda = 1$ and $M = 1$. Thus, (4.3.57) implies the existence of $s > 0$ depending only on the data and δ_{P_0} such that

$$|\partial_{\boldsymbol{\tau}_{\mathrm{w}}}\varphi(\boldsymbol{\xi}) - \partial_{\boldsymbol{\tau}_{\mathrm{w}}}\varphi(P_0)| \le C_1|\boldsymbol{\xi}-P_0|^{\alpha} \qquad \text{for all } \boldsymbol{\xi}\in\overline{\Omega\cap B_s(P_0)}.$$

Combining this with (16.6.72), we obtain

$$|D\varphi(\boldsymbol{\xi}) - D\varphi(P_0)| \le C_1|\boldsymbol{\xi} - P_0|^\alpha \qquad \text{for all } \boldsymbol{\xi} \in \overline{\Omega \cap B_s(P_0)}. \tag{16.6.76}$$

3. Now we show the $C^{1,\alpha}$–regularity of φ up to P_0 by using Proposition 4.3.11. We show that the assumptions of Proposition 4.3.11 are satisfied with the constants depending only on $(\rho_0, \rho_1, \gamma, \theta_{\rm w}^*, \delta_{P_0})$.

Domain $\Omega \cap B_s(P_0)$ is between the two curves $\Gamma_{\rm shock} \cap B_s(P_0)$ and $\Gamma_{\rm wedge} \cap B_s(P_0)$. Since $\Gamma_{\rm wedge}$ is a straight segment, it satisfies (4.3.85) with $M = 0$. Also, $\Gamma_{\rm shock}$ satisfies (4.3.85) by Corollary 8.2.14 which holds by Proposition 15.2.1. This can be seen by using assertion (i) of Corollary 8.2.14 and (8.2.24) with $P_0 = P_1$ and $\mathbf{e} = \boldsymbol{\nu}_{\rm w}$ to obtain (4.3.85) depending only on (ρ_0, ρ_1, γ).

Next, we check condition (4.3.86). Recall that $\varphi(P_0) = \varphi_2(P_0)$ and $D\varphi(P_0) = D\varphi_2(P_0)$ for subsonic admissible solutions by (15.1.3). Also, from the continuous dependence of the parameters of state (2) on $\theta_{\rm w} \in [\theta_{\rm w}^{\rm d}, \theta_{\rm w}^{\rm s}]$, it follows that $\|D\varphi_2^{(\theta_{\rm w})}\|_{C^2(B_1(P_0))} \le C(\rho_0, \rho_1, \gamma)$ for any $\theta_{\rm w} \in [\theta_{\rm w}^{\rm d}, \theta_{\rm w}^{\rm s}]$. From (16.6.76), we have

$$|\varphi(\boldsymbol{\xi}) - \varphi_2(\boldsymbol{\xi})| \le C[d(\boldsymbol{\xi})]^{1+\alpha}, \qquad |D\varphi(\boldsymbol{\xi}) - D\varphi_2(\boldsymbol{\xi})| \le C[d(\boldsymbol{\xi})]^\alpha \tag{16.6.77}$$

for any $\boldsymbol{\xi} \in \overline{\Omega \cap B_s(P_0)}$, where $d(\boldsymbol{\xi}) = |\boldsymbol{\xi} - P_0|$. Furthermore, angle $\zeta^{(\theta_{\rm w})}$ between line $S_1^{(\theta_{\rm w})} = \{\varphi_1 = \varphi_2^{(\theta_{\rm w})}\}$ and $\Gamma_{\rm wedge}$ is in $(0, \frac{\pi}{2})$ and depends continuously on $\theta_{\rm w} \in [\theta_{\rm w}^{\rm d}, \theta_{\rm w}^{\rm s}]$ so that $\hat{\zeta} = \min_{\theta_{\rm w} \in [\theta_{\rm w}^{\rm d}, \theta_{\rm w}^{\rm s}]} \zeta^{(\theta_{\rm w})} \in (0, \frac{\pi}{2})$. Then we have

$$\operatorname{dist}(\boldsymbol{\xi}, \Gamma_{\rm wedge}) \ge d(\boldsymbol{\xi}) \sin\hat{\zeta} \qquad \text{for any } \boldsymbol{\xi} \in S_1^{(\theta_{\rm w})} \text{ and } \theta_{\rm w} \in [\theta_{\rm w}^{\rm d}, \theta_{\rm w}^{\rm s}].$$

Thus, recalling that $\varphi = \varphi_1$ on $\Gamma_{\rm shock}$ and using (16.6.46) and (16.6.77), we have

$$\operatorname{dist}(\boldsymbol{\xi}, \Gamma_{\rm wedge}) \ge d(\boldsymbol{\xi}) \sin\hat{\zeta} - C[d(\boldsymbol{\xi})]^{1+\alpha} \qquad \text{for all } \boldsymbol{\xi} \in \Gamma_{\rm shock} \cap B_s(P_0).$$

Reducing s further if necessary, depending only on constant C in the above inequality, we obtain (4.3.86) with $M = \frac{2}{\sin\hat{\zeta}}$.

We write equation (2.2.11) in terms of φ:

$$(c^2 - \varphi_{\xi_1}^2)\varphi_{\xi_1\xi_1} - 2\varphi_{\xi_1}\varphi_{\xi_2}\varphi_{\xi_1\xi_2} + (c^2 - \varphi_{\xi_2}^2)\varphi_{\xi_2\xi_2} - 2c^2 + |D\varphi|^2 = 0, \tag{16.6.78}$$

where $c^2 = c^2(|D\varphi|^2, \varphi)$ is given by (1.14). This equation is satisfied in $\Omega \cap B_s(P_0)$. The boundary conditions are (16.6.4) on $\Gamma_1 = \Gamma_{\rm shock} \cap B_{s'/C}(P_0)$ and (16.6.8) on $\Gamma_2 = \Gamma_{\rm wedge} \cap B_{s'/C}(P_0)$.

Then the equation satisfies the regularity properties in (4.3.49)–(4.3.50) and also (4.3.92) from the explicit expressions of (a_{ij}, a) which are polynomials in $(\mathbf{p}, z)$, but independent of $\boldsymbol{\xi}$. Also, the ellipticity in (4.3.53) on φ holds in Ω by (16.6.2). The boundary conditions satisfy (4.3.87), since (16.6.8) follows from its explicit expression, and (16.6.4) follows from Lemma 16.6.3.

The nondegeneracy in (4.3.88) is satisfied for (16.6.4) by (16.6.7) and, for (16.6.8), we explicitly calculate that $|D_{\mathbf{p}} g^{\rm w}(\mathbf{p})| = |\boldsymbol{\nu}_{\rm w}| = 1$. Finally, (4.3.89) holds by (16.6.76).

Therefore, we apply Proposition 4.3.11(i) to obtain

$$\|\varphi\|_{C^{1,\beta}(\Omega\cap B_{s/2}(P_0))} \le C,$$

where $\beta \in (0,1)$ and C depend only on the constants in the estimates discussed above, and hence on $(\rho_0, \rho_1, \gamma, \theta_{\rm w}^*, \delta_{P_0})$. With this, using (16.6.33), we have

$$\|f_{\mathbf{e}}\|_{C^{1,\beta}([0,\frac{s}{2}])} \le C.$$

Thus, we can apply Proposition 4.3.11(ii) with $\sigma = \beta$ to obtain

$$\|\varphi\|^{(-1-\alpha),\{P_0\}}_{2,\beta,\Omega\cap B_{s/4}(P_0)} \le C.$$

Then, using (16.6.33) again, we have

$$\|f_{\mathbf{e}}\|_{C^{1,\alpha}([0,\frac{s}{4}])} \le C.$$

Now, we apply Proposition 4.3.11(ii) with $\sigma = \alpha$ and (16.6.37) with $\frac{s}{8}$ instead of $\frac{s}{2}$. From this, applying (16.6.33), we obtain (16.6.38) (with $\frac{s}{8}$). Proposition 16.6.11 is proved. □

16.6.4 Global estimate

Using Proposition 16.6.11, we can obtain the global estimate for the wedge angles $\theta_{\rm w} \in (\theta_{\rm w}^{\rm c}, \frac{\pi}{2})$, where $\theta_{\rm w}^{\rm c}$ is from Definition 15.7.3.

Corollary 16.6.12. *Let $\delta_{P_0} \in (0,1)$. Let $\theta_{\rm w}^* \in (\theta_{\rm w}^{\rm c}, \frac{\pi}{2})$. Let α from Lemma 10.5.2 be extended to all $\theta_{\rm w} \in [\theta_{\rm w}^*, \frac{\pi}{2})$ by Proposition 15.7.5. Then there exists C depending only on the data and $(\delta_{P_0}, \theta_{\rm w}^*)$ such that, if $\theta_{\rm w} \in (\theta_{\rm w}^{\rm c}, \frac{\pi}{2})$ satisfies (16.6.1) and φ is an admissible solution of* **Problem 2.6.1** *with the wedge angle $\theta_{\rm w}$, then*

$$\begin{aligned}
&\varphi^{\rm ext} \in C^{1,\alpha}(\overline{\Omega^{\rm ext}}) \cap C^\infty(\overline{\Omega^{\rm ext}} \setminus \{P_0, P_0^-, P_3\}),\\
&\|\varphi\|^{(-1-\alpha),\{P_0,P_3\}}_{2,\alpha,\Omega} \le C, \\
&D^m(\varphi - \varphi_2)(P_0) = 0 \qquad \text{for } m = 0, 1.
\end{aligned} \tag{16.6.79}$$

Furthermore, the shock function $f_{O_1,\rm sh}$ for $\Gamma^{\rm ext}_{\rm shock}$, introduced in Corollary 10.1.3, satisfies

$$\begin{aligned}
&f_{O_1,\rm sh} \in C^{1,\alpha}([\theta_{P_0}, \theta_{P_0^-}]) \cap C^\infty((\theta_{P_0}, \theta_{P_0^-})),\\
&\|f_{O_1,\rm sh}\|^{(-1-\alpha),\{\theta_{P_0},\theta_{P_0^-}\}}_{2,\alpha,(\theta_{P_0},\theta_{P_0^-})} \le C.
\end{aligned} \tag{16.6.80}$$

Proof. Using Proposition 16.6.11, we follow the proof of Corollary 11.4.7 by using Proposition 16.3.2 and the other results in this chapter, which extend the results of the previous chapters to $\theta_{\rm w} \in [\theta_{\rm w}^{\rm d}, \frac{\pi}{2}]$, in order to obtain the estimate:

$$\|\varphi\|^{(-1-\alpha),\{P_0,P_3\}}_{2,\alpha,\Omega} \le C\left(\|\varphi\|^{(-1-\alpha),\{P_0\}}_{2,\alpha,\Omega\cap B_s(P_0)} + \|\varphi\|^{(-1-\alpha),\{P_3\}}_{2,\alpha,\Omega\setminus B_{s/2}(P_0)}\right).$$

This estimate, together with (15.1.3), implies (16.6.79), where the regularity of φ^{ext} in $C^\infty(\overline{\Omega^{\text{ext}}} \setminus \{P_0, P_0^-, P_3\})$ is obtained similarly to the argument for the corresponding part in Corollary 11.4.7.

Then (16.6.79) implies (16.6.80) by Lemma 10.1.1 with Proposition 15.7.1. □

16.6.5 Some uniform estimates for all the supersonic and subsonic reflection solutions away from the detachment angle.

Using the regularity results proved above, we can express Γ_{shock} near P_0 as a graph in the (x, y)–coordinates considered in §11.1 for any wedge angle $\theta_{\text{w}} \in (\theta_{\text{w}}^{\text{d}}, \frac{\pi}{2})$, including the strictly subsonic–away-from-sonic case. This extends Proposition 16.1.4.

Moreover, we show that the shock functions in the (x, y)–coordinates have the uniform estimates in $C^{1,\alpha}$ near P_1 and in $C^{2,\alpha}$ away from P_1 for all the wedge angles $\theta_{\text{w}} \in (\theta_{\text{w}}^{\text{d}}, \frac{\pi}{2})$, where we have used the convention that $P_1 = P_4 = P_0$ and $\overline{\Gamma_{\text{sonic}}} = \{P_0\}$ for the subsonic wedge angles.

We note that, by now, we have considered these coordinates near Γ_{sonic}. However, from its definition in (11.1.1)–(11.1.2), the (x, y)–coordinates are smooth coordinates in $\mathbb{R}^2 \setminus \mathcal{O}_2$. From the continuous dependence of state (2) and P_0 on θ_{w}, there exists $\varepsilon > 0$ such that $|P_0{}^{(\theta_{\text{w}})}\mathcal{O}_2^{(\theta_{\text{w}})}| \geq 100\varepsilon$ for any $\theta_{\text{w}} \in [\theta_{\text{w}}^{\text{d}}, \theta_{\text{w}}^{\text{s}}]$. We thus consider $\Gamma_{\text{shock}} \cap B_\varepsilon(P_0)$ in the (x, y)–coordinates. Specifically, we extend (16.1.10)–(16.1.13) to include all $\theta_{\text{w}} \in (\theta_{\text{w}}^{\text{d}}, \theta_{\text{w}}^{\text{s}}]$ in the following way:

Corollary 16.6.13. *For any $\theta_{\text{w}}^* \in (\theta_{\text{w}}^{\text{d}}, \frac{\pi}{2})$, there exist $\varepsilon_0 > 0$, $\omega > 0$, $\alpha \in (0, 1)$, and C depending only on $(\rho_0, \rho_1, \gamma, \theta_{\text{w}}^*)$ such that, for any admissible solution φ with $\theta_{\text{w}} \in [\theta_{\text{w}}^*, \frac{\pi}{2}]$, there is a shock function $\hat{f} \in C^{(-1-\alpha),\{x_{P_1}\}}_{2,\alpha,(x_{P_1}, x_{P_1}+\varepsilon_0)}$ so that, for each $\varepsilon \in (0, \varepsilon_0]$, region Ω_ε defined by (16.1.3) satisfies the properties in (16.1.10)–(16.1.13) and*

$$\frac{1}{C} \leq -\partial_y(\varphi_1 - \varphi) \leq C \qquad \text{in } \Omega_{\varepsilon_0}, \tag{16.6.81}$$

and moreover,

$$\begin{aligned} &\|\hat{f}\|^{(-1-\alpha),\{x_{P_1}\}}_{2,\alpha,(x_{P_1}, x_{P_1}+\varepsilon_0)} \leq C, \\ &\hat{f}(x_{P_1}) = \hat{f}_0(x_{P_1}), \quad \hat{f}'(x_{P_1}) = \hat{f}_0'(x_{P_1}), \end{aligned} \tag{16.6.82}$$

where $\hat{f}_0$ is the function from (11.2.8) *in Lemmas* 11.2.2 *and* 16.1.2.

Proof. We divide the proof into two steps.

1. We first show (16.1.10)–(16.1.13) and (16.6.81). In the light of Proposition 16.1.4, which implies that (16.1.10)–(16.1.13) and Lemma 11.2.7 hold for any wedge angle satisfying (16.1.7), it remains to consider only the subsonic reflection solutions with the wedge angles for which (16.6.1) holds. To prove the assertion in that case, we note that, by Lemma 16.1.2 and Proposition 11.2.2,

$$\partial_y(\varphi_1 - \varphi_2)(P_1) \le -\frac{2}{M} \qquad \text{for all } \theta_{\rm w} \in [\theta_{\rm w}^{\rm d}, \frac{\pi}{2}]$$

with uniform $M > 0$. Then (16.6.79) implies that there exists $\varepsilon_0 > 0$ depending only on the data and $\theta_{\rm w}^*$ such that

$$\partial_y(\varphi_1 - \varphi) \le -\frac{1}{M} \qquad \text{on } \overline{\Omega} \cap B_{\varepsilon_0}(P_1) \text{ for all } \theta_{\rm w} \in [\theta_{\rm w}^*, \frac{\pi}{2}]. \tag{16.6.83}$$

With this, since $\varphi_1 > \varphi$ in Ω with $\varphi_1 = \varphi$ on $\Gamma_{\rm shock} \subset \partial\Omega$, we obtain that the properties in (16.1.10)–(16.1.13) follow from (16.6.79). Note also that (16.6.81) follows from (16.6.79) and (16.6.83).

2. Let δ_{P_0} be the constant from Proposition 16.4.6. Then estimate (16.6.82) for all the supersonic reflection solutions is obtained from:

(a) Estimate (16.3.3) with constant C defined in Proposition 16.3.2 for $\delta = \frac{\delta_{P_0}}{2}$, by using Remark 11.4.2, for the wedge angles satisfying (16.3.5) with $\delta = \frac{\delta_{P_0}}{2}$;

(b) Estimate (16.4.60) in Proposition 16.4.6 for the wedge angles satisfying (16.4.1).

Let δ_{P_0} be the constant in Proposition 16.5.3. Then estimate (16.6.82) for all the subsonic and sonic reflection solutions with $P_1 = P_0$ follows from:

(a) Estimate (16.5.18) for the wedge angles satisfying (16.5.1);

(b) Estimate (16.6.37) by using (16.6.83) and (15.1.3) for the wedge angles satisfying (16.6.1).

□

We also note that estimate (11.2.23) can be extended to any admissible solution for $\theta_{\rm w} \in [\theta_{\rm w}^*, \frac{\pi}{2}]$ as follows:

Lemma 16.6.14. *For any $\theta_{\rm w}^* \in (\theta_{\rm w}^{\rm d}, \frac{\pi}{2})$, there exist $\varepsilon_0 > 0$ and C depending only on $(\rho_0, \rho_1, \gamma, \theta_{\rm w}^*)$ such that, if φ is any admissible solution with $\theta_{\rm w} \in [\theta_{\rm w}^*, \frac{\pi}{2})$,*

$$|D_{(x,y)}\psi| \le Cx \qquad \text{in } \Omega_{\varepsilon_0}, \tag{16.6.84}$$

where Ω_ε is defined by (16.1.3).

Proof. By Proposition 16.1.4, the assertion is true for all the admissible solutions with any wedge angle $\theta_{\rm w} \in \mathcal{I}_{\rm w}(\delta_{P_0}) \cap [\theta_{\rm w}^*, \frac{\pi}{2})$, where we have used the notation in (16.1.7).

Therefore, it remains to consider the subsonic reflection case with the wedge angles $\theta_{\rm w} \in (\theta_{\rm w}^*, \theta_{\rm w}^{\rm s})$ for which (16.6.1) holds. Since δ_{P_0} in Proposition 16.1.4

is positive and depends only on the data, it follows that there exists $\hat{d} > 0$ depending only on the data and $\theta_{\rm w}^*$ such that $x_{P_0(\theta_{\rm w})} \geq \hat{d}$ for any $\theta_{\rm w} \in (\theta_{\rm w}^*, \theta_{\rm w}^{\rm s})$ satisfying (16.6.1). Therefore, by (16.1.10), $x \geq \hat{d}$ in Ω_{ε_0} for any admissible solution with such a wedge angle. Then (16.6.84) for such an admissible solution ψ follows from the L^∞–estimate of $D\psi$ in Ω, given by (16.6.79) in Corollary 16.6.12. □

Chapter Seventeen

Existence of Global Regular Reflection-Diffraction Solutions up to the Detachment Angle

17.1 STATEMENT OF THE EXISTENCE RESULTS

In this chapter, we prove the following existence assertion:

Proposition 17.1.1. *Let $\gamma > 1$ and $\rho_1 > \rho_0 > 0$. Let $\theta_{\rm w}^{\rm c} \in [\theta_{\rm w}^{\rm d}, \frac{\pi}{2})$ be the corresponding critical wedge angle, defined in Definition* 15.7.4. *Then, for any $\theta_{\rm w} \in (\theta_{\rm w}^{\rm c}, \frac{\pi}{2})$, there exists an admissible solution of* **Problem 2.6.1**.

The rest of this chapter is a proof of Proposition 17.1.1. To achieve this, it suffices to prove the existence of admissible solutions with $\theta_{\rm w} \in [\theta_{\rm w}^*, \frac{\pi}{2})$ for each $\theta_{\rm w}^* \in (\theta_{\rm w}^{\rm c}, \frac{\pi}{2})$. Thus, throughout this chapter, we fix $\theta_{\rm w}^* \in (\theta_{\rm w}^{\rm c}, \frac{\pi}{2})$.

Note that, if $\theta_{\rm w}^* > \theta_{\rm w}^{\rm s}$, the existence of admissible solutions with $\theta_{\rm w} \in [\theta_{\rm w}^*, \frac{\pi}{2})$ follows from Proposition 12.1.1. Thus, we focus essentially on the case:

$$\theta_{\rm w}^{\rm c} < \theta_{\rm w}^* \le \theta_{\rm w}^{\rm s},$$

although we do not assume this in this chapter (the argument works for Case $\theta_{\rm w}^* > \theta_{\rm w}^{\rm s}$ as well, which is essentially reduced to the proof of Proposition 12.1.1).

17.2 MAPPING TO THE ITERATION REGION

17.2.1 Mapping into the iteration region

We follow the construction in §12.2 for all $\theta_{\rm w} \in (\theta_{\rm w}^{\rm d}, \frac{\pi}{2})$, with only notational changes, by employing the convention that $P_0 = P_1 = P_4$ and $\overline{\Gamma_{\rm sonic}} = \{P_0\}$ for the subsonic wedge angles and the fact that all the estimates of Chapters 8–11 are extended to all $\theta_{\rm w} \in (\theta_{\rm w}^{\rm d}, \frac{\pi}{2})$ in Chapters 15–16. We now sketch the construction.

Using Definitions 11.2.1 and 16.1.1, we can find $\delta_* > 0$ such that (12.2.2) holds for any $\theta_{\rm w} \in [\theta_{\rm w}^{\rm d}, \frac{\pi}{2}]$. Fix this $\delta_* > 0$. Then, in the same way as in §12.2, for each wedge angle $\theta_{\rm w} \in (\theta_{\rm w}^{\rm d}, \frac{\pi}{2})$, we define S_{1,δ_*} by (12.2.1), point $\hat{P}_1$ by (12.2.3) in which we use circle $\partial B_{|\mathcal{O}_2 P_4|}(\mathcal{O}_2)$ instead of $\partial B_{c_2}(\mathcal{O}_2)$ (note that $|\mathcal{O}_2 P_4| = c_2$ for supersonic/sonic wedge angles, while $|\mathcal{O}_2 P_4| < c_2$ for subsonic wedge angles), arc $\Gamma_{\rm sonic}^{(\delta_*)}$ as the smaller arc $\hat{P}_1 P_4$ of $\partial B_{|\mathcal{O}_2 P_4|}(\mathcal{O}_2)$ (which is an arc, rather than one point, even for subsonic wedge angles), and $\mathcal{Q}$ and $\mathcal{Q}_{\rm bd}$ by (12.2.5). Then (12.2.6) holds for all $\theta_{\rm w} \in (\theta_{\rm w}^{\rm d}, \frac{\pi}{2}]$ by the same proof.

Moreover, to define $\mathcal{D}_{\delta_*,\varepsilon}$ for any wedge angle $\theta_{\rm w} \in [\theta_{\rm w}^{\rm d}, \frac{\pi}{2}]$, we modify (12.2.8) as follows: Let ε_0 and ε_1 be the constants from Corollary 16.6.13 and Lemma 16.1.2, respectively. For each $\varepsilon \in (0, \varepsilon_0]$, denote

$$\begin{aligned}\mathcal{D}_{\delta_*,\varepsilon}^{(\theta_{\rm w})} &:= \{\varphi_2 < \varphi_1 + \delta_*\} \cap \Lambda \cap \mathcal{N}_{\varepsilon_1}(\Gamma_{\rm sonic}^{(\delta_*)}) \cap \{x_{P_1} < x < x_{P_1} + \varepsilon\} \\ &\equiv \mathcal{Q} \cap \mathcal{N}_{\varepsilon_1}(\Gamma_{\rm sonic}^{(\delta_*)}) \cap \{x_{P_1} < x < x_{P_1} + \varepsilon\},\end{aligned} \tag{17.2.1}$$

where φ_2 and $\Gamma_{\rm sonic}^{(\delta_*)}$ correspond to the wedge angle $\theta_{\rm w}$. Then there exists a function $\hat{f}_{0,\delta_*} \in C^\infty([x_{P_1}, x_{P_1} + \varepsilon_0])$ such that

$$\begin{aligned}&\mathcal{D}_{\delta_*,\varepsilon_0} = \{(x,y) \,:\, x_{P_1} < x < x_{P_1} + \varepsilon_0,\;\; \theta_{\rm w} < y < \hat{f}_{0,\delta_*}(x)\}, \\ &S_{1,\delta_*} \cap \mathcal{N}_{\varepsilon_1}(\Gamma_{\rm sonic}^{(\delta_*)}) \cap \{x_{P_1} < x < x_{P_1} + \varepsilon_0\} \\ &\qquad = \{(x,y) \,:\, x_{P_1} < x < x_{P_1} + \varepsilon_0,\;\; y = \hat{f}_{0,\delta_*}(x)\}.\end{aligned} \tag{17.2.2}$$

We recall that, in the (x, y)–coordinates, $x_{P_1} = x_{P_4} = 0$ and $y_{P_1} > y_{P_4} = \theta_{\rm w}$ for supersonic wedge angles, while $x_{P_1} = x_{P_4} > 0$ and $y_{P_1} = y_{P_4} = \theta_{\rm w}$ for subsonic wedge angles.

Lemma 17.2.1. *Lemma* 12.2.2 *holds for any wedge angle* $\theta_{\rm w} \in [\theta_{\rm w}^{\rm d}, \frac{\pi}{2}]$, *with constants* $\delta > 0$, $C > 0$, *and* $\varepsilon_0 > 0$ *depending only on* (ρ_0, ρ_1, γ) *and the following notational changes that affect only subsonic/sonic wedge angles:* $P_0 = P_1 = P_4$ *for subsonic wedge angles; assertion* (iv) *in Lemma* 12.2.2 *should be replaced by the following (which applies to all the wedge angles* $\theta_{\rm w} \in [\theta_{\rm w}^{\rm d}, \frac{\pi}{2}]$, *since* $x_{P_4} = 0$ *for supersonic/sonic wedge angles):*

(iv) $F_1(P) = (x_P - x_{P_4}, y_P - y_{P_4})$ *for all* $P \in \mathcal{D}_{\delta_*, 3\varepsilon_0/4}^{(\theta_{\rm w})}$ *under the notation in* (17.2.1). *In particular,*

$$F_1(P_1) = (0, y_{P_1} - y_{P_4}), \quad F_1(P_4) = \mathbf{0},$$
$$F_1(\overline{\Gamma_{\rm sonic}}) = \{s = 0,\, 0 \le t \le y_{P_1} - y_{P_4}\}.$$

Also, $F_1(\mathcal{Q}_{\rm bd} \setminus \mathcal{D}_{\delta_*, 3\varepsilon_0/4}^{(\theta_{\rm w})}) = F_1(\mathcal{Q}_{\rm bd}) \cap \{s \ge \frac{3\varepsilon_0}{4}\}$.

Proof. The proof of Lemma 12.2.2 applies, with the following change in several places that refer to the (x, y)–coordinates: x should be replaced by $x - x_{P_4}$ (which does not cause any change for the supersonic angles considered in Lemma 12.2.2). Moreover, estimate (16.6.81) is used in the proof of (iv) instead of the estimate of $\partial_y(\varphi_1 - \varphi)$ in (11.2.37). □

Next, we have

Lemma 17.2.2. *For any $\theta_{\rm w}^* \in (\theta_{\rm w}^{\rm d}, \frac{\pi}{2})$, Lemma 12.2.3 holds for any admissible solutions with the wedge angle $\theta_{\rm w} \in [\theta_{\rm w}^*, \frac{\pi}{2})$ and the normal reflection for $\theta_{\rm w} = \frac{\pi}{2}$, with constants ε_0, ε_1, ε, $\delta > 0$ and $C \geq 1$ depending only on $(\rho_0, \rho_1, \gamma, \theta_{\rm w}^*)$, and constants $C(\hat{\varepsilon})$ (resp. $C(\alpha)$) depending, in addition, on $\hat{\varepsilon}$ (resp. α), given the following changes:*

- *Instead of (12.2.30), there exists $\alpha \in (0,1)$ such that*

$$\|\mathfrak{g}_{\rm sh}\|^{(-1-\alpha),\{0\}}_{2,\alpha,(0,\varepsilon_0)} \leq C, \quad \mathfrak{g}_{\rm sh}(0) = \mathfrak{g}_{S_1}(0), \quad \mathfrak{g}'_{\rm sh}(0) = \mathfrak{g}'_{S_1}(0). \tag{17.2.3}$$

We note that (17.2.3) is equivalent to (12.2.30) with norm $\|\cdot\|^{(1+\alpha),(\rm par)}_{2,\alpha,(0,\varepsilon)}$ instead of $\|\cdot\|^{(\rm par)}_{2,\alpha,(0,\varepsilon)} \equiv \|\cdot\|^{2,(\rm par)}_{2,\alpha,(0,\varepsilon)}$.

- *Instead of (12.2.31), for any $s \in (0, \hat{s}(\theta_{\rm w}))$ and $\theta_{\rm w} \in [\theta_{\rm w}^*, \frac{\pi}{2})$,*

$$\begin{aligned} &\min(\mathfrak{g}_{\rm sh}(0) + \frac{s}{M}, \frac{1}{M}) \\ &\qquad \leq \mathfrak{g}_{\rm sh}(s) \leq \min(\mathfrak{g}_{\rm sh}(0) + Ms, \eta^{(\theta_{\rm w})}(s) - \frac{1}{M}), \end{aligned} \tag{17.2.4}$$

where $\mathfrak{g}_{\rm sh}(0) = t_{P_1}(\theta_{\rm w}) \geq 0$, and $M \geq 1$ depends only on $(\rho_0, \rho_1, \gamma, \theta_{\rm w}^*)$, and function $\eta^{(\theta_{\rm w})}(\cdot)$ is from (12.2.10).

Proof. We follow the proof of Lemma 12.2.3 by using Corollary 16.6.13 and the estimates of Chapters 15–16 which extend the estimates of Chapters 8–11 to all $\theta_{\rm w} \in (\theta_{\rm w}^{\rm d}, \frac{\pi}{2})$. In particular, (17.2.3) follows directly from (16.6.82), since $\mathfrak{g}_{S_1}(s) = \hat{f}_0(s + x_{P_4}) - \theta_{\rm w}$ and $\mathfrak{g}_{\rm sh}(s) = \hat{f}(s + x_{P_4}) - \theta_{\rm w}$.

To prove (17.2.4), we note that, from (16.1.12)–(16.1.13), (16.6.82) in Corollary 16.6.13, and Lemma 17.2.1(iv), it follows that there exist $\hat{\varepsilon} \in (0, \varepsilon)$ and $M \geq 1$ depending only on $(\rho_0, \rho_1, \gamma, \theta_{\rm w}^*)$ such that

$$Ms + t_{P_1} \geq \mathfrak{g}_{\rm sh}(s) \geq \frac{1}{M}s + t_{P_1} \qquad \text{for all } s \in [0, \hat{\varepsilon}].$$

Recall that $t_{P_1} = \mathfrak{g}_{\rm sh}(0) \geq 0$. Now we employ Lemma 15.7.4(iv) (with $r = \frac{\hat{\varepsilon}}{2}$) and the proof of the upper bound in (12.2.31) (see the proof of Lemma 12.2.3) which works in the present case without change to conclude the proof of (17.2.4). □

Now we define a mapping into the iteration region $Q^{\rm iter} = [0,1]^2$. We first note

Lemma 17.2.3. *Lemma 12.2.4 holds for any $\theta_{\rm w} \in (\theta_{\rm w}^{\rm d}, \frac{\pi}{2}]$. In particular, $\theta_{\rm w}^{\rm s}$ is replaced by $\theta_{\rm w}^{\rm d}$ in (12.2.35) and in the definition of set $\mathcal{B}$ after (12.2.38).*

Proof. We argue in a way similar to Lemma 12.2.4, with only notational changes. Equation (12.2.39) in the (x,y)–coordinates, in the neighborhood $\mathcal{D}_{\delta_*,\varepsilon_0}$ defined now by (17.2.1), takes the form:

$$\tilde{\varphi}_{20}(x,y) := \begin{cases} \zeta_1(x-x_{P_1})\varphi_0(x,y)+(1-\zeta_1(x-x_{P_1}))\varphi_2(x,y) & \text{for } (x,y)\in\mathcal{D}_{\delta_*,\varepsilon_0}, \\ \varphi_0 & \text{otherwise}, \end{cases}$$

where we have used the expression of φ_2 in the (x,y)–coordinates in the first equality. With this, using Lemma 17.2.1(iv) and $x_{P_1} = x_{P_4}$, we follow the argument for Lemma 12.2.4 with the only changes that $(\varphi_1-\varphi_2)_{|(x,y)=(s,t+\theta_{\rm w})}$ is replaced by $(\varphi_1-\varphi_2)_{|(x,y)=(s+x_{P_1},t+\theta_{\rm w})}$ in (12.2.40) and in the estimate of $\frac{\partial I_2}{\partial t}$. Note that all of the changes affect only the subsonic angles, since $x_{P_1}=0$ in the supersonic/sonic case.

We also note that the wedge angles near $\frac{\pi}{2}$ are supersonic, so that Step 2 of the proof of Lemma 12.2.4 is unchanged. □

Let $\theta_{\rm w}\in(\theta_{\rm w}^{\rm d},\frac{\pi}{2})$, and let φ be an admissible solution. Then (12.2.24) holds by Lemma 17.2.2, which determines function $\mathfrak{g}_{\rm sh}(s)$. We define the map:

$$F_{(2,\mathfrak{g}_{\rm sh})}:\ \tilde{\Omega}_\infty=(0,\hat{s}(\theta_{\rm w}))\times(0,\infty)\mapsto\mathbb{R}^2$$

by (12.2.41). The map is well-defined by (17.2.4). However, for the subsonic/sonic wedge angles with $\mathfrak{g}_{\rm sh}(0)=t_{P_1}=0$, the map becomes singular as $s\to 0+$. Furthermore, noting that region $\tilde{\Omega}=F_1(\Omega)$ is of form (12.2.24), we find that $F_{(2,\mathfrak{g}_{\rm sh})}$ is one-to-one on $\tilde{\Omega}$, $F_{(2,\mathfrak{g}_{\rm sh})}(\tilde{\Omega})=Q^{\rm iter}$, and its inverse $F^{-1}_{(2,\mathfrak{g}_{\rm sh})}: Q^{\rm iter}\mapsto\tilde{\Omega}$ is given by (12.2.43). We note that the regularity of $F^{-1}_{(2,\mathfrak{g}_{\rm sh})}(\cdot)$ on $\overline{Q^{\rm iter}}$ is determined by the regularity of $\mathfrak{g}_{\rm sh}(\cdot)$ on $[0,\hat{s}(\theta)]$ for any $\theta_{\rm w}^*\in(\theta_{\rm w}^{\rm d},\frac{\pi}{2})$, even for the subsonic/sonic angles, *i.e.*, $F^{-1}_{(2,\mathfrak{g}_{\rm sh})}$ does not become singular as $s\to 0+$. However, in the subsonic/sonic case, $F^{-1}_{(2,\mathfrak{g}_{\rm sh})}$ does not extend to a one-to-one map on $\overline{Q^{\rm iter}}$ since $F^{-1}_{(2,\mathfrak{g}_{\rm sh})}(0,t)=(0,0)$ for any $t\in[0,1]$.

Now we can define function u on $Q^{\rm iter}$ by (12.2.44).

17.2.2 Uniform estimates of admissible solutions mapped into the iteration region

In order to state the extension of Proposition 12.2.5, *i.e.*, to give the uniform estimates for function u on $Q^{\rm iter}$ defined by (12.2.44) with an admissible solution φ, in the case of all the wedge angles including the subsonic/sonic case, we need to examine the properties of function u near $\{s=0\}$ in the case of subsonic and sonic wedge angles in more detail. Recall that $\mathfrak{g}_{\rm sh}(0)=0$ in this case, so that transform (12.2.41) becomes singular near $\{s=0\}$.

Suppose that φ is an admissible solution, and u on $Q^{\rm iter}$ is defined by (12.2.44). Using the property of Lemma 12.2.2(iv) extended to the subsonic reflection case in the form given in Lemma 17.2.1, we can use the (x,y)–coordinates on $\Omega\cap\mathcal{D}_{\varepsilon_0/10}$ for any $\theta_{\rm w}\in(\theta_{\rm w}^{\rm d},\frac{\pi}{2})$. For any $\boldsymbol{\xi}\in\Omega_{\varepsilon_0/10}$, set $(s,t)=F_{(2,\mathfrak{g}_{\rm sh})}\circ F_1(\boldsymbol{\xi})$. Then the (x,y)–coordinates of $\boldsymbol{\xi}$ satisfy

$$(x-x_{P_4},y-\theta_{\rm w})=(hs,\mathfrak{g}_{\rm sh}(hs)t)\qquad\text{for all } s\in(0,\frac{\varepsilon_0'}{10}),\ t\in(0,1).\tag{17.2.5}$$

Here

$$h:=\hat{s}(\theta_{\rm w}),\tag{17.2.6}$$

and ε_0' is defined similarly to (12.2.45), but with $\theta_{\rm w}^{\rm d}$ instead of $\theta_{\rm w}^{\rm s}$:

$$\varepsilon_0':=\frac{\varepsilon_0}{\max_{\theta_{\rm w}\in[\theta_{\rm w}^{\rm d},\frac{\pi}{2}]}\hat{s}(\theta_{\rm w})},\tag{17.2.7}$$

where ε_0 is from Lemma 17.2.1, and we have used the continuity of $\hat{s}(\cdot)$ on $[\theta_{\rm w}^{\rm s},\frac{\pi}{2}]$. Then $\varepsilon_0'\in(0,\frac{1}{2})$ depends only on the data.

Now, using (12.2.35) that holds by Lemma 17.2.3, we find that, for $\psi=\varphi-\varphi_2$ expressed in the (x,y)–coordinates and u given by (12.2.44), we have

$$\psi(x,y)=u(\frac{x-x_{P_4}}{h},\frac{y-\theta_{\rm w}}{\mathfrak{g}_{\rm sh}(x-x_{P_4})})\qquad\text{for all } (x,y)\in\Omega_{\varepsilon_0/10}.\tag{17.2.8}$$

Proposition 16.6.11 and Corollary 16.6.12 imply that $\psi\in C^{(-1-\alpha),\{P_0\}}_{2,\alpha,\Omega_\varepsilon}$ with $(\psi,D\psi)(P_0)=(0,\mathbf{0})$ for any subsonic/sonic reflection solution, for uniform $\varepsilon>0$, where $P_0=P_1=P_4$. Moreover, after possibly increasing M, we can write (17.2.4) in the subsonic/sonic case, *i.e.*, when $\mathfrak{g}_{\rm sh}(0)=0$, in the form:

$$\frac{s}{M}\le\mathfrak{g}_{\rm sh}(s)\le Ms\qquad\text{on } s\in(0,h),\tag{17.2.9}$$

where we have used the notation in (17.2.6).

Now we discuss the properties of functions $u(s,t)$ obtained by (17.2.8) from functions $\psi\in C^{(-1-\alpha),\{P_0\}}_{k,\alpha,\Omega}$ with $(\psi,D\psi)(P_0)=(0,\mathbf{0})$.

Let $h>0$, $x_{P_4}\in\mathbb{R}$, and $\theta_{\rm w}\in\mathbb{R}$, and let $\mathfrak{g}_{\rm sh}(\cdot)$ be a function on $[0,h]$ satisfying (17.2.9). This determines the following map $\mathbf{x}:\ \mathbb{R}^2\mapsto\mathbb{R}^2$:

$$(x,y)=\mathbf{x}(s,t)\qquad\text{defined by (17.2.5).}\tag{17.2.10}$$

Then map $\mathbf{x}$ is one-to-one on $(0,\frac{\varepsilon_0'}{10})\times(0,1)$, but not one-to-one on $[0,\frac{\varepsilon_0'}{10})\times(0,1)$. That is, map $\mathbf{x}$ loses its one-to-one property on the side: $\{s=0\}\times[0,1]$.

For $\varepsilon\in(0,h)$, define

$$\Omega_\varepsilon:=\mathbf{x}(Q_{\varepsilon'}),\qquad P_0=P_1=P_4:=\mathbf{x}(0,0),\tag{17.2.11}$$

with $\varepsilon':=\frac{\varepsilon}{h}$ and $Q_{\varepsilon'}:=Q^{\rm iter}\cap\{s<\varepsilon'\}$. Then Ω_ε is of the same form as in (16.1.10) with $\hat{f}(x)=\theta_{\rm w}+\mathfrak{g}_{\rm sh}(x-x_{P_4})$, where we recall that $P_4=(x_{P_4},\theta_{\rm w})$.

For $k\in\{1,2\}$ and $\alpha'\in(0,1)$ satisfying $k+\alpha'\ge1+\alpha$, denote

$$\hat{C}^{(-1-\alpha),\{P_0\}}_{k,\alpha',\Omega_\varepsilon}:=\{v\in C^{(-1-\alpha),\{P_0\}}_{k,\alpha',\Omega_\varepsilon}\ :\ v(P_0)=0,\ Dv(P_0)=\mathbf{0}\}.\tag{17.2.12}$$

We now describe the space of functions u on $Q_{\varepsilon'}$, obtained by (17.2.8) with $\psi \in \hat{C}^{(-1-\alpha),\{P_0\}}_{k,\alpha',\hat{\Omega}}$. For that, we define the following norm: For $\alpha \in (0,1)$ and points $\mathbf{s} = (s,t)$ and $\tilde{\mathbf{s}} = (\tilde{s},\tilde{t})$, denote

$$\delta_\alpha^{(\text{subs})}(\mathbf{s},\tilde{\mathbf{s}}) := \left(|s-\tilde{s}|^2 + \left(\max(s,\tilde{s})\right)^2|t-\tilde{t}|^2\right)^{\frac{\alpha}{2}}. \tag{17.2.13}$$

Then, for $\sigma > 0$, $\alpha \in (0,1)$, and a nonnegative integer m, define

$$\begin{aligned}
\|u\|^{(\sigma),(\text{subs})}_{m,0,Q_\varepsilon} &:= \sum_{0\le k+l\le m} \sup_{\mathbf{s}\in Q_\varepsilon} \left(s^{k-\sigma}|\partial_s^k\partial_t^l u(\mathbf{s})|\right),\\
[u]^{(\sigma),(\text{subs})}_{m,\alpha,Q_\varepsilon} &:= \sum_{k+l=m} \sup_{\mathbf{s},\tilde{\mathbf{s}}\in Q_\varepsilon, \mathbf{s}\ne\tilde{\mathbf{s}}} \Bigg(\min(s^{\alpha+k-\sigma}, \tilde{s}^{\alpha+k-\sigma})\\
&\qquad\qquad \times \frac{|\partial_s^k\partial_t^l u(\mathbf{s}) - \partial_s^k\partial_t^l u(\tilde{\mathbf{s}})|}{\delta_\alpha^{(\text{subs})}(\mathbf{s},\tilde{\mathbf{s}})}\Bigg),\\
\|u\|^{(\sigma),(\text{subs})}_{m,\alpha,Q_\varepsilon} &:= \|u\|^{(\sigma),(\text{subs})}_{m,0,Q_\varepsilon} + [u]^{(\sigma),(\text{subs})}_{m,\alpha,Q_\varepsilon},
\end{aligned} \tag{17.2.14}$$

where $Q = Q^{\text{iter}}$, $Q_\varepsilon := Q \cap \{s < \varepsilon\}$, and k and l are nonnegative integers.

Denote $C^{m,\alpha}_{\sigma,(\text{subs})}(Q_\varepsilon) := \{u \in C^m(Q_\varepsilon) : \|u\|^{(\sigma),(\text{subs})}_{m,\alpha,Q_\varepsilon} < \infty\}$.

We first note that, in (17.2.13), we can use $\min(s,\tilde{s})$ instead of $\max(s,\tilde{s})$, and the resulting norm (17.2.14) is equivalent, with the constant depending only on (m,α):

Lemma 17.2.4. *If $s, \tilde{s}, t, \tilde{t} \in (0,1)$, then*

$$|s-\tilde{s}|^2 + \left(\max(s,\tilde{s})\right)^2|t-\tilde{t}|^2 \le 5\left(|s-\tilde{s}|^2 + \left(\min(s,\tilde{s})\right)^2|t-\tilde{t}|^2\right). \tag{17.2.15}$$

Proof. Let $s, \tilde{s}, t, \tilde{t} \in (0,1)$. Assume that $\tilde{s} \le s$. Then we prove (17.2.15) by considering two separate cases.

If $s \le 2\tilde{s}$, then, recalling that $\tilde{s} \le s$, we see that $\max(s,\tilde{s}) \le 2\min(s,\tilde{s})$, so that (17.2.15) holds.

In the opposite case, recalling that $\tilde{s} \le s$, we obtain that $0 < \tilde{s} < \frac{s}{2}$, so that

$$s - \tilde{s} > \frac{s}{2} = \frac{1}{2}\max(s,\tilde{s}) \ge \frac{1}{2}\max(s,\tilde{s})|t-\tilde{t}|,$$

which implies

$$5|s-\tilde{s}|^2 \ge |s-\tilde{s}|^2 + \left(\max(s,\tilde{s})\right)^2|t-\tilde{t}|^2.$$

This means that (17.2.15) holds in this case as well. □

Now we prove the first main fact about spaces $C^{m,\alpha}_{\sigma,(\text{subs})}(Q_\varepsilon)$.

Lemma 17.2.5. *Let $M, M_1 \geq 1$, $x_{P_4} \in \mathbb{R}$, and $\theta_{\mathrm{w}} \in \mathbb{R}$. Let $h \in (\frac{1}{M_1}, M_1)$ and $\varepsilon \in (0, \frac{h}{2})$. Let $m \in \{1,2\}$ and $\alpha \in [0,1)$. Assume that a function $\mathfrak{g}_{\mathrm{sh}}(s)$ on $(0,\varepsilon)$ satisfies* (17.2.9) *on $(0,\varepsilon)$ and*

$$\|\mathfrak{g}_{\mathrm{sh}}\|^{(-1-\alpha),\{0\}}_{m,\alpha,(0,\varepsilon)} \leq M. \tag{17.2.16}$$

Let ψ be a function on Ω_ε. Define a function u on $Q_{\varepsilon'}$ by

$$u(s,t) = \psi\big(x_{P_4} + hs,\ \theta_{\mathrm{w}} + \mathfrak{g}_{\mathrm{sh}}(hs)t\big),$$

so that (17.2.8) *holds. Then $\psi \in \hat{C}^{(-1-\alpha),\{P_0\}}_{m,\alpha,\Omega_\varepsilon}$ if and only if $u \in C^{m,\alpha}_{1+\alpha,(\mathrm{subs})}(Q_{\varepsilon'})$, and*

$$\frac{1}{C}\|\psi\|^{(-1-\alpha),\{P_0\}}_{m,\alpha,\Omega_\varepsilon} \leq \|u\|^{(1+\alpha),(\mathrm{subs})}_{m,\alpha,Q_{\varepsilon'}} \leq C\|\psi\|^{(-1-\alpha),\{P_0\}}_{m,\alpha,\Omega_\varepsilon},$$

where C depends only on (M, M_1), and we have also used the notations in (17.2.10)–(17.2.11).

Proof. The proof consists of three steps.

1. From (17.2.12) and structure (16.1.10) of Ω_ε with $\hat{f}(x) = \mathfrak{g}_{\mathrm{sh}}(\frac{x-x_{P_4}}{h})$ satisfying (17.2.9), it is easy to see that, for $m = 1,2$, and $\alpha \in [0,1)$, space $\hat{C}^{(-1-\alpha),\{P_0\}}_{m,\alpha,\Omega_\varepsilon}$ can be characterized as the set of all $v \in C^{(-1-\alpha),\{P_0\}}_{m,\alpha,\Omega_\varepsilon}$ for which $\|v\|'^{,(-1-\alpha)}_{m,\alpha,\Omega_\varepsilon} < \infty$, where

$$\begin{aligned}
&\|v\|'^{,(-1-\alpha)}_{m,0,\Omega_\varepsilon} := \sum_{0\leq|\beta|\leq m} \sup_{\mathbf{x}\in\Omega_\varepsilon} \Big(|x - x_{P_4}|^{|\beta|-1-\alpha}|D^\beta v(\mathbf{x})|\Big),\\
&[v]'^{,(-1-\alpha)}_{m,\alpha,\Omega_\varepsilon}\\
&\quad := \sum_{|\beta|=m} \sup_{\mathbf{x},\tilde{\mathbf{x}}\in\Omega_\varepsilon,\mathbf{x}\neq\tilde{\mathbf{x}}} \Big(\min(|x-x_{P_4}|^{m-1}, |\tilde{x}-x_{P_4}|^{m-1})\frac{|D^\beta v(\mathbf{x}) - D^\beta v(\tilde{\mathbf{x}})|}{|\mathbf{x}-\tilde{\mathbf{x}}|^\alpha}\Big),\\
&\|v\|'^{,(-1-\alpha)}_{m,\alpha,\Omega_\varepsilon} := \|v\|'^{,(-1-\alpha)}_{m,0,\Omega_\varepsilon} + [v]'^{,(-1-\alpha)}_{m,\alpha,\Omega_\varepsilon},
\end{aligned} \tag{17.2.17}$$

where $\mathbf{x} = (x,y)$, $\tilde{\mathbf{x}} = (\tilde{x},\tilde{y})$, and $\beta = (\beta_1,\beta_2)$ is a multi-index. Moreover, norms $\|\cdot\|^{(-1-\alpha),\{P_0\}}_{m,\alpha,\Omega_\varepsilon}$ and $\|\cdot\|'^{,(-1-\alpha)}_{m,\alpha,\Omega}$ are equivalent on $\hat{C}^{(-1-\alpha),\{P_0\}}_{m,\alpha,\Omega_\varepsilon}$, with constant C depending only on (M, M_1), where only the case that $m = 1,2$, and $\varepsilon < h$ is considered.

Then it suffices to show that

$$\frac{1}{C}\|\psi\|'^{,(-1-\alpha)}_{m,\alpha,\Omega_\varepsilon} \leq \|u\|^{(1+\alpha),(\mathrm{subs})}_{m,\alpha,Q_{\varepsilon'}} \leq C\|\psi\|'^{,(-1-\alpha)}_{m,\alpha,\Omega_\varepsilon} \tag{17.2.18}$$

with C depending only on (M, M_1). In the argument below, the universal constant C depends only on these parameters.

2. We note that the first and second derivatives of ψ are expressed through the derivatives of u by

$$\begin{aligned}
\psi_x &= \frac{1}{h}u_s - t\frac{\mathfrak{g}'}{\mathfrak{g}}u_t, \qquad \psi_y = \frac{1}{\mathfrak{g}}u_t,\\
\psi_{xx} &= \frac{1}{h^2}u_{ss} - 2\frac{t\mathfrak{g}'}{h\mathfrak{g}}u_{st} + \Big(\frac{t\mathfrak{g}'}{\mathfrak{g}}\Big)^2 u_{tt} - t\Big(\frac{\mathfrak{g}''}{\mathfrak{g}} - 2\frac{(\mathfrak{g}')^2}{\mathfrak{g}^2}\Big)u_t,\\
\psi_{xy} &= \frac{1}{h\mathfrak{g}}u_{st} - \frac{t\mathfrak{g}'}{\mathfrak{g}^2}u_{tt} - \frac{\mathfrak{g}'}{\mathfrak{g}^2}u_t, \qquad \psi_{yy} = \frac{1}{\mathfrak{g}^2}u_{tt},
\end{aligned} \tag{17.2.19}$$

where $\mathfrak{g}(\cdot)$ is used for $\mathfrak{g}_{\rm sh}(\cdot)$, and $(D,D^2)\psi = (D,D^2)\psi(x,y)$, $(D,D^2)u = (D,D^2)u(\frac{x-x_{P_4}}{h}, \frac{y-\theta_{\rm w}}{\mathfrak{g}(x-x_{P_4})})$, and $(\mathfrak{g},\mathfrak{g}') = (\mathfrak{g},\mathfrak{g}')(x-x_{P_4})$. Note that, using map (17.2.10) and denoting by $\mathbf{s}(\cdot)$ its inverse, then $(D,D^2)u = (D,D^2)u(\mathbf{s}(\mathbf{x}))$.

From (17.2.19), using (17.2.9) on $(0,\varepsilon)$ and $\|\mathfrak{g}_{\rm sh}\|^{(-1-\alpha),\{0\}}_{m,\alpha,(0,\varepsilon)} \le M$, we directly obtain

$$\frac{1}{C}\|\psi\|^{',(-1-\alpha)}_{m,0,\Omega_\varepsilon} \le \|u\|^{(1+\alpha),({\rm subs})}_{m,0,Q_{\varepsilon'}} \le C\|\psi\|^{',(-1-\alpha)}_{m,0,\Omega_\varepsilon}.$$

Indeed, let $m=1$ in order to fix the setting. Then $u \in C^{1,0}_{1+\alpha,({\rm subs})}(Q_{\varepsilon'})$ implies that, for $(s,t) \in (0,\varepsilon') \times (0,1)$,

$$|u(s,t)| \le Ls^{1+\alpha}, \quad |u_s(s,t)| \le Ls^{\alpha}, \quad |u_t(s,t)| \le Ls^{1+\alpha},$$

where $L = \|u\|^{(1+\alpha),({\rm subs})}_{1,0,Q_{\varepsilon'}}$. By (17.2.19) (where the arguments in the functions are specified after the equation), and using the lower bound $\mathfrak{g}(s) \ge \frac{s}{M}$ in (17.2.9), we have

$$|\psi_y| = |\frac{1}{\mathfrak{g}}u_t| \le CL(x-x_{P_4})^{\alpha}.$$

Similarly, using (17.2.16) in addition to the previous properties, we have

$$|\psi_x| = |\frac{1}{h}u_s - t\frac{\mathfrak{g}'}{\mathfrak{g}}u_t| \le CL(x-x_{P_4})^{\alpha}.$$

Also, from (17.2.8), $|\psi| \le CL(x-x_{P_4})^{1+\alpha}$. Combining these estimates, we have

$$\|\psi\|^{',(-1-\alpha)}_{m,0,\Omega_\varepsilon} \le C\|u\|^{(1+\alpha),({\rm subs})}_{m,0,Q_{\varepsilon'}} \qquad \text{for } m=1.$$

For $m=2$, the proof is similar, on account of using the expressions of the second derivatives in (17.2.19). Also, the estimates:

$$\|u\|^{(1+\alpha),({\rm subs})}_{m,0,Q_{\varepsilon'}} \le C\|\psi\|^{',(-1-\alpha)}_{m,0,\Omega_\varepsilon} \qquad \text{for } m=1,2$$

are obtained similarly by noting from (17.2.19) that $(D,D^2)u$ can be expressed through $(D,D^2)\psi$ and using the upper bound that $|\mathfrak{g}(s)| \le Ms$ in (17.2.9).

3. It remains to show the estimates of the difference-quotient terms:

$$[\psi]^{',(-1-\alpha)}_{m,\alpha,\Omega_\varepsilon} \le C\|u\|^{(1+\alpha),({\rm subs})}_{m,\alpha,Q_{\varepsilon'}}, \quad [u]^{(1+\alpha),({\rm subs})}_{m,\alpha,Q_{\varepsilon'}} \le C\|\psi\|^{',(-1-\alpha)}_{m,\alpha,\Omega_\varepsilon} \tag{17.2.20}$$

for $m = 1, 2$. We show the details of the estimate for one of the terms of the expression of $[u]^{(1+\alpha),(\mathrm{subs})}_{m,\alpha,Q_{\varepsilon'}}$ in (17.2.14) with $m = 2$: $k = 0$ and $l = 2$. That is, we show that

$$\sup_{\mathbf{s},\tilde{\mathbf{s}}\in Q_{\varepsilon'},\mathbf{s}\neq\tilde{\mathbf{s}}} \Big(\min(s^{-1}, \tilde{s}^{-1}) \frac{|u_{tt}(\mathbf{s}) - u_{tt}(\tilde{\mathbf{s}})|}{\delta_{\alpha}^{(\mathrm{subs})}(\mathbf{s}, \tilde{\mathbf{s}})} \Big) \le C \|\psi\|^{\prime,(-1-\alpha)}_{2,\alpha,\Omega_{\varepsilon}}. \tag{17.2.21}$$

To prove this, we change the variables from $\mathbf{s} = (s, t)$ to $\mathbf{x} = (x, y)$ on the left-hand side of (17.2.21) by transformation (17.2.5), use (17.2.19) to replace u by ψ, and then estimate the resulting expression. In order to do this, we first show that the denominators in the difference quotients in (17.2.14) and (17.2.17) are comparable to each other under the change of variables (17.2.5).

Claim 17.2.6. *Denote by* $\mathbf{x}(\mathbf{s})$ *the transformation:* $Q_{\varepsilon'} \to \Omega_{\varepsilon}$, *given by* (17.2.5) *and* (17.2.10). *Then there exists* C *depending only on* (M, M_1) *such that*

$$\frac{1}{C} \delta_1^{(\mathrm{subs})}(\mathbf{s}, \tilde{\mathbf{s}}) \le |\mathbf{x}(\mathbf{s}) - \mathbf{x}(\tilde{\mathbf{s}})| \le C \delta_1^{(\mathrm{subs})}(\mathbf{s}, \tilde{\mathbf{s}}).$$

This can be seen as follows: For $\mathbf{s} = (s, t)$ and $\tilde{\mathbf{s}} = (\tilde{s}, \tilde{t})$, from (17.2.5),

$$|\mathbf{x}(\mathbf{s}) - \mathbf{x}(\tilde{\mathbf{s}})|^2 = h^2 (s - \tilde{s})^2 + \big(\mathfrak{g}(hs)t - \mathfrak{g}(h\tilde{s})\tilde{t}\big)^2.$$

We may assume that

$$\tilde{s} = \min(s, \tilde{s}). \tag{17.2.22}$$

Since $\mathfrak{g}(hs)t - \mathfrak{g}(h\tilde{s})\tilde{t} = t\big(\mathfrak{g}(hs) - \mathfrak{g}(h\tilde{s})\big) + \mathfrak{g}(h\tilde{s})(t - \tilde{t})$, and $s, t, \tilde{s}, \tilde{t} \in (0, 1)$, then we can employ (17.2.16) and the upper bound: $|\mathfrak{g}(s)| \le Ms$ in (17.2.9) to obtain

$$\begin{aligned} |\mathbf{x}(\mathbf{s}) - \mathbf{x}(\tilde{\mathbf{s}})|^2 &= h^2 (s - \tilde{s})^2 + \big(t(\mathfrak{g}(hs) - \mathfrak{g}(h\tilde{s})) + \mathfrak{g}(h\tilde{s})(t - \tilde{t})\big)^2 \\ &\le C\big((s - \tilde{s})^2 + \tilde{s}^2 (t - \tilde{t})^2\big) \le C\big(\delta_1^{(\mathrm{subs})}(\mathbf{s}, \tilde{\mathbf{s}})\big)^2. \end{aligned}$$

Thus, the upper bound of $|\mathbf{x}(\mathbf{s}) - \mathbf{x}(\tilde{\mathbf{s}})|$ in the claim is proved.

To prove the lower bound of $|\mathbf{x}(\mathbf{s}) - \mathbf{x}(\tilde{\mathbf{s}})|$, we consider two cases.

In the case that $|s - \tilde{s}| \ge \frac{s}{2M^2} |t - \tilde{t}|$, we estimate

$$\begin{aligned} |\mathbf{x}(\mathbf{s}) - \mathbf{x}(\tilde{\mathbf{s}})| &= \Big(h^2 (s - \tilde{s})^2 + \big(\mathfrak{g}(hs)t - \mathfrak{g}(h\tilde{s})\tilde{t}\big)^2 \Big)^{\frac{1}{2}} \\ &\ge \frac{1}{C} |s - \tilde{s}| \ge \frac{1}{C_1} \delta_1^{(\mathrm{subs})}(\mathbf{s}, \tilde{\mathbf{s}}), \end{aligned}$$

where we have used the condition of this case and (17.2.22) in the last inequality.

In the opposite case, when $|s - \tilde{s}| \le \frac{s}{2M^2} |t - \tilde{t}|$, we use

$$\mathfrak{g}(hs)t - \mathfrak{g}(h\tilde{s})\tilde{t} = \tilde{t}\big(\mathfrak{g}(hs) - \mathfrak{g}(h\tilde{s})\big) + \mathfrak{g}(hs)(t - \tilde{t})$$

and the lower bound: $\mathfrak{g}(s) \geq \frac{s}{M}$ in (17.2.9) to estimate:

$$\begin{aligned}|\mathbf{x}(\mathbf{s}) - \mathbf{x}(\tilde{\mathbf{s}})| &\geq \left|\tilde{t}(\mathfrak{g}(hs) - \mathfrak{g}(h\tilde{s})) + \mathfrak{g}(hs)(t - \tilde{t})\right| \\ &\geq \left|\mathfrak{g}(hs)(t - \tilde{t})\right| - \left|\mathfrak{g}(hs) - \mathfrak{g}(h\tilde{s})\right| \\ &\geq \frac{hs}{M}|t - \tilde{t}| - Mh|s - \tilde{s}| \\ &\geq \frac{hs}{2M}|t - \tilde{t}| \geq \frac{1}{C}\delta_1^{(\mathrm{subs})}(\mathbf{s}, \tilde{\mathbf{s}}),\end{aligned}$$

where we have used the condition of this case and (17.2.22) in the last two inequalities. Now Claim 17.2.6 is proved.

We continue the proof of (17.2.21).

Let $\mathbf{s}, \tilde{\mathbf{s}} \in Q_{\varepsilon'}$ and $\mathbf{s} \neq \tilde{\mathbf{s}}$. Let $\mathbf{x} = \mathbf{x}(\mathbf{s})$ and $\tilde{\mathbf{x}} = \mathbf{x}(\tilde{\mathbf{s}})$, where $\mathbf{x}(\cdot)$ is given by (17.2.5) and (17.2.10). Let $\mathbf{s}(\cdot)$ be the inverse of $\mathbf{x}(\cdot)$. Then $\mathbf{x}, \tilde{\mathbf{x}} \in \Omega_\varepsilon$ and $\mathbf{x} \neq \tilde{\mathbf{x}}$. We change the variables from (s,t) to (x,y) on the left-hand side of (17.2.21), substitute $u_{tt}(\mathbf{s}(\mathbf{x})) = \mathfrak{g}^2(x - x_{P_4})\psi_{yy}(\mathbf{x})$ by (17.2.19), and employ also Claim 17.2.6 to replace the denominator in the difference quotient. Then, assuming without loss of generality that $\tilde{s} \leq s$, which implies that $0 < \tilde{x} - x_{P_4} \leq x - x_{P_4}$ so that $|x - \tilde{x}| \leq x - x_{P_4}$, and employing the upper bound in (17.2.9) and (17.2.16), we have

$$\begin{aligned}&\min(s^{-1}, \tilde{s}^{-1})\frac{|u_{tt}(\mathbf{s}) - u_{tt}(\tilde{\mathbf{s}})|}{\delta_\alpha^{(\mathrm{subs})}(\mathbf{s}, \tilde{\mathbf{s}})} \\ &\leq C(x - x_{P_4})^{-1}\frac{|\mathfrak{g}^2(x - x_{P_4})\psi_{yy}(\mathbf{x}) - \mathfrak{g}^2(\tilde{x} - x_{P_4})\psi_{yy}(\tilde{\mathbf{x}})|}{|\mathbf{x} - \tilde{\mathbf{x}}|^\alpha} \\ &\leq C(x - x_{P_4})^{-1}\Big(|\psi_{yy}(\mathbf{x})|\frac{|\mathfrak{g}^2(x - x_{P_4}) - \mathfrak{g}^2(\tilde{x} - x_{P_4})|}{|\mathbf{x} - \tilde{\mathbf{x}}|^\alpha} \\ &\qquad\qquad + \mathfrak{g}^2(\tilde{x} - x_{P_4})\frac{|\psi_{yy}(\mathbf{x}) - \psi_{yy}(\tilde{\mathbf{x}})|}{|\mathbf{x} - \tilde{\mathbf{x}}|^\alpha}\Big) \\ &\leq C(x - x_{P_4})^{-1}\Big(\|\psi\|^{',(-1-\alpha)}_{2,0,\Omega_\varepsilon}(x - x_{P_4})^{\alpha - 1}(x - x_{P_4})\frac{|x - \tilde{x}|}{|\mathbf{x} - \tilde{\mathbf{x}}|^\alpha} \\ &\qquad\qquad + (\tilde{x} - x_{P_4})^2[\psi]^{',(-1-\alpha)}_{2,\alpha,\Omega_\varepsilon}(x - x_{P_4})^{-1}\Big) \\ &\leq C\|\psi\|^{',(-1-\alpha)}_{2,\alpha,\Omega_\varepsilon}.\end{aligned}$$

Since this is true for each $\mathbf{s}, \tilde{\mathbf{s}} \in Q_{\varepsilon'}$ such that $\mathbf{s} \neq \tilde{\mathbf{s}}$, this proves (17.2.21). The other terms in the expression of $[u]^{(1+\alpha),(\mathrm{subs})}_{m,\alpha,Q_{\varepsilon'}}$ in (17.2.14) are estimated similarly. This proves the second estimate in (17.2.20).

The first estimate in (17.2.20) is proved similarly, *i.e.*, for each term in the expression of $[\psi]^{',(-1-\alpha)}_{m,\alpha,\Omega_\varepsilon}$ in (17.2.17), we change the variables from (x,y) to (s,t) by (17.2.5) and (17.2.10), express the derivatives of ψ through the derivatives of u by (17.2.19), replace the denominator $|\mathbf{x} - \tilde{\mathbf{x}}|^\alpha$ with $\delta_\alpha^{(\mathrm{subs})}(\mathbf{s}, \tilde{\mathbf{s}}) = \big(\delta_1^{(\mathrm{subs})}(\mathbf{s}, \tilde{\mathbf{s}})\big)^\alpha$ by using Claim 17.2.6, and then perform the estimates by using the lower bound in (17.2.9). Now Lemma 17.2.5 is proved. □

Corollary 17.2.7. *Let $\varepsilon' \in (0,1)$, $\alpha, \hat{\alpha} \in [0,1)$ with $\alpha < \hat{\alpha}$, and $m \in \{1,2\}$. Assume that there exists $M > 0$ such that $\|u_k\|^{(1+\hat{\alpha}),(\mathrm{subs})}_{m,\hat{\alpha},Q_{\varepsilon'}} \le M$ holds for a sequence of functions $\{u_k\}_{k=1}^{\infty}$ on $Q_{\varepsilon'}$. Then there exists a subsequence $\{u_{k_j}\}_{j=1}^{\infty}$ converging in norm $\|\cdot\|^{(1+\alpha),(\mathrm{subs})}_{m,\alpha,Q_{\varepsilon'}}$.*

Proof. Let $h = 1$, $x_{P_4} = \theta_{\mathrm{w}} = 0$, $\varepsilon = h\varepsilon' = \varepsilon'$, and $\mathfrak{g}_{\mathrm{sh}}(s) = s$ on $(0,\varepsilon)$. This defines map $\mathbf{x}(\cdot)$ by (17.2.10), and Ω_ε and P_0 by (17.2.11). Now, using Lemma 17.2.5, the assertion follows from the compactness of $C^{(-1-\hat{\alpha}),\{P_0\}}_{m,\hat{\alpha},\Omega_\varepsilon}$ in $C^{(-1-\alpha),\{P_0\}}_{m,\alpha,\Omega_\varepsilon}$. □

We also note that one side of the estimates in Lemma 17.2.5 holds in the general case of the conditions in (17.2.4). Specifically, let $\mathfrak{g}_{\mathrm{sh}}(\cdot)$ be a function on $[0,h]$ satisfying

$$\min(\mathfrak{g}_{\mathrm{sh}}(0) + \frac{s}{M}, \ \delta_{\mathrm{sh}}) \le \mathfrak{g}_{\mathrm{sh}}(s) \le \mathfrak{g}_{\mathrm{sh}}(0) + Ms, \tag{17.2.23}$$

where $\mathfrak{g}_{\mathrm{sh}}(0) \ge 0$ and $\delta_{\mathrm{sh}} > 0$ satisfy $\delta_{\mathrm{sh}} \ge \mathfrak{g}_{\mathrm{sh}}(0)$.

We fix $x_{P_4} \in \mathbb{R}$ and use map (17.2.10). For $\varepsilon \in (0,h)$, denote

$$\begin{aligned} &\Omega_\varepsilon := \mathbf{x}(Q_{\varepsilon'}) \qquad \text{with } \varepsilon' := \frac{\varepsilon}{h} \ \text{ and } \ Q_{\varepsilon'} := Q^{\mathrm{iter}} \cap \{s < \varepsilon'\}, \\ &P_1 := \mathbf{x}(0,1) = (x_{P_4}, \theta_{\mathrm{w}}), \ \ P_4 := \mathbf{x}(0,0) = (x_{P_4}, \theta_{\mathrm{w}} + \mathfrak{g}_{\mathrm{sh}}(0)), \\ &\Gamma_{\mathrm{sonic}} := \partial\Omega_\varepsilon \cap \{x = x_{P_4}\} = \mathbf{x}(\{0\} \times (0,1)). \end{aligned} \tag{17.2.24}$$

Note that these notations are consistent with (17.2.11) when $\mathfrak{g}_{\mathrm{sh}}(0) = 0$. Also, Ω_ε is of the form as in (16.1.10) with $\hat{f}(x) = \theta_{\mathrm{w}} + \mathfrak{g}_{\mathrm{sh}}(x - x_{P_4})$.

Similarly to (17.2.12) for $k \in \{1,2\}$ and $\alpha' \in (0,1)$ satisfying $k + \alpha' \ge 1 + \alpha$, we denote

$$\hat{C}^{(-1-\alpha),\{\Gamma_{\mathrm{sonic}}\}}_{k,\alpha',\Omega_\varepsilon} := \{v \in C^{(-1-\alpha),\{\Gamma_{\mathrm{sonic}}\}}_{k,\alpha',\Omega_\varepsilon} \ : \ v = 0, \ Dv = \mathbf{0} \ \text{ on } \Gamma_{\mathrm{sonic}}\}. \tag{17.2.25}$$

Lemma 17.2.8. *Let $M, M_1, M_2 \ge 1$, $x_{P_4} \in \mathbb{R}$, and $\theta_{\mathrm{w}} \in \mathbb{R}$. Let $h \in (\frac{1}{M_1}, M_1)$, $\delta_{\mathrm{sh}} \ge \frac{1}{M_2}$, and $\varepsilon \in (0, \frac{h}{2})$. Let $m \in \{1,2\}$ and $\alpha \in [0,1)$. Assume that a function $\mathfrak{g}_{\mathrm{sh}}(s)$ on $(0,\varepsilon)$ satisfies (17.2.16) and (17.2.23). Let u be a function on $Q_{\varepsilon'}$, and define function ψ on Ω_ε by (17.2.8). Then $u \in C^{m,\alpha}_{1+\alpha,(\mathrm{subs})}(Q_{\varepsilon'})$ implies that $\psi \in \hat{C}^{(-1-\alpha),\{\Gamma_{\mathrm{sonic}}\}}_{m,\alpha,\Omega_\varepsilon}$ with*

$$\|\psi\|^{(-1-\alpha),\{\Gamma_{\mathrm{sonic}}\}}_{m,\alpha,\Omega_\varepsilon} \le C\|u\|^{(1+\alpha),(\mathrm{subs})}_{m,\alpha,Q_{\varepsilon'}},$$

where C depends only on (M, M_1, M_2), and we have also used the notations in (17.2.10)–(17.2.11) and (17.2.24).

Proof. We follow the proof of Lemma 17.2.5.

First, we note that $x = x_{P_4}$ on $\Gamma_{\rm sonic}$, which implies that, by an argument similar to that for Lemma 17.2.5 for $m = 1, 2$, and $\alpha \in [0, 1)$, space $\hat{C}^{(-1-\alpha),\{\Gamma_{\rm sonic}\}}_{m,\alpha,\Omega_\varepsilon}$ can be characterized as the set of all $v \in C^{(-1-\alpha),\{\Gamma_{\rm sonic}\}}_{m,\alpha,\Omega_\varepsilon}$ for which $\|v\|'^{,(-1-\alpha)}_{m,\alpha,\Omega_\varepsilon} < \infty$, where norm $\|v\|'^{,(-1-\alpha)}_{m,\alpha,\Omega_\varepsilon}$ is defined by (17.2.17).

Moreover, norms $\|\cdot\|^{(-1-\alpha),\{\Gamma_{\rm sonic}\}}_{m,\alpha,\Omega_\varepsilon}$ and $\|\cdot\|'^{,(-1-\alpha)}_{m,\alpha,\Omega}$, for $m = 1, 2$, are equivalent to $\hat{C}^{(-1-\alpha),\{\Gamma_{\rm sonic}\}}_{m,\alpha,\Omega_\varepsilon}$, with constant C depending only on (M, M_1).

Thus, it suffices to show

$$\|\psi\|'^{,(-1-\alpha)}_{m,\alpha,\Omega_\varepsilon} \le C\|u\|^{(1+\alpha),(\rm subs)}_{m,\alpha,Q_{\varepsilon/h}} \tag{17.2.26}$$

with C depending only on (M, M_1, M_2).

For that, we first consider the case that $\varepsilon \le \frac{M}{M_2}$. In this case, for any $s \in (0, \varepsilon)$,

$$\delta_{\rm sh} \ge \frac{1}{M_2} \ge \frac{s}{M},$$

so that, from (17.2.23),

$$\mathfrak{g}_{\rm sh}(s) \ge \min(\mathfrak{g}_{\rm sh}(0) + \frac{s}{M}, \delta_{\rm sh}) \ge \min(\mathfrak{g}_{\rm sh}(0) + \frac{s}{M}, \frac{s}{M}) \ge \frac{s}{M}.$$

Then we follow the proof of estimate (17.2.26) in Lemma 17.2.5 without change, by noting that, for the proof of that estimate, only the lower bound $\mathfrak{g}_{\rm sh}(s) \ge \frac{s}{M}$ in (17.2.9) has been used, and this estimate holds in our present setting as shown above. Thus, we obtain (17.2.26) with C depending only on (M, M_1) when $\varepsilon \le \frac{M}{M_2}$.

If $\varepsilon > \frac{M}{M_2}$, we first note that (17.2.26) is already proved with $\tilde{\varepsilon} = \frac{M}{M_2}$ instead of ε. Next, from (17.2.16) and (17.2.23), transform (17.2.9) is one-to-one on $Q_{\varepsilon/h} \setminus Q_{\tilde{\varepsilon}/(2h)}$ onto $\Omega_\varepsilon \setminus \Omega_{\tilde{\varepsilon}/2}$, with $\mathbf{x}(\cdot) \in C^{m,\alpha}(\overline{Q_{\varepsilon/h} \setminus Q_{\tilde{\varepsilon}/(2h)}}; \mathbb{R}^2)$ and $\mathbf{x}^{-1}(\cdot) \in C^{m,\alpha}(\overline{\Omega_\varepsilon \setminus \Omega_{\tilde{\varepsilon}/2}}; \mathbb{R}^2)$. Moreover, norms $\|\cdot\|'^{,(-1-\alpha)}_{m,\alpha,\Omega_\varepsilon\setminus\Omega_{\tilde{\varepsilon}/2}}$ and $\|\cdot\|^{(1+\alpha),(\rm subs)}_{m,\alpha,Q_{\varepsilon/h}\setminus Q_{\tilde{\varepsilon}/(2h)}}$ are equivalent to the standard $C^{m,\alpha}$–norms in the closures of the corresponding domains, with the constants depending only on $(\varepsilon, \tilde{\varepsilon})$, and hence on (M, M_1, M_2), since $\frac{M_1}{2} \ge \varepsilon \ge \tilde{\varepsilon} = \frac{M}{M_2}$. Then

$$\|\psi\|'^{,(-1-\alpha)}_{m,\alpha,\Omega_\varepsilon\setminus\Omega_{\tilde{\varepsilon}/2}} \le C\|u\|^{(1+\alpha),(\rm subs)}_{m,\alpha,Q_{\varepsilon/h}\setminus Q_{\tilde{\varepsilon}/(2h)}}$$

with C depending only on (M, M_1, M_2). Combining this with (17.2.26) for $\tilde{\varepsilon} = \frac{M}{M_2}$, we obtain (17.2.26) for ε with C depending only on (M, M_1, M_2). ☐

We will also need the following:

Lemma 17.2.9. *If* $\mathfrak{g}_{\rm sh}(0) > 0$ *in the conditions of Lemma* 17.2.8, *then*

$$|\partial_y\psi(x,y)| \le \|u\|^{(1+\alpha),(\rm subs)}_{1,\alpha,Q_{\varepsilon'}}(x - x_{P_4})^{\frac{1}{2}+\alpha} \qquad \text{for all } (x,y) \in \Omega_r,$$

where $r = \min\{\mathfrak{g}^2_{\rm sh}(0),\ \varepsilon\}$.

Proof. By (17.2.14) and (17.2.19), we have

$$\psi_y(x,y) = \frac{1}{\mathfrak{g}_{\rm sh}(x-x_{P_4})} u_t\Big(\frac{x-x_{P_4}}{h}, \frac{y-\theta_{\rm w}}{\mathfrak{g}_{\rm sh}(x-x_{P_4})}\Big),$$

$$|u_t(s,t)| \le \|u\|^{(1+\alpha),({\rm subs})}_{1,\alpha,Q_{\varepsilon'}} s^{1+\alpha}.$$

Also, $\mathfrak{g}_{\rm sh}(s) \ge \mathfrak{g}_{\rm sh}(0)$ for any $s \in (0,r)$ by (17.2.23). Thus, for $(x,y) \in \Omega_r$, using that $0 \le x - x_{P_4} \le r \le \mathfrak{g}^2_{\rm sh}(0)$, we have

$$\begin{aligned} |\psi_y(x,y)| &\le \|u\|^{(1+\alpha),({\rm subs})}_{1,\alpha,Q_{\varepsilon'}} \frac{(x-x_{P_4})^{1+\alpha}}{\mathfrak{g}_{\rm sh}(0)} \\ &\le \|u\|^{(1+\alpha),({\rm subs})}_{1,\alpha,Q_{\varepsilon'}} \frac{\sqrt{r}}{\mathfrak{g}_{\rm sh}(0)} (x-x_{P_4})^{\frac{1}{2}+\alpha} \\ &\le \|u\|^{(1+\alpha),({\rm subs})}_{1,\alpha,Q_{\varepsilon'}} (x-x_{P_4})^{\frac{1}{2}+\alpha}. \end{aligned}$$

□

In the next two lemmas, we show the relation between spaces $C^{m,\alpha}_{\sigma,({\rm subs})}(Q_{\varepsilon'})$ and $C^{m,\alpha}_{\sigma,({\rm par})}(Q_{\varepsilon'})$.

Lemma 17.2.10. *Let $\varepsilon' \in (0,1)$, $\alpha \in (0,1)$, $\sigma \ge 0$, and m be a nonnegative integer. If $u \in C^{m,\alpha}_{\sigma,({\rm subs})}(Q_{\varepsilon'})$, then $u \in C^{m,\alpha}_{\sigma,({\rm par})}(Q_{\varepsilon'})$, where the parabolic norms are with respect to $\{s=0\}$, i.e., (x,y) in (4.6.2) are (s,t) in $Q_{\varepsilon'}$. Moreover,*

$$\|u\|^{(\sigma),({\rm par})}_{m,\alpha,Q_{\varepsilon'}} \le \|u\|^{(\sigma),({\rm subs})}_{m,\alpha,Q_{\varepsilon'}}.$$

Proof. The assertion follows directly from the expressions of these norms in (4.6.2) (with (s,t) instead of (x,y)) and (17.2.14), since the weights in the corresponding terms are larger than those in the expression of $\|\cdot\|^{(\sigma),({\rm par})}_{m,\alpha,Q_{\varepsilon'}}$ (*i.e.*, $s^{k+\frac{l}{2}-\sigma} \le s^{k-\sigma}$ and $s^{\alpha+k+\frac{l}{2}-\sigma} \le s^{\alpha+k-\sigma}$ since $s \in (0,1)$ and $l \ge 0$), and also $\delta^{({\rm par})}_\alpha(\mathbf{s},\tilde{\mathbf{s}}) \ge \delta^{({\rm subs})}_\alpha(\mathbf{s},\tilde{\mathbf{s}})$ from their definitions (4.6.1) and (17.2.13) for $s, \tilde{s} \in (0,1)$. □

In the next lemma, we restrict the values of α to $(0,\frac{1}{3}]$.

Lemma 17.2.11. *Let $\varepsilon' \in (0,1)$ and $\alpha \in (0,\frac{1}{3}]$. If $u \in C^{2,0}_{2,({\rm par})}(Q_{\varepsilon'})$, then $u \in C^{1,\alpha}_{1+\alpha,({\rm subs})}(Q_{\varepsilon'})$ with*

$$\|u\|^{(1+\alpha),({\rm subs})}_{1,\alpha,Q_{\varepsilon'}} \le 4\|u\|^{(2),({\rm par})}_{2,0,Q_{\varepsilon'}}, \qquad (17.2.27)$$

where the parabolic norms are defined with respect to $\{s=0\}$.

Proof. By (4.6.2) (with (s,t) instead of (x,y)) and (17.2.14), we have the following explicit expressions of the components of the norms in (17.2.27):

$$\begin{aligned}\|u\|^{(1+\alpha),(\mathrm{subs})}_{1,0,Q_{\varepsilon'}} &=\|s^{-1-\alpha}u\|_\infty + \|s^{-\alpha}u_s\|_\infty + \|s^{-1-\alpha}u_t\|_\infty,\\ \|u\|^{(2),(\mathrm{par})}_{2,0,Q_{\varepsilon'}} &=\|s^{-2}u\|_\infty + \|s^{-1}u_s\|_\infty + \|s^{-\frac{3}{2}}u_t\|_\infty + \|u_{ss}\|_\infty \\ &\quad + \|s^{-\frac{1}{2}}u_{st}\|_\infty + \|s^{-1}u_{tt}\|_\infty,\end{aligned} \tag{17.2.28}$$

where we have used $\|s^p v\|_\infty := \sup_{(s,t)\in Q_{\varepsilon'}}\big(s^p|v(s,t)|\big)$.

Since $\alpha\in(0,\frac{1}{3}]$ and $s\in(0,1)$, it follows that

$$\|u\|^{(1+\alpha),(\mathrm{subs})}_{1,0,Q_{\varepsilon'}} \le \|u\|^{(2),(\mathrm{par})}_{2,0,Q_{\varepsilon'}}. \tag{17.2.29}$$

Thus, it remains to estimate the terms of $[u]^{(1+\alpha),(\mathrm{subs})}_{1,\alpha,Q_{\varepsilon'}}$ in (17.2.14). We first estimate the term with $k=1$ and $l=0$, which is

$$\sup_{\mathbf{s},\tilde{\mathbf{s}}\in Q_{\varepsilon'},\mathbf{s}\neq\tilde{\mathbf{s}}} \frac{|u_s(\mathbf{s})-u_s(\tilde{\mathbf{s}})|}{\delta^{(\mathrm{subs})}_\alpha(\mathbf{s},\tilde{\mathbf{s}})}. \tag{17.2.30}$$

From (17.2.13), using that $s,\tilde{s},t,\tilde{t}\in(0,1)$, we obtain that, for $\alpha\in(0,\frac{1}{3}]$,

$$\begin{aligned}&\sqrt{|s-\tilde{s}|}\le|s-\tilde{s}|^\alpha\le\delta^{(\mathrm{subs})}_\alpha(\mathbf{s},\tilde{\mathbf{s}}),\\ &\sqrt{\max(s,\tilde{s})|t-\tilde{t}|}\le\big(\max(s,\tilde{s})|t-\tilde{t}|\big)^\alpha\le\delta^{(\mathrm{subs})}_\alpha(\mathbf{s},\tilde{\mathbf{s}}).\end{aligned} \tag{17.2.31}$$

Then, using (17.2.28), we find that, for $\mathbf{s},\tilde{\mathbf{s}}\in Q_{\varepsilon'}$,

$$\begin{aligned}|u_s(\mathbf{s})-u_s(\tilde{\mathbf{s}})| &\le |u_s(s,t)-u_s(\tilde{s},t)|+|u_s(\tilde{s},t)-u_s(\tilde{s},\tilde{t})|\\ &\le \max_{s'\in[s,\tilde{s}]}|u_{ss}(s',t)|\,|s-\tilde{s}|+\max_{t'\in[t,\tilde{t}]}|u_{st}(\tilde{s},t')|\,|t-\tilde{t}|\\ &\le \|u\|^{(2),(\mathrm{par})}_{2,0,Q_{\varepsilon'}}\big(|s-\tilde{s}|+\sqrt{\tilde{s}}|t-\tilde{t}|\big)\\ &\le \|u\|^{(2),(\mathrm{par})}_{2,0,Q_{\varepsilon'}}\big(\sqrt{|s-\tilde{s}|}+\sqrt{\tilde{s}}\sqrt{|t-\tilde{t}|}\big)\\ &\le 2\|u\|^{(2),(\mathrm{par})}_{2,0,Q_{\varepsilon'}}\delta^{(\mathrm{subs})}_\alpha(\mathbf{s},\tilde{\mathbf{s}}),\end{aligned}$$

where we have used (17.2.31) in the last inequality. Therefore, term (17.2.30) is estimated from above by $2\|u\|^{(2),(\mathrm{par})}_{2,0,Q_{\varepsilon'}}$.

The term in semi-norm $[u]^{(1+\alpha),(\mathrm{subs})}_{1,\alpha,Q_{\varepsilon'}}$ in (17.2.14) with $k=0$ and $l=1$ is

$$\sup_{\mathbf{s},\tilde{\mathbf{s}}\in Q_{\varepsilon'},\mathbf{s}\neq\tilde{\mathbf{s}}}\Big\{\big(\max(s,\tilde{s})\big)^{-1}\frac{|u_t(\mathbf{s})-u_t(\tilde{\mathbf{s}})|}{\delta^{(\mathrm{subs})}_\alpha(\mathbf{s},\tilde{\mathbf{s}})}\Big\}. \tag{17.2.32}$$

To estimate (17.2.32), we choose $\mathbf{s},\tilde{\mathbf{s}}\in Q_{\varepsilon'}$ with $s\le\tilde{s}$ (without loss of generality) and consider the following two cases:

The first case is when $0 < \frac{\tilde{s}}{2} \le s \le \tilde{s}$ and $|t-\tilde{t}| \le \sqrt{\tilde{s}}$. Using these conditions and their consequence: $|s-\tilde{s}| < \tilde{s}$, we find that, for any $\alpha \in [0, \frac{1}{3}]$,

$$\begin{aligned}|u_t(\mathbf{s}) - u_t(\tilde{\mathbf{s}})| &\le \max_{s'\in[s,\tilde{s}]} |u_{st}(s',t)|\,|s-\tilde{s}| + \max_{t'\in[t,\tilde{t}]} |u_{tt}(\tilde{s},t')|\,|t-\tilde{t}| \\ &\le \|u\|^{(2),(\mathrm{par})}_{2,0,Q_{\varepsilon'}} \left(\sqrt{\tilde{s}}|s-\tilde{s}| + \tilde{s}|t-\tilde{t}|\right) \\ &\le \|u\|^{(2),(\mathrm{par})}_{2,0,Q_{\varepsilon'}} \left(\tilde{s}^{\frac{7}{6}}|s-\tilde{s}|^{\frac{1}{3}} + \tilde{s}^{\frac{4}{3}}|t-\tilde{t}|^{\frac{1}{3}}\right) \\ &\le 2\tilde{s}\,\|u\|^{(2),(\mathrm{par})}_{2,0,Q_{\varepsilon'}} \delta^{(\mathrm{subs})}_{\alpha}(\mathbf{s},\tilde{\mathbf{s}}),\end{aligned}$$

where we have used that $\delta^{(\mathrm{subs})}_{\alpha}(\mathbf{s},\tilde{\mathbf{s}}) \ge |s-\tilde{s}|^{\frac{1}{3}}$ and $\delta^{(\mathrm{subs})}_{\alpha}(\mathbf{s},\tilde{\mathbf{s}}) \ge \tilde{s}^{\frac{1}{3}}|t-\tilde{t}|^{\frac{1}{3}}$ if $\alpha \in [0, \frac{1}{3}]$.

In the opposite case, either $0 < s \le \frac{\tilde{s}}{2}$ or $|t-\tilde{t}| \ge \sqrt{\tilde{s}}$. If $0 < s \le \frac{\tilde{s}}{2}$, then $\delta^{(\mathrm{subs})}_{1}(\mathbf{s},\tilde{\mathbf{s}}) \ge |\tilde{s}-s| \ge \frac{\tilde{s}}{2}$. If $|t-\tilde{t}| \ge \sqrt{\tilde{s}}$, then $\delta^{(\mathrm{subs})}_{1}(\mathbf{s},\tilde{\mathbf{s}}) \ge \tilde{s}|\tilde{t}-t| \ge \tilde{s}^{\frac{3}{2}}$. Thus, in both cases, we obtain that $\delta^{(\mathrm{subs})}_{1}(\mathbf{s},\tilde{\mathbf{s}}) \ge \frac{\tilde{s}^{\frac{3}{2}}}{2}$, which implies that $\delta^{(\mathrm{subs})}_{\alpha}(\mathbf{s},\tilde{\mathbf{s}}) \ge \frac{\sqrt{\tilde{s}}}{2}$ for any $\alpha \in (0, \frac{1}{3}]$. Using that and (17.2.28), we estimate:

$$\begin{aligned}|u_t(\mathbf{s}) - u_t(\tilde{\mathbf{s}})| &\le |u_t(\mathbf{s})| + |u_t(\tilde{\mathbf{s}})| \le \|u\|^{(2),(\mathrm{par})}_{2,0,Q_{\varepsilon'}}(\tilde{s}^{\frac{3}{2}} + s^{\frac{3}{2}}) \\ &\le 2\tilde{s}^{\frac{3}{2}}\|u\|^{(2),(\mathrm{par})}_{2,0,Q_{\varepsilon'}} \le 4\tilde{s}\|u\|^{(2),(\mathrm{par})}_{2,0,Q_{\varepsilon'}} \delta^{(\mathrm{subs})}_{\alpha}(\mathbf{s},\tilde{\mathbf{s}}).\end{aligned}$$

Combining the estimates obtained in the two cases above, we see that (17.2.32) is estimated from above by $4\|u\|^{(2),(\mathrm{par})}_{2,0,Q_{\varepsilon'}}$. Combining with (17.2.29), we complete the proof. □

Now we can state an extension of Proposition 12.2.5 to the case of angles $\theta^*_{\mathrm{w}} \in (\theta^{\mathrm{d}}_{\mathrm{w}}, \frac{\pi}{2})$. We use ε'_0 defined by (17.2.7).

Proposition 17.2.12. *Let $\theta^*_{\mathrm{w}} \in (\theta^{\mathrm{d}}_{\mathrm{w}}, \frac{\pi}{2})$. There exist $M > 0$ and $\bar{\alpha} \in (0, \frac{1}{3}]$ depending on the data and θ^*_{w} such that, for any admissible solution φ for $\theta_{\mathrm{w}} \in [\theta^*_{\mathrm{w}}, \frac{\pi}{2}]$, function u defined by* (12.2.44) *satisfies that, for any $\alpha \in [0, \bar{\alpha}]$,*

$$\|u\|^{(-1-\alpha),\{(1,0)\}}_{2,\alpha,Q^{\mathrm{iter}}\setminus\overline{Q}_{\varepsilon'_0/10}} + \|u\|^{(1+\alpha),(\mathrm{par})}_{2,\alpha,Q_{\varepsilon'_0}} + \|u\|^{(1+\alpha),(\mathrm{subs})}_{1,\alpha,Q_{\varepsilon'_0}} \le M, \tag{17.2.33}$$

where $Q_{\varepsilon'_0} := Q^{\mathrm{iter}} \cap \{s < \varepsilon'_0\}$, and norm $\|\cdot\|^{(1+\alpha),(\mathrm{par})}_{2,\alpha,A}$ is stated with respect to $\{s = 0\}$.

Proof. In this proof, the universal constant C depends only on the data and θ^*_{w}.

We first consider the supersonic wedge angles $\theta_{\mathrm{w}} \in [\theta^*_{\mathrm{w}}, \frac{\pi}{2}]$. From Corollary 16.4.8, (12.2.26) with $\hat{\varepsilon} = \frac{\varepsilon_0}{2}$, and (17.2.3)–(17.2.4) in Lemma 17.2.2, we have

$$\|u\|^{(-1-\bar{\alpha}),\{(1,0)\}}_{2,\bar{\alpha},Q^{\mathrm{iter}}\setminus\overline{Q}_{\varepsilon'_0/10}} + \|u\|^{(2),(\mathrm{par})}_{2,\bar{\alpha},Q_{\varepsilon'_0}} \le C, \tag{17.2.34}$$

by using the C^3–regularity of map F_1^{-1} and the explicit form (12.2.43) of map $F^{-1}_{(2,\mathfrak{g}_{\rm sh})}$. We note that $F^{-1}_{(2,\mathfrak{g}_{\rm sh})}$ does not become singular as $t_{P_1}(\theta_{\rm w}) \to 0+$ in (17.2.4), *i.e.*, estimate (17.2.34) holds with uniform constant C for all the supersonic wedge angles $\theta_{\rm w} \in [\theta_{\rm w}^*, \frac{\pi}{2}]$. Using this estimate and Lemmas 4.1.2, 4.6.4, and 17.2.11, we obtain (17.2.33) for all the supersonic wedge angles $\theta_{\rm w} \in [\theta_{\rm w}^*, \frac{\pi}{2}]$ and all $\alpha \in [0, \bar{\alpha}]$.

Now we consider the subsonic/sonic wedge angles $\theta_{\rm w} \in [\theta_{\rm w}^*, \frac{\pi}{2}]$. We note that estimate (16.5.26) in Corollary 16.5.26 is stronger than estimate (16.6.79) in Corollary 16.6.12, which implies that every subsonic/sonic reflection solution for a wedge angle $\theta_{\rm w} \in [\theta_{\rm w}^*, \frac{\pi}{2}]$ satisfies (16.6.79). Then we use Lemma 17.2.5, which can be applied since (17.2.4) holds with $t_{P_1}(\theta_{\rm w}) = 0$ for every subsonic/sonic reflection solution, *i.e.*, (17.2.9) holds. Using (17.2.3), (12.2.26) with $\hat{\varepsilon} = \frac{\varepsilon_0}{2}$ in Lemma 17.2.2, (16.6.79), and Lemma 17.2.5, we have

$$\|u\|^{(-1-\bar{\alpha}),\{(1,0)\}}_{2,\bar{\alpha},Q^{\rm iter}\setminus\overline{Q}_{\varepsilon_0'/10}} + \|u\|^{(1+\bar{\alpha}),({\rm subs})}_{2,\bar{\alpha},Q_{\varepsilon_0'}} \le C.$$

From this, using Lemmas 4.1.2, 4.6.4, and 17.2.10, we obtain (17.2.33) for all the subsonic/sonic wedge angles $\theta_{\rm w} \in [\theta_{\rm w}^*, \frac{\pi}{2}]$ and all $\alpha \in [0, \bar{\alpha}]$. □

17.2.3 Mapping from the iteration region

Next, for a function u on $Q^{\rm iter}$, we determine the corresponding domain Ω and function φ.

We extend the construction in §12.2.3, Definition 12.2.6, and Lemma 12.2.7 to the wedge angles $\theta_{\rm w} \in (\theta_{\rm w}^{\rm d}, \frac{\pi}{2}]$ with the changes cited below, which mostly affect only the case of subsonic/sonic angles (besides some estimates in the supersonic-close-to-sonic case).

Fix $\theta_{\rm w}^* \in (\theta_{\rm w}^{\rm d}, \frac{\pi}{2})$. Since $P_1 = P_4$ in the subsonic/sonic case and $P_1 \ne P_4$ in the supersonic case, (12.2.47) now takes the form:

$$\inf_{\partial Q^{(\theta_{\rm w})}_{\rm bd}\cap F_1^{-1}(\{s=0\})} (\varphi_1 - \tilde{\varphi}_2^{(\theta_{\rm w})}) < 0 \le \sup_{\partial Q^{(\theta_{\rm w})}_{\rm bd}\cap F_1^{-1}(\{s=0\})} (\varphi_1 - \tilde{\varphi}_2^{(\theta_{\rm w})}), \quad (17.2.35)$$

with the equality on the right if and only if $\theta_{\rm w}$ is a subsonic/sonic angle. This follows from the calculation before (12.2.47).

Now Definition 12.2.6 applies to the present situation, *i.e.*, for $\theta_{\rm w}^* \in (\theta_{\rm w}^{\rm d}, \frac{\pi}{2})$, without change. Note that we still have the strict inequalities in (12.2.49) for all $s^* \in (0, \hat{s}(\theta_{\rm w})]$ for any wedge angle $\theta_{\rm w} \in (\theta_{\rm w}^*, \frac{\pi}{2})$, which includes the subsonic/sonic angles. In particular,

$$\mathfrak{S} \text{ and } \mathfrak{S}(\theta_{\rm w}) \quad \text{are defined by (12.2.50)–(12.2.51).}$$

Then $\mathfrak{S}(\theta_{\rm w})$ is non-empty for any $\theta_{\rm w} \in [\theta_{\rm w}^*, \frac{\pi}{2}]$ by (12.2.37) with Lemma 17.2.3, (17.2.35), and Lemma 12.2.2(i) via Lemma 17.2.1. Furthermore, since (12.2.50) includes the conditions that (u, Du) vanish on $\{s = 0\}$, it follows from (17.2.35) that, for any $(u, \theta_{\rm w}) \in \mathfrak{S}$, the functions defined by Definition 12.2.6(ii) satisfy

$$\mathfrak{g}_{\rm sh}^{(u,\theta_{\rm w})}(0) = \mathfrak{h}_{\rm sh}^{(u,\theta_{\rm w})}(0) = t_{P_1}(\theta_{\rm w}), \tag{17.2.36}$$

where we recall that $t_{P_1} = t_{P_4} = 0$ for the subsonic/sonic angle $\theta_{\rm w}$, and $t_{P_1} > t_{P_4} = 0$ for the supersonic angle $\theta_{\rm w}$. Also, the admissible solutions satisfy (17.2.4). For this reason, instead of the set in (12.2.55), we define that, for $M \geq 1$ and $\delta_{\rm sh} > 0$,

$$\hat{\mathfrak{S}}_{M,\delta_{\rm sh}}^{(\alpha)} := \{(u, \theta_{\rm w}) \in \mathfrak{S} \; : \; (17.2.23) \text{ and } (12.2.54) \text{ hold}\}. \tag{17.2.37}$$

Then we have

Lemma 17.2.13. *Let $\theta_{\rm w}^* \in (\theta_{\rm w}^{\rm d}, \frac{\pi}{2})$. All the assertions of Lemma* 12.2.7 *remain true, except assertion* (viii) *which now takes the following form:*

(viii′) *Let $\alpha \in (0,1)$, $\sigma \in (1,2]$, and M, $\delta_{\rm sh} > 0$. Let $(u, \theta_{\rm w}) \in \mathfrak{S}$ with $\theta_{\rm w} \in [\theta_{\rm w}^*, \frac{\pi}{2}]$, and let $(u, \theta_{\rm w})$ satisfy* (12.2.64) *and* (17.2.23). *Then*

(a) *For every $\varepsilon \in (0, \frac{\hat{s}(\theta_{\rm w})}{2})$, there exists $C_{\varepsilon,M}$ depending on the data and $(\theta_{\rm w}^*, \alpha, \delta_{\rm sh}, \varepsilon, M)$ such that*

$$\|\mathfrak{F}_{(u,\theta_{\rm w})}^{-1}\|_{2,\alpha,\Omega(u)\setminus\mathcal{D}_\varepsilon}^{(-1-\alpha),\Gamma_{\rm sym}} \leq C_{\varepsilon,M}, \tag{17.2.38}$$

$$\|\varphi - \tilde{\varphi}_2^{(\theta_{\rm w})}\|_{2,\alpha,\Omega\setminus\mathcal{D}_\varepsilon}^{(-1-\alpha),\Gamma_{\rm sym}} \leq C_{\varepsilon,M}; \tag{17.2.39}$$

(b) *Moreover, if $\theta_{\rm w}$ is supersonic and satisfies $\frac{|D\varphi_2(P_0)|}{c_2}(\theta_{\rm w}) \geq 1 + \delta$ for some $\delta > 0$, then* (12.2.69)–(12.2.71) *hold for $(u, \theta_{\rm w})$, with $\hat{C}$ depending only on the data and $(\theta_{\rm w}^*, \alpha, \sigma, \delta_{\rm sh}, \delta)$, and with $\hat{C}_0$ depending only on the data and δ.*

Proof. We first note that all the assertions of Lemma 12.2.7 (except assertion (viii)) remain true, since all the conditions of these assertions in Lemma 12.2.7 are satisfied in the present case, and their proofs can be repeated without change by using Lemmas 12.2.2 and 12.2.4 held in the present case by Lemmas 17.2.1 and 17.2.3, respectively.

Thus, it remains to prove assertion (viii′). Fix $\varepsilon \in (0, \frac{\hat{s}(\theta_{\rm w})}{2})$. Then, by (17.2.23),

$$\min\{\frac{\varepsilon}{M}, \delta_{\rm sh}\} < \mathfrak{g}_{\rm sh}^{(u,\theta_{\rm w})}(s) < M\hat{s}(\theta_{\rm w}) + \max_{\theta_{\rm w}\in[\theta_{\rm w}^*, \frac{\pi}{2}]} t_{P_1}(\theta_{\rm w}) \qquad \text{on } (\varepsilon, \hat{s}(\theta_{\rm w})),$$

where we have used the continuity of $t_{P_1}(\theta_{\rm w}) = y_{P_1}(\theta_{\rm w}) - \theta_{\rm w}$ as a function of $\theta_{\rm w}$ on $[\theta_{\rm w}^*, \frac{\pi}{2}]$. Then condition (12.2.53) holds on $(\varepsilon, \hat{s}(\theta_{\rm w}))$ with constant b depending only on the data and $(M, \delta_{\rm sh}, \varepsilon)$. Also, as we have discussed above, the assertion of Lemma 12.2.7(vii) holds in the present case so that, from (12.2.66), for any ε,

$$\|\mathfrak{g}_{\rm sh}^{(u,\theta_{\rm w})}\|^{(-1-\alpha),\{\hat{s}(\theta_{\rm w})\}}_{2,\alpha,[\varepsilon,\,\hat{s}(\theta_{\rm w})]} \le C_{\varepsilon,M},$$

where $C_{\varepsilon,M}$ depends only on the data and $(\theta_{\rm w}^*, \alpha, M, \delta_{\rm sh}, \varepsilon)$. Using these two properties of $\mathfrak{g}_{\rm sh}^{(u,\theta_{\rm w})}$, estimate (17.2.38) follows from the explicit expressions (12.2.41) and (12.2.52) by using Lemma 12.2.2, which holds by Lemma 17.2.1.

Since (12.2.64) holds, then, for any $\varepsilon' \in (0, \frac{1}{2})$,

$$\|u\|^{(-1-\alpha),\{1\}\times(1,0)}_{2,\alpha,Q^{\rm iter}\cap\{s>\varepsilon'\}} \le C(\varepsilon')M,$$

where $C(\varepsilon')$ depends only on ε'. This, combined with (17.2.38), implies (17.2.39) by using the explicit formula given in Definition 12.2.6(v) and Lemma 17.2.3.

Finally, if $\theta_{\rm w}$ is supersonic and satisfies $\frac{|D\varphi_2(P_0)|}{c_2}(\theta_{\rm w}) \ge 1+\delta$, then $t_{P_1} \ge \frac{1}{C(\delta)} > 0$, where $C(\delta)$ depends only on the data and δ. Thus, (17.2.23) implies that condition (12.2.53) holds on $(0, \hat{s}(\theta_{\rm w}))$ with constant b depending only on δ and the data. Therefore, the proof of Lemma 12.2.7(viii) implies that (12.2.69)–(12.2.71) hold with $\hat{C}$ and $\hat{C}_0$ depending only on the parameters described in the formulation of this lemma. □

17.3 ITERATION SET

Now we define an iteration set to construct global regular reflection solutions for the wedge angles between $\frac{\pi}{2}$ and the detachment angle $\theta_{\rm w}^{\rm d}$, *i.e.*, $\theta_{\rm w} \in (\theta_{\rm w}^{\rm d}, \frac{\pi}{2})$. This extends the previous definition of the iteration set in Definition 12.3.2 to allow for proving the existence of both subsonic and supersonic reflection solutions depending on $\theta_{\rm w}$: If the wedge angle $\theta_{\rm w} \in (\theta_{\rm w}^{\rm d}, \frac{\pi}{2})$ is such that

$$\frac{|D\varphi_2|^2}{c^2(|D\varphi_2|, \varphi_2)} > 1 \quad \text{or} \quad \frac{|D\varphi_2|^2}{c^2(|D\varphi_2|, \varphi_2)} \le 1 \qquad \text{at } P_0,$$

then we can construct a supersonic or subsonic/sonic reflection solution for $\theta_{\rm w}$.

Below, we use constants ε_0 in Lemma 17.2.1 and ε_0' defined by (17.2.7).

We use the notations in (i)–(ii) and (iv) of Definition 12.3.1, extended to all $\theta_{\rm w} \in (\theta_{\rm w}^{\rm d}, \frac{\pi}{2})$ with the use of notation (17.3.2), and introduce some additional notations in the definition below.

Definition 17.3.1. (i) *For* $\alpha \in (0,1)$, *define*

$$\|u\|^{(**)}_{2,\alpha,Q^{\rm iter}} = \|u\|^{(-1-\alpha),\{1\}\times(0,1)}_{2,\alpha,Q^{\rm iter}\setminus\overline{Q}_{\varepsilon_0'/10}} + \|u\|^{(1+\alpha),({\rm par})}_{2,\alpha,Q_{\varepsilon_0'}} + \|u\|^{(1+\alpha),({\rm subs})}_{1,\alpha,Q_{\varepsilon_0'}}, \tag{17.3.1}$$

where we have used the norms defined by (4.6.2) *with respect to* $\{s=0\}$ *and by* (17.2.14), *and* $Q_\sigma := Q^{\rm iter} \cap \{s<\sigma\}$. *Denote by* $C^{2,\alpha}_{(**)}(Q^{\rm iter})$ *the set of all* $C^2(Q^{\rm iter})$*-functions with finite norm* (17.3.1).

(ii) *For* $\theta_{\rm w} \in [\theta_{\rm w}^{\rm d}, \frac{\pi}{2}]$ *and* $\sigma \in (0, \varepsilon_0]$, *denote*

$$\mathcal{D}_\sigma^{(\theta_{\rm w})} := \{\varphi_2 < \varphi_1\} \cap \Lambda \cap \mathcal{N}_{\varepsilon_1}(\Gamma_{\rm sonic}^{(\delta_*)}) \cap \{x_{P_1} < x < x_{P_1} + \sigma\}, \qquad (17.3.2)$$

$$\mathcal{D}_{0,\sigma}^{(\theta_{\rm w})} := \{\varphi_2 < \varphi_1\} \cap \Lambda \cap \mathcal{N}_{\varepsilon_1}(\Gamma_{\rm sonic}^{(\delta_*)}) \cap \{x_{P_1} < x < \sigma\}, \qquad (17.3.3)$$

where δ_* *and* $\Gamma_{\rm sonic}^{(\delta_*)}$ *have been defined at the beginning of* §17.2.1, *and* ε_1 *is from Lemma* 16.1.2. *We also write* $(\mathcal{D}_\sigma, \mathcal{D}_{0,\sigma})$ *for* $(\mathcal{D}_\sigma^{(\theta_{\rm w})}, \mathcal{D}_{0,\sigma}^{(\theta_{\rm w})})$.

Remark 17.3.2. *From* (17.3.2)–(17.3.3), *it follows that, for any* $\sigma \in (0, \varepsilon_0]$,

(a) $\mathcal{D}_\sigma \equiv \mathcal{D}_\sigma^{(\theta_{\rm w})}$ *coincides with the set defined in* (17.2.1) *for* $\varepsilon = \sigma$;

(b) $\mathcal{D}_{0,\sigma} \subset \mathcal{D}_\sigma$ *for all the wedge angles;*

(c) $\mathcal{D}_{0,\sigma} = \mathcal{D}_\sigma$ *for all the supersonic/sonic wedge angles;*

(d) $\mathcal{D}_{0,\sigma} = \emptyset$ *for all the subsonic wedge angles satisfying* $x_{P_0} > \sigma$;

(e) *From Proposition* 16.1.3, *specifically from* (16.1.3) *and* (16.1.5)–(16.1.6), *it follows that, for any admissible solution,*

$$\Omega_\sigma = \Omega \cap \mathcal{D}_\sigma \qquad \textit{for all } \sigma \in (0, \varepsilon_0). \qquad (17.3.4)$$

We will use the following compactness property of space $C^{2,\alpha}_{(**)}(Q^{\rm iter})$:

Lemma 17.3.3. *Let* $0 \le \beta < \alpha < 1$. *Then* $C^{2,\alpha}_{(**)}(Q^{\rm iter})$ *is compactly embedded into* $C^{2,\beta}_{(**)}(Q^{\rm iter})$.

This follows directly from Lemma 4.6.3, Corollary 17.2.7, and the standard results on the compactness in the Hölder spaces.

Now we are ready to define the iteration set. We choose below a constant $\alpha \in (0, \frac{\bar\alpha}{2}]$, where $\bar\alpha$ is the constant in Proposition 17.2.12. Let the small constants $\delta_1, \delta_2, \delta_3, \varepsilon, \lambda > 0$ and large constant $N_1 > 1$ be fixed and chosen below. We always assume that $\varepsilon < \varepsilon_0$, where ε_0 is from Lemma 17.2.1.

Definition 17.3.4. *Let* $\theta_{\rm w}^* \in (\theta_{\rm w}^{\rm d}, \frac{\pi}{2})$. *The iteration set* $\mathcal{K} \subset C^{2,\alpha}_{(**)}(Q^{\rm iter}) \times [\theta_{\rm w}^*, \frac{\pi}{2}]$ *is the set of all* $(u, \theta_{\rm w})$ *satisfying the following:*

(i) $(u, \theta_{\rm w})$ *satisfy the estimates:*

$$\|u\|^{(**)}_{2,\alpha,Q^{\rm iter}} < \eta_1(\theta_{\rm w}), \qquad (17.3.5)$$

where $\eta_1 \in C(\mathbb{R})$ *is defined by*

$$\eta_1(\theta_{\rm w}) = \begin{cases} \delta_1 & \textit{if } \frac{\pi}{2} - \theta_{\rm w} \le \frac{\delta_1}{N_1}, \\ N_0 & \textit{if } \frac{\pi}{2} - \theta_{\rm w} \ge \frac{2\delta_1}{N_1}, \\ \textit{linear} & \textit{if } \frac{\pi}{2} - \theta_{\rm w} \in (\frac{\delta_1}{N_1}, \frac{2\delta_1}{N_1}), \end{cases}$$

with $N_0 = \max\{10M, 1\}$ *and constant* M *in* (17.2.33).

(ii) $(u, \theta_w) \in \mathfrak{S}$, *where we have used* (12.2.50). *Then* $\mathfrak{g}_{sh}$, $\Omega = \Omega(u)$, $\Gamma_{shock} = \Gamma_{shock}(u)$, *and* $\varphi = \varphi^{(u)}$ *are defined in Definition* 12.2.6 *and* §17.2.3, *and* $\Gamma_{sym} = \Gamma_{sym}(u)$ *is defined in Lemma* 12.2.7(iii) *with Lemma* 17.2.13.

(iii) *The properties of* Γ_{shock}:

$$\mathrm{dist}(\Gamma_{shock}, B_{c_1}(\mathcal{O}_1)) \geq \frac{1}{N_5} \tag{17.3.6}$$

with $N_5 = 2C$, *where* C *is from Proposition* 15.5.3. *Moreover, we require*

$$\begin{aligned}&\min(\mathfrak{g}_{sh}(0) + \frac{s}{N_2}, \frac{1}{N_2}) < \mathfrak{g}_{sh}(s)\\ &< \min(\mathfrak{g}_{sh}(0) + N_2\, s,\ \eta^{(\theta_w)}(s) - \frac{1}{N_2})\end{aligned} \tag{17.3.7}$$

for any $s \in (0, \hat{s}(\theta_w))$ *and* $\theta_w \in [\theta_w^*, \frac{\pi}{2})$, *where* $\mathfrak{g}_{sh}(0) = t_{P_1}(\theta_w) \geq 0$, $N_2 = 2M$, M *is from* (17.2.4), *and function* $\eta^{(\theta_w)}(\cdot)$ *is from* (12.2.10).

(iv) φ *satisfies*

$$\psi > \eta_2(\theta_w) \qquad \text{in } \overline{\Omega} \setminus \mathcal{D}_{\varepsilon/10}, \tag{17.3.8}$$

$$|\partial_x \psi(x,y)| < \eta_3(\theta_w)\, x \qquad \text{in } \overline{\Omega} \cap (\mathcal{D}_{\varepsilon_0} \setminus \mathcal{D}_{0,\varepsilon/10}), \tag{17.3.9}$$

$$|\partial_y \psi(x,y)| < N_3 x \qquad \text{in } \overline{\Omega} \cap (\mathcal{D}_{\varepsilon_0} \setminus \mathcal{D}_{0,\varepsilon/10}), \tag{17.3.10}$$

$$|(\partial_x \psi,\, \partial_y \psi)| < N_3 \varepsilon \qquad \text{in } \overline{\Omega} \cap \mathcal{D}_{\varepsilon}, \tag{17.3.11}$$

$$\|\psi\|_{C^{0,1}(\Omega)} < N_4, \tag{17.3.12}$$

$$\partial_{\mathbf{e}_{S_1}}(\varphi_1 - \varphi) < -\eta_2(\theta_w) \qquad \text{in } \overline{\Omega} \setminus \mathcal{D}_{\varepsilon/10}, \tag{17.3.13}$$

$$\partial_{\xi_2}(\varphi_1 - \varphi) < -\eta_2(\theta_w) \qquad \text{in } \overline{\Omega} \setminus \mathcal{N}_{\varepsilon/10}(\Gamma_{sym}), \tag{17.3.14}$$

$$\partial_{\boldsymbol{\nu}}(\varphi_1 - \varphi) > \mu_1 \qquad \text{on } \overline{\Gamma_{shock}}, \tag{17.3.15}$$

$$\partial_{\boldsymbol{\nu}}\varphi > \mu_1 \qquad \text{on } \overline{\Gamma_{shock}}, \tag{17.3.16}$$

where $\mathcal{D}_\sigma$ *and* $\mathcal{D}_{0,\sigma}$ *are defined by* (17.3.2) *and* (17.3.3), *respectively, functions* $\eta_2, \eta_3 \in C(\mathbb{R})$ *are defined by*

$$\eta_2(\theta_w) = \delta_2 \min(\frac{\pi}{2} - \theta_w - \frac{\delta_1}{N_1^2}, \frac{\delta_1}{N_1^2}),$$

$$\eta_3(\theta_w) = \begin{cases} \frac{2-\mu_0}{1+\gamma} & \textit{if } \frac{|D\varphi_2(P_0)|}{c_2}(\theta_w) \geq 1 - \frac{\delta_{P_0}}{2}, \\ N_3 & \textit{if } \frac{|D\varphi_2(P_0)|}{c_2}(\theta_w) \leq 1 - \delta_{P_0}, \\ \textit{linear} & \textit{if } \frac{|D\varphi_2(P_0)|}{c_2}(\theta_w) \in (1 - \delta_{P_0}, 1 - \frac{\delta_{P_0}}{2}) \end{cases}$$

for δ_{P_0} *in Proposition* 16.1.4, *and constants* ε_0, μ_k, N_3, *and* N_4 *are chosen as follows:*

(a) ε_0 *is from Corollary* 16.6.13;

(b) $\mu_0 = \frac{\delta}{2}$, *where* δ *is from Lemma* 11.2.5, *extended by Proposition* 16.1.4;

(c) $\mu_1 = \frac{\min\{\hat{\delta},\hat{\delta}_1\}}{2}$, *where* $\hat{\delta}$ *and* $\hat{\delta}_1$ *are the constants from Corollary* 10.1.2, *extended by Proposition* 15.7.1;

(d) $N_3 = 10C$, *where* C *is the constant from estimate* (16.6.84) *in Lemma* 16.6.14.

In (17.3.9)–(17.3.11), *we have used* φ *and* φ_2 *expressed in the* (x,y)-*coordinates* (11.1.1)–(11.1.2).

(v) *Uniform ellipticity in* $\overline{\Omega}\setminus\mathcal{D}_{\varepsilon/10}$: *For any* $\boldsymbol{\xi}\in\overline{\Omega}\setminus\mathcal{D}_{\varepsilon/10}$,

$$\frac{|D\varphi|}{c(|D\varphi|^2,\varphi)}(\boldsymbol{\xi}) < \max(1-\frac{\hat{\zeta}}{2}, \ \frac{|D\varphi_2(P_1)|}{c_2}(\theta_{\rm w}) - \lambda\,{\rm dist}(\boldsymbol{\xi},\overline{\Gamma_{\rm sonic}}))$$

with $\lambda = \frac{\tilde{\mu}}{2}$, *where* $\tilde{\mu}$ *and* $\hat{\zeta}$ *are from Proposition* 15.6.2, *and* $\overline{\Gamma_{\rm sonic}} = \{P_0\}$ *for the subsonic/sonic wedge angles. Note that the formula above combines both cases in Proposition* 15.6.2 *with relaxed constants, which allows us to achieve a strict inequality for admissible solutions.*

(vi) *Bounds on the density:*

$$\rho_{\min} < \rho(|D\varphi|^2,\varphi) < \rho_{\max} \qquad \text{in } \overline{\Omega}$$

with $\rho_{\max} > \rho_{\min} > 0$ *defined by* $\rho_{\min} = \frac{a}{2}\rho_1$ *and* $\rho_{\max} = 2C$, *where* $a = \left(\frac{2}{\gamma+1}\right)^{\frac{1}{\gamma-1}}$ *and* C *are the constants from Lemma* 9.1.4 *extended by Proposition* 15.2.2.

(vii) *Consider the boundary value problem* (12.3.25)–(12.3.29) *determined by* $(u,\theta_{\rm w})$, *with the equation and boundary condition on* $\Gamma_{\rm shock}$ *defined below in* §17.3.2 *and the boundary condition on* $\Gamma_{\rm sonic}$ *understood as a one-point condition at* P_0 *for the subsonic/sonic wedge angles. This problem has a solution* $\hat{\psi}\in C^2(\Omega(u,\theta_{\rm w}))\cap C^1(\overline{\Omega}(u,\theta_{\rm w}))$ *with the following properties: Let*

$$\hat{\varphi} := \hat{\psi} + \varphi_2 \qquad \text{in } \Omega(u,\theta_{\rm w}),$$

and let function $\hat{u}$ *be defined by* (12.2.44) *from function* $\hat{\varphi}$, *where map* $F_{(2,\mathfrak{g}_{\rm sh})}$ *is determined by function* $\mathfrak{g}_{\rm sh}$ *that corresponds to* $(u,\theta_{\rm w})$, *i.e.*,

$$\hat{u} := (\hat{\varphi}-\tilde{\varphi}_2)\circ F_1^{-1}\circ F_{(2,\mathfrak{g}_{\rm sh})}^{-1} \qquad \text{on } Q^{\rm iter}. \tag{17.3.17}$$

Then $\hat{u}$ *satisfies*

$$\hat{u}\in\mathfrak{S}(\theta_{\rm w}), \tag{17.3.18}$$

$$\|\hat{u}-u\|^{(**)}_{2,\alpha/2,Q^{\rm iter}} < \delta_3. \tag{17.3.19}$$

Remark 17.3.5. *The choice of constants* $(\mu_0, \mu_1, \alpha, \varepsilon, \lambda, \delta_1, \delta_2, \delta_3)$ *and* N_k, *for* $k = 0, \dots, 5$, *is made below to keep only the following dependence: Constants* (μ_0, N_3, N_4) *are fixed above depending on the data; constants* $(\mu_1, N_0, N_2, N_5, \lambda)$ *are fixed above depending on the data and* $\theta_{\rm w}^*$; *constant* α *is fixed later depending only on the data and* $\theta_{\rm w}^*$; *and the other constants, in addition to the dependence on the data and* $\theta_{\rm w}^*$, *have the following dependence: small* ε *depends on* α, *small* δ_1 *depends on* (α, ε), *large* N_1 *depends on* δ_1, *small* δ_2 *depends on* $(\delta_1, N_1, \varepsilon)$, *and then* $\delta_3 > 0$ *is chosen as small as needed, depending on all the other constants.*

Remark 17.3.6. *The difference between Definitions* 12.3.2 *and* 17.3.4 *is that the different norm is used in conditions* (i) *and* (vii), *and conditions* (iii) *and* (v) *and condition* (17.3.9) *of Definition* 17.3.4 *are generalized from conditions* (iii) *and* (v) *and condition* (12.3.7) *of Definition* 12.3.2 *to take into account the cases of supersonic/subsonic reflections. Otherwise, the conditions in Definitions* 12.3.2 *and* 17.3.4 *are of the same structure with the constants adjusted to the corresponding cases.*

Remark 17.3.7. *The following version of Remark* 12.3.4 *holds in the present case: Let* $(u, \theta_{\rm w}) \in C^{2,\alpha}_{(**)}(\mathcal{Q}^{\rm iter}) \times [\theta_{\rm w}^*, \frac{\pi}{2}]$ *satisfy conditions* (i)–(iii) *of Definition* 17.3.4, *and let* $\Omega = \Omega(u, \theta_{\rm w})$, $\Gamma_{\rm shock} = \Gamma_{\rm shock}(u, \theta_{\rm w})$, *and* $\varphi = \varphi^{(u,\theta_{\rm w})}$ *be as in Definition* 12.2.6. *Also, let* $\mathfrak{g}_{\rm sh}$ *be the function determined by* $(u, \theta_{\rm w})$ *in Definition* 12.2.6(i). *Then, from* (17.3.5) *and Lemma* 12.2.7(vii) *that holds by Lemma* 17.2.13, $\mathfrak{g}_{\rm sh}$ *satisfies* (12.2.66) *with* $\sigma = 1 + \alpha$ *and* $M = C\eta_1(\theta_{\rm w})$, *where* C *depends only on the data and* α. *From that, we have*

$$\begin{aligned}
&\|\mathfrak{g}_{\rm sh}\|^{(-1-\alpha),\{0,\,\hat{s}(\theta_{\rm w})\}}_{2,\alpha,[0,\,\hat{s}(\theta_{\rm w})]} \le C\eta_1(\theta_{\rm w}) \le CN_0,\\
&(\mathfrak{g}_{\rm sh} - \mathfrak{g}_{S_1})(0) = (\mathfrak{g}_{\rm sh}' - \mathfrak{g}_{S_1}')(0) = 0,
\end{aligned} \tag{17.3.20}$$

where N_0 *is from Definition* 17.3.4(i). *In particular, using that* F_1 *and* F_1^{-1} *are* C^3*–maps, we conclude that* $\Gamma_{\rm shock}$ *is a* $C^{1,\alpha}$*–curve up to its endpoints. Furthermore, using Lemma* 17.2.1(iv), *we have*

$$\Gamma_{\rm shock} \cap \mathcal{D}_{\varepsilon_0/10} = \{(x, f_{\rm sh}(x)) \;:\; x_{P_1} < x < x_{P_1} + \frac{\varepsilon_0}{10}\}$$

in the (x, y)*–coordinates, and*

$$\begin{aligned}
&\|f_{\rm sh} - \hat{f}_0\|^{(1+\alpha),({\rm par})}_{2,\alpha,(0,\varepsilon_0)} < C\eta_1(\theta_{\rm w}),\\
&(f_{\rm sh} - \hat{f}_0)(0) = (f_{\rm sh} - \hat{f}_0)'(0) = 0,\\
&(12.3.23)\ \ \textit{holds},
\end{aligned} \tag{17.3.21}$$

where $\hat{f}_0$ *is from the expression of* S_1 *as a graph in the* (x, y)*–coordinates, and constant* C *in* (12.3.23) *depends only on the data.*

Furthermore, arguing as in Remark 12.3.4, *we see that* (12.3.18) *holds with* $M_{\rm dom}$ *depending only on the data and is independent of the parameters in the definition of* $\mathcal{K}$.

Remark 17.3.8. *Let* $(u, \theta_{\rm w})$, Ω, $\Gamma_{\rm shock}$, *and* φ *be as in Remark* 17.3.7. *Using* (17.3.5) (*with norm* (17.3.1)) *and* (17.3.7), *we combine Lemma* 17.2.8 (*applied with* $\varepsilon = \varepsilon_0'$) *and* (17.2.39) (*applied with* $\varepsilon = \frac{\varepsilon_0'}{10}$), *employ Lemma* 12.2.4 *held by Lemma* 17.2.3, *choose* $\delta_1 \le \mu$ *for* μ *in* (12.2.34), *recall that* $N_0, N_1 \ge 1$, *and use* (12.3.18) *held by Remark* 17.3.7 *to obtain*

$$\|\psi\|_{1,\alpha,\Omega} < C\eta_1(\theta_{\rm w}) \le \hat{N}_0, \qquad (\psi, D\psi) = (0, \mathbf{0}) \quad \text{on } \overline{\Gamma_{\rm sonic}}, \tag{17.3.22}$$

where C *and* $\hat{N}_0$ *depend on the data and* $(\theta_{\rm w}^*, \alpha)$, *and we have used notation* $\overline{\Gamma_{\rm sonic}} := \{P_0\}$ *for the subsonic/sonic wedge angles.*

Moreover, from (17.3.5) *and* (17.3.7), *we can apply Lemma* 17.2.13(viii′) *to obtain* (17.2.39), *and then we use* (12.3.18) *to obtain that, for each* $\tau \in (0, \varepsilon_0)$,

$$\|\varphi\|^{(-1-\alpha),\Gamma_{\rm sym}}_{2,\alpha,\Omega\setminus\mathcal{D}_\tau} + \|\psi\|^{(-1-\alpha),\Gamma_{\rm sym}}_{2,\alpha,\Omega\setminus\mathcal{D}_\tau} \le C, \tag{17.3.23}$$

where C *depends only on the data and* $(\theta_{\rm w}^*, \alpha, \tau)$.

Remark 17.3.9. *From* (17.3.5) *and Lemma* 17.2.9 (*applied with* $\varepsilon = \varepsilon_0'$), *we see that, if* $(u, \theta_{\rm w}) \in C^{2,\alpha}_{(**)}(Q^{\rm iter}) \times [\theta_{\rm w}^*, \frac{\pi}{2}]$ *satisfies conditions* (i)–(iii) *of Definition* 17.3.4, *and* $\theta_{\rm w}$ *is a supersonic angle, i.e.,* $\frac{|D\varphi_2(P_0)|}{c_2}(\theta_{\rm w}) > 1$, *then*

$$|\partial_y \psi(x, y)| \le Cx^{\frac{1}{2}+\alpha} \qquad \text{for all } (x, y) \in \Omega \cap \mathcal{D}_r, \tag{17.3.24}$$

where $r = \min\{\mathfrak{g}_{\rm sh}^2(0), \varepsilon_0\}$, C *depends only on the data and* $(\theta_{\rm w}^*, \alpha)$, *and we have used that* $\mathfrak{g}_{\rm sh}(0) > 0$ *and* $x_{P_4} = 0$ *for the supersonic wedge angles.*

Remark 17.3.10. *Let* $\delta \in (0, 1)$. *Let* $(u, \theta_{\rm w}) \in C^{2,\alpha}_{(**)}(Q^{\rm iter}) \times [\theta_{\rm w}^*, \frac{\pi}{2}]$ *satisfy conditions* (i)–(iii) *of Definition* 17.3.4 *and* $\frac{|D\varphi_2(P_0)|}{c_2}(\theta_{\rm w}) \ge 1 + \delta$. *Then, using part* (b) *of Lemma* 17.2.13(viii′) *and arguing as in Remark* 12.3.5, *we find that* (12.3.19) *holds for* $\delta^* = \alpha$, *with constant* $N_0^*(\delta)$ *depending only on the data and* $(\theta_{\rm w}^*, \alpha, \delta)$.

Remark 17.3.11. *If* $(u, \theta_{\rm w}) \in C^1(\overline{Q^{\rm iter}}) \times [\theta_{\rm w}^*, \frac{\pi}{2}]$ *satisfies the conditions in Definition* 17.3.4(ii) *and the corresponding* ψ *satisfies* (17.3.22), *then, arguing as in Remark* 12.3.6, *we see that* (12.3.22)–(12.3.23) *hold.*

Remark 17.3.12. *Arguing as in Remarks* 12.3.7–12.3.8, *and employing* (17.3.15) *instead of* (vi), *and Definition* 17.3.4(vi) *instead of Definition* 12.3.2(vi), *we find that the assertions of Remarks* 12.3.7–12.3.8 *hold in the present case.*

17.3.1 Closure of the iteration set

Similarly to Definition 12.3.9, we introduce the following notations:

Definition 17.3.13. (i) $\mathcal{K}^{\rm ext}$ *is the set of all* $(u, \theta_{\rm w}) \in C^{2,\alpha}_{(**)}(Q^{\rm iter}) \times [\theta_{\rm w}^*, \frac{\pi}{2}]$ *which satisfy conditions* (i)–(vi) *of Definition* 17.3.4;

(ii) $\overline{\mathcal{K}}$ *and* $\overline{\mathcal{K}^{\rm ext}}$ *are the closures of* $\mathcal{K}$ *and* $\mathcal{K}^{\rm ext}$ *in* $C^{2,\alpha}_{(**)}(Q^{\rm iter})\times[\theta_{\rm w}^*,\frac{\pi}{2}]$, *respectively*;

(iii) *For each* $\mathcal{C}\in\{\mathcal{K},\mathcal{K}^{\rm ext},\overline{\mathcal{K}},\overline{\mathcal{K}^{\rm ext}}\}$ *and each* $\theta_{\rm w}\in[\theta_{\rm w}^*,\frac{\pi}{2}]$,

$$\mathcal{C}(\theta_{\rm w}):=\{u\ :\ (u,\theta_{\rm w})\in\mathcal{C}\}\subset C^{2,\alpha}_{(**)}(Q^{\rm iter}).$$

We show that Lemma 12.3.10 holds in the present setting.

Lemma 17.3.14. *The sets introduced in Definition* 17.3.13 *have the following properties*:

(i) $\overline{\mathcal{K}^{\rm ext}}\subseteq\mathfrak{S}$; *that is, if* $(u,\theta_{\rm w})\in\overline{\mathcal{K}^{\rm ext}}$, *then* $(u,\theta_{\rm w})$ *satisfies condition* (ii) *of Definition* 17.3.4;

(ii) *If* $(u,\theta_{\rm w})\in\overline{\mathcal{K}^{\rm ext}}$, *then* $(u,\theta_{\rm w})$ *satisfies conditions* (i) *and* (iii)–(vi) *of Definition* 17.3.4 *with the nonstrict inequalities in the estimates*;

(iii) *The properties in Remarks* 17.3.6–17.3.12 *hold with the nonstrict inequalities for any* $(u,\theta_{\rm w})\in\overline{\mathcal{K}^{\rm ext}}$.

Proof. We divide the proof into three steps.

1. To prove (i), we follow the argument in Step 1 of the proof of Lemma 12.3.10. Then, using (17.3.7) instead of (12.3.5), it follows that, for each $(u,\theta_{\rm w})\in\mathcal{K}^{\rm ext}$, we have the following inequality instead of (12.3.24):

$$\inf_{\mathcal{Q}_{\rm bd}^{(\theta_{\rm w})}(s)}(\varphi_1-\tilde{\varphi}_2^{(\theta_{\rm w})})+\frac{\delta s}{N_2}\le u(\frac{s}{\hat{s}(\theta_{\rm w})},1)\le\sup_{\mathcal{Q}_{\rm bd}^{(\theta_{\rm w})}(s)}(\varphi_1-\tilde{\varphi}_2^{(\theta_{\rm w})})-\frac{\delta}{N_2}\tag{17.3.25}$$

for any $s\in(0,\hat{s}(\theta_{\rm w}))$. Then, arguing similarly to Step 1 of the proof of Lemma 12.3.10 and employing Lemma 12.2.7(vi) to the present case by Lemma 17.2.13, we conclude that (17.3.25) holds for each $(u,\theta_{\rm w})\in\overline{\mathcal{K}^{\rm ext}}$. This implies (12.2.49) (with the strict inequalities) for any $(u,\theta_{\rm w})\in\overline{\mathcal{K}^{\rm ext}}$, and hence assertion (i).

2. To show assertion (ii), we argue as in Step 2 of the proof of Lemma 12.3.10. The only difference is that Lemma 12.2.7(viii) cannot be applied in the present situation (since its conditions cannot be satisfied when $\theta_{\rm w}$ is a subsonic/sonic wedge angle). Instead, we apply (17.2.38) from Lemma 17.2.13(viii′). Then we have

$$(D\varphi\circ\mathfrak{F})_{(u^{(i)},\theta_{\rm w}^{(i)})}\to(D\varphi\circ\mathfrak{F})_{(u,\theta_{\rm w})},\quad(D\psi\circ\mathfrak{F})_{(u^{(i)},\theta_{\rm w}^{(i)})}\to(D\psi\circ\mathfrak{F})_{(u,\theta_{\rm w})}$$

in $C^\alpha(\overline{Q^{\rm iter}}\cap\{(s,t)\ :\ s\ge\delta\})$ for any $\delta\in(0,\frac{1}{2})$. Combining this with the other convergence and mapping properties shown in Step 2 of the proof of Lemma 12.3.10, and following that argument, we obtain assertion (ii).

3. Assertion (iii) follows from (i)–(ii). □

17.3.2 The equation and boundary conditions for the iteration

In this section, we define explicitly equation (12.3.25) and the boundary condition (12.3.26) on Γ_{shock} for $(u, \theta_{\text{w}}) \in \overline{\mathcal{K}^{\text{ext}}}$. We also discuss some properties of Problem (12.3.25)–(12.3.29) mapped onto the iteration region Q^{iter}.

In a fashion similar to the notational convention in Remark 12.3.11, we now have the following:

Remark 17.3.15. *In the rest of §17.3.2 and its subsections, we will always consider the nonstrict inequalities in the estimates in conditions* (i) *and* (iii)–(vii) *of Definition* 17.3.4, *and in Remarks* 17.3.6–17.3.12, *as in Lemma* 17.3.14(ii)–(iii).

Fix $(u, \theta_{\text{w}}) \in \overline{\mathcal{K}^{\text{ext}}}$ and consider the corresponding $(\Omega, \Gamma_{\text{shock}}, \Gamma_{\text{sonic}}, \varphi, \psi)$. We will define equation (12.3.25) such that

(i) For the supersonic/sonic wedge angles, the equation is strictly elliptic in $\overline{\Omega} \setminus \overline{\Gamma_{\text{sonic}}}$ with elliptic degeneracy on the sonic arc $\overline{\Gamma_{\text{sonic}}}$, which is one point for sonic wedge angles;

(ii) For the subsonic wedge angles, the equation is strictly elliptic in $\overline{\Omega}$ with the ellipticity constant depending on $\frac{|D\varphi_2(P_0)|}{c_2}(\theta_{\text{w}})$, and its regularity near P_0 is sufficiently high;

(iii) For a fixed point $\hat{\psi} = \psi$, equation (12.3.25) coincides with the original equation (2.2.11) written in terms of ψ;

(iv) The equation depends continuously on θ_{w} in a sense specified below.

Similarly to §12.4, we define equation (12.3.25) in $\Omega \cap \mathcal{D}_{\varepsilon_{\text{eq}}}$ and $\Omega \setminus \mathcal{D}_{\varepsilon_{\text{eq}}/10}$ separately, and then combine them over region $\mathcal{D}_{\varepsilon_{\text{eq}}} \setminus \mathcal{D}_{\varepsilon_{\text{eq}}/10}$ by using a cutoff function, where $\varepsilon_{\text{eq}} \in (0, \frac{\varepsilon_0}{2})$ is determined depending only on the data, ε_0 is from Lemma 17.2.1, and $\mathcal{D}_\varepsilon$ is defined by (17.3.2). Also, from now on, we always assume that ε in Definition 17.3.4 satisfies (12.4.1).

We use below the potential flow equation in the form of (12.4.2) in terms of ψ. Now we first define the iteration equation.

17.3.2.1 The iteration equation in Ω away from $\overline{\Gamma_{\text{sonic}}}$

We continue to consider $(u, \theta_{\text{w}}) \in \overline{\mathcal{K}^{\text{ext}}}$ and the corresponding (Ω, ψ). We also recall that the notational convention $\overline{\Gamma_{\text{sonic}}} := \{P_0\}$ is used for subsonic/sonic wedge angles.

In $\Omega \setminus \mathcal{D}_{\varepsilon_{\text{eq}}/10}$, we define the equation in the same way as in §12.4.1, by (12.4.5). This defines coefficients $(A^1_{ij}, A^1_i) = (A^1_{ij}, A^1_i)(\boldsymbol{\xi})$ in $\Omega \setminus \mathcal{D}_{\varepsilon_{\text{eq}}/10}$, with $A^1_{12} = A^1_{21}$ and $A^1_i \equiv 0$. In terms of equation (12.4.2),

$$A^1_{ij}(\boldsymbol{\xi}) = A^{\text{potn}}_{ij}(D\psi(\boldsymbol{\xi}), \psi(\boldsymbol{\xi}), \boldsymbol{\xi}).$$

Condition (v) of Definition 17.3.4 and (12.4.1) imply that the equation is uniformly elliptic in $\Omega \setminus \mathcal{D}_{\varepsilon_{\text{eq}}/10}$, with the ellipticity constant depending only on the data and $\varepsilon_{\rm eq}$. Also, from (17.3.23), coefficients (A^1_{ij}, A^1_i) are in $C^{(-\alpha),\Gamma_{\rm sym}}_{1,\alpha,\overline{\Omega}\setminus\mathcal{D}_{\varepsilon_{\rm eq}/10}}$, and (12.4.6) holds with C depending only on the data and $(\theta^*_{\rm w}, \varepsilon_{\rm eq})$, and hence only on the data and $\theta^*_{\rm w}$, once we fix $\varepsilon_{\rm eq}$ depending only on the data.

17.3.2.2 The iteration equation in Ω near $\overline{\Gamma_{\rm sonic}}$

Now we define equation (12.3.25) in region $\Omega\cap\mathcal{D}_{\varepsilon_{\rm eq}}$, *i.e.*, near $\overline{\Gamma_{\rm sonic}}$. In the $\boldsymbol{\xi}$–coordinates, the equation is of form (12.4.20) with (12.4.21), whose coefficients $(A^2_{ij}, A^2_i) = (A^2_{ij}, A^2_i)(D\hat{\psi}, \boldsymbol{\xi})$ are defined separately for the supersonic up to the subsonic-close-to-sonic case and for the sufficiently strictly subsonic case, and are then combined by a cutoff function depending on $\theta_{\rm w}$.

In the supersonic reflection case, we define the coefficients in the same way as in §12.4.2. Working in the (x, y)–coordinates (11.1.1)–(11.1.2), we define equation (12.4.18) with coefficients $(\hat{A}_{ij}, \hat{A}_i)(\mathbf{p}, x, y)$ by (12.4.14)–(12.4.15), and then change the coordinates to $\boldsymbol{\xi}$ in order to obtain (12.4.20)–(12.4.21) with coefficients $(A^2_{ij}, A^2_i)(\mathbf{p}, \boldsymbol{\xi})$. The properties of this equation in the supersonic reflection case are collected in Lemma 12.4.2. Now we note that this can be extended to the subsonic wedge angles which are sufficiently close to and include the sonic angle.

Lemma 17.3.16. *Lemma* 12.4.1 *holds in the present case, i.e., for any* $(u, \theta_{\rm w}) \in \mathfrak{S}$ *satisfying* (17.3.9)–(17.3.11) *with* $\theta_{\rm w} \in [\theta^*_{\rm w}, \frac{\pi}{2})$.

Proof. All the properties used in the proof of Lemma 12.4.1 have been extended to the present case; in particular, see (17.3.21) and Lemma 17.2.13. □

Lemma 17.3.17. (a) *There exist* $\lambda_1 > 0$, $\varepsilon_{\rm eq} \in (0, \frac{\varepsilon_0}{2})$, *and* $N_{eq} \geq 1$ *depending only on the data such that all the assertions of Lemma* 12.4.2 *(with* (v) *in a modified form, given below) hold for any* $(u, \theta_{\rm w}) \in \overline{\mathcal{K}^{\rm ext}}$ *with* $\dfrac{|D\varphi_2(P_0)|}{c_2}(\theta_{\rm w}) \geq 1 - \dfrac{\delta_{P_0}}{2}$, *where* δ_{P_0} *is from Proposition* 16.1.4.

Assertion (v) *of Lemma* 12.4.2 *now has the following form:*

(v′) *For any* $(u, \theta_{\rm w}) \in \overline{\mathcal{K}^{\rm ext}}$ *with the supersonic wedge angle* $\theta_{\rm w}$,

$$|D_y(\hat{A}_{11}, \hat{A}_{12})(\mathbf{p}, x, y)| \leq \hat{C}\sqrt{x} \qquad \text{in } \Omega\cap\mathcal{D}_r, \tag{17.3.26}$$

where $r = \min\{\mathfrak{g}^2_{\rm sh}(0),\ \varepsilon_{\rm eq}\}$, *and* $\hat{C}$ *depends on the data and* $(\theta^*_{\rm w}, \alpha)$. *Moreover, for any* $\delta \in (0, \frac{\delta_{P_0}}{2})$, *there exists* $\hat{C}_\delta$ *depending only on the data and* $(\theta^*_{\rm w}, \alpha, \delta)$ *such that, for any* $(u, \theta_{\rm w}) \in \overline{\mathcal{K}^{\rm ext}}$ *with* $\dfrac{|D\varphi_2(P_0)|}{c_2}(\theta_{\rm w})$ $\geq 1 + \delta$,

$$|D_y(\hat{A}_{11}, \hat{A}_{12})(\mathbf{p}, x, y)| \leq \hat{C}_\delta\sqrt{x} \qquad \text{in } \Omega\cap\mathcal{D}_{\varepsilon_{\rm eq}}. \tag{17.3.27}$$

(b) *For subsonic wedge angles,* $D^k_{\mathbf{p}}(\hat{A}_{ij}, \hat{A}_i)(\mathbf{p}, \cdot, \cdot)$ *are in* $C^{1,\alpha}(\overline{\Omega \cap \mathcal{D}_{\varepsilon_{\rm eq}}})$ *for* $k = 0, 1, 2$*. More specifically, for any* $\delta \in (0, \frac{\delta_{P_0}}{2})$*, there exists* C_δ *depending on the data and* δ *such that, for any* $(u, \theta_{\rm w}) \in \overline{\mathcal{K}^{\rm ext}}$ *with* $\frac{|D\varphi_2(P_0)|}{c_2}(\theta_{\rm w}) \in [1 - \frac{\delta_{P_0}}{2}, 1 - \delta]$*, the coefficients satisfy*

$$\sup_{\mathbf{p}\in\mathbb{R}^2} \|(D^k_{\mathbf{p}}\hat{A}_{ij}, D^k_{\mathbf{p}}\hat{A}_i)(\mathbf{p}, \cdot, \cdot)\|_{C^{1,\alpha}(\overline{\Omega\cap\mathcal{D}_{\varepsilon_{\rm eq}}})} \le C_\delta \qquad \text{for } k = 0, 1, 2. \tag{17.3.28}$$

Similar estimates hold for $(A^2_{ij}, A^2_i)(\mathbf{p}, \cdot)$ *in the* $\boldsymbol{\xi}$*-coordinates.*

(c) *Let* $\theta_{\rm w} \in [\theta^*_{\rm w}, \frac{\pi}{2})$ *satisfy* $\frac{|D\varphi_2(P_0)|}{c_2}(\theta_{\rm w}) \in [1 - \frac{\delta_{P_0}}{2}, 1)$ *and*

$$x^{(\theta_{\rm w})}_{P_0} \ge \frac{\varepsilon}{10}. \tag{17.3.29}$$

Then, if $(u, \theta_{\rm w}) \in \overline{\mathcal{K}^{\rm ext}}$*, and* (Ω, ψ) *are determined by* $(u, \theta_{\rm w})$*, the corresponding operator* $\hat{\mathcal{N}}_2$ *from* (12.4.18) *satisfies* (12.4.29) *in* $\Omega \cap \mathcal{D}_{\varepsilon_{\rm eq}}$*. In particular, in* $\Omega \cap \mathcal{D}_{\varepsilon_{\rm eq}}$ *in the* $\boldsymbol{\xi}$*-coordinates,*

$$c_2 A^2_{ij}(D\psi(\boldsymbol{\xi}), \boldsymbol{\xi}) = A^{\rm potn}_{ij}(D\psi(\boldsymbol{\xi}), \psi(\boldsymbol{\xi}), \boldsymbol{\xi}), \qquad A^2_i(D\psi(\boldsymbol{\xi}), \boldsymbol{\xi}) = 0.$$

Proof. We first show assertion (a). The proof of Lemma 12.4.2 uses only properties (12.3.7)–(12.3.9) and the properties of the weak state (2), specifically the continuous dependence of (u_2, c_2) on $\theta_{\rm w} \in [\theta^{\rm s}_{\rm w}, \frac{\pi}{2}]$ and the positivity of $c_2(\theta_{\rm w})$ (hence the uniform positive lower bound of c_2 for all $\theta_{\rm w} \in [\theta^{\rm s}_{\rm w}, \frac{\pi}{2}]$).

In the present case, if $\frac{|D\varphi_2(P_0)|}{c_2}(\theta_{\rm w}) \ge 1 - \frac{\delta_{P_0}}{2}$, condition (17.3.9) is of form (12.3.7), where we have used the explicit definition of $\eta_3(\theta_{\rm w})$ in Definition 17.3.4(iv). Conditions (17.3.10)–(17.3.11) clearly are of the same form as (12.3.8)–(12.3.9). The properties of the weak state (2) mentioned above hold for all $\theta_{\rm w} \in [\theta^{\rm d}_{\rm w}, \frac{\pi}{2}]$, by Theorem 7.1.1. Also, with Lemma 17.3.16, the results in the lemma follow from the proof of Lemma 12.4.2, quoted verbatim (with the updated range of wedge angles, now $[\theta^{\rm d}_{\rm w}, \frac{\pi}{2}]$).

Furthermore, to show assertion (v′), we use Remarks 17.3.9–17.3.10 to show that the additional condition in Lemma 12.4.2(v) holds in the regions of (17.3.26) and (17.3.27), respectively. Noting also the dependence of the constants in Remarks 17.3.9–17.3.10, we obtain (v′) by repeating the proof of Lemma 12.4.2(v).

Now we show assertion (b). Recall that $\zeta_1 \in C^\infty(\mathbb{R})$ satisfies (12.4.10)–(12.4.11). Since $x^{(\theta_{\rm w})}_{P_0} > 0$ for subsonic wedge angles by (16.1.1), and $x \in (x^{(\theta_{\rm w})}_{P_0}, x^{(\theta_{\rm w})}_{P_0} + \varepsilon_{\rm eq})$ for any $(x, y) \in \Omega \cap \mathcal{D}_{\varepsilon_{\rm eq}}$ by (17.3.2), it follows that, for each $\mathbf{p} \in \mathbb{R}^2$ fixed, $(x, \zeta_1(\frac{p_1}{x}), \zeta_1(\frac{p_2}{x}))$ are in $C^\infty(\overline{\Omega \cap \mathcal{D}_{\varepsilon_{\rm eq}}})$ as functions of (x, y). Then the explicit expressions in (12.4.14)–(12.4.15) and (11.1.5) of $(\hat{A}_{ij}, \hat{A}_i)$, combined with (17.3.22), imply

$$(\hat{A}_{ij}, \hat{A}_i, D^k_{\mathbf{p}}\hat{A}_{ij}, D^k_{\mathbf{p}}\hat{A}_i)(\mathbf{p}, \cdot, \cdot) \in C^{1,\alpha}(\overline{\Omega \cap \mathcal{D}_{\varepsilon_{\rm eq}}}) \qquad \text{for } k = 1, 2, \dots.$$

Next, we show (17.3.28). We first note that, from the continuous dependence of the parameters of the weak state (2) on $\theta_{\rm w}$, it follows that there exists $c_1(\delta) > 0$ depending on the data and δ such that $x^{(\theta_{\rm w})}_{P_0} \geq c_1(\delta)$ if $\frac{|D\varphi_2(P_0)|}{c_2}(\theta_{\rm w}) \in [1 - \frac{\delta_{P_0}}{2}, 1 - \delta]$. Then, for any $(u, \theta_{\rm w}) \in \overline{\mathcal{K}^{\rm ext}}$ with such $\theta_{\rm w}$, we find that $x \in (c_1(\delta), \varepsilon_{\rm eq})$ for any $(x, y) \in \Omega \cap \mathcal{D}_{\varepsilon_{\rm eq}}$. Using (12.4.33) and calculating similar to (12.4.34), we obtain that, for $m = 0, 1, 2, \dots$, $K > 0$, and $q \in \mathbb{R}$,

$$|D^m_q \zeta_1(\frac{q}{Kx})| \leq C(K, \delta, \varepsilon_{\rm eq}) \qquad \text{for any } (x, y) \in \Omega \cap \mathcal{D}_{\varepsilon_{\rm eq}}.$$

Then, again using the explicit expressions of $(\hat{A}_{ij}, \hat{A}_i)$ and (17.3.22), we obtain (17.3.28).

Finally, we show assertion (c). Using Remark 17.3.2(d), we note that condition (17.3.29) implies that (17.3.9)–(17.3.10) now hold in $\Omega \cap \mathcal{D}_{\varepsilon_0}$, which lead to (12.4.28). Thus, (12.4.29) holds in $\Omega \cap \mathcal{D}_{\varepsilon_{\rm eq}}$, which is shown by repeating the proof of Lemma 12.4.2(viii). This implies assertion (c). ☐

The properties of the equation in Lemmas 12.4.2 and 17.3.17, shown above, follow from the properties of the potential flow equation in the region where the solution is close to sonic (*i.e.*, $\frac{|D\varphi|}{c(|D\varphi|^2,\varphi)}$ is close to one), and the fact that state (2) (hence φ for $(u, \theta_{\rm w}) \in \overline{\mathcal{K}^{\rm ext}}$) is close to sonic in $\Omega \cap \mathcal{D}_{\varepsilon_{\rm eq}}$ for the wedge angles satisfying the conditions of Lemma 17.3.17.

On the other hand, for the subsonic wedge angles satisfying

$$\frac{|D\varphi_2(P_0)|}{c_2}(\theta_{\rm w}) \leq 1 - \delta \qquad \text{for some } \delta > 0,$$

ψ is not close to sonic in $\Omega \cap \mathcal{D}_{\varepsilon_{\rm eq}}$ in general, so that we cannot define the iteration equation based on these properties. Instead, we use the fact that the potential flow equation is uniformly elliptic in $\Omega \cap \mathcal{D}_{\varepsilon_{\rm eq}}$, with the uniform ellipticity constant (depending on δ) for all $(u, \theta_{\rm w}) \in \overline{\mathcal{K}^{\rm ext}}$ such that $\frac{|D\varphi_2(P_0)|}{c_2}(\theta_{\rm w}) \leq 1 - \delta$. In order to do that, we follow the way that we have used to define the iteration boundary condition on $\Gamma_{\rm shock}$ in (12.5.47), which takes a simpler form in the present case due to the quasilinear structure of the potential flow equation.

We first note the following:

Lemma 17.3.18. *In the present case, Lemma* 12.5.6 *holds for any wedge angle* $\theta_{\rm w} \in [\theta^*_{\rm w}, \frac{\pi}{2}]$. *Specifically, fix* $\tilde{\sigma}_1 \in (0, 1)$. *Then, for each* $(u, \theta_{\rm w}) \in \overline{\mathcal{K}^{\rm ext}}$, *there exists* $v \equiv v^{(u,\theta_{\rm w})} \in C^4(\overline{\Omega}^{(u,\theta_{\rm w})})$ *such that assertions* (i)–(ii) *of Lemma* 12.5.6 *hold with* $C(\tilde{\sigma}_1)$ *in* (12.5.40) *depending only on the data and* $(\theta^*_{\rm w}, \tilde{\sigma}_1, \alpha)$ *in Definition* 17.3.4.

Proof. In the present case, $(u, \theta_{\rm w}) \in \overline{\mathcal{K}^{\rm ext}} \subset C^1(\overline{Q^{\rm iter}}) \times [\theta^*_{\rm w}, \frac{\pi}{2}]$, and it satisfies the conditions of Definition 17.3.4(ii), (17.3.22), and (17.3.20) with $(\hat{N}_0, C)$ depending on the data and $(\theta^*_{\rm w}, \alpha)$.

Thus, the only condition of Lemma 12.5.6 which is not satisfied now is (12.3.5). Instead, property (17.3.7) is now satisfied.

In the proof of Lemma 12.5.6, condition (12.3.5) is used only to obtain the existence of $\delta_0 > 0$ and $\hat{C}$ depending only on $\min_{[\theta^*_{\rm w}, \frac{\pi}{2}]} \hat{s}(\cdot)$ (hence on the data) and $(\hat{N}_0, N_2)$ such that (12.5.43) holds.

In the present case, we can follow the proof of (12.5.43) by using (17.3.7) instead of (12.3.5), for which the argument works without change. Thus, we obtain the existence of $\delta_0 > 0$ and $\hat{C}$ depending only on the data and $\theta^*_{\rm w}$ such that (12.5.43) holds. Then the assertions of Lemma 12.5.6 hold with $C(\tilde{\sigma}_1)$ in (12.5.40) depending only on the data and $(\theta^*_{\rm w}, \alpha)$ for each $\tilde{\sigma}_1$. □

Now we define the iteration equation of form (12.4.20) in $\Omega \cap \mathcal{D}_{\varepsilon_{\rm eq}}$ (where $\varepsilon_{\rm eq}$ will be adjusted below), with coefficients $A^{2,({\rm subs})}_{ij}(D\hat{\psi}, \boldsymbol{\xi})$ and $A^{2,({\rm subs})}_{i} \equiv 0$, for any $(u, \theta_{\rm w}) \in \overline{\mathcal{K}^{\rm ext}}$ such that $\dfrac{|D\varphi_2(P_0)|}{c_2}(\theta_{\rm w}) \le 1 - \delta_M$ for some $\delta_M \in (0,1)$.

Fix $(u, \theta_{\rm w}) \in \overline{\mathcal{K}^{\rm ext}}$. Let $\eta \in C^\infty(\mathbb{R})$ be a cutoff function such that $\eta \equiv 1$ on $(-\infty, 1)$, $\eta \equiv 0$ on $(2, \infty)$, and $0 \le \eta \le 1$ on $\mathbb{R}$. Fix $\tilde{\sigma}_1 \in (0,1)$, and denote

$$\eta_{\tilde{\sigma}_1}(t) = \eta(\frac{t}{\tilde{\sigma}_1}).$$

Let $\tilde{\sigma}_1 \in (0,1)$ be defined later, depending only on the data. Then we define $(A^{2,({\rm subs})}_{ij}, A^{2,({\rm subs})}_{i})(\mathbf{p}, \boldsymbol{\xi})$ for $(\mathbf{p}, \boldsymbol{\xi}) \in \mathbb{R}^2 \times \overline{\Omega \cap \mathcal{D}_{\varepsilon_{\rm eq}}}$ by

$$A^{2,({\rm subs})}_{ij}(\mathbf{p}, \boldsymbol{\xi}) = \eta_{\tilde{\sigma}_1} A^{({\rm potn})}_{ij}(\mathbf{p}, \psi(\boldsymbol{\xi}), \boldsymbol{\xi}) + (1 - \eta_{\tilde{\sigma}_1}) A^{({\rm potn})}_{ij}(Dv(\boldsymbol{\xi}), \psi(\boldsymbol{\xi}), \boldsymbol{\xi}), \tag{17.3.30}$$

for $\eta_{\tilde{\sigma}_1} = \eta_{\tilde{\sigma}_1}(|\mathbf{p} - Dv(\boldsymbol{\xi})|)$, where $v(\boldsymbol{\xi})$ is defined in Lemma 12.5.6 for $(u, \theta_{\rm w})$ and $\tilde{\sigma}_1$, and $A^{({\rm potn})}_{ij}(\mathbf{p}, z, \boldsymbol{\xi}), j = 1, 2$, are the coefficients of the potential flow equation (12.4.2).

Lemma 17.3.19. *There exist $\lambda_s > 0$ (depending only on the data) and C (depending only on the data and $\theta^*_{\rm w}$) such that, if ε and δ_1 in Definition* 17.3.4 *are small (depending only on the data), then, for any $(u, \theta_{\rm w}) \in \overline{\mathcal{K}^{\rm ext}}$ with $\theta_{\rm w}$ satisfying $\dfrac{|D\varphi_2(P_0)|}{c_2}(\theta_{\rm w}) \le 1 - \dfrac{\delta_{P_0}}{4}$, coefficients $A^{2,({\rm subs})}_{ij}$ defined by* (17.3.30) *with $\tilde{\sigma}_1 = \sqrt{\delta_1}$ satisfy the following properties:*

(i) $A^{2,({\rm subs})}_{ij}(\mathbf{p}, \boldsymbol{\xi}) = A^{({\rm potn})}_{ij}(\mathbf{p}, \psi(\boldsymbol{\xi}), \boldsymbol{\xi})$ *for any* $(\mathbf{p}, \boldsymbol{\xi}) \in \mathbb{R}^2 \times \overline{\Omega \cap \mathcal{D}_{\varepsilon_{\rm eq}}}$ *with* $|\mathbf{p} - D\psi(\boldsymbol{\xi})| < \dfrac{\sqrt{\delta_1}}{2}$;

(ii) *For any* $(\mathbf{p}, \boldsymbol{\xi}) \in \mathbb{R}^2 \times (\Omega \cap \mathcal{D}_{\varepsilon_{\rm eq}})$,

$$\left| A^{2,({\rm subs})}_{ij}(\mathbf{p}, \boldsymbol{\xi}) - A^{2,({\rm subs})}_{ij}(D\psi(\boldsymbol{\xi}), \boldsymbol{\xi}) \right| \le C\sqrt{\delta_1}; \tag{17.3.31}$$

(iii) $D^k_{\mathbf{p}} A^{2,(\mathrm{subs})}_{ij}(\mathbf{p},\cdot) \in C^{1,\alpha}(\overline{\Omega \cap \mathcal{D}_{\varepsilon_{\mathrm{eq}}}})$ *for* $k = 0, 1, 2$, *and for any* $\mathbf{p} \in \mathbb{R}^2$, *with*

$$\|D^k_{\mathbf{p}} A^{2,(\mathrm{subs})}_{ij}(\mathbf{p},\cdot)\|_{C^{1,\alpha}(\overline{\Omega\cap\mathcal{D}_{\varepsilon_{\mathrm{eq}}}})} \le C;$$

(iv) *Uniform ellipticity holds: For any* $\boldsymbol{\xi} \in \Omega \cap \mathcal{D}_{\varepsilon_{\mathrm{eq}}}$ *and* $\mathbf{p}, \mu \in \mathbb{R}^2$,

$$\lambda_s|\mu|^2 \le \sum_{i,j=1}^{2} A^{2,(\mathrm{subs})}_{ij}(\mathbf{p},\boldsymbol{\xi})\mu_i\mu_j \le \lambda_s^{-1}|\mu|^2;$$

(v) *If* $(u, \theta_{\mathrm{w}}) \in \overline{\mathcal{K}^{\mathrm{ext}}}$, *and* $A^{2,(\mathrm{subs})}_{ij}(\mathbf{p},\boldsymbol{\xi})$ *are determined by* (u,θ_{w}), *then, for the corresponding* (Ω,ψ),

$$A^{2,(\mathrm{subs})}_{ij}(D\psi(\boldsymbol{\xi}),\,\boldsymbol{\xi}) = A^{(\mathrm{potn})}_{ij}(D\psi(\boldsymbol{\xi}),\,\psi(\boldsymbol{\xi}),\,\boldsymbol{\xi}) \qquad \text{in } \Omega.$$

Proof. Assertion (i) follows from (12.5.39) with $\tilde{\sigma}_1 = \sqrt{\delta_1}$ if $\delta_1 \le \frac{1}{4}$.

We now show (ii). In this argument, the universal constant C depends only on the data and θ^*_{w}. From (12.5.39) with $\tilde{\sigma}_1^2 = \delta_1 \le \frac{1}{4}$, we find that $\eta_{\tilde{\sigma}_1}(|D\psi(\boldsymbol{\xi}) - Dv(\boldsymbol{\xi})|) = 1$. Thus, we have

$$A^{2,(\mathrm{subs})}_{ij}(D\psi(\boldsymbol{\xi}),\boldsymbol{\xi}) = A^{(\mathrm{potn})}_{ij}(D\psi(\boldsymbol{\xi}),\psi(\boldsymbol{\xi}),\boldsymbol{\xi}) \qquad \text{for all } \boldsymbol{\xi} \in \Omega\cap\mathcal{D}_{\varepsilon_{\mathrm{eq}}}. \tag{17.3.32}$$

With this, we have

$$A^{2,(\mathrm{subs})}_{ij}(\mathbf{p},\boldsymbol{\xi}) - A^{2,(\mathrm{subs})}_{ij}(D\psi(\boldsymbol{\xi}),\boldsymbol{\xi}) = I_1 + I_2,$$

where

$$I_1 = \eta_{\tilde{\sigma}_1}\big(A^{(\mathrm{potn})}_{ij}(\mathbf{p},\psi(\boldsymbol{\xi}),\boldsymbol{\xi}) - A^{(\mathrm{potn})}_{ij}(D\psi(\boldsymbol{\xi}),\psi(\boldsymbol{\xi}),\boldsymbol{\xi})\big),$$
$$I_2 = (1-\eta_{\tilde{\sigma}_1})\big(A^{(\mathrm{potn})}_{ij}(Dv(\boldsymbol{\xi}),\psi(\boldsymbol{\xi}),\boldsymbol{\xi}) - A^{(\mathrm{potn})}_{ij}(D\psi(\boldsymbol{\xi}),\psi(\boldsymbol{\xi}),\boldsymbol{\xi})\big),$$

with $\eta_{\tilde{\sigma}_1} = \eta_{\tilde{\sigma}_1}(|\mathbf{p} - Dv(\boldsymbol{\xi})|)$. Then I_1 can be nonzero only in $|\mathbf{p} - Dv(\boldsymbol{\xi})| \le 2\tilde{\sigma}_1$, which implies that $|\mathbf{p} - D\psi(\boldsymbol{\xi})| \le 3\tilde{\sigma}_1$ by (12.5.39). From (12.4.4), combined with (12.3.18) (which holds by Remark 17.3.7) and (17.3.22), we obtain that $|I_1| \le C\tilde{\sigma}_1 = C\sqrt{\delta_1}$. Similarly,

$$|I_2(\boldsymbol{\xi})| \le C|D(v-\psi)(\boldsymbol{\xi})| \le C\tilde{\sigma}_1^2 = C\delta_1.$$

Now (ii) is proved.

Assertion (iii) follows from (12.4.4), combined with (12.3.18) (which holds by Remark 17.3.7) and (17.3.22).

Assertion (iv) can be seen as follows: Since $P_1 = P_0$ in the subsonic reflection case, the condition that $\frac{|D\varphi_2(P_0)|}{c_2}(\theta_{\mathrm{w}}) \le 1 - \frac{\delta_{P_0}}{4}$, combined with (17.3.32) and the conditions of Definition 17.3.4(v)–(vi), implies that (iv) holds for $(\mathbf{p},\boldsymbol{\xi}) =$

$(D\psi(\boldsymbol{\xi}), \boldsymbol{\xi})$ for any $\boldsymbol{\xi} \in \Omega \cap \mathcal{D}_{\varepsilon_{\rm eq}}$, with λ_s depending only on the data and δ_{P_0}, and hence only on the data. Now (17.3.31) with small δ_1, depending only on the data and $\theta_{\rm w}^*$, implies assertion (iv) with $\frac{\lambda_s}{2}$ for all $(\mathbf{p}, \boldsymbol{\xi}) \in \mathbb{R}^2 \times (\Omega \cap \mathcal{D}_{\varepsilon_{\rm eq}})$.

Assertion (v) follows from (i). □

Now we combine coefficients (A_{ij}^2, A_i^2) with $A_{ij}^{2,(\rm subs)}$ (recall that $A_i^{2,(\rm subs)} \equiv 0$) such that the resulting coefficients are equal to (A_{ij}^2, A_i^2) in the supersonic reflection case and to $(A_{ij}^{2,(\rm subs)}, 0)$ in the *subsonic-away-from-sonic* case, and the coupling is continuous with respect to $\theta_{\rm w}$, so that property (iii) of Lemma 12.7.1 holds in the present case (to be shown later).

Let $\zeta \in C^\infty(\mathbb{R})$ satisfy

$$\zeta(s) = \begin{cases} 0 & \text{if } s \le 1 - \frac{\delta_{P_0}}{2}, \\ 1 & \text{if } s \ge 1 - \frac{\delta_{P_0}}{4}, \end{cases} \qquad \text{and} \quad \zeta'(s) \ge 0 \quad \text{on } \mathbb{R}. \tag{17.3.33}$$

Given $(u, \theta_{\rm w}) \in \overline{\mathcal{K}^{\rm ext}}$, for each $\mathbf{p} \in \mathbb{R}^2$ and $\boldsymbol{\xi} \in \Omega \cap \mathcal{D}_{\varepsilon_{\rm eq}}$, define

$$\begin{aligned} &A_{ij}^3(\mathbf{p}, \boldsymbol{\xi}) = c_2 \zeta A_{ij}^2(\mathbf{p}, \boldsymbol{\xi}) + (1 - \zeta) A_{ij}^{2,(\rm subs)}(\mathbf{p}, \boldsymbol{\xi}), \\ &A_i^3(\mathbf{p}, \boldsymbol{\xi}) = c_2 \zeta A_i^2(\mathbf{p}, \boldsymbol{\xi}), \end{aligned} \tag{17.3.34}$$

where $\zeta = \zeta(\frac{|D\varphi_2(P_0)|}{c_2}(\theta_{\rm w}))$; see Remark 12.4.4 regarding coefficient c_2 in the terms of (A_{ij}^2, A_i^2).

We state the properties of (A_{ij}^3, A_i^3) in the following lemma:

Lemma 17.3.20. *There exist $\lambda_1 > 0$, $\varepsilon_{\rm eq} \in (0, \frac{\varepsilon_0}{2})$, $N_{\rm eq} \ge 1$ depending on the data, $C > 0$ depending on the data and $\theta_{\rm w}^*$, and $\lambda(\delta) > 0$ for each $\delta \in (0, \frac{\delta_{P_0}}{2})$ depending on the data and δ with $\delta \mapsto \lambda(\delta)$ being non-decreasing such that the following hold: Let ε and δ_1 in Definition 17.3.4 be small, depending only on the data. Then, for any $(u, \theta_{\rm w}) \in \overline{\mathcal{K}^{\rm ext}}$, coefficients (A_{ij}^3, A_i^3) defined by (17.3.34) with (17.3.33) satisfy the following properties:*

(i) *If $\frac{|D\varphi_2(P_0)|}{c_2}(\theta_{\rm w}) \ge 1 - \frac{\delta_{P_0}}{4}$, then $(A_{ij}^3, A_i^3) = (A_{ij}^2, A_i^2)$. Moreover, these functions satisfy the properties in Lemma 17.3.17;*

(ii) *If $\delta \in (0, \frac{\delta_{P_0}}{2})$ and $\frac{|D\varphi_2(P_0)|}{c_2}(\theta_{\rm w}) \le 1 - \delta$, then*

  (a) *Coefficients A_{ij}^3 are uniformly elliptic with constant $\lambda(\delta)$: For any $\boldsymbol{\xi} \in \Omega \cap \mathcal{D}_{\varepsilon_{\rm eq}}$ and $\mathbf{p}, \mu \in \mathbb{R}^2$,*

$$\lambda(\delta)|\mu|^2 \le \sum_{i,j=1}^2 A_{ij}^3(\mathbf{p}, \boldsymbol{\xi}) \mu_i \mu_j \le \frac{1}{\lambda(\delta)}|\mu|^2;$$

(b) (A^3_{ij}, A^3_i) *satisfy* (17.3.28) *with constant* C_δ *depending on the data and* $(\theta^*_{\rm w}, \delta)$.

The proof of this lemma directly follows from definition (17.3.34) with (17.3.33), by using Lemmas 17.3.17 and 17.3.19.

17.3.2.3 Combined iteration equation in Ω

Next, we combine the equations introduced above by defining the coefficients of (12.3.25) in Ω, similar to (12.4.42) with (A^3_{ij}, A^3_i) instead of (A^2_{ij}, A^2_i):

$$\begin{aligned} A_{ij}(\mathbf{p},\boldsymbol{\xi}) &= \hat{\zeta}_2^{(\varepsilon_{\rm eq})}(\boldsymbol{\xi})A^1_{ij}(\boldsymbol{\xi}) + \big(1-\hat{\zeta}_2^{(\varepsilon_{\rm eq})}(\boldsymbol{\xi})\big)A^3_{ij}(\mathbf{p},\boldsymbol{\xi}),\\ A_i(\mathbf{p},\boldsymbol{\xi}) &= \big(1-\hat{\zeta}_2^{(\varepsilon_{\rm eq})}(\boldsymbol{\xi})\big)A^3_i(\mathbf{p},\boldsymbol{\xi}), \end{aligned} \tag{17.3.35}$$

where $A^1_{ij}(\boldsymbol{\xi})$ are defined by (12.4.5), $(A^3_{ij}, A^3_i)(\mathbf{p},\boldsymbol{\xi})$ are defined by (17.3.34) with (17.3.33), and $\hat{\zeta}_2^{(\varepsilon_{\rm eq})}$ is defined by (12.4.37)–(12.4.38). This equation has the properties, similar to those in Lemma 12.4.5 and Corollary 12.4.6.

Lemma 17.3.21. *There exists* $\varepsilon_{\rm eq} \in (0, \frac{\varepsilon_0}{2})$ *depending only on the data such that, if* $\varepsilon \in (0, \frac{\varepsilon_{\rm eq}}{2})$ *in Definition* 17.3.4, *then, for any* $(u, \theta_{\rm w}) \in \overline{\mathcal{K}^{\rm ext}}$, *equation* (12.3.25) *with coefficients* (17.3.35) *satisfies assertions* (i)–(vii) *of Lemma* 12.4.5 *(with a modification in* (v)–(vi) *detailed below), with* $\lambda_0 > 0$ *depending only on the data and* $\theta^*_{\rm w}$, *and with* C *depending on the data and* $(\theta^*_{\rm w}, \alpha)$. *The modified properties* (v)–(vi) *are:*

(v′) $(A_{ij}, A_i)(\mathbf{p},\boldsymbol{\xi}) = (A^3_{ij}, A^3_i)(\mathbf{p},\boldsymbol{\xi})$ *for any* $(\mathbf{p},\boldsymbol{\xi}) \in \mathbb{R}^2 \times (\Omega \cap \mathcal{D}_{\varepsilon_{\rm eq}/2})$. *Thus, the coefficients in* (17.3.35) *in* $\mathbb{R}^2 \times (\Omega \cap \mathcal{D}_{\varepsilon_{\rm eq}/2})$ *satisfy Lemma* 17.3.20;

(vi′) $(A_{ij}, A_i)(\mathbf{p},\boldsymbol{\xi}) = (A^3_{ij}, A^3_i)(\mathbf{p},\boldsymbol{\xi})$ *for any* $(\mathbf{p},\boldsymbol{\xi}) \in \mathbb{R}^2 \times (\Omega \cap \mathcal{D}_{\varepsilon_{\rm eq}/2})$.

Moreover, assertion (viii) *of Lemma* 12.4.5 *is replaced by the following:*

(viii′) *Let* ψ *be a fixed point; that is,* $\hat{\psi} = \psi$ *satisfies equation* (12.3.25) *in* Ω. *Then*

(a′) (12.3.25) *coincides with the potential flow equation* (12.4.2) *in* $\Omega \setminus \mathcal{D}_{0,\varepsilon/10}$, *where we have used the notation in* (17.3.3):

$$A_{ij}(D\psi,\boldsymbol{\xi}) = A^{\rm (potn)}_{ij}(D\psi,\psi,\boldsymbol{\xi}), \quad A_i(D\psi,\boldsymbol{\xi}) = 0. \tag{17.3.36}$$

In particular, if $x_{P_0} \geq \frac{\varepsilon}{10}$, (12.3.25) *coincides with the potential flow equation* (12.4.2) *in* Ω, *i.e.,* (17.3.36) *holds in* Ω.

(b′) *If* $\dfrac{|D\varphi_2(P_0)|}{c_2}(\theta_{\rm w}) \leq 1 - \dfrac{\delta_{P_0}}{2}$, (12.3.25) *coincides with the potential flow equation in* Ω, *i.e.,* (17.3.36) *holds in* Ω.

(c′) *Assume that* (12.4.7) *holds for* $\hat{\psi} = \psi$. *Assume also that* ε *(in the iteration set) is small, depending on the data and the constants in* (12.4.7). *Then assertions (b) and (c) of Lemma* 12.4.5(viii) *hold for any* $(u, \theta_{\rm w}) \in \overline{\mathcal{K}^{\rm ext}}$ *with* $\theta_{\rm w}$ *satisfying* $\frac{|D\varphi_2(P_0)|}{c_2}(\theta_{\rm w}) \geq 1 - \frac{\delta_{P_0}}{4}$.

Proof. The conditions in (12.4.43) hold in the present case with M depending on the data and $(\theta_{\rm w}^*, \alpha)$ by (17.3.23) for $\tau = \frac{\varepsilon_{\rm eq}}{2}$. Then all the conditions of Lemma 12.4.5 are satisfied, and its assertions with modifications given above follow in the present case by repeating the proof of Lemma 12.4.5 and using the properties of (A^3_{ij}, A^3_i) in Lemma 17.3.20.

We comment only on property (b′) of assertion (viii′). From its definition in §17.3.2.1, coefficients A^1_{ij} for $(u, \theta_{\rm w}) \in \overline{\mathcal{K}^{\rm ext}}$ satisfy $A^1_{ij}(\boldsymbol{\xi}) = A^{\rm (potn)}_{ij}(D\psi, \psi, \boldsymbol{\xi})$ in Ω, where we have used the form of (12.4.2) for the potential flow equation. Also, $A^{2,{\rm (subs)}}_{ij}(D\psi(\boldsymbol{\xi}), \boldsymbol{\xi}) = A^{\rm (potn)}_{ij}(D\psi(\boldsymbol{\xi}), \psi(\boldsymbol{\xi}), \boldsymbol{\xi})$ by Lemma 17.3.19(v). Since $A^3_{ij}(\mathbf{p}, \boldsymbol{\xi}) = A^{2,{\rm (subs)}}_{ij}(\mathbf{p}, \boldsymbol{\xi})$ and $A^3_i(\mathbf{p}, \boldsymbol{\xi}) = 0$ in $\mathbb{R}^2 \times \Omega$ by (17.3.34) for $\theta_{\rm w}$ satisfying $\frac{|D\varphi_2(P_0)|}{c_2}(\theta_{\rm w}) \leq 1 - \frac{\delta_{P_0}}{2}$, then property (b′) of assertion (viii′) follows from (17.3.35). □

We have the following analogue of Lemma 12.4.7.

Lemma 17.3.22. *Let* $\varepsilon_{\rm eq} \in (0, \frac{\varepsilon_0}{2})$ *be as in Lemma* 17.3.21, *and let* $\delta_{P_0} > 0$ *be from Proposition* 16.1.4. *Then, for every* $\delta_{\rm e} \in (0, \frac{\delta_{P_0}}{2}]$, *there exists* $\varepsilon_{\delta_{\rm e}}$ *depending on the data and* $(\theta_{\rm w}^*, \delta_{\rm e})$ *such that, if* $\varepsilon \in (0, \varepsilon_{\delta_{\rm e}}]$ *in Definition* 17.3.4, *the following hold: Let* $(u, \theta_{\rm w}) \in \overline{\mathcal{K}}$ *be such that* $\hat{\psi} = \psi$ *in* Ω *for function* $\hat{\psi}$ *introduced in Definition* 17.3.4(vii). *Assume also that, in the* (x, y)*-coordinates,*

$$|\psi_x| \leq \frac{2 - \frac{\mu_0}{5}}{1+\gamma} x, \ |\psi_y| \leq N_3 x \qquad in\ \Omega \cap \mathcal{D}_{0,\varepsilon/4}\ \ if\ \frac{|D\varphi_2(P_0)|}{c_2}(\theta_{\rm w}) \geq 1 - \delta_{\rm e}. \tag{17.3.37}$$

Then ψ *satisfies the potential flow equation* (12.4.2) *in* Ω. *Moreover, equation* (12.4.2) *is strictly elliptic for* ψ *in* $\overline{\Omega} \setminus \overline{\Gamma_{\rm sonic}}$.

Proof. Fix $(u, \theta_{\rm w}) \in \overline{\mathcal{K}}$ satisfying the conditions above. We divide the proof into two steps.

1. We first show that, if $\varepsilon > 0$ is chosen small in Definition 17.3.4, ψ satisfies the potential flow equation (12.4.2) in Ω.

If $\theta_{\rm w}$ satisfies $\frac{|D\varphi_2(P_0)|}{c_2}(\theta_{\rm w}) \leq 1 - \frac{\delta_{P_0}}{2}$, then, by assertion (b′) in Lemma 17.3.21(viii′) and since $\hat{\psi} = \psi$ satisfies (12.3.25), it follows that ψ satisfies the potential flow equation (12.4.2).

Next, we can choose $\delta_{\rm e} \in (0, \frac{\delta_{P_0}}{2})$. Then there exists $\varepsilon_{\delta_{\rm e}} > 0$ small, depending on the data and $(\theta_{\rm w}^*, \delta_{\rm e})$, such that

$$x_{P_0}^{(\theta_{\rm w}^*)} \geq \varepsilon_{\delta_{\rm e}} \qquad \text{if } \frac{|D\varphi_2(P_0)|}{c_2}(\theta_{\rm w}) \leq 1 - \delta_{\rm e}.$$

Let $\varepsilon \in (0, \varepsilon_{\delta_e}]$. With this choice, if θ_w satisfies

$$\frac{|D\varphi_2(P_0)|}{c_2}(\theta_w) \in (1 - \frac{\delta_{P_0}}{2}, 1 - \delta_e),$$

then, using assertion (a′) of Lemma 17.3.21(viii′), we come to the same conclusion as that in the previous case.

Finally, if θ_w satisfies $\frac{|D\varphi_2(P_0)|}{c_2}(\theta_w) \geq 1 - \delta_e$, we combine condition (17.3.37) with (17.3.9) (in which $\eta_3(\theta_w) = \frac{2-\mu_0}{1+\gamma}$ from the condition of the present case, with $\delta_e \in (0, \frac{\delta_{P_0}}{2})$) and (17.3.10) to obtain

$$|\partial_x \psi(x,y)| \leq \frac{2-\mu_0}{1+\gamma} x, \quad |\partial_y \psi(x,y)| \leq N_3 x \qquad \text{in } \Omega \cap \mathcal{D}_{\varepsilon_{eq}}. \tag{17.3.38}$$

From this, arguing as in the proof of (12.4.29) in Lemma 12.4.2, we find that (12.4.29) holds. Thus, in the $\boldsymbol{\xi}$–coordinates,

$$c_2 A_{ij}^2(D\psi, \boldsymbol{\xi}) = A_{ij}^{(\text{potn})}(D\psi, \psi, \boldsymbol{\xi}), \quad A_i^2(D\psi, \boldsymbol{\xi}) = 0 \qquad \text{in } \Omega \cap \mathcal{D}_{\varepsilon_{eq}}.$$

Combining this with Lemma 17.3.19(v) and using $A_{ij}^1(\boldsymbol{\xi}) = A_{ij}^{(\text{potn})}(D\psi, \psi, \boldsymbol{\xi})$ from its definition, we obtain from (17.3.35) that (17.3.36) holds in $\Omega \cap \mathcal{D}_{\varepsilon_{eq}}$. Then, using again that $\hat{\psi} = \psi$ satisfies (12.3.25), we obtain that ψ satisfies the potential flow equation (12.4.2).

Therefore, for any $\theta_w \in (\theta_w^*, \frac{\pi}{2})$, if the conditions of this lemma hold and ε_{δ_e} is chosen as above, then (17.3.36) holds in Ω, and ψ satisfies equation (12.4.2).

2. The result that equation (12.4.2) is elliptic for ψ follows from both (17.3.36) in Ω and the fact that equation (12.3.25) is strictly elliptic for ψ in $\overline{\Omega} \setminus \overline{\Gamma_{\text{sonic}}}$ by Lemma 12.4.5(i) (which holds by Lemma 17.3.21). □

17.3.2.4 Boundary conditions for the iteration

Recall that, by Proposition 16.2.1, operators $\mathcal{M}_0(D\psi, \psi, \boldsymbol{\xi})$ and $\mathcal{M}_1(D\psi, \psi, \boldsymbol{\xi})$, defined by (11.3.2) and (11.3.23), and function $F(z, \boldsymbol{\xi})$ in (11.3.20) are defined for all $\theta_w \in (\theta_w^d, \frac{\pi}{2})$ and satisfy the properties as in §11.3.

Then all the constructions of §12.5 hold without change for all $\theta_w \in (\theta_w^d, \frac{\pi}{2})$. In particular, definition (12.5.15) and Lemma 12.5.1 hold without change if $\theta_w \in (\theta_w^d, \frac{\pi}{2})$, and $(u, \theta_w) \in C^{2,\alpha}_{(**)}(Q^{\text{iter}}) \times [\theta_w^*, \frac{\pi}{2}]$ satisfies conditions (i)–(ii) and (vi) of Definition 17.3.4 and (17.3.15).

Furthermore, we note that Lemma 12.5.6 holds in the present case by Lemma 17.3.18. Then, for each $(u, \theta_w) \in \overline{\mathcal{K}^{\text{ext}}}$, we define the iteration boundary condition by (12.5.47).

Lemma 17.3.23. *If parameters $(\varepsilon, \delta_1, \delta_3, \frac{1}{N_1})$ of the iteration set in Definition 17.3.4 and $\delta_{\mathcal{K}} > 0$ are all small, depending only on the data and θ_w^*, then the following holds: Let $(u, \theta_w) \in \overline{\mathcal{K}^{\text{ext}}}$ be such that there exists $(u^\#, \theta_w^\#) \in \overline{\mathcal{K}}$*

satisfying (12.5.31) *with* $(u, \theta_{\rm w})$. *Define function* $\mathcal{M}_{(u,\theta_{\rm w})}(\cdot)$ *by* (12.5.47) *with* $\tilde{\sigma}_1 = \sqrt{\delta_1}$. *Then all the assertions of Lemma* 12.5.7 *hold for* $\mathcal{M}_{(u,\theta_{\rm w})}(\cdot)$ *with* C *depending only on the data,* $\delta_{\rm bc}$ *depending only on the data and* $\theta_{\rm w}^*$, *and* C_α *depending only on the data and* $(\theta_{\rm w}^*, \alpha)$.

Proof. If $(u^\#, \theta_{\rm w}^\#) \in \overline{\mathcal{K}}$ satisfies (12.5.31) with $(u, \theta_{\rm w})$ for constant $\delta_{\mathcal{K}}$, there exists $(\tilde{u}^\#, \tilde{\theta}_{\rm w}^\#) \in \mathcal{K}$ satisfying (12.5.31) with $(u, \theta_{\rm w})$ for constant $2\delta_{\mathcal{K}}$. We use $(\tilde{u}^\#, \tilde{\theta}_{\rm w}^\#)$ below instead of $(u^\#, \theta_{\rm w}^\#)$. Then we note that all the conditions of Lemma 12.5.7 are satisfied.

Indeed, the requirements of Lemma 12.5.2 are satisfied for $(u, \theta_{\rm w}) \in \overline{\mathcal{K}^{\rm ext}}$ by conditions (ii) and (v)–(vi) of Definition 17.3.4, (17.3.11), and (17.3.15)–(17.3.16), and the dependence of the constants in the conditions are specified in Definition 17.3.4.

Also, $(\tilde{u}^\#, \tilde{\theta}_{\rm w}^\#) \in \mathcal{K}$ satisfies the conditions of Lemma 12.5.3 on $(u^\#, \theta_{\rm w}^\#)$ there, with $b > 0$ depending only on the data and $\theta_{\rm w}^*$, which follows from (17.3.7) for $(\tilde{u}^\#, \tilde{\theta}_{\rm w}^\#)$. Indeed, from Lemma 12.2.2(ii), which holds by Lemma 17.2.1, we see that $|P_2 - P_3| \geq \frac{1}{C}\mathfrak{g}_{\rm sh}(\hat{s}(\theta_{\rm w}))$ so that $|P_2 - P_3| \geq \frac{1}{C_1} =: b$, by (17.3.7) and the uniform lower bound of $\hat{s}(\theta_{\rm w})$ for $\theta_{\rm w} \in [\theta_{\rm w}^{\rm d}, \frac{\pi}{2}]$.

Furthermore, Lemma 12.5.6 holds for $(u, \theta_{\rm w}) \in \overline{\mathcal{K}^{\rm ext}}$ by Lemma 17.3.18, with $C(\tilde{\sigma}_1)$ in (12.5.40) depending only on the data and $(\theta_{\rm w}^*, \alpha, \tilde{\sigma}_1)$.

Thus, we have checked that all the conditions of Lemma 12.5.7 are satisfied. We have also specified the dependence of the constants in the conditions.

Furthermore, using Proposition 16.2.1, we can check that the proofs of Lemmas 12.5.2–12.5.3 and 12.5.7 and Corollary 12.5.4 work without change in the sonic/subsonic reflection case.

Then, if parameters $(\varepsilon, \delta_1, \delta_3, \frac{1}{N_1})$ of the iteration set in Definition 17.3.4 and $\delta_{\mathcal{K}} > 0$ are chosen to satisfy the smallness conditions of Lemma 12.5.7 for the constants in its conditions determined above, we complete the proof by employing Lemma 12.5.7, taking into account the dependence of the constants in the conditions. □

Therefore, we also obtain the following:

Corollary 17.3.24. *Corollary* 12.5.8 *holds without change in the present case, i.e., for any* $\theta_{\rm w}^* \in (\theta_{\rm w}^{\rm d}, \frac{\pi}{2})$.

17.3.2.5 The boundary value problem in the iteration region $Q^{\rm iter}$

Let $(u, \theta_{\rm w}) \in \overline{\mathcal{K}^{\rm ext}}$, and let (Ω, ψ) be determined by $(u, \theta_{\rm w})$.

Similarly to §12.7.1, we rewrite Problem (12.3.25)–(12.3.29) for an unknown function $\hat{\psi}$ in Ω (with $\overline{\Gamma_{\rm sonic}} = \{P_0\}$ in the subsonic case) as a boundary value problem for function $\hat{u}$ in $Q^{\rm iter}$ determined by (12.3.15). Then we obtain that the problem for $\hat{u}$ is of form (12.7.2)–(12.7.6), and Lemmas 12.7.1–12.7.2 hold. The proofs work without change, where, in Step 1 of the proof of Lemma 12.7.1, we use the C^∞–smoothness with respect to $\theta_{\rm w} \in [\theta_{\rm w}^{\rm d}, \frac{\pi}{2}]$ of function

$\zeta(\frac{|D\varphi_2(P_0)|}{c_2}(\theta_{\rm w}))$ in (17.3.34), in addition to the other properties used there. Also, when extending the proofs to the subsonic/sonic case, it is important that, in Lemma 12.7.1, we assert only the interior continuity of $(\mathcal{A}_{ij}, \mathcal{A}_i)(\mathbf{p}, s, t)$ and $f(s,t)$ with respect to variables $(s,t) \in Q^{\rm iter}$, and the convergence of the sequence of such functions on compact subsets of the interior of $Q^{\rm iter}$. Similarly, for the boundary conditions, in Lemma 12.7.2, we have established the continuity and convergence of the ingredients of the boundary conditions within the relative interiors of the corresponding boundary segments (continuity) and their compact subsets (convergence). Therefore, the singularity near $\Gamma_{\rm sonic}$ of map $\mathfrak{F}^{-1} : \Omega \mapsto Q^{\rm iter}$ does not affect the proofs.

17.4 EXISTENCE AND ESTIMATES OF SOLUTIONS OF THE ITERATION PROBLEM

We now solve the iteration problem (12.3.25)–(12.3.29) for each $(u, \theta_{\rm w}) \in \overline{\mathcal{K}^{\rm ext}}$, *i.e.*, establish the existence and estimates of solutions.

Since $(u, \theta_{\rm w}) \in \overline{\mathcal{K}^{\rm ext}}$ is considered, then, throughout this section, when referring to the conditions of Definition 17.3.4, we always assume the nonstrict inequalities as described in Remark 17.3.15.

The outline of the argument is the following:

(i) We first show the existence of solutions $\hat{\psi}$ of the iteration problem for all $(u, \theta_{\rm w}) \in \overline{\mathcal{K}^{\rm ext}}$, by using Propositions 4.7.2 and 4.8.7. It remains to obtain appropriate estimates of solutions $\hat{\psi}$ in four different cases.

(ii) For the *supersonic-away-from-sonic* angles, *i.e.*, $\theta_{\rm w}$ satisfying (16.3.5) for any given $\delta > 0$, we follow the argument in §12.7.2. The estimates then depend on δ.

(iii) For the *supersonic-near-sonic* angles, *i.e.*, $\theta_{\rm w}$ satisfying

$$\frac{|D\varphi_2(P_0)|}{c_2}(\theta_{\rm w}) \in (1, 1 + \hat{\delta}_{P_0}) \tag{17.4.1}$$

with appropriate $\hat{\delta}_{P_0} > 0$, we follow the argument in §16.4, changing from the free boundary problem to the fixed boundary problem, *i.e.*, the iteration problem (12.3.25)–(12.3.29).

(iv) For the *subsonic-near-sonic* angles, *i.e.*, $\theta_{\rm w}$ satisfying

$$\frac{|D\varphi_2(P_0)|}{c_2}(\theta_{\rm w}) \in (1 - \hat{\delta}_{P_0}, 1], \tag{17.4.2}$$

we similarly follow a fixed boundary version of the argument in §16.5 to obtain the *a priori* estimates of solutions.

(v) For the *subsonic-away-from-sonic* angles, *i.e.*, $\theta_{\rm w}$ satisfying

$$\frac{|D\varphi_2(P_0)|}{c_2} \le 1 - \hat{\delta}_{P_0}, \tag{17.4.3}$$

we use the estimates of solutions given in Proposition 4.8.7.

(vi) Finally, we combine all these estimates into one estimate for $\hat{u}$ in $Q^{\rm iter}$ for any $(u, \theta_{\rm w}) \in \overline{\mathcal{K}^{\rm ext}}$, similarly to Proposition 17.2.12.

Constant $\hat{\delta}_{P_0}$ in the conditions above is determined below, depending only on the data and $\theta_{\rm w}^*$.

We sketch some details of this procedure below. For this, we first prove a fixed boundary version of Lemma 16.4.3; that is, we now assume only one condition on $\Gamma_{\rm shock}$ instead of two, but assume *a priori* a higher regularity of $\Gamma_{\rm shock}$, where we note that only the Lipschitz regularity (16.1.13) of $\Gamma_{\rm shock}$ is assumed in Lemma 16.4.3.

Lemma 17.4.1. *In the conditions of Lemma* 16.4.3, *we replace the boundary condition* (16.4.36) *by*

$$\|\hat{f}\|_{C^{(-1-\beta),\{0\}}_{2,\beta,(0,2\varepsilon)}} \le M \qquad \text{for some } \beta \in (0,1). \tag{17.4.4}$$

Then the conclusion of Lemma 16.4.3 *holds without change.*

Proof. Note that this lemma is an extension of Theorem 4.7.4 to a more general scaling, in the same way as Lemma 16.4.3 is an extension of the estimates in Proposition 11.4.3 to a more general scaling.

We follow the proof of Lemma 16.4.3 to consider Cases (i)–(iii) in rectangles (16.4.46). In Cases (i)–(ii), the argument works without change. In particular, we obtain the same estimates for both the ingredients of the rescaled equation and the rescaled solution $\psi^{(z_0)}$ defined by (16.4.44).

Thus, we focus now on Case (iii). Since ψ satisfies (16.4.35) on $\Gamma_{\rm shock}$, function (16.4.44) satisfies the rescaled boundary condition (16.4.52) on $\Gamma^{(z_0)}_{\rm shock}$, defined by (16.4.53). By assumption, b satisfies (11.3.41)–(11.3.42) on $\Omega_{2\varepsilon}$. Then we combine these properties with (16.4.29)–(16.4.30) to conclude that $b^{(z_0)}$ on $\Gamma^{(z_0)}_{\rm shock}$ satisfies (11.4.28).

Using condition (17.4.4) and repeating the argument between (4.7.39) and (4.7.40) in the proof of Theorem 4.7.4, we obtain

$$\|D_S F^{(z_0)}\|_{C^{1,\alpha}([-\frac{1}{2},\frac{1}{2}])} \le C(M,\alpha).$$

This is estimate (11.4.34) with α instead of β. Then, repeating the argument in the proof of Proposition 11.4.3 after equation (11.4.34), we obtain (11.4.36) (with α instead of β) in the present case. This completes the proof. □

17.4.1 Existence of solutions of the iteration problem

We now show that Proposition 12.7.3 holds in the present case, *i.e.*, for any $\theta_{\rm w} \in (\theta_{\rm w}^{\rm d}, \frac{\pi}{2})$. We use $\bar{\alpha}$ from Proposition 17.2.12.

Proposition 17.4.2. *Let parameters $(\varepsilon, \delta_3, \frac{1}{N_1})$ of the iteration set in Definition 17.3.4 and $\delta_{\mathcal{K}} > 0$ be all small, depending only on the data and $\theta_{\rm w}^*$. Let δ_1 be small, depending on the data and $(\theta_{\rm w}^*, \alpha)$, for which (δ_1, α) are from Definition 17.3.4. Then there exist $\hat{\alpha} \in (0, \bar{\alpha})$ depending only on the data and $\theta_{\rm w}^*$, and $C \geq 1$ depending only on the data and $(\theta_{\rm w}^*, \alpha)$ and, for each $s \in (0, \varepsilon_0)$, there exists $C_s \geq 1$ depending only on the data and $(\theta_{\rm w}^*, s, \alpha)$ such that, for each $(u, \theta_{\rm w}) \in \overline{\mathcal{K}^{\rm ext}}$ satisfying (12.7.13) with some $(u^\#, \theta_{\rm w}^\#) \in \overline{\mathcal{K}}$, there is a unique solution $\hat{\psi} \in C^2(\Omega) \cap C^1(\overline{\Omega} \setminus \overline{\Gamma_{\rm sonic}}) \cap C(\overline{\Omega})$ of Problem (12.3.25)–(12.3.29), determined by $(u, \theta_{\rm w})$, satisfying (12.7.14)–(12.7.16).*

Moreover, for every $\delta_{\rm el} \in (0,1)$, there exist $\alpha_2 \in (0, \frac{1}{2})$ depending on the data and $(\theta_{\rm w}^*, \delta_{\rm el})$, and $C_{\delta_{\rm el}}$ depending on the data and $(\theta_{\rm w}^*, \alpha, \delta_{\rm el})$ such that, if $\theta_{\rm w}$ satisfies $\frac{|D\varphi_2(P_0)|}{c_2}(\theta_{\rm w}) \leq 1 - \delta_{\rm el}$, then

$$\|\hat{\psi}\|_{2,\alpha_2,\Omega}^{(-1-\alpha_2),\ \{P_0\}\cup\overline{\Gamma_{\rm sym}}} \leq C_{\delta_{\rm el}}. \tag{17.4.5}$$

Proof. The proof consists of two steps.

1. In the first step, we consider the wedge angles $\theta_{\rm w} \in [\theta_{\rm w}^*, \frac{\pi}{2})$ satisfying

$$\frac{|D\varphi_2(P_0)|}{c_2}(\theta_{\rm w}) \geq 1 - \delta_{\rm e}, \tag{17.4.6}$$

where $\delta_{\rm e} \in (0, \frac{\delta_{P_0}}{4})$ will be determined below, depending only on the data and $\theta_{\rm w}^*$. Note that all the supersonic wedge angles $\theta_{\rm w}$ satisfy (17.4.6).

Fix $(u, \theta_{\rm w}) \in \overline{\mathcal{K}^{\rm ext}}$ with $\theta_{\rm w}$ satisfying (17.4.6), and assume that it satisfies (12.7.13) with some $(u^\#, \theta_{\rm w}^\#) \in \overline{\mathcal{K}}$. We show the existence of solutions for Problem (12.3.25)–(12.3.29) determined by $(u, \theta_{\rm w}) \in \overline{\mathcal{K}^{\rm ext}}$. We follow the proof of Proposition 12.7.3.

We first note that, by Lemma 17.2.1 and Lemma 12.2.2(ii), transform $F_1(\cdot)$ is C^3 with C^3–inverse, endowed with the uniform estimates in these norms for $\theta_{\rm w} \in [\theta_{\rm w}^{\rm d}, \frac{\pi}{2}]$. Thus, we make the change of coordinates $(s,t) = F_1(\boldsymbol{\xi})$ and consider the problem in the (s,t)–variables in domain $F_1(\Omega)$. We will employ Proposition 4.7.2 to obtain the existence of solutions. Therefore, we need to check its conditions.

By (17.2.4) and Lemma 17.2.1,

$$\mathfrak{g}_{\rm sh}(0) = t_{P_1}(\theta_{\rm w}) = y_{P_1} - y_{P_4}.$$

Then $\mathfrak{g}_{\rm sh}(0) > 0$ for the supersonic wedge angles, and $\mathfrak{g}_{\rm sh}(0) = 0$ for the subsonic/sonic wedge angles. From (12.7.17) and (17.3.7), it follows that $F_1(\Omega)$ is of the structure of (4.5.1) and (4.5.3) with $h = \hat{s}(\theta_{\rm w})$, $f_{\rm bd} = \mathfrak{g}_{\rm sh}$, and $t_0 = \mathfrak{g}_{\rm sh}(0)$.

Since $\delta_{\rm e} \in (0, \frac{\delta_{P_0}}{4})$, then, from Lemma 17.3.17(a), Lemma 17.3.20(i), and Lemma 17.3.21, we see that equation (12.3.25) is of the same form and properties as in Proposition 12.7.3, where we note that Lemma 12.4.2(v) is *not* used in Proposition 12.7.3. Also, the boundary condition is of form (12.3.26) with the same properties as in Proposition 12.7.3 by Lemma 17.3.23.

Next, we rewrite Problem (12.3.25)–(12.3.29) in the coordinates $(s,t) = F_1(\boldsymbol{\xi})$ and show that it satisfies the properties listed after the problem. We first discuss property (i) under this change of variables. We use property (iv) of Lemma 17.2.1. Noting that $x_{P_1} = x_{P_4}$, we substitute $(x,y) = (s + x_{P_1}, t + y_{P_4})$ into (i) to obtain that, for any $(s,t) \in F_1(\Omega) \cap \{s < \varepsilon_{\rm eq}\}$ and $\mathbf{p}, \boldsymbol{\kappa} \in \mathbb{R}^2$,

$$\lambda_1 |\boldsymbol{\kappa}|^2 \le \sum_{i,j=1}^{2} \hat{A}_{ij}(\mathbf{p}, s + x_{P_1}, t + y_{P_4}) \frac{\kappa_i \kappa_j}{(s + x_{P_1})^{2 - \frac{i+j}{2}}} \le \frac{1}{\lambda_1} |\boldsymbol{\kappa}|^2. \tag{17.4.7}$$

This condition is of the same form as condition (4.5.90) for $(x_1, x_2) = (s,t)$ with

$$\delta = x_{P_1}. \tag{17.4.8}$$

Note that $x_{P_1}(\theta_{\rm w}) = 0$ for supersonic/sonic wedge angles $\theta_{\rm w}$, and $x_{P_1}(\theta_{\rm w}) > 0$ for subsonic wedge angles $\theta_{\rm w}$, by Proposition 16.1.3. Thus, in $F_1(\Omega) \cap \{s < \varepsilon_{\rm eq}\}$, the equation is degenerate elliptic for the supersonic/sonic wedge angles $\theta_{\rm w}$, and uniformly elliptic for the subsonic wedge angles $\theta_{\rm w}$.

With this, following the argument for Proposition 12.7.3, we see for Problem (12.3.25)–(12.3.29), rewritten in coordinates $(s,t) = F_1(\boldsymbol{\xi})$ in domain $F_1(\Omega)$, that all the assumptions of Proposition 4.7.2 are satisfied and that the dependence of the parameters in the conditions of Proposition 4.7.2 on $\delta_{\mathcal{K}}$ and the parameters of the iteration set is the same as those in Proposition 12.7.3, except that δ is determined by (17.4.8):

α is the same as α in the iteration set;
$(\varepsilon, \kappa, \lambda)$ depend only on the data and $\theta_{\rm w}^*$;
$\beta = \frac{3}{4}$;
$\sigma = C_0 \sqrt{\delta_1}$, with $C_0 > 0$ depending on the data;
M depends only on the data and $(\theta_{\rm w}^*, \alpha)$;
$h, t_1, t_2 \in (\frac{1}{C}, C)$ and $M_{\rm bd} \le C$, with C depending only on $(\rho_0, \rho_1, \gamma, \theta_{\rm w}^*)$.

Now, applying Proposition 4.7.2 and changing the variables by $\boldsymbol{\xi} = F_1^{-1}(s,t)$, we obtain the existence of $\bar{\alpha} \in (0, \frac{1}{8})$ (depending only on the data and $\theta_{\rm w}^*$) and $\delta_0, \hat{\delta}_1 > 0$ (depending only on the data and $(\theta_{\rm w}^*, \alpha)$) such that, if $\delta_1 \in (0, \hat{\delta}_1)$ in the iteration set and $x_{P_1} < \delta_0$, there exists a unique solution of Problem (12.3.25)–(12.3.29) determined by $(u, \theta_{\rm w})$, and (12.7.14)–(12.7.16) hold.

Next we note that the continuous dependence of $\frac{|D\varphi_2(P_0)|}{c_2}$ and x_{P_1} on $\theta_{\rm w} \in [\theta_{\rm w}^{\rm d}, \frac{\pi}{2}]$ and the fact that $x_{P_1}(\theta_{\rm w}) = 0$ if $\frac{|D\varphi_2(P_0)|}{c_2}(\theta_{\rm w}) \ge 1$ (supersonic/sonic wedge angles $\theta_{\rm w}$) imply that there exists $\delta_{\rm e} \in (0, \frac{\delta_{P_0}}{8})$ such that (17.4.6) implies that $x_{P_1} < \delta_0$. Taking the supremum of such δ_e, we obtain that $\delta_e \in (0, \frac{\delta_{P_0}}{4})$ depending only on the data and (θ_w^*, α), since δ_0 depends only on these parameters. Now the proposition is proved for the wedge angles satisfying (17.4.6).

2. It remains to consider the wedge angles $\theta_w \in [\theta_w^*, \frac{\pi}{2})$ satisfying

$$\frac{|D\varphi_2(P_0)|}{c_2}(\theta_w) \leq 1 - \delta_e, \tag{17.4.9}$$

where $\delta_e > 0$ has been fixed in Step 1. We again make the change of variables $(s,t) = F_1(\boldsymbol{\xi})$ and consider the problem in the (s,t)–variables in domain $F_1(\Omega)$.

In this case, the assertion follows from a direct application of Proposition 4.8.7. Indeed, by Lemma 17.3.21, we can apply Lemma 17.3.20(ii) with $\delta = \delta_e$ and Lemma 12.4.5(i), which imply (4.8.13) with (δ, λ) depending only on the data and (θ_w^*, α), since δ_e depends only on these parameters. Also, from Lemma 17.3.20(ii) with $\delta = \delta_e$, we obtain (17.3.28) with $\delta = \delta_e$, which implies (4.8.14) with M depending only on the data and (θ_w^*, α), since δ_e depends only on these parameters. To summarize, we have shown (4.8.13)–(4.8.14) with the constants depending only on the data and (θ_w^*, α).

The other conditions of Proposition 4.8.7 are checked in the same way as we have done for the conditions of Proposition 4.7.2 in Step 1, and the constants in these conditions have the same dependence on the parameters of the iteration set and $\delta_{\mathcal{K}}$ as in Step 1. Now we apply Proposition 4.8.7 to obtain the existence of solutions, if $\sigma > 0$ depends only on parameters $(\kappa, \lambda, \delta, M, \alpha, \varepsilon)$ of the conditions of Proposition 4.8.7. Translating in terms of the parameters of the iterations set and using the dependence of the parameters given in Step 1 (as discussed above) give the condition that δ_1 is small, depending only on the data and (θ_w^*, α).

Also, estimates (12.7.14)–(12.7.16) and (17.4.5) follow from Lemma 4.8.6.

□

Next, we have

Lemma 17.4.3. *For any $(u, \theta_w) \in \overline{\mathcal{K}^{\rm ext}}$, the results of Lemmas* 12.7.4–12.7.5 *and* 12.7.7*, as well as Corollary* 12.7.6*, hold in the present case with only the following notational changes: Space $C^{2,\alpha}_{*,1+\delta^*}(Q^{\rm iter})$ is replaced by space $C^{2,\alpha}_{**}(Q^{\rm iter})$, and the references to the definitions, conditions, and results in* §12.3 *and* §12.7 *are replaced by the references to the corresponding statements in* §17.3 *and* §17.4.

Proof. The proofs of Lemmas 12.7.4–12.7.5 and 12.7.7 are the same by using Proposition 17.4.2. More specifically, in extending the proof of Lemma 12.7.4(i) to the present case, we note that Lemma 12.2.4 and (12.2.68) hold in the present case by Lemmas 17.2.3 and 17.2.13.

The proof of Lemma 12.7.5 works in the present situation with only notational changes as described in the statement of this lemma.

Corollary 12.7.6 is extended to the present case by the same argument with only notational changes and by using Lemma 12.7.5 which is already extended to the present case.

In extending the proof of Lemma 12.7.7 to the present case, we note that all the subsonic/sonic wedge angles correspond to the case that has been considered in Step 1 of the proof. Then we note that the argument in Step 1 uses the property in Lemma 12.2.2(iv) which now holds in the form given in Lemma 17.2.1(iv), and the fact that the equation is strictly elliptic in $\Omega \cap \mathcal{D}_{\varepsilon_{\rm p}}$ and the condition on $\Gamma_{\rm shock} \cap \mathcal{D}_{\varepsilon_{\rm p}}$ is oblique and homogeneous. Also, we use that, from Corollary 12.7.6 extended to the present case, any $(u, \theta_{\rm w}) \in \overline{\mathcal{K}}$ satisfies Definition 17.3.4(vii) with the nonstrict inequality in (17.3.19). □

17.4.2 Estimates of solutions of the iteration problem for the supersonic wedge angles

In this section, we consider the supersonic wedge angles $\theta_{\rm w}$, *i.e.*, satisfying

$$\frac{|D\varphi_2(P_0)|}{c_2}(\theta_{\rm w}) > 1. \tag{17.4.10}$$

We first extend Lemma 12.7.8 to this case, which now takes the following form:

Lemma 17.4.4. *If the parameters in Definition* 17.3.4 *are chosen as in Lemma* 12.7.7 *(through Lemma* 17.4.3*), and if* $\varepsilon > 0$ *is further reduced if necessary depending on the data, then, for any* $(u, \theta_{\rm w}) \in \overline{\mathcal{K}^{\rm ext}}$ *with the supersonic wedge angle* $\theta_{\rm w}$*, estimate* (12.7.27) *holds with* C *depending only on the data and* $(\theta_{\rm w}^*, \alpha)$.

Proof. In extending the proof of Lemma 12.7.8 to the present case, we use the estimates of Proposition 17.4.2 and the fact that, in the supersonic case, the iteration equation and boundary condition on $\Gamma_{\rm shock}$ are the same as in §12.3.3, as we have shown in Step 1 of the proof of Proposition 17.4.2. We note that Lemma 12.4.2(v) is not used in the proof of Lemma 12.7.8. We also note that estimate (12.4.17) is true in the present case with C depending only on the data by Lemma 17.3.16. □

Next we need to extend Proposition 12.7.9 to the present case. However, this is not straightforward because the proof of Proposition 12.7.9 is obtained by employing Theorem 4.7.4, so that it uses the properties of Definition 12.3.2(iii) and Lemma 12.4.2(v) to satisfy the conditions of Theorem 4.7.4. Both properties hold for each supersonic wedge angle, but the constants degenerate as $\frac{|D\varphi_2(P_0)|}{c_2}(\theta_{\rm w}) \to 1+$. Thus, we consider separately the cases of *supersonic-away-from-sonic* and *supersonic-close-to-sonic* wedge angles.

We first consider the case of the *supersonic-away-from-sonic* wedge angles.

Proposition 17.4.5. *If the parameters in Definition* 17.3.4 *are chosen as in Lemma* 17.4.4*, then, for any* $\delta \in (0,1)$ *and* $\sigma \in (0,1)$*, there exist* $\hat{\varepsilon}_{\rm p} \in (0, \frac{\varepsilon_{\rm p}}{2}]$ *(with* $\varepsilon_{\rm p} := \frac{1}{2}\min\{\varepsilon_{\rm eq}, \varepsilon_{\rm bc}\}$*) and* C*, depending only on the data and* $(\theta_{\rm w}^*, \alpha, \sigma, \delta)$*,* *such that the following holds: For any $(u^{\#}, \theta_{\rm w}^{\#}) \in \overline{\mathcal{K}}$, there is $\delta_{u^{\#},\theta_{\rm w}^{\#}} > 0$ small so that, for any $(u, \theta_{\rm w}) \in \overline{\mathcal{K}^{\rm ext}}$ satisfying* (12.7.22) *and*

$$\frac{|D\varphi_2(P_0)|}{c_2}(\theta_{\rm w}) \geq 1 + \delta, \tag{17.4.11}$$

solution $\hat{\psi} \in C^2(\Omega) \cap C^1(\overline{\Omega} \setminus \overline{\Gamma_{\rm sonic}}) \cap C(\overline{\Omega})$ of Problem (12.3.25)–(12.3.29) *determined by $(u, \theta_{\rm w})$ satisfies* (12.7.28).

Proof. The continuous dependence of $\frac{|D\varphi_2(P_0)|}{c_2}(\theta_{\rm w})$, y_{P_1}, and y_{P_4} on $\theta_{\rm w} \in [\theta_{\rm w}^{\rm d}, \frac{\pi}{2}]$ implies that, for each $\delta > 0$, there exists $C_\delta > 0$ depending on the data and δ such that, if $\theta_{\rm w}$ satisfies (17.4.11),

$$(y_{P_1} - y_{P_4})(\theta_{\rm w}) \geq \frac{1}{C_\delta}.$$

With this, using Lemma 17.2.1(iv) and (17.2.4), we conclude that, for each $(u, \theta_{\rm w}) \in \overline{\mathcal{K}^{\rm ext}}$ with $\theta_{\rm w}$ satisfying (17.4.11),

$$\mathfrak{g}_{\rm sh}(0) = (y_{P_1} - y_{P_4})(\theta_{\rm w}) \geq \frac{1}{C_\delta}.$$

Now condition (iii) of Definition 17.3.4 implies that the property of Definition 12.3.2(iii) holds with $N_2 = \frac{1}{C_\delta} > 0$. Also, Lemma 17.3.20(i) and (17.3.27) imply that property (v) of Lemma 12.4.2 now holds with the constant depending only on the data and $(\theta_{\rm w}^*, \alpha, \delta)$. Then we follow the proof of Proposition 12.7.9 to obtain (12.7.28) with $\varepsilon_{\rm p}$ and C depending on the data and $(\theta_{\rm w}^*, \alpha, \sigma, \delta)$. ☐

Now we consider the case of the *supersonic-close-to-sonic* wedge angles.

Proposition 17.4.6. *If the parameters in Definition* 17.3.4 *are chosen as in Lemma* 17.4.4, *then, for any $\sigma \in (0,1)$, there exist $\hat{\delta}_{P_0} > 0$ depending only on the data and $(\theta_{\rm w}^*, \sigma)$, $\hat{\varepsilon}_{\rm p} \in (0, \frac{\varepsilon_{\rm p}}{2}]$ (with $\varepsilon_{\rm p} := \frac{1}{2}\min\{\varepsilon_{\rm eq}, \varepsilon_{\rm bc}\}$), and C depending only on the data and $(\theta_{\rm w}^*, \alpha, \sigma)$ such that, for any $(u^{\#}, \theta_{\rm w}^{\#}) \in \overline{\mathcal{K}}$, there is $\delta_{u^{\#},\theta_{\rm w}^{\#}} > 0$ small so that, for any $(u, \theta_{\rm w}) \in \overline{\mathcal{K}^{\rm ext}}$ satisfying* (12.7.22) *and* (17.4.1), *solution $\hat{\psi} \in C^2(\Omega) \cap C^1(\overline{\Omega} \setminus \overline{\Gamma_{\rm sonic}}) \cap C(\overline{\Omega})$ of Problem* (12.3.25)–(12.3.29) *determined by $(u, \theta_{\rm w})$ satisfies* (12.7.28).

Proof. We follow the proof of Proposition 16.4.6 with the following change: We use the fixed boundary estimates – Theorem 4.7.4 and Lemma 17.4.1, instead of the corresponding free boundary estimates – Proposition 11.4.3 and Lemma 16.4.3. Below we show that the conditions of the corresponding results are satisfied in the present case.

We will fix $\hat{\varepsilon}_{\rm p} \in (0, 1]$ later. Fix $(u, \theta_{\rm w}) \in \overline{\mathcal{K}^{\rm ext}}$ satisfying the conditions above with corresponding $(\Omega, \Gamma_{\rm shock}, \mathfrak{g}_{\rm sh}, \psi)$.

1. We show the algebraic growth of ψ as in Proposition 16.4.2 with $m = 4$ by using Lemma 16.4.2. We write the equation and condition of $\Gamma_{\rm shock} \cap \mathcal{D}_{\hat{\varepsilon}_{\rm p}}$ as a linear homogeneous equation and boundary condition, and check the conditions of Lemma 16.4.2.

Domain $\Omega \cap \mathcal{D}_{\hat{\varepsilon}_{\rm p}}$ in the (x, y)–coordinates is of the structure as in (11.2.38)–(11.2.39) with $f_{\rm sh}(x) = \mathfrak{g}_{\rm sh}(x) + y_{P_4}$ by Definition 12.2.6(iv), Lemma 17.2.1(iv) with $x_{P_4} = 0$ and $t_{P_1} = y_{P_1} - y_{P_4} > 0$ in the supersonic case (by (16.1.11)–(16.1.12)), and (17.3.7). Also, using (11.2.10) and (12.3.23) with C depending only on the data (which holds by Remark 17.3.7), recalling that $\hat{\varepsilon}_{\rm p} \leq 1$, and reducing ε depending on the data, we see that (11.2.40) holds for all $x \in [0, \hat{\varepsilon}_{\rm p}]$. Moreover, (16.4.12) holds by (17.3.7), and $b_{\rm so} = \mathfrak{g}_{\rm sh}(0)$.

Using Lemma 17.3.20(i) and Lemma 17.3.17(a) in the supersonic case, we obtain that Lemma 12.4.2 holds for equation (12.3.25) in the (x, y)–coordinates in $\Omega \cap \mathcal{D}_{\hat{\varepsilon}_{\rm p}}$. Then the strict ellipticity and conditions (16.4.5) in $\Omega \cap \mathcal{D}_{\hat{\varepsilon}_{\rm p}}$ follow from (12.4.22) with $\hat{A}_{12} = \hat{A}_{21}$. Similarly, (16.4.7) follows from (12.5.55). The homogeneity of (16.4.6) follows from (12.5.56), and the obliqueness follows from Lemma 12.5.7(iv), where we have used Lemma 17.3.23.

Note that the constants in (11.2.40), (12.5.55), and (16.4.5) depend only on the data and $\theta_{\rm w}^*$. Thus, we obtain that the corresponding constants $\hat{\varepsilon}, b_{\rm so}^* > 0$ and $\hat{C} > 0$ in Lemma 16.4.2 depend only on the data and $\theta_{\rm w}^*$. We set $\hat{\varepsilon}_{\rm p} = \min\{\hat{\varepsilon}, 1\}$. Also, there exists $\hat{\delta}_{P_0} > 0$ such that $\mathfrak{g}_{\rm sh}(0) = (y_{P_1} - y_{P_4})(\theta_{\rm w}) \leq b_{\rm so}^*$ if $\theta_{\rm w}$ satisfies (17.4.1). We fix this $\hat{\varepsilon}_{\rm p}$. Then we have

$$0 \leq \psi(x, y) \leq Cx^4 \qquad \text{in } \Omega_{\hat{\varepsilon}_{\rm p}} \cap \{x > \frac{b_{\rm so}^2(\theta_{\rm w})}{10}\}, \tag{17.4.12}$$

where $b_{\rm so} = \mathfrak{g}_{\rm sh}(0)$, C depends only on the data and $\theta_{\rm w}^*$, and the lower bound follows from Lemma 12.7.7.

2. We follow Step 1 of the proof of Proposition 16.4.6. Then we estimate ψ in $\Omega_{b_{\rm so}^2}$. We can assume that $b_{\rm so}^2 < \hat{\varepsilon}_{\rm p}$; otherwise, we replace $b_{\rm so}^2$ by $\hat{\varepsilon}_{\rm p}$ in the following argument.

We will use Theorem 4.7.4 with $(x_1, x_2) = (x, y)$, $l = b_{\rm so} \equiv \mathfrak{g}_{\rm sh}(0)$, and $r = 2l^2$. Therefore, we check its conditions.

As we discussed in Step 1, for the equation, we can use Lemma 12.4.2, except for (v), which is replaced by Lemma 17.3.17(v′). For the conditions on $\Gamma_{\rm shock}$, we use Lemma 12.5.7, in which (12.5.56) implies the homogeneity of this condition on $\Gamma_{\rm shock} \cap \mathcal{D}_{\hat{\varepsilon}_{\rm p}}$.

Then we check the conditions of Theorem 4.7.4, as in the proof of Proposition 12.7.10, and conclude that they are satisfied with the constants depending on the data and $(\theta_{\rm w}^*, \alpha)$; in particular, (4.7.14) holds by (17.3.26).

Let $\sigma \in (0, 1)$. Then r_0, determined in Theorem 4.7.4 with $\alpha = \sigma$ by the constants discussed above, depends only on the data and $(\theta_{\rm w}^*, \sigma)$. We reduce $\hat{\varepsilon}_{\rm p}$ so that $\mathfrak{g}_{\rm sh}(0) = (y_{P_1} - y_{P_4})(\theta_{\rm w}) \leq \sqrt{r_0}$ if $\theta_{\rm w}$ satisfies (17.4.1). Then $\hat{\varepsilon}_{\rm p}$ depends only on the data and $(\theta_{\rm w}^*, \sigma)$.

Now $l^2 = \mathfrak{g}_{\rm sh}^2(0) \leq r_0$. Applying Theorem 4.7.4 with $\alpha = \sigma$, we obtain

$$\|\psi\|_{2,\sigma,\Omega_{b_{\rm so}^2}}^{(\rm par)} \leq C, \tag{17.4.13}$$

where C depends only on the data and $(\theta_{\mathrm{w}}^*, \alpha, \sigma)$.

3. Now we follow Step 2 of the proof of Proposition 16.4.6. Thus, we estimate ψ in $\Omega_\varepsilon \cap \{\frac{b_{\mathrm{so}}^2}{5} < x < \hat{\varepsilon}_{\mathrm{p}}\}$. We use Lemma 17.4.1, with the same parameters as in Lemma 16.4.3 which have been used in Step 2 of the proof of Proposition 16.4.6. Therefore, we need to show that the conditions of Lemma 17.4.1 are satisfied.

As in Step 2 of the proof of Proposition 16.4.6, we use

$$h(x) = \frac{x}{10k}, \qquad g(x) = \frac{x^4}{10k}.$$

The growth of the solution is estimated in (17.4.12).

Using (17.3.7), we check that (16.4.63) holds with $b_{\mathrm{so}} = \mathfrak{g}_{\mathrm{sh}}(0)$ and $k = N_2$ by reducing $\hat{\varepsilon}_{\mathrm{p}}$ if necessary, depending only on the data and θ_{w}^*. Then it follows that rectangles (16.4.64)–(16.4.65) satisfy (11.4.8) and (16.4.66). This implies condition (16.4.41).

Using the properties of the equation and the condition on Γ_{shock} discussed in Steps 1–2, we obtain that ellipticity (16.4.33) and the first line in (16.4.34) follow from Lemma 12.4.2(i) with $A_{12} = A_{21}$, where the last property holds by Lemma 12.4.2(ii). To check the other conditions in (16.4.34), we note that, in the condition on $D_y A_{11}(\mathbf{p}, z, x, y)$, the right-hand side is $\frac{Mx}{h(x)}$, which is equal to $10MN_2$ in the present case. Now the conditions in (16.4.34) (from the second line) follow from Lemma 12.4.2(ii). Constants (λ, M) in (16.4.33)–(16.4.34) then depend only on the data.

Also, using (12.5.55), the condition on $\Gamma_{\mathrm{shock}} \cap \mathcal{D}_{\hat{\varepsilon}_{\mathrm{p}}}$ can be written in the form of (16.4.35). Using Lemma 12.5.7(i) and (12.5.55)–(12.5.56), we show that this condition satisfies the required properties (11.3.41)–(11.3.42) with the constants depending only on the data and $(\theta_{\mathrm{w}}^*, \alpha)$.

Therefore, all the conditions of Lemma 17.4.1 are satisfied, whose constants depend only on the data and $(\theta_{\mathrm{w}}^*, \alpha)$.

Applying Lemma 17.4.1, we obtain (16.4.68) (with $\hat{\varepsilon}_{\mathrm{p}}$ instead of ε) with notation (16.4.67). Then, arguing as in the rest of the proof of Proposition 16.4.6, we obtain (12.7.28). □

Now we obtain the global estimate of supersonic reflection solutions in the norm introduced in Definition 12.3.1(iv).

Proposition 17.4.7. *If parameters ε and $\frac{1}{N_1}$ of the iteration set in Definition 17.3.4 are small – depending on the data and θ_{w}^*, δ_1 is small – depending on the data and $(\theta_{\mathrm{w}}^*, \alpha)$, and δ_3 is small – depending on the data and $(\theta_{\mathrm{w}}^*, \delta_2)$, then, for any $(u^\#, \theta_{\mathrm{w}}^\#) \in \overline{\mathcal{K}}$, there exists $\delta_{u^\#, \theta_{\mathrm{w}}^\#} > 0$ so that, for any $(u, \theta_{\mathrm{w}}) \in \overline{\mathcal{K}^{\mathrm{ext}}}$ satisfying (12.7.22) and (17.4.10) (i.e., θ_{w} is supersonic), solution $\hat{\psi}$ of (12.3.25)–(12.3.29) determined by (u, θ_{w}) satisfies $\hat{\psi} \in C^{2,\hat{\alpha}}_{*,2}(\Omega)$ with*

$$\|\hat{\psi}\|^{*,(2)}_{2,\hat{\alpha},\Omega} \le C, \tag{17.4.14}$$

where $\hat{\alpha} \in (0, \frac{1}{8})$ is the constant determined in Proposition 17.4.2 (*depending on the data and $\theta^*_{\rm w}$*), *and C depends only on the data and* $(\theta^*_{\rm w}, \alpha)$.

Proof. Fix $\sigma = \hat{\alpha}$. This fixes constant $\hat{\delta}_{P_0}$ in Proposition 17.4.6. Since $\hat{\alpha}$ in Proposition 17.4.2 depends on the data and $\theta^*_{\rm w}$, then $\hat{\delta}_{P_0}$ fixed above also depends only on the data and $\theta^*_{\rm w}$.

Choose $\delta = \hat{\delta}_{P_0}$ in Proposition 17.4.5.

The choices made above fix constant $\hat{\varepsilon}_{\rm p}$ for $\alpha = \hat{\alpha}$ in Propositions 17.4.5 and 17.4.6, depending only on the data and $\theta^*_{\rm w}$.

Fix $(u, \theta_{\rm w})$ satisfying the conditions of this proposition. Thus, $\theta_{\rm w}$ satisfies either (17.4.1) or (17.4.11) with $\delta = \hat{\delta}_{P_0}$.

Then we complete the proof by combining estimate (12.7.16) (for $s = \frac{\hat{\varepsilon}_{\rm p}}{10}$) from Proposition 17.4.2 with estimate (12.7.28) (for $\sigma = \hat{\alpha}$), obtained from either Proposition 17.4.5 (for $\delta = \hat{\delta}_{P_0}$) or Proposition 17.4.6, depending on $\theta_{\rm w}$. For this, we have employed the fact that, since $(\hat{\alpha}, \hat{\varepsilon}_{\rm p}, \hat{\delta}_{P_0})$ depend only on the data and $\theta^*_{\rm w}$, the constants in all the estimates quoted above depend only on the data and $(\theta^*_{\rm w}, \alpha)$. □

17.4.3 Estimates of solutions of the iteration problem for the subsonic/sonic wedge angles

In this section, we consider the subsonic/sonic wedge angles $\theta_{\rm w}$, *i.e.*, satisfying

$$\frac{|D\varphi_2(P_0)|}{c_2}(\theta_{\rm w}) \in (0, 1]. \tag{17.4.15}$$

We first consider the case of sonic and subsonic-close-to-sonic wedge angles, and derive a local estimate near P_0. We use constant δ_{P_0} determined in Proposition 16.1.4 depending on the data.

Proposition 17.4.8. *If the parameters in Definition* 17.3.4 *are chosen as in Lemma* 12.7.7 (*through Lemma* 17.4.3), *then there exist $\hat{\delta}_{P_0} \in (0, \frac{\delta_{P_0}}{4}]$ and $\hat{\varepsilon}_{\rm p} \in (0, \frac{\varepsilon_{\rm p}}{2}]$ depending only on the data and $\theta^*_{\rm w}$ and, for any $\sigma \in (0, 1)$, there exists $C > 0$ depending only on the data and $(\theta^*_{\rm w}, \alpha, \sigma)$ such that the following estimate holds: For any $(u^\#, \theta^\#_{\rm w}) \in \overline{\mathcal{K}}$, there is $\delta_{u^\#, \theta^\#_{\rm w}} > 0$ small so that, for any $(u, \theta_{\rm w}) \in \overline{\mathcal{K}^{\rm ext}}$ satisfying* (12.7.22) *and* (17.4.2), *solution $\hat{\psi} \in C^2(\Omega) \cap C^1(\overline{\Omega} \backslash \overline{\Gamma_{\rm sonic}}) \cap C(\overline{\Omega})$ of Problem* (12.3.25)–(12.3.29) *determined by $(u, \theta_{\rm w})$ satisfies*

$$\|\hat{\psi}\|_{C^{2,\sigma}(\overline{\Omega_{\hat{\varepsilon}_{\rm p}}})} \le C, \qquad D^m \hat{\psi}(P_0) = 0 \ \ \textit{for } m = 0, 1, 2. \tag{17.4.16}$$

Proof. The proof is obtained by following the argument for Proposition 16.5.3 and using the fixed boundary estimate (Lemma 17.4.1) instead of the corresponding free boundary estimates (Lemma 16.4.3), in a similar way to the proof of Proposition 17.4.6.

We first show the algebraic growth of the solutions by using Lemma 16.5.1 with $m = 5$ and choosing $\hat{\delta}_{P_0}$ and $\hat{\varepsilon}_{\rm p}$ sufficiently small, where we check its conditions as in Step 1 of the proof of Proposition 17.4.6, and the corresponding constants depend only on the data and $\theta_{\rm w}^*$. The choice of small positive $\hat{\delta}_{P_0}$ and $\hat{\varepsilon}_{\rm p}$, depending on the data and $\theta_{\rm w}^*$, is then performed as follows: $\hat{\varepsilon}_{\rm p}$ is constant ε determined in Lemma 16.5.1, and then $\hat{\delta}_{P_0}$ is determined as the largest constant such that condition (17.4.2) for $\theta_{\rm w}$ implies that $x_{P_1}(\theta_{\rm w}) \le \frac{\hat{\varepsilon}_{\rm p}}{10}$. This implies that $\hat{\delta}_{P_0} > 0$ by the continuous dependence of $x_{P_1}(\theta_{\rm w})$ on $\theta_{\rm w} \in [\theta_{\rm w}^{\rm d}, \frac{\pi}{2}]$ by noting that $x_{P_1}(\theta_{\rm w}) = 0$ for the supersonic/sonic wedge angles $\theta_{\rm w}$.

We apply Lemma 17.4.1 in the same setting as we have used for Lemma 16.4.3 in Step 2 in the proof of Proposition 16.5.3, *i.e.*, with

$$h(x) = \frac{1}{10k}d_{\rm so}(x) = \frac{1}{10k}(x - x_{P_1}),$$
$$g(x) = \frac{1}{10k}d_{\rm so}^5(x) = \frac{1}{10k}(x - x_{P_1})^5.$$

We check the conditions of Lemma 17.4.1 as in Step 3 of the proof of Proposition 17.4.6, where we only note the following: In (16.4.34), the condition on $D_yA_{11}(\mathbf{p}, z, x, y)$ has the right-hand side that is $\frac{Mx}{h(x)}$, which is $\frac{10MN_2x}{x - x_{P_1}} \ge 10MN_2$ in the present case. Thus, the L^∞–bound of $D_yA_{11}(\mathbf{p}, z, x, y)$ is sufficient to satisfy this condition, and this bound holds by Lemma 12.4.2(ii). Then, applying Lemma 17.4.1 (instead of Lemma 16.4.3) and arguing as in the rest of the proof of Proposition 16.5.3, we obtain (17.4.16). □

Now we consider all the subsonic/sonic wedge angles and derive a global estimate in Ω.

Proposition 17.4.9. *If the parameters in Definition* 17.3.4 *are chosen as in Lemma* 12.7.7 *(through Lemma* 17.4.3*), there exist* $\hat{\alpha} \in (0, \frac{1}{3})$ *depending only on the data and* $\theta_{\rm w}^*$*, and* $C > 0$ *depending only on the data and* $(\theta_{\rm w}^*, \alpha)$ *such that the following estimate holds: For any* $(u^\#, \theta_{\rm w}^\#) \in \overline{\mathcal{K}}$*, there exists* $\delta_{u^\#, \theta_{\rm w}^\#} > 0$ *small so that, for any* $(u, \theta_{\rm w}) \in \overline{\mathcal{K}^{\rm ext}}$ *satisfying* (12.7.22) *with subsonic/sonic wedge angle* $\theta_{\rm w}$*, i.e., with* (17.4.15)*, solution* $\hat{\psi} \in C^2(\Omega) \cap C^1(\overline{\Omega} \setminus \overline{\Gamma_{\rm sonic}}) \cap C(\overline{\Omega})$ *of Problem* (12.3.25)*–*(12.3.29) *determined by* $(u, \theta_{\rm w})$ *satisfies*

$$\|\hat{\psi}\|_{2,\hat{\alpha},\Omega}^{(-1-\hat{\alpha}),\,\{P_0\}\cup\overline{\Gamma_{\rm sym}}} \le C, \qquad (\hat{\psi}, D\hat{\psi})(P_0) = (0, \mathbf{0}). \tag{17.4.17}$$

Proof. Let $\hat{\delta}_{P_0} \in (0, \frac{\delta_{P_0}}{4}]$ and $\hat{\varepsilon}_{\rm p} \in (0, \frac{\varepsilon_{\rm p}}{2}]$ be the constants determined in Proposition 17.4.8, depending only on the data and $\theta_{\rm w}^*$.

Let $(u, \theta_{\rm w}) \in \overline{\mathcal{K}^{\rm ext}}$ satisfy the conditions of this proposition. If $\theta_{\rm w}$ satisfies (17.4.2), we apply first estimate (12.7.16) (for $s = \frac{\hat{\varepsilon}_{\rm p}}{10}$) which holds by Proposition 17.4.2, with $\hat{\alpha}$ determined in Proposition 12.7.3 (adjusted to the wedge angle range $\theta_{\rm w} \in [\theta_{\rm w}^{\rm d}, \frac{\pi}{2})$), depending only on the data. Then we apply Proposition 17.4.8 with $\sigma = \hat{\alpha}$. Combining these estimates, we obtain (17.4.17), with $\hat{\alpha}$ chosen as above, depending only on the data, and C depending on the data and $(\theta_{\rm w}^*, \alpha)$.

Otherwise, recalling that $\theta_{\rm w}$ is subsonic or sonic, it follows that $\theta_{\rm w}$ satisfies $\frac{|D\varphi_2(P_0)|}{c_2}(\theta_{\rm w}) \leq 1 - \hat{\delta}_{P_0}$. Then we apply (17.4.5) for $\delta_{\rm el} = \hat{\delta}_{P_0}$. Thus, we obtain (17.4.17), in which $(\hat{\alpha}, C)$ are (α_2, C_δ) from (17.4.5) corresponding to $\delta_{\rm el} = \hat{\delta}_{P_0}$. Therefore, $\hat{\alpha}$ depends only on the data and $\theta_{\rm w}^*$ (since $\hat{\delta}_{P_0}$ depends on these parameters), and C depends on α, in addition to the data and $\theta_{\rm w}^*$.

Combining the two cases considered above, we conclude the proof. □

Next, we combine the estimates obtained above for $\hat{\psi}$ for all the cases into one uniform estimate for function $\hat{u}$ on $Q^{\rm iter}$.

Corollary 17.4.10. *Assume that the conditions of Propositions* 17.4.7 *and* 17.4.9 *hold, and choose* $\alpha = \frac{\hat{\alpha}}{2}$ *in Definition* 17.3.4 *where* $\hat{\alpha}$ *is the smaller one of* $\hat{\alpha}$ *in Propositions* 17.4.7 *and* 17.4.9. *Then, for any* $(u^\#, \theta_{\rm w}^\#) \in \overline{\mathcal{K}}$, *there exists* $\delta_{u^\#, \theta_{\rm w}^\#} > 0$ *so that, for any* $(u, \theta_{\rm w}) \in \overline{\mathcal{K}^{\rm ext}}$ *satisfying* (12.7.22), *solution* $\hat{u}$ *of Problem* (12.7.2)–(12.7.6) *determined by* $(u, \theta_{\rm w})$ *satisfies*

$$\|\hat{u}\|^{(**)}_{2,\alpha,Q^{\rm iter}} < C, \tag{17.4.18}$$

where C *depends only on the data and* $\theta_{\rm w}^*$.

Proof. We argue in a manner similar to the proof of Proposition 17.2.12. In this proof, the universal constant C depends only on the data and $\theta_{\rm w}^*$.

We first consider the case of the supersonic wedge angles $\theta_{\rm w} \in [\theta_{\rm w}^*, \frac{\pi}{2}]$. Then we map the estimate of Proposition 17.4.7 into the iteration region $Q^{\rm iter}$ by using $\alpha = \frac{\hat{\alpha}}{2}$, (17.3.7), (17.3.20), the C^3–regularity of map F_1^{-1}, and the explicit form (12.2.43) of map $F^{-1}_{(2,\mathfrak{g}_{\rm sh})}$. Here we note that $F^{-1}_{(2,\mathfrak{g}_{\rm sh})}$ does not become singular as $t_{P_1}(\theta_{\rm w}) \to 0+$ in (17.2.4), *i.e.*, these estimates are uniform for all the supersonic wedge angles $\theta_{\rm w} \in [\theta_{\rm w}^*, \frac{\pi}{2}]$. Then we obtain

$$\|\hat{u}\|^{(-1-\alpha),\{1\}\times(0,1)}_{2,\alpha,Q^{\rm iter}\setminus\overline{Q}_{\varepsilon_0'/10}} + \|\hat{u}\|^{(2),({\rm par})}_{(2),\alpha,Q_{\varepsilon_0'}} \leq C.$$

From this estimate, using Lemmas 4.1.2, 4.6.4, and 17.2.11, we obtain (17.4.18) for all the supersonic wedge angles $\theta_{\rm w} \in [\theta_{\rm w}^*, \frac{\pi}{2}]$.

Now we consider the case of the subsonic/sonic wedge angles $\theta_{\rm w} \in [\theta_{\rm w}^*, \frac{\pi}{2}]$. We employ Proposition 17.4.9 and Lemma 17.2.5, which can be applied since (17.3.7) holds with $t_{P_1}(\theta_{\rm w}) = 0$, *i.e.*, (17.2.9) holds, for every subsonic/sonic wedge angle $\theta_{\rm w} \in [\theta_{\rm w}^*, \frac{\pi}{2}]$. Then, from (17.3.20) and (17.4.17), using Lemma 17.2.5, we have

$$\|\hat{u}\|^{(-1-\alpha),\{1\}\times(0,1)}_{2,\alpha,Q^{\rm iter}\setminus\overline{Q}_{\varepsilon_0'/10}} + \|\hat{u}\|^{(1+\alpha),({\rm subs})}_{2,\alpha,Q_{\varepsilon_0'}} \leq C.$$

From this, using Lemmas 4.1.2, 4.6.4, and 17.2.10, we obtain (17.4.18) for all the subsonic/sonic wedge angles $\theta_{\rm w} \in [\theta_{\rm w}^*, \frac{\pi}{2}]$. □

17.5 OPENNESS OF THE ITERATION SET

We first show that $\mathcal{K}^{\rm ext}$ is open.

Lemma 17.5.1. *If δ_2 is small, depending only on the data and (δ_1, N_1), then $\mathcal{K}^{\rm ext}$ is relatively open in $C^{2,\alpha}_{(**)}(Q^{\rm iter}) \times [\theta^*_{\rm w}, \frac{\pi}{2}]$.*

Proof. The proof is along the lines of the proof of Lemma 12.8.1. However, there are some new features related to the present case of subsonic/sonic wedge angles. We only sketch the argument here. The universal constant C in this argument depends only on (ρ_0, ρ_1, γ) and $\theta^*_{\rm w}$.

It suffices to show that each of conditions (i)–(vi) of Definition 17.3.4, together with condition (i), determines an open subset of $C^{2,\alpha}_{(**)}(Q^{\rm iter}) \times [\theta^*_{\rm w}, \frac{\pi}{2}]$. Now we prove this in the following five steps:

1. Condition (i) defines a relatively open set, since function $\eta_1(\theta_{\rm w})$ is continuous;

2. Condition (ii) uses set $\mathfrak{S}$ defined by (12.2.50), that is, by the inequalities in (12.2.49). We first note that all the three terms in the inequalities in (12.2.49) are continuous functions of $(s^*, \theta_{\rm w})$. For the middle term $u(\frac{s^*}{\hat{s}(\theta_{\rm w})}, 1)$, this continuity follows from the continuity and positive lower bound of $\hat{s}(\cdot)$ and the fact that $u \in C^{2,\alpha}_{(**)}(Q^{\rm iter})$. For the right and left terms, such a continuity follows from the fact that the parameters of state (2) depend continuously on $\theta_{\rm w}$. Also, since $u(0, \cdot) = 0$ by (12.2.50), then

(a) In the supersonic case, the strict inequalities in (17.2.35) hold. This implies that the strict inequalities in (12.2.49) hold for all $s^* \in [0, \hat{s}(\theta_{\rm w})]$. Since the $C^{2,\alpha}_{(**)}(Q^{\rm iter})$–norm is stronger than the $C(\overline{Q^{\rm iter}})$–norm, the assertion follows.

(b) In the subsonic/sonic case, we have an equality on the right in (17.2.35). Thus, we cannot extend (12.2.49) to all $s^* \in [0, \hat{s}(\theta_{\rm w})]$, so that an extra argument is needed. We note first that, for any $\theta_{\rm w}$, inclusion $(u, \theta_{\rm w}) \in \mathfrak{S}$ satisfying Definition 17.3.4(i) implies that (17.3.20) holds. We also use (17.2.4) which holds by Lemma 17.2.2. This, with (17.2.35) in the subsonic/sonic case and (17.2.36), implies that $\mathfrak{g}_{\rm sh}(0) = 0$ and $\mathfrak{g}'_{\rm sh}(0) > \frac{1}{C}$. Then

$$\mathfrak{g}_{\rm sh}(s) \geq \frac{s}{2C} \qquad \text{on } [0, \tilde{\varepsilon}], \tag{17.5.1}$$

where $\tilde{\varepsilon}(N_0) \in (0, \varepsilon_0)$. If $(\tilde{u}, \tilde{\theta}_{\rm w})$ satisfies that $\|u - \tilde{u}\|^{(**)}_{2,\alpha,Q^{\rm iter}} + |\theta_{\rm w} - \tilde{\theta}_{\rm w}| \leq \sigma$, we show that (12.2.49) holds for $(\tilde{u}, \tilde{\theta}_{\rm w})$ if $\sigma > 0$ is small, depending on $(u, \theta_{\rm w})$. First we note that, by (17.3.1), $\|u - \tilde{u}\|^{(1+\alpha),(\rm par)}_{2,\alpha,Q_{\varepsilon'_0}} \leq \sigma$, so that

$$(u - \tilde{u})(0, t) = 0, \quad |\partial_s(u - \tilde{u})(s, t)| \leq \sigma s^\alpha \qquad \text{if } (s, t) \in Q_{\varepsilon'_0}.$$

Also, for any $s \in [0, \tilde{\varepsilon}]$, we have

$$\left| \frac{s}{\hat{s}(\tilde{\theta}_{\rm w})} - \frac{s}{\hat{s}(\theta_{\rm w})} \right| \leq Cs|\theta_{\rm w} - \tilde{\theta_{\rm w}}| \leq Cs\sigma$$

for C depending only on the data. Furthermore, $(u, \theta_{\rm w})$ satisfies (17.3.5). Then we obtain

$$\begin{aligned}&\Big|\tilde{u}(\frac{s}{\hat{s}(\tilde{\theta}_{\rm w})},1)-u(\frac{s}{\hat{s}(\theta_{\rm w})},1)\Big|\\&\le\Big|(\tilde{u}-u)(\frac{s}{\hat{s}(\tilde{\theta}_{\rm w})},1)\Big|+\Big|u(\frac{s}{\hat{s}(\tilde{\theta}_{\rm w})},1)-u(\frac{s}{\hat{s}(\theta_{\rm w})},1)\Big|\le C\sigma s^{1+\alpha}\end{aligned}$$

for C depending only on the data and α. With Definition 12.2.6(i)–(ii), this can be written as

$$\Big|\tilde{u}(\frac{s}{\hat{s}(\tilde{\theta}_{\rm w})},1)-(\varphi_1-\tilde{\varphi}_2^{(\theta_{\rm w})})((F_1^{(\theta_{\rm w})})^{-1}(s,\mathfrak{g}_{\rm sh}(s)))\Big|\le C\sigma s^{1+\alpha}.$$

Combining this with (12.2.37) in Lemma 17.2.3 and (17.5.1), and choosing σ small depending on the data, we have

$$\tilde{u}(\frac{s}{\hat{s}(\tilde{\theta}_{\rm w})},1)<(\varphi_1-\tilde{\varphi}_2^{(\theta_{\rm w})})((F_1^{(\theta_{\rm w})})^{-1}(s,0))-\frac{1}{\hat{C}}s \tag{17.5.2}$$

for any $s\in(0,\tilde{\varepsilon})$. Now we need to replace $\theta_{\rm w}$ by $\tilde{\theta}_{\rm w}$ on the right-hand side of the last estimate. For this, we note from Lemma 12.2.2(iii) (which holds by Lemma 17.2.1) and (12.2.38) in Lemma 17.2.3 that, for $\sigma\le 1$,

$$\Big|D(\varphi_1-\tilde{\varphi}_2^{(\theta_{\rm w})})((F_1^{(\theta_{\rm w})})^{-1}(q,0))-D(\varphi_1-\tilde{\varphi}_2^{(\tilde{\theta}_{\rm w})})((F_1^{(\tilde{\theta}_{\rm w})})^{-1}(q,0))\Big|\le C\sigma$$

for any $q\in[0,1]$, where C depends on the data and $\theta_{\rm w}$. Also, we note that $(F_1^{(\theta_{\rm w})})^{-1}(\mathbf{0})=P_4{}^{(\theta_{\rm w})}$. Moreover, $(\varphi_1-\varphi_2^{(\theta_{\rm w})})(P_4{}^{(\theta_{\rm w})})$ is positive if $\theta_{\rm w}$ is supersonic, and is zero, otherwise. Also, $\varphi_2^{(\theta_{\rm w})}(P_4{}^{(\theta_{\rm w})})=\tilde{\varphi}_2^{(\theta_{\rm w})}(P_4{}^{(\theta_{\rm w})})$ by (12.2.35) in Lemma 17.2.3. Since, in our case, $\theta_{\rm w}$ is subsonic or sonic, and $\tilde{\theta}_{\rm w}$ may be of any type, we have

$$(\varphi_1-\tilde{\varphi}_2^{(\theta_{\rm w})})(P_4{}^{(\theta_{\rm w})})\le(\varphi_1-\tilde{\varphi}_2^{(\tilde{\theta}_{\rm w})})(P_4{}^{(\tilde{\theta}_{\rm w})}).$$

Combining these observations, we have

$$\begin{aligned}&(\varphi_1-\tilde{\varphi}_2^{(\theta_{\rm w})})((F_1^{(\theta_{\rm w})})^{-1}(s,0))\\&=(\varphi_1-\tilde{\varphi}_2^{(\theta_{\rm w})})(P_4{}^{(\theta_{\rm w})})+\int_0^s\partial_q(\varphi_1-\tilde{\varphi}_2^{(\theta_{\rm w})})((F_1^{(\theta_{\rm w})})^{-1}(q,0))\,dq\\&\le(\varphi_1-\tilde{\varphi}_2^{(\tilde{\theta}_{\rm w})})(P_4{}^{(\tilde{\theta}_{\rm w})})\\&\quad+\int_0^s\partial_q(\varphi_1-\tilde{\varphi}_2^{(\tilde{\theta}_{\rm w})})((F_1^{(\tilde{\theta}_{\rm w})})^{-1}(q,0))\,dq+C\sigma s\\&=(\varphi_1-\tilde{\varphi}_2^{(\tilde{\theta}_{\rm w})})((F_1^{(\tilde{\theta}_{\rm w})})^{-1}(s,0))+C\sigma s.\end{aligned}$$

Then, from (17.5.2), using (12.2.38) in Lemma 12.2.4 and reducing σ if necessary, we have

$$\tilde{u}(\frac{s}{\hat{s}(\tilde{\theta}_{\rm w})},1)<(\varphi_1-\tilde{\varphi}_2^{(\tilde{\theta}_{\rm w})})((F_1^{(\tilde{\theta}_{\rm w})})^{-1})(s,0)\le\sup_{\mathcal{Q}_{\rm bd}^{(\tilde{\theta}_{\rm w})}(s)}(\varphi_1-\tilde{\varphi}_2^{(\tilde{\theta}_{\rm w})}),$$

where we have used the notation in (12.2.48). Combining this with (12.2.49) for u and (17.2.35) (where the left inequality is strict), recalling that $\|u - \tilde{u}\|^{(**)}_{2,\alpha,Q^{\rm iter}} \le \sigma$, and reducing σ depending on $(u,\theta_{\rm w})$, we obtain (12.2.49) for $\hat{u}$.

3. Condition (iii) defines a relatively open set.

The proof that (17.3.6) defines a relatively open set is as in Lemma 12.8.1. The proof of (17.3.7) repeats the argument of the previous step.

4. Conditions (iv)–(vi) define a relatively open set.

We note first that, using (17.2.38) instead of (12.2.69)–(12.2.70) in the proof of (12.8.1), we obtain that, for every $\tilde{\varepsilon} \in (0, \frac{1}{10})$,

$$\begin{aligned}
&\|(D\varphi \circ \mathfrak{F})_{(u,\theta_{\rm w})} - (D\varphi \circ \mathfrak{F})_{(\tilde{u},\tilde{\theta}_{\rm w})}\|_{L^\infty(\overline{Q^{\rm iter}}\cap\{s\ge\tilde{\varepsilon}\})} \\
&+ \|(D\psi \circ \mathfrak{F})_{(u,\theta_{\rm w})} - (D\psi \circ \mathfrak{F})_{(\tilde{u},\tilde{\theta}_{\rm w})}\|_{L^\infty(\overline{Q^{\rm iter}}\cap\{s\ge\tilde{\varepsilon}\})} \\
&\le C(\tilde{\varepsilon})\big(\|u-\tilde{u}\|_{C^1(\overline{Q^{\rm iter}})} + |\theta_{\rm w} - \tilde{\theta}_{\rm w}|\big),
\end{aligned} \tag{17.5.3}$$

where $C(\tilde{\varepsilon})$ depends only on the data and $(\theta_{\rm w}^*, \tilde{\varepsilon})$.

If $(u,\theta_{\rm w}) \in \mathcal{K}^{\rm ext}$, then, as proved above, if $(\tilde{u},\tilde{\theta}_{\rm w})$ is sufficiently close to $(u,\theta_{\rm w})$ in the norm of $C^{2,\alpha}_{(**)}(Q^{\rm iter})$, $(\tilde{u},\tilde{\theta}_{\rm w}) \in \mathfrak{S}$ so that $\tilde{\Omega} = \Omega^{(\tilde{u},\tilde{\theta}_{\rm w})}$, $\Gamma^{(\tilde{u},\tilde{\theta}_{\rm w})}_{\rm shock}$, $\tilde{\varphi} = \varphi^{(\tilde{u},\tilde{\theta}_{\rm w})}$, and $\tilde{\psi} = \psi^{(\tilde{u},\tilde{\theta}_{\rm w})}$ are well-defined. Also, $(\tilde{u},\tilde{\theta}_{\rm w})$, taken sufficiently close to $(u,\theta_{\rm w})$, satisfies

$$\|u\|^{(**)}_{2,\alpha,Q^{\rm iter}} < N_0. \tag{17.5.4}$$

We prove properties (iv)–(vi) for $(\tilde{u},\tilde{\theta}_{\rm w})$.

Since (17.3.8) and (17.3.13) are in domain $\Omega \setminus \mathcal{D}_{\varepsilon/10}$, then, using (17.5.3) instead of (12.8.1), we prove these properties for $(\tilde{u},\tilde{\theta}_{\rm w})$ by the same argument as in Lemma 12.8.1 (the argument for properties (iv)–(vi) there).

Now we prove (17.3.9) for $(\tilde{u},\tilde{\theta}_{\rm w})$. The right-hand side satisfies that, for any $\theta_{\rm w} \in [\theta_{\rm w}^*, \frac{\pi}{2}]$,

$$\eta_3(\theta_{\rm w})\,x \ge \frac{2-\mu_0}{10(1+\gamma)}\varepsilon \qquad \text{in } \tilde{\Omega} \cap (\mathcal{D}_{\varepsilon_0} \setminus \mathcal{D}_{0,\varepsilon/10}).$$

On the other hand, since $(\tilde{u},\tilde{\theta}_{\rm w}) \in \mathfrak{S}$ and satisfies (17.5.4), $\tilde{\psi}$ satisfies (17.3.22). Thus, $|D\tilde{\psi}(x,y)| \le C(x - x_{P_1})^\alpha$, where C depends only on the data and $\theta_{\rm w}^*$. Then (17.3.9) holds for $\tilde{\psi}$ in $\tilde{\Omega} \cap (\mathcal{D}_{\tilde{\varepsilon}} \setminus \mathcal{D}_{0,\varepsilon/10})$, where $\tilde{\varepsilon} \in (0, \varepsilon_0)$ is small, depending on the data and $\theta_{\rm w}^*$.

Remark 17.5.2. *We note that region $\tilde{\Omega} \cap (\mathcal{D}_{\tilde{\varepsilon}} \setminus \mathcal{D}_{0,\varepsilon/10})$ may be empty. For example, this happens in the supersonic reflection case (for which $\mathcal{D}_{0,\varepsilon/10} = \mathcal{D}_{\varepsilon/10}$); see Remark 17.3.2 when $\tilde{\varepsilon} \le \frac{\varepsilon}{10}$. In fact, it is expected that $\tilde{\varepsilon} \ll \frac{\varepsilon}{10}$ (that can be seen from the procedure of determining $\tilde{\varepsilon}$). On the other hand, $\tilde{\Omega} \cap (\mathcal{D}_{\tilde{\varepsilon}} \setminus \mathcal{D}_{0,\varepsilon/10})$ is definitely non-empty for the subsonic wedge angles $\tilde{\theta}_{\rm w}$ satisfying $x_{P_1}(\tilde{\theta}_{\rm w}) \ge \frac{\varepsilon}{10}$. In fact, the argument has been done mostly in this case, since $\tilde{\Omega} \cap (\mathcal{D}_{\varepsilon_0} \setminus \mathcal{D}_{0,\varepsilon/10}) = \tilde{\Omega} \cap (\mathcal{D}_{\varepsilon_0} \setminus \mathcal{D}_{\hat{\varepsilon}})$ in the opposite case, where $\hat{\varepsilon} = \frac{\varepsilon}{10} - x_{P_1}(\tilde{\theta}_{\rm w}) > 0$.*

In the remaining region $\tilde{\Omega} \cap (\mathcal{D}_{\varepsilon_0} \setminus \mathcal{D}_{\tilde{\varepsilon}})$, we can use (17.5.3) so that we can follow the proof in Lemma 12.8.1. Now (17.3.9) is proved for $(\tilde{u}, \tilde{\theta}_{\rm w})$.

The arguments for properties (17.3.10)–(17.3.13) and (v)–(vi) are similar.

5. It remains to show (17.3.14)–(17.3.16) for $(\tilde{u}, \tilde{\theta}_{\rm w})$, provided that it is sufficiently close in the norm of $C^{2,\alpha}_{(**)}(Q^{\rm iter})$ to $(u, \theta_{\rm w}) \in \mathcal{K}^{\rm ext}$. We first show (17.3.14) for $(\tilde{u}, \tilde{\theta}_{\rm w})$. Rewrite it as

$$\partial_{\xi_2}(\varphi_1 - \varphi_2^{(\tilde{\theta}_{\rm w})}) + \partial_{\xi_2}\tilde{\psi} < -\eta_2(\tilde{\theta}_{\rm w}) \qquad \text{in } (\overline{\Omega} \setminus \mathcal{N}_{\varepsilon/10}(\Gamma_{\rm sym}))^{(\tilde{u},\tilde{\theta}_{\rm w})},$$

where $\partial_{\xi_2}(\varphi_1 - \varphi_2^{(\tilde{\theta}_{\rm w})})$ is independent of $\boldsymbol{\xi}$. Since $\partial_{\xi_2}(\varphi_1 - \varphi_2^{(\theta_{\rm w})}) = -v_2^{(\theta_{\rm w})}$ is negative for any $\theta_{\rm w} \in [\theta_{\rm w}^{\rm d}, \frac{\pi}{2})$ and continuous with respect to $\theta_{\rm w} \in [\theta_{\rm w}^{\rm d}, \frac{\pi}{2}]$ by the continuous dependence of the parameters of the weak state (2) on $\theta_{\rm w}$, we obtain

$$\partial_{\xi_2}(\varphi_1 - \varphi_2^{(\theta_{\rm w})}) \le -\frac{1}{C} \qquad \text{for all } \theta_{\rm w} \in [\theta_{\rm w}^{\rm d}, \tfrac{\pi}{2} - \tfrac{\delta_1}{2N_1^2}].$$

Then, using that $\eta_2(\theta_{\rm w}) < 0$ for $\theta_{\rm w} \in [\frac{\pi}{2} - \frac{\delta_1}{N_1^2}, \frac{\pi}{2}]$ and $\eta_2(\cdot) \in C([\theta_{\rm w}^{\rm d}, \frac{\pi}{2}])$, and choosing δ_2 small – depending only on the data, we have

$$g(\theta_{\rm w}) := -\eta_2(\theta_{\rm w}) - \partial_{\xi_2}(\varphi_1 - \varphi_2^{(\theta_{\rm w})}) \ge \frac{1}{C} \qquad \text{for all } \theta_{\rm w} \in [\theta_{\rm w}^{\rm d}, \frac{\pi}{2}],$$

where $C > 0$ depends only on the data.

From this, following the proof of (17.3.9), we show that there exists $\tilde{\varepsilon}$ depending only on the data and $(\theta_{\rm w}^*, g(\theta_{\rm w}))$ such that, for any $(\tilde{u}, \tilde{\theta}_{\rm w})$ that is sufficiently close to $(u, \theta_{\rm w})$ in the norm of $C^{2,\alpha}_{(**)}(Q^{\rm iter})$, (17.3.14) holds for $\tilde{\psi}$ in $(\Omega \cap \mathcal{D}_{\tilde{\varepsilon}})^{(\tilde{u},\tilde{\theta}_{\rm w})}$. In $(\Omega \setminus (\mathcal{D}_{\tilde{\varepsilon}} \cup \mathcal{N}_{\varepsilon/10}(\Gamma_{\rm sym})))^{(\tilde{u},\tilde{\theta}_{\rm w})}$, we can employ (17.5.3) so that we can repeat again the proof in Lemma 12.8.1. Now (17.3.14) is proved for $(\tilde{u}, \tilde{\theta}_{\rm w})$.

For the remaining properties (17.3.15)–(17.3.16), the proofs are similar to the ones for (17.3.14), since we can write these properties for $(\tilde{u}, \tilde{\theta}_{\rm w})$ as

$$\partial_{\boldsymbol{\nu}}(\varphi_1 - \varphi_2^{(\tilde{\theta}_{\rm w})}) + \partial_{\boldsymbol{\nu}}\tilde{\psi} > \mu_1 \qquad \text{on } \overline{\Gamma_{\rm shock}},$$
$$\partial_{\boldsymbol{\nu}}\varphi_2^{(\tilde{\theta}_{\rm w})} + \partial_{\boldsymbol{\nu}}\tilde{\psi} > \mu_1 \qquad \text{on } \overline{\Gamma_{\rm shock}}.$$

Also, if $(u, \theta_{\rm w}) \in \mathcal{K}^{\rm ext}$ in which $\theta_{\rm w}$ is a subsonic or sonic wedge angle, then, from (17.3.22) and (17.3.15)–(17.3.16), we obtain

$$\partial_{\boldsymbol{\nu}}(\varphi_1 - \varphi_2^{(\theta_{\rm w})})(P_0) > \mu_1 \qquad \text{on } \overline{\Gamma_{\rm shock}},$$
$$\partial_{\boldsymbol{\nu}}\varphi_2^{(\theta_{\rm w})}(P_0) > \mu_1 \qquad \text{on } \overline{\Gamma_{\rm shock}}.$$

Now we follow the previous argument by using the fact that (12.8.2) also holds in the present situation, which can be seen by checking that all the properties used in the proof of (12.8.2) hold in the present case by Lemmas 17.2.1 and 17.2.13. □

Now we are ready to prove the main result of this section.

Proposition 17.5.3. *If parameters* $(\varepsilon, \delta_1, \frac{1}{N_1})$ *of the iteration set in Definition* 17.3.4 *are small – depending on the data and* $\theta_{\rm w}^*$, δ_2 *is small – depending on the data and* (δ_1, N_1), δ_3 *is small – depending on the data and* $(\theta_{\rm w}^*, \delta_2)$, *and* $\alpha = \frac{\hat{\alpha}}{2}$ *for* $\hat{\alpha}$ *determined in Corollary* 17.4.10, *then the iteration set* $\mathcal{K}$ *is relatively open in* $C^{2,\alpha}_{(**)}(Q^{\rm iter}) \times [\theta_{\rm w}^*, \frac{\pi}{2}]$.

Proof. We follow the proof of Proposition 12.8.2, replacing $C^{2,\alpha}_{*,1+\delta^*}(Q^{\rm iter})$ by space $C^{2,\alpha}_{(**)}(Q^{\rm iter})$ and using the results of §17.4, specifically Proposition 17.4.2 and Corollary 17.4.10, instead of the corresponding results of §12.7.2.

Also, we use Lemma 17.3.3 to show the compactness of $C^{2,\alpha}_{(**)}(Q^{\rm iter}) \times [\theta_{\rm w}^*, \frac{\pi}{2}]$ in $C^{2,\frac{\alpha}{2}}_{(**)}(Q^{\rm iter}) \times [\theta_{\rm w}^*, \frac{\pi}{2}]$. Specifically, the argument in the sentence after (12.8.5) is now replaced by the following: Since sequence $\{\hat{u}^{(i)}\}$ is bounded in $C^{2,\alpha}_{(**)}(Q^{\rm iter})$ by Corollary 17.4.10, there exists a subsequence $\{\hat{u}_{i_j}\}$ converging in $C^{2,\frac{\alpha}{2}}_{(**)}(Q^{\rm iter})$. □

17.6 ITERATION MAP AND ITS PROPERTIES

We follow the construction of §13.1–§13.3, via replacing space $C^{2,\alpha}_{*,1+\delta^*}(Q^{\rm iter})$ by space $C^{2,\alpha}_{(**)}(Q^{\rm iter})$ and using the results of §17.4, specifically Proposition 17.4.2 and Corollary 17.4.10, instead of the corresponding results of §12.7.2.

Now we follow §13.1 to describe some points that require clarification or updating. The constants in this argument depend only on the data and $\theta_{\rm w}^*$.

Remark 17.6.1. *Let* $(u, \theta_{\rm w}) \in \overline{\mathcal{K}}$. *Then, by Lemma* 17.3.14 *and* 17.4.3 *with Corollary* 12.7.6, $(u, \theta_{\rm w})$ *satisfies all the conditions of Definition* 17.3.4 *and Remarks* 17.3.6–17.3.12 *with the nonstrict inequalities in the estimates. We will consider such nonstrict inequalities when referring to these properties in* §17.6–17.9.

Let $(u, \theta_{\rm w}) \in \overline{\mathcal{K}}$. Then $\mathfrak{g}_{\rm sh}$ satisfies (17.3.20). Also, (Ω, φ) is determined for $(u, \theta_{\rm w})$ as in Definition 17.3.4(ii), and $(\hat{\varphi}, \hat{u})$ is determined by $(u, \theta_{\rm w})$ as in Definition 17.3.4(vii).

Define $(v, \hat{v}, v_1)$ by (13.1.1). We first note that $F_{(2,\mathfrak{g}_{\rm sh})} = \mathfrak{F}^{-1}_{(u,\theta_{\rm w})} \circ (F_1^{(\theta_{\rm w})})^{-1}$ by (12.2.52). Following the calculation in (13.1.2) and using Lemma 17.2.1(iv), Lemma 12.2.2(ii), Lemma 17.2.8, (17.2.38), and (17.3.19), we have

$$\begin{aligned} \|v - \hat{v}\|_{1,\frac{\alpha}{2},F_1(\Omega)} &= \|(u - \hat{u}) \circ F_{(2,\mathfrak{g}_{\rm sh})}\|_{1,\frac{\alpha}{2},F_1(\Omega)} \\ &\le C\|(u - \hat{u}) \circ \mathfrak{F}^{-1}_{(u,\theta_{\rm w})}\|_{1,\frac{\alpha}{2},\Omega} \\ &\le CM\|u - \hat{u}\|^{(**)}_{2,\frac{\alpha}{2},Q^{\rm iter}} \\ &< CM\delta_3 \end{aligned} \tag{17.6.1}$$

with constant C depending only on the data and $\theta_{\rm w}^*$. This estimate replaces (13.1.2) in the present case. We also note that the structure of $F_1(\Omega)$ is as in (13.1.3) in the present case, with $\mathfrak{g}_{\rm sh}(0)=0$ for the subsonic/sonic wedge angle $\theta_{\rm w}$. Then we follow the argument after (13.1.3) until (13.1.5).

Since $(u,\theta_{\rm w})\in\overline{\mathcal{K}}$, it follows that $\hat\psi=\varphi-\varphi_2^{(\theta_{\rm w})}$ satisfies (17.4.17) in the subsonic-sonic case, and (17.4.14) in the supersonic case. We note that (17.4.14) implies

$$\|\hat\psi\|_{2,\hat\alpha,\Omega}^{(-1-\hat\alpha),\,\overline{\Gamma_{\rm sonic}}\cup\overline{\Gamma_{\rm sym}}}\le C,\qquad (\hat\psi,D\hat\psi)=(0,\mathbf{0})\ \text{ on }\overline{\Gamma_{\rm sonic}}.\tag{17.6.2}$$

The last estimate also holds in the subsonic-sonic case, when $\overline{\Gamma_{\rm sonic}}=\{P_0\}$, since it coincides with (17.4.17) in that case. Thus, (17.6.2) holds for any $(u,\theta_{\rm w})\in\overline{\mathcal{K}}$. Also, since the definition of $\hat v$ in (13.1.1) can be written as $\hat v=(\hat\psi+\varphi_2-\tilde\varphi_2)\circ F_1^{-1}$, then, using Lemma 12.2.2(ii), and (12.2.35) and (12.2.38) from Lemma 17.2.3, it follows that $\hat v\in C_{2,\hat\alpha,F_1(\Omega)}^{(-1-\hat\alpha),F_1(\overline{\Gamma_{\rm sonic}})\cup F_1(\overline{\Gamma_{\rm sym}})}$ with

$$\begin{aligned}&\|\hat v\|_{2,\hat\alpha,F_1(\Omega)}^{(-1-\hat\alpha),F_1(\overline{\Gamma_{\rm sonic}})\cup F_1(\overline{\Gamma_{\rm sym}})}\le C,\\&(\hat v,D\hat v)=(0,\mathbf{0})\qquad \text{on } F_1(\overline{\Gamma_{\rm sonic}}).\end{aligned}\tag{17.6.3}$$

The structure of domain $F_1(\Omega)$ indicated in (13.1.3) allows us to apply Theorem 13.9.8 to extend $\hat v$ defined on $F_1(\Omega)$ to a function

$$\mathcal{E}_{\mathfrak{g}_{\rm sh}}^{(\hat s(\theta_{\rm w}))}(\hat v)\in C_{2,\hat\alpha,\mathcal{D}^{\rm ext}}^{(-1-\hat\alpha),\overline{\Gamma_{\rm sonic}^{\rm ext}}\cup\overline{\Gamma_{\rm sym}^{\rm ext}}},$$

where

$$\begin{aligned}&\mathcal{D}^{\rm ext}=\{(s,t)\ :\ 0<s<\hat s(\theta_{\rm w}),\ 0<t<(1+\sigma)\mathfrak{g}_{\rm sh}(s)\},\\&\overline{\Gamma_{\rm sonic}^{\rm ext}}=\partial\mathcal{D}^{\rm ext}\cap\{(s,t)\ :\ s=0\},\\&\overline{\Gamma_{\rm sym}^{\rm ext}}=\partial\mathcal{D}^{\rm ext}\cap\{(s,t)\ :\ s=\hat s(\theta_{\rm w})\},\end{aligned}\tag{17.6.4}$$

and $\sigma\in(0,1)$ depends only on $\mathrm{Lip}[\mathfrak{g}_{\rm sh}]$, and hence on the data. Specifically, by Theorem 13.9.8(ii)–(iii) and (17.6.3),

$$\begin{aligned}&\|\mathcal{E}_{\mathfrak{g}_{\rm sh}}^{(\hat s(\theta_{\rm w}))}(\hat v)\|_{2,\hat\alpha,\mathcal{D}^{\rm ext}}^{(-1-\hat\alpha),\overline{\Gamma_{\rm sonic}^{\rm ext}}\cup\overline{\Gamma_{\rm sym}^{\rm ext}}}\le C,\\&(\hat v,D\hat v)=(0,\mathbf{0})\qquad \text{on } \overline{\Gamma_{\rm sonic}^{\rm ext}}.\end{aligned}\tag{17.6.5}$$

This, combined with (13.1.5), implies that there exists $\zeta>0$ depending only on the data and $\theta_{\rm w}^*$ such that

$$\frac{\partial(v_1-\mathcal{E}_{\mathfrak{g}_{\rm sh}}^{(\hat s(\theta_{\rm w}))}(\hat v))}{\partial t}\le-\frac{1}{2C}\tag{17.6.6}$$

on $\mathcal{D}^{\rm ext}\cap\{(s,t)\ :\ 0<s<\hat s(\theta_{\rm w}),\ |t-\mathfrak{g}_{\rm sh}(s)|\le\zeta\}$. Using (17.3.7) and (17.6.4), we have

$$\{(s,t)\ :\ 0<s<\hat s(\theta_{\rm w}),\ |t-\mathfrak{g}_{\rm sh}(s)|\le\min(\hat\sigma s,\ \hat\delta_{\rm sh})\}\subset\mathcal{D}^{\rm ext},\tag{17.6.7}$$

where $\hat{\sigma} = \frac{\sigma}{N_2} > 0$ and $\hat{\delta}_{\rm sh} = \frac{\sigma}{2N_2} > 0$ depend only on the data and $\theta_{\rm w}^*$.

Also, starting from (17.3.22) and following the argument for deriving (17.6.3) from (17.6.2), we obtain

$$\|v\|_{1,\alpha,F_1(\Omega)} \le C, \qquad (v, Dv) = (0, \mathbf{0}) \quad \text{on } F_1(\overline{\Gamma_{\rm sonic}}). \tag{17.6.8}$$

By (17.6.3) and (17.6.8), we have

$$(v - \hat{v}, D(v - \hat{v})) = (0, \mathbf{0}) \qquad \text{on } F_1(\overline{\Gamma_{\rm sonic}}).$$

Then, using (17.6.1), we have

$$|(v - \hat{v})(s,t)| \le C\delta_3 s^{1+\frac{\alpha}{2}} \qquad \text{in } F_1(\overline{\Omega}).$$

From this, since $v = v_1$ on $F_1(\Gamma_{\rm shock})$ from the fact that $\varphi = \varphi_1$ on $\Gamma_{\rm shock}$ combined with (13.1.1), we have

$$|(v_1 - \mathcal{E}_{\mathfrak{g}_{\rm sh}}^{(\hat{s}(\theta_{\rm w}))}(\hat{v}))(s,t)| = |(v_1 - \hat{v})(s,t)| \le C\delta_3 s^{1+\frac{\alpha}{2}} \qquad \text{on } F_1(\Gamma_{\rm shock}). \tag{17.6.9}$$

From (17.6.6)–(17.6.7) and (17.6.9), reducing δ_3 depending only on the data and $\theta_{\rm w}^*$, we conclude that there exists a unique function $\hat{\mathfrak{g}}_{\rm sh}(s)$ on $(0, \hat{s}(\theta_{\rm w}))$ such that

$$\begin{aligned} &\mathcal{N}_\varsigma(F_1^{(\theta_{\rm w})}(\Gamma_{\rm shock})) \cap \{\mathcal{E}_{\mathfrak{g}_{\rm sh}}^{(\hat{s}(\theta_{\rm w}))}(\hat{v}) = v_1\} \\ &= \{(s, \hat{\mathfrak{g}}_{\rm sh}(s)) \ : \ 0 < s < \hat{s}(\theta_{\rm w})\} \subset F_1^{(\theta_{\rm w})}(\mathcal{Q}_{\rm bd}^{(\theta_{\rm w})}). \end{aligned} \tag{17.6.10}$$

Moreover, from (17.6.3) and (17.6.6) with smooth v_1 (with uniform $C^3(\overline{\mathcal{D}^{\rm ext}})$–estimates),

$$\begin{aligned} &\|\hat{\mathfrak{g}}_{\rm sh}\|_{2,\hat{\alpha},(0,\hat{s}(\theta_{\rm w}))}^{(-1-\hat{\alpha}),\{0,\hat{s}(\theta_{\rm w})\}} < C, \\ &(\hat{\mathfrak{g}}_{\rm sh} - \mathfrak{g}_{S_1})(0) = (\hat{\mathfrak{g}}_{\rm sh}' - \mathfrak{g}_{S_1}')(0) = 0, \end{aligned} \tag{17.6.11}$$

where we have used the definition of $\hat{v}$ in (13.1.1) to obtain the properties of $\hat{\mathfrak{g}}_{\rm sh}$ at $s = 0$. Now, using (17.6.1) and (17.6.6), we have

$$\begin{aligned} &\|\hat{\mathfrak{g}}_{\rm sh} - \mathfrak{g}_{\rm sh}\|_{1,\hat{\alpha}/2,(0,\hat{s}(\theta_{\rm w}))} \le C\delta_3, \\ &(\hat{\mathfrak{g}}_{\rm sh} - \mathfrak{g}_{\rm sh})(0) = (\hat{\mathfrak{g}}_{\rm sh}' - \mathfrak{g}_{\rm sh}')(0) = 0. \end{aligned} \tag{17.6.12}$$

From (17.6.12), further reducing δ_3, we obtain that $\hat{\mathfrak{g}}_{\rm sh}$ satisfies (17.3.7) with $2N_2$ instead of N_2. Now we define $\tilde{u}$ by (13.1.13).

Lemma 17.6.2. *If $(u, \theta_{\rm w}) \in \overline{\mathcal{K}}$, then $\tilde{u}$ defined by (13.1.13) satisfies*

$$\|\tilde{u}\|_{2,\hat{\alpha},Q^{\rm iter}}^{(**)} \le C, \tag{17.6.13}$$

where C depends only on the data and $\theta_{\rm w}^*$.

Proof. Since $(u, \theta_{\rm w}) \in \overline{\mathcal{K}}$, it follows that $\hat{\psi} = \varphi - \varphi_2^{(\theta_{\rm w})}$ satisfies (17.4.14) in the supersonic case and (17.4.17) in the subsonic-sonic case.

Then, for the supersonic wedge angle $\theta_{\rm w}$, we start by following the proof of Lemma 13.1.1. First, using (17.4.14) instead of (12.7.29), we obtain (13.1.6). From this, using also (17.6.11) and Theorem 13.9.5, we obtain (13.1.8). Also, we note that (17.6.11) is equivalent to (13.1.15) with $\delta^* = \hat{\alpha}$. Then, following the proof of Lemma 13.1.1, we obtain (13.1.14) with C depending only on the data and $\theta_{\rm w}^*$. From this, using the expression of the norm in (13.1.14) given in Definition 12.3.1(ii), and employing Lemma 17.2.11, we conclude (17.6.13) for the supersonic wedge angle $\theta_{\rm w}$.

For the subsonic/sonic wedge angle $\theta_{\rm w}$, we employ (17.6.5). Also, we use the fact that (17.6.12) with sufficiently small δ_3 implies that $\hat{\mathfrak{g}}_{\rm sh}$ satisfies (17.3.7) with $2N_2$ instead of N_2. This, combined with $\mathfrak{g}_{\rm sh}(0) = 0$ for the subsonic/sonic wedge angle $\theta_{\rm w}$ so that $\hat{\mathfrak{g}}_{\rm sh}(0) = 0$ by (17.6.12), shows that $\hat{\mathfrak{g}}_{\rm sh}$ satisfies (17.2.9) with $M = 2N_2$. Then, from (17.6.5), (17.6.11), and Definition 12.2.6(v), applying Lemma 17.2.5 with $m = 2$ and $\varepsilon' = \varepsilon_0'$ for ε_0' from (17.2.7), we have

$$\|\tilde{u}\|^{(1+\hat{\alpha}),(\rm subs)}_{2,\hat{\alpha},Q_{\varepsilon_0'}} \le C.$$

Applying Lemma 17.2.10, we have

$$\|\tilde{u}\|^{(1+\hat{\alpha}),(\rm par)}_{2,\hat{\alpha},Q_{\varepsilon_0'}} \le \|\tilde{u}\|^{(1+\hat{\alpha}),(\rm subs)}_{2,\hat{\alpha},Q_{\varepsilon_0'}} \le C.$$

Also, using (17.6.5), (17.6.11), and the explicit expression (12.2.43) in (13.1.13), we have

$$\|\tilde{u}\|^{(-1-\hat{\alpha}),\{1\}\times(0,1)}_{2,\hat{\alpha},Q^{\rm iter}\setminus\overline{Q}_{\varepsilon_0'/10}} \le \|\mathcal{E}^{(\hat{s}(\theta_{\rm w}))}_{\mathfrak{g}_{\rm sh}}(\hat{v})\|^{(-1-\hat{\alpha}),\overline{\Gamma^{\rm ext}_{\rm sym}}}_{2,\hat{\alpha},\mathcal{D}^{\rm ext}\cap\{s>\frac{\varepsilon_0'}{10}\}} \le C.$$

The last two displayed estimates imply (17.6.13) in the subsonic-sonic case. □

Using Lemma 17.6.2, we have

Definition 17.6.3. *The iteration map* $\mathcal{I} : \overline{\mathcal{K}} \mapsto C^{2,\alpha}_{(**)}(Q^{\rm iter})$ *is defined by*

$$\mathcal{I}(u, \theta_{\rm w}) = \tilde{u},$$

where $\tilde{u}$ *is from* (13.1.13). *For each* $\theta_{\rm w} \in [\theta_{\rm w}^*, \frac{\pi}{2}]$, *we define map* $\mathcal{I}^{(\theta_{\rm w})} : \overline{\mathcal{K}}(\theta_{\rm w}) \mapsto C^{2,\alpha}_{(**)}(Q^{\rm iter})$ *by*

$$\mathcal{I}^{(\theta_{\rm w})}(u) = \mathcal{I}(u, \theta_{\rm w}).$$

Then Lemma 13.1.3, with space $C^{2,\alpha}_{*,1+\delta^*}(Q^{\rm iter})$ replaced by $C^{2,\alpha}_{(**)}(Q^{\rm iter})$, holds in the present case; its proof is independent of the detailed properties of the spaces in the definition of $\mathcal{I}^{(\theta_{\rm w})}$ and now works without change.

17.7 COMPACTNESS OF THE ITERATION MAP

We continue to follow the notational convention of Remark 17.6.1. Below we use $\hat{\alpha}$ that is the smaller one of $\hat{\alpha}$ in Propositions 17.4.7 and 17.4.9, and define $\alpha = \frac{\hat{\alpha}}{2}$ as in Corollary 17.4.10, both depending only on the data and $\theta_{\rm w}^*$. Then

$$\alpha' := \hat{\alpha} - \alpha = \frac{\hat{\alpha}}{2} > 0.$$

Now we show the continuity and compactness.

Lemma 17.7.1. *The iteration map* $\mathcal{I} : \overline{\mathcal{K}} \subset C^{2,\alpha}_{(**)}(Q^{\rm iter}) \times [\theta_{\rm w}^*, \frac{\pi}{2}] \mapsto C^{2,\alpha}_{(**)}(Q^{\rm iter})$ *is continuous and compact with*

$$\|\mathcal{I}^{(\theta_{\rm w})}(u)\|^{(**)}_{2,\alpha+\alpha',Q^{\rm iter}} \le C. \tag{17.7.1}$$

Proof. We divide the proof into two steps.

1. We first show the continuity of $\mathcal{I}$. We follow the proof of Lemma 13.2.2, replacing space $C^{2,\alpha}_{*,1+\delta^*}(Q^{\rm iter})$ by space $C^{2,\alpha}_{(**)}(Q^{\rm iter})$ and using the results of §17.4 instead of the corresponding results of §12.7.2. However, there are new features introduced by the singular nature of the map from $\Omega^{(u,\theta_{\rm w})}$ to $Q^{\rm iter}$ for the subsonic wedge angle $\theta_{\rm w}$. Then our proof is as follows:

Let $(u^{(j)}, \theta_{\rm w}^{(j)}) \in \overline{\mathcal{K}}$ converge to $(u, \theta_{\rm w}) \in \overline{\mathcal{K}}$ in the norm of $C^{2,\alpha}_{(**)}(Q^{\rm iter}) \times [\theta_{\rm w}^*, \frac{\pi}{2}]$. Denote

$$(\Gamma^{(j)}_{\rm shock}, \Omega^{(j)}) := (\Gamma_{\rm shock}(u^{(j)}, \theta_{\rm w}^{(j)}), \Omega(u^{(j)}, \theta_{\rm w}^{(j)})),$$
$$(\Gamma_{\rm shock}, \Omega) := (\Gamma_{\rm shock}(u, \theta_{\rm w}), \Omega(u, \theta_{\rm w})).$$

From Definition 12.2.6(i)–(iii) and (12.2.59) that holds by Lemma 17.2.13, we have

$$\mathfrak{h}^{(j)}_{\rm sh} \to \mathfrak{h}_{\rm sh} \qquad \text{in } C^{1,\frac{\alpha}{2}}([0,1]). \tag{17.7.2}$$

Fix a compact set $K \Subset F_1^{(\theta_{\rm w})}(\Omega)$. Then $K \subset F_1(\Omega) \cap \{s \ge s_K\}$, where $s_K > 0$ depends only on the data and K. Thus, from (17.3.7), there exists C_K depending on the data and K such that, for any $(u^\#, \theta^\#) \in \overline{\mathcal{K}}$,

$$\frac{1}{C_K} < \mathfrak{g}^\#_{\rm sh}(s) < C_K \qquad \text{for all } (s,t) \in K. \tag{17.7.3}$$

Combining this with (17.7.2) and using (12.2.41), we have

$$F_{(2,\mathfrak{g}^{(j)}_{\rm sh})} \to F_{(2,\mathfrak{g}_{\rm sh})} \qquad \text{in } C^{1,\frac{\alpha}{2}}(K). \tag{17.7.4}$$

In particular, there exists a compact subset $\hat{K} \Subset Q^{\rm iter}$ such that $F^{(j)}_{(2,\mathfrak{g}_{\rm sh})}(K) \subset \hat{K}$ for all j. Then, using that $\hat{v} = \hat{u} \circ F_{(2,\mathfrak{g}_{\rm sh})}$, $\hat{v}^{(j)} = \hat{u}^{(j)} \circ F_{(2,\mathfrak{g}^{(j)}_{\rm sh})}$, and $\hat{u}^{(j)} \to \hat{u}$

in C^2 on compact subsets of Q^{iter} by Lemma 12.7.5 (which holds by Lemma 17.4.3), we have

$$\hat{v}^{(j)} \to \hat{v} \qquad \text{in } C^{1,\frac{\alpha}{2}}(K) \text{ for all } K \Subset F_1(\Omega), \tag{17.7.5}$$

where we have used the property that $K \Subset F_1(\Omega)$ is an arbitrary set in the argument above.

Next we consider functions $\hat{\mathfrak{h}}_{\text{sh}}^{(j)}$ and $\hat{\mathfrak{h}}_{\text{sh}}$, determined by functions $\hat{\mathfrak{g}}_{\text{sh}}^{(j)}$ and $\hat{\mathfrak{g}}_{\text{sh}}$ in (17.6.10) by the formulas from Definition 12.2.6(ii), and show that

$$\hat{\mathfrak{h}}_{\text{sh}}^{(j)} \to \hat{\mathfrak{h}}_{\text{sh}} \qquad \text{in } C^{(-1-\alpha),\{0,1\}}_{2,\alpha,(0,1)}. \tag{17.7.6}$$

For that, we note that, for $k = 1, 2, \dots$, each $\psi^{(j)}$ satisfies (17.3.23) with $\tau = \frac{\varepsilon_0}{k}$. Then, writing v in (13.1.1) as $v := (\psi + \varphi_2 - \tilde{\varphi}_2) \circ F_1^{-1}$ and applying this to $v^{(j)}$ for each j, we obtain that, for each j,

$$\|v^{(j)}\|^{(-1-\alpha),F_1^{\theta_{\text{w}}^{(j)}}(\Gamma_{\text{sym}})}_{2,\alpha,F_1^{\theta_{\text{w}}^{(j)}}(\Omega^{(j)})\cap\{s>\frac{\varepsilon_0}{k}\}} \le C_k \qquad \text{for } k = 1, 2, \dots,$$

where C_k depends only on the data and (θ_{w}, k). Restricting to a smaller region, we find that, for each j,

$$\|v^{(j)}\|_{2,\alpha,F_1^{\theta_{\text{w}}^{(j)}}(\Omega^{(j)})\cap\{\frac{\varepsilon_0}{k}<s<\hat{s}(\theta_{\text{w}}^{(j)})-\frac{\varepsilon_0}{k}\}} \le C_k \qquad \text{for } k = 1, 2, \dots.$$

Combining this with (17.7.2) (which holds in the present case as we have discussed above), and using the continuity of $\hat{s}(\cdot)$, we can apply Theorem 13.9.5(iii), which shows that, for

$$w_j(s,t) := \mathcal{E}^{(\hat{s}(\theta_{\text{w}}^{(j)}))}_{\mathfrak{g}_{\text{sh}}^{(j)}}(\hat{v}^{(j)})\Big(\frac{\hat{s}(\theta_{\text{w}})}{\hat{s}(\theta_{\text{w}}^{(j)})}s, t\Big),$$

we obtain that, for each $k = 1, 2, \dots$,

$$w_j \to \mathcal{E}^{(\hat{s}(\theta_{\text{w}}))}_{\mathfrak{g}_{\text{sh}}}(\hat{v}) \qquad \text{in } C^{2,\alpha}\big(\mathcal{D}^{\text{ext}}_{\sigma/2} \cap \{\frac{\varepsilon_0}{k} < s < \hat{s}(\theta_{\text{w}}) - \frac{\varepsilon_0}{k}\}\big), \tag{17.7.7}$$

where

$$\mathcal{D}^{\text{ext}}_{\sigma/2} = \{(s,t) \,:\, 0 < s < \hat{s}(\theta_{\text{w}}),\ 0 < t < (1 + \frac{\sigma}{2})\mathfrak{g}_{\text{sh}}(s)\},$$

and $\sigma > 0$ is from (17.6.4). From (17.3.7), (17.7.2), and (17.6.12) applied for each j and the limiting function, it follows that, reducing δ_3 depending only on the data and θ_{w}^*, then, for $k = 1, 2, \dots$,

$$\{(s, \hat{\mathfrak{g}}_{\text{sh}}^{(j)}(\frac{\hat{s}(\theta_{\text{w}}^{(j)})}{\hat{s}(\theta_{\text{w}})}s)) \,:\, \frac{\varepsilon_0}{k} < s < \hat{s}(\theta_{\text{w}}) - \frac{\varepsilon_0}{k}\} \subset \mathcal{D}^{\text{ext}}_{\sigma/2} \qquad \text{for } j \ge N(k),$$

and the same holds for $\{(s, \hat{\mathfrak{g}}_{\text{sh}}(s)) \,:\, \frac{\varepsilon_0}{k} < s < \hat{s}(\theta_{\text{w}}) - \frac{\varepsilon_0}{k}\}$. Then, from (17.6.6) and further reducing δ_3 (depending on ζ, and hence on the data and

$\theta_{\rm w}^*$) in (17.6.12) applied for each j and the limiting function, the convergence in (17.7.7) for each $k = 1, 2, \dots$ implies that $\hat{\mathfrak{h}}_{\rm sh}^{(j)} \to \hat{\mathfrak{h}}_{\rm sh}$ in $C^2([\frac{1}{m}, 1 - \frac{1}{m}])$ for each $m = 1, 2, \dots$. In particular, $\hat{\mathfrak{h}}_{\rm sh}^{(j)} \to \hat{\mathfrak{h}}_{\rm sh}$ pointwise on (0,1). Then, using (17.6.11) for each j and recalling that $\alpha < \hat{\alpha}$, we obtain (17.7.6).

From (17.6.13) for $\tilde{u}^{(j)}$ and $\tilde{u}$, using (17.7.6)–(17.7.7), we obtain that $\tilde{u}^{(j)} \to \tilde{u}$ pointwise in the open region $Q^{\rm iter}$. Then, using (17.6.13) for each $\tilde{u}^{(j)}$ and Lemma 17.3.3 with $\alpha < \hat{\alpha} = \alpha + \alpha'$, we find that $\tilde{u}^{(j)} \to \tilde{u}$ in $C^{2,\alpha}_{(**)}(Q^{\rm iter})$. This shows the continuity of the iteration map.

2. Now we prove the compactness of $\mathcal{I}$ and (17.7.1). Lemma 17.6.2 and Definition 17.6.3 imply (17.7.1), where $\alpha + \alpha' = \hat{\alpha}$. Then Lemma 17.3.3 with $\alpha < \hat{\alpha}$ and the continuity proved in Step 1 imply the compactness of the iteration map $\mathcal{I} : \overline{\mathcal{K}} \mapsto C^{2,\alpha}_{(**)}(Q^{\rm iter})$. □

17.8 NORMAL REFLECTION AND THE ITERATION MAP FOR $\theta_{\rm w} = \frac{\pi}{2}$

We recall that the normal reflection $\varphi = \varphi_2$ in Ω for $\theta_{\rm w} = \frac{\pi}{2}$ corresponds to

$$u^{(\rm norm)} \equiv 0 \qquad \text{in } Q^{\rm iter},$$

by (12.2.34) and (12.2.44).

The results in §13.3 and their proofs hold without change in the present case. Then we have

Proposition 17.8.1. *For any $(u, \frac{\pi}{2}) \in \overline{\mathcal{K}}$,*

$$\mathcal{I}^{(\frac{\pi}{2})}(u) = u^{(\rm norm)} \equiv 0.$$

17.9 FIXED POINTS OF THE ITERATION MAP FOR $\theta_{\rm w} < \frac{\pi}{2}$ ARE ADMISSIBLE SOLUTIONS

We continue to follow the notational convention in Remark 17.6.1. The main result of this section is the following:

Proposition 17.9.1. *If the parameters in Definition* 17.3.4 *are chosen so that δ_1 and ε are small, depending on the data and $\theta_{\rm w}^*$, and if $N_1 \geq 8$, then the following holds: Let $(u, \theta_{\rm w}) \in \overline{\mathcal{K}}$ for $\theta_{\rm w} \in [\theta_{\rm w}^*, \frac{\pi}{2})$, and let u be a fixed point of map $\mathcal{I}^{(\theta_{\rm w})}$. Let φ be determined by $(u, \theta_{\rm w})$ as in Definition* 12.2.6(v). *Then φ is an admissible solution of* **Problem 2.6.1** *in the sense of Definition* 15.1.1 *if $\theta_{\rm w}$ is supersonic, and in the sense of Definition* 15.1.2 *if $\theta_{\rm w}$ is subsonic or sonic.*

Before proving Proposition 17.9.1, we prove the following preliminary property:

Lemma 17.9.2. *In the conditions of Proposition* 17.9.1, *reducing* ε *further depending on the data and* $\theta_{\rm w}^*$ *if necessary, then the following holds: Let* $(u, \theta_{\rm w}) \in \overline{\mathcal{K}}$ *and let* u *be a fixed point of map* $\mathcal{I}^{(\theta_{\rm w})}$. *Let* $\theta_{\rm w}$ *be subsonic or sonic, i.e.,* $\frac{|D\varphi_2(P_0)|}{c_2}(\theta_{\rm w}) \leq 1$. *Let* (Ω, ψ) *be determined by* $(u, \theta_{\rm w})$. *Then* ψ *satisfies the potential flow equation* (12.4.2) *in* Ω. *Moreover, equation* (12.4.2) *is strictly elliptic for* ψ *in* $\overline{\Omega} \setminus \overline{\Gamma_{\rm sonic}}$.

Proof. Let $\hat{\delta}_{P_0} \in (0, \frac{\delta_{P_0}}{4}]$ be the constant determined in Proposition 17.4.8. Let $\varepsilon_{\rm eq} \in (0, \frac{\varepsilon_0}{2})$ be the constant determined in Lemma 17.3.22 for $\delta_{\rm e} = \frac{\hat{\delta}_{P_0}}{2}$. We choose $\varepsilon \in (0, \varepsilon_{\rm eq}]$.

If $\theta_{\rm w}$ satisfies that $\frac{|D\varphi_2(P_0)|}{c_2}(\theta_{\rm w}) \in [1 - \frac{\hat{\delta}_{P_0}}{2}, 1]$, we can apply Proposition 17.4.8 to obtain (17.4.16) with $\sigma = \hat{\alpha}$ for ψ, since $\hat{\psi} = \psi$. From the choice of σ, it follows that C in (17.4.16) depends on the data and $\theta_{\rm w}^*$. Choosing ε sufficiently small, depending on the data and $\theta_{\rm w}^*$, we find from (17.4.16) that

$$|\psi_x| \leq \frac{2 - \frac{\mu_0}{5}}{1+\gamma} x, \; |\psi_y| \leq N_3 x \quad \text{in } \Omega \cap \mathcal{D}_{0,\varepsilon/2} \text{ if } \frac{|D\varphi_2(P_0)|}{c_2}(\theta_{\rm w}) \in [1 - \frac{\hat{\delta}_{P_0}}{2}, 1].$$

Then, applying Lemma 17.3.22 with $\delta_{\rm e} = \frac{\hat{\delta}_{P_0}}{2}$ and using assertion (b$'$) of Lemma 17.3.21(viii$'$) to handle the wedge angles satisfying that $\frac{|D\varphi_2(P_0)|}{c_2}(\theta_{\rm w}) \leq 1 - \frac{\delta_{P_0}}{2}$, we conclude the proof for all the fixed points with the wedge angles $\theta_{\rm w}$ satisfying that $\frac{|D\varphi_2(P_0)|}{c_2}(\theta_{\rm w}) \leq 1$. □

We follow the argument in §13.4 with the adjustments specified below.

We recall that $\alpha = \frac{\hat{\alpha}}{2}$ is fixed as defined in Corollary 12.7.11, where $\hat{\alpha}$ is defined in Proposition 12.7.10. Thus, α depends only on the data and $\theta_{\rm w}^*$.

We note that, in order to prove Proposition 17.9.1, it suffices to show the following properties of φ:

$$\varphi_2 \leq \varphi \leq \varphi_1 \quad \text{in } \Omega, \tag{17.9.1}$$

$$\partial_{\mathbf{e}_{S_1}}(\varphi_1 - \varphi) \leq 0 \text{ and } \partial_{\xi_2}(\varphi_1 - \varphi) \leq 0 \quad \text{in } \Omega, \tag{17.9.2}$$

$$|\psi_x| \leq \frac{2 - \frac{\mu_0}{5}}{1+\gamma} x \quad \text{in } \Omega \cap \mathcal{D}_{\varepsilon/4} \text{ if } \frac{|D\varphi_2(P_0)|}{c_2}(\theta_{\rm w}) > 1. \tag{17.9.3}$$

The argument that properties (17.9.1)–(17.9.3) imply Proposition 17.9.1 repeats the corresponding argument after (13.4.1)–(13.4.3) by using Definition 17.3.4 instead of Definition 12.3.2 with obvious adjustments for the structure of subsonic-sonic solutions, and space $C^{2,\alpha}_{*,1+\delta^*}(\Omega)$ instead of space $C^{2,\alpha}_{(**)}(\Omega)$ in item (e) of the argument. We only give the modified version of item (c):

(c') φ satisfies the potential flow equation (2.2.8) in Ω, which is elliptic for φ in $\overline{\Omega} \setminus \overline{\Gamma_{\text{sonic}}}$.

Indeed, by Lemma 17.9.2, this is also true when θ_{w} is subsonic or sonic, *i.e.*, $\frac{|D\varphi_2(P_0)|}{c_2}(\theta_{\text{w}}) \le 1$. It remains to consider the supersonic wedge angle θ_{w} satisfying that $\frac{|D\varphi_2(P_0)|}{c_2}(\theta_{\text{w}}) > 1$. In that case, by Proposition 17.4.7, estimate (17.4.14) holds. Since $\hat{\psi} = \psi$ in the present case, we find from (17.4.14) that $|\psi_y| \le Cx^{\frac{3}{2}}$ in $\Omega \cap \mathcal{D}_{\varepsilon_0}$, with C depending only on the data and θ_{w}^*. Choosing ε small, we have

$$|\psi_y| \le N_3 x \qquad \text{in } \Omega \cap \mathcal{D}_{\varepsilon/4}.$$

Combining this with (17.9.3) and applying Lemma 17.3.22, we obtain that φ satisfies the potential flow equation (2.2.8) in Ω, and equation (2.2.8) is elliptic for φ in $\overline{\Omega} \setminus \overline{\Gamma_{\text{sonic}}}$.

Then we prove (17.9.1)–(17.9.3) in the remaining part of this section.

We first notice the following fact.

Lemma 17.9.3. *If $\varepsilon > 0$ in the definition of $\mathcal{K}$ is sufficiently small, depending only on the data and θ_{w}^*, then, for any $(u, \theta_{\text{w}}) \in \overline{\mathcal{K}}$ such that u is a fixed point of map $\mathcal{I}^{(\theta_{\text{w}})}$ and θ_{w} is a supersonic wedge angle, the corresponding function ψ satisfies assertions* (a)–(c) *in Lemma* 12.4.5(viii).

Proof. Fix $(u, \theta_{\text{w}}) \in \overline{\mathcal{K}}$ for which u is a fixed point and θ_{w} is a supersonic wedge angle. Let $\hat{\psi}$ be the unique solution of (12.3.25)–(12.3.29) defined by (u, θ_{w}).

By Lemma 13.1.3, $u = \hat{u}$. Using Definition 12.2.6(v) and (12.3.15), then $\varphi = \hat{\varphi}$ so that $\psi = \hat{\psi}$. From this, since θ_{w} is supersonic, estimate (17.4.14) of Proposition 17.4.7 implies that ψ satisfies (12.4.7). Now the assertion follows from assertions (a′) and (c′) of Lemma 17.3.21(viii′). □

Lemma 17.9.4. *Let $(u, \theta_{\text{w}}) \in \overline{\mathcal{K}}$ be such that u is a fixed point of map $\mathcal{I}^{(\theta_{\text{w}})}$. Then the corresponding function ψ satisfies* (13.4.4).

Proof. From Propositions 17.4.7 and 17.4.9, for any fixed point $(u, \theta_{\text{w}}) \in \overline{\mathcal{K}}$, $\psi = \hat{\psi}$ satisfies $\psi \in C^{1,\alpha}(\overline{\Omega}) \cap C^{2,\alpha}(\overline{\Omega} \setminus (\overline{\Gamma_{\text{sonic}}} \cup \overline{\Gamma_{\text{sym}}}))$, where $\overline{\Gamma_{\text{sonic}}} = \{P_0\}$ in the subsonic/sonic case. Then we can follow the proof of Lemma 13.4.3. □

17.9.1 $\varphi \ge \varphi_2$ for fixed points

Lemma 17.9.5. *The result of Lemma* 13.4.4 *holds without change.*

Proof. We can follow the proof of Lemma 13.4.4, since Lemmas 13.1.3 and 12.7.7 are available (the last one from Lemma 17.4.3). □

17.9.2 Directional monotonicity for $\varphi_1 - \varphi$ for fixed points

We first note that estimate (13.4.24) is available in the present case by (17.3.22). Also, estimate (12.3.22) is available by (17.3.21).

We first show that the result of Lemma 13.4.6 holds in the present case.

Lemma 17.9.6. *If the parameters in Definition* 17.3.4 *are chosen so that* δ_1 *and* ε *are small, depending only on the data and* $\theta_{\rm w}^*$*, and if* $N_1 \geq 8$*, then, for every* $(u, \theta_{\rm w}) \in \overline{\mathcal{K}}$ *such that* $\theta_{\rm w} \in [\theta_{\rm w}^*, \frac{\pi}{2})$ *and* u *is a fixed point of map* $\mathcal{I}^{(\theta_{\rm w})}$*, the corresponding function* φ *satisfies*

$$\partial_{\mathbf{e}_{S_1}}(\varphi_1 - \varphi) \leq 0 \qquad in\ \Omega.$$

Proof. Let $(u, \theta_{\rm w}) \in \overline{\mathcal{K}}$ be a fixed point of the iteration map. Then $\psi = \hat{\psi}$ as we have shown above.

By Lemmas 17.9.2 and 17.9.3, ψ satisfies an equation of structure (12.4.45) in Ω.

Then we follow the proof of Lemma 13.4.6 by using estimates (17.3.21)–(17.3.22) instead of (12.3.22) and (13.4.24). Also, the proof of Lemma 13.4.6 uses that $(\psi, D\psi) = (0, \mathbf{0})$ on $\overline{\Gamma_{\rm sonic}}$. In the present case, this holds by (17.3.22), where $\overline{\Gamma_{\rm sonic}} = \{P_0\}$ for the subsonic/sonic wedge angles so that the proof works without change.

The only point we need to verify is the following: In Step 2 of the proof of Lemma 13.4.6, we have shown that $P_1\mathcal{O}_2$ and $\boldsymbol{\nu}_{S_1}$ are not parallel to each other. The proof uses the fact that P_1 lies on the sonic circle $\partial B_{c_2}(\mathcal{O}_2)$ of state (2). In the present case, that proof works in the supersonic case.

Now consider the subsonic/sonic case. Then $P_1 = P_0$, so that we need to show that, for the subsonic/sonic wedge angle $\theta_{\rm w}$, vectors $P_0\mathcal{O}_2$ and $\boldsymbol{\nu}_{S_1}$ are not parallel to each other. Since $D\varphi_k(P_0) = P_0\mathcal{O}_k$ for $k = 1, 2$, and S_1 is the shock line between the uniform states φ_1 and φ_2 on which they satisfy the Rankine-Hugoniot conditions, then, from (6.1.6) in Lemma 6.1.2 applied with $\varphi^- = \varphi_1$ and $\varphi^+ = \varphi_2$, we find that $\mathcal{O}_1\mathcal{O}_2 \perp S_1$. However, if $P_0\mathcal{O}_2$ is parallel to $\boldsymbol{\nu}_{S_1}$, then $P_0\mathcal{O}_2 \perp S_1$ so that $\mathcal{O}_1\mathcal{O}_2 \perp S_1$ implies that $P_0\mathcal{O}_2$ is parallel to $\mathcal{O}_1\mathcal{O}_2$, which means that these lines coincide. From that, since $\mathcal{O}_1 = (u_1, 0) \notin \{\xi_2 = \xi_1 \tan\theta_{\rm w}\}$, while both $\mathcal{O}_2$ and P_0 lie on $\Gamma_{\rm wedge} \subset \{\xi_2 = \xi_1 \tan\theta_{\rm w}\}$, it follows that $\mathcal{O}_2 = P_0$ so that $\mathcal{O}_2 \in S_1$. This contradicts (6.1.5). Therefore, $P_0\mathcal{O}_2$ and $\boldsymbol{\nu}_{S_1}$ are not parallel to each other.

The rest of the proof of Lemma 13.4.6 works without change. □

Next we show that the result of Lemma 13.4.9 holds in the present case.

Lemma 17.9.7. *If the parameters in Definition* 17.3.4 *are chosen so that* δ_1 *and* ε *are small, depending on the data and* $\theta_{\rm w}^*$*, and if* $N_1 \geq 8$*, then, for every* $(u, \theta_{\rm w}) \in \overline{\mathcal{K}}$ *such that* $\theta_{\rm w} \in [\theta_{\rm w}^*, \frac{\pi}{2})$ *and* u *is a fixed point of map* $\mathcal{I}^{(\theta_{\rm w})}$*, the corresponding function* φ *satisfies*

$$\partial_{\xi_2}(\varphi_1 - \varphi) \leq 0 \qquad in\ \Omega.$$

Proof. We argue in the same way as in the proof of Lemma 17.9.6, by showing that the proof of Lemma 13.4.9 works in the present case.

Then it suffices to check the argument in Step 2 of the proof of Lemma 13.4.9, which shows that $(1,0)$ is not parallel to $P_1\mathcal{O}_2$ for all the wedge angles $\theta_{\rm w} \in [\theta_{\rm w}^*, \frac{\pi}{2}]$. That argument works without change in the supersonic reflection case.

In the subsonic/sonic reflection case, $P_1 = P_0$ so that we need to show that, for the subsonic/sonic wedge angle $\theta_{\rm w}$, vectors $P_0\mathcal{O}_2$ and $(1,0)$ are not parallel to each other. This immediately follows from the fact that segment $P_0\mathcal{O}_2$ lies on $\Gamma_{\rm wedge} \subset \{\xi_2 = \xi_1 \tan\theta_{\rm w}\}$ with $\theta_{\rm w} \in [\theta_{\rm w}^*, \theta_{\rm w}^{\rm s}] \subset (0, \frac{\pi}{2})$.

The rest of the proof of Lemma 13.4.9 works without change. □

17.9.3 Removing the cutoff in the equation for supersonic fixed points

It remains to show that (17.9.3) holds. Thus, we focus on the supersonic wedge angle $\theta_{\rm w}$. We then have the following analogue of Lemmas 13.4.10–13.4.11.

Lemma 17.9.8. *If the parameters in Definition* 12.3.2 *are chosen so that the conditions of Proposition* 17.4.7 *are satisfied, and if ε is further reduced if necessary, depending on the data and $\theta_{\rm w}^*$, then the following holds: Let $(u, \theta_{\rm w}) \in \overline{\mathcal{K}}$ be such that $\theta_{\rm w} \in [\theta_{\rm w}^*, \frac{\pi}{2})$ is a supersonic wedge angle, i.e., satisfying* (17.4.10), *and let u be a fixed point of map $\mathcal{I}^{(\theta_{\rm w})}$. Then the corresponding function $\psi = \varphi - \varphi_2$ satisfies*

$$|\psi_x| \le \frac{2 - \frac{\mu_0}{5}}{1+\gamma}x \qquad \text{in } \Omega \cap \{x \le \frac{\varepsilon}{4}\}. \tag{17.9.4}$$

Proof. We follow the argument in the proof of Lemmas 13.4.10–13.4.11 to establish the upper and lower bounds of ψ_x in $\Omega \cap \{x \le \frac{\varepsilon}{4}\}$. We note that the supersonic reflection solutions satisfy the same iteration equation as in Lemmas 13.4.10–13.4.11, which follows from Lemma 17.3.21(v′), Lemma 17.3.20(i), and Lemma 17.3.17(a). Also, the supersonic reflection solutions satisfy the same iteration boundary condition on $\Gamma_{\rm shock}$ as in Lemmas 13.4.10–13.4.11, by Lemma 17.3.23. Furthermore, the monotonicity property used in the proof of Lemma 13.4.11 is extended to the present case in Lemma 17.9.6.

Thus, the argument works without change for each supersonic wedge angle $\theta_{\rm w}$ so that (17.9.4) is obtained for sufficiently small ε. We need to check that a uniform constant $\varepsilon > 0$ can be chosen for all the fixed points of the iteration map, corresponding to the supersonic wedge angles. In Lemmas 13.4.10–13.4.11, the smallness of ε depends only on constant C in estimate (12.7.29) in Proposition 12.7.10. In the present case, a similar estimate (17.4.14) as in Proposition 17.4.7, with C depending only on the data and $\theta_{\rm w}^*$, holds for any supersonic wedge angle. Therefore, ε can be chosen uniformly in the present case. □

Corollary 17.9.9. *Corollary* 13.4.13 *holds in the present case.*

Proof. We follow the proof of Corollary 13.4.13. By Lemma 17.9.8, we conclude that φ satisfies the potential flow equation (2.2.8) in Ω. Then we follow the argument for Lemma 8.3.2. This argument uses the property that $\varphi_1 - \varphi = \varphi_1 - \varphi_2 \geq 0$ on $\overline{\Gamma_{\rm sonic}}$. This property holds in the present case for the supersonic, sonic, and subsonic wedge angles, where $\overline{\Gamma_{\rm sonic}} = \{P_0\}$ in the last two cases, and $\varphi_1 = \varphi_2 = \varphi$ at P_0 in the subsonic/sonic cases. The rest of the argument follows without change. □

17.9.4 Completion of the proof of Proposition 17.9.1

Note that

- (17.9.1) follows from Lemma 17.9.5 and Corollary 17.9.9;
- (17.9.2) follows from Lemmas 17.9.6 and 17.9.7;
- (17.9.3) follows from Lemma 17.9.8.

Then the proof of Proposition 17.9.1 is completed.

17.10 FIXED POINTS CANNOT LIE ON THE BOUNDARY OF THE ITERATION SET

We assume that the parameters of the iteration set satisfy the conditions of Proposition 17.9.1. Then the fixed points are admissible solutions of **Problem 2.6.1** in the sense of Definitions 15.1.1–15.1.2.

We first note the following preliminary results:

Lemma 17.10.1. *The assertions of Lemmas* 12.6.1–12.6.2 *hold in the present setting, where* (12.6.1) *is replaced by*

$$\|u - u^{(\rm norm)}\|^{(**)}_{2,\alpha,Q^{\rm iter}} < \mu. \tag{17.10.1}$$

Proof. We first follow the proof of Lemma 12.6.1, replacing Definition 12.3.2 by Definition 17.3.4. Then we follow the proof of Lemma 12.6.2, replacing spaces $C^{2,\alpha}_{*,1+\delta^*}(Q^{\rm iter})$ and $C^{2,\bar\alpha}_{*,2}(Q^{\rm iter})$ by spaces $C^{2,\alpha}_{(**)}(Q^{\rm iter})$ and $C^{2,\bar\alpha}_{(**)}(Q^{\rm iter})$, respectively, and using Lemma 17.3.3, to obtain (17.10.1). □

Now we prove the main result of this section.

Lemma 17.10.2. *If the parameters in Definition* 17.3.4 *are chosen to satisfy the conditions of Proposition* 17.9.1, N_1 *is further increased (depending on the data and* $\theta_{\rm w}^*$*), and* δ_2 *is chosen sufficiently small (depending on* (ε, δ_1)*), then, for any* $\theta_{\rm w} \in [\theta_{\rm w}^*, \frac{\pi}{2}]$,

$$\mathcal{I}(u, \theta_{\rm w}) \neq u \qquad \text{for all } (u, \theta_{\rm w}) \in \partial\mathcal{K},$$

where boundary $\partial\mathcal{K}$ is considered relative to space $C^{2,\alpha}_{(**)}(Q^{\rm iter}) \times [\theta_{\rm w}^*, \frac{\pi}{2}]$; in particular, both $(u, \frac{\pi}{2})$ and $(u, \theta_{\rm w}^*)$ may lie in the interior of $\mathcal{K}$.

Proof. We follow the proof of Lemma 13.5.1 by using the results of this chapter (instead of the corresponding results of Chapter 13) and the fact that the constants in the estimates in Definition 17.3.4 are chosen such that the admissible solutions satisfy these estimates with the strict inequalities by the *a priori* estimates cited when these constants are introduced in Definition 17.3.4. □

17.11 PROOF OF THE EXISTENCE OF SOLUTIONS UP TO THE CRITICAL ANGLE

From Proposition 17.8.1 and Theorem 3.4.7(i), it follows that the fixed point index of map $\mathcal{I}^{(\frac{\pi}{2})}$ on set $\overline{\mathcal{K}}(\frac{\pi}{2})$ is nonzero. Specifically,

$$\mathbf{Ind}(\mathcal{I}^{(\frac{\pi}{2})}, \overline{\mathcal{K}}(\frac{\pi}{2})) = 1.$$

From Proposition 17.5.3, and Lemmas 17.7.1 and 17.10.2, we see that the conditions of Theorem 3.4.8 are satisfied, which implies

$$\mathbf{Ind}(\mathcal{I}^{(\theta_{\rm w})}, \overline{\mathcal{K}}(\theta_{\rm w})) = 1 \qquad \text{for all } \theta_{\rm w} \in [\theta_{\rm w}^*, \frac{\pi}{2}].$$

Then, from Theorem 3.4.7(ii), a fixed point of map $\mathcal{I}^{(\theta_{\rm w})}$ on domain $\overline{\mathcal{K}}(\theta_{\rm w})$ exists for any $\theta_{\rm w} \in [\theta_{\rm w}^*, \frac{\pi}{2}]$. By Proposition 17.9.1, the fixed points are admissible solutions. Thus, the admissible solutions exist for all $\theta_{\rm w} \in [\theta_{\rm w}^*, \frac{\pi}{2}]$. This holds for any $\theta_{\rm w}^* \in (\theta_{\rm w}^{\rm c}, \frac{\pi}{2})$. Then the proof of Proposition 17.1.1 is completed.

17.12 PROOF OF THEOREM 2.6.6: EXISTENCE OF GLOBAL SOLUTIONS UP TO THE DETACHMENT ANGLE WHEN $u_1 \leq c_1$

We now prove Theorem 2.6.7. The existence follows from Proposition 17.1.1. The regularity follows from the *a priori* estimates in the supersonic/subsonic cases, and Corollary 16.4.8, 16.5.4, and 16.6.12.

The property that the corresponding $(\rho, \Phi)(t, \mathbf{x})$ is a weak solution of **Problem 2.2.1** in the sense of Definition 2.3.3 satisfying the entropy condition is proved in the same way as in §13.7, where we note that, for the subsonic and sonic regular reflections, the argument works without changes.

17.13 PROOF OF THEOREM 2.6.8: EXISTENCE OF GLOBAL SOLUTIONS WHEN $u_1 > c_1$

We now prove Theorem 2.6.9. We follow the argument in §13.8 by using Proposition 17.1.1 and Corollaries 16.4.8, 16.5.4, and 16.6.12 to prove the existence and regularity for $\theta_{\mathrm{w}} > \theta_{\mathrm{w}}^{\mathrm{c}}$. If $\theta_{\mathrm{w}}^{\mathrm{c}} > \theta_{\mathrm{w}}^{\mathrm{d}}$, we argue as in §13.8, by employing Lemma 15.7.4(iii) and Proposition 15.4.3 (instead of Proposition 9.4.8) with the corresponding straightforward (and mostly notational) changes in the argument, and the regularity results near Γ_{sonic} or the reflection point, Corollary 16.4.7, and Propositions 16.5.3 and 16.6.11.

Part V

Connections and Open Problems

Chapter Eighteen

The Full Euler Equations and the Potential Flow Equation

In this chapter, we first analyze the system of full Euler equations and its planar shock-front solutions, and formulate the shock reflection-diffraction problem into an initial-boundary value problem for the system. Then we employ the self-similarity of the problem to reformulate the initial-boundary value problem into a boundary value problem for a system of first-order nonlinear PDEs of composite-mixed hyperbolic-elliptic type. We also present the local theory and von Neumann's conjectures for shock reflection-diffraction. Finally, we discuss the role of the potential flow equation in the shock reflection-diffraction problem even in the realm of the full Euler equations.

18.1 THE FULL EULER EQUATIONS

As described in Chapter 1, the full Euler equations for compressible fluids in $\mathbb{R}^3_+ = \mathbb{R}_+ \times \mathbb{R}^2$, $t \in \mathbb{R}_+ := (0, \infty)$ and $\mathbf{x} \in \mathbb{R}^2$, are of the following form:

$$\begin{cases} \partial_t \rho + \nabla_{\mathbf{x}} \cdot (\rho \mathbf{v}) = 0, \\ \partial_t(\rho \mathbf{v}) + \nabla_{\mathbf{x}} \cdot (\rho \mathbf{v} \otimes \mathbf{v}) + \nabla_{\mathbf{x}} p = 0, \\ \partial_t(\frac{1}{2}\rho|\mathbf{v}|^2 + \rho e) + \nabla_{\mathbf{x}} \cdot \big((\frac{1}{2}\rho|\mathbf{v}|^2 + \rho e + p)\mathbf{v}\big) = 0, \end{cases} \tag{18.1.1}$$

where ρ is the density, $\mathbf{v} = (u, v)$ the fluid velocity, p the pressure, and e the internal energy. Two other important thermodynamic variables are temperature θ and entropy S. Also, $\mathbf{a} \otimes \mathbf{b}$ denotes the tensor product of vectors $\mathbf{a}$ and $\mathbf{b}$.

For a polytropic gas,

$$p = (\gamma - 1)\rho e, \qquad e = c_v \theta, \qquad \gamma = 1 + \frac{R}{c_v}, \tag{18.1.2}$$

or equivalently,

$$p = p(\rho, S) = \kappa \rho^\gamma e^{S/c_v}, \qquad e = e(\rho, S) = \frac{\kappa}{\gamma - 1}\rho^{\gamma-1} e^{S/c_v}, \tag{18.1.3}$$

where $R > 0$, $c_v > 0$, $\gamma > 1$, and $\kappa > 0$ are some constants given.

System (18.1.1) can be written as the following general form as a hyperbolic system of conservation laws (*cf.* Dafermos [100], Glimm-Majda [139], and Lax [171]):

$$\partial_t \mathbf{u} + \nabla_{\mathbf{x}} \cdot \mathbf{f}(\mathbf{u}) = 0, \qquad \mathbf{x} \in \mathbb{R}^2, \tag{18.1.4}$$

where $\mathbf{u} = (\rho, \rho\mathbf{v}, \frac{1}{2}\rho|\mathbf{v}|^2 + \rho e)^\top$ and $\mathbf{f}(\mathbf{u}) : \mathbb{R}^4 \mapsto \mathbb{R}^4$ is a nonlinear mapping.

Definition 18.1.1 (Weak Solutions). *A function* $(\mathbf{v}, p, \rho) \in L^\infty_{\rm loc}(\mathbb{R}_+ \times \mathbb{R}^2)$ *is called a weak solution of* (18.1.1) *in a domain* $\mathcal{D} \subset \mathbb{R}_+ \times \mathbb{R}^2$*, provided that*

$$\int_{\mathcal{D}} \Big(\mathbf{u}\, \partial_t \zeta + \mathbf{f}(\mathbf{u}) \cdot \nabla_{\mathbf{x}} \zeta\Big) d\mathbf{x}\, dt = 0 \tag{18.1.5}$$

for any $\zeta \in C_0^1(\mathcal{D})$.

As physically required, we focus on entropy solutions – the weak solutions satisfying the entropy condition. To motivate the entropy condition, we note that, for smooth solutions, system (18.1.1) implies the following conservation property of entropy:

$$\partial_t(\rho S) + \nabla_{\mathbf{x}} \cdot (\rho S \mathbf{v}) = 0; \tag{18.1.6}$$

that is, any smooth solution of (18.1.1) satisfies (18.1.6). Moreover, from (18.1.3), one can check that the function: $\mathbf{u} \mapsto -\rho S$ is convex. Then we define that a weak solution $(\mathbf{v}, \rho, p)$ satisfies the entropy condition if

$$\partial_t(\rho S) + \nabla_{\mathbf{x}} \cdot (\rho S \mathbf{v}) \geq 0 \tag{18.1.7}$$

in the distributional sense:

$$\int_{\mathcal{D}} \Big(\rho S\, \partial_t \zeta + \rho S \mathbf{v} \cdot \nabla_{\mathbf{x}} \zeta\Big) d\mathbf{x}\, dt \leq 0 \tag{18.1.8}$$

for any $\zeta \in C_0^1(\mathcal{D})$ with $\zeta \geq 0$.

In the study of a piecewise smooth weak solution of (18.1.2) with jump for $(\mathbf{v}, p, \rho)$ across an oriented surface $\mathcal{S}$ with unit normal $\mathbf{n} = (n_t, \mathbf{n_x}), \mathbf{n_x} = (n_1, n_2)$, in the $(t, \mathbf{x})$–coordinates, the requirement of the weak solution of (18.1.2) in the sense of Definition 18.1.1 implies the *Rankine-Hugoniot conditions* across $\mathcal{S}$:

$$\begin{aligned}
&[\rho]n_t + [\rho\mathbf{v}] \cdot \mathbf{n_x} = 0,\\
&[\rho\mathbf{v}]n_t + [\rho\mathbf{v} \otimes \mathbf{v}] \cdot \mathbf{n_x} + [p]\mathbf{n_x} = 0,\\
&[\frac{1}{2}\rho|\mathbf{v}|^2 + \rho e]n_t + [(\frac{1}{2}\rho|\mathbf{v}|^2 + \rho e + p)\mathbf{v}] \cdot \mathbf{n_x} = 0,
\end{aligned} \tag{18.1.9}$$

where $[w]$ denotes the jump of quantity w across the oriented surface $\mathcal{S}$. Conditions (18.1.9) can be rewritten/simplied in several ways (*cf.* [38, 236]).

Given any direction $\boldsymbol{\eta} \in \mathbb{S}^1$, *i.e.*, $|\boldsymbol{\eta}| = 1$, the characteristic equation of system (18.1.1) for eigenvalues λ in the direction $\boldsymbol{\eta}$ is

$$\rho^2(\lambda - \mathbf{v} \cdot \boldsymbol{\eta})^2\big((\lambda - \mathbf{v} \cdot \boldsymbol{\eta})^2 - c^2\big) = 0,$$

where $c = \sqrt{\frac{\gamma p}{\rho}}$ is the sonic speed. This implies that the four eigenvalues in the direction $\boldsymbol{\eta}$ are

$$\lambda_2 = \lambda_3 = \mathbf{v} \cdot \boldsymbol{\eta} \qquad \text{(repeated)}, \tag{18.1.10}$$

and

$$\lambda_j = \mathbf{v} \cdot \boldsymbol{\eta} + (-1)^j c, \qquad j = 1, 4. \tag{18.1.11}$$

Thus, system (18.1.1) is always hyperbolic in the unbounded domain $\{p > 0, \rho > 0\}$ in the phase space $(\mathbf{v}, p, \rho)$, but not strictly hyperbolic.

The right eigenvectors $\mathbf{r}_j$ corresponding to λ_j, $j = 1, 2, 3, 4$, are respectively:

$$\mathbf{r}_2 = (-\eta_2, \eta_1, 0, 0)^\top, \qquad \mathbf{r}_3 = (-\eta_2, \eta_1, 0, 1)^\top, \tag{18.1.12}$$

and

$$\mathbf{r}_j = (\boldsymbol{\eta}, (-1)^j \rho c, (-1)^j \rho c^{-1})^\top, \qquad j = 1, 4, \tag{18.1.13}$$

so that

$$\nabla_{(\mathbf{v},p,\rho)} \lambda_j \cdot \mathbf{r}_j \equiv 0, \qquad j = 2, 3, \tag{18.1.14}$$

and

$$\nabla_{(\mathbf{v},p,\rho)} \lambda_j \cdot \mathbf{r}_j = \frac{\gamma + 1}{2} \neq 0, \qquad j = 1, 4. \tag{18.1.15}$$

This means that the characteristic fields corresponding to $\lambda_j, j = 2, 3$, are always linearly degenerate (for which corresponding vortex sheets and entropy waves may be formed), while the characteristic fields corresponding to $\lambda_j, j = 1, 4$, are always genuinely nonlinear (for which corresponding shock waves and rarefaction waves may be formed). For more on this, see Dafermos [100], Glimm-Majda [139], and Lax [171]; also [35, 72, 181, 286].

For any fixed direction $\boldsymbol{\eta} \in \mathbb{S}^1$, since $\lambda_j = \mathbf{v} \cdot \boldsymbol{\eta}, j = 2, 3$, are repeated eigenvalues with two linearly independent right eigenvectors, there are two different types of discontinuity waves in the $(t, \mathbf{x})$–coordinates with a state $(\mathbf{v}_0, p_0, \rho_0)$ that is connected to possible other states $(\mathbf{v}, p, \rho)$ in the phase space given by

$$[\mathbf{v}] \cdot \boldsymbol{\eta} = 0, \tag{18.1.16}$$

$$[p] = 0, \tag{18.1.17}$$

forming a two-dimensional manifold in the phase space, where we have denoted that $[w] := w - w_0$ for any quantity w across the discontinuity wave from now on. Notice that the discontinuity waves are surfaces with co-dimension one in the $(t, \mathbf{x})$–coordinates across which the physical state $(\mathbf{v}, p, \rho)$ has a jump. This yields two fundamental waves:

Vortex sheets: Planar waves $\lambda_2 t - \boldsymbol{\eta} \cdot \mathbf{x} = 0$ in the $(t, \mathbf{x})$–coordinates connect two physical states $(\mathbf{v}_0, p_0, \rho_0)$ and $(\mathbf{v}, p, \rho)$ with

$$[\mathbf{v} \cdot \boldsymbol{\eta}^\perp] \neq 0, \quad \mathbf{v} \cdot \boldsymbol{\eta} = \mathbf{v}_0 \cdot \boldsymbol{\eta} = 0, \quad [p] = [S] = 0.$$

Entropy waves: Planar waves $\lambda_3 t - \boldsymbol{\eta} \cdot \mathbf{x} = 0$ connect two states $(\mathbf{v}_0, p_0, \rho_0)$ and $(\mathbf{v}, p, \rho)$ with

$$[S] \neq 0, \quad [\mathbf{v}] = 0, \quad [p] = 0.$$

The other nonlinear waves are rarefaction waves and shock waves.

Rarefaction waves: Given $\boldsymbol{\eta} \in \mathbb{S}^1$ and a state $(\mathbf{v}_0, p_0, \rho_0)$, any state $(\mathbf{v}, p, \rho) \in \mathcal{R}_j(\mathbf{v}_0, p_0, \rho_0; \boldsymbol{\eta}), j = 1, 4$, of a plane rarefaction wave connecting to state $(\mathbf{v}_0, p_0, \rho_0)$ can be expressed by

$$\mathcal{R}_j(\mathbf{v}_0, p_0, \rho_0; \boldsymbol{\eta}) : \begin{cases} [S] = 0, \\ [\mathbf{v}] = (-1)^j \frac{2c_0}{\gamma-1}\big((\frac{\rho}{\rho_0})^{\frac{\gamma-1}{2}} - 1\big)\boldsymbol{\eta}, \end{cases} \qquad j = 1, 4. \tag{18.1.18}$$

Eliminating $\boldsymbol{\eta}$, all possible states $(\mathbf{v}, p, \rho)$ that connect to $(\mathbf{v}_0, p_0, \rho_0)$ form a wave-cone $\mathcal{R}_j(\mathbf{v}_0, p_0, \rho_0)$ in the phase space as follows:

$$\mathcal{R}_j(\mathbf{v}_0, p_0, \rho_0) : \begin{cases} (\mathbf{v} - \mathbf{v}_0)^2 = (\frac{2}{\gamma-1})^2(c - c_0)^2, \\ p\rho^{-\gamma} = p_0\rho_0^{-\gamma}, \end{cases} \qquad j = 1, 4. \tag{18.1.19}$$

It is easy to check that, along $\mathcal{R}_j(\mathbf{v}_0, p_0, \rho_0; \boldsymbol{\eta})$,

$$(-1)^j \frac{d\lambda_j}{d\rho} > 0 \qquad \text{for } j = 1, 4.$$

This shows that $(\mathbf{v}, p, \rho)$ can be connected to $(\mathbf{v}_0, p_0, \rho_0)$ by a first or fourth family of planar centered rarefaction wave if and only if

$$(\mathbf{v}, p, \rho) \in \mathcal{R}_1(\mathbf{v}_0, p_0, \rho_0), \qquad \rho < \rho_0, \tag{18.1.20}$$

or

$$(\mathbf{v}, p, \rho) \in \mathcal{R}_4(\mathbf{v}_0, p_0, \rho_0), \qquad \rho > \rho_0. \tag{18.1.21}$$

Shock waves: Given a state $(\mathbf{v}_0, p_0, \rho_0)$, a planar shock-front $\sigma_j t = \boldsymbol{\eta} \cdot \mathbf{x}$ for $j = 1, 4$, in the $(t, \mathbf{x})$–coordinates connecting a physical state $(\mathbf{v}, p, \rho)$ to state $(\mathbf{v}_0, p_0, \rho_0)$ requires that $(\mathbf{v}, p, \rho) \in \mathcal{S}_j(\mathbf{v}_0, p_0, \rho_0; \boldsymbol{\eta})$ be determined by the Rankine-Hugoniot conditions of the form:

$$\mathcal{S}_j(\mathbf{v}_0, p_0, \rho_0; \boldsymbol{\eta}) : \begin{cases} \frac{p}{p_0} = \frac{\rho - \mu^2\rho_0}{\rho_0 - \mu^2\rho}, \\ \mathbf{v} - \mathbf{v}_0 = (-1)^j(\rho - \rho_0)\sqrt{\frac{1}{\rho_0\rho}\frac{p-p_0}{\rho-\rho_0}}\,\boldsymbol{\eta} \end{cases} \tag{18.1.22}$$

with the shock speed:

$$\sigma_j = \mathbf{v}_0 \cdot \boldsymbol{\eta} + (-1)^j\sqrt{\frac{\rho}{\rho_0}\frac{p - p_0}{\rho - \rho_0}}, \qquad j = 1, 4, \tag{18.1.23}$$

where $\mu^2 = \frac{\gamma-1}{\gamma+1}$. Then state $(\mathbf{v}, p, \rho)$ must be on the circular cone:

$$\mathcal{S}_j(\mathbf{v}_0, p_0, \rho_0) : \begin{cases} (\mathbf{v} - \mathbf{v}_0)^2 = \frac{1}{\rho\rho_0}(p - p_0)(\rho - \rho_0), \\ \frac{p}{p_0} = \frac{\rho - \mu^2\rho_0}{\rho_0 - \mu^2\rho}. \end{cases} \tag{18.1.24}$$

It can be checked that any state $(\mathbf{v}, p, \rho) \in \mathcal{S}_j(\mathbf{v}_0, p_0, \rho_0)$ satisfies the Lax geometric entropy inequalities:

$$\lambda_j(\mathbf{v}, p, \rho; \boldsymbol{\eta}) < \sigma_j(\rho; \mathbf{v}_0, p_0, \rho_0, \boldsymbol{\eta}) < \lambda_j(\mathbf{v}_0, p_0, \rho_0; \boldsymbol{\eta})$$

if and only if

$$(-1)^j(\rho - \rho_0) < 0, \qquad j = 1, 4, \tag{18.1.25}$$

which is the entropy condition for shock waves $\mathcal{S}_j(\mathbf{v}_0, p_0, \rho_0), j = 1, 4$. That is, $(\mathbf{v}, p, \rho)$ can be connected to $(\mathbf{v}_0, p_0, \rho_0)$ by a planar shock-front of the first or fourth field if and only if

$$(\mathbf{v}, p, \rho) \in \mathcal{S}_1(\mathbf{v}_0, p_0, \rho_0), \qquad \rho > \rho_0, \tag{18.1.26}$$

or

$$(\mathbf{v}, p, \rho) \in \mathcal{S}_4(\mathbf{v}_0, p_0, \rho_0), \qquad \rho < \rho_0. \tag{18.1.27}$$

18.2 MATHEMATICAL FORMULATION I: INITIAL-BOUNDARY VALUE PROBLEM

Using (18.1.22)–(18.1.25), we can formulate the problem of shock reflection-diffraction by a wedge for the full Euler equations (18.1.1) in $\mathbb{R}^3_+$ in the following way:

Problem 18.1 (Initial-Boundary Value Problem). *Seek a solution of system* (18.1.1) *satisfying the initial condition at* $t = 0$:

$$(\mathbf{v}, p, \rho) = \begin{cases} (0, 0, p_0, \rho_0), & |x_2| > x_1 \tan\theta_{\mathrm{w}}, x_1 > 0, \\ (u_1, 0, p_1, \rho_1), & x_1 < 0, \end{cases} \tag{18.2.1}$$

and the slip boundary condition along the wedge boundary:

$$\mathbf{v} \cdot \boldsymbol{\nu} = 0, \tag{18.2.2}$$

where $\boldsymbol{\nu}$ *is the outward normal to the wedge boundary, and states* (0) *and* (1) *satisfy*

$$u_1 = \sqrt{\frac{(p_1 - p_0)(\rho_1 - \rho_0)}{\rho_0\rho_1}}, \quad \frac{p_1}{p_0} = \frac{\rho_1 - \mu^2\rho_0}{\rho_0 - \mu^2\rho_1}, \qquad \rho_1 > \rho_0. \tag{18.2.3}$$

That is, given (ρ_0, p_0, ρ_1) *and* $\gamma > 1$, *the other variables* (u_1, p_1) *are uniquely determined by* (18.2.3). *In particular, the incident-shock Mach number* $\mathcal{M}_I := \frac{u_1}{c_1}$ *is determined by*

$$\mathcal{M}_I^2 = \frac{2(n_I - 1)^2}{(\gamma + 1)\big(n_I - \mu^2\big)}, \tag{18.2.4}$$

where $c_1 = \sqrt{\frac{\gamma p_1}{\rho_1}}$ *is the sonic speed of state* (1), *and* $n_I = \frac{\rho_1}{\rho_0}$, *which means that the incident-shock strength can be represented by either the incident-shock Mach number* $\mathcal{M}_I$ *or the ratio* n_I *of the densities of the incident shock; see* Fig. 18.1.

This problem is a two-dimensional lateral Riemann problem with physical Riemann data.

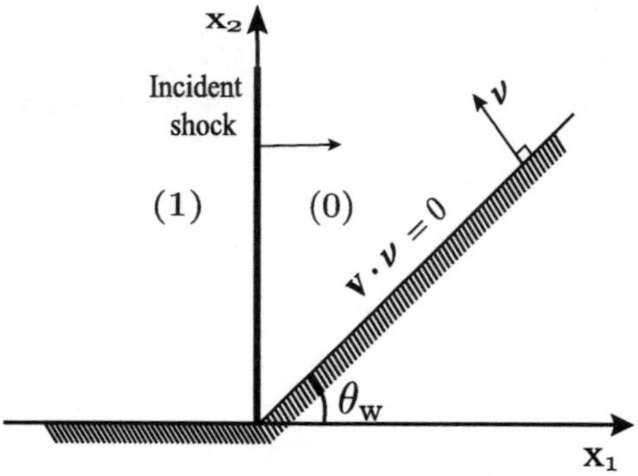

Figure 18.1: Initial-boundary value problem

18.3 MATHEMATICAL FORMULATION II: BOUNDARY VALUE PROBLEM

Notice that the initial-boundary value problem (**Problem 18.1**) is invariant under the self-similar scaling:

$$(t, \mathbf{x}) \longmapsto (\alpha t, \alpha \mathbf{x}) \qquad \text{for any } \alpha \neq 0.$$

Therefore, we seek self-similar solutions:

$$(\mathbf{v}, p, \rho)(t, \mathbf{x}) = (\mathbf{v}, p, \rho)(\boldsymbol{\xi}), \qquad \boldsymbol{\xi} = \frac{\mathbf{x}}{t}.$$

Then the self-similar solutions are governed by the following system:

$$\begin{cases} \operatorname{div}(\rho \mathbf{U}) + 2\rho = 0, \\ \operatorname{div}(\rho \mathbf{U} \otimes \mathbf{U}) + Dp + 3\rho \mathbf{U} = 0, \\ \operatorname{div}\big((\frac{1}{2}\rho|\mathbf{U}|^2 + \frac{\gamma p}{\gamma - 1})\mathbf{U}\big) + 2(\frac{1}{2}\rho|\mathbf{U}|^2 + \frac{\gamma p}{\gamma - 1}) = 0, \end{cases} \tag{18.3.1}$$

where $\mathbf{U} = \mathbf{v} - \boldsymbol{\xi} = (U, V)$ is the pseudo-velocity and $D = (\partial_{\xi_1}, \partial_{\xi_2})$ is the gradient with respect to the self-similar variables $\boldsymbol{\xi} = (\xi_1, \xi_2)$.

Similarly, system (18.3.1) can be written as the following general form as a system of balance laws:

$$\operatorname{div}\mathbf{A}(\mathbf{w}) + \mathbf{B}(\mathbf{w}) = 0, \tag{18.3.2}$$

where $\mathbf{w} = (\mathbf{U}, p, \rho)$ is unknown, while $\mathbf{A} : \mathbb{R}^4 \mapsto (\mathbb{R}^4)^2$ and $\mathbf{B} : \mathbb{R}^4 \mapsto \mathbb{R}^4$ are given nonlinear mappings.

The eigenvalues of system (18.3.1) are

$$\lambda_2 = \lambda_3 = \frac{V}{U} \text{ (repeated)}, \qquad \lambda_j = \frac{UV + (-1)^j c\sqrt{|\mathbf{U}|^2 - c^2}}{U^2 - c^2}, \quad j = 1, 4,$$

where $c = \sqrt{\frac{\gamma p}{\rho}}$ is the sonic speed.

When the flow is pseudo-subsonic, *i.e.*, $|\mathbf{U}| < c$, eigenvalues $\lambda_j, j = 1, 4$, become complex so that the system consists of two transport-type equations and two nonlinear equations of mixed hyperbolic-elliptic type. Therefore, system (18.3.1) is, in general, of *composite-mixed hyperbolic-elliptic type.*

Definition 18.3.1 (Weak Solutions). *A function $\mathbf{w} \in L^\infty_{\mathrm{loc}}(\Omega)$ for an open region $\Omega \subset \mathbb{R}^2$ is a weak solution of* (18.3.1) *in Ω, provided that*

$$\int_\Omega \Big(\mathbf{A}(\mathbf{w}(\boldsymbol{\xi})) \cdot \nabla\zeta(\boldsymbol{\xi}) + \mathbf{B}(\mathbf{w}(\boldsymbol{\xi}))\,\zeta(\boldsymbol{\xi})\Big) d\boldsymbol{\xi} = 0 \tag{18.3.3}$$

for any $\zeta \in C^1_0(\Omega)$.

As before, we focus on entropy solutions – the weak solutions satisfying the entropy condition. Rewriting condition (18.1.7) and its weak form (18.1.8) in the self-similar variables $\boldsymbol{\xi}$ in terms of functions $(\mathbf{U}, \rho, p)$, we arrive at the following definition: A weak self-similar solution $(\mathbf{U}, \rho, p)$ of (18.3.1) satisfies the entropy condition if

$$\mathrm{div}(\rho S \mathbf{U}) + 2\rho S \geq 0 \tag{18.3.4}$$

in a weak sense; that is,

$$\int_\Omega \Big(\rho S \mathbf{U} \cdot \nabla\zeta - 2\rho S \zeta\Big) d\boldsymbol{\xi} \leq 0 \tag{18.3.5}$$

for any $\zeta \in C^1_0(\Omega)$ with $\zeta \geq 0$. We note that, if a weak self-similar solution $(\mathbf{U}, \rho, p)$ of (18.3.1) satisfies the entropy condition (18.3.5), then the corresponding solution $(\mathbf{v}, \rho, p)$ of (18.1.1) in the $(t, \mathbf{x})$–coordinates satisfies the entropy condition (18.1.8).

In the study of a piecewise smooth weak solutions of (18.3.1) with jump for $(\mathbf{U}, p, \rho)$ across an oriented surface $\mathcal{S}$ with unit normal $\boldsymbol{\nu} = (\nu_1, \nu_2)$, in the $\boldsymbol{\xi}$–coordinates, the requirement of the weak solution of (18.3.1) in the sense of (18.3.3) implies the *Rankine-Hugoniot conditions* across $\mathcal{S}$:

$$\begin{aligned}
&[\rho \mathbf{U}] \cdot \boldsymbol{\nu} = 0,\\
&[\rho \mathbf{U} \otimes \mathbf{U}] \cdot \boldsymbol{\nu} + [p]\boldsymbol{\nu} = 0,\\
&[(\frac{1}{2}\rho|\mathbf{U}|^2 + \frac{\gamma p}{\gamma - 1})\mathbf{U}] \cdot \boldsymbol{\nu} = 0.
\end{aligned} \tag{18.3.6}$$

Note that the first line in (18.3.6) implies that $\rho\mathbf{U}\cdot\boldsymbol{\nu}$ is well-defined on $\mathcal{S}$ by taking a limit to $\mathcal{S}$ from either side. Then (18.3.6) can be simplified to the following:

$$\begin{aligned}
&[\rho\mathbf{U}]\cdot\boldsymbol{\nu}=0,\\
&(\rho\mathbf{U}\cdot\boldsymbol{\nu})[\mathbf{U}\cdot\boldsymbol{\nu}]+[p]=0,\\
&(\rho\mathbf{U}\cdot\boldsymbol{\nu})[\mathbf{U}\cdot\boldsymbol{\tau}]=0,\\
&(\rho\mathbf{U}\cdot\boldsymbol{\nu})[\frac{1}{2}\rho|\mathbf{U}|^2+\frac{\gamma p}{\gamma-1}]=0,
\end{aligned}\tag{18.3.7}$$

where $\boldsymbol{\tau}$ is the unit tangent vector on $\mathcal{S}$.

Then, for a given state $(\mathbf{U}_0,p_0,\rho_0)$, the states of the discontinuity waves connecting to this state in the self-similar coordinates are the following:

Vortex sheets:

$$[p]=0,\quad \mathbf{U}\cdot\boldsymbol{\nu}=\mathbf{U}_0\cdot\boldsymbol{\nu}=0,\quad [\mathbf{U}\cdot\boldsymbol{\tau}]\neq 0,\quad [S]=0$$

with speed $\sigma_2=\frac{V}{U}=\frac{V_0}{U_0}$, where $\boldsymbol{\nu}$ and $\boldsymbol{\tau}$ are the unit normal and tangent vectors on the waves.

Entropy waves:

$$[p]=0,\quad [\mathbf{U}]=0,\quad [S]\neq 0$$

with the same speed $\sigma_3=\sigma_2$.

Shock waves $\mathcal{S}_j(\mathbf{U}_0,p_0,\rho_0), j=1,4$, are the curves determined by

$$\begin{cases}
\frac{[U]}{[V]}=-\sigma_j,\\
[p]=\frac{2\rho_0c_0^2}{(\gamma+1)(\rho_0-\mu^2\rho_1)}[\rho],\\
[V]=\frac{2c_0^2}{(\gamma+1)(\rho_0-\mu^2\rho_1)(U_0\sigma_j-V_0)}[\rho]
\end{cases}\tag{18.3.8}$$

with the shock speeds:

$$\sigma_j=\frac{U_0V_0+(-1)^j\sqrt{\bar{c}^2(|\mathbf{U}_0|^2-\bar{c}^2)}}{U_0^2-\bar{c}^2}\tag{18.3.9}$$

for $\bar{c}^2=\frac{\rho}{\rho_0}\frac{[p]}{[\rho]}$.

The Rankine-Hugoniot conditions may be rewritten in a different form. For a shock connecting two states $(\mathbf{U}_0,p_0,\rho_0)$ and $(\mathbf{U},p,\rho)$, let L and N be the tangent and normal components of pseudo-velocity $\mathbf{U}$ to the discontinuity wave with $|\mathbf{U}|^2=L^2+N^2$. Then the Rankine-Hugoniot conditions are equivalent to

$$[\rho N]=0,\tag{18.3.10}$$

$$[\rho LN]=0,\tag{18.3.11}$$

$$[\rho N^2+p]=0,\tag{18.3.12}$$

$$[\rho N(h+\frac{N^2+L^2}{2})]=0,\tag{18.3.13}$$

where $h = \frac{\gamma p}{(\gamma-1)\rho}$.

If $N_0 = 0$, (18.3.10) implies that $N = 0$, so that (18.3.11) and (18.3.13) are satisfied, and (18.3.12) is equivalent to $[p] = 0$. This discontinuity wave corresponds to either a vortex sheet or an entropy wave.

If $N_0 \neq 0$, then $N \neq 0$ from (18.3.10) so that (18.3.11) and (18.3.13) become

$$[L] = 0, \tag{18.3.14}$$

$$[h + \frac{N^2}{2}] = 0. \tag{18.3.15}$$

In this case, when the pseudo-flow passes through the wave, the tangential pseudo-velocity L is continuous, while the normal pseudo-velocity N experiences a jump.

If the pseudo-flow passes the wave from the front with state $(\mathbf{U}_0, p_0, \rho_0)$ to the back with state $(\mathbf{U}, p, \rho)$, the entropy condition for a shock is

$$\rho_0 < \rho. \tag{18.3.16}$$

From (18.3.10)–(18.3.12), it follows that

$$N_0^2 = \frac{\rho}{\rho_0} \frac{p - p_0}{\rho - \rho_0}, \tag{18.3.17}$$

$$N^2 = \frac{\rho_0}{\rho} \frac{p - p_0}{\rho - \rho_0}. \tag{18.3.18}$$

Substituting these into (18.3.15), we obtain

$$m := \frac{p}{p_0} = \frac{n - \mu^2}{1 - \mu^2 n}, \tag{18.3.19}$$

where

$$n := \frac{\rho}{\rho_0} > 1, \qquad \mu^2 := \frac{\gamma - 1}{\gamma + 1}.$$

Then (18.3.17)–(18.3.18) imply that

$$\frac{N_0^2}{c_0^2} = \frac{n}{\gamma} \frac{m - 1}{n - 1}, \qquad \frac{N^2}{c^2} = \frac{1}{\gamma n} \frac{m - 1}{n - 1},$$

that is,

$$\frac{|\mathbf{U}_0|^2}{c_0^2} \cos^2 \tau = \frac{n}{\gamma} \frac{m - 1}{n - 1}, \qquad \frac{|\mathbf{U}|^2}{c^2} \cos^2(\tau + \delta) = \frac{1}{\gamma n} \frac{m - 1}{n - 1}, \tag{18.3.20}$$

where τ is the angle from the normal of the shock-front to velocity $\mathbf{U}_0$ in front of the wave, and δ is the angle from velocity $\mathbf{U}_0$ to $\mathbf{U}$ behind the wave; see Fig. 18.2.

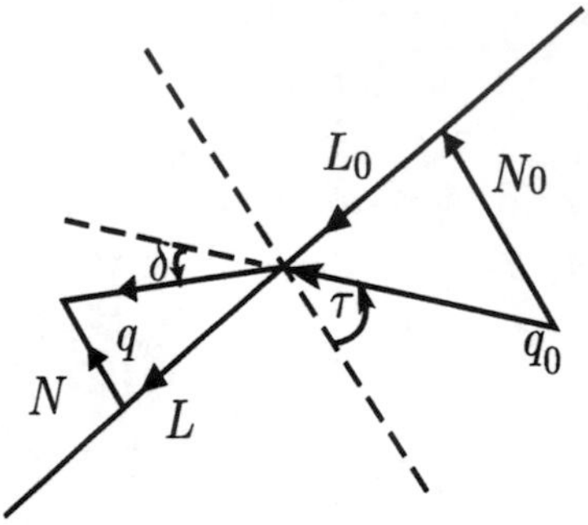

Figure 18.2: The pseudo-flow passes the shock from the front to the back

Let

$$z = \tan\tau. \tag{18.3.21}$$

Then

$$\cos^2\tau = \frac{1}{1+z^2},$$

hence

$$\tan(\tau+\delta) = \frac{L}{N} = \frac{nL_0}{N_0} = nz,$$

which implies

$$\cos^2(\tau+\delta) = \frac{1}{1+n^2z^2}. \tag{18.3.22}$$

Substituting (18.3.21)–(18.3.22) into (18.3.20), we have

$$M_0^2 = \frac{2(1+z^2)n}{(\gamma+1)(1-\mu^2 n)}, \tag{18.3.23}$$

$$M^2 = \frac{2(1+z^2n^2)}{(\gamma+1)(n-\mu^2)}, \tag{18.3.24}$$

where M_0 and M are the Mach numbers of the pseudo-flow in front of and behind the shock, respectively.

As in the potential flow case, the problem is symmetric with respect to the ξ_1–axis. Thus, it suffices to consider the problem in half-plane $\xi_2 > 0$ outside the following half-wedge:

$$\Lambda := \{\xi_1 < 0, \xi_2 > 0\} \cup \{\xi_2 > \xi_1 \tan\theta_{\rm w},\ \xi_1 > 0\}.$$

Then the initial-boundary value problem (**Problem 18.1**) in the $(t, \mathbf{x})$–coordinates can be formulated as the following boundary value problem in the self-similar coordinates $\boldsymbol{\xi}$:

Problem 18.2 (Boundary Value Problem). *Seek a solution to system* (18.3.1) *satisfying the slip boundary condition on the wedge boundary and the matching condition on the symmetry line* $\xi_2 = 0$:

$$\mathbf{U}\cdot\boldsymbol{\nu} = 0 \qquad \text{on } \partial\Lambda = \{\xi_1 \le 0, \xi_2 = 0\} \cup \{\xi_1 > 0, \xi_2 \ge \xi_1 \tan\theta_{\mathrm{w}}\},$$

and the asymptotic boundary condition as $|\boldsymbol{\xi}| \to \infty$:

$$(\mathbf{U}+\boldsymbol{\xi}, p, \rho) \longmapsto \begin{cases} (0,0,p_0,\rho_0), & \xi_1 > \xi_1^0, \xi_2 > \xi_1 \tan\theta_{\mathrm{w}}, \\ (u_1, 0, p_1, \rho_1), & \xi_1 < \xi_1^0, \xi_2 > 0, \end{cases}$$

where

$$\xi_1^0 = u_1 + \sqrt{\frac{\rho_0(p_1 - p_0)}{\rho_1(\rho_1 - \rho_0)}} = \sqrt{\frac{\rho_1(p_1 - p_0)}{\rho_0(\rho_1 - \rho_0)}}. \tag{18.3.25}$$

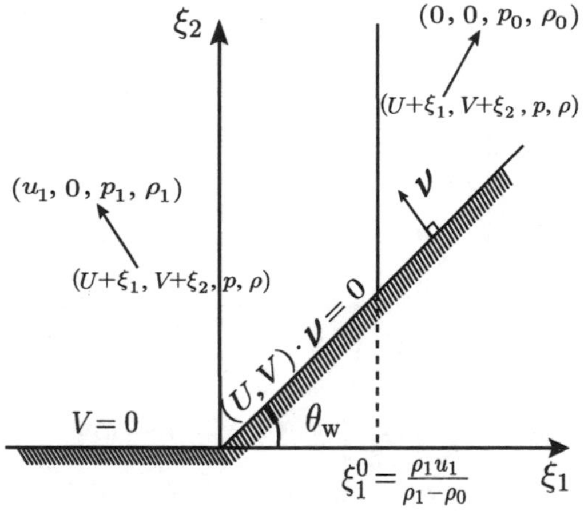

Figure 18.3: Boundary value problem in the unbounded domain Λ

Two of the main features of this problem are the wedge corner along the solid boundary and the jump of the asymptotic boundary data, which are not conventional for the unbounded boundary value problems. It is expected that the solutions of **Problem 18.2** contain all possible patterns of shock reflection-diffraction configurations as observed in physical and numerical experiments; *cf.* [12, 99, 139, 143, 166, 205, 206, 243, 259] and the references cited therein.

Observe that, for any solution with bounded physical variables $(\mathbf{u}, p, \rho)$, system (18.3.1) must be hyperbolic when $|\boldsymbol{\xi}|$ is large enough so that $|\mathbf{U}| > c$, that is, system (18.3.1) is always hyperbolic in the far field and becomes composite-mixed elliptic-hyperbolic around the wedge corner in the $\boldsymbol{\xi}$–coordinates.

18.4 NORMAL REFLECTION

The simplest case of **Problem 18.2** is when the wedge angle $\theta_{\rm w} = \frac{\pi}{2}$. In this case, **Problem 18.2** simply becomes the normal reflection problem, for which the incident shock normally reflects, and the reflected-diffracted shock becomes a plane. It can be shown that there exist a *unique* state $(\bar{p}_2, \bar{\rho}_2)$ with $\bar{\rho}_2 > \rho_1$ and a *unique* location of the reflected shock:

$$\bar{\xi}_1 = -\frac{\rho_1 u_1}{\bar{\rho}_2 - \rho_1} \qquad \text{with} \quad u_1 = \sqrt{\frac{(\bar{p}_2 - p_1)(\bar{\rho}_2 - \rho_1)}{\rho_1 \bar{\rho}_2}} \tag{18.4.1}$$

such that state (2) $= (-\boldsymbol{\xi}, \bar{p}_2, \bar{\rho}_2)$ is subsonic inside the sonic circle with its center at the origin and radius $c_2 = \sqrt{\frac{\gamma \bar{p}_2}{\bar{\rho}_2}}$, and supersonic outside the sonic circle (see Fig. 18.4).

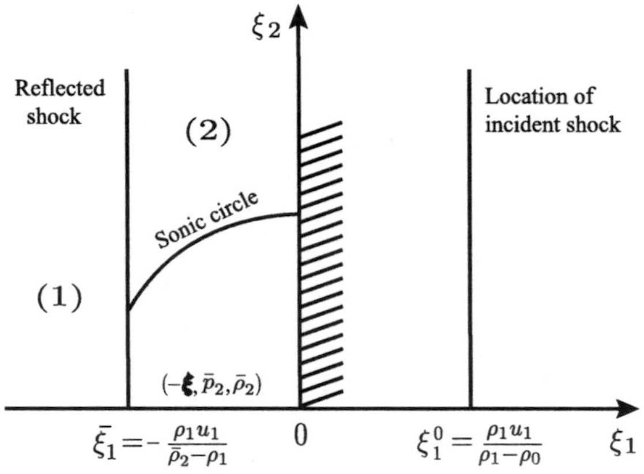

Figure 18.4: The normal reflection solution

In this case,

$$\mathcal{M}_I^2 = \frac{2(n_2 - 1)^2}{1 - \mu^2 n_2}, \qquad \frac{\bar{p}_2}{p_1} = \frac{n_2 - \mu^2}{1 - \mu^2 n_2}, \tag{18.4.2}$$

and $n_2 = \frac{\bar{\rho}_2}{\rho_1} > 1$ is the unique root of

$$\big(2c_1^2 + (\gamma - 1)u_1^2\big)n_2^2 - \big(4c_1^2 + (\gamma + 1)u_1^2\big)n_2 + 2c_1^2 = 0,$$

that is,

$$n_2 = \frac{4c_1^2 + (\gamma + 1)u_1^2 + u_1\sqrt{16c_1^2 + (\gamma + 1)^2 u_1^2}}{2\big(2c_1^2 + (\gamma - 1)u_1^2\big)}, \tag{18.4.3}$$

where $c_1 = \sqrt{\frac{\gamma p_1}{\rho_1}}$ is the sonic speed of state (1).

In other words, given $\mathcal{M}_I, p_0, \rho_0$, and $\gamma > 1$, state (2) $= (-\boldsymbol{\xi}, \bar{p}_2, \bar{\rho}_2)$ is uniquely determined through (18.2.3)–(18.2.4) and (18.4.1)–(18.4.3).

18.5 LOCAL THEORY FOR REGULAR REFLECTION NEAR THE REFLECTION POINT

The necessary condition for regular reflection-diffraction configurations to exist is the existence of two-shock configurations (one is the incident shock and the other is a reflected shock) formed locally around point P_0 (see Fig. 18.5).

Theorem 18.5.1 (Local Theory). *There exist a unique detachment angle $\theta_{\rm w}^{\rm d} = \theta_{\rm w}^{\rm d}(M_I, \gamma) \in (0, \frac{\pi}{2})$ and a unique sonic angle $\theta_{\rm w}^{\rm s} = \theta_{\rm w}^{\rm s}(\mathcal{M}_I, \gamma) \in (0, \frac{\pi}{2})$ with $\theta_{\rm w}^{\rm d} < \theta_{\rm w}^{\rm s}$ such that*

(i) *There are two states (2), $(\mathbf{U}_2^a, p_2^a, \rho_2^a)$ and $(\mathbf{U}_2^b, p_2^b, \rho_2^b)$, such that*

$$|\mathbf{U}_2^a| > |\mathbf{U}_2^b| \quad \textit{and} \quad |\mathbf{U}_2^b| < c_2^b$$

if and only if $\theta_{\rm w} \in (\theta_{\rm w}^{\rm d}, \frac{\pi}{2})$;

(ii) *In particular,*

$$|\mathbf{U}_2^a| > c_2^a$$

if and only if the wedge angle $\theta_{\rm w} \in (\theta_{\rm w}^{\rm s}, \frac{\pi}{2})$,

where $c_2^a = \sqrt{\frac{\gamma p_2^a}{\rho_2^a}}$ and $c_2^b = \sqrt{\frac{\gamma p_2^b}{\rho_2^b}}$ are the sonic speeds, and $\mathcal{M}_I = \frac{u_1}{c_1}$ is the incident-shock Mach number that can be expressed by $n_I = \frac{\rho_1}{\rho_0}$ and $\gamma > 1$ through (18.2.4).

Theorem 18.5.1 directly follows from Theorems 18.5.2–18.5.3 below. As indicated in Theorem 18.5.1, the detachment angle $\theta_{\rm w}^{\rm d}$ and the sonic angle $\theta_{\rm w}^{\rm s}$ can be expressed explicitly in terms of $n_I = \frac{\rho_1}{\rho_0}$ and $\gamma > 1$, equivalently, the incident-shock Mach number $\mathcal{M}_I = \frac{u_1}{c_1}$ and $\gamma > 1$, as indicated in Theorems 18.5.2–18.5.3.

Theorem 18.5.2. *The necessary and sufficient condition for the existence of the two-shock configuration at P_0 is that the wedge angle $\theta_{\rm w}$ is larger than the detachment angle $\theta_{\rm w}^{\rm d}$ determined by*

$$\theta_{\rm w}^{\rm d} = \frac{\pi}{2} - \operatorname{arccot}\Big(\frac{1}{n_I}\sqrt{\frac{2}{3}\sqrt{b^2+3c}\cos(\frac{\beta}{3}) + \frac{b}{3} - 1}\Big), \tag{18.5.1}$$

where

$$n_I = \frac{\rho_1}{\rho_0}, \qquad a = \frac{2}{1-\mu^2}(n_I - 1)(n_I - \mu^2), \qquad b = 1 + \frac{2\mu^2}{1-\mu^2}a,$$
$$d = (\frac{1}{n-1} + \frac{\mu^2}{n-\mu^2})a^2, \qquad \beta = \arccos\Big(\frac{27a^2 + 2b^2 + 9bd}{2(b^2+3d)^{1/3}}\Big) \in (0, \frac{\pi}{2}).$$

Proof. We divide the proof into four steps.

1. From (18.3.8), the two states (u_0, v_0, p_0, ρ_0) and (u_1, v_1, p_1, ρ_1) can be connected by the incident shock I located at $\xi_1 = \xi_1^0$ in the self-similar plane $\boldsymbol{\xi} = (\xi_1, \xi_2)$:

$$\mathcal{S}_0 : \begin{cases} \xi_1^0 = u_1 + \sqrt{\frac{\rho_0(p_1-p_0)}{\rho_1(\rho_1-\rho_0)}} = \sqrt{\frac{\rho_1(p_1-p_0)}{\rho_0(\rho_1-\rho_0)}}, \\ \frac{u_1}{\rho_1-\rho_0} = \sqrt{\frac{1}{\rho_0\rho_1}\frac{p_1-p_0}{\rho_1-\rho_0}}, \\ \frac{p_1}{p_0} = \frac{\rho_1-\mu^2\rho_0}{\rho_0-\mu^2\rho_1}, \\ p_1 > p_0 \Longleftrightarrow \rho_1 > \rho_0 \Longleftrightarrow u_1 > 0, \end{cases} \tag{18.5.2}$$

where $\mu^2 = \frac{\gamma-1}{\gamma+1}$.

The necessary condition for the existence of the regular reflection-diffraction configuration is that $\mathcal{S}_0$ intersects with the wedge boundary at point P_0 and produces a reflected shock $\mathcal{S}_1$.

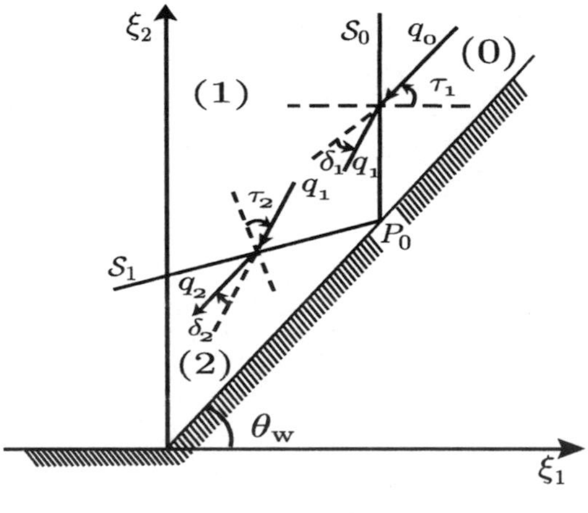

Figure 18.5: Two-shock reflection configuration

2. In Fig. 18.5, τ_i denotes the angle between the normal of the shock-front and the velocity on the shock, and δ_i is the angle of deflection of the flow across the shock, where $i = 1$ and 2 correspond to the incident and reflected shock, respectively.

For convenience, we denote

$$m_i = \frac{p_i}{p_{i-1}} > 0, \quad n_i = \frac{\rho_i}{\rho_{i-1}}, \quad z_i = \tan\tau_i, \quad M_i = \frac{|\mathbf{U}_i|}{c_i}, \qquad i = 1, 2,$$

where $\tau_1 = \theta_{\mathrm{w}}$. Then we have

$$\begin{cases} m_i = \frac{n_i - \mu^2}{1-\mu^2 n_i}, \\ \tan\delta_i = \frac{(n_i-1)z_i}{n_i z_i^2+1}, \\ M_{i-1}^2 = \frac{(1-\mu^2)(z_i^2+1)n_i}{1-\mu^2 n_i}, \\ M_i^2 = \frac{(1-\mu^2)(z_i^2 n_i^2+1)}{n_i-\mu^2}, \\ 1 < n_i < \frac{1}{\mu^2}, \quad 0 < z_1 < \infty. \end{cases} \tag{18.5.3}$$

From the boundary condition, we find that $\delta_1 + \delta_2 = 0$, that is,

$$\frac{\tan\delta_1 + \tan\delta_2}{1 - \tan\delta_1 \tan\delta_2} = 0. \tag{18.5.4}$$

Denoting $(n, z) := (n_1, z_1)$ as independent parameters, we now examine the case that there exists a solution satisfying

$$1 < n_2 < \frac{1}{\mu^2} \tag{18.5.5}$$

in domain $\{(n,z) \,:\, 1 < n < \frac{1}{\mu^2}, 0 < z < \infty\}$.

From (18.5.3)–(18.5.4), we have

$$n_2 = \frac{n^2 z^2 + 1}{n(z_2^2+1) + \mu^2(n^2 z^2 - z_2^2)}. \tag{18.5.6}$$

Substituting (18.5.6) into (18.5.4), we have

$$(z_2 - nz)\big((n-\mu^2)(nz^2+1)z_2^2 + (1-\mu^2)z(n^2z^2+1)z_2 + (n-1)(\mu^2 nz^2+1)\big) = 0. \tag{18.5.7}$$

Then we conclude that either

$$z_2 = nz \tag{18.5.8}$$

or

$$z_2 = \frac{-(1-\mu^2)z(n^2z^2+1) \pm \sqrt{\Theta(n,z^2)}}{2(n-\mu^2)(nz^2+1)}, \tag{18.5.9}$$

where

$$\Theta(n,z^2) = (1-\mu^2)^2 z^2 (n^2z^2+1)^2 - 4(n-1)(n-\mu^2)(nz^2+1)(\mu^2 nz^2+1). \tag{18.5.10}$$

Substituting (18.5.8) into (18.5.6), we have

$$n_2 = \frac{1}{n} < 1,$$

which is not the solution we want, since it does not satisfy the entropy condition.

On the other hand, (18.5.9) has provided two real solutions of z_2. Now we examine the issue of which one is a physical solution. This requires the physical stability: State (2), *i.e.*, $(\mathbf{u}_2, p_2, \rho_2)$, should tend to the constant state $(\mathbf{u}_0, p_0, \rho_0)$ when $n \to 1$, and shocks I and R through $(\xi_1^0, \xi_1^0 \tan\theta_{\mathrm{w}})$ should tend to two characteristic lines. Since $\tau_1 = -\tau_2$, then

$$z_2(n,z)|_{n=1} = -z < 0.$$

Substituting $n = 1$ into (18.5.9), we have

$$z_2|_{n=1} = \frac{-z(z^2+1) \pm z(z^2+1)}{2(z^2+1)} = \begin{cases} 0 & \text{for `` + '' sign,} \\ -z & \text{for `` − '' sign.} \end{cases}$$

Based on this, we must take z_2 with the " − " sign in (18.5.9), that is,

$$z_2 = \frac{-(1-\mu^2)z(n^2z^2+1) - \sqrt{\Theta(n,z^2)}}{2(n-\mu^2)(nz^2+1)}. \tag{18.5.11}$$

Then the necessary and sufficient condition for the existence of a real root z_2 is

$$\Theta(n, z^2) \geq 0. \tag{18.5.12}$$

3. We now check (18.5.5) under the assumption that (18.5.12) holds. Substituting (18.5.6) into (18.5.5), it requires that

$$\mu^2(n-\mu^2)(z_2^2+1)+\mu^4(n^2z^2+1) < \mu^2(n^2z^2+1) < \mu^2(n^2z^2+1)+(n-\mu^2)(z_2^2+1).$$

The second inequality above holds automatically. The first one is equivalent to

$$z_2^2+1 < \frac{(1-\mu^2)(z^2n^2+1)}{n-\mu^2}. \tag{18.5.13}$$

By (18.5.11), inequality (18.5.13) is equivalent to

$$\begin{aligned}&\sqrt{\Theta(n,z^2)} \\ &< -(1-\mu^2)z(n^2z^2+1) + 2(n-\mu^2)(nz^2+1)\sqrt{\frac{(1-\mu^2)(z^2n^2+1)}{n-\mu^2} - 1},\end{aligned} \tag{18.5.14}$$

that is,

$$\begin{aligned} J(n,z) := {}& (1-\mu^2)(n^2z^2+1)(nz^2+1) - (n-\mu^2)(nz^2+1) \\ & - (1-\mu^2)(n^2z^2+1)z\sqrt{\frac{(1-\mu^2)(z^2n^2+1)}{n-\mu^2} - 1} \\ & + (n-1)(\mu^2nz^2+1) > 0. \end{aligned} \tag{18.5.15}$$

Since

$$\sqrt{\frac{(1-\mu^2)(z^2n^2+1)}{n-\mu^2}-1} < nz,$$

we have

$$\begin{aligned} J(n,z) > {} & (1-\mu^2)(n^2z^2+1)(nz^2+1)-(n-\mu^2)(nz^2+1) \\ & -(1-\mu^2)(n^2z^2+1)nz^2+(n-1)(\mu^2nz^2+1)=0, \end{aligned} \tag{18.5.16}$$

which implies that (18.5.13) holds. This shows that, under assumption (18.5.12), (n_2, z_2) determined by (18.5.6) and (18.5.11) is the solution that we seek.

4. Rewrite $\Theta(n, z^2)$ as

$$\Theta(n,Z)=\frac{(1-\mu^2)^2}{n^2}\Big(Z^3-(1+\frac{2\mu^2a}{1-\mu^2})Z^2-(\frac{1}{n-1}+\frac{\mu^2}{n-\mu^2})a^2Z-a^2\Big)$$

with $Z = n^2z^2 + 1$ and

$$a=\frac{2}{1-\mu^2}(n-1)(n-\mu^2). \tag{18.5.17}$$

Then (18.5.12) is equivalent to

$$F(Z)=Z^3-bZ^2-dZ-a^2\geq 0 \tag{18.5.18}$$

with

$$b=1+\frac{2\mu^2}{1-\mu^2}a, \qquad d=\Big(\frac{1}{n-1}+\frac{\mu^2}{n-\mu^2}\Big)a^2. \tag{18.5.19}$$

Let $Z = y + \frac{b}{3}$. Then (18.5.18) becomes

$$f(y)=y^3+ky+l\geq 0, \tag{18.5.20}$$

where

$$k=-d-\frac{1}{3}b^2<0, \qquad l=-a^2-\frac{2}{27}b^3-\frac{1}{3}bd<0.$$

The discriminant for $f(y) = 0$ is

$$\frac{l^2}{4}+\frac{k^3}{27}=\frac{a^3}{54(1-\mu^2)^3}\Delta(n),$$

where

$$\begin{aligned} \Delta(n) := {} & -16\mu^4n^4+\big(40\mu^4+24\mu^2-16\big)n^3-\big(37\mu^4+38\mu^2-27\big)n^2 \\ & -\big(\mu^6-17\mu^4-9\mu^2+9\big)n-\frac{(1-\mu^2)^4(n+1)}{n-\mu^2}-\mu^2(1-\mu^2)^2. \end{aligned} \tag{18.5.21}$$

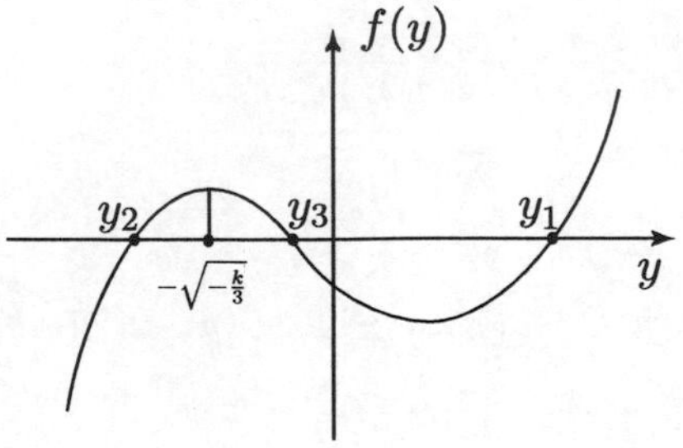

Figure 18.6: The behavior of function $f(y)$

From

$$\Delta(1) < 0, \quad \Delta'(1) < 0, \quad \Delta''(1) < 0, \quad \Delta'''(1) < 0, \quad \Delta^{(4)} < 0,$$

we have

$$\Delta(n) < \Delta(1) < 0.$$

It follows that $f(y) = 0$ has three real roots:

$$y_{1,2,3} = \sqrt[3]{-\frac{l}{2} + \frac{a}{3}\sqrt{-\frac{a}{6}\frac{\Delta(n)}{1-\mu^2}}\, i} + \sqrt[3]{-\frac{l}{2} - \frac{a}{3}\sqrt{-\frac{a}{6}\frac{\Delta(n)}{1-\mu^2}}\, i}.$$

That is,

$$y_1 = \frac{2}{3}\sqrt{b^2+3d}\,\cos(\frac{\beta}{3}), \tag{18.5.22}$$

$$y_2 = \frac{2}{3}\sqrt{b^2+3d}\,\cos(\frac{\beta+2\pi}{3}), \tag{18.5.23}$$

$$y_3 = \frac{2}{3}\sqrt{b^2+3d}\,\cos(\frac{\beta+4\pi}{3}), \tag{18.5.24}$$

where

$$\beta = \arccos(\frac{27a^2+2b^2+9bd}{2(b^2+3d)\sqrt{b^2+3d}}) \in (0, \frac{\pi}{2}). \tag{18.5.25}$$

Then we have

$$y_1 > 0 > y_3 > y_2.$$

It follows that

$$f(y) \geq 0 \qquad \text{for } y \geq y_1 \text{ or } y_2 \leq y \leq y_3;$$

see Fig. 18.6.

Furthermore, we find that $f(1-\frac{b}{3}) < 0$ and $1-\frac{b}{3} > -\sqrt{-\frac{k}{3}}$, so that

$$y_3 < 1 - \frac{b}{3} < y_1.$$

Therefore, we conclude from $Z = n^2 z^2 + 1 > 1$ that

$$f(y) \geq 0 \qquad \text{if and only if } \; y \geq y_1.$$

That is, the necessary and sufficient condition for the existence of the two-shock configuration at P_0 is that the wedge angle $\theta_{\rm w}$ is larger than the detachment angle $\theta_{\rm w}^{\rm d}$ determined by (18.5.1). This completes the proof. □

Now we determine the sonic angle $\theta_{\rm w}^{\rm s}$.

Theorem 18.5.3. *The weak state (2) at point P_0 is supersonic if and only if the wedge angle $\theta_{\rm w}$ is larger than the sonic angle $\theta_{\rm w}^{\rm s}$ determined by*

$$\theta_{\rm w}^{\rm s} = \arctan\big(\frac{1}{n}\sqrt{Z_0 - 1}\big) \tag{18.5.26}$$

with

$$Z_0 = \frac{1}{2}\big(\sqrt{Z_1} + \sqrt{Z_2} + \sqrt{Z_3}\big) + \frac{1}{4}(b+1), \tag{18.5.27}$$

where

$$Z_1 = \sqrt[3]{-\frac{s}{2} + \sqrt{\frac{s^2}{4} + \frac{r^3}{27}}} + \sqrt[3]{-\frac{s}{2} + \sqrt{\frac{s^2}{4} - \frac{r^3}{27}}} - \frac{2}{3}a_1 + \frac{1}{4}(b+1)^2,$$

$$Z_2 = \omega\sqrt[3]{-\frac{s}{2} + \sqrt{\frac{s^2}{4} + \frac{r^3}{27}}} + \omega^2\sqrt[3]{-\frac{s}{2} + \sqrt{\frac{s^2}{4} - \frac{r^3}{27}}} - \frac{2}{3}a_1 + \frac{1}{4}(b+1)^2,$$

$$Z_3 = \omega^2\sqrt[3]{-\frac{s}{2} + \sqrt{\frac{s^2}{4} + \frac{r^3}{27}}} + \omega\sqrt[3]{-\frac{s}{2} + \sqrt{\frac{s^2}{4} - \frac{r^3}{27}}} - \frac{2}{3}a_1 + \frac{1}{4}(b+1)^2,$$

and

$$\begin{aligned}
\omega &= -\frac{1}{2} + \frac{\sqrt{3}}{2}i,\\
a_1 &= b - d - \frac{\mu^2(n-1)^2}{(1-\mu^2)^2},\\
b_1 &= d - a^2 - \frac{\big(\mu^2(n^2-2)+n\big)(n-1)^2}{(1-\mu^2)^2},\\
d_1 &= a^2 - \frac{(n-\mu^2)(n+1)(n-1)^3}{(1-\mu^2)^2},\\
r &= -4d_1 - (b+1)b_1 - \frac{1}{3}a_1^2,\\
s &= \frac{8}{3}a_1 d_1 - \frac{1}{3}(b+1)a_1 b_1 - (b+1)^2 d_1 - b_1^2 - \frac{2}{27}a_1^3.
\end{aligned}$$

Proof. From (18.5.3), we have

$$
\begin{aligned}
M_2^2 - 1 &= \frac{1}{n_2 - \mu^2}\left((1-\mu^2)n_2^2 z_2^2 + 1 - n_2\right) \\
&= \frac{1}{(n_2-\mu^2)\big((n-\mu^2)z_2^2 + n(\mu^2 n z^2 + 1)\big)^2} G(z_2^2), \qquad (18.5.28)
\end{aligned}
$$

where

$$
\begin{aligned}
G(z_2^2) := {}& (n-\mu^2)^2 z_2^4 \\
&+ \big((1-\mu^2)(n^2z^2+1)^2 + 2n(n-\mu^2)(\mu^2 n z^2 + 1) \\
&\quad - (n-\mu^2)(n^2z^2+1)\big) z_2^2 \\
&+ n^2(\mu^2 n z^2 + 1)^2 - n(n^2z^2+1)(\mu^2 n z^2 + 1). \qquad (18.5.29)
\end{aligned}
$$

The equation: $G(z_2^2) = 0$ has two distinct roots: $r_- < 0 < r_+$. We then see that

$$(z_2^2 - r_+)G(z_2^2) > 0 \qquad \text{for } z_2^2 \neq r_+.$$

Thus, the weak state (2) is supersonic if and only if

$$z_2^2 \geq r_+.$$

Notice that equation: $z_2^2 = r_+$ is equivalent to the following equation:

$$nz^2\Theta = (n-1)^2(\mu^2 n z^2 + 1)(z^2+1), \qquad (18.5.30)$$

which is from (18.5.9).

We rewrite (18.5.30) as

$$H(Z) := (Z-1)h(Z) - \frac{(n-1)^2 n^3}{(1-\mu^2)^2} = 0, \qquad (18.5.31)$$

where $Z = 1 + n^2 z^2$ and

$$h(Z) := Z^3 - bZ^2 - \Big(d + \frac{\mu^2(n-1)^2}{(1-nu^2)^2}\Big) - a^2 - \frac{(n-1)^2\big(\mu^2(n^2-1)+n\big)}{(1-\mu^2)^2}. \qquad (18.5.32)$$

Similarly to the argument for Theorem 18.5.2, we conclude that the equation:

$$(Z-1)h(Z) = 0$$

has two roots: 1 and $\bar{Z} \geq 1$.

Since $\frac{(n-1)^2 n^3}{(1-\mu^2)^2} \geq 0$, there exists a unique solution $Z_0 > \bar{Z} \geq 1$ of (18.5.31); the other solutions are less than 1. Then the solution of equation (18.5.31) is (18.5.27), which implies that

$$z_0 = \frac{1}{n}\sqrt{Z_0 - 1} \qquad (18.5.33)$$

is the solution of (18.5.30). Furthermore, we have

$$\begin{aligned}(z_2)_z = &- \frac{1}{2(n-\mu^2)(nz^2+1)^2\sqrt{\Theta}} \\ &\times \Big((1-\mu^2)(n^3z^4+3n^2z^2-nz^2+1)\sqrt{\Theta} \\ &\quad + (1-\mu^2)^2 z(nz^2+1)(n^2z^2+1)^2 \\ &\quad + 2nz^3(1-\mu^2)^2(n-1)(n^2z^2+1) \\ &\quad + 4nz(1-\mu^2)(n-1)(n-\mu^2)(nz^2+1)\Big) < 0.\end{aligned}$$

Combining $z_2 < 0$ together with (18.5.30) yields

$$(z_2^2)_z = 2z_2(z_2)_z > 0.$$

Thus, we have

$$(z_2^2 - r_+)(z - z_0) > 0 \qquad \text{when } z \neq z_0.$$

It follows that

$$G(z_2^2) \begin{cases} > 0 & \text{for } z > z_0, \\ = 0 & \text{for } z = z_0, \\ < 0 & \text{for } \tan\theta_w^d < z < z_0. \end{cases}$$

Then, for

$$\theta_w^s = \arctan z_0 = \arctan(\frac{1}{n}\sqrt{Z_0 - 1}),$$

we conclude

$$M_2^2 - 1 \begin{cases} > 0 & \text{for } \theta_w > \theta_w^s, \\ = 0 & \text{for } \theta_w = \theta_w^s, \\ < 0 & \text{for } \theta_w^d < \theta_w < \theta_w^s. \end{cases}$$

This completes the proof. □

Fig. 18.7 shows the results in Theorems 18.5.2–18.5.3 for $\gamma = 1.4$ (air). In Fig. 18.7, the incident angle $\alpha = \frac{\pi}{2} - \theta_w$ is a function of the ratio of pressures, $\frac{p_0}{p}$, the thick curve is the curve of the detachment incident angle $\alpha^d = \frac{\pi}{2} - \theta_w^d$ for the existence of the two-shock configuration, and the dashed curve is the curve that provides the relative weak outflow behind the reflected-diffracted shock to be sonic. In the region between these two curves, state (2) is subsonic. In the region under the dashed curve, state (2) is supersonic.

18.6 VON NEUMANN'S CONJECTURES

As in the potential flow case, for the wedge angle $\theta_w \in (0, \frac{\pi}{2})$, different patterns of reflected-diffracted configurations may occur for the full Euler case. Several

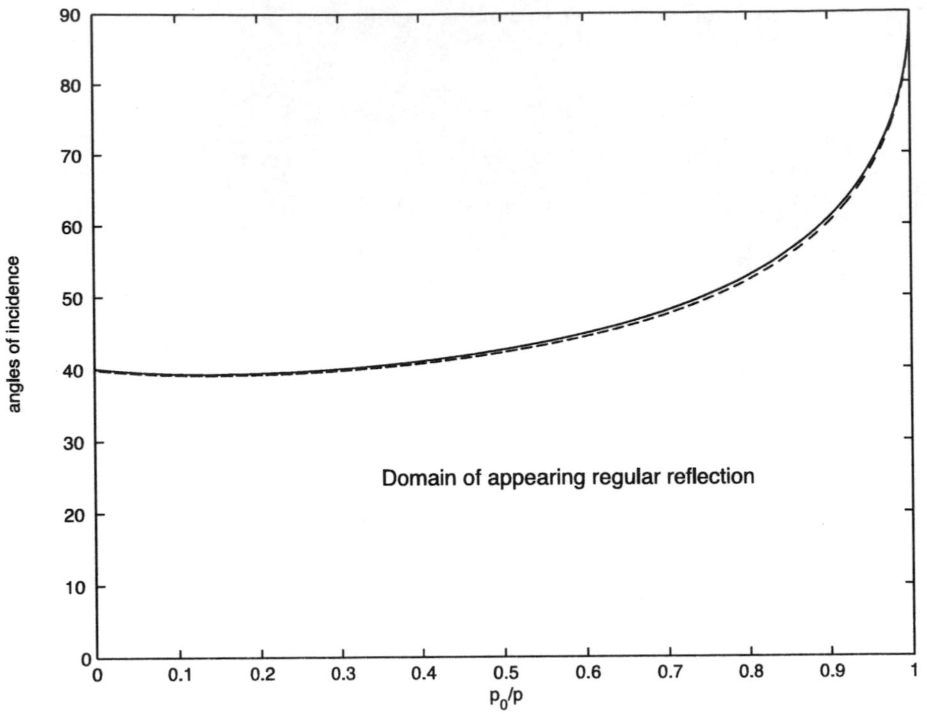

Figure 18.7: The sonic conjecture vs the detachment conjecture $\theta_{\rm w}^{\rm s} > \theta_{\rm w}^{\rm d}$ when $\gamma = 1.4$. Courtesy of Sheng-Yin [241].

criteria and conjectures have been proposed for the existence of different configurations for the patterns (*cf.* Ben-Dor [12]). As indicated in Chapter 2, one of the most important conjectures made by von Neumann [267, 268] in 1943 is the *detachment conjecture*, which states that the regular reflection-diffraction configuration may exist globally whenever the two-shock configuration (one is the incident shock and the other is a reflected shock) exists locally around point P_0 (also see Figs. 2.3–2.6).

Similarly, von Neumann's detachment conjecture for the full Euler equations can be stated as follows:

von Neumann's Detachment Conjecture: *There exists a global regular reflection-diffraction configuration whenever the wedge angle $\theta_{\rm w}$ is in $(\theta_{\rm w}^{\rm d}, \frac{\pi}{2})$, as shown in Figs.* 2.3–2.6. *That is, the existence of state* (2) *implies the existence of a regular reflection-diffraction solution of* **Problem 18.2**.

It is clear that the regular reflection-diffraction configuration is impossible without a local two-shock configuration at the reflection point P_0 on the wedge, and it is expected that the Mach reflection-diffraction configurations occur when the wedge angle $\theta_{\rm w}$ is smaller than the detachment angle $\theta_{\rm w}^{\rm d}$. Similarly, when the wedge angle $\theta_{\rm w}$ is larger than the detachment angle $\theta_{\rm w}^{\rm d}$, the local theory indicates that there are two possible states for state (2), and there has long been a debate as to which one is physically admissible; see Courant-Friedrichs [99], Ben-Dor [12], and the references cited therein.

Since the reflection-diffraction problem is not a local problem, we take a different point of view, as in §7.5, for the potential flow case. The selection of state (2) should be determined by not only the local feature of the problem but also the global features of the problem, and more precisely, by the stability/continuity of the configuration with respect to the wedge angle $\theta_{\rm w}$ as $\theta_{\rm w} \to \frac{\pi}{2}$.

Stability/Continuity Criterion to Select the Correct State (2): *Since the solution is uniquely determined when the wedge angle* $\theta_{\rm w} = \frac{\pi}{2}$, *it is necessary that the global regular reflection-diffraction configurations should converge to the unique normal reflection solution when* $\theta_{\rm w} \to \frac{\pi}{2}$, *provided that such global configurations can be constructed.*

As for the potential flow equation, we employ this stability criterion to conclude that the choice for state (2) should be $(U_2^a, V_2^a, p_2^a, \rho_2^a)$, which is a weaker state (*i.e.*, the jump of the corresponding shock is smaller), if such a global configuration can be constructed.

In general, $(U_2^a, V_2^a, p_2^a, \rho_2^a)$ may be supersonic or subsonic. If it is supersonic, the characteristic propagation speeds are finite, and state (2) is completely determined by the local information: state (1), state (0), and the location of point P_0. That is, any information from the reflection region, especially the disturbance at corner P_3, cannot travel towards the reflection point P_0. However, if it is subsonic, the information can reach P_0 and interact with it, potentially creating a new type of shock reflection-diffraction configurations. In particular, it is clear that no supersonic reflection configurations exist beyond the sonic angle. This argument motivated the following conjecture:

von Neumann's Sonic Conjecture: *There exists a supersonic regular reflection-diffraction configuration when* $\theta_{\rm w} \in (\theta_{\rm w}^{\rm s}, \frac{\pi}{2})$ *for* $\theta_{\rm w}^{\rm s} > \theta_{\rm w}^{\rm d}$, *as shown in Figs.* 2.3 *and* 2.5. *This is, the supersonicity of the weak state* (2) (*i.e.*, $|(U_2^a, V_2^a)| > c_2^a$) *at* P_0 *implies the existence of a supersonic regular reflection-diffraction solution of* **Problem 18.2**.

There has been a long debate in the literature as to whether there still exists a global regular reflection-diffraction configuration beyond the sonic angle up to the detachment angle; see Ben-Dor [12] and the references cited therein. As shown in Fig. 18.7, the difference between the wedge angles of the sonic conjecture and the detachment conjecture is only fractions of a degree; a resolution has greatly challenged even sophisticated modern numerical and laboratory experiments. In Part IV, we have rigorously proved that the global regular reflection-diffraction configuration does exist beyond the sonic angle up to the detachment angle for potential flow.

Similarly, to solve von Neumann's conjectures, we can also reformulate **Problem 18.2** into the following free boundary problem:

Problem 18.3 (Free Boundary Problem). *For $\theta_{\rm w} \in (\theta_{\rm w}^{\rm d}, \frac{\pi}{2})$, find a free boundary (curved reflected-diffracted shock) $\Gamma_{\rm shock}$ and a vector function $(\mathbf{U}, p, \rho)$ defined in region Ω, as shown in Figs. 2.3–2.6, such that φ satisfies:*

(i) *Equation* (18.3.1) *in Ω;*

(ii) *On the free boundary $\Gamma_{\rm shock}$,*

$$\mathbf{U}\cdot\boldsymbol{\tau} = \mathbf{U}_1\cdot\boldsymbol{\tau}, \quad \rho\mathbf{U}\cdot\boldsymbol{\nu} = \rho_1\mathbf{U}_1\cdot\boldsymbol{\nu}, \quad p + \rho(\mathbf{U}\cdot\boldsymbol{\nu})^2 = p_1 + \rho_1(\mathbf{U}_1\cdot\boldsymbol{\nu})^2,$$

$$\frac{\gamma p}{(\gamma-1)\rho} + \frac{1}{2}(\mathbf{U}\cdot\boldsymbol{\nu})^2 = \frac{\gamma p_1}{(\gamma-1)\rho_1} + \frac{1}{2}(\mathbf{U}_1\cdot\boldsymbol{\nu})^2,$$

$$\rho > \rho_1,$$

where $\boldsymbol{\nu}$ and $\boldsymbol{\tau}$ are the unit normal and tangent vectors of the free boundary $\Gamma_{\rm shock}$;

(iii) $(\mathbf{U}, p, \rho) = (\mathbf{U}_2, p_2, \rho_2)$ *on $\Gamma_{\rm sonic}$ in the supersonic reflection case as shown in Figs. 2.3 and 2.5, and at P_0 in the subsonic reflection case as shown in Figs. 2.4 and 2.6;*

(iv) $\mathbf{U}\cdot\boldsymbol{\nu} = 0$ *on $\Gamma_{\rm wedge}\cup\Gamma_{\rm sym}$,*

where $\boldsymbol{\nu}$ is the interior unit normal to Ω on $\Gamma_{\rm wedge}\cup\Gamma_{\rm sym}$.

If the free boundary can be written as $\Gamma_{\rm shock} = \{\xi_2 = f_{\rm sh}(\xi_1)\}$ with $f_{\rm sh}(\xi_1^0) = \xi_1^0\tan\theta_{\rm w}$, then the free boundary conditions can be written as

$$\frac{df_{\rm sh}}{d\xi_1} = \frac{U_1V_1 - \bar{c}_1\sqrt{|\mathbf{U}_1|^2 - \bar{c}_1^2}}{U_1^2 - \bar{c}_1^2} = -\frac{u-u_1}{v-v_1},$$

$$\frac{p}{p_1} = \frac{\rho - \mu^2\rho_1}{\rho_1 - \mu^2\rho},$$

$$p - p_1 = -\rho_1\Big(U_1\frac{u-u_1}{v-v_1} + V_1\Big)(v - v_1),$$

$$\rho > \rho_1,$$

where $\bar{c}_1^2 = \frac{\rho}{\rho_1}\frac{p-p_1}{\rho-\rho_1}$ and $\mu^2 = \frac{\gamma-1}{\gamma+1}$.

The boundary condition (U, V, p, ρ) along the sonic circle is the Dirichlet boundary condition, so that (U, V, p, ρ) is continuous across the sonic circle.

Similarly, we notice that the key obstacle to the existence of regular reflection-diffraction configurations is a new additional possibility that, for some wedge angle $\theta_{\rm w}^{\rm c} \in (\theta_{\rm w}^{\rm d}, \frac{\pi}{2})$, the reflected-diffracted shock P_0P_2 may attach to the wedge vertex P_3, as observed by experimental results (*cf.* [263, Fig. 238]).

On the other hand, solutions with an attached shock do not exist when the initial data of **Problem 18.1** (or equivalently, the parameters of state (1) in **Problem 18.2**) satisfy $\mathcal{M}_I \le 1$, or equivalently, $u_1 \le c_1$, since the front state of the shock must be supersonic. Moreover, in this case, the regular reflection-diffraction solution of **Problem 18.2** should exist for each $\theta_{\rm w} \in (\theta_{\rm w}^{\rm d}, \frac{\pi}{2})$, as von Neumann conjectured [267, 268].

In the other case, $\mathcal{M}_I > 1$, there is a possibility that there exists $\theta_{\rm w}^{\rm c} \in (\theta_{\rm w}^{\rm d}, \frac{\pi}{2})$ such that a regular reflection-diffraction solution with an attached shock of **Problem 18.2** exists for each $\theta_{\rm w} \in (\theta_{\rm w}^{\rm c}, \frac{\pi}{2})$ in the sense that a solution with $P_2 = P_3$ exists for $\theta_{\rm w} = \theta_{\rm w}^{\rm c}$.

Since $\mathcal{M}_I$ is a function of $n_I = \frac{\rho_1}{\rho_0}$ for fixed $\gamma > 1$, determined by (18.2.4), the condition that $\mathcal{M}_I \le 1$ or $\mathcal{M}_I > 1$ is equivalent to the corresponding condition on n_I; that is to say, there exists $n^* > n_I$ such that

(i) $\mathcal{M}_I < 1$ for any $n_I \in (1, n^*)$;

(ii) $\mathcal{M}_I = 1$ for $n_I = n^*$;

(iii) $\mathcal{M}_I > 1$ for any $n_I \in (n^*, \infty)$.

Thus, Case (i) corresponds to the case of a relatively weaker incident shock, while Case (iii) corresponds to the case of a relatively stronger incident shock.

In fact, the regime between angles $\theta_{\rm w}^{\rm s}$ and $\theta_{\rm w}^{\rm d}$ is very narrow and only fractions of a degree apart; see Fig. 18.7 from Sheng-Yin [241], where the solid curve is the detachment angle curve with respect to $\frac{p_0}{p_1}$ and the dash curve is the sonic angle curve with respect to $\frac{p_0}{p_1}$. This shows that the difference between the two conjectures is only fractions of a degree. Therefore, a complete understanding of the behavior of the two different configurations requires further rigorous mathematical analysis.

18.7 CONNECTIONS WITH THE POTENTIAL FLOW EQUATION

In this section, we discuss the role of the potential flow equation for shock reflection-diffraction, even in the realm of the full Euler equations.

Under the Hodge-Helmoltz decomposition:

$$\mathbf{U} = D\varphi + \mathbf{W}, \qquad \mathrm{div}\mathbf{W} = 0$$

for some function φ and vector $\mathbf{W}$, the Euler equations (18.3.1) become

$$\mathrm{div}(\rho D\varphi) + 2\rho = -\mathrm{div}(\rho \mathbf{W}), \tag{18.7.1}$$

$$D\big(\frac{1}{2}|D\varphi|^2 + \varphi\big) + \frac{Dp}{\rho} = (D\varphi + \mathbf{W}) \cdot D\mathbf{W} + (D^2\varphi + I)\mathbf{W}, \tag{18.7.2}$$

$$(D\varphi + \mathbf{W}) \cdot D\omega + (1 + \Delta\varphi)\,\omega = -\mathrm{curl}\,\big(\frac{Dp}{\rho}\big), \tag{18.7.3}$$

$$(D\varphi + \mathbf{W}) \cdot DS = 0, \tag{18.7.4}$$

$$\mathrm{div}\,\mathbf{W} = 0, \tag{18.7.5}$$

where $\omega = \operatorname{curl}\mathbf{W} = \operatorname{curl}\mathbf{U}$ is the vorticity of the fluid, $S = c_v \ln(p\rho^{-\gamma})$ the entropy, and D the gradient with respect to the self-similar variables $\boldsymbol{\xi}$.

When $\omega = 0, S = const.$, and $\mathbf{W} = 0$ on a curve Γ transverse to the fluid direction, we conclude from (18.7.4) that, in domain Ω_E formed by the fluid trajectories past Γ:

$$\frac{d}{dt}\boldsymbol{\xi} = (D\varphi + \mathbf{W})(\boldsymbol{\xi}),$$

we have

$$S = const.$$

This implies

$$p = const.\, \rho^\gamma \qquad \text{in } \Omega_E,$$

so that

$$\operatorname{curl}\big(\frac{Dp}{\rho}\big) = const.\ \operatorname{curl}(D\rho^{\gamma-1}) = 0 \qquad \text{in } \Omega_E.$$

Notice that, if $\mathbf{U} = D\varphi + \mathbf{W} \in C^{0,1}$, the theory of ordinary differential equations implies that Ω_E has an open interior, as long as curve Γ has a relatively open interior.

Then equation (18.7.3) becomes

$$(D\varphi + \mathbf{W}) \cdot D\omega + (1 + \Delta\varphi)\omega = 0,$$

which implies that

$$\omega = 0, \qquad i.e., \ \operatorname{curl}\mathbf{W} = 0 \qquad \text{in } \Omega_E.$$

This yields that $\mathbf{W} = const.$ in Ω_E, since $\operatorname{div}\mathbf{W} = 0$. Then we conclude

$$\mathbf{W} = 0 \qquad \text{in } \Omega_E,$$

since $\mathbf{W}|_\Gamma = 0$, which yields that the right-hand side of equation (18.7.2) vanishes.

By scaling, we finally conclude that any solution of system (18.7.1)–(18.7.4) in domain Ω_E is determined by the potential flow equation (2.2.8) with (2.2.9) for self-similar solutions.

In our case, when $\theta_{\rm w} \in (\theta_{\rm w}^{\rm s}, \frac{\pi}{2})$, the weak state (2) is supersonic and satisfies

$$\omega = 0, \quad \mathbf{W} = 0, \quad S = S_2. \tag{18.7.6}$$

If our solution $(\mathbf{U}, p, \rho)$ is $C^{0,1}$ in domain $P_0P_1P_2P_3$ (as shown in Figs. 2.3 and 2.5) and the tangential derivative of U is continuous across the sonic arc $\Gamma_{\rm sonic}$, (18.7.6) still holds along $\Gamma_{\rm sonic}$ on the side of Ω. Therefore, we conclude by asserting the following theorem:

Theorem 18.7.1. *Let* $(\mathbf{U}, p, \rho)$ *be a solution of* **Problem 18.2** *such that* $(\mathbf{U}, p, \rho)$ *is* $C^{0,1}$ *in the open domain* $P_0P_1P_2P_3$ *and the tangential derivative of* $\mathbf{U}$ *is continuous across the sonic arc* $\Gamma_{\rm sonic}$. *Let* Ω_E *be the subdomain of* Ω *formed by the*

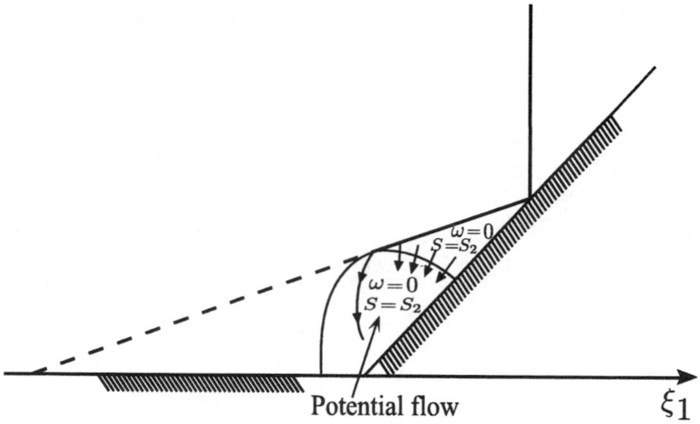

Figure 18.8: The potential flow equation coincides with the full Euler equations in an important domain Ω_E

fluid trajectories past the sonic arc Γ_{sonic}. *Then, in* Ω_E, *the full Euler equations* (18.7.1)–(18.7.4) *coincide with the potential flow equation* (2.2.8) *with* (2.2.9) *for self-similar solutions, that is, equation* (2.2.8) *with* (2.2.9) *is exact in domain* Ω_E *for* **Problem 18.2**; *see Fig.* 18.8.

Remark 18.7.2. *The domains like* Ω_E *also exist in several Mach reflection-diffraction configurations.* Theorem 18.7.1 *applies to such domains whenever solution* $(\mathbf{U}, p, \rho)$ *is* $C^{0,1}$ *and the tangential derivative of* $\mathbf{U}$ *is continuous. In fact,* Theorem 18.7.1 *indicates that, for any solution* φ *of the potential flow equation* (2.2.8) *with* (2.2.9), *the* $C^{1,1}$*–regularity of* φ *and the continuity of the tangential component of the velocity field* $\mathbf{U} = D\varphi$ *are optimal across the sonic arc* Γ_{sonic} *in general.*

Remark 18.7.3. *The importance of the potential flow equation* (1.5) *with* (1.4) *in the time-dependent Euler flows was also observed by Hadamard* [146] *through a different argument. Moreover, as indicated in* §2.1, *for the solutions containing a weak shock, especially in many aerodynamic applications, the potential flow model* (2.2.8)–(2.2.9) *and the full Euler flow model* (18.7.1)–(18.7.4) *match each other well up to the third order of the shock strength. For more on this, see Bers* [16], *Glimm-Majda* [139], *and Morawetz* [221].

Remark 18.7.4. *Since two eigenvalues of system* (18.7.1)–(18.7.4) *are always real and repeated, with two linearly independent eigenvectors, system* (18.7.1)–(18.7.4) *can be decomposed of two transport-type equations and two nonlinear equations of mixed hyperbolic-elliptic type. While we have developed essential techniques in Parts* I–IV *to handle two mixed-type nonlinear equations with similar difficulties as for the potential flow equation* (2.2.8) *with* (2.2.9), *there may be an additional new difficulty for the two coupled transport-type equations in the* *other domain* $\Omega \setminus \Omega_E$ *containing the stagnation points* $\mathbf{U} = 0$ *at which pressure* p *and vorticity* ω *may have additional singularity* (*cf. Serre* [236]).

Notes: Theorem 18.5.2 was rigorously proved in Chang-Chen [31] based on the work by Bleakney-Taub [18] and von Neumann [267, 268], while Theorem 18.5.3 is due to Sheng-Yin [241]. Theorem 18.7.1 and related analysis for the role of the potential flow equation are due to Chen-Feldman [55].

Chapter Nineteen

Shock Reflection-Diffraction and New Mathematical Challenges

As demonstrated in Chapters 1–18, the analysis of shock reflection-diffraction has advanced the development of new mathematical approaches, techniques, and ideas that will be useful for solving further problems involving shock reflection and/or diffraction, as well as other related problems involving similar analytical difficulties. Nevertheless, further understanding of these phenomena requires the development of ever more powerful new mathematics.

19.1 MATHEMATICAL THEORY FOR MULTIDIMENSIONAL CONSERVATION LAWS

As we have indicated, shock reflection-diffraction configurations are a fundamental class of structural patterns of solutions of the two-dimensional Riemann problem for the Euler equations for compressible fluids. In addition, the Euler equations are core fundamental nonlinear PDEs in the multidimensional hyperbolic systems of conservation laws, and the Riemann solutions are expected to be the global attractors and determine the local structure of general entropy solutions of the Cauchy problem (the initial value problem). Therefore, any further understanding of the shock reflection-diffraction configurations will advance our understanding of the behavior of general entropy solutions of multidimensional hyperbolic systems of conservation laws, which plays an important role in the multidimensional mathematical theory of hyperbolic conservation laws.

In this book, we have established the global existence, regularity, and structural stability of regular reflection-diffraction configurations for potential flow up to the detachment angle or the critical angle. Further fundamental mathematical problems include the following:

Problem 19.1. *Establish a mathematical theory for shock reflection-diffraction for the two-dimensional isentropic and/or full Euler equations.*

As seen in Chapters 1–18, we have understood the mathematical difficulties relatively well for the transonic shocks, the Kelydsh degeneracy near the sonic arc, the singularities of solutions near the wedge vertex and the corner between the transonic shock and the sonic arc for the nonlinear second-order elliptic equations (*cf.* Theorem 18.7.1), as well as a one-point singularity at the corner

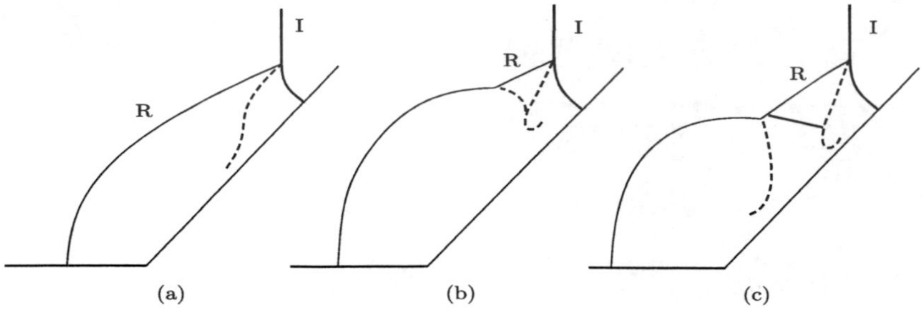

Figure 19.1: Different types of Mach reflection-diffraction configurations

between the reflected shock and the wedge at the reflection point for the transition from the supersonic to subsonic reflection configurations when the wedge angle decreases across the sonic angle up to the detachment angle or the critical angle. On the other hand, the following two new features of the problem for the isentropic and/or full Euler equations still need to be understood:

(i) Solutions of transport-type equations with rough coefficients and stationary transport velocity;

(ii) Estimates of the vorticity of the pseudo-velocity.

Indeed, the calculations in Serre [236] have shown difficulties in estimating vorticity $\omega = \nabla \times \mathbf{U}$. It is possible that the vorticity will have some singularities in the region, perhaps near the wedge boundary and/or corner. In fact, even for potential flow, the second derivatives of the velocity potential, *i.e.*, the first derivatives of the velocity, may blow up at the wedge corner.

Problem 19.2. *Analyze Mach reflection-diffraction configurations for the two-dimensional isentropic and/or full Euler equations.*

When the wedge angle $\theta_{\rm w}$ decreases across the detachment angle $\theta_{\rm w}^{\rm d}$, or the incident-shock strength decreases, there is a transition from the Regular Reflection (RR) to a Mach Reflection (MR). The MR configurations include the Simple Mach Reflection and become increasingly complex when they include the Complex Mach Reflection, the Double Mach Reflection, the von Neumann Reflection, and the Guderley Reflection, among others, as the wedge angle $\theta_{\rm w}$ decreases; see Fig. 19.1, as well as Figs. 1.2–1.6, for the types of the Mach Reflection and the Guderley Reflection. A detailed analysis based on the shock polar can be found in [12, 13, 31, 151], while [96, 159, 230, 233, 257, 258, 259] and the references cited therein provide supportive numerical experiments.

There are several additional difficulties for the Mach reflection-diffraction configurations, including:

(i) *A priori* understanding of the location of the triple-point where the incident shock and the reflected-diffracted shock meet;

(ii) At the triple-point of the Mach reflection, an additional shock called a Mach stem connects this point to the wedge. Since a pure three-shock pattern

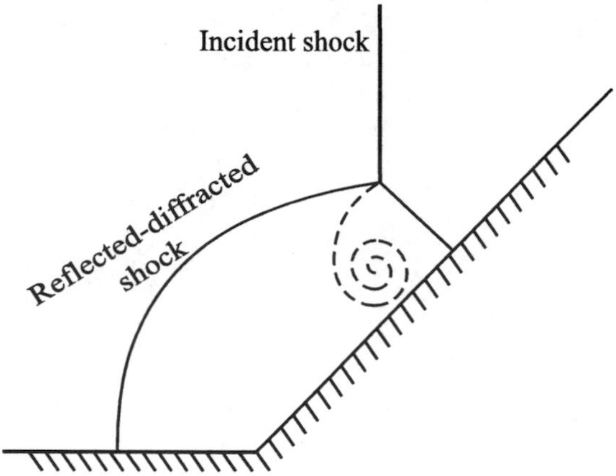

Figure 19.2: Vorticity waves in the Mach reflection-diffraction configurations

is not normally possible, as indicated in Courant-Friedrichs [99] and Serre [236], a fourth wave – vortex sheet, or other types of waves, must originate from this triple-point generally. See Fig. 19.2 for a possible vorticity wave formed by the vortex sheet originating from the triple-point in the Mach reflection configuration as observed experimentally. The stability of vortex sheets is a very sensitive issue, as indicated by Artola-Majda [2, 3]; for example, such waves are known to be nonlinearly unstable dynamically in general, unless the jump of the tangential velocity exceeds $2\sqrt{2}c$ (see [97, 98]). The question is whether the self-similarity stabilizes such vortex sheets. In particular, it is important to identify the right function space for the vorticity function $\omega = \nabla \times \mathbf{U}$, especially to examine whether the vortex sheet for the vorticity wave is a chord-arc with the form:

$$z(s) = z_0 + a(s) \int_0^s e^{ib(\tau)} d\tau, \qquad b \in \mathrm{BMO},$$

or a similar form.

Problem 19.3. *Study the shock reflection-diffraction configurations for the higher dimensional potential flow equations and isentropic/full Euler equations.*

When a planar shock hits an n-dimensional cone, $n \geq 3$, higher dimensional shock reflection-diffraction phenomena occur. Then the supersonic part of the reflection-diffraction region, corresponding to state (2) in the two-dimensional case, is no longer a constant physical state, and the reflected-diffracted shock is no longer a straight cone. In particular, for the higher dimensional potential flow equation, the pseudo-potential function is no longer a quadratic function and the supersonic part of the reflected-diffracted shock should also be treated as a free boundary. The supersonic state is determined by the *degenerate* second-order hyperbolic equations, which has to be solved first.

Another additional difficulty is that the boundary of the physical domain has a more delicate singularity at the origin, *i.e.*, the cone vertex. It is important to understand how the pseudo-velocity vanishes in the subsonic domain in the self-similar coordinates. Further points to understand include the degeneracy of solutions across the sonic surface (in the RR case) and the singularity of solutions near the curved wedge-like corner between the transonic shock and the sonic surface in the higher dimensional case.

It is also important to study the problem for higher dimensional isentropic and/or full Euler equations.

19.2 NONLINEAR PARTIAL DIFFERENTIAL EQUATIONS OF MIXED ELLIPTIC-HYPERBOLIC TYPE

As we have known from the previous chapters, nonlinear PDEs of mixed elliptic-hyperbolic type arise naturally from the shock reflection-diffraction problem. Such mixed equations also arise naturally in many other fundamental problems in fluid mechanics, differential geometry, elasticity, relativity, calculus of variations, and related areas. The mathematical approaches and techniques developed in this book will be useful for solving a variety of nonlinear PDEs involving similar analytical difficulties. Nevertheless, further new mathematical ideas, techniques, and approaches need to be developed for solving these PDEs.

Problem 19.4. *Study further elliptic free boundary problems for nonlinear equations of mixed type, and the corresponding nonlinear degenerate elliptic equations and systems.*

From our solution of the regular shock reflection-diffraction problem for the potential flow equation, the physical problem has been formulated into a free boundary problem for a nonlinear second-order degenerate elliptic equation with ellipticity degenerating on a part of the fixed boundary (the sonic curve), and mathematical techniques have been developed for this specific form of elliptic degeneracy. Our analysis has the following features:

(i) The problem is global. In particular, even the local regularity estimates for the nonlinear degenerate elliptic equation near the sonic arc require some global properties of the solutions, including the positivity of ψ in §11.4. Similarly, in the subsonic reflection case, the estimates for the uniformly elliptic equation in the domain with free boundary near the reflection point in §16.6 also rely on a global property – the monotonicity cone of $\varphi_1 - \varphi$. Furthermore, the use of global properties is essential in the other estimates; for example, the estimates in §11.4 do not hold definitely if ψ is negative or changes its sign. For more on this, see the estimates in §9.5.

(ii) In order to obtain the elliptic solution of a mixed-type nonlinear equation, the formulation of the free boundary problem should match with the nonlinear equation naturally.

(iii) Our regularity estimates near the sonic arc, where the equation is degenerate elliptic, are obtained by using a comparison function (which utilizes the Keldysh-type elliptic degeneracy and depends on the detailed structure of the self-similar potential flow equation), combined with rescaling techniques whose scaling is determined by both the equation and the structure of the domain. Indeed, we have developed three different scalings; see §11.4 and §16.4–§16.5.

Since nonlinear degenerate elliptic problems for PDEs of mixed type arise in many applications, it is important to obtain a deeper understanding of the different types of elliptic degeneracy and the natural global settings for such problems.

Problem 19.5. *Study further nonlinear degenerate hyperbolic equations and systems.*

Nonlinear degenerate hyperbolic equations/systems also arise in many shock reflection/diffraction problems, especially in the high dimensional case with sonic boundaries, as described in Problem 19.3. There are two types of degeneracy near the degenerate sets: hyperbolic degeneracy (Tricomi-type) and parabolic degeneracy (Keldysh-type). The prototype for linear degenerate hyperbolic equations is the celebrated Euler-Poisson-Darboux equation (*cf.* [37, 38, 41]), which has been successfully applied in solving some important one-dimensional nonlinear problems involving nonlinear degenerate hyperbolic systems (*cf.* [38, 61, 62, 63, 64] and the references cited therein). The degeneracy behavior of self-similar flow for the two-dimensional Euler equations near the vacuum has been exhibited through the characteristic decompositions in Zheng [289]; see also [182, 180, 161]. Further understanding of nonlinear degenerate hyperbolic equations and systems in the multidimensional case deserves special attention.

Problem 19.6. *Develop further new ideas, techniques, and approaches to analyze nonlinear PDEs of mixed elliptic-hyperbolic type*

The study of degenerate hyperbolic equations and degenerate elliptic equations as proposed in Problems 19.4–19.5 will be very helpful for developing further new ideas for nonlinear PDEs that change the type from hyperbolic to elliptic. The unification of the mathematical theory for these two classes of nonlinear PDEs is at the cutting edge of modern mathematical research, and has important implications for the parts of the wider scientific community which use nonlinear PDEs. Despite this, nonlinear mixed-type PDEs still remain poorly understood and are exceedingly challenging from both the theoretical and numerical point of view. The mathematics of nonlinear mixed-type PDEs is largely uncharted territory. Any advance in this direction can be expected to fertilize and provide deeper insights into the theory of nonlinear PDEs in general, to lead to efficient new algorithms for the computation of solutions, and to impact upon developments in other branches of mathematics.

The mathematical methods expected for further development may include weak convergence methods, entropy methods, energy estimate techniques, positive symmetrization techniques, and measure-theoretical methods, among others. These methods are largely independent of the types of PDEs under consideration (*cf.* [54, 69, 70, 115, 126]) and will therefore work well within a unified framework of nonlinear mixed-type PDEs.

19.3 FREE BOUNDARY PROBLEMS AND TECHNIQUES

As presented in this book, the shock reflection-diffraction problem can be formulated as free boundary problems involving nonlinear PDEs of mixed elliptic-hyperbolic type. Similarly, many other transonic problems can also be formulated as free boundary problems involving nonlinear PDEs of mixed type (*cf.* [38, 58] and the references cited therein). The understanding of these transonic problems requires a complete mathematical solution of the corresponding free boundary problems for nonlinear mixed PDEs, such as what we have done for the shock reflection-diffraction problem in this book. These problems are not only longstanding open problems in fluid mechanics, but also fundamental in the mathematical theory of multidimensional conservation laws; see [58, 139, 236, 286] and the references cited therein. In this sense, we have to understand these free boundary problems in order to fully understand global entropy solutions to multidimensional hyperbolic systems of conservation laws.

Problem 19.7. *Develop further free boundary techniques for solving transonic shock problems.*

As we have shown, the shock reflection-diffraction problem has been formulated as a one-phase free boundary problem for nonlinear degenerate elliptic PDEs. Similarly, other shock reflection/diffraction problems such as the Prandtl-Meyer problem for supersonic flow impinging onto solid wedges can be formulated as one-phase free boundary problems involving other mathematical difficulties (*cf.* [5, 6]). Similar one-phase, or even two-phase, free boundary problems also arise in many other transonic flow problems, such as steady transonic flow problems including transonic nozzle flow problems (*cf.* [7, 38, 53, 179]), steady transonic flows past obstacles (*cf.* [38, 43, 47, 85]), and the local stability of Mach configurations (*cf.* [81, 82]), as well as higher dimensional versions of Problem 19.3 (shock reflection-diffraction by a solid cone). Some of these involve various types of degeneracy of free boundaries (*i.e.*, the strength of jump tends to zero at some points or singularities along the free boundary). In Chapters 1–18, we have discussed recently developed mathematical approaches and techniques for solving some of these free boundary problems. On the other hand, the analysis of other types of transonic problems demands further development of mathematical ideas, techniques, and approaches.

Problem 19.8. *Develop mathematical techniques for continuous transition between the supersonic and subsonic phases.*

Consider, for example, transonic flow problems involving supersonic bubbles in subsonic flow (*cf.* [95, 159, 217, 218, 219, 257, 258, 259]) and steady transonic flows past obstacles and nozzles (*cf.* [7, 38, 43, 47, 53, 85, 179]). Some recent experimental and numerical results have shown that supersonic bubbles may occur even in the shock reflection-diffraction configurations, such as the Guderley reflection-diffraction configurations (*cf.* [243, 257, 258, 259]). The common feature of these problems is the involvement of the continuous transition between the supersonic and subsonic phases. Many of these problems can be formulated as free boundary problems for nonlinear PDEs of mixed elliptic-hyperbolic type, whose free boundary is the phase boundary between the elliptic and hyperbolic phases, which degenerate at least on some part of the transition set. There have not yet emerged mathematical techniques and approaches efficient enough to tackle such free boundary problems. One of the strategies is to deal first with some fundamental, concrete physical transonic problems involving supersonic bubbles in subsonic flow.

Problem 19.9. *Develop mathematical techniques to understand the behavior of solutions near a point or set where a free boundary meets a fixed boundary, especially when the free and/or fixed boundaries are transition boundaries between the hyperbolic and elliptic phases.*

As we have presented in this book, the free boundary problem formulated from the shock reflection-diffraction problem involves the corner between the free boundary and a fixed boundary that is a degenerate sonic curve for the nonlinear PDE. In this book, as well as in earlier papers [4, 54], we have developed mathematical techniques for handling the optimal regularity of both the free boundary and the corresponding solution at corner P_1 on Fig. 2.3. The reflected shock P_0P_2 is C^2 near P_1, so that the curved free boundary Γ_{shock} matches at P_1 not only the slope but also the curvature of the straight reflected shock P_0P_1. The corresponding solution in Ω is $C^{1,1}$ near P_1, but not C^2 up to P_1. Our techniques have been developed based on the features of the underlying nonlinear PDE. It would be interesting to develop further the mathematical techniques and approaches for understanding the behavior of solutions near the point or set where a free boundary meets a fixed boundary, especially when the free and/or fixed boundaries are transition boundaries between the hyperbolic and elliptic phases for general second-order PDEs of mixed type.

19.4 NUMERICAL METHODS FOR MULTIDIMENSIONAL CONSERVATION LAWS

The main challenge in designing numerical algorithms for hyperbolic conservation laws is that weak solutions are not unique, and multidimensional waves such as vortex sheets and entropy waves are often sensitive with perturbations, so that the numerical schemes should be consistent with the Clausius inequality (the entropy inequality), as well as other physical requirements to capture the sensitivity of multidimensional waves. Excellent numerical schemes should also be numerically simple, robust, fast, and low cost, and have sharp oscillation-free resolutions and high accuracy in the domains where the solution is smooth. It is also desirable that the schemes capture vortex sheets, vorticity waves, and entropy waves, as they appear in the shock reflection-diffraction configurations, and are coordinate invariant, among other features. The main difficulty in calculating fluid flows with discontinuities is that it is very hard to predict, even in the process of a flow calculation, when and where new discontinuities arise and interact. Moreover, tracking the discontinuities, especially their interactions, is numerically burdensome (see [20, 100, 150, 174]).

Problem 19.10. *Develop more efficient numerical methods for further understanding shock reflection-diffraction phenomena.*

Since its fundamental importance in fluid mechanics and the mathematical theory of multidimensional conservation laws with rich structure, the shock reflection-diffraction problem has become a standard test problem for multidimensional numerical methods; see [12, 104, 105, 120, 133, 134, 135, 137, 138, 139, 141, 149, 151, 152, 159, 160, 170, 174, 177, 221, 232, 236, 240, 256, 257, 258, 259, 264, 273, 282] and the references cited therein. On the other hand, the sensitivity with respect to the wedge angle and the strength of the incident shock, and the complexity of the shock reflection-diffraction configurations urgently ask for more efficient numerical methods.

Problem 19.11. *Develop more efficient numerical methods for multidimensional conservation laws.*

Shock waves, vorticity waves, and entropy waves are, by nature, fundamental discontinuity waves, and they arise in supersonic or transonic gas flow, or from a very sudden release (explosion) of chemical, nuclear, electrical, radiate, or mechanical energy in a limited space. Tracking these discontinuities and their interactions, especially when and where new waves arise and interact in the motion of gases, is one of the main motivations for numerical computation for the gas dynamics equations.

An efficient numerical approach is shock capturing algorithms for general hyperbolic conservation laws with form (1.10) or (1.11). Modern numerical ideas of shock capturing for computational fluid dynamics date back to 1944 when von Neumann first proposed a new numerical method, a centered difference scheme, to treat the hydrodynamical shock problem, for which numerical calculations showed oscillations on the mesh scale (see Lax [173]). In 1950, von Neumann's dream of capturing shocks was first realized when he and Richtmyer [271] introduced the ingenious idea of adding a numerical viscous term of the same size as the truncation error into the hydrodynamic equations. Their numerical viscosity guarantees that the scheme is consistent with the Clausius inequality, *i.e.*, the entropy inequality. The shock jump conditions – the Rankine-Hugoniot conditions – are satisfied, provided that the Euler equations of gas dynamics are discretized in conservation form. Then oscillations are eliminated by the judicious use of the artificial viscosity; the solutions constructed by this method converge uniformly, except in a neighborhood of shocks where they remain bounded and are spread out over a few mesh intervals.

In the one-dimensional case, further examples of success include the Lax-Friedrichs scheme (1954), the Glimm scheme (1965), the Godunov scheme (1959) and related high order shock capturing schemes; for example, van Leer's MUSCL (1981), Colella-Wooward's PPM (1984), Harten-Engquist-Osher-Chakravarthy's ENO (1987), the more recent WENO (1994, 1996), the Lax-Wendroff scheme (1960) and its two-step version, the Richtmyer scheme (1967), as well as the MacCormick scheme (1969). See also [66, 71, 100, 120, 141, 178, 256, 261] and the references cited therein.

In the multidimensional case, one approach is to generalize directly the one-dimensional methods for solving multidimensional problems; such an approach has led to several useful numerical methods including semi-discrete methods and Strang's dimension-dimension splitting methods.

Observe that multidimensional effects do play a significant role in the behavior of the solution locally, as shown in the shock reflection-diffraction configurations. Also note that the approach that only solves one-dimensional Riemann problems in the coordinate directions clearly lacks the use of all the multidimensional information. The development of fully multidimensional methods requires a good mathematical theory in order to understand the multidimensional behavior of entropy solutions, especially the shock reflection-diffraction configurations that have been analyzed in this book. However, further understanding of the shock reflection-diffraction configurations requires more efficient numerical methods. Current efforts in this direction include employing more information about the multidimensional behavior of solutions, determining the direction of primary wave propagation and applying wave propagation in other directions, and using transport techniques, upwind techniques, finite volume techniques, relaxation techniques, and kinetic techniques from the microscopic level; see [33, 169, 203, 256], [120, 139, 141, 178, 261], and the references cited therein.

Other useful methods for calculating sharp fronts for gas dynamics equations include front-tracking algorithms [91, 138] and level set methods [223, 239].

Bibliography

[1] Alt, H. W., Caffarelli, L. A. and Friedman A. (1985). Compressible flows of jets and cavities, *J. Diff. Eqs.* **56**: 82–141.

[2] Artola, M. and Majda, A. (1987). Nonlinear development of instability in supersonic vortex sheets, I: The basic kink modes, *Phys. D.* **28**: 253–281.

[3] Artola, M. and Majda, A. (1989). Nonlinear development of instability in supersonic vortex sheets, II: Resonant interaction among kink modes, *SIAM J. Appl. Math.* **49**: 1310–1349.

[4] Bae, M., Chen, G.-Q. and Feldman, M. (2009). Regularity of solutions to regular shock reflection for potential flow, *Invent. Math.* **175**: 505–543.

[5] Bae, M., Chen, G.-Q. and Feldman, M. (2013). Prandtl-Meyer reflection for supersonic flow past a solid ramp, *Quart. Appl. Math.* **71**: 583–600.

[6] Bae, M., Chen, G.-Q. and Feldman, M. (2017). *Prandtl-Meyer Reflection Configurations, Transonic Shocks, and Free Boundary Problems*, Research Monograph, *In preparation.*

[7] Bae, M.-J. and Feldman, M. (2011). Transonic shocks in multidimensional divergent nozzles, *Arch. Rational Mech. Anal.* **201**: 777–840.

[8] Ball, J. M. and Carstensen, C. (1999). Compatibility conditions for microstructures and the austenite-martensite transition, *Mater. Sci. Eng. A.* **273–275**: 231–236.

[9] Ball, J. M. and James, R. D. (1987). Fine phase mictures as minimizers of energy, *Arch. Rational Mech. Anal.* **100**: 13–52.

[10] Ball, J. M., Kirchheim, B. and Kristensen, J. (2000). Regularity of quasiconvex envelopes, *Calc. Var. Partial Diff. Eqs.* **11**: 333–359.

[11] Bazhenova, T. V., Gvozdeva, L. G. and Nettleton, M. A. (1984). Unsteady interaction of shock waves, *Prog. Aero. Sci.* **21**: 249–331.

[12] Ben-Dor, G. (1991). *Shock Wave Reflection Phenomena*, Springer-Verlag: New York.

[13] Ben-Dor, G. (2001). Oblique shock wave reflections, In: *Handbook of Shock Waves* (Eds. G. Ben-Dor, O. Igra, and T. Elperin), Vol. 2, pp. 67–179, Academic Press: San Diego.

[14] Benzoni-Gavage, S. and Serre, D. (2007). *Multidimensional Hyperbolic Partial Differential Equations: First-Order Systems and Applications*, The Clarendon Press and Oxford University Press: Oxford.

[15] Bernoulli, D. (1738). *Hydrodynamica*, Argentorati.

[16] Bers, L. (1958). *Mathematical Aspects of Subsonic and Transonic Gas Dynamics*, John Wiley & Sons, Inc.: New York; Chapman & Hall, Ltd.: London.

[17] Bitsadze, A. V. (1964). *Equations of the Mixed Type*, Macmillan Company: New York.

[18] Bleakney, W. and Taub, A. H. (1949). Interaction of shock waves, *Rev. Modern Phys.* **21**: 584–605.

[19] Bressan, A. (2000). *Hyperbolic Systems of Conservation Laws: The One-dimensional Cauchy Problem*, Oxford University Press: Oxford.

[20] Bressan, A., Chen, G.-Q., Lewicka, M. and Wang D.-H. (2011). *Nonlinear Conservation Laws and Applications*, IMA Volume in Mathematics and Its Applications, **153**, Springer-Verlag: New York.

[21] Caffarelli, L. A. (1987). A Harnack inequality approach to the regularity of free boundaries, I: Lipschitz free boundaries are $C^{1,\alpha}$, *Rev. Mat. Iberoamericana*, **3**: 139–162.

[22] Caffarelli, L. A. (1989a). A Harnack inequality approach to the regularity of free boundaries, II: Flat free boundaries are Lipschitz, *Comm. Pure Appl. Math.* **42**: 55–78.

[23] Caffarelli, L. A. (1989b). A Harnack inequality approach to the regularity of free boundaries, III: Existence theory, compactness, and dependence on X, *Ann. Scuola Norm. Sup. Pisa Cl. Sci.* (4), **15**: 583–602.

[24] Caffrelli, L. A., Jerison, D. and Kenig, C. (2002). Some new monotonicity theorems with applications to free boundary problems, *Ann. of Math.* **155**: 369–404.

[25] Caffarelli, L. and Salsa, S. (2005). *A Geometric Approach to Free Boundary Problems*, American Mathematical Society: Providence, RI.

[26] Canić, S., Keyfitz, B. L. and Kim, E. H. (2000). Free boundary problems for the unsteady transonic small disturbance equation: Transonic regular reflection, *Meth. Appl. Anal.* **7**: 313–335.

[27] Canić, S., Keyfitz, B. L. and Kim, E. H. (2002). A free boundary problem for a quasilinear degenerate elliptic equation: regular reflection of weak shocks, *Comm. Pure Appl. Math.* **55**: 71–92.

[28] Canić, S., Keyfitz, B. L. and Lieberman, G. M. (2000). A proof of existence of perturbed steady transonic shocks via a free boundary problem, *Comm. Pure Appl. Math.* **53**: 484–511.

[29] Cauchy, A.-L. (1827). Sur les relations qui existent dans l'état d'équilibre d'un corps solide ou fluide, entre les pressions ou tensions et les forces accélératrices, *Exercises de Mathématiques*, **2**: 108–111.

[30] Cauchy, A.-L. (1829). Sur l'équilibre et le mouvenment intérieur des corps considérés comme des masses continues, *Exercises de Mathématiques*, **4**: 293–319.

[31] Chang, T. and Chen, G.-Q. (1986). Diffraction of planar shock along the compressive corner, *Acta Math. Scientia*, **6**: 241–257.

[32] Chang, T., Chen, G.-Q. and Yang, S. (1992). 2-D Riemann problem in gas dynamics and formation of spiral, In: *Nonlinear Problems in Engineering and Science–Numerical and Analytical Approach* (*Beijing, 1991*), pp. 167–179, Science Press: Beijing.

[33] Chang, T., Chen, G.-Q. and Yang, S. (1995). On the Riemann problem for two-dimensional Euler equations I: Interaction of shocks and rarefaction waves, *Discrete Contin. Dynam. Systems*, **1**: 555–584.

[34] Chang, T., Chen, G.-Q. and Yang, S. (2000). On the Riemann problem for two-dimensional Euler equations II: Interaction of contact discontinuities, *Discrete Contin. Dynam. Systems*, **6**: 419–430.

[35] Chang, T. and Hsiao, L. (1989). *The Riemann Problem and Interaction of Waves in Gas Dynamics*, Longman Scientific & Technical: Harlow; and John Wiley & Sons, Inc.: New York.

[36] Chapman, C. J. (2000). *High Speed Flow*, Cambridge University Press: Cambridge.

[37] Chen, G.-Q. (1997). Euler-Poisson-Darboux equations and hyperbolic conservation laws. In: *Nonlinear Evolutionary Partial Differential Equations*, pp. 11–25, AMS/IP Studies in Advanced Mathematics, Vol. **3**, International Press: Cambridge, Massachusetts.

[38] Chen, G.-Q. (2005). Euler equations and related hyperbolic conservation laws, Chapter 1, *Handbook of Differential Equations, Evolutionary Equations*, Vol. **2**, Eds. C. M. Dafermos and E. Feireisl, Elsevier: Amsterdam, The Netherlands.

[39] Chen, G.-Q. (2008). On nonlinear partial differential equations of mixed type, In: *Proceedings of the 4th International Congress of Chinese Mathematicians*, Eds. L. Jin, K.-F. Liu, L. Yang, and S.-T. Yau, Higher Education Press: Beijing.

[40] Chen, G.-Q. (2015a). Gas dynamics equations: Computation. In: *Encyclopedia of Applied and Computational Mathematics*, Ed. B. Engquist, Springer-Verlag: Berlin and Heidelberg. arXiv:1205.4433.

[41] Chen, G.-Q. (2015b). The Tricomi equation. In: *Equations, Laws and Functions of Applied Mathematics*, the Princeton Companion to Applied Mathematics, Ed. N. J. Higham, III. 30, pp. 170, Princeton University Press. arXiv:1311.3338.

[42] Chen, G.-Q., Chen, J. and Feldman, M. (2007). Transonic shocks and free boundary problems for the full Euler equations in infinite nozzles, *J. Math. Pures Appl.* **88**: 191–218.

[43] Chen, G.-Q., Chen, J. and Feldman, M. (2016). Transonic flows with shocks past curved wedges for the full Euler equations, *Discrete Contin. Dyn. Syst.* **36**: 4179–4211.

[44] Chen, G.-Q., Chen, J. and Song, K. (2006). Transonic nozzle flows and free boundary problems for the full Euler equations, *J. Diff. Eqs.* **229**: 92–120.

[45] Chen, G.-Q., Clelland, J., Slemrod, M., Wang, D. and Yang, D. (2017). Isometric embedding via strongly symmetric positive systems, *Asian J. Math.* (in press). arXiv:1502.04356.

[46] Chen, G.-Q., Dafermos, C. M., Slemrod, M. and Wang, D. (2007). On two-dimensional sonic-subsonic flow, *Commun. Math. Phys.* **271**: 635–647.

[47] Chen, G.-Q. and Fang, B.-X. (2009). Stability of transonic shock-fronts in steady potential flow past a perturbed cone, *Discrete Conti. Dynamical Systems*, **23**: 85–114.

[48] Chen, G.-Q. and Fang, B.-X. (2017). Stability of transonic shocks in steady supersonic flow past multidimensional wedges, *Adv. Math.* **314**: 493–539.

[49] Chen, G.-Q. and Feldman, M. (2003). Multidimensional transonic shocks and free boundary problems for nonlinear equations of mixed type, *J. Amer. Math. Soc.* **16**: 461–494.

[50] Chen, G.-Q. and Feldman, M. (2004a). Steady transonic shocks and free boundary problems in infinite cylinders for the Euler equations, *Comm. Pure Appl. Math.* **57**: 310–356.

[51] Chen, G.-Q. and Feldman, M. (2004b). Free boundary problems and transonic shocks for the Euler equations in unbounded domains, *Ann. Scuola Norm. Sup. Pisa Cl. Sci.* (*5*), **3**: 827–869.

[52] Chen, G.-Q. and Feldman, M. (2005). Potential theory for shock reflection by a large-angle wedge, *Proc. Nat. Acad. Sci. U.S.A.* **102**: 15368–15372.

[53] Chen, G.-Q. and Feldman, M. (2007). Existence and stability of multidimensional transonic flows through an infinite nozzle of arbitrary cross-sections, *Arch. Rational Mech. Anal.* **184**: 185–242.

[54] Chen, G.-Q. and Feldman, M. (2010a). Global solutions to shock reflection by large-angle wedges for potential flow, *Ann. of Math.* **171**: 1019–1134.

[55] Chen, G.-Q. and Feldman, M. (2010b). Shock reflection-diffraction and multidimensional conservation laws, In: *Hyperbolic Problems: Theory, Numerics and Applications, Proc. Sympos. Appl. Math.* **67**: 25–52, Amer. Math. Soc.: Providence, RI.

[56] Chen, G.-Q. and Feldman, M. (2011). Shock reflection-diffraction and nonlinear partial differential equations of mixed type, In: *Complex Analysis and Dynamical Systems IV, Contemporary Mathematics*, **554**: 55–72, Amer. Math. Soc.: Providence, RI.

[57] Chen, G.-Q. and Feldman, M. (2012). Comparison principles for self-similar potential flow, *Proc. Amer. Math. Soc.* **140**: 651–663.

[58] Chen, G.-Q. and Feldman, M. (2015). Free boundary problems in shock reflection/diffraction and related transonic flow problems, In: *Free Boundary Problems and Related Topics.* Theme Volume: Phil. Trans. R. Soc. **A373**: 20140276, The Royal Society: London.

[59] Chen, G.-Q., Feldman, M., Hu, J. and Xiang, W. (2017). Loss of regularity of solutions of the shock diffraction problem by a convex corned wedge for the potential flow equation, Preprint, arXiv:1705.06837.

[60] Chen, G.-Q., Feldman, M. and Xiang, W. (2017). Convexity of self-similar transonic shock waves for potential flow, Preprint.

[61] Chen, G.-Q. and Kan, P. T. (1995). Hyperbolic conservation laws with umbilic degeneracy, I. *Arch. Rational Mech. Anal.* **130**: 231–276.

[62] Chen, G.-Q. and Kan, P. T. (2001). Hyperbolic conservation laws with umbilic degeneracy, II. *Arch. Rational Mech. Anal.* **160**: 325–354.

[63] Chen, G.-Q. and LeFloch, P. G. (2000). Compressible Euler equations with general pressure law, *Arch. Rational Mech. Anal.* **153**: 221–259.

[64] Chen, G.-Q. and LeFloch, P. G. (2003). Existence theory for the isentropic Euler equations, *Arch. Rational Mech. Anal.* **166**: 81–98.

[65] Chen, G.-Q. and Li, T.-H. (2008). Well-posedness for two-dimensional steady supersonic Euler flows past a Lipschitz wedge, *J. Diff. Eqs.* **244**: 1521–1550.

[66] Chen, G.-Q. and Liu, J.-G. (1997). Convergence of difference schemes with high resolution for conservation laws, *Math. Comp.* **66**: 1027–1053.

[67] Chen, G.-Q., Shahgholian, H. and Vázquez, J.-V. (2015). Free boundary problems: The forefront of current and future developments, In: *Free Boundary Problems and Related Topics.* Theme Volume: Phil. Trans. R. Soc. **A373**: 20140285, The Royal Society: London.

[68] Chen, G.-Q., Slemrod, M. and Wang, D. (2008). Vanishing viscosity method for transonic flow, *Arch. Rational Mech. Anal.* **189**: 159–188.

[69] Chen, G.-Q., Slemrod, M. and Wang, D. (2010a). Isometric embedding and compensated compactness, *Commun. Math. Phys.* **294**: 411–437.

[70] Chen, G.-Q., Slemrod, M. and Wang, D. (2010b). Weak continuity of the Gauss-Codazzi-Ricci system for isometric embedding, *Proc. Amer. Math. Soc.* **138**: 1843–1852.

[71] Chen, G.-Q. and Toro, E.-F. (2004). Centered difference schemes for nonlinear hyperbolic equations, *J. Hyper. Diff. Eqs.* **1**: 531–566.

[72] Chen, G.-Q. and Wang, Y.-G. (2012). Characteristic discontinuities and free boundary problems for hyperbolic conservation laws, In: *Nonlinear Partial Differential Equations, The Abel Symposium 2010*, Chapter 5, pp. 53–82, H. Holden and K. H. Karlsen (Eds.), Springer-Verlag: Heidelberg.

[73] Chen, G.-Q. and Yuan, H. (2009). Uniqueness of transonic shock solutions in a duct for steady potential flow, *J. Diff. Eqs.* **247**: 564–573.

[74] Chen, G.-Q., Zhang, Y. and Zhu, D. (2006). Existence and stability of supersonic Euler flows past Lipschitz wedges, *Arch. Rational Mech. Anal.* **181**: 261–310.

[75] Chen, S.-X. (1996). Linear approximation of shock reflection at a wedge with large angle, *Comm. Partial Diff. Eqs.* **21**: 1103–1118.

[76] Chen, S.-X. (1998a). Asymptotic behavior of supersonic flow past a convex combined wedge, *Chinese Ann. of Math.* **19B**: 255–264.

[77] Chen, S.-X. (1998b). Supersonic flow past a concave double wedge, *Sci. China Mathematics*, **41A**: 39–47.

[78] Chen, S.-X. (1998b). Global existence of supersonic flow past a curved convex wedge, *J. Partial Diff. Eqs.* **11**: 73–82.

[79] Chen, S.-X. (2001). Existence of stationary supersonic flows past a point body, *Arch. Rational Mech. Anal.* **156**: 141–181.

[80] Chen, S.-X. (2005). Stability of transonic shock fronts in two-dimensional Euler systems, *Trans. Amer. Math. Soc.* **357**: 287–308.

[81] Chen, S.-X. (2006). Stability of a Mach configuration, *Comm. Pure Appl. Math.* **59**: 1–35.

[82] Chen, S.-X. (2008). Mach configuration in pseudo-stationary compressible flow, *J. Amer. Math. Soc.* **21**: 63–100.

[83] Chen, S.-X. (2012). E-H type Mach configuration and its stability, *Commun. Math. Phys.* **315**: 563–602.

[84] Chen, S.-X. (2015). Global existence and stability of a stationary Mach reflection, *Sci. China Mathematics*, **58**: 11–34.

[85] Chen, S.-X. and Fang, B. (2007). Stability of transonic shocks in supersonic flow past a wedge, *J. Diff. Eqs.* **233**: 105–135.

[86] Chen, S.-X., Hu, D. and Fang, B. (2013). Stability of the E-H type regular shock refraction, *J. Diff. Eqs.* **254**: 3146–3199.

[87] Chen, S.-X. and Li, D. (2014). Cauchy problem with general discontinuous initial data along a smooth curve for 2-D Euler system, *J. Diff. Eqs.* **257**: 1939–1988.

[88] Chen, S.-X. and Qu, A. (2012). Two-dimensional Riemann problems for Chaplygin gas, *SIAM J. Math. Anal.* **44(3)**: 2146–2178.

[89] Chen, S.-X., Xin, Z. and Yin, H. (2002). Global shock waves for the supersonic flow past a perturbed cone, *Commun. Math. Phys.* **228**: 47–84.

[90] Chen, S.-X. and Yuan, H. (2008). Transonic shocks in compressible flow passing a duct for three-dimensional Euler systems, *Arch. Rational Mech. Anal.* **187**: 523–556.

[91] Chern, I.-L., Glimm, J., McBryan O., Plohr B. and Yaniv S. (1986). Front tracking for gas dynamics, *J. Comp. Phys.* **62**: 83–110.

[92] Chorin, A. J. (1994). *Vorticity and Turbulence*, Springer-Verlag: New York.

[93] Ciarlet, P. G. and Mardare, C. (2010). Existence theorems in intrinsic nonlinear elasticity, *J. Math. Pures Appl.* **94**: 229–243.

[94] Clausius, R. (1865). Über verschiedene für die Anwendung bequeme Formen der Hauptgleichungen der mechanischen Wärmatheorie, *Annalen der Physik und Chemie*, **125**: 353–400.

[95] Cole, J. D. and Cook, L. P. (1986). *Transonic Aerodynamics*, North-Holland: Amsterdam.

[96] Colella, P. and Henderson, L. F. (1990). The von Neumann paradox for the diffraction of weak shock waves, *J. Fluid Mech.* **213**: 71–94.

[97] Coulombel, J. F. and Secchi, P. (2004). Stability of compressible vortex sheet in two space dimensions, *Indiana Univ. Math. J.* **53**: 941–1012.

[98] Coulombel, J. F. and Secchi, P. (2008). Nonlinear compressible vortex sheets in two space dimensions, *Ann. Sci. Ecole Norm. Sup.* **41**: 85–139.

[99] Courant, R. and Friedrichs, K. O. (1948). *Supersonic Flow and Shock Waves*, Springer-Verlag: New York.

[100] Dafermos, C.-M. (2016). *Hyperbolic Conservation Laws in Continuum Physics*, 4th Edition, Springer-Verlag: Berlin.

[101] Dafermos, C.-M. and Feireisl, E. (2004–06). *Handbook of Differential Equations*, Vols. 1–3, Elsevier: Amsterdam.

[102] Dafermos, C.-M. and Pokorny, M. (2008–09). *Handbook of Differential Equations*, Vols. 4–5, Elsevier: Amsterdam.

[103] Daskalopoulos, P. and Hamilton, R. (1999). The free boundary in the Gauss curvature flow with flat sides, *J. Reine Angew. Math.* **510**: 187–227.

[104] Dem'yanov, A. Yu. and Panasenko, A. V. (1981). Numerical solution to the problem of the diffraction of a plane shock wave by a convex corner, *Fluid Dynamics* **16**: 720–725 (Translated from the original Russian).

[105] Deschambault, R L. and Glass, I. I. (1983). An update on non-stationary oblique shock-wave reflections: actual isopycnics and numerical experiments, *J. Fluid Mech.* **131**: 27–57.

[106] Ding, X., Chen, G.-Q. and Luo P. (1989). Convergence of the fractional step Lax-Friedrichs scheme and Godunov scheme for isentropic gas dynamics, *Commun. Math. Phys.* **121**: 63–84.

[107] Dong, G. (1991). *Nonlinear Partial Differential Equations of Second Order*, Transl. Math. Monographs, **95**, Amer. Math. Soc.: Providence, RI.

[108] Elling, V. (2009). Counter exammples to the sonic criterion, *Arch. Rational Mech. Anal.* **194** (3): 987–1010.

[109] Elling, V. (2012). Non-existence of strong regular reflections in self-similar potential flow, *J. Diff. Eqs.* **252**: 2085–2103.

[110] Elling, V. and Liu, T.-P. (2005). The ellipticity principle for self-similar potential flows, *J. Hyper. Diff. Eqs.* **2**: 909–917.

[111] Elling, V. and Liu, T.-P. (2008). Supersonic flow onto a solid wedge, *Comm. Pure Appl. Math.* **61**: 1347–1448.

[112] Euler, L. (1755). Principes géenéraux du mouvement des fluides, *Mém. Acad. Sci. Berlin* **11**: 274–315.

[113] Euler, L. (1759). Supplément aux recherches sur la propagation du son, *Mém. Acad. Sci. Berlin* **15**: 210–240.

[114] Euler, L. (1759, 1761). De principiis motus fluidorum, *Novi Comm. Acad. Petrop.* **xiv** & **vi**.

[115] Evans, L. C. (1990). *Weak Convergence Methods for Nonlinear Partial Differential Equations*, CBMS-RCSM, 74, Amer. Math. Soc.: Providence, RI.

[116] Evans, L. C. (2010). *Partial Differential Equations*, Amer. Math. Soc.: Providence, RI.

[117] Fang, B.-X. (2006). Stability of transonic shocks for the full Euler system in supersonic flow past a wedge, *Math. Methods Appl. Sci.* **29**: 1–26.

[118] Feldman, M. (2001). Regularity of Lipschitz free boundaries in two-phase problems for fully nonlinear elliptic equations, *Indiana Univ. Math. J.* **50**: 1171–1200.

[119] Ferrari, C. and Tricomi, F. G. (1968). *Transonic Aerodynamics*, Academic Press: New York, English Transl. of *Aerodinamica Transonica*, Cremonese (1962).

[120] Fey, M. and Jeltsch, R. (1999). *Hyperbolic Problems: Theory, Numerics, and Applications*, I, II, International Series of Numerical Mathematics 130, Birkhäuser Verlag: Basel.

[121] Finn, R. and Gilbarg, D. (1957). Three-dimensional subsonic flows, and asymptotic estimates for elliptic partial differential equations, *Acta Math.* **98**: 265–296.

[122] Francheteau, J. and Métivier, G. (2000). *Existence de Chocs Faibles pour des Systèmes Quasi-linéaires Hyperboliques Multidimensionnels*, Astérisque, **268**, SMF: Paris.

[123] Freistühler, H. and Szepessy, A. (2001). *Advances in the Theory of Shock Waves*, Birkhäuser: Boston.

[124] Friedman, A. (1988). *Variational Principles and Free-Boundary Problems*, 2nd Edition. Robert E. Krieger Publishing Co., Inc.: Malabar, FL [First edition, John Wiley & Sons, Inc.: New York, 1982].

[125] Friedlander, S. and Serre, D. (2002-07). *Handbook of Mathematical Fluid Dynamics*. Elsevier: Amsterdam.

[126] Friedrichs, K. O. (1958). Symmetric positive linear differential equations, *Comm. Pure Appl. Maths.* **11**: 333–418.

[127] Gamba, I. M. and Morawetz, C. S. (1996). A viscous approximation for a 2-D steady semiconductor or transonic gas dynamic flow: existence theorem for potential flow, *Comm. Pure Appl. Math.* **49**: 999–1049.

[128] Gamba, I., Rosales, R. R. and Tabak, E. G. (1999). Constraints on possible singularities for the unsteady transonic small disturbance (UTSD) equations, *Comm. Pure Appl. Math.* **52**: 763–779.

[129] Gibbs, J. W. (1873). A method of geometrical representation of the thermodynamic properties of substances by means of surfaces, *Trans. Connecticut Acad.* **2**: 382–404.

[130] Gilbarg, D. and Hörmander, L. (1980). Intermediate Schauder estimates, *Arch. Rational Mech. Anal.* **74**(4): 297–318.

[131] Gilbarg, D. and Trudinger, N. (1983). *Elliptic Partial Differential Equations of Second Order,* 2nd Edition, Springer-Verlag: Berlin.

[132] Glass, I. I. (1974). *Shock Waves and Man*, University of Toronto Press: Toronto.

[133] Glaz, H. M., Colella, P., Glass, I. I. and Deschambault, R. L. (1985a). A numerical study of oblique shock-wave reflection with experimental comparisons, *Proc. Roy. Soc. Lond.* **A398**: 117–140.

[134] Glaz, H. M., Colella, P., Glass, I. I. and Deschambault, R. L. (1985b). A detailed numerical, graphical, and experimental study of oblique shock wave reflection, *Lawrence Berkeley Laboratory Report*, **LBL-20033**.

[135] Glaz, H. M., Walter, P. A., Glass, I. I. and Hu, T. C. J. (1986). Oblique shock wave reflections in SF_6: A comparison of calculation and experiment, *AIAA J. Prog. Astr. Aero.* **106**: 359–387.

[136] Glimm, J. (1991). Nonlinear waves: overview and problems. In: *Multidimensional Hyperbolic Problems and Computations* (Minneapolis, MN, 1989), IMA Vol. Math. Appl. **29**: 89–106, Springer: New York.

[137] Glimm, J., Ji, X.-M., Li, J.-Q., Li, X.-L., Zhang, P., Zhang, T. and Zheng, Y.-X. (2008). Transonic shock formation in a rarefaction Riemann problem for the 2-D compressible Euler equations, *SIAM J. Appl. Math.* **69**: 720–742.

[138] Glimm, J., Klingenberg, C., McBryan, O., Plohr, B., Sharp, D. and Yaniv, S. (1985). Front tracking and two-dimensional Riemann problems, *Adv. Appl. Math.* **6**: 259–290.

[139] Glimm, J. and Majda, A. (1991). *Multidimensional Hyperbolic Problems and Computations*, Springer-Verlag: New York.

[140] Glimm, J., Kranzer, H. C., Tan, D. and Tangerman, F. M. (1997). Wave fronts for Hamilton-Jacobi equations: the general theory for Riemann solutions in $\mathbb{R}^n$, *Commun. Math. Phys.* **187**: 647–677.

[141] Godlewski, E. and Raviart, P. (1996). *Numerical Approximation of Hyperbolic Systems of Conservation Laws*, Springer-Verlag: New York.

[142] Gu, C.-H. (1962). A method for solving the supersonic flow past a curved wedge (in Chinese), *Fudan Univ. J.* **7**: 11–14.

[143] Guderley, K. G. (1962). *The Theory of Transonic Flow*, Translated from the German by J. R. Moszynski, Pergamon Press: Oxford-London-Paris-Frankfurt; Addison-Wesley Publishing Co. Inc.: Reading, Mass.

[144] Guès, O., Métivier, G., Williams, M. and Zumbrun, K. (2005). Existence and stability of multidimensional shock fronts in the vanishing viscosity limit, *Arch. Rational Mech. Anal.* **175**: 151–244.

[145] Guès, O., Métivier, G., Williams, M. and Zumbrun, K. (2006). Navier-Stokes regularization of multidimensional Euler shocks, *Ann. Sci. École Norm. Sup.* **(4) 39**: 75–175.

[146] Hadamard, J. (1903). *Leçons sur la Propagation des Ondes et les Équations de l'Hydrodynamique*, Hermann: Paris (Reprinted by Chelsea 1949).

[147] Han, Q. and Hong, J.-X. (2006). *Isometric Embedding of Riemannian Manifolds in Euclidean Spaces*, Amer. Math. Soc.: Providence, RI.

[148] Harabetian, E. (1987). Diffraction of a weak shock by a wedge, *Comm. Pure Appl. Math.* **40**: 849–863.

[149] Hindman, R. G., Kutler, P. and Anderson, D. (1979). A two-dimensional unsteady Euler-equation solver for flow regions with arbitrary boundaries, *AIAA Report* **79-1465**.

[150] Holden, H. and Risebro, N. H. (2002). *Front Tracking for Hyperbolic Conservation Laws*, Springer-Verlag: New York.

[151] Hornung, H. G. (1986). Regular and Mach reflection of shock waves, *Ann. Rev. Fluid Mech.* **18**: 33–58.

[152] Hornung, H. G., Oertel, H. and Sandeman, R. J. (1979). Transition to Mach reflexion of shock waves in steady and pseudosteady flow with and without relaxation, *J. Fluid Mech.* **90**: 541–560.

[153] Hugoniot, H. (1887). Sur la propagation du mouvement dans les corps et spécialement dans les gaz parfaits, I. *J. Ecole Polytechnique* **57**: 3–97.

[154] Hugoniot, H. (1889). Sur la propagation du mouvement dans les corps et spécialement dans les gaz parfaits, II. *J. Ecole Polytechnique* **58**: 1–125.

[155] Hunter, J. K. (1988). Transverse diffraction of nonlinear waves and singular rays, *SIAM J. Appl. Math.* **48**: 1–37.

[156] Hunter, J. (1989). Hyperbolic waves and nonlinear geometrical acoustics, In: *Transactions of the 6th Army Conference on Appl. Math. and Computing*, Boulder CO., pp. 527–569.

[157] Hunter, J. (1997). Nonlinear wave diffraction. In: *Geometrical Optics and Related Topics (Cortona, 1996), Progr. Nonlinear Diff. Eqs. Appl.* **32**, pp. 221–243, Birkhäuser Boston: Boston, MA.

[158] Hunter, J. and Keller, J. B. (1984). Weak shock diffraction, *Wave Motion* **6**: 79–89.

[159] Hunter, J. K. and Tesdall, A. (2002). Self-similar solutions for weak shock reflection, *SIAM J. Appl. Math.* **63**: 42–61.

[160] Ivanov, M. S., Vandromme, D., Formin, V. M., Kudryavtsev, A. N., Hadjadj, A. and Khotyanovsky, D. V. (2001). Transition between regular and Mach reflection of shock waves: new numerical and experimental results, *Shock Waves* **11**: 199–207.

[161] Ji, X. and Zheng, Y. (2013). Characteristic decouplings and interactions of rarefaction waves of 2D Euler equations, *J. Math. Anal. Appl.* **406**: 4–14.

[162] Keller, J B. and Blank, A. A. (1951). Diffraction and reflection of pulses by wedges and corners, *Comm. Pure Appl. Math.* **4**: 75–94.

[163] Keyfitz, B. L. and Warnecke, G. (1991). The existence of viscous profiles and admissibility for transonic shocks, *Comm. Partial Diff. Eqs.* **16**: 1197–1221.

[164] Kinderlehrer, D. and Nirenberg, L. (1977). Regularity in free boundary problems, *Ann. Scuola Norm. Sup. Pisa Cl. Sci. (4)* **4**: 373–391.

[165] Kirchhoff, G. (1868). Über den Einfluss der Wärmeleitung in einem Gase auf die Schallbewegung, *Ann. Physik* **134**: 177–193.

[166] Korobeinikov, V. P. (1986). *Nonstationary Interactions of Shock and Detonation Wave in Gases*, Nauka: Moscow, USSR (in Russian).

[167] Krehl, V. P. and van der Geest, M. (1991). The discovery of the Mach reflection effect and its demonstration in an auditorium, *Shock Waves* **1**: 3–15.

[168] Kristensen, J. (1999). On the non-locality of quasiconvexity, *Ann. Inst. H. Poincaré Anal. Non Linéaire* **16**: 1–13.

[169] Kurganov, A. and Tadmor, E. (2002). Solution of two-dimensional Riemann problems for gas dynamics without Riemann problem solvers, *Numer. Methods Partial Diff. Eqs.* **18**: 584–608.

[170] Kutler, P. and Shankar, V. (1977). Diffraction of a shock wave by a compression corner, Part I: Regular reflection, *AIAA J.* **15**: 197–203.

[171] Lax, P. D. (1973). *Hyperbolic Systems of Conservation Laws and the Mathematical Theory of Shock Waves*, CBMS-RCSM, SIAM: Philiadelphia.

[172] Lax, P. D. (1986a). Hyperbolic systems of conservation laws in several space variables, In: *Current Topics in Partial Differential Equations*, pp. 327–341, Kinokuniya: Tokyo.

[173] Lax, P. D. (1986b). On dispersive difference schemes, *Physica D* **18**: 250–254.

[174] Lax, P. D. (2000). Mathematics and computing, In: *Mathematics: Frontiers and Perspectives*, pp. 417–432, Arnold, V.I., Atiyah, M., Lax, P. and Mazur, B. (Eds.), Amer. Math. Soc.: Providence, RI.

[175] Lax, P. D. and Liu, X.-D. (1998). Solution of two-dimensional Riemann problems of gas dynamics by positive schemes, *SIAM J. Sci. Comput.* **19**: 319–340.

[176] LeFloch, Ph. G. (2002). *Hyperbolic Systems of Conservation Laws: The Theory of Classical and Nonclassical Shock Waves*, Birkhäuser-Verlag: Basel.

[177] LeVeque, R. J. (1992). *Numerical Methods for Conservation Laws*, Second Ed., Birkhäuser-Verlag: Basel.

[178] LeVeque, R. J. (2002). *Finite Volume Methods for Hyperbolic Problems*, Cambridge University Press: Cambridge.

[179] Li, J., Xin, Z. and Yin, H. (2003). Transonic shocks for the full compressible Euler system in a general two-dimensional de Laval nozzle, *Arch. Rational Mech. Anal.* **207**: 533–581.

[180] Li, J., Yang, Z. and Zheng, Y. (2011). Characteristic decompositions and interactions of rarefaction waves of 2-D Euler equations, *J. Diff. Eqs.* **250**: 782–798.

[181] Li, J., Zhang, T. and Yang, S. (1998). *The Two-Dimensional Riemann Problem in Gas Dynamics*, Longman (Pitman Monographs 98): Essex.

[182] Li, J. and Zheng, Y. (2009). Interaction of rarefaction waves of the two-dimensional self-similar Euler equations, *Arch. Rational Mech. Anal.* **193**: 623–657.

[183] Li, T.-T. (1980). On a free boundary problem, *Chinese Ann. Math.* **1**: 351–358.

[184] Lieberman, G. M. (1985a). The Perron process applied to oblique derivative problems, *Adv. Math.* **55**: 161–172.

[185] Lieberman, G. M. (1985b). Regularized distance and its applications, *Pacific J. Math.* **117**: 329–352.

[186] Lieberman, G. M. (1986a). Intermediate Schauder estimates for oblique derivative problems, *Arch. Rational Mech. Anal.* **93**: 129–134.

[187] Lieberman, G. M. (1986b). Mixed boundary value problems for elliptic and parabolic differential equations of second order, *J. Math. Anal. Appl.* **113**: 422–440.

[188] Lieberman, G. M. (1986c). Regularity of solutions of nonlinear elliptic boundary value problems, *J. Reine Angew. Math.* **369**: 1–13.

[189] Lieberman, G. M. (1987a). Local estimates for subsolutions and supersolutions of oblique derivative problems for general second order elliptic equations, *Trans. Amer. Math. Soc.* **304**: 343–353.

[190] Lieberman, G. M. (1987b). Hölder continuity of the gradient of solutions of uniformly parabolic equations with conormal boundary conditions, *Ann. Mat. Pura Appl. (4)* **148**: 77–99.

[191] Lieberman, G. M. (1987c). Two-dimensional nonlinear boundary value problems for elliptic equations, *Trans. Amer. Math. Soc.* **300**: 287–295.

[192] Lieberman, G. M. (1988a). Hölder continuity of the gradient at a corner for the capillary problem and related results, *Pacific J. Math.* **133**: 115–135.

[193] Lieberman, G. M. (1988b). Oblique derivative problems in Lipschitz domains. II. Discontinuous boundary data, *J. Reine Angew. Math.* **389**: 1–21.

[194] Lieberman, G. M. (1989). Optimal Hölder regularity for mixed boundary value problems, *J. Math. Anal. Appl.* **143**: 572–586.

[195] Lieberman, G. M. (1999). Nonuniqueness for some linear oblique derivative problems for elliptic equations, *Comment. Math. Univ. Carolinae* **40(3)**: 477–481.

[196] Lieberman, G. M. (1996). *Second Order Parabolic Differential Equations*, World Scientific Publishing Co. Inc.: River Edge, NJ.

[197] Lieberman, G. M. and Trudinger, N. S. (1986). Nonlinear oblique boundary value problems for nonlinear elliptic equations, *Trans. Amer. Math. Soc.* **295**: 509–546.

[198] Lien, W.-C. and Liu, T.-P. (1999). Nonlinear stability of a self-similar 3-dimensional gas flow, *Commun. Math. Phys.* **204**: 525–549.

[199] Lighthill, M. J. (1949). The diffraction of a blast I, *Proc. Roy. Soc. London* **198A**: 454–470.

[200] Lighthill, M. J. (1950). The diffraction of a blast II, *Proc. Roy. Soc. London* **200A**: 554–565.

[201] Lin, F. H. and Wang, L. (1998). A class of fully nonlinear elliptic equations with singularity at the boundary, *J. Geom. Anal.* **8**: 583–598.

[202] Liu, T.-P. (2000). *Hyperbolic and Viscous Conservation Laws*, CBMS-NSF Regional Conf. Series in Appl. Math. **72**, SIAM: Philadelphia.

[203] Liu, X. D. and Lax, P. D. (1996). Positive schemes for solving multidimensional hyperbolic systems of conservation laws, *J. Comp. Fluid Dynamics* **5**: 133–156.

[204] Lock, G. D. and Dewey, J. M. (1989). An experimental investigation of the sonic criterion for transition from regular to Mach reflection of weak shock waves, *Exp. in Fluids* **7**: 289–292.

[205] Lyakhov, V. N., Podlubny, V. V. and Titarenko, V. V. (1989). *Influence of Shock Waves and Jets on Elements of Structures*, Mashinostroenie: Moscow (in Russian).

[206] Mach, E. (1878). Über den verlauf von funkenwellenin der ebene und im raume, *Sitzungsber. Akad. Wiss. Wien* **78**: 819–838.

[207] Majda A. and Thomann, E. (1987). Multidimensional shock fronts for second order wave equations, *Comm. Partial Diff. Eqs.* **12**: 777–828.

[208] Majda, A. (1983a). The stability of multidimensional shock fronts, *Mem. Amer. Math. Soc.* **41**, no. 275.

[209] Majda, A. (1983b). The existence of multidimensional shock fronts, *Mem. Amer. Math. Soc.* **43**, no. 281.

[210] Majda, A. (1984). *Compressible Fluid Flow and Systems of Conservation Laws in Several Space Variables*, Springer-Verlag: New York.

[211] Majda, A. (1991). One perspective on open problems in multi-dimensional conservation laws, In: *Multidimensional Hyperbolic Problems and Computations (Minneapolis, MN, 1989), IMA Vol. Math. Appl.* **29**, pp. 217–238, Springer: New York.

[212] Majda, A. and Bertozzi, A. (2002). *Vorticity and Incompressible Flow*, Cambridge University Press: Cambridge.

[213] Métivier, G. (1990). Stability of multi-dimensional weak shocks, *Comm. Partial Diff. Eqs.* **15**: 983–1028.

[214] Michael, J. H. (1981). Barriers for uniformly elliptic equations and the exterior cone condition, *J. Math. Anal. Appl.* **79**: 203–217.

[215] Michael, J. H. (1977). A general theory for linear elliptic partial differential equations, *J. Diff. Eqs.* **23**: 1–29.

[216] Mises, R. V. (1958). *Mathematical Theory of Compressible Fluid Flow*, Academic Press: New York.

[217] Morawetz, C. S. (1956). On the non-existence of continuous transonic flows past profiles I, *Comm. Pure Appl. Math.* **9**: 45–68.

[218] Morawetz, C. S. (1957). On the non-existence of continuous transonic flows past profiles II, *Comm. Pure Appl. Math.* **10**: 107–131.

[219] Morawetz, C. S. (1958). On the non-existence of continuous transonic flows past profiles III, *Comm. Pure Appl. Math.* **11**: 129–144.

[220] Morawetz, C. S. (1985). On a weak solution for a transonic flow problem, *Comm. Pure Appl. Math.* **38**: 797–818.

[221] Morawetz, C. S. (1994). Potential theory for regular and Mach reflection of a shock at a wedge, *Comm. Pure Appl. Math.* **47**: 593–624.

[222] Morawetz, C. S. (1994). On steady transonic flow by compensated compactness, *Meth. Appl. Anal.* **2**: 257–268.

[223] Osher, S. and Fedkiw, R. (2003). *Level Set Methods and Dynamic Implicit Surfaces*, Springer-Verlag: New York.

[224] Otway, T. H. (2015). *Elliptic-Hyperbolic Partial Differential Equations: A Mini-Course in Geometric and Quasilinear Methods*, Springer-Verlag: New York.

[225] Otway, T. H. (2012). *The Dirichlet Problem for Elliptic-Hyperbolic Equations of Keldysh Type*, Springer-Verlag: Berlin-Heidelberg.

[226] Rankine, W. J. M. (1870). On the thermodynamic theory of waves of finite longitudinal disturbance, *Phil. Trans. Royal Soc. London*, **160**: 277–288.

[227] Rayleigh, Lord (J. W. Strutt) (1910). Aerial plane waves of finite amplitude, *Proc. Royal Soc. London* **84A**: 247–284.

[228] Reichenbach, H. (1983). Contribution of Ernst Mach to fluid mechanics, *Ann. Rev. Fluid Mech.* **15**: 1–28.

[229] Riemann, B. (1860). Über die Fortpflanzung ebener Luftvellen von endlicher Schwingungsweite, *Gött. Abh. Math. Cl.* **8**: 43–65.

[230] Sakurai, A., Henderson, L. F., Takayama, K., Walenta, Z. and Collela, P. (1989). On the von Neumann paradox of weak Mach reflection, *Fluid Dyn. Research* **4**: 333–345.

[231] Schaeffer, D. G. (1976). Supersonic flow past a nearly straight wedge, *Duke Math. J.* **43**: 637–670.

[232] Schneyer, G. P. (1975). Numerical simulation of regular and Mach reflection, *Phy. Fluids* **18**: 1119–1124.

[233] Schulz-Rinne, C. W., Collins, J. P. and Glaz, H. M. (1993). Numerical solution of the Riemann problem for two-dimensional gas dynamics, *SIAM J. Sci. Comput.* **14**: 1394–1414.

[234] Serre, D. (1995). Perfect fluid flow in two independent space variables: Reflection of a planar shock by a compressive wedge, *Arch. Rational Mech. Anal.* **132**: 15–36.

[235] Serre, D. (1996). Spiral waves for the two-dimensional Riemann problem for a compressible fluid, *Ann. Fac. Sci. Toulouse Math. (6)* **5**: 125–135.

[236] Serre, D. (2007). Shock reflection in gas dynamics. In: *Handbook of Mathematical Fluid Dynamics*, Vol. **4**, pp. 39–122, Eds: S. Friedlander and D. Serre, Elsevier: North-Holland.

[237] Serre, D. (2009). Multidimensional shock interaction for a Chaplygin gas, *Arch. Rational Mech. Anal.* **191**: 539–577.

[238] Serre, D. (2000). *Systems of Conservation Laws*, Vols. 1–2, Cambridge University Press: Cambridge.

[239] Sethian, J. A. (1999). *Level Set Methods and Fast Marching Methods: Evolving Interfaces in Computational Geometry, Fluid Mechanics, Computer Vision, and Materials Science*, 2nd Edition. Cambridge University Press: Cambridge.

[240] Shankar, V., Kutler, P. and Anderson, D. (1978). Diffraction of a shock wave by a compression corner, Part II: Single Mach reflection, *AIAA J.* **16**: 4–5.

[241] Sheng, W.-C. and Yin, G. (2009). Transonic shock and supersonic shock in the regular reflection of a planar shock, *Z. Angew. Math. Phys.* **60**: 438–449.

[242] Shiffman, M. (1952). On the existence of subsonic flows of a compressible fluid, *J. Rational Mech. Anal.* **1**: 605–652.

[243] Skews, B. and Ashworth, J. (2005). The physical nature of weak shock wave reflection, *J. Fluid Mech.* **542**: 105–114.

[244] Smirnov, M. M. (1978). *Equations of Mixed Type*, Translated from the Russian. Translations of Mathematical Monographs **51**, American Mathematical Society: Providence, RI.

[245] Smoller, J. (1994). *Shock Waves and Reaction-Diffusion Equations*, Springer-Verlag: New York.

[246] Song, K. (2003). The pressure-gradient system on non-smooth domains, *Comm. Partial. Diff. Eqs.* **28**: 199–221.

[247] Song, K., Wang, Q. and Zheng, Y. (2015). The regularity of semihyperbolic patches near sonic lines for the 2-D Euler system in gas dynamics, *SIAM J. Math. Anal.* **47(3)**: 2200–2219.

[248] Srivastava, R. S. (1995). Starting point of curvature for reflected diffracted shock wave, *AIAA J.* **33**: 2230–2231.

[249] Srivastava, R. S. (2007). Effect of yaw on the starting point of curvature for reflected diffracted shock wave (subsonic and sonic cases), *Shock Waves* **17**: 209–212.

[250] Stefan, J. (1889). Über die Theorie der Eisbildung, insbesondere über die Eisbildung im Polarmeere, Sitzungsberichte der Österreichischen Akademie der Wissenschaften, *Mathematisch-Naturwissenschaftliche Klasse, Abteilung 2, Mathematik, Astronomie, Physik, Meteorologie und Technik* **98**: 965–983.

[251] Stein, E. M. (1970). *Singular Integrals and Differentiability Properties of Functions*, Princeton Mathematical Series **30**, Princeton University Press: Princeton, NJ.

[252] Stein, E. M. (1993). *Harmonic Analysis: Real-Variable Methods, Orthogonality, and Oscillatory Integrals*, Princeton Mathematical Series **43**, Princeton University Press: Princeton, NJ.

[253] Stokes, G. G. (1848). On a difficulty in the theory of sound, *Philos. Magazine, Ser. 3*, **33**: 349–356.

[254] Sturtevant, B. and Kulkarny, V. A. (1976). The focusing of weak shock waves, *J. Fluid. Mech.* **73**: 1086–1118.

[255] Tabak, E. and Rosales, R. (1994). Focusing of weak shock waves and the von Neumann paradox of oblique shock reflection, *Phys. Fluids*, **6**: 1874–1892.

[256] Tadmor, E., Liu, J.-G. and Tzavaras, A. (2009). *Hyperbolic Problems: Theory, Numerics and Applications*, Parts I, II, Amer. Math. Soc.: Providence, RI.

[257] Tesdall, A. M. and Hunter, J. K. (2002). Self-similar solutions for weak shock reflection, *SIAM J. Appl. Math.* **63**: 42–61.

[258] Tesdall, A. M., Sanders, R. and Keyfitz, B. (2006). The triple point paradox for the nonlinear wave system, *SIAM J. Appl. Math.* **67**: 321–336.

[259] Tesdall, A. M., Sanders, R. and Keyfitz, B. (2008). Self-similar solutions for the triple point paradox in gasdynamics, *SIAM J. Appl. Math.* **68**: 1360–1377.

[260] Timman, R. (1964). Unsteady motion in transonic flow, In: *Symposium Transsonicum (IUTAM, Aachen, Sept. 1962)*, pp. 394–401, ed. K. Oswatitsch, Springer-Verlag: Berlin.

[261] Toro, E. (2009). *Riemann Solvers and Numerical Methods for Fluid Dynamics: A Practical Introduction*, 3rd Edition. Springer-Verlag: Berlin.

[262] Trudinger, N. (1985). On an interpolation inequality and its applications to nonlinear elliptic equations, *Proc. Amer. Math. Soc.* **95**: 73–78.

[263] Van Dyke, M. (1982). *An Album of Fluid Motion*, The Parabolic Press: Stanford.

[264] Vasil'ev, K. and Kraiko, A. (1999). Numerical simulation of weak shock diffraction over a wedge under the von Neumann paradox conditions, *Comput. Math. & Math. Phys.* **39**: 1335–1345.

[265] von Karman, T. (1947). The similarity law of transonic flow, *J. Math. Phys.* **26**: 182–190.

[266] von Neumann, J. (1943). Theory of shock waves, *Progress Report*, U.S. Dept. Comm. Off. Tech. Serv. No. **PB32719**, Washington, DC.

[267] von Neumann, J. (1943). Oblique reflection of shocks, *Explo. Res. Rep.* **12**, Navy Department, Bureau of Ordnance, Washington, DC.

[268] von Neumann, J. (1945). Refraction, intersection, and reflection of shock waves, *NAVORD Rep.* **203-45**, Navy Department, Bureau of Ordnance, Washington, DC.

[269] von Neumann, J. (1963). *Collected Works*, Vol. **6**, Pergamon: New York.

[270] von Neumann, J. (2010). Discussion on the existence and uniqueness or multiplicity of solutions of the aerodynamical equation [Reprinted from MR0044302], *Bull. Amer. Math. Soc.* (*N.S.*) **47**: 145–154.

[271] von Neumann, J. and Richtmyer, R. D. (1950). A method for the numerical calculation of hydrodynamical shocks, *J. Appl. Phys.* **21**: 380–385.

[272] Wang, P.-Y. (2000). Regularity of free boundaries of two-phase problems for fully nonlinear elliptic equations of second order. I. Lipschitz free boundaries are $C^{1,\alpha}$, *Comm. Pure Appl. Math.* **53**: 799–810.

[273] Woodward, P. and Colella, P. (1984). The numerical simulation of two-dimensional fluid flow with strong shocks, *J. Comp. Phys.* **54**: 115–173.

[274] Woodward, P. (1985). Simulation of the Kelvin-Helmholtz instability of a supersonic slip surface with the piecewise-parabolic method (PPM). In: *Numerical Methods for the Euler Equations of Fluid Dynamics* (*Rocquencourt, 1983*), pp. 493–508, SIAM: Philadelphia, PA.

[275] Wu, X.-M. (1956). *Equations of Mathematical Physics*, Higher Education Press: Beijing (in Chinese).

[276] Xie, C. and Xin, Z. (2007). Global subsonic and subsonic-sonic flows through infinitely long nozzles, *Indiana Univ. Math. J.* **56**: 2991–3023.

[277] Xin, Z. and Yin, H. (2005). Transonic shock in a nozzle: two-dimensional case, *Comm. Pure Appl. Math.* **58**: 999–1050.

[278] Yang, G.-J. (1989). *The Euler-Poisson-Darboux Equations*, Yunnan University Press: Yunnan (in Chinese).

[279] Yuan, H. (2006). On transonic shocks in two-dimensional variable-area ducts for steady Euler system, *SIAM J. Math. Anal.* **38**: 1343–1370.

[280] Yuan, H. (2007). Transonic shocks for steady Euler flows with cylindrical symmetry, *Nonlinear Analysis* **66**: 1853–1878.

[281] Zabolotskaya, G. I. and Khokhlov, R. V. (1969). Quasi-plane waves in the nonlinear acoustics of confined beams, *Sov. Phys.-Acoustics* **15**: 35–40.

[282] Zakharian, A. R., Brio, M., Hunter, J. K. and Webb, G. M. (2000). The von Neumann paradox in weak shock reflection, *J. Fluid Mech.* **422**: 193–205.

[283] Zeidler, E. (1986). *Nonlinear Functional Analysis and Its Applications I: Fixed-Point Theorems*, Translated from the German by P. R. Wadsack, Springer-Verlag: New York.

[284] Zhang, T. and Zheng, Y. (1990). Conjecture on the structure of solutions of the Riemann problem for two-dimensional gas dynamics, *SIAM J. Math. Anal.* **21**: 593–630.

[285] Zhang, Y.-Q. (2003). Steady supersonic flow past an almost straight wedge with large vertex angle, *J. Diff. Eqs.* **192**: 1–46.

[286] Zheng, Y. (2001). *Systems of Conservation Laws: Two-Dimensional Riemann Problems*, Birkhäuser: Boston.

[287] Zheng, Y. (2003). A global solution to a two-dimensional Riemann problem involving shocks as free boundaries, *Acta Math. Appl. Sin. Engl. Ser.* **19**: 559–572.

[288] Zheng, Y. (2006). Two-dimensional regular shock reflection for the pressure gradient system of conservation laws, *Acta Math. Appl. Sin. Engl. Ser.* **22**: 177–210.

[289] Zheng, Y. (2009). The compressible Euler system in two space dimensions, In: *Nonlinear Conservation Laws, Fluid Systems and Related Topics*, pp. 301–390, Ser. Contemp. Appl. Math. CAM **13**, World Sci. Publishing: Singapore.

Index